WATER RESOURCES HANDBOOK

Related Titles of Interest in Water and Water Resources

WATER RESOURCES HANDBOOK

Larry W. Mays Editor in Chief

Department of Civil and Environmental Engineering
Arizona State University
Tempe, Arizona

McGRAW-HILL

New York San Francisco Washington, D.C. Auckland Bogotá
Caracas Lisbon London Madrid Mexico City Milan
Montreal New Delhi San Juan Singapore
Sydney Tokyo Toronto

Library of Congress Cataloging-in-Publication Data

Water resources handbook / Larry W. Mays, editor-in-chief.
    p.   cm.
    Includes index.
    ISBN 0-07-041150-6
    1. Water-supply. I. Mays, Larry W.
  TD345.H257   1996
  333.91'15—dc20   96-3165
                                     CIP

McGraw-Hill

A Division of The McGraw-Hill Companies

1 2 3 4 5 6 7 8 9 DOC/DOC 9 0 1 0 9 8 7 6

ISBN 0-07-041150-6

The sponsoring editor for this book was Larry S. Hager, the editing supervisor was Virginia Carroll, and the production supervisor was Suzanne W. B. Rapcavage. It was set in Times Roman by North Market Street Graphics.

Printed and bound by R. R. Donnelley & Sons Company.

INTERNATIONAL EDITION
Copyright © 1996. Exclusive rights by McGraw-Hill for manufacture and export. This book cannot be re-exported from the country to which it is consigned by McGraw-Hill. The International Edition is not available in North America.

When ordering this title, use ISBN 0-07-114517-6.

This book is printed on acid-free paper.

CONTENTS

Part 2   Water Resource Quality (Natural Systems)

Chapter 8. Water Quality   8.3

Part 4   Water Resources Excess Management

Part 5   Water Resources for the Future

Chapter 29. Global Climate Change: Effect on Hydrologic Cycle   29.3

Chapter 30. Ecological Effects of Global Climate Change
on Freshwater Ecosystems with Emphasis on Streams and Rivers   30.1

Chapter 31. Energy and Water   31.1

Chapter 32. Water-Use Management: Permit
and Water-Transfer Systems   32.1

Chapter 33. Decision Support Systems (DSS) for Water-Resources Management

CONTRIBUTORS

Robert B. Ambrose, Jr. *US EPA National Exposure Research Lab* (CHAP. 14)

Lawrence A. Baker *Arizona State University* (CHAP. 9)

Thomas O. Barnwell, Jr. *US EPA National Exposure Research Lab* (CHAP. 14)

Franklin L. Burton *Burton Environmental Engineering* (CHAP. 19)

Steven C. Chapra *University of Colorado* (CHAPS. 10, 33)

M. Hanif Chaudhry *Washington State University* (CHAP. 2)

Zaid K. Chowdhury *Malcolm Pirnie Inc.* (CHAP. 17)

Albert J. Clemmens *U.S. Water Conservation Laboratory* (CHAP. 22)

James Crook *Black & Veatch* (CHAP. 21)

Benedykt Dziegielewski *Southern Illinois University-Carbondale* (CHAP. 23)

J. Wayland Eheart *University of Illinois at Urbana-Champaign* (CHAP. 32)

Stuart G. Fisher *Arizona State University* (CHAP. 30)

David Ford *Consulting Hydrologic Engineer* (CHAPS. 27, 28)

Michael J. Graham *Batelle Pacific Northwest Laboratories* (CHAP. 11)

Nancy B. Grimm *Arizona State University* (CHAP. 30)

T. J. Grizzard *Virginia Tech* (CHAP. 26)

Douglas Hamilton *Consulting Engineer* (CHAP. 28)

Robert W. Hinks *Arizona State University* (CHAP. 24)

Benjamin F. Hobbs *The Johns Hopkins University* (CHAP. 31)

Hydrologic Engineering Center *U.S. Army Corps of Engineers* (CHAP. 27)

Marvin E. Jensen *Consultant* (CHAP. 22)

D. F. Kibler *Virginia Tech* (CHAP. 26)

Dennis P. Lettenmaier *University of Washington* (CHAP. 29)

G. V. Loganathan *Virginia Tech* (CHAP. 26)

Daniel P. Loucks *Cornell University* (CHAP. 15)

Jay R. Lund *University of California at Davis* (CHAPS. 31, 32)

David Maidment *University of Texas at Austin* (CHAP. 23)

Joseph F. Malina, Jr. *University of Texas at Austin* (CHAP. 8)

Larry W. Mays *Arizona State University* (CHAPS. 1, 6, 24)

Gregory McCabe *U.S. Geological Survey* (CHAP. 29)

Steve C. McGutcheon *US EPA National Exposure Research Lab* (CHAP. 14)

F. Blaine Metting *Batelle Pacific Northwest Laboratories* (CHAP. 11)

Richard L. Mittelstadt *U.S. Army Corps of Engineers* (CHAP. 31)

Clay L. Montague   *University of Florida* (CHAP. 12)

David H. Moreau   *University of North Carolina at Chapel Hill* (CHAP. 4)

Eva M. Opitz   *Planning and Management Consultants, Ltd.* (CHAP. 23)

Rene F. Reitsma   *University of Colorado* (CHAP. 33)

John A. Replogle   *U.S. Water Conservation Laboratory* (CHAP. 22)

Curtis J. Richardson   *Duke University* (CHAP. 13)

Eugene Z. Stakhiv   *U.S. Army Corps of Engineers* (CHAP. 29)

Kenneth M. Strzepek   *University of Colorado* (CHAP. 33)

A. Dan Tarlock   *Chicago-Kent College of Law* (CHAP. 5)

George Tchobanoglous   *University of California–Davis* (CHAP. 20)

James M. Thomas   *U.S. Geological Survey* (CHAP. 11)

Yeou-Koung Tung   *University of Wyoming* (CHAPS. 6, 7)

Thomas M. Walski   *Wilkes University* (CHAP. 18)

George H. Ward, Jr.   *University of Texas at Austin* (CHAP. 12)

Garret P. Westerhoff   *Malcolm Pirnie Inc.* (CHAP. 17)

Joseph R. Williams   *US EPA Natural Risk Management Lab* (CHAP. 14)

William Yeh   *University of California–Los Angeles* (CHAP. 16)

Ben Chie Yen   *University of Illinois at Urbana-Champaign* (CHAP. 25)

Robert A. Young   *Colorado State University* (CHAP. 3)

Edith A. Zagona   *University of Colorado* (CHAP. 33)

PREFACE

Water affects the life of every human on Earth: too little or too much can be detrimental to our livelihood; the quality affects our health and well-being; and the beauty can be found in the many streams, rivers, lakes, estuaries, and oceans. If you ask a child what water means to them, they might write a poem about water similar to what my twelve-year-old son, Travis, wrote, called "Water, Water, Everywhere":

> Calm and glistening
> Sparkles like many crystals
> Clear and beautiful
> You drink it every day
> Sparkling and clear water

Unfortunately, as we all know, there is much more to the poem that can be written, not so optimistically.

In her book *Silent Spring,* Rachel Carson presented some selected quotes in the front pages. The first is by Albert Schweitzer: "Man has lost the capacity to foresee and to forestall. He will end by destroying the Earth." The second is by E. B. White:

> I am pessimistic about the human race because it is too ingenious for its own good. Our approach to nature is to beat it into submission. We would stand a better chance of survival if we accommodated ourselves to this planet and viewed it appreciatively instead of skeptically and dictatorially.

In the last chapter of *The Other Road,* Rachel Carson wrote:

> The road we have long been traveling is deceptively easy, a smooth superhighway on which we progress with great speed, but at its end lies disaster. The other fork of the road—the one "less traveled by"—offers our last, our only chance to reach a destination that assures the preservation of our earth.

After having thought about these many times I am optimistic that the human race has the ability to foresee and is not too ingenious for its own good, that we will not beat nature into submission. We will survive as we continue to learn how to accommodate ourselves to this planet and view it appreciatively. Hopefully, we have taken, and will continue to take, the other fork of the road in the development and utilization of our water resources on Earth. Let the effort in developing this handbook be a contribution to that process of learning how to accommodate ourselves on this planet.

During the past few decades, water resources and the related fields have undergone a revolution in scientific development and in methods available for analysis and design. The advent of the computer has changed our entire thought process and ability to understand and model complex processes as well as archive and analyze voluminous amounts of data. Such abilities were simply impossible only a few decades ago. A major emphasis of this book is to present the state of the art of the many topics related to water resources.

First and foremost, this handbook is intended to be a reference book for those wishing to expand their knowledge of water resources. The handbook can be used as a companion along the pathway of learning about water resources in general or about special topics in water resources. It can be used to begin the journey and/or to continue the journey, as a road map to other publications and specific topics beyond the intent of this handbook. The handbook should serve as a reference to engineers, planners, economists, attorneys, managers, hydrologists, designers, policy makers, geologists, political scientists, biologists, educators, ecologists, limnologists, geographers, public administrators, resource developers, environmentalists, soil scientists, and others.

The field of water resources is diverse, resulting in a wide range of subject matter that should be covered in a water resources handbook. The handbook subject matter selected reflects my personal perception of water resources. The chapters were therefore chosen to reflect what I feel are the most directly related topics to the widest range of readers with an interest in water resources. Space limitations, however, did not allow for an all-inclusive treatment of all fields related to water resources. Examples include many related topics in the biological sciences, physical sciences, social sciences, and engineering.

The handbook comprises five sections: Principles for Water Resources, Chapters 1 to 7; Water Quality of Natural Systems, Chapters 8 to 14; Water Resource Supply Systems, Chapters 15 to 23; Water Resources Excess Management, Chapters 24 to 28; and Water Resources for the Future, Chapters 29 to 33. My intent was that these five sections cover the major topics of a water resources handbook, at least within the framework in which I have thought about water resources for many years. The individual chapters were then chosen to cover the topics I felt were most important for these five sections. I sincerely hope you find what you need in this handbook or at least the information to lead you in the right direction on your pathway of learning about water resources.

Larry W. Mays

ACKNOWLEDGMENTS

I must first acknowledge the authors who made this handbook possible. It has been a sincere privilege to have worked with such an excellent group of dedicated people. They are all experienced professionals who are among the leading experts in their respective fields. References to material in this handbook should be attributed to the respective chapter authors.

Each chapter was reviewed by professionals in their particular area of expertise. I would like to thank each of those people for their time and effort. You have certainly been a major contribution to the completion of this handbook.

During the past twenty years of my academic career I have received help and encouragement from so many people that it is not possible to name them all. These people represent a wide range of universities, research institutions, government agencies, and professions. To all of you I express my deepest thanks.

I would like to acknowledge the support that I have received from Arizona State University. Even though I have served as Chair of the Department of Civil and Environmental Engineering during this time, my efforts were made possible through the hard work and dedication of Debbie Trimmels and Ethel Bruce of my staff, who not only helped on this handbook but also kept the department going. A special thanks goes to Professor Paul Ruff, the associate chair of the department, who kept things going in my absence and who covered me in meetings when I could not attend. A very special thanks goes to Dr. Guihua Li, who helped in many ways, including work on the many figures and permissions.

I appreciate the advice and encouragement of Larry Hager of McGraw-Hill. Last of all, I would like to give a special thanks to Ginny Carroll and team of North Market Street Graphics in Lancaster, Pennsylvania, who were in charge of the handbook production.

This handbook has been part of a personal journey that began over forty years ago when I was a young boy with a love of water. Books are companions along the journey of learning. I hope that you will be able to use this handbook in your own journey of learning about water resources. Have a happy and wonderful journey.

I would like to dedicate this handbook to humanity and human welfare.

Larry W. Mays

ABOUT THE EDITOR

Larry W. Mays, Ph.D., P.E., P.H., is professor and chair of the Civil and Environmental Engineering Department at Arizona State University. He was formerly director of the Center for Research in Water Resources at the University of Texas at Austin, where he held an Engineering Foundation Endowed Professorship.

A registered professional engineer in several states and a registered professional hydrologist, Dr. Mays has served as principal investigator on numerous water resource research projects sponsored by federal, state, and local government agencies. He is a member of the American Society of Civil Engineers and many other professional organizations, including the Universities Council on Water Resources, for which he has served as president.

Dr. Mays has published extensively in the water resources literature. Among his previous books, he is coauthor of *Applied Hydrology* and *Hydrosystems Engineering and Management,* both published by McGraw-Hill.

P · A · R · T · 1

PRINCIPLES FOR WATER RESOURCES

CHAPTER 1
WATER RESOURCES: AN INTRODUCTION

Larry W. Mays

Department of Civil and Environmental Engineering
Arizona State University
Tempe, Arizona

1.1 HISTORICAL PERSPECTIVES ON WATER RESOURCES

Water resources cannot be studied without studying humanity. To quote Chief Seattle, chief of the Suquamish tribe, who lived across Puget Sound from the site of the city that later arose in Seattle's name, "Man did not weave the web of life, he is merely a strand in it. Whatever he does, to the web, he does to himself." The first images of the surface of the Earth as seen from the moon over two decades ago helped us visualize Earth as a unit, an integrated set of systems—land masses, atmosphere, oceans, and the plant and animal kingdom. These images have also helped us to realize that threats to one of these integrated systems could harm them all.

1.1.1 Water Resources Development in Antiquity

Humans have spent most of their history as hunting and food gathering beings. Only in the last 9000 to 10,000 years have we discovered how to raise crops and tame animals. Such revolution probably first took place in the hills to the north of present day Iraq and Syria. From there the agricultural revolution spread to the Nile and Indus valleys. During this agricultural revolution, permanent villages took the place of a wandering existence. About 6000 to 7000 years ago, farming villages of the Near and Middle East became cities. Farmers learned to raise more food than they needed, allowing others to spend time making things useful to their civilization. People began to invent and develop technologies, including how to transport and manage water for irrigation.

The first successful efforts to control the flow of water were made in Egypt and Mesopotamia. Remains of these prehistoric irrigation works still exist. In ancient Egypt the construction of canals was a major endeavor of the Pharaohs and their servants, beginning in Scorpio's time. One of the first duties of provincial governors was the digging and repair of canals, which were used to flood large tracts of land while the Nile was flowing high. The land was checkerboarded with small basins, defined by a system of dikes. Problems of the uncertainty of the Nile flows were recognized. During very high flows the dikes were washed away and villages were flooded, drowning thousands. During low flows the land did not receive water and no crops could grow. In many places, where fields were too high to receive water from the canals, water was drawn from the canals or the Nile directly by the *swape* or *shaduf.* These consisted of a bucket on the end of a cord which hung from the long end of a pivoted boom, counterweighted at the short end (de Camp, 1963). The building of canals continued in Egypt throughout the centuries.

The Sumerians in southern Mesopotamia built city walls and temples, and dug canals that were the world's first engineering works. It is also of interest that these people, from the beginning of recorded history, fought over water rights. Irrigation was extremely vital to Mesopotamia, Greek for "the land between the (Tigris and Euphrates) rivers." An ancient Babylonian curse was, "May your canal be filled with sand" (de Camp, 1963). Their ancient laws even dealt with canals and water rights. The following, from about −VI*, illustrates such a law (de Camp, 1963):

> The gentleman who opened his wall for irrigation purposes, but did not make his dyke strong and hence caused a flood and inundated a field adjoining his, shall give grain to the owner of the field on the basis of those adjoining.

Flooding problems were more serious in Mesopotamia than in Egypt because the Tigris and Euphrates carried several times more silt per unit volume of water than the Nile. This resulted in rivers rising faster and changing their courses more often in Mesopotamia.

Both the Mesopotamian irrigation system and that in the Egyptian Delta were of the basin type, which were opened by digging a gap in the embankment and closed by placing mud back into the gap. Water was hoisted using the swape, as in Egypt. Laws in Mesopotamia not only required farmers to keep their basins and feeder canals in repair, but also required everyone to help with hoes and shovels in times of flood, or when new canals were to be dug or old ones repaired (de Camp, 1963). Some canals may have been used for 1000 years before they were abandoned and others were built. Even today, 4000 to 5000 years later, the embankments of the abandoned canals are still present. These canal systems, in fact, supported a denser population than lives there today. Over the centuries, the agriculture of Mesopotamia began to decay because of the salt in the alluvial soil. Then in 1258, the Mongols conquered Mesopotamia and destroyed the irrigation systems.

The Assyrians also developed extensive public works. Sargon II invaded Armenia in −714, discovering the *ganat* (Arabic name) or *kariz* (Persian name). This is a tunnel used to bring water from an underground source in the hills down to the foothills. Sargon destroyed the area in Armenia, but brought the concept back to Assyria. This method of irrigation spread over the Near East into North Africa over the centuries and is still used. Sargon's son Sennacherib also developed water works

---

* Centuries are indicated by Roman numerals preceded by + or − according to whether they are centuries of the Christian era or B.C.; hence −VI refers to the sixth century B.C. Years are treated likewise, with Arabic instead of Roman numerals; for example −714 is 714 B.C. and +714 is 714 A.D. This system of dates has been used by George Sarton in *History of Science,* by Joseph Needham in *Science and Civilization in China,* and by L. Sprague de Camp in *Ancient Engineer.*

by damming the Tebitu River and using a canal to bring water to Nineveh, where the water could be used for irrigation without hoisting devices. During high water in the spring, overflows were handled by a municipal canebrake that was built to develop marshes used as game preserves for deer, wild boar, and birch breeding areas. When this system was outgrown, a new 30-mile canal was built, with an aqueduct which had a layer of concrete or mortar under the upper layer of stone to prevent leakage.

The Greeks were the first to show the connection between engineering and science, although they borrowed ideas from the Egyptians, the Babylonians, and the Phoenicians. During the Hellenistic period, Ktesibius (–285– –247), who lived in Alexandria, had several inventions including the force pump, the hydraulic pipe organ, the musical keyboard, the metal spring, and the water clock. Shortly after Ktesibius, Philen of Byzantium also had many inventions, one of which was the water wheel. One application of this water wheel was a bucket-chain water hoist powered by an undershot waterwheel. This water hoist may have been the first recorded case of using the energy of running water for practical use. Probably the greatest Hellenistic engineer, and one of the greatest intellects of all time, was Archimedes of Syracuse (–287– –212), a contemporary of Philan. Archimedes founded the ideas of hydrostatics and buoyancy. The Hellenistic kings began to build public bath houses.

The early Romans devoted much of their time to useful public works projects. They built roads, harbor works, aqueducts, temples, forums, town halls, arenas, baths, and sewers. The prosperous early-Roman bourgeois typically had a dozen-room house, with a square hole in the roof to let rain in, and a cistern beneath the roof to store the water. Many aqueducts were built by the Romans; however, they were not the first to build these. King Sennacherio built aqueducts, as did both the Phoenicians and the Helenes. The Romans and Helenes needed extensive aqueduct systems for their fountains, baths, and gardens. They also realized that water transported from springs was better for their health than river water, and did not require lifting the water to street level as did river water. Roman aqueducts were built on elevated structures to provide the needed slope for water flow. Knowledge of pipe making—using bronze, lead, wood, tile, and concrete—was in its infancy, and the difficulty of making good large pipes was a hinderance. Most Roman piping was made of lead, and even the Romans recognized that water transported by lead pipes is a health hazard.

The water source for a typical water-supply system of a Roman city was a spring or a dug well, usually with a bucket elevator to raise the water. If the well water was clear and of sufficient quantity, it was conveyed to the city by aqueduct. Also, water from several sources was collected in a reservoir, then conveyed by aqueduct or pressure conduit to a distributing reservoir (*castellum*). Three pipes conveyed the water—one to pools and fountains, the second to the public baths for public revenue, and the third to private houses for revenue to maintain the aqueducts (Rouse and Ince, 1957).

Irrigation was not a major concern, because of the terrain and the intermittent rivers. Romans did, however, drain marshes to obtain more farm land and because they were concerned about the bad air, or "harmful spirits," rising from the marshes, which they thought caused disease (de Camp, 1963). The disease-carrying mechanism was not the air, or spirits, but the malaria-carrying mosquito. Empedocles, the leading statesman of Acragas in Sicily during the Persian War (–V), drained the local marshes of Selinus to improve the people's health (de Camp, 1963). He also theorized that all matter is made of four elements: Earth, air, fire, and water.

The fall of the Roman Empire extended over a 1000-year transition period called the Dark Ages. During this period, the concepts of science related to water resources probably retrogressed. After the fall of the Roman Empire, water and sanitation—

indeed, public health—declined in Europe. Historical accounts tell of incredibly unsanitary conditions—polluted water, human and animal wastes in the streets, and water thrown out of windows onto passersby. Various epidemics ravaged Europe. During the same period, Islamic cultures, on the periphery of Europe, had religiously mandated high levels of personal hygiene, along with highly developed water supplies and adequate sanitation systems.

Next came the Renaissance. Some modest advances in the utilization of water occurred in the fourteenth century—waterpower was used in sawmills, a rather complicated pump was developed, and well borings were recorded for the first time.

Early Chinese history contains a great deal of fable and very little recorded history, making it impossible to trace the development of water resources very far into the past. The emperor Yu devised systems of dikes for flood protection that lasted several thousand years. Irrigation reached an advanced level in China at least 3000 years ago. Li built a system of canals, set monuments to record water stages, and established rules of operation (–II) (Rouse and Ince, 1957).

1.1.2 Early Development in Water Science

The following passage from Buras (1972) eloquently summarizes the history of water science:

> Man's quest for a better use of the available water is as old as mankind itself. His understanding of the phenomena connected with the occurrence of water in nature is relatively recent, and until about 150 years ago it was rather limited. Nevertheless, man has tried to offer plausible explanations about the natural hydrological processes, at a level of sophistication commensurate with the general *niveau* of scientific development at the time.

During the Renaissance, a gradual change occurred from purely philosophical concepts toward observational science. Leonardo da Vinci (1452–1519) made the first systematic studies of velocity distribution in streams, using a weighted rod held afloat by an inflated animal bladder. By releasing the rod at different points in the stream's cross section, Leonardo traced the velocity distribution across the channel. The 8000 existing pages of Leonardo's notes (MacCurdy, 1939) contain more entries concerning hydraulics than any other subject (Frazier, 1974).

The French Huguenot scientist Bernard Palissy (1510–1589) showed that rivers and springs originate from rainfall, thus refuting an age-old theory that streams were supplied directly by the sea. The French naturalist Pierre Perrault (1608–1680) measured runoff, and found it to be only a fraction of rainfall. He recognized that rainfall is a source for runoff, and correctly concluded that the remainder of the precipitation is lost by transpiration, evaporation, and diversion. Evangelista Torricelli (1608–1647) related the barometric height of mercury to the weight of the atmosphere, and the form of a liquid jet to the trajectory of free fall. Blaise Pascal (1623–1662) clarified principles of the barometer, hydraulic press, and pressure transmissibility. Isaac Newton (1642–1727) explored various aspects of fluid resistance (inertial, viscous, and wave) and discovered jet contraction.

Hydraulic measurements and experiments flourished during the eighteenth century. New hydraulic principles were discovered, such as the Bernoulli (1700–1782) equation and Chezy's (1718–1798) formula, and better instruments were developed, including the tipping-bucket rain gauge and the current meter. Henri de Pitot (1695–1771) constructed a device to indicate water velocity through a differential head. Leonard Euler (1707–1783) first explained the role of pressure in fluid flow, and formulated the basic equation of motion and the so-called Bernoulli theorem. Hydrogen was discovered in 1766 by an English scientist, Henry Cavendish. Oxygen was discovered independently by two chemists, Carl Scheele of Sweden and Joseph Priestley of England in the 1770s.

Concepts of hydrology advanced during the nineteenth century. Dalton (1802) established a principle for evaporation, the theory of capillary flow was described by the Hagen-Poiseuille (1839) equation, and the rational method for determining peak flood flow was proposed by Mulvaney (1850). Darcy (1856) developed the law of porous media flow, Rippl (1883) presented a diagram for determining reservoir storage requirements, and Manning (1891) proposed an open-channel flow formula. Hydraulics research continued in the nineteenth century, with Louis Marie Henri Navier (1785–1836) extending the equations of motion to include molecular forces. Jean-Claude Barre de Saint-Venant wrote prolifically in many fields on hydraulics. Jean Louis Poiseuille (1799–1867) studied the resistance of flow through capillary tubes. Henry Philibert Gaspard Darcy (1803–1858) performed tests on filtration and pipe resistance. Others, such as Dupuit, Weisbach, Froude, Francis, Kutter, Stokes, Breese, Kirchoff, Lord Kelvin, Pelton, Reynolds, and Boussinesq, advanced the knowledge of fluid flow and hydraulics during the nineteenth century.

In England in the 1840s waterborne sewage was introduced on a large scale, as a result of recommendations of a parliamentary commission charged to find alternatives to cesspools, privies, buckets, and earth closets used in urban areas. The advances and progressive improvements that have occurred over time have been in response to new scientific discoveries and to previous failures. The use of separate sanitary and storm sewers was recommended in 1842 in England, and in the 1870s in the United States.

The beneficial effects of sewerage became much clearer during the close of the nineteenth century. By sewering certain towns in England, the death rate from pulmonary diseases alone was reduced by 50 percent. People began to realize that in densely populated areas the disposal of solid and liquid refuse was a serious problem. The Mosaic regulations (Deut. xxiii, 12–13, Old Testament) could no longer be enforced, and to store the filth of a city within the city is simply to invite disease and death (Staley and Pierson, 1899). Julius Adams (1880) in the introduction to his book wrote:

> Although Sanitary Science, as we understand the term, is of modern growth, and it is only within the last thirty or forty years that the subject has engaged the attention of scientific men, yet the remains of ancient works exhibit the fact, that so soon as the progress of civilization tended to concentrate population within centres of comparatively limited areas, if not his health, was man himself; and the best method for the prompt removal of noxious refuse from the vicinity of his dwelling, which to the primitive inhabitant was a subject scarcely worthy of a thought, soon forced itself upon his attention as a subject above all others deserving of his careful consideration.

At the beginning of the twentieth century quantitative hydrology was basically the application of empirical approaches to solve practical hydrological problems. Gradually, hydrologists replaced empiricism with rational analysis of observed data. Green and Ampt (1911) developed a physically-based model for infiltration; Hazen (1914) introduced frequency analysis of flood peaks and water storage requirements; Richards (1931) derived the governing equation for unsaturated flow; Sherman devised the unit hydrograph (1932); Horton (1933) developed infiltration theory and a description of drainage basin form (1945); Gumbel proposed the extreme value law for hydrologic studies (1941); and Hurst (1951) demonstrated that hydrologic observations may exhibit sequences of low or high values that persist over many years.

1.1.3 Recent Developments in Water Science

During the last six decades "the evolution of hydrologic science has been in the direction of ever-increasing space and time scales, from small catchments to large river basins to the earth system, and from storm event to seasonal cycle to climatic trend" (National Research Council [NRC], 1991).

The United Nations sponsored the International Hydrologic Decade, from 1965 to 1974, as a result of the need for international cooperation to effectively use transnational water resources, and for broad-scale international cooperation to acquire hydrologic data. A primary benefit of this program was to raise consciousness about the regional- and global-scale problems and about human impact on the hydrologic cycle. The realization of human impact has evolved into contemporary views of the interactive role of humans in the hydrologic cycle, in that human activity has become an integral and inseparable part of the hydrologic cycle. The quality of water as it moves through the hydrologic cycle is as important as the quantity, and, in fact, the quality of water can influence important quantity fluxes of the hydrologic cycle (NRC 1991).

The United Nations sponsored the International Drinking Water Supply and Sanitation Decade (1981–1990), which had the goal of providing sanitation services and access to safe drinking water to those without these services. Unfortunately, over a decade later after enormous effort, expense, and progress, 1800 million people in the world are still without access to sanitation services, and nearly 1300 million lack access to clean water (Gleick, 1993).

A somewhat landmark event was the 1992 United Nations Conference on Environment and Development (Earth Summit) held in Rio de Janeiro. This conference focused the attention of world governments and the international public on the new and continuing threats to the global environment. The conference served to highlight the seriousness of the continuing threats to the biosphere, and how massive and complex an undertaking it will be to overcome the threats. The agreements achieved at the Earth Summit are certainly steps in the right direction.

1.2 A PERSPECTIVE ON EARTH AND THE UNIVERSE

1.2.1 The Beginning

Earth is a place that is much different from the rest of the universe. Our present understanding is that the universe began from an explosion unlike anything that we could imagine. This theory of the early universe is basically what we refer to as the *big bang* theory, or what astronomers refer to as the *standard model*. According to Weinberg (1993), at about one-hundredth of a second into the explosion, the temperature of the universe was about a hundred thousand million $(10^{11})°C$. At this temperature none of the components of matter—molecules, atoms, or even the nuclei of atoms—could have held together. The matter spreading out in this explosion consisted of the various elementary particles, such as electrons, positrons, neutrons, and photons. After about one-tenth of a second, the temperature dropped to thirty thousand million $(3 \times 10^{10})°C$ and continued to decrease. At about 14 seconds, the temperature dropped to three thousand million degrees, which was cool enough so that electrons and positrons were destroyed faster than their recreation from photons and neutrinos. By the end of three minutes the temperature reached one thousand million degrees, which was cool enough for the protons and neutrons to start forming into complex nuclei, such as those of hydrogen (deuterium) and helium. By this time (end of the first three minutes), the universe consisted mostly of light, neutrons and antineutrons (Weinberg, 1993). As time went on, the matter continued to rush apart, steadily becoming cooler and less dense, so that after a few hundred thousand years, temperatures were cool enough for electrons to join with nuclei to form atoms of hydrogen and helium. Under the influence of gravitation, these gases began to form clumps, which ultimately condensed to form the galaxies and stars of the present universe (Weinberg, 1993). Our universe indeed is still expanding. Hawking (1988) points out that "time had a beginning of the big bang in the sense that earlier times simply would not be defined." The space-time scales presented in Table 1.1 provide a perspective on the vast ranges of space and time in our universe.

TABLE 1.1   Distance-Time Scales

Light-years	Meters	Distances	Years	Seconds	Times
		? ? ? ? ? ? ? ?			? ? ? ? ? ? ? ?
	10^{27}			10^{18}	Age of universe
		Edge of universe			Age of earth
10^9			10^9		
	10^{24}			10^{15}	
10^6		To nearest neighbor galaxy			Earliest humans
	10^{21}		10^6		
		To center of our galaxy		10^{12}	Age of pyramids
10^3			10^3		
	10^{18}				Age of U.S.
		To nearest star		10^9	Life of a human
1			1		
	10^{15}			10^6	
		Radius of orbit of Pluto			One day
	10^{12}			10^3	Light goes from sun to earth
		To the sun		1	One heart beat
	10^9			10^{-3}	Period of sound wave
		To the moon		10^{-6}	Period of radiowave
	10^6			10^{-9}	Light travels one foot
		Height of a Sputnik		10^{-12}	Period of molecular rotation
	10^3			10^{-15}	Period of atomic vibration
		Height of a TV antenna tower		10^{-18}	Light crosses an atom
	1	Height of a child		10^{-21}	
	10^{-3}				Period of nuclear vibration
		A grain of salt		10^{-24}	Light crosses a nucleus
	10^{-6}				? ? ? ? ? ? ? ?
		A virus			
	10^{-9}				
		Radius of an atom			
	10^{-12}				
	10^{-15}	Radius of a nucleus			
		? ? ? ? ? ? ? ?			

Source: Feynman, et al., 1963.

According to Carl Sagan (1980) and Stephen Hawking (1988), there are some hundred billion galaxies, each on the average having a hundred billion stars. Earth revolves around the sun, which is a star only eight light-minutes away. The next nearest star, called Proxima Centauri, is about four light years away. A *light-year* is defined as the distance light travels in one year (at 186,000 mi/s, about 23 million miles). Most of the stars that we see are a few hundred light-years away. The visible stars that we see at night are concentrated into a band called the Milky Way galaxy, which contains some 400 billion stars. Each of the stars may have planets. Sagan (1980) suggests that there may be as many planets as stars, 10 billion (10^{22}). According to Hawking (1988), our galaxy is about one hundred thousand light-years across and is slowly rotating. The stars in the Milky Way galaxy's spiral orbit about the center of the galaxy approximately once every several hundred million years. Andromeda is the large galaxy nearest to our own.

Galaxies are embedded into clusters of galaxies as our sun is embedded in a cluster of stars. This Local Group of galaxies is several million light years across and consists of some 20 galaxies. As Feynman, et al. (1963) have noted, stars attract each other, as do the galaxies forming the clusters, illustrating that the laws of gravity extend out forever, and inversely as the square of distance. Scientists have concluded that the universe is expanding, as the distance between the different galaxies is growing. Scientists feel that this is one of the great intellectual revolutions of the twentieth century. To put things in perspective, our Earth is such a small part of our own solar system, which in turn is such a small minute part of our galaxy, which in turn is such a small minute part of the universe.

The origin of Earth as a planet began as debris from a stellar explosion condensed about a nucleus. As the mass expanded its gravitational force grew, compressing the mass even further. Earth was initially very hot, and without an atmosphere. As the earth cooled, an atmosphere developed that contained no oxygen, but other gases that could sustain primitive forms of life. A chance combination of atoms grew into large structures (macromolecules) that could assemble other atoms in the oceans into similar structures, reproducing themselves and multiplying. This process of evolution led to the development of more and more complicated organisms that could self-produce. The first primitive forms of life consumed various materials and released oxygen, which eventually changed the atmosphere to its present day composition. This composition allowed the development of higher forms of life, such as fish, reptiles, mammals, and ultimately humans.

1.2.2 Earth–Sun Relationship

As Carl Sagan (1980) wrote,

> Welcome to the planet Earth—a place of blue nitrogen skies, oceans of liquid vapor, cool forests and soft meadows, a world positively rippling with life.

The sun is continually shedding part of its mass by radiating waves of electromagnetic energy and high-energy particles into space. The sun's constant emission represents, in the long run, almost all the energy available to the earth (except for a small portion emanating from radioactive decay of earth minerals). The amount of solar energy is affected by four factors: Solar output, sun–earth distance, sun angle, and length of day. Solar energy originates from nuclear reactions within the sun's hot core, and is transmitted to the sun's surface by radiation and hydrogen convection. Visible solar radiation (light) comes from the cooler outer layer, called the *photosphere*. Temperatures rise in the outer chromosphere and corona, which is continually expanding into space. Outflowing hot gases (plasma) from the sun are referred to as *solar wind*. The solar wind interacts with the earth's magnetic field and upper atmosphere. The sun behaves as a *blackbody*, meaning that it both absorbs all energy received and in turn radiates energy at the maximum rate possible.

Earth has two principal motions: Revolution in an orbit around the sun and rotation on its axis. The earth rotates from west to east, and the path of its orbit around the sun is in the same direction. During the elliptical orbit of the earth, its distance to the sun varies, with a mean distance of 149.6 million kilometers (93 million miles). The revolution of the earth is illustrated in Fig. 1.1 showing the *perihelion* and the *aphelion* which occur on January 3 and July 5, respectively. Perihelion is the shortest distance (147 million km, 91.5 million mi) and aphelion is the longest distance of the earth from the sun (152 million km, 94.5 million mi). One might expect the earth to be warmest when it is closest to the sun; however, the opposite is true for the Northern Hemisphere. Winter in the Northern Hemisphere coincides with the perihelion and summer with the aphelion. This variation of distance between the earth and the sun has only a very minor effect (<3.5 percent) on the variation of seasons, (i.e., receipt of solar energy).

An explanation of the seasons is presented in Fig. 1.1. As the earth revolves around the sun it is inclined at an angle of 66°35′ from the vertical throughout the revolution. During part of the orbit the Southern Hemisphere is pointed toward the sun, and during the other part of the orbit the Northern Hemisphere is pointed toward the sun. The *sun angle* is the angle at which solar radiation hits the earth's atmosphere and surface. The greatest angle possible is 90° at 23°27′N latitude, as shown in Fig. 1.1, and the smallest angle is 0°.

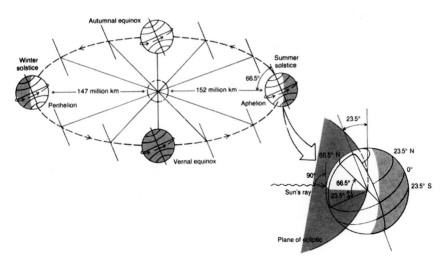

FIGURE 1.1 The revolution of the earth and the seasons. Note that the angle of inclination of the earth's axis is the same in all seasons. *(Marsh, 1987.)*

1.2.3 Energy for Earth

The primary source of energy that drives the earth's processes is the sun, which supplies energy and solar radiation. A secondary source of energy (about 0.002 percent) is the earth's interior, in the form of heat and movement of rocks. Figure 1.2 shows that the total flow of energy toward the earth's surface is equivalent to approximately 172,032,000,000 MW/s, 172 billion from the sun and 32 million from secondary sources. Of the solar radiation that reaches the earth's atmosphere, approximately 60.2 billion MW/s is reflected from the atmosphere; 30.1 billion MW/s is absorbed by the atmosphere; and 81.7 billion MW/s is absorbed by the earth. Basically the 111.83 billion, 30.1 billion, and secondary sources from the earth (0.032 billion MW/s) provide the energy available to drive the processes at the earth's surface. This 111.83 billion MW/s allows photosynthesis and plant growth, induces winds which generate water waves, and evaporates water which becomes rainfall. This 111.83 billion MW/s is also emanated from the earth's surface and atmosphere, along with the 60.2 billion MW/s reflected from the atmosphere, so that the total leaving is 172.032 billion MW/s. The conservation of energy states that there can be no absolute loss of energy within a system; it may change form, as from light to heat, or it may be stored, but it cannot be created or destroyed.

The amount of incoming solar radiation possible for a given latitude is a function of the sun angle and the duration of daylight. Figure 1.3 illustrates the variation in total daily solar radiation at the outer edge of the atmosphere as a function of time of year and latitude of location. Figure 1.4 illustrates the average annual solar energy received at the earth's surface. Values of solar radiation in the equatorial zone are lower than those in the subtropics, because of the more pronounced effect of cloud cover in the equatorial zone.

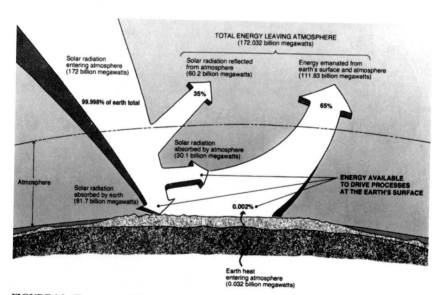

FIGURE 1.2   Energy to and from the earth's surface in megawatts per second. All but 0.002 percent comes from the sun. *(Marsh, 1987.)*

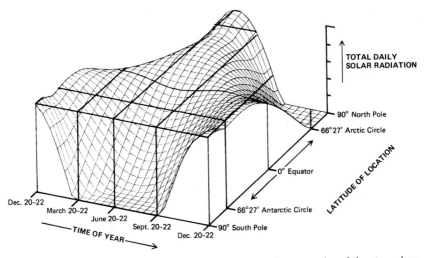

FIGURE 1.3 The variation in total daily solar radiation at the outer edge of the atmosphere. *(Marsh, 1987.)*

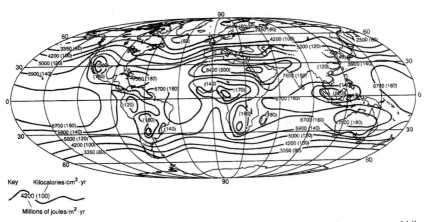

FIGURE 1.4 The worldwide distribution of solar radiation in millions of joules per year and kilocalories per square centimeter per year. Note that the values in the equatorial zone are lower than those in the subtropics, mainly because of differences in cloud cover. *(Marsh, 1987.)*

Solar radiation is absorbed by the atmosphere, the land, and the oceans, and provides energy for the earth's air, water, and biological systems. As previously described, however, this energy is returned to space to maintain an energy balance. The energy from the sun is classified by the *electromagnetic spectrum,* wherein types of radiation are classified by wavelength, because the energy actually travels in the form of electromagnetic waves. Four categories (Marsh, 1987) are: (1) very short wavelength (<0.15 μm) includes gamma rays, x rays, and some ultraviolet radiation, (2) relatively short wavelength (0.15 to 3.0–4.0 μm) includes ultraviolet, visible light, and near infrared radiation, (3) relatively long wavelength (3.0–4.0 to 100 μm) which is infrared radiation, and (4) very long wavelength (>100 μm) which includes radio waves, television waves, and microwaves. Solar radiation mostly includes the categories 2 and 3, which are termed *shortwave* and *longwave,* respectively.

The balance of radiation, taking into account the shortwave and the longwave radiation, is the *net radiation* and defines the radiation flux for a geographic area over a time period. During mornings, net radiation (incoming – outgoing) is typically positive, as the flux of shortwave radiation increases and the landscape heats up. During the late afternoon and evening, it is typically negative, as the shortwave flux declines and heated surfaces emit a larger amount of longwave radiation. The major components of the atmospheric energy balance are the net radiation R, ground (or water) heat flux H_G, sensible heat flux H_S, and latent heat flux H_L, with the energy balance at any time being $R + H_G + H_S + H_L = 0$.

Sensible heat flux is the heat exchange between the surface (ground or water) and the overlying air by conduction and convection. *Latent heat flux* is the heat exchange between the earth's surface and the overlying atmosphere, where moisture is the transfer vehicle. The uneven temporal and spatial distribution of radiation leads to the uneven distribution of temperature in the earth–atmosphere system. Meteorologic and hydrologic processes occur in the redistribution of energy throughout the earth–atmosphere system. The energy transport involves large transfers of mass in the earth–atmosphere system. These transport processes determine the climate and weather processes as we understand them. Water has a tendency to store the heat it receives and land quickly returns it to the atmosphere.

1.2.4   Atmospheric and Ocean Circulation

Atmospheric circulation on Earth is a very complex process which is influenced by many factors. Major influences are differences in heating between low and high altitudes, the earth's rotation, and heat and pressure differences associated with land and water. The general circulation of the atmosphere is due to latitudinal differences in solar heating of the earth's surface and inclination of its axis, distribution of land and water, mechanics of the atmospheric fluid flow, and the Coriolis effect. In general, the atmospheric circulation is thermal in origin and is related to the earth's rotation and global pressure distribution. If the earth were a nonrotating sphere, atmospheric circulation would appear as in Fig. 1.5. Air would rise near the equator and travel in the upper atmosphere toward the poles, then cool, descend into the lower atmosphere, and return toward the equator. This is called *Hadley circulation.*

The rotation of the earth from west to east changes the circulation pattern. As a ring of air about the earth's axis moves toward the poles, its radius decreases. In order to maintain angular momentum, the velocity of air increases with respect to the land surface, thus producing a westerly air flow. The converse is true for a ring of air moving toward the equator—it forms an easterly air flow. The effect producing these changes in wind direction and velocity is known as the *Coriolis force.*

The actual pattern of atmospheric circulation has three cells in each hemisphere, as shown in Fig. 1.6. In the *tropical cell,* heated air ascends at the equator, proceeds toward the poles at upper levels, loses heat, and descends toward the ground at latitude 30°. Near the ground it branches, one branch moving toward the equator and the other toward the pole. In the *polar cell,* air rises at 60° latitude and flows toward the poles at upper levels, then cools, and flows back to 60° near the earth's surface. The *middle cell* is driven frictionally by the other two; its surface flows toward the pole, producing prevailing westerly air flow in the midlatitudes.

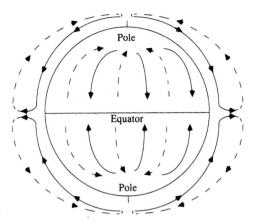

FIGURE 1.5   One-cell atmosphere circulation pattern for a nonrotating planet. *(Chow, et al., 1988.)*

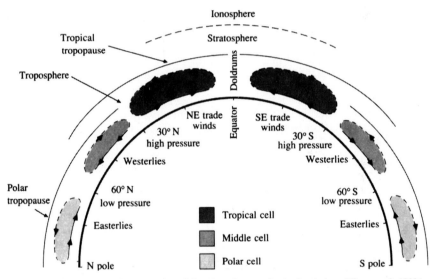

FIGURE 1.6   Latitudinal cross section of the general atmospheric circulation. *(Chow, et al., 1988.)*

The uneven distribution of ocean and land on the earth's surface, coupled with their different thermal properties, creates additional spatial variation in atmospheric circulation. The annual shifting of the thermal equator due to the earth's revolution around the sun causes a corresponding oscillation of the three-cell circulation pattern. With a larger oscillation, exchanges of air between adjacent cells can be more frequent and complete, possibly resulting in many flood years. Also, monsoons may advance deeper into such countries as India and Australia. With a smaller oscillation, intense high pressure may build up around 30° latitude, thus creating extended dry periods. Since the atmospheric circulation is very complicated, only the general pattern can be identified.

The atmosphere is divided vertically into various zones. The atmospheric circulation described above occurs in the *troposphere,* which ranges in height from about 8 km at the poles to 16 km at the equator. The temperature in the troposphere decreases with altitude at a rate varying with the moisture content of the atmosphere. For dry air the rate of decrease is called the *dry adiabatic lapse rate* and is approximately 9.8°C/km. The *saturated adiabatic lapse rate* is less, about 6.5°C/km, because some of the vapor in the air condenses as it rises and cools, releasing heat into the surrounding air. These are average figures for lapse rates that can vary considerably with altitude. The *tropopause* separates the troposphere from the *stratosphere* above. Near the tropopause, sharp changes in temperature and pressure produce strong narrow air currents known as *jet streams* with velocities ranging from 15 to 50 m/s (30 to 100 mi/h). They flow for thousands of kilometers, and have an important influence on air-mass movement.

The oceans exert an important control on global climate. Because water bodies have a high volumetric heat capacity, the oceans are able to retain great quantities of heat. Through wave and current circulation, the oceans redistribute heat to considerable depths and even large areas of the oceans. Redistribution is east–west or west–east, and is also across the midaltitudes from the tropics to the subartic, enhancing the overall poleward heat transfer in the atmosphere. Waves are predominately generated by wind. Ocean circulation is illustrated in Figure 1.7.

Oceans have a significant effect on the atmosphere; however, an exact understanding of the relationships and mechanisms involved are not known. The correlation between ocean temperatures and weather trends and midlatitude events has not been solved. One trend is the growth and decline of a warm body of water in the equatorial zone of the eastern Pacific Ocean, referred to as El Niño (meaning "The Infant" in Spanish, alluding to the Christ Child, because the effect typically begins around Christmas). The warm body of water develops and expands every five years or so off the coast of Peru, initiated by changes in atmospheric pressure resulting in a decline of the easterly trade winds. This reduction in wind reduces resistance, causing the eastwardly equatorial countercurrent to rise. As El Niño builds up, the warm body of water flows out into the Pacific and along the tropical west coast of the Americas, displacing the colder water of the California and Humboldt currents. One of the interesting effects of this weather variation is the South Oscillation, which changes precipitation patterns—

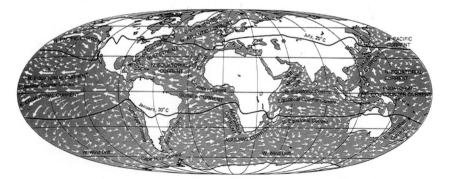

FIGURE 1.7   The actual circulation of the oceans. Major currents are shown with heavy arrows. *(Marsh, 1987.)*

resulting in drier conditions where there would normally be substantial precipitation, and in wetter conditions in areas of normally little precipitation.

1.2.5 Global Climates

The global climate must be viewed as operating within a complex atmosphere-land-ocean-ice system. Climate classification can be made in the form of a genetic classi-fication, as that proposed by Strahler (1969). He considers the three major climates as: (1) low-latitude climates, which are controlled by equatorial and tropical air masses, (2) middle-latitude climates, which are controlled by both tropical and polar air masses, and (3) high-latitude climates, which are controlled by polar and artic air masses. These are subdivided into 15 climatic regions as shown in Fig. 1.8.

1.3 WATER IN THE EARTH ATMOSPHERE SYSTEM

1.3.1 Origin of Water

Venus, Earth, and Mars all have atmospheres with solar-forced circulations. Earth's atmosphere is made up mainly of nitrogen and oxygen, which is controlled by bio-logical processes. The atmospheres on Venus and Mars both have carbon dioxide, controlled by abiotic processes. The clouds on each of these planets, however, have far different constituents—Venus has sulfuric acid, Earth has water, and Mars has dust.

Two classes of theories, evolutionary and genetic, have been used to explain water on Earth. Genetic theory contends that the chemical equilibrium of accreting gas and dust in the solar nebula led to the formation of solid constituents rich in hydrated minerals in Venus, Mars, and Earth. The water in these minerals, and other volatiles, were released to varying degrees over time in the formation of planetary atmospheres. The source of water was the outgassing of water vapor from the earth's interior through the extrusion of material by volcanoes and ocean upwellings over geological time. Once released from the earth's interior, the *juvenile water* con-densed, because the combined temperature and pressure at the earth's surface were ideal for water to exist in liquid form. Venus and Mars had different results. Higher accretion temperatures and tectonic activities on Venus led to outgassing followed by irreversible photodissociation of any water into hydrogen, which escaped to space, and oxygen, which reacted with surface elements. Carbon dioxide created a runaway greenhouse effect, resulting in a dry surface with a temperature of 464°C (National Research Council, 1991). Outgassing on Mars has been limited by lower accretion temperatures and no tectonic activity. There is, however, evidence of sur-face erosion by flowing liquid, possibly water, the source of which is unknown. Mars's atmosphere is thin and cold (−53°C), which has led to seasonal polar caps of frozen carbon dioxide and the possibility of extensive frozen subsurface water (National Research Council, 1991).

The evolutionary theory contends that the planets began with similar volatiles, and that subsequent events led to their current composition. Some of these events may have even included meteorite impacts (National Research Council, 1991).

Earth probably once had a carbon dioxide atmosphere that was reduced by unique processes, such as biological processes. Probably the most unique thing about

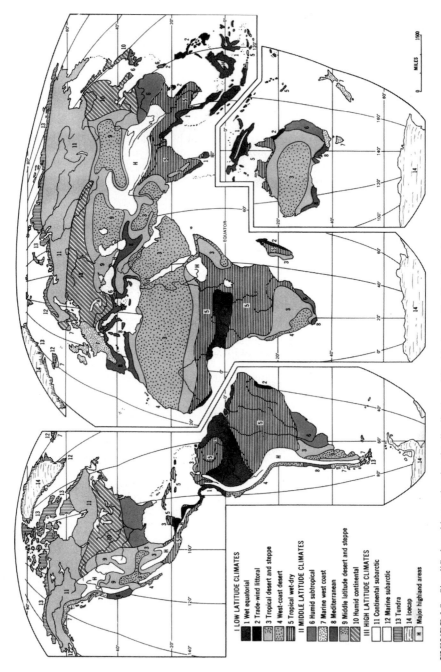

FIGURE 1.8 Simplified world map showing the distribution of Strahler's genetic climatic regions. *(Strahler, 1969.)*

I LOW LATITUDE CLIMATES
1 Wet equatorial
2 Trade-wind littoral
3 Tropical desert and steppe
4 West-coast desert
5 Tropical wet-dry

II MIDDLE LATITUDE CLIMATES
6 Humid subtropical
7 Marine west coast
8 Mediterranean
9 Middle latitude desert and steppe
10 Humid continental

III HIGH LATITUDE CLIMATES
11 Continental subarctic
12 Marine subarctic
13 Tundra
14 Icecap
   Major highland areas

the earth–atmosphere system is the ability for all three phases of water (solid, liquid, and vapor) to coexist, which is certainly unique among the terrestrial planets. Figure 1.9 illustrates the planetary positions on the phase diagram of water.

1.3.2 What Is Water?

The water molecule is a unique combination of hydrogen and oxygen atoms, with electrons being shared between them as shown in Fig. 1.10. The symmetry of the distribution of electrons leaves one side of each molecule with a positive charge, resulting in an electrostatic attraction between molecules. Water molecules can form four such relatively weak hydrogen bonds. The hydrogen, or polar, bonds of water molecules are much weaker than the covalent bonds between hydrogen and oxygen within the molecule. These polar bonds cause water molecules to cluster in tetrahedral patterns, as shown in Fig. 1.11 for ice. In the solid state, the tetrahedral arrangement of the bonding produces a tetrahedral crystalline structure. In the fluid state, increases in temperature weaken the hydrogen bonding.

Ice processes heat energy from the vibration of atoms and molecules in the fixed structure. As ice warms, the vibrations increase to the point where the tetrahedral structure breaks down and the ice melts. The molecules of the liquid phase are closer than in the solid state, as illustrated in Fig. 1.11, making water slightly more dense than ice at its melting point. Molecules of water in the liquid phase vibrate faster as temperature rises. Once the vibrations are great enough, some molecules are thrown off (or escape) the liquid surface, forming a gaseous or vapor phase, called evaporation. This evaporation consumes a large amount of energy, called *latent heat of vaporization*. The phase changes for water are: (1) *evaporation*—liquid to vapor, (2) *condensation*—vapor to liquid, (3) *sublimation*—vapor to solid or solid to vapor,

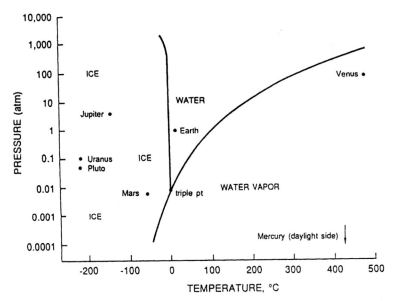

FIGURE 1.9 Planetary positions on the phase diagram of water. (*National Research Council, 1991.*)

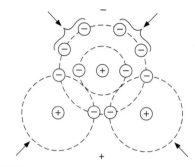

FIGURE 1.10   The Water Molecule. *(After Sutcliffe, 1968.)*

(4) melting—solid to liquid, and (5) freezing—liquid to solid.

The physical properties of water are unique compared to substances with similar molecular mass. Water has the highest specific heat of any known substance, which means that temperature change within it occurs very slowly. Water has a high viscosity and a high surface tension compared to most common liquids, which is caused by the hydrogen bonding. This produces capillary rise of water in soils and causes rain to form into spherical droplets. Physical properties of water in the solid and liquid phases vary with temperature. In these states the variation in density differs more significantly than in most liquids. Water in the gaseous phase (*water vapor*) exerts a partial pressure in the atmosphere, referred to as its *vapor pressure*. In the atmosphere above a liquid water sur-

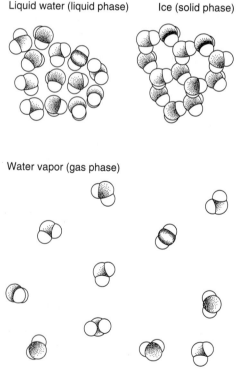

FIGURE 1.11   The three states of water. *(After Sutcliffe, 1968.)*

face, water molecules are constantly being exchanged between the air and the water. For a drier atmosphere, the rate of uptake of molecules is greater than the rate of return to the surface. At a state of equilibrium, when the number of molecules leaving the surface is equal to the number arriving, saturation of the vapor pressure of air has been reached. Additional water molecules to the air are balanced by deposition on the water surface. Table 1.2 lists some common physical constants of pure water. The latent heat of vaporization is about 8 times larger than is necessary to melt ice, and about 600 times larger than its heat capacity (the energy necessary to raise water temperature by 1°C). Evaporation is then the dominant component of energy balance in the hydrologic cycle. About 23 percent of the solar radiation reaching the earth is absorbed by evaporating water (Maidment, 1993). The latent heat of water is larger than for any other liquid.

1.3.3 Earth's Hydrologic Cycle

The National Research Council (1991) report defines the *hydrologic cycle* as "the pathway of water as it moves in its various phases through the atmosphere, to the Earth, over and through the land, to the ocean, and back to the atmosphere" as shown in Fig. 1.12. During the cycle, which has no beginning or end, a single water molecule may assume various states, returning to the hydrologic pathway as new chemical compounds are mixed with various solid and liquid substances. As shown in the figure, water evaporates from the oceans and the land surface to become part of the atmosphere; water vapor is transported and lifted in the atmosphere until it condenses and precipitates on the land or oceans; precipitated water may be intercepted by vegetation, become overland flow over the ground surface, infiltrate into the ground, flow through the soil as subsurface flow, and discharge into streams as surface runoff. Large amounts of the intercepted water and surface runoff returns to the atmosphere through evaporation. Infiltrated water may percolate deeper to recharge groundwater, and later emerge in springs, or seepage into streams, to form surface runoff. Finally, this water may flow out to the sea or evaporate into the atmosphere. Throughout this cycle, water may take on many quality aspects.

The hydrologic cycle can also be viewed on a global scale, as shown in Fig. 1.13. Our knowledge of the amount of water in space and in the earth's mantle is very limited. There is evidence that space and the earth's mantle both exchange water with the primary crustal, ice, the atmosphere, and the ocean. The hydrologic cycle can also be viewed as a global geophysical process, as shown in Fig. 1.14. Water vapor and methane molecules are diffused into space, causing loss of the hydrogen in water. These hydrogen atoms subsequently escape by photochemistry. The addition of

TABLE 1.2 Physical Constants of Pure Water

Specific heat, 15°C	4.18 J g^{-1} deg^{-1}
Latent heat of melting	334.4 J g^{-1}
Latent heat of vaporization, 15°C	2462 J g^{-1}
Surface tension	7340 mN m^{-2} cm^{-1}
Tensile strength	1418.5 kN m^{-2} cm^{-2}
Melting point, 1013 mb	0°C
Boiling point, 1013 mb	100°C

Source: Sutcliffe, 1968.

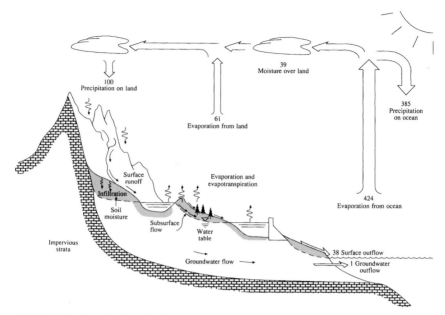

FIGURE 1.12   Hydrologic cycle with global annual average water balance given in units relative to a value of 100 for the rate of precipitation on land. *(Chow, et al., 1988.)*

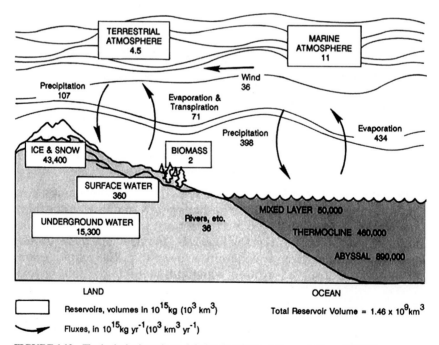

FIGURE 1.13   The hydrologic cycle at global scale *(National Research Council, 1986.)*

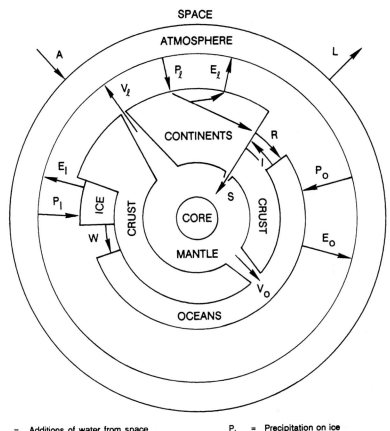

FIGURE 1.14 The hydrologic cycle as a global geophysical process. Enclosed areas represent storage reservoirs for the earth's water, and the arrows designate the transfer fluxes between them. *(National Research Council, 1991.)*

water from space is a controversial issue. Volcanic activity vents water vapor to the atmosphere and liquid vapor to the ocean. Water recirculates on a geological time scale by the subduction of water-containing crustal material.

This fascinating phenomena called the hydrological cycle is being changed by human activities. We have begun to realize the effects on nature and our environment brought about by these changes in the hydrologic cycle. This realization is changing our contemporary views of the interactive role of people in the hydrologic cycle. Figure 1.15 illustrates the classical and modern view points of the role of people in the hydrologic cycle. Human activities are an integral and inseparable part of

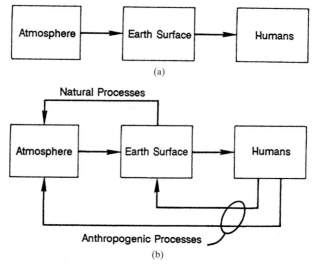

FIGURE 1.15 The role of humans in the hydrologic cycle. (*a*) classical viewpoint; (*b*) modern viewpoint. (*National Research Council, 1982.*)

the hydrologic cycle. One of our most important realizations is that the quality of water in this cycle is of as much concern as the quantity.

1.4 AVAILABILITY OF WATER ON EARTH

Figure 1.16 illustrates the variation in average annual precipitation for the world's land areas. Data on global water resources are presented in Table 1.3. The oceans contain 96.5 percent of the water on Earth, whereas freshwater reserves are only 2.53 percent (or 35 million km³) of the total 1.384 billion km³. A large fraction of the freshwater (24 million km³, or 68.7 percent) is ice and permanent snow cover in the Antarctic and Arctic region. The main sources of water for human consumption, freshwater lakes and rivers, contain on the average about 90,000 km³ of water (0.26 percent of the total global freshwater reserves). The atmosphere contains only 12,900 km³, which is 0.001 percent of the total water, or 0.04 percent of the freshwater.

It is also of interest to review the data for annual runoff and water consumption by physiographic and economic regions of the world, as listed in Table 1.4. The total water withdrawn for use in 1990 was 9.3 percent of the total surface runoff. Unrecoverable consumptive use was 5.2 percent. By year 2000, these values could be 11.6 percent and 6.5 percent respectively (Shiklomanov, 1993).

The dynamics of actual water availability in different regions of the world is particularly interesting in understanding the water balance on Earth. As illustrated in Table 1.5, during the 30-year period from 1950 to 1980, the actual level of per capita water supply decreased rather significantly in many regions of the world, due to population increases. Significant impacts were in North Africa, North China and Mongolia, Central Asia, and Kazakhstan. In addition to the regions listed above, by year 2000 low water availability per capita is anticipated in central and southern Europe,

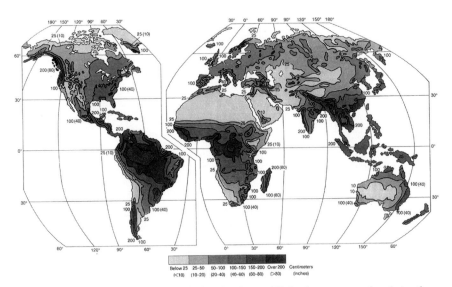

FIGURE 1.16 Average annual precipitation for the world's land areas, excepting Antarctica. *(Marsh, 1987.)*

TABLE 1.3 Water Reserves on Earth

	Distribution area, 10^3 km^2	Volume, 10^3 km^3	Layer, m	Percentage of global reserves	
				Of total water	Of fresh-water
World ocean	361,300	1,338,000	3,700	96.5	—
Groundwater	134,800	23,400	174	1.7	—
Freshwater		10,530	78	0.76	30.1
Soil moisture		16.5	0.2	0.001	0.05
Glaciers and permanent snow cover	16,227	24,064	1,463	1.74	68.7
Antarctic	13,980	21,600	1,546	1.56	61.7
Greenland	1,802	2,340	1,298	0.17	6.68
Arctic islands	226	83.5	369	0.006	0.24
Mountainous regions	224	40.6	181	0.003	0.12
Ground ice/permafrost	21,000	300	14	0.022	0.86
Water reserves in lakes	2,058.7	176.4	85.7	0.013	—
Fresh	1,236.4	91	73.6	0.007	0.26
Saline	822.3	85.4	103.8	0.006	—
Swamp water	2,682.6	11.47	4.28	0.0008	0.03
River flows	148,800	2.12	0.014	0.0002	0.006
Biological water	510,000	1.12	0.002	0.0001	0.003
Atmospheric water	510,000	12.9	0.025	0.001	0.04
Total water reserves	510,000	1,385,984	2,718	100	—
Total freshwater reserves	148,800	35,029	235	2.53	100

Source: Shiklomanov, 1993.

TABLE 1.4 Annual Runoff and Water Consumption by Continents and by Physiographic and Economic Regions of the World

Continent and region	Mean annual runoff mm	Mean annual runoff km³/yr	Aridity index, R/LP	Water consumption, km³/yr 1980 Total	1980 Irretrievable	1990 Total	1990 Irretrievable	2000 Total	2000 Irretrievable
Europe	310	3,210	0.6	435	127	555	178	673	222
North	480	737	0.7	9.9	1.6	12	2.0	13	2.3
Central	380	705	1.4	141	22	176	28	205	33
South	320	564	0.7	132	51	184	64	226	73
European USSR (North)	330	601	1.5	18	2.1	24	3.4	29	5.2
European USSR (South)	150	525		134	50	159	81	200	108
North America	340	8,200	0.8	663	224	724	255	796	302
Canada and Alaska	390	5,300	1.5	41	8	57	11	97	15
United States	220	1,700	1.2	527	155	546	171	531	194
Central America	450	1,200		95	61	120	73	168	93
Africa	150	4,570	8.1	168	129	232	165	317	211
North	17	154	2.5	100	79	125	97	150	112
South	68	349	2.2	23	16	36	20	63	34
East	160	809	2.5	23	18	32	23	45	28
West	190	1,350	2.5	19	14	33	23	51	34
Central	470	1,909	0.8	2.8	1.3	4.8	2.1	8.4	3.4

Asia	330	14,410	2.2	1,910	1,380	2,440	1,660	3,140	2,020
North China and Mongolia	160	1,470	1.3	395	270	527	314	677	360
South	490	2,200	2.7	668	518	857	638	1,200	865
West	72	490	0.7	192	147	220	165	262	190
Southeast	1,090	6,650	3.1	461	337	609	399	741	435
Central Asia and Kazakhstan	70	170	0.9	135	87	157	109	174	128
Siberia and Far East	230	3,350	1.2	34	11	40	17	49	25
Trans-Caucasus	410	77		24	14	26	18	33	21
South America	660	11,760	0.6	111	71	150	86	216	116
Northern area	1,230	3,126	0.7	15	11	23	16	33	20
Brazil	720	6,148	1.3	23	10	33	14	48	21
West	740	1,714	2.0	40	30	45	32	64	44
Central	170	812		33	20	48	24	70	31
Australia and Oceania	270	2,390	4.0	29	15	38	17	47	22
Australia	39	301	0.6	27	13	34	16	42	20
Oceania	1,560	2,090		2.4	1.5	3.3	1.8	4.5	2.3
Land area (rounded off)		44,500		3,320	1,450	4,130	2,360	5,190	2,900

Source: Shiklomanov, 1993.

the southern European part of the former Soviet Union, Southeast Asia, and West, East, and South Africa (Shiklomanov, 1993). The very high natural nonuniformity in the distribution of water supply throughout the earth is increasing with time, as a result of the extremely rapid rate of human economic activities and population change. Data for year 2000 presented in Table 1.5 does not consider the possible anthropogenic global-scale climatic changes through the year 2000.

1.5   INGREDIENTS FOR WATER RESOURCES

The management of water resources can be subdivided into three broad categories: (1) *water-supply management,* (2) *water-excess management,* and (3) *environmental restoration.* All modern multipurpose water resources projects are designed and built for water-supply management and/or water-excess management. In fact,

TABLE 1.5   Dynamics of Actual Water Availability in Different Regions of the World

Continent and region	Area, 10^6 km²	Actual water availability, 10^3 m³/yr per capita				
		1950	1960	1970	1980	2000
Europe	10.28	5.9	5.4	4.9	4.6	4.1
North	1.32	39.2	36.5	33.9	32.7	30.9
Central	1.86	3.0	2.8	2.6	2.4	2.3
South	1.76	3.8	3.5	3.1	2.8	2.5
European USSR (North)	1.82	33.8	29.2	26.3	24.1	20.9
European USSR (South)	3.52	4.4	4	3.6	3.2	2.4
North America	24.16	37.2	30.2	25.2	21.3	17.5
Canada and Alaska	13.67	384	294	246	219	189
United States	7.83	10.6	8.8	7.6	6.8	5.6
Central America	2.67	22.7	17.2	12.5	9.4	7.1
Africa	30.10	20.6	16.5	12.7	9.4	5.1
North	8.78	2.3	1.6	1.1	0.69	0.21
South	5.11	12.2	10.3	7.6	5.7	3.0
East	5.17	15.0	12	9.2	6.9	3.7
West	6.96	20.5	16.2	12.4	9.2	4.9
Central	4.08	92.7	79.5	59.1	46.0	25.4
Asia	44.56	9.6	7.9	6.1	5.1	3.3
North China and Mongolia	9.14	3.8	3.0	2.3	1.9	1.2
South	4.49	4.1	3.4	2.5	2.1	1.1
West	6.82	6.3	4.2	3.3	2.3	1.3
Southeast	7.17	13.2	11.1	8.6	7.1	4.9
Central Asia and Kazakhstan	2.43	7.5	5.5	3.3	2.0	0.7
Siberia and Far East	14.32	124	112	102	96.2	95.3
Trans-Caucasus	0.19	8.8	6.9	5.4	4.5	3.0
South America	17.85	105	80.2	61.7	48.8	28.3
North	2.55	179	128	94.8	72.9	37.4
Brazil	8.51	115	86	64.5	50.3	32.2
West	2.33	97.9	77.1	58.6	45.8	25.7
Central	4.46	34	27	23.9	20.5	10.4
Australia and Oceania	8.59	112	91.3	74.6	64.0	50.0
Australia	7.62	35.7	28.4	23	19.8	15.0
Oceania	1.34	161	132	108	92.4	73.5

Source: Shiklomanov, 1993.

throughout human history all water resources projects have been designed and built for one or both of these categories. A *water resources system* is a system for redistribution, in space and time, the water that is available to a region to meet societal needs (Plate, 1993). Water can be utilized from *surface water systems,* from *groundwater systems,* or from *conjunctive/ground surface water systems.* When discussing water resources, we must consider both the quantity and the quality aspects. As pointed out earlier in this chapter, the hydrologic cycle must be defined in terms of both the water quantity and the water quality. Because of the very complex water issues and problems that we face today, many fields of study are involved in the solution of these problems. These include the biological sciences, engineering, physical sciences, and social sciences. Figure 1.17 attempts to present the wide diversity of disciplines involved in water resources.

1.6 FUTURE DIRECTIONS FOR WATER RESOURCES

1.6.1 Water Development

As we approach the twenty-first century, we are questioning the viability of our patterns of development, industrialization and resources usage. We are now beginning to discuss the goals of attaining an equitable and sustainable society in the international community. Looking into the future, a new set of problems face us, including the rapidly growing population in developing countries, uncertain impacts of global climate change, possible conflicts over shared freshwater resources, thinning of the

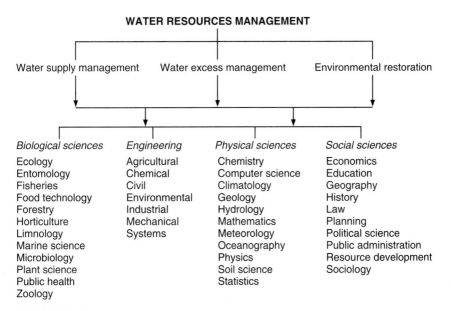

FIGURE 1.17   Ingredients for water resources management.

ozone layer, destruction of rain forests, threats to wetlands, farmland, and other renewable resources, and many others.

During the 1980s, the United Nations sponsored the International Water Supply and Sanitation Decade. Data in Table 1.6 present the water supply and sanitation coverage for developing regions for 1980 and 1990. The striking fact is that four-fifths of the world's population is covered by this table, including 100 percent of the population of developing countries. Many factors interfere with the improvement of water supplies and the provision of sanitation services in developing countries. Some constraints for developing countries include an insufficient number of trained professionals, insufficient funding, inadequate operation and maintenance, ineffective logistics, inadequate cost-recovery framework, inappropriate institutional framework, insufficient health education effects, intermittent water service, and lack of planning and design.

The future population will have a direct and very significant impact on future water availability, use, and quality. By year 2000 the total population on Earth may exceed 6 billion people, and by year 2050, possibly 10 billion, according to estimates by the United Nations. In discussing the impacts of population it must be recognized that the total population and growth rates are very different in developing countries and developed countries. For developed countries, the rate is under 1 percent per year, whereas it exceeds 2 percent per year in developing countries, and in some parts of Africa, Asia, and the Middle East, it exceeds 3 percent per year (Gleick, 1993). The consequence is that over 90 percent of future population increases will be in developing countries, where access to clean water, sanitation services, and other amenities for a satisfactory quality of life are inadequate.

Over the past decade, the total population in urban areas grew tremendously, due to massive migrations to the larger urban areas in developing countries. Most of these urban areas have never had adequate clean water and sanitation services. There is no doubt that during the next decade, as the population-growth rate in these urban areas in developing countries increases, the situation will worsen. Because the total amount of freshwater is fixed, these growing populations will continuously reduce the water available per capita (see Table 1.5).

It is interesting to note that throughout the world, poor rural women spend 60 to 90 hours per week gathering wood, collecting water, preparing food, and caring for children (Gleick, 1993). The United Nations has shown that the education of women can improve child health, and often leads to improved availability of water and sanitation. Educating people about family planning and public health can also be effective in tackling the future problems of water availability. The world's population must be stabilized, as it cannot continue to grow indefinitely. At the same time, we must work to reduce the enormous suffering caused by what already exists.

Al Gore (1992), in his book *Earth in the Balance,* proposed a Global Marshall Plan which includes the following five goals to save the global environment:

1. Stabilizing the world population

2. The rapid creation and development of environmentally appropriate technologies, also referred to as a Strategic Environment Initiative (SEI)

3. A comprehensive and ubiquitous change in the economic rules of the road by which we measure the impact of our decisions on the global environment—a new global economics

4. The negotiation and approval of a new generation of international agreements

5. The establishment of a cooperative plan for educating the world's citizens about our global environment

TABLE 1.6 Water Supply and Sanitation Coverage for Developing Regions, 1980 and 1990

Region/sector	1980				1990			
	Population, 10^6	Percent coverage	Number served, 10^6	Number unserved, 10^6	Population, 10^6	Percent coverage	Number served, 10^6	Number unserved, 10^6
Africa								
Urban water	119.77	83	99.41	20.36	202.54	87	176.21	26.33
Rural water	332.83	33	109.83	223.00	409.64	42	172.06	237.59
Urban sanitation	119.77	65	77.85	41.92	202.54	78	160.01	42.53
Rural sanitation	332.83	18	59.91	272.92	409.64	26	106.51	303.13
Latin America and the Caribbean								
Urban water	236.72	82	194.11	42.61	324.08	87	281.95	42.13
Rural water	124.91	47	58.71	66.20	123.87	62	76.80	47.07
Urban sanitation	236.72	78	184.64	52.08	324.08	79	256.02	68.06
Rural sanitation	124.91	22	27.48	97.43	123.87	37	45.83	78.04
Asia and the Pacific								
Urban water	549.44	73	401.09	148.35	761.18	77	586.11	175.07
Rural water	1,823.30	28	510.52	1,312.78	2,099.40	67	1,406.60	692.80
Urban sanitation	549.44	65	357.14	192.30	761.18	65	494.77	266.41
Rural sanitation	1,823.30	42	765.79	1,057.51	2,099.40	54	1,133.68	965.72
Western Asia (Middle East)								
Urban water	27.54	95	26.16	1.38	44.42	100	44.25	0.17
Rural water	21.95	51	11.19	10.76	25.60	56	14.34	11.26
Urban sanitation	27.54	79	21.76	5.78	44.42	100	44.42	0.00
Rural sanitation	21.95	34	7.46	14.49	25.60	34	8.70	16.90
Totals for these regions								
Urban water	933.47	77	720.77	212.70	1,332.22	82	1,088.52	243.70
Rural water	2,302.99	30	690.25	1,612.74	2,658.51	63	1,669.79	988.72
Urban sanitation	933.47	69	641.39	292.08	1,332.23	72	955.22	377.00
Rural sanitation	2,302.99	37	860.64	1,442.35	2,658.51	49	1,294.72	1,363.79

Source: Gleick, 1993.

These five goals are all interrelated, so they should be pursued simultaneously. An integrating goal would be "the establishment, especially in the developing world, of the social and political conditions most conducive to the emergence of sustainable societies."

1.6.2 Research Directions

The Carnegie Commission on Science, Technology, and Government (1992) defined environmental research as directed to maintaining environmental quality, including monitoring, testing, evaluation, prevention, mitigation, assessment, and policy analysis. Their definition includes:

- Investigations designed to understand the structure and function of the biosphere, and the impact that human activities have on it
- Research to understand the conditions necessary to support human existence without destroying the resource base
- Research to define the properties and adverse effects of toxic substances on human health and the environment
- The development of technologies to monitor pollutants and their impacts
- The development of pollution-control technologies
- The economic and social research directed at understanding the many complex, interrelated factors that influence environmental quality

   The Commission concluded that the present research and development system in the United States has basically been a "catch up, clean up" dominated approach. They felt that the research and development system is diffuse, reactive, and focused on short-range, end-of-the-pipe solutions. The mechanism to coordinate and integrate the research products are weak. In the future there must be more effort put into: (1) environmental biology, (2) interdisciplinary studies, (3) understanding ecological processes, (4) understanding the interrelation of land, water, and biota in landscapes, and (5) there needs to be better integration of economic, social, and political studies of environmental issues with the natural sciences.
   The Committee on Opportunities in Hydrologic Sciences of the NRC (1991) developed priority categories of scientific opportunity under the premises that: "(1) the largest potential for such contribution lies in the least explored scales and in making the linkages across scales, and (2) hydrologic science is currently data-limited." The unranked research areas of highest priority are: (1) chemical and biological components of the hydrologic cycle, (2) scaling of dynamic behavior, (3) land surface–atmosphere interaction, (4) coordinated global-scale observation of water reservoirs and the fluxes of water and energy, and (5) hydrologic effects of human activity.
   The research area of chemical and biological components of the hydrologic cycle includes:

- Understanding the interaction between ecosystems and the hydrologic cycle
- Understanding the pathways of water through soil and rock through the use of aqueous geochemistry to reveal the historical states for climate research, and to reconstruct the erosional history of continents

- Combining efforts in aquatic chemistry, microbiology, and physics of flow to reveal solute transformation, biochemical functioning, and the mechanism for both contamination and purification of soils and water

Scaling of dynamic behavior involves research:

- To quantify predictions of large-scale hydrologic processes under the three-dimensional heterogeneity of natural systems, which are orders of magnitude larger in scale than idealized one-dimensional laboratory conditions
- To quantify the inverse problem by disaggregating conditions at large scale to obtain small scale information, e.g. in the parameterization of subgrid-scale processes in climate models

Understanding land surface–atmosphere interactions has become somewhat urgent, because of the accelerating human-induced changes in land surface characteristics globally, on issues ranging from the mesoscale upward to continental scales. A better understanding of the following are needed:

- Our knowledge of the time and space distribution of rainfall, soil moisture, groundwater recharge, and evapotranspiration
- Knowledge of the variability and sensitivity of local and regional climates to alterations in land surface properties

Coordinated global-scale observation of water reservoirs and the fluxes of water and energy is needed for a better understanding of the state and variability of the global water balance. Two programs that will help in this effort are the World Climate Data Program (WCDP) to assemble historical and current data, and the World Climate Research Program (WCRP), which is planning a global experimental program to place future observations on a sound and coordinated effort, called the Global Energy and Water Cycle Experiment (GEWEX).

The Global Energy and Water Cycle Experiment (GEWEX), proposed to begin in the late 1990s, is designed to verify large-scale hydrologic models and to validate global-scale satellite observations. This initiative of the World Climate Research Program addresses four scientific objectives:

1. Determine water and energy fluxes by global measurements of observable atmosphere and surface properties.
2. Model the hydrologic cycle and its effects on the atmosphere and ocean.
3. Develop the ability to predict variations of global and regional hydrologic processes and water resources and their response to environmental change.
4. Foster the development of observing techniques, and data management and assimilation systems suitable for operational applications to long-range weather forecasting and to hydrologic and climate predictions.

A central goal of the GEWEX program is to develop and improve modeling of hydrologic processes, and to integrate surface and ground water processes on the catchment scale into fully interactive global land–atmosphere models.

Hydrologic effects of human activity research should focus on the quantitative forecasts of anthropogenic hydrologic change, which is largely indistinguishable from the temporal variability of the natural system.

In summary, as asked by the Carnegie Commission (1992), "Can scientists and engineers generate the kind of large-scale and highly focused effort that took us into space and apply it to developing the understanding necessary to protect our global environment?" An international effort will be required to meet the environmental challenges that we face today. Obtaining a sustainable development will require a wide range of research advances.

REFERENCES

Adams, J., *Sewers and Drains for Populous Districts,* Van Nostrand, New York, 1880.

Barry, R. G., and R. J. Charley, *Atmosphere, Weather, and Climate,* 6th ed., Routledge, London and New York, 1992.

Buras, N., *Scientific Allocation of Water Resources,* Elsevier, New York, 1972.

Carnegie Commission on Science, Technology, and Government, *Environmental Research and Development, Strengthening the Federal Infrastructure,* Task Force on the Organization of Federal Government R & D Program, New York, 1992.

Chow, V. T., D. R. Maidment, and L. W. Mays, *Applied Hydrology,* McGraw-Hill, 1988.

deCamp, L. S., *The Ancient Engineers,* Dorset Press, New York, 1963.

Dalton, J., "Experimental Essays on the Constitution of Mixed Gases; on the Force of Steam or Vapor from Waters and Other Liquids, Both in a Torricellian Vacuum and in Air; on Evaporation; and on the Expansion of Gases by Heat," *Mem. Proc. Man. Lit. Phil. Soc.,* 5:535–602, 1802.

Darcy, H., *Les Fontaines Publiques de la Ville de Dijon,* V. Dalmont, Paris, 1856.

Frazier, A. H., "Water Current Meters," *Smithsonian Studies in History and Technology,* no. 28, Smithsonian Institution Press, Washington, D.C., 1974.

Gleick, P. (ed.), *Water in Crisis: A Guide to the World's Fresh Water Resources,* Oxford University Press, Oxford, New York, 1993.

Gore, A., *Earth in the Balance,* Houghton Mifflin, New York, 1992.

Green, W. H., and G. A. Ampt, "Studies on Soil Physics," *J. Agric. Sci.,* vol. 4, pt. 1, pp. 1–24, 1911.

Feynman, R. P., R. B. Leighton, and M. Sands, *The Feynman Lecture Notes on Physics, Vol. I.,* Addison-Wesley, Reading, Mass., 1963.

Hagen, G. H. L., "Über die Bewegung des Wassers in Engen Cylindrischen Rohren," *Poggendorfs Annalen der Physik und Chemie,* 16, 1839.

Hawking, S. H., *A Brief History of Time: From the Big Bang to Black Holes,* Bantam Books, New York, 1988.

Hazen, A., "Storage to be Provided in Impounding Reservoirs for Municipal Water Supply," *Trans. Am. Soc. Civ. Eng.,* 77:1539–1640, 1914.

Horton, R. E., "The Role of Infiltration in the Hydrologic Cycle," *Trans. Am. Geophys. Union,* 14:446–460, 1933.

Hurst, H. E., "Long-Term Storage Capacity of Reservoirs," *Trans. Am. Soc. Civ. Eng.,* 116, paper no. 2447, pp. 770–799, 1951.

MacCurdy, E., *The Notebooks of Leonardo da Vinci,* vol. 1, Reynal and Hitchcock, New York, 1939.

Maidment, D. R. (ed.), *Handbook of Hydrology,* McGraw-Hill, New York, 1993.

Manning, R., "On the Flow of Water in Open Channels and Pipes," *Trans. Inst. Civ. Eng. Ireland,* 20:161–207, 1891; supplement 24:179–207, 1895.

Marsh, W. M., *Earthscape: A Physical Geography,* John Wiley, New York, 1987.

Marsh, W. M., and J. Dozier, *Landscape: An Introduction to Physical Geography,* John Wiley, New York, 1986.

Mulvaney, T. J., "On the Use of Self-Registering Rain and Flood Gauges in Making Observations of the Relations of Rainfall and of Flood Discharges in a Given Catchment," *Proc. Inst. Civ. Eng. Ireland,* 4:18–31, 1850.

National Research Council, *Opportunities in the Hydrologic Sciences,* National Academy Press, Washington, D.C., 1991.

National Research Council, *Scientific Basis of Water-Resource Management,* National Academy Press, Washington, D.C., 1982.

National Research Council, *Global Change in the Geosphere-Biosphere,* National Academy Press, Washington, D.C., 1986.

Needham, J., *Science and Civilization in China, Vol. I,* Cambridge University Press, Cambridge, England, 1954.

Plate, E. J., "Sustainable Development of Water Resources: A Challenge to Science and Engineering," *Water International,* International Water Resources Association, 18,(2):84–94, June 1993.

Richards, L. A., "Capillary Conduction of Liquids through Porous Mediums," *Physics. A Journal of General and Applied Physics,* American Physical Society, Minneapolis, Minn., vol. 1, pp. 318–333, July–Dec. 1931.

Rippl, W., "Capacity of Storage Reservoirs for Water Supply," *Minutes of Proceedings, Institution of Civil Engineers,* 70:270–278, 1883.

Rouse, H., and S. Ince, *History of Hydraulics,* Dover, New York, 1957.

Sagan, C., *Cosmos,* Random House, New York, 1980.

Sarton, G., *A History of Science,* Harvard University Press, Cambridge, Mass., 1952–59.

Sherman, L. K., "Streamflow from Rainfall by the Unit-Graph Method," *Eng. News Rec.,* 108:501–505, 1932.

Shiklomanov, I., "World Fresh Water Resources," Chap. 2 in P. Gleick (ed.), *Water in Crisis,* Oxford University Press, Oxford, New York, 1993.

Strahler, A. N., *Physical Geography,* 3d ed., John Wiley, New York, 1969.

Stahley, C., and G. S. Pierson, *The Separate System of Sewage,* Van Nostrand, New York, 1899.

Sutcliffe, J., *Plants and Water,* Edward Arnold, London, 1968.

Weinberg, S., *The First Three Minutes,* 2d ed., Basic Books, New York, 1993.

White, I. D., D. N. Mottershead, and S. J. Harrison, *Environmental Systems,* 2d ed., Chapman and Hall, London, 1992.

CHAPTER 2
PRINCIPLES OF FLOW OF WATER

M. Hanif Chaudhry

Department of Civil and Environmental Engineering
Washington State University
Pullman, Washington

2.1 INTRODUCTION

In this chapter, principles of flow of water necessary for the analysis of water resource projects are briefly reviewed. A number of commonly used terms are first defined. Closed-conduit, open-channel and subsurface flows are then discussed. The data compiled from various sources which practicing engineers will find useful are presented. The material in this chapter has been intentionally kept free of advanced mathematics and fluid mechanics concepts. For additional coverage of closed conduits, see Chap. 18; for open channels, see Chap. 25; and for subsurface flow, refer to Chaps. 11 and 16.

2.2 DEFINITIONS

2.2.1 Fluid Properties

The *density* ρ of a fluid (Roberson and Crowe, 1993) is its mass per unit volume, while the *specific weight* γ is its weight per unit volume. The density and specific weight are related by the equation

$$\gamma = \rho g \qquad (2.1)$$

in which g = acceleration due to gravity. In SI units, ρ is expressed in kg/m^3, and in customary English units, it is expressed in slugs/ft^3. The units for specific weight in SI units are N/m^3 and in English units are lbs/ft^3.

The *specific gravity* of a substance is the ratio of its mass density to that of pure water at standard conditions. Physicists use 4°C (39.2°F) as the standard temperature, while engineers ordinarily use 15°C (60°F). The density of water at 4°C is 1000 kg/m³, and it changes slightly with change in temperature.

An *ideal fluid* may be defined as one in which there is no friction, i.e., viscosity is zero. In a *real fluid*, shear force exists whenever motion takes place, thus producing fluid friction. An ideal fluid does not exist in reality, but the concept is useful in simplifying many analyses. The *viscosity* of a fluid is a measure of its resistance to shear or angular deformation. If du/dy is the velocity gradient and τ is the shearing stress between any two thin sheets of fluid, then

$$\tau = \mu \frac{du}{dy} \tag{2.2}$$

The coefficient μ is called the *absolute* or *dynamic viscosity* and its units are poise. The *kinematic viscosity* v is defined as the dynamic viscosity μ divided by the mass density ρ, i.e.,

$$v = \frac{\mu}{\rho} \tag{2.3}$$

and is expressed in stokes (v is called kinematic viscosity, since no force units are involved).

At a gas–liquid interface, forces develop at the liquid surface due to molecular attraction, and this causes the liquid surface to behave like a stretched membrane. The intensity of this molecular attraction per unit length is called *surface tension.* In SI units, surface tension is expressed in N/m and in customary English units, in lb/ft.

The *compressibility* of water is characterized by the bulk modulus of elasticity K which is defined as the ratio of relative change in volume due to a differential change in pressure. The SI units for K are N/m² and the customary English units are lb/in² (psi).

2.2.2 Pressure

Fluid pressure p is the force exerted on a unit area. Commonly used English units for pressure are lbs/in² (psi), lbs/ft², feet of water, and inches of mercury. In SI units, pressure is expressed in pascal, equal to 1 N/m². Neglecting the pressure on the surface of a liquid, the pressure at depth h is

$$p = \gamma h \tag{2.4}$$

The pressure at a point is equal in all directions.

The pressure measured by using atmospheric pressure as the datum is called *gauge pressure.* However, if it is measured above absolute zero, then it is termed absolute pressure. When the pressure is less than the atmospheric pressure, it is termed vacuum, and its gauge value is the amount by which it is below the atmospheric pressure. A perfect vacuum corresponds to absolute zero pressure. Figure 2.1 illustrates the relationship between these pressures.

The total force F exerted by a fluid on a plane area A is the product of the area and the pressure at its centroid, i.e.,

$$F = \gamma \bar{h} A \tag{2.5}$$

in which $\bar{h} =$ depth of the fluid over the centroid (Fig. 2.2).

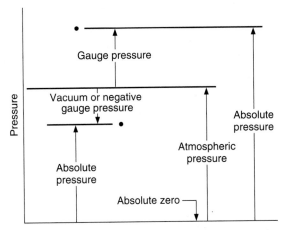

FIGURE 2.1   Relationship between various pressures.

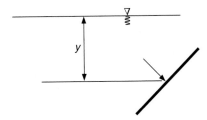

FIGURE 2.2   Hydrostatic pressure.

2.2.3   Steady and Uniform Flows

If all conditions (velocity, pressure) at a point in a flowing fluid remain constant with respect to time, the flow is called *steady flow;* if they vary with time, the flow is called *unsteady flow* (Roberson and Crowe, 1993; Chaudhry, 1993, 1987; Roberson, et al., 1988; Chaudhry and Yevjevich, 1980).

If the flow velocity is the same at every point in space at a particular instant of time, the flow is called *uniform flow.* The steady and uniform flows may exist independent of each other. Thus, the following four combinations are possible: Steady uniform, steady nonuniform, unsteady uniform, and unsteady nonuniform.

2.2.4   Rate of Discharge

The quantity of liquid flowing per unit time across any section is called the rate of discharge (commonly called discharge or flow). The following units are used to express the rate of discharge: Cubic feet per second (cfs), cubic meters per second (m³/s), million gallons per day (mgd), and gallons per minute (gpm). The conversion factors between these units are:

1 m³/s = 35.3 cfs

1 cfs = 449 gpm (American gallons)

1 million gal/day = 694 gpm

If the flow velocity v varies across the cross section, then flow

$$Q = \int_{A} v\,dA = VA \qquad (2.6)$$

in which v = velocity through an infinitesimal area dA, and V = mean velocity over the entire cross-sectional area A.

2.2.5 Continuity Equation

For steady flow between two sections, we can write

$$Q = A_1 V_1 = A_2 V_2 \tag{2.7}$$

in which the subscripts refer to variables at a particular section (Fig. 2.3). This equation is valid if there is no inflow or outflow between the two sections.

2.2.6 Momentum Equation

The momentum equation may be written for a volume of water between two cross sections in one-dimensional flow as

$$\sum F_x = \rho Q(V_{out} - V_{in}) \tag{2.8}$$

in which $\sum F_x$ = vectorial sum of the component of all the external forces acting on the water in the x-direction; V_{out} = flow velocity in the x-direction at the downstream cross section; V_{in} = flow velocity in the x-direction at the upstream cross section; and the x-direction is along the conduit axis or along the channel bottom, considered positive in the downstream direction.

2.2.7 Energy Equation

In this section, we first develop an expression for a coefficient (Chaudhry, 1993) so that the mean flow velocity may be used to compute the velocity head, $V^2/(2g)$.

2.2.7.1 Kinetic Energy.   Let us assume that the velocity distribution across a section is as shown in Fig. 2.4. The mass of fluid flowing through an area dA per unit time is $(\gamma/g)v\,dA$, in which v = flow velocity through an area dA. Then, the flow of kinetic energy per unit time through this area is $(\gamma/g)v\,dA\ v^2/2 = (\gamma/2g)v^3 dA$. If the

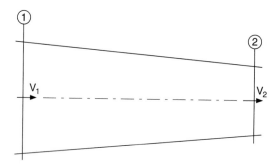

FIGURE 2.3 Notation for continuity equation.

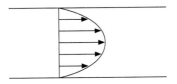

FIGURE 2.4   Velocity distribution in a pipe.

velocity distribution is known, then total kinetic energy flowing through the section per unit time is

$$\frac{\gamma}{2g} \int_A v^3 dA \qquad (2.9)$$

By using the mean flow velocity V and a coefficient α, we can write that the total energy per unit weight is $V^2/(2g)$. Since the flow across the entire section is γAV, we can write

$$\text{Total kinetic energy transmitted} = (\gamma AV)\left(\alpha \frac{V^2}{2g}\right) = \gamma \alpha A \frac{V^3}{2g} \qquad (2.10)$$

Hence it follows from Eqs. (2.9) and (2.10) that

$$\frac{\gamma}{2g} \int_A v^3 dA = \gamma \alpha A \frac{V^3}{2g}$$

or

$$\alpha = \frac{1}{AV^3} \int_A v^3 dA \qquad (2.11)$$

As the average of the sum of the cubes of numbers is greater than the cube of the average, the value of α is greater than 1. The value of α for flow in circular pipes flowing full with parabolic velocity distribution is equal to 2 for laminar flow, and normally ranges between 1.03 to 1.06 for turbulent flow. Since usually the value of α is not precisely known, it is not commonly used, and the kinetic energy of fluid per unit weight is taken equal to $V^2/2g$.

The *law of conservation of energy* is well known. When applied to a flowing fluid, it states that the total energy at a downstream section is equal to the total energy at an upstream section, minus the head losses between the two sections. If there is a turbo-machine, such as a pump or a turbine, between the two sections, then the energy added or extracted by the machine has to be taken into consideration. Referring to Fig. 2.5, we can write

$$z_1 + \frac{p_1}{\gamma} + \frac{V_1^2}{2g} = z_2 + \frac{p_2}{\gamma} + \frac{V_2^2}{2g} + h_f \qquad (2.12)$$

in which p = pressure; z = height above datum; h_f = head losses between sections 1 and 2; and the subscripts 1 and 2 refer to the quantities for sections 1 and 2, respectively. Note that in Eq. (2.12) we have assumed that the kinetic-energy coefficient is equal to unity.

2.2.8   Laminar and Turbulent Flows

In laminar flow, the particles move in definite paths, and the fluid appears to move by sliding of laminations of infinitesimal thickness relative to the adjacent layers. The resistance to flow is produced by viscous shear of fluid particles and varies as the first power of the flow velocity.

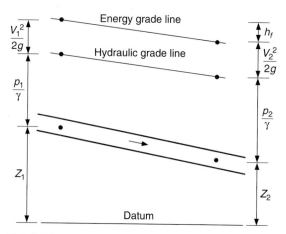

FIGURE 2.5 Notation for energy equation.

In turbulent flow, particles follow irregular and erratic paths, and no two particles have identical or similar motions. A distinguishing characteristic of turbulence is its irregularity and an absence of definite flow patterns. Experiments show that frictional resistance in turbulent flows varies with approximately second power of velocity.

2.2.9 Hydraulic and Energy Grade Lines

Let us consider flow in a pipe as shown in Fig. 2.5. If a series of piezometers were erected along the pipe, then liquid would rise in them to various levels. The line drawn through the top of the liquid columns is called the *hydraulic grade line*. The vertical distance of the energy grade line above any specified datum is equal to total head or total energy.

2.3 CLOSED-CONDUIT FLOW

The analysis of closed conduit flow becomes necessary for the planning, design, and operation of water resource projects. Any flow-conveying structure with a closed top and flowing full is called a closed conduit (Chaudhry and Yevjevich, 1980). Such analyses involve both steady and unsteady flow.

2.3.1 Steady Flow

Steady flow in closed conduits is well-understood and there is a sufficient amount of information available to analyze these flows. The calculation of head losses in closed conduits is needed to determine an optimum size during design and to determine limitations and constraints for operation.

2.3.2 Hydraulic Transients

Whenever the flow conditions are changed from one steady state to another, the intermediate stage flow is called *transient flow*. The magnitude of pressure waves produced in these flows may cause the resulting pressures to exceed the design pressures, resulting in rupturing or collapsing of the conduit walls.

As a rough rule of thumb, the pressure change ΔH caused by an *instantaneous* change in flow velocity ΔV may be computed from the expression $\Delta H = 100\Delta V$, where ΔH is in ft (m) of water and ΔV is in ft/s (m/s). Since changes in the flow velocity on actual projects are usually gradual, pressure change is less than that given by this expression if the time during which the flow changes is greater than twice the transit time of the system. The transit time is the wave travel time from the boundary where the flow is changed to the end boundary where the waves are reflected back.

Spectacular failures have occurred when a system was not designed or operated to account for these pressures (Chaudhry, 1993, 1987; Roberson, et al., 1988; Chaudhry and Yevjevich, 1980).

Head losses may be classified as friction and form losses (Roberson and Crowe, 1993; Chaudhry, 1993; Chaudhry, 1987; Roberson, Cassidy, and Chaudhry, 1988; Chaudhry and Yevjevich, 1980). *Friction losses* result from shear forces between the fluid and the boundary containing the fluid. *Form losses* are due to eddies generated by changes in the geometry of the containing vessels, such as change in cross-sectional area, and bends in the pipeline profile. Expressions and experimental data for computing the friction and form losses are presented in the following sections.

2.4 FRICTION LOSSES

In this section, we discuss friction losses in circular and in noncircular conduits.

2.4.1 Circular Conduits

Various formulas, such as Chezy, Darcy-Weisbach, Manning, Hazen-William, and Scobey, have been proposed for establishing the relationship between the friction losses, physical characteristics of a conduit, and the flow parameters of these formulas. Of these, the Darcy-Weisbach formula (Roberson and Crowe, 1993; Chaudhry, 1993, 1987; Roberson, et al., 1988; Chaudhry and Yevjevich, 1980; Davis and Sorenson, 1969; ASCE, 1965) is scientifically based, the friction factor is dimensionless, no fractional powers are involved, and the formula applies to both laminar and turbulent flows. According to the Darcy-Weisbach formula

$$h_f = \frac{fLV^2}{D\,2g} \tag{2.13}$$

where h_f = total friction losses
f = friction factor
L = conduit length
D = inside diameter of the conduit
V = mean flow velocity
g = acceleration due to gravity

The friction factor f depends upon the Reynolds number R_e and the relative roughness k/D, where k represents the average nonuniform roughness of the conduit. For laminar flow ($R_e < 2000$)

$$f = \frac{64}{R_e} \qquad (2.14)$$

in which $R_e = DV/v$, and v = kinematic viscosity.

Based on Nikuradse's (Nikuradse, 1932) experiments, von Karman and Prandtl (Roberson and Crowe, 1993; ASCE, 1965; USBR, 1977; von Karman, 1930) proposed the following equations for friction factor in turbulent flow:

Smooth pipe $\qquad \dfrac{1}{\sqrt{f}} = 2 \log_{10} R_e \sqrt{f} - 0.8 \qquad (2.15)$

Rough pipe $\qquad \dfrac{1}{\sqrt{f}} = 2 \log_{10} \dfrac{r_o}{k} + 1.74 \qquad (2.16)$

in which r_o = pipe radius.

A smooth pipe is defined as having small irregularities compared with the thickness of the boundary layer. Rough pipes are significant in that the irregularities of the walls are sufficient to break up the laminar boundary layer, with the result that a turbulent flow is fully developed. Eqs. (2.15) and (2.16) are unsatisfactory in the transition zone between the smooth and rough pipe regions (Fig. 2.6). Colebrook and White (1939) showed that this is caused by the fact that the resistance to flow for uniform sand roughness is different from that for an equivalent nonuniform roughness, such as exists in commercial pipe. They proposed the following semi-empirical formula, which is asymptotic to both smooth and rough pipe equations (Eqs. [2.15] and [2.16]):

$$\frac{1}{\sqrt{f}} - 2 \log_{10} \frac{r_o}{k} = 1.74 - 2 \log_{10}\left[1 + 18.7 \frac{r_o/k}{R_e \sqrt{f}} \right] \qquad (2.17)$$

Moody (1944), utilizing Prandtl-Karman experimental data, Colebrook-White function, and experiments on commercial pipe, plotted a diagram between f and R_e as shown in Fig. 2.6. This diagram, called the Moody diagram, is now widely used. To use this diagram, one should know the roughness of the conduit surface, for which no practical and satisfactory method is available at present. Herein, the uniform sand-grain roughness is designated by k while the nonuniform roughness, as found in commercial pipes, is referred to as the *rugosity* and is designated by ϵ.

The determination of rugosity is very difficult, because the protuberances in conduits vary in size, pattern, and spacing. The surfaces may be uniformly fine grained, medium grained, or coarse grained with irregularly spaced pits, depressions, and protuberances, etc. As an infinite number of combinations are possible, ϵ may be estimated from the experience charts (USBR, 1977) presented in Fig. 2.7.

Table 2.1 lists the friction factors determined from prototype tests conducted by the U.S. Army Corps of Engineers Waterways Experiment Station (ASCE, 1965; USBR, 1977). Investigations reported by the U.S. Bureau of Reclamation show that a continuous-interior steel pipe can be both smoother and rougher than a concrete pipe. In large conduits, the friction factor may not change from year to year, depending upon the maintenance of the conduit surface, as it does in the case of small pipes. For example, the friction factor for a 34-year-old penstock was found to be compa-

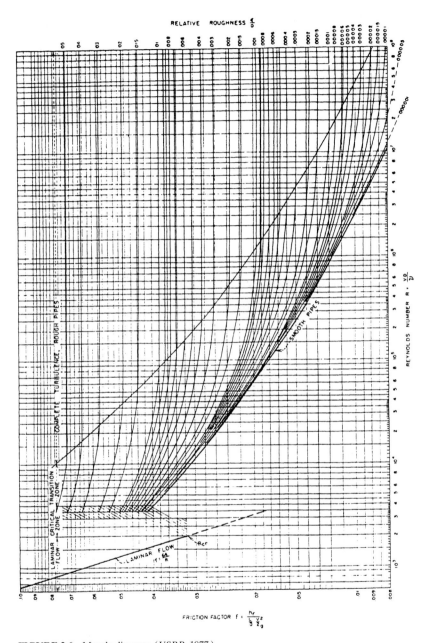

FIGURE 2.6 Moody diagram. (*USBR, 1977.*)

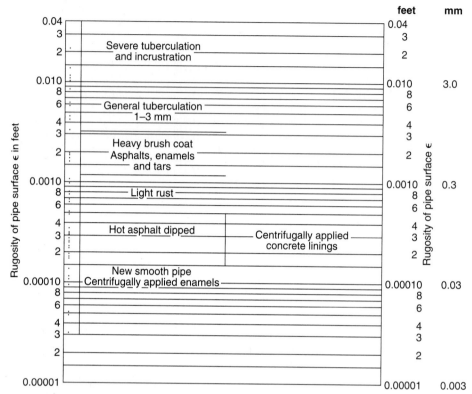

FIGURE 2.7 Rugosity values for various pipes: (*a*) butt-welded, steel pipe (continuous interior). (*USBR, 1977.*)

rable to that of a new pipe. Unless the water flowing through conduits had significant quantities of abrasive material, the friction factor of well-constructed concrete pipes did not change with age. In some cases, algae growth was reported to have little effect on the carrying capacity of concrete pipes until it reduced the cross-sectional area, while in the San Diego Aqueduct, a thin film of algae decreased the carrying capacity by about 10 percent during summer.

Theoretically speaking, the value of the friction factor lies between the smooth-pipe curve and a constant value of approximately 0.054 for rough surfaces. If the wall irregularities are sufficiently large so as to produce noticeable expansion and contraction losses, such as in unlined tunnels, the factor may be greater than 0.054. For example, $f = 0.10$ was obtained (Elder, 1958) for the unlined portion of the Apalachia Tunnel.

Substantial savings in cost may be achieved if large and long conduits are designed to provide a smooth surface, and consequently a reduced cross-sectional area. The cost of producing a smooth surface should be compared with the savings resulting from a reduced cross-sectional area. Conservative values should be used in

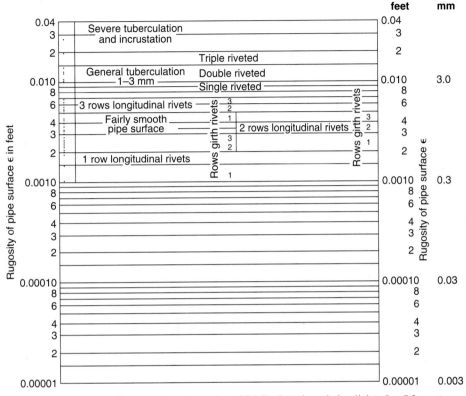

FIGURE 2.7   Rugosity values for various pipes: (*b*) fully riveted, steel pipe (joints 6 to 8 ft apart, longitudinal seams riveted). (*USBR, 1977.*)

design, in view of the scarcity of field data and the difficulty of achieving precision during field tests.

2.4.2   Unlined Rock Tunnels

Based on prototype observations, Rahm (1953, 1958) has proposed the following expression for the friction factor of unlined tunnels:

$$f = 0.00275\delta \tag{2.18}$$

in which the value of δ was determined as follows: Cross-sectional areas were taken every 50 ft along the tunnel length, and after excluding the upper and lower 1 percent for practical reasons, these data were plotted on an ordinary normal-distribution logarithmic diagram. A straight line was drawn (Fig. 2.8) approximating the curve, and passing through the point corresponding to the mean cross-sectional area of the tunnel and a frequency of 50 percent. The slope of the line represents the

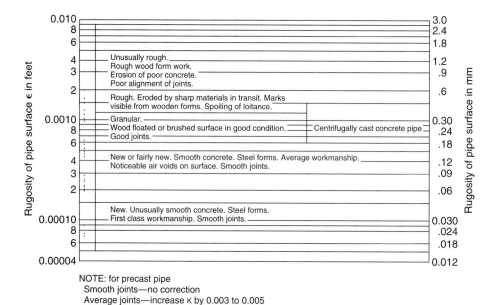

FIGURE 2.7 Rugosity values for various pipes: (*c*) concrete pipe. (*USBR, 1977.*)

variation in cross-sectional area of the tunnel, and can be expressed in terms of percentage inclination δ by

$$\delta = \frac{A_{99} - A_1}{A_1} 100 \tag{2.19}$$

in which A_{99} and A_1 are the cross-sectional areas corresponding to a frequency of 99 and 1 percent, respectively.

Furthermore, it has been found that

$$\delta = \frac{200b_m}{R} \tag{2.20}$$

in which b_m = average excess over break in inches, and R = hydraulic radius. This expression indicates that f is proportional to b_m and thus emphasizes the desirability of accurate drilling to keep b_m as low as possible. Smoothness of the rock surface can be controlled significantly by a close placing of the contour drill holes, exact parallel drilling, and careful blasting. Heggstad states that in Norway, the use of *smooth blasting* has resulted in reducing the tunnel size by 20 to 30 percent over rough-blasted tunnels. Model tests (Thomas and Whitham, 1964) show that:

1. Because of the sawtooth effect, an upstream direction of tunnel-driving yields a lower friction factor (equivalent to a reduction of 0.006 in Manning *n*). This difference was with sharp teeth. However, when the teeth were beveled, the difference was not so definite.

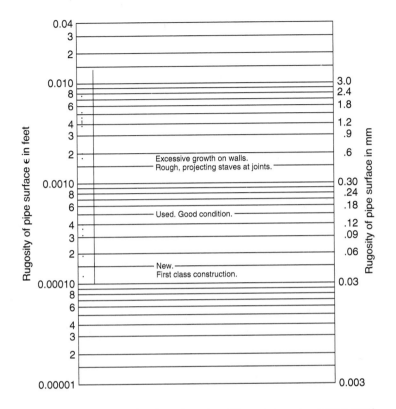

FIGURE 2.7 Rugosity values for various pipes: (*d*) wood-stave pipe. (*USBR, 1977.*)

TABLE 2.1 Observed Resistance Coefficients

	Coefficient	
Surface character	*f*	*n*
(a) $R = 10^8$ (Approximate)		
Concrete, wood forms, joints ground (Oahe)	0.0068	0.0098
Concrete, steel forms (Denison)	0.0072	0.0103
Steel, coal tar (Ft. Randall)	0.0085	0.0114
(b) $R = 10^7$ (Approximate)		
Steel, vinyl (Ft. Randall)	0.0075	0.0107
Concrete, wood forms (Enid)	0.0130	0.0125
Concrete, wood forms (Pine Flat)	0.0132	0.0115
Concrete, wood forms, roughened with use (Pine Flat)	0.0181	0.0135

Source: ASCE, 1965.

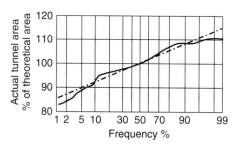

FIGURE 2.8 Distribution of actual tunnel area. (*USBR, 1977.*)

2. For a mechanically-bored tunnel, the combined effect of rough surface and wavy profile would produce a Manning n of about 0.0195. The wavy profile is produced by slight changes in the direction of the machine because of varying rock hardness, and by larger changes in direction caused by changes in machine alignment as the side thrusters are retracted and advanced.

Prototype tests at the tailrace tunnel of the Pirttikoski power plant (Ponni, et al., 1962) in Finland, and at Mammoth Pool power tunnel (Davis and Sorenson, 1969) in California, have confirmed the validity of Rahm's method. Table 2.2 (ASCE, 1965) lists the friction coefficients for a wide range of tunnel sizes.

2.5 FORM LOSSES

It is rarely possible to provide a straight, uniform conduit for conveying water from one point to another. Normally, between the source and the destination, a variety of deviations from such an ideal passage become necessary. These departures from uniformity, whether in the form of partial obstacles, changes in section, branches, or bends, impose an additional energy loss on the flowing water. Because they generally represent only a small part of the total, such losses are commonly referred to as *minor or secondary losses.* Nevertheless, they can be an important consideration, and their effect should always be taken into consideration.

Generally, form losses are the result of fully-developed turbulence, and thus can be expressed in terms of the nominal velocity head $V^2/(2g)$. However, since the velocity distribution may be entirely disturbed as it traverses the loss-producing section, important changes in boundary layer may also be expected. These will be reflected in a modification of the wall resistance over that part of the conduit where the regime is disturbed—usually a distance of 20 diameters or more.

A practicing engineer is usually interested in the net contribution of the form losses, and is not especially concerned with the exact mechanism involved. Thus, the results of laboratory and prototype tests have customarily been reported in the form of coefficients (Rouse, 1950; King, 1954; Addison, 1954; Idel'chik, 1966; Miller, 1971; Creager and Justin, 1955; U.S. Army Corps of Engineers, 1977; Wunderlich, 1963; Brown, 1958) to be applied to the velocity head at the section in question, with the coefficient including (unless otherwise stated) the subsidiary effects previously noted. This approach is more convenient than the alternative practice of converting

TABLE 2.2 Unlined Tunnel Friction Coefficients

No.	Tunnel name*	Location	Type of rock	Design area, in ft²	Average driven area in ft²	Percentage over-break	n	f	Variation in cross-sectional area, δ
1	Cresta	California	Granite	578	656	13.5	0.035	0.075	27
2	West Point	California	Granite	180	222	23.5	0.033	0.080	NA
3	Bear River	California	Granite	82	93	14.0	0.028	0.066	NA
4	Balch	California	Granite	144	169	17.5	0.032	0.079	NA
5	Haas	California	Granite	151	184	21.9	0.030	0.068	26
6	Cherry	California	Granite	133	150	12.5	0.034	0.090	NA
7	Jaybird	California	Granite	177	195	10	0.032	0.077	NA
8	Apalachia	Tennessee	Quartzite & slate	380	431	13.5	0.038	0.095	NA
9	Alfta	Sweden	Granite-gneiss	323	364	12.7	0.036	0.086	29
10	Harspranget	Sweden	Granite	2045	2195	7.4	0.032	0.052	24
11	Jarpstrommen	Sweden	Slate	1130	1230	9.8	0.029	0.048	20
12	Krokstrommen	Sweden	Granite	970	1090	12.9	0.029	0.048	17
13	Porjus I	Sweden	Granite-gneiss	538	618	14.8	0.034	0.073	27
14	Porjus II	Sweden	Granite-gneiss	538	662	23.0	0.030	0.055	19
15	Selsfore	Sweden	Slate w/granite	753	865	15.0	0.044	0.114	39
16	Sillre	Sweden	Gneiss	54	71	32.0	0.034	0.102	37
17	Sunnerstaholm	Sweden	Granite-gneiss	323	386	19.7	0.039	0.104	37
18	Tasan	Sweden	Gneiss	183	185	1.2	0.033	0.081	33
19	Eucumbene-Tumut	Australia	36% Granite 64% Metam. Sedim.	400	445	11.2	0.029	0.054	NA
20	Tooma-Tumut	Australia	Granite	125	153	22.4	0.031	0.074	NA
21	Murrumbidgee-Eucumbene	Australia	10% Granite 90% Metam. Sedim.	100	127	27.0	0.036	0.104	NA
22	Big Creek 2	California	Granite	113	130	15.0	0.037[†]	NA	NA
23	Big Creek 3	California	Granite	434	515	18.7	0.035[‡]	NA	NA
24	Big Creek 8	California	Granite	357	400	12.0	0.038[†]	NA	NA
25	Ward	California	Granite	211	274	29.9	0.036[†]	NA	NA
26	Big Creek 4	California	Granite	409	462	13.0	0.030[‡]	NA	NA
27	Mammoth Pool	California	Granite	336	367	9.2	0.029[‡]	NA	NA

* Tunnels 1 to 5 are owned by Pacific Gas and Electric Co., 6 by the City of San Francisco, 7 by Sacramento Municipal Utility District, 8 by the TVA, 9 to 18 by Swedish State Power Board, 19 to 21 by the Snowy Mountains Hydro-Electric Authority, and 22 to 27 by the Southern California Edison Co.
[†] Based on gross head loss observations including special losses at adits, bends, lined sections, etc.
[‡] Based on net head loss observations on unlined sections only, or else on gross head loss measurements corrected by deducting special losses.
Source: ASCE, 1965.

minor losses to equivalent conduit length, and treating them as additional frictional resistance.

The following sections are intended to indicate what is available to the engineer for estimating the probable minor losses for a given conduit configuration. For this purpose, it is convenient to classify these losses into two groups: (1) at or near the intake, and (2) along the conduit length.

2.5.1 Intake Losses

The *intake losses* include the entrance loss, trash rack loss, and head gate loss. It should be noted that an accurate experimental determination of the intake losses depends on the conduit being of sufficient length to permit a uniform friction gradient to be established, based on fully developed turbulence. A comprehensive treatment of the hydrodynamic principles involved and an analysis of the velocity and pressure distributions for various shapes of conduit inlets has been given by Rouse (1950).

2.5.2 Entrance Loss

In addition to the pressure drop representing the velocity head, a loss is incurred at the entrance to a conduit analogous to that in a short tube. The value of the coefficient K_e in the expression

$$h_L = K_e \frac{V^2}{2g} \tag{2.21}$$

depends largely on the geometry of the entrance. Representative published values are listed in Table 2.3.

2.5.3 Trash Rack Loss

Creager and Justin (1955) give the following equation for these losses:

$$K_t = 2.45 - 0.45R - R^2 \tag{2.22}$$

in which R = the ratio of net to gross area at the rack section. For a typical value of $R = 0.65$, the resulting K value is 0.74, to be applied to the velocity head for the net area. Figure 2.9 shows the variation of K_t with R for the bars of several different shapes. In this figure, h = head loss through the rack, in feet, and V = the velocity at the section without the rack, in feet per second. Generally, the velocity is low at the

TABLE 2.3 Representative K Values

Type of entrance	Rouse (1950)	King (1954)	Addison (1954)	Creager and Juston (1955)
Inward projecting	1.0	0.78	0.70	0.56–0.93
Sharp-cornered	0.50	0.50	0.50	0.56
Slightly rounded	—	0.23	—	0.23
Bell-mouthed	0.01–0.05	0.04	0.04	0.06

Source: ASCE, 1965.

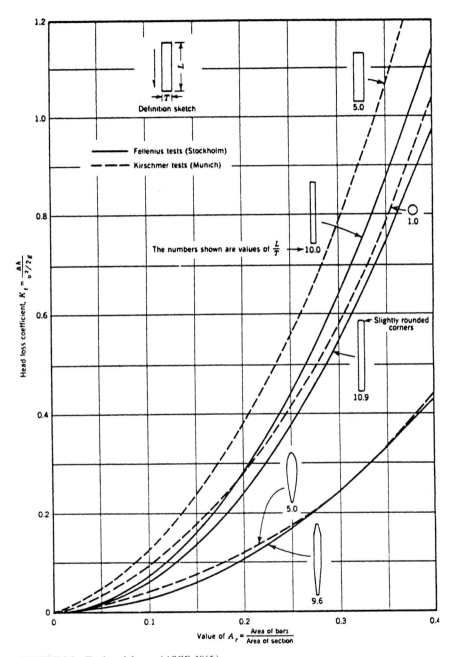

FIGURE 2.9   Trash rack losses. (*ASCE, 1965.*)

rack section and hence the head loss. Obstruction by trash, however, can substantially increase the losses.

2.5.4 Head Gate Loss

No special head loss need be assumed for a gate located at the conduit inlet and not interfering with the entrance flow conditions. Where the opening is farther downstream but continuous with the flow passage, a coefficient K_g less than 0.1 may be used to account for the loss caused by the eddies in a gate slot. For gates mounted such that the flow is contracted, the coefficient depends on the shape of the lip and the degree of suppression of the jet. Because of the analogy with flow through an orifice, the coefficient is usually reported as the coefficient of discharge C in

$$Q = CA \sqrt{2gH} \tag{2.23}$$

in which A = area of gate opening, and H = the difference in head acting on the gate. The value of C is readily converted to the equivalent K value by

$$K = \frac{1}{C^2} - 1 \tag{2.24}$$

K is then applied to the velocity at the gate section.

King (1954) has reported various model and prototype data, indicating a range for C from 0.62 to 0.83 (i.e., K = 2.6 to 0.45) relatively independent of the percentage of gate opening. He also gives an expression relating C to the differential head, and another relating it to the width of opening. The U.S. Army Corps of Engineers Waterways Experiment Station (WES) has related C to gate opening for certain shapes of lip, and has suggested a design curve in which C varies from 0.73 to 0.80 (K = 0.87 to 0.56) as the opening increases from 10 to 80 percent.

The U.S. Bureau of Reclamation (USBR, 1977) summarized various reported data, as listed in Table 2.4. A comprehensive treatment of the subject is given by Wunderlich (1963). The coefficients are given in terms of the gate opening, type of discharge, and gate design.

2.5.5 Expansion Loss

A sudden expansion or enlargement incurs a head loss that may be theoretically expressed as

$$h_L = \frac{(V_1 - V_2)^2}{2g} \tag{2.25}$$

TABLE 2.4 Values of C and K

Geometry of gate	Discharge coefficient, C			Loss coefficient, K		
	Max	Min	Avg	Max	Min	Avg
Unsuppressed contraction	0.70	0.60	0.63	1.80	1.00	1.50
Bottom and sides suppressed	0.81	0.68	0.70	1.20	0.50	1.00
Corners rounded	0.95	0.71	0.82	1.00	0.10	0.50

Source: ASCE, 1965.

in which V_1 and V_2 are the velocities in the smaller and larger conduits, respectively. King (1954) states that the experimentally derived formula by Archer

$$h_L = 0.01705(V_1 - V_2)^{1.919} \qquad (2.26)$$

gives more satisfactory results than that of Eq. (2.25). He provides a table based on this relationship, which gives the coefficient K to be applied to the velocity head in the smaller pipe.

Gradual expansions or transitions entail a lesser head loss (because the turbulent eddies produced by the transition are suppressed) and may be calculated from

$$h_L = K\frac{(V_1 - V_2)^2}{2g} \qquad (2.27)$$

in which K depends on the angle of flare or divergence. Brown (1958) gives values of K ranging from 0.20 to 2.07 for flare angles of 2° to 90°. King's corresponding values are considerably lower, ranging from 0.03 to 0.67. He also provides a table based on these coefficients, that converts K to the more usual form (i.e., in terms of the larger velocity head). Creager and Justin (1955) also express K in the latter form, giving

$$K = \left(1 - \frac{a_1}{a_2}\right)^2 \sin \theta \qquad (2.28)$$

in which a_1 and a_2 = the smaller and larger areas, and θ = one-half of the flare angle.

It should be noted that all the treatments mentioned exclude the friction loss within the enlargement section. When this is considered, Addison (1954) states that the optimum taper is 1 in 10, resulting in an expression for head loss as

$$h_L = 0.14\frac{(V_1 - V_2)^2}{2g} \qquad (2.29)$$

2.5.6 Contraction Loss

A sudden contraction incurs a loss coefficient K of as much as 0.5 for pronounced differences in area. Rouse (1950), King (1954), and Creager and Justin (1955) provide data showing the variation of K with the ratio of diameter. The head loss in gradual contractions is usually negligible.

2.5.7 Bend Loss

The loss at a bend results from a distortion of the velocity distribution, thereby causing additional shear stresses in the fluid. The bend loss is calculated from

$$h_b = K_B \frac{V^2}{2g} \qquad (2.30)$$

in which h_b = head loss in the bend, K_B = bend-loss coefficient, and V = velocity in the pipe.

Customarily, the bend-loss coefficient does not include the effect of normal friction loss in the length of pipe involved. It is generally accepted that the principal parameters affecting the loss coefficient K are the deflection angle of the bend, the ratio of bend radius to conduit diameter, and the Reynolds number. The latter is usually either not considered or is regarded as unimportant for the ranges reported.

Graphs and curves giving K_B as a function of these parameters are given by Rouse (1950), King (1954), Creager and Justin (1955), and the U.S. Army Corps of Engineers (Fig. 2.10). These researchers indicate that a radius to diameter ratio of 4 to 6 is the optimum, yielding a K_B value of approximately 0.15 for a 90° bend. For smaller deflection angles, the value of K is smaller.

Rouse (1950) and Addison (1954) indicate that the bend loss can be materially reduced by providing a grid of deflecting vanes extending across the flow passage. This is seldom done, however, because these losses are usually small.

2.5.8 Valve Losses

For circular gate valves in the fully open position, the loss coefficients given by some references are listed in Table 2.5. For partial opening of this type of valve, the loss increases rapidly until free or shooting flow occurs. Coefficients for this condition are plotted in Figs. 2.11 and 2.12. The basic equation used in Fig. 2.12 is

$$K_v = \frac{H_L}{V^2/2g} \tag{2.31}$$

in which K_v = the valve-loss coefficient, H_L = head loss through the valve, and V = average velocity in the pipe.

In Fig. 2.11, the values are calculated from

$$Q = CA \sqrt{2gH_e} \tag{2.32}$$

in which C = valve discharge coefficient, A = area based on the nominal valve diameter, and H_e = the energy head measured to the centerline of the conduit immediately upstream of the valve.

In Fig. 2.12, the data are for valves having the same diameter as the pipe, and for the downstream pipe flowing full. The data in Fig. 2.11 are based on tests conducted by the U.S. Bureau of Reclamation (1977) on free-flow, 8-in diameter gate valves, located at the downstream end of a conduit having the same nominal diameter as that of the valve.

A butterfly valve, even in the fully open position, obstructs the flow due to disk thickness. Creager suggests $K = t/d$, in which t = the disk thickness and d = the conduit diameter, with K to be applied to the velocity head based on the gross area. Mahon (1957) observes that $K = 0.25$ is a commonly reported value. Boyd (1958) notes that the loss can be reduced by the use of a converging section on the downstream side of the disk. Figure 2.13 shows the loss in terms of a modified discharge coefficient, and shows, as would be expected, a rapid increase in loss for partial openings. The following equation is used in this figure:

$$Q = C_q D^2 \sqrt{g\Delta H} \tag{2.33}$$

where
Q = discharge, ft³/s
C_q = discharge coefficient
D = valve diameter, ft
ΔH = pressure drop across the valve, in ft of water

The head loss in a needle or plunger valve is treated briefly in Creager and Justin (1955). The coefficients for Howell-Bunger valves, flap gates, and reverse tainter valves are shown on the U.S. Army Corps of Engineers (1977) design criteria.

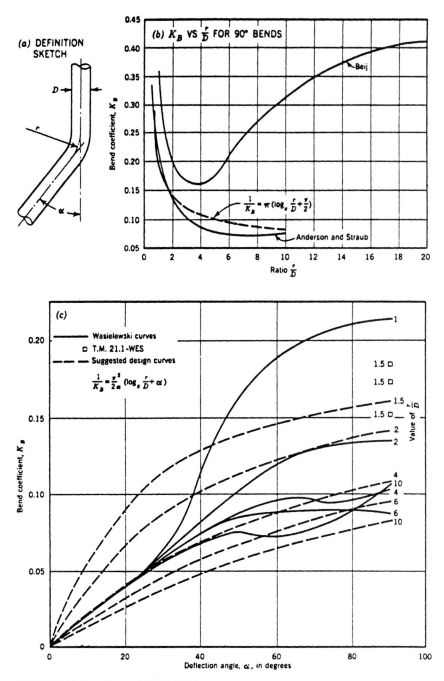

FIGURE 2.10 Bend losses. (*ASCE, 1965.*)

TABLE 2.5 Circular Gate Valve Loss Coefficients

	Diameter of pipe, in	K_v
WES	to 18	0.18
Rouse	Not stated	0.19
King	12	0.07
Brown	24	0
Creager	Not stated	Less than 0.1

Source: ASCE, 1965.

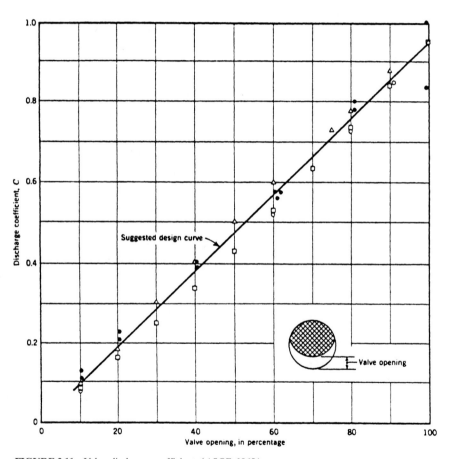

FIGURE 2.11 Valve-discharge coefficient. (*ASCE, 1965.*)

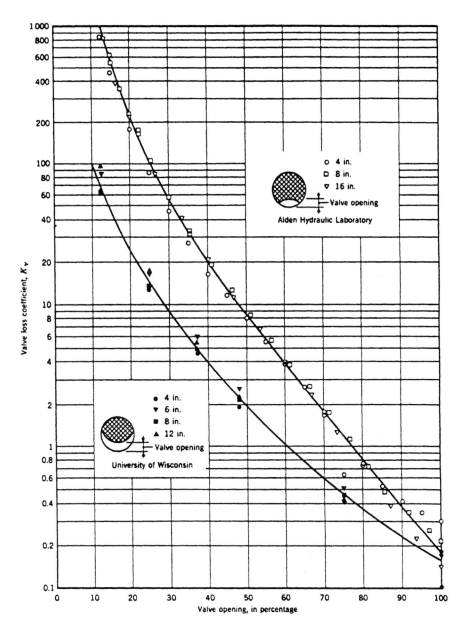

FIGURE 2.12 Valve-loss coefficient. (*ASCE, 1965.*)

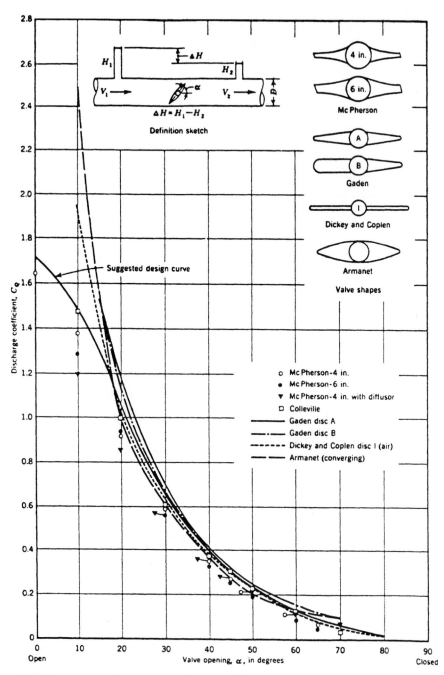

FIGURE 2.13   Butterfly valve, loss coefficient. (*ASCE, 1965.*)

2.5.9 Wyes and Bifurcations

These configurations occur frequently in water supply and cooling-water systems, and in hydroelectric power plants. Although there has been a considerable amount of research, much of it is either theoretical or based on laboratory tests only. There is a paucity of systematic and codified data suitable for direct application to proto-type installations. The problem is compounded because of the possibility of many combinations of sizes, angles, and fabrication methods.

Mososnyi (1957) noted that the hydraulic losses at wye pieces are governed by the angle of bifurcation and by the ratio of cross-sectional areas. Diversion through a branch of part of the flow in a pipeline having a constant cross section will result in the velocities in the main pipe downstream from the point of bifurcation being suddenly reduced. The hydraulic phenomena thus corresponds to that in a sudden expansion. A sudden change in the flow velocity and flow direction results in separation phenomena that may be reduced to minimal values by properly shaped corners and appropriate decrease in cross-sectional areas. He finds the loss coefficients for appropriately proportioned wyes to vary from 0.25 to 0.50, in contrast to those with improperly shaped wyes, in which the coefficients may be appreciably higher. In fact, the coefficient with 90° bifurcations may be as high as 2.0.

Based on the discharge delivered by the main conduit to the point of bifurcation as $Q = Q_m + Q_b$, Mosonyi developed Table 2.6 giving values of the loss coefficients for various ratios of Q_b/Q and Q_m/Q_b, and for bifurcations at 90° and at 30°, in which the subscripts m and b denote the main penstock and branch pipe, respectively. He observed that the local losses in the main conduit may assume negative values for small discharge ratios. This does not constitute a violation of the energy laws, because the increase originates from a decrease in velocity. Such negative loss coefficients are shown in Table 2.6 as zero. Rouse (1950) shows the necessity of including correction coefficients for several velocity heads if the true energy balance is to be described.

A carefully conceived and executed test program (Muller, 1949) was conducted at the Lucendro Power Station in Switzerland to determine the head losses in a section of the 2.2-m diameter welded steel penstock containing two 55° wye junctions. For $Q_m = Q_b$, the loss coefficient for each wye averaged nearly $K = 2.2$ (based on the velocity head in the main penstock). A minimum K of approximately 0.2 was achieved when approximately 40 percent of the main penstock flow was diverted through a branch. For the entire section, total head loss was distinctly minimized when the flow was evenly divided between the two branches. The results confirmed earlier data obtained by the hydraulic models. Ruus (1970) reported head loss coefficients for wyes and manifolds.

TABLE 2.6 Coefficients of Head Loss at Bifurcations

Discharge ratio for $Q = Q_m + Q_b$		Loss coefficients for bifurcations of			
		90°		Acute (30°) angles	
Q_b/Q	Q_m/Q_b	K_m	K_b	K_m	K_b
0.10	9	0.03	0.89	0	0.78
0.25	3	0	0.88	0	0.63
0.50	1	0	0.91	0	0.44
0.75	1/3	0.18	1.06	0.16	0.36
0.90	1/9	0.28	1.20	0.26	0.41

Source: ASCE, 1965.

2.5.10 Miscellaneous Junctions

The head loss coefficients for several tees, acute- and obtuse-angled elbows, and other configurations are given (King, 1954; Addison, 1954; Idel'chik, 1966; Miller, 1971; Creager and Justin, 1955). More detailed data on various junctions are furnished by Freeman (1942).

2.5.11 Exit Loss

When a conduit empties into a reservoir, the condition is analogous to the sudden enlargement discussed previously. Unless a gradually tapered transition is provided, the entire velocity head is lost in the formation of eddies.

2.6 OPEN-CHANNEL FLOW

2.6.1 Definitions

The flow in a conduit having free surface is termed *open-channel flow*. The free surface is usually subjected to atmospheric pressure. Note that the conduit itself may be open or closed at the top; it is the top surface which must be free for the flow to be designated as open-channel flow (Chaudhry, 1993; Chow, 1959).

A channel is said to be *prismatic* if it has constant cross section and a constant bottom slope. The depth of flow y is the vertical distance of the lowest point of a channel section from the free surface, and the depth of flow section d is the depth of flow normal to the flow direction, as shown in Fig. 2.14. Since the bottom slope is

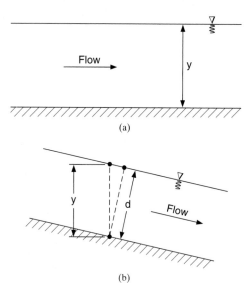

FIGURE 2.14 Depth of flow and depth of flow section: (*a*) small bottom slope; (*b*) steep bottom slope.

usually small, y and d are assumed to be equal. For steep slopes, this assumption is not valid, and appropriate adjustments become necessary.

The *stage* is the elevation or vertical distance of the free surface above a datum. The *wetted perimeter P* is the length of the line of intersection of the channel wetted surface with a cross-sectional plane normal to the direction of flow. The water area A is the cross-sectional area of the flow normal to the direction of flow. The *hydraulic radius R* is the ratio of the water area to its wetted perimeter, i.e.,

$$R = \frac{A}{P} \tag{2.34}$$

Open-channel flow is said to be *uniform* if the depth of flow is the same at every section of the prismatic channel. The flow is varied if the flow depth changes along the length of the channel. The varied flow may be classified as *rapidly* or *gradually varied*. The flow is rapidly varied if the depth changes abruptly over a comparatively short distance, such as in a hydraulic jump; otherwise it is gradually varied, e.g., a flood wave without a bore formation.

The Froude number, F_r is given by the equation

$$F_r = \frac{V}{\sqrt{gD}} \tag{2.35}$$

in which V = flow velocity, D = hydraulic depth = A/T, A = cross-sectional area normal to flow, and T = top water-surface width.

If $F_r = 1$, i.e., $V = \sqrt{gD}$, flow is called *critical flow* and the flow velocity is called *critical velocity;* if $F_r > 1$, i.e., $V > \sqrt{gD}$, the flow is called *supercritical flow;* and if $F_r < 1$, i.e., $V < \sqrt{gD}$, the flow is called *subcritical flow*. Since critical velocity V is equal to the celerity of a small gravity wave (Chaudhry, 1993) the above definitions have special significance. In supercritical flow, the flow velocity is greater than the wave celerity. Hence, a disturbance propagates only in the downstream direction. In subcritical flow, however, a disturbance propagates both in the upstream and downstream directions, since the celerity is more than the flow velocity.

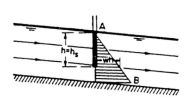

FIGURE 2.15 Pressure distribution in parallel flow. (*Chow, 1959.*)

In a channel having small bottom slope, the pressure at any point is directly proportional to the depth of the point below the free surface, and is equal to hydrostatic pressure corresponding to this depth. In other words, the pressure distribution is hydrostatic (Fig. 2.15). This is valid for uniform flow in which the streamlines have neither curvature nor divergence, and thus have no acceleration normal to the flow direction. Similarly, for practical purposes, gradually varied flow may be assumed to behave as uniform flow, since the rate of change of the flow depth with distance is small, and the streamlines have slight curvature. Hence, the pressure distribution in both uniform and gradually varied flows may be considered as hydrostatic.

If the streamlines have substantial curvature, then the flow is called *curvilinear flow*. Depending upon the curvature, a centrifugal force acts in the downward or upward direction, and thus the pressure may be higher or lower than hydrostatic. The centrifugal force acts in the upward direction in a convex flow, and therefore the

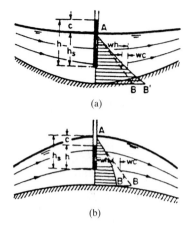

(a)

(b)

FIGURE 2.16 Pressure distribution in curvilinear flow: (a) concave flow; (b) convex flow. (*Chow, 1959.*)

pressure is less than hydrostatic (see Fig. 2.16). The increase or decrease in pressure c due to curvature in a curvilinear flow may be computed from

$$c = \frac{d}{g}\frac{V^2}{r} \qquad (2.36)$$

where d = depth of flow
        r = radius of curvature
        V = flow velocity

Note that c is positive for concave flow, zero for parallel flow, and negative for convex flow.

For a channel having steep slope, the unit pressure may be computed from the following equation (for notation, see Fig. 2.17):

$$h = y \cos^2 \theta = d \cos \theta \qquad (2.37)$$

2.6.2 Energy Equation

Applying the law of conservation of energy to channel flow shown in Fig. 2.18, we obtain

$$z_1 + d_1 \cos \theta + \frac{V_1^2}{2g} = z_2 + d_2 \cos \theta + \frac{V_2^2}{2g} + h_f \qquad (2.38)$$

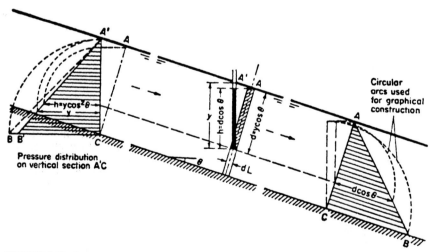

FIGURE 2.17 Pressure distribution in channels having steep slopes. (*Chow, 1959.*)

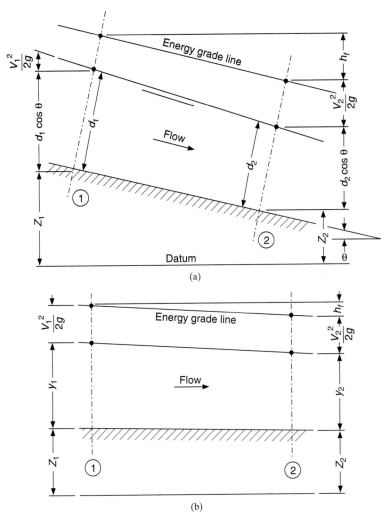

FIGURE 2.18 Notation for energy equation: (*a*) channel having steep slope; (*b*) channel having small slope.

In this equation, subscripts 1 and 2 refer to the section number and h = head loss between these two sections. Note that the channel may have steep slope and the kinetic-energy coefficient is assumed to be equal to unity. If the channel slope is small, then

$$z_1 + y_1 + \frac{V_1^2}{2g} = z_2 + y_2 + \frac{V_2^2}{2g} + h_f \tag{2.39}$$

Equations (2.38) and (2.39) are valid for parallel as well as for gradually varied flow.

Specific energy E in a channel section is defined as the energy per unit weight of water at the section measured above the channel bottom. Hence for a channel having steep slope

$$E = d \cos \theta + \frac{V^2}{2g} \tag{2.40}$$

and for a channel having small bottom slope

$$E = y + \frac{V^2}{2g} \tag{2.41}$$

or

$$E = y + \frac{Q^2}{2gA^2} \tag{2.42}$$

The specific energy at critical flow is a minimum. Hence, $dE/dy = 0$. Differentiating Eq. (2.42) with respect to y, equating it to zero and noting that $dA/dy = T$, we obtain

$$\frac{V^2}{2g} = \frac{D}{2} \tag{2.43}$$

Equation (2.43) states that for critical flow, the velocity head is equal to one-half the hydraulic depth.

Various empirical formulas have been proposed to compute uniform flow in open channels. Notable among these formulas are Chezy and Manning. The latter has been used widely in the English speaking countries. In this formula

$$V = \frac{C_o}{n} R^{2/3} S^{1/2} \tag{2.44}$$

where V = flow velocity
n = Manning coefficient
R = hydraulic radius
S = slope of the energy-grade line

Since the flow is assumed to be uniform, $S = S_o$ in which S_o = bottom slope. In English units, $C_o = 1.49$, V is in ft/s, and R is in ft; in SI units, $C_o = 1$, V is in m/s and R is in m. The Manning coefficient has the same value in both system of units if a proper value for C_o is included.

Table 2.7 (from Chow, 1959) lists n values for various flow surfaces. The value of n mainly depends upon the type of flow surface; however, other factors affect the value of n, such as size and shape of channel, stage and discharge, and suspended material and bed load.

For a given channel, i.e., specified n and S_o, and for a given discharge Q, there is only one depth at which flow will be uniform. This depth, called *normal depth*, is determined by solving the following equation:

$$Q = \frac{C_o}{n} A R^{2/3} S_o^{1/2} \tag{2.45}$$

TABLE 2.7 Values of the Roughness Coefficient n

(Boldface figures are values generally recommended in design)

Type of channel and description	Minimum	Normal	Maximum
A. Closed conduits flowing partly full			
A-1. Metal			
a. Brass, smooth	0.009	**0.010**	0.013
b. Steel			
1. Lockbar and welded	0.010	0.012	0.014
2. Riveted and spiral	0.013	0.016	0.017
c. Cast iron			
1. Coated	0.010	0.013	0.014
2. Uncoated	0.011	0.014	0.016
d. Wrought iron			
1. Black	0.012	0.014	0.015
2. Galvanized	0.013	0.016	0.017
e. Corrugated metal			
1. Subdrain	0.017	0.019	0.021
2. Storm drain	0.021	**0.024**	0.030
A-2. Nonmetal			
a. Lucite	0.008	0.009	0.010
b. Glass	0.009	**0.010**	0.013
c. Cement			
1. Neat, surface	0.010	0.011	0.013
2. Mortar	0.011	0.013	0.015
d. Concrete			
1. Culvert, straight and free of debris	0.010	0.011	0.013
2. Culvert with bends, connections, and some debris	0.011	**0.013**	0.014
3. Finished	0.011	0.012	0.014
4. Sewer with manholes, inlet, etc., straight	0.013	0.015	0.017
5. Unfinished, steel form	0.012	0.013	0.014
6. Unfinished, smooth wood form	0.012	**0.014**	0.016
7. Unfinished, rough wood form	0.015	0.017	0.020
e. Wood			
1. Stave	0.010	0.012	0.014
2. Laminated, treated	0.015	0.017	0.020
f. Clay			
1. Common drainage tile	0.011	0.013	0.017
2. Vitrified sewer	0.011	0.014	0.017
3. Vitrified sewer with manholes, inlet, etc.	0.013	0.015	0.017
4. Vitrified subdrain with open joint	0.014	0.016	0.018
g. Brickwork			
1. Glazed	0.011	0.013	0.015
2. Lined with cement mortar	0.012	0.015	0.017
h. Sanitary sewers coated with sewage slimes, with bends and connections	0.012	0.013	0.016
i. Paved invert, sewer, smooth bottom	0.016	0.019	0.020
j. Rubble masonry, cemented	0.018	0.025	0.030
B. Lined or built-up channels			
B-1. Metal			
a. Smooth steel surface			
1. Unpainted	0.011	**0.012**	0.014
2. Painted	0.012	0.013	0.017
b. Corrugated	0.021	0.025	0.030
B-2. Nonmetal			
a. Cement			
1. Neat, surface	0.010	0.011	0.013
2. Mortar	0.011	0.013	0.015

TABLE 2.7 Values of the Roughness Coefficient *n* *(Continued)*

(Boldface figures are values generally recommended in design)

Type of channel and description	Minimum	Normal	Maximum
b. Wood			
1. Planed, untreated	0.010	0.012	0.014
2. Planed, creosoted	0.011	0.012	0.015
3. Unplaned	0.011	0.013	0.015
4. Plank with battens	0.012	0.015	0.018
5. Lined with roofing paper	0.010	0.014	0.017
c. Concrete			
1. Trowel finish	0.011	**0.013**	0.015
2. Float finish	0.013	0.015	0.016
3. Finished, with gravel on bottom	0.015	0.017	0.020
4. Unfinished	0.014	0.017	0.020
5. Gunite, good section	0.016	0.019	0.023
6. Gunite, wavy section	0.018	0.022	0.025
7. On good excavated rock	0.017	0.020	
8. On irregular excavated rock	0.022	0.027	
d. Concrete bottom float finished with sides of			
1. Dressed stone in mortar	0.015	0.017	0.020
2. Random stone in mortar	0.017	0.020	0.024
3. Cement rubble masonry, plastered	0.016	0.020	0.024
4. Cement rubble masonry	0.020	0.025	0.030
5. Dry rubble or riprap	0.020	0.030	0.035
e. Gravel bottom with sides of			
1. Formed concrete	0.017	0.020	0.025
2. Random stone in mortar	0.020	0.023	0.026
3. Dry rubble or riprap	0.023	0.033	0.036
f. Brick			
1. Glazed	0.011	**0.013**	0.015
2. In cement mortar	0.012	**0.015**	0.018
g. Masonry			
1. Cemented rubble	0.017	0.025	0.030
2. Dry rubble	0.023	0.032	0.035
h. Dressed ashlar	0.013	0.015	0.017
i. Asphalt			
1. Smooth	0.013	0.013	
2. Rough	0.016	0.016	
j. Vegetal lining	0.030	—	0.500
C. Excavated or dredged			
a. Earth, straight and uniform			
1. Clean, recently completed	0.016	0.018	0.020
2. Clean, after weathering	0.018	**0.022**	0.025
3. Gravel, uniform section, clean	0.022	0.025	0.030
4. With short grass, few weeds	0.022	0.027	0.033
b. Earth, winding and sluggish			
1. No vegetation	0.023	0.025	0.030
2. Grass, some weeds	0.025	0.030	0.033
3. Dense weeds or aquatic plants in deep channels	0.030	0.035	0.040
4. Earth bottom and rubble sides	0.028	0.030	0.035
5. Stony bottom and weedy banks	0.025	0.035	0.040
6. Cobble bottom and clean sides	0.030	0.040	0.050
c. Dragline-excavated or dredged			
1. No vegetation	0.025	0.028	0.033
2. Light brush on banks	0.035	0.050	0.060
d. Rock cuts			
1. Smooth and uniform	0.025	0.035	0.040
2. Jagged and irregular	0.035	0.040	0.050

TABLE 2.7 Values of the Roughness Coefficient *n* *(Continued)*

(Boldface figures are values generally recommended in design)

Type of channel and description	Minimum	Normal	Maximum
e. Channels not maintained, weeds and brush uncut			
1. Dense weeds, high as flow depth	0.050	0.080	0.120
2. Clean bottom, brush on sides	0.040	0.050	0.080
3. Same, highest stage of flow	0.045	0.070	0.110
4. Dense brush, high stage	0.080	0.100	0.140
D. Natural streams			
D-1. Minor streams (top width at flood stage <100 ft)			
a. Streams on plain			
1. Clean, straight, full stage, no rifts or deep pools	0.025	0.030	0.033
2. Same as above, but more stones and weeds	0.030	0.035	0.040
3. Clean, winding, some pools and shoals	0.033	0.040	0.045
4. Same as above, but some weeds and stones	0.035	0.045	0.050
5. Same as above, lower stages, more ineffective slopes and sections	0.040	0.048	0.055
6. Same as 4, but more stones	0.045	0.050	0.060
7. Sluggish reaches, weedy, deep pools	0.050	0.070	0.080
8. Very weedy reaches, deep pools, or floodways with heavy stand of timber and underbrush	0.075	0.100	0.150
b. Mountain streams, no vegetation in channel, banks usually steep, trees and brush along banks submerged at high stages			
1. Bottom: gravels, cobbles, and few boulders	0.030	0.040	0.050
2. Bottom: cobbles with large boulders	0.040	0.050	0.070
D-2. Flood plains			
a. Pasture, no brush			
1. Short grass	0.025	0.030	0.035
2. High grass	0.030	0.035	0.050
b. Cultivated areas			
1. No crop	0.020	0.030	0.040
2. Mature row crops	0.025	0.035	0.045
3. Mature field crops	0.030	0.040	0.050
c. Brush			
1. Scattered brush, heavy weeds	0.035	0.050	0.070
2. Light brush and trees, in winter	0.035	0.050	0.060
3. Light brush and trees, in summer	0.040	0.060	0.080
4. Medium to dense brush, in winter	0.045	0.070	0.110
5. Medium to dense brush, in summer	0.070	0.100	0.160
d. Trees			
1. Dense willows, summer, straight	0.110	0.150	0.200
2. Cleared land with tree stumps, no sprouts	0.030	0.040	0.050
3. Same as above, but with heavy growth of sprouts	0.050	0.060	0.080
4. Heavy stand of timber, a few down trees, little undergrowth, flood stage below branches	0.080	0.100	0.120
5. Same as above, but with flood stage reaching branches	0.100	0.120	0.100
D-3. Major streams (top width at flood stage >100 ft). The *n* value is less than that for minor streams of similar description, because banks offer less effective resistance.			
a. Regular section with no boulders or brush	0.025	—	0.060
b. Irregular and rough section	0.035	—	0.100

Source: Chow, 1959.

The flow depth along the channel having gradually varied flow may be computed by integrating the following differential equation:

$$\frac{dy}{dx} = \frac{S_o - S_f}{1 - (Q^2 T)/(gA^3)} \tag{2.46}$$

Various numerical procedures have been used in the past to integrate this equation. Of these, fourth-order Runge-Kutta method has been used (Chaudhry, 1993) for solving this equation on a digital computer. Gradually, varied flow profiles may be computed by solving the energy equation between two consecutive cross sections. For this purpose, the standard-step method (Chow, 1959) dating back to manual computations is employed in the widely used computer program HEC-2, developed by Hydrologic Engineering Center, U.S. Army Corps of Engineers (1990).

Figure 2.19 (from Chow, 1959) shows various flow profiles. To compute these profiles, we begin with the known depth at the control section, and then proceed in the upstream direction if the flow is subcritical, or proceed in the downstream direction if the flow is supercritical.

2.7 GROUNDWATER FLOW

2.7.1 Definitions

The water that occurs beneath the earth's surface is termed *groundwater*. In general, this definition extends to all such water resident within the pore spaces contained in subsurface soil and rock. In terms of classification, this water can be grouped by depth into the *vadose* and *saturated zones* (Freeze and Cherry, 1979; Hathhorn, 1994). The vadose zone is in turn comprised of an upper unsaturated region and a lower capillary fringe. Characteristically, the pore water within the vadose zone is subject to pressures which are less than atmospheric. In general, this negative pressure decreases with depth from a minimum at the surface to a point where it is again zero, marking a surface known as the *water table*, the physical lower boundary to the vadose zone. Within the vadose zone, the saturation (i.e., the fraction of voids filled with water) acts inversely proportional to the pressure, as governed under capillary theories. In the unsaturated zone, the pores are filled with a combination of gas (usually air) and liquid (usually water). At depth, the saturation approaches nearly 100 percent within the capillary fringe.

Beneath the vadose zone, the hydrogeology is completely saturated. Its composition includes a general layering of aquifers and aquicludes. By formal definition, an *aquifer* is any saturated hydrogeologic unit that contains significant quantity of water, and sufficient permeability to transmit that water. The terms *significant* and *sufficient* here usually refer to the economic viability of developing that water for human use. By contrast, an *aquiclude* is a unit which contains significant water, but does not possess the necessary permeability for development. The sequence of aquifers (with depth) is typically led by a *phreatic unit*, that is, an aquifer whose upper boundary is the water table (also known as an *unconfined aquifer*). Beneath this upper unconfined unit, there exists a sequence of layered confined (or semiconfined) units. The terminology *confined* is derived from the observation that these lower units typically possess internal piezometric heads which are capable of lifting their water above their upper physical boundaries. That physical confinement is

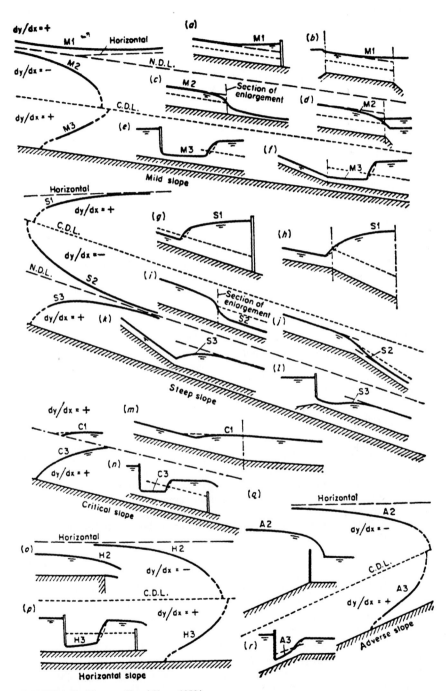

FIGURE 2.19 Flow profiles. (*Chow, 1959.*)

owed to the presence of an aquiclude, thereby trapping the positive pore water pressure generated by the weight of soil, rock, and water above (i.e., overburden).

Quantitatively, the void volume in a porous medium is described by the *porosity*

$$n = \frac{V_v}{V_t} \tag{2.47}$$

in which V_v is the volume of void per unit total volume, V_t. The water in that void volume is defined by the *volumetric moisture content*

$$\theta = \frac{V_w}{V_t} \tag{2.48}$$

or *saturation* (as a percent)

$$S = \frac{V_w}{V_t} \times 100 \tag{2.49}$$

where V_w is the volume of water present. This water is thought to be made up of a mobile and an immobile fraction. The immobile fraction is attached to solid grains, due to molecular adhesion, and is quantified as a *residual moisture content*

$$\theta_r = \frac{V_w^*}{V_t} \tag{2.50}$$

where V_w^* is the volume of immobile water, usually some 1 to 5 percent of the porosity. The mobile fraction, in turn, is able to move freely under gravity or other pressure gradients present. In these latter terms, one speaks of an *effective or kinematic moisture content, or porosity.*

2.7.2 Darcy Law

Following the traditional view of fluid mechanics, the energy of a subsurface fluid particle can be quantified in Bernoulli terms as the sum of the pressure head, elevation, and velocity head. However, the actual velocity of groundwater is usually very small, on the order of about a meter per day or less. Thus, the velocity head can be ignored, and Eq. (2.12) is reduced to a sum of the remaining terms, i.e., $p/g + z$, which is referred to as the *piezometric head h*. Based on this, we may write the energy equation for the flow of water in a tube filled with porous material between any two sections 1 and 2 (Fig. 2.20) as:

$$z_1 - z_2 = h_1 - h_2 = h_f \tag{2.51}$$

Since the average flow velocity is very small, the flow is laminar and the head loss may be taken as proportional to velocity V, i.e.,

$$h_f = \frac{LV}{K} \tag{2.52}$$

in which L = length and K = constant of proportionality, commonly referred to as *hydraulic conductivity*. The value of K is a function of the size and shape of the voids and the viscosity of the water. For example, K has high values for sand and gravel, but low values for clay and rocks.

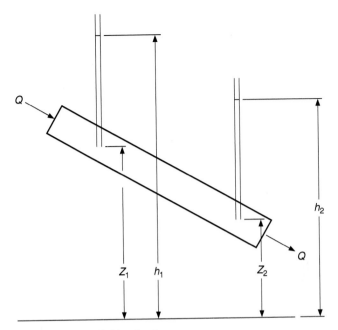

FIGURE 2.20 Definition sketch.

It follows from Eqs. (2.51) and (2.52) that

$$V = -K \frac{h_2 - h_1}{L} \tag{2.53}$$

or, we may write in differential form as

$$V = -K \frac{dh}{dl} \tag{2.54}$$

This is referred to as *Darcy's law*. Note that V is based on the total cross-sectional area of the tube, and in the literature on groundwater flow is usually referred to as *specific discharge* (or *Darcy velocity*, or *Darcy flux*), and the derivative term dh/dl as the *hydraulic gradient i*. A dimensional analysis of Eq. (2.53) shows that K has the dimensions of velocity.

An alternative form of Darcy's law is

$$Q = -K \frac{dh}{dl} A \tag{2.55}$$

or

$$Q = -KiA \tag{2.56}$$

Typical values of K are listed in Table 2.8.

In more formal terms, one notes that V and $-dh/dl$ in Eq. (2.54) are vectors while that of K are second-order tensors. This generalization allows for a complete description of flow in *anisotropic* (i.e., directional dependent) *media*.

TABLE 2.8 Porosity and Hydraulic Conductivity for Various Materials

Material	Porosity	Hydraulic conductivity (m/day)
Soil	0.55	10^{-3}–5
Clay	0.50	10^{-7}–10^{-4}
Sand	0.25	0.06–120
Gravel	0.20	100–7000
Limestone	0.20	10^{-4}–5000
Sandstone	0.11	10^{-5}–0.5
Basalt	0.11	10^{-8}–1000
Granite	0.001	10^{-8}–5

Source: Roberson, et al., 1988.

In order to model groundwater flow, one must account for the conservation of mass

$$\frac{1}{\rho}\frac{\partial}{\partial t}\rho n = \frac{\partial V}{\partial x} + \frac{\partial V}{\partial y} + \frac{\partial V}{\partial z} \tag{2.57}$$

where the term ρn quantifies the mass of water in a bulk volume of porous medium and x, y, and z are the Cartesian coordinates. The term on the left-hand side of Eq. (2.57) represents the storage, while that on the right-hand side the net influx. In applications, however, the storage term must be related to h and Darcy's law inserted for the flux. In doing this, one may write

$$S^{*}\frac{\partial h}{\partial t} = \frac{\partial}{\partial x}\left(K_{x}\frac{\partial h}{\partial x}\right) + \frac{\partial}{\partial y}\left(K_{y}\frac{\partial h}{\partial y}\right) + \frac{\partial}{\partial z}\left(K_{z}\frac{\partial h}{\partial z}\right) \tag{2.58}$$

where S^{*} is an appropriate storage for either confined or unconfined systems, and $K_{x}, K_{y},$ and K_{z} are the hydraulic conductivities in the x, y, and z directions. It is common, however, to use this relation subject to depth averaging. Thus, Eq. (2.58) for a confined system becomes

$$S\frac{\partial h}{\partial t} = \frac{\partial}{\partial_{x}}\left(T_{x}\frac{\partial h}{\partial_{x}}\right) + \frac{\partial}{\partial_{y}}\left(T_{y}\frac{\partial h}{\partial y}\right) \tag{2.59}$$

where T_{x} and T_{y} are the transmissivities along the x and y directions. For an unconfined system, it takes the form

$$S_{y}\frac{\partial h}{\partial t} = \frac{\partial}{\partial x}\left(K_{x}h\frac{\partial h}{\partial x}\right) + \frac{\partial}{\partial y}\left(K_{y}h\frac{\partial h}{\partial y}\right) \tag{2.60}$$

Here, S and S_{y} are known as the *confined storativity* and *unconfined specific yield*, each term used to quantify the volume of water released per unit head decline per unit area of aquifer in their respective systems.

2.8 TRANSPORT PROCESSES

In order to investigate environmental concerns, water resource specialists are usually called upon to study the transport of various substances in water. These substances, referred to as *constituents,* may be contaminants, pollutants, artificial tracers, or other materials being transported in water. The motion and spreading of a constituent may be due to advection, diffusion, and dispersion. In this section, we shall first define commonly used terms and then present the governing differential equation of transport of a constituent in a fluid.

2.8.1 Definitions

The amount of substance in water is specified by the concentration C which is defined as the mass of substance per unit volume of water. A constituent is said to be *conservative* if it does not decay, is not adsorbed or absorbed, and does not undergo chemical, biological, or nuclear transformation.

The transport of a constituent due to bulk motion of the fluid is called *advection.* Dispersion caused entirely by the motion of the fluid is referred to as *mechanical dispersion,* and that mainly due to concentration gradient is called *diffusion.* A combination of diffusion and mechanical dispersion is called *hydrodynamic dispersion.* The spreading of the constituent and its resulting dilution is due to hydrodynamic dispersion.

In order to illustrate these concepts, let us consider uniform, laminar flow through a pipe. The velocity distribution in this flow at a cross section is parabolic. Let a substance be introduced uniformly across the pipe cross section. Due to higher flow velocity at the center of the pipe, the substance will be carried to a greater distance near the center than near the walls. Thus, the material will be dispersed due to nonuniform velocity distribution (Fig. 2.21).

To illustrate different processes, let us consider steady uniform flow through a pipe. Let the flow velocity be U and let the concentration of a constituent be initially zero. Let us assume that at time t_o, we introduce at the upstream end of the pipe a constituent such that concentration C_o is maintained at the pipe entrance. Let us designate the concentration at any location in the pipe by C. In order to plot the results in nondimensional form, we will use relative concentration, $C_r = C/C_o$. The time variation of C_r will plot as a step function, as shown in Fig. 2.22a. If the constituent is conservative and there is no dispersion and diffusion, then the constituent will propagate as plug flow, and at time t_f will have moved to location 1 as shown by the dotted line. However, due to dispersion and diffusion, the relative concentration at

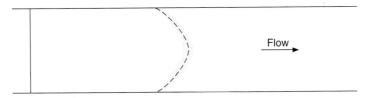

FIGURE 2.21 Dispersion in pipe flow.

the outlet will be as shown in Fig. 2.22b. The constituent front will first appear at time t_1. If we plot the relative concentration at different times as the front moves through the pipe, it will appear as shown in Fig. 2.22c. Due to mechanical dispersion and molecular diffusion, some of the constituent particles move faster than the average flow velocity, while others move slower.

The mass of diffusing constituent per unit time passing through a given cross section in a stationary fluid is proportional to the concentration gradient. This is known as *Fick's First Law* and may be expressed as

$$F = -D \frac{dC}{dx} \tag{2.61}$$

where
- F = mass flux per unit time per unit area $[M/L^2T]$
- D = diffusion coefficient $[L^2/T]$
- C = constituent concentration $[M/L^3]$
- dC/dx = concentration gradient

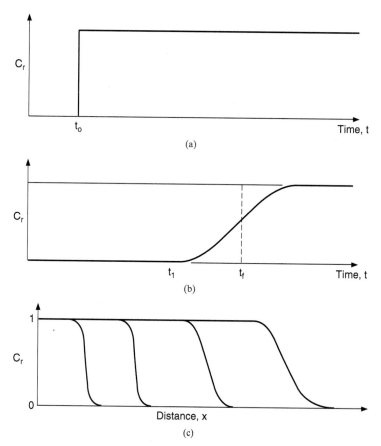

FIGURE 2.22 Dispersion in one-dimensional flow. *(After Freeze and Cherry, 1979.)*

Fick's law is based on molecular transport, and states that a substance tends to equalize its distribution; i.e., it flows from a zone of high concentration to a zone of low concentration.

Transport of constituents due to fluid motion is called *advection*. The advection equation for a conservative substance may be written as:

$$\frac{\partial C}{\partial t} + U \frac{\partial C}{\partial x} = 0 \tag{2.62}$$

in which U = mean fluid velocity.

The *Peclet number* is the ratio of diffusion to advection over the characteristic length L. A small Peclet number indicates that the transport of a substance (Liggett, 1994) is mainly due to diffusion.

2.8.1.1 Governing Equation.   It is necessary for the constituent within an elemental volume to satisfy the law of conservation of mass, i.e.,

Net rate of change of mass of constituent = efflux of constituent
out of the element − influx of constituent into the element
± loss or gain of constituent due to reactions

We may combine the different transport processes to obtain the following equation for one-dimensional flow:

$$\frac{\partial C}{\partial t} + U \frac{\partial C}{\partial x} = \frac{\partial}{\partial x} \left(D \frac{\partial C}{\partial x} \right) + RC + S \tag{2.63}$$

in which R is the reaction rate and S is the source term. This equation is called *advection-dispersion equation*. Note that this form of mass conservation is valid for transport in pipes, open channels, and groundwater. The main difference is in the manner in which the dispersion is quantified in each system, along with the fact that partitioning may take place in groundwater due to the presence of solid particles.

The dispersion coefficient D for pipe flow may be determined from the following equation (Holly, 1975):

$$D = 10.1 R_o u_* \tag{2.64}$$

where R_o = pipe radius
u_* = shear velocity = $\sqrt{\tau_o/\rho}$
τ_o = shear stress at the wall
ρ = mass density of the fluid

For dispersion in waterways, the following equation (Holly, 1985) may be used to estimate D:

$$D = 5.93 u_* h \tag{2.65}$$

in which h = flow depth. Field measurements (Fischer, et al., 1979) indicate that the coefficient 5.93 in this equation varies from 8.6 to 7500.

In groundwater, one speaks of longitudinal and transverse dispersion, i.e.,

$$D_L = \alpha_L V \tag{2.66}$$

$$D_T = \alpha_T V \tag{2.67}$$

in which α_L and α_T are the longitudinal and transverse dispersivities, and V is the seepage velocity.

The U.S. Environmental Protection Agency suggests the following expressions (U.S. EPA, 1986):

$$\alpha_L = 0.1x_r \tag{2.68}$$

$$\alpha_T = 0.33\alpha_L \tag{2.69}$$

where x_r is the transport distance from the source.

REFERENCES

Addison, H., *A Treatise on Applied Hydraulics,* Chapman and Hall Ltd., London, 1954.

Amer. Soc. of Civ. Engrs., (ASCE) "Factors Influencing Flow in Large Conduits," Report of the Task Force on Flow in Large Conduits, *Jour. Hyd. Div.,* Nov. 1965, pp. 123–152.

Boyd, L. M., "Large Hydro-Electric Valves," Canadian Electrical Assn., Hydr. Power Sect. Meeting, January, 1958.

Brown, J. G., *Hydroelectric Engineering Practice,* Blackie and Son Ltd., London, 1958.

Chaudhry, M. H., *Open-Channel Flow,* Prentice Hall, Englewood Cliffs, N.J., 1993.

Chaudhry, M. H., *Applied Hydraulic Transients,* 2d ed., Van Nostrand Reinhold, New York, 1987.

Chaudhry, M. H., and V. Yevjevich (eds.), *Closed-Conduit Flow,* Water Resources Publications, Fort Collins, Colo., 1980.

Chow, V. T., *Open-Channel Hydraulics,* McGraw-Hill, New York, 1959.

Colebrook, C. F., and C. M. White, "Turbulent Flow in Pipes with Particular Reference to the Transition Region Between Smooth and Rough Pipe Laws," *Journal,* Institution of Civil Engineers, London, vol. 11, 1939, p. 133.

Creager, W. P., and J. D. Justin, *Hydroelectric Handbook,* 2d ed., John Wiley, New York, 1955.

Davis, C. V., and K. E. Sorenson, *Handbook of Applied Hydraulics,* 3d ed., McGraw-Hill, New York, 1969.

Elder, R. A., "Friction Measurements in Apalachia Tunnel," *Trans.,* ASCE, 123:1249, 1958.

Fischer, H. B., E. J. List, R. C. Y. Koh, J. Imberger, and N. H. Brooks, *Mixing in Inland and Coastal Waters,* Academic Press, New York, 1979.

Freeman, J. R., "Experiments Upon the Flow of Water in Pipes and Pipe Fittings," Amer. Soc. Mech. Engrs., New York, 1942.

Freeze, R. A., and J. A. Cherry, *Groundwater,* Prentice Hall, Englewood Cliffs, N.J., 1979.

Hathhorn, W. E., Private Communications with M. H. Chaudhry, 1994.

Holly, F. M., "Dispersion in Rivers and Coastal Waters—1. Physical Principles and Dispersion Equations," in *Developments in Hydraulic Engineering—3,* P. Novak (ed.), Elsevier Applied Science Publishers, London, 1985, p. 18.

Holly, F. M., "Two-Dimensional Mass Dispersion in Rivers," Hydrology Paper No. 78, Colorado State University, Fort Collins, Colo., 1975.

Idel'chik, I. E., *Handbook of Hydraulic Resistance Coefficients of Local Resistance and of Friction,* A. Barouch (trans.), Israel Program for Scientific Translation, 1966.

King, H. E., *Handbook of Hydraulics,* 4th ed., McGraw-Hill, New York, 1954.

Liggett, J. A., *Fluid Mechanics,* McGraw-Hill, New York, 1994.

Mahon, R. L., "Hydraulic Butterfly Valves," *Civil Engineering,* ASCE, 27:698 1957.

Miller, D. S., *Internal Flow, A Guide to Losses in Pipes and Duct Systems,* British Hydromechanics Research Association, Cranfield, Bedford, U.K., 1971.

Moody, L. F., "Friction Factors for Pipe Flow," *Trans.,* Amer. Soc. Mech. Engrs., 66:672, 1944.

Mosonyi, E., *Water Power Development,* Publishing House of the Hungarian Academy of Sciences, Budapest, 1957.

Muller, W., "Friction Losses in the High-Pressure Pipe Line and Distributing system of the Lucendro Power Station, Switzerland," *Sulzer Technical Review,* (4):2, 1949.

Nikuradse, J., "Gesetzmassigkeiten der Turbulenten Stromung in Glatten Rohren," *VDI Forschungsheft* 356, 1932 (in German).

Ponni, K., H. Sistonen, and E. Voipio, "Surge Chamber of the Low-Head Power Plant at Pirttikoski," paper 96 112, 6th World Power Conference, Melbourne, Australia, Oct. 1962.

Rahm, L., "Friction Losses in Swedish Rock Tunnels," *Water Power,* Dec. 1958, p. 457.

Rahm, L., "Flow Problems with Respect to Intakes and Tunnels of Swedish Hydroelectric Power Plants," *Bulletin No. 36,* Institution of Hydraulics at the Royal Institution of Tech., Stockholm, 1953.

Roberson, J. A., J. J. Cassidy, and M. H. Chaudhry, *Hydraulic Engineering,* Houghton Mifflin, Boston, Mass., 1988.

Roberson, J. A., and C. T. Crowe, *Engineering Fluid Mechanics,* 5th ed., Houghton Mifflin, Boston, Mass., 1993.

Rouse, H., (ed.), *Engineering Hydraulics,* John Wiley, New York, 1950.

Ruus, E., "Head Losses in Wyes and Manifolds," *Jour. Hydraulic Div.,* ASCE, 96:593–607, March 1970.

Thomas, H. H., and L. S. Whitham, "Tunnels for Hydroelectric Power in Tasmania," *Jour. Power Div.,* ASCE, 90:11–28, Oct. 1964.

U.S. Army Corps of Engineers (USACE), "HEC-2 Water Surface Profiles, User's Manual," Hydrologic Engineering Center, Davis, Cal., 1990.

U.S. Army Corps of Engineers (USACE), "Hydraulic Design Criteria," Waterways Experiment Station, Vicksburg, Miss., 1977.

U.S. Bureau of Reclamation, (USBR) "Friction Factors for Large Conduits Flowing Full," *Engineering Monograph No. 7,* July 1977.

U.S. Bureau of Reclamation, (USBR) *Design of Small Dams,* Denver, Colo., 1977.

U.S. Environmental Protection Agency (U.S. EPA), *Federal Register,* United States Printing Office, 51(9), 1986, p. 1652.

Von Karman, T., "Mechanische Aehnlichkeit und Turbulenz," *Proceedings, 3d International Congress for Applied Mechanics,* vol. 1, Stockholm, 1930, p. 85.

Wunderlich, W., "Die Grndablasse an Talsperren (Deep Sluices in Dams)," *Die Wasserwirtschaft,* March and April, 1963.

CHAPTER 3
WATER ECONOMICS

Robert A. Young
Professor Emeritus
Department of Agricultural and Resource Economics
Colorado State University
Ft. Collins, Colorado

3.1 INTRODUCTION AND OVERVIEW

It is fitting that attention be turned to the subject of managing water. The water resource in its varied forms supplies important benefits to humankind, both commodity benefits and environmental values. However, throughout the world, significant water management problems abound, and in many areas are rapidly becoming worse (Clarke, 1993; Gleick, 1993). Growing populations and incomes are imposing ever-increasing demands for water for agricultural, industrial, and residential uses on limited surface and groundwater supplies. These same forces add to the pollution discharged to the world's waterways, and to the encroachment of human activities onto lowlands vulnerable to flooding, or onto important natural ecosystems.

The subject of water resource management seems to be viewed by many of its practitioners and the public (not to mention the organizers of this volume) to be mainly an interesting hydrologic-engineering problem made inconveniently messy by the mysterious and unpredictable activities of humankind. One task of this chapter is to demonstrate an alternative perspective: That the significant challenges of water management as we approach the twenty-first century are much more issues of "people-coordination" than of physical or technical water management, and that economics and other social sciences have an important role to play in addressing these issues.

Economists study the ways in which individuals and societies respond to the scarcity of means available for achieving a multiplicity of wants. Ideally, economic analysis is designed to anticipate and assess the impact of alternative policies over the longer term and on all affected parties, not only on those immediately affected. Water, and the resources required to both exploit and protect it, are increasingly scarce; hence, it is in the interest of the public to apply economic criteria to water management decisions.

Two principal theses underlie the presentation in this chapter. First, the outward manifestations of water problems—shortages, pollution, conflicts over entitlements to use water, environmental degradation—are but symptoms. To a great extent, the actual problems arise from underlying economic policy failures: Water is underpriced, and its uses are underregulated and/or beset with counterproductive incentives. Second, while human behavior may be inherently more difficult to understand than physical or biological relationships, sufficient regularities can be found in aggregate human interactions with the water resource to help in describing recurring and systematic patterns of water use, diagnosing the sources of water-related problems, and prescribing improved water allocations and the institutional arrangements for achieving them.

This chapter presents for the noneconomist and the nonspecialist in water economics an interpretive survey of how economists currently approach problems arising from managing water, including allocation, pricing, pollution control and natural hazard management. The breadth of the subject matter, combined with the limits of space, dictate that the style is more a literature review than the "how to" character of analyses found in other chapters of this volume. A previous survey (Young and Haveman, 1985) emphasized earlier concerns addressed by water economists, principally the design and application of economic feasibility (cost-benefit) tests for investments in water supply structures. (See also Kneese, 1988). This chapter focuses primarily on literature and issues receiving the most attention during the past ten years—water allocation, water quality, and floodplain management—although interest in these issues arose and significant work occurred much earlier.

A proposition stated earlier is that water management problems have become primarily people-coordination problems. Accordingly, this chapter stresses economic approaches to design of institutional arrangements (i.e., interrelated sets of organizations and rules or laws) which serve to coordinate the activities of people who use or benefit from water resources, so as to achieve the maximum value from scarce water, environmental, and other resources.

The balance of the chapter is divided, as are the activities of economists themselves, into two major categories. *Positive* economics, taken up in Sec. 3.3, is concerned with observable facts and recurring relationships—it seeks to describe, explain, and predict economic phenomena. For example, what are the effects of changing prices, incomes, policies, or technologies on water consumption patterns? Or, what role does water play in regional economic growth? *Normative* economics is concerned not only with matters of fact, but with criteria for policy and questions of optimal policy. For example, *should* a particular water supply project be undertaken? Are markets *preferable* to a government administrative agency in accommodating changing patterns of demand for water? *Should* pollution be discouraged, and if so, with what type of policies? Normative economics employs the empirical studies of positive economics, and combines them with value judgments reflecting notions about the ideal society to derive policy recommendations. The concluding portions of the chapter, Secs. 3.4 to 3.8, address normative analyses, focusing on three specific issues: Water allocation, water quality management, and flood hazard management. The remainder of this introductory section introduces some of most important concepts used by economists in the study of water allocation and management.

3.1.1 The Economist's Approach

A few key concepts identify the economist's particular view on the way an economy functions, and how policies should be designed. (See Rhoads, 1985, for a sympathetic review and critique.) An extract from Gary Becker's Nobel Lecture (1993, pp. 385–386) provides a starting point.

It is a *method* of analysis, not an assumption about particular motivations. . . . The analysis assumes that individuals maximize welfare *as they conceive it,* whether they be selfish, altruistic, loyal, spiteful or masochistic. Their behavior is forward-looking, and is assumed to be consistent over time. In particular, they try as best as they can to anticipate the uncertain consequences of their actions. . . . While this approach to behavior builds on an expanded theory of individual choice, it is not mainly concerned with individuals. It uses theory at the micro level as a powerful tool to derive implications at the group or macro level. [Emphasis in the original.]

One of the most important economic concepts is that of *opportunity costs,* which refers to the benefits foregone when a scarce resource is used for one purpose instead of its next best alternative use. It is important to recognize that spending and regulatory decisions that use scarce resources impose costs in the form of foregone alternatives—that is, opportunities that can no longer be undertaken.

Another significant concept identified with the economic approach is *marginalism.* In the context of resource allocation decisions, marginalism emphasizes the importance of considering incremental gains relative to incremental costs. Rather than setting spending decisions on ranking of problems by their seriousness, spending should be prioritized on the basis of the marginal potential gains relative to incremental costs.

Closely linked with marginalism are the notions of *diminishing returns* and *resource substitutibility.* Diminishing returns refers to the fact that, on the producers' side, increases in the use of a given input (when all other inputs are held constant) lead to decreasing increments of product. Similarly, for consumers, additions to consumption yield decreasing increments of utility or satisfaction. Resource substitutibility means that consumers and producers are not limited to fixed proportions of resource use in their consumption or production activities. Changing relative prices, or scarcities, make it attractive to substitute plentiful resources for scarce ones. Farmers may take more care and expend more labor in crop irrigation under scarce water conditions than they would under conditions of relative plenty, or householders might replace inefficient plumbing fixtures as water prices rise.

Water was involved in one of the most famous intellectual conundrums in the history of economic thought: The water–diamond paradox. This problem was resolved in the eighteenth century by what came to be known as the distinction between value in use and value in exchange. Although its price is low, water has enormous value *in use* to humans, because it is necessary to existence. Diamonds, in contrast, are not at all essential, but have high value *in exchange* (on the market). Final resolution of the paradox came with the additional distinction between total and marginal values. The total utility of water clearly exceeds that of diamonds. However, the marginal utility of diamonds is greater than the marginal utility of water. Because diamonds are scarce (the marginal costs of acquiring more are high) and the marginal utility for diamonds is high, diamonds are priced higher than is water.

Another important idea is that *incentives matter.* Based on the belief that individuals act to maximize welfare as they see it, economists expect that the individual producer or consumer will adjust behavior when incentives change. Since the time of Adam Smith over two centuries ago, an emphasis on designing institutions so as to make private interests more consistent with public goals has been a main attribute of the economist's approach. Paying attention to private interests in designing government programs will frequently permit important goals to be achieved more cheaply.

Economics has been characterized as the study of *unintended consequences* of human action in that part of the social system encompassing production, exchange, and consumption of goods and services (O'Driscoll, 1977). Economics goes beyond direct observation. The nonspecialist can recognize the immediate effect of policy decisions: A policy of holding water prices below costs makes the resource less expensive and improves the economic well-being of some consumers. In this case, other consumers must pay part of the costs. An investment in a reliable, high-quality water supply may provide increased employment and income to the regional economy. However, economists attempt to also address the hidden impacts of these policies, and illuminate considerations not readily recognized as aspects of these problems. Low-cost water will lead to overuse and waste of the resource, while the financing of an investment in water supply implies foregone employment and income elsewhere in the economy.

Nonmarketed goods and services valuation is an important aspect of environmental and resource economics. Economists recognize that people value things—including many important services of the earth's water supply—that they do not purchase through a market, or that they may value for reasons independent of their own purchase and use. Further, not everything that reduces utility—such as pollution—is costed in markets. Although practitioners of the dismal science are sometimes equated with Oscar Wilde's cynic (who knows the price of everything and the value of nothing), environmental economists in fact spend much of their professional efforts attempting to estimate the public's value (often called a *shadow price*) for nonmarketed goods and services. The modern economic paradigm assumes that values of goods and services rest on underlying demand and supply relationships, that are usually, but not always, reflected in market prices. Economics is not just the study of markets, but more generally, the study of preferences and human behavior (Hanemann, 1994).

The principal strengths of the economic approach to rational policymaking are its focus on assessing the consequences—both beneficial and adverse—of policy actions and its attempts to be sensitive to the particular facts of decision situations. By expressing consequences in terms of a common denominator of money value, it provides a method of resolving tradeoffs among competing and valued ends, including taking account of the economic costs (foregone benefits) of achieving those ends.

It would be pleasing to be able to assure the reader that resource economists all speak with the same voice on how economics is applied to research and policy issues. As with the discipline taken as a whole, and indeed, in common with other social sciences, resource economists exhibit a spectrum of methodological perspectives, as well as diverse ideological views, on the appropriate role for private and government entities in managing natural and environmental resources. Moreover, there are important differences across the profession regarding the uses and limits of economics as a policy tool (see Randall, 1985 for a full discussion). In contrast to the perceived mainstream focus on markets and prices, *institutionalists,* from a policy stance skeptical of market mechanisms, have emphasized the importance of studying the distributive effects of economic institutions. Many of their concerns, however, have been coopted into mainstream environmental and resource economics practice. Some economists in the institutionalist tradition reject the mainstream's primary emphasis on economic efficiency as a criterion for policy analysis, emphasizing the importance of other values, such as income distribution (e.g., Bromley, 1991). Others have focused on the limitations of market allocation systems for dealing with potential long-term environmental problems (Costanza, et al., 1990). On the other hand are the *Individualists* or *Austrians,* who emphasize the role of individual liberties as well as economic efficiency, and urge decentralization, property rights, and markets for resolving water and environmental problems (e.g., Anderson and Leal, 1991).

3.2 WHY IS WATER POLICY DESIGN SO DIFFICULT? ECONOMIC AND RELATED CONSIDERATIONS

Water *is* indeed different. A number of special characteristics distinguish water from most other resources or commodities, and pose significant challenges for the design and selection of water allocation and management institutions. These unique characteristics are considered here under four headings: Water supply, water demand, social attitudes, and legal-political considerations.

3.2.1 Physical and Hydrologic Characteristics

3.2.1.1 Mobility.   Water, usually a liquid, tends to flow, evaporate, and seep as it moves through the hydrologic cycle. Mobility presents problems in identifying and measuring specific units of the resource. Due to its physical nature, and for other reasons, water is what economists call a *high-exclusion cost* resource, implying that the exclusive property rights which are the basis of a market or exchange economy are relatively difficult and expensive to establish and enforce.

3.2.1.2 Uncertainty in Supply.   Water supplies, although generally renewable, are typically relatively variable and unpredictable in time, space, and quality. Local water availability usually changes systematically throughout the seasons of the year (with climatic variations) and over longer cyclical swings. Forecasts of significant global climate change—from both natural and human causes—raise concerns about longer-term supply trends. Problems for humankind are encountered at the extremes of the probability distributions of supply (floods and droughts). Too much or too little water can, of course, yield adverse effects on human societies. Flooding from excess rainfall or snowmelt is an important hazard in many areas, imposing significant costs, and most governments have undertaken programs for flood control. At the opposite extreme, droughts can have a major negative impact on an economy, particularly those relying heavily on agriculture.

3.2.1.3 Solvent Properties.   Water is a nearly universal solvent which, together with plentiful supply, creates an inexpensive capacity for absorbing wastes and pollutants, and for diluting them and transporting them to less-adverse locations. Managing the assimilative capacity of the hydrologic system is, then, understood as the management of a scarce collective or public asset. In many situations, water quality considerations are increasingly as important as direct use and other public benefits.

3.2.1.4 Pervasive Interdependency Among Users.   The physical nature of water, combined with supply variability, causes a unique but unpredictable degree of inter-relationship among water users. Water is rarely completely consumed (i.e., lost to evaporation) in the course of human consumption or production activities. So-called "water uses" generally result in return flows to an aquifer or stream. In crop irrigation, for example, it is not unusual to find that fifty percent or more of water diverted returns, in the form of surface or subsurface drainage, to the hydrologic system, while an even larger proportion is typically returned from municipal and industrial withdrawals. Other, particularly downstream, users are greatly affected (for good or ill) by the quantity, quality, and timing of releases or return flows by upstream users. These interdependencies lead to effects called *externalities,* (or *spillover* or *third party* effects), which are uncompensated side effects of individual activities. In such cases, the full costs of economic activity are not recognized in individual producer or consumer decisions, and outcomes for the society will be suboptimal.

3.2.1.5 Site-Specificity of Water Problems.   Because of water supply variations, and also due to localized demand-side considerations, problems with water resources are typically rather site-specific. The relative supplies of surface and groundwater at any location depend, of course, on climatic variations (rainfall and snowpack), as well as on the available aquifer storage. Water demand and quality issues are likewise specific to population size and economic development level. The implication of these facts is that water management problems tend to be specific to areas, and policy treatment often needs to be adapted to local conditions.

3.2.1.6 Economies of Large Size.   The capture, storage, and delivery of water— especially surface water—exhibits economies of large size (falling unit costs). When costs decline over the range of existing demands, a single supplying entity is the most economically efficient organizational arrangement. This is a classical natural monopoly situation—the least-cost supply is with a single producing organization. Accordingly, public regulation or ownership is often invoked to avoid monopolistic pricing. (Groundwater seems to present a different story, as most size economies are achieved at relatively small outputs. Moreover, such size economies as are observed may be offset by increased pumping costs, and rising third-party spillover costs, due to water-table drawdown.)

3.2.1.7 Distinctive Attributes of Groundwater Supplies.   Groundwater aquifers are an important source of water throughout the world. An aquifer is defined as a geologic formation actually, or potentially, containing water in its pores or voids, which can be removed economically and used as a water supply. Several differences in supply attributes from surface water can be identified for groundwater, including a slow rate of flow and extra difficulties in accurately knowing the potential yield and quality of an aquifer. This point will be elaborated upon somewhat in Section 3.5.

3.2.2 Water Demand—Characteristics from Users' Perspectives

3.2.2.1 Preliminary Remarks on Water Demand.   People obtain many types of value and benefits from water. Because each benefit type usually calls for specialized management approaches, it will be useful to classify the types of value into five classes. These are: (1) commodity benefits, (2) waste assimilation benefits, (3) public and private aesthetic and recreational values, (4) species and ecosystem preservation, and (5) social and cultural values. The first three of these are treated here as economic considerations, because they are characterized by increasing scarcity, and the associated problems of allocation among competing uses to maximize economic value. The final pair, preservation and sociocultural issues, are discussed separately as noneconomic values.

It may be most useful to begin by recognizing that the economic characteristics of water demand varies across the continuum from *rival* to *nonrival* goods. A good or service is said to be *rival* in consumption, if one person's use in some sense precludes or prevents uses by other individuals or businesses. Goods that are rival in consumption are the types that are amenable to supply and allocation by market or quasi-market processes, and are often called *private* goods. The opposite end of the continuum is occupied by goods that are *nonrival* in consumption, meaning that one person's use does not preclude enjoyment by others. Goods that are nonrival are often called *public* or *collective* goods. Because nonpayers cannot be easily excluded, private firms will not find it profitable to supply nonrival goods. Water for agricultural or industrial use tends toward the rival end, while the aesthetic value of a beautiful stream is nonrival.

The significance of nonrivalry can be better understood by noting its association with high exclusion costs (Schmid, 1989). *Exclusion cost* refers to the resources required to keep those not entitled from using the good or service. Water is frequently a high-exclusion-cost good because of its physical nature: When the service exists for one user, it is difficult to exclude others. In such cases, it is hard to limit the use of the good to those who have helped pay for its costs of production. (The refusal of some beneficiaries to pay their share of the provision of a public good, from whose benefits they cannot be excluded, is called the *free rider* problem. To circumvent the problem, public goods must normally be financed by general taxes, rather than by specific charges.)

3.2.2.2 *Variety and Economic Characteristics of Water Uses.*

The first type of benefit is the *commodity benefit:* Those derived from personal drinking, cooking, and sanitation, and those contributing to productive activities on farms and in businesses and industries. What are here called *commodity values* are distinguished by the fact of being rival in use, meaning that one person's use of a unit of water necessarily precludes use of that unit by others. Commodity uses tend to be private goods or services.

Some additional distinctions will be helpful in continuing the discussion of commodity-type uses. Those uses of water which normally take place away from the natural hydrologic system may be called *withdrawal* (or *offstream*) uses. Since they typically involve at least partial consumption (evaporation), they may further be distinguished as *consumptive* uses. Other economic commodity values associated with water may not require it to leave the natural hydrologic system. This group may be labelled *instream* water uses—hydroelectric power generation and waterways transportation are important examples. Since instream uses often involve little or no physical loss, they are also called *nonconsumptive* uses. Although instream uses do not "consume" much water, in the sense of evaporating it to the atmosphere, they do often require a change in the time and/or place of availability—as with releases from a hydropower reservoir—and therefore exhibit some aspects of the rivalness of a private good.

The second general class of economic benefit of water use is the value of waste disposal. Bodies of water are significant assets because of their assimilative capacity, meaning that they can carry away wastes, dilute them, and, for some substances, aid in processing wastes into less undesirable forms. The assimilative capacity of water is closer to being a public or collective (rather than private) value, because of the difficulty in excluding dischargers from utilizing these services.

A third type of economic benefit from water is its value for recreation, aesthetics, and fish and wildlife habitat. Once regarded as luxury goods inappropriate for governmental concern, these benefits are increasingly important. The citizens of developed countries more and more often choose water bodies for recreational activities. In developing nations, as income and leisure time grow, water-based recreation is also increasingly important to the citizens, and often provides a basis for attracting the tourist trade. As is waste assimilation, recreational and aesthetic values are nearer the public-good end of the spectrum. Enjoyment of an attractive water body does not necessarily deny similar enjoyment to others. However, congestion at special sites, such as waterfalls, may adversely affect total enjoyment of the resource. Significant instream values are also found as habitat for wildlife and fish.

The economic value of water can also extend to *nonuse* values. In addition to valuation of goods and services which are actually used or experienced, it is observed that people are willing to pay for environmental services they will neither use nor experience. Nonuse values are benefits received from knowing that a good exists, even though the individual may not ever directly experience the good. Voluntary contribution toward preserving an endangered fish species is an illustrative example. Many resource economists argue that nonuse values should be included with use values, so as to more accurately measure total environmental values.

Some environmentalists object to policies which acknowledge the commodity aspects of water, because they fear this will lead to sacrifice of important public benefits. However, it will likely be more fruitful to recognize both the commodity and environmental characteristics of water demand, and design policies with this duality in mind.

3.2.2.3 Water Is a Low-Valued Commodity. The economic value per unit weight or volume of water tends to be relatively low, placing water among commodities which economists call *bulky*. Capital and energy costs for transportation, lifting, and storage tend to be high relative to economic value at the point of use. (For example, in crop irrigation, much of the water applied may yield direct economic value—profit after production costs—of less than US$0.04 per ton.) Extensive water-conserving technologies (closed conduits, recycling, and metering) as well as incentives for conservation (marketable property rights, increasing-block pricing) are presently found only where water is recognized as scarce and valuable. (Although water may be low valued, it nevertheless may be underpriced relative to cost of supply or opportunity costs.)

3.2.2.4 Variability in Demand. As on the supply side, variability is also important on the demand side. Agricultural needs oscillate, responding to temperature and rainfall patterns over seasons of the year and over longer cycles. Residential and industrial water uses also vary depending on daily, weekly, and seasonal considerations. Both storage and conveyance systems and management institutions must be prepared to satisfy peak loads in high demand periods.

3.2.3 Social Attitudes Toward Water

3.2.3.1 Conflicting Social Cultural Values. Because water is essential to life, and because clean water and sanitation are essential to health, market allocation mechanisms are often rejected in favor of regulatory approaches. The Dublin Conference on Water and Environment in January 1992 asserted as one of its guiding principles for action that "... it is vital to recognize first the basic right of all human beings to have access to clean water and sanitation at an affordable price." The significance of water for life receives further emphasis in arid regions, where crop irrigation is essential to production of the other staff of life, food.

Moreover, many view water as contributing special cultural, religious and social values, and prefer not to have water treated as an economic commodity. Goals other than economic efficiency play an unusually large role in selecting water management institutions. Boulding (1980) has observed that "the sacredness of water as a symbol of ritual purity exempts it somewhat from the dirty rationality of the market." Some cultures or religions (i.e., Islam) proscribe water allocation by market forces.

However, focus on the necessity for life as the basis for design of social institutions tends to obscure the fact that in modern societies only a tiny fraction of water consumption actually is used for drinking and for preservation of human life. Most water is used for convenience, comfort, and aesthetic pleasure. In the arid western United States, residential water withdrawal frequently exceeds 400 liters per capita per day, up to nearly half of which may be applied to irrigate lawns and gardens, with most of the remainder for flushing toilets, bathing, and washing cars (Gleick, 1993).

3.2.4 Legal-Political Considerations

A number of considerations for water policy design fall on the border between economics and political science, or what is sometimes called *political economy.*

3.2.4.1 Transactions Costs Versus the Relative Scarcity of Water. The term *transactions costs* refers to the resources required to establish, operate, and enforce a resource allocation, management, or regulatory system. Transactions costs are also termed ICE costs, because they comprise the costs of obtaining *information* (such as knowledge about the needs and attitudes of other participants), *contracting* costs (resources required to reach agreements), and *enforcement* costs (the expense of enforcing contracts and public laws and regulations). Given the supply and demand characteristics of water noted earlier, transactions costs for water management and allocation tend to be high relative to its value. Where water is plentiful relative to demand, water laws tend to be simple and only casually enforced. Where water is scarce, more elaborate management systems have emerged. In many regions, water supplies are only now becoming scarce enough to require formal management systems. Increased resource scarcity and technological advances, which reduce the transactions cost of monitoring and enforcing regulations, both act to encourage innovations in allocative institutions, so as to economize on the scarce resource.

3.2.4.2 The Cumulative Impact of Many Small Decisions. A related point is that water policy makers must often confront the problem aptly termed the "tyranny of small decisions" by Alfred Kahn (1965). Even though each individual act of water use, taken alone, might have a negligible impact, the sum total can be of major importance. One example is found in the rapid spread of tubewells for irrigation in south Asia. Any one of these small wells would have little effect on the total groundwater supply, but in total, some aquifers are being rapidly depleted. Another case is nonpoint pollution from chemicals carried by runoff from farmers' fields or from forest harvest. Effective public regulation of many small, scattered decision-makers is exceedingly difficult, but increasingly necessary.

3.2.4.3 The "Common Pool" Aspect of Water Resources. *Common pool natural resources* are defined by two characteristics (Gardner, et al., 1990). The first is rivalry, or subtractibility, meaning that a unit of resource withdrawn by one individual is not fully available to other potential users. The second is that the costs to a government entity of excluding potential beneficiaries from exploiting the resource is relatively high. Water and other fugitive or mobile resources, such as petroleum, wildlife, or migratory wildfowl, can be examples of common pool resources.

Common pool problems, or dilemmas, arise when individually rational resource-use decisions bring about a result that is not optimal when considered from the perspective of the exploiters as a group, i.e., of society. The roots of the problems associated with common pools are found in the inadequate economic and institutional framework within which the resource is exploited. Common pool resources have been typically utilized in an open access framework, within which resource ownership is according to a rule of capture. When no one owns the resource, users have no incentive to conserve for the future, or to consider the foregone benefits to others, and the self-interest of individual users leads them to over-rapid and/or excessive exploitation. The characteristics of the economic institutions governing their use is the fundamental issue in managing common pool resources.

Gardner, et al. (1990) specify the following additional conditions as necessary to produce a common pool resource dilemma. First, there are many appropriators, or users, withdrawing the resource. Second, the actions of the individual users, given the particular situation with respect to the resource itself, the characteristics of users, demand for the resource, and extraction technology, bring about suboptimal outcomes from the group's viewpoint. Finally, there must exist institutionally feasible strategies for collective management of the resource that are more efficient than the current situation. (See also Ostrom, et al., 1994.) Both surface and ground water have often been utilized under open access rules, leading to various forms of suboptimality.

To sum up, we see that water is truly an unusual resource, and for numerous physical, social, political, and economic reasons, presents special challenges to getting the incentives right. Before taking up the problems of water policy design, the discussion turns to positive economic analysis applied to water.

3.3 POSITIVE ECONOMICS OF WATER: EMPIRICAL RELATIONSHIPS AND MEASUREMENTS

To perform empirical descriptive measurements of important relationships, to explain and understand the regularities beneath the apparently haphazard occurrences of economic life, and to predict the effects of changes in the society and its policies on water use, are the purposes of positive economic analysis. With regard to the water resource, the regularities that are of primary interest here are: (1) those socioeconomic factors that affect the amount of water consumed, (2) the responsiveness of water use to price and other variables, and (3) the relationship between water supply and regional economic growth. This section will examine empirical evidence on the general patterns of water consumption and on the factors which influence water demand and supply.

Descriptive statistics illuminate broad water use and consumption patterns. In the United States, crop irrigation is the major user of water, accounting for 42 percent of withdrawals and 84 percent of consumption in 1990. The domestic-commercial category represented 11 percent of withdrawals and 7 percent of consumption, while industry took 8 and 5 percent, and thermoelectric power accounted for 39 and 4 percent of the same categories. Water withdrawal and consumption patterns elsewhere in the world reflect climate, degree of economic development, and other factors, but as in the United States, crop irrigation represents the major consumptive use of water in the world. (Space limits preclude the display of the national and worldwide data on water withdrawals and consumption by sector. See Solley, et al., 1993, or Rogers, 1993, for summaries for the United States, and Gleick, 1993, for both a global overview and detailed data on water use.)

3.3.1 Measuring Demand for Water

Human uses of water are conveniently divided into withdrawal (offstream) and in-place (instream) categories. *Withdrawal uses,* those for which water resources are diverted from their natural bodies of water, include agriculture, industry, commercial, and residential purposes. In-place uses include values for recreation, fish and wildlife habitat, hydroelectric power, waste load assimilation, and the like. A further cross-classification identifies whether the demand is from use and the price of water. Demand is the willingness of users or consumers to pay for goods and services, as that willingness to pay varies (usually inversely) with the amount being purchased. Water demand is very site-specific, varying with a range of natural and socioeconomic factors. The demand relationship is represented graphically by the familiar demand curve, or algebraically as:

$$Q_W = Q_W(P_W, P_a, P, Y, \mathbf{Z}) \tag{3.1}$$

where Q_W refers to the individual's level of consumption of water in a specified time period; P_W refers to the price of water; P_a denotes the price of an alternative water source; P refers to an average price index representing all other goods and services; Y is the consumer's income, and $\mathbf{Z}$ is a vector representing other factors, such as climate and consumer preferences.

Similarly, a model for producers' demands for water can be obtained from the theory of a cost-minimizing producer:

$$Q_W = Q_W(P_W, P_i, P_a, X, \mathbf{S}) \tag{3.2}$$

where, as before, P_W and P_a represent the price of water from the given system and from an alternative source; P_i represents a vector of prices of inputs (capital, labor, and materials); X stands for the quantity of product to be produced; and $\mathbf{S}$ represents a vector of other factors, such as technology and climate. (See Kindler and Russell, 1984; Munasinghe, 1992, Chap. 7; or Spulber and Sabbaghi, 1993, for more complete developments of these models).

In the everyday language of water management, *demand* is often used synonymously with *requirement*. However, these two ideas should be distinguished. If the quantity used is the same no matter what the price, the term *requirement* is appropriate. While the human body requires some minimum daily amount of water for survival, the true requirement for survival is a very small fraction of the amount we normally use—almost all water uses are for production, sanitation, or convenience in daily living. Both consumers and producers can normally change their patterns of water use if price or scarcity makes it in their interest to do so. Households can decrease the frequency or duration of water using activities (i.e., bathing or lawn watering), install water-saving plumbing fixtures or change their outdoor landscaping, while producers might similarly adopt water-efficient technologies or altered production processes.

Forecasting of water use into the distant future is fraught with difficulties. The simplistic extrapolation of trends in per capita "requirements" in water system planning has resulted in many cases in which future water use was greatly overestimated. Rogers (1993, Fig. 6.1) compares actual 1990 withdrawals in the United States with several authoritative forecasts made in the 1970s which foresaw large growth in water use by the 1990s and beyond. In reality, by 1990 the amount of freshwater withdrawal had actually declined from its 1975 level. (See also Solley, et al., 1993.) Some of that decline was doubtless due to increased application of economic rationing mechanisms, such as metering combined with pricing schemes designed to confront customers with more of the costs of their water-use decisions. Bower, et al. (1984) note that similar errors of overestimating water demand were made in Europe, citing the increasing use of volumetric charging mechanisms and the response of industrial water users to water quality regulations, which had the incidental effect of reducing withdrawals because of increased recycling.

3.3.1.1 Measuring Water Users' Response to Price and Other Factors: Withdrawal Uses. The economic study of regularities in consumer use of water begins with response to prices and incomes. Economic analysts usually apply either of two broad approaches to modeling water user behavior (Kindler and Russell, 1984). One approach is statistical, making inferences from actual observations on quantities consumed, together with the corresponding data on prices, incomes, climatic factors and so on. An abstract demand function is formulated, which asserts a hypothesized connection between water consumption (the variable to be explained) and factors influencing the dependent variable. Parameters of demand equations in this approach are inferred via statistical inference, usually multiple regression techniques. (Statistical inference applied to economic phenomena is referred to as *econometrics.*) Data to estimate the function can come from repeated observations over time on the same activity (time series data), or simultaneous observation of many activities during the same time period (cross-sectional data). Sufficient numbers of accurate observations for developing reliable water-demand functions have been difficult to come by, because of the limited number of observable cases where water is volumetrically priced, although the proportion of water suppliers using meters has increased greatly in recent years. Demand equation estimates based on cross-sectional data are more frequent in the literature, because of the limited records for time series, but time series are also attempted. Recent water-demand studies using the statistical method are Schneider and Whitlach (1991), Griffin and Chang (1992), and Renzetti (1993). See Chap. 23 for details.

The second approach employs mathematical optimization combined with engineering design, and is usually applied to producers' water use (in agriculture and industry). Studies are performed to identify a number of feasible water-using unit processes, together with alternative designs, and alternative inputs, operating conditions, and associated costs for each. This information is incorporated into a mathematical programming model with an objective function formulated to select the combination of processes which maximizes firm profits or minimizes costs. The demand function is approximated by solving the model for numerous water prices, and recording the amount of water required at each price. Solutions are often repeated under alternative assumptions about product prices, resource constraints, and technology, to determine sensitivity to alternative model specifications and to learn more about the nature of response patterns.

The mathematical programming approach has long been used to model irrigators' response to price, cost, or policy changes. Early models provided only for omission of crops with the lowest return per unit water in response to increased price or scarcity. Subsequent approaches (e.g., Bernardo, et al., 1986) incorporated choices in the number, timing, and amounts of irrigation application (based on highly detailed agronomic simulations of crop response to water stress), and choices among irrigation application technologies. Howitt, Watson, and Adams (1980), and Dudley (1988), represent other variations. (See Boggess, Lacewell, and Zilberman, 1993, for more on theory and practice regarding agricultural water demand studies.)

Economic models of industrial water use decisions might include variables representing alternative water or product prices, public policy constraints (as for water quality), varying technologies or production processes, varying prices of other factor inputs, varying product mix, and/or output level. Water price may need to reflect not only the basic cost of intake water, but of treatment costs prior to use, and disposal costs (including waste treatment). The mathematical optimization approach to industrial water demand was pioneered by Russell for the case of petroleum refining (see review in Russell, 1984) and has been applied to electricity generation by Stone and Whittington (1984).

3.3.1.2 Elasticities of Demand for Water.   The responsiveness of consumers' purchases to varying price is an important concept to microeconomists. A measure of this responsiveness, the *elasticity of demand* with respect to a demand-determining variable, is defined as the percentage by which the quantity taken changes in response to a one-percent change in the variable. The most frequently used elasticity concept is *price elasticity,* which is defined as the percentage change in quantity taken if price is changed one percent. The *income elasticity* of demand is a related concept: It measures the percentage change in quantity demanded associated with a one-percent change in consumer income.

The *price elasticity of demand* for water measures the willingness of consumers to give up water use in the face of rising prices, or conversely, the tendency to use more as price falls. In one sense, price elasticity reflects the availability of opportunities for water conservation, or for substituting other goods or services for water. For example, water-saving toilets and shower heads reduce the amount of water needed by a typical family. Sprinkler irrigation systems lower the water needed to irrigate an area of crops. The larger the range of conservation opportunities, and the smaller the expenditures required to use them, the more price-elastic the demand. It is important to distinguish between short-run and long-run price elasticities. A change in price is not likely to have a major immediate effect, but we would expect that the long-run price elasticity, when sufficient time has passed to allow adjustments, will be greater than in the short run.

Empirical estimates of the price elasticity of residential water demand in the United States usually report that demand is price-inelastic, meaning that a rise in price results in a less than proportional reduction in water taken. This also means that a rise in price will increase total revenue of the vendor. The estimates vary, depending on such factors as the season, local conditions, whether the adjustment period is long-run or short-run, and the form of function hypothesized to represent the data. Almost all estimates of long-run price elasticity of residential water demand in the United States seem to fall between -0.3 and -0.7, meaning that other factors held constant, a 10 percent increase in price would lead to a 3 to 7 percent decrease in amount purchased. Demand for water used solely in homes is found to be even more price inelastic, around -0.2, but when outdoor uses (watering lawns and gardens) are considered in isolation, they are found to show an elastic response, especially in regions where frequent rain may be expected to serve as a substitute (Howe, 1983). Schneider and Whitlach (1991) present one of the most comprehensive water-demand studies available, and also provide an extensive survey of the previous literature. They analyzed a very large data set (some 30 years of individual accounts from a number of communities supplied by the City of Columbus, Ohio, water system) and derived price elasticity estimates for each of 5 sectors, as well as for the total of all metered-demand accounts. For residential demands, their long-run model yielded price elasticity estimates of -0.26, while short-run adjustments (1 year) were quite inelastic at -0.12. For total metered demand, including residential, commercial, industrial, government, and schools, their analysis came up with long-run price elasticities of -0.50 and short-run of -0.14. Griffin and Chang (1991) found from a sample of Texas counties that demand is somewhat more inelastic in winter (about -0.3) than in summer (about -0.4). Lyman (1992) compared peak with off-peak demands from a small Idaho city, finding a quite elastic response to peak prices (-2.6 to -3.3), while long-run off-peak price elasticity was about -0.7.

Due to the difficulty in obtaining observation sets with a wide enough range of incomes, empirical estimates of the income elasticity of water demand are not plentiful. As expected, those estimated in the United States tend to show that water is a normal good, with consumption reflecting an income consumption curve that increases with income, but at a decreasing rate with higher incomes. Income elasticity is found to be small but positive; an increase in income brings about a less-than-proportional increase in quantity consumed. Schneider and Whitlach (1991), for example, report a value of 0.21 for their Columbus, Ohio study, and other estimates tend to fall in the 0.30 to 0.50 range. An increasingly important income elasticity for water planners is that for recreational services of the water resource. Walsh (1986, p. 267) reports that the income elasticity for general expenditures on recreation in the United States is relatively large, in excess of 1.0, while for specific water-based activities (swimming, boating, fishing) it ranges from 0.3 to 0.5. The rise in incomes in Western economies then is seen to lead to a rapid rise in water-based recreational activities, and accounts for much of the increasing public concern with access to high-quality water for recreation, and for the competition for in-place uses against traditional offstream water-use sectors.

Municipal and private water agency managers sometimes contend that water use is completely unresponsive to price increases. However, negative effects of price increases on use tend to be offset to a degree by positive income effects. Moreover, once the price increases upon which these conclusions are based are adjusted for inflation, it has sometimes been found that little—or often, no—real price increase actually occurred. Martin and Kulakowski (1991), for a case study of Tucson, Arizona, conclude that prices must be raised by the rate of inflation each year, plus approximately the rate of change in real per capita income, just to maintain a constant rate of per capita water use.

Empirical studies of residential water demand have recently been completed for a number of developing countries. Demand-function studies based on metered water are rarely feasible, of course, but some important and interesting aspects of demand can be measured. The World Bank Water Demand Research Team (1993) reports on studies of the willingness to pay for improved levels and types of rural water supply facilities in Latin America, Africa, and South America. Both direct questioning (contingent valuation) and indirect inference methods were employed. Income elasticity estimates were quite low, generally in a range around 0.10. However, in some African locations, rural villagers were found to be spending nearly 10 percent of their income on water in the dry season, but the more frequent finding was that willingness to pay for improved quality or convenience was only a few percent of income. The characteristics of existing supply alternatives (cost, quality, reliability) were crucial determinants of willingness to pay.

Turn now to elasticity measures for water as an intermediate good, in agriculture and industry. Herrington (1987) summarizes a number of earlier modeling studies on irrigation demand, which show demand to be quite price-inelastic at average price levels, but more responsive at higher water charges. Moore, Gollehon, and Negri (1994) report from U.S. Census data that output elasticities of irrigation water production are highly inelastic for 13 irrigated crops in the 17 western U.S. states, implying that marginal reductions in water supply would have relatively small effects on crop production in that region, and conversely, demand would be responsive to price. Only a few studies have examined industrial water demand with econometric procedures. As with residential use, industrial water use is reported to be somewhat inelastic (Schneider and Whitlach, 1991). Renzetti (1993) reports that from a large sample of Canadian industries, the average price elasticity for intake water was −0.38, while individual industry estimates ranged from −0.15 to −0.59. Renzetti also confirms that recirculation of water is a substitute for both water intake and water discharge, and concludes from this that economic incentives, such as effluent fees, are likely to encourage both reduced water intake and increased recirculation.

3.3.1.3 *Demand for Water-Related Public Benefits.*

In those cases where water yields a public or collective good (nonrival in consumption), neither diversion for production nor purchase prices exist, and special data collection and demand evaluation methods must be adopted. These cases are often associated with recreation or aesthetic enjoyment of water in its natural surroundings, or with water quality improvement. Several different approaches to measuring benefits and costs of instream and public benefits are commonly employed. See Freeman (1993) for an authoritative and exhaustive treatment. Cropper and Oates (1992) and Braden and Kolstad (1991) also provide particularly useful reviews of the issues. Walsh, Johnson, and McKean (1992) distill generalizations regarding the findings of two decades of earlier empirical studies.

Two general lines of approach have been developed, both based on user surveys of actual or hypothetical behavior. The first broad approach is called the *observed indirect* approach. Relying on observations of actual expenditure choices made by consumers, the method infers net willingness to pay from the differences in expenditures observed with varying levels of environmental amenities. If the use of water-based recreational services influences the demand for any marketed commodity, observations on purchasing behavior related to the marketed commodity can be analyzed to derive information on the willingness to pay for the environmental amenity.

The *travel cost* method is the most widely used example of the observed indirect method. The underlying assumption of the travel cost approach is that observable recreationist behavior, as related to increasing costs of travel, reflect the changes in demand for the activity which would occur if prices were actually charged. The travel cost approach involves two steps: The first is to estimate the individual recreationist's demand for the resource, and the second is to statistically derive the relevant resource demand curve. A sample of recreationists must be contacted and questioned regarding their actual expenditures. The samples are designed so as to question respondents from a range of travel distances, whose expenditures will vary with distances traveled.

The second type of observed indirect method is *hedonic pricing*. This method rests on the assumption that the price of a marketed good depends on its different characteristics, and the contribution of alternative characteristics can be identified statistically. The hedonic approach most often analyzes real property (land) sales price data exhibiting differing but measurable environmental characteristics (e.g., varying water supplies or water qualities). This technique has received limited application in water-resource planning to date. Among the few examples, d'Arge and Shogren (1989) found that the value of recreational second-home properties on a lake in northwest Iowa were depressed $13.50 per square foot, as compared to similar properties on a cleaner lake. Torell, Libbin, and Miller (1990), measured the value of groundwater for irrigation in the southern High Plains (in Colorado, Kansas, and New Mexico). Driscoll, Dietz, and Alwang (1994) estimated the effect of flood risk on property values in Roanoke, Virginia.

For services whose benefits cannot be linked to a marketed good, another approach employing user surveys, the *contingent valuation method* (CVM) is frequently employed. This form of direct questioning approach relies on sample surveys, in which respondents report or reveal their values (in monetary terms) to hypothetical future situations. In principle, the CVM seeks to measure the money willingly paid (or received) which is necessary to maintain the level of satisfaction (utility) to a person experiencing an increase (or decrease) in his or her level of environmental services. This amount of compensation reflects willingness to pay (WTP) or willingness to accept (WTA) compensation for the change. Personal interviews or mail questionnaires ask WTP or WTA for hypothetical changes in amount, reliability, or quality of water supplies. Questionnaires implementing the CVM usually contain four basic elements: First, an explanation of the rules and structure of the hypothetical market in which the good or service is being sold; second, a clear description of the good and how it is to be made available; third, a question to elicit monetary values from the respondent; and fourth, questions which validate acceptance and comprehension of the questionnaire format, and obtain social and economic characteristics to help explain variations in responses. Photographs, maps, tables, or diagrams are often used to clarify the hypothesized changes (Mitchell and Carson, 1989).

The reliability of the CVM method depends crucially on the design of the questionnaire. The hypothetical change in water supply situations must be carefully described, and the rules of the hypothetical market game must be clearly spelled out. Critics have questioned the reliability of responses regarding hypothetical situations never before experienced, or even considered, by the interviewee (Diamond and Hausman, 1994). However, the method appears to yield plausible and consistent results when carefully applied, and represents the only way to estimate WTP or WTA for nonobservable experiences, or nonuse values, which might be enhanced or diminished by policy changes. Cummings, Ganderton, and McGuckin (1994) represent an example of improved technique, showing from a sample of New Mexico residents that consideration of substitute goods and services have an important effect on empirical CVM measurements. See Hanemann (1994) for a thoughtful defense of the CVM procedure and its application to policy.

There have been numerous applications of CVM to water resource management issues. Carson (1991) reports on a survey of California voters regarding their WTP for water supply reliability, finding that median annual household WTP to avoid shortages was $83 for a mild shortage, up to $258 for the most severe case. Mitchell and Carson (summarized by Carson, 1991) attempted to measure the value of achieving each of several degrees of improvement in water quality (boatable, fishable, and swimmable) at the national level. Estimated annual national benefits of nationwide swimmable-quality waters were in the $20 billion range, which was not enough to justify the cost of attaining that standard. Cummings, Ganderton, and McGuckin (1994) report an average WTP for preservation of endangered fish species of $3.42 per household. Loomis (1987a), Ward (1987), Hansen and Hallam (1991), and Duffield, Neher, and Brown (1992) are examples of applications of CVM to valuation of instream flows. These studies tend to report that instream flow values at certain times and places may exceed those for offstream use in agriculture, suggesting that a shift in incentives might lead to improvement in net economic benefits. A few applications have been made in developing countries. Whittington, et al. (1992) studied the demand for improved sanitation services in Ghana.

Contingent valuation has also been applied to a number of cases involving nonuse benefits. (Recall that nonuse values are benefits received from knowing that a good exists, even though the individual may not ever directly experience the good. Preserving an endangered species is an example.) Loomis, et al. (1991) surveyed households in California to determine WTP to protect and expand wetland, as well as to reduce contamination of wildlife habitat, reporting that California households would pay $254 per annum in additional taxes for an increase in wetland acreage with an associated 40 percent increase in bird population.

3.3.2 Estimating the Economic Value of Water

Measurements of the benefits (economic value) and cost associated with water policy options are one of the most important activities in positive water economics.

(Some might include this topic in the normative economics section, because economic values are designed to fit into the underlying value judgments of benefit-cost analysis. However, because value estimates are most often performed with empirical data and related analytic techniques, and are subject to standards of refutation and criticism consistent with empirical science, the subject is placed here under the heading of positive analysis.)

For applied benefit-cost analysis (BCA), benefits and costs must be expressed in monetary terms, by applying the appropriate prices to each physical unit of input and product. Three types of estimates are employed. First, an important source of the prices used for BCA are the result of observing market activities. So long as the observed markets are functioning properly (reflecting numerous buyers and sellers, and the absence of monopoly powers and external costs), the use of market prices is appropriate. Second, when markets exist, but are not competitive, the prices observed are not proper reflections of underlying preferences or costs, and need to be adjusted (for example, when agricultural commodity prices are controlled by government regulation, or when minimum wage rates are set above market clearing prices). And third, in many cases it will be necessary to estimate prices for impacts which are not registered in any market (such as the value of water used for power generation, environmental preservation, or recreation). The latter two cases are most typical in cost-benefit analysis for water resource planning; in these instances the prices employed are called *accounting prices* or *shadow prices*.

In applied BCA, the terms *benefit* and *cost* take on a narrow technical economic interpretation. The prices used in BCA are interpreted as expressions of WTP for a particular good or service by individual consumers, producers, or units of government. Therefore, the discussion of methods of water-demand analysis for consumer and producer commodity benefits and public good benefits in Sec. 3.3.1 is applicable. The definition of value, or benefits, as WTP is obvious for market prices, since the equilibrium market price represents the willingness to pay at the margin of potential buyers of the good or service. But WTP is also the theoretical basis on which shadow prices are calculated for nonmarketed goods. The assertion that WTP should be the measure of value, or cost, follows from the principle that public policy should be based on the aggregation of individual preferences. WTP represents the total value of an increment of project output, i.e., the demand for that output. WTP is the individual's best offer to purchase the increment. Therefore, benefits are defined as any positive effect, material or otherwise, for which identifiable impacted parties are willing to pay. [Critics of certain applications of BCA (e.g., Bromley, 1991), observing that WTP is dependent on the existing distribution of income, caution against any unquestioning application of the technique for public investment decisions.] Particularly useful discussions of theory and practice of shadow pricing of environmental resources are found in Braden and Kolstad (1991), and Freeman (1993). Applied methods of valuing water in withdrawal and intermediate good uses have not received the intensive critical scrutiny which has been devoted to determining environmental and nonuse values, and a number of frequently used approaches tend to overestimate willingness to pay for intermediate good values (Young and Gray, 1986).

Gibbons (1986) summarizes the empirical water demand and valuation literature up to the early 1980s, and updates and reproduces numerous empirical estimates of marginal water values in alternative use sectors for the United States. Her analysis, too extensive to summarize succinctly, indicates that value estimates vary widely, depending on local conditions of supply and demand, type or sector of water use, and the conceptual framework employed. Comparing marginal values between sectors, such as required for assessing economic efficiency of intersectoral allocations, requires adjustments to express values in commensurate terms of form (treated or raw water), place, and time. For example, studies of residential water demand yield estimates of very high values per unit volume relative to those for other withdrawal uses, such as crop irrigation or industrial cooling. However, once the residential demand estimates are adjusted to develop a derived demand for raw water at the water source, the premium of residential value over the other sectors becomes much smaller (but does not vanish).

The idea of costs in economic analysis is also based on the willingness to pay concept; in this case, the WTP for foregone benefits or opportunity costs. Economic costs are the value of the opportunities foregone because of the commitment of resources to a project, or the willingness to pay to avoid detrimental effects. In cases where the market for an input is biased due to government intervention, or absent altogether, a shadow price based on the opportunity cost principle will be developed in the analysis. Particularly knotty issues are the shadow prices for labor (wages) and capital (interest rate, also called the social rate of discount). See Schmid (1989), Ward and Deren (1992), or Pearce and Warford (1993) for detailed treatments of these topics. Measuring damage costs or damage functions for analysis of water pollution and flood control policies is an increasingly important problem. (See, for example: Adams and Crocker, 1991; Crocker, Forster, and Shogren, 1991; and Shabman, 1994.) Costs tend to be more site-specific than benefit estimates. Frederick (1994) provides a useful survey of the costs of alternative sources of water supply, including water development projects, recycling, and desalination, in the United States.

3.3.3 Measuring the Impacts of Water Developments on Regional Economies

The conventional wisdom in many parts of the world is that water resource development can serve as an effective driving force for regional economic development. Water-supply developments are thought to yield significant direct employment and income impacts in the project region. For example, a frequently cited figure in the popular agricultural press in the United States is that agricultural sales carry a multiplier of seven, meaning that each dollar of farm sales generates seven dollars worth of further sales in the nonfarm economy. The perception held by members of the U.S. Congress of large regional impacts from water developments is said to be the basis for the Regional Economic Development account, which was required to be measured in the U.S. Water Resources Council's *Principles and Guidelines for Water and Related Land Resource Implementation Studies* (1983). The same hypothesis, asserting important economic linkages between water usage and regional economies, is expressed in connection with reduced water use, often in the context of debating market transfers of water from agricultural to urban uses, or assessing the effects of droughts.

Both econometric and regional interindustry analyses have been applied to measure these impacts. Young and Haveman (1985) survey earlier econometric and descriptive evidence from throughout the world, reporting rather limited evidence in support of a strong water–regional growth linkage. Mann, et al. (1987) developed an econometric approach to measuring the contribution of groundwater irrigation from the Ogallala Aquifer to the economy of the northern High Plains, finding statistically significant but empirically limited employment effects. Howe, et al. (1990) employed an input-output model to assess the employment effects on the Colorado economy of a specific transfer of water rights from over 40,000 acres of irrigated land to urban uses, finding almost imperceptible effects on the state's agricultural employment level. Berck, Robinson, and Goldman (1991) present an even more theoretically sophisticated regional economic model for water policy analysis, a computable general equilibrium model of agricultural water use in an important California agricultural area, the southern San Joaquin Valley. In contrast to the fixed-price assumptions of input-output models, product demand functions are incorporated into the CGE model. From their analysis of the economic effects of transferring water from agriculture, they report that, at most, about 5000 agricultural jobs would be lost if as much as 20 percent of the region's irrigation water supply were sold. Also using a combination of interindustry and computable general equilibrium models, Brookshire, McKee, and Watts (1993) projected regional economic impacts of a proposed program for flow maintenance for endangered species protection in the upper Colorado River Basin in the southwestern United States. They reported net *gains* to the regional economy, largely because the foregone benefits of reduced upper-basin agricultural uses would be more than offset by gains in hydropower and lower-basin offstream uses. The evidence from these analyses suggests that in the capital-intensive U.S. farming sector, little employment and income impact are found from adding water to, or withdrawing water from, agricultural uses. While water is a necessary condition for economic activity, the evidence suggests that augmenting water supplies is not sufficient alone to generate regional economic growth, nor, in most cases, will marginal reductions in available water have major adverse regional impacts.

3.3.4 Other Empirical Studies

In addition to conventional demand and price analyses, economists are turning to empirical studies of the behavior and operation of water management institutions, often in collaboration with other social science disciplines. Saliba and Bush (1987) examined a series of cases in the southwestern United States, describing how water market institutions function and identifying trends in prices. For the same region, Colby (1990) estimated the transactions costs (fees charged by hydrologists, attorneys, and engineers) of water transfers, and identified factors influencing those costs, such as size of the transfer, state regulations governing the transaction, and the general scarcity and value of water in the region. Transactions costs averaged about $90 per acre foot (which was about 6 percent of the average sales price), but the figure varied considerably from state to state, according to water transfer regulations. Tang (1992) investigated the range of incentives and constraints developed for self-managed irrigation organizations in a number of developing nations, to ascertain those factors contributing most significantly to efficiency and productivity. Ostrom and Gardner (1993) studied the prospects for overcoming asymmetric interests in self-managed irrigation systems with a model of strategic interaction among water users, testing the model with a regression analysis of data from Nepal.

3.4 NORMATIVE ECONOMICS AND WATER POLICY

3.4.1 Normative Economics

For at least two reasons, formal policy analysis is unavoidably normative. First, questions of "ought" identify a basis for dissatisfaction with the current state of affairs, and define a policy problem. Second, normative criteria are necessary for identifying an "improved" policy. Policy analysis presupposes some ethical principle, or principles, which can provide a standard of evaluation for existing and proposed policies. The normative branch of economics is *welfare economics,* which combines value judgments regarding the nature of the desirable organization of society with positive studies of empirical economic regularities to develop policy recommendations.

Selecting value criteria is probably the most important aspect of normative analysis. [Although the strong positivist distinction between positive and normative studies (on the ground of the supposed ability to accept or reject factual statements solely on the basis of empirical evidence) has not withstood philosophical critiques, the distinction remains useful.] Normative economics of the mainstream variety is based on a variant of Utilitarian ethics, which can roughly be characterized as holding that policies should be decided on the criterion of the "greatest good for the greatest number." The value judgments underlying modern welfare economics are attributed to the early twentieth century Italian scholar, Vilfredo Pareto. Three normative propositions are the basis for Paretian welfare economics (Maler, 1985). First, each individual is the best judge of his or her own welfare. Second, the welfare of society is based on the welfare of its individual citizens. Third, if the welfare of one individual increases and the welfare of no other individual decreases, the welfare of society increases. This last proposition is called a *Pareto improvement.* When no possible increase in any individual's welfare is possible without diminishing satisfactions for some other person, a *Pareto optimum* is said to be reached.

However, in a complex society, nearly all policy changes would lower the welfare of some while improving the welfare of others, so that few proposals would gain approval under the strict Paretian concept. Welfare theorists circumvented this difficulty with the *compensation test:* If the gainers from the proposed change could fully compensate those experiencing losses, and still enjoy a net gain, the change would be judged acceptable. In practice, compensation is not practicable in many cases, so the test of acceptability becomes one of a *potential* Pareto improvement, determining if the gainers could in principle compensate losers, whether or not compensation actually occurs.

3.4.2 Applied Welfare Economics

Benefit-cost analysis (BCA) is the term used to label the practical application of the welfare economics test for potential Pareto improvement. Empirical economic methods are employed to predict whether a proposed policy initiative would produce beneficial effects in excess of adverse effects, both measured in commensurate monetary terms (Randall, 1986; Smith, 1986). Beneficial effects are those which produce positive utility, or remove anything that causes disutility, while costs are reductions in desired things, or increases in undesired impacts. Methods for measuring the value (economic benefits) and costs of water-related policy proposals are discussed briefly in Sec. 3.4.3. More generally, see any of the numerous texts on applied benefit-cost analysis (e.g., Pearce and Nash, 1983; Schmid, 1989).

The sketch of the overall conceptual framework for welfare economics and BCA should not be left without mentioning the major critiques of the approach. A serious drawback to the model is that a potential Pareto improvement treats all affected individuals equally, and is therefore consistent with making the poor poorer while the rich become richer. Put another way, efficient allocations are not necessarily just or fair. A related criticism is that measures of benefits and costs, as conventionally derived, are dependent on the existing distribution of wealth, and hence may serve to enshrine the status quo. Critics from within and without the profession question whether preferences can be measured by existing techniques. More fundamental challenges come from philosophers, some of whom contest the Utilitarian basis of BCA, particularly the dominant roles of preference satisfaction and wealth maximization. An important alternative ethical criterion, tracing its origins to Lockean social contract theory, is based not on outcomes, but on individual rights and duties. The focus of rights-based ethical theory is on process and the actions that are allowed. Certain actions may be prohibited if they infringe on individual rights and freedoms. Yet another critique challenges the anthropocentric character of the major normative economic and political decision criteria, advancing a theory of the intrinsic value of nature and the environment. (For elaboration on these points, see Randall, 1986, and the papers in Gillroy and Wade, 1992.)

3.4.3 Economics and Multiple Objective Evaluation

Most mainstream economic policy analyses rely primarily on potential Pareto improvement (economic efficiency) as the normative criterion by which to to assess policy options. However, public water-policy decisions are typically made in light of multiple, rather than single, objectives, and evaluation procedures should reflect this larger perspective.

Several approaches to multiple objective decision analysis have been developed (Chankong and Haimes, 1983), many of which have been applied to issues of water policy. One approach is to maintain a formal optimization procedure. Weights must be assigned to each of the objectives, to indicate the relative importance of each. A variant is to optimize one objective, subject to constraints on the others. However, selection of weights (or of constraint levels) tends to be rather difficult, and often exhibits a high degree of arbitrariness. Some efforts at multiobjective evaluation have focused on the mathematics of the process, but have given inadequate thought to the normative problems of specifying objectives, and the appropriate weighting procedures. This technocratic approach has received little endorsement outside the ranks of academic researchers.

The major alternative approach is to treat the analysis as one of displaying information to the political decision process, rather than trying to optimally solve the decision problem at the analytic stage. Information on impacts relating to the various objectives is displayed, with the task of weighting left to the bureaucratic or legislative decision makers. Economic evaluation, in this framework, calculates the present value of net benefits according to textbook rules for BCA, and the balancing of social objectives is performed elsewhere. Economists can further assist with the broader analysis by determining and displaying how the economic impacts are distributed, according to categories such as income groups, political subdivisions, regions, age groups, or water-use sectors. One example of the information display approach to multiobjective evaluation is found in the OECD handbook *Management of Water Projects* (1985). Water policy impacts for three classes of objectives are to be evaluated: Economic, Social, and Environmental. The handbook provides broad instructions for evaluating each of these considerations and displaying the impact measures. (Distributive economic measures, such as income and employment impacts, are elements in the Social Impact Indicators list.) A similar approach was adopted earlier by the U.S. Water Resources Council, whose latest version is *Principles and Guidelines* (1983). Analyses of impacts relating to two major objectives, National Economic Development (i.e., economic efficiency), and Environmental Quality, were required. Regional Economic Development and Other Social Effects were optional accounts to be assessed, if needed, in specific situations.

3.5 BROAD POLICY OPTIONS FOR MANAGING WATER: MARKETS VERSUS GOVERNMENTS

Before addressing specific water management policy issues, it will be useful to review and assess broad policy options for managing water resources. As with other resources, a number of approaches are possible for ameliorating or solving problems arising in water system management, and economic analysis can contribute to choices among these. In this section, the broad policy options of market versus government are addressed, and the strengths and limitations of each are reviewed.

3.5.1 Markets

The voluntary exchange of commodities or services at agreed-upon prices provides the basis for markets. *Price* provides the key signal to coordinate the economic activities of people dispersed in space and/or time, and not necessarily even aware of each other's existence, so as to satisfy consumers' wants. Low prices relative to costs of production signal producers to shift resources to other, more profitable, activities. Conversely, high prices relative to production costs indicate profitable opportunities, and encourage increases in output. Producers' activities are coordinated with consumers' wants as each entity (firm, household) pursues its own self-interest.

Several questions must be answered by any economic system. First, what goods and services are to be produced? Second, what resources and technologies are used in producing them? And third, who is to enjoy the use of products? The adoption of the market system to answer these questions is based on the premises that the personal wants of individuals should decide the employment of resources in production, distribution, and exchange, and that individuals are themselves the best judges of their own wants (consumer sovereignty).

3.5.1.1 Advantages of Private Markets.

A market system that has many producers and consumers, who are well informed, motivated by individual self-interest, and individually own and control resources, is called a *competitive market.* Such an idealized system can be shown to have certain desirable properties. One such desirable attribute is that the system will yield the maximum *economic efficiency* in resource allocation—it will produce the maximum-valued bundles of goods and services, given the endowment of resources, the available technology, the preferences of consumers, and the distribution of purchasing power. (More formally, economic efficiency in resource allocation is attained when no individual could be made better off without making someone else worse off—a condition, as noted earlier, termed Pareto optimality.) Individual producers and consumers, acting in their own self-interest, will, in accordance with Adam Smith's "invisible hand," arrive at an allocation of resources that cannot be improved upon. Firms, encouraged by prospective profit, buy inputs as cheaply as possible, combine them in the most efficient form, and produce those things that have the highest value relative to cost. Consumers' tastes and preferences influence their expenditure patterns, thereby encouraging firms to produce the commodities people want. Prices are bid up for the commodities most desired, and producers allocate resources in the direction of greatest profits. The firms most successful in the process, producing desired goods most efficiently, are rewarded by profit, and the unsuccessful are eliminated, so production occurs at least cost.

A further desirable property of the market system is its ability to accommodate change in conditions of production and patterns of consumption. New knowledge and technology are rapidly reflected in the prices that producers are willing to accept for their products. On the consumer side, changes in income and preference are soon reflected in expenditure patterns. Hence, a market system yields maximum satisfaction in not only a static but a dynamic context.

3.5.1.2 Limitations of Markets: The Theory of Market Failure.

The precise preconditions of the idealized competitive market may not always be present under real-world conditions. Mixed capitalistic systems are based on the presumption that, for most goods and services, the allocation resulting from market processes sufficiently approximates the idealized system. Where this is not the case, regulatory processes or public production are provided to allocate resources.

The theory of *market failure* (Wolf, 1988; Cullis and Jones, 1987) identifies the shortcomings of markets. Market failure is said to arise when the incentives facing individuals and groups in the economy encourage behaviors that do not meet the appropriate efficiency or equity criteria. Several classifications of conditions that lead to market failure are found. They usually include externalities, public or collective goods, increasing returns, market imperfections, and distributional inequity.

Externalities, also called *spillover effects,* are uncompensated side effects of individual activities. In such cases, the full costs of economic activity are not recognized in individual producer and/or consumer decisions, and market outcomes will be suboptimal. Modern economies, with their high material consumption, highly technical production processes, and ever-increasing concentration of populations into large cities, are fraught with externalities. Examples are readily found in the case of water. One occurs in the form of wastewater discharges detrimental to downstream water users, a cost unreflected in upstream water users' allocation decisions. *Public* or *collective goods* are nonrival in consumption, in the sense that one person's consumption doesn't preclude use by others. It is difficult or costly to exclude nonpayers, and, therefore, for the private producer to appropriate all the benefits provided. Water-storage projects often provide environmental or recreational amenities, or flood control benefits, of a public-good nature. *Decreasing cost industries* refers to the case where average costs are falling throughout the range of market demand. In the presence of decreasing costs throughout the range of production, no combination of several firms can produce at as low a cost as can one supplier, so one firm is able to take over the entire market and become what is called a *natural monopoly.* (Water supply, such as an urban water-supply system, or canal irrigation, represent examples of such natural monopolies, since more than one (competitive) supplier would present much higher distribution costs.) A socially inefficient production situation will result, since unregulated monopolists will restrain production, charge excessive prices, and have little incentive to innovate. However, in the presence of increasing returns, the lowest-cost mode is by a single producer, so society is likely to be benefited by regulating, or perhaps owning, the monopoly, rather than attempting to encourage competitive suppliers. *Market imperfections* occur when actual characteristics of information, mobility, and price depart significantly from those assumed by the competitive model. The economy will produce below its capacity when information concerning market opportunities is not equally available to all producers, when factors of production cannot move quickly in response to market information, and when prices or interest rates fail to reflect opportunity costs and relative scarcities. *Distributional inequity* arises because the unfettered market often concentrates income and wealth in the hands of the most successful minority, while leaving little to those possessing lesser abilities, or suffering ill-chance in business or health.

Market failures can be corrected, or at least reduced, by introducing appropriate incentive structures. Financial inducements, such as taxes or subsidies, are one approach. Modified or new rights and duties, such as tradeable water rights, are another example. These approaches are discussed in more detail in subsequent sections.

3.5.2　Government

Charles Wolf (1988) and others (see Cullis and Jones, 1987) have also catalogued, in considerable detail, the advantages and disadvantages of government hierarchies for allocation of resources. Whereas private sector hierarchies are subject to market tests of profitability in order to survive and prosper, and to government regulation of market failures, government hierarchies face different goals and performance requirements.

3.5.2.1　*Advantages of Public Sector Production and Allocation.*　Wolf (1988) makes the following positive case for government activities in the economy. The advantages of public sector production and allocation activities are, in a sense, the opposite side of the coin from private market allocation. They include the following:

Regulating externalities.　Collective action to protect third parties from unwanted costs (for example, regulating pollution by cities, industries and agriculture) is a necessary role of the public sector.

Regulating or supplanting natural monopoly.　The supplier of water has, as a natural monopoly, literal power to impose exorbitant costs, or even economic ruin, on those served. Advocates argue that public regulation, or more often, public supply, can, in principle, avoid the possible undesirable effects of a private, profit-oriented monopoly.

Providing or protecting public or collective goods.　Because private entrepreneurs cannot exclude beneficiaries, and therefore capture a return on investment in public goods, the market system will not supply this type of need. Health, public safety, and education are the most common examples. In the case of water and natural resources, governments can preserve unique or beautiful natural environments by creating national parks.

Reflecting broader social goals.　Although the private market is seen as an efficient engine for producing the maximum-valued bundle of goods and services, public action may incorporate a broader range of social goals. Primary among these might be the amelioration of inequalities in income and wealth among members of the society, and perhaps among political subdivisions or regions, via tax and subsidy policies. Protection of environmental resources is another role.

3.5.2.2　*Limitations of Public Sector Resource Allocation: The Theory of Non-market Failure.*　A long-ongoing debate over the advantages of capitalist versus socialist organization of an economy provides the basis for the main critique of production and resource allocation by governments. The mechanisms which promote efficient allocation of resources and innovation in markets, are less effective or even absent in government. Market advocates emphasize the information efficiency of the price system, when compared to government control of resources. On a related point, some economists hypothesize that citizens of a democracy are "rationally ignorant" of the consequences of voting choices, because of the cost of being fully informed relative to their potential individual gain.

Wolf (1988) has formulated a more elaborate model of *nonmarket failure* that focuses on those performance incentives in public agencies that result in divergence from socially preferable outcomes in terms of allocative efficiency and distributional equity criteria. A general lesson here is that even in the presence of market failures, nonmarket solutions may not necessarily be superior to suboptimal market approaches.

Among problems with public hierarchies, Wolf lists the following:

"Products" are hard to define. The outputs of nonmarket activities are often difficult to define in practice, and hard to measure independently of the inputs that produced them. Floodwater management, or amenity benefits of water storage reservoirs, are examples of water system outputs that are hard to measure.

Evidence of quality is elusive. When consumer preferences transmitted by market prices are missing, it is difficult to know if public performance is improving or deteriorating.

No single performance measure. Because there is no single bottom line for evaluating performance, the public cannot effectively determine the value of public action. Hence, there is seldom a reliable mechanism for terminating unsuccessful programs.

Public agents may have private goals. Wolf refers to the internal goals of an organization as *internalities*. These, in addition to the agency's public purposes, provide the motivations, rewards, and penalties for individual performance. Such internal goals are characteristic of any large organization, private or public, but the problem of performance measures noted above makes public inefficiencies less likely to be terminated. Specific examples of counterproductive internal goals include budget maximization, overly expensive high-tech solutions, and outright nonperformance of duties. In the first case, when profit is not available as a performance measure, the budget often serves as a proxy. Agency heads are often, for example, provided with staff and perquisites according to the size of their budget, reinforcing the incentive distortion. Second, high technology solutions, or "technical quality," may become an agency goal. Illustrating with a water management example, sprinkler or drip irrigation systems may be recommended when less technically advanced but more reliable methods are more economical. State of the art control systems may be installed, even though less elaborate or less technically advanced structures are functional and cost-effective. Finally, agency personnel may be persuaded, by gifts or other inducements, to violate operating rules for a favored few.

Public actions may cause external effects. Public agencies, as well as private firms and households, can be a major source of third-party spillover effects. For example, military weapons production has polluted surface and groundwater bodies. Salinity and/or waterlogging of downslope lands from inappropriately managed public canal irrigation system projects represent another example.

Power may be inequitably distributed. However noble the intent, public sector responsibilities may not be exercised scrupulously or competently. Yet the monopoly control of water supplies by public agencies provides certain individuals with so much power over the economic welfare of water users that procedures to protect those of limited influence must be of prime importance.

In addition to the elements just discussed, which largely relate to the limitations of bureaucracies, the government failure model also addresses—in economic terms—voter behavior and the success of interest groups. Because obtaining information on public policy issues costs effort, the rational citizen tends to be poorly informed about the effects of any particular policy. Potential policies which provide significant benefits to a small segment of society may encourage that segment to pursue their interests by lobbying the government, even if no larger social purposes are served. When excessive costs of such policies can be diffused among the public at large, the public's incentives to oppose inefficient policies is small and each voter's stake in the outcome is insufficient to warrant action. This process, by which well-organized interest groups can obtain concentrated local benefits while dispersing costs to the public treasury, is called *rent-seeking* in the public choice economics literature (the term *rent* in this context referring to unearned income.) Water development projects often provide textbook examples of how this interest-group process operates (see, e.g., Mayhew and Gardner, 1994).

3.6   WATER PRICING AND ALLOCATION ISSUES

Broadly speaking, *allocation* refers to the apportionment of water among users within and between use sectors. The central condition of contemporary water management is that all water users cannot have all the water they might want. How water is shared is at the heart of water resource allocation.

Judged by the standard embodied in BCA—achieving the maximum return from water and associated capital, labor and other resources—economists see considerable room for improvement. (Economists define problems of water allocation in terms of failure to achieve the optimum, in the Pareto sense.) These suboptimalities are attributed to a number of factors, including improper pricing, lack of (or constraints on) markets to reflect opportunity costs in alternative uses of water (Howe, Schurmeier, and Shaw, 1986; R. A. Young, 1986; Wahl, 1989), and inappropriate incentives incorporated in related policies—such as underpriced energy, or overpriced agricultural commodities (Frederick, 1991; M. D. Young, 1992).

Several types of water allocation issues are economically important, including apportionment among individual users, among water use sectors (including offstream and instream), over time and space dimensions, and from alternative sources (such as surface or groundwater). In this section, only the first two will be treated. (Although allocation may require engineering facilities to store and transport water, the focus here is on the economics of the institutional aspects of the topic.) For each of these issues, allocation to individuals and allocation between sectors, two broad topics are discussed. One concerns the conditions for the economic optimum allocation of water. The second, and more complex, issue is the selection of institutional arrangements for allocating water. Institutional arrangements for allocation can be grouped into price-based or quantity-based instruments.

3.6.1   Allocation of Water to Individual Users

3.6.1.1   Optimal Quantity of Water to Allocate to the Individual User. The question of the optimal amount of water to allocate to individual users will be taken up first. The assumption at this stage is that water is supplied to each individual user entity (firm or household) by an existing supply agency. The agency's problem may be viewed as devising a means to divide the available water among its clientele, while recovering all or part of expenditures. The society's problem is somewhat more complex, including achievement of economic efficiency (reflecting external and other nonmarket costs and benefits) and equity.

Figure 3.1, which portrays conventional demand and marginal cost curves, will help illustrate the important concepts. The horizontal axis represents the quantity of water used per unit time, while the vertical axis is price or cost in monetary units. The curve MB (for marginal benefit, or demand) reflects the user's (either consumer's or producer's) marginal willingness to pay for additional units of water use per unit time. Following the principles of diminishing marginal returns (for producers) or diminishing marginal utility (for consumers), marginal willingness to pay falls as additional units are taken. The other curve MC (marginal cost) expresses the incremental cost of supply.

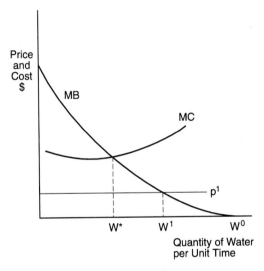

FIGURE 3.1 Optimal allocation of water.

The optimal quantity to allocate to this user is the amount labelled W^*, which is the point where marginal cost and marginal benefit are equated. At that point, the marginal willingness to pay for an additional unit exactly equals the opportunity cost (willingness to pay for foregone alternatives) of supplying that unit. Any allocation greater than W^* will involve use of marginal units whose worth to the user is less than the incremental cost of supply, and net total benefits will be less than maximum. Conversely, use of units of water less than W^* have value to the user exceeding their marginal cost, and net total benefit could be increased by allocating more units to the user.

This discussion emphasizes only economic efficiency as the criterion for allocation. Formal approaches to incorporating equity considerations into water allocation have begun to appear. (See Sampath, 1991, for example.)

3.6.1.2 Prices and Charges as Mechanisms for Allocating Water. The price to individuals for water seldom is set by market forces; rather, charges are set by a publicly owned water supply agency or a regulated private water utility. Water prices (rates, in public utility jargon) have both efficiency and equity impacts, as well as influencing agency revenues.

Marginal Cost Pricing. Economists typically advocate *marginal cost* pricing for public utilities, because this approach will yield the most economically efficient allocation of water and associated resources. Marginal cost represents the incremental cost of supplying the good or service. As we saw in the discussion of Fig. 3.1, when prices are set at marginal cost, the rational purchaser will take incremental units so long as his or her marginal willingness to pay (demand) exceeds the charge. Any consumption greater than that level will involve units whose value is less than the incremental cost of supply, while at use below that level, benefits can be increased by increasing water use. Marginal cost pricing therefore achieves an economically efficient level of water use. The components of marginal cost calculations are important (Saunders, Warford, and Mann, 1983; Herrington, 1987). To perform the desired rationing, marginal costs must represent longer-term capital and operating costs, and measure potential future costs for the next major lump of capacity. A long-term investment plan is therefore required.

In the context of marginal cost pricing of water, opportunity costs, user costs and external costs should also be taken into account (in addition to conventional capital and operating costs). The major relevant opportunity cost is the value of water in alternative offstream and instream uses. Figure 3.2 illustrates this form of suboptimality. The curve MC' represents the marginal cost, in this case, including an external cost. The optimal quantity, incorporating external cost, is W^{**}. Once again, the degree of nonoptimality is measured by the difference between the quantities taken with versus without considering the externality.

User cost is a term for a form of future opportunity cost, referring to the present value of foregone future benefits of present use of a nonrenewing resource. The concept is particularly applicable to groundwater extraction. For example, Martin, et al.

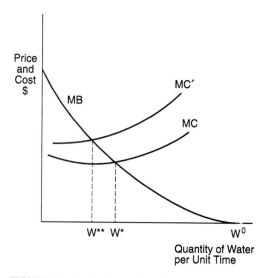

FIGURE 3.2 Optimal allocation of water in the presence of externalities.

(1984) argue that incorporating the user cost of extracting nonrenewing groundwater for the City of Tucson, Arizona, is necessary to signal the scarcity of water in that desert environment. This policy would add about 60 percent to present city water rates.

Rate schedules, in practice, normally fail to correctly reflect the economic concept of marginal costs, focusing instead on pricing only for recovering historical or embedded costs. Herrington (1987) recounts a number of objections from water managers regarding the application of marginal cost pricing. One obstacle is that, in practice, the long-term planning exercises required for marginal cost pricing are difficult to accomplish. This leads to a further claim that, because demand for water is so unresponsive to price, the effort needed to determine marginal costs is unwarranted. As Herrington observes, the lack of a long-term plan with estimates of future incremental costs is an imprudent management strategy, and is not of itself a compelling reason for avoiding a pricing policy which reflects opportunity, user, and external costs. (Also, as noted in Sec. 3.3, numerous empirical studies throughout the world have shown that imposition of a measuring-metering program, combined with some sort of volumetric charging structure, results in significant long-run curtailment of water consumption.)

Water resource projects often present even more complex pricing problems. Marginal cost pricing in decreasing cost industries may imply revenue shortfalls. Also, for multiple purpose projects, one or more public goods (which cannot be directly priced) may be among the outputs. Budgetary constraints may require that a predetermined fraction of costs be recovered efficiently. For such contexts, Freeman and Norris (1988) draw on Ramsey pricing rules to determine the efficient division of a multiple-purpose water project's cost among the beneficiaries of marketable outputs and the taxpayers.

Joint Cost Allocation.   Another issue in cost recovery analyses for multipurpose water projects is the allocation of joint or common costs to use sectors. (Note that the term *allocation* here refers to components of costs, not to water itself.) Production of goods and services by a public enterprise often results in the joint output of several goods, some of the costs of which (both fixed or variable) are separable (attributable unambiguously to beneficiaries), and some are joint or nonseparable (those which contribute simultaneously to all outputs). When it is necessary to allocate these nonseparable costs of production among the beneficiaries, in order to implement a cost recovery or pricing scheme, the problem arises of determining a fair and efficient allocation. Charging beneficiaries for the separable costs of supplying their benefits is straightforward. However, finding a formula for allocation of common costs, such as dams and reservoirs, is less easy, and because it involves multiple criteria, has no simple straightforward solution. One solution is to allocate joint costs according to the *proportional use of capacity,* a distribution of costs based on physical use of water, or alternatively, on reservoir storage capacity. This approach and variants of it has failed in practice, both because the measure of "use" is not obvious for some beneficiary classes (flood control, recreation), but more importantly, because it often allocated costs in excess of potential gains to some beneficiaries (such as irrigators). This latter drawback is overcome by the most commonly used approach: The *separable costs–remaining benefits* (SCRB) method. The SCRB method allocates joint costs using the ratio of *remaining benefits* (the lesser of benefits or cost of least-cost alternative minus separable costs) to the *total remaining benefits* (total benefits minus separable costs) to allocate joint costs. The SCRB procedure retains other limitations (see H. P. Young, et al., 1982), but adjustments can ameliorate most of these. Game theoretic concepts have also been employed to illuminate problems of joint cost allocation in both private sector and public sector accounting: See, for example, the essays in H. P. Young (1985). Driessen and Tijs (1985) discuss the relationship between game-theoretic methods and the more traditional approaches, concluding that for many practical applications, the game-theoretic and the SCRB approaches yield the same solutions. An alternative approach is provided by Freeman and Norris (1988), who come at the cost allocation question by treating joint cost allocation as a process within a Ramsey pricing model.

Finally, it is worth noting that economists have encountered considerable difficulty in persuading policy makers and public utility regulators that economic efficiency is the only appropriate criterion for pricing in public water supply and in regulated public utility industries. Howe (1988) recounts a discouraging history of pricing and cost-sharing policy by the U.S. Government. For six federal irrigation projects examined, prices range from about one-fifth to nearly one-twenty-fifth of the full cost of providing the water. Rather than serving as a means of efficiently determining which water services to deliver, and rationing access to public projects, federal cost-recovery policy emerges as a successful effort to shield beneficiaries from paying costs. In the context of regulated public utilities, it is clear that rate-setting bodies are concerned primarily with recovering actual expenditures made in behalf of customers—economists' concepts of opportunity costs, user costs, or external costs are regarded as difficult to quantify, and even of questionable legitimacy. Based on long experience with public utility regulation, Zajac (1985) contends that applied rate setting involves a balancing of social goals, and economic efficiency in resource allocation is, in practice, largely overshadowed by perceived considerations of justice and fairness between suppliers and customers. However, understanding the regulators' point of view does not necessarily require one to endorse it. Focus on "equitable" treatment of present customers can cause serious misallocation in the efficiency sense. Moreover, ignoring external and user costs itself brings about its own form of inequity, that imposed on those eventually forced to experience the inefficiencies or externalities.

This framework can be drawn upon to illustrate a number of suboptimalities or policy failures. Policy problems are conventionally defined as failure to achieve some desired state of affairs. For economists, most suboptimalities are failures of one type or another to achieve an economically efficient (Pareto optimal) use of resources.

Flat Rate Charges.   Referring once again to Fig. 3.1, consider the effect of a flat rate pricing scheme, one in which charges are unrelated to volume of water taken. Under a flat rate scheme, the marginal charge is zero at all quantities. The rational, self-interested user will take water so long as marginal benefits are positive; i.e., to point w^0 on the horizontal axis. The difference between w^* and w^0 (the range of water use where marginal cost exceeds marginal value to the user) measures the degree of economic waste or suboptimality. Also in Fig. 3.1, the line p' represents another common pricing scheme, a uniform volume charge. The charge is drawn to reflect price below long run marginal cost, leading to the suboptimal allocation at w'.

In practice, the most commonly employed pricing policy for water sets a flat rate charge, although this policy is rapidly giving way to metered systems by urban suppliers in the developed nations. Such charges are designed primarily to equitably recover costs, rather than achieve optimal resource allocation. By definition, flat rates do not charge according to a measure of the volume received, although some proxy for volume is usually the basis for the charge. In agriculture, the area irrigated is the most frequent basis for the charge. In residential settings in the industrialized world, flat rate charges have been based on number of residents, number of rooms, number and type of water-using fixtures, or on measures of property value.

Flat rates provide no incentive for rationing water to the optimal amount of use; hence they find little support among economists. Returning to Figure 3.1, we see that a flat rate without any additional supply constraint would allow the user to take water so long as marginal net benefit remains positive, labelled w^0. The amount taken would considerably exceed the economically efficient allocation. However, flat rate schemes are easily understood by consumers, are simple to administer, and assure the supplying entity of adequate and stable revenue. The high cost of installing and monitoring meters has been suggested as a reason for staying with the flat rate approach, an argument which is convincing only in cases where water is plentiful and supply costs are low.

However, a number of factors have contributed to an increasingly widespread adoption of volumetric pricing schemes. These factors include the rising scarcity of water and of capital for developing new supplies, and technological advances which have lowered the costs of meters and meter reading. Additional factors are the recognition that user charges represent the most equitable means of financing water supply, together with evidence (reviewed in Sec. 3.3) that pricing does indeed restrict water use. In many places, water is only now becoming scarce enough to justify the tangible and intangible costs of establishing formal pricing systems. Flat rates could satisfy repayment requirements in the absence of serious shortages. However, when the signals of scarcity of water (and of costs of the capital, labor, and energy) are absent, pressures arise for structural solutions to satisfy incorrectly perceived water "needs." The inevitability of increasingly scarce water supplies suggests the eventual desirability of adopting rate schemes which reflect the opportunity costs of water and the resources needed to supply it.

Space limits preclude a more detailed discussion of these issues. See Spulber (1989) for a thorough treatment of the theoretical issues arising in decreasing cost industries. Herrington (1987) and Munasinghe (1992, Chap. 8) provide competent analyses of applied water pricing. Small and Carruthers (1991) and Sampath (1992) review issues relating to pricing of irrigation water.

3.6.1.3 Allocating Water to Individuals by Quantity-Based Mechanisms. In many instances, particularly with irrigation water, the cost of measurement and administration, and the uncertainty of water supply, discourages use of pricing as an allocative procedure. In principle, a quantity-based mechanism (sometimes called a *quota*) could be designed so as to achieve the same optimal allocation derived in Figure 3.1. Several criteria are important in specifying quota or entitlement systems (Young and Haveman, 1985). Security of tenure, affording protection against legal and physical uncertainties, is one consideration. Another attribute is certainty—the rules of water use must be easy to discover and to understand. Third, a desirable system should minimize the possibility that water users would impose uncompensated costs on third parties. Finally, the rules must be consistently and fairly enforced.

Irrigation water allocation presents extra difficulties, particularly the frequently uncertain water supply, the numerous smallholders, the tendency for large conveyance losses to be incurred between the water source and the ultimate customers, and the difficulties in communicating demand and supply information. A number of quantity-based approaches for allocating water to individual irrigation users have been applied. In the western United States, the prior-appropriation doctrine of "first in time—first in right" is found (Tarlock, this volume, Chap. 5). Alternative quota mechanisms for irrigation systems have been studied by economists, building on the path-breaking simulation modeling work of Maass and Anderson (1978). Bowen and Young (1986) modeled a number of alternative irrigation water allocation and cost-recovery methods (including both pricing and quota systems) with linear programming for a case study in Egypt, accounting for administrative costs, conveyance losses, and government commodity price distortions, concluding that under existing conditions, where water supply is not constraining, a water quota system combined with flat land taxes provides the best compromise among allocative and cost-recovery mechanisms. Johnson (1990) studied irrigation water allocation in Indonesia with an updating of the Maass-Anderson approach.

Paterson (1989) and Dudley (1992) outline a proposal for *capacity sharing* for managing reservoir water supplies, which has attracted considerable attention in Australia. In contrast with the widely used concept of release sharing, which assigns entitlements in terms of delivered water, capacity sharing allocates explicit shares of storage capacity, inflow, and losses—and hence of stored water—to end users. The fully worked-out proposal provides for rights that are explicit, enforceable, exclusive, and exchangeable. An important aspect of the concept is that rights are assigned not only for releases, but for both beginning and ending carryovers of water. Thus, the system facilitates the marketing of water already in the system, and the rights to future water in the long term. Proponents of capacity sharing contend that this approach can better satisfy the requirements for an economically efficient allocation of water among competing sectors and across a number of years, such as in the event of long-term drought.

3.6.2 Allocating Water Between and Among Sectors

3.6.2.1 The Optimal Allocation of Water Between and Among Sectors. From the economic perspective, the optimal allocation of limited water in a given geographic area (river basin or other defined region) is that which maximizes the net economic benefits of the region's water. A multisector net benefit function is formulated, which represents the discounted present value of the streams of future benefits in each of the relevant sectors, minus corresponding costs. (Approaches to net benefit measurement for nonmarketed benefits and costs for water planning were sketched in Sec. 3.3.) The net benefit function is maximized when the net marginal benefits per unit of water used are equal in all use sectors. This latter proposition represents an instance of what is commonly known as the *equimarginal principle*.

The equimarginal principle can be illustrated graphically for a simplified two-sector case, as in Figure 3.3. The horizontal axis represents a fixed supply of developed water in the river basin or region. The vertical axes are in money terms. The curves represent the dollar value of marginal net benefits in two alternative withdrawal use sectors, such as agriculture (MB_a) and urban demands (MB_u). The curve which slopes from upper left to lower right of the graph is a conventional marginal benefit function reflecting diminishing returns. The other curve is also a marginal benefit function, but drawn in reverse (from the right-hand axis) for purposes of this illustration. The point of intersection (W^*) represents the optimal balance between use sectors, because from any alternative point, reallocation toward W^* would provide gains to the region.

The regional net benefit functions must be defined and measured with care, when a number of potential pitfalls are possible. One rule is that marginal benefits in competing sectors must be measured in commensurate terms. For example, net economic benefits should be measured for the same location, quality (degree of treatment and processing), and planning horizon. It is incorrect, although frequently seen even in some of the technical literature, to express marginal benefits in the urban sector in terms of price charged for water treated and delivered to a retail consumer, and compare them with the marginal value of a completely different economic commodity, raw river water delivered to a farmer's intake point. A preferable approach would express both demand sectors in raw water terms, by determining the derived demand

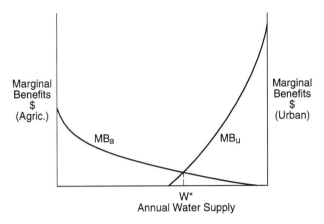

FIGURE 3.3   Optimal intersectoral allocation of water.

of urban users for raw water (which requires accounting for transportation, storage, treatment, and delivery costs). Another consideration is whether use should be measured in terms of withdrawals, or as evaporative consumption. When reallocation is considered only between offstream uses, both marginal benefit measures should be in the same units. However, when instream uses are introduced, the problem becomes much more complex, and a hydrologic-based model, which can reflect spatial and temporal considerations, is a desirable basis for analysis (e.g., Booker and Young, 1984).

3.6.2.2 Institutional Arrangements for Intersectoral Water Allocation and Reallocation.

Changing and growing demands for water, particularly in the more arid regions, bring about a need for more responsive allocative institutions. Rising populations and incomes have brought about increasing demands for water in urban uses, and for meeting recreational and amenity values. Proposals for major new water projects confront several realities: Sites with low capital and operating costs are already utilized; there are increased future costs for energy and capital, and a rising public concern over the potential foregone environmental benefits of water developments. The largest user of water is crop irrigation, particularly in arid regions, although its value at the margin is relatively low. Competition for water in traditional uses arises from residential, industrial, and commercial uses, and from instream uses such as power generation, waste load dilution, recreation, and fish and wildlife habitat. These factors combine to encourage search for water supplies from existing uses, whose economic value is less than the cost of developing new supplies. The relative scarcity of water is an important consideration influencing the form of water-allocation institutions. Given the high information, contracting, and enforcement costs required to establish and enforce a water allocation system, where water is plentiful relative to demand, laws governing water use and allocation tend to be uncomplicated and casually enforced. More elaborate systems have only evolved where water is scarce and valuable, and where inefficiencies and inequities prompt action by community and state.

In response to these changes, a need arises for institutional arrangements to facilitate transfers of water use, while protecting affected interests. A transfer of water rights may involve a change in type of use, a change in location, or in point of diversion. The main focus of study has been the case where water rights are permanently sold in a willing buyer–willing seller transaction (Saliba and Bush, 1987; Howe, et al., 1986; Young, 1986). The advantages of market transfers were made in detail in Sec. 3.3. Briefly recapping, markets provide price signals which encourage resources to move from lower- to higher-valued uses, and economic efficiency is enhanced. Present users are assured that their water right is transferred only on terms to which they willingly agree.

The primary objections to reallocation by markets arise from the extent and nature of external costs, those uncompensated effects imposed on third parties not participating directly in the transaction. Proposals for solutions to these problems range from increasing public regulation of transactions, so as to protect affected third parties, to assignment of responsibility to an administrative agency. Because the case against administrative allocation of resources was addressed in detail in Sec. 3.5.1, no further discussion will be presented here. However, the principal objections to market transfers will now be taken up.

To introduce the issue, some concepts are appropriate. Economists classify external effects as either technological or pecuniary. Technological externalities are real changes in production or consumption opportunities available to third parties, and generally involve some physical or technical linkage among the parties (such as degraded water quality). This type of externality represents a change in welfare, and should be reflected in evaluation of the economic efficiency effects of policies or projects. The methods for measuring nonmarketed impacts described in Sec. 3.3 are appropriate. *Pecuniary impacts* (often referred to as secondary economic impacts in the water-planning literature) are those reflected in changes in incomes or prices (such as effected by increased purchases of goods and services in a regional economy). Secondary economic impacts likely represent income distribution impacts. From the larger perspective of nation or state, secondary impacts registered on a specific locality are likely to be offset by similar, but more difficult to isolate, effects on income of opposite sign elsewhere (Sugden and Williams, 1978). Economic convention therefore suggests that secondary impacts not be taken into account in economic evaluations.

Technological or real external costs can take the form of changes in quantity, timing, and quality of water available for other uses. A principal focus of concern is on the return flows, that portion of the water not consumed by evaporation, and normally available downstream to other water users. Transfer of an entire diversionary right to an alternative location would adversely affect existing downstream water users. This is a case of a technological externality, and warrants preventive policy. The usual solution is to distinguish between the right to divert and the right to transfer, the latter being limited to consumptive use (U.S. National Research Council, 1992). When transferred rights are restrained to historical consumptive use, return flows remain available for traditional users, and this externality is largely prevented.

The issue of policy for secondary economic impacts of water transfer is more contentious. As implied above, economists are inclined to the view that local secondary costs associated with water rights transfers are pecuniary impacts not of much concern in economic efficiency analysis. Moreover, such impacts are likely to be more than offset by secondary gains elsewhere—as was the finding of Howe, et al., 1990, in a case study—and thus can be ignored from the larger perspective. Also, the regional economic linkages (secondary employment and income) with agricultural water resource use tend to be empirically less significant than the public appears to believe (see Sec. 3.3.2). More generally, growing economies must adopt processes which allow for declining industries to be displaced by those which can make more productive use of resources. Public policy provides a general safety net to help affected individuals accommodate such change. However, influential voices argue that agriculture represents a special case, and that area-of-origin protection policies should be instituted to protect farming communities from these third-party economic impacts, and from associated social and environmental damages. (See U.S. National Research Council, 1992, for further discussion.)

Several other mechanisms for inter- and intrasectoral transfers are possible. One form is a *water lease,* in which an agreement is reached to temporarily transfer a fixed amount of water for a limited time period. Such transactions are not uncommon, both within the agricultural sector and, to a lesser extent, between agricultural and other user groups. A related mechanism is a *water bank,* which is a formal arrangement for pooling temporary offerings of numerous individual right-holders, for lease to other potential users with more urgent or valuable demands. Water banks usually involve a change in purpose and place of use of water within the same year, but the banked water may also be placed in surface or underground storage in anticipation of future needs (U.S. National Research Council, 1992). Part of the rental price usually goes to cover administrative expenses. Some banking arrangements require a review, to assure protection of the local public interest, before implementation. Another method of temporary water transfer is the dry year option, or water-supply option contracts. Water users operating under the prior appropriation doctrine of water law in the western United States, who have sufficient rights to meet their needs in all but the driest years, may enter into a long-term contingent contract or option agreement with senior water-right holders, to acquire the senior rights only during contractually-specified drought conditions. Michelsen and Young (1993) show that under certain conditions this approach can be less expensive than permanent purchase of seldom-used senior water rights, although only a few such agreements have been reached. In the third-world urban domestic water demand context, Crane (1994) reports that extending permission to households with city water supplies to resell to those without such supplies increased water use and lowered costs, having the effect of extending the city's water-supply network.

3.6.3 Allocating Groundwater

Groundwater allocation presents several differences in supply attributes from surface water that warrant separate treatment. One important difference is rate of flow. While groundwater moves in response to inflows or outflows, as does surface water, it typically moves at a much slower rate, perhaps only a few centimeters per hour. A second characteristic is the volume of water as a proportion of the total volume of the aquifer. The water-bearing capacity of a formation is but a fraction of the total volume (usually less than 15 percent). The final difference noted here refers to the limited information typically available regarding aquifer characteristics, and the high incremental cost of further knowledge. Normally, specialized skills and measuring instruments are required to determine the size and economic characteristics of an aquifer, such as the depth to water, depth to the geologic bottom, as well as the speed of flow and the water-bearing capacity referred to earlier. Information on these parameters is typically meager, being derived from analysis of scattered boreholes, so the incremental cost of gaining further knowledge is relatively high.

A fairly extensive literature addresses the economics of managing groundwater as a nonrenewing resource. Once again, the analysis focuses on two issues: The optimal rate of use over a long planning period, and the appropriate institutional structure to encourage individual pumpers to achieve this optimum. Groundwater has often been left as an open access resource, and over-rapid exploitation and external costs (land subsidence, intrusion of poor quality water) are two of the major inefficiencies which occur in the absence of adequate collective management. In the most advanced analyses, the optimal allocation is derived by dynamic optimization procedures, which balance diminishing returns to present period use against the effects of increased pumping costs and the discounted future benefits of future uses. Decision rules specifying the optimal rate of pumping for each year of a long-term horizon are derived. A tax on pumping to internalize the externality is one proposal for achieving optimal temporal allocation of groundwater. However, pumping quotas or permits, particularly marketable ones, appear to be the preferable approach. (See R. A. Young, 1992, for a discussion of optimal groundwater management, and a review of alternative institutional arrangements for groundwater allocation). Although a number of economic studies have concluded that groundwater management shows little economic advantage over an open access regime in the United States (see Provencher and Burt, 1994, for a survey), these sanguine results are mainly for large aquifers where withdrawals do not greatly exceed recharge, or where groundwater is used primarily to supplement surface sources. It is the writer's opinion that from a worldwide perspective, problems of groundwater allocation—over-rapid exploitation, water quality degradation, and subsidence—are among the most important and serious, as well as the most intractable, of water management problems presently facing water policy makers.

3.7 ECONOMICS OF WATER AND ENVIRONMENTAL QUALITY MANAGEMENT

3.7.1 Introductory Remarks

A number of water policy issues fall under the general heading of environmental quality management. Among economists, perhaps the most attention has been directed at the problem of managing water pollution (both point- and nonpointsource), and that emphasis is reflected here. Additional topics touched upon are the economics of nature preservation, and integrated watershed management.

In the modern world, it must be recognized that environmental pollution is an inevitable and pervasive consequence of the production and consumption activities of humankind. Raw materials are extracted from the environment and processed into consumer goods by the production sector. Some wastes (*residuals*) from the production process are returned to the environment (for example, waste chemicals from petroleum refineries discharged to rivers). Similarly, the household sector returns unwanted byproducts of consumption activities to the environment through sewers, to the air, or to solid waste receiving sites. The *materials balance principle,* derived from basic laws of physics regarding the conservation of matter, asserts that over the long run, the mass of residuals discharged to the environment by consumers and producers will equal the mass of materials originally extracted from the environment to make consumption goods. In consequence, the environment is a scarce resource, equally important as an assimilator of residuals as it is as a source of materials. (See Pearce and Turner, 1990, Chap. 2, for further discussion.)

One important implication of the materials balance principle is that control of residuals discharges to the environment must be balanced. This means that managing discharges into watercourses must be taken in conjunction with management of waste disposal to the atmosphere and to landfills. Residuals must end up somewhere, as mass or energy. Reducing the amount of wastes discharged into water may not solve the overall problem of society if the wastes are sent elsewhere in the environment, such as to the atmosphere by burning, or by being dumped on or under the landscape.

The deleterious effect of residuals discharges is increasingly becoming an issue. Adverse effects on human health, and reduction in the quality of life, are political issues in most nations. Any economy which fails to recognize that the environment is a scarce resource, and permits it to be a free good, will fail to achieve an economically efficient allocation of environmental resources. When there are no costs or constraints for using scarce environmental services as a sink for residuals, producers and consumers have no, or inadequate, incentives to control their discharges to the environment. Governments are being called upon to remedy these unhealthy and unpleasant conditions.

Water pollution problems are classified as originating from either point or nonpoint sources. The first type, point-source pollution, refers to cases where there is a readily identifiable pipe or ditch through which the pollutant is transported to the receiving water body. Regulation and monitoring of compliance with regulations can focus on the point of discharge. Nonpoint sources are those for which there exists no identifiable single source of pollutant discharge. They are most often associated with human uses of land, such as farming and construction. Soil particles eroded from farm or forest lands, which subsequently wash into lakes or streams, represent a primary example. Fertilizers and pesticides can be carried off the soil surface or percolate into groundwater deposits. However, much nonpoint-source pollution arises from nature itself, examples being organic matter from forests, and dissolved solids from both surface and underground contact with sedimentary rock formations.

Two types of policy issues related to water-quality management are addressed here. The first concerns what level of water quality should be targeted by public policy. The second issue is what package of policy instruments should be adopted to achieve the desired standard for point- and nonpoint-source pollution.

3.7.2 The Economically Optimal Level of Water Quality

Policy makers face a paradoxical situation with regard to water quality: The public desires near-zero waste elimination, but this would be enormously expensive, or even impossible, unless important industries are closed down. Recognizing the need for tradeoffs, as in other instances of resource scarcity, economists have proposed an economic-efficiency-based model to aid in balancing these concerns.

Figure 3.4 illustrates the model of optimum pollution control. Consider the case of a damaging pollutant or residual being discharged to a river. Assume that the adverse effects of possible concentrations of this pollutant are known, and that monetary values can be placed on these effects. The damaging discharges can be reduced by treatment or process changes at a known cost. The decision variable of interest is residuals concentration, shown on the horizontal axis. Treatment cost T rises (moving from right to left) as the residual concentration falls. The willingness to pay of beneficiaries is assumed to be the damages avoided by the treatment. Damages avoided D fall (also moving from right to left) as residual concentration is reduced. For initial units of improvement, the cost of initial increments of improved water quality are likely to be fairly small, while the water user's (receptor's) willingness to pay (damages avoided) will likely be fairly high. The total cost of disposing of residuals in the water is the sum of damages and treatment costs (given by the curve $T + D$). The optimum level of pollution control is found at the minimum of total cost, the point Q^*. This is the point which economic analysis suggests should be the target, or optimal, level of environmental quality. Neither emitter nor receptor will be entirely pleased with this compromise position, the receptor preferring treatment to the threshold of damages Q^1, while the emitter prefers not to incur costs of treatment, Q^2.

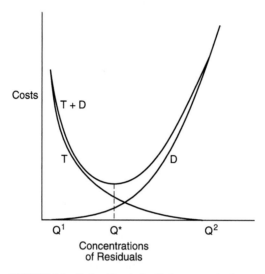

FIGURE 3.4 Optimal level of pollution concentration.

Note that this model is subject to versions of the criticisms of applied utilitarianism noted in Sec. 3.4. The BCA model places damage costs to receptors on an equal footing with the costs of treatment by polluters. The model implies that cost should be borne by the least-cost avoider, whether that entity is the source of, or is damaged by, the pollutant. Within the range where the incremental damages to receptors are less than the incremental costs of treatment (to the left of Q^* in Fig. 3.4), the model does not require the polluter to treat to higher standard. An alternative view, expressed by individualists among economists advocating a *Pareto safety rule* (no uncompensated third-party damage; e.g., Buchanan, 1977), and environmental philosophers arguing for a criterion of *avoid doing harm* (Goodin, 1992), contends that that public pollution control policy should focus on minimizing all damages to third parties.

3.7.3 Influencing Dischargers to Control Point Source Water Pollution

A pollution control agency faces the situation that the actions of many independent and individual actors (producers, consumers, and government agencies) affect environmental quality. Technologies and product mixes will differ among actors. The discharge, rather than treatment, of wastes transfers costs to others distant in time and space, so the agency must induce the actors to take steps which are likely to be contrary to their perceived self-interest. Moreover, the agency cannot costlessly know what levels of pollutants are being discharged at any given time by each of the actors. Several approaches to controlling pollution are possible. These include direct regulation, economic incentives, and persuasion. The relative effectiveness, or desirability, of these alternatives must be appraised not only from an economic efficiency perspective, but in terms of other factors, including the distribution of costs and benefits across the economy, the ease of monitoring and enforcement, and flexibility in the face of economic change. (See Bohm and Russell, 1985; or Pearce and Warford, 1992, for detailed analyses of the advantages and disadvantages of alternative pollution-control policy instruments.)

3.7.3.1 Direct Regulatory Approaches.

The most frequently used approach to pollution control is by government-imposed standards or targets for ambient water quality. These standards are implemented by setting discharge levels for each type of pollutant, for each polluter or polluter class. The environmental agency monitors compliance, and is often empowered to impose penalties on violators. Several different approaches to specifying target discharge levels have been tried: (1) limits on maximum rates of discharge from a pollution source, (2) specified percentage removal of pollutants from emissions, and (3) requirement to implement some technology package ("best available technology"). Many believe that the regulatory approach has proven to be an effective and equitable means for reducing pollution discharges and raising environmental quality. The bargaining process that precedes the selection of exact control packages has provided flexibility, and encourages voluntary compliance by polluters. Regulatory agencies and legislatures are both comfortable with the approach.

However, economists have criticized direct regulatory approaches (e.g., Baumol and Oates, 1988). One objection is that there is an inadequate concern for the costs of environmental improvement. Modeling studies suggest that direct regulations provide more costly solutions to pollution control than do other means. In particular, incentive-based approaches are asserted to be more likely to find the pollution discharge rate that minimizes the total cost to society. More general objections arise from government failure, discussed in Sec. 3.5. Regulations may be poorly conceived, arbitrary, and manipulated for purposes unrelated to their original intent.

3.7.3.2 Economic Incentive Approaches.

A general alternative approach to pollution control regulation is decentralized incentives and disincentives. A key assumption of the incentives approach is that social policy regarding pollution control should encourage the selection of the set of pollution control options which yields the least cost to society, where pollution damage costs, as well as residuals treatment costs, are both considered.

The *effluent charge* (also called an *emissions or pollution tax*), based on the *polluter pays* principle, is one such option. The regulatory authority imposes a fee or tax on each unit of the contaminant discharged. The charge might be set so as to represent the economic value of the damages to third parties caused by the pollutant. (Economists refer to this as *internalizing the external costs*, which encourages polluters toward achievement of economic efficiency.) The unit charge would likely rise with increased levels of discharge. Polluters would, under the emissions tax approach, be free to respond to the charge as they choose. Firms with low unit costs of reducing pollution relative to the charge would presumably take steps to lower discharges, so long as unit costs are less than the charge. Others might find it cheaper to pay the tax than make the necessary pollution control expenditures. These incentives encourage reducing pollution by the least-cost methods available, but all dischargers would share the costs of abatement to some degree. Moreover, all firms would find it in their interest to seek changes in process technologies, and/or in treatment of discharges, which reduce the social costs of coping with the problem of residuals disposal. In effect, the polluter is faced with an incentive to economize on the use of scarce environmental assimilative capacity.

Even following several decades of advocacy by economists, effluent taxes are seldom selected as pollution-control policy instruments. Objections to the approach are of several types. The principal protests come from polluting industries themselves, who object to a charge for environmental services previously received free, and point to adverse profit and employment impacts. Their concerns are also likely based on a fear that taxes might go beyond the stage of taxing to attain the optimal degree of discharge, to punitive taxation for pollution itself. Also, the tax may raise the production costs of firms, not only by the amount of the tax, but also by the costs of treatment to avoid part of the tax, although this latter concern could be ameliorated in part by setting the tax to begin only above certain pollution discharge levels. An administrative drawback is the difficulty faced by the authority in accurately measuring the external costs or damages (the *damage function*). Environmentalists express the fear that a tax, by putting a "price" on the environment, sends the message to polluters that the environment is merely another economic commodity to be bought and sold (e.g., Kelman, 1986).

Marketable effluent permits are another variation on the decentralized, incentive-based approach (M. D. Young, 1992). A central authority, perhaps for a specific river basin, would decide on the total permissible pollution discharges consistent with general water-quality policy guidelines. Permits equivalent to that total would be issued, subject to the requirement that no discharges are allowed without a permit. The initial issue of permits might be on the basis of historical discharges (or a fraction thereof) or via an auction to the highest bidder. Once issued, the permits would be marketable, and the opportunity costs of discharge versus treatment would influence behavior. The authority could fine-tune water quality by withdrawing permits, or by issuing more. Private interests (e.g., environmental or recreational associations) could buy permits for permanent retirement, to improve water quality for aesthetic purposes above the level chosen by the basin authority.

In principle, the opportunity costs embodied in the permit system would create the same incentives on polluters to reduce emissions, and find low-cost abatement technologies, as would the effluent tax. Polluters whose cost of emission reduction is relatively expensive would buy additional permits, while those with less costly opportunities for reducing discharges could implement these opportunities, and profit by selling some permits. Potential new polluting industries would need to enter the permit market, buying permits and replacing other dischargers. The price of permits would also signal to potential new entrants the scarcity of environmental quality relative to other basins, and to technological in-plant abatement opportunities.

In addition to the economic efficiency (cost-effectiveness) argument, a principal advantage in implementing the tradeable permit approach is that it represents a natural extension to the pollution permit system already existing in many nations. Objections are similar to those voiced concerning effluent taxes. Critics of the idea have worried about one firm cornering the permit market, so as to reduce competition for its products. Other concerns have centered, from the emitters' side, on the potential impacts on their profits, and hence, over the longer term, on the net worth and market value of their firm. Environmental interests are dubious about a policy which grants rights to what they regard as morally reprehensible activities (although, as we have seen from the mass balance principle, the reduction of water pollution will tend to increase problems of residual discharges to the air and the land, or require large amounts of energy and other resources to abate discharges).

When the assumption of perfect knowledge is relaxed, the equivalence of taxes and tradable permits may disappear. According to what has become known as the Weitzman theorem, where marginal cost and benefit functions are imperfectly known, the expected welfare gain may differ. The choice of policy instrument then depends on the relative steepness of the marginal cost and benefit curves (see Cropper and Oates, 1992, for a review).

Economic incentive policies for water pollution management have seen little actual application as yet. Surveys of pollution control in the OECD nations (de Savornin Lohman, 1994) show that environmental charges, for the most part, were not applied to induce reduced polluting behavior, but for financing specific environmental regulation expenditures. De Savornin Lohman proposes a number of reasons—including uncertain environmental effectiveness, cost effectiveness, distributional impacts, and the contexts of public policymaking and administration—why economic incentives are not more frequently adopted.

Cognitive (voluntary) approaches to pollution control, using education, moral suasion, and technical assistance, have been tried in some countries, but progress is elusive. They are attractive because of their low economic and political cost. Several factors appear to account for their limited success. Private costs of the necessary changes in production and consumption practices are often nontrivial, while private gains may be negligible and not at all obvious. Because of the uncertain linkage between changed private production decisions and improved water quality (often at distant locations), individuals are little inclined to try new approaches in the absence of monetary disincentives (Bohm and Russell, 1985).

3.7.4 Nonpoint-Source Pollution Control Options

As progress is made with point-source pollution control, attention turns to the selection and implementation of control options for nonpoint-source pollution (NPSP). Nonpoint sources are now the major source of water pollution in the United States: They are thought to be responsible for almost all suspended solids, and most oxygen-demanding loadings, nutrients, and bacteria counts. Control of NPSP presents special difficulties. Sources are difficult to identify, monitoring emissions presents daunting challenges, and the fates of pollutants in the environment are uncertain (Tomasi, Segerson, and Braden, 1994; Malik, Larsen, and Ribaudo, 1994).

The great variety of sources, and the variety of nonpoint-source pollutants, contribute to the difficulty. The primary source of nonpoint pollutants is the agricultural sector. Runoff from farms and forests may carry with it suspended solids and sediments, dissolved solids and chemicals (fertilizers—particularly nitrogen and phosphorus—and pesticides). Drainage waters from irrigated lands in arid areas carry with them dissolved mineral solids (salinity), pesticides, and nutrients. Urban storm drainage, leakage from buried fuel tanks, and subsurface and surface mining are other major contributors. Other substances often occurring in diffuse source runoff are oxygen-demanding organic matter, petroleum products, heavy metals, and fecal bacteria.

Nonpoint-source pollution is also characterized by its episodic and random nature. Occasional heavy rainfall or snowmelt events (over which the source has no control) typically are the trigger, in contrast to the more even flow of discharges by point sources. These characteristics of source type and timing imply that a variety of control technologies may be required for effective abatement of this type of water pollution.

Another aspect of complexity of nonpoint control arises from the nature of the human activities from which it originates. For example, the pollution arising from a farmer's land depends not only on the rainfall patterns and the land characteristics (slope and soil texture), but on numerous previous land use and production decisions, including choice of crops, tillage practices, and pesticide and fertilizer use. The farmer's production choices are, in turn, influenced by market prices for inputs and products, as well as by government price and income support programs. There is increasing support for the view that an indirect source of pollution problems in agriculture in both the United States and Europe is government price support and subsidy policies, which make intensive agricultural production overly attractive (M. D. Young, 1992; Scheierling, 1994). Successful policy interventions must change those aspects of land-use decisions which are the source of diffuse pollutants.

As with point-source pollution control, policy options for nonpoint pollution control can be classed as cognitive (voluntaristic), regulatory, or incentive-based. As with point-source pollution, education, moral suasion, and technical assistance have seen limited success. (See Harrington, et al., 1985, for a good succinct discussion of nonpoint control policies. Chapters by Segerson, 1990, on incentive policies, and by Anderson, et al., 1990, on regulatory approaches to agricultural nonpoint pollution control, provide more detailed discussions).

Regulatory approaches call for specific actions by those regulated, or prohibitions on certain behaviors leading to water-quality degradation. One approach, for example, is the use of design standards (Harrington, et al., 1985). Such standards specify actions to be taken (such as a management plan for sediment control), or actions prohibited (such as avoiding certain cropping practices on highly erodible lands). Performance standards, in contrast, place limits on the rate of pollution discharge to the water body. In this case, interference with land use practices would be only in response to observed violations (Anderson, et al., 1990).

Neither technique is without limitations. Regulations of the design standard type are easier to enforce. However, they may be unnecessarily costly, because their general application may impose costs on some who contribute little to the problem. In contrast, performance standards, at least in principle, focus more directly on the pollutant source. However, performance standards are difficult to enforce in practice because accurate measurement of discharges (particularly from small farms) are nearly impossible, creating a continuing dispute over actual sources of pollutants.

The major alternative to regulatory policies are various *incentive-based approaches*. Policies in this group can include several specific mechanisms, including taxes, subsidies, and trading policies (Segerson, 1990; Tomasi, Segerson, and Braden, 1994; Zilberman and Marra, 1993).

Taxes or fees could be levied on either inputs or pollution outputs. For example, extra charges have been imposed on agricultural fertilizers in Sweden, with proceeds used to fund water-quality monitoring. Higher costs are expected to reduce fertilizer application rates and therefore water pollution. (However, experience suggests that taxes are unlikely to be set high enough to significantly affect land use, because of the adverse income effects.) Alternatively, charges might be levied on pollution itself, by imposing an "effluent charge" such as was discussed for point-source pollution. The technical and administrative complexity of setting fees for numerous farmers, precisely linked to the damages caused by their effluents, is mind-boggling. No example of successful implementation of this form of taxation has come to the writer's attention.

Subsidies—payments to potential sources of pollution for actions to reduce the pollution—could be made for actual reduction in pollution loadings, for adopting appropriate land use practices, or for making investments which reduce loadings. Subsidies to prevent soil erosion (and the associated productivity losses) have a long history in the United States and elsewhere. Subsidies are the most politically attractive (to polluters) of the available options. In contrast to other approaches, which impose costs on the emitting source and spread benefits over the society at large, subsidy costs are spread over the general population, and gains are offered to the land user. A drawback is that paying polluters to avoid polluting activities is objectionable to some groups. Also, payments must be made to some who would adopt proper practices anyway, and targeting specific instances where changing practices is most conducive to improved water quality is important. (See Lichtenberg, Strand, and Lessley, 1993, for a more detailed discussion of subsidies for nonpoint-source pollution control. Braden, Netusil, and Kosubud (1994) analyze the potential role of "green payments" to reward farmers for practices which yield environmental benefits.)

A trading policy for emissions could be adopted in situations where costs of abatement differ sharply across potential emitters. Those with high abatement costs could purchase emissions rights, and so reduce total costs of abatement. Costs, in this instance, are borne primarily by the polluters. Shortle and Adler (1994) analyze three incentive structures, including a mixed system of taxes and tradable input permits for the nonpoint pollution-control problem, in which emissions are stochastic and nonobservable, and there is differential information about cost of control. Letson, Crutchfield, and Malik (1993) studied the potential of emissions trading for controlling agricultural emissions to coastal waters in the United States, finding that the potential for application is limited and implementation would be extremely difficult and complex.

Finally, outright purchases of actual water rights and/or land use rights provides another approach. The public agency acquires rights to part or all of the land use rights from polluting lands, and manages the land to assure water quality. Purchase of tropical forest lands by either public or private agencies has been undertaken to preserve first-growth forests, but water-quality improvements are a side-benefit. Again, costs are borne primarily by beneficiaries, rather than by the land users whose practices are actually responsible for pollution.

Considerable progress is evident in integrated physical-biological-economic modeling to evaluate alternative nonpoint-source pollution-control mechanisms. See, for example, the analyses by Taylor, Adams, and Miller (1992) on nitrogen effluent control in Oregon, and Dinar, et al. (1993) on problems of irrigation drainage in California.

3.7.5 Other Issues Related to Water Quality

Other important issues only recently receiving attention are monitoring and enforcement, and the potential role of liability laws and the courts.

In a world of limited regulatory agency budgets, how much resources should be spent for monitoring and enforcement of pollution discharge regulations? The environmental quality control literature has largely tended to ignore considerations of monitoring and enforcement, assuming that polluters will comply with whatever regulatory system is in place, regardless of their own self-interest. However, when the probability of detection of violations is nonzero and agents of the polluting organization can decide whether or not to risk detection, problems of monitoring and enforcement arise. The analysis of these problems has examined the differential implications of the choice of policy instruments (charges versus standards), the implications of marginal penalty structures, and the determination—in setting penalties for violators—of whether merely negligence or strict liability is imposed on violators. See Russell, Harrington, and Vaughn (1986) for a discussion and analysis of these issues.

A related topic is the increasing use of the courts—by private interests in addition to public regulation—together with liability laws to improve environmental quality. As the standard problem of point-source pollution has come under control, issues arising from accidental pollution discharges, particularly those from toxic substances, have received increasing attention. Private lawsuits for damages can augment the public agency regulation of water pollution by obtaining court injunctions against polluting behavior or by filing lawsuits claiming tort liability for infractions of pollution standards. These potential actions provided further disincentives to polluting behavior. (See the papers collected in Tietenberg, 1992, for an analysis of the role of environmental liability law.)

3.7.6 Integrated Management of Water Supplies, Water Quality, and Watersheds

Water supply and water quality have typically been considered independently of one another. It is increasingly clear that integrating the quantity and quality aspects of water management is essential. Spulber and Sabbaghi (1994) have developed an elaborate formal economic framework which integrates quantity and quality considerations. Their principal organizing concepts are quality-graded demand and supply functions for water. A discrete set of water supply and demand functions is envisaged, each representing a defined water quality. This important extension from the traditional approach—of ignoring water quality—permits simultaneous representation of both the quantity and quality dimensions in analyses of intersectoral allocation issues. While the approach is theoretically quite attractive, no real-world empirical implementation is attempted, and it remains to be seen if this will be a fruitful approach in practical policy analyses. Booker and Young (1994) reported an alternative approach to integrating water quantity and quality. A combined hydrologic-economic optimization model was developed for the Colorado River basin in the southwestern United States. The main water quality problem is with dissolved mineral solids (salinity), which arises naturally in runoff from the sedimentary rock formations in the watershed, and secondarily from irrigation drainage waters. Salinity damage functions, for each subregion and type of use, quantify the economic effects of quality degradations. The hydrologic model routes salinity downriver, and the economic optimization model balances water quantity (including hydropower) and quality considerations.

Because a hydrologic unit is a natural entity for managing water supply and quality, interest has been renewed in the broader perspective for managing water supplies and environmental quality. Beginning in the 1930s the United States government developed a planning and implementation strategy for managing river basins for multiple-purpose outputs. Emphasis was on water supply for irrigation, municipal and industrial water uses, floodplain management, and power generation. By the 1960s, a concern for water quality became evident. Growing problems of water pollution from nonpoint sources (particularly sediments and nutrients) brought out the importance of incorporating land-use decisions into water management.

Integrated watershed management initially was employed to describe planning approaches which focused simply on allocation and supply among competing water-using sectors. More recently, the term has come to encompass the selection among water and land-use policy options according to multiple policy criteria. Interdisciplinary approaches are essential. For example, Easter, Dixon, and Hufschmidt (1986) present a conceptual framework emphasizing an interdisciplinary, multiple-objective planning approach which integrates land use and water management. The concepts are illustrated by case studies from Asia. Munasinghe (1992, Chap. 2) also conceptualizes integrated water resources planning as involving multisectoral, multiobjective water policy.

3.8 ECONOMIC EVALUATION OF POLICIES FOR FLOODPLAIN MANAGEMENT

Throughout the world, floods each year bring about considerable damages to property and disruption of economic activity, as well as causing injuries and deaths. By some measures, floods cause more damage and deaths to humankind than do any other natural hazard (Rodda, 1995). In the United States, floods have caused an estimated average damages of $2 billion per year over the past 30 years, and damages from the Mississippi Valley floods of 1993 exceeded $12 billion. Even so, on the average, flood damages represent a minuscule impact on the U.S. economy (Shabman, 1988).

So as to make the best use of valuable floodplains and to reduce losses to their citizens, governments expend resources to change flow regimes and adopt policies to influence behavior of floodplain occupants. During recent years, it appears that economists in the United States have taken relatively less interest in the policy issues posed by flooding than they have in water allocation and water quality issues, although a continuing program is evident in the United Kingdom (Penning-Rowsell, Parker, and Harding, 1986; Penning-Rowsell and Fordham, 1994).

Several elements of the theory market limitations discussed in Sec. 3.4 provide the economic justification for public policy to mitigate damages from natural hazards such as floods (Milliman, 1983). Basinwide actions designed to reduce damages from floods are classic instances where markets would be inadequate in achieving optimal resource allocation. The extent of flood hazard in downstream areas is influenced by land use and water-channeling decisions made upstream, often in another state or nation, so public coordination at the regional, national, and, often, international level is essential for efficient mitigation of these externalities. Another justification for public intervention is the public's imperfect knowledge regarding the actual risks of low-probability/high-consequence flood events. A public agency may be better able to evaluate tradeoffs regarding risks than the individuals that experience them. More generally, public flood management programs produce benefits which tend to be nonrival in consumption and exhibit high exclusion costs; once flood control services are produced, all floodplain occupants have access to those services, and individuals residing or conducting business on a floodplain cannot be readily excluded from enjoying floodplain management benefits.

The following discussion will address several issues: the economic optimum of public spending directed toward floodplain management, the institutional arrangements for moving toward that optimum, the approaches taken by economists in understanding the behavior of floodplain occupants, and methods of measuring benefits of flood hazard mitigation.

3.8.1 The Optimal Allocation of Resources to Floodplain Management

Economic evaluation of floodplain management policies was given impetus in the United States by language in the Flood Control Act of 1936, which asserts that "benefits to whomsoever they may accrue" must exceed the costs. Therefore, the economic approach follows the general BCA principles (discussed in Sec. 3.4.2) of balancing beneficial effects against adverse effects, focusing on the tradeoffs associated with purchasing additional risk reduction. Distinctive aspects include a probabilistic element in the evaluation of benefits, reflecting the uncertainty of flood events. Also—as in the case of the pollution control model discussed in Sec. 3.7—benefits of flood damage mitigation are measured in terms of willingness to pay to *avoid damages.* The basic economic model of natural hazard management hypothesizes that a rational, fully informed floodplain occupant would be willing to pay up to the present discounted expected (probability-weighted) value of losses, in order to avoid potential losses. The expected losses typically decline rapidly with initial levels of expenditure, but level off, as additional damage reduction is more difficult to attain. Flood hazard adjustments are subject to diminishing expected marginal returns, because they protect against the most frequent events. Mitigation costs tend to rise at an increasing rate with increased degrees of mitigation. Costs of protection tend also to be nonlinear. In the case of structural adjustments to floods (such as dams and levees) the volume (hence costs) is a power of the height, so additional protection is obtained at disproportionate expense.

As always, the selection of the objective function is crucial in how problems are defined and remedies evaluated. Analyses by noneconomists sometimes emphasize minimizing flood damages. Such an approach does not consider the costs of damage mitigation. An economically efficient level of expenditure on flood hazard mitigation is that which minimizes the sum of costs and expected damages, in present value terms. An equivalent formulation is to maximize the sum of discounted net benefits, when benefits are measured as the difference between flood losses with versus without protection. (The evaluation is, of course, site specific, depending on both hydrologic conditions and the nature and density of present and prospective human activity on the floodplain.)

This emphasis on tradeoffs differs from conventional technical risk analysis as applied to natural hazard mitigation policy. Technical approaches select a risk level roughly reflecting the potential severity of adverse effects, and design projects to satisfy the selected degree of risk (Renn, 1992). For example, it is sometimes suggested that standards for flood loss reduction be set to protect to the one percent annual chance ("100-year") flood.

The technical approach treats all affected areas and parties equitably, but it ignores economic efficiency considerations. Under technical flood risk standards, flood control policies are not subjected to systematic comparison of costs of mitigation with the expected losses averted. Therefore, large expenditures may sometimes be made which have little prospect for a corresponding reduction of damages. In contrast, the economic approach goes beyond the identification of the probability of some adverse event to the measurement of the disutility of such events to humans. It calls for an explicit assessment of potential damages to human floodplain residents (usually expressed in monetary units) as related to degree of flood control or mitigation. The economic approach to optimal protection from flood hazards can claim several advantages in addition to incorporating nonphysical (social) implications of risky events. It provides a common denominator in which risks and costs of amelioration can be compared in the same units, and different types of damages can be measured with the same unit.

An important implication of the economic optimum level of adjustment (but also of the technical standards approach) is that not all flood risks are prevented. The economic criterion is also consistent with increasing damages through time, as the real value of floodplain activities grows with the economy. Nonspecialists sometimes mistakenly diagnose policy failure when damages rise in proportion to the degree of protection afforded or increase over time in "protected" floodplains.

3.8.2   Measurement of Economic Costs and Benefits of Flood Mitigation

Estimation of costs of floodplain management policy evaluation presents few concerns unique to this issue. Two topics will be mentioned here. One has to do with adequately accounting for certain opportunity costs or external costs of floodplain management policies. For example, evaluations of public flood hazard mitigation policies have sometimes not adequately accounted for foregone nonmarketed benefits, such as the value of wetlands drained (Penning-Rowsell, Parker, and Harding, 1986). External costs have also been overlooked. Stavins and Jaffe (1990) found that in the Mississippi Valley, public flood management policies unintentionally, and contrary to federal policy, contributed to a substantial amount of private conversion of forested wetlands to agricultural uses.

A second issue concerns the risk of failure of high-hazard dams. While dam failure is not limited to flood control structures, the issue has much in common with flood hazard mitigation and has been analyzed in this context. Conventional water-project evaluation has tended to ignore the risk of structural failure brought about by events which exceed the design limits of the dam. Failure of a dam would likely result in several types of costs, including damage to property downstream, and income losses, as well as the foregone benefits of hydropower, irrigation, or flood control. The sum of these costs, weighted by the probability of failure, should be accounted for in water-project evaluation, or flood control structures are underdesigned and their benefits overstated (Cochrane, 1989). Ellingwood, et al. (1993), attempt an estimate of the cost of dam failures based on several actual cases.

The methods for estimating the economic benefits of flood risk reduction are similar to those used in other contexts, but because of the problems of the public's imperfect knowledge of flood probabilities and likely damages, and the potential for intangible impacts including the risk of death, their application presents a number of difficult and contentious aspects. Howe and Cochrane (1992) provide a conceptual discussion of the process of estimating probable damage functions for natural hazards, including floods.

The principal technique for estimating urban flood risk reduction benefits has been the *property damage avoided* (PDA) *approach,* which reflects the present value of real (inflation-free) expected property damages avoided by the project or policy. The replacement and repair costs to buildings and other property and structures, with and without the flood hazard, is estimated for each of a number of river flow levels. The estimated annual benefit for a given flow level is the difference between repair costs. Each flow is weighted by its probability of occurrence, and the benefits over all flows summed to estimate the expected benefit for each year. The annual benefits must be estimated for each year of the planning period, incorporating predicted changes in economic activity on the floodplain over time. More detailed descriptions of the approach can be found in the U.S. Water Resources Council (1983), or Penning-Rowsell, et al., (1992, Chap. 5).

Shabman (1994a), Howe and Cochrane (1992), and others note that the PDA method has been criticized for not incorporating nonproperty effects, such as individual and community disruption, medical expenses, productivity losses, and pre-flood anxiety. Another critic of the conventional approach is James (1994), who argues that, in the case of Bangladesh, a focus on property damages ignores the point that the primary benefits of flood damage reduction policies are "the intangible values of better lives for people who escape poverty." Shabman (1994a) provides probably the first comparison of alternative flood risk reduction benefit measures for the same locality. Estimates using PDA, land price analysis, and contingent valuation (CVM) methods were each developed for a portion of Roanoke, Virginia. It was found that benefit estimates varied significantly across techniques. In particular, the CV approach yielded willingness to pay estimates which did not systematically increase with an increasing probability of inundation.

Other particular valuation issues arise with assigning values to potential loss of life and to psychological effects of flood events. Although deaths from floods are a major risk for few of the population, the topic of valuing risks to life cannot be ignored in economic evaluations. The issue is controversial, because many disapprove of the idea of valuing life as if it were a commodity. Risk analysts acknowledge that society is unwilling to place a value on saving a particular individual from certain death. However, society does value safety or, conversely, reductions in risks to life. A number of approaches, including both market and nonmarket techniques, can be used to infer the value society places on reducing risks to life and health (Viscusi, 1986).

3.8.3 Policy Options for Flood Hazard Mitigation

The economic approach to flood damage mitigation policy owes much to the writings of the geographer Gilbert White and his associates (e.g., Burton, Kates, and White, 1993). White's natural hazards paradigm stresses the linkages between human use of the environment and the uncertain events flowing from the processes of natural systems. The interaction of extreme events with human activities produces hazards and influences responses to them. Acknowledging that floods are acts of nature, but contending that flood losses are largely due to acts of humans, White began urging the need for altered policies nearly 50 years ago. Strict emphasis on structural approaches to *flood control* should be superseded by a broader concept of *floodplain management.* Individuals should be encouraged to take action to reduce flood damages to complement structural approaches by public agencies. White proposed regulatory policies (such as zoning and other land-use controls) and economic incentives (subsidies and insurance) designed to influence floodplain occupants to take location and construction decisions to mitigate potential flood damages.

Criticisms of the limited cost effectiveness of the government's (primarily Corps of Engineers) structural approaches led to a reconsideration in the 1960s. In 1966, the Presidential Task Force on Federal Flood Control Policy recommended reorientation toward nonstructural, *people-management* policies, such as land-use controls and subsidized insurance for floodplain occupants. Legislation in 1968 established a nationwide federally subsidized flood insurance program, contingent upon local governments implementing land-use controls. The flood insurance concept was intended to deter risky site decisions by confronting floodplain occupants with the expected costs of their actions. (See Johnston and Associates, 1992, for a detailed account of the U.S. experience with floodplain management.)

On the whole, nonstructural approaches to floodplain management have been less effective in mitigating damages than its proponents in the social sciences had hoped. Several significant flood events occurred during the years following the 1968 legislation. It was observed that at-risk homeowners had little interest in purchasing flood insurance, even after the policy was revised to heavily subsidize purchasers and assure that insurance was in their interest in actuarial terms. Similarly, individuals seldom adopt loss prevention measures, such as floodproofing of structures, to protect themselves against flood damage. Lord (1994) further notes the problems that arose from the federal governments's adoption of the standard of a one percent probability of occurrence of a flood in any year for delineating Special Flood Hazard Areas for the flood insurance program. The standard had the unintentional effect of encouraging both the public and local government agencies that flood risk was zero outside the zone, a proposition that is incorrect even if the zoning exercise was correct, which, given the difficulties, it may not be. Green, Parker, and Penning-Rowsell (1993) from U.K. and French experience further conclude that benefits of flood warnings have typically been overstated. The assumptions regarding both reliability of the warnings and the response of the public appear to have been overestimated.

Nonstructural flood damage mitigation policies implemented by the U.S. government were studied by economists, psychologists, and other social scientists to assess individual and group behavior in response to natural and technological hazards. A team of psychologists and economists (prominent among whom were psychologists Paul Slovic and Baruch Fischoff) initiated a series of field surveys and controlled laboratory experiments to analyze the factors influencing individual responses to low-probability/high-consequence events. These studies concluded that most individuals are not accustomed to thinking in probability terms and, particularly in the context of low-probability events, tend not to rely on statistical data as the basis for making tradeoffs between benefits and costs. Rather, people seem to base their actions on past experience and advice from acquaintances. However, low-probability events such as floods, by their nature, do not provide individuals with much experience on which to make decisions on how to respond. Even experts with extensive access to data often disagree sharply about the likelihood of hazards. Psychologists find that individuals seem to adopt rules of thumb that permit them to ignore potentially disastrous occurrences of low-probability events, by assuming they fall below a threshold worthy of concern. (See Kunreuther, 1992, for a review by a central participant in these studies.) O'Grady and Shabman (1994) emphasize the additional problems of setting a discount (interest) rate in the context of long-term plans. The conclusions of this line of analysis are that few people make the mental computations implied by the discounted expected utility model of choice in the face of risk, and that an alternative, richer model of behavior will be more fruitful for flood policy analysis.

The concept of *moral hazard* provides a second behavioral explanation for the limited success of incentive policies for mitigating flood damages. Moral hazard—the term originated in the economic literature on insurance—refers to risks to an insurer arising from the behavior of the insured. Moral hazard arises when the probability of an event insured against is to some degree under the control of the insured, but the insurer cannot readily monitor the insured's actions to price the hazard. The insurance market fails in this case because the policy's very availability alters the incentives to the insured and the probabilities on which the insurer must rely. Moral hazards range from simple carelessness to outright fraud. Lichtenberg (1994) adapts the moral hazard concept to characterize the frequently observed phenomenon of further floodplain encroachment following public installation of flood management structures. When floods recur, the encroachment yields increased damages and thus partially offsets the expected benefits of the structures. People seem to assume that structures will subject them to a negligible flood risk and develop land behind the structures, increasing the actual damages incurred in rare but inevitable flood events.

Both Lichtenberg (1994) and Shabman (1994b) emphasize the perverse incentives operating in floodplains in the United States. Liberal postdisaster aid from the federal government ensures that those investing in floodplains never bear the full costs of the actions. Homeowners, businesses, and local governments are eligible for subsidized loans and grants to rebuild damaged structures as well as tax write-offs for losses. The federal government has further subsidized a large share of the costs of structural flood protection, creating a preference in favor of structural approaches and leading to excessive development on floodplains. Reformed policies, which confront those choosing to locate in the floodplain with the full costs of their activities—either from experiencing the loss in the event of flooding or from actuarially fair insurance rates—are needed.

Methodological investigations of the economic benefits of flood hazard reduction and further analysis of incentive systems for encouraging individual floodplain occupants toward mitigating flood damages will be fruitful areas for further research. Proposals introduced into the U.S. Congress in early 1995 call for more emphasis on an economic approach to risk analysis of both human-source and natural hazards. If benefit-cost considerations become a requirement of hazard analyses, progress on monetized measures of health and safety considerations will be required. Because of the difficulties experienced by people in evaluating low-probability/high-consequence events such as floods, and the problem of moral hazard, it is equally clear that these aspects of flood damage mitigation policy will warrant increased attention of researchers.

3.9 CONCLUDING REMARKS

This chapter has emphasized the theme that most water-management issues facing the world, rather than being purely physical or technical in nature, include significant demand-side or people-coordination components. Resource and environmental economists have made considerable progress in positive water economics—understanding how people respond to economic variables and incentives in their use of and interaction with water resources. Similar progress has been made on the normative side—in improving techniques for evaluating policies and in combining positive analysis and normative criteria to assess alternative policy proposals. Much, however, remains to be done. Improvements in actual water policy at times seem glacially slow, while in the meantime in many parts of the world, water is allocated to less valued uses, water quality continues to decline, groundwater basins are overexploited, and floods and droughts take an unnecessarily severe toll. Water-policy analysis is an inherently interdisciplinary problem. Increased collaboration among the varied scientific disciplines and more focus on working with the responsible government agencies on the real problems will be the path to improving our understanding of problems and providing better prescriptions for their solution.

Acknowledgments

I thank H. C. Cochrane, S. L. Gray, S. M. Scheierling, and several anonymous referees for helpful suggestions on earlier drafts.

REFERENCES

Adams, R., and T. Crocker, "Materials Damages," in J. B. Braden and C. D. Kolstad (eds.), *Measuring the Demand for Environmental Quality,* Elsevier Science Publishers, Amsterdam, 1991, pp. 271–301.

Anderson, T. L., and D. R. Leal, *Free Market Environmentalism,* Westview Press, Boulder, Colo., 1991.

Anderson, G., et al., in J. B. Braden, and S. B. Lovejoy (eds.), *Agriculture and Water Quality: International Perspectives,* Lynne Reiner Publishers, Boulder, Colo. and London, 1990.

Baumol, W. J., and W. E. Oates, *The Theory of Environmental Policy,* 2d ed., Cambridge University Press, Cambridge and New York, 1988.

Becker, G. S. "The Economic Way of Looking at Behavior" (Nobel lecture), *Journal of Political Economy* 101 (3): 385–409, 1993.

Berck, P., S. Robinson, and G. Goldman, "The Use of Computable General Equilibrium Models to Assess Water Policies," in A. Dinar, and D. Zilberman (eds.), *The Economics and Management of Water and Drainage in Agriculture,* Kluwer Aademic Publishers, Boston, 1991.

Bohm, P., and C. F. Russell, "Comparative Analysis of Policy Instruments," in A. V. Kneese and J. L. Sweeney (eds.), *Handbook of Natural Resource and Energy Economics,* vol. 1, Elsevier, Amsterdam, 1985, pp. 395–460.

Boggess, W., R. Lacewell, and D. Zilberman, "Economics of Water Use in Agriculture," in G. A. Carlson, D. Zilberman, and J. A. Miranowski (eds.), *Agricultural and Environmental Resource Economics,* Oxford University Press, New York, 1993, pp. 319–391.

Booker, J. E., and R. A. Young, "Modeling Intrastate and Interstate Markets for Colorado River Water Resources," *Jour. Environ. Econ. and Management* 26(1): 66–87, 1994.

Boulding, K. E., "The Implications of Improved Water Allocation Policy," in M. Duncan (ed.), *Western Water Resources: Coming Problems and Policy Alternatives,* Westview Press, Boulder, Colo., 1980.

Bower, B. T., J. Kindler, C. S. Russell, and W. R. D. Sewell, "Water Demand," in J. Kindler and C. S. Russell (eds.), *Modeling Water Demands,* Academic Press, London, 1984.

Braden, J. B., and S. B. Lovejoy (eds.), *Agriculture and Water Quality: International Perspectives.* Lynne Reiner Publishers, Boulder, Colo., and London, 1990.

Braden, J. B., and C. D. Kolstad (eds.), *Measuring the Demand for Environmental Quality.* Elsevier, Amsterdam, 1991.

Braden, J. B., N. R. Netusil, and R. F. Kosobud, "Incentive-Based Nonpoint Source Pollution Abatement in a Revised Clean Water Act," *Water Resources Bulletin* 30 (5): 781–791, 1994.

Bromley, D. W., *Environment and Economy: Property Rights and Public Policy.* Basil Blackwell, Oxford, 1991.

Brookshire, D. S., M. McKee, and G. Watts, *Draft Economic Analysis of Proposed Critical Habitat Designation in the Colorado River Basin for the Razorback Sucker, Humpback Chub, Colorado Squawfish and Bonytail,* Report to U.S. Fish and Wildlife Service, University of New Mexico, Albuquerque, 1993.

Burton, Ian, R. W. Kates, and G. F. White, *The Environment as Hazard,* 2d ed., Guilford Press, New York, 1993.

Carson, R. T and K. M. Martin, "Measuring the Benefits of Freshwater Quality Changes: Techniques and Empirical Findings," in A. Dinar and D. Zilberman (eds.), *The Economics and Management of Water and Drainage in Agriculture,* Kluwer Academic Publishers, Boston, 1991, pp. 389–410.

Clarke, R., *Water: The International Crisis.* MIT Press, Cambridge, Mass., 1993.

Colby, B. C., "Transaction Costs and Efficiency in Western Water Allocation," *Am. Jour. Agric. Econ.* 72(5): 1185–1192, 1990.

Costanza, Robert, H. E. Daly, and J. A. Bartholomew, "Goals, Agenda and Policy Recommendations for Ecological Economics," in R. Costanza (ed.), *Ecological Economics: The Science and Management of Sustainability,* Columbia University Press, New York, 1991.

Crane, R., "Water Markets, Market Reform and the Urban Poor: Results from Jakarta, Indonesia," *World Development* 22(1): 71–84, 1994.

Crocker, T. D., B. A. Forster, and J. F. Shogren, "Valuing Potential Groundwater Protection Benefits," *Water Resources Research* 27(1): 1–6, 1991.

Cropper, M. L., and W. E. Oates, "Environmental Economics: A Survey," *Journal of Economic Literature* 30(2): 675–740, 1992.

Cummings, R. G., P. T. Ganderton, and T. McGuckin, "Substitution Effects in Contingent Valuation Estimates," *Am. J. Agric. Econ.,* 76(2): 205–214, 1994.

de Savornin Lohman, A. P., "Economic Incentives in Environmental Policy: Why are They White Ravens?," in H. Opschoor and K. Turner (eds.), *Economic Incentives and Environmental Policies: Principles and Practices.* Kluwer Academic Publishers, Dordrecht, 1994, pp. 55–67.

Diamond, P. A., and J. A. Hausman, "Contingent Valuation: Is Some Number Better Than No Number?" *J. Econ. Perspectives* 8(4): 45–64, 1994.

Dinar, A., E. T. Loehman, M. P. Aillery, M. R. Moore, R. E. Howitt, and S. A. Hatchett, "Regional Economic Modeling and Incentives to Control Drainage Pollution," in C. S. Russell and J. F. Shogren (eds.), *Theory, Modelling and Experience in Management of Nonpoint-Source Pollution,* Kluwer Academic Publishers, Boston, 1993.

Dinar, A., and D. Zilberman (eds.), *The Economics and Management of Water and Drainage in Agriculture,* Kluwer Academic Publishers, Boston, 1991.

Driessen, T. S. H., and S. H. Tijs, "The Cost Gap Method and Other Cost Allocation Methods for Multipurpose Water Projects," *Water Resources Research,* 21 (10), 1985.

Driscoll, P., B. Dietz, and J Alwang, "Welfare Analysis When Budget Constraints Are Nonlinear: The Case of Flood Hazard Reduction," *J. Env. Econ. and Mgmt.* 26(2): 181–199, 1994.

Dudley, N. J., "A Single Decision-Maker Approach to Irrigation Reservoir and Farm-Management Decision-Making," *Water Resources Research* 24(4): 633–640, 1988.

Dudley, N. J., "Water Allocation by Markets, Common Property and Capacity Sharing: Companions or Competitors," *Natural Resources Journal* 32(4): 757–778, 1992.

Duffield, J. W., C. J. Nehrer and T. C. Brown, "Recreation Benefits of Instream Flow: Application to Montana's Big Hole and Bitterroot Rivers," *Water Resources Research* 28 (9): 2169–2181, 1992.

Easter, K. W., J. A. Dixon, and M. M. Hufschmidt (eds.), *Watershed Resources Management: An Integrated Framework,* Westview Press, Boulder, Colo., 1986.

Easter, K. W., "Differences in Transactions Costs of Strategies to Control Agricultural Offsite and Undersite Costs," in C. S. Russell and J. F. Shogren (eds.), *Theory, Modelling and Experience in Management of Nonpoint-Source Pollution,* Kluwer Academic Publishers, Boston, 1993.

Ellingwood, B., R. B. Corotis, J. Boland, and N. P. Jones, "Assessing Costs of Dam Failure," *J. Water Resource Planning and Management* 119(1): 64–82, 1994.

Fisher, A. C., and J. V. Krutilla, "The Economics of Nature Preservation," in A. V. Kneese and J. L. Sweeney (eds.), *Handbook of Natural Resource Economics,* vol. I, Elsevier, Amsterdam, 1985, pp. 165–189.

Frederick, K. D., "Water Resources: Increasing Demand and Scarce Supplies," in K. D. Frederick and R. A. Sedjo (eds.), *America's Renewable Resources: Historical Trends and Current Challenges,* Resources for the Future, Washington, D.C., 1991.

Frederick, K. D., "Long-Term Water Costs," in S. Durden and R. Patrick (eds.), *Proceedings, Fifth National Conference on Water: Our Next Crisis,* Academy of Natural Sciences of Philadelphia, 1994, pp. 303–324.

Freeman, A. M. III, "The Ethical Basis for the Economic View of the Environment," in D. VanDe Veer and C. Pierce (eds.), *People, Penguins and Plastic Trees: Basic Issues in Environmental Ethics,* Wadsworth, Belmont, Cal., 1986.

Freeman, A. M. III, "Water Pollution Policy," in P. R. Portney (ed.), *Public Policies for Environmental Protection,* Resources for the Future, Washington, D.C., 1991.

Freeman, A. M. III, *The Measurement of Environmental and Resource Values: Theory and Methods,* Resources for the Future, Washington, D.C., 1993.

Freeman, A. M. III, and J. C. Norris, "Quasi-Optimal Pricing for Cost Recovery in Multiple Purpose Water Projects," in V. K. Smith (ed.), *Environmental Resources and Applied Welfare Economics: Essays in Honor of John V. Krutilla,* Resources for the Future, Washington, D.C., 1988, pp. 119–134.

Gardner, R., E. Ostrom, and J. M. Walker, "The Nature of Common Pool Resources," *Rationality and Society* 2:335–358, 1990.

Gibbons, D. C., *The Economic Value of Water,* Resources for the Future, Washington, D.C., 1986.

Gillroy, J. M., and M. Wade (eds.), *The Moral Dimensions of Public Policy Choice: Beyond the Market Paradigm,* University of Pittsburgh Press, Pittsburgh, 1992.

Gleick, P. H. (ed.), *Water in Crisis: A Guide to the World's Fresh Water Resources,* Oxford University Press, New York, 1993.

Goodin, R. E., "Ethical Principles for Environmental Protection" (original, 1983), reprinted in J. M. Gillroy and M. Wade (eds.), *The Moral Dimensions of Public Policy Choice: Beyond the Market Paradigm,* University of Pittsburgh Press, Pittsburgh, 1992, pp. 411, 429.

Green, C. H., D. J. Parker, and E. C. Penning-Rowsell, 1993. "Designing for Failure," in P. A. Merriman and C. W. A. Browitt (eds.), *Natural Disasters: Protecting Vulnerable Communities,* Thomas Telford, London, 1993, pp. 78–91.

Griffin, R. C. and C. Chang, "Seasonality in Community Water Demand," *Western J. Agric. Econ.,* 16(2): 207–217, 1991.

Griffin, R. C. and S.-H. Hsu, "The Potential for Water Market Efficiency When Instream Flows Have Value," *Am. J. Agric. Econ.,* 75(2): 292–303, 1993.

Hamilton, J. R., and N. K. Whittlesey, 1986. "Energy and Limited Water Resource Competition," in N. K. Whittlesey (ed.), *Energy and Water Management in Western Irrigated Agriculture,* Westview Press, Boulder, Colo., 1986, pp. 309–327.

Hanemann, W. M., "Valuing the Environment through Contingent Valuation," *J. Econ. Perspectives,* 8(4): 19–43, 1994.

Hansen, L. T. and A. Hallam, "National Estimates of the Recreational Value of Streamflow," *Water Resources Research* 27(2): 167–175, 1991.

Harrington, W., A. J. Krupnick, and H. M. Peskin, "Policies for Nonpoint-Source Pollution Control," *Journal of Soil and Water Conservation* 40:27–33, 1985.

Herrington, P. R., *Pricing of Water Services,* OECD Publications, Organisation for Economic Cooperation and Development, Paris, 1987.

Howe, C. W., "Impact of Price on Residential Water Demand: New Insights," *Water Resources Research* 18(5): 713–716, 1983.

Howe, C. W., "Public Intervention Revisited: Is Venerability Vulnerable?" in V. K. Smith (ed.), *Environmental Resources and Applied Welfare Economics: Essays in Honor of John V. Krutilla,* Resources for the Future, Washington, D.C., 1988, pp. 191–209.

Howe, C. W., D. R. Schurmeier, and W. D. Shaw, "Innovative Approaches to Water Allocation: The Potential for Water Markets," *Water Resources Research* 22(4): 439–445, 1986.

Howe, C. W., J. K. Lazo, and K. R. Weber, "The Economic Impacts of Agriculture to Urban Water Transfers on the Area of Origin: A Case Study of the Arkansas River Valley in Colorado," *Am. Jour. Agric. Econ.* 72(5): 1200–1204, 1990.

Howe, C. W., and H. C. Cochrane, *Guidelines for the Uniform Definition, Identification and Measurement of Economic Damages from Natural Hazard Events,* Special Publication 28, Institute of Behavioral Science, University of Colorado, Boulder, Colo., 1992.

James, L. D., "Flood Action: An Opportunity for Bangladesh," *Water International* 19(2): 61–69, 1994.

Johnston, L. R., Associates, *Floodplain Management in the United States: An Assessment Report,* (2 vols.), Federal Interagency Floodplain Management Task Force, FEMA, Washington, D.C., 1992.

Kahn, A. E., "The Tyranny of Small Decisions: Market Failure, Imperfections and the Limits of Economics," *Kyklos* 19:23–47, 1966.

Kelman, S., "Cost-Benefit Analysis: An Ethical Critique," in D. VanDe Veer and C. Pierce (eds.), *People, Penguins and Plastic Trees: Basic Issues in Environmental Ethics,* Wadsworth, Belmont, Cal., 1986, pp. 242–249.

Kindler, J., and C. S. Russell (eds.), *Modeling Water Demands,* Academic Press, London, 1984.

Kneese, A. V., "Three Decades of Water Resources Research: A Personal Perspective," in V. K. Smith (ed.), *Environmental Resources and Applied Welfare Economics: Essays in Honor of John V. Krutilla,* Resources for the Future, Washington, D.C., 1988, pp. 45–55.

Kunreuther, H., "A Conceptual Framework for Managing Low-Probability Events," in S. Krimsky and D. Golding (eds.), *Social Theories of Risk,* Praeger Publishers, Westport, Conn., 1992, pp. 301–337.

Letson, D., S. Crutchfield, and A. Malik, "Point/Nonpoint-Source Trading for Controlling Pollutant Loadings to Coastal Waters: A Feasibility Study," in C. S. Russell and J. F. Shogren (eds.), *Theory, Modeling and Experience in the Management of Nonpoint-Source Pollution,* Kluwer Academic Publishers, Boston, 1993.

Libby, L. W., and W. G. Boggess, "Agriculture and Water Quality: Where are We and Why?," in J. B. Braden and S. B. Lovejoy (eds.), *Agriculture and Water Quality: International Perspectives,* Lynne Reiner Publishers, Boulder, Colo., and London, 1990, pp. 9–38.

Lichtenberg, E., "Sharing the Challenge: An Economist's View," *Water Resources Update* 97:39–43, 1994.

Lichtenberg, E., I. E. Strand, Jr., and B. V. Lessley, "Subsidizing Agricultural Nonpoint-Source Pollution Control: Targeting Cost Sharing and Technical Assistance," in C. S. Russell and J. F. Shogren (eds.), *Theory, Modeling and Experience in the Management of Nonpoint-Source Pollution,* Kluwer Academic Publishers, Boston, 1993.

Loomis, J. B., "The Economic Value of Instream Flow: Methodology and Benefit Estimates for Optimum Flows," *Jour. Environmental Management* 24(2): 169–179, 1987a.

Loomis, J. B., "Balancing Public Trust Resources of Mono Lake and Los Angeles' Water Right," *Water Resources Research* 23(8): 1440–1456, 1987b.

Loomis, J. B., M. Hanemann, B. Kanninen, and T. Wegge, "Willingness to Pay to Protect Wetland and Reduce Wildlife Contamination from Agricultural Drainage," in A. Dinar and D. Zilberman (eds.), *The Economics and Management of Water and Drainage in Agriculture,* Kluwer Academic Publishers, Boston, 1991.

Lord, W. B., "Flood Hazard Delineation: The One Percent Standard," *Water Resources Update* 95 (Spring): 36–39, 1994.

Lyman, R. A., "Peak and Off-Peak Residential Water Demand," *Water Resources Research* 28(9): 2159–2167, 1992.

Malik, A. S., B. A. Larsen, and M. Ribaudo, "Economic Incentives for Agricultural Nonpoint Source Pollution Control," *Water Resources Bulletin* 30(3): 471–480, 1994.

Martin, W. E., H. M. Ingram, N. K. Laney, and A. H. Griffin, *Saving Water in a Desert City,* Resources for the Future, Washington, D.C., 1984.

Martin, W. E., and S. Kulakowski, "Water Price as a Policy Variable in Managing Water Use: Tucson, Arizona," *Water Resources Research,* 27(2): 157–166, 1991.

Mayhew, S., and B. D. Gardner. "The Political Economy of Early Federal Reclamation in the West" in T. L. Anderson and P. J. Hill, eds. *The Political Economy of the American West.* Rowman and Littlefield Publishers, Lanham, Maryland, 1994.

Michelsen, A. M., and R. A. Young, "Optioning Agricultural Water Rights for Urban Water Supplies During Drought," *American J. Agric. Econ.* 75(4): 1010–1020, 1993.

Milliman, J. W., "An Agenda for Economic Research on Flood Mitigation," in S. A. Changnon, Jr., R. J. Schicht, and R. G. Semonin, *A Plan for Research on Floods and their Mitigation in the United States,* Illinois State Water Survey Division, Champaign, Ill., 1983.

Mitchell, R. C., and R. C. Carson, *Using Surveys to Value Public Goods: The Contingent Valuation Method,* Resources for the Future, Washington, D.C., 1989.

Moore, M. R., N. R. Gollehon, and D. H. Negri, "Alternative Forms of Production Functions of Irrigated Crops," *J. Agr. Econ. Research* 44(3): 16–32, 1994.

Munasinghe, M., *Water Supply and Environmental Management: Developing World Applications,* Westview Press, Boulder, Colo., 1992.

Nieswiadomy, M. L., "Estimating Urban Residential Water Demand: Effects of Price Structure, Conservation and Education," *Water Resources Journal* 28(3): 609–615, 1992.

O'Grady, K., and L. Shabman, "Uncertainty and Time Preference in Shore Protection," in Y. Y. Haimes, D. A. Moser, and E. Z. Stakhiv (eds.), *Risk-Based Decision Making in Water Resources VI,* American Society of Civil Engineers, New York, 1994, pp. 136–154.

Organisation for Economic Cooperation and Development (OECD), *Management of Water Projects: Decision-making and Investment Appraisal,* OECD Publications, Paris, 1985.

———, *Environmental Policy: How to Apply Economic Instruments,* OECD Publications, Paris, 1991.

———, *Market and Government Failures in Environmental Management: Wetlands and Forests,* OECD Publications, Paris, 1992.

Ostrom, E., and R. Gardner, "Coping with Asymmetries in the Commons: Self-Governing Irrigation Systems Can Work," *J. Econ. Perspectives* 7(4): 93–112, 1993.

Ostrom, E., R. Gardner, and J. Walker, *Rules, Games and Common-Pool Resources,* Univ. of Michigan Press, Ann Arbor, 1994.

Paterson, J., "Rationalised Law and Well-Defined Water Rights for Improved Water Management," in Organisation for Economic Cooperation and Development (OECD), *Renewable Natural Resources: Economic Incentives for Improved Management,* OECD, Paris, 1989, pp. 43–64.

Pearce, D. W., and R. K. Turner, *Economics of Natural Resources and the Environment,* Johns Hopkins University Press, Baltimore, Md., 1990.

Pearce, D. W., and J. J. Warford, *World Without End: Economics, Environment and Sustainable Development,* Oxford University Press, New York, 1993.

Penning-Rowsell, E. C., D. J. Parker, and D. M. Harding, 1986. *Floods and Drainage: British Policies for Hazard Reduction, Agricultural Improvement and Wetland Conservation,* Allen and Unwin, London, 1986.

Penning-Rowsell, E. C., C. H. Green, et al., *The Economics of Coastal Management: A Manual of Benefit Assessment Techniques,* Bellhaven, London, 1992.

Penning-Rowsell, E. C. and M. Fordham (eds.), *Floods Across Europe: Flood Hazard Assessment, Modelling and Management,* London, Middlesex University Press, 1994.

Provencher, B., and O. Burt, "A Private Property Rights Regime for the Commons: The Case for Groundwater," *Amer. J. Agr. Econ.* 76(4): 875–888, 1994.

Randall A., "Methodology, Ideology and The Economics of Policy: Why Resource Economists Disagree," *Am. J. Agric. Econ.* 67:1022–1029, 1985.

Randall, A., "Valuation in a Policy Context," in D. W. Bromley (ed.), *Natural Resource Economics: Policy Problems and Contemporary Analysis,* Kluwer-Nijhoff, Boston, 1986, pp. 163–199.

Renn, O., "Concepts of Risk: A Classification," in S. Krimsky and D. Golding (eds.), *Social Theories of Risk,* Praeger Publishers, Westport, Conn., 1992, pp. 53–79.

Renzetti, S., "Estimating the Structure of Industrial Water Demands: The Case of Canadian Manufacturing," *Land Economics* 68(4): 396–404, 1992.

Rhoads, S. S., *The Economist's View of the World: Governments, Markets and Public Policy,* Cambridge University Press, Cambridge, 1985.

Rodda, J. C., "Whither World Water?" *Water Resources Bulletin* 31(1): 1–7, 1995.

Rogers, P., *America's Water: Federal Roles and Responsibilities,* Cambridge, Mass., MIT Press, 1993.

Russell, C. S. "Programming Models for Regional Water Demand Analysis," in J. Kindler and C. S. Russell (eds.), *Modeling Water Demands,* Academic Press, London, 1984.

Russell, C. S., W. Harrington, and W. J. Vaughan, *Enforcing Pollution Control Laws,* Johns Hopkins University Press, Baltimore, Md., 1986.

Russell, C. S., and J. F. Shogren (eds.), *Theory, Modeling and Experience in the Management of Nonpoint-Source Pollution,* Kluwer Academic Publishers, Boston, 1993.

Saliba, Bonnie Colby, and D. B. Bush, *Water Markets in Theory and Practice: Market Transfers and Public Policy,* Westview Press, Boulder, Colo., 1987.

Sampath, R. K., "On Some Aspects of Irrigation Distribution in India," *Land Economics* 66(4): 448–463, 1990.

Sampath, R. K., "Issues in Irrigation Pricing in Developing Countries," *World Development* 20(7): 967–977, 1992.

Saunders, R. J., J. J. Warford, and P. C. Mann, "Alternative Definitions of Marginal Cost," in G. M. Meier (ed.), *Pricing Policy for Development Management,* Johns Hopkins University Press, Baltimore, 1983, pp. 203–207.

Scheierling, S. M., "Agricultural Water Pollution: The Challenge of Integrating Agricultural and Environmental Policies. Lessons from the European Community Experience," World Bank Technical Paper, 1994.

Schmid, A. A., *Benefit-Cost Analysis: A Political Economy Approach,* Westview Press, Boulder, Colo., 1989.

Schneider, M. L., and E. E. Whitlach, "User-Specific Water Demand Elasticities," *Jrnl. Water Resources Planning and Management* 17(1): 52–73, 1991.

Segerson, K., "Incentive Policies," in J. B. Braden and S. B. Lovejoy (eds.), *Agriculture and Water Quality: International Perspectives,* Lynne Reiner Publishers, Boulder, Colo., and London, 1990.

Shabman, L., "The Benefits and Costs of Flood Control: Reflections on the Flood Control Act of 1936," in H. Rosen and M. Reuss (eds.), *The Flood Control Challenge: Past, Present and Future,* Public Works Historical Society, Chicago, 1988, pp. 109–123.

Shabman, L., "Measuring the Benefits of Flood Risk Reduction," in Y. Y. Haimes, D. A. Moser, and E. Z. Stakhiv (eds.), *Risk-Based Decision Making in Water Resources VI,* New York, American Society of Civil Engineers, 1994, pp. 122–135.

Shabman, L., "Responding to the 1993 Flood: The Restoration Option," *Water Resources Update* 95 (Spring), 1994b, pp. 26–30.

Shortle, J. S., and D. G. Abler, 1994. "Incentives for Nonpoint Pollution Control," in C. Dosi and T. Tomasi (eds.), *Nonpoint Source Pollution Regulation: Issues and Analysis,* Kluwer Academic Publishers, Dordrecht, 1994, pp. 137–149.

Small, L. E., and I. Carruthers, *Farmer-Financed Irrigation: The Economics of Reform,* Cambridge University Press, Cambridge, 1991.

Smith, V. K., "A Conceptual Overview of the Foundations of Benefit-Cost Analysis," in J. D. Bentkover, V. T. Covello, and J. Mumpower (eds.), *Benefit Assessment: The State of the Art,* Reidel, Boston, 1986, pp. 13–34.

Smith, V. K., and W. Desvouges, *Measuring Water Quality Benefits,* Kluwer Nijhoff, Norwell, Mass., 1986.

Solley, W. B., R. R. Pierce, and H. A. Perlman, "Estimated Use of Water in the United States, 1990," Circular 1081, U.S. Geological Survey, Washington, D.C., 1993.

Spulber, Nicolas, *Regulation and Markets,* MIT Press, Cambridge, Mass., 1989.

Spulber, Nicolas, and Asghar Sabbaghi, *Economics of Water Resources: From Regulation to Privatization,* Kluwer Academic Publishers, Norwell, Mass., 1994.

Stavins, R. N., and A. B. Jaffe, "Unintended Impacts of Public Investments on Private Decisions: The Depletion of Forested Wetlands," *Am. Econ. Review* 80(3): 337–352, 1990.

Stone, J. C., and D. Whittington, "Industrial Water Demands," in J. Kindler and C. S. Russell (eds.), *Modeling Water Demands,* Academic Press, London, 1984.

Sugden, R., and A. Williams, *The Principles of Practical Cost-Benefit Analysis,* Oxford University Press, Oxford, 1978.

Tang, S. Y., *Institutions and Collective Action: Self Governance in Irrigation,* ICS Press, San Francisco, 1992.

Thomas, J. F., and G. J. Syme, "Estimating Residential Price Elasticity of Demand for Water: A Contingent Valuation Approach," *Water Resources Research* 24(11): 1847–1857, 1988.

Tietenberg, T. H. (ed.), *Innovation in Environmental Policy: Recent Developments in Environmental Enforcement and Liability,* Edward Elgar Publishing, Brookfield, Vt., 1992.

Tomasi, T., K. Segerson, and J. Braden, "Issues in the Design of Incentive Schemes for Nonpoint Source Pollution Control," in C. Dosi and T. Tomasi (eds.), *Nonpoint Source Pollution Regulation: Issues and Analysis,* Kluwer Academic Publishers, Dordrecht, 1994, pp. 1–37.

U.S. National Research Council, Board on Water Science and Technology, *Water Transfers in the West: Efficiency, Equity and the Environment,* National Academy Press, Washington, D.C., 1992.

U.S. Water Resources Council, *Economic and Environmental Principles and Guidelines for Water and Related Land Resource Implementation Studies,* Government Printing Office, Washington, D.C., 1983.

Viscusi, W. K., "The Valuation of Risks to Life and Health: Guidelines for Policy Analysis," in J. D. Bentkover, V. T. Covello, and J. Mumpower (eds.), *Benefit Assessment: The State of the Art,* Reidel, Boston, 1986, pp. 193–210.

Wahl, R. W., *Markets for Federal Water: Subsidies, Property Rights and the Bureau of Reclamation,* Resources for the Future, Washington, D.C., 1989.

Walsh, R. G., *Recreation Economic Decisions: Comparing Benefits and Costs,* Venture Publishing, State College, Pa., 1986.

Walsh, R. G., D. M. Johnson, and J. R. McKean, "Benefits Transfer of Outdoor Recreation Demand Studies: 1968–1988," *Water Resources Research* 28(3): 707–713, 1992.

Ward, F., "Economics of Water Allocation to Instream Uses: Evidence from a New Mexico Wild River," *Water Resources Research* 23(3): 381–392, 1987.

Ward, W. A., and B. J. Deren, *The Economics of Project Analysis: A Practitioner's Guide,* World Bank, Washington, D.C., 1991.

Whittington, D., et al., "Household Demands for Improved Sanitation Services in Kumasi, Ghana: A Contingent Valuation Study," *Water Resources Research* 29(6): 1539–1560, 1992.

Wolf, C., Jr., *Markets or Governments: Choosing Between Imperfect Alternatives,* MIT University Press, Cambridge, Mass., 1988.

World Bank, *Water Resources Management: A World Bank Policy Paper,* International Bank for Reconstruction and Development, Washington, D.C., 1993.

World Bank Water Demand Research Team, "The Demand for Water in Rural Areas: Determinants and Policy Implications," *World Bank Research Observer* 8(1): 47–70, 1993.

Young, H. P., N. Odaka, and T. Hashimoto, "Cost Allocation in Water Resources Development," *Water Resources Research* 18(3): 463–475, 1982.

Young, H. P. (ed.), *Cost Allocation: Methods, Principles and Practices,* Elsevier, Amsterdam, 1985.

Young, M. D., *Sustainable Investment and Resource Use: Equity, Environmental Integrity and Economic Efficiency,* UNESCO, Paris, 1992.

Young, R. A., "Why Are There So Few Transactions Among Water Users?," *Am. J. Agric. Econ.* 68:1143–1151, 1986.

Young, R. A., "Managing Aquifer Overexploitation: Economics and Policies," in Ian Simmers, F. Villaroya, and L. F. Rebollo (eds.), *Selected Papers on Aquifer Overexploitation,* Heise Publishers, Hannover, Germany, 1992.

Young, R. A., and S. L. Gray, "Input-Output Models, Economic Surplus and the Evaluation of State or Regional Water Plans," *Water Resources Research* 21(12): 1819–1823, 1985.

Young, R. A., and R. H. Haveman, "Economics of Water Resources: A Survey," in *Handbook of Natural Resources and Energy Economics,* vol. II, A. V. Kneese and J. L. Sweeney (eds.), Elsevier Science Publishers, Amsterdam, 1985.

Zajac, E. E., "Perceived Economic Justice: The Example of Public Utility Regulation," in H. P. Young (ed.), *Cost Allocation: Methods, Principles and Practices,* Elsevier, Amsterdam, 1985.

Zilberman, D., and M. Marra, "Agricultural Externalities," in G. A. Carlson, D. Zilberman, and J. A. Miranowski (eds.), *Agricultural and Environmental Resource Economics,* Oxford University Press, New York, 1993.

CHAPTER 4
PRINCIPLES OF PLANNING AND FINANCING FOR WATER RESOURCES IN THE UNITED STATES

David H. Moreau, Professor
Department of City and Regional Planning
University of North Carolina at Chapel Hill

4.1 INTRODUCTION

Principles for planning and financing of water resources in the United States have evolved from a very complex and dynamic set of circumstances. Complexities arise from many sources. Responsibilities for water management are distributed across federal, state, and local governments and the private sector. Hydrologic systems are highly variable both in time and space, and the same resource is often managed to serve several purposes simultaneously. Pollution of the resource comes from a wide array of sources, and tracing the transport, fate, and effects of harmful substances on human health and the environment is demanding on existing knowledge.

Existing mixes of institutions and policies and the principles they reflect are manifestations of the history of initiatives undertaken in the country to address water problems as they arose on the national agenda. First it was the need for navigation, then public water supply, then flood control, irrigation, hydroelectric power, and finally, water quality. Understanding the present array of organizations and principles under which they operate is difficult without a knowledge of their origins.

Yet, national water resource management policy is not static, and any effort to describe principles today is but a snapshot of a rapidly changing environment. Just within the period from 1965 to 1990, fundamental shifts have occurred in water policies, planning, and financing. Within that period an entirely new structure for water resource planning was created and dismantled. During that period the national construction program for reservoirs declined from an all-time high to a crawl, and financial arrangements were completely overhauled. During that same period, basic changes were made in the way the United States plans, manages, and finances water-quality programs. Many of these changes, especially those relating to planning and financing, are still very much in a state of flux. Thus, an understanding of the existing situation is enriched by a knowledge of recent history and current trends.

Only a few principles and the complexities behind them can be covered in a single chapter. These principles are put in historical perspective because it is from that viewpoint that the dynamics of policy formation are revealed. In recent years, much of water policy in the United States has been rooted in state and local government, but limited time and space restrict consideration of those changes to general observations and acknowledgement of their importance.

The chapter begins with a listing of the eight principles to be discussed and moves to an examination of each of those principles. A few observations are offered in the concluding section.

4.2   PRINCIPLES

Principles discussed in this chapter are:

1. River basins and watersheds within basins are logical spatial units for water resource planning. Groups of basins and watersheds may be appropriate when demands and impacts on the resource extend across multiple units.

2. All potential uses should be considered when planning for the resource, and to the extent that it is prudent to do so, facilities should be used for multiple purposes.

3. Land and ecological resources which are significant to water should be incorporated in the planning process.

4. Planning should identify those allocations and reallocations that result in the highest valued uses of the resource.

5. A sufficient number of alternatives should be formulated and evaluated to enhance the search for economically efficient and environmentally sound alternatives and to display the tradeoffs among competing objectives that decision makers must make in selecting a plan.

6. A well-developed set of objectives, criteria, and procedures is essential to formulation and evaluation of alternatives.

7. Financial arrangements to implement the plan should require beneficiaries of the plan to pay in proportion to the cost of services that they derive from the plan, to the extent that such arrangements are not inconsistent with other principles of equity and efficient use of the resource.

8. To the extent possible, organizational arrangements for water and related land resources should be integrated across all purposes within the planning unit.

4.3   RIVER BASINS AND WATERSHEDS

During the last quarter of the nineteenth century several of these principles for water management evolved, at least in part in response to the growing demand for water-based services. During the period 1875 to 1900, population in the United States increased from about 45 million to about 76 million, an increase of 70 percent. Much of that growth was occurring in the flood-prone Mississippi Valley and the arid

West where water problems were shared over large regions. Urbanization was also increasing at a very rapid pace, with the number of cities with populations in excess of 50,000 rising from 30 to 78.

4.3.1   Origins

One of the earliest principles to evolve was the use of river basins or watersheds as the basic unit for planning. Debate over the need for basinwide planning was triggered in part by a heated controversy over proper flood-control strategies in the 1850s, but it was not until the end of the nineteenth century that the principle was clearly articulated. In 1850, Congress directed that a study be undertaken to protect low-lying lands along the Mississippi River. Captain A. A. Humphreys of the U.S. Army Corps of Engineers (USACE) was given responsibility to prepare the report, and he later requested help from Lieutenant H. L. Abbot. Humphreys and Abbot completed their report in 1861 and recommended a "levees only" strategy to control flooding in the Mississippi Valley. They also devoted a considerable portion of their report to an attempt to discredit 1849 and 1851 reports by a well-known civil engineer, Charles Ellet, who had recommended an alternative strategy that included a system of headwater reservoirs in addition to downstream levees. Ellet's recommendation was bitterly attacked by Humphreys and Abbot, and their views tended to dominate flood-control policy on the Mississippi until the devastating floods of 1927 (Kemper, 1949, pp. 37–57).

A more explicit statement of the principle came from leaders of the conservation movement during the administration of President Theodore Roosevelt. Among those leaders were Frederick H. Newell, an assistant to Major John Wesley Powell in the U.S. Geological Survey (USGS); Gifford Pinchot, referred to by many as America's first trained forester, who emerged as a prominent advisor to Roosevelt; and W. J. McGee, a self-educated anthropologist and geologist, who also had been an assistant to Powell in the USGS and has been described as the chief theorist of the conservation movement (Hays, 1959, p. 108).

When Roosevelt became President, Pinchot and Newell held positions of influence in his administration. One of the earliest issues on which their mettle was tested was a conflict over water power. The President and his advisors were concerned that federal land and water policies were resulting in virtual giveaways of immensely valuable resources to powerful private corporations whom they feared would establish monopolies with little public benefit. A particular issue arose when private power interests proposed to build a dam on the Tennessee River at Muscle Shoals, Alabama. Roosevelt was persuaded to veto a permit for that project, arguing that any such development should await more careful planning for all of the river in a manner that would be of greatest benefit to the public (McGeary, 1960, p. 74).

McGee, writing in an article in 1907, argued that each stream is an interrelated system in which control of any part will affect to some extent any other part. He also felt that the work of the Reclamation Service, then headed by Newell, best exemplified the practice of river development. Pinchot, McGee, and Newell found many forums to express their views on water management (McGeary, 1960, p. 94; Hays, 1959, p. 105). They recommended a national commission on waterways, a recommendation that Roosevelt made a reality when he established the Inland Waterways Commission (IWC) in 1908. All three were appointed to the commission along with Senator Francis Newlands of Nevada, a strong advocate of irrigation.

Several basic principles were articulated in Roosevelt's letter to the commission. Among them was that river basins are the logical unit for planning. His letter said (IWC, 1908, p. IV):

Each river system, from its headwaters in the forest to its mouth on the coast, is a single unit and should be treated as such. Navigation of the lower reaches of a stream can not be fully developed without the control of flows and low waters by storage and drainage. Navigable channels are directly concerned with the protection of source waters and with soil erosion, which takes materials for bars and shoals from the richest portion of our farms. The uses of a stream for domestic and municipal water supply, for power, and in many cases for irrigation, must also be taken into account.

4.3.2 Practice: 1927–1960

River-basin planning was given an impetus when the Federal Water Power Act was passed in 1920. That act established a regular process for licensing of private power projects, but for several years Congress did not provide the Federal Power Commission with funds for planning (Shad, 1979, p. 13). In 1925, Congress recognized the need for planning to guide the FPC's licensing process and directed that agency and the Corps of Engineers to prepare estimates of planning costs for those streams on which power development appeared to be feasible. Those streams were listed in House Document 308 in 1927, and the Rivers and Harbors Act of 1927 directed the Corps of Engineers to undertake surveys and plans for them. Over 200 reports were prepared as a result of that action, and these so-called *308 reports* established the first set of comprehensive river-basin development plans for the nation (Senate Select Committee, 1959, p. 13).

In the 1940s and 1950s river basins continued to be used as the basic spatial unit for planning under so-called comprehensive plans prepared under the dominant leadership of one or more federal agencies. In its 1960 report, the Senate Select Committee on Water Resources, chaired by Senator Robert Kerr of Oklahoma, recommended creation of a planning structure that would bring the several water-related federal agencies under common leadership and elevate the role of states. Recommendations of the Kerr Committee were largely realized with passage of the Water Resources Planning Act of 1965, which created a cabinet-level Water Resources Council. River-basin planning and management were given a prominent place by the Council, but the Council also recognized that to address some water problems, especially those of large urban areas that drew supplies from multiple basins, planning could not be limited to single basins. It was given authority under Title II to establish river-basin commissions for basins, groups of basins, and other areas to coordinate federal, state, local, and private sector planning. River-basin commissions were given responsibilities to recommend priorities for data collection, investigations, planning, and projects. Each state and each federal agency with jurisdiction in areas covered by commissions were to be represented on the governing body, the chair of which was to be appointed by the President. Six such commissions were created, including those for New England, the Ohio River basin, the Great Lakes basin, the Missouri River basin, the Pacific Northwest, and the Souris-Red-Rainy basins. Planning was also to be conducted by the Council in areas not covered by commissions.

4.3.3 Water Quality

River basins have also been commonly accepted as logical spatial units for water-quality planning. North Carolina, like many other states, began its comprehensive pollution-control program in 1951 by undertaking pollution surveys, classifying streams, and preparing pollution control plans for each of 17 major basins in the state. When the Federal Water Pollution Control Act Amendments (PL92-500) were passed in 1972, states were required under Section 303 to prepare implementation plans by river basin and subbasins within basins.

PL92-500 added a significant new dimension to the array of planning activities, however, with its incentives for metropolitan-level planning. Minervini (1979) pointed out that Senator Edmund Muskie of Maine, the principal author and advocate of PL92-500, was also highly influential in his advocacy of areawide comprehensive planning for all federally assisted programs affecting metropolitan areas. He had been influential in passage of the Demonstration Cities and Metropolitan Development Act of 1966 and wrote the Intergovernmental Cooperation Act of 1968. Under the 1965 Water Quality Act, the Senate had encouraged a 10-percent bonus for construction grants that conformed to comprehensive regional development plans, and that incentive was substantially increased in Section 208 of PL92-500. That provision authorized $300 million for 100-percent federally funded grants to metropolitan agencies for the preparation of areawide waste-treatment plants.

Despite the key role that was envisioned for areawide planning when the act was being formulated, it played a relatively minor part when the act was implemented. The U.S. Environmental Protection Agency (EPA), the federal agency charged with protection of water quality, chose to follow the path established by the states, with primary emphasis on state program planning under Section 106 and basin-level implementation plans under Section 303(e).

4.3.4 Watersheds

Agricultural interests were not satisfied with the emphasis on large scale, multiple purpose planning that emerged from the 308 plans. They argued that farmers were being damaged by floods on small watersheds that were being ignored by planning at the basin scale. The Department of Agriculture (USDA) had been granted authority by the Flood Control Act of 1936 to undertake investigations in those watersheds for which planning by the Corps of Engineers had been authorized, and the Flood Control Act of 1944 authorized USDA to undertake land treatment and other nonstructural measures to reduce flooding and erosion. It was not until 1952, however, that a bill was introduced to authorize USDA to construct facilities in small watersheds. Objections from the USACE and the Bureau of Reclamation (USBR) were overcome in 1954, when Public Law 566 was enacted, and the Small Watershed Program was established by the Soil Conservation Service. Initially, this program was limited to watersheds of 250,000 acres (390 mi^2) or less, to structures containing no more than 5000 acre-ft of storage, and federal assistance was limited to surveys, planning, and engineering services. Amendments in 1956 expanded federal assistance and increased the size of eligible structures to 25,000 acre-ft, so long as storage allocated to flood control did not exceed 5000 acre-ft (Holmes, 1972, p. 28).

Watersheds have also been rediscovered by EPA through its Watershed Protection Approach, a program initiated in 1991. Although that program is somewhat eclectic and is not highly structured, it is responsive to recommendations of diverse groups calling for nationwide watershed management. Cases cited by EPA under this program range from a small 20-mi^2 watershed that would fit comfortably under USDA's small watershed program to an area covering portions of seven states in the Lower Mississippi Valley (USEPA, 1991; 1992).

4.4   COMPREHENSIVE PLANNING AND
MULTIPLE PURPOSE FACILITIES

In addition to the use of river basins as planning units, leaders of the conservation movement also advocated the principle of multiple-purpose development. Roosevelt's letter to the Inland Waterways Commission stated that

> ... works designed to control our waterways have thus far usually been undertaken for a single purpose (navigation, power, irrigation, protection from floods, supply water for domestic and manufacturing purposes). ... the time has come for merging local projects and uses of the inland waters in a comprehensive plan for the benefit of the entire country. Such a plan should consider and include all uses to which streams may be put, and should bring together and coordinate the points of view of all users of water.

Two concepts are embedded in this principle. First, all facilities should be designed to serve multiple purposes, and second, planning should be sufficiently comprehensive so as to include all uses to which streams may be put. It is a principle founded in concern for economic efficiency. Marginal investments to projects that serve a single purpose can yield substantial benefits to other purposes.

Although this principle is appealing and has been widely advocated, practice still lags far behind. Numerous obstacles have hindered achievement of fully comprehensive planning, including debates over appropriate roles of government in planning and management, conflicts between federal and nonfederal roles, and interagency competition. Many of these obstacles arose from fundamental debates on matters of policy about federal and state investments in navigation, about federal investments in flood control, irrigation, and hydroelectric power, public versus private roles in public water supplies, and federal versus state authorities in management of water quality. Several of these issues are traced in the sections that follow.

4.4.1   Navigation

Use of rivers for navigation has been a matter of national policy since the country was established. Even before adoption of the Constitution, George Washington was involved in attempts to settle a conflict between states bordering the Potomac River, and some have attributed the calling of the Constitutional Convention in 1787 to the need to resolve that dispute (Shad, 1979; Fox, 1964). Early leaders were mindful of the complex array of waterway tariffs that had stifled the flow of commerce in Europe.

The Ordinance of 1787 for governance of the Northwest Territory is cited by Hull and Hull (1967, p. 3), however, as the cornerstone of a free waterway policy in the United States. That ordinance, formulated and adopted by the last Congress under the Articles of Confederation, established a government for the territory north of the Ohio River and east of the Mississippi. Framers of the ordinance wrote in Article IV that "... the navigable waters leading into the Mississippi and St. Lawrence Rivers ... shall be common highways and forever free ... without any tax, impost or duty. ..." Within two years that ordinance was adopted without change as one of the first acts of Congress under the new Constitution in 1789.

Precedents had been established in the Treaties of Paris of 1763 and 1783. In 1763, when France and Spain ceded lands east of the Mississippi River to Great Britain, the three countries agreed that the Mississippi River would remain free of tolls from its source to the sea. Following the Revolutionary War, the United States and Great Britain agreed in 1783 to reaffirm the freedom of commerce along the Mississippi River to all citizens of the two countries.

Establishing authority over the use of waterways did not mean that the federal government would take all initiatives to overcome barriers to navigation and control destructive floods. Early advocates of federal financing of improvements to navigation included George Washington, Alexander Hamilton, James Madison, John Jay, and John C. Calhoun (Hull and Hull, 1967, pp. 9–11), but despite active leadership in both the executive and legislative branches of government, federal investment in the early years was stymied by the War of 1812 and strict interpretations of the Constitution. The issue of federal undertakings to improve the nation's waterways was settled in 1824, when the Supreme Court ruled in the famous *Gibbons v. Ogden* case that federal participation in internal improvements of waterways is justified under the commerce clause of the Constitution.

Even after the issue of federal authority to invest in internal improvements was settled, federal investments in water resources remained at a modest level. The private sector took the early initiative to improve navigation by building two major canals—the Santee, opened in 1800, connecting the port at Charleston, South Carolina to the Santee River, and the Middlesex Canal in 1803, connecting the port of Boston with the Merrimack River. Joint ventures between private corporations and states were also common. Virginia, New York, and other states actively sought participation in private ventures. Virginia and Maryland purchased nearly one-half of the shares offered by the Potomac Company, and Virginia was heavily invested in the James River and Dismal Swamp Companies (Goodrich, 1960, p. 20).

The need for capital soon overwhelmed the capacity of private investors, however, and states began to take the lead, many with financially disastrous outcomes. New York State invested heavily in development of the Erie, Champlain, and Oswego Canals among others. In 1908, the Inland Waterways Commission reported that 4469 miles of canals had been built in the United States to that time, most of it from 1820 to 1840, often referred to as the Great Canal Era. A financial crisis in 1837, competition from railroads, and mismanagement caused many of these ventures to yield revenues far short of expenses, leading many to require public subsidies or default on loans. Over 2200 miles had been abandoned by 1890. By 1908, there were only 2189 miles of canals still in operation, their ownership distributed as shown in Table 4.1.

4.4.2 Flood Control

As the country began its westward expansion, especially with settlement of the Lower Mississippi River Valley, the devastating effects of flooding became a matter

TABLE 4.1 Canals in Operation in the United States in 1908

Ownership	Number	Mileage	Percent of mileage
United States government	17	194	9
States	12	1359	62
Private	16	636	29
Total	45	2189	100

Source: Data compiled from Report of the Inland Waterways Commission, 1908.

of increasing national concern. In 1810, only 129,000 people lived in Arkansas, Louisiana, Mississippi, and Missouri. Within 40 years that number increased to over 2 million, and by 1850, 1 out of every 12 citizens lived in one of those four states (*Historical Statistics of the United States: Colonial Times to 1970*). Many of them were engaged in agricultural activities in alluvial valleys of the Mississippi and its tributaries, where periodic floods destroyed their crops.

Early efforts to reduce damage from floods were undertaken by individual land owners and territorial governments. Initially, settlements tended to locate along rivers, one plantation deep, and territorial tradition required each plantation to build its share of a continuous line of levees. In some instances, assistance was provided from local and state sources, occasionally with federal aid. Louisiana and Mississippi adopted territorial traditions into law in 1816 and 1819, respectively, by requiring frontholders to build levees. If they did not, parishes and counties were given authority to construct them and sue the landowner for the expense. As settlements began on interior lands, however, there was increasing demand by frontholders that levees be paid for from general taxes. In the 1840s and 1850s, Louisiana, Mississippi, and Arkansas established levee boards within each parish or county to oversee levee construction and to levy a general land tax to finance their construction (Harrison, 1961, pp. 60–63).

During this period, political leaders in the Mississippi Valley states continued to call for assistance from the federal government, claiming that federal lands were benefiting from private and local investments in levee construction. The federal response is particularly interesting in light of present debates over wetlands. As early as 1826, Congress had debated granting wetlands to the states for the purpose of "reclaiming" those lands for agricultural purposes. As Harrison noted, however, many members of Congress were unsure of the character and extent of the "wet and overflowed lands," and thought that the granting of these lands to the states should await further surveys. It was not until 1849 and 1850 that Congress took steps, through the Swamp Land Acts, to transfer much of these lands to the states, on the condition that funds from the sale of these lands be used to build additional levees and drainage works. Four states—Louisiana, Arkansas, Missouri, and Mississippi—acquired about 24 million acres of land through this process, with its incentive to convert them to agriculture (Harrison, 1961, p. 68).

Whether or not the federal government should make direct investments in flood-control projects remained a subject of considerable debate until passage of the Flood Control Act of 1936. One flurry of activity occurred after the flood of 1874, when many levees along the Mississippi River remained in poor condition from damage suffered during the Civil War. For years Congress had rejected flood control bills as being unconstitutional, but over the period 1874 to 1879, advocates argued that navigation and flood control were one and the same problem. A bill to establish a permanent Mississippi River Commission (MRC) to prepare and implement plans for improvements to navigation and flood control was introduced in 1879, but opponents saw it as a first step toward a major federal program to reclaim flood-prone lands throughout the Mississippi Valley. Their position was that while navigation had been clearly established as a legitimate federal activity under the constitution, it was being used as a guise for federal expenditures for flood control.

Advocates of the MRC prevailed, but it had jurisdiction only in the Mississippi Valley. Other special situations were subsequently authorized. USBR began including flood control and other purposes in its projects in 1906. When the USACE prepared the 308 river-basin plans to guide licensing of hydroelectric dams, it required private investors to include flood control, as well as several other purposes, in their projects. It was not until the Flood Control Act of 1936, however, that the federal government was authorized to construct dams and other works primarily for the purpose of flood control.

In the 1960s, increased attention was given to floodplain management as a complement to dams and levees. That initiative is discussed later in this chapter.

4.4.3 Public Water Supply

Continued movement of jobs and people from rural areas to concentrated urban centers has had important impacts on water supplies, wastewater disposal, floodplain management, and other aspects of water management. When the new Constitution was signed in 1789, only two cities had populations over 25,000, none over 50,000. Less than one percent of the entire population lived in urban places. By 1860, there were 16 cities with over 50,000 people, and nearly 20 percent of the population of the country lived in urban places. That trend continued into the twentieth century. By 1960, 113 million, or 63 percent of the nation's population, were living in metropolitan areas. Twenty years later, the metropolitan population had increased by more than 50 percent, and its share of the national total had jumped to 75 percent (*Historical Statistics of the United States: Colonial Times to 1970*).

Withdrawals to satisfy needs of urban residents have followed growth in population, reaching 38.5 bgd in 1990 (USGS, 1993). Publicly-supplied water was the only one of four major categories of use, as defined by USGS, that experienced continual growth from 1960 to 1990. As shown in Fig. 4.1, withdrawal for this purpose is substantially less than for cooling of thermoelectric power plants and irrigation, but water for this purpose is limited to high-quality sources. Rising public concern about that quality was translated into legislative action in the form of the Safe Drinking Water Act (SDWA), passed by Congress in 1974, which for the first time established national drinking water standards for all public supplies.

Private water companies were the first to respond to needs of urban residents, but those companies more often than not gave way to municipal ownership. In the

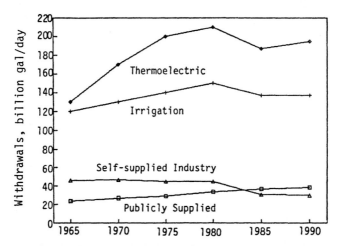

FIGURE 4.1 Withdrawals for water use in the United States. (*W. B. Solley, R. R. Pierce, and H. A. Perlman,* Estimated Use of Water in the United States in 1990, *United States Geological Survey, USGPO, Washington, D.C., 1993.*)

period 1795 to 1800, Massachusetts chartered 18 private companies, most of which were organized to serve small towns. Similar activity was occurring in other states, especially New York, New Jersey, and South Carolina. One of the most successful of the private ventures was the Baltimore Water Company, but many private companies (including the one in Baltimore) could not keep pace with demands for service. After an abortive joint venture with a private canal company, the City of Philadelphia began supplying its own water in 1798. In New York City, a private water company chartered in 1799 was more interested in using its profits to go into the banking business than meeting the city's growing demand for water. After several decades of conflict, the city took over the system in 1834. Relations between Boston and the Boston Aqueduct Company were apparently more amicable, but the outcome was much the same as in New York. The city began delivery of water from its new Cochituate Aqueduct in 1848 (much more detailed discussions of these events are given by Blake, 1956).

Similar stories have been told of other cities. By 1860, 12 of the 16 cities over 50,000 were served by municipally-owned systems. The principle of local government ownership and management of public water supplies was well-established. Blake (1956, p. 268) quoted an 1844 statement by the Committee on Public Health of the American Medical Association, which declared:

> ... an abundant supply of water is so intimately connected with the health of a city that the municipal authorities should rank this among the most important of their public duties.... The public welfare is too deeply interested in their faithful performance, safely to permit them to pass into the hands of incorporated companies.

Local governments continue to be the prime deliverers of public water in the United States. Approximately 85 percent of all production for public supplies is by utilities owned and operated by local governments, the balance being supplied largely by investor-owned companies (Immerman, 1986). Data from the *Census of Governments* indicate that in 1987, local governments accounted for 99.4 percent of all governmental expenditures for water supply, with municipalities and special districts accounting for over 90 percent of that total.

Capital expenditures by local government for water and related sewer services, shown in Fig. 4.2, are substantially greater than federal expenditures for water resource development. Expenditures for capital outlays for water-supply facilities has been about $6 billion a year (1987 dollars) since 1987, and capital expenditures for sewer service, which is now largely a local government responsibility, was $7 billion in 1990.

4.4.4 Irrigation

While urbanization was becoming more of a fact of life in the East in the latter half of the nineteenth century, the West was just being settled. Its population in 1860 was 619,000, or 2.0 percent of the U.S. population (these numbers exclude the sparsely populated areas of Arizona, Idaho, Montana, and Wyoming that had not joined the Union). By 1900 that population had increased sevenfold to 4.3 million (*Historical Statistics of the United States: Colonial Times to 1970*).

Much of that growth was encouraged by federal land policy. The Homestead Act of 1862 initiated an era in which public lands were given in 160-acre parcels to individuals, if they would agree to establish residence on the land and meet specified cul-

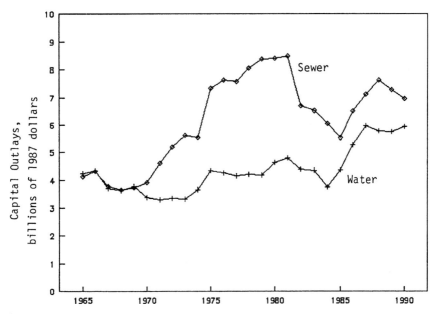

FIGURE 4.2 Capital outlays by local governments for water and wastewater facilities in constant 1987 dollars. (*Various reports of the U.S. Bureau of the Census.*)

tivation requirements. Irrigation was essential to the success of these ventures, and development of supply systems for this purpose began with private entrepreneurs. Sites at which small-scale irrigation could be profitably developed were limited in number, however, and private development of larger areas was hampered by limited expertise in soil science, limited capital, and a decentralized patchwork of distribution systems that could not capture economies of scale. It was not long before there was a call for federal financial assistance for western water development, but Congress turned a deaf ear for decades. In 1866, Congress declared that irrigation was to follow local customs, laws, and court decisions, thereby subjecting irrigation to state laws. Like the Swamp Lands Act earlier, the Desert Lands Act of 1877 created incentives to individuals in four states by offering 640-acre tracts of federally owned land at give-away prices if buyers would irrigate within three years. That incentive was ineffective, however, and after the irrigation boom–bust period of 1887 to 1893, private investors became disillusioned. Western political leaders then increased the intensity of their campaign for more federal assistance (Hays, 1959, p. 9). Congress responded in part by passing the Carey Act in 1894. This time, each state was granted up to one million acres to do whatever was necessary to get settlers to occupy, cultivate, and irrigate these lands.

The Carey Act likewise brought only a modest response. The original law and its amendments in later years led to applications to the Department of the Interior by eligible states for approximately 8.5 million acres of land, but residency and irrigation criteria were met on only 1.1 million acres (Golze, 1961, p. 19). That did not satisfy irrigation boosters like Francis G. Newlands, a prominent figure at the first National Irrigation Congress (NIC) in 1891, who was elected to the U.S. House of Representatives in 1892. By 1898, NIC, together with several national associations of commercial and industrial interests, gained sufficient support to get both national political parties to include irrigation financing as part of their platforms (Hays, 1959, pp. 9–11).

In 1901, Newlands proposed the Reclamation Fund, through which the federal government would finance the construction of irrigation facilities with proceeds from the sale of public lands. He proposed in that legislation that only family-owned farms could benefit from projects financed by the Reclamation Fund, and he compromised his original 80-acre limit on farm size, settling for a limit of 160 acres. When Theodore Roosevelt became President in 1902, he took an interest in irrigation and was able to persuade reluctant Republicans to support Newland's bill. It passed in 1902, and Roosevelt established the Reclamation Service within USGS to administer the program, and appointed Frederick Newell director (Hays, 1959, pp. 11–14). Several years later, the Bureau of Reclamation was created as an independent agency within the Department of the Interior to administer the program.

The Reclamation program may have been most visible, but it did not dominate the irrigation scene. Data in Fig. 4.3, compiled by Wahl (1989, p. 17), indicate that the USBR has never accounted for more than one-quarter of the irrigated acreage in the 17 western states.

4.4.5 Electric Power

Just as the push for major investments in irrigation facilities was beginning to mature in the West, a new technology that would have a major impact on U.S. water resources was emerging. Development of lighting systems in the 1880s created a market for electric power (Rudolph and Ridley, 1986, pp. 24–36) that led to an explosive growth in demand for hydroelectric dams and for huge quantities of water to cool thermoelectric generating plants. As illustrated in Fig. 4.4, in 1920 the installed capacity of electricity utilities was under 13,000 MW. In 1990, it was 735,000 MW

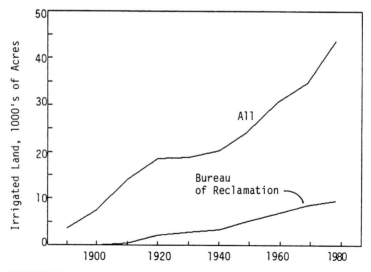

FIGURE 4.3   Irrigated lands in the United States. (*Wahl, 1989.*)

(*Historical Statistics of the United States*). Hydroelectric facilities played an important part in the early growth of the industry. From 1920 to 1950 hydroelectric plants generated 25 to 30 percent of all electricity. After 1950, hydroelectric plants became less important in terms of total output, but they remain a very important part of the mix of technologies for generating peaking power. Steam and nuclear plants accounted for 87 percent of total capacity in 1990, and as shown in Fig. 4.1, they withdraw more water than any other water-using sector.

The electric power industry as a whole is dominated by the private sector, with investor-owned utilities holding 77 percent of all generating capacity in the United States, 83 percent of all nonhydro capacity. Government has developed the largest share of water-based generating capacity. Of the 12 percent of total capacity for which water is now the prime mover, governments own 66 percent and the private sector 34 percent. Federal agencies hold 68 percent of government-owned capacity; state and local governments own the balance (Edison Electric Institute, 1990).

4.4.6 Water Quality

Quite unlike water resource planning and management programs, which have been dominated by the federal government since the beginning of the country, water pollution-control–quality-management programs were creations of state governments. Although the federal government took over the leadership role in the 1960s, these programs are marked much more by a federal–state partnership than has existed in water resource development activities.

Massachusetts is generally credited with establishing the first pollution-abatement program and setting a model for other states. In 1886, the legislature

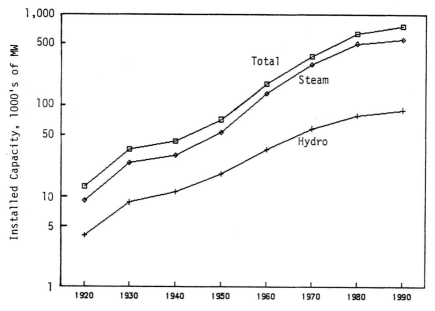

FIGURE 4.4   Installed generating capacity of electric utilities.

passed "An Act to Protect the Purity of Inland Waters" and provided funding for a full-time engineering staff under the board of health (Fair, 1950). By 1905 more than 40 states had laws regarding water pollution, some of them quite weak. Only 8 of those 40 states had enforcement capabilities (Tarr, et al., 1980, p. 72). Even in those states with enforcement capabilities, there were numerous exceptions.

Enactment of more effective statutory authority to control pollution followed an uneven pace from state to state. As of 1939, almost every state had vested some power to manage water pollution upstream of public water supplies in its state board of health (National Resources Committee, 1939, Appendix I), but the states in which that authority was made explicit may have numbered less than 20. In many other states, the authority was stated in very general terms, subsumed under very broad delegations of authority to protect public health.

By 1962, considerable advances had occurred at the state level, but the pattern of unevenness persisted. A survey of state water-pollution control programs (McKee and Wolf, 1963) indicated that at least 20 states had reasonably comprehensive programs in 1962, but about half of the states had no water quality standards or only minimal guidelines.

As the issue of environmental quality, and more particularly issues of water and air quality, moved to a more prominent position on the national agenda in the 1960s, the absence of such basic information as a census of industrial wastes made it appear that inadequate action was being taken at the state level to achieve clean water. Pressure mounted to put the federal government in a more dominant leadership position.

In the early 1960s Congress was dissatisfied with the pace of progress toward pollution control, and that discontent was led by Senator Muskie, chairman of the newly created Senate Subcommittee on Air and Water Pollution (Davies, 1970, p. 43). By 1963, Muskie had formulated legislation that would have: (1) transferred administration of pollution control from the Public Health Service to a new Federal Water Pollution Control Administration, still within the Department of Health, Education, and Welfare (HEW) and (2) established a federal floor on water-quality standards for the states. That legislation passed Congress as the Water Quality Act of 1965. Implementation of the act required each state to adopt water-quality standards at least as strict as the federal standards and to develop implementation plans for achieving them. Those plans had to be approved by the Secretary of HEW.

The Water Quality Act of 1965 was just the beginning of federal dominance. Amidst the clamor of the environmental movement for greater action, the Federal Water Pollution Control Act Amendment of 1972 (PL92-500) was passed, and, with modest changes in 1977, 1982, and 1987, it has been the framework for water-pollution control in the United States for over 20 years. That law, now referred to as the Clean Water Act (CWA), strikes a precarious balance between federal and state responsibilities in managing water quality, but it has provided little impetus for more comprehensive water resource management.

4.4.7   Protection and Restoration

The Clean Water Act protects water quality for designated uses—water supply, recreation, fish and wildlife—and where streams and other water bodies have been degraded in quality, the CWA can be used to restore water quality and related ecological systems to support designated uses. Although the CWA has been expanded by court decisions and rulemaking to protect certain resources, most notably wetlands, it was not designed to keep streams in near-natural condition. Other programs have addressed that purpose more directly.

Central to this effort was the Wild and Scenic Rivers Act (WSRA) passed in 1968. Amidst the rapid pace of dam construction in the 1950s (discussed later in this chapter), a group of naturalists and planners inside and outside the Department of the Interior began to formulate the concept of preserving selected streams as "wild" rivers. With the advent of the environmental movement in the 1960s and support from the Kennedy and Johnson administrations, the time was ripe to move legislation through Congress. A bill to designate the Saint Croix River was introduced by Senator Gaylord Nelson of Wisconsin in 1963, but the first national river to be established was the Ozark National Scenic Riverways in 1964. Sixteen other wild and scenic river bills were introduced before the WSRA was passed in 1968. The original act designated parts of 12 rivers and listed 27 others for study. As of 1992, the National Wild and Scenic River System included 151 officially designated streams, but those designations include many named tributary streams. An unofficial count listed 314 rivers and named tributaries and 11 other streams with related designations. A total of 11,277 miles of streams were included in the system (Palmer, 1993, pp. 10–25, 97–98).

In several cases, the USACE has made substantial outlays to restore previously modified river systems to a more natural condition. Two examples for which the National Research Council has provided some details are the Kissimmee Riverine-Floodplain System in Florida and the Atchafalaya Basin in Louisiana (National Research Council, 1992). The Endangered Species Act is also leading to changes in water management, particularly in the Pacific Northwest. Restoration of wild salmon runs and protection of other species has led to proposals for significant changes in management of the Columbia and Snake Rivers (Volkman, 1992).

4.4.8 Groundwater

Restoration has been a key feature of planning for groundwater since the discovery of a number of well-publicized incidents of contamination brought this once-neglected resource to national attention. Declining groundwater levels have also attracted attention in several states.

A national overview of groundwater availability (USGS, 1984, pp. 36–45) showed groundwater withdrawals increasing from about 35 bgd in 1950 to 88 bgd in 1980, an annual average increase of 3.1 percent. About two-thirds of those withdrawals were concentrated in only eight states, particularly California and several midwestern states that have drawn down the High Plains Aquifer. Resulting drawdowns in water levels and reduced pressures in artesian aquifers have drawn considerable national attention.

Far greater publicity and concern has been focused on health and environmental threats of groundwater contamination, largely from chemical wastes. Since World War II, production of materials now classified as hazardous wastes has spiralled. Approximately 65,000 commercially available chemicals have been identified and regulated by EPA under authority of the Toxic Substances Control Act. EPA estimated in its Toxics-Release Inventory that 18 billion pounds of chemicals were released to the environment in 1987 (USEPA, 1990, p. 7). As limits on emissions of pollutants to the atmosphere and discharge to surface waters became more restrictive, some generators of hazardous wastes chose to dispose of them in poorly designed, and poorly operated, landfills and lagoons. A 1977 EPA study found that groundwater contamination was the most common form of damage resulting from improper disposal of hazardous materials (USEPA, 1977). Disposal of solid wastes from municipalities also posed a serious threat to groundwater quality.

Consequences of these disposal practices began to emerge in the late 1960s and 1970s. Among the most widely publicized cases was the Hooker Chemical and Plastics Corporation's dump of highly toxic substances at its Love Canal property, near Niagara Falls, New York, during the years 1942 to 1952. Residents near that site began to notice some effects of the dump in the late 1950s, and concerns about potential adverse health effects continued to grow over the next 15 years. By the late 1970s it had become a state and national issue (Levine, 1982, pp. 1–26). Similar cases, including the "Valley of the Drums" near Shepardsville, Kentucky, and the Velsicol Chemical sites near Memphis, Tennessee, have been reported in the media and reviewed in both popular and technical literature (Epstein, Brown, and Pope, 1982).

Even before many of these cases reached a level of national notoriety, Congress began working on legislation to protect groundwater quality. A modest program to protect aquifers that are the sole source of community water supplies was established under the Safe Drinking Water Act of 1974. The Solid Waste Act of 1970 strongly encouraged states to close open dumps, but it was not until 1974 that Congress began to address the issue of hazardous wastes. Representative Paul Rogers, chair of the Health Subcommittee of the House Commerce Committee, favored a comprehensive approach to solid-waste management, one that would establish a strong federal role in solid waste, as well as hazardous wastes. The Nixon Administration wanted to leave solid-waste management in the hands of states, reserving federal initiatives to hazardous wastes. Rogers was able to craft legislation in 1975 with relatively modest opposition after he compromised on the issue of federal intrusion into management of municipal solid waste.

The Resource Conservation and Recovery Act (RCRA), as the bill became known, passed in 1976 when many members of Congress and the Ford Administration underestimated its significance (Epstein, Brown, and Pope, 1982, pp. 182–194). Among other provisions, it directed EPA to: (1) create a cradle-to-grave manifest system for the transport, storage, treatment, and disposal of hazardous wastes; (2) establish a permit system for owners of disposal facilities; and (3) set standards for construction and operation of those systems.

The problem of cleaning up abandoned chemical dumps, however, was not solved by RCRA. The act covered new waste management activities and those active at the time of the legislation. The Comprehensive Environmental Response Compensation and Liability Act (CERCLA), commonly known as Superfund, passed in November 1980, in the waning days of the Carter administration. The Superfund created by this legislation was a $1.6 billion pot of money to be used to clean up spills of hazardous materials and abandoned waste disposal sites if no responsible party could be found or responsible parties were unwilling to pay. Financing was provided by a tax on oil and a list of specific chemicals. The legislation also empowered the government to sue responsible parties for the cost of clean up.

Other legislation, including amendments to RCRA (1984), SDWA (1986), CERCLA (1986), and CWA (1987), has been added to strengthen groundwater protection programs. EPA's implementation strategy began to take shape in 1980, and its Groundwater Protection Strategy was published in 1984 (USEPA, 1984). That strategy stressed: (1) enhancement of state capabilities to manage groundwater, (2) improved management of sources of contamination, and (3) creation of a consistent and rational policy for protection and cleanup.

Pursuant to that strategy, EPA reported in 1990 that states had received approximately $7 million per year since 1985 for development of groundwater management strategies. All of the states had prepared strategies; 33 had passed additional legislation, 22 had developed formal groundwater classification systems, and 11 others had "informal or implicit" systems (USEPA, 1990, p. 9). Unfortunately, only a small number of states have developed comprehensive water resource plans that give adequate consideration to groundwater resources.

Omission of groundwater in state plans is not the only failure to achieve the principle of multipurpose comprehensive planning. The large number of purposes discussed in this section, each with its own policy and institutional framework involving different levels of government and the private sector, makes integrated, comprehensive planning and management difficult and complex to achieve. Efforts to create organizational arrangements to undertake that task are discussed later in this chapter.

4.5 PLANNING FOR WATER AND RELATED LAND RESOURCES

Some efforts toward restoration reflect the growing importance of the principle that water resource planning cannot be carried out independently of that for closely related land resources and ecological systems. This principle began to evolve even before those more contemporary issues emerged. One of the early advocates of relating land and water management was Major John Wesley Powell. In his famous *Report on the Lands of the Arid Regions of the United States* (1878), Powell recommended several policies that related land development to water management, including:

1. Because water is not sufficiently abundant to irrigate all lands in the region, irrigation should be reserved for those lands that are best suited for it.
2. Homesteads should be defined topographically, not rectilinearly, and irrigation should be permitted only on frontlands, while use of backlands would be restricted to pasture.
3. The federal government should undertake surveys to identify those lands that are most appropriate for irrigation.
4. Development should not be left to settlers or the private sector.

During the conservation movement, the importance of land–water relationships became more widely recognized. River basins began to be viewed not only as unified hydrologic systems, but also as enormous land and water complexes, in which both public and private planning interests must be reconciled (Fox, 1964, p. 67).

A particular controversy sparked much debate between foresters and engineers in the latter part of the nineteenth century and the early part of this century. Forests were known to play some role in controlling the flow of water to flood-prone streams, and forest management was being touted as a means of controlling floods. Many engineers, including a Special Committee of the American Society of Civil Engineers, acknowledged possible connections between forests and flooding, but they argued that those effects were not sufficiently predictable to justify inclusion of forest management as a part of flood control plans (White, 1957). Today, those effects are much more predictable in particular locations, through the use of computer-based hydrologic models.

Control of erosion from land-disturbing activities, especially agriculture, is another important result of the well-established linkages between land and water management. Significant milestones that have marked recognition of this relationship include: (1) establishment of the Soil Conservation Service in 1933 under the leadership of Hugh Bennett, (2) passage of the Flood Control Act of 1936 which authorized the Secretary of Agriculture to make investigations of runoff and soil erosion relating to flood control, and (3) the Small Watershed Program established in 1954 (White, 1957).

The need to integrate water resource planning with that of land resources is reflected in the titles of federal documents governing water planning. Senate Document 97, published in 1961, was entitled "Principles and Standards for Planning and Evaluating Water and Related Land Resources." That same phrase was incorporated in the title of the Water Resource Council's principles and standards in 1973.

Probably no activity has recognized the need for integrated water- and land-use planning more than efforts to plan and manage water quality. Erosion from agriculture, forestry, and urban areas were among the nonpoint sources recognized by amendments to the Federal Water Pollution Control Act (FWPCA) in 1972. EPA now ranks agriculture as the leading source of water quality impairment (USEPA, 1994a). Siltation, nutrients, and pesticides are listed among the leading causes of degradation in rivers, lakes, and estuaries. Other nonpoint sources transported by runoff from land surfaces, particularly from urban areas and resource extraction sites, are also listed among leading sources of degradation. Because of the dispersed nature of these sources and large volumes of runoff, it is impractical to collect and treat contaminants from these sources in a central treatment facility. They must be managed through a variety of land-use management techniques that either control the location and intensity of pollutant activities or require appropriate on-site management of potential contaminants.

In more recent years, effects of land development on the quality of drinking water has received considerable attention, causing a number of urban areas and some states to adopt special watershed protection programs (Miller, et al., 1981; Robbins, et al., 1992). A special problem is protecting public water supplies from surface sources that are not being treated by filtration plants. Under the Surface Water Treatment Rule promulgated by EPA in 1989, watersheds from which such supplies are drawn are subject to especially stringent land-use regulations.

Yet another impetus to integrate water and related land resource planning is the relatively recent attempt at ecological restoration. Interactions between land and water systems often result from complex ecological systems, particularly those involving wetlands. In its study of restoration of aquatic ecosystems, the National Research Council (NRC) reported on a number of cases to provide some details about restoration efforts (NRC, 1992). Unfortunately, that study failed to identify a viable institutional mechanism and planning arrangements for carrying out integrated restoration programs involving intermediate to large-scale water resource systems.

4.6 ALLOCATION AND REALLOCATION
TO HIGHEST VALUED USES

Water transfers have become more commonplace in western states in recent years, and changes in statutes have occurred to better accommodate this practice (e.g., see Argent, 1989; Jensen, 1987; Howe, 1989; Wahl, 1989; Wahl and Osterhoudt, 1986). These transfers usually involve the displacement of lower-valued agricultural uses by urban uses, but other types of transactions have also occurred, notably the transfer of water saved by reducing conveyance losses in canals from the Imperial Irrigation District (IID) to the Metropolitan Water District of Southern California (MWD). Considerable attention has been given to water markets as a means of facilitating such exchanges, particularly in western states, where water laws are more compatible with exchanges of water rights of this type than in eastern states. Most western-state water law is derived from the doctrine of prior appropriation (first in time, first in right), whereas eastern water law, evolving from the riparian doctrine, closely ties rights to use water to ownership of lands over and next to which it flows.

Wahl (1989, pp. 155–56) concluded from his analysis of federal policy that considerable latitude for voluntary water transfers existed under reclamation law in 1989, but that additional legislation was needed to remove certain ambiguities in the uses to which transferred water can be put. Muys (1993) asserts that the USBR has taken a hands-off policy to voluntary transfers, allowing transfers to occur when they comply with state law. He also reviewed changes in legislation governing the federally owned Central Valley Project in California, in which policy on transfers was made explicit, allowing mutually agreeable transfers so long as they satisfied a list of conditions relating to repayment for the project, environmental conditions, third-party effects, and other factors.

Transfers of water in eastern states to meet growing urban demands have also drawn attention, but many of the more notable cases are not mutually agreeable, and the absence of a market makes it difficult to arrive at exchanges that are satisfactory to all parties. One case now under investigation is a proposed reallocation of storage in Lake Lanier in Georgia to increase the supply of drinking water to Atlanta and the surrounding metropolitan area (American Water Resources Association, 1992). The transfer would ostensibly come from storage now allocated to hydroelectric power. In an attempt to settle a growing dispute between those who would benefit from this reallocation and downstream interests in Alabama and Florida who allege that they would be damaged by such a transfer, Congress authorized and funded a comprehensive study for the Alabama-Coosa-Tallapoosa and Apalachicola-Chattahoochee-Flint River basins. One goal of that study is to recommend a formal coordination mechanism for basinwide management of the resource, but it remains to be seen what form that mechanism will take and if it can resolve the issues that are put before it.

4.7 CONSIDERATION OF ALTERNATIVES

Throughout much of the history of water management in the United States, planning has concentrated on building structures to expand supplies, control floods, and treat wastewater. Only relatively recently has much attention been given to nonstructural alternatives that reduce demand, reduce exposure of people and property to flooding, reduce wastewater flows, and prevent pollution. Consideration of these options often leads to more economically efficient and less environmentally damaging plans. Even when structural measures are most appropriate, alternative configurations of structures, construction of projects in multiple stages, and alternative types of structures can result in very different beneficial and adverse effects of economic efficiency, environmental quality, and other objectives by which plans are to be evaluated. Only by formulating and evaluating a variety of alternatives can planners display the kinds of tradeoffs among competing objectives that responsible decision makers must consider in selecting among alternatives.

That a variety of alternatives should be formulated and evaluated as a basic step in any water resource planning process is accepted as a general principle. This discussion is focused, however, on two important cases where nonstructural measures are gaining favor as compliments to structural measures. The first case examines the historical emphasis on dam construction in the United States and the emerging roles of floodplain management and conservation. The second case examines conservation and pollution prevention as alternatives to traditional emphasis on construction of new and expanded wastewater treatment plants.

4.7.1 Dam Construction

Past emphasis on structural solutions and recent shifts are illustrated very well by the history of dam construction. As of 1910, there were less than 300 large reservoirs in the United States that had a normal (nonflood control) capacity of at least 5000 acre-ft or a maximum capacity (including flood control) of at least 25,000 acre-ft. (See Table 4.2.) Combined normal storage in all of those reservoirs was less than 14 million acre-ft (maf). In 1988 there were over 2700 reservoirs in that size class, and their combined normal storage was 467 maf (Ruddy and Hitt, 1990). Major growth in these facilities began in the 1920s, and as shown in Fig. 4.5, that growth did not slow until the 1970s. Thus, it seems appropriate to refer to those years as the Great Dam Era just as many refer to the period 1830–1860 as the Great Canal Era.

In fact, there are now over 66,000 dams in the country, but most of these facilities are relatively small. Approximately 98 percent of the normal capacity of all reservoirs in the country is in large reservoirs (Langbien, 1982).

Federal and nonfederal organizations (states, special districts, municipal and county governments, and the private sector) were all significant actors in reservoir development. Nonfederal projects far outnumber federal projects, about 2.5 to 1, but federal reservoirs, particularly those owned by USACE, USBR, and the Tennessee Valley Authority, are much larger on average than those owned by nonfederal interests. In fact, most very large reservoirs in the country are owned by one of these three agencies. They own only 29 percent of all reservoirs, but those projects include 68 percent of all normal capacity (Ruddy and Hitt, 1990).

The Great Dam Era came to a close after 1965, however. Sharp declines in construction rates for new reservoirs were experienced after that date, reflecting changes in attitude toward environmental values, increased costs of dam and reservoir construction, a diminishing number of attractive reservoir sites, and other regulatory programs, such as dam safety.

The downturn in new construction is also reflected in expenditure data for USACE and USBR projects, as shown in Fig. 4.6. Data in that diagram are for con-

TABLE 4.2   Large Reservoirs in the United States, 1988

Ownership	Approximate number	Normal capacity, MAF	Average capacity, acre-ft
Nonfederal	1,945	148	77,000
Federal	780	319	409,000
Corps of Engineers	467	172	370,000
Bur. of Reclamation	214	134	625,000
Tenn Valley Auth.	35	10.5	300,000
Total	2,728	467	486,000

Note: Numbers do not necessarily add to totals because of rounding and selected approximations.
Source: Ruddy and Hitt, 1990.

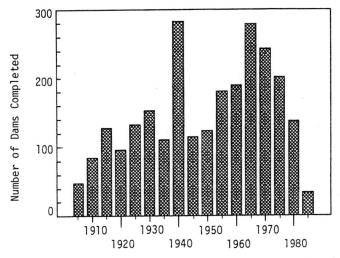

FIGURE 4.5 Number of large dams completed in successive five-year periods. (*Ruddy and Hitt, 1990.*)

struction and rehabilitation of water and power projects, in constant 1987 dollars. They show very sharp increases in spending by the Corps just after World War II, reaching a peak in 1967. After that, spending dropped dramatically, especially during Ronald Reagan's first term as President. USBR spending peaked in real dollar spending in 1950, then settled down to relatively stable levels.

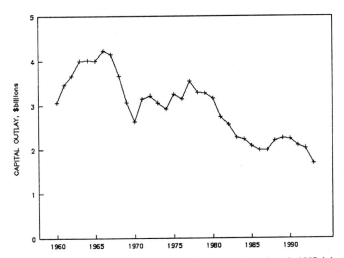

FIGURE 4.6 Capital outlays for federal water resource projects in 1987 dollars. (*Historical Tables, Budget of the United States, 1992.*)

4.7.2 Floodplain Management

Despite these large investments in reservoirs and other flood control works, average annual flood damages continued to rise through the 1950s. Gilbert White, University of Chicago Professor of Geography in the 1940s and 1950s, wrote a series of monographs on floodplain occupancy and adjustment to floods. He, probably more than any other student of the subject, pointed to the fact that as the frequency of flood events is reduced by building control structures, protected lands become more attractive for urban development. When floods do occur, however, property damage is considerably higher, resulting in higher average damages. Before 1970, only anecdotal information about prior floods was available in most communities. Other property owners were aware of potential risks but chose to gamble against the odds. When devastating floods did occur, government could not, and still cannot, resist pleas for financial relief from victims, including individual property owners and communities that are ravaged by these events.

After a series of flood disasters in the early 1960s, White headed a Bureau of the Budget Task Force on Federal Flood Control Policy. When the Task Force issued its report, *A Unified National Program for Managing Flood Losses*, it called for a broader perspective on flood control that embraced management of floodplains. The report, published as House Document 465, recommended, among other actions, that:

1. Flood-prone areas be delimited by appropriate federal agencies.

2. Uniform techniques be established for defining flood frequencies, and flood forecasting techniques be improved.

3. Incentives in federal grants assure that state and local governments pay due attention to flood hazards.

4. A feasibility study of flood insurance programs be undertaken.

5. Federal agencies be authorized to include floodproofing, land acquisition, and relocation in flood management programs.

Congress directed the Department of Housing and Urban Development (HUD) to study the feasibility of a national flood insurance program in 1965, even before the Bureau of the Budget report had been made final. The HUD task force, chaired by Marion Clawson, a senior economist at Resources for the Future, saw flood insurance as having two desirable effects: It would tend to steer development away from flood hazards, and it would provide relief to victims. This task force recommended that a national flood insurance program be established, with federal subsidies to provide the difference between reasonable premiums to be paid by property owners and full actuarial costs. It also recommended that local governments be encouraged to restrict development in floodplains.

Recommendations of the task force were translated into policy in 1968, when Congress passed the National Flood Insurance Act. Modifications to the law were necessary in 1969 and 1973, to overcome local government hesitancies to adopt sufficiently strict land-use regulations. The Flood Disaster Protection Act of 1973 included a provision that made adoption of those controls a prerequisite for federal financial assistance.

An assessment of these efforts was published in 1992 by the Federal Interagency Flood Plain Management Task Force (FIFMTF). The report points to a long list of achievements—more widespread public perception of the hazard; improved knowledge, standards, and technology; an extensive body of judicial decisions; and well-established development standards. Measuring program effectiveness remains an elusive task, however. The report points to the lack of ". . . consistent, reliable data about program activities and their impacts" as a principal complication (FIFMTF, 1992, pp. 60–61). It notes that susceptibility to flooding in the country is being reduced at individual sites and in local communities through a variety of land-use controls and emergency-preparedness activities. Evidence reviewed by the task force led it to conclude, however, that overall vulnerability has either increased or remained the same because of the large amount of preexisting vulnerable development, numerous exceptions in state and local policies, or the inability of governments at all levels to respond quickly to new spurts of development activity.

Following the disastrous 1993 floods in the Mississippi-Missouri River basin, a review committee was established in January 1994 to investigate the causes and consequences of that flood, evaluate the performance of existing floodplain and watershed management programs, and make recommendations for appropriate changes. Among the committee's numerous findings was that initial estimates of those actually covered by flood insurance ranged from below 10 percent up to 20 percent of insurable buildings in identified flood hazard areas in the Midwest. For the nation as a whole, the range is 20 to 30 percent. The committee recommended taking more vigorous steps to market the program and reducing postdisaster support for those who are eligible but do not purchase adequate coverage (Interagency Floodplain Management Review Committee, 1994, pp. 131–134).

4.7.3 Conservation

Just as floodplain management has been elevated in importance relative to the building of flood control dams, conservation or demand management has received much greater attention in recent years as an option for bringing urban supply and demand into balance. Previously, conservation was little-used except in drought emergencies. Although there are no comprehensive data on the use of conservation and its effectiveness, there is considerable evidence of its growing importance.

Much of the leadership promoting conservation has come from state and local governments. By April 1992, at least 16 states had adopted revised plumbing codes requiring water-efficient fixtures to be installed in new homes and businesses (Eddy, 1992). The Energy Policy and Conservation Act, passed by Congress in October 1992, established national standards governing the manufacture of plumbing fixtures. These actions affect only the installation of new fixtures, however, and if conservation programs address only new construction, many years will pass before even modest effects of conservation will be observed. California, through its Department of Water Resources, has been among the few states actively promoting an aggressive conservation program, prompted in large part by the prolonged drought from 1987 through 1992.

Seattle is one of several cities that have undertaken aggressive programs to manage demand. In 1993 it had underway a program to: (1) distribute conservation kits to all single-family residential customers, (2) pay for replacement of leaking toilets in high-consumption residential and commercial buildings, (3) conduct water audits for multiple family buildings and industries, (4) initiate water-use patrols to promote efficiency in lawn irrigation, (5) modify the price structure, and (6) start a research and development program on water reclamation and reuse (Dieteman, 1993).

Many California cities had little choice but to take aggressive steps toward conservation. In 1991 the governor of California directed all local governments to prepare plans to reduce consumption by 50 percent (*US Water News,* April 1991). The Metropolitan Water District of Southern California then instituted a policy in response to that directive, requiring urban customers to reduce consumption by 30 percent. An aggressive conservation program was also undertaken by the East Bay Municipal Utility District, serving the metropolitan area on the east side of San Francisco Bay. That program, which has many features in common with Seattle's, has a special emphasis on landscape irrigation practices, as well as plumbing fixtures. It provides financial incentives to existing customers who modify lawn irrigation practices to meet standards and to existing customers who exceed code requirements for showers and toilets.

4.8 EVALUATION

Water planners face a challenging task in evaluating which of several alternatives best serves the public interest or whether any alternative makes the public better or worse off than doing nothing. Objectives and criteria for the formulation and evaluation of alternatives are subject to change with shifts in social values. Measuring the beneficial and adverse effects of investments to manage water resources is often difficult, especially when some inputs required to make those investments and the services that they deliver are not readily quantifiable in terms of economic gain and loss. Despite these difficulties, however, it is accepted as a matter of principle that the public is better served when planners can reduce the enormous detail about beneficial and adverse effects of alternatives to a relatively small number of indicators of social value. Only then is it possible to have a meaningful assessment of the trade-offs among conflicting values that must be made in reaching decisions.

4.8.1 Early Initiatives

Development of clear statements of principles, standards, and procedures for measuring beneficial and adverse effects is essential to guide the evaluation process. Considerable effort has been expended to develop appropriate theory and procedures to support the process, and much of that has occurred—and continues to occur—amidst controversy. In the absence of formal procedures for evaluation, however, decision makers are likely to resort to vote trading for projects to serve their own or other special interests in what has become known as log-rolling and pork barrel politics. Such was the case when Congress began funding projects in 1826 when planning and evaluation was virtually absent. Annual rivers-and-harbors bills included both authorizations and appropriations for a variety of individual projects, but many of those projects had never been subjected to field surveys or feasibility studies (Shad, 1974).

Economic development has always been one of the basic objectives for water resource investments in the United States. Albert Gallatin, Secretary of the Treasury under Thomas Jefferson, was directed by Congress in 1807 to prepare a plan for improving navigation by removing obstructions and building canals. That plan, detailed in Gallatin's 1808 *Report on Roads and Canals,* laid out a series of roads and canals. The grid of east–west and north–south routes, many of which were later built, envisioned a transportation network that would provide major arteries for travel and commerce, connecting most areas of the nation as it existed at the time.

Not only did Gallatin propose a series of specific water resource improvements, he offered the rudiments of an investment criterion. One measure of benefit, he argued, is the annual savings in cost of moving goods over the alternative, unimproved condition—if that amount exceeds the interest on capital invested to make the improvement, the difference is a net gain to the nation, even if tolls are not sufficient to cover costs. He also argued that benefits will occur if new commerce is generated by the proposed improvements. On the question of financing, Gallatin argued that, if sufficient capital is available to the private sector, improvements could be left to individuals, but because that is not the case for large-scale capital projects, the government is justified in financing such projects (Gallatin, 1808).

Although formal statements of objectives and criteria were not developed until the 1940s, concepts developed during the conservation movement were strongly based in principles of economic efficiency. Leaders of the movement were driven by the need to make wise use of resources rather than a desire to preserve them. Their great crusade was about efficient use of resources to generate wealth for present and future generations. W. J. McGee expressed great confidence in the application of science to control nature for the benefit of humankind. According to Hays (1959, p. 127):

> The apostles of the gospel of efficiency subordinated the aesthetic to the utilitarian. Preservation of natural scenery and historic sites, in their scheme of things, remained subordinate to increasing industrial productivity.

4.8.2 Benefit Cost Analysis

A perusal of several of the 308 plans suggests that in the 1920s and early 1930s, the Corps of Engineers was using a crude form of benefit-cost analysis to evaluate plans and individual projects within plans. However, adoption of formal criteria for evaluating projects is generally traced to language in the Flood Control Act of 1936. The oft-quoted language in Section 1 of that legislation states:

> . . . the Federal Government should improve or participate in the improvement of navigable waters . . . for flood control purposes if the benefits to whomsoever they may accrue are in excess of the estimated costs, and if the lives and social security of people are otherwise adversely affected.

There was not another mention of the benefit-cost standard in that act, and Congressional intent when that language was included is not clear (Dorfman, 1976). Furthermore, the bill went on to authorize a long list of projects, many of which may well have failed to meet the test so boldly stated in Section 1.

Regardless of Congressional intent, the benefit-cost standard became entrenched in federal water planning for several decades. In 1946, FIARBC established a Subcommittee on Benefits and Costs and charged it to formulate principles for estimating benefits and costs for water projects that would be mutually acceptable to participating agencies. The subcommittee's report was delivered in May 1950, accepted by the Committee, and published as *Proposed Practices for Economic Analysis of River Basin Projects,* a document which became known as the *Green Book* and which was revised in 1958. Although that document acknowledged that public policy may be influenced by factors other than economic considerations, it argued that maximization of benefits less cost is a ". . . fundamental requirement for the formulation and economic justification of projects and programs." It differentiated between *primary* and *secondary* benefits and between *tangible* and *intangible* effects. Primary benefits were defined as the value of products and services that flowed directly from a project, while secondary benefits were defined as the macroeconomic benefits of regional employment and expenditures that could be attributed to the project. Secondary benefits were effectively excluded from consideration in project analysis.

The Green Book recognized limitations on economic evaluation imposed by deficiencies in market prices, including: (1) some private market prices may not adequately reflect values of resources from a public viewpoint and (2) some values, like improvements in public health, protection of scenic values, and prevention of loss of life, are not fully measurable in monetary terms. The argument was advanced that, lacking any other alternative, nonmarket effects should be evaluated in economic terms to the fullest extent possible. Intangible effects, those not susceptible to monetary evaluation, were to be described and not "overlooked." Principles for measuring benefits for various project purposes were also included.

Debate over the appropriate uses and misuses of benefit-cost analysis was intense even in the 1950s and 1960s. Some of the issues were (Holmes, 1972, p. 32):

1. Overestimation of benefits and underestimation of costs by agency planners seeking to justify projects

2. Use of secondary benefits by the USBR to justify projects, when, at least after the depression years of the 1930s, these benefits were considered to be quite small

3. Inadequate treatment of intangible benefits and costs

4. Failure to evaluate planning alternatives, especially nonstructural alternatives for flood control, alternative farming practices for irrigated lands, and private alternatives to public projects

4.8.3 Principles and Standards

The Kerr Committee recommended in 1960 that one of the duties of the proposed water resources council be to develop principles, standards, and procedures for evaluating water projects. President Kennedy did not wait for passage of the bill before taking action, however. In 1961, a panel of consultants, chaired by Maynard Hufschmidt, was appointed by the Bureau of the Budget to offer opinions on these matters. One of the primary contributions of the report from that group was the articulation of "a broader and more fundamental statement of objectives" as a prerequisite for principles and standards. The panel went on to argue that the all-embracing objective is to maximize national welfare, but that national welfare consists of multiple components that cannot be captured by a single measure. They discussed several of those components, among them: (1) increasing national income, (2) equitable distribution of income and the means for doing so, and (3) preservation of aesthetic and cultural values. Their view of the multidimensional nature of national welfare led them to recommend that standards and criteria be framed in terms of multiple objectives rather than the single objective of national income. A number of other issues, including the discount rate and secondary benefits, were also addressed by the panel. They recommended an interest or discount rate that synthesized the Administration's social discount rate and the opportunity cost of capital. Previous policies that tended to overly restrict the use of secondary benefits were questioned, but the consultants noted that careful analysis of secondary benefits on each dimension of national welfare was required before they could be included as project benefits (Bureau of the Budget, 1961).

Work by the panel of consultants was reflected in a subsequent review by the President's Water Resources Council, resulting in adoption of a new set of standards and procedures for water resource planning, published in 1962 as Senate Document

97 of the Second Session of the 82d Congress. It directed that full consideration should be given to several objectives, namely national economic development and development in each region of the country, preservation of natural resources of the nation, and social well-being of people. It directed that "All viewpoints—national, regional, State, and local—shall be fully considered and taken into account—," that planning should be multipurpose in character, that nonstructural as well as structural measures should be considered, and that river basins are usually the most appropriate geographical unit for planning. It allowed the use of secondary benefits, requiring that they be clearly identified in planning reports. It also carried forward the prior tangible-intangible benefits distinction with language similar to that in the Green Book and similar directives to quantify benefits and costs in ". . . comparable economic terms to the fullest extent possible."

When the Water Resources Planning Act was passed in 1965, one of the first tasks to be undertaken by the new Water Resources Council (WRC) was development of its *Proposed Principles and Standards for Planning Water and Related Land Resources,* published in December 1971. The final version, commonly referred to as the *Principles and Standards,* or *P&S,* was published in September 1973. There is little doubt that this process was strongly influenced by enactment of the National Environmental Policy Act (NEPA) in 1969, which put the full authority of the federal government behind establishment of environmental quality as a major national objective. NEPA also required federal agencies to prepare environmental impact statements on all actions that might have a significant effect on the environment. When WRC issued its *Proposed Principles and Standards for Planning Water and Related Land Resources* in 1971, it unveiled what Dorfman (1976) viewed as radical changes in the benefit-cost standard. The proposed *Principles and Standards* became final in 1973, with only modest changes from the version proposed in 1971.

WRC adopted a multiple-objective framework for its P&S, an elaboration of what had been recommended by the panel of consultants in 1961. Two objectives were recognized, namely national economic development (NED) and environmental quality (EQ). The P&S also placed considerable emphasis on the formulation of alternative plans, each to be evaluated with respect to the two objectives. Other effects of alternatives on regional development and social well-being were also to be evaluated. Thus, two objectives were to guide the formulation of alternative plans, but beneficial and adverse effects of each plan were to be evaluated in four separate accounts, one of each of the objectives, one for regional development, and one for social well-being.

Components of NED included the value of increased output of goods and services and the value of external economies. The EQ objective was seen as including: (1) protection or enhancement of areas of natural beauty and human enjoyment; (2) protection or enhancement of especially valuable natural resources and ecological systems; (3) enhancement of the quality of water, air, and land resources; and (4) avoidance of irreversible commitments of resources to future uses.

The regional development account picked much of what had been referred to as secondary benefits in the Green Book and Senate Document 97. Regional development was defined as consisting of regional income effects and effects on the number of jobs, stability of employment, population distribution, and the regional environment. Social well-being included effects on: (1) real income distribution; (2) life, health, and safety; (3) educational, cultural, and recreational resources; and (4) emergency preparedness.

Dorfman (1976) saw in the P&S three fundamental shifts from prior policy. First, benefit-cost analysis was demoted from its decisive role in water resource planning. While acknowledging that the social welfare function by which public investments could be properly evaluated was not fully known, he argued that it has at least the two components recognized in the P&S—increasing national income and environmental quality. Second, explicit recognition was given to the fact that no single measurement can be made of the social desirability of a project. He argued that the final judgment about investments in water projects involves a weighing of beneficial and adverse effects on more than one objective—judgments that constitute a political choice. Third, he viewed the instructions to display the distribution of income effects on classes or groups as a rejection of the aggregative concept implied in the language of the Flood Control Act of 1936.

While the issue of appropriate objectives occupied much of the debate, other important principles were addressed in the P&S. Under the heading of "General Evaluation Principles," the document addressed several key matters, including:

1. Beneficial and adverse effects of projects should be measured *with and without* proposed investments.
2. The *discount rate* should be based on the cost of borrowing money.
3. Evaluation should be based on consideration and comparison of proposed investments with other *alternatives.*
4. *Risk and uncertainty* should be considered, and methods and strategies for treating them should be explicit.
5. Alternatives should be subjected to *sensitivity analysis* to show effects of possible errors in data and projected future conditions.

In the process of formulating plans, P&S stressed the importance of:

1. Evaluation of existing and expected conditions without a plan (the do-nothing option)
2. Formulation of alternative plans that reveal tradeoffs which must be made among competing objectives and display effects of externally imposed constraints

President Carter's aggressive use of benefit-cost analysis and other elements of P&S was a source of irritation to congressional advocates of water projects. When he came to that office in 1977, Carter drew special attention to 60 water projects previously authorized by Congress. He directed USACE and USBR to review 19 projects for possible deauthorization on the grounds that they had been justified on questionable economic and environmental grounds. White House staff later identified another 14 questionable projects, and they were added to what became known as the "hit list."

4.8.4 Recent Changes

When Ronald Reagan became President, the Water Resources Council and its P&S were easy targets for an administration bent on reducing the size of government. The Council was given no budget, and P&S was rescinded. James Watt, President Reagan's Secretary of the Interior, was a strong proponent of water projects, and in his statement to the House Subcommittee on Appropriations for Energy and Water Development in 1983, he stated that repeal of P&S was another of his "significant accomplishments" because they were "so onerous that much needed water resources plans and projects could not be developed." He had persuaded President Reagan to use his authority under the Water Resources Planning Act to replace P&S with the nonbinding *Economic and Environmental Principles and Guidelines for*

Water and Related Land Resources Implementation Studies, published in March 1983. Aside from the fact that the *Principles and Guidelines* no longer had the authority of formal rules, the primary shift was back toward a dominant role for benefit-cost analysis. The language in the P&G is that

> The federal objective . . . is to contribute to national economic development consistent with protecting the nation's environment . . .

Although the P&G retained all four of the accounts established in the P&S, the NED account is the only one that is required. Most other principles found in the P&S survived.

Improvements in evaluation methods continue to occur. NEPA and P&S have spurred a substantial effort since 1970 to develop formal methods for environmental assessment. Nichols (1988) reviewed 15 methods that were developed by the early 1980s. They range from simple checklists of potential environmental effects to sophisticated weighting schemes, but all are based on considerable subjective judgments of persons making the analysis. At least two methods have received attention in the literature and in practice in recent years. One is the Habitat Evaluation Procedure developed by the U.S. Fish and Wildlife Service, a technique for estimating requirements to mitigate adverse effects of water resource and other types of development projects. The other is the Wetland Evaluation Technique (Adamus, et al., 1987), developed with the support of the Corps of Engineers to assist that agency in evaluating applications for dredge and fill permits under Section 404 of the Clean Water Act.

Advances have also been made in recent years in estimating economic values of damage to natural resources and the benefits of water-quality improvements. Development of techniques to estimate economic damage to natural resources has been prompted, at least in part, by the Comprehensive Environmental Response, Compensation, and Liability Act of 1980 (CERCLA), commonly known as Superfund. Federal and state governments were authorized by that act to seek compensation for damages to public natural resources, and a significant body of literature has developed to determine a rational basis for determining appropriate amounts of compensation (see Kopp and Smith, 1993).

Parallel developments have also been made in efforts to estimate economic benefits of improvements in water quality, and other aspects of environmental quality (Hanley and Spash, 1993). Among recent literature on this subject is an effort by Carson and Mitchell (1993) to estimate the public's willingness to pay for various levels of clean water. Using techniques similar to those used for estimating damage to natural resources, the authors estimated benefits for the national water quality program. Although the validity of particular estimation procedures is still being judged by resource economists, these techniques represent substantial advances in economic evaluation of investments to improve environmental quality.

4.9 FINANCING

From the viewpoint of society as a whole, merits of a water resource plan may be determined independently of who pays for it. Beneficial and adverse effects on the national economy or the environment can—and, in some cases, should—be counted without regard to either the distribution of those who benefit most from the plan or the distribution of those who are most burdened by its costs. That was the viewpoint of Congress when it passed the 1936 Flood Control Act, but as discussed earlier, that view was somewhat modified when the Water Resource Council's *Principles and Standards* required a display of the distribution of benefits and costs.

Significant changes in the practice of financing investments have since occurred. Financing of several sectors of water management has moved closer to the principle that costs should be borne in direct proportion to benefits received. Several practical difficulties and some policy objectives limit the rigor with which that principle can be applied. Some difficulties include: (1) limitations of measuring either benefits or costs; (2) problems of clearly identifying who benefits and to what extent they benefit from a particular plan; (3) allocating costs of a plan to benefitting groups and individuals, particularly when there are joint costs among complementary purposes and projects; and (4) problems of designing tariffs that achieve the objectives, especially when benefits cannot be excluded from those who choose not to pay.

There are also legitimate policy objectives that are not consistent with the benefitor-pays principle. Society has chosen, in some instances, to extend services and other benefits to those who cannot afford to pay. The federal government has also used financial incentives as a means to achieve other ends, such as settlement of the West, economic development in depressed areas, and enhancement of water quality.

Despite these difficulties and other policy objectives, water-financing policy has clearly shifted toward the benefitor-pays principle through such changes as federal cost-sharing policy, flood insurance, reduced subsidies for wastewater treatment facilities, and increasing reliance on user fees. That shift is but part of a larger trend of moving a greater part of public-service financing from the federal to state and local government and to users. Nixon's "New Federalism" strategy was to give local governments greater discretion in the expenditure of federal funds—to move toward block grants instead of traditional categorical grants. This policy was predicated on an assumption that the total amount of federal funds would continue to increase. Under the Carter administration, funding became more constant, and under Reagan the size of the pie began to shrink. The Republican majority in Congress, resulting from the 1992 elections, supported more drastic reductions.

Accompanying the decline in federal aid was a decided shift among the several sources from which state and local governments collect their own revenues (what the Bureau of the Census refers to as own-source revenues, to distinguish them from intergovernmental transfers like federal grants). The most important of those sources are general taxes and service charges, such as those for water supply, sewer, electricity, and liquor stores. Since 1970, the proportion of own-source revenues derived from service charges has been increasing relative to general taxes, reflecting the user-pays principle of government finance. These trends are shown graphically by municipal general revenue data in Fig. 4.7. Federal sources contributed only 4.4 percent of local government general revenue in 1968. That share peaked at about 16 percent in 1978, and by 1990 had dropped back to about its 1968 level. The percentage of municipalities' own-source general revenues derived from user charges jumped from 26 percent in 1968 to 40 percent in 1986.

4.9.1 Federal Cost Sharing Policy

When appropriations for water projects slowed in the 1970s, the real water project issue before Congress may not have been benefits and costs or environmental effects, but who was going to pay for water projects (Caulfield, 1984). The National Water Commission (NWC), in 1973, was among the first to suggest changes in federal cost-sharing formulas. That organization was created by an act of Congress in

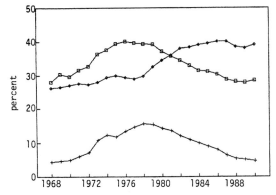

FIGURE 4.7 Selected statistics of local government finances. (*U.S. Bureau of the Census.*)

1968, with a five-year life to review and recommend changes in federal water resource policies. Unlike the Water Resources Council, which was an administrative agency governed by cabinet secretaries or their representatives, NWC was an advisory group of seven members appointed by the President, all of whom served part-time, and none of whom were officials or employees of the federal government (NWC, 1973, p. ix).

When NWC issued its final report in 1973, it found (to no one's surprise) that existing joint financing by federal agencies and nonfederal partners was useful and appropriate, but that cost sharing policies were "grossly inconsistent and lead to inefficiencies and inequities" at both federal and nonfederal levels (NWC, 1973, p. 494). NWC took a dim view of federal subsidies in general. Although it did not advocate eliminating all subsidies, it recommended that, to the extent practicable and administratively feasible, identifiable beneficiaries should bear appropriate shares of capital, operating, and maintenance costs through appropriate pricing or user charges. NWC recommended recovery of all costs for municipal and industrial water supplies, new irrigation projects, hydroelectric power facilities, and wastewater treatment plants (after a 10-year period). The commission also recommended recovery of operation and maintenance costs from users of inland waterways, a sharp departure from the long-standing "free waterways" policy that had stood since the writing of the Constitution. If NWC's recommendations had been followed, costs for flood control, drainage, and shoreline and hurricane protection would have been recovered through local governments with the power to assess benefitted properties (NWC, 1973, pp. 497–498).

Pursuant to recommendations of NWC, Congress directed the Water Resources Council, in Section 80(c) of the 1974 Water Resources Development Act, to review cost-sharing policies. When WRC fulfilled that charge with its 1975 report, it reviewed alternative policies but did not recommend any particular action, leaving the options to Congress (Reuss, 1991, pp. 32–33).

Senator Domenici of New Mexico took the initiative to establish waterway user charges in 1977, calling for the administration to develop a user-fee system to recover one-half of the construction cost and all of the operation and maintenance costs for inland waterways facilities through tolls and license fees. The House responded by rejecting the Domenici bill and proposing a fuel tax, which the administration estimated would require 40 cents per gallon. After much wrangling within Congress, and between Congress and the White House, a Domenici-sponsored compromise was enacted. That law imposed a 4-cent fuel tax that would increase gradually to 12 cents, far below the amount needed to fully recover costs (Reuss, 1991, p. 55).

The slowdown in new funding that resulted from President Carter's handling of water projects created a substantial backlog when President Reagan took office in 1981. David Stockman, Reagan's Director of the Office of Management and Budget, took a very active role in pressing the administration's view that existing cost-sharing policies were unacceptable. However, officials within the executive branch could not agree on what was an acceptable policy.

A flurry of congressional activity in 1983 and 1984 came to naught as authorizing and appropriations committees within each house could not agree. Finally, a compromise bill was produced in June 1985, and Stockman concurred. Senator Dole was given substantial credit for bringing together the various parties. The bill was titled the Water Resources Development Act of 1986 (WRDA-86).

Stakes in the outcome of this bill were high. It authorized expenditure of $16 billion for water projects, three-fourths of which would be paid by the federal government. It changed cost-sharing policies for flood control, beach-erosion control, and recreation. Local sponsors now have to provide 25 to 35 percent of the cost of flood-control facilities, whereas the federal government had previously picked up all of those costs. Local governments now have to pay 50 percent of the cost of projects that are attributable to recreational use, and nonfederal partners have to pay 35 to 50 percent of beach-erosion control. A subtle but important change is that local sponsors now have to share 50–50 the cost of feasibility studies. Where it was once possible to get preliminary plans and evaluations for projects of questionable value at no expense to local governments, projects must now be sufficiently attractive to nonfederal sponsors to convince them to pay for feasibility studies early in the planning process. With these changes, many projects are being more fully subjected to market-type reviews than ever before.

User fees on inland waterway transportation were also an important part of the legislation. Senator Mark Hatfield's ad valorem tax, proposed in 1983, survived at the rate of 0.04 percent on imports and exports. The tax on fuel consumed by users of waterways was increased from the prevailing 10 to 20 cents per gallon over a 10-year period. These rates still left the federal government providing about 60 percent of the cost of operating, maintaining, and replacing the system (Reuss, 1991, p. 167).

4.9.2 Financing of Wastewater Treatment Facilities

Like the funding of water projects, financing of wastewater treatment facilities has changed dramatically over the past 25 years. When Congress passed the Federal Water Pollution Control Act Amendments in 1972, the majority believed that one of the primary obstacles to achieving clean water was the financial burden that construction of wastewater treatment plants would place on local governments. Before 1972, the federal government was contributing less than $1 billion annually to the construction of wastewater treatment plants. Soon after President Nixon's environmental message in February 1970, the administration introduced a bill calling for federal construction grants of $2 billion a year. By the time PL92-500 passed in 1972, authorized federal expenditures had been tripled to $6 billion a year, and the federal share of construction costs had been increased to 75 percent.

This unprecedented level of federal funding for waste treatment facilities amounted to one of the largest public works programs in the history of the country, second only to the interstate highway program. By infusing this large amount of capital, the national rate of expenditure for pollution-control facilities was increased, but by focusing only on federal expenditures, the picture of total expenditures was distorted. Expenditures by local governments for water and sewer services, shown in Fig. 4.2, indicate that by 1972, when PL92-500 was passed, annual expenditures by local governments for sewers were already more than $4 billion (constant 1987 dollars). When the stream of federal subsidies, shown in Fig. 4.8, began to flow to local coffers in large amounts, overall spending increased but not by the amount of federal subsidies. As federal contributions increased, expenditures from local government own-source revenues declined sharply. The federal share of local government outlays for sewer service over the period 1972 to 1985, estimated from the curves in Figs. 4.2 and 4.8, came to about 50 percent. State contributions rose to over $0.5 billion by the mid 1980s, accounting for over 10 percent of local expenditures (Moreau, 1988).

Federal subsidies are now on the decline, however, and they have been following that trend since 1977. Total spending dropped in real terms from 1980 to 1985, but it has been growing since, as local governments take up the slack.

As local governments get less from the federal and state governments, they have shifted a greater share of their costs to user fees. A comparison of local government revenue and expenditure data with a somewhat idealized revenue–expenditure model indicates that water services have been nearly fully financed by user-fees for many years. The data show clearly that sewer services in the United States are moving in that direction, away from a system financed in large part by grants and local tax revenue (McCullough, Moreau, and Linton, 1993).

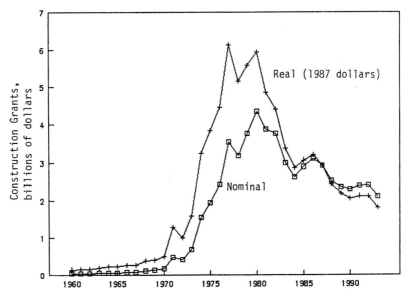

FIGURE 4.8 Federal Clean Water Act construction grants. (*Nominal data from Historic Tables of the U.S. Budget; adjusted for inflation using GNP implicit price deflators.*)

4.10 INTEGRATED MANAGEMENT

Throughout much of the history of water management in the United States, water resource planners have recognized the need for organizational arrangements to coordinate and integrate the work of multiple federal agencies, states, local government, and the private sector. Actual arrangements have been marked by several different models, fluctuating from decentralized planning by the several construction agencies, to efforts at coordinating activities through interagency committees, to the establishment of separate agencies to plan and coordinate federal and state activities. Only planning within separate agencies survived for a very long period.

4.10.1 Early Efforts

An attempt was made during the conservation movement to establish a unified agency for water resources planning. Following publication of the report of the Inland Waterways Commission, Senator Newlands introduced a bill in 1907 to carry out the commission's recommendations. He proposed a permanent agency to investigate problems, authorize projects, oversee construction, and coordinate the work of other federal water resource agencies. Members of the governing board of the agency would have been appointed by the President, and financing would have been through appropriations to a special waterway fund. That proposal engendered strong, often bitter opposition from the Corps of Engineers, and it was received coolly by members of Congress, who resisted delegation of authority over water projects to an executive agency. Despite support from Roosevelt and near-unanimous support in the House of Representatives, the proposal died in the Senate in December, 1908. Newlands' proposal was revived in 1911, and it continued to gain popularity among interest groups until it was finally passed in 1917. However, Congressional supporters of the Corps of Engineers effectively blocked appointments to the commission until after Newlands died in 1919, and the Federal Water Power Act of 1920 repealed authorization of the commission (Hays, 1959, pp. 109–114, 219–239). Without that coordinating body, Congress turned to a single agency, the USACE, to prepare the 308 plans to guide Federal Power Commission decision making.

During the 1930s, water resource planning was a part of Franklin D. Roosevelt's initiatives to lead the nation out of its most severe depression. It was also part of Roosevelt's efforts to establish national planning in the executive branch of government: a progression of "alphabet" agencies starting with the Mississippi Valley Committee (MVC) of the Public Works Administration and the National Planning Board in 1933; a cabinet-level committee called the President's Committee on Water Flow (PCWF) and the National Resources Board (NRB) in 1934; the National Resources Committee (NRC) in 1935; and the National Resources Planning Board (NRPB) in 1939. Congress grew leery of these efforts to concentrate national planning in the executive branch and abolished the NRPB in 1943, including specific language that forbade the transfer of those functions to other agencies (Holmes, 1972, pp. 13–22).

With the demise of NRPB, several agencies agreed to form a voluntary interagency mechanism to coordinate water resource planning and development, one that was to last for more than 20 years. In 1943, the Departments of War, Interior, and Agriculture, and the Federal Power Commission agreed to form the Federal Interagency River Basin Committee (FIARBC), a committee that operated without benefit of Congressional authority or staff. It was reorganized under the Eisenhower administration in 1954 as the Interagency Committee on Water Resources, and the Departments of Health, Education, and Welfare, Commerce, and Labor became participants in its activities (Holmes, 1972, pp. 38–39; Shad, 1979).

4.10.2 Water Resources Council

The Kerr Committee addressed the need for better coordination and integration in 1960. It recommended several changes in water planning, declaring that, among other activities, the Federal Government should (Holmes, 1979, pp. 40–41):

1. Prepare, in cooperation with the states, comprehensive development and management plans for all major river basins, giving full recognition to streamflow regulation, outdoor, recreation, and fish and wildlife development.
2. Make grants to states over a 10-year period to develop their capabilities to participate in comprehensive planning.
3. Support a coordinated research program to increase available supplies of water and to promote efficiency of use in agriculture and industry.
4. Prepare biennial assessments of water supply and demand for each region of the country.

When the Water Resources Planning Act was passed in 1965, it implemented most of these recommendations and created an organization not unlike that proposed by Senator Newlands some 50 years before. The Water Resources Council (WRC) was charged with encouraging the conservation, development, and utilization of water and related land resources in the United States on a comprehensive, coordinated basis. Specific responsibilities included:

1. Preparation of biennial assessments (only two of which were ever conducted)
2. Maintenance of river basin plans
3. Establishment, subject to presidential approval, of principles, standards, and procedures for planning and evaluation of water and related land resources
4. Recommendation to the president of any actions concerning water policy and water resource projects

Another significant activity of WRC was its incentives to states to develop their own planning capabilities, an area of state government that had been neglected for many years. Title III of WRPA granted WRC authority to make 50-percent matching grants to states. Some see this as its most lasting contribution.

President Carter pushed centralized planning and formal analysis to its breaking point. He proposed that WRC be empowered to review and comment on all water projects before they were sent to Congress for authorization. Needless to say, Congressional reaction to Carter's handling of water projects was hostile. That hostility was particularly strong among members from western states where many of the targeted projects were located, but it was sufficiently widespread that the House actually voted to abolish the WRC in 1978, and the Senate voted to eliminate it from the budget. Only compromise gave it a reprieve (Reuss, 1991, p. 61).

No compromise could save WRC from the Reagan administration. When it came to office in 1981, WRC was one of the first casualties. The new administration argued that the council was an ineffective institution that fell far short of managing the nation's water resources and assessing national needs. Reagan created a Cabinet

Council on Natural Resources and the Environment to address policy issues in these fields. His administration took the position that by shifting deliberations on policy from WRC (which met rarely, and which cabinet secretaries almost never attended) to the new cabinet council, water issues were being given elevated status and would be handled with greater administrative efficiency. All staff were terminated at the end of Fiscal Year 1982 (they were offered positions in other federal agencies), and continuing obligations of WRC were dispersed to other organizations in the government. All funding for WRC activities, including grants to states and river basin commissions, was eliminated after 1983 (Bahr, 1993). Since then, planning at the federal level has generally declined, and what planning has occurred has tended to follow the single agency model.

4.10.3 Role of the States

States have taken an increasingly active role in water planning and management, and their role has become more visible as the federal role has diminished. Some of these changes have directly resulted from government policy, while others are more indirect consequences, reflecting greater state and local authority in policy arenas that have been elevated in priority. Probably the most direct effect came from the state grant program authorized by the Water Resources Planning Act. Statistics of employment in state and local water programs relative to that in federal agencies from 1960 to 1980 offers strong evidence of the growing importance of nonfederal roles (Schmandt, et al., 1988, p. 12). Certainly the Water Resources Development Act of 1986, which brought substantial changes in cost-sharing formulas, created much more of a partnership in project planning than ever before.

Other forces that tended to increase state roles were less direct. Schooler and Ingram (1982) argue that relative roles of state and federal governments are established by priorities given to the issues. Federal interests tended to dominate prior to 1970, but as water policy moved to address issues of environmental quality and allocation of scarce resources, the action moved toward those arenas in which states had the dominant authority. Schooler and Ingram base their arguments largely on state power over water resource allocation, but equally important is the traditional "state primacy" role that has been incorporated in much of the environmental legislation covering water resources—the Clean Water Act, the Safe Drinking Water Act, and the Resource Conservation and Recovery Act, to name a few of the more important ones.

At least two surveys of state programs have been reported (WRC, 1980; Viessman and Biery-Hamilton, 1986), but drawing valid generalizations from these assessments is difficult. Water problems are highly variable from one region of the country to another, and states tend to devise policies to address particular problems. Furthermore, surveys of state programs are largely based on statutes, legislative and administrative directives, and other formal actions. There is little in these surveys to evaluate how well and to what extent these mandates have been fulfilled. Many mandates may have gone unfunded or underfunded.

The WRC report found that three states—Delaware, Florida, and Washington—had mandated comprehensive approaches to planning and management of water and related land resources, operating through a single agency, with the authority of a permit system for withdrawals. In Florida, however, water quality was managed through a separate state agency. WRC found that 32 additional states had some form of legislatively authorized comprehensive planning, but the extent of integration between quantity and quality and comprehensiveness was variable. They noted that the trend was toward establishment of some form of permit system for withdrawals, but few other generalizations were possible. Descriptions of planning and management for quantity and quality were dealt with in separate sections of the report.

Schmandt, et al. (1988) took a closer look at a few states. They found that in the western states of Arizona, California, and Texas, improved management of existing sources was of primary concern. These states had experienced high rates of growth and considerable pressure on limited resources. Findings from these western states were in contrast to those for Florida, North Carolina, and Wisconsin, where the major water-management issue was quality. A general finding was that states have diverse water problems, and they respond to particular water problems with particular and often innovative policies. Regional water management districts were part of strategies in Florida and Arizona.

Comprehensive planning in the six states had an uneven history. A state water-use plan developed in Florida contained many worthwhile goals but was lacking in legislative policy and financial mechanisms. Effective water planning in Texas was stymied by preoccupation with procedural rather than substantive policies. In California, where there is a long history of comprehensive planning, the authors found well-crafted supply plans but no comprehensive water-use plan.

4.10.4 Recent Proposals

Several organizations have been suggesting a variety of decentralized models to fill the void left by collapse of the Water Resources Council. Voices of the states, groups representing the water industry, and environmental groups are being heard more frequently and with greater influence. Among proposals that have been offered are the so-called "Park City Principles," developed by the Western Governor's Association (WGA) and the Western States Water Council (WSWC) in 1991, and subsequently reviewed by the Interstate Conference on Water Policy (ICWP) in its 1992 National Policy Roundtable; and the 1990 policy statement by the American Water Works Association (AWWA, 1992). WGA-WSWC principles as adapted by ICWP: (1) stress the primary role of state government in a broad range of management activities; (2) encourage the use of holistic or systematic approaches in "problemsheds"; (3) favor a policy framework that is responsive to economic, social, and environmental values, that is flexible yet predictable, and is adaptable; (4) encourage decentralization of authority and accountability within a national policy framework; (5) promote negotiation and market-like processes; (6) encourage participation by all levels of government and the public; (7) recognize freshwater as a fundamental integrating ingredient in natural resource management; and (8) encourage adoption of freshwater sustainability as a guiding principle for future water-resource management (Light, 1992).

Another set of recommendations has come from the Interagency Floodplain Management Review Committee (IFMRC), established by the Clinton Administration in 1994. IFMRC was created to review flood events and needs for policy changes following extensive flooding in the Upper Mississippi Valley in 1993. To enhance organizational arrangements for national floodplain management, IFMRC recommended, among other things, that: (1) the Water Resources Council be revitalized, (2) river basin commissions be revised to reflect current needs, and (3) the role of state governments in the process be increased (IFMRC, 1994, pp. 73–89).

4.11 CONCLUDING COMMENTS

Most recommendations from these policy advocates reflect the basic principles that have evolved from experience in water resource planning in the United States as outlined and discussed in this chapter. They confirm the use of river basins and watersheds as logical geographical units for planning; they include comprehensive, integrated resource planning; they call for consideration of a wide array of alternative management options; they advocate integration of water and related land resource planning; and they embody concepts of multiple-objective policies and planning.

Some of these recommendations also reflect the continuing debate over the extent to which policy, planning, and financing will be centralized or decentralized. Importance of the role of states is affirmed by all, but one stresses the primacy of states within a consistent national policy framework. Achievement of that goal would be difficult, however, because consistency declines as authority is decentralized. At the other end of the spectrum are the voices supporting restoration of the Water Resources Council and river basin commissions, both symbols of a centralized federal model. This tension in national water management has been present from the beginning, and it is likely to remain as a part of the larger national debate over the size and authority of the federal government versus that of state governments.

REFERENCES

Adamus, P. R., et al., "Wetland Evaluation Technique (WET); Vol. II: Methodology," Draft Technical Report Y-87, USACE Waterways Experimental Station, Vicksburg, Miss., 1987, pp. 9–10, 15–24, 28–31, 34–35, 38–42, B-18.

American Water Resources Association, "Issues in Water Resources—Reallocation of Water Storage in Lake Lanier," *Hydata,* 11(1), 1992.

Argent, Gala, "Water Marketing: Driven by Low Supplies," *Western Water,* Sacramento, Cal., Water Education Foundation, May/June, 1989, pp. 4–11.

Bahr, Thomas G., Director, Water Resources Research Institute, New Mexico State University, and formerly, Director, Office of Water Policy, U.S. Department of the Interior, 1993. Personal Communication.

Blake, Nelson Manford, *Water for the Cities,* Syracuse, N.Y., Syracuse University Press, 1956.

Bureau of the Budget, *Standards and Criteria for Formulating and Evaluating Federal Water Resources Development,* Report of Panel of Consultants, Washington, D.C., June 30, 1961.

Carson, Richard T., and Robert C. Mitchell, "The Value of Clean Water: The Public's Willingness to Pay for Boatable, Fishable, and Swimmable Quality Water," *Water Resources Research,* 29(7): 2445–2454, 1993.

Caulfied, Henry P., Jr., "Fulfilling the Promises of the Water Resources Planning Act," *25th Annual Meeting of the Interstate Conference on Water Problems,* Pittsburgh, Pa., 1984.

Council on Environmental Quality, *Environmental Quality,* First Annual Report, U.S. Government Printing Office, 1970.

Davies, J. Clarence, III, *The Politics of Pollution,* New York, Pegasus, 1970.

Dieteman, Alan, Personal Communication with the Director of Conservation for the Seattle Water Department, 1993.

Dorfman, Robert, "Forty Years of Cost-Benefit Analysis," Discussion Paper No. 498, Harvard Institute of Economic Research, Cambridge, Mass., Harvard University Press, 1976.

Eddy, Natalie, *Small Flows,* a publication of the National Small Flows Clearing, Morgantown, W.Va., West Virginia University, April, 1992.

Edison Electric Institute, *Statistical Yearbook of the Electric Utility Industry/1990,* No. 58, 1991.

Epstein, Samuel S., Lester O. Brown, and Carl Pope, *Hazardous Waste in America,* San Francisco, Cal., Sierra Club Books, 1982.

Fair, Gordon M., "Sanitary Engineering in a Changing World," *Sewage and Industrial Wastes,* 22(1): 11–16, 1950.

Federal Interagency Floodplain Management Task Force, *Floodplain Management in the United States: An Assessment Report,* Vol. 1, Summary Report, Natural Hazards Research and Applications Information Center, Boulder, Colo., University of Colorado, 1992.

Federal Inter-Agency Committee, *Proposed Practices for Economic Evaluation of River Basin Projects,* Report of the Subcommittee on Benefits and Costs, 1950.

Fox, Irving K., "Review and Interpretation of Experiences in Water Resources Planning," in C. E. Kindsvater (ed.), *Organization and Methodology for River Basin Planning,* Water Research Center, Georgia Institute of Technology, 1964.

Golze, Alfred R., *Reclamation in the United States,* Caldwell, Ida., Caxton Printers, 1961.

Goodrich, Carter, *Government Promotion of American Canals and Railroads 1800–1890,* New York, N.Y., Columbia University Press, 1960.

Hanley, Nick, and Clive L. Spash, *Cost-Benefit Analysis and the Environment,* Brookfield, Vt., Edgar Elgar, 1993.

Harrison, Robert W., *Alluvial Empire,* Pioneer Press, Little Rock, Ark., 1961.

Hays, Samuel P., *Conservation and the Gospel of Efficiency,* Cambridge, Mass., Harvard University Press, 1959.

Holmes, Beatrice H., *A History of Federal Water Resources Programs, 1800–1960,* U.S. Department of Agriculture, Miscellaneous Publication No. 1233, 1972.

Holmes, Beatrice H., *History of Federal Water Resources Programs and Policies, 1961–70,* U.S. Department of Agriculture, Miscellaneous Publication No. 1379, 1979.

Howe, Charles W., "The Increasing Importance of Water Transfers and the Need for Institutional Reforms," *Water Resources Update,* Carbondale, Ill., Universities Council on Water Resources, no. 79, Spring 1989.

Hull, William J., and Robert W. Hull, *The Origin and Development of the Waterways Policy of the United States,* Washington, D.C., National Waterways Conference, 1967.

Immerman, Frederick W., *Financial Descriptive Summary: 1986 Survey of Community Water Systems,* U.S. Environmental Protection Agency, Office of Drinking Water, 1986.

Inter-Agency Committee on Water Resources, *Proposed Practices for Economic Analysis of River Basin Projects,* Report by the Subcommittee on Evaluation Standards, Washington, D.C., U.S. Government Printing Office, 1958.

Interagency Floodplain Management Review Committee, *Sharing the Challenge: Floodplain Management into the 21st Century,* Washington, D.C., U.S. Government Printing Office, 1994.

Jensen, Ric, "The Texas Water Market," *Texas Water Resources,* College Station, Tex., Texas Water Resources Institute, vol. 13, no. 2, 1987.

Kemper, J. P., *Rebellious River,* Boston, Mass., Bruce Humphries, 1949.

Kopp, Raymond J., and V. Kerry Smith, *Valuing Natural Assets—The Economics of Natural Resource Damage Assessment,* Washington, D.C., Resources for the Future, 1993.

Langbein, W. B., *Dams, Reservoirs, and Withdrawals for Water Supply—Historic Trends,* U.S. Geological Survey, Open-File Report No. 82-256, 1982.

Levine, Adeline Gordon, *Love Canal: Science, Politics, and People,* Lexington, Mass., Lexington Books, 1982.

Lieber, Harvey, *Federalism and Clean Waters: The 1972 Water Pollution Control Act,* Lexington, Mass., Lexington Books, 1975.

Light, Steven (ed.), *National Water Policy Roundtable,* Washington, D.C., Interstate Conference on Water Policy, 1992.

McCullough, James S., David H. Moreau, and Brenda L. Linton, "Financing Waste Water Services in Developing Countries," Research Triangle Park, N.C., Research Triangle Institute, 1993.

McGeary, M. Nelson, *Gifford Pinchot: Forester-Politician,* Princeton, N.J., Princeton University Press, 1960.

McKee, Jack. E., and Harold W. Wolf, *Water Quality Criteria,* 2d Ed., Sacramento, Cal., California State Water Quality Control Board, 1963.

Miller, T. L., R. J. Burby, E. J. Kaiser, and D. H. Moreau, *Protecting Drinking Water Supplies Through Watershed Management: A Casebook for Devising Local Programs,* Center for Urban and Regional Studies, Chapel Hill, N.C., University of North Carolina, 1981.

Moreau, David H., "State Financing of Water Resource Projects," *Proceedings of the 1988 Woodlands Conference: New State Roles: Environment, Resources and the Economy,* Woodlands, Tex., Center for Growth Studies, 1989.

Muys, Jerome C., "National Water Policy in Transition: The Role of Water Marketing," *Texas Water Resources Forum,* Austin, Tex., 1989.

National Research Council, *Restoration of Aquatic Ecosystems,* Washington, D.C., National Academy Press, 1992.

National Resources Committee, "Water Pollution Control in the United States," 3d Report of the Special Advisory Committee on Water Pollution, Washington, D.C., U.S. Government Printing Office, 1939.

National Water Commission, "Water Policies for the Future," Final Report to the President and to the Congress, Washington, D.C., U.S. Government Printing Office, 1973.

Robert C. Nichols, "A Review and Analysis of Fourteen Environmental Assessment Methods," in E. Hyman, B. Stiftel, D. Moreau, and R. Nichols (eds.), *Combining Facts and Values in Environmental Impact Assessment,* 1988, pp. 115–223.

Palmer, Tim, *The Wild and Scenic Rivers of America,* Washington, D.C., Island Press, 1993.

President's Water Resources Council, "Policies, Standards, and Procedures in the Formulation, Evaluation, and Review of Plans for Use and Development of Water and Related Land Resources," Senate Document No. 97, 87th Congress, 2d Session, Washington, D.C., U.S. Government Printing Office, 1962.

Reuss, Martin, *Reshaping National Water Politics: The Emergence of the Water Resources Development Act of 1986,* Institute for Water Resources, U.S. Army Corps of Engineers, IWR Policy Study 91-Ps-1, 1991.

Robbins, R. W., J. L. Glicker, D. M. Bloem, and B. M. Niss, *Effective Watershed Management for Surface Water Supplies,* Denver, Colo., American Water Works Association Research Foundation, 1991.

Ruddy, Barbara C., and Kerie J. Hitt, "Summary of Selected Characteristics of Large Reservoirs in the United States and Puerto Rico, 1988," Open-File Report 90-163, Denver, Colo., U.S. Geological Survey, 1990.

Rudolph, Richard, and Scott Ridley, *Power Struggle—The Hundred-Year War Over Electricity,* New York, Harper and Row, 1986.

Schmandt, Jurgen, Ernest T. Smerdon, and Judith Clarkson, *State Water Policies—A Study of Six States,* New York, Praeger, 1988.

Schooler, Dean, and Helen Ingram, "Water Resources Development," *Policy Studies Review,* 1(2): 243–254, 1982.

Senate Select Committee on Water Resources, *Water Resource Activities in the United States,* Committee Print no. 2, 86th Congress, 1st Session, 1959.

Shad, Theodore M., "Water Resources Planning—Historical Development," *Journal of the Water Resources Planning and Management Division, Proceedings of the American Society of Civil Engineers,* 105(WRI): 9–25, March 1979.

Tarr, Joel A., James McCurley, and Terry F. Yosie, "The Development and Impact of Urban Wastewater Technology: Changing Concepts of Water Quality Control, 1850–1930," in Martin V. Melosi (ed.), *Pollution and Reform in American Cities, 1870–1930,* Austin, Tex., University of Texas Press, 1980.

U.S. Environmental Protection Agency, "Waste Disposal Practices and Their Effects on Ground Water," Report to Congress prepared by the Office of Water Supply and the Office of Solid Waste Management, 1977.

U.S. Environmental Protection Agency, "Ground-Water Protection Strategy," 1984.

U.S. Environmental Protection Agency, "Progress in Ground-Water Protection and Restoration," Office of Water, EPA 440/6-90-001, 1990.

U.S. Geological Survey, "National Water Summary 1983—Hydrologic Events and Issues," Water Supply Paper No. 2250, 1984.

U.S. Environmental Protection Agency, "The Watershed Protection Approach Framework Document," Office of Wetlands, Oceans, and Watersheds, October 1991.

U.S. Environmental Protection Agency, "The Watershed Approach—Annual Report 1992," Office of Water, EPA A840-S-93-001, 1993.

U.S. Water Resources Council, *State of the States: Water Resources Planning and Management, Fiscal Year 1981 Update,* Washington, D.C., 1981.

U.S. Water Resources Council, "Establishment of Principles and Standards," *Federal Register,* vol. 38, no. 174 (September 10, 1973), pp. 24778–24869.

Viessman, Warren, Jr., and Gay M. Biery-Hamilton, "An Analysis of State Water Resources Planning Processes in the United States," Report to the five Florida Water Management Districts, Department of Environmental Engineering Sciences, University of Florida, 1986.

Volkman, John M., "Making Room in the Ark," *Environment,* 34(4): 18–20, 37–43, 1992.

Wahl, Richard W., *Markets for Federal Water: Subsidies, Property Rights, and the Bureau of Reclamation,* Washington, D.C., Resources for the Future, 1989.

Wahl, Richard W., and Frank H. Osterhoudt, "Voluntary Transfers of Water in the West, " *National Water Summary—1985,* U.S. Geological Survey, Water Supply Paper 2300, 1986.

Water Resources Activities in the United States, Committee Print No. 2, Senate Select Committee on National Water Resources, 86th Congress, 1st Session, October, 1959.

White, Gilbert F., "A Perspective of River Basin Development," *Law and Contemporary Problems,* Durham, N.C., Duke University, 22(2) (Spring): 157–187, 1957.

CHAPTER 5
WATER LAW

A. Dan Tarlock
Chicago-Kent College of Law
Chicago, Illinois

5.1 FUNCTION OF WATER LAW

5.1.1 Creation of Private Property Rights in Scarce Resources

Water law has two basic, related functions: (1) the creation of correlative private property rights in scarce resources, and (2) the imposition of public interest limitations on private use.* The first function requires the recognition of individual rights to use a portion of a common supply and their simultaneous limitation to protect the interests of similarly situated users; and the second requires the imposition of additional limitations on the exercise of these rights to promote more general public values. There are three models of legal regimes to allocate common resources, such as streams and aquifers, and to resolve disputes among competing water users. These models are not exclusive; they may exist simultaneously. However, they represent something of a regulatory progression as society has come to appreciate the scarcity of water resources, the quantity and quality interdependencies among uses, and the importance of water resources for the promotion of biodiversity. The models are: (1) a *tort approach,* (2) an *exclusive property-rights regime,* and (3) a *managed property-rights regime.* However, for any of these schemes to function effectively, the creation of a property-rights regime and the delineation of limitations on the exercise of those rights, is necessary.

The tort model assumes that water resources are essentially free goods and that the societal interest in water is limited to the adjudication of infrequent disputes among limited classes of users. Individuals are allowed to use water without the need for any kind of prior approval, to the point that another user suffers a demonstrable harm such as diminished flow or pollution. This is the *common law* or *riparian rights model.* Under this model, the common law of nuisance, which imposes a reasonable-

---

* The standard modern comprehensive treatments of water law are David H. Getches, *Water Law in a Nutshell* (2d ed., 1990); Wells A. Hutchins, *Water Rights Laws in the Nineteen Western States* (completed by Harold R. Ellis and J. Peter DeBraal, 1971); National Water Commission, *A Summary Digest of State Water Laws* (Richard L. Dewsnup and Dallin Jensen, eds., 1973); A. Dan Tarlock, *Law of Water Rights and Resources* (1988) and *Water and Water Rights* (Robert Beck, ed., 1991).

ness qualification on the enjoyment of property, forms the basis for water law doctrines which allow injured users to seek after-the-fact relief for their injuries. The exclusive property rights model assumes that all available resources should be used to the point of exhaustion, but that each user should know in advance the extent of his right, so that the risk of interference with other users will be minimized. Absolute exclusivity is impossible to achieve, because water resources are almost always shared among competing claimants, but maximum exclusivity is a feasible objective for a common property-rights regime. The doctrine of prior appropriation, developed in the western United States, has attempted to develop maximum feasible exclusive individual rights to the maximum extent possible in surface and, in many states, groundwater resources. Management involves the administrative allocation and distribution of water to meet multiple objectives. These objectives are generally achieved within a legislative framework, which defines the limits of administrative discretion, and subject to the constraints on constitutionally protected property rights. The managed property-rights model superimposes public interest limitations on a base of individual exclusive rights. Management has been superimposed on both riparian and appropriative regimes, but management takes place within a property-rights framework.

5.1.2   Public Interest Limitations on Private Use

Water has both commodity and community value. These dual values arise from the fact that all uses of a stream or aquifer are related. The chain of claimants with either a property right or a recognized interest often extends throughout a stream or aquifer system. Thus, unlike other forms of property such as land, the use of water has always been limited to protect both the interests of other users and of the community. The concept of public interest is an evolving one. Originally, public interest limitations played almost no role in water law, but advances in scientific understanding and changing social values have increased its role (DuMars and Minnis, 1989). Public interest can encompass the conservation of nonrenewable groundwater resources to prevent or limit mining, the maximization of net returns on a use,* the maintenance of instream flows for fish and wildlife preservation, the limitation of consumptive uses to enhance water quality, and other values such as community stability. Property rights in water are subject to public interest limitations. States have greater power to regulate water use, compared to dry land, because the use of water can directly affect the enjoyment of other users. Water rights have always been more limited compared to other forms of property rights.[†]

The basis of the public interest limitation is the concept of correlative rights. Water rights have long been classified as *correlative*. Correlative property rights are individual property rights subject to limitations to protect the equal or "correlative" rights of similarly situated users. For example, an artesian groundwater user may be required to cap the well to maintain the pressure in the aquifer to protect the equal rights of other users to a fair pressure level.[‡] The need to protect correlative rights

---

* Water allocation generally seeks to promote the efficient allocation of resources, unless there are compelling equity constraints. Thus, water should be allocated to a given use to the point that marginal cost equals marginal value. See Bonnie Colby Saliba and David B. Bush, *Water Markets in Theory and Practice,* (1987).

[†] The extent of the state's regulatory power is the subject of intense debate. See, e.g., Joseph L. Sax, "The Constitution, Property Rights and the Future of Water Law," 61 Colo. L. Rev. 257 (1990) and "Point/Counterpoint, Long's Peak Report: Reforming National Water Policy," 24 Envtl. L. 125 (1994).

[‡] *Lindsley v. National Carbonic Gas Co.,* 220 U.S. 61 (1911).

was the original basis for the regulation of water rights and remains the core justification for public interest limitations. Water rights may be limited to advance the following related public objectives:

1. *The Maintenance of Correlative Rights.* Water rights are often limited to guarantee that all similarly situated users have equal access to the resource.

2. *Conservation of a Limited Supply of a Nonrenewable Resource.* This rationale is closely related to the protection of correlative rights, but the objective of conservation limitations has historically been to maximize the value of a nonrenewable resource over some period of time, rather than simply to do equity among similar users. The right to use nonrenewable resources such as aquifers is often limited to prolong the life of an aquifer. Conservation objectives can vary from regulations designed to sustain an aquifer perpetually by limiting use to safe yield to regulations which specify the conditions under which mining can occur.

3. *The Promotion of Efficiency.* The promotion of efficiency is the economic rationale for conservation. Many western state administrative regimes have long included the power to deny new appropriations to protect the public interest, and this power has gradually been extended to applications to transfer water rights. The power to deny new applications when water is available or to permit a transfer when there is no demonstrated injury to third-party rights is seldom actually applied, and when it is applied, the purpose is generally to favor more efficient or less efficient projects.* In recent years, the concept of efficiency has been extended to the evaluation of broader social costs, such as environmental values. Thus, appropriations can be denied if the state concludes that, on balance, the environmental damage would exceed the value of the project for which a water right is sought.[†]

4. *The Promotion of Public Values.* The public interest can include the promotion of public values, such as environmental quality, or the maintenance of local water-dependent communities. State environmental quality assessment acts have expanded the opportunity to raise broader economic and social objections to new proposed water diversions. Little-NEPAs (National Environmental Protection Act) apply to new appropriation or related applications and require that the state agency consider a variety of factors, such as impacts on sensitive ecosystems, beyond the traditional statutory criteria for granting new water rights.[‡]

5.2 WATER LAW SYSTEMS

5.2.1 Land-Based Systems: Riparian Rights and Groundwater "Ownership"

The common law of water rights is a land-based system. A riparian right is a right to use a portion of the flow of a watercourse that arises by virtue of ownership of land bordering a stream and lakes (Dewsnup and Jensen, 1974). The word *riparian* is derived from the Latin *ripa*, which means river bank. Riparian rights exist *jure nature* because the riparian's land "has the advantage of being washed by the stream" (Gould, 1900). The common law viewed riparian rights as one of the cluster

---

* The leading case of *Tanner v. Bacon,* 136 P.2d 957 (Utah 1943) subordinated a prior application to protect the feasibility of a larger Bureau of Reclamation project.

[†] e.g., *Shokal v. Dunn,* 707 P.2d 441 (Idaho 1985).

[‡] e.g., *Stempel v. Department of Water Resources,* 508 P.2d 166 (Wash. 1973).

of rights incident to land that included the rights to lateral and subjacent support. It is interesting to speculate why the common law limited the right to use water to owners of land bordering on a stream. A modern American legal historian has suggested that this reflects the pre-nineteenth century attitude that land was not a productive resource, but something to be enjoyed for its own sake (Freyfogle, 1986; Horwitz, 1977). It is rational for a primarily agricultural society to define the "natural" rights incident to land rights as those attributes of land that preserved its natural condition, and to be less concerned with promoting widespread access to resources. However, the limitation to land ownership proved unsuited for an urban society, and a variety of means have been developed to allow water be used beyond stream and watershed systems, although there is a revival of interest in the riparian system because of its greater protection of instream uses (Butler, 1987).

Riparian rights attach only to riparian land. To qualify as riparian land, there must be contact with a watercourse capable of supporting riparian rights.* Riparian rights depend on ownership of the bank of a watercourse, and not on ownership of its bed.† The contact must occur during periods of ordinary flow.‡ Land left dry as the result of a change in the course of a river due to a flood may no longer be riparian.§ Any contact with the highest line of ordinary flow is sufficient; there is no minimum ratio between the size of land and the length of frontage.¶ Riparian land may also expand and contract according to the laws of accretion, avulsion, and reliction, which allocate land title in response to shifting streams and changing lake levels.**

Riparian rights have been constantly questioned, but they endure, probably because their bark is worse than their bite and they continue to provide a reasonable basis for allocation of water-rich areas. Modifications have occurred when the doctrine is perceived as inconsistent with economic efficiency. The merits of riparian rights were extensively debated in California in the late nineteenth and early twentieth centuries. Downstream users were afraid that the doctrine would block access to water and contribute to the monopolization of the resource. California and other dual-system states solved this problem by allowing the appropriation of surplus water, water beyond that used by riparians. Similar concerns have been expressed in other parts of the country, but in practice the doctrine has not blocked access to water by major users. Municipalities have exercised the power of eminent domain to condemn water rights outside their territorial limits and transfer water to areas of

---

* *Clippinger v. Birge,* 14 Wash. App. 976, 547 P.2d 871 (1966) (land separated from lake by road and strip owned by another party was not riparian); *Monroe Carp Pond Co. v. River Raisin Paper Co.,* 240 Mich. 279. 215 N.W. 325 (1917). *Cf. Cabal v. County of Kent,* 72 Mich. App. 523, 250 N.W.2d 191 (1976).

† e.g., *Bouris v. Largent,* 94 Ill. App. 2d 251, 236 N.E. 2d 15 (1968). See F. Maloney, J. Plager, and F. Baldwin, *Water Law and Administration: The Florida Experience,* § 21.6 (1968). *Cf. United States v. Antanum Irrigation Dist.,* 236 F.2d 321, 325 (9th Cir. 1956), *cert. denied,* 352 U.S. 998 (1957), awarded federal Indian reserved rights, which are analogous to riparian rights, in a stream described as the northern boundary of the reservation. "It is true that the Yakima treaty described the Antanum as the north boundary of the reservation, whereas the boundary" of the reservation in *Winters v. United States,* 207 U.S. 564 (1983), "was described as beginning at a point in the middle of the main channel of the Milk River. But a tract of land bounded by a nonnavigable stream is deemed to extend to the middle of the stream."

‡ *Wholey v. Caldwell,* 108 Cal. 95, 41 P.31 (1895); *In the Matter of the Determination of the Ordinary High Water Mark and the Outlet Elevation for Beaver Lake,* 466 N.W.2d 163 (S.D. 1991) (remand to determine if slough and lake joined under ordinary and continuous conditions rather than at times of extraordinary high water).

§ *El Paso County Water Improvement Dist. No. 1 v. City of El Paso,* 133 F. Supp. 894 (W.D. Tex. 1955), *aff'd in part,* 243 F.2d 927 (5th Cir.), *cert. denied,* 355 U.S. 820 (1957).

¶ *Charnock v. Higuerra,* 111 Cal. 478, 44 P.171 (1896); *Town of East Troy v. Flynn,* 169 Wis. 2d 330, 485 N.W.2d 415 (Wis. App. 1992).

** For an introduction to the complicated law of riparian boundary disputes see Ronald W. Tank, *Legal Aspects of Geology* (1983).

demand, and it is becoming easier to sever water rights from riparian land in many states (Harnsberger, 1969; Ziegler, 1959).

Groundwater was also initially allocated by the ownership of overlying surface land, but this has created greater problems because no riparian sharing limitations were imposed upon use. Thus, the common law does not effectively restrain use. At common law, the right to use groundwater is an incident of surface ownership. Two rules developed to allocate the resource. Under the *absolute ownership* rule, an overlying land owner can use as much water as can be pumped, unless the purpose is malicious.* Many courts modified absolute ownership by adopting the *reasonable use* rule. A pumper may still use as much as can be pumped, so long as the use is for a productive purpose and is confined to the overlying land. The practical effect of this rule is to require municipal pumpers to compensate injured farmers. Neither of these rules prevents rapid exploitation or prior use. In many states, the common law has been supplemented by legislation that authorizes a state agency to impose pumping limitations on large pumpers in short-term droughts.

Many western states attempted to solve this problem by applying the law of prior appropriation, discussed in Sec. 5.2.2, to groundwater. Other states have applied the common law of riparian rights to groundwater through the adoption of the California correlative rights rule, or have adopted the Restatement of Torts (Second) approach. The correlative rights rule applies riparian surface law rules to overlying pumpers in a basin,† and Restatement of Torts (Second) provides large-scale pumpers may be liable if "the withdrawal of groundwater unreasonably causes harm to a proprietor of neighboring land . . ."‡ Prior appropriation is difficult to apply to groundwater, and priorities are seldom enforced. For example, most states have rejected a senior "right to lift" because it would freeze pressure levels and discourage subsequent use.§ Juniors have a right to lower pressure to a "reasonable" level. For this reason, states with exhaustible supplies have imposed a variety of regulatory regimes to stabilize or roll back pumping levels in basins where withdrawals exceed safe yield.¶

5.2.2 Use-Based Systems: Prior Appropriation and Permit Systems

Prior appropriation is a use-based rather than land-based system of property rights. The system was developed in the mining camps of California to allocate water for placer mining, and spread throughout the West because it was thought to promote irrigation economies. Appropriate rights are not tied to the locus of the use of the water. They apply to direct flow diversions, and to the storage of water for subsequent release. No eastern state currently follows the prior appropriation, but the administration of permit schemes in states such as Florida creates a de facto priority schedule. Water may be used any place to which it can be transported within a state.** A water right is perfected by diverting water and applying it to a beneficial use. Rights are allocated by priority. In times of shortage, there is no prorate curtail-

---

* e.g., *Wiggins v. Brazil Coal & Clay Co.,* 440 N.E.2d 495 (Ind. 1983).

† The leading case is *Katz v. Walkinshaw,* 74 P. 766 (Cal. 1903).

‡ e.g., Cline v. American Aggregates Corp., 474 N.E.2d 324 (Ohio 1984).

§ e.g., Wayman v. Muarry City Cor., 23 Utah 2d 97, 458 P.2d 861 (1969).

¶ See A. Dan Tarlock, James N. Corbridge, and David H. Getches, *Water Resource Management,* Chap. 5 (4th ed., 1992).

** Ironically, many states have imposed statutes which prohibit or restrict the export of water across state lines. Export prohibitions are unconstitutional, but statutes which prefer in- to out-of-state users may be constitutional.

ment. Junior rights must cut back, so that senior right-holders will obtain the full amount of their rights. Holders of senior rights are entitled to take the full amount of their rights, regardless of the comparative efficiencies of junior and senior uses. Most western states have also applied prior appropriation to groundwater, but the large pumping states of California, Nebraska, and Texas have not.

Appropriative water rights are the opposite of riparian rights. There need be no relationship between the source of water and the locus of use. Los Angeles, for example, enjoys water appropriated on the Colorado and Owens Rivers, hundreds of miles from the city. However, an appropriator must have access to water to perfect an appropriation. The early appropriations were on public land, and the federal government provided access by the Act of 1866, which sanctioned the custom of prior appropriation based on trespass.* Two federal statutes permit the acquisition of right of way for reservoirs and ditches used pursuant to a state-granted water right.[†] This same thinking has been carried over to cases dealing with appropriations founded on trespasses on private lands, although there is no common law privilege to trespass to exercise a public right. Courts have often stated that a water right cannot be perfected by a trespass on private land,[‡] but courts often do not penalize a trespassing appropriator, to promote widespread access and to prevent the monopolization of the resource. Acts that resulted from illegal entries can qualify the trespasser for a water right.[§] Colorado has held that an applicant for a conditional water right may trespass on the lands of another to take a first step to appropriate water, and then condemn the necessary easement as authorized by state law.[¶]

Access to water open to appropriation can generally be acquired by eminent domain, if necessary. To prevent de facto riparianism, western states passed statutes permitting a water-rights claimant to condemn the necessary rights of way to bring the water from the stream to the place of use.** These statutes were presumptively unconstitutional, because in the nineteenth century the courts limited the power of eminent domain to public uses.[††] A private appropriator did not meet the use by the public standards; the Supreme Court sustained a Utah statute because of the "peculiar condition of the soil or climate" in the arid West.[‡‡] Nebraska refused to follow

---

* e.g., Stookey v. Green, 53 Utah 311, 178 P. 586 (1919); Peterson v. Wood, 71 Utah 77, 262 P. 828 (1927).

[†] 43 U.S.C. §§ 661 and 946–49 provide for the reservation of an interest on unpatented federal lands for reservoirs and ditches to be used in connection with a state-granted water. Section 661 creates an easement, *Laughlin v. State Board of Control,* 21 Wyo. 99, 128 P. 517 (1912), but §§ 946–949 have been said by the United States Supreme Court, *Kern River Co. v. United States,* 227 U.S. 147 (1921), and the Wyoming Supreme Court, *Johnson Irrigation Co. v. Ivory,* 46 Wyo. 221, 24 P.2d 1053 (1933), to have created a limited fee. Montana construes both acts to create easements. *E.E. Eggebrecht, Inc. v. Waters,* 704 P.2d 422 (Mont. 1985).

[‡] e.g., *In re the General Determination of the Rights to the Use of Surface and Ground Waters of the Payette River Drainage Basin,* 107 Idaho 221, 687 P.2d 1348 (1984); *Geary v. Harper,* 92 Mont. 242, 12 P.2d 276 (1932).

[§] e.g., *State ex rel. Ham, Yearsley & Ryrie v. Superior Court,* 70 Wash. 442, 126 P.945 (1912); *Bassett v. Swensen,* 51 Idaho 256, 5 P.2d 722 (1931).

[¶] *Bubb v. Christensen,* 200 Colo. 21, 610 P.2d 1343 (1980).

** Ariz.: Ariz. Rev. Stat. Ann § 12-1111(16).
   Colo.: Colo. Rev. Stat. § 37-86-102.
   Idaho: Idaho Code § 42-1102.
   Nev.: Nev. Rev. Stat. § 533.050.
   N.M.: N.M. Stat. Ann. § 72-1-5.
   S.D.: S.D. Codified Laws §§ 46-8-1-18.
   Utah: Utah Code Ann. § 78-34-1(5).
   Wyo.: Wyo. Stat. § 1-26-401.

[††] See Comment, "The Public Use Limitation on Eminent Domain: An Advance Requiem," 58 Yale L.J. 599 (1949).

[‡‡] *Clark v. Nash,* 198 U.S. 361 (1905). The need to divert waters from their natural channels to obtain the fruits of the soil was characterized as "so universal and imperious that it claims recognition of the law" in *Yunker v. Nichols,* 1 Colo. 551 (1872).

the Supreme Court's lead because it was insufficiently arid,* although this decision is an exception to the willingness of the courts to allow public use to be defined by custom. Eminent domain may be used to transport water so long as the use is beneficial; beneficial uses are presumed public uses.† States have also authorized the use or enlargement of common ditches‡ and temporary entries to gain information or to maintain a ditch.§ Once an appropriative right is perfected, the appropriator may use a natural stream to transport it to a distant locus of use.¶ The Supreme Court now equates public use with public purpose,** so the constitutionality of these statutes seems beyond doubt.

5.3 WATER USE RULES

5.3.1 Acquisition of Rights

Riparian rights are an incident of property ownership. The right is acquired when riparian land is acquired. The right does not depend on use. Water rights are generally described as *real property rights.*†† A riparian right is an incorporeal rather than corporeal right, because one cannot possess the flow of a stream; one can only use the water.‡‡ They may be acquired either through the purchase of riparian land, or by the severance of riparian rights from the land. An agreement to allow the use of waters subject to riparian rights is generally classified as an easement rather than a personal covenant or as a profit a prendre.§§ Interests in water are subject to the law of eminent domain,¶¶ and public projects that unduly interfere with the exercise of riparian rights may constitute takings.*** Likewise, the enjoyment of riparian rights is subject to police power restrictions and the protection of the Fifth and Fourteenth Amendments. For example, there are cases that must uphold the power of cities to limit the surface use of lakes to prevent pollution††† and those that find that this restriction of riparian use is a taking.‡‡‡

---

* *Vetter v. Broadhurst,* 100 Neb. 356, 160 N.W. 109 (1916).
† *Kaiser Steel Corp. v. W.S. Ranch Co.,* 81 N.M. 414, 467 P.2d 986 (1970) (coal mining).
‡ Cal.: Cal. Water Code § 1800.
N.M.: N.M. Stat. Ann § 75-1-5.
Utah: Utah Code Ann. § 73-1-7.
§ *Stoll v. MacPherson Duck Club, Ltd.,* 43 Colo. App. 377, 607 P.2d 1019 (1979).
¶ *Rock Creek Ditch & Flume Co. v. Miller,* 93 Mont. 248, 17 P.2d 1074 (1933); *Cascade Town Co. v. Empire Water & Power Co.,* 181 F. 1011 (D. Colo. 1910).
** *Hawaii Housing Authority v. Midkiff,* 467 U.S. 229 (1984).
†† *Mayor v. Commissioners,* 7 Pa. 348 (1857); *Nevada Co. v. Kidd,* 37 Cal. 282 (1869); *Willow River Chief v. Wade,* 100 Wis. 86, 76 N.W. 273 (1989). See Note, "The Constitutional Sanctity of a Property Interest in a Riparian Right," 1969 Wash. U. L.Q. 327.
‡‡ The Vermont Supreme Court speaking of a "perpetual right to tap" a spring and brook, *Davidson v. Vaugh,* 114 Vt. 243 44 A.2d 144, 147 (1945), stated:

> [O]ur rule is that the grant of a right to take water from a stream or spring conveys a right in the land itself, and is something more than an easement; it is an interest partaking of the nature of a profit a prendre. *Clement v. Rutland Country Club,* 94 Vt. 63, 66, 108 A. 843; *Lawrie v. Silsby,* 76 Vt. 240, 251, 56 A. 1106, 104 Am.St.Rep. 927; *Village of Brattleboro v. Yauvey,* 101 Vt. 314, 318, 143 A. 295.

§§ *Johnson v. Armour and Co.,* 69 N.D. 769, 291 N.W. 113 (1940) (downstream riparian who contracted with upstream riparian to allow waste discharges subjected the stream to an easement that bound his successors in interest).
¶¶ *Jeter v. Vinton-Roanoke Water Co.,* 114 Va. 769, 76 S.E. 921 (1913).
*** *United States v. Gerlach Live Stock Co.,* 339 U.S. 725 (1950).
††† *People v. Hulbert,* 131 Mich. 156, 92 N.W. 211 (1902); *Bino v. Hurley,* 273 Wis. 10, 76 N.W.2d 571 (1956).
‡‡‡ *State v. Morse,* 84 Vt. 387, 80 A. 189 (1911); *State v. Heller,* 123 Conn. 492, 196 A. 337 (1937).

Riparian rights ordinarily do not attach to artificial watercourses.* The reason is the expectation of the shore landowners. As a court explained, denying relief to a resort owner who was located on a bay of the Mississippi River that had been dammed as part of a drainage project, and was subsequently lowered for drainage reasons, "[a]ppellant acquired no vested right in a natural body of water by establishing her business adjacent to the lake *after* the bay had . . . changed. . . . She had no common law right to the continuance of the bay in its new, artificial state, and had no legally protected interest in the maintenance of the water at any particular level.† However, because expectation is the basis for the denial of riparian rights in artificial bodies of water, it can also be the basis for the recognition of rights in these waters; the particular character of the water body and circumstances surrounding its use may give rise to reasonable expectations. Thus, the simple rule that there are no riparian rights in artificial watercourses is often not a good guide to the results in the actual cases.

Abutting landowner interests in the level of artificial waters can be created by express or implied easements. Implied appurtenant easements to an artificial lake may thus arise by dedication.‡ The sale of lots with reference to a plat which depicts a lake, oral representations about prospective lake use made to purchasers, and the representation of lake use opportunities in advertising, will support a finding that a landowner impliedly dedicated easements to all lot purchasers in a subdivision to maintain a lake at the level which it existed at the time lots were first sold.§

Appropriative rights were originally acquired by posting a notice on a stream and digging a ditch. To perfect a water right, the appropriator must complete the necessary diversion works to put the water to a beneficial use within a reasonable period of time. If this is done, the priority date "relates" back to the commencement of the project. If the project is not completed with due diligence, the right dates only from the completion of the project. Today, in all states except Colorado, appropriative rights are assigned and administered by a state agency. However, the prepermit judge-made law forms the basis for the administration of the system. An appropriator must apply to a state agency for a permit or water license. The appropriator must show that: (1) unappropriated water is available, (2) the proposed appropriation will not interfere with prior rights, and (3) the appropriation is in the public interest. The resulting right is specific to quantity, time and place of use is subject to a water duty—the maximum amount allowed for a specific crop.

The essence of a priority system is that prior rights are superior to subsequent ones and that water be constantly applied to beneficial use. Thus, to acquire a new appropriative right, a claimant or permit applicant must show that there is unappropriated water available. Availability is generally measured by a "normal water year."¶ Under this standard, water is available for appropriation even if the right cannot be satisfied every year, but water is obviously unavailable for appropriation if only one-half of the decree holders can usually be served during periods of peak

---

* *Anderson v. Bell,* 433 So. 2d 1202 (Fla. 1983); *Thompson v. Enz,* 379 Mich. 667, 679, 143 N.W.2d 473, 480 (1967); *Crenshaw v. Graybeal,* 597 So 2d 650 (Miss. 1992); *Ours v. Grace Property, Inc.,* 412 S.E.2d 490 (W. Va. 1991); and *United States v. 1,629.6 Acres of Land, More or Less,* 503 F.2d 764 (3d Cir. 1974). See Corbridge, "Surface Rights in Artificial Watercourses," 24 Natural Resources J. 887 (1984).

† *Wood v. South River Drainage Dist.,* 422 S.W.2d 33 (Mo. 1967).

‡ *Black v. Williams* 417 So. 2d 911 (Miss. 1982); *Our v. Grace Property, Inc.,* 412 S.E.2d 490 (W. Va. 1991) (easement to construct flood control dam did not include power to transfer the right to use the lake to third parties).

§ *Shear v. Stevens Bldg. Co.,* 418 S.E.2d 841 (N.C. App. 1992).

¶ *Dovel v. Dobson,* 122 Idaho 59, 831 P.2d 527 (1992).

demand.* But, water use on most streams is like the federal budget, and the right to use water is subject to a limitation not found with dry-land rights.

Beneficial use is the basis, measure, and limit of appropriative rights. To promote the maximum use by the maximum number of users, courts imposed three historic limitations on the enjoyment of appropriative rights. First, a right must be used to be held; unused rights are subject to abandonment or statutory forfeiture. Second, the use must not be wasteful. Waste was originally defined by community custom, but more recent decisions and administrative practices impose a higher standard, although it falls short of technological or economic efficiency. Third, the use must be for a beneficial purpose. This is a flexible standard, as it encompasses new uses such as instream flows, so it has, in effect, become a restatement of the antiwaste prohibition.

The importance of the beneficial use limitation is illustrated by the rules for the determination of available water. No one really knows how much water is actually being put to beneficial use by how many people. Paper claims and decrees may fully appropriate the stream, but this may not represent the amount actually dedicated to service existing rights. A prior appropriator's right extends only to the amount of water put to beneficial use, so the amount of firm water rights may often be less than the amount of paper claims.[†] "The decreed amount may be prima facie evidence of an appropriator, entitlement . . . , but such evidence may be rebutted by showing actual historical beneficial use. Beneficial use is not a concept which is considered only at the time an appropriation is obtained. The concept represents a continuing obligation which must be satisfied in order for the appropriation to remain viable."[‡] Most states do not simply add up permit amounts to decide if unappropriated water is available.[§] For example, Colorado holds that water is available for appropriation on paper overappropriated streams if the junior appropriator can show that there will be no injury to senior water-rights holders.[¶] Colorado's scheme of water-rights administration does not contemplate an automatic cessation of junior rights each time a senior makes a call.[**] In contrast, Texas does add up permit amounts to determine availability. Water availability is determined by the sum of certified filings and permits applicable to the river, less any rights cancelled by proceedings brought by the state.[††] The theory is that the definition promotes the centralized permit system intended by the legislature. Because water permits have been more accurately described as licenses to appropriation, not adjudication,[‡‡] the Texas approach has been criticized because it ignores the realities of western water administration.[§§]

---

* *St. Johns Irrigation & Ditch Co. v. Arizona State Water Comm.*, 127 Ariz. 350, 621 P.2d 37 (1980) (writ of prohibition may issue to prevent state agency from exercising jurisdiction over new applications).

[†] e.g., *Jones v. Warmsprings Irrig. Dist.*, 162 Or. 186, 91 P.2d 542 (1939). See note 3, supra.

[‡] *Basin Elec. Power Coop. v. State Bd. of Control*, 578 P.2d 557 (Wyo. 1978).

[§] e.g., *Temescal Water Co. v. Department of Public Works*, 44 Cal. 2d 90, 280 P.2d 1 (1955), and *Benz v. Water Resources Comm'n*, 94 Or. App. 73, 764 P.2d 594 (1988), which approved a state policy to issue a permit to anyone who is willing to use whatever unappropriated water may become available, to promote the maximum beneficial use of water.

[¶] *Southeastern Colo. Conservancy Dist. v. Rich*, 625 P.2d 977 (Colo. 1981).

[**] For an interesting example of this doctrine see *Central Platte Natural Resources District v. State of Wyoming*, 245 Neb. 439, 513 N.W.2d 847 (1994) which holds that undiverted surface and pumped groundwater constitute water available for instream flow appropriation.

[††] *Lower Colo. River Auth. v. Texas Dep't of Water Resources*, 689 S.W.2d 873 (Tex. 1984). *Accord Office of the State Eng'r v. Morris*, 819 P.2d 203 (Nev. 1991). See Note, "Unappropriated Water: The Amount of Water Remaining After Taking into Account All Existing Permits at Their Recorded Levels," 16 Tex. Tech. L. Rev. 989 (1985).

[‡‡] *Wyoming v. Colorado*, 259 U.S. 419, 488, *modified and reh'g denied*, 260 U.S. 1 (1922), *vacated*, 353 U.S. 953 (1957).

[§§] e.g., *Benz v. Water Resources Comm'n*, 94 Or. App. 73, 764 P.2d 594 (1988) (junior appropriation on ungauged stream allowed over objections of senior appropriators that illegal diversions would not be accurately monitored by state).

5.3.2   Use of Right

Water rights are different from property rights in other resources, such as land or hard rock minerals, because they are subject to limitations to protect the interests of other users and the public generally. They are of necessity incomplete property rights. All natural resources have two basic characteristics: (1) the maximum amount or stock of the resource is fixed by geologic events that occurred long before we attached value to the resource, or (2) if the amount of the resource changes, we have no or very limited control over the rate of change. Water resources can be either stock or flow resources.* An aquifer being depleted in excess of the annual rate of recharge is a stock resource, and a surface stream is a flow resource. In either case, the geographical distribution of the resource and basic fairness demands that the resource be shared. Not only do similarly situated claimants have all claims to the source, but the fixed or variable quantity of the source gives society a strong interest in how the resource is allocated among different users over time.[†] In addition, it is impossible to define the amount of the resource that will constantly be available for consumption. Any claim to the resource, therefore, must be somewhat temporary, or subject to diminution. In short, it is difficult to define exclusive property rights in water compared to land, because of the contingent nature of the ability to enjoy the resource.

The special characteristics of water have led courts to adopt the Roman and civil law classifications of water, and to define all water rights—riparian and appropriative—in terms of the right to use, rather than ownership of the corpus of the water. Riparian and appropriative water rights are *usufructuary property rights,* which Roman law defined as the use of a res by a nonowner that left the character of the res essentially unchanged. The concept is used to determine the major incidents of the right, and increasingly, to define the limitations of the exercise of the right. The major uses of the usufruct concept in the resolution of conflicts are:

1. *Description of the Basic Right.*   The leading American water law decision, *Tyler v. Wilkinson,* accepted as settled the proposition that "strictly speaking, he [a water right holder] has no property in the water itself; but a simple use of it, while it passes along."[‡]

2. *Water Rights Are Separate from the Soil.*   The Colorado Supreme Court described an appropriative water right as a usufruct to support its holding that the common law of riparian rights never applied in Colorado. "Instead of being a mere incident to the soil, it rises, when appropriated, to the dignity of a distinct usufructuary estate, or right of property,"[§] but the same principle applies to riparian rights. Such rights are said to arise from ownership of land bordering a stream, not from ownership of the bed of a stream, and may be severed from riparian land.[¶]

3. *Riparians and Appropriators Have a Duty to Share Among Themselves.*   Because riparian rights are usufructuary, a riparian "has only the right to enjoy the advantage of a reasonable use of the stream as it flows through the land, subject to a

---

  * Economists view these two special conditions as the distinguishing feature of natural resources. McInerney, "Natural Resource Economics: The Basic Analytical Principles" 30, 31 in *The Economy of Environmental and Natural Resources Policy* (J. Butlin, ed., 1981).
  † Economists have devoted considerable energy to calculating the optimal rate of resource depletion. See Chapter 6, § 6.02, infra.
  ‡ 24 Fed. Cas. 472 (C.C.D.R.I. 1827).
  § *Coffin v. Left Hand Ditch Co.,* 6 Colo. 443 (1882).
  ¶ *Belvedere Development Corp, v. Division of Administration, State Dep't of Transportation,* 413 So. 2d 847 (Fla. App. 1982).

like right belonging to other riparian proprietors."* Appropriators do not have to share like riparians, but courts have developed a number of rules to limit appropriators, so that water will be available for other appropriators. The best-known limitation is the previously discussed beneficial use rule, which enjoins wasteful practices.

4. *Water Rights Are Subject to Greater Exercises of the Police Power Compared to Land.* The exercise of water rights have been limited by legislation that extinguishes unexercised common law rights, limits the nonconsumptive use of water to further a public interest, or even retroactively limits the enjoyment of vested rights. In each instance, water-right holders argue that the legislature has taken property without due process of law. The issue is a complicated one. Water-right holders sometimes prevail, but courts have sustained statutes that extinguish unexercised rights, and recently have held that the "public trust" applies retroactively to vested water rights. To justify these results, courts stress the usufructuary nature of the water right to support their conclusion that a water-right holder's expectations of continued exclusive enjoyment (of consumptive and nonconsumptive rights)[†] are always less than a dry surface owner's.[‡]

The *doctrine of beneficial use* limits the use of appropriative rights. Riparian rights have also been limited by the incorporation of the doctrine of reasonableness. There are two theories of riparian rights: The *natural flow* and *reasonable use* rules. In theory, the difference is substantial. The natural flow theory allocates the flow equally to all users. No user can divert or retain the flow, except the last user of the stream. In contrast, the reasonable use rule allows reasonable diversions and retentions. For all practical purposes, few if any states follow the natural flow theory today, although the issue may be open in Alabama (Apolinsky, 1959). Natural flow language survives in some eastern states, but it is always mixed with reasonable use language. One of the last explicit natural flow cases[§] held that downstream riparians could enjoin defendant from diverting water from a river when defendant substantially diminished the flow of the stream, because "[t]he plaintiffs, as riparian owners along the Noroton River, are entitled to the natural flow of the water of the running stream through or along their land, in its accustomed channel, undiminished in quantity or unimpaired in quality." However, the result would be the same under the reasonable use theory. When the consequences of the natural flow theory are severe, courts are likely to reject it. For example, a federal district court refused to recognize a right to the undiminished flow of the Hackensack River in favor of a downstream water company against an upstream city.[¶]

The natural flow theory began to break down by the mid-nineteenth century because it was too rigid. Courts never applied the natural flow theory to prevent all impoundments and consumptive uses, but a combination of the restrictive rules on

---

* *Crawford Co. v. Hathaway,* 67 Neb. 325, 93 N.W. 781 (1903).

† *Brusco Towboat Co. v. State,* By and Through Straub, 30 Ore. App. 509, 567 P.2d 1037, 1042 (1977) ("right [to erect structures on submerged and submersible land] should be treated as a usufruct rather than as a property interest").

‡ e.g., *United States v. State Water Resources Control Board,* 182 Cal. App. 3d 82, 227 Cal. Rptr. 161 (1986), review denied. See *Aerojet Gen. Corp. v. San Mateo County Super. Ct.,* 209 Cal. App. 3d 973, 257 Cal. Rptr. 621 (1st Dist. 1989), *supplemented on denial of reh'g,* 211 Cal. App. 3d 216, 258 Cal. Rptr. 684, of the use of the usufructuary nature of a water right. The court rejected the polluter's argument that the state, as opposed to "a traditional fee simple property owner," could not sue under CERCLA for surface water environmental damages, because as trustee for the public it had a mere usufructuary right.

§ *Collens v. New Canaan Water Co.,* 155 Conn. 477, 234 A.2d 825 (1967). See also *Smith v. Morgantown,* 187 N.C. 801, 123 S.E. 88 (1924).

¶ *Hackensack Water Co. v. Village of Nyack,* 289 F. Supp. 671 (S.D.N.Y. 1968).

impoundments and the possibility that an impoundment or consumptive use would not be recognized led to the modification of the natural flow theory. In its place, courts adopted the reasonable use rule to prevent the monopolization of the use of a stream.* A leading and still widely quoted case[†] reasoned that the natural flow theory was inconsistent with the correlative rights of each riparian to use the watercourse, and replaced it with a rule that allowed one riparian to interfere with another's use whether "under all the circumstances of the case the use of the water by one is reasonable and consistent with a correspondent enjoyment of right by another."[‡]

Courts have developed a number of catch-all standards or statements of relevant factors. A multifactor standard announced by the Minnesota Supreme Court is often quoted:

> regard must be had to the subject-matter of the use; the occasion and manner of its application; the object, extent, necessity, and duration of the use; the nature and size of the stream; the kind of business to which it is subservient; the importance and necessity of the use claimed by one party, and the extent of injury to the other party. . . .

The Restatement of Torts (Second) Section 850A retains the classic multi-factor test, but adds factors such as the practicability of mutual mitigation and "the protection of existing values of water uses, land, investments and enterprises . . ."

5.3.3 Loss of Right

Both riparian and appropriative rights can be lost if the holder fails to take steps to protect them. Appropriative rights are easier to lose than riparian rights. Riparian rights, in theory, do not depend on use, and thus they cannot be lost by nonuse. There are three important qualifications to this sweeping principle. First, a subsequent use can wholly or partially divest a prior use. However, under the Restatement of Torts (Second), discussed earlier, priority is a factor in determining the reasonableness of a use. Thus, it is harder for a subsequent use to bump a prior use. Second, riparian rights can be lost by prescription.[§] An upper riparian who takes more than a reasonable share, and thereby interferes with a downstream proprietor's use for a substantial period of time, acquires a prescriptive right against the lower riparian. Third, in California, which recognizes both appropriative and riparian rights, a subsequently asserted riparian right may be subordinated to existing appropriative rights.[¶]

Loss of water rights is much more common in appropriative systems. An appropriative right depends on its continuous application to beneficial use, so a right can be lost by nonuse. Other appropriators may challenge a right on the theory that it has been lost by nonuse. There are two theories of loss: Abandonment and forfeiture. *Abandonment* is a common law that requires an intent to relinquish a property right, followed by some act of relinquishment. Intent to abandon and actual relinquishment are two separate jural acts, but a substantial period of nonuse raises a presumption of abandonment. The holder of right must rebut the presumption by showing that there was no intention to abandon the right, and that the nonuse was justified by economic, financial, or administrative obstacles.

---

* The transition from the natural flow to the reasonable use rule is traced in Plager, "Some observations On the Law of Water Allocation as a Variable in Industrial Site Location," 1968 Wis. L. Rev. 673.
  † *Dumont v. Kellogg*, 29 Mich. 420 (1874).
  ‡ *Id.* at 423.
  § *Pabst v. Finmand*, 211 P.2d 11 (1922).
  ¶ *In re Waters of Long Valley Creek Stream System*, 25 Cal.3d 339, 599 P.2d 656 (1979).

Forfeiture is a more draconian theory of water right loss. Forfeiture is a statutory procedure which terminates water rights that are not used for a specific period, usually five years. There are, however, a number of defenses to forfeiture similar to abandonment. Statutes sometimes recognize exemptions such as drought or active military service. Originally, appropriative rights could be lost by prescription. However, to protect the integrity of state permit systems, most states have precluded the acquisition of prescriptive rights by legislation or judicial decision.

5.3.4 Transfer of a Right

Water rights are property rights, and thus they may be transferred. There is considerable confusion of the transferability of riparian rights, but both nonconsumptive and consumptive rights may be transferred. The transfer of consumptive rights entails considerable risk, because the right remains subject to the reasonable use claims of other riparians. Appropriative rights have long been transferred, and the transfer of existing water rights is now a major reallocation strategy in some areas of the country. Appropriative rights transfers are subject to two levels of constraints, which are, in all states except Colorado, applied by the state water agency (MacDonnell, 1990). First, the correlative nature of the right has long been recognized by the rule that no transfer will be allowed if junior rights will be injured. Second, as the pace of water marketing increases, states may impose new constraints on water transfers, to protect areas of origin and environmental values.

5.3.5 Water Quality Rights

Water rights have traditionally been thought of as confined to a given quantity of water, but there is often a relationship between the flow of a stream and its water quality. Historically, both water law and federal and state pollution law have ignored this relationship, but many states are beginning to integrate these two dimensions of water use. There is a modest tradition to do this. Both the law of riparian rights and prior appropriation include a quality dimension to the right; a water use has an action, based on common law nuisance, against dischargers whose discharges cause an interference with a beneficial or reasonable use. To close the artificial separation of quantity and quality, western states increasingly assert the power to deny new appropriations if the new use will impair a prior user's supply (Getches, MacDonnell, and Rice, 1991). California has integrated quantity and quality management, and courts have interpreted this integration to require that all water rights be exercised in a manner which does not cause the deterioration of water quality.*

5.4 PUBLIC WATER RIGHTS

5.4.1 Public Trust Rights

Private water rights are often subject to superior public ones. The source of public rights is the public trust, a judicial doctrine which limits the use of the beds and waters of navigable lakes and rivers for "nontrust" purposes. The public trust is lim-

---

* *United States v. State Water Resources Control Board,* 182 Cal.APp.3d 82, 227 Cal.Rptr. 161 (1st Dist. 1986).

ited to navigable waters. *Navigability* is a technical legal term; the federal definition has progressed from tidal waters to highways of commerce.* Many states have expanded the federal definition to streams and lakes capable of supporting public recreation, because states may constitutionally decide whether federally nonnavigable waters within their borders are subject to the public trust.† The basic trust principle is that states, as owners of most submerged lands, are the trustees of the beds and waters of navigable lakes, rivers, and their nonnavigable tributaries, to ensure that these resources are used consistent with the trust. This power can be retained by the state or delegated to local governments, as it has been in many states.

The core trust ideas are that lake and river bottoms should only be used, in modern terms, for water-related purposes, and that all citizens should have access to these resources. Originally, the trust protected the use of rivers and lakes for commercial navigation. Structures which interfered with navigation were always a public nuisance. The leading case of *Illinois Central Railroad v. Illinois*‡ held that the trust prevents states from alienating trust lands if trust values will be substantially impaired, and this case is the basis for judicial review of the state grants of trust lands for private use. The public trust does not preclude all use of lakes and rivers for nonenvironmental purposes, but it imposes two major limitations on public activities that threaten the integrity of trust resources. First, the trust contains substantive environmental standards, which prohibit many harmful activities. Second, the trust effectively places the burden on public agencies to justify an environmentally harmful activity. The trust is primarily a judicial doctrine, and the standards are formulated and enforced by state courts. A few states, most notably Michigan and Minnesota, have enacted legislation which adopts the public trust as a standard to review administrative actions.

The trust now protects the recreational use of navigable waters, and has the potential to dedicate these waters to the maintenance of ecological integrity. The leading precedent is *National Audubon Society v. Superior Court.*§ This case was brought to enjoin the City of Los Angeles from diverting water, authorized by vested water rights, from the tributaries of Mono Lake, because the lowered lake levels threatened to destroy the lake's ecosystem. Los Angeles relied on its 1940 state water licenses, but the Court held that: "[o]nce the state has approved an appropriation, the public trust imposes a duty of continuing supervision over the taking and use of the appropriated water. In exercising its sovereign power to allocate water resources in the public interest, the state is not confined by past allocation decisions which may be incorrect in light of current knowledge or inconsistent with current needs." Three years later, an intermediate California appellate court relied on *National Audubon* to hold that California water-pollution law requires that the State Water Resources Control Board has a duty to establish water quality standards necessary to maintain a fresh–salinity balance in the San Francisco Bay Delta, without taking into account the impact of the standards on vested water rights. The rationale was that the public trust doctrine gives the state the power to modify prior vested rights to ensure that the exercise of the water right does not impair the public trust values of the Delta.¶ *National Audubon* has not been directly applied outside of California.

---

* *United States v. Appalachian Power Co.,* 311 U.S. 377 (1940).
† e.g., *Arkansas v. McIlroy,* 268 Ark. 227, 595 S.W.2d 659 (1980), *cert. denied* 449 U.S. 843 (1981).
‡ 146 U.S. 387 (1892).
§ 33 Cal.3d 419, 189 Cal.Rptr. 346, 658 P.2d 709, *cert. denied,* 464 U.S. 977 (1983).
¶ *United States v. State Water Resources Control Board,* 182 Cal.App.3d 82, 227 Cal.Rptr. 161 (1986).

5.4.2   Indian and Public Land Reserved Rights

Native American water rights are a continuing source of controversy among the federal government, the tribes, and the states. Native American tribes have a special class of water rights that adhere to treaty and executive order reservations. Native American water rights have characteristics of both appropriative and riparian rights, and are superior to most state-created rights. The distinguishing feature of all aboriginal peoples is that their identity is tied to a specific geographic location. Indian reservations are the remnants of the society that existed in North America before European discovery and conquest. These reservations sometimes represent true aboriginal homelands; in other cases, they represent lands on which tribes were resettled in the nineteenth century. All reservations may potentially claim federal reserved Indian water rights to support reservation uses either because they are pre-existing or aboriginal rights reserved by the treaty creating the reservation, or they were granted by the federal government, as owner of all public lands including Indian Country.*

The Supreme Court first recognized tribal water rights in the case of *Winters v. United States*,[†] and the rights are popularly referred to as *Winters rights*. The case arose when the federal government sued to enjoin non-Indian users from interfering with irrigation on the Fort Belknap Reservation, created between 1874 and 1888, on the Milk River in Montana. The non-Indians claimed prior rights under Montana law, but the Supreme Court held that the tribe had superior rights which dated from the creation of the reservation, either because the rights were confirmed by treaty or granted by the treaty. *Winters* rights are quasi-riparian because the right is based on land ownership, not, as in the case with appropriative rights, on the application of water to beneficial use, but they are also appropriative, because the right has a priority date. The usual priority date is the date of creation of the reservation (Royster, 1994). Since most reservations were created to clear the way for non-Indian settlement, this date is sufficient to give the tribe a right superior to most state-created rights. True aboriginal rights based on immemorial practices would, of course, be superior to any state-created right.

Until the 1960s, tribal rights were asserted by the federal government under its trust responsibility. As a result, Winters rights were generally only claimed to supported existing or planned tribal irrigation needs. *Winters* rights are now asserted directly by the tribes and tribal–state tensions have risen. Tribes assert rights to large amounts of water long allocated by state law, to the use of water for irrigation and nonirrigation purposes, and for the right to lease the water for nonreservation uses. *Winters* was decided at the height of the assimilation era in American Indian policy, and this history influenced the Court's definition of the scope of the water right. The basic idea behind assimilation was to break up the reservations, and turn the Indians into yeomen farmers like their non-Indian settler neighbors. Thus, the Supreme Court spoke only of the use of the water for irrigation and irrigation has been the historic measure of the right. In 1963, the Supreme Court held that the right entitled the tribes to all the water necessary to irrigate the *"practicable irrigable acreage"* on the reservation.[‡] This standard requires that the land must be: (1) capable of irrigation, and (2) at a reasonable cost.[§] This standard may be met with irrigation efficiencies less than for a comparable non-Indian project.

---

* The proposition, much disputed today, that Indian tribes have no title against the federal government was established in *Johnson v. M'Intosh*, 21 U.S. 543 (1823).

[†] 207 U.S. 564 (1908).

[‡] *Arizona v. California*, 373 U.S. 546 (1963).

[§] *In re General Adjudication of All Rights to Use Water in the Big Horn River System*, 753 P.2d 76 (Wyo. 1988), *aff'd sub. nom. Wyoming v. United States*, 492 U.S. 496 (1989).

Practicable irrigable acreage gives many tribes large blocks of paper water rights. It is unlikely that much of this water will actually be used on the reservation to which it is attached. Many tribes wish to transfer the rights for use both on and off the reservation, but many western states and water users oppose transfers. The legal rationale is that *Winters* rights were created to turn the Indians into a pastoral people, and thus the right is appurtenant to the reservation and the water is limited to irrigation and domestic use. The Indian role in transfers is not easy to predict, because tribes can both further and impede transfers. Indians have inchoate claims to a great deal of water through recent and pending Indian water-rights settlements. The legal power of tribes to transfer water remains in doubt. The power to lease to non-Indians is often asserted but has never been judicially sanctioned. The transfer of tribal land, and probably water, requires Congress' consent under the Nonintercourse Act of 1790, and this may apply to leases as well as permanent title transfers (Royster, 1994).

Many tribes want to use water for nonirrigation uses. Courts have recognized *Winters* rights for instream flows and fisheries, but the idea has not been universally accepted. A major Wyoming Supreme Court opinion has held that *Winters* does not apply to groundwater, and that *Winters* rights cannot be used for fishery maintenance.

The federal government may also assert reserved rights to carry out the water-related purposes of withdrawn public land units. These rights are limited by an important Supreme Court doctrine that basically limits them to national parks and monuments. Most non-Indian reserved rights claims are based on the implied rather than the express intent of Congress in withdrawing public land from entry. In a case denying reserved rights for national forests, the Court developed a high threshold test: (1) there must be strong evidence of implied intent, (2) the water must be for the primary, not secondary, purpose of the reservation, and (3) the right is limited to the minimum amount of water necessary to carry out the purpose of the withdrawal.*

5.4.3 Regulatory Water Rights

Regulatory water rights are de facto rather than de jure proprietary rights that arise because of federal and state regulatory programs. *Regulatory property rights* refer to the impact on state water rights from federal programs, which require flow releases that may be inconsistent with state water law. The two most important federal programs that can supersede state water law are Section 404 of the Clean Water Act, and the Federal Endangered Species Act. Prior to the 1970s, the federal government generally asserted only proprietary rights. In addition to reserved rights, discussed in Sec. 5.4.2, the federal government sometimes asserted a navigation servitude. All users of navigable waters are subject to a servitude, which prevents the acquisition of compensable rights against the federal government when the river is developed for navigation or hydroelectric power.

Programs such as the Clean Water Act, Federal Power Act of 1920, and the Endangered Species Act of 1973, have the potential to require that large quantities of water be released from federal reservoirs or left in streams to fulfill the federal objectives. These decisions may preempt state water allocation law. For example, the Endangered Species Act applies to both new and existing federal water projects and

---

* *United States v. New Mexico,* 438 U.S. 696 (1978).

to federally licensed projects. A federal court recently required the Federal Energy Regulatory Commission (FERC) to consider wildlife mitigation measures in its annual operating licenses, pending a full relicensing (Whooping Crane Trust, 1989). The effect on vested state water rights remains unclear. No court has held that a vested water right must give way to fish, but the potential is there. In 1988, the Endangered Species Act was reauthorized without carving out a special exception for western water rights, and the act has been interpreted to allow the federal government to deny the necessary federal permits to enjoy state water rights,* to require that the federal government dedicate previously dedicated reservoir blocks to the protection of endangered species,† and to enjoin state water-rights holders from continuing diversions that harm endangered species.‡ State regulatory water rights may exist as well as similar statutes or under assertions of the "public trust."

5.5 FEDERAL–STATE RELATIONS

5.5.1 Federal–State Conflicts

Water planning, management, and allocation responsibility is shared between the federal government and the states. During the nineteenth century, federal and state interests were generally complementary, and thus there were few jurisdictional or legal conflicts, but federal power was limited in purpose and geographic scope. Water allocation and management is not an enumerated Constitutional power, but the commerce clause was accepted as a source of federal power to control navigation, and until the twentieth century, the assumption was that federal power was limited to commercially navigable rivers. Accordingly, the federal government confined itself to the promotion of inland navigation (Holmes, 1972), and to the disposition of the public domain to the states and to private individuals. This minimal federal role changed at the end of the nineteenth century due to reliable scientific information about the consequences of rapid resource exploitation. Science provided the foundation for the progressive conservation movement, which was premised on the need for a strong federal role in natural resources policy (Hays, 1959). To conserve resources, the progressive conservation agenda included federal retention of public lands, the construction of multiple-purpose water resources projects, and federal financial support for irrigation projects. With the exception of flood-control projects, most federal efforts were concentrated in the western United States, because federal public land disposition policy was inadequate for arid lands (see Chap. 4).

The major legal consequences of the progressive conservation movement were delayed until the New Deal, when the modern era of large multiple-purpose water projects began. Federal–state conflicts were initially minimal because the federal government did not seek to displace state allocation institutions and law. The opposite was true. Congress mandated that the two major federal construction water agencies, the United States Army Corps of Engineers and the Bureau of Reclamation, must acquire the necessary water rights under state law and operate the federal projects in a manner consistent with state law. Congress imposed a similar mandate on the Federal Power Commission (now the Federal Energy Regulatory Commission), which was created in 1920 to license private hydroelectric projects on naviga-

---

* *Riverside Irrigation District v. Andrews,* 758 F.2d 508 (10th Cir. 1985).
† *Truckee Water Conservancy District v. Clark,* 741 F.2d 257 (9th Cir. 1984).
‡ *United States v. Glenn-Colusa Irrigation District,* 788 F.Supp. 1126 (E.D.Cal. 1992).

ble waters. Power follows the purse, however, and federal policies began to conflict with state policies. To resolve these conflicts, the Supreme Court eventually rendered a series of decisions that validated the expanding federal water resources allocation, and preempted inconsistent state water laws, but has now began to return to the original vision of the federal–state balance. The strong preemption decisions were tied to the construction or operation of multiple-purpose flood control and reclamation projects, and their immediate relevance has diminished as the era of the construction of federal water projects to promote regional growth has ended.* However, the federal government's management of public lands for environmental purposes, and the growing importance of federal environmental regulatory programs which impact the use of water, such as the Clean Water Act and the Endangered Species Protection Act, continue to produce federal–state conflicts.

From a strictly legal point of view, there are no federal–state conflicts over water management. The federal government either has the constitutional power to control the use of water or it does not. If the federal government has the power, Article VI, Clause 2 of the Constitution, the supremacy clause, displaces state law. If the federal government lacks the constitutional power to regulate, state law is the exclusive source of authority. Federal–state disputes have classically revolved around three statutory construction problems: (1) To what extent did Congress intend to exercise federal power? (2) To what extent did Congress intend to preempt the field and exclude state regulation? and (3) To what extent did Congress provide for continuing, coordinate or subordinate state power? The last inquiry is particularly important because Congress has generally deferred to state water law in the operation of federal programs. Section 8 of the Reclamation Act is the model for shared federal–state responsibility in the use of water. It provides "[t]hat nothing in this Act shall be constructed as affecting or intended to affect in any way interfere with the laws of any State or Territory relating to the control, appropriation, use or distribution of water used in irrigation . . ."

5.5.1.1 *The Bureau of Reclamation.*

The basic idea between the Reclamation Act of 1902 was simple. The federal government would provide financing for the construction of small reclamation projects to serve family farmers. Water rights would be acquired and distributed in accordance with state law. To limit the distribution of water to bona fide yeomen, as opposed to speculators, the delivery of water was limited to 160 acres in single ownership, or 320 acres for land owned by a husband and wife. The history of federal reclamation law is one of a constant adjustment to the reality of irrigation. In most areas, land was too costly to return enough to repay the projects. Reclamation gradually evolved from a self-financing to a federal subsidy program, financed by the generation of hydroelectric power. Reclamation program subsidies take the form of interest-free or ability-to-pay repayment obligations (Wahl, 1989). The scale of federal projects also became larger than originally envisioned, and in states such as California, substantial blocks of water were delivered to "excess" lands.

The Supreme Court's reclamation jurisprudence has evolved through three phases: (1) the denial of federal power, (2) the confirmation of federal power and the displacement of state law, and (3) a presumption of state supremacy. The constitutionality of the Reclamation Act of 1902 arose indirectly in the first equitable apportionment litigation, when the United States petitioned to intervene to urge the

---

* Moreau (see Chap. 4) traces the shift from federal dominance caused by the emergence of water quality, the maintenance of environmental quality, urban water supply, and the augmentation of federal flood control policy with nonstructural alternatives as primary water policy issues in the 1960s.

Supreme Court to adopt a uniform rule of prior appropriation. Instead, the Court dismissed the petition, because the federal government had no power to do so. Reclamation of arid lands was not among the enumerated powers of the Constitution.* Federal power was dramatically expanded by the Supreme Court during the New Deal, and this expansion led to the conclusion that the federal government need no longer rely on the fiction of navigation improvement to allocate water. In 1950, the Supreme Court held that the spending power authorized federal reclamation projects.†

Federal law did not displace state law until a seminal 1958 decision, arising out of California's historic resistance to the Reclamation Act of 1958. Prior to this case, reclamation law was premised on the assumption that the federal government was simply a carrier and distributor of project water, and thus had to acquire all water rights needed for a federal project pursuant to state law. The Bureau held these rights in trust for the project beneficiaries, and thus had to operate the project in conformity with state laws.‡ To maintain deliveries to excess lands, farmers in California convinced the state Supreme Court that the excess lands provision of the Reclamation Act of 1902 violated state law. This was a dubious conclusion, but even if the California Supreme Court was correct, federal law controls, because Section 5 of the Reclamation Act clearly and unequivocally imposes a federal policy on the delivery of water to project beneficiaries. Section 8 was never intended to repeal this express federal policy, and the Supreme Court held that federal law supersedes state law.§ But the Supreme Court went beyond this simple holding, and characterized Section 8 as merely a just compensation rule. The Bureau of Reclamation merely had to follow state law to determine the amount of money due if it interfered with any vested state rights.

This reading of Section 8 became the new reclamation law. Five years later, the Court held that a California law which recognized a preference for areas of origin was preempted by the Reclamation Act, although—unlike the 160-acre limitation—there was no clear conflict between federal policy and state law.¶ State law served only to define the "property interests, if any, for which compensation must be made." The New Deal presumption that a broader and exclusive federal sphere was necessary to carry out federal objectives was fully articulated in 1963 in the epic litigation over the Colorado River. To the surprise of all parties, the court held that the Secretary of the Interior had the discretion to allocate the river, in times of shortage, as he or she saw fit, state law and an interstate compact not withstanding.** This presumption was reversed in 1978, and Reclamation law restored the original intent of Congress. The leading case of *California v. United States*†† holds that state law controls the operation of federal reclamation projects unless it conflicts with an explicit congressional directive.‡‡

5.5.1.2 FERC. The Federal Energy Regulatory Commission, the successor to the Federal Power Commission, is also subject to a statute which protects state water

---

* *Kansas v. Colorado,* 206 U.S. 46 (1907).
† *United States v. Gerlach Livestock Co.,* 339 U.S. 725 (1950).
‡ *Nebraska v. Wyoming,* 325 U.S. 589 (1945).
§ *Ivanhoe Irrigation District v. McCracken,* 357 U.S. 275 (1958).
¶ *City of Fresno v. California,* 372 U.S. 627 (1963).
** *Arizona v. California,* 373 U.S. 546 (1963).
†† 438 U.S. 645 (1978).
‡‡ The basic conflict was whether the state of California could impose releases to support downstream whitewater rafting until the federal government could demonstrate that all water was needed to fulfill project purposes. California prevailed on this ground. *United States v. California,* 694 F.2d 1171 (9th Cir. 1982).

law from federal interference. However, the Supreme Court has consistently held that a FERC license preempts state laws which interfere with a FERC license. FERC has the power to issue licenses for hydroelectric projects on navigable streams which are best adapted to the comprehensive development of the river. The relationship between state law and the Federal Power Act came before the Court in 1946, and the Court read Section 27 to require deference to states only for the purpose of defining compensable water rights. Any further deference to state law would defeat the objective of the act, which the Court erroneously defined as "the comprehensive development of the water resources of the nation."[*] In 1990, the Court reaffirmed this decision, and held that California could not impose minimum flow license conditions because they interfered with FERC's comprehensive planning authority.[†] The states, however, have found a partial end-run around the wall of preemption which shields FERC licenses from state regulation. Section 401 of the Clean Water Act requires that all federal projects and licenses obtain a certification from the state pollution authority that the proposed activity will not interfere with state water quality standards. The Supreme Court has interpreted this statute broadly to apply to activities which discharge pollutants into streams as well as the construction of hydroelectric dams which decrease stream flows, noting that the relationship between water quantity and quality is an artificial one.[‡]

5.5.2 Equitable Apportionment Among States

Interstate rivers must be shared among riparian states. States qua states have water rights in interstate rivers and lakes under the doctrine of equitable apportionment. By virtue of their quasi-sovereign constitutional status, all riparian states have an equal right to use interstate waters which border them. This right includes both the right to consume a fair share of the water, and the right to be free from pollution.

Equitable apportionment is a judicial doctrine based on the basic rule of international law that all sovereign nations are equal. The constitutional basis of the doctrine is Article III of the U.S. Constitution. Article III gives the Supreme Court original jurisdiction to adjudicate disputes among states. Original jurisdiction lawsuits are the only constitutional avenue of relief for states, absent an express agreement, such as interstate compact, to vindicate their quasi-sovereign interests in interstate waters. In 1907 the Supreme Court rationalized original jurisdiction suits between states for the apportionment of interstate waters, on the theory that the limited powers of the states in a federal system requires a federal common law of interstate waters (adapted from Tarlock, 1985). States may assert their quasi-sovereign interests in interstate waters within their borders by suing *parens patriae,* to represent the interests of all the citizens of the state, in the Supreme Court's original jurisdiction. In brief, a state must sue to vindicate the interests of a large number of its citizens to the use of an interstate stream. It can not sue to vindicate the rights of an individual water-right holder against a similarly situated one in another state. State courts offer relief for simple cross-boundary water-rights disputes.

To obtain an equitable apportionment, a state must file an original action in the Supreme Court, and the Court must accept jurisdiction. It is not easy to convince the Court to accept jurisdiction. Historically, the Court has been reluctant to exercise its

---

[*] *First Iowa-Hydro-Electric Cooperative v. Federal Power Commission,* 328 U.S. 152 (1946).
[†] *California v. Federal Energy Regulatory Commission,* 495 U.S. 490 (1990).
[‡] *PUD 1 of Jefferson County and City of Tacoma v. Washington Department of Ecology,* 114 S.Ct. (1990) U.S. (1994).

jurisdiction without a high showing of relatively immediate injury by a state affected by a use or a proposed use. Thus, the standards for apportionment have remained general, and as a consequence, it remains difficult for a state to estimate the amount of its future share or its ability to stop a diversion in another state in advance of litigation. The principle barrier that a state seeking to protect its future use of a river must overcome is the doctrine of ripeness. A court's primary role is to administer justice *after* an individual has suffered harm. Courts are much less willing to *prevent* injury, preferring to wait and see what actually happens. For example, the Chicago diversion produced the Court's first major equitable apportionment ripeness decision. *Missouri v. Illinois** was the first case to set a high standard of injury as a prerequisite to Supreme Court relief. Missouri sued to protect the health of residents of St. Louis and other riparian cities from the sewage sent by Chicago into the Mississippi. Missouri invoked the common law rule that a riparian had a right to the flow of a stream unimpaired in quality and quantity. To dismiss Missouri's suit, a higher standard of proof than would be applied to a suit for equitable relief between private parties was articulated: "Before this Court ought to intervene the case should be of serious magnitude, clearly and fully proved, and the principle applied should be one which the Court is prepared deliberately to maintain against all considerations on the other side."† Relief was not warranted on the facts.

The Court applied *Missouri v. Illinois* in 1931, when it dismissed Connecticut's attempt to prevent a Massachusetts transbasin diversion planned to benefit Boston.‡ Connecticut relied on the strict common law rule that all uses outside of the watershed are per se unreasonable, but the Court found at least three reasons to dismiss the action. Connecticut, the lower riparian state, failed to prove any injury, and thus the case arguably fell within the more "modern" common law rule that only transwatershed diversions that actually cause injury to downstream riparians are actionable.

Ripeness is not an absolute barrier to an equitable apportionment, because states have interests beyond those of private parties. State police power over water resources extends to the protection of future allocation options. A state suing *parens patriae* is not simply another rightholder claiming that the proposed diversion will interfere with the exercise of a prior right. The police power encompasses the power to anticipate the risks to existing and future in-state users presented by out-of-state diversions, and to try to minimize them before actual harm occurs by asserting that the diversion exceeds the diverter's fair share of the river. This must be balanced against the competing demands of other states, because if this logic were carried to its conclusion, downstream states would have a virtual veto over an upstream state's diversions. This would destroy the principle of equality among states announced in *Kansas v. Colorado*. Justice William O. Douglas' dissenting opinion in *Nebraska v. Wyoming*§ illustrates the liberal view of ripeness. Justice Douglas took the occasion to articulate his view, over a strong dissent, that if all claims, perfected or not, on a stream exceed the dependable flow, then a conflict exists and injury should be presumed:

> What we have then is a situation where three States assert against a river, whose dependable natural flow during the irrigation season has long been over-appropriated, claims based not only on present uses but on projected additional uses as well. The various statistics with which the record abounds are inconclusive in showing the existence or extent of actual damage to Nebraska. But we know that deprivation of water in arid

---

* 200 U.S. 496 (1906).
† *Id.* at 521.
‡ *Connecticut v. Massachusetts,* 282 U.S. 660 (1931).
§ 325 U.S. 589 (1945).

or semi-arid regions cannot help but be injurious. If this were an equity suit to enjoin threatened injury, the showing made by Nebraska might possibly be insufficient. But *Wyoming v. Colorado* . . . indicates that where the claims to the water of a river exceed the supply a controversy exists appropriate for judicial determination.*

Justice Douglas' definition of ripeness is a limited extension of the concept. It would not help any state seeking an equitable apportionment to fend off future diversion proposals. There would have to be, at a minimum, a diversion project that posed some immediate threat to the interests of other states.

5.5.2.1 *Apportionment Standards.*   Once the Court accepts original jurisdiction and appoints a master to take the evidence, the issue becomes what law to apply. The Court initially rejected local (state) law as the basis for an apportionment, then accepted it as *the* apportionment basis among states that followed the same law, and finally downgraded local law to a "guiding principle." Fair allocation rather than consistency with locally generated expectations ultimately became the touchstone of equitable apportionment. Although the Court has never been very precise about the source of the law of equitable apportionment, its early decision makes it clear that the grant of original jurisdiction requires a federal common law and a federal statutory law to prevent one state from using its law to gain an unfair advantage over another.[†] The use of local law as a basis for allocation is thus not compelled by the Constitution, and it might be unconstitutional to rely exclusively on local law. Local law may, however, serve as a source of principles to apply since a federal common law must, of course, examine the most relevant sources of substantive law, and in our federal system that will generally be state law.[‡] In end, local water is the reliable guide to the apportionment standards that the Court will follow.

In the 1920s and 1930s, the Court seemed to adopt a simple rule or set of rules: Appropriation among prior appropriation jurisdictions, and riparian rights among riparian jurisdictions.[§] But this was an overly simplistic assumption, in light of the decisions that the Court was rendering. The cases consistently made it clear that the Court was not bound by local law, and could refuse to apply local laws in situations where it determined that other states would be deprived of water to protect excess claims. The Court first refused to follow the strict common law rules of riparian rights in situations where proposed consumptive uses would have been flatly prohibited. For example, in *Connecticut* it refused to adopt the strict common law of riparian rights as the sole basis for an equitable apportionment when it impedes the fair distribution of resources.[¶] The restrictive rules of water usage inherent in the common law violate that principle of equality of access which the Court has consistently tried to follow. Finally, in *Nebraska v. Wyoming,*** the Court adopted the current open-ended standard of equitable apportionment:

> So far as possible those established uses should be protected though strict application of the priority rule might jeopardize them. Apportionment calls for the exercise for an

---

* *Id.* at 618.

[†] See *Wyoming v. Colorado*, 259 U.S. 419 (1922).

[‡] See Tarlock, *supra* note 28, at 394–402.

[§] See *Wyoming v. Colorado*, 259 U.S. 419 (1922); *New Jersey v. New York*, 283 U.S. 336 (1931).

[¶] Four years later, the Court applied its high standards of injury to a familiar western water law doctrine, and dismissed a suit by Washington against Oregon because the former's call would be futile. *Washington v. Oregon*, 297 U.S. 517 (1936).

** 32 U.S. 589 (1945).

informed judgment of consideration of many factors. Priority of appropriation is the guiding principle. But physical and climatic conditions, the consumptive use of water in the several sections of the river, the character and rate of return flows, the extent of established uses, the availability of storage water, the practical effect of wasteful uses on downstream areas, the damage to upstream areas as compared to the benefits to downstream areas if a limitation is imposed on the former—these are all relevant factors.*

The doctrine seems to contain the root principle that "a State may not preserve solely for its own inhabitants natural resources located within its own borders."[†] In short, the Supreme Court has now linked the policies underlying equitable apportionment to those underlying the dormant, or negative commerce clause.

The Court has recently announced a decision in an equitable apportionment case that has significant implications for state efforts to prevent water raids.[‡] The opinion can be read as a reaffirmation of *Nebraska v. Wyoming* or as an indication that the Court is willing to break new ground to increase the conditions that must be met before a state can refuse to share common resources. *Colorado v. New Mexico* arose from an attempt by the state of Colorado to bump the priorities of a marginal irrigation district in New Mexico on an interstate stream to benefit a new industrial development upstream in Colorado. The special master rejected a strict application of prior appropriation and balanced the equities in favor of the new, upstream use. In its initial opinion, the Court agreed with the master in theory, and added a new standard to the law of equitable apportionment. If the upstream state can demonstrate by clear and convincing evidence that the equities favor bumping existing uses, it may be able to prevail, because "an important consideration is whether the existing users could offset the diversion by reasonable conservation measures to prevent waste."[§] The case was remanded to the master for more specific findings to determine if the new conservation duty should be invoked.

The requirement that a state take reasonable conservation measures to preserve its priority has not historically been part of either the law of prior appropriation or riparian rights in any meaningful way. Many argue that the law should encourage conservation, and *Colorado v. New Mexico I* broke new ground by applying the argument that conservation duties must be incorporated into the law of equitable apportionment. The radical nature of the decision was apparently too much for a majority of the Court.[¶] *Colorado v. New Mexico II* returned to the rule of *Wyoming v. Colorado,* and protected New Mexico's priority. Writing for the majority, Justice O'Connor concluded that Colorado had failed to demonstrate by clear and convincing evidence that the diversion should be allowed. The Court had announced this standard in the previous opinion, but Justice O'Connor tightened the standard by requiring a guarantee that the diverting state will bare all of the risks of an erroneous diversion. On one level, the Court's placement of the burden of bumping existing uses of the initiating state returns the law of equitable apportionment to pre–*Nebraska v. Wyoming* standards. On another level, however, the decision opens up new possibilities for states wanting to claim water for new uses and places new risks on states seeking to protect existing uses. *Colorado v. New Mexico II* did not reject the possibility that the conservation duty imposed in *Colorado v. New Mexico I* might be applied in more appropriate cases. In the end, Colorado seems to have

---

* *Id.* at 618.
[†] *Idaho ex rel Evans v. Oregon,* 462 U.S. 1017, 1023 (1983).
[‡] *Colorado v. New Mexico,* 459 U.S. 176 (1982) [hereinafter cited as *Colorado v. New Mexico I*].
[§] *Id.* at 188.
[¶] See *Colorado v. New Mexico,* 463 U.S. 1204 (1984) [hereinafter cited as *Colorado v. New Mexico II*].

lost because it had not done the comprehensive planning necessary to justify the new use. The lesson of the cases is that states must undertake extensive studies to justify the water uses claimed in an equitable apportionment action and must make some showing that they are taking steps to manage their sources efficiently. As previously mentioned, the Court has recently joined the law of equitable apportionment with the policies underlying the dormant or negative commerce clause. The end result is that states may not only have to justify the need for the water, but now must show that they have, and are, taking reasonable steps to allocate available supplies fairly and efficiently.

The law of equitable apportionment has been supplemented by the use of interstate compacts to allocate waters and the power of Congress to apportion interstate streams. The ripeness barrier and the vague quantification standards have led the Court to express a preference for interstate compacts, agreements among states, rather than adjudications as means of sharing interstate waters. Interstate compacts may allocate waters in advance of an actual conflict among competing states and the states are not bound by any particular sharing formula. For example, in 1922 the Upper Basin states entered into the Colorado River Compact to reserve a share of the river against the Lower Basin, and thus precluded California, and to a lesser extent Arizona, from preempting future upstream uses by perfecting rights based on actual use. Each basin was given the right to 7.5 acre-ft per year. As does the doctrine of equitable apportionment, interstate compacts create state water rights. Compact rights are an inherent limitation on the use and enjoyment of all water rights. Thus, a state may be required to curtail its own users to satisfy its compact obligations to another state.* Originally, states assumed that equitable apportionment or compact shares were dedicated to use within the states, but uneven growth rates challenge this assumption. There is an increasing interest in interstate water marketing, including the sale or lease of a state's unused share to users in another state.

Congress may also directly apportion interstate rivers. The source of this power is the commerce clause, and Congress' power was confirmed in *Arizona v. California.*[†] The Court held the Boulder Canyon Project Act, which authorized the Lower Colorado River Storage Project, apportionment of the Lower Basin's share among Arizona, California, and Nevada, and delegated to the Secretary of Interior the power to decide how to allocate the states' shares in times of shortage. There is substantial doubt that Congress did in fact intend to supercede state law, and this expansive power has seldom been invoked by Congress. When Congress has apportioned an interstate stream, it has generally ratified an agreement worked out by the states and other interested users.[‡]

REFERENCES

Apolinsky, "The Development of Riparian Law in Alabama," 12 Ala. L. Rev. 155, 170 (1959).

Butler, Lynda, "Allocating Consumptive Water Rights in a Riparian Jurisdiction: Defining the Relationship Between Public and Private Interests," 47 U.Pitt. L. Rev. 95 (1987).

---

* *Hinderlider v. La Plata River and Cherry Creek Ditch Co.,* 304 U.S. 92 (1938).

[†] 373 U.S. 546 (1963). See generally Charles J. Meyers, "The Colorado River," 19 Stan. L, Rev. 1 (1966).

[‡] e.g., Truckee-Carson-Pyramid Lake Water Rights Settlement Act, Pub.L. No. 101-618, 104 Stat. 3294 (1990).

Dewsnup, R., and D. Jensen, eds., *A Summary-Digest of State Water Laws 32,* National Water Commission, 1974.

DuMars, Charles T., and Michele Minnis, "New Mexico Water Law: Determining Public Welfare Values in Water Rights Allocation," 31 Ariz. L. Rev. 817 (1989).

Freyfogle, "*Lux v. Haggin* and the Common Law Burdens of Water Law," 57 U. Colo. L. Rev. 485 (1986).

Getches, David, Lawrence MacDonnell, and Teresa Rice, *Controlling Water Use: The Unfinished Business of Water Quality Protection* (1991).

Gould, J., *Treatise on the Law of Waters,* (3d ed.), § 148, 1900.

Harnsberger, "Eminent Domain and Water Law," 48 Neb. L. Rev. 325, 366–69 (1969).

Hays, Samuel P., *Conservation and the Gospel of Efficiency: The Progressive Conservation Movement 1890–1920,* Harvard Univ. Press, 1959.

Holmes, Beatrice H., *A History of Federal Water Resources Programs, 1800–1960,* U.S. Department of Agriculture, Misc. Pub. 1233, 1972.

Horwitz, M., *The Transformation of American Law, 1786–1860* (1977).

MacDonnell, Lawrence J., "Transferring Water Uses in the West," 43 Okla. L. Rev. 119 (1990).

Royster, Judith V., "A Primer on Indian Water Rights: More Questions Than Answers," 30 Tulsa L.J. 61 (1994).

Tarlock, A. Dan, *The Law of Equitable Apportionment Revisited, Updated and Restated,* 56 Colo. L. Rev. 381 (1985).

Wahl, Richard W., *Markets for Federal Water: Subsidies, Property Rights, and the Bureau of Reclamation,* Resources for the Future, Inc., 1989.

Ziegler, "Acquisition and Protection of Water Supplies by Municipalities," 57 Mich. L. Rev. 349 (1959).

CHAPTER 6
SYSTEMS ANALYSIS

Larry W. Mays
Department of Civil and Environmental Engineering
Arizona State University
Tempe, Arizona

Yeou-Koung Tung
Wyoming Water Research Center and Statistics Department
University of Wyoming
Laramie, Wyoming

6.1 SYSTEMS CONCEPT

6.1.1 What Is Optimization?

The major types of water resources problems that must be solved for various hydrosystems include (Buras, 1972):

1. Determination of the optimal scale of project development
2. Determination of the optimal dimensions of system components
3. Determination of the optimal operation of the system

Water resource problems deal with both design and analysis. *Analysis* is concerned with determining the behavior of an existing system or a trial system that is being designed. In many cases, study of the system behavior is to determine operation of the system or the response of a system under specified inputs. The *design problem* is to determine the sizes of system components. As an example, the design of a reservoir system determines the size and location of reservoirs. Analysis of a reservoir system is the process of determining operation policies for the reservoir system. Operation of the reservoir system is required to check the design. In other words, a design is formulated and followed by an analysis to see if it performs according to specifications. If a design satisfies the specifications, then an acceptable design is found. New designs can be formulated and then analyzed.

Conventional procedures for design and analysis are iterative trial and error. The effectiveness of conventional procedures are dependent upon an engineer's intuition, experience, skill, and knowledge of the hydrosystem under investigation.

Therefore, conventional procedures are highly related to the human element which could lead to inefficient design and analysis of complex systems. Conventional procedures are typically based on simulation models in a trial and error process. Sometimes a simulation model is applied iteratively in an attempt to arrive at an optimal solution.

Optimization procedures eliminate the trial and error process of changing a design and resimulating with each new design change. Instead, an optimization model automatically changes the design parameters. An optimization procedure has mathematical expressions that describe the system and its response to the system inputs for various design parameters. These mathematical expressions are constraints in the optimization model. In addition, constraints are used to define the limits of the design variables, and performance is evaluated through the use of objective functions which could be to minimize cost.

An advantage of the conventional process is that the engineer's experience and intuition are used in making changes in the system or to change or make additional specifications. But the conventional procedure can lead to nonoptimal, or uneconomical, designs and operation policies. Also, the conventional procedure can be very time consuming and labor intensive. An optimization procedure requires one to explicitly identify the design variables, the objective function or measure of system performance to be optimized, and the constraints for the system. In contrast to the conventional procedure, the optimization procedure is more organized, using appropriate mathematical approaches to select the decisions.

An optimization problem in water resources may be formulated in a general framework with an objective function to

$$\text{Optimize } f(\mathbf{x}) \qquad\qquad (6.1)$$

subject to constraints

$$\mathbf{g}(\mathbf{x}) = \mathbf{0} \qquad\qquad (6.2)$$

and bound constraints on the decision variables

$$\underline{\mathbf{x}} < \mathbf{x} < \overline{\mathbf{x}} \qquad\qquad (6.3)$$

where $\mathbf{x}$ is a vector of n decision variables $(x_1, x_2, \ldots, x_n)$, $\mathbf{g}(\mathbf{x})$ is a vector of m equations called constraints, and $\underline{\mathbf{x}}$ and $\overline{\mathbf{x}}$ represent vectors of the lower and upper bounds, respectively, on the decision variables. In general, the objective equation (6.1) is to be maximized or minimized. Maximizing $f(\mathbf{x})$ is equivalent to minimizing $-f(\mathbf{x})$ or vice versa.

Every optimization problem has two essential parts: the objective function and the set of constraints. The *objective function* describes the performance criteria of the system. *Constraints* describe the system or process that is being designed or analyzed, which can be of two forms: equality constraints and inequality constraints. A *feasible solution* to an optimization problem is a set of values of the decision variables that simultaneously satisfies the constraints. The *feasible region* is the region of feasible solutions defined by the constraints. An *optimal solution* is a set of values of the decision variables that satisfies the constraints and provides an optimal value of the objective function.

Depending upon the nature of the objective function and the constraints, an optimization problem can be classified as: (1) linear vs. nonlinear, (2) deterministic vs. probabilistic, (3) static vs. dynamic, (4) continuous vs. discrete, and (5) lumped parameter vs. distributed parameter. *Linear programming* problems consist of a linear objective function where all constraints are linear. *Nonlinear programming* problems have nonlinear constraints and/or the objective function. *Deterministic* problems consist of parameters that can be assigned fixed values, whereas *probabilistic* problems consist of parameters that are considered as random variables. *Static* problems do not explicitly consider the variable time aspect, whereas *dynamic* problems do consider variable time. Static problems are referred to as *mathematical programming* problems, and dynamic problems are often referred to as *optimal control* problems, which involve difference or differential equations. *Continuous* problems have variables that can take on continuous values, whereas with *discrete* problems, the variables must take on discrete values. Typically, discrete problems are posed as *integer programming* problems, in which the variables must be integer values. A *lumped* problem considers the parameters and variables to be homogeneous throughout the system, whereas *distributed* problems must take into account detailed variations of system behavior from one location to another.

The method of optimization used depends on: (1) the type of objective function, (2) the type of constraints, and (3) the number of decision variables. Table 6.1 lists some general steps to solve an optimization problem. Some problems may not require that the engineer follow the steps in the exact order, but each step should be considered in the process. The overall objective in optimization is to determine a set of values of the decision variables that satisfies the constraints and provides the optimal response to the objective function.

6.1.2 Single-Objective versus Multiple-Objective Optimization

The solutions to a growing number of problems facing water resource professionals today are becoming more complex. In most water resource problems, the decision-making process is cultivated by the desire to achieve several goals simultaneously, and many of them could be *noncommensurate* and conflicting. In such circumstances, improvement of some objectives cannot be obtained without the sacrifice of others. Therefore, the ideological theme of optimality in the single-objective context

TABLE 6.1 Six Steps Used to Solve Optimization Problems

---

1. Analyze the process itself so that the process variables and specific characteristics of interest are defined, i.e., make a list of all of the variables.

2. Determine the criterion for optimization and specify the objective function in terms of the above variables together with coefficients. This step provides the performance model (sometimes called the economic model when appropriate).

3. Develop, via mathematical expressions, a valid process or equipment model that relates the input-output variables of the process and associated coefficients. Include both equality and inequality constraints. Use well-known physical principles (mass balances, energy balances), empirical relations, implicit concepts, and external restrictions. Identify the independent and dependent variables to get the number of degrees of freedom.

4. If the problem formulation is too large in scope:
   a. Break it up into manageable parts, or
   b. Simplify the objective function and model

5. Apply a suitable optimization technique to the mathematical statement of the problem.

6. Check the answers, and examine the sensitivity of the result to changes in the problem coefficients and the assumptions.

Source: Edgar and Himmelblau, 1988.

is no longer appropriate. Instead, the goal of optimality in the single-objective framework is replaced by the concept of *noninferiority* in the multiple-objective analysis. More discussion of the multiobjective problem is given in Sec. 6.7.

6.2 LINEAR PROGRAMMING (LP)

Linear programming models have been extensively applied to optimal resource allocation problems. As the name implies, LP models have two basic characteristics; i.e., both the objective function and constraints are linear functions of the decision variables. The general form of an LP model can be expressed as

$$\text{Max (or min)} \quad x_o = \sum_{j=1}^{n} c_j x_j \tag{6.4}$$

subject to

$$\sum_{j=1}^{n} a_{ij} x_j = b_i, \quad \text{for} \quad i = 1, 2, \ldots, m \tag{6.5}$$

$$x_j \geq 0, \quad \text{for} \quad j = 1, 2, \ldots, n \tag{6.6}$$

where the c_j's are the *objective function coefficients*, the a_{ij}'s are the *technological coefficients*, and the b_i's are the right-hand side (RHS) coefficients. In matrix form, an LP model can be concisely expressed as

$$\text{max (or min)} \quad x_o = \mathbf{c}^T \mathbf{x} \tag{6.7}$$

subject to

$$\mathbf{A}\mathbf{x} \leq \mathbf{b} \tag{6.8}$$

$$\mathbf{x} \geq \mathbf{0} \tag{6.9}$$

where $\mathbf{c}$ is an n by 1 column vector of objective function coefficients, $\mathbf{x}$ is an n by 1 column vector of decision variables, $\mathbf{A}$ is an m by n matrix of technological coefficients, $\mathbf{b}$ is an m by 1 column vector of the RHS coefficients, and the superscript T represents the transpose of a matrix or a vector. Excellent textbooks on LP include: Gass (1985), Hillier and Lieberman (1990), Taha (1987), and Winston (1987).

Four basic assumptions are implicitly built into LP models:

1. *Proportionality Assumption.*   This implies that the contribution of the jth decision variable to the system effectiveness $c_j x_j$ and its usage of the various resources $a_{ij} x_j$ are directly proportional to the value of the decision variable.

2. *Additivity Assumption.*   This assumption means that, at a given level of activity $(x_1, x_2, \ldots, x_n)$, the total usage of resources and contribution to the overall system effectiveness are equal to the sum of the corresponding quantities generated by each activity conducted by itself.

3. *Divisibility Assumption.*   Activity units can be divided into any fractional level, so that noninteger values for the decision variables are permissible.

4. *Deterministic Assumption.*   All parameters in the model are known without uncertainty. The effect of parameter uncertainty on the result can be investigated by conducting sensitivity analysis.

6.2.1 Simplex Method

A well-known algorithm for solving an LP problem is the *simplex method* which consists of the following three steps:

1. *Initialization Step.*   Start at a feasible extreme point.

2. *Iterative Step.*   Move to a better adjacent feasible extreme point.

3. *Stopping Rule.*   Stop iterations when the current feasible extreme point is better than all its adjacent extreme points.

In linear algebra, the system of equations is called *indeterminant* if there are more unknowns than equations. For problems with n unknowns and m equations, where $n > m$, there is no unique solution to the problem. The possible solutions to an indeterminant system of equations can be obtained by letting $(n - m)$ unknowns equal zero and solving for the remaining m unknowns. The solutions so obtained are called the *basic solutions*. In theory, there will be, at maximum, a total of $n!/(m!(n - m)!)$ basic solutions to such problems, if all exist. Geometrically, each basic solution represents the intersection point of each pair of constraint equations, an *extreme point*.

The $(n - m)$ decision variables set to zero are called the *nonbasic variables* while the remaining m decision variables, whose values are obtained by solving a system of m equations with m unknowns, are called the *basic variables*. The solution of a basic variable that is feasible is called the *basic feasible solution* corresponding to a feasible extreme point in the feasible space.

The simplex method solves an LP model by exploiting the feasible extreme points. The algorithm searches for the optimum of an LP model by abiding by two fundamental conditions: (1) the *optimality condition* and (2) the *feasibility condition*. The *optimality condition* ensures that no inferior solutions (relative to the current solution point) are ever encountered. The *feasibility condition* guarantees that, starting with a basic feasible solution, only basic feasible solutions are enumerated during the course of computation.

6.2.2 New Methods of Solution

The simplex algorithm searches the optimality of an LP problem by examining the feasible extreme points. The path of optimum seeking is made along the boundary of the feasible region. The algorithm, since its conception (Dantzig, 1963), has been widely used and considered as the most efficient approach for solving LP problems. Only recently have two new methods for solving LP problems been developed, which have an entirely different algorithmic philosophy (Khatchian, 1979; Karmarkar, 1984).

In contrast to the simplex algorithm, *Khatchian's ellipsoid method* and *Karmarkar's projective scaling method* seek the optimum solution to an LP problem by moving through the interior of the feasible region. A schematic diagram illustrating the algorithmic differences between the simplex and the two new algorithms is shown in Fig. 6.1. Khatchian's ellipsoid method approximates the optimum solution of an LP problem by creating a sequence of ellipsoids (an ellipsoid is the multidimensional analog of an ellipse) that approach the optimal solution. Both Khatchian's method and Karmarkar's method have been shown to be *polynomial time algorithms*. This means that the time required to solve an LP problem of size n by the two new methods would take at most αn^β where α and β are two positive numbers.

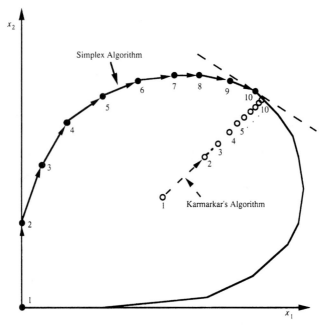

FIGURE 6.1   Search path for simplex algorithm and Karmarkar's algorithm. *(Mays and Tung, 1992.)*

On the other hand, the simplex algorithm is an *exponential time algorithm* in solving LP problems. This implies that, in solving an LP problem of size n by the simplex algorithm, there exists a positive number γ such that for any n the simplex algorithm would find its solution in a time of at most $\gamma 2^n$. For a large enough n (with positive α, β, and γ), $\gamma 2^n > \alpha n^\beta$. This means that, in theory, the polynomial time algorithms are superior to exponential time algorithms for large LP problems.

6.2.3   Integer Programming

Many water resource engineering problems involve decision variables that must be integer-valued. Typical examples are those representing a discrete number of objects, such as pumps and culvert barrels, and those representing quantification of codes, such as whether to accept or reject a particular water development alternative or alternatives. Mathematical model formulation for an integer programming problem is identical to Eqs. (6.7) to (6.9) for *integer linear program* or to Eqs. (6.17a–c) for *integer nonlinear program*. An integer program can be *mixed* or *pure*, depending on whether some or all its decision variables are restricted to be integer-valued.

Several algorithms have been developed for solving integer programming problems, and the ones that are mature in commercial application are those for integer linear programming problems. However, unlike the linear programming algorithms, the computation performance of integer programming algorithms is not uniformly efficient, and is erratic (Taha, 1987). This is especially true when the size of an integer programming problem gets larger. The main reason is the fact that the feasible space of an integer programming problem is the effect of round-off error resulting from the use of the digital computer for problem solving. Reference books on integer programming optimization can be found elsewhere (Garfinkel and Nemhauser, 1972; Salkin, 1975; Taha, 1975; Nemhauser and Wolsey, 1988).

Methods for solving an integer programming problem can be categorized as *cutting methods* and *search methods*. The cutting methods, developed by Gomory (Taha, 1995), are primarily for integer programming problems which include the *fractional algorithm* for the pure integer programs and the *mixed algorithm* for mixed integer programs. In essence, the methods solve an integer programming problem by first treating all integer variables as if they are continuous variables. If all the integer variables are integer-valued, the optimal solutions are found. Otherwise, a sequence of secondary constraints is added systematically to the original problem until the resulting optimal solution satisfies the integer conditions. The added secondary constraints, in effect, cut off portions of the original feasible space that do not contain feasible integer points.

The search methods originate from the idea to enumerate all feasible integer solutions. By incorporating some efficient and smart tests, the required number of enumerations to reach the optimum can be reduced to a minimum. Among the search methods, the *branch-and-bound algorithm,* originally developed by Land and Doig (1960), is the most widely used. This algorithm, unlike the cutting methods, applies directly to both pure and mixed integer programming problems. It also starts solving the problem by ignoring the restriction of integer. For those integer-constrained variables with fractional values, one of them is selected to branch the original problem into two subproblems which are subsequently solved. The branching process eliminates the feasible space that does not contain feasible integers. The process can be repeated until the integrality of all integer-constrained variables is satisfied. However, the process can be very time consuming and impractical. As a result, the concept of bounding is introduced based on the concept that, if the continuous optimum solution of a subproblem yields a worse objective function value than the one associated with the best available integer solution, it does not deserve further exploration. In this case the subproblem is said to be *fathomed* and may henceforth be deleted. The computation efficiency of the branch-and-bound algorithm depends on the selection of good branching variables and the sequence in which the subproblem is evaluated. Unfortunately, there is no definite optimal rule for this.

6.2.4 Network Methods

Many water resource planning and management problems are concerned with allocating water among various locations at different times. Some such allocations are directional, as dictated by flow direction. Figure 6.2 is a typical example in which the system consists of several reservoirs connected by a network of channels. Problems of this nature can often be represented by a network involving nodes (reservoirs) and arcs (channels). Decision variables are the releases from the reservoirs at different times. Network problems can be categorized into one of four models: (1) network minimization model, (2) shortest-route model, (3) maximum-flow model, and (4) minimum-cost capacitated network model (Taha, 1987; Hillier and Lieberman, 1990).

Constraints in the mathematical programming model for a network problem are largely made up of the flow continuity equations. Therefore, the constraint matrix is generally very sparse, and its elements are either +1 or −1. Although most network

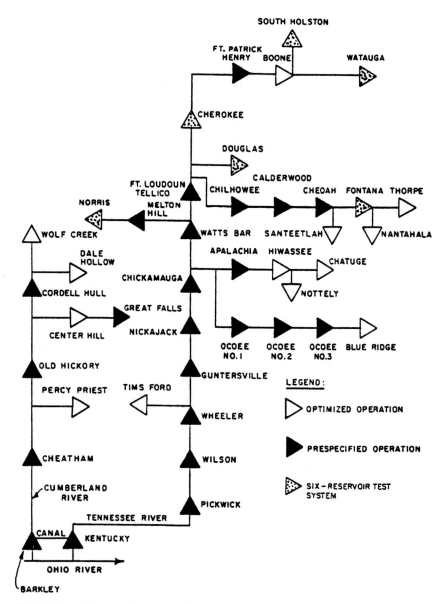

FIGURE 6.2   Schematic diagram of multireservoir system. *(Wunderlich, 1991.)*

problems can be formulated as linear programming problems, they are not solved by
the linear programming algorithms. Instead, highly efficient algorithms have been
developed to take advantage of the special structure of network optimization models (Jensen and Barnes, 1980). Nonlinearity in network optimization presents a challenge ahead.

6.3 NONLINEAR PROGRAMMING

6.3.1 Unconstrained Nonlinear Optimization

This section describes the basic concepts of unconstrained nonlinear optimization, including the necessary and sufficient conditions of a local optimum. Further, unconstrained optimization techniques for univariate and multivariate problems are described. Understanding unconstrained optimization procedures is important because these techniques are the fundamental building blocks in many of the constrained nonlinear optimization algorithms.

The problem of *unconstrained minimization* can be stated as

$$\min_{\mathbf{x} \in E^n} \; f(\mathbf{x}) \tag{6.10}$$

in which $\mathbf{x}$ is a vector of n decision variables $\mathbf{x} = (x_1, x_2, \ldots, x_n)^T$ defined over the entire Euclidean space E^n. Since the feasible region is infinitely extended without bound, the optimization problem does not contain any constraints.

Assume that $f(\mathbf{x})$ is a nonlinear function and twice differentiable; it could be convex, concave, or a mixture of the two over E^n. In the one-dimensional case, the objective function $f(\mathbf{x})$ could behave as Fig. 6.3, consisting of peaks, valleys, and inflection

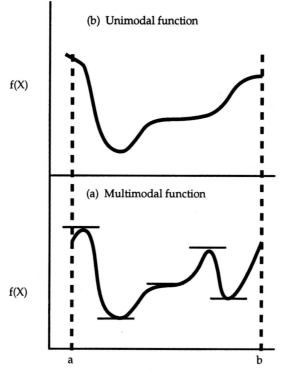

FIGURE 6.3 Definition of unimodal functions.

points. The *necessary conditions* for a solution to Eq. (6.10) at an optimal solution $\mathbf{x}^*$ are: (1) the gradient vector $\nabla \mathbf{f}(\mathbf{x}^*) = \mathbf{0}$ and (2) the Hessian matrix $\nabla^2 \mathbf{f}(\mathbf{x}^*) = \mathbf{H}(\mathbf{x}^*)$ is strictly positive definite. The *sufficient conditions* for an unconstrained minimum are: (1) $\nabla \mathbf{f}(\mathbf{x}^*) = \mathbf{0}$ and (2) $\nabla^2 \mathbf{f}(\mathbf{x}^*) = \mathbf{H}(\mathbf{x}^*)$ is strictly positive definite.

In theory, the solution to Eq. (6.10) can be obtained by solving the following system of n nonlinear equations with n unknowns,

$$\nabla \mathbf{f}(\mathbf{x}^*) = \mathbf{0} \tag{6.11}$$

The approach has been viewed as indirect in the sense that it backs away from the original problem of minimizing $\mathbf{f}(\mathbf{x})$. Furthermore, an iterative numerical procedure is required to solve the system of nonlinear equations which tends to be computationally inefficient.

By contrast, the preference is given to those solution procedures which directly attack the problem of minimizing $\mathbf{f}(\mathbf{x})$. Direct solution methods, during the course of iteration, generate a sequence of solution points in E^n that terminate or converge to a solution to Eq. (6.10). Such methods can be characterized as *search procedures*.

In general, all search algorithms for unconstrained minimization consist of two basic steps. The first step is to determine the *search direction* along which the objective function value decreases. The second step is called a *line search* (or *one dimensional search*) to obtain the optimum solution point along the search direction determined by the first step. Mathematically, minimization for the line search can be stated as

$$\min_{\beta} f(\mathbf{x}^0 + \beta \mathbf{d}) \tag{6.12}$$

in which $\mathbf{x}^0$ is the current solution point, $\mathbf{d}$ is the vector indicating the search direction, and β is a scalar, $0 < \beta < \infty$, representing the step size whose optimal value is to be determined. There are many search algorithms whose differences primarily lie in the way the search direction $\mathbf{d}$ is determined.

Due to the very nature of search algorithms, it is likely that different starting solutions might converge to different local minima. Hence, there is no guarantee of finding the global minimum by any search technique applied to solve Eq. (6.10) unless the objective function is a convex function over E^n.

In implementing search techniques, specification of convergence criteria or stopping rules is an important element that affects the performance of the algorithm and the accuracy of the solution. Several commonly used stopping rules in an optimum seeking algorithm are

$$\|\mathbf{x}^k - \mathbf{x}^{k+1}\| < \epsilon_1 \tag{6.13a}$$

$$\frac{\|\mathbf{x}^k - \mathbf{x}^{k+1}\|}{\|\mathbf{x}^k\|} < \epsilon_2 \tag{6.13b}$$

$$|f(\mathbf{x}^k) - f(\mathbf{x}^{k+1})| < \epsilon_3 \tag{6.13c}$$

$$\left| \frac{f(\mathbf{x}^k) - f(\mathbf{x}^{k+1})}{f(\mathbf{x}^k)} \right| < \epsilon_4 \tag{6.13d}$$

in which superscript k is the index for iteration; ϵ represents the tolerance or accuracy requirement; $\|\mathbf{x}\|$ is the length of the vector $\mathbf{x}$; and $\|$ is the absolute value. The specification of the tolerance depends on the nature of the problem and on the accu-

racy requirement. Too small a value of ϵ (corresponding to a high accuracy requirement) could result in excessive iterations, wasting computer time. On the other
hand, too large a value of ϵ could make the algorithm terminate prematurely at a
nonoptimal solution.

Referring to Eq. (6.10), the solution of these types of problems can be stated in
an algorithm involving the following basic steps:

Step 0   Select an initial starting point $\mathbf{x}^{k=0} = (x_0^1, x_0^2, \ldots, x_0^n)$.

Step 1   Determine a search direction, $\mathbf{d}^k$.

Step 2   Find a new point $\mathbf{x}^{k+1} = \mathbf{x}^k + \beta^k \mathbf{d}^k$ where β^k is the step size, a scalar, which
          minimizes $f(\mathbf{x}^k + \beta^k \mathbf{d}^k)$.

Step 3   Check convergence criteria such as Eq. (6.13a–d) for termination; if not
          satisfied, set $k = k + 1$ and return to Step 1.

Various unconstrained multivariate methods differ in the way the search directions are determined. The recursive line search for an unconstrained minimization
problem in Step 2 above is expressed as

$$\mathbf{x}^{k+1} = \mathbf{x}^k + \beta^k \mathbf{d}^k \tag{6.14}$$

Table 6.2 lists the equations for determining the search direction for four basic
groups of methods: Descent methods, conjugate direction methods, quasi-Newton
methods, and Newton's method. The simplest ones are the steepest descent methods,
while the Newton methods are the most computationally intensive.

In the *steepest descent method* the search direction is $-\nabla \mathbf{f}(\mathbf{x})$ because $\nabla \mathbf{f}(\mathbf{x})$ points
in the direction of the maximum rate of increase in objective function value, there-
fore a negative sign is associated with the gradient vector in Eq. (6.14) for a minimization problem. The recursive line search equation for the steepest descent
method, then, is written as

$$\mathbf{x}^{k+1} = \mathbf{x}^k - \beta^k \nabla \mathbf{f}(\mathbf{x}^k) \tag{6.15}$$

Using *Newton's method,* the recursive equation for a line search is

$$\mathbf{x}^{k+1} = \mathbf{x}^k - \mathbf{H}^{-1}(\mathbf{x}^k) \nabla \mathbf{f}(\mathbf{x}^k) \tag{6.16}$$

Although Newton's method converges faster than most other algorithms, the major
disadvantage is that it requires inverting the Hessian matrix in each iteration, which
is a cumbersome task, especially when the number of decision variables is large.

The *conjugate direction methods* and *quasi-Newton's methods* are intermediate
between the steepest descent and Newton's methods. The conjugate direction
methods are motivated by the need to accelerate the typically slow convergence of
the steepest descent method. Conjugate direction methods, as can be seen in Table
6.2, define the search direction by utilizing the gradient vector of the objective function of the current iteration and the information on the gradient and search direction of the previous iterations. The motivation of quasi-Newton methods is to avoid
inverting the Hessian matrix as required by Newton's method. These methods use
approximations to the inverse Hessian with a different form of approximation for
the different quasi-Newton methods. Detailed descriptions and theoretical development can be found in textbooks such as Luenberger (1984); Fletcher (1980);
Dennis and Schnable (1983); Gill, Murray, and Wright (1981); and Edgar and Himmelblau (1988).

TABLE 6.2 Computation of Search Directions*

Search Direction	Definition of Terms

Steepest Descent
$\mathbf{d}^{k+1} = -\nabla f(\mathbf{x}^{k+1})$

Conjugate gradient methods
(1) Fletcher-Reeves

$\mathbf{d}^{k+1} = -\nabla f(\mathbf{x}^{k+1}) + a_1\mathbf{d}^k$ $\qquad a_1 = \dfrac{\nabla^T f(\mathbf{x}^{k+1})\nabla f(\mathbf{x}^{k+1})}{\nabla^T f(\mathbf{x}^k)\nabla f(\mathbf{x}^k)}$

$\qquad\qquad\qquad\qquad\qquad\qquad\qquad \mathbf{d}^0 = -\nabla f(\mathbf{x}^0)$

(2) Polak-Ribiere

$\mathbf{d}^{k+1} = -\nabla f(\mathbf{x}^{k+1}) + a_2\mathbf{d}^k$ $\qquad a_2 = \dfrac{\nabla^T f(\mathbf{x}^{k+1})\mathbf{Y}^{k+1}}{\nabla^T f(\mathbf{x}^k)\nabla f(\mathbf{x}^k)}$

$\qquad\qquad\qquad\qquad\qquad\qquad\qquad \mathbf{Y}^{k+1} = \nabla f(\mathbf{x}^{k+1}) - \nabla f(\mathbf{x}^k)$

$\qquad\qquad\qquad\qquad\qquad\qquad\qquad \mathbf{d}^0 = -\nabla f(\mathbf{x}^0)$

(3) 1-step Broyden-Fletcher-Goldfard-Shanno (BFGS)

$\mathbf{d}^{k+1} = -\nabla f(\mathbf{x}^{k+1}) + a_3(a_4\mathbf{S}^{k+1} + a\mathbf{Y}^k)$ $\qquad a_3 = \dfrac{1}{(\mathbf{S}^{k+1})^T\mathbf{Y}^{k+1}}$

$$a_4 = -\left(1 + \frac{(\mathbf{Y}^{k+1})^T\mathbf{Y}^{k+1}}{(\mathbf{S}^{k+1})^T\mathbf{Y}^{k+1}}\right)(\mathbf{S}^{k+1})^T\nabla f(\mathbf{x}^{k+1})$$

$$+(\mathbf{Y}^{k+1})^T\nabla f(\mathbf{x}^{k+1})$$

$$a_5 = (\mathbf{S}^{k+1})^T\nabla f(\mathbf{x}^k)$$

$$\mathbf{S}^{k+1} = \mathbf{x}^{k+1} - \mathbf{x}^k$$

$$\mathbf{d}^0 = -\nabla f(\mathbf{x}^0)$$

Quasi-Newton methods
(1) Davidon-Fletcher-Powell (DFP) Method (variable metric method)[†]

$\mathbf{d}^{k+1} = \mathbf{G}^{k+1}\nabla f(\mathbf{x}^{k+1})$ $\qquad \mathbf{G}^{k+1} = \mathbf{G}^k + \dfrac{\mathbf{S}^k(\mathbf{S}^k)^T}{(\mathbf{S}^k)^T\mathbf{Y}^k} - \dfrac{\mathbf{G}^k\mathbf{Y}^k(\mathbf{G}^k\mathbf{Y}^k)^T}{(\mathbf{Y}^k)^T\mathbf{G}^k\mathbf{Y}^k}$

(2) Broyden-Fletcher-Goldfarb-Shanno (BFGS) method[†]

$\mathbf{d}^{k+1} = \mathbf{G}^{k+1}\nabla f(\mathbf{x}^{k+1})$ $\qquad \mathbf{G}^{k+1} = \mathbf{G}^k + \left(\dfrac{1 + (\mathbf{Y}^k)^T\mathbf{G}^k\mathbf{Y}^k}{(\mathbf{Y}^k)^T\mathbf{S}^k}\right)\dfrac{\mathbf{S}^k(\mathbf{S}^k)^T}{(\mathbf{S}^k)^T\mathbf{Y}^k}$

$$-\frac{\mathbf{Y}^k(\mathbf{S}^k)^T\mathbf{G}^k + \mathbf{G}^k\mathbf{S}^k(\mathbf{Y}^k)^T}{(\mathbf{Y}^k)^T\mathbf{S}^k}$$

(3) Broyden Family
$\mathbf{G}^\phi = (1 - \phi)\mathbf{G}^{\text{DFP}} + \phi\mathbf{G}^{\text{BFGS}}$

Newton method
$\mathbf{d}^{k+1} = -\mathbf{H}^{-1}(\mathbf{x}^k)\nabla f(\mathbf{x}^k)$

* Formulas for other search directions can be found in Luenberger (1984).
[†] $\mathbf{G}^\phi = \mathbf{I}$ (identity matrix).
Source: Mays and Tung, 1992.

6.3.2 Constrained Optimization: Optimality Conditions

Consider the general nonlinear programming (NLP) problem with the nonlinear objective:

$$\text{Minimize } f(\mathbf{x}) \tag{6.17a}$$

subject to

$$g_i(\mathbf{x}) = 0 \qquad i = 1, \ldots, m \tag{6.17b}$$

and

$$\underline{x}_i \leq x_i \leq \overline{x}_i \qquad j = 1, 2, \ldots, n \tag{6.17c}$$

in which Eq. (6.17c) is a bound constraint for the jth decision variable x_j with $\underline{x}_j$ and $\overline{x}_j$ being the lower and upper bounds, respectively.

Unlike an unconstrained problem, the feasible space of a constrained optimization problem is not infinitely extended. As a result, the solution that satisfies the optimality condition of the unconstrained optimization problem is not guaranteed to be feasible in constrained problems. In other words, a local optimum for a constrained problem might be located on the boundary or a corner of the feasible space at which the gradient vector is not equal to zero. Therefore, modifications to the optimality conditions for unconstrained problems are necessary.

The most important theoretical results for the nonlinear constrained optimization are the *Kuhn-Tucker conditions*. These conditions must be satisfied at any constrained optimum, local or global, of any LP and NLP problems. They form the basis for the development of many computational algorithms.

Without losing generality, consider an NLP problem stated by Eqs. (6.17a–b) with no boundary constraints. Note that constraint Eq. (6.17b) is all equality constraints. Under this condition, the *Lagrange multiplier method* converts a constrained NLP problem into an unconstrained one by developing an augmented objective function, called the *Lagrangian*. The *Lagrangian function* $L(\mathbf{x}, \boldsymbol{\lambda})$ is defined as

$$L(\mathbf{x}, \boldsymbol{\lambda}) = f(\mathbf{x}) + \boldsymbol{\lambda}^T \mathbf{g}(\mathbf{x}) \tag{6.18}$$

in which $\boldsymbol{\lambda}$ is the vector of *Lagrange multipliers* and $\mathbf{g}(\mathbf{x})$ is a vector of the constraint equation. Algebraically, Eq. (6.18) can be written

$$L(x_1, \ldots, x_n; \lambda_1, \ldots, \lambda_m) = f(x_1, \ldots, x_n) + \sum_{i=1}^{m} \lambda_i g_i(x_1, \ldots, x_n) \tag{6.19}$$

For a minimization problem, $L(\mathbf{x}, \boldsymbol{\lambda})$ is the objective function, $m + n$ variables, to be minimized. The necessary and sufficient conditions for the optimal solution, $\mathbf{x}^*$ are:

1. $f(\mathbf{x}^*)$ is convex and $g(\mathbf{x}^*)$ is convex in the vicinity of $\mathbf{x}^*$

2. $\dfrac{\partial L(\mathbf{x}^*)}{\partial x_j} = \dfrac{\partial f}{\partial x_j} + \sum_{i=1}^{m} \lambda_i \dfrac{\partial g_i}{\partial x_j} = 0 \qquad j = 1, \ldots, n \tag{6.20a}$

3. $\dfrac{\partial L}{\partial \lambda_i} = g_i(x) = 0 \qquad i = 1, \ldots, m \tag{6.20b}$

4. λ_i is unrestricted-in-sign $\qquad i = 1, \ldots, m \tag{6.20c}$

Solving equations (6.20a) and (6.20b) simultaneously provides the optimal solution.

Lagrange multipliers have an important interpretation in optimization problems. For a given constraint, the corresponding multiplier indicates the sensitivity of the optimal objective function value due to a unit change in the right-hand side of the constraint, that is,

$$\frac{\partial f}{\partial b_i}\bigg|_{\mathbf{x}\,=\,\mathbf{x}^*} = \lambda_i$$

The λ_i's are called *dual variables* or *shadow prices*.

6.3.2.1 Kuhn-Tucker Conditions.   Equations (6.20a–b) form the optimality conditions for an optimization problem involving only equality constraints. The Lagrange multipliers associated with the equality constraints are unrestricted-in-sign. Using the Lagrange multiplier method, the optimality conditions for the following generalized NLP problem can be derived.

$$\text{minimize } f(\mathbf{x})$$

subject to

$$g_i(\mathbf{x}) = 0 \qquad i = 1, \ldots, m$$

and

$$\underline{x}_j \le x_j \le \overline{x}_j \qquad j = 1, \ldots, n$$

In terms of the Lagrangian method, the above nonlinear minimization problem can be written as

$$\min L = f(\mathbf{x}) + \boldsymbol{\lambda}^T \mathbf{g}(\mathbf{x}) + \underline{\boldsymbol{\lambda}}^T(\mathbf{x} - \underline{\mathbf{x}}) + \overline{\boldsymbol{\lambda}}^T(\mathbf{x} - \overline{\mathbf{x}}) \tag{6.21}$$

in which $\boldsymbol{\lambda}$, $\underline{\boldsymbol{\lambda}}$, and $\overline{\boldsymbol{\lambda}}$ are vectors of Lagrange multipliers corresponding to constraints $\mathbf{g}(\mathbf{x}) = \mathbf{0}$, $\underline{\mathbf{x}} - \mathbf{x} \le \mathbf{0}$, and $\mathbf{x} - \overline{\mathbf{x}} \le \mathbf{0}$, respectively. The Kuhn-Tucker conditions for the optimality of the above problem are

$$\nabla_x L = \nabla_x f + \boldsymbol{\lambda}^T \nabla_x \mathbf{g} - \underline{\boldsymbol{\lambda}} + \overline{\boldsymbol{\lambda}} = 0 \tag{6.22a}$$

$$g_i(\mathbf{x}) = 0 \qquad i = 1, 2, \ldots, m \tag{6.22b}$$

$$\underline{\lambda}_j(\underline{x}_j - x_j) = \overline{\lambda}_j(x_j - \overline{x}_j) = 0 \qquad j = 1, 2, \ldots, n \tag{6.22c}$$

The m basic variables in theory can be expressed in terms of the $n - m$ nonbasic variables as $\mathbf{x}_B(\mathbf{x}_N)$. Assume that constraints $\mathbf{g}(\mathbf{x}) = \mathbf{0}$ are differentiable and the *m basis matrix* $\mathbf{B}$ can be obtained as

$$\mathbf{B} = \left[\frac{\partial \mathbf{g}(\mathbf{x})}{\partial \mathbf{x}_B} \right]$$

which is nonsingular such that there exists a unique solution of $\mathbf{x}_B(\mathbf{x}_N)$. *Nonsingular* means that the determinant of the basis matrix $\mathbf{B}$ is not equal to 0.

The objective called a *reduced objective* can be expressed in terms of the nonbasic variables as

$$F(\mathbf{x}_N) = f(\mathbf{x}_B(\mathbf{x}_N), \mathbf{x}_N) \tag{6.23}$$

The original NLP problem is transformed into the following *reduced problem*

$$\text{minimize } F(\mathbf{x}_N) \tag{6.24a}$$

subject to

$$\underline{\mathbf{x}}_N \le \mathbf{x}_N \le \overline{\mathbf{x}}_N \tag{6.24b}$$

which can be solved by an unconstrained minimization technique with slight modification to account for the bounds on nonbasic variables. *Generalized reduced gradient algorithms,* therefore, solve the original problem (6.17a–c) by solving a sequence of reduced problems (6.24a–b) using unconstrained minimization algorithms.

6.3.2.2 Generalized Reduced Gradient Methods.   Consider solving the reduced problem (6.24a–b) starting from an initial feasible point $\mathbf{x}^0$. To evaluate $F(\mathbf{x}_N)$ by Eq. (6.23), the values of the basic variables $\mathbf{x}_B$ must be known. Except for a very few cases, $\mathbf{x}_B(\mathbf{x}_N)$ cannot be determined in closed form; however, it can be computed for any $\mathbf{x}_N$ by an iterative method which solves a system of m nonlinear equations with the same number of unknowns as equations. A procedure for solving the reduced problem starting from the initial feasible solution $\mathbf{x}^{k\,=\,0}$ is

Step 0   Start with initial feasible solution $\mathbf{x}^{k\,=\,0}$ and set $\mathbf{x}_N^{\,k} = \mathbf{x}^{k\,=\,0}$

Step 1   Use $\mathbf{x}_N^k$ to determine the corresponding values of $\mathbf{x}_B$ by an iterative method for solving m nonlinear equations $\mathbf{g}(\mathbf{x}_B(\mathbf{x}_N^k), \mathbf{x}_N^k) = \mathbf{0}$.

Step 2   Determine the search direction $\mathbf{d}^k$ for the nonbasic variables by an uncontained minimization scheme.

Step 3   Choose a step size for the line search scheme, β^k, by solving the one-dimensional search problem

$$\text{minimize } F(\mathbf{x}_N^k + \beta \mathbf{d}^k)$$

with $\mathbf{x}$ restricted so that $\mathbf{x}_N^k + \beta \mathbf{d}^k$ satisfies the bounds on $\mathbf{x}_N$. This one-dimensional search requires repeated applications of Step 1 to evaluate F for the different β values.

Step 4   Test the current point $\mathbf{x}^k = (\mathbf{x}_B^k, \mathbf{x}_N^k)$ for optimality, if not optimal set $k = k + 1$ and return to Step 1.

6.3.2.3 Penalty Function Methods.   Because unconstrained problems are much easier to solve, the idea of penalty function methods is to transform constrained NLP problems into a sequence of unconstrained optimization problems by adding one or more functions of the constraints to the objective function and to delete the constraints. Using a penalty function, a constrained nonlinear programming problem is transformed to an unconstrained problem as

$$\left.\begin{array}{l}\text{minimize } f(\mathbf{x}) \\ \text{subject to } \mathbf{g}(\mathbf{x})\end{array}\right\} \Rightarrow \text{minimize } L[\,f(\mathbf{x}), \mathbf{g}(\mathbf{x})]$$

where $L[\,f(\mathbf{x}), \mathbf{g}(\mathbf{x})]$ is a *penalty function.* Various forms of penalty functions have been proposed which can be found elsewhere (McCormick, 1983; Gill, Murray, and Wright, 1981). The penalty function is minimized by stages for a series of values of parameters associated with the penalty function. For many of the penalty functions, the Hessian of the penalty function becomes increasingly *ill-conditioned* (i.e., the function value is extremely sensitive to a small change in the parameter value) as the solution approaches the optimum. This section briefly describes a penalty function method called the *augmented Lagrangian method.*

The augmented Lagrangian method adds a quadratic penalty term to the Lagrangian function [Eq. (6.18)], to obtain

$$L_A(\mathbf{x}, \lambda, \psi) = f(\mathbf{x}) + \sum_{i=1}^{m} \lambda_i \mathbf{g}_i(\mathbf{x}) + \frac{\psi}{2} \sum_{i=1}^{m} \mathbf{g}_i^2(\mathbf{x}) = \mathbf{f}(\mathbf{x}) + \lambda^T \mathbf{g}(\mathbf{x}) + \frac{\psi}{2}\, \mathbf{g}(\mathbf{x})^T \mathbf{g}(\mathbf{x}) \quad (6.25)$$

where ψ is a positive penalty parameter. Some desirable properties of Eq. (6.25) are discussed by Gill, Murray, and Wright (1981).

For ideal circumstances, $\mathbf{x}^*$ can be computed by a single unconstrained minimization of the differentiable function [Eq. (6.25)]. However, in general, λ^* is not available until the solution has been determined. An augmented Lagrangian method, therefore, must include a procedure for estimating the Lagrange multipliers. Gill, Murray, and Wright (1981) present the following algorithm:

Step 0   Select initial estimates of the Lagrange multipliers $\lambda^{k=0}$, the penalty parameter ψ, an initial point $\mathbf{x}^{k=0}$. Set $k = k + 1$ and set the maximum number of iterations as J.

Step 1   Check to see if $\mathbf{x}^k$ satisfies optimality conditions or if $k > J$. If so, terminate the algorithm.

Step 2   Minimize the augmented Lagrangian function, $L_A(\mathbf{x}, \lambda, \psi)$, in Eq. (6.25). Procedures to consider unboundedness must be considered. The best solution is denoted as $\mathbf{x}^{k+1}$.

Step 3   Update the multiplier estimate by computing λ^{k+1}.

Step 4   Increase the penalty parameter ψ if the constraint violations at $\mathbf{x}^{k+1}$ have not decreased sufficiently from those at $\mathbf{x}^k$.

Step 5   Set $k = k + 1$ and return to Step 1.

6.4 DYNAMIC PROGRAMMING

6.4.1 Dynamic Programming Concept

Dynamic programming (*DP*) transforms a sequential or multistage decision problem that may contain many interrelated decision variables into a series of single-stage problems, each containing only one or a few variables. In other words, the DP technique decomposes an N-decision problem into a sequence of N separate, but interrelated, single-decision subproblems. Decomposition is very useful in solving large, complex problems by decomposing a problem into a series of smaller subproblems and then combining the solutions of the smaller problems to obtain the solution of the whole model-composition. Using decomposition, a problem is solved more efficiently, resulting in significant computational savings. As a rule of thumb, computations increase exponentially with the number of variables, but only linearly with the number of subproblems. Books that deal with dynamic programming are Cooper and Cooper (1981), Denardo (1982), and Drefus and Law (1977).

Exhaustive enumeration possesses three main shortcomings: (1) it would become impractical if the number of alternative combinations is large, (2) the optimal course of action cannot be verified, even if it is obtained in the early computation, until all the combinations are examined, and (3) infeasible combinations (or solutions) cannot be eliminated in advance. Dynamic programming can overcome the shortcomings of an exhaustive enumeration procedure using the following concepts:

1. The problem is decomposed into subproblems, and the optimal alternative is selected for each subproblem sequentially so that it is never necessary to enumerate all combinations of the problem in advance.

2. Because optimization is applied to each subproblem, nonoptimal combinations are automatically eliminated.

3. The subproblems should be linked together in a special way so that it is never possible to optimize over infeasible combinations.

Referring to Fig. 6.4, the basic elements in the DP formulation are introduced as follows.

1. *Stages n* are the points of the problem where decisions are to be made. If a decisionmaking problem can be decomposed into N subproblems, there will be N stages in the DP formulation.

2. *Decision variables* d_n are courses of action to be taken for each stage. The number of decision variables d_n in each stage is not necessarily equal to one.

3. *State variables* S_n are variables describing the state of a system at any stage n. A state variable can be discrete or continuous, finite or infinite. Referring to Fig. 6.4, at any stage n there are input states S_n and output states S_{n+1}. The state variables of the system in a DP model have the function of linking succeeding stages so that, when each stage is optimized separately, the resulting decision is automatically feasible for the entire problem. Furthermore, it allows one to make optimal decisions for the remaining stages without having to check the effect of future decisions for decisions previously made.

4. *Stage return* r_n is a scalar measure of the effectiveness of decisionmaking in each stage. It is a function of the input state, the output state, and the decision variables of a particular stage. That is, $r_n = r(S_n, S_{n+1}, d_n)$.

5. *Stage transformation* t_n is a single-valued transformation which expresses the relationships between the input state, the output state, and the decision. In general, through the stage transformation, the output state at any stage n can be expressed as the function of the input state and the decision as

$$S_{n+1} = t_n(S_n, d_n) \qquad (6.26)$$

The basic features that characterize all DP problems are as follows:

1. The problem is divided into stages, with decision variables at each stage.

2. Each stage has a number of states associated with it.

3. The effect of the decision at each stage is to produce return, according to the stage return function, and to transform the current state variable into the state variable for the next stage, through the state transformation function.

4. Given the current state, an optimal policy for the remaining stages is independent of the policy adopted in previous stages. This is called *Bellman's principle of optimality,* which serves as the backbone of dynamic programming.

5. The solution begins by finding the optimal decision for each possible state in the last stage (called the *backward recursive*) or in the first stage (called the *forward*

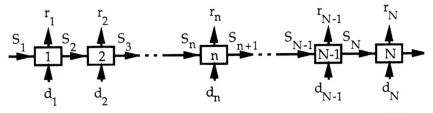

FIGURE 6.4 Sequential representation of serial dynamic programming problems.

recursive). A forward algorithm computationally advances from the first to the last stage, whereas a backward algorithm advances from the last stage to the first.

6. A recursive relationship that identifies the optimal policy for each state at any stage n can be developed, given the optimal policy for each state at the next stage, $n + 1$. This backward recursive equation, referring to Fig. 6.4, can be written as

$$f_n^*(S_n) = \underset{d_n}{\text{opt}} \left\{ r_n(S_n, d_n) \ o \ f_{n+1}^*(S_{n+1}) \right\}$$

$$= \text{opt}\{r_n(S_n, d_n) \ o \ f_{n+1}^*[t_n(S_n, d_n)]\} \qquad (6.27)$$

where o represents an algebraic operator which can be $+, -, \times,$ or $\div$ whichever is appropriate to the problem. The recursive equation for a forward algorithm is stated as

$$f_n^*(S_n) = \text{opt}\{r_n(S_n, d_n) \ o \ f_{n-1}^*(S_{n-1})\} \qquad (6.28)$$

An increase in the number of discretizations and/or state variables would increase the number of evaluations of the recursive formula and core memory requirement per stage. This problem of rapid growth of computer time and core memory requirement associated with multiple state variable DP problems is referred to as the *curse of dimensionality*. From the problem solving viewpoint, the problem of increased computer time is of much less concern than that of the increased computer storage requirement. An increase in required computer core memory might result in exceeding the available storage capacity of a particular computer facility in use, and the problem cannot be solved. On the other hand, an increase in computer time requires one to be more patient for the final result. Therefore, the rapid growth in computer memory requirements associated with multiple state variable problems can make the difference between solvable and unsolvable problems.

6.4.2 Discrete Differential Dynamic Programming

Discrete differential dynamic programming (DDDP) is an iterative DP procedure which is specially designed to overcome some of the difficulties of the DP approach previously mentioned. DDDP uses the same recursive equation as DP to search among the discrete states in the state-stage domain. Instead of searching over the entire state-stage domain for the optimum, as is the case for DP, DDDP examines only a portion of the state-stage domain saving computer time and memory (Chow, et al., 1975). This optimization procedure is solved through iterations of trial states and decisions to search for the optimum for a system subject to the constraints that the trial states and decisions should be within the respective *admissible domain*, i.e., feasible in the state and decision spaces.

In DDDP, the first step is to assume a trial sequence of admissible decisions called the *trial policy*, and the state vectors of each stage are computed accordingly. This sequence of states within the admissible state domain for different stages is called the *trial trajectory*. An alternative to the above procedure is first to assume a trial trajectory and then use it to compute the trial policy. Several states, located in the neighborhood of a trial trajectory, can be introduced to form a band called a *corridor* around the trial trajectory (see Fig. 6.5). It is a common practice to discretize the state space into uniform increments, called the *state increment*, where the total number of discretizations, referred to as *grid points* or *lattice points*, for each state variable is the same. The decisions, therefore, have to be made with respect to the method of discretizing the state variables. The interval of the decision variable is dependent on the corresponding interval of the state variable.

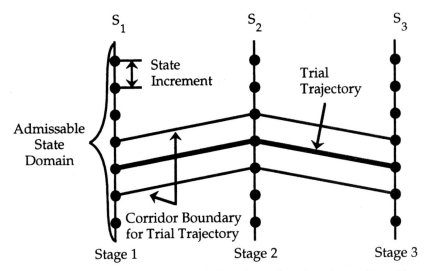

FIGURE 6.5 Formation of corridor around trial trajectory for a three-stage problem. *(Mays and Tung, 1992.)*

The traditional DP approach is applied in a DDDP problem to the states within the corridor using the recursive relationship for an improved trajectory to define a new policy, or set of decisions, within the introduced corridor. The improved trial trajectory is then adopted as the new trajectory to form a new corridor. This process of corridor formation, optimization with respect to the states within the corridor and trace-back to obtain an improved trajectory for the entire system is called an *iteration*.

This procedure is repeated beyond some iteration k which produces a corridor providing a system return f_k such that further iterations with this corridor size will produce a difference in system return $f_k - f_{k-1}$ less than a specified tolerance. At this point in the optimization procedure, the size of the state increment can be reduced to set up a new corridor in which the states or lattice points are closer together. A smaller corridor is formed around the improved trajectory for the last iteration completed. The iterations continue reducing the size of the state increment through the system accordingly until a specified minimum corridor size is reached.

The criterion used to determine when the magnitude of the state increment size should be reduced is based on the relative change of the optimal objective function value of the previous iteration, i.e.,

$$|f_k - f_{k-1}|/f_{k-1} \leq \epsilon \qquad (6.29)$$

where ϵ is the specified tolerance level. Figure 6.6 is a flowchart showing the general algorithm of the DDDP procedure.

Although the use of DDDP partially circumvents the problem associated with the curse of dimensionality of the regular DP, it introduces other considerations involved with choosing the initial trajectory, changing the lattice point or state spacing, and converging to a local optimum instead of the global optimum. The major factors affecting the performance of DDDP include:

1. The number of lattice points
2. The initial state increment along with the number of lattice points which determine corridor width

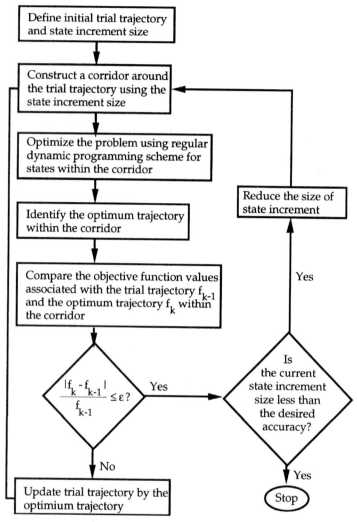

FIGURE 6.6 General algorithm for discrete differential dynamic programming. *(Mays and Tung, 1992.)*

3. The initial trial trajectory used to establish the location of the corridors within the state space for the first iteration

4. The reduction rate of the state increment size which determines the corridor width for various iterations

These factors are somewhat dependent upon one another for expedient and efficient use of DDDP for water resources problems. The choice of the initial trial trajectory and the initial corridor width are interdependent. For example, if a small corridor width is chosen in conjunction with an initial trial trajectory which is far from the optimal region, unnecessary iterations are required to move the trajectory into the optimal or near-optimal region. It is possible for DDDP, under these circumstances, to converge to a solution which is far from the optimal. The number of lattice points and the initial state increment together determine the initial corridor width. Using a combination of a small number of lattice points and a small initial state increment could also result in local optimal solutions far from the optimal.

The effect of choosing a poor initial trajectory can be reduced if a large number of lattice points and/or large initial state increments are used. In essence, the better the initial trajectory, the smaller the number of lattice points and initial state increments, or simply, the smaller the initial corridor width can be used. Very small state increments with more lattice points can also be used to establish a small corridor width. This can result in improved convergence; however, increasing the number of lattice points increases the computation time and storage requirements. Computation time can be improved by increasing the reduction rate of the state increment at each iteration; however, too large a reduction rate of the state increment may miss the optimal region, thus not providing the optimal solution. Choosing a large initial state increment and a large reduction rate of the state increment size may be advantageous. However, when the initial state increment is too large, resulting in large corridor widths, unnecessary computations are performed in regions of the state space far from the optimal.

6.4.3 Differential Dynamic Programming

This section presents the differential dynamic programming (DDP) method for discrete-time optimal control problems. The term DDP used by Jacobson and Mayne (1970) broadly refers to stagewise nonlinear programming procedures. Earlier works that basically developed DDP procedures for unconstrained discrete-time control problems include Bellman and Dreyfus (1962), Mayne (1966), Gershwin and Jacobson (1970), Dyer and McReynolds (1970), Jacobson and Mayne (1970), Yakowitz and Rutherford (1984), and Yakowitz (1989). Ohno (1978) and Murray and Yakowitz (1981) contributed to DDP techniques for constrained optimal control problems.

The basic *optimal control problem* is stated as follows

$$\min_{u} Z = \sum_{t=1}^{T} f_t(\mathbf{x}_t, \mathbf{u}_t; t) \tag{6.30}$$

subject to

$$\mathbf{x}_{t+1} = \mathbf{g}_t(\mathbf{x}_t, \mathbf{u}_t; t) \qquad t = 1, \ldots, T \tag{6.31}$$

The objective of DDP is to minimize a quadratic approximation instead of solving the actual control problem using the algorithm approach defined in Fig. 6.7. Yakowitz and Rutherford (1984) pointed out the following properties of DDP for unconstrained problems:

1. DDP overcomes the curse of dimensionality in that the computational requirements grow as $N^3 T$ and memory requirements as NnT, where n and N are, respectively, the state and control variable dimensions, and T is the number of decision times.

2. Under lenient conditions, DDP is globally convergent.

3. No discretization of state or control spaces is required.

4. The convergent rule of the DDP algorithm is quadratic for control problems in which the Hessian matrix of the objective function is convex in a neighborhood of the solution.

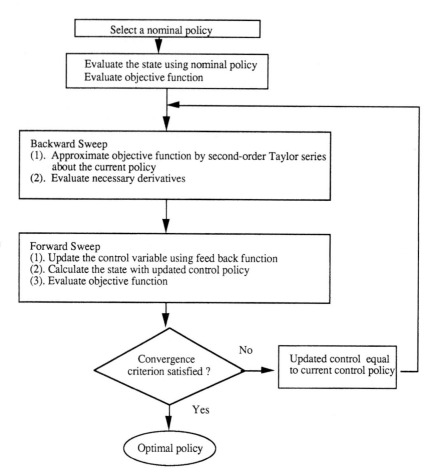

FIGURE 6.7   Flowchart of the unconstrained DDP method.

6.4.3.1 Algorithm Definition.   Define the current or known control policy as $\bar{\mathbf{u}}_t$ for $t = 1, \ldots, T$ and the current or known state trajectory $\bar{\mathbf{x}}_t$ for $t = 1, \ldots, T + 1$. The initial state is $\mathbf{x}_1$. For any function $W(\mathbf{x}, \mathbf{u})$ defined by the control and state variables, let $QP(W(\mathbf{x}, \mathbf{u}))$ denote the linear and quadratic path of the Taylor's series expansion of $W(\ )$ about $(\bar{\mathbf{u}}, \bar{\mathbf{x}})$. The quadratic path for time T where the DDP backward recursion begins is

$$L(\mathbf{x}, \mathbf{u}; T) = QP((f(\mathbf{x}, \mathbf{u}; T)) = \frac{1}{2} \delta\mathbf{x}^T\left(\frac{\partial^2 f}{\partial\mathbf{x}^2}\right)\delta\mathbf{x} + \delta\mathbf{x}^T\left(\frac{\partial^2 f}{\partial\mathbf{x}\partial\mathbf{u}}\right)\delta\mathbf{u}$$

$$+ \frac{1}{2}\delta\mathbf{u}^T\left(\frac{\partial^2 f}{\partial\mathbf{u}^2}\right)\delta\mathbf{u} + \left(\frac{\partial f}{\partial\mathbf{u}}\right)^T\delta\mathbf{u} + \left(\frac{\partial f}{\partial\mathbf{x}}\right)^T\delta\mathbf{x} \qquad (6.32)$$

where $\delta\mathbf{x} = (\mathbf{x} - \bar{\mathbf{x}}_T)$ and $\delta\mathbf{u} = (\mathbf{u} - \bar{\mathbf{u}}_T)$ are state and input perturbations, $(\partial f/\partial \mathbf{u})$ and $(\partial f/\partial \mathbf{x})$ are the gradients and $(\partial^2 f/\partial \mathbf{x}^2)$, $(\partial^2 f/\partial \mathbf{x}\partial \mathbf{u})$, and $(\partial^2 f/\partial \mathbf{u}^2)$ are the Hessian of $f(\mathbf{x}, \mathbf{u}; T)$ which are evaluated at $\bar{\mathbf{x}}_T$ and $\bar{\mathbf{u}}_T$. Equation (6.32) can be presented in a more compact form as

$$L(\mathbf{x}, \mathbf{u}; T) = \delta\mathbf{x}^T\mathbf{A}_T\delta\mathbf{x} + \delta\mathbf{x}^T\mathbf{B}_T\delta\mathbf{u} + \delta\mathbf{u}^T\mathbf{C}_T\delta\mathbf{u} + \mathbf{D}_T\delta\mathbf{u} + \mathbf{E}_T\delta\mathbf{x} \qquad (6.33)$$

The idea of DDP is to minimize the quadratic approximation instead of the actual control problem value function, thereby obtaining an amenable function which is at the expense of involving truncation error. A necessary condition that a control $\mathbf{u}^*$ minimizes $L(\mathbf{x}, \mathbf{u}; T)$ is

$$\nabla_u L(\mathbf{x}, \mathbf{u}; T) = 2\mathbf{C}_T\delta\mathbf{u} + \mathbf{B}_T\delta\mathbf{x} + \mathbf{D}_T = \mathbf{0} \qquad (6.34)$$

Then the optimal control $\mathbf{u}^*$ can be found from Eq. (6.34), under the assumption that $\mathbf{C}_T$ is nonsingular, as

$$\delta\mathbf{u}(\mathbf{x}; T) = (\mathbf{u}^* - \bar{\mathbf{u}}_T) \qquad (6.35a)$$

$$= -\left(\frac{1}{2}\right)\mathbf{C}_T^{-1}(\mathbf{D}_T + \mathbf{B}_T\delta\mathbf{x}) \qquad (6.35b)$$

$$= \alpha_T + \beta_T\delta\mathbf{x} \qquad (6.35c)$$

where

$$\alpha_T = \left(-\frac{1}{2}\right)\mathbf{C}_T^{-1}\mathbf{D}_T$$

and

$$\beta_T = \left(-\frac{1}{2}\right)\mathbf{C}_T^{-1}\mathbf{B}_T$$

The optimal value function is

$$F(\mathbf{x}; T) = \min_u Z = \min f(\mathbf{x}_T, \mathbf{u}_T; T) \qquad (6.36)$$

which is approximated by the quadratic as

$$V(\mathbf{x}; T) = L((\mathbf{x}, \mathbf{u}(\mathbf{x}, T); T))$$

$$= L(\mathbf{x}, \bar{\mathbf{u}}_T + \alpha_T + \beta_T\delta\mathbf{x}; T) \qquad (6.37)$$

The following quadratic $V(\mathbf{x}; T)$ is determined by substituting equation (6.35) into equation (6.33),

$$V(\mathbf{x}; T) = \delta\mathbf{x}^T\mathbf{P}_T\delta\mathbf{x} + \mathbf{Q}_T\delta\mathbf{x} \qquad (6.38)$$

where

$$\mathbf{P}_T = \mathbf{A}_T - \left(\frac{1}{4}\right)\mathbf{B}_T^T\mathbf{C}_T^{-1}\mathbf{B}_T \qquad (6.39)$$

$$\mathbf{Q}_T^T = -\left(\frac{1}{2}\right)\mathbf{D}_T^T\mathbf{C}_T^{-1}\mathbf{B}_T + \mathbf{E}_T^T \qquad (6.40)$$

as long as $\mathbf{C}_T$ is nonsingular.

The DDP backward recursion procedure is performed for $t = T, T - 1, \ldots, 1$ using the quadratic

$$L(\mathbf{x}, \mathbf{u}; t) = QP\{f(\mathbf{x}, \mathbf{u}; t) + V[\mathbf{g}(\mathbf{x}, \mathbf{u}; t); t + 1]\} \qquad (6.41)$$

where $V(\mathbf{x}; t+1)$ is the quadratic approximate optimal return function defined as

$$V(\mathbf{x}; t+1) = (\delta\mathbf{x})^T \mathbf{P}_{t+1} \delta\mathbf{x} + \mathbf{Q}_{t+1} \delta\mathbf{x} \qquad (6.42)$$

Similar to equation (6.33) the quadratic can be expressed as

$$L(\mathbf{x}, \mathbf{u}; t) = \delta\mathbf{x}^T \mathbf{A}_t \delta\mathbf{x} + \delta\mathbf{u}^T \mathbf{B}_t \delta\mathbf{x} + \delta\mathbf{u}^T \mathbf{C}_t \delta\mathbf{u} + \mathbf{D}_t^T \delta\mathbf{u} + \mathbf{E}_t^T \delta\mathbf{x} \qquad (6.43)$$

The coefficients $\mathbf{A}_t, \mathbf{B}_t, \mathbf{C}_t, \mathbf{D}_t^T$ and $\mathbf{E}_t^T$ are

$$\mathbf{A}_t = \frac{1}{2}\left(\frac{\partial^2 f}{\partial \mathbf{x}^2}\right)_t + \left(\frac{\partial \mathbf{g}}{\partial \mathbf{x}}\right)_t^T \mathbf{P}_{t+1}\left(\frac{\partial \mathbf{g}}{\partial \mathbf{x}}\right)_t + \frac{1}{2}\sum_{i=1}^{n}(\mathbf{Q}_{t+1})_i\left(\frac{\partial^2 \mathbf{g}}{\partial x_i^2}\right) \qquad (6.44)$$

$$\mathbf{B}_t = \left(\frac{\partial^2 f}{\partial \mathbf{x}\partial \mathbf{u}}\right)_t + 2\left(\frac{\partial \mathbf{g}}{\partial \mathbf{x}}\right)_t^T \mathbf{P}_{t+1}\left(\frac{\partial \mathbf{g}}{\partial \mathbf{x}}\right)_t + \frac{1}{2}\sum_{i=1}^{n}(\mathbf{Q}_{t+1})_i\left(\frac{\partial^2 \mathbf{g}}{\partial x_i\partial u_i}\right) \qquad (6.45)$$

$$\mathbf{C}_t = \frac{1}{2}\left(\frac{\partial^2 f}{\partial \mathbf{u}^2}\right)_t + \left(\frac{\partial \mathbf{g}}{\partial \mathbf{x}}\right)_t^T \mathbf{P}_{t+1}\left(\frac{\partial \mathbf{g}}{\partial \mathbf{x}}\right)_t + \frac{1}{2}\sum_{i=1}^{n}(\mathbf{Q}_{t+1})_i\left(\frac{\partial^2 \mathbf{g}}{\partial u_i^2}\right) \qquad (6.46)$$

$$\mathbf{D}_t^T = \left(\frac{\partial f}{\partial \mathbf{u}}\right)_t + \mathbf{Q}_{t+1}^T\left(\frac{\partial \mathbf{g}}{\partial \mathbf{x}}\right)_t \qquad (6.47)$$

$$\mathbf{E}_t^T = \left(\frac{\partial f}{\partial \mathbf{u}}\right)_t + \mathbf{Q}_{t+1}^T\left(\frac{\partial \mathbf{g}}{\partial \mathbf{x}}\right)_t \qquad (6.48)$$

The first-order derivatives of $f(\mathbf{x}, \mathbf{u}; t)$ in the above equations are components of the gradient of $f(\mathbf{x}, \mathbf{u}; t)$; the second-order derivatives are the components of the Hessian of $f(\mathbf{x}, \mathbf{u}; t)$; the first-order derivatives of $g(\mathbf{x}, \mathbf{u}; t)$ are components of the Jacobian of $g(\mathbf{x}, \mathbf{u}; t)$; and the second-order derivatives of $g(\mathbf{x}, \mathbf{u}; t)_i$ for $1 \le i \le n$ are the blocks of the Hessian matrices of the coordinates of $g(\mathbf{x}, \mathbf{u}; t)$. All derivatives are evaluated about the current states and controls.

The first-order necessary condition for the optimality is

$$\nabla_u L(\mathbf{x}, \mathbf{u}; t) = \mathbf{0} \qquad (6.49)$$

so that the minimizing strategy for the quadratic $L(\mathbf{x}, \mathbf{u}; t)$ is

$$\delta\mathbf{u}(\mathbf{x}; t) = \boldsymbol{\alpha}_t + \boldsymbol{\beta}_t(\mathbf{x} - \bar{\mathbf{x}}_t) \qquad (6.50)$$

where

$$\boldsymbol{\alpha}_t = -\frac{1}{2}\mathbf{C}_t^{-1}\mathbf{D}_t \qquad (6.51)$$

and

$$\boldsymbol{\beta}_t = -\frac{1}{2}\mathbf{C}_t^{-1}\mathbf{B}_t \qquad (6.52)$$

The approximating polynomial for the optimal return function is

$$V(\mathbf{x}; t) = L(\mathbf{x}, \mathbf{u}(\mathbf{x}, t); t)$$
$$= (\delta\mathbf{x})^T \mathbf{P}_t(\mathbf{x} - \bar{\mathbf{x}}_t) + \mathbf{Q}_t^T \delta\mathbf{x} \qquad (6.53)$$

where

$$\mathbf{P}_t = \mathbf{A}_t - \frac{1}{4}\,\mathbf{B}_t^T\mathbf{C}_t^{-1}\mathbf{B}_t \qquad (6.54)$$

$$\mathbf{Q}_t = -\frac{1}{2}\,\mathbf{D}_t\mathbf{C}_t^{-1}\mathbf{B}_t + \mathbf{E}_t^T \qquad (6.55)$$

These equations are necessary for the DDP backward recursion. α_t and β_t for $1 \le t \le T$ must be stored for use in the forward sweep. The forward sweep determines the successor DDP policy by successively selecting controls according to the rule $\mathbf{u}\,(\mathbf{x}^*; t)$ and then calculating the successor state at each time so that $\mathbf{u}_1^* = \mathbf{u}(\mathbf{x}_1^*; 1)$ and $\mathbf{x}_2^* = \mathbf{g}(\mathbf{x}_1^*, \mathbf{u}_1^*; 1)$. Then the following is for $t = 2, \ldots, T$

$$\mathbf{u}_t^* = \mathbf{u}(\mathbf{x}_t^*; t) + \bar{\mathbf{u}}_t \qquad (6.56)$$

and

$$\mathbf{x}_{t+1}^* = \mathbf{g}(\mathbf{x}_t^*, \mathbf{u}_t^*; t) \qquad (6.57)$$

For the next DDP iteration the DDP successor control is the current control sequence $\bar{\mathbf{u}}$.

6.5   OPTIMAL CONTROL

6.5.1   Concepts

Linear quadratic (*LQ*) problems find an optimal control law which minimizes a *quadratic performance index* subject to a *dynamic linear system*. Since the quadratic performance index is a function of states and controls at all times, the LQ problem is to find an optimal control law such that the states and controls are as close as possible to the desired states and controls, respectively. There are several methods of solving the posed problem. One way is to use the DP technique to derive an analytical optimal control law which is a linear function of the current state. Books on optimal control theory include: Anderson and Moore (1990); Athans and Falb (1966); Auslander, et al. (1974); Brogan (1974); Dyer and McReynolds (1970); Jacobson (1980); Luus (1967); and Maybeck (1979, 1982).

If the desired states are zero, the LQ problem is a *linear quadratic regulator* (*LQR*) problem which regulates the states to zero. If the desired states are not zero, the LQ problem is a *linear quadratic tracker* (*LQT*) problem which forces the system to track the desired states at all times. For the LQR problem (or LQT) problem, if the dynamic linear system equation is deterministic, the LQR is called *LQRD* (or LQTD for LQT); if the dynamic linear system equation is stochastic, the LQR is called *LQRS* (or LQTS for LQT). In essence, the LQR problem is a special case of the LQT problem.

The combination of LQ problems with a Kalman-Bucy filter (Maybeck, 1982) leads to the *linear quadratic-Gaussian* (*LQG*). The *Kalman-Bucy filter* gives and optimum state estimate by minimizing the mean square error about the state. Since the optimal control law for LQ problems is a linear function of the current state, if the current state is directly known without uncertainty, the optimal control law can directly apply. If the current state is not directly known, the current state can be estimated by the Kalman-Bucy filter. The optimal control law applies by using the estimate of the state by the Kalman-Bucy filter.

Consider a continuous dynamic system defined by the following *state-space model* as

$$\dot{\mathbf{x}} = \mathbf{f}(\mathbf{x}(t), \mathbf{u}(t); t) \tag{6.58}$$

in which $\mathbf{x}(t)$ and $\mathbf{u}(t)$ are the vectors of state and control of the system, respectively. The problem is, at a given initial state, to use $\mathbf{u}(t)$ to guide the system such that the following *performance index is* minimized,

$$J = F(\mathbf{x}(t_f), t_f) + \int_{t_o}^{t_f} L(\mathbf{x}(t), \mathbf{u}(t), t)\, dt \tag{6.59}$$

in which F and L are real and scalar-valued functions, t_o and t_f are the initial and terminal times, respectively. The minimal cost from time t_o to the terminal time t_f is defined as

$$g(\mathbf{x}(t), t_f - t) = \min_{\mathbf{u}(t)} \left\{ F(\mathbf{x}(t_f), t) + \int_t^{t_f} L(\mathbf{x}(t), \mathbf{u}(t), t)\, dt \right\} \tag{6.60}$$

which is sometimes called the *optimal cost-to-go*.

The well-known Hamilton-Jacoby-Bellman (or *Hamilton-Jacobi equation*) is

$$\frac{\partial g}{\partial t_r} = \min_{\mathbf{u}(t)} \left\{ L(\mathbf{x}(t), \mathbf{u}(t), t) + [\nabla_x g]^T \dot{\mathbf{x}} \right\} \tag{6.61}$$

with the boundary condition

$$g(\mathbf{x}(t), t_r)|_{t_r = 0} = F(\mathbf{x}(t_f), t_f) \tag{6.62}$$

The *optimal control* is the one which minimizes the term in the pair of brackets on the right-hand side of the Hamilton-Jacobi-Bellman equation. If there is no constraint on $\mathbf{u}(t)$, the necessary condition for optimality can be derived by taking derivatives,

$$\frac{\partial L}{\partial \mathbf{u}} + \frac{\partial \mathbf{f}}{\partial \mathbf{u}} \nabla_x g = \mathbf{0} \tag{6.63}$$

Consider a linear dynamic system

$$\dot{\mathbf{x}}(t) = \mathbf{A}(t)\mathbf{x}(t) + \mathbf{B}(t)\mathbf{u}(t) \tag{6.64}$$

with loss function

$$L = [\mathbf{x}(t) - \mathbf{x}_D(t)]^T \mathbf{P}(t)[\mathbf{x}(t) - \mathbf{x}_D(t)] + \mathbf{u}(t)^T \mathbf{R}(t)\mathbf{u}(t)$$

and

$$F = [\mathbf{x}(t_f) - \mathbf{x}_D(t_f)]^T \mathbf{H}(t)[\mathbf{x}(t_f) - \mathbf{x}_D(t_f)] \tag{6.65}$$

in which $\mathbf{x}_D$ denotes the desired value of the state variable and $\mathbf{P}(t)$, $\mathbf{R}(t)$, and $\mathbf{H}(t)$ are weighting matrices.

The derivatives $\partial \mathbf{f}/\partial \mathbf{u}$ and $\partial \mathbf{L}/\partial \mathbf{u}$, respectively, are

$$\frac{\partial \mathbf{f}}{\partial \mathbf{u}} = \mathbf{B}^T(t) \tag{6.66}$$

$$\frac{\partial \mathbf{L}}{\partial \mathbf{u}} = 2\mathbf{R}(t)\mathbf{u}(t) \tag{6.67}$$

The optimal control policy for $\mathbf{u}(t)$ is

$$\mathbf{u}^*(t) = -\frac{1}{2}\mathbf{R}^{-1}\mathbf{B}^T\nabla_x g(\mathbf{x}(t), t_r) \tag{6.68}$$

In order to obtain $g(\mathbf{x}(t), t_f - t)$ for this linear system with quadratic loss function, the Hamilton-Jacobi-Bellman differential equation for $g(\mathbf{x}(t), t_f - t)$ is solved after substitution of the optimal control policy. In order to solve this equation, the classic method is to assume a quadratic function form for $g(\mathbf{x}(t), t_f - t)$ which has some unknown coefficients. Then substitute it into the Hamilton-Jacobi-Bellman equation to obtain the so-called Riccati-type equation, which is then solved for the unknown coefficients in the quadratic function (Brogan, 1974).

6.5.2 Linear Quadratic Regulator (LQR)

6.5.2.1 *Linear Quadratic Regulator LQR with Deterministic System Equation (LQRD).* The purpose of LQRD is to guide a linear system during a fixed period such that the states of the system are as close as possible to the zero states with the least control energy. It is to find an optimal law which minimizes a quadratic performance index such as

$$J(\mathbf{x}(0)) = \mathbf{x}(N)^T\mathbf{P}(N)\mathbf{x}(N) + \sum_{k=0}^{N-1}[\mathbf{x}(k)^T\mathbf{P}(k)\mathbf{x}(k) + \mathbf{u}^T(k)\mathbf{R}(k)\mathbf{u}(k)] \tag{6.69}$$

subject to a dynamic and time-variant linear system

$$\mathbf{x}(k+1) = \mathbf{A}(k)\mathbf{x}(k) + \mathbf{B}(k)\mathbf{u}(k) \tag{6.70}$$

$$\mathbf{x}(0) = \mathbf{x}_o \tag{6.71}$$

in which T is the transpose of a matrix; the vector $\mathbf{x}(k)$ of $(m \times 1)$ is the state at time k; the vector $\mathbf{u}(k)$ of $(n \times 1)$ is the control law applied at time k; the $(m \times m)$ weighting matrix $\mathbf{P}(k)$ is symmetric positive semidefinite; the weighting matrix $\mathbf{R}(k)$ of $(n \times n)$ is symmetric positive definite; the matrices $\mathbf{A}(k)$ of $(m \times m)$ and $\mathbf{B}(k)$ of $(m \times n)$ describe the system; k is the time sequence from 0 to N; and $\mathbf{x}_o$ is a given initial state. Eq. (6.70) can be called a *system transition equation* or transition equation.

An analytic solution of the optimal control law can be derived (Jacobson et al., 1980) by using dynamic programming

$$\mathbf{u}^*(k) = \mathbf{L}(k)\mathbf{x}(k) \tag{6.72}$$

in which

$$\mathbf{L}(k) = -(\mathbf{R}(k) + \mathbf{B}^T(k)\mathbf{K}_c(k+1)\mathbf{B}(k))^{-1}\mathbf{B}^T(k)\mathbf{K}_c(k+1)\mathbf{A}(k) \tag{6.73}$$

in which the matrix $\mathbf{K}_c(k)$ satisfies the recursive Riccati equation,

$$\mathbf{K}_c(k) = \mathbf{A}^T(k)\mathbf{K}_c(k+1)[\mathbf{I} - \mathbf{B}(k)(\mathbf{B}^T(k)\mathbf{K}_c(k+1)\mathbf{B}(k)$$
$$+ \mathbf{R}(k))\text{-}\mathbf{B}^T(k)\mathbf{K}c(k+1)]\mathbf{A}(k) + \mathbf{P}(k) \tag{6.74}$$

with the terminal condition

$$\mathbf{K}_c(N) = \mathbf{Q}(N) \tag{6.75}$$

The minimized performance index is given by

$$J^*(\mathbf{x}(0)) = \mathbf{x}^T(0)\mathbf{K}_c(0)\mathbf{x}(0) \tag{6.76}$$

The corresponding minimized cost-to-go performance index is

$$J^*(\mathbf{x}(k)) = \mathbf{x}^T(k)\mathbf{K}_c(k)\mathbf{x}(k) \qquad (6.77)$$

The result shows that the optimal control law is a linear function of the current state. If the current state is known without uncertainty, the optimal control law can directly apply. This fact accounts for the great popularity of the LQ problem. It can be seen that the transition equation is deterministic.

6.5.3 Linear Quadratic Gaussian (LQG)

When the state can only be directly observed with uncertainty and the system transition equation uncertainty is also considered, the LQG problem can be formulated and solved on the basis of the *separation theorem* which is an important concept in modern linear control theory. Essentially, LQG can be divided into two parts, the *estimator* part and the *actuator* part (Bertsekas, 1987). The estimator part is an optimal solution of the problem of estimating the state $\mathbf{x}(k)$ assuming no control takes place, while the actuator part is an optimal solution of the control problem assuming perfect state information prevails. The estimator can be achieved by using the Kalman-Bucy filtering which estimates the optimal state by minimizing the mean squared error of the estimated state. It can be shown that this estimate is also the expected value of the state condition on the current and previous measurements.

6.6 INTERFACING OPTIMIZERS WITH PROCESS SIMULATORS

6.6.1 Problem Formulation

A general process optimization problem in water resources may be formulated in an NLP framework in terms of state or dependent y variables and control or independent x variables:

$$\text{minimize } f(\mathbf{y}, \mathbf{x}) \qquad (6.78)$$

subject to process (simulation) equations

$$\mathbf{g}(\mathbf{y}, \mathbf{x}) = \mathbf{0} \qquad (6.79)$$

and additional constraints for design, operation, budget, etc. on the dependent $\mathbf{y}$ and independent $\mathbf{x}$ variables:

$$\underline{\mathbf{w}} \le \mathbf{w}(\mathbf{y}, \mathbf{x}) \le \overline{\mathbf{w}} \qquad (6.80)$$

The process simulation equations for hydrosystems applications basically consist of the governing physical equations that simulate a physical process such as conservation of mass, energy, and momentum. These equations are typically large in number, sparse, and nonlinear expressions of the governing partial differential equations. Conceptually, the simplest of the real-world problems cannot be solved in this manner as a result of their size. The existing NLP codes cannot solve such large, sparse problems. An alternative approach is to use the appropriate process simulator to solve the constraint equation [Eq. (6.79)] each time the objective and constraints are evaluated in the optimizer. The major advantage of such an approach is the reduced size of the nonlinear optimization problem, so that only a small subset of the constraint equations is treated by the optimizer. The basic idea is that the optimizer only handles the following reduced problem:

$$\text{minimize } f(\mathbf{x}(\mathbf{y}), \mathbf{y}) \tag{6.81}$$

subject to

$$\underline{\mathbf{w}} \le \mathbf{w}(\mathbf{y}, \mathbf{x}) \le \overline{\mathbf{w}} \tag{6.82}$$

as opposed to the much larger problem defined by Eqs. (6.78) to (6.80).

State (or dependent) variables and the control (or independent) variables are implicitly related through the simulator. In essence, the simulator equations are used to express the states in terms of the controls yielding a much smaller nonlinear optimization problem. The reduced gradient $[\partial F/\partial \mathbf{y}]$ where $F(\mathbf{y}) = f(\mathbf{x}(\mathbf{y}), \mathbf{y})$ is required by the optimization routines. The reduced problems are solved by combining augmented Lagrangian and reduced gradient procedures. Techniques for computing gradients of the reduced problem functions were developed by Lasdon and Waren (1978):

Step 1   Use the simulator to solve process Eq. (6.79)

Step 2   Solve for π, the Lagrange multiplier vector, using the linear difference equations

$$\pi^T = \left[\frac{\partial \mathbf{g}}{\partial \mathbf{x}} \right]^{-1} \left[\frac{\partial f}{\partial \mathbf{x}} \right]^T \tag{6.83}$$

Step 3   Evaluate the reduced gradient

$$\left[\frac{\partial F}{\partial \mathbf{y}} \right]^T = \left[\frac{\partial f}{\partial \mathbf{y}} \right]^T - \pi^T \left[\frac{\partial \mathbf{g}}{\partial \mathbf{y}} \right] \tag{6.84}$$

In equations (6.83) and (6.84), all elements of $\partial f/\partial \mathbf{x}$, $\partial \mathbf{g}/\partial \mathbf{x}$, $\partial f/\partial \mathbf{y}$ and $\partial \mathbf{g}/\partial \mathbf{y}$ are evaluated at some solution $(\overline{\mathbf{x}}, \overline{\mathbf{y}})$ and Eq. (6.84) yields the reduced gradient, $\partial F/\partial \mathbf{y}$, at $\overline{\mathbf{y}}$. The Lagrange multiplier π has its usual shadow price interpretation.

Solution of the simulator for the state variables, given the control variables from the optimizer, does not guarantee that the state variables satisfy the bound constraint in the optimizer. As a result, the bounds on the state variables are combined with the objective function into a penalty-like function called an augmented Lagrangian (Fletcher, 1975). This approach forces the state bounds to be satisfied and also reduces the number of constraints seen by the optimizer. The *augmented Lagrangian* is:

$$\text{minimize } L(\mathbf{y}, \mu, \sigma) = F(\mathbf{y}) + \frac{1}{2} \sum_i \sigma_i \min\left[0, c_i - \frac{\mu_i}{\sigma_i} \right]^2 + \frac{1}{2} \sum_i \frac{\mu_i}{\sigma_i} \tag{6.85}$$

where i is the index for each bound constraint; s_i and m_i are the penalty weights and Lagrange multipliers for the ith bound; and c_i is the violation of the bounds either above the upper or below the minimum defined as $c_i = \min (y_i - \underline{y}_i, \overline{y}_i - y_i)$.

6.7 MULTIOBJECTIVE PROGRAMMING

6.7.1 Multiobjective Concepts

Multiobjective programming deals with problems involving several objectives that are noncommensurate and conflicting with each other. Among the objective functions involved, there is no single one whose importance is overwhelmingly dominant over all others. Under this circumstance, the ideological theme of optimality in the single-objective context is no longer appropriate. The solution to a multiobjective problem is a best compromisable solution, according to the decisionmaker's preference, among the objectives and the noninferior solutions to the problem.

A *noninferior solution* to a multiobjective programming problem is a feasible solution to which there is no other feasible solution that will yield an improvement in one objective without causing degradation to at least one other objective (Cohon, 1978). The collection of such inferior solutions allows the assessment of tradeoff among conflicting objectives. Such tradeoff is often measured by the negative of the slope of the noninferior set, called the *marginal rate of transformation (MRT)*, representing the *feasible tradeoff* between objectives.

To obtain the solution to a multiobjective programming problem, the preference of the decisionmaker among the conflicting objectives must be known. Information concerning the decisionmaker's preference is commonly called the *utility function* which is a function of the objective function values. Geometrically, the utility function can be depicted as a series of *indifference curves*. The utility of a decisionmaker will be the same for combinations of solutions that fall on the same indifference curve. The negative slope of an indifference curve is the *marginal rate of substitution (MRS)* indicating the decisionmaker's *desired tradeoff* among different objectives. The best compromised solution to a multiobjective programming problem is a unique set of alternatives which possess the property of maximizing the decisionmaker's utility and are elements of the noninferior solution set. Referring to Fig. 6.8, the best compromised solution occurs at the point where the indifference curve and noninferior solution set are tangent. In other words, at the best compromised solution, MRT equals to MRS.

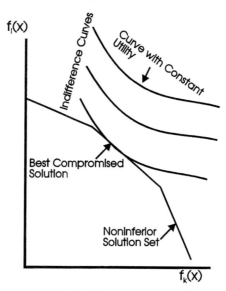

FIGURE 6.8 Noninferior solution and indifference curves.

Mathematically, multiobjective programming problems can be expressed in terms of *vector optimization* as

$$\text{maximize } \mathbf{f}(\mathbf{x}) = [f_1(\mathbf{x}), f_2(\mathbf{x}), \ldots, f_K(\mathbf{x})] \qquad (6.86a)$$

subject to

$$\mathbf{g}(\mathbf{x}) = \mathbf{0} \qquad (6.86b)$$

in which $\mathbf{f}(\mathbf{x})$ is a vector of K objective functions, $\mathbf{g}(\mathbf{x}) = \mathbf{0}$ are vectors of constraints, and $\mathbf{x}$ is a vector of decision variables.

6.7.2   Types of Methods

There are many methods proposed to derive the solution to a multiobjective programming problem. Basically, they can be classified into two categories (Cohon, 1978): *Generating techniques* and *techniques incorporating knowledge of decisionmaker's preference.* The purpose of generating techniques is to provide decisionmakers with information about the inferior solution set or feasible tradeoff to the multiobjective problem. System analysts play the role of information providers and do not actively engage in decisionmaking. On the other hand, techniques in the other category explicitly incorporate the decisionmaker's preference to search for the best compromised solution. Detailed descriptions of various techniques for solving multiobjective problems can be found elsewhere (Cohon, 1978; Goicoechea, et al., 1982; Chankong and Haimes, 1983; and Steuer, 1986). In the following sections, some frequently used techniques in multiobjective water resource management are briefly discussed.

6.7.3   Noninferior Solution Generating Techniques

Two commonly used generating techniques are the *weighting method* and the *constraint method.* The weighting method casts the multiobjective programming problem, Eqs. (6.86a–b), into

$$\text{maximize } \sum_{k=1}^{K} w_k f_k(\mathbf{x}) \qquad (6.87a)$$

subject to

$$\mathbf{g}(\mathbf{x}) = \mathbf{0} \qquad (6.87b)$$

where w_k is the weight on the kth objective. The weighted method is based on the fact that the tradeoff between two different objectives $f_{k'}(\mathbf{x})$ and $f_{k''}(\mathbf{x})$ is represented by the respective weights as $-w_k/w_{k'}$. The optimal solution to Eqs. (6.87a–b) is noninferior if it is unique under strictly positive-valued weights. Through systematic assignment of positive-valued weights to each objective function and solving Eqs. (6.87a–b) repeatedly, the noninferior solution set can be derived.

The constraint method arbitrarily selects one objective and places all other objective functions into the constraint set as

$$\text{maximize } f_h(\mathbf{x}) \qquad (6.88a)$$

subject to

$$f_k(\mathbf{x}) \geq \underline{f}_k \qquad \text{for all} \qquad k \neq h \tag{6.88b}$$

$$\mathbf{g}(\mathbf{x}) = \mathbf{0} \tag{6.88c}$$

in which $f_h(\mathbf{x})$ is the selected objective and $\underline{f}_k$ is the specified lower bound to be achieved by the kth objective. Note that the specified lower bound $\underline{f}_k$ should be that the optimal solution to Eqs. (6.88a–c) exits. The noninferior solution set to the multiobjective programming problem, Eqs. (6.86a–b) can be obtained by systematically varying $\underline{f}_k$ and solving Eqs. (6.88a–c) accordingly.

Realize that both the weighted method and constraint method derive the inferior solution set to the multiobjective programming problem by repeatedly solving a single objective problem. This computational burden can be greatly reduced if appropriate parametric programming algorithms can be applied. For more detailed descriptions on the theory and algorithm of the two methods and other generating techniques, readers are referred to Cohon (1978) and Chankong and Haimes (1983).

6.7.4   Goal Programming

The *goal programming* procedure for a multiobjective programming problem is to identify an r when the goal is underachieved, overachieved, or both. For each objective, two types of goals can be used in a goal programming formulation, depending on the nature of the goal: *Two-sided goal* or *one-sided goal*. Referring to Fig. 6.9, a two-sided goal considers a penalty when the goal is not met exactly by the problem solution, whereas a one-sided goal incurs a penalty only when the specified goal is underachieved or overachieved, but not both, by the solution. For example, for the objective of profit maximization, the goal type is one-sided because the penalty would not occur if the solution results in a higher benefit than the specified goal; for the objective of maintaining the groundwater table at a fixed level, the goal type is a two-sided one because undesirable consequences would occur if such a level is not maintained.

In the goal programming formulation, the multiobjective problem, equation (6.86), is transformed into

$$\text{minimize} \sum_{k=1}^{K} (w_k^+ d_k^+ + w_k^- d_k^-) \tag{6.89a}$$

subject to

$$f_k(\mathbf{x}) - d_k^+ + d_k^- = G_k \qquad \text{for} \qquad k = 1, 2, \ldots, K \tag{6.89b}$$

$$\mathbf{g}(\mathbf{x}) = \mathbf{0} \tag{6.89c}$$

in which d_k^+ and d_k^- are the additional nonnegative decision variables representing, respectively, the deviations due to overachieving and underachieving the specified goal G_k for the kth objective, $f_k(\mathbf{x})$; w_k^+ and w_k^- are weighting factors associated with two types of deviation, respectively; $\mathbf{x}$ and $\mathbf{g}(\mathbf{x})$ is the vector of the original decision variables and constraints, respectively. The two nonnegative decision variables d_k^+ and d_k^- satisfy the condition that both will not be simultaneously greater than zero. The weighting factors w_k^+ and w_k^- can be used to indicate the relative importance of different types of deviations for the various objective functions considered in the problem. The solution technique for solving a goal programming problem depends on the nature of the original objective functions $f_k(\mathbf{x})$, and the constraints $\mathbf{g}(\mathbf{x})$.

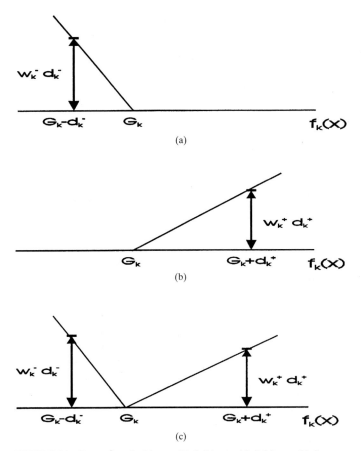

FIGURE 6.9   Types of goals: (*a*) one-sided; (*b*) one-sided; (*c*) two-sided.

The objective function of the goal programming formulation given above assumes that goals for all objectives are two-sided. Modifications should be made according to the goal type associated with each objective. For an objective function with a one-sided goal for which only underachievement is to be penalized, the term $w_k^+ d_k^+$ associated with the objective should be removed. Similarly, the term $w_k^- d_k^-$ is removed for the objective to which only overachievement is undesirable. It should be noted that the two deviation variables d_k^+ and d_k^- represent the status of the objective function with respect to its goal G_k. Therefore, regardless of the goal type, they must remain in the constraint equations corresponding to the objectives.

6.7.5   Utility Function Methods

Utility function techniques are based on the *multiattribute utility theory* (*MAUT*) which was designed to evaluate the *utility function* representing the preference of a decisionmaker in terms of multiple attributes. In the context of the MAUT framework, the objectives of a multiobjective programming problem are attributes and decision variables are alternatives of choice. Therefore, the utility function of a decisionmaker for a multiobjective programming problem can be viewed as a superobjective function to which the problem strives to maximize. Using the MAUT, the multiobjective programming model, Eqs. (6.86a–b), can be expressed as

$$\text{maximize } u[f_1(\mathbf{x}), f_2(\mathbf{x}), \ldots, f_K(\mathbf{x})] \tag{6.90a}$$

subject to

$$\mathbf{g}(\mathbf{x}) = \mathbf{0} \tag{6.90b}$$

in which $u[\ ]$ is the utility function depending on the values of objective functions. The decisionmaker prefers a feasible solution $\mathbf{x}_1$ to $\mathbf{x}_2$ if $u[f_1(\mathbf{x}_1), f_2(\mathbf{x}_1), \ldots, f_K(\mathbf{x}_1)] > u[f_1(\mathbf{x}_2), f_2(\mathbf{x}_2), \ldots, f_K(\mathbf{x}_2)]$.

The crust of solving a multiobjective programming problem using MAUT is the development of a utility, $u[\ ]$, which adequately represents the preference behavior of the decisionmaker for the multiobjective problem under consideration. For practicality reasons, the MAUT avoids direct assessment of $u[f_1, f_2, \ldots, f_K]$ because deriving a utility function for a single attribute, $u_k(f_k)$, is difficult enough. Many multiattribute utility functions used in practice are of the following general form

$$u[f_1, f_2, \ldots, f_K] = h[u_1(f_1), u_2(f_2), \ldots, u_K(f_K)] \tag{6.91}$$

which indicates that the utility function can be decomposed in terms of uniattribute utility functions. Decomposition reduces the dimensionality of the problem, and hence greatly simplifies the task of assessing the utility function. Zeleny (1982) lists several commonly used forms for utility decomposition.

Two fundamental concepts are essential in the development of utility functions: *Preferential independence* and *utility independence*. Zeleny (1982) offers the following definitions for the two. A pair of attributes f_a and f_b is preferentially independent of attribute f_c if the tradeoff between f_a and f_b is not affected by the level of f_c. Attribute f_a is utility independent of attribute f_b when conditional preferences for lotteries on f_a, given f_b, do not depend on the level of f_b. After information about the decisionmaker's preference is extracted, it is important to verify the compliance of the two independence conditions. Zeleny provides an excellent elaboration on the subject of multiattribute utility measurement.

Since the value of the utility function represents the degree of decisionmaker's preference, then, the decisionmaker would be indifferent to different combinations of objective function values that yield the same utility value. Therefore, geometric representation of constant utility in the objective function space is the indifference curve or surface. On an indifference curve, the change in utility value is zero, that is,

$$du = \sum_{k=1}^{K} \left(\frac{\partial u}{\partial f_k} \right) df_k \tag{6.92}$$

From Eq. (6.92), the tradeoff between the objectives f_k and f_1 can be computed as

$$\text{MRS}_{k1} = -\frac{\partial f_1}{\partial f_k} = -\frac{\partial u/\partial f_k}{\partial u/\partial f_1} \qquad \text{for} \qquad k \neq 1 \tag{6.93}$$

where MRS_{k1} is the marginal rate of substituting f_k for f_1.

6.7.6 Surrogate Worth Tradeoff Technique

The surrogate worth tradeoff (SWT) method is described by Haimes and Hall (1974); Haimes, et al. (1975); and Haimes (1977). Because of the difficulty in defining the utility function for real-life problems, the method utilizes the approximated utility based on the values of objective functions and the MRS.

Using the SWT method, the multiobjective programming problem in Eq. (6.86) is written as

$$\text{maximize } f_1(\mathbf{x}) \tag{6.94a}$$

subject to

$$f_k(\mathbf{x}) \geq f_k^o \quad \text{for all} \quad k \neq 1 \tag{6.94b}$$

$$\mathbf{g}(\mathbf{x}) = \mathbf{0} \tag{6.94c}$$

in which f_k^o are prespecified values for the various objective functions. Since the Lagrange multiplier associated with a constraint indicates the rate of change of the objective function with respect to the constraint's RHS (Sec. 6.3.2), therefore, the Lagrange multiplier λ_k corresponding to the constraint equation [Eq. (6.94b)] represents the tradeoff between the objective functions $f_k(\mathbf{x})$ and $f_1(\mathbf{x})$. Of interest in a multiobjective problem are those negative-valued λ's. Haimes and Hall (1974) show that the values of Lagrange multipliers λ_k for constraint equation Eq. (6.94b) is the MRT between $f_k(\mathbf{x})$ and $f_1(\mathbf{x})$, MRT_{k1}, indicating the amount of $f_1(\mathbf{x})$ that can be gained by sacrificing one unit of $f_k(\mathbf{x})$. In general, the MRT_{k1} is a non-decreasing function of f_k^o.

When tradeoff between $f_k(\mathbf{x})$ and $f_1(\mathbf{x})$ exists than $\lambda_k < 0$. To avoid the difficult task of assessing the true utility function, the SWT method requires the decisionmaker to use an ordinal scale to represent a preference according to the relative difference between the feasible tradeoff MRT_{k1} and the decisionmaker's desirable tradeoff MRS_{k1} to sacrifice $f_k(\mathbf{x})$ for gaining $f_1(\mathbf{x})$. The value assigned to the MRT_{k1} is called the *surrogate worth* SW_{k1} which, in turn, depends on f_k^o. In the SWT method, the *surrogate worth function* $\text{SW}_{k1}(f_k^o)$ is applied to a monotonic decreasing function of f_k^o. The assignment of surrogate worth value can be made using

$$\text{SW}_{k1} = \begin{cases} > \min(0, 10) & \text{if} \quad \text{MRS}_{k1} > \text{MRT}_{k1} \\ > \max(-10, 0) & \text{if} \quad \text{MRS}_{k1} < \text{MRT}_{k1} \end{cases} \tag{6.95}$$

with +10 indicating complete favor for substituting $f_k(\mathbf{x})$ for $f_1(\mathbf{x})$ and −10 indicating complete disfavor. The best compromised solution for $f_k(\mathbf{x})$ can be obtained when $\text{SW}_{k1} = 0$. This can be achieved by solving Eq. (6.94a–c) for $\lambda_k = \text{MRT}_{k1}$ under different f_k^o's in Eq. (6.94b) and requesting the corresponding surrogate worth from the decisionmaker (see Fig. 6.10).

The process can be repeated by considering different objective functions from which the best compromised objective function values, f_k^* for all $k \neq 1$, can be obtained. Then, the solution to the entire multiobjective problem can be obtained by solving Eqs. (6.94a–c) with f_k^o in Eq. (6.94b) replaced by f_k^*.

6.8 SELECTED APPLICATIONS IN WATER RESOURCES

Since the existing literature is too extensive to be exhaustively listed, this section only provides some sample references applying different optimization techniques to

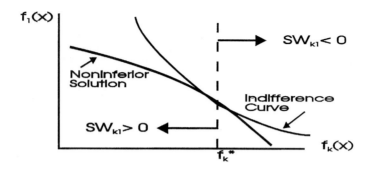

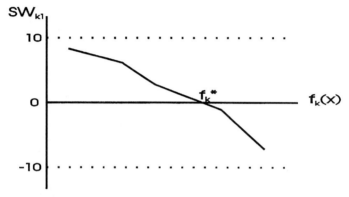

FIGURE 6.10   Identification of best compromised solution by surrogate worth tradeoff method.

a variety of water resource management, planning, and design problems. The applications are categorized according to the optimization technique used and the type of problem addressed. Tables 6.3 to 6.11, respectively, list applications of linear programming, nonlinear programming, dynamic programming, differential dynamic programming, multiobjective programming, linear quadratic control techniques, and applications of interfacing optimizers with simulators to various water resource problems, including reservoir operation and management, irrigation and agricultural management, conjunctive use management, water distribution systems, floodplain management, urban drainage systems, water quality management, groundwater management, and hydrologic and hydraulic modelings.

TABLE 6.3 List of Sample Applications of Linear Programming to Water Resource Problems

Application type	References
Conjunctive use	Paudyal and Das Gupta (1990); Tyagi (1988); Kaushal, et al. (1985); Chaturvedi and Chaube (1985); Louie, et al. (1982).
Water distribution systems	Kessler, et al. (1990); Ormsbee and Kessler (1990); Swamee and Shamir (1990); Kessler and Shamir (1989); Fujiwara and Dey (1988); Fujiwara, et al. (1987); Morgan and Goulter (1985); Bhave (1983a,b); Khanjani and Busch (1983); Quindry, et al. (1981); Shamir (1979).
Floodplain management	Weisz and Day (1974).
Groundwater management	Piper, et al. (1989); Chau (1988); Lindner, et al. (1988); Gorelick (1988); Atwood and Gorelick (1985); Schafer (1984); Tyagi and Narayana (1984); Willis and Liu (1984); Helweg and Arora (1983); Heidari (1982); Aguado, et al. (1974).
Urban drainage	Neugebauer and Schilling (1991); Elimam, et al. (1989); Yeh, et al. (1980).
Agriculture/irrigation	Buller, et al. (1991); Chaudhry and Young (1989); Batie, et al. (1989); Kaushal, et al. (1985); Tyagi and Narayana (1984); Howitt (1983); Moore, et al. (1974).
Hydrological modeling	Loaiciga and Church (1990); Lall and Olds (1987); Wang and Yu (1986); Bardsley (1984); Han and Rao (1980).
Reservoir operation and management	Reznicek and Simonovic (1992); Reznicek, et al. (1991); Tao and Lennox (1991); Lognathan and Bhattacharya (1990); Diaz and Fontane (1989); Tollow (1989); Kuczera (1989); Bhaskar and Whitlatch (1987); Lele (1987); Martin (1987); Moy, et al. (1986); Pogge, et al. (1985); Yeh (1985).
Water quality management	Lall and Lin (1991); Walker, et al. (1990); Ponnambalam, et al. (1990); Hayes (1989); Eheart and Park (1989); Bose, et al. (1988); Fontaine and Lesht (1987); McBean (1987); Valocchi and Eheart (1987); Jenq, et al. (1984); Olufeagba and Flake (1981).
Estuary management	Martin (1987).

TABLE 6.4 List of Sample Applications of Integer Programming to Water Resource Problems

Application type	References
Hydropower	Trezos (1991)
Drought management	Martin (1991)

TABLE 6.5 List of Sample Applications of Network Optimization to Water Resource Problems

Application type	References
Reservoir operation	Sabet and Creel (1991); Afshar, et al. (1991)

TABLE 6.6 List of Sample Applications of Nonlinear Programming to Water Resource Problems

Application type	References
Conjunctive use	Kiashyap and Chandra (1982); Rydzewski and Rashid (1981); Matsukawa, et al. (1992).
Water distribution systems	Ostfelt and Shamir (1993 a, b); Brion and Mays (1991); Lansey and Basnet (1991); Duan, et al. (1990); Lansey and Mays (1989); Lansey, et al. (1989); Chiplunkar, et al. (1986); Rowell and Barnes (1982); Whitlatch and Asplund (1981).
Groundwater management	Liu (1988); Schafer (1984); Larson, et al. (1977).
Urban drainage	Musselman and Talavage (1979); Lott (1976).
Reservoir operation and management	Sabet and Creel (1991); Cantiller (1986); Rosenthal, et al. (1983); Ford, et al. (1981).
Water quality management	Nix and Heaney (1988); Gorelick, et al. (1984); Loucks (1978); Spofford (1976); McNamara (1976); Bundgaard-Nielsen and Hwang (1976).
Estuary management	Tung, et al. (1990).
Water treatment plant	Wu and Chu (1991).
Hydropower	Tejada-Guibert, et al. (1990).

TABLE 6.7 List of Sample Applications of Dynamic Programming to Water Resource Problems

Application type	References
Conjunctive use	Onta, et al. (1991); Becklynck, et al. (1983).
Water distribution systems	Lansey and Awumah (1994); Martin (1990); Zessler and Shamir (1989); Flynn and Marino (1989); Ormsbee, et al. (1989).
Floodplain management	Hopkins, et al. (1982); Morin, et al. (1981); Hopkins, et al. (1976).
Groundwater management	Cleveland and Yeh (1991); Jones, et al. (1987).
Urban drainage	Kuo, et al. (1991); Li and Matthew (1990); Han and Rao (1987); Nouh (1987); Han, et al. (1980); Wenzel, et al. (1979); Froise and Burges (1978); Mays, et al. (1976); Mays and Bedient (1982).
Water supply systems	Martin (1987); Buras (1985).
Agriculture/irrigation	Tsakiris and Kiountouzis (1984); Jensen, et al. (1980); Rhenals and Bras (1981)
Hydrological/hydraulic modeling	Zavesky and Goodman (1988).
Reservoir operation and management	Braga Jr., et al. (1991); Piccardi and Soncini-Sessa (1991); Vedula and Mohan (1990); Kuo, et al. (1990); Druce (1990); Kelman, et al. (1990); Alaouze (1989); Liango (1988); Whitlatch, et al. (1988); Foufoula-Georgiou and Kitanidis (1988); Yeh (1976); Maidment and Chow (1981).
Water quality management	Joshi and Modak (1989, 1987); Stedinger and Bell-Graf (1979); Shih and Meier (1972).
Environmental	Liebman and Lynn (1966).

TABLE 6.8 List of Sample Applications of Differential Dynamic Programming to Water Resources Problems

Application type	References
Groundwater management	Chang, et al. (1992); Culver and Shoemaker (1992); Jones, et al. (1987); Liao and Shoemaker (1991); Makinde-Odusola and Marino (1989); Whiffen and Shoemaker (1993).
Reservoir operation and management	Murray and Yakowitz (1979), Carriaga and Mays (1995a, 1995b).
Estuary management	Li and Mays (1995).

TABLE 6.9 List of Sample Applications of Multiobjective Programming to Water Resources Problems

Application type	References
Groundwater management	Duckstein, et al. (1994); Chang, et al. (1992); Culver and Shoemaker (1992); Jones, et al. (1987); Liao and Shoemaker (1991); Makinde-Odusola and Marino (1989); Whiffen and Shoemaker (1993).
Reservoir operation and management	Crawley and Dandy (1993); Loganatham and Bhattacharya (1990); Can and Houck (1984).
Estuary management	Mao and Mays (1994).

TABLE 6.10 List of Sample Applications of Linear Quadratic Control to Water Resources Problems

Application type	References
Reservoir operation and management	Hooper, et al. (1991); Georgakakos and Marks (1987); Wasimi and Kitanidis (1983).
Estuary management	Zhao and Mays (1995a, b).

TABLE 6.11 List of Sample Applications of Interfacing Optimizers with Simulators

Application type	References
Water distribution system	Brion and Mays (1991); Lansey and Mays (1989); Cullinane, et al. (1992).
Groundwater management	Gorelick, et al. (1984); Wanakule, et al. (1986); See Chap. 16
Reservoir operation and management	Unver and Mays (1990).
Estuary management	Bao and Mays (1994a, b).

REFERENCES

Afshar, A., M. A. Marino, and A. Abrishamchi, "Reservoir Planning for Irrigation District," *Journal of Water Resources Planning and Management,* ASCE, 117(1): 74–85, 1991.

Aguado, F., I. Remson, M. F. Pikul, and W. A. Thomas, "Optimum Pumping for Aquifer Dewatering," *J. Hydraulic Division,* ASCE, 100: 860–877, 1974.

Alaouze, C. M., "Reservoir Releases to Uses with Different Reliability Requirements," *Water Resources Bulletin,* 25(6): 1163–1168, 1989.

Anderson, B., and J. B. Moore, *Optimal Control of Linear Quadratic Methods,* Prentice-Hall, Englewood Cliffs, N.J., 1990.

Athans, M., and P. L. Falb, *Optimal Control,* McGraw-Hill, New York, 1986.

Atwood, D. F., and S. M. Gorelick, "Optimal Hydraulic Containment of Contaminated Ground Water," *Proceedings of the Fifth National Symposium and Exposition on Aquifer Restoration and Ground Water Monitoring,* The Fawcett Center, Columbus, Ohio, May 21–24, 1985, pp. 328–344.

Auslander, D. M., Y. Takahashi, and M. J. Rabins, *Introduction to Systems and Control,* McGraw-Hill, New York, 1974.

Bao, Y., and L. W. Mays, "Optimization of Freshwater Inflows to the Lavaca-Tres Palacios Estuary," *Journal of Water Resources Planning and Management,* ASCE, 120(2): 218–236, March/April 1994a.

Bao, Y., and L. W. Mays, "New Methodology for Optimization of Freshwater Inflows to Estuaries," *Journal of Water Resources Planning and Management,* ASCE, 120(2): 199–217, March/April 1994b.

Bardsley, W. E., "Conservative Estimation of Groundwater Volumes: Application of Linear Programming to Tritium Data," *Journal of Hydrology,* 67 (1–4): 183–193, 1984.

Batie, S. S., R. A. Kramer, and W. E. Cox, "Economic and Legal Analysis of Strategies for Managing Agricultural Pollution of Groundwater," Virginia Polytechnic Institute and State University, Blacksburg, Va., Dept. of Agricultural Economics, Report, October 1989.

Becklynck, J., M. Besbes, P. Combes, P. Hubert, and G. Marsily, "Managing the Integration of Groundwater and Surface Water Resources: A Case Study of Supplying Potable Water to the Lille Metropolitan Area, Ground Water in Water Resources Planning," *Proceedings* of a Symposium, Koblenz, West Germany, Aug. 28–Sept. 3, 1983, IAHS Publication no. 142, vol. II. : 685–695.

Bellman, R., and S. Dreyfus, *Applied Dynamic Programming,* Princeton University Press, Princeton, N.J., 1962.

Bertsekas, D. P. *Dynamic Programming: Deterministic and Stochastic Models,* Prentice-Hall, Inc., Englewood Cliffs, N.J., 1987.

Bhaskar, N. R., and E. E. Whitlatch, "Comparison of Reservoir Linear Operation Rules Using Linear and Dynamic Programming," *Water Resources Bulletin,* 23(6): 1027–1036, 1987.

Bhave, P. R., "Optimization of Gravity-Fed Water Distribution Systems: Theory," *Journal of Environmental Engineering,* ASCE, 109 (1): 189–205, 1983a.

Bhave, P. R., "Optimization of Gravity-Fed Water Distribution Systems: Application," *Journal of Environmental Engineering,* ASCE, 109(2): 383–395, 1983b.

Bose, S. K., P. Ray, and B. K. Dutta, "Water Quality Management of the Hooghly Estuary—A Linear Programming Model," *Water Science and Technology,* 20(6/7): 235–242, 1988.

Braga Jr., B. P. F., W. W-G. Yeh, L. Becker, and M. T. L. Barrow, "Stochastic Optimization of Multiple-Reservoir-System Operation," *Journal of Water Resources Planning and Management,* ASCE, 117(4): 471–481, 1991.

Brion, L. M., and L. W. Mays, "Methodology for Optimal Operation of Pumping Stations in Water Distribution Systems," *Journal of Hydraulic Engineering,* ASCE, 117(11): 1551–1571, 1991.

Brogan, W. L., *Modern Control Theory,* Quantum Publishers, New York, 1974.

Buller, O., H. L. Manges, L. R. Stone, and J. R. Williams, "Modeled Crop Water Use and Soil Water Drainage," *Agricultural Water Management*, 19(2): 117–134, 1991.

Bundgaard-Nielsen, M., and C. L. Hwang, "A Review on Decision Models in Economics of Regional Water Quality Management," *Water Resources Bulletin*, 12(3): 461–479, 1976.

Buras, N., "Application of Mathematical Programming in Planning Surface Water Storage," *Water Resources Bulletin*, AWRA, 21(6): 1013–1020, 1985.

Buys, J. D., "Dual Algorithms for Constrained Optimization Problems," Ph.D. thesis, University of Leiden, Netherlands, 1972.

Can, E. K., and M. H. Houck, "Real-Time Reservoir Operations by Goal Programming," *Journal of Water Resources Planning and Management*, ASCE, 110(3): 297–309, 1984.

Cantiller, R. R., "Nonlinear Approach to the Determination of Reservoir Operating Rules," Ph.D. dissertation, Purdue University, Dept. of Civil Engineering, Lafayette, Ind., 1986.

Carriaga, C. C., and L. W. Mays, "Optimal Control Approach for Sedimentation Control in Alluvial Rivers," *Journal of Water Resources Planning and Management*, ASCE, 121(6): 408–417, Nov./Dec. 1995a.

Carriaga, C. C., and L. W. Mays, "Optimization Modeling for Sedimentation in Alluvial Rivers," *Journal of Water Resources Planning and Management*, ASCE, 121(3): 251–259, Nov./Dec. 1995b.

Chang, L.-C., "The Application of Constrained Optimal Control Algorithms to Groundwater Remediation," Ph.D. dissertation, Cornell University, Ithaca, N.Y., 1990.

Chang, L.-C., C. A. Shoemaker, and P. L.-F. Liu, "Application of a Constrained Optimal Control Algorithm to Groundwater Remediation," *Water Resources Research*, 28(12): 3157–3173, 1992.

Chankong, V., and Y. Y. Haimes, *Multiobjective Decisionmaking: A Theory and Methodology*, Elsevier Science Publishing, New York, 1983.

Chaturvedi, M. C., and U. C. Chaube, "Irrigation System Study in International Basin," *Journal of Water Resources Planning and Management*, ASCE, 111(2): 137–148, 1985.

Chau, T. S., "Analysis of Sustained Ground-Water Withdrawals by the Combined Simulation-Optimization Approach," *Groundwater*, 26(4): 454–463, 1988.

Chaudhry, M. A., and R. A. Young, "Valuing Irrigation Water in Punjab Province, Pakistan: A Linear Programming Approach," *Water Resources Bulletin*, AWRA, 25(5): 1055–1061, 1989.

Chiplunkar, A. V., S. L. Mehndiratta, and P. Khanna, "Looped Water Distribution System Optimization for Single Loading," *Journal of Environmental Engineering*, ASCE, 112(2): 264–279, 1986.

Cleveland, T. G., and W. W-G. Yeh, "Optimal Configuration and Scheduling of Ground-Water Tracer Test," *Journal of Water Resources Planning and Management*, ASCE, 117(1): 37–51, 1991.

Cohon, J. L., *Multiobjective Programming and Planning*, Academic Press, New York, 1978.

Cooper, L. L., and M. W. Cooper, *Introduction to Dynamic Programming*, Pergamon Press, Elmsford, N.Y., 1981.

Crawley, P. D., and G. C. Dandy, "Optimal Operation of Multiple-Reservoir System," *Journal of Water Resources Planning and Management*, ASCE, 119(1): 1–17, 1993.

Cullinane, M. J., K. Lansey, and L. W. Mays, "Optimization—Availability Based Design of Water Distribution Networks," *Journal of Hydraulic Engineering*, ASCE, 118(3): 420–441, Mar. 1992.

Culver, T. B., "Dynamic Optimal Control of Groundwater Remediation with Management Periods: Linearized and Quasi-Newton Approaches," Ph.D. dissertation, Cornell University, Ithaca, N.Y., 1991.

Culver, T., and C. Shoemaker, "Dynamic Optimal Control for Groundwater Remediation with Flexible Management Periods," *Water Resources Research*, 28(3): 629–641, 1992.

Culver, T. B., and C. A. Shoemaker, "Dynamic Optimal Control for Groundwater Remediation with Flexible Management Periods," *Water Resources Research*, AGU, 28(3): 629–641, 1992.

Denardo, E. V., *Dynamic Programming Theory and Applications,* Prentice-Hall, Englewood Cliffs, N.J., 1982.

Dennis, J. E., and R. B. Schnable, *Numerical Methods for Unconstrained Optimization,* Prentice-Hall, Englewood Cliffs, N.J., 1983.

Diaz, G. E., and D. G. Fontane, "Hydropower Optimization via Sequential Quadratic Programming," *Journal of Water Resources Planning and Management,* ASCE, 115(6): 715–734, 1989.

Dreyfus, S., and A. Law, *The Art and Theory of Dynamic Programming,* Academic Press, New York, 1977.

Druce, D. J., "Incorporating Daily Flood Control Objectives Into a Monthly Stochastic Dynamic Programing Model for a Hydroelectric Complex," *Water Resources Research,* AGU, 26(1): 5–11, 1990.

Duan, N., L. W. Mays, and K. E. Lansey, "Optimal Reliability-Based Design of Pumping and Distribution Systems," *Journal of Hydraulic Engineering,* ASCE, 116(2): 249–268, 1990.

Duckstein, L., W. Treichel, and S. El Magnouni, "Ranking Ground-Water Management Alternative by Multicriterion Analysis," *Journal of Water Resources Planning and Management,* ASCE, 120(4): 546–565, 1994.

Dyer, P., and S. McReynolds, *The Computational Theory of Optimal Control,* Academic Press, New York, 1970.

Edgar, T. F., and D. M. Himmelblau, *Optimization of Chemical Processes,* McGraw-Hill, New York, 1988.

Eheart, J. W., and H. Park, "Effects of Temperature Variation on Critical Stream Dissolved Oxygen," *Water Resources Research,* 25(2): 145–151, 1989.

Elimam, A. A., C. Charalambous, and F. H. Ghobrial, "Optimum Design of Large Sewer Networks," *Journal of Environmental Engineering,* ASCE, 15(6): 1171–1190, 1989.

Fletcher, R., *Practical Methods of Optimization, Vol. 1, Unconstrained Optimization,* John Wiley, New York, 1980.

Flynn, L. E., and M. A. Marino, "Aqueduct and Reservoir Capacities for Distribution Systems," *Journal of Water Resources Planning and Management,* ASCE, 115(5): 547–565, 1989.

Fontaine, T. D., and B. M. Lesht, "Improving the Effectiveness of Environmental Management Decisions with Optimization and Uncertainty Analysis Techniques," in: *Systems Analysis in Water Quality Management.* Pergamon Press, New York, 1987, pp. 31–41.

Ford, D. T., R. Garland, and C. Sullivan, "Operation Policy Analysis: Sam Rayburn Reservoir," *Journal of the Water Resources Planning and Management Division,* ASCE, 107(WR2): 339–350, 1981.

Froise, S., and S. J. Burges, "Least-Cost Design of Urban-Drainage Networks," *Journal of Water Resources Planning and Management,* ASCE, 104(WR1): 75–92, 1978.

Foufoula-Georgiou, E., and P. K. Kitanidis, "Gradient Dynamic Programming for Stochastic Optimal Control of Multidimensional Water Resources Systems," *Water Resources Research,* 24(8): 1345–1359, 1988.

Fujiwara, O., and D. Dey, "Method for Optimal Design of Branched Networks on Flat Terrain," *Journal of Environmental Engineering,* 114(6): 1464–1475, 1988.

Fujiwara, O., B. Jenchaimahakoon, and N. C. P. Edirisinghe, "Modified Linear Programming Gradient Method for Optimal Design of Looped Water Distribution Networks," *Water Resources Research,* 23(6): 977–982, 1987.

Garfinkel, R. S., and G. L. Nemhauser, *Integer Programming,* John Wiley, New York, 1972.

Gass, S. I., *Linear Programming: Methods and Application,* McGraw-Hill, New York, 1985.

Georgakakos, A. P., and D. H. Marks, "A New Method for the Real-Time Operation of Reservoir Systems," *Water Resources Research,* 23(7): 1376–1390, 1987.

Gershwin, S., and D. Jacobson, "A Discrete-Time Differential Dynamic Programming Algorithm with Application to Optimal Orbit Transfer," *AIAA J.,* 8: 1616–1626, 1970.

Gill, P. E., W. Murray, and M. H. Wright, *Practical Optimization,* Academic Press, London and New York, 1981.

Gorelick, S. M., *Review of Groundwater Management Models, Efficiency in Irrigation: The Conjunctive Use of Surface and Groundwater Resources,* Winrock International, Arlington, Va., 1988, pp. 103–121.

Gorelick, S. M., C. I. Voss, P. E. Gill, W. Murray, M. A. Saunders, and M. M. Wright, "Aquifer Reclamation Design: The Use of Containment Transport Simulation Combined with Nonlinear Programming," *Water Resources Research,* AGU, 20(4): 415–427, 1984.

Hall, W., and W. Butcher, "Optimal Timing of Irrigation," *Journal of Irrigation and Drainage Division,* ASCE, 94(2): 267–275, 1968.

Han, J., and A. R. Rao, "Optimal Design of Storm Drain Systems," *Computational Hydrology '87,* Lighthouse Publications, Mission Viejo, Cal., 1987, pp. C4–C8.

Han, J., and A. R. Rao, "Optimal Parameter Estimation and Investigation of Objective Functions of Urban Runoff Models," Purdue University Water Resources Research Center, Lafayette, Ind., Technical Report no. 135, 1980.

Han, J., A. R. Rao, and M. H. Houck, "Least Cost Design of Urban Drainage Systems," Purdue University Water Resources Research Center, Lafayette, Ind., Technical Report no. 138, 1980.

Hayes, D. F., "Systems Approach to Water Quality Management During Drought Periods in the Cumberland River Basin," *Proceedings of the Nineteenth Mississippi Water Resources Conference,* Water Resources Research Institute, Mississippi State University, Mississippi State, Miss., 1989.

Heidari, M., "Application of Linear System Theory and Linear Programming to Groundwater Management in Kansas," *Water Resource Bulletin,* 18: 1003–1012, 1982.

Helweg, O., and S. Arora, "Controlling Ground Water Degradation in the Tulare Lake Basin," *Water Resources Bulletin,* 19(5): 837–844, 1983.

Hillier, F. S., and G. J. Lieberman, *Introduction to Operation Research,* 5th ed., McGraw-Hill, New York, 1990.

Himmelblau, D. M., *Applied Nonlinear Programming,* McGraw-Hill, New York, 1972.

Hooper, E. R., A. P. Georgakakos, and D. P. Lettenmaier, "Optimal Stochastic Operation of Salt River Project, Arizona," *Journal of Water Resources Planning and Management,* ASCE, 117(5): 566–587, 1991.

Hopkins, L. D., E. D. J. Brill, J. C. Liebman, and H. G. J. Wenzel, "Flood Plain Management Through Allocation of Land Uses—A Dynamic Programming Model," Illinois Water Resources Center, Urbana, Ill., *Research Report* no. 117, 1976.

Hopkins, L. D., E. D. J. Brill, and B. Wong, "Generating Alternative Solutions for Dynamic Programming Models of Water Resources Problems," *Water Resources Research,* 18(4): 782–790, 1982.

Howitt, R. E., "Calibration of Large-Scale Economic Models of the Agricultural Sector and Reliable Estimation of Regional Derived Demand for Water," University of California, Davis, California Water Resources Center Report, 1983.

Jacobson, D. H., D. H. Martin, M. Pachter, and T. Geveci, "Extensions of Linear-Quadratic Control Theory," in A. V. Balakrishnan, and M. Thoma, eds., *Lecture Notes in Control and Information Sciences,* Springer-Verlag, 1980.

Jacobson, D., and D. Mayne, *Differential Dynamic Programming,* Elsevier, New York, 1970.

Jenq, T-R., M. L. Granstrom, S-F. Hsueh, and C. G. Uchrin, "Phosphorus Management LP Model Case Study," *Water Resources Bulletin,* 20(4): 511–520, 1984.

Jensen, P., and J. W. Barnes, *Network Flow Programming,* Wiley, New York, 1980.

Jensen, P. A., H. W. Chu, and D. D. Cochard, "Network Flow Optimization for Water Resources Planning With Uncertainties in Supply and Demand," University of Texas, Center for Research in Water Resources, Technical Report, CRWR-172, 1980.

Jones, L., R. Willis, and W. W. G. Yeh, "Optimal Control of Nonlinear Groundwater Hydraulics Using Differential Dynamic Programming," *Water Resources Research,* 23(11): 2097–2106, 1987.

Jones, L. C., R. Willis, and W. W. Yeh, "Optimal Control of Nonlinear Groundwater Hydraulics Using Differential Dynamic Programming," *Water Resource Research,* 23(11): 2097–2217, 1987.

Joshi, V., and P. Modak, "Heuristic Algorithms for Waste Load Allocation in a River Basin," *Water Science and Technology*, 21(8/9): 1057–1064, 1989.

Joshi, V., and P. Modak, "Reliability Based Discrete Differential Dynamic Programming Model for River Water Quality Management under Financial Constraint," *Systems Analysis in Water Quality Management*, Pergamon Press, New York, 1987, pp. 15–22.

Karmarkar, N., "A New Polynomial-Time Algorithm for Linear Programming," *Combinatorica*, 4(4): 373–395, 1984.

Kaushal, M. P., S. D. Khepar, and S. N. Panda, "Saline Groundwater Management and Optimal Cropping Pattern," *Water International*, 10(2): 86–91, 1985.

Kelman, J., J. R. Stedinger, L. A. Cooper, E. Hsu, and S-Q. Yuan, "Sampling Stochastic Dynamic Programming Applied to Reservoir Operation," *Water Resources Research*, 26(3): 447–454, 1990.

Kessler, A., and U. Shamir, "Analysis of the Linear Programming Gradient Method for Optimal Design of Water Supply Networks," *Water Resources Research*, 25(7): 1469–1480, 1989.

Kessler, V., L. Ormsbee, and U. Shamir, "Methodology for Least-Cost Design of Invulnerable Water Distribution Networks," *Civil Engineering Systems*, 7(1): 20–28, 1990.

Khanjani, M. J., and J. R. Busch, "Optimal Irrigation Distribution Systems with Internal Storage," *Transactions of the American Society of Agricultural Engineers*, 26(3): 743–747, 1983.

Khatchian, L., "A Polynomial Algorithm in Linear Programming," *Soviet Mathematics*, Doklady 20, 1979, pp. 191–194.

Kiashyap, D., and S. Chandra, "Distributed Conjunctive Use Model for Optimal Cropping Pattern, Optimal Allocation of Water Resources," IAHS Publication No. 135, *Proceedings* of a Symposium held at the First Scientific General Assembly of the IAHS, Exeter, England, July 19–30, 1982, pp. 377–384.

Kuczera, G., "Fast Multireservoir Multiperiod Linear Programing Models," *Water Resources Research*, 25(2): 169–176, 1989.

Kuo, J-T., B. C. Yen, and G-P. Hwang, "Optimal Design for Storm Sewer System with Pumping Stations," *Journal of Water Resources Planning and Management*, ASCE, 117(1): 11–27, 1991.

Kuo, J. T., N. S. Hsu, W. S. Chu, S. Wan, and Y. J. Lin, "Real-Time Operation of Tanshui River Reservoirs," *Journal of Water Resources Planning and Management*, ASCE, 116(3): 349–361, 1990.

Lall, U., and J. Olds, "Parameter Estimation Model for Ungaged Streamflows," *Journal of Hydrology*, 92(3–4): 245–262, 1987.

Lall, U., and Y. C. Lin, "Groundwater Management Model for Salt Lake County, Utah with Some Water Rights and Water Quality Considerations," *Journal of Hydrology*, 123(3–4): 367–393, 1991.

Land, A., and A. Doig, "An Automatic Method of Solving Discrete Programming Problems," *Econometrica*, 28(3), 1960.

Lansey, K. E., and K. Awumah, "Optimal Pump Operations Considering Pump Switches," *Journal of Water Resources Planning and Management*, ASCE, 120(1): 17–35, 1994.

Lansey, K. E., and C. Basnet, "Parameter Estimation for Water Distribution Networks," *Journal of Water Resources Planning and Management*, ASCE, 117(1): 126–144, 1991.

Lansey, K. E., N. Duan, L. W. Mays, and Y-K. Tung, "Water Distribution System Design under Uncertainties," *Journal of Water Resources Planning and Management*, ASCE, 115(5): 630–645, 1989.

Lansey, K. E., and L. W. Mays, "Optimization Model for Water Distribution System Design," *Journal of Hydraulic Engineering*, ASCE, 115(10): 1401–1418, 1989.

Lapidus, L., and R. Luus, *Optimal Control of Engineering Processes*, Blaisdell Publishing Company, Waltham, Mass. 1967.

Larson, S. P., T. Maddock III, and S. S. Papadopulos, "Optimization Techniques Applied to Ground-Water Development," *Proceedings* of Symposium on the Optimal Development and Management of Groundwater, Birmingham, England, July 1977, International Association of Hydrogeology, pp. E57–E66.

Lasdon, L. S., R. L. Fox, and M. W. Ratner, "Nonlinear Optimization Using the Generalized Reduced Gradient Method," *Revue Francaise d'Automatique, Informatique et Recherche Operationnelle,* V-3, November 1974, pp. 73–104.

Lasdon, L. S., and A. D. Waren, "Generalized Reduced Gradient Software for Linearly and Nonlinearly Constrained Problems," in *Design and Implementation of Optimization Software,* H. J. Greenberg(ed.), Sijthoff and Noordhoff, Rockville, Md., 1978, pp. 363–397.

Lele, S. M., "Improved Algorithms for Reservoir Capacity Calculation Incorporating Storage-Dependent Losses and Reliability Norm," *Water Resources Research,* 23(10): 1819–1823, 1987.

Li, G., and R. G. S. Matthew, "New Approach for Optimization of Urban Drainage Systems," *Journal of Environmental Engineering,* 116(5): 927–944, 1990.

Li, G., and L. W. Mays, "Differential Dynamic Programming for Estuarine Management," *Journal of Water Resources Planning and Management,* ASCE, 121(6): 455–463, Nov./Dec. 1995.

Liango, Q., "Reliability Constrained Markov Decision Programming and Its Practical Application to the Optimization of Multipurpose Reservoir Regulation, Computational Methods in Water Resources," in *Numerical Methods for Transport and Hydrologic Processes,* vol. 2, Elsevier, New York, 1988, pp. 417–422.

Liao, L. Z., and C. A. Shoemaker, "Convergence in Unconstrained Discrete-Time Differential Dynamic Programming," *IEEE Trans. Automatic Control,* 36(6): 692–706, 1991.

Liebman, J. S., L. S. Lasdon, L. Schrage, and A. Waren, *Modeling and Optimization with GINO,* Scientific Press, Palo Alto, Cal., 1986.

Liebman, J., and W. Lynn, "The Optimal Allocation of Stream Dissolved Oxygen," *Water Resources Research,* 2(3): 581–591, 1966.

Lindner, W., K. Lindner, and G. Karadi, "Optimal Groundwater Management in Two-Aquifer Systems," *Water Resources Bulletin,* 24(1): 27–33, 1988.

Liu, C. C. K., "Solute Transport Modeling in Heterogeneous Soils: Conjunctive Application of Physically Based and System Approaches," *Journal of Contaminant Hydrology,* 3(1): 97–111, 1988.

Loaiciga, H. A., and R. L. Church, "Linear Programs for Nonlinear Hydrologic Estimation," *Water Resources Bulletin,* 26(4): 645–656, 1990.

Loganathan, G. V., and D. Bhattacharya, "Goal-Programming Techniques for Optimal Reservoir Operations," *Journal of Water Resources Planning and Management,* ASCE, 116(6): 820–838, 1990.

Lott, D. L., "Optimization Model for the Design of Urban Flood-Control Systems," Texas University at Austin, Center for Research in Water Resources, Technical Report, CRWR-141, 1976.

Loucks, D. P., "Planning Comprehensive Water Quality Protection Systems," American-Soviet Symposium on Use of Mathematical Models to Optimize Water Quality Management (Kharkov and Rostov-on-Don, USSR; December 9–16, 1975), Environmental Research Laboratory, U.S. Environmental Protection Agency, Gulf Breeze, Fla., Report no. EPA-600/9-78-024, 1978: 1–33.

Loucks, D. P., "Planning Comprehensive Water Quality Protection Systems, American-Soviet Symposium on Use of Mathematical Models to Optimize Water Quality Management" (Kharkov and Rostov-on-Don, USSR; December 9–16, 1975), Environmental Research Laboratory, U.S. Environmental Protection Agency, Gulf Breeze, Fla., Report no. EPA-600/9-78-024, 1978: 1–33.

Louie, P. W. F., W. W.-G. Yeh, and N.-S. Hsu, "An Approach to Solving a Basin-Wide Water Resources Management Planning Problem with Multiple Objectives," California Water Resources Center, Contribution no. 184, 1982.

Luenberger, D. G., *Introduction to Linear and Nonlinear Programming,* Addison-Wesley, Reading, Mass., 1984.

Maidment, D. R., and V. T. Chow, "Stochastic State Variable Dynamic Programming for Reservoir System Analysis," *Water Resources Research,* 17(6): 1578–1584, 1981.

Makinde-Odusola, B. A., and M. A. Marino, "Optimal Control of Groundwater by the Feedback," *Water Resources Research,* 25(6): 1341–1352, 1989.

Martin, Q. W., "Linear Water-Supply Pipeline Capacity Expansion Model," *Journal of Hydraulic Engineering,* 116(5): 675–690, 1990.

Martin, Q. W., "Estimating Freshwater Inflow Needs for Texas Estuaries by Mathematical Programming," *Water Resources Research,* 23(2): 230–238, 1987.

Martin, Q. W., "Hierarchical Algorithm for Water Supply Expansion," *Journal of Water Resources Planning and Management,* 113(5): 677–695, 1987.

Martin, Q. W., "Optimal Daily Operation of Surface-Water Systems," *Journal of Water Resources Planning and Management,* 113(4): 453–470, 1987.

Martin, Q. W., "Drought Management Plan for Lower Colorado River in Texas," *Journal of Water Resources Planning and Management,* ASCE, 117(6): 645–661, 1991.

Mao, N., and L. W. Mays, "Goal Programming Models for Determining Freshwater Inflows to Bays and Estuaries," *Journal of Water Resource Planning and Management,* ASCE, 120(3): 316–329, May/June 1994.

Matsukawa, J., B. A. Finney, and R. Willis, "Conjunctive Use Planning in Mad River Basin, California," *Journal of Water Resources Planning and Management,* ASCE, 118(2): 115–132, 1992.

Maybeck, P. S., *Stochastic Models, Estimation, and Control,* vol. 1, Academic Press, New York, 1979.

Maybeck, P. S., *Stochastic Models, Estimation, and Control,* vol. 3, Academic Press, New York, 1982.

Mayne, D., "A Second-Order Gradient Method for Determining Optimal Trajectories of Non-Linear Discrete-Time Systems," *Internat. J. Control,* 3: 85–95, 1966.

Mays, L. W., and P. B. Bedient, "Model for Optimal Size and Location of Detention," *Journal of the Water Resources Planning and Management Division,* 108(WR3): 270–285, 1982.

Mays, L. W., and Y. K. Tung, *Hydrosystems Engineering and Management,* McGraw-Hill, New York, 1992.

Mays, L. W., H. G. Wenzel, and J. C. Liebman, "Model for Layout and Design of Sewer Systems," *Journal of the Water Resources Planning and Management Division,* ASCE, 102(WR2): 385–405, 1976.

McBean, E. A., "Developing Management Positions for Acidic Emission Reduction Negotiations," *Systems Analysis in Water Quality Management.* Pergamon Press, New York, 1987, pp. 51–59.

McCormick, G. P., *Nonlinear Programming: Theory, Algorithms, and Applications,* John Wiley, New York, 1983.

McNamara, J. R., "An Optimization Model for Regional Water Quality Management," *Water Resources Research,* 12(2): 125–134, 1976.

Moore, C. V., H. H. Snyder, and P. Sun, "Effects of Colorado River Water Quality and Supply on Irrigated Agriculture," *Water Resources Research,* 10(2): 137–144, 1974.

Morgan, D. R., and I. C. Goulter, "Optimal Urban Water Distribution Design," *Water Resources Research,* 21(5): 642–652, 1985.

Morin, T. L., W. L. J. Meier, and K. S. Nagaraj, "Optimal Mix of Adjustments to Floods," Purdue University Water Resources Research Center, Lafayette, Ind., Technical Report no. 139, 1981.

Moy, W. S., J. L. Cohon, and C. S. ReVelle, "Programming Model for Analysis of the Reliability, Resilience, and Vulnerability of a Water Supply Reservoir," *Water Resources Research,* 22(4): 489–498, 1986.

Murray, M., and S. J. Yakowitz, "The Application of Optimal Control Methodology to Nonlinear Programming Problems," *Math. Programming,* vol. 21, 1981, pp. 331–347.

Murray, D. M., and S. J. Yakowitz, "Constrained Differential Dynamic Programming and its Application to Multireservoir Control," *Water Resource Research,* 15(5): 1017–1027, 1979.

Murtaugh, B. A., and M. A. Saunders, "Large-Scale Linearly Constrained Optimization," *Mathematical Programming,* 14: 41–72, 1978.

Musselman, K. J., and J. J. Talavage, "Interactive Multiple Objective Optimization," Purdue University Water Resources Research Center, Lafayette, Ind., Technical Report no. 121, 1979.

Nemhauser, G. L., and L. A. Wolsey, *Integer and Combinatorial Optimization,* Wiley, New York, 1988.

Neugebauer, K., W. Schilling, and J. Weiss, "Network Algorithm for the Optimum Operation of Urban Drainage Systems," *Water Science and Technology,* 24(6): 209–216, 1991.

Nix, S. J., and J. P. Heaney, "Optimization of Storm Water Storage-Release Strategies," *Water Resources Research,* 24(11): 1831–1838, 1988.

Nouh, M., "Storm Sewer Design Sensitivity Analysis Using ILSD-2 Model," *Journal of Water Resources Planning and Management,* 113(1): 151–158, 1987.

Ohno, K., "A New Approach of Differential Dynamic Programming for Discrete Time Systems," *IEEE Trans. Automat. Control,* AC-23: 37–47, 1978.

Olufeagba, B. J., and R. H. Flake, "Modelling and Control of Dissolved Oxygen in an Estuary," *Ecological Modelling,* 14(1): 79–94, 1981.

Onta, P. R., A. Das Gupta, and R. Harboe, "Multistep Planning Model for Conjunctive Use of Surface-Water and Ground-Water Resources," *Journal of Water Resources Planning and Management,* 117(6): 662–678, 1991.

Ormsbee, L. E., T. M. Walski, D. V. Chase, and W. W. Sharp, "Methodology for Improving Pump Operation Efficiency," *Journal of Water Resources Planning and Management,* 115(2): 148–164, 1989.

Ormsbee, L., and A. Kessler, "Optimal Upgrading of Hydraulic-Network Reliability," *Journal of Water Resources Planning and Management,* 116(6): 784–802, 1990.

Ostfelt, A., and U. Shamir, "Optimal Operation of Multiquality Networks. II: Steady Condition," *Journal of Water Resources Planning and Management,* ASCE, 119(6): 645–662, 1993a.

Ostfelt, A., and U. Shamir, "Optimal Operation of Multiquality Networks. II: Unsteady Condition," *Journal of Water Resources Planning and Management,* ASCE, 119(6): 663–684, 1993b.

Paudyal, G. N., and A. Das Gupta, "Irrigation Planning by Multilevel Optimization," *Journal of Irrigation and Drainage Engineering,* 116(2): 273–291, 1990.

Piccardi, C., and R. Soncini-Sessa, "Stochastic Dynamic Programming for Reservoir Optimal Control: Dense Discretization and Inflow Correlation Assumption Made Possible by Parallel Computing," *Water Resources Research,* 27(5): 729–741, 1991.

Piper, S., W. Y. Huang, and M. Ribaudo, "Farm Income and Ground Water Quality Implications from Reducing Surface Water Sediment Deliveries," *Water Resources Bulletin,* 25(6): 1217–1230, 1989.

Pogge, E. C., Y. S. Yu, and G. T. Wang, "Study Of Multireservoir Operation With Minimum Desirable Flow Constraints," Kansas Water Resources Research Institute, Manhattan, Kan., Contribution no. 251, 1985.

Ponnambalam, K., E. A. McBean, and T. E. Unny, "Impacts of Meteorological Variations on Acid Rain Abatement Decisions," *Journal of Environmental Engineering,* 116(6): 1063–1075, 1990.

Quindry, G. E., E. D. Brill, and J. C. Liebman, "Optimization of Looped Water Distribution Systems," *Journal of the Environmental Engineering Division,* 107(EE4): 665–679, 1981.

Reznicek, K. K., S. P. Simonovic, and C. R. Bector, "Optimization of Short-Term Operation of a Single Multipurpose Reservoir—A Goal Programming Approach," *Canadian Journal of Civil Engineering,* 18(3): 397–406, 1991.

Reznicek, K. K., and S. P. Simonovic, Issues in Hydropower Modeling Using GEMSLP Algorithm," *Journal of Water Resources Planning and Management,* ASCE, 118(1): 54–70, 1992.

Rhenals, A. E., and R. L. Bras, "The Irrigation Scheduling Problem and Evapotranspiration Uncertainty," *Water Resources Research,* 17(5): 1328–1338, 1981.

Rockafellar, R. T., "A Dual Approach to Solving Nonlinear Programming Problems by Unconstrained Optimization," *Math. Programming,* 5: 354–373, 1973a.

Rockafellar, R. T., "The Multiplier Method of Hestenes and Powell Applied to Convex Programming," SIAM, *J. Control and Optimization,* 12: 268–285, 1973b.

Rockafellar, R. T., "Augmented Lagrangian Multiplier Functions and Duality in Nonconvex Programming," SIAM, *J. Applied Math,* 12: 555–562, 1974.

Rosenthal, R. E., M. Hanscom, and R. S. Dembo, "Nonlinear Programming Methods in Reservoir System Management," Tennessee University, Knoxville, Tenn., Research Report no. 93, 1983.

Rosenthal, R. E., "Scheduling Reservoir Releases for Maximum Hydropower Benefit by Nonlinear Programming on a Network," University of Tennessee, College of Business Administration, Knoxville, Tenn., Working Paper no. 43, 1977.

Rowell, W. F., and J. W. Barnes, "Obtaining Layout of Water Distribution Systems," *Journal of the Hydraulics Division,* ASCE, 108(HY1): 137–148, 1982.

Rubin, Y., *Hierarchical Method for the Design of Water Allocation and Water Distribution Networks Based on Graph-Theory, Irrigation and Water Allocation,* International Association of Hydrological Sciences Press, Institute of Hydrology, Wallingford, Oxfordshire, UK, IAHS Publication no. 169, 1987, pp. 207–220.

Rydzewski, J. R., and HA-H. Rashid, "Optimization of Water Resources for Irrigation in East Jordan," *Water Resources Bulletin,* 17(3): 367–372, 1981.

Sabet, H., and C. L. Creel, "Network Flow Modeling of Oroville Complex," ASCE, *Journal of Water Resources Planning and Management,* 117(3): 301–320, 1991.

Salkin, H. M., *Integer Programming,* Addison-Wesley, Reading, Mass., 1975.

Schafer, J. M., "Determining Optimum Pumping Rates for Creation of Hydraulic Barriers to Ground Water Pollutant Migration," *Proceedings* of the Fourth National Symposium and Exposition on Aquifer Restoration and Ground Water Monitoring, May 23–25, 1984, The Fawcett Center, Columbus, Ohio, pp. 50–60.

Shamir, U., "Optimization in Water Distribution Systems Engineering," *Math. Programming,* no.(11): 65–75, 1979.

Shih, C. S., and W. L. J. Meier, "Integrated Management of Quantity and Quality of Urban Water Resources," *Water Resources Bulletin,* 8(5): 1006–1017, 1972.

Spofford, W. O. J., "Integrated Residuals Management: A Regional Environmental Quality Management Model," Resources for the Future, Washington, D.C., Reprint 130, 1976.

Stedinger, J. R., and J. M. Bell-Graf, "Systems Analysis," Literature Review, *Journal of the Water Pollution Control Federation,* 51(6): 1092–1098, 1979.

Swamee, P. K., and A. K. Sharma, "Reorganization of Water-Distribution System," *Journal of Environmental Engineering,* 116(3): 588–600, 1990.

Taha, A. T., *Operations Research: An Introduction,* Macmillan, New York, 1987.

Taha, H., *Integer Programming: Theory, Applications, and Computations,* Academic Press, New York, 1975.

Tao, T., and W. C. Lennox, "Reservoir Operations by Successive Linear Programming," *Journal of Water Resources Planning and Management,* 117(2): 274–280, 1991.

Tejada-Guibert, A., J. R. Stedinger, and K. Staschus, "Optimization of Value of CVP's Hydropower Production," *Journal of Water Resources Planning and Management,* ASCE, 116(1): 52–70, 1990.

Tennessee Valley Authority, Water Resource Management Methods Staff, "Testing of Different Optimization Methods on a Reservoir Subsystem," Knoxville, Tenn., Report B-16, 1977.

Texas Water Development Board, "System Simulation for Management of a Total Water Resource," Report 118, 1970.

Tollow, A. J., "Operation of Water Supply Reservoirs by 'Control Bands' Derived by Simulation," *Hydrological Sciences Journal,* 34(4): 449–463, 1989.

Trezos, T., "Integer Programming Application for Planning of Hydropower Production," *Journal of Water Resources Planning and Management,* ASCE, 117(3): 340–351, 1991.

Tsakiris, G., and E. Kiountouzis, "Optimal Intraseasonal Irrigation Water Distribution," *Advances in Water Resources,* 7(2): 89–92, 1984.

Tung, Y. K., Y. Bao, L. Mays, and G. Ward, "Optimization of Freshwater Inflow to Estuaries," *Journal of Water Resources Planning and Management,* 116(A) July/Aug. 1990.

Tyagi, N. K., "Managing Salinity through Conjunctive Use of Water Resources," *Ecological Modelling,* 40(1): 11–24, 1988.

Tyagi, N. K., and V. V. D. Narayana, "Water Use Planning for Alkali Soils under Reclamation," *Journal of Irrigation and Drainage Engineering,* 110(2): 192–207, 1984.

Unver, O. I., and L. W. Mays, "Model for Real-Time Optimal Flood Control Operation of a Reservoir System," *Water Resources Management,* Kluver Academic Publishers, vol. 4, 1990, pp. 21–46.

Valocchi A. J., and J. W. Eheart, "Incorporating Parameter Uncertainty into Groundwater Quality Management Models," in *Systems Analysis in Water Quality Management,* Pergamon Press, New York, 1987.

Vedula, S., and S. Mohan, "Real-time Multipurpose Reservoir Operation: A Case Study," *Hydrological Sciences Journal,* 35(4): 447–462, 1990.

Walker, D. J., D. T. Noble, and R. S. Magleby, "Effective Cost-Share Rates and the Distribution of Social Costs in the Rock Creek, Idaho, Rural Clean Water Project," *Journal of Soil and Water Conservation,* 45(4): 477–479, 1990.

Wanakule, N., L. W. Mays, and L. S. Lasdon, "Optimal Mangement of Large-Scale Aquifers: Methodology and Application," *Water Resources Research,* 22(4): 447–466, 1986.

Wang, G.-T., and Y.-S. Yu, "Estimation of Parameters of the Discrete, Linear, Input-Output Model," *Journal of Hydrology,* 85(1/2): 15–30, 1986.

Wasimi, S. A., and P. K. Kitanidis, "Real-Time Forecasting and Daily Operation of a Multi-reservoir System During Floods by Linear Quadratic Gaussian Control," *Water Resources Research,* 19(6): 1511–1522, 1983.

Weisz, R. N., and J. C. Day, "A Methodology for Planning Land Use and Engineering Alternatives for Flood Plain Management: The Flood Plain Management System Model," U.S. Army Engineer Institute for Water Resources, Fort Belvoir, Va., IWR Paper 74-p2, 1974.

Wenzel, H. G. J., B. C. Yen, and W. H. Tang, "Advanced Methodology for Storm Sewer Design—Phase II," Water Resources Center, University of Illinois, Research Report no. 140, 1979.

Whiffen, G. J., and C. A. Shoemaker, "Nonlinear Weighted Feedback Control of Groundwater Remediation Under Uncertainty," AGU, *Water Resources Research,* 29(9): 3277–3289, 1993.

Whitlatch, E. E., and P. L. Asplund, "Capital Cost of Rural Water Distribution Systems," *Water Resources Bulletin,* 17(2): 310–314, 1981.

Whitlatch, E. E., N. R. Bhaskar, and M. J. Martin, "Operating Policies for Reservoir Pumped-Water Interties," *Journal of Water Resources Planning and Management,* 114(6): 687–703, 1988.

Willis, R., and P. Liu, "Optimization Model for Ground-Water Planning," *Journal of Water Resources Planning and Management,* 110(3): 333–347, 1984.

Winston, W. L., *Operations Research: Applications and Algorithms,* PWS Publisher, Boston, Mass., 1987.

Wu, M-Y., and W-S. Chu, "System Analysis of Water Treatment Plant in Taiwan," *Journal of Water Resources Planning and Management,* ASCE, 117(5): 536–548, 1991.

Wunderlich, W. O., "Chapter 11: System Planning and Operation," in *Hydropower Engineering Handbook,* J. S. Gulliver, and R. E. A. Arndt, eds., McGraw-Hill, New York, 1991.

Yakowitz, S., and B. Rutherford, "Computational Aspects of Discrete-Time Optimal Control," *Applied Mathematics and Computation,* 15: 29–45, 1984.

Yakowitz, S., "Algorithms and Computational Techniques in Differential Dynamic Programming," *Control and Dynamic Systems,* Academic Press, New York, 31, 1989, pp. 75–91.

Yeh, W. W.-G., "Optimization of Real Time Daily Operation of a Multiple Reservoir System," Report no. UCLA-ENG-7628, 1976.

Yeh, W. W.-G., "Reservoir Management and Operations Models: A State-of-the-Art Review," *Water Resources Research,* 21(12): 1797–1818, 1985.

Yeh, W. W.-G., L. Becker, D. Toy, and A. L. Graves, "Central Arizona Project: Operations Model," *Journal of the Water Resources Planning and Management Division,* ASCE, 106(WR2): 521–540, 1980.

Zavesky, R. R., and A. S. Goodman, "Water-Surface Profiles Without Energy Loss Coefficients," *Water Resources Bulletin,* 24(6): 1281–1287, 1988.

Zessler, U., and U. Shamir, "Optimal Operation of Water Distribution Systems," *Journal of Water Resources Planning and Management,* 115(6): 735–752, 1989.

Zhao B., and L. W. Mays, "Calibration and Control for Estuary Systems Under Uncertainty," submitted to *Water Resources Management,* Kluwer Academic Publishers, 1995a.

Zhao, B., and L. W. Mays, "Estuary Management by Stochastic Linear Quadratic Optimal Control," *Journal of Water Resources Planning and Management,* ASCE, 121(5): 382–391, Sep./Oct. 1995b.

CHAPTER 7
UNCERTAINTY AND RELIABILITY ANALYSIS

Yeou-Koung Tung
Wyoming Water Resources Center and Statistics Department
University of Wyoming
Laramie, Wyoming

7.1 INTRODUCTION

7.1.1 Uncertainties in Water Resources Engineering

In the analysis and design of water resources engineering systems, uncertainties arise in various aspects including, but not limited to, hydraulic, hydrologic, structural, environmental, and socioeconomic aspects. Of particular interest to water resources engineers are those related to hydraulic and hydrologic uncertainties. Uncertainties attribute mainly to our lack of a perfect understanding with regard to the phenomena and processes involved, and a perfect knowledge of how to determine parameter values for processes that are fairly well understood. *Uncertainty* could simply be defined as the occurrence of events that are beyond our control (Mays and Tung, 1992). Yen and Ang (1971) classified uncertainties into two types: *Objective uncertainties* associated with any random process or deducible from statistical samples and *subjective uncertainties* for which no quantitative factual information is available.

More specifically, in water resources engineering analyses and design, uncertainties can arise from several sources (Yen, et al., 1986) such as *natural uncertainties, model uncertainties, parameter uncertainties, data uncertainties,* and *operational uncertainties.* Natural uncertainty is associated with the inherent randomness of natural processes such as the occurrence of precipitation and flood events. In water resources engineering, models ranging from simple empirical equations to sophisticated computer simulations are used. Model uncertainty reflects the inability of the model or design technique to accurately represent the system's true physical behavior. Consequently, using an imperfect model for prediction could result in error, compromising the performance of the engineering design. Parameter uncertainties result from the inability to accurately quantify the inputs to and parameters of a model. Parameter uncertainty can be caused by changes in operational conditions of hydraulic structures, inherent variability of inputs and parameters in time and space, and lack of sufficient data. Data uncertainties include: (1) measurement errors, (2) inconsistency and nonhomogeneity of data, (3) data handling and transcription errors, and (4) inadequate representativeness of data sample due to time and space limitations. Operational uncertainties include those associated with construction, manufacture, deterioration, maintenance, and human factors.

In general, uncertainty due to inherent randomness of physical processes cannot be eliminated. Other uncertainties such as those associated with the lack of complete knowledge of the process, models, parameters, and data can be reduced through research, data collection, and careful manufacture.

Several expressions have been used to describe the degree of uncertainty of a parameter, a function, a model, or a system. The most complete and ideal description of uncertainty is the probability density function (defined in Sec. 7.2.3) of the quantity subject to uncertainty. However, in most practical problems such a probability function cannot be derived or found precisely.

Another measure of the uncertainty of a quantity is to express it in terms of a *confidence interval.* Use of confidence intervals has two drawbacks: (1) The uncertain quantities may not be normally distributed as assumed in the conventional procedures to determine the confidence interval and (2) no means is available to directly combine confidence intervals of individual contributing random components to give the confidence interval of the system as a whole.

A useful alternative is to use the *statistical moments* associated with a quantity subject to uncertainty. In particular, the second-order moment (see Sec. 7.2.4) is a measure of the dispersion of a random variable.

The existence of various uncertainties (including inherent randomness of natural processes) is the main contributor to potential failure of hydraulic structures. Knowledge of statistical information describing the uncertainties is essential in reliability analysis. Therefore, uncertainty analysis is a prerequisite for reliability analysis.

The main objective of uncertainty analysis is to identify the statistical properties of a model output as a function of stochastic input parameters. Uncertainty analysis provides a formal and systematic framework to quantify the uncertainty associated with the model output. Furthermore, it offers the designer insight regarding the contribution of each stochastic input parameter to the overall uncertainty of the model output. Such knowledge is essential to identify the "important" parameters to which more attention should be given, in order to have a better assessment of their values, and accordingly, to reduce the overall uncertainty of the output.

7.1.2 Reliability of Water Resources Systems

All hydraulic structures placed in a natural environment are subject to various external stresses. The *resistance* or *strength* of a hydraulic structure is its ability to accomplish the intended mission satisfactorily, without failure, when subject to *load* of demands or external stresses. Failure occurs when the resistance of the system is exceeded by the load.

Reliability analysis can be applied to various types of engineering problems such as design of engineering facilities; evaluation of the safety of existing structures; inspection, repair and other maintenance operations; routing, distribution and allocation, and other planning and management operations; data sampling and measurement network design; real-time prediction and long-term forecasting; and comparison and assessment of techniques and procedures.

The *reliability* of a hydraulic structure is the probability of safety p_s that the load does not exceed the resistance of the structure,

$$p_s = P\,[L \leq R] \tag{7.1}$$

in which P[   ] denotes probability. Conversely, the failure probability p_f can be expressed as

$$p_f = P\,[L > R] = 1 - p_s \tag{7.2}$$

Failure of hydraulic structures can be classified as *structural failure* and *performance failure* (Yen and Ang, 1971; Yen, et al., 1986). Structural failure involves damage or change of the structure or facility, hindering its ability to function as desired. On the other hand, performance failure does not necessarily involve structural damage, but the performance limit of the structure is exceeded and undesirable consequences occur. Generally, the two types of failure are related. Some hydraulic structures, such as dams and levees, are designed on the concept of structural failure, whereas other hydraulic structures, such as sewers and water-supply networks, are designed on the basis of performance failure.

A common practice for measuring the reliability of a water resources engineering structure is the *return period* or *recurrence interval*. The return period is defined as the long-term average (or expected) time between two successive failure-causing events. Flood frequency analysis using annual maximum flow series is a typical example of this kind. The main disadvantage of using return period is that reliability is measured only in terms of time of occurrence of loads, without considering their interactions with resistance (Melchers, 1987).

Two other types of reliability measures are frequently used in engineering practice to consider the relative magnitudes of resistance and anticipated load (called the *design load*). One measure is the *safety margin* (SM) defined as the difference between the resistance and the anticipated load, that is,

$$SM = R - L. \tag{7.3}$$

The other measure is the *safety factor* (SF) which is the ratio of resistance to load as

$$SF = \frac{R}{L} \tag{7.4}$$

Several types of safety factors and their applications to hydraulic engineering design are discussed by Yen (1979).

7.1.3 Overview of Uncertainty and Reliability Analysis Methods

Uncertainty analysis techniques involve different levels of mathematical complexity and data requirements. The appropriate technique to be used depends on the nature of the problem at hand including the availability of information, model complexity, and type and accuracy of results desired.

Several useful analytical approaches for uncertainty analysis including derived distribution techniques and integral transform techniques are described in Sec. 7.3. Although these analytical approaches are rather restrictive in practical applications, they are, however, powerful tools for deriving complete information about a stochastic process in some situations. Also, Sec. 7.3 describes several approximation techniques that are particularly useful for problems involving complex functions which cannot be dealt with analytically. They are primarily developed to estimate the statistical moments about the underlying random processes.

There are two basic probabilistic approaches to evaluate the reliability of a hydraulic structure or system. The most direct approach is a statistical analysis of data of past failure records for similar structures. The other approach is reliability analysis which considers and combines the contribution of each potentially influencing factor. The former is a lumped system approach requiring no knowledge about the behavior of the facility or structure and its load and resistance interaction. However, in many cases this direct approach is impractical because: (1) The sample size may be too small to be statistically reliable, especially for low-probability/high-consequence events, (2) the sample may not be representative of the structure or of the population, and (3) the physical conditions of a hydraulic structure or system may vary with respect to time, that is, the system is *nonstationary*.

There are two major steps in reliability analysis: (1) Identify and analyze the uncertainties of each of the contributing factors, and (2) combine the uncertainties of the stochastic factors to determine the overall reliability of the structure. The second step, in turn, may also proceed in two different ways: (1) directly combining the uncertainties of all the factors, and (2) separately combining the uncertainties of the factors belonging to different components or subsystems to first evaluate the respective subsystem reliability and then combine the reliabilities of the different components or subsystems to yield the overall reliability of the structure.

7.2 REVIEW OF PERTINENT PROBABILITY AND STATISTICAL THEORIES

In probability theory, an *experiment* represents, in general, the process of observing random phenomena. A phenomenon is considered *random* if the outcome of an observation cannot be predicted with absolute precision. The totality of all possible outcomes of an experiment is called the *sample space*. An *event* is any subset of outcomes contained in the sample space. Therefore, an event may be an empty (or null) set, subset of the sample space, or the sample space itself. If two events contain no common elements in the sets, then they are *mutually exclusive* or *disjoint events*. If the occurrence of one event depends on the occurrence of another event, then these are called *conditional events*.

Probability is a numeric measure of the likelihood of the occurrence of an event. In general, the probability of occurrence of an event can be assessed objectively through observations of the occurrence of the event or subjectively on the basis of experience and judgment.

7.2.1 Rules of Probability Computations

In probability computations, the three basic axioms of probability are intuitively understandable: (1) $P(A) \geq 0$, (2) $P(S) = 1$ with S being the sample space, and (3) If A and B are mutually exclusive events, then $P(A \cup B) = P(A) + P(B)$. Relaxing the requirement of mutual exclusiveness in axiom (3), the probability of the union of N events can be written as

$$P\left(\bigcup_{i=1}^{N} A_i\right) = \sum_{i=1}^{N} P(A_i) - \sum_{i<j}\sum P(A_i, A_j)$$

$$+ \sum_{i < j < k} \sum \sum P(A_i, A_j, A_k) - \cdots$$

$$+ (-1)^N P(A_1, A_2, \ldots, A_N) \tag{7.5}$$

in which $P(A_i, A_j) = P(A_i \cap A_j)$. If all events involved are mutually exclusive, all but the first summation on the right-hand side of Eq. (7.5) vanish.

If two events are *statistically independent,* this implies that the occurrence of one event has no influence on the occurrence of the other event. The probability of joint occurrence of N independent events is

$$P\left(\bigcap_{i=1}^{N} A_i \right) = P(A_1) \cdot P(A_2) \cdots P(A_N) = \prod_{i=1}^{N} P(A_i) \tag{7.6}$$

The probability that a conditional event occurs is called the *conditional probability.* The conditional probability $P(A \mid B)$ can be computed as

$$P(A \mid B) = \frac{P(A, B)}{P(B)} \tag{7.7}$$

in which $P(A \mid B)$ is the probability of occurrence of event A given that event B occurred. The probability of joint occurrence of N events (not necessarily independent) can be evaluated as

$$P\left(\bigcap_{i=1}^{N} A_i \right) = P(A_1) \cdot P(A_2 \mid A_1) \cdot P(A_3 \mid A_2, A_1) \cdots P(A_N \mid A_{N-1}, A_{N-2}, \cdots, A_2, A_1) \tag{7.8}$$

Sometimes, the event E occurs along with several attribute events A_i that are mutually exclusive ($A_i \cap A_j = \varnothing$, for $i \neq j$) and *collectively exhaustive* ($A_1 \cup A_2 \cup \cdots \cup A_N) = S$. The probability of occurrence of event E, regardless of the attributes, can be computed as

$$P(E) = \sum_{i=1}^{N} P(E, A_i) = \sum_{i=1}^{N} P(E \mid A_i) P(A_i) \tag{7.9}$$

which is called the *total probability theorem.*

7.2.2 Random Variables and Their Distributions

In analyzing statistical characteristics of water resources system performance, many events of interest can be defined by the related random variables. A *random variable* is a real-valued function that is defined on the sample space. A random variable can be discrete or continuous.

The *cumulative distribution function* (*CDF*) or simply *distribution function* (*DF*) of a random variable X is defined as

$$F(x) = P(X \leq x) \tag{7.10}$$

The CDF $F(x)$ is a nondecreasing function of its argument. Furthermore, $\lim_{x \to -\infty} F(x) = 0$ and $\lim_{x \to +\infty} F(x) = 1$.

For a discrete random variable X, the *probability mass function* (*PMF*) of X is defined as

$$p(x) = P(X = x) \tag{7.11}$$

The PMF of any discrete random variable must satisfy two conditions: (1) $p(x_i) \geq 0$ for all x_i's, and (2) $\Sigma_{\text{all } i} \, p(x_i) = 1$.

For a continuous random variable, the *probability density function (PDF)* $f(x)$ is defined as

$$f(x) = \frac{dF(x)}{dx} \tag{7.12}$$

in which $F(x)$ is the CDF of X. Similar to the discrete case, any PDF of a continuous random variable must satisfy two conditions: (1) $f(x) \geq 0$, and (2) $\int f(x) \, dx = 1$.

When problems involve multiple random variables, the *joint distribution* and *conditional distribution* are used. For example, in the design and operation of a flood-control reservoir, one often has to consider the flood peak and flood volume simultaneously. In such cases, one would need to develop a *joint PDF* of flood peak and flood volume. For the purpose of illustration, the discussions are limited to problems involving two continuous random variables. For discrete random variables, one simply replaces the integrals by summations.

The joint PDF of two continuous random variables X and Y is denoted as $f_{X,Y}(x, y)$ and the corresponding joint CDF is

$$F_{X,Y}(x, y) = \int_{-\infty}^{x} \int_{-\infty}^{y} f_{X,Y}(x, y) \, dx \, dy \tag{7.13}$$

Two random variables X and Y are statistically independent if and only if $f_{X,Y}(x,y) = f_X(x)f_Y(y)$ and $F_{X,Y}(x, y) = F_X(x)F_Y(y)$. Consequently, a problem involving multiple independent random variables is, in effect, a univariate problem in which each individual random variable can be treated separately.

7.2.3   Statistical Properties of Random Variables

In probability and statistics, the term *population* describes the complete assemblage of all the values representative of a particular random process. A *sample* is any subset of the population. Further, *parameters* in a statistical model are quantities that are descriptive of a population. *Sample statistics* or simply *statistics* are quantities calculated on the basis of sample observations.

Descriptors that are commonly used to assess statistical properties of a random variable are those showing the central tendency, dispersion, and asymmetry of a distribution. These descriptors are related to the *statistical moments* of the random variable. Without losing generality, the following discussion will consider continuous random variables. For discrete random variables, the integral sign is replaced by the summation sign and the PDF by the PMF.

The rth moments of a random variable X about any reference point $X = x_0$ is defined, for the continuous case, as

$$E[(X - x_0)^r] = \int_{-\infty}^{\infty} (x - x_0)^r f(x) \, dx \tag{7.14}$$

where $E[\ \ ]$ is a *statistical expectation operator.*

Two types of statistical moments are commonly used: *moments about the origin* where $x_0 = 0$ and the *central moments* where $x_0 = \mu$ with μ being the mean of the random variable defined in Eq. (7.17). The rth central moment is denoted as $\mu_r = E[(X - \mu)^r]$ while the rth moment about the origin is denoted as $\mu'_r = E[X^r]$. Through the binomial expansion, the central moments μ_r can be obtained from the moments about the origin μ'_r as

$$\mu_r = \sum_{i=0}^{r} (-1)^i \,_rC_i \, \mu^i \, \mu'_{r-i} \tag{7.15}$$

where $\,_rC_i = r!/[i! \, (r-i)!]$. Conversely, the moments about the origin μ'_r can be obtained from the central moments μ_r as

$$\mu'_r = \sum_{i=0}^{r} \,_rC_i \, \mu^i \, \mu_{r-i} \tag{7.16}$$

The central tendency is commonly measured by the *expectation* which is defined as

$$E[X] = \mu = \int_{-\infty}^{\infty} x f(x) \, dx \tag{7.17}$$

This expectation is also known as the *mean* of a random variable. Two operational properties of the expectation are useful:

1. The expectation of the sum of several random variables equals the sum of the individual expectations,

$$E\left(\sum_{i=1}^{N} a_i X_i \right) = \sum_{i=1}^{N} a_i E[X_i] \tag{7.18}$$

2. If $X_1, X_2, \ldots, X_N$ are statistically independent random variables, then

$$E\left(\prod_{i=1}^{N} X_i \right) = \prod_{i=1}^{N} E[X_i] \tag{7.19}$$

Two other types of measures of central tendency are sometimes used in practice. They are the *median* and *mode*. The median of a random variable is the value that splits the distribution into two equal halves. The mode is defined as the value of the random variable at which the PDF is maximum.

The variability of a random variable is measured by the *variance,* which is defined as

$$\mathrm{Var}[X] = \sigma^2 = E[(X - \mu)^2] = \int_{-\infty}^{\infty} (x - \mu)^2 f(x) \, dx \tag{7.20}$$

The variance is the second-order central moment. The positive square root of variance is called the *standard deviation,* which is often used as the measure of the degree of uncertainty associated with a random variable. To compare the degree of uncertainty of two random variables of different units, a dimensionless measure $\Omega = \sigma/\mu$, called the *coefficient of variation,* is useful.

Two important operational properties of the variance are:

1. $\mathrm{Var}[X] = E[X^2] - E^2[X]$ (7.21)
2. When $X_1, X_2, \ldots, X_N$ are independent random variables,

$$\text{Var}\left(\sum_{i=1}^{N} a_i X_i\right) = \sum_{i=1}^{N} a_i^2 \sigma_i^2 \tag{7.22}$$

where a_i is a constant and σ_i is the standard deviation of the random variable X_i.

The asymmetry of the PDF of a random variable is measured by the *skew coefficient* γ, defined as

$$\gamma = \frac{\mu_3}{\mu_2^{1.5}} = \frac{E[(X-\mu)^3]}{\sigma^3} \tag{7.23}$$

The skew coefficient is dimensionless and is related to the third central moment. The sign of the skew coefficient indicates the extent of symmetry of the probability distribution about its mean. If $\gamma = 0$, the distribution is symmetric about its mean; if $\gamma > 0$, the distribution has a long tail to the right; if $\gamma < 0$, the distribution has a long tail to the left. In practice, statistical moments higher than three are rarely used because their accuracies decrease rapidly when estimated from the small samples.

When a problem involves two dependent random variables, the degree of linear dependence between the two can be measured by the *correlation coefficient* $\rho(X,Y)$ which is defined as

$$\rho(X,Y) = \frac{\text{cov}(X,Y)}{\sigma_X \sigma_Y} \tag{7.24}$$

where $\text{cov}(X,Y)$ is the covariance between random variables X and Y defined as

$$\text{cov}(X,Y) = E[(X-\mu_X)(Y-\mu_Y)] = E(XY) - \mu_X \mu_Y \tag{7.25}$$

It can be easily shown that $\text{cov}(Z_1, Z_2) = \rho(Z_1, Z_2)$ with Z_1 and Z_2 being the standardized random variables. Figure 7.1 graphically illustrates several cases of correlation coefficient. If the two random variables X and Y are independent, then $\rho(X,Y) = \text{cov}(X,Y) = 0$. However, the reverse is not necessarily true because $\rho(X,Y)$ does not reflect nonlinear dependence between X and Y. If the random variables involved are not independent, Eq. (7.22) can be generalized as

$$\text{Var}\left(\sum_{i=1}^{N} a_i X_i\right) = \sum_{i=1}^{N} a_i^2 \sigma_i^2 + 2 \sum_{i=1}^{N-1} \sum_{j=i+1}^{N} a_i a_j \text{cov}[X_i, X_j] \tag{7.26}$$

7.2.4 Univariate Probability Distributions

This section summarizes two discrete distributions and several commonly used univariate continuous distributions.

7.2.4.1 Binomial Distribution. The *binomial distribution* is a discrete distribution applicable to a random process with only two possible outcomes. In water resources systems, the state of components or subsystems can be classified as either functioning or failed. Consider a system involving a total of n independent components with each component having two possible states, functioning or failed. If the probability of each component being failed is p, then the probability of having x number of failed components in the system can be computed as

$$p(x) = {}_nC_x p^x q^{n-x} \qquad x = 0, 1, 2, \ldots, n \tag{7.27}$$

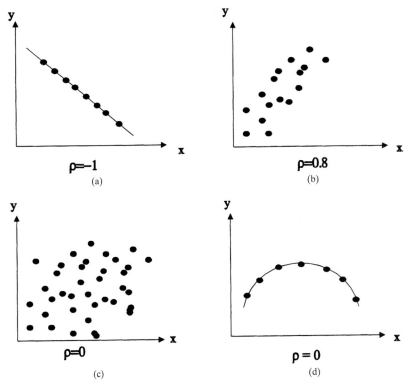

FIGURE 7.1 Different cases of correlation between two random variables: (*a*) perfectly linearly correlated in opposite direction; (*b*) strongly linearly corrected in positive direction; (*c*) uncorrelated in linear fashion; (*d*) perfectly correlated in nonlinear fashion, but uncorrelated linearly.

where $q = 1 - p$, the probability of being in the functioning state for each component. A random variable X having a binomial distribution with parameters n and p has the expectation $E(X) = np$ and variance $\text{Var}(X) = npq$.

7.2.4.2 Poisson Distribution. When the number of system components $n \to \infty$ and $p \to 0$ while $np = \nu = $ constant, the binomial distribution becomes a *Poisson distribution* with the PMF defined as

$$p(x) = e^{-\nu}\nu^x/x! \qquad x = 0, 1, 2, \ldots \qquad (7.28)$$

where the parameter $\nu > 0$ represents the mean of a discrete random variable X having a Poisson distribution. For a Poisson random variable, the mean and the variance are identical to ν.

The Poisson distribution has been applied widely in modeling the number of occurrences of a random event within a specified time or space interval. Equation (7.28) can be modified as

$$p(x) = e^{-\lambda t}(\lambda t)^x/x! \qquad x = 0, 1, 2, \ldots \qquad (7.29)$$

in which the parameter λ is the average rate of occurrence of the random event in an interval $(0, t)$.

7.2.4.3 Normal Distribution. The *normal distribution* is also called the *Gaussian distribution*. A normal random variable having the mean μ and variance σ^2 has the PDF

$$f(x \mid \mu, \sigma^2) = \frac{1}{\sqrt{2\pi}\,\sigma} \exp\left[-\frac{1}{2}\left(\frac{x-\mu}{\sigma}\right)^2\right] \quad \text{for} \quad -\infty < x < \infty \quad (7.30)$$

The normal distribution is bell-shaped and symmetric with respect to $x = \mu$. Therefore, the skew coefficient of a normal random variable is zero. A random variable which is the sum of independent normal random variables is also a normal random variable with mean and variance that can be computed by Eqs. (7.18) and (7.22), respectively.

The normal distribution sometimes provides a viable alternative to approximate the probability of a random variable which is not normally distributed. An important theorem relating to the sum of independent random variables is the *central limit theorem*, which can be loosely stated as: The distribution of the sum of a number of independent random variables, regardless of their individual distribution, can be approximated by a normal distribution as long as none of the variables has a dominant effect on the sum. The larger the number of random variables involved in the summation, the better the approximation.

Probability computations for normal random variables are made by first transforming to a *standardized variable Z* as

$$Z = (X - \mu)/\sigma \quad (7.31)$$

in which Z has a mean of zero and a variance of one. The PDF of Z, called the *standard normal distribution*, is

$$\phi(z) = \frac{1}{\sqrt{2\pi}}\, e^{-z^2/2} \quad -\infty < z < \infty \quad (7.32)$$

Computations of probability for a normal random variable can be made as

$$P(X \le x) = P[(X-\mu)/\sigma \le (x-\mu)/\sigma] = P[Z \le z] = \Phi(z) \quad (7.33)$$

where $\Phi(z)$ is the CDF of the standard normal random variable Z. A table of standard normal probability is shown in Table 7.1. Several algebraic formulas are available for fairly accurate approximation of the normal probability (Abramowitz and Stegun, 1972; Stedinger, et al., 1993).

7.2.4.4 Lognormal Distribution. The *lognormal distribution* is a commonly used continuous distribution when random variables cannot be negative-valued. A random variable X is lognormally distributed if its logarithmic transform $Y = \ln(X)$ has a normal distribution with mean $\mu_{\ln X}$ and variance $\sigma^2_{\ln X}$. The PDF of a lognormal random variable is

$$f(x \mid \mu_{\ln X}, \sigma^2_{\ln X}) = \frac{1}{\sqrt{2\pi}\,\sigma_{\ln X}\, x} \exp\left[-\frac{1}{2}\left(\frac{\ln(x) - \mu_{\ln X}}{\sigma_{\ln X}}\right)^2\right] \quad x > 0 \quad (7.34)$$

Statistical properties of a lognormal random variable on the original scale can be computed from those of the log-transformed variable. To compute the statistical moments of X from those of $\ln(X)$, the following formulas are useful.

TABLE 7.1 Cumulative Standard Normal Probability, $\Phi(z) = P(Z \le z)$.

z	0.00	0.01	0.02	0.03	0.04	0.05	0.06	0.07	0.08	0.09
−3.4	0.0003	0.0003	0.0003	0.0003	0.0003	0.0003	0.0003	0.0003	0.0003	0.0002
−3.3	0.0005	0.0005	0.0005	0.0004	0.0004	0.0004	0.0004	0.0004	0.0004	0.0003
−3.2	0.0007	0.0007	0.0006	0.0006	0.0006	0.0006	0.0006	0.0005	0.0005	0.0005
−3.1	0.0010	0.0009	0.0009	0.0009	0.0008	0.0008	0.0008	0.0008	0.0007	0.0007
−3.0	0.0013	0.0013	0.0013	0.0012	0.0012	0.0011	0.0011	0.0011	0.0010	0.0010
−2.9	0.0019	0.0018	0.0017	0.0017	0.0016	0.0016	0.0015	0.0015	0.0014	0.0014
−2.8	0.0026	0.0025	0.0024	0.0023	0.0023	0.0022	0.0021	0.0021	0.0020	0.0019
−2.7	0.0035	0.0034	0.0033	0.0032	0.0031	0.0030	0.0029	0.0028	0.0027	0.0026
−2.6	0.0047	0.0045	0.0044	0.0043	0.0041	0.0040	0.0039	0.0038	0.0037	0.0036
−2.5	0.0062	0.0060	0.0059	0.0057	0.0055	0.0054	0.0052	0.0051	0.0049	0.0048
−2.4	0.0082	0.0080	0.0078	0.0075	0.0073	0.0071	0.0069	0.0068	0.0066	0.0064
−2.3	0.0107	0.0104	0.0102	0.0099	0.0096	0.0094	0.0091	0.0089	0.0087	0.0084
−2.2	0.0139	0.0136	0.0132	0.0129	0.0125	0.0122	0.0119	0.0116	0.0113	0.0110
−2.1	0.0179	0.0174	0.0170	0.0166	0.0162	0.0158	0.0154	0.0150	0.0146	0.0143
−2.0	0.0228	0.0222	0.0217	0.0212	0.0207	0.0202	0.0197	0.0192	0.0188	0.0183
−1.9	0.0287	0.0281	0.0274	0.0268	0.0262	0.0256	0.0250	0.0244	0.0239	0.0233
−1.8	0.0359	0.0352	0.0344	0.0336	0.0329	0.0322	0.0314	0.0307	0.0301	0.0294
−1.7	0.0446	0.0436	0.0427	0.0418	0.0409	0.0401	0.0392	0.0384	0.0375	0.0367
−1.6	0.0548	0.0537	0.0526	0.0516	0.0505	0.0495	0.0485	0.0475	0.0465	0.0455
−1.5	0.0668	0.0655	0.0643	0.0630	0.0618	0.0606	0.0594	0.0582	0.0571	0.0559
−1.4	0.0808	0.0793	0.0778	0.0764	0.0749	0.0735	0.0722	0.0708	0.0694	0.0681
−1.3	0.0968	0.0951	0.0934	0.0918	0.0901	0.0885	0.0869	0.0853	0.0838	0.0823
−1.2	0.1151	0.1131	0.1112	0.1093	0.1075	0.1056	0.1038	0.1020	0.1003	0.0985
−1.1	0.1357	0.1335	0.1314	0.1292	0.1271	0.1251	0.1230	0.1210	0.1190	0.1170
−1.0	0.1587	0.1562	0.1539	0.1515	0.1492	0.1469	0.1446	0.1423	0.1401	0.1379
−0.9	0.1841	0.1814	0.1788	0.1762	0.1736	0.1711	0.1685	0.1660	0.1635	0.1611
−0.8	0.2119	0.2090	0.2061	0.2033	0.2005	0.1977	0.1949	0.1922	0.1894	0.1867
−0.7	0.2420	0.2389	0.2358	0.2327	0.2296	0.2266	0.2236	0.2206	0.2177	0.2148
−0.6	0.2743	0.2709	0.2676	0.2643	0.2611	0.2578	0.2546	0.2514	0.2483	0.2451
−0.5	0.3085	0.3050	0.3015	0.2981	0.2946	0.2912	0.2877	0.2843	0.2810	0.2776
−0.4	0.3446	0.3409	0.3372	0.3336	0.3300	0.3264	0.3228	0.3192	0.3156	0.3121
−0.3	0.3821	0.3783	0.3745	0.3707	0.3669	0.3632	0.3594	0.3557	0.3520	0.3483
−0.2	0.4207	0.4168	0.4129	0.4090	0.4052	0.4013	0.3974	0.3936	0.3897	0.3859
−0.1	0.4602	0.4562	0.4522	0.4483	0.4443	0.4404	0.4364	0.4325	0.4286	0.4247
−0.0	0.5000	0.4960	0.4920	0.4880	0.4840	0.4801	0.4761	0.4721	0.4681	0.4641
0.0	0.5000	0.5040	0.5080	0.5120	0.5160	0.5199	0.5239	0.5279	0.5319	0.5359
0.1	0.5398	0.5438	0.5478	0.5517	0.5557	0.5596	0.5636	0.5675	0.5714	0.5753
0.2	0.5793	0.5832	0.5871	0.5910	0.5948	0.5987	0.6026	0.6064	0.6103	0.6141
0.3	0.6179	0.6217	0.6255	0.6293	0.6331	0.6368	0.6406	0.6443	0.6480	0.6517
0.4	0.6554	0.6591	0.6628	0.6664	0.6700	0.6736	0.6772	0.6808	0.6844	0.6879
0.5	0.6915	0.6950	0.6985	0.7019	0.7054	0.7088	0.7123	0.7157	0.7190	0.7224
0.6	0.7257	0.7291	0.7324	0.7357	0.7389	0.7422	0.7454	0.7486	0.7517	0.7549
0.7	0.7580	0.7611	0.7642	0.7673	0.7704	0.7734	0.7764	0.7794	0.7823	0.7852
0.8	0.7881	0.7910	0.7939	0.7967	0.7995	0.8023	0.8051	0.8078	0.8106	0.8133
0.9	0.8159	0.8186	0.8212	0.8238	0.8264	0.8289	0.8315	0.8340	0.8365	0.8389
1.0	0.8413	0.8438	0.8461	0.8485	0.8508	0.8531	0.8554	0.8577	0.8599	0.8621
1.1	0.8643	0.8665	0.8686	0.8708	0.8729	0.8749	0.8770	0.8790	0.8810	0.8830
1.2	0.8849	0.8869	0.8888	0.8907	0.8925	0.8944	0.8962	0.8980	0.8997	0.9015
1.3	0.9032	0.9049	0.9066	0.9082	0.9099	0.9115	0.9131	0.9147	0.9162	0.9177
1.4	0.9192	0.9207	0.9222	0.9236	0.9251	0.9265	0.9278	0.9292	0.9306	0.9319

TABLE 7.1 Cumulative Standard Normal Probability, $\Phi(z) = P(Z \le z)$ (Continued)

z	0.00	0.01	0.02	0.03	0.04	0.05	0.06	0.07	0.08	0.09
1.5	0.9332	0.9345	0.9357	0.9370	0.9382	0.9394	0.9406	0.9418	0.9429	0.9441
1.6	0.9452	0.9463	0.9474	0.9484	0.9495	0.9505	0.9515	0.9525	0.9535	0.9545
1.7	0.9554	0.9564	0.9573	0.9582	0.9591	0.9599	0.9608	0.9616	0.9625	0.9633
1.8	0.9641	0.9649	0.9656	0.9664	0.9671	0.9678	0.9686	0.9693	0.9699	0.9706
1.9	0.9713	0.9719	0.9726	0.9732	0.9738	0.9744	0.9750	0.9756	0.9761	0.9767
2.0	0.9772	0.9778	0.9783	0.9788	0.9793	0.9798	0.9803	0.9808	0.9812	0.9817
2.1	0.9821	0.9826	0.9830	0.9834	0.9838	0.9842	0.9846	0.9850	0.9854	0.9857
2.2	0.9861	0.9864	0.9868	0.9871	0.9875	0.9878	0.9881	0.9884	0.9887	0.9890
2.3	0.9893	0.9896	0.9898	0.9901	0.9904	0.9906	0.9909	0.9911	0.9913	0.9916
2.4	0.9918	0.9920	0.9922	0.9925	0.9927	0.9929	0.9931	0.9932	0.9934	0.9936
2.5	0.9938	0.9940	0.9941	0.9943	0.9945	0.9946	0.9948	0.9949	0.9951	0.9952
2.6	0.9953	0.9955	0.9956	0.9957	0.9959	0.9960	0.9961	0.9962	0.9963	0.9964
2.7	0.9965	0.9966	0.9967	0.9968	0.9969	0.9970	0.9971	0.9972	0.9973	0.9974
2.8	0.9974	0.9975	0.9976	0.9977	0.9977	0.9978	0.9979	0.9979	0.9980	0.9981
2.9	0.9981	0.9982	0.9982	0.9983	0.9984	0.9984	0.9985	0.9985	0.9986	0.9986
3.0	0.9987	0.9987	0.9987	0.9988	0.9988	0.9989	0.9989	0.9989	0.9990	0.9990
3.1	0.9990	0.9991	0.9991	0.9991	0.9992	0.9992	0.9992	0.9992	0.9993	0.9993
3.2	0.9993	0.9993	0.9994	0.9994	0.9994	0.9994	0.9994	0.9995	0.9995	0.9995
3.3	0.9995	0.9995	0.9995	0.9996	0.9996	0.9996	0.9996	0.9996	0.9996	0.9997
3.4	0.9997	0.9997	0.9997	0.9997	0.9997	0.9997	0.9997	0.9997	0.9997	0.9998

Source: Devore, 1987.

$$\mu_X = \exp\left(\mu_{\ln X} + \frac{\sigma_{\ln X}^2}{2}\right) \tag{7.35}$$

$$\sigma_X^2 = \mu_X^2 \left[e^{\sigma_{\ln X}^2} - 1\right] \tag{7.36}$$

$$\gamma_X = \Omega_X^3 + 3\Omega_X \tag{7.37}$$

Equation (7.37) indicates that lognormal distributions are always positively skewed. Conversely, the statistical moments of $\ln X$ can be computed from those of X by

$$\mu_{\ln X} = \frac{1}{2}\ln\left[\frac{\mu_X^2}{1 + \Omega_X^2}\right] = \ln(\mu_X) - \frac{1}{2}\sigma_{\ln X}^2 \tag{7.38}$$

$$\sigma_{\ln X}^2 = \ln(1 + \Omega_X^2) \tag{7.39}$$

Since the sum of normal random variables is normally distributed, the product of independent lognormal random variables is also lognormally distributed. In cases that two lognormal random variables are correlated with a correlation coefficient $\rho_{X,Y}$ in the original space, then the covariance terms in the log-transformed space must be included in determining the variance. From $\rho_{X,Y}$, the correlation coefficient in the log-transformed space can be computed as

$$\rho_{\ln X, \ln Y} = \frac{\ln(1 + \rho_{X,Y}\Omega_X\Omega_Y)}{\sqrt{\ln(1 + \Omega_X^2)\ln(1 + \Omega_Y^2)}} \tag{7.40}$$

7.2.4.5 *Gamma Distribution and Its Variations.* The *gamma distribution* is another rather versatile continuous distribution for a positive-valued random variable. The two-parameter gamma distribution has the PDF defined as

$$f(x|\alpha, \beta) = \frac{\beta}{\Gamma(\alpha)} \, (\beta x)^{\alpha - 1} e^{-\beta x} \qquad \text{for} \qquad x > 0 \tag{7.41}$$

in which $\alpha > 0$ and $\beta > 0$ are parameters and $\Gamma(\ )$ is a gamma function. When $\alpha = 1$, the gamma distribution reduces to the *exponential distribution* with the PDF defined as

$$f(x|\beta) = \beta \exp(-\beta x) \qquad \text{for} \qquad x > 0 \tag{7.42}$$

The exponential distribution is widely used for describing the life span of various electronic and mechanical components. It plays an important role in reliability mathematics using time-to-failure analysis (see Sec. 7.6).

Two variations of the gamma distribution are frequently used in hydrologic frequency analysis, namely, *Pearson* and *log-Pearson type 3 distributions.* The log-Pearson type 3 distribution is recommended for use by the U.S. Interagency Advisory Committee on Water Data (1982) as the standard distribution for flood frequency analysis. The Pearson type 3 distribution is a three-parameter gamma distribution with the third parameter being the lower bound of the random variable whereas the log-Pearson type 3 distribution is simply a three-parameter log-gamma distribution. Stedinger, et al. (1993) provide a good summary of these two distributions.

7.2.4.6 *Extremal Distributions.* Water resources engineering reliability analysis often focuses on the statistical characteristics of extreme events. The work on statistics of extremes was pioneered by Fisher and Tippett (1928) and expanded by Gumbel (1958). Three types of asymptotic distributions of extremes have been derived based on the different characteristics of the underlying distribution (Haan, 1977):

Type I—Parent distributions are unbounded in the direction of extremes and all statistical moments exist.

Type II—Parent distributions are unbounded in the direction of extremes but all moments do not exist. Thus type II is of little use in practical engineering applications.

Type III—Parent distributions are bounded in the direction of the desired extreme.

Type I extremal distribution is sometimes referred to as a *Gumbel distribution,* a *Fisher-Tippett distribution,* or a *double exponential distribution.* The CDF and PDF of type I extremal distribution have, respectively, the following forms

$$F(x|x_o, \beta) = \exp\left[-\exp \mp \left(\frac{x - x_o}{\beta}\right)\right] \qquad -\infty < x, x_o < \infty \tag{7.43a}$$

$$f(x|x_o, \beta) = \frac{1}{\beta} \exp\left[\mp\left(\frac{x - x_o}{\beta}\right) - \exp \mp \left(\frac{x - x_o}{\beta}\right)\right] \qquad -\infty < x, x_o < \infty \tag{7.43b}$$

in which $\beta \geq 0$; $-$ applies to the largest extreme whereas $+$ is for the smallest extreme.

Type III extremal distribution is widely used for modeling the smallest extreme. The resulting distribution is also known as the *Weibull distribution* having a PDF defined as

$$f(x|x_o, \alpha, \beta) = \frac{\alpha}{\beta}\left(\frac{x - x_o}{\beta}\right)^{\alpha - 1} \exp{-\left(\frac{x - x_o}{\beta}\right)^{\alpha}}, x \geq x_o, \alpha, \beta > 0 \qquad (7.44)$$

The Weibull distribution is very versatile and can take many shapes. The CDF of the Weibull distribution can be derived as

$$F(x|x_o, \alpha, \beta) = 1 - \exp{-\left(\frac{x - x_o}{\beta}\right)^{\alpha}} \qquad (7.45)$$

Recently, the *generalized extreme value distribution (GEV)* has been used in hydrologic frequency analysis. The GEV distribution provides an expression that encompasses all three types of extreme value distributions. Detailed descriptions of the GEV distribution are provided by Stedinger, et al. (1993).

7.2.5 Multivariate Probability Distributions

Multivariate probability distributions are extensions of univariate probability distributions which jointly account for more than one random variable. *Bivariate* and *trivariate* distributions are special cases when two and three random variables, respectively, are involved. The fundamental bases of multivariate probability distributions are described in Sec. 7.2.3. There are several ways to construct a multivariate distribution, and detailed descriptions can be found in Johnson and Kotz (1976). In this section, two commonly used multivariate distributions, namely, multivariate normal and multivariate lognormal, are described.

7.2.5.1 Multivariate Normal Distribution.

The *multivariate normal* PDF for N correlated normal random variables is defined as

$$f(\mathbf{x}) = \frac{|\mathbf{C}^{-1}|^{1/2}}{(2\pi)^{N/2}} \exp\left[-\frac{1}{2}(\mathbf{x} - \mathbf{\mu})^T \mathbf{C}^{-1}(\mathbf{x} - \mathbf{\mu})\right] \qquad -\infty < \mathbf{x} < \infty \qquad (7.46)$$

in which $\mathbf{\mu} = (\mu_1, \mu_2, \ldots, \mu_N)$, an $N \times 1$ vector of means $\mathbf{C} = E[(\mathbf{X} - \mathbf{\mu})(\mathbf{X} - \mathbf{\mu})^T]$ is an $N \times N$ covariance matrix and T represents the transpose. In terms of the standard normal random variables $Z_i = (X_i - \mu_i)/\sigma_i$ the *standardized multivariate normal PDF* can be expressed as

$$\phi(\mathbf{z}|\mathbf{R}) = \frac{|\mathbf{R}^{-1}|^{1/2}}{(2\pi)^{N/2}} \exp\left(-\frac{1}{2}\mathbf{z}^T \mathbf{R}^{-1}\mathbf{z}\right) \qquad -\infty < \mathbf{z} < \infty \qquad (7.47)$$

in which $\mathbf{R} = E[\mathbf{Z}\mathbf{Z}^T]$ is a $N \times N$ correlation matrix for the multivariate normal variables.

Evaluation of the probability of multivariate normal random variables involves multidimensional integration as

$$\Phi(\mathbf{z}|\mathbf{R}) = P(Z_1 \leq z_1, Z_2 \leq z_2, \ldots, Z_N \leq z_N|\mathbf{R})$$

$$= \int_{-\infty}^{z_1} \int_{-\infty}^{z_2} \cdots \int_{-\infty}^{z_N} \phi(\mathbf{z}|\mathbf{R}) \, dz \qquad (7.48)$$

Accurate evaluation for $\Phi(\mathbf{z}|\mathbf{R})$ is generally difficult, and is often resolved by approximations. Several practical methods have been proposed to compute the bounds of multivariate normal probability based on univariate or bivariate normal probabilities (Rackwitz, 1978; Ditlevsen, 1979, 1982, 1984; Bennett and Ang, 1983).

7.2.5.2 *Multivariate Lognormal Distribution.* The *bivariate log-normal* PDF is

$$f(x_1, x_2) = \frac{1}{2\pi x_1 x_2 \, \sigma_1' \sigma_2' \sqrt{1 - \rho'^2}} \; e^{-Q'/[2(1 - \rho'^2)]} \qquad \text{for} \qquad x_1, x_2 > 0 \qquad (7.49)$$

in which

$$Q' = \frac{[\ln (x_1) - \mu_1']^2}{\sigma_1'^2} + \frac{[\ln (x_2) - \mu_2']^2}{\sigma_1'^2} - 2\rho' \frac{[\ln (x_1) - \mu_1'][\ln (x_2) - \mu_2']}{\sigma_1' \sigma_2'}$$

where μ' and σ' are the mean and standard deviation of log-transformed random variables, subscripts 1 and 2 indicate the random variables X_1 and X_2, respectively, and ρ' is the correlation coefficient of the two log-transformed random variables. After log-transformation is made, properties of multivariate lognormal random variables follow exactly as the multivariate normal case.

7.3 METHODS FOR UNCERTAINTY ANALYSIS

From the previous discussions of uncertainty in water resources systems modeling and analysis, it is not difficult to find that many quantities of interest are functionally related to several variables which are subject to varying levels of uncertainty. The methods applicable for uncertainty analysis are dictated by the information available on the stochastic variables involved and the functional relationships among the variables. In principle, it would be ideal to derive the exact probability distribution of the model output as a function of those of the stochastic variables. In this section, some analytical and approximate methods for uncertainty analysis are described.

7.3.1 Analytical Methods

7.3.1.1 *Derived Distribution Method.* The *derived distribution method* determines the distribution of a random variable W which is related to another random variable X, whose PDF and/or CDF are known, through the functional relationship $W = g(X)$. The CDF of W can be obtained as

$$H_W(w) = P[W \le w] = F_X[g^{-1}(w)] \qquad (7.50)$$

where $g^{-1}(w)$ represents the inverse function of g. The PDF of W, $h_W(w)$, then, can be obtained according to Eq. (7.12) by taking the derivative of $H_W(w)$ with respect to w.

In the case that the functional relation between X and W is either strictly increasing or strictly decreasing, the PDF of W can be derived directly from the PDF of X as

$$h_W(w) = f_X(x) \, |dx/dw| \qquad (7.51)$$

in which $|dx/dw|$ is called the *Jacobian*. This derived distribution method is also called the *transformation of variables technique*.

Consider a multivariate case in which N random variables $W_1, W_2, \ldots, W_N$ are related to N other random variables $X_1, X_2, \ldots, X_N$ through a system of N equations as

$$W_i = g_i (X_1, X_2, \ldots, X_N), \qquad \text{for} \qquad i = 1, 2, \ldots, N$$

When the functions $g_1(.), g_2(.), \ldots, g_N(.)$ satisfy a monotonic relationship, the joint PDF of random variables W can be directly obtained from the following equation

$$h(w_1, w_2, \ldots, w_N) = f(x_1, x_2, \ldots, x_N) \, |\mathbf{J}| \qquad (7.52)$$

where $f(x_1, x_2, \ldots, x_N)$ and $h(w_1, w_2, \ldots, w_N)$ are the joint PDFs of Xs and Ws, respectively; and $|\mathbf{J}|$ is the absolute value of the determinant of the $N \times N$ *Jacobian matrix*

$$|\mathbf{J}| = \begin{vmatrix} \dfrac{\partial x_1}{\partial w_1} & \dfrac{\partial x_1}{\partial w_2} & \cdots & \dfrac{\partial x_1}{\partial w_N} \\ \cdot & \cdot & \cdots & \cdot \\ \cdot & \cdot & \cdots & \cdot \\ \cdot & \cdot & \cdots & \cdot \\ \dfrac{\partial x_N}{\partial w_1} & \dfrac{\partial x_N}{\partial w_2} & \cdots & \dfrac{\partial x_N}{\partial w_N} \end{vmatrix} \qquad (7.53)$$

In theory, the above concept of deriving the PDF of a random variable as a function of the PDFs of other random variables is simple and straightforward. However, the success of implementing such procedures largely depends on the functional relation, the form of the PDFs involved, and the analyst's mathematical skill. It is in common situations in which analytical derivations are virtually impossible.

7.3.1.2 Integral Transform Techniques. *Integral transformation techniques* originally found their applications in univariate statistical analysis. The well-known integral transforms are *Fourier transform, Laplace transform,* and *exponential transform.* One useful transform technique, but less known to water resources engineering, is the *Mellin transform* (Epstein, 1948; Park, 1987).

The Fourier transform of a PDF leads to the *characteristic function* while the Laplace and exponential transforms yield the *moment generating function.* The Laplace transform is the same as the exponential transform except that the Laplace transform is only applicable to nonnegative random variables. Once the characteristic function or moment generating function is derived, statistical moments of a random variable can be computed. One major advantage of integral transforms is that, if such transforms of a PDF exist, the relationship between the PDF and its integral transform is unique (Kendall, et al., 1987).

In dealing with a multivariate problem in which a random variable is a function of several random variables, the convolution property of these integral transforms becomes analytically powerful, especially when the stochastic variables in the model are independent. Fourier and exponential transforms are powerful in treating the sum and difference of random variables, while the Mellin transform is attractive to the quotient and product of random variables (Griffin, 1975; Springier, 1979).

The Fourier transform of a PDF, $f(x)$, is defined as

$$\mathscr{F}_f(s) = \int_{-\infty}^{\infty} e^{isx} f(x) \, dx = E\left[e^{isx}\right] \qquad (7.54)$$

where s is the parameter in the transformed space and $i = \sqrt{-1}$. The resulting Fourier transform $\mathscr{F}(s)$ of a PDF is called the characteristic function. The characteristic function of a random variable is unique and always exists for all values of the argument s. Therefore, given a characteristic function of a random variable, its probability density function can be uniquely determined through the inverse transform. The characteristic functions of some commonly used PDFs are shown in Table 7.2. Using the characteristic function, the rth-order moment about the origin of the random variable X can be obtained as

$$E[X^r] = \mu'_r = \frac{1}{i^r} \left[\frac{d^r \, \mathcal{F}_f(s)}{d \, s^r} \right]_{s=0} \tag{7.55}$$

The Fourier transform is particularly useful when random variables are related linearly. In such cases, the convolution property of the Fourier transform can be applied to derive the characteristic function of the resulting random variable. More specifically, consider that $W = X_1 + X_2 + \cdots + X_N$ and all X's are independent random variables with known PDF, $f_i(x_i), i = 1, 2, \ldots, N$. The characteristic function of W then can be obtained as

$$\mathcal{F}_W(s) = \mathcal{F}_1(s) \, \mathcal{F}_2(s) \cdots \mathcal{F}_N(s) \tag{7.56}$$

which is the product of the characteristic function of each individual random variable X. The resulting characteristic function for W can be used in Eq. (7.55) to obtain the statistical moments of any order for the random variable W.

The *Laplace/exponential transforms* of a PDF $f(x)$ are defined, respectively, as

$$\mathcal{L}_f(s) = \int_0^\infty e^{sx} f(x) \, dx \tag{7.57}$$

$$\mathcal{E}_f(s) = \int_{-\infty}^\infty e^{sx} f(x) \, dx \tag{7.58}$$

The transformed functions given by Eqs. (7.57) and (7.58) of a PDF are called moment generating functions. The moment generating functions of commonly used PDFs are shown in Table 7.2. Similar to the characteristic function, statistical moments of a random variable X can be derived, from its moment generating function, as

$$E[X^r] = \mu'_r = \left[\frac{d^r \, \mathcal{E}_f(s)}{d \, s^r} \right]_{s=0} \tag{7.59}$$

Basically, the moment generating function has properties similar to the characteristic function such as the convolution properties described by Eq. (7.58). However, it possesses two deficiencies: (1) the moment generating function of a random variable may not always exist for all distribution functions and all values of s and (2) the correspondence between a PDF and moment generating function may not necessarily be unique.

The *Mellin transform* of a function $f(x)$, for $x > 0$, is defined as (Griffin, 1975; Springier, 1979)

$$M_X(s) = M[f(x)] = \int_0^\infty x^{s-1} f(x) \, dx \qquad x > 0 \tag{7.60}$$

where $M_X(s)$ is the Mellin transform of the function $f(x)$. A one-to-one correspondence exists between $M_X(s)$ and $f(x)$. The Mellin transform provides an alternative way to find the moments of any order of a nonnegative random variable.

The Mellin transform of the convolution of the PDFs associated with two independent random variables in a product form is equal to the product of the Mellin transform of two individual PDFs as

TABLE 7.2 Characteristic Functions of Some Commonly Used Distributions

Distribution function	PDF	Range	Characteristic function	Moment generating
Binomial	$p(x) = {}_nC_x p^x q^{n-x}$	$x = 0, 1, 2, \ldots, n$	$(q + pe^{is})^n$	$(q + pe^s)^n$
Poisson	$p(x) = e^{-v}v^x/x!$	$x = 0, 1, 2, \ldots$	$\exp\{v(e^{is} - 1)\}$	$\exp\{v(e^s - 1)\}$
Normal	$f(x) = \dfrac{1}{\sqrt{2\pi}\,\sigma}\exp\left[-\dfrac{1}{2}\left(\dfrac{x-\mu}{\sigma}\right)^2\right]$	$-\infty < x < \infty$	$\exp\{i\mu s - 0.5s^2\,\sigma^2\}$	$\exp\{\mu s + 0.5\sigma^2\,s^2\}$
Gamma	$f(x) = \dfrac{\beta}{\Gamma(\alpha)}(\beta x)^{\alpha-1}e^{-\beta x}$	$x > 0$	$\left(\dfrac{\beta}{\beta - is}\right)^\alpha$	$\left(\dfrac{\beta}{\beta - s}\right)^\alpha$
Extreme-value I	$f(x) = \dfrac{1}{\beta}\exp\left[\pm\left(\dfrac{x - x_o}{\beta}\right) - \exp\pm\left(\dfrac{x - x_o}{\beta}\right)\right]$	$-\infty < x < \infty$	$e^{i\alpha s}\Gamma(1 \mp i\beta s)$	$e^{\alpha s}\Gamma(1 \mp \beta s)$

$$M_W(s) = M[f(w)] = M[g(x)*h(y)] = M_X(s) \, M_Y(s) \qquad (7.61)$$

in which $*$ is the convolution operator. Equation (7.61) can be extended to a general case involving more than two independent random variables.

From this convolution property of the Mellin transform and its relationship to statistical moments, one can immediately see the advantage of the Mellin transform as a tool for obtaining the moments of a random variable which is related to other random variables in a multiplicative fashion. In addition to the convolution property, which is of primary importance, the Mellin transform also has several useful properties as summarized in Table 7.3. Furthermore, the Mellin transform of some commonly used PDFs are tabulated in Table 7.4. Tung (1990) applied the Mellin transform to uncertainty analysis of hydraulic and hydrologic problems.

TABLE 7.3 Mellin Transform of Products and Quotients of Random Variables (Park, 1987)

Random variable	PDF given	$M_W(s) =$
$W = X$	$f(x)$	$M_X(s)$
$W = X^b$	$f(x)$	$M_X(bs - b + 1)$
$W = 1/X$	$f(x)$	$M_X(2 - s)$
$W = XY$	$f(x), g(y)$	$M_X(s)M_Y(s)$
$W = X/Y$	$f(x), g(y)$	$M_X(s)M_Y(2 - s)$
$W = aX^bY^c$	$f(x), g(y)$	$a^{s-1}M_X(bs - b + 1)M_Y(cs - c + 1)$

a, b, c: constants; X, Y, W: random variables.
Source: Park, 1987.

TABLE 7.4 Mellin Transforms for Some Commonly Used Probability Density Functions

Probability	PDF	Range	Mellin transform
Uniform	$f(x) = \dfrac{1}{b-a}$	$a \le x \le b$	$\dfrac{b^s - a^s}{s(b-a)}$
Exponential	$f(x) = \beta \exp(-\beta x)$	$x > 0$	$\beta^{1-s}\,\Gamma(s)$
Gamma	$f(x) = \dfrac{\beta}{\Gamma(\alpha)}(\beta x)^{\alpha-1}e^{-\beta x}$	$x > 0$	$\dfrac{\beta^{1-s}\,\Gamma(\alpha+s-1)}{\Gamma(\alpha)}$
Triangular	$f(x) = \dfrac{2(x-a)}{(c-a)(b-a)}$	$a \le x \le b$	
	$f(x) = \dfrac{2(c-x)}{(c-a)(c-b)}$	$b \le x \le c$	$\dfrac{2}{(c-a)s(s+1)}\left[\dfrac{c(c^s-b^s)}{c-b} - \dfrac{a(b^s-a^s)}{b-a}\right]$
Standard beta	$f(x) = \dfrac{\Gamma(\alpha+\beta)}{\Gamma(\alpha)\Gamma(\beta)}x^{\alpha-1}(1-x)^{\beta-1}$	$0 \le x \le 1$	$\dfrac{\Gamma(\alpha+\beta)\Gamma(\alpha+s-1)}{\Gamma(\alpha)\Gamma(\alpha+\beta+s-1)}$
Standard normal	$f(x) = \dfrac{1}{\sqrt{2\pi}}e^{-x^2/2}$	$-\infty < x < \infty$	$2^{(s-3)/2}\,\Gamma(s/2)$

Although the Mellin transform is useful for uncertainty analysis, it possesses one drawback which should be pointed out: Under some combinations of distribution and functional form, the resulting transform may not be analytic for all values of s. This could occur especially when quotients or variables with negative exponents are involved. Under such circumstances, other transforms such as the Laplace or Fourier transform could be used to find the moments.

7.3.2 Approximation Methods

Most of the models or design procedures used in water resources engineering analyses are nonlinear and highly complex. This basically limits analytical derivation of the statistical properties of model output using the methods described in Sec. 7.3.1. Engineers frequently have to resort to methods that yield approximations to the statistical properties of uncertain model output. In the following, three methods that are useful for uncertainty analysis are described.

7.3.2.1 First-Order Variance Estimation Method. The *first-order variance estimation (FOVE) method* is also called the *variance propagation method* (Berthouex, 1975). It estimates uncertainty in terms of the variance of system output, which is evaluated on the basis of statistical properties of the system's stochastic variables. The method approximates the function involving stochastic variables by the Taylor series expansion.

Consider that a hydraulic or hydrologic design quantity W is related to N stochastic variables X as

$$W = g(\mathbf{X}) = g(X_1, X_2, \ldots, X_N) \tag{7.62}$$

where $\mathbf{X}$ is an N-dimensional column vector of stochastic variables. The Taylor series expansion of the function $g(\mathbf{X})$ with respect to a selected point $\mathbf{X} = \mathbf{x}_o$ in the parameter space can be expressed as

$$W = g(\mathbf{x}_o) + \sum_{i=1}^{N} \left(\frac{\partial g}{\partial X_i} \right)_{\mathbf{x}_o} (X_i - x_{io}) + \frac{1}{2} \sum_{i=1}^{N} \sum_{j=1}^{N} \left(\frac{\partial^2 g}{\partial X_i \partial X_j} \right)_{\mathbf{x}_o} (X_i - x_{io})^2 + \epsilon \tag{7.63}$$

in which ϵ represents the higher-order terms. The partial derivative terms are called the *sensitivity coefficients*, each represents the rate of change of model output W with respect to a unit change of each variable around $\mathbf{x}_o$. Dropping the second- and higher-order terms, the preceding equation can be expressed, in matrix/vector form, as

$$W \approx g(\mathbf{x}_o) + \mathbf{s}_o^{T \cdot} (\mathbf{X} - \mathbf{x}_o) \tag{7.64}$$

in which $\mathbf{s}_o$ is the vector of sensitivity coefficients evaluated at $\mathbf{X} = \mathbf{x}_o$. The mean and variance of W, by the first-order approximation, can be expressed, respectively, as

$$E[W] \approx g(\mathbf{x}_o) + \mathbf{s}_o^{T \cdot} (\boldsymbol{\mu} - \mathbf{x}_o) \tag{7.65}$$

and

$$\mathrm{Var}[W] \approx \mathbf{s}_o^T \mathbf{C}(\mathbf{X}) \mathbf{s}_o \tag{7.66}$$

in which $\boldsymbol{\mu}$ and $\mathbf{C}(\mathbf{X})$ are the vector of means and covariance matrix of stochastic variables $\mathbf{X}$, respectively.

The common practice of the first-order variance method is to take the expansion point $x_o = \mu$ and the mean and variance of W can be estimated as

$$E[W] \approx g(\mu) \tag{7.67}$$

and

$$\text{Var}[W] \approx s^T C(X) s \tag{7.68}$$

in which s is an N-dimensional vector of sensitivity coefficients evaluated at $x_o = \mu$. When all the stochastic variables are independent, the variance of model output W can be approximated as

$$\text{Var}[W] \approx s^T D s = \sum_{i=1}^{N} s_i^2 \sigma_i^2 \tag{7.69}$$

in which $D = \text{diag}(\sigma_1^2, \sigma_2^2, \ldots, \sigma_N^2)$, a diagonal matrix of variances of the stochastic variables. The ratio $s_i^2 \sigma_i^2 / \text{var}[W]$ indicates the proportion of overall uncertainty in the model output contributed by the uncertainty associated with variable X_i.

In general, $E[g(X)] \neq g(E[X])$ unless $g(X)$ is a linear function of X. Improvement of the accuracy can be made by incorporating higher-order terms in the Taylor series expansion. However, as the higher-order terms are included, the mathematical complication, and the required information, increase rapidly. The method can be expanded to include the second-order term for estimating the mean to account for the presence of correlation between stochastic variables.

The first two moments are used in uncertainty analysis for practical engineering design. The FOVE method does not require knowledge of the PDF of stochastic variables which simplifies the analysis. However, this advantage is also a disadvantage of the method because it is insensitive to the distributions of stochastic variables in the uncertainty analysis.

The computational effort associated with the FOVE method largely depends on how the sensitivity coefficients are calculated. For simple analytical functions the computation of the derivatives is a trivial task. However, for functions that are complex and/or implicit in the form of computer programs, or charts and figures, the task of computing the derivatives could become cumbersome. In such cases point estimation (PE) techniques can be used to circumvent the problems.

7.3.2.2 Rosenblueth's Probabilistic Point Estimation Method.

Rosenblueth's point estimation (PE) method was first developed for handling stochastic variables that are symmetric (Rosenblueth, 1975), which is later extended to treat nonsymmetric random variables (Rosenblueth, 1981). Referring to Fig. 7.2, the basic idea of Rosenblueth's PE method is to approximate the original PDF or PMF of the random variable X by assuming that the entire probability mass of X is concentrated at two points x_- and x_+. The four unknowns, that is, the locations of x_- and x_+ and the corresponding probability masses p_- and p_+, are determined in such manner that the first three moments of the original random variable X are preserved. The solutions for $x_-, x_+, p_-,$ and p_+ are

$$x_- = \mu - z_- \sigma \tag{7.70a}$$

$$x_+ = \mu + z_+ \sigma \tag{7.70b}$$

$$p_+ = \frac{z_-}{z_+ + z_-} \tag{7.70c}$$

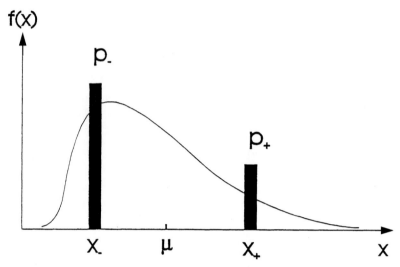

FIGURE 7.2   Schematic diagram of Rosenblueth's PE method in univariate case.

$$p_- = 1 - p_+ \tag{7.70d}$$

where

$$z_+ = \frac{\gamma}{2} + \sqrt{1 + \left(\frac{\gamma}{2}\right)^2} \tag{7.70e}$$

$$z_- = z_+ - \gamma \tag{7.70f}$$

with γ being the skew coefficient of the stochastic variable X.

When the distribution of the random variable X is symmetric ($\gamma = 0$), $z_- = z_+ = 1$ and $p_- = p_+ = 0.5$. This implies that, with a symmetric random variable, the two points are located at one standard deviation to either side of the mean, with equal probability mass assigned at the two locations.

For problems involving N stochastic variables, the two points for each variable are computed according to Eqs. (7.70a–f) and permutated producing a total of 2^N possible points of evaluation in the parameter space. The rth moment of $W = g(\mathbf{X}) = g(X_1, X_2, \ldots, X_N)$ about the origin can be approximated as

$$E(W^r) \approx \sum p_{(\delta 1, \delta 2, \ldots, \delta N)} \, w^r_{(\delta 1, \delta 2, \ldots, \delta N)} \tag{7.71}$$

in which subscript δ_i is a sign indicator that can only be $+$ or $-$ for representing the stochastic variable X_i having the value of $x_{i+} = \mu_i + z_{i+}\,\sigma_i$ or $x_{i-} = \mu_i - z_{i-}\,\sigma_i$, respectively; $p_{(\delta 1, \delta 2, \ldots, \delta N)}$ can be determined as

$$p_{(\delta 1, \delta 2, \ldots, \delta N)} = \prod_{i=1}^{N} p_{i, \delta i} + \sum_{i=1}^{N-1} \left(\sum_{j=i+1}^{N} \delta_i\, \delta_j\, a_{ij} \right) \tag{7.72}$$

in which

$$a_{ij} = \frac{\rho_{ij}/2^N}{\sqrt{\prod_{i=1}^{N}\left[1 + \left(\frac{\gamma_i}{2}\right)^2\right]}} \qquad (7.73)$$

where ρ_{ij} is the correlation coefficient between stochastic variables X_i and X_j. The number of terms in the summation of Eq. (7.71) is 2^N which corresponds to the total number of possible combinations of + and − for all N stochastic variables.

For each term of the summation in Eq. (7.71), the model has to be evaluated once at the corresponding point in the parameter space. This indicates a potential drawback of Rosenblueth's PE method as it is applied to practical problems. When N is small, the method is practical for performing uncertainty analysis. However, for moderate or large N, the number of required function evaluations of $g(\mathbf{X})$ could be too numerous to implement practically, even on the computer. To circumvent this shortcoming, Harr (1989) developed an alternative PE method that reduces the 2^N function evaluations required by Rosenblueth's method down to $2N$.

7.3.2.3 Harr's Probabilistic Point Estimation Method.

Harr's point estimation method (Harr, 1989) reduces the required function evaluations by PE methods. The method utilizes the first two moments (that is, the mean and variance) of the random variables involved and their correlations. Skew coefficients of the random variables are ignored by the method. Hence, the method is appropriate for treating random variables that are symmetric. For problems involving only a single random variable, Harr's PE method is identical to Rosenblueth's method with zero skew coefficient. The theoretical basis of Harr's PE method is built on the orthogonal transformation using eigenvalue-eigenvector decomposition.

Orthogonal transformation is an important tool for treating problems involving correlated random variables. The procedure maps correlated random variables from their original space to a new domain in which they become uncorrelated. Hence, the analysis is greatly simplified. Consider N multivariate random variables $\mathbf{X} = (X_1, X_2, \ldots, X_N)^T$ having a mean vector $\boldsymbol{\mu} = (\mu_1, \mu_2, \ldots, \mu_N)^T$ and covariance matrix $\mathbf{C}(\mathbf{X})$. The vector of correlated standardized random variables $\mathbf{X}' = \mathbf{D}^{-1/2}(\mathbf{X} - \boldsymbol{\mu})$, that is $\mathbf{X}' = (X'_1, X'_2, \ldots, X'_N)^T$ with $\mathbf{X}'_i = (X_i - \mu_i)/\sigma_i$ for $i = 1, 2, \ldots, N$, would have mean vector of zero, $\mathbf{0}$, and the covariance matrix equal to the correlation matrix $\mathbf{R}(\mathbf{X})$ with D, being an $N \times N$ diagonal matrix of variances of the stochastic variables.

Orthogonal transformation can be made by several ways (Young and Gregory, 1973; Golub and Van Loan, 1989). One frequently used approach is the *eigenvalue-eigenvector factorization* by which the correlation matrix $\mathbf{R}(\mathbf{X})$ is decomposed as

$$\mathbf{R}(\mathbf{X}) = \mathbf{C}(\mathbf{X}') = \mathbf{V}\,\boldsymbol{\Lambda}\,\mathbf{V}^T \qquad (7.74)$$

in which $\mathbf{V}$ is an $N \times N$ eigenvector matrix consisting of N eigenvectors as $\mathbf{V} = (\mathbf{v}_1, \mathbf{v}_2, \ldots, \mathbf{v}_N)$ with $\mathbf{v}_i$ being the ith eigenvector and $\boldsymbol{\Lambda} = \text{diag}(\lambda_1, \lambda_2, \ldots, \lambda_N)$ is a diagonal eigenvalue matrix. Frequently, the eigenvectors $\mathbf{v}$'s are normalized such that its norm is equal to unity, that is, $\mathbf{v}^T\mathbf{v} = 1$. Using the eigenvector matrix $\mathbf{V}$, the following transformation can be made

$$\mathbf{U} = \mathbf{V}^T\mathbf{X}' \qquad (7.75)$$

The resulting transformed stochastic variables $\mathbf{U}$ have the mean $\mathbf{0}$ and covariance matrix $\boldsymbol{\Lambda}$ indicating that $\mathbf{U}$ are uncorrelated because their covariance matrix $\mathbf{C}(\mathbf{U})$ is a diagonal matrix $\boldsymbol{\Lambda}$. Hence, each new stochastic variable U_i has the standard deviation equal to $\sqrt{\lambda_i}$, for all $i = 1, 2, \ldots, N$.

For multivariate problems involving N stochastic variables, Harr's PE method selects its points for function evaluation which are located at the intersections of the N eigenvector axes and a hypersphere of radius $\sqrt{N}$ centered at the origin in the transformed $\mathbf{U}$-space. Referring to Fig. 7.3, the corresponding points in the original parameter space can be obtained as

$$\mathbf{x}_{i\pm} = \boldsymbol{\mu} \pm \sqrt{N}\, \mathbf{D}^{1/2}\, \mathbf{v}_i \qquad i = 1, 2, \ldots, N \tag{7.76}$$

in which $\mathbf{x}_{i\pm}$ is the vector of coordinates of the N stochastic variables in the parameter space on the ith eigenvector $\mathbf{v}_i$.

Based on the $2N$ points determined by Eq. (7.76), the function values at each of the $2N$ points can be computed. From that the rth moment of the function $W = g(\mathbf{X})$ can be calculated according to

$$\overline{w}_i^r = \frac{w_{i+}^r + w_{i-}^r}{2} = \frac{g^r(\mathbf{x}_{i+}) + g^r(\mathbf{x}_{i-})}{2} \qquad i = 1, 2, \ldots, N; r = 1, 2, \ldots \tag{7.77}$$

$$E[W^r] = \mu_r'(W) = \frac{\displaystyle\sum_{i=1}^{N} \lambda_i\, \overline{w}_i^r}{N} \qquad r = 1, 2, \ldots \tag{7.78}$$

7.4 LOAD-RESISTANCE INTERFERENCE RELIABILITY ANALYSIS

In numerous hydrologic and hydraulic problems, uncertainties in data and in theory, including design and analysis procedures, warrant a probabilistic treatment of the problems. The failure associated with a hydraulic structure is the result of the combined effect from inherent randomness of external load and various uncertainties involved in the analysis, design, construction, and operational procedures.

Failure of an engineering system occurs when the load L (external forces or demands) on the system exceeds the resistance R (strength, capacity, or supply) of the system. The definitions of reliability and failure probability, Eqs. (7.1) and (7.2), are applicable to component reliability, as well as total system reliability. In hydraulic and hydrologic analyses, the resistance and load are frequently functions of a number of stochastic variables, that is, $L = g(\mathbf{X}_L)$ and $R = h(\mathbf{X}_R)$. Accordingly, the reliability is a function of the stochastic variables involved as

$$p_s = P\left[g(\mathbf{X}_L) \le h(\mathbf{X}_R)\right] \tag{7.79}$$

Computations of reliability using Eq. (7.79) do not consider the time-dependence of loads as if $\mathbf{X}_L$ and $\mathbf{X}_R$ are stationary random variables. This is referred to as the *static reliability model*, which is generally applied when the performance of the system subject to a single worst load event is of concern.

However, a hydraulic structure is expected to serve its designed function over an expected period of time. Engineers are frequently interested in knowing the reliability of the structure over its expected service life. In such circumstances, elements of service duration, randomness of occurrence of loads, and possible change of resistance characteristics over time must be considered. Reliability models incorporating

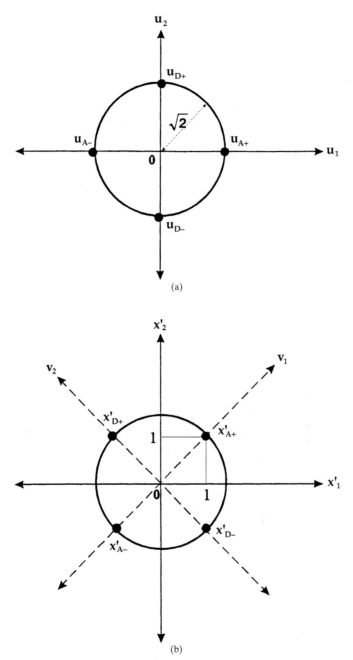

FIGURE 7.3   Schematic diagram of Harr's PE method for bivariate problems: (*a*) selected points in eigen-space; (*b*) corresponding selected points in original standardized space.

these elements are called *time-dependent reliability models* (Kapur and Lamberson, 1977; Tung and Mays, 1980b; Wen, 1987).

7.4.1 Reliability Performance Measures

In reliability analysis, Eq. (7.79) is often expressed in terms of a performance function $W(\mathbf{X}) = W(\mathbf{X}_L, \mathbf{X}_R)$ as

$$p_s = P\,[W(\mathbf{X}_L, \mathbf{X}_R) \ge 0] = P\,[W(\mathbf{X}) \ge 0] \tag{7.80}$$

In reliability analysis, the system state is divided into a safe (satisfactory) set defined by $W(\mathbf{X}) \ge 0$ and a failure (unsatisfactory) set defined by $W(\mathbf{X}) < 0$. The boundary that separates the safe set and failure set is a surface defined by $W(\mathbf{X}) = 0$ and this surface is called the *failure surface* or *limit-state surface*. Since the performance function $W(\mathbf{X})$ defines the condition of the system, it is sometimes called the *system state function*.

The performance function $W(\mathbf{X})$ can be expressed differently as (Tung and Yen, 1993)

$$W_1(\mathbf{X}) = R - L = h(\mathbf{X}_R) - g(\mathbf{X}_L) \tag{7.81}$$

$$W_2(\mathbf{X}) = (R/L) - 1 = [h(\mathbf{X}_R)/g(\mathbf{X}_L)] - 1 \tag{7.82}$$

$$W_3(\mathbf{X}) = \ln\,(R/L) = \ln\,[h(\mathbf{X}_R)] - \ln\,[g(\mathbf{X}_L)] \tag{7.83}$$

Equation (7.81) is identical to the notion of safety margin, while Eqs. (7.82) and (7.83) are based on safety factor representations.

The *reliability index* β is another frequently used reliability indicator, which is defined as the reciprocal of the coefficient of variation of the performance function $W(\mathbf{X})$ as

$$\beta = \mu_W/\sigma_W \tag{7.84}$$

in which μ_W and σ_W are the mean and standard deviation of the performance function, respectively. From Eq. (7.84), assuming an appropriate PDF for the random performance function $W(\mathbf{X})$, the reliability can then be computed as

$$p_s = 1 - F_W\,(0) = 1 - F_{W'}\,(-\beta) \tag{7.85}$$

in which $F_W(\cdot)$ is the CDF of the performance variable W and W' is the standardized performance variable defined as $W' = (W - \mu_W)/\sigma_W$. The expression of reliability p_s for eight different distributions of $W(\mathbf{X})$ are given in Table 7.5. In practice, the normal distribution is commonly used for $W(\mathbf{X})$, in which case the reliability can be computed as

$$p_s = 1 - \Phi(-\beta) = \Phi(\beta) \tag{7.86}$$

in which $\Phi(\cdot)$ is the standard normal CDF.

7.4.2 Direct Integration Method

From Eqs. (7.1) and (7.2), the computation of reliability requires knowledge of probability distributions of load and resistance, or the performance function, W. In terms of the joint PDF of the load and resistance, the reliability can be expressed as

TABLE 7.5 Reliability Formulas for Selected Distributions

Distribution of W	PDF, $f_W(w)$	Mean, μ_w	Coefficient of variation, Ω_W	Reliability $p_s = P(W \geq 0)$
Normal	$\dfrac{1}{\sigma\sqrt{2\pi}}\left[e^{-\frac{1}{2}\left(\frac{w-\mu}{\sigma}\right)^2}\right]; -\infty < w < \infty$	μ	σ/μ	$\Phi\left(\dfrac{\mu}{\sigma}\right)$
Lognormal	$\dfrac{1}{\sqrt{2\pi}w\sigma_{\ln w}}\left[e^{-\frac{1}{2}\left(\frac{\ln(w)-\mu_{\ln w}}{\sigma_{\ln w}}\right)^2}\right], w \geq 0$	$\mu_w = e^{\mu_{\ln w} + (\sigma_{\ln w}^2/2)}$	$\sqrt{e^{\sigma_{\ln w}^2}-1}$	$\Phi\left(\dfrac{\mu_{\ln w}}{\sigma_{\ln w}}\right)$
Exponential	$\beta e^{-\beta(w-w_0)}, w \geq w_0$	$\dfrac{1}{\beta}+w_0$	$\dfrac{1}{1+\beta w_0}$	$e^{-\beta(w-w_0)}$
Gamma	$\dfrac{\beta^\alpha(w-\xi)^{\alpha-1}e^{-\beta(w-\xi)}}{\Gamma(\alpha)}, w \geq \xi$	$\dfrac{\alpha}{\beta}+\xi$	$\dfrac{\sqrt{\alpha}}{\alpha+\beta\xi}$	$1-I[-\lambda a, k]$
Beta	$\dfrac{1}{B(\alpha,\beta)}\dfrac{(w-a)^{\alpha-1}(b-w)^{\beta-1}}{(b-a)^{\alpha+\beta+1}}, a \leq w \leq b$	$a+\dfrac{\alpha}{\alpha+\beta}(b-a)$	$\sqrt{\dfrac{\alpha\beta}{\alpha+\beta+1}}\dfrac{(b-a)}{(\alpha+\beta)\mu_w}$	$1-\dfrac{B_u(\alpha,\beta)}{B(\alpha,\beta)}; u=\dfrac{a}{a-b}$
Triangular	$\dfrac{2}{b-a}\left(\dfrac{w-a}{m-a}\right)$ for $a \leq w \leq m$ $\dfrac{2}{b-a}\left(\dfrac{b-w}{b-m}\right)$ for $m \leq w \leq b$	$\dfrac{a+m+b}{3}$	$\left[\dfrac{1}{2}-\dfrac{ab+am+bm}{6\mu_w^2}\right]^{1/2}$	$1-\dfrac{(w-a)^2}{(b-a)(m-a)}$, for $a \leq w \leq m$ $\dfrac{(b-w)^2}{(b-a)(b-m)}$, for $m \leq w \leq b$
Uniform	$\dfrac{1}{b-a}, a \leq w \leq b$	$\dfrac{a+b}{2}$	$\dfrac{1}{\sqrt{3}}\dfrac{b-a}{b+a}$	$\dfrac{b-w}{b-a}$

Source: Yen, et al., 1986.

$$p_s = \int_{r_1}^{r_2} \left[\int_{\ell_1}^{r} f_{R,L}(r,\ell) \, d\ell \right] dr = \int_{\ell_2}^{\ell_2} \left[\int_{\ell}^{r_2} f_{R,L}(r,\ell) \, dr \right] d\ell \qquad (7.87)$$

in which r and ℓ are dummy arguments for the resistance and load, respectively; (r_1, r_2) and (ℓ_1, ℓ_2) are the lower and upper bounds for the resistance and load, respectively. This computation of reliability is called the *load-resistance interference*.

When the load and resistance are independent, Eq. (7.87) can be reduced to

$$p_s = E_R[F_L(R)] = 1 - E_L[F_R(L)] \qquad (7.88)$$

in which $E_R[F_L(R)]$ is the expectation of the CDF of load over the possible range of resistance. A schematic diagram illustrating the load-resistance interference, when load and resistance are independent random variables, is shown in Fig. 7.4.

The method of direct integration requires the PDFs of the load and resistance or the performance function be known or derived. This information is seldom available in practice, especially for the joint PDF, because of the complexity of hydrologic and hydraulic models used in design. Explicit solution of direct integration can be obtained for only a few PDFs. Yen, et al. (1986) listed the solution of the failure probability $p_f = 1 - p_s$ for eight different distributions of the performance function W. For most PDFs numerical integration may be necessary. When using numerical integration (including Monte Carlo simulation described in Sec. 7.5), difficulty may be encountered when one deals with a multivariate problem.

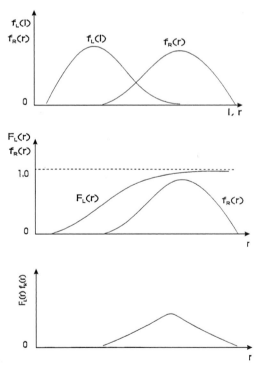

FIGURE 7.4 Schematic diagram of load-resistance interference.

7.4.3   Mean-Value First-Order Second-Moment (MFOSM) Method

The *MFOSM method* for reliability analysis employs the FOVE method to estimate the statistical moments of the performance function $W(\mathbf{X})$. Once the mean and standard deviation of $W(\mathbf{X})$ are obtained, the reliability index β_{MFOSM} is computed as

$$\beta_{\text{MFOSM}} = \frac{W(\boldsymbol{\mu})}{\sqrt{\mathbf{s}^T \, \mathbf{C}(\mathbf{X}) \, \mathbf{s}}} \tag{7.89}$$

from which the reliability is computed according to Eqs. (7.85) or (7.86).

Yen and Ang (1971), Ang (1973), and Cheng, et al. (1986b) indicated that, provided $p_s < 0.99$, reliability is not greatly influenced by the choice of distribution for W, and the assumption of a normal distribution is quite satisfactory. However, for reliability higher than this value (e.g., $p_s = 0.999$), the shape of the tails of a distribution becomes critical. In such cases accurate assessment of the distribution of $W(\mathbf{X})$ should be made to evaluate the reliability or failure probability. The MFOSM method has been used widely in various hydraulic structural and facility designs, such as storm sewers (Tang and Yen, 1972; Tang, et al., 1975; Yen and Tang, 1976; Yen, et al., 1976), culverts (Yen, et al., 1980; Tung and Mays, 1980b), levees (Tung and Mays, 1981a; Lee and Mays, 1986), floodplains (McBean, et al., 1984), and open channel hydraulics (Huang, 1986).

The application of the MFOSM method is simple and straightforward. However, it possesses certain weaknesses in addition to the difficulties with accurate estimation of extreme failure probabilities as mentioned above. These weaknesses include:

1. Inappropriate choice of the expansion point: In reliability computation, one should be concerned with those points in the parameter space that fall on the failure surface or limit-state surface.

2. Inability to handle distributions with large skew coefficient: This is mainly due to the fact that MFOSM method incorporates only the first two moments of the random variables involved.

3. Generally poor estimation of the mean and variance of highly nonlinear functions: This is because the first-order approximation of a highly nonlinear function is not accurate.

4. Sensitivity of the computed failure probability to the formulation of the performance function W.

5. Inability to incorporate available information on probability distributions of the stochastic variables affecting the load and resistance.

From the above arguments, the general rule of thumb is not to rely on the result of the MFOSM method if any of the following conditions exist: (1) high accuracy requirements for the estimated reliability or risk, (2) high nonlinearity of the performance function, and (3) many skewed random variables are involved in the performance function.

7.4.4   Advanced First-Order Second-Moment (AFOSM) Method

The main thrust of the *AFOSM method* is to reduce the error of the MFOSM method associated with the nonlinearity and noninvariability of the performance function, while keeping the simplicity of the first-order approximation. In the AFOSM method, the expansion point $\mathbf{x}_* = (\mathbf{x}_L^*, \mathbf{x}_R^*)$ is located on the failure surface defined by the limit-state equation $W(\mathbf{x}) = 0$ which defines the boundary that separates the system performance from being unsatisfactory (unsafe) or being satisfactory (safe), that is,

$$W(\mathbf{x})\begin{cases} > 0 & \text{system performance is satisfactory (or safe region)} \\ = 0 & \text{failure surface} \\ < 0 & \text{system performance is unsatisfactory (or failure region)} \end{cases}$$

Among all the possible values of $\mathbf{x}$ that fall on the limit-state surface $W(\mathbf{x}) = 0$, one is more concerned with the combination of stochastic variables that would yield the lowest reliability or highest risk. The point on the failure surface with the lowest reliability is the one having the shortest distance to the point where the means of the stochastic variables are located. This point is called the *design point* by Hasofer and Lind (1974) or the *most probable failure point* (Shinozuka, 1983).

Consider that $\mathbf{X} = (X_1, X_2, \ldots, X_N)^T$ are N uncorrelated stochastic variables having a vector mean $\boldsymbol{\mu}$ and covariance matrix $\mathbf{D}$. The original stochastic variables X can be standardized into $\mathbf{X}'$ according to Eq. (7.27). Hence, the design point in $\mathbf{x}'$-space is the one that has the shortest distance from the failure surface $W(\mathbf{x}') = 0$ to the origin $\mathbf{x}' = \mathbf{0}$. Such a point can be found by solving

$$\text{minimize } \|\mathbf{x}'\| = (\mathbf{x}'^T \mathbf{x}')^{1/2} \tag{7.90a}$$

subject to

$$W(\mathbf{x}') = 0. \tag{7.90b}$$

This constrained nonlinear minimization problem can be solved by the Lagrangian multiplier method (see Sec. 6.3.2) and the design point $\mathbf{x}'_*$ can be obtained as

$$\mathbf{x}'_* = -\xi_* \|\mathbf{x}'_*\| \nabla_{\mathbf{x}'} W(\mathbf{x}'_*) \tag{7.91}$$

Furthermore, from Eq. (7.91) the distance between the origin $\mathbf{x}' = \mathbf{0}$ to the design point $\mathbf{x}'_*$ can be obtained as

$$\mathbf{x}'_* = -\frac{\nabla_{\mathbf{x}'} W(\mathbf{x}'_*)}{\|\nabla_{\mathbf{x}'} W(\mathbf{x}'_*)\|} \|\mathbf{x}'_*\| = -\|\mathbf{x}'_*\| \boldsymbol{\alpha}_* \tag{7.92}$$

in which $\boldsymbol{\alpha}_* = \nabla_{\mathbf{x}'} W(\mathbf{x}'_*)/\|\nabla_{\mathbf{x}'} W(\mathbf{x}'_*)\|$ is a unit vector eminating from the design point $\mathbf{x}'_*$ and pointing toward the origin (see Fig. 7.5). The elements of $\boldsymbol{\alpha}_*$, α_{i*}, are called the *directional derivatives* representing the value of the cosine angle between the gradient vector $\nabla_{\mathbf{x}'} W(\mathbf{x}')$ and axes of the standardized variables. Geometrically, Eq. (7.97) shows that the vector $\mathbf{x}'_*$ is perpendicular to the tangent hyperplane passing through the design point. The shortest distance can be expressed as

$$|\mathbf{x}'_*| = -\boldsymbol{\alpha}_*^T \mathbf{x}'_* = \frac{-\sum_{i=1}^{N} \left(\dfrac{\partial W(\mathbf{x}')}{\partial x_i'} \right)_* x_{i*}'}{\sqrt{\sum_{j=1}^{N} \left(\dfrac{\partial W(\mathbf{x}')}{\partial x_i'} \right)_*^2}} \tag{7.93}$$

Recall that $X_i = \mu_i + \sigma_i X_i'$, for $i = 1, 2, \ldots, N$. By the chain rule in calculus, Eq. (7.93) can be written, in terms of the original variables $\mathbf{X}$, as

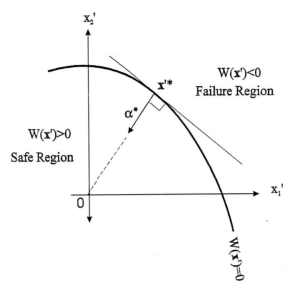

FIGURE 7.5 Characteristics of design point in standardized spaces.

$$|\mathbf{x}'_*| = \frac{\sum_{i=1}^{N} \left(\frac{\partial W(\mathbf{x})}{\partial x_i} \right)_* (\mu_i - x_{i*})}{\sqrt{\sum_{j=1}^{N} \left(\frac{\partial W(\mathbf{x})}{\partial x_i} \right)_*^2 \sigma_i^2}} \tag{7.94}$$

in which $\mathbf{x}_*$ is the point in $\mathbf{x}$-space corresponding to the design point $\mathbf{x}'_*$ in $\mathbf{x}'$-space which can be easily determined as $\mathbf{x}_* = \mu + \mathbf{D}^{1/2}\mathbf{x}'_*$.

Refer to the first-order approximation of the performance function $W(\mathbf{X})$. Taking the expansion point $\mathbf{x}_o = \mathbf{x}_*$, the expected value and variance of the performance function $W(\mathbf{X})$ can be approximated as

$$\mu_W \approx \mathbf{s}_*^T (\mu - \mathbf{x}_*) \tag{7.95}$$

$$\sigma_W^2 \approx \mathbf{s}_*^T \mathbf{C}(\mathbf{X}) \, \mathbf{s}_* \tag{7.96}$$

in which $\mathbf{s}_* = (s_{1*}, s_{2*}, \dots, s_{N*})^T$ is a vector of sensitivity coefficients of the performance function $W(\mathbf{X})$ evaluated at $\mathbf{x}_*$; and μ and $\mathbf{C}(\mathbf{X})$ are the mean vector and covariance matrix of the stochastic variables, respectively. If the stochastic variables are uncorrelated, Eq. (7.96) reduces to

$$\sigma_W^2 = \sum_{i=1}^{N} s_{i*}^2 \, \sigma_i^2 \tag{7.97}$$

in which σ_i is the standard deviation of the ith stochastic variable. The standard deviation of the performance function $W(\mathbf{X})$ can alternatively be expressed in terms of the directional derivatives as

$$\sigma_W = \sum_{i=1}^{N} \alpha_{i*} \, s_{i*} \, \sigma_i \tag{7.98}$$

where α_{i*} is the directional derivative for the ith stochastic variable at the expansion point $\mathbf{x}_*$.

$$\alpha_{i*} = \frac{s_{i*} \, \sigma_i}{\sqrt{\displaystyle\sum_{j=1}^{N} s_{j*}^2 \, \sigma_j^2}} \tag{7.99}$$

Equation (7.99) is identical to the one defined in Eq. (7.93). With the mean and standard deviation of the performance function $W(\mathbf{x})$ computed at $\mathbf{x}_*$, the AFOSM reliability index β_{AFOSM} can be determined as

$$\beta_{\text{AFOSM}} = \frac{\mu_W}{\sigma_W} = \frac{\displaystyle\sum_{i=1}^{N} s_{i*} \, (\mu_i - x_{i*})}{\displaystyle\sum_{i=1}^{N} \alpha_{i*} \, s_{i*} \, \sigma_i} \tag{7.100}$$

Equation (7.100) indicates that the AFOSM reliability index β_{AFOSM} is identical to the shortest distance from the origin to the design point in the standardized parameter space. β_{AFOSM} is also called the *Hasofer-Lind reliability index*.

Once β_{AFOSM} is computed, the reliability can be estimated as $p_s = \Phi(\beta_{\text{AFOSM}})$. Since $\beta_{\text{AFOSM}} = \|\mathbf{x}'_*\|$, the sensitivity of β_{AFOSM} with respect to the uncorrelated standardized random variables is

$$\nabla_{\mathbf{x}'} \beta_{\text{AFOSM}} = \nabla_{\mathbf{x}'} \|\mathbf{x}'_*\| = \frac{\mathbf{x}'_*}{|\mathbf{x}'_*|} = -\alpha_* \tag{7.101}$$

Equation (7.101) shows that $-\alpha_{i*}$ represents the rate of change in β_{AFOSM} due to a one standard deviation change in stochastic variable X_i, that is,

$$-\alpha_{i*} = \left[\frac{\partial \beta_{\text{AFOSM}}}{\partial X'_i} \right]_{\mathbf{x}'_*} = \left[\frac{\partial \beta_{\text{AFOSM}}}{\partial X_i} \right]_{\mathbf{x}*} \sigma_i \tag{7.102}$$

It can also be easily shown that

$$\left(\frac{\partial p_s}{\partial X'_i} \right)_{\mathbf{x}'_*} = -\alpha_{i*} \, \Phi(\beta_{\text{AFOSM}}); \quad \left(\frac{\partial p_s}{\partial X_i} \right)_{\mathbf{x}*} = -\sigma_i \, \alpha_{i*} \, \Phi(\beta_{\text{AFOSM}}) \tag{7.103}$$

for $i = 1, 2, \ldots, N$.

In the case that $\mathbf{X}$ are independent normal stochastic variables, Hasofer and Lind (1974) proposed a recursive equation for determining $\mathbf{x}_*$.

$$\mathbf{x}_{(r+1)} = \mu + \mathbf{Ds}_{(r)} \frac{[\mathbf{x}_{(r)} - \mu]^T \mathbf{s}_{(r)} - W(\mathbf{x}_{(r)})}{\mathbf{s}_{(r)}^T \mathbf{Ds}_{(r)}} \quad \text{for} \quad r = 1, 2, \ldots \tag{7.104}$$

Based on Eq. (7.104), the Hasofer-Lind algorithm for the AFOSM reliability analysis can be outlined as follows:

Step 1   Select an initial trial solution $\mathbf{x}_{(r)}$.

Step 2   Compute $W(\mathbf{x}_{(r)})$ and the corresponding sensitivity coefficient vector $\mathbf{s}_{(r)}$.

Step 3   Revise solution point $\mathbf{x}_{(r+1)}$ according to Eq. (7.104).

Step 4   Check if $\mathbf{x}_{(r)}$ and $\mathbf{x}_{(r+1)}$ are sufficiently close. If yes, compute the reliability index β_{AFOSM} according to Eq. (7.103) and the corresponding reliability $p_s = \Phi(\beta_{\mathrm{AFOSM}})$, then, stop; otherwise, update the solution point by letting $\mathbf{x}_{(r)} = \mathbf{x}_{(r+1)}$ and return to Step 2.

Step 5   Compute the sensitivity of reliability with respect to changes in stochastic variables.

Due to the nature of nonlinear optimization, the algorithm AFOSM-HL does not necessarily coverge to the true design point associated with the minimum reliability index. Therefore, Madsen, et al. (1986) suggest that different initial trial points be used and the smallest reliability index be selected to compute the reliability.

Ang and Tang (1984) also proposed an iterative procedure for finding the design point. The core of their updating procedure relies on the fact that, according to Eq. (7.100), the following relationship should be satisfied

$$\sum_{i=1}^{N} s_{i*} \left(\mu_i - x_{i*} - \alpha_{i*}\, \beta_*\, \sigma_i \right) = 0 \qquad (7.105)$$

Since variables X_i's are random and uncorrelated, Eq. (7.105) defines the failure point within the first-order context. Hence, Eq. (7.100) can be decomposed into

$$x_{i*} = \mu_i - \alpha_{i*}\, \beta_*\, \sigma_i \qquad \text{for} \qquad i = 1, 2, \ldots, N \qquad (7.106)$$

Ang and Tang (1984) present the following iterative procedure to locate the design point $\mathbf{x}_*$ and the corresponding reliability index β_{AFOSM} under the condition that stochastic variables are independent normal random variables. The algorithm involves the following steps:

Step 1   Select an initial point $\mathbf{x}_{(r)}$ in the parameter space. For practicality, the point μ is a viable starting point.

Step 2   At the selected point $\mathbf{x}_{(r)}$, compute the mean of the performance function $W(\mathbf{X})$ by

$$\mu_W = W(\mathbf{x}_{(r)}) + \mathbf{s}_{(r)}^T \left(\mu - \mathbf{x}_{(r)} \right) \qquad (7.107)$$

and the variance according to Eq. (7.91).

Step 3   Compute the corresponding reliability index β according to Eq. (7.84).

Step 4   Compute the values of directional derivative α_i for all $i = 1, 2, \ldots, N$, according to Eq. (7.99).

Step 5   Revise the location of expansion point $\mathbf{x}_{(r+1)}$ according to Eq. (7.112) using β obtained from Steps 4 and 5.

Step 6   Check if the revised expansion point $\mathbf{x}_{(r+1)}$ differs significantly from the previous trial expansion point $\mathbf{x}_{(r)}$. If yes, use the revised expansion point as the trial point by letting $\mathbf{x}_{(r)} = \mathbf{x}_{(r+1)}$ and go to Step 2 for another iteration. Otherwise, the iterative procedure is considered complete and the latest reliability index β_{AFOSM} is to be used in Eq. (7.86) to compute the reliability p_s.

Similar to the Hasofer-Lind algorithm, the above Ang-Tang algorithm does not guarantee convergence to the design point for all problems.

The task to determine the critical failure point $\mathbf{x}^*$ that minimizes the reliability is equivalent to minimizing the value of the reliability index β. Cheng, et al. (1986a) proposed the following optimization model to identify the design point.

$$\text{minimize } |W(\mathbf{x})| \text{ subject to Eq. (7.106)} \qquad (7.108)$$

The approach by Cheng, et al. (1986a) is a viable tool when the iterative algorithms of Ang-Tang and Hasofer-Lind do not converge.

The AFOSM method described above is suitable for the case that all stochastic variables in the load and resistance functions are independent, normal random variables. In reality, stochastic variables in a performance function may be nonnormal and correlated. In the following, the treatment of stochastic variables that are nonnormal and correlated are discussed.

7.4.4.1   Treatment of Nonnormal Stochastic Variables.

When nonnormal random variables are involved, it is advisable to transform them into equivalent normal variables. Rackwitz (1976) proposed an approach which transforms a nonnormal distribution into an equivalent normal distribution, so that the value of the CDF of the transformed equivalent normal distribution *is* the same as that of the original nonnormal distribution at the failure point $\mathbf{x}_*$, that is,

$$F_i(x_{i*}) = \Phi\left(\frac{x_{i*} - \mu_{i*N}}{\sigma_{i*N}}\right) \qquad (7.109)$$

in which $F_i(x_{i*})$ is the CDF of the stochastic variable X_i having a value at x_{i*}; and μ_{i*N} and σ_{i*N} are the mean and standard deviation of normal equivalent for the ith stochastic variable at $X_i = x_{i*}$. From Eq. (7.109), the following equations are obtained

$$\mu_{i*N} = x_{i*} - z_{i*}\,\sigma_{i*N} \qquad (7.110)$$

$$\sigma_{i*N} = \frac{\Phi(z_{i*})}{f_i(x_{i*})} \qquad (7.111)$$

in which $z_{i*} = \Phi^{-1}[F_i(x_{i*})]$ is the standard normal quantile.

To incorporate the normal transformation for nonnormal but uncorrelated stochastic variables, the iterative algorithms described previously for the AFOSM method can be modified. In the Hasofer-Lind algorithm, replace the updating equation in Step 3 by

$$\mathbf{x}_{(r+1)} = \mu_N + \mathbf{D}_N\,\mathbf{s}_{(r)}\,\frac{(\mathbf{x}_{(r)} - \mu_N)^T\,\mathbf{s}_{(r)} - W(\mathbf{x}_{(r)})}{\mathbf{s}_{(r)}^T\,\mathbf{D}_N\,\mathbf{s}_{(r)}} \qquad (7.112)$$

In the Ang-Tang algorithm, compute the directional derivatives in Step 4 according to Eq. (7.104) substituting the σ_{i*N}'s for the corresponding standard deviations of nonnormal stochastic variable σ_i's. Then, in Step 5, use β and α_{iN} to revise the location of the expansion point $\mathbf{x}_{(r+1)}$ according to

$$x_{i(r+1)} = \mu_{iN} - \alpha_{iN}\,\beta\,\sigma_{iN} \qquad \text{for} \qquad i = 1, 2, \ldots, N \qquad (7.113)$$

7.4.4.2   Treatment of Correlated Stochastic Variables.

When some of the stochastic variables involved in the performance function are significantly correlated, transformation of correlated variables to uncorrelated ones is made before the step of normal transformation is taken. Consider that the stochastic variables in the performance function are multivariate normal random variables with the vector of mean μ and a covariance matrix $\mathbf{C}(\mathbf{X})$. Following the orthogonal transformation described in Sec. 7.3.2, the directional derivatives of the performance function in the transformed, uncorrelated $\mathbf{Y}$-space $\tilde{\alpha}$ are computed as

$$\tilde{\alpha}_i = \frac{\tilde{s}_i \, \tilde{\sigma}_i}{\sqrt{\sum_{j=1}^{N} \tilde{s}_j^2 \, \tilde{\sigma}_j^2}} = \frac{\tilde{s}_i \, \sqrt{\lambda_i}}{\sqrt{\sum_{j=1}^{N} \tilde{s}_j^2 \, \lambda_j}} \qquad (7.114)$$

in which $\tilde{s}_{i*}$ is the sensitivity coefficient of the performance function $W(\mathbf{X})$ with respect to the transformed stochastic variable Y_i evalutated at $\mathbf{Y} = \mathbf{y}_*$ which could be obtained as

$$\tilde{s}_{i*} = \left(\frac{\partial W(\mathbf{X})}{\partial \mathbf{Y}_i} \right)_{y*} = \sigma_{i*} \, \mathbf{s}_*^T \, \mathbf{v}_i \qquad (7.115)$$

in which $\mathbf{s}_*$ is the vector of sensitivity coefficients of the performance function with respect to the original stochastic variables evaluated at $\mathbf{X} = \mathbf{x}_*$; and $\mathbf{v}_i$ is the corresponding eigenvector in matrix $\mathbf{V}$ associated with the correlation matrix.

For the most general condition where stochastic variables are correlated and nonnormal, the following algorithm can be applied.

Step 1   Decompose the correlation matrix $\mathbf{R}(\mathbf{X})$ to find the eigenvector matrix $\mathbf{V}$ and eigenvalues λ.

Step 2   Select an initial point $\mathbf{x}_{(r)}$ in the original parameter space.

Step 3   At the selected point $\mathbf{x}_{(r)}$ compute the mean and variance of the performance function $W(\mathbf{X})$ according to Eqs. (7.107) and (7.96), respectively.

Step 4   Compute the corresponding reliability index β according to Eq. (7.84).

Step 5   Compute the mean μ_{iN} and standard deviation σ_{iN} of the normal equivalent using Eqs. (7.110) and (7.111) for the nonnormal stochastic variables.

Step 6   Compute sensitivity coefficients $\tilde{s}_i$ in the transformed domain according to Eq. (7.115) for all $i = 1, 2, \ldots, N$ with σ_i replaced by σ_{iN}.

Step 7   Compute the directional derivatives $\tilde{\alpha}$ according to Eq. (7.114).

Step 8   Using β and $\tilde{\alpha}_i$ obtained from Steps 4 and 7, compute the location of expansion point $\mathbf{y}_{(r+1)}$ in the transformed domain according to

$$y_{i,(r+1)} = -\tilde{\alpha}_i \beta \sqrt{\lambda_i} \qquad \text{for} \qquad i = 1, 2, \ldots, N \qquad (7.116)$$

Step 9   Convert the obtained expansion point $\mathbf{y}_{(r+1)}$ in the transformed domain in Step 7 back to the original parameter space as

$$\mathbf{x}_{(r+1)} = \mu_N + \mathbf{D}_N^{1/2} \, \mathbf{V} \mathbf{y}_{(r+1)} \qquad (7.117)$$

in which μ_N is the vector of means of the normal equivalent at solution point $\mathbf{x}$ and $\mathbf{D}_N$ is the diagonal matrix of variances of the normal equivalent.

Step 10   Check if the revised expansion point $\mathbf{x}_{(r+1)}$ differs significantly from the previous trial expansion point $\mathbf{x}_{(r)}$. If yes, use the revised expansion point as the new trial point by letting $\mathbf{x}_{(r)} = \mathbf{x}_{(r+1)}$; go to Step 3 for another iteration. Otherwise, the iterative procedure is considered complete and the latest reliability index β is used in Eq. (7.86) to compute the reliability p_s.

The use of AFOSM reliability index removes the problem of lack of invariance associated with the MFOSM reliability index. This allows placing different designs on the same common ground for comparing their relative reliabilities using β_{AFOSM}.

A design with a higher value of β_{AFOSM} would be associated with a higher reliability and lower failure probability. However, such conclusions are valid only if the curvatures of failure surfaces at the design points for different designs are similar.

One drawback of the preceding Ang-Tang algorithm is the potential inconsistency between the orthogonally transformed variables $\mathbf{U}$ and the normal-transformed space in computing the directional derivatives in Steps 6 and 7. This is because the eigenvalues and eigenvectors associated with $\mathbf{R(X)}$ will not be identical to those in the normal-transformed variables. To correct this inconsistency, Der Kiureghian and Liu (1986) developed a normal-transformation that preserves the marginal probability contents and the correlation structure of the multivariate nonnormal random variables.

Based on the marginal PDFs of two stochastic variables X_i and X_j, and their correlation coefficient ρ_{ij}, Der Kiureghian and Liu (1986) apply Nataf's bivariate distribution model (Nataf, 1962)

$$\rho_{ij} = \int_{-\infty}^{\infty} \int_{-\infty}^{\infty} \left(\frac{x_i - \mu_i}{\sigma_i} \right) \left(\frac{x_j - \mu_j}{\sigma_j} \right) \Phi_{ij}(z_i, z_j, \rho_{ij}^*) \, dz_i \, dz_j \qquad (7.118)$$

in which Z_i and Z_j are the two correlated standard normal random variables having the equivalent marginal cumulative probabilities corresponding to a pair of random variables X_i and X_j in the original space; ρ_{ij}^* is the correlation coefficient between Z_i and Z_j; ρ_{ij} is the correlation coefficient between X_i and X_j; μ_i and σ_i are, respectively, the mean and standard deviation of X_i; Φ_{ij} is the bivariate standard normal PDF with zero means, unit standard deviations, and correlation coefficient ρ_{ij}^*.

Two conditions are inherently considered in the bivariate distribution model of Eq. (7.118): (1) the normal-transformation satisfies Eq. (7.109), and (2) the value of the correlation coefficient in the normal space lies between -1 and 1. For a pair of nonnormal stochastic variables X_i and X_j, with known means μ_i and μ_j, standard deviations σ_i and σ_j, and correlation coefficient ρ_{ij}, Eq. (7.118) can be applied to solve for ρ_{ij}^*. To avoid the required computation for solving ρ_{ij}^* in Eq. (7.118), Der Kiureghian and Liu (1986) developed a set of semiempirical formulas for ten marginal distributions commonly used in reliability computations as

$$\rho_{ij}^* = T_{ij} \cdot \rho_{ij} \qquad (7.119)$$

in which T_{ij} is a transformation factor depending on the marginal distributions and correlation of the two random variables considered.

7.5 RELIABILITY ANALYSIS: TIME-TO-FAILURE ANALYSIS

In previous sections, evaluations of reliability are based on the analysis of interaction between loads on the system and the resistance of the system. A system would perform its intended function satisfactorily within a specified time period if its capacity exceeds the loading. Instead of considering detailed interactions of resistance and loading over time, a system or its components can be treated as a black box or a lumped-parameter system, and their performances are observed over time. This reduces the reliability analysis to a one-dimensional problem involving time as the only variable. In such cases, the *time-to-failure* (*TTF*) of a system or a component of the system is the random variable. It should be pointed out that the term *time* could be used in a more general sense. In some situations other physical scale measures, such as distance or length, may be appropriate for system performance evaluation.

The time-to-failure analysis is particularly suitable for assessing the reliability of systems and/or components which are repairable. For a system that is repairable after its failure, the time period it would take to have it repaired back to the operational state is uncertain. Therefore, the *time-to-repair* (*TTR*) is also a random variable.

For a repairable system or component, its service life can be extended indefinitely if repair work can restore the system as if it was new. Intuitively, the probability of a repairable system available for service is greater than that of a nonrepairable system. This section will focus on characteristics of failure, repair, and availability of repairable systems by time-to-failure analysis.

7.5.1 Failure Density, Failure Rate, and Mean-Time-To-Failure

Any system will fail eventually; it is just a matter of time. Due to the presence of many uncertainties that affect the operation of a physical system, the time that the system fails to satisfactorily perform its intended function is a random variable.

The *failure density function* is the probability distribution that governs the time occurrence of failure. The failure density function serves as the common thread in the reliability assessments in time-to-failure analysis. The reliability of a system or a component within a specified time interval $[0, t]$ can be expressed as

$$p_s(t) = P(\text{TTF} > t) = \int_t^\infty f(\tau)\, d\tau \qquad (7.120)$$

in which TTF is the random time-to-failure having $f(t)$ as the failure density function. The reliability $p_s(t)$ represents the probability that the system experiences no failure within $[0, t]$. The failure probability or *unreliability* can be expressed as $p_f(t) = 1 - p_s(t)$. Note that unreliability $p_f(t)$ is the probability that a component or a system would experience its first failure within the time interval $(0, t)$. As the age of the system t increases, the reliability $p_s(t)$ decreases, whereas the unreliability $p_f(t)$ increases. Conversely, the failure density function can be obtained from reliability or unreliability as

$$f(t) = -\frac{d\,[p_s(t)]}{dt} = \frac{d\,[p_f(t)]}{dt} \qquad (7.121)$$

The time-to-failure is a continuous, non-negative random variable by nature. Many distribution functions described in Sec. 7.2.5 are appropriate for modeling the stochastic nature of time-to-failure. Among them, the exponential distribution, Eq. (7.42), is perhaps the most widely used. Besides its mathematical simplicity, the exponential distribution has been found, both phenomenologically and empirically, to adequately describe the time-to-failure distribution for components, equipment, and systems involving components with mixtures of life distributions.

The *failure rate* is defined as the number of failures occurring per unit time in a time interval $(t, t + \Delta t)$ per unit of the remaining population at time t. The *instantaneous failure rate* $m(t)$ can be obtained as

$$m(t) = \lim_{\Delta t \to 0} \left[\frac{N_F(\Delta t)/\Delta t}{N(t)} \right] = \frac{f(t)}{p_s(t)} \qquad (7.122)$$

where $N_F(\Delta t)$ is the number of failures in a time interval $(t, t + \Delta t)$; and $N(t)$ is the number of failures from the beginning up to time t. This instantaneous failure rate is also called the *hazard function* or *force of mortality function* (Pieruschka, 1963).

The failure rate for many systems or components has a bathtub shape as shown in Fig. 7.6 in that three distinct life periods can be identified. They are *early life* (or *infant mortality*) *period, useful life period,* and *wearout life period.* In the early life period, quality failures and stress-related failures dominate, with little contribution from wearout failures. During the useful life period, all three types of failures contribute to the potential failure of the system or component, and the overall failure rate remains more or less constant over time. In the later part of life, the overall failure rate increases with age, because wearout failures and stress-related failures are the main contributors, and wearout becomes a more and more dominating factor for the failure of the system with age.

The reliability can be directly computed from the failure rate as

$$p_s(t) = \exp\left(\int_0^t m(\tau)\, d\tau\right) \tag{7.123}$$

Substituting Eq. (7.123) into Eq. (7.122), the failure density function $f(t)$ can be expressed, in terms of failure rate, as

$$f(t) = m(t)\left[\exp\left(\int_0^{-t} m(\tau)\, d\tau\right)\right] \tag{7.124}$$

In general, the reliability of a system or a component is strongly dependent on its age. This can be mathematically expressed by the conditional probability as

$$p_s(\xi|t) = \frac{p_s(t+\xi)}{p_s(t)} \tag{7.125}$$

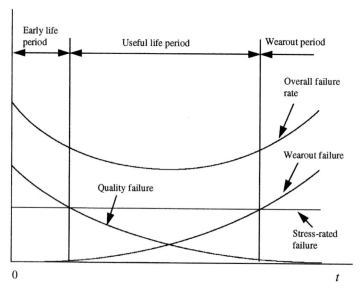

FIGURE 7.6   Shape of typical hazard function and its components.

in which t is the age of the system and up to that point the system has not failed, $p_s(\xi|t)$ is the reliability over a new mission period ξ, having successfully operated over a period of t.

A commonly used reliability measure of system performance is the *mean-time-to-failure (MTTF)* which is the expected time-to-failure. The MTTF can be mathematically defined as

$$\mathrm{MTTF} = E(\mathrm{TTF}) = \int_0^\infty \tau f(\tau)\, d\tau \qquad (7.126)$$

The mean-time-to-failure for some failure density functions is given in Table 7.6.

7.5.2 Repairable Systems

For repairable water resources systems, such as pipe networks, pump stations, and storm runoff drainage structures, failed components within the system can be repaired or replaced so that the system can be put back into service. The time required to have the failed system repaired is uncertain, and consequently, the total time required to restore the system from its failure to operational state is a random variable.

Like the time-to-failure, the random time-to-repair (TTR) has the *repair density function* $g(t)$ describing the random characteristics of the time required to repair a failed system when failure occurs at time zero. The *repair probability* $G(t)$ is the probability that the failed system can be restored within a given time period $[0, t]$,

$$G(t) = P[\mathrm{TTR} \le t] = \int_0^t g(\tau)\, d\tau \qquad (7.127)$$

The repair probability $G(t)$ is one measure of maintainability (Kapur, 1988).

The *repair rate* $r(t)$, similar to the failure rate, is the conditional probability that the system is repaired per unit time, given that the system failed at time zero and is still not repaired at time t. The quantity $r(t)dt$ is the probability that the system is repaired during the time interval $(t, t + dt)$ given that the system fails at time t. The relation between the repair density function, repair rate, and repair probability is

$$r(t) = \frac{g(t)}{1 - G(t)} \qquad (7.128)$$

Given a repair rate $r(t)$, the repair density function and the maintainability can be determined, respectively, as

$$g(t) = r(t) \exp\left(-\int_0^t r(\tau)\, d\tau\right) \qquad (7.129)$$

$$G(t) = 1 - \exp\left(-\int_0^t r(\tau)\, d\tau\right) \qquad (7.130)$$

The *mean-time-to-repair (MTTR)* is the expected value of time-to-repair of a failed system, that is,

$$\mathrm{MTTR} = \int_0^\infty t g(t)\, dt \qquad (7.131)$$

TABLE 7.6 Mean-Time-to-Failure for Some Failure Density Functions

Distribution	PDF, $f(t)$	Reliability, $p_s(t)$	Failure rate, $m(t)$	MTTF
Normal	$\dfrac{1}{\sqrt{2\pi}\sigma_T}\exp\left[-\dfrac{1}{2}\left(\dfrac{t-\mu_T}{\sigma_T}\right)^2\right]$	$\Phi\left(\dfrac{t-\mu_T}{\sigma_T}\right)$	$\dfrac{f(t)}{\Phi\left(\dfrac{t-\mu_T}{\sigma_T}\right)}$	μ_T
Lognormal	$\dfrac{1}{\sqrt{2\pi}\sigma_{\ln t}\,t}\exp\left[-\dfrac{1}{2}\left(\dfrac{\ln(t)-\mu_{\ln T}}{\sigma_{\ln T}}\right)^2\right]$	$\Phi\left(\dfrac{\ln(t)-\mu_{\ln T}}{\sigma_{\ln T}}\right)$	$\dfrac{f(t)}{\Phi\left(\dfrac{\ln(t)-\mu_{\ln T}}{\sigma_{\ln T}}\right)}$	$\exp\left(\mu_{\ln T}+\dfrac{\sigma_{\ln T}^2}{2}\right)$
Exponential	$\beta e^{-\beta t}$	$e^{-\beta t}$	β	$\dfrac{1}{\beta}$
Rayleigh	$\dfrac{t}{\beta^2}\exp\left[-\dfrac{1}{2}\left(\dfrac{t}{\beta}\right)^2\right],\beta>0$	$\exp\left[-\dfrac{1}{2}\left(\dfrac{t}{\beta}\right)^2\right]$	$\dfrac{t}{\beta^2}$	$1.253\,\beta$
Gamma	$\dfrac{\beta}{\Gamma(\alpha)}(\beta t)^{\alpha-1}e^{-\beta t}$	$\displaystyle\int_t^\infty f(\tau)\,d\tau$	$\dfrac{f(t)}{p_s(t)}$	$\dfrac{\alpha}{\beta}$
Gumbel	$e^{\pm y}e^{\mp e^{\pm y}}; y=\dfrac{t-t_o}{\beta}$	$1-e^{-e^{\pm y}}$	$\dfrac{f(t)}{p_s(t)}$	$x_o\pm0.577\,\beta$
Weibull	$\dfrac{\alpha}{\beta}\left(\dfrac{t-t_o}{\beta}\right)^{\alpha-1}e^{-\left(\dfrac{t-t_o}{\beta}\right)^\alpha}$	$e^{\left(\dfrac{t-t_o}{\beta}\right)^\alpha}$	$\dfrac{\alpha(t-t_o)^{\alpha-1}}{\beta^\alpha}$	$t_o+\beta\Gamma\left(1+\dfrac{1}{\alpha}\right)$
Uniform	$\dfrac{1}{b-a}$	$\dfrac{b-t}{b-a}$	$\dfrac{1}{b-t}$	$\dfrac{a+b}{2}$

The MTTR measures the elapsed time required to perform the maintenance operation and is used to estimate the downtime of a system. It is also a measure for maintainability of a system. Various types of maintainability measures are derivable from the repair density function (Kraus, 1988).

The MTTR is a proper measure of the mean life span of a nonrepairable system. For a repairable system, a more representative indicator for the fail-repair cycle is the *mean-time-between-failure* (*MTBF*) which is the sum of MTTR and MTTR, that is,

$$MTBF = MTTF + MTTR \qquad (7.132)$$

The *mean-time-between-repair* (*MTBR*) is the expected value of the time between two consecutive repairs, and it is equal to MTBF.

7.5.3 Availability and Unavailability

A repairable system experiences a repetition of repair-to-failure and failure-to-repair processes during its service life. Hence, the probability that a system is in operating condition at any given time t for a repairable system is different than that of a nonrepairable system. The term *availability* $A(t)$ is generally used for repairable systems, to indicate the probability that the system is in operating condition at any given time t. On the other hand, reliability $p_s(t)$ is appropriate for nonrepairable systems, indicating the probability that the system has been *continuously* in its operating state starting from time zero up to time t.

Availability can also be interpreted as the percentage of time that the system is in operating condition within a specified time period. In general, the availability and reliability of a system satisfies the following inequality relationship

$$0 \le p_s(t) \le A(t) \le 1 \qquad (7.133)$$

with the equality holding for nonrepairable systems. The reliability of a system decreases monotonically to zero as the age of the system increases, whereas the availability decreases but converges to a positive probability. This is shown in Fig. 7.7.

The compliment to availability is the *unavailability* $U(t)$, which is the probability that a system is in the failed condition at time t, given it is in operating condition at time zero. In other words, unavailability is the percentage of time that the system is not available for the intended service in time period $(0,t)$, given it is operational at time zero. Availability, unavailability, and unreliability satisfy the following relationships:

$$A(t) + U(t) = 1.0 \qquad (7.134)$$

$$0 \le U(t) \le p_f(t) \le 1 \qquad (7.135)$$

For a nonrepairable system, the unavailability is equal to the unreliability, that is, $U(t) = p_f(t)$.

Determination of availability or unavailability of a system requires full accounting of the failure and repair processes. The basic elements that describe such processes are failure density function $f(t)$ and the repair density function $g(t)$. Consider a system with a constant failure rate λ and a constant repair rate η. The availability and unavailability, respectively, are

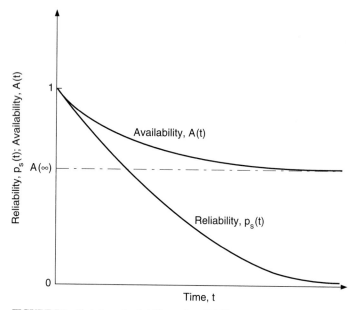

FIGURE 7.7   Variation of reliability and availability with time.

$$A(t) = \frac{\eta}{\lambda + \eta} + \frac{\lambda}{\lambda + \eta} \, e^{-(\lambda + \eta)t} \qquad\qquad (7.136)$$

$$U(t) = \frac{\lambda}{\lambda + \eta} \left[1 - e^{-(\lambda + \eta)t} \right] \qquad\qquad (7.137)$$

As the time approaches infinity ($t \to \infty$) the system reaches its stationary condition. Then, the stationary availability $A(\infty)$ and unavailability $U(\infty)$ are

$$A(\infty) = \frac{\eta}{\lambda + \eta} = \frac{1/\lambda}{1/\lambda + 1/\eta} = \frac{\text{MTTF}}{\text{MTTF} + \text{MTTR}} \qquad\qquad (7.138)$$

$$U(\infty) = \frac{\lambda}{\lambda + \eta} = \frac{1/\eta}{1/\lambda + 1/\eta} = \frac{\text{MTTR}}{\text{MTTF} + \text{MTTR}} \qquad\qquad (7.139)$$

Other properties for a system with constant failure rate and repair rate are summarized in Table 7.7.

7.6   MONTE CARLO SIMULATION

Simulation is a process of replicating the real world based on a set of assumptions and conceived models of reality (Ang and Tang, 1984). Because the purpose of a simulation model is to duplicate reality, it is a useful tool for evaluating the effect of different designs on system performance. The Monte Carlo procedure is a

TABLE 7.7 Summary of Constant Rate Model

Repairable systems	Nonrepairable systems
Failure process	
$m(t) = \lambda$	$m(t) = \lambda$
$p_s(t) = e^{-\lambda t}$	$p_s(t) = e^{-\lambda t}$
$p_f(t) = 1 - e^{-\lambda t}$	$p_f(t) = 1 - e^{-\lambda t}$
$f(t) = \lambda e^{-\lambda t}$	$f(t) = \lambda e^{-\lambda t}$
$\text{MTTF} = 1/\lambda$	$\text{MTTF} = 1/\lambda$
Repair process	
$r(t) = \eta$	$r(t) = 0$
$G(t) = 1 - e^{-\eta t}$	$G(t) = 0$
$g(t) = \eta e^{-\eta t}$	$g(t) = 0$
$\text{MTTR} = 1/\eta$	$\text{MTTR} = \infty$
Dynamic behavior of whole process	
$U(t) = \dfrac{\lambda}{\lambda + \eta}\left[1 - e^{-(\lambda + \eta)t}\right]$	$U(t) = 1 - e^{-\lambda t} = p_f(t)$
$A(t) = \dfrac{\eta}{\lambda + \eta} + \dfrac{\lambda}{\lambda + \eta}\left[1 - e^{-(\lambda + \eta)t}\right]$	$A(t) = e^{-\lambda t} = p_s(t)$
$w(t) = \dfrac{\lambda\eta}{\lambda + \eta} + \dfrac{\lambda^2}{\lambda + \eta}\left[1 - e^{-(\lambda + \eta)t}\right]$	$w(t) = f(t) = \lambda e^{-\lambda t}$
$\gamma(t) = \dfrac{\lambda\eta}{\lambda + \eta}\left[1 - e^{-(\lambda + \eta)t}\right]$	$\gamma(t) = 0$
$W(0, t) = \dfrac{\lambda\eta}{\lambda + \eta}t + \dfrac{\lambda^2}{(\lambda + \eta)^2}\left[1 - e^{-(\lambda + \eta)t}\right]$	$W(0, t) = p_f(t)$
$\Gamma(0,t) = \dfrac{\lambda\eta}{\lambda + \eta}t - \dfrac{\lambda\eta}{(\lambda + \eta)^2}\left[1 - e^{-(\lambda + \eta)t}\right]$	$\Gamma(0, t) = 0$
Stationary values of whole process	
$U(\infty) = \dfrac{\lambda}{\lambda + \eta} = \dfrac{\text{MTTR}}{\text{MTTF} + \text{MTTR}}$	$U(\infty) = 1$
$A(\infty) = \dfrac{\eta}{\lambda + \eta} = \dfrac{\text{MTTF}}{\text{MTTF} + \text{MTTR}}$	$A(\infty) = 0$
$w(\infty) = \dfrac{\lambda\eta}{\lambda + \eta} = \dfrac{1}{\text{MTTF} + \text{MTTR}}$	$w(\infty) = 0$
$\gamma(\infty) = \dfrac{\lambda\eta}{\lambda + \eta} = w(\infty)$	$\gamma(\infty) = 0$

Source: Adapted from Henley and Kumamoto, 1981.

numerical simulation to reproduce random variables preserving the specified distributional properties.

7.6.1   Generation of Random Numbers

The most commonly used techniques to generate a sequence of pseudorandom numbers are those that employ some form of recursive computation. In principle, such recursive formulas are based on calculating the residual modulo of some integers of a linear transformation. The process of producing a random number sequence is completely deterministic. However, the generated sequence would appear to be uniformly distributed and independent.

Congruential methods for generating M random numbers are based on the fundamental congruence relationship which can be expressed as (Lehmer, 1951)

$$X_i = \{aX_{i-1} + c\} \,(\text{mod } m) \qquad i = 1, 2, \ldots, M \tag{7.140}$$

in which a is the multiplier, c is the increment, and m is an integer-valued modulus. Several forms of congruential algorithms are based on Eq. (7.140). Applying Eq. (7.140) to generate a random number sequence requires the specification of a, c, and m along with X_0, called the *seed*. Once the sequence of random number Xs is generated, the random number from the unit interval $[0, 1]$ can be obtained as

$$U_i = \frac{X_i}{m} \qquad i = 1, 2, \ldots, M \tag{7.141}$$

The process of generating uniform random numbers is the building block in Monte Carlo simulations.

7.6.2   Classifications of Generation Algorithms

Algorithms to generate random variates from a specified distribution can be categorized into three types: *CDF-inverse method, acceptance-rejection (AR) method,* and *variable transformation method.* The CDF-inverse method is based on the fact that the CDF $F_X(x)$ of a random variable X is a nondecreasing function with respect to its value and $0 \leq F_X(x) \leq 1$.

For the great majority of continuous probability distributions applied in water resources engineering analysis, $F_X(x)$ is a strictly increasing function of x. Hence, there exists a unique relationship between $F_X(x)$ and u, that is, $u = F_X(x)$, as shown in Fig. 7.8. It can be shown that, if U is a standard uniform random variable defined over the unit interval $[0, 1]$, the following relation holds,

$$X = F_X^{-1}(U) \tag{7.142}$$

To efficiently apply the CDF-inverse method for generating random numbers, an explicit expression between X and U is required. When analytical forms of the CDF inverse are not available, the CDF-inverse method requires solving

$$u = \int_{-\infty}^{x} f_X(t)\, dt \tag{7.143}$$

for x from the known u which could be inefficient.

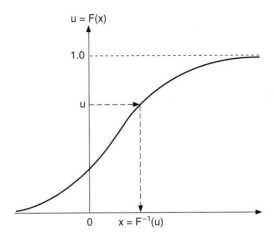

FIGURE 7.8 CDF-inverse method for random variate generation.

An alternative approach is the *acceptance-rejection approach*. Its basic idea is to replace the original $f_X(x)$ by an appropriate PDF $h_X(x)$, from which random numbers can be easily and efficiently produced. The generated random number from $h_X(x)$ then is subject to testing before it is accepted as a random number from the original $f_X(x)$. This approach for generating random numbers has become popular.

The *variable transformation method* generates variates for the random variable of interest based on its known statistical relationship with other random variables whose variates can easily be produced. Therefore, the variable transformation method is sometimes effective for generating random variates from a complicated distribution based on variates produced from simple distributions.

7.6.3 Generation of Random Variates for Some Commonly Used Distributions

Several books have been written for generating univariate random numbers (Rubinstein, 1981; Dagpunar, 1988; Law and Kelton, 1991). A number of computer programs are available in the public domain. For example, STATLIB is a system for distribution of statistical software which is available through e-mail address: stalib@lib.stat.cmu.edu. Press, et al. (1989) present computer programs for Monte Carlo simulation which are available at low cost. There is other proprietary computer software frequently used for random number generation. An example of such software is the International Mathematical and Statistical Library (IMSL), which is available on most mainframe computer systems.

To generate multivariate random variates, the joint PDF is required. Compared with univariate random variate generators, algorithms for multivariate random variates are much more restricted to a few joint distributions, such as multivariate normal, multivariate lognormal (Parrish 1990), multivariate gamma (Ronning, 1977), and others (Johnson, 1987). If the multivariate random variables involved are correlated with a mixture of marginal distributions, the joint PDF is difficult to formulate. Chang, et al. (1994) proposed a practical multivariate Monte Carlo simulation procedure for problems involving mixtures of non-normal random variables by utilizing the semi-empirical formulas derived by Der Kiureghian and Liu (1985) and Liu and Der Kiureghian (1986).

7.6.4 Monte Carlo Integration

In reliability analysis, computations of system and/or component reliability and other related quantities involve integration operations. A simple example is the time-to-failure analysis in which the reliability of a system within a time interval $(0, t)$ is obtained from Eq. (7.120). A more complex example of reliability computation is by the load-resistance interference in that the reliability is computed by Eq. (7.80).

For cases of integration on one or two dimensions and the integrands are well-behaved (e.g., no discontinuity), conventional numerical integration methods, such as trapezoidal approximation or Simpson's rule, are efficient and accurate. Gould and Tobochnik (1988) show that, in general, if the error for one-dimensional integration is $O(k^{-a})$, the error with a d-dimensional integration would be $O(k^{-a/d})$. This indicates that the accuracy of conventional numerical integration schemes decreases rapidly as the dimension of integration increases. For multiple integrals, the Monte Carlo method becomes a more suitable numerical technique for integration.

Consider the following integration problem

$$G = \int_a^b g(x)\,dx \qquad a \le x \le b \qquad (7.144)$$

The *sample-mean method* is a simple Monte Carlo integration procedure based on the idea that the integral Eq. (7.144) can alternatively be expressed as

$$G = \int_a^b \left[\frac{g(x)}{f_X(x)} \right] f_X(x)\,dx \qquad a \le x \le b \qquad (7.145)$$

in which $f_X(x) \ge 0$ is a PDF defined over $a \le x \le b$. The transformed integral given by Eq. (7.145) is equivalent to the computation of expectation of $g(X)/f_X(X)$, namely,

$$G = E_X \left[\frac{g(X)}{f_X(X)} \right] \qquad (7.146)$$

with X being a random variable having a PDF $f_X(x)$ defined over $a \le x \le b$. The estimation of $E_X[g(X)/f_X(X)]$ by Monte Carlo method is

$$\hat{G} = \frac{1}{M} \sum_{i=1}^M \frac{g(x_i)}{f_X(x_i)} \qquad (7.147)$$

in which x_i is the random variate generated according to $f_X(x)$ and M is the number of random variates produced.

The sample-mean Monte Carlo integration involves the following steps:

1. Select $f_X(x)$ defined over the region of the integral from which M random variates are generated.
2. Compute $g(x_i)/f_X(x_i)$ for $i = 1, 2, \ldots, M$.
3. Calculate the sample average based on Eq. (7.147) as the estimate for G.

7.6.5 Variance Reduction Techniques

Since the Monte Carlo integration is a sampling procedure, the results obtained inevitably involve sampling errors which decrease as the sample size increases. Increasing sample size, for achieving higher precision, generally means an increase in computer time for generating random variates and data processing. *Variance reduction techniques* aim at obtaining high precision for the Monte Carlo simulation results without having to substantially increase the required sample size. Several variance reduction techniques are briefly described as follows.

7.6.5.1 Importance Sampling Technique. The *importance sampling technique* concentrates the distribution of sampling points in the part of the domain that is most "important" for the task rather than spreading the sampling points out evenly (Marshall, 1956). Rubinstein (1981) shows that the error associated with the sample-mean method can be considerably reduced if $f_X(x)$ is chosen to have a similar shape as $|g(x)|$. However, in practical implementations of this technique, consideration must be given to the tradeoff between the desired error reduction and the difficulties of sampling from $f_X(x)$, especially when $|g(x)|$ is not well-behaved.

7.6.5.2 Antithetic-Variates Technique. Consider that Θ_1 and Θ_2 are two unbiased estimators of a parameter θ. The two estimators can be combined together to form another estimator in the following fashion

$$\Theta_a = \frac{1}{2}(\Theta_1 + \Theta_2) \tag{7.148}$$

The new estimator Θ_a is also unbiased and has a variance as

$$\text{Var}(\Theta_a) = \frac{1}{4}[\text{Var}(\Theta_1) + \text{Var}(\Theta_2) + 2\text{cov}(\Theta_1, \Theta_2)] \tag{7.149}$$

From Eq. (7.149) one realizes that the variance associated with Θ_a could be reduced if Monte Carlo simulation can generate random variates which result in strongly negative correlation between Θ_1 and Θ_2.

In a Monte Carlo simulation, the estimators Θ_1 and Θ_2 are functions of generated random variates which, in turn, are related to the standard uniform random variates. Therefore, Θ_1 and Θ_2 are functions of the two standard uniform random variables U_1 and U_2. The objective to produce negative $\text{cov}[\Theta_1(U_1), \Theta_2(U_2)]$ can be achieved by producing U_1 and U_2 which are negatively correlated. A simple approach that generates negatively correlated uniform random variates with minimal computation is to let $U_1 = 1 - U_2$.

7.6.5.3 Correlated Sampling Technique. Correlated sampling techniques are especially effective for variance reduction when the primary objective of the simulation study is to evaluate small changes in system performance or to compare the difference in system performance between two specific designs (Rubinstein, 1981; Ang and Tang, 1984). For example, consider two designs A and B for the same system involving a vector of N random variables $\mathbf{X} = (X_1, X_2, \ldots, X_N)$ which could be correlated with a joint PDF $f_X(\mathbf{x})$ or independent, each with a marginal PDF, $f_i(x_i)$, $i = 1$, $2, \ldots, N$. The performance of the system under the two designs can be expressed as

$$\Theta_A = g(\mathbf{a}, \mathbf{X}); \Theta_B = g(\mathbf{b}, \mathbf{X}) \tag{7.150}$$

in which $g(\ )$ is a function defining the system performance; and **a** and **b** are vectors of design parameters for designs A and B, respectively. Since the two performance measures Θ_A and Θ_B are dependent on the same random variables through the same performance function $g(\ )$, their estimators would be positively correlated. In this case, independently generating two sets of random variates, according to their probability laws for designs A and B, would still result in positive correlation between $\hat{\Theta}_A$ and $\hat{\Theta}_B$. However, an increase in correlation between $\hat{\Theta}_A$ and $\hat{\Theta}_B$ leading to further reduction in $\text{var}(\hat{\Delta}_\Theta)$ can be achieved by using a common set of standard uniform random variates for both designs A and B.

7.6.5.4 Stratified Sampling Technique.   *Stratified sampling* is a well established area in statistical sampling (Cochran, 1966). In spirit, it is similar to importance sampling. The basic difference between the two techniques is that the stratified sampling technique takes more observations at regions that are more "important" whereas importance sampling chooses an optimal PDF. Variance reduction by the stratified sampling technique is achieved through taking more samples at important subregions.

7.6.5.5 Latin Hypercubic Sampling Technique.   The *Latin hypercubic sampling technique* is a special method under the realm of stratified sampling which selects random samples of each random variable over its range in a stratified manner (KcKay 1988). The technique divides the plausible range of each random variable into K equal-probability intervals for each stochastic variable involved. Within each interval, a random variate is generated resulting in K random variates for each stochastic variable. Then, the generated random sequences are randomly permutated for all random variables. The integral represented by Eq. (7.144) is estimated by

$$\hat{G} = \frac{1}{K} \sum_{k=1}^{K} g(\mathbf{x}_k) \qquad (7.151)$$

7.7 SYSTEM RELIABILITY

Most systems involve many subsystems and components whose performances affect in turn the performance of the system as a whole. The reliability of the entire system is affected not only by the reliability of individual subsystems and components, but also the interaction and configuration of the subsystems and components. Furthermore, water resources systems involve multiple failure modes, that is, there are several potential modes of failure in which the occurrence of any or a combination of such failure modes constitute the system failure (Ang and Tang, 1984). Due to the fact that different failure modes might be defined over the same stochastic variables space, the failure modes are correlated in general.

For a complex system involving many subsystems, components, and contributing stochastic variables, it is generally difficult, if not impossible, to directly assess the reliability of the system. Instead, the system is divided into subsystems, components, and modes of operation, and the reliability of each is quantified.

7.7.1 Overview of System Reliability Evaluation

Evaluation of system reliability requires knowing what constitutes the system being in a failed or satisfactory state. Such knowledge is essential for system classification and dictates the methods to be used for system reliability determination.

From the reliability computation viewpoint, the classification of the system primarily depends on how the system performance is affected by its components or its modes of performance. A multiple-component system that requires that all of its components to perform satisfactorily for satisfactory performance of the entire system is called a *series system*. Similarly, a single-component system involving several modes of operation is also viewed as a series system, if satisfactory performance of the system requires satisfactory performance of all its different modes of operation.

A second basic type of system is the *parallel system*. A parallel system is characterized by the property that the system would serve its intended purpose as long as at least one of its components or modes of operation performs satisfactorily.

For most real-life problems, systems are *complex systems* in which the components are arranged as a mixture of in-series and in-parallel, or in the form of a loop. In dealing with a complex system, the general approach is to reduce the system configuration, based on its component arrangement or modes of operation, to a simpler system for which the analysis can be performed easily. However, this goal may not be achieved for all cases, necessitating the development of a special procedure.

7.7.2 Reliability of Simple Systems

7.7.2.1 Series Systems. A series system requires that all of its components or modes of operation perform satisfactorily to ensure a satisfactory operation of the entire system. In the context of load-resistance interference, the failure probability of a series system involving M components or modes of operation can be expressed as

$$p_{f,\text{sys}} = P\left[\bigcup_{i=1}^{M} (W_i < 0)\right] = P\left[\bigcup_{i=1}^{M} (Z_i < -\beta_i)\right] \tag{7.152}$$

in which W_i is the performance function for the ith component or mode of operation.

In the framework of time-to-failure analysis, the reliability of a series system, when the performance of individual components is independent of each other, is

$$p_{s,\text{sys}}(t) = P\left(\bigcap_{i=1}^{M} F_i'\right) = \prod_{i=1}^{M} p_{s,i}(t) \tag{7.153}$$

in which $p_{s,i}(t)$ is the reliability of the ith component over the time interval $[0, t]$. Similarly, the availability of a series system involving M independent components is

$$A_{\text{sys}}(t) = \prod_{i=1}^{M} A_i(t) \tag{7.154}$$

in which $A_{\text{sys}}(t)$ and $A_i(t)$ are the availabilities of the system and the ith component, respectively, at time t.

For the special case of an exponential failure density function, the reliability of a series system involving M independent components is

$$p_{s,\text{sys}}(t) = \prod_{i=1}^{M} e^{-\lambda_i t} = \exp\left[-\left(\sum_{i=1}^{M} \lambda_i\right)t\right] \tag{7.155}$$

where λ_i is the constant failure rate of the ith component. Assuming an exponential repair function for each independent component, the availability for a series system is

$$A_{\text{sys}}(t) = \prod_{i=1}^{M} \left[\frac{\eta_i}{\lambda_i + \eta_i} + \frac{\lambda_i}{\lambda_i + \eta_i} e^{-(\lambda_i + \eta_i)t}\right] \tag{7.156}$$

in which η_i is the constant repair rate for the ith component. The stationary system availability is

$$A_{\text{sys}}(\infty) = \prod_{i=1}^{M} \left(\frac{\eta_i}{\lambda_i + \eta_i} \right) = \prod_{i=1}^{M} \left(\frac{\text{MTTF}_i}{\text{MTTR}_i + \text{MTTF}_i} \right) \tag{7.157}$$

in which MTTR_i and MTTF_i are, respectively, the mean-time-to-repair and mean-time-to-failure of the ith component.

7.7.2.2 Parallel Systems. For a parallel system, the entire system will perform satisfactorily if one or more of its components or modes of operation function satisfactorily. In the framework of load-resistance interference for different modes of operation, the failure probability of a parallel system is

$$p_{f,\text{sys}} = P\left[\bigcap_{i=1}^{M} (W_i < 0) \right] = P\left[\bigcap_{i=1}^{M} (Z_i < -\beta_i) \right] \tag{7.158}$$

In the framework of time-to-failure analysis, the failure probability of a parallel system involving M independent components can be computed as

$$p_{f,\text{sys}}(t) = \prod_{i=1}^{M} p_{f,i}(t) \tag{7.159}$$

If each component has an exponential failure density function with parameter λ_i, for $i = 1, 2, \ldots, M$, the system failure probability is

$$p_{f,\text{sys}}(t) = \prod_{i=1}^{M} \left(1 - e^{-\lambda_i t} \right) \tag{7.160}$$

In a case where all components have an identical failure rate, that is, $\lambda_1 = \lambda_2 = \cdots = \lambda_M = \lambda$, the MTTF of the system is

$$\text{MTTF} = \frac{1}{\lambda} \sum_{i=1}^{M} \frac{1}{i} \tag{7.161}$$

The unavailability of a parallel system involving M independent components is

$$U_{\text{sys}}(t) = \prod_{i=1}^{M} U_i(t) \tag{7.162}$$

Under the condition of independent, exponential repair functions for the M components, the unavailability of a parallel system is

$$U_{\text{sys}}(t) = \prod_{i=1}^{M} \frac{\lambda_i}{\lambda_i + \eta_i} \left[1 - e^{-(\lambda_i + \eta_i)t} \right] \tag{7.163}$$

and the stationary system unavailability is

$$U_{\text{sys}}(\infty) = \prod_{i=1}^{M} \frac{\lambda_i}{\lambda_i + \eta_i} = \prod_{i=1}^{M} \left(\frac{\text{MTTR}_i}{\text{MTTR}_i + \text{MTTF}_i} \right) \tag{7.164}$$

7.7.2.3 K-Out-of-M Parallel Systems. This is a parallel system of M components for which the system will function if K ($K < M$) or more components function.

This type of system is also called a *partially redundant system*. The general reliability formula for this system is rather cumbersome. For components having identical reliability functions, that is, $p_{s,i}(t) = p_s(t)$, the system reliability, when component performances are independent, is

$$p_{s,\text{sys}}(t) = \sum_{j=K}^{M} {}_{M}C_j \, [p_s(t)]^j \, [1 - p_s(t)]^{M-j} \tag{7.165}$$

in which ${}_{M}C_j$ is the binomial coefficient. Furthermore, if the failure density function is an exponential distribution, the system reliability can be expressed as

$$p_{s,\text{sys}}(t) = \sum_{j=K}^{M} {}_{M}C_j \, [e^{-\lambda t}]^j \, [1 - e^{-\lambda t}]^{M-j} \tag{7.166}$$

The availability of the system can be obtained from substituting component availability for component reliability in Eq. (7.166).

7.7.2.4 Standby Redundant Systems.

A *standby redundant system* is a parallel system in which only one component or subsystem is in operation. It is a special case of a K-out-of-M system with $K = 1$. If the operating component fails, then another component is operated. This type of system is different from the parallel system described previously where all the components are operating because standby units do not operate. The reliability for a system with M components out of which $M - 1$ units are on standby, is the probability that at most $M - 1$ components fail. Applying the Poisson distribution, this probability can be expressed by

$$p_{s,\text{sys}}(t) = \sum_{i=0}^{M-1} \frac{(\lambda t)^i \, e^{-\lambda t}}{i!} \tag{7.167}$$

The above equation is valid under the following assumptions: The switch functions perfectly, the units are identical, the component failure rates are constant, the standby units are as good as new, and the unit failures are statistically independent. The mean-time-to-failure is

$$\text{MTTF} = \int_0^\infty p_{s,\text{sys}}(t) \, dt = \int_0^\infty \sum_{i=0}^{M-1} \frac{(\lambda t)^i e^{-\lambda t}}{i!} \, dt = \frac{M}{\lambda} \tag{7.168}$$

Since the system's operation is the result of a relay of a series of components, the system MTTF is the sum of the MTTF of the individual components.

7.7.3 Methods for Computing Reliability of Complex Systems

Many practical water resources systems, such as water distribution systems, have neither series nor parallel configuration. Evaluation of reliability for such complex systems is generally difficult. For some systems, it is possible to combine components into groups in such a manner that it would appear as in-series or in-parallel. For other systems, special techniques need to be developed which requires a certain degree of insight and ingenuity from engineers. This section describes some of the potentially useful techniques for water resources system reliability evaluation.

7.7.3.1 State Enumeration Method.

The *state enumeration method* lists all possible mutually exclusive states of the system components that define the state of the entire system. In general, for a system containing M components in which each can be classified into N operating states, there will be N^M possible states for the entire system. Once all the possible system states are enumerated, the states that result in successful system operation are identified, and the probability of the occurrence of each successful state is computed. The last step is to sum all of the successful state probabilities which yield the system reliability.

7.7.3.2 Path Enumeration Method. This is a very effective method for system reliability evaluation. A *path* is defined as a set of components or modes of operation which leads to a certain state of the system. In system reliability analysis, the system states of interest are those of failed state and operational state. Under this category, the *tie-set analysis* and *cut-set analysis* are the two well-known techniques.

Cut-Set Analysis. The *cut-set* is defined as a set of system components or modes of operation which, when failed, causes failure of the system. Cut-set analysis is powerful for evaluating system reliability for two reasons: (1) It can be easily programmed on digital computers for fast and efficient solution of any general system configuration, especially in the form of a network and (2) the cut-sets are directly related to the modes of system failure and therefore identify the distinct and discrete ways in which a system may fail.

The cut-set method utilizes the minimum cut-set for calculating the system failure probability. The *minimum cut-set* is a set of system components which, when all fail, causes failure of the system, but when any one component of the set does not fail, the system does not fail. A minimum cut-set implies that all components of the cut-set must be in the failure state to cause system failure. Therefore, the components or modes of operation involved in the minimum cut-set are effectively connected in parallel, and each minimum cut-set is connected in series. Consequently, the failure probability of a system can be expressed as

$$p_{f,\text{sys}} = P\left[\bigcup_{i=1}^{I} C_i\right] = P\left[\bigcup_{i=1}^{I}\left(\bigcap_{j=1}^{J_i} F_{ij}\right)\right] \tag{7.169}$$

in which C_i is the ith of the total I minimum cut-sets, J_i is the total number of components or modes of operation in the ith minimum cut-set, and F_{ij} represents the failure event associated with the jth component or mode of operation in the ith minimum cut-set.

Tie-Set Analysis. As the complement of a cut-set, a *tie-set* is a minimal path of the system in which system components or modes of operation are arranged in series. Consequently, a tie-set fails if any of its components or modes of operation fail. All tie-sets are effectively connected in parallel, that is, the system will be in the operating state if any of its tie-sets are functioning. The main disadvantage of the tie-set method is that failure modes are not directly identified. Direct identification of failure modes is sometimes essential if a limited amount of a resource is available to place emphasis on a few dominant failure modes.

7.7.3.3 Conditional Probability Approach. The *conditional probability approach* starts with a selection of key components and modes of operation whose states (operational or failed) would decompose the entire system into simple series and/or parallel subsystems for which the reliability or failure probability of subsystems can be easily evaluated. Then, the reliability of the entire system is obtained by combining those of the subsystems using the conditional probability rule:

$$p_{s,\text{sys}} = p_{s|F_i'} \cdot p_{s,i} + p_{s|F_i} \cdot p_{f,i} \tag{7.170}$$

in which $p_{s|F_i'}$ and $p_{s|Fi}$ are the conditional system reliabilities given that component i is operational F_i', and failed F_i, respectively; and $p_{s,i}$ and $p_{f,i}$ are the reliability and failure probabilities of the ith component, respectively.

Except for very simple and small systems, a nested conditional probability operation is inevitable. Efficient evaluation of system reliability of a complex system hinges entirely on the proper selection of key components which generally is a difficult task when the system is large. Furthermore, the method cannot be easily adopted to computerized problem solving.

7.7.3.4 Fault Tree Analysis. Conceptually, *fault-tree analysis* traces from a system failure backward, searching for possible causes of the failure. A *fault tree* is a logical diagram representing the consequence of component failures (basic or primary failures) on system failure (top failure or top event). The major objective of fault

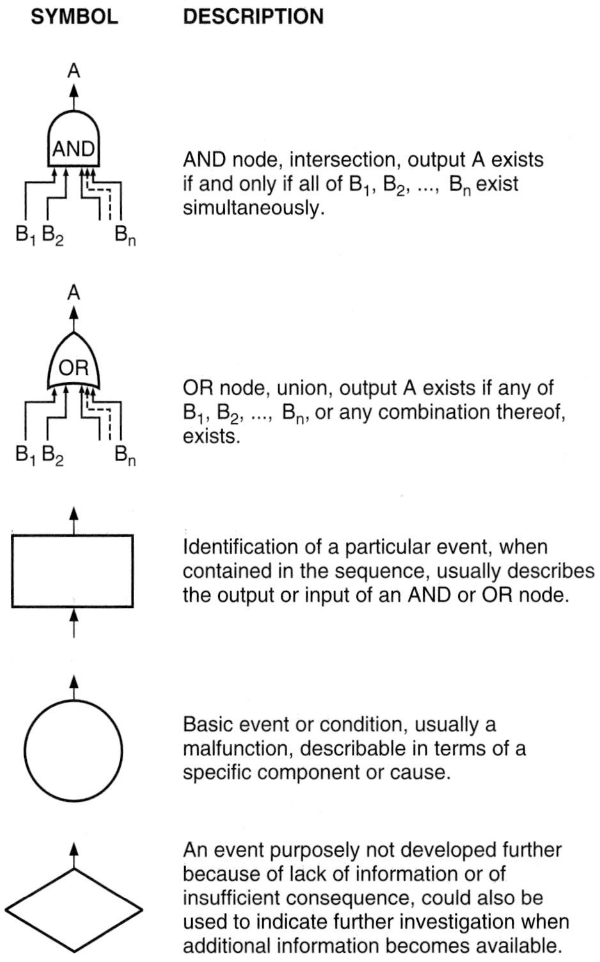

SYMBOL   DESCRIPTION

AND node, intersection, output A exists if and only if all of B_1, B_2, ..., B_n exist simultaneously.

OR node, union, output A exists if any of B_1, B_2, ..., B_n, or any combination thereof, exists.

Identification of a particular event, when contained in the sequence, usually describes the output or input of an AND or OR node.

Basic event or condition, usually a malfunction, describable in terms of a specific component or cause.

An event purposely not developed further because of lack of information or of insufficient consequence, could also be used to indicate further investigation when additional information becomes available.

FIGURE 7.9 Some gate and event symbols used in fault tree.

tree construction is to represent the system condition, which may cause system failure, in a symbolic manner. In other words, the fault tree consists of event sequences that lead to system failure. There are actually two types of building blocks: *gate symbols* and *event symbols*. Gate symbols connect events according to their causal relation such that there may be one or more input events but only one output event. Figure 7.9 shows the two commonly used gate symbols. Three types of commonly used event symbols are also shown in Fig. 7.9. A simple fault tree for the failure of an existing dam is shown in Fig. 7.10.

Henley and Kumamoto (1981) present heuristic guidelines for constructing fault trees:

1. Replace abstract events by less abstract events.
2. Classify an event into more elementary events.
3. Identify distinct causes for an event.
4. Couple trigger event with "no protection actions."
5. Find cooperative causes for an event.
6. Pinpoint component failure events.
7. Develop component failure.

To evaluate the fault tree, one should always start from the minimal cut-sets which, in essence, are *critical paths*. Basically, the fault tree evaluation comprises two distinct processes: (1) the determination of the logical combination of events that cause the top event (failure) expressed in minimal cut-sets and (2) the numerical evaluation of the failure probability. Several computer codes are available for generating cut-sets. For example, MOCUS (Fussell, et al., 1974) was developed to obtain minimal cut-sets from fault trees.

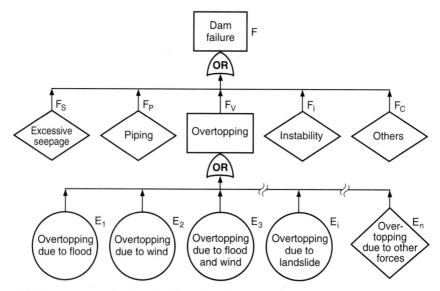

FIGURE 7.10 Simple fault tree for failure of existing dams. (*Cheng, et al., 1982.*)

The system availability $A_s(t)$ is the probability that the top event does not occur at time t, which is the probability of the system operating successfully when the top event is an OR combination of all system hazards. System reliability $p_{s,\text{sys}}(t)$ is the probability that the top event does not occur over time interval $(0, t)$. System reliability requires continuation of the nonexistence of the top event, and its value is less than or equal to the availability.

7.8 RISK-BASED DESIGN OF WATER RESOURCES SYSTEMS

All water resources engineering problems involve a number of interconnected and interrelated components. The analysis should take such interactions into account so that the overall behavior of the system is properly modeled. Water resources systems analysis using optimization techniques is described in Chap. 6 of this book. In this section, emphasis is placed on how to incorporate uncertainties for optimal water resources system designs.

Reliability analysis can be applied to the design of various hydraulic structures with or without considering *risk costs,* which are the costs associated with the failure of hydraulic structures or systems. Table 7.8 lists some applications of reliability analysis of hydraulic structures that have appeared in the literature. The risk-based least-cost design of hydraulic structures promises to be, potentially, the most significant application of reliability analysis.

7.8.1 Risk-Based Design

The *risk-based design* of hydraulic structures integrates the procedures of economic, uncertainty, and reliability analyses in design practice. Engineers using a risk-based design procedure consider tradeoffs among various factors such as risk, economics, and other performance measures in hydraulic structure design. When risk-based design is embedded into an optimization framework, the combined procedure is called *optimal risk-based design.*

7.8.1.1 Basic Concept. Because the cost associated with the failure of a hydraulic structure cannot be predicted from year to year, a practical way to quantify it is to use an expected value on the annual basis. The *total annual expected cost* (*TAEC*) is the sum of the annual installation cost and annual expected damage cost which can be mathematically expressed as

$$\text{TAEC}(\mathbf{x}) = \text{FC}(\mathbf{x})*\text{CRF} + E(D|\mathbf{x}) \qquad (7.171)$$

in which FC is the first or total installation cost, which is function of the size and configuration $\mathbf{x}$ of the hydraulic structure considered; CRF is the *capital recovery factor,* which brings the present worth of the installation costs to an annual basis; and $E(D|\mathbf{x})$ is the annual expected damage cost associated with the structural failure.

Referring to Fig. 7.11, as the structural size increases, the annual installation cost increases while the annual expected damage cost associated with failure decreases. The optimal risk-based design determines the optimal structural size, configuration, and operation such that the annual total expected cost is minimized. Mathematically, the optimal risk-based design problem can be stated as:

TABLE 7.8 Some Applications of Reliability Analysis in Hydrosystem Design

Method	Hydrosystem	References
Direct integration	Levees Dams	Plate & Ihringer (1986) Meon (1992)
MFOSM	Storm sewers	Tang and Yen (1972) Tang et al. (1975) Yen and Tang (1976) Yen and Jun (1984)
	Culverts	Mays (1979) Yen, et al. (1980) Tung and Mays (1980)
	Levees	Tung and Mays (1981a,b) Lee and Mays (1986)
	Bridges Open channels	Tung and Mays (1982) Huang (1986) Yeh, et al. (1993)
AFOSM	Sewers Dams Sea dikes and barriers Freeboard Bridge scour Runoff modeling	Melching and Yen (1986) Cheng, et al. (1982, 1993) Vrijling (1987, 1993) Cheng, et al. (1986a) Yen and Melching (1991) Melching, et al. (1990) Melching (1992)
	Groundwater	Sitar, et al. (1987) Jang and Sitar (1990)
Monte Carlo simulation	Bridge scour	Johnson (1992) Chang, et al. (1994)
	Water distribution network	Bao and Mays (1990)
	Open channel	Mizumura and Ouazar (1992)
Probabilistic PE methods	Open channel	Chang, et al. (1993) Tung and Yeh (1993)
Time-to-failure analysis	Water distribution systems	Mays (1989)
Availability analysis	Water distribution systems	Mays (1989) Culinane, et al. (1992)
	Dams	Tang and Yen (1991, 1993) Kuo, et al. (1992)
Fault tree analysis	Dams Sea dikes and barriers Bridge scour	Cheng, et al. (1982) Vrijling (1987, 1993) Yen and Melching (1991)
System reliability analysis	Sewer systems	Yen, et al. (1976)
	Water distribution systems	Tung (1985) Mays (1989) Duan, et al. (1990) Bouchart and Goulter (1992)
	Dams	Cheng, et al. (1982) Yen and Melching (1991)
	Storm surge barriers	Vrijling (1993)

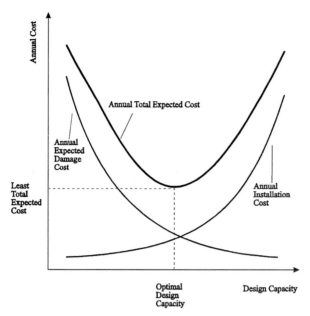

FIGURE 7.11 Optimal risk-based design concept.

$$\text{Minimize TAEC}(\mathbf{x}) = \text{FC}(\mathbf{x}) * \text{CRF} + E(D|\mathbf{x}) \qquad (7.172a)$$

subject to

$$g_i(\mathbf{x}) = 0 \qquad i = 1, 2, \ldots, m \qquad (7.172b)$$

where $g_i(\mathbf{x}) = 0$ are constraints representing the design specifications that must be satisfied. The solution to Eqs. (7.172a) to (7.172b) could be obtained through the use of appropriate optimization algorithms.

7.8.1.2 Evaluation of Annual Expected Damage Cost. In the optimal risk-based design of hydraulic structures, the thrust of the exercise is to evaluate $E(D|\mathbf{x})$ as the function of the PDFs of loading and resistance, damage function, and the types of uncertainty considered. The annual expected damage cost in the conventional risk-based hydraulic design can be computed as

$$\mathbf{E}_1(D|\mathbf{x}) = \int_{q_c^*}^{\infty} D(q|q_c^*)\, f(q)\, dq \qquad (7.173)$$

where q_c^* is the deterministic flow capacity of a hydraulic structure subject to random loads following a PDF $f(q)$; and $D(q|q_c^*)$ is the damage function corresponding to the magnitude of the load q and the hydraulic structural capacity q_c^*. Due to the complexity of the damage function and the form of the PDF of loads, the integration of Eq. (7.173), in most real-life applications, is carried out numerically.

The conventional risk-based hydraulic design considers only the inherent hydrologic uncertainty due to the random occurrence of loads. It does not consider hydraulic and economic uncertainties. Furthermore, the probability distribution of the load to the water resources system is assumed known, which is generally not the case in reality.

Incorporation of Hydraulic Uncertainty. As described in Sec. 7.1.2, uncertainties also exist in the process of determining the flow-carrying capacity of the hydraulic structure. In other words, q_c is a quantity subject to uncertainty. From the uncertainty analysis of q_c, the statistical properties of q_c can be estimated. Hence, to incorporate the uncertainty features of q_c in the risk-based design, the annual expected damage can be calculated as

$$E_2(D) = \int_0^\infty \left[\int_{q_c}^\infty D(q \mid q_c)\, f(q)\, dq \right] g(q_c) = \int_0^\infty E_1(D \mid q_c)\, g(q_c)\, dq_c \qquad (7.174)$$

in which $g(q_c)$ is the PDF of random flow carrying capacity q_c.

Incorporation of Hydrologic Parameter Uncertainty. Since the occurrence of load is random by nature, the statistical properties such as the mean, standard deviation and skewness of the distribution calculated from a finite sample are also subject to sampling errors. Therefore, the estimated load of a specified return period q_T is also a random variable associated with its probability distribution (see Fig. 7.12), instead of being a single-valued quantity presented by its "average" as commonly done in practice. The expected damage corresponding to the load of a T-year return period can be expressed as

$$E(D_T \mid q_c^*) = \int_{q_c^*}^\infty D(q_T \mid q_c^*)\, h(q_T)\, dq_T \qquad (7.175)$$

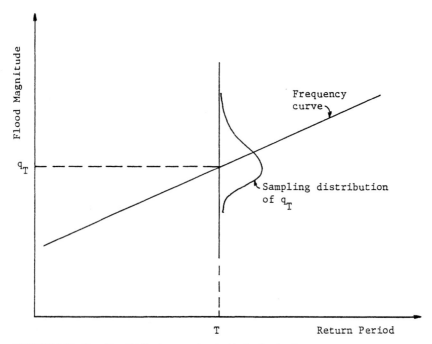

FIGURE 7.12 Sampling distribution associated with the flood estimator.

where $E(D_T|q_c^*)$ is the expected damage corresponding to a T-year load event given a known flow capacity of the hydraulic structure q_c^*; $h(q_T)$ is the sampling PDF of the estimated load with a T-year return period. Sampling distributions for a T-year event of some distributions in hydrologic frequency analysis have been developed by Stedinger (1983) and Chowdhury and Stedinger (1991).

To combine the inherent hydrologic uncertainty, represented by the PDF of the annual load event $f(q)$, and the hydrologic parameter uncertainty, represented by the sampling PDF for a load of a given return period $h(q_T)$, the annual expected damage cost can be written as

$$E_3(D|q_c^*) = \int_{q_c^*}^{\infty} \left[\int_{q_c^*}^{\infty} D(q_T|q_c^*)\, h(q_T|q)\, dq_T \right] f(q)\, dq \qquad (7.176)$$

Incorporation of Combined Hydrologic and Hydraulic Uncertainties. To include hydrologic inherent and parameter uncertainties along with the hydraulic uncertainty associated with the flow carrying capacity, the annual expected damage cost can be written as

$$E_4(D) = \int_0^{\infty} \left[\int_{q_c}^{\infty} \left[\int_{q_c}^{\infty} D(q_T, q_c)\, h(q_T)\, dq_T \right] f(q)\, dq \right] g(q_c)\, dq_c$$

$$= \int_0^{\infty} E_3(D)\, g(q_c)\, dq_c \qquad (7.177)$$

From the preceding formulations for computing annual expected damage in the risk-based design of hydraulic structures, one realizes that the mathematical complexity increases as more uncertainties are considered. However, to obtain an accurate estimation of annual expected damage associated with structural failure would require the consideration of all uncertainties, if such can be practically done. Otherwise, the annual expected damage would, in most cases, be underestimated, leading to inaccurate optimal design. Numerical investigations by Tung (1987) indicate that, without providing full account of uncertainties in the analysis, the resulting annual expected damage is significantly underestimated, even with a 75-year-long flood record.

7.8.2 Optimization of Water Resources Systems by Chance-Constrained Methods

In all fields of science and engineering, the decision-making process is dependent on several parameters describing system behavior and characteristics. More often than not, at least one of these system parameters cannot be assessed with certainty. In a system optimization model, if some of the coefficients in the constraints are random variables, compliance with the constraints, under a given set of solutions, cannot be ensured with certainty. Due to the random nature of the constraint coefficients, a certain likelihood always exists that constraints will be violated.

The basic idea of *chance-constrained methods* is to find the solution to an optimization problem such that the constraints will be complied with at a specified reliability. Chance-constrained formulations have been applied to groundwater management (Tung, 1986), reservoir operations (Loucks, et al., 1981), waste load allocation (Tung and Hathhorn, 1990), water distribution systems (Lansey, et al., 1989). This section briefly describes the properties of chance-constrained models.

Refer to the general nonlinear optimization problem as stated in Eq. (6.17) and consider a constraint $g(\mathbf{x}) \leq b$ with $\mathbf{x}$ being a vector of decision variables. In general, decision variables $\mathbf{x}$ in an optimization model are controllable or have little uncertainty. Suppose that some of the parameters on the left-hand side of the constraint $g(\mathbf{x})$ and/or the right-hand side (RHS) coefficient b are subject to uncertainty. Under this circumstance, the chance-constrained formulation expresses the original constraint in a probabilistic format as

$$P[g(\mathbf{x}) \leq b] \geq \alpha \qquad (7.178)$$

where $P[\ ]$ is the probability and α is the specified reliability for constraint compliance. Since the above chance-constrained formulation involves probability, it is not mathematically operational for algebraic solution. For this reason, the so-called *deterministic equivalent* must be derived. There are three cases in which the random elements in Eq. (7.178) could occur: (1) Only elements in $g(\mathbf{x})$ are random, (2) only the RHS b is random, and (3) both $g(\mathbf{x})$ and b are random.

The simplest case is case (2) where only the RHS coefficient b is random. The constraint can be rewritten as

$$P[g(\mathbf{x}) \leq B] \geq \alpha \qquad (7.179)$$

where B is a random RHS coefficient associated with a CDF, $F_B(\ )$. The deterministic equivalent of the original chance-constraint Eq. (7.179) is

$$g(\mathbf{x}) \leq b_{1-\alpha} \qquad (7.180)$$

where $b_{1-\alpha}$ is the $(1-\alpha)$th quantile of the random RHS coefficient B satisfying $P[B \leq b_{1-\alpha}] = 1-\alpha$. If the RHS coefficient B is a normal random variable with mean μ_B and standard deviation σ_B, Eq. (7.180) can be written as

$$g(\mathbf{x}) \leq \mu_B + z_{1-\alpha}\,\sigma_B \qquad (7.181)$$

with $z_{1-\alpha}$ being the $(1-\alpha)$th standard normal quantile.

In the third case, both the left-hand side and right-hand side of a constraint are random. They are denoted as $G(\mathbf{x})$ and B, respectively. The chance-constraint can then be expressed as

$$P[G(\mathbf{x}) - B \leq 0] \geq \alpha \qquad (7.182)$$

The deterministic equivalent of Eq. (7.182) can be derived as

$$F_{G(\mathbf{x})-B}^{-1}(\alpha) \geq 0 \qquad (7.183)$$

where $F_{G(\mathbf{x})-B}^{-1}(\alpha)$ is the inverse of the CDF of random $G(\mathbf{x}) - B$ taken on the value of α.

As a special case, consider a linear programming (LP) formulation as

$$\text{Maximize } \mathbf{c}^T\mathbf{x}$$

subject to

$$\mathbf{Ax} \leq \mathbf{b}$$

$$\mathbf{x} \geq 0$$

in which $\mathbf{A}$ is the technological coefficients matrix; $\mathbf{b}$ is the vector of RHS coefficients; and $\mathbf{c}$ is the vector of objective function coefficients. Suppose that the elements in **A** and **b** are subject to uncertainty. By imposing a reliability restriction α on the system constraints, the constraints in the above LP model can be transformed into the following chance-constrained formulation:

$$P[\mathbf{Ax} \le \mathbf{b}] \ge \alpha \tag{7.184}$$

In a chance-constrained LP model, the elements in **A, b,** and **c** can be random variables. When the objective function coefficient c_j's are random variables, it is common to replace them by their expected values. Consider the general case where elements A_{ij} and B_i are simultaneously random variables. Assume that random technological coefficients and random RHS coefficients are correlated within a constraint, and these coefficients are uncorrelated between constraints. The chance-constraint format can be written as

$$P\left[\sum_{j=1}^{n} A_{ij}x_j - B_i \le 0 \right] \ge \alpha_i \tag{7.185}$$

The deterministic equivalent of Eq. (7.185) can be derived, assuming a normal distribution for the random terms in Eq. (7.185), as

$$\sum_{j=1}^{n} E[A_{ij}]x_j + \Phi^{-1}(\alpha_i)\ \sqrt{\mathbf{x}^T\mathbf{C}_i\mathbf{x} + 2\sum_{j=1}^{n} x_j\text{cov}(A_{ij}, B_i) + \sigma_{B_i}^2} \le E(B_i),\, i = 1, 2, \ldots, n \tag{7.186}$$

where $E[A_{ij}]$ is an expectation of the random technological coefficient A_{ij}, $\mathbf{C}_i$ is an $n \times n$ covariance matrix of n random technological coefficients $(A_{i1}, A_{i2}, \ldots, A_{in})$ in the ith constraint, and $\text{cov}(A_{ij}, B_i)$ is the covariance between the random technological coefficient A_{ij} and the random RHS coefficient B_i for the ith constraint.

REFERENCES

Abramowitz, M., and I. A. Stegun, eds., *Handbook of Mathematical Functions with Formulas, Graphs and Mathematical Tables,* 9th ed., Dover, New York, 1972.

Ang, A. H.-S., "Structural Risk Analysis and Reliability-Based Design," *Journal of Structural Engineering Division,* ASCE, 99(9):1891–1910, 1973.

Ang, A. H.-S., and W. H. Tang, *Probability Concepts in Engineering Planning and Design, Vol. II: Decision, Risk, and Reliability,* John Wiley, New York, 1984.

Bao, Y., and L. W. Mays, "Model for Water Distribution System Reliability," *Journal of Hydraulic Engineering,* ASCE, 116(9):1119–1137, 1990.

Berthouex, P. M. "Modeling Concepts Considering Process Performance, Variability, and Uncertainty," in T. M. Keinath and M. P. Wanielista, eds., *Mathematical Modeling for Water Pollution Control Processes,* Ann Arbor Science, Ann Arbor, Mich., 1975, pp. 405–439.

Bouchart, F., and I. Goulter, "Selecting Valve Location to Optimize Water Distribution Network Reliability," in J. T. Kuo and G. F. Lin, eds., *Stochastic Hydraulics '92, Proceedings 6th IAHR International Symposium, Taipei,* Water Resources Publications, Littleton, Colo., 1992, pp. 155–162.

Chang, C. H., J. C. Yang, and Y. K. Tung, "Sensitivity and Uncertainty Analyses of a Sediment Transport Model: A Global Approach," *Journal of Stochastic Hydrology and Hydraulics,* 7(4): 299–314, 1993.

Chang, C. H., Y. K. Tung, and J. C. Yang, "Monte Carlo Simulation for Correlated Variables with Marginal Distributions," *Journal of Hydraulic Engineering,* ASCE, 120(2): 313–331, 1994.

Cheng, S. T., B. C. Yen, and W. H. Tang, "Overtopping Risk for an Existing Dam," *Hydraulic Engineering Series,* no. 37, Department of Civil Engineering, University of Illinois at Urbana-Champaign, Ill., 1982.

Cheng, S. T., B. C. Yen, and W. H. Tang, "Wind Induced Overtopping Risk of Dams," in B. C. Yen, ed., *Stochastic and Risk Analysis in Hydraulic Engineering,* Water Resources Publications, Littleton, Colo., 1986a, pp. 48–58.

Cheng, S. T., B. C. Yen, and W. H. Tang, "Sensitivity of Risk Evaluation to Coefficient of Variation," in B. C. Yen, ed., *Stochastic and Risk Analysis in Hydraulic Engineering,* Water Resources Publications, Littleton, Colo., 1986b, pp. 266–273.

Cheng, S. T., B. C. Yen, and W. H. Tang, "Stochastic Risk Modeling of Dam Overtopping," in B. C. Yen and Y. K. Tung, eds., *Reliability and Uncertainty Analyses in Hydraulic Design,* ASCE, 1993, pp. 123–132.

Chowdhury, J. U., and J. R. Stedinger, "Confidence Interval for Design Floods with Estimated Skew Coefficient," *Journal of Hydraulic Engineering,* 117(7):811–831, 1991.

Cochran, W., *Sampling Techniques,* 2d ed., Wiley, New York, 1966.

Cullinane, M. J., K. E. Lansey, and L. W. Mays, "Optimization-Availability-Based Design of Water Distribution Networks," *Journal of Hydraulic Engineering,* ASCE, 118(3):420–441, 1992.

Dagpunar, J., *Principles of Random Variates Generation,* Oxford University Press, New York, 1988.

Der Kiureghian, A., and P. L. Liu, "Structural Reliability Under Incomplete Probability Information." *Journal of Engineering Mechanics,* ASCE 112(1):85–104, 1985.

Devore, J. L., *Probability and Statistics for Engineering and Sciences,* 2d ed., Brooks/Cole, Monterey, Cal., 1987.

Duan, N., L. W. Mays, and K. E. Lansey, "Optimal Reliability-Based Design of Pumping and Distribution Systems," *Journal of Hydraulic Engineering,* ASCE, 116(2):249–268, 1990.

Epstein, B., "Some Application of the Mellin Transform in Statistics," *Annals of Mathematical Statistics,* 19:370–379, 1948.

Fisher, R. A., and L. H. C. Tippett, "Limiting Forms of the Frequency Distribution of the Largest or Smallest Member of a Sample," *Proceedings of the Cambridge Philosophical Society,* 24: 180–190, 1928.

Fussell, J. B., E. B. Henry, and N. H. Marshall, "MOCUS—A Computer Program to Obtain Minimal Cut Sets from Fault Trees," ANCR-1156, 1974.

Giffin, W. C., *Transform Techniques for Probability Modeling,* Academic Press, New York, 1975.

Golub, G. H., and C. F. Van Loan, *Matrix Computations,* 2d edition, Johns Hopkins University Press, Baltimore, Md. 1989.

Gould, H., and J. Tobochnik, *An Introduction to Computer Simulation Methods: Applications to Physical Systems, Part 2,* Addison-Wesley, Reading, Mass., 1988.

Gumbel, E. J., *Statistics of Extremes,* Columbia University Press, New York, 1958.

Haan, C. T., *Statistical Methods in Hydrology,* Iowa State University Press, 1977.

Harr, M. E., "Probabilistic Estimates for Multivariate Analyses," *Applied Mathematical Modelling,* 13: 313–318, 1989.

Hasofer, A. M. and N. C. Lind, "Exact and Invariant Second-Moment Code Format," *Journal of Engineering Mechanics Div.,* ASCE, 100(EM1):111–121, 1974.

Henley, E. J., and H. Kumamoto, *Reliability Engineering and Risk Assessment,* Prentice-Hall, Englewood Cliffs, N. J., 1981.

Huang, K. Z., "Reliability Analysis of Hydraulic Design of Open Channel," in B. C. Yen, ed., *Stochastic and Risk Analysis in Hydraulic Engineering,* Water Resources Publications, Littleton, Colo., 1986.

Interagency Advisory Committee on Water Data, *Guidelines for Determining Flood Flow Frequency,* Bulletin 17B, U.S. Department of Interior, U.S. Geological Survey, Office of Water Data Coordination, Reston, Va., 1982.

Jang, Y. S. and N. Sitar, "Reliability Analysis of Contaminant Transport Through Clay Liners," Department of Civil Engineering, Report no. UCB/GT-90/04, 1990.

Johnson, N. L. and S. Kotz, *Distributions in Statistics: Continuous Multivariate Distributions,* John Wiley, New York, 1976.

Johnson, M. E., *Multivariate Statistical Simulation,* John Wiley, New York, 1987.

Johnson, P. A., "Reliability-Based Pier Scour Engineering," *Journal of Hydraulic Engineering,* ASCE, 118(10): 1344–1358, 1992.

Kapur, K. C., and L. R. Lamberson, *Reliability in Engineering Designs,* John Wiley, New York, 1977.

Kapur, K. C., "Chapter 19—Mathematical and Statistical Methods and Models in Reliability and Life Studies," in W. G. Ireson and C. F. Crombs, Jr., eds., *Handbook of Reliability Engineering and Management,* McGraw-Hill, New York, 1988.

Kendall, M., A. Stuart, and J. K. Ord, *Kendall's Advanced Theory of Statistics, Vol. 1: Distribution Theory,* 5th Edition, New York, Oxford University Press, 1987.

Kraus, J. W., "Chapter 15—Maintainability and Reliability," in W. G. Ireson and C. F. Coombs, Jr., eds., *Handbook of Reliability Engineering and Management,* New York, McGraw-Hill, 1988.

Kuo, J. T., B. C. Yen, J. I. Lin, G. P. Hwang, and S. K. Hsu, "Dam Safety Inspection Scheduling for Shihmen Reservoir," in J. T. Kuo and G. F. Lin, eds., *Stochastic Hydraulics '92, Proceedings 6th IAHR International Symposium, Taipei,* Water Resources Publications, Littleton, Colo., 1992, pp. 99–106.

Lansey, K. E., N. Duan, L. W. Mays, and Y. K. Tung, "Water Distribution System Design under Uncertainties," *Journal of Water Resources Planning and Management,* 115(5): 630–645, 1989.

Law, A. M. and W. D. Kelton, *Simulation Modeling and Analysis,* McGraw-Hill, New York, 1991.

Lee, H. L., and L. W. Mays, "Hydraulic Uncertainties in Flood Levee Capacity," *Journal of Hydraulic Engineering,* ASCE, 112(10):928–934, 1986.

Lehmer, D. H., "Mathematical Methods in Large-Scale Computing Units," *Ann. Comp. Lab.,* Harvard University, 26: 141–146, 1951.

Liu, P. L., and A. Der Kiureghian, "Multivariate Distribution Models with Prescribed Marginals and Covariances," *Probabilistic Engineering Mechanics,* 1(2):105–112, 1986.

Loucks, D. P., J. R. Stedinger, and D. A. Haith, *Water Resources Systems Planning and Analysis,* Prentice-Hall, Englewood Cliffs, N.J., 1981.

Madsen, H. O., S. Krenk, and N. C. Lind, *Methods of Structural Safety,* Prentice-Hall, Englewood Cliffs, N.J., 1986.

Marshall, A. W., "The Use of Multi-Stage Sampling Schemes in Monte Carlo Computations," in M. A. Meyer, ed., *Symposium on Monte Carlo Methods,* John Wiley, New York, 1956.

Mays, L. W., "Optimal Design of Culverts Under Uncertainties," *Journal of Hydraulic Division,* ASCE, 105(5):443–460, 1979.

Mays, L. W., ed., *Reliability Analysis of Water Distribution Systems,* ASCE, New York, 1989.

Mays, L. W., and Y. K. Tung, *Hydrosystems Engineering and Management.* McGraw-Hill, New York, 1992.

McBean, E. A., J. Penel, and K.-L. Siu, "Uncertainty Analysis of a Delineated Floodplain," *Canadian Journal of Civil Engineers,* 11:387–395, 1984.

McKay, M. D., "Sensitivity and Uncertainty Analysis Using a Statistical Sample of Input Values," in Y. Ronen, ed., *Uncertainty Analysis,* CRC Press, Boca Raton, Fla., 1988.

Melchers, R. E., *Structural Reliability Analysis and Prediction,* Ellis Horwood Limited, Chichester, England, 1987.

Melching, C. S., "An Improved First-Order Reliability Approach for Assessing Uncertainties in Hydrologic Modeling," *Journal of Hydrology,* 132: 157–177, 1992.

Melching, C. S., H. G. Wenzel, Jr., and B. C. Yen, "A Reliability Estimation in Modeling Watershed Runoff with Uncertainties," *Water Resources Research,* 26: 2275–2286, 1990.

Meon, G., "Overtopping Probability of Dams under Flood Load," in J. T. Kuo and G. F. Lin, eds., *Stochastic Hydraulics '92, Proceedings, 6th IAHR International Symposium, Taipei,* Water Resources Publications, Littleton, Colo., 1992, pp. 99–106.

Mizumura, K., and D. Ouazar, "Stochastic Characteristics of Open Channel Flow," in J. T. Kuo and G. F. Lin, eds., *Stochastic Hydraulics '92, Proceedings 6th IAHR International Symposium, Taipei,* Water Resources Publications, Littleton, Colo., 1992, pp. 99–106.

Park, C. S., "The Mellin Transform in Probabilistic Cash Flow Modeling," *The Engineering Economist,* 32(2):115–134, 1987.

Parrish, R. S., "Generating Random Deviates From Multivariate Pearson Distributions." *Computational Statistics & Data Analysis,* 9:283–296, 1990.

Pieruschka, E., *Principles of Reliability,* Prentice-Hall, Englewood Cliffs, N.J., 1963.

Plate, E. J., and J. Ihringer, "Failure Probability of Flood Levees on a Tidal River," in B. C. Yen, ed., *Stochastic and Risk Analysis in Hydraulic Engineering,* Water Resources Publications, Littleton, Colo., 1986, pp. 48–58.

Press, W. H., B. P. Flannery, S. A. Teukolsky, and W. T. Vetterling, *Numerical Recipes—The Art of Scientific Computing (Fortran Version),* Cambridge University Press, New York, 1989.

Rackwitz, R., "Practical Probabilistic Approach to Design," *Bulletin* 112, Comite European du Beton, Paris, 1976.

Ronning, G., "A Simple Scheme for Generating Multivariate Gamma Distributions with Non-negative Covariance Matrix," *Technometrics,* 19(2):179–183, 1977.

Rosenblueth, E., "Point Estimates for Probability Moments," *Proceedings, National Academy of Science,* 72(10):3812–3814, 1975.

Rosenblueth, E., "Two-Point Estimates in Probabilities," *Applied Mathematical Modelling,* 5:329–335, 1981.

Rubinstein, R. Y., *Simulation and the Monte Carlo Method,* John Wiley, New York, 1981.

Shinozuka, M., "Basic Analysis of Structural Safety," *Journal of Structural Engineering Div.,* ASCE, 109(3):721–740, 1983.

Sitar, N., J. D. Cawlfield, and A. Der Kiureghian, "First Order Reliability Approach to Stochastic Analysis of Subsurface Flow and Contaminant Transport," *Water Resources Research,* 23(5): 794–804, 1987.

Springer, M. D., *The Algebra of Random Variables,* John Wiley, New York, 1979.

Stedinger, J. R., "Confidence Intervals for Design Events," *Journal of Hydraulic Engineering,* 109(1):13–27, 1983.

Stedinger, J. R., R. M. Vogel, and E. Foufoula-Georgiou, "Chapter 18: Frequency Analysis of Extreme Events," in D. R. Maidment, ed., *Handbook of Hydrology,* McGraw-Hill, New York, 1992.

Tang, W. H., and B. C. Yen, "Hydrologic and Hydraulic Design under Uncertainties," *Proceedings, International Symposium on Uncertainties in Hydrologic and Water Resources Systems,* Tucson, Ariz., 2:868–882, 1972.

Tang, W. H., L. W. Mays, and B. C. Yen, "Optimal Risk-Based Design of Storm Sewer Networks," *Journal of Environmental Engineering Division,* ASCE, 103(3):381–398, 1975.

Tang, W. H., and B. C. Yen, "Dam-Safety Inspection Scheduling," *Journal of Hydraulic Engineering,* ASCE, 117(2):214–229, 1991.

Tang, W. H., and B. C. Yen, "Probabilistic Inspection Scheduling for Dams," in B. C. Yen and Y. K. Tung, eds., *Reliability and Uncertainty Analyses in Hydraulic Design,* ASCE, 1993, pp. 107–122.

Tung, Y. K., "Evaluation of Water Distribution Network Reliability," *Hydraulics and Hydrology in Small Computer Age, Proceedings of the Hydraulic Division Specialty Conference,* ASCE, Orlando, Fla., August 12–17, 1985.

Tung, Y. K., "Groundwater Management by Chance-Constrained Model," *Journal of Water Resources Planning and Management,* 112:1–19, 1986.

Tung, Y. K., "Effects of Uncertainties on Optimal Risk-Based Design of Hydraulic Structures," *Journal of Water Resources Planning and Management,* 113(5):709–722, 1987.

Tung, Y. K., "Mellin Transform Applied to Uncertainty Analysis in Hydrology/Hydraulics," *Journal of Hydraulic Engineering,* ASCE. 116(5):659–674, 1990.

Tung, Y. K. and W. E. Hathhorn, "Stochastic Waste Load Allocation," *Journal of Ecological Modelling,* 51: 29–46, 1990.

Tung, Y. K., and L. W. Mays, "Optimal Risk-Based Design of Hydraulic Structures," *Technical Report CRWR-171,* Center for Research in Water Resources, University of Texas, Austin, Tex., 1980b.

Tung, Y. K., and L. W. Mays, "Risk Models for Levee Design," *Water Resources Research,* AGU, 17(4):833–841, 1981a.

Tung, Y. K., and L. W. Mays, "Optimal Risk-Based Design of Flood Levee Systems," *Water Resources Research,* AGU, 17(4):843–852, 1981b.

Tung, Y. K., and K. C. Yeh, "Evaluation of Safety of Hydraulic Structures Affected by Migrating Pit," *Journal of Stochastic Hydrology and Hydraulics,* 7(2):131–145, 1993.

Tung, Y. K. and B. C. Yen, "Some Recent Progress in Uncertainty Analysis for Hydraulic Design," in B. C. Yen and Y. K. Tung, eds., *Reliability and Uncertainty Analyses in Hydraulic Design,* ASCE, 1993, pp. 17–34.

Tung, Y. K., and L. W. Mays, "Risk Models for Levee Design," *Water Resources Research,* AGU, 17(4):833–841, 1981a.

Vrijling, J. K., "Probabilistic Design of Water Retaining Structures," in L. Duckstein and E. J. Plate, eds., *Engineering Reliability and Risk in Water Resources,* Martinus Nijhoff, Dordrecht, The Netherlands, 1987, pp. 115–134.

Vrijling, J. K., "Development in Probabilistic Design of Flood Defenses in the Netherlands," in B. C. Yen and Y. K. Tung, eds., *Reliability and Uncertainty Analyses in Hydraulic Design,* ASCE, 1993, pp. 133–178.

Wen, Y. K., "Approximate Methods for Nonlinear Time-Variant Reliability Analysis," *Journal of Engineering Mechanics,* ASCE, 113(12):1826–1839, 1987.

Yeh, K. C., and Y. K. Tung, "Uncertainty and Sensitivity of a Pit Migration Model," *Journal of Hydraulic Engineering,* ASCE, 119(2):262–281, 1993.

Yen, B. C., "Safety Factor in Hydrologic and Hydraulic Engineering Design," in E. A. McBean, K. W. Hipel and T. E. Unny, eds., *Reliability in Water Resources Management,* Water Resources Publications, Littleton, Colo., 1979, pp. 389–407.

Yen, B. C. and A. H.-S. Ang, "Risk Analysis in Design of Hydraulic Projects," in C. L. Chiu, ed., *Stochastic Hydraulics, Proceedings of First International Symposium,* University of Pittsburgh, Pittsburgh, Pa., 1971, pp. 694–701.

Yen, B. C., S. T. Cheng, and W. H. Tang, "Reliability of Hydraulic Design of Culverts," *Proceedings, International Conference on Water Resources Development,* IAHR Asian Pacific Division 2d Congress, Taipei, Taiwan, 2:991–1001, 1980.

Yen, B. C., S. T. Cheng, and C. S. Melching, "First-Order Reliability Analysis," in B. C. Yen, ed., *Stochastic and Risk Analysis in Hydraulic Engineering,* Water Resources Publications, Littleton, Colo., 1986, pp. 1–36.

Yen, B. C., and W. H. Tang, "Risk-Safety Factor Relation for Storm Sewer Design," *Journal of Environmental Engineering Division,* ASCE, 102(2):509–516, 1976.

Yen, B. C., and B. H. Jun, "Risk Consideration in Design of Storm Drains," *Proceedings, 3d IAHR/IAWPRC International Conference on Urban Storm Drainage,* Chalmers University of Technology, Goteborg, Sweden, 2:695–704, 1984.

Yen, B. C., and C. S. Melching, "Reliability Analysis Methods for Sediment Problems," *Proceedings, 5th Federal Interagency Sediment Conference,* Las Vegas, 2:9.1–9.8, March 1991.

Yen, B. C., and Y. K. Tung, "Some Recent Progress in Reliability Analysis for Hydraulic Design," in B. C. Yen and Y. K. Tung, eds., *Reliability and Uncertainty Analyses in Hydraulic Design,* ASCE, 1993, pp. 35–80.

Yen, B. C., H. G. Wenzel, Jr., L. W. Mays, and W. H. Tang, "Advanced Methodologies for Design of Storm Sewer Systems," Research Report, no. 112, Water Resources Center, University of Illinois at Urbana-Champaign, Ill., August 1976.

Young, D. M., and R. T. Gregory, *A Survey of Numerical Mathematics, Vol. II,* Addison-Wesley, Reading, Mass., 1973.

P · A · R · T · 2

WATER RESOURCES QUALITY (NATURAL SYSTEMS)

CHAPTER 8
WATER QUALITY

Joseph F. Malina, Jr., Ph.D., P.E., DEE

C.W. Cook Professor
Environmental Engineering
Civil Engineering Department
The University of Texas
Austin, Texas

8.1 WATER QUALITY

8.1.1 Introduction

A safe and potable water supply is critical to the survival of human life as we know it. Humans and all flora and fauna need water to survive. Society in general requires water for maintenance of public health, fire protection, cooling water for electricity generation, use in industrial processes, irrigation of agriculture lands, and navigation.

8.1.2 Water Characteristics and the Hydrologic Cycle

The hydrologic cycle plays an important role in the quality of water. Precipitation of water washes particulate materials and gases from the air. Particles may have organic or inorganic chemical compounds attached. Some chemicals washed by rainfall are sulfuric acid (acid rain), and nitrates. Runoff resulting from precipitation may carry inorganic chemicals which dissolve into the water runoff flow over different geological formation and soils; e.g., limestone dissolves in runoff, resulting in hard water. Runoff from agricultural lands carries large amounts of silt and suspended solids, as well as pesticides, into streams, lakes, and other surface water bodies.

The myriad of chemicals, solids, and other debris from urban areas affects the quality of receiving streams. Evaporation of water from oceans, lakes, and other impoundments tends to increase the salinity and concentration of conservative substances in surface waters.

Infiltration and percolation of surface runoff into groundwater aquifers can markedly affect the quality of the groundwater. Nitrogenous compounds and phosphorus in various forms are washed from the soil and transported into the aquifer, where these compounds can undergo different chemical and biological transformations. Petroleum products and synthetic organic chemicals also find their way into aquifers. Water quality in streams and surface waters throughout the world varies considerably.

Physical, chemical, and biological characteristics of different sources of water are summarized in Table 8.1. Analytical procedures used to analyze natural water samples for most of the biological, chemical, and physical parameters used to assess water quality are presented in detail in *Standard Methods* (1995) and will not be repeated in this chapter.

TABLE 8.1　Physical, Chemical, and Biological Characteristics of Various Water Sources

Characteristics	Typical surface water	Typical groundwater	Domestic wastewater (U.S.)
Physical			
Turbidity, NTU	—	—	—
Solids, total, mg/L	—	—	700
Suspended, mg/L	>50	—	200
Settleable, mL/L	—	—	10
Volatile, mg/L	—	—	300
Filterable (dissolved), mg/L	<100	>100	500
Color, (Color units)	—	—	—
Odor, number	—	—	Stale
Temperature, °C	0.5–30	2.7–25	10–25
Temperature, °F	33–86	37–77	50–77
Chemical: inorganic matter			
Alkalinity, mg/L as $CaCO_3$	<100	>100	>100
Hardness, mg/L as $CaCO_3$	<100	>100	—
Chlorides, mg/L	50	200	<100
Calcium, mg/L	20	150	—
Heavy metals, mg/L	—	0.5	—
Nitrogen, mg/L	<10	<10	40
Organic, mg/L	5	—	15
Ammonia, mg/L	—	—	25
Nitrate, mg/L	<5	5	0
Phosphorus, total, mg/L	—	—	12
Sulfate, mg/L	—	—	—
pH, (pH units)	—	6.5–8	6.5–8.5
Chemical: organic matter			
Total organic carbon (TOC), mg/L	<5	—	150
Fats, oils, greases, mg/L	—	—	100
Pesticides, mg/L	<0.1	—	—
Phenols, mg/L	<0.001	—	—
Surfactants, mg/L	<0.5	<0.5	—
Chemical: Gases			
Oxygen, mg/L	7.5	7.5	<1.0
Biological			
Bacteria, MPN/100mL	<2000	<100	10^8–10^9
Viruses, plaque forming units (pfu)	<10	<1	10^2–10^4

Source:　Adapted from National Interim Primary Drinking Regulations (1975).

Return flows from agricultural land (runoff), power plants (cooling water), industrial facilities (discharge of treated process effluents, cooling waters, and storm water runoff), and urban areas (treated municipal wastewater and storm water runoff) also affect the quality of water in receiving streams, and may indirectly affect groundwater quality.

8.2 PHYSICAL CHARACTERISTICS OF WATER

Water quality is related to a safe and healthy source of water. Therefore a water supply should not be offensive to the senses of touch, sight, smell, and taste. Physical characteristics of water are determined qualitatively by these senses, e.g., temperature by touch; color, floating debris, turbidity, and suspended solids by sight; and taste and odor by smell and taste.

8.2.1 Temperature

The temperature of freshwater normally varies from 0 to 35°C (32 to 95°F), depending on the source, depth, and season. Saline water temperature in oceans varies with depth. Temperature of ocean water near the surface varies from approximately 21°C (70°F) down to 2.3°C (36°F), depending on the season of the year and the latitude. The temperature of deep ocean water (i.e., at depths >2000 ft [610 m]) is uniformly approximately 2.3°C (36°F). Water temperature in groundwater aquifers depends on the latitude, the time of the year, and the depth of the aquifer below the surface of the ground. The ambient temperature of the soil will affect the temperature of the groundwater.

The temperature of water affects some of the important physical properties and characteristics of water, such as density, specific weight, viscosity, surface tension, thermal capacity, enthalpy, vapor pressure, specific conductivity and conductance, salinity, and solubility of dissolved gases (e.g., oxygen and carbon dioxide).

Chemical and biological reaction rates increase with increasing temperature. Reaction rates are usually assumed to double for an increase in temperature of 10°C (50°F).

The effects of changes in temperature on water quality are most noticeable in lakes. Water quality in lakes at a given latitude changes dramatically with the season of the year. A detailed discussion of the effects of water temperature and temperature changes on lake stratification and turnover, as well as the effects of these conditions on water quality, is presented in Chap. 9.

8.2.2 Color

Color in water is primarily a concern of water quality for aesthetic reasons. Colored water gives the appearance of being unfit to drink, even though the water may be perfectly safe for public use. On the other hand, color can indicate the presence of organic substances, such as algae or humic compounds. More recently, color has been used as a quantitative assessment of the presence of potentially hazardous or toxic organic materials in the water.

Decaying vegetation results in brown-colored water. This color is caused by suspended and dissolved tannins, which are released by biodegradation of vegetation.

Algae give water an apparent green color. Water can leach tannins and vegetable color from plants and/or from organic detritus resulting, in a yellow to brown color in the water. Some of the color-causing components may be particulate matter, which results in an apparent color. However, some of the tannins and vegetable color are true colors. Red and/or yellow color in natural waters may also be derived from mineral sediments. Color may be considered to be true color if it is caused by dissolved compounds and substances. True color caused by dissolved organic substances may be removed by the addition of powdered activated carbon (PAC) to the water, with subsequent removal of the PAC by flocculation, sedimentation, and filtration. Granular activated carbon columns can also be used to remove true color from water. Apparent color is caused by particulate solids (e.g., algae) and/or turbidity. Apparent color can be removed by filtration.

8.2.3 Taste and Odor

Taste and odor are human perceptions of water quality. Human perception of taste includes sour (hydrochloric acid), salt (sodium chloride), sweet (sucrose), and bitter (caffeine). Relatively simple compounds produce sour and salty tastes. However, sweet and bitter tastes are produced by more complex organic compounds. Humans detect many more types of odors than tastes. Odor thresholds of some organic compounds in water are summarized in Table 8.2.

Taste and odor problems are more common in surface water than in groundwater. Organic materials discharged directly to water, such as falling leaves, runoff, and agricultural drainage, frequently are sources of tastes and odor-producing compounds released during biodegradation.

Some bacteria and fungi responsible for the decomposition of organic material in natural water also cause odors and tastes in the water. Bacteria and fungi associated with tastes and odors in natural water are listed in Table 8.3.

Algae also are a principal cause of taste and odors in some surface water. Tastes and odors associated with algae are presented in Table 8.4. Some protozoa also can contribute to problems with odors and tastes in natural water. Some of the more

TABLE 8.2 Odor Thresholds of Various Substances in Water

Compound	Threshold odor concentration, (mg/L)
2-Octanol	0.13
Styrene	0.05
Ethylbenzene	0.1
Naphthalene	0.007
p-Dichlorobenzene	0.15
Chloroform	20.0
Nonanal	0.001
Methyl sulfide	0.003
Geosmin	0.000005
Methylisoborneol (MIB)	0.000005
Trichloroethene	2.6
Tetrachloroethene	2.8
Dichloromethane	24.0
Toluene	0.14

Source: James M. Montgomery (1985).

TABLE 8.3 Bacteria in Surface Water Storage Facilities and Associated Odors

Organism	Specific organism	Associated taste and odor
Bacteria	*Beggiatoa*	Decayed, very offensive
	Crenothrix	Decayed, very offensive
	Leptothrix	Medicinal with chlorine
	Sphaerotilus natans	Decayed, very offensive
	Thitothrix	Sulfur

Source: Adapted from Tchobanoglous and Schroeder (1985).

common protozoa associated with odors in natural water are listed in Table 8.5. Along with the resulting odor commonly detected.

Actinomycetes (mold-like filamentous bacteria) produce severe musty odors and tastes as a result of the bacterial degradation of algae and products of algal metabolism. Actinomycetes are frequently associated with blue green algae mats floating on water surfaces. The two compounds produced by actinomycetes which cause tastes and odors in water are geosmin (musty odor) and methylisoborneol (MIB) (Silvey, 1964; 1968).

Odors and tastes caused by dissolved organic substances in natural water may be removed by the addition of powdered activated carbon (PAC) to the water, with subsequent removal of the PAC by flocculation, sedimentation, and filtration. Granular activated carbon columns can also be used to remove odors and tastes in natural water.

Most odors and tastes in groundwater supplies are caused by bacterial actions in aquifers, or by salts and minerals which dissolve in the groundwater percolating through underground mineral deposits. Many groundwater supplies contain hydrogen sulfide (characterized as a rotten egg odor), which is produced during the anaerobic bacterial conversion of organic sulfur, elemental sulfur, sulfates, and sulfites to hydrogen sulfide (H_2S). H_2S can be detected as odor in water at a concentration of less than 100 mg/L (0.0001/mg/L). Concentrations of H_2S in water greater than 0.1 to 0.5 mg/L are offensive to smell. Hydrogen sulfide can be detected qualitatively at low concentrations in air (less than 0.5 parts per billion). Data presented in Table 8.6 summarize some of the information available that deals with the detection of hydrogen sulfide.

The hydrogen sulfide concentration in the atmosphere is a function of the concentration of hydrogen sulfide in water, which also is a function of the temperature and pH of the water. The effects of temperature on solubility are shown in Table 8.7, and the effects of pH on the solubility of hydrogen sulfide are illustrated in Table 8.8.

These data indicate that the mole fraction (percentage) of hydrogen sulfide in water is dependent on the pH of the water, assuming that no other reactions with the sulfide are taking place, such as precipitation of sulfide as heavy metal sulfides under reduced environmental conditions. The percentage of hydrogen sulfide which is ionized (HS^-) increases as the pH increases, i.e., the percentage of un-ionized hydrogen sulfide (H_2S) decreases as the pH increases. Therefore, at pH = 6, 90.1 percent of the H_2S dissolved in the water is available for stripping from the wastewater. This percentage of available H_2S for stripping decreases to 22.5 percent of the dissolved hydrogen sulfide at pH = 7.5, and the potential for hydrogen sulfide odor is reduced.

Hydrogen sulfide may be removed by air-stripping the gas from the water and capturing the stripped gases. Granular activated carbon columns can also be used to remove hydrogen sulfide from natural water.

TABLE 8.4 Algae in Surface Water Storage Facilities and Associated Odors

Organism	Specific organism	Associated taste and odor
Chlorophyceae (green)	*Cladophora*	Septic
	Closterium	Grassy
	Dictyosphaerium	Grassy, nasturtium, fishy
	Eudorina	Fishy
	Gloecystis	Septic
	Hydrodictyon	Septic
	Pandorina	Fishy
	Scenedesmus	Grassy
	Spirogyra	Grassy
	Staurastrum	Grassy
	Ulothrix	Grassy
	Volvox	Fishy
Cyanophyceae (blue-green)	*Anabaena*	Moldy, grassy, septic
	Anacystis	Grassy, septic (sweet)
	Aphanizomenon	Moldy, grassy, septic
	Coelosphaerium	Grassy (sweet)
	Cylindrosphermum	Grassy, septic
	Gloeocapsa	Red, septic
	Microcystis	Grassy, septic
	Oscillatoria	Grassy, musty, spicy
	Rivularia	Moldy, grassy, musty
Diatomaceae (usually brown)	*Asterionella*	Aromatic, geranium, fishy
	Cyclotella	Geranium, fishy
	Diatoma	Aromatic
	Fragilaria	Geranium, musty
	Melosira	Geranium, musty
	Meridion	Aromatic, spicy
	Stephanodiscus	Geranium, fishy
	Synedra	Earthy, musty
	Tabellaria	Geranium, fishy

Source: Adapted from Tchobanoglous and Schroeder (1985).

TABLE 8.5 Protozoa in Surface Water Storage Facilities Associated with Odors

Organism	Specific organism	Associated taste and odor
Protozoa	*Bursaria*	Irish moss, salt marsh, fishy
	Ceratium	Fish, septic (bitter), red-brown color
	Chlamydomonas	Musty, grassy (sweet)
	Cryptomonas	Candied violets (sweet)
	Dinobryon	Aromatic, violets, fishy
	Glenodinium	Fishy
	Mallomonas	Aromatic, violets, fishy
	Peridinium	Fishy, like clamshells (bitter)
	Synura	Cucumber, muskmelon, fishy (bitter)
	Uroglena	Fishy, oily, cod-liver oil

Source: Adapted from Tchobanoglous and Schroeder (1985).

TABLE 8.6   Olfactory Detection of Hydrogen Sulfide

Concentration, part per million by volume	Detectability
<0.00021	Olfactory detection threshold
0.00047	Olfactory recognition threshold
0.5 to 30	Strong odor
10 to 50	Headache, nausea, and eye, nose, and throat irritation

Source: ASCE (1989).

TABLE 8.7   Solubility of Hydrogen Sulfide

Temperature, °C	Hydrogen sulfide, mg/L
10	5160
15	4475
20	3925
25	3470
30	3090
40	2520

Source: ASCE (1989).

TABLE 8.8   Solubility of Hydrogen Sulfide as a Function of pH and Percentage of Hydrogen Sulfide

pH	Solubility, mg/L	Percentage of H_2S (%)	Percentage of HS^- (%)
6	3,840	90.1	9.9
7	7,270	47.7	52.3
7.5	15,400	22.5	77.5
8	41,800	8.3	91.7
8.5	124,000	2.8	97.2
9	390,000	0.9	99.1

Source: ASCE (1989).

Ferrous iron and manganese may also cause unpalatable water. Taste- and odor-producing compounds in groundwater are caused by human activities such as chemical dumping, sanitary landfills, mining, industrial waste disposal, and agriculture. In situ biological degradation and in situ chemical oxidation have been used to destroy organic chemicals causing odors and tastes in groundwater.

8.2.4 Turbidity

Turbidity is a measure of the light-transmitting properties of water and is comprised of suspended and colloidal material. Turbidity is expressed as nephelometric turbidity units (NTU). Turbidity is important for aesthetic and health reasons. Turbidity is associated with microorganisms. Viruses and bacteria become attached to particulate material. These attached potential pathogens can be protected by the particulate material from the bactericidal and viricidal effects of chlorine, ozone and other disinfecting agents. Particulate material and turbidity also add to the chlorine demand of natural surface water. The organic compounds attached to particulate material and the organic constituents of the turbidity in water have also been indicted as precursors of chlorinated products in water supplies.

8.2.5 Solids

The *total solids* content of water is defined as the residue remaining after evaporation of the water and drying the residue to a constant weight at 103 to 105°C (217.5 to 221°F). The organic fraction, or *volatile solids* content, is considered to be related to the loss of weight of the residue remaining after evaporating the water and after ignition of the residue at a temperature of $500 \pm 50°C$ ($932 \pm 122°C$). The organic or volatile solids will oxidize at this temperature and will be driven off as gas. The inorganic residue, or fixed solids, remains as inert ash. Solids also can be classified as settleable solids, nonfilterable (suspended) solids, and filterable solids. Settleable solids (e.g., silt and heavy organic solids) will settle under the influence of gravity. Suspended and filterable solids are also classified based on particle size and the retention of suspended solids on standard glass-fiber filters.

8.2.5.1 Suspended Solids.   Natural water samples are filtered through a standard glass-fiber filter using an appropriate filter support and vacuum apparatus to determine the *suspended solids* concentration. The standard glass-fiber filters are usually 2.2 to 12.5 cm (0.86 to 4.92 in) in diameter and may be classified by the average diameter of the pores or openings, ranging from 0.45 μm, which can retain bacterial cells, up to average pore size of 1.2 μm. Therefore, it is essential to be aware of the pore size of the glass-fiber filters or the type of filter used (e.g., Whatman grade 934AH; Gelman type A/E; Millipore type AP40; E-D Scientific Specialties grade 161; or other equivalent filters) to understand the reported data. Suspended solids may be organic or inorganic materials originating from a wide variety of sources, such as decaying vegetation algae, solids discharged by industries and municipalities, urban and agricultural runoff, and physical degradation of geological formations.

8.2.5.2 Filterable Solids.   *Filterable solids* consist of dissolved solids and colloidal particles which pass through the standard glass-fiber filters used to determine the suspended solids concentration. Filterable solids can be considered to be total dissolved solids (TDS). The colloidal fraction consists of particulate matter with an approximate size range of 0.001 to 1 μm (usually >0.45 but <1.0 μm). The dissolved solids are organic and inorganic molecules and ions, which are introduced to natural water as a result of weathering of rocks and soils and the leaching of organic compounds from decomposing plant and animal residues and detritus.

A summary of the physical characteristics of natural water in streams and rivers throughout the world is presented in Table 8.9. These data provide an insight into the wide variability of natural water quality. The quality of natural water sources used for recreational purposes, agricultural irrigation, and municipal and industrial supplies should be established in terms of the specific water-quality parameters which most affect the possible use of the water.

TABLE 8.9 Water Quality of Streams and Rivers Physical Characteristics

Water quality parameter	Typical value	Observed ranges
Physical characteristics		
Temperature, °C	Variable	0–30
Color, color units	1–10	0–500
Turbidity, NTU		0–3
Specific conductance, µS/cm at 25°C	70	40–1500
Total dissolved solids, mg/L	73–89	5–317
Suspended solids, mg/L	10–110	0.3–50,000
Total solids, mg/L		20–1000

Source: Adapted from McCutcheon (1993); Livingstone (1963); Hem (1971).

8.3 CHEMICAL CHARACTERISTICS OF WATER

The chemical characteristics of natural water are a reflection of the soils and rocks with which the water has been in contact. In addition, agricultural and urban runoff, and municipal and industrial wastewater after treatment, impact the water quality. Microbial and chemical transformations also affect the chemical characteristics of water.

8.3.1 Inorganic Minerals

Runoff from precipitation causes erosion and weathering of geological formations, rocks, and soils as the runoff travels to surface-water bodies, or infiltrates and percolates to groundwater aquifers. Inorganic chemical characteristics of groundwater and surface waters will be different. During this period of contact with rocks and soils the water dissolves inorganic minerals, which enter the natural waters. For example, runoff flowing over limestone formations will dissolve calcium carbonate ($CaCO_3$). Many the inorganic minerals may dissociate, to varying degrees, to cations (positively charged ions) and anions (negatively charged species). $CaCO_3$ may dissociate to calcium cations (Ca^{+2}) and carbonate anions (CO_3^-).

The use of water for domestic purposes also results in the addition of mineral constituents to receiving streams. In this way the inorganic quality of the receiving water is affected by the addition of return flows from municipal systems. The data presented in Table 8.10 indicate the concentrations of cations and anions introduced into receiving streams by treated municipal wastewater after a single municipal use.

8.3.1.1 Major Cations in Water. *Cations* found in natural water include calcium, magnesium, sodium, and potassium and the major anions include chloride, sulfate, carbonate, bicarbonate, fluoride, and nitrate. The concentrations of the major cations and anions exceeds 1 mg/L. The combination of anions and cations in water is controlled by various chemical reactions, such as dissolving of minerals into water, precipitation of solid minerals, and ionization of some constituents. In addition, cations and anions which may be adsorbed to organic and inorganic suspended solids may be desorbed or released, depending on the chemical condition in the water. Sorption is the attachment of cations and/or anions to a solid surface by electrochemical or thermodynamic forces.

TABLE 8.10 Typical Mineral Contribution to Water
Resulting from Domestic Use

Constituent	Increment added, mg/L	
	Range	Typical
Cations		
Ammonium (NH_4^{+4})	5–30	15
Calcium (Ca^{+2})	15–40	20
Magnesium (Mg^{+2})	10–30	10
Potassium (K^+)	7–15	10
Sodium (Na^+)	30–80*	50*
Anions		
Bicarbonate (HCO_3^-)	40–180	100
Carbonate (CO_3^{-2})	0–10	—
Chloride (Cl^-)	20–50*	40*
Nitrate (NO_3^- - N)	5–20	10
Phosphate (PO_4^{-3})	8–20	12
Sulfate (SO_4^{-2})	10–30	20
Other Data		
Hardness as $CaCO_3$	75–225	90
Alkalinity as $CaCO_3$	30–160	80
Total dissolved solids	140–480	300

* Excluding the additions from domestic water softeners.
Source: Adapted from Tchobanoglous and Schroeder (1985).

Typical concentrations of major ions in the classic "world average" river are presented in Table 8.11. These values are calculated concentrations, based on the theoretical concentrations of various constituents in waters of the rivers on each
continent, if each of the rivers was sampled and the concentrations were weighted to
reflect the relative volume of discharge at the source (Livingstone, 1963).

Calcium is the most prevalent cation in water, and second inorganic ion to bicarbonate in most surface waters. Calcium concentrations of up to 300 mg/L or higher

TABLE 8.11 Typical Concentrations of Major Ions in the Classic "World
Average" River

Constituent	Concentration, mg/L	Cations, meq/L	Anions meq/L
Cations			
Ca^{+2}	15	0.750	—
Mg^{+2}	4.1	0.342	—
Na^+	6.3	0.274	—
K^+	2.3	0.059	—
Anions			
HCO_3^-	58.4	—	0.958
SO_4^{-2}	11.2	—	0.233
Cl^-	7.8	—	0.220
NO_3^-	1	—	0.017
Sum	106.1	1.425	1.428

Source: Adapted from Livingstone (1963).

have been reported; however, calcium concentrations of 40 to 120 mg/L are more common. Calcium is found as Ca^{+2} dissolved in water.

The principal concern about calcium is related to the fact that calcium is the primary constituent of water hardness. Calcium precipitates as $CaCO_3$ in iron and steel pipes. A thin layer of $CaCO_3$ (scale) can help inhibit corrosion of the metal; however, excessive accumulation of $CaCO_3$ in boilers, hot water heaters, heat exchangers, and associated piping affects heat transfer and could lead to plugging of the piping.

Magnesium is not as abundant in rock as calcium; therefore, although magnesium salts are more soluble than calcium, less magnesium is found in groundwater and surface water. In natural waters magnesium is found predominantly as the free ion Mg^{+2}.

Other constituents in natural water in concentrations of 1 mg/L or higher include aluminum, boron, iron, manganese, phosphorus, organic carbon, ammonia, nitrite, organic nitrogen nitrates, and dissolved gases (carbon dioxide, dissolved oxygen, and hydrogen sulfide), among others. Sodium and potassium are commonly found as free ions Na^+ or K^+. The concentrations of sodium and potassium ions in natural water usually are low.

8.3.1.2 Major Anions in Water.   Biacarbonate (HCO_3^-) is the principal anion found in natural water. One source of bicarbonate ions in natural water is the dissociation of carbonic acid (H_2CO_3) which is formed when carbon dioxide (CO_2) from the atmosphere, or from animal (e.g., fish) and bacterial respiration, dissolves in water. Bicarbonate ions are important in the carbonate system, which provides a buffer capacity to natural water and is responsible in a great measure for the alkalinity of water.

In addition to bicarbonates (HCO_3^-), anions commonly found in natural water are chlorides (Cl^-), sulfates (SO_4^{-2}), and nitrates (NO_3^-). These anions are released during the dissolution and dissociation of common salt deposits in geologic formations. Nitrates (NO_3^-) are frequently found in groundwater as the result of the bacteriological oxidation of nitrogenous materials in soils. Other anions found in water include fluorides (F^-), carbonates (CO_3^{-2}), and phosphates (PO_4^{-3}).

Therefore, at one end of the spectrum natural waters can be classified as containing calcium bicarbonate ($CaCO_3$), or hard water. On the other end of the spectrum are the waters considered to be soft, which contain sodium (Na^+) as the principal cation and chloride (Cl^-) as the major anion.

Minor cations and anions found in water may be categorized as metallic ions, non-metallic anions, alkali metals, and alkaline earths. Typical concentrations of these ions and elements are observed in natural water in μg/L and are listed in Tables 8.12, 8.13, and 8.14, respectively, as well as the significance of these elements in water supplies.

8.3.2 Carbonate Equilibrium

The principal anion in natural water is bicarbonate (HCO_3^-). The bicarbonate ion is a key species in the CO_2^- carbonate-bicarbonate system. The carbonate-bicarbonate system is probably the most important chemical system in natural waters. The carbonate system provides the buffering capacity essential for maintaining the pH of natural water systems in the range required by bacteria and other aquatic species. The interrelationship between pH and the carbonate system is discussed in Sec. 8.3.3.

The carbonate system includes the following species: CO_2, H_2CO_3, HCO_3^-, CO_3^{-2}, OH^- and H^+. Frequently Ca^{+2} and $CaCO_3$ are included in the carbonate system, since Ca^{+2} is second in abundance to HCO_3^- in natural waters. The solution of carbon diox-

TABLE 8.12   Minor and Trace Elements Metallic Elements

Constituent	Concentration in natural water, µg/L	Chemistry	Significance in water supplies
Other metallic elements			
Titanium	8.6 Median river <1.5 Median public water	One of the 10 most abundant elements, but quite insoluble. Highly resistant to corrosion.	
Vanadium	<70	Occurs as anions and cations of V^{+3}, V^{+4}, and V^{+5}	May concentrate in vegetation.
Chromium	5.8 Median river 0.43 Median public water	Cr^{+3} and Cr^{+6} stable in surface waters $Cr_2O_7^{-2}$ and CrO_4^{-2} stable in groundwaters	Industrial pollutant.
Molybdenum	0.35 Median river 1.4 Median public water	Rare element. Concentrations in water reflect ambient mineral content	Accumulated by vegetation. Forage crops may become toxic.
Cobalt	ND–1.0	+2 and +3 valence states. Adsorption and complexation control concentration	Essential in nutrition in small quantities.
Nickel	10	Found generally at +2 oxidation state. Adsorbed by manganese + iron oxides	
Copper	10	Solubility limited to 64 µg/L at pH = 7.0 and 6.4 µg/L at pH = 8.0 by cupric oxide or hydroxy carbonate minerals	Utilized in water treatment and metal fabrication. Used to inhibit algae growth in reservoirs. Essential for nutrition of flora and fauna.
Silver	0.1–0.3	Solubility predicts concentration of 0.1–10 µg/L in dilute aerated water.	Has been used as disinfectant.
Gold	ND–trace	Very low solubility.	
Zinc	10	Probability controlled by availability. Natural water does not contain levels dictated by solubility.	Widely used in industry. Sources: wastes, galvanized pipes, cooling water treatment, etc.
Cadmium	ND–10	Relatively rare. Used in electroplating.	Toxic. Presence may indicate industrial contamination.
Mercury	ND–<10	Relatively rare in rocks or water.	Highly toxic. Sources: pollution from mining, industry, or metallurgical works.
Lead	1–10	Solubility limited by cerussite, a lead carbonate or $PbSO_4$, in oxidized systems, to ~2 µg/L	Older plumbing systems contain lead, which may dissolve at low pH.

Source: Data from NAS (1977); Livingstone (1963); Turekian (1971); and Hem (1971).

TABLE 8.13   Minor and Trace Elements Nonmetallic Anions

Constituent	Concentration in natural water, µg/L	Chemistry	Significance in water supplies
Nonmetals (anions in natural waters)			
Arsenic	0–1000	Arsenate spp. $H_2AsO_4^-$ and $HAsO_4^2$ equilibrium forms in natural waters. $HAsO_2$ (aq) may be found in reducing environment. Solution controlled by metal arsenates, e.g., Cu^{+2}. Adsorption also important.	Used in industry in some herbicides and pesticides. Lethal in animals above 20 mg/lb. Long-term ingestion of 0.21 mg/L reported to be poisonous.
Selenium	0.2	Selenite form in natural waters SeO_3^{-2}. Low solubility.	Taken up by vegetation.
Bromine	20	Present in natural waters as Br^-. Present in rainwater.	
Iodine	0.2–2	Cycling controlled through biochemical process.	Essential nutrient in higher animals. Has been used to seed clouds.

Source: Data from NAS (1977); Livingstone (1963); Turekian (1971); and Hem (1971).

TABLE 8.14   Minor and Trace Elements Alkali Metals and Alkaline Earths

Constituent	Concentration in natural water µg/L	Chemistry	Significance in water supplies
Alkali metals			
Lithium	0.001–0.3		Not toxic to plants, at concentration in irrigation waters.
Rubidium	0.0015		
Cesium	0.05–0.02*		
Alkaline earths			
Beryllium	0.001–1	Found as a constituent of beryl. Typically is particulate.	Highly toxic, but occurs at very low concentration.
Strontium	0.6 Median river water 0.11 Median public water	Can replace Ca and Mg in igneous rocks. Concentration in natural water is less than solubility.	See radionuclides Sec. 8.3.6.
Barium	0.043 Median public water	Concentration controlled by $BaSO_4$ solubility $K_{sr} \sim 10^{-10}$	Ingestion of soluble barium salts can be fatal. Normal water concentrations have no effect.

* Observed in six analyses of rivers in Japan.
Source: Data from NAS (1977); Livingstone (1963); Turekian (1971); and Hem (1971).

ide (CO_2) into natural water causes the formation of carbonic acid (H_2CO_3). The H_2CO_3 dissociates to bicarbonate (HCO_3^-) and hydrogen (H^+) ions. In turn HCO_3^- can dissociate and produce carbonate (CO_3^{-2}) and hydrogen (H^+) ions. The hydrogen ion (H^+) concentration in water controls the pH of the solution. The pH of water is defined as the negative logarithm of the [H^+], where [H^+] is the hydrogen ion concentration expressed in moles per liter (mol/L).

$$pH = -\log [H^+] = \log \frac{1}{[H^+]} \qquad (8.1)$$

The pH is the negative power to which 10 must be raised to equal the hydrogen ion concentration, or

$$[H^+] = 10^{-pH} \qquad (8.2)$$

In a neutral solution [H^+] is 10^{-7}, or pH = 7. At greater hydrogen ion concentrations the pH is lower, and for greater hydroxide ion concentrations the pH increases. The pH range is from 0 (extremely acidic) to 14 (extremely basic).

The [H^+] affects the pH which is defined as:

$$pH = \log \frac{1}{[H^+]} \qquad (8.3)$$

Therefore, the pH of the water controls which species is predominant. Water molecules HOH (commonly written (H_2O) dissociate, or ionize to H^+ and OH^- ions.

$$2\, H_2O \underset{\leftarrow}{\overset{\rightarrow}{}} H_3O^+ + OH^- \qquad (8.4)$$

or written in a simpler form

$$H_2O \underset{\leftarrow}{\overset{\rightarrow}{}} H^+ + OH^- \qquad (8.5)$$

The product of [H^+] and [OH^-], in mol/L, is constant.

$$[H^+]\,[OH^-] = K_w = 1 \times 10^{-14} \qquad (8.6)$$

where K_w is the ion-product constant of water. If the hydrogen ion concentration is 10^{-4} mol/L, the [OH^-] concentration must be $10^{-14}/10^{-4} = 10^{-10}$ mol/L. Since 10^{-10} is smaller than 10^{-4}, the solution is acidic. A large amount of hydrogen ions (H^+) in water makes the water acid, and a lack of hydrogen ions makes the water basic. A basic solution has predominance of hydroxide ions (OH^-). The dissociation reactions of carbonic acid (H_2CO_3) and of the bicarbonate ion (HCO_3^-) can be written as the following equations:

$$H_2CO_3 \underset{\leftarrow}{\overset{\rightarrow}{}} HCO_3^- + H^+$$

$$k_1 = 10^{-6.35} \text{ @ } 25°C \qquad (8.7)$$

$$HCO_3^- \underset{\leftarrow}{\overset{\rightarrow}{}} CO_3^{-2} + H^+$$

$$k_2 = 10^{-10.3} \text{ @ } 25°C \qquad (8.8)$$

These equations can be used to define the relative distribution of carbonate species as a function of pH. Changes in pH can have drastic effects on the species present in the carbonate system.

The total carbonate species in water include carbonic acid $[H_2CO_3]$, bicarbonate ions $[HCO_3^-]$, and carbonate ions $[CO_3^{-2}]$. Therefore, the fraction of carbonic acid $[H_2CO_3]$ present may be expressed as α_0 and written as the fraction:

$$\alpha_0 = \frac{[H_2CO_3]}{[H_2CO_3] + [HCO_3^-] + [CO_3^{-2}]} \tag{8.9}$$

Similarly, the fraction of bicarbonate ions $[HCO_3^-]$ and carbonate ions $[CO_3^{-2}]$ among the total carbonate species may be expressed as α_1 and α_2, respectively and written as the following fractions:

$$\alpha_1 = \frac{[HCO_3^-]}{[H_2CO_3] + [HCO_3^-] + [CO_3^{-2}]} \tag{8.10}$$

$$\alpha_2 = \frac{[CO_3^{-2}]}{[H_2CO_3] + [HCO_3^-] + [CO_3^{-2}]} \tag{8.11}$$

The reciprocals of these expressions for α_0 may be written as $1/\alpha_0$

$$\frac{1}{\alpha_0} = 1 + \frac{[HCO_3^-]}{[H_2CO_3]} + \frac{[CO_3^{-2}]}{[H_2CO_3]} \tag{8.12}$$

This expression may be expressed in terms of the dissociation constants k_1 and k_2, respectively, for the dissociation of carbonic acid to bicarbonate ions and for the dissociation of bicarbonate to carbonate ions. These equilibrium constants k_1 and k_2 may be expressed, respectively, in terms of the ratio of bicarbonate ions to carbonic acid and the ratio of carbonate to bicarbonate ion species as:

$$k_1 = \frac{[HCO_3^-][H^+]}{[H_2CO_3]} \quad \text{and} \quad \frac{k_1}{[H^+]} = \frac{[HCO_3^-]}{[H_2CO_3]} \tag{8.13}$$

and

$$k_2 = \frac{[CO_3^{-2}][H^+]}{[HCO_3^-]} \quad \text{and} \quad \frac{k_2}{[H^+]} = \frac{[CO_3^{-2}]}{[HCO_3^-]} \tag{8.14}$$

The ratio of carbonate ions to carbonic acid may be written by combining these expressions as

$$\frac{k_1 k_2}{[H^+]^2} = \frac{[CO_3^{-2}]}{[H_2CO_3]} \tag{8.15}$$

Substituting these ratios into the expression of the reciprocal of $1/\alpha_0$ above yields

$$\frac{1}{\alpha_0} = 1 + \frac{K_1}{[H^+]} + \frac{K_1 K_2}{[H^+]^2} \tag{8.16}$$

and solving for α_0 yields the following

$$\alpha_0 = \frac{[H^+]^2}{[H^+]^2 + [H^+]K_1 + K_1 K_2} \tag{8.17}$$

Similarly, one can develop the following expressions:

$$\alpha_1 = \frac{K_1[H^+]}{[H^+]^2 + K_1[H^+] + K_1K_2} \tag{8.18}$$

$$\alpha_2 = \frac{K_1K_2}{[H^+]^2 + K_1[H^+] + K_1K_2} \tag{8.19}$$

These relationships clearly indicate the significance of pH on the distribution of carbonate species and the concentration of $[OH^-]$ in natural water. The effects of pH on α_0, α_1, and α_2 are illustrated in Table 8.15, and the relationship among pH and fraction of carbonate species in water expressed in terms of α_0, α_1, and α_2 is shown in Figure 8.1.

8.3.3   pH and Alkalinity

Water quality data reported for streams and rivers throughout the world, expressed in terms of pH, carbon dioxide, alkalinity, and acidity, are presented in Table 8.16. A review of the carbonate system illustrates the effect of changes in pH on the car-

TABLE 8.15   Relationship of pH and α_0, α_1, and α_2

pH	α_0	α_1	α_2
1	1.0000	0.0000	0.0000
1.5	1.0000	0.0000	0.0000
2	1.0000	0.0000	0.0000
2.5	0.9999	0.0001	0.0000
3	0.9996	0.0004	0.0000
3.5	0.9986	0.0014	0.0000
4	0.9955	0.0045	0.0000
4.5	0.9861	0.0139	0.0000
5	0.9572	0.0428	0.0000
5.5	0.8762	0.1238	0.0000
6	0.6911	0.3089	0.0000
6.5	0.4143	0.5856	0.0001
7	0.1827	0.8169	0.0004
7.5	0.0660	0.9326	0.0014
8	0.0218	0.9737	0.0045
8.5	0.0087	0.9798	0.0115
9	0.0021	0.9533	0.0446
9.5	0.0005	0.7712	0.2283
10	0.0002	0.6811	0.3187
10.5	0.0000	0.4032	0.5968
11	0.0000	0.1761	0.8239
11.5	0.0000	0.0633	0.9367
12	0.0000	0.0209	0.9791
12.5	0.0000	0.0067	0.9933
13	0.0000	0.0021	0.9979
13.5	0.0000	0.0006	0.9994
14	0.0000	0.0002	0.9998

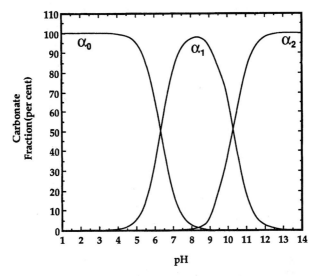

FIGURE 8.1 Relationship among pH and α_0, α_1, and α_2.

bonate species present in the water. pH is also an extremely important variable because it is the controlling factor determining the solubility of most metals, and also because most microorganims can only survive within a narrow pH range.

pH is important in natural waters and in water treatment. Aquatic organisms are sensitive to pH changes and require a pH of 6 to 9.

In water treatment proper chemical treatment, including disinfection, requires pH control. Water treatment also frequently requires the addition of chemicals which can change the pH; therefore, measurement of the pH alone is not sufficient for process control—the pH must be complemented with some measure of the capacity of the water to resist changes in pH. A measure of this capacity to resist changes in pH is the alkalinity.

Carbonates and bicarbonates are responsible primarily for buffering, and for the acid-neutralizing capacity of water. *Alkalinity* is defined as the capacity of natural water to neutralize acid added to it. *Total alkalinity* is the amount of acid required to reach a specific pH, viz. pH = 4.3 to 4.8.

Total alkalinity can be approximated by alkalinity $\approx [HCO_3^-] + 2[CO_3^{-2}] + [OH^-] - [H^+]$.

TABLE 8.16 Water Quality of Streams and Rivers pH, Carbon Dioxide, Alkalinity, Acidity

Water quality parameter	Typical value, mg/L	Observed ranges, mg/L
pH (pH units)	4.5–8.5	1–9
Carbon dioxide (CO_2)	0–5	0–50
Alkalinity (as $CaCO_3$)	150	5–250
Acidity (as $CaCO_3$)		2.8–23.3

Source: Adapted from McCutcheon (1993); Livingstone (1963); Hem (1971).

Total alkalinity in natural waters includes the following components:

$$\text{Hydroxide alkalinity} = [OH^-]$$

$$\text{Bicarbonate alkalinity} = [HCO_3^-]$$

$$\text{Carbonate alkalinity} = [CO_3^{-2}]$$

Alkalinity is the acid-neutralizing capacity of a sample of water. Alkalinity, expressed in units of eq/L or mg/L as $CaCO_3$, is a measure of the amount of acid (in equivalents) required to reduce the pH of one liter of the water to pH $\approx$ 4.3 (more properly to the H_2CO_3 endpoint, which is pH $\approx$ 4.3).

The carbonate system (H_2CO_3–HCO_3^-–CO_3^{-2}) usually is the only acid/base system present which transfers protons in the pH range 4 to 10 (besides water itself), alkalinity can usually be thought of as the amount of acid it takes to convert the bicarbonate (HCO_3^-) and carbonate (CO_3^{-2}) to the carbonic acid form (H_2CO_3). The simultaneous transfer of $[H^+]$ in water must also be taken into account:

$$\text{Alkalinity} = 2[CO_3^{-2}] + [HCO_3^-] + [OH^-] - [H^+] \tag{8.20}$$

$$= 2\,\alpha_2\,C_T + \alpha_1\,C_T + [OH^-] - [H^+] \tag{8.21}$$

where

$$C_T = [H_2CO_3] + [HCO_3^-] + [CO_3^{-2}] \tag{8.22}$$

$$= \text{total carbonate}$$

$$\text{Total alkalinity} = (2\alpha_2 + \alpha_1)C_T + \frac{K_w}{[H^+]} - [H^+] \tag{8.23}$$

where

$$K_w = 10^{-14} = [H^+][OH^-]$$

pH and alkalinity of a water which can be measured in a laboratory are the only information required to calculate the total alkalinity (C_T). The total alkalinity includes three components:

$$\text{Hydroxide alkalinity} = [OH^-]$$

$$\text{Bicarbonate alkalinity} = [HCO_3^-]$$

$$\text{Carbonate alkalinity} = 2[CO_3^{-2}]$$

If the pH and total alkalinity are measured, the concentration of the various components of alkalinity can be calculated. The total alkalinity C_T may be calculated using the values of α_1 and α_2 determined for the pH of the water. These values can be used to calculate the:

$$\text{Hydroxide alkalinity} = \frac{K_w}{[H^+]} \tag{8.24}$$

$$\text{Bicarbonate alkalinity} = \alpha_1\,C_T \tag{8.25}$$

and

$$\text{Carbonate alkalinity} = 2\alpha_2\,C_T \tag{8.26}$$

A simplified short cut to approximate the species of alkalinity in a water sample requires determination of two points on the alkalinity titration curve (Fig. 8.2). The first point is the *phenolphthalein alkalinity*, i.e., the amount of strong acid (in eq/L) required to change color of the water from pink to clear (colorless) when a small amount of a phenolphthalein reagent is put into the water sample. This color change occurs at approximately pH = 8.3. Continuing the titration to pH = 4.3, which is the H_2CO_3 endpoint, yields the total alkalinity.

Two transformations occur while reaching the phenolphthalein endpoint P (first endpoint), at pH = 8.3, namely:

$$CO_3^{-2} \rightarrow HCO_3^-$$

and

$$OH^- \rightarrow H_2O$$

at pH = 8.3 the $[OH^-] = 10^{-5.7}$ and $[OH^-] \approx 0$.

One transformation occurs going from the first to the second endpoint (methyl orange endpoint), pH ≈ 4.3, total alkalinity T, namely:

$$HCO_3^- \rightarrow H_2CO_3$$

The HCO_3^- represents the HCO_3^- originally present plus the HCO_3^- formed by transfer from CO_3^{-2}. The values of each of the three forms of alkalinity can be determined using the relative values of the phenolphthalein alkalinity P and the total alkalinity T, expressed as either eq/L or mg/L $CaCO_3$ as presented in Table 8.17.

One simplification assumes that $[OH^-]$ and $[HCO_3^-]$ cannot exist together in the same sample, which is not totally correct. Therefore, five possible situations can occur:

1. Hydroxide $[OH^-]$ only.

   High pH, pH > 10.

   Titration usually essentially is complete at the phenolphthalein end point.

   $[OH^-]$ alkalinity = phenolphthalein alkalinity.

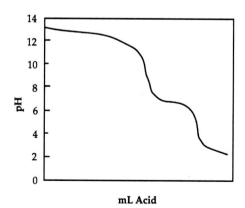

FIGURE 8.2 Typical titration curve illustrating the acid requirements to neutralize pH during alkalinity titration.

TABLE 8.17 Relationship Among Titration Results and Hydroxyl, Bicarbonate and Carbonate Alkalinity

Titration results	OH^- alkalinity	HCO_3^- alkalinity	CO_3^{-2} alkalinity
$P = 0$	0	T	0
$T = 2P$	0	0	T
$T > 2P$	0	$T - 2P$	$2P$
$T = P$	T	0	0

Source: Sawyer, McCarty, and Parkin (1994)

2. Carbonate $[CO_3^{-2}]$ only.

   pH > 8.5.

   Phenolphthalein end point equals exactly one-half the total titration.

   $[CO_3^{-2}]$ alkalinity = total alkalinity.

3. Hydroxide $[OH^-]$ plus carbonate $[CO_3^{-2}]$.

   High pH, usually pH > 10.

   Titration from phenolphthalein end point to the methyl orange end point represents one-half of the carbonate $[CO_3^{-2}]$ alkalinity.

   $[CO_3^{-2}]$ alkalinity = 2 (titration from phenolphthalein end point to the methyl orange end point).

   $[OH^-]$ alkalinity = total alkalinity – $[CO_3^{-2}]$ alkalinity.

4. Carbonate $[CO_3^{-2}]$ plus bicarbonate $[HCO_3^-]$.

   pH > 8.3, and pH < 11.

   Titration to phenolphthalein end point represents one-half of the $[CO_3^{-2}]$ alkalinity.

   $[CO_3^{-2}]$ alkalinity = 2(titration to phenolphthalein end point) $[HCO_3^-]$ alkalinity = Total alkalinity – $[CO_3^{-2}]$ alkalinity.

5. Bicarbonate $[HCO_3^-]$ only.

   pH ≦ 8.3

   $[HCO_3^-]$ alkalinity = total alkalinity.

8.3.4 Acidity

Acidity is the "quantitative capacity of aqueous media to react with hydroxyl ions." Water, mineral acids, carbonic acids, and others contribute to acidity. Titration with a strong base (e.g., NaOH) to defined end points (pH = 4.3 and pH = 8.3; methyl orange and phenolphthalein) is used to determine acidity. Acidity indicates the corrosiveness of acidic water on steel, concrete, and other materials. Mineral acidity is caused by metal salts which ionize in water. Sources of mineral acidity include mine drainage, leachate from tailings or ore dumps, fly ash, and some organic acids.

8.3.5 Indicators of Water Quality (Inorganics)

Water quality characteristics frequently are used to indicate trends in quality changes. Some of the inorganic parameters include hardness, total dissolved solids, conductivity, and sodium adsorption ratio. The interrelationship among these parameters can indicate the addition of salts to natural water sources.

8.3.5.1 Hardness. *Hardness* (expressed as mg/L $CaCO_3$) observed for streams and rivers throughout the world ranges between 1 to 1000 mg/L as $CaCO_3$. Typical concentrations are 47 to 74 mg/L $CaCO_3$. Hardness reflects the composite measure of the polyvalent cation concentration in water. Calcium and magnesium ions (Ca^{+2} and Mg^{+2}) are the primary constituents of hardness. Hardness expressed as mg/L $CaCO_3$ is used to classify waters from "soft" to "very hard." This classification is summarized in Table 8.18.

Hardness is an indicator to industry of potential precipitation of calcium carbonates in cooling towers or boilers, interference with soaps and dyes in cleaning and textile industries, and with emulsifiers in photographic development.

8.3.5.2 Total Dissolved Solids. *Total dissolved solids* (TDS) are a measure of salt dissolved in a water sample after removal of suspended solids. TDS is the residue remaining after evaporation of the water. TDS concentrations in water in arid regions or in water subjected to agricultural runoff can be much higher.

For example, the TDS concentration in southern California is 700 to 800 mg/L, and the TDS concentration exceeds 1000 mg/L as the river discharges into the Baja California. The EPA recommends a maximum TDS concentration in drinking water supplies of 500 mg/L. The average increase in TDS concentration for water issued for domestic purposes is approximately 300 mg/L.

The TDS load carried in streams throughout the world has been estimated by Livingstone (1963) and Alekin and Brazhnikova (1961). Data reported by Livingstone (1963) are presented in Table 8.19. The estimates reported by Alekin and Brazhnikova (1961) indicate a higher water flux (36.45×10^{12} m^3/yr) and lower TDS concentration (88 mg/L) resulting in a lower TDS flux (3.488×10^{12} kg/yr).

8.3.5.3 Conductivity. The concentration of TDS is related to electrical conductivity (EC; μmhos/cm) or specific conductance. The *conductivity* measures the capacity of water to transmit electrical current. The conductivity increases as the concentration of TDS increases. The conductivity is a relative term, and the relationship between the TDS concentration and conductivity (μmhos/cm) is unique to a given water sample and in a specific TDS concentration range. TDS and conductivity affect the water sample and the solubility of slightly soluble compounds and gases in water (e.g., $CaCO_3$ and O_2). In general, the corrosivity of the water increases as TDS and EC increase, assuming other variables are kept constant.

8.3.5.4 Sodium Adsorption Ratio. The *sodium adsorption ratio* (SAR) is used to evaluate the hazard in irrigation waters caused by sodium. The SAR relates the

TABLE 8.18 Relationship of Hardness Concentration and Classification of Natural Waters

Hardness as mg/L $CaCO_3$	Classification
0–60	Soft
61–120	Moderately hard
121–180	Hard
>180	Very hard

TABLE 8.19 Estimates of the Total Runoff of the World and the Flux of Dissolved Solids

	Livingstone (1963)			Alekin and Brazhnikova (1961)		
		Dissolved solids			Dissolved solids	
	Water flux, 10^{12} m^3/yr	Concentration, mg/L	Flux, 10^{12} kg/yr	Water flux, 10^{12} m^3/yr	Concentration including organics, mg/L	Flux, 10^{12} kg/yr
North America	4.55	142	0.646	6.43	89	0.572
Europe	2.50	182	0.455	3.00	101	0.303
Asia	11.05	142	1.570	12.25	111	1.360
Africa	5.90	121	0.715	6.05	96	0.581
Australia	0.32	59	0.019	0.61	176	0.107
South America	8.01	69	0.552	8.10	71	0.575
Total	32.33	120	3.957	36.45	88	3.498

concentration of sodium ions to the concentration of magnesium and calcium ions. The *SAR* is defined as:

$$\text{SAR} = \frac{\text{Na}^+}{\sqrt{(\text{Ca}^{+2} + \text{Mg}^{+2})/2}} \quad \text{in meq/L}$$

The proper ratio of sodium ions to calcium and magnesium ions in irrigation water results in irrigated soil which is granular in texture, easily worked, and permeable. With increasing proportions of sodium as the SAR increases, soil tends to become less permeable and more difficult to work.

8.3.6 Radionuclides

Radionuclides in water are classified according to the type of energy released: (1) alpha radiation (positively charged helium nuclei), (2) beta radiation (electrons), or (3) gamma radiation (electromagnetic energy). Each type of radiation has different health effects. Alpha particles travel at velocities up to 10^7 m/sec. When ingested, the relatively massive alpha particles can be very damaging to body tissue. Beta particles travel at about the speed of light, penetrate to greater depth because of the smaller mass, and create less damage. Gamma radiation penetrates deeply, but has limited effects at low levels.

The units of radiation measurement are curies (Ci) or rems. [Ci = 3.7 × 10^{10} nuclear transformations per second; picocurie (pCi) = 10^{-12}Ci]. A rem is the radiation dose producing the same biological effect (rem = absorbed dose × quality factor).

General properties of some of the more common naturally occurring or synthetically produced radiation are found in elements presented in Table 8.20. Natural radiation is found in elements in the earth's crust [potassium-40 (^{40}K)]. Another source of natural radiation results from cosmic ray bombardment in the atmosphere [tritium (^{3}H) and carbon-14 (^{14}C)]. Other high-atomic-weight, naturally occurring isotopes found in natural water include uranium-238, thorium-232, uranium-235, and breakdown products, radium-226 and radium-228.

Nuclear fission from weapons testing, radiopharmaceuticals, and nuclear fuel processing and use are sources of radiation caused by human activity. Contribution from weapons testing has decreased since 1963, but radionuclides used in medicine

TABLE 8.20 Radioactivity in Water

Nuclide	Half-life	Average concentration in natural water	Significance in water
Naturally occurring Uranium-238 (α)*	4.5×10^9 yr	0.02–200 µg/L[†]	Probably produces greatest amount of radioactivity in water (USGS).
Radium-226 (α)	1.6×10^3 yr	0.01–0.1 pCi/L	Susceptible to removal via flocculation or softening.
Radon-222 (α)	3.82 days	1 pCi/L surface water >1000 pCi/L aquifer	Gas formed by decay of radium-226.
Uranium-235 (α)	7×10^8 yr	0.02–200 µg/L[†]	
Radium-223 (α)	11.4 days		
Thorium-232 (α)	NA[‡]	Extensively measured	
Radium-223 (β)[§]	5.8 yr	0.005–0.1 pCi/L	Decay yields alpha-emitting daughters.
Radium-224 (α)	3.6 days		
Radon-220 (α)	<1 min		
Tritium (^{3}H) (β)	12.3 yr	10–25 pCi/L	Cosmic ray bombardment of rain.
Carbon-14	5730 yr	6 pCi/g C[§]	Cosmic ray bombardment of CO_2.
Potassium-40 (β)	1.3×10^9 yr	<4 pCi/L	Constant 0.0118% of total potassium.
Rubidium-87 (β)	5×10^{11} yr		
Polonium-210 (α)	138 days		
Nuclear fission products			Weapons testing.
Strontium-90 (β)	30 yr	0.5 pCi/L	
Cesium-137 (β)	30 yr	0.1 pCi/L	
Iodine-131	8.1 days	NA	
Tritium (H^3)	12.5 yr	10–25 pCi/L	
Carbon-14	5730 yr	6 pCi/L	
Pharmaceuticals			
Iodine-125	60 days	Variable periodic effluent releases	Could affect thyroid gland.
Iodine-131	8.1 days		
Technetium-99m	6.0 h		
Nuclear fuel			
Tritium (H^3)	12.5 yr	10–25 pCi/L	
Plutonium-239	24,360 yr		Released at ~1 mCi/yr from liquid storage sites.
Iodine-129	1.7×10^7 yr	<0.01 pCi/L	

* Alpha-emitting nuclide (Radium-226 and radium-228) Drinking water contributes 0.01–0.05 pCi/day or 2% to total daily intake of alpha radiation.
[†] Represents the concentration of all uranium in water, primarily ^{238}U.
[‡] Beta emitting nuclide, consisting of electrons or positrons.
[§] That is, with 1 mg C/L would have ~0.006 pCi ^{14}C.
Source: NAS (1977) and Hem (1971).

and energy production have increased; therefore, source control has kept contributions to waters relatively constant, although production has increased.

The primary importance of radionuclides in water is protection of human health and safety. ^{3}H and ^{14}C are used as tracers and age determinants. The data presented in Table 8.21 indicate that the body dose that accrues from drinking water compared to natural background radiation is low (estimated to be less than 0.24 percent by NAS); however, EPA policy assumes that potential harm exists from any level of radiation.

8.3.7 Organic Materials

Organic compounds in water also affect water quality. Organic chemicals cause disagreeable tastes and odors in drinking water. Organic chemicals are made up of car-

TABLE 8.21 Annual Radiation Dose by Source

Source	Annual dose (mrem)
Cosmic rays	44
Terrestrial radiation	
External to the body	80
Internal to the body	18
Total natural sources	102
Hypothetical water supply in the United States	0.244

Source: Adapted from NAS (1977).

bon (C) and hydrogen (H), as well as nitrogen (N) and oxygen (O). Organic compounds are derived from living organisms as well as from industrial sources. A wide assortment of organic compounds are produced in the chemical and petrochemical industries. Organic compounds also may contain sulfur (S), phosphorus (P), fluorine (F), chlorine (Cl), bromine (Br), and iodine (I). The strong carbon–carbon bonds in organic compounds distinguish these materials from inorganic compounds.

In today's industrialized society, water purveyors must be aware that water can be toxic to the consumers. Vinyl chloride, benzene, and other organic contaminants are known carcinogenic agents, while chloroform is a cancer-suspect agent. These anthropogenic materials generally are present in water supplies in low concentration. However, the question that needs to be answered is: At what level do trace organic contaminants exert an impact on human health?

8.3.7.1 Natural Organics. Naturally occurring organic material in water reflects the degradation of organic materials (carbohydrates and proteins) to humic end products, which accumulate in soils until runoff carries them into water. The organic material in soils originates with plant and animal degradation products. These degradation products condense and polymerize into fulvic and humic acids, to kerogen, and finally coal. Chemical and/or microbial processes cause the transformations by first attacking functional groups and aliphatic side chains. Condensation and polymerization of various reactive groups results in larger, more aromatic molecules, which decrease in solubility until kerogen or humin is produced. The end products are not soluble in acid or alkali and are resistant to biodegradation and chemical reaction.

8.3.7.2 Man-made Organics. Synthetic organic compounds include a broad variety of aliphatic and aromatic compounds. Many manufactured organic compounds may be found at very low concentrations in natural water. Isolation, identification, and evaluation of health effects of these synthetic organics at low concentration are lacking. Data on the sources, concentrations, and health effects of 129 synthetic organics found or likely to be found in drinking water supplies have been compiled and analyzed (NAS, 1977). These compounds include pesticides, industrial chemicals, plasticizers, polychlorinated biphenyls (PCBs), polynuclear aromatics (PNAs), and halogenated methane and ethane derivatives.

8.3.7.3 Measurement of Organics in Water—Organic Carbon. Organics in water can be expressed in terms of *total organic carbon* (TOC), especially at small concentrations. The TOC is the difference between total carbon (TC) and inorganic carbon (IC). Analytically, the carbonaceous compounds are thermally oxidized to carbon dioxide in the presence of a catalyst. The concentration of carbon dioxide in the carrier gas is quantitatively measured, using an infrared analyzer.

Typical concentrations of organic matter observed in natural water in streams and rivers throughout the world are presented in Table 8.22. Some discrepancy among the observed range of data for dissolved organic carbon (DOC) and total organic carbon (TOC) may be attributable to the various sources used to compile the data. In some cases TOC was reported but no DOC data was included. In other instances only DOC data were reported.

8.3.8 Biological Oxygen Demand

Biochemical oxygen demand (BOD), the most widely used parameter, is a measure of the amount of oxygen used by indigenous microbial population in water in response to the introduction of degradable organic material. The 5-day BOD (BOD_5) is most widely used. Typical concentrations of BOD_5 reported for streams and rivers throughout the world are <2 to 15 mg/L and the observed range is <2 to 65 mg/L. As a reference, the effluent limitations established by the U.S. EPA for biologically treated municipal wastewater is 30 mg/L, and many states require municipal effluents to contain an average BOD_5 of 20 mg/L, with concentrations as low as 5 mg/L in water-quality sensitive water segments.

The BOD_5 of natural waters is related to the dissolved oxygen concentration, which is measured at zero time and after 5 days of incubation at 20°C. The difference is the dissolved oxygen used by the microorganisms in the biochemical oxidation of organic matter. The BOD_5 can be calculated as $BOD_5 = D_0 - D_1$ in which BOD_5 is in mg/L and D_0 and D_1 are the dissolved oxygen concentrations (mg/L) at time 0 and 5 days, respectively. If the BOD_5 of natural water exceeds the dissolved oxygen concentration at saturation (≈ 9 mg/L), the sample of natural water must be diluted with aerated dilution water and the BOD_5 is calculated as

$$BOD_5 = \frac{D_0 - D_1(100)}{P\,(\%)} \tag{8.27}$$

in which P is the percent of sample (natural water) used. The BOD_5 test can also be used to characterize municipal and industrial wastewaters. However, in these tests, in addition to dilution water, acclimated seed organisms, nutrients, and the presence or absence of toxic substances must be considered. In standard BOD_5 analyses the samples are incubated at 20°C. However, if the samples are incubated at some other temperature, the rate coefficient can be adjusted using the relationship

$$K_5 - K_{20}\theta^{(T-20)} \tag{8.28}$$

TABLE 8.22   Water Quality of Streams and Rivers Physical Characteristics

Water quality parameter	Typical value, mg/L	Observed ranges, mg/L
Inorganic carbon (IC) (C)	50	5–250
Total organic carbon (TOC) (C)	1–10	0.01–40
Dissolved organic carbon (DOC) (C)	1–6	0.3–32
Volatile organic carbon (VOC) (C)	0.05	
Total organic matter	2–20	0.02–80

Source: Adapted from McCutcheon (1993); Livingstone (1963); Hem (1971).

where $\theta = 1.047$. The mathematical expression for BOD may be developed. The BOD remaining at time t is

$$L_t = (10^{-kt}) \tag{8.29}$$

and y is the amount of BOD that has been exerted (satisfied) at time t

$$y = L - L_t \tag{8.30}$$

where L = ultimate BOD (mg/L). Therefore

$$y = L\ (1 - 10^{-kt}) \tag{8.31}$$

t is incubation time in days.

Nitrification of ammonia in the BOD test can yield erroneous results. Ammonia can be produced biologically during the hydrolysis of proteins. The biooxidation of ammonia results in the formation of partially oxidized nitrogen species, nitrite, which subsequently can be biologically oxidized to nitrate. These biochemical reactions are illustrated in the generalized reactions as follows:

$$NH_3 + \frac{3}{2}O_2 \xrightarrow[Nitrobacter]{nitrite\text{-}forming\ bacteria} HNO_2 \tag{8.32}$$

$$HNO_2 + \frac{1}{2}O_2 \xrightarrow[Nitrobacter]{nitrate\text{-}forming\ bacteria} HNO_3 + H_2O \tag{8.33}$$

$$NH_3 + 2O_2 \rightarrow HNO_3 + H_2O \tag{8.34}$$

The oxygen required to oxidize ammonia completely to the nitrate form may be considered to be the nitrogenous oxygen demand (NOD). Theoretically the NOD of ammonia is 3.765 g O_2 per g NH_3 oxidized to HNO_3. However, the NOD of ammonia-nitrogen (NH_3-N) is 4.57 g O_2 per g N oxidized.

Carbonaceous biochemical oxygen demand (CBOD) may be measured by inhibiting nitrification in the BOD test using chemical inhibitors such as methylene blue, thiourea and allylthiourea, 2-chlor-6 (trichloromethyl) pyridine, or proprietary products. Detailed procedures for the BOD test and for suppression of the nitrification reaction in the BOD test are listed in the latest edition of *Standard Methods* (1995).

8.3.9 Chemical Oxygen Demand

The *chemical oxygen demand* (COD) test of natural water yields the oxygen equivalent of the organic matter that can be oxidized by a strong chemical oxidizing agent in an acidic medium. The COD observed in natural streams and rivers is <2 mg/L to 100 mg/L. Potassium dichromate is the oxidizing chemical used in the COD test in North America; however, potassium permanganate is the oxidizing compound in other parts of the world. Silver sulfate is added as a catalyst and to minimize the interference of chloride on the COD test. Mercuric sulfate is also added to inhibit interferences of metals on the oxidation of organic compounds. The reaction of the dichromate with organic matter is presented here in a general way:

$$\text{Organic matter } (C_aH_bO_c) + Cr_2O_7^{-2} + H^- \xrightarrow[heat]{catalyst} 2Cr^{+3} + CO_2 + H_2O \tag{8.35}$$

8.3.10 Dissolved Gases

8.3.10.1 Gas Transfer. The principal transfer of gas in natural water is the transfer of oxygen from the atmosphere to the water. However, gas transfer also is used to strip hydrogen sulfide (H_2S), ammonia (NH_3) and volatile organic compounds from water. In both processes, material is transferred from one bulk phase to another across a gas–liquid interface. For example, oxygen is transferred from the bulk gaseous phase (atmosphere) across the gas–liquid interface into the bulk liquid phase (water). In the case of stripping *volatile organic compounds* (VOC) from liquid, the VOC is transferred from the bulk liquid phase (water) across the liquid–gas interface into the bulk gaseous phase (atmosphere). Mass transfer occurs in the direction of decreasing concentration.

The rate of transfer across the air–water interface depends on the equilibrium concentration and the turbulence at the interface. The dissolved substance must be transported from the bulk fluid to the gas–liquid interface, transferred across the interface, and finally transferred into the bulk of the new phase.

8.3.10.2 Solubility of Gases. The equilibrium of each phase, concentration of gases or volatile compounds dissolved in water depends on the temperature, the type of gas or volatile compound, and the partial pressure of the gas or volatile compound adjacent to the water. The relationship between the partial pressure of the gas in the atmosphere above the water and the concentration of the gas or volatile compound in the water is described by *Henry's law*:

$$P_g = Hx_g \qquad (8.36)$$

where

P_g = partial pressure of gas, atm

H = Henry's law constant

x_g = equilibrium mole fraction of dissolved gas

$$= \frac{\text{mol gas } (n_g)}{\text{mol gas } (n_g) + \text{mol water } (n_w)}$$

The *Henry's law constant* is a function of the type of gas or volatile organic compound, temperature of the bulk liquid, and constituents of the liquid (water). Values of Henry's law constant for various gases which are slightly soluble in water at 20°C (68°F) are listed in Table 8.23.

Thermodynamic parameters which are required to adjust Henry's law constants for different temperatures also are included in Table 8.21. The effects of temperature on Henry's constant may be determined using the relationship:

$$\log H = \frac{-\Delta H}{RT + {}^\circ K} \qquad (8.37)$$

ΔH (kcal/kmol) is heat absorbed in the evaporation of 1 mol of component from solution at constant temperature and pressure, T is temperature in °K, K is an empirical constant, and $R = 1.987$ kcal °K – kmol (gas constant).

8.3.10.3 Dissolved Oxygen. Dissolved oxygen is important in natural water, because dissolved oxygen is required by many microorganisms and fish in aquatic systems. Dissolved oxygen also establishes an oxic environment, in which oxidized

TABLE 8.23   Henry's Law Constants Selected Gases that are Slightly Soluble in Water

Gas	Formula	Henry's constant at 20°C, atm[b]	Temperature dependence ΔH, kcal/kmol $\times 10^{-3}$	K
Oxygen	O_2	4.3×10^4	1.45	7.11
Nitrogen	N_2	8.6×10^4	1.12	6.85
Methane	CH_4	3.8×10^4	1.54	7.22
Ozone	O_3	5.0×10^3	2.52	8.05
Carbon dioxide	CO_2	1.51×10^2	2.07	6.73
Hydrogen sulfide	H_2S	5.15×10^2	1.85	5.88
Chlorine	Cl_2	5.85×10^2	1.74	5.75
Chlorine dioxide	ClO_2	54	2.93	6.76
Sulfur dioxide	SO_2	38	2.40	5.68
Ammonia	NH_3	0.76	3.75	6.31

Source: James M. Montgomery (1985).

forms of many constituents in water are predominant. Typical dissolved oxygen concentrations reported for natural water in streams and rivers throughout the world are 3 to 9 mg/L, which is the concentration of dissolved oxygen in fresh water at saturation at 20°C (68°F). The observed range of dissolved oxygen concentrations reported worldwide for streams and rivers is 0 mg/L (anoxic conditions) and 19 mg/L (supersaturated conditions). Supersaturated conditions are caused by algal blooms and usually occur late in the afternoon in the summer in temperate climates. Under nighttime conditions bacteria and algae consume the dissolved oxygen and the respiration of these microorganisms results in anoxic or anaerobic conditions. Under anoxic conditions, or periods of zero dissolved oxygen in the water, reduced forms of chemical species are formed and frequently lead to the release of undesirable odors until oxic or aerobic conditions develop. Henry's law constants for selected volatile organic compounds (VOC) are presented in Table 8.24. Thermodynamic parameters required for determining the Henry's law constant also are listed in Table 8.24.

TABLE 8.24   Henry's Law Constants Selected VOC's That Are Slightly Soluble in Water

Gas	Formula	Henry's constant at 20°C, atm[b]	Temperature dependence ΔH, kcal/kmol $\times 10^{-3}$	K
Vinyl chloride	CH_2CHCl	3.55×10^5	—	—
Carbon tetrachloride	CCl_4	1.29×10^3	4.05	10.06
Tetrachloroethylene	C_2Cl_4	1.10×10^3	4.29	10.38
Trichloroethylene	$CCHCl_3$	550	3.41	8.59
Benzene	C_6H_6	240	3.68	8.68
Chloroform	$CHCl_3$	170	4.00	9.10
Bromoform	$CHBr_3$	35	—	—
Pentachlorophenol	C_6OHCl_5	0.12	—	—

Source: James M. Montgomery (1985).

8.4 MICROBIOLOGICAL CHARACTERISTICS OF WATER

The biological characteristics of water are related to the transmission of disease by pathogenic organisms in water, to the development of tastes and odors in surface water and groundwater, and to corrosion and biofouling of cooling system heat transfer surfaces and water supply facilities. The impact of microorganisms in water on humans and on water quality can be best understood if the basic concepts of microbiology are discussed.

8.4.1 Classification of Organism

The principal groups of microorganisms in natural water include protists, plants, and animals. These organisms are presented in Table 8.25.

The nomenclature for classification of microorganisms is based on groups of increasing size:

$$\text{Kingdom} \rightarrow \text{Phylum} \rightarrow \text{Class} \rightarrow \text{Order} \rightarrow \text{Family} \rightarrow \text{Genus} \rightarrow \text{Species}$$

The scientific name of any organism includes the genus and the species. For example, the scientific name of human beings is *Homo sapiens;* i.e., humans belong to the species *sapiens* and to the genus *Homo.* Humans also belong to the family Hominidae, the order Primate, the class Mamalia, the phylum Chordata, and finally the kingdom Animal.

8.4.2 The Nature of Biological Cells

Living cells contain a cell membrane which forms the outer boundary of the cell. Bacteria, algae, fungi, and plants also have a cell wall outside the membrane. The cell wall defines the characteristic shape of the cell. Cell mobility is provided by flagella or some hair-like appendages called *cilia.* The cell membrane contains cytoplasm,

TABLE 8.25   Simplified Classification of Microorganisms in Water

Kingdom	Representative members	Cell classification
Animal	Crustaceans Worms Rotifers	Eucaryotic cells (containing a nucleus enclosed within a well-defined nuclear membrane)
Plant	Rooted aquatic plants Seed Plants Ferns Mosses	
Protista Higher	Protozoa Algae Fungi (molds and yeasts)	Procaryotic cells (nucleus not enclosed in a true nuclear membrane)
Lower	Blue-green algae Bacteria	

Source: Adapted from Tchobanoglous and Schroeder (1985).

which includes various organelles and a colloidal suspension of proteins, carbohydrates, and other complex organic matter. The cell membrane regulates the passage of water and nutrient materials (substrate) into and out of the cell. The cytoplasm also contains the nucleus (or nuclear area), nucleic acids, deoxyribonucleic acid (DNA), the genetic material vital to reproduction, and ribonucleic acid (RNA), which is important in the synthesis of proteins.

8.4.3 Nutritional Requirements of Organisms

Microorganisms require a source of energy, carbon, and other nutrients to synthesize new cells. In addition, macronutrients such as nitrogen, and phosphorus, micronutrients such as sulfur, potassium, calcium, and magnesium, and trace elements must be available.

8.4.4 Energy and Carbon Requirements

Organisms also use organic matter as a source of energy. Organisms may be classified according to the source. Carbon required for cell synthesis may be derived from carbon dioxide by *autotrophic* (chemoautotrophic) *bacteria*, or from the carbon in organic matter by *heterotrophic* (chemoheterotrophic) *microbes*. Energy is available to heterotrophic organisms from the oxidation or reduction of the organic substrate during biodegradation of the organic material. Autotrophic organisms use energy released from the oxidation or reduction of inorganic compounds, while phototrophic microbes derive energy from sunlight through photosynthetic reactions. A general classification of microbes based on source of energy and carbon is presented in Table 8.26.

8.4.5 Oxygen Requirements

Oxygen is a major constituent of cellular matter (e.g., carbohydrates, polysaccharides, lipids, proteins, amino acids, and nucleic acids) and also plays an important role in cell growth and metabolism. Organisms which require molecular oxygen (dissolved O_2) for metabolism are *obligate aerobes,* and those organisms for which molecular oxygen is toxic are *obligate anaerobes.* These anaerobes derive the oxygen needed for cell synthesis and energy from the biodegradation of chemical compounds. Other organisms are *facultative.* Facultative microbes prefer dissolved oxygen, if available; however facultative organisms also can derive oxygen and energy from the degradation of organic substrates.

TABLE 8.26 General Classification of Microorganisms Based on Sources of Energy and Carbon

Classification	Energy source	Carbon source	Representative organisms
Photoautotrophs	Light	CO_2	Higher plants, algae, photosynthetic bacteria
Photoheterotrophs	Light	Organic matter	Photosynthetic bacteria
Chemoautotrophs	Inorganic matter	CO_2	Bacteria
Chemoheterotrophs	Organic matter	Organic matter	Bacteria, fungi, protozoa, animals

8.4.6 Environmental Effects on Organisms

The environment also affects the growth and reproduction of microorganisms. Specifically, environmental factors include: (1) the type and composition of substrate, (2) pH, (3) substance's oxygen, (4) temperature, (5) light, and (6) toxic or inhibitory substances.

8.4.7 Microorganisms in Water

The principal microorganisms of concern in water include bacteria, fungi, algae, protozoa, worms, rotifers, crustaceans, and viruses. Some of the physical and biological characteristics of organisms important to water quality considerations are summarized in Table 8.27. The sizes of the microorganisms range from 10^{-2} μm for viruses up to 10^5 μm for some of the helminths (worms).

8.4.7.1 Bacteria. *Bacteria* are single-cell protists which may be characterized according to morphology as: spheroid (coccus), rod (bacillus), curved rod (vibrio), or spiral (spirillum). A typical coccus is a sphere about 1 to 3 μm in diameter. Two common configurations of cocci are *Staphylococcus,* which is an amorphous grouping of many cocci organisms, and *Streptococcus,* which is a long chain of cocci organisms resembling a string of pearls. Bacillus varies in size from 0.3 to 1.5 μm in diameter and from 1.0 to 10.0 μm in length. Organisms commonly found in human excrement are about 0.5 μm in diameter and about 2 μm long. *Vibrios* vary in size from 0.6 to 1.0 μm in diameter and from 2 to 6 μm in length. Spirillum may be as long as 50 μm.

8.4.7.2 Fungi. *Fungi* are aerobic, multicellular, nonphotosynthetic heterotrophic, eucarotic protists. Most fungi are *saprophytes* which degrade dead organic matter. Fungi grow in low-moisture areas, and are tolerant of low-pH environments. Fungi release carbon dioxide and nitrogen during the breakdown of organic material. Fungi are essential to the decomposition of vegetation and woody materials, which are resistant to bacterial degradation.

8.4.7.3 Algae. *Algae* are autotrophic, photosynthetic organisms. A classification of algae is presented in Table 8.28. Algae take on the color of the pigment which is the catalyst for photosynthesis. In addition to chlorophyll (green), different algae have different pigments such as carotenes (orange), phycocyanin (blue), phyco-

TABLE 8.27 Physical and Biological Characteristics of Micrornanisms of Water Quality Importance

Organism	Size, mm	Surface charge	Shape	Environmentally resistant stage
Viruses	10^{-2}–10^{-1}	Negative	Variable	Viron
Bacteria	10^{-1}–10^1	Negative	Rod, coccoid, spiral comma	Spores, cystlike
Blue-green algae	10^0	Negative	Coccoid, filamentous	Cysts
Green algae	10^0–10^2	Negative	Colloid, pennatic	Spores, cysts
Protozoa	10^0–10^2	Negative	Variable	Cysts
Fungi	10^0–10^2	Negative	Filamentous, coccoid	Spores
Helminths	10^0–10^5	Negative	Variable	Eggs

Source: Adapted from James M. Montgomery (1985).

erythrin (red), fucoxanthin (brown), and xanthophylls (yellow). Various colored algae observed in nature contain a combination of pigments. Algae derive carbon from CO_2 in water, and energy for cell synthesis is available through photosynthesis, using chlorophylls, carotenoids, and phycobilins as the photosynthetic catalysts. Oxygen is a product of the photosynthesis. Algae utilize oxygen for respiration in the absence of light. Algae and bacteria have a symbiotic relationship in aquatic systems. Algae produce oxygen which is used by the bacterial population.

Excessive algal growth (algal blooms) can result in supersaturated oxygen conditions in the daytime and anaerobic conditions at night. Algae also affect water quality, since some algae cause tastes and odors in natural water. In addition, actinomycetes, which grow associated with algal mats resulting from blooms, also contribute tastes and odors to water.

Some algae can obtain carbon from the bicarbonates and carbonates in natural waters. The pH of the water generally increases as the bicarbonates are used. Some algae also can fix nitrogen from the atmosphere and may grow in spite of control of nitrogen sources.

8.4.7.4 Protozoa. *Protozoa*, single-cell animals without cell walls, are aerobic or facultative heterotrophs. Some protozoa species are parasitic in a host organism. A classification of protozoa commonly associated with natural water are presented in Table 8.29.

TABLE 8.28   Classification of Major Algal Groups

Group	Descriptive name	Aquatic environment
Chlorophyta	Green algae	Fresh and saltwater
Chrysophta	Diatoms, golden-brown algae	Fresh and saltwater
Cryptophta	Cryptomonads	Saltwater
Euglenophyta	Euglenea	Freshwater
Phaeophyta	Brown algae	Saltwater
Pyrrhophyta	Dinoflagellates	Fresh and saltwater
Rhodophyta	Red algae	Fresh and saltwater
Xanthophyta	Yellow-green algae	Fresh and saltwater

Source: Adapted from Palmer (1962).

TABLE 8.29   Classification of Major Groups of Protozoa

Class	Mode of motility	Typical members	
		Name	Remarks
Ciliata	Cilia (usually) multiple	*Paramecium*	Free-swimming
Mastigophora	Flagella (one or more)	*Giardia lamblia*	Causative agent for giardiasis
Sarcodina	Pseudopodia (some with flagella)	*Entamoeba histolytica*	Causative agent for dysentery
Sporozoa	Creeping, often nonmotile flagella at some stages	*Plasmodium vivax*	Causative agent for malaria

Giardia lamblia is a flagellate protozoa which causes the intestinal disease giardiasis. Symptoms of giardiasis include diarrhea, nausea, indigestion, flatulence, bloating, fatigue, and appetite and weight loss. This disease can be chronic without proper treatment. It occurs throughout the world and is contracted by drinking contaminated water.

8.4.7.5 Worms and Rotifers. A number of worms and rotifers are of importance to water quality. Rotifers are aerobic chemoheterotrophs which prey on bacteria. Many worms are aquatic parasites. Flatworms of the class *Trematoda* are known as flukes, and the *Cestoda* are tapeworms. *Nematodes* of public health concerns are *Trichinella,* which causes trichinosis; *Necator,* which causes hookworm; *Ascaris,* which causes common roundworm infestation; and *Filaria,* which causes filariasis. These nematodes are present in contaminated water sources.

8.4.7.6 Crustaceans. *Crustaceans* also are aerobic chemoheterotrophs that feed on bacteria and algae. These hard-shelled, multicellular animals are a source of food for fish.

8.4.7.7 Viruses. *Viruses* are parasitic organisms containing DNA or RNA in a coating of protein. Reported data on the survival of viruses in natural surface waters are summarized in Table 8.30.

The environment provided in natural water is not conducive for the long-term survival of viruses. However, virus longevity is enhanced by organic particulate material, onto which the virus can become attached.

The virion particles invade living cells and the viral genetic material redirects cell activities toward production of new viral particles. A large number of viruses are released to infect other cells when the infected cell dies. Viruses are host-specific, and a specific type of virus attacks only one type of organism. Common waterborne viral diseases are infectious hepatitis and gastroenteritis.

8.4.7.8 Waterborne Diseases. Many bacteria, viruses, and protozoa are causative organisms for some of the more virulent diseases transmitted to humans directly through water and indirectly through contaminated food. A summary of waterborne diseases in which bacteria are species involved is presented in Table 8.31.

TABLE 8.30 Survival of Viruses in Natural Waters

Agent	Initial concentration, mL^{-1}	Type of water	Temperature, °C	Time for 100% loss of infectivity, days
Poliovirus 2	10^4	River	16–20	29–35
Coxsackie virus B3	10^4	River	16–20	29–35
Echovirus 7	10^4	River	16–20	29–35
Coxsackie virus A4	0.32	River	4–8	>150
Coxsackie virus A4	32	River	20–22	>45
Echovirus 7	$10^{2.3}$	Well	20	66
Echovirus 7	$10^{2.3}$	Well	10	113
Coxsackie virus B3	$10^{2.3}$	Well	20	66
Echovirus 7	10^3	River	4–6	90

Source: Adapted from James M. Montgomery (1985) from Berger (1983).

TABLE 8.31 Bacteria of Public Health Significance in Water

Genus	Species	Host(s)	Disease	Transmission	Occurrence
Salmonella	Several hundred serotypes: S. typhi, S. Enteritides, S. Typhimurium	Found in human and animal guts, polluted water, contaminated foods.	Typhoid fever from S. typhi, acute diarrhea and cramping; typhoid fever can be fatal.	Transmitted through water or food processed with contaminated water.	Worldwide, common in areas of Far and Middle East, Eastern Europe, Central and South American, Africa.
Shigella	Four species: S. Sonnei, S. Flexneri, S. Boydii, S. Dysenteriae	Rarely found in hosts other than humans.	Shigellosis, or bacillary dysentery, causes fever and bloody diarrhea.	Transmitted via contaminated food, polluted water and person to person.	Worldwide endemic in tropical populations; in United States moderately endemic in lower economic areas.
Leptospira	Common human isolates include:	Found in infected humans and animals. Excreted via urine.	Leptospirosis: acute infections of kidney, liver, central nervous system.	Transmitted through skin abrasions or mucous membranes to blood stream; may come from contact with animal carriers and/or polluted water.	Worldwide; occupational hazard to rice field workers, farmers, sewer workers, and miners.
Pasturella	P. tularensis	Humans and wild animals, especially rabbits.	Tularemia: causes chills and fever, with an ulcer at site of infection, swollen lymph nodes.	Handling of infected animals, arthropod bites (deer fly, wood tick), drinking contaminated water.	North America, Europe, USSR, Japan.
Vibrio	V. cholerae	Humans.	Cholera: an acute intestinal disease causing vomiting diarrhea, dehydration, can be fatal.	Contaminated water, person-to-person contact	Endemic in parts of India, East Pakistan; Philippines, Indonesia, Thailand, Hong Kong, Korea.
Enteric pathogenic E. coli	Various serotypes	Warm-blooded animals.	Diarrhea, urinary infections.	Sewage contaminated water or food; person to person.	Worldwide.
Yersinia	Y. enterocolitica, Y. pseudotuberculosis	Birds and mammals.	Diarrhea, fever, vomiting, anorexia, acute abdominal pain, abscesses, septicemia.	Animal to human, person to person, contaminated water.	Worldwide.

TABLE 8.31   Bacteria of Public Health Significance in Water (*Continued*)

Genus	Species	Host(s)	Disease	Transmission	Occurrence
Mycobac-terium	*M. tuberculosis, M. balnei, M. bovis*	Humans, diseased cattle.	Pulmonary or skin tuberculosis.	Typically airborne, but sewage-contaminated water also is a route.	Worldwide.

Source: Adapted from James M. Montgomery (1985).

Typical diseases associated with water are listed Table 8.32. The common name of the disease, causative agent, and symptoms also are included.

Outbreaks of waterborne diseases in the United States for the period 1971 to 1981 are summarized in Table 8.33. During that period bacterial gastroenteritis was most prevalent, accounting for more that 62.5 percent of the outbreaks and 52.7 percent of the cases, followed by *Giardiasis* outbreaks, which made up 16.3 percent of the observed incidence and 26.3 percent of the cases reported.

TABLE 8.32   Typical Diseases Associated with Water

Category and method of contraction	Disease	Causative agent	Symptoms
Waterborne: ingesting contaminated water	Amebiasis, (amoebic dysentery)	Protozoan (*Entamoeba histolytica*)	Diarrhea with bleeding, abscesses of liver and small intestine
	Shigellosis (dysentery)	Bacteria (*Shigella,* 4 spp.)	
	Cholera	Bacteria (*Vibrio cholerae*)	Extremely heavy diarrhea, dehydration, high death rate
	Gastroenteritis	Virus (enteroviruses, parvovirus, rotavirus)	Mild to severe diarrhea
	Giardiasis	Protozoan (*Giardia lamblia*)	Mild to severe diarrhea, nausea, indigestion, flatulence
	Infectious hepatitis	Virus (hepatitis A virus)	Jaundice, fever
	Leptospirosis (Weil's disease)	Bacteria (*Leptospira*)	Jaundice, fever
	Salmonellosis	Bacteria (*Salmonella,* ~1700 spp.)	Fever, nausea, diarrhea
	Typhoid fever	Bacteria (*Salmonella typhosa*)	High fever, diarrhea, ulceration of small intestine
Water-washed: washing with contaminated water	Shigellosis (dysentery)	Bacteria (*Shigella*)	Mild to severe diarrhea
	Scabies	Mite	Skin ulcers
	Trachoma	Virus	Eye inflammation, partial or complete blindness
Water based: worm infections water one stage in cycle	Filariasis	Worm	Blocking lymph nodes, permanent damage to tissue
	Guinea worm	Worm	Arthritis of joints
	Schistosomiasis	Worm (schistosomes)	Tissue damage and blood loss in bladder and intestinal venous drainage

Sources: Tchobanoglous (1985), as adapted from Hawkes (1971) and Salvato (1982).

TABLE 8.33 Outbreaks of Waterborne Diseases in the United States 1971 to 1981

Disease	Outbreaks	Percent of total*	Illnesses	Percent of total*
Gastroenteritis				
Unidentified agent	192	62.5	39,845	52.7
Giardiasis	50	16.3	19,863	26.3
Shigellosis	25	8.1	5,448	7.2
Salmonellosis	8	2.6	1,150	1.5
Hepatitis A	16	5.2	463	0.5
Campylobacter diarrhea	4	1.3	3,902	5.2
Viral Gastroenteritis	10	3.2	3,147	4.2
Vibrio cholerae	1	0.3	17	0.02
Rotavirus	1	0.3	1,761	2.3
Total	307		75,596	

* Percentages do not total 100% because of rounding.
Source: Adapted from Craun (1983) in James M. Montgomery (1985).

Many outbreaks of waterborne diseases probably are not recognized; therefore, their incidence are not reported. A total of 34 outbreaks were reported in 17 states for the two-year period 1991 and 1992. (Moore, 1994). These outbreaks reportedly affected 17,464 individuals. In the majority (23/34 or 67.6 percent) of the outbreaks, acute gastroenteritis illness was reported to have been the cause of 13,367 cases, but the causative agent could not be identified. The causative agents in 7 of the remaining 11 outbreaks were protozoan parasites *Giardia* (4 outbreaks; 123 cases), and *Cryptosporidium* (3 outbreaks; 3551 cases). Hepatitis A and *Shigella sonnei* each were identified as the agent in one case, affecting 10 and 150 individuals, respectively. Chemicals were involved in two outbreaks affecting 263 illnesses.

These outbreaks of waterborne diseases reflect on the quality of the water sources, as well as on the ineffectiveness of water treatment and distribution systems providing drinking water in the states reporting outbreaks. However, considering the fact that no waterborne outbreak or illness was reported in 33 states, water treatment facilities in the United States are quite efficient in providing a pathogen-free water supply to protect the health of the public.

Assay and confirmation of the presence of the causative agent of waterborne diseases are lengthy and time consuming. In lieu of specific analyses, coliform organisms have been used to determine the biological characteristics of natural waters.

8.4.7.9 *The Use of* **Escherichia coli** *as an Indicator Organism.* The coliform group of bacteria are aerobic and/or facultative gram-negative, nonspore-forming, rod-shaped bacteria that ferment lactose to gas: *Escherichia coli* is commonly used as an indicator organism. This organism is present in the intestine of warm-blooded animals, including humans. Therefore the presence of *E. coli* in water samples indicates the presence of fecal matter and of the possible presence of pathogenic organisms of human origin. The concentration of indicator organisms is reported in MPN/100 mL (MPN = most probable number) or in CFU/100 mL (CFU = colony forming units).

E. coli has been used and continues to be considered a good indicator organism for the following reasons:

1. *E. coli* are found in large numbers in human excrement.

2. *E. coli* are relatively easy to detect and assay.

3. The data lend themselves to statistical analyses.

4. *E. coli* responds the same as pathogens to changes in the aquatic environment.

Recently, problems related to the use of *E. coli* as an indicator organism include:

1. *E. coli* is not a single species.

2. *Proteus* and *Aerobacter* are normally found outside the human intestinal tract in soil.

3. Other organisms found in water have the ability to ferment lactose.

4. *E. coli* identical to those in humans are in the intestinal tract of other warm-blooded animals. In spite of these short comings, *E. coli* continues to be used as an indicator organism.

Other enteric organisms which are also considered indicator organisms are fecal streptococci (*Sreptococcus faecalis*) and clostridia (*Clostridium perfringens*). These organisms are more resistant to changes in the environment than *E. coli*. Therefore, fecal streptococci are used in conjunction with the coliform test as an indication of human fecal contamination. Estimated per capita contribution of indicator organisms is presented in Table 8.34. Ducks, swine, and sheep all contribute higher numbers of fecal indicator organisms that humans. Therefore, using *E. coli* and *F. streptococcus* as indicator organisms may not obviate problems with the inability of indicator organisms to positively identify the presence of pathogenic organisms in natural water or in water supplies. The epidemiological follow-up of recently reported outbreaks of waterborne diseases indicates that bacterial indicator organisms may not be effective in identifying the presence of protozoan parasitic or viral causative agents.

Protecting society from waterborne diseases, especially in water supplies, is essential. Epidemic waterborne diseases cause emotional and economic losses in communities affected. Indicator organisms are usually used to warn of the threat of waterborne disease. Therefore, if significant numbers of coliforms are detected in a water supply, detailed testing is needed to identify what bacterial species or other causative agents associated with the human intestinal tract are present.

TABLE 8.34   Estimated Per Capita Contribution of Indicator Microorganisms from Humans and Some Animals

| Animal | Average indicator organisms, number/g feces | | Average contribution, per capita/24 hr | | |
	Fecal coliform 10^6	Fecal streptococci 10^6	Fecal coliform 10^6	Fecal streptococci 10^6	Ratio FC/FS*
Human	13.0	3.0	2,000	450	4.4
Chicken	1.3	3.4	240	620	0.4
Cow	0.23	1.3	5,400	31,000	0.2
Duck	33.0	54.0	11,000	18,000	0.6
Pig	3.3	84.0	8,900	230,000	0.04
Sheep	16.0	38.0	18,000	43,000	0.4
Turkey	0.29	2.8	130	1,300	0.1

* FC/FS = Fecal coliform/fecal streptococci.
Source: From Tchobanoglous and Schroeder (1985).

8.4.8 Bioassays

Bioassays are designed to evaluate the responses of aquatic organisms to water constituents in natural waters. Acute toxicity is assessed in short-term tests. Intermediate term bioassays estimate the effects on the life stages of organisms with long life cycles. Long-term flow-through tests assess the maximum allowable toxicant concentration (MATC) which does not produce harmful effects with continuous exposure.

Fish are usually the test organisms. The species used should be indigenous or representative of species in the receiving water and should be sensitive to environmental changes. Typically, short-term bioassays using fish are conducted by adding various toxicants or pollutants to an aliquot of natural water in several aquaria, each having 10 to 20 fish. Survival of the test fish is recorded with time. Bioassay data are reported in terms of the median lethal concentration (LC50), which is defined as the concentration of the contaminant resulting in death of 50 percent of the test fish after 24, 48, and 96 hours. Procedures of the bioassay tests using fish and other organisms are given in *Standard Methods* and expanded by the U.S. EPA.

8.5 BIOLOGICAL CHARACTERISTICS

The microbial consortium present in natural waters is responsible for the chemical transformations that occur in aquatic ecosystems. All classifications of organisms are involved in recycling of elements. A schematic diagram of an *aquatic ecosystem* is presented in Fig. 8.3. Carbonaceous, nitrogenous, phosphorous, and sulfurous compounds are converted into simpler forms finally to the mineralized forms.

In all aquatic ecosystems, the producers are the algae and rooted plants which convert sunlight to synthesize new plant material. Aquatic animals consume the plants and produce detritus. Dead plants and animals add to the detritus, which the decomposers (bacteria and fungi) convert to simple inorganic substances such as ammonia, sulfate, carbon dioxide, and water. The nitrogen, phosphorus, and sulfur cycles are fundamental to recycling of these elements in the aquatic ecosystem.

In a typical aquatic ecosystem, plant and animal tissues are formed. Plant and animal materials are composed of carbon, hydrogen, oxygen, nitrogen, phosphorus, and sulfur. These elements are building blocks for carbohydrates, lipids, proteins, phospholipids, and nucleic acids. These carbonaceous, nitrogenous, phosphorous, and sul-

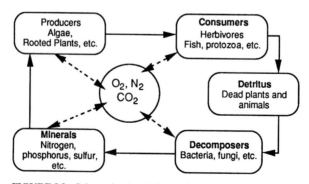

FIGURE 8.3   Schematic of typical aquatic ecosystem.

furous substances undergo decomposition upon the death of the plants and animals. Decomposition is the result of bacterial and fungal degradation of complex organic materials into simpler compounds, finally resulting in mineralization of the substance to elemental form.

Schematic diagrams illustrating the nitrogen, carbon, and sulfur cycles in aerobic [environment contains dissolved oxygen and oxidizing conditions, indicated by a positve oxidation reduction potential (ORP = +200 to 300 mv)] and anaerobic [characterized by no dissolved oxygen, and reducing conditions (ORP = −100 to 500 mv)] environments are presented in Figs. 8.4 and 8.5, respectively.

Typical concentrations and ranges of nitrogen and phosphorus in streams and rivers throughout the world are presented in Table 8.35.

8.5.1 Nitrogen Cycle

Nitrogen is a constituent of protein and nucleic acids and is required by organisms in greatest quantity after carbon and oxygen. Organic nitrogen, ammonia (NH_3), nitrite (NO_2^-), nitrate (NO_3^-), and nitrogen gas (N_2) are important nitrogen-containing compounds in aquatic systems. The atmosphere is the reservoir for nitrogen. Some bacteria and algae fix nitrogen from the atmosphere and incorporate this nitrogen into cell material.

Organic nitrogen degrades to ammonia via bacterial ammonification of organic matter. Ammonia undergoes nitrification in a two-step bacterial conversion of ammonia to nitrite and to nitrate. Nitrate is utilized in the production of organic nitrogen in the form of bacteria and plants. A portion of the nitrates are returned to the atmosphere by anaerobic bacterial denitrification.

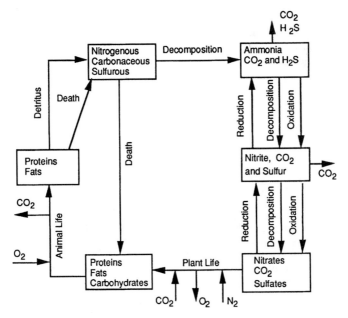

FIGURE 8.4 Aerobic cycle in aquatic ecosystems.

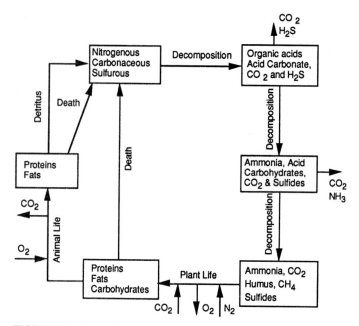

FIGURE 8.5 Anaerobic cycle in aquatic ecosystems.

TABLE 8.35 Water Quality of Streams and Rivers Nitrogen and Phosphorus

Water quality parameter	Typical value, mg/L	Observed ranges, mg/L
Total nitrogen (N)	0.1–10	0.004–>100
Organic nitrogen	0.1–9	<0.2–20
Ammonia (NH$_3$-N)	0.01–10	<0.01–45
Nitrite (NO$_2$-N)	0.01–0.5	<0.002–10
Nitrate (NO$_3$-N)	0.23	0.01–250
Nitrogen gas (N$_2$)	0–18.4	
Total phosphorus (P)	0.02–6	0.01–30
Orthophosphate (PO$_4$-P)	0.01–0.5	<0.01–14

8.5.2 Phosphorus Cycle

Phosphorus also undergoes environmental cycles involving organic and inorganic forms. All living matter contains phosphorus. Orthophosphate (PO_4^{-3}) is the only form of phosphorus that is used readily by most plants and microorganisms. The phosphorus cycle involves: (1) the bacterial conversion of organic phosphorus, and (2) the microbial conversion of inorganic phosphorus to organic compounds. Microorganisms convert insoluble phosphorus compounds [$Ca_2(HPO_4)_2$] into soluble forms, principally PO_4^{-3}. Bacterial decomposers degrade the organic phosphorus in dead plant and animal tissue and in detritus to PO_4^- which subsequently is incorporated into plant and animal tissue in the aquatic environment, and the cycle continues.

Phosphorus concentrations in many natural water environments is low, and algal growth is limited. Introduction of large amounts of phosphorus in partially treated wastewater and irrigation return flows frequently results in algal blooms.

8.5.3 Sulfur Cycle

Sulfate occurs in most natural waters as organic sulfur, hydrogen sulfide (H_2S), elemental sulfur (S^0), and sulfate SO_4^{-2}. Sulfate is taken up by plants, which in turn are consumed by animals. Dead plants and animals and detritus are decomposed by bacteria to hydrogen sulfide, which in turn, under aerobic conditions, is oxidized to elemental sulfur. Sulfide is also oxidized to sulfur and sulfate by photosynthetic sulfur bacteria and blue-green algae. The reduction of sulfate to hydrogen sulfide is carried out by a small group of anaerobic sulfate-reducing bacteria.

Hydrogen sulfide is toxic to many organisms, and also is the source of odors in natural water. Hydrogen sulfide also can combine and precipitate heavy metals such as iron, zinc, and cobalt. These metal elements are required for bacterial growth; therefore high levels of H_2S may inhibit growth.

8.5.4 Ecological Relationships

The *nutrient cycles* are related to eutrophication and acid rain. Impurities such as toxics and growth stimulators affect the ecosystem. Floating algae, or toxic-producing blue-green algae (microcystis), are included. In arid and semiarid areas in the southwestern parts of the United States, *microcystis* produce toxins which cause illness in cattle and humans and can be fatal.

Heavy metals (e.g., Zn^{+2}) are required in trace amounts for plant and animal growth, but are toxic at concentrations approaching 0.1 to 1.0 g/m^3. Anions (e.g., Cl^-) have minimal ecological effect over concentration ranges of several orders of magnitude.

Ammonia occurs in equilibrium with NH_4^+ in water. However, NH_3 may be lethal at 0.1 mg/L. Carbon dioxide is in equilibrium with carbonic acid (H_2CO_3), bicarbonate ions (HCO_3^-), and carbonate ions (CO_3^{-2}), which is affected by pH. The carbonate equilibrium condition is significant as the buffering characteristic of water.

Biostimulants are chemical elements that stimulate biological growth. Nitrogen (NH_4^+ and NO_3^-) and phosphate (PO_4^{-3}) are the most common biostimulants, but any growth-limiting nutrient can act in this manner.

Natural waters may be classified as *oligotrophic, mesotrophic,* and *eutrophic.* Oligotrophic waters contain low concentrations of essential nutrients (viz, nitrogen, phosphorus, and iron) and organic matter. Life forms, but few plants, are present in small numbers in oligotrophic waters. Water is considered to be mesotrophic as the amounts of nutrients and organic matter increase. An abundance and diversity of animals and plants at all trophic levels (food chain) also characterize mesotrophic conditions. Eutrophic water bodies contain large amounts of nutrients and organic matter, which support substantial plant growth.

Lake Superior in the Great Lakes, Lake Tahoe in California and Nevada, and Crater Lake in Oregon are examples of oligotrophic waters. The nutrient cycles and the natural processes of growth, death, and decay, as well as input of nutrients from return flows, increase the nutrient concentrations, and the waters become increasingly biologically productive. As the body of water ages fewer species are present, but the concentrations of algae continue to increase. Diurnal fluctuations in dissolved oxygen are common. This process of aging is called *eutrophication*. The eutrophication process is natural; however, human activities accelerate the rate.

8.6   WATER QUALITY MANAGEMENT

Water quality management requires an understanding of the physical, chemical, and biological characteristics of water and the interactions among the physical, chemical, and biological processes, transformations, and conversions. The water quality requirements associated with various beneficial uses of water also must be understood in detail.

Environmental and/or water pollution control agencies in each state usually establish beneficial uses for surface water and groundwater in the state. Beneficial uses include: municipal water supply, industrial water supply, recreation, agricultural irrigation, power, navigation, and protection and enhancement of fish and wildlife. The beneficial use usually is based on the quality of the water body in question, existing and projected sources of pollution, acceptable alternate water resources, historical use patterns, and existing treatment systems for pollution abatement.

Quality of water for public water supplies is the highest, followed by water used for contact recreation. The next highest water quality varies with the geographical location and the economic sector of importance in the area, i.e., agriculture versus heavy industry. Power generation has other water quality requirements. Water quality concerns for navigation are not very important, provided the water is not corrosive and flammable.

Once a beneficial use has been designated, water quality criteria are developed to protect this use. These criteria frequently are based solely on existing data and scientific judgment, without technical considerations or economic feasibility. Criteria can lead to the promulgation of water quality standards, which carry legal authorization (e.g., standards for safe drinking water).

Historically, the Public Health Service (PHS) prepared standards which were strictly applied to suppliers of water to interstate commerce, and were intended to protect the health of the public on commercial trains, airplanes, buses, and similar vehicles. The PHS standards were based on the ability to detect bacteria and the use of chlorine to disinfect water. These standards specified a maximum permissible limit for plate counts for a coliform organism.

The PHS standards were recognized informally as water quality criteria and were adopted and/or adapted as standards by many state or local regulatory agencies. Over the years in the United States, the number of regulated constituents of water has continued to increase and maximum permissible concentrations have decreased. These trends are the result of improved analytical procedures and more sensitive instrumentation, coupled with the availability of toxicological data on various contaminants in water.

Water quality goals define desired water quality and establish pollutant and contaminant concentrations much more stringent than standards. Goals also can address aspects or constituents not covered by standards, or focus on overall water quality objectives (e.g., one goal of PL92-500 was zero discharge of pollutants into navigable streams of the United States).

Water quality management efforts must take into consideration the impact of various federal regulations on the quality of surface water and groundwater. The overall improvement in water quality in streams and rivers in the United States brought about by enforcement of PL-92-500, the Water Pollution Control Amendments of 1972 (U.S. Congress, 1973) and subsequent amendments is quite dramatic. Although the goal of zero discharge of pollutants has not been achieved, effluent limitations on industrial and municipal point sources (return flows) have markedly reduce the pollution load discharged into the streams and rivers of the United States.

Currently, water pollution control agencies are directing attention at appropriate treatment technologies for the control of nonpoint source pollutants, specifically urban, industrial, and highway runoff. Nonpoint sources of water pollution (storm water runoff from agricultural lands and runoff from streets and highways in municipalities and urban industrial centers) are drawing more attention, because the impact of runoff on receiving water bodies is much more noticeable since control of point sources of water pollution, coupled with more stringent effluent limitations, have markedly improved the quality of industrial and municipal discharges. These return flows are important resources in many parts of the United States, and the quality of effluent discharges and storm water runoff markedly affect water-quality management strategies and programs.

Other federal regulations, namely, the Resource Conservation and Recovery Act (RCRA), focus on the control of discharge of pollutants in the leachates from the improper disposal of industrial and hazardous wastes into groundwater and surface water sources. Superfund regulation, along with RCRA, should be effective in cleaning up groundwater resources which are polluted with leachates from land disposal facilities, or from direct but inadvertent infiltration of accidentally spilled hazardous and/or toxic chemicals into groundwater aquifers. Water-quality management calls for methods of predicting changes in water quality due to environmental changes; and to the methods of modification of water quality.

8.6.1 Water Quality Criteria Development

Water quality criteria for various beneficial uses, including drinking water, have been addressed in numerous documents. In 1952, the California State Water Pollution Control Board published *Water Quality Criteria*, in which scientific and technical information dealing with water quality for various beneficial uses was discussed. These criteria were revised in 1963 by McKee and Wolf, and in 1971 the California State Water Resources Control Board published a revision of these criteria (McKee and Wolf, 1963; 1971). These documents were used extensively throughout the United States and in many countries throughout the world.

Federal agencies have developed water quality criteria documents in response to the Federal Water Pollution Control Amendments of 1972 (1973) and various amendments, as well as to the Safe Drinking Water Act (1974) and amendments and promulgation of rules. These documents include:

- *Water Quality Criteria,* National Technical Advisory Committee to the Secretary of the Interior, reprinted by the EPA. ("Green" Book) USEPA, 1972.
- *Water Quality Criteria,* prepared by the National Academy of Sciences and the National Academy Engineering for the EPA. ("Blue" Book) NAS, 1973.
- *Quality Criteria for Water,* published by the EPA. ("Red" Book) USEPA, 1976.

In spite of efforts to control the quality of water sources and of drinking water supplies, variations among the quality of water supplies continues to reflect the local water supplies.

TABLE 8.36 Primary Drinking Water Standards Maximum
Contaminant Levels*

Contaminant	Maximum contaminant level, mg/L
Inorganic chemicals	
Asbestos	7 MFL[†]
Arsenic	0.05 mg/L
Barium	2.0
Cadmium	0.005
Chromium	0.1
Lead	0.05
Mercury	0.002
Nitrate (as N)	10
Nitrite (as N)	1
Selenium	0.05
Silver	0.05
Fluoride[†]	4
Pesticides/PCB	
Endrin	0.0002
Lindane	0.004
Methoxychlor	0.04
Toxaphene	0.003
2,4-D	0.07
2,4,5-TP Silvex	0.05
Ethylene dibromide	0.00005
Heptachlor	0.0004
PCB's	0.0005
Chlordane	0.002
Volatile Organic Chemicals	
Benzene	0.005
Carbon tetrachloride	0.005
1,1-Dichloroethylene	0.007
Trichloroethylene	0.005
Total trihalomethanes	0.1
Turbidity	1–5 NTU
Coliform bacteria	1/100 mL (mean)
Radiological	
Gross alpha	15 pCi/L
Radium 226 and 228	5
Gross beta	50
Tritium	20,000
Strontium 90	8

* Safe Drinking Water Act (1991 update).
[†] MFL = Million fibers per liter longer than 10 μm.

   The National Academy of Sciences (1977, 1980) developed quantitative criteria and established principles for the assessment of risk and safety of drinking water containing chemical constituents:

- Effects observed in animals when properly qualified are applicable to humans.
- No technique is available to establish a threshold for long-term effects of toxic agents.

TABLE 8.37 National Secondary Drinking Water Regulations (1979)

Constituent	Maximum contaminant level	Effect on water quality
Chloride	250 mg/L	Salty taste
Color	15 color units	Objectionable appearance
Copper	1 mg/L	Undesirable taste
Corrosivity	Noncorrosive	Stains; dissolution of metals; economic loss
Foaming agents	0.5 mg/L	Undesirable appearance
Iron	0.3 mg/L	Bitter, astringent taste; stained laundry
Manganese	0.05 mg/L	Impaired taste; discolored laundry
Odor	3 threshold odor number	Undesirable smell
pH	6.5–8.5	Corrosivity; impaired taste
Sulfate	250 mg/L	Detectable taste; laxative at high levels
Total dissolved solids	500 mg/L	Objectionable taste
Zinc	5 mg/L	Undesirable taste; milky appearance at high levels

- Exposure of experimental animals to high doses of toxic agents is necessary and is a valid technique for identifying possible carcinogenic effects on humans.

- Toxic agents and other materials should be assessed in terms of risk to human health rather than as safe or unsafe.

Federal and state water quality regulation and standards will continue to evolve to support water quality goals and to protect the health and welfare of the public. Water quality criteria will continue to be developed to protect the beneficial uses identified for specific water bodies. The interaction and interrelationship among beneficial use, water quality criteria, standards, and goals will continue to be more and more important aspects of water resources quality management, as pressures on limited water resources continue to increase.

8.6.2 Drinking Water Supply Regulations

Current guidelines for evaluating the suitability of a surface water or groundwater source for use as a public water supply are the regulations mandated by the U.S. Environmental Protection Agency, Title 40, Parts 141 and 143 of the Safe Drinking Water Act (1974). Two types of standards, primary and secondary, must be established. Primary standards are designed to protect human health, and secondary standards are directed at constituents of water which cause odors, negatively affect the appearance of water, or adversely affect public welfare.

Primary standards may be performance requirements (maximum permissible levels or MCLs) or treatment requirements. The MCLs listed in Table 8.36 are based on a 1991 update of the rules promulgated at that time.

Secondary drinking water regulations are presented in Table 8.37. The Safe Drinking Water Act regulations continue to evolve. New rules are proposed, discussed at public hearings, commented on by interested parties, and revised as necessary before final promulgation. These regulations are directed at the quality of water at the point of use, and water quality is usually monitored at the treatment plant.

These drinking water regulations also impact the management of water quality in natural water resources, which may have multiple demands imposed upon them by different users. Therefore, water-quality management must take into consideration the requirements of the Safe Drinking Water Act, as well as the restrictions and limitations of the various federal and state regulations and statutes which affect the quality of surface and underground water resources.

REFERENCES

Alekin, O. A., and L. V. Brazhnikova, "The Discharge of Soluble Matter from Dry Land of the Earth," *Gidrokhim. Materialy,* 32:12–24, 1961.

American Public Health Association, American Water Works Association, and Water Environment Federation, *Standard Methods for the Examination of Water and Wastewater,* 19th ed., American Public Health Association, New York, 1995.

American Society of Civil Engineers, *Sulfide in Wastewater Collection and Treatment Systems,* ASCE, Manuals and Reports on Engineering Practice, no. 69, American Society of Civil Engineers, New York, 1989.

Amoore, J., J. Johnston, and M. Rubin, "The Stereochemical Theory of Odor," *Scientific American,* 42, February, 1964.

Berger, P. S., and Y. Argaman, "Assessment of Microbiology and Turbidity Standards of Water," U.S. Environmental Protection Agency 570/9-83-001, July 1983.

Craun, G. F., S. C. Waltrip, and A. F. Hammonds, "Waterborne Outbreaks in the United States 1971–81," U.S. Environmental Protection Agency, MERL, 1983.

Hawkes, H. A., *Microbial Aspects of Pollution,* Academic Press, London, 1971.

Hem, J. D., "Study and Interpretation of the Chemical Characteristics of Natural Water," Geological Survey, Water Supply Paper 1473, U.S. Government Printing Office, Washington, D.C., 1971.

Livingstone, D. A., "Chemical Composition of Rivers and Lakes, Data of Geochemistry," 6th ed., Professional Paper 440-G, Chap. G., U.S. Geological Survey, Washington, D.C., 1963.

McCutcheon, S. C., *Quality Modeling: Biogeochemical Cycles in Rivers,* CRC Press, Boca Raton, Fla., 1993.

McGauhey, P. H., *Engineering Management of Water Quality,* McGraw-Hill, New York, 1968.

McKee, J. E., and H. W. Wolf, *Water Quality Criteria,* 2d ed., Publication No. 3-A, California State Water Quality Control Board, Sacramento, Cal., 1963.

McKee, J. E., and H. W. Wolf, *Water Quality Criteria,* 2d ed., Publication No. 3-A, California State Water Quality Control Board, Sacramento, Cal., 1971.

Montgomery, J. M., *Water Treatment Principles and Design,* Wiley-Interscience, New York, 1985.

Moore, A. C., B. L. Herwaldt, G. F. Craun, R. L. Calderon, A. K. Highsmith, and D. D. Juranek, "Waterborne Disease in the United States 1991 and 1992," *Journal American Water Works Association,* 86:87–99, 1994.

Moore, B. E., B. P. Sagik, and J. F. Malina, Jr., "Viral Association with Suspended Solids," *Water Research,* 9:197–203, 1975.

National Academy of Sciences and Engineering, "Water Quality Criteria 1972," National Academy of Science, Ecological Research Series, USEPA Report R3-73-033, Washington, D.C., 1973.

National Academy of Sciences, Safe Drinking Water Committee, *Drinking Water and Health,* National Academy of Sciences Printing and Publishing Office, Washington, D.C., 1977.

National Academy of Sciences, Safe Drinking Water Committee, *Drinking Water and Health,* vol. 2 and 3, National Academy of Sciences Printing and Publishing Office, Washington, D.C., 1980.

National Academy of Sciences, Safe Drinking Water Committee, *Drinking Water and Health,* vol. 4, National Academy of Sciences Printing and Publishing Office, Washington, D.C., 1982.

National Academy of Sciences, Safe Drinking Water Committee, *Drinking Water and Health,* vol. 5, National Academy of Sciences Printing and Publishing Office, Washington, D.C., 1983.

National Interim Primary Drinking Water Regulations, *Federal Register,* part IV, Washington D.C., 1975, pp. 59566–89.

Palmer, C. M., *Algae in Water Supplies,* U.S. Department of Health, Education, and Welfare, Office of Water Supply, Cincinnati, Ohio, 1962.

Salvatto, J. A., *Environmental Engineering and Sanitation,* 3d ed., Wiley Interscience, New York, 1982.

Sawyer, C. N., P. L. McCarty, and G. F. Parkin, *Chemistry for Environmental Engineers,* McGraw-Hill, New York, 1994.

Silvey, J., W. Glaze, A. Hendricks, D. Henley, and J. Matlock, "Gas Chromatographic Studies on Taste and Odor in Water," *Journal American Water Works Association* 60:440, 1968.

Silvey, J. K. G., and A. W. Roach, "Studies on Microbiotic Cycles in Surface Waters," *Journal American Water Works Association,* 56(1):60–72, 1964.

Stumm, W., and J. Morgan, *Aquatic Chemistry,* 2d ed., Wiley-Interscience, New York, 1981.

Symons, J., *Water Quality Behavior in Reservoirs,* Federal Water Pollution Control Administration Department of Health, Education, and Welfare and Department of Interior Government Printing Office, Washington, D.C., 1969.

Tchobanoglous, G., and E. D. Schroeder, *Water Quality,* Addison-Wesley, Reading, Mass., 1985.

Turekian, K. K., "Rivers, Tributaries, and Estuaries," in D. W. Hood, ed., *Impingement of Man on the Ocean,* John Wiley, New York, 1971.

U.S. Congress, *Water Pollution Control Act, Amendments of 1972,* PL 92-500, Washington, D.C., 1973.

U.S. Congress, *Safe Drinking Water Act,* PL 93-523, Washington, D.C., 1974.

U.S. Environmental Protection Agency, "Water Quality Criteria," EPA R3-73-033, Washington, D.C., 1972.

U.S. Environmental Protection Agency, "Quality Criteria for Water," Washington, D.C., 1976.

U.S. Environmental Protection Agency, "Waterborne Transmission of Giardiasis," EPA-600/9-79-001, Washington, D.C., 1979.

U.S. Environmental Protection Agency, "Interim Primary Drinking Water Regulations; Amendments," *Fed. Reg.,* 45 (168): 57332–57357, August 27, 1980.

U.S. Environmental Protection Agency, "Quality Criteria for Water," EPA-440/5-86-001, U.S. EPA, Office of Water Regulations and Standards, Washington D.C., 1987.

Wetzel, R. G., *Limnology,* W. B. Saunders, Philadelphia, 1975.

CHAPTER 9
LAKES AND RESERVOIRS

Lawrence A. Baker
Department of Civil and Environmental Engineering
Arizona State University
Tempe, Arizona

9.1 INTRODUCTION

9.1.1 Importance of Lakes and Reservoirs

There are an estimated 131,600 lakes in the United States (EPA, 1990). They are an invaluable ecological resource that also serve many human needs. Lakes and reservoirs provide 68 percent of the water used by the nation's largest utilities (>50,000 customers; NRC, 1992). Most of the surface water used for agricultural irrigation in the arid western United States comes from impounded rivers. About 3 percent of our nation's total energy production comes from hydroelectric power, mostly provided through reservoir releases (Kaufman and Franz, 1993). Lakes and reservoirs enhance our lives by providing opportunities for swimming, boating, and fishing. In the United States, fishing alone has an estimated annual economic impact of $28 billion (Fisheries, 1988).

9.1.2 Objectives

The first half of this chapter examines the general physical and chemical characteristics of lakes and reservoirs, with particular emphasis on chemical mass balances and biogeochemical processes that control the fate and transport of pollutants. Because many of our aquatic systems are impaired in one way or another, the second half of this chapter addresses specific environmental problems and restoration approaches. This is a short overview of a very broad topic, so I have tried to develop a useful reference list for the reader. A few that find a place on the bookshelf of many aquatic scientists include the following: limnology (Hutchinson, 1957; Wetzel, 1983; Goldman and Horne, 1983; Thornton, et al., 1990), lake restoration (Cooke, et al., 1993; NRC, 1992), water-quality modeling (Thomann and Mueller, 1987; Reckhow and Chapra, 1983), aquatic chemistry (Stumm and Morgan, 1981; Snoeyink and Jenkins, 1980; Morel and Hering, 1993; Brezonik, 1994), and engineering hydrology (Linsley, et al., 1982).

9.2   PHYSICAL CHARACTERISTICS

9.2.1   Vertical Stratification

The density of water varies with temperature, with a maximum at 4°C. Because of the dependency of density on temperature, lakes are often vertically stratified. This stratification is often distinct, resulting in an upper, warm *epilimnion,* and a lower, cool *hypolimnion.* These two layers are separated by the *metalimnion,* a region in which the temperature changes sharply. The term *thermocline* refers to the position at which the temperature gradient is largest. The following description of lake stratification was developed from Wetzel (1983); more refined classifications are found in Hutchinson (1957) and Ruttner (1963).

9.2.1.1   Patterns of Stratification.   In temperate regions, most deep lakes exhibit a *dimictic* pattern of stratification. Following the spring ice-out, dimictic lakes typically mix completely. During this *turnover* (mixis) the lake is isothermal from top to bottom. Almost immediately after turnover the surface layers warm, and within a few weeks stable stratification is established. In the fall, as the surface layers cool and winter storms begin, the lake experiences a fall turnover, again becoming isothermal. After ice cover forms, dimictic lakes develop an inverse temperature stratification, with the warmest (4°C) water at the bottom and cooler water above.

   Monomictic lakes experience only one stratification period. Warm monomictic lakes occur in areas where temperatures do not drop below 4°C. These are common in the southeast United States, where warm summers promote a summer stratification period for deeper lakes. Since ice does not form in the winter, these lakes are completely mixed during the mild winter. In contrast, cold monomictic lakes occur in regions where water temperatures never exceed 4°C, primarily in the Arctic or at high elevations. These stratify in the winter under ice, but are well-mixed throughout the summer.

   Polymictic lakes circulate frequently or are continuously mixed. Warm polymictic lakes occur in tropical areas and are common in Florida. These may stratify temporarily during the dog days of August, but the stratification is weak and quickly broken up during storms. Many shallow tundra lakes are cold polymictic lakes, with surface temperatures never exceeding 4°C. *Oligomictic lakes* are also found primarily in tropical areas. They are stratified most of the time, but the stratification is broken up at long and irregular intervals.

   Finally, *meromictic lakes* exhibit stable stratification (years to thousands of years) because of salinity gradients (*chemoclines*). In these lakes, a saline lower layer is found below a less saline layer. Some important examples of meromictic lakes are Lake Mono in California, and the Red Sea in Israel.

9.2.1.2   Chemical Consequences of Stratification.   Vertical stratification has a profound effect on the chemical and biological characteristics of lakes. Although particles can pass downward through a thermocline or chemocline, the diffusion of dissolved substances is greatly restricted when there are sharp temperature or chemical gradients. In strongly stratified lakes there is therefore very little exchange of water or solutes between the epilimnion and the hypolimnion. In most stratified lakes, algal production occurs primarily in the epilimnion; the hypolimnion is "fed" by a rain of sedimenting organic particles. Because decomposition is occurring and there is no contact with the atmosphere, dissolved oxygen (DO) usually becomes depleted in the hypolimnion.

   For a typical dimictic lake, the entire water column mixes to the surface during turnover. The result is a well-oxygenated water column with DO levels near saturation, generally around 8 to 12 mg/L, depending on the temperature. Following the development of stratification, the epilimnion generally remains well oxygenated. In *eutrophic lakes,* the epilimnion can become supersaturated during the day as the result of photosynthetic activity, and slightly undersaturated at night as the result of high respiration rates, but the departure from saturation in either direction is usually only a few mg/L.

In the hypolimnion, decomposition usually predominates over photosynthesis, and there is a net consumption of oxygen during the stratification period. The extent of oxygen depletion depends upon the flux of organic matter into the hypolimnion and its volume. In deep, *oligotrophic lakes* the depletion can be small, but in eutrophic lakes oxygen levels sometimes fall to zero. For many eutrophic lakes, low oxygen levels limit the suitability of the hypolimnion as fish habitat. This can be especially problematic in warm climates, where the fish seek cool water but cannot live there because oxygen is depleted.

The onset of anoxic conditions dramatically changes the biogeochemical environment. Some common events include the production of hydrogen sulfide (H_2S), the reduction of iron and manganese, the production of methane, and the release of iron-bound phosphorus from sediments. The products of these anaerobic reactions accumulate in the hypolimnion, with generally undesirable effects. Hydrogen sulfide is especially problematic, because it is toxic to aquatic organisms, smells, and is highly corrosive. In reservoirs, high levels of hydrogen sulfide can severely damage sluice gates (S. L. Ashby, U.S. Corps of Engineers, personal communication). The discharge of anaerobic reservoir waters can result in levels of hydrogen sulfide and ammonia that are toxic to fish living downstream. The presence of reduced iron and manganese is a problem with drinking water, because both metals impart taste and cause staining problems. Finally, nutrients released during the decomposition process accumulate in the hypolimnion. During the fall turnover, decomposition endproducts are distributed throughout the lake. The mixing of nutrients into the photic zone often results in algal blooms during the spring and fall turnover. When very large quantities of hydrogen sulfide and ammonia have accumulated in the hypolimnion, turnover can result in toxic concentrations of these chemicals throughout the water column, causing fish kills.

9.2.2 Hydrology

9.2.2.1 Residence Time. Lakes and reservoirs receive water from streams, direct precipitation, and subsurface seepage. Water is lost by evaporation, streamflow, and subsurface seepage. The rate of water exchange is usually expressed as a *hydraulic residence time* (Θ_H), which has units of time. Generally, Θ_H is calculated as the ratio between volume and outflow:

$$\Theta_H = \frac{V}{Q_{\text{out}}} \tag{9.1}$$

where V = volume
        Q_{out} = outflow

Since the outflow from many lakes is smaller than the inflow due to evaporation losses, a residence time calculated on the basis of inflow will usually yield a smaller value. For mass balance models (Sec. 9.3), Θ_H is always computed on the basis of outflow. Authors of limnological studies often do not state how Θ_H is calculated, but should in order to avoid ambiguity. Hydrologic residence times for lakes and reservoirs vary from a few weeks to hundreds of years (Table 9.1).

TABLE 9.1 Hydraulic Residence Times for Several Groups of Lakes and Reservoirs in the United States

Values based on outflow unless stated otherwise

Lake or group of lakes	Area, ha	Drainage area: lake area	Overflow rate, m/yr	Residence time, yr
Northeast NSWS drainage lakes ($n = 544$)[a]	18	14	—	0.23
Upper Midwest drainage lakes ($n = 258$)[a]	22	15	—	0.53
Western NSWS drainage lakes ($n = 532$)[a]	5	24	—	0.29
Southeastern U.S. reservoirs ($n = 116$)[b]	8,570	439	24	0.4
U.S. Corp of Engineers' reservoirs ($n = 107$)[c]	3,450	93	18.6	0.37
Seepage lakes in Florida and the Upper Midwest[d]	~5–10	0	0.08–0.43	9–113
Lake Superior[e]	8.2×10^6	1.5	0.82	191
Lake Michigan[e]	5.8×10^6	2.0	0.90	99
Lake Huron[e]	6.0×10^6	2.2	2.8	23
Lake Erie[e]	2.6×10^6	3.0	10.5	2.6
Lake Ontario[e]	1.9×10^6	3.6	11.6	7.9
Lake Washington, WA[f]	8.8×10^3	—	13.3	2.5

[a] NSWS = National Surface Water Survey, a statistically-based survey of low alkalinity lakes throughout the United States. Values represent medians. Residence times for drainage lakes were computed on the basis of inflow (Baker, et al., 1990).
[b] Soballe, et al. (1992).
[c] Thornton, et al. (1980), cited from Cooke, et al. (1993).
[d] Based on water budgets for about a dozen seepage lakes (Baker, et al., 1988; 1991a).
[e] Thomann and Mueller (1987).
[f] Barnes and Schell (1972).

For many modeling applications in which chemical loss terms are calculated as velocities, the *overflow rate* or *water loading rate* q_s is used:

$$q_s = \frac{Q_{out}}{A} \tag{9.2}$$

where q_s has units of L/T and A is the area. The overflow rate is typically on the range of a few tenths of meters to a few meters per year (Table 9.1).

9.2.2.2 Hydrologic Types. The hydrology of various types of North American lakes is reviewed by Winter and Woo (1990). Lakes can be classified according to their sources and losses of water, which in turn strongly influence their geochemical character. *Drainage lakes,* the most common lake type, have surface inlets and outlets. Usually, but not always, most of the water input is derived from the catchment, and precipitation to the lake surface comprises a minor source of water. In drainage lakes with short hydraulic residence times, the major ion composition of lakewater closely resembles that of the input streams. Hydraulic residence times for drainage lakes vary from a few weeks for small mountain lakes to several hundred years for some very large lakes (Table 9.1).

Bartsch and Gakstatter (1978; in Thomann and Mueller, 1987) developed a statistical relationship to predict the hydraulic residence time from the ratio of the drainage area (DA) to the lake area (LA) for lakes in the northeastern United States:

$$\log_{10}\Theta_H = 4.077 - 1.177 \log\left(\frac{DA}{LA}\right) \tag{9.3}$$

where Θ_H is in years. Since this is purely an empirical relationship, it cannot be used outside the northeast.

Reservoirs typically have larger watershed:lake area ratios and are more rapidly flushed than natural lakes (Table 9.1). These characteristics can translate into higher loadings of sediment and nutrients for reservoirs than for natural lakes. They also tend to be more elongated and have more complex shapes than natural lakes. Despite higher nutrient loadings, turbidity limits algae growth in some reservoirs, and the growth of algae in very rapidly flushed reservoirs may be limited by cell washout (Soballe, et al., 1992). Reservoirs that experience major drawdowns have little development of littoral biological communities and have planktonic-based food webs. Elongated reservoirs nearly always exhibit longitudinal changes in turbidity; because of this, concentrations of metals, nutrients, and other particle-bound chemicals also tend to decrease along the length of the reservoir. Relationships between hydrology and the chemical and biological characteristics of reservoirs are reviewed by Soballe, et al. (1992) and Thornton, et al. (1990).

Closed basins have surface inlets but no outlets. Many discharge to groundwater and are hydrologically similar to drainage lakes. For lakes that have no *outseepage* (groundwater recharge), water leaves only by evaporation. These lakes have a $\Theta_H = \infty$. Because solutes are added and never removed, concentrations of normally conservative solutes build up until they precipitate as salts. Closed basin, saline lakes are fairly common in the western United States. An example is Mono Lake, California, which has a salinity of 90 g/L, three times that of the ocean (NRC, 1987). The biota of extremely saline lakes is generally dominated by a very few species (brine shrimp; brine flies; no fish) that are sometimes present in massive numbers.

Seepage lakes lack channeled inlets or outlets. These are common in dune systems, in karst topographies, and in glaciated regions, where they form in potholes. These can be further subdivided into *groundwater recharge lakes*, which receive most of their water input from precipitation and lose water to the groundwater system, and *groundwater flow-through lakes*, which receive a substantial portion of their total inflow from groundwater. For some seepage lakes in the Upper Midwest and Florida, precipitation can contribute more than 90 percent of the total water input. They have received considerable study in recent years because they are particularly susceptible to acidification by acidic deposition (Eilers, et al., 1983; Baker, et al., 1986a). These lakes tend to be dilute; their chemical composition is controlled to a large extent by within-lake processes (Brezonik, et al., 1987). In contrast, the chemistry of groundwater flow-through lakes tends to reflect that of the groundwater system. These two types of seepage lakes are visually identical; classification is based on chemical composition or hydrology. Water budgets for about a dozen seepage lakes in Florida and the Upper Midwest have been developed in the course of acidification studies. Their hydraulic residence times range from a few years to over 100 years (Baker, et al., 1991a).

9.3 HYDROLOGIC AND CHEMICAL BUDGETS

Aquatic scientists frequently construct hydrologic and chemical budgets of lakes for several objectives: (1) to elucidate the mechanisms of pollutant transport and degradation, (2) to develop mathematical models, (3) to predict the effect of pollutant control options, and (4) to evaluate the efficacy of pollution control measures after they are implemented. Because there is little written on the general aspects of chemical mass balances, and because mass balances are often developed rather uncritically, this section focuses on the methodology of developing chemical mass balances and on the errors that one is likely to incur along the way. For a thorough review of some of the problems associated with water budgets, the reader is referred to the work of Winter (1981).

9.3.1   Mathematical Balance for Water and Chemicals

Mathematical balances for water and chemicals are widely used in the analysis of the fate and transport of pollutants in lakes. A generalized *chemical budget* for a lake-sediment system can be written (using the nomenclature of Thomann and Mueller, 1987):

$$V\frac{dC_T}{dt} = W_T - \Phi C_{T1} + K_f A(f_{d2}C_{T2}/\Phi - f_{d1}C_{T1}) - K_{d1}f_{d1}V_1 C_{T1}$$

$$+ k_1 A \frac{C_g}{H_e - f_{d1}C_{T1}} - v_s A f_{p1}C_{T1} + v_u A f_{p2}C_{T2} \quad (9.4)$$

$$\frac{VdC_{T2}}{dt} = -K_f A(f_{d2}C_{T2}/\Phi - f_{d1}C_{T1}) - K_{d2}f_{d2}V_2 C_{T2} + v_s A f_{p1}C_{T1} - v_u A f_{p2}C_{T2}$$

$$- v_d A f_{p2}C_{T2} \quad (9.5)$$

where C_{Tx} = total concentration in lake ($x = 1$) and sediment ($x = 2$), M/L^3
         f_{px} = fraction of C_T in the particulate form
         f_{dx} = fraction of C_T in the dissolved form
         W_T = loading to lake, M/T
         k_1 = overall mass transfer coefficient for volatilization, L/T
         C_g = concentration in gas phase, M/L^3
         H_e = Henry's Law constant, M$_T$/(L$_g^3 \cdot$ M$_T \cdot$ L$_w^3$)
         K_{dx} = dissolved chemical overall decay rate, L/T
         v_s = settling velocity of solids, L/T
         v_u = solids resuspension velocity, L/T
         v_d = sedimentation (burial) velocity, L/T
         K_f = sediment-water dissolved chemical diffusion rate, L/T
         Φ = porosity of sediment

The loading W_T is in turn comprised of precipitation, stream loadings, inseepage, and aerosol deposition to the lake surface. Outputs include stream outputs and groundwater seepage. The direction of the volatilization flux depends upon the concentration of the volatile species in the atmosphere relative to its concentration in the lake (see *Volatilization* in Sec. 9.3.1.1).

9.3.1.1   *Components of Input-Output Budgets.*   *Atmospheric Inputs.*   Chemical inputs from the atmosphere comprise two general components: wet precipitation (rainfall) and dry deposition. The latter includes deposition of aerosols (particles) and the adsorption of gases (such as SO_2 and HNO_3) directly to the lake surface. Although often neglected in chemical budgets, atmospheric loadings can be an important component of the chemical budget of lakes. This is particularly true for lakes with relatively small watershed:lake ratios (Tisue and Fingleton, 1987; Swack-hammer and Eisenreich, 1991; Messer and Brezonik, 1978; Baker, et al., 1986a).

The least expensive approach to estimating precipitation volumes for lakes is to use one of a number of algorithms to interpolate precipitation values from the

National Weather Service (NWS) network. Errors incurred in this approach can be very high, especially when the network density is low and in mountainous regions; errors are higher for short periods of time than for longer periods (Winter, 1981). Chemical measurements are not made in the NWS network, so estimates of chemical loading also require an estimate of the chemical composition of rainfall. Useful reviews on atmospheric deposition rates include Brezonik (1975), deposition rates for nutrients; Stoddard (1994), nitrogen inputs throughout North America; Jeffries and Snyder (1981), deposition of metals; and Baker, et al. (1991a), gradients in deposition of major ions across North America. Multiplying literature values of chemical composition with interpolated rainfall will lead to very high uncertainties and should be used only when chemical inputs from precipitation are unlikely to comprise a major portion of the total chemical input.

For more accurate estimates of atmospheric loadings, on-site monitoring is required. Two types of instruments are available for this purpose. *Bulk deposition collectors* collect both rainfall and dry deposition using a collector that is continuously open. *Wet/dry collectors*, such as the type manufactured by the Aero-Chemetrics Corp., are comprised of two buckets and a movable lid. The lid is activated by a moisture sensor. One bucket (the dry bucket) is exposed when it is not raining and the other (wet bucket) is exposed when it is raining. The National Acid Deposition Program (NADP) uses this type of collector, so there is a substantial data base for atmospheric deposition of major ions. Brezonik, et al. (1983) reported that the deposition of major ions at collocated bulk and wet/dry collectors generally differed by less than 20 percent.

Atmospheric deposition collectors are beset with a number of operational problems. The lids of the wet/dry collectors do not always open and close, so the collection efficiency on the wet side is usually less than 100 percent. Consequently, volume collected in the wet bucket is normalized to precipitation volume as determined by a standard rain gauge located at the site. Because of this, and the fact that bulk deposition collectors seem to collect the same input, there seems to be little advantage to using wet/dry collectors instead of the much simpler bulk collectors. Problems associated with wet/dry collectors are discussed by Gatz, et al. (1988) and Baker (1991c).

Bulk collectors and wet/dry collectors substantially underestimate gas deposition to wet surfaces (Dasch, 1983; Lewis, 1983). This is particularly problematic for SO_2 and various nitrogen species. An approach for estimating gaseous inputs of these species follows.

Volatilization. Certain chemicals, including a variety of organic compounds (DDT and PCBs) and some inorganic compounds (e.g., H_2S, NH_3, and CH_4), move across the air–water interface. The volatilization rate r_c depends upon the degree of saturation of the compound in water:

$$r_c = k_1 \left(\frac{C_{Tl} - C_g}{H_e} \right) \tag{9.6}$$

When a compound is oversaturated the direction of movement is from the lake to the atmosphere. Conversely, when $C_{Tl} < C_g/H_e$, the direction of gas transfer is from the atmosphere to the lake. When $C_{Tl} = 0$,

$$r_c = \frac{-k_1 C_g}{H_e} \tag{9.7}$$

Under these conditions, k_1/H_e is a constant, which is sometimes called the *deposition velocity* v_d. Thus, for chemicals that undergo rapid transformation (e.g., $SO_2 \rightarrow SO_4^{2-}$)

in water, the mass transfer rate is often represented as the product of the atmospheric gas concentration and a deposition velocity:

$$r_c = v_d C_g \tag{9.8}$$

Voldner, et al. (1986) review the literature on nitrogen and sulfur dry deposition. We have used the following estimates of v_d for gases in several lake chemical budgets: SO_2, 0.5 cm/s; HNO_3, 1 cm/s; NO_2, 0.1 cm/s (Baker, 1991c), but the accuracy of these estimates is unknown. There is no good method for measuring gas adsorption rates to lake surfaces; modeling studies are extremely complex and well outside the scope of almost any limnological study.

Evaporation. The most common method of estimating evaporation from lakes, and the only method that can be used in reservoir design, is to measure *pan evaporation*. Pan evaporation rates are adjusted to lake evaporation rates using a pan coefficient. Problems with this approach (summarized from Winter, 1981) include: (1) temporal and regional variations in pan coefficients; (2) variability of evaporation rates among different types of pans, which can account for as much as 30 percent variation among types; (3) the need for daily maintenance of the pans; and (4) uncertainties in extrapolating from measurement sites to lake sites. Winter suggests that the commonly used pan coefficient of 0.7 be used for annual estimates only. A summary of annual pan coefficients is found in Linsley, et al. (1982). Figure 9.1 could be used as a rough estimate of evaporation from shallow lakes.

The mass transfer method is based on the concept that the rate of evaporation is a product of the difference between the saturation vapor pressure and the measured vapor pressure and a mass transfer coefficient.

$$E = (e_o - e_a)(a + bv) \tag{9.9}$$

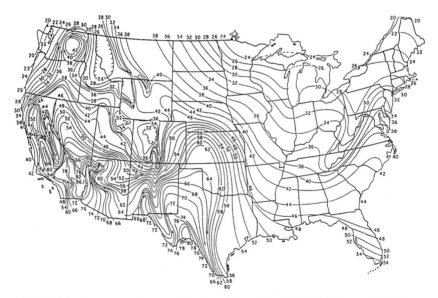

FIGURE 9.1   Average annual evaporation from shallow lakes, in in/yr. (*Linsley, et al., 1982.*)

where e_o = saturation vapor pressure, calculated from the temperature of the water surface; e_a = vapor pressure of the air at a fixed height above the water surface; v = wind velocity at a fixed height; and a and b are empirically determined coefficients. Harbeck (1962; in Linsley, et al., 1982) found that for reservoirs up to 120 km², $a = 0$ and

$$b = 0.29A^{-0.05} \tag{9.10}$$

where A = area, km²; the wind velocity is measured at 2 m above the surface; and e_a is measured in the unmodified (upwind) air. This method requires relatively standard meteorological measurements: windspeed, relative humidity, and temperature of the lake and air.

Winter (1981) considered the energy budget method to be the most accurate method of estimating evaporation, but this method has not been widely used because most radiometers measure only shortwave radiation; both longwave and shortwave net radiation is needed for energy budget calculations (see Linsley, et al., 1982).

Subsurface Seepage. Subsurface seepage is sometimes neglected in water balance calculations, but in many circumstances subsurface inflows or outflows can be very important. Even when seepage flows are relatively small, chemical inputs from seepage can be large if the groundwater is enriched in a constituent of interest. In the Upper Midwest, for example, groundwater inflows that comprise only a few percent of total water input can provide sufficient alkalinity to buffer against acidification from acid deposition (Baker, et al., 1991a). In the past, the most common approach for estimating net seepage inputs (Q_{gw}) to lakes has been to compute seepage as the residual term in a water balance:

$$\frac{dV}{dt} = = Q_i + Q_p - Q_e - Q_o + Q_{gw} \tag{9.11}$$

where Q_i = stream inflows
Q_p = precipitation to the lake surface
Q_e = evaporation losses
Q_o = stream outflows

All are in units of L^3/T. There are two problems with this approach. First, it determines only *net seepage flow*. This is a problem for lakes that have both inseepage and outseepage. For example, in McCloud Lake, Florida, groundwater flowed to the lake from all sides during the first half of 1980 and from the lake to the groundwater in the second half of the year, when a multiyear drought was ended by a period of high precipitation (Baker, et al. 1986a). In other lakes, groundwater moves inward through one end of the lake and lakewater flows outward to the groundwater system on the other (see Winter, 1986). Estimates of net seepage would not convey this information. Second, this approach aggregates all errors incurred in the water budget into the seepage term, making it highly unreliable (Winter, 1981). Errors in seepage inputs by this method can exceed 100 percent.

A second common method of measuring seepage flows in lakes is the *flow net approach* (e.g., Winter, 1986). In this approach, a network of piezometers is placed throughout the watershed near the lake to determine hydraulic gradients; flows are then computed by Darcy's law. From a hydrologic standpoint, the main problem with this approach is the determination of hydraulic conductivities. For chemical mass balances, another limitation is that one must normally assume that concentrations of lakewater and groundwater are representative of outflows and inflows, respectively. This may be unrealistic because the composition of water may change significantly while passing through the lake sediment.

A third approach, which is becoming more popular, is to measure seepage directly across the lake bottom using either *seepage meters* or *hydraulic potentiometers* (reviewed by Rosenberry, 1990). Seepage meters, first developed by Lee (1977), are most often made of 55-gal drums cut in half. A tube with a stopcock is installed in the top, to which is attached a baggie filled with water. After placing the seepage meter into the lake sediments, the stopcock is opened. After a period of time (days to weeks), the baggie is removed and weighed. Losses of water from the baggie are used to compute outseepage; gains of water in the baggie are considered inseepage. To make lakewide estimates of seepage movement requires a network of seepage meters (we used 20 in a study of a 5-ha lake in Florida; Baker, et al., 1986a). In small lakes, the highest flows occur in the littoral zone; flows decline exponentially with distance from the shoreline (Rosenberry, 1990). The main disadvantage of this approach is that it is time-consuming and expensive. It has two major advantages: (1) the location of inflows and outflows can be determined accurately, and (2) during periods of inseepage, samples can be collected for chemical analysis, thereby providing a basis for direct computation of chemical loadings.

Hydraulic potentiometers measure the difference in head between the lake surface and the underlying groundwater. Using measured or estimated values for the hydraulic conductivity, the gradient, dH/dt, is used to estimate flows using Darcy's law. A design was published by Winter, et al. (1988).

Stream Chemical Loadings. Stream inputs are the dominant source of water and chemicals to most lakes and reservoirs. There is often substantial error in stream chemical loadings, and the methodology of loading calculation should be considered carefully. Four methods are widely used: (1) simple methods, in which the product of flow and concentration is determined for discrete time intervals; (2) regression methods; (3) complex methods that make greater utilization of flow variations; and (4) estimation methods. Several investigators have examined loading methodologies for the common monitoring situation in which flow is measured continuously and concentration data is collected at regular intervals, e.g., biweekly or monthly (Dolan, et al., 1981; Stevens and Smith, 1978; Young, et al., 1988). Factors to consider in selecting a methodology include: (1) the amount of data available and, in particular, whether instantaneous flow has been or will be measured, (2) the flashiness of the stream, (3) the degree of accuracy needed, and (4) funds available for chemical monitoring.

Simple methods of computing stream loading are made by simply multiplying flow times concentration. This is probably the most commonly used method, but it is subject to considerable error unless an enormous amount of data is available (e.g., daily measurements of concentration and instantaneous flow) or the system has very stable flows. More often, this method is used for monitored streams with continuous flow measurement and regular, infrequent, concentration measurements. Because a major portion of the loading occurs within a small period of high flows, this approach tends to underrepresent chemical inputs during high flows. In a Monte Carlo analysis of alternative loading methods in which the true loading of the Grand River in Michigan was known from daily flow and concentration measurements, Dolan, et al. (1981) found that loadings calculated from daily flow and average monthly concentration had a standard deviation of 11 percent of the true loading. Thus, an annual loading calculated by this approach would be within 11 percent of the true loading only 67 percent of the time.

A second approach for regularly monitored streams having instantaneous flow measurements is to develop statistical relationships between flow and concentration. These regression equations are then used to compute concentration and loadings on days when concentration is not measured. This method also takes advantage of the abundance of flow data that is available for gauged streams.

An extension of this approach is used for very flashy systems in which storms represent the major flow. In this method, a number of storms (typically 4 to 12) are sampled at numerous points throughout the storm to generate event loadings. Regression analysis is used to develop a relationship between total event flow and total event chemical loadings. Flow is measured continuously, allowing the calculation of total storm flow for each event; the regression equation is then used to infer chemical loading. Baseflow, which represents a small portion of the total loading, is sampled at regular intervals to determine baseflow loading. This approach is nearly always used to analyze stormwater loadings.

Finally, several investigators have used *Beale's ratio estimator* to calculate loadings. In this method the estimated loading is calculated as:

$$\mu_y = \mu_x \frac{m_y}{m_x} - \left[\frac{1 + S_{xy}/nm_xm_y}{1 + S_x^2/nm_x^2} \right] \tag{9.12}$$

where μ_y = estimated load
   μ_x = mean daily flow for the year
   m_y = mean daily loadings for the days on which concentrations were determined
   m_x = mean daily flow for days on which concentrations were determined
   n = number of days on which concentrations were determined

$$S_{xy} = \frac{1}{(n-1)} \sum_{i=j}^{n} x_i y_j - nm_x m_y$$

$$S_x^2 = \frac{1}{(n-1)} \sum_{i=j}^{n} x_i^2 - nm_x^2$$

where x_i = individual measured flows
   y_i = daily loading for each day on which concentration was determined

This approach makes maximum use of flow information and is best suited for situations with an abundance of flow data but little concentration data. Dolan, et al. (1981) and Young, et al. (1988) have advocated use of this approach, finding it superior relative to the simple methods or the regression method. Further improvement occurs with postsampling stratification into high and low flow periods.

An advantage of the ratio method over regression methods is that it does not depend upon a flow–concentration relationship. This is important because a significant regression relationship between flow and concentration may not exist. One disadvantage of the ratio method compared to the regression method is that the ratio method cannot be used to estimate loadings during periods when concentration is not measured (breaks in the data). Here the regression approach has an advantage: Once a flow–concentration relationship is developed for a river, it can be used to predict loadings during years when no concentration data is collected. In an analysis of arsenic loadings in several Arizona rivers, we found that Beale's ratio method and the regression method yielded nearly identical loadings for a six-year period of record. However, annual loadings computed by the regression approach are probably more accurate, because the regression approach makes full use of the information contained in the entire data set (Qureshi, 1995).

In designing (or redesigning) monitoring networks, several investigators have concluded that sampling frequency should reflect the importance of high flows in the overall annual budget. Reckhow and Chapra (1983) have suggested that the frequency of stream chemical sampling be stratified according to flow, not time. This sage advice is rarely heeded.

Estimation methods are used when stream loadings cannot readily be determined. Estimation methods are most commonly used in eutrophication studies, in which the nonpoint source loadings of phosphorus and nitrogen are estimated by the use of *export coefficients*. In this method, land export coefficients unique for specific land uses are used to compute watershed inputs of nutrients. The total loading is:

$$L = \Sigma A_i \beta_i + \text{point source loadings} \tag{9.13}$$

where A_i = area in land use i, ha
β_i = export coefficient, kg/ha · yr

Export coefficients are usually determined for single-use watersheds (Table 9.2) but we have had some success with a stepwise backward regression approach to determine export coefficients for 40 multiple-use watersheds in Florida (Baker, et al., 1981). Point source loadings are determined separately. Gakstatter, et al. (1978) compiled data on nutrient loadings from several hundred municipal wastewater treatment plants; Reckhow and Chapra (1983) include a discussion of phosphorus inputs from various sources. The error incurred with estimation methods is very large. They are therefore most useful for preliminary calculations and for designing sampling strategies.

Changing Storage. *Lake volume* is determined by developing a depth–volume relationship. Typically, this is done by running multiple sonar transects across a lake bottom to determine depths at various locations. The depths are then mapped, and the area between isopleths is used to compute the volume of each stratum. Total storage for a constituent is computed as the product of the concentration of that constituent within a stratum and the volume of the stratum. One problem with this approach is that the position of the boat must be accurately known. Traditionally, this is done by starting at a landmark and moving across the lake at a known speed and precise direction. Recently Gubala, et al. (in press) have developed a computerized system that incorporates the global positioning system, a research-grade fathometer, and the geographic information system for developing lake bathymetric maps. The system is capable of developing bathymetric maps far faster than the conventional approach.

The degree to which errors in storage are important depends upon the residence time of the chemical in question. Errors in chemical storage will generally be larger

TABLE 9.2 Export Coefficients for Several Water Quality Constituents, by Watershed Type, kg/ha · yr

	Urban	Agriculture	Forest	Precipitation
Phosphorus*	0.5–5.0	0.1–3.0	0.02–0.45	0.15–0.6
	(0.8–3.0)	(0.4–1.7)	(0.15–0.3)	(0.20–0.50)
Nitrogen[†]	8–10	0.8–75	1–8	3–30[‡]
Suspended solids[†]	300–2,500	5–8,000	2–900	—
Lead[†]	0.2–0.6	0.003–0.09	0.01–0.05	0.12–3.5[§]

* From Table 8.13 of Reckhow and Chapra (1983). Values are the entire range of observations; values in parentheses are typical values.

[†] Range of land use export coefficients from Novotny and Chesters (1981) cited in EPA (1991) except as noted.

[‡] Brezonik (1975).

[§] Jeffries and Snyder (1981); data from North America.

for chemicals that have residence times longer than Θ_H (as the result of in-lake sources or dry deposition) than for chemicals that have residence times shorter than Θ_H (the result of in-lake sinks). Errors in chemical storage are also likely to be larger for lakes with rapidly changing depth (e.g., a fluctuating reservoir) than for a lake with a stable water level.

9.3.1.2 *Measurements of Internal Fluxes.*   Input-output budgets reveal only net sinks and sources of chemicals and provide little or no knowledge about the transformations involved. The development of internal chemical budgets can provide additional information about the fate and transport of pollutants within the lake or reservoir system.

Sedimentation.   Settling rates of particles in lakes and reservoirs are not accurately predicted by Stoke's law. Difficulties arise in determining shape factors for individual algae species and in assigning a density, because algae regulate their buoyancy. Furthermore, Stoke's law does not account for turbulence and flow velocities (Jorgenson, 1989). Sedimentation velocities can be estimated within a broad range using literature values (tabulated in Jorgenson, 1989), but gross sedimentation is often measured directly using sediment traps. *Sediment traps* are usually an array of narrow tubes with capped bottoms that are placed in a support frame. The trap is suspended at a specified depth and left in place for periods of a few days up to several weeks. Trap material is then filtered or centrifuged and analyzed for dry weight, elemental composition, and (occasionally) particle characteristics. There is considerable debate over how sediment traps should be constructed and maintained in order to mimic natural conditions, and there is no standard protocol for their use. Important variables include the duration of collection period, the use of preservatives, the width and the width:length ratio of the tubes, and the placement of traps (Blomqvist and Kofoed, 1981; Hakanson and Jansson, 1983).

Diffusion Across the Sediment–Water Interface.   There is often substantial movement of solutes across the sediment–water interface. These fluxes occur for two reasons. First, catabolic processes within the sediments consume oxygen, produce nutrients and methane, reduce sulfate and nitrate, and decompose some organic toxins. Second, partition coefficients for metals and organic compounds are lower for particles in sediments than for particles suspended in water. This results in desorption of adsorbed chemicals in the sediment environment (Thomann and Mueller, 1987). Both of these mechanisms create chemical gradients across the sediment–water interface that result in a movement of solutes from the region of higher concentration to the region of lower concentration.

The most common method of measuring sediment consumption of oxygen and production of nutrients is to measure concentration changes in sediment chambers. Both laboratory and in situ chambers have been used; these are operated either in a batch mode (usually with a stirrer) or as flow-through systems. These are usually made of Plexiglass and have surface areas of <0.25 m^2. They may be equipped with sampling ports, chemical probes (especially for oxygen), and sometimes a stirring mechanism. Changes in concentration over time (typically <24 h) are used to compute areal flux rates:

$$F = \frac{dC}{dt}\frac{V}{A} \tag{9.14}$$

where F = areal flux, $M/L^2 - T$
A = cross-sectional area of the chamber, L^2
dC/dt = change in concentration with time, $M/L^3 - T$

A number of operational conditions greatly effect measured flux rates from chamber experiments. These include: (1) stirring speed or flow; (2) buildup or depletion of chemicals, especially depletion of oxygen, within the chamber; and (3) length of the measurement period. Sediment chambers have been the most common method for measuring sediment oxygen demand (SOD) and for measuring short-term fluxes of nutrients. They are less useful for chemicals with lower flux rates, because the chambers often go anoxic during the measurement period. Some useful studies include James (1974), Belanger (1980), Bowman and Delfino (1980), and Sonzogni, et al. (1977).

There has been an increasing use of gradient techniques to measure flux rates across the sediment–water interface. In this technique, chemical gradients near the sediment–water interface are measured using porewater equilibrators (Hesslein, 1976) or by extracting porewater from sectioned sediment cores. *Fick's first law* is used to compute fluxes:

$$F = \frac{dC}{dZ} D_c \tag{9.15}$$

where dC/dZ = concentration gradient at the sediment–water interface, M/L^4, and D_c = effective diffusion coefficient, L^2/T. The *effective diffusion coefficient* is computed as $D_c = \Phi \times D_b$ where Φ is the porosity of the sediment (typically $0.7 < \Phi < 0.95$) and D_b is the bulk diffusion coefficient. Values of D_b for some common chemicals are found in Li and Gregory (1974). The valid use of Fick's first law depends upon several assumptions: (1) there is no advection of water through the sediment, (2) there is no bioturbation, and (3) the reaction is occurring in the sediment below the interface (see Berner, 1980). Although many investigators have used single porewater profiles to estimate annual fluxes of solutes across the sediment–water interface, studies by Rudd, et al. (1990) and Sherman, et al. (1994) show that chemical fluxes across the sediment–water interface vary in time and space.

Sediment Accumulation.   Measurements of sediment chemical accumulation are done by collecting intact cores. These are segmented; the segments are then dated and analyzed for the chemical constituents of interest. Sediments less than 150 years old are dated by ^{210}Pb analysis (Oldfield and Appleby, 1984). Sediment ages are often verified by examining cultural markers (e.g., charcoal from forest fires or the occurrence of ragweed pollen coinciding with early agricultural development). Concomitant measurements of bulk density and dry weight within each core stratum are used to compute dry mass sedimentation rates. Chemical accumulation rates are calculated as the product of the dry mass sedimentation rate and the chemical concentration within each stratum.

Sediment cores have been widely used to document historical changes in the water quality and biota of lakes (Sec. 9.5.3.2). Sediment cores can be used to quantify lakewide accumulation of chemicals. By integrating data from multiple cores, Engstrom, et al. (1994) computed recent and preindustrial lakewide accumulation rates of mercury in a dozen lakes in the midwestern United States, concluding that atmospheric input rates have increased by approximately 3- to 4-fold in the past 100 years. The determination of net chemical sinks by sediment coring has the advantage over the conventional method of calculating net sinks as the difference between inputs and outputs in that it is a direct measurement. This approach is valid only for chemicals that have negligible (or known) degradation and volatilization losses and do not migrate through the sediments.

Sediment cores can also be used to determine recycling rates. In this method, recycling of chemicals at the sediment–water interface is computed by subtracting surficial accumulation (deposition) with accumulation at depth (burial) (Armstrong, et al., 1987). Linked with measurements of gross sedimentation, Armstrong, et al. used this method to compute recycling of nitrogen phosphorus, silica, and PCBs in Crystal Lake, Wisconsin (Fig. 9.2).

9.3.1.3   Checks on Mass Balances.   Ideally, one would have information on the statistical uncertainty associated with each measured input or output from a lake; the magnitude of potential errors could then readily be determined by statistical error propagation methods. Unfortunately, this type of information is almost never available. Monte Carlo methods are sometimes used to estimate cumulative errors, but in using this approach the bounds of uncertainty need to be specified. Even if statistical uncertainties on measurements were known, there is the additional problem that a measurement technique can be nonrepresentative (e.g., improper location of a bulk deposition collector).

One of the most common ways to check the reliability of the hydrologic balance is to complete a chloride budget and examine the magnitude of the chloride sink. The operational assumption is that chloride is conservative; nonzero sinks are taken as an indication of potential error in the water budget. Although chloride is often conservative, this assumption should not be accepted uncritically (see discussion in Baker, et al., 1991a).

A second approach is to use sensitivity analysis to evaluate the effect of potential errors. For example, if one is concerned with measured seepage inputs, one could examine the effect of doubling or halving the measured seepage input. As with the Monte Carlo method, this assumes that the investigator has some idea of the potential magnitude of the uncertainty.

A third approach is to develop water or chemical budgets in which all components are measured, and the error is computed as a residual. Ideally, the computed

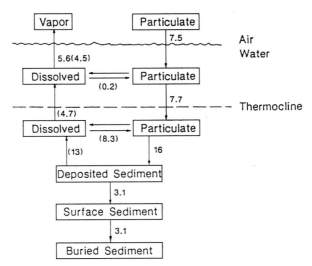

FIGURE 9.2   Deposition and recyling of PCBs in Crystal Lake, Wisconsin. Arrows denote fluxes ($\mu g/m^2 \cdot yr$); values in parentheses were calculated by difference. (*Armstrong, et al., 1987.*)

residual error is smaller than any of the measured inputs or outputs, although this of course is not always the case. This approach is commonly used for evaluating water budgets.

Finally, one can develop an overdetermined chemical budget. Because one is often interested in net sinks, sediment chemical accumulation measured directly by coring analysis can be used as a check on the sink determined from an input/output budget. This approach was used to develop a sulfur budget for Little Rock Lake, Wisconsin. In this study, the lakewide sulfate sink computed from measured inputs and outputs was compared with the sink computed by calculating sulfur accumulation in multiple cores throughout the depositional zone (Baker, et al., 1989; Fig. 9.3). The relatively close correspondence between the two approaches provided confidence that errors in measured inputs and outputs were relatively small (Baker, et al., 1989).

9.4 BIOGEOCHEMICAL PROCESSES OF POLLUTANTS

The effect of pollutants on the biota of aquatic systems depends in large measure upon their fate and transport. Biogeochemical transformations of a few important pollutants are presented here. Regrettably, there is no general reference on the subject of biogeochemical cycling in aquatic systems, although several limnology books (Wetzel, 1983; Goldman, 1983) describe the basics for major nutrients.

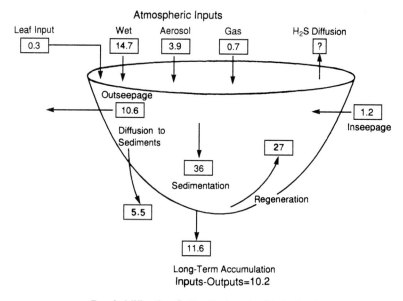

Pre-Acidification Sulfur Budget for Little Rock Lake.

FIGURE 9.3 Sulfur budget for Little Rock Lake, Wisconsin. Fluxes are in mmole/m² · yr. Net sinks were estimated from inputs minus outputs and by sediment accumulation measured in four cores in the depositional zone. (*Baker, et al., 1989.*)

9.4.1 Nitrogen

Nitrogen is sometimes the limiting nutrient to algal growth, comprising 1 to 10 percent of the dry mass of algae. The process of nitrification, which converts NH_4^+ to NO_3^-, is important because it consumes oxygen and can cause substantial oxygen depletion in aquatic systems. Free ammonia (NH_3) is toxic to fish and other aquatic organisms at concentrations that are sometimes encountered in hypereutrophic systems (Effler, et al., 1990). Nitrate is toxic to humans, the basis of a drinking water maximum contaminant level (MCL) of 10 mg/L, but this level is rarely seen in lakes.

9.4.1.1 Components. *Nitrogen Fixation.* For most lakes, the majority of nitrogen enters the lake in the form of NH_4^+, NO_3^-, or organic nitrogen. Many blue-green algae and certain bacteria can also fix nitrogen gas from the atmosphere, forming organic nitrogen in the process. Species capable of doing this include several blue-green algae of the genera *Aphanizomenon, Anabaena, Gleotrichia, Nodularia,* and *Nostoc* and certain bacteria (Wetzel, 1983). The contribution of *nitrogen fixation* to the total annual nitrogen budget is highly variable, ranging from <1 to 90 percent (summarized by Goldman and Horne, 1983). The contribution of fixation tends to be small for lakes that have high external N inputs. For Lake Mendota, Wisconsin, a typical midwestern eutrophic lake, nitrogen fixation accounts for about 5 to 10 percent of the total nitrogen input (Fig. 9.4). Fixation by blue-green algae is inhibited when inorganic nitrogen is available. Nitrogen-fixing blue-green algae often form blooms when the supply of combined nitrogen is low, commonly during midsummer, because their ability to fix N gives them a competitive advantage over non-N-fixing algae. Nitrogen fixation also occurs in the nodules of some wetland species, such as *Alnus* (alder).

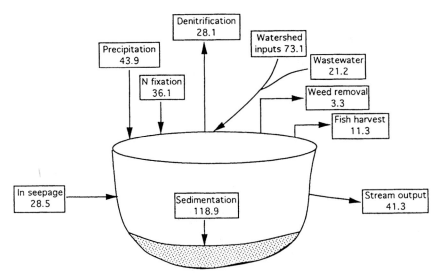

FIGURE 9.4 An estimated nitrogen budget for Lake Mendota. Fluxes are in kg · N/yr. Adapted from Table 12-12 of Wetzel (1983) based on data from several studies.

Nitrogen fixation is most commonly measured by the *acetylene reduction technique*. In this method, acetylene is introduced into a water sample. The acetylene mimics N_2 and is converted to ethylene, which is readily measured by gas chromatography.

Denitrification.　*Denitrification* is an anaerobic process in which NO_3^- is used as an electron acceptor in respiration, resulting in the formation of nitrogen gas (N_2). Denitrification occurs in the anoxic hypolimnia and sediments of lakes and in adjacent wetlands. A wide variety of bacteria can reduce nitrate. In a review of denitrification studies Messer and Brezonik (1983) found that losses of N by denitrification accounted for 0 to 81 percent of input nitrogen in the 20 or so lakes for which denitrification contributions had been determined. Biogeochemical considerations suggest that denitrification should be more important in shallow, eutrophic lakes than in deeper, less productive lakes. Detailed reviews of denitrification in aquatic systems are found in Brezonik (1977) and Seitzinger (1988; 1990). Details of several analytical approaches for measuring denitrification are found in several chapters of Revsbech and Sorenson (1990).

The most common method for measuring denitrification is the *acetylene inhibition* (or blockage) *technique*, but absolute confirmation requires isotope techniques (conversion of $^{15}NO_3^-$ to $^{15}N_2$). Denitrification at the sediment–water interface can also be studied by examining vertical gradients of nitrate (Messer and Brezonik, 1983; Rudd, et al., 1986). For determining whole-lake denitrification rates, Messer and Brezonik (1978) developed a mass balance approach predicated on the fact that denitrification will cause a decrease in the N:P ratio of the sediments relative to that of lake inputs. Several techniques for measuring denitrification are compared by Messer and Brezonik (1983) in their study of Lake Okeechobee, Florida.

Nitrification.　*Nitrification* is the process by which ammonium (NH_4^+) is converted to nitrate in the presence of oxygen. It is carried out by several specialized group of bacteria (primarily *Nitrosomonas* and *Nitrobacter*) which obtain energy from the oxidation process. The two-step process occurs as follows:

$$NH_4^+ + 3/2O_2 \rightarrow NO_3^- + 2H^+ + H_2O \tag{9.16}$$

$$NO_2^- + 1/2O_2 \rightarrow NO_3^- \tag{9.17}$$

An important aspect of nitrification is that it can reduce oxygen concentrations. This is especially important in rivers, where the nitrogenous oxygen demand can play an important role in oxygen dynamics. In lakes, it is not uncommon to observe a buildup of ammonium under the ice during the first few months of winter that is followed by an increase in nitrate and concomitant decline in NH_4^+ as nitrification proceeds. The nitrification process is one of the more sensitive biogeochemical processes with respect to pollutant effects. A buildup of NO_2^-, an intermediate process in the oxidation, is often taken as a sign of toxicity (Rheinheimer, 1971). Although nitrification has been widely considered to be inhibited by pH values <5 to 6, this point has been debated in acidification studies (Rudd et al., 1988; Sampson et al., 1994).

Assimilation.　Both NH_4^+ and NO_3^- are utilized by algae, and several species can utilize organic nitrogen species. Half saturation constants K_s for uptake of both species have been determined for about a dozen species. Goldman and Horne (1983) assert that K_s values increase with cell size and with increasing nutrient concentrations, with values ranging from 3 to 1000 µg/L.

Mineralization.　Most of the nitrogen in algal cells occurs as proteins, so the first step in mineralization is the release of NH_4^+ from the amino group of proteins. A wide variety of bacteria carry out the primary mineralization step.

Volatilization.　The pK_a (acid dissociation constant) for ammonium is 9.3. At higher pH levels, volatile free ammonia (NH_3) predominates; at lower pH values the nonvolatile NH_4^+ species predominates. Volatilization is probably significant only in lakes with high pH values and elevated total ammonium concentrations, conditions that would occur primarily in hypereutrophic systems. Bouldin, et al. (1974) reported NH_3 losses of 2 to 38 percent per day from experimental ponds. For very small lakes, ammonia-caused fish kills may be far more common than reported, because very high NH_3 levels would generally be transient: By the time local agencies sampled a lake to determine the cause of a fish kill, NH_3 levels might well be below toxic levels due to plant uptake, declining pH (conversion of NH_3 to NH_4^+), or volatilization.

9.4.1.2 Nitrogen Balance for Lake Mendota. Figure 9.4 is a nitrogen balance for Lake Mendota, a typical Midwestern eutrophic lake located in Madison, Wisconsin with a drainage area:lake area ratio of 18:1. Note that precipitation inputs plus nitrogen fixation account for more nitrogen input than stream watershed inputs. Major outputs are sedimentation (58 percent), denitrification (14 percent), stream outlets (20 percent), and fish and weed harvesting (7 percent). Because nitrogen budgets at this degree of refinement have been compiled for only a few lakes, it is impossible to make broad generalizations about the relative importance of the sources and sinks of nitrogen in other lakes.

9.4.2 Phosphorus

Phosphorus is of considerable interest in the management of lakes and reservoirs because it is the nutrient that is most commonly limiting to plant growth. Algae normally have a P content of 0.1 to 1 percent of dry weight. The N:P ratio of algae in the idealized *Redfield equation* is 16:1 (molar ratio; 7:1 weight ratio). Using bioassay data Chiaudani and Vighi (1974) determined that N:P ratios >23:1 (10:1 as a weight ratio) are phosphorus-limited. In a nonrandom survey of 49 lakes in the United States, Miller, et al. (1974) determined from algal bioassays that 71 percent were phosphorus-limited.

Phosphorus exists in several forms in aquatic environments. Soluble (or *filtrable*) phosphorus occurs as orthophosphate (PO_4^{-3}), as condensed polyphosphates (from detergents), and as various organic species. A sizable fraction of the soluble P pool (30 to 90 percent; see Wetzel, 1983) is in the organic form, originating from the decay or excretion of nucleic acids and algal storage products (e.g., Minear, 1972; Herbes, et al., 1975; Manny and Minear, 1994). Particulate phosphorus occurs in the organic form (particulate organic P) as a part of living organisms and detritus, and in the inorganic form in the form of minerals such as apatite.

Although most of the input-output phosphorus modeling (Sec. 9.5.3) is based on total phosphorus, a substantial amount of the phosphorus that enters lakes is probably not available to algae growth. *Bioavailable* species include orthophosphorus, polyphosphorus, most soluble organic phosphorus, and a portion of the particulate fraction. In a review of phosphorus bioavailability, Sonzogni, et al. (1982) concluded that bioavailable P in Great Lakes tributaries generally did not exceed 60 percent of total P and suggested that eutrophication models take bioavailability into account.

9.4.2.1 Processes. Phosphorus undergoes no redox reactions in aquatic systems, nor are any common species volatile, so the biogeochemical pathways for phosphorus are much simpler than for nitrogen.

Sedimentation. Most lakes retain a significant portion of input phosphorus. The main mechanism for retaining phosphorus is simple sedimentation of P-containing particles. Particulate phosphorus can originate from the watershed (*allochthonous material*) or be formed within the lake (*autochthonous material*). Soluble phosphorus is also removed from the water column by coprecipitation with iron and manganese hydroxides. Shaffer and Armstrong (1994) developed a detailed breakdown of the various sedimentary forms of phosphorus in Lake Michigan (Fig. 9.5).

Recycling. Much of the phosphorus that enters a lake is recycled, and a large fraction of the recycling occurs at the sediment–water interface. In a classic study, Mortimer (1941; 1942) demonstrated that recycling of phosphorus is linked to iron recycling. According to this paradigm, soluble phosphorus in the water column is removed by adsorption onto iron hydroxides, which precipitate under aerobic conditions. When the hypolimnion becomes anaerobic, the iron becomes reduced, freeing up phosphorus. This is consistent with a fairly general observation that P release rates are nearly an order of magnitude higher under anaerobic conditions than under aerobic conditions. For example, Holdren and Armstrong (1980) showed that phosphorus release rates were greater in calcium poor (iron-rich) sediments than from calcium-rich sediments and that the anaerobic:aerobic ratio of P release was higher for calcium-poor sediments than for calcium-rich sediments. Recently Gatcher (1988) and Bostrom, et al. (1988) have disputed the putative iron-phosphorus linkage, pointing to the significance of microbial metabolism and calcium carbonate precipitation and solubilization in controlling phosphate cycling in sediments.

Factors that control phosphorus recycling rates are not well understood, although it is clear that oxygen status, phosphorus speciation, temperature, and pH are important variables. Modelers generally represent P release from sediments as a zeroth order term, with one constant for aerobic conditions (typically <1 mg/m$^2 \cdot$ day) and a higher constant for anaerobic conditions.

In addition to direct regeneration from sediments, *macrophytes* (large aquatic plants) often play a significant role in phosphorus recycling. Macrophytes with

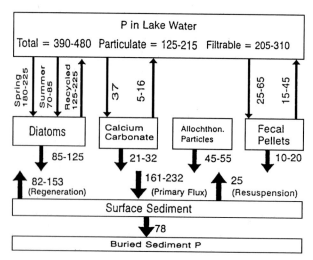

FIGURE 9.5 Phosphorus sedimentation and regeneration in southern Lake Michigan. Values in boxes are areal concentrations, mg/m^2; values on arrows are fluxes mg/m$^2 \cdot$ yr. Areal concentrations are based on the mean depth of 85 m. (*Shafer and Armstrong, 1994.*)

highly developed root systems derive most of their phosphorus from the sediments (e.g., Bristowe and Whitcombe, 1971; Denny, 1972). Regeneration to the water column can occur by excretion or through decay, effectively "pumping" (McRoy and Bardsdate, 1972) phosphorus from the sediments. Of the two mechanisms, decay is probably much more important in lakes (Bristowe and Whitecombe, 1971; Bole and Allen, 1978; Smith, 1978). The decay of macrophytes during the late summer and fall may provide sufficient phosphorus to promote algal blooms in small lakes with extensive macrophyte beds (Landers, 1982). Growth rates for several species of macrophytes have been summarized by McNabb and Tierney (1972), Grace and Wetzel (1978), and Boyd and Hess (1970). Crude models to represent macrophyte growth and recycling of P have been proposed for lakes by Baker (1984) and for rivers by Thomann and Mueller (1987), but standard eutrophication models do not contain realistic representations of macrophyte growth and their role in phosphorus regeneration. The internal loading generated by recycling is particularly important in shallow, eutrophic lakes. In several cases where phosphorus inputs have been reduced to control algal blooms, regeneration of phosphorus from P-rich sediments has slowed the rate of recovery. One of best examples of this phenomenon is Shagwa Lake, Minnesota (Larsen, et al., 1981).

9.4.3 Toxic Organic Chemicals

In the past three decades there has been considerable interest in the fate and effects of toxic organic chemicals, much of it originating from the fish and wildlife mortality caused by the indiscriminate use of organochlorine pesticides and PCBs. Organochlorine pesticides and PCBs have been banned or had their use restricted in the 1970s and early 1980s, but many are highly persistent in the environment and still present at levels of concern (Sec. 9.6.1). The focus in this section will be on high-molecular-weight, persistent, toxic compounds, since these have received the most attention as contaminants of surface waters. For a much more detailed treatment the reader might start with the recent compilation of review papers by Jones (1991).

9.4.3.1 Processes. Partitioning. Because many organic chemicals adsorb onto the surfaces of particles, sedimentation is a primary mechanism by which many organic compounds are removed from the water column of lakes. The degree to which organic compounds bind to particles is described by the *partition coefficient K_d* :

$$K_d = \frac{r}{C_d} \qquad (9.18)$$

where r is the concentration of the compound on particles in solution at equilibrium, in µg/kg or mg/kg, and C_d is the concentration of the compound in the water at equilibrium, in µg/L or mg/L. The units of K_d are L/kg. Values of K_d are directly related to octanol-water partition coefficients, which are in turn inversely related to solubility. Highly soluble (*hydrophilic*) compounds tend to have low K_{ow} values and are not readily adsorbed; conversely, weakly soluble (*hydrophobic*) compounds have high K_{ow} and have a stronger tendency to be adsorbed to particles. From laboratory experiments Karikoff, et al. (1979) developed the relationship:

$$K_d = 0.617 f_{oc} K_{ow} \qquad (9.19)$$

where f_{oc} is the weight fraction of organic carbon. One can readily predict the distribution of an organic chemical in the water column with a knowledge of the suspended solids concentration, the carbon content of the suspended material, the total concentration of the compound, and its K_{ow}:

$$C_p = \frac{C_T m K_d}{(1 + m K_d)} \qquad (9.20)$$

where C_p = concentration of compound in the particulate form
          C_T = total (water + particulate) concentration
          m = suspended solids concentration

It follows that:

$$C_d = C_T - C_p \qquad (9.21)$$

Note that the fraction of compound in the dissolved form depends upon both the distribution coefficient and the suspended solids concentration (Fig. 9.6). K_{ow} values for a variety of organic compounds are tabulated in Lyman, et al. (1990).

Volatilization.   Many organic compounds are volatile and can move across the air–water interface. The rate of volatilization depends upon the concentration of the compound in solution relative to its equilibrium concentration and an overall volatilization constant k_1 (Eq. 9.6). The value of k_1 depends upon both the properties of the lake surface (wind speed and water velocity) and the chemical (Henry's law constant). Mass transfer coefficients have been determined for only a handful of chemicals. Values of k_1 for other chemicals are based on extrapolation based on molecular weight (Brezonik, 1994).

Organic chemicals can move in either direction across the air–water interface, depending upon the degree of saturation of the chemical in the water. For example, since the ban of PCBs in 1979, the net direction of the movement of PCBs across the air–water interface in Lake Michigan has reversed. Whereas the lake once adsorbed PCBs from the atmosphere, it may now a source of PCBs to the atmosphere (Swackhammer and Eisenreich, 1991).

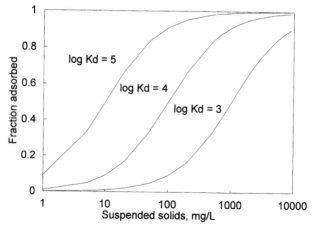

FIGURE 9.6   Relationship between the distribution coefficient and the fraction of a chemical adsorbed to particles [$C_p/C_T = K_d m/(1 + K_d m)$]; units are defined in text.

Degradation. Organic contaminants may be degraded by both biotic and abiotic processes. *Degradation* generally refers to any alteration of the initial chemical, which sometimes yields relatively stable intermediate endproducts. *Mineralization* is the complete degradation of an organic chemical to inorganic endproducts (e.g., CO_2, H_2O, and various other inorganic ions). Both anaerobic and aerobic metabolism are important in the degradation of organic contaminants. In addition, abiotic degradation can occur by direct or indirect *photodegradation* and by *hydrolysis.* A brief introduction to the degradation of organic contaminants is found in Hemond and Fechner (1994); a far more detailed presentation is found in Schwarzenbach, et al. (1993). Rate constants for various organic degradation processes are summarized in Lyman, et al. (1990). Degradation constants k, in units of T^{-1}, are often expressed in terms of half-lives $t_{1/2}$, where $t_{1/2} = 0.69/k$.

Bioaccumulation. High-molecular-weight compounds often accumulate in the tissues of living organisms at concentrations far higher than occur in the water, a process generally referred to as *bioaccumulation.* Bioaccumulation occurs by two mechanisms. *Bioconcentration* is the direct uptake of chemical by aquatic organisms through gills and by adsorption. *Bioconcentration factors* (BCFs) are calculated as a ratio of the concentration in the organism to the concentration in the water:

$$BCF = \frac{\text{concentration in organism, mg/kg}}{\text{concentration in water, mg/L}} \qquad (9.22)$$

Bioconcentration factors are directly related to octanol-water partition coefficients, which in turn are inversely related to solubility (reviewed by Barron, 1990). *Biomagnification* generally refers to the process by which pollutants become more concentrated as one organism (e.g., an algae) is consumed by another (e.g., a zooplankton). Bioconcentration is generally believed to be the more important process in aquatic systems. Biomagnification is the dominant process in terrestrial food chains and is important in terrestrial–aquatic linkages (e.g., piscivorous birds).

Overall bioaccumulation factors (calculated in the same manner as BCFs) can exceed 10^6. Figure 9.7 shows the bioaccumulation of DDT in the Lake Ontario ecosystem.

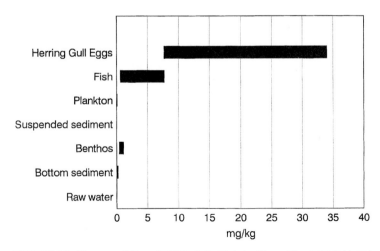

FIGURE 9.7 Bioaccumulation of DDT in Lake Ontario. (*From Allan, 1989, Table 11.*)

9.4.3.2 Pesticides in the Great Lakes. The Great Lakes have been a focus of much of the research on the environmental behavior of organochlorine compounds. In summarizing the mass balances for PCBs, Swackhammer and Eisenreich (1991) found that most (58 to 90 percent) of the total PCB input for the upper lakes (Superior, Michigan, and Huron) comes from the atmosphere. Although atmospheric input rates are similar for the lower lakes, river inputs make major contributions, so that atmospheric inputs are relatively less important (<15 percent of total input). The major sink for PCBs are volatilization and sedimentation. Volatilization is the dominant sink for PCBs in all five lakes. Because removal processes are very active, the chemical residence time of PCBs in the Great Lakes is only 1 to 17 percent of the water residence times.

9.4.4 Trace Metals

The primary practical interest in trace metals is their toxicity, although many are also micronutrients (e.g., Mn, Zn, Co, and Cu). A number of natural sources of trace metals can lead to elevated concentrations (e.g., geothermal springs; see Hem, 1970), but the major cause for toxic levels of metals in lakes is anthropogenic activities. Lead in most aquatic systems is probably derived mostly from the use of leaded gasolines (levels have declined dramatically since leaded gasolines have been phased out); urban stormwater is highly enriched from Pb, Cu, and Zn (EPA, 1983a); and mine drainage is enriched in various metals (Herlihy, et al., 1990; Moore, 1994). A penultimate source of mercury to many lakes in the Upper Midwest is atmospheric deposition (Swain, et al., 1992). Arsenic is quite common from natural sources and is also a contaminant of mine drainage (Welch, et al., 1988; Baker, et al., 1994).

9.4.4.1 Processes. *Complexation and Precipitation.* In all but the most acidic aquatic systems, the major fraction of most trace metals is bound as either inorganic or organic complexes. Many metal complexes are insoluble, resulting in removal of the metal from the water column. Under oxic conditions, the most prevalent inorganic metal complexes are those of hydroxides and carbonates; additionally, aluminum forms tight complexes with fluoride. Under saline conditions, sulfate and chloride complexes can be important. Under anaerobic conditions, metal solubility is often controlled by sulfide complexes. Some inorganic complexation is not sensitive to redox conditions, whereas other metal complexes are redox-sensitive either directly (oxidation and reduction of the metal ion) or indirectly (e.g., by reactions with sulfide).

A widely studied non-redox-sensitive complexation process of ecological importance is the complexation of aluminum in low-alkalinity waters. Aluminum hydroxide complexes are: $Al(OH)^{+2}$, $Al(OH)_2^+$, $Al(OH)_3)_s$, and $Al(OH)_4^-$. Aluminum solubility is minimal at around pH 5.5 to 6.0; $Al(OH)_2^+$ and free aluminum (Al^{+3}) predominate at lower pH values. Solubilization of soil gibbsite $[Al(OH)_3]$ and other soil minerals has resulted in elevated aluminum concentrations in acidic lakes and streams in the eastern United States that have become acidified by acid deposition (Fig. 9.8). Concentrations of inorganic monomeric aluminum in low-pH surface waters throughout the United States are controlled by both chemical solubility and hydrological interactions (Herlihy, et al., in preparation).

Several metals (e.g., As and Cr) undergo redox reactions that control their solubilities (reviewed by Kuhn, et al., 1994). The solubilities of many other metals are controlled indirectly by the presence of sulfide. Sulfide–metal complexes are extremely insoluble, so the formation or oxidation of sulfide plays a major role in the solubility of metals, especially in sediments. Sulfide complexation also plays a major role in controlling the toxicity of metals to sediment biota (DiToro, et al., 1991).

Moore (1994) describes a *redox pump* mechanism that controls the solubility of metals in a mine-contaminated reservoir in Montana.

Finally, trace metals are often scavenged from the water column of lakes by coprecipitation, generally by hydroxides of manganese or iron. This often is a seasonal process. Under anoxic conditions in a stratified lake, reduced species (Mn^{+2}, Fe^{+2}) are produced. When the lake turns over, these metals are oxidized, and trace metals are adsorbed onto the resulting Mn or Fe hydroxides, which precipitate. Figure 9.9 illustrates a typical seasonal pattern of manganese and iron in relation to oxygen at the bottom of a eutrophic lake.

In addition to forming inorganic complexes, trace metals form complexes with natural organic acids (fulvic and humic acids). This organic complexation greatly affects the toxicity and environmental behavior of metals. For example, organic complexation exerts considerable control over the phytotoxicity of copper (Gatcher, et al., 1978; McKnight, 1981) and the toxicity of aluminum to fish (Baker and Schofield, 1982; Baker and Christensen, 1990). Organic complexes also increase the solubility of metals.

Detailed discussions of metal complexation are found in Snoeyink and Jenkins (1980), Stumm and Morgan (1981), and Hem (1970). Several computer programs are available to predict metal complexation and precipitation (e.g., MINTEQA2; Allison, et al., 1991).

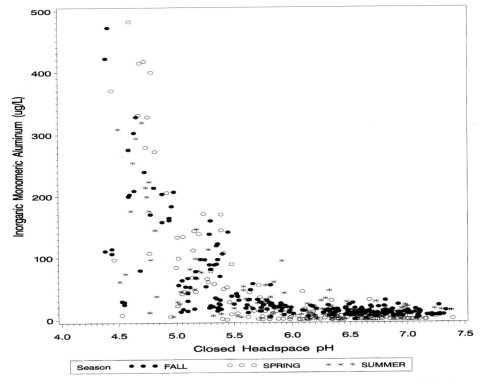

FIGURE 9.8 Relationship between monomeric aluminum and pH for lakes in the EPA's National Surface Water Survey. (*Baker, et al., 1990.*)

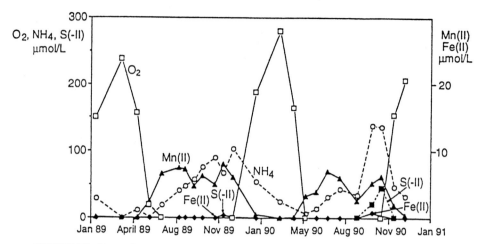

FIGURE 9.9   Seasonal pattern of manganese, sulfide, iron, and ammonia at 30 m in eutrophic Lake Greifen, Switzerland. (*Kuhn, et al., 1994.*)

9.4.4.2 Adsorption.   Because metals are often bound in one way or the other to particles, the fate of metals is often intimately linked to the behavior of particles. A common modeling approach to predict the retention of metals in lakes and reservoirs is to use a partition coefficient to describe the extent of association of the particular metal with particulates; metal removal rates are then tied directly to the sedimentation rate of suspended particles (Ambrose, et al., 1988a; Thomann and Mueller, 1987). Partition coefficients for metals are often in the range 10^4 to 10^5 but are highly variable among sites; site-specific computation of metal k_d values is recommended for modeling studies (Thomann and Mueller, 1987).

9.5 ASSESSMENT METHODS

In the past two decades there has been considerable emphasis on environmental analysis of aquatic systems. Some general goals of environmental assessment include:

- Is there an existing problem with water quality or the biotic community that requires management action?
- What management strategies are likely to be successful at restoring a desirable condition?
- Is there a trend of deterioration in condition and, if so, can a cause be established?
- Conversely, have ongoing water quality management practices been efficacious?

9.5.1 Assessment of Current Condition

9.5.1.1 Sampling and Analytical Methods.   Chemical laboratory methods for most water quality studies are well standardized (APHA, 1992; EPA, 1983b). A useful reference on field limnological procedures is Likens and Wetzel (1990), and a standard reference for fisheries field studies is Nielsen and Johnson (1983). Although considerable attention has been given to the standardization and quality assurance of laboratory methods over the past decade, limnological field methods remain nonstandardized and highly variable.

Mass balances, an important component of many lake environmental assessments, were discussed in Sec. 9.3. This section focuses on two other aspects of lake environmental assessment: the use of chemical and biological indicators and methods of estimating chemical toxicity.

9.5.1.2 Condition Indicators.

For years ecologists have sought the ideal indicator of ecosystem condition, but the state of science of ecological indicators is still fairly primitive, especially with regard to lakes. *Condition indicators* are of three types: (1) biological indicators, developed on the basis of biological assemblages, (2) chemical indicators, and (3) composite indicators.

A variety of biological metrics have been proposed to measure the biological condition of surface waters. Among the first were biomass and productivity. These parameters respond readily to changes in nutrient inputs but are fairly resilient to toxic stress, unless it is severe. Measures of community structure include species richness (numbers of species per sample), diversity indices, which incorporate information on the number of species and the number of individuals in each species group, and biotic indices (Ford, 1989). *Biotic indices* are potentially the most useful, since they are generally designed as measures of specific types of stress. In the past, most have been linear indices in which species or functional groups are assigned weighting factors based on their environmental tolerances. These have reached a higher stage of development in streams than in lakes. Examples of biological indices in streams include numerous indices of organic enrichment (reviewed in Ford, 1989) and Karr's Index of Biological Integrity (Karr, 1987; Miller, et al., 1988; Hughes and Gammon, 1987). To date, the most useful biological indices in lakes have been based on the composition of diatom communities, reviewed by Dixit, et al. (1992). Most modern indices are based on multivariate statistical techniques (e.g., Charles, et al., 1994). The development of improved biological indicators is a goal of EPA's nascent Ecological Monitoring and Assessment Program (Paulsen, et al., 1991).

Some very useful chemical and composite indicators have been developed for eutrophication and acidification. Several researchers have developed *trophic state indices* (TSIs). Carlson (1977) proposed three TSI scales, one based upon total phosphorus, one on chlorophyll *a,* and one on Secchi disk transparency. The first two were related to Secchi disk transparency by regression equations and all three were normalized to a 0 to 100 scale. Kratzer and Brezonik (1981) introduced a nitrogen-based TSI to be used in N-limited lakes and proposed averaging the TSI values for Secchi disk, chlorophyll *a,* and the lower of the phosphorus or nitrogen TSI values to give a composite TSI (TSI_{ave}). Other indices have been developed that use macrophytes, hypolimnetic dissolved oxygen, primary production, and other variables (reviewed by Brezonik, 1984, and Cooke, et al., 1993). None of these has been universally accepted; the disparate approaches used by various state agencies for estimating lake trophic conditions is summarized by Larsen, et al. (1991).

Several diagnostic ratios have been developed to indicate lake chemical status. The N:P ratio is a useful indicator of nutrient limitation, with ratios (wt:wt) <10:1 indicating N limitation and values >20:1 indicating P limitation. In the realm of lake acidification, $SO_4^{-2}{:}C_b$ (eq:eq; C_b = sum of base cations) ratios >1.0 have been used to implicate atmospheric deposition of sulfuric acid as a cause of lake acidification (Munson and Gherini, 1991).

9.5.2 Evaluation of Toxic Effects

9.5.2.1 Experimental Approaches.   Laboratory bioassays to determine the acute toxicity of a chemical to an aquatic organism are fairly straightforward. The methodology is well developed and standardized for a variety of organisms, including fish, aquatic invertebrates, zooplankton, and algae (APHA, 1992). Methodologies to determine overall toxicity of a particular water are also well developed, but determining which constituent is causing the toxicity is more problematic. The vast majority of information on toxic effects of chemicals in aquatic systems is in the form of mortality responses. The most common expression of mortality is the concentration at which 50 percent mortality occurs over a specific time period, most commonly 96 hours. This is called the *96 hour LC-50* (LC = lethal concentration). There has recently been a growing emphasis on developing toxicity data for more sensitive life stages (e.g., larvae and eggs) and reproduction (Giesy and Graney, 1989; Gentile, et al., 1991).

A major challenge of ecotoxicology is to extrapolate from short-term mortality-based bioassays to natural conditions (Cairns, 1989). For a given species and chemical, ratios of chronic:acute toxicity (ACR) are reasonably consistent; Kenaga (1982; cited in Giesy and Graney, 1989) found that the ACR was approximately 25 for 93 percent of the chemicals studied. Extrapolating from one chemical to other chemicals within a given class, or from one species to another, is more problematic (Giesy and Graney, 1989; Gentile, et al., 1991).

Seeking a higher level of realism, some ecologists have turned to *in situ mesocosms* or even whole-lake experiments to determine the effects of toxic substances, lake acidification, or biological manipulations. In situ mesocosms are intermediate between laboratory bioassays and whole-lake experiments in simulating nature. These are typically plastic cylinders that vary in diameter from 1 to many m. They may be open to the sediments or sealed at the bottom. A useful design presented by Landers (1979) can easily be scaled up to a diameter of 4 to 5 m. Some problems with mesocosms are replicability, difficulty in controlling the entry of fish, leakage, and animal damage (McQueen and Post, 1986; Perry, et al., 1986). The degree to which a mesocosm simulates a lake environment probably decreases with time. As a practical matter, we have found that the construction and installation of 10 4-m littoral mesocosms open to the sediment takes about 200 working hours.

Whole-lake experiments are less common because of high costs, problems with controlling public access, land use restrictions, and environmental regulations. Nevertheless, multiyear, whole-lake experiments provide insights on the effects of perturbations on reproduction, fish behavior, biogeochemical cycling, and other aspects of lake ecology that would be difficult, if not impossible, to deduce from laboratory experimentation. Whole-lake experiments also captivate the public, appealing in part because the experimental results are often highly visible, e.g., the famous photo of Dave Schindler's eutrophication experiments of the early 1970s (Schindler, 1975). The statistical design of whole-lake experiments is reviewed by Carpenter (1989).

9.5.2.2 Biomarkers.   *Biomarkers* are "indicators of variation in cellular or biochemical components of processes, structures, or function that are measurable in biological systems or samples" (definition of the National Academy of Science, cited in Gentile, et al., 1991). Biomarkers range from rather general indicators (lipid content and RNA:DNA ratios) to indicators of specific chemicals (e.g., disruption of the chloride cells in fish gills by addition of acid to lakewater). The problem with biomarkers is that we cannot yet infer ecological effects from biomarker data (Gentile, et al., 1991).

9.5.2.3   Tissue Analysis.   A common approach for assessing the impact of toxic pollutants in lakes is to measure concentrations of toxicants in the tissues of fish or other organisms. Results are typically expressed as mg of toxicant per kg of tissue. Comparisons among studies are sometimes limited by differences in the parts of the organism analyzed; some common parts include the whole body, edible flesh, and specific organs (especially the liver). Toxicant concentrations are highly variable among tissues within a given individual, among individuals of differing sizes (large fish generally have higher toxicant levels than small fish), and among species (species with high fat content usually have higher toxicant levels than do species with lower fat contents).

A common assessment endpoint is the Food and Drug Administration's *action limits* for food consumption by humans. Another important concern, especially for toxicants that bioaccumulate, is the health of piscivorous mammals and birds. An excellent source of information on the interpretation of contaminant levels in fish and wildlife is the U.S. Fish and Wildlife Service's Contaminant Hazards Reviews series, available through the USFWS's Patuxent Wildlife Research Center in Laurel, Maryland.

9.5.3   Analysis of Water Quality Trends

There has been a growing interest in analyzing trends in water quality, partly inspired by a desire to evaluate the effects of water-pollution policies over the past two decades. Methodologies to analyze water-quality trends have also improved, particularly in the areas of statistical trend analysis and paleolimnology. These advances have made trend analysis a practical tool for limnological assessment.

Section 9.5.3.1 examines the topic of retrospective analysis of water quality and biological data collected at various points in time on a time scale of 5 to 50 years. Section 9.5.3.2 looks at the use of sediment records to infer historical change, a branch of aquatic science called *paleolimnology*. Paleolimnological analysis does not rely upon the prior collection of water quality data and allows one to reconstruct limnological histories that began prior to human influence.

9.5.3.1   Retrospective Analysis of Limnological Data.   Many investigators have examined historical records of water chemistry, fish populations, and other parameters to develop inferences about ecological changes in lakes and reservoirs. What would appear to be a fairly straightforward task is generally not; valid interpretation of historical data is fraught with difficulty. Discussion of the various types of problems encountered in developing a retrospective analysis follows.

Methodological Changes.   A common problem with historical water-quality data is that the analytical methods for virtually all chemical constituents and even many physical parameters have changed, sometimes several times, during the time period of interest. This would not necessarily be a problem if changes in methodology were carefully documented, overlap studies were conducted, and rigorous quality assurance procedures were followed. Regrettably, this is almost never the case, especially for data more than 10 to 20 years old. In several cases, key conclusions regarding historical changes reported by one group have been challenged on purely methodological grounds by another group. Two notable cases include an analysis of pH and alkalinity changes in Adirondack lakes (Asbury, et al., 1989 versus Kramer, et al., 1986) and an analysis of silica trends in Lake Michigan (summarized in Schelske and Carpenter, 1992).

Changes in taxonomy sometimes obsfucate the interpretation of biological trends. This problem is sometimes solved by the fact that biological samples are easily preserved, and many biologists save *type specimens* for posterity. The result of this is that errors in taxonomy made by earlier investigators can sometimes be corrected by later investigators.

Documentation and Quality Assurance.   Starting around the mid-1970s, many agencies became aware of problems with the quality of water-quality data. Since then there have been enormous improvements in the quality of data collected, although improvements in laboratory analysis have generally been greater than improvements in field methods (Hren, et al., 1990). An example of the slop incurred in earlier studies is illustrated by phosphorus trends reported by various agencies for Lake Griffin, Florida (Preston and Brezonik, 1985). As can be seen in Fig. 9.10, the expenditure of what must have been several hundred thousand dollars has resulted in no useful information regarding trends in phosphorus levels in this lake. This is almost certainly not an isolated case.

Type of Data.   A third problem with historical data is that it often is inappropriate, too sparse, and nonrandom. For example, in an analysis of historical data that had been collected in Ohio and Colorado, Hren, et al. (1990) found that 5 percent or less of the samples were analyzed for priority pollutants, one-fourth or less of the sample stations had 10 or more samples collected during the period 1980 to 1984, and most of data collection stations were in small areas with known or suspected water-quality problems. Systematic monitoring for biological information in lakes is almost nonexistent.

Statistical Techniques.   In the past decade there have been major advances in the statistical treatment of water-quality trends. Of particular concern for analyzing water-quality trends is the large seasonality of the data, which makes statistical trend analysis difficult. Loftis, et al. (1989) recently reviewed seven statistical trend tests for EPA's Long-Term Monitoring Project, a monitoring network of several dozen lakes throughout the United States. For annual sampling, Loftis, et al. recommended the Kendall-tau test and for quarterly sampling they recommended either the seasonal Kendall test or an analysis of covariance procedure.

Natural Variability in Water Quality.   Water quality managers often do not appreciate the extent to which natural interyear hydrologic fluctuations alter water quality. An example of the interrelationship between hydrology and lake chemistry is the

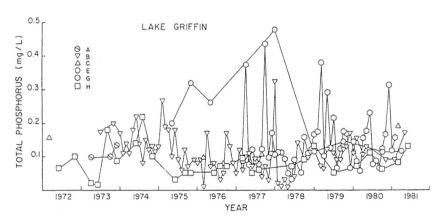

FIGURE 9.10   Phosphorus trends in Lake Griffin, Florida, based on analysis by six agencies (A–H). (*Preston and Brezonik, 1985.*)

drought-driven acidification trend observed in Nevins Lake, Michigan (Webster, et al., 1990; Fig. 9.11). For this small seepage lake, the acid neutralizing capacity (ANC) was normally around 200 µeq/L. The ANC dropped suddenly during the early 1980s. Webster, et al. concluded that the lake normally received well-buffered groundwater and that this input was considerably reduced during the drought of the mid-1980s, causing the lake to become rapidly acidified. In the late 1980s precipitation rates increased; renewed groundwater input caused an increase in the lake's ANC.

Lake nutrient budgets also exhibit large interyear variations. In a six-year study of Mirror Lake, New Hampshire (surface area = 15 ha; watershed:lake area = 5.7), the coefficient of variation in annual chemical loadings was 5 to 15 percent. The ratio of highest annual loading:lowest annual loading ranged from 1.4 to 2.2 (Na^+ and Cl^-

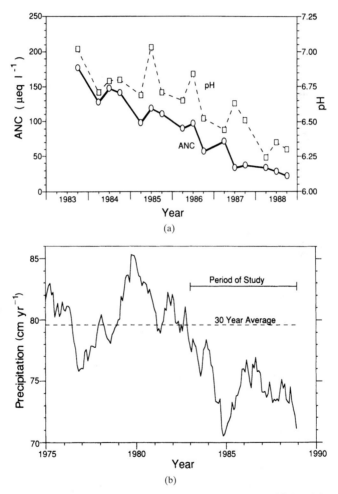

FIGURE 9.11 Rapid acidification of Nevins Lake, Michigan. (*a*) Trends in pH and ANC; (*b*) five-year running means of annual precipitation at nearby Grand Marais. The dashed line shows the 30-year mean. (*Webster, et al., 1990.*)

are excluded because they exhibited a systematic increase during the study period due to highway construction; Likens, et al., 1977). Several studies in Florida reveal that for small streams in agricultural areas, variations in annual loadings of N and P can exceed a factor of 2 (Stewart, et al., 1978; Campbell, 1978). Spooner, et al. (1988) developed an approach for determining the minimum detectable change (MDC) in phosphorus concentrations that could be revealed by a sampling program designed to show the effect of a non-point-source pollution control program in a 110,000-acre watershed. Because of natural variations in hydrology, the MDC was 10 to 59 percent over nine years, even with biweekly sampling.

In addition to normal hydrologic fluctuations, catastrophic events can alter the interpretation of water quality trends. Notably, landslides and hurricanes can cause extraordinary loadings of sediment for short periods (Meade and Parker, 1985; Grant and Wolff, 1991).

9.5.3.2 *Paleolimnological Analysis of Lake Sediments.* During the past 10 years there has been considerable development in paleolimnological methods to the point that paleolimnology has become a practical lake management tool. The essence of paleolimnology is that lake sediments are archives of historical information. The basic methodology involves collecting undisturbed sediment cores, typically to a depth of around 1 meter for histories of 100 years or less. These cores are extruded with a piston, and the sediments from each section (usually at 1 cm intervals) are analyzed individually. For periods of 100 years or less, core segments are dated by lead-210 analysis. From this analysis the date at which the sediment layer was deposited can be estimated, usually within a few years. Ideally, the date of select core segments is verified by independent analysis. Common verification methods include the date of a peak in cesium-137 between 1953 and 1963 (the period of open atmospheric nuclear bomb testing) and the first appearance of ragweed pollen, an indicator of agricultural development.

Depending upon the need, core segments can be analyzed for physical characteristics, chemical constituents, and biological remains. Dozens of paleolimnological studies of lakes have documented the onset of cultural eutrophication, deposition-induced acidification, changes in sedimentation rates associated with agricultural development and urbanization, changes in the accumulation of metals, nutrients, sulfur, and organic chemicals, the occurrence of forest fires, the disappearance of fish, and other ecological changes in lakes. Key reviews include Dixit, et al. (1992), Charles and Hites (1987), Charles, et al. (1989; 1994), and Engstrom, et al. (1994).

One of the major developments in paleolimnology has been the development of quantitative methods that utilize diatoms to make inferences about changes in lake chemistry. Diatoms are particularly useful as indicators of ecological change because they are ubiquitous and numerous, there are many species of diatoms, many of which have narrow ecological tolerances, and they are usually well preserved in sediments (Charles, et al., 1989). The identification of diatoms requires considerable expertise, and ecological indices based on diatoms are still evolving, hence diatom-based inferences of changes in lake environments remain in the province of taxonomic experts.

Figure 9.12 shows the paleolimnological history of Harvey's Lake, Vermont (Engstrom, et al., 1985). The sediment record reflects the cultural history of the watershed: the operation of a lumber mill between 1780 and 1945, early agricultural development in the 1780 to 1945 period, and rapid eutrophication since 1945.

9.5.3.3 *Causality.* The underlying cause of a particular environmental trend is often unclear or complex. One might, for example, be looking for a relationship

between phosphorus loading and algal production. In most cases there will be a number of obsfucating factors that muddy the relationship between an environmental trend of interest (e.g., eutrophication) and a putative cause (e.g., loadings from municipal wastewater treatment plants). The role of natural hydrologic variations was discussed earlier. Another important factor that is often overlooked is the role of direct biological manipulations such as fish stocking, fish removal (by rotenone treatment), and the introduction of exotic species. For example, in an analysis of historical fisheries data in the Adirondack Mountains, Baker, et al. (1993) found that acidification was an important, but not necessarily the dominant, factor influencing losses of fish populations. Fisheries rehabilitation, the introduction or invasion of exotic species, and changes in stocking policy also played a major role in fish population losses. Schelske and Carpenter (1992) reviewed the many factors that have altered the Lake Michigan ecosystem over the past century.

9.5.4 Models of Lakes and Reservoirs

Chemical models have become a key tool in the evaluation of lake management alternatives. Chapter 14 deals with models of aquatic systems, so the intent here is to focus on the simplest of models, which also happen to be the most widely used.

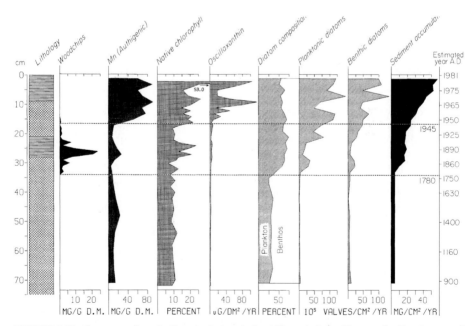

FIGURE 9.12 Summary of a paleolimnological analysis of Harvey's Lake, Vermont, by Engstrom, et al. (1985). Data show the occurrence of a lumber mill (appearance of woodchips in sediments), an increase in the concentration of authigenic manganese, indicating low oxygen status, the appearence of blue green algae (indicated by the pigment Oscillaxanthin), an increase in the overall abundance of algae (an increase in native chlorophyll), a decline in the ratio of planktonic:benthic diatoms (indicating decreased water clarity), and an increased sedimentation (indicating increased erosion rates).

9.5.4.1 The CSTR Model for Lakes. The application of CSTR (*continuously stirred tank reactor*) or blackbox models to lake eutrophication (Dillon and Rigler, 1974a; Vollenweider, 1976 and dozens of subsequent papers) has forever changed limnology. For a given chemical, the CSTR model in its simplest form (shown for constant volume) includes an input, an output, and a single sink:

$$V \frac{dC}{dt} = W - CQ_o - kCV \tag{9.23}$$

where
V = volume, L^3
W = chemical loading, M/T
C = lake concentration, M/L^3
k = first order decay constant, T^{-1}
Q_o = outflow, L^3/T

At steady state ($dC/dt = 0$), Eq. 9.23 can be solved for concentration:

$$C = \frac{W}{Q_o + kV} \tag{9.24}$$

Dividing the numerator and denominator of Eq. 9.24 by area results in the following:

$$C = \frac{L}{q + v} \tag{9.25}$$

where
L = areal loading rate, $M/L^2 - T$
$Q_o/A = q$ = overflow rate, L/T
and
$Vk/A = v_i$ = an apparent settling velocity, L/T

This modeling approach is commonly used to predict phosphorus concentrations in lakes. A commonly used value for the phosphorus settling velocity v_p is 10 m/yr, but Thomann and Mueller (1987) show that v_p values are highly variable. Their compilation of values showed that v_p values for natural lakes span one order of magnitude (3 to 30 m/yr); v_p values for reservoirs were even more variable. These data suggest that site-specific calibration is needed for accurate prediction.

Kamoto (1966), Dillon and Rigler (1974b), Jones and Bachman (1976), and others have found a reasonably consistent statistical relationship between the average concentration of total phosphorus at spring turnover and average summertime chlorophyll *a* concentrations. A commonly used relationship is (Dillon and Rigler, 1974b):

$$\log (\text{chl } a)_{\text{summer}} = -1.14 + 0.1.583 \log (P_{\text{sp}}) \tag{9.26}$$

where $(\text{chl } a)_{\text{summer}}$ = average summer chlorophyll concentration, mg/m^3, and P_{sp} = concentration of total phosphorus during spring turnover, also in mg/m^3. There is considerable variation in the chlorophyll–phosphorus relationship among groups of lakes, so Eq. 9.26 should be regarded as a rough approximation. Nutrient limitation, climate, and other factors affect this relationship, so a regional calibration of chlorophyll–phosphorus relationships to be used for management purposes is advised.

Equations 9.25 and 9.26 can be melded together to produce an equation that predicts chlorophyll *a* concentrations from phosphorus loading. This linkage is widely used in practical lake management. Additional details and limitations of this approach are described in several texts (Reckhow and Chapra, 1983; Cooke, et al., 1993; Thomann and Mueller, 1987).

One-box CSTR models with single loss terms have also been developed for nitrogen in Florida lakes (Baker, et al., 1985), sulfate in acid-sensitive lakes (Baker, et al., 1986b; Kelly, et al., 1987), suspended sediments in the Great Lakes (Thomann and Mueller, 1987), and several other constituents.

9.5.4.2 Complex Models. Sophistication of lake models is gained by adding process and hydrodynamic complexity. For many pollutants, a single sink term is inadequate for describing in-lake transformations. When this is the case, several loss terms are used, each representing a single process. Complex models may include terms for chemical volatilization, adsorption onto sediments and subsequent deposition, recycling at the sediment–water interface, biological degradation, and chemical degradation (Thomann and Mueller, 1987).

The most common level of hydrodynamic complexity is the use of a two-box model in which the epilimnion and hypolimnion are modeled as separate compartments. For dendritic reservoirs, a common modification is to represent the system as a series or network of interconnected CSTRs. Three-dimensional water quality modeling for lakes is relatively rare. Useful texts on lake modeling include Thomann and Mueller (1987), Reckhow and Chapra (1983), and Jorgensen and Gromiec (1989). Ambrose, et al. (1988b) reviewed EPA's water quality models. Water quality models are reviewed in Chap. 14.

9.5.5 Regional Assessments

Although limnologists have been involved in *synoptic* studies for many years, beginning with the classic studies of Birge and Juday in the early 1900s, most of these have been nonrandom surveys conducted within fairly small areas, or were oriented toward specific problems (e.g., EPA's National Eutrophication Survey). Statistically haphazard surveys have been useful in the development of limnological concepts, but cannot be considered a quantitative assessment tool. The development of statistically-based, assessment-oriented regional analysis began with EPA's National Surface Water Survey (NSWS), conducted in the mid-1980s with the purpose of quantifying the numbers of acidic and low-pH lakes in areas of the United States susceptible to acidification by acid deposition. The NSWS had five essential characteristics: (1) the target population of lakes and streams was explicitly defined, (2) sampling was done using a stratified random design, (3) chemical conditions were sampled within a defined index period, (4) it employed uniform sampling and analytical procedures, and (5) it had a documented quality assurance program. The result of the NSWS and related regional modeling studies was one of the most comprehensive, defensible, and expensive environmental assessments ever conducted (NAPAP, 1990).

The EPA and other federal agencies have continued to develop statistically based surveys through the Ecological Monitoring and Assessment Program (EMAP). The Surface Water component of EMAP (EMAP-Surface Waters) is under development (EPA, 1991; Paulsen, et al., 1991). The broad goal of EMAP-Surface Waters is to provide a data base that can be tapped by a variety of investigators interested in the environmental characteristics of lakes and streams. By design, it is intended to provide information on trends in water quality throughout the United States. One of the main problems encountered by EMAP is the lack of suitable indicators of environmental quality for lakes (see Sec. 9.5.1.2).

One recent impetus for expanding the role of regional limnological studies has come by way of the Clean Water Act Amendments of 1987, which direct states to

adopt quantitative biological criteria by 1995. One approach for developing biological criteria has been to use an *ecoregion* framework, in which the condition of a lake
or stream is compared with reference sites within a more or less homogeneous
ecoregion. This approach has been used to develop legal standards for dissolved
oxygen for Arkansas streams, biological integrity for Ohio streams, and phosphorus
criteria for Minnesota lakes (summarized in Hughes and Larsen, 1988; Fig. 9.13) and
is being developed as a basis for setting legal biological standards in several other
states. Problems with this approach include: (1) the establishment of baseline or reference conditions in areas in which there are few remaining undisturbed water bodies, (2) the general lack of development of biological indices for lakes, and (3)
variability of biological conditions as a function of elevation, lake size, and other limnological characteristics within an ecoregion. A useful approach for developing biological criteria in lakes would be to use paleolimnogical studies to establish
reference conditions. The development of regional monitoring programs and biological criteria for aquatic systems is discussed by Gallant, et al. (1989), Suter (1990),
Hughes and Larsen (1988), and Miller, et al. (1988).

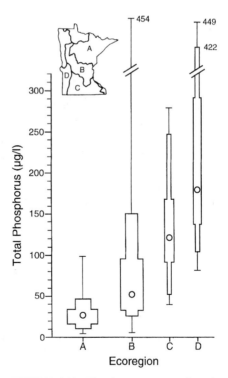

FIGURE 9.13 Phosphorus concentrations in
reference lakes in five ecoregions of Minnesota.
The data show that natural phosphorus levels
decline as one moves from south to north.
(*Hughes and Larson, 1988.*)

9.6 ENVIRONMENTAL PROBLEMS OF LAKES AND RESERVOIRS

9.6.1 Types of Problems

Although our focus for protecting aquatic ecosystems over the past twenty years has been chemical quality, the ecological condition of lakes and reservoirs depends not only upon chemical quality but also upon the condition of the habitat and the introduction of *exotic species*—species that have been deliberately introduced into a lake or arrive there through biological invasions. Our failure to address the problems of physical and biological perturbations has been highlighted by NRC (1992). This review of the major types of environmental problems experienced by lakes and reservoirs follows the outline of NRC (1992).

9.6.1.1 Eutrophication.   The term *eutrophication* has been used in two disparate ways. Traditionally, the term eutrophication was used to describe the slow filling of lakes, but the term is more commonly used to describe the array of biological, chemical, and physical changes that occur in response to increases in nutrient inputs, generally as the result of anthropogenic activities. To make the matter more confusing, the term *eutrophic* is used to describe nutrient-rich lakes in general, regardless of whether the condition has been a long-standing one or a recent development. As a practical matter, lake trophic status is often based upon algal abundance and nutrient levels (Table 9.3).

Ubiquitous characteristics of eutrophic lakes include high nutrient levels accompanied by large populations of *phytoplankton* (algae) and low transparency. Bluegreen algae tend to predominate in highly eutrophic lakes; these are considered undesirable because some of them form unsightly surface blooms and are responsible for undesirable tastes and odors. For stratified lakes, high levels of algal production in the epilimnion can result in depletion of oxygen in the hypolimnion which, as noted earlier (Sec. 9.2.1.2), has undesirable water-quality consequences. The buildup of free ammonia in the epilimnion of hypereutrophic lakes can be toxic to fish (Effler, et al., 1990). The littoral zones of many eutrophic lakes are choked with aquatic weeds which inhibit boating and swimming. Although modest eutrophication probably leads to a more productive and diverse fishery, excessive eutrophication results in shifts in fish communities that are culturally undesirable in the United States. For a coldwater fishery dominated by salmonids, eutrophication results in a shift to a warmwater fishery dominated by species such as bass, perch, carp, and bullheads, all relatively tolerant of enriched conditions.

Although there was a substantial controversy over limiting nutrients during the 1970s (see the collection of papers in Likens, 1972; Schindler, 1975), there is little

TABLE 9.3   Trophic State Delineation from the EPA's National Eutrophication Survey

Trophic state	Chlorophyll *a*, μg/L	Total P, μg/L	Secchi disk tranparency, m
Oligotrophic	<7	<10	>3.7
Mesotrophic	7–12	10–20	2.0–3.7
Eutrophic	>12	>20	<2.0

Source: Reckhow and Chapra (1983).

doubt that the major limiting nutrients for algal growth in temperate lakes are phosphorus and nitrogen. Of the two major limiting nutrients, phosphorus is probably more commonly limiting (Sec. 9.4.2). Accordingly, most lake restoration activities have been focused on controlling phosphorus inputs or limiting recycling in lakes (Sec. 9.8).

9.6.1.2 Hydrologic and Physical Alterations.

Hydrologic and physical alterations are a greater problem with rivers and streams than for lakes, but water-level alterations can have a profound effect upon a lake ecosystem. Many species of fish (e.g., northern pike) depend upon emergent vegetation or gravel beds for spawning, and water-level stabilization can eliminate spawning habitat (NRC, 1992). Many flood control reservoirs experience dramatic changes in water level that prevent the development of anything that resembles a natural littoral area.

Hydrologic diversions can greatly alter the chemical and biological conditions in lakes and reservoirs. The diversion of freshwater inflows from Lake Mono, California, for use as municipal water supply has been a major concern because of its effect on the salinity of the lake (NRC, 1987). The diversion of the Bear River to Bear Lake, Utah, previously a closed basin, has resulted in a profound shift in the unique chemistry of the lake (Nunan, 1972) and has caused eutrophication through the addition of nutrients (Lamarra, et al., 1984). Hydrologic diversions are sometimes employed to reduce symptoms of eutrophication (Sec. 9.7.1.2).

The construction of reservoirs has a profound impact on downstream rivers (reviewed by Petts, 1984). Reservoirs can be major sediment traps (Fig. 9.14); sediment transport along the lower Colorado and the Missouri rivers has been greatly reduced as the result of reservoir construction in the mid-1900s (Meade and Parker, 1985). The release of cooler hypolimnetic waters has been considered a benefit in some southern streams, allowing the development of a trout fishery where none was possible before. On the other hand, the release of clear, cool waters has imperiled native species of the lower Colorado River that are adapted to warm, silt-rich

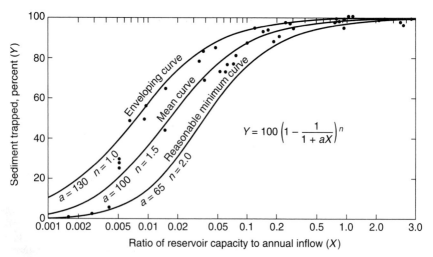

FIGURE 9.14 Trap efficiency of normal ponded reservoirs as a function of the reservoir volume: annual inflow. (*Linsley, et al., 1982.*)

waters. The release of hypolimnetic waters with low dissolved oxygen concentrations and elevated levels of ammonia and sulfide can impair the downstream biota.

9.6.1.3 Siltation. Sedimentation rates in most large lakes and reservoirs, typically on the order of <1 to a few cm/yr, are often not sufficient to cause filling at an appreciable rate. Exceptions are reservoirs that receive very high sediment loadings and shallow lakes in urban or agricultural areas with high erosion rates. Linsley, et al. (1982) outline a general approach for estimating the time of filling for reservoirs. In addition to the problem of filling, high turbidity levels can limit the productivity of algae and macrophytes by reducing light transmission.

9.6.1.4 Exotic Species. One of the most important causes of degradation of natural biological communities in the United States has been the spread of exotic species, either by deliberate introduction or by accidental invasion. Many species of aquatic organisms have become established in the United States in the past century, sometimes with dramatic consequences. *Water hyacinth* (*Eichornia crassipes*), introduced at the St. Louis World Fair, now clogs waterways throughout the southeast, and *water milfoil* (*Myriophyllum spicatum*) is a perennial problem in lakes throughout much of the United States. The deliberate introduction of carp and the accidental introduction of the sea lamprey have had profound influences on the fisheries of U.S. surface waters. More recently, the accidental introduction of the zebra mussel has wrecked havoc on water-supply systems along the Great Lakes, with an estimated economic loss of $5 billion by the year 2000 (Giacomo and Randell, 1993). The introduction of exotic species is considered one of the major factors contributing to the extinctions of fish in the United States (Miller, et al., 1989). In Arizona alone, there are nearly twice as many exotic species (~60) as native species (~30, of which 15 are endangered; P. Marsh, Southwest Center for Environmental Research, personal communication).

9.6.1.5 Acidification. Widespread acidification of surface waters is caused by mine drainage and by acidic deposition. In addition to surface waters that have been made acidic through human activities, there are many naturally acidic surface waters. Most of these have high levels of dissolved organic carbon or are found in areas of geothermal activity (Baker, et al., 1991b).

The EPA's NSWS showed that nearly 7 percent of the lakes and 3 percent of the total stream length in surveyed areas of the eastern United States were *acidic,* defined as having an acid-neutralizing capacity <0 μeq/L (corresponding to pH values of roughly 5.0 to 5.5). These numbers understate the biological impact of acidification because: (1) sampling was conducted in the fall rather than in the spring (episodic acidification occurs in many lakes during snowmelt and rainstorms), and (2) biological damage occurs in lakes that did not meet the NSWS criteria of "acidic" (Baker, et al., 1991b).

Among the acidic lakes, nearly 75 percent were deemed *deposition-dominated;* 23 percent were deemed *organic-dominated.* The chemical signature of the deposition-dominated acidic lakes (SO_4^{-2}:base cation ratios usually >1.0) and paleolimnological analysis showed that significant acidification of lakes has occurred in the Adirondack Mountains and New England. Some acidification has probably occurred in the eastern Upper Midwest and Florida, but not in the western United States (NAPAP, 1990). Typically, lakes susceptible to acidification are small (median size = 10 ha) lakes in forested watersheds. In the Northeast, most acidic lakes are rapidly flushed drainage lakes, but in Florida and the Upper Midwest most acidic lakes are seepage lakes. Losses of fish populations have been carefully documented for Adirondack lakes

(Baker, et al., 1993); biological impacts of deposition-induced acidification have probably also occurred in the Upper Midwest, the Atlantic Coastal Plain, New England, and Florida (NAPAP, 1990). Biological effects of acidification were reviewed by Baker and Christensen (1991).

In 1990, Congress passed amendments to the Clean Air Act that will result in a 50 percent decline in sulfur dioxide emissions. Modeling studies conducted during the NAPAP show that significant recovery is expected in northeast lakes and that further acidification in Appalachian streams will be limited (NAPAP, 1990). Liming of lakes has been very widely used to ameliorate the effects of acidification in Scandinavia and, to a lesser extent, in the United States and Canada. Liming methodology is reviewed in Olem (1990).

9.6.1.6 Toxic Contamination.   National interest in toxic organic compounds was piqued by Rachel Carson's *Silent Spring* (1962). In those days, indiscriminate use of pesticides was causing fairly widespread mortality among nontarget organisms, including bald eagles, fish-eating ducks, robins, and fish. Among the most harmful organic contaminants in lakes are the organochlorine pesticides—DDT, lindane, chlordane, toxaphene, and a few others, plus polychlorinated biphenyls (PCBs), which were used for a variety of purposes.

These compounds have two characteristics that make them among the most insidious pollutants: they bioaccumulate and are highly persistent in the environment. Although the production and use of organochlorine compounds in the United States was phased out in the late 1970s through the early 1980s, they are still found in many aquatic environments because: (1) they degrade extremely slowly, and (2) there are continued inputs to lakes from erosion of contaminated soil and atmospheric transport from contaminated land areas (Allen, 1989; Standley and Hites, 1991). Rappaport, et al. (1984) concluded that inputs of fresh DDT in Upper Midwest peat bogs occurred by atmospheric transport from Mexico, where the pesticide is still in use. Data from 100 U.S. sites in the National Contaminant Biomonitoring Program show that concentrations of several organochlorine compounds have declined slowly since the 1970s (Schmitt, et al., 1990).

A number of trace metals are highly toxic to fish and other aquatic organisms. Metal contamination of aquatic environments is most often associated with mining activity or urban runoff. Leaching of metals from abandoned coal mines in the Appalachian region and abandoned metals mines in the West is a widespread problem (Herlihy, et al., 1990; Moore and Luoma, 1990). By one estimate, there are approximately 180,000 acres of acidic mine drainage-impacted lakes in the United States (Klinemann and Hedin, 1990; cited in NRC, 1992).

A second major source of metals contamination is urban runoff. In an extensive survey of urban runoff in 22 cities, the EPA (1983a) concluded that copper, lead, and zinc in urban runoff pose a significant threat to aquatic life. Median concentrations of 34 µg/L for copper, 144 µg/L for lead, and 160 µg/L for zinc exceeded EPA's criteria for the protection of aquatic life (EPA, 1983a). The ecological effect of metals in urban runoff has received relatively little study.

The metal that has received the most attention in recent years is mercury. Although mercury was used with casual disregard for many years (e.g., as a pesticide), dispersal of mercury into the environment is now tightly regulated. The major source of mercury in many lakes today probably is atmospheric deposition.

About 50 to 75 percent of global mercury emissions are the result of anthropogenic mercury emissions (EPRI, 1994); mercury deposition rates in the northern temperate zone are 3 to 4 times higher currently than in preindustrial times (Swain, et al., 1992). Mercury bioaccumulates in fish. Although there is little evidence of mercury-induced damage to fish in the Upper Midwest, many Upper Midwest lakes contain some large fish that exceed the 0.5 mg/kg action limit of the World Health Organization (Gloss, et al., 1990). At these levels, deleterious effects to fish-eating mammals and birds would be expected (Eisler, 1987).

In the United States, fishing bans or consumption advisories imposed because of elevated concentrations of toxic organic compounds and mercury were in effect in 1988 for 721 lakes, totalling 2.5 million acres. Twenty-two states reported fishing restrictions associated with PCBs, 17 had restrictions associated with chlordane, and seven reported bans due to DDT (EPA, 1990).

9.6.2 National Aquatic Assessment

Under Section 305(b) of the Clean Water Act, EPA is directed to conduct a biennial assessment of water quality in the United States. These assessments provide a glimpse at the causes and sources of water quality problems in lakes throughout the United States. Although the vast majority (95 percent) of lakes and reservoirs meet the *swimmable and fishable goal* of the Clean Water Act, only about 74 percent fully supported their *designated uses* as determined by the states (Table 9.4). These designated uses include categories such as drinking-water supply, contact recreation, coldwater fishery, and warmwater fishery.

The methodology of the 305(b) reports used to derive Table 9.4 is far from ideal for a number of reasons: (1) the fraction of assessed resources is small (41 percent of the total lake acreage) and certainly not random, (2) beneficial uses are politically determined, (3) states do not use a consistent assessment methodology, and (4) lakes assessed in one reporting period are not necessarily the same ones assessed in a later reporting period (EPA, 1990). EPA may eventually use results from their EMAP program (Sec. 9.5.5) as a basis for the 305(b) reports, greatly enhancing our ability to track environmental trends in lakes.

Despite these weaknesses, the 305(b) reports provide considerable insights regarding water-quality problems at the national level. First, the "traditional" pollutants—excessive nutrients, siltation, and organic loading—appear to be the major causes of water-quality impairment in lakes and reservoirs; toxic pollutants are much less important. Second, agricultural runoff was identified by EPA as the

TABLE 9.4 Water Quality Impairment of Designated Uses for Lakes in the United States

	Acres of lake surface
Total resource	38,400,000
Resource within 40 assessed states	22,347,961
Assessed lake acreage	16,314,012
Assessed lakes meeting use designation, as percent of assessed resource:	
Full supporting	73.7
Threatened*	17.8
Partially supporting	16.6
Not supporting	9.8

* The "threatened" category is a subset of "fully supporting" in the 305(b) report. Because of this, the percentages for the four categories add up to more than 100%.
Source: EPA (1990).

largest single contributor of pollution to lakes and reservoirs, affecting about half of the impaired lakes. These findings and several related studies (NRC, 1989; GAO, 1990; EPA, 1989; Water Quality 2000, 1993) have led policy analysts to suggest that greater emphasis be placed on nonpoint sources of pollutants.

9.7 RESTORATION OF LAKES AND RESERVOIRS

The term *restoration* has been used rather broadly in the realm of lake management and in various environmental laws. The recent NRC report *Restoration of Aquatic Systems* (NRC, 1992) delineated the distinction between *restoration* and related activities defining restoration as "the reestablishment of predisturbance aquatic functions and related physical, chemical and biological characteristics." NRC recognized that restoration can never be perfect; nevertheless restoration is a broader goal than *rehabilitation,* which is often used in the context of visual improvements to an ecosystem, such as improving water clarity. This recognition is important because many of the lake management activities deemed "restoration" are in fact narrow in their goals and oriented primarily toward improving human utilization.

9.7.1 Source Control of Pollutants

Although declining ecological conditions in lakes can have a variety of causes, chemical pollution remains high on the list. Many lake restoration efforts involve direct manipulations of the lake itself, but successful lake management activities to mitigate the effects of chemical pollutants have nearly always involved a reduction in pollutant inputs and, particularly, reductions of nutrients. The outcome of several major efforts to reduce pollutant inputs to lakes is discussed below.

9.7.1.1 Organochlorine Pesticides.   Organochlorine pesticide residues in fish in the Great Lakes were well above FDA action limits prior to the banning of these chemicals during the 1970s through the early 1980s. In the Great Lakes (Fig. 9.15), as well as in other aquatic systems throughout the country (Schmidt, et al., 1990), concentrations of these contaminants have been slowly declining. Limnological factors affecting the rate of decline of organochlorine pesticides include hydraulic residence time, suspended solids concentration, and productivity (Allan, 1989). For many lakes, continued inputs of organochlorine compounds from groundwater (leaching of contaminated industrial or agricultural sites), erosion of soils from contaminated sites, or long-range atmospheric transport has slowed the decline (Allan, 1989; Schmidt, et al., 1990).

9.7.1.2 Phosphorus.   Reduction of phosphorus inputs to lakes is the most common approach for reversing the eutrophication of lakes. Most efforts to reduce phosphorus inputs to lakes have been done either by diverting wastewater effluent from the lake or reducing phosphorus concentrations in wastewater effluents through advanced treatment or phosphorus detergent bans. A classic case study of the effect of diversion is Lake Washington, Seattle, where diversion of wastewater effluent from the lake (100 percent diversion) resulted in decreased phosphorus concentrations and reduced algal abundance (Fig. 9.16). For the Great Lakes, reduction of phosphorus inputs from municipal wastewater treatment plants by 51 to 67 percent has resulted in marked reductions in algal abundances, especially in the lower lakes (CEQ, 1990).

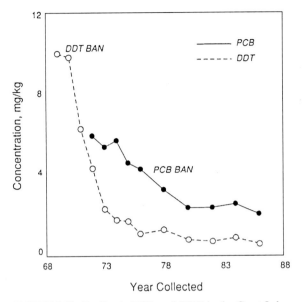

FIGURE 9.15   Decline in PCBs and DDT in the Great Lakes. (*NRC, 1992.*)

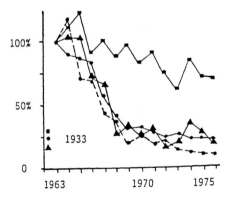

FIGURE 9.16 Reduction in phosphorus and chlorophyll concentrations in Lake Washington following a complete diversion of wastewater. Circles and dashed line = chlorophyll; circles with solid line = total phosphorus; triangles with solid line = phosphate; squares with solid line = nitrate. (*Cooke, et al., 1993; data of T. Edmundson, 1978.*)

Reductions in phosphorus loadings are not always accompanied by immediate reductions in algal abundance and other symptoms of eutrophication. As noted earlier, internal loading from phosphorus-rich sediments can impede the rate of recovery, particularly for shallow lakes. Case studies of phosphorus reduction by wastewater diversion and tertiary treatment are reviewed in Cooke, et al. (1993).

Reductions in point-source inputs are often not sufficient to achieve the desired level of nutrient control. For many lakes, the reversal of eutrophication also requires the control of non-point-source inputs. Unfortunately there is little demonstrated long-term success at source control of phosphorus from diffuse sources. Where reduction of non-point-source phosphorus inputs has been successful, it has generally been associated with the control of feedlot runoff or the installation of detention basins, wetlands, or other interception devices to reduce the movement of sediment. Smolen, et al. (1988) reviewed the status of 20 agricultural non-point-source projects and identified some of the problems that have limited the success of these projects; Logan (1987) discusses the potential for reductions of non-point-source loadings of phosphorus to Lake Erie. *Best management practices* for reducing pollutant loadings from urban areas are reviewed by EPA (1992).

9.7.1.3 Lake Acidification. Although liming can mitigate the problem (Olem, 1990), most countries that have experienced a widespread lake acidification problem have based their mitigation strategy upon the reduction of sulfur dioxide emissions from coal-burning power plants. In most instances where large reductions in emissions have occurred, acidic lakes have experienced an increase in pH and alkalinity (NAPAP, 1990).

9.7.2 In-Lake Physical and Chemical Restoration Techniques

Physical and chemical treatment techniques are often used in conjunction with reductions in nutrient inputs to decrease the response time that would occur with nutrient reduction alone. In-lake management practices are generally practical only for relatively small lakes with high recreational or aesthetic value. Some of the more common techniques are discussed here. For thorough treatments of lake restoration the reader is referred to Cooke, et al. (1993) and NRC (1992).

9.7.2.1 Dilution and Flushing. Dilution of the inflow water with the addition of low-P water lowers the P concentration of the inflow and increases the water loading, with the net effect of decreasing lakewater P concentrations. The degree of reduction can be deduced from the general P-loading model (Eq. 9.25). Several projects that have utilized dilution used water with relatively low phosphorus concentrations (to keep the new P loading as low as possible), but it is theoretically possible to reduce lake phosphorus concentrations by simply increasing the water loading rate alone (same P concentration). In addition to the effect of reducing P concentrations, reductions in algae growth may also occur under extreme water loadings (Θ_H < 10 days) because of cell washout. The limiting factor in utilizing dilution to reduce algal abundance is an inexpensive supply of dilution water. Lake dilution and flushing has been successful in Moses Lake and Green Lake in Washington State (reviewed by Cooke, 1993).

9.7.2.2 Sediment Treatments. In small lakes, sediments are sometimes dredged and removed. This is often done simply to increase the depth and thereby enhance recreational or navigational uses. Sediment dredging has also been used to remove toxic materials, to reduce the growth of aquatic plants, and to reduce phosphorus regeneration (Peterson, 1981). Removal of toxic materials is uncommon in small lakes, but has been done in several rivers contaminated with PCBs and other persistent organic toxins. For macrophyte control, the goal is to create a water depth great enough to prevent colonization of macrophytes through light limitation. The Wisconsin DNR (reported in Cooke, et al., 1993) found that the maximum depth of plant growth (Y) is related to *Secchi disk transparency* by the equation:

$$Y = 0.83 + 1.22\, X \tag{9.27}$$

where X is the Secchi disk transparency, in m (from Cooke, et al., 1993). Although other environmental factors affect the distribution of macrophytes, nuisance macrophytes rarely occur in water depths >3 m.

Several approaches can be used to decide how much sediment needs to be dredged to reduce internal nutrient cycling. For lakes that have recently become eutrophic, one approach is to simply remove the layer of nutrient-rich sediments that has accrued since nutrient enrichment began. For shallow lakes, Hanson and Stefan (1984) proposed dredging to a depth that allows stable stratification, thereby preventing mixing of nutrients in the water column during the summer. They used a lake model (MINLAKE) to establish dredging design depths for a group of lakes in Fairmont, Minnesota. For some shallow lakes, dredging can increase algal growth by reducing sediment resuspension and thereby increasing light penetration (Hanson and Stefan, 1984).

A major factor in determining the cost of dredging is the availability of a suitable site for disposing of dredge spoils. Dredging costs vary tremendously, in most cases from $3,000 to $60,000 per hectare (1991 dollars, Cooke, et al., 1993).

On occasion, sediments are treated to reduce the flux of pollutants from polluted sediments to the overlying water. One of the more popular methods of sediment treatment is the application of alum (aluminum sulfate) to reduce the regeneration of phosphorus. Alum added to the surface of a lake will immediately result in coprecipitation of phosphorus and improved clarity, but the main objective is to form a stable floc at the sediment–water interface that will reduce the regeneration of phosphorus for an extended period of time. The traditional method of dosing is to add as much alum as possible without lowering the pH below 6 (Kennedy and Cooke, 1982; Cooke, et al., 1993), a dose well above the stoichiometric requirement for removing phosphorus from the water column. Higher doses will lower the pH and likely result in toxic aluminum concentrations. The effective lifetime of alum treatments ranges from a few years to over 10 years (Garrison and Knauer, 1984). The mixing characteristics of the lake and the density of sediments are important determinants for success. Alum is best considered as an adjunct to nutrient reduction that can hasten the reversal of eutrophication problems; success is limited in cases where nutrient loadings have not been reduced.

A variation on this theme is *sediment oxidation*, in which calcium nitrate is added to sediments. The basic concept is that the nitrate will be used as an electron acceptor for bacteria, thereby preventing reducing conditions and the release of phosphorus (Ripl and Lindmark, 1978; reported in Cooke, et al., 1993). Although this technique has been used with some success, it should work only when anaerobic release of phosphorus is an important part of the overall phosphorus budget (Noonan, 1986).

9.7.2.3 Aeration. *Aeration of lakes* is widely practiced, primarily as a remediation step to prevent problems associated with anoxia. Aeration of the hypolimnia of lakes is used to maintain high DO levels for fish and to prevent the buildup of H_2S, manganese, and iron in drinking water supply reservoirs. There are many types of aerators; the most common type of *hypolimnetic aerator* is a partial lift system (Fig. 9.17). Although hypolimnetic aeration should at least in some circumstances reduce the buildup of phosphorus and thereby reduce algal problems, there is little empirical evidence of this (Cooke, et al., 1993).

In regions where ice forms in winter, aeration is also used to prevent winterkill or to provide open water habitat for waterfowl. The objective here is usually to produce a well-oxygenated refuge rather than to aerate the entire water column. Although the goal of preventing winterkill is generally achieved, a major problem associated with winter aeration is that it often causes holes in the ice. In Minnesota, several deaths have occurred when people wandered near aerators and fell through the ice, leading the state to develop a permit system to ensure safety (Baker and Swain, 1989).

9.7.2.4 Artificial Circulation. Complete vertical circulation of lakes can be accomplished by injecting air near the bottom of the lakes. *Artificial circulation* is a popular technique for reducing symptoms of eutrophication (Pastorek, et al., 1981). In theory, artificial circulation should decrease light availability, but observations regarding changes in algal abundance following circulation are mixed. The most consistently observed responses are increased DO levels, slightly decreased pH, and oxidation of iron and manganese. A shift in algal dominance from blue-greens to green algae is often observed. An extensive review of lake aeration and artificial circulation is provided by Pastorek, et al. (1981), also summarized in Cooke, et al. (1993).

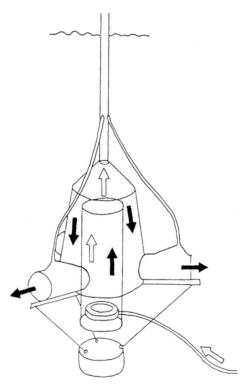

FIGURE 9.17 Schematic of an Atlas Copco partial airlift hypolimnetic aerator. (*Cooke, et al., 1986.*)

9.7.3 Biological Manipulations

9.7.3.1 Macrophyte Control. The control of overly abundant macrophytes is one of the most widely practiced lake-management efforts. *Macrophyte control* can be accomplished by mechanical harvesting, herbicide treatments, sediment covers, and lake drawdown. Harvesting and herbicide treatments are the most common methods. Macrophyte control is generally a remedial operation, intended to improve boating or the aesthetic qualities of a lake. Most mechanical harvesters are one-step operations in which the cut material is moved by a conveyor belt to a barge, allowing the harvested biomass to be removed from the lake. Utilization of harvested macrophytes is usually not economically feasible. The side effects of *macrophyte harvesting* are generally considered to be minimal, but herbicide treatments of large areas without subsequent removal of decaying plant material can lead to algal blooms or oxygen depletion (Carpenter and Adams, 1978; Nichols and Keeney, 1973).

Harvesting of macrophytes may lead to improved trophic status by reducing internal cycling of nutrients. This is most likely to occur in shallow lakes with extensive macrophyte beds and low external loadings, conditions that would best apply to small urban lakes. Peterson, et al. (1974) and Carpenter and Adams (1978) concluded that even extensive and repeated macrophyte harvesting would not reduce phosphorus levels if external nutrient inputs were high.

Lake drawdown has also been used to control macrophytes. The effect of lake drawdown varies with species and with timing. Some species, such as softstem bullrush (*Scirpus validus*) require mud flats to germinate, but then survive well with large fluctuations in water level. Cooke, et al. (1993) summarize the observed response to lake drawdown. The response of fish to lake drawdown is variable. Lake drawdown has the additional advantage of desiccating and compacting sediments, which may result in reduced nutrient cycling (reviewed by Cooke, et al., 1993).

In the southern United States, grass carp have been used to control macrophytes. Although they do control macrophyte growth, the wisdom of their introduction is questionable. In most states introduced grass carp must be sterile, but the species has already spread throughout parts of the southeastern United States.

9.7.4 Integration of Fisheries and Water Quality Objectives

Lake management efforts directed by water quality agencies and implemented by engineering agencies are often detached from fisheries management efforts; often neither includes the goal of a balanced, natural ecosystem. NRC (1992) has pointed to the need for a better understanding of the interactive effects of management efforts directed towards aquatic plants, fisheries, and phytoplankton.

The term *biomanipulation* (Shapiro, et al., 1975) has been used to include a variety of biological controls for controlling eutrophication. Most efforts under this heading have been directed toward removing first-level predators, promoting the development of larger herbivorous zooplankton populations (often *Daphnia*). This can be accomplished by complete fish removal, but most efforts at biomanipulation have involved the introduction of large predators who feed on zooplanktivores. This approach has worked, at least over the short term, in several whole-lake experiments (reviewed by NRC, 1993; Fig. 9.18).

Cooke, et al. (1993) suggest that the term *biomanipulation* might also include the control of detritovores which tend to promote recycling of nutrients, at least in shal-

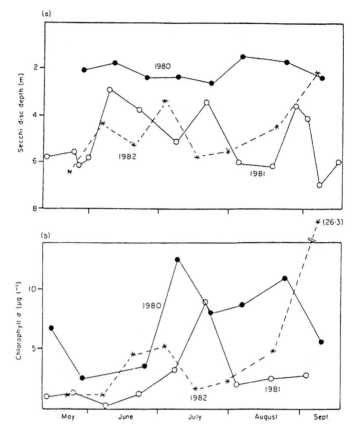

FIGURE 9.18 Biomanipulation of Round Lake, Minnesota. The entire fish population was killed with rotenone in the fall of 1980, then restocked with a high density of piscivorous fish (walleye and largemouth bass) and smaller numbers of bluegill and channel catfish. The experiment was successful at reducing algae levels for the first two years. By the third year, bluegill populations had increased, resulting in greater predation of the herbivorous zooplankton, and algae populations increased. (*Shapiro and Wright, 1984.*)

low, littoral areas. The long-standing fisheries management practice of removing rough fish as a prelude towards *restoration* (with trout, of course) of lakes may have the effect of reducing nutrient regeneration and thereby reducing algae blooms.

Finally, there has been a modest effort toward "restructuring" littoral zones with the purpose of facilitating multiple uses (Engel, 1984; 1987). Engel's proposed management practices include *management zoning,* selective mechanical harvesting to create boat channels and an *edge effect* for fish, and modification of the aquatic plant community by harvesting undesirable species and transplanting of desirable species.

In summary, although limnologists have recognized that interactions among trophic levels are a determinant of algal production and water quality, our ability to predict the effect of biological manipulations on lakewater quality is still poor.

REFERENCES

Allan, R. J., "Factors Affecting Sources and Fate of Persistent Toxic Organic Chemicals: Examples from the Laurentian Great Lakes," in: A. Boudou and F. Ribeyre, eds., *Aquatic Ecotoxicology* (2 vols.), CRC Press, Boca Raton, Fla., 1989.

Allison, J. D., D. S. Brown, and K. J. Novo-Gradac, MINTEQA2/PRODFA2, "A Geochemical Assessment Model for Environmental Systems: Version 3.0 User's Manual," U.S. Environmental Protection Agency, Office of Research and Development, EPA/600/3/91/021, 1991.

Ambrose, R. B., Jr., R. A. Wool, J. P. Connolly, and R. W. Schanz, "WASP4, A Hydrodynamic and Water Quality Model—Model Theory, User's Guide, and Programmer's Guide," U.S. Environmental Protection Agency Research Lab, Athens, Ga., EPA/600/3-87/039, 1988a.

Ambrose, R. B., Jr., J. P. Connolly, E. Southerland, T. O. Barnwell, Jr., and J. L. Schnoor, "Waste Allocation Simulation Models," *J. Water Pollution Control Fed.*, 60:1646–1655, 1988b.

APHA, *Standard Methods for the Examination of Water and Wastewater*, 18th ed., American Public Health Association, American Water Works Association, and Water Pollution Control Federation, Washington, D.C., 1992.

Armstrong, D. E., J. P. Hurley, D. L. Swackhammer, and M. M. Shafer, "Cycles of Nutrient Elements, Hydrophobic Organic Compounds, and Metals in Crystal Lake," in R. A. Hites and S. J. Eisenreich, eds., *Sources and Fates of Aquatic Pollutants*, Advances in Chemistry Series no. 216, American Chemical Society, Washington, D.C., 1987, pp. 491–518.

Asbury, C. E., F. A. Vertucci, M. D. Mattson, and G. E. Likens, "Acidification of Adirondack Lakes," *Environ. Sci. Technol.*, 23:362–365, 1989.

Baker, J. P., and C. L. Schofield, "Aluminum Toxicity to Fish in Acidic Waters," *Water Air Soil Pollut.*, 18:289–309, 1982.

Baker, J. P., and S. Christensen, "Effects of Acidification on Aquatic Biota," in D. F. Charles, ed., *Acidic Deposition and Aquatic Ecosystems: Regional Case Studies*, Springer-Verlag, New York, 1991, pp. 83–106.

Baker, J. P., W. J. Warren-Hicks, J. Gallagher, and S. W. Christensen, "Fish Population Losses from Adirondack Lakes: The Role of Surface Water Acidity and Acidification," *Water Resources Research* 29:861–874, 1993.

Baker, L. A., P. L. Brezonik, and C. K. Kratzer, "Nutrient Loading-Trophic State Relationships for Florida Lakes," Florida Water Resources Center Publication no. 56, Gainesville, Fla., 1981.

Baker, L. A., "A Model for Phosphorus Cycling in Lakes," Internal Memorandum no. 100, St. Anthony Falls Hydraulics Lab, University of Minnesota, 1984.

Baker, L. A., "Regional Estimates of Dry Deposition," Appendix B in *Acidic Deposition and Aquatic Ecosystems*, edited by Don Charles, Springer-Verlag, New York, pp. 645–652, 1991.

Baker, L. A., P. L. Brezonik, and C. K. Kratzer, "Nutrient Loading-Trophic State Relationships for Florida Lakes," Florida Water Resources Center Publication no. 56, Gainesville, Fla., 1981.

Baker, L. A., P. L. Brezonik, and C. Kratzer, "Nutrient Loading Models for Florida Lakes," in J. F. Taggart and L. M. Moore, eds., *Lake and Reservoir Management*, vol. 1, North American Lake Management Society, Washington, D.C., 1985, pp. 253–258.

Baker, L. A., P. L. Brezonik, and E. S. Edgerton, "Sources and Sinks of Ions in a Softwater, Acidic Lake," *Water Resources Res.*, 22:715–722, 1986a.

Baker, L. A., P. L. Brezonik, and C. D. Pollman, "Model of Internal Alkalinity Generation in Softwater Lakes: Sulfate Component," *Water, Air, Soil Pollution*, 30:89–94, 1986b.

Baker, L. A., and E. Swain, "Lake Management in Minnesota," *Lake and Reservoir Management*, 5:1–10, 1989.

Baker, L. A., N. R. Urban, and P. L. Brezonik, "Sulfur Cycling in Little Rock Lake," in W. J. Cooper and E. S. Saltzman, eds., *Biogenic Sulfur in the Environment*, ACS Symposium Series no. 393, American Chemical Society, Washington, D.C., 1989, pp. 79–100.

Baker, L. A., P. R. Kaufmann, A. T. Herlihy, and J. M. Eilers, "Current Acid-Base Status of Surface Waters," State-of-Science Report no. 9, National Acid Precipitation Assessment Program Integrated Assessment, 722 Jackson Place, Washington, D.C., 1990.

Baker, L. A., J. M. Eilers, R. B. Cook, P. R. Kaufmann, and A. T. Herlihy, "Interregional Comparison of Surface Water Chemistry and Biogeochemical Processes," in Don Charles, ed., *Acidic Deposition and Aquatic Ecosystems,* Springer-Verlag, New York, 1991a, pp. 567–614.

Baker, L. A., A. Herlihy, and P. Kaufmann, "Acidic Lakes and Streams in the United States: Role of Acidic Deposition," *Science* 252:1151–1154, 1991b.

Baker, L. A., T. Qureshi, and L. Farnsworth, "Sources of Arsenic in the Verde and Salt Rivers, Arizona," *Proc. 67th Annual Conference and Exposition, Water Environment Federation,* paper no. AC942402, Water Environment Federation, Alexandria, Va., 1994.

Barron, M. G., "Bioconcentration," *Environmental Science and Technol.,* 24:1612–1618, 1990.

Belanger, T. V., "Benthic Oxygen Demand in Lake Apopka, Florida," *Water Research,* 15:267–274, 1980.

Berner, R. A., *Early Diagenesis: a Theoretical Approach,* Princeton University Press, Princeton, N.J., 1980.

Blomqvist, S., and C. Kofoed, "Sediment Trapping—A Subaquatic in situ Experiment," *Limnology and Oceanography,* 26:585–589, 1981.

Bole, J. B., and J. R. Allen, "Uptake of Phosphorus from Sediments by Aquatic Plants, *M. spicatum,* and *H. verticulatus,*" *Water Research,* 12:353, 1978.

Bostrom, B., J. M. Anderson, S. Fleischer, and M. Jansson, "Exchange of Phosphorus across the Sediment–Water Interface," *Hydrobiologia,* 170:229–244, 1988.

Bouldin, D. R., R. L. Johnson, C. Burda, and Chun-Wei Kao, "Losses of Inorganic Nitrogen from Aquatic Systems," *J. Environ. Quality* 3:107–114, 1974.

Bowman, G. T. and J. J. Delfino, "Sediment Oxygen Demand Techniques: A Review and Comparison of Laboratory and *in situ* Systems," *Water Research,* 14:491–499, 1980.

Boyd, C. E., and L. W. Hess, "Factors Influencing Shoot Production and Mineral Nutrient Levels in *Typha latifolia,*" *Ecology,* 51:296–300, 1970.

Brezonik, P. L., "Nutrients and Other Biologically Active Substances in Atmospheric Precipitation," in *Proc. First Spec. Symp. on Atmos. Contr. to Chemistry of Lake Waters, Internat. Assoc. Great Lakes Res.,* Sept. 28–Oct. 1, 1975.

Brezonik, P. L., "Denitrification in Natural Waters," *Prog. Water Techn.,* 8:373–392, 1977.

Brezonik, P. L., C. D. Hendry, Jr., E. S. Edgerton, R. L. Schulze, and T. L. Crisman, "Acidity, Nutrients, and Minerals in Atmospheric Precipitation over Florida: Depositional Patterns, Mechanisms, and Ecological Effects," EPA 600/3-83-004 (NTIS PB-83 165-837), U.S. Environmental Protection Agency, Washington, D.C., 1983.

Brezonik, P. L., "Trophic State Indices: Rationale for Multivariate Approaches," "Lake and Reservoir Management," EPA 440/5/84-001, U.S. Environmental Protection Agency, Washington, D.C., 1984, pp. 441–445.

Brezonik, P. L., *Chemical Kinetics and Process Dynamics in Aquatic Systems,* Lewis Publishers, Ann Arbor, Mich., 1994.

Brezonik, P. L., L. A. Baker, and T. E. Perry, "Mechanisms of Alkalinity Generation in Acid-Sensitive Softwater Lakes," in R. A. Hites and S. J. Eisenreich, eds., *Sources and Fates of Aquatic Pollutants,* Advances in Chemistry Series no. 216, American Chemical Society, Washington, D.C., 1987, pp. 229–261.

Bristow, J. M., and M. Whitcombe, "The Role of Roots in the Nutrition of Aquatic Vascular Plants," *Am. J. Bot.,* 58:8–13, 1971.

Brock, T. D., *A Eutrophic Lake: Lake Mendota, Wisconsin,* Ecological Series no. 55, Springer-Verlag, New York, 1985.

Cairns, J. Jr., "Applied Ecotoxicology and Methodology," in A. Boudou and F. Ribeyre, eds., *Aquatic Ecotoxicology,* vol. II, CRC Press, Boca Raton, Fla., 1989, pp. 275–290.

Campbell, K. L., "Nutrient Loads in Streamflow from Sandy Soils in Florida," *Trans. ASAE,* 22:1115–1120, 1979.

Carlson, R. E., "A Trophic State Index for Lakes," *Limnol. Oceanogr.,* 22:361–369, 1977.

Carpenter, S. R., "Replication and Treatment Strength in Whole-Lake Experiments," *Ecology,* 70:453–463, 1989.

Carpenter, S. R., and M. S. Adams, "Macrophyte Control by Harvesting and Herbicides: Implications for Phosphorus Cycling in Lake Wingra, Wisconsin," *J. Aquatic Plant Management,* 16:20–23, 1978.

Carson, R., *Silent Spring,* 25th Anniversity ed. (1987), Houghton Mifflin, Boston, 1962.

Charles, D. F., and J. P. Smol, "Long-Term Chemical Changes in Lakes: Quantitative Inferences from Biotic Remains in the Sediment Record," in L. A. Baker, ed., *Environmental Chemistry of Lakes and Reservoirs,* Advances in Chemistry Series no. 237, American Chemical Society, Washington, D.C., 1994, pp. 3–31.

Charles, D. F., R. W. Battarbee, I. Renberg, H. van Dam, and J. P. Smol, "Paleoecological Analysis of Lake Acidification Trends in North America and Europe Using Diatoms and Chrysophytes," in S. A. Norton, S. E. Lindberg, and A. L. Page, eds., *Acid Precipitation, vol. 4: Soils, Aquatic Processes, and Lake Acidification,* Springer-Verlag, New York, 1989, pp. 207–276.

Charles, M. J., and R. A. Hites, "Sediments as Archives of Environmental Pollution Trends," in *Sources and Fates of Aquatic Pollutants,* Advances in Chemistry Series no. 216, American Chemical Society, Washington, D.C., 1987, pp. 364–389.

Chiaudani, G., and M. Vighi, "The N:P Ratio and Tests with *Selanastrum* to Predict Eutrophication in Lakes," *Water Research,* 8:1063–1069, 1974.

Cooke, G. D., E. B. Welch, S. A. Peterson, and P. R. Newroth, *Lake and Reservoir Restoration,* Butterworth, New York, 1986.

Cooke, G. D., E. B. Welch, S. A. Peterson, and P. R. Newroth, *Restoration and Management of Lakes and Reservoirs,* 2d ed., Lewis Publishers, Ann Arbor, Mich., 1993.

Council on Environmental Quality, *Environmental Quality: 20th Annual Report,* Executive Office of the President, Washington, D.C., 1990.

Dasch, J. M., "A Comparison of Surrogate Surfaces for Dry Deposition Collection," in H. R. Pruppache, R. G. Semonin, and W. G. N. Slinn, eds., *Precipitation Scavenging, Dry Deposition, and Resuspension,* Elsevier, New York, 1983, pp. 883–900.

Denny, P., "Sites of Nutrient Adsorption in Aquatic Macrophytes," *J. Ecology,* 60:819–829, 1972.

Dillon, P. J., and F. H. Rigler, "A Test of a Simple Nutrient Budget Model Predicting the Phosphorus Concentration in Lake Water," *J. Fish. Res. Board Canada,* 31:1771–1778, 1974a.

Dillon, P. J., and F. H. Rigler, "The Phosphorus-Chlorophyll Relationship in Lakes," *Limnol. Oceanogr.,* 19:767–773, 1974b.

Ditoro, D. M., J. D. Mahony, D. J. Hansen, K. J. Scott, A. R. Carlson, and G. T. Ankley, "Acid Volatile Sulfide Predicts the Acute Toxicity of Cadmium and Nickel in Sediments," *Environ. Sci. Technol.,* 26:96–101, 1991.

Dixit, S. S., J. P. Smol, J. C. Kingston, and D. F. Charles, "Diatoms: Powerful Indicators of Environmental Change," *Environmental Science and Technology,* 26:23–33, 1992.

Dolan, D. M., A. K. Yui, and R. D. Geist, "Evaluation of River Load Estimation Methods for Total Phosphorus," *J. Great Lakes Res.,* 7:207–214, 1981.

Effler, S. W., C. M. Brooks, M. T. Auer, and S. M. Doerr, "Free Ammonia and Toxicity Criteria in a Polluted Urban Lake," *Res. J. Water Pollution Control Fed.,* 62:771–779, 1990.

Eilers, J. M., G. E. Glass, K. E. Webster, and J. A. Rogalla, "Hydrologic Control of Lake Susceptibility to Acidification," *Can. J. Fisheries Aquatic Sci.,* 40:1896–1904, 1983.

Eisler, R., "Mercury Hazards to Fish, Wildlife, and Invertebrates: A Synoptic Review," U.S. Fish and Wildlife Survey Biol. Report 85(1.10), 1987.

Electric Power Research Institute, "Mercury Atmospheric Processes: A Synthesis Report," Prepared by the Expert Panel on Mercury Atmospheric Processes, March 11–18, Tampa, Fla., Electric Power Research Institute, Palo Alto, Cal., EPRI/TR-104214, 1994.

Engel, S., "Restructuring of Littoral Zones: A Different Approach to an Old Problem," in *Lake and Reservoir Management, Proc. North Am. Lake Management Society,* Washington, D.C., EPA 440/5-84-001, 1984, pp. 463–466.

Engel, S., "The Restructuring of Littoral Zones," *Lake and Reservoir Management,* 3:235–42, 1987.

Engstrom, D. R., E. B. Swain, and J. C. Kingston, "A Palaeolimnological Record of Human Disturbance from Harvey's Lake, Vermont: Geochemistry, Pigments, and Diatoms," *Freshwater Biology* 15:261–288, 1985.

Engstrom, D. R., E. B. Swain, T. A. Henning, M. E. Brigham, and P. L. Brezonik, "Atmospheric Mercury Deposition to Lakes and Watersheds: A Quantitative Reconstruction from Multiple Cores," in L. A. Baker, ed., *Environmental Chemistry of Lakes and Reservoirs*, edited by Advances in Chemistry Series no. 237, American Chemical Society, Washington, D.C., 1994, pp. 33–36.

Fisheries, "National Recreational Fisheries Policy," *Fisheries*, March–April, 1988, p. 35.

Ford, J. "The Effects of Chemical Stress on Aquatic Species Composition and Community Structure," in S. A. Levine, M. A. Harwell, J. R. Kelly, and K. D. Kimball, eds., *Ecotoxicology: Problems and Approaches*, Springer Advanced Texts in Life Sciences, Springer-Verlag, New York, 1989, pp. 100–144.

Gakstatter, J. H., M. O. Allum, S. E. Dominguez, and M. R. Crouse, "A Survey of Phosphorus and Nitrogen Levels in Treated Municipal Wastewater," *Journal of the Water Pollution Control Federation*, 50:718–722, 1978.

Gallant, A. L., T. R. Whittier, D. P. Larsen, J. M. Omernik, and R. M. Hughes, "Regionalization as a Tool for Managing Environmental Resources," U.S. EPA Environmental Research Laboratory, Corvallis, Ore., EPA/600/3-89/060, 1989, 152 pp.

Garrison, P. J., and D. R. Knauer, "Long-Term Evaluation of Three Alum-Treated Lakes," *Lake and Reservoir Management, Proceedings of the Third Annual Conference of the North American Lake Management Society*, U.S. Environmental Protection Agency, Washington, D.C., EPA 440/5/84-001, 1984, pp. 513–518.

Gatcher, R., J. S. Davis, and A. Mares, "Regulation of Copper Availability to Phytoplankton by Macromolecules in Lake Water," *Env. Sci. Technol.*, 12:1416–1421, 1978.

Gatcher, R., J. S. Meyer, and A. Mares, "Contribution of Bacteria to Release and Fixation of Phosphorus in Lake Sediments," *Limnol. Oceanogr.*, 33:1542–1558, 1988.

Gatz, D. F., V. C. Bowersox, and J. Su, "Screening Criteria for NADP Dry-Bucket Sample Data," Presentation at the 81st Annual Meeting of the Air Pollution Control Association, June 19–24, 1988, Reprint no. 838, Illinois State Water Survey, Champaign, Ill., 1988.

Gentile, J. H., W. H. van der Schalie, and W. P. Wood, "Summary Report on Issues in Ecological Risk Assessment," U.S. Environmental Protection Agency, Office of Research and Development, Washington, D.C., EPA/625/3-91-018, 1991.

Giacomo, R. S., and N. G. Randell, "Invasion of the Zebra Mussels," *Civil Engineering*, May, 1993, pp. 56–58.

Giesy, J. P., and R. L. Graney, "Recent Developments in and Intercomparisons of Acute and Chronic Bioassays and Bioindicators," in M. Munawar, G. Dixon, C. I. Mayfield, T. Reynoldson, and M. H. Sadar, eds., *Environmental Bioassay Techniques and their Application*, Kluwer Academic Publishers, 1989.

Gloss, S. P., T. M. Grieb, C. T. Driscoll, C. L. Schofield, J. P. Baker, D. Landers, and D. B. Porcella, "Mercury Levels in Fish from the Upper Pennisula of Michigan (ELS Subregion 2B) in Relation to Acidity," EPA Office of Acid Deposition, Environmental Monitoring, and Quality Assurance, EPA/600/3-90/068, 1990.

Goldman, C. R., and A. J. Horne, *Limnology*, McGraw-Hill, New York, 1983.

Government Accounting Office, "Greater EPA Leadership Needed to Reduce Nonpoint Source Pollution," GAO/RCED-91-10. Washington, D.C., 1990.

Grace, J. B., and R. G. Wetzel, "The Production Biology of Eurasion Milfoil (*Myriophyllum spicatum* L.): A Review," *J. Aquat. Plant Management*, 16:1–11, 1983.

Grant, G. E., and A. L. Wolff, "Long-Term Patterns of Sediment Transport after Timber Harvest, Western Cascade Mountains, Oregon, USA," in D. E. Walling, and N. E. Peters, eds., *Sediment and Stream Water Quality in a Changing Environment: Trends and Perspectives, Proc. IAHS Symposium*, Vienna, August, 1991.

Gubala, C. P., C. Branch, N. Roundy, and D. Landers, "Automated Global Positioning Charting of Environmental Attributes: A Limnologic Case Study," *Science of the Total Environment*, in press.

Hakanson, L., and M. Jansson, *Principles of Lake Sedimentology,* Springer-Verlag, New York, 1983.

Hanson, M. J., and H. G. Stefan, "Shallow Water Quality Improvement by Dredging," in *Lake and Reservoir Management: Practical Applications, Proc. 4th Ann. Conf. and Int. Symp., North Am. Lake Management Society,* Washington, D.C., 1984, pp. 162–171.

Hem, J. D., *Study and Interpretation of the Chemical Characteristics of Water,* U.S. Geological Supply Paper 1473, U.S. Government Printing Office, Washington, D.C., 1970.

Hemond, H. F., and E. J. Fechner, *Chemical Fate and Transport in the Environment,* Academic Press, New York, 1994.

Herbes, S. E., H. E. Allen, and K. H. Mancy, "Enzymatic Characterization of Soluble Organic Phosphorus in Lake Water," *Science,* 187:432–434, 1975.

Herlihy, A. T., P. R. Kaufmann, M. E. Mitch, and D. D. Brown, "Regional Estimates of Acid Mine Drainage Impact on Streams in the Mid-Atlantic and Southeastern United States," *Water Air Soil Pollut.,* 50:91–107, 1990.

Hesslein, R. H., "An *in situ* Sampler for Close Interval Porewater Studies," *Limnol. Oceanogr.,* 21:912–914, 1976.

Holdren, G. C., and D. E. Armstrong, "Factors Affecting Phosphorus Release from Intact Lake Sediment Cores," *Environ. Sci. Technol.,* 14:79–87, 1980.

Hren, J., C. J. O. Childress, J. M. Norris, T. H. Chaney, and D. N. Myers, "Regional Water Quality: Evaluation of Data for Assessing Conditions and Trends," *Environ. Science and Technol.,* 24:1122–1127, 1990.

Hughes, R. M., and G. R. Gammon, "Longitudinal Changes in Fish Assemblages and Water Quality in the Willamette River, Oregon," *Trans. Am. Fisheries Soc.,* 116:196–209, 1987.

Hughes, and D. P. Larsen, "Ecoregions: An Approach to Surface Water Protection," *Journal of the Water Pollution Control Federation,* 60:486–493, 1988.

Hutchinson, G. E., *A Treatise on Limnology I: Geography, Physics, and Chemistry,* John Wiley, New York, 1957.

James, A., "The Measurement of Benthal Respiration," *Water Research,* 8:955–959, 1974.

Jeffries, D. S., and W. R. Snyder, "Atmospheric Deposition of Heavy Metals in Central Ontario," *Water, Air, Soil Poll.,* 15:127–152, 1981.

Jones, K. C., *Organic Contaminants in the Environment,* Environmental Management Series, Elsevier, New York, 1991.

Jones, J. R., and R. W. Bachman, "Prediction of Phosphorus and Chlorophyll Levels in Lakes," *Journal Water Poll. Control Fed.,* 48:2176–2182, 1976.

Jorgenson, S. E., "Sedimentation," in S. E. Jorgensen, and M. J. Gromiec, eds., *Mathematical Submodels in Water Quality Systems,* Elsevier, New York, 1989, pp. 109–124.

Jorgenson, S. E., and M. J. Gromiec, *Mathematical Submodels in Water Quality Systems,* Developments in Environmental Modeling no. 14, Elsevier, New York, 1989.

Kaufman, D., and C. M. Franz, *Biosphere 2000: Protecting our Global Environment,* Harper-Collins, New York, 1993.

Kamoto, M., "Primary Production by Phytoplankton Community in Some Japanese Lakes and its Dependence upon Lake Depth," *Arch. Hydrobiol.,* 62:1–28, 1966.

Karickhoff, S. W., D. S. Brown, and T. A. Scott, "Sorption of Hydrophobic Pollutants on Natural Sediments," *Water Research,* 13:241–248, 1979.

Karr, J. R., "Biological Monitoring and Environmental Assessment: A Conceptual Framework," *Environmental Management,* 11:249–256, 1987.

Kelly, C. A., J. W. M. Rudd, R. H. Hesslein, C. T. Driscoll, S. A. Gherini, and R. E. Hecky, "Prediction of Biological Acid Neutralization in Acid-Sensitive Lakes," *Biogeochemistry,* 3:129–140, 1987.

Kennedy, R. H., and G. D. Cooke, "Control of Lake Phosphorus with Aluminum Sulfate: Dose Determination and Application Techniques," *Water Resources Bull.,* 18:389–395, 1982.

Kramer, J. R., A. W. Andren, R. A. Smith, A. H. Johnson, R. B. Alexander, and G. W. Oehlert, "Streams and Lakes," in *Acid Deposition: Long-Term Trends,* National Research Council, National Academy Press, Washington, D.C., 1986, pp. 231–299.

Kratzer, C. R., and P. L. Brezonik, "A Carlson-Type Trophic State Index for Nitrogen in Florida Lakes," *Water. Resources Bull.,* 17:713–715, 1981.

Kuhn, A., C. A. Johnson, and L. Sigg, "Cycles of Trace Elements with a Seasonally Anoxic Hypolimnion," in L. A. Baker, ed., *Environmental Chemistry of Lakes and Reservoirs,* Advances in Chemistry Series no. 237, American Chemical Society, Washington, D.C., 1994, pp. 473–497.

Lamarra, V. A., V. D. Adams, C. Thomas, R. Herron, P. Birdsey, V. Kollock, and M. Pitts, "A Historical Perspective and Present Water Quality Conditions in Bear Lake, Utah-Idaho," in "Lake and Reservoir Management," EPA 440/84-001, U.S. Environmental Protection Agency, Washington, D.C., 1984.

Landers, D. H., "A Durable, Reusable Enclosure System that Compensates for Changing Water Level," *Limnol. Oceanogr.,* 24:991–994, 1979.

Landers, D. H., "Effects of Senescing Aquatic Macrophytes on Nutrient Chemistry and Chlorophyll *a* of Surrounding Waters," *Limnol. Oceanogr.,* 27:428–439, 1982.

Larsen, D. P., D. W. Schultz, and K. W. Malueg, "Summer Internal Phosphorus Supplies in Shagwa Lake, Minnesota," *Limnology and Oceanogr.,* 26:740–753, 1981.

Larsen, D. P., D. L. Stevens, A. R. Selle, and S. G. Paulsen, "Environmental Monitoring and Assessment Program, EMAP-Surface Waters: A Northeast Lakes Program," *Lake and Reservoir Management,* 7:1–11, 1991.

Lee, D. R., "A Device for Measuring Seepage Flux in Lakes and Estuaries," *Limnol. Oceanogr.,* 22:155–163, 1977.

Lewis, W. M., "Collection of Airborne Materials by a Water Surface," *Limnology and Oceanography,* 28:1242–1246, 1983.

Li, Y., and S. Gregory, "Diffusion of Ions in Sea Water and in Deep-Sea Sediments," *Geochem. Cosmochim. Acta,* 38:703–714, 1974.

Likens, G. E., ed., *Nutrients and Eutrophication: The Limiting Nutrient Controversy, Proceedings of the Symp. on Nutrients and Eutrophication,* W.K. Kellog Biological Station, Feb. 11–12, 1971, Am. Society of Limnology and Oceanography, Lawrence, Kan., 1972.

Likens, G. E., and R. G. Wetzel, *Limnological Analysis,* Springer-Verlag, New York, 1990.

Likens, G. E., J. S. Eaton, N. M. Johnson, and R. S. Pierce, "Flux and Balance of Water and Chemicals," in G. E. Likens, ed., *An Ecosystem Approach to Aquatic Ecology: Mirror Lake and its Environment,* Springer-Verlag, New York, 1977.

Linsley, R. K. Jr., M. A. Kohler, and J. L. H. Paulhus, *Hydrology for Engineers,* 3d ed., McGraw-Hill, New York, 1982.

Loftis, J. C., R. C. Ward, R. D. Phillips, and C. H. Taylor, "An Evaluation of Trend Detection Techniques for Use in Water Quality Monitoring Programs," U.S. Environmental Protection Agency, Corvallis Environmental Research Laboratory, EPA/600/3-98/037, 1989.

Logan, T. J., "Diffuse (Nonpoint) Source Loading of Chemicals to Lake Erie," *J. Great Lakes Res.,* 13:649–658, 1987.

Lyman, W. J., W. F. Reehl, and D. H. Rosenblatt, *Handbook of Chemical Property Estimation Methods,* American Chemical Society, Washington, D.C., 1990.

Manny, M. A., and R. A. Minear, "Organic Phosphorus in the Hydrosphere: Characterization via ^{31}P FT-NMR," in L. A. Baker, ed., *Environmental Chemistry of Lakes and Reservoirs,* Advances in Chemistry Series no. 237, American Chemical Society, Washington, D.C., 1994, pp. 161–191.

McKnight, D., "Chemical and Biological Processes Controlling the Response of a Freshwater Ecosystem to Copper Stress: A Field Study of the $CuSO_4$ Treatment of Mill Pond Reservoir, Burlington, Massachusetts," *Limnol. Oceanogr.,* 26:518–531, 1981.

McNabb, C. D., and D. P. Tierney, "Growth and Mineral Accumulation of Submerged Vascular Hydrophytes in Pleioeutrophic Environments," Technical Report no. 26, Institute of Water Research, Michigan State University, East Lansing, 1972.

McQueen, D. J., and J. R. Post, "Enclosure Experiments: The Effects of Planktivorous Fish," in G. Redfield, J. F. Taggart, and L. Moore, eds., *Lake and Reservoir Management, Vol. II., Proc. 5th Ann. Conf. Internat. Symp. on Applied Lake and Watershed Management,* North American Lake Management Soc., Washington, D.C., 1986, pp. 313–318.

McRoy, C. P., and R. J. Bardsdate, "Phosphorus Cycling in an Eelgrass (*Zostera marina* L.) Ecosystem," *Limnol. Oceanogr.,* 17:48–67, 1972.

Meade, R. H., and R. S. Parker, *Sediments in Rivers of the United States,* National Water Summary, 1984, Water Supply Paper 2275, U.S. Geological Survey, Reston, Va., 1985.

Messer, J. J., and P. L. Brezonik, "Denitrification in the Sediments of Lake Okeechobee," *Verh. Internat. Limnol.,* 20:2207–2216, 1978.

Messer, J. J., and P. L. Brezonik, "Comparison of Denitrification Rate Estimation Techniques in a Large, Shallow Lake," *Water Research,* 17:631–640, 1983.

Miller, D. L., P. M. Leonard, R. M. Hughes, J. R. Karr, P. B. Moyle, L. H. Schrader, B. A. Thompson, R. A. Daniels, K. D. Fausch, G. A. Firzhugh, J. R. Gammon, D. B. Halliwell, P. L. Angermeier, and D. J. Orth, "Regional Applications of an Index of Biotic Integrity for Use in Water Resource Management," *Fisheries,* 13:12–20, 1988.

Miller, R. R., J. D. Williams, and J. E. Williams, "Extinctions of North American Fishes During the Past Century," *Fisheries,* 14:22–38, 1989.

Miller, W. E., T. E. Maloney, and J. C. Greene, "Algal Productivity in 49 Lake Waters as Determined by Algal Assays," *Water Research,* 8:667–679, 1974.

Minear, R. A., "Characterization of Naturally Occurring Dissolved Organophosphate Compounds," *Environmental Science and Technol.,* 6:431–437, 1972.

Moore, J. N., "Contaminant Mobilization Resulting from Redox Pumping in a Metal-Contaminated River-Reservoir System," in L. A. Baker, ed., *Environmental Chemistry of Lakes and Reservoirs,* Advances in Chemistry no. 293, American Chemical Society, Washington, D.C., 1994, pp. 451–471.

Moore, J. N., and S. Luoma, "Hazardous Wastes from Large-Scale Metal Extraction," *Environ. Sci. Technol.,* 24:1278–1285, 1990.

Morel, F. M. M., and J. G. Hering, *Principles and Applications of Aquatic Chemistry,* John Wiley, New York, 1993.

Mortimer, C. H., "The Exchange of Dissolved Substances Between Mud and Water in Lakes (Parts I and II)," *J. Ecology,* 29:280–329, 1941.

Mortimer, C. H., "The Exchange of Dissolved Substances Between Mud and Water in Lakes (Parts III and IV)," *J. Ecology,* 30:147–201, 1942.

Munson, R. K., and S. A. Gherini, "Hydrochemical Assessment Methods for Analyzing the Effects of Acidic Deposition of Surface Waters," in *Acidic Deposition and Aquatic Ecosystems,* Don Charles, ed., Springer-Verlag, New York, 1991, pp. 9–34.

National Acid Precipitation Assessment Program, "1990 Integrated Assessment Report," Washington, D.C., 1990.

National Research Council, *The Mono Basin Ecosystem: Effects of Changing Water Level,* Mono Basin Ecosystem Committee, National Academic Press, Washington, D.C., 1987.

National Research Council, *Alternative Agriculture,* Committee on the Role of Alternative Farming Methods in Modern Production Agriculture, National Academy Press, Washington, D.C., 1989.

National Research Council, *Restoration of Aquatic Ecosystems,* Committee on Restoration of Aquatic Ecosystems, National Academy Press, Washington, D.C., 1992.

Nichols, D. S., and D. R. Keeney, "Nitrogen and Phosphorus Release from Decaying Water Milfoil," *Hydrobiologia,* 42:509–525, 1973.

Nielsen, L. A. and D. L. Johnson, eds., *Fisheries Techniques,* Southern Printing Company, Blacksburg, Va., 1983.

Noonan, T. A., "Water Quality in Long Lake, Minnesota, Following Riplox Treatment," in G. Redfield, J. F. Taggart, and L. M. Moore, eds., *Lake and Reservoir Management, Vol. 2, Proc. 5th Annual Conf. and Internat. Symp.,* N. Am. Lake Management Society, Washington, D.C., 1986, pp. 131–137.

Novotny, V., and G. Chesters, *Handbook of Nonpoint Pollution: Sources and Management,* Van Nostrand Reinhold, New York, 1981.

Nunan, R. L., "Effect of Bear River Storage on Water Quality in Bear Lake, Utah," M.S. thesis, Utah State University, Logan, 1972.

Oldfield, F., and P. G. Appleby, "Empirical Testing of ^{210}Pb-Dating Models for Lake Sediments," in E. Y. Haworth and J. W. F. Lund, eds., *Lake Sediments and Environmental History,* University of Minnesota Press, 1984, pp. 94–124.

Olem, H., "Liming Acidic Surface Waters," NAPAP Report no. 15, National Acid Precipitation Assessment Program, Washington, D.C., 1990.

Pastorek, R. A., T. C. Ginn, and M. W. Lorenzen, "Evaluation of Aeration/Circulation as a Lake Restoration Technique," U.S. Environmental Protection Agency Environmental Research Lab, EPA-600/3-81-014, 1981.

Paulsen, S. G., et al., "EMAP-Surface Waters Monitoring and Research Strategy, Fiscal Year 1991," U.S. Environmental Protection Agency, Office of Research and Development, EPA/600/3-91/022, 1991.

Perry, T. E., L. A. Baker, and P. L. Brezonik, "Comparison of Sulfate Reduction Rates Measured in Laboratory Microcosms, Field Mesocosms, and *in situ* in Little Rock Lake, Wisconsin," in G. Redfield, J. F. Taggart, and L. M. Moore, eds., *Lake and Reservoir Management, Vol. 2,* North American Management Soc., Washington, D.C., 1986, pp. 309–312.

Peterson, S. A., W. L. Smith, and K. W. Malueg, "Full-Scale Harvest of Aquatic Plants: Nutrient Removal from a Eutrophic Lake," *J. Water Poll. Control Fed.,* 46:697–707, 1974.

Peterson, S. A., "Sediment Removal as a Lake Restoration Technique," U.S. Environmental Protection Agency, Corvallis Environmental Research Lab, Corvallis, Ore., EPA 600/3-81-013, 1981.

Petts, G. E., *Impounded Rivers: Perspectives for Ecological Management,* Wiley-Interscience, New York, 1984.

Preston, S. D., and P. L. Brezonik, "Water Quality in the Oklawaha Chain of Lakes: A Case Study on Problems and Limitations in Compiling Long-Term Data Bases," in *Lake and Reservoir Management, Proc. 4th Annu. Conf. Int. Symp. N. Am. Lake Manag. Soc.,* Oct. 16–19, 1984, McAfee, N.J., N. Am. Lake Manag. Soc., Washington, D.C., 1985, pp. 101–107.

Qureshi, T., "Sources of Arsenic in the Verde and Salt Rivers, Arizona," M.S. thesis, Arizona State University, 1995.

Ramade, F., "The Pollution of the Hydrosphere by Global Contaminants and its Effects on Aquatic Ecosystems," in A. Boudou and F. Ribeyre, eds., *Aquatic Ecotoxicology* (2 vols.), CRC Press, Boca Raton, Fla., 1989.

Rappaport, R. A., N. R. Urban, P. D. Capel, J. B. Baker, B. Looney, and S. J. Eisenreich, " 'New' DDT Inputs to North America: Atmospheric Deposition," *Chemosphere,* 14:1167–1173, 1984.

Reckhow, K. H., and S. C. Chapra, *Engineering Approaches for Lake Management* (2 vols.), Butterworth, Boston, 1983.

Revsech, N. P., and J. Sorenson, *Denitrification in Soils and Sediments,* Plenum Press, New York, 1990.

Rheinheimer, G., *Aquatic Microbiology,* John Wiley, New York, 1971.

Rosenberry, D. O., "Inexpensive Groundwater Monitoring Methods for Determining Hydrologic Budgets of Lakes and Wetland," in *Proc. 1990 Conf., North Am. Lake Management Soc.,* North Am. Lake Management Soc., Washington, D.C., 1990, pp. 123–313.

Rudd, J. W. M., C. A. Kelly, V. St. Louis, R. H. Hesslein, A. Furutani, and M. H. Holoka, "Microbial Consumption of Nitric and Sulfuric Acids in North Temperate Lakes," *Limnol. Oceanogr.,* 31:1267–1280, 1986.

Rudd, J. W. M., C. A. Kelly, D. W. Schindler, and M. S. Turner, "Disruption of the Nitrogen Cycle in Acidified Lakes," *Science,* 240:1515–1517, 1988.

Rudd, J. W. M., C. A. Kelly, D. W. Schindler, and M. A. Turner, "A Comparison of the Acidification Efficiencies of Nitric and Sulfuric Acids in Lake Sediments," *Limnol. Oceanogr.,* 31:1281–1291, 1990.

Ruttner, F., *Fundamentals of Limnology,* trans. D. G. Frey and F. E. J. Frey, University of Toronto Press, Toronto, 1963.

Sampson, C., P. L. Brezonik, and E. P. Weir, "Effects of Acidification on Chemical Composition and Chemical Cycles in a Seepage Lake: Inferences from a Whole-Lake Experiment," in L. A. Baker, ed., *Environmental Chemistry of Lakes and Reservoirs,* ACS Advances in Chemistry no. 237, American Chemical Society, Washington, D.C., 1994, pp. 121–159.

Scheider, W. A., J. J. Moss, and P. J. Dillon, "Measurement and Uses of Hydraulic and Nutrient Budgets," in *Lake Restoration, Proceedings of a National Conference,* August 22–24, 1978, Minneapolis, EPA 440/5-79-001, 1979.

Schelske, C. L., and S. R. Carpenter, "Restoration Case Studies: Lake Michigan," in *Restoration of Aquatic Ecosystems,* National Research Council Committee on Restoration of Aquatic Ecosystems: Science, Technology, and Public Policy, National Academy Press, Washington, D.C., 1992, pp. 380–392.

Schindler, D. W., "Whole-Lake Eutrophication Experiments with Phosphorus, Nitrogen, and Carbon," *Verh. Internat. Verein. Limnol.,* 19:3221–3231, 1975.

Schmitt, C. J., J. J. Zajicek, and P. H. Peterman, "National Pesticide Monitoring Program: Residues of Organochlorine Chemicals in U.S. Freshwater Fish, 1976–84," *Arch. Environ. Contam. Toxicol.,* 19:748–781, 1990.

Schwarzenbach, R. P., P. M. Gschwend, and D. M. Imboden, *Environmental Organic Chemistry,* Wiley, New York, 1993.

Scientific Advisory Board, "Reducing Risk: Setting Priorities and Strategies for Environmental Protection," SAB-EC-90-021, U.S. Environmental Protection Agency, Washington, D.C., 1990.

Seitzinger, S. P., "Denitrification in Freshwater and Coastal Marine Ecosystems: Ecological and Geochemical Significance," *Limnol. Oceanogr.,* 33:702–724, 1988.

Seitzinger, S. P., "Denitrification in Sediments," in N. P. Revsbech and J. Sorensen, eds., *Denitrification in Soils and Sediments,* Plenum Press, New York, 1990, pp. 301–322.

Shafer, M. M., and D. E. Armstrong, "Mass Fluxes and Recycling of Phosphorus in Lake Michigan: Role of Major Particle Phases in Regulating the Annual Cycle," in L. A. Baker, ed., *Environmental Chemistry of Lakes and Reservoirs,* Advances in Chemistry Series no. 237, American Chemical Society, Washington, D.C., 1994, pp. 285–322.

Shapiro, J., V. Lamarra, and M. Lynch, "Biomanipulation: An Ecosystem Approach to Lake Restoration," in P. L. Brezonik and J. L. Fox, eds., *Water Quality Management Through Biological Control,* Department of Environmental Engineering Sciences, University of Florida, Gainesville, 1975.

Shapiro, J., and E. B. Swain, "Lessons from the Silica 'Decline' in Lake Michigan," *Science* 221:457–459, 1983.

Shapiro, J., and D. I. Wright, "Lake Restoration by Biomanipulation: Round Lake, Minnesota, the First Two Years," *Freshwater Biology,* 14:371–383, 1984.

Sharpley, A. N., J. K. Syers, and P. O. O'Connor, "Phosphorus Inputs into a Stream Draining an Agricultural Watershed," *Water, Air, Soil Poll.,* 6:39–52, 1976.

Sherman, L. A., L. A. Baker, P. L. Brezonik, and E. P. Weir, "Sediment Porewater Dynamics of Little Rock Lake, Wisconsin: Geochemical Processes and Seasonal and Spatial Variability," *Limnol. and Oceanogr.,* 39:1155–1171, 1994.

Slinn, S. A., and W. G. N. Slinn, "Predictions for Particle Deposition on Natural Waters," *Atmos. Environ.,* 14:1013–1016, 1980.

Smith, C. S., "Phosphorus Uptake by Roots and Shoots of *Myriophyllum spicatum* L.," Ph.D. dissertation, Department of Botany, University of Wisconsin, Madison, 1978.

Smith, R. A., R. B. Alexander, and M. G. Wolman, "Water Quality Trends in the Nation's Rivers," *Science,* 235:1608–1615, 1987.

Smolen, M. D., F. J. Humenik, J. Spooner, S. L. Brichford, and R. P. Maas, "NWQEP 1987 Annual Report: Status of Agricultural Nonpoint Source Projects," U.S. Environmental Protection Agency, Office of Water, Washington, D.C., 1988.

Snoeyink, V. L., and D. Jenkins, *Water Chemistry,* John Wiley, New York, 1980.

Soballe, D. M., B. L. Kimmel, R. H. Kennedy, and R. F. Gaugush, "Reservoirs," in *Biodiversity of the Southeastern United States: Aquatic Communities,* John Wiley, New York, 1992, pp. 421–474.

Sonzogni, W. C., D. P. Larsen, K. W. Malueg, and M. D. Schuldt, "Use of Large Submerged Chambers to Measure Sediment-Water Interactions," *Water Research,* 11:461–464, 1977.

Sonzogni, W. C., S. C. Chapra, D. E. Armstrong, and T. J. Logan, "Bioavailability of Phosphorus to Lakes," *J. Environmental Quality,* 11:555–563, 1982.

Spooner, J., S. L. Brichford, R. P. Maas, M. D. Smolen, D. A. Dickey, G. Ritter, and E. Flaig, "Determining the Statistical Sensitivity of the Water Quality Monitoring Program in the Taylor-Nubbin Slough, Florida Project," *J. Lake and Reservoir Management* 4:113–124, 1988.

Standley, L. J., and R. A. Hites, "Chlorinated Organic Contaminants in the Atmosphere," in K. C. Jones, ed., *Organic Contaminants in the Environment,* Elsevier Applied Science, New York, 1991, pp. 1–32.

Stevens, R. J., and R. V. Smith, "A Comparison of Discrete and Intensive Sampling for Measuring the Loads of Nitrogen and Phosphorus in the River Main, County Antrim," *Water Res.,* 12:823–830, 1978.

Stewart, E. H., L. H. Allen, Jr., and D. V. Calvert, "Water Quality of Streams on the Upper Taylor Creek Watershed, Okeechobee County, Florida," *Proc. Soil and Crop Science Soc. Florida,* 37:117–120, 1978.

Stoddard, J. L., "Long-Term Changes in Watershed Retention of Nitrogen: Its Causes and Consequences," in L. A. Baker, ed., *Environmental Chemistry of Lakes and Reservoirs,* ACS Advances in Chemistry no. 237, American Chemical Society, Washington, D.C., 1994, pp. 223–284.

Stumm, W., and J. J. Morgan, *Aquatic Chemistry: An Introduction Emphasizing Chemical Equilibria in Natural Waters,* John Wiley, New York, 1981.

Suter, G., "Endpoints for Regional Ecological Risk Assessment," *Environmental Management,* 14:9–23, 1990.

Swackhammer, D. L., and S. J. Eisenreich, "Processing of Organic Chemicals in Lakes," in K. C. Jones, ed., *Organic Contaminants in the Environment: Environmental Pathways and Effects,* Elsevier Applied Science, New York, 1991, pp. 33–86.

Swain, E. B., D. R. Engstrom, M. E. Brigham, T. A. Henning, and P. L. Brezonik, "Increasing Rates of Mercury Deposition in Midcontinental North America," *Science,* 257:784–787, 1992.

Taylor, C. H., and J. C. Loftis, "Testing for Trends in Lake and Ground Water Quality Time Series," *Water Resources Bull.,* 25:715–726, 1989.

Thomann, R. V., and J. A. Mueller, *Principles of Surface Water Quality Modeling and Control,* HarperCollins, New York, 1987.

Thornton, K. W., B. L. Kimmel, and F. E. Payne, *Reservoir Limnology: Ecological Perspectives,* Wiley-Interscience, New York, 1990.

Tisue, T., and D. Fingleton, "Atmospheric Inputs and the Dynamics of Trace Elements in Lake Michigan," in R. A. Hites, and S. J. Eisenreich, eds., *Sources and Fates of Aquatic Pollutants,* Advances in Chemistry Series no. 216, American Chemical Society, Washington, D.C., 1987, pp. 105–125.

U.S. Environmental Protection Agency, "Results of the Nationwide Urban Runoff Program, Vol. 1 - Final Report," WH-554, Water Planning Division, Washington, D.C., 1983a.

U.S. Environmental Protection Agency, "Methods for the Chemical Analysis of Water and Wastes," Environmental Support Laboratory, Cincinnati, Ohio, EPA 600/4-79-020, 1983b.

U.S. Environmental Protection Agency, "Nonpoint Sources: Agenda for the Future," WH-556, Office of Water, Washington, D.C., 1989.

U.S. Environmental Protection Agency, "National Water Quality Inventory: 1988 Report to Congress," Washington, D.C., EPA 440-4-90-003, 1990.

U.S. Environmental Protection Agency, "EMAP Monitor, January 1991," Office of Research and Development, Washington, D.C., 1991.

U.S. Environmental Protection Agency, "Environmental Impacts of Stormwater Discharges," Office of Water, EPA 841-R-92-001, 1992.

Voldner, E. C., L. A. Barrie, and A. Sirois, "A Literature Review of Dry Deposition of Oxides of Sulfur and Nitrogen with Emphasis on Long-Range Transport Modeling in North America," *Atmos. Environ.,* 20:2101–2123, 1986.

Vollenweider, R. A., "Advances in Defining Critical Loading Levels for Phosphorus in Lake Eutrophication," *Mem. 1st Ital Idrobiol.,* 33:53–83, 1976.

Water Quality 2000, "Water Quality 2000: Phase II Report, Problem Identification," Water Quality 2000, 601 Wythe Street, Alexandria, Va., 1990.

Webster, K., A. D. Newell, L. A. Baker, and P. L. Brezonik, "Climatically Induced Rapid Acidification of a Softwater Seepage Lake," *Nature,* 347:374–376, 1990.

Welch, A. H., M. S. Lico, and J. L. Hughes, "Arsenic in Groundwater of the Western United States," *J. Ground Water,* May–June, 1988.

Wetzel, R. G., 1983. *Limnology,* Saunders College Publishing, Harcourt Brace Jovanovich College Publishers, New York, 1983.

Whitfield, P. H., "Selecting a Method for Estimating Substance Loadings," *Water Resources Bull.,* 18:203–211, 1982.

Williams, J. E., J. E. Johnson, D. A. Hendrickson, S. Contreras-Balderas, J. D. Williams, M. Navvarro-Mendoza, D. E. McAllister, and J. E. Deacon, "Fishes of North America Endangered, Threatened, or of Special Concern, 1989," *Fisheries,* 14:2–21, 1989.

Winter, T. C., "Uncertainties in Estimating the Water Balance of Lakes," *Water Resources Bulletin,* 17:82–115, 1981.

Winter, T. C., "Effect of Ground-Water Recharge on Configuration of the Water Table Beneath Sand Dunes and on Seepage in the Sandhills of Nebraska, U.S.A.," *J. Hydrology,* 86:221–237, 1986.

Winter, T. C., J. W. LaBaugh, and D. O. Rosenberry, "The Design and Use of a Hydraulic Potentiometer for Direct Measurement of Differences in Hydraulic Head Between Groundwater and Surface Water," *Limnol. Oceanogr.,* 33:1209–1214, 1988.

Winter, T. C., and M. K. Woo, "Hydrology of Lakes and Wetlands," In M. G. Wolman and H. C. Riggs, eds., *Surface Water Hydrology,* Geological Society of America, vol. 0–1, Boulder, Colo., 1990.

Young, T. C., J. V. DePinto, and T. Heidtke, "Factors Affecting the Efficiency of Some Estimators of Fluvial Total Phosphorus Load," *Water Resources Res.,* 24:1535–1540, 1988.

CHAPTER 10

RIVERS AND STREAMS

Steven C. Chapra
Professor, CADSWES
Civil, Environmental and Architectural Engineering
University of Colorado
Boulder, Colorado

Because they provide a convenient source of water and transportation, rivers have been the site of urban development from ancient times. Along with their other uses, rivers have also served as a receptacle for the large quantities of waste generated by urban populations. Consequently, river systems downstream from cities were among the first settings where scientists and engineers confronted environmental problems.

Due to its historical and continued significance, the urban wastewater problem (Fig. 10.1) will serve as the starting point for the present review. In the following description, we will focus on both the problem and the management schemes developed to control it. This will be followed by a discussion of a number of river water-quality problems that have emerged in recent times and which require new directions.

10.1 CLASSICAL WASTEWATER PROBLEMS

Beginning in the late nineteenth century, civil engineers devoted a great deal of effort to the design of urban water and wastewater systems (Fig. 10.1). The development of such systems was stimulated by the awareness that waterborne pathogens cause disease. Consequently, engineers became involved in the design and construction of water treatment plants, distribution networks, and wastewater collection systems. The design of these projects was fairly straightforward because the objectives were so well defined. The goals were to deliver an adequate quantity of potable water to the urban populace and to carry off their wastes safely.

In contrast, the question of what to do with the waste was a more ambiguous proposition. At first, municipalities discharged raw sewage directly to receiving waters. As it became apparent that such action would ultimately transform rivers into large sewers, the need for wastewater treatment plants became evident. However, such treatment could range from relatively inexpensive sedimentation to costly physical/chemical treatment. In the extreme case, the latter could actually result in an effluent that was more pristine than the receiving water. Clearly, both extremes were unacceptable. Consequently, some design goal had to be established that would protect the environment adequately but economically.

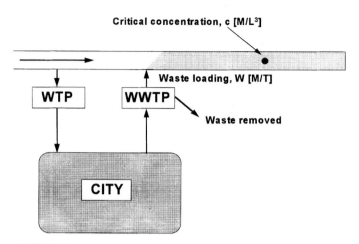

FIGURE 10.1 An urban water-wastewater system. A water treatment plant (WTP) purifies river water for human consumption. A wastewater treatment plant (WWTP) removes pollutants from sewage to protect the receiving water.

The establishment of this design goal was the start of stream water-quality analysis. It was decided that waste treatment should be designed to produce an effluent that induced an acceptable level of water quality in the receiving water. However, to determine this proper level of treatment, it was necessary to define water quality. To understand how this was done, we must first understand how sewage affects a river.

10.1.1 Stream Dissolved Oxygen

If a stream is unpolluted, dissolved oxygen levels above an urban waste discharge will be near saturation. In addition, there will usually be a diverse community of plants and animals interacting in what is a healthy, natural ecosystem.

As depicted in Fig. 10.2, such a healthy ecosystem can be viewed as a balanced cycle of life and death. Powered by the sun, autotrophic* organisms (primarily plants) convert simple inorganic nutrients into more complex organic molecules. In this *photosynthesis* process, solar energy is stored as chemical energy in the organic molecules. In addition, oxygen is liberated and carbon dioxide is consumed. The organic matter then serves as an energy source for heterotrophic organisms (bacteria and animals) in the reverse process of *respiration* and *decomposition*. This returns the organic matter to the simpler inorganic state. During breakdown, oxygen is consumed, and carbon dioxide is liberated.

This production/decomposition cycle can be used to understand the environment in a stream below a wastewater discharge (Fig. 10.3). If the stream is unpolluted, dissolved oxygen concentrations above the discharge will be near saturation. The introduction of the sewage will elevate the levels of both dissolved and solid organic matter. This has three impacts. First, the solid matter makes the water turbid and

---

* The term *autotrophic* refers to organisms (like plants and some bacteria) that do not depend on other organisms for nutrition. In contrast, *heterotrophic* organisms (such as animals and most bacteria) subsist on matter derived from other organisms.

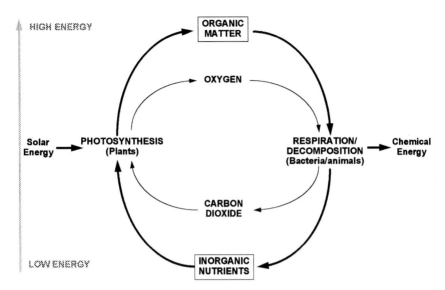

FIGURE 10.2   The natural cycle of organic production and decomposition.

unsightly. Thus, light cannot penetrate, and plant growth is suppressed. Second, the solids settle downstream from the sewage outfall and create sludge beds that can emit noxious odors. Third, the organic matter provides food for heterotrophic organisms. Consequently, the right side of the cycle in Fig. 10.2 begins to become overemphasized. Large populations of decomposer organisms break down the organic matter in the water and in the process deplete the dissolved oxygen. In addition, breakdown of organic matter takes place in the sludge bed, and a sediment oxygen demand supplements oxygen depletion in the water.

The fundamental reactions governing the process can be expressed by

$$\underset{\text{organic carbon}}{C_6H_{12}O_6} + \underset{\text{oxygen}}{6O_2} \xrightarrow{\text{respiration/decomposition}} \underset{\text{carbon dioxide}}{6CO_2} + \underset{\text{water}}{6H_2O} \quad (10.1)$$

In this simple characterization, the many different organic molecules in sewage are represented by the sugar glucose, $C_6H_{12}O_6$. As the glucose is decomposed, oxygen is utilized by the bacteria. At the same time, carbon dioxide and water are produced.

Although such a formula provides a broad description of the chemical changes, the actual reactions are much more complex and involve many more chemical constituents. However, because it would be unrealistic to characterize all these constituents, the first environmental scientists took a different approach. They disregarded the composition of the sewage and instead devised a simple experimental measurement. A quantity of sewage was placed in a closed bottle along with clean water and seed bacteria. The resulting oxygen depletion was then measured over time. In this way, they could directly assess the amount of oxygen required by the microorganisms to decompose the sewage. The resulting quantity was dubbed *biochemical oxygen demand* or *BOD.*

Now, aside from depletion, a counteracting force acts to replenish oxygen in the stream. As dissolved oxygen levels drop, atmospheric oxygen enters the water across the air–water interface to compensate for the oxygen deficit. At first, oxygen deple-

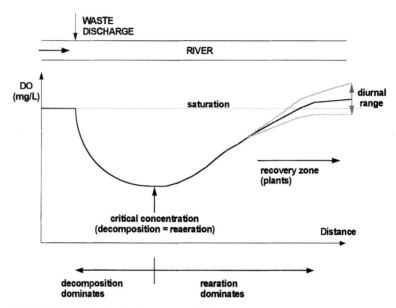

FIGURE 10.3 The dissolved oxygen "sag" that occurs below sewage discharges into streams.

tion dwarfs this *reaeration*. However, as the organic matter is assimilated, the rate of depletion diminishes as the reaeration rate increases. Consequently, there comes a point at which the depletion and the reaeration will be in balance. At this point, a lowest or *critical* level of oxygen will be reached (Fig. 10.3). Beyond this critical level, the reaeration process dominates and oxygen levels begin to rise again. Thus, a DO "sag" is created in the river below the sewage discharge.

Along with decomposition, the water becomes clearer as much of the solid matter from the discharge settles to the bottom. In addition, inorganic nutrients liberated during the decomposition process are high. Consequently, the reaction (Eq. 10.1) reverses and plant photosynthesis begins to dominate in the recovery zone. Thus, the left side of the cycle (Fig. 10.2) can become overemphasized. As the result of plant growth, carbon dioxide and water are utilized, and oxygen is produced. Consequently, large diurnal (that is, over the daily cycle) oxygen swings can occur in the recovery zone. In the late afternoon, supersaturated conditions can occur as plant photosynthesis is at its peak. Conversely, in early morning, depletion can be exacerbated by respiration. For poorly buffered waters, large swings in pH can also occur due to the creation and utilization of carbon dioxide.

Beyond the chemical changes, sewage also leads to significant effects on the biota. As in Fig. 10.4b and c, molds and bacteria dominate near the discharge. The bacterial populations flourish until the BOD and/or the oxygen is exhausted. In addition, the bacteria themselves provide a food source for a succession of organisms, consisting of ciliates, rotifers, and crustaceans.

The diversity of higher organisms decreases drastically in the degradation and active decomposition zones below the discharge. At the same time, the total number of organisms increases (Fig. 10.4d and e). As recovery ensues, these trends are reversed.

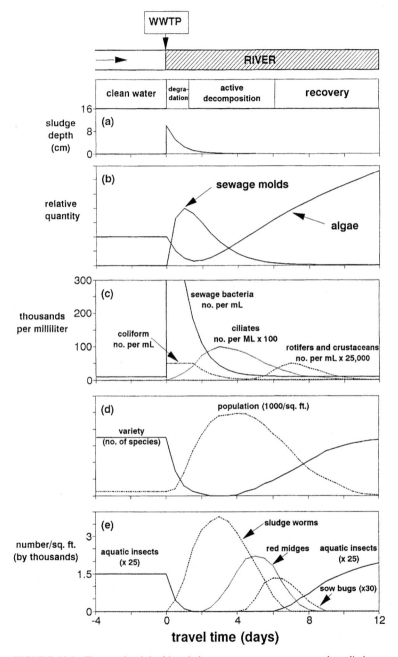

FIGURE 10.4 The trends of the biota below a wastewater treatment plant discharge into a river. (*Redrawn from Bartsch and Ingram, 1967.*)

It should be noted that whereas the patterns depicted in Figs. 10.3 and 10.4 are generally correct, the specific magnitudes shown depend on a number of factors, including the amounts of BOD and oxygen in the discharge, and the characteristics of the water body, such as flow rate, depth, and reaeration. In order to be useful for management, mathematical models have been developed to quantify these interactions.

10.1.2 Dissolved Oxygen Modeling

Now that we have provided an overview of the chemical environment below an urban sewage outfall, we can outline the modeling approaches that have been developed to quantify the phenomenon. The earliest such models were developed by Streeter and Phelps (1925) for the Ohio River. Since that time, additions have been made to what has been called the *Streeter-Phelps model.*

The following pair of differential equations provide mass balances for BOD and oxygen deficit in a stream segment with constant flow and geometry

$$U \frac{dL}{dx} = -k_r L \tag{10.2}$$

$$U \frac{dD}{dx} = -k_a D + k_d L \tag{10.3}$$

where U = average stream velocity, m/d
         L = CBOD concentration, mg/L
         x = distance downstream from the discharge, m
         k_r = total removal rate of BOD by decomposition and settling, d^{-1}
         D = dissolved oxygen deficit, mg/L
         k_a = reaeration rate, d^{-1}
         k_d = removal rate of BOD by decomposition, d^{-1}

Note that the deficit is related to the oxygen concentration c by

$$c = c_s - D \tag{10.4}$$

where c_s = saturation concentration for dissolved oxygen, mg/L.

Also, observe that Eqs. (10.2) and (10.3) are written for a steady state condition. That is, they are designed to simulate the spatial distribution of oxygen below a treatment plant. Variations in time are not considered in this simple representation. Further, they ignore longitudinal dispersion and ignore sources and sinks of oxygen such as photosynthesis, respiration, and sediment oxygen demand.

Equations (10.2) and (10.3) can be solved for the BOD and dissolved oxygen concentration below a single point source of sewage as

$$L = L_o e^{-(k_r/U)x} \tag{10.5}$$

and

$$c = c_s - D_o e^{-(k_a/U)x} - \frac{k_d L_o}{k_a - k_r} \left(e^{-(k_r/U)x} - e^{-(k_a/U)x} \right) \tag{10.6}$$

where L_o and D_o are the concentrations of BOD and oxygen deficit at the mixing point where the waste is discharged. These can be calculated by a simple mass balance (Fig. 10.5). For example, in the case of BOD,

$$L_o = \frac{Q_r L_r + W}{Q_r + Q_w} \qquad (10.7)$$

where Q_r is the river flow rate upstream of the mixing point, L_r is the river BOD, W is the point source of BOD, and Q_w is the point source flow rate.

Over the years, Eqs. (10.5) and (10.6) have been expanded to incorporate more realism and complexity into the Streeter-Phelps approach. For example, a more detailed solution for deficit can be developed as

$$D = D_o e^{-(k_a/U)x} \qquad \text{initial deficit}$$

$$+ L_o \left[\frac{k_d}{k_a - k_r} \left(e^{-(k_r/U)x} - e^{-(k_a/U)x} \right) \right] \qquad \text{initial BOD (includes point loading)}$$

$$- \frac{(p_a - R)}{k_a} \left(1 - e^{-(k_a/U)x} \right) \qquad \text{photosynthesis/respiration}$$

$$+ \frac{S_b'}{k_a} \left(1 - e^{-(k_a/U)x} \right) \qquad \text{sediment oxygen demand} \qquad (10.8)$$

where p_a and R are the plant photosynthesis and respiration rates, respectively, and S_b' is the rate of sediment oxygen demand, mg/L/d.

Over the years, the basic framework of the Streeter-Phelps model has been extended in other ways. Some major advances include:

- As outlined in Eq. (10.8) the model is set up to evaluate a single point source discharging into a river reach with constant parameters. That is, the flow, depth, reaction rates, and other parameters are assumed to be constant over the stretch of river being modeled. Investigators such as O'Connor (1967) have shown how the equation can be extended by applying the equation to a system in a piecewise

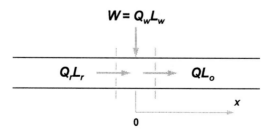

FIGURE 10.5 Mass balance for a point source discharged into a plug-flow system. This figure shows a mass balance for BOD. Note that instantaneous lateral and vertical mixing is assumed to occur in the dashed segment. This segment is also assumed to be infinitesimally thin so that no degradation occurs in the mixing zone.

fashion. By linking each piece with a mass balance, the approach can be employed to evaluate multiple sources and systems with spatially-varying parameters.

- The model assumes that all river water moves downstream at a single constant velocity—that is, as *plug flow*. In fact, shear forces cause some waters to move slower than others. This can cause longitudinal mixing or *dispersion* of the constituents carried by the flow. O'Connor (1960) broadened the approach to incorporate dispersion.

- The system is characterized as being one-dimensional. That is, it is assumed that the river is completely mixed laterally and vertically. Thus, point sources are assumed to mix instantaneously as in Eq. (10.7). In addition, vertical and lateral gradients are not simulated. For wide or deep rivers, this may not be valid. Although the approach is almost always applied in this fashion for rivers, Thomann (1963) developed a finite-difference expression of the model that allows it to be applied to two- and three-dimensional systems.

- Zero-order kinetics (that is, constants) were originally used to characterize plant activity and sediment oxygen demand. As will be addressed further on in this review, neither of these mechanisms is actually that simple. However, when evaluating the effects of primary and secondary treatment of point sources, such crude approximations were deemed adequate.

Although these modifications have been made, the classic Streeter-Phelps model still retains some fundamental simplifying assumptions. These include steady state and linearity. As described next, these characteristics have influenced the management schemes based on the approach.

10.1.3 Dissolved Oxygen Management

Now that we have a basic understanding of urban stream pollution and the Streeter-Phelps framework, we can outline the schemes that have been devised to manage the problem. These can be divided into two classes: effluent-based and quality-based approaches.

10.1.3.1 Effluent-Based Approaches. Effluent approaches are based on setting limits on the concentration of point sources. These limits are often related to available treatment technology. For example, the BOD and total suspended solids in untreated wastewater in the United States is approximately 200 to 250 mg/L. Conventional secondary treatment (primary sedimentation and secondary biological treatment) typically removes about 85 to 90 percent of both these constituents. Hence, an effluent standard of about 30 mg/L might be promulgated.

This approach was first instituted in the 1960s and 1970s when efforts were made to mandate secondary treatment across the United States. In the intervening years, two other variants of the effluent-based approach have occurred. The first is the idea of *zero discharge*. This goal was included in the 1972 Water Pollution Control Act (Public Law 92-500). It was intended to completely eliminate the urban wastewater problem by essentially forbidding municipalities to discharge their wastes to receiving waters. Although it would have obviously solved the water-quality problem, it was flawed because of its prohibitive cost. These costs related to the great expense that would have been incurred to remove all constituents from the wastewater. In addition, considerable cost would also be associated with the disposal of the large quantities of solid and gaseous wastes that would be generated. As a consequence of the costs, the zero-discharge goal was never realized.

The second variant of the effluent-based approach has arisen in the management of pollutants such as metals and toxic organics. In this case, the idea is to set effluent levels at or below the concentrations desired in the environment. The rationale behind this approach is that the discharge level would then represent an upper bound on the resulting environmental levels. Although this method moderates the zero-discharge idea by allowing some discharge, it also has flaws. As with zero discharge, it can also be prohibitively expensive. In addition, a greater shortcoming relates to the fact that in some cases water-quality standards might still be violated. For example, it is known that certain contaminants concentrate in bottom sediments and the food chain. Thus, just because effluent limits are set at desired water concentrations, there is no guarantee that levels in the sediments and biota would not exceed standards.

The effluent-based approach served a purpose by providing a legal framework for universal institution of secondary wastewater treatment. However, as mentioned above, its extension to zero discharge is economically unfeasible. Further, its use to mandate levels of treatment somewhere between secondary and total removal is also flawed in that it does not ensure that the receiving water will be adequately protected. Clearly, as originally envisioned by Streeter and Phelps, a cause–effect linkage between loadings and response is required. Thus, water-quality approaches provide an alternative.

10.1.3.2 Quality-based Approaches.

Although it espoused a goal of zero discharge, the 1972 Water Pollution Control Act included a special designation for stream segments where effluent-based standards did not result in quality that complied with water-quality standards. For such cases, the quality of the receiving water becomes the basis for the magnitude of the effluent. Thus, the law returns to Streeter and Phelps' original idea.

Effective implementation of such an approach is based on a two-step process. First, stream standards are specified. This often relates to protection of aquatic organisms such as fish. For oxygen, common standards might range from 4 to 6 mg/L. Then, a model is implemented to estimate the waste loading required to meet the standard. The approach is shown in Fig. 10.6.

Although this approach might seem straightforward, there are additional issues that must be addressed before it can be implemented. The first relates to the time scale over which the standard is to be met. In traditional stream-quality management, a steady state corresponding to a summer, low-flow condition was chosen. Conditions during this period were assumed to be the most critical for many water

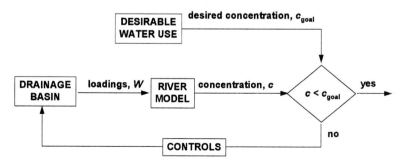

FIGURE 10.6 The water-quality management approach.

bodies. Hence, if standards were maintained during summer low-flow conditions, it was assumed they would be exceeded at all other times.*

The second issue relates to multiple sources. For cases where a single source impacts the river, the approach is straightforward. As depicted in Fig. 10.3, there will exist a critical concentration (the critical deficit) which is totally governed by the single point source. Consequently, there will be a unique level of waste reduction that brings this critical concentration to the required standard.

For multiple sources, the relationship is not so straightforward. In particular, the critical concentration will usually be governed by the combined effect of several sources. Further, it is possible that after the critical level is raised to the standard at one point, a critical level at another point in the stream will still violate the standard. This new critical concentration will also be dictated by several sources. Thus, because there is no unique combination of treatment levels that can be derived, unless additional constraints are devised, subjective choices must be made.

There have been several approaches to remedy such ambiguity. The first is the idea of *uniform treatment.* In this approach, the fractional treatment levels for all point sources are ratcheted up equally until the standard is met at all locations on the stream. Although this scheme is the easiest to implement in a legal sense ("all treatment plants are created equal"), it has the flaw that it is suboptimal. That is, it will result in overtreatment, and hence excess cost.

Several variants of uniform treatment have been proposed. The *transferable permit* approach starts by using a uniform treatment approach as the basis for issuing discharge permits to individual polluters. Thereafter, the polluters are free to trade permits. This approach was first developed for air pollution. The approach makes sense where basinwide quality is being controlled, since all sources contribute to overall basin quality. Consequently, it could provide a workable solution for water-quality problems in well-mixed systems like certain lakes and bays. However, for distributed systems such as streams, where quality varies with location, it could lead to violation of standards.

An alternative approach involves incorporating treatment costs into the scheme. In such *least-cost allocations,* the expense of removing waste from each point source is quantified. Then optimization methods such as linear programming are employed to meet the water-quality standards at the least cost. This results in expenditures being made where they yield the most improvement.

A third general alternative, called *zoned-uniform treatment,* takes an intermediate approach. In this case, cost data are employed to set optimum treatment levels. However, these levels are determined for clusters or zones of discharges, rather than for individual point sources.

In summary, a rational basis for managing the urban wastewater problem exists. Unfortunately, both least-cost and zoned treatment approaches have not been widely adopted to date. However, in an era of tightening budgets and increased environmental concerns, such economically efficient waste reduction schemes could become more attractive in the future. We will return to this topic at the end of this chapter.

10.1.4   Bacteria and Other Pathogens

As mentioned at the beginning of this chapter, the initial impetus for water-quality control arose because of concerns over waterborne diseases. The primary agents of

---

* For some cold climates, winter low-flow conditions, in which ice cover retards reaeration, may also be critical for dissolved oxygen.

these diseases are called *pathogens*. These are disease-producing organisms that grow and multiply within the host. Some pathogens enter the human body through the skin. More commonly they are ingested along with drinking water.

Pathogens can be divided into categories. The most common groups associated with water pollution are summarized in Table 10.1.

Although the organisms listed in Table 10.1 are the cause of waterborne diseases, their concentrations are often very difficult to measure. Consequently, classical water-quality management and modeling has focused on the levels of *indicator organisms*. These are typically groups of organisms that are convenient to measure and which are abundant in human and animal waste. If they are present, it is assumed that pathogens may also be present.

There are three major types of indicator bacteria. These are:

1. *Total coliform (TC).* A large group of anaerobic, gram-negative, non-spore-forming, rod-shaped bacteria that ferment lactose with gas formation within 48 hours at 35°C. They exist in both polluted and unpolluted soils and occur in the feces of warm-blooded animals. *Escherichia coli* (or *E. coli*) and *Aerobacter aerogenes* are common members of the group that occur in organisms and soils, respectively.

2. *Fecal coliform (FC).* A subset of TC that come from the intestines of warm-blooded animals. Thus, because they do not include soil organisms, they are preferable to TC. They are measured by running the standard total coliform test at an elevated temperature (44.5°C). As a general rule of thumb, the FC is about 20 percent of TC (Kenner 1978). However, the ratio is highly variable.

TABLE 10.1 Some Waterborne Pathogenic Organisms

Category	Description	Species and groups
Bacteria	Microscopic, unicellular organisms that lack a fully-defined nucleus and contain no chlorophyll.	*Vibrio cholarae* *Salmonella* *Shigella* *Legionella enteropathogenic E. coli*
Viruses	A large group of submicroscopic (10 to 25 nm) infectious agents. They are composed of a protein sheath surrounding a nucleic acid core, and thus contain all the information required for their own reproduction. However, they require a host in which to live.	Hepatitis A Enteroviruses Polioviruses Echoviruses Coxsackieviruses Rotaviruses
Protozoa	Unicellular animals that reproduce by fission.	*Giardia lambia* *Entamoeba histolytica* *Cryptosporidium* *Naegleria fowleri*
Helminths (intestinal worms)	Intestinal worms and wormlike parasites.	Nematodes *Schistoma haematobium*
Algae	Large group of nonvascular plants. Certain species produce toxins, which if consumed in large quantities may be harmful.	*Anabaena flos-aquae* *Microcystis aeruginosa* *Aphanizomenon flos-aquae*

3. *Fecal streptococci (FS).* These include several varieties of streptococci which originate from humans (*Streptococcus faecalis*) as well as domesticated animals such as cattle (*Streptococcus bovis*) and horses (*Streptococcus equinus*).

Although the TC measurement has traditionally been the most widely used indicator of contamination, its use is problematic because of the presence of nonfecal coliform bacteria. Consequently, emphasis is shifting more to fecal coliforms and fecal streptococci.

Further, the ratio of FC to FS (FC/FS) is often used to determine whether contamination is due to human or animal sources. In general, an FC/FS > 4 is often taken to indicate human contamination, whereas FC/FS < 1 is interpreted as originating from other warm-blooded animals. However, the ratio should be used with care because of differential die-off of FC and FS. Thus, as the distance from a sewage outfall increases the ratio can change, and naive interpretation of the ratio could be misleading.

The distribution of an indicator organism such as total coliform can be modeled in a fashion similar to BOD,

$$U \frac{dB}{dt} = -k_b B \tag{10.9}$$

where B = the concentration of bacteria, number/100 mL; and k_b = the loss rate, d^{-1}, which for total coliform bacteria can be represented as (Mancini, 1978; Thomann and Mueller, 1987)

$$k_b = \underbrace{(0.8 + 0.02S)1.07^{T-20}}_{\left(\begin{smallmatrix} \text{natural} \\ \text{mortality} \end{smallmatrix}\right)} + \underbrace{\frac{\alpha I_o}{k_e H}(1 - e^{-k_e H})}_{\text{(light)}} + \underbrace{F_p \frac{v_s}{H}}_{\text{(settling)}} \tag{10.10}$$

where
k_b = the total loss rate, d^{-1}
S = salinity, ppt or g/L
T = temperature, °C
α = a proportionality constant ($\cong 1$)
I_o = surface light energy, langley hr^{-1}
k_e = an extinction coefficient, m^{-1}
H = depth, m
F_p = the fraction of the bacteria that are attached to settling particles
v_s = the settling velocity of the particles, m/d

If all its parameters are constant, Eq. 10.9 can be solved for

$$B = B_o e^{-(K_b/U)x} \tag{10.11}$$

where B_o = the initial concentration of bacteria at the mixing point, number/mL. Thus, as in Fig. 10.7, the model would predict an exponential decrease in bacterial concentration downstream from the source. Note that the total loss rates calculated by Eq. (10.10) are usually on the order of 1 d^{-1}. Consequently, high concentrations of bacteria will tend to be located directly downstream of the discharge point.

Much has been done to diminish the threat posed by disease-carrying organisms in rivers and streams. However, bacteria and viruses still pose problems related to disease transmission and interference with uses of water for recreation. In addition,

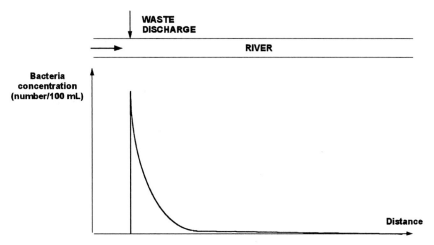

FIGURE 10.7   The distribution of bacteria below a point source discharge into a stream. In contrast to dissolved oxygen (compare with Fig. 10.3), the critical concentration occurs at the mixing point. In addition, because of the rapid die-off rate of indicator bacteria, the resulting contamination will be more localized.

most undeveloped and developing countries still experience great problems related to pathogens. Consequently, they are still a highly relevant component of stream water quality.

10.2   MODERN STREAM POLLUTION PROBLEMS

The description of BOD/dissolved oxygen and bacteria presented earlier represents the paradigm that defined stream water-quality management from 1925 through the early 1970s. Over the past 20 years, further study and observations have broadened our concept of stream water quality. The following sections outline other problems and constituents that must be addressed by modern stream water-quality management schemes.

10.2.1   Nitrogen

Figure 10.8 depicts the nitrogen cycle in natural waters. As can be seen, the cycle affects the water's oxygen level. In addition, several other water-quality problems occur. These impacts are reviewed in the subsequent paragraphs.

10.2.1.1   Nitrification.   In addition to carbonaceous BOD, nitrogen compounds in wastewater also have an impact on a river's oxygen resources (Fig. 10.8). Sewage nitrogen can be broadly broken down into organic nitrogen compounds (such as proteins and urea) and ammonia. With time, the organic nitrogen compounds are hydrolyzed to create additional ammonia. Autotrophic bacteria then assimilate the ammonia and create nitrite (NO_2^-) and nitrate (NO_3^-).

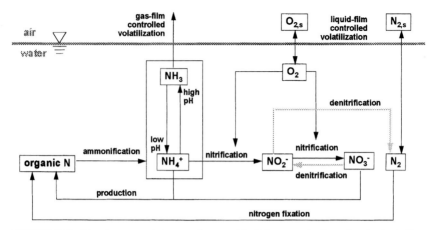

FIGURE 10.8 The nitrogen cycle in natural waters. The light arrows indicate that the denitrification reactions take place under anaerobic conditions.

The conversion of ammonia to nitrate is collectively called *nitrification*. It can be represented by a series of reactions (Gaudy and Gaudy, 1980). In the first, bacteria of the genus *Nitrosomonas* convert the ammonium ion (NH_4^+) to nitrite,

$$NH_4^+ + 1.5O_2 \rightarrow 2H^+ + H_2O + NO_2^- \tag{10.12}$$

In the second, bacteria of the genus *Nitrobacter* convert nitrite to nitrate,

$$NO_2^- + 0.5O_2 \rightarrow NO_3^- \tag{10.13}$$

The entire process consumes 4.57 g of oxygen per gram of nitrogen oxidized.

As with the decomposition of carbonaceous matter, nitrification will cause a dissolved oxygen sag in the stream. The entire process, as depicted in Fig. 10.9, shows the collective effect of organic carbon and nitrogen decomposition on a stream's oxygen resources. In Fig. 10.9a, the carbonaceous BOD is elevated at the point where the sewage enters the stream. It then decreases as the CBOD is decomposed and settles. The effect of decomposition in tandem with reaeration results in an oxygen sag.

In Fig. 10.9b, nitrification also induces a sag. However, because of the sequential nature of the nitrification reactions, the sag's minimum occurs farther downstream than for the carbon induced depletion in Fig. 10.9a.

Finally, Fig. 10.9c shows the total depletion that results from the combined effect of carbon and nitrogen. Note that the resulting DO profile is worse than the individual sags from carbon and nitrogen alone, and the critical DO level occurs at a distance somewhere between the carbon and the nitrogen cases.

Early attempts to simulate the impact of nitrification used a nitrogenous BOD (NBOD) to characterize the process. In recent years, a more mechanistic approach that attempted to model organic nitrogen, ammonia, and nitrate explicitly has been developed (Fig. 10.10). The mass balance equations for the model are (compare with Eqs. [10.2] and [10.3])

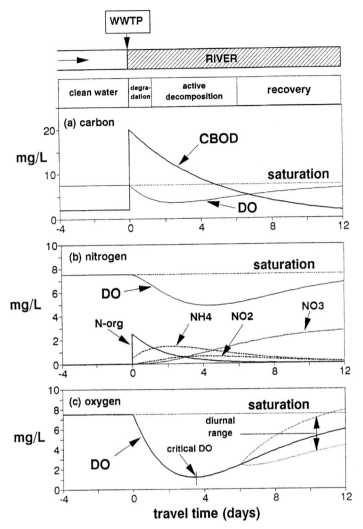

FIGURE 10.9 The trends of (*a*) carbon, (*b*) nitrogen, and (*c*) oxygen below a wastewater treatment plant discharge into a river.

$$-U \frac{dN_o}{dx} = -k_{oa}N_o \tag{10.14}$$

$$-U \frac{dN_a}{dx} = k_{oa}N_o - k_{ai}N_a \tag{10.15}$$

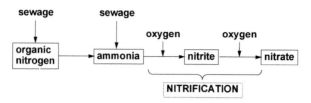

FIGURE 10.10   The decomposition of nitrogen compounds below a wastewater treatment plant discharge. Models have been developed to simulate this process based on a series of first-order reactions.

$$-U \frac{dN_i}{dx} = k_{ai}N_a - k_{in}N_i \tag{10.16}$$

$$-U \frac{dN_n}{dx} = k_{in}N_i \tag{10.17}$$

where the subscripts o, a, i, and n denote organic, ammonium, nitrite, and nitrate, respectively, and the k's are first-order conversion rates, d^{-1}. An oxygen deficit balance can also be written as

$$-U \frac{dD}{dx} = r_{oa}k_{ai}N_a + r_{oi}k_{in}N_i - k_aD \tag{10.18}$$

where the r's are oxygen-to-nitrogen ratios. These equations can be solved to develop the types of nitrogen and oxygen profiles observed in Fig. 10.9*b*.

10.2.1.2   *Nitrate Pollution.* As depicted in Fig. 10.9*b*, the ultimate result of the nitrification process is nitrate. Unfortunately, in sufficiently high concentrations, nitrate in drinking water can cause serious and occasionally fatal effects to infants. The problem can become especially critical in agricultural regions where nonpoint sources of nitrate supplement high levels due to nitrification from point sources. Although the model framework in Fig. 10.10 was originally developed to simulate the effect of nitrification on dissolved oxygen, it can also serve as a model of nitrate build up.

10.2.1.3   *Ammonia Toxicity.* Ammonia exists in two forms in natural water: ammonium ion (NH_4^+) and ammonia gas (NH_3). Whereas the former is innocuous at the levels encountered in most natural waters, the un-ionized form is toxic to fish. The interrelationship between the two forms is governed by the following equilibrium reaction,

$$NH_4^+ \rightleftarrows NH_3 + H^+ \tag{10.19}$$

At high pH (and to a lesser extent at high temperatures), this reaction shifts to the right. For example, at moderate temperatures ($\cong 20°C$) and pH levels above 9, upwards of 20 percent of the total ammonia will be in the un-ionized form (Fig. 10.11).

Although the pH of natural waters is usually lower than 9, there are some cases where the pH is high enough so that harmful levels of un-ionized ammonia occur. The most noteworthy instance occurs in the recovery zone below a secondary sewage discharge into a shallow stream. For such cases, suspended solids will be low and ammonia will be high. Thus, profuse plant growth can occur. In the late afternoon, such plant photosynthesis can deplete carbon dioxide from the water and, for weakly buffered waters, induce a large increase in pH. The late afternoon is also coincidentally the time of day when temperatures are highest. Consequently, the high pH and temperature can result in elevated levels of un-ionized ammonia (Fig. 10.12).

10.2.1.4   *Eutrophication.*   Aside from its other characteristics, nitrogen serves as an essential nutrient for plant growth. We will reserve detailed discussion of the eutrophication phenomenon until the next section. However, there are two aspects of nitrogen as a nutrient that bear mentioning. First, algae use both ammonium and nitrate as a source of the nutrient. They prefer ammonium, but are capable of generating an enzyme that allows them to utilize nitrate when ammonium levels are low.

Second, certain blue-green algae are capable of fixing free nitrogen. This gives them a competitive advantage in situations where other forms of nitrogen are in short supply. This state of affairs can sometimes occur when advanced treatment includes nitrogen removal. In such cases, blue-green algae can become dominant. This is especially important because some of the more common nitrogen-fixing species contribute to scum formation and taste and odor problems.

10.2.1.5   *Summary.*   In summary, the nitrogen problem in streams is multifaceted. First, ammonia can cause oxygen depletion via nitrification. If this occurs, one of the byproducts is nitrate, which is itself a pollutant. Further, depending on stream temperature and pH, the ammonia can manifest itself in an un-ionized form which is toxic to aquatic organisms. Finally, both ammonia and nitrate are essential nutrients for photosynthesis. Thus, they can stimulate excessive plant growth which, as described next, constitutes a water-quality problem in its own right.

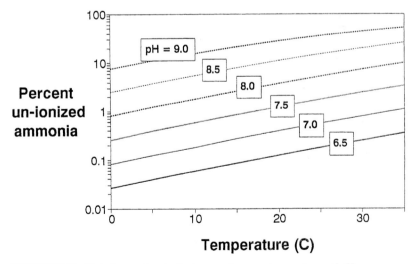

FIGURE 10.11   Plot of percent un-ionized ammonia versus temperature and pH.

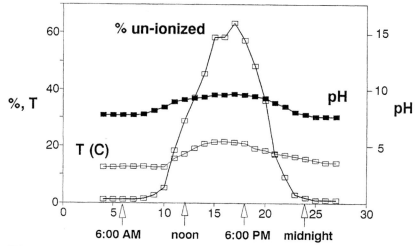

FIGURE 10.12   Plot of temperature, pH and percent un-ionized ammonia versus time of day in Boulder Creek, Colorado.

10.2.2   Stream Eutrophication

Eutrophication refers to the overstimulation of plant growth due to the discharge of excess nutrients to surface waters. In general, most study of eutrophication has focused on estuaries and standing waters (that is, lakes and impoundments) rather than streams. However, as secondary treatment has become routinely implemented across the United States, more attention is being directed towards the problem of plant growth in rivers. This is especially true for agriculturalized and urbanized basins, where nutrient contributions from runoff can be substantial.

In general, stream eutrophication can have a number of deleterious effects on a stream or river. First, the profuse growth of plants decrease water clarity and some species form unsightly scums. Second, certain species of algae cause taste and odor problems in drinking water. Third, certain blue-green algae can be toxic when consumed by animals (Table 10.1). Fourth, eutrophication can alter the species composition of a river ecosystem. Native biota may be displaced as the environment becomes more productive. Finally, the nutrients can indirectly affect other aspects of stream chemistry. For example, the uptake and release of carbon dioxide by plants can alter the system's pH.

The primary controllable nutrients causing eutrophication are nitrogen and phosphorus. As described in the previous section, both ammonium and nitrate are available for plant growth. For phosphorus, dissolved reactive phosphorus (orthophosphate) is considered to be readily available.

A rough rule of thumb for assessing which nutrient is limiting relates to the nitrogen-to-phosphorus ratio. In general, an N/P ratio less than 7 to 10 suggests that nitrogen is limiting. Conversely, higher levels imply that phosphorus will limit plant growth. As displayed in Table 10.2, sewage is generally enriched in phosphorus. Therefore, streams dominated by wastewater effluents tend to be nitrogen limited. Similarly, estuaries tend to be deficient in nitrogen and, hence, are usually nitrogen limited. In contrast, those systems subject to phosphorus removal and non-point-source input are generally phosphorus limited.

Although Table 10.1 provides general patterns, individual streams must be assessed on a case by case basis. Consequently, mathematical models have been developed to attempt to simulate the eutrophication process.

A simple representation of such a model is depicted in Fig. 10.13. Three mass-balance equations are written for a single limiting nutrient, a phytoplankton and a zooplankton group.

$$-U \frac{dP}{dx} = G_m(T)\phi_\ell \frac{n}{k_{sn}+n} P - C_z(T)ZP - k_{dp}(T) \tag{10.20}$$

$$-U \frac{dZ}{dx} = \epsilon C_z(T)ZP - k_{dz}(T)Z \tag{10.21}$$

$$-U \frac{dn}{dx} = -G_m(T)\phi_\ell \frac{n}{k_{sn}+n} P + (1-\epsilon)C_z(T)ZP + k_{dz}(T)Z + k_{dp}(T)P \tag{10.22}$$

where P, Z, and n = phytoplankton, zooplankton, and nutrient concentrations, respectively. All these concentrations are written in terms of mg nutrient/m^3. Thus, stoichiometric conversions do not have to be taken into account. The other parameters are: $G_m(T)$ = temperature-dependent phytoplankton maximum growth rate, d^{-1}; ϕ_ℓ = an attenuation coefficient due to light [dimensionless number between zero (total light limitation) and 1 (no limitation) that is dependent on surface light, water clarity, and water depth]; k_{sn} = half-saturation constant for nutrient limitation, mg/m^3; $C_g(T)$ = zooplankton grazing rate, m^3 mg^{-1} d^{-1}; ϵ = zooplankton grazing efficiency [dimensionless number from 0 to 1]; $k_{dp}(T)$ and $k_{dz}(T)$ = respiration rates for phytoplankton and zooplankton, d^{-1}, respectively.

Although they represent a gross simplification, Eqs. (10.20) to (10.22) illustrate the fundamental way in which eutrophication modeling differs from classic Streeter-Phelps oxygen and bacteria models. These are:

- The eutrophication computations are intrinsically time variable. A steady-state is inappropriate because plant biomass continually changes as light, temperature and grazing pressure varies over the course of the year. Consequently, the usual time scale is the annual cycle. In addition, because of problems such as ammonia toxicity, diurnal models are beginning to be developed.

TABLE 10.2 N/P Ratios for Point, Nonpoint and Marine Waters

Source type	TN/TP*	IN/IP[†]	Limiting nutrient
Raw sewage	4	3.6	Nitrogen
Activated sludge	3.4	4.4	Nitrogen
Activated sludge plus nitrification	3.7	4.4	Nitrogen
Activated sludge plus phosphorus removal	27.0	22.0	Phosphorus
Activated sludge plus nitrogen removal	0.4	0.4	Nitrogen
Activated sludge plus nitrogen and phosphorus removal	3.0	2.0	Nitrogen
Nonpoint sources	28	25	Phosphorus
Marine waters		2	Nitrogen

* TN/TP = total nitrogen/total phosphorus.
[†] IN/IP = inorganic nitrogen/inorganic phosphorus.
Source: Thomann and Mueller 1987; Omernik 1977; and Goldman, et al., 1973.

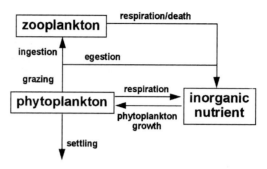

FIGURE 10.13 A simple representation of the interactions simulated in a nutrient/food-chain model.

• Nonlinear, time-dependent kinetics are employed to simulate the food-chain interactions. Although linear, constant-parameter food-chain models are sometimes adequate for descriptive purposes, nonlinear, time-dependent kinetics are necessary to simulate the effects of remedial measures such as nutrient load reductions. Several mechanisms contribute to this increase in model complexity. First, predator-prey interactions between the plants and animals are modeled with nonlinear Lotka-Volterra equations. Second, substrate and food utilization by both plants and animals are formulated as nonlinear saturating kinetics. Third, the light limitation of the plants is influenced by self-shading. Finally, model parameters are dependent on temporally varying factors such as light and temperature.

As noted above, the representation in Fig. 10.13 is an oversimplification. In fact, nutrient/food-chain models often include several limiting nutrients and a much more complex food chain (Fig. 10.14). Further, because photosynthesis and respiration can impact dissolved oxygen levels, eutrophication mass balances have been coupled with BOD/oxygen deficit balances to provide a more mechanistic framework for assessing stream oxygen resources. Computer models are available to implement such frameworks (Brown and Barnwell, 1987).

Before concluding this section, it must be stressed that this is a very simple introduction to eutrophication modeling that is intended to outline major features of the approach. The reader is directed to other references for a more complete description of the approach's strengths and shortcomings (Thomann and Mueller [1987] and Chapra [1996] provide overviews).

In addition, it should be noted that the aforementioned description applies to all natural waters (that is, rivers, estuaries, and lakes). However, specific application to rivers leads to the need for further refinements. For example, because seasonal simulations are required, the system's hydrology must usually be described. Further, a distinction must be made between free-floating and fixed plants. In lakes, although both exist, most emphasis has been placed on floating plants such as phytoplankton. For shallow streams, either one or the other or both can exist, and should be simulated.

10.2.3 Sediment Oxygen Demand (SOD)

As in Eq. 10.8, sediment oxygen demand (SOD) was specified as a zero-order source term in classical Streeter-Phelps models. In some cases, the SOD rate was measured

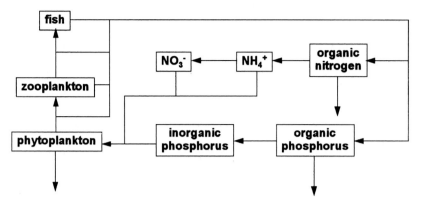

FIGURE 10.14 A more complex representation of the interactions simulated in a nutrient/food-chain model. This version, in contrast to Fig. 10.13, includes two nutrients and an expanded food chain.

directly. More often, it was arbitrarily chosen to calibrate the model to fit observed oxygen profiles. Once treatment was installed, the approach becomes problematic because it is necessary to decide how the SOD changes following treatment.

The two most common approaches were to: (1) leave the SOD unchanged, or (2) assume linearity and lower it in direct proportion to the load reduction. The situation deteriorated further for multiple-source problems.

Such arbitrary methods were usually justified under the assumption that a truly mechanistic understanding of the process was not necessary when making crude assessments of the effects of primary and secondary treatment. Although this assumption is probably questionable in itself, it becomes even more tenuous as models are used to evaluate advanced waste treatment.

Experimental observations bear this out. Early on, investigators recognized that the relationship between SOD and the organic content of sediments was nonlinear (Fig. 10.15). For example, Fair, et al. (1941) performed experiments that suggested a square root relationship between SOD and sediment organic matter concentration. Subsequent work supported this observation (for example, Gardiner, et al., 1984).

In a landmark paper, Di Toro, et al. (1990) have developed a model of the process that mechanistically arrives at the same square-root relationship. In essence, they show that the relationship can be explained by the fact that decomposition occurs in two sediment zones: an aerobic and an anaerobic layer (Fig. 10.16). In the aerobic surface layer, organic carbon is decomposed according to Eq. (10.1). In the deeper anaerobic zone, further decomposition occurs by methanogenesis according to

$$C_6H_{12}O_6 \quad \rightarrow \quad 3CO_2 \quad + \quad 3CH_4 \tag{10.23}$$

organic carbon   carbon dioxide   methane

The methane diffuses back up to the aerobic zone where it is oxidized as in

$$3CH_4 \quad + \quad 6O_2 \quad \rightarrow \quad 3CO_2 \quad + 6H_2O \tag{10.24}$$

methane   oxygen   carbon dioxide   water

Now, when the organic carbon content of the sediment gets high, the amount of methane produced in the anaerobic sediments can exceed its solubility, and bubbles

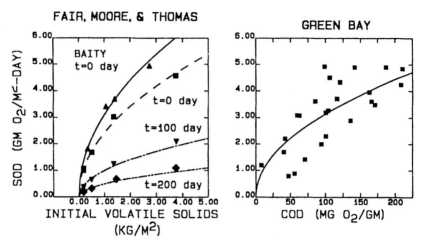

FIGURE 10.15   Sediment oxygen demand versus initial area volatile solids (*Fair, et al., 1941, and Baity, 1938*) and surface sediment COD (*Gardiner, et al., 1984*). (*Reprinted from Di Toro, et al., 1990.*)

form. These bubbles migrate upward due to their buoyancy and, consequently, represent a loss of organic carbon that does not exert a sediment oxygen demand. This loss partly explains the observed square root relationship between SOD and sediment organic carbon content.

   In addition, Di Toro demonstrated how the two-zone scheme affects nitrogen. For this case, decomposition of organic nitrogen inputs generate ammonia in the sediments. In the aerobic zone, nitrification serves as an oxygen-demanding reaction

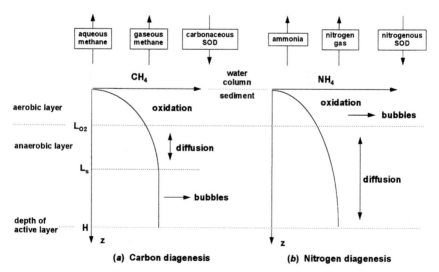

FIGURE 10.16   A schematic diagram of Di Toro's SOD model. (*Redrawn from Di Toro, et al., 1990.*)

which exerts an SOD. In the anaerobic zone, ammonia is created via ammonification of organic N. Thus, a gradient is created which feeds additional ammonia to the aerobic layer where it is also subject to nitrification. Ammonia that is not converted to nitrate diffuses back into the water via diffusion. Finally, the nitrate created by the nitrification process is converted to nitrogen gas via denitrification.

By providing a means to calculate SOD as a function of external loadings, the Di Toro framework is a major improvement over prior schemes. Predictions should be more realistic than those obtained with the two alternatives used by earlier modelers—fixed SOD or linear decrease in proportion to load reductions. The former is overly conservative whereas the latter is overly optimistic. By allowing a more physically realistic middle course, the Di Toro approach should provide a basis for better management decisions. To date, the Di Toro's framework has been applied to Chesapeake Bay (Cerco and Cole, 1993). In the coming years it will undoubtedly be incorporated into stream oxygen models to provide a more rigorous characterization of SOD.

10.2.4 Organic Toxicants and Metals

To this point, our discussion has focused on conventional pollutants. Now we will turn to so-called toxic pollutants. The major differences between conventional and toxic pollution relate to four areas:

1. *Natural vs. Alien.* The conventional pollution problem typically deals with the natural cycle of organic production and decomposition (recall Fig. 10.2). The discharge of sewage adds both organic matter and inorganic nutrients to a water body. The decomposition of the organic matter by bacteria can result in oxygen depletion. The inorganic nutrients can stimulate excess plant growth. In both cases, the problem involves an overstimulation of the natural processes governing the waterbody. In contrast, many toxins do not occur naturally. Examples are pesticides and other synthetic organics. For such alien pollutants, the problem is one of poisoning or interference with natural processes.

2. *Aesthetics vs. Health.* Although it would be an overstatement to contend that conventional pollution deals solely with aesthetics, a strong case can be made that the mitigation of "visual" pollution has been a prime motivation for conventional waste treatment. In contrast, the toxic substance issue is almost totally dominated by health concerns. Most toxic remediation focuses on the contamination of: (1) drinking water and (2) aquatic food stuffs.

3. *Few vs. Many.* Conventional water-quality management deals with on the order of about 10 pollutants. In contrast, there are thousands of organic chemicals that could potentially be introduced into our natural waters. If even a fraction of these prove toxic, the sheer numbers of potential toxicants will have a profound effect on the resulting control strategies. Further, it is difficult to obtain detailed information on the factors governing their transport and fate on an individual basis. The study of problems such as dissolved-oxygen depletion and eutrophication is facilitated by the fact that they involve a few chemicals. In contrast, toxicant modeling is complicated by the vast number of contaminants that are involved.

4. *Single Species vs. Solid/Liquid Partitioning.* In most cases, conventional water-quality models usually treat the pollutant as a single species. Consequently, its strength in the water body is measured by a single concentration—for example, dissolved oxygen deficit. In contrast, the transport, fate, and ecosystem impact of

toxicants is intimately connected with how they partition or associate with solid matter in and below the water body. Thus, toxic substance analysis must distinguish between dissolved and particulate forms. The distinction between dissolved and particle-associated toxicant has an impact on transport and fate in the sense that certain mechanisms differentially impact the two forms. For example, volatilization only acts on the dissolved component. The distinction also has importance from the standpoint of assessing ecosystem impact. For example, if biotic contamination is being assessed, toxicant mass normalized to biotic mass is a better metric of contamination than toxicant mass normalized to water volume.

The four points outlined above have implications for both the modeling and the management of toxicants. From a modeling perspective, it means that along with the system's hydrology, the transport and fate of solids must be characterized (Thomann and Di Toro, 1983; Chapra and Reckhow, 1983; O'Connor, 1988). In particular, the association of toxicants with settling and resuspending particles represents a major mechanism controlling their transport and fate in natural waters. Further, small organic particles, such as phytoplankton and detritus, can be ingested and passed along to higher organisms (Thomann, 1981). Such food-chain interactions have led the modelers to view nature's organic carbon cycle as more than an end in itself. Rather, the food chain is viewed as a conveyer and concentrator of contaminants.

To date, most toxicant modeling has been focused on estuaries and lakes. However, some initial efforts have been made to address rivers and streams (see O'Connor, 1988). Because of the emphasis on solids, mass balances are written for both the stream and its underlying sediments (Fig. 10.17). The following steady-state mass balances illustrate the major features of such models

$$-U \frac{dc_1}{dx} = -\frac{v_s}{H_1} F_{p1}c_1 - \frac{v_v}{H_1} F_{d1}c_1 - k_1 c_1 + \frac{v_r}{H_1} c_2 \tag{10.25}$$

$$0 = \frac{v_s}{H_2} F_{p1}c_1 - k_2 c_2 - \frac{v_r}{H_2} c_2 - \frac{v_b}{H_2} c_2 \tag{10.26}$$

where the subscripts 1 and 2 designate the water and sediments, respectively. The other parameters are: c = the concentration of the toxic substance, mg/m^3; v_s = settling velocity of particulate matter, m/d; v_r = resuspension velocity of the bottom sediments, m/d;

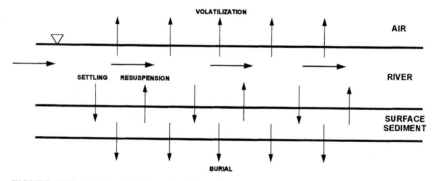

FIGURE 10.17 A schematic of a toxic substance model for a stream. Note that in addition to the transfers shown in this diagram, toxicants can also be lost from the system by decomposition reactions.

v_b = a burial velocity; H = depth, m; k = a decay coefficient, d^{-1}; F_p = the fraction of the toxicant associated with solid matter; and F_d = the fraction of the toxicant in dissolved form. Note that the fractions associated with particles and dissolved are themselves a function of the stream's suspended solids concentration and a partition coefficient,

$$F_{d1} = 1 - F_{p1} = \frac{1}{1 + K_d m} \tag{10.27}$$

where K_d = the partition coefficient that quantifies the toxicant's association with solid matter, $m^3\ g^{-1}$; and m = the suspended solids concentration in the stream. Also note that the sediment concentration can be expressed on a mass-specific basis. This is a more meaningful expression than the volume-specific concentration in the water column and is the usual way in which sediment toxicant standards are expressed.

A solution of Eqs. (10.25) and (10.26) for a point source is depicted in Fig. 10.18. The result indicates a decay as the substance is removed by volatilization and decomposition. In addition, the underlying sediments track on the water concentration. For both cases, the highest concentrations occur at the mixing point.

From a management perspective, toxics differ from conventional pollutants in two fundamental ways. First, because of public and ecosystem health concerns, the assimilative capacity approach may not be the best approach. In some cases, effluent

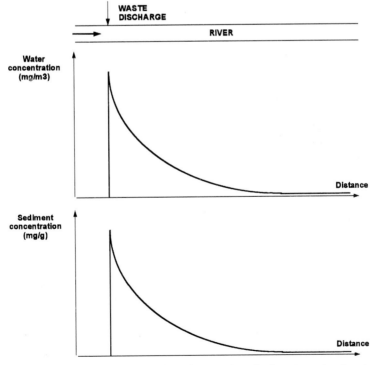

FIGURE 10.18 The distribution of a toxic contaminant in the water and sediments below a point-source discharge into a stream. Note that the sediment concentration is expressed on a mass-specific basis.

limitations might provide a better alternative. Second, the plethora of toxics means that it might be impractical to model each contaminant individually. Thus, alternative approaches such as screening (Chapra, 1991) might be considered.

10.2.5 Particulate and Dissolved Solids

The fate and transport of particulate and dissolved solids has been studied for many years. In contrast to other areas of water-quality analysis, these models have usually developed within the field of water resources rather than environmental engineering. The reason for this is that these problems were among the first of the nonpoint problems. Whereas most early environmental engineers focused on the urban wastewater problem, water-resource engineers concentrated on subjects such as hydrology that viewed the drainage basins as a system. Thus, they often confronted nonpoint pollution.

In addition, both the fate and transport of particulate and dissolved solids (at least at the first level of approximation) are much less governed by chemical and biological processes than for other environmental problems. In many cases, they are dictated by the physics of transport, a strong focus area of the water-resource engineers. Finally, in the United States, much of the early environmental modeling dealt with the urban pollution problems associated with large eastern cities. In contrast, many water-resource engineers worked in the water-scarce West, where problems like erosion and salinity were more prominent.

10.2.5.1 Particulate Solids. There is a substantial body of information connected with the simulation of sediment transport in streams. In the past, this field has dealt with the physical aspects of sediment transport. In particular, it focused on how stream hydrology scours, suspends, transports, and deposits sediments.

Today, this perspective is being broadened to include biological and chemical aspects of sediment transport. Thus, there is great potential for integrating sediment transport modeling with water-quality analyses. Three problems that provide attractive settings for such integration are:

1. *Biotic Habitat.* The life cycles of many organisms, such as fish, are strongly dependent on the bottom sediments of riverine systems. For example, many species require specific substrate types to successfully reproduce. In addition, a bottom sediment-based food chain supplies nutrition to many organisms. Thus, protection of resources such as fisheries can be supported by models and understanding from the field of sediment transport. Concern over endangered species provides added stimulus for such efforts.

2. *Toxicant Transport and Fate.* As outlined in the previous section, the fate of toxicants is intimately tied to the fate of solids in aquatic systems. In particular, many contaminants concentrate in the bottom sediments of natural waters. Therefore, models of toxic substances in streams require an adequate characterization of sediment transport.

3. *Sediment Oxygen Demand.* It has been observed that streams receiving sewage input sometimes experience periods of severe oxygen depletion following short periods of high flow. One explanation is that enriched bottom sediments are resuspended and induce a short-term oxygen deficit. Sediment transport mechanisms could provide a means to rationalize the simulation of such phenomena.

10.2.5.2 *Dissolved Solids.* The build-up of dissolved solids in natural waters can sometimes constitute a severe water-quality problem. Such increases are usually connected with agricultural return water. As depicted in Fig. 10.19, water can be applied to fields repeatedly as a river flows through agricultural land. On each pass through a field, the water picks up dissolved constituents. The process is exacerbated by the fact that water is lost via evapotranspiration (that is, the loss of water to the atmosphere through evaporation and plants). In addition, if the river has been impounded, evaporation losses can compound the problem by further concentrating the salt solution. Because many such constituents are conservative (that is, they are not removed by decomposition, volatilization, and sedimentation), they can build up in substantial concentrations. Beyond a certain point, such water can have an adverse affect on subsequent use for human and industrial consumption, as well as further agricultural use.

O'Connor (1976) has developed some analytical solutions that can be useful in characterizing such systems. In addition, he has also shown how the levels of dissolved solids can be simulated in impoundments (O'Connor, 1989). This work is significant to stream modeling, because salinity problems often occur in drainage basins where impoundments and rivers form a network.

Computer-oriented models presently exist to simulate levels of dissolved solids in some large Western river basins, such as that of the Colorado River (Schuster, 1987). These models contrast with conventional water-quality frameworks in that a dynamic water budget is required. However, in most cases, although levels in impoundments are computed, salt concentrations in the connecting rivers are not simulated. Further, the models are often system-specific. Consequently, there is a need for a general modeling framework to compute salt budgets in river systems.

10.2.6 Temperature

The mathematical modeling of the transport and fate of heat in natural waters has been the subject of extensive study. Edinger, et al. (1974) provide an excellent and comprehensive report of this research. Thomann and Mueller (1987) and Chapra (1996) have summarized the fundamental approach as it relates to water-quality modeling.

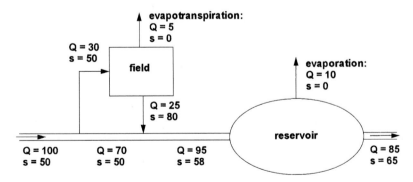

FIGURE 10.19 Simple salt budget for a river stretch subject to irrigation and impoundment. Notice how irrigation of the field increases the salt concentration by the combined effect of leaching and evapotranspiration. For the reservoir, additional concentration occurs due to water losses from evaporation.

Most of this work has been oriented toward evaluating cooling-water discharges and has dealt with systems such as cooling ponds and large rivers. Today, there is a broadening of interest in temperature modeling of streams that goes beyond the effects of point sources of heat such as power plants. There are several problem contexts:

- There is growing interest in the diurnal temperature variations of shallow, turbulent streams that are commonly found in upland regions. Although these systems are sometimes subjected to anthropogenic heat loads, their response to natural forcing functions is also of interest. Physical modifications, such as channelization and riparian-zone denudation, can have a pronounced effect on their thermal regimes. Remedial measures have been proposed to reverse these effects (Oswald and Roth, 1988). Mathematical models could prove useful in the evaluation of these modifications. In addition, diurnal temperature variations are relevant to the modeling of the fate of pollutants in such systems. For example, as mentioned previously, ammonia toxicity is sensitive to diurnal temperature variations.

- Temperature effects on biota, some of which are threatened or endangered, are directing attention toward heat management. Aside from the drainage basin modifications noted above, reservoir releases can also have an impact on temperatures in tail water streams.

- Biological and chemical transformations are sensitive to temperature. Therefore, in order to characterize these other problems adequately, there may be cases where an accompanying analysis of heat would be needed.

For all these problems, the extensive body of theory concerning the fate of heat provides a basis for temperature models of streams. In many of these contexts, this will mean that the river's hydrology will need to be simulated, because changing water levels and flows can have a pronounced effect on stream temperature.

10.3 THE FUTURE

In this review, I have attempted to show how the urban wastewater problem and dissolved oxygen defined early attempts to model and manage stream water quality. Then, a number of areas representing current and future directions were outlined.

These modern areas can be divided into two general classes. First, some involve a broadening of the classical perspective. For example, efforts to improve characterizations of nitrogen, eutrophication, and SOD all reflect extensions of the earlier Streeter-Phelps theory. Second, new areas such as toxics, solids, and temperature represent more of a departure from the classical perspective.

All the modern developments clearly point to the need for more sophisticated models. However, beyond efforts to upgrade models via inclusion of better mechanisms, there are wider implications related to the general scope of stream water-quality management. These are:

- *Broader Definition of Stream Water Quality.*   The original focus on oxygen must be broadened to encompass other water-quality parameters. Some of these, such as plant biomass and temperature, have already been incorporated into existing stream model frameworks (Brown and Barnwell, 1987). Others, such as un-ionized ammonia, toxic substances, and solids, should also be included.

- *Drainage Basin System Perspective.* The fundamental organizing principle of new modeling and management frameworks must be the drainage basin system itself (Fig. 10.20). From this perspective, the land and the water can be viewed as a whole: the land serving as the source of pollution and the water serving as the transmission network.

- *Point/Nonpoint Sources.* Nonpoint sources must be explicitly included in the new framework. This has two implications. First, the steady-state approaches employed for traditional wasteload allocations should be supplemented by a time-variable perspective because nonpoint sources (in particular, those associated with runoff) are intrinsically dynamic. Second, geographical information must be a key element of the model's database in order to adequately assess the land-related factors governing nonpoint sources.

- *Time-Variable Water Quality.* Aside from considering nonpoint sources, time-variable approaches are required to address problems such as eutrophication and ammonia toxicity, which have seasonal and diurnal dynamics. Beyond simulation models, this means that the idea of water-quality standards must be recast to reflect a dynamic perspective. Further, it will also require a rethinking of traditional steady-state management schemes. For example, what is the best approach to optimize water-quality control dynamically?

- *Quantity/Quality Interactions.* In the past, the hydrology and hydraulics used for water-quality simulations have typically been viewed as model inputs rather than as integral parts of the water-quality modeling process. The incorporation of a time-variable approach means that real hydrology should be integrated into the new modeling framework. Among other things, this opens the possibility that, for controlled basins, flow augmentation can be viewed as an alternative to additional treatment.

- *Modeling Environments and Computing.* All the above-mentioned features have one common aspect: they will all require increased information management

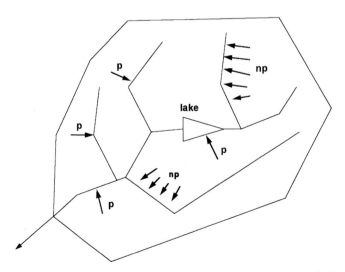

FIGURE 10.20   A river basin system showing point (p) and nonpoint (np) sources of pollution.

capabilities well beyond those of presently available water-quality modeling tools. Consequently, an effective computing environment will be an essential element for their effective integration and implementation.

10.4 ECONOMICS/COST-EFFECTIVENESS

Although we have attempted to note the management aspects of water-quality control, the present review has emphasized modeling. Obviously, such models are but one facet, albeit an important one, in the overall control process. As a final note, we believe that economic trends could have a strong effect on stream water-quality management in the coming decades.

In essence, strong economic pressures should stimulate interest in cost-effective yet robust solutions for pollution abatement over the coming decades. Growing budget deficits and a burgeoning national debt mean that hard choices must be made regarding future public and private expenditures for water-quality control. This is especially true because the financing of public-works projects such as waste treatment have increasingly moved from federal to state and local levels over the past decade. Hence, as perceived pollution-control expenditures come closer to the individual taxpayer, effective expenditures of public funds should gain public support over the coming years.

This notion is reinforced by the fact that in the United States, the least expensive point-source treatment options have already been implemented. As depicted in Fig.

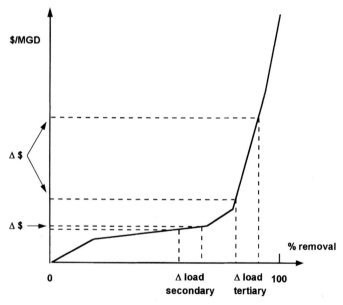

FIGURE 10.21 Capital construction costs versus degree of treatment for municipal wastewater treatment. Note that most decisions relating to tertiary waste treatment presently deal with high percent removals. Consequently, a faulty decision carries a much higher economic penalty today than in earlier years, when primary and secondary waste treatment were dominant.

10.21, treatment currently deals with the steepest part of the cost curve for point sources. Further, controls of nonpoint or diffuse sources are typically more expensive than point-source controls.

These trends will have two impacts on stream water-quality management and modeling. First, from the perspective of a new water-quality control paradigm, this means that economic factors should probably be integrated into decision-support frameworks. At a minimum, some cost-benefit accounting and optimization capabilities should be incorporated explicitly.

Second, they mean that there is a much greater incentive for sound decisions today because of the associated costs. As depicted in Fig. 10.21, a modeling mistake today incurs a much larger economic penalty than in the past. For this reason, it is critical that effort be expended to increase scientific understanding and expand and enhance water-quality modeling capabilities for rivers and streams over the coming decades.

REFERENCES

Bartsch, A. F., and W. F. Ingram, *Biology of Water Pollution,* U.S. Dept. of Interior, Water Pollution Control Administration, 1967.

Brown, L. C., and T. O. Barnwell, Jr., *The Enhanced Stream Water Quality Models QUAL2E and QUAL2E-UNCAS.* EPA/600/3-87/007. U.S. Environmental Protection Agency, Athens, Ga., 1987.

Cerco, C. F., and T. Cole, "Three-Dimensional Eutrophication Model of Chesapeake Bay," *J. Environ. Engr. ASCE,* 119(6):1006–1025, 1993.

Chapra, S. C., *Surface Water Quality Modeling,* McGraw-Hill, New York, 1996.

Chapra, S. C., "A Toxicant Loading Concept for Organic Contaminants in Lakes," *J. Environ. Engr.* 117(5):656–677, 1991.

Chapra, S. C., and R. P. Canale, *Numerical Methods for Engineers,* 2d ed. McGraw-Hill, New York, 1988.

Chapra, S. C., and K. H. Reckhow, *Engineering Approaches for Lake Management, Vol. 2: Mechanistic Modeling,* Butterworth, Woburn, Mass., 1983.

Chen, C. W., "Concepts and Utilities of Ecological Models," *J. San. Engr. Div. ASCE,* 96(SA5):1085–1086, 1970.

Deininger, R. A., "Water Quality Management—The Planning of Economically Optimal Control Systems," *Proc. of the First Annual Meeting of the Am. Water Resources Assoc.,* 1965.

Di Toro, D. M., "Recurrence Relations for First-Order Sequential Reactions in Natural Waters," *Water Resour. Res.,* 8(1):50–57, 1972.

Di Toro, D. M., R. V. Thomann, and D. J. O'Connor, "A Dynamic Model of Phytoplankton Population in the Sacramento-San Joaquin Delta," in R. F. Gould, ed., *Advances in Chemistry Series 106: Nonequilibrium Systems in Natural Water Chemistry,* American Chemical Society, Washington, D.C., 1971.

Di Toro, D. M., P. R. Paquin, K. Subburamu, and D. A. Gruber, "Sediment Oxygen Demand Model: Methane and Ammonia Oxidation," *J. Environ. Engr. ASCE,* 116(5):945–986, 1990.

Fair, G. M., E. W. Moore, and H. A. Thomas, "The Natural Purification of River Muds and Pollutional Sediments," *Sewage Works J.,* 13(2):270–307; 13(4):756–779; 13(6):1209–1228, 1941.

Gardiner, R. D., M. T. Auer, and R. P. Canale, "Sediment Oxygen Demand in Green Bay (Lake Michigan)," *Proc. 1984 Specialty Conf.,* ASCE, 514–519, 1984.

Gaudy, A. F., and E. T. Gaudy, *Microbiology for Environmental Scientists and Engineers,* McGraw-Hill, New York, 1980.

Goldman, J. C., D. R. Tenore, and H. I. Stanley, "Inorganic Nitrogen Removal from Wastewater: Effect on Phytoplankton Growth in Coastal Marine Waters," *Science*, 180:955–956, 1973.

Kenner, B. A., "Fecal Streptococcal Indicators," in G. Berg, ed., *Indicators of Viruses in Water and Food,* Ann Arbor Science Publishers, Ann Arbor, Mich., 1978.

Loucks, D. P., C. S. Revelle, and W. R. Lynn, "Linear Programming Models for Water Pollution Control," *Management Science,* 14(4):B166–B181, 1967.

Mancini, J. L., "Numerical Estimates of Coliform Mortality Rates under Various Conditions," *J. Water Poll. Control Fed.,* 50(11), 1978.

O'Connor, D. J., "Oxygen Balance of an Estuary," *J. San. Engr. Div. ASCE,* 86(SA3):35–55, 1960.

O'Connor, D. J., "The Temporal and Spatial Distribution of Dissolved Oxygen in Streams," *Water Resour. Res.,* 3(1):65–79, 1967.

O'Connor, D. J., "The Concentration of Dissolved Solids and River Flow," *Water Resour. Res.,* 12(2):279–294, 1976.

O'Connor, D. J., "Models of Sorptive Toxic Substances in Freshwater Systems. III: Streams and Rivers," *J. Envir. Engr.,* 114(3):552–574, 1988.

O'Connor, D. J., "Seasonal and Long-Term Variations of Dissolved Solids in Lakes and Reservoirs," *J. Environ. Engr. ASCE,* 115(6):1213–1234, 1989.

Omernik, J. M., "Non-Point Source-Stream Nutrient Level Relationships: A Nationwide Study, Corvallis ERL," ORD. USEPA, Corvallis, Oreg., EPA-600/3-77-105, 1977.

Oswald, T., and C. Roth, "Make Us a River," *Rod and Reel,* May/June:28, 1988.

Reckhow, K. H., and S. C. Chapra, *Engineering Approaches for Lake Management, Vol. 1: Data Analysis and Empirical Modeling,* Butterworth, Woburn, Mass., 1983.

Riley, G. A., "Factors Controlling Phytoplankton Population on Georges Bank," *J. Mar. Res.,* 6:104–113, 1946.

Schuster, R. J., *Colorado River Simulation System Documentation: System Overview,* U.S. Bureau of Reclamation, Denver, Colo., 1987.

Streeter, H. W., and E. B. Phelps, *A Study of the Pollution and Natural Purification of the Ohio River, III. Factors Concerning the Phenomena of Oxidation and Reaeration,* U.S. Public Health Service, Pub. Health Bulletin No. 146, February, 1925, reprinted U.S. Dept. of Health, Education, and Welfare, Public Health Administration, 1958.

Thomann, R. V., "Mathematical Model for Dissolved Oxygen," *J. San. Engr. Div. ASCE,* 89(SA5):1–30, 1963.

Thomann, R. V., "Equilibrium Model of Fate of Microcontaminants in Diverse Aquatic Food Chains," *Can. J. Flsh. Aquat. Sci.,* 38:280–296, 1981.

Thomann, R. V., and J. A. Mueller, *Principles of Surface Water Quality Modeling and Control,* Harper & Row, New York, 1987.

Thomann, R. V., and M. J. Sobel, "Estuarine Water Quality Management and Forecasting," *J. San. Engr. Div. ASCE,* 90(SA5):9–36, 1964.

Velz, C. J., "Deoxygenation and Reoxygenation," *Proc. Am. Soc. Civ. Engr.,* 65(4):677–680, 1938.

Velz, C. J., "Factors Influencing Self-Purification and Their Relation to Pollution Abatement," *Sewage Works Journal,* 19(4):629–644, 1947.

CHAPTER 11
GROUNDWATER

Michael J. Graham
Battelle Pacific Northwest Laboratories
Richland, Washington

James M. Thomas
U.S. Geological Survey, Water Resources Division
Carson City, Nevada

F. Blaine Metting
Battelle Pacific Northwest Laboratories
Richland, Washington

Groundwater quality is an area of increasing focus and importance as society strives to protect, and in some cases restore, this precious resource. In this chapter we will explore the conditions of natural groundwater and learn how the study of chemistry can give insights into the groundwater flow system. Two areas of anthropogenic contamination are covered. The first, contamination by chemicals, is an analysis of various point and nonpoint sources of pollution. The second is an overview of contamination of groundwater by microorganisms.

11.1 QUALITY OF NATURAL GROUNDWATER

The study of natural groundwater quality can provide important insights into the nature of the resource. By evaluating the natural chemical and isotopic compositions of groundwater, inferences can be made of the reactions that produce natural water chemistry and the recharge, movement, mixing, and discharge of groundwater.

11.1.1 Natural Groundwater Quality

Groundwater in natural systems generally contains less than 1000 mg/L dissolved solids, unless groundwater has (1) encountered a highly soluble mineral, such as gypsum, (2) been concentrated by evapotranspiration, or (3) been geothermally heated. In natural systems, groundwater generally acquires dissolved constituents by dissolution of aquifer gases, minerals, and salts. Thus, soil zone and aquifer gases and the most soluble minerals and salts in an aquifer generally determine the chemical composition of groundwater in an aquifer.

11.1.1.1 Major Ion Chemistry. Most groundwater is recharged through a soil zone which contains partial pressures of carbon dioxide (CO_2) gas that are higher than atmospheric. Thus, recently recharged groundwater generally contains high inorganic carbon concentrations. Carbon forms carbonic acid (H_2CO_3), bicarbonate ions (HCO_3^-), and carbonate ions (CO_3^{-2}). As pH increases above about 4.5, HCO_3^- increases in concentration, and near neutral pH, dissolved carbon is present mainly as the HCO_3^- ion. Above a pH of about 8.3, CO_3^{-2} becomes increasingly more important as the pH increases. Cations dissolved in the water that balance the electrical charge of the carbon anions depend primarily on the mineral composition of the aquifers.

Aquifers composed of carbonate soil and rock contain highly soluble calcite and dolomite. Thus, waters in these types of aquifers contain primarily calcium (Ca^{+2}), magnesium (Mg^{+2}), and HCO_3^- ions. For example, groundwater in the carbonate rock aquifers of southern Nevada contain predominantly Ca^{+2}, Mg^{+2}, and HCO_3^- ions (Table 11.1; Winograd and Pearson, 1976). Carbonate rock aquifers often contain interbedded evaporite deposits, which contain highly soluble minerals such as gypsum, resulting in a groundwater with sulfate (SO_4^{-2}) being a major ion. For example, the Madison Aquifer in Montana, South Dakota, and Wyoming contains groundwater with predominately Ca^{+2}, Mg^{+2}, HCO_3^-, and SO_4^{-2} ions (Table 11.1; Plummer, et al., 1990). Aquifers consisting of silicate soil and rock contain a wide range of silicate minerals. Thus, groundwater in these aquifer types have chemical compositions that range widely in major ion composition depending on the most soluble minerals in the aquifer. Aquifers composed of granitic soils and rock generally contain biotite, hornblende, K-feldspar, muscovite, plagioclase feldspar, and quartz, which dissolve to form a water containing mainly Ca^{+2}, Na^+, and HCO_3^- ions (Table 11.1; Garrels and Mackenzie, 1967; Nordstrom, et al., 1989).

Aquifers consisting of rhyolitic soils and rock generally contain volcanic glass and feldspar minerals as their most soluble components. Thus, water chemistry in these aquifers is generally dominated by Na^+ and HCO_3^- ions, with lesser amounts of Ca^{+2} (White, 1979; Thomas, et al., 1989).

Descriptions of major ion water chemistries associated with different aquifer rock types are given by Hem (1985, pp. 189–202); Drever (1986, Chap. 9); Mazor (1991, pp. 23–25); and Morel and Hering (1993, Chap. 5). Mazor (1991, p. 24) presents a summary of rock type and typically associated groundwater chemistry in his Table 3.1.

11.1.1.2 Minor and Trace Element Chemistry. Minor and trace element compositions of natural groundwater depend on the availability of minor and trace elements in easily soluble phases or on sorption sites, and the redox state of the water in the aquifer. Some trace elements are also present in significant amounts in the atmosphere from anthropogenic sources, and these elements can be introduced into groundwater when precipitation recharges an aquifer.

Minor element concentrations, such as fluoride (F^-), bromide (Br^-), and strontium (Sr^{+2}), generally are derived from the dissolution of minerals or salts that contain these elements in small amounts, or from minerals that are in small abundance. For example, calcite generally contains some Sr^{+2} substituted for Ca^{+2} in the calcite structure, so when calcite dissolves a small amount of Sr^{+2} is released to the water, as well as Ca^{+2} and CO_3^{-2}. Trace element concentration is related to the availability of

TABLE 11.1 Examples of Natural Groundwater Chemistry*

Site	Temperature	pH	Ca	Mg	Na	K	HCO_3	SO_4	Cl	SiO_2
Carbonate rock aquifer										
Cold Creek Spring[a]	10	7.6	1.72	0.64	0.07	0.01	4.79	0.09	0.03	0.11
Corn Creek Spring[a]	21	7.7	1.14	1.34	0.27	0.05	4.67	0.18	0.20	0.28
Indian Springs[a]	26	7.4	1.15	0.92	0.18	0.03	3.97	0.16	0.10	0.20
Carbonate rock aquifer with evaporite mineral										
Lewiston Big Spring[b]	10.6	7.58	1.87	1.15	0.10	0.02	3.31	1.46	0.05	—
Hanover Flowing Well[b]	20.4	7.63	2.10	1.19	0.12	0.03	3.42	1.77	0.04	—
Vanek Warm Spring[b]	19.6	7.40	3.25	1.65	0.16	0.03	3.53	3.44	0.07	—
Landusky Spring[b]	20.4	7.24	6.50	4.08	1.79	0.24	3.81	10.11	0.54	—
Granitic rock aquifer										
Ephemeral Springs[c]	—	6.2	0.08	0.03	0.13	0.03	0.33	0.01	0.01	0.27
Perennial Springs[c]	—	6.8	0.26	0.08	0.26	0.04	0.90	0.03	0.03	0.41
Shallow Well 2[d]	8.0	7.89	0.85	0.19	0.54	0.04	2.34	0.09	0.10	0.21
Rhyolitic rock aquifer										
Lower Indian Spring[e]	21.0	7.9	0.15	0.04	2.48	0.04	2.07	0.18	0.42	0.80
Crystal Springs[e]	24.0	7.7	0.55	0.15	2.18	0.09	2.33	0.23	0.59	0.75
S. Brown Well[f]	10.6	6.8	0.50	0.16	1.02	0.10	1.54	0.14	0.23	0.93
N. Brown Well[f]	14.7	7.5	0.87	0.23	2.09	0.17	2.46	0.54	0.71	1.03

* All analyses are in millimoles per liter and temperature in degrees Celsius.
[a] Winograd and Pearson, 1976.
[b] Plummer, et al., 1990.
[c] Garrels and MacKenzie, 1967.
[d] Nordstrom, et al., 1989.
[e] White, 1979.
[f] Thomas, et al., 1989.

the trace element in soluble phases or on sorption sites; complexation of the trace element with ions or organic matter dissolved in the water; and the redox state of water in the aquifer (Drever, 1988, Chap. 15; Morel and Hering, 1993, Chaps. 6 and 7). For example, an element such as uranium (U) has to be present in a mineral, or be adsorbed on coatings of mineral surfaces or organic matter, to be available to enter groundwater. In addition to being present as simple ions, most trace elements combine with other ions or dissolved organic matter to form complex species. Thus, the total concentration of a trace element also depends on the availability of other dissolved constituents. The solubility of minerals containing U can be increased by several orders of magnitude by U complexation with carbon, phosphate, and fluoride, which would result in greatly increased total dissolved U concentration (Langmuir, 1978). The presence or absence of oxygen also is an important factor influencing trace element concentration. Most trace elements have more than one valence state, so they can be present as oxidized or reduced species. Dissolved iron (Fe) is rapidly removed from groundwater in the presence of oxygen by iron oxide formation. This reaction is reversible, so the depletion of oxygen in an aquifer that contains iron oxide coatings on mineral grains produces a water with high Fe concentration (Chapelle, 1993; Vroblesky and Chapelle, 1994).

11.1.1.3 Isotope Chemistry. The natural isotopic composition of groundwater depends on the isotopic composition of recharge water and geochemical reactions involving the recharge water and materials within the aquifer that contain the isotope element of interest. Deuterium and oxygen-18 are part of the water molecule, so they make excellent tracers of groundwater movement. In addition, they have great natural variability in groundwater, which enhances their usefulness as groundwater tracers. Deuterium and oxygen-18 compositions of precipitation that recharges aquifers varies greatly because of the fractionation of the isotopes. Variability of deuterium and oxygen-18 compositions in precipitation is the result of: (1) temperature, (2) seasonal, (3) altitude, (4) rainout, (5) continental, and (6) latitudinal effects (Dansgaard, 1964; Fritz and Fontes, 1980; Gat and Gonfiantini, 1981; Mazor, 1991; Coplen, 1993).

Dansgaard (1964) showed that the deuterium and oxygen-18 composition of precipitation depends on the temperature of condensation at which precipitation is formed (temperature effect). Thus, precipitation during the cooler winter is isotopically lighter (more negative) than warmer summer precipitation (seasonal effect) (Dansgaard, 1964; Siegenthaler and Oeschger, 1980). Deuterium and oxygen-18 values also decrease in precipitation with increasing altitude as demonstrated by Siegenthaler and Oeschger (1980) (altitude effect). During a storm, deuterium and oxygen-18 compositions of the rain become continually more depleted (more negative) as the amount of rainfall increases (rainout effect) (Yurtsever and Gat, 1981). Deuterium and oxygen-18 also decrease away from ocean sources (continental effect) and with increasing latitude away from the equator (latitudinal effect) (Sonntag, et al., 1979; Yurtsever and Gat, 1981). The differences in isotopic composition of groundwater resulting from these effects can be used to help determine source areas, flow paths, and mixing of groundwater, as will be discussed in Sec. 11.1.2.3.

The natural isotopic composition of an element dissolved in groundwater depends on the isotopic composition of the element in precipitation that recharges an aquifer and geochemical reactions involving the element within an aquifer. For example, the carbon-13 composition of carbon dissolved in groundwater in an aquifer results from the initial carbon-13 composition of carbon dissolved in precipitation, dissolution of soil zone CO_2, dissolution and precipitation of minerals containing carbon, exsolution of CO_2 gas, and oxidation/reduction reactions involving carbon. Atmospheric CO_2 has a $\delta^{13}C$ of about -7 permil (Pearson and Friedman, 1970), so dissolution of atmospheric CO_2 produces a water with a $\delta^{13}C$ of approximately -7 permil. Soil zone CO_2 is primarily from root respiration and decomposition of organic matter and has a $\delta^{13}C$ composition of about -25 permil in temperate climates and about -15 permil in arid climates (Coplen, 1993). Thus, dissolution of soil zone CO_2 results in groundwater with a $\delta^{13}C$ composition that is heading toward -25 or -15 permil. Marine calcite and dolomite generally have a $\delta^{13}C$ composition of about 0 permil (Mazor, 1991), so dissolution of these minerals results in a $\delta^{13}C$ that heads toward 0 permil. When dissolved inorganic carbon is removed from the groundwater by mineral precipitation or gas exsolution, carbon-13 is fractionated because of the mass difference between carbon-12 and carbon-13.

Thus, the precipitation of calcite, or exsolution of CO_2 gas, results in the $\delta^{13}C$ of the water becoming heavier (more positive) because carbon-12 is more easily liberated from carbon dissolved in groundwater than carbon-13. The amount of fractionation of carbon-13 is pH and temperature dependent (Wigley, et al., 1978). The oxidation of organic matter produces CO_2 or CH_4 gas that has a light $\delta^{13}C$ composition, so oxidation of organic matter with subsequent dissolution of the gas produced results in a water that is becoming isotopically lighter in $\delta^{13}C$ (Baedecker and Back, 1979). The carbon-13 composition of carbon dissolved in groundwater is usually the result of several of the above listed processes, and therefore the $\delta^{13}C$ groundwater composition can be used to determine the types and amounts of natural geochemical processes that are producing the groundwater chemistry.

Other elements dissolved in groundwater, such as S, Sr, and U, also have isotopes that change due to geochemical processes and can be used to determine natural geochemical processes in aquifers.

11.1.2 Natural Groundwater Chemistry and Flow Systems

Natural groundwater chemistry of flow systems can be used to determine: (1) geochemical reactions that produce observed water chemistry and changes in observed water chemistry, (2) groundwater flow paths, (3) groundwater mixing, and (4) groundwater ages and flow rates.

11.1.2.1 Major Ion Chemistry. Major ion chemistry can be used to determine geochemical reactions, flow paths, and mixing of groundwater by defining the chemical evolution of natural groundwater using mass balance and thermodynamic calculations (Plummer, et al., 1983). Geochemical reactions that produce changes in groundwater chemistry can be described by a set of mass transfer reactions for changes in ion mass between the groundwater and aquifer materials. The mass transfer reactions have to be chosen for phases—minerals or gases—that have been identified in the aquifer, and the reactions have to be thermodynamically favorable for the reactions to take place (Plummer and Back, 1980; Plummer, et al., 1983). Thus, a water has to be below saturation with respect to a phase (mineral or gas) and the phase has to be present to dissolve, or a water has to be above saturation with respect to a phase for a mineral to precipitate or a gas to exsolve from a water. Ion exchange, if clay minerals with ions on exchange sites are present, and mixing of chemically different groundwaters can also be included in mass transfer calculations. In addition, isotope data, such as carbon-13 and sulfur-34, can be used in the mass transfer calculations to constrain plausible geochemical reactions. That is, calculated mass transfers of elements have to accurately predict the isotope value measured in the groundwater for the geochemical model to be valid. A unique set of geochemical reactions that describes changes in water chemistry does not exist, but a choice of phases based on observed mineralogic and gas data will result in a realistic geochemical model if the geochemical reactions are thermodynamically feasible.

An example of the use of major ion chemistry to define geochemical reactions and flow paths is given for the Floridan aquifer (Plummer, et al., 1983). In central Florida, groundwater flows from an area of high potentiometric head near Polk City down gradient to Wauchula (Fig. 11.1). A geochemical reaction model that defines the changes in natural water chemistry along this flow path includes dolomite $[CaMg(CO_3)_2]$ dissolution, gypsum $[CaSO_4 \cdot 2H_2O]$ dissolution, unsaturated zone CO_2 gas dissolution along the first half of the flow path, followed by organic carbon $[CH_2O]$ oxidation along the second half of the flow path, ferric iron hydroxide $[FeOOH]$ dissolution, calcite $[CaCO_3]$ precipitation, and pyrite $[FeS_2]$ formation.

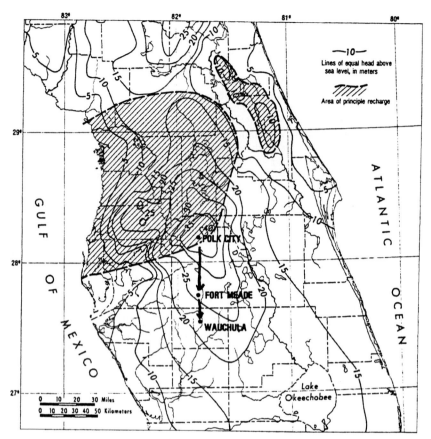

FIGURE 11.1 Groundwater flow in the Floridan aquifer (*Plummer et al., 1983*).

The amounts of mass transfer needed to describe the changes in water chemistry along the flow path from Polk City to Wauchula are, in mmol/kg:

Polk City water $+ 0.96CaMg(CO_3)_2 + 1.68CaSO_4 \cdot 2H_2O + 0.17CH_2O + 0.53CO_2$

$$+ 0.03FeOOH - 1.84CaCO_3 - 0.03FeS_2 = \text{Wauchula water}$$

Thermodynamic calculations for the water chemistries show that the proposed geochemical model is thermodynamically acceptable for the main phases involved in mass transfer. Gypsum is below saturation in both waters and dolomite is below saturation in the initial water, so they would dissolve. Dolomite would also dissolve in the Wauchula water, even though it is slightly above saturation, because of dedolomitization (the incongruent dissolution of dolomite with calcite precipitation driven by the irreversible dissolution of gypsum). The waters are above saturation with respect to calcite, so it would precipitate (Plummer, et al., 1983). Chemical analyses for Fe dissolved in the water is lacking, so mineral saturation could not be calculated for these phases, but the redox state of water in the aquifer indicates the small amounts of FeOOH dissolution and FeS_2 formation would most likely occur (Plummer, et al., 1983).

This example presents a valid geochemical model, which in addition to identifying probable geochemical reactions that describe the observed changes in water chemistry and the amounts of mass transfer of the elements between the groundwater and the aquifer, also shows that the proposed flow path is geochemically viable.

Major ion chemistry can be used to determine the mixing of chemically different waters. The percentages of chemically different waters that mix to produce a water that is chemically distinct from the original mixing waters can be calculated from major ion ratios. Nonreactive ions, such as Cl^-, which are not generally involved in geochemical reactions, can be used to calculate percentages of mixing waters. Whereas ions like Ca^{+2}, which are more likely to be involved in geochemical reactions, are less likely to give valid mixing ratios. Mazor, et al. (1973) and Mazor (1991) present examples of how to calculate percentages of mixing waters when geochemical reactions after mixing are minor. A mixture of waters can also undergo geochemical reactions during and after mixing. The percent of each water involved in mixing can usually be determined using a conservative ion like Cl^- and accounting for the geochemical reactions producing changes in water chemistry in addition to mixing by using the mass transfer and thermodynamic approach that was previously described. Glynn (1991) presents an example of mixed waters that have undergone geochemical reactions.

11.1.2.2 Minor and Trace Element Chemistry. Natural minor and trace element concentrations can also be used to help determine flow paths and mixing of groundwater and define geochemical reactions involving trace constituents. Geochemical reactions involving trace elements are commonly redox reactions, so trace elements can often be used to determine the redox state of groundwater in an aquifer.

The trace element selenium (Se) was used in a shallow aquifer to help evaluate flow paths and mixing of groundwater in the western San Joaquin Valley, California (Deverel and Fio, 1991). Selenium concentrations were greater than 700 µg/L in most groundwater collected from depths of greater than 6 m below land surface and in the vicinity of drain 2 (Deverel and Fio, 1991; Fig. 11.2). Groundwater less than 6 m below land surface in the remainder of the study area contained an average 108 µg/L of Se. This difference in Se concentration indicates that deep groundwater (>6 m) containing high concentrations of Se was flowing upward and discharging into drain 2. Groundwater in the aquifer was shown to be flowing both horizontally and vertically and mixing before being discharged to a drain system (Deverel and Fio, 1991). Groundwater discharge was shown to be a mixture of about 30 percent deep (>6 m) and 70 percent shallow (<6 m) groundwater for a shallow drain (1.8 m below land surface) and about 60 percent deep and 40 percent shallow groundwater for a deep drain (2.7 m below land surface). When water was applied on top of the flow system for irrigation, flow in the shallow groundwater system changed such that only shallow groundwater was discharged into the shallow drain, and discharge of deep groundwater into the deep drain decreased from about 60 to 30 percent of the total flow.

Iron dissolved in groundwater in the Middendorf aquifer in South Carolina changed in concentration from about 0.5 µg/L in the oxygenated recharge area to commonly greater than 1000 µg/L in a zone 15 to 55 km down gradient of the recharge area (Chapelle and Lovely, 1992). This change in dissolved Fe concentration was the result of the aquifer becoming anaerobic and Fe reducing bacteria reducing Fe^{+3} oxyhydroxides, thus releasing Fe^{+2} to the groundwater. The lack of dissolved oxygen and sulfides in the water allowed Fe^{+2} concentration, and hence total dissolved Fe, to increase to greater than 1000 µg/L.

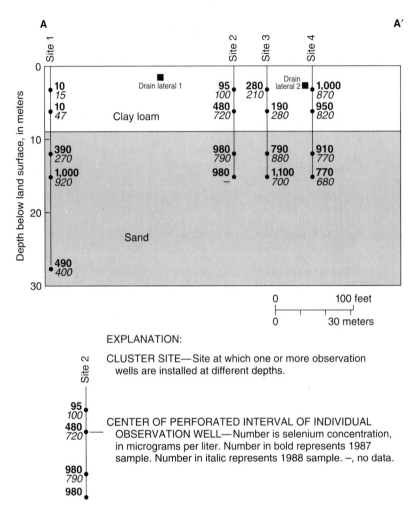

FIGURE 11.2   Selenium in groundwater in the San Joaquin Valley, California (*Deveral and Fio, 1991*).

11.1.2.3   Isotope Chemistry.   Isotopes are a very powerful tool for understanding natural groundwater flow systems. Deuterium and oxygen-18 can be used as finger-prints to track parcels of water that are isotopically different. This difference in iso-topic composition can be used to determine source areas, flow paths, and mixing of groundwaters. Winograd and Friedman (1972) used deuterium data to show that water discharging from large springs in southern Nevada and the Death Valley area of California is from several sources (Fig. 11.3). Water recharged in mountains in southern Nevada and groundwater flowing into southern Nevada from north-central Nevada flows beneath topographic divides of several valleys for tens of miles to discharge at large springs. The deuterium composition of these waters is different because of different recharge areas, so the isotopic data were the key to determining source areas and flow paths of water discharging at these large springs. Water dis-

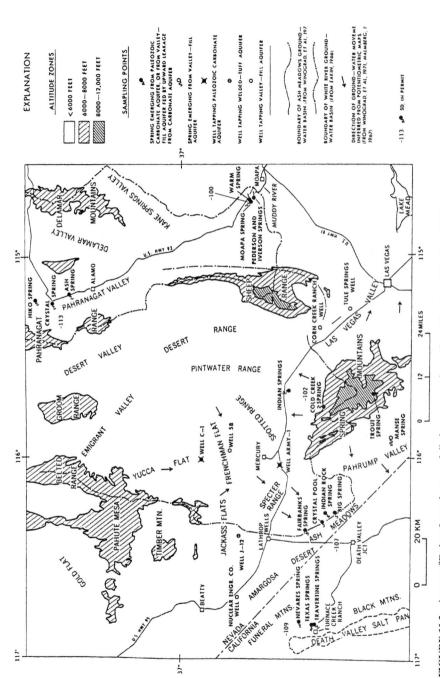

FIGURE 11.3 Average δD data for recharge and discharge waters in southern Nevada and the Death Valley area of California (*Winograd and Friedman, 1972*).

11.9

charging at Ash Meadows springs had a median deuterium value (–107 permil), between that of recharge to the mountains in southern Nevada (–102 permil) and water flowing into southern Nevada from the north (–113 permil). Thus, by using isotopic data in conjunction with hydrologic and geologic information, water discharging from Ash Meadows Springs was determined to be a mixture of these waters. Furthermore, the percentage of the two waters discharging from the springs could be calculated.

Isotopes of elements dissolved in groundwater can also be used to determine geochemical reactions, flow paths, and groundwater ages, and hence, flow rates. For example, carbon isotope data have been used to determine geochemical reactions that produce changes in natural water chemistry and to estimate the age and flow rates of groundwater (Borole, et al., 1979; Plummer, et al., 1983; Mazor, et al., 1985; Phillips, 1989; Plummer, 1991). In the previous example of the Floridan aquifer, carbon and sulfur isotope data were included in the geochemical modeling to help determine the possible geochemical reactions in the flow system and to calculate the age and flow rate of groundwater in the aquifer (Plummer, et al., 1983). By constraining the geochemical reactions with isotopic data, some of the mass transfer models that describe the processes producing changes in natural groundwater chemistry were invalidated. The carbon isotope data were important in showing that the flow system was initially open to soil zone CO_2, and later closed to CO_2 with organic carbon oxidation becoming important. Including the carbon isotope data in the mass transfer model results in a calculated groundwater age of 11,800 years, assuming the soil zone CO_2 contains 50 percent modern carbon-14 (Plummer, et al., 1983, Table 9). This groundwater age corresponds to an average flow rate of 5.8 m/yr between Polk City and Wauchula (Fig. 11.1). This geochemically calculated flow rate is similar to a flow rate calculated with a hydrologic model developed for the Floridan aquifer (Hanshaw, et al., 1965). Thus, the flow path and geochemical reactions that describe the changes in natural groundwater chemistry are reasonable.

Isotope data have also been used to identify oxidation/reduction reactions in natural groundwater systems. For example, sulfate reduction has been identified in aquifers in Canada, Florida, and Texas (Pearson and Rightmire, 1980; Rye, et al., 1981; Robertson and Schiff, 1994). In a sandy forested recharge area in south-central Canada, SO_4^{-2} concentrations of 6 to 27 mg/L at approximately 10 m below land surface drop below 0.5 mg/L at a depth of about 20 m. Over this same interval, the sulfur-34 value of SO_4^{-2} changes from 4 to 9 permil to 20 to 40 permil, indicating SO_4^{-2} has been biogenically reduced. The $^{234}U/^{238}U$ ratio of U isotopes and $^{87}Sr/^{86}Sr$ ratio of Sr isotopes dissolved in groundwater have been used to help identify geochemical reactions and to delineate source areas, flow paths, and mixing of groundwater in aquifers (Osmond, 1980; Peterman, et al., 1992). The ratio of these isotopes are fingerprints much like deuterium and oxygen-18 and can be used the same way. For example, Sr isotopes have been used for interpretation of water sources to Ash Meadows springs (Peterman, et al., 1992). The Sr isotope data (Fig. 11.4) indicate that water discharging from Ash Meadows springs is a mixture of waters, as did the deuterium data (Fig. 11.3). The Sr isotope data also provide additional information about geochemical reactions within the flow system. They indicate that water in the flow system has either encountered and reacted with noncarbonate rocks, or that a small amount of water is flowing into the carbonate rock aquifers from a noncarbonate aquifer source.

Isotopes of other elements, such as nitrogen, lithium, and boron, can also be used to help determine sources and flow paths of groundwater and geochemical reactions in natural groundwater flow systems.

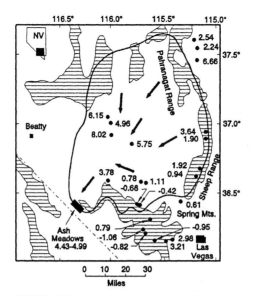

FIGURE 11.4 Strontium isotope data for southern Nevada. Strontium isotope values given as $\delta^{87}Sr = [(^{87}Sr/^{86}Sr/0.70920) - 1] \cdot 1000$. (*Peterman, et al., 1992.*)

11.1.3 Summary of Natural Groundwater Quality

In summary, natural groundwater chemistry is important for understanding geochemical reactions that produce the chemical composition of groundwater and for determining source areas, flow paths, and mixing of groundwater. Understanding the natural geochemical system is required before the effects of aquifer contamination can be evaluated. When the natural water chemistry of a flow system is evaluated, and the processes that produce the water chemistry are understood, then remediation of a contaminant problem, whether intrinsically or engineered, can be efficiently applied.

11.2 ANTHROPOGENIC CONTAMINATION

Groundwater contamination is a great concern. It may take decades for contaminated groundwater to naturally return to a useable resource, given the slow rates of groundwater movement and recharge. The restoration of groundwater systems through engineered treatment is also slow and costly. Groundwater contamination emanates from point sources, such as waste disposal in landfills, and nonpoint or distributed sources, such as agricultural application of pesticides. In the following sections, we will give an overview of these problems.

11.2.1 Nature of the Problem

In the United States, groundwater is supplied to over 40 percent of the population served by public water utilities. Accounting for private systems, which are dominantly groundwater, over half of the population relies upon groundwater for drinking water (Solley, et al., 1988). Groundwater is the source of about one-third of U.S. irrigation water. Total use of groundwater for 1985 was estimated at 73 billion gallons per day (USGS, 1990), more than double the usage in 1950.

With the increase in population and society's increasing disposal of solid and liquid wastes to the environment, contamination of groundwater resources is a growing problem. For example, the electric utility industry produces approximately 75 million tons of solid waste annually, which may double by the turn of the century (Murarka and McIntosh, 1987). An estimated 20.5 million tons of fertilizers were applied to crops during 1988 to 1989 (USDA, 1989).

Groundwater contamination is a concern in almost every state. Forty or more states report that their groundwater quality is threatened by agricultural activity, septic tanks, and/or underground storage tanks (USEPA, 1990). In the largest monitoring study conducted in the United States (EPA's National Pesticide Survey, 1992), residues of one or more pesticides were detected in about 10.4 percent of wells in community water systems and in about 4.2 percent of rural domestic wells. The study concluded that less than 1 percent of all drinking-water wells in the United States exceed a health-based limit for pesticides.

11.2.2   Types of Groundwater Contamination

Groundwater has been contaminated by a wide range of substances. The Office of Technology Assessment (OTA) compiled an extensive list of substances that have been found in groundwater (1984). These substances include approximately 175 organic chemicals, 50 inorganic chemicals, biological organisms, and radionuclides (App. 11.A). Also listed in App. 11.A are examples of the uses of the substances, which provides insights to the sources of the contaminants.

Since the OTA report of 1984, much more data have been collected on contaminated groundwater. The National Research Council (National Academy of Sciences, 1994) recently compiled a list of the 25 most frequently detected groundwater contaminants at hazardous waste sites (Table 11.2).

Nitrates from fertilizers and animal wastes are prevalent contaminants of groundwater, as well as pesticides from agricultural activities. National surveys of this type of nonpoint source groundwater pollution have been made by the U.S. Environmental Protection Agency and the U.S. Geological Survey (USEPA, 1990; 1992; Puckett, 1994).

11.2.3   Mechanisms for Groundwater Contamination

Groundwater can be contaminated by localized releases from sources such as hazardous waste disposal sites, municipal landfills, surface impoundments, underground storage tanks, gas and oil pipelines, back-siphoning of agricultural chemicals into wells, and injection wells. Groundwater can also become contaminated by substances released at or near the soil surface in a more dispersed manner, including pesticides, fertilizers, septic tank leachate, and contamination from other nonpoint sources (National Academy of Sciences, 1993). An illustration of various mechanisms of groundwater contamination is given in Fig. 11.5 (from Fetter, 1993) and discussed in detail in following sections.

Unless a substance is placed directly into an aquifer (e.g., via an injection well), it has to pass through the unsaturated (vadose) zone, which has been referred to as "the

TABLE 11.2 Most Frequently Detected Groundwater Contaminants
at Hazardous Waste Sites

Rank	Compound	Common sources
1	Trichloroethylene	Dry cleaning; metal degreasing
2	Lead	Gasoline (prior to 1975); mining; construction material (pipes); manufacturing
3	Tetrachloroethylene	Dry cleaning; metal degreasing
4	Benzene	Gasoline; manufacturing
5	Toluene	Gasoline; manufacturing
6	Chromium	Metal plating
7	Methylene chloride	Degreasing; solvents; paint removal
8	Zinc	Manufacturing; mining
9	1,1,1-Trichloroethane	Metal and plastic cleaning
10	Arsenic	Mining; manufacturing
11	Chloroform	Solvents
12	1,1-Dichloroethane	Degreasing; solvents
13	1,2-Dichloroethene, trans-	Transformation product of 1,1,1-trichloroethane
14	Cadmium	Mining; plating
15	Manganese	Manufacturing; mining; occurs in nature as oxide
16	Copper	Manufacturing; mining
17	1,1-Dichloroethene	Manufacturing
18	Vinyl chloride	Plastic and record manufacturing
19	Barium	Manufacturing; energy production
20	1,2-Dichloroethane	Metal degreasing; paint removal
21	Ethylbenzene	Styrene and asphalt manufacturing; gasoline
22	Nickel	Manufacturing; mining
23	Di(2-ethylhexyl)phthalate	Plastics manufacturing
24	Xylenes	Solvents; gasoline
25	Phenol	Wood treating; medicines

Source: National Research Council (1994).

buffer between human activity and the groundwater sources" (Goldshmid, 1984). Figure 11.6 (from Last, et al., 1994) shows the movement of carbon tetrachloride from a source through the vadose zone to groundwater. The movement of substances through the vadose zone is influenced by the properties of the substance (quantity, volatility, and density), the properties of the geologic material (physical heterogeneity and hydraulic properties), reactions between the substance and the subsurface environment (sorption, ion exchange, geochemical and microbial transformation), and the hydrologic setting (quantity and time distribution of precipitation, evaporation and transpiration). Thus, not all contamination sources actually result in groundwater contamination. EPA estimates that groundwater contamination has occurred at about 80 percent of the nation's Superfund sites (USEPA, 1993).

11.2.4 Sources of Contamination

The Office of Technology Assessment (1984) cataloges sources of groundwater contamination into six general categories (Table 11.3).

Category II sources (those designed to store, treat, and/or dispose of substances) are common sources of groundwater contamination, receiving national attention

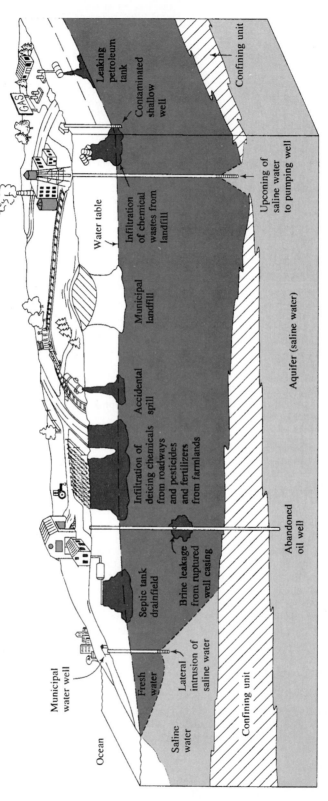

FIGURE 11.5 Mechanisms of groundwater contamination (*Fetter, 1993*).

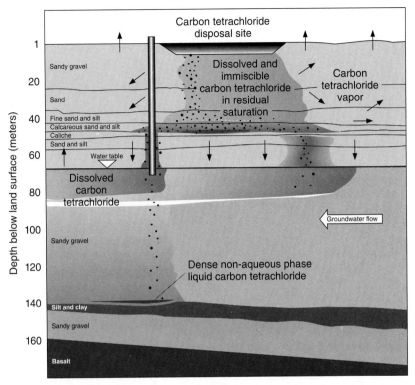

FIGURE 11.6 Conceptual model of carbon tetrachloride contamination in the vadose zone and groundwater (*Last, et. al., 1994*).

since the discovery of problems at Love Canal. These are point sources of contamination. The most prevalent sources of groundwater contamination are from Category IV sources, primarily agricultural practices. These are nonpoint sources of contamination.

11.2.4.1 Point Sources. In 1980, Congress passed the Comprehensive Environmental Response, Compensation, and Liability Act (CERCLA). CERCLA, commonly referred to as the Superfund Act, made groundwater cleanup a national priority. A few years later Congress amended the Resource Conservation and Recovery Act (RCRA) of 1974, further expanding the nation's groundwater restoration program. CERCLA legislation established a multibillion-dollar fund to pay for restoration of abandoned hazardous waste sites and provided authorization for EPA to sue parties responsible for the contamination in order to recover costs. RCRA focuses on cleanup of contamination at active Category II sources. EPA estimates that there are 2000 CERCLA sites and 1500 to 3500 RCRA sites where groundwater contamination may have occurred (USEPA, 1993). The Office of Technology Assessment estimates that there are from 300,000 to 400,000 leaking underground storage tanks (1989). Several states have enacted laws similar to RCRA and CERCLA. There are an estimated 20,000 to 40,000 hazardous waste sites that fall under this legislation (USEPA, 1993; Office of Technology Assessment, 1989).

TABLE 11.3 Source Categories for Groundwater Contamination with Examples

Source category	Examples
I. Sources designed to discharge substances	Septic tanks, cesspools Injection wells
II. Sources designed to store, treat, and/or dispose of substances	Landfills, dumps, abandoned hazardous waste sites Above-ground and underground storage tanks Radioactive waste disposal sites
III. Sources designed to retain substances during transport	Pipelines Material transport and transfer
IV. Sources discharging substances as a consequence of other planned activities	Irrigation Pesticide/fertilizer applications Farm animal wastes
V. Sources providing a conduit for contaminated water to enter aquifers	Production wells Monitoring wells
VI. Naturally occurring sources whose discharge is created and/or exacerbated by human activity	Groundwater/surface water interactions Saltwater intrusion

Source: Modified from Office of Technology Assessment (1984).

The Department of Energy (DOE) and the Department of Defense (DOD) subsequently established environmental programs to bring their facilities into compliance with RCRA and CERCLA requirements. The DOE Environmental Restoration and Waste Management Program (now referred to as Environmental Management, or EM) has identified over 4000 hazardous and radioactive waste sites across the complex. DOD's Installation Restoration Program has identified over 7000 hazardous waste sites. The largest point source groundwater contamination problems, and most costly from a remediation standpoint, come from the DOE and DOD installations. Examples follow to illustrate the nature of these problems.

Department of Energy. The Department of Energy and its predecessor agencies produced materials and parts for nuclear warheads. Wastes were generated at every step in the process, from research, to production, to testing. Across the DOE complex there are 0.85 million cubic meters of stored waste (including high-level waste, mixed hazardous, and radioactive low-level waste) and 3.1 million cubic meters of buried waste (USDOE, 1995). In addition to these stored wastes, operations resulted in significant contamination of soil and groundwater by hazardous and radioactive wastes.

The Hanford site in Richland, Washington (Fig. 11.7; Freshley and Graham, 1988) is considered one of DOE's largest groundwater contamination problems. Tritium and nitrates from the reprocessing of fuels have contaminated tens of square miles of the aquifer underlying the site (Freshley and Thorne, 1992). The spread of the tritium plume with time is shown in Fig. 11.8 (USDOE, 1995). Since tritium and nitrate are conservative tracers (moving with the groundwater), the spread of these plumes provided a means to calibrate model estimates of travel times (Freshley and Thorne, 1992). Other chemicals and radionuclides have also been disposed at the Hanford site and are found in the groundwater (Dirkes, et al., 1994). These substances are less prevalent in the groundwater due to limited disposal, attenuation, and/or degrada-

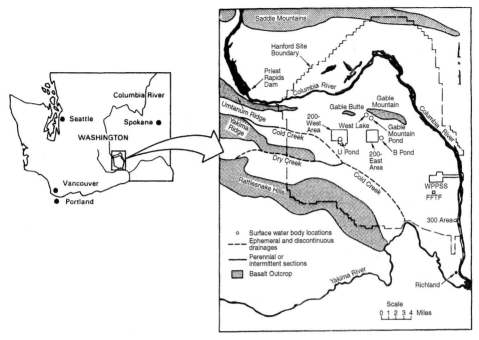

FIGURE 11.7   Location of the Hanford site in Richland, Washington (*Freshley and Graham, 1988*).

tion in the vadose zone and the aquifer. Figure 11.9 shows the distribution of carbon tetrachloride in groundwater under one area of the site. Carbon tetrachloride is immiscible in water but exhibits a relatively high solubility and high mobility in groundwater. Mobilization can also occur through vapor transport, providing a continued source of groundwater contamination from free-phase carbon tetrachloride resident at depth in the vadose zone. Thus even after the surface contamination is removed, expansion of the plume in groundwater can occur. Technologies to remediate this problem have been tested and demonstrated as part of the DOE's remediation program. The current estimate to remediate contamination at all DOE sites is $230 billion, projected to take 75 years.

Department of Defense.   The Department of Defense has diverse and widespread groundwater contamination problems. Most of DOD's groundwater problems are from chemicals, such as degreasing agents, or fuel and fuel oil from transportation activities. The Eielson Air Force Base in Alaska provides a good example of the type of groundwater problems found at DOD installations (and many industrial complexes as well). Eielson AFB is located about 40 km southeast of Fairbanks, Alaska. The base was constructed during World War II, with major expansions in 1954 to provide tactical air support forces. In 1989, Eielson AFB was listed by the EPA on the National Priorities List with six operable units containing a total of 64 potential source areas of contamination. An evaluation of Operable Unit 1 (completed in 1994) included investigations of eight source areas contaminated with petroleum, oil, and lubricants (Gilmore, et al., 1994). The investigation identified primary contaminants of concern that included fuels and fuel-related contaminants (diesel; benzene, toluene, ethylbenzene, and xylene; total petroleum

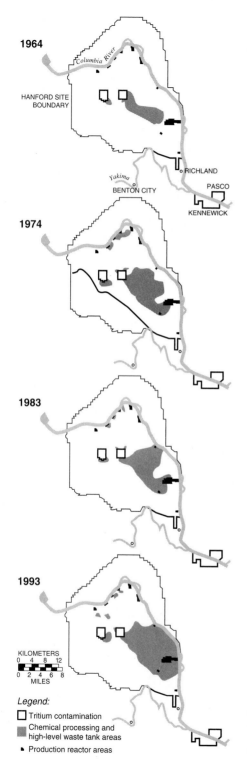

FIGURE 11.8 Spreading tritium contamination at the Hanford Site in Washington. The shaded areas on these maps show how tritium contamination in concentrations above safe drinking-water standards has spread over time (*DOE, 1995*).

11.18

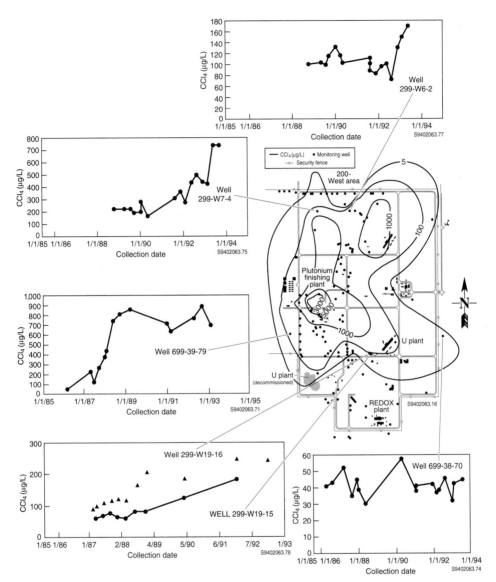

FIGURE 11.9 Distribution of carbon tetrachloride in groundwater under the 200-West area of the Hanford site (*Dirkes, et al., 1994*).

hydrocarbons; polycyclic aromatic hydrocarbons), maintenance-related solvents and cleaners (volatile chlorinated hydrocarbons such as trichloroethylene), polychlorinated biphenyls, and dichlorodiphenyltrichloroethane (DDT). The origins of these contaminants included leaks from storage tanks, drums and piping, and spills. Ongoing operations and past sitewide practices also contributed to the contamination. These practices included spraying mixed oil and solvent wastes on unpaved roads and aerial spraying of DDT. The DOD is now evaluating which areas pose the highest risk and the most effective methods to remediate the contamination found to be above acceptable limits. Similar investigations are ongoing at most of the DOE installations.

11.2.4.2 Nonpoint Sources.   Nonpoint, or Category IV, sources of groundwater contamination contribute to more widespread, or regional, contamination of groundwater than point sources, although the levels of contamination are typically less acute. Alley (1993) provides a comprehensive overview of regional water quality issues. The Clean Water Act, also known as the Federal Water Pollution Control Act as amended by Public Law 92-500 in 1972, addresses many non-point-source pollution issues. Under this legislation, each state is directed to establish objectives for controlling non-point-source pollution. Congress is in the process of developing amendments to reauthorize the Clean Water Act.

Nitrates from fertilizers and animal wastes are the most pervasive types of groundwater contamination. Pesticides also contribute significantly to groundwater contamination. These two examples of non-point-source contamination are examined in more detail in the following sections. Other nonpoint sources of contamination include sediment runoff from agricultural fields; acid deposition from the atmosphere; percolation of atmospheric pollutants; detergents, oils, and metals carried in urban runoff; deicing salt applications; and mining and mine drainage.

Nitrates.   Agricultural use of nitrogen reached 12 million tons by 1985, a fourfold increase from 1960 (USDA, 1987). Each year, farm animals in the United States produce millions of tons of manure that contains an estimated 6.5 million tons of nitrogen. Agricultural nonpoint sources of nitrate pollution account for over 90 percent of the total nitrogen added to the environment. Atmospheric deposition of nitrate, small relative to agricultural inputs, may be another significant nonpoint source in those western watersheds devoted to forestry or in remote headwater areas (Puckett, 1994). Nitrate levels in groundwater across major portions of the United States have elevated, commensurate with this increase in use. In the early 1980s, approximately one-third of all wells in southern Delaware exceeded the drinking-water standard for nitrate, due in part to agricultural fertilizer applications (Ritter and Chirnside, 1984). The U.S. Geological Survey, through the National Water Quality Assessment (NAQUA) Program, and numerous states have conducted studies and surveys on nitrate contamination of groundwater. In 1991, a NAQUA study of herbicides and nitrate in groundwater was conducted for the mid-continental United States (Kolpin and Burkart, 1991). Nitrate contamination was found to be nonuniform in the near-surface aquifers in the region. Nitrates equal to or exceeding 3.0 mg/L was found in 29 percent of the samples, and 6 percent had NO_3 concentrations exceeding the drinking-water standard of 10 mg/L (Burkart and Kolpin, 1993). The study found that unconsolidated aquifers were more susceptible to nitrate contamination, due to greater recharge rates and shorter flow paths than the bedrock aquifers sampled.

A study of the processes affecting the distribution of nitrate in the subsurface environment in an agricultural region of eastern Washington sheds more light on the problem (Geyer, et al., 1992). In this area, large masses of agricultural nitrate have been transported to depths up to 16 m in percolating and recharging groundwater. Nitrate concentrations were affected by biological denitrification (or microbial breakdown), plant uptake, biological and chemical incorporation into soil organic matter, and transport in flowing water. Groundwater concentrations were highest at the bottom of slopes, where the vadose zone is relatively thin. However, large masses of nitrate reside in upslope locations deep in the vadose zone, posing a long-term threat to groundwater quality. This study illustrates the importance of the vadose zone in the assessment of threats to groundwater quality.

Pesticides. The first reported instances of groundwater contamination by pesticides occurred in 1979 when dibromochloropropane (DBCP) and aldicarb were detected in California and New York, respectively. Subsequently, there was widespread detection of ethylene dibromide (EDB) in groundwater in California, Florida, Georgia, and Hawaii. In July 1983, 10 drinking-water supply wells in central Oahu were closed due to DBCP and EDB contamination. By 1988, pesticides had been detected in the groundwater of more than 26 states (USEPA, 1988). These findings set forth several studies by the EPA and the Department of Agriculture (see for example USDA, 1987; USEPA, 1992). As a result, the use of some pesticides (such as DBCP and EDB) has been suspended, and application practices have been modified to mitigate contamination of the groundwater. A review of agricultural approaches to reduce contamination of groundwater by pesticides and other agrichemicals was published by the Office of Technology Assessment in 1990 (Office of Technology Assessment, 1990).

The following two examples demonstrate that non-point-source groundwater contamination by pesticides is a complex and long-term problem. Even though DBCP was banned in the late 1970s, it is still detected in many aquifers underlying agricultural regions. DBCP was detected in about one-third of over 3000 wells in the San Joaquin Valley sampled from 1975 to 1988 (Fig. 11.10) (Domagalski and Dubrovsky, 1991). The highest concentrations generally were found in areas of high use. Although in the western part of the valley, concentrations are lower even in areas of high use. This is probably related to the greater presence of fine-grained sediments in the western part of the valley, where transport through the soil would be slower than in coarse-grained sediments, thus allowing more time for degradation of the pesticide in the vadose zone. However, pesticide migration through fine-grained soils has shown to be enhanced by the presence of fractures in clays (Jorgensen and Fredericia, 1992). Mobile bacteria can also facilitate groundwater transport of pesticides (Lindqvist and Enfield, 1992).

A study of atrazine contamination of groundwater via groundwater–surface water interactions demonstrates another mechanism for non-point-source pollution (Duncan, et al., 1991). Drawdown from a well field located near the Platte River in Lincoln, Nebraska induces recharge of the aquifer from the river, particularly during the summer months. Water samples were collected from a set of monitoring wells located at varying distances from the river (Fig. 11.11). Atrazine in the groundwater was attributed to this induced recharge. Concentrations of the herbicide in groundwater correlate with the high concentrations in the river during the summer, when induced recharge is the highest. The concentrations in groundwater decrease with distance from the river due to dilution.

Results from this study are consistent with a larger study of atrazine in the groundwater in the midcontinental United States (Burkart and Kolpin, 1993). The frequency of herbicide detection was related to the proximity of wells to streams in this region. The frequency of detection in samples from wells within 30 m of a stream was more than double that for samples from wells more than 30 m from a stream. This was attributed to hydrologic connections between the aquifers and the streams.

11.2.5 Summary and Conclusions

The National Research Council, in its assessment of the vulnerability of groundwater to contamination (1993), established a first law of groundwater vulnerablility: *All*

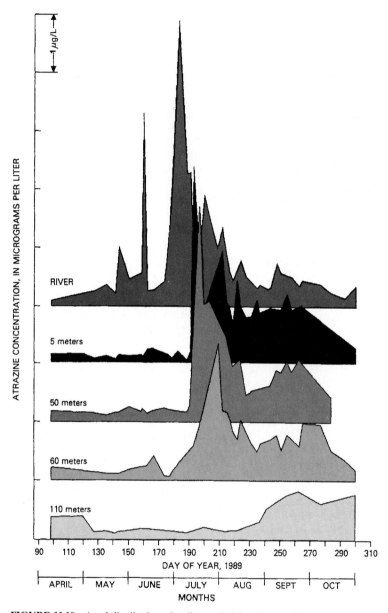

FIGURE 11.10   Areal distribution of wells sampled for dibromochloropropane in the San Joaquin Valley, California, 1975 to 1988 (*after Domagalski and Dubrovsky, 1991*).

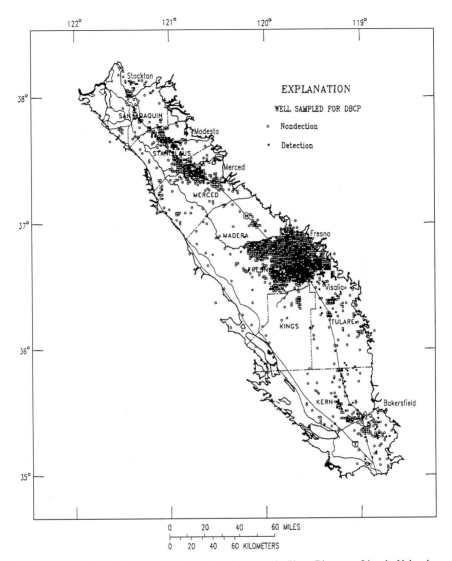

FIGURE 11.12 Atrazine concentation in wells adjacent to the Platte River near Lincoln, Nebraska. Approximate distances of wells from the river are indicated at the left-hand side of each graph (*Duncan, et al., 1991*).

groundwater is vulnerable. That is, an aquifer or portion of an aquifer can only be judged to be more or less vulnerable to contamination than other aquifers or other portions of a given aquifer. We have seen that groundwater can be contaminated through a variety of mechanisms by numerous contaminants. We are just beginning to characterize and clean up some of the large sources of point-source contamination, as in the example of DOE's cleanup program that may continue for 75 or more years. There is widespread uncertainty associated with the task of subsurface remediation; many approaches have proven ineffective (National Academy of Sciences, 1994). Non-point-source pollution is an even more confounding issue. Global expenditures for end-of-pipe pollution control, waste disposal, and remedial clean-up of pollution was estimated at $200 billion in 1990, with growth to $300 billion annually by the year 2000 (OTA, 1994). The potential cost of remediating soil and groundwater contamination in the United States may be as large as $750 billion (National Academy of Sciences, 1994).

Consequently, government and industry are being driven to policies and practices that minimize the production of pollutants that contaminate our environment. The practices of waste minimization, material recycling, and product end-of-life recycling are captured under the rubric of *industrial ecology*. Industrial ecology addresses issues that will impact future production, use, and disposal technologies (Patel, 1992). Proper implementation of the concept should provide for increased protection of groundwater and reduce significantly the resources devoted to remediation of groundwater in the future.

11.3 MICROBIOLOGICAL POLLUTION OF GROUNDWATER

11.3.1 Introduction to Subsurface Microbiology

Prior to scientific exploration over the past decade and a half, it was not known that microbial communities of varying complexity are common and integral components of aquifers and other subsurface environments. Despite occasional outbreaks of disease from contaminated drinking water, little effort was directed at understanding interactive geochemical, physical, and microbiological phenomena and how these natural processes influence groundwater quality. Recent research in deep microbiology has confirmed the existence of a microbial ecosystem throughout the subsurface that relies on sources of organic (such as lignin-like materials) or inorganic (metals and hydrogen) electron donors under largely nutrient-limited conditions (Ghiorse and Wilson, 1988; Stevens and McKinley, 1995). The significance of subsurface microbiology is in understanding and potentially reversing or otherwise controlling groundwater contamination by toxic and hazardous chemicals and pathogens.

Microbial pathogens, including viruses (colloidal dimensions), bacteria (≈ 1 to 10μ) and protozoa (≈ 10 to 100μ) are introduced to groundwater from domestic and municipal sewage systems, sludge amendment of soils, and from feedlots and dairies. *Enteric bacteria* (inhabitants of the gastrointestinal tract of mammals) are the most common pathogens in groundwater and include various species and strains of *Salmonella, Escherichia, Shigella,* and *Vibrio,* among others. They are more common than *obligate pathogens* (viruses and many protozoan parasites) by virtue of an opportunistic nature that allows them to persist under nutrient-sufficient conditions. Viruses do not replicate outside the host, so their persistence in the subsurface is determined by physical mechanisms that affect colloidal behavior (i.e., temperature and sorption/desorption phenomena) and the influence of the geochemical environment on integrity of the structural elements of the viral particle (i.e., pH and osmotic potential). Parasitic protozoa, such as *Giardia* and *Entamoeba,* are less common because of reduced transport potential, as a consequence of their larger size relative to matrix pore sizes (Keswick, 1984). Diseases caused by these microorganisms and known to have originated from contaminated groundwater include gastroenteritis (enterotoxigenic *E. coli*), shigellosis, cholera, typhoid fever, viral hepatitis, and giardiasis (Bitton and Gerba, 1984; Craun, 1984).

11.3.2 Scope of the Problem/Sources of Groundwater Pathogen Contamination

As many as half of all waterborne disease outbreaks in the United States are caused by contaminated groundwater that is either untreated or suffers interruption of the treatment process for a period of time. The remainder of disease outbreaks are associated with drinking water from surface sources. (Bitton and Harvey, 1992; Craun, 1984). About half of contamination events and subsequent disease outbreaks are caused by use of untreated groundwater and half are due to deficiencies in water-treatment systems where existing or potential contamination is recognized (Craun, 1984).

Point sources of microbiological pollution of groundwater include land disposal of sewage, sewage sludge and solid wastes, septic tanks and cesspools, leaking sewer systems, and agricultural practices (i.e., feedlots, dairies, swine, and poultry operations). From the public health perspective, contamination from domestic and municipal septic and sewage systems are the most significant sources. Nonpoint source microbiological contamination of the subsurface, such as urban runoff, is probably much less important.

From a volume standpoint, septic tank/soil adsorption treatment systems are the largest single source of waste disposal to soil and subsurface systems. These systems employ preliminary anaerobic waste treatment in the tank followed by aerobic treatment and filtration by upper soil horizons. Effective treatment that minimizes contamination of groundwater is related to: (1) depth to groundwater, (2) whether unsaturated or saturated flow conditions predominate, (3) the physical nature of the matrix (e.g., particle and pore size distributions and presence of preferred flow paths), and (4) bacterial properties, including survivability and metabolic and surface properties that directly influence adsorption and transport in the subsurface. The design of septic and drainage systems takes these factors into account and, when properly installed, can effectively minimize movement of pathogens downward to groundwater in most cases (Hagedorn, 1984).

Bacterial and viral transport in the subsurface is influenced by adhesion properties of the cells with respect to the porous matrix. Cell and viral particle surfaces usually have a net negative charge and can be bound electrostatically to surfaces through cation bridging and physical forces in the diffuse double layer surrounding soil particles. Many bacteria also produce extracellular slimes or biosurfactants that can play a significant role in retardation or movement. Physical filtering is important for larger microorganisms and explains much of what has been observed and reported for distribution of fecal coliforms and potential pathogens near point sources of pollution, such as septic tank drain fields.

11.3.3 Microbial Ecology of the Subsurface

Understanding the microbial ecology of natural subsurface communities contributes to determining how anthropogenic contamination occurs and might be dealt with. Native microbial communities in the subsurface are to some degree phylogenetically distinct from surface microorganisms, including pathogens. In all likelihood, pathogens found in the subsurface have been transported downward from different sources of pollution, whereas native populations include organisms transported from the surface in addition to those that have evolved in place over geologic time. Research with subsets of bacterial isolates in a Subsurface Microbial Culture Collection, supported by the U.S. Department of Energy at Florida State University and the Oregon Graduate Institute, has identified as native to the subsurface strains related to *Arthrobacter globiformis, Micrococcus luteus, Terrabacter tumescens, Acinetobacter calcoaceticus, Alcaligenes eutrophus* and *Sphingomonas* (Balkwill, 1993). Of the *Sphingomonas* isolates studied to date, at least one-third harbor plasmids (extrachromosomal DNA), many of unusually large size that may reflect a unique mechanism for adaptation to the subsurface and are known in surface pathogens to code for antibiotic resistance (Fredrickson, et al., in press). The discovery of the first obligately anaerobic *Bacillus* (*B. infernus*) also supports the hypothesis that native subsurface communities are unique in the biosphere (Boone, et al., 1995).

The structure and function of native microbial communities reflect physical and geochemical conditions, much as for surface habitats. The occurrence and range of activity is primarily reflective of water and substrate availability, with numbers at or below detection limits only in locations devoid of water. Factors including pH, temperature, pressure, salinity, and oxidation/reduction potential do not uniquely limit microbial populations compared to surface habitats; however, subsurface communities are usually less structurally complex and include smaller numbers of organisms (Ghiorse and Wilson, 1988).

A range of pathogenic microorganisms has been observed in groundwater, including viruses, bacteria, fungi, protozoa, and helminths. However, pathogenic microorganisms are probably not native to subsurface environments and so their presence and importance in public health depends on transport, survival, and retention by pores and particulate properties of the solid matrix. In this regard, there are four major factors controlling these properties: (1) the nature of the porous medium, including physical, geochemical, and biological heterogeneities, (2) climate (primarily rainfall), (3) the hydrologic regime (saturated versus unsaturated conditions), and (4) the nature of the microorganisms (Gerba and Bitton, 1984). Transport to the subsurface is favored in areas of high rainfall, and within the subsurface by fractures (preferred pathways) in the geological matrix, by a large percentage of pore diameters greater than the size of the microorganism, and by saturated flow conditions. Smaller pathogens (viruses and smaller bacteria) move further and faster than larger types. As with the native microbiota, persistence of bacterial pathogens in the subsurface is largely controlled by availability of water and nutrients, with temperature being particularly important to survival of viruses (Bitton and Harvey, 1992).

The study of subsurface microbial ecology is made difficult by the need to take measures to avoid contamination during sampling and by the cost of drilling and sampling the subsurface. Most of the technical hurdles have largely been overcome through the use of aseptic sampling and sample handling procedures and the use of various tracer techniques (Colwell, et al., 1992). Inert chemical, gaseous, and particulate tracers have been used in quality assurance sampling protocols to provide indices of contamination by, respectively, soil gases, groundwater and colloids (including viruses), and cellular microorganisms. More commonly, tracers are used to derive information about groundwater velocities and microbial transport in studies designed to better understand subsurface microbial ecology (Bitton and Harvey, 1992).

Methods of indirect detection of microorganisms in subsurface groundwater and sediment samples include traditional plating methods (i.e., bacterial/fungal plate counts, viral plaque-forming units) and enrichment techniques to isolate specific microbial groups (e.g., enteric and/or coliform bacteria). Assays for specific enzyme activities (e.g., hemolytic activity) have also been used as indirect evidence of the presence of groups of microorganisms. More direct methods for interrogating samples that rely on molecular biological methods to detect the presence, potential activity, and ongoing activities of culturable and nonculturable microorganisms include various DNA or RNA fingerprinting methods which are based on chemical hybridization of genes, gene segments, or RNA gene products with homologous material in the sample. For nucleic acids present in minute amounts, the polymerase chain reaction (PCR) has been used to amplify and detect molecules specific to enteroviruses and bacteria in groundwater (Abbaszadegan, et al., 1993).

APPENDIX 11.A   SUBSTANCES KNOWN TO OCCUR IN GROUNDWATER WITH EXAMPLES OF USES

Contaminant	Examples of uses
Aromatic Hydrocarbon	
Acetanilide	Intermediate manufacturing, pharmaceuticals, dyestuffs
	Detergents
Alkyl benzene sulfonates	Detergents
Aniline	Dyestuffs, intermediate, photographic chemicals, pharmaceuticals, herbicides, fungicides, petroleum refining, explosives
Anthracene	Dyestuffs, intermediate, semiconductor research
Benzene	Detergents, intermediate, solvents, antiknock gasoline
Benzidine	Dyestuffs, reagent, stiffening agent in rubber compounding
Benzyl alcohol	Solvent, perfumes and flavors, photographic developer inks, dyestuffs, intermediate
Butoxymethylbenzene	NA
Chrysene	Organic synthesis, coal tar by-product
Creosote mixture	Wood preservatives, disinfectants
Dibenz[a.h.]anthrancene	NA
Di-butyl-*p*-benzoquinone	NA
Dihydrotrimethylquinoline	Rubber antioxidant
4,4-Dinitrosodiphenylamine	NA
Ethylbenzene	Intermediate, solvent, gasoline
Fluoranthene	Coal tar by-product
Fluorene	Resinous products, dyestuffs, insecticides, coal tar by-product
Fluorescein	Dyestuffs
Isopropyl benzene	Solvent, chemical manufacturing
4,4'-methylene-*bis*-2-chloroaniline (MOCA)	Curing agent for polyurethanes and epoxy resins
Methylthiobenzothiazole	NA
Napthalene	Solvent, lubricant, explosives, preservatives, intermediate, fungicide, moth repellant
o-Nitroaniline	Dyestuffs, intermediate, interior paint pigments, chemical manufacturing

APPENDIX 11.A   SUBSTANCES KNOWN
TO OCCUR IN GROUNDWATER
WITH EXAMPLES OF USES (Continued)

Contaminant	Examples of uses
Nitrobenzene	Solvent, polishes, chemical manufacturing
4-Nitrophenol	Chemical manufacturing
n-Nitrosodiphenylamine	Pesticides, retarder of vulcanization of rubber
Phenanthrene	Dyestuffs, explosives, synthesis of drugs, biochemical research
n-Propylbenzene	Dyestuff, solvent
Pyrene	Biochemical research, coal tar by-product
Styrene (vinyl benzene)	Plastics, resins, protective coatings, intermediate
Toluene	Adhesive solvent in plastics, solvent, aviation and high-octane blending stock, dilutent and thinner, chemicals, explosives, detergents
1,2,4,-Trimethylbenzene	Manufacture of dyestuffs, pharmaceuticals, chemical manufacturing
Xylenes (m,o,p)	Aviation gasoline, protective coatings, solvent, synthesis of organic chemicals, gasoline
Oxygenated hydrocarbons	
Acetic acid	Food additives, plastics, dyestuffs, pharmaceuticals, photographic chemicals, insecticides
Acetone	Dyestuffs, solvent, chemical manufacturing, cleaning and drying of precision equipment
Benzophenone	Organic synthesis, odor fixative, flavoring, pharmaceuticals
Butyl acetate	Solvent
n-Butyl-benzylphthalate	Plastics, intermediate
Di-n-butyl phthalate	Plasticizer, solvent, adhesives, insecticides, safety glass, inks, paper coatings
Diethyl phthalate	Chemical manufacturing, solvent, analytical chemistry, anesthetic, perfumes
Diisopropyl ether	Plastics, explosives, solvent, insecticides, perfumes
2,4-Dimethyl-3-hexanol	Pharmaceuticals, plastics, disinfectants, solvent, dyestuffs, insecticides, fungicides, additives to lubricants and gasolines
Di-n-octyl phthalate	Plasticizer for polyvinyl chloride and other vinyls
1,4-Dioxane	Solvent, lacquers, paints, varnishes, cleaning and detergent preparations, fumigants, paint and varnish removers, wetting agent, cosmetics
Ethyl acrylate	Polymers, acrylic paints, intermediate
Formic acid	Dyeing and finishing, chemicals, manufacture of fumigants, insecticides, solvents, plastics, refrigerants
Methanol (methyl alcohol)	Chemical manufacturing, solvents, automotive antifreeze, fuels

APPENDIX 11.A   SUBSTANCES KNOWN
TO OCCUR IN GROUNDWATER
WITH EXAMPLES OF USES *(Continued)*

Contaminant	Examples of uses
Methylcyclohexanone	Solvent, lacquers
Methyl ethyl ketone	Solvent, paint remover, cements and adhesives, cleaning fluids, printing, acrylic coatings
Methylphenyl acetamide	NA
Phenols (e.g., *p*-tert-butylphenol)	Resins, solvent, pharmaceuticals, reagent, dyestuffs and indicators, germicidal paints
Phthalic acid	Dyestuffs, medicine, perfumes, reagent
2-Propanol	Chemical manufacturing, solvent, deicing agent, pharmaceuticals, perfumes, lacquers, dehydrating agent, preservatives
2-Propyl-1-heptanol	Solvent
Tetrahydrofuran	Solvent
Varsol	Paint and varnish thinner
Hydrocarbons with specific elements (e.g., N, P, S, CI, Br, I, F)	
Acetyl chloride	Dyestuffs, pharmaceuticals, organic preparations
Alachlor (Lasso)	Herbicides
Aldicarb (sulfoxide and sulfone; Temik)	Insecticide, nematocide
Aldrin	Insecticides
Atrazine	Herbicides, plant growth regulator, weed-control agent
Benzoyl chloride	Medicine, intermediate
Bromacil	Herbicides
Bromobenzene	Solvent, motor oils, organic synthesis
Bromochloromethane	Fire extinguishers, organic synthesis
Bromodichloromethane	Solvent, fire extinguisher fluid, mineral and salt separations
Bromoform	Solvent, intermediate
Carbofuran	Insecticide, nematocide
Carbon tetrachloride	Degreasers, refrigerants and propellants, fumigants, chemical manufacturing
Chlordane	Insecticides, oil emulsions
Chlorobenzene	Solvent, pesticides, chemical manufacturing
Chloroform	Plastics, fumigants, insecticides, refrigerants and propellants
Chlorohexane	NA
Chloromethane (methyl chloride)	Refrigerants, medicine, propellants, herbicide, organic synthesis
Chloromethyl sulfide	NA
2-Chloronaphthalene	Oil: plasticizer, solvent for dyestuffs, varnish gums and resins, waxes; wax: moisture-, flame-, acid-, and insect-proofing and fibrous materials; moisture- and flame-proofing of electrical cable; solvent (see oil)
Chlorpyrifos	NA
Chlorthal-methyl (DCPA, or Dacthal)	Herbicide

APPENDIX 11.A   SUBSTANCES KNOWN
TO OCCUR IN GROUNDWATER
WITH EXAMPLES OF USES (Continued)

Contaminant	Examples of uses
p-Chlorophenyl methylsulfone	Herbicide manufacture
Chlorophenylmethyl sulfide	Herbicide manufacture
Chlorophenylmethyl sulfoxide	Herbicide manufacture
o-Chlorotoluene	Solvent, intermediate
p-Chlorotoluene	Solvent, intermediate
Cyclopentadine	Insecticide manufacture
Dibromochloromethane	Organic synthesis
Dibromochloropropane (DBCP)	Fumigants, nematocide
Dibromodichloroethylene	NA
Dibromoethane (ethylene dibromide, EDB)	Fumigant, nematocide, solvent, waterproofing preparations, organic synthesis
Dibromomethane	Organic synthesis, solvent
Dichlofenthion (DCFT)	Pesticides
o-Dichlorobenzene	Solvent, fumigants, dyestuffs, insecticides, degreasers, polishes, industrial odor control
p-Dichlorobenzene	Insecticides, moth repellant, germicide, space odorant, intermediate, fumigants
Dichlorobenzidine	Intermediate, curing agent for resins
Dichlorocyclooctadiene	Pesticides
Dichlorodiphenyldichloroethane (DDD, TDE)	Insecticides
Dichlorodiphenyldichloroethylene (DDE)	Degradation product of DDT, found as an impurity in DDT residues
Dichlorodiphenyltrichloroethane (DDT)	Pesticides
1,1-Dichloroethane	Solvent, fumigants, medicine
1,2-Dichloroethane	Solvent, degreasers, soaps and scouring compounds, organic synthesis, additive in antiknock gasoline, paint and finish removers
1,1-Dichloroethylene (vinylidiene Chloride)	Saran (used in screens, upholstery, fabrics, carpets, etc.), adhesives, synthetic fibers
1,2-Dichloroethylene (*cis* and *trans*)	Solvent, perfumes, lacquers, thermoplastics, dye extraction, organic synthesis, medicine
Dichloroethyl ether	Solvent, organic synthesis, paints, vanishes, lacquers, finish removers, drycleaning, fumigants
Dichloroiodomethane	NA
Dichloroisopropylether (=*bis*-2-chloroisopropylether)	Solvent, paint and varnish removers, cleaning solutions
Dichloromethane (methylene chloride)	Solvent, plastics, paint removers, propellants, blowing agent in foams
Dichloropentadiene	NA
2,4-Dichlorophenol	Organic synthesis
2,4-Dichlorophenoxyacetic acid (2,4-D)	Herbicides
1,2-Dichloropropane	Solvent, intermediate, scouring compounds, fumigants, nematocide, additive for antiknock fluids

APPENDIX 11.A SUBSTANCES KNOWN
TO OCCUR IN GROUNDWATER
WITH EXAMPLES OF USES (Continued)

Contaminant	Examples of uses
Dicyclopentadiene (DCPD)	Insecticide manufacture
Dieldrin	Insecticides
Diiodomethane	Organic synthesis
Diisopropylmethyl phosphonate (DIMP)	Nerve gas manufacture
Dimethyl disulfide	NA
Dimethylformamide	Solvent, organic synthesis
2,4-Dinotrophenol (Dinoseb, DNBP)	Herbicides
Dithiane	Mustard gas manufacture
Dioxins (e.g., TCDD)	Impurity in the herbicide 2,4,5-T
Dodecyl mercaptan (lauryl mercaptan)	Manufacture of synthetic rubber and plastics, pharmaceuticals, insecticides, fungicides
Endosulfan	Insecticides
Endrin	Insecticides
Ethyl chloride	Chemical manufacturing, anesthetic, solvent, refrigerants, insecticides
Bis-2-ethylexylphthalate	Plastics
Di-2-ethylexylphthalate	Plasticizers
Fluorobenzene	Insecticide and larvicide intermediate
Fluoroform	Refrigerants, intermediate, blowing agent for foams
Heptachlor	Insecticides
Heptachlorepoxide	Degradation product of heptachlor, also acts as an insecticide
Hexachlorobicycloheptadiene	NA
Hexachlorobutadiene	Solvent, transformer and hydraulic fluid, heat-transfer liquid
α-Hexachlorocyclohexane (=Benzenehexachloride, or α-BHC)	Insecticides
β-Hexachlorocyclohexane (β-BHC)	Insecticides
γ-Hexachlorocyclohexane (γ-BHC, or Lindane)	Insecticides
Hexachlorocyclopentadiene	Intermediate for resins, dyestuffs, pesticides, fungicides, pharmaceuticals
Hexachloroethane	Solvent, pyrothechnics and smoke devices, explosives, organic synthesis
Hexachloronorbornadiene	NA
Isodrin	Intermediate compound in manufacture of Endrin
Kepone	Pesticides
Malathion	Insecticides
Methoxychlor	Insecticides
Methyl bromide	Fumigants, pesticides, organic synthesis
Methyl parathion	Insecticides
Oxathine	Mustard gas manufacture
Parathion	Insecticides
Pentachlorophenol (PCP)	Insecticides, fungicides, bactericides, algicides, herbicides, wood preservative
Phorate (Disulfoton)	Insecticides

APPENDIX 11.A   SUBSTANCES KNOWN
TO OCCUR IN GROUNDWATER
WITH EXAMPLES OF USES (Continued)

Contaminant	Examples of uses
Polybrominated biphenyls (PBBs)	Flame retardant for plastics, paper, and textiles
Polychlorinated biphenyls (PCBs)	Heat-exchange and insulating fluids in closes system
Prometon	Herbicides
RDX (Cyclonite)	Explosives
Simazine	Herbicides
Tetrachlorobenzene	NA
Tetrachloroethanes (1,1,1,2 and 1,1,2,2)	Degreasers, paint removers, varnishes, lacquers, photographic film, organic synthesis, solvent, insecticides, fumigants, weed killer
Tetrachloroethylene (or perchloroethylene, PCE)	Degreasers, drycleaning, solvent, drying agent, chemical manufacturing, heat-transfer medium, vermifuge
Toxaphene	Insecticides
Triazine	Herbicides
1,2,4-Trichlorobenzene	Solvent, dyestuffs, insecticides, lubricants, heat-transfer medium (e.g., coolant)
Trichloroethanes (1,1,1 and 1,1,2)	Pesticides, degreasers, solvent
1,1,2,-Trichloroethylene (TCE)	Degreasers, paints, drycleaning, dyestuffs, textiles, solvent, refrigerant and heat exchange liquid, fumigant, intermediate, aerospace operations
Trichlorofluoromethane (Freon 11)	Solvent, refrigerants, fire extinguishers, intermediate
2,4,6-Trichlorophenol	Fungicides, herbicides, defoliant
2,4,5-Tricholophenoxyacetic acid (2,4,5,-T)	Herbicides, defoliant
2,4,5-Trichlorophenoxypropionic acid (2,4,5-TP or Silvex)	Herbicides and plant growth regulator
Trichlorotrifuoroethane	Dry-cleaning, fire extinguishers, refrigerants, intermediate, drying agent
Trinitrotoluene (TNT)	Explosives, intermediate in dye stuffs and photographic chemicals
Tris-(2,3-dibromopropyl) phosphate	Flame retardant
Vinyl chloride	Organic synthesis, polyvinyl chloride and copolymers, adhesives
Other hydrocarbons	
Alkyl sulfonates	Detergents
Cyclohexane	Organic synthesis, solvent, oil extraction
1,3,5,7-Cyclooctatetraene	Organic research
Dicyclopentadiene (DCPD)	Intermediate for insecticides, paints and varnishes, flame retardants
2,3-Dimethylhexane	NA
Fuel oil	Fuel, heating
Gasoline	Fuel
Jet fuels	Fuel
Kerosene	Fuel, heating solvent, insecticides

APPENDIX 11.A SUBSTANCES KNOWN TO OCCUR IN GROUNDWATER WITH EXAMPLES OF USES *(Continued)*

Contaminant	Examples of uses
Lignin	Newsprint, ceramic binder, dyestuffs, drilling fuel additive, plastics
Methylene blue activated substances (MBAS)	Dyestuffs, analytical chemistry
Propane	Fuel, solvent, refrigerants, propellants, organic synthesis
Tannin	Chemical manufacturing, tanning, textiles, electroplating, inks, pharmaceuticals, photography, paper
4,6,8-Trimethyl-nonene	NA
Undecane	Petroleum research, organic synthesis
Metals and cations	
Aluminum	Alloys, foundry, paints, protective coatings, electrical industry, packaging, building and construction, machinery and equipment
Antimony	Hardening alloys, solders, sheet and pipe, pyrotechnics
Arsenic	Alloys, dyestuffs, medicine, solders, electronic devices, insecticides, rodenticides, herbicide, preservative
Barium	Alloys, lubricant
Beryllium	Structural material in space technology, inertial guidance systems, additive to rocket fuels, moderator and reflector of neutrons in nuclear reactors
Cadmium	Alloys, coatings, batteries, electrical equipment, fire-protection systems, paints, fungicides, photography
Calcium	Alloys, fertilizers, reducing agent
Chromium	Alloys, protective coatings, paints, nuclear and high-temperature research
Cobalt	Alloys, ceramics, drugs, paints, glass, printing, catalyst, electroplating, lamp filaments
Copper	Alloys, paints, electrical wiring, machinery, construction materials, electroplating, piping, insecticides
Iron	Alloys, machinery, magnets
Lead	Alloys, batteries, gasoline additive, sheet and pipe, paints, radiation shielding
Lithium	Alloys, pharmaceuticals, coolant, batteries, solders, propellants
Magnesium	Alloys, batteries, pyrotechnics, precision instruments, optical mirrors
Manganese	Alloys, purifying agent
Mercury	Alloys, electrical apparatus, instruments, fungicides, bactericides, mildew proofing, paper, pharmaceuticals
Molybdenum	Alloys, pigments, lubricant

APPENDIX 11.A SUBSTANCES KNOWN TO OCCUR IN GROUNDWATER WITH EXAMPLES OF USES *(Continued)*

Contaminant	Examples of uses
Nickel	Alloys, ceramics, batteries, electroplating, catalyst
Palladium	Alloys, catalyst, jewelry, protective coatings, electrical equipment
Potassium	Alloys, catalyst
Selenium	Alloys, electronics, ceramics, catalyst
Silver	Alloys, photography, chemical manufacturing, mirrors, electronic equipment, jewelry, equipment, catalyst, pharmaceuticals
Sodium	Chemical manufacturing, catalyst, coolant, nonglare lighting for highways, laboratory reagent
Thallium	Alloys, glass, pesticides, photoelectric applications
Titanium	Alloys, structural materials, abrasives, coatings
Vanadium	Alloys, catalysts, target material for x rays
Zinc	Alloys, electroplating, electronics, automotive parts, fungicides, roofing, cable wrappings, nutrition
Nonmetals and anions	
Ammonia	Fertilizers, chemical manufacturing, refrigerants, synthetic fibers, fuels, dyestuffs
Boron	Alloys, fibers and filaments, semiconductors, propellants
Chlorides	Chemical manufacturing, water purification, shrink-proofing, flame retardants, food processing
Cyanides	Polymer production (heavy duty tires), coatings, metallurgy, pesticides
Fluorides	Toothpastes and other dentrifices, additive to drinking water
Nitrates	Fertilizers, food preservatives
Nitrites	Fertilizers, food preservatives
Phosphates	Detergents, fertilizers, food additives
Sulfates	Fertilizers, pesticides
Sulfites	Pulp production and processing, food preservatives
Microorganisms	
Bacteria (coliform)	
Giardia	
Viruses	
Radionuclides	
Cesium-137	Gamma radiation source for certain foods
Chromium-51	Diagnosis of blood volume, blood cell life, and cardiac output

APPENDIX 11.A SUBSTANCES KNOWN TO OCCUR IN GROUNDWATER WITH EXAMPLES OF USES (Continued)

Contaminant	Examples of uses
Cobalt-60	Medical diagnosis, therapy, leak
Iodine-131	detection, tracers (e.g., to study efficiency of mixing pulp fibers, chemical reactions, and thermal stability of additives to food products), measuring film thicknesses
Iron-59	Medicine, tracer
Lead-210	NA
Phosphorus-32	Tracer, medical treatment, industrial measurements (e.g., tire-tread wear and thickness of films and ink)
Plutonium-238, -243	Energy source, weaponry
Radium-226	Medical treatment, radiography
Radium-228	Naturally occurring
Radon-222	Medicine, leak detection, radiography, flow rate measurement
Ruthenium-106	Catalyst
Scandium-46	Tracer studies, leak detection, semiconductors
Strontium-90	Medicine, industrial applications (e.g., measuring thicknesses, density control)
Thorium-232	Naturally occurring
Tritium	Tracer, luminous instrument dials
Uranium-238	Nuclear reactors
Zinc-65	Industrial tracers (e.g., to study wear in alloys, galvanizing, body metabolism, function of oil additives in lubricating oils)
Zirconium-95	NA

Source: Modified from Office of Technology Assessment, 1984.

REFERENCES

Abbaszazdegan, M., M. S. Huber, C. P. Gerba, and I. L. Pepper, "Detection of Enteroviruses in Groundwater with the Polymerase Chain Reaction," *Applied and Environmental Microbiology*, 59:1318–1324, 1993.

Alley, W. M., ed., *Regional Ground-Water Quality*, Van Nostrand Reinhold, New York, 1993.

Baedecker, M. J., and W. Back, "Hydrogeological Processes and Chemical Reactions at a Landfill," *Ground Water*, 17:429–437, 1979.

Balkwill, D. L., "DOE Makes Subsurface Cultures Available," *ASM News*, 59:504–506, 1993.

Bitton, G., and C. P. Gerba, "Groundwater Pollution Microbiology: The Emerging Issue," in G. Bitton and C. P. Gerba, eds., *Groundwater Pollution Microbiology*, John Wiley, New York, 1984, pp. 1–7.

Bitton, G., and R. W. Harvey, "Transport of Pathogens Through Soils and Aquifers," in R. Mitchell, ed., *Environmental Microbiology*, Wiley-Liss, New York, 1992, pp. 103–124.

Boone, D. R., Y. Liu, Z. Zhao, et al., "*Bacillus infernus* sp. nov., an Fe(III)- and Mn(IV)-Reducing Anaerobe from the Deep Terrestrial Subsurface," *International Journal of Systematic Bacteriology*, 45:441–448, 1995.

Borole, D. V., S. K. Gupta, S. Krishnaswami, et al., "Uranium Isotopic Investigations and Radio-carbon Measurements of River-Groundwater Systems, Sabaramati Basin, Gujarat, India," in *Isotope Hydrology 1978, IAEA,* Vienna, 1:181–201, 1979.

Burkart, M. R., and D. W. Kolpin, "Hydrologic and Land-Use Factors Associated with Herbicides and Nitrate in Near-Surface Aquifers," *Journal of Environmental Quality,* 22:646–656, 1993.

Chapelle, F. H., *Ground-Water Microbiology and Geochemistry,* John Wiley, New York, 1993.

Chapelle, F. H., and D. R. Lovely, "Competitive Exclusion of Sulfate Reduction by Fe(III)-Reducing Bacteria: A Mechanism for Producing Discrete Zones of High-Iron Ground Water," *Groundwater,* 30(1):29–36, 1992.

Collwell, F. S., G. J. Stormberg, T. J. Phelps, et al., "Innovative Techniques for Collection of Saturated and Unsaturated Subsurface Basalts and Sediments for Microbiological Characterization," *Journal of Microbiological Methods,* 15:279–292, 1992.

Coplen, T. B., "Uses of Environmental Isotopes," in W. M. Alley, ed., *Regional Ground-Water Quality,* Van Nostrand Reinhold, New York, 1993.

Craun, G. F., "Health Aspects of Groundwater Pollution," in G. Bitton and C. P. Gerba, eds., *Groundwater Pollution Microbiology,* John Wiley, New York, 1984, pp. 135–179.

Dansgaard, W., "Stable Isotopes in Precipitation," *Tellus,* 16:436–469, 1964.

Deverel, S. J., and J. L. Fio, "Groundwater Flow and Solute Movement to Drain Laterals, Western San Joaquin Valley, California 1. Geochemical Assessment," *Wat. Resour. Res.,* 27(9): 2233–2246, 1991.

Dirkes, R. L., R. W. Hanf, and R. K. Woodruff, *Hanford Site Environmental Report for Calendar Year 1993,* PNL-9823, Battelle Pacific Northwest Laboratory, Richland, Washington, 1994.

Domagalski, J. L., and N. M. Dubrovsky, *Regional Assessment of Nonpoint-Source Pesticide Residues in Ground Water, San Joaquin Valley, California,* U.S. Geological Survey Water-Resources Investigations Report 91-4027, Sacramento, Cal., 1991.

Drever, J. I., *The Geochemistry of Natural Waters,* 2d ed., Prentice-Hall, Englewood Cliffs, N.J., 1988.

Duncan, D., D. T. Pederson, T. R. Sheperd, and J. D. Carr, "Atrazine Used as a Tracer of Induced Recharge," *Ground Water Monitoring Review* 11(4):144–150, 1991.

Fetter, C. W., *Contaminant Hydrogeology,* Macmillan, New York, 1993.

Fredrickson, J. K., D. L. Balkwill, G. R. Drake, et al., "Aromatic-Degrading *Sphingomonas* Isolates from the Deep Subsurface," *Applied and Environmental Microbiology,* 61:1917–1922.

Freshley, M. D., and M. J. Graham, *Estimation of Ground-Water Travel Time at the Hanford Site: Description, Past Work, and Future Need,* PNL-6328, Pacific Northwest Laboratory, Richland, Washington, 1988.

Freshley, M. D., and P. D. Thorne, *Ground-Water Contribution to DOE From Past Hanford Operations,* PNWD-1974 HEDR, Battelle Pacific Northwest Laboratories, Richland, Washington, 1992.

Fritz, P., and J. Ch. Fontes, *Handbook of Environmental Isotope Geochemistry, Vol. 1, The Terrestrial Environment, A,* Elsevier, Oxford, U.K., 1980.

Garrels, R. M., and F. T. Mackenzie, "Origin of the Chemical Compositions of Some Springs and Lakes," *Advances in Chemistry Series,* no. 67, Amer. Chem. Soc., 1967, Chap. 10, pp. 222–242.

Gerba, C. P., and G. Bitton, "Microbial pollutants: Their Survival and Transport Pattern to Groundwater," in G. Bitton and C. P. Gerba, eds., *Groundwater Pollution Microbiology,* John Wiley, New York, 1984, pp. 65–88.

Geyer, D. J., C. K. Keller, J. L. Smith, and D. L. Johnstone, "Subsurface Fate of Nitrate as a Function of Depth and Landscape Position in Missouri Flat Creek Watershed, U.S.A.," *Journal of Contaminant Hydrology,* 11:127–147, 1992.

Ghiorse, W. C., and J. T. Wilson, "Microbial Ecology of the Terrestrial Subsurface," *Advances in Applied Microbiology,* 33:107–173, 1988.

Gilmore, T. J., R. M. Fruland, T. L. Liikala, et al., *Final Remedial Investigation Report Operable Unit 1 Eielson Airforce Base, Alaska,* PNL-9817, Pacific Northwest Laboratory, Richland, Wash., 1994.

Glynn, P. D., "Effect of Impurities in Gypsum on Contaminant Transport at Pinal Creek, Arizona," in G. E. Mallard and D. A. Aronson, eds., *U.S. Geological Survey Toxic Substances Hydrology Program—Proceedings of the Technical Meeting,* Monterey, California, March 11–15, 1991, Water-Resources Investigations Report 91-4034, 1991, pp. 466–474.

Goldshmid, J., "Introductory Comments," in B. Yaron, G. Dagan, and J. Goldschmid, eds., *Pollutants in Porous Media,* Springer-Verlag, Berlin, 1984, pp. 208–211.

Hagedorn, C., "Microbiological Aspects of Groundwater Pollution Due to Septic Tanks," in G. Bitton and C. P. Gerba, eds., *Groundwater Pollution Microbiology,* John Wiley, New York, 1984, pp. 181–193.

Hanshaw, B. B., W. Back, and M. Rubin, "Radiocarbon Determinations for Estimating Groundwater Flow Velocities in Central Florida," *Science,* 148:494–495, 1965.

Hem, J. D., *Study and Interpretation of the Chemical Characteristics of Natural Water,* U.S. Geological Survey Water Supply Paper 2254, 1985.

Jelinski, L. W., R. E. Graedel, R. A. Laudise, et al., "Industrial Ecology: Concepts and Approaches," *Proceedings of the National Academy of Sciences,* Washington, D.C., 89:793–797, 1992.

Jorgensen, P. R., and J. Fredericia, "Migration of Nutrients, Pesticides, and Heavy Metals in Fractured Clayey Till," *Geotechnique* 42(1):66–77, 1992.

Keswick, B. H., "Sources of Groundwater Pollution," in G. Bitton and C. P. Gerba, eds., *Groundwater Pollution Microbiology,* John Wiley, New York, 1984, pp. 40–64.

Kolpin, D. W., and M. R. Burkart, *Work Plan for Regional Reconnaissance for Selected Herbicides and Nitrate in Ground Water of the Mid-Continental United States, 1991,* U.S. Geological Survey Open-File Report 91-59, 1991.

Langmuir, D., "Uranium Solution-Mineral Equilibria at Low Temperatures with Applications to Sedimentary Ore Deposits," *Geochim. Cosmochim. Acta,* 42:547–569, 1978.

Last, G. V., R. E. Lewis, D. E. Beaver, et al., "Computer Enhanced Conceptual Model of Carbon Tetrachloride Contamination at the Hanford Site," *Geological Society of America with Abstracts,* vol. 26, no. 7, Annual Meeting, 1994.

Lindqvist, R., and C. G. Enfield, "Biosorption of Dichlorodiphenyltrichoroethane and Hexachlorobenzene in Groundwater and Its Implications for Facilitated Transport," *Applied and Environmental Microbiology* 58(7):2211–2218, 1992.

Mazor, E., *Applied Chemical and Isotopic Groundwater Hydrology,* Halsted Press, New York, 1991.

Mazor, E., A. Kaufman, and I. Carmi, "Hammet Gader (Israel): Geochemistry of a Mixed Thermal Spring Complex," *J. Hydrol.,* 18:289–303, 1973.

Mazor, E., F. D. Vautaz, and F. C. Jaffe, "Tracing Groundwater Components by Chemical, Isotopic and Physical Parameters, Example: Schinziach, Switzerland," *J. of Hydrol.,* 76:233–246, 1985.

Morel, F. M. M., and J. G. Hering, *Principles and Applications of Aquatic Chemistry,* John Wiley, New York, 1993.

Murarka, I. P., and D. A. McIntosh, *Solid Waste Environmental studies (SWES): Description, Status, and Available Results,* Electric Power Research Institute, EPRI EA-5322-SR, Palo Alto, Cal., 1987.

National Academy of Sciences, *Ground Water Vulnerability Assessment: Predicting Relative Contamination Potential Under Conditions of Uncertainty,* National Academy Press, Washington, D.C., 1993.

National Academy of Sciences, *Alternatives for Ground Water Cleanup,* National Academy Press, Washington, D.C., 1994.

National Research Council, *Ground Water Vulnerability Assessment, Contamination Potential Under Conditions of Uncertainty,* National Academy Press, Washington, D.C., 1993.

Nordstrom, D. K., J. W. Ball, R. J. Donahoe, and D. Whittemore, "Groundwater Chemistry and Water-Rock Interactions Stripa," *Geochim. Cosmochim. Acta,* 53:1727–1740, 1989.

Office of Technology Assessment, *Protecting the Nation's Groundwater from Contamination,* U.S. Congress, OTA O-276, 2 vols., Washington, D.C., 1984.

Office of Technology Assessment (OTA), *Coming Clean: Superfund Problems Can Be Solved,* PB90-142209, National Technical Information Service, Springfield, Va., 1989.

Office of Technology Assessment, *Beneath the Bottom Line: Agricultural Approaches to Reduce Agrichemical Contamination of Groundwater,* OTA-F418, Washington, D.C., 1990.

Office of Technology Assessment, *Industry, Technology and the Environment: Competitive Challenges and Business Opportunities,* OTA-ITE-586, Washington, D.C., 1994.

Osmond, J. K., "Uranium Disequilibrium in Hydrologic Studies," in P. Fritz, and J. Ch. Fontes, eds., *Handbook of Environmental Isotope Geochemistry, Vol. 1, The Terrestrial Environment, A,* Elsevier, Oxford, U.K., 1980, pp. 259–282.

Paces, T., "Chemical Characteristics and Equilibration in Natural Water-Felsic Rock-CO_2 System," *Geochim. Cosmochim. Acta,* 36:217–240, 1972.

Patel, C. K. N., "Industrial Ecology," *Proceedings of the National Academy of Sciences,* Washington, D.C., 89:798–799, 1992.

Pearson, F. J., Jr., and I. Friedman, "Sources of Dissolved Carbonate in an Aquifer Free of Carbonate Minerals," *Wat. Resour. Res.,* 6:1775–1781, 1970.

Pearson, F. J., Jr., and C. T. Rightmire, "Sulphur and Oxygen Isotopes in Aqueous Sulphur Compounds," in P. Fritz, and J.Ch. Fontes, eds., *Handbook of Environmental Isotope Geochemistry, Vol. 1, The Terrestrial Environment, A,* Elsevier, Oxford, U.K., 1980, pp. 227–258.

Peterman, Z. E., J. S. Stuckless, S. A. Mahan, B. D. Marshall, E. D. Gutentag, and J. S. Downey, "Strontium Isotope Characterization of the Ash Meadows Ground-Water System, Southern Nevada, USA," in Y. K. Kharaka, and A. S. Maest, eds., *Water-Rock Interaction,* vol. 1, A.A. Balkema, Rotterdam and Brookfield, 1992, pp. 825–829.

Phillips, F. M., M. K. Tansey, L. A. Peeters, S. Cheng, and A. Long, "An Isotopic Investigation of Groundwater in the Central San Juan Basin, New Mexico: Carbon 14 Dating as a Basis for Numerical Flow Modeling," *Wat. Resour. Res.,* 25(10):2259–2273, 1989.

Plummer, L. N., and W. Back, "The Mass Balance Approach: Applications to Interpreting the Chemical Evolution of Hydrologic Systems," *Am. J. Sci.,* 280:130–142, 1980.

Plummer, L. N., J. F. Busby, R. W. Lee, and B. B. Hanshaw, "Geochemical Modeling of the Madison Aquifer in Parts of Montana, Wyoming, and South Dakota," *Wat. Resour. Res.,* 26(9): 1981–2014, 1990.

Plummer, L. N., D. L. Parkhurst, and D. C. Thorstenson, "Development of Reaction Models for Groundwater Systems," *Geochim. Cosmochim. Acta,* 47:665–685, 1983.

Puckett, L. J., *Nonpoint and Point Sources of Nitrogen in Major Watersheds of the United States,* U.S. Geological Survey, 1994.

Ritter, U. F., and A. E. M. Chirnside, "Impact of Land Use on Groundwater Quality in Southern Delaware," *Ground Water,* 22:38–47, 1984.

Roberston, W. D., and S. L. Schiff, "Fractionation of Sulphur Isotopes During Biogenic Sulphate Reduction Below a Sandy Forested Recharge Area in South-Central Canada," *J. of Hydrol.,* 158:123–134, 1994.

Rye, R. O., W. Back, B. B. Hanshaw, et al., "The Origin and Isotopic Composition of Dissolved Sulfide in Groundwater from Carbonate Aquifers in Florida and Texas," *Geochim. Cosmochim. Acta,* 45:1941–1950, 1981. ·

Siegenthaler, U., and H. Oeschger, "Correlation of ^{18}O in Precipitation with Temperature and Altitude," *Nature,* 285:314–317, 1980.

Solley, W. B., C. F. Merk, and R. R. Pierce, *Estimated use of Water in the United States in 1985,* U.S. Geological Survey Circular 1004, 1988.

Sonntag, C., E. Klitzsch, E. P. Lohnert, et al., "Paleoclimatic Information from Deuterium and Oxygen-18 in Carbon-14 Dated North Saharian Groundwaters," in *Isotope Hydrology 1978,* vol. 2, IAEA, Vienna, 1979, pp. 569–581.

Stevens, T. O., and J. P. McKinley, "Lithoautotrophic Microbial Ecosystems in Deep Basalt Aquifers," *Science,* 270:450–454, 1995.

Thomas, J. M., A. H. Welch, and A. M. Preissler, "Geochemical Evolution of Ground Water in Smith Creek Valley—a Hydrologic Ally Closed Basin in Central Nevada, U.S.A.," *Appl. Geochem.,* 4:493–510, 1989.

U.S. Department of Agriculture, "The Magnitude and Cost of Groundwater Contamination from Agricultural Chemicals, A National Perspective," Staff Report AGES870318, U.S. Department of Agriculture, Environmental Research Service, Washington, D.C., 1987.

U.S. Department of Agriculture, "Agricultural Resources: Inputs, Situation and Outlook," U.S. Department of Agriculture, Economic Research Service, Washington, D.C., 1989.

U.S. Department of Energy, "Closing the Circle on the Splitting of the Atom, The Environmental Legacy of Nuclear Weapons Production in the United States and What the Department of Energy is Doing About It," U.S. Department of Energy, Office of Environmental Management, Washington D.C., 1995.

U.S. Environmental Protection Agency, "Pesticides in Ground Water Data Base: Interim Report," U.S. Environmental Protection Agency, Washington, D.C., 1988.

U.S. Environmental Protection Agency, "National Water Quality Inventory: 1988 Report to Congress," EPA-440-4-90-003, Washington, D.C., 1990.

U.S. Environmental Protection Agency, "National Pesticide Survey: Update and Summary of Phase II Results," EPA 570/9-91-020, EPA, Office of Pesticide Programs, Washington, D.C., 1992.

U.S. Environmental Protection Agency, "Cleaning Up the Nation's Waste Sites: Markets and Technology Trends", EPA 542-R-92-012, EPA, Office of Solid Waste and Emergency Response, Washington, D.C., 1993.

U.S. Geological Survey, "National Water Summary 1987—Hydrologic Events and Water Supply and Use," U.S. Geological Survey Water-Supply Paper 2350, Washington, D.C., 1990.

Vroblesky, D. A., and F. H. Chapelle, "Temporal and Spatial Changes of Terminal Electron-Accepting Processes in a Petroleum Hydrocarbon-Contaminated Aquifer and the Significance for Contaminant Biodegradation," *Wat. Resour. Res.,* 30(5):1561–1570, 1994.

White, A. F., *Geochemistry of Ground Water Associated with Tuffaceous rocks, Oasis Valley, Nevada,* U.S. Geol. Sur. Prof. Pap. 712-E, Washington, D.C., 1979.

Wigley, T. M. L., L. N. Plummer, and F. J. Pearson, Jr., "Mass Transfer and Carbon Isotope Evolution in Natural Water Systems," *Geochim. Cosmochim. Acta,* 42:1117–1139, 1978.

Winograd, I. J., and F. J. Pearson, Jr., "Major Carbon 14 Anomaly in a Regional Carbonate Aquifer: Possible Evidence for Megascale Channeling, South Central Great Basin," *Wat. Resour. Res.,* 12(6):1125–1143, 1976.

Yurtsever, Y., and J. R. Gat, "Atmospheric Waters," in J. R. Gat, and R. Gonfiantini, eds., *Stable Isotope Hydrology, Deuterium and Oxygen-18 in the Water Cycle,* IAEA, Vienna, 1981, pp. 103–142.

CHAPTER 12
ESTUARIES

George H. Ward, Jr.
Center for Research in Water Resources
The University of Texas at Austin
Austin, Texas

Clay L. Montague
Department of Environmental Engineering Sciences
University of Florida
Gainesville, Florida

12.1   INTRODUCTION

This chapter addresses a particular class of water body, which has occupied a central role in the development of water-resource management, the estuary. A rather imprecise description of an estuary is "a watercourse in which seawater and freshwater intermix," which encompasses a range of possibilities: from the seaward reaches of rivers to coastal embayments, from deltas with broad, shallow tidal flats to deep, windswept fjords.

Estuaries contain some of the most productive areas of the world, with respect to both ecology and economics. This is not a coincidence. At the coastal land–water interface, natural subsidies of production accrue to both nature and humanity. For humanity, the land–water interface is advantageous for development, affording access both to the sea and the interior. While this statement is true for the coastline in general, the estuary affords a special attraction, its sheltered morphology offering a natural protective harbor, with access to freshwater supplies in its upper reaches, and a bounty of fish and game. Many of the great population centers of the world are established on estuarine harbors. In developed countries, coastal zones are attractive for their natural beauty and recreational opportunities. Estuaries subsidize heavy industry. Costs of transporting raw materials and finished goods over water are much less than those of other forms of transportation, and the sheltered land–water interface of an estuary is a natural transfer zone. Heavy industry is also subsidized by the large volume of water available in estuaries, as a source of process water and for dilution of wastes. In modern society, the estuary function of a rich food supply remains important. The land–water interface is the base for the commercial fishing industry, and is especially intimately connected to the estuaries.

Within the estuary that supports all of this socioeconomic activity is a rich ecological system consisting of large quantities of productive plants and animals. These include intertidal marshes and fringing growths of mangroves, subtidal seagrasses and other vegetation, oyster reefs, clam banks, worm flats, and many species of fish and birds. Only a few of these have direct economic importance in the coastal fishery, but most contribute by processing wastes, maintaining water quality, and providing habitat and food for the few that are commercially harvested. Like economic production, ecological production is subsidized in estuaries. Work that must be done by plants and animals elsewhere is done for them in estuaries by the convergence and discharge of many natural energies. The estuary is the receptor of the nutrients and sediments carried to the coast by rivers. Estuarine plants and animals benefit from this riverborne subsidy. The flood and ebb of the tides, the action of wind and waves, and the salinity gradient from the head of the estuary to the ocean inlet together provide considerable circulation of water. This supplies nutrients for, and removes waste products of, sedentary plants and animals, and provides transport from place to place for plankton, larvae, and juveniles, as well as for many larger motile animals.

Many of the concepts and quantitative techniques for addressing water-resource problems in estuaries have been imported from other watercourses. For example, the methods of one-dimensional transport analysis employed in riverine systems have been extended to estuaries. These include application of the momentum equations (e.g., the St. Venant equation) for surges and floodwaves in estuaries, and application of the advection-diffusion equation with kinetics (e.g., the Streeter-Phelps equation) to the evaluation of waste loads in an estuary. Modeling, in particular, has become an important element of the management toolbox in estuaries. The kinetics of constituents carried in solution or suspension in an estuary may be rendered more complex due to the higher energy of its hydrodynamics or the intermixing with seawater, but their conceptual and analytical treatment is basically the same as other aquatic systems: Estuarine chemistry is a subset (albeit perhaps complicated) of aquatic chemistry. Fishery management in an estuary is similar to that in oceanic or limnological systems, with the added complexity of migratory species and compressed gradients of water chemistry and habitats. Wetlands are essential features of the estuarine ecosystem, but from a management point of view are handled little differently from their analogs in a riparian environment.

Thus, a complete treatment of the estuarine environment would require inclusion of the freshwater and marine processes that interact in the estuary. Not only would this far exceed the space available in this handbook, it would also substantially overlap material presented in other chapters. Our approach, instead, is to present a précis of the estuarine environment, directing the reader to sources in the literature for more detailed treatment, and to focus on those aspects of estuarine management that are unique to this environment. Matters of (for example) general analysis of waste load impacts in a surface watercourse, of connections between streamflow and transport processes, or of modern modeling methods, are all addressed in other chapters of this handbook, and are not treated here. From this standpoint, it seems to us that management of estuaries is generally differentiated from that of other aquatic systems by the following:

- Hydrodynamic processes are more complex, and must be given explicit attention whenever transport of waterborne constituents is an issue.

- When physiochemical behavior of a waterborne constituent is a concern, the special properties of the estuary environment must be considered, including the seaward salinity gradient, variable buffering capacity, high turbidity, and estuarine micropopulaces.

- The ecological system is more frequently the ultimate objective of management, and must be addressed explicitly (rather than implicitly, e.g., through satisfaction of water-quality criteria).

The focus of this chapter is on how the special properties of the estuary must be accommodated when any of these considerations applies. There are several comprehensive texts on estuaries which the reader should consult for detailed information, cited in the appropriate sections following. As a comprehensive survey, the proceedings of the Jekyll Island conference (Lauff, 1967) remains a basic source. Also, the proceedings of the biennial conferences of the Estuarine Research Federation published by Academic (e.g., Cronin, 1975) have become standard references. The books by Day, et al. (1989), Knox (1986), and Mann (1974) provide good introductions to estuarine ecology.

12.1.1 Special Features of the Estuarine Environment

The estuary defies precise definition, but is generally considered to include the following properties (Ketchum, 1951; Stommel, 1951; Pritchard, 1952; 1967a; Cameron and Pritchard, 1965; Officer, 1976; Orlando, et al., 1993):

- Coastal waterbody
- Semienclosed
- Free connection to open sea
- Influx of sea water[*]
- Freshwater influx[†]
- Small to intermediate scale[‡]

The important inference is that an estuary is a complex watercourse that is transitional between a purely riverine system and one that is purely marine. Therefore, an estuary is governed by hydrographic processes that are both riverine and marine (e.g., floods and tides, respectively). It is also subject to processes that are unique to the estuarine environment, originating from the interaction of marine and riverine influences, and its semienclosed morphology. Estuaries tend to be broad, well-circulated systems. There is usually a clear zonation in morphology and habitats with distance from the sea, tending from being deep, saline, and well-aerated near the main inlet to the sea, to shallow, brackish, and poorly flushed in the upper reaches. The principal hydrographic features of an estuary are: (1) morphology and bathymetry, (2) hydrology, (3) tides, (4) meteorology, and (5) density currents. The hydrographic characteristics of an estuary, or a segment of an estuarine system, can often be judged by determining the relative influence of these factors.

Morphology, the shape of the estuary, is a reflection of the processes forming and maintaining the system, usually on a geological time and space scale. *Bathymetry* refers to the submerged physiography of the estuary, i.e., its depths and patterns of deep and shoal areas within the system. The mouth, or inlet to the sea, is one of the

---

[*] "... containing a measurable quantity of sea salt" (Pritchard, 1952)

[†] "... measurable dilution of seawater by land water drainage" (Ketchum, 1951)

[‡] The Gulf of Mexico, Mediterranean Sea, and Baltic, for example, are not considered to be estuaries, though they satisfy the other criteria. Among oceanographers, these larger systems are referred to as *inland, marginal, enclosed* or *confined* seas.

fundamental morphological controls, since it determines the exchange with the sea. Littoral sand supply and riverine sediment loads further establish patterns of shoal areas that are sculpted and shaped by waves and currents. Many estuaries include extensive deltaic regions created by the sediment load of rivers.

By definition, an estuary includes a riverine inflow. Inflow affects the hydrography of the estuary by establishing a gradient of salinity across the system, and further influences water quality by its associated influx of constituents of terrestrial origin, frequently including human waste loads. Both the magnitude and time sequence of inflow are important in the overall estuarine hydrography. Some systems have highly time-variable inflows, ranging over several orders of magnitude, while others have relatively steady inflows. When there is a prominent seasonality in the freshwater inflow, the character of the estuary can change greatly from the low-flow to the high-flow season. (An extreme example is the estuaries of India, e.g., the Vellar, which shift with the monsoon from pure freshwater systems to pure seawater.) It is important to recognize that the inflow feature of an estuary enlarges the geographical area of concern to encompass the entire watershed of the feeding rivers, which may entail a completely different hydroclimatology than that of the coastal region in which the estuary is located.

The *tide* is, of course, the most obvious marine influence on estuary hydrography. The ocean margin tide range itself is greatly variable around the earth, from a few centimeters to several meters, and may also vary in its basic periodicity from semidiurnal (12.4 hours) to diurnal (24.8 hours), frequently with longer period intertidal variability. As the tide propagates into the estuary it is generally attenuated and lagged by the frictional energy loss associated with the shallow, constricted watercourse, though there may be regions whose morphology amplifies the tidal range.

Because estuaries are semienclosed coastal systems, they tend to be responsive to meteorological forcing, of which the wind is the most important agent. Even fjords are subject to pronounced wind forcing (e.g., Dooley, 1979), but this is especially true of those estuaries that are broad and relatively shallow. Wind generates steep shortcrested waves that can become efficacious mixing agents. On a larger scale, wind can generate autonomous circulations (*gyres*) within the estuary. Suddenly varying winds can induce *wind tides* by effecting a tilt in the water surface across the estuary and an abrupt water level differential between the estuary and the adjacent sea. Estuaries are additionally subject to the variety of meteorological conditions peculiar to the coastal zone, including tropical storms and sea–land breezes.

The *density current* is perhaps the least obvious and most poorly understood of the principal estuarine controls, but it is a basic element of estuarine circulation. This is the current generated by a *horizontal* difference in density. In an estuary, this arises from the horizontal salinity gradient, more saline water being denser than fresher water. Essentially, the density current is the flow of denser water displacing lighter water, but modulated by mixing processes and the shape of the estuary. In a longitudinal estuary, this is manifested as a mean circulation directed upstream in the lower layer and downstream in the upper. In broader systems, such as lagoons and bays, the density current can be manifested as a flow directed upstream in the deeper sections of the estuary, compensated by a seaward return flow in the shallower sections. The density current is the prime vehicle for salinity intrusion and often establishes a dynamic equilibrium with the steady freshwater inflow from the river.

While estuaries provide many subsidies that enhance both ecological and economic production, their environments are also very dynamic. For survival of biota, salinity, light, water depth, temperature, and dissolved oxygen are among the most influential environmental parameters, each of which is subject to wide variation in the estuary. Tides and winds produce large changes in water level, which leave intertidal organisms alternately submerged and exposed to air, and alter the light available to plants and animals on the bottom. Excursions in river flow, wind, and sea level create especially wide swings in salinity. The variations in environment found at any one spot in estuaries can be difficult for sedentary species to withstand; they have therefore evolved special adaptations. In fact, relatively few such sedentary species live in estuaries, though those that do, like oysters and marsh grasses, may be very productive because of the energy subsidies they receive.

Although motile animals may pass through a wide range of environments when going from place to place, their mobility allows them to avoid extremes that must be withstood by sedentary organisms. Hence, more kinds of motile animals live in estuaries, and a wide variety of seasonal migrants visit to take advantage of the productive food and cover. The transient population, in particular, whose migration into the estuary is linked to specific life cycle stages, is a major feature of the estuarine ecosystem. Some species are anadromous, like salmon, immigrating to the estuary to spawn. Others, like shrimp and many finfish, spawn offshore, and the young (larvae, postlarvae, or juveniles) immigrate into the estuary to develop in the protected, nutrient-rich habitats. This nursery function of estuaries is essential for many commercial species that are in fact harvested in the open sea. Ecological management of estuaries often reduces to a focus on habitat: the combination of water quality, bed characteristics (including sediment), and flora that support general types of aquatic communities.

12.1.2   An Overview of Water Resources Management in Estuaries

This chapter is addressed primarily to the scientist or engineer involved in the management of estuaries. Management, in this context, involves gaining and exercising some degree of physical and political control over the major environmental influences of the estuary in order to achieve a prescribed objective. At the largest scale this includes control over the estuary boundary conditions: the quantity of freshwater inflow and the size and shape of the inlet throat. It can include development of the estuary for navigation and shipping, operation of petrochemical wells, installation of protective structures for storm surges, and use of estuary water for industrial supply. Control over water quality and habitat within the estuary can also be important for many objectives, such as preserving estuary fisheries while allowing waste discharges.

Because estuaries are the physical interface of watersheds and the ocean, they typically exist in a complex political environment consisting of numerous public agencies and private interests either promoting specialized needs or imposing, perhaps inadvertently, impacts on the estuary. With management control comes conflict among objectives, and the usual management problem is to achieve a compromise. This requires quantitative methods, based upon cause-and-effect relations.

12.1.2.1   Analysis and Modeling in Estuarine Management.   The management of estuaries requires being able to determine the effect on estuary circulation or constituent concentrations, or upon elements of the estuary dependent upon these, such as biological communities, that results from a specific event or external control. For some of these, especially those caused by natural events, the determination of effects can be based entirely upon data collection and analysis. In many management situations, however, a human activity must be evaluated *before* it is implemented; these require some sort of predictive capability. The standard methodology is to apply a predictive *model.* Models assume a variety of forms, including scaled hydraulic (i.e., physical) models, laboratory cultured ecosystems (microcosms), and statistical regressions. The attributes that all of these types of models share are that each is (1) a simplified depiction of reality, and (2) quantitative. In the present context we apply the term strictly to some sort of mathematical formulation of a physical relationship.

In this chapter, the generic term *analysis* is employed to describe the quantitative procedures by which the behavior of aspects of estuarine systems is determined, and with which predictions of that behavior are made for various scenarios. This includes both the prosecution of scientific research and the evaluation of management alternatives. In this same generic sense, there are two general categories of techniques: statistical analysis based upon historical measurements, and deterministic formulas. The first relies solely upon the available data base, and seeks to extract relations between the dependent variable and what are thought to be the controlling variables. The second develops functional relations between the dependent variable and its controls by inference from basic laws of physics. In practice, the analytical methodology (*model*) will be a mix of the two, depending upon the nature of (and knowledge of) the relevant processes, the complexity of the mathematical relations, and the available data. The estuarine environment poses no novelty in application of these general methodologies, and their presentation properly belongs within the scope of other chapters in this handbook or in other references in the professional literature. Therefore, in this chapter models and model features are employed for their conceptual value, but no detailed treatment is given. On the other hand, we attempt to make note of those features of estuaries that dictate special analytical requirements.

In estuaries, a preponderance of management problems reduces to the concentration of waterborne indicator parameters. Analysis (i.e., modeling) of those parameters begins with compilation of available data. The data density in space must be sufficient to determine the principal gradients, and the sampling interval in time determines the temporal variability that can be resolved in the data. This is one respect in which the estuarine environment differs from other aquatic systems, in that the spatial complexity and great temporal variability impose special demands on both the analytical methodology and the data base. The task of greatest urgency facing an estuarine manager usually is access to field data. The foundation of data collection should be measurements of tides and salinities within the estuary, but this must be extended to include water-quality parameters of engineering or ecological importance. Good data on inflow from the watersheds are also needed.

Because of the range of time-space scales that operate in a typical estuary, a long period of data is often necessary to perform a comprehensive analysis. The resources of a single data collection program rarely are adequate for characterizing salinity in an estuary, so some effort may be justified in seeking out and combining results from several sampling programs, or in implementing special-purpose intensive data-collection campaigns. Frequently, the analyst is faced with the situation of having data in only a few regions or even at a single station, so a realistic appraisal must be made of what analyses the data base will or will not support. The important point is that a resource of historical data is indispensable to evaluate operational problems and devise appropriate management strategies.

The concentration of any constituent is governed by transport processes (including mixing) and kinetic processes, so the model must include a determination of hydrodynamic transports as well as a mass balance of the water-quality constituents. This is true whether the watercourse is a river, lake, aquifer, or estuary. For an estuary, however, the complex geometry and complicated hydrodynamics make model formulation especially difficult. For this reason, the special topic of estuary modeling has long received concentrated attention, and there is an extensive literature on the subject (e.g., Ward and Espey, 1971). Because any model is a simplification of nature based upon various assumptions, it is necessary to establish that the model achieves its intended purpose by comparing its predictions with actual measured data. This is the process of model verification (see, e.g., Thomann and Barnwell, 1980; Ditmars, et al., 1987).

The general causal controls on estuary water quality are shown schematically in Fig. 12.1. In this diagram, oblong boxes represent external factors which control elements of either circulation or the concentration of indicator parameters, the rectangles specific classes of processes, and the circles variables in those processes that are of independent significance. This figure represents reality. The question in model selection and development is: For the specific estuary problem of concern, how can the model simplify this complex reality and still depict the constituent concentra-

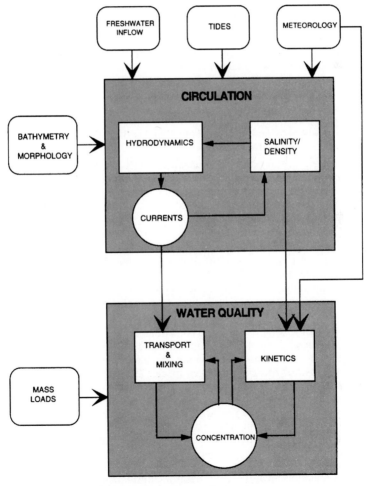

FIGURE 12.1 Principal controls on water quality in estuary.

tions to an adequate accuracy? There are several traditional strategies and assumptions for simplifying the model formulation, as follow:

- *Reduction of Dimensionality.* Estuaries are three-dimensional entities, with a complex spatial variability. The mathematical equations representing variation of water quality therefore also must be three-dimensional, which creates great difficulty in their solution. If only larger scale variation is of concern, and there is a prominent dimension in the estuary morphology, the dimensionality of the equations may be reduced by averaging over the lesser important dimensions. A broad, shallow lagoon may be averaged in the vertical and treated with a two-dimensional (horizontal) geometry. A river-channel estuary may be averaged over the cross section and treated as a one-dimensional longitudinal system. Of course, this may entail including terms to parameterize dispersion due to current variations across the section (see Sec. 12.3.4).

- *Long-Term Temporal Averaging.* If detailed time variability is not of concern, the model may address variables averaged over a longer period of time, thereby simplifying both the input-data burden and the mathematical solution. One particularly important judgment is whether *intratidal variability* must be treated. If intertidal (or *residual*) variation is sufficient, then the model can address circulations and concentrations assumed to be averaged over several tidal cycles. This considerably simplifies the model. Of course, this may entail including terms to parameterize mixing by tidal dispersion (Sec. 12.3.4). The extreme instance of long-term temporal averaging is to assume a dynamic equilibrium, i.e., a steady state. This may be appropriate for well-defined seasons in which the river inflow becomes steady, provided there is some evidence that the real system becomes stable (in the tidally averaged sense).

- *Elimination of Meteorological Variability.* Since meteorological events are essentially random on a long-term basis, their effects on circulation and kinetics can be absorbed into the noise (i.e., experimental error), and removed from further consideration. This is an especially attractive assumption in conjunction with long-term temporal averaging.

- *Separation of Salinity in the Hydrodynamics.* The simultaneous solution for currents and salinity creates many mathematical complexities, since the two equations are coupled in the nonlinear terms. If the estuary can be assumed homogeneous in density, then the hydrodynamic solution is greatly simplified. Salinity is then treated as a passive tracer and modeled separately. Of course, this may entail including terms to parameterize the transports effected by density currents, usually assumed to be dispersion-type terms (Sec. 12.3.4).

- *Simplification of Kinetics.* For many constituents, the kinetic terms may be quite complex. Dissolved oxygen, for example, is subject to losses due to a complex of aerobic organisms that stabilize various organic carbon forms, and to aerobic nitrifiers that oxidize inorganic nitrogen species. These are frequently lumped into a carbonaceous BOD (biochemical oxygen demand) term and a nitrogenous BOD term, each assumed to be first-order with a rate coefficient that is, at most, temperature dependent (see Sec. 12.4.3). Nutrients appear in a number of chemical complexes with kinetic interconversions. This is frequently suppressed by using a single measure, such as the total elemental concentration, and assuming first-order kinetics.

Model formulation must be based upon a careful analysis of the management problem, identifying the space-time scales of importance and the factors controlling the estuary response at those scales. The unit of measure is the tidal excursion, the horizontal distance moved by a parcel of water on the flood tide. If the zone of degraded water quality is located within a tidal excursion of the point of waste load, then the model time resolution must be intratidal, and capable of detailed spatial resolution, at least in the vicinity of the waste load. On the other hand, if the zone of degraded water quality is distant from the point of waste load by several or many tidal excursions, then an intertidal analysis will probably be sufficient, with only large-scale spatial depiction.

The fact that these conceptual simplifications have been cast in the context of model application might lead one to infer that they are some way invoked by modeling. Working back from field observations requires analogous considerations. Once a water-quality problem occurs, the specific causes must be established before management strategies can be adopted. The same type of water quality problem, e.g., low dissolved oxygen (DO), can be caused by several different hydrographic processes operating on different space-time scales. It is easy to jump to conclusions about the cause of a problem without adequate resolution of time-space variation. Some water-quality limitations may be controlled by larger scale transport and kinetic processes, and be amenable to management only by controlling where discharges are sited. This emphasizes the need for advance regional planning of the development of an estuary for maximum benefits and minimum impacts, based upon information about the hydrography and water quality of the system. In estuaries, like other aquatic systems, repairing one problem can cause another. The classic dilemma of implementing waste treatment to reduce turbidity in a watercourse, only to have the watercourse turn green, can be manifested in a much more complex way in an estuary, due to the great number of physical and biochemical interactions.

12.1.2.2   Some Example Generic Management Problems

Wastewater Discharges.   One of the common management problems of an estuary (like many other watercourses that receive effluents) is that of evaluation of water quality as influenced by one or several waste discharges. The motivation of the analysis can be to diagnose the source of degraded water quality, to improve that quality by imposing limits on the existing dischargers, or to predict the effects on water quality of a proposed discharge. A broad class of discharges is characterized by low volumes of flow (in comparison to, say, freshwater inflow) but high concentrations of potential contaminants. These include domestic wastewater, effluent from sewage treatment plants, and process water or wastewater from industrial operations.

Concern with domestic wastewater is primarily as a source of oxygen-demanding organics, nutrients (especially nitrogen and phosphorus), and pathogens. Nutrient discharges are especially significant to estuaries because the highly productive estuary biota have a capability for uptake and cycling of nutrients that exceeds that of normal surface waters. As an example of the sorts of impacts that may be entailed in an estuary, nutrients from a domestic effluent usually stimulate the growth of phytoplankton, increasing turbidity. This can dramatically reduce the growth of bottom vegetation, if already light-limited, and may impair its functions of sediment stabilization (Frostick and McCave, 1979) and habitat for juvenile stages of fish and invertebrates. Estuaries commonly have extensive intertidal marsh areas; in some estuaries these have been exploited to receive domestic discharges. The nutrients and fresher water stimulate marsh vegetation instead of phytoplankton, which helps assimilate some of the nutrients before they enter the open estuary waters.

Effluents containing industrial wastes entail an additional problem of toxins, especially metals, such as mercury, lead, and zinc. Though discharged in small concentrations, these can become concentrated in the bodies of animals and plants, generally increasing up the food chain (biomagnification). A type of industrial discharge of importance in some regions is that from pulp mills, which produce an organically rich effluent that may also contain toxicants. Chlorine bleach used to whiten paper mixes with the organic extracts from the processed pulpwood to create a host of chlorinated hydrocarbons, many of which are toxic, carcinogenic, or mutagenic. Dioxin and other polychlorinated biphenyls (PCBs) are of particular concern because of their great toxicity. Interestingly, the heavy organic load may not be especially problematic if sited appropriately. Estuaries contain numerous organisms capable of assimilating and converting the organic matter in the pulp-mill wastes.

One high-volume industrial discharge frequently encountered in estuaries is the heated cooling-water returns from steam–electric power plants. Because of the high-volume rate of discharge, the resulting plume of elevated temperature can become enormous, hundreds of hectares in extent. During the 1960s and 1970s these were the center of management and public attention because of perceived "thermal pollution." This concern has abated somewhat, at least in estuaries, so long as the discharge is sited to allow avoidance and not encroach upon sedentary communities or critical habitats. However, in some systems, they can be an attractive nuisance if they provide warm, but unreliable, winter havens for biota. If the plant shuts down, the temperature shock may not be tolerated by animals in the plume. This is a concern for manatees found in the Crystal River, Florida, power plant cooling-water discharge canal. To avoid recirculation of the plume into the intake, a power-plant discharge is often separated by great distances from the source, or placed on the opposite side of a peninsula. Because power plants circulate large volumes of water (20 to 50 m^3/s), they can become significant elements in the circulation of the estuary, perhaps transporting water into a region of very different chemical composition or salinity.

Assimilative Capacity. An important management strategy for estuaries that receive high loads of wastes is an analysis of *assimilative capacity,* i.e., the determination of how much waste load can be assimilated by an estuary, its *carrying capacity.* In large estuaries with multiple waste load discharges, this is carried one step further, to form the basis for so-called *waste load allocations* (Southerland, et al., 1984), in which specific waste load limits are imposed on individual dischargers to maintain a lower bound on water quality throughout the receiving watercourse.

The general procedure of assimilative capacity determination is indicated schematically in Fig. 12.2. The process starts with the specification of:

- Critical conditions, i.e., that combination of external controls that maximizes impacts of the waste load on water quality; for example, low river flow and high temperatures.

- The concentration level that determines acceptable water quality (*threshold of impact* in Fig. 12.2). This is often a criterion concentration or regulatory standard.

A suitable mathematical model is used to determine the concentration that results from a given level of waste load. This model is indicated by the shaded boxes of Fig. 12.2, emphasizing that for an estuary there is both a hydrodynamic and a mass balance aspect of the modeling (addressed in Secs. 12.3 and 12.4). The waste load magnitudes are then adjusted until the predicted concentration nowhere exceeds the threshold of impact. This waste load value is the greatest that can be discharged within the specified threshold, and is, therefore, the assimilative capacity for the system.

Such a carrying capacity analysis requires a considerable amount of preparatory work before the procedure of Fig. 12.2 can actually be performed. The following tasks, which are nontrivial, must have been accomplished:

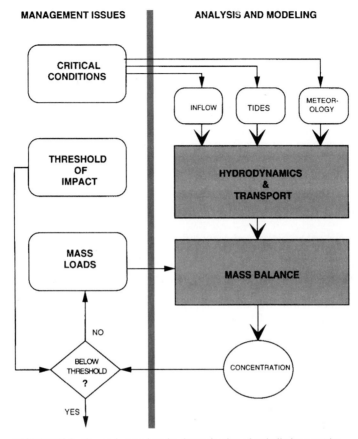

FIGURE 12.2 General procedure for determination of assimilative capacity.

- Specification of which parameters will form the basis of water-quality evaluation
- Definition of the parameter values corresponding to acceptable water quality (i.e., the threshold of impact)
- Development and verification of a model for the specified parameters that is appropriate for the estuary of concern
- Establishment of the combination of external conditions that are critical for water quality

In all of the preceding, for simplicity the procedure is presented as though it would be applied to the estuary in toto. In fact, the assimilative capacity determination is a strong function of position in the estuary. There will be areas in almost any estuary that are well circulated and subject to regular water-mass replacement, that will generally have a high degree of assimilative capacity. There will also be areas that are poorly circulated with frequent stagnation (dead zones), which will have a low assimilative capacity. The location of the hypothetical waste load relative to well circulated or poorly circulated zones, and relative to other existing waste loads, is important to the ability of the estuary to assimilate that waste load. As noted in preceding sections, the water-quality parameters defining the problem and the operational time-space scales must be carefully specified in order to analyze the problem and establish appropriate management actions. A comprehensive treatment of modeling in water-resource management, including estuaries, and of the prosecution of wasteload allocations is given by Thomann and Mueller (1987).

Freshwater Inflow and Salinity Intrusion.   A defining criterion for an estuary is the freshwater inflow. This governs salinity, contributes a major sediment load (for maintaining marshes and mud flats in the face of erosion and rising sea levels), and provides nutrients and essential trace elements derived from drainage of the watershed and weathering of rock formations and soils, as well as an influx of contaminants from upstream cities. With the increasing development of watersheds, the demands of water supply, and the construction of dams and diversions, there is an increasing possibility of modifications in the inflow to the estuary. Decisions made anywhere in the watershed that alter the quality or quantity of freshwater discharge at the coast can potentially influence the ecological and economic development in estuaries.

An important aspect of freshwater inflow is its volume of flow and the resulting patterns of salinity. This is certainly true for inflow from rivers and peripheral drainageways. In some areas, groundwater discharge into estuaries also affects salinity regimes, and occasionally may add significant amounts of nutrients. Freshwater inflow establishes a gradient of salinity across the estuary, from zero at the source point of inflow, to oceanic values near the estuary entrance. This gradient is displaced, steepened, or flattened by variations in inflow. At one extreme, intense flood events (freshets) can flush the estuary, perhaps rendering it fresh (e.g., Schroeder, 1977). At the other extreme, prolonged freshwater inflow deficit leads to seawater intrusion throughout the estuary and increasing salinities (perhaps augmented by evaporation deficit). As suggested previously, salinity exerts strong hydrodynamic controls on the estuary because it in effect drives the density structure. Therefore, in the regions of large horizontal and/or vertical gradients of salinity, circulation and turbulence are most strongly affected by density structure; these regions are especially important in characterizing the response of salinity structure to inflow.

Not only can the amount of freshwater flow affect the salinity-gradient-driven circulation patterns in estuaries, but also it can influence the regime of salinity variation. The location of the zone of greatest salinity fluctuation and the frequency and amplitude of fluctuations within that zone, especially the seasonal pattern in inflow, are often more important than the mean salinity in determining distribution and abundance of submerged vegetation (Montague and Ley, 1993), or salinity cues for migrating organisms at critical life-cycle stages. Many estuaries have a regional climatology of alternating wet and dry seasons, to which the estuarine community is adapted. Also, the finer scales of the inflow time signal affect the response of the estuary, especially in salinity extrusion events (e.g., Lepage and Ingram, 1988).

A dramatic example of the effect of augmenting freshwater inflow is the case of Charleston Harbor, South Carolina. In 1941, the flow of the Santee River was diverted into Charleston Harbor, increasing its mean freshwater inflow by two orders of magnitude, from ~3 m^3/s to ~400 m^3/s. In the first decade following this diversion, annual maintenance dredging of the harbor channels increased from a modest 6×10^4 m^3 to 1.8×10^6 m^3, with a concomitant increase in expense. This was associated not only with the increased fluvial sediment load, but also a fundamental change in the estuary circulation (Simmons, 1966): a more prominent density current due to the higher salinity gradient, and an increased effectiveness in trapping the fine sediments (see Sec. 12.4.2). In 1985, 70 percent of the river flow was diverted back into the Santee (away from the estuary), resulting in a mean flow of ~130 m³/s, after which the tidal amplitude increased, salinity stratification diminished, the longitudinal salinity gradient was displaced landward, and sedimentation decreased (Bradley, et al., 1990; Kjerfve and Magill, 1990).

Freshwater inflows also transport minerals, nutrients (Armstrong, 1982) and detritus (Darnell, 1967) into the estuary, and can have a direct effect on its ecological communities apart from their influence on hydrography. These factors may be incapable of direct formulation due to the complexity of processes of delivery, uptake, and assimilation into organisms and communities, but their influence is evident in statistical associations between inflow and elements of the ecosystem (e.g., Montagna and Kalke, 1992) or fisheries (e.g., Copeland, 1966; Armstrong, 1982; Longley, 1994). Moreover, if the constituents transported by freshwater inflow emerge as the management focus, either in association with or apart from the consideration of volume and timing of inflows, then the concern may not simply be the river flows per se, but also the factors within the watershed that are the ultimate source of dissolved and suspended matter. (This merges into the next management problem topic.) Importance of inflows to estuarine productivity can entail direct conflicts between demand for freshwater in upland regions, and the need to preserve estuarine productivity in coastal regions, that have both environmental and legal dimensions. This becomes especially problematic in semiarid or water-stressed basins (Kaiser and Kelly, 1987). Three decades ago, B. J. Copeland (1966) commented, "With continuation of man's activities in allowing less and less fresh water downstream to the estuary, man may have to pave the estuarine areas and sell them for real estate." His optimism is as applicable today.

Non-Point-Source Loading. In this category is included the set of problems falling under the rubric of non-point-source pollution, though we would expand it to encompass nonpoint sources of constituents other than pollutants. This includes:

- Diffuse sources (i.e., poorly defined or widely distributed in space) around the estuary periphery, such as runoff from local drainage areas, and groundwater interflow

- Highly transient loading, generally in association with storm runoff or precipitation washout

- Areal sources, such as atmospheric deposition on the estuary surface

- Riverine sources, i.e., distant loads (either point or nonpoint) transported into the estuary from upstream, in the process of which the constituent is subject to dispersion and detention, intermixing with other sources, and chemical transformation

Because the individual pollutant sources are small and numerous, perhaps infinite, they are not amenable to control in the same way as point sources; therefore nonpoint-source pollution has become an increasingly complex and formidable aspect of water-quality management.

The estuary environment poses two somewhat novel aspects. First, the specific marine influence of the coastal zone may have to be considered. This includes the sea as a source of aerosols and condensation nuclei, as well as its effect on climatology in the coastal zone, particularly precipitation processes and mesoscale circulations, notably the seabreeze. Second, and more importantly, the entire estuary watershed may have to be addressed in the budgets of pollutants and other constituents, including the effects of urbanization, agriculture, and land management. Increasingly, water-quality studies of estuaries must be extended great distances inland to document the ultimate source of nutrients and contaminants, e.g., Wolfe, et al. (1991), Correll, et al. (1992), Hager and Schemel (1992), Jaworski, et al. (1992), Nienhuis (1992), and Scudlark and Church (1993). Not only does this entail a much larger study domain, but there are also multiple pathways by which the constituents enter the estuary, some of which involve physicochemical or biochemical transformations.

Navigation and Shoreline Development. The natural environment of the estuary offers benefits to human settlement with access to the sea, so estuarine coastal development is often associated with shipping. While estuaries are natural harbors, their physical configuration must frequently be modified to better accommodate seagoing traffic. The civil works involved include enlarging or creation of inlets, protection of inlets from waves, sediment, and adverse currents by jettying, stabilization of the shoreline by bulkheading and revetment, dredging of deep-draft channels and the associated movement and disposal of dredged sediments, protection of channels by groins and dikes, and construction and dredging of harbors.

The most significant morphological alterations to an estuary are often the result of navigational improvements. These civil works can fundamentally modify exchange between estuary and sea, and the trajectory and intensity of currents within the estuary; therefore, they have the potential to alter circulation processes of the estuary. One common navigational action is channelization, which can result in increased salinity intrusion due to greater water depths and the associated density current. This in turn can modify patterns of sediment transport and deposition. Quantitative documentation of such effects from historical channelization may be difficult, however, because navigation channels are usually dredged in incrementally greater depths over many years. Even if a sufficiently long period of record of salinity measurements exists, which is itself rare, it can be difficult to discriminate the channel effects from the other sources of salinity variability (e.g., Kjerfve and Magill, 1990). One exception is the case of Matagorda Bay on the Texas coast. Unlike most of the other large bays on the Gulf of Mexico coast, Matagorda Bay had no deep-draft channel until 1963, when an 11-m project was dredged. This was a deep-draft channel suddenly imposed where no channel had previously existed. Analysis of salinity data from before and after the dredging disclosed that salinities in the bay increased around 5 ppt due to the channel (Ward, 1983). There are also conditions in which channelization can alter the response of the estuary to meteorological forcing (e.g., Hearn, et al., 1987).

Other modifications associated with shoreline development include construction of bridges and causeways, filling and reclamation, and permanent alterations in the character of the nearshore environment. Disposal of dredge spoil and land creation by diking and filling replace the original environments, which are often valuable fringe habitats such as shallow bay bottom or salt marsh. Bulkheading and pier installation are also responsible for fundamental modification of the nearshore habitat, converting the original sloping, rugose, reticulated and vegetated nearshore zone to an abrupt rectilinear barrier. An instructive compilation of various levels of shoreline and peripheral development, and the resulting impacts on the estuarine environment, is the series of research reports issued by the Council for Scientific and Industrial Research of South Africa (e.g., Heydorn and Tinley, 1982).

Inlet Management. The exchange with the sea is another defining characteristic of estuaries, and the estuary entrance is a key mediator of that exchange. Many deliberate modifications to estuary inlets have the specific objective of altering access to the sea, but some modifications for other purposes inadvertently alter this exchange. Among the former are: (1) shipping improvements, such as channel dredging and jettying, which deepen an inlet, or protect it from shoaling and wave action, for the purpose of passage of oceangoing vessels, (2) barriers, to protect the estuary from marine influences such as storm surges and salinity intrusion, and (3) bar structure removal and sand bypassing, to stabilize the inlet and prevent shoaling, implemented for a variety of reasons from navigation to maintenance of the estuary salinity within favorable limits (e.g., Kassner and Black, 1982). Among the latter are revetting and bulkheading for erosion control, and any disruption to the littoral transport processes along the shoreline, such as construction of groins and seawalls, perhaps distant from the affected inlet. Probably the most extreme modification to the estuary's interaction with the sea is the complete creation or closure of an inlet. The premier example is the Delta Project of the Netherlands, the construction of barriers across many of the important inlets to the estuaries of the rivers Rhine, Meuse and Scheldt.

Much of the productivity and biological importance of estuaries stems from the passage of organisms between the sea and the protected estuarine environment. This passage occurs exclusively through the estuary inlet, so it follows that inlet modifications have the potential to affect biological communities both within and external to the estuary. Indeed, one objective of inlet creation is to provide fish passes, to facilitate utilization of an estuary by diadromous fish and shellfish (Carothers and Innis, 1960). It has become increasingly apparent that the morphology of the inlet and its hydrodynamic behavior are important constraints on the ability of species to use an inlet for access to the estuary (Darnell, 1979). Passage of essentially planktonic forms, i.e., larvae and some juveniles, may be completely passive (Pietrafesa and Janowitz, 1988), but even feeble swimmers are capable of directed migration. These organisms apparently select favorable currents to carry them through the inlet, employing their limited motility for vertical migration or movement between lateral shallows and the current zone (Cronin and Forward, 1982; Leggett, 1984; Miller, et al., 1984). Such morphological factors as a range of depths and bottom types are necessary for these organisms to negotiate the swift currents of an inlet. Influences on the transport of larval fishes and invertebrates from offshore spawning grounds into juvenile nurseries in estuaries have therefore been a recent concern in inlet projects (Mehta and Montague, 1991).

From the management point of view, *inlet manipulation* must be recognized as an action with fundamental potential consequences on a range of estuarine processes. Alterations of the width, depth, and length of the inlet, its internal physiography, and adjacent shoreline can affect the tidal energy, salt, and sediment entering the estuary, the detention of freshwater within the estuary, and use of the inlet by migrating fishes and other marine species. Moreover, these effects can be felt beyond the geographical bounds of the estuary. The hydrodynamic behavior of an inlet impacts the movement and distribution of sand along the adjacent shore as well. Removal of the inlet bar structure may result in modifications of the transport and sedimentary processes within the estuary, but also along the downdrift beaches. The imposition of a barrier such as a jetty can be expected to disrupt littoral sand transport for a great distance down the coast. Erosion, a chronic problem in many coastal regions, including the United States, is not only problematic to the owners of beachfront property, but can also interfere with littoral-zone organisms (for instance, success of sea turtle nesting, Montague, 1993). And, of course, modification of the effectiveness of an inlet as a biological access can alter recruitment and year-class productivity in both the estuary and offshore fishery.

An example of inlet modification is that of Matagorda Bay, noted previously. For various reasons, the route of its 11-m channel included transecting the barrier island with an artificial landcut, rather than bringing the channel through the natural tidal inlet of Pass Cavallo. This energetic inlet had a long history of a rapacious appetite for ships, beginning in 1685 with La Salle's flagship *l'Aimable* and continuing to the present, a consequence of its extensive and vagarious bar structure. The Corps of

Engineers anticipated the new landcut to have a propensity for scour instability, because of tidally induced water-level differential across the barrier island. The Corps therefore was prepared with stone riprap to revet the landcut when it was opened to the sea in September 1963. The instability was more dramatic than expected: Swift currents began spontaneously scouring and widening the landcut. The stone riprap was snapped up like peanuts. The landcut became an emergency, and the Corps revetted the full length of both banks, using much heavier stonework, to finally control the lateral enlargement of the landcut. However, it continued to scour downward, finally stabilizing at depths of 15 to 20 m, nearly double the project depth.

Because of this intensive scour, the landcut has proved to be self-maintaining, and has never required dredging. Moreover, it has captured half the original tidal prism of Pass Cavallo (Ward, 1982; van de Kreeke, 1985). Since 1963, Pass Cavallo, the natural inlet, has begun shoaling, and as of this writing has lost over half of its original cross section. As the pass shoals, more of the tidal prism is usurped by the landcut. The long-term consequences may be more than hydraulic. Pass Cavallo was a broad, geomorphologically complex inlet, presenting a range of bottom types, current speeds and depths to the varied swimming capabilities of migrating organisms. The landcut, in contrast, is a deep, U-shaped inlet with uniform, swift currents. Field data from both inlets indicate that the landcut is much less utilized by diadromous species. If Pass Cavallo shoals to closure, and is replaced by the landcut as the principal tidal pass for the bay, recruitment in Matagorda Bay could be significantly impaired.

Accidental Spills. The concentration of shipping and heavy industries in estuarine zones creates the threat of accidental releases (spills) of large volumes of hazardous chemical or petroleum compounds. The danger of such spills and the potential vulnerability of the estuarine ecosystem demand special management consideration of spill response. This lies beyond the scope of the present chapter, but some comment is warranted. From the strategic viewpoint, the manager should determine in advance:

- Zones of exposure (e.g., industrial sites, heavy ship traffic routes, pipeline locations, and well operations)

- Most probable sites for accidents (channel turns and intersections, loading docks, and pipeline crossings)

- Locations of sensitive habitats (e.g., marshes and shallow protected waters) or biological communities (e.g., oyster reefs and spawning grounds), including, since the *Exxon Valdez* oil spill of 1989 in Alaska, an advance determination of monetary value of coastal ecological resources

- Predicted transport processes under probable hydrometeorological scenarios

This information will guide planning in personnel training, response procedures, and the storage and maintenance of control and recovery equipment. Once an accidental spill is under way, however, the situation becomes tactical, and management must focus on control of the source, confinement and recovery of the spilled material, and deployment of protective structures (booms, skimmers, and so forth) around threatened critical habitats. In coastal and estuarine regions, special hydrographic skills are needed to quickly interpret spill movement and advise management. In the United States, the National Ocean Service of the National Oceanic and Atmospheric Administration has established a cadre of trained staff to provide these services wherever such accidents may occur:

Hazardous Material Response Team
7600 Sand Point Way NE
Seattle, Washington 98115-0070
Telephone: 206-526-6311

12.2 MORPHOLOGY

12.2.1 Principal Types of Estuaries

Morphology is frequently used as a basis for classification of estuaries. Some of the
principal classes are:

River channel estuary   The downstream reach of a river that flows directly into a
marine system, also referred to as a *tributary estuary*. There is little change in
cross-sectional area or depth as the mouth is approached. This is probably the
simplest form that an estuary can assume. One geomorphological feature frequently associated with the river channel estuary is an extensive delta.

Coastal plain estuary   Also referred to as a *drowned river valley* estuary, the submerged lower segment of a river and its associated relict floodplain. It is therefore dendritic with a prominent longitudinal dimension. Some authors include
river channel estuaries in this class.

Fjord   Extremely deep systems that have been scoured by glaciers, encompassing such geomorphic classes as fjords, fjards, and firths. The inflow is small in comparison to the estuary volume and lower-salinity water tends to lie in a thin layer
at the surface. Because of their characteristic formation, these systems are found
at higher latitudes.

Lagoon   A broad embayment, generally shallow, and poorly connected to the
ocean only through narrow inlets. Frequently, the estuary is separated from the
ocean by a barrier island complex, which means that lagoons are typically found
on gently sloping shelves with high littoral transport, such as the Atlantic and
Gulf of Mexico coasts of North America.*

Tectonobay   Estuaries formed by tectonic processes, for example, faulting and
vulcanism. They are characterized by high relief, irregular shapes and, frequently,
regions of moderate depth. San Francisco Bay in California is an example.

This classification offers a convenient means for the analyst to anticipate estuary
hydrographic and transport behavior based upon an inspection of its morphology.
Like any classification system, each class represents an idealized abstraction. Usually,
a real-world estuary will be a composite of several of these idealized morphological
classes. Lagoonal systems can be found within tectonobays and coastal plain systems,
for example, and coastal plain or river channel systems can be found within lagoons
and tectonobays. Also, the river channel and fjord classes would not even be regarded
as estuaries by some workers, and bays would be placed in a parallel companion category (but our view is more egalitarian). There are additional coastal aquatic systems
that do not conveniently fit these classifications, such as bights and the plumes of

---

* Pritchard (1952) referred to this class as bar-built estuaries. The class defined here is intended to be
somewhat broader, including sounds and coastal lagoons (see also Kjerfve, 1986).

major rivers, that are occasionally referred to as estuaries. For example, the zone referred to as the estuary of the Amazon is in fact the coastal inner continental shelf adjacent to the river mouth (e.g., Sholkovitz, 1993). While exhibiting many hydrographic and biochemical features of an estuary, it is certainly not semienclosed. (Lest one think that there is unanimity in what does and does not constitute an estuary, consider the essays of Herdendorf, 1990; Odum, 1990; and Schubel and Pritchard, 1990.)

The type specimen for the coastal plain estuary is Chesapeake Bay, Fig. 12.3, certainly one of the most important estuaries in North America and the subject of continuous and intensive study (e.g., Smith, et al., 1992). The dendritic pattern of tributaries and subtributaries, resulting from flooding by sea-level rise of the fluvially scoured river valleys, should especially be noted. Several major rivers conflow into the Chesapeake, each of which is an estuarine system in its own right, including the James, York, Potomac, Patuxent, and Susquehanna. An example of a fjord system, shown in Fig. 12.4, is the Vestfjord, the upper segment of the great Oslofjord (Gade, 1970). The great depths and orientations of the bedform structures are evidence of glacial scour. Galveston Bay, on the Gulf of Mexico coast, Fig. 12.5, is a lagoonal system, relatively shallow, and nearly landlocked behind a system of barrier islands. The shape of the bay is arcuate due to wind-wave erosion. The principal inflow is the Trinity River, whose delta is apparent in the northern reach of the system. Another Gulf of Mexico lagoonal system is Mobile Bay, Fig. 12.6, which is much more dominated by freshwater inflow than the Galveston system. An additional prominent example of this morphological class is the Albemarle-Pamlico Sound on the U.S. Atlantic coast (Roelofs and Bumpus, 1953; Inman and Dolan, 1989). The Golfo de Fonseca on the Pacific coast of Central America, Fig. 12.7, is an example of a tectonobay. Flanked by volcanic peaks, its bathymetry is irregular and varied, including submerged pinnacles, tidally scoured channels (including the notorious Dyer Strait), and extensive deltaic systems with mangrove swamps.

There are other systems of estuary classification, most notably based upon salinity stratification (Pritchard, 1955), upon density structure and gravitational circulation (Hansen and Rattray, 1966), and upon a combination of physiography, coastal geology, and climate (Fairbridge, 1980), but in our view the classification presented is most serviceable from the resource-management perspective. (Kennish, 1986, provides a literature summary of a variety of estuary classification approaches.) Perhaps a better approach to geomorphological classification is to recognize that the shape of the estuary is determined by *both* the geological structure of the coast, including its geological history, and the hydrodynamic environment, that is, waves, currents, marine bathymetry, and character of the bed sediments. Such a classification scheme—which has not yet been developed—would combine the two, such as a combination of the tectonic approach of Inman and Nordstrom (1971) with the process-based approach of Sheppard (1973).

12.2.2 Connection to the Sea: Mouths, Inlets, and Channels

The estuary mouth is in many respects the principal environmental control point for an estuary. This opening controls both the influence of the sea (tides, salts, and littoral sediments) and the river (freshwater, riverborne nutrients, and fluvial sediments). The estuary mouth is also the access for diadromous species, the migratory fishes, crabs, shrimps, turtles, and other animals that are central to estuary and shelf ecology. While the nature of the connection to the sea is somewhat implicit in the morphological classifications given earlier, it is of value to identify specific types since their properties affect the hydrography and water quality of the estuary.

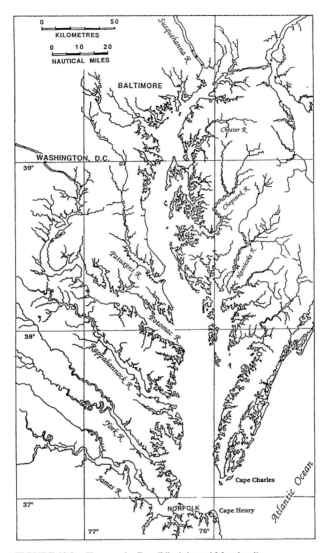

FIGURE 12.3   Chesapeake Bay (Virginia and Maryland).

There are two broad categories of *estuary mouth geometry*: (1) the convergent-section (or horn), and (2) the contraction-expansion (or contraction, or venturi) inlet.* The convergent-section mouth exhibits diminishing width more or less monotonically with distance upstream. This includes the classical funnel-shaped river mouth, such as the Chesapeake (Fig. 12.3), as well as bight-like bays, such as Golfo

---

* Coastal morphologists distinguish these by the respective terms *estuary* and *inlet,* a usage that is in obvious conflict with the present context.

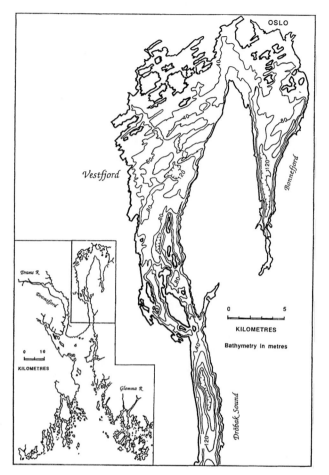

FIGURE 12.4 Vesfjord, upper reach of the Oslofjord (inset).

de Fonseca (Fig. 12.7), and many fjords. Opposite poles of this category of estuary mouth would be the deltaic channel of nearly constant cross section, and the coastline bight, such as Apalachee Bay (Florida), both of which press the limits of what is conventionally meant by an estuary. An approximate representation of the convergence of width is $W = W_o \exp\{-x/R\}$, for x measured from the coastline, in which R is a length measuring the reciprocal of the rate of convergence, i.e., the distance up the estuary required for the width to reduce to $1/e$ of the opening at the coastline. Shape parameters are R and W_o/R.

The contraction inlet includes the classical barrier-island tidal pass, of which the main entrances to Galveston Bay (Fig. 12.5) are examples, as well as bedrock-constrained inlets on hard-rock coasts such as the Golden Gate of San Francisco Bay. Inlets in sandy beaches are an important topic in coastal engineering (see USCE, 1984, for a summary). Shape parameters that have emerged from coastal engineering studies include A_c/A_b, L/W, and h/D, where A_c = cross-sectional area of

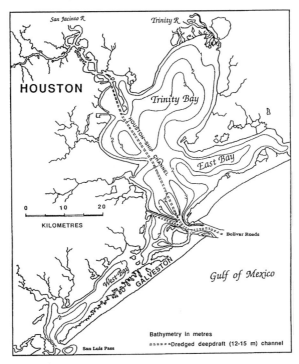

FIGURE 12.5   Galveston Bay (Texas).

inlet throat; A_b = surface area of the estuary behind the inlet; L = length of inlet; W = width of inlet throat; h = amplitude of ocean tide at inlet entrance; and D = controlling depth of the inlet. Additional inlet morphological parameters include the number of inlets affording access to the estuary, the degree of offset of the inlet entrance relative to the centroid of the estuary (or to its thalweg, the locus of greatest depths), seasonal variation in water levels and river flow, and the relative magnitudes of the inlet and the estuary volumes.

Estuary entrances can also be distinguished according to whether or not depth constrictions are present in the mouth. Fjords generally have an entrance constricted by a sill of glacial deposits. Some coastal plain estuaries have extensive shoals and bars deposited in the entrance reach, while others are swept clear by tidal currents. Inlets on sand beaches have extensive bar systems, both inside and outside the inlet (USCE, 1984; Boothroyd, 1978; and the case studies in Cronin, 1975, vol. 2). In some estuaries, under certain hydrological and hydrographic conditions, the entrance deposits shoal to the point of practically eliminating exchange between the sea and estuary; several examples of these *choked* or *blind* estuaries are found on the cape coast of South Africa (Heydorn and Tinley, 1980).

Which entrance configuration obtains for a specific estuary is determined by local geology, tides and tidal currents, river flow and fluvial sediment load, littoral longshore drift and wave climate (see Fairbridge, 1980), and the response of the inlet to rare cataclysmic events such as hurricanes, tsunamis, or intense midlatitude storms. The concern in the context of estuarine management is, however, not the

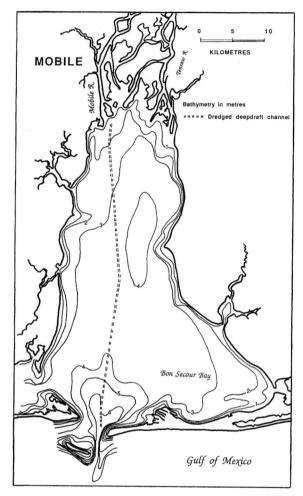

FIGURE 12.6   Mobile Bay (Alabama).

mode of formation, but rather the implication of entrance morphology for exchange dynamics of the estuary, especially the propagation of tides, development of density currents, and extent of retention of waters within the estuary boundaries.

12.3   HYDRODYNAMICS AND TRANSPORT PROCESSES

Practical estuary management is *always* a compromise between resources of data and analytical tools, and limitations of funding and time. This is especially true for hydrodynamics, for which proper measurements are expensive and rarely available

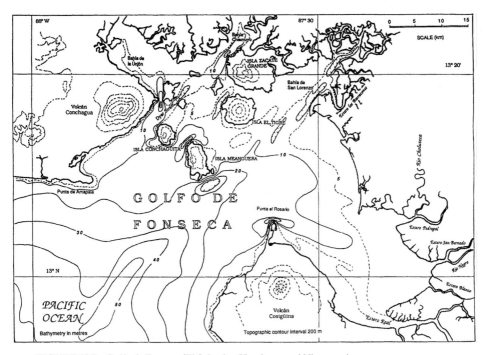

FIGURE 12.7   Golfo de Fonseca (El Salvador, Honduras, and Nicaragua).

from past surveys, and for which analytical methodologies are complex and costly. Moreover, hydrodynamics per se rarely represents the management endpoint, but rather an intermediate step toward that endpoint; for instance, in defining the transport of a contaminant. The nature of the management problem dictates space-time scales of concern, and therefore the extent to which the analysis can be simplified and data requirements minimized. In a new situation, e.g., an estuary that has received little prior study, or in which a novel management parameter (a recently identified toxicant, or exposure to a spill from a new industrial operation) is involved, space-time simplification should be implemented interactively with data analysis and modeling. Occasionally, management experience from another related estuary can be imported to shorten the process. Sometimes, lamentably, the press of time or a limitation on resources forces assumptions about the system, whose validity—established post facto, if at all—is dependent upon the judgement of the analyst.

The *fluid state of an estuary* is defined by current velocity (both speed and direction), water depth, and density. Variation in these parameters in space and time is induced by the combination of forcings to which an estuary is subjected. The usual analytical approach is to identify the predominant controls that force estuary motion, so that a preliminary inference about probable hydrodynamic responses can be made. These inferences are made quantitative by analysis of measured data, application of a model, or both. In this section, we first summarize the primary hydrodynamic controls involved in estuarine processes, then address the resultant circulation and transport processes, and those considerations underlying space-time simplification. For readers with interest and need for comprehensive treatments, we recommend Ippen (1966), Dyer (1973; 1979), and Officer (1976).

12.3.1 Tides

Tide refers generally to the movement of water—the currents and associated excursions of water level—resulting from the differential gravitational accelerations of the earth–moon–sun system. (There are other, nonastronomical sources of water movement to which the term *tide* is colloquially attached, but these usages are excluded here.) *Tide* is also applied specifically to the rise and fall of water level, as differentiated from *tidal current,* the intended usage usually clear from context.

12.3.1.1 Astronomical Forcing of the Oceanic Tide. The fact that tides are forced by gravitational interactions among astronomical bodies means that tidal analysis has a considerable advantage over the analysis of periodic behavior of a system of unknown forcing. The latter is approached by the spectral analysis of its measured response through application of the Fourier-Stieltjes transform. But in the case of the former, the frequencies that are expected to be present in the signal can be identified a priori, and specific techniques applied to the extraction of these signals from the measured tide record. These expected frequencies are determined from the theory of the *equilibrium tide,* which predicts the form assumed by a shallow layer of water (equal to the mass of the ocean) on a frictionless and uniform earth that would balance the gravitational forces instantaneously at each point on the surface of the planet. The major astronomical periodicities controlling the equilibrium tide are listed in Table 12.1, in which σ_i denotes the corresponding frequency (*speed*) in degrees per hour.

Each of the two highest frequency components σ_0 and σ_1 emerges in the computation as a squared harmonic, which can be rewritten as a sum of harmonics of the fundamental frequency and its double. Therefore, the equilibrium tide has frequencies arising from the rotation of the earth, σ_0 and $2\sigma_0$ with respect to the sun (solar day), and σ_1 and $2\sigma_1$ with respect to the moon (lunar day). This is the mathematical expression of the physical feature of the equilibrium tide of having a water-depth maximum on both the side of the earth toward the astronomical body and on the side opposite, so as the earth rotates, a point on its surface registers the passage of two maxima per day (lunar or solar). The longer period components appear in the equilibrium tide as modulators of the short period harmonic, i.e., of the form

$$(1 + a \cos \sigma_i t) \cos \sigma_k t = \cos \sigma_k t + \frac{a}{2} \left[\cos (\sigma_k + \sigma_i)t + \cos (\sigma_k - \sigma_i)t\right]$$

TABLE 12.1 Principal Astronomical Frequencies

Name	Frequency designation	Period, mean solar time
Solar day	σ_0	24 h
Lunar day	σ_1	24.84 h
Tropical month	σ_2	27.32 days
Tropical year	σ_3	365.24 days
Lunar perigee rotation	σ_4	8.85 yr
Lunar nodal retrogression	σ_5	18.6 yr

so that sidebands of sum $(\sigma_k + \sigma_i)$ and difference $(\sigma_k - \sigma_i)$ frequencies are generated. Moreover, the sum and difference frequencies of the longer-period components interacting among themselves behave as modulators as well. These harmonics can be expressed in turn as a function of a small number of astronomical parameters, e.g., the celestial longitudes of the moon, sun, lunar perigee, and lunar ascending node, all of which are capable of prediction by astronomical methods.

Because of its proximity, the tidal forcing of the moon is over twice that of the sun, so the high frequency component $2\sigma_1$ is dominant. The tide therefore has a basic periodicity of 12.4 hours with a subordinate periodicity of 12.0 hours. The superposition of these is a low-frequency variation of $2(\sigma_0 - \sigma_1) \cong 1.016°/h$, or a period of about 354.3 hours. This is the fortnightly spring-neap cycle, in which the sun–moon–earth system passes from alignment (*syzygy*) to quadrature and back, marked by the phase of the moon. The declination of the moon or sun (i.e., its elevation angle above the equatorial plane) induces an inequality in the tidal maxima and minima during each rotation of the earth. The more important is the declination of the moon, passing through a complete cycle in about 27.2 days (the *nodical month*), i.e., a frequency of $\sigma_2 + \sigma_5$. The maximum absolute value of the declination (*great declination*) creates the greatest inequality and therefore the greatest 24.8-h range of tide, referred to as a tropic tide. Zero declination results in a minimal inequality, the equatorial tide. The 27.2-day lunar-declination cycle includes a positive and negative maximum declination, so there are two occurrences of tropical and equatorial tides during the cycle. The beat frequency $2[(\sigma_0 - \sigma_1) - (\sigma_2 + \sigma_5)] \cong 2.1°/day$ defines the 173-day period in which the spring-neap and tropical-equatorial cycles come into phase, maximizing their combined range.

The larger harmonic components (*partial tides*), their periods and frequencies, and conventional designations are given in Table 12.2. Hydraulic analyses generally focus on one semidiurnal component M_2 and one diurnal component K_1 to characterize the propagation of the tide. (While these components have a precise period, there is an informal patois in which the M_2—or K_1—component is referred to as *the* semidiurnal—or diurnal—tide.) The ratios of the amplitudes of the dominant components are frequently used as a convenient index for classifying tides, for example,

$$F = \frac{K_1 + O_1}{M_2 + S_2}$$

defines a *semidiurnal* tide if $F < 0.25$, *diurnal* if $F > 3.0$, and *mixed* otherwise.

The real ocean tide is considerably different from the equilibrium tide, being additionally affected by friction, inertia, bathymetry, and the barriers to flow of continents. The net effect is that the ocean behaves as a filter, suppressing or enhancing different frequencies in the tidal signal, a filter whose characteristics

TABLE 12.2 Predominant Components of the Equilibrium Tide, Mean Solar Hours

Semidiurnal			Diurnal		
Designation	Frequency, °/h	Period, h	Designation	Frequency, °/h	Period, h
M_2	28.984	12.42	K_1	15.041	23.93
S_2	30.000	12.00	O_1	13.943	25.82
N_2	28.440	12.66	P_1	14.959	24.07
K_2	30.082	11.97	Q_1	13.399	26.87

vary from place to place and which unfortunately can be determined only by direct observation. As noted above, knowing what frequencies can be expected allows empirical determination of their amplitudes and phases, and their dependency upon the controlling astronomical variables, from which tidal predictions are prepared. This requires operation of tide gauges at the desired locations over a sufficient duration of time to encompass the important tidal periodicities. (This duration ranges from a few months to 20 years, depending upon the dominant components at that location and the required accuracy; see Pugh, 1987. Practical tidal analysis considers the complementary question of what constituents can be resolved from a record of limited duration.) Strong signals of meteorological origin, e.g., the effect of daily solar heating, the annual climatological cycle, and the development of a daily seabreeze, are included in the measured water levels and therefore are also incorporated into the coefficients for that periodicity. For example, the seabreeze and daily solar heating frequently augment the K_1 and P_1 components.

A *tide gauge* is a device for measuring the detailed time variation of water level at or near the coastline. Traditionally, float-type gauges have been employed for this purpose, geared to a drum-type recorder. In recent years, a variety of different sensors have come into use, such as *bubblers* (in which the rate of bubble production depends upon orifice pressure), *laser* or *acoustic pulse transmitters* (in which the time of passage from a signal source to the water surface is converted to water-surface elevation), and piezometric pressure transducers (which generate an electric signal in proportion to the pressure, hence the depth of the overlying water). Also, *digital data logging* has virtually replaced the older analog recorders, and the use of telemetry links to a central recording station, allowing real-time data acquisition and display, is becoming common. Despite these technological advances, the number of tide gauges operating on a coastline is still limited, due to the expense of each gauge, and problems of access, installation, maintenance and security. The historical method has been to establish a small number of permanent gauges, for which extensive records are maintained and full harmonic analyses performed. Tidal predictions are published for these *reference stations*. The relations for tidal extrema between other, nonreference (secondary) stations and the nearest appropriate reference station are determined from shorter, less reliable gauge records at the former, and are provided as an addendum to the tide predictions.

There is an extensive literature on the venerable subject of tides, yielding however much detail the reader may be masochistic enough to desire. General principles are treated in oceanography texts (e.g., Proudman, 1953; Defant, 1961; and Neumann and Pierson, 1966). Harmonic analysis and prediction are described by Pugh (1987) and Godin (1972), which contain references to the basic mathematical literature. A more general hydraulic analysis is given by Dronkers (1964). The U.S. practice of tidal analysis is summarized by Shureman (1958), and those aspects of pertinence to U.S. coastal and navigation engineering, especially as engaged by the U.S. Corps of Engineers, are presented by Pillsbury (1956). An advanced reference, with much modern material, is the volume edited by Parker (1991). Tidal predictions and associated information (datums, archival tide records, harmonic constants, manuals, and coastal pilots) are distributed in the United States by the National Ocean Service:

> Tidal Analysis and Prediction
> National Ocean Service, NOAA
> 1305 East West Highway
> Silver Spring, Maryland 20910

12.3.1.2   *Tides in Estuaries.* To a first approximation, the tide may be viewed as
being forced in the open ocean from which it propagates as a progressive long wave
into the shallower coastal regions. The effects of the shallower water are to retard the
speed of the wave as the square root of depth, and to increase the amplitude inversely
as the fourth root of depth. Depending upon the geometry of the coastal zone, partial
wave reflection may occur as well. The effect of frictional drag is to dissipate energy
and to produce higher harmonics of the frequencies present in the waveform. Thus
the tide in shallower coastal areas can be considerably distorted from that of the open
ocean. It is still considered to be characterized by harmonics deriving from astronomical frequencies (with integral multiples due to nonlinearities).

The primary results of tidal dynamics as they apply to tidal motions in an estuary
(see Lamb, 1932; Ippen, 1966; Dronkers, 1986; Uncles, 1988) may be summarized as
follows:

- The tide propagates as a long wave, i.e., one whose wavelength λ is large compared to the mean water depth D.
- The speed (*celerity*) of the propagating wave is approximately $C_o = (gD)^{1/2}$, g
  denoting acceleration of gravity, and its corresponding wavelength $\lambda = C_o T$ for
  period T.
- In the open ocean, the tidal wave is progressive. For a wave of harmonic form, surface elevation ζ relative to the mean is given by

$$\zeta = H \cos 2\pi \left(ft - \frac{x}{\lambda} \right)$$

  in which H is wave amplitude, x is distance in the direction of propagation and
  f is frequency (cycles per unit time).
- The current associated with the progressive wave is given by

$$u = H \frac{C_o}{D} \cos 2\pi \left(ft - \frac{x}{\lambda} \right)$$

  i.e., the tidal elevation and the tidal current are in phase.
- In an enclosed coastal basin of uniform depth in communication with the ocean,
  tides are forced by propagation of the ocean tide into the basin, rather than by
  direct astronomical forcing, referred to as a *co-oscillating* tide.
- The presence of barriers to flow allows wave reflection, so that the tide and tidal
  current consist of the superposition of two oppositely propagating progressive
  waves of amplitudes H and H_r, respectively,

$$\zeta = H \left[(1 + r) \sin \frac{2\pi x}{\lambda} \cos 2\pi ft - (1 - r) \cos \frac{2\pi x}{\lambda} \sin 2\pi ft \right]$$

$$u = H \frac{C_o}{D} \left[(1 - r) \sin \frac{2\pi x}{\lambda} \cos 2\pi ft - (1 + r) \cos \frac{2\pi x}{\lambda} \sin 2\pi ft \right]$$

  where $r = H_r/H$, the reflection coefficient.
- As the barriers to flow (i.e., the shoreline) are approached, the reflection coefficient increases. This induces a phase lag between tide and tidal current, with current leading water level.

- In the upper reaches of a co-oscillating basin, the reflection coefficient approaches unity, and the tide becomes a standing wave with current leading water level by 90°. For a standing wave, the water level rises and falls simultaneously throughout the basin.

- The effect of friction is to dampen the amplitude of the tide and retard its propagation. Friction substantially diminishes the reflection coefficient.

- Most coastal plain and river channel estuaries exhibit a diminishing cross section with distance upstream. This convergence of cross section increases the tidal amplitude. The behavior of the tide range with distance upstream is the result of opposing influences of amplification due to cross-section convergence and dampening due to friction.

- An additional effect of friction, specifically its nonlinearity, in combination with cross-section convergence is the generation of overtides, at multiples of the basic tidal period. For estuaries in which the semidiurnal tide M_2 is prominent, the most important overtide is usually half that period, denoted M_4. Intertidal storage in estuaries also contributes to development of overtides (Friedrichs and Aubrey, 1988).

- Two degrees of freedom in the horizontal plane, as would be afforded in the open ocean, a broad bay, or large complex fjord, allow the development of rotary currents, in which the direction of the tidal current varies through the full 360° with the tide. Rotary currents can be caused by the astronomical tidal forcing itself, by bathymetry, and by the rotation of the earth. The last is generally predominant, resulting in a *cum sole* turning of the tidal current (clockwise in the northern hemisphere).

- The families of *cotidal lines* (isopleths of synchronous tides, i.e., equal phase) and *corange lines* (isopleths of equal tidal range) generally form an intersecting net. If friction is negligible, these families are orthogonal. As r varies from $r = 0$ (progressive wave) to $r = 1$ (standing wave), the cotidal lines range from aligning perpendicular to the direction of maximum current to aligning parallel.

Tide Observations and Analysis. Any analysis of tidal behavior of an estuary must be founded upon measured tides, usually available from a public agency. The objective of operating tide gauges in an estuary is to determine the range of tide and how it varies with position, most often for navigational purposes, but also for determining limitations on shoreline development and ownership (Shalowitz, 1962), and for issuing warnings of extremes in water level. The minimum tide gauge distribution for this purpose is a gauge that measures the oceanic tide, placed either at the mouth of the estuary or on the ocean shoreface, and a gauge measuring the tide in the inland reach of the estuary. For irregular estuaries, especially those that are large enough that internal tide propagation can become complex, more inland gauges will be needed. Since such a network of "tide" gauges in fact measures water-level variation, meteorological responses of the estuary as well as astronomical tides would be obtained in the process. A careful analysis can allow separation of the two, and determination of the response of the estuary to wind and atmospheric pressure.

As a general rule, tide range diminishes with distance into the estuary, due to frictional dissipation as the tide wave propagates. However, as summarized previously, the morphology of the estuary can amplify the tidal wave so as to create internal regions of higher tides. Two examples are shown in Figs. 12.8 and 12.9, from the Delaware and the St. Lawrence. The Delaware is a classic example of a coastal plain

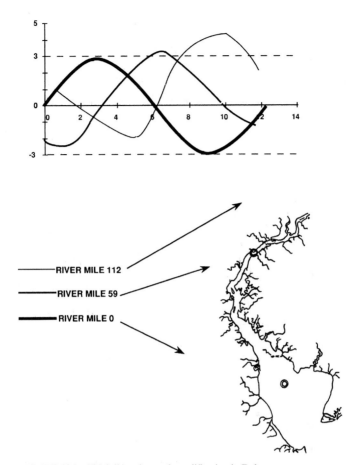

FIGURE 12.8   Tidal distortion and amplification in Delaware estuary.

estuary. The St. Lawrence is a fjord, but the St. Lawrence channel per se is shallower and receives a large inflow from the river, and therefore exhibits features more like a coastal plain system. In both cases, there is a region in the midzone of the estuary in which the tidal range is amplified. In both cases, there is a lag in propagation and distortion of shape of the tide as it propagates upstream.

The value of tide gauging in estuary management is to determine tidal water exchange and movement. Therefore, the problem is to infer currents from observations of tides. Even with a satisfactory network of tide gauges, inference of currents must be founded upon dynamic analysis. The hydrodynamics of tidal propagation in an estuary is primarily acceleration of the fluid forced by the hydrostatic head of varying water-surface elevation. Depending upon the estuary morphology, this acceleration is further modified by a combination of frictional dissipation and the local component of the earth's rotation. The principal tidal phenomena in most estuaries can be analyzed by assuming the water homogeneous and by treating the

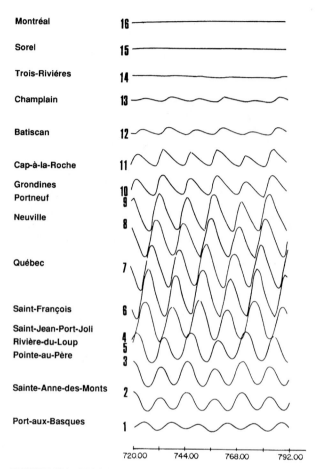

Montréal	16
Sorel	15
Trois-Riviéres	14
Champlain	13
Batiscan	12
Cap-à-la-Roche	11
Grondines **Portneuf**	10 9
Neuville	8
Québec	7
Saint-François	6
Saint-Jean-Port-Joli **Rivière-du-Loup** **Pointe-au-Père**	4 5 3
Sainte-Anne-des-Monts	2
Port-aux-Basques	1

720.00   744.00   768.00   792.00

FIGURE 12.9 Tidal distortion and amplification in the St. Lawrence estuary. Estuary mouth is at bottom. (*Godin, 1979.*)

vertical-mean currents. The model embodying these assumptions is the so-called *shallow water equations:*

$$\frac{\partial u}{\partial t} + u\frac{\partial u}{\partial x} + v\frac{\partial u}{\partial y} = -g\frac{\partial h}{\partial x} + fv + \frac{1}{\rho D}\left(\tau_{sx} - \tau_{bx}\right)$$

$$\frac{\partial v}{\partial t} + u\frac{\partial v}{\partial x} + v\frac{\partial v}{\partial y} = -g\frac{\partial h}{\partial y} - fu + \frac{1}{\rho D}\left(\tau_{sy} - \tau_{by}\right) \tag{12.1}$$

$$\frac{\partial h}{\partial t} + \frac{\partial u}{\partial x} + \frac{\partial v}{\partial y} = 0$$

where u,v = vertical-means of x and y component current
         h = instantaneous water level
         f = coriolis parameter
         τ_s, τ_b = stresses at the estuary surface and bed
         ρ = water density

This type of model, rather than the simpler one-dimensional form, is mandated when the estuary is large enough that rotary currents and complex trajectories can develop. These equations can only be solved numerically, for which there is a considerable body of literature (e.g., Weiyan, 1992), and there exist several computer programs of general applicability. The proper formulation of bed stress is a complex and subtle matter in itself, whose review is beyond the scope of this brief treatment. A *quadratic stress* is usually employed for most coastal and estuary problems (i.e., $\tau_b \propto \mathbf{u} \mid \mathbf{u} \mid$), with coefficients developed in the theory of open-channel uniform flow, a practice founded more on tradition than physics.

One index to the intensity of tidal currents is the *tidal excursion*, the distance that a parcel of water is carried by the incoming tide. For a harmonic nonrotary tidal current, this is

$$\xi = \int_0^{T/2} U_o \sin \sigma t \, dt = \frac{T U_o}{\pi}$$

Tidal excursion is clearly a function of location in the estuary and of the frequencies in the tide. A bulk measure of tidal transport is the *tidal prism*, the volume crossing a section of the estuary on the incoming tide. For tidal flow across a section of area A, the prism is $\Pi = A \, \xi$.

One-Dimensional Tidal Propagation.   In the case of a predominantly longitudinal estuary that can be treated one-dimensionally, the analysis is amenable to a simpler approach. The governing equations become

$$\frac{\partial u}{\partial t} + u \frac{\partial u}{\partial x} = -g \frac{\partial \zeta}{\partial x} - \frac{g \, u \, |u|}{C^2 D} \tag{12.2a}$$

$$\frac{A}{D} \frac{\partial \zeta}{\partial t} + \frac{\partial A u}{\partial x} = 0 \tag{12.3}$$

in which u now denotes the section-mean current over the cross section of area A and mean depth D; C is the coefficient of de Chézy; ζ is the instantaneous water level (relative to the mean level) averaged over the width; and the surface stress is neglected. It can be useful to study such tidal behavior by simplified models amenable to a spreadsheet or even hand calculation, some of which are summarized here (see, e.g., Pillsbury, 1959, and Dronkers, 1964). These methods universally assume sufficiently gradual changes along the longitudinal axis, so that the vexatious nonlinear term is negligible in comparison to the head gradient.

As a first approximation, friction may be neglected, so Eq. (12.2a) simplifies to:

$$\frac{\partial u}{\partial t} = -g \frac{\partial \zeta}{\partial x} \tag{12.2b}$$

implying

$$\frac{\partial^2 \zeta}{\partial t^2} = g \, D \, \frac{\partial^2 \zeta}{\partial x^2}$$

which is the wave equation for a waveform with celerity $C_o = \sqrt{gD}$. For a single tidal harmonic of frequency σ (radians per unit time, $\sigma = 2\pi f$) propagating in the positive x direction with celerity c,

$$\zeta(x,t) = H \cos \sigma\left(t - \frac{x}{c}\right) \equiv H \cos(\sigma t - \kappa x)$$

where *wavenumber* $\kappa \equiv 2\pi/\lambda = \sigma/c$, λ denoting the wavelength. Since $u = (C_o/D)\,\zeta$, as noted above, the extrema in current speed coincide with extrema in tidal stage, a property that is diagnostic for a progressive wave. One boundary condition is always applied, namely the tidal variation at the estuary entrance. Taking $x = 0$ here, with x directed up the estuary, this is $\zeta = H_o \cos(\sigma t - \varphi_o)$.

If the estuary has constant cross section and terminates at $x = L$ at a head of tide (e.g., a barrier, dead-end basin, or fall line), where the boundary condition $u(L) = 0$ applies, then

$$\zeta = \frac{H_o \cos(\sigma t - \varphi_o) \cos \kappa(L - x)}{\cos(\kappa L)}$$

$$u = -\frac{(gH_o/c) \sin(\sigma t - \varphi_o) \sin \kappa(L - x)}{\cos(\kappa L)}$$

This is a case of perfect reflection at the barrier $x = L$, so that $r = 1$, and ζ can be resolved into upstream and downstream propagating waves of equal amplitude and celerity $c = C_o$. The composite waveform is stationary, stage lags current by 90° (a property diagnostic for a standing wave), and nodes are at $\kappa(L - x) = (1 + 2k)\pi/2$, k an integer. At the extrema, at $\kappa(L - x) = k\pi$, the range is $2H_o/\cos(\kappa L)$. This range becomes infinite if $L = (1 + 2k)\pi/2\kappa = (1 + 2k)\,\lambda/4$, a situation of resonance in which the channel-estuary length L is "tuned" to the tidal wavelength. As an example, for the principal lunar frequency $\sigma = 2\sigma_1$ and a depth of $D = 10$ m, the tidal wavelength $\lambda = \sqrt{gD}\,T \approx 450$ km, so the minimum channel length to exhibit resonance would be about 110 km. (A tidal extremum occurs at the reflecting barrier $x = L$, so this resonance condition is equivalent to there being a node at the entrance $x = 0$.) These results are modified substantially if the geometry of the estuary varies with distance along its axis. For example, for an estuary with constant depth and an exponentially converging width, say $W = W_o \exp\{-x/R\}$, the amplitude of a wave progressing up the estuary is found to increase as $\exp\{x/2R\}$ and that of a wave progressing down the estuary to decrease as $\exp\{-x/2R\}$. Moreover the wavenumber is decreased by $\kappa^2 = \sigma^2/gD - (2R)^{-2}$.

With friction included, the tidal behavior changes. For a uniform cross section, a progressive wave is found to decay in amplitude (e.g., Ippen, 1966), so that

$$\zeta = H_o \exp\{-\mu x\} \cos(\sigma t - \kappa x)$$

$$u = \frac{H_o \sigma}{D}\,(\mu^2 + \kappa^2)^{-1/2} \exp\{-\mu x\} \cos(\sigma t - \kappa x + \varphi)$$

With a reflected wave, H declines and H_r increases in amplitude with distance up the estuary, so that r increases. If the estuary is sufficiently long (or sufficiently shallow) the composite tide changes from progressive to standing with distance up the estuary. Wavenumber κ is increased relative to the frictionless value, $\kappa^2 = \sigma^2/gD + \mu^2$. With a variable geometry, the tidal response is even more complex. A converging cross section (such as the exponential configuration considered above) increases the tidal range while friction decreases it. Whether the tide range increases or decreases therefore depends upon the relative influence of the two factors. For the exponentially converging section $W = W_o \exp \{-x/R\}$, the total tide variation is given by

$$\zeta = \frac{H_o}{2} \exp \left\{ -\frac{x}{2R} \right\} [\exp \{\mu x\} \cos (\sigma t + \kappa x) + \exp \{-\mu x\} \cos (\sigma t - \kappa x)] \quad (12.4a)$$

and the variation of amplitude by

$$H/H_o = \exp \left\{ -\frac{x}{2R} \right\} [(\cos 2\kappa x + \cosh 2\mu x)]^{1/2} \quad (12.4b)$$

(see Lamb, 1932; Harleman, 1966; Dronkers, 1986. An exponential width convergence that exactly balances friction, so that the tidal range remains constant, is referred to as an *ideal* estuary; see Pillsbury, 1956).

As a rough estimate, for an estuary of relatively simple geometry, these equations can be used to determine the variation of tide and tidal range. The coefficient μ, referred to as the damping parameter or damping modulus, is an empirical parameter that typically ranges 0.001 to 0.1 km^{-1}. If tidal records are available from two or more sites, values of μ can be fitted to match the observed ranges. If a tide record is available at only one position along the estuary, a typical value can be assumed. This is not an accurate procedure, however, because μ is not governed strictly by physical properties such as frictional roughness, but also by wavelength (itself influenced by geometry and friction) and the local reflection coefficient, which means that μ is a strong function of x and of tidal wave mechanics. Moreover, Eqs. (12.4 a and b) prove to be excessively sensitive to the precise values and longitudinal distribution of μ.

Once ζ is determined, the associated expression for u can be obtained. For the channel estuary (of constant cross section) with constant μ and κ,

$$u_{\max} = U_o = \frac{\cosh 2\mu x - \cos 2\kappa x}{\mu \sinh \mu x \cos \kappa x + \kappa \cosh \mu x \sin \kappa x}$$

from which tidal excursion ξ and tidal prism $\Pi = A\xi$ can be computed. Rather complicated closed-form expressions (in terms of exponential and trigonometric functions) are also obtainable for converging estuaries of various geometries, such as the exponential form considered above (see Ippen, 1966, and citations therein).

Fitting a simple geometric form to an estuary and utilizing an exponential damping to model frictional attenuation of the tide are both artifices that seek to retain the simplicity of simple harmonic tidal waves. These can be relaxed and the harmonic tide form retained by discretizing the longitudinal axis of the estuary, as follows.

Step 1. Subdivide the estuary into longitudinal segments of length Δx_i, at the ends of which tidal records are available. Based upon water level and mean current defined at the ends of each segment, the linearized momentum equation is approximated by:

$$\frac{1}{g} \left(\frac{\partial u_i}{\partial t} \right) + \frac{(\zeta_{i+1} - \zeta_i)}{\Delta x_i} + \frac{u_i |u_i|}{(C^2 D)} = 0$$

where u_i = section-mean current at midpoint of segment i
ζ_i, ζ_{i+1} = water levels (relative to mean) at lower and upper ends of segment i

C is estimated from knowledge of bottom texture (e.g., Henderson, 1966) and bed-forms, and typically ranges 40 to 90 √m/s in estuaries.

Step 2. Assume a harmonic form for the tide:

$$\zeta_i = H_i \cos{(\sigma t - \varphi_i)}$$

where H_i = amplitude of tide at lower end of segment i
φ_i = phase of tide at lower end of segment i

As before, $\sigma = 2\pi f$ is the frequency of the principal tidal component. The values of H_i and φ_i are found by inspection of the tidal records at the ends of the segment. The surface-head term is also harmonic

$$\zeta_{i+1} - \zeta_i = \mathcal{H}_i \cos{(\sigma t - \Phi_i)} \tag{12.5}$$

Compute its amplitude $\mathcal{H}_i$ and phase Φ_i from:

$$\tan \Phi_i = \frac{H_{i+1} \sin{(\varphi_{i+1})} - H_i \sin{(\varphi_i)}}{H_{i+1} \cos{(\varphi_{i+1})} - H_i \cos{(\varphi_i)}} \tag{12.6a}$$

$$\mathcal{H}_i = \frac{H_{i+1} \cos{(\varphi_{i+1})} - H_i \cos{(\varphi_i)}}{\cos \Phi_i} \tag{12.6b}$$

with the quadrant of Φ_i determined from the signs of

$$\mathcal{H}_i \sin \Phi_i = H_{i+1} \sin{(\varphi_{i+1})} - H_i \sin{(\varphi_i)}$$

$$\mathcal{H}_i \cos \Phi_i = H_{i+1} \cos{(\varphi_{i+1})} - H_i \cos{(\varphi_i)}$$

Step 3. Assume a harmonic form for the tidal current:

$$u_i = U_i \cos{(\sigma t - \psi_i)} \tag{12.7}$$

from which the acceleration is

$$\partial u_i / \partial t = -\sigma U_i \sin{(\sigma t - \psi_i)}$$

and the friction term is approximately

$$\frac{u_i |u_i|}{(c^2 D)} = \frac{8\,U_i^2}{3\pi c^2 D} \cos{(\sigma t - \psi_i)}$$

so that the momentum equation becomes

$$-\frac{\sigma U_i}{g} \sin{(\sigma t - \psi_i)} + S_i \cos{(\sigma t - \Phi_i)} + \frac{8\,U_i^2}{3\pi c^2 D} \cos{(\sigma t - \psi_i)} = 0$$

where we define $S_i \equiv \mathcal{H}_i / \Delta x_i$. Expressions for U_i and ψ_i are found by setting $\sigma t = 0$ and $\sigma t = -\pi/2$, and cross-multiplying by $\sin \psi_i$ and $\cos \psi_i$, viz.

$$S_i \sin{(\Phi_i - \psi_i)} = \sigma U_i / g \tag{12.8a}$$

$$S_i \cos{(\Phi_i - \psi_i)} = -\frac{8\,U_i^2}{3\pi c^2 D} \tag{12.8b}$$

Since these equations involve only nonnegative quantities, $\Phi_i - \psi_i$ is an angle between $90°$ and $180°$. Compute $\Phi_i - \psi_i$ from:

$$\cos(\Phi_i - \psi_i) = p - \sqrt{1 + p^2}$$

$$p = \frac{3\pi c^2 \sigma^2 D}{16\, S_i g^2} \tag{12.9}$$

Compute U_i from either of Eqs. (12.8a) or (12.8b) depending upon how close $\Phi_i - \psi_i$ is to either $90°$ or $180°$.

This procedure depends upon having tidal records at the ends of each reach. Of course, tide gauges are rarely distributed so conveniently, so observations will usually provide tide records from gauges at best widely separated along the channel length. These gauges could be used to define subreaches of the estuary, but these may prove to be too long for the difference approximations used above to be trustworthy. It is usually better to use shorter subreaches, at the ends of which tide range and phase are estimated. The computation of current at these intermediate reaches is then compromised in accuracy. If there is a means of reliably computing current in any segment of the estuary (see following), this can be extended to the adjacent segments by the continuity equation, which is the final step in the procedure:

Step 4. From the continuity equation Eq. (12.2b):

$$A_i u_i = A_{i-1} u_{i-1} - a_i \frac{\partial \zeta_i}{\partial t} \tag{12.10}$$

and the assumed harmonic variation of the tide,

$$A_i u_i = A_{i-1} u_{i-1} + a_i H_i \sigma \sin(\sigma t - \varphi_i)$$

where A_i = cross-sectional area at midpoint of segment i

a_i = surface area between midpoints of segments $i - 1$ and i

Working from a segment of known velocity, say

$$u_o = U_o \cos(\sigma t - \psi_o)$$

successive application of Eq. (12.10) yields the current phase and amplitude at segment i

$$\tan \psi_i = \frac{U_o A_o \sin \psi_o + \sum\limits_{k=o}^{i-1} a_k H_k \sigma \sin \varphi_k}{U_o A_o \cos \psi_o + \sum\limits_{k=o}^{i-1} a_k H_k \sigma \cos \varphi_k}$$

$$U_i = \frac{U_o A_o \cos \psi_o + \sum\limits_{k=o}^{i-1} a_k H_k \sigma \cos \varphi_k}{A_i \cos \psi_i}$$

with the quadrant of ψ_i determined from the signs of

$$U_i A_i \sin \psi_i = U_o A_o \sin \psi_o + \sum\limits_{k=o}^{i-1} a_k H_k \sigma \sin \varphi_k$$

$$U_i A_i \cos \psi_i = U_o A_o \cos \psi_o + \sum\limits_{k=o}^{i-1} a_k H_k \sigma \cos \varphi_k$$

The accuracy of this computation can be improved by applying the basic procedure in various successive approximations. For example, if observed tide records are available only at the head and entrance of an estuary, the current at the central reach of the estuary is computed by assuming a straight-line water slope S_i, i.e., by carrying out Steps 1 to 3 assuming the estuary to be represented by a single reach. Then Step 4 is applied, computing from this central reach where current is now known to both the estuary head and entrance. The resulting current amplitudes and phases are used in the equations of Step 3 to determine the surface-head at each segment, which are then summed over the segments to compute the tide variation at the ends of the estuary. This is compared to the given tide records, the slope at the central reach readjusted and the computation repeated.

If there is only a tide gauge at the estuary entrance, but the inland reach of the estuary terminates at a head of tide, the current is specified here to be zero, a first-approximation tidal record is assumed (for convenience, the same amplitude and phase as given at the entrance), and Step 4 is applied to compute current amplitude and phase at the remaining sections. These are used to determine surface-head amplitude and phase, using the equations of Step 3, and summed to compute the entrance channel tide. This is compared to the entrance record given, the assumed tide at the estuary head readjusted, and the computation repeated.

Similarly, the tides and currents computed by this procedure can be used as first approximations to estimate the distortions introduced by the nonlinear advection term $u\partial u/\partial x$ or to accommodate higher frequencies (overtides) in the computation. This general method was employed for many years by the U.S. Corps of Engineers for tidal computations in estuaries (see, e.g., Pillsbury, 1956).

The attractiveness of these calculations is that they exploit the assumption of a harmonic form to simplify the arithmetic. Several approximations must be introduced to maintain this simplicity. For a specific estuary, these approximations may be dubious: There may be more than one dominant frequency in the tide, a simple damping parameter may not be appropriate, or the estuary geometry may be too complex to be depicted by simple algebraic functions. While these can be accommodated in principle by more complicated analyses, or in the case of the discretized segments, further subdividing the estuary, the computational labor increases quickly, undermining the benefits of a "simple" hand calculation. In these cases, it is recommended that a fully general numerical model be employed instead. For a longitudinal estuary, such a model is easily implemented, and its generality and ease of application justify its use. General numerical solutions suitable for a digital computer are presented in numerous references, e.g., Dronkers (1964), Harleman (1971), and Vreugdenhil (1989).

Cubature.   One additional computation, which can often be applied to a set of tide records to determine the tidal prism and the associated currents, is the so-called *method of cubature.* This is a direct application of the continuity equation Eq. (12.3) integrated from a tide station at x_1 (say) to a tide station at x_2,

$$\Delta Q_1 \equiv Q_1 - Q_2 = a\, \partial\, \bar{\zeta}/\partial t$$

where a is the surface area of the estuary between x_1 and x_2, and Q_i is the instantaneous flow at station x_i. The mean tidal variation $\bar{\zeta}$ over the section from x_1 to x_2 is the crucial quantity that must be approximated by the average of the records at x_1 and x_2, which requires that they be close enough that the water surface can be assumed planar between them. The rate of change of mean water level $\partial\, \bar{\zeta}/\partial t$ is computed by finite difference approximation (for which a 30-min time increment is typical). The flow at x_1 is then found by summing the incremental discharges up to a station where flow is known (e.g., a head of tide), $Q_1 = \Sigma\Delta Q_i$. Practical application requires close attention to selection of a representative tidal cycle for the calculation and how the time differencing and spatial averaging are performed (see Pillsbury, 1956). The tidal prism at x_1 is determined by integrating in time from the time of slack current before flood t_{so} to the time of slack current before ebb t_{s1} at x_1,

$$\Pi_1 = \sum_{i=1}^{n-1} a_i [\overline{\zeta}_i(t_{s1}) - \overline{\zeta}_i(t_{so})] + Q_n$$

The *component tidal prism* for each section is $a_i [\overline{\zeta}_i(t_{s1}) - \overline{\zeta}_i(t_{so})]$. The difference $[\overline{\zeta}_i(t_{s1}) - \overline{\zeta}_i(t_{so})]$ looks like a tidal range (and if stations i and $i+1$ are close enough it may be approximated by the mean of their tidal ranges). Much violence has been done to this equation by interpreting it as the sum of the tidal range times surface area for area segments between tide stations. In fact, substantial error will be incurred if computed this way, since the time lag in propagation of the tidal wave is neglected. The time integration, it should be noted, is carried out from slack water to slack water *at station 1*, which may not correspond to slack waters at station 2. This method of cubature is best adapted to a longitudinal estuary with tide stations along its length. However, it is often extended to broad, irregular estuaries. In this application, the sections must be bounded by nonflow boundaries (shorelines, jetties, dikes, and so forth) and by cotidal lines.

12.3.2 Hydrology and Freshwater Inflows

One of the defining properties of an estuary is that within its boundaries seawater is intermixed with freshwater. The sources of this freshwater—and more generally the budget of freshwater—are referred to collectively as the *hydrology of the estuary.* The physical bounds of estuarine hydrology are the watershed draining into the estuary, the water surface and overlying atmosphere, the bed, and aquifers in direct or indirect contact with estuarine waters, in order of generally decreasing importance. All of the considerations and techniques of the science of hydrology (Eagleson, 1970; Chow, et al., 1988; Maidment, 1993) apply therefore to estuarine hydrology. In most estuaries, the freshwater inflow is the only source of throughflow, therefore the volume of the inflow averaged over a sufficiently long period is also the volume debouching to the sea. On a more dynamic basis, inflow hydrographs can completely displace water in a region of an estuary, and if applied in sufficient volume over a short duration can displace the entirety of the estuary water volume (e.g., Schroeder, 1977). Inflows in estuaries are major sources of nutrients, sediments and other waterborne physicochemical constituents (see Sec. 12.4). Large estuaries can have multiple sources of inflows, entering the system through various drainageways, each governed by the hydrological processes operating within its own watershed. The resulting estuary response can therefore be a complex function of space.

Special considerations in estuary hydrology can apply to the following:

- Characteristic seasonal patterns of hydrology
- Specific time patterns of inflow, especially flood-hydrograph morphology, quickflow spikes, and drought
- Spatial distribution in the estuary of the sources of inflow, and its relation to estuary morphology
- Long-term alterations in inflow volume, especially due to anthropogenic alterations of the watershed
- Development of runoff and inflow from peripheral coastal watersheds

Because salinity is a direct measure of the ratio of seawater to freshwater, analysis of estuary hydrology usually proceeds closely with the analysis of its salinity (treated in the following section).

The first concern of the analyst is appraisal of the data resources available, i.e., the streamflow gauges operated within the estuary watershed, the extent of coverage of the principal drainageways, period of records, archival practices and formats, etc. To the extent that gauged streamflow data is unavailable, methods of synthetic hydrology may be necessitated to estimate inflows (see Maidment, 1993, and references therein). Even when the inflowing streams are satisfactorily gauged, there will be a limit to how far downstream such a gauge can be situated, due to the corrupting effects of tides on the measurement of stage. Consequently, the lowermost segment of a watershed is rarely gauged (the exception being when the drainageway passes through a critical-flow transition, such as a fall line, weir or headgate), and the inflow contributed by this ungauged portion will have to be estimated separately. Moreover, because the ungauged segment lies within the coastal zone, it may be governed by a different hydroclimatology than the inland section of the watershed, for which simple methods of drainage-area extrapolation would not be appropriate. For many estuaries, during some parts of the year, the additional effect of precipitation and evaporation directly on the water surface must be determined. Again, it may be inappropriate to extrapolate data acquired at a terrestrial station, due to the different hydroclimatology operating in the coastal setting.

Fortunately, streamflow and meteorological observations have become two of the most ubiquitously measured environmental parameters, even in third-world countries, due to the efforts of national and international agencies. Moreover, the digital revolution has led to a wide accessibility of these data sets. (See Mosley and McKerchar, 1993, for information on accessing streamflow archives.) Usually, the analyst will find that the most satisfactory information available for a specific estuary is its freshwater inflow. While this may require a considerable effort of data manipulation to properly characterize the inflow regime of the estuary, at least the basic data exist. Unfortunately, this is frequently not the case for the other hydrodynamic controls.

12.3.3 Salinity

Salinity is a fundamental estuary measurement, for several reasons:

- Salinity can be measured in the field quickly and inexpensively.*
- Because the salinity of freshwater is nearly zero, the value of salinity at a point in an estuary can be interpreted as a measure of the proportion of seawater at that point (unless evaporation is large enough to be important in the salt budget).
- Salinity is virtually conservative, and therefore can be used as a water tracer.

---

* In oceanography, salinity must be determined to a high level of precision, 4 or 5 significant figures, because it varies proportionately so little in the sea. In an estuary, however, salinity ranges from 0 to >30 ppt, and is "noisy," being influenced by random small-scale displacements of water. The necessary accuracy of measurement is much lower, therefore, and less expensive methods are quite capable of providing useful data. Mangelsdorf (1967) commented that the (then-current) oceanographic precision of ±0.003 ppt would be wasted in an estuary "unless the position of the station were specified to within 40 feet, the time specified to within 90 seconds, and the depth specified to a quarter of an inch!"

- In an estuary, the variation of water density is dominated by salinity, and is rarely affected by temperature. (This is the reverse of the usual situation in freshwater lakes and in the ocean.) Therefore, salinity is an important indicator of hydrodynamic processes affected or controlled by density, such as turbulence and density currents.
- Salinity is one of the key variables determining habitat, due to the varying osmoregulatory capabilities of estuarine organisms and their ability or inability to survive in waters of a given salinity.

Because salinity is a natural water tracer, and because a more extensive data base is likely to be available than for other parameters, the first step in studying circulation processes in an estuary is its analysis. This is treated in Sec. 12.3.4 in the context of transport processes. In the present section, we summarize some of the basic features of estuary salinity, and the associated hydrographic parameters of temperature and density.

12.3.3.1 Salinity and Density Structure.

In an estuary, salinity has both horizontal and vertical structure. Both ultimately result from introduction of freshwater inflows, which establish a gradient of salinity across the estuary from fresh to oceanic, though their specific time-space configuration is influenced by many other hydrodynamic processes of circulation and mixing (see Sec. 12.3.4). One implication is related to the data base from which salinity structure is to be analyzed: The distribution of sampling points must be capable of resolving both vertical and horizontal structure. This means numerous horizontal stations are needed, sufficient to map the principal gradients of salinity through the system, and at each station vertical profiles, to depict the stratification of salinity. Also, the sampling interval must be sufficiently dense in time to resolve the important scales of variability.

Two examples of horizontal salinity structure are given in Figs. 12.10 and 12.11, displaying *isohalines* constructed from a network of sampling stations distributed through the estuary, for, respectively, Albemarle-Pamlico Sound on the U.S. East Coast and Galveston Bay on the U.S. Gulf Coast. Both are lagoonal-type estuaries, though with respectable water depths (for lagoons). The locations of *salinity gradient zones* are governed by the relative position of the inlets to the ocean and the sources of freshwater inflow. Large gradients in salinity are associated with the tongue of high salinity intruded from the ocean inlet (such as shown in Fig. 12.10) and the plume of freshwater at the point of river inflow (for example, the Trinity River in upper Galveston Bay, Fig. 12.11).

Spatial structure in estuary temperature can result either from temperature differences between seawater and freshwater, or from differential heating and cooling. In most estuaries in temperate latitudes, the former is by far the more important. In lagoonal and coastal plain estuaries, the intensity of vertical mixing prevents formation of a warmer surface layer and a thermocline. Whatever vertical stratification in temperature occurs in these systems is generally in association with salinity stratification, hence a reflection of inflow and oceanic temperature differences, rather than surface heat absorption. In fjords, in contrast, with their great depths and small relative volume of inflow, a warm-season thermocline can form. In most situations, even fjords, in which freshwater inflow is important in establishing a salinity gradient zone, temperature distribution will be dominated by mixing rather than thermodynamics. (An extreme example is the temperature structure of the upper St. Lawrence; Pelletier and Lebel, 1979.) The general domination of the estuary density structure by salinity is due to two facts: (1) the high range of salinity (0 to 35 ppt) in comparison to the small range in temperature (rarely more than about 5°C), and (2)

FIGURE 12.10   Salinity distribution (surface) in Albemarle-Pamlico Sound (North Carolina), April conditions. (*After Pietrafesa, et al., 1986.*)

the greater change in density per unit change in salinity. Fig. 12.12 displays isopleths of density versus salinity and temperature, based upon the International Equation of State 1980 (Pond and Pickard, 1983) from which the relative importance of either can be judged from known ranges in the system of interest.

12.3.3.2 Density Currents.   The existence of a horizontal density gradient is responsible for one of the more dramatic and often unexpected circulation regimes of the estuary: the *estuarine density current* (in the older literature, the salinity current). Consider a longitudinal estuary of constant cross section. Its density-driven circulation is shown schematically in Fig. 12.13, the classical two-layer density-current circulation. This circulation is basically an intrusion of denser, more saline water in the lower layer, compensated by an outflow of fresher water in the upper layer. This is an *intertidal circulation:* It is exposed only if vertical current profiles taken over one or several tidal cycles are averaged, as suggested by Fig. 12.13(*a*). While a density current will tend to produce vertical salinity stratification, because of the differential transport of fresher water downstream near the surface and more saline water upstream at depth, it is important to emphasize that it is the *horizontal* gradient in salinity that drives it. In fact, a density current can exist even though the salinity is unstratified in the vertical.

If we assume a tidal-mean steady state with an imposed horizontal density gradient in which turbulent stress is balanced in the horizontal by the pressure gradient, the equations of motion (laterally averaged) reduce to:

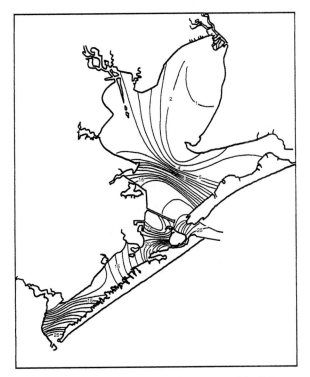

FIGURE 12.11 Salinity distribution (vertical-mean) in Galveston Bay, during freshet of Trinity River, July 16, 1968.

$$\frac{\partial}{\partial z} N_z \frac{\partial u}{\partial z} = \frac{1}{\rho} \frac{\partial p}{\partial x}$$

$$-g = \frac{1}{\rho} \frac{\partial p}{\partial z}$$

$$\frac{\partial u}{\partial x} + \frac{\partial w}{\partial z} = 0$$

in which p is pressure; z is directed upward from the seabed; w is vertical-component current; and N_z is the diffusion coefficient ("viscosity") for momentum, also a function of z. For this simplified system, w and p can be eliminated by integration, and the three equations combined into one for tidal-mean current $u(z)$. If the horizontal density gradient $\partial\rho/\partial x = \Lambda \partial s/\partial x$ ($\Lambda \approx 0.815$ for s in ppt and ρ in kg m^{-3}) is given, the form of the vertical profile of current $u(z)$ is governed by that of the eddy viscosity N_z. We introduce a parameter

$$\theta \equiv \frac{g D^3 \dfrac{\partial \log \rho}{\partial x}}{N_z v_f}$$

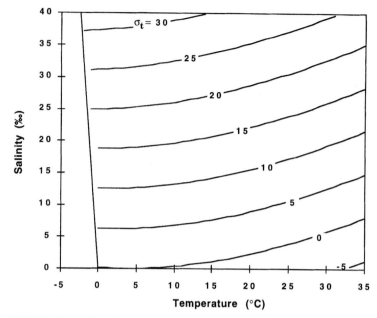

FIGURE 12.12 Water density as function of salinity and temperature for estuarine ranges, based on International Equation of State 1980. Density $\rho(\text{kg m}^{-3}) = \sigma_t + 1000$.

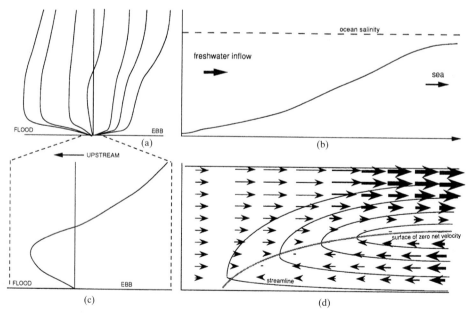

FIGURE 12.13 Schematic of tidal-mean density current structure in longitudinal estuary: (*a*) Vertical profiles of current during tidal cycle; (*b*) Longitudinal profile of vertical-mean salinity; (*c*) Vertical profile of tidal-mean current; (*d*) Tidal-mean circulation in vertical-longitudinal section.

in which $\overline{N_z}$ is the vertical mean of N_z, *log* denotes the natural logarithm, and $v_f \equiv Q/A$ ($v_f < 0$ for x positive from the mouth to the head). This number is nonnegative and dimensionless, and (as will be seen) measures the intensity of the density current. Typical estuarine values of the parameters range:

N_z: $1–100 \times 10^{-4}$ m^{2}s^{-1}

D: $1–100$ m

$\partial s/\partial x$: $0.01–1$ppt km^{-1} hence $\dfrac{\partial \log \rho}{\partial x}$: $1–100 \times 10^{-8}$ m^{-1}

v_f: $10^{-3}–10^{-1}$ ms^{-1}

so that θ ranges 10 to 10^6, typically on the order of 10^3.

For simple mathematical forms of N_z, closed-form expressions for $u(z)$ can be obtained that are instructive, albeit theoretical. Consider first constant $N_z = \overline{N_z}$. With boundary conditions $u(0) = 0$ at the seabed, $\partial u/\partial z = 0$ at the surface (zero stress) and $\int_o^D u(z)dz = v_f D$, the profile of $u(z)$ is:

$$u(z) = v_f \theta \left[\left(3\theta^{-1} - \frac{1}{8} \right)\left(\frac{z}{D} \right) + \left(\frac{5}{16} - \frac{3}{2}\theta^{-1} \right)\left(\frac{z}{D} \right)^2 - \frac{1}{6}\left(\frac{z}{D} \right)^3 \right]$$

For $\theta > 24$, a reversal in $u(z)$ occurs, with upstream flow in the lower layer, whose level becomes asymptotic to $0.58\,D$ as θ increases beyond 10^3. The approximate flow in the density current Q_s (computed by integrating current over the section from the bottom up to the reversal level) is given by $Q_s/Q = 9.4 \times 10^{-3}\,\theta$, which demonstrates:

- The density current can drive a flow at least on the order of, and typically 1 to 2 orders of magnitude greater than, the freshwater throughflow.
- The density current is driven directly by the horizontal gradient in salinity and is diminished by the intensity of vertical mixing.
- The density current varies as the cube of water depth (for a given value of horizontal salinity gradient), therefore is greatly enhanced by estuary depth.

The vertical turbulent excursions, and therefore the vertical fluxes, are limited by the presence of barriers to flow, especially the bed and surface. Assuming these excursions are proportional to the distance from the barrier (the traditional *mixing-length* model; see Kent and Pritchard, 1959) leads to a parabolic profile for the diffusivity: $N_z = 6\,\overline{N_z}\,z(D - z)\,D^{-2}$. The resulting current profile is log-linear:

$$u(z) = v_f \left[\frac{\theta}{12}\frac{(z - z_o)}{D} + \frac{\left(\dfrac{\theta}{24} - 1 \right)}{\log\left(\dfrac{z_o}{D} \right) + 1} \log \frac{z}{z_o} \right]$$

where z_o is the roughness height, the elevation above the bed at which the mean current vanishes (e.g., Tennekes and Lumley, 1972). Example profiles for both the constant-N_z and parabolic-N_z forms, for a value of $\theta = 10^3$, are shown in Fig. 12.14. The flow of the density current is proportional to θ for $\theta > 200$, so the same conclusions as before apply. The parabolic-N_z profile, i.e., log-linear current profile, better reflects the physics than does the constant-N_z (with a frictional logarithmic boundary layer at the seabed, and a zero-stress condition at the surface that does not impose a separate condition on $\partial u/\partial z$). Moreover, tidal-mean profiles in real estuaries often exhibit this log-linear form, e.g., Pritchard and Kent (1953), Pritchard (1967b), Bowden and Gilligan (1971), van de Kreeke and Robaczewska (1989), and Wallis, et al. (1989).

While these numerical results postulate a tidal-mean steady state, the physical processes of interaction of a horizontal pressure gradient with vertical turbulent flux apply as well to nonsteady processes of salinity intrusion and extrusion. Fig. 12.15 shows the vertical distribution of salinity along the main axis of Chesapeake Bay, displaying the response to a freshet. In the upper panel, the influx of freshwater has led to diluted salinities in the shallower depths. Salinity is reintruding from the sea, much faster at depth due to the density current, leading to the more pronounced stratification in the lower panel. The movement of the 20-ppt isohaline during the 22-day interval between these two sections should be noted.

Estuaries with substantial width exhibit much more complex density currents than the simple channel system considered above, with circulation in both the horizontal and vertical planes. In Chesapeake Bay, for example, the upstream layer in the deep axis extends through a greater depth than would be expected for a constant cross-section channel, with greater downstream flows in the lateral shallower sections. A system such as Galveston Bay (Figs. 12.5 and 12.11), a shallow lagoonal system with a deep narrow ship channel dredged through it, can exhibit a density current with net upstream flow throughout the depth in the channel and a compensating return flow to the sea in the shallow areas to either side (Ward, 1980). Upstream flow throughout the depth along the estuary axis with downstream flow in the lateral sections also occurs whenever the estuary widens and the central axis has sufficient depth. An example is the lower St. Lawrence around Rimouski, where the

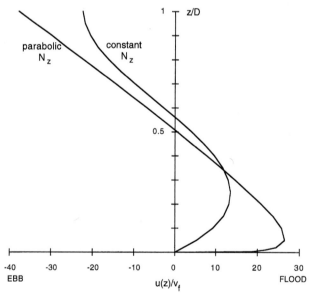

FIGURE 12.14 Theoretical vertical profiles of residual current, for constant and parabolic profiles of N_z.

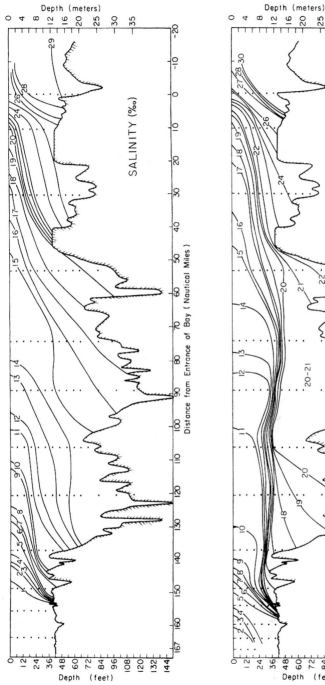

FIGURE 12.15 Longitudinal-vertical section of salinity along axis of Chesapeake Bay, April 11 (*above*) and May 8 (*below*), 1968. (*Seitz, 1971.*)

channel begins to expand towards its mouth at the Gulf, and a central deep of 300 m exists (El-Sabh, 1979).

12.3.3.3 *Salinity Analysis.* Analysis of salinity in an estuary begins with compilation of available data, and, as noted earlier, a realistic appraisal must be made of what analyses the data base will or will not support. One obvious independent variable is freshwater inflow, and a first step is to examine the association between the two. Probably no analysis has provoked as much frustration in estuarine studies, however, as the statistical dependency of salinity upon inflow, because there is an intuitive cause and effect relationship that refuses to emerge from the statistics. Salinity in an estuary is dependent upon freshwater inflow: without inflow, the salinities would eventually acquire oceanic values. The fallacy is to conclude from this that there is a direct association between a given level of inflow and the salinity at a point in the bay. The nature of the problem is illustrated by the salinity data of Fig. 12.16, showing the association of salinities in the center of Galveston Bay (Fig. 12.5) with gauged flow of the Trinity River, the principal source of inflow. While there is a discernible downward slope in the relation, as we would expect, the variance of salinity encompasses nearly the entire estuarine range, independent of the level of inflow. This high variance is a quantitative demonstration of the complexity of the response of salinity in the bay to many factors, only one of which is freshwater inflow. Other hydrographic mechanisms, such as tides, meteorology, and internal circulations, as well as the boundary value at the estuary entrance, govern the internal transports of waters of different salinities in the bay, and dictate how freshwater influences salinity.

There are several scales of time variation in the salinity signal, ranging from short-term tidal and meteorological to long-term seasonal and multiannual. Most data programs of the water-quality monitoring type, including those from which the data of Fig. 12.16 were obtained, sample on too infrequent a basis to resolve the shorter-term variability. The tidal variation, for instance, is virtually unsampled. Thus the data of Fig. 12.16 represent random sampling of these shorter time scales, which is nonresolvable and an intrinsic source of variance. An additional source of variance is due to the dynamic response of salinity. First, there is a lag between the freshwater signal as measured at an inflow gauge and its effect somewhere in the estuary. In addition to this lag, salinity in the bay responds more as an integrator of freshwater inflow, i.e., with a longer time scale of variation than that of the inflow itself, and is affected by the operative physical processes, e.g., tidal excursion, antecedent salinity gradients, and semipermanent circulation patterns. Salinity intrusion takes place by mixing by tidal currents and advection by density currents, and intrusion into the upper reaches of an estuary generally requires a longer time, perhaps on the order of weeks to months. Salinity extrusion due to a freshet, on the other hand, is basically a mechanism of displacement by freshwater, and occurs rather rapidly. It is not surprising, therefore, that there is no unique relation between salinity and inflow, but analyses like Fig. 12.16 rather are random samplings of a range of responses.

While many estuaries are dominated in the shorter time scales by tidal variations, this is not always true. Fig. 12.17 shows a time series of salinity measured at hourly intervals at a fixed station over a one-month period in a shallow lagoonal estuary on the Texas coast. At the beginning of this month, salinities neared oceanic values. During this period, the freshwater inflow increased sharply by two orders of magnitude and displaced the water in the estuary. A high-gradient zone formed at the leading edge of the freshwater. The greatest tidal variation in salinity during this period occurs when the monitoring station was encompassed within the high-gradient zone, whereupon the modest tidal excursion (about 1 km) becomes capable of substan-

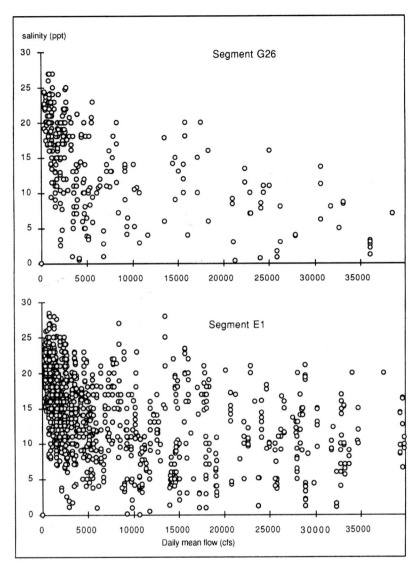

FIGURE 12.16 Salinity (surface) in two segments in central Galveston Bay versus Trinity River flow. (*Ward and Armstrong, 1992.*)

tially altering the salinity. From this figure, one would conclude that tidal variation is relatively unimportant except when large concentration gradients are manifest, and that the system is most responsive to large variations in inflow.

Analysis of salinity in an estuary requires delineating its spatial structure, and identifying its temporal variations and their sources of forcing. A recent analysis of salinity in U.S. estuaries by the National Ocean Service attempts a qualitative display of salin-

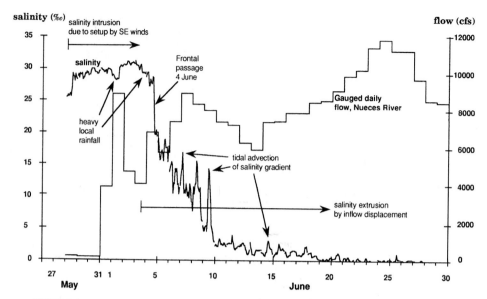

FIGURE 12.17   Salinity in Nueces Bay (Corpus Christi Bay system, Texas), June 1987.

ity time-space variability and the factors which force that variability in a form suitable for estuary intercomparison (Orlando, et al., 1993; 1994). Summary matrices have been constructed, based upon objective analyses of salinity data, that display the relative importance of different time scales of variability and the physical factors thought to force that variability, an example of which for Galveston Bay (Figs. 12.5, 12.11, and 12.16) is shown in Fig. 12.18. This project will eventually develop such matrices (to the extent permitted by the historical data base) for all of the major U.S. estuaries.

12.3.4   Transports by Circulation and Mixing

As noted in the introduction, management concerns in estuaries frequently center on the concentration of waterborne indicators, including pollutants and planktonic organisms. For many indicators, concentrations are affected greatly by transport processes, i.e., by current velocity and parameters closely related to velocity. The mass conservation equation for concentration c of a waterborne parameter is:

$$\frac{\partial c}{\partial t} = -\left\{ u\frac{\partial c}{\partial x} + v\frac{\partial c}{\partial y} + w\frac{\partial c}{\partial z} \right\} + \left\{ \frac{\partial}{\partial x}K_x\frac{\partial c}{\partial x} + \frac{\partial}{\partial y}K_y\frac{\partial c}{\partial y} + \frac{\partial}{\partial z}K_z\frac{\partial c}{\partial z} \right\} + \left\{ \sum S_i \right\}$$

$$(12.11)$$

| Local time change | = | Advective transport | + | Diffusive transport | + | Sum of sources and sinks |

This expresses the local time change of c as the sum of transport and sources, the last term in braces on the right being the sum of all sources (which includes sinks as

	Time Scale of Salinity Response				
Mechanism	Hours	Days to Weeks	Months to Seasons	Year to Year	Episodic
Freshwater Inflow		L **S**	H **D**	H **D**	
Tides		LIT **M**			
Wind		LIT **S**			
Density Currents		LIT **M**	LIT **S**		
Shelf Processes			LIT **S**	LIT **M**	
	UNKNOWN	LOW	MEDIUM	MEDIUM	UNKNOWN
	Effect on Salinity Variability				

Salinity Variability	Importance of Mechanism	Assessment Reliability
Very High => 21 ppt High = 11-20 ppt Medium = 6-10 ppt Low = 3-5 ppt Very Low = < 2 ppt	D - dominant S - secondary M - minor	H - high M - moderate L - low LIT - Literature Only

Relative importance of mechanism

Assessment Reliability → H   **D**

FIGURE 12.18 Salinity variability matrix for Galveston Bay. (*Orlando, et al., 1993.*)

well), and the other two terms in braces being transports. The first of these is *advective transport,* and the second *turbulent diffusion.* (An idealized model has already been slipped in, that turbulent flux is proportional to and opposite the local concentration gradient, the proportionality K_x, K_y, K_z referred to as eddy diffusivities for *c*. Derivation of this equation and underlying considerations are treated in standard references such as Neumann and Pierson, 1966.)

12.3.4.1 Advection and Space-Time Scales of Motion.

In an estuary, advective transport is one of the primary processes governing the concentration distribution of waterborne substances. Therefore characterization of currents is an indispensable preliminary to analyzing concentration distribution. Sufficiently comprehensive direct measurements of current velocity is rarely economically possible, however. Instead, the standard procedure is to employ measurements of related variables, usually tides and salinity, in such a way that inferences can be made about circulation processes. For many estuaries, these may be the only quantitative data available. Of course, there must also be available measurements of the predominant controls, especially ocean tides, meteorology, and river inflows.

The crucial decision in the analysis is identifying the space and time scales of concern. This is foremost determined by the nature of the management problems to be addressed. Is the concern one addressing the overall estuary, such as gross loadings of nutrients and sediments, or is it focused in the immediate region of a waste discharge, say of a highly reactive and quickly decaying pollutant? Are long-term average conditions of uppermost concern, or the extremes encountered during floods or droughts? Does the problem focus on a target organism, which may be present in the estuary only in certain seasons or under certain hydrographic conditions? This decision is also determined by the actual space-time scales on which the controlling processes operate, and by the capabilities of the data and analytical model for resolving those scales.

In the time domain, fluctuations occur on a range of periodicities, governed by the external controls on the estuary. Fig. 12.19 is a schematic of how energy might be distributed in an estuary (see Monin, et al., 1977), which for specificity is assumed to be freely exchanging with the sea and located on the tropical Pacific coast of North America. (These assumptions affect the relative amplitudes of the tidal frequencies, the periodicity of intertidal motions, seasonality of the freshwater inflow, and the prominence of El Niño.) The second-to-minute variations in currents and density due to turbulence and waves are rarely important in analyzing circulation or water quality, but they play a major role in diffusion and mixing, which cannot be neglected. In an estuary the tidal signal is a particularly strong component of fluctuations. The meteorological sources of variation can be numerous, depending upon the climatology of the estuary. The 24-hour seabreeze signal can reinforce or partially cancel the 24.8-hour tidal frequency, depending upon their relative phasing. In midlatitudes, frontal passages exert a fluctuation with several days periodicity (not reflected in Fig. 12.19, of course). In lower latitudes, tropical disturbances become a major source of variability over a wide range of frequencies. The seasonal signal (semiannual or annual) is probably the most prominent *periodicity* arising from meteorological effects, and would be even stronger in a midlatitude estuary than indicated in Fig. 12.19. In an estuary in which there is a great seasonal variation in inflows, such as those on the Atlantic and Gulf of Mexico coasts of North America, this can result in a significant annual signal.

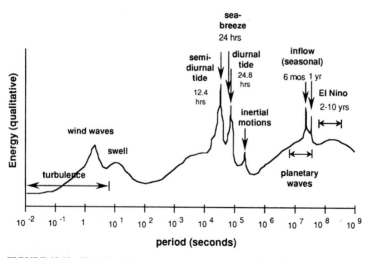

FIGURE 12.19   Hypothetical energy spectrum in a tropical Pacific estuary.

The spatial scale of motion is intimately related to time scale, because the observed fluctuation in time is frequently associated with displacements of water masses of different properties (an example of which was given in Fig. 12.17). In addition, spatial structure is a reflection of the morphology and bathymetry of the estuary: Internal shoals and barriers, large variations in depth, and multiple inflow sources create local water-property differentiation, whereas open, uniform, well-mixed systems tend to be more homogeneously graded in properties.

Clearly, one of the prominent types of motion in an estuary is tidal. To first approximation, water movement in an estuary can be viewed as an oscillatory tidal motion superposed on larger-scale longer-term circulations. A key decision in analyzing estuary circulations is therefore the importance of intratidal detail, that equal to or finer than the tidal periodicity: i.e., whether tidal motion must be explicitly considered or can be averaged out. (A closely related consideration regards interdiurnal time scales.) The space scale associated with the tidal period is the tidal excursion. If the time resolution is intertidal, then the spatial resolution should not be finer than the tidal excursion. Conversely, spatial structure resolved on a scale of several-to-many tidal excursions is addressed on an intertidal basis.

12.3.4.2 Space-Time Averages and Dispersion.
By far, the more common strategy in estuary management is to treat the longer-term intertidal state of the estuary, by considering averages of properties over several tidal cycles, or measurements with a sampling interval much greater than the tidal periodicity. The longer-term intertidal circulation is also referred to as *residual*. One of the major components of residual circulation in an estuary is the density current, addressed earlier in the context of salinity. There are other sources of residual circulation in an estuary, including wind (Schroeder and Wiseman, 1986; Vieira, 1986; Sanford and Boicourt, 1990), freshwater inflow and nonlinear effects of the tide (Nihoul and Ronday, 1975; Uncles and Jordan, 1979). A useful collection of papers at an advanced level treating residual circulations is that edited by Cheng (1990).

Of course, tides and other shorter-term fluctuations cannot be disposed of so easily. If point measurements are made on an occasional basis, for example, water samples are drawn at a network of stations over a day's period, without any regard to stage of tide, the intratidal fluctuations will contribute an element of variance ("noise") in the data. More importantly, while one would like to be able to work with a transport equation exactly like Eq. (12.11) except in which all quantities are tidal averages, the influence of intratidal frequencies of motion on transport processes cannot be eliminated by averaging. For example, assume u and c to be each a simple harmonic fluctuation of frequency $\sigma = 2\pi/T$ superposed on tidal-mean values of u_o and c_o, with c phase-lagging u by φ, and consider the first advective term in Eq. (12.11):

$$-u\frac{\partial c}{\partial x} = -[u_o + U\cos(\sigma t + \varphi)]\frac{\partial}{\partial x}(c_o + C\cos\sigma t)$$

which averaged over the period T is

$$-u_o\frac{\partial c_o}{\partial x} - \frac{1}{2}U\frac{\partial C}{\partial x}\cos\varphi$$

The average advective transport is the advection "of the mean" plus a term whose size depends upon the phasing of u and c and upon the amplitude of their fluctuations. It is easy to see that there are conditions in which the second term equals or dominates the first.

This transport that is additional to that determined from tidal means of u and c is referred to as *tidal dispersion*. It is especially important in concert with physiographic perturbations, which develop secondary circulations in association with the ebbing and flooding tidal current. Geyer and Signall (1992) propose that its relative importance increases as the ratio of the tidal excursion to the characteristic scale of the perturbation. A common treatment of these nonlinear flux terms in estuary modeling is to postulate that tidal dispersion behaves as a diffusive flux of the tidal-mean concentration $\underline{c}$, i.e., that the flux is down the tidal-mean gradient, hence of the form $-T_x \partial\underline{c}/\partial x$. With this assumption, the tidal average of Eq. (12.11) assumes exactly the same mathematical form as the instantaneous equation, except that the coefficients in the diffusive transport terms are $(T_x + K_x)$, $(T_y + K_y)$, and $(T_z + K_z)$.

Another complexity in the treatment of estuarine transport is the explicit consideration of vertical profiles of current and concentration. In the preponderance of estuaries the width to depth ratio is large, and there is usually greater concern with variation from place to place in the estuary than with variation with depth in the water column. Moreover, in modeling based upon the mass-conservation equation Eq. (12.11) the vertical terms are particularly troublesome, both in their mathematical formulation and in the computational resources they demand. (For example, to resolve vertical profiles by values at three levels in the depth, which is a rather minimal resolution, the requisite computer storage for a numerical model is clearly tripled.) Many management problems can be treated just as well with vertical-mean values of the concentration. Again, it is desirable to have an equation exactly like Eq. (12.11) but with the z-dependencies eliminated and all variables replaced by their vertical means. And, again, this desire is thwarted by the nonlinear advective terms. Replace each variable by the sum of its vertical mean from surface to bottom and the profile departure from that mean, e.g., $u(x,y,z,t) = \bar{u}(x,y,t) + u'(x,y,z,t)$, and consider again the first advective term in Eq. (12.20):

$$-u\frac{\partial c}{\partial x} = -(\bar{u} + u')\frac{\partial}{\partial x}(\bar{c} + c')$$

which averaged over the vertical becomes

$$-\bar{u}\frac{\partial \bar{c}}{\partial x} - \overline{u'\frac{\partial c'}{\partial x}}$$

As was the case with tidal averaging, the vertical-mean advective transport is the advection "of the mean" plus a term dependent upon the structure in the vertical profiles of u and c. This additional transport in a vertical-averaged formulation is referred to as *shear dispersion*. As with tidal dispersion, it is common to postulate that shear dispersion behaves as a diffusive flux of the vertical-mean concentration $\bar{c}$, i.e. that the flux is down the gradient of vertical-mean concentration, hence $-V_x \partial \bar{c}/\partial x$.

For both tidal- and vertical-mean quantities, the form of the mass-conservation equation is

$$\frac{\partial c}{\partial t} = -\left\{u\frac{\partial c}{\partial x} + v\frac{\partial c}{\partial y}\right\} + \left\{\frac{1}{D}\left(\frac{\partial}{\partial x}DE_x\frac{\partial c}{\partial x} + \frac{\partial}{\partial y}DE_y\frac{\partial c}{\partial y}\right)\right\} + \left\{\sum S_i\right\} \quad (12.12)$$

in which overbars and underbars are omitted and the various diffusion terms are combined into single gradient terms with the E's referred to simply as dispersion coefficients (Fischer, et al., 1979; Holley and Jirka, 1986). For a longitudinal estuary, it is usually desirable to suppress the lateral y dependence as well, representing the dependent variables as cross section means (e.g., West, et al., 1990), in which case the dispersion term includes the effects of cross-channel variation (such as secondary flows and lateral concentration distributions) in addition to vertical variation, and in addition to tides if the equation represents the tidal average:

$$\frac{\partial c}{\partial t} = -\left\{\frac{Q}{A}\frac{\partial c}{\partial x}\right\} + \left\{\frac{1}{A}\frac{\partial}{\partial x}AE\frac{\partial c}{\partial x}\right\} + \left\{\sum S_i\right\} \qquad (12.13)$$

In many respects, the parameterization of the dispersion terms in these averaged equations is misleading, since it suggests that the dispersion coefficient is a physical parameter that is in principle measurable, and, once determined, can be assumed to be fixed function of space. In fact, it is not directly measurable but must be inferred from observed distributions of waterborne parameters, and may vary with hydrological and hydrographical conditions, as well as depend on the specific substance whose concentration is measured.

Unfortunately, terminological inconsistency is rampant. Dispersion coefficients are referred to variously in the literature as diffusion, eddy diffusion and exchange coefficients, and may have dimensions of L^2T^{-1}, L^4T^{-1}, L^2, or $M(LT)^{-1}$. In this chapter, we reserve the term *eddy diffusion* for the effective down-gradient flux resulting from small-scale space-time fluctuations associated with turbulence, a terminology consistent with that of physical oceanography. We apply the term *dispersion* to the effective flux resulting from a large-scale average of advective transport, specifically that in excess of the flux given by the formal product of the average current and average concentration. Therefore, dispersion is associated with systematic correlated variation in concentration and current over the average, either in time or space, or both. We use *diffusion* to refer to a specific mathematical parameterization form: flux as the product of the negative of the concentration gradient in space and a coefficient. Holley (1969) proposed a different convention, that *diffusion* refer to the excess effective advective flux when a time average is employed, and *dispersion* when a space average, a proposal which has much to recommend it. The best advice to be given the analyst is to carefully examine the *use* of the transport equation in question. How the values of u and c that go into the advection term $u\partial c/\partial x$ are determined, e.g., whether they are section means, vertical means, lateral means, and/or tidal means, defines the residual transport that is absorbed into the *dispersion* term.

The derivation of the averaged equations Eqs. (12.12) or (12.13) involves two key steps:

- The velocity and concentration are each written as the sum of its mean value and a departure from that mean.

- The collection of the means of the product of departures are replaced by a diffusion-type term with a dispersion coefficient.

This derivation assumes nothing specific to an estuary, but in fact may apply to any surface watercourse. Indeed, equations of exactly the same form have been applied to transport processes in streams, lakes, and oceans. In particular, Eq. (12.13) has been long used in modeling dispersion in streams. Typical longitudinal dispersion coefficients in streams range 1 to 50 m²/s (e.g., Fischer, 1968). Example values of longitudinal dispersion coefficients from a sampling of estuaries are given in Table 12.3. It is immediately clear that dispersion coefficients are much larger in estuaries than in streams. Moreover, dispersion varies greatly along the longitudinal axis of an estuary. Generally, progressing from the head of the estuary to the sea, dispersion increases in the tidal reach of a river, and becomes even larger in the brackish and saline reaches of the estuary. This is due to the greater transverse and vertical circulations associated with salinity intrusion and with the greater widths and depths (and therefore secondary circulations) encountered with distance down the estuary towards its mouth.

Much can be learned about transport processes, and their space-time scales, by analysis of salinity. Moreover, development of a transport model for waterborne constituents in an estuary almost always begins with modeling salinity. Salinity is virtually conservative so $\Sigma S_i = 0$ in Eq. (12.11) and its distribution in the estuary hinges on transport processes. Dispersion coefficients must be inferred from observed distributions, for which salinity is a traditional parameter. Its historical role in estuary modeling has been a "calibrator" variable in this respect.

The fact that salinity controls water density in an estuary provides additional insight into the nature of dispersive transport. In order to solve a mass conservation model, say Eq. (12.11), directly for c, one must have given the advective transport variables (u,v,w) and the diffusive parameters (K_x,K_y,K_z). The former would presumably come from a companion solution of the hydrodynamic equations. In this procedure c is treated as a passive tracer and modeled by a feed-forward process of first determining the component currents, given tides and inflows, then computing c. Salinity, however, is not a passive tracer, because salinity governs density, and density exerts a major control on hydrodynamics. Properly, therefore, the hydrodynamic model should include density-driven horizontal acceleration terms, for example the x-component would be

$$\frac{\partial u}{\partial t} + u\frac{\partial u}{\partial x} + v\frac{\partial u}{\partial y} + w\frac{\partial u}{\partial z} = -g\frac{\partial h}{\partial x} - \frac{g}{\rho}\int_z^h \frac{\partial \rho}{\partial x}\,dz + fv - \tau_{bx}$$

The density gradient results in a pressure gradient acceleration, the second term on the right, in addition to that resulting from the slope of the water surface. If c denotes salinity, with a relation $\rho(c)$ (which is practically linear), the solution of Eq. (12.11) would now be coupled with that of the hydrodynamic equations. This is a nontrivial alteration, rendering the mathematical problem nonlinear, and imposing substantial computational demands on its numerical solution.

TABLE 12.3 Longitudinal Dispersion Coefficients for Selected Estuaries, Determined from Salinity Distributions

Estuary	D^*, m	s^*, ppt	$\partial s/\partial x$, ppt/km	$E, m^2/s$	Source
Bay of Fundy	40	31	0.008	300	Billen and Smitz (1978)
Upper Chesapeake Bay	10	8	0.2	500	Boicourt (1969)
Cook Inlet (1969)	35	24	0.062	200	Matthews and Rosenberg
Upper Delaware	7	1		120	Paulson (1970)
Eems	5	15	0.3	200	Eggink (1967)
Gironde	3	3		235	Dyer (1978)
Humber	7	24	0.45	171	Gameson (1982)
Patuxent	5	12	0.24	365	Owen (1969)
Patuxent	4	12	0.13	75	Ulanowicz and Flemer (1978)
Upper San Francisco Bay	6	16	0.71	144	Glenne (1966)
Scheldt	10	8	0.2	100	Wollast (1978)
Severn	10	28		200	Uncles and Radford (1980)
Sungai Berbok	10	24	0.4	80	Ong, et al. (1991)
Volkerak	10	20	0.4	250	van der Kreeke (1990)
Yaquina Bay	7	22	1.8	140	Burt and Marriage (1957)

* Typical value for reach over which E was computed.

If salinity is treated as a passive tracer, with a feed-forward model, as indicated diagrammatically in Fig. 12.20(*a*), the currents from the hydrodynamic model will not include the effect of density-driven transport, and this flux must be incorporated into the dispersion coefficients. Any other truly passive tracer would have to be modeled the same way, because the currents would be the same, i.e., not including density effects. On the other hand, if the effect of salinity on density is directly incorporated through the pressure-gradient acceleration, salinity and hydrodynamics become coupled, as then the currents will directly include the effect of density acceleration, indicated in Fig. 12.20(*b*), and the currents so computed will include density-driven transport. The dispersion terms can be restricted to the dispersive fluxes they are supposed to represent, including turbulence, shear dispersion and tides.

But there is another, more subtle aspect to the effect of salinity on transport, viz., the influence of vertical density gradients on the intensity of turbulence and there-

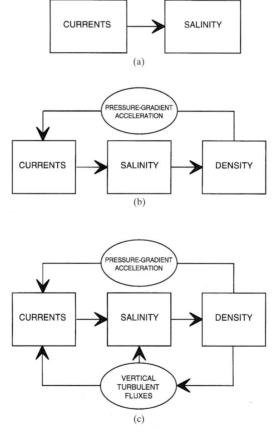

FIGURE 12.20   Alternative strategies of coupling salinity transport and hydrodynamics: (*a*) Feedforward model; (*b*) Coupled hydrodynamic/salinity model with feedback for density currents; (*c*) Coupled hydrodynamic/salinity model with feedback for density currents and turbulent fluxes.

fore on the diffusivities (K_x, K_y, K_z). A vertical density gradient creates a dissipation of turbulent energy in working against gravity. The more complete hydrodynamic-salinity model takes the form of Fig. 12.20(c), with salinity feedback added as well in the turbulent flux terms. Modeling of turbulent fluxes in a density-stratified fluid is especially problematic. Analysis of field data and heuristic reasoning suggests the eddy flux should be an inverse function of Richardson number (e.g., Kent and Pritchard, 1959) or Brunt-Väisälä frequency (e.g., Ward, 1977), both of which are directly related to $\partial\rho/\partial z$. More promising is to improve the eddy flux parameterization by the use of higher-order closures, e.g., Mellor and Yamada (1982), which is the basis for the Princeton estuary model (Oey, et al., 1985). This adds additional non-linear coupling and mathematical complexity in the model, and more computer demands for its numerical solution.

Nowadays, the analyst faced with practical estuarine management problems has many software resources to rely upon, and will rarely—if ever—need to personally contend with translating "squiggles and del-byes" into numerical results. The above formulations should, however, serve to caution the analyst that estuary transport processes are complex, and one does not achieve simplification without sacrifice. The choice of a model, especially entailing space and time averages, does not reduce the burden of quantifying transport but merely transfers it from one type of depiction to another. The correct procedure is:

- Determine the space and time scales necessary for accurately treating the management problem of concern.

- Establish the relative importance of different transport processes in the estuary.

- Employ a formulation (a model) that exploits simplifications appropriate to that system.

12.3.4.3 *Flushing.*

Flushing.   Frequent reference is made to the *flushing time* of an estuary. This is defined as $\tau = V/Q_f$ where V is the volume of the estuary and Q_f the long-term average river inflow (see, e.g., Officer, 1976). This is a concept that has been imported into the estuary from lakes and rivers. Also referred to as *renewal time, replacement time,* and *residence time,* this is the time required for the freshwater inflow to replace the volume of water in the estuary. It is directly related to the degree of dilution with "new, uncontaminated" water. Used for this purpose in water quality assessments, flushing time is a useful parameter for lakes and rivers, in which the river inflow is virtually the only supply of new water.

For management purposes in an estuary, the parameter has little utility, for two reasons. First, dilution varies strongly as a function of position in the estuary. A single number attempting to characterize the entire system is useful only for gross, relative comparisons between estuaries (such as those presented by Officer and Ryther, 1977), not for any absolute characterization of the estuary's ability to assimilate specific waste loads. Second, and more importantly, there are other mechanisms of dilution and water replacement operating in an estuary in addition to river inflow. Most importantly, these are tides, meteorological flushing, and the influx of seawater driven by density currents (e.g., Kranenburg, 1986; Goodrich, 1988). As an example, in longitudinal estuaries (river channel estuaries), the flow associated with the density current circulation (see Fig. 12.13) is nominally an order of magnitude greater than the river flow that maintains the salinity gradient.

Sometimes this concept of flushing time is generalized by accounting for the fraction f of freshwater implied by the estuary-mean salinity s, $f = (s_o - s)/s_o$, so that $T = fV/Q$ (see Ketchum, 1951a; Officer, 1976), s_o denoting ocean salinity. This can be further extended by explicit consideration of tidal exchange. Van de Kreeke (1988) computes residence time as

$$\tau = \frac{V}{\epsilon \Pi} \left[\frac{(s_o - s)}{(s_o - s_e)} \right]$$

where Π is the tidal prism for the estuary, and s_e the salinity exiting the estuary. Here ϵ is the fraction of the ebb volume that is not drawn back into the estuary on the next flooding tide, i.e., the proportion of "new" seawater in the tidal prism, a parameter that is not readily available for a given estuary. A better approach to determine flushing based upon tidal exchange is as a function of position. This entails subdividing the estuary into segments based upon the length of the tidal excursion, and computing the flushing time of each segment based upon renewal by tidal influx. (This method is due to Ketchum, 1951b; it and related methods are reviewed by Pritchard, 1952, and Officer, 1976.) However, even accommodating tidal exchange in this way does not account for the additional dilution afforded by the other processes, including density currents and internal circulations. We observe that the effort in doing this sort of analysis for real-world estuaries begins to approach that of a complete transport model, which would be more accurate in any event and more directly applicable to management problems. Zimmerman (1976; 1988) gives a more rigorous summary of the definition and computation of residence times appropriate to estuaries (see also Dronkers and Zimmerman, 1981).

12.4 WATER QUALITY

Water *quality* is a broad term, referring in general to any quantitative parameter or suite of parameters, taken collectively, that can serve as an indicator for a potential use of water or aquatic habitat in an estuary. *Use* includes the uses of nature as well as human activities. In this context, *quality* would range from physical properties such as viscosity and current velocity to communities of large organisms of the estuary, and would extend in space to include the atmospheric and terrestrial environs. We adopt a narrower definition, consistent with common practice in water-resource management, that quality is defined by physical, chemical and microbiological constituents in suspension or in solution in the estuary waters, particularly those that delimit biological use. We also broaden this definition to include constituents present in the estuary sediments accessible to organisms or capable of exposure and resuspension by currents. *Environmental quality* might be a term more appropriate than *water quality* but the weight of convention prevails. A useful treatment of estuarine water and sediment quality may be found in the review monograph edited by Olausson and Cato (1980); more specific references are cited later.

12.4.1 Water Quality Indicators

Being based upon substances in solution or suspension, an evaluation of quality requires: (1) numerical magnitudes of the substance as a function of space and time in the estuary (or in the region of the estuary of concern), and (2) quantification of the levels that impose an impairment of biological use. By far, the most important consideration in estuary water-quality management obtains when the substance is a contaminant whose presence in the estuary is due to anthropogenic loadings. But there are circumstances in which the manager must consider adverse levels of naturally occurring indicators as well, especially to properly interpret apparent degradation of estuarine habitat, alterations in species populations, or organism mortality. In this chapter we are primarily concerned with determination of the magnitudes of the substance in the estuary, involving a combination of data collection and analysis, and modeling. Quantification of the threshold levels of impairment entails setting criteria and standards, usually based upon biological responses.

12.4.1.1 Mass Budgeting. The concentration distribution in space and time is governed by the mass-balance equation, either Eq. (12.11) or a space/time-averaged version that has the formal appearance of Eqs. (12.11), (12.12) or (12.13). The predominance and complexity of the transport terms in an estuary were described in Sec. 12.3.4, and we assume that the proper analysis has been performed to quantify these terms. Application of the mass-balance equation to a specific constituent requires:

- Specification of the source and sinks of the constituent, ΣS_i in Eq. (12.11) et seq.
- Establishment of boundary conditions, especially definition of any mass loads within the interior or periphery of the watercourse

These would be required for any watercourse, including an estuary, and therefore in themselves present no novelty. The simplest implementation of the mass-balance equation Eq. (12.11) et seq. for an estuary is the so-called *box model,* in which the estuary is subdivided into large segments for which average values of physical parameters, flows, and concentrations are determined, and between which transport fluxes are computed. The box model may be developed rigorously by integrating Eq. (12.11) over the volumes or surfaces of the elements, or less rigorously by using finite-difference-type approximations of Eq. (12.11). The success of a box model depends upon: (1) defining the segments strategically, to take advantage of physiographic and hydrodynamic constraints on flow to best approximate the segment means and boundary fluxes, and (2) having an adequate base of data. For example, for a longitudinal coastal-plain system with a density-current circulation, a two-layer box system would be employed, with the layer boundary taken as the surface of zero residual motion, i.e., the current reversal level in Fig. 12.13 (see, e.g., Doering, et al., 1990; O'Connor and Lung, 1981; Lung and O'Connor, 1985; Hager and Schemel, 1992; Prego, 1993). Frequently in estuary analysis, and in particular in box models, the observed distribution of conservative parameters, such as salinity, is used to estimate transport terms, which are then applied to other parameters.

The idea of applying Eq. (12.11) to large physically-defined segments of a complex fluid system is an old one, and can be traced to early work by oceanographers (e.g., the salt-budget method of Knudsen). The interested reader is directed to the formal reviews by Zimmerman (1976) and Officer (1980). In the present context, we regard box models as a special kind of discrete model, i.e., a finite-approximation solution to the mass-balance equation, which is treated elsewhere in this handbook. The advantages of a box model over more sophisticated numerical transport models are:

- The box model is conceptually simpler, therefore appeals to intuition.
- It can be formulated like a budget of accounts, and therefore offers a convenient framework for delineating the magnitude and distribution of loads and of high concentrations, and for assessing the adequacy of field data.
- The box model is computationally simple, capable of hand calculation or of implementation in spreadsheet software.

The disadvantages of a box model are that it may not admit adequate resolution to treat the management problem of concern, the nature and implications of simplifying assumptions may be lost (such as a long-term steady state, or the difference approximations employed in computing transports), and it may give the user a meretricious illusion of precision.

With respect to the kinetics terms in the mass-balance equation, many substances are subjected to a first-order sink of form $S_i = -Kc$, in which K^{-1} is the decay time constant (e^{-1}-folding or relaxation time), or equivalently, $S_i = -0.693\,(T_{1/2})^{-1}c$, $T_{1/2}$ denoting the aquatic halflife of the substance. Some substances are subjected to a zeroth-order source/sink, one in which the rate of supply/decay is independent of the concentration of the substance. An example of the combination of the two is the rate of transfer across the water surface of a gas of low solubility,

$$S_a = K_a(c_s - c)$$

where K_a is a gas transfer coefficient per unit volume of water and c_s is the saturation concentration for the gas. In a mass-balance equation for c, this represents the sum of a zeroth-order source and first-order sink. (This is for a gas of greater concentration in the atmosphere than in the water. For the opposite, the sign of S_a would be reversed.)

It should be remarked that only radionuclides can be rigorously considered to follow a first-order process. For water-quality indicators of management concern, including all of those addressed here, first-order kinetics should be regarded as a simplified mathematical model representing a much more complex array of physiochemical and biochemical reactions. For example, in the gas transfer term given above, the concentration c strictly should be evaluated within a thin film at the surface, and the K_a depicts a complex process involving turbulent replacement both within this thin film and within the overlying atmosphere. In practice, however, c would be taken to be a section mean, or a vertical mean, over an appreciable depth (perhaps the total depth of the watercourse), and K_a would at best be related to some bulk atmospheric variable, such as wind speed at standard anemometer height. Such approximations are justified by the fact that the complexity of the individual kinetics terms generally does not arise so much from their mathematical forms as it does their dependency upon other water-quality parameters, which necessitates coupling multiple mass-balance equations.

Boundary conditions are applied at each *port* to the estuary, i.e., at each point of inflow and at each opening to the sea. In addition, mass loads must be specified at the points of discharges containing the parameter of concern. The specification of boundary conditions and the formulation of kinetic processes of a parameter are not completely independent. The magnitude of the source/sink terms implies a time scale of reaction that the substance exhibits and therefore its intrinsic "memory," which can be converted to a distance. For the first-order reaction, K^{-1} is the obvious measure of time scale. If K^{-1} is on the order of several periods of the fundamental tide cycle, then the reaction will affect tidal-mean concentrations over a distance of several or many tidal excursions. (Conversely, for evaluating the behavior of the parameter on shorter time-space scales, say within a tidal excursion of a load point, or over a time period less than a tidal cycle, the reaction may be neglected and the parameter treated as quasi-conservative.) On the other hand, for highly reactive parameters, so that K^{-1} is less than, say, a few hours, the analysis need only be carried out within a short distance (less than a tidal excursion) of the load point, but may have to explicitly treat tidal and similar short-term current variation.

12.4.1.2   Salinity as a Kinetic Indicator.   The quintessential water-quality variable in an estuary is salinity. Because of the fundamental role played by salinity in
delineating circulation and mixing in an estuary, its analysis was treated earlier in
Sec. 12.3.3. In the present context, it should be re-emphasized that salinity is one of
the controlling determinants of habitat and is a natural water-mass tracer, so analysis of estuarine water quality begins with this parameter. (Of almost equal importance is the concentration of suspended solids, which is addressed separately in the
following section.)

Analysis of the relation between salinity and other water-quality parameters can
be useful to determine the extent of similarities in distribution, hence transport, and
to detect nonconservative behavior in the water-quality parameter. A qualitative
diagnostic device is a graphical depiction of the concentration of the water-quality
constituent of concern as ordinate versus salinity of the water sample as abscissa,
interpretation of which is suggested by the sketches of Fig. 12.21. A series of water
samples from various points in the estuary will plot as a straight line on such a graph
if the following are true:

• Oceanic salinity, out from the entrance to the estuary, is stable, i.e., undisturbed by
  river plumes or shelf processes.

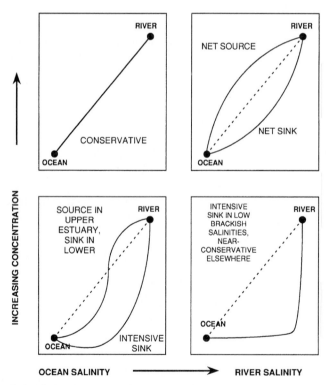

FIGURE 12.21   Interpretation of mixing diagram versus salinity in terms
of source or sink processes. (*After Peterson, et al., 1975.*)

- The constituent has well-defined values in both river and seawater (i.e., relatively constant over at least the same time scale as salinity intrusion into the estuary).
- There are no additional boundary fluxes (besides the river and the sea) into the estuary of either freshwater or saltwater, or of constituent mass loads.
- Neither the constituent nor salinity is subject to a net source or sink within the estuary.

Under these conditions, the constituent concentration will be determined by mixing of seawater and river water in exactly the same proportion as salinity, hence the straight-line relation. Departure from this straight line is used to infer the operation of a source or sink, as suggested by Fig. 12.21. This has proven valuable in appraising the behavior of substances that are brought to the sea by runoff and inflow, such as boron, silicon, and fluorides (e.g., Liss and Pointon, 1973; Kullenberg and Gupta, 1973), as well as metals, nutrients, and sediments. This method has the considerable advantage of being independent of position in the estuary, so sample-station location does not enter the analysis, and independent of variations in the space-time structure of salinity. One must be cautious, however, of the assumptions delineated above. Departure from a straight line means that one of these assumptions is violated, not necessarily that the constituent is nonconservative. For example, if net precipitation-evaporation is significant, then variation of salinity will not be determined simply by mixing and a nonlinear correlation may result. Also, the occurrence of a straight-line relation does not mean the constituent is free of internal kinetics, but rather that its sources and sinks balance, which can be a very different matter.

12.4.1.3 Dissolved Oxygen, BOD and Nutrients. The earliest water-quality problem addressed analytically for estuaries, like that of flowing streams, was the sag in dissolved oxygen (DO) due to injection of an organic, oxygen-demanding load. As in freshwater streams, the impact on DO is quantified by the biochemical oxygen demand (BOD). Like the stream, the estuary is usually addressed by a version of the Streeter-Phelps equations, coupling BOD into the sink term for DO, but in which a longitudinal dispersion term is added with coefficient considerably greater than that of a flowing stream (e.g., O'Connor, 1960; 1965). BOD is computed first then fed forward into the calculation for DO. In this case, BOD L has a first order sink $-K_rL$ which also is a zeroth-order sink for dissolved oxygen (or a source for dissolved oxygen deficit, as the problem is usually formulated; see Thomann and Mueller, 1987). As with rivers and streams, some analytical value is derived from treating the carbonaceous component of the BOD separately from oxidation of nitrogen compounds, incorporating the latter into the DO mass-balance equation as additional sink terms. This strategy is particularly useful in evaluating wasteloads with high concentrations of nitrogen compounds.

Departing for a moment from the estuarine focus of this chapter, we remark that BOD as a water-quality indicator remains problematic. There is a multiplicity of factors that can affect the BOD of a water parcel, whether it be a laboratory BOD bottle on a shelf or a moving parcel of water embedded within the flow of a natural watercourse, including:

- Types of bacteria present in the water
- Initial quantities of each type of bacteria present
- Multiplication (or growth) rates for each type of bacteria present
- Chemical characteristics of the substrate, i.e., the oxidizable organic constituents within the water

- Concentrations of the oxidizable constituents
- Constituents which act as an inhibitor or a stimulant for the bacterial metabolism
- Environmental parameters, most notably pH and temperature
- Other aerobic organisms in the water, notably protozoa (which predate on the bacteria) and plankton (which respire)

The heterotrophic bacteria indigenous to a natural watercourse, especially an estuary, comprise a numerous, complex, and diverse community of organisms with a wide range of specific metabolic capabilities. In a natural system, the nutrient source will be made up of many organic carbon constituents, some more labile than others, and different species of bacteria will attack that component of the substrate most appropriate for an energy source in terms of availability and the specific biochemical abilities of the species. Further, waste loads not only provide a rich and varied source of organic carbon, but also represent in themselves significant sources of bacteria. In addition, there are chemautotrophs, which derive their cell carbon from inorganic carbon compounds. Of particular concern are the nitrifying chemautotrophs, *Nitrosomonas* spp. and *Nitrobacter* spp., the principal organisms responsible for the oxidation of inorganic nitrogen.

The analyst may be faced with a variety of potentially inconsistent measurements of oxygen depletion in a laboratory bottle, all described as BOD, such as dilution-series 5-day BOD with cultured seed, dilution-series 20-day BOD with cultured seed, dilution-series 3-day BOD with natural seed, and nitrogen-suppressed 20-day BOD with or without cultured seed, and lack a meaningful way to interconvert from one to the other. From the standpoint of oxygen-budget analysis, the ultimate BOD is the desired measure. This is frequently well approximated by a long-period (say, 20-day) incubation. If laboratory BODs are to be used for rate determinations, it is mandatory that the samples utilize the natural bacterial community, i.e., natural seed. Dilution should be avoided, since this is a source of error itself. In many studies employing BOD, the phenomenon has been encountered of increasing BOD (per unit volume) as the sample is subjected to greater dilutions. This has been attributed to toxicity in the natural water and to differential stimulation of bacterial species. (A possible explanation for the phenomenon lies in the Monod equation for bacterial growth, formally equivalent to Michaelis-Menten kinetics. A varying substrate concentration implies that the rate constant for BOD exertion is a nonlinear function of the dilution factor, which contradicts the basic assumption of the dilution approach, that the BOD depletion is simply proportional to dilution.)

If the manager can influence laboratory methodology, we recommend that an oxygen depletion series be carried out on undiluted water samples with only natural seed, based upon sequential DO measurements (at, say, daily intervals) with the sample being reaerated as necessary to maintain an aerobic environment. With modern self-stirring electrometric probes, the effort and expense are minimal, the data can be used for extrapolation to the BOD_u asymptote, multiple stages of BOD exertion can be identified, and rate coefficients estimated directly.

Because of the great biological productivity of estuaries, the management of nutrient loads to these systems is frequently accorded a high priority, especially nitrogen and phosphorus. For modest influxes of N and P, the relatively high phosphorus concentrations of seawater render many estuaries nitrogen-limited, the reverse of the usual situation in freshwater systems, so management concerns in these estuaries focus on nitrogen (e.g., Rudek, et al., 1991). However, heavily loaded estuaries may have such high concentrations of nitrogen and phosphorus that they must both be addressed. Biological uptake by phytoplankton and macrophytes is a major sink. The most rigorous means of incorporating this sink is a coupled analysis of primary producers (see Di Toro, et al., 1977, and also the review of Fransz, et al., 1991), but often a simpler course is taken of specifying a first-order sink whose rate coefficient is determined empirically. In watercourses in which large areas of anoxic waters occur, especially adjacent to the bottom, denitrification may become a source of nitrogen. (Apart from the problem of time/space mapping of zones of depleted oxygen, this kinetic process is not specific to estuaries nor does it entail formulations unique to the estuarine environment.) An additional nutrient of potential management importance is silicon. Silicon is important in an estuary due to requirements of the diatom community, which often dominates the phytoplankton. Modification in the riverine inflow can alter the supply of silicon and therefore the population of algae, the base of the food chain in estuaries.

12.4.1.4 Pathogens and Toxics.

The importance of aquatic bacteria to the quality of a watercourse was exemplified above by the oxygen budget. Bacteria play many other important roles in estuarine waters and sediments (Wood, 1972; Stevenson and Erkenbrecher, 1976). Bacteria are the chief agents of the decay of organic detritus, and thereby mediate the transfer of energy from plants through detritus to the rest of the estuarine food chain. Bacteria also account for the high sediment oxygen demand of estuaries and the resulting anaerobic zones. (Estuaries are important sites for the anaerobic portions of the global nitrogen and sulfur cycles.) Bacteria may also stabilize sediments in estuaries (Montague, 1986; Montague, et al., 1993). Rarely does the manager have to confront the distribution of bacteria in an estuary directly. Rather, the action of bacteria is implicit in the kinetic terms of other water-quality indicators, and their effects are generally determined empirically. For example, for the dissolved oxygen balance described earlier, the effect of bacteria in the stabilization of organic matter is parameterized and measured by the BOD. The ecological functions of bacteria are further described in Sec. 12.5.

The exception—the bacteria whose distribution may have to be directly confronted in estuary management—is the case of pathogens. A few species of bacteria are pathogenic to humans and therefore represent a public-health problem. In estuaries, as in freshwater systems, *E. coli,* an intestinal bacterium, is used as an indicator of the potential presence of human pathogens of enteric origin. For this reason the distribution of coliforms in the water column is a determinant for limiting water uses, and this distribution is capable of analysis and prediction by mass-balance methods. (Coliform concentrations in receptor organisms, such as shellfish, are also important, especially from the public-health standpoint.) Moreover, management of enteric bacteria focuses on source control, based on considerations analogous to other pollutant-source management. Increasing attention is being given those bacteria that are indigenous to the aquatic environment, estuaries included, which are opportunistic pathogens, such as *Clostridium botulinum* and *Aeromonas hydrophila* (Grimes, 1991). One specific to estuaries, especially of concern in shellfish management, is *Vibrio vulnificus.*

Both metals and organic toxicants comprise a class of contaminants that can be an issue in any aquatic system. In estuaries, they assume special importance, because:

- The loads to estuaries are frequently greater than to inland waterways, due to riverborne influxes ultimately debouching into the estuary, but also to the additional loads from heavy concentrations of population, industry and shipping in estuarine areas.

- Many of these contaminants have a close association with suspended solids, due either to the originating load or the hydroscopic character of the compound, which can result in high accumulations in the estuary.

- The more productive and energetic ecosystem of the estuary leads to greater rates of assimilation and accumulation in biota than encountered in other aquatic systems.
- Accumulation of these pollutants in fish and shellfish can impact large-harvest coastal fisheries, either through mortality of the organism, or through human consumption of contaminated tissue.

Analysis of distribution of these constituents is addressed through mass-balance methods, i.e., Eq. (12.11) et seq. Apart from the transport terms, and the possible need to include suspended solids in the analysis (see Sec. 12.4.2), the methodology does not differ materially from application to other types of watercourses. (A useful development of the fundamental equations for sorptive substances is given by O'Connor, 1988.) Many of the analytical methods and the quantification of kinetic properties have originated from studies in estuaries because it was in these systems that water-quality problems were first manifested.

A major class of organic toxicants is the chlorinated hydrocarbons (or organochlorines). This includes pesticides and industrial chemicals such as polychlorinated biphenyls (PCBs). It can also include compounds produced inadvertently in any industrial process in which chlorine is brought into contact with hydrocarbons. An example of recent concern is the production of cyclodienes, especially dieldrin, through the use of chlorine in paper bleaching, which renders any kraft paper mill a potential discharger. Another example is the formation of hexachlorobenzene (HCB) during electrolysis of salt water, which implicates any industrial process using carbon electrodes as a potential source.

Perhaps the most famous organochlorine compound is dichlorodiphenyltrichloroethane (DDT), and it is important in estuary water quality both because it is a long-lived toxic with a history of extensive application, and because it is a model for the behavior of other, less-studied organochlorines. While DDT concentrations in estuaries (and other aquatic systems) have been declining in recent years in the United States due to governmental control, in other countries it remains a major estuary contaminant. As a technical product DDT is comprised of as many as 14 analogs and isomers, the most important of which are p,p'-DDT and o,p'-DDT. The relative proportion of the two is a function of the proportion in the initial source and of their relative kinetics in the receiving water. The former is about 70 percent p,p'-DDT and 20 percent o,p'-DDT in technical grade DDT (Reutergårdh, 1980; Buechel, 1983). (A rule-of-thumb of a 3:1 ratio of p,p'-DDT to o,p'-DDT seems to be current.) Both forms are hydroscopic and sorb readily to fine particulates (Crompton, 1985), especially to humics (Pierce, et al., 1974), and o,p'-DDT degrades at a rate somewhat faster than p,p'-DDT (Reutergårdh, 1980). Since DDT and its residues (mainly DDE and DDD), like many organochlorines, adsorb to particles, they can be attached to bacteria and phytoplankton, as well as inorganic sediments. The rate of actual uptake of DDT residues by the phytoplankton will therefore underestimate the quantities potentially entering the food web through grazing on phytoplankton.

Another class of hydrocarbons that can be of special importance in estuaries is petroleum hydrocarbons, associated with crude oil or its refined products, which we collectively refer to here as *oil*. These hydrocarbons generally are a mixture, with large proportions of cycloalkanes, aromatics, and polynuclear aromatics (PAHs) (see NAS, 1975, and Carlberg, 1980). Most of these are toxic, comprising an environmental problem in even dilute concentrations. But the physical properties of oil in heavy concentrations pose a biological danger, apart from its toxicity, in interfering with biological functions of an organism.

Oil is generally brought to mind in the problem of management of hazardous substance spills. As noted in Sec. 12.1.2.6, much of spill management is strategic planning and tactical response. Within the scope of this chapter, the ultimate impacts of a spill are the resulting distribution and fate of the spilled substance. This is approached from a mass-balance perspective, exactly like the other parameters considered here. The chief difference is the transient element of the analysis:

- A point load delivered instantaneously or over a short duration of time
- The operative hydrographic and hydrometeorological processes immediately during and following the loading episode
- The trajectories from the load point to the zones of stranding or efflux from the estuary
- The sources and sinks to which the substance is subjected over these trajectories

12.4.2 Sediments and Their Role in Estuarine Water Quality

The term *sediments* is used broadly to refer to solid particles of greater density than water but capable of being carried, perhaps briefly, in suspension by fluid flow. Such sediment can be organic or mineral, and in the water column is manifested as *suspended solids* or *particulate matter* (terms which in this chapter are used interchangeably). A related parameter is turbidity, which refers to interference with the passage of light and is therefore an indirect indicator of the concentration of suspended matter in the water.

Sediment is a normal, ubiquitous component of the estuarine environment. Unlike many freshwater systems, estuaries are typically turbid, and the suspended particulates have a significant and indispensable function in their biochemistry and ecology. (Those whose experience is confined to freshwater systems should take careful note of these statements.) Moreover, the bed sediments have an equal—if not greater—function, both as a continuing supply of suspended sediments and as a seat for the benthal element of the ecological community. Estuarine sediment transport processes may be considered to fall into three classes:

1. *Littoral processes,* due to coastal marine sediment mechanics, including beach and shoreline erosion, and interception of the longshore transport by the estuary inlet/mouth hydrodynamic circulations.
2. *Fluvial processes,* associated with sediments of terrestrial origin brought into the estuary by rivers and peripheral runoff.
3. Interior reworking of sediment deposits.

Therefore estuarine sediment transport encompasses the major fields of sedimentary processes, with nuances added by the peculiarities of the estuarine environment. By now the reader has become accustomed to our whining about scope and space and no doubt anticipates this necessarily superficial summary. (Among the voluminous literature on the subject, especially for fluvial systems, representative textbooks are Graf, 1971, and Yalin, 1976. A review of the processes affecting cohesive sediments in estuaries is given by Mehta, et al., 1989.)

12.4.2.1 Mobilization and Transport of Sediment. Fundamental properties of sediment affecting its transport are its density (relative to that of water), which drives gravitational settling, and its grain size and shape, which govern the ability of the fluid to resist settling by exerting horizontal and upward forces. Quiescent sediment over which water flows is resistive to movement up to a threshold value of the stress exerted by the flowing fluid (specifically, its component parallel to the surface of the sediment). Once this threshold is exceeded, sediment grains are mobilized from the bed, and are carried by the fluid. Several modes of motion are involved, characterized by the degree to which the particles are lifted from the bed, viz., rolling (or creeping), saltation, and suspension. The former two can organize themselves into spatially coherent bedforms (ripples and dunes) through a complex interaction with the flowing fluid. The mass transport due to creeping and saltation is collectively referred to as bedload, in contradistinction to suspended load. Clearly fluid velocity is controlling, especially its magnitude and duration.

Littoral sediments tend to be coarse silts and sands (mean grain diameters from $\frac{1}{32}$ to 2 mm), frequently quartz, more or less rounded, with a specific gravity of about 2.7. The stress threshold for grain mobilization τ_c is a strong function of grain size d, practically linear within the turbulent regime: $\tau_c \approx 41 \, d$ for d in cm and τ_c in dynes cm^{-2}. Once resuspension from the bed begins, its rate is approximately proportional to the excess stress $(\tau - \tau_c)$. When $\tau < \tau_c$, deposition occurs, governed by the fall velocity w_s, which under turbulent conditions is proportional to $\sqrt{d}$. Resuspension and settling are not, therefore, symmetric processes with respect to current speed or bed stress.

In the coastal environment the primary mechanism for resuspension is waves, specifically velocities associated with orbital motions and the complex wave transformations within the surf zone. Also, the interaction of waves with the nearshore profile forces net movements both perpendicular (onshore-offshore) and parallel (longshore) to the coastline, resulting in a net transport of the sediments in suspension. The complexity of the processes have necessitated an empirical approach in relating longshore transport directly to characteristics of waves at the seaward limit of the surf zone (see Komar, 1976 and USCE, 1984). Littoral sediments are transported into the vicinity of the estuary entrance, where they are entrained into the tidal currents and carried into the estuary on the flooding tide.

For those estuaries with a contraction entrance (see Sec. 12.2.2), this littoral sediment entrainment into tidal currents can result in dramatic patterns of sandbars and shoals due to the dynamics of the inlet. The tidal current velocity is increased by confluence in flowing through the inlet, so that bed stress increases and exceeds the critical value for even coarse sediments. These sediments are therefore easily transported through the inlet (and deeper tidal channels may be scoured as well). The difluence of flow and reduction in current speed at the exit of the inlet diminishes stress below critical for all but the finest particles, leading to a zone of sediment deposition. This zone within the inlet, formed by difluence of the flood tidal current, is the flood bar, and the corresponding zone on the exterior of the inlet is the ebb bar. The former is generally more extensive than the latter, due to the relatively sheltered environment of the estuary compared to that of the open coast. Examples and descriptions of the genesis of these and other estuarine sand bedforms are provided in Boothroyd (1978); see also Inman and Dolan (1989).

Sediments carried into the estuary by rivers and peripheral runoff tend to have much higher proportions of clays and colloids, i.e., muds, compared to those of littoral origin, and also to contain varying proportions (10 to 50 percent) of organics, especially humics. The muds are not only finer particles than the littoral silts and sands, ranging 0.5 to 4 µm, but are also mineralogically different, e.g., illites, chlorites, and smectites. The particles are platelike flakes and electrostatically charged. As they collide, they aggregate to form flocs that are much larger than the clay particles per se. The organic material present can enhance and stabilize flocculation through bacterial films and organic binding (Nowell, et al., 1981; Montague, 1986).

Because flocs are geometrically complex, open agglomerations, they behave as a water-sediment mass with lower effective density than the individual flakes (effective densities range 1.1 to 1.9 g cm^{-3}) but greater effective diameter, by orders of magnitude. The net effect is a marked increase in fall velocity over that implied by the individual grain sizes. The magnitude of this fall velocity is clearly dependent upon the rate of flocculation, for which there are two important controls: the concentration of suspended solids per se (hence the opportunity for collision and further aggregation) and any factors which alter the tendency for particles to attract and bind. With respect to the former, there is a clear increase in fall velocity w_s with a power of TSS concentration C^k, k in the range 1 to 2, up to a concentration C on the order of 10^4 mg/L (Mehta, 1986). Higher concentrations result in "hindered settling" in which w_s decreases roughly as $(1 - C)^5$ (see Mehta, 1986; Dyer, 1986). With respect to the latter, the presence of ions in the water decreases the electrostatic repulsion between particles and enhances the formation of flocs. Salinity is therefore generally regarded to have a strong influence on flocculation and settling in estuaries (Dyer, 1986), but the quantitative dependence remains unclear. Some researchers find greatly enhanced sedimentation in the brackish salinities (3 to 10 ppt), while others find a near-linear increase over the entire range of salinity from 0 to 30 ppt. Meade (1968) in reviewing data from the U.S. Atlantic coast found no decline of suspended sediment with increasing salinity that could not be accounted for by simple dilution. Burt (1986) found no dependency on salinity at all in his measurements in the Thames, which he attributed to the long time available for flocculation in the estuary, compared to the short times of laboratory experiments upon which the empirical salinity dependence is generally based. Eisma, et al. (1991) determined from studies in the Ems, Rhine, and Gironde that hydrodynamics and organic binding, especially carbohydrates, are the principal factors causing flocculation, and that salinity plays very little part (see also Eisma, 1986).

For the sands typical of littoral sediments, discussed earlier, the bed sediments are noncohesive, and scour or deposition depends upon bed stress τ (which varies as the square of velocity) relative to a critical value τ_c. For clay flocs, in contrast, the bed sediments are cohesive. Deposition occurs if the bed stress is less than a critical value for deposition τ_d, and scour occurs if the bed stress exceeds a critical value τ_s; these critical stresses are not even approximately equal, τ_s exceeding τ_d by about an order of magnitude. Deposition rate proceeds roughly as $(1 - \tau/\tau_d)C$, but the proportionality depends upon both fall velocity and the time duration over which flocculation takes place (Mehta, 1986; Dyer, 1986). While reliable values of τ_c (for littoral sediments) have been established in the literature and depend almost entirely upon grain size, both τ_d and τ_s are complex functions of the sedimentary characteristics of the bed, are site-specific, and only a few determinations in estuaries have been published.

The third class of estuarine sediment transport processes enumerated previously is the internal reworking of sediments. This refers generally to the scour and erosion of sediments from relict deposits, whose riverine or littoral origin and original deposition belong to the geological past. These may be noncohesive sand bars or cohesive well-consolidated clays, and are exposed to currents either on the bottom of the estuary or along its shoreline. Because of their age and the different hydrological/hydrographic conditions prevailing at their initial deposition, they may differ from the sediments of modern littoral or fluvial origin. Consequently, the parameters governing thresholds and rates of erosion may also be different. From the standpoint of the estuarine sediment budget, probably their most important property is as a sediment load different in its location from either the tidal inlets or the river mouths, and governed by different hydrometeorological and hydrographic factors than those controlling the littoral or fluvial sources. For example, windwave erosion of relict sand deposits (ancient strandplain dunes or barrier islands) around the shoreline of the estuary will depend upon the direction and fetch of wind, and may involve episodic slumping along retreating scarps. Exposed relict clay layers (such as ancient river deltas) will probably be tight, firm, dense clays that tend to erode in lumps and be transported in bedload.

12.4.2.2 Sediment Mass Budgets. Mass budget analysis and modeling of suspended matter usually considers the suspended solids totalled over a range of grain sizes, such as muds (i.e., clays and finer) or TSS (retained on 0.45 μm filter). Riverine loads, influxes at the mouth or inlets to the estuary, and peripheral sources such as beach erosion, are specified as boundary conditions. For the source and sink terms, scour (erosion) is zeroth-order in suspended sediment concentration, and deposition (settling) is first-order, with coefficients that are functions of sediment characteristics, the near-bed currents, and perhaps other factors. One of the more important hydrographic controls is tidal currents that can exceed the critical stress for resuspension at the race of the current (i.e., the maxima in current speed), leading to a time pattern of suspension and settling at twice the dominant tidal frequency.

If turbulent mixing is sufficiently vigorous that suspended sediment is vertically homogeneous, the vertical-mean mass budget equation Eq. (12.12) or section-mean equation Eq. (12.13) may directly apply. Vertical homogeneity is rarely the case, however; due to the settling processes, suspended particulates increase in concentration downward. If there is a density current with even slight vertical shear, also usually the case in the presence of a horizontal salinity gradient, there will clearly be a differential transport, greater upstream in the lower layer (with higher concentration) than downstream in the upper layer (with lower concentration). The combined effect of vertical settling and the long-term mean differential motion due to the density current is that sediments converge in the salinity intrusion zone, such as diagrammed in Fig. 12.22. In some estuaries, the zone of sediment convergence is concentrated near the inland limit of salinity intrusion, creating a distinct estuarine turbidity maximum. An example is shown in Fig. 12.23 from the principal channel of the San Joaquin delta region of San Francisco Bay. Other examples of estuaries with a prominent and well-defined turbidity maximum include the St. Lawrence, Columbia, Savannah, Weser, Gironde, Chesapeake Bay, Delaware, and Tamar. The location, extent, and concentration of the turbidity maximum are strong functions of hydrographic and hydrological controls (e.g., d'Anglejan, 1980; Geyer, 1993). Also, stimulated flocculation of river sediments as they are transported into the region of salinity intrusion will enhance the turbidity maximum.

There are two important implications for the estuary from this convergence of sediment. First, estuaries tend to be collectors—sinks—for sediment. Even when an estuary does not have a pronounced zone of maximum turbidity, due to variability and complexity of its hydrography, the processes of settling and density current transport into its interior produce a convergence of both littoral and fluvial sediments that is widespread within the estuary. The usual situation is that an estuary is slowly filling with sediments. There are examples of rivers having completely filled their estuaries with sediment in the recent geological past, i.e., the sediment accumulation exceeding the rate of sea-level rise, such as the now-nonexistent bays of the Brazos and Colorado Rivers on the Texas coast and that of the Yangtze in China.

Second, TSS is a prime example of the failure of the dispersion coefficient parameterization in the mass-transport equation. The reason is that the flux $\overline{u'\partial c'/\partial x}$ (see Sec. 12.3.4) is not directed down the concentration gradient (as is assumed by writing it $-E_x \partial \, \overline{c}/\partial x$) but rather *up* the concentration gradient. To carry out a mass budget of TSS, one must explicitly include the density current transports. The simplest

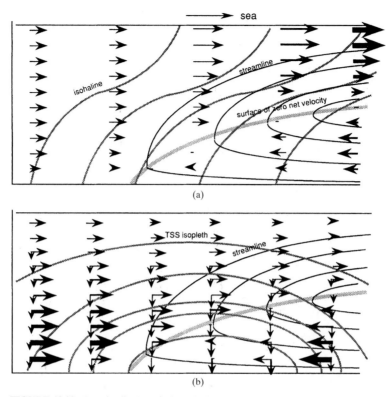

FIGURE 12.22 Longitudinal-vertical residual transport of suspended sediment in rectilinear estuary driven by density current (see Fig. 12.13), showing convergence into the turbidity maximum: (*a*) Tidal-mean circulation in vertical-longitudinal section; (*b*) Tidal-mean
suspended sediment transport.

approach would be a two-layer box model. Owen (1977) summarizes a two-layer
coupled hydrodynamic-salinity-sediment model appropriate for a longitudinal estuary with sediment sources at the mouth and head, and settling within the estuary,
solved by finite differences. O'Connor and Lung (1981) employ essentially the same
two-layered model, but by assuming that the horizontal gradient of salinity is given
a priori, develop an analytical solution for current and TSS.

It is clear that grain-size analyses can yield important insights into the processes
of deposition and resuspension of sediments in an estuary. Moreover, many constituents have an affinity for adsorption to surfaces, for which suspended sediments
are a prime target. This is especially true for the muds, due to their chemical (and
biochemical) nature, which stimulates binding, and to their large flocs, which offer a
considerable surface area per unit volume. Grain-size determinations can frequently
clarify the efficacy of sediments for adsorption. Unfortunately, grain-size determinations are often overlooked as a routine aspect of sediment analysis. If the manager is
in a position to influence data-collection practices, he should encourage some—even
rudimentary—textural determinations whenever sediments or suspended sediments
are sampled.

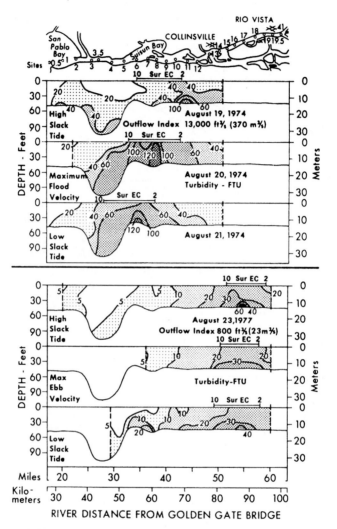

FIGURE 12.23 Longitudinal-vertical profiles of turbidity in Carquinez Strait region, upper San Francisco Bay. The brackets mark the zone of surface conductivity ranging 2 to 10 mmhos/cm (salinity 1 to 6 ppt). (*Arthur and Ball, 1979.*)

In assembling a data base for suspended matter, the analyst will frequently find a wealth of turbidity measurements. These can be used to estimate TSS; while the estimate is rough, they can be useful in constructing long-term trends or identifying zonation in suspended matter through the estuary. The traditional method of viewing a candle flame through a vertical tube containing the water sample motivated the definition of the *Jackson Turbidity Unit* (JTU), see APHA (1989). The modern nephelometer is now the preferred alternative (e.g., Lamont, 1981; Kirk, 1983), and the measurement is reported in *nephelometric turbidity units* (NTUs), defined to be numerically about the same as JTUs. The depth of the Secchi disc has for many years been the limnologist's and oceanographer's standard means for field measurement of turbidity (Hutchinson, 1957). Unfortunately, the relation between *Secchi depth* (SD) and conventional measures of turbidity is murky at best, a deep analysis of which is given by Preisendorfer (1986). As a first approximation, there is reason to expect an inverse relation between suspended solids and Secchi depth, TSS = b/SD, and a direct linear relation between suspended solids and turbidity, TSS = aT, with constants in the ranges of $10 < b < 50$ and $0.2 < a < 0.5$, for SD in meters, TSS in mg/L, and T in NTU (e.g., Jones and Willis, 1956; Holmes, 1970; Di Toro 1978; Effler, 1988).

12.4.3 Kinetics in the Estuarine Environment

Application of the mass-balance equation Eq. (12.11), as noted earlier, is not unique to estuaries, but the nature of the estuary system dictates some important differences in the formulation of the equation. The complexity of estuary hydrodynamics and density structure requires special treatment of the transport terms. Boundary conditions may also be complicated, reflecting both riverine and marine influences as well as internal loads. An additional factor which may differentiate the estuarine mass-balance equation from those of other watercourses is special forms of the source and sink terms. These may arise from the following properties of the estuary:

- The hydrometeorological and hydrographic regime of the estuary, including complex bathymetry and physiography, and internal circulations
- Salinity and its variation from freshwater to seawater concentrations
- The varied and productive biotic communities, especially the microflora
- High concentrations of sediments, especially fine-grained clay flocs

12.4.3.1 Dissolved Oxygen Kinetics. A good example of the need for special kinetic formulations in an estuary is the dissolved oxygen budget. Solubility of DO is governed by temperature, as in freshwater systems, but also by salinity. The dependence of saturation concentration C_s is given approximately by the Fair-Geyer (1954) expression:

$$C_s = \frac{5(100 - \text{Cl})}{T + 35}$$

where C_s is in mg/L, Cl is chlorinity in ppt (= salinity/1.81), and T is temperature in °C. (The coefficients were re-evaluated from those of Fair and Geyer, 1954, using the recent data in APHA, 1989.)

The surface reaeration term also requires modification of the coefficient for application to an estuary. The O'Connor-Dobbins (1958) formula,

$$K_a = [D_L \, U \, D^{-3}]^{1/2}$$

though originally proposed for flowing streams, has found wide application in estuaries, especially those of coastal-plain type that have a morphological similarity to large rivers. Here $D_L \approx 2.1 \times 10^{-9}$ m^{2}s^{-1} is the molecular diffusivity of oxygen in water. The current velocity U is that which is dominant in the region of interest, the throughflow velocity in a stream, but usually the mean tidal current in an estuary. (By this is meant the tidal mean of the absolute value of the current.) This equation is strictly applicable only to those systems in which turbulence is developed by the flow of the fluid over the bed. For narrow, sheltered estuaries, it would therefore be appropriate. Many estuaries are, however, broad, open systems in which wind plays an equal or greater part in aerating the water. In these systems, a dependency more like

$$K_a = a \, W^{1/2} \, D^{-1}$$

is suggested, where W is wind speed, in m/s at standard anemometer height (10 m), and $a \approx 4 \times 10^{-6}$ (m/s)$^{1/2}$, based upon observations in the laboratory, large lakes and the oceans (e.g., Kanwisher, 1963; Eloubaidy and Plate, 1972; Liss, 1973; Peng, et al., 1974; Mattingly, 1977; Broecker, et al., 1978). Observations in estuaries are scarce, due to the difficulty of the procedures. Recent measurements (Downer, et al., 1991) suggest that this relation may be too high by about 20 percent, at least for shallow bays.

The biological terms in the oxygen balance may require modification as well for application in an estuary. An explicit *sediment oxygen demand* term may have to be used to reflect the frequently vigorous respiration of benthal microflora. Generally, rate coefficients in estuarine waters for BOD and nitrifiers are on the same order as those in freshwater systems. (There is some indication of a reduction in metabolic rates with increasing salinity; Mudryk and Donderski, 1991.) On the other hand, photosynthesis and respiration by the phytoplankton community are often large enough that they must be added, as zeroth-order terms dependent upon light penetration and nutrient concentrations (see Di Toro, et al., 1977; Thomann and Mueller, 1987). A recent comprehensive review of dissolved oxygen kinetics edited by Smith, et al. (1992), though strictly addressing the Chesapeake, is of general applicability to other estuaries.

12.4.3.2 Nutrients. As indicated in Sec. 12.4.1, nutrients and their management are of particular importance in estuaries. We consider silicon, nitrogen, and phosphorus. The disposition of these nutrients in an estuary is displayed qualitatively by the generalized (generic) diagrams of Figs. 12.24 et seq., in which the principal transfers of the nutrient are shown between its major pools. With these, as almost all biochemical constituents of estuarine waters, a distinction must be made between the occurrence of the constituent in solution (dissolved) and in association with suspended solids (particulate). (The distinction is operational, based upon passing the sample through a filter, usually 0.45 μm but occasionally coarser; e.g., Eaton, 1979.) These diagrams are intended to display the important transfer paths, but not the relative magnitudes of the pools. The dominant transfers are indicated by bold arrows, and those that are occasionally important, or that are of secondary but nonnegligible magnitude, by broken arrows.

For silicon, the important pool is that in solution, since dissolved silicon is available for uptake by microorganisms. (This is a comparative statement. Biotic availability of an element depends upon its chemical form, i.e., whether it occurs as an inorganic ion, or inorganic or organic complexes, and specifically which ones.) River waters are rich sources of dissolved silicon, typically two orders of magnitude above the concentrations in seawater. There is only one sink of any consequence of dissolved silicon in the estuary: biotic uptake. The rate of biotic uptake seems to be highly variable, as much as 30 percent of the source load over the entire estuarine reach in some cases, but low enough in others that Si is conservative and directly correlated with salinity. There is little evidence for regeneration (Liss and Pointon, 1973; Aston, 1980). The two primary controls on relative biotic uptake are phytoplankton metabolism (in turn driven by insolation and light penetration) and river flow. Since river flow is the dominant source of Si, a low flow implies a low load, so the uptake becomes large by comparison and is capable of reducing Si concentration. This

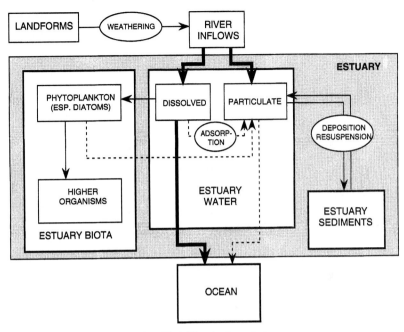

FIGURE 12.24 Schematic of simplified silicon budget of estuary.

effect is clearly demonstrated by the data of Peterson, et al. (1975) from San Francisco Bay, shown in Fig. 12.25, showing silicon-salinity relations through the saline intrusion reach for 13 different sampling runs over a range of inflows.

The mass transfers of nitrogen, shown in Fig. 12.26, are not only more complex than those of silicon, but also differ in several fundamental respects. First, unlike silicon, there are substantial anthropogenic sources of nitrogen, from both wastewater and from runoff containing nitrogen fertilizers. These additional loads can potentially increase the dissolved nitrogen supply and flora biomass, much of which ultimately enters the sediment pool. Second, the effects of solids (i.e., the particulate and sediment pools) on the solution pool are greater than is the case for silicon. There are sources of dissolved nitrogen from both the bed sediments (especially from vegetation in the shallow areas of the estuary) and from the suspended solids (by desorption), but these are of secondary importance to the inflow and wastewater sources. An added complexity, suppressed in the diagram of Fig. 12.26, is that there are several interconverting species of nitrogen involved (McCarty, et al., 1970), most important of which are the inorganic forms, ammonia, nitrites, and nitrates.

The situation for phosphorus is more complex yet, as shown by Fig. 12.27. There are multiple sources of phosphorus, both natural and anthropogenic, and there are several interconverting species (McCarty, et al., 1970). The oceanic source is more important than is the case for either silicon or nitrogen. Phosphorus has an affinity for fine-grain particulates, especially the clay flocs, so interacts much more with the suspended sediments in the estuary. (Lebo and Sharp, 1993, report a peak in total phosphorus associated with the estuary turbidity maximum in the Delaware.) The high uptakes and regeneration of dissolved phosphorus by the biota should be par-

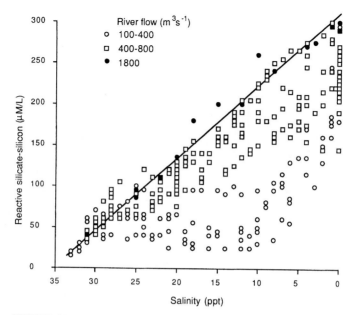

FIGURE 12.25 Silicon versus salinity in San Francisco Bay, 1971 to 1973. (*Data from Peterson, et al., 1975.*)

ticularly noted. This high turnover of phosphorus (i.e., low value of the ratio of water-column mass to rate of utilization, in dimensions of time) is well known in other aquatic systems (e.g., Wetzel, 1975), and operates in estuaries as well (Lebo and Sharp, 1993), and in fact the kinetic rate coefficients from freshwater systems seem to be equally applicable in estuaries. This high turnover rate means that the dissolved phosphorus pool is generally very small, because it is taken up by flora almost as soon as it is available. While there is substantial exchange with the particulate pool, sediments generally act as a net sink for phosphorus, especially in the deeper estuaries. However, many estuaries exhibit a summer phosphorus maximum, which has been suggested to result from phosphate release from sediments under anoxic conditions (Taft and Taylor, 1976).

12.4.3.3 Metals and Organic Toxicants. A generalized (generic) transfer diagram for trace metals and sorptive organics is shown in Fig. 12.28. The central feature is the role of the particulates and sediment pools as both a source of bioavailable dissolved forms and direct biological uptake due to particle feeders, and a sink through sorption and either settling or transport to the ocean with the finer clays and colloids (Duinker, 1980; Domagalski and Kuivila, 1993; Olsen, et al., 1993). In any aquatic system, suspended matter and sediments act as carriers for these trace constituents; in an estuary, sediment transport is especially important. The greater the turbidity of an estuary, the more likely dissolved trace constituents are to behave nonconservatively in comparison to salinity as a tracer. Adsorption is a primary mechanism for the transfer of organics and metals from solution to suspension, and settling is a primary mechanism for the loss of total concentration (dis-

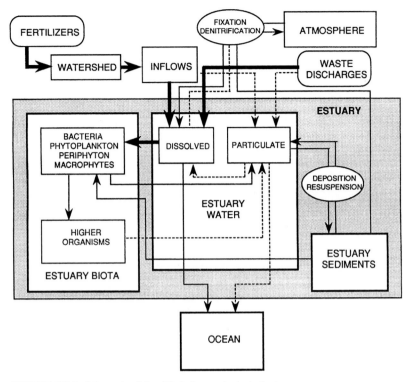

FIGURE 12.26   Schematic of simplified nitrogen budget of estuary.

solved plus particulate) from the water column. The effect of particulates is to act as scavengers for the solution pool. Trace organics and metals in the bed sediments can be reintroduced into the water column by the physical processes of scour, resuspension, and bioturbation, and into the dissolved phase by desorption, whose rate depends upon pH, redox potential, and salinity (i.e., increasing with salinity).

For those estuaries in which the principal metals load is from the head of the estuary (the Rhine being the classic example), there is generally a seaward decline in trace metal concentrations in both bed sediments and suspended particulates. This is considered to be the combined result of greater mobilization of sediment metals in the lower estuary, from the decay of organic matter enhancing solution of particulate metals by providing ligands, in combination with higher tidal velocities and resulting resuspension, and a gradual increase in proportion of uncontaminated muds of littoral origin relative to the contaminated muds of fluvial origin. Concentrations of iron and manganese are important indicators for the efficacy of the adsorption-deposition sink for metals, because their oxides greatly enhance the adsorption of other metals. When the trace constituent does behave nonconservatively, the kinetic rates involved, either from inorganic conversions or biological uptake, are generally much smaller than, say, nitrogen. Moreover, concentration in the higher trophic levels is a frequent consequence (see Sec. 12.5.3). A comprehensive survey of trace metals in estuaries may be found in Salomons and Förstner

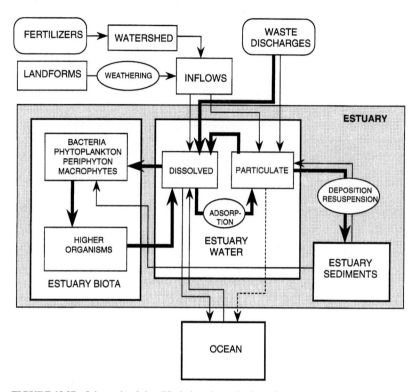

FIGURE 12.27   Schematic of simplified phosphorus budget of estuary.

(1984). The reviews by Reutergårdh (1980) and Livingston (1976) on organochlorines and Förstner (1980) on metals are also recommended. The analyst should bear in mind that while most metals data are reported in elemental form, the kinetics, i.e., stability and reaction rates, depend upon the specific complexes involved (e.g., van den Berg, 1993).

While there are natural sources for trace metals, human activity has greatly increased the loads both from the watershed and from direct industrial discharges to the watercourse (Förstner, 1980). For most trace organics, the natural sources are practically negligible, and anthropogenic sources represent the bulk of the loadings, though these can originate both from the watershed and from direct discharges to the watercourse. For many trace metals and organochlorines, a potentially important additional source is deposition from the atmosphere. Anthropogenic loading, especially of trace metals, can follow this transfer path if large industrial or urban sources occur upwind and proximate to the watercourse, a circumstance frequently characteristic of urbanized estuaries. This can also increase the loading from river flow and runoff by deposition on the watershed and subsequent drainage into the estuary. Atmospheric loading of pesticides originates in agricultural applications, perhaps great distances from the estuary. Harder, et al. (1980) report significant rainfall deposition of toxaphene in the tidal distributaries of North Inlet estuary (South Carolina) and Young, et al. (1976) report both dry deposition and precipitation loads of DDT and PCBs in the southern California coastal area.

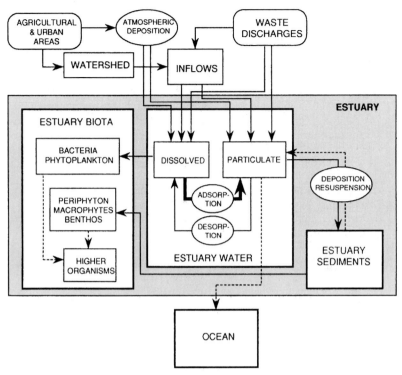

FIGURE 12.28 Schematic of simplified budget of generic heavy metal or sorptive organics in estuary.

12.4.4 Analytic Approach

The analysis of distribution and fate of nutrients and of toxic organics and metals is therefore quite complex in an estuary, a complexity aggravated by the different physicochemical forms the substance can assume, difficult and expensive laboratory analytical procedures of highly variable precision, and usually a sparse data base. The schematics of Figs. 12.24 to 12.28 are simplifications and generalizations, which will probably not apply directly to any specific estuary. For example, both the time and space dimensions are suppressed in these diagrams. Nutrients especially exhibit characteristic seasonal variation (even silicon; see, e.g., Peterson, et al., 1986), and the dominant transfers will change with the season. In some estuaries seasonal variation has been observed in metals as well (e.g., Boyden, et al., 1979). Clearly, seasonality of freshwater inflow is an important determinant of seasonal behavior in these constituents. In space, the loads, transports and kinetics will all vary with position in the estuary. Also, there may be important transfer paths operating in a specific system, not depicted on the schematic. The atmospheric pathway has been found to be important for nutrient loading in some estuaries. In some of the U.S. Atlantic coast estuaries (Jaworski, et al., 1992; Scudlark and Church, 1993), atmospheric deposition has been found to represent a significant load of nitrogen, both on the watershed and directly to the estuarine water surface. In some estuaries, birds have been identified as important transfer pathways for nitrogen and phosphorus (Bildstein, et al., 1992), and in others dense shellfish beds (Dame, et al., 1991).

The analyst is advised to proceed as follows:

- Adopt a time scale of many tidal cycles, i.e., a long-term average, since the sources (loads) are expected to be diffuse.
- Identify and quantify the principal loads of the substance of concern to the estuary.
- Determine the linearity of the substance in solution relative to salinity, and determine reaches of systematic source or sink processes.
- Determine the distribution of both dissolved and particulate concentrations in the estuary waters; in particular, evaluate variation with river flow and other hydrographic controls, and establish the spatial variation and its relation to known load sites.
- Determine the distribution of the substance in the sediments of the estuary; evaluate the spatial patterns relative to known load sites.
- Couple the analysis with a mass budget of suspended sediments in the estuary; use grain-size as an independent measure if data exist.

Often an occurrence of biological toxicity or an episode of tissue contamination (and potential human health impacts) is the motivator for concern with a specific organic or metal substance. The nature of the organism so affected can possibly provide additional spatio-temporal information on the probable path of contamination. Similarly, companion data on productivity or algal biomass (including nuisance blooms) can provide the same information for nutrients.

If each of these steps can be accomplished, a rudimentary mass budget should be capable of closure for the estuary or large subreaches. If the data set is adequate, a mass-balance model, i.e., Eq. (12.11) et seq., should be the goal. This can take the form of a box model, as briefly described in Sec. 12.4.1, in which the compartments are defined to discriminate the principal transfer paths in space. Unfortunately, the usual situation is that the mass-budgeting process proves to be data-limited. Water sample analyses in particular tend to be noisy, because these are occasional point samples drawn from a range of tidal, meteorological, and hydrological variation, characteristic of estuaries. It is usually better under these circumstances to emphasize the sediment analysis, as sediments offer a kind of integrated response to the overlying water, and may yield intelligible information even when sampled infrequently. Though data limitation may frustrate a mass budget, by attempting to carry out the preceding steps, the analyst is in a better position to recommend specific data collection procedures.

12.5 ECOLOGY

12.5.1 Features of Estuarine Ecology

12.5.1.1 Ecological Productivity and Production Subsidies. Estuaries are among the most ecologically productive areas in the world—generally higher in net primary production than eutrophic lakes, croplands, and tropical rain forests. Major categories of estuarine primary producers are given in Table 12.4. *Spatially integrated net primary production* (gross production less plant respiration) in estuaries generally lies within the range from 300 to 3000 grams of organic carbon

TABLE 12.4 Main Primary Producers of Estuaries

Producer	Production, $gC\ m^{-2}y^{-1}$
Microalgae	~270 (35–550)
Phytoplankton	
Periphyton	
Macrophytes	~1500 (700–3000)
Intertidal	
Salt marshes	
Mangroves	
Subtidal	
Seagrasses	
Macroalgae	

Source: Boynton, et al., 1982; Montague and Wiegert, 1990.

fixed per square meter per year, $gC\ m^{-2}y^{-1}$. Typically estuarine productivity is far greater than that of the uplands or nearshore continental shelf immediately adjacent to the estuary.

The exceptional production in estuaries can be attributed to two general phenomena: the continual movement of water, and the trapping of nutrients. Water movement subsidizes plant and animal growth by distributing nutrients and seeds, removing metabolic wastes, and providing passive transportation. The energy savings are then put into additional production (Schelske and Odum, 1961; Odum, 1980). Tides, salinity gradients, river discharge, runoff, and winds all contribute to water movement (see Sec. 12.3). Nutrients that stimulate both plant and animal production are brought from within the entire river watershed in river water. Quantities of sediment are also transported to estuaries by river water (Meade, 1982; Meade and Parker, 1985). Here nutrients tend to remain—trapped both in vegetation and in accumulations of muddy sediments (see Sec. 12.4.2). The estuarine circulation pattern greatly retards the escape of sediments through the mouths of estuaries, and brings in additional nutrients from the nearshore continental shelf. Most estuaries are experiencing a rising sea level, which continually increases the capacity for nutrient-rich sediment accumulation.

Several sources of particles contribute to estuarine muds. Besides fluvial and littoral particles brought in by estuarine circulation, new particles form and settle spontaneously through flocculation. Significant to estuarine productivity and perhaps even more important to sedimentation is the biological production of particles in estuaries. Populations of phytoplankton and bacteria take up dissolved nutrients, converting them into new living particles. Likewise, submerged and intertidal vegetation converts inorganic nutrients to biomass, their own leaves, stems, roots, blades, and holdfasts. Some produce vast quantities of calcium carbonate sediment in tropical estuaries (e.g., Stockman, et al., 1967). All eventually die and decompose into tiny particles of organic detritus that settle with silt and clay in quiet waters, and there continue to contribute to the richness of the nutrient environment of estuaries.

The wide range of production values for different estuaries (300 to 3000 $gC\ m^{-2}y^{-1}$) depends in large part on the fraction of the total estuarine area covered by intertidal and subtidal vegetation such as salt marshes, mangroves, and seagrass beds. Productivity in these ecosystems (~1500 $gC\ m^{-2}y^{-1}$) is much greater than the planktonic systems of open water (~270 $gC\ m^{-2}y^{-1}$). It is important to realize, however, that quantity of primary production is not the only factor that determines the quantity of animal production in an estuary. Per unit mass, the lesser quantity of microalgal production supports considerably more animal production (Montague, et al., 1987). Nevertheless, the ratio of the surface area of an estuary to its volume can be a good predictor of estuarine primary production (Welsh, et al., 1982). Broad shallow estuaries, which have extensive intertidal zones and well-lit subtidal zones, support more intertidal and subtidal macrophytes, so are more productive than deep, steep-sided estuaries (in which phytoplankton habitat is more prevalent). Other factors, such as latitude, tidal range, and river discharge, contribute to variation in production from estuary to estuary (Day, et al., 1989).

Although estuaries are in general highly productive, not all regions within an estuary are productive. Low production can occur in areas where water circulation is very poor, or perhaps in areas of frequently alternating extremes of salinity (Montague and Ley, 1993). At the landward fringe of intertidal marshes may be areas that are barren because the soil salinity is so high (>150 ppt). Some quiet waters do not contain vegetation despite sufficient light because the shallow waters become too warm, or oxygen remains too low at night to allow vascular plant survival.

12.5.1.2 Stresses in the Estuarine Environment.

Along with the subsidies to production in estuaries come significant stresses. Because of the fluctuating water levels and variations in river discharge and local weather, the estuarine environment is highly variable in key physiological variables: salinity, dissolved oxygen, temperature, light, and water level. An implication of the dynamic estuarine environment is one of significant energetic tradeoffs for animal survival and evolution in estuaries. The energy gained by remaining in rich feeding grounds may just balance that needed to power the physiological mechanisms required to do so.

Many estuarine species have physiological mechanisms to withstand the stresses. The most prevalent species of estuarine animals and plants are *euryhaline* (able to withstand a broad range of salinity). Others are *eurythermal*. Many animals have adaptations for tolerating short periods of low oxygen. Most estuarine algae have accessory pigments that allow better use of available wavelengths of light under turbid conditions. For a review of various mechanisms for adapting to the estuarine environment, see Newell (1976). Nevertheless, the ability to withstand a broad range and long duration of stress has its limits—not only in amplitude but also in the frequency of alternating between extremes, timing, and suddenness of change. Hence, algal blooms and die-offs of animals and plants are common occurrences even in pristine estuaries. In developed estuaries, these events are exacerbated by pollution, becoming more frequent and intense, and causing farther-reaching economic damage and public concern.

Several strategies for survival in estuaries are employed by resident and transient organisms including tolerance, avoidance, and boom-and-bust. Sedentary organisms must withstand extremes or die. Only a few species have evolved with all the necessary physiological mechanisms required for this, however. Large swimming animals can avoid harmful extremes simply by leaving unsuitable habitat, but this may require that they continually swim away from shifting water masses of detrimental water quality, occasionally having to leave favorable feeding grounds. Compared to sedentary animals and plants, many more species of transient organisms use estuarine habitat. The boom-and-bust strategy is common among tiny, rapidly colonizing and reproducing organisms such as species of single-celled algae and bacteria. These can colonize an area rapidly and reproduce when conditions are favorable. Algal blooms are an example. While many individual cells may die when unsuitable conditions occur, a few encystic survivors can repopulate rapidly when favorable conditions return. A large number of species of single-celled algae occur in estuaries, some of which are able to dominate in blooms for extended periods.

12.5.1.3 *Estuarine Trophic Chains and Food Webs.*

Ecologists recognize three types of trophic chains in estuaries defined by the form of food or by how it is consumed at the base of the chain: grazing, detritus, and osmotic. Animals in estuaries, such as crabs, shrimp, fish, oysters, and worms, are produced in a rather complex food web that occurs at the intersection of these trophic chains (Fig. 12.29). At the base of grazing chains are living plants—ranging in size from phytoplankton to mangroves—that are consumed whole or in part by herbivores. Most phytoplankton is consumed alive—both by herbivorous zooplankton, which in turn are consumed largely by predatory zooplankton such as larval fishes and crabs, and by filter-feeding worms and bivalves on the bottom, which are fed upon by predatory snails, worms, echinoderms, and fishes. Conversely, only about 10 percent of intertidal and subtidal vascular plants and macroalgae are consumed alive. In the intertidal zone, the herbivores are mostly insects, which give rise to a terrestrial-like grazing trophic chain that includes spiders and songbirds. In beds of seagrasses, the principal grazers are often sea urchins.

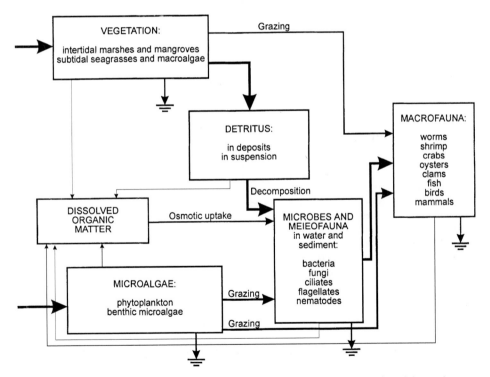

FIGURE 12.29 The rich production of estuarine animals arises at the intersection of the grazing, detritus, and osmotic trophic chains.

Technically, dead biomass of any kind forms the base of detritus trophic chains, but vascular plant material, with its difficult to decompose ligno-cellulose structure, comprises the bulk of the nonliving particulate organic matter in estuaries. With few exceptions, this cannot be assimilated by animals. Vascular plant detritus must first be decomposed by fungi and bacteria, which in turn can be consumed by animals. This extra step in the detritus trophic chain consumes much of the energy available in the detritus. Hence, animal production per unit plant production is much lower in detritus trophic chains than in grazing trophic chains, perhaps by as much as ten times (Barnes, 1980; Montague, et al., 1987).

Estuarine vascular plant detritus is produced in intertidal marshes, mangroves, and subtidal seagrass beds, and is imported from land in rivers and runoff. The relative amount from each source of course varies from estuary to estuary. In shallow estuaries, an individual particle of detritus is more likely to have been produced within estuarine intertidal and subtidal vegetation, while in deep estuaries (e.g., fjords) detritus may be more of terrestrial origin. Exceptions to this rule occur when large rivers or canal drainage dominate shallow estuaries.

Dissolved organic matter forms the base of the osmotic trophic chain. Dissolved organics leach from all living things—including bacteria, fungi, algae, vascular plants, and animals—both while alive and after death. In general, the first beneficiaries of the most readily usable (labile) forms of organic matter are bacteria. Although nearly all living things can assimilate glucose, for example, bacteria, with their small size and high surface to volume ratio, quickly absorb the majority of labile organics nearly as soon as they appear in water. Some dissolved organics may also be used by microalgae when they are not producing enough organics by photosynthesis. What remains to be measured as dissolved organic matter in water is in general the less usable (refractory) forms. These may photodegrade into more labile forms over time, which are then quickly absorbed by bacteria.

Most juvenile and adult estuarine animals are omnivorous. These in turn are eaten by a smaller number of predatory animals, including popular estuarine sport fish (e.g., drums and flounder), a variety of wading birds, certain birds of prey (e.g., osprey), and at least one marine mammal (bottlenose dolphin). Deposit feeders, such as fiddler crabs, certain snails, sea cucumbers, and many species of polychaete worms, ingest whole mud or sand. Filter feeders, such as clams, mussels, oysters, and certain polychaete worms, filter all particles within a size range from water. Both deposit and filter feeders consume whatever their digestive enzymes allow during the residence time of the material in the gut. Some omnivores (e.g., fiddler crabs and oysters) sort and separate the more inorganic fraction from the more organic fraction, attempting to digest the latter and egesting the former as pseudofeces. Others (e.g., certain sea cucumbers) simply swallow everything.

The material digested by these omnivores primarily consists of tiny living organisms: bacteria, fungi, microalgae, microfauna (<50 µm), and meiofauna (50 to 500 µm). These tiny organisms form a complex microbial food web that combines detritus, microalgal grazing, and osmotic trophic chains of estuaries. This microbial web then gives rise to the vast majority of the familiar animals of estuaries that are popular with snorkelers, anglers, and beachcombers.

Gause's competitive exclusion principle (Gause, 1936; Hardin, 1960) implies that two species competing for the same resource cannot both survive for long. The large variety of estuarine omnivores seem to share the same resource and defy this principle. They can share the resource, however, only because they feed on different size fractions, digest different compounds (leaving organically rich feces for other omnivores), and feed at different times and locations depending on their tolerance of environmental conditions (which change frequently in estuaries). Hence, many omnivores can coexist in estuaries without directly competing for the same resource for very long (Montague, et al., 1981).

12.5.2 Major Estuarine Habitats

12.5.2.1 Open Water Habitat. Open water is the medium that connects all estuarine habitats and is itself a habitat for plankton, neuston, and nekton (Fig. 12.30). Plankton are suspended in the water and are defined primarily by their inability to swim against adverse water currents. Neuston are similar in this respect, but occupy the surface film at the air–water interface. Nekton on the other hand are vagile swimmers—able to swim against the current to change location (e.g., most fishes, shrimp, and swimming crabs).

Plankton range in size from tiny viruses and bacteria to large jellyfish (megaplankton). They include plants (phytoplankton) and animals (zooplankton). Zooplankton include species that spend their entire life cycle as plankton (holoplankton); however, most other marine creatures (including most fishes and invertebrates) spend the larval stage of their life cycle as plankton (meroplankton).

The phytoplankton, which account for open-water primary production, consist of a mix of many species from several taxonomic divisions of plants. Species of diatoms, dinoflagellates, bluegreens, and green algae comprise the larger "net" phytoplank-

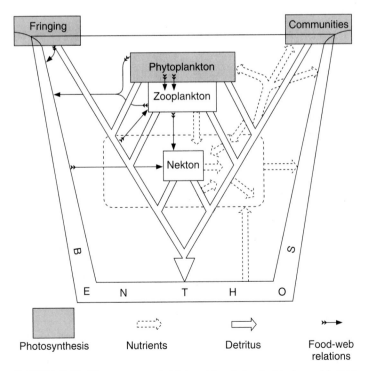

FIGURE 12.30 The open water habitat and its relation to fringing and bottom habitats in estuaries. (*Barnes, 1980.*)

ton (those retained on a typical 63-μm mesh phytoplankton net). The phytoplankton that pass through the net are sometimes referred to as nanoplankton. They are less well known, but consist of smaller species and individuals of the net phytoplankton plus a variety of others, including species of chrysophytes and coccolithophores. Although these smaller algae do not account for a lot of phytoplankton biomass, they account for perhaps half of the open-water gross primary production (Pomeroy, 1974). This follows the general rule that smaller organisms have higher rates of production and respiration per unit biomass.

The total amount of phytoplankton in an estuary is determined by the balance between cell division and cell death of the entire collection of species that coexist in estuarine water. While each species has particular responses to a variety of stimuli, light, nutrients, and temperature are major factors that affect the productivity of the entire community of species. Grazing by herbivorous zooplankton and filter-feeding bivalves is usually the most important determinant of cell death, though in deep estuaries, sinking or being swept by vertical currents down and out of the photic zone for an extended period may be a significant cause of phytoplankton death.

Since all phytoplankton contain chlorophyll-*a*, the amount of this green pigment in a volume of water is often used to estimate phytoplankton biomass. Chlorophyll-*a* is the light-absorbing photosynthetic pigment common to all plants. It differentially absorbs red light, thereby yielding its reflective green color, and activates the process of photosynthesis. Phytoplankton community productivity is expressed by the community assimilation number, i.e., the carbon converted from inorganic to organic form via photosynthesis per unit of chlorophyll-*a* per unit of time by the entire collection of phytoplankton.

Light Effects on Phytoplankton.   The response of community assimilation number to light is roughly linear at low light intensities, but reaches saturation at higher levels. An equation that has found wide use in phytoplankton ecology is:

$$A = A_{max} (1 - e^{-\alpha I})$$
(12.14)

where A is the assimilation number at light level I, A_{max} is the maximum assimilation number under ideal light conditions, and α is the logarithm of the fractional decrease in A below A_{max} per unit increase in I. The α of Eq. (12.14) equals the initial slope of the curve (at I just above 0) divided by A_{max}. Other authors have used the symbol α to represent the initial slope in a similar equation (e.g., Platt, et al., 1980), but the form given here is preferred for its simplicity.

Very high levels of light may actually depress the assimilation number, a phenomenon known as photoinhibition. This effect is not thoroughly understood, but has been demonstrated to result from exposure to ultraviolet radiation. Equation (12.14) can be multiplied by the following photoinhibition term:

$$Inh = e^{-\beta I}$$
(12.15)

where β is the fractional logarithmic reduction in assimilation number per unit increase in light I. When this term is added, assimilation number increases from zero at zero light to a maximum at an optimal light. This maximum is somewhat below A_{max} of Eq. (12.14), however, which is defined only in the absence of photoinhibition. It is not necessary that the wavelength of light be the same in Eq. (12.15) as in Eq. (12.14), as long as this distinction is specified, of course. Equation (12.14) may use photosynthetically active radiation, for example (I_{par}, that is, 400 to 700 nm wavelength), while Eq. (12.15) can refer to ultraviolet radiation (I_{uv}, 200 to 400 nm; Cullen, et al., 1992).

Because phytoplankton absorb light, their density determines the amount of light available at depth below the surface. The extinction of light with depth in estuaries generally follows the *Beer-Lambert law:*

$$I_z = I_o e^{-kz} \tag{12.16}$$

where I_z is the light at depth z; I_o is the light at the surface; and k is the extinction coefficient (the logarithmic fraction of light absorbed per unit depth). The extinction coefficient is determined by the absorption of light by water, dissolved substances, and suspended particles, especially the phytoplankton themselves. There is no simple relation between extinction coefficient and the more conventional measures of turbidity, though to a rough approximation k is proportional to turbidity in NTU (e.g., Holmes, 1970; Effler, 1988). The various wavelengths of light are differentially absorbed both in pure water and in water with other light-absorbing constituents. Both red and ultraviolet wavelengths are absorbed most. Photoinhibition attenuates with the absorption of ultraviolet light, but ironically, since chlorophyll-*a* differentially absorbs red light, photosynthesis also attenuates. Blue and green light penetrates to greatest depth. Light color can stimulate different activities in phytoplankton. Carbohydrate synthesis, for example, is more prevalent in white light, while protein synthesis is stimulated in green light (Wallen and Geen, 1971). Most phytoplankton have one or more accessory pigments (e.g., carotenoids) that absorb the more deeply penetrating colors of light. These accessory pigments allow greater photosynthesis at depth than would be possible without them.

Although the Beer-Lambert equation Eq. (12.16) is approximately correct for well-mixed layers of estuarine waters, the absorption of light may not always decline smoothly with depth across the halocline, or where masses of phytoplankton form into discrete horizontal bands. Some phytoplankton can position themselves vertically in the water column by swimming with flagella (e.g., dinoflagellates), or through adding buoyancy by producing lipid (e.g., diatoms). If the vertical mixing rate is low relative to these abilities, as may occur when wind and tidal energy are low, discrete bands of one species may form when they position themselves en masse at depths with favorable conditions. In shallow well-mixed estuaries, however, vertical mixing is usually sufficient to prevent banding.

Conversely, in deep vigorously mixed estuaries, phytoplankton production may suffer because the depth of mixing is below a critical depth for their survival. Phytoplankton are constantly mixed from the well-lit upper parts of the water column to less-well-lit portions beneath. Phytoplankton can be mixed below the point where light is just sufficient for photosynthesis to balance respiration (known as the compensation point). If the cells have stored enough photosynthate while they were in the upper zone they can survive a period of relative darkness below the compensation point. The critical depth is the depth at which this stored photosynthate is depleted. If mixing transports phytoplankton below this depth, they will not survive. Each species can be characterized by its own critical depth. Determining a critical depth for the entire community is problematic because a natural selection for species with different light requirements occurs when the mixing depth changes. The dominant species at the time of the determination may not be those that dominate after the mixing depth changes. Nevertheless, the concept may explain unexpectedly low rates of photosynthesis in deep estuaries with seemingly plenty of light and nutrients.

Nutrient Effects on Phytoplankton.   The response of phytoplankton assimilation number to individual nutrients (e.g., nitrogen, phosphorus, and silicon) is similar to that of light: a linear response under low levels, but reaching a saturation level at higher levels. A popular equation for representing this relationship is:

$$A = A_{\max} \frac{N}{k_m + N} \tag{12.17}$$

where N is the concentration of a particular nutrient in the water and k_m is the concentration of nutrient N that yields an assimilation number A equal to half the maximum assimilation number $A_{\max}$. This equation has been used to describe enzyme kinetics, and seems to fit laboratory data well. The analogy of nutrient uptake to the formation of an enzyme–substrate complex in biochemistry is also appealing, as is the ease of linearization of this equation with an inverse transformation, though other saturation equations, e.g., Eq. (12.14), may be suitably adapted, and have also been widely used.

Although saturation to high levels of nutrients is the norm in laboratory studies of single species, the picture is somewhat complicated in estuarine waters where many species of phytoplankton coexist. Addition of nutrients (e.g., from a sewage outfall) above the saturation limit for an initially dominant species may simply result in a shift in dominance to species that can grow faster at the higher concentrations. Ultimately as the new dominant species bloom, nutrient saturation will still occur if another factor (e.g., light) becomes so severely limiting to production throughout the estuary that continued additions of the nutrient have no further effect.

The nutrients that most often become limiting to production in estuaries are generally either nitrogen or phosphorus, two nutrients used in abundance by plants. In productive estuaries this limitation is not because the absolute concentrations of nitrogen and phosphorus are small, but rather because they are out of proportion to each other. Stoichiometric requirements for these major nutrients are often assumed to follow approximately those found in North Atlantic phytoplankton and water by Redfield (Redfield, 1958). *Redfield's ratio,* as it is known, implies a balanced nutrient medium of 16 atoms of nitrogen to every atom of phosphorus. Large departures from this ratio (greater than a factor of 2) indicate an excess of one of these nutrients and a limitation to overall production by the other. Although nitrogen is brought into estuaries by rivers and runoff, it can also be fixed from the atmosphere by nitrogen-fixing bacteria and returned to the atmosphere by denitrifying bacteria (see Fig. 12.26). The anaerobic zones required for these processes are readily available in estuarine muds. In estuaries with high N:P ratios, nitrogen fixation may exceed denitrification, and vice versa.

Phosphorus, on the other hand, has no gaseous atmospheric pool and arrives in most estuaries primarily in river water and runoff, often sorbed to suspended particles of clay (Fig. 12.27). These particles settle with fine organic particles in marshes and mud flats. Phosphate is released from the clay particles in anaerobic muds and can leak into overlying water, where it can be incorporated into phytoplankton. Estuaries that receive little clay from land-derived sediment discharges are prone to being phosphorus-poor. Florida Bay, at the southern tip of the Florida mainland, is an example. Here the fine sediments are mostly calcium carbonate marl rather than clay. Phosphorus import into this region is low, and unlike clay, marl tends to chemically bind and sequester phosphate, making it difficult to leach out of sediments into overlying water. In Florida Bay N:P ratios may exceed 300. In contrast, the estuaries of the coast of Georgia are phosphate-rich. Nitrogen is often in short supply and, when added, stimulates production. The red clay of the Georgia piedmont carries enormous quantities of phosphorus to the coast, where it supports considerable primary production. Moreover, the sediments already there hold enough phosphorus to support this high level of primary production for several hundred years (Pomeroy, et al., 1972).

Temperature Effects on Phytoplankton. Within a limited range of temperature, the *Van't Hoff relationship* has been used to represent temperature effects in organisms and biochemical reactions, and even communities of many species. This equation predicts an exponential rise in the rates of biochemical reactions with temperature, parameterized by Q_{10}, the factor by which the rate or collection of rates increase with every 10°C increase in temperature. This equation can be applied to the phytoplankton assimilation number as follows:

$$A = A' \, Q_{10} \frac{T - T'}{10} \tag{12.18}$$

where T' is the temperature, °C at which A equals A'. The value of Q_{10} is often close to 2, meaning that a doubling of assimilation number occurs for every 10°C increase in temperature.

Assimilation number drops when temperature is above and below the range over which Eq. (12.18) is valid. Under these circumstances, individuals may suffer physiological damage, or even death. However, at some temperatures, certain species may suffer while others still flourish. At the ecosystem level this means a replacement of species and continued phytoplankton production. Hence the function of phytoplankton production has a broader temperature range than does the structure of the phytoplankton community that performs this function. This is most certainly true of the ecosystem response to every other environmental factor as well (e.g., light and nutrients). The community of estuarine phytoplankton is diverse, which adds a measure of functional stability as community structure changes in response to environment.

Grazing by Zooplankton and Bivalves. An important control on phytoplankton biomass is grazing by herbivorous zooplankton, bivalves, and certain polychaete worms. Herbivorous copepods such as *Acartia tonsa* dominate the zooplankton community in estuaries. Filter-feeding bivalves include clams, oysters, and mussels. Filter-feeding polychaetes include the parchment tube worm, *Chaetopterus*. Grazing not only reduces phytoplankton biomass, but removes all types of suspended particles from the water, wraps them in a much denser fecal pellet, and sends them to the bottom. Side effects are that the estuarine water is clearer and the bottom sediments richer. The zooplankton and bivalves assimilate perhaps 40 percent or so of the energy contained in the phytoplankton and other particles. The rest begins to decay on the estuarine floor.

12.5.2.2 Intertidal Vegetation (Marshes and Mangroves).

At the edges of the estuary is the intertidal zone, a unique habitat that is alternately flooded and exposed as the tide rises and falls. The horizontal extent of the intertidal zone is determined by the range of the tides divided by the topographic slope through the zone. The horizontal extent of the intertidal zone in estuaries is nearly everywhere covered by rooted plants. Only pounding waves, erosive currents, or hard-rock substrates prevent vegetation from covering the intertidal zone. The type of intertidal vegetation depends in large measure on the number of days per year below freezing, the regularity and depth of inundation by flood waters, and soil salinity. The variety of marshes in Florida estuaries are representative, since that state spans such a great change in latitude. Schematic profiles of typical marshes of Florida are given in Fig. 12.31.

In the tropics, where freezes do not occur, all but the highest intertidal elevations are covered with tall mangrove trees. Light is dim beneath the trees. Mangroves shade out shorter types of vegetation. The coverage by mangroves declines

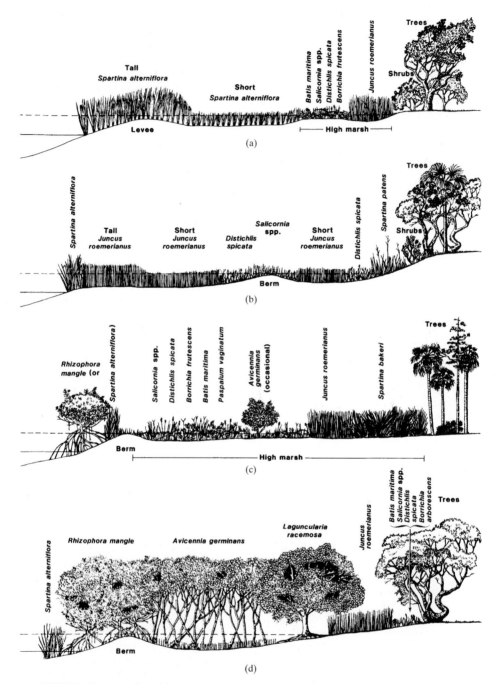

FIGURE 12.31   Profiles of intertidal vegetated habitats in Florida: (*a*) Northeastern; (*b*) Northwestern; (*c*) Indian River Lagoon (eastern); (*d*) Southwestern. (*Montague and Wiegert, 1990.*)

at higher latitudes because of freeze damage. In Florida, mangrove trees occur in intertidal zones exposed, on average, less than 5 days per year below freezing, with proportionately greater coverage in areas with fewer freezes (Montague, unpublished data).

Where mangroves do not grow, a variety of low-growing intertidal grasses, sedges, rushes, and other plants occupy intertidal zones (Montague and Wiegert, 1990). Below mean low water, where the marsh is regularly flooded and drained, the vegetation is usually dominated by a single species. In the eastern and southern United States, for example, these areas are covered by nearly monotypic stands of saltmarsh cordgrass (*Spartina alterniflora*). Irregularly flooded intertidal zones predominate in estuaries with little lunar tide. In such estuaries wind and annual changes in sea level are the primary determinants of flooding and drainage cycles. In the eastern and southern United States, irregularly flooded intertidal zones are often dominated by black needlerush (*Juncus roemerianus*). Brackish intertidal zones may be covered by saltmarsh bulrush (*Scirpus robustus*), though a variety of other species may dominate in various locations. Intertidal zones above mean high water contain a variety of so-called highmarsh plants, many of which are especially adapted to hot, often dry, and salty soils. Here salts often accumulate to high levels because of poor circulation of water due to infrequent flooding, and high evaporation rates when saltwater does inundate the area. In the extreme, salt may build to levels so high that no rooted plants can survive. This results in barren salt pans in the highest intertidal elevations. An interesting exception is the bluegreen algae community that thrives on mud flats in the Laguna Madre of Texas. These flats are frequently inundated by wind tides. The algal community (*Microcoleus, Anacystis, Schizothrix,* and other marine bluegreens as well as marine diatoms) forms a leathery mat as much as several centimeters thick, that survives dormant during the flats' prolonged exposure to air.

Some highmarsh plants can form extensive colonies and make distinctive bands dominated by a different species at different elevations. Such zonation is characteristic of all intertidal zones and is related to the amount of exposure to air, water, salt, aquatic, and terrestrial consumers, and at the edge, competition for space by adjacent dominant organisms. The distinctiveness of the edge of these bands may relate either to a sudden change in soil conditions, or to intense competition between the plant species.

The stems and leaves of intertidal vegetation slow tidal currents, dampen waves, and trap particles contained in flood waters. Much of the fine clay sediment entering estuaries from rivers and runoff settle in intertidal marshes. Marshes with sufficient sediment supply can grow vertically as sea level rises. With even more sediment, marshes can expand laterally until only a narrow, serpentine tidal creek is left to carry tidal waters in and out. As the constriction of the tide progresses with the expansion of marshes, current velocities increase, presumably until an equilibrium is reached between tidal hydraulic erosion and the tendency of marshes to expand. Subsequently, levees of muddy sediment may form in the marsh at the edges of tidal creeks. In estuaries with little input of fine sediments, intertidal vegetation cannot expand, but may simply migrate with the intertidal zone as sea level rises or falls.

Primary production of intertidal vegetation far exceeds that of phytoplankton, perhaps because the roots are surrounded by nutrient rich soils while the leaves are bathed in full sunlight. Nevertheless, production can be stimulated experimentally in intertidal marsh, which helps to illustrate the controls on primary production there. Vegetation near the edges of tidal creeks is much taller and may be two to three times as productive as that in the center portions of the marsh. The movement of the tides has been suggested as the reason for this (Schelske and Odum, 1961; Steever, et al., 1976; Turner, 1976). The circulation of water subsidizes production by recharging sediment nutrients and removing metabolic wastes. Many types of experimental manipulations performed in areas of short *Spartina alterniflora* have resulted in enhanced growth. These include adding freshwater, adding nitrogen, adding oxygen, increasing soil drainage, and adding fiddler crab burrows. These results generally support the concept of the tidal subsidy hypothesis. Tides create circulation of interstitial water, which helps to replenish depleted nutrient reserves, restore oxygen levels, and remove accumulated salts in the vicinity of plant roots. Fiddler crab burrows assist this process by providing avenues into sediment for tidal water to circulate. Moreover, fiddler crabs excrete nitrogen in burrow water, which injects a needed nutrient in close proximity to plant roots (Montague, 1982). On a regional basis, production of intertidal vegetation positively correlates with tidal range and negatively with latitude (Steever, et al., 1976; Turner, 1976).

Unlike for phytoplankton, direct grazing of intertidal vegetation by herbivorous animals is usually not a major cause of biomass reduction. That being said, some intertidal marshes have been intentionally grazed by cattle, and wintering geese are known to eat large quantities of saltmarsh vegetation in the middle Atlantic states, leaving vast holes in the marsh called *eat-outs*. On average, however, only about 10 percent of the net production of intertidal rooted plants is consumed alive, mostly by herbivorous insects. The rest dies and decays mostly on the marsh surface, thereby starting the detritus trophic chain in intertidal sediments.

The ecologically productive intertidal zone affects estuarine water as water floods and ebbs with the tides. The nutrient, organic, oxygen, and biotic content of the water changes. However, these changes are difficult to measure directly because of the disparity between the large scale required to detect the impact and the small scale of available sampling procedures. Reviews of intertidal exchange phenomena are given by Nixon (1980) and Montague, et al. (1987).

Tidal creeks are fundamentally important to the interaction of intertidal ecosystems with estuarine water (Montague and Wiegert, 1990). These are the access sites of roaming marine animals looking for food. While the inner portions of the marsh are good habitat from the standpoint of food and protection from predation, when the tide falls, aquatic organisms must head for the creeks or remain stranded. The edges of creeks are the sites of initial and greatest energy dissipation and deposition of suspended sediments. They are also feeding sites for a multitude of wading birds.

Intertidal marshes are often said to export large quantities of detritus via tidal creeks. However, some marshes import more detritus than they export. In some cases an equivalent amount of organic matter is imported as is exported, but the form has changed from particles to dissolved organics. This inconsistency of the relationship between intertidal marshes and estuarine water can be attributed in part to the intertidal sediments, which contain a community of decomposers, and to a combination of many other factors that can influence the exchange process in a given setting. Because the marsh is very productive and because most of that production dies and begins to decay in place, intertidal sediments contain a vast capacity to decompose detritus in the form of numerous specialized bacteria and fungi. Detritus produced elsewhere that enters the marsh on incoming tides can be decomposed there, along with that produced by intertidal plants. Therefore, depending in part on the relative amount of detritus produced elsewhere, marshes may be net importers or net exporters of detritus. Other processes are important, including the decomposability of the detritus itself. Rapidly decomposing detritus is more likely to disappear into the intertidal food web before it can be transported out of the marsh. Marshes that have considerable throughflow of ground water or surface water are more likely to export detritus and other materials. Likewise, marshes with high densities of creeks and those that experience high tidal ranges are more likely to export. Marshes in areas of more frequent and intense storms are also more likely to export detritus and other materials.

The value to estuarine ecological production of intertidally produced detritus, however, can be realized without any detrital export. Animals that arise from the intertidal detritus–algae trophic chain (e.g., fiddler crabs) are themselves food for many transient birds, fish, and even mammals that feed in the intertidal zone and associated tidal creeks. The presence of the detritus creates a rich, attainable food supply in marsh. The valuable export from intertidal zones may be in the form of larger animals rather than as suspended particles of detritus.

A stockpile of detritus can buffer the supply of food to the detritus trophic chain in times of low detrital inputs. Detritus builds because of the relatively long time required for it to completely decompose. The quantity of slowly decomposing organic matter that settles with clays in estuarine muds ensures the maintenance of anaerobic zones just below the mud surface. Anaerobic zones are important in phosphorus release, nitrogen fixation and denitrification, and sulfur cycling. Hence, estuarine nutrient cycles are assisted by the accumulation of detritus from all sources, not the least of which are intertidal zones. Moreover, the vast accumulation of biomass in intertidal vegetation holds nutrients for some time before they are released back to the estuary. This effectively buffers the nutrient levels in the estuary where intertidal vegetation accounts for a large fraction of the total estuarine area.

Intertidal vegetation, in estuaries where it is prevalent, undoubtedly reduces the damage to onshore structures by damping waves, slowing currents, attenuating the storm surge, and absorbing some of the wind energy. The full evaluation of this potential remains to be calculated, however. Likewise, by holding sediments in place, rooted vegetation conceivably reduces the rate of refilling of dredged channels and basins and hence lowers maintenance costs, though the amount of the cost savings has not been evaluated to our knowledge. Nevertheless, these are two additional public benefits of intertidal vegetation, along with the habitat they contribute to a variety of vertebrates, and the buffer they provide to food supply, nutrient cycles, and water quality.

12.5.2.3 Submersed Vegetation (Seagrasses, Certain Macroalgae). Only a few species of seagrasses and macroalgae exist around the world. Submerged beds of vegetation are often dominated by nearly monotypic stands of one species, with some zonation of species occurring close to shore (Fig. 12.32). The dominants change with light, temperature, and salinity. Among marine species, subtropical communities have the greatest diversity. Turtle grass (*Thalassia testudinum*), manatee grass (*Syringodium filiforme*), and shoal grass (*Halodule wrightii*) are the three most commonly dominant seagrasses in the Gulf of Mexico and Caribbean. In addition, a variety of green, red, and brown macroalgae may cover the bottom (Littler, et al., 1989). Along the Atlantic coast of the United States, the dominant seagrass is eelgrass (*Zostera marina*). Species of *Zostera* dominate at high latitudes around the world. In brackish waters, other plants dominate, especially widgeon grass (*Ruppia maritima*), musk "grass" (actually a green alga, *Chara hornemannii*), and ribbon-leaf grass (*Valisneria americanum*), among others, as waters become fresher.

The areal extent of dense beds of subtidal vegetation depends primarily on the slope of the bottom between two extremes: low-tide exposure to air at the shallow side, and the high-tide availability of light at the deep side. The critical depth for the development of beds of subtidal vegetation is where roughly 15 percent of surface light remains (Kenworthy and Haunert, 1991). Hence, subtidal vegetation beds

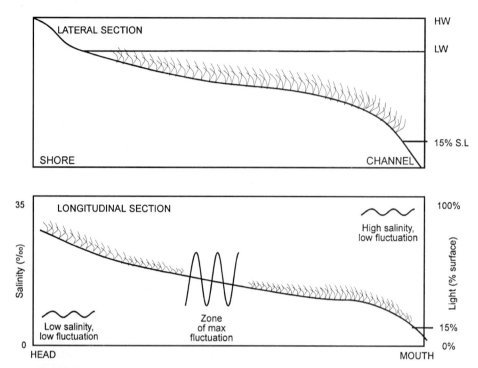

FIGURE 12.32   Profile of typical sub-tropical seagrass meadow: (*top*) Lateral section with elevations indicated [HW = high water, LW = low water, 15% S.L. = critical depth for seagrass bed development (15% of surface light remains)] (*bottom*) Longitudinal section with salinity, salinity fluctuation, and critical depth for seagrass bed development indicated.

occur to greater depths in less turbid water. Estuaries with dense concentrations of phytoplankton or suspended clays cannot also support seagrasses to a great depth.

Broad flat shallows in areas of low tidal range (e.g., Gulf coast of Florida) can have rich beds of seagrass because light penetrates to the bottom despite some turbidity in the water, but exposure to air is uncommon. Conversely, the coast of Georgia supports no seagrass beds, because tidal range is very great and the water too turbid to allow any area between low water and the critical depth for seagrasses at high water.

These simple criteria of exposure to air and limitation by light generally assume a layer of sediment is present of sufficient thickness for the establishment of a sufficient root system for seagrasses or holdfast system for macroalgae. Seagrasses, for example, do not grow on rock; however, a variety of macroalgae can form dense colonies on rock. Erosion from rapid currents in channels or instability of sediments caused by waves can prevent macrophyte beds. Erosion, instead of a lack of light, sometimes limits the expansion of such beds at the deep end. Likewise at the shallow side, exposure to wave energy may limit macrophyte beds. Ponded water at low tide may have high temperature by day or low oxygen at night that limits macrophyte expansion in shallow areas. A minimum current velocity is required to replenish depleted nutrients and remove waste products in microzones around leaves and roots. Areas of continuously near-stagnant conditions do not support productive communities of subtidal vegetation. Also, areas of rapidly changing salinity extremes apparently do not support productive subtidal vegetation, presumably because communities are continually in a state of species replacement (Montague and Ley, 1993). Nevertheless, the simple criteria of exposure to air and limitation by light are often sufficient for a rough assessment and can be used for a management parameter as long as the other possibilities are kept in mind to explain discrepancies.

Nutrient availability is a special problem for subtidal vegetation. Unless plants can rapidly grow completely to the surface and shade out competitors, the success of beds of subtidal vegetation often rests on keeping nutrients out of the overlying water. Nutrients in the water stimulate phytoplankton, which absorb light. Although bottom vegetation can also take nutrients from the water, the higher surface to volume ratio of the phytoplankton allows them to compete more successfully. However, bottom plants, especially seagrasses, can also take nutrients from sediments, where phytoplankton cannot unless sediments are resuspended. Interestingly, the presence of bottom vegetation, once established, helps to keep nutrients in the sediments where they can accumulate over time and repeatedly recycle to the bottom vegetation.

A sudden die-off of seagrasses can result in a release of nutrients to the water. This in turn stimulates phytoplankton, which reduces light. Hence a secondary die-off of seagrasses can begin. This secondary die-off may be much farther reaching than the initial die-off, as additional nutrients are released and sediment destabilizes, further exacerbating the death of subtidal vegetation. Eventually vast areas of bottom vegetation may die, as the overlying water turns green with phytoplankton. Such a situation occurred in the formerly very productive turtlegrass beds of western Florida Bay during the late 1980s and early 1990s. This scenario also occurred in the "wasting disease" that struck eelgrass beds on both sides of the North Atlantic in the 1930s. Although sudden die-offs can occur from acute changes in the environment, a chronic influx of nutrients (e.g., from sewage) can stimulate phytoplankton production and result in extensive loss of bottom vegetation over time. Such a shift in habitat can alter the success of local fisheries, and will usually be met with great public concern.

The functions and values attributed to intertidal vegetation in Sec. 12.5.4.2 can be repeated for subtidal vegetation in estuaries where rooted seagrasses or large macroalgae cover the bottom. Area for area, these plants are as productive as intertidal vegetation as long as the water is clear and other environmental conditions (salinity, temperature, and turbulence) remain within limits of tolerance. Submerged plants are even better habitat for smaller marine animals than intertidal vegetation because water always surrounds the leaves and stems. Many more animals can remain in the habitat at low tide. Submerged macrophytes do not reduce wind speeds, but they can effectively attenuate storm surges, wave energy, and currents. Good coverage of the bottom by vegetation also effectively holds sediments (undoubtedly reducing channel maintenance costs), but as for intertidal vegetation, the potential for this has not been evaluated.

12.5.2.4 *Mud and Sand.*

Wherever conditions are unsuitable for the growth of intertidal or subtidal vegetation, exposed mud or sand occurs. Although superficially these areas may appear barren, inside the sediment are numerous worms, crustacea and mollusks (Fig. 12.33). Mud accumulates in quiet waters. Particles of clay, marl, and organic detritus which make up mud are not massive enough to settle in areas of much water movement. Underneath the mud may be sand or rock. Sand accumulates (or remains exposed) in areas of higher current or waves. The grain size of deposited sand is a function of the kinetic energy of water movement. Greater

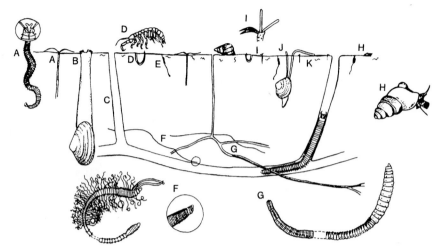

FIGURE 12.33   Some common organisms of a New England mud flat. Suspension feeder: B = *Mya arenaria* (bivalve). Deposit feeders: A = *Polydora ligni* (polychaete); C = *Nereis virens* (polychaete); D = *Corophium* spp. (amphipod); E = *Tharyx* sp. (polychaete); F = *Lumbrinereis tenuis* (polychaete); G = *Heteromastus filiformis* (polychaete); H = *Hydrobia totteni* (gastropod); I = *Streblospio benedicti* (polychaete); J = *Macoma balthica* (bivalve); K = oligochaete (omnivore); L = *Ilyanassa obsoleta* (gastropod). (*Whitlatch, 1982.*)

movement results in larger sizes since finer grains are winnowed out. Mud is richer in nutrients, detritus, and microbes than sand and is consumed by a variety of deposit feeders. The anaerobic zone is much shallower over mud than over sand, occurring within a centimeter or two of the sediment. This is because of: (1) greater oxygen demand from the greater accumulation of organic matter and bacteria, (2) lower diffusivity of oxygen owing to the smaller pore spaces between tiny particles of mud, and (3) lower advection in areas where mud can settle.

Muddy sediment forms much faster than indicated either by the slow rate of settling of the tiny mud particles, or by their input from rivers. Filter-feeding animals continually sweep any fine particles from the water in an incessant effort to consume the nutritive ones. The collection of filter feeders is annually capable of removing many times the total amount of suspended matter, and in fact causes sedimentation that far exceeds the total fluvial input of sediments (Biggs and Howell, 1984; Dame, et al., 1980)! Much of what is removed, however, is not assimilated by the filter-feeding animals. Instead it is packaged and egested as denser feces and pseudofeces (rejecta), which fall to the bottom, contributing to the nutrient content of estuarine sediments.

The detritus food chain is in full operation in estuarine sediments. Once the oxygen is used up in the decomposition of organic matter, other oxidizing agents are substituted, and decomposition continues, as illustrated in Fig. 12.34 (Fenchel and Riedl, 1970). Nitrate reduction allows further use of organic matter, and the growth of certain bacteria. When neither oxygen nor nitrate is available, iron and sulfate reduction allows further organic use and the growth of special iron and sulfate reducing bacteria (with the resulting accumulation of black anaerobic muds and the characteristic low-tide smell of sulfide). Once these oxidizing agents are no longer available, deep in the anaerobic sediments, even carbonate ions can be reduced to

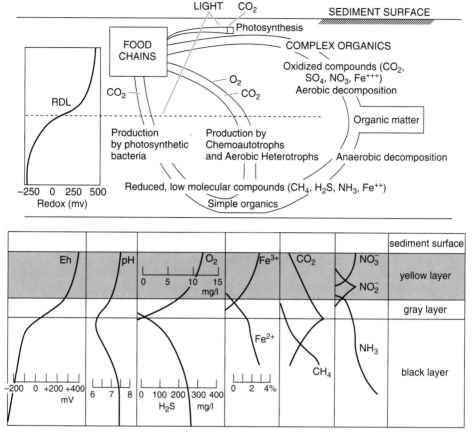

FIGURE 12.34   Anaerobic processes in estuarine sediments and a profile of sediment chemistry. (*Fenchel and Riedl, 1970, as cited and redrawn in Day, et al., 1989.*)

consume organic matter, which results in methane formation (swamp gas) and the growth of methanogenic bacteria.

Fermentation also occurs in the anaerobic zone. It results in the conversion of large, complex organic molecules to simpler forms, such as acetate and alcohol. Without these simpler organics, the processes of nitrate, sulfate, and carbonate reduction would be much slower, because the microbes involved can apparently use only these simple organic compounds. Because of the abundance of sulfate in well-oxygenated seawater, sulfate reduction in estuarine sediments may be limited more by the availability of simple organics than by a lack of sulfate.

The sulfide, ammonia, and methane produced by various bacteria in estuarine sediments can rise through the mud and appear in the better oxygenated surface layer of sediments. These reduced gases release energy when oxidized. Not surprisingly, other special bacteria exist that can use the energy to fix inorganic carbon from the air (like plants do with solar energy) in a process called chemosynthesis. Chemosynthesis can be measured as the rate of uptake of inorganic carbon in the dark (dark fixation). It may amount to 10 percent or so of the total uptake of inorganic carbon in estuaries.

The operation of these various processes by which the energy in organic matter is converted to biomass in sediments results over time in the nearly complete use of even difficult-to-decompose detritus. The resulting bacteria—which have converted perhaps 20 percent of the energy available in the detritus into their own cells (80 percent is used in their metabolism)—are available for consumption by protozoans, meiofauna, and deposit feeders, thus initiating the first stages of the detritus trophic chain. Larger animals on mud flats include numerous species of abundant polychaete worms and a variety of crustaceans and mollusks. The most abundant, such as polychaetes of the genus *Nereis,* are deposit feeders. A small number of predatory worms also occur in the mud. Several different predatory species use various appendages as weapons and a variety of hunting techniques to capture the deposit feeders and each other. All of these worms, mollusks, and crustacea of mud flats are food for fishes that pass in search of food.

Mud and sand that receive sufficient light contain microalgae, especially diatoms, and blue-green algae. Where diatoms are especially abundant, the mud or sand may reflect a golden-brown sheen. Purple photosynthetic sulfur bacteria, which occupy well-lit anaerobic zones, may form brilliant purple bands just below the sediment surface. The collection of photosynthetic microbes is generally about one-tenth as productive as rooted vegetation, but as for phytoplankton, most of this production is consumed live by deposit feeders. Hence it is a valuable form of energy for estuarine animals and, along with the microbes associated with decomposing detritus, comprises the food for numerous deposit-feeding worms, shrimp, crabs, and fishes.

12.5.2.5 *Rock and Other Hard Surfaces.* Hard surface has been said to be the most limiting resource in the sea. Any hard submerged surface is quickly invaded by animals and plants (Fig. 12.35). Some estuaries contain natural rocks and accumulations of shell, but nearly all contain quantities of pilings, boat hulls, bulkhead walls, jetties, revetments, groins, platforms, and multitudes of bottles and cans. Hard surfaces are good habitat for filter-feeding or water-straining animals such as oysters, tunicates, sponges, and barnacles, which wait for food to arrive in moving water. The planktonic larvae of these animals usually occur in excess of available settlement sites.

The structure provided both by hard objects and the animals attached to them attract other animals. Some come to hide, others to feed. All of these animals excrete nutrients. Where light is sufficient, these nutrients stimulate the production of algae of various sizes. The algae in turn provide food for herbivorous animals and additional cover for other animals. Soon a fully developed hard surface ecosystem appears.

Intertidal hard surfaces, like other intertidal zones, exhibit distinctive zonation that has been an ecological curiosity for many years. The zones arise primarily from the differential tolerance of various species to exposure to air at low tide. Few organisms can tolerate long periods of drought. The topmost animal band usually consists of one species of balanoid barnacle of the genus *Chthamalus.* Below are other species, with diversity increasing with less exposure. In zones above the topmost animal band are often colorful bands of lichens, fungi, and algae (Stephenson and Stephenson, 1972). Sharp boundaries between zones may be caused by intense competition for space at the interface (Connell, 1961), or by sharp changes in inundation frequency with elevation, which result from complex tidal patterns (Doty, 1946).

Many workers have been tempted to use specific zones as a quick way to identify elevations around an estuary for legal purposes. Unfortunately, the elevation of bands of the same species are affected by several physical and biological factors that

INTERTIDAL ZONATION

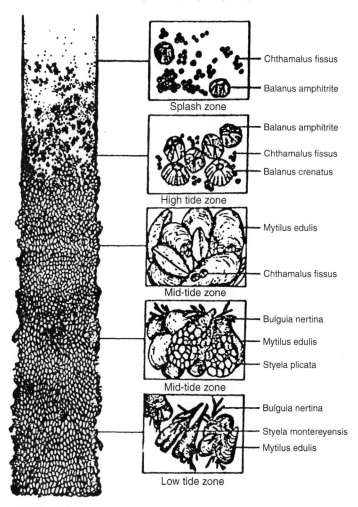

FIGURE 12.35 A typical hard surface estuarine ecosystem. (*Reish, 1968; as cited in Odum, et al., 1974.*)

change from place to place, as illustrated in Fig. 12.36 (Stephenson and Stephenson, 1972). Chief among these is wave height and sea spray. Parts of the estuary with greater windward fetch experience higher waves and more sea spray. Therefore, moisture reaches higher elevations in these cases. This causes an expansion of each biological zone to higher elevations unrelated to mean water level.

Well-developed hard-surface communities provide habitat for the early juvenile stages of many fishes and small shrimps. These in turn are food for larger fish that are attracted to these areas. This attractive value of hard surfaces is well utilized for fishering, and in some estuaries, artificial reefs are in demand.

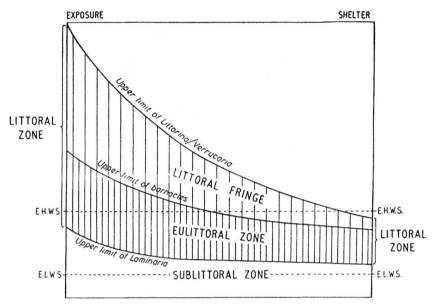

FIGURE 12.36   Effect of exposure to waves on biotic bands on intertidal hard surfaces. (*Lewis, 1964.*)

12.5.2.6   Consumer Reefs.   A characteristic of estuaries is a patchwork of *consumer reefs*. These include oyster reefs, mussel reefs, clam banks, and worm reefs. All contain extremely dense accumulations of filter feeders (oysters, clams, mussels, and sabellariid polychaete worms). They eat phytoplankton and other suspended organic particles and can considerably clarify the overpassing water. The dominant species provides food and cover for a variety of less abundant secondary species. It also produces huge numbers of planktonic larvae. The larvae settle on hard surfaces, and some have a known preference for settlement on or near the shells of their own species. This propensity is responsible for the development of such huge masses, given sufficient food supply, and the availability of mature larvae that are ready to settle. The latter depends on the survival of larvae and the transport by currents to appropriate settlement sites. Larval survival is highly variable and often unsuccessful, which, in an evolutionary sense, accounts for the huge numbers of larvae that are produced.

Often only one size of individuals is prevalent amongst the dominant consumer at a given location. In other locations all sizes coexist. Circumstances adequate for survival may not be suitable for settlement. Conditions suitable for settlement may be ephemeral. When both have occurred, a bed of clams, oysters, or worms will consist of a single age cohort. Sometimes only young and adult stages are present, perhaps a result of approaching the density limit imposed by the food supply in the water. Oyster reefs can serve to illustrate this hypothesis (Fig. 12.37). Unharvested oyster reefs often contain large adult oysters as well as recently settled oyster spat. Oyster reefs can be started quickly on old oyster shell because larval oysters preferentially settle there. If food is plentiful, these young oysters will grow and the reef can expand. If not, they cannot survive competition with the larger oysters, so medium sizes may be absent. If food has been plentiful for a long time, all sizes should be represented and the reef has the potential to expand in size until demand

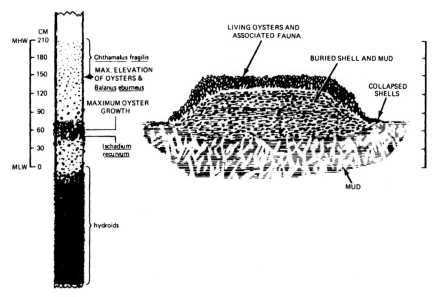

FIGURE 12.37 Oyster reef, an example of a consumer reef. Note maximum growth of oysters occurs somewhat below maximum elevation of reef. (*Bahr and Lanier, 1981.*)

for food equals supply (and medium sizes disappear). Because these filter feeders are so efficient, however, oysters along the edges of a large, shallow reef obtain more food than those in the center. In many cases, the centers of oyster reefs contain far more dead shells, while the expanding edges contain multiple sizes of oysters.

Chronic disease, predation, harvest by humans, and extreme fluctuations of the environment can prevent or substantially slow the expansion of a consumer reef to the limit dictated by the food supply. Acute episodes can completely reset reef development. As long as food is plentiful, planktonic larvae are consistently produced somewhere within the range of transport by water currents, and they can eventually survive and find an appropriate place to settle, so that the return of fully developed consumer reefs will simply be a matter of time.

12.5.3 Estuarine Management Strategies

There are two broad strategies for addressing ecological performance of estuaries in management: *functional habitats* and *target species.* Other quantitative measures, for example, nutrient turnover, productivity, biomass, and community indices such as diversity, which have found some management utility in streams and lakes, have not been as successful in estuary ecology management, perhaps because of the greater mutability and resilience of estuaries. Generally, of these two strategies, the functional habitat approach is more useful for planning and forecasting the effects of proposed modifications, while the target species approach is more diagnostic, seeking to account for observed responses in the organisms.

As its name implies, the habitat approach focuses on the extent and health of specific habitats within the estuary, the philosophy being that these habitats contribute in a variety of ways to the maintenance and productivity of various estuarine populations. The principal types of estuarine habitats were recounted in the preceding section. In a specific estuary, subtypes of these principal habitats may have to be characterized, e.g., a particular seagrass meadow, a tidal *S. alterniflora* peripheral marshstand, species-demarcated mangrove flats, or a brackish *Rangia*-dominated secondary bay. In any case, the physical and chemical parameters necessary for the maintenance of the habitat must be quantified, for example:

- Tidal range
- Ranges of salinity, temperature, and DO
- Substrate characteristics
- Turbidity and light penetration
- Meteorological water-level variations
- Freshwater drainage
- Nutrient supply
- Sediment supply
- Wave exposure
- Currents and throughflow

Changes in any of these are then used to project the response of the habitat.

The target species approach, in contrast, identifies a specific organism whose response to natural or anthropogenic factors is to be determined. This may be either a *key* species, which has intrinsic public importance, usually as a commercial or recreational fishery, or an *indicator* species, which is selected because it typifies elements of the ecosystem or responds sensitively to some factor in the estuarine environment.

Defined in this way, the target species approach subsumes fishery management. The topic of fisheries and recruitment is one of the most intensely studied of all estuarine issues. For this chapter, we only touch upon basic concepts that may be useful to estuarine managers and water-quality engineers unfamiliar with fishery ecology and management. For a more complete introduction, the interested reader is referred to the chapters on nekton and fisheries (and references therein) in Day, et al. (1989). Also, the reader is referred to the current Estuarine Living Marine Resources (ELMR) Program of NOAA, which is an ambitious compilation of data on life histories and occurrence of 135 fish and invertebrate species in 118 estuaries in the contiguous United States (e.g., Monaco, et al., 1990; Emmett, et al., 1991; Nelson, 1992; see also Bulger, et al., 1993).

The vast majority of commercially and recreationally important species of fish and shellfish use estuaries at some point in their life history. Two-thirds of the 3 million metric tons landed in the United States can be considered in this category. This phenomenon relates to characteristics of both the fishery organism and the fisher. Firstly, it is less expensive both in fuel and labor to fish close to shore, so coastal fishes are more prevalent in the world catch. Secondly, those species in greatest demand are those that can be caught in abundance at a relatively large size. Because large estuarine fishes and invertebrates begin life as planktonic larvae and grow to larger adults, those surviving to adulthood have experienced suitable conditions for survival over a considerable amount of space and time (Fig. 12.38). Larvae are often spawned offshore and drift back toward shore over a period of weeks. Once in the longshore currents, they may be swept through inlets into estuaries on incoming

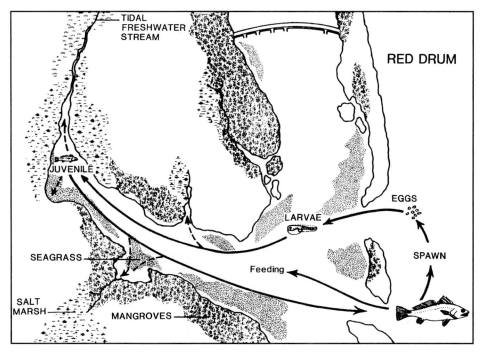

FIGURE 12.38   Life cycle of red drum. Note the use of a wide range of estuarine habitats. (*Lewis, et al., 1985.*)

tides. Here they remain through postlarval, juvenile, subadult and adult stages. Adults return to offshore spawning grounds during their spawning season.

A central class of information needed by the manager is the preference and tolerance limits of each organism to its physical and chemical environment. This may vary with life stage, and becomes especially complex for diadromous organisms, since their life cycle ranges over a variety of conditions. The set of estuarine habitats, which together span a variety of scales and locations, are used sequentially by such animals as they grow. Hence, the larger estuarine animals are good integrators of estuarine conditions in both space and time, so their sustained abundance indicates an ecologically healthy estuary. Nevertheless, even an estuary that contains an abundance of fishes may have problems with the fishery. The fishery will be inadequate if landings cannot meet demand, if landings exceed demand, if specific species can no longer be caught, if catch per unit effort has declined, or if the safety of fishes for human consumption is questionable.

Moreover, survival in estuaries is not guaranteed. The environment can suddenly change from favorable to unfavorable, and numerous predators are abundant. Many more individuals die in estuarine habitats than grow to adulthood. Fishery management often consists of providing habitat and limiting catch after problems with the fishery are noted. Ability to predict and plan preventative management protocols is hampered by the variability of the environment, the lack of adequate data, and the traditional independence and understandable skepticism of the fisher (which can translate into difficulty convincing fishers to reduce catch during good years on the chance that catch will be more consistently good in later years). As stocks become depleted, the entire fishing industry becomes a boom-and-bust activity. Oscillations in fishery organisms are followed by oscillations in fishing success, boat building, loans, loan defaults, and perhaps even welfare payments, a pattern that is predictable in simple simulation models, but difficult to manage.

Survival of a given year-class of fish and invertebrates is highly variable. Successful animals can offset this by repeatedly producing large quantities of offspring, each with little parental investment, in what is referred to by ecologists as a bet-hedging reproductive strategy. Adult females of larger species each produce billions of eggs every year. This attests to the tremendous likelihood of death before adulthood, since over the long run, a population needs only to replace itself with one adult for each adult that dies. In fact, in most years very few or no subadults exist to mature into adults. Rarely is this dismal picture offset by conditions for good survival of a given year-class throughout all stages of the life cycle. This type of pattern results in two common phenomena experienced by fishers. First, the uniformity of size in a single season's catch, and second, the memory of a year when the catch was truly exceptional. For the survival of a fish population, one such year need occur only rarely to offset the hardships of many previous years. This also means that the lack of recent good years of fishing for one or a few species may not be a bellwether of a deteriorating environment caused by coastal development, though this cannot be ruled out either. However, if the entire set of estuarine dependent fish and invertebrates is consistently failing through a variety of years, a deteriorating coastal environment may be the cause. By monitoring a large enough subset of commercial, game, and nongame fish and wildlife, the status of estuarine health can be better assessed.

An indicator organism may or may not enjoy a public perception of importance, but is employed in estuarine management as an index or a biosentinel. Organisms which concentrate toxins in their tissues are serviceable as indicators of the extent of pollution by that toxin. Filter or deposit feeders and top carnivores are particularly valuable in this respect (and may, therefore, also be important as recreational or commercial fisheries). Toxins appear in estuaries both naturally (as in toxic dinoflagellate blooms) and through human activity. Small quantities of toxins taken up by animals in foods, or absorbed through the gills or skin, will concentrate in tissues if they cannot be broken down and eliminated as fast as they are absorbed. If this happens at each step in a trophic chain, the tissues of long-lived carnivorous animals can become lethal to their predators, including unsuspecting people eating a seafood dinner. Of course long before this occurs, many animals and plants may have suffered sublethal effects, which lower production and abundance.

Ideally, *indicator organisms* are those which can serve as early warnings (a miner's canary) of deteriorating conditions caused by toxins or other pollutants in the environment. This might be manifested by the accumulation of toxins in tissues, by changes in internal anatomy or physiology (so-called biomarkers), or by changes in the population of the organisms in the estuary. This latter includes both the disappearance of plants or animals and the appearance of particular organisms associated with specific types of pollution.

Appropriate indicators are those that offer some practical monitoring advantages as well as represent some theoretical rationale for their ability to indicate ecological damage in a deteriorating estuarine environment. Some general criteria may be specified: (1) ease of monitoring and identification, preferably with nondestructive methods (so that the necessary data density can be obtained without the collection procedure altering the results), (2) not so sensitive that natural variation in its abundance or production swamps a steady change with deteriorating conditions, and

(3) not so insensitive that changes go unnoticed. Relatively long-lived sedentary plants and animals, such as rooted vegetation, filter feeders, and deposit feeders may tolerate widely ranging conditions and hence may be too insensitive. Fish that integrate all estuarine conditions as mentioned above may be too sensitive. Landings of commercially harvested organisms are relatively easy to monitor, but since fishing grounds may contain animals from various estuaries, it is difficult to attribute even long-term changes in landings to conditions in a particular estuary.

The goal is to have a set of indicators that provide a high information content per unit of monitoring effort. Finding good indicators of ecological conditions is difficult in any environment, but in one as naturally variable as an estuary, the problem of separating natural blooms and die-offs from those induced by deteriorating estuarine conditions is daunting. If appropriate indicators can be found, data on their occurrence and relevant physiology must be collected consistently, intensively in space and time, and for a long duration in order to establish their status and trends.

Acknowledgments

The authors acknowledge the University of Texas (GHW) and the University of Florida (CLM) for providing facilities and equipment needed for completion of the manuscript. This publication was partially supported by Grant # NA89AA-D-SG139 from the National Oceanic and Atmospheric Administration through the National Sea Grant College Program, and by Grant # BSC-8914585 from the National Science Foundation, to one of the authors (GHW). The views expressed herein do not necessarily reflect those of NOAA or any of its subagencies, nor of NSF.

REFERENCES

American Public Health Association, *Standard Methods for the Examination of Water and Wastewater, 17th Ed.,* APHA, Washington, D.C., 1989.

d'Anglejan, B., "On the Advection of Turbidity in the Saint Lawrence Middle Estuary," *Estuaries,* 4(1):2–15, 1980.

Armstrong, N. E., "Responses of Texas Estuaries to Freshwater Inflows," in V. Kennedy, ed., *Estuarine comparisons,* Academic Press, New York, 1982, pp. 103–120.

Arthur, J., and M. Ball, "Factors Influencing the Entrapment of Suspended Material in the San Francisco Bay-Delta Estuary," in T. Conomos, ed., *San Francisco Bay: The Urbanized Estuary,* Amer. Assn. Adv. Sci., Pacific Div., San Francisco, 1979, pp. 143–174.

Aston, S. R., "Nutrients, Dissolved Gases, and General Biogeochemistry in Estuaries," in E. Olausson and I. Cato, eds., *Chemistry and Biogeochemistry of Estuaries* John Wiley, Chichester, U.K., 1980, pp. 233–262.

Bahr, L., and W. Lanier, "The Ecology of Intertidal Oyster Reefs of the South Atlantic Coast: A Community Profile," Report FWS/OBS-81/15, Office of Biological Services, U.S. Fish and Wildlife Service, Washington, D.C., 1981.

Barnes, R. S. K., "The Unity and Diversity of Aquatic Systems," in R. Barnes and K. Mann, eds., *Fundamentals of Aquatic Ecosystems,* Blackwell Scientific Publications, Boston, 1980, pp. 5–23.

Biggs, R., and B. Howell, "The Estuary as a Sediment Trap: Alternate Approaches to Estimating its Filtering Efficiency," in V. Kennedy, ed., *The Estuary as a Filter,* Academic Press, New York, 1984, pp. 107–129.

Bildstein, K., E. Blood, and P. Frederick, "The Relative Importance of Biotic and Abiotic Vectors in Nutrient Transport," *Estuaries,* 15(2):147–157, 1992.

Billen, G., and J. Smitz, "Mathematical Model of Water Quality in a Highly Polluted Estuary," in J. Nihoul, ed., *Hydrodynamics of Estuaries and Fjords,* Elsevier, Amsterdam, 1978, pp. 55–62.

Boicourt, W., "A Numerical Model of the Salinity Distribution in Upper Chesapeake Bay," Tech Rep. 54, Chesapeake Bay Institute, Johns Hopkins University, Baltimore, Md., 1969.

Boothroyd, J. C., "Mesotidal Inlets and Estuaries," in R. Davis, ed., *Coastal Sedimentary Environments* Springer-Verlag, New York, 1978, pp. 287–360.

Bowden, K., and R. Gilligan, "Characteristic Features of Estuarine Circulation as Represented in the Mersey Estuary," *Limnology and Oceanography,* 16(3):490–502, 1971.

Boyden, C., S. Aston, and I. Thornton, "Tidal and Seasonal Variations of Trace Elements in Two Cornish Estuaries," *Est. Coast. Mar. Sci.,* 9(3):303–317, 1979.

Boynton, W., W. Kemp, and C. Keefe, "A Comparative Analysis of Nutrients and Other Factors Influencing Estuarine Phytoplankton Production," in V. Kennedy, ed., *Estuarine Comparisons,* Academic Press, New York, 1982, pp. 69–90.

Bradley, P., B. Kjerfve, and J. Morris, "Rediversion Salinity Change in the Cooper River, South Carolina: Ecological Implications," *Estuaries,* 13(4):373–379, 1990.

Broecker, et al., "The Influence of Wind on CO_2-Exchange in a Wind-Wave Tunnel," *J. Marine Res.,* 36:595–610, 1978.

Buechel, K., ed., *Chemistry of Pesticides,* John Wiley, New York, 1983.

Bulger, A., B. Hayden, M. Monaco, D. Nelson, and M. McCormick-Ray, "Biologically-Based Estuarine Salinity Zones Derived from a Multivariate Analysis," *Estuaries,* 16(2):311–322, 1993.

Burt, T. M., "Field Settling Velocities of Estuary Muds," in A. Mehta, ed., *Estuarine Cohesive Sediment Dynamics,* Springer-Verlag, New York, 1986, pp. 126–150.

Burt, W., and L. Marriage, "Computation of Pollution in the Yaquina River Estuary," *Sewage and Industrial Wastes,* 29(12):1385–1389, 1957.

Cameron, W., and D. Pritchard, "Estuaries," *The Sea,* vol. 2, John Wiley, New York, 1965, pp. 306–324.

Carlberg, S. R., "Oil Pollution of the Marine Environment," in E. Aoausson and I. Cato, eds., *Chemistry and Biogeochemistry of Estuaries,* John Wiley, New York, 1980, pp. 367–402.

Carothers, H., and H. Innis, "Design of Inlets for Texas Coastal Fisheries," *J. Waterways and Harbors,* ASCE, 86 (WW3), 1960.

Cheng, R. T., ed., *Residual Currents and Long-Term Transport,* Springer-Verlag, New York, 1990.

Chow, V., D. Maidment, and L. Mays, *Applied Hydrology,* McGraw-Hill, New York, 1988.

Connell, J. H., "The Influence of Interspecific Competition and Other Factors on the Distribution of the Barnacle *Chthamalus stellatus,"* *Ecology,* 42:710–723, 1961.

Copeland, B. J., "Effects of Decreased River Flow on Estuarine Ecology," *J. Water Pol. Control Fed.,* 38(11):1831–1839, 1966.

Crompton, T., *Determination of Organic Substances in Water,* John Wiley, New York, 1985.

Cronin, L. E., ed., *Estuarine Research, 1, Chemistry, Biology and the Estuarine System; 2, Geology and Engineering,* Academic Press, New York, 1975.

Cronin, T., and R. Forward, "Tidally Timed Behavior: Effects on Larval Distributions in Estuaries," in V. Kennedy, ed., *Estuarine Comparisons,* Academic Press, New York, 1982, pp. 505–520.

Cullen, J., P. Neale, and M. Lesser, "Biological Weighting Function for the Inhibition of Phytoplankton Photosynthesis by Ultraviolet Radiation," *Science,* 258:646–650, 1992.

Dame, R., N. Dankers, T. Prins, H. Jongsma, and A. Smaal, "The Influence of Mussel Beds on Nutrients in the Western Wadden Sea and Eastern Scheldt Estuaries," *Estuaries,* 14(2):130–138, 1991.

Dame, R., R. Zingmark, H. Stevenson, and D. Nelson, "Filter Feeder Coupling Between the Estuarine Water Column and Benthic Subsystems," in V. Kennedy, ed., *Estuarine Perspectives,* Academic Press, New York, 1980, pp. 521–526.

Darnell, R. M., "Organic detritus in relation to the estuarine ecosystem," in G. Lauff, ed., *Estuaries,* American Association for the Advancement of Science, Pub. No. 83, Washington, D.C., 1967, pp. 376–382.

Darnell, R. M., "The Pass as a Physically-Dominated, Open Ecological System," in R. Livingston, ed., *Ecological Processes in Coastal and Marine Systems,* Plenum Press, New York, 1979, pp. 383–393.

Day, J., C. Hall, W. Kemp, and A. Yáñez-Arancibia, *Estuarine Ecology,* John Wiley, New York, 1989.

Ditmars, J., E. Adams, K. Bedford, and D. Ford, "Performance Evaluation of Surface Water Transport and Dispersion Models," *J. Hydr. Engr.,* ASCE, 113(8):961–980, 1987.

Di Toro, D. M., "Optics of Turbid Estuarine Waters: Approximations and Applications," *Water Res.,* 12:1059–1068, 1978.

Di Toro, D., R. Thomann, D. O'Connor, and J. Mancini, "Estuarine Phytoplankton Biomass Models—Verification Analyses and Preliminary Applications," in E. Goldberg, I. McCave, J. O'Brien, and J. Steele, eds., *The Sea,* vol. 6, John Wiley, New York, 1977, pp. 969–1020.

Doering, P., C. Oviatt, and M. Pilson, "Control of Nutrient Concentrations in the Seekink-Providence River Region of Narragansett Bay, Rhode Island," *Estuaries,* 13(4):418–430, 1990.

Domagalski, J., and K. Kuivila, "Distribution of Pesticides and Organic Contaminants Between Water and Suspended Sediment in San Francisco Bay, California," *Estuaries,* 16(3A):512–520, 1993.

Dooley, H. D., "Factors Influencing Water Movements in the Firth of Clyde," *Est. Coast. Mar. Sci.,* 9(5):631–641, 1979.

Doty, M. S., "Critical Tide Factors That Are Correlated With the Vertical Distribution of Marine Algae and Other Organisms Along the Pacific Coast," *Ecology,* 27:315–328, 1946.

Downer, C., E. Holley, and G. Ward, "Tracer Gas Method for Shallow Bays," Report CRWR 233, Center for Research in Water Resources, The University of Texas at Austin, 1991.

Dronkers, J. J., *Tidal Computations in Rivers and Coastal Waters,* North-Holland, Amsterdam, 1964.

Dronkers, J. J., "Tidal Asymmetry and Estuarine Morphology," *Neth. J. Sea Res.,* 20(2/3): 117–131, 1986.

Dronkers, J., and J. Zimmerman, "Some Principles of Mixing in Tidal Lagoons," *Oceanologica Acta,* 4(SP):107–117, 1982.

Duinker, J. C., "Suspended Matter in Estuaries: Adsorption and Desorption Processes," in E. Olausson, and I. Cato, eds., *Chemistry and biogeochemistry of estuaries,* John Wiley, New York, 1980, pp. 121–151.

Dyer, K. R., *Estuaries: A Physical Introduction,* John Wiley, London, 1973.

Dyer, K. R., "The Balance of Suspended Sediment in the Gironde and Thames Estuaries," in B. Kjerfve, ed., *Estuarine Transport Processes,* University of South Carolina Press, Columbia, S.C., 1978, pp. 135–145.

Dyer, K. R., ed., *Estuarine Hydrography and Sedimentation, a Handbook,* Cambridge University Press, Cambridge, 1979.

Dyer, K. R., *Coastal and Estuarine Sediment Dynamics,* John Wiley, Chichester, U.K., 1986.

Eagleson, P. S., *Dynamic Hydrology,* McGraw-Hill, New York, 1970.

Eaton, A., "Removal of 'Soluble' Iron in the Potomac River Estuary," *Est. Coast. Mar. Sci.,* 9(1):41–49, 1979.

Effler, S. W., "Secchi Disc Transparency and Turbidity," *J. Env. Engr., ASCE,* 114(6):1436–1447, 1988.

Eggink, H. J., "Predicted Effects of Future Discharges of Industrial Wastes into the Eems Estuary," *Adv. Water Pollution Research,* 3:1–27, 1967.

Eisma, D., "Flocculation and De-Flocculation of Suspended Matter in Estuaries," *Netherlands J. Sea Research,* 20(2/3):183–199, 1986.

Eisma, D., P. Bernard, G. Cadée, V. Ittekkot, J. Kalf, R. Laane, J. Martin, W. Mook, A. van Put, and T. Schuhmacher, "Suspended-Matter Particle Size in Some West-European Estuaries; Part II: A Review on Floc Formation and Break-Up," *Neth. J. Sea Res.,* 26(3):215–220, 1991.

Eloubaidy, A., and E. Plate, "Wind Shear Turbulence and Reaeration Coefficient," *J. Hydr. Engr. Div., ASCE,* 98(HY1):153–170, 1972.

El-Sabh, M. I., "The Lower St. Lawrence Estuary as a Physical Oceanographic System," *Naturaliste Canadien,* 106(1):55–73, 1979.

Emmett, R., S. Stone, S. Hinton, and M. Monaco, "Distribution and Abundance of Fishes and Invertebrates in West Coast Estuaries, 2: Species Life History Summaries," ELMR Rep. no. 8, Strategic Environmental Assessments Div., National Ocean Service, NOAA, Silver Spring, Md., 1991.

Fair, G. M., and J. C. Geyer, *Water Supply and Waste-Water Disposal,* John Wiley, New York, 1954.

Fairbridge, R. W., "The Estuary: its Definition and Geodynamic Cycle," in E. Olausson and I. Cato, eds., *Chemistry and Biogeochemistry of Estuaries* John Wiley, New York, 1980, pp. 1–35.

Fenchel, T., and R. Riedl, "The Sulfide System: A New Biotic Community Underneath the Oxidized Layer of Marine Sand Bottoms," *Marine Biology,* 7:255–268, 1970.

Fischer, H. B., "Dispersion Predictions in Natural Streams," *J. Sanitary Engr. Div., ASCE,* 94(SA5):927–943, 1968.

Fischer, H., E. List, R. Koh, J. Imberger, and N. Brooks, *Mixing in Inland and Coastal Waters,* Academic Press, New York, 1979.

Förstner, U., "Inorganic Pollutants, Particularly Heavy Metals in Estuaries," in E. Olausson and I. Cato, eds., *Chemistry and Biogeochemistry of Estuaries,* John Wiley, New York, 1980, pp. 308–348.

Fransz, H. J. Mammaerts, and G. Radach, "Ecological Modelling of the North Sea," *Neth. J. Sea Res.,* 28(1/2):67–140, 1991.

Friedrichs, C., and D. Aubrey, "Non-Linear Tidal Distortion in Shallow Well-Mixed Estuaries: A Synthesis," *Estuarine, Coastal and Shelf Science,* 27:521–545, 1988.

Frostick, L., and I. McCave, "Seasonal Shifts of Sediment Within an Estuary Mediated by Algal Growth," *Est. Coast. Mar. Sci.,* 9(5):569–576, 1979.

Gade, H. G., "Hydrographic Investigations in the Oslofjord, a Study of Water Circulation and Exchange Processes," Report 24 (3 vols.), Geophysical Institute, University of Bergen, Norway, 1970.

Gameson, A. L., ed., *The Quality of the Humber Estuary, 1961–1981,* Humber Estuary Committee, Yorkshire Water Authority, U.K., 1982.

Gause, G. F., "The Principles of Biocenology," *Quart. Rev. of Biol.,* 11:320–396, 1936.

Geyer, W. R., "The Importance of Suppression of Turbulence by Stratification on the Estuary Turbidity Maximum," *Estuaries,* 16(1):113–125, 1993.

Geyer, W., and R. Signall, "A Reassessment of the Role of Tidal Dispersion in Estuaries and Bays," *Estuaries,* 15(2):97–108, 1992.

Glenne, B., "Diffusive Processes in Estuaries," SERL Rep. 66-6, College of Engineering, University of California at Berkeley, 1966.

Godin, G., *The Analysis of Tides,* University of Toronto Press, Toronto, 1972.

Godin, G., "La Marée Dans le Golfe et l'Estuaire du Saint-Laurent," *Naturaliste Canadien,* 106(1):105–121, 1979.

Goodrich, D. M., "On Meteorologically Induced Flushing in Three U.S. East Coast Estuaries," *Estuarine, Coastal and Shelf Science,* 26:111–121, 1988.

Graf, W. H., *Hydraulics of Sediment Transport,* McGraw-Hill, New York, 1971.

Grimes, D. J., "Ecology of Estuarine Bacteria Capable of Causing Human Disease: A Review," *Estuaries,* 14(4):345–360, 1991.

Hager, S., and L. Schemel, "Sources of Nitrogen and Phosphorus to Northern San Francisco Bay," *Estuaries,* 15(1):40–52, 1992.

Hansen, D., and M. Rattray, "New Dimensions in Estuary Classification," *Limn. Oceanogr.,* 11(3):319–326, 1966.

Harder H., E. Christensen, J. Matthews, and T. Bidleman, "Rainfall Input of Toxaphene to a South Carolina Estuary," *Estuaries,* 3(2):142–147, 1980.

Hardin, G., "The Competitive Exclusion Principle," *Science,* 131:1292–1297, 1960.

Harleman, D. R., "Tidal Dynamics in Estuaries," in A. Ippen, ed., *Estuary and Coastline Hydrodynamics,* McGraw-Hill, New York, 1966, pp. 522–545.

Harleman, D. R., "One-Dimensional Models," in G. Ward and W. Espey, eds., *Estuarine Modeling: An Assessment,* 1971, pp. 34–89.

Hearn, C., J. Hunter, and M. Heron, "The Effects of a Deep Channel on the Wind-Induced Flushing of a Shallow Bay or Harbor," *J. Geophys. Res.,* 92(C4):3919–3924, 1987.

Henderson, F., *Open Channel Flow,* Macmillan, New York, 1966.

Herdendorf, C. E., "Great Lakes Estuaries," *Estuaries* 13(4):493–503, 1990.

Heydorn, A. E. F., and K. L. Tinley, *"Estuaries of the Cape; Part 1, Synopsis of the Cape Coast,"* Research Report 380, Council for Scientific & Industrial Research, National Research Institute for Oceanology, Stellenbosch, South Africa, 1980.

Holland, A. F., ed., "Near Coastal Program Plan for 1990: Estuaries," Doc. no. EPA/600/4-90/033, Environmental Research Laboratory, U.S. Environmental Protection Agency, Narragansett, R.I., 1990.

Holley, E. R., "Unified View of Diffusion and Dispersion," *J. Hydraulics Div. ASCE,* 95(HY2):612–631, 1969.

Holley, E., and G. Jirka, "Mixing in Rivers," Tech. Rep. E-86-11, Environmental Laboratory, U.S. Army Waterways Experiment Station, Vicksburg, Miss., 1986.

Holmes, R. W., "The Secchi Disk in Turbid Coastal Waters," *Limn. Oceanogr.,* 15:688–694, 1970.

Hutchinson, G. E., *A Treatise on Limnology,* vol. 1, John Wiley, New York, 1957.

Inman, D. and C. Nordstrom, "On the Tectonic and Morphologic Classification of Coasts," *J. Geol.,* 79(1):1–21, 1971.

Inman, D., and R. Dolan, "The Outer Banks of North Carolina: Budget of Sediment and Inlet Dynamics Along a Migrating Barrier System," *J. Coastal Res.,* 5(2):193–237, 1989.

Ippen, A. T., ed., *Estuary and Coastline Hydrodynamics,* McGraw-Hill, New York, 1966.

Ippen, A. T., "Tidal Dynamics in Estuaries," in A. Ippen, ed., *Estuary and Coastline Hydrodynamics,* McGraw-Hill, New York, 1966, pp. 493–522.

Jaworski, N., P. Groffman, A. Keller, and J. Prager, "A Watershed Nitrogen and Phosphorus Balance: The Upper Potomac River Basin," *Estuaries,* 15(1):83–95, 1992.

Jones, D., and M. Willis, "The Attenuation of Light in Sea and Estuarine Waters in Relation to the Concentration of Suspended Solid Matter," *J. Mar. Biol. Assn. U.K.,* 35:431–44, 1956.

Kaiser, R., and S. Kelly, "Water Rights for Texas Estuaries," *Texas Tech Law Review,* 18(4):1121–1156, 1987.

Kanwisher, J., "On the Exchange of Gases Between the Atmosphere and the Sea," *Deep Sea Res.,* 10:195–207, 1963.

Kassner, J., and J. Black, "Efforts to Stabilize a Coastal Inlet: A Case Study of Moriches Inlet, New York," *Shore and Beach,* April 1982, pp. 21–29.

Kennish, M. J., *Ecology of Estuaries, I: Physical and Chemical Aspects,* CRC Press, Boca Raton, Fla., 1986.

Kent, R., and D. Pritchard, "A Test of Mixing-Length Theories in a Coastal Plain Estuary," *J. Mar. Res.,* 18(1):62–72, 1959.

Kenworthy, W., and D. Haunert, "The Light Requirements of Seagrasses: Proceedings of a Workshop to Examine the Capability of Water Quality Criteria, Standards and Monitoring Programs to Protect Seagrasses," NOAA Technical Memorandum NMFS-SEFC-287, National Oceanic and Atmospheric Administration, Washington, D.C., 1991.

Ketchum, B. H., "The Flushing of Tidal Estuaries," *Sewage & Industrial Wastes,* 23:198–209, 1951a.

Ketchum, B. H., "The Exchange of Fresh and Salt Water in Tidal Estuaries," *J. Mar. Res.,* 10:18–38, 1951.

Kirk, J. T., *Light and Photosynthesis in Aquatic Ecosystems,* Cambridge University Press, London, 1983.

Kjerfve, B., "Comparative Oceanography of Coastal Lagoons," in D. Wolfe, ed., *Estuarine Variability,* Academic Press, New York, 1986, pp. 63–81.

Kjerfve, B., and K. Magill, "Salinity Changes in Charleston Harbor 1922–1987," *J. Waterway, Port, Coastal, and Ocean Engr., ASCE,* 116(2):153–168, 1990.

Knox, G. A., *Estuarine Ecosystems: A Systems Approach,* vols. 1 and 2, CRC Press, Boca Raton, Fla., 1986.

Komar, P. D., *Beach Processes and Sedimentation,* Prentice-Hall, Englewood Cliffs, N.J., 1976.

Kranenburg, C., "A Time Scale for Long-Term Salt Intrusion in Well-Mixed Estuaries," *J. Phys. Oceanogr.,* 16(7):1329–1331, 1986.

Kullenberg, B., and R. Sen Gupta, "Fluoride in the Baltic," *Geochim. et Cosmochim. Acta,* 37(5):1327–1337, 1973.

Lamb, H., *Hydrodynamics,* 6th ed. University Press, Cambridge, 1932. (Republished Dover, New York, 1945.)

Lamont, I. M., ed., "Measurement of Color and Turbidity," in *Water Research Topics,* vol. 1, Stevenage Laboratory, Hertfordshire, 1981.

Lauff, G. H., ed., *Estuaries,* American Association for the Advancement of Science, Pub. no. 83, Washington, D.C., 1967.

Lebo, M. and J. Sharp, "Distribution of Phosphorus Along the Delaware, an Urbanized Coastal-Plain Estuary," *Estuaries,* 16(2):290–301, 1993.

Leggett, W. C., "Fish Migrations in Coastal and Estuarine Environments: A Call for New Approaches to the Study of an Old Problem," in J. McCleave, G. Arnold, J. Dodson, and W. Neill, eds., *Mechanisms of Migration in Fishes,* Plenum, New York, 1984, pp. 159–178.

Lepage, S., and R. Ingram, "Estuarine Response to a Freshwater Pulse," *Estuarine, Coastal and Shelf Science,* 26:657–667, 1988.

Lewis, J. R., *The Ecology of Rocky Shores,* The English Universities Press, London, 1964.

Lewis, R., R. Gilmore, D. Crewz, and W. Odum, "Mangrove Habitat and Fishery Resources in Florida," in W. Seaman, ed., *Florida Aquatic Habitat and Fishery Resources,* Florida Chapter of the American Fisheries Society, Kissimmee, Fla., 1985, pp. 281–336.

Liss, P. S., "Process of Gas Exchange Across an Air-Water Interface," *Deep-Sea Res.,* 20:221–238, 1973.

Liss, P., and M. Pointon, "Removal of Dissolved Boron and Silicon During Estuarine Mixing of Sea and River Waters," *Geochim. et Cosmochim. Acta,* 37(6):1493–1498, 1973.

Littler, D., M. Littler, K. Bucher, and J. Norris, *Marine Plants of the Caribbean: A Field Guide From Florida to Brazil,* Smithsonian Institution Press, Washington, D.C., 1989.

Livingston, R. J., "Dynamics of Organochlorine Pesticides in Estuarine Systems: Effects on Estuarine Biota," in M. Wiley, ed., *Estuarine Processes,* vol. 1, Academic Press, New York, 1976, pp. 507–522.

Longley, W. L., ed., *Freshwater Inflows to Texas Bays and Estuaries: Ecological Relationships and Methods for Determination of Needs,* Texas Water Development Board and Texas Parks and Wildlife Department, Austin, Texas, 1994.

Lung, W., and D. O'Connor, "Two-Dimensional Mass Transport in Estuaries," *J. Hydr. Engr., ASCE,* 110(10):1340–1357, 1985.

Maidment, D. R., ed., *Handbook of Hydrology,* McGraw-Hill, New York, 1993.

Mangelsdorf, P. C., "Salinity Measurements in Estuaries," in G. Lauff, ed., *Estuaries,* Publ. no. 83, AAAS, Washington, D.C., 1967.

Mann, K. H., *Ecology of Coastal Waters: A Systems Approach,* University of California Press, Berkeley, 1982.

Matthews, J, and D. Rosenberg, "Numeric Modeling of a Fiord Estuary," Report R-69-4, Institute of Marine Science, University of Alaska, College, Ask., 1969.

Mattingly, G. E., "Experimental Study of Wind Effects on Reaeration," *J. Hydr. Engr. Div. ASCE,* 103(HY3):311–323, 1977.

McCarty, P., W. Echelberger, J. Hem, D. Jenkins, G. Lee, J. Morgan, R. Robertson, R. Schmidt, J. Symons, M. Trexler, and J. van Wazer, "Chemistry of Nitrogen and Phosphorus in Water," *J. American Water Works Assoc.,* 62:127–140, 1970.

McPhearson, R. M., "The hydrography of Mobile Bay and Mississippi Sound, Alabama," *J. Marine Sci. Alabama,* 1(2):1–83, 1970.

Meade, R. H., "Relations Between Suspended Matter and Salinity in Estuaries of the Atlantic Seaboard, U.S.A.," in *Geochemistry, Precipitation, Evaporation, Soil-moisture, Hydrometry,* Pub. no. 78, International Association of Scientific Hydrology, 1968, pp. 96–109.

Meade, R. H. "Sources, Sinks, and Storage of River Sediment in the Atlantic Drainage of the United States," *Journal of Geology,* 90:235–252, 1982.

Meade, R., and R. Parker, "Sediment in Rivers of the United States," in *National Water Summary 1984: Hydrologic Events, Selected Water-Quality Trends, and Ground-Water Resources,* Water-supply Paper 2275, United States Geological Survey, Reston, Va., 1985, pp. 49–60.

Mehta, A. J., "Characterization of Cohesive Sediment Properties and Transport Processes in Estuaries," in A. Mehta, ed., *Estuarine Cohesive Sediment Dynamics,* Springer-Verlag, New York, 1986, pp. 290–325.

Mehta, A., E. Hayter, W. Parker, R. Krone, and A. Teeter, "Cohesive Sediment Transport, I: Process Description," *J. Hydr. Engr, ASCE,* 115(8):1076–1093, 1989.

Mehta, A. and C. Montague, "A Brief Review of Flow Circulation in the Vicinity of Natural and Jettied Inlets: Tentative Observations on Implications for Larval Transport at Oregon Inlet, North Carolina," UFL/COEL/MP-91/03, Coastal and Oceanographic Engineering Department, University of Florida, Gainesville, 1991.

Mellor, G., and T. Yamada, "Development of a Turbulence Closure Model for Geophysical Fluid Problems," *Rev. Geophys. Space Phys.,* 20:851–875, 1982.

Miller, J., J. Reed, and L. Pietrafesa, "Patterns, Mechanisms and Approaches to the Study of Migrations of Estuarine-Dependent Fish Larvae and Juveniles," in J. McLeave, G. Arnold, J. Dodson, and W. Neill, eds., *Mechanisms of Migration in Fishes,* Plenum, New York, 1984, pp. 209–225.

Monaco, M. D. Nelson, R. Emmett, and S. Hinson, "Distribution and Abundance of Fishes and Invertebrates in West Coast Estuaries, 1: Data Summaries," ELMR Rep. no. 4, Strategic Environmental Assessments Div., National Ocean Service, NOAA, Silver Spring, Md., 1990.

Monin, A., V. Kamenkovich, and V. Kort, *Variability of the Oceans* (English trans.), John Wiley, New York, 1977.

Montagna, P., and R. Kalke, "The Effect of Freshwater Inflow on Meiofaunal and Macrofaunal Populations in the Guadalupe and Nueces Estuaries, Texas," *Estuaries* 15(3):307–326, 1992.

Montague, C. L., "The Influence of Fiddler Crab Burrows and Burrowing on Metabolic Processes in Salt Marsh Sediments," in V. Kennedy, ed., *Estuarine Comparisons,* Academic, New York, 1982, pp. 283–301.

Montague, C. L., "Influence of Biota on Erodibility of Sediments," in A. Mehta, ed., *Estuarine Cohesive Sediment Dynamics,* Springer-Verlag, New York, 1986, pp. 251–269.

Montague, C. L., "Ecological Engineering of Inlets in Southeastern Florida: Design Criteria for Sea Turtle Nesting Beaches," *Journal of Coastal Research SI,* 18:267–276, 1993.

Montague, C., S. Bunker, E. Haines, M. Pace and R. Wetzel, "Aquatic Macroconsumers," in L. Pomeroy and R. Wiegert, eds., *The Ecology of a Salt Marsh,* Springer-Verlag, New York, 1981, pp. 69–85.

Montague, C., and J. Ley, "A Possible Effect of Salinity Fluctuation on Abundance of Benthic Vegetation and Associated Fauna in Northeastern Florida Bay," *Estuaries,* 16:703–717, 1993.

Montague, C., M. Paulic, and T. Parchure, "The Stability of Sediments Containing Microbial Communities: Initial Experiments with Varying Light Intensity," in A. Mehta, ed., *Nearshore and Estuarine Cohesive Sediment Transport,* American Geophysical Union, Washington, D.C., 1993, pp. 348–359.

Montague, C., and R. Wiegert, "Salt Marshes," in R. Myers and J. Ewel, eds., *Ecosystems of Florida,* University of Central Florida Press, Orlando, 1990, pp. 481–516.

Montague, C., A. Zale V, and H. Percival, "Ecological Effects of Coastal Marsh Impoundments: A Review," *Environmental Management,* 11:743–756, 1987.

Mosley, M., and A. McKerchar, "Streamflow," in Maidment, ed., *Handbook of Hydrology,* 1993, pp. 8.1–8.39.

Mudryk, Z., and W. Donderski, "Effect of Sodium Chloride on the Metabolic Activity of Halophilic Bacteria Isolated from the Lake Gardno Estuary," *Estuaries,* 14(4):495–496, 1991.

National Academy of Sciences, *Petroleum in the Marine Environment,* National Academy of Sciences Press, Washington, D.C., 1975.

Nelson, D. M., ed., "Distribution and Abundance of Fishes and Invertebrates in Gulf of Mexico Estuaries, 1: Data summaries," ELMR Report no. 10, Strategic Environmental Assessments Division, NOS/NOAA, Silver Spring, Md., 1992.

Neumann, G., and W. Pierson, *Principles of Physical Oceanography,* Prentice-Hall, Englewood Cliffs, N.J., 1966.

Newell, R. C., *Adaptation to Environment,* Butterworths, London, 1976.

Nienhuis, P. H., "Eutrophication, Water Management and the Functioning of Dutch Estuaries and Coastal Lagoons," *Estuaries,* 15(4):538–548, 1992.

Nihoul, J., and F. Ronday, "The Influence of the 'Tidal Stress' on the Residual Circulation," *Tellus,* 27(5):484–489, 1975.

Nixon, S. W., "Between Coastal Marshes and Coastal Waters—A Review of Twenty Years of Speculation and Research on the Role of Salt Marshes in Estuarine Productivity and Water Chemistry," in P. Hamilton and K. MacDonald, eds., *Estuarine and Wetland Processes,* Plenum, New York, 1980, pp. 437–525.

Nowell, A., P. Jumars, and J. Eckman, "Effects of Biological Activity on the Entrainment of Marine Sediments," *Mar. Geol.,* 42:133–153, 1981.

O'Connor, D. J., "Oxygen Balance of an Estuary," *J. Sanitary Engr. Div. ASCE,* 86(SA3):35–55, 1960.

O'Connor, D. J., "Estuarine Distribution of Nonconservative Substances," *J. Sanitary Engr. Div. ASCE,* 91(SA1):23–42, 1965.

O'Connor, D. J., "Models of Sorptive Toxic Substances in Freshwater Systems, Parts I, II, III," *J. Env. Engr. Div., ASCE,* 114(3):507–574, 1988

O'Connor, D., and W. Dobbins, "Mechanism of Reaeration in Natural Streams," *Transactions ASCE,* 123:641–684, 1958.

O'Connor, D., and W. Lung, "Suspended Solids Analysis of Estuarine Systems," *J. Env. Engr. Div., ASCE,* 107(EE1):101–120, 1981.

Odum, E. P., "The status of three ecosystem-level hypotheses regarding salt marsh estuaries: tidal subsidy, outwelling, and detritus-based food chains," in V. Kennedy, ed., *Estuarine Perspectives,* Academic, New York, 1980, pp. 485–495.

Odum, W. E., "The Lacustrine Estuary Might be a Useful Concept," *Estuaries,* 15(3):506–507, 1990.

Odum, H., B. Copeland, and E. McMahon, *Coastal Ecological Systems of the United States,* The Conservation Foundation, Washington, D.C., 1974.

Oey, L.-Y., G. Mellor, and R. Hires, "A Three-Dimensional Simulation of the Hudson-Raritan Estuary," *J. Phys. Oceanogr.,* 15:1676–1720, 1985.

Officer, C. B., *Physical Oceanography of Estuaries (and Associated Coastal Waters),* John Wiley, New York, 1976.

Officer, C. B., "Box Models Revisited," in P. Hamilton and K. MacDonald, eds., *Estuary and Wetland Processes,* Plenum, New York, 1980, pp. 65–114.

Officer, C., and J. Ryther, "Secondary Sewage Treatment versus Ocean Outfalls: An Assessment," *Science,* 197:1056–1060, 1977.

Olausson, E., and I. Cato, eds., *Chemistry and Biogeochemistry of Estuaries,* John Wiley, New York, 1980.

Olsen, C., I. Larsen, P. Mulholland, K. von Damm, J. Grebmeier, L. Schaffner, R. Diaz, and M. Nichols, "The Concept of an Equilibrium Surface Applied to Particle Sources and Contaminant Distributions in Estuary Sediments," *Estuaries,* 16(3B):683–696, 1993.

Ong, J., W. Gong, C. Wong, Z. Din, and B. Kjerfve, "Characterization of a Malaysian Mangrove Estuary," *Estuaries,* 14(1):38–48, 1991.

Orlando, S. P., L. P. Rozas, G. H. Ward, and C. J. Klein, *Salinity Characteristics of Gulf of Mexico Estuaries,* National Ocean Service, National Oceanic and Atmospheric Administration, Silver Spring, Md., 1993.

Orlando, S., P. Wendt, C. Klein, M. Pattillo, K. Dennis and G. Ward, *Salinity Characteristics of South Atlantic Estuaries,"* National Ocean Service, National Oceanic and Atmospheric Administration, Silver Spring, Md., 1994.

Owen, M. W., "Problems in the Modeling of Transport, Erosion, and Deposition of Cohesive Sediments," in E. Goldberg, I. McCave, J. O'Brien and J. Steele, eds., *The Sea,* vol. 6, John Wiley, New York, 1977, pp. 515–537.

Owen, W., "A study of the physical hydrography of the Patuxent River and its estuary," Tech. Rep. 53, Chesapeake Bay Institute, Johns Hopkins University, Baltimore, Md., 1969.

Parker, B. B., ed., *Tidal Hydrodynamics,* John Wiley, New York, 1991.

Paulson, R. W., "Variation of the Longitudinal Dispersion Coefficient in the Delaware River Estuary as a Function of Freshwater Inflow," *Water Res. Res.,* 6(2):516–526, 1970.

Pelletier, E., and J. Lebel, "Hydrochemistry of Dissolved Inorganic Carbon in the St. Lawrence Estuary (Canada)," *Est. Coast. Mar. Sci.,* 9(6):785–795, 1979.

Peng, et al., "Surface Radon Measurements in the North Pacific Ocean Station Papa," *J. Geophys. Res.,* 79(12):1772–1780, 1974.

Peterson, D., T. Conomos, W. Broenkow, and E. Scrivani, "Processes Controlling the Dissolved Silica Distribution in San Francisco Bay," in L. Cronin, ed., *Estuarine Processes,* vol. 1, Academic, New York, 1975, pp. 153–187.

Peterson, D., D. Cayan and J. Festa, "Interannual Variability in Biogeochemistry of Partially Mixed Estuaries: Dissolved Silicate Cycles in Northern San Francisco Bay," in D. Wolfe, ed., *Estuarine Variability,* Academic, New York, 1986, pp. 123–138.

Pierce, R., C. Olney, and G. Felbeck, "*pp'*-DDT Adsorption to Suspended Particulate Matter in Sea Water," *Geochim. Cosmochim. Acta,* 38(7):1061–1073, 1974.

Pietrafesa, L., G. Janowitz, T.-Y. Chao, R. Weisberg, F. Askari, and E. Noble, "The Physical Oceanography of Pamlico Sound," Publ. UNC-WP-86-5, North Carolina Sea Grant Program, North Carolina State University, Raleigh, N.C., 1986.

Pietrafesa, L. and G. Janowitz, "Physical Oceanographic Processes Affecting Larval Transport Around and Through North Carolina Inlets," *American Fisheries Society Symposium,* 3:34–50, 1988.

Pillsbury, G. B., *Tidal Hydraulics,* U.S. Corps of Engineers Waterways Experiment Station, Vicksburg, Miss., 1956.

Platt, T., C. Gallegos, and W. Harrison, "Photoinhibition of Photosynthesis in Natural Assemblages of Marine Phytoplankton," *J. Mar. Res.,* 38:687–701, 1980.

Pomeroy, L. R., "The Ocean's Food Web, a Changing Paradigm," *BioScience,* 24:499–503, 1974.

Pomeroy, L., L. Shenton, R. Jones, and R. Reimold, "Nutrient Flux in Estuaries," in G. Likens, ed., *Nutrients and Eutrophication,* Special Symposium 1, American Society of Limnology and Oceanography, 1972, pp. 274–291.

Pond, S., and G. Pickard, *Introductory Dynamical Oceanography,* 2d ed., Pergamon, Oxford, 1983.

Prego, R., "General Aspects of Carbon Biogeochemistry in the Ria of Vigo, Northwestern Spain," *Geochim. Cosmochim. Acta,* 57(9):2041–2052, 1993.

Preisendorfer, R. W., "Secchi Disc Science: Visual Optics of Natural Waters," *Limn. Oceanogr.,* 31(5):909–926, 1986.

Pritchard, D. W., "Estuarine Hydrography," in *Advances in Geophysics,* vol. 1, Academic, New York, 1952, pp. 243–280.

Pritchard, D. W., "Estuarine Circulation Patterns," *Proceedings ASCE,* 81(Separate no. 717), 1955.

Pritchard, D. W., "What is an Estuary: Physical Viewpoint," in G. Lauff, ed., *Estuaries,* American Association for the Advancement of Science, Pub. no. 83, Washington, D.C., 1967a, pp. 3–5.

Pritchard, D. W., "Observations of Circulation in Coastal Plain Estuaries," in G. Lauff, ed., *Estuaries,* American Association for the Advancement of Science, Pub. no. 83, Washington, D.C., 1967b, pp. 37–44.

Pritchard, D., and R. Kent, "The Reduction and Analysis of Data from the James River Operation Oyster Spat," Tech. Rep. VI (Ref. 53-12), Chesapeake Bay Institute, The Johns Hopkins University, Baltimore, Md., 1953.

Pugh, D. T., *Tides, Surges and Mean Sea-Level,* John Wiley, Chichester, U.K., 1987.

Redfield, A. C., "The Biological Control of Chemical Factors in the Environment," *Am. Scientist,* 46:205–221, 1958.

Reish, D. J., *Marine Life of Alamitos Bay,* Seaside Printing Co., Long Beach, Cal., 1968.

Reutergårdh, L., "Chlorinated Hydrocarbons in Estuaries," in E. Aoausson and I. Cato, eds., *Chemistry and Biogeochemistry of Estuaries,* John Wiley, New York, 1980, pp. 349–365.

Roelofs, E., and D. Bumpus, "The Hydrography of Pamlico Sound," *Bull. Mar. Sci. Gulf & Caribbean,* 3(3):181–205, 1953.

Rudek, J., H. Paerl, M. Mallin, and P. Bates, "Seasonal and Hydrological Control of Phytoplankton Nutrient Limitation in the Lower Neuse River estuary, North Carolina," *Mar. Ecol. Prog. Ser.,* 75:133–142, 1991.

Salomons, W., and U. Förstner, *Metals in the Hydrocycle,* Springer-Verlag, Berlin, 1984.

Sanford, L., and W. Boicourt, "Wind-Forced Salt Intrusion into a Tributary Estuary," *J. Geophys. Res.,* 95(C8):13,357–13,371, 1990.

Schelske, C., and E. Odum, "Mechanisms Maintaining High Productivity in Georgia Estuaries," *Proc. Gulf and Caribbean Fisheries Inst.,* 14:75–80, 1961.

Schroeder, W. W., "The Impact of the 1973 Flooding of the Mobile River System on the Hydrography of Mobile Bay and East Mississippi Sound," *Northeast Gulf Science,* 1(2):68–76, 1977.

Schroeder, W., and W. Wiseman, "Low-Frequency Shelf-Estuarine Exchange Processes in Mobile Bay and Other Estuarine Systems on the Northern Gulf of Mexico," in D. Wolfe, ed., *Estuarine Variability,* Academic, New York, 1986, pp. 355–367.

Schubel, J., and D. Pritchard, "Great Lakes Estuaries—Phooey," *Estuaries,* 15(3):508–509, 1990.

Scudlark, J., and T. Church, "Atmospheric Input of Inorganic Nitrogen to Delaware Bay," *Estuaries,* 16(4):747–759, 1993.

Seitz, R. C., "Temperature and Salinity Distributions in Vertical Sections Along the Longitudinal Axis and Across the Entrance of the Chesapeake Bay," Graph. Summ. No. 5, Ref. 71-7, Chesapeake Bay Institute, The Johns Hopkins University, Baltimore, Md., 1971.

Shalowitz, A., *Shore and Sea Boundaries,* vols. 1 and 2, Publication 10-1, Coast and Geodetic Survey, U.S. Department of Commerce, Government Printing Office, Washington, D.C., 1962.

Shepard, F. P., *Submarine Geology,* Harper & Row, New York, 1973.

Sholkovitz, E. R., "The Geochemistry of Rare Earth Elements in the Amazon River Estuary," *Geochim. Cosmochim. Acta,* 57(10):2181–2190, 1993.

Shureman, P., *Manual of Harmonic Analysis and Prediction of Tides,* Sp. Pub. 98, U.S. Coast & Geodetic Survey, Government Printing Office, Washington, D.C., 1958, reprinted 1976.

Simmons, H. B., "Tidal and Salinity Model Practice," in Ippen, ed., *Estuary and Coastline Hydrodynamics,* McGraw-Hill, New York, 1966, pp. 711–731.

Smith, D., M. Leffler, and G. Mackiernan, eds., *Oxygen Dynamics in the Chesapeake Bay: A Synthesis of Recent Research,* Maryland Sea Grant Program, University of Maryland, College Park, Md., 1992.

Southerland, E., R. Wagner, and J. Metcalfe, *Technical Guidance Manual for Performing Waste Load Allocations: Book III, Estuaries,* Environmental Protection Agency, Washington, D.C., 1984.

Steever, E., R. Warren, and W. Wiering, "Tidal Energy Subsidy and Standing Crop Production of *Spartina alterniflora*," *Est. Coastal Mar. Sci.*, 4:473–478, 1976.

Stephenson, T., and A. Stephenson, *Life Between Tidemarks on Rocky Shores*, W. H. Freeman, San Francisco, 1972.

Stevenson, L., and C. Erkenbrecher, "Activity of Bacteria in the Estuarine Environment," in M. Wiley, ed., *Estuarine Processes*, vol. 1, Academic, New York, 1976, pp. 381–394.

Stockman, K., R. Ginsberg, and E. Shinn, "The Production of Lime Mud by Algae in South Florida," *J. Sed. Petrol.*, 37:633–648, 1967.

Stommel, H., "Recent Developments in the Study of Tidal Estuaries," Ref. No. 51-33, Woods Hole Oceanographic Institution, 1951.

Taft, J., and W. Taylor, "Phosphorus Dynamics in Some Coastal Plain Estuaries," in M. Wiley, ed., *Estuarine Processes*, vol. 1, Academic Press, New York, 1976, pp. 79–89.

Tennekes, H., and J. Lumley, *A First Course in Turbulence*, MIT Press, Cambridge, Mass., 1972.

Thomann, R., and T. Barnwell, eds., "Workshop on Verification of Water Quality Models," Report EPA-600/9-80-016, Environmental Protection Agency, Washington, D.C., 1980.

Thomann, R., and J. Mueller, *Principles of Surface Water Quality Modeling and Control*, Harper & Row, New York, 1987.

Turner, R. E., "Geographic Variations in Salt Marsh Macrophyte Production: A Review," *Contr. mar. sci.*, 20:47–68, 1976.

Ulanowicz, R., and D. Flemer, "A Synoptic View of a Coastal Plain Estuary," in J. Nihoul, ed., *Hydrodynamics of Estuaries and fjords*, Elsevier, Amsterdam, 1978, pp. 1–26.

Uncles, R. J., "Tidal Dynamics of Estuaries," in B. Kjerfve, ed., *Hydrodynamics of Estuaries*, vol. 1, Boca Raton, Fla., CRC Press, 1988, pp. 59–73.

Uncles, R., and M. Jordan, "Residual Fluxes of Water and Salt at Two Stations in the Severn Estuary," *Estuarine and Coastal Marine Science*, 9:287–302, 1979.

Uncles, R., and P. Radford, "Seasonal and Spring-Neap Tidal Dependence of Axial Dispersion Coefficients in the Severn—A Wide, Vertically Mixed Estuary," *J. Fluid Mech.*, 98(4):703–722, 1980.

U.S. Corps of Engineers, *Shore Protection Manual*, 4th ed., Coastal Engineering Research Center, Waterways Experiment Station, Department of the Army. U.S. Government Printing Office, Washington, D.C., 1984.

van de Kreeke, J., "Stability of Tidal Inlets—Pass Cavallo, Texas," *Estuarine, Coastal & Shelf Science*, 21(1):33–43, 1985.

van de Kreeke, J., "Dispersion in Shallow Estuaries," in B. Kjerfve, ed., *Hydrodynamics of Estuaries*, vol. 1, CRC Press, Boca Raton, Fla., 1988, pp. 27–39.

van de Kreeke, J., "Longitudinal Dispersion of Salt in the Volkerak Estuary," in R. Cheng, ed., *Residual Currents and Long-Term Transport*, Springer-Verlag, New York, 1990, pp. 151–164.

van de Kreeke, J., and K. Robaczewska, "Effect of Wind on the Vertical Circulation and Stratification in the Volkerak Estuary," *Neth. J. Sea Res.*, 23(3):239–253, 1989.

van den Berg, C. M. G., "Complex Formation and the Chemistry of Selected Trace Elements in Estuaries," *Estuaries*, 16(3A):512–520, 1993.

Vieira, M. E. C., "The Meteorologically Driven Circulation in Mid-Chesapeake Bay," *J. Mar. Res.*, 44:473–493, 1986.

Vreugdenhil, C., *Computational Hydraulics*, Springer-Verlag, Berlin, 1989.

Wallen, D., and G. Geen, "The Nature of the Photosynthate in Natural Phytoplankton Populations in Relation to Light Quality," *Marine Biology*, 10:157–168, 1971.

Wallis, S., J. Crowther, J. Curran, D. Milne, and J. Findlay, "Consideration of a One Dimensional Transport Model of the Upper Clyde Estuary," in M. Palmer, ed., *Advances in Water Modelling and Measurement*, BHRA, The Fluid Engineering Centre, Cranfield, Bedford, UK, 1989, pp. 23–41.

Ward, G. H., "Formulation and Closure of a Model of Tidal-Mean Circulation in a Stratified Estuary," in L. Cronin, ed., *Estuarine Processes*, vol. 2, Academic Press, New York, pp. 365–378.

Ward, G. H., "Hydrography and Circulation Processes of Gulf Estuaries," in P. Hamilton and K. MacDonald, eds., *Estuary and Wetland Processes*, Plenum, New York, 1980, pp. 183–215.

Ward, G. H., "Pass Cavallo, Texas: Case Study of Tidal-Prism Capture," *J. Waterway, Port, Coastal, Ocean Div., ASCE*, 108(WW4):513–525, 1982.

Ward, G. H., "The Effect of Deepdraft Ship Channels on Salinity Intrusion in Shallow Bays," *Ports '83*, Am. Soc. Civil Engrs., 1983, pp. 532–542.

Ward, G., and N. Armstrong, "Ambient Water and Sediment Quality of Galveston Bay: Present Status and Historical Trends," Report GBNEP-22, Galveston Bay National Estuary Program, Webster, Tex., 1992.

Ward, G., and W. Espey, eds., "Estuarine Modeling: An Assessment," Report 16070 DZV 02/71, Environmental Protection Agency, Washington, D.C., 1971.

Weiyan, T., *Shallow Water Hydrodynamics*, Elsevier, Amsterdam, 1992.

Welsh, B., R. Whitlatch, and W. Bohlen, "Relationship Between Physical Characteristics and Organic Carbon Sources as a Basis for Comparing Estuaries in Southern New England," in V. Kennedy, ed., *Estuarine Comparisons*, Academic, New York, 1982, pp. 53–67.

West, J., R. Uncles, J. Stephens, and K. Shiono, "Longitudinal Dispersion Processes in the Upper Tamar Estuary," *Estuaries*, 13(2):118–124, 1990.

Wetzel, R. G., *Limnology*, W.B. Saunders, Philadelphia, 1975.

Whitlatch, R. B., "The Ecology of New England Tidal Flats: A Community Profile," Report FWS/OBS-81/01, Biological Services Program, U.S. Fish and Wildlife Service, Washington, D.C., 1982.

Wolfe, D., R. Monahan, P. Stacey, D. Farrow and A. Robertson, "Environmental Quality of Long Island Sound: Assessment and Management Issues," *Estuaries*, 14(3):224–236, 1991.

Wollast, R., "Modelling of Biological and Chemical Processes in the Scheldt Estuary," in J. Nihoul, ed., *Hydrodynamics of Estuaries and Fjords*, Elsevier, Amsterdam, 1978, pp. 63–77.

Wood, E. J. F., "Ecology of Bacteria in Estuarine Systems," in B. Nelson, ed., *Environmental Framework of Coastal Plain Estuaries*, GSA Memoir 133, Geological Society of America, Boulder, Colo., 1972, pp. 237–246.

Yalin, M. S., *Mechanics of Sediment Transport*, Pergamon, Oxford, 1976.

Young, D., D. McDermott, and T. Heezen, "Aerial Fallout of DDT in Southern California," *Bull. Environ. Contam. Tox.*, 16:604–611, 1976.

Zimmerman, J. T. F., "Mixing and Flushing of Tidal Embayments in the Western Dutch Wadden Sea, Part I: Distribution of Salinity and Calculation of Mixing Time Scales," *Neth. J. Sea Res.*, 10(2):149–191, 1976.

Zimmerman, J. T. F., "Mixing and Flushing of Tidal Embayments in the Western Dutch Wadden Sea, Part II: Analysis of Mixing Processes," *Neth. J. Sea Res.*, 10(4):397–439, 1976.

Zimmerman, J. T. F., "Estuarine Residence Times," in B. Kjerfve, ed., *Hydrodynamics of Estuaries*, vol. 1 CRC Press, Boca Raton, Fla., 1988, pp. 75–84.

CHAPTER 13
WETLANDS

Curtis J. Richardson
Professor and Director
Duke University Wetland Center
Durham, North Carolina

13.1 INTRODUCTION

Wetlands comprise approximately 6 percent ($8.5 \text{ km}^2 \times 10^3$) of the world's land surface and are found in every climate from the tropics to the frozen tundra (Maltby and Turner, 1983). The generic term wetland is now used worldwide and includes specific ecosystems known regionally in North America as bogs, bottomlands, fens, floodplains, mangroves, marshes, mires, moors, muskegs, playas, peatlands, pocosins, potholes, reedswamps, sloughs, swamps, wet meadows, and wet prairies. *Freshwater wetlands* comprise more than 95 percent of the total area of wetlands in North America with the peatlands (mires) of Canada, Alaska, and the lower 48 states making up the vast majority of area at 150,000,000 ha, 49,400,000 ha, and 10,240,000 ha, respectively (Bord na Mona, 1984). Today wetlands cover nearly 55 million ha in the United States with states like California and Iowa suffering over 90 percent in losses and many states, like Florida and North Carolina, having reduced their wetlands by nearly 50 percent since 1790 (Dahl and Johnson, 1991) (Fig. 13.1). The intensive conversion of wetlands to agriculture, forestry, and urban areas has resulted in the loss of 53 percent of all wetland habitats in the conterminous United States during the period from 1780 to 1980 (Dahl, 1990).

Wetland ecosystems represent the transition between terrestrial and aquatic systems and have water at or near the surface of the ground for much of the year. The native plants and animals living in wetlands are uniquely adapted to live under conditions of intermittent flooding, lack of oxygen (*anoxia*), and harsh (often toxic) conditions of reduced chemical species (e.g., H_2S rather than SO_4). Key ecosystem functions on the landscape include water and nutrient storage, chemical transformations of N, P, S, and C, and high primary productivity, with storage of vast amounts of carbon as peat due to low decomposition rates. These functions are lost once the wetland is drained. At the population level wetlands function as wildlife habitats, maintaining unique species and increased biodiversity (Richardson, 1994).

The great diversity of wetlands have one thing in common: Their formation, processes and characteristics are controlled by water. To understand what characterizes

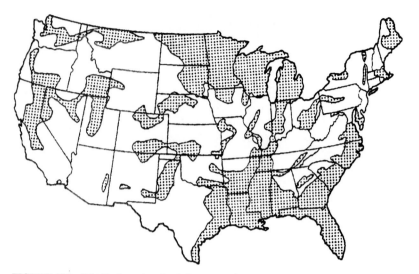

FIGURE 13.1 Distribution of wetlands in conterminous United States. (*National Wetlands Inventory, U.S. Fish and Wildlife Service, St. Petersburg, Fla.*)

a wetland and the role of wetland ecosystems on the landscape, we will first define wetlands and then review how wetlands are formed. It is also necessary to analyze the relative proportion of three water sources—precipitation, ground water discharge, and lateral surface flow—inputs that influence the type of wetland one finds on the landscape. A brief discussion outlining wetland functions is followed by: (1) a categorization of the types of wetlands that exist on the landscape, (2) an analysis of how their biogeochemical functions influence water quality, and (3) how hydrologic processes affect water quantity on the landscape. The hydrologic effects of wetland drainage, conversion to agriculture, forestry, or peat mining is presented as a case study for wetlands of the southeastern United States.

13.2   WETLANDS DEFINED

Wetlands are not easily defined. It is a collective term used to describe a great diversity of ecosystems worldwide whose formation and existence is dominated by water. Permanently flooded deep water areas (generally greater than two meters) are not considered wetlands. Wetlands form part of a continuous water gradient from the drier upland to the continuously wet open-water ecosystems on the landscape. Thus, the upper and lower limits of a wetland are somewhat arbitrary, a fact causing considerable problems in the legal determination of what is or is not defined as a wetland area regulated under the U.S. Army Corps of Engineers (COE) Section 404 of the 1977 Clean Water Act. Historically, wetlands were defined by specialists like botanists and foresters, who focused on plants adapted to flooding and/or saturated soil conditions. A hydrologist's definition emphasized the position of the water table relative to the ground surface over time. Wildlife biologists focused on waterfowl, wading birds, and game and fish species habitats. No formal or universally recognized definition of wetlands existed until the U.S. Fish and Wildlife Service (FWS) in 1979, after years of review, prosed a comprehensive definition (Cowardin, et al., 1979). The wetland definition used as the criteria for a new National Wetlands definition is as follows:

> Wetlands are lands transitional between terrestrial and aquatic systems where the water table is usually at or near the surface or the land is covered by shallow water. Wetlands must have one or more of the following three attributes: (1) at least periodically, the land supports predominantly hydrophytes (water-loving plants), (2) the substrate is predominately undrained hydric soil (wet soils), and (3) the substrate is nonsoil and is saturated with water or covered by shallow water at some time during the growing season of each year.

This definition is widely accepted today by scientists, land use planners, and state managers alike, and has been adopted internationally in many countries. This definition, although comprehensive and useful for the purposes of classification and mapping of wetlands, does not lend itself to the legal definition of wetlands that must be used to determine whether or not a wetland is a *jurisdictional wetland* for purposes of a dredge and fill permit issued by the COE under section 404 of the Clean Water Act of 1977. In 1989, four federal agencies (COE, Environmental Protection Agency [EPA], Soil Conservation Service, and the FWS) collaborated and developed an "accepted" jurisdictional wetland definition which, unlike the FWS definition, is more restrictive since it must meet all three attributes. In simple terms, a jurisdictional wetland must have hydric or waterlogged soils, fifty percent or more of the dominant plants must be wetland plants, and the water table must be found within 30 cm of the ground surface for seven days during the growing season. The interagency legal definition of wetlands is as follows:

> The term "wetlands" means those areas that are inundated or saturated by surface or ground water at a frequency and duration sufficient to support, and that under normal circumstances do support, a prevalence of vegetation typically adapted for life in saturated soil conditions. Wetlands generally include swamps, marshes, bogs, and similar areas (EPA, 40 CFR 230.3 and CE, 33 CFR 328.3).

Current proposed legislation in the U.S. Congress (1995) has ignored the National Academy of Sciences report on wetland delineations and may change this definition in the near future (NRC, 1995).

13.3 WETLAND TYPES

Part of the confusion about the vast number of wetland types is due to the fact that ecologically similar types of wetlands are referred to by many different names throughout the world. For example, a *swamp* in North America refers to a wetland with trees or shrubs, but in Europe they are called *carrs*. The *pocosins* (Algonquin Indian word for *swamp on a hill*) of the southeastern coastal plains of the United States are really nothing more than shrub bogs. Likewise, a *bottomland* refers to a floodplain wetland, generally along a stream; and both are often called *riparian wetlands,* even though this term represents only the area of streamside zone influence. The *billabongs* of Australia are known as *lagoons* or *backswamps* elsewhere. *Mire* comes from the Old Norse *myrr,* and is used to describe peatlands in Europe. In Canada, peatlands are often called *muskegs,* another Algonquin Indian term, and in Germany and England they are called *moors.* The use of local and regional names greatly increases the confusion about the true number of wetland types.

A classification of distinct wetland types characterized according to their sources of water, nutrients, ecological similarities, and topographic position on the landscape clarifies wetland types. The key to ordering these systems is based on their dominant sources of water. For classification purposes, water inflows can be simplified to inputs from: (1) precipitation, (2) ground water discharge, and (3) surface and near-surface inflow (e.g., tides, overbank flow from stream channels, or overland flow) (Fig. 13.2). A particular wetland may have only one dominant source of water input or a combination of sources. The relative position of major wetland types, when scaled according to the relative importance of water sources, shows that bogs and pocosins receive their water almost exclusively from rainfall (*ombrotrophic*) and as such are nutrient-poor (*oligotrophic*). By contrast, fens, seeps, and some marsh types are controlled by groundwater inputs. These wetlands are nourished by minerals from the ground (*minerotrophic*) and are often more nutrient-rich (*eutrophic*). The surface-flow dominated wetland types are very diverse and would include swamps, salt marshes, and wetlands found along the

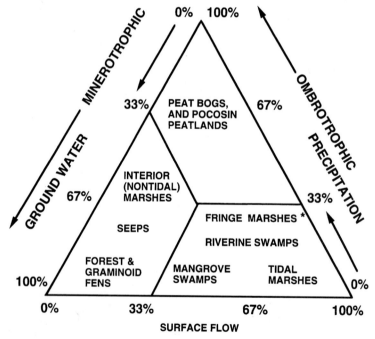

FIGURE 13.2 Relative importance of water source to major wetland types. The percentage of water contributions to a wetland from precipitation, groundwater, or surface flow creates different types of wetlands. Ombrotrophic bogs lack groundwater contributions and upstream inflow. Minerotrophic wetlands, such as fens, receive water and minerals from groundwater. * = Includes both lacustrine and estuarine marshes. (*Modified from Brinson, 1993.*)

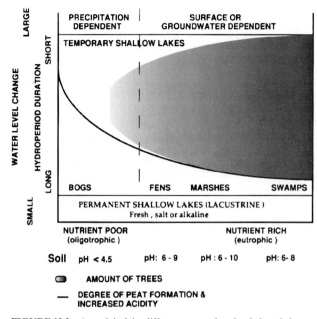

FIGURE 13.3 A model of the different types of wetlands found along a gradient of nutrients (*x* axis) and water regimes (*y* axis). The swamps are dominated by trees under both long and short hydroperiods. The water-level changes within wetland types vary from permanently flooded or long hydroperiods and small water-level changes (lacustrine marshes) to those with short hydroperiods and large water-level changes (tropical swamp forests). (*Modified from Gopal, et al., 1990.*)

fringe of lakes and streams. The relative position of these wetland types to each other on the landscape is best displayed when scaled out along axes of both water regime and nutrients (Fig. 13.3). The nutrient content of the vegetation and soils is lowest in the bogs, and increases along the fen, marsh, and swamp gradient. By contrast, the amount of peat formation decreases along this gradient due to increased decay (decomposition) of organic matter in enriched sites with higher pH. The biomass of trees in proportion to shrubs in bogs is generally low, but forest species increase to become dominant in the swamps. The duration of *hydroperiod* (seasonal pattern of water level at or near the surface) ranges from permanently flooded in the *lacustrine* (shallow lake) *wetland* to periods ranging from 6 to 9 months in the bogs and as little as two weeks in some swamp systems. The seasonal magnitude of water-level change ranges from as small as 0.5 meters in bogs to over 7 meters in tropical swamp forest systems. Thus, the amount of rainfall versus surface or groundwater can be used to separate the general types of wetlands, bog, fens, marshes, and swamps. The amount of peat formation and plant and tree productivity are related to nutrient content and acidity.

The modern *FWS classification of wetlands* divides wetland and deepwater habitats into five *ecological systems*: (1) marine, (2) estuarine, (3) riverine, (4) lacustrine, and (5) palustrine, plus a number of subsystems and classes (Fig. 13.4). Freshwater wetlands comprise the last three systems and make up over 90 percent of the world's wetlands. The *marine system* consists of open ocean and its associated coastline. Mostly deepwater habitat, marine wetlands are limited to intertidal areas, rocky shores, beaches, and some coral reefs with salinities exceeding 30 ppt. The *estuarine system* is more closely associated with land, and comprises salt and brackish tidal marshes, mangroves, swamps, and intertidal mud flats, as well as bays, sounds, and coastal rivers where ocean-derived salts measure less than 0.5 ppt. The *riverine system* is limited to *lotic* (flowing) freshwater river and stream channels and is mainly a deepwater habitat. The *lacustrine* (lake) *system* includes wetlands situated in *lentic* (nonflowing) water bodies such as lakes, reservoirs, and deep pond habitats where trees, shrubs, and emergent plants do not make up more than 30 percent areal coverage in areas less than 2 meters in depth. The *palustrine system* comprises the vast majority (over 90 percent) of the world's inland marshes, bogs, mires, and swamps and does not include any deepwater habitats.

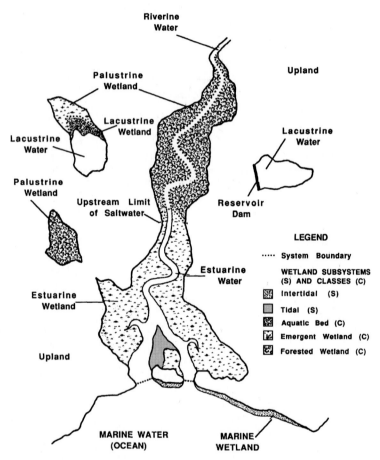

FIGURE 13.4 The five major wetland systems of the U.S. Fish and Wildlife Service classification system are displayed along with several subsystems and classes. The general boundary and relationships among systems are shown. (*Tiner, 1985.*)

13.4 WETLAND FORMATION

The climatic processes and water conditions which formed the swampy environments of the Carboniferous period are evident through fossil analysis and the fuel (e.g., coal) we burn in today's industrial society. Compared with mountains and most rivers, our present wetlands are a young and dynamic ecosystem. Most of the current peat-based wetlands date from the past 12,000 years (postglacial), although wetland habitats surely exist from the Pleistocene ice ages (2 million to 10,000 years ago) near lakes and former glaciers. Any process that produces a hollow or depression in the landscape and holds sufficient water may result in wetland formation. Wetlands are found in deserts near springs and in high rainfall or runoff areas of the mountains. Many of the northern bogs and prairie potholes of the midwestern United States and Canada were formed in depressions left by buried ice blocks (*kettle holes*) from retreating glaciers 10,000 years ago. Along rivers and streams, periodic flooding lays down layers of silt and mud (*alluvial deposits*) along the banks and floodplains, creating bottomlands or swamp forests. Beavers also play a vital role in creating thousands of acres of new wetlands each year.

Wetland formation can happen suddenly, as when a major flood of the Mississippi created new wetlands along the Atchafalaya delta in 1973, or when debris dams up local rivers or streams. Much slower formation processes are at work in the Arctic, where only the upper frozen layer of ice melts in the summer. Because of *permafrost* (permanently frozen ground), cool conditions, and low evaporation, the small amount of annual rainfall and ice melt creates waterlogged soils and, in turn, the world's most extensive peatlands, primarily in northern Russia, Canada, and Alaska. Humans have also contributed greatly to the formation of wetlands in Great Britain and Europe by cutting the forests, which caused the water table to rise and form the present day peatlands. Peat extraction for fuel in the Netherlands and building of fish ponds in Bohemia during the Middle Ages created human-made wetlands. The secret to successful creation of functioning wetlands on the landscape, whether human-made or natural, is primarily dependent on maintaining the hydrological integrity (i.e., maintaining the volume of water at the surface [hydroperiod] and the seasonal water level patterns [*hydropatterns*]).

Wetlands are usually found in depressions or along rivers, lakes, and coastal waters, where they are subjected to periodic flooding (Fig. 13.5a). They also can occur on slopes adjacent to groundwater seeps or can cover an entire landscape like large areas of Ireland, northern Minnesota, and coastal North Carolina, where precipitation levels exceed evapotranspiration (ET) and blocked drainage creates peat soils, which can be over 90 percent water by volume. *Surface water depressions* receive precipitation and overland flow (Fig. 13.5b). Losses occur through ET and downward seepage into the water table. *Groundwater depressions* are in contact with the water table, so they receive groundwater plus rainfall and overland flow (Fig. 13.5c). *Seepage wetlands* occur on slopes where groundwater flows near the surface (Fig. 13.5d). They differ from groundwater depressions in that they have an outlet. The size of these wetlands depends directly on the amount of groundwater and overload discharge into them. *Overflow wetlands* receive water from river flooding, lake water, or even tidal influences (Fig. 13.5e). The water level in these wetlands closely follows the water levels or flooding frequency of the water source. Clearly, the most important factor controlling the type of wetland found on the landscape is the sources of water input: precipitation, surface flow, or groundwater. Nutrient chemistry, pH, and salinity also greatly influence the type of wetland plant communities found within each wetland.

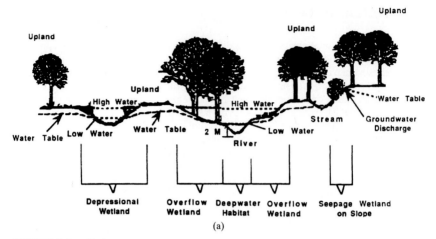

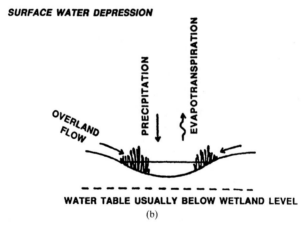

FIGURE 13.5a   Wetlands, deepwater habitats, and uplands on a landscape. Depressional, overflow, and seepage wetland types are due to distinct types of hydrologic input and topographic position on the landscape. (*Modified from Tiner, 1984.*)

FIGURE 13.5b–e   Four major hydrologic input regimes that characterize wetlands with regard to water source, water table, and land forms. (*Modified from Novitzki, 1979.*)

13.5   WETLAND FUNCTIONS VERSUS VALUES

The *value* of a wetland is an estimate, usually subjective, of the worth, merit, quality, or importance of a particular ecosystem or portion thereof. The term imposes an *anthropocentric* (human) focus, which connotes something of use or desirable to *Homo sapiens*. Values ascribed to many wetlands include providing habitats for fishing, hunting, waterfowl, timber harvesting, wastewater assimilation, water quality, and flood control, to name a few. However, these perceived values directly arise

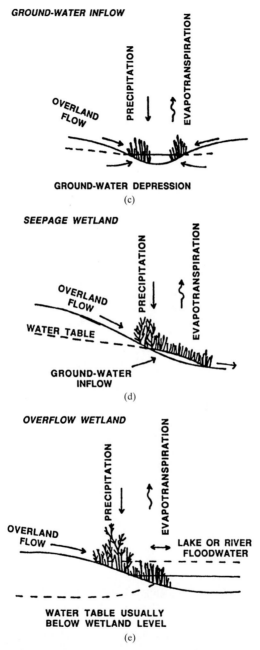

FIGURE 13.5*b–e (Continued)* Four major hydrologic input regimes that characterize wetlands with regard to water source, water table, and land forms. (*Modified from Novitzki, 1979.*)

from the ecological functions within the wetlands. For example, wetlands under *anaerobic* conditions (no oxygen) process nitrate to N_2O and release nitrogen as a gas to the atmosphere. This is an ecological function of wetlands. This function has value on the landscape, in that wetlands can be used to remove nitrogen from agricultural or municipal wastewater runoff (see Sec. 13.5.4.3). This is a value to society that is based directly on the ecological function of nitrogen cycling within the wetland ecosystem which will be lost if the wetland is drained.

A general listing of the functions and values often attributed to natural wetlands is given in Table 13.1 (Richardson, 1994; 1995). The wetland functions are placed in five ecosystem-level categories with specific examples listed below. The functions and values are placed in separate categories to clarify the differences between processes that wetland systems perform and the values that society extracts from these functions. *Wetland values* are derived directly from functions, and the specific relationships are noted by numbers following the values given in Table 13.1. The most important or direct relationship is listed first. *Wetland functions* that are potentially altered or destroyed when a certain value is obtained from the wetland can be directly assessed from this relationship. For example, wastewater treatment is a function of biogeochemical cycling and biological productivity (functions 3 and 2 in Table 13.1) that is directly affected by any impact or harvesting technique that reduces cycling rates or the net primary productivity of the site. Drainage could also seriously affect biogeochemical cycling (3), decomposition (4), or other functions. The flood storage value of a site is obviously primarily related to the hydrologic flux (1), but may also be seriously impacted by a reduction in the biological productivity (2) of the site. Moreover, any activity that significantly impacts the hydrologic function (1) of the wetland in question may well result in loss of the wetland itself or a loss in the ability of the wetland to perform flood control, flood storage, and sediment and erosion control, as well as water quality, water supply, or even timber production values on the landscape.

A brief comparison and analysis of the productivity, decomposition, and wildlife functions of native wetlands follows. The importance of each function and value derived from these processes depends on the wetland type, since not all functions and values are found to the same degree in each wetland. Special emphasis has been given to both the hydrologic and biogeochemical functions in this chapter, since they relate directly to the main focus of this volume. Moreover, an understanding of these processes greatly enhances one's ability to manage or restore wetlands and construct wetlands for water quality or wastewater treatment. Excellent volumes exist on the use of wetlands for wastewater treatment and the topic will not be covered in detail in this chapter (see Godfrey, et al., 1985; Reddy and Smith, 1987; Hammer, 1989; Moshiri, 1993).

13.5.1 Primary Productivity

Wetlands have some of the highest reported plant-growth rates of any ecosystem in the world. Their rates greatly exceed grasslands, cultivated lands, and most forests (Fig. 13.6). Cattail, salt marsh, mangrove swamps, and reed grass marshes are the most productive and more than double the above- and below-ground annual plant biomass production (annual increase in plant organic matter fixed from sunlight by photosynthesis) stored in sedge marshes or bogs, fens, and muskegs. Grasslands and croplands of the United States produce only one-fourth of the fiber of the most productive wetlands, whereas upland forests produce about one-half. Only the

TABLE 13.1 Attributes Generally Reported as Functions and Values of Wetland Ecosystems

---

Wetland functions
1. Hydrologic flux and storage
   a. Aquifer (groundwater) recharge to wetland and/or discharge from the ecosystem
   b. Water storage reservoir and regulator
   c. Regional stream hydrology (discharge and recharge)
   d. Regional climate control (evapotranspiration export, large scale atmospheric losses of H_2O)
2. Biological productivity
   a. Net primary productivity
   b. Carbon storage
   c. Carbon fixation
   d. Secondary productivity
3. Biogeochemical cycling and storage
   a. Nutrient source or sink on the landscape
   b. C, N, S, P, etc. transformations (oxidation/reduction reactions)
   c. Denitrification
   d. Sediment and organic matter reservoir
4. Decomposition
   a. Carbon release (global climate impacts)
   b. Detritus output for aquatic organisms (downstream energy source)
   c. Mineralization and release of N, S, C, etc.
5. Community/wildlife habitat
   a. Habitat for species (unique and endangered)
   b. Habitat for algae, bacteria, fungi, fish, shellfish, wildlife, and wetland plants
   c. Biodiversity
Wetland values
   1. Flood control (conveyance), flood storage (1,2)*
   2. Sediment control (filter for waste) (3,2)
   3. Wastewater treatment system (3,2)
   4. Nutrient removal from agricultural runoff and wastewater systems (3,2)
   5. Recreation (5,1)
   6. Open space (1,2,5)
   7. Visual—cultural (1,5)
   8. Hunting (fur-bearers, beavers, muskrats) (5,2)
   9. Preservation of flora and fauna (endemic, refuge) (5)
  10. Timber production (2,1)
  11. Shrub crops (cranberry and blueberry) (2,1)
  12. Medical (streptomycin) (5,4)
  13. Education and research (1–5)
  14. Erosion control (1,2,3)
  15. Food production (shrimp, fish, ducks) (2,5)
  16. Historical, cultural, and archaeological resources (2)
  17. Threatened, rare, endangered species habitat (5)
  18. Water quality (3,1,4)
  19. Water supply (1)
  20. Global carbon storage (4,2)

Source: Richardson, 1994.

* Wetland values that are directly related to wetland functions (1–5) or those functions that can be adversely affected by the overutilization of values. The order of the numbers suggests which primary function is most directly or first affected.

tropical rain forest comes close to matching the primary productivity of the mangrove swamps or cattail and salt marshes. This biomass serves as food for a great
multitude of aquatic and terrestrial animals, as well as for many endemic species
that exist only in wetlands. Mammals such as moose, caribou, and muskrat graze on
marsh plants. Waterfowl depend heavily on marsh plant seeds. When the plants die,
they decay and form plant fragments called *detritus*. This organic material forms the
base of the aquatic food chain, which supports zooplankton (single-celled floating
animals) and, in turn, shrimp, clams, and fish. Detritus from wetlands is the food
source for many aquatic insects important to commercial species like salmon,
shrimp, crabs, and the majority of nonmarine aquatic animals. In 1991, three million
tons of fish were landed in the United States; two million tons were from species
that spend part of their life cycle (estuarine dependent) in estuarine ecosystems.
Wetlands thus provide important habitat and food for many juvenile species of
shrimp, crayfish, crabs, and fish.

13.5.2 Decomposition

Another unique feature of wetlands compared to terrestrial ecosystems is that dead
plants and animals do not decompose as fast, especially in acid peatlands. They follow more the aquatic production/decomposition cycle (see Fig. 10.2, this volume).
The decay rate in peatlands is only one-half to one-quarter the rate of more aerobic uplands. When leaves, small stems, and roots die they normally are shredded by
macroinvertebrates, earthworms, nematodes, and microbes. The complex organic
matter, which is 45 percent carbon by weight, usually decays to CO_2 and H_2O within

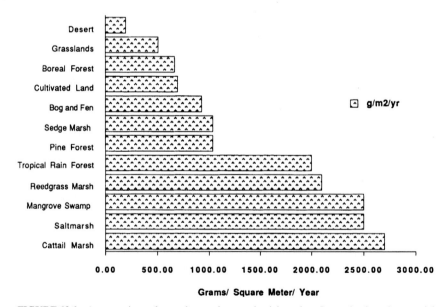

FIGURE 13.6   A comparison of net primary plant productivity values for wetlands and terrestrial
ecosystems. All values are in g m^{-2} yr^{-1} dry weight for above- and below-ground growth. (*Richardson,
1979; Maltby, 1986.*)

one year in terrestrial ecosystems, and the mineral elements are released for new growth. This is not the case in wetlands. The slow decay rates in wetlands are not only due to a lack of oxygen; low pH, shortages of calcium, and low soil temperatures often contribute. These anaerobic conditions eliminate most of the decomposer organisms like filamentous fungi, and the existing anaerobic bacteria obtain energy by inefficient processes like fermentation. Very reduced soils (i.e., no NO_3 or SO_4 present) release important greenhouse gases, like methane and hydrogen, as "marsh gas." Normal metabolism is so greatly reduced in wetlands that many of these ecosystems slowly build up peat (undecomposed organic matter and minerals) at 1 to 2 mm/yr on average (Craft and Richardson, 1993). Moreover, peatlands store vast quantities of carbon and help balance the global carbon cycle (Gorham, 1991). Not only have peat-forming organisms and local biota been preserved in this peat over thousands of years, but pollen and archaeological artifacts, including human remains, have been found buried almost intact. One preserved human body known as the Tollund Man, over 2000 years old, was discovered in a peat bog in Denmark in 1950 with skin, whiskers, and fingernails intact. Analysis of pollen and macrofossils from peats all over the world have allowed scientists to establish historical vegetation changes, climatic variations, and human impacts on the landscape as far back as 275,000 years. Undisturbed wetlands are a time machine to past conditions.

13.5.3 Community/Habitat

Wetlands are important year-round habitats for hundreds of bird species, amphibians, reptiles, and mammals, especially in warmer climates. It has been estimated that 150 types of birds and 200 fish species are wetland-dependent. Furbearers like the North American muskrat (*Ondatra zibethicus*) depend on wetland habitats and food like cattail roots, while other species like otter (*Lutra lutra*) and mink (*Luterola lutreola*) heavily utilize these communities. In tropical wetlands alligators, crocodiles, and hippopotamus (*Alligator, Crocodilus, Hippopotamus,* and other genera) feed voraciously on animals and plants. The American crocodile, an endangered species, is now found only in mangroves and coastal waters of Florida. The rare bog turtle (*Clemmys muhlenbergi*) and Pine Barrens tree frog (*Hyla andersoni*) exist only in a few acid bog and swamp habitats and are threatened by the draining of those wetlands. Rare orchids (*Habenaria spp.*) and unusual insectivorous plants such as sundews (*Drosera spp.*), Venus flytraps (*Dionea muscipula*), and pitcher plants (*Sarracenia spp.*) exist only in wetland habitats.

Wetlands also provide important breeding grounds, overwintering areas, and feeding grounds for millions of migratory waterfowl and other birds. Sixty to 70 percent of the 10 to 12 million ducks that breed annually utilize the prairie pothole region of the United States and Canada alone. The bottomland hardwood forests and marshes of the southeast are the winter home of millions of ducks that wing their way south along the meandering rivers and adjacent forested riparian wetlands. These bottomland hardwood forests also provide habitat for neotropical migrant species. There are a diverse number of habitats within each wetland based on the depth, amount of water, and type of plants adapted to each zone. The diversity of wetland types as well as the variation within each wetland thus provides essential habitats for animals and plants. Additionally, many upland animal species depend on wetlands for water and food, especially during drought periods. Thus, maintaining the hydrologic function of wetlands on the landscape is the key to maintaining their importance in the biosphere.

13.5.4 Biogeochemical Cycling

One of the key ecological functions of wetlands is their ability to store, transform, and cycle nutrients. Wetlands maintain among the widest range of oxidized and reduced chemical states of all ecosystems, due to their periodic flooding and drying cycles. The unique shift from aerobic (free oxygen present) to anaerobic (no free oxygen) soil conditions provides these systems with the ability to process PO_4, NO_3, SO_4, and C and release gases (N_2O, S, H_2S, CH_4, CO_2) to the atmosphere, thus affecting biosphere-level problems such as acid rain, global warming, and greenhouse gases.

Specifically, these ecosystems maintain unique biogeochemical processes that allow them to transform some elements (such as denitrification of inorganic nitrogen and reduction of sulfate), act as a sink or source for others (e.g., carbon), and function as sediment filtering systems in the landscape (Deevey, 1970; Patrick and Reddy, 1976; Boto and Patrick, 1979; Bartlett, et al., 1979; Armentano, 1980; Tilton, et al., 1976; Tourbier and Pierson, 1976; Richardson and Nichols, 1985; Richardson, 1989; Richardson and Craft, 1993; Correll, 1986; Mitsch and Gosselink, 1993). These studies, although often conflicting, do suggest that wetlands have ecological value on a regional-landscape basis that is directly related to their ability to transform, store, or cycle nutrients.

The words *filter* and *sink* are often used interchangeably in ecosystem level studies, but it is clear from an analysis of their etymology and definitions that these terms are not synonymous. A *filter* is any device in which carbon, sand, or similar material is held, and through which liquid is passed to remove suspended impurities or solids. A *sink*, as defined by Webster, is a basin or receptacle for receiving, storing, and carrying off materials; a low-lying area where water collects. Thus, by definition a dissolved nutrient that is filtered by the wetland will pass through the ecosystem. The term *sink* should be utilized when we wish to convey the idea that the ecosystem is storing the element or material and that output or net yield of material is less than input of that material. It is important to keep in mind that a system functioning as a net sink can still release a significant mass of material to downstream ecosystems. This is especially true for wetlands which often receive high loading rates from upstream areas in the watershed. The term *source* should be utilized when the ecosystem output exceeds inputs of the particular material in question. I suggest that an annual time frame be utilized when determining whether or not an ecosystem is a sink or source on the landscape. It is important to note, however, that seasonal differences will often occur and a system may switch from being a source to a sink or vice versa. Finally, the word *transformer* is defined as the thing that changes the form or outward appearance; to change the condition, nature or function of. . . . Wetlands would be defined as *nutrient transformers* when processes within the ecosystem result in significant change in the valence state of an element, change from inorganic to organic forms of a nutrient (or vice versa), or convert from liquid to gaseous phase. It should be noted from these definitions that a wetland can simultaneously function as a transformer and either a sink or source on the landscape.

With these words defined we can now look at the ecological function of wetlands in terms of their ability to filter materials, transform nutrients, and function as sources or sinks. The incomplete database for many elements from different wetland types prevented me from comparing many of these systems and elements. I have therefore focused primarily on nitrogen, phosphorus, and carbon. The factors controlling biogeochemical cycling and the internal processes regulating elemental fluxes are discussed in other chapters of this volume, in Mitsch and Gosselink (1993), and in Schlesinger (1991), and are not repeated in detail here.

13.5.4.1 Wetlands as Filters on the Landscape. The best evidence that wetlands function as effective filters comes from studies of wastewater and urban runoff additions to these ecosystems (Boto and Patrick, 1979; Bastian and Reed, 1979; Godfrey, et al., 1985; Cooper, et al., 1987; Hammer, 1989; Moshiri, 1993; Barling and Moore, 1994). Values for *suspended solids (SS) removal* range from 60 to 90 percent for the wastewater case studies reviewed by Reed, et al. (1980). An artificial cattail marsh system in Listowel, Canada, was found to retain from 57 to 93 percent (4-year average for 5 different marsh system designs) of SS (Herskowitz, 1986). A cattail marsh was found to retain 97 percent of nonvolatile SS and 76 percent of volatile SS from urban runoff in Minneapolis/St. Paul, Minnesota (Brown, 1984). Boto and Patrick (1979) indicate that with the removal of suspended sediment comes the added benefit of heavy metal and toxicant removal, which are often adsorbed on sediment and then buried within the system. Cooper, et al. (1986) utilized Cs-137 sediment analysis and found that 80 percent of sediment loss from cultivated fields in North Carolina was deposited in riparian areas above the floodplain swamp. The remaining 20 percent was found in the swamp. About 50 percent of the Cs-137 sediment in the riparian areas was found within 100 m of the export point from the fields. These few studies all indicate that freshwater wetlands are very efficient filters in the landscape. There is a desperate need for more quantitative sediment analysis (i.e., Cooper type) in order to quantify the filtering capacity for most freshwater wetlands, especially under higher loading levels.

13.5.4.2 Wetlands as Transformers. To estimate nutrient transformation potential for freshwater wetlands, the annual biogeochemical cycles for nitrogen, phosphorus, and carbon for a *bog, fen,* and *swamp ecosystem* are presented. An analysis of the cycles allows us to quantify if a significant proportion of these elements are transformed prior to storage or export from these ecosystems.

A classic example of these transformations is shown by the role of wetlands in the nitrogen cycle (Fig. 13.7). *Nitrogen transformations* are a complex assortment of microbially mediated processes and chemical reactions strongly influenced by the *redox status* (degree of reduced or anaerobic conditions) of the wetland soils. Most flooded wetland soils have a thin, oxidized layer at the surface caused by proximity to the air or higher dissolved oxygen in overlying floodwater. The reduced layer below dominates the soil and lower water column processes and prevents upland plants, bacteria, and animals from invading these ecosystems due to a lack of oxygen for normal metabolism. Multiple biotic and abiotic transformations take place in these layers and involve seven valence states (+5 to -3) and the formation of such oxidized nitrogen species as nitrate (NO_3^-), nitrite (NO_2^-) and nitrogen dioxide gas (N_2O) as well as the reduced ammonium ion (NH_4^+). Organic N is mineralized in both the reduced and oxidized layers to NH_4^+. The thin oxidized zone is where NH_4^+ is oxidized to NO_3^- by bacteria (*nitrification*). In the reduced layer, NH_4^+ is stable and may be adsorbed to sediment or utilized by plants and microbes for growth. Any NO_3^- that diffuses downward in this zone is *denitrified* (microbially converted) to N_2 or N_2O and released to the atmosphere as a gas. Thus, wetlands provide one of the key ecosystems processes that keeps N in global circulation. Their most unique function may be related to their ability as ion transformers, since these systems maintain the widest range of oxidation/reduction reactions on the landscape.

Bowden (1987), in a review of the N cycle in wetlands, states that sediments are the single largest pool of N in wetlands (100s to 1000s g N m^{-2}) and that nitrogen fixation is an important supplemental input to some wetlands (<1 to 3 g N m^{-2} yr^{-1}). He also reported that both nitrification (<0.1 to 10 g N m^{-2} yr^{-1}) and denitrification to ammonia (= 0.5 g N m^{-2} yr^{-1}) probably operate below their potential due to anaero-

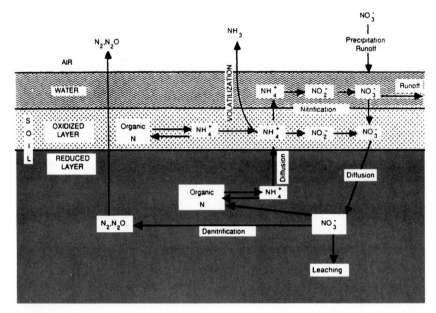

FIGURE 13.7   Nitrogen transformations in wetlands. (*Modified from Faulkner and Richardson, 1989.*)

bic conditions. Seasonal variations in these processes are unknown, but the existing data suggest that N loss from natural wetlands is limited to a small proportion of the mineralizable nitrogen and a fraction of the total nitrogen in the system. Recent studies in Sweden have pointed out the importance of vegetation uptake of N in riparian wetlands, as well as the retention of ions due to sediment deposits (Ambrus and Lowrance, 1991; Cooper, 1990; Hanson, et al., 1994; Barling and Moore, 1994; Vought, et al., 1994). However, they found the main mechanism for N removal to be denitrification, as did Jacks, et al. (1994) in a forested system in southern Sweden. Gilliam (1994) reported that riparian areas in North Carolina reduced NO_3 in shallow groundwater from agricultural fields from 15 mg/L to 1 to 2 mg/L in the first 10 to 15 m of the riparian areas.

The effects of *nitrogen loadings* (e.g., inorganic N from wastewater runoff) on wetland output of N indicate that significant shifts can take place in the transformation and storage of N (also see Sec. 13.5.4.3). For example, Kelly and Harwell (1985) noted that no excess water outflow of N above background would exist up to an input rate of 4.4 g N m^{-2} yr^{-1}, but a 21 percent increase in output would occur at loadings of 8.8 g N m^{-2} yr^{-1}, and asymptotically approach 43 percent of input at loading rates exceeding 30 g N m^{-2} yr^{-1}. This suggests that high N processing capacity (such as adsorption and denitrification) occurs in wetlands, since outputs under these increased loading rates appear to be highly regulated (Richardson and Nichols, 1985).

An annual biogeochemical cycle completed for P in a fen ecosystem in central Michigan revealed that permanent storage of P ranged between 0.2 and 0.5 g m^{-2} yr^{-1} (Richardson, 1985; Richardson and Marshall, 1986). The annual plant uptake of P ranged from 0.7 to 0.9 g m^{-2} of which 35 percent was recycled in the ecosystem via litterfall (Fig. 13.8). The pool of *dissolved inorganic phosphorus (DIP)* in the soil is utilized by plants and microorganisms and receives low-level inputs from rainfall, leaching, root exudation, and decomposition. The amount of DIP available for biotic uptake is very low in comparison to annual uptake requirements. A rapid turnover in the DIP compartment was indicated from ^{32}P studies (Richardson and Marshall, 1986) and suggests that the exchangeable P pool is the main contributor of P. The inputs to the DIP pool in the fen during the growing season are rapidly (<1 h) removed from the water column by the microbial component of the ecosystem and transformed to high-energy P (Richardson, 1985). Phosphorus soil adsorption capacity in the fen is very high (>2500 µg/g) and is controlled by high aluminum (2295 ± 1804 µg/g) and iron content (4924 ± 1980 µg/g) in the peat. The soil organic pool is the main long-term P sink but is not readily available for plant growth or microbial uptake. The inputs of P via rainfall and runoff are not adequate to meet plant uptake requirements, which suggests that P transformations and exchange are essential for maintenance of plant growth (Fig. 13.8).

Macrophytes are mainly responsible for the recycling of P in the fen through root uptake from the soil; i.e., plants are the source of P for the water rather than the reverse. Decomposition, annual release of P by microorganisms, and litterfall mineralization are also key transformations that contribute to annual plant uptake requirements. Microbial uptake and soil adsorption were found to limit the amount of DIP made available for plant growth during the growing season (Richardson, 1985). The long-term P storage capacity of this fen is primarily as organic P, with low concentrations of inorganic P cycling annually.

Richardson and Craft (1993) reported a value of 0.44 g m^{-2} yr^{-1} of P retention for a portion of the northern Everglades which had received TP loading of nearly 60 t/yr for two decades from agricultural runoff. Qualls and Richardson (1995) report that this area transforms inorganic P and stores nearly 70 percent of it as organic P and

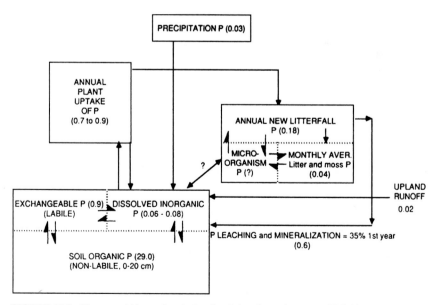

FIGURE 13.8 The annual biogeochemical cycle of phosphorus in a central Michigan fen. Uptake, storage and transfer rates of P are shown (in brackets) and are reported in g m^{-2} yr^{-1}. (*Modified from Richardson and Marshall, 1986.*)

that calcium is responsible for most of the soil sorption and precipitation in this alkaline wetland (Richardson and Vaithiyanathan, 1995). Thus, P storage in wetlands is low, especially in peat-based wetlands (Richardson, 1985), but the rate of transformation from inorganic P (ortho-P) to organic or complexed P is high (Richardson and Craft, 1993).

A detailed C budget for a swamp forest in North Carolina was completed by Mulholland in 1980; transformations, storage, and export of C are presented in Fig. 13.9. The swamp is partitioned into three living components (trees and saplings, shrub and herbaceous plants, and algae) and two detritus components (water column and swamp floor). Most of the organic C in the swamp (22,000 g m^{-2}) was found in the vegetation (63 percent) and the remainder in the swamp litter, logs, and organic soil (to 25 cm depth) (Fig. 13.9). The total input of 902 g C m^{-2} yr^{-1} was 77 percent from primary productivity and 21 percent from C in stream flow (Mulholland, 1980). Annual internal recycling of 352 g C m^{-2} comprised 77 percent litterfall, 17 percent macrolitter, and 6 percent stemflow and throughfall (Fig. 13.9). Output of C from the swamp was mostly by litter respiration (45 percent) during low-water periods and by stream flow during high water levels (40 percent). Aquatic, water column, and anaerobic respiration accounted for the remaining 15 percent of C loss. Input exceeds output by 361 g C m^{-2} yr^{-1} with all but 38 g C m^{-2} of this stored in annual above-ground net wood increment (Mulholland, 1980). Thus, even a flow-through system like a swamp functions as a sink for C. An analysis of water column data indicates that this component functions as the flow-through or transport system. Allochthonous hydrologic inputs of C were 196 g C m^{-2} yr^{-1} or 76 percent of the total water column input. Hydrologic outputs accounted for 89 percent of the C output with respiration accounting for the remainder (Mulholland, 1980). The ecosystem was efficient at transforming nearly 63 percent of the total detrital inputs to CO_2 via metabolic processes. Mulholland's data indicate that the primary C cycle in swamps is from inorganic CO_2 to organic C via primary productivity, and back to the gaseous state via respiration. This process is similar in both terrestrial and wetland ecosystems. However, other processes involving the transformation of organic matter into CH_4 and the export of organic C, mostly as dissolved and colloidal humic materials via the hydrologic cycle, is of greater importance in wetland ecosystems. Bridgham, et al. (1995) have noted, for example, methane flux rates from wetlands as high as 1.3 g C m^{-2} yr^{-1} for bogs, and Roulet, et al. (1992) reported values of 3.6 g C m^{-2} yr^{-1} for shrub swamps in southern Ontario.

The major fluxes and storage values previously given for biogeochemical cycles of N, P, and C in three different wetland types cannot be uniformly taken as typical rates for all wetlands. What is evident is that wetlands function as effective transformers (e.g., primary reducer systems on the landscape), and can be either a sink or source depending on the valence state and form of the element in question. Outputs are regulated by hydrologic conditions, soil adsorption and storage, and biotic uptake and recycling.

Watershed data (i.e., mass of nutrients in outflow or input/output budgets) are available for some wetland types. I have used this information in the next section to address the question of whether or not freshwater wetlands function primarily as a source or sink for N, P, K, Ca, Mg, Na, C, and S.

13.5.4.3 *Wetlands a Sink or Source for Nutrients?*

A comparison of the principal storage compartments for nutrients in wetlands reveals that the majority of nutrients are stored in the peat and litter. For N, P, and K more than 55 percent of the total amount present is in peat and litter; plants are second in importance, and the water compartment contains the smallest amount (Verhoeven, 1986; Richard-

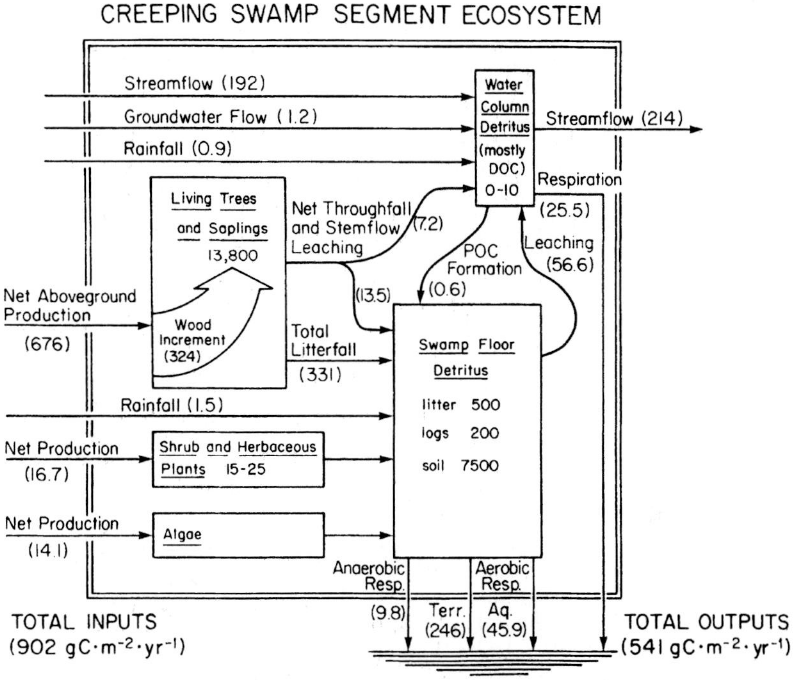

CREEPING SWAMP SEGMENT ECOSYSTEM

FIGURE 13.9 A model of organic carbon flux in a swamp ecosystem in coastal North Carolina, USA. Carbon standing stocks are reported as g C m^{-2} and all fluxes as g C m^{-2} yr^{-1}. (*Mulholland, 1980.*)

son and Craft, 1993; Qualls and Richardson, 1995). Potassium, a highly mobile ion in wetlands, is proportionally more abundant in water and plants than N and P since the K in peat is in a highly exchangeable form (Richardson, et al., 1978). Of note is the fact that more than 95 percent of the N and P in peatlands is found in the bound organic form and is not readily available for biotic uptake and recycling (Richardson, et al., 1978; Qualls and Richardson, 1995). For example, the upper 20 cm of peat soil in a fen in Michigan averaged 6830, 242, 215, and 3577 kg/ha of total N, P, K, and Ca, respectively (Richardson, et al., 1978). This compartment represented over 97 percent of the total N, P, and Ca and 82 percent of the K. A fractionation of the forms of phosphorus in soils from nutrient-enriched areas of the Everglades compared to unenriched areas revealed that 50 to 70 percent of the P was stored in organic forms at both sites and that Fe and Al controlled less than 10 percent of the P, while Ca controlled nearly 15 percent of sorption (Qualls and Richardson, 1995).

The annual storage rate of organic N and P in wetlands is low. Only a small portion of the total vegetative production is accumulated. Rates of peat accumulation in wetlands in Canada, Ireland, and Finland range from about 10 to 100 g dry matter m^{-2} yr^{-1} (Reader and Stewart, 1972; Parkarinen, 1975; Tolonen 1979). Hemond (1980) estimates net peat accumulation in Thoreau's Bog in Massachusetts to be 180 g m^{-2} yr^{-1}. Rates for peat accumulation in the Everglades were 188 g m^{-2} yr^{-1} in unenriched areas and 322 g m^{-2} yr^{-1} in enriched areas (Craft and Richardson, 1993).

Ranges of 1.0 to 2.6 percent N and 0.05 to 0.12 percent P in organic soils are typical (Stanek, 1973; Richardson, et al., 1978; Westman, 1979). Thus the rate at which N and P are accumulated in peat ranges from 0.10 to 4.7 g N m^{-2} yr^{-1} and 0.005 to 0.22 g P m^{-2} yr^{-1} in moderate climates, and possibly up to 10.0 g N m^{-2} yr^{-1} or 0.50 g P m^{-2} yr^{-1} in warm, highly productive areas. Schlesinger (1978) estimated annual accumulations of 3.8 g of N and 0.15 g of P m^{-2} in Georgia's Okefenokee Swamp. The annual accumulation rate of P in a central Michigan fen was estimated to be 0.2 to 0.5 g m^{-2} yr^{-1} (Richardson and Marshall, 1986). Recent studies in the Florida Everglades have shown that P retention capacity is less than 1.0 g P m^{-2} yr^{-1} and averages 0.44 g P m^{-2} yr^{-1} in areas receiving agricultural runoff and is near 0.2 g m^{-2} yr^{-1} in unenriched areas (Richardson and Craft, 1993). These accumulation rates suggest a limited N and P storage capacity for wetlands if peat accumulation is the main biogeochemical mechanism suggested for processing and storage.

A meaningful comparison of input-output nutrient budgets for diverse ecosystem types requires that one take into account the differences in inputs (i.e., only rainfall in perched bogs versus rainfall plus runoff in swamp, marsh, or fen ecosystems), climatic and geological differences, as well as amounts of artificial nutrient loadings (such as fertilizer runoff). It should also be realized that input-output budgets do not, for the most part, take into account gaseous losses from the ecosystem processes (e.g., denitrification), and thus wetlands are often incorrectly depicted as sinks when in fact they should be recognized as transformers. Exports from input-output diagrams should therefore be labeled as nongaseous exports when these transformations are not accounted for. Differences in measurement technique also lead to wide variations in the data sets and are noted in the budgets presented in Table 13.2. (*Note:* g m^{-2} × 10 = kg ha^{-1}).

A comparison of gross watershed nutrient budgets for a range of wetland types and an upland hardwood forest shows that wetlands retain or process most of the entering NO_3-N and NH_4-N and consequently, they do not release large amounts of these inorganic forms of N to downstream waters (Table 13.2). The terrestrial ecosystem by comparison is also a net retainer of NO_3-N but releases 84 percent of input. The exact proportion of uptake and adsorption of NO_3-N and NH_4-N versus N transformations (e.g., denitrification and gaseous release) cannot be determined from gross budgets, but many studies report high denitrification rates in wetlands (Bartlett, et al., 1979; Hemond, 1983). Bowden (1987) does not agree with these findings and suggests that denitrification rates are based on potential denitrification rates, not actual denitrification, and thus the importance of this component of the N cycle in wetlands may be greatly overestimated. However, a good indication of N transformation capacity by wetlands is shown by the removal of 96 percent of the high inputs of NO_3-N from fertilizer into an Iowa marsh (Table 13.2). Davis, et al. (1981) suggest that the main N removal mechanism in this marsh was denitrification. Lowrance, et al. (1984) have shown that a riparian forest (stream-side bottomland hardwood swamp) in Georgia was able to denitrify 61 percent of the 52 kg ha^{-1} yr^{-1} of N input. Studies by Vought, et al. (1994) in Sweden concur with the concept that N is primarily lost via denitrification in wetland buffer strips; interestingly, they report that nitrate is lost within the first 10 to 20 m of the riparian zone.

The total output for NO_3-N is low for bog ecosystems when compared to upland forest or heavily fertilized marsh (Table 13.2). The swamp ecosystem is the only other wetland system to show a high output of NO_3-N, and this is attributable to the high flow rates through the ecosystem and the high amount of nonpoint runoff in the large watershed included in this study. The net loss was, however, minimal. The amounts of output of NH_4-N are generally low for all ecosystems except the blanket bog in Ireland (Table 13.2).

Input-output diagrams were utilized to analyze whether wetland ecosystems are a source or sink for N, P, K, and Ca (Figs. 13.10a–d). A solid diagonal line is used to show where inputs equal outputs. Wetland sites above the line are a source (i.e., in nongaseous form) of the specific nutrient and those below are functioning as a sink or, in the case of N and S, may be releasing it as a gas (e.g., denitrification of N). The export of total nitrogen (TN) is much higher from several of the wetlands than from upland forest ecosystems (Table 13.2, Fig. 13.10a). The blanket bog, Iowa marsh, and Florida swamp ecosystems showed a net loss of organic N, especially when compared to upland forests. The exception to this trend is the Minnesota bog, which is partially surrounded by mineral soils (Verry and Timmons, 1982). It should be noted that the bog site in England is an eroding bog, which contributes to its higher output levels of TN, and the marsh and swamp received high N inputs from their watersheds. The pocosin systems (ombrotrophic bogs in the southeastern United States) are net retainers of TN. The upland forest sites proved to be the most efficient retainers of TN input (Table 13.2; Fig. 13.10a). Collectively, (Table 13.2) the input-output data indicate that raised bogs and some swamps are a sink for TN, but wetlands can be a source of TN under high input levels or eroding conditions. As noted earlier, these ecosystems are efficient transformers (inorganic N to organic N and denitrification), but they can also export inorganic N under higher loading rates.

All of the systems measured were net retainers of dissolved P (DP) (Table 13.2). The only ecosystems that showed a net loss of total phosphorus (TP) were most of the bogs and the Florida swamp (Table 13.2; Fig. 13.10b). Only 30 percent of the 5.6 kg ha^{-1} yr^{-1} of P input from agricultural runoff and rainfall to a bottomland swamp in Georgia was retained by the ecosystem (Lowrance, et al., 1984). Of interest is the fact that the wetland outputs are an order of magnitude higher than yields reported from the upland hardwood forest. The average output of TP for wetlands is 0.4 kg ha^{-1} yr^{-1} versus 0.2 kg ha^{-1} yr^{-1} for upland terrestrial ecosystems (Richardson, 1985). These data suggest that wetlands do not cycle P as tightly as uplands and are not as effective as P sinks. The exception to this may be the swamp ecosystems which apparently retain a higher percentage of dissolved P input. Soil P retention is controlled by adsorption which is regulated by extractable amorphous aluminum and iron content in acid soils and Ca in alkaline soils (Richardson, 1985; Richardson and Vaithiyanathan, 1995; Qualls and Richardson, 1995). Swamps have been shown to have higher levels of Al and Fe than peatland systems, and P retention is directly related to these elements (Richardson, 1985; Richardson and Marshall, 1986).

Potassium and sodium are two of the most mobile elements in ecosystems (Table 13.2). For the most part, both wetlands and terrestrial ecosystems lose more K (Table 13.2; Fig. 13.10c) and Na than they receive from rainfall and runoff (i.e., geologic inputs have not been included in most wetland budgets and current higher outputs must reflect earlier storage of these elements). The higher levels of Na loss by pocosins, Finnish bogs, and the eroding bog in England reflect their high input levels from the nearby oceans. Calcium losses (Table 13.2; Fig. 13.10d) are very high from the eroding bog because of water contact with the underlying limestone rock in the watershed (Crisp, 1966). The wetlands in North Carolina and the bogs in Ireland and Minnesota do show a net retention of Ca, but output levels average over 10 kg ha^{-1} yr^{-1}. Magnesium losses are also high for most of the ecosystems, especially those systems in contact with the mineral soil. A bottomland swamp forest in Georgia retained only 39 percent of 53 kg ha^{-1} yr^{-1} of Ca input; 23 percent of 20 kg ha^{-1} yr^{-1} of Mg input; 7 percent of 105 kg ha^{-1} yr^{-1} of Cl input; and 6 percent of 23 kg ha^{-1} yr^{-1} of K input (Lowrance, et al., 1984).

This database is limited, but the evidence is that wetland output for Ca is almost equal to input but is below that of forests which have a much higher mineral sub-

TABLE 13.2 A Comparison of Nutrient Budgets for Wetland Ecosystems and a Terrestrial Hardwood Forest

Research Site / Ecosystem type / Dominant vegetation and soil	Budget	Nutrients, kg ha⁻¹ yr⁻¹										Annual rainfall
		NO_3-N	NH_4-N	TN	DP	TP	K	Ca	Mg	Na	C	
Minnesota, USA[a]												
Perched bog	Input	2.0	2.2	12.7	0.4	1.2	11.1	20.9	5.9	4.2	—	80
Black spruce	Output	0.3	0.7	6.4	0.2	0.5	6.1	9.1	3.8	2.4	—	
Organic	Net gain or loss	+1.7	+1.5	+6.3	+0.2	+0.7	+5.0	+11.8	+2.1	+1.8	—	
Pennine, England[b]												
Blanket bog	Input	8.2*		8.2	0.6	0.6	3.1	9.0	—	25.5	—	213
Sphagnum-calluna	Output	2.9		17.7	0.4	0.9	11.0	58.3	—	45.5	41,878	
Organic	Net gain or loss	+5.3		-9.5	+0.2	-0.3	-7.9	-49.3	—	-20.0	-41,800	
Finland national average[c]												
Bogs	Input	—	—	6	—	0.1	2.4	2	1	2	—	57
Organic	Output	—	—	2	—	0.3	4.6	12	4	6	—	
	Net gain or loss	—	—	+4	—	-0.2	-2.2	-10	-3	-4	—	
Glenamoy, Ireland[d]												
Blanket bog	Input	6.0	18.1	—	0.7	—	18.1	23.2	—	—	—	139
Carex-shrub	Output	4.5	6.7	—	0.2†	—	4.7	22.0	—	—	—	
Organic	Net gain or loss	+1.5	+11.4	—	+0.5	—	+13.4	+1.2	—	—	—	
Edinburgh, Scotland[e]												
Blanket bog	Input	—	—	6.1‡	—	0.1	1.9	3.0	—	—	—	100
calluna-eriophorum	Output	—	—	18.4	—	0.2	2.0	4.6	—	—	—	
Organic	Net gain or loss	—	—	-12.3	—	-0.1	-0.1	-1.6	—	—	—	
Iowa, USA[f]												
Prairie pothole marsh	Input	209.5	9.1	23.7§	3.5	4.1	—	—	—	—	302¶	94
Typha-sparganium	Output	29.8	2.0	28.5	2.8	3.6	—	—	—	—	516	
	Net gain or loss	+179.7	+7.1	-4.8	+0.7	+0.5	—	—	—	—	-214	

Florida, USA[g] Swamp Cypress-gum Alluvial	Input	26.4	3.6	64.2	2.3	4.7	—	—	—	—	—	—
	Output	26.5	2.7	68.7	2.1	5.3	—	—	—	—	—	—
	Net gain or loss	-0.1	+0.9	-4.5	+0.2	-0.6	—	—	—	—	—	
North Carolina, USA[h] Swamp Red maple-ash gum Clay-silt	Input**	2.7	2.4	3.1	0.5	0.6	1.1	9.0	2.2	9.0	902[j]	112
	Output	0.2	0.1	0.5	0.01	0.02	2.6	7.8	2.6	8.3	541	
	Net gain or loss	+2.5	+2.3	+2.6	+0.49	+0.58	-1.5	+1.2	-0.4	+0.7	+361	
North Carolina, USA[i] Pocosin Pond pine-shrub Mineral	Input	3.0	4.9	14.2	—	0.8	1.9	12.6	3.9	18.7	2	116
	Output	0.4	0.6	3.6	—	0.2	2.4	11.4	6.8	29.6	94	
	Net gain or loss	+2.6	+4.3	+10.6	—	+0.6	-0.5	+1.2	-2.9	-10.9	-92	
North Carolina, USA[j] Pocosin Pond pine-shrub Organic	Input	3.0	4.7	13.7	—	1.2	1.9	13.6	2.5	13.3	2	103
	Output	0.1	0.6	6.0	—	0.4	2.8	7.6	2.9	20.1	263	
	Net gain or loss	+2.9	+4.1	+7.7	—	+0.8	-0.9	+6.0	-0.4	-6.8	-261	
New Hampshire, USA[k] Hardwood forest Beech-maple Sandy loam	Input	19.0	2.9	20.7	—	0.04[††]	0.9	2.2	0.6	1.6	1484	130
	Output	16.1	0.3	4.0	—	0.02	2.4	13.9	3.3	7.5	12	
	Net gain or loss	+2.9	+2.6	+16.7	—	+0.02	-1.5	-11.7	-2.7	-5.9	+1472	

[a] Verry and Timmons (1982)
[b] Crisp (1966)
[c] Viro (1953)
[d] Burke (1975)
[e] Malcolm and Cuttle (1983)
[f] Davis, et al. (1981)
[g] Elder (1985)
[h] Kuenzler, et al. (1977)
[i] Skaggs, et al. (1980)

[j] Kuenzler, et al. (1980). Values include biomass accretion and respiration losses for C.
[k] Likens, et al. (1977)
* Note: NO_3-N + NH_4-N. Data for this site includes eroding peat; C calculated as 45% of peat loss.
† Dissolved P comprised > 80% total P.
‡ Average of 3 years of data
§ Kjeldahl N and total P were not measured in precipitation; fertilizer levels of N were high upstream.
¶ Soluble organic carbon only
** Average of 1975–76 data. Inputs are only given in terms of rainfall.
†† PO_4-P only; C from this site includes biomass accretion.

Source: Richardson, 1989.

13.23

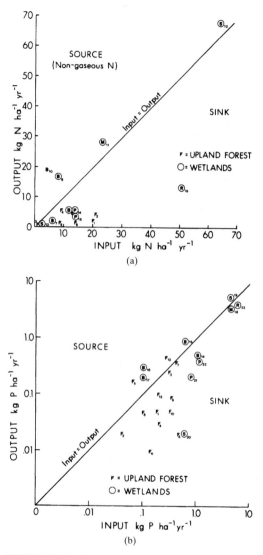

FIGURE 13.10*a–d* Input-output budgets for wetland and
forested ecosystems. (F = forested uplands; B = bogs; S =
swamps; P = pocosins; M = marshes; R = riparian.) Numbers
indicate the location of each data set. (*Likens, et al., 1977
for F, B, S, and P; Lowrence, et al., 1984 for R.*)

strate input due to weathering. Both wetlands and forests do not retain K or Na
effectively and are a source for downstream systems.

The large variation in carbon budgets shown for ecosystems in Table 13.2 is due
to the fact that some researchers included biomass accretion and some only measured soluble carbon inputs. It is clear that the outputs of C from freshwater wet-

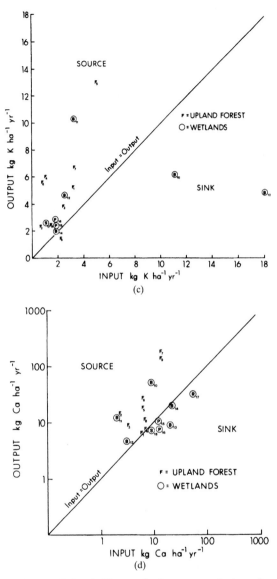

FIGURE 13.10*a–d (Continued)* Input-output budgets for wetland and forested ecosystems. (F = forested uplands; B = bogs; S = swamps; P = pocosins; M = marshes; R = riparian.) Numbers indicate the location of each data set. (*Likens, et al., 1977 for F, B, S, and P; Lowrence, et al., 1984 for R.*)

lands greatly exceed the yields from terrestrial ecosystems. The large output of C from the English bog is due to the high amount of erosion. The inputs to this system are unknown, but it is safe to say that outputs exceed 41,000 kg ha^{-1} yr^{-1}. The high exports of C from marsh, swamp, and pocosin bogs suggest that waters downstream from wetlands are subjected to high loadings of organic matter (i.e., wetlands are a source of dissolved organic carbon, humics, and fulvics) and that freshwater wetlands, like saltwater wetlands, are net exporters of C. A direct comparison of the organic carbon export from swamps and upland ecosystems as a function of the annual runoff shows that C export from swamp watersheds is more than five times that for uplands (Mulholland and Kuenzler, 1979; Fig. 13.11). There is a direct but different relationship (note slope differences on Fig. 13.11) between C export and runoff in swamplands versus uplands. The larger export of C per unit of runoff in swamplands was attributed to the combined effects of increased leaching of dissolved organic compounds from organic-rich soils and the concentrating effects of increased evapotranspiration in wetlands versus uplands (Mulholland and Kuenzler, 1979). The importance of soluble C and detritus material output from freshwater wetlands to downstream secondary productivity and/or water quality (e.g., increased biological oxygen demand) has not been fully quantified.

Only a few ecosystem-level studies have been done on sulfur retention and release. Terrestrial ecosystems, in contrast to wetlands, release much of the SO$_4$ received as precipitation (Likens, et al., 1977; Bayley and Schindler, 1986). Urban, et al., (1986) reported that a bog in northern Minnesota retained 56 percent of low-level SO$_4$ atmospheric inputs. Hemond (1980) calculated that nearly 77 percent of the high levels of atmospheric input of SO$_4$ were sequestered by a bog in New England. A fen system in Canada was found to retain an average of 51 percent of the

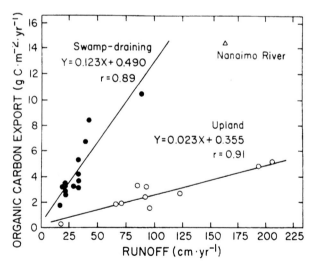

FIGURE 13.11　The annual export of organic carbon as a function of annual runoff for various upland and swamp ecosystems and the Nanaimo River watershed in Canada. The Nanaimo River watershed comprises both wetland and upland ecosystems. Regression equations are given for both uplands and swamp-draining systems. Carbon export can be predicted from runoff estimates X. (*Mulholland, 1980.*)

108 meq m^{-2} yr^{-1} of SO$_4$ additions (including acid treatments) over a 4-year period (Bayley, et al., 1986). However, retention varied from a low of 22 percent during dry summer periods, when O$_2$ diffused into the peat and reduced sulfides were oxidized to SO$_4$, to a high of 73 percent when the fen remained anaerobic. Wieder and Lang (1988) also observed that SO$_4$ was removed from acid mine drainage by a sphagnum bog in West Virginia. Wieder (1985) reported a significantly higher export of SO$_4$ during dry periods, and that on average 81 percent of the S stored in the bogs was as carbon-bonded S. These data suggest that wetlands function as effective sinks for SO$_4$ so long as anaerobic conditions are maintained. They become a source of SO$_4$ to downstream areas during dry periods. This change in the SO$_4$ source/sink function follows the change in soil oxidative status and suggests that wetlands are also effective transformers of S.

13.5.5 Hydrologic Function

The *hydrologic function of wetlands* on the landscape is determined by type (e.g., bog, fen, or swamp), topographical position (i.e., proximity to large or small streams or lakes), overall size, and connection to groundwater. Many wetlands, including bogs, pocosins, peatlands, and marshes, function as water pumps on the landscape, losing approximately two-thirds of their annual water by ET and leaving 30 percent or less for annual runoff or groundwater recharge (Fig. 13.12). The proportion of water inflow compared to rainfall varies greatly, with swamps receiving most of their inputs by stream flow. They also gain and lose some water to the shallow aquifer beneath them. Stream flow out of swamps comprises the biggest loss, although water can be held for months and released more slowly than in areas where the vegetation is removed. Bogs receive most of their water by rain and lose almost none to groundwater. The large portion of annual runoff from perched bogs (i.e., water table elevated above the regional water table) is in spring after winter snowmelt or after the bog peat has filled with rain due to low winter ET rates. The peat soils of these systems act much like a sponge on the landscape, holding and slowly releasing vast quantities of rainfall when the peat is drier, and rapidly giving up water when the sponge is full. Many small bogs have no flowing outlets and lose almost all their water by ET. Because of their size, often 20,000 ha or more, pocosins slowly lose water by surface and subsurface runoff; it is distributed over vast areas to the surrounding lakes, rivers, and estuaries. *Fen systems* are flow-through systems with runoff often exceeding ET, especially during ice-thaw periods in the spring. The annual flow from fens connected to groundwater is very uniform through the year.

Wetlands can be an important source or sink of regional surface and groundwater supplies kilometers from the actual wetland itself. Recent studies have shown that the relationship between the regional water table and wetlands is more complex and dynamic than once thought, and that seasonal reversals of groundwater flow among wetlands, and between wetlands and lakes, often occur (Winter, 1989). This suggests that the drainage of a wetland in one area may have serious consequences for water conditions in lakes, groundwater, other wetlands, or downstream areas that are not readily evident without detailed hydrologic studies. It has been demonstrated that peak stream flows are only 20 percent as large in basins with 40 percent lake and wetland area as in similar basins with no lake or wetland area (Novitzki, 1979). Recent studies have shown a relationship between the amount of wetland area on the landscape and stormwater flow peaks. These data suggest that when less than 10 percent of the watershed is in wetlands, significant peak flows occur (Johnston 1994). The effects of land-use changes on the hydrologic functions of wetlands

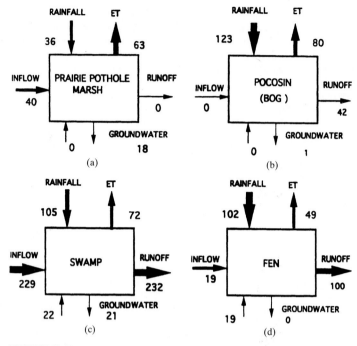

FIGURE 13.12   Annual water budgets for: (*a*) a prairie pothole marsh in North Dakota; (*b*) a pocosin bog in North Carolina; (*c*) an alluvial cypress swamp in southern Illinois; and (*d*) a rich fen in North Wales, U.K. All water-flow values are reported in cm yr^{-1}. (*Mitsch and Gosselink, 1993; Richardson, 1994.*)

are presented next as a case study for the southeastern United States. This study is based on analysis recently published in *Wetlands* (Richardson and McCarthy, 1994).

13.6   THE EFFECTS OF LAND-USE PRACTICES ON WETLAND HYDROLOGIC FUNCTIONS: A SOUTHEASTERN CASE STUDY

The way in which wetlands are managed or altered has a major impact on the hydrologic fluxes and routes of water loss from ecosystems. Many wetlands, including the marshes, bogs, pocosins, and peatlands of the United States, function as water pumps on the landscape, losing over two-thirds of their annual water input by evapotranspiration (ET); leaving only 30 percent or less of annual input for runoff or groundwater recharge (Richardson and Gibbons, 1993). An earlier watershed analysis of water budgets and flow-duration curves by Daniel (1981) for coastal peatlands showed that construction of drainage channels in wetlands, without further alterations in land or vegetation cover, resulted in increased peak flows and low flows at the expense of midrange flows, but not total annual flows. Skaggs, et al. (1991) found peak runoff rates from developed fields (130 ha) in eastern North Carolina were two to four times natural wetlands (25 L/s versus 100 L/s), the exact ratio of runoff depending on the size of the experimental unit, land use, and antecedent soil water conditions. Skaggs, et al. (1991) also reported that the higher peak outflows occurred sooner on developed sites. In addition, with drainable porosity of coastal wetland soils being 1 to 7 percent (Skaggs, 1978), water tables could rise 200 mm with only a 10-mm infiltration into soil from rainfall (Skaggs, et al., 1991). Moreover, drainage modifications coupled with a radical change in vegetation from forest to agriculture resulted in an increase in total runoff due to a decrease in canopy interception and a shorter residence time for water in the watershed. Precipitation falling on wetlands can be intercepted by the plant canopy, with the rate of interception varying with canopy density, intensity and duration of rainfall, season of the year, and other climate conditions (Kozlowski, 1983; McCarthy and Skaggs, 1992). Rutter, et al. (1971) estimated a 20 to 40 percent interception of precipitation in conifer plantations, while McCarthy (1990) reported values of 10 to 30 percent for a drained forested pine plantation watershed on the coast of North Carolina. Interception of rainfall by agricultural plants is seasonally dependent (i.e., fallow versus covered fields) as well as crop specific (corn versus soybeans or wheat).

Forest management practices in southern plantation pine forests typically consist of site preparation, planting, thinning, fertilization, and harvesting (Campbell and Hughes, 1981). Agricultural practices include extensive drainage and site preparation, as well as annual harvesting, and fertilizer and pesticide additions (Barnes, 1981). In both land uses on flat, poorly drained soils, some form of water management (e.g., free drainage or controlled drainage) is often practiced. An extensive set of experimental field trials since the late 1970s has been used to test and refine the water management model DRAINMOD, which has been used to predict hydrologic responses for various land-use and wetland alterations on the coast of North Carolina (Skaggs, 1978; Skaggs, et al., 1980; Gregory, et al., 1984; McCarthy, 1990; and Amatya, 1993). For an intensive review of the DRAINMOD model algorithms, sensitivity analyses and management alternatives tested, see the papers noted. In this case study analysis, a field tested model (DRAINMOD) was used to: (1) estimate the effects of wetland drainage and soil and vegetation removal on hydrologic response and (2) compare natural wetland water losses with losses resulting from land clearing, agriculture, forestry, and peat-mining activities on deep peat wetlands. An earlier and more complete version of this analysis is given in Richardson and McCarthy (1994).

13.6.1 Modeling Approach

The computer model DRAINMOD was developed as a tool to facilitate the design of agricultural water-management systems (Skaggs, 1978; 1980). The model is based on physical processes and requires inputs describing site surface conditions, soil properties, vegetation, and meteorology. It was designed for application to shallow-water-table mineral soils, but has been modified and tested for peat soils by Gregory, et al. (1984), and forestry applications (McCarthy, 1990; 1992; Amatya, 1993). The model has been specifically field tested and calibrated against watershed runoff for pocosin peatlands subjected to agriculture and peat mining in Dare and Hyde Counties in North Carolina (Skaggs 1980; Gregory, et al., 1984; Gale and Adams, 1984), and forest management activities on drained mineral soils (McCarthy, 1990; Amatya, 1993). The model has been described in detail in Skaggs (1978; 1980) and Gregory, et al. (1984); thus only the salient features are presented as they pertain to the simulation of wetland development conditions.

13.6.2 DRAINMOD: Model

This field-scale model has the capability of simulating, on a day to day, hour by hour basis, the surface runoff, subsurface drainage, evapotranspiration, soil water content, and water table position using climatological and soil property input data under both natural and assigned water-management plans (Gregory, et al., 1984). The model is based on a soil-water balance for a column of soil which extends from the impermeable layer to the surface, where the water balance for a given time frame is as follows:

$$V_a = D + ET + DS - F$$

where V_a is the change in the air volume; D is the drainage from the column; ET is the actual evapotranspiration; DS is the deep seepage; and F is the infiltration entering the section in time t. All values have units of cm^3/cm^2 or cm (Skaggs, 1978; Skaggs, et al., 1980). The surface runoff and surface storage are computed with the following water balance equation:

$$RO = P - F - S$$

where RO is the surface runoff, P is the precipitation, F is the infiltration, and S is the change in surface storage during time t. All values are in cm and the time increment used in calculation is 1 h. For a more complete description of the model and details of field tests and validation see Skaggs, et al. (1980) and Gregory, et al. (1984). The specific conditions modeled in the analysis are given in Richardson and McCarthy (1994).

13.6.3 Natural Conditions on Peat Soils

13.6.3.1 *Precipitation* **P** *and Potential Evapotranspiration (PET)*. The average annual rainfall for the Hyde County region of coastal North Carolina is 123 cm over the last 20 years (Wiser, 1975). The average monthly distribution of rainfall for that period is shown in Fig. 13.13. On average, the wettest months are July and August. Simulated PET exceeds or matches rainfall only during June but is close to rainfall in both July and August. Average monthly rainfall compared to PET, when integrated over the whole year, indicates that 80 cm of the average annual 123-cm input is discharged from vegetated pocosin sites as ET. Thus, nearly 66 percent of annual rainfall leaves a natural pocosin ecosystem as ET, and 34 percent leaves as runoff, as groundwater recharge, or is stored in the peat.

13.6.3.2 *Groundwater Discharge (GWD)*. The organic histosols of the North Carolina coastal plain in Hyde County are underlain by poorly permeable subsurface layers of clay and sand (Foutz, 1983). The predicted flow rate of deep seepage to the Castle Hayne aquifer was determined to be only 0.043 cm/yr (Foutz, 1983). Badr (1978) also studied this region and reported that surface-water flow from the organic soils to groundwater was only 0.02 cm/yr. Groundwater discharge rates representative of the coastal region are very low and have been estimated to be approximately 1 cm/yr (Heath, 1975; Daniel, 1981). Groundwater discharges are thus estimated to be <1 percent of the annual water budget of regional wetlands and play an insignificant role in the annual water flux.

13.6.3.3 *Runoff (RO) and Water Storage*. A simulation of runoff for a natural (mature) pocosin ecosystem on deep peat indicates that runoff is highest (>7 cm)

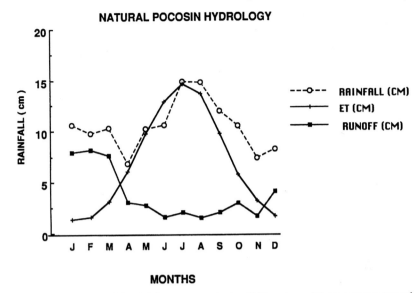

NATURAL POCOSIN HYDROLOGY

MONTHS

FIGURE 13.13 Monthly rainfall, evapotranspiration (ET), and runoff from a mature natural pocosin site on deep peat in Hyde County, N.C. Values represent monthly averages from a 20-year simulation with DRAINMOD. (*Richardson and McCarthy, 1994.*)

during the winter months and lowest during the summer months (<2.5 cm) (Fig. 13.13). Given the previous average hydrologic rainfall and ET values, mean annual runoff from natural pocosin areas can be estimated as:

$$RO = P - ET - GWD$$

$$RO = 123 \text{ cm} - 80 \text{ cm} - 1 \text{ cm}$$

$$RO = 42 \text{ cm/yr}$$

By comparison Skaggs, et al. (1991) calculated an average annual average outflow for a natural pocosin area on a Portsmouth sandy loam near Wilmington, N.C. to be 36.5 cm/yr. However, the annual variation in rainfall and ET over several decades shows that runoff from peatlands can vary considerably from average conditions during periods of extreme rainfall or drought (Fig. 13.14). Following several years of high rainfall and reduced soil storage capacity, runoff nearly reached 75 cm/yr in the mid 1950s, while runoff dropped to near 10 cm/yr following the lower rainfall periods and higher ET of the 1970s. By comparison, drained pocosin, forest plantation, and agriculture on the same Portsmouth soil discharged on average 39.2, 35.6, and 42.1 cm/yr over the 32-year period from 1950 to 1982. Skaggs reported runoff ranging from 8.4 to 65.9 cm/yr for the Portsmouth sites over the 32-year simulation. Thus, high year-to-year variations in runoff can be expected on both peat and mineral soils, depending on climatic variability.

A predicted water-table depth profile for peat soils in Hyde County in 1977, considered representative of an average year, shows that the water table remains above or near the surface until day 85 (Fig. 13.15). Over the next 100 days it remains 10 to

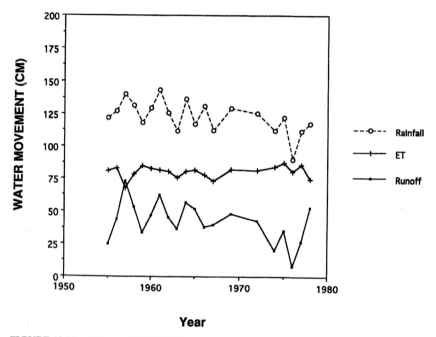

FIGURE 13.14 A 20-year DRAINMOD simulation of annual rainfall, evapotranspiration (ET), and runoff for a mature natural area with peat soils in coastal North Carolina. (*Richardson and McCarthy, 1994.*)

15 cm below the surface. From July until October the water-table depth dropped to nearly 100 cm below the surface, with the exception of one major rainfall event which drove the water table up to within 20 cm of the surface. This rapid water-level response is due to the limited pore space available in hydric soils (Skaggs, 1978; 1980). By the end of the year, the water table was within a few centimeters of the surface. Similar seasonal patterns were found in the mineral soils of the Portsmouth soils (Skaggs, et al., 1991).

13.6.4 Runoff from Altered Peat Soils

The annual runoff for a 404-ha block of land in Hyde County, N.C. under natural, disturbed conditions where all the vegetation was removed from the peatlands, during mining activities where 30 cm of peat surface was removed, under agriculture, and under managed forestry (both early years 1 to 3 and closed canopy conditions) is shown in Fig. 13.16. The conditions used in these simulations are given in Richardson and McCarthy (1994).

This 404-ha area was utilized as a test-size watershed for the normalization of field measurements and model analysis. The total runoff for any size area with comparable drainage, soils, and plant cover can then be estimated and compared. The highest annual average simulated runoff was 63 cm/yr (2.5×10^6 m^3/yr), which was

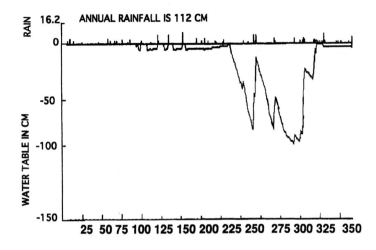

FIGURE 13.15   A DRAINMOD simulation of water-table depth for a natural pocosin site on deep peat with mature natural vegetation in coastal North Carolina, in 1977. (*Richardson and McCarthy, 1994.*)

the water discharge projected during the period of active mining on 404 ha of wetlands land. Average annual outflow would exceed natural conditions by 50 percent.

Runoff was simulated for a 20-year period for 30 cm intervals of peat removal (i.e., removal of 30, 60, 90, and 120 cm of peat) during each mining phase (Fig. 13.16). Thus, the model indicates that continued peat removal would not result in any increased runoff above initial measured increases at the first 30 cm of removal. This is due to the similarity in hydraulic conductivity conditions below the 30-cm level and a similar ET value for disturbed peatlands (Gregory, et al., 1984). The next-highest level of simulated runoff was 47.5 cm/yr (1.9×10^6 m^3/yr) from land that is classified as disturbed (i.e., ditched and vegetation partly or totally removed). Runoff for disturbed landscapes exceeds natural conditions by 13 percent annually.

Runoff was simulated for a 20-year period on reclaimed peat areas with a land cover characterized by row crop agriculture with a wheat/soybean rotation. The average annual runoff was 45.7 cm/yr (1.8×10^6 m^3/yr) (Fig. 13.17). This output exceeded natural levels of runoff by 9 percent. Simulated runoff from pine plantations for initial conditions during the first 3 to 5 years of growth (low cover percentage and reduced ET) and during increased biomass and closed canopy conditions (i.e., increased ET) were 42.7 cm/yr (1.7×10^6 m^3/yr), and 34.1 cm/yr (1.3×10^6 m^3/yr), respectively. Average early pine plantation runoff was only 2 percent higher than natural runoff levels, but by year 7 forestry runoff was reduced 19 percent below natural wetland projected runoff. The general trends of agriculture land use increasing runoff above natural conditions on Portsmouth sandy loam soils, and forest plantations decreasing runoff compared to natural systems was also shown by Skaggs, et al. (1991). However, agricultural runoff was 15 percent greater than natural outflow, and pine plantations only reduced outflow by 3 percent. The different results for each land-use scenario relate directly to the variations in ET and water interception among cover types as well as the physical difference of peat versus sandy loam soils.

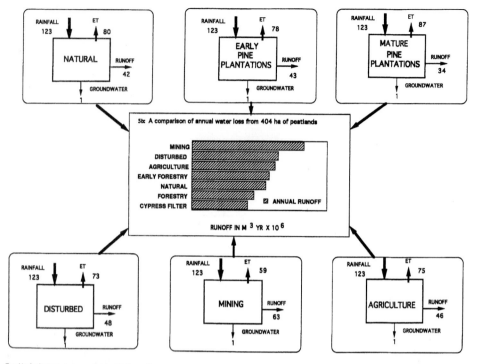

5a: Hydrologic inputs and outputs in cm/yr

FIGURE 13.16 Hydrologic inputs and outputs by land-use type in cm/yr. *Note:* Output values were rounded to nearest whole number and may not exactly match inputs. (*Richardson and McCarthy, 1994.*)

13.6.5 Monthly Runoff Comparisons by Land Use Type

A comparison of monthly runoff rates for each block of land for natural, agriculture, mining, and silviculture land use is shown in Fig. 13.17. The highest simulated runoff (>9 cm/month) for all land-use types occurred during the winter months and the lowest runoff (<1 cm/month) was during the summer months. The highest runoff in the winter months was from mining and agriculture land uses. However, runoff from mining and agriculture was only 1.2 cm above natural site rates during this period. The lowest runoff, <2.5 cm/month for most of the year, was in the silviculture-reclaimed areas. This suggests that reclamation of mined areas with forestry would result in a significant reduction in total monthly and annual runoff. The next-best land-use type in terms of reduced runoff was natural pocosin areas (Fig. 13.17). The highest predicted runoff during the summer months was from the mining sites, where ET was significantly reduced. Agriculture sites also exhibited higher output than natural sites, especially during the fall months. The relatively high runoff during mining would require that outflow in excess of natural conditions be managed in most state water programs due to mining discharge regulations. By contrast, agricultural runoff is not currently regulated in most states.

20 YEAR SIMULATION OF MONTHLY RUNOFF

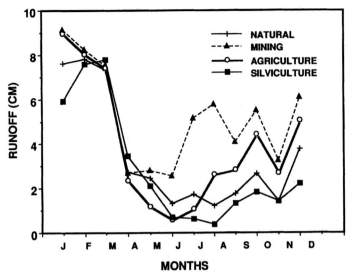

FIGURE 13.17   A comparison of simulated monthly runoff levels for natural, mining, agriculture, and silviculture land use. (*Richardson and McCarthy, 1994.*)

13.6.6   Storm Events by Land-Use Types

13.6.6.1   Flow Duration Curves.   While the data presented in Fig. 13.17 are useful for depicting expected monthly runoff, these data provide little indication of peak flows for various land use conditions. One technique which may be used to estimate the frequency of flows at a given point is the flow duration curve (Dunne and Leopold, 1978). The flow duration curves presented in Fig. 13.18 indicate the proportion of the time that a flow rate might be expected to equal or exceed a certain value. High flow periods are an important consideration in water management, since periods of high flow are likely to result in soil erosion and increased nutrient loss. Therefore, it is important to develop a quantitative relationship which shows the cumulative frequency of high flow periods.

Data for the flow duration curves were obtained by using DRAINMOD to simulate 20 years of daily unit area runoff for each of the scenarios listed in the preceding section. The daily unit area runoff was averaged over the 20-year period of record on a daily basis, yielding 365 observations of average daily runoff. Average daily runoff in cm/day was then converted to flow, in L/s, for a 404-ha tract of land. The described procedure provided data to simulate expected canal flow rates (Gregory, et al., 1984).

After obtaining average daily flow rates, data were sorted in ascending order and a frequency histogram was developed. Flow duration curves presented in Fig. 13.18 depict the recurrence interval of exceeding a certain flow for natural, mining, agriculture, and forest management conditions. Flow volumes above 275 L/s occur roughly 2.5 percent of the time, regardless of the condition simulated. Hence, for cer-

A COMPARISON OF FLOW DURATION CURVES

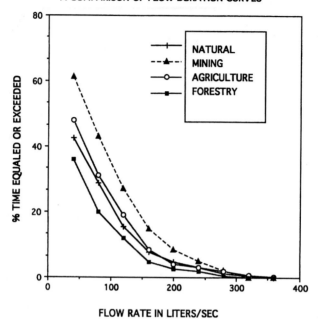

FIGURE 13.18 Flow duration curves for four land-use types on peat-lands in coastal North Carolina. Data simulated over a 20-year period by DRAINMOD. (*Richardson and McCarthy, 1994.*)

tain periods, high flows may be expected even under natural conditions. Mining conditions displayed higher flows than all other flows simulated. As an example, outflows of 195 L/s occurred 10 percent of the time. By comparison, flow volumes of 130 L/s would be exceeded only 10 percent of the time when the area is utilized for silviculture. One can postulate that reasons for the increased flow during the mining phase include removal of natural vegetation, resulting in decreased evapotranspiration, and removal of soil layers high in hydraulic conductivity.

Also, the flows under natural conditions and during the agricultural phase are coincident for flows which may be expected less than 10 percent of the time; that is, the high flow periods. For more typical rates of flow, that is, flows that may be equaled or exceeded 20 to 50 percent of the time, flow from agricultural blocks of land only slightly exceeds flow under natural conditions. In addition, at the points where flow during the agricultural phase exceeds that under natural conditions, flows will be under 160 L/s.

Of all the scenarios that were simulated using DRAINMOD, for any given probability of exceedence, flow under forested conditions was the lowest. As an example, consider flows which may be expected to be exceeded 40 percent of the time. Under natural, premined conditions, flows of just under 50 L/s may be expected. Under forested conditions, flows of just under 30 L/s may be equaled or exceeded 40 percent of the time.

To summarize, under all conditions simulated, even natural, there is a certain (albeit small) probability that high flows may be expected from wetlands. Under forest management there will be a net reduction in freshwater runoff from the altered wetlands. Peak flows will also be greatly reduced. Agricultural conversion of wetlands also increases annual and peak runoff. Mining activities greatly increase runoff at the expense of ET.

13.7 SUMMARY AND CONCLUSIONS

It is clear that ecologists need to do a better job of quantifying wetland functions and relating them to human values, so that we can at least determine what we will lose when we alter or develop wetlands. The key ecological question to be addressed under any development scenario is whether wetland functions have been significantly altered. For example, if water storage from a wetland was the important process on the landscape, then the question becomes: What will it cost to replace that function? Moreover, can the wetland function be replaced at all? Replacement costs are often far higher than the cost of selecting another area or avoiding the wetland functional loss. The billions of dollars of property and crop loss along the Mississippi and Missouri Rivers during the record summer floods of 1993 is a case in point. States like Iowa and Missouri, which have developed nearly 90 percent of their wetlands, suffered record flooding levels due to a lack of water storage capacity in natural wetlands as well as uncontrolled development in the floodplains themselves. Once we understand all their ecological functions, these former wastelands will become ever-more-valuable wetlands. Specific ideas that emerge from this review are as follows:

- Wetlands are among the most productive ecosystems in the world. They provide valuable habitat for many species, and they are an essential ecosystem on the landscape in terms of their unique functions.

- The terms *filter* and *sink* are often used interchangeably in the literature. These terms are not synonymous. This suggests that ecosystem scientists have introduced confusion by suggesting that wetlands may function as nutrient filters on the landscape, when in fact they mean that wetlands may function as nutrient sinks or traps.

- Wetlands do filter 60 to 90 percent of the suspended solids from wastewater-addition studies, and riparian areas have been shown to retain 80 percent of sediment runoff from adjacent agriculture lands.

- Wetlands do function as effective transformers (i.e., primary oxidation/reduction systems on the landscape) of N, P, and C. Their role in sulfur transformations is also important, but few studies exist on S cycling in freshwater wetlands (Deevey, 1970; Weider and Lang, 1986). These ecosystems do have the unique characteristic of transforming and releasing significant quantities of N_2 and some H_2S gas to the atmosphere.

- Wetlands can function as either a sink or source, depending on the element in question, type of wetland, loading level, season of the year, and whether or not the ecosystem is aggrading. Wetlands are not efficient sinks for either K or Na. Wetlands, especially bogs, marshes, and fens with organic soils, do not retain P as efficiently as forested systems and are often a source rather than a sink. They do often contain vast stores of nutrients, i.e., N, P, K, and C, but these have been slowly collected over hundreds—and in the case of peatlands, thousands—of years. Bogs appear to be a sink for TN and carbon-bonded S.

- Aggrading wetlands are a net sink for C but they export a significant amount of C via hydrologic outflow, especially when compared to terrestrial ecosystems. Also, the nutrient load carried with organic material in outflow waters can make up the vast majority of elemental losses from wetlands, but has not been measured by many researchers.

- The greatest function of wetlands may be as transformers, since these ecosystems maintain the widest range of oxidation/reduction reactions of any ecosystem on the landscape.

- Generally, peat-based wetlands like pocosins or the Everglades lose 66 percent of annual rainfall as ET, with 34 percent leaving as diffuse runoff. Development for agriculture, forestry, and peat mining will result in a shift in the hydrologic output from ET to point-source runoff.

- Agriculture, land-clearing, and peat-mining simulations suggest that these land use activities will increase annual runoff over natural undrained wetlands by 9, 13, and 50 percent respectively. Closed-canopy forest pine plantations will reduce runoff by 19 percent over natural conditions.

A technical challenge lies ahead in developing watershed management methodology and techniques that consider long-term hydrologic behavior, spatial arrangement of wetlands on the landscape, and the effects of wetland development on hydrologic functions. This article hopefully provides some insights, as well as estimates of the changes in wetland hydrologic and biogeochemical functions that can be expected under varying land-use practices.

Acknowledgments

This chapter is a revised, greatly enhanced, and updated version of a recently published chapter on freshwater wetlands which appeared in *Encyclopedia of the Environment,* vol. 3, Academic Press. In addition, the section on hydrology and biogeochemical analysis were abstracted and updated from earlier versions published in *Wetlands* and the volume *Freshwater Wetlands and Wildlife,* published by DOE in 1989.

REFERENCES

Amatya, D. M., "Hydrologic Modeling of Drained Forested Lands," Ph.D. dissertation, North Carolina State University, Raleigh, N.C., 1993.

Ambus, P., and R. Lowrance, "Comparison of Denitrification in Two Riparian Soils," *Soil Science Society of America,* 55:994–997, 1991.

Armentano, T. V., "Drainage of Organic Soils as a Factor in the World Carbon Cycle," *Biosci.,* 30:825–830, 1980.

Badr, A. W., "The Physical Properties of Some North Carolina Organic Soils and the Effect of Land Development on These Properties," M.S. thesis, Department of Biological and Agricultural Engineering, North Carolina State University, Raleigh, N.C., 1978.

Barling, R. D., and I. D. Moore, "Role of Buffer Strips in Management of Waterway Pollution: A Review," *Environmental Management,* 18(4):543–558, 1994.

Barnes, S., "Agricultural Adaptability of Wet Soils of the North Carolina Coastal Plain," in C. J. Richardson, ed., *Pocosin Wetlands,* Hutchinson Ross, Stroudsburg, Pa., 1981.

Bartlett, M. S., L. C. Brown, N. B. Hanes, and N. H. Nickerson, "Denitrification in Freshwater Soil," *J. Environ. Qual.,* 111:460–464, 1979.

Bastian, R. K., and S. C. Reed, "Aquaculture Systems for Wastewater Treatment," EPA 430/90-80-006, U.S. Environmental Protection Agency, Washington, D.C., 1979.

Bayley, S. E., and D. W. Schindler, "Sources of Alkalinity in Precambrian Shield Watersheds under Natural Conditions and After Fire or Acidification, Effects of Acid Deposition on Forests, Wetlands, and Agricultural Ecosystems," in T. Hutchinson and K. Meema, eds., *Proceedings of the Workshop,* Springer-Verlag, New York, 1986.

Bayley, S. E., R. S. Behr, and C. A. Kelly, "Retention and Release of S from a Freshwater Wetland," *Water, Air, and Soil Pollution,* 31:101–114, 1986.

Bord na Mona, "Fuel Peat in Developing Countries," Study report prepared for the World Bank, Dublin, Ireland, 1984.

Boto, K. G., and W. H. Patrick, Jr., "Role of Wetlands in the Removal of Suspended Sediments," in P. E. Greeson, Jr., R. Clark, and J. E. Clark, eds., *Wetland Function and Values: The State of Our Understanding,* American Water Resources Association, Minneapolis, Minn., 1979.

Bowden, W. B., "The Biogeochemistry of Nitrogen in Freshwater Wetlands," *Biogeochemistry,* 4:313–348, 1987.

Bridgham, S. D., C. A. Johnston, J. Pastor, and K. Updegraff, "Potential Feedbacks of Northern Wetlands on Climate Change," *BioScience,* 45(4):262–274, 1995.

Brinson, M. M., "A Hydrogeomorphic Classification for Wetlands," Technical Report WRP-DE-4, U.S. Army Corps of Engineers, Waterways Experiment Station, Vicksburg, Miss., 1993.

Brown, C. A., "Morphology and Biology of Cypress Trees," K. C. Ewel and H. T. Odum, eds., in *Cypress Swamps,* Univ. of Florida, Press, Gainesville, 1984, pp. 16–24.

Burke, W., "Fertilizer and Other Chemical Losses in Drainage Water from a Blanket Bog," *Irish J. Agric. Res.,* 14:163–178, 1975.

Campbell, R. G., and J. H. Hughes, "Forest Management Systems in North Carolina Pocosins," in C. J. Richardson, ed., *Pocosin Wetlands,* Hutchinson Ross, Stroudsburg, Pa., 1981, pp. 199–213.

Cooper, A. B., "Nitrate Depletion in the Riparian Zone and Stream Channel of a Small Headwater Catchment," *Hydrobiologia,* 202:13–26, 1990.

Cooper, J. R., J. W. Gilliam, R. B. Daniels, and W. P. Robarge, "Riparian Areas as Filters for Agricultural Sediment," *Soil Sci. Soc. Am. J.,* 51:416–420, 1987.

Cooper, J. R., J. W. Gilliam, and T. C. Jacobs, "Riparian Areas as a Control of Nonpoint Pollutants," in D. L. Correll, ed., *Watershed Research Perspectives,* Smithsonian Institution Press, Washington, D.C., 1986, pp. 166–192.

Correll, D. L., ed., *Watershed Research Perspectives,* Smithsonian Institution Press, Washington, D.C., Smithsonian Environmental Research Center, Edgewater, Md., 1986.

Cowardian, L. M., V. Carter, F. C. Golet, and E. T. LaRoe, "Classification of Wetlands and Deepwater Habitats of the United States," U.S. Fish & Wildlife Service Pub. FWS/OBS-79/31, Washington, D.C., 1979.

Craft, C. B., and C. J. Richardson, "Peat Accretion and N, P, and Organic C Accumulation in Nutrient-Enriched and Unenriched Everglades Peatlands," *Ecological Applications,* 3(3):446–458, 1993a.

Craft, C. B., and C. J. Richardson, "Peat Accretion and Phosphorus Accumulation along a Eutrophication Gradient in the Northern Everglades," *Biogeochemistry,* 22:133–156, 1993b.

Crisp, D. T., "Input and Output of Minerals for an Area of Penine Moorland: The Importance of Precipitation, Drainage, Peat Erosion, and Animals," *J. Appl. Ecol.,* 44:327–348, 1966.

Dahl, T. E., "Wetlands Losses in the United States, 1780s to 1980s," U.S. Department of the Interior, Fish and Wildlife Service, Washington, D.C., 1990.

Dahl, T. E., and C. E. Johnson, "Wetlands Status and Trends in the Conterminous United States Mid-1970s to mid-1980s," U.S. Department of Interior, Fish and Wildlife Service, Washington, D.C., 1991.

Daniel, C. C., III, "Hydrology, Geology and Soils of Pocosins: A Comparison of Natural and Altered Systems," in C. J. Richardson, ed., *Pocosin Wetlands,* Hutchinson Ross, Stroudsburg, Pa., 1981, pp. 69–108.

Davis, C. B., J. L. Baker, A. G. van der Valk, and C. E. Beer, "Prairie Pothole Marshes as Traps for Nitrogen and Phosphorus in Agricultural Runoff," in B. Richardson, ed., *Selected Proc. of the Midwest Conf. on Wetland Values and Management,* Freshwater Soc. Pub., St. Paul, Minn., 1981, pp. 153–164.

Deevey, E. S., Jr., "In Defense of Mud," *Bull. of the Ecol. Soc. of America,* 51:5–8, 1970.

Dunne, T., and L. B. Leopold, *Water in Environmental Planning,* W. H. Freeman, New York, 1978.

Elder, J. F., "Nitrogen and Phosphorus Speciation and Flux in a Large Florida River-Wetland System," *Water Resour. Res.,* 21:724–732, 1985.

Faulkner, S. P., and C. J. Richardson, "Physical and Chemical Characteristic of Freshwater Wetland Soils," in D. A. Hammer, ed., *Constructed Wetlands for Wastewater Treatment: Municipal, Industrial and Agricultural,* Lewis Publishers, Chelsea, Mich., 1989, pp. 41–72.

Foutz, T. L., "Effects of Peat Mining on Ground Water Presses," M.S. thesis, North Carolina State University, Raleigh, N.C., 1983.

Gale, J. A., and D. A. Adams, "Cumulative Impacts of Peat Mining Final Project Report," Coastal Energy Impact Program Report no. 40, Raleigh, N.C., 1984.

Gilliam, J. W., "Riparian Wetlands and Water Quality," *J. of Environmental Quality,* 23:896–900, 1994.

Godfrey, P. J., E. R. Kaynor, S. Pelczarski, and J. Benforado, eds., *Ecological Considerations in Wetlands Treatment of Municipal Wastewaters,* Van Nostrand Reinhold, New York, 1985.

Gopal, B., et al., "Definition and Classification," in B. C. Pattern, et. al., eds., *Wetlands and Shallow Continental Water Bodies,* vol. 1, SPB Academic Publishing, The Hague, Netherlands, 1990, pp. 9–15.

Gopal, B., and V. Masing, "Biology and Ecology," in B. C. Patten, ed., *Wetland and Shallow Continental Water Bodies,* SPB Academic Publishing, The Hague, Netherlands, 1990, pp. 91–239.

Gopal, B., R. E, Turner, R. G. Wetzel, and D. F. Whighman, eds., *Wetlands-Ecology and Management,* National Institute of Ecology and International Scientific Publications, Jaipur, India, 1982.

Gorham, E., "Northern Peatlands: Role in the Carbon Cycle and Probable Responses to Climatic Warming," *Ecological Applications,* 1:182–195, 1991.

Greeson, P. E., J. R. Clark, and J. E. Clark, eds., *Wetland Functions and Values: The State of Our Understanding,* American Water Resources Association, Minneapolis, Minn., 1978.

Gregory, J. D., R. W. Skaggs, R. G. Broadhead, R. H. Culbreath, J. R. Bailey, and T. Foutz, "Hydrologic and Water Quality Impacts of Peat Mining in North Carolina," UNC Water Resources Research Institute Report no. 214, Raleigh, N.C., 1984.

Hammer, D. A., *Constructed Wetlands for Wastewater Treatment,* Lewis, Chelsea, Mich., 1989.

Hanson, G. C., P. M. Groffman, and A. J. Gold, "Denitrification in Riparian Wetlands Receiving High and Low Groundwater Nitrate Inputs," *J. of Environmental Quality,* 23:917–922, 1994.

Heath, R. C., "Hydrology of the Albemarle-Pamlico Region, North Carolina. A Preliminary Report on the Impact of Agricultural Developments," U.S. Geological Survey, Water Resource Investigations 9-75, 1975.

Hemond, H. F., "Biogeochemistry of Thoreau's Bog, Concord, Massachusetts," *Ecol. Monogr.,* 50:507–526, 1980.

Hemond, H. F., "The Nitrogen Budget of Thoreau's Bog," *Ecol.,* 64:99–109, 1980.

Herskowitz, T., "Listowel Artificial Marsh Project," Ministry of Canada Project no. 128RR, Ontario, Canada, 1986.

HISARS, "Hydrologic Information and Retrieval Systems Reference Manual," in E. H. Wiser, ed., North Carolina Technical Bulletin no. 215, Raleigh, N.C., 1975.

Jacks, G., A. Joelsson, and S. Fleischer, "Nitrogen Retention in Forest Wetlands," *Ambio,* 23(6):358–362, 1994.

Johnston, C. A., "Cumulative Impacts to Wetlands," *Wetlands,* 14:49–55, 1994.

Kelly, J. R., and M. A. Harwell, "Comparisons of the Processing of Elements by Ecosystems I: Nutrients," in P. J. Godfrey, E. R. Kaynor, S. Pelczarski, and J. Benforado, eds., *Ecological Considerations in Wetlands Treatment of Municipal Wastewater,* Van Nostrand Reinhold, New York, 1985, pp. 137–157.

Kozlowski, T. T., *Water Deficits and Plan Growth,* vol. III: *Additional Woody Plants,* Academic Press, New York, 1983.

Kuenzler, E. J., P. J. Mulholland, L. A. Ruley, and R. P. Sniffen, "Water Quality in North Carolina Coastal Plain Streams and Effects of Channelization," Water Resour. Res. Inst. Rep., 127:1–160, University of North Carolina, 1977.

Kuenzler, E. J., P. J. Mulholland, L. A. Yarbro, and L. A. Smock, "Distributions and Budgets of Carbon, Phosphorus, Iron and Manganese in a Floodplain Swamp Ecosystem," Water Resour. Res. Inst. Rep., 157:1–234, University of North Carolina, 1980.

Likens, G. E., F. H. Borman, R. S. Pierce, J. S. Eaton, and N. M. Johnson, *Biogeochemistry of a Forested Ecosystem,* Springer-Verlag, New York, 1977.

Lowrance, R., R. Todd, J. Fail., Jr., Hendrickson, Jr., R. Leonard, and L. Asmussen, "Riparian Forests as Nutrient Filters in Agricultural Watersheds," *Biosci.,* 34:374–377, 1984.

Malcolm, D. C., and S. P. Cuttle, "Fertilizers Applied to Drained Peat 1: Nutrient Losses in Drainage," *Forestry,* 56:155–176, 1983.

Maltby, E., *Waterlogged Wealth,* Earthscan, London, 1986.

Maltby, E., and R. E. Turner, "Wetlands of the World," *Geog. Mag.,* 55:12–17, 1983.

McCarthy, E. J., "Modification, Testing and Application of a Hydrologic Model for a Drained Forest Watershed," doctoral dissertation, North Carolina State University, Raleigh, N.C., 1990.

McCarthy, E. J., and R. W. Skaggs, "Simulation and Evaluation of Water Management Systems for a Pine Plantation Watershed," *Southern J. Appl. For.,* 16(1):44–52, 1992.

Mitsch, W. J., and J. G. Gosselink, *Wetlands,* 2d ed., Van Nostrand Reinhold, New York, 1993.

Moshiri, G. A., *Constructed Wetlands for Water Quality Improvement,* CRC Press, Boca Raton, Fla., 1993.

Mulholland, P. J., "Organic Carbon Cycling and Export," in E. J. Kuenzler, P. J. Mulholland, L. A. Yarbro, and L. A. Smock, eds., "Distribution and Budgets of Carbon, Phosphorus, Iron and Mananese in a Floodplain Swamp Ecosystem, Water Resour. Res. Rep.," 157:15–89, University of North Carolina, 1980.

Mulholland, P. J., and E. J. Kuenzler, "Organic Carbon Export from Upland and Forested Wetland Watersheds," *Limnol. Oceanogr.,* 24:960–966, 1979.

National Research Council, *Wetlands: Characteristics and Boundaries,* National Academy of Sciences, Washington, D.C., 1995.

Novitzki, R. P., "Hydrologic Characteristics of Wisconsin's Wetlands and Their Influence on Floods, Stream Flow, and Sediment," in P. E. Greeson, J. R. Clark, and J. E. Clark, eds., *Wetland Functions and Values: The State of Our Understanding,* American Water Resource Association, Minneapolis, Minn., 1979, pp. 377–388.

Pakarinen, P., "Bogs as Peat-Producing Ecosystems," *Intern. Peat Soc. Bull.,* 7:51–54, 1975.

Patrick, W. H., Jr., and K. R. Reddy, "Nitrification-Denitrification Reactions in Flooded Soils and Water Bottoms: Dependence on Oxygen Supply and Ammonium Diffusion," *J. Environ. Qual.,* 5:469–472, 1976.

Qualls, R. G., and C. J. Richardson, "Forms of Soil Phosphorus Along a Nutrient Enrichment Gradient in the Northern Everglades," *Soil Science,* 160:183–198, 1995.

Reader, R. J., and J. M. Stewart, "The Relationship between Net Primary Production and Accumulation for a Peatland in Southeastern Manitoba," *Ecol.,* 53:1024–1037, 1972.

Reddy K. R., and W. H. Smith, *Aquatic Plants for Water Treatment and Resource Recovery,* Magnolia, Orlando, Fla, 1987.

Richardson, C. J., "Primary Productivity Values in Freshwater Wetlands," in P. E. Greeson, J. R. Clark, and J. E. Clark, eds., *Wetland Functions and Values: The State of Our Understanding,* American Water Resources Assoc., Minneapolis, Minn., 1979, pp. 131–145.

Richardson, C. J., *Pocosin Wetlands,* Hutchinson Ross, Stroudsburg, Pa., 1981.

Richardson, C. J., "Pocosins: Ecosystem Processes and the Influence of Man," in C. J. Richardson, ed., *Pocosin Wetlands,* Hutchinson Ross, Stroudsburg, Pa., 1981, pp. 135–154.

Richardson, C. J., "Mechanisms Controlling Phosphorus Retention Capacity in Freshwater Wetlands," *Science,* 228:1424–1427, 1985.

Richardson, C. J., "Reclamation of areas of oil and gas activity in the Big Cypress National Preserve," Report to Energy, Mining and Minerals Division, National Park Service, Denver, Colo., 1986.

Richardson, C. J., "Freshwater Wetlands: Transformers, Filters, or Sinks?," in R. R. Sharitz, and J. W. Gibbons, eds., *Freshwater Wetlands and Wildlife,* CONF-8603101, DOE Symposium Series no. 61, U.S. Department of Energy, 1989, pp. 1989.

Richardson, C. J., "Ecological Functions and Human Values in Wetlands: A Framework for Assessing Forestry Impacts," *Wetlands,* 14(1):1–9, 1994.

Richardson, C. J., "Wetlands Ecology," in *Encyclopedia of Environmental Biology,* vol. 3, Academic Press, New York, 1995, pp. 535–550.

Richardson, C. J., and C. B. Craft, "Effective Phosphorus Retention in Wetlands: Fact or Fiction?," in G. A. Moshiri, ed., *Constructed Wetlands for Water Quality Improvement,* Lewis, Boca Raton, Fla., 1993, pp. 271–282.

Richardson, C. J., and D. S. Nichols, "Ecological Analysis of Wastewater Management Criteria in Wetland Ecosystems," in P. J. Godfrey, E. R. Kaynor, S. Pelczarski, and J. Benforado, eds., *Ecological Considerations in Wetlands Treatment of Municipal Wastewater,* Van Nostrand Reinhold, New York, 1985, pp. 351–391.

Richardson, C. J., and E. J. McCarthy, "Effect of Land Development and Forest Management on Hydrologic Response in Southeastern Coastal Wetlands: A Review," *Wetlands,* 14:56–71, 1994.

Richardson, C. J., and J. A. Davis, "Natural and Artificial Wetland Ecosystems: Ecological Opportunities and Limitations," in K. E. Reddy, ed., "Aquatic Plants for Wastewater Treatment and Resource Recovery," University of Florida Press, Gainesville, Fla., 1986, pp. 819–854.

Richardson, C. J., and L. W. Gibbons, "Pocosins, Carolina Bays and Mountain Bogs," in B. Martin, et al., eds., *Biodiversity of the Southern United States: Terrestrial Communities,* John Wiley, New York, 1993, pp. 257–310.

Richardson, C. J., and P. Vaithiyanathan, "P Sorption Characteristics of the Everglades Soils Along an Eutrophication Gradient," *Soil Science Society of America,* 59:1570–1578, 1995.

Richardson, C. J., and P. E. Marshall, "Processes Controlling the Movement, Storage, and Export of Phosphorus in a Fen, Peatland," *Ecol. Monogr.,* 56:279–302, 1986.

Richardson, C. J., D. L. Tilton, J. A. Kadlec, J. P. M. Chamie, and W. A. Wentz, "Nutrient Dynamics of Northern Wetland Ecosystems," in R. E. Good, D. F. Whigham, and R. L. Simpson, eds., *Freshwater Wetlands: Ecological Processes and Management Potential,* Academic Press, New York, 1978, pp. 217–241.

Richardson, C. J., R. E. Evans, and D. Carr, "Pocosins: An Ecosystem in Transition," in C. J. Richardson, ed., *Pocosin Wetlands,* Hutchinson Ross, Stroudsburg, Pa., 1981, pp. 3–19.

Richardson, C. J., W. A. Wentz, J. P. M. Chamie, J. A. Kadlec, and R. H. Kadlec, "Background Ecology and the Effects of Nutrient Additions on a Central Michigan Wetland," in M. W. Lefor, W. C. Kennard, W. and T. B. Helfsolt, eds., *Proc. of the 3rd Wetlands Conf.,* Inst. Water Res., University of Connecticut, 1976, pp. 34–72.

Roulet, N. T., R. Ash, and T. R. Moore, "Low Boreal Wetlands as a Source of Atmospheric Methane," *J. Geophys. Res.,* 97:3739–3749, 1992.

Rutter, A. J., K. A. Krshaw, P. C. Robbins, and A. J. Morton, "A Predictive Model of Rainfall Interception in Forests. I. Derivation of the Model from Observations in a Plantation of Corsican Pine," *Agric. Meterol.,* 9(5/6):367–384, 1971.

Schlesinger, W. H., "Community Structure, Dynamics, and Nutrient Cycling in the Okefenokee Cypress Swamp Forest," *Ecol. Monogr.,* 48:43–65, 1978.

Schlesinger, W. H., *Biogeochemistry: An Analysis of Global Change,* Academic Press, San Diego, Calif., 1991.

Skaggs, R. W., "A Water Management Model for Shallow Water Table Soils," UNC Water Resources Research Institute Report no. 134, Raleigh, N.C., 1978.

Skaggs, R. W., "DRAINMOD reference report: Methods for Design and Evaluation of Drainage Water Management Systems for Soils with High Water Tables," Soil Conservation Service, Fort Worth, Tex., 1980.

Skaggs, R. W., J. W. Gilliam, and R. O. Evans, "A Computer Simulation Study of Pocosin Hydrology," *Wetlands*, 11:399–416, 1991.

Skaggs, R. W., J. W. Gilliam, T. J. Sheets, and J. S. Barnes, "Effects of Agricultural Land Development on Drainage Waters in North Carolina Tidewater Regions," UNC Water Resources Research Institute Report no. 159, Raleigh, N.C., 1980.

Stanek, W., "Classification of Muskeg," in N. W. Radforth, and C. O. Brawner, eds., *Muskeg and the Northern Environment in Canada*, University Toronto Press, Toronto, Canada, 1973, pp. 31–62.

Stowell, R., R. Ludwig, J. Colt, and G. Tchobanoglous, "Concepts in Aquatic Treatment System Design," *J. Env. Div., Processing of the ASCE*, 107, no. EE5, 1981.

Tchobanoglous, G., and E. D. Schroeder, *Water Quality*, Addition-Wesley, Reading, Mass., 1985.

Teal, J. M., and M. Teal, *Life and Death of the Salt Marsh*, Little, Brown, Boston, Mass., 1969.

Terry, T. A., and J. H. Hughes, "The Effects of Intensive Management on Planted Loblolly Pine (*Pinus taeda L.*) Growth on Poorly Drained Soils of the Atlantic Coastal Plain," in B. Bernier, and C. H. Winget, eds., *Forest Soils and Forest Land Management, Proc. 4th N. Am. For. Soils Conf.*, Les Presses de l'Universite Laval, Quebec, 1975, pp. 351–357.

Thornthwaite, C. W., "An Approach toward a Rational Classification of Climate," *Geological Review*, 38:55–94, 1948.

Thornthwaite, C. W., and J. R. Mather, "Instructions and Tables for Computing Potential Evapotranspiration and the Water Balance," Drexel Institute of Technology, *Climatology*, 10(3):185–311, 1957.

Tilton, D. L., R. H. Kadlec, and C. J. Richardson, eds., "Freshwater Wetlands and Sewage Effluent Disposal," *Proc. of a National Symp. May 10–11, 1976*, University of Michigan, Ann Arbor, Mich., 1976.

Tiner, R. W. Jr., *Wetlands of the United States: Current Status and Recent Trends*, U.S. Fish and Wildlife Service, Washington, D.C., 1984.

Tiner, R. W. Jr., *Wetlands of New Jersey*, National Wetlands Inventory, U.S. Fish and Wildlife Service, Newton Corner, Mass., 1985.

Tolonen, K., "Peat as a Renewable Resource," *Proc. of the Intern. Peat Soc. on Classification of Peat and Peatlands*, Sept 17–21, Hyytiala, Finland, 1979, pp. 282–296.

Tourbier, J., and R. W. Pierson, *Biological Control of Water Pollution*, University of Pennsylvania Press, Philadelphia, Pa., 1976.

U.S. Army Corps of Engineers, "Corps of Engineers Wetlands Delineation Manual," Technical Report Y-87-1, U.S. Army Engineer Waterways Experiment Station, Vicksburg, Miss., 1987.

U.S. Environmental Protection Agency, "Freshwater Wetlands for Wastewater Management Handbook," EPA 904/9-85-135, Region IV, Atlanta, Ga., 1985.

Verhoeven, J. T. A., "Nutrient Dynamics in Minerotrophic Peat Mires," *Aquatic Botany*, 25:117–137, 1986.

Verry, E. S., "Streamflow Chemistry and Nutrient Yields from Upland-Peatland Watersheds in Minnesota," *Ecol.* 56:1149–1157, 1975.

Verry, E. S., and D. R. Timmons, "Waterborne Nutrient Flow through an Upland-Peatland Watershed in Minnesota," *Ecol.*, 63:1456–1467, 1982.

Viro, P. J., "Loss of Nutrients and the Natural Nutrient Balance of the Soil in Finland," *Comm. Inst. For. Fenn.*, 42:1–50, 1953.

Weller, M. W., *Freshwater Marshes: Ecology and Wildlife Management*, 2d ed., University of Minnesota Press, Minneapolis, Minn., 1987.

Westman, C. J., "Climate-Dependent Variation in the Nutrient Content of the Surface Peat Layer from Sedge Pine Swamps," *Proc. of the Intern. Peat Soc. Symp. on Classification of Peat and Peatlands*, Sept. 17–21, Hyytiala, Finland, 1979, pp. 160–170.

Wetzel, R. G., *Limnology,* 2d ed., W. B. Saunders, Philadelphia, Pa., 1983.

Wieder, R. K., "Peat and Water Chemistry at Big Bog Run, a Peatland in the Appalachian Mountains of West Virginia, U.S.A.," *Biogeochemistry,* 1:277–302, 1985.

Wieder, R. K., and G. E. Lang, "Fe, Al, Mn, and S Chemistry of *Sphagnum* Peat in Four Peatlands with Different Metal and Sulfur Input," *Water Air Soil Poll.,* 29:309–320, 1986.

Wieder, R. K., and G. E. Lang, "Cycling of Inorganic and Organic Sulfur in Peat from Big Run Bog, West Virginia," *Biogeochemistry,* 5:221–242, 1988.

Wieder, R. K., and G. E. Lang, "Influence of Wetlands and Coal Mining on Stream Water Chemistry," *Water, Air and Soil Pollution,* 23:381–396, 1984.

Winter, Thomas C., "Hydrologic Studies of Wetlands in the Northern Prairie," in Arnold Van der Valk, ed., *Northern Prairie Wetlands,* Iowa State University Press, Ames, Iowa, 1989, pp. 16–54.

Wiser, E. H., "HISARS—Hydrologic Information and Retrieval Systems Reference Manual," North Carolina Technical Bulletin no. 215, Raleigh, N.C., 1975.

CHAPTER 14

COMPUTER MODELS FOR WATER-QUALITY ANALYSIS

Robert B. Ambrose, Jr., P.E.
Thomas O. Barnwell, Jr.
Steve C. McCutcheon, Ph.D., P.E.
US EPA National Exposure Research Lab
Ecosystems Research Division, Athens, Georgia

Joseph R. Williams
US EPA National Risk Management Lab
Subsurface Protection and Remediation Division
Ada, Oklahoma

14.1 INTRODUCTION

Computer models are used extensively in the analysis of complex water-quality issues and problems. To aid practicing hydrologists and engineers, we summarize models that handle traditional water-quality indicators, including dissolved oxygen, biochemical oxygen demand (BOD), nutrients, and phytoplankton, as well as toxicants such as heavy metals and organic chemicals. We do not cover hydrologic, hydrodynamic, or sediment transport models, except as they are incorporated within a water-quality modeling framework. We will confine our review to simulation models, not decision models that use linear or nonlinear optimization techniques to select among management alternatives.

Software reviews summarizing the state of the art of various modeling fields are being produced regularly. Many of these reviews will contain more detailed information than we can present here, and are recommended to the reader. Wurbs (1993) provides a thorough review of water-management models, including groundwater, watershed, stream hydraulics, and river/reservoir water quality. Villars, et al. (1993) review surface water, groundwater, and soil models of chemical concentrations. Donigian and Huber (1991) provide a detailed summary and review of urban and nonurban non-point-source models, with references to many other reviews. McCutcheon (1989) provides detailed information on transport and water-quality models for streams and rivers. Orlob (1992) and Stefan, et al. (1989) overview surface water quality modeling. Van der Heijde, et al. (1993) overview and summarize groundwater models. Williams, et al. (1994) cover vadose zone models.

Model development is dynamic and ongoing. Some models are being continuously refined, upgraded, and integrated with other models and support software, while other models remain static and are eventually abandoned. Some of the information here will quickly become outdated. The reader is cautioned to use this review as general background on the applied state of the art in the mid-1990s, and as a starting point in choosing appropriate water-quality models.

This chapter does not attempt an exhaustive compilation of existing models. Even if this could be done fairly, the list would very quickly become dated as new models are developed and old models are refined. We describe a representative sample of the best models that are used for routine applications, along with a few new models that promise to improve the applied state of the art. We limit our analysis to operational models with documentation, support, and experience. Most are in the public domain, although a few must be purchased from private vendors.

Implementation of an off-the-shelf model or method will be easiest if the model can be characterized as operational in the sense of:

1. *Documentation.* This should include a user's manual that explains model theory and numerical procedures, data needs, data input format, and so forth. Documentation most often separates the many computerized procedures found in the literature from models that can be accessed and easily used by others.

2. *Support.* This is sometimes provided by the model developer but often by a federal agency, such as the U.S. Army Corps of Engineers Hydrologic Engineering Center (HEC), Waterways Experiment Station (WES), or the U.S. Environmental Protection Agency (EPA).

3. *Experience.* Every model must be used a first time, but it is best to rely on a model with a proven track record.

The models described in the following sections are all operational according to these criteria. New methods and models are constantly under development and should not be neglected simply because they lack one of these characteristics, but the user should be aware of potential difficulties if any characteristic is lacking.

14.1.1   Relevant Model Features

There are many ways to categorize water-quality models for review. Some major features include dimensionality, temporal domain, water-quality variables, and water-quality processes. Our first division will be among models that simulate non-point-source runoff, surface water systems, and groundwater. Within these major subject areas, we examine several major categories separately. Non-point-source models are divided into urban and nonurban. Surface water models are divided into water quality and mixing zones. Groundwater models are divided into vadose (unsaturated) zone and saturated zone. Some models are more general than others and fit into more than one category. Their capabilities are discussed as they apply to the different categories.

At the present time, most model applications are run on DOS-based 386/486 microcomputers. Some advanced users install the models on engineering work stations that allow multitasking. Few models are presently configured for the WINDOWS environment.

14.1.2   Sources and Acquisition Procedures

Model availability is an important issue in the water-quality profession. Various governmental and private software outlets have been in operation since the late 1980s and early 1990s. Each operates under its own policies regarding distribution and support; at times, policies change due to programmatic initiatives or budgetary constraints. The reader is encouraged to contact these organizations to obtain current software listings and acquisition and support policies.

The U.S. Environmental Protection Agency (EPA) has maintained the Center for Exposure Assessment Modeling (CEAM) in Athens, Georgia, since the mid-1980s. Water quality and exposure assessment software is available free of charge to all requestors through diskette exchange, the CEAM electronic bulletin board system, and the Internet. User assistance is provided by telephone for many of the distributed models. Inquiries should be directed to

Manager, CEAM
U.S. Environmental Protection Agency
960 College Station Road
Athens, GA 30605-2720
706-546-3549

Models may be downloaded directly from the CEAM bulletin board at (706) 546-3402 (1200-14400, n-8-1). Internet access is gained via anonymous ftp at node earth1.epa.gov, or through World Wide Web browsers (e.g., Mosaic, Netscape) via universal resource locator (URL) file http://ftp.epa.gov/epa_ceam/wwwhtml/ceam_home.html.

EPA has also maintained the Center for Subsurface Modeling Support (CSMoS) in Ada, Oklahoma since the early 1990s. Saturated and unsaturated subsurface models are available free of charge to all requestors through diskette exchange, the CSMoS electronic BBS, and the Internet. User assistance is provided by telephone. Inquiries should be directed to

Manager, CSMoS
R.S. Kerr Environmental Research Laboratory
919 Kerr Research Drive
P.O. Box 1198
Ada, OK 74820
405-436-8586

Models may be downloaded directly from the CSMoS bulletin board at (405) 436-8506 (1200-9600, n-8-1). Internet access is gained via anonymous ftp at node ftp.epa.gov/pub/gopher/ada/models or through World Wide Web browsers via URL file http://www.epa.gov/ada/kerrlab.html.

The U.S. Army Corps of Engineers develops and distributes water quality models through the Waterways Experiment Station (WES) in Vicksburg, Mississippi, and the Hydrologic Engineering Center (HEC) in Davis, California. WES models are presently available to the public free of charge. Inquiries should be directed to

Water Quality and Contaminant Modeling Branch
Environmental Laboratory
U.S. Army Engineer Waterways Experiment Station
3909 Halls Ferry Rd.
Vicksburg, MS 39180
601-634-3785

HEC models are available to the public through registered private firms, which offer varying degrees of support at varying prices. A list may be obtained from

HEC
609 Second Street
Davis, CA 95616-4687
916-756-1104

The International Ground Water Modeling Center (IGWMC) distributes and supports a large number of subsurface models developed by the USGS, EPA, other agencies, universities, and private firms. Many of the models in the IGWMC inventory are available for a nominal handling fee. A software catalogue is available upon request at

IGWMC
Colorado School of Mines
Golden, CO 80401-1887
303-273-3103

The U.S. Geological Survey has developed a number of widely-used ground water models (Appel and Reilly, 1988). Most of these models are available from

USGS
NWIS Program Office
437 National Center
Reston VA 22092
703-648-5695

Model documentation, reports, and circulars can be obtained from

USGS Map Distribution
Box 25286, MS 306
Denver Federal Center
Denver, CO 80225
303-236-7477

USGS hydrologic models can be obtained by contacting

USGS
Hydrologic Analysis Support Section
415 National Center
Reston, VA 22092
703-648-5313

Two European model software centers are referenced in this chapter. The Danish Hydraulic Institute offers a series of generalized hydraulic and water resources engineering models, along with support and training. Inquiries should be sent to

Danish Hydraulic Institute
Agern Alle 5
DK-2970
Horsholm, Denmark
phone +45 45 76 95 55

Wallingford Software, the software house within HR Wallingford, offers software packages for hydraulic and water management studies, along with customer support and training. For more information, contact

UK Sales and Marketing Office
Wallingford Software
HR Wallingford Limited
Howbery Park
Wallingford
Oxfordshire
OX10 8BA United Kingdom
phone +44 491 35381

14.2 NON-POINT-SOURCE MODELS

Stormwater runoff from all categories of land use—urban, agricultural cropland, pasture, and forest—is involved in many water-quality problems. Models can be used to help in setting total maximum daily loads (TMDLs) for basins or watersheds, in registering new pesticides or assessing the risk of current pesticide use, or in assessing the risk from hazardous-waste sites. But by no means should it be assumed that every water-quality problem requires a water-quality modeling effort. Some problems related to flooding are mostly hydraulic in nature. In some instances, local or state regulations may prescribe a nominal solution that excludes water-quality analysis. Other problems may be resolved through the use of measured data without the need to model.

14.2.1 Urban Runoff-Quality Models

Five models—USGS, HSPF, STORM, SWMM, and AUTO-QI—collectively make up the best choice of full-scale simulation models for urban areas. At least two European models are also available that simulate water quality. Finally, there are many models well known in the hydrologic literature, such as those developed by the HEC and SCS, that might be useful in the hydrologic and sediment-related aspects of water-quality studies but that do not simulate water quality directly. A general comparison of model attributes is given in Table 14.1. This table includes the EPA Statistical Method since with the publication of the recent FHWA study, it can be considered a formalized procedure (Driscoll et al., 1989). It is most often implemented as a spreadsheet procedure.

DR3M-QUAL. A version of the USGS Distributed Routing Rainfall Runoff Model that includes quality simulation, DR3M-QUAL is a continuous- and event-simulation model available from the U.S. Geological Survey (Alley and Smith, 1982a; 1982b). Runoff generation and subsequent routing use the kinematic wave method. Quality is simulated using buildup and washoff functions, with settling of solids in storage units dependent on particle size distribution. The model has been used in some of the NURP studies that were conducted by the USGS (Alley, 1986). DR3M-QUAL Fortran 77 source code and documentation are available for a nominal fee from the USGS Hydrologic Analysis Support Section.

HSPF. The Hydrological Simulation Program—FORTRAN (HSPF) is a continuous-simulation model for upland watersheds. HSPF is the culmination of

TABLE 14.1 Comparison of Urban Model Attributes

Model attribute	DR3M-QUAL	HSPF	Statistical[a]	STORM	SWMM	AUTO-QI
Sponsoring agency	USGS	EPA	EPA	HEC	EPA	State of Illinois
Simulation type[b]	C,SE	C,SE	N/A	C	C,SE	C,SE
No. pollutants	4	10	Any	6	10	5
Rainfall/runoff analysis	Y	Y	N[c]	Y	Y	Y
Sewer system flow routing	Y	Y	N/A	N	Y	N
Full, dynamic flow routing equations	N	N	N/A	N	Y[d]	N/A
Surcharge	Y[e]	N	N/A	N	Y[d]	N/A
Regulators, overflow structures	N	N	N/A	Y	Y	N/A
Special solids routines	Y	Y	N	N	Y	N
Storage analysis	Y	Y	Y[f]	Y	Y	N
Treatment analysis	Y	Y	Y[f]	Y	Y	Y
Suitable for screening(S), design (D)	S,D	S,D	S	S	S,D	S,D
Available on microcomputer	N	Y	Y[g]	N	Y	Y
Data, personnel requirements[h]	Medium	High	Medium	Low	High	Medium
Overall model complexity[i]	Medium	High	Medium	Medium	High	Medium

[a] EPA procedure.
[b] C = continuous simulation; SE = single event simulation.
[c] Runoff coefficient used to obtain runoff volumes.
[d] Full dynamic equations and surcharge calculations only in Extran Block of SWMM.
[e] Surcharge simulated by storing excess inflow at upstream end of pipe. Pressure flow not simulated.
[f] Storage and treatment analyzed analytically.
[g] FHWA study, Driscoll, et al. (1989).
[h] General requirements for model installation, familiarization, data requirements, etc. To be interpreted only very generally.
[i] Reflection of general size and overall model capabilities. Note that complex models may still be used to simulate very simple systems with attendant minimal data requirements.
Source: Donigian and Huber (1991).

hydrologic routines that originated with the Stanford Watershed Model in 1966 and eventually incorporated many non-point-source modeling efforts of the EPA Athens Environmental Research Laboratory (Bicknell, et al., 1993). It includes urban water-quality algorithms similar to SWMM. This model has been widely used for nonurban non-point-source modeling and is described additionally in that section of this report. Additional guidelines for application are provided by Donigian, et al. (1984). HSPF is available from EPA CEAM.

STORM. The first significant use of continuous simulation in urban hydrology came with the Storage, Treatment, Overflow, Runoff Model (STORM), developed by the Corps of Engineers Hydrologic Engineering Center (HEC, 1977; Roesner, et al., 1974). HEC also provides application guidelines (Abbott, 1977). STORM uses a simple runoff coefficient, SCS and unit hydrograph methods for generation of hourly runoff depths from hourly rainfall inputs. The buildup and washoff formulations are used for simulation of six prespecified pollutants. STORM source code (not executable) is available directly from HEC.

SWMM. The original version of the Storm Water Management Model (SWMM) was developed as a single-event model specifically for the analysis of combined sewer overflows, but its scope has vastly broadened since the original release. Version 4 (Huber and Dickinson, 1988; Roesner, et al., 1988) of the model performs both continuous and single-event simulation throughout the whole model, can simulate backwater, surcharging, pressure flow, and looped connections in its Extran Block, and has a variety of options for quality simulation, including traditional buildup and washoff formulations as well as rating curves and regression techniques. A bibliography of SWMM usage is available (Huber, et al., 1985) that contains many references to case studies. SWMM is available from EPA CEAM.

Automated Q-ILLUDAS (AUTO-QI). The Illinois Urban Drainage Area Simulator (ILLUDAS) evolved from the British Road Research Laboratory Model. ILLUDAS is a continuous- and event-simulation model that uses time-area methods for generation of runoff coupled with Horton or SCS infiltration on pervious areas. Its simplicity and early metric option have resulted in ILLUDAS being widely used. Quality has been formally included in a recent release of the model, now called AUTO-QI (Terstriep, et al., 1990). This release includes an interface to the ARC/INFO Geographical Information System. This software is available from the Illinois State Water Survey, Champaign, Illinois.

14.2.1.1 Two European Models.
The U.S. models (except, perhaps, AUTO-QI) discussed here do not take full advantage of graphics and other user-friendly capabilities of microcomputers. Two well known and commercially available European models, MOUSE and Wallingford, are excellent examples of application of the full power of the microcomputer when used in conjunction with recent programming languages and graphics hardware and software.

MOUSE. The Danish Hydraulic Institute, I. Kruger AS, PH-Consult Aps and EMOLET have produced the continuous-simulation MOUSE (Modeling of Urban Sewers) model. Included in the package are modules for generation of runoff from rainfall using either a time-area method or a nonlinear reservoir model; a sewer routing module using alternative computational levels corresponding to kinematic, diffusive, and dynamic wave theory; and a model for computation and analysis of combined sewer overflows. MOUSE is available for a fee from Danish Hydraulic Institute.

WALLINGFORD Procedure. The Wallingford Procedure (Price, 1991) is a continuous-simulation set of models maintained by Hydraulics Research Ltd. in the United Kingdom. It consists of a cluster of modules, including runoff generation from rainfall (WASSP), simple and fully dynamic sewer routing (WALLRUS and SPIDA, respectively), and a quality routine (MOSQITO). The group of Wallingford models is available for a fee from Wallingford Software.

14.2.2 Nonurban Runoff-Quality Models and Methods

In this section we summarize the primary nonurban runoff-quality models; as noted earlier, additional details on each model are provided in Donigian and Huber, 1991. Summaries are presented here for HSPF, CREAMS/GLEAMS, ANSWERS,

AGNPS, PRZM, and SWRRB. Table 14.2 shows a comparison of selected model attributes and capabilities.

HSPF. The Hydrological Simulation Program—FORTRAN (HSPF) (Bicknell, et al., 1993) is a comprehensive package for continuous simulation of watershed hydrology and water quality for both conventional and toxic organic pollutants. HSPF incorporates watershed scale models into a basin-scale analysis framework that includes fate and transport in one-dimensional stream channels. It is the only comprehensive model of watershed hydrology and water quality that allows the integrated simulation of land and soil contaminant runoff processes with instream hydraulic, water temperature, sediment transport, nutrient, and sediment-chemical interactions. The hydrologic routines originated with the Stanford Watershed Model. Erosion calculations are based on detachment processes driven by the energy of rainfall. The runoff quality capabilities include both simple relationships (i.e., empirical buildup/washoff and constant concentrations) and detailed soil process options (i.e., leaching, sorption, soil attenuation, and soil nutrient transformations). HSPF is available from EPA CEAM.

CREAMS/GLEAMS. Chemicals, Runoff, and Erosion from Agricultural Management Systems (CREAMS) was developed by the USDA—Agricultural Research Service (Knisel, 1980; Leonard and Ferreira, 1984) for the analysis of agricultural best-management practices for pollution control. CREAMS is a continuous-simulation field-scale model that uses separate hydrology, erosion, and chemistry submodels connected together by pass files. The hydrology is generally calculated using the SCS Curve Number approach, although a detailed hydrology option is

TABLE 14.2 Comparison of Nonurban Model Attributes

Attribute	AGNPS	ANSWERS	CREAMS	HSPF	PRZM	SWRRB
Sponsoring agency	USDA	Purdue	USDA	EPA	EPA	USDA
Simulation type	C,SE	SE	C,SE	C,SE	C	C
Rainfall/runoff analysis	Y	Y	Y	Y	Y	Y
Erosion modeling	Y	Y	Y	Y	Y	Y
Pesticides	Y	N	Y	Y	Y	Y
Nutrients	Y	Y	Y	Y	N	Y
User-defined constituents	N	N	N	Y	N	N
Soil processes						
Pesticides	N	N	Y	Y	Y	Y
Nutrients	N	N	Y	Y	N	Y
Multiple land type capability	Y	Y	N	Y	N	Y
Instream water quality simulation	N	N	N	Y	N	N
Available on microcomputer	Y	Y	Y	Y	Y	Y
Data and personnel requirements	M	M/H	H	H	M	M
Overall model complexity	M	M	H	H	M	M/H

Y = yes; N = no; M = Moderate; H = High; C = Continuous; SE = Storm Event
Source: Donigian and Huber, 1991.

available, requiring short time interval rainfall data. Daily water balance, erosion, and sediment yield are estimated at the edge of the field. Plant nutrients and pesticides are simulated, and storm load and average concentrations of sediment-associated and dissolved chemicals are determined in the runoff, sediment, and percolation through the root zone (Leonard and Knisel, 1984). User-defined management activities, including aerial spraying, soil incorporation of pesticides, animal waste management, and agricultural best-management practices can be simulated by CREAMS. Calibration is not specifically required, but is usually desirable.

Groundwater Loading Effects of Agricultural Management Systems (GLEAMS) was developed by the USDA Agriculture Research Service (Leonard, et al., 1987) to utilize the management-oriented physically based CREAMS model and incorporate a component for vertical flux of pesticides. GLEAMS is the vadose zone component of the CREAMS model. GLEAMS and CREAMS can be obtained from

USDA-ARS
Southeast Watershed Research Lab
P.O. Box 946
Tifton, GA 31793

ANSWERS. Areal Nonpoint Source Watershed Environment Response Simulation (ANSWERS) was developed at the Agricultural Engineering Department of Purdue University (Beasley and Huggins, 1981). It is an event-based, distributed parameter gridded model capable of predicting the hydrologic and erosion response of agricultural watersheds. Within each one- to four-hectare element, the model simulates the processes of interception, infiltration, surface storage, surface flow, subsurface drainage, and sediment drainage, detachment, transport, and deposition. Nutrients are simulated using correlation relationships between chemical concentrations, sediment yield, and runoff volume. ANSWERS can be obtained from

University of Georgia
Department of Agricultural Engineering
Coastal Plains Experiment Station
P.O. Box 748
Tifton, GA 31793

AGNPS. Agricultural Nonpoint Source Pollution Model (AGNPS) is a continuous and event-based watershed model developed by the USDA Agriculture Research Service (Young, et al., 1986) to obtain uniform and accurate estimates of runoff quality, with primary emphasis on nutrients and sediments, and to compare the effects of various pollution-control practices that could be incorporated into the management of watersheds. The Modified Universal Soil Loss Equation is used for predicting soil erosion, and a unit hydrograph approach used to represent the flow in the watershed. The methods used to predict nitrogen and phosphorus yields from the watershed and individual cells were developed by Frere, et al. (1980) and are also used in CREAMS. AGNPS can be obtained from

USDA-ARS
North Central Research Laboratory
Norris, MN 56267

PRZM. Pesticide Root Zone Model (PRZM) was developed at the U.S. EPA Environmental Research Laboratory in Athens, Georgia by Carsel, et al. (1984) and has been updated in the PRZM2 release by Mullins, et al., 1993. It is a one-dimensional, field-scale compartmental model that can be used for continuous simulations of chemical movement in the unsaturated zone within and immediately below the plant root zone. The hydrology component is based on the Soil Conservation Service curve number procedure and the Universal Soil Loss Equation, respectively. Pesticide applications on soil or on the plant foliage are modeled in the chemical transport simulation. PRZM2 is an integral part of the Pesticide Assessment Tool for Rating Investigations of Transport (PATRIOT, Imhoff, et al., 1993), which is designed to provide rapid analyses of groundwater vulnerability to pesticides on a regional, state, or local level. PATRIOT is composed of: (1) a pesticide fate and transport model (PRZM2), (2) a comprehensive database, (3) an interface that facilitates database exploration, (4) a directed sequence of interaction that guides the user in providing all the necessary information to perform alternative model analyses, and (5) user-selected methods of summarizing and visualizing model results. PRZM and PATRIOT are available from EPA CEAM.

SWRRB. Simulator for Water Resources in Rural Basins (SWRRB) was developed by Williams, et al. (1985), and Arnold, et al. (1989) for evaluating basin-scale water quality. SWRRB is a continuous-simulation model that operates on a daily time step and simulates weather, hydrology, crop growth, sedimentation, and nitrogen, phosphorous, and pesticide movement. The model was developed by modifying the CREAMS model for application to large, complex, rural basins. Surface runoff is calculated using the Soil Conservation Service Curve Number technique. Sediment yield is computed for each basin by using the Modified Universal Soil Loss Equation. The crop growth model computes total biomass each day during the growing season as a function of solar radiation and leaf area index. The Pesticide Runoff Simulator (PRS) was developed for the U.S. EPA Office of Pesticides and Toxic Substances by Computer Sciences Corporation (1980) to simulate pesticide runoff and adsorption into the soil on small agricultural watersheds. PRS is based on SWRRB. The model includes a built-in weather generator based on temperature, solar radiation, and precipitation statistics. SWRRB can be obtained from

USDA-ARS
Attn: Dr. Jeffrey G. Arnold
808 East Blackland Road
Temple, TX 76502
817-770-6502

14.2.3 Discussion

The models discussed briefly here do not represent all of the modeling options available for runoff-quality simulation, but they are certainly the most notable, most widely used, and most operational. Selection from among these models is often made on the basis of personal preference and familiarity, in addition to needed model capabilities. For urban modeling, various in-house versions of STORM are still used by consultants—even though the official HEC version has not been updated since 1977—because these versions have been adapted to the needs of the firms and because STORM has proven to provide useful continuous simulation results. The USGS DR3M-QUAL model has been perhaps the least used by persons outside the USGS, but it has worked satisfactorily in several applications. Support for both STORM and DR3M-QUAL is minimal. CREAMS has been used most extensively for field-scale agricultural runoff modeling because of its agricultural origins and ties to the agricultural research community.

HSPF and SWMM are the most widely used of the models, with the nod to SWMM if urban hydrology and hydraulics must be simulated in detail and to HSPF for mixed land use. On the other hand, the water quality routines in HSPF for sediment erosion, pollutant interactions, and surface water quality are superior, and the capability to efficiently handle all types of land uses and pollutant sources (including urban and agriculture, point and nonpoint), is a definite advantage when needed for large complex basins. Both models are somewhat overwhelming to the novice user in terms of size, but only the components of interest of either model need be used in a given study, and the catchment schematization can often be coarse for purposes of simulation of water quality at the outlet. In general, considerable judgment and experience are required to use these computer programs effectively.

Continuing model development and testing within the agricultural research community will likely lead to further enhancements of agricultural models like CREAMS, SWRRB, and AGNPS. In fact, USDA continues to support a wide range of model development work in individual research facilities. SWRRB development appears to be focusing on a level of complexity between HSPF and the detailed field-scale models that are limited to small areas. Most of these efforts, however, still focus primarily on agricultural areas, with limited abilities to be used in large, complex multi-land-use basins.

Regression, spreadsheet, statistical methods, and loading functions are most useful as screening tools. Indeed, if the statistical method or EPA's screening procedures indicate that there should be no water-quality problem (as defined by exceedance of a specified concentration level with a specified frequency), then more detailed water-quality simulation may not be required at all. Simple techniques are identified in the Model Compendium (U.S. EPA, 1992). If sensitivity analyses and worst-case evaluations further support these conclusions, detailed water-quality modeling will probably not be needed.

14.3   SURFACE-WATER-QUALITY MODELS

Surface water bodies are valuable resources, providing habitat for fish and wildlife, and supporting human populations with drinking water, commercial fisheries, recreational opportunities, wastewater receptors, and navigational routes. Surface water quality can be impacted by many human activities, including land use, such as farming and urban development, discharge of municipal and industrial wastewater, and direct use, such as dredging for navigation, or impoundment for water supply or flood control. Environmental assessments are often used to manage human activities to preserve or enhance water quality. Surface-water-quality models are a central component of many environmental assessments, including waste load allocations and watershed TMDLs, risk assessments for pesticides and hazardous-waste sites, and operation of reservoirs.

A primary division for surface-water-quality models can be made between far-field models, which simulate water quality in entire sections of a water body, and mixing zone models, which simulate near-field dilution processes. Far-field water-quality models can be further subdivided on the basis of water body type, i.e., streams and rivers, lakes and reservoirs, and estuaries (including tidal rivers). Because some models can be applied to more than one water body type, they are summarized together in the water quality section.

14.3.1 Mixing Zone Models

There are five basic sources of information on mixing zone modeling. Fischer, et al. (1979) and Holley and Jirka (1986) review the theories of mixing in surface waters. EPA (1985) provides a brief overview of mixing calculations and how these can be used to meet waste discharge regulations. Finally, there are two public domain models supported by the U.S. EPA available to calculate mixing and to formulate mixing zones. These include Baumgartner, et al. (1993) and Jirka and colleagues (Doneker and Jirka, 1990; Akar and Jirka, 1991; Jones and Jirka, 1991; Jirka and Hinton, 1993).

The PLUMES model interface and model manager (Baumgartner, et al., 1993), interfaces with two initial dilution plume models, RSB and UM. PLUMES initiates two far-field mixing algorithms beyond the zone on initial dilution calculated with RSB and UM. In addition, PLUME incorporates the flow classification used by the more elaborate CORMIX expert system for mixing-zone modeling. This flow classification scheme leads to an easy linkage between the PLUMES and CORMIX systems to cover a wide range of mixing conditions.

The PLUMES system is intended to simulate plumes discharged into deeper freshwater or marine waters where buoyancy dominates over local flow conditions. Simulations cover buoyant or dense plumes, and single sources or multiple-source diffuser outfalls in many different configurations.

CORMIX Version 2.10 is a comprehensive system for modeling diverse types of aquatic pollutant discharges of conventional or toxic pollutants into all types of receiving water bodies, including streams, rivers, lakes, reservoirs, estuaries, and coastal waters. The user can easily switch among these design solutions for any site with given ambient conditions. Version 2.10 allows for nonconservative pollutant types with first-order reaction processes and/or surface heat loss, calculates plume travel times, and considers the effect of wind on plume mixing. The prediction of plume geometry and dilution characteristics within a receiving water's initial mixing zone are made so that compliance with regulatory constraints may be judged. The system also predicts discharge plume behavior at distances beyond the initial mixing zone.

Both PLUMES and CORMIX are available from EPA CEAM.

14.3.2 Water-Quality Models

In this section we summarize eleven water-quality models that are documented and available to the public. These models cover a range of capabilities from conventional water quality to toxicants, and from steady one-dimensional to dynamic multidimensional solutions.

QUAL2E. The Enhanced Stream Water Quality Model QUAL2E and QUAL2E-UNCAS (Brown and Barnwell, 1987) permits simulation of several conventional water-quality constituents in a branching stream system under steady conditions. The model uses a finite difference solution to the one-dimensional, longitudinal advective-dispersive mass transport and reaction equation. QUAL2E predicts DO, CBOD, temperature, and phytoplankton dynamics as affected by nutrients and organic material. It comes with a pre- and postprocessor and software that facilitates sensitivity and uncertainty analysis. QUAL2E is available from EPA CEAM.

SMPTOX3. The Simplified Method Program—Variable Complexity Stream Toxics Model (SMPTOX3) is an interactive computer program for performing waste load allocations for toxics (LimnoTech, 1990). This steady, one-dimensional, longitudinal model calculates suspended solids and dissolved and particulate toxicant concentrations in the water column and stream bed resulting from point-source discharges into streams and rivers. The model is imbedded in a user-friendly shell that handles data input, graphical output, sensitivity analysis, and uncertainty analysis. SMPTOX3 is available from EPA CEAM.

HSPF. The Hydrological Simulation Program—FORTRAN (HSPF) (Bicknell, et al., 1993) is a comprehensive package for simulation of watershed hydrology and water quality for both conventional and toxic organic pollutants. HSPF incorporates watershed scale models into a basin-scale analysis framework that includes fate and transport in one-dimensional stream channels. Several hydrodynamic routing options are available. The conventional water-quality routines predict DO, CBOD, temperature, and phytoplankton dynamics as affected by nutrients and organic material. The toxicant routines combine organic chemical process kinetics with sediment balance algorithms to predict dissolved and sorbed chemical concentrations in the upper sediment bed and overlying water column. HSPF is available from EPA CEAM.

WASP5. The Water Quality Analysis Simulation Program, WASP5 (Ambrose, et al., 1987) is a generalized compartment modeling framework for simulating water quality and contaminant fate and transport in surface waters. It is dynamic and can be applied in one, two, or three dimensions. Two major subcomponent models are provided with WASP5. The toxics subcomponent, TOXI5, combines organic chemical process kinetics with simple sediment balance algorithms to predict dissolved and sorbed chemical concentrations in the sediment bed and overlying water column. The dissolved oxygen/eutrophication subcomponent EUTRO5 predicts DO, CBOD, and phytoplankton dynamics as affected by nutrients and organic material. Transport can be driven by specified flows or by coupling to external hydrodynamic models. WASP5 is available from EPA CEAM.

EXAMSII. The Exposure Analysis Modeling System EXAMSII (Burns, 1990) is an interactive modeling system that allows a user to specify and store the properties of chemicals and ecosystems, modify either via simple commands, and conduct rapid evaluations and error analyses of the probable aquatic fate of synthetic organic chemicals. The solution can be steady-state or dynamic, and the model can be applied in one, two, or three dimensions. EXAMS combines chemical loadings, transport, and transformation into a set of differential equations using the law of conservation of mass as an accounting principle. The model is imbedded in an interactive, command-driven shell that handles data input and graphical or tabular output. EXAMSII is available from EPA CEAM.

CE-QUAL-RIV1. The U.S. Army Engineer Waterways Experiment Station's hydrodynamic and water-quality model for streams, CE-QUAL-RIV1, accommodates one-dimensional, branching systems with multiple hydraulic control structures. The model simulates DO, BOD, and phytoplankton dynamics as affected by nutrients and organic material in highly unsteady systems. CE-QUAL-RIV1 is available from WES.

CE-QUAL-W2. The U.S. Army Engineer Waterways Experiment Station's hydrodynamic and water-quality model CE-QUAL-W2 provides dynamic two-dimensional (longitudinal-vertical) simulations for branching, stratified waterbodies such as reservoirs and narrow estuaries. The hydrodynamic and water-quality modules are directly coupled, so that predicted flows are influenced by variable water density calculated from temperature, salinity, and solids concentrations. The model can simulate DO and phytoplankton dynamics as affected by nutrients and organic material. CE-QUAL-W2 is available from WES.

CE-QUAL-ICM. The U.S. Army Engineer Waterways Experiment Station originally developed CE-QUAL-ICM as an integrated compartment model for application to Chesapeake Bay. Transport can either be specified by the user or imported from coupled multidimensional hydrodynamic models. The model features advanced phytoplankton-nutrient balance routines and a sediment diagenesis submodel for predicting sediment oxygen demand and nutrient fluxes. This model is under active development, and the capability to simulate toxicants is planned. CE-QUAL-ICM is available from WES.

HEC5Q.   The U.S. Army Engineer Hydrologic Engineering Center developed HEC-5Q to simulate the operation of river-reservoir systems in basinwide studies. The stream component is a one-dimensional unsteady model driven by a selection of available hydrologic routing models. The reservoir component is a one-dimensional vertical model. The model simulates DO, BOD, a selection of conservative and nonconservative constituents, and contains a phytoplankton option. HEC5Q is available from HEC.

SALMON-Q.   HR Wallingford assembled SALMON-Q from a suite of programs to model unsteady flow, siltation, and water quality in one-dimensional channel networks, including rivers and estuaries. SALMON-Q incorporates the modules in a user-friendly graphics interface. Water quality parameters include DO, BOD, nutrients, and phytoplankton. SALMONQ is available for a fee from Wallingford Software.

MIKE11.   The Danish Hydraulic Institute developed MIKE-11 for the simulation of unsteady flow, sediment transport, and water quality in one-dimensional channel networks, including rivers and estuaries. MIKE-11 is operated through an interactive menu system that supports modules for hydrology, hydrodynamics, cohesive and noncohesive sediment transport, and water quality. Water-quality parameters include DO, BOD, nutrients, phytoplankton, zooplankton, bottom vegetation, and metals. MIKE11 is available for a fee from the Danish Hydraulics Institute.

14.3.3   Comparison of Water-Quality Model Attributes

Tables 14.3 and 14.4 summarize important model attributes. The first major category in Table 14.3 is the water body type. Some models specialize in rivers, including QUAL2E, SMPTOX3, HSPF, CE-QUAL-RIV1, SALMON-Q, and MIKE-11. Other models are more general, though perhaps less efficient for particular types of water bodies. EXAMS, WASP5, and CE-QUAL-ICM are "box" or "compartment" models that can be applied to all types of water bodies. CE-QUAL-W2 was developed as a reservoir model but has been applied to deep, slow-moving rivers as well as estuaries. HEC5Q is used for river basin evaluation, including rivers and reservoirs.

The next major category, dimensionality, determines how a model can spatially represent real water bodies. Most river and stream models are one-dimensional. All that are listed here can be applied to branching systems. Two dimensional models can cover either longitudinal and lateral (x/y) or longitudinal and vertical (x/z) dimensions. CE-QUAL-W2 is a true x/z model. The three box models listed here can be applied in one, two, or three dimensions. The user should be aware, however, that the underlying equations are not inherently multidimensional in form.

The third major category, time, determines how a model can represent water body dynamics. Steady-state models, such as QUAL2E and SMPTOX3, predict concentrations that do not vary in time. These are useful primarily for rivers under low-flow design conditions. Both dynamic and quasidynamic solutions predict concentrations that vary with time. Quasidynamic solutions allow some major forcing functions, such as flow, loading, or solar effects on photosynthesis, to vary with time. Typically, one or more forcing functions will be held constant or allowed to

TABLE 14.3 Surface Water Quality Models—Basic Information

Model	QUAL2E	SMPTOX3	HSPF	WASP5	EXAMS	CEQUALRIV1	CEQUALW2	CEQUALICM	HECSQ	SALMONQ	MIKE11
Water body type											
Stream, river	Y	Y	Y	Y	Y	Y	Y	Y	Y	Y	Y
Lake, reservoir	N	N	N	Y	Y	N	Y	Y	Y	N	N
Estuary	Y	N	N	Y	N	Y	Y	Y	N	N	N
Dimension											
1-D, Branching	Y	Y	Y	Y	Y	Y	Y	Y	Y	Y	Y
2-D, X/Y	N	N	N	Y	Y	N	N	Y	N	N	N
2-D, X/Z	N	N	N	Y	Y	N	N	Y	N	N	N
3-D, Box	N	N	N	Y	Y	N	N	Y	N	N	N
Time											
Steady	Y	Y	N	N	Y	N	N	N	N	N	N
Quasidynamic	Y	N	Y	Y	Y	N	N	N	N	Y	Y
Dynamic	N	N	Y	Y	N	Y	Y	Y	Y	Y	Y
Hydrodynamics											
Input	Y	Y	Y	Y	Y	N	N	N	N	N	Y
Simulated	N	N	Y	Y	N	Y	Y	Y	Y	Y	Y
Control struct.	N	N	Y	N	N	Y	Y	Y	Y	Y	Y
Transport											
Advection	Y	Y	Y	Y	Y	Y	Y	Y	Y	Y	Y
Dispersion	Y	Y	N	Y	Y	Y	Y	Y	Y	Y	Y
Benthic exchange	N	Y	Y	Y	Y	N	Y	Y	N	Y	Y
Loading											
Input, steady	Y	Y	Y	Y	Y	Y	Y	Y	Y	Y	Y
Input, variable	N	N	Y	Y	N	N	N	N	N	Y	Y
Simulated	N	N	Y	Y	Y	N	N	N	N	Y	Y
Other											
Preprocessor	Y	Y	N	Y	Y	N	N	N	N	Y	Y
Postprocessor	Y	Y	Y	Y	Y	N	N	N	N	Y	Y

vary slowly. Dynamic solutions allow all important forcing functions to vary with time. This would include daily changes in inflow and loading, or hourly changes in a tidal boundary.

The fourth major category is hydrodynamics. All water quality models require information on the movement of water. Some provide this information in the input dataset. These models are steady or quasidynamic in nature. QUAL2E, SMPTOX3, and EXAMS require input information on flow or velocity; HSPF, WASP5, CE-QUAL-ICM, and MIKE11 can either accept input flows, or be linked to simulated hydrodynamics. Some water quality models, such as HSPF and CE-QUAL-W2 provide for the hydrodynamic calculations internally. Most, however, require linkage to an external hydrodynamics file generated by a separate hydrodynamics program. This may provide additional flexibility, allowing linkage with several different hydrodynamics programs or generated files. Often, however, the linkage to external programs is not as smooth or transparent as an internal linkage. Some linked water-quality/hydrodynamic models include processes to simulate control structures, such as weirs or low-head dams in rivers, or selective withdrawal from reservoirs. These can be more easily applied to highly regulated or engineered bodies of water.

The fifth major category deals with transport. All models mentioned in this section simulate advection, and all but HSPF include dispersion. Most 1-D riverine simulations do not require dispersion, because most rivers are not highly dispersive and the model network and solution technique introduce some degree of numerical dis-

TABLE 14.4   Surface Water Quality Models—Variables and Processes

Model		QUAL2E	SMPTOX3	HSPF	WASP5	EXAMS	CEQUALRIV1	CEQUALW2	CEQUALICM	HEC5Q	MIKE11	SALMONQ
Chemical processes	First-order decay	Y	Y	Y	Y	Y	Y	Y	Y	Y	Y	Y
	Process kinetics	N	N	Y	Y	Y	N	N	N	N	N	N
	Daughter products	N	N	Y	Y	Y	N	N	N	N	N	N
	Sorption	N	Y	Y	Y	Y	N	N	N	N	Y	N
Sediment processes	Input rates	N	Y	Y	Y	N	Y	Y	Y	Y	Y	Y
	Noncohesive processes	N	N	Y	N	N	N	N	N	N	Y	N
	Cohesive processes	N	N	Y	N	N	N	N	N	N	Y	N
Water quality processes	Temperature	Y	N	Y	N	N	Y	Y	Y	Y	Y	Y
	Salinity	Y	N	Y	N	N	Y	Y	Y	Y	Y	Y
	Bacteria	N	N	Y	N	N	Y	Y	Y	Y	Y	Y
	DO-BOD	Y	N	Y	Y	N	Y	Y	N	Y	Y	Y
	DO-Carbon balance	N	N	N	N	N	N	Y	Y	N	N	N
	Nitrogen cycle	Y	N	Y	Y	N	Y	Y	Y	Y	Y	Y
	Phosphorus cycle	Y	N	Y	Y	N	Y	Y	Y	Y	Y	Y
	Silicon cycle	N	N	N	N	N	N	Y	Y	N	Y	Y
	Phytoplankton	Y	N	Y	Y	N	Y	Y	3	Y	Y	Y
	Zooplankton	N	N	Y	N	N	N	N	Y	N	Y	N
	Benthic algae	N	N	Y	N	N	N	N	N	N	Y	Y
	Simulate SOD	N	N	N	Y	N	N	Y	Y	N	N	N

persion. Accurate dispersion coefficients are usually required for simulations of lakes, reservoirs, and estuaries. Most models require the user to manipulate the calculational time step to minimize numerical dispersion; special efforts in reducing numerical dispersion in CE-QUAL-W2 have resulted in improved temperature calibrations with minimal parameter tuning. Transport to and from benthic compartments is an important component of toxic chemical models, such as SMPTOX3, HSPF, WASP5, and EXAMS. More recently, benthic exchange has been included as part of conventional water-quality simulations, as with CE-QUAL-ICM and, at the user's discretion, with WASP5.

The sixth major category is pollutant loading. All models allow the user to input steady pollutant loads. Some allow the specification of variable loads from the input dataset. Finally, a few models provide internal or external linkage to non-point-source loading simulations. HSPF features internal linkage with pervious and impervious land modules (see discussion in Sec. 14.2). WASP5, on the other hand, provides for external linkage to many loading models through a formatted non-point-source loading file. MIKE11 and SALMONQ come with linkage to urban wastewater modules, and CE-QUAL-ICM was linked with HSPF for application to Chesapeake Bay.

The last category in Table 14.3 summarizes software features and availability. Many of the models come with preprocessors to aid the user in specifying input data, and graphical postprocessors to aid in the interpretation of simulations. The scope and style of support software vary from tightly integrated packages to loosely connected modules that may link to commercial software. The organizations cited for model availability—CEAM, WES, HEC, Wallingford Software, and the Danish Hydraulics Institute—are summarized in Sec. 14.1.2.

Table 14.4 summarizes variables and water-quality processes employed by surface-water-quality models. The first major category, chemical processes, determines whether and how a model can be used to simulate toxicants. Some models here deal explicitly with toxicants in a simple manner, including SMPTOX3, HSPF, WASP5, and MIKE11. Other models have the capability to simulate toxicants in a more complex fashion. These include HSPF, WASP5, and EXAMS. Similar algorithms are being added to CE-QUAL-ICM.

Almost all models include a state variable that degrades at a user-specified first-order rate. Some models that specialize in the simulation of organic chemicals will include process kinetics. These algorithms predict chemical degradation rates from chemical properties in combination with various environmental forcing functions, such as pH, light, and temperature. Most of the chemical process models can track the daughter products of the transformation reactions.

In addition to degradation, toxicant models should contain algorithms to handle sorption to suspended and benthic solids. The simplest technique is equilibrium linear sorption, which is characterized by a simple partition coefficient. This coefficient may be specified directly or calculated internally for organic chemicals from such chemical and environmental properties as octonol–water partition coefficient and sediment organic carbon content. At high concentrations, the sorption process becomes nonlinear with respect to the chemical concentrations. None of the models listed here explicitly contains nonlinear sorption algorithms. Sorption kinetics are allowed as an option in HSPF, and may be handled indirectly in WASP5 or EXAMS using transformation reactions along with partition coefficients.

The second major category in Table 14.4, sediment processes, is most important for models that simulated toxicants. Most of the models that simulate sediment or solids as a state variable employ a simple mass-balance approach with user input deposition, resuspension, and burial rates. Some models, such as HSPF and MIKE11, contain algorithms to predict cohesive and/or noncohesive sediment deposition and resuspension based upon flow and sediment characteristics. There are many models that specialize in solids transport, but do not include water-quality variables or processes. These sediment transport models are not listed here.

The final major category covers water-quality processes. Most of the conventional pollutant water-quality models listed here simulate temperature using simple heat-balance techniques. In WASP5 and EXAMS, temperature is provided as user input and is not simulated. Salinity is another variable that is explicitly included in most models. It is used both as a tracer to calibrate transport, and as a component of the DO saturation calculations. Although salinity is not explicitly included in WASP5 or EXAMS, it can be simulated using a chemical state variable with no decay or sorption. Bacteria, such as total or fecal coliform, is included in many models. Those that do not simulate bacteria explicitly can nevertheless use a chemical state variable with first-order decay and no sorption to simulate bacterial dynamics. The user must then be careful to properly specify bacterial loadings and boundary conditions, given the different analytical units reported for bacteria (i.e., counts/100 mL versus mg/L).

Most conventional pollutant models handle dissolved oxygen (DO) linked to carbonaceous biochemical oxygen demand (CBOD), as well as to nitrification, phytoplankton growth and respiration, and sediment oxygen demand (SOD). CE-QUAL-ICM calculates the DO budget using organic carbon rather than CBOD. The nitrogen and phosphorus cycles are treated explicitly by those models addressing eutrophication. Most models treat the nutrient cycles in a relatively simple manner. The phosphorus cycle, for example, contains inorganic phosphorus (or orthophosphate), phytoplankton phosphorus, and nonliving organic phosphorus. The nitrogen cycle contains ammonium nitrogen, nitrite nitrogen, nitrate nitrogen, phytoplankton nitrogen, and nonliving organic nitrogen. Some models, such as CE-QUAL-ICM, employ more complex nutrient cycles that simulate organic nitrogen and phosphorus in various forms, such as particulate and dissolved, labile and refractory. Some also treat the silicon cycle in order to handle the growth of diatoms.

Most eutrophication models here treat phytoplankton as a single variable. CE-QUAL-ICM handles three phytoplankton classes separately. Zooplankton are included as a variable in two models here—CE-QUAL-ICM and MIKE11. Benthic algal routines have recently been added to some models in experimental versions; MIKE11 and SALMONQ include benthic algae in their standard release. Sediment oxygen demand (SOD), along with benthic nutrient flux, is almost always specified by the user at present. Although WASP5 has primitive capabilities to simulate benthic nutrients and SOD, only CE-QUAL-ICM offers this capability in a fully developed form.

14.3.4 Discussion

Simple analytical water-quality models have been employed since the 1920s. More complex, computerized water-quality models have now had more than 25 years of use. Many of the models listed here have evolved over a long period, and have stood the test of time. An excellent example is QUAL2E, which has been a worldwide standard for conventional water-quality analysis in streams. Other models are of recent origin and employ promising new scientific algorithms. Good examples of these are the CORMIX expert system for mixing-zone analysis and CE-QUAL-ICM for conventional water-quality interactions in water column and benthic sediment.

Toxicant pollutant models have been operational over the past decade. Some models, like EXAMS, have emphasized organic chemical reactions, and have been used primarily for pesticide analysis in simple aquatic systems. Models such as HSPF and WASP5 have imbedded EXAMS chemical kinetics in transport frameworks that simulate flow, sediment transport, and chemical dynamics in more complex aquatic systems. HSPF handles watersheds with internal coupling of land-surface and river-reach modules. WASP5 features external linkage to hydrodynamic and runoff models. SPMTOX3 has simplified the chemical kinetics and transport algorithms to give a more easily used but less predictive modeling system for toxicants.

The series of models produced by the U.S. Army Corps of Engineers have emphasized dynamic flow and transport in regulated rivers and reservoirs. CE-QUAL-RIV1, CE-QUAL-W2, and HEC5Q contain a solid description of conventional water-quality interactions. CE-QUAL-ICM, designed to link with external multidimensional hydrodynamic models, contains the most complex water-quality dynamics.

Several emerging trends are apparant in surface-water-quality modeling. One key area is facilitating realistic simulations by linking models within graphical shells and decision-support systems. These shells can not only link models with databases but also with other models through standardized file transfers. In particular, we are now seeing more routine linkage of water-quality models with three-dimensional hydrodynamics programs. Linked watershed–water body modeling systems are also being produced. An important trend in conventional pollutant modeling is the replacement of the benthic boundary with direct simulation of sediment oxygen demand and benthic nutrient fluxes.

14.4 GROUNDWATER-QUALITY MODELS

Groundwater is an important resource in many areas for its use as a source of drinking water and irrigation water. In addition, groundwater often affects the condition of important surface water resources. In many areas, groundwater may be threatened by the leaching of pesticides and other agricultural chemicals and the leaching of industrial chemicals from hazardous-waste sites. Because of the importance of this resource, and because the degradation of groundwater cannot be easily reversed, the assessment of threats to groundwater quality from human activities is often required. Groundwater-quality models are increasingly used as part of these assessments.

14.4.1 Vadose Zone Models

Several models (see Table 14.5) have been developed to simulate contaminant fate in soils and the underlying vadose, or unsaturated zone. These have been used to analyze the problem of leaching from agricultural fields or hazardous waste sites. All the vadose zone models covered here are one-dimensional, vertical in orientation. More complex multidimensional, saturated/unsaturated models are covered in the next section.

RITZ. The RITZ model (Nofziger, et al., 1988) was developed to simulate the transport and fate of hydrocarbons from a single waste application under different land-treatment scenarios. RITZ is a one-dimensional transport and fate model, utilizing steady-state water movement and linear partitioning of contaminant. The waste, containing the contaminant of interest, is assumed to be incorporated uniformly into the upper layer of the vadose zone. Although the waste can contain oily materials, the model assumes that the oil will not migrate. The contaminant and oil are subject to first-order decay. The model estimates the amount of the contaminant that is lost to the atmosphere by volatilization, and the concentration of contaminant migrating to groundwater is calculated. RITZ is available from EPA CSMoS.

VIP. VIP (Stevens, et al., 1989) is an extension of the RITZ model developed for evaluating the fate of a hazardous substance in the unsaturated zone following land application of oily wastes. The model simulates vadose zone processes including volatilization, degradation, adsorption/desorption, advection, and dispersion. Four physical phases in the vadose zone are considered, including water, oil, soil grains, and soil-pore air. VIP is available from EPA CSMoS.

CMLS. The Chemical Movement in Layered Soils (CMLS) model (Nofzinger and Hornsby, 1988) estimates the depth of the peak concentration of a nonpolar organic chemical as a function of time after application. The model calculates the relative amount of the chemical in the soil profile as a function of time. Chemicals are assumed to move only in the liquid phase in response to soil water movement. The change in depth of the chemical depends upon the quantity of water moving

TABLE 14.5 Unsaturated Groundwater Quality Models

Model		RITZ	VIP	CMLS	HYDRUS	MOFAT	MULTIMED	VLEACH	SESOIL	PRZM2	VADOFT
Time	Steady	N	N	N	N	N	Y	Y	N	N	N
	Quasidynamic	Y	Y	Y	N	N	Y	Y	Y	N	N
	Dynamic	N	N	N	Y	Y	N	N	N	Y	Y
Flow	Input	N	N	Y	N	N	Y	N	N	N	N
	Simulated	Y	Y	N	Y	Y	Y	N	Y	Y	Y
	Darcian flow	N	N	Y	Y	Y	Y	N	Y	N	Y
	Variable soil properties	N	N	Y	Y	Y	Y	N	Y	Y	Y
	Hysteresis	N	N	N	Y	N	N	N	Y	N	N
	Water retention function	N	N	N	Y	Y	Y	N	N	N	Y
	Evapotranspiration	Y	Y	Y	Y	N	N	N	Y	Y	N
	Root water uptake	N	N	Y	Y	N	N	N	Y	Y	N
Transport	Advection	Y	Y	Y	Y	Y	Y	Y	Y	Y	Y
	Dispersion—water phase	N	Y	N	Y	Y	Y	N	Y	Y	Y
	Dispersion—air phase	N	N	N	N	N	N	Y	N	N	N
Chemical processes	First-order decay	Y	Y	N	Y	N	Y	N	Y	Y	Y
	Process kinetics	N	N	N	N	N	Y	N	N	Y	Y
	Volatilization	Y	Y	N	N	Y	N	Y	Y	Y	N
	Daughter products	N	N	N	N	N	Y	N	N	Y	Y
	Linear sorption	Y	N	Y	Y	Y	Y	Y	N	Y	Y
	Nonaqueous phase	Y	Y	N	N	Y	N	N	N	N	N
	Initial contaminant soil	N	Y	N	Y	Y	N	Y	Y	Y	Y

past the chemical, properties of the chemical, and selected soil properties. CMLS is available from

University of Florida
IFAS Software Support
203 Building 120
Gainesville, FL 32611

HYDRUS. HYDRUS simulates one-dimensional variably saturated water flow and solute transport in porous media (Kool and van Genuchten, 1991). The solution of the flow problem considers the effects of root water uptake and hysteresis in the unsaturated soil hydraulic properties. The solute transport equation incorporates the effects of ionic or molecular diffusion, hydrodynamic dispersion, linear or non-linear equilibrium adsorption, and first-order decay. The boundary conditions for the flow and transport equations may be constant or time-varying. The code utilizes fully implicit, Galerkin-type linear finite element solutions of the governing flow and transport equations. HYDRUS is available from

USDA Soil Salinity Laboratory
4500 Glenwood Drive
Riverside, CA 92501

MOFAT. MOFAT is a two-dimensional, multiphase, multicomponent finite element model (Katyal, et al., 1991). Flow and transport of three fluid phases (water, NAPL, and gas) are simulated, along with up to five partitionable (equilibrium) components. MOFAT has the flexibility to handle the conditions expected in a field scenario; notably, the flow of a gas phase can be modeled directly, and its impact on volatilization can be assessed. This flexibility results in a large data requirement. MOFAT is available from EPA CSMoS.

MULTIMED. The Multimedia Exposure Assessment Model (MULTIMED) simulates the movement of contaminants leaching from a waste disposal facility. The model consists of a number of modules that predict contaminant concentrations at a receptor due to transport in the subsurface, surface water, or air. The model includes options for directly specifying infiltration rates to the unsaturated and saturated zones, or a MULTIMED module can be used to estimate infiltration rates. MULTIMED is available from EPA CEAM.

VLEACH. VLEACH is a one-dimensional finite difference model that simulates the transport of chemicals through the vadose zone to the water table via aqueous advection and gaseous diffusion. The code can simulate leaching in a number of distinct "polygons," or vertical stacks of cells that may differ in soil properties, recharge rate, depth to water, or initial conditions. This model does not consider chemical or biological degradation processes. VLEACH is available from the EPA CSMoS and the IGWMC.

SESOIL. The Seasonal Soil Compartment Model (SESOIL) is a one-dimensional, finite difference flow and transport model designed for long-term hydrologic, sediment, and pollutant fate simulations (Hetrick and Scott, 1993). Designed as a screening-level tool for evaluation of the movement of contaminants through the vadose zone, SESOIL input data requirements are not very detailed, and the model is relatively simple to use. SESOIL is available from IGWMC and from EPA OTS as part of the GEMS modeling system.

PRZM2. The Pesticide Root Zone Model (PRZM2) is a one-dimensional, dynamic, compartmental model that can be used to simulate chemical movement in the unsaturated zone within and immediately below the plant root zone (Carsel, et al., 1984; Mullins, et al., 1993). The original PRZM water-balance algorithms, based on the SCS curve number hydrology, are linked with the VADOFT unsaturated zone model. PRZM2 is linked with a comprehensive database and interface in the Pesticide Assessment Tool for Rating Investigations of Transport (PATRIOT). PATRIOT is designed to provide rapid analyses of groundwater vulnerability to pesticides on a regional, state, or local level (Imhoff, et al., 1993). PRZM2 and PATRIOT are available from EPA CEAM.

14.4.2 Saturated Zone Models

This section summarizes multidimensional groundwater-quality models (see Table 14.6) that can be applied to the saturated zone. Many of the models cited here include equations for variable saturation, and can be applied to combined saturated/unsaturated groundwater systems.

AT123D. The Analytical Transient 1,2,3-Dimensional Model (AT123D) provides an analytical one-, two-, or three-dimensional solution to flow and transport in

TABLE 14.6 Saturated Groundwater Quality Models

	Model	AT123D	MULTIMED	RANDOMWALK	MOC	SUTRA	SWIFT-II	FASTCHEM	3DLEWASTE
Subsurface system	Unsaturated	N	Y	N	N	Y	N	Y	Y
	Saturated, unconfined	Y	Y	Y	N	Y	Y	Y	Y
	Saturated, confined	Y	Y	Y	Y	Y	Y	Y	Y
Dimension	1-D	Y	Y	N	N	N	N	N	N
	2-D	Y	Y	Y	Y	Y	N	Y	N
	3-D	Y	Y	N	N	N	Y	N	Y
Time	Steady	Y	Y	Y	Y	Y	Y	Y	Y
	Quasidynamic	Y	Y	N	N	N	N	N	N
	Dynamic	N	N	Y	Y	Y	Y	Y	Y
Flow	Input	N	Y	N	N	N	N	N	N
	Simulated	Y	N	Y	Y	Y	Y	Y	Y
	Darcian flow	Y	Y	Y	Y	Y	Y	Y	Y
	Fractured flow	N	N	N	N	N	Y	N	N
	Heterogeneous aquifer	N	N	Y	Y	Y	Y	Y	Y
	Variable density flow	N	N	N	N	Y	Y	N	N
Transport	Advection	Y	Y	Y	Y	Y	Y	Y	Y
	Dispersion— water phase	Y	Y	Y	Y	Y	Y	Y	Y
Chemical processes	First-order decay	Y	Y	Y	Y	Y	Y	Y	Y
	Organic process kinetics	N	Y	N	N	N	N	N	N
	Geochemical reactions	N	N	N	N	N	N	Y	N
	Daughter products	N	Y	Y	N	N	N	N	N
	Linear sorption	Y	Y	Y	Y	Y	Y	Y	Y
	Nonlinear sorption	N	N	Y	Y	Y	N	Y	N
	Temperature/ heat balance	N	N	N	N	Y	Y	N	N

a homogeneous aquifer (Yeh, 1981). While flows are steady, the model can handle either constant or time-variable chemical loadings to the groundwater. Chemical processes include linear sorption and first-order decay. AT123D is available from IGWMC and from EPA OTS as part of the GEMS modeling system.

MULTIMED. The Multimedia Exposure Assessment Model (MULTIMED) simulates the movement of contaminants leaching from a waste disposal facility. The model consists of a number of modules that predict contaminant concentrations at a receptor due to transport in the subsurface, surface water, or air. Steady saturated flows are input by the user. MULTIMED calculates the mixing of a leachate plume from the vadose zone into a homogeneous confined aquifer, and the three-dimensional advection and dispersion downgradient. Chemical processes include linear sorption, hydrolysis, and first-order decay. MULTIMED is available from EPA CEAM.

RANDOM WALK. RANDOM WALK (Prickett, et al., 1981) is a two-dimensional contaminant transport model for heterogeneous, confined, or unconfined aquifers. Transport is driven by flows from the Prickett-Lonnquist Aquifer Simulation Model PLASM. Advection and dispersion are handled using particle in cell and random walk techniques. Variable-shaped chemical source areas can be continuous or slug release. First-order chemical reactions are simulated. RANDOM WALK is available from the IGWMC or from

Director of Computer Service
Illinois State Water Survey
Box 5050, Station A
Champaign, IL 61820

MOC. The U.S. Geological Survey Method of Characteristics (MOC) is a two-dimensional finite difference solute transport and dispersion model (Konikow and Bredehoeft, 1978) for heterogeneous, anisotropic confined aquifers. Implicit unsteady flow calculations are uncoupled from solute transport. Advection is solved using the method of characteristics. A two-step explicit procedure is used to solve for the effects of dispersion and dilution. Chemical reactions include first-order decay, equilibrium sorption using a choice of isotherms, and ion exchange. Software is available from the USGS and the IGWMC.

SUTRA. The U.S. Geological Survey Saturated-Unsaturated Transport (SUTRA) model is a two-dimensional model for variably saturated subsurface environments (Voss, 1984). SUTRA simulates fluid, heat, and solute movement with a hybrid finite element and integrated finite difference scheme. The model solution can be configured in an areal or a cross-sectional network. Chemical boundary conditions, sources, and sinks can be time variable. Chemical reactions include first-order decay, sorption, and production. SUTRA is available from the USGS and the IGWMC.

SWIFT-II. The Sandia Waste Isolation, Flow, and Transport Model (SWIFT-II) is a three-dimensional coupled flow and transport model for saturated porous and fractured media (Reeves, et al., 1986). SWIFT-II employs a finite difference scheme to simulate fluid, heat, brine, and trace solute movement. Fractured zones may be represented by a dual porosity or a discrete fracture approach. Chemical reactions include first-order decay and linear sorption. SWIFT-II is available from Argonne National Laboratory.

FASTCHEM. The Fly Ash and FGD Sludge Transport and Geochemistry (FASTCHEM) model is a package of six computer codes to simulate the two-dimensional transport of inorganic chemicals through unsaturated-saturated groundwater systems (Hostetler, et al., 1989). FASTCHEM uses a two-dimensional finite element scheme to calculate steady flow through a heterogeneous subsurface environment. Inorganic chemicals are advected and dispersed through a series of one-dimensional streamtubes derived from the flow model. An equilibrium speciation model is used to calculate aqueous speciation, solubility, ion exchange, and absorption reactions. FASTCHEM is available from the Electric Power Research Institute and Battelle Pacific Northwest Laboratories.

3DFEMWATER/3DLEWASTE. The Three-Dimensional Finite Element Model of Water Flow Through Saturated-Unsaturated Media (3DFEMWATER) and the Three-Dimensional Lagrangian-Eulerian Finite Element Model of Waste Transport Through Saturated-Unsaturated Media (3DLEWASTE) are related numerical codes that can be used together to model flow and transport in three-dimensional, variably saturated porous media under transient conditions with multiple distributed and point sources/sinks (Yeh, et al., 1992). Chemical reactions include first-order decay and linear sorption. The complexity of the 3DFEMWATER/3DLEWASTE numerical models requires that they be used by experienced numerical modelers with a strong background in hydrogeology. 3DFEMWATER/3DLEWASTE is available from EPA CEAM.

14.4.3 Discussion

The use of saturated zone models has proliferated over the last 20 years, with applications for many different site types and hydrogeologic settings. The use of unsaturated zone models, however, has been primarily focused in the theoretical and research applications during this same time period. It has only been within the last 5 to 6 years that unsaturated zone models have been applied in nonresearch settings, and significant experience and education are required for their appropriate application to real-world problems. These models have been primarily used for the investigation of water movement in unsaturated systems, without much regard to the contaminant processes that are occurring. A considerable amount of research is still being conducted to test the principles and theory being utilized by these models, especially in the area of contaminant transport and fate.

With the increased emphasis on remediation of hazardous-waste sites, management of pesticide use, and remediation of leaking underground storage tanks, the use of unsaturated zone models has become of great interest to managers and decision makers. This is especially true for the application of models to determine soil cleanup levels, remediation goals, and pollutant action levels. A basic problem has become apparent—there is a general lack of knowledge about the physical, chemical, and hydraulic processes of the unsaturated zone. Not only are contaminants in this system being driven by hydraulic gradients, preferential flow paths, and gravity (primary characteristics of a saturated system), they are also affected by capillary pressure gradients, vapor concentration gradients, and other processes. For a science that has primarily dealt with water movement in soils and, in more recent times, pesticide retardation and fate, to address many of the organic and inorganic pollutants is a tremendous leap to the edge of scientific understanding.

It is then appropriate to question the routine use of these groundbreaking modeling efforts in management and policy decision making when the scientific review process has not had adequate opportunity to reach a credible consensus. Thus it becomes important that models selected have adequate review, and have been shown to be applicable to the problem for which they are being considered. The models that do not meet these minimum requirements are likely to result in lengthy technical and policy-driven debates, and are likely to be disallowed when their technical soundness might otherwise be acceptable.

A renewed interest in the acceptance of saturated and unsaturated zone models has developed within the last few years due to the greater need for justifying the models' applications. Two primary factors are generally being addressed when assessing the use of models: (1) model testing and credibility and (2) parameter variability effects on model outcomes.

Parameter variability issues are becoming more important for understanding how to interpret the results of both saturated and unsaturated zone models. These systems are naturally heterogeneous, and cannot always be treated as homogeneous and isotropic. Most models, however, must make these assumptions to simplify the numerical and analytical techniques used to simulate these systems. Therefore, to understand the mechanisms and the outcomes of contaminant fate and transport in subsurface systems, the model user must deal with the parameter variability issues.

Tremendous advances are occurring in the application of models to hazardous-waste site remediation and waste management. These advances will continue as long as the viability and the limitations of the models are not overlooked, and adequate model testing is not sidestepped to advance a policy concept. Model testing can build credibility and contribute much to the application of models. Proper testing and consideration of parameter variability can provide a foundation for model application to support policy-based and remediation-based decision making.

REFERENCES

Abbott, J., "Guidelines for Calibration and Application of STORM," Training Document no. 8, Hydrologic Engineering Center, Corps of Engineers, Davis, Calif., 1977.

Akar, P. J, and G. H. Jirka, "CORMIX2: An Expert System for Hydrodynamic Mixing Zone Analysis of Conventional and Toxic Submerged Multipoint Discharges," EPA/600/3-91/073, U.S. Environmental Protection Agency, Athens, Ga., 1991.

Alley, W. M., and P. E. Smith, "Distributed Routing Rainfall-Runoff Model Version II," USGS Open File Report 82-344, Gulf Coast Hydroscience Center, NSTL Station, Miss., 1982a.

Alley, W. M., and P. E. Smith, "Multi-Event Urban Runoff Quality Model," USGS Open File Report 82-764, Reston, Va., 1982b.

Ambrose, R. B., T. A. Wool, J. P. Connolly, and R. W. Schanz, "WASP4, A Hydrodynamic and Water Quality Model," EPA/600/3-87/039, U.S. Environmental Protection Agency, Athens, Ga., 1987.

Appel, C. A., and T. E. Reilly, "Selected Reports that Include Computer Programs Produced by the U.S. Geological Survey for Simulation of Ground-Water Flow and Quality," Water Resources Investigations Report 87-4271, U.S. Geological Survey, Reston, Va., 1988.

Arnold, J. G., J. R. Williams, A. D. Nicks, and N. B. Sammons, "SWRRB, A Basin Scale Simulation Model for Soil and Water Resources Management," Texas A&M Press, 1989.

Baumgartner, D. L., W. E. Frick, and P. J. W. Roberts, "Dilution Models for Effluent Discharges," 2d ed., EPA/600/R-93/139, Newport, Oreg., 1993.

Beasley, D. B., and L. F. Huggins, "ANSWERS Users Manual," EPA-905/9-82-001, U.S. Environmental Protection Agency, Region V, Chicago, Ill., 1981.

Bicknell, B. R., J. C. Imhoff, J. L. Kittle, A. S. Donigian, and R. C. Johanson, "Hydrological Simulation Program FORTRAN (HSPF): User's Manual for Release 10," EPA-600/R-93/174. U.S. Environmental Protection Agency, Athens, Ga., 1993.

Brown, L. C., and T. O. Barnwell, "The Enhanced Stream Water Quality Model QUAL2E and QUAL2E-UNCAS: Documentation and User Manual," EPA-600/3-87/007, U.S. Environmental Protection Agency, Athens, Ga., 1987.

Burns, L. A., "Exposure Analysis Modeling System: User's Guide for EXAMSII Version 2.94," EPA/600/3-89/084, U.S. Environmental Protection Agency, Athens, Ga., 1990.

Computer Science Corporation, "Pesticide Runoff Simulator User's Manual," U.S. Environmental Protection Agency, Office of Pesticides and Toxic Substances, Washington, D.C., 1980.

Danish Hydraulic Institute, "MOUSE Users Guide and Documentation," Danish Hydraulic Institute, Agern Alle 5, DK-2970 Hørsholm, Denmark, 1990.

Doneker, R. L., and G. H. Jirka, "CORMIX1: An Expert System for Mixing Zone Analysis of Conventional and Toxic Single Port Aquatic Discharges," EPA/600/3-90/012, U.S. Environmental Protection Agency, Athens, Ga., 1990.

Donigian, A. S., Jr., J. C. Imhoff, B. R. Bicknell, and J. L. Kittle, Jr., "Application Guide for Hydrological Simulation Program FORTRAN (HSPF)," EPA-600/3-84-065, U.S. Environmental Protection Agency, Athens, Ga., 1984.

Donigian, A. S. Jr., and W. C. Huber, "Modeling of Nonpoint Source Water Quality in Urban and Non-urban Areas," EPA/600/3-91/039, U.S. Environmental Protection Agency, Athens, Ga., 1991.

Driscoll, E. D., P. E. Shelley, and E. W. Strecker, "Pollutant Loadings and Impacts from Highway Stormwater Runoff," Vol. I, Design Procedure (FHWA-RD-88-006); Vol. III, "Analytical Investigation and Research Report" (FHWA-RD-88-008), Office of Engineering and Highway Operations R & D, Federal Highway Administration, McLean, Va., 1989.

Feldman, A. D., "HEC Models for Water Resources System Simulation: Theory and Experience," *Advances in Hydroscience,* vol. 12, Academic Press, New York, 1981, pp. 297–423.

Fischer, H. B., E. J. List, R. C. Y. Koh, J. Imberger, and N. H. Brooks, *Mixing in Inland and Coastal Waters,* Academic Press, New York, 1979.

Frere, M. H., J. D. Ross, and L. J. Lane, "The Nutrient Submodel," in W. G. Knisel, ed., "CREAMS: A Field Scale Model for Chemicals, Runoff, and Erosion from Agricultural Management Systems," U.S. Department of Agriculture, Conservation Research Report no. 26, 1980, pp. 65–86.

Hetrick, D. M., and S. J. Scott, "The New SESOIL User's Guide," Publication no. PUBL-SW-200, Wisconsin Department of Natural Resources, Madison, Wis., 1993.

Holley, E. R., and G. H. Jirka, "Mixing in Rivers," Tech. Report E-86-11, U.S. Army Waterways Experiment Station, Vicksburg, Miss., 1986.

Holst, R. W., and L. L. Kutney, "U.S. EPA Simulator for Water Resources in Rural Basins," Exposure Assessment Branch, Hazard Evaluation Division, Office of Pesticide Programs, U.S. Environmental Protection Agency (draft), 1989.

Hostetler, C. J., R. L. Erikson, J. S. Fruchter, and C. T. Kincaid, "FASTCHEM Package," vols. 1–5, EPRIEA-5870, Project 2485-2, Pacific Northwest Laboratories, Richland, Wash., 1989.

Huber, W. C., and R. E. Dickinson, "Storm Water Management Model User's Manual, Version 4," EPA/600/3-88/001a (NTIS PB88-236641/AS), U.S. Environmental Protection Agency, Athens, Ga., 1988.

Huber, W. C., Heaney, J. P., and B. A. Cunningham, "Storm Water Management Model (SWMM) Bibliography," EPA/600/3-85/077 (NTIS PB86-136041), U.S. Environmental Protection Agency, Athens, Ga., 1985.

Hydrologic Engineering Center, "Storage, Treatment, Overflow, Runoff Model, STORM, User's Manual," Generalized Computer Program 723-S8-L7520, Corps of Engineers, Davis, Calif., 1977.

Imhoff, J. C., P. R. Hummel, J. L. Kittle, Jr., and R. F. Carsel, "PATRIOT—A Methodology and Decision Support System for Evaluating the Leaching Potential of Pesticides," U.S. Environmental Protection Agency, Athens, Ga., 1993.

Jirka, G. H., and S. W. Hinton, "User's Guide for the Cornell Mixing Zone Expert System (CORMIX)," Technical Bulletin 624, National Council of the Paper Industry for Air and Stream Improvement, New York, 1992.

Jones, G. R., and G. H. Jirka, "CORMIX3: An Expert System for the Analysis and Prediction of Buoyant Surface Discharges," Technical Report of the School of Civil and Environmental Engineering, Cornell University, Ithaca, N.Y., 1991.

Konikow, L. F., and J. D. Bredehoeft, "Computer Program of Two-Dimensional Solute Transport and Dispersion in Ground Water," "Techniques of Water Resources Investigations," Book 7, Chap. C2, U.S. Geological Survey, 1978.

Knisel, W., ed., "CREAMS: A Field-Scale Model for Chemicals, Runoff, and Erosion from Agricultural Management Systems," U.S. Department of Agriculture, Conservation Research Report no. 26, 1980.

Knisel, W. G., R. A. Leonard, and F. M. Davis, "Agricultural Management Alternatives: GLEAMS Model Simulations," *Proceedings of the Computer Simulation Conference,* July 24–27, Austin, Tex., 1989, pp. 701–706.

Leonard, R. A., W. G. Knisel, and D. A. Still, "GLEAMS: Groundwater Loading Effects of Agricultural Management Systems," *Trans. of the ASAE,* 30(5):1403–1418, 1987.

Limno-Tech, Inc. (LTI), "Simplified Method Program—Two Phase Toxics with Bed Interactions (SMPTOX3) User's Manual," Prepared for U.S. EPA Region IV, Atlanta, Ga., supported by U.S. EPA Assessment and Watershed Protection Division, Washington, D.C., 1990.

McCutcheon, S. C., "Transport and Surface Exchange in Rivers," in *Water Quality Modeling,* Vol. I, CRC Press, Boca Raton, Fla., 1989.

Mills, W. B., D. B. Porcella, M. J. Ungs, S. A. Gherini, K. V. Summers, Lingfung Mok, G. L. Rupp, and G. L. Bowie, "Water Quality Assessment: A Screening Procedure for Toxic and Conventional Pollutants in Surface and Ground Waters" (Revised 1985), EPA-600/6-85-002a and b, vols. I and II, U.S. Environmental Protection Agency, Athens, Ga., 1985.

Mullins, J. A., R. F. Carsel, J. E. Scarbrough, and A. M. Ivery, "PRZM-2, A Model for Predicting Pesticide Fate in the Crop Root and Unsaturated Zones: Users Manual for Release 2.0," EPA/600/R-93/046, U.S. Environmental Protection Agency, Athens, Ga., 1993.

Orlob, G. T., "Water Quality Modeling for Decision Making," *Journal of Water Resources Planning and Management,* ASCE, 118(3), May 1992.

Price, R. K., "Product Brief and Technical Notes for WALLRUS and SPIDA," Hydraulics Research Limited, Wallingford, Oxford, U.K., 1991.

Prickett, T. A., R. G. Naymik, and C. G. Lonnquist, "A Random-Walk Solute Transport Model for Selected Groundwater Quality Evaluations," Bulletin 65, Illinois State Water Survey, 1981.

Reeves, M., D. S. Ward, N. D. Johns, and R. M. Cranwell, "Theory and Implementation for Swift-II, The Sandia Waste-Isolation Flow and Transport Model for Fractured Media, Release 4.84," NUREG/CR-3328, SAND83-1159, Sandia National Laboratories, 1986.

Roesner, L. A., J. A. Aldrich, and R. E. Dickinson, "Storm Water Management Model User's Manual Version 4: Addendum I, EXTRAN," EPA/600/3-88/001b (NTIS PB88-236658/AS), U.S. Environmental Protection Agency, Athens, Ga., 1988.

Salhotra, A. M., P. Mineart, S. Sharp-Hansen, and T. Allison, "MULTIMED, the Multimedia Exposure Assessment Model for Evaluating the Land Disposal of Wastes—Model Theory," EPA/600/R-93/081, U.S. Environmental Protection Agency, Athens, Ga., 1993.

Scarbrough, J., "3DFEMWATER/3DLEWASTE: Numerical Codes for Delineating Wellhead Protection Areas in Agricultural Regions Based on the Assimilative Capacity Criterion," EPA/600/R-92/223, U.S. Environmental Protection Agency, Athens, Ga., 1992.

Stefan, H. G., R. B. Ambrose, and M. S. Dortch, "Formulation of Water Quality Models for Streams, Lakes, and Reservoirs: Modeler's Perspective," Miscellaneous Paper E-89-1, U.S. Army Corps of Engineers, Waterways Experiment Station, Vicksburg, Miss., 1989.

Terstriep, M. L., M. T. Lee, E. P. Mills, A. V. Greene, and M. R. Rahman, "Simulation of Urban Runoff and Pollutant Loading from the Greater Lake Calumet Area—Part 1, Theory and Development; Part 2, AUTO-QI User's Manual," Illinois State Water Survey, Champaign, Ill., 1990.

U.S. Environmental Protection Agency, "Results of the Nationwide Urban Runoff Program, Vol. I, Final Report," NTIS PB84-185552, U.S. Environmental Protection Agency, Washington, D.C., 1983.

U.S. Environmental Protection Agency, "Technical Support Document for Water Quality-based Toxics Control," U.S. EPA Office of Water, Washington, D.C., 1985.

U.S. Environmental Protection Agency, "Compendium of Watershed-Scale Models for TMDL Development," EPA841-R-92-002, U.S. EPA Office of Water, Washington, D.C., 1992.

van der Heidje, P. K. M., and O. A. Elnawawy, "Compilation of Ground-Water Models," EPA/600/R-93/118, U.S. Environmental Protection Agency, 1993.

Villars, M., M. Gerath, and D. Galya, "Review of Mathematical Models for Health Risk Assessment: III. Chemical Concentrations in Surface Water, Groundwater and Soil," *Environ. Software,* 8:135–155, 1993.

Voss, C. I., "A Finite-Element Simulation Model for Saturated-Unsaturated, Fluid-Density-Dependent Ground-Water Flow with Energy Transport or Chemically-Reactive Single-Species Solute Transport," U.S. Geological Survey, 1984.

Williams, J. R., A. D. Nicks, and J. G. Arnold, "Simulator for Water Resources in Rural Basins," *ASCE J. Hydraulic Engineering,* 111(6):970–986, 1985.

Williams, J. R., and H. D. Berndt, "Sediment Yield Prediction Based on Watershed Hydrology," *Transactions of the ASAE,* 20(6):1100–1104, 1977.

Williams, J. R., and A. D. Nicks, "CREAMS Hydrology Model—Option 1," in V. P. Singh, ed., "Applied Modeling in Catchment Hydrology, Proceedings of the International Symposium on Rainfall-Runoff Modeling," Water Resources Publications, Littleton, Colo., 1982, pp. 69–86.

Wurbs, R. A., "Water Management Models," IWR Report 93-NDS-16, Institute for Water Resources, U.S. Army Corps of Engineers, 1993.

Yeh, G. T., "AT123D: Analytical Transient One-, Two-, and Three-Dimensional Simulation of Waste Transport in the Aquifer System," Publication no. 1439, ORNL-5602, Oak Ridge National Laboratory, Environmental Sciences Division, Oak Ridge, Tenn., 1981.

Yeh, G. T., S. Sharp-Hansen, B. Lester, R. Strobl, and J. Scarbrough, "3DFEMWATER/3DLE-WASTE: Numerical Codes for Delineating Wellhead Protection Areas in Agricultural Regions Based on the Assimilative Capacity Criterion," EPA/600/R-92/223, U.S. Environmental Protection Agency, Athens, Ga., 1992.

Young, R. A., C. A. Onstad, D. D. Bosch, and W. P. Anderson, "Agricultural Nonpoint Source Pollution Model: A Watershed Analysis Tool," Agriculture Research Service, U.S. Department of Agriculture, Morris, Minn., 1986.

P · A · R · T · 3

WATER RESOURCES SUPPLY SYSTEMS

CHAPTER 15
SURFACE WATER RESOURCE SYSTEMS

Daniel P. Loucks
School of Civil and Environmental Engineering
Cornell University
Ithaca, New York

15.1 SURFACE WATER RESOURCE SYSTEMS: AN INTRODUCTION

It is common practice when studying or writing about water-resource systems to distinguish between surface waters and groundwaters, and between water quantity and water quality. The organization of this handbook illustrates this tradition. In the real world, however, these components of an integrated water-resources system may interact with and influence each other. This writer's charge is to outline some approaches to the modeling of surface-water-quantity systems, i.e., systems that are designed and operated to control, regulate, and, in general, manage the flows and storage of water that we can see on the surface of a watershed, river basin, or region. To the extent such surface water systems are responsive to groundwater–surface water interactions, and to water quality–quantity interactions, these other aspects of water-resource systems must also be included in surface water system models, at least in some approximate manner. The challenge, of course, is how to do this, recognizing the differences in the appropriate time and space scales for describing the processes that apply to these different components.

Each water-resource *system* is a set of interdependent water bodies and structures, each impacting on the state and performance of others, and together contributing to the overall performance of the system. Those involved in the design and operation of each structure, be it a single- or multipurpose reservoir, a diversion canal, a control structure affecting the input or output of a natural lake or wetland area, a hydropower plant, a groundwater withdrawal or recharge pumping plant, a water or wastewater treatment plant, or a flood control levee, must examine the impacts resulting from those individual components on the other components in the system. *Integrated water-resource systems planning and management* focuses not only on the performance of individual components, but also on the performance of the entire system of components.

An integrated view of water-resource systems cannot be compartmentalized into either surface water or groundwater and either water quantity or water quality just because the respective time and space scales make the modeling or study of such divisions convenient. The modeling approaches described in this chapter will attempt to address this interaction, at least to the extent appropriate when primarily focusing on the management of surface-water-quantity systems.

Models of water-resource systems must describe the essential features of such systems and the interdependencies among system components in the detail needed for the particular decisions being considered. They provide the means of predicting the expected overall system performance, given a set of assumed design and operating parameters and policies. Whether they use formal optimization procedures or not, models are used to simulate, i.e., to evaluate what might happen in the future if certain actions or decisions or policies were to be implemented. All model results are based on assumptions—assumptions about an uncertain future. Given a set of hydrologic, wasteload, design- and operating-policy assumptions, models can help predict what impacts might be associated with those assumptions. Hence, models can help identify what assumptions are important, and what assumptions are not so important, with respect to particular management objectives or goals. There is no value collecting more precise data if its precision will not affect the decisions being considered. With the help of models, humans can focus their attention on what models can not do, i.e., identify the best set of assumptions upon which to base their decisions.

Since this chapter is to be a description of surface-water systems modeling, the modeling approach used to frame this discussion will be a descriptive one, i.e., a *descriptive simulation modeling approach* (which may include prescriptive optimization). It is not the intent here to advocate any specific simulation modeling approach. The approaches used here simply serve as a means to describe at least some of the essential features of different simulation models of multicomponent surface-water-resource systems that interact with groundwater and water-quality systems.

The level of detail needed to model surface and groundwater and water quantity and quality can vary depending on the system and objectives to be served. Descriptive models of integrated water-resource systems involving surface and groundwaters and water quantity and quality are typically less detailed than descriptive models of only one of these four components. Since the focus in this chapter is mainly on surface water quantity, the level of detail in modeling groundwater and water-quality components is less than what it would usually be if these components were of primary interest. More detailed groundwater and water-quality approaches than those presented in this chapter are found in other chapters of this handbook.

15.2 DESCRIPTIVE MODELS AND MODEL USE

15.2.1 Why Model?

Structures and facilities used to control water quantity and/or quality do so by altering the distribution of water quantity and/or quality over time and space. Alternative designs and operating policies of systems that alter the distribution of water quantity and quality over time and space will generate different sets of multiple impacts. Models can help identify some of these spatial and temporal impacts, whether in areas as small as a catchment containing a single stream and a single water user or in regions encompassing multiple river basins containing multiple multipurpose reservoirs, water-supply diversions, groundwater pumping and recharge wells, and water and wastewater treatment plants. Some of these impacts may be beneficial, others may not. Furthermore, since the distribution of both benefits and costs over affected populations will vary, issues of equity are important. Finally, different individuals may evaluate the benefits and costs of these impacts differently.

The use of computer models of water-resource systems is one way of identifying effective design parameters and operating policies of system components and of estimating the probabilities of possible impacts resulting from various alternative system designs and operating policies. There are few, if any, better ways of obtaining this information, short of actually implementing a particular system and observing its performance. The latter option is clearly expensive and risky, especially if the system does not perform very well.

15.2.2 Evaluating Alternative Designs and Operating Policies

Given a set of assumptions regarding system design and operation, models can be developed to assess system performance. This is done by keeping track of where water is, where it goes, and what its quality is, for a fixed sequence of hydrologic and wasteload inputs during a fixed sequence of time periods. If applicable, the amount of power and energy generated and how much energy is used for pumping may also be determined. If the hydrologic and wasteload inputs are representative of what might occur in the future, the model results should be indicative of what one can expect to observe, at least in a statistical sense, in the future. Through multiple model runs under different hydrologic and wastewater-loading assumptions, individuals can test, modify, and evaluate various designs and operating policies in a systematic search for the ones that perform best.

15.2.3 Evaluating Sensitivity of Assumptions

It is obvious that the results of any computer model of a particular system design and operating policy are always very dependent on the assumptions incorporated into the model. The assumptions that must be incorporated into any model are numerous, and are often based on unknown or uncertain information. These assumptions typically include future hydrology, future user demands and consumption rates, changing facility design and operation, and numerous values of parameters of functions that define how the system works, e.g., the production of hydroelectric energy, the flow of groundwater, and the fate and transport of pollutants. It is impossible to know all of this information with any degree of precision.

It is not the purpose of a model to help determine which of numerous credible assumptions are best. Models can not replace human judgement. They can, however, be used to estimate what may happen given any set of assumptions. Thus they can be used to test how sensitive the performance of any particular system may be to various uncertain inputs and parameter values. This is a major, and often overlooked, benefit of modeling. There is no need to spend considerable time and money collecting more precise data on some aspect of the system being simulated if, through sensitivity analyses, one can determine that having more accurate data will not make much difference in the value of various system performance measures considered important. Through models, one can determine which assumptions substantially affect the performance of the system, and hence which assumptions may be worthy of additional study.

15.2.4 Models for Planning and Management

There are many kinds of water-resources models that can be used to address various planning, management, and operation issues. Some of these models are designed to examine in considerable detail the hydrologic or hydraulic processes that take place in watersheds, in water bodies, or in structures built to control or treat water. Others are designed to identify and evaluate potentially attractive structural and management alternatives prior to their detailed design or analysis. The descriptive modeling approaches outlined in this chapter are primarily designed to assist those involved in planning, management, and operation. First we will consider just planning and management.

Models for planning and management are typically based on mass balances of water quantities and water-quality constituents. In addition to mass balances throughout a system in each of multiple time steps, any model of a particular system needs to include the particular processes that take place in that system depending on the duration of the time steps being modeled. For example, if the discrete time-step durations are shorter than the time required for water to travel throughout the system, keeping track of where the water is within each simulated time step may be necessary. On the other hand, if the simulation time-step duration is larger than the time-of-flow through the system being modeled, flow routing may be not be necessary. Other phenomena that may be necessary to model could include seepage, evaporation, and consumption.

Models may also have to take into account the possible interaction between ground and surface waters, and between water quantity and water quality, again as applicable depending on the system being simulated. Physical and biochemical processes that take place in surface and groundwater systems usually involve quite different time and space scales. The level of detail needed in any particular model will depend in part on the purpose of the modeling exercise or study, the complexity of the system being modeled, and the available data. In this chapter the focus is on surface-water systems, but one cannot adequately model surface-water systems that interact with groundwater systems, or model quantity regulation if it is partly based on quality, without modeling, at least in some approximate way, the interacting groundwater and water-quality components of an entire water-resource system.

Water-resource systems models can be either generic or site specific. *Generic models,* as opposed to *site-specific models,* are designed to be applicable for many water-resource systems. There is no model now, nor is it likely that there will be a model developed in the future, that will be applicable to all systems anywhere in the world. If any generic model cannot address the issues of concern or the questions being asked, i.e., if it cannot provide the information needed for planning the development, or effective management, of a particular water-resource system, then an alternative site-specific model should be developed and used.

Fig. 15.1 broadly outlines a modeling process applicable to the planning and management of many water-resources systems. Obviously, how well each step of the process is performed influences the applicability and reliability of the model results and the necessary number of iterations through the modeling process.

Beginning in Box A of Fig. 15.1, it is important to note that a model developer and user should have a thorough understanding of both the objectives of the modeling study and the capabilities and limitations of the models being developed and/or used. Careful consideration of model study objectives and model capabilities and limitations will reduce the time and expenditure required for these multiple model runs.

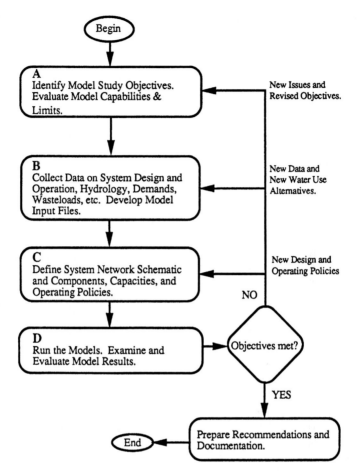

FIGURE 15.1 Modeling procedure for planning and management.

The process of data collection and analysis and conversion for model use is likely to continue as long as models are being used to study or analyze the system. All the steps between Boxes B and D in Fig. 15.1 are likely to be repeated many times in any study examining and evaluating design and operating policy alternatives. It is generally most efficient to perform the tasks outlined in Box B as completely as possible early in the study. Even so, it is possible that supplemental data will have to be obtained and analyzed throughout the study.

Activities in Box B of Fig. 15.1 must be done with care. The development of input data may well be the most difficult and labor intensive part of a model application. Errors in input data will certainly affect the accuracy of the model output. The user must develop data that represents the possible range of important supply, demand, and water-quality events, as applicable for the particular system being simulated.

The outcome of this modeling process should be a better understanding of the multiple impacts, over time and space, that are likely to result from various system component designs and operating policies, as well as a better understanding of how various data and assumptions affect those impacts. The latter information is helpful in planning specific data-collection and analysis programs.

15.2.5 Models for Operation

Models are also useful for *real-time operation*. In this mode, models are used to obtain estimates of effective operating decisions over time. Models can define operating decisions at each decision site that take into account the current and expected future state of the entire system, not just a part of the current state as is usually the case when applying *operating policies* (reservoir release, groundwater pumping, and diversion allocation). Often such policies are based only on the state of the system at the site where a decision must be made.

An example may help to illustrate this point. Consider a site on a river where water is diverted upstream of the reservoir. Most diversion policies will define the amount of water to withdraw from the stream as a function of the flow in that stream at the withdrawal or diversion site. The amount diverted will not be a function of the storage in the downstream reservoir, or of the need for water in future periods downstream of that reservoir. Having models that can look at the impacts of particular decisions over the entire system, and into future time periods, permits a more integrated system-wide operating policy to be defined for the upstream diversion. Of course, it is an interesting institutional question as to whether adaptive system-wide operating policies based on model results can be implemented. Can, for example, the current relative priorities of various users be identified in the political process and then incorporated into these simulation models—and even if so, will the model results be useful in the debate on how to operate the system?

Real-time operating models can also be used to help manage extreme events, such as floods and droughts, that occur over relatively short time periods. They can also be used to help manage a complex system on a continuing basis. These two different types of model applications will require models having different spatial and temporal resolutions.

Operating models are used sequentially. They are continually updated and rerun to get the most current estimates of what operating decisions should be made for each component, as applicable, in each future decision period. While only those decisions in the current period are of interest, the impact of these decisions in the future must be considered. To ensure that the current operating decisions included in the model results are not *myopic*, i.e., based only on the current state of the system and not what it might be in the near or distant future, the model must include a sufficient number of future time periods. The number of periods that should be included in the model can be determined by finding the future period in which whatever happens, it does not affect the operating decisions of the current period. Alternatively, desired storage volume and flow targets may be defined and set for some specified period in the future. Given a set of existing storage volumes and assumed future natural inflows, the operating policies for regulating storage volumes and flows are set so as to meet these future system-state targets.

Information concerning the current state of the system, the current and future inflows, wasteloads, costs and benefits, objectives or demand targets (or the relative importance of various possible objectives or demand targets), and any current and/or future design and/or operating constraints, must be continually updated each time the model is run. Fig. 15.2 illustrates the process of using models for real-time management.

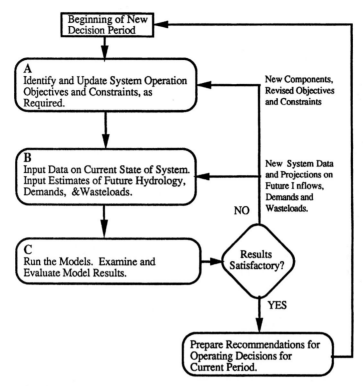

FIGURE 15.2 Modeling procedure for real-time operation.

The modeling process involves a continuing series of model data updates and model solutions. At the beginning of each decision period the assumptions and data in the model or models are updated, based on current flows, storage volumes, and quality conditions as appropriate, and on the best estimates of future conditions. This is represented by Boxes A and B in Fig. 15.2. The models are then solved (box C in Fig. 15.2) to obtain the most effective set of decisions for the current period and for a number of periods into the future. Only the current period's values are of interest. If they are considered satisfactory, they are then implemented. Then at the beginning of the next period the model's database is again updated and the modeling process is repeated. This sequential modeling process continues on into the future.

Most models that simulate this process of real-time operation usually employ an optimization routine for computing the operating decisions in each decision time period. Optimization replaces fixed predefined operating policies. Unless optimization is used, or is to be used, in the real world, the results from such optimization-based simulation models are not realistic. Such simulation models will be assuming operating policies based on optimization, when in fact optimization is not used in practice. It is not a trivial, if even possible, task to define operating policies from the results of optimization-based simulations. They—and any other modeling approach for that matter—cannot predict changes in system objectives and constraints that may result from surprises, i.e., events not foreseen or forecast.

Simulation models containing optimization in place of predefined operating policies include HEC-PRN (HEC, 1993), the Acres Simulation Model (Sigvaldason, 1976), MODSIM3 (Labadie, 1984) and WATNET (Kuczera, 1990). Also see Hooper, et al. (1991); Grygier and Stedinger (1985); Kuo, et al. (1990); Randall, et al. (1995); Shane (1980); and Yeh, et al. (1992).

Simulation models of water-resource systems, whether used for planning or for real-time management, merely provide information. They should not be expected to dictate actual decisions. Humans must still assume responsibility for any decisions made, since they are the ones making and entering the assumptions and data into the models that affect model results. Models cannot be used to determine which assumptions and data are best. They can only help identify the impacts of those assumptions and data. Models used for planning, management, and real-time operation can not substitute for the people responsible for recommending and making design and operating decisions.

15.2.6 User Interface Requirements

To be useful for examining the impacts of various design and operating policy decisions and for testing the effects of numerous assumptions, simulation models must be easy to use. Data input must be done in a manner that makes all assumptions clear and mistakes easy to detect and correct. All input data must be easy to modify for sensitivity analyses. The different operations of the model and program must meet the needs of a variety of users and be easy to control (see, for example, Keys and Palmer, 1993; Palmer, 1993). The model output must be able to be presented in a variety of ways, and each way must be easy to execute and understand. In addition, if a model is to be used for studying a wide variety of water-resource systems, the model's data input and editing features, its operation, and its output display capabilities must be readily adaptable and applicable to those different systems (see, for example, Fedra, 1992; Pabst, 1993).

15.3 SYSTEM STRUCTURE, CONFIGURATION, AND DATA

There are two general options for describing a surface-water system to be modeled. One is based on sets of *contiguous cells,* the other on a *node-link network.* Water flows from each cell to adjacent cells, or from each node to adjacent nodes, through connecting links. The rate or amount of flow depends on the water available and on the physics built into the cells, or nodes and links. The *cell-based approach* is reasonable when area-wide hydrologic processes must be modeled and the direction of flow between cells may vary due to differences in heads over time. These flows between cells can include not only surface-water flows, but also groundwater flows, having both horizontal and vertical components. Hence the cells can represent three-dimensional hydrologic systems. Nodes of a node-link network can also represent various subsurface zones or aquifers, or portions of aquifers, in three-dimensional space.

15.3.1 Cellular Network Structure

The area to be simulated can be divided into cells. These cells need not be uniform in size or shape, but they should be arranged to leave no gaps or areas not included in a cell. Each cell can be considered homogeneous with respect to area, or if three-dimensional, with respect to each vertical zone in each cell. What must be described (modeled) and simulated is the water flows to and from each zone in each cell, both vertically and horizontally.

Computational as well as physical considerations often dictate the type, shape, and sizes of the cellular structure. Figure 15.3 illustrates three options, an irregular rectangular cell network, an irregular triangular cell network (TIN), and an irregular polygon cell network. The more rigid rectangular cell network representation has features that facilitate the computation of flows between cell zones, whereas the more flexible irregular polygon network representation has the greatest ability to capture more accurately the spatial characteristics of the system. Each method permits an increase in the number of cells, and hence a reduction in individual cell size, in areas where more spatial detail is desired or warranted (Taylor and Behrens, 1995; Razavian, et al., 1990).

Figure 15.3 illustrates plane views of alternative cell networks. These cells may need to be extended into the third dimension if groundwater processes are to be simulated. Figure 15.4 illustrates how different subsurface zones in several three-dimensional triangular cells, for example, might be defined. It also identifies some of the hydrologic processes that can affect the water balance in these cell zones. These zones and processes could be included in a simulation model, as appropriate.

The hydrologic processes that are included in any simulation model depend on the model developer. The processes needed in any simulation model depend on the area or region being modeled and the level of detail required for the decisions being made. Descriptions of hydrologic processes, including those listed in Fig. 15.4, are found in almost any hydrologic text or handbook (e.g., Maidment, 1993; also see Fiering, et al., 1982, and Chaps. 24, 26, 27 and 28 of this handbook)

15.3.2 Node-Link Network Structure

Most water-resource systems can also be represented by schematic node-link networks. Examples include those discussed or incorporated into models by Andreu, et al. (1991); Basson, et al. (1994); Bureau of Reclamation (1987); CADSWES (1992); Eichert (1992); Doland and DeLuca (1991); HEC (1985); Kuczera and Diment (1987); Loucks, et al. (1994); Stockholm Environment Institute (1993); and Strzepek, et al., (1979; 1989), to mention only a few. The nodes of the network can represent aquifers, gauge sites, consumption sites, natural lakes, reservoirs, wetland areas, confluences, diversions, or simply monitoring sites. A single node may be a combination

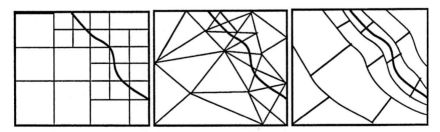

FIGURE 15.3 Three alternative irregular discretizations of geographic areas for hydrologic simulation modeling: rectangular, triangular, and polygon.

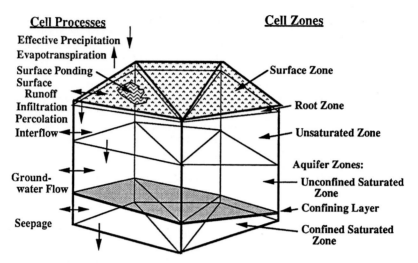

FIGURE 15.4 A three-dimensional cell network, defining various subsurface zones and hydrologic processes, as appropriate.

of some of these node types. Any storage node can be divided into substorage nodes, connected by bidirectional links. The unidirectional or bidirectional links of the network can represent river reaches, diversions, and area-wide water transfers, say between aquifers, between aquifers and surface-water components, and between wetlands and other components of the system. Any river reach or diversion link may contain a hydropower plant or pumping station (e.g., for pumped-storage hydroelectric plants or for aquifers subjected to pumping and/or artificial recharge).

Figures 15.5 and 15.6 illustrate a relatively simple water-resource system and its node-link network representation. This particular system does not have all the features or system components that could be included in a model network. It merely illustrates the process of representing an existing system as a node-link network.

An alternative way of representing the system shown in Fig. 15.5 would be to omit the diversion to the town, as shown in Fig. 15.6, and define consumption at the reservoir as a function of the reservoir release. This function would be the same as the town's diversion allocation function, i.e., a function defining the amount of water diverted to, and consumed by, the town given the total outflow from the reservoir. The simulated results would be the same. Under this alternative scheme, one less node and link would be required.

Although not illustrated in Fig. 15.5 and 15.6, multiple diversions upstream of a particular demand site of interest may be represented by a single diversion when their individual diversions are not of importance to the downstream demand site.

At each node, of whatever type, mass balances must be performed in each model time step. The initial storage volume plus the total inflow less the discharge and losses and consumption, if any, equals the final or end-of-period storage volume. If the node is a nonstorage node (i.e., its storage volume capacity is zero) then its initial and final storage volumes are zero, and there are no evaporation or seepage losses.

Flows between geographic locations (nodes) are usually represented by links. The direction of flow in links depends on the difference in storage (or groundwater

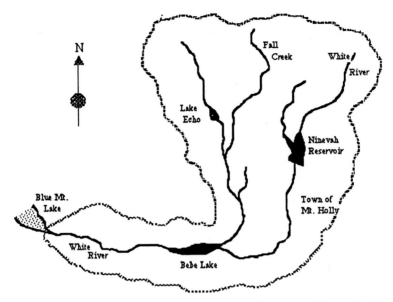

FIGURE 15.5 A sample water resource system that can be represented by a node-link network.

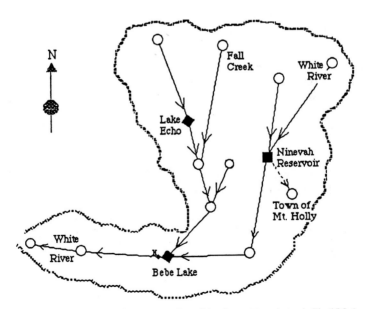

FIGURE 15.6 A network representation of the river system shown in Fig. 15.5. A hydropower plant is shown just downstream of Bebe Lake. All nodes are sites where inflows can occur and where flow and quality data can be monitored and recorded.

pressure) heads. In situations where this does not change over time, unidirectional links may be defined. Otherwise bidirectional links or two directional links in opposite directions may be appropriate.

Hydropower plants and/or pumping stations can be associated with any link. Hydropower plants may be fixed-head run-of-river plants or variable-head plants if their upstream nodes are reservoirs (see Chap. 31). Heads for power and energy production are computed as the difference in elevation between the upstream (reservoir) and either the downstream (tail-water) heads or the turbine elevation, whichever is less. If the difference is negative, then pumping requiring energy consumption can be computed, but only if a pumping plant has been defined for that link in the simulation model. If hydropower and pumping are assigned to a link, pumped-storage operations may also be modeled.

In each link of a water-resource system network, a mass balance of quantity and quality constituents must take place. What enters each network link component must either remain or exit. What remains may be due to flow routing in river reaches or diversion canals. What exits may include some type of loss (evaporation, seepage, decay, or transformation).

All water-resource system simulation models are based on the principle of mass balance. What actually happens or can happen in any particular node or link component will depend on the type of component and possibly on the state of adjacent components.

15.3.3   Model Time Periods

Simulation models of water-resource systems are typically based on discrete time periods. The number and duration of these discrete time periods will depend on the system being simulated, the available data, possibly the computational time required, and—most importantly—the questions being addressed, i.e., the purpose of the simulation. In each discrete time period input data, including precipitation or streamflows, are usually read from input data files, and the simulation proceeds to account for these new inputs, along with existing system flows and storage volumes, in a manner defined by the system's operating policy. Exogenous events (such as precipitation, inflows from runoff, evaporation rates, demands, and water-quality standards) that take place during, or apply to, the simulation time periods are usually assumed to remain constant throughout each time period, but may vary over different time periods. Endogenous events (like streamflows, reservoir releases and lake outflows, energy production or consumption, and storage volumes) can change or vary during the simulation time period.

Figure 15.7 illustrates the relationships among the different types of discrete time periods that can be defined in simulation models. This particular illustration is of a multi-year simulation model, divided into (not necessarily equal) within-year time periods. Each within-year period can be further divided into equal simulation time steps.

If time is discretized as shown in Fig. 15.7, input data and operating policies are usually defined for each discrete within-year period, and the model output is usually expressed as either average values or beginning and ending values for each within-year period over a sequence of years. Dividing each within-year period into discrete simulation time steps is one way to increase the accuracy of the simulation, especially when short time events (e.g., water-quality processes) are being modeled or when decisions (e.g., reservoir releases) are being made at regular intervals within each within-year period. An alternative to defining multiple simulation time steps

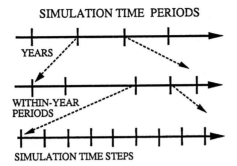

FIGURE 15.7 Types of time periods that can be included in simulation models.

within each within-year period is to implement event-based simulation that iterates over space within each within-year period until sufficient accuracy is achieved (CADSWES, 1992; Behrens and Loucks, 1994).

15.3.4 Hydrologic and Water-Quality Data

Simulation models using the cellular network representation often simulate the rainfall-runoff processes that generate stream flows. Precipitation is the forcing function that creates changes in the amounts of water available at different locations throughout a region being simulated. Models using the node-link network representation are often driven by time series of unregulated streamflows (runoff) at various gauge (flow measuring and recording) sites. In such cases, it is convenient, and therefore common, to assume that flows and quality constituents entering the system enter at the nodes of the node-link network schematic. Hence such models of water-resource systems require one or more time-series sequences of unregulated flows at all node sites. These may be computed from unregulated flow data at designated gauge sites.

The estimation of natural unregulated flow data is always difficult, given the lack of adequate records of all past events that may have altered the flows that were actually measured at various gauge stations. There can also be measurement errors, but these are often insignificant in comparison to the uncertainty regarding human-made events or actions that may have altered natural flows. Furthermore, what is of interest today is what those natural flows would have been given current and future land uses. Predictions of these data are very uncertain. Add to this the additional uncertainty regarding normal weather variations, variations that may be changing due to changes in our climate, i.e., the possibility that the spatial and temporal probability distributions of past precipitation and runoff may not be representative of what may occur in the future. Clearly, precipitation and streamflow estimations are just that—estimations. Whatever hydrologic data sequences are used, whether they are the actual measured (and perhaps deregulated) historical data or are synthetically generated data, we know with certainty they will not occur in the future. Hence different plausible hydrologic sequences should be used to test the sensitivity of these assumed future hydrologic data in any modeling application.

Having estimates of the natural unregulated flows at each node site, the incremental natural inflows to the system between adjacent node sites can be estimated. Incremental natural inflows at each node site are the differences between the natural flows at that site and the natural flows at all adjacent node sites upstream of that site.

Incremental inflows may have pollutant concentrations from nonpoint pollution sources, and if water quality is being simulated, these concentrations must be defined for each quality constituent being modeled, along with point-source quantities and constituent concentrations.

15.3.5 Operating Policies and Data

In addition to hydrologic and water-quality inputs, operating policy information is needed before any simulation can proceed. Operating policies are best defined as input data rather than as subroutines in the simulation program. This facilitates the analysis of different operating policies, one of the main purposes and advantages of simulation modeling. (See, for example, CADSWES, 1992; Eichert, 1993; Ford, 1990; HEC, 1985; Loucks, et al., 1994, and Chap. 28 of this handbook.)

Operating policies are usually defined for each within-year period. These policies for each within-year period can vary over successive years, perhaps in response to changing demands, but most models assume fixed policies from one year to the next. Operating policies need to be defined where and whenever a decision must be made, i.e., where and when there is a choice of action. Such situations include the allocation of flows to consumption sites or to each of multiple outflowing diversion canals or river reaches, releases from reservoirs, hydropower production, pumped storage, groundwater pumping (withdrawals and artificial recharge) and wastewater treament.

15.3.6 Link Allocation and Node Consumption Functions

Consumption can occur at any point in a surface-water system, but for modeling purposes it is often convenient to assume consumption can only occur at a node. *Consumption* can be defined as a function of the node outflow. Any amount consumed is lost from the system. If a portion of an allocation to a demand site is returned to the system, then the consumption must be defined as only that portion of the allocation that is not returned.

If a surface-water node serves as the upstream or inflow node of multiple links transporting water from that node to other surface water nodes, then link-flow allocation functions need to be defined, preferably in the model input data set. These link-flow allocation functions can either reflect policy decisions or natural physical conditions. They define the amounts of the total node outflow (excluding any consumption and/or loss at the node) that enter each of the multiple outflowing links. These allocations can be defined as functions of the total node outflow, as illustrated in Fig. 15.8, or as functions of demand targets or target deficits at nodes further downstream. *Demand targets* are the amounts of flow or storage desired at some point in time and space. They are also part of the input data. Demand-driven allocations would be constrained by the amount of water available to be allocated.

The allocation functions may differ in each within-year period, but in all cases they must preserve continuity of flow. No more water can enter all the outflowing links than is available, and all available water should be allocated.

15.3.7 Supply-Based Reservoir Release and Storage Target Functions

There are various ways to define reservoir release policies. Most typical are policies based on the storage volume in the reservoir, or in a group of reservoirs, and the par-

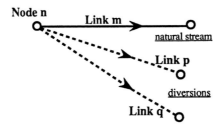

Node n	Outflow Allocations to		
Outflow	Link m	Link p	Link q
0	0	0	0
10	10	0	0
30	10	5	15
50	10	20	20
90	45	25	20

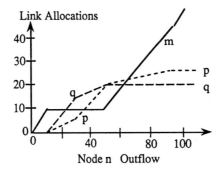

FIGURE 15.8 Link allocation functions for the two diversion links (dashed lines) and a main channel link carrying water from the node n.

ticular within-year time period (Kaczmarek and Kindler, 1982). These policies are often called *reservoir release rules* or *rule curves.* They can either be defined by specifying an ideal storage volume target in each within-year time period that the operator tries to maintain, or by dividing up the storage capacity into various zones in each within-year time period and specifying a release associated with each storage zone in each within-year period. There may also be some constraints as to the minimum and maximum desired releases, regardless of the release-rule policy.

Figure 15.9 illustrates a storage-zone based reservoir release rule.

15.3.8   Single Independent Reservoir Releases

Release policies for independent single reservoirs can be specified in the form of traditional storage- and time-dependent reservoir release rules, similar in form to that shown in Figs. 15.9 and 15.10. These release rules for a reservoir at node n are

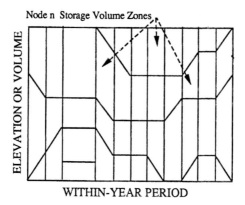

FIGURE 15.9 Reservoir release rule based on storage volumes. Storage capacity is divided into discrete zones and releases are defined for each zone in each within-year period. The reservoir operator has only to know what zone the current storage volume is in and the time period to know what to release.

based on a number of predefined storage zones in each within-year time period t. Each storage zone z includes a range of storage volumes. For each zone z, a policy can specify the minimum [$BS_{mn}(n,z,t)$ and $ES_{mn}(n,z,t)$] and maximum [$BS_{mx}(n,z,t)$ and $ES_{mx}(n,z,t)$] storage volumes at the beginning (B) and end (E) of the zone's within-year period t. The policy must also specify the zone's releases, $BR_{mn}(n,z,t)$, $ER_{mn}(n,z,t)$, $BR_{mx}(n,z,t)$, and $ER_{mx}(n,z,t)$, associated with these minimum and maximum storage volumes at the beginning and end of the within-year period t.

The reservoir releases $BR_{mn}(n,z,t)$, $ER_{mn}(n,z,t)$, $BR_{mx}(n,z,t)$, and $ER_{mx}(n,z,t)$ associated with the four corners of each storage zone z (as shown in Fig. 15.10) can be any nonnegative values. There is no required relationship among these four "corner" releases in each zone z, or among these and other corner releases in other zones located above, below, or to either side of zone z. For example, release $BR_{mn}(n,z,t)$ can be greater than $BRmx(n,z,t)$; however this is not usual practice. Normally it makes more sense to reduce releases as the storage volume decreases.

If constant and equal releases at the beginning and end of the within-year period in a particular zone are specified, as is often done, then $BR_{mn}(n,z,t) = ER_{mn}(n,z,t) = BR_{mx}(n,z,t) = ER_{mx}(n,z,t)$. The release will be constant for all storage volumes within the storage zone. Most reservoir operating rules specify constant releases for each storage zone, and reduced constant releases in decreasing storage zones.

The release in any simulation time step can be computed based on an interpolation of the four corner releases over both storage and time within the applicable storage zone and within-year period. This interpolation is illustrated in Fig. 15.10.

The storage zone z containing the actual initial storage volume $S(n,tt)$ can be identified by checking if the initial storage $S(n,tt)$ is within the storage volume ranges $S_{min}(n,z,t)$ and $S_{max}(n,z,t)$ of each zone z of the applicable within-year period t. Once the applicable zone z has been identified, the reservoir release targets $R_{min}(n,z,tt)$ and $R_{max}(n,z,tt)$ associated with the interpolated and applicable storage volume zone boundaries $S_{min}(n,z,tt)$ and $S_{max}(n,z,tt)$ can be calculated. It is then possible to interpolate the daily release $R(n,tt)$ for the simulation time step tt.

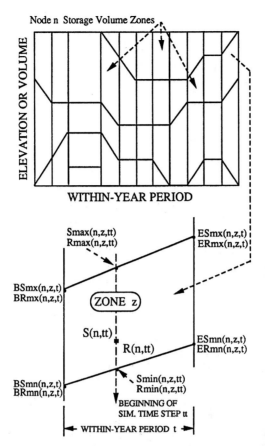

FIGURE 15.10 Reservoir *n* release rule showing release calculation in simulation time step *tt* within a within-year period *t* based on initial storage $S(n,tt)$, in a zone *z* and the maximum and minimum storage volumes for that zone at the beginning and ending of the period. Associated with these four storage maximum and minimum zone volumes are four corresponding releases. The release $R(n,tt)$ in each simulation time step *tt* is an interpolated value based on these four releases.

As previously stated, if all release values $BR_{mn}(n,z,t)$, $ER_{mn}(n,z,t)$, $BR_{mx}(n,z,t)$, and $ER_{mx}(n,z,t)$ in any zone *z* are the same, then the release is fixed at that value for any storage volume $S(n,tt)$ in the zone. This condition is typical of many reservoir release rules (HEC, 1985; Kaczmarek and Kindler, 1982).

15.3.9 Multiple Interdependent Reservoir Releases

For multiple reservoirs being operated as a group, i.e., interdependently, a supply-driven reservoir release rule typically will be a function of the total combined storage in the group of reservoirs. The reservoir group release rule, similar to that shown in Fig. 15.10, will usually be defined at the most downstream reservoir of the group and will apply to the total storage volume in the reservoir group. The release rule may also define a flow that is to be achieved at a site downstream of all of the reservoirs in the group.

The releases from all reservoirs in a reservoir group not located at the release-rule site can be defined based on specified reservoir storage targets or balancing functions. These balancing functions, as illustrated in Fig. 15.11, can differ for each

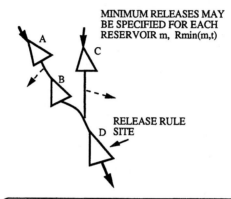

RESERVOIR GROUPS:

MINIMUM RELEASES MAY
BE SPECIFIED FOR EACH
RESERVOIR m, Rmin(m,t)

RELEASE RULE
SITE

Total Storage Volume	Target Storage Volume in Reservoir			
	A	**B**	**C**	**D**
10	5	0	5	0
20	10	0	10	0
60	30	5	20	5
100	30	25	25	20
200	30	45	60	65

FIGURE 15.11 Multiple reservoir storage targets, each expressed as functions of the total storage volume in all reservoirs of the group. These functions can be defined for all reservoirs in a group not a reservoir release-rule site. The target storage volumes are defined by interpolations between the volumes listed in the table. They can vary in each within-year period.

within-year period t. They define the desired or target storage in each upstream reservoir m of the group as a function of the current total reservoir storage in the entire group. The multireservoir release policy can attempt to keep all reservoirs in balance (at their target storage volumes) during the simulation by releasing in each simulation time step some fraction of the water in excess of the target storage volume. The particular fraction can be a function of the total reservoir group release from the most downstream reservoirs.

In each simulation time step each reservoir group release can be computed using the group release rule based on the total storage volume in the group. Knowing the total storage volume at the beginning of each simulation time step and the total reservoir group release permits the estimation of the end-of-simulation-time-step storage volume of the group. Using the storage target functions, the end-of-time-step storage volume for each reservoir m in the group can be calculated. The releases from each reservoir m in the group not at the release-rule site will be either the excess storage over this target storage volume, or a specified nonnegative minimum release.

These releases or outflows may be increased, if possible, due to downstream accumulated demand deficits, as will be discussed next.

15.3.10   Demand-Based Reservoir Release Functions

The reservoir-release policy just described is based on supply. Releases can also be based on downstream demand deficits. A *demand deficit* is the difference between a desired flow or storage volume and the actual flow or storage volume. Releases to meet demand deficits can have priority over releases based on available supply. Supply-based releases can be in addition to any releases based on demand.

Any accumulated deficit in meeting a demand can, if the simulation model includes such a policy, cause additional releases from designated upstream reservoirs. The amount of those releases can depend on the extent of the deficit and the number of decision periods (simulation time steps) remaining in the within-year period being simulated.

Natural lakes, by definition, are not controlled. Their outflows are a function of their storage volumes. Hence they cannot be used as upstream source sites for meeting downstream demand target deficits. If they are controllable, or if pumping from a natural lake can occur, then this natural lake may be modeled as a reservoir site whose outflow can be defined, or at least influenced, by an operating policy.

15.3.11   Aquifer Pumping Functions

If the node is a groundwater aquifer, the user must define any water withdraw or artificial recharge pumping policy. Such policies may be functions of the state of the surface-water system and/or demands or demand deficits at potential consumption sites.

15.3.12   Hydropower or Pumping Functions

Hydropower and/or pumping plants can be simulated, as applicable. Part of the data required to define the characteristics and operation of facilities where hydropower production or pumping energy consumption occurs includes the energy production constants that convert daily flows and head to energy. These constants include the plant and penstock efficiencies. While the energy production constants actually vary with head (because the plant and penstock efficiencies vary with head) they are often assumed to be constant for planning purposes. (See Chap. 31.)

Energy production constants can be derived based on the power equation:

$$\text{Power (kW)} = 9.81 \cdot \text{flow (m}^3\text{/s)} \cdot \text{head (m)} \cdot \text{energy production efficiency}$$

$$= 0.08467 \cdot \text{flow (ft}^3\text{/s)} \cdot \text{head (ft)} \cdot \text{efficiency}$$

If these flows and heads are maintained for the number of hours in a model time step, say Δt hours, then the value of energy produced in that time step, expressed in kilowatt hours, is the power multiplied by Δt hours.

$$\text{Energy (kWh)} = \text{power (kW)} \cdot \Delta t$$

This applies when the flow is expressed in cubic meters per second and the duration of the period Δt is expressed in hours. If the flow is expressed in units of cubic meters per time period, then the cubic meters per second flow term gets replaced by the cubic meters per period flow term divided by the number of seconds in the period:

$$\text{Flow (m}^3\text{/s)} = \frac{\text{flow (m}^3\text{/period)}}{\{\Delta t \cdot 3600 \text{ s/period}\}}$$

Hence,

$$\text{Energy (kWh)} = \frac{9.81 \cdot \text{flow (m}^3\text{/period)} \cdot \text{head(m)} \cdot \text{efficiency}}{(3600 \text{ s/h})}$$

If energy is to be expressed in units of kilowatt hours, and the units of flow is thousands of cubic meters per simulation time period, then the kilowatt-hours produced in the period is:

$$\text{Energy (kWh)} = \frac{9.81 \cdot \text{flow (1000 m}^3\text{/period)} \cdot \text{head (m)} \cdot \text{efficiency}}{3.600}$$

$$= 2.725 \cdot \text{flow (1000 m}^3\text{/period)} \cdot \text{head (m)} \cdot \text{efficiency}$$

$$= 2725 \cdot \text{flow (million m}^3\text{/period)} \cdot \text{head (m)} \cdot \text{efficiency.}$$

In these two cases, the energy production constant for the applicable link is 2.725 or 2725 times the plant production efficiency, depending on whether the flow is expressed as thousands or millions of cubic meters per period, respectively, and assuming energy is expressed as kilowatt-hours.

These constants apply for any simulation time-step duration. However, other values will apply if other units of energy, flow and/or head are used. For example, using the fact that there are about 3.28084 feet in a meter:

$$\text{Energy (kWh)} = 0.02352 \cdot \text{flow (1000 ft}^3\text{/period)} \cdot \text{head (ft)} \cdot \text{efficiency}$$

Energy consumption constants for turbines using energy to pump flow against a head are greater than energy production constants for turbines generating energy from flow falling over a given head. For any hydropower plant also serving as a pumping station, the energy consumption constant equals the energy production constant divided by the product of the production efficiency and the pumping efficiency. Since the energy production constant is the product of a constant (that converts head times flow to energy) times the production efficiency, the energy consumption constant is simply the conversion constant divided by pumping efficiency. Thus for any pumping station on a link l, the energy consumption constant for flows expressed in cubic meters per simulation time period and head expressed in meters is:

$$\text{Energy consumption constant} = \frac{2.725}{\text{pumping efficiency}}$$

The energy consumption constant for flows expressed in cubic feet per simulation time period and head expressed in feet is:

$$\text{Energy consumption constant} = \frac{0.02352}{\text{pumping efficiency}}$$

15.4   SIMULATION MODELING PROCEDURES

There are a variety of simulation modeling procedures. Three main types of simulation procedures are sequential, event-driven, and those that use optimization (see Chap. 6) instead of predefined operating policies. Each type has its advantages and limitations. Each type moves through time in discrete steps. These will be called simulation time steps. Clearly, as the duration of the simulated time steps decreases, the time required for simulating a fixed amount of time will increase.

Each type of simulation involves a number of passes or loops through various data sets. These data sets include the hydrologic flow sequences, the various time periods being considered in the simulation, and the network of cells or of nodes and links. The three types of simulation approaches distinguish themselves in how they simulate the network of cells or of nodes and links.

As illustrated in Fig. 15.12, the simulation can begin by first identifying a hydrologic sequence. For each of the possible multiple hydrologic sequences (or replicates), the simulation can proceed through the sequence of successive within-year time periods, from one year to the next, through all years of the simulation. Within each within-year period, the simulation can proceed from one simulation time step to another.

15.4.1   A Sequential Simulation Procedure

In each simulation time step of each within-year period, a sequential simulation proceeds through the system network a number of times to determine the flows, storage volumes, evaporation and seepage losses, and hydropower production and quality at each node and link, as applicable. Figure 15.13 shows one possible approach for defining and arranging separate simulation subroutines or loops for simulating a node-link network (after Loucks, et al., 1994).

In a procedural simulation, the sequence of simulation events within each simulation time step, such as defined in Fig. 15.13, is specified at or near the beginning of the program code. To illustrate, consider the simulation of a series of cells or nodes and links. The processes that occur in each cell, or in each node or link, can be defined in a subroutine whose arguments can include all the data needed to perform the subroutine's functions and all the data produced by those functions. For a river reach, these data could include, as applicable, the inflow, the parameter values for flow routing and for computing evaporation and seepage losses, and the names of the objects that provide the reach inflow and receive the reach outflow. The subroutine arguments could also include the output data generated by the procedures or methods incorporated into the reach subroutine. This output data will likely be the input data for one or more other subroutines.

In a procedural simulation, one typically programs the simulation to begin at the beginning of a headwater reach (i.e., a reach having no upstream reaches whose outflow becomes part of the inflow to the reach of interest) and proceeds sequentially in the direction of flow until all river reaches have been simulated. Of course, an actual simulation may be somewhat more complex, especially if the direction of flow

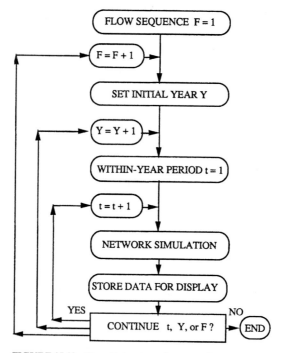

FIGURE 15.12   Overall structure of many surface-water simulation programs. The order of flow sequence replicate, year and within-year time period can be changed to simulate all flow replicates before moving to another within-year period t. In any event, the network simulation sequence is performed after the flows and within-year periods are defined.

in some reaches depends on the flow or if a reservoir release, diversion allocation, or reach flow is determined, in part, by downstream system states and demands as well as upstream supplies, but essentially the simulation sequence is fixed for a given system. The sequence is determined, in part, by the geographic relationships of the system components (cells or nodes and links) and possibly the type of components, regardless of the flow, quality, or hydropower data used in the simulation. That sequence is known (e.g., Fig. 15.13) prior to any simulation and is independent of the data being simulated.

In this procedural example only one subroutine is needed for each process being modeled. If, for example, all the reaches have the same functions or methods to perform, then the reach subroutine can be used as many times as there are reaches, each time with a different set of input argument values pertaining to the particular reach being simulated. The simulation manager or shell sees to it that the data sets are supplied in the appropriate order. Once all system components have been simulated, the simulation proceeds to the next time step.

15.4.2   An Event-Driven Simulation Procedure

In an event-based simulation, the sequence of simulation events within each simulation time step, such as defined in Fig. 15.13, does not exist. In event-based simulation,

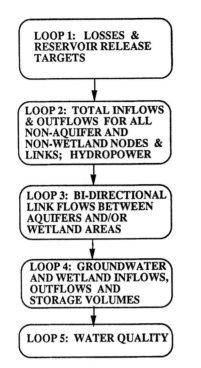

FIGURE 15.13 Sequence of calculations that could be used to simulate a network of nodes and links in each simulation time step. Once completed, the simulation moves to the next simulation time step and the loops are repeated.

each component of the water-resource system reacts to changes in its state when those changes occur and not in some predefined order. Changes in system states, say from the addition of inflow or deviations from meeting various storage or flow demand targets, may well cause any one of many of the loops or subroutines identified in Fig. 15.13 to be invoked for a particular component, depending on what is needed at a particular site or group of sites. The sequence of methods that will be executed will depend on the state of the system and the hydrologic, water quality, and operating policy input data. The invoking or execution of these loops or other methods becomes data- or event-dependent rather than order-dependent within each simulation time step (CADSWES, 1992; Behrens and Loucks, 1994).

While there is no unique way of defining an event-driven procedure, the simulation approach could proceed as follows. Let each component of the water-resource system have its own data base. At the beginning of each simulation time step, all exogenous data, such as natural incremental inflows and predefined demand targets and operating rules, are entered into the data bases of all appropriate cells, nodes, or links. Once a cell, node, or link has sufficient data so that its processes or operations can be performed, those processes or operations are performed. This will result in additional data that are needed by other components before they can perform their processes and operations. Eventually all components will have the data they need to perform their simulations, and all will be in balance with each other; i.e., mass balances will be maintained throughout the system. Once that condition is met, the simulation program can move to the next simulation time period.

For example, consider a simple two component system—an upstream reservoir and a downstream demand site. Once the reservoir with a known initial storage volume has its total inflow, evaporation, seepage, and operating policy data it can determine its release and final storage volume. Its release becomes the inflow to the downstream demand site that has a demand target. The downstream component now has its inflow and can perform its processes or operations to determine if its demands are met. Suppose it does not have sufficient inflow to meet its flow demand and hence needs additional inflow. This need for additional inflow translates to a need for additional releases from the upstream reservoir. This may cause a change in the reservoir operating policy. This change in reservoir data will cause it to simulate its processes again in an attempt to provide the additional releases required by the downstream demand site. If additional releases occur, this changes the inflows at the demand site, which in turn causes the downstream demand site to again simulate its processes.

This iterative event-driven procedure is repeated until the desired inflow at the demand site equals the natural incremental inflow at that demand site plus the release from the upstream reservoir. Whether or not the downstream demand target is met, the two components are in balance. Assuming nothing happens during the current simulation period to change their data as other system components are being simulated, they will await any changes in their data that may occur during the next or subsequent simulation time periods. If there are losses of flow between the upstream reservoir and the downstream demand site, an additional river reach component must enter and become a part of this iterative event-driven simulation procedure.

This is a simplified description of one approach to event-driven simulation within each simulation time period. Within each time period it is neither possible nor necessary to determine the sequence of component simulations prior to knowing the data that are being simulated. The determination of simulation order will be made by the components themselves, depending on the input data provided by the user and by the simulation of other components in the system. The only program control needed is an event manager so that two or more event-driven simulations do not try to occur simultaneously in the same processor. In each simulation time period, the event-driven processes can iterate until there is sufficient accuracy and until there is agreement among all component input and output data. The direction of simulation can be upstream or downstream, or in both directions, as needed.

15.4.3   A Network Optimization Procedure

In contrast to the sequential and event-driven simulation approaches, some simulation models use a network optimization procedure to allocate the flows and storage volumes throughout a network of nodes and links in each simulation time step (see, for example, Kuczera, 1989; 1993; Labadie, et al., 1984; Sigvaldason, 1976). The optimization algorithm replaces the specific operating policy functions that are either defined in the input data or built into a simulation model. The algorithm also replaces the procedures used to implement any specified operating policy. Replacing prespecified operating policies with sets of decisions based on optimization can result in improved system performance. Optimization can identify sets of decisions that are based on a much broader view of the impacts of those decisions, over time and space, and hence can more efficiently accomplish specified system operating objectives (Wurbs, 1991; Yeh, 1986).

When simulation models are used for planning and management, model results based on optimization solutions must be able to identify implementable operating policies. This may be difficult to do unless the model has been calibrated to conform to some predefined operating policy—one that can be implemented in practice without the aid of optimization. However, when such models are used for operation, on a periodic basis and with updated data describing the current system state and expected future inflows and demands each time the model is run, the model results can identify a more integrated set of operating decisions for any particular set of hydrologic inputs. These allocation decisions can be based on the state of all system components and the impact of any decision on those components now and in the future. Figure 15.2 illustrates a real-time operation framework in which such simulation-optimization models can be used.

The optimization procedure or algorithm usually incorporated into simulation models for defining operating decisions takes advantage of the node-link network representation of the water-resource system. A network flow programming (optimization) algorithm is usually applied to the network flow and storage volume allocation problem. The flows in each river reach or diversion link and/or storage volumes in each lake, reservoir, wetland, or aquifer storage node have costs or penalties that are (usually piecewise-linear) functions of the flow or storage volume. These cost or penalty functions are defined in such a way as to identify the priority of water allocations over space and time. The model solution identifies the flows and volumes in each flow link and storage node that minimize the total sum of all costs or penalties.

If a network optimization algorithm is to be used as the optimization procedure, each storage node of a node-link network can be divided into storage zones, each zone having a different cost or penalty. Storage zones are used to equate storage ranges of equal desirability among all the storage reservoirs in the system. It allows for the drawing down of reservoir storage in a predefined manner, if physically feasible, and for the allocation of water supplies among multiple reservoirs, again if physically feasible. This helps to achieve *reservoir balancing,* i.e., reducing the chance that some reservoirs may be full while other reservoirs may be at their dead storage volumes unless such a condition is desired.

Similarly, each flow link can be expanded to become multiple links, each of which has an upper bound on the flow contained within that link, and a cost per unit flow assigned to the link. The unknown variables in the optimization are the storage node volumes and link flows. These unknown variables are multiplied by the unit costs or penalties associated with those flows or volumes, i.e., the unit costs or penalties associated with the applicable link or storage zone, and then summed to get the total cost or penalty which the optimization is trying to minimize. Network optimization essentially solves the following model for each node n and link l in each period t.

$$\text{Minimize } \Sigma_{nt} \text{ penalty}_{nt}(\text{storage\_volume}_{nt}) + \Sigma_{lt} \text{ penalty}_{lt}(\text{link\_outflow}_{lt})$$

Subject to:

Mass balance at each node n in each period t:

$$\text{Storage\_volume}_{n,t+1} = \text{storage\_volume}_{nt} + \text{node\_inflow}_{nt} - \text{node\_outflow}_{nt}$$
$$- \text{evaporation\_seepage\_losses}_{nt} - \text{consumption}_{nt}$$

Mass balance at each link l in each period t:

$$\text{Storage\_volume}_{l,t+1} = \text{storage\_volume}_{lt} + \text{link\_inflow}_{lt} \text{ link\_outflow}_{lt}$$
$$- \text{evaporation\_seepage\_losses}_{lt}$$

Capacity constraints at each node n in each period t:

$$\text{Storage\_volume}_{nt} \leq \text{maximum capacity}_{nt}$$

Network continuity for all nodes n and periods t:

$$\Sigma \text{ link\_outflow}_{lt} + \text{incremental\_flow}_{lt} = \text{node\_inflow}_{nt}$$

where the sum is over all links l whose outflow node is n.

$$\text{node\_outflow}_{nt} \leq \Sigma \text{ link\_inflow}_{lt}$$

where the sum is over all links l whose inflow node is n and the inequality applies only if all outflowing links are diversion (nonnatural) links.

The *penalty functions* included in the objective function may appear as illustrated in Fig. 15.14. These functions associated with each node and link are defined so as to set priorities among different water users or purposes over space and time.

The optimization model may include additional constraints such as for defining flow routing in links, if applicable (which would result in link storage volumes as shown earlier), defining evaporation and seepage losses as functions of storage volumes or flows, and for defining other physical or operating policy relationships pertaining to the particular system being modeled. Additional constraints may also be needed to linearize penalty and other nonlinear functions, to define integer variables, upper and lower bounds on certain variables, to include water quality, and to define approximate energy production or consumption.

15.5 SIMULATION PROCESSES AND ASSUMPTIONS

15.5.1 Storage Node Evaporation and Seepage Losses

Surface-water bodies are usually subject to evaporation and seepage losses, and groundwater aquifers may be subject to seepage losses. These losses can be expressed as functions of the storage volumes in those water bodies at the beginning and ending of each simulation time period. These losses can be calculated from elevation–surface area or storage volume–surface area functions and evaporation rates, and from storage volume–seepage loss functions.

The difficulty in calculating evaporation and seepage calculations in discrete time periods is that these processes affect the final volumes or flows, which in turn affect the evaporation and seepage losses. Hence an iteration process can be implemented for increased accuracy. However, if the simulation time periods are relatively short, so that there is not much difference between initial and final conditions in a time period, evaporation and seepage can often be approximated by basing these losses only on initial conditions in each simulation time period. There is probably more inaccuracy in the data on evaporation and seepage than in the loss of precision from

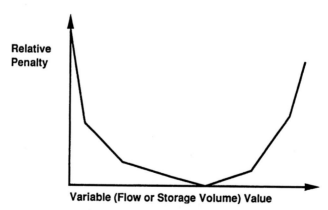

FIGURE 15.14   A penalty function associated with a simulated variable value at a particular node or link.

ignoring end-of-period flow or storage conditions when calculating evaporation and seepage, especially if the time periods are relatively short.

Note that evaporation and seepage losses represent quantities lost to the system. This is not the case if the seepage contributes to the inflow or volume in another system component. For example, seepage from a surface-water reservoir could add to the volume in a groundwater aquifer, and water from a groundwater aquifer could be leaking into another adjacent aquifer. If such aquifers are included in the simulation model, a link between the two applicable nodes should be defined. The seepage should be included as part of the outflowing node's outflow that, after being allocated to the appropriate link, becomes part of another node's inflow.

15.5.2 Surface-Water Node Inflows and Aquifer Natural Recharges

In each simulated time step the natural incremental inflow between successive surface-water nodes and the natural recharge of groundwater aquifer nodes, if they occur, must be calculated. To compute the natural incremental inflows, the natural flows themselves must be known or computed.

The natural unregulated flow at each surface-water node in each simulation time step is usually some function of the recorded (and deregulated, as required) gauge site flows at various gauge sites. Knowledge of rainfall-runoff patterns and land topography are helpful in defining the relationships between ungauged natural flows in ungauged sites and the natural flows at gauge sites. If the area is relatively homogeneous, with no pronounced topographic features that would cause more rainfall and runoff in some areas compared to others, the usual procedure for flow estimation is to compute the runoff per unit area based on the gauge flows, and apply the same runoff per unit area to the ungauged sites. If such a procedure does not apply, then one should use whatever procedures do apply, and these procedures may differ within the region being modeled. Gauge flow data are usually contained in external files and are read in as required during the simulation.

The natural unregulated incremental flow associated with any link representing a natural channel is the difference between the unregulated flows at the two end nodes of the link. If there are no nodes upstream of a particular node then that node is the initial node of a river or stream. At this node the incremental flow is the same as the total (usually natural or unregulated) flow at that site. The total incremental flow entering a node is the sum of the incremental flows associated with each incoming link.

Natural incremental flows can be negative. A negative incremental flow reflects the loss in natural flow due to evaporation and seepage or unaccounted diversions. Note that these losses are independent of actual (possibly partially regulated) flows in the link. To compute the losses based on actual flows in any link, the model input data can include a link-flow loss function for that link. If such a loss function is used, the natural incremental flows based on gauge flows would have to be adjusted so that they do not include any losses. In this case all incremental flows would be nonnegative.

The total surface-water inflow into a surface-water node equals the incremental unregulated flow plus the sum of the outflows from each of the node's incoming links. The total inflow of any surface-water node must be nonnegative.

If the node is a groundwater aquifer, its natural recharge must be estimated, and these estimates are usually based on the areal extent of the aquifer and the amount of effective precipitation that percolates to the aquifer per unit area of the watershed.

15.5.3 Groundwater Flows

Groundwater flows among aquifers and among aquifers and surface-water components are often represented by flows in bidirectional links connecting the aquifer to other system components. These flows are either based on the physics of groundwater flow (see Chap. 2) or on a pumping policy. In both cases these groundwater flows can be defined as a function of the pressure heads of the total amount of water available at each end of the link.

If one of the nodes connected to an aquifer by a groundwater link is a surface-water node, the water available at that node is its storage, if any, plus its inflow. If an aquifer is connected to another aquifer via a groundwater link, then the transfer of water depends only on the storage volumes (heads) in those two aquifers. Groundwater pumping policies can be defined for withdraws or for artificial recharge.

15.5.4 Reservoir Releases

As previously described, the outflow or release from each surface-water reservoir is specified by its release rule or operation policy. Reservoir releases are typically based on current reservoir storage volumes, on downstream demand target deficits, on a specified minimum required release, or on target storage volumes in the case of multiple reservoirs operated as a group, as applicable. Release rules can be based on the current and expected future state of any or all components, but such policies are not typically defined and implemented except for very complex systems (see, for example, CADSWES, 1992; and Bureau of Reclamation, 1987).

Releases made to meet downstream demand deficits may take priority over the releases made based on storage volumes. If the total release based on the reservoir operating policy and any demand deficits is less than the minimum required release for the simulation time period, then an additional release can be made, if possible, to meet the minimum release requirements for the simulation time period. If the available storage is insufficient to make this release, then some rule is needed to determine just how much should be released. If the remaining available storage is released, the reservoir will be at its lowest regulated level. This lowest regulated storage volume is usually the dead storage volume, which through evaporation and seepage can decrease.

The reservoir release policies based on multiple storage release zones, storage target balancing functions, and prorated demand target deficits, were discussed and defined in the previous section on simulation operating policies. Other policies are discussed in Hufschmidt and Fiering (1966) and Kaczmarek and Kindler (1982).

15.5.5 Natural Lake Discharges

The discharge from each natural lake is obtained from the lake-storage-discharge function. By definition, the discharges from natural lakes are not controlled. The greater the lake storage volume, the greater will be the discharge from the lake. Discharges from natural lakes can, but need not, vary among different within-year periods.

15.5.6 Multiple and Diversion Link Flow Allocations

Simulation models not using optimization need to have policies defining link-flow allocations in situations where the flow can enter multiple links. Link-flow allocations may be supply-driven or demand-driven based on downstream prorated demand target deficits. The total allocation of water in any constructed diversion link may have to be constrained to not exceed its maximum specified flow capacity.

Normally only one link of multiple links transporting water from any node represents the natural channel. If two or more outgoing links representing natural channels exist (i.e., they have not been designated diversion links), then each should have allocation functions defined and each will be allocated amounts defined by those functions. If the allocation functions for multiple natural channels have not been defined, there must be some procedure built into the model to decide how much of the available flow will be allocated to each link.

If no outgoing natural channel has been designated (i.e., if the node is a terminal node or if all the outgoing links are diversions and their allocations do not equal the amount available for allocation), any node outflow in excess of the sum of allocated link flows could be lost to the system.

15.5.7 Surface Flow Routing

Flow routing procedures may be appropriate for some simulation models of some systems. Routing should be implemented whenever flow travel times could be greater than the duration of the time period being simulated. There are a number of different routing procedures that could be included in any simulation model (see Maidment, 1993, and Chaps. 24 and 25 of this handbook). A four-parameter routing model that ignores backwater flow, but otherwise fits a wide range of time series flow data, requires the division of the applicable reach into subreaches. The number of subreaches is one of the four parameters that needs to be determined. Each subreach i of a link l can be assumed to be a lake whose outflow or release, outflow(i,l,t), in each time period t is a function of its inflow, inflow(i,l,t), and its initial storage volume, volume(i,l,t). This routing function is assumed to have the following form:

$$\text{outflow}(i,l,tt) = [c1(l,t) \cdot \text{volume}(i,l,t) + c2(l,t) \cdot \text{inflow}(i,l,t)]^{c3(l,t)}$$

where the parameters $c1(\ )$, $c2(\ )$ and $c3(\ )$ are the other three parameters for each link l and period t that need to be determined. The values of $c1(\ )$ and $c2(\ )$ can range from 0 to 1, and $c3(\ )$ must be greater than 0 and no greater than 1.

If both the expression inside the square brackets $[\cdot]$ and the exponent $c3(\ )$ are less than one, the computed outflow will be larger than the value of the expression within the brackets. To insure against this, the above expression can be modified to become:

$$\text{outflow}(i,l,t) = \min \{[c1(l,t) \cdot \text{volume}(i,l,t) + c2(l,t) \cdot \text{inflow}(i,l,t)],$$

$$[c1(l,t) \cdot \text{volume}(i,l,t) + c2(l,t) \cdot \text{inflow}(i,l,t)]^{c3(l,t)}\}$$

The inflow into each link segment i of link l in time period t is the release from the immediate upstream link segment $i-1$, less any losses that may occur:

$$\text{inflow}(i,l,t) = \text{outflow}(i-1,l,t) - \text{loss}(i-1,l,t)$$

The daily volume of flow lost in the entire link due to seepage and evaporation for any daily link inflow can be approximated by defining a negative unregulated incremental flow, or it can be obtained from the link flow loss function. If the latter loss function is used, the loss of flow in any subreach would equal the total link loss based on the subreach inflow divided by the number of subreaches.

The ending storage volume$(i,l,t+1)$ in each subreach segment i is the initial storage in the next time period.

$$\text{volume}(i,l,t+1) = \text{volume}(i,l,t) + \text{inflow}(i,l,t) - \text{outflow}(i,l,t)$$

The number of subreaches i in link l and the value of the coefficients $c1(l,t)$, $c2(l,t)$, and $c3(l,t)$ can be estimated by calibration procedures. To calibrate the flow routing function for any reach (link), time series of inflows and outflows can be used. Given the inflow series, users can compare the computed outflow series with the measured (or otherwise determined) outflow series. Simulation models can be used to do this through trial and error, or users can apply search techniques, such as genetic algorithms, to find the best (calibrated) values of the four parameters.

The default values of $c1(l,t)$, $c2(l,t)$ and $c3(l,t)$ routing coefficients can be set to 1.0. These default values for $c1(l,t)$, $c2(l,t)$ and $c3(l,t)$ will equate the inflow to the outflow regardless of the number of link segments. No delayed routing will occur. Link volumes will equal zero. The flow will be instantaneous, as is usually assumed for mass-balance simulations when the time-of-flow through the system is less than the within-year period being simulated.

If $c3(l,t)$ is 1, then the routing results from a series of linear lakes. Furthermore, if $c1(l,t)$ and $c3(l,t)$ are 1 and $c2(l,t)$ is 0, then any water entering the reach link l that is not lost due to evaporation and seepage will take the number of simulation time steps equal to the number of link segments to reach the end of the link. This represents a form of routing where the link outflows equal the link inflows some specified number of previous simulation time periods. The outflows are time lagged inflows, and the duration of the time lags are independent of the magnitude of the flow.

If $c1(l,t)$ and/or $c2(l,t)$ are less than 1, then inputs to the link will be dampened out over space as well as delayed over time as they proceed through the link. It is likely $c1(l,t)$ and $c3(l,t)$ will have values very close, if not equal, to 1, and that $c2(l,t)$ will be close, or equal, to either 0 or 1. Calibrations of actual flow data suggest that there are usually several different sets of parameter values that obtain very good fits of even highly variable time-series outflow data associated with a given time series of inflow data.

15.5.8 Hydropower and Energy Production

Hydropower facilities can be an important component of a water-resource system (see Chap. 31). The output of a hydropower plant is measured in terms of power and energy. Hydropower plants are operated to produce power and energy, often supplementing that produced by other sources, e.g., thermal or nuclear plants. Hydropower power and energy is usually sold to a power utility, and it is the utility that often dictates just how much power and energy is needed and when it is needed. The demand for power and energy usually varies over each day, and over each season. (For further discussion of demand loads as well as for additional detail on hydropower systems, see Basson, et al., 1994).

Hydropower plants may be located on any surface-water link. If a hydropower plant is located on a particular link, all the flow entering the link can be assumed to be available for the production of energy; however, the energy that may be produced will depend on the installed plant capacities (less any allowance for maintenance) and plant factors. The plant factor for each simulation period is the fraction of the period hydropower can be produced. Both power and energy can be produced in each simulation time period.

Energy can be produced on any surface-water link l containing a hydropower plant only if there is flow above the minimum required by the turbines for energy production in the link, and only if the head at the hydropower plant is greater than 0. The head is the difference between the water-surface elevation at the upstream node minus the turbine elevation or downstream node storage volume elevation, whichever is less.

If the hydroelectric plant is a run-of-river plant, the water storage levels are assumed to be constant and the resulting head is usually the rated head. The rated head is a constant value representing the head for which the turbines were designed.

In each simulation time step, the energy produced is the minimum of what could be produced with the given flow through the turbines and the storage head; what could be produced given the power plant capacity and plant factor; or 0 if the flow does not equal or exceed the minimum flow necessary for energy production.

The power available at each hydroelectric plant is a function of the existing head, the rated head, and plant capacity at the rated head.

$$\text{System power}(t) = \Sigma_l \frac{(\text{plant\_capacity}(l) \cdot \text{head}(l,t)}{\max \{\text{head}(l,t), \text{rated head}(l)\})}$$

summed over all hydropower links l

Reliable system power $= \min$ (system power(t) for all time steps t)

Hydropower systems are just one component of a multipurpose water-resource system, and they are often just one component of an electrical power generation and distribution system. Hence, considerably more detail can be included in a model that includes hydropower. For a more thorough discussion of hydropower systems modeling, see Basson, et al. (1994) and Chap. 31 of this handbook.

15.5.9 Pumping and Energy Consumption

Pumping stations can be located on any link. Pumping can occur on groundwater links if withdrawals or artificial recharge are being implemented. The only reason to designate a groundwater link a pumping link and to enter storage-head data associated for an aquifer is to compute and record energy consumption associated with pumping against a negative net head. The actual flow, or volume of water pumped, in any groundwater link will be determined by the groundwater link flows based on available storage volumes or water quantities at the end nodes of the link.

Hydroelectric power plants can serve as pumping stations on any surface-water link, but pumping will only occur if users have designated the link a pumping (as well as a hydropower) link. Pumping will be assumed to occur on any link if the flow in the link is increasing in elevation, i.e., if the head, head(l,tt), is negative. If the consumption of energy due to pumping is to be calculated and recorded, pumping capacity has to be defined for the link. The direction and amount of flow in the link need not be affected by the presence or absence of a pumping station.

Some of the same data required for hydropower plants are also required for pumping stations, but with several differences. In place of energy production constants, the simulation model will require energy consumption constants. Plant factors indicating the fraction of time hydroelectric energy can be produced, minimum flow requirements for energy production, and power plant capacities, are not needed when calculating energy consumption from pumping operations.

If the head on a pumping link is negative, pumping will be assumed to occur. Maximum pumping capacities may restrict the amount of flow that can be pumped. If the entire link flow cannot be pumped, the model must be programmed to know what to do with the amount that cannot be pumped. Pumped flows on groundwater links, specified by groundwater link-flow policies, should also not exceed the pumping capacities on those links.

15.5.10   Water-Quality Concentrations

The discharge of waterborne wastes into the surface or groundwaters may be assumed to occur at any node of the node-link network. If the simulation model is data driven, the names of all modeled wastewater constituents, the initial concentrations of all of these constituents in the nodes and links, and the constituent concentrations in the natural incremental flow entering any or all nodes of the network are inputs to the model. The time-series water-quality data can be contained in a separate file that can be read during the simulation.

The prediction of any quality constituent concentration in any component of the network depends in part on the particular system component and on the flows and storage volumes, if any, in that component. The flows and storage volumes themselves are often independent (determined without regard to) the water quality, but this need not be the case. Water-quality constituent loadings (inputs) are usually assumed constant throughout each simulation time period. See Chaps. 8, 9, 10, 11, and 14 for more details on water quality aspects of water resource systems and modeling.

15.5.11   Mass Inputs

Water-quality constituent inputs at each node n are defined by the concentrations $WC(i,n,t)$ of each constituent i in the uncontrolled flows entering each node in each period t. These inputs can be contained in an external file which is read during the simulation. They may vary from one simulation period to another, but are usually assumed constant within each simulation period.

The mass discharges $W(i,n,t)$ of waste constituents i in simulation time step t entering a node n are the wasteload concentrations $WC(i,n,t)$ of those wastes times the natural incremental flow $NIQ(n,t)$ entering that node.

$$W(i,n,t) = NIQ(n,t) \cdot WC(i,n,t)$$

In each simulation time step t, the total mass $NM(i,n,t)$ of each constituent i entering any node n is calculated by adding the mass discharge $W(i,n,t)$ to all the masses [link outflows $LOQ(l,t)$ times the end-of-link concentrations $LOC(i,l,t)$] from the incoming links.

$$NM(i,n,t) = W(i,n,t) + \Sigma_l LOQ(l,t) \cdot LOC(i,l,t)$$

where the sum is over all links l whose flow enters node n.

The incoming node concentration is the incoming mass $NM(i,n,t)$ divided by the incoming flow $QI(n,t)$.

$$NC(i,n,t) = \frac{NM(i,n,t)}{QI(n,t)}$$

For non-storage nodes, the input mass equals the output mass. Complete mixing is assumed to occur. Thus the outgoing concentration $NC(i,n,t)$ of constituent i at a nonstorage node n in simulation time step t is the same as the initial node concentration $NC(i,n,t)$, which of course equals the concentration entering each outflowing link.

$$NC(i,n,t) = LIC(i,l,t) \text{ in all links } l \text{ whose inflow node is nonstorage node } n$$

The mass of constituent i entering a link $LIM(i,l,t)$ is the link inflow concentration $LIC(i,l,t)$ multiplied by the link inflow $LIQ(l,t)$.

$$\text{Link incoming mass} = LIM(i,l,t) = LIC(i,l,t) \cdot LIQ(l,t) \text{ for each link } l.$$

15.5.12 Processes

The simulation of water quality in storage nodes representing natural lakes (Chap. 9), reservoirs (Chap. 9), wetlands (Chap. 13), or groundwater aquifers (Chap. 11), and in links representing stream reaches or surface water diversions, is usually based on the flows and storage volumes resulting from the water-quantity simulation. As discussed earlier, link and node inflows and outflows are typically assumed constant over each simulated time step but can change from one time step to another. The storage volumes in the node may change within the simulation time step. Link storage volumes in each simulation time step are usually those computed from flow routing if flow routing is implemented. If flow routing is not performed, link volumes can be based on time-of-flow functions that define the time of flow $TF(l)$, usually in days, from the upstream to the downstream end, for any given daily flow $LQ(l,t)$ for each reach link l.

$$\text{Link volume} = \text{average flow} \cdot \text{time of flow} = LQ(l,t) \cdot TF(l)$$

If flow routing has not been performed and time of flow functions have not been entered, link volumes are based on the positive-valued initial link volumes defined by the user. If these volumes are zero, then the water-quality constituent concentrations will usually be set to 0.

Each storage node or link can contain a series of storage elements. Storage elements are depicted in Fig. 15.15. The instantaneous rate of change in mass of a water quality constituent i in a storage element m equals the incoming mass less the outgoing mass less the decay or consumption of that constituent mass plus the growth or increase in mass of that constituent i resulting from the transformation of other quality constituents j in the storage element.

The mass loading rate $M(i,m)$ per day of a constituent i entering the first storage volume element m of a link or node during a simulation time step includes the mass load discharges, if any, from the watershed. The incoming mass in any storage volume element also includes the discharges from immediate upstream elements. The outgoing mass from storage volume element m is the product of the element's daily outflow $Q^{out}(m)$, and the element's volume concentration, $conc(i,m)$.

The total growth, decay, or consumption of constituent i in a storage volume element during a simulation time step is governed by the rate constants $K(i,j,t)$ for the particular storage volume element and by the stoichiometric transformation constants $T(j,i)$. Assuming the growth rate constants are >0 and the decay rate constants are <0,

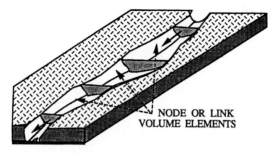

In Each Volume Element:

Time Rate of Change of MASS of CONSTITUENT i =

MASS INPUT - MASS in OUTFLOW

+ GROWTH - DECAY or CONSUMPTION

+ TRANSFORMATION OF CONSTITUENTS j

FIGURE 15.15 Water-quality volume element and processes governing the mass and concentration of a constituent i in the volume element. The transformation of other constituents j may contribute to the mass of constituent i.

$$d\frac{[\text{conc}(i,m)\cdot\text{vol}(m)]}{dt} = M(i,m) - \{Q^{out}(m)\cdot\text{conc}(i,m)\} + \{\Sigma_j\, K(i,j,t)\cdot\text{conc}(i,m)\cdot\text{vol}(m)\}$$

$$- \{\Sigma_{j\,\neq\,i}\,[K(j,i,t)\cdot T(j,i)\cdot\text{vol}(m)\cdot\text{conc}(j,m)]\}$$

In this equation, the rate constants:

$K(i,j,t)$ = rate constant for growth (>0), decay, or consumption (<0) of constituent i as it is influenced by the presence of constituent j.

$K(i,i,t)$ (usually <0) = the rate constant associated with settling, volatilization, or other factors that result in a loss or gain of constituent i that does not involve a transformation to other constituents being modeled.

These rate constants are usually in units of day^{-1}. The values of all the rate constants $K(i,j,t)$ are dependent not only on the simulation time period t but also on the particular storage node or link. Some simulation models permit users to assign default values to all node rate constants and/or to all link rate constants, and these default values will apply to all nodes and links that have not had their individual node- or link-specific rate constants entered.

The stoichiometric transformation constants are dimensionless:

$T(j,i)$ = the mass of constituent i created by a decrease of a unit of mass of constituent j. Note that $T(j,i) = 1/T(i,j)$.

This differential equation defining the rate of change in the mass of a constituent i in a storage volume element can be approximated by the finite difference equation for each simulation time step t. For the moment we will omit the index m indicating the particular storage volume element and assume the duration of the simulation time step t is expressed in days. To simplify the notation a little let $C(i,t)$ and $V(t)$ represent initial concentration and volume, respectively, $M(i,t)$ the mass loading (input) of constituent i, and $QO(t)$ represent outflow, during the simulation time step t.

$$C(i,t+1) \cdot [V(t+1) - V(t)] + V(t+1) \cdot [C(i,t+1) - C(i,t)]$$

[the change in mass during simulation time step t]

$$= \left\{ M(i,t) - QO(t) \cdot \frac{[C(i,t) + C(i,t+1)]}{2} \right\} \cdot \Delta t$$

[average mass inflow less outflow during t]

$$+ \left\{ \Sigma_j K(i,j,t) \cdot \frac{[V(t) \cdot C(i,t) + V(t+1) \cdot C(i,t+1)]}{2} \right\} \cdot \Delta t$$

[growth or decay of constituent i due to constituent j during t]

$$- \left\{ \Sigma_{j \neq i} K(j,i,t) \cdot T(j,i) \cdot \frac{[V(t) \cdot C(j,t) + V(t+1) \cdot C(j,tt+1)]}{2} \right\} \cdot \Delta t$$

[transformation of average mass of constituent j to constituent i during t]

Rearranging terms provides a set of simultaneous equations that can be solved to predict the concentrations of each constituent in each storage element m at the end of each simulation time step t.

$$C(i,tt+1) = \{M(i,t) \cdot \Delta t + C(i,t) \cdot [V(t+1) + (\Sigma_j 0.5 K(i,j,t) \cdot V(t) - 0.5 QO(t)) \cdot \Delta t]$$

$$- \Sigma_{j \neq i} [0.5 K(j,i,t) \cdot T(j,i) \cdot (V(t) \cdot C(j,t) + V(t+1) \cdot C(j,t+1)) \cdot \Delta t] \}/\{2V(t+1)$$

$$- V(t) + 0.5 [QO(t) - \Sigma_j K(i,j,t) \cdot V(t+1)] \cdot \Delta t\}$$

In the preceding equation, the end-of-time-step concentrations, the $C(i,t+1)$ and $C(j,t+1)$ terms, are unknown at the beginning of each simulation time step t.

To avoid the need to solve simultaneous equations in each simulation time step, one can base the change in constituent masses due to growth, decay and transformation on only the initial constituent concentrations rather than average concentrations in each simulation time step. Errors caused by using only the initial concentrations as a basis for growth, decay, and transformation in each simulation time step may be reduced (but not completely eliminated) by adjusting the values of the rate constants. This compensating adjustment will occur in the calibration process.

All the terms on the right-hand side of the following equations are now known at the beginning of each simulation time step.

$$C(i,t+1) = \{M(i,t) \cdot \Delta t + C(i,t) \cdot [V(t+1) + (\Sigma_j K(i,j,t) \cdot V(t) - 0.5 QO(t)) \cdot \Delta t]$$

$$- \Sigma_{j \neq i} [K(j,i,t) \cdot T(j,i) \cdot V(t) \cdot C(j,t) \cdot \Delta t]\}/\{2 V(t+1) - V(t) + 0.5 [QO(t) \cdot \Delta t]\}$$

These equations for each constituent i in each node and link storage volume element can be used to predict all storage element concentrations $C(i,t+1)$ at the end of each simulation time step t. They assume changing, but well mixed, storage volumes in each volume element. Dispersion is complete within each volume element but not over all volume elements.

The mixing that takes place in each volume element does not take place instantaneously. If it did, some portion of any input of a constituent discharged into the river at some upstream discharge node site would travel downstream through the remainder of the system in a single time step. To eliminate this problem, the concentration of the inflow into a volume element in time step t can be assumed to be the average concentration $0.5 \cdot \{C(i,t-1) + C(i,t)\}$ in the upstream volume element in the previous time step $t-1$. Thus the mass input of a given constituent i to a volume element in time step t, $M(i,t)$, is

$$M(i,t) = QO(t) \cdot [0.5 \cdot \{C(i,t-1) + C(i,t)\}]$$

where the terms on the right-hand side apply to the immediate upstream volume element. The simulation time step duration is therefore assumed to be the time it takes for complete mixing to occur in each volume element.

15.5.13 Limiting Growth Factors

The growth and decay rate constants in the above equations for predicting end-of-simulation time-step concentrations $C(i,t+1)$ of each constituent i being modeled can be modified to account for the effects of limiting concentrations of multiple constituents on the rate of growth of some specified constituents, and the rate of consumption or decay of other constituents. Examples might include the growth of algae from the consumption of nutrients such as nitrate or phosphate. Algal growth could be limited by the smallest nutrient (e.g., nitrate or phosphate) concentration, or by the available concentration of the constituent (e.g., algae) itself that is feeding on these nutrients.

In such cases the minimum of all the appropriate terms $K(i,j,t) \cdot C(i,t)$ and $K(j,i,t) \cdot T(j,i) \cdot C(j,t)$ involving the appropriate nutrients j and the feeder constituent i must be determined. Once this minimum value has been determined, the values of each $K(i,j,t)$ and $K(j,i,t)$ involved in that selected group must be adjusted so that each product $[K(i,j,t) \cdot C(i,t)]$ and $[K(j,i,t) \cdot T(j,i) \cdot C(j,t)]$ expression within the group will equal the same minimum value.

To illustrate, consider the situation where constituent i grows as it feeds on constituent j. If the (i,j) relationship is defined, then the positive $K(i,j,t)$ and $T(i,j)$ values must be determined and entered into the database. The growth expression $[K(i,j,t) \cdot C(i,tt)]$ will appear in the equation for constituent i, and the consumption expression $[-K(i,j,t) \cdot T(i,j) \cdot C(i,tt)]$ will appear in the equation for constituent j.

$$\Delta[C(i,t)] = \{K(i,j,t) \cdot C(i,t)\} \cdot \Delta t$$

$$\Delta[C(j,t)] = -\{K(i,j,t) \cdot T(i,j) \cdot C(i,t)\} \cdot \Delta t$$

Both equations assume there is an adequate amount of constituent j to achieve the growth $\Delta[C(i,t)]$ of constituent i [specified by the expression $K(i,j,t) \cdot C(i,t) \cdot \Delta t$], by consuming an amount $\Delta[C(j,t)]$ of constituent j [specified by the expression $K(i,j,t) \cdot T(i,j) \cdot C(i,t) \cdot \Delta t$]. If there is not a sufficient amount of constituent j available, both the growth of constituent i and the consumption of constituent j will be reduced.

For each i,j relationship requiring the specification of the $K(i,j,t)$ rate constant and the $T(i,j)$ transformation constant, the following adjustment procedure can be performed.

$$K'(i,j,t) = \min \left\{ \frac{C(j,tt)}{[K(i,j,t) \cdot T(i,j) \cdot C(i,tt) \cdot \Delta t(t)], 1.0} \right\} \cdot K(i,j,t)$$

where the $K'(i,j,t)$ on the left-hand side is the adjusted value of the original $K(i,j,t)$ on the right-hand side of the equation. If any of the terms in the product $K(i,j,t) \cdot T(i,j) \cdot C(i,t) \cdot \Delta t$ is zero then the value of the rate constant $K(i,j,t)$ does not change.

This adjustment procedure applies to all growth rate constants, i.e., $K(i,j,t) > 0$ or $-K(j,i,t) > 0$. Hence if a negative $K(j,i,t)$ rate constant were defined instead of a positive $K(i,j,t)$, then the growth expression $[-K(j,i,t) \cdot T(j,i) \cdot C(j,t)]$ for constituent i will appear in the equation for constituent i, and the consumption expression $[K(j,i,t) \cdot C(j,tt)]$ for constituent j will appear in the equation for constituent j. The adjustment procedure is the same.

$$K'(j,i,t) = \min \left\{ \frac{C(i,t)}{[T(j,i) \cdot C(j,t) \cdot K(j,i,t) \cdot \Delta t], 1.0} \right\} \cdot K(j,i,t)$$

This automatic adjustment procedure insures consistency within multiple water-quality constituent equations, and insures against negative concentrations.

The next adjustment of rate constant $K(\ )$ values takes place over all expressions included in a minimization group. For all $K(i,j,t)$ terms in the minimization group included in the equation for constituent i,

$$K'(i,j,t) = \min \frac{\{\text{all expressions involving } i \text{ in the group}\}}{C(i,t)}$$

Similarly, for all $K(j,i,t)$ terms in the group contained in the equation for constituent i,

$$K'(j,i,t) = \min \frac{\{\text{all expressions involving } i \text{ in the group}\}}{[T(j,i) \cdot C(j,t)]}$$

The primary purpose of a minimization group within an equation for a particular constituent (say constituent a) is to insure consistency in cases where there are multiple constituents (say a, b, and c), and the constituent concentration that is most limiting (the smallest of the expressions involving a, b, or c) determines the growth of the particular constituent a, and hence the consumption of the other constituents b and c.

This rate constant adjustment operation satisfies two important conditions. The first is that no constituent concentration will become negative. The second is that the consumption of each nutrient by any constituent will be limited by the growth of that constituent, which in turn is limited by the most critical (usually smallest) nutrient concentration. For example, consider algae feeding on nitrate and phosphate. If there is no nitrate, assume there can be no algal growth. In this obvious case, nitrate is the limiting nutrient. The equation for predicting the change in phosphorus concentration will not include any nitrate terms, but it will include adjusted rate constants $K(\ )$ involving algae and phosphorus used to predict the reduction of phosphorus due to algae growth. In this example these adjusted $K(\ )$ values will be zero since there can be no algal growth (due to a lack of nitrate) even though the algal biomass and phosphate concentrations are positive and not limiting.

The original differential equation for the prediction of end-of-time-step constituent concentrations can be rewritten to include this consideration of limiting nutrients or limiting concentrations of the consuming constituent.

$$d \frac{[\text{conc}(i,m) \cdot \text{vol}(m)]}{dt} = M(i,m) - \{Q^{\text{out}}(m) \cdot \text{conc}(i,m)\}$$

mass loading rate   mass outflow

$$+ \min \{|K(i,j,t) \cdot \text{conc}(i,m)| \text{ or } |K(j,i,t) \cdot T(j,i) \cdot \text{conc}(j,m)|\} \cdot \text{vol}(m)$$

over all applicable (user-specified) constituents j involved in the growth of constituent i

$$+ \{\Sigma_j K(i,j,t) \cdot \text{conc}(i,m) \cdot \text{vol}(m)\}$$

over all user-specified constituents j involved in the decay or consumption [i.e., $K(i,j,t) < 0$] of constituent i

$$- \{\Sigma_{j \neq i} [K(j,i,t) \cdot T(j,i) \cdot \text{vol}(m) \cdot \text{conc}(j,m)]\}$$

over all user-specified constituents j involved in the increase in constituent i and not included in the above minimum group [$K(j,i,t) < 0$].

Before any of these equations are solved, the $K(\ )$ rate adjustment procedure must be carried out. It is based on all of the terms included in all of the minimum groups in all of the equations. This permits the adjusted value of any $K(\ )$ included in the minimum group of one equation to be determined by its inclusion in a more critical minimum group in another equation.

15.5.14   Dispersion

Dispersion in a storage node or link is dependent on the number of defined fully mixed storage volume elements in each system component (node or link), as illustrated in Fig. 15.15. The number of volume elements in each node and link is a dispersion calibration parameter. Each storage volume element is assumed to be completely mixed and hence of uniform concentrations. Increasing the number of storage volume elements decreases the amount of dispersion assumed to take place in the node or link component.

For example, if there is only one storage element defined for a particular node or link, the concentration of each constituent at the end of each simulation time step will be the same throughout the entire node or link component. If there are 20 different storage elements, there can be 20 different discrete average concentrations of each constituent over the spatial extent of that water body at the beginning and at the end of each simulation time period.

15.5.15   Durations of Simulation Time Steps

The duration of the simulation time step Δt should be no greater than the smallest detention time of all of the volume elements.

$$\text{Detention time of an element volume} = \frac{\text{volume}}{\text{outflow}}$$

If during a simulation the detention time in any storage node or subnode volume element or in any link or sublink volume element in any simulation period is less than the simulated time step, the simulation results may be in error. Simulation model users should reduce the duration of the simulation time step to be no greater than the shortest detention time of any storage volume element.

15.5.16   Limitations

While this linear (first-order) growth, decay, consumption, and transformation structure of the water-quality model presented here permits considerable flexibility for simulating multiple interdependent water-quality constituents, it is not appropriate for light or temperature. If temperature is being modeled, all temperature-dependent parameters should be sensitive to changes in temperature. Light together with temperature could be the limiting factor governing growth of biomass. Adding equations for modeling temperature and light are not very difficult; getting data for them is often much more difficult.

The decay rate constant $K(i,i,t)$ for dissolved oxygen deficit (DOD) ($i = $ DOD) is the reaeration rate constant. This rate constant can vary, in reality, with flow, and of course with atmospheric pressure, wind, and other factors. In addition, the predicted dissolved oxygen deficit cannot exceed the dissolved oxygen saturation concentration. When the DOD concentration is equal or more than the saturation concentration, the model will fail to predict other oxygen-dependent constituent concentration values if their rate constants are set for aerobic conditions. Users of models should always check the output to see if it conforms to the input assumptions.

Sediment transport and the interaction of sediments with certain pollutants is becoming increasingly important to model. Sediment transport and pollutant interaction is one of our continuing research challenges.

15.5.17   Model Calibration

All rate constants and the number of storage volume elements for each storage node or link for each within-year time period must be defined. The appropriate values of these parameters may be determined from a model calibration study using measured (real) time-series data or data from other calibrated and verified models of particular nodes or links. Model calibration and verification are necessary before any credence can be placed on model results.

15.6   CONCLUSIONS

This chapter has outlined some approaches for modeling surface-water-quantity systems, i.e., systems that are designed and operated to control, regulate, and, in general, manage water that flows and is stored on the surface of the ground. The emphasis has been on models used for evaluating alternative plans and management policies for the integrated operation of system components. Much more detailed modeling is required for the design of hydraulic structures. To the extent that surface-water system components are responsive to water stored under the surface of the ground, and are managed to achieve water quality as well as water quantity goals or targets, the interaction of surface-water components with both groundwater and water-quality components may also have to be considered. Some simplified ways of doing this have been identified, recognizing that the focus is primarily on surface-water systems. Other chapters in this book address groundwater and water-quality modeling in a much more detailed manner.

Models of water-resource systems attempt to describe the essential features of such systems and the interdependencies among system components in the detail needed for the particular decisions being considered and issues being addressed. These could include how to handle long-term droughts, or how to meet stated regulations and priorities governing water use over time and space, especially under different and changing hydrologic supply and demand scenarios.

Models that are intended to be generic, i.e., applicable to a number of water-resource systems, are increasingly becoming data driven. The assumptions involving the types and configuration of components, their processes, their operations, and their interactions with other components are defined in the input data rather than in the software code. The software is simply a manager of these data. It provides the graphical user interface that facilitates the interactive use of these data and the understanding of the results of any analysis of these data. This chapter has focused on the data and assumptions, rather than on the interface of descriptive simulation models.

Data-driven simulation models enable their users to address many *what if* questions. The primary purpose of modeling is to predict what might happen if some change is made or occurs in the existing system. Any prediction of future events or impacts resulting from possible changes in the system must be based on a set of assumptions about what supplies, demands, and goals will be in the future. Since no one knows, this information has to be assumed. Having models available to test the outcome of *what if* assumptions may help reduce the risks or vulnerability inherent in such systems as well as identify opportunities for change that could increase the overall performance of the system.

The development and use of descriptive models of water-resource systems requires an understanding of the systems being modeled and of the reasons for developing and/or using the models. This understanding is necessary in order to define the appropriate level of spatial detail and the appropriate number and durations of time periods to be modeled. This understanding is needed to identify those components that must be considered and the processes that take place in each of those components. It is needed to identify how each component interacts with other components and what options are available for changing the design and/or operation of each of those components. An understanding is also needed of the limitations of any modeling approach, and the assumptions behind the use of any particular simulation or optimization modeling procedure.

This chapter has outlined several modeling approaches, with an emphasis on the integration of multiple system components into a single descriptive model designed for the study of alternative plans and management policies. No matter what modeling approach is used, it is likely that it will need to be modified in time, to address new issues, to incorporate new understandings or greater detail, and to meet new needs. Given the premise that whatever model is developed, be it generic or basin specific, at some point it will need to be modified, a model architecture or structure is needed that will facilitate this change as well as facilitate the adaptation of new technology. Given the current rate of technology change, this presents a considerable challenge.

This chapter is merely an overview on the modeling of surface-water-resource systems. Its emphasis has been on the modeling of water and some of its main attributes; it has not been on how one can estimate the important economic or social impacts resulting from various flows, storage volumes, energy production or consumption, and possible pollutant concentrations. Those who wish to have more detail on various modeling approaches as well as on the estimation of economic and social impacts are encouraged to consult other references on these subjects, including many of those listed here.

REFERENCES

Andreu, J., J. Capilla, and E. Sanchis, "AQUATOOL: A Computer-Assisted Support System for Water Resources Research Management Including Conjunctive Use," in D. P. Loucks and J. R. da Costa, eds., *Decision Support Systems: Water Resources Planning,* NATO ASI Series G, vol. 26, Springer-Verlag, Berlin, 1991, pp. 363–356.

Basson, M. S., R. B. Allen, G. G. S Pegram, and J. A. van Rooyen, *Probabilistic Management of Water Resource and Hydropower Systems,* Water Resources Publications, Highlands Ranch, Colo., 1994.

Behrens, J. S., and D. P. Loucks, *Event-Driven Simulation of River Systems,* Cornell University, Ithaca, N.Y., 1994.

Bureau of Reclamation, "Colorado River Simulation System Documentation, System Overview," US Department of the Interior, Denver, Colo., 1987.

Center for Advanced Decision Support for Water and Environmental Systems, "River Simulation System (RSS) User's Manual, Programming Guide, and Tutorial," University of Colorado, Boulder, Colo., 1992.

Dolan, L. S., and D. K. DeLuca, "Use of a Hydrologic Model in a Basin-Wide Water Allocation Proceeding," *Water Resources Bulletin,* 29(1):107–117, 1993.

Eichert, B. S., "Program Description for Eichert's Engineering HEC-5 Package: H5A," El Macero, Calif., 1992.

Fedra, K., "Advanced Computer Applications," *Options,* International Institute for Applied Systems Analysis, Laxenburg, Austria, December, 1992, pp. 4–14.

Fiering, M. B., et al., *Scientific Basis of Water Resource Management, Studies in Geophysics,* National Research Council, National Academy Press, Washington D.C., 1982.

Ford, D. T., "Reservoir Storage Reallocation Analysis with PC, *Journal of Water Resources Planning and Management,* 116(3):402–416, 1990.

Hooper, E. R., A. P. Georgakakos, and D. P. Lettenmaier, "Optimal Stochastic Operation of Salt River Project," *Journal of Water Resources Planning and Management,* 117(5):566–587, 1991.

Hufschmidt, M. F., and M. B Fiering, *Simulation Techniques for Design of Water-Resource Systems,* Harvard University Press, Cambridge, Mass., 1966.

Hydrologic Engineering Center, "Water Supply Simulation Using HEC-5," Training Document no. 20, U.S. Army Corps of Engineers, Davis, Calif., 1985.

Hydrologic Engineering Center, "HEC-PRN Simulation Model, User's Manual," U.S. Army Corps of Engineers, Davis, Calif., 1993.

Hydrological Modelling Unit, "Integrated Quality-Quantity Model IQQM, Reference Manual," Department of Water Resources, New South Wales, Australia, 1994.

Grygier, J. C., and J. R. Stedinger, "Algorithm for Optimizing Hydropower System Operations," *Water Resources Research,* 21(1):1–10, 1985.

Kaczmarek, Z., and J. Kindler, *The Operation of Multiple Reservoir Systems,* IIASA Collaborative Proceedings Series, CP-82-S3, International Institute for Applied Systems Analysis, Laxenburg, Austria, 1982.

Keys, A. M. and R. N. Palmer, "The Role of Object-Oriented Simulation Models in Drought Preparedness Studies," *Proceedings of the 20th Conference on Water Resources Planning and Management Division,* ASCE, New York, 1993, pp. 479–482.

Kuczera, G., and G. A. Diment, "A General Water Supply Simulation Model: WASP," *Journal of Water Resources Planning and Management, ASCE,* 114(4):365–382, 1987.

Kuczera, G., "Fast Multi-Reservoir Multi-Period Linear Programming Models," *Water Resources Research,* 25(2):169–176, 1989.

Kuczera, G., "Network Linear Programming Codes in Water Supply Headworks Modeling," *Journal of Water Resources Planning and Management, ASCE,* 119(3):412–417, 1993.

Kuczera, G., "WATNET, Generalized Water Supply Simulation Using Network Linear Programming," Department of Civil Engineering and Surveying, University of Newcastle, New South Wales, Australia, 1990.

Kuo, J-T., N-S. Hsu, W-S. Chu, S. Wan, and Y-J. Lin, "Real-Time Operation of Tanshui River Reservoirs," *Journal of Water Resources Planning and Management,* 116(3):349–361, 1990.

Labadie, J. W., A. M. Pineda, and D. A. Bode, "Network Analysis of Raw Supplies under Complex Water Rights and Exchanges: Documentation for Program MODSIM3," Colorado Water Resources Institute, Fort Collins, Colo., 1984.

Loucks, D. P., P. N. French, and M. R. Taylor, "Interactive River-Aquifer Simulation, Program Description and Operation," School of Civil and Environmental Engineering, Cornell University, Ithaca, N.Y., 1994.

Maidment, D. R., ed., *Handbook of Hydrology,* McGraw Hill, New York, 1993.

Pabst, A. F., "Next Generation HEC Catchment Modeling," *Proceedings, ASCE International Symposium on Engineering Hydrology,* July 25–30, San Francisco, Calif., 1993.

Palmer, R. N., A. M. Keys, and S. Fisher, "Empowering Stakeholders through Simulation in Water Resources Planning," *Proceedings of the 20th Conference on Water Resources Planning and Management Division, ASCE,* New York, 1993, pp. 451–454.

Randall, D., G. W. Link, L. Cleland, C. S. Kuehne, and D. P. Sheer, "A Water Supply Planning Simulation Model Using Mixed Integer Linear Programming 'Engine'," Water Resources Management, Columbia, Md., 1995.

Razavian, D., A. S. Bleed, R. J. Supalla, and N. R. Gollehon, "Mustistage Screening Process for River Basin Planning," *Journal of Water Resources Planning and Management*, 116(3): 323–334, 1990.

Shane, R. M., and K. C. Gilbert, "Weekly Time Step Reservoir System Scheduling Model," TVA Report, Water Systems Development Branch, Tennessee Valley Authority, Norris, Tenn., 1980.

Sigvaldason, O. T., "A Simulation Model for Operating a Multipurpose Multireservoir System," *Water Resources Research,* 12(2):263–278, 1976.

Stockholm Environment Institute, *WEAP, A Computerized Water Evaluation and Planning System, User Guide,* Boston Center Tellus Institute, Boston, Mass., 1993.

Strzepek, K. M., "User's Manual for MIT River Basin Simulation Model," Massachusetts Institute of Technology, Civil Engineering Department, Report no. 242, Cambridge, Mass., 1979.

Strzepek, K. M., L. Garcia, and T. Over, "MITSIM2.1 River Basin Simulation Model: User Manual," Center for Advanced Decision Support for Water and Environmental Systems, University of Colorado, Boulder, Colo., 1989.

Taylor, M. R., and J. S. Behrens, "Object Oriented Hydrologic Simulation," Resources Planning Associates, Ithaca, N.Y., 1995.

Wurbs, R. A., "Optimization of Multiple-Purpose Reservoir System Operations: A Review of Modeling and Analysis Approaches," Research Document no. 34, Hydrologic Engineering Center, U.S. Army Corps of Engineers, Davis, Calif., 1991.

Yeh, W. W-G, "Reservoir Management and Operation Models: A State-of-the-Art Review," *Water Resources Research,* 21(12):1797–1818, 1986.

Yeh, W. W-G, L. Becker, S-Q. Hua, D-P. Wen, and J-M. Liu, "Optimization of Real-Time Hydrothermal System Operation," *Journal of Water Resources Planning and Management,* 118(6):636–653, 1992.

CHAPTER 16
GROUNDWATER SYSTEMS

William W-G. Yeh
Department of Civil and Environmental Engineering
UCLA
Los Angeles, California

16.1 INTRODUCTION

Groundwater models are physically based mathematical models derived from Darcy's law and the law of conservation of mass. Various established solution techniques utilizing either finite difference or finite element approximations, or a combination of both, are available for solving the governing equation of the model, provided that model parameters and initial and boundary conditions are properly specified. Groundwater models are used as tools for decision making in the management of a water-resources system. The accuracy of model prediction depends, to a great extent, on the reliability of the estimated parameters as well as on the accuracy of prescribed initial and boundary conditions. Hydrogeological information about an aquifer is useful in the estimation of initial and boundary conditions. However, parameters used in deriving the governing equation are not directly measurable from the physical point of view, and, in practice, model parameters are required to be estimated (identified) from historical input-output observations (data) using an inverse procedure of parameter identification. See Chap. 2 for fundamental groundwater definitions.

The importance of *parameter identification* in groundwater modeling is well understood. Unfortunately, because data is limited in both quantity and quality, the inverse problem is often ill-posed. The ill-posedness is generally characterized by the nonuniqueness and instability of the inverse solution.

This chapter reviews inverse solution procedures for parameter identification, groundwater-management models, and optimal experimental design methodologies. The review focuses on water supply as opposed to water quality. The fact is that water-supply and water-quality issues are inseparable in the management of a groundwater system. We will address these issues when appropriate.

We will use a typical two-dimensional groundwater-flow model to illustrate the three areas being reviewed. Consider an unsteady flow in an inhomogeneous, isotropic, and confined aquifer for which the governing equation can be expressed as

$$\frac{\partial}{\partial x}\left(T\frac{\partial h}{\partial x}\right) + \frac{\partial}{\partial y}\left(T\frac{\partial h}{\partial y}\right) = S\frac{\partial h}{\partial t} \pm Q \tag{16.1}$$

subject to the following initial and boundary conditions:

$$h(x,y,0) = f_0(x,y) \qquad [(x,y) \in \Omega] \tag{16.2a}$$

$$h(x,y,t) = f_1(x,y,t) \qquad [(x,y) \in \partial\Omega_1] \tag{16.2b}$$

$$T\frac{\partial h}{\partial n} = f_2(x,y,t) \qquad [(x,y) \in \partial\Omega_2] \tag{16.2c}$$

where $h(x,y,t)$ = hydraulic head
$T(x,y)$ = transmissivity
S = storage coefficient
$Q(x,y)$ = sink (or source) term
t = time
Ω = flow region
$\partial\Omega$ = boundary offlow region $(\partial\Omega_1 \cup \partial\Omega_2 = \partial\Omega)$
$\partial / \partial n$ = normal derivative to boundary
f_0, f_1, f_2 = known functions

If the storage coefficient S is assumed to be known and initial and boundary conditions are provided, the inverse problem seeks to identify the transmissivity distribution $T(x,y)$. We will use the identification of $T(x,y)$ to illustrate the inverse solution procedures. However, in practice, we may face a much more complex inverse problem that must also determine the storage coefficient, missing initial or boundary condition, and any other aquifer properties that might control the flow of groundwater. Nevertheless, in principle, the procedures used to identify $T(x,y)$ can also be used to identify other parameters. Eq. (16.1) subject to (16.2) can be solved numerically by either finite difference and/or finite element approximations.

If space variables x and y are discretized by finite difference approximations while keeping time variable t continuous, and using an integrating method (Varga, 1962), the following system of ordinary differential equations results (Yeh and Yoon, 1976):

$$S\frac{d\mathbf{h}}{dt} = -\mathbf{B(T)h} + \boldsymbol{\gamma} \tag{16.3}$$

where $\mathbf{h} = [h_1(t),\cdots,h_N(t)]^T$
$\mathbf{B(T)}$ = positive definite banded symmetric matrix of order N
$\boldsymbol{\gamma}$ = N-dimensional vector stemming from the sink (or source) term and boundary conditions
N = total number of interior nodes resulting from the discretization of x and y
T = transpose of a vector when used as a superscript

If the *Crank-Nicolson scheme* is used to approximate the time derivative, the resulting set of linear algebraic equations can be represented as

$$S\frac{\mathbf{h}^{m+1} - \mathbf{h}^m}{\Delta t} = -\mathbf{B(T)}\frac{\mathbf{h}^{m+1} + \mathbf{h}^m}{2} + \boldsymbol{\gamma}^{m+1/2} \tag{16.4}$$

$$\left[\mathbf{I} + \frac{\Delta t}{2S}\mathbf{B(T)}\right]\mathbf{h}^{m+1} = \left[\mathbf{I} - \frac{\Delta t}{2S}\mathbf{B(T)}\right]\mathbf{h}^m + \frac{\Delta t}{S}\boldsymbol{\gamma}^{m+1/2} \tag{16.5}$$

where $\quad \Delta t$ = time increment
$\mathbf{h}^m = \mathbf{h}(m \cdot \Delta t)$
$\gamma^{m + 1/2} = \gamma[(m + \frac{1}{2}) \cdot \Delta t]$
m = index used to denote time increment
$\mathbf{I}$ = identity matrix

Solutions obtained by solving Eq. (16.5), with appropriate initial and boundary conditions and some estimated values of the parameters, provide head variations in the flow region.

For identification purposes, $T(x,y)$ is generally parameterized, by either an interpolation or zonation procedure, to reduce it to a finite dimensional form. Under the zonation procedure, the aquifer domain is divided into a finite number of zones, with each zone characterized by a constant parameter value. The number of parameters to be identified is equal to the number of zones. Under the interpolation procedure, the aquifer domain is divided into a finite number of elements connected at nodes. Each node is associated with a chosen local basis function. The unknown $T(x,y)$ is then approximated by a linear combination of the basis functions, where the number of parameters to be identified is equal to the number of the unknown nodal transmissivity values.

16.2 INVERSE PROBLEM OF PARAMETER IDENTIFICATION

The problem of parameter identification in distributed parameter systems has been studied extensively during the last three decades. The term *distributed parameter system* implies that the response of the system is governed by a partial differential equation and parameters embedded in the equation are spatially dependent. A review of the inverse problem of parameter identification in groundwater hydrology was presented by Yeh (1986) and Carrera (1988). The inverse problem of parameter identification concerns the optimal determination of the parameters by observing the output of the dependent variable in the spatial and time domains. The number of observations is finite and limited, whereas the spatial domain is continuous. For an inhomogeneous aquifer the dimension of the parameter is theoretically infinite. In practice, spatial variables are approximated by either a finite-difference or finite-element scheme while the aquifer system is subdivided into several subregions with each subregion characterized by a constant parameter. The reduction of the number of parameters from the infinite dimension to a finite dimensional form is called *parameterization*. As mentioned earlier, parameterization can be achieved by either the zonation method (Coats, et al., 1970; Emsellem and de Marsily, 1971; Yeh and Yoon, 1976; and Cooley, 1977; 1979) or the interpolation method (Distefano and Rath, 1975; Yoon and Yeh, 1976; and Yeh and Yoon, 1981).

There are two types of errors associated with the inverse problem: (1) the system modeling error, as represented by a performance criterion, such as the least squares error, and (2) the error associated with parameter uncertainty. For a given set of data in which the information contained is finite, an increase in parameter dimension (the number of unknown parameters associated with parameterization) will generally improve the system modeling error, but the parameter uncertainty error will start to increase at some point and grow rapidly from that point. It has been found, e.g., Yeh and Yoon (1981), that the optimum level of parameterization depends on the quantity and quality of data (observations).

16.2.1 Ill-Posedness

The inverse problem is often ill-posed. The *ill-posedness* is generally characterized by the nonuniqueness and instability of the identified parameters. The instability of the inverse solution stems from the fact that small errors in heads will cause serious errors in the identified parameters.

Chavent (1974) studied the uniqueness problem in connection with parameter identification in distributed parameter systems. As was pointed out by Chavent, the uniqueness problem has a great practical importance, because in the case of nonuniqueness, the identified parameters will differ according to the initial estimate of the parameters, and there will be no reason for the estimated parameters to be close to the "true" parameters. As a consequence, the responses of model and system may differ for inputs different from those that have been used for identification. Chavent studied the uniqueness problem for two situations: (1) the case of constant parameters; and (2) the case of spatially distributed parameters. In situation 1, i.e., constant parameters, there are generally more measurements than unknowns, so that, in general, the inverse problem is unique. In situation 2, i.e., distributed parameters, if measurements are made only at a limited number of locations in the spatial domain, the inverse problem is always nonunique.

The uniqueness problem in parameter identification is intimately related to identifiability. The notion of *identifiability* addresses the question of whether it is at all possible to obtain unique solutions of the inverse problem for unknown parameters of interest in a mathematical model, from data collected in the spatial and time domains. Kitamura and Nakagiri (1977) formulated the parameter identification problem as the one-to-one property of the inverse problem, i.e., the one-to-one property of mapping from the space of system outputs to the space of parameters. However, the uniqueness of such a mapping is extremely difficult to establish and often nonexistent. They defined the identifiability as follows: "We shall call an unknown parameter 'identifiable' if it can be determined uniquely in all points of its domain by using the input-output relation of the system and the input-output data." Kitamura and Nakagiri also obtained some results for parameter identifiability or nonidentifiability for a system characterized by a linear, one-dimensional parabolic partial differential equation.

Another definition of identifiability was given by Chavent (1979b), which is suited to the identification process using the output least squares error criterion. If such criterion is used for solving the inverse problem of parameter identification, the parameter is said to be output least squares identifiable if and only if a unique solution of the optimization problem exists and the solution depends continuously on observations. Chavent (1983) presented a weaker sufficient condition for output least squares identification.

Identifiability is usually not achievable in the case where data is only available at a limited number of locations in the spatial domain. In view of the various uncertainties involved in groundwater modeling, a groundwater model can only be used to approximate the behavior of an aquifer system. If a small, prescribed error is allowed in prediction, Yeh and Sun (1984) developed an extended identifiability criterion which can be used for designing an optimum pumping test to assist parameter identification. The extended identifiability is called the δ-*identifiability*, which is based on the concept of weak uniqueness.

16.2.2 Classification of Parameter Identification Methods

There are only two types of error criteria that have been used in the formulation of the inverse problem. It is generally accepted that the inverse procedures can be classified under either the equation error criterion or the output error criterion (Yeh, 1986).

16.2.2.1 Equation Error Criterion.

(Direct method as classified by Neuman, 1973.) When equation error criterion is employed for parameter estimation, it requires an explicit formulation of the unknown parameters. Yeh (1986) has shown that when the governing equation, Eq. (16.1), is discretized by finite-difference approximations, and that observations (observed and interpolated) are substituted for all nodes in the flow domain in the discretized finite-difference equations, the resulting system of algebraic equations in terms of the unknown nodal transmissivities are:

$$\mathbf{A}_m \mathbf{T}_g = \mathbf{b}_m + \boldsymbol{\epsilon}_m \qquad m = 1, 2, \ldots, N \tag{16.6}$$

where $\mathbf{A}_m$ = coefficient matrix, a function of $\mathbf{h}$
$\mathbf{T}_g$ = transmissivity vector containing transmissivity values at all nodal points
N = total number of time steps
$\mathbf{b}_m$ = column vector, a function of $\mathbf{h}$
m = index for time
$\boldsymbol{\epsilon}_m$ = error term

Note that to account for the lack of equality, an error term $\boldsymbol{\epsilon}_m$ has been added to the equation. The error term contains observation and interpolation errors.

In a more compact matrix form, this becomes

$$\mathbf{A}\mathbf{T}_g = \mathbf{b} + \boldsymbol{\epsilon} \tag{16.7}$$

where $\mathbf{A} = [\mathbf{A}_1^T, \mathbf{A}_2^T, \cdots, \mathbf{A}_N^T]^T$
$\mathbf{b} = [\mathbf{b}_1^T, \mathbf{b}_2^T, \cdots, \mathbf{b}_N^T]^T$
$\boldsymbol{\epsilon} = [\boldsymbol{\epsilon}_1^T, \boldsymbol{\epsilon}_2^T, \cdots, \boldsymbol{\epsilon}_N^T]^T$

T is a transpose operator when used as a superscript. It should be noted that whether finite difference or finite element approximations are used as the numerical method for solving the governing equation, the resulting equation error will always have the form of Eq. (16.7) The advantage of this formulation is that Eq. (16.7) is linear, and $\mathbf{T}_g$ can be determined by minimizing the equation error $\boldsymbol{\epsilon}$.

From Eq. (16.7), the least squares error (or residual sum of squares) can be expressed by

$$\boldsymbol{\epsilon}^T\boldsymbol{\epsilon} = (\mathbf{A}\mathbf{T}_g - \mathbf{b})^T(\mathbf{A}\mathbf{T}_g - \mathbf{b}) \tag{16.8}$$

Minimizing the least squares error, the transmissivity vector can be estimated as

$$\hat{\mathbf{T}}_g = (\mathbf{A}^T\mathbf{A})^{-1}\mathbf{A}^T\mathbf{b} \tag{16.9}$$

where $\hat{\mathbf{T}}_g$ is the estimated transmissivity vector of $\mathbf{T}_g$. Note that the solution of Eq. (16.9) implicitly assumes homoscedasticity and lack of correlation among residuals. The solution is also highly dependent on the level of discretization used in the numerical solution of the governing equation. Another disadvantage is that solution of Eq. (16.9) is generally unstable in the presence of noise. Parameterization is generally required to ensure a stable solution.

The graphical method developed by Stallman (1956) for the identification of the regional transmissivity distribution is a direct inverse technique. The method relies on the construction of a flow net and, hence, can only be applied to a two-dimensional, steady-state situation. It requires distributed head observations (or estimated) in the entire flow domain and a transmissivity value at some point along each streamtube of the constructed flow net. Applications of this technique include Nelson (1960; 1961); Hunt and Wilson (1974); Hawkins and Stevens (1983); and Rice and Gorelick (1985).

16.2.2.2 Output Error Criterion. (Indirect method as classified by Neuman, 1973.) For modeling purposes, the objective is to determine $T(x,y)$ from a limited number of observations of $h(x,y,t)$ scattered in the spatial and time domains so that a certain criterion is optimized. If the classical least squares error is used to represent the output error, the objective function to be minimized is

$$\min_{T(x,y)} J = [\mathbf{h}_D - \mathbf{h}_D^*]^T [\mathbf{h}_D - \mathbf{h}_D^*] \qquad (16.10)$$

where $\mathbf{h}_D$ is the vector of calculated heads at observation wells, based on some estimated values of parameters, and $\mathbf{h}_D^*$ is the vector of observed heads.

For identification purposes, $T(x,y)$ is again parameterized by either a zonation or interpolation method.

The Gauss-Newton algorithm has proven to be an effective algorithm to perform the minimization indicated in Eq. (16.10). The original or the modified version of the algorithm has been used by many researchers in the past in solving the inverse problem; e.g., Jacquard and Jain (1965); Jahns (1966); Thomas, et al. (1972); Gavalas, et al. (1976); Yoon and Yeh (1976); Cooley (1977; 1982); and Hill (1992). The popularity of the algorithm stems from the fact that it does not require the calculation of the Hessian matrix as is required by the Newton method, and the rate of convergence is superior when compared to the classical gradient searching procedures. The algorithm is basically developed for unconstrained minimization. However, constraints such as upper and lower bounds, the most common type in parameter identification, are easily incorporated in the algorithm with minor modifications. The algorithm starts with a set of initial estimates of parameters and converges to a local optimum. If the objective function is convex, the local optimum would be the global optimum. Due to the presence of observation noise, the inverse problem is usually nonconvex, and hence only a local optimum can be assured in the minimization.

Let $\overline{\mathbf{T}}$ be a vector of parameters that contains $[T_1, T_2, \cdots, T_L]$. The algorithm generates the following parameter sequence for an unconstrained minimization problem:

$$\overline{\mathbf{T}}^{k+1} = \overline{\mathbf{T}}^k - \rho^k \mathbf{d}^k \qquad (16.11)$$

with

$$\mathbf{A}^k \mathbf{d}^k = \mathbf{g}^k \qquad (16.12)$$

where $\mathbf{A}^k = [\mathbf{J}_D(\overline{\mathbf{T}}^k)]^T [\mathbf{J}_D(\overline{\mathbf{T}}^k)], (L \times L)$
  $\mathbf{g}^k = [\mathbf{J}_D(\mathbf{T}^k)]^T [\mathbf{h}_D(\mathbf{T}^k) - \mathbf{h}_D^*], (L \times 1)$
  $\mathbf{J}_D$ = Jacobian matrix of head with respect to $\overline{\mathbf{T}}, (M \times L)$
  ρ^k = step size, (scalar)
  $\mathbf{d}^k$ = Gauss–Newton direction vector, $(L \times 1)$
  M = number of observations
  L = parameter dimension

The step size ρ^k, a scalar, can be determined by a quadratic interpolation scheme such that $\mathbf{J}(\overline{\mathbf{T}}^{k+1}) < \mathbf{J}(\overline{\mathbf{T}}^k)$, or simply by a trial-and-error procedure. Occasionally, the direction matrix $[\mathbf{J}_D^T \mathbf{J}_D]$ may become ill-conditioned. Corrections must be made in order for the algorithm to continue, and the methods suggested by Levenberg (1944) and Marquardt (1963) are a modification of the Gauss-Newton direction. As stated earlier, the basic Gauss-Newton algorithm does not handle constraints. If constraints are imposed on the parameters, such as the upper and lower bounds, the Gauss-Newton algorithm can be allied with a gradient projection technique (Yoon and Yeh, 1976).

The elements of the Jacobian matrix are represented by the sensitivity coefficients,

$$\mathbf{J}_D = \begin{bmatrix} \dfrac{\partial h_1}{\partial T_1} & \dfrac{\partial h_1}{\partial T_2} & \cdots & \dfrac{\partial h_1}{\partial T_L} \\[2ex] \dfrac{\partial h_2}{\partial T_1} & \dfrac{\partial h_2}{\partial T_2} & \cdots & \dfrac{\partial h_2}{\partial T_L} \\[2ex] \cdot & & & \\ \cdot & & & \\ \cdot & & & \\ \dfrac{\partial h_M}{\partial T_1} & \dfrac{\partial h_M}{\partial T_2} & \cdots & \dfrac{\partial h_M}{\partial T_L} \end{bmatrix} \qquad (16.13)$$

The transpose of the Jacobian matrix is

$$\mathbf{J}_D^T = \begin{bmatrix} \dfrac{\partial h_1}{\partial T_1} & \dfrac{\partial h_2}{\partial T_1} & \cdots & \dfrac{\partial h_M}{\partial T_1} \\[2ex] \dfrac{\partial h_1}{\partial T_2} & \dfrac{\partial h_2}{\partial T_2} & \cdots & \dfrac{\partial h_M}{\partial T_2} \\[2ex] \cdot & & & \\ \cdot & & & \\ \cdot & & & \\ \dfrac{\partial h_1}{\partial T_L} & \dfrac{\partial h_2}{\partial T_L} & \cdots & \dfrac{\partial h_M}{\partial T_L} \end{bmatrix} \qquad (16.14)$$

In solving the inverse problem, an efficient method must be used in the calculation of the sensitivity coefficients. We will now focus our attention on the techniques developed for calculating the sensitivity coefficients.

16.2.3 Computation of Sensitivity Coefficients

Sensitivity coefficients, the partial derivatives of head with respect to each of the parameters, play an important role in the solution of the inverse problem. In the Gauss-Newton algorithm, elements of the Jacobian matrix are represented by the sensitivity coefficients, $\partial h_i / \partial T_l$, $i = 1, \cdots, M$; $l = 1, \cdots, L$. If $\mathbf{h}$ is the head vector, the sensitivity coefficients are $\partial \mathbf{h} / \partial T_l$, $l = 1, \cdots, L$. Literature review indicates that three methods have been used in the past in the calculation of sensitivity coefficients. We will summarize these methods as follows.

16.2.3.1 *Influence Coefficient Method.*   The *influence coefficient method* (Becker and Yeh, 1972) uses the concept of parameter perturbation. The *l*th row of $\mathbf{J}_D^T$ is approximated by

$$\frac{\partial h_i}{\partial T_l} \approx \frac{h_i(\overline{\mathbf{T}} + \Delta T_l e_l) - h_i(\overline{\mathbf{T}})}{\Delta T_l} \qquad i = 1,\cdots,M \qquad (16.15)$$

where ΔT_l is a small increment of T_l and e_l is the *l*th unit vector. The values of $h(\overline{\mathbf{T}})$ and $h(\overline{\mathbf{T}} + \Delta T_l e_l)$ are obtained by solving the governing equation (by simulation), subject to the imposed initial and boundary conditions. The method requires perturbing each parameter one at a time. If there are L parameters to be identified, the governing equation has to be solved (simulated) $(L + 1)$ times for each iteration in the nonlinear least squares minimization to numerically produce the sensitivity coefficients. The numerical representation of $\mathbf{J}_D^T$ Eq. (16.14) is called the influence coefficient matrix (Becker and Yeh, 1972). The elements of the influence coefficient matrix, represented by a_{li} are numerical approximations of the sensitivity coefficients,

$$
\begin{array}{ccccc}
 & h_1 & h_2 & \cdots & h_M \\
T_1 & a_{11} & a_{12} & \cdots & a_{1M} \\
T_2 & a_{21} & a_{22} & \cdots & a_{2M} \\
T_L & a_{L1} & a_{L2} & \cdots & a_{LM}
\end{array}
\qquad (16.16)
$$

Each element in the matrix represents the ratio of change in the head to the change in a particular parameter. The value of ΔT_l is a small increment of T_l by which parameter T_l is perturbed. The appropriate value of ΔT_l is usually determined on a trial-and-error basis. Bard (1974) has suggested some guidelines in choosing the value of ΔT_l.

16.2.3.2 *Sensitivity Equation Method.*   In this approach, a set of sensitivity equations are obtained by taking the partial derivatives with respect to each parameter in the governing equation and initial and boundary conditions. After taking the partial derivatives, the following set of *sensitivity equations* result (Yeh, 1986):

$$\frac{\partial}{\partial x}\left[\mathbf{T}\frac{\partial\left(\frac{\partial h}{\partial T_l}\right)}{\partial x}\right] + \frac{\partial}{\partial y}\left[\mathbf{T}\frac{\partial\left(\frac{\partial h}{\partial T_l}\right)}{\partial y}\right] = S\frac{\partial\left(\frac{\partial h}{\partial T_l}\right)}{\partial t} \qquad (16.17)$$

$$+\left[-\frac{\partial}{\partial x}\left(\frac{\partial\mathbf{T}}{\partial T_l}\frac{\partial h}{\partial x}\right) - \frac{\partial}{\partial y}\left(\frac{\partial\mathbf{T}}{\partial T_l}\frac{\partial h}{\partial y}\right)\right] \qquad l = 1,\cdots,L$$

The associated initial and boundary conditions are

$$\frac{\partial h(x,y,0)}{\partial T_l} = 0 \qquad l = 1,\cdots,L$$

$$\frac{\partial h(x,y,t)}{\partial T_l} = 0 \qquad l = 1,\cdots,L \qquad (16.18)$$

$$\mathbf{T}\frac{\partial\left(\frac{\partial h}{\partial T_l}\right)}{\partial n} = -\frac{\partial\mathbf{T}}{\partial T_l}\frac{\partial h}{\partial g} \qquad l = 1,\cdots,L$$

The numerical values of $\partial h/\partial x$ and $\partial h/\partial y$ are obtained from the solution of the governing equation. If we replace $(\partial h/\partial T_l)$ by h and consider the term

$$\left[-\frac{\partial}{\partial x}\left(\frac{\partial \mathbf{T}}{\partial T_l}\frac{\partial h}{\partial x}\right) - \frac{\partial}{\partial y}\left(\frac{\partial \mathbf{T}}{\partial T_l}\frac{\partial h}{\partial y}\right)\right]$$

as Q, the set of sensitivity equations would be of the same form as that of the governing equation. Hence the solution method used for solving the governing equation can be used to solve the set of sensitivity equations. The number of simulation runs required to generate the sensitivity coefficients per iteration is $(L + 1)$, which is the same as that of the influence coefficient method.

16.2.3.3 Variational Method.

The *variational method* was first used to solve the inverse problem of parameter identification by Jacquard and Jain (1965) and then by Carter, et al. (1974; 1982) associated with finite difference schemes. Sun and Yeh (1985) extended the method to the case of a finite element scheme. Following Carter, et al. (1974), the sensitivity coefficients can be computed by the following equation:

$$\frac{\partial h^{(j)}}{\partial T_l^{(i)}} = \int\int_{(\Omega_i)}\int_0^t \nabla q'\,(x,y,t-\tau)\nabla h(x,y,\tau)d\tau\,dx\,dy \qquad j=1,2,\cdots,N_o \qquad i=1,2,\cdots,N_n$$

$$(16.19)$$

where (Ω_i) is the exclusive subdomain of node i as defined by Sun and Yeh; ∇ is the gradient operator; $h(x,y,t)$ is the solution of the governing equation; N_o is the total number of observation wells; N_n is the total number of nodes used in the numerical solution; $q'(x,y,t)$ is the time derivative of $q(x,y,t)$, which is the solution of the following set of adjoint equations:

$$\frac{\partial}{\partial x}\left[\mathbf{T}\frac{\partial q}{\partial x}\right] + \frac{\partial}{\partial y}\left[\mathbf{T}\frac{\partial q}{\partial y}\right] = S\frac{\partial q}{\partial t} + G_j(x,y)H(t) \qquad (16.20)$$

subject to the following initial and boundary conditions:

$$q(x,y,0) = 0 \qquad (x,y) \in \Omega$$

$$q(x,y,t) = 0 \qquad (x,y) \in \partial\Omega_1 \qquad (16.21)$$

$$\frac{\partial q}{\partial n}(x,y,t) = 0 \qquad (x,y) \in \partial\Omega_2$$

where

$$G_j(x,y) = \frac{1}{P_j} \qquad (x,y) \in (\Omega_j)$$

$$G_j(x,y) = 0 \qquad \text{otherwise} \qquad (16.22)$$

$$H(t) = 0 \qquad t \leq 0$$

$$H(t) = 1 \qquad t > 0$$

P_j is the area of subdomain (Ω_j).

Note that the adjoint equation for $q(x,y,t)$, Eq. (16.20), has the same form as that of the governing equation for $h(x,y,t)$, Eq. (16.1), and hence the same numerical scheme can be used to solve h and q. By solving the governing equation one time only and solving the adjoint equation for each observation well, all sensitivity coefficients $[(\partial h^{(j)} / \partial T_l^{(i)})]$ ($j = 1,2,\cdots,N_0; i = 1,2,\cdots,N_n$) can be produced. Hence the number of simulation runs required to calculate the sensitivity coefficients per iteration is $(N_0 + 1)$, as compared to $(L + 1)$, which is required by either the influence coefficient method or the sensitivity equation method.

Comparing the previously mentioned three methods in the calculation of sensitivity coefficients, it is clear that the variational method would be advantageous if $L > N_0$, the case where the number of parameters to be identified is greater than the number of observation wells. On the other hand, if $N_0 > L$, the influence coefficient and sensitivity equation methods are preferred. The influence coefficient method is easy to implement, and the perturbation value should be appropriately chosen so that the numerical approximation of the sensitivity coefficient is valid. However, the sensitivity and variational methods are intrinsically much more accurate. The need for an efficient method for calculating the sensitivity coefficients in solving inverse problems has also been pointed out by Dogru and Seinfeld (1981), and Sykes, et al. (1985).

16.2.4 Parameter Uncertainty, Parameter Structure, and Optimum Parameter Dimension

Parameter identification in a distributed parameter system should, in principle, include the determination of both the parameter structure and its value. If zonation is used to parameterize the unknown parameters, parameter structure is represented by the number and shape of zones. On the other hand, if the finite element method is used for parameterization, parameter structure concerns the number and location of nodal values of parameters. Emsellem and de Marsily (1971) were among the first to consider the problem of optimal zoning pattern. Yeh and Yoon (1976) suggested a systematic procedure based upon a statistical criterion for the determination of an optimum zoning pattern. Sun and Yeh (1985) formulated a general parameter structure identification problem in terms of combinatorial optimization and solved it by a simple one-at-a-time searching procedure. They clearly pointed out that optimum parameter values cannot be separated from the optimum parameter structure. That is, if the parameter structure is incorrect, the identified parameter values will also be incorrect. Shah, et al. (1978) showed the relationship between the optimal dimension of parameterization and observations in considerable depth. The necessity to limit the dimension of parameterization has been further studied by Yeh and Yoon (1981); Yeh, et al. (1983); and Kitanidis and Vomvoris (1983). The dimension of parameterization is directly related to the quantity and quality of data (observations). In practice, the number of observations is limited, and observations are corrupted with noise. Without controlling parameter dimension, instability often results (Yakowitz and Duckstein, 1980). If instability occurs in the inverse problem solution, parameters will become unreasonably small (sometimes negative, which is physically impossible) and/or large, if parameters are not properly constrained. In the constrained minimization, instability is characterized by the fact that during the solution process parameter values are bouncing back and forth between the upper and lower bounds. Reduction of parameter dimension can make the inverse solution stable. As mentioned earlier, as the number of zones (in the zonation case) is increased, the modeling error (least squares) decreases while the parameter uncertainty error at some point will start to increase. A tradeoff of the two types of errors can then be made, from which an optimum parameter dimension can be determined. A standard procedure is to gradually increase the parameter dimension, starting from the lowest dimension, i.e., the homogeneous case, and calculate the two types of errors for each parameterization. The error in parameter uncertainty can be represented by a norm of the covariance matrix of the estimated parameters (Yeh and Yoon, 1976; Shah, et al., 1978).

An approximation of the covariance matrix of the estimated parameters in nonlinear regression can be represented by the following form (Bard, 1974; Yeh and Yoon, 1976; 1981; Shah, et al., 1978; Yeh, 1986):

$$\text{Cov}\,(\hat{\mathbf{T}}) = \frac{J(\hat{\mathbf{T}})}{M - L}\,[\mathbf{A}(\mathbf{T})]^{-1} \tag{16.23}$$

where $J(\hat{\mathbf{T}})$ = least squares error
M = number of observations
L = parameter dimension
$\mathbf{A} = [J_D^T J_D]$
J_D = Jacobian matrix of $\mathbf{h}$ with respect to $\mathbf{T}$

A norm of the covariance matrix has been used to represent the error in parameter uncertainty. Norms, such as trace, spectral radius (maximum eigenvalue), and determinant have been used in the literature. Equation (16.23) also assumes homoscedasticity and uncorrelated errors. This assumption is generally not satisfied and the actual covariance may be much higher than that given by Eq. (16.23).

The covariance matrix of the estimated parameters also provides information regarding the reliability of each of the estimated parameters. A well-estimated parameter is generally characterized by a small variance as compared to an insensitive parameter that is associated with a large variance. By definition, the correlation matrix of the estimated parameters is

$$\mathbf{R} = \begin{bmatrix} \dfrac{c_{11}}{(c_{11}c_{11})^{1/2}} & \cdots & \dfrac{c_{1L}}{(c_{11}c_{LL})^{1/2}} \\[2ex] \cdot & & \cdot \\ \cdot & & \cdot \\ \cdot & & \cdot \\[1ex] \dfrac{c_{L1}}{(c_{LL}c_{11})^{1/2}} & \cdots & \dfrac{c_{LL}}{(c_{LL}c_{LL})^{1/2}} \end{bmatrix} \tag{16.24}$$

where c_{ij}'s are elements of the covariance matrix of the estimated parameters. The more sensitive the parameter, the closer and quicker the parameter will converge. A correlation analysis of the estimated parameters would indicate the degree of interdependence among the parameters with respect to the objective function. Correlation of parameters is called the *collinearity problem*. Such problems can cause slow rate of convergence in minimization and in most cases result in nonoptimal parameter estimates. A more rigorous treatment of the collinearity problem is to use the more sophisticated statistical techniques, such as ridge regression (Cooley, 1977) and the method of principal components.

16.2.5 Statistical and Stochastic Inverse Procedures

The inverse procedures just mentioned assume that model parameters are deterministic and, after parameterization, can be represented by a set of constant parameters. A disadvantage of the deterministic approach is that it cannot capture the smaller-scale variability of a distributed parameter. A different approach is to represent the model parameter, e.g., transmissivity, as a random function with certain underlying properties. The inverse problem then seeks to identify the constants that are imbedded in the random function.

Stochastic methods have been introduced for solving the inverse problem in groundwater modeling. Neuman and Yakowitz (1979) and Clifton and Neuman (1982) proposed the use of kriging to generate the distributed prior estimation of log transmissivity and its covariance matrix estimation. The prior estimation was then imposed on the original least squares minimization problem of fitting head observations either as a constraint or as another objective in the framework of a multiobjective optimization. The general form of the weighted objective function can also be derived from the *maximum likelihood estimation* (MLE) (Carrera and Neuman, 1986a).

The composite least squares criterion proposed by Neuman and Yakowitz is

$$J = [\mathbf{h}^* - \mathbf{f}(\overline{\mathbf{T}})]^T \mathbf{V}_h^{-1} [\mathbf{h}^* - \mathbf{f}(\overline{\mathbf{T}})] + \lambda (\mathbf{T}^* - \overline{\mathbf{T}})^T \mathbf{V}_T^{-1} (\mathbf{T}^* - \overline{\mathbf{T}}) \qquad (16.25)$$

where $\mathbf{T}^*$ = prior estimate of $\overline{\mathbf{T}}$
$\mathbf{V}_T$ = known symmetric positive definite matrix
$\mathbf{V}_h$ = known matrix, symmetric and positive definite
λ = unknown positive parameter
$\mathbf{f}(\overline{\mathbf{T}})$ = model solution
$\mathbf{h}^*$ = observed head

The observed head $\mathbf{h}^*$ and the prior estimates of transmissivity $\mathbf{T}^*$ are related to true head $\mathbf{h}$ and true transmissivity $\overline{\mathbf{T}}$ by

$$\mathbf{h}^* = \mathbf{h} + \boldsymbol{\epsilon}$$

$$\mathbf{T}^* = \overline{\mathbf{T}} + \mathbf{v} \qquad (16.26)$$

where $\boldsymbol{\epsilon}$ and $\mathbf{v}$ are noise vectors and

$$E(\boldsymbol{\epsilon}) = 0$$

$$\text{Var}(\boldsymbol{\epsilon}) = \sigma_h^2 \mathbf{V}_h$$

$$E(\mathbf{v}) = 0 \qquad (16.27)$$

$$\text{Var}(\mathbf{v}) = \sigma_T^2 \mathbf{V}_T$$

It is assumed that $\mathbf{V}_h$ and $\mathbf{V}_T$ are known, but σ_h and σ_T do not enter the computations. The second term in the composite objective function provides a smoothing effect in the minimization. Neuman and Yakowitz proposed two methods, called *cross-validation* and *comparative residual analysis*, to select the optimum value of λ.

Kitanidis and Vomvoris (1983) presented a *geostatistical approach* to parameter identification for a one-dimensional steady flow problem. Their method consists of two main steps: (1) the structure of the parameter field is identified, i.e., mathematical representations of the variogram and the trend are selected and their parameters are established; and (2) kriging is applied to provide minimum variance and unbiased point estimates of hydrogeological parameters using all information available. In their approach, it is assumed that several point measurements of head and transmissivities (in logarithms) are available. In effect, parameterization is achieved by representing the hydrogeological parameters as a random field which can be characterized by the variogram and trend with a small number of parameters. The reduction of parameter dimension has resulted in stable inverse problem solutions, even with the presence of errors.

A stochastic inverse method was used in a two-dimensional steady flow problem by Hoeksema and Kitanidis (1984) and Dagan (1985). In the former, a finite difference numerical model of groundwater flow was used to relate the head and transmissivity variability and cokriging was used to estimate the unknown transmissivity field. Calculating the cross-covariance matrix required $(N + 1)$ simulation runs, where N is the number of grids. In the later, an analytical technique and Gaussian conditional mean are used in place of kriging to calculate the covariance matrix under some simplified assumptions. The analytical method was further extended to the case of quasi-steady flow by Dagan and Rubin (1988), subject to additional limitations. The stochastic inverse approach has also been applied to some simplified field problems (Hoeksema and Kitanidis, 1985a, b; Rubin and Dagan, 1987a, b; Dagan and Rubin, 1988).

Recently, Sun and Yeh (1992) have extended the stochastic inverse approach to a general transient groundwater-flow problem. It was assumed that the aquifer is confined, the storage coefficient is constant in the entire aquifer, and the parameter to be identified is K. The log hydraulic conductivity $Y = \log K$ is taken to be a normally distributed field (Freeze, 1975; Hoeksema and Kitanidis, 1985a). Also, the random field Y is characterized by a constant mean and an isotropic, exponential covariance (Dagan, 1985; Hoeksema and Kitanidis, 1985b; Wagner and Gorelick, 1989):

$$E[Y] = \mu_Y$$

$$\mathrm{Cov}_{YY}(x_i, x_j) = \sigma_Y^2 \exp\left(-\frac{d_{ij}}{l_Y}\right)$$

where σ_Y^2 = log hydraulic conductivity variance
l_Y = log hydraulic conductivity correlation scale
d_{ij} = distance between points x_i and x_j

The hydraulic conductivity can thus be estimated by identifying the statistical parameters μ_Y, σ_Y^2 and l_Y.

First, the adjoint state equations were derived for the *stochastic partial differential equation (SPDE)* relating head and log K perturbations and used to calculate all cross-covariance matrices. The SPDE for the expected head H is

$$S\frac{\partial H}{\partial t} = e^F\left[\frac{\partial}{\partial x}\left(m\frac{\partial H}{\partial x}\right) + \frac{\partial}{\partial x}\left(m\frac{\partial H}{\partial y}\right)\right] + Q \tag{16.28}$$

where H = expectation of the head random field, h
F = expectation of the log hydraulic conductivity random field, Y
x,y = space variables
S = storage coefficient
Q = sink/source term
m = thickness of the aquifer

and the SPDE for perturbations is

$$S\frac{\partial g}{\partial t} = e^F\left[\frac{\partial}{\partial x}\left(m\frac{\partial g}{\partial x}\right) + \frac{\partial}{\partial y}\left(m\frac{\partial g}{\partial y}\right)\right] + e^F m\left[\frac{\partial f}{\partial x}\frac{\partial H}{\partial x} + \frac{\partial f}{\partial y}\frac{\partial H}{\partial y}\right] + \left(S\frac{\partial H}{\partial t} - Q\right)f \tag{16.29}$$

where h and f are zero mean perturbations such that $Y = F + f$ and $h = H + g$. The corresponding adjoint state equation is

$$S\frac{\partial \varphi}{\partial t} + e^F\left[\frac{\partial}{\partial x}\left(m\frac{\partial \varphi}{\partial x}\right) + \frac{\partial}{\partial y}\left(m\frac{\partial \varphi}{\partial y}\right)\right] = \frac{\partial h}{\partial g} \qquad (16.30)$$

where φ is the adjoint state variable and R is a user-chosen performance function. The Jacobian can then be calculated using the variational method with only $(N_0 + 1)$ simulation runs, where N_0 is the total number of observation wells. Once the Jacobian is obtained, the measurement covariance matrices can be approximated to the first order. MLE was then used to estimate the statistical parameters. The utilization of head observations of all observation times simultaneously produced much better results than the sequential estimation approach used by Dagan and Rubin (1988).

The log K field can be estimated by applying either Gaussian conditional mean estimate or cokriging and using the statistical parameters and observation data (Dagan, 1985; Hoeksema and Kitanidis, 1985b; de Marsily, 1986; Dagan and Rubin, 1988). For cokriging, all head observations at different observation times were again used simultaneously.

16.3 GROUNDWATER MANAGEMENT MODELS

Groundwater production from an aquifer is traditionally developed to satisfy municipal, industrial, or agricultural water demands to a full or partial extent. The following planning problems are associated with the management of groundwater supply (Willis and Yeh, 1987):

1. The determination of the optimal pumping pattern, i.e., the location of all pumping wells in the system, and the magnitude and duration of pumping necessary to satisfy the given water targets.
2. The timing and staging of well-field development, i.e., the capacity expansion problem.
3. The design of surface storage and transport facilities to distribute the groundwater supply to the water demands in the basin.

The physical basis of these planning models includes the governing equations of the groundwater system. The state variable is the head, and decision variables may include pumping and recharge rates. The numerical solution of the planning model, which relates the state and decision variables, is found by finite difference and/or finite element approximations, provided that model parameters have been properly estimated.

Gorelick, in his review paper (1983), classified groundwater hydraulic management models into the embedding and response matrix approaches. In the *embedding approach,* finite difference and/or finite element approximations of the governing equation, Eq. (16.5), are treated as part of the constraint set of an optimization model. If the objective function and constraints are all linear, the formulated optimization model can be solved by linear programming (LP). In the *response matrix approach,* the governing equation along with the approximate initial and boundary conditions are used to determine the influence of pumping and/or recharge at selected locations in the flow region. The influence coefficients are predicated on an initial policy which consists of a set of the estimated values of the decision variables used in the management model. These influence coefficients form the response matrix which is used to replace Eq. (16.5) in the optimization model. The decision variables in a linear, mixed integer, or quadratic program include pumping or recharge rates at certain given locations. The response matrix has a much smaller dimension than the discretized governing equation. Therefore, a considerable amount of computer time and storage can be saved in the optimization. However, iterative solution of the optimization problem is generally required because the response surface of the governing equation is not necessarily linear. At each new iteration, the response matrix should be updated using the optimized values of the decision variables.

In the case of a nonlinear system, such as transient unconfined flow, nonlinear programming (NLP) and differential dynamic programming (DDP) have been used to solve the management models. Jones, et al. (1987) developed a generalized DDP algorithm to effectively alleviate the inherent dimensionality problem in connection with the solution of large-scale, nonlinear groundwater-optimization models.

16.3.1 Linear Programming (LP) Models

LP has been one of the most widely used techniques in water-resources management. It is concerned with solving a special type of problem: one in which all relations among the variables are linear, both in constraints and in the objective function to be optimized. (See Chap. 6.)

A typical LP model is

$$\min_{\mathbf{x}} Z = \mathbf{c}^T \mathbf{x} \qquad (16.31)$$

subject to

$$\mathbf{Ax} \geq \mathbf{b} \qquad (\mathbf{x} \geq 0) \qquad (16.32)$$

in which $\mathbf{c} = n$-dimensional vector of cost coefficients
$\mathbf{x} = n$-dimensional vector of decision variables
$\mathbf{b} = m$-dimensional vector of right-hand sides
$\mathbf{A} = m \times n$ matrix of constraint ("technological") coefficients
T = transpose operation

LP has been used extensively in groundwater-management models that are based upon either the embedding or the response matrix approach. For a confined aquifer system, the governing equation of groundwater flow is linear; hence, the resulting set of finite difference (or finite element) equations is also linear, as represented by Eq. (16.5). For a linear objective function, the management model can be solved by a standard LP package. Deninger (1970) used the response matrix approach in maximizing water production from a well field. LP was used to maximize the total well discharge, with constraints limiting drawdowns. Aquado and Remson (1974) used the embedding approach in the control of groundwater hydraulics. The objective was the maximization of hydraulic heads at specific locations with constraints on head gradients and pumping rates. LP was used to solve the optimization problem. For a quadratic objective function subject to a set of linear constraints, the problem lends itself to a quadratic program that can be solved by LP modified with restricted entry. In the case of nonlinear equations, Willis (1984) used the technique of quasilinearization (Bellman and Kalaba, 1965) to linearize the nonlinear equations, and solutions were obtained by solving a series of linear programs.

LP has been used as a valuable tool in the optimization of the planning and management of groundwater systems. Although objective functions as well as some of the constraints are often nonlinear, various linearization techniques (such as piecewise linearization and first-order Taylor series expansion) allied with an iterative scheme have been successfully used to obtain solutions. The essential advantages of LP include the following: (1) it can accommodate relatively high dimensionality with comparative ease; (2) universal optima are obtained; (3) no initial policy is needed; and (4) standard computer codes are readily available.

16.3.2   Mixed-Integer and Quadratic Programming Models

Rosenwald and Green (1974) used mixed integer programming allied with a response matrix to determine the optimum location of wells in an oil reservoir system so that the production-demand curve was met as closely as possible. An external finite difference model was used to generate a transient response matrix which described pressure changes caused by pumping. The solution of the nonlinear problem of minimizing pumping costs was presented by Maddock (1972). A response matrix was used to represent the drawdown at a point as the superposition of drawdowns from each well. A quadratic programming algorithm was used for solution.

16.3.3   Differential Dynamic Programming (DDP) Models

DP is especially suited for solving the resources allocation problem or the multistage decision-making problem. The popularity and success of DP can be attributed to the fact that the nonlinear and stochastic features that characterize a large number of water-resources systems can be translated into a DP formulation.

When the returns are independent and additive and the objective is to maximize the total return, a typical backward recurrence relation is

$$f_n(\mathbf{x}_n) = \max_{\mathbf{d}_n} \; [R_n(\mathbf{x}_n, \mathbf{d}_n) + f_{n-1}(\mathbf{x}_{n-1})] \tag{16.33}$$

where
- $\mathbf{x}$ = state variable
- $\mathbf{d}$ = decision variable
- R = return function
- n = stage
- $\mathbf{x}_{n-1} = t_n(\mathbf{x}_n, \mathbf{d}_n)$ = state-to-stage transformation equation
- $f_0(\mathbf{x}_0)$ = given for all terminal states

The state and decision variables are generally vectors.

Application of classical DP to groundwater management problems is limited because of the dimensionality problem resulting from the large number of state variables associated with groundwater modeling. The state variable vector is the distributed head (x, y, t) and the decision variable vector is the distributed pumping/recharge (x, y, t). Jones et al. (1987) formulated a groundwater management problem in terms of optimal control, and presented a modified *differential dynamic programming* (DDP) algorithm for solving the resulting large-scale, nonlinear groundwater optimization model. See Chap. 6 for more details on DDP.

DDP is an iterative procedure based on the concept developed in classical DP for solving models with a quadratic objective function subject to a set of linear constraints. The DDP method can handle a large number of state variables without discretizing both the state and decision variables and, consequently, reduces the dimensionality problem. In Jones et al. (1987), Taylor series expansion was used to provide a quadratic approximation of the nonlinear objective function and linear expressions for the nonlinear equations. A quadratic programming algorithm was used to obtain stage-to-stage solutions. A different definition was used by Culver and Shoemaker (1992) for the method developed by Jones et al. (1987); they called it the *successive approximation linear quadratic regulator* (SALQR) method since the problem presented by Jones et al. (1987) requires linearization of the dynamics in the optimization step. If dynamics (set of finite difference equations) are linear, SALQR and DDP are identical.

Jones, et al. (1987) solved some large-scale, nonlinear groundwater-management models using a generalized DDP. They found that: (1) the computational requirements for solving DDP are at least an order of magnitude less than under conventional embedding-optimization models; (2) central memory requirements of the DDP model are far less than under conventional groundwater-optimization models; and (3) the effects of pumping and upper bounds on heads can be effectively controlled using quadratic approximations about the current nominal solutions.

16.3.4 Nonlinear Programming (NLP) Models

In general, a nonlinear programming problem can be stated as:

$$\min_{\mathbf{x} \in X} f(\mathbf{x}) \tag{16.34}$$

subject to

$$\mathbf{g}(\mathbf{x}) \geq 0 \tag{16.35}$$

where $\mathbf{x}$ = n-dimensional vector of decision variables
     $f(\mathbf{x}), \mathbf{g}(\mathbf{x})$ = real-valued and m vector-valued given functions, respectively

NLP is a formulation of the most general mathematical programming, and can effectively handle a nonseparable objective function and/or nonlinear constraints. The constraint includes the hydraulic response equations, Eq. (16.5), upper and lower bounds of the decision and state variables, and water-demand requirements. The objective function of the model could reflect the cost or loss in not meeting water targets or the overall net economic benefit. Although NLP formulation is the most general, rate of convergence of the algorithm and computer requirements are major obstacles in the solution of practical groundwater problems. If the objective function and constraints are separable and are convex functions of the decision variables, separable convex programming methods can be used to solve the original problem. MINOS, a commercially available software package (Murtagh and Saunders, 1982) is designed to solve large-scale nonlinear programming problems with sparse constraint matrices. For problems that do not have sparse constraint matrices, an alternative package is NPSOL (Gill, et al., 1986).

16.3.5 Simulation Models

Simulation is a modeling technique that is used to approximate the behavior of a system on the computer, representing all the characteristics of the system by mathematical relationships. It is an effective tool for studying the management of a complex water-resource system by incorporating the experience and judgment of the planner or designer into the model. Simulation models have been successfully used by various practitioners. However, in recent years a tendency has developed toward incorporating an optimization scheme into a simulation model to perform a certain degree of optimization. As a result, the distinction made between optimization and simulation is often obscured. It has become quite common to have a few optimization routines nested in a simulation model.

The *optimization-groundwater simulation system* developed by Wanakule, et al. (1986) consists of the generalized reduced-gradient model and a groundwater-simulation model. Finite difference was used in the simulation, and the alternating direction implicit scheme was used to solve the finite difference equations. The approach presented by Wanakule, et al. (1986) is closely related to the work by Gorelick, et al. (1984) in combining simulation and optimization. It is an improved version of the embedding approach. Wanakule, et al. (1986) viewed the overall problem as one of discrete time optimal control, where variables describing the aquifer system were divided into state and control variables. Through the simulation model, it was possible to express the state variable (head) as an implicit function of the control variable (pumping). Thus, the model constraints could be conceptually eliminated, yielding a smaller reduced problem involving only the control variable. Methods using the NLP algorithm, such as the generalized reduced-gradient method, necessitates a large number of functional evaluations using the simulation model. Hence, large computation times are still required in the solution process.

Recent papers on the use of optimization in groundwater hydraulic management include Danskin, et al. (1985); Noel and Howitt (1982); Casola, et al. (1986); Makinde-Obusola and Marino (1989); Dougherty and Marryott (1991); Karatzas and Pinder (1993); and Ahlfeld and Heidari (1994). Systems analysis techniques have also been used for the management of groundwater quality and aquifer remediation studies. These include Willis (1976; 1979); Gorelick, et al. (1979; 1982); Gorelick and Remson (1982); Ahlfeld et al. (1988a; 1988b); Shoemaker, et al. (1989); Andricevic and Kitanidis (1990); and Culver and Shoemaker (1993).

16.3.6 Stochastic Management Models

The management models discussed so far are predominantly deterministic, i.e., model parameters as well as inflow (recharge) and demand are assumed to be deterministic. However, virtually all hydrologic parameters are uncertain. Although sensitivity analysis can be used to assess parameter uncertainty, the procedure does not explicitly account for parameter uncertainty in the management model, and as a consequence, it may not lead to satisfactory results.

Tung (1986) presented a simple multiple-period stochastic groundwater-management model for controlling the drawdowns in a homogenous confined aquifer system. The methodology accounts for the random characteristics of transmissivity and the storage coefficient. The stochastic management model was formulated by transforming model constraints containing random variables (transmissivity and storage coefficient) into deterministic constraints using the chance-constrained method (Charnes, et al., 1958). The hydraulic response matrix method (Maddock, 1972) was used, with the response coefficients generated using the Cooper-Jacob equation for the idealized aquifer system. An iterative LP solution procedure was developed to account for the nonlinear constraints and convergence. Tung (1986) did not consider the spatial variability in model parameters. A simulation-regression-management model that explicitly accounts for parameter uncertainty was presented by Wagner and Gorelick (1987). The approach used is to incorporate the groundwater-simulation model as part of a nonlinear chance-constrained stochastic optimization model. The effects of uncertainty on management decisions due to the unknown spatial variability of hydraulic conductivity were further analyzed by Wagner and Gorelick (1989). They developed a procedure which explicitly incorporates this uncertainty for the optimal design of aquifer remediation strategies.

Andricevic (1990) developed a sequential algorithm for a two-dimensional groundwater hydraulic management problem that explicitly accounts for the uncertainty in the hydraulic conductivity field. The focus of the study is on the groundwater prediction errors arising from the initial imperfect knowledge of the spatially variable hydraulic conductivity field.

16.4 OPTIMAL EXPERIMENTAL DESIGN FOR PARAMETER IDENTIFICATION

The optimization of an experimental design for parameter identification deals with the selection of a set of experiment conditions (decision variables) such that a specified criterion (performance measure) is optimized subject to a given set of constraints. In a groundwater system, the experimental conditions that can be controlled include the number and location of pumping (or recharge) wells, the number and location of observation wells, the pumping and recharge rates, and the sampling frequency. The constraints frequently encountered include the following: cost, reliability of the estimated parameters, allowable drawdowns at selected locations, time duration of the experiment, maximum pumping and recharge rates, and allowable time interval between two consecutive measurements. To apply the optimal design theory in practice requires that: (1) a criterion (performance measure) be established so that different experiments can be compared; and (2) an algorithm be developed for optimizing the established criterion over a set of possible experimental designs. Hsu and Yeh (1989) proposed an iterative solution procedure for the optimal experimental design for parameter identification. The criteria that have been established for optimal experimental design can be divided into the following two general categories.

16.4.1 Classical Criteria Derived from Linear Statistical Models

Steinberg and Hunter (1984) presented an excellent review of the classical criteria used in linear statistical models governed by a linear algebraic equation that relates the dependent variable to several independent variables. Let us assume that the experimental data can be represented by the following linear model

$$\mathbf{y} = \mathbf{X}\boldsymbol{\beta} + \boldsymbol{\epsilon} \tag{16.36}$$

where $\mathbf{y}$ = vector of observations
$\mathbf{X}$ = matrix of independent variables (predictor variables)
$\boldsymbol{\beta}$ = vector of parameters to be estimated
$\boldsymbol{\epsilon}$ = vector of experimental errors

It is also assumed that the errors are uncorrelated and have constant variance σ^2. The least squares estimate for parameter vector $\boldsymbol{\beta}$ is

$$\hat{\boldsymbol{\beta}} = (\mathbf{X}^T\mathbf{X})^{-1}\,\mathbf{X}^T\mathbf{y} \tag{16.37}$$

and the covariance matrix of the estimated parameters is

$$\text{Cov}\,(\hat{\boldsymbol{\beta}}) = \sigma^2\,(\mathbf{X}^T\mathbf{X})^{-1} \tag{16.38}$$

It can be shown that the variance of the estimated response at X is

$$d(\mathbf{X}) = \sigma^2\mathbf{X}(\mathbf{X}^T\mathbf{X})^{-1}\,\mathbf{X}^T \tag{16.39}$$

The most popular optimality criteria are:

1. *D-optimality.* A design is said to be D-optimal if it minimizes the determinant of the covariance matrix of the estimated parameters. Since σ^2 is a constant, it is sufficient to use $(\mathbf{X}^T\mathbf{X})^{-1}$ in the minimization.

2. *A-optimality.* A design is said to be A-optimal if it minimizes the trace of the covariance matrix of the estimated parameters.

3. *E-optimality.* A design is said to be E-optimal if it minimizes the maximal eigenvalue of the covariance matrix of the estimated parameters.

4. *G-optimality.* A design is said to be G-optimal if it minimizes max $d(\mathbf{X})$, where the maximum is taken over all possible vectors $\mathbf{X}$ of predictor variables.

5. *I_λ-optimality.* A design is said to be I_λ-optimal if it minimizes $\int d(\mathbf{X})\lambda\,(d\mathbf{X})$ where λ is a probability measure on the space of predictor variables. This criterion, which is sometimes called average integrated variance, also belongs to a more general class of L-optimality criteria discussed by Fedorov (1972).

St. John and Draper (1975) reviewed the D-optimal design criterion and presented algorithms for constructing such designs. Hsu and Yeh (1989) formulated the optimal experimental design problem of a groundwater-flow system. The objective was to minimize the experimental cost subject to a set of constraints, including parameter reliability. They used the A-optimal reliability criterion to solve several parameter identification problems where the unknown parameter vector was transmissivity. Nishikawa and Yeh (1989) used D-optimality as a reliability criterion to generate optimal pumping test designs for the purpose of identifying the unknown transmissivity in a hypothetical aquifer system.

The inherent difficulty in using one of the classical criteria for experimental design is that designs are predicated on unknown parameters that their values are to be estimated. In practice, initial estimates of parameters are determined from prior information and, based upon the initial estimates, an optimal design is carried out and data collected. The inverse problem of parameter identification is solved to update the parameter estimates, and the design of the experiment is repeated. This procedure is called sequential design, and its convergence property has been investigated by Nishikawa and Yeh (1989) and Cleveland and Yeh (1990).

In mass transport, sampling strategies for parameter identification have been studied by Wagner and Gorelick (1987) and Knopman and Voss (1987). Cleveland and Yeh (1991) developed a forward DP algorithm for the optimal configuration and scheduling of an aquifer tracer test whose data were to be used to estimate the transport parameters by maximizing a measure of the information matrix (inverse of the covariance matrix) without exceeding a budget.

The importance of experimental design in connection with the planning of monitoring networks has been recognized for some time. Recent papers include Carrera, et al. (1984); Chu, et al. (1987); Meyer and Brill (1988); Knopman and Voss (1988); and Loaiciga (1989).

16.4.2   Criteria Based on the Concept of Extended Identifiability

The δ-*identifiability criterion* presented by Yeh and Sun (1984) represents the first attempt in the development of an extended identifiability criterion for experimental design for parameter estimation. The criterion emphasizes the reliability of model prediction rather than seeking the uniqueness of the identified parameters. A δ-identifiable pumping test is an experiment that produces sufficient data to guarantee that the parameter estimates of the groundwater model yield predictions that are sufficiently accurate for the overall management objective.

In mathematical terms, the criterion can be stated as follows:

$$\min_{\mathbf{T},\mathbf{T}^0} \left\{ \frac{1}{M} \sum_{i=1}^{M} [h_i(\mathbf{T}) - h_i(\mathbf{T}^0)]^2 \right\}^{1/2} \geq \delta \tag{16.40}$$

subject to

$$\| \mathbf{P}(\mathbf{T}) - \mathbf{P}(\mathbf{T}^0) \| \geq \epsilon \tag{16.41}$$

$$(\mathbf{T},\mathbf{T}^0) \in \mathbf{R} \tag{16.42}$$

where h_i = calculated head value from the groundwater simulation model corresponding to the ith observation
   $\mathbf{T}^0$ = true value of parameter vector $\mathbf{T}$
   M = total number of observations
   δ = small positive number representing the criterion of identifiability
   $\mathbf{P}$ = vector representing the predicted objectives
   ϵ = acceptable error for the overall management objective
   R = admissible set of parameters
   $\| \ \|$ = a prescribed norm

McCarthy and Yeh (1990) applied this concept to obtain a minimum-cost pumping test for a hypothetical aquifer where the uncertain parameter vector was transmissivity. They showed that the δ-identifiability could be achieved by limited quantity and quality of data.

16.4.3   Systematic Approach for Experimental Design, Parameter Identification, and Management Decision

A recent paper by Sun and Yeh (1990) presented a methodology that linked experimental design, parameter identification, and management decision into a systematic procedure that encompassed all three of these essential stages in groundwater modeling and management. They generalized the extended identifiability concept into the following three criteria:

1. *Interval Identifiability (INI).*   The uniqueness of the inverse problem is relaxed if the identified parameters are not "too far" from the true parameters. The INI is defined as follows:

For a parameter $\mathbf{p}_0 \in Ad$ and given weighting matrices $\mathbf{C}_p$ and $\mathbf{C}_D$, if there is a design D and a number $\delta > 0$, such that

$$\| \mathbf{u}_D(\mathbf{p}) - \mathbf{u}_D(\mathbf{p}_0) \|_{\mathbf{C}_p} < \delta \qquad \text{implies}$$

$$\| \mathbf{p} - \mathbf{p}_0 \|_{\mathbf{C}_p} < 1 \qquad \text{for any} \qquad \mathbf{p} \in Ad \tag{16.43}$$

then $\mathbf{p}_0$ is said to be δ-interval identifiable with respect to design D and weighting matrices $\mathbf{C}_p$ and $\mathbf{C}_D$. The notation $\|\bullet\|$ represents the generalized least squares norm, p is the parameter vector, and $\mathbf{u}_D$ is the model output.

2. *Prediction Equivalence Identifiability (PEI).* The PEI is a generalization of the extended identifiability of Yeh and Sun (1984). For PEI it is assumed that there is a prediction project E with prediction vector $\mathbf{g}_E$ and an accuracy requirement described by weighting matrix $\mathbf{C}_E$. $\mathbf{C}_D$ is a given weighting matrix for the observation space. For a parameter $p_0 \in Ad$, if there exists a design D and a number $\delta > 0$, such that

$$\| \mathbf{u}_D(\mathbf{p}) - \mathbf{u}_D(\mathbf{p}_0)\|_{C_D} < \delta \qquad \text{implies}$$

$$\| \mathbf{g}_E(\mathbf{p}) - \mathbf{g}_E(\mathbf{p}_0)\|_{C_E} < 1 \qquad \text{for any} \qquad \mathbf{p} \in Ad \qquad (16.44)$$

then the parameter $\mathbf{p}_0$ is said to be δ-prediction equivalence identifiable with respect to design D and weighting matrices $\mathbf{C}_E$ and $\mathbf{C}_D$.

3. *Management Equivalence Identifiability (MEI).* Analogous to the PEI, the MEI is defined as follows: Assume that there is a management objective J with the optimal decision vector $\mathbf{q}_J$ and an accuracy requirement described by weighting matrix $\mathbf{C}_J$. $\mathbf{C}_D$ is a given weighting matrix for the observation space. For a parameter $\mathbf{p}_0 \in Ad$, if there exists a design D and a number $\delta > 0$, such that

$$\| \mathbf{u}_D(\mathbf{p}) - \mathbf{u}_D(\mathbf{p}_0)\|_{C_D} < \delta \qquad \text{implies}$$

$$\| \mathbf{q}_J(\mathbf{p}) - \mathbf{q}_J(\mathbf{p}_0)\|_{C_J} < 1 \qquad \text{for any} \qquad \mathbf{p} \in Ad \qquad (16.45)$$

then parameter $\mathbf{p}_0$ is said to be δ-management equivalence identifiable with respect to design D and weighting matrices $\mathbf{C}_E$ and $\mathbf{C}_D$. The sufficiency of a design D for MEI can be confirmed by solving the following optimization problem:

$$\min F(\mathbf{p}, \mathbf{p}_0) = \| \mathbf{u}_D(\mathbf{p}) - \mathbf{u}_D(\mathbf{p}_0)\|_{\hat{C}_D}^2 \qquad (16.46)$$

subject to

$$\| \mathbf{q}_J(\mathbf{p}) - \mathbf{q}_J(\mathbf{p}_0)\|_{\hat{C}_J}^2 \geq 1, \qquad \mathbf{p} \in \mathbf{P}_{ad}, \mathbf{p}_0 \in \mathbf{P}_{ad}$$

If the purpose of building a model is to find an optimal decision vector for a given management objective rather than using the model for prediction, then it is only necessary to consider the error of the optimal decision variables caused by the uncertainty of the identified parameters. The design criterion based on the MEI criterion produces reliable designs for model applications, while the experimental cost is minimized. The best compromise is sought to balance the management risk and the cost of data collection.

Additionally, Sun and Yeh (1990) presented a systematic procedure for finding a reliable design using the aforementioned extended identifiability criteria.

16.5 CONJUNCTIVE USE OF SURFACE WATER AND GROUNDWATER

Sound management of water resources depends on an integrated plan that considers simultaneously all planning aspects: allocating surface water and groundwater, artificially recharging groundwater, and preventing undesirable effects from conjunctive use of surface water and groundwater. With proper management, conjunctive use of the total water resources of an area can increase the efficiency, reliability, and cost-effectiveness of water use, particularly in regions with spatial or temporal imbalances of water demands and natural supplies (Louie, et al., 1984; Willis and Yeh, 1987). *Conjunctive use,* as applied to Southern California water use, means integrating imported surface-water supplies with existing groundwater reserves through sound management (Kendall and Sienkiewich, 1988). For example, supplemental water available during the winter months from the State Water Project or the Colorado River—water that would otherwise be lost to the ocean—would be spread for recharge in local groundwater basins. Such surplus flow would be combined with local supplies for use during sustained dry periods. The importance of artificial groundwater recharge in Southern California was also pointed out by Madrid (1988). Bouwer (1988) emphasized that design and management of recharge systems are site-specific and should be adopted to local conditions of water quality, climate, soil, hydrogeology and environmental constraints. Accurate simulation models are essential for planning and operation of a water-resources system (Willis and Yeh, 1987). Additional literature dealing with conjunctive use of surface and groundwater includes Burt (1964); Aron and Scott (1971); Maddock (1974); Noel and Howitt (1982); and Wildermuth, et al. (1988).

16.5.1 Multiobjective Analysis

Essential to a sound basin-wide water resources management practice is a unified plan that simultaneously considers all planning aspects. Such basin-wide planning has long been recognized as a management problem with multiple objectives (Louie, et al., 1984; Yeh, et al., 1992).

Over the years, various solution techniques for solving the multiple-objective problems have been developed. Cohon and Marks (1975) conducted a comprehensive review of these techniques and have categorized them into two groups: *generating techniques,* which completely identify the set of noninferior solutions, and *preference-oriented techniques,* which are based on the interaction with the expressed preference of the decision-maker prior to or during the progress of the analysis.

The representative works under the generating techniques category are given by Gass and Saaty (1955); Maass, et al. (1962); Marglin (1967); Major (1969); Miller and Byers (1973); Cohon and Marks (1973); and by Cohon, et al. (1978). Techniques which incorporate decision maker's preferences are typified by Charnes and Cooper (1961); Geoffrion (1967); and Haimes and Hall (1974).

In a basin-wide planning problem involving three objectives, Louie, et al. (1984) demonstrated that the use of the constraint method for multiobjective optimization and the application of the influence coefficient method (Becker and Yeh, 1972) for linking the simulation model (groundwater flow and mass transport) with the optimization model can be successful for finding good solutions for planning purposes. The constraint method of solution chooses one of the objectives as a scalar objective to be optimized and adds the other objectives to the constraint set. Parametric variation of the objectives in the constraint set from their lower bounds to upper bounds traces out an approximation of the noninferior set (*Pareto optimum*). Pareto optimality is defined as follows: A Pareto optimal alternative is one for which increasing the optimum level of satisfaction of one of the objectives can be done only at the expense of decreasing the optimum level of another objective. Therefore, Pareto optimality also quantifies the tradeoff among the objectives. In theory, if each objective is parameterized K times, the number of required scalar optimal solutions may be as high as K^{P-1}, where P is the total number of objectives. In practice, a satisfactory approximation of the noninferior set can often be traced out with considerably fewer solutions.

The three objectives considered by Louie, et al. (1984) are: (1) to meet increasing and fluctuating future water demands of the basin, (2) to optimize the conjunctive use of surface water and groundwater, and (3) to minimize the undesirable effects of groundwater overdraft. To quantify the objectives, mathematical representations of each objective must be formulated.

16.5.1.1 Water-Supply Management Objective (Objective 1).
One of the most important objectives in basin-wide planning is to meet the water demands in the area at lowest overall cost. Although in actuality the cost structure may be more complicated, a linear cost structure was used by Louie, et al. (1984) to simplify the cost objective function

$$Z_{TC}(\mathbf{q}) = \sum_{i=1}^{S}\sum_{j=1}^{U} C_{i,j}^{s}q_{i,j}^{s} + \sum_{j=1}^{U}\sum_{k=1}^{T} C_{j,k}^{t}q_{j,k}^{t} + \sum_{k=1}^{T}\sum_{l=1}^{D} C_{k,l}^{d}q_{k,l}^{d} \tag{16.47}$$

where Z_{TC} = the total cost incurred in supplying a quantity of $q_{i,j}^{s}$ from source i to user j with a unit cost $C_{i,j}^{s}$, in transmitting a quantity $q_{j,k}^{t}$ of wastewater from user j to treatment plant k at a unit cost of $C_{j,k}^{t}$, and in transporting the treated water of quantity $q_{k,l}^{d}$ from treatment plant k to disposal site l at a unit cost of $C_{k,l}^{d}$
   S = number of supply sources (surface water and groundwater)
   U = number of water use groups
   T = number of treatment plants
   D = number of disposal sites

All of the unit costs pertain only to the energy and operation and maintenance costs; no capital investment is considered because all facilities are assumed to be in existence.

16.5.1.2 Water-Quality Objective (Objective 2).
The water-quality objective is expressed in terms of the levels of water-quality constituents. Clearly, the objective of water-quality control is to minimize the deviation between the simulated concentration levels and the limiting levels for the operational areas. The deviation used by Louie, et al. (1984) is a one-sided measure, i.e., the deviation would be a positive value if the computed level is greater than the limiting level; otherwise, it would be zero. For this objective, several functions can be formulated, e.g., the sum or weighted sum of the deviations from all the subareas; another is the maximum deviation selected from among the subareas. For the latter case, the objective can be expressed as

$$Z_{WQ} = \max_{a}(D_1, D_2, \cdots, D_a, \cdots, D_N) \tag{16.48}$$

in which

$$D_a = C_a - CL_a \quad \text{if} \quad C_a > CL_a; \quad 0 \quad \text{if} \quad C_a \le CL_a \tag{16.49}$$

where C_a = computed concentration level for operational area a
   CL_a = limiting (allowable) level of concentration for area a
   N = total number of operational areas

Therefore, Z_{WQ} is defined as the maximum deviation from allowable levels, which is to be minimized.

16.5.1.3 *Prevention of Undesirable Groundwater Overdraft (Objective 3).*
Another objective is the prevention of undesirable overdraft of groundwater basins. This objective can be achieved by maintaining water levels in the basin by controlling the amount of extractions and recharges. Therefore, the objective function can be expressed as the next extraction from (or net recharge into) the basin as

$$Z_0 = \sum_{i \in E} q_i - \sum_{j \in R} q_j \tag{16.50}$$

where q_j = the quantity recharged from the jth source (e.g., from treatment plant)
q_i = the quantity extracted from the ith groundwater source
R = the set of recharge sources
E = the set of groundwater sources

In a conjunctive use study for the Santa Ana basin in Southern California, the objective function used by Yeh, et al. (1992) consists of the following three objectives: (1) the minimization of the total costs, expressed as a summation of capital costs, operational costs, and imported water costs on an annualized basis as linear functions of the decision variables, (2) the minimization of the total dissolved solids (TDS) in the basin, and (3) the minimization of the total inorganic nitrogen. They also used the constraint method to determine the tradeoffs among the three objectives.

16.5.2 Multiobjective Optimization

Clearly it is desirable to minimize all of the objectives concurrently; however, these three objectives are conflicting in the sense that the optimal plans for the individual objectives are different. Thus, the cost will invariably go up if better water quality is to be maintained and higher groundwater tables are to be kept; furthermore, poorer water quality will be observed if greater amounts of reclaimed water are recharged into the groundwater basin.

The familiar formulation for a multiple objective optimization problem involving the three aforementioned objectives can be stated as

$$\min_{q \in F} Z = (Z_1, Z_2, Z_3) \tag{16.51}$$

subject to

$$\mathbf{q} \in F: \mathbf{g}^1(\mathbf{q}) \geq 0 \tag{16.52}$$

where $Z_1(\mathbf{q})$ = Z_{TC} (objective function expressing the total operational cost incurred)
$Z_2(\mathbf{q})$ = Z_{WQ} (objective function expressing the maximum deviation between the resulting and allowable water quality levels)
$Z_3(\mathbf{q})$ = Z_0 (objective function indicating overdraft conditions)
$\mathbf{q}$ = $(q_1, q_2, \cdots, q_i, \cdots, q_K)$ a vector of decision (planning) variables expressing the quantities of water transferred between the sources, users, treatment plants, and disposal sites, in which K represents the total number of connections
F = $\{\mathbf{q} | \mathbf{q}_u \leq \mathbf{q} \geq \mathbf{q}_l; \mathbf{q} \geq 0\}$; $\mathbf{q}_u$ = upper bound on $\mathbf{q}$; $\mathbf{q}_l$ = lower bound on $\mathbf{q}$; and
$\mathbf{g}^1(\mathbf{q})$ = a set of constant functions

The solution procedure is described by the flow diagram (Fig. 16.1) and followed by the detailed descriptions given here:

Step 1. All the necessary input data and elements associated with the groundwater model, the model representing the network of water supply, wastewater treatment and disposal facilities, and the constraints and cost coefficients for the optimization procedures are processed at this step.

Step 2. A linear program is solved to obtain the optimal solution to the first objective problem—minimizing the total variable cost for water supply, water transport, treatment, and disposal. This is done with no regard to the second and third objectives of the problem. The optimization problem for the first objective can be written as

$$\min_{q \in F} Z_1(\mathbf{q}) \tag{16.53}$$

subject to

$$
\left.
\begin{array}{ll}
\displaystyle\sum_{i=1}^{S} q_{ij}^s \geq \text{demand } j & \forall j = 1,2,3,\cdots,U \\[2ex]
\displaystyle\sum_{j=1}^{U} q_{ij}^s \leq \text{supply } i & \forall i = 1,2,3,\cdots,S \\[2ex]
\displaystyle\sum_{i=1}^{S} q_{ij}^s = \sum_{k=1}^{T} q_{jk}^t & \forall j = 1,2,3,\cdots,U \\[2ex]
\displaystyle\sum_{j=1}^{U} q_{jk}^t (1 - \text{Tlut}_{jk}) \leq \text{TPC}_k & \forall k = 1,2,3,\cdots,T \\[2ex]
\displaystyle\sum_{j=1}^{U} q_{jk}^t (1 - \text{Tlut}_{jk}) = \sum_{l=1}^{D} q_{kl}^d & \forall k = 1,2,3,\cdots,T \\[2ex]
\displaystyle\sum_{k=1}^{T} q_{kl}^d (1 - \text{Tltd}_{kl}) \leq \text{DC}_l & \forall l = 1,2,3,\cdots,D
\end{array}
\right\} \; g^1(q) \tag{16.54}
$$

where Tlut_{jk} = transmission loss between user j and treatment plant k
Tltd_{kl} = loss between treatment plant k and disposal site l
TPC_k = capacity of the kth treatment plant
DC_l = capacity of the lth disposal site

Step 3. Prior to obtaining the optimal solution for the second objective problem (water quality control), linkage between the optimization procedure and the simulation model has to be established, since water quality can only be evaluated through the simulation model. The influence coefficient method (Becker and Yeh, 1972) is used to serve that purpose. The conceptual idea of the method is based on the Taylor series expansion of the second objective function; its first order approximation can be written as

$$D_a^1 = D_a^0 + \sum_{k=1}^{P} \frac{\partial D_a}{\partial q_k}\bigg|_{q^0} \Delta q_k, \qquad \forall a = 1,2,\cdots,N \tag{16.55}$$

where P = number of variables
q_k = the kth decision variable; $q_k \in q = (q^s, q^t, q^d)$

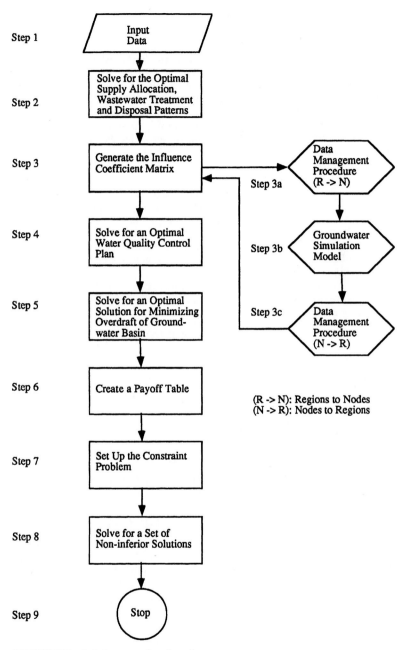

FIGURE 16.1 Solution procedure flow diagram.

$$\Delta q_k = q_k^{\,1} - q_k^{\,0}$$

$\partial D_a / \partial q_k \big|_{q^0}$ = the influence coefficient defined as the change in the concentration deviation in area a due to a unit change in the kth variable evaluated around the initial value q^0

All of these computations for the influence coefficients are carried out by executing Steps 3a through 3c as shown in the flow diagram.

Step 3a.   Taking into account the water demand requirements, supply distribution systems, treatment plants, and disposal sites, the water is routed through the system and then distributed into the nodal inputs for the simulation model. This process is handled by the data-management procedure, which processes the data from the regional level to the nodal level.

Step 3b.   Solve the groundwater-simulation model to obtain solutions of flow (head and velocity) and mass transport (concentration).

Step 3c.   Since the management options are exercised on the regional level (operational areas), the nodal values (water quality) must be converted back to the regional level. This reverse process can be a simple averaging, weighted averaging, or taking the maximum value of the nodal values to represent the regional value. The maximum nodal value is used to reflect a more stringent water-quality-control policy for each subarea in this study.

Step 4.   Obviously, the objective in water-quality control is to minimize the deviation between the simulated concentrations and the allowable limits in the regions. Thus, the second objective problem can be expressed as

$$\min_{q \in F} Z_2 \left[\text{or} \min_{q \in F} (\max_a D_a) \right] \tag{16.56}$$

subject to

$$g^1(q) \geq 0 \tag{16.57}$$

Equivalently, the minimax LP formulation can be converted to the following LP problem:

$$\min_{q \in F} D \tag{16.58}$$

subject to

$$D_a^0 + \sum_{k=1}^{P} \frac{\partial D_a}{\partial q_k} \bigg|_{q^0} \Delta q_k \leq D_a \forall a; \qquad D_a \leq D \forall a;$$

$$g^1(q) \geq 0; \qquad D_a, D \geq 0 \tag{16.59}$$

Note that this problem is solved without consideration of the first and third objectives.

Step 5.   The optimization problem involving the third objective is solved at this step. The problem can simply be written as

$$\min_{q \in F} \left(\sum_{i \in E} q_i - \sum_{j \in R} q_j \right) \tag{16.60}$$

subject to

$$g^1(q) \geq 0$$

Step 6.   After the three optimization problems have been solved individually, a payoff table can be constructed as shown in Table 16.1.

TABLE 16.1 Payoff Table: Three Optimization Problems

	Z_1	Z_2	Z_3
$Z(q^1)$	$Z_1(q^1)$	$Z_2(q^1)$	$Z_3(q^1)$
$Z(q^2)$	$Z_1(q^2)$	$Z_2(q^2)$	$Z_3(q^2)$
$Z(q^3)$	$Z_1(q^3)$	$Z_2(q^3)$	$Z_3(q^3)$

Step 7. With the payoff table constructed (Table 16.1), the original multiple objective problem can be converted to a constraint problem (Cohon and Marks, 1973). For example, selecting Z_2 as the objective function, the constraint problem can be expressed as

$$\min_{q \in F} Z_2 \tag{16.61}$$

subject to

$$\mathbf{g}^1(\mathbf{q}) \geq 0 \tag{16.62}$$

$$Z_i \leq L_i + \frac{\gamma}{\tau - 1}(H_i - L_i); \qquad i = 1 \text{ and } 3 \tag{16.63}$$

where L_i = minimum value of the ith column of the payoff table
   H_i = maximum value of the ith column of the payoff table
   γ = incremental index from 0 to $\tau - 1$
   τ = total number of increments from L_i to H_i

Step 8. The constraint problem is solved by a parametric linear programming procedure.
Step 9. The original problem is completed at this step, resulting in a noninferior set of solutions.

16.6 CONCLUSIONS

The development of groundwater-simulation models in the early 1970s provided groundwater planners with quantitative techniques for analyzing alternative groundwater pumping or recharge strategies. Costs and benefits are developed for each planning or management alternative so that optimization techniques can be applied to the groundwater systems. Groundwater-optimization models provide optimal groundwater planning or design alternatives in the context of each system's objective and constraints.

The accuracy of a simulation model is dependent, to a certain extent, on the accuracy of the inverse solution, which in turn is determined by the quantity and quality of data used for solving the inverse problem of parameter identification. Optimization of experimental design for parameter identification is an important part of groundwater planning and management. The classical criteria derived from linear statistical theory suffer from the fact that designs are predicated on parameters that themselves are to be estimated. Experimental design based on the concept of extended identifiability relaxes the uniqueness requirement of the inverse solution, yet still produces reliable designs that satisfy the management objective. A key advancement in groundwater planning and management has been the unification of experimental design, parameter identification, and management decision into a single systematic procedure.

REFERENCES

Aboufirassi, M., and M. A. Marino, "Kriging of Water Levels in the Souss Aquifer" (Morocco), *Math. Geol.*, 15(4):537–551, 1983.

Aboufirassi, M., and M. A. Marino, "Cokriging of Aquifer Transmissivities from Field Measurements of Transmissivity and Specific Capacity," *Math. Geol.*, 16(1):19–35, 1984a.

Aboufirassi, M., and M. A. Marino, "A Geostatistically-Based Approach to Identification of Aquifer Transmissivity in Yolo Basin" (California), *Math. Geol.*, 16,(2):125–137, 1984b.

Ahlfeld, D. P., and M. Heidari, "Applications of Optimal Hydraulic Control to Groundwater Systems," *J. Water Resour. Plng. and Mgmt.*, ASCE, 120(3):350–365, 1994.

Ahlfeld, D. P., J. M. Mulvey, G. F. Pinder, and E. F. Wood, "Contaminated Groundwater Remediation Design Using Simulation, Optimization, and Sensitivity Theory, 1, Model Development," *Water Resour. Res.*, 24(3):431–441, 1988a.

Ahlfeld, D. P., J. M. Mulvey, G. F. Pinder, and E. F. Wood, "Contaminated Groundwater Remediation Design Using Simulation, Optimization, and Sensitivity Theory, 2, Analysis of a Field Site," *Water Resour. Res.*, 24(3):443–452, 1988b.

Alley, W. M., E. Aquado, and I. Remson, "Aquifer Management under Transient and Steady-State Conditions," *Water Resour. Bull.*, 12(3):963–972, 1976.

Andricevic, R., "A Real-Time Approach to Management and Monitoring of Groundwater Hydraulics," *Water Resour. Res.*, 26(11):2747–2755, 1990.

Andricevic, R., and P. K. Kitanidis, "Optimization of Pumping Schedule in Aquifer Remediation under Uncertainty," *Water Resour. Res.*, 26(5):875–885, 1990.

Anger, G., ed., *Inverse and Improperly Posed Problems in Differential Equations*, Akademie-Verlag, Berlin, 1979.

Aquado, E., and I. Remson, "Groundwater Hydraulics in Aquifer Management," *J. Hydr. Div.*, ASCE, 100(1):103–118, 1974.

Aquado, E., and I. Remson, "Groundwater Management with Fixed Charges," *J. Water Res. Plng. and Mgmt. Div.*, ASCE, 106(WR2):375–382, 1980.

Aron, G., and V. H. Scott, "Dynamic Programming for Conjunctive Water Use," *J. Hydr. Div.*, ASCE, 97(HY5):705–720, 1971.

Bard, Y., *Nonlinear Parameter Estimation*, John Wiley, New York, 1974.

Bastin, G., and M. Gevers, "Identification and Optimal Estimation of Random Fields from Scattered Point-Wise Data," *Automatica*, 21(2):139–155, 1985.

Beale, E. M. L., "Confidence Regions in Non-linear Estimation," *J. R. Stat. Soc.*, Ser. B, 22:41–76, 1960.

Bear, J., *Dynamics of Fluids in Porous Media*, Elsevier, New York, 1972.

Beck, A., *Parameter Estimation in Engineering Science*, John Wiley, New York, 1977.

Becker, L., and W. W-G. Yeh, "Identification of Parameters in Unsteady Open-Channel Flows," *Water Resour. Res.*, 8(4):956–965, 1972.

Bellman, R., and R. Kabala, *Quasilinearization and Nonlinear Boundary-Value Problems*, Elsevier, New York, 1965.

Birtles, A. B., and E. H. Morel, "Calculation of Aquifer Parameters from Sparse Data," *Water Resour. Res.*, 15(4):832–844, 1979.

Bouwer, H., "Systems for Artificial Recharge of Groundwater," in A. I. Hohnson and J. Finlayson, eds., *Proceedings of the International Symposium on Artificial Recharge of Groundwater*, August 23–27, ASCE, Anaheim, Calif., 1988, pp. 2–12.

Bredehoef, J. D., and R. A. Young, "The Temporal Allocation of Ground-Water: A Simulation Approach," *Water Resour. Res.,* 6(1):3–21, 1970.

Bruch, J. C., Jr., C. M. Lam, and T. M. Simundich, "Parameter Identification in Field Programs," *Water Resour. Res.,* 10(1):73–79, 1974.

Burt, O. R., "The Economics of Conjunctive Use of Ground and Surface Water," *Hilgardia,* University of California, Davis, 1964.

Burt, O. R., "Temporal Allocation of Groundwater," *Water Resour. Res.,* 3(1):45–56, 1967.

California Department of Water Resources, "Planned Utilization of Groundwater Basins, San Gabriel Valley, Appendix C: Economics," *Bulletin,* 104-2, 1966.

California State Water Resources Control Board, Santa Ana Region 8, "Water Quality Control Plan Report," parts I and II, vol. 1, Calif., 1975.

Carrera, J., "State of the Art of the Inverse Problem Applied to Flow and Solute Transport Equations," in E. Custodio, A. Gurgui, and J. P. Lobo Ferreira, eds., *Groundwater Flow and Quality Modeling,* D. Reidel, Hingham, Mass., 1988, pp. 549–583.

Carrera, J., and S. P. Neuman, "Estimation of Aquifer Parameters under Transient and Steady-State Conditions, 1, Maximum Likelihood Method Incorporating Prior Information," *Water Resour. Res.,* 22(2):199–210, 1986a.

Carrera, J., and S. P. Neuman, "Estimation of Aquifer Parameters under Transient and Steady-State Conditions, 2, Uniqueness, Stability, and Solution Algorithms," *Water Resour. Res.,* 22(2):211–227, 1986b.

Carrera, J., and S. P. Neuman, "Estimation of Aquifer Parameters under Transient and Steady-State Conditions, 3, Application to Synthetic and Field Data," *Water Resour. Res.,* 22(2):228–242, 1986c.

Carrera, J., E. Usnoff, and F. Szidarovsky, "A Method for Optimal Observation Network Design for Groundwater Management," *J. Hydrol.,* 73($\frac{1}{2}$):147–163, 1984.

Carter, R. D., L. F. Kemp, Jr., and A. C. Pierce, "Discussion of Comparison of Sensitivity Coefficient Calculation Methods in Automatic History Matching," *Soc. Pet. Eng. J.,* 22(2):205–208, 1982.

Carter, R. D., L. F. Kemp, Jr., A. C. Pierce, and D. L. Williams, "Performance Matching with Constraints," *Soc. Pet. Eng. J.,* 14(2):187–196, 1974.

Casola, W. H., R. Narayanan, C. Duffy, and A. B. Bishop, "Optimal Control Model for Groundwater Management," *J. Water Res. Plng. and Mgmt.,* ASCE, 112(2):183–197, 1986.

Chang, S., and W. W-G. Yeh, "A Proposed Algorithm for the Solution of the Large-Scale Inverse Problem in Groundwater," *Water Resour. Res.,* 12,(3):365–374, 1976.

Charnes, A., and W. W. Cooper, *Management Models and Industrial Applications of Linear Programming,* vol. 1, John Wiley, New York, 1961.

Charnes, A., W. W. Cooper, and G. H. Symonds, "Cost Horizons and Certainty Equivalents: An Approach to Stochastic Programming of Heating Oil," *Management Science,* 4(3):225–263, 1958.

Chavent, G., "Identification of Functional Parameters in Partial Differential Equations," in R. E. Goodson and M. Polis, eds., *Identification of Parameters in Distributed Systems,* American Society of Mechanical Engineers, New York, 1974, pp. 31–48.

Chavent, G., "About the Stability of the Optimal Control Solution of Inverse Problems," in G. Anger, ed., *Inverse and Improperly Posed Problems in Differential Equations,* Akademie-Verlag, Berlin, 1979a, pp. 45–58.

Chavent, G., "Identification of Distributed Parameter System: About the Output Least Square Method, Its Implementation, and Identifiability," in R. Isermann, ed., *Identification and System Parameter Estimation,* vol. 1, Pergamon, New York, 1979b, pp. 85–97.

Chavent, G., "Local Stability of the Output Least Square Parameter Estimation Technique," *Math. Appl. Comp.,* 2(1):3–22, 1983.

Chavent, G., M. Dupuy, and P. Lemonnier, "History Matching by Use of Optimal Theory," *Soc. Pet. Eng. J.,* 15(1):74–86, 1975.

Chen, W. H., G. R. Gavalas, J. H. Seinfeld, and M. L. Wasserman, "A New Algorithm for Automatic Historic Matching," *Soc. Pet. Eng. J.,* 14(6):593–608, 1974.

Chu, W. S., E. W. Strecker, and D. P. Lettenmaier, "An Evaluation of Data Requirements for Groundwater Contaminant Transport Modeling," *Water Resour. Res.,* 23(3):408–424, 1987.

Cleveland, T. G., and W. W-G. Yeh, "Sampling Network Design for Transport Parameter Identification," *J. Water Resour. Plng. and Mgmt.,* ASCE, 116(6):764–783, 1990.

Cleveland, T. G., and W. W-G. Yeh, "Optimal Configuration and Scheduling of Groundwater Tracer Tests," *J. Water Resour. Plng. and Mgmt.,* ASCE, 117(1):37–51, 1991.

Clifton, P. M., and S. P. Neuman, "Effects of Kriging and Inverse Modeling on Conditional Simulation of the Avra Valley Aquifer in Southern Arizona," *Water Resour. Res.,* 18(4):1215–1234, 1982.

Coats, K. H., J. R. Dempsey, and J. H. Henderson, "A New Technique for Determining Reservoir Description from Field Performance Data," *Soc. Pet. Eng. J.,* 10(1):66–74, 1970.

Cohon, J. L., *Multipleobjective Programming and Planning,* Academic Press, New York, 1978.

Cohon, J. L., R. L. Church, and D. P. Sheer, "Generating Multiobjective Trade-Offs: An Algorithm for Bicriterion Problems," *Water Resour. Res.,* 15(5):1001–1010, 1979.

Cohon, J. L., and D. M. Marks, "Multiobjective Screening Models and Water Resource Investment," *Water Resour. Res.,* 9(4):826–836, 1973.

Cohon, J. L., and D. M. Marks, "A Review and Evaluation of Multiobjective Programming Techniques," *Water Resour. Res.,* 11(2):208–220, 1975.

Cooley, R. L., "A Method of Estimating Parameters and Assessing Reliability for Models of Steady State Ground Flow, 1, Theory and Numerical Properties," *Water Resour. Res.,* 13(2):318–324, 1977.

Cooley, R. L., "A Method for Estimating Parameters and Assessing Reliability for Models of Steady State Groundwater Flow, 2, Application of Statistical Analysis," *Water Resour. Res.,* 15(3):603–617, 1979.

Cooley, R. L., "Incorporation of Prior Information on Parameters into Nonlinear Regression Groundwater Flow Models, 1, Theory," *Water Resour. Res.,* 18(4):965–976, 1982.

Cooley, R. L., "Incorporation of Prior Information on Parameters into Nonlinear Regression Groundwater Flow Models, 2, Applications," *Water Resour. Res.,* 19(3):662–676, 1983.

Cooley, R. L., and P. J. Sinclair, "Uniqueness of a Model of Steady-State Groundwater Flow," *J. Hydrol.,* 31:245–269, 1976.

Cressie, N., "Fitting Variogram Models Using Weighted Least Squares," *J. Math. Geol.,* 17:693–702, 1985.

Culver, T. B., and C. A. Shoemaker, "Dynamic Optimal Control for Groundwater Remediation with Flexible Management Periods," *Water Resour. Res.,* 28(3):629–641, 1992.

Culver, T. B., and C. A. Shoemaker, "Optimal Control for Groundwater Remediation by Differential Dynamic Programming with Quasi-Newton Approximations," *Water Resour. Res.,* 29(4):823–831, 1993.

Dagan, G., "Stochastic Modeling of Groundwater Flow by Unconditional and Conditional Probabilities, 1, Conditional Simulation and the Direct Problem," *Water Resour. Res.,* 18(4):813–833, 1982.

Dagan, G., "Stochastic Modeling of Groundwater Flow by Unconditional and Conditional Probabilities: The Inverse Problem," *Water Resour. Res.,* 21(1):65–72, 1985.

Dagan, G., and Y. Rubin, "Stochastic Identification of Recharge, Transmissivity, and Storativity in an Aquifer Transient Flow: A Quasi-Steady Approach," *Water Resour. Res.,* 24(10): 1698–1710, 1988.

Danskin, W. R., and S. M. Gorelick, "A Policy Evaluation Tool: Management of a Multiaquifer System Using Controlled Stream Recharge," *Water Resour. Res.,* 21(11):1731–1747, 1985.

de Coursey, D. G., and W. M. Snyder, "Computer Oriented Method of Optimizing Hydrologic Model Parameters," *J. Hydrol.,* 9:34–53, 1969.

de Marsily, G., *Quantitative Hydrogeology,* Academic, San Diego, Calif., 1986.

Delhomme, J. P., "Spatial Variability and Uncertainty in Groundwater Flow Parameters: A Geostatistical Approach," *Water Resour. Res.,* 15(2):269–280, 1979.

Deninger, R. A., "Systems Analysis of Water Supply Systems," *Water Resour. Bull.,* 6(4):573–579, 1970.

DiStefano, N., and A. Rath, "An Identification Approach to Subsurface Hydrological Systems," *Water Resour. Res.,* 11(6):1005–1012, 1975.

Dogru, A. H., and J. H. Seinfeld, "Comparison of Sensitivity Coefficient Calculation Methods in Automatic History Matching," *Soc. Pet. Eng. J.,* 21(5):551–557, 1981.

Dougherty, D. E., and R. A. Marryott, "Optimal Groundwater Management. 1. Simulated Annealing," *Water Resour. Res.,* 27(10):2493–2509, 1991.

Douglas, J., Jr., "Alternating Direction Methods for Three Space Variables," *Num. Math.,* 4:41–63, 1962.

Emsellem, Y., and G. de Marsily, "An Automatic Solution for the Inverse Problem," *Water Resour. Res.,* 7(5):1264–1283, 1971.

Fedorov, V. V., *Theory of Optimal Experiments,* W. J. Studden and E. M. Klimko, trans. and ed., Academic Press, New York, 1972.

Flores, E. Z., A. L. Gutjahr, and L. W. Gelhar, "A Stochastic Model for the Operation of a Stream-Aquifer System," *Water Resour. Res.,* 14(1):30–38, 1975.

Freeze, R. A., "A Stochastic-Conceptual Analysis of One-Dimensional Groundwater Flow in Nonuniform Homogeneous Media," *Water Resour. Res.,* 11(5):725–741, 1975.

Frind, E. O., and G. F. Pinder, "Galerkin Solution of the Inverse Problem for Aquifer Transmissivity," *Water Resour. Res.,* 9(5):1397–1410, 1973.

Garay, H. L., Y. Y. Haimes, and P. Das, "Distributed Parameter Identification of Groundwater Systems by Nonlinear Estimation," *J. Hydrol.,* 30:47–61, 1976.

Gass, S., and T. Saaty, "The Computation Algorithm for the Parametric Objective Function," *Naval Research Logistics Quarterly,* 2(1&2):39–45, 1955.

Gavalas, G. R., P. C. Shah, and J. H. Seinfeld, "Reservoir History Matching by Bayesian Estimation," *Soc. Pet. Eng. J.,* 16(6):337–350, 1976.

Gelhar, L. W., "Stochastic Subsurface Hydrology from Theory to Applications," *Water Resour. Res.,* 22(9):135S–145S, 1986.

Geoffrion, A. M., "Solving Bicriterion Mathematical Programs," *Operations Research,* 15(1):39–54, 1967.

Gill, P. E., W. Murray, M. A. Sanders, and M. H. Wright, "Users Guide for NPSOL (Version 4.0): A FORTRAN Package for Nonlinear Programming," *Tech. Rep. SOL,* 86(2), Systems Optimization Laboratory, Stanford University, Stanford, Calif., 1986.

Gorelick, S. M., "A Model for Managing Sources of Groundwater Pollution," *Water Resour. Res.,* 18(4):773–781, 1982.

Gorelick, S. M., "A Review of Distributed Parameter Groundwater Management Modeling Methods," *Water Resour. Res.,* 19(2):305–319, 1983.

Gorelick, S. M., B. Evans, and I. Remson, "Identifying Sources of Groundwater Pollution: An Optimization Approach," *Water Resour. Res.,* 19(3):779–790, 1983.

Gorelick, S. M., and I. Remson, "Optimal Dynamic Management of Groundwater Pollutant Resources," *Water Resour. Res.,* 18(1):71–76, 1982.

Gorelick, S. M., I. Remson, and R. W. Cottle, "Management Model of a Groundwater System with a Transient Pollutant Source," *Water Resour. Res.,* 15(5):1243–1249, 1979.

Gorelick, S. M., C. I. Voss, P. E. Gill, M. Murray, M. A. Sanders, and M. M. Wright, "Aquifer Reclamation Design: the Use of Contaminant Transport Simulation Combined with Nonlinear Programming," *Water Resour. Res.,* 20(4):415–427, 1984.

Guvanasen, V., and R. E. Volker, "Identification of Distributed Parameters in Groundwater Basins," *J. Hydrol.,* 36:279–293, 1978.

Haimes, Y. Y., and W. A. Hall, "Multiobjective in Water Resource System Analysis: The Surrogate Worth Tradeoff Method," *Water Resour. Res.,* 10(4):615–624, 1974.

Haimes, Y. Y., R. L. Perrine, and D. A. Wismer, "Identification of Aquifer Parameters by Decomposition and Multilevel Optimization," *Water Resour. Cent. Contr.*, vol. 123, University of California, Los Angeles, 1968.

Hawkins, D. B., and D. B. Stephens, "Groundwater Modeling in a Southwestern Alluvial Basin," *Ground Water*, 21(6):733–739, 1983.

Hefez, E., V. Shamir, and J. Bear, "Identifying the Parameters of an Aquifer Cell Model," *Water Resour. Res.*, 11(6):993–1004, 1975.

Heidari, M., "Application of Linear System Theory and Linear Programming to Groundwater Management in Kansas," *Water Resour. Bull.*, 18(6):1003–1012, 1982.

Helweg, O. J., and J. W. Labadie, "Linked Models for Managing River Basin Salt Balance," *Water Resour. Res.*, 13(2):329–336, 1977.

Hill, M. C., "A Computer Program (MODFLOWP) for Estimating Parameters of a Transient, Three-Dimensional, Ground-Water Flow Model Using Nonlinear Regression," U.S. Geological Survey, Open-File Report 91-484, Denver, Colo., 1992.

Hoeksema, R. J., and P. K. Kitanidis, "An Application of the Geostatistical Approach to the Inverse Problem in Two-Dimensional Groundwater Modeling," *Water Resour. Res.*, 20(7):1003–1020, 1984.

Hoeksema, R. J. and P. K. Kitanidis, "Analysis of the Spatial Structure of Properties of Selected Aquifers," *Water Resour. Res.*, 21(4):563–572, 1985a.

Hoeksema, R. J., and P. K. Kitanidis, "Comparison of Gaussian Conditional Mean and Kriging Estimation in the Geostatistical Solution of the Inverse Problem," *Water Resour. Res.*, 21(6):825–836, 1985b.

Hsu, N. S., and W. W-G. Yeh, "Optimum Experimental Design for Parameter Identification in Groundwater Hydrology," *Water Resour. Res.*, 25(5):1025–1040, 1989.

Hughes, J. P., and D. P. Lettenmaier, "Data Requirement for Kriging: Estimation and Network Design," *Water Resour. Res.*, 17(6):1641–1650, 1981.

Hunt, B. W., and D. D. Wilson, "Graphical Calculation of Aquifer Transmissivity in Northern Canterbury, New Zealand," *J. Hydrol. (N. Z.)*, 13(2):66–81, 1974.

Illangasekare, T. H., and H. J. Morel-Seytoux, "Stream-Aquifer Influence Coefficients for Simulation and Management," *Water Resour. Res.*, 18(1):168–176, 1982.

Illangasekare, T. H., and H. J. Morel-Seytoux, "Design of a Physically-Based Distributed Parameter Model for Arid-Zone Surface-Groundwater Management," *J. Hydrol.*, 74:213–232, 1984.

Illangasekare, T. H., and H. J. Morel-Seytoux, "A Discrete Kernel Model for Conjunctive Management of a Stream-Aquifer System," *J. of Hydrol.*, 85:319–330, 1986.

Irmay, S., "Piezometric Determination of Inhomogeneous Hydraulic Conductivity," *Water Resour. Res.*, 16(4):691–694, 1980.

Jackson, D. R., and G. Aron, "Parameter Estimation in Hydrology: The State of the Art," *Water Resour. Bull.*, 7(3):457–471, 1971.

Jacquard, P., and C. Jain, "Permeability Distribution from Field Pressure Data," *Soc. Pet. Eng. J.*, 5(4):281–294, 1965.

Jahns, H. O., "A Rapid Method for Obtaining a Two-Dimensional Reservoir Description from Well Pressure Response Data," *Soc. Pet. Eng. J.*, 6(4):315–327, 1966.

Jones, L., R. Willis, and W. W-G. Yeh, "Optimal Control of Nonlinear Groundwater Hydraulics Using Differential Dynamic Programming," *Water Resour. Res.*, 23(11):2097–2106, 1987.

Journel, A. G., and J. C. Huijbregts, *Mining Geostatistics*, Academic, Orlando, Fla., 1978.

Judge, G. J., W. E. Griffiths, R. C. Hill, and T-C. Lee, *The Theory and Practice of Econometrics*, John Wiley, New York, 1980.

Kalman, R. E., "A New Approach to Linear Filtering and Prediction Problems," *Trans. ASME J. Basic Eng.*, 82:35–45, 1960.

Karatzas, G. P., and G. F. Pinder, "Groundwater Management Using Numerical Simulation and the Outer Approximation Method for Global Optimization," *Water Resour. Res.*, 29(10): 3371–3378, 1993.

Kashyap, D., and S. Chandra, "A Nonlinear Optimization Method for Aquifer Parameter Estimation," *J. Hydrol.*, 57:163–173, 1982.

Kendall, D. R., and A. Sienkiewich, "Conjunctive Use Opportunities in Southern California," in A. I. Johnson and D. J. Finlayson, eds., *Proceedings of the International Symposium on Artificial Recharge of Groundwater*, August 23–27, ASCE, Anaheim, Calif., 1988, pp. 385–395.

Kitamura, S., and S. Nakagiri, "Identifiability of Spatially-Varying and Constant Parameters in Distributed Systems of Parabolic Type," *SIAM J. Contr. Optimiz.*, 15(5):785–802, 1977.

Kitanidis, P. K., "Parametric Estimation of Covariances of Regionalized Variables," *Water Resour. Bull.*, 23(4):557–567, 1987.

Kitanidis, P. K., and E. G. Vomvoris, "A Geostatistical Approach to the Inverse Problem in Groundwater Modeling (Steady State) and One-Dimensional Simulations," *Water Resour. Res.*, 19(3):677–690, 1983.

Kleinecke, D., "Use of Linear Programming for Estimating Geohydrologic Parameters of Groundwater Basins," *Water Resour. Res.*, 7(2):367–375, 1971.

Kleinecke, D., "Comments on 'An Automatic Solution for the Inverse Problem' by Y. Emsellem and G. de Marsily," *Water Resour. Res.*, 8(4):1128–1129, 1972.

Knopman, D. S., and C. I. Voss, "Behavior of Sensitivities in the One-Dimensional Advection-Dispersion Equation: Implications for Parameter Estimation and Optimal Design," *Water Resour. Res.*, 23(2):253–272, 1987.

Knopman, D. S., and C. I. Voss, "Discrimination Among One-Dimensional Models of Solute Transport in Porous Media: Implications for Sampling Design," *Water Resour. Res.*, 24(11): 1859–1876, 1988.

Kruger, W. D., "Determining a Real Permeability Distribution by Calculations," *T. Soc. Pet. Eng.*, 222(7):691–696, 1961.

Kubrusly, C. S., "Distributed Parameter System Identification, a Survey," *Int. J. Contr.*, 26(4):509–535, 1977.

Labadie, J. W., "Decomposition of a Large Scale Nonconvex Parameter Identification Problem in Geohydrology," Rep. ORC 72-73, Oper. Res. Cent., University of California, Berkeley, 1972.

Levenberg, K., "A Method for the Solution of a Certain Nonlinear Problems in Least Squares," *Q. Appl. Math.*, 2:164–168, 1944.

Lin, A. C., and W. W-G. Yeh, "Identification of Parameters in an Inhomogeneous Aquifer by Use of the Maximum Principle of Optimal Control and Quasilinearization," *Water Resour. Res.*, 10(4):829–838, 1974.

Loaiciga, H. A., "An Optimization Approach for Groundwater Quality Monitoring Network Design," *Water Resour. Res.*, 25(8):1771–1782, 1989.

Loaiciga, H. A., and M. A. Marino, "Groundwater Management and the Inverse Problem," *Stochastic Hydrol. and Hydr.*, 1(3):161–168, 1987a.

Loaiciga, H. A., and M. A. Marino, "Parameter Estimation in Groundwater: Classical, Bayesian, and Deterministic Assumptions and Their Impact on Management Policies," *Water Resour. Res.*, 23(6):1027–1035, 1987b.

Loaiciga, H. A., and M. A. Marino, "Error Analysis and Stochastic Differentiability in Subsurface Flow Modeling," *Water Resour. Res.*, 25(12):2897–2902, 1990.

Louie, P. W. F., W. W-G. Yeh, and N. S. Hsu, "An Approach to Solving a Basin-Wide Water Resources Management Planning Problem with Multiple Objectives," California Water Resources Center Contribution Report, University of California, Davis, Calif., July, 1982.

Louie, P. W. F., W. W-G. Yeh, and N. S. Hsu, "Multiobjective Water Resources Management Planning," *J. Water Resour. Plng. and Mgmt.*, ASCE, 110(1):39–56, 1984.

Lovell, R. E., L. Duckstein, and C. C. Kisiel, "Use of Subjective Information in Estimation of Aquifer Parameters," *Water Resour. Res.*, 8(3):680–690, 1972.

Maass, A., M. M. Hufschmidt, R. Dorfman, H. A. Thomas, Jr., S. A. Marglin, and G. M. Fair, *Design of Water Resources Systems*, Harvard University Press, Cambridge, Mass., 1962.

Maddock, T. III., "Algebraic Technological Function from a Simulation Model," *Water Resour. Res.*, 8(1):129–134, 1972.

Maddock, T. III, "The Operation of a Stream-Aquifer System under Stochastic Demands," *Water Resour. Res.,* 10(1):1–10, 1974.

Madrid, C., "Artificial Ground Water Recharge in Southern California," in A. I. Johnson and D. J. Finlayson, ed., *Proceedings of the International Symposium on Artificial Recharge of Groundwater,* August 23–27, ASCE, Anaheim, Calif., 1988, pp. 378–384.

Major, D. C., "Benefit-Cost Ratios for Projects in Multiple Objective Investment Programs," *Water Resour. Res.,* 5(6):1174–1178, 1969.

Makinde-Obusola, B. A., and M. A. Marino, "Optimal Control of Groundwater by the Feedback Method of Control," *Water Resour. Res.,* 25(6):1341–1352, 1989.

Mantoglou, A., and J. L. Wilson, "The Turning Bands Method for Simulation of Random Fields Using Line Generation by a Spectral Method," *Water Resour. Res.,* 18(5):1379–1394, 1982.

Marglin, S. A., *Public Investment Criteria,* MIT Press, Cambridge, Mass., 1967.

Marino, M. A., and W. W-G. Yeh, "Identification of Parameters in Finite Leaky Aquifer Systems," *J. Hydraul. Div., ASCE,* 99(HY2):319–336, 1973.

Marquardt, D. W., "An Algorithm for Least Squares Estimation of Nonlinear Parameters," *SIAM, J.* 11:431–441, 1963.

Martensson, K., "Least Square Identifiability of Dynamic Systems," Tech. Rep. RB 7344, Department of Electrical Engineering, University of Southern California, Los Angeles, 1973.

Matheron, G., "The Intrinsic Random Functions and Their Applications," *Adv. Appl. Probab.,* 5:438–468, 1973.

McCarthy, J. M., and W. W-G. Yeh, "Optimal Pumping Test Design for Parameter Estimation and Prediction in Groundwater Hydrology," *Water Resour. Res.,* 26(4):779–791, 1990.

McElwee, C. D. and M. A. Yukler, "Sensitivity of Groundwater Models with Respect to Variations in Transmissivity and Storage," *Water Resour. Res.,* 14(3):451–459, 1978.

McElwee, C. D., "Sensitivity Analysis and the Groundwater Inverse Problem," *Groundwater,* 20(6):723–735, 1982.

McKinney, D. C., and D. P. Loucks, "Network Design for Predicting Groundwater Contamination," *Water Resour. Res.,* 28(1):133–147, 1992.

McLaughlin, D. B., "Investigation of Alternative Procedures for Estimating Groundwater Basin Parameters," Report Prepared for the Office of Water Research and Technology, U.S. Department of the Interior, Water Resources Engineering, Walnut Creek, Calif., 1975.

McLaughlin, D. B., "Hanford Groundwater Modeling—A Numerical Comparison of Baysian and Fisher Parameter Estimation Techniques," Rockwell Hanford Operations, Energy Systems Group, Rockwell International, Richland, Wash., 1979.

McLaughlin, D. B., and E. F. Wood, "A Distributed Parameter Approach for Evaluating the Accuracy of Groundwater Model Predictions, 1, Theory," *Water Resour. Res.,* 24(7): 1037–1047, 1988a.

McLaughlin, D. B., and E. F. Wood, "A Distributed Parameter Approach for Evaluating the Accuracy of Groundwater Model Predictions, 2, Application to Groundwater Flow," *Water Resour. Res.,* 24(7):1048–1060, 1988b.

Meyer, P. D., and E. D. Brill, Jr., "A Method for Locating Wells in a Groundwater Monitoring Network under Conditions of Uncertainty," *Water Resour. Res.,* 24(8):1277–1282, 1988.

Miller, W. L., and D. M. Byers, "Development and Display of Multiple Objective Project Impacts," *Water Resour. Res.,* 9(1):11–20, 1973.

Mishra, S., and J. C. Parker, "Parameter Estimation for Coupled Unsaturated Flow and Transport," *Water Resour. Res.,* 25(3):385–396, 1989.

Murtagh, B. A., and M. A. Saunders, "A Projected Lagrangian Algorithm and Its Implementation for Sparse Nonlinear Constraints," *Mathematical Programming Study 16,* North-Holland, Amsterdam, The Netherlands, 1982.

Navarro, A., "A Modified Optimization Method of Estimating Aquifer Parameters," *Water Resour. Res.,* 13(6):935–939, 1977.

Nelson, R. W., "In-place Measurement of Permeability in Heterogeneous Media, 1, Theory of a Proposed Method," *J. Geophys. Res.,* 65(6):1753–1758, 1960.

Nelson, R. W., "In-place Measurement of Permeability in Heterogeneous Media, 2, Experimental and Computational Considerations," *J. Geophys. Res.,* 66(8):2469–2478, 1961.

Nelson, R. W., "Conditions for Determining a Real Permeability Distribution by Calculation," *Soc. Pet. Eng. J.,* 2(3):223–224, 1962.

Nelson, R. W., "In-place Determination of Permeability Distribution for Heterogeneous Porous Media through Analysis of Energy Dissipation," *Soc. Pet. Eng. J.,* 8(1):33–42, 1968.

Nelson, R. W., and W. L. McCollum, "Transient Energy Dissipation Methods of Measuring Permeability Distributions in Heterogeneous Porous Materials," Rep. CSC 691229, Water Resources Division, U.S. Geological Survey, Washington, D.C., 1969.

Neuman, S. P., "Calibration of Distributed Parameter Groundwater Flow Models Viewed as a Multiple-Objective Decision Process under Uncertainty," *Water Resour. Res.,* 9(4):1006–1021, 1973.

Neuman, S. P., "Role of Subjective Value Judgment in Parameter Identification," in G. C. Vansteenkiste, ed., *Modeling and Simulation of Water Resources Systems,* North-Holland, Amsterdam, 1975.

Neuman, S. P., "A Statistical Approach to the Inverse Problem of Aquifer Hydrology, 3, Improved Solution Method and Added Perspective," *Water Resour. Res.,* 16(2):331–346, 1980.

Neuman, S. P., G. E. Fogg, and E. A. Jacobson, "A Statistical Approach to the Inverse Problem of Aquifer Hydrology, 2, Case Study," *Water Resour. Res.,* 16(1):33–58, 1980.

Neuman, S. P., and S. Yakowitz, "A Statistical Approach to the Inverse Problem of Aquifer Hydrology, 1, Theory," *Water Resour. Res.,* 15(4):845–860, 1979.

Nishikawa, T., and W. W-G. Yeh, "Optimal Pumping Test Design for the Parameter Identification of Groundwater Systems," *Water Resour. Res.,* 25(7):1737–1747, 1989.

Noel, J. E., and R. E. Howitt, "Conjunctive Multibasin Management: An Optimal Control Approach," *Water Resour. Res.,* 18(4):753–763, 1982.

Nutbrown, D. A., "Identification of Parameters in a Linear Equation of Groundwater Flow," *Water Resour. Res.,* 11(4):581–588, 1975.

Pierce, A., "Unique Identification of Eigenvalues and Coefficients in a Parabolic Problem," *SIAM J. Contr. Optimiz.,* 17(4):494–499, 1979.

Pinder, G. F., J. D. Bredehoeft, and H. H. Cooper, Jr., "Determination of Aquifer Diffusivity from Aquifer Response to Fluctuation in River Stage," *Water Resour. Res.,* 5(4):850–855, 1969.

Ponzini, G., and A. Lozej, "Identification of Aquifer Transmissivity: The Comparison Model Method," *Water Resour. Res.,* 18(3):597–622, 1982.

Rajaram, H., and D. McLaughlin, "Identification of Large-Scale Spatial Trends in Hydrologic Data," *Water Resour. Res.,* 26(10):2411–2424, 1990.

Rice, W. A., and S. M. Gorelick, "Geological Inference from 'Flow Net' Transmissivity Determination: Three Case Studies," *Water Resour. Bull.,* 21(6):919–930, 1985.

Rosenwald, G. W., and D. W. Green, "A Method for Determining the Optimum Location of Wells in a Reservoir Using Mixed-Integer Programming," *Soc. Pet. Eng. J.,* 14(1):44–54, 1974.

Rowe, P. P., "An Equation for Estimating Transmissivity and Coefficient of Storage from River Level Fluctuation," *J. Geophys. Res.,* 65(10):3419–3424, 1960.

Rubin, Y., and G. Dagan, "Stochastic Identification of Transmissivity and Effective Recharge in Steady Groundwater Flow, 1, Theory," *Water Resour. Res.,* 23(7):1185–1192, 1987a.

Rubin, Y., and G. Dagan, "Stochastic Identification of Transmissivity and Effective Recharge in Steady Groundwater Flow, 2, Case Study," *Water Resour. Res.,* 23(7):1193–1200, 1987b.

Rubin, Y., and G. Dagan, "Stochastic Analysis of Boundaries Effects on Head Spatial Variability in Heterogeneous Aquifers, 1, Constant Head Boundary," *Water Resour. Res.,* 24(10):1689–1697, 1988.

Sadeghipour, J., and W. W-G. Yeh, "Parameter Identification of Groundwater Aquifer Models: A Generalized Least Squares Approach," *Water Resour. Res.,* 20(7):971–979, 1984.

Sagar, B., "Galerkin Finite Element Procedure for Analyzing Flow through Random Media," *Water Resour. Res.*, 14(6):1035–1044, 1978.

Sagar, B., S. Yakowitz, and L. Duckstein, "A Direct Method for the Identification of the Parameters of Dynamic Nonhomogeneous Aquifers," *Water Resour. Res.*, 11(4):563–570, 1975.

Samper, F. J., and S. P. Neuman, "Estimation of Spatial Covariance Structures by Adjoint State Maximum Likelihood Cross Validation, 1, Theory," *Water Resour. Res.*, 25(3):351–362, 1989.

Schwartz, J., "Linear Models for Groundwater Management," *J. Hydrol.*, 28(24):377–392, 1976.

Shah, P. C., G. R. Gavalas, and J. H. Seinfeld, "Error Analysis in History Matching: The Optimum Level of Parameterization," *Soc. Pet. Eng. J.*, 18(3):219–228, 1978.

Shoemaker, C., L. C. Chang, L. Z. Liao, and P. Liu, "Supercomputers and Optimal Control of Groundwater Quality," in *Proc. 16th Annual Conf. on Water Resour. Plng. and Mgmt.*, May 21–25, ASCE, 1989, pp. 129–132.

Silvey, S. D., *Optimal Design—An Introduction to the Theory for Parameter Estimation*, Chapman and Hall, New York, 1980.

Slater, G. E., and E. J. Durer, "Adjustment of Reservoir Simulation Models to Match Field Performance," *Soc. Pet. Eng. J.*, 11(3):295–305, 1971.

Smith, P. J., and B. S. Piper, "A Non-Linear Optimization Method for the Estimation of Aquifer Parameters," *J. Hydrol.*, 39:255–271, 1978.

St. John, R. C., and N. R. Draper, "D-Optimality for Regression Designs: A Review," *Technometrics*, 17(1):15–23, 1975.

Stallman, R. W., "Numerical Analysis of Regional Water Levels to Define Aquifer Hydrology," *Trans. AGU*, 37(4):451–460, 1956.

Steinberg, D. M., and W. G. Hunter, "Experimental Design: Review and Comment," *Technometrics*, 26(2):71–97, 1984.

Sun, N. Z., and W. W-G. Yeh, "A Proposed Upstream Weight Numerical Method for Simulating Pollutant Transport in Groundwater," *Water Resour. Res.*, 19(6):1489–1500, 1983.

Sun, N. Z., and W. W-G. Yeh, "Identification of Parameter Structure in Groundwater Inverse Problem," *Water Resour. Res.*, 21(6):869–883, 1985.

Sun, N. Z., and W. W-G. Yeh, "Coupled Inverse Problem in Groundwater Modeling, 1, Sensitivity Analysis and Parameter Identification," *Water Resour. Res.*, 26(10):2507–2525, 1990a.

Sun, N. Z., and W. W-G. Yeh, "Coupled Inverse Problems in Groundwater Modeling, 2, Identifiability and Experimental Design," *Water Resour. Res.*, 26(10):2527–2540, 1990b.

Sun, N. Z., and W. W-G. Yeh, "A Stochastic Inverse Solution for Transient Groundwater Flow: Parameter Identification and Reliability Analysis," *Water Resour. Res.*, 28(12):3269–3280, 1992.

Swain, L. A., "Predicted Water-Level and Water-Quality Effects of Artificial Recharge in the Upper Coachella Valley, California Using a Finite-Element Digital Model," *Water-Resources Investigations 77-29*, prepared in cooperation with the Desert Water Agency and the Coachella Valley County Water District, 1978.

Sykes, J. F., J. L. Wilson, and R. W. Andrews, "Sensitivity Analysis for Steady State Groundwater Flow Using Adjoint Operators," *Water Resour. Res.*, 21(3):359–371, 1985.

Tang, D. H., and G. F. Pinder, "Simulation of Groundwater Flow and Mass Transport under Uncertainty," *Adv. Water Resour.*, 1(1):25–30, 1977.

Tang, D. H., and G. F. Pinder, "A Direct Solution to the Inverse Problem in Groundwater Flow," *Adv. Water Resour.*, 2(2):97–99, 1979.

Theil, H., "On the Use of Incomplete Prior Information in Regression Analysis," *Am. Stat. Assoc. J.*, 58(302):401–414, 1963.

Theil, H., *Principles of Econometrics*, John Wiley, New York, 1971.

Theis, C. V., "The Relation between the Lowering of the Piezometric Surface and the Rate and Duration of Discharge of a Well Using Groundwater Storage," *Trans. AGU*, 16:519–524, 1935.

Thomas, L. K., L. J. Hellums, and G. M. Reheis, "A Nonlinear Automatic History Matching Technique for Reservoir Simulation Models," *Soc. Pet. Eng. J.*, 12(6):508–514, 1972.

Townley, L. R., and J. L. Wilson, "Computationally Efficient Algorithms for Parameter Estimation and Uncertainty Propagation in Numerical Models of Groundwater Flow," *Water Resour. Res.,* 21(12):1851–1860, 1985.

Trescott, P. C., G. F. Pinder, and S. P. Larson, "Finite-Difference Model for Aquifer Simulation in Two Dimensions with Results of Numerical Experiments," *Techniques of Water-Resources Investigations of the U.S.G.S.,* Book 7, Chap. C1, 1976.

Tucciarelli, T., and G. Pinder, "Optimal Data Acquisition Strategy for the Development of a Transport Model for Groundwater Remediation," *Water Resour. Res.,* 27(4):577–588, 1991.

Tung, Y. K., "Groundwater Management by Chance-Constrained Model," *J. Water Resour. Plng. and Mgmt.,* ASCE, 112(1):1–19, 1986.

Van Geer, F. C., C. B. M. Te Stroet, and Y. Zhou, "Using Kalman Filtering to Improve and Quantify the Uncertainty of Numerical Groundwater Simulations, 1, The Role of System Noise and Its Calibration," *Water Resour. Res.,* 27(8):1987–1994, 1991.

Varga, R. S., *Matrix Iterative Analysis,* Prentice-Hall, Englewood Cliffs, N.J., 1962.

Vermuri, V., J. A. Dracup, R. C. Erdmann, and N. Vermuri, "Sensitivity Analysis Method of System Identification and Its Potential in Hydrologic Research," *Water Resour. Res.,* 5(2): 341–349, 1969.

Vermuri, V., and W. J. Karplus, "Identification of Nonlinear Parameters of Groundwater Basin by Hybrid Computation," *Water Resour. Res.,* 5(1):172–185, 1969.

Wagner, B. J., and S. M. Gorelick, "Optimal Groundwater Quality Management Under Parameter Uncertainty," *Water Resour. Res.,* 23(7):1162–1174, 1987.

Wagner, B. J., and S. M. Gorelick, "Reliable Aquifer Remediation in the Presence of Spatially Variable Hydraulic Conductivity: From Data to Design," *Water Resour. Res.,* 25(10): 2211–2225, 1989.

Wanakule, N., L. W. Mays, and L. S. Lasdon, "Optimal Management of Large-Scale Aquifers: Methodology and Applications," *Water Resour. Res.,* 22(4):447–465, 1986.

Wasserman, M. L., A. S. Emanuel, and J. H. Seinfeld, "Practical Applications of Optimal-Control Theory to History-Matching Multiphase Simulator Models," *Soc. Pet. Eng. J.,* 15(4):347–355, 1975.

Water Resources Engineers, Inc., "An Investigation of Salt Balance in the Upper Santa Ana River Basin," report to the California State Water Resources Control Board and the Water Quality Control Board, Santa Ana Region, March, 1969.

Wildermuth, M. J., D. J. Ringel, P. Overeynder, and D. H. Johnson, "Optimizing Limited Water Resources through Conjunctive Use," in A. I. Johnson and D. J. Finlayson, eds., *Proceedings of the International Symposium on Artificial Recharge of Groundwater,* August 23–27, ASCE, Anaheim, California, 1988, pp. 332–341.

Willis, R., "Optimal Groundwater Quality Management: Well Injection of Waste Water," *Water Resour. Res.,* 12(1):47–53, 1976.

Willis, R., "A Planning Model for the Management of Groundwater Quality," *Water Resour. Res.,* 15(6):1305–1312, 1979.

Willis, R., "A Unique Approach to Regional Groundwater Management," in J. S. Rosenshein and G. D. Bennet, eds., *Groundwater Hydraulics,* American Geophysical Union, Washington, D.C., 1984.

Willis, R., and W. W-G. Yeh, *Groundwater Systems Planning and Management,* Prentice-Hall, Englewood Cliffs, N.J., 1987.

Wilson, J. L., and M. Dettinger, "State Versus Transient Parameter Estimation in Groundwater Systems," paper presented at Specialty Conference on Verification of Mathematical and Physical Models in Hydraulic Engineering, Aug. 9–11, ASCE, University of Maryland, College Park, 1978.

Wismer, D. A., R. L. Perrine, and Y. Y. Haimes, "Modeling and Identification of Aquifer Systems of High Dimension," *Automatica,* 6:77–86, 1970.

Yakowitz, S., and L. Duckstein, "Instability in Aquifer Identification: Theory and Case Studies," *Water Resour. Res.,* 16(6):1045–1064, 1980.

Yakowitz, S., and P. Noren, "On the Identification of Inhomogeneous Parameters in Dynamic Linear Partial Differential Equations," *J. Math. Anal. Appl.*, 53:521–538, 1976.

Yeh, W. W-G., "Aquifer Parameter Identification," *J. Hydraul. Div.*, ASCE., 10(HY9): 1197–1209, 1975a.

Yeh, W. W-G., "Optimal Identification of Parameters in an Inhomogeneous Medium with Quadratic Programming," *Soc. Pet. Eng. J.*, 15(5):371–375, 1975b.

Yeh, W. W-G., "Review of Parameter Identification Procedures in Groundwater Hydrology: the Inverse Problem," *Water Resour. Res.*, 22(2):95–108, 1986.

Yeh, W. W-G., "On the Ill-Posedness of the Inverse Problem," in *Proc. XXII Congress, Int. Assoc. for Hydr. Res., Topics in Hydr. Modeling*, Aug. 31–Sep. 4, International Association for Hydraulic Research, 1987, pp. 397–401.

Yeh, W. W-G., "Systems Analysis in Ground-Water Planning and Management," ASCE, 118(3):224–237, 1992.

Yeh, W. W-G., and L. Becker, "Linear Programming and Channel Flow Identification," *J. Hydraul. Div.*, ASCE, 99(HY11):2013–2021, 1973.

Yeh, W. W-G., and L. Becker, "Multiobjective Analysis of Multireservoir Operations," *Water Resour. Res.*, 18(5):1326–1336, 1982.

Yeh, W. W-G., L. Becker, W-K. Liang, N. Z. Sun, M. Davert, and M. C. Jeng, "Development of a Multi-Objective Optimization Model for Water Quality Management Planning in the Upper Santa Ana Basin," prepared for the Santa Ana Watershed Project Authority, June 1992.

Yeh, W. W-G., and G. W. Tauxe, "A Proposed Technique for Identification of Unconfined Aquifer Parameters," *J. Hydrol.*, 12:117–128, 1971.

Yeh, W. W-G., and G. W. Tauxe, "Quasilinerization and the Identification of Aquifer Parameters," *Water Resour. Res.*, 7(2):375–381, 1971.

Yeh, W. W-G., and G. W. Tauxe, "Optimal Identification of Aquifer Diffusivity Using Quasilinearization," *Water Resour. Res.*, 7(4):955–962, 1971.

Yeh, W. W-G., and N. Z. Sun, "An Extended Identifiability in Aquifer Parameter Identification and Optimal Pumping Test Design," *Water Resour. Res.*, 20(12):1837–1847, 1984.

Yeh, W. W-G., and N. Z. Sun, "Variational Sensitivity Analysis, Data Requirements, and Parameter Identification in a Leaky Aquifer System," *Water Resour. Res.*, 26(9):1927–1938, 1990.

Yeh, W. W-G., and Y. S. Yoon, "A Systematic Optimization Procedure for the Identification of Inhomogeneous Aquifer Parameters," in Z. A. Saleem, ed., *Advances in Groundwater Hydrology*, American Water Resources Association, Minneapolis, Minn., 1976, pp. 72–82.

Yeh, W. W-G., and Y. S. Yoon, "Aquifer Parameter Identification with Optimum Dimension in Parameterization," *Water Resour. Res.*, 17(3):664–672, 1981.

Yeh, W. W-G., Y. S. Yoon, and K. S. Lee, "Aquifer Parameter Identification with Kriging and Optimum Parameterization," *Water Resour. Res.*, 19(1):225–233, 1983.

Yeh, W. W-G., and N. Z. Sun, "An Extended Identifiability Approach for Experimental Design in Groundwater Modeling," *Computational Methods in Water Resources X*, vol. 2, Peters, et al., eds., Kluwer Academic Publishers, 1994, pp. 899–906.

Yoon, Y. S., and W. W-G. Yeh, "Parameter Identification in an Inhomogeneous Medium with the Finite-Element Method," *Soc. Pet. Eng. J.*, 16(4):217–226, 1976.

Yziquel, A., and J. C. Bernard, "Automatic Computing of a Transmissivity Distribution Using Only Piezometric Heads," in C. A. Brebbia, W. G. Gray, and G. F. Pinder, eds., *Finite Elements in Water Resources*, Pentech, London, 1978, pp. 1.157–1.185.

Zhou, Y., C. B. M. Te Stroet, and F. C. Van Geer, "Using Kalman Filtering to Improve and Quantify the Uncertainty of Numerical Groundwater Simulations, 2, Application to Monitoring Network Design," *Water Resour. Res.*, 27(8):1995–2006, 1991.

CHAPTER 17
WATER TREATMENT SYSTEMS

Garret P. Westerhoff
Zaid K. Chowdhury
Malcolm Pirnie Inc.
White Plains, New York

17.1 WATER TREATMENT GOAL

The goal of a water-treatment system is to provide potable water to the consumer at a reasonable cost. *Potable water,* although the term continues to have new meaning with the advance of science, refers to a drink which is pleasant to taste and safe to human health. A *water-treatment system* may be defined as the well-planned integration of a water source, treatment facilities, and a system of conveying the treated water to the consumer. The entire system must be operated in a way to develop and maintain customer confidence in the water quality and the reliability of the supply. Watershed activities usually have a significant impact on the quality of the raw-water supply to the treatment facility. The transmission and distribution system conveying treated water to the consumer has a significant impact on the quality of the treated water at the consumer's tap.

In this book, components of a complete water-treatment system are presented as separate chapters by different authors. This should not lead the reader to conclude that the need for an integrated approach is, or should be, disregarded in actual application. Water is susceptible to a variety of sources of contamination, directly affecting a water utility's ability to provide potable water to the consumer at a reasonable cost. Water quality and quantity must be maintained to ensure that adequate resources are available to meet the needs of present and future generations. In order to provide a protected, reliable supply, a comprehensive resource-management program must be developed and implemented.

This chapter provides a brief overview of water-treatment systems with particular emphasis on the engineering aspects of a water-treatment facility. The parameters that affect water quality and the operation of a treatment system are summarized in this section. Placement of a water-treatment facility in the broader concept of a complete water-treatment system is discussed briefly in the subsequent sections followed by a discussion of major issues involved in planning for water-treatment systems. Objectives of design of a treatment facility, the appropriate criteria for design, and descriptions of major water-treatment processes are also provided in the subsequent sections of this chapter.

17.1.1 Historical Perspective on Water Quality

Water-quality issues facing the drinking-water industry can be addressed and reasonable protection of drinking-water supplies can be attained only through appropriate planning, engineering, scientific discovery, and consumer involvement. Nowadays, there is a rise in public interest and expectations about drinking water. This heightened public concern is prompting more stringent water-quality regulations with the passage of time. Water-quality standards are presently in a period of evolving change in an attempt to balance concerns about infectious organisms and the potential chronic effects of a variety of organic and inorganic contaminants. The issues involved in water quality, primarily those addressing the protection of public health, have evolved over time and will continue to become increasingly more complex as discussed in the following:

- Earlier in this century, attention was focused primarily on prevention of infectious diseases, such as cholera, and on aesthetic quality. Diseases of concern at that time were usually associated with a latency period of several days to weeks. Only a handful of contaminants was regulated to satisfy this objective.

- In the 1970s, the consequences of the chemical revolution began to be recognized and attention turned to chronic diseases caused by chemicals which are *carcinogenic* (cancer causing), *mutagenic* (causing gene mutation), and/or *teratogenic* (causing birth defects) when ingested. Associated latency periods for these concerns are decades long. As a result, a large number of contaminants are beginning to be regulated at very low levels.

- Historically, groundwater was considered to be of high quality, requiring little or no treatment. During the 1980s it became apparent that a significant number of groundwater supplies were under the influence of surface water resulting in microbiological concerns similar to those for surface waters. More extensive analysis revealed the presence of contaminations from human-made (synthetic) organic chemicals as well as isolated concerns for inorganic contamination. As a result, groundwater systems are being regulated similarly to surface-water systems.

- Bacteria, viruses, and other pathogens in water supplies, including the protozoa *Giardia lamblia* and *Cryptosporidium oocyst,* are still a concern despite the widespread use of disinfectants. *Cryptosporidium,* first recognized as a cause of human illness in 1976, is one of several waterborne pathogens still challenging the water industry today. In 1993 a single outbreak of cryptosporidiasis in Milwaukee reportedly affected over 370,000 people. To cope with these challenges, regulations are being continually revised to more stringent levels.

- Over the past several years disinfection, the practice that brought epidemic proportions of waterborne diseases under control, has been challenged as creating contaminants with potential chronic health effects. As a result, *disinfection by-products* (DBPs) were regulated. These regulations are continually made more stringent as more DBPs are identified and analytical methods are developed and refined.

In the United States the agency currently responsible for the development of water-quality standards for potable use is the U.S. Environmental Protection

Agency (USEPA), created in 1970. Prior to the formation of the EPA, water-quality regulations were promulgated by the U.S. Public Health Service (USPHS). Many other countries have their own set of standards such as European standards (Knoppert, 1980). International water-quality standards are set by the World Health Organization (WHO, 1984).

The first set of U.S. water-quality regulations was adopted by the USPHS in 1914 (AWWA, 1990) and was limited to bacteriological quality. Although the 1914 standards were applicable to interstate carriers, many states adopted them. These standards were revised in 1925, 1942, 1946, and 1962. Under the 1962 standards 28 water-quality constituents, including inorganic and microbiological parameters, were regulated (USPHS, 1962). As a result of several studies (USPHS, 1970; Symons, 1974; and Consumer Reports, 1974) indicating the relatively poor quality of water supply in the United States public awareness was raised which resulted in congressional action. The Congress passed the Safe Drinking Water Act (SDWA) in 1974 (PL 93-523, 1974). Under this act EPA promulgated enforcable (primary) water-quality standards in 1975; they were amended in 1977 and 1979. Nonenforcable (secondary) water-quality standards were promulgated in 1979.

In 1986, Congress revised the original 1974 SDWA and the SDWA amendments of 1986 were signed into law (PL 99-339). These amendments required EPA to set additional water-quality standards for maintaining the safety of the nation's drinking-water supply. EPA's rulemaking efforts are leading to the regulation of an increased number of parameters and an evolving set of water-quality standards. Although EPA is lagging in the implementation schedule for water-quality standards as proposed in the SDWA amendments, the number of regulated parameters has increased significantly since 1962 when the first significant set of rules was promulgated.

Some of the newly proposed regulations under the 1986 SDWA amendments include regulations limiting the concentration of several synthetic and volatile contaminants in potable water supplies; *Total Coliform Rule* (TCR), dealing with the microbiological quality of water in the distribution system; *Surface Water Treatment Rule* (SWTR), dealing with proper disinfection of surface waters; and the *Lead and Copper Rule*, dealing with corrosion of these two metals in water-distribution systems. Several other regulations are currently being developed by the EPA under the SDWA amendments of 1986. These regulations include the *Ground Water Disinfection Rule* (GWDR), requiring proper disinfection of groundwater systems; *Disinfectants-Disinfection By-product Rule* (D/DBP), controlling the concentration of disinfectants and disinfection by-products in the water-distribution system; and the *Arsenic Rule*, limiting the concentration of arsenic in the water-distribution system. For a complete reference on the current status of regulations, readers are encouraged to consult EPA and AWWA publications such as *SDWA Advisor: Regulatory Update Service*, published and updated on a regular basis by AWWA (Pontius, 1994).

17.1.2 Contaminants Affecting Water Quality

Source-water contaminants that have the greatest significance in determining the quality of potable water can be grouped into three categories:

1. Microbiological/particulate
2. Chemical
3. Natural organic matter (NOM)

Microbiological contamination, which results in the spread of waterborne disease, continues to pose a threat to human health. Filtration and disinfection are the primary defenses against this threat. Microbiological control cannot be sacrificed as part of a strategy to meet any of the requirements to protect consumer health against chronic effects of chemical contaminants. In fact, the thrust of a recent water-quality standard developed by the EPA (the SWTR) is to improve the effectiveness of disinfection practices by ensuring that the primary objective of producing a microbiologically safe water is being achieved within the water-treatment facilities. Current rulemaking efforts are considering more stringent regulations in this regard to safeguard the consumer health, especially in areas where the microbiological quality of source water is particularly poor.

Chemical contamination by a large variety of both inorganic and organic chemicals is also of concern to human health. Some metals, such as lead, chromium, mercury, silver, and arsenic, and nonmetals such as nitrate have long been recognized as significant, but others, particularly synthetic organic chemicals such as pesticides and herbicides, have only entered the environment in large quantities since the early decades of the twentieth century. Another type of organic chemical contaminant in drinking water—disinfection by products—was present since the advent of chemical disinfection of water, but was only recently identified as a concern to human health. Many of the chemical contaminants found in treated water today are suspected to be mutagenic, carcinogenic, and/or teratogenic when ingested even in very small quantities over long periods of time.

Natural organic matter (NOM) is present in all water supplies. NOM is primarily derived from vegetative and organic decomposition in the watershed (e.g., rotting leaves), in soils (e.g., decaying plants and organisms), and in water bodies (e.g., decaying aquatic plants and organisms). In some water sources the effluent from biological wastewater treatment facilities can contribute significantly to the total concentration of NOM. Due to the natural removal processes during percolation through the soil medium the concentration of NOM is normally much less in groundwater when compared with surface sources. NOM can cause various interferences in water treatment such as:

- React with the chemical disinfectant (e.g., chlorine and ozone) to form disinfection by-products such as trihalomethanes (THM)
- Exert a demand for chemical coagulants (e.g., alum and ferric chloride) and reduce their efficiency to remove particulate contaminants such as microbiological particles and soil or clay particles
- Bind with other chemical contaminants (e.g., metals) and act as a carrier of these contaminants through the water-treatment and distribution system.

In many cases, the concentration of NOM [as represented by dissolved organic carbon (DOC) concentration] is the single most important parameter to consider in water-treatment design; e.g., the design of a granular activated carbon (GAC) adsorption process. In addition to interfering with some treatment processes and determining the bounds of other treatment processes, NOM can affect the aesthetic quality (e.g., color) of drinking-water supplies.

17.1.3 System Components

The overall system which can provide protection to the consumer can be considered to have three principal components: source of supply, treatment system, and distribution system (see Fig. 17.1).

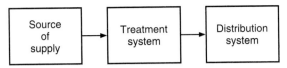

FIGURE 17.1   System components.

Through most of this century, attention has focused on the treatment facility component, primarily on filtration and disinfection, to economically produce a water safe from infectious disease. As the complexity of the issues associated with delivering high-quality water to the consumer increases, it is becoming clear that water-supply planning and engineering must better integrate all three system components in a multiple-barrier approach.

17.1.4   Multiple-Barrier Concept

Essentially, the *multiple-barrier concept* means providing multiple or redundant opportunities to assure water quality. In other words, this approach consists of raising more than one barrier against the passage of contaminants from the source to the consumer. In this manner, if one barrier fails, the system as a whole still performs satisfactorily. Historically, source protection was provided as the first protective barrier. The multiple-barrier approach was then developed further to include implementation of multiple points of disinfection within the treatment facility. Nowadays, treatment systems strive to provide multiple levels of protection at various parts of a water-supply system.

Each major component of the total system provides some level of redundancy to the subsequent component. For example, maintenance and protection of the source provides some redundancy to the treatment component, and proper design, operation, maintenance, and control of the treatment facilities serve to protect the transmission and distribution system. The system as a whole operating with these multiple levels of barriers acts together to assure protection of consumers' health.

Considerations for the overall arrangement of the multiple barriers often result in a more economical approach compared to concentrating efforts on specific treatment components. The design of a water-treatment system should consider anticipated future standards in addition to the existing regulatory requirements, consumers' demand for better water quality, the need for redundancy, and the risk or vulnerability of introducing other sources of contamination. In addition, system security and protection from sabotage should be incorporated into the system design. In this manner, multiple barriers of protection are again imposed within the treatment component of a water-supply system.

17.1.4.1   *Treatment Requirements and Process Redundancy.*   The new water-quality standards in the United States recognize the need for a multiple-barrier concept within treatment systems. For example, one of the recent water-quality standards calls for removal as well as *inactivation* (rendering microbes inactive) of microbiological contaminants, whereas one or the other process could suffice, if operated efficiently. According to this regulation, the Surface Water Treatment Rule, water systems with filtration are required to provide primary disinfection for microbiological inactivation, even though filtration provides a substantial degree of removal. Once the finished water enters the transmission and distribution system, another

water-quality standard (the Total Coliform Rule) requires the maintenance of a disinfectant residual to ensure continued protection of the water quality. Treatment processes must be designed to ensure that this multiple-barrier approach is achieved.

More attention needs to be paid to minimize the natural organic content of the source water, both in the watershed and at the treatment facility. Within treatment facilities, systems must be put in place and operations modified to reduce the NOM concentration by converting the dissolved portion to the particulate phase and thereby providing effective NOM removal. Renewed attention must also be paid to the design and operation of distribution systems to protect the quality of water delivered to the customer. For source waters vulnerable to contamination by a broad spectrum of synthetic organic chemicals, barriers are needed to minimize the possibility of organics reaching the consumer. Where these contaminants cannot be adequately prevented from reaching the source water, a barrier within the treatment component may be necessary. Such a barrier can be provided by using a dissolved NOM removal process such as activated carbon adsorption.

17.1.4.2 Redundant Equipment. Often when microbiological contamination occurs, it can be associated with equipment failure at treatment and/or disinfection facilities. Providing equipment redundancy and back-up power supplies can significantly reduce the chance of microbiological contamination. Some of the newly proposed regulations in the United States make it a requirement to provide such equipment redundancy. Based on an EPA study of over 20 existing unfiltered surface supplies, essentially all existing systems failed to provide adequate redundancy of equipment and power. Providing such facilities will create an additional barrier and ensure greater confidence that the supply will be adequately treated and disinfected prior to entry into the transmission system.

17.1.4.3 Qualified Operators. Qualified operators with a clear understanding of the treatment processes, process objectives, and systems operation provide a necessary and vital component of protection, and hence, another barrier. When failures occur or system modifications are made, the availability of properly educated and experienced plant operators can provide assurances that the water is adequately treated and is safe for the consumers. As water treatment moves forward, it becomes more complex in response to new concerns and regulations. Operating staffs will require expanded skill development and, in some cases, upgrading to meet the more sophisticated requirements.

17.1.4.4 Quality Assurance/Quality Control. Standard operating procedures which address all process and operational components should be developed to aid the operators in maintaining a minimum level of water quality. Regular review of system operating practices will ensure that objectives of quality assurance/quality control are being met. In addition, a rigorous quality-control program will help to identify a problem early, reducing or eliminating the risk of providing lesser-quality finished water.

17.2 TYPES OF TREATMENT SYSTEMS

Types and levels of contaminants present in source waters vary depending on the type of sources considered. Based upon the types of sources, water-treatment systems can be broadly classified into two basic categories: *surface-water-treatment sys-*

tems and *groundwater-treatment systems.* Elements of these two types of treatment systems reflect the general characteristics of the two types of source water. In general, surface waters contain higher levels of suspended contaminants, whereas groundwaters contain higher levels of dissolved contaminants. Surface waters represent a more oxygenated environment compared to groundwater, which is often found in anoxic condition. Surface-water sources are readily impacted by climatological changes on the surface of the earth, whereas groundwaters are affected more gradually. Surface and groundwater sources cannot always be differentiated based upon these characteristics, as some surface waters may resemble the characteristics of groundwaters, e.g., surface waters available in upstate New York.

Because of the high degree of variability in the types and concentrations of contaminants in surface-water sources, surface-water-treatment facilities must be designed with a greater degree of flexibility to handle ranges of water-quality conditions; consequently, they often require more attentive operation compared to groundwater systems. Groundwater systems on the other hand, because source-water quality is less variable, need less flexibility in design, as well as a lesser degree of operator attentiveness.

17.2.1 Treatment of Surface Water

Surface-water sources are readily impacted by weather conditions such as rain and other forms of precipitation. If a treatment system receives water from a river, the degree of variation in water quality is expected to be greater compared to a system receiving water from a lake or a reservoir, which often attenuate fluctuations in raw-water quality as a result of the holding time within the water body. For lakes and reservoirs, however, seasonal water-quality variations are often observed as temperature and sunlight conditions change during the year.

Surface-water-treatment systems must be designed to accommodate seasonal and temporal variations in water quality. For some source waters, however, the variations are smaller and the system design can be modified to accommodate less flexibility. For the highly variable surface-water source, a common treatment scheme is to use a "conventional" treatment process. Figure 17.2 shows a schematic description of a conventional treatment process. Descriptions of each component of the treatment process are provided in Sec. 17.6. In a conventional process chemical conditioning of the suspended contaminants is achieved by the coagulation/flocculation

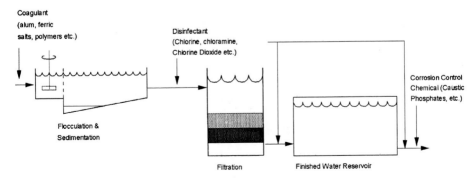

FIGURE 17.2 Schematic of conventional water treatment system.

steps, followed by the physical-separation steps of sedimentation and filtration. Disinfection of the water using a chemical disinfectant is often achieved at multiple locations, prior to coagulation, after sedimentation, and after filtration. Due to concerns for disinfection by-products, it is prudent to delay the disinfection process until the cleanest water possible is achieved, i.e., settled or filtered water.

Higher quality surface water treatment systems (i.e., low turbidity and low TOC) which do not face large variations in water quality sometimes eliminate the sedimentation step. A process flow schematic for these treatment systems is shown in Fig. 17.3. This flow schematic is termed a *direct filtration scheme*. A direct filtration treatment scheme is usually designed for optimal removal of suspended contaminants and is not amenable to the modified water treatment objective of removing higher levels of dissolved contaminants by coagulation; it also may not be suitable for reducing color, taste, and odor.

Some surface water sources contain high concentrations of polyvalent metal ions (e.g., calcium and magnesium) which imparts high hardness to the water. Plants treating these waters require the removal of hardness (softening) through the use of chemical precipitation or membrane filtration. In these plants, chemical precipitation may replace or be placed in addition to the coagulation step shown in Fig. 17.2. If a membrane softening step is desired, this step is often situated after the filtration process. Some other surface water systems require an additional barrier to organic contaminants, both synthetic and natural. For these systems, an additional treatment step involving GAC adsorption can be incorporated after the filters.

Source waters often contain significant concentrations of biodegradable contaminants, including some fraction of the NOM present in source waters. It is important for a water treatment system to remove these biodegradable contaminants to reduce the biological regrowth problem within the water distribution system. In addition, biological treatment often provides an inexpensive avenue for the reduction of NOM and disinfection by-products.

Although biological treatment of drinking water is a relatively modern concept, traditional treatment processes such as slow sand filtration can be quite effective for the removal of biodegradable contaminants. Recent application of the biological treatment of drinking water includes the use of biologically active rapid filters. These filters can be constructed with the traditional sand and anthracite media or sand and GAC media. Use of GAC adsorption to remove SOCs and NOM can also serve as an effective process for the removal of biodegradable contaminants.

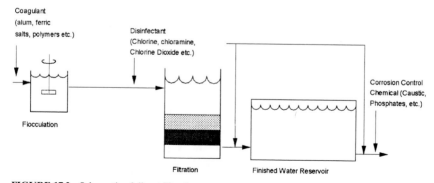

FIGURE 17.3   Schematic of direct filtration water treatment system.

17.2.2 Treatment of Groundwater

It is difficult to generalize groundwater treatment schemes, since treatment systems for groundwater are often designed with specific objectives for the removal of site-specific contaminants. Groundwater sources may be high in dissolved contaminants such as metal ions causing hardness, organic solvents such as TCE (in contaminated sites), or complex natural organic compounds (e.g., colored water aquifers). Suspended contaminants in groundwater are often present in very low concentrations which generally are not a concern for groundwater treatment.

Most groundwater systems are designed to remove dissolved contaminants, and this may include chemical or membrane softening for hardness removal, oxidation or membrane filtration for color removal, air stripping or advanced oxidation for volatile contaminants removal, and ion exchange for the removal of specific ions such as nitrates and arsenic-bearing anions. Removal of color-causing NOM from groundwater through oxidation has the drawback of producing known and unknown oxidation by-products (disinfection by-products). Highly colored groundwater such as that found in the Biscayne Aquifer, Florida, can also be treated with membrane processes or, to a lesser degree, by optimized coagulation.

Some groundwaters, because of the anoxic environment in the aquifer, are rich in odor-producing compounds such as hydrogen sulfide, which can also be removed through air stripping. The air stripping process for the removal of volatile contaminants results in a contaminated off-gas stream which requires further treatment before it is released to the atmosphere. Contaminated off-gas can be treated with water scrubbers, gaseous phase GAC adsorption, chemical oxidation, or thermal destruction.

17.2.3 Treatment System Size Considerations

Surface water treatment facilities are, in general, designed to serve larger populations, although many small surface water treatment facilities exist to serve small communities. Smaller communities generally depend on groundwater treated at the source for specific contaminant removal. Larger treatment facilities are often designed using a more conventional approach and are often operated continuously with constant operator attendance. Smaller systems tend to be more varied in nature to satisfy specific treatment objectives and are designed to operate intermittently with less operator intervention.

The majority of water customers receive water from large treatment systems in the United States, even though the number of small treatment systems far exceeds the number of larger treatment systems. Because smaller treatment systems may lack the necessary operator training and financial resources to satisfy stringent water quality goals, adequate treatment and operation of small systems is a serious concern for the water industry.

The capacity of a treatment system is dictated by the water demands of the community, and many systems are designed to satisfy future water demands. Depending on the source of water and the topography of the distribution system, several treatment systems may be designed to feed into a common distribution system. Many treatment systems are suitable for modular construction in which additional process units are constructed as the demand for treated water increases. Various statistical and empirical methods have been developed for projecting future water demands of a community. Details of water demand analyses are presented in Chap. 23 of this book.

17.3 DEVELOPING A WATER QUALITY MASTER PLAN

Although water treatment technologies are available to obtain very high levels of water quality, none offers "magic" solutions that apply to all waters for all water quality problems. Neither do all utilities face the same problems or have the same approach towards water treatment. Many source waters are of high quality and need little more than well-operated filtration and protection of the watershed. At the other extreme, some utilities, with poorer source water quality, struggle to meet the existing water quality standards. The development of new and more stringent water quality standards put these utilities in even more challenging situations. To satisfy the water quality standards of today and tomorrow and to provide the consumer with the best possible water, these utilities need to engage in careful planning with a thorough understanding of the regulatory directions, available water quality, and treatment options. The logical and prudent approach for most utilities is to develop a system-specific *Water Quality Master Plan*. A well-developed Water Quality Master Plan can provide the guidance necessary for a logical and progressive program of changes which can be implemented in a phased approach to meet water quality objectives.

While each utility's approach will vary, the elements of an integrated Water Quality Master Plan typically include consideration of regulatory impacts; establishment of a water quality objective; technical, operational, and management audits; treatability studies; a public participation program; assessment of monitoring and analytical needs; and a cost analysis. These steps of a Water Quality Master Plan are briefly described as follows.

17.3.1 Water Quality Objective

Planning begins with a review of the current water quality standards and the probable impacts of new regulations being developed. In the United States, a number of new regulations are being developed in response to the 1986 SDWA Amendments. Of particular significance is the probable impact of these regulations on a specific system. The regulatory requirements establish a minimum threshold for the development of site-specific water quality objectives.

17.3.2 Technical Audit

This is a review of the capabilities of an existing water treatment and distribution system to meet the water quality objectives. A technical audit includes identification of potential problem areas and development of a program to investigate any identified problem areas. A technical audit is a comparison of the present performance of the water system to:

- Current and probable regulations
- Specific water quality objectives and goals
- Current daily operational practices.

Depending on the specific needs of a water system, a technical audit can include a variety of evaluations such as:

- Water source and watershed management
- Water treatment system(s)
- Distribution system
- Laboratory capabilities
- Water quality monitoring programs

The goal of the technical audit is to identify areas where improvements are needed to meet established water quality objectives and to provide a course of action. This might include:

- Identifying alternatives to improve water quality
- Establishing the need to acquire additional data if available information is not sufficient to evaluate alternatives
- Developing a plan for near-term and long-term actions to conform to the established water treatment philosophy and meet the established water quality objectives
- Developing budgets and schedules for any required actions

17.3.3 Treatability Studies

Based on the results of the technical audit, treatability studies may be necessary to evaluate site-specific application of alternatives available to improve water quality. The objectives of treatability studies are to confirm and quantify the potential benefits of a treatment alternative and to develop design criteria for cost comparisons and subsequent implementation. These studies can include bench, pilot, and plant-scale testing, as previously discussed.

17.3.4 Focused Management Audit

This is a review of the management practices of the utility, focusing on the question of how the organization may benefit from changing or improving its present operating practices across a number of functional areas in response to the changes being imposed by the new regulations. A management audit may include a review of operation, maintenance, and laboratory functions. Of particular importance are staffing, training, and internal communications.

17.3.5 Public Involvement Initiatives

Drinking water regulations place considerable emphasis on public notification and are, in major part, being driven by a rise in public interest, perceptions, and expectations about water and water quality. Customers must be brought into the decision-making processes surrounding water to shape the basic philosophy of a water treatment system. Therefore, this element includes development of a positive and proactive program for involving consumers and the general public in water system decisions. It also establishes an information base for use in preparing public notices and presentations to regulatory commissions, elected officials, and consumers.

17.3.6 Monitoring Requirements/Laboratory Needs

The monitoring requirements of the new regulations should be carefully assessed to determine the probable need to expand an existing monitoring program. Alternatives available for analysis should be evaluated to determine the feasibility of performing all analyses in-house or with the use of a contract laboratory service.

17.3.7 Economic Implications

Utilities must develop ranges of costs associated with the preceding activities. This may include creating an outline of financing and implementing alternatives along with evaluating impacts on water rates.

17.4 DESIGN OBJECTIVES

Effective and economical removal of microbiological, chemical, and naturally occurring contaminants from a source water in a water treatment system can be achieved only through a properly designed and properly operated treatment system. In addition to providing a source of potable water, a properly designed water treatment system is expected to build and maintain adequate multiple barriers against contaminants that affect the source water quality and the quality of water supplied to the consumer. The objective of designing a water treatment system is to provide the most efficient means for constructing these barriers. Types of contaminants that affect a particular source water as well as the water quality requirements for the treated water need to be considered during the process of developing an effective and economical treatment system design.

17.4.1 Economical Removal of Contaminants

The contaminants that affect the quality of source waters can be microbiological, chemical, or naturally occurring as discussed in Sec. 17.1. These contaminants may be present in one of two physical states: dissolved or particulate. Removal of particulate contaminants may be achieved through a physical separation process (e.g., in a sedimentation basin and/or in granular media filter), whereas the removal of dissolved contaminants is generally achieved in two stages: conversion from the dissolved to particulate phase, followed by a physical separation of the particulate phase. There are exceptions to this two-step process in which a direct removal of dissolved contaminants is achieved, e.g., adsorption on *granular activated carbon* (GAC), *ion exchange, air stripping* of dissolved contaminants, and separation of ionic forms of contaminants through *membrane systems*, such as *reverse osmosis* membranes.

Many different types of treatment processes are available to effectively remove various contaminants. Some processes are more expensive than others, while some are more effective in removing certain types of contaminants. A prudent design approach is to put the least expensive treatment processes to maximum use while minimizing the use of more expensive treatment processes. For example, both coagulation and GAC adsorption will effectively remove dissolved organic contaminants, although GAC is more expensive than coagulation. A treatment system could be designed with coagulation followed by GAC adsorption rather than a GAC adsorption system alone. This sequential process takes advantage of the less expensive coagulation process to extend the useful life of the GAC system while producing a higher-quality water. The challenge facing a design engineer is to develop the most cost-effective design for removing contaminants from source waters without compromising water quality standards and goals.

17.4.2 Removal of Particulate Contaminants

Particulate contaminants in a source water are generally of concern for public health and aesthetic reasons as well as for potential interference with disinfection processes. Although microbiological contaminants often do not create an aesthetic problem, these organisms posing health risks may seek shelter among the aesthetically unpleasant particulate contaminants. Removal of particulate contaminants, therefore, removes a large portion of the microbiological contaminants. For this reason, drinking water regulations in the United States provide credits for microbiological contaminant removal for processes that efficiently remove particulate matter.

Given sufficient time, most particulate contaminants will eventually separate from water as a result of natural forces. The time frame for such action, however, is very long and not practical within a treatment system. In a treatment system, the process is expedited through chemical conditioning of the suspended contaminants, allowing for faster settling and efficient removal in a granular media filter. The design objective of effectively removing nearly all particulate matter is achieved by providing minimal turbidity in treated water as mandated by the prevailing water quality standards.

17.4.3 Removal of Dissolved Contaminants

Dissolved contaminants present in a source water can be synthetic or natural. In either case, removal can be effected by first converting dissolved contaminants to a particulate phase and subsequently removing the particulates, or alternately by directly removing the dissolved contaminants. The types of processes that are appropriate for removal are specific to the contaminants of concern. Some dissolved contaminants (such as arsenic) are amenable to removal through an initial chemical conversion to a particulate phase, while others such as trichloroethylene are amenable to conversion to a gaseous phase by air stripping. Still others such as a portion of the natural organic matter and certain synthetic organic contaminants are amenable to removal through adsorption on GAC.

For many dissolved contaminants, a multiple-step removal process is the most suitable and economical. For example, if it is desired that a treatment system remove most of the dissolved organic carbon from a source water, the best process may be a combination of optimized coagulation followed by GAC adsorption. The design objective for the removal of dissolved contaminants is to make the best use of the most feasible technologies and judiciously choose the advanced technologies to satisfy the additional removal requirements.

17.4.4 Elements of Design

An appropriate combination of physical and chemical processes can effectively reduce the concentration of contaminants in a source water to very low levels for potable consumption. The objective of design of a treatment system is to make the best and most economical use of available technologies to meet the desired level of water quality. This objective can be achieved with a thorough knowledge of water chemistry and the current treatment technologies, an understanding of the present and future water quality requirements, and proper engineering evaluation of alternative treatment processes.

17.5 DESIGN CRITERIA

The initial step in establishing design criteria for a new water treatment system or for adding new processes to an existing water treatment system is the development of a design philosophy and objectives. Once the design philosophy has been established, specific project objectives can be established. Critical to the objectives are the quantity and quality requirements to be met by the new design.

The design philosophy should be developed, taking into consideration the views of the facility owner and operators, regulatory requirements, the consumers, and the designer. A clear, concise philosophy will provide a beacon for all those involved in the development and operation of the facilities. The most important element of a design philosophy beyond the need for minimum reliability at a reasonable cost is to address the maximum possible reduction of the water's organic content and the production of a microbiologically safe water with minimum chronic health effects.

17.5.1 Water Supply

Issues associated with both the quantity and quality aspects of water resources and water supply to a treatment system are addressed in Part 2 of this Handbook and are not be repeated here. Suffice it to say that detailed knowledge of the water quality elements and their variability over time, seasons, and various meteorological events forms the basis for many water treatment decisions. Organic content of source water is a very important water quality parameter which often dictates the design of a specific unit process. Due to the lack of regulatory requirements to monitor organic contents of source waters, organic content and characteristics of many source waters are unknown.

17.5.2 Establishing Water Quality Objectives

The establishment of water quality objectives for a treatment system is based on using water quality standards and pertaining regulations as minimum criteria. Many facilities, recognizing that regulations change over time and responding to their customer's desires, are establishing their own specifically developed objectives to satisfy a local need or provide a desired level of responsible water supply management. These additional objectives might include softening and taste and odor reduction/control as local water quality objectives, and more reliably reducing organisms and further limiting disinfectant residuals as public health–related issues.

17.5.3 Evaluation of Alternative Treatment Processes

Evaluation of treatment technologies requires an understanding of the technologies themselves, an extensive knowledge of water chemistry, and treatability studies on candidate water supplies. Significant information can and should be obtained from studies which have been conducted previously. In most cases, it is not prudent to proceed with final selection of a treatment scheme without at least some degree of treatability studies. In selecting a treatment process based on *treatability studies*, treatment cost should be considered, since, ultimately, it is financing the construction that determines the fate of a treatment system.

17.5.3.1 Types of Treatability Studies. Treatability studies can be grouped into five categories:

- Paper evaluation
- Bench-scale studies

  Batch operation

  Continuous flow
- Small-scale pilot tests (continuous flow)

  Simulate individual unit process

  Simulate complete treatment process
- Large-scale pilot tests (continuous flow)
- Prototype pilot studies (continuous flow)

17.5.3.2 Treatability Study Program. Much can be learned through bench-scale batch type studies at significantly less expense than a continuous-flow pilot plant. For this reason, bench-scale studies should be undertaken first to screen alternatives, establish the process chemistry, and as an aid in developing the testing and evaluation program for any subsequent continuous-flow pilot studies. A paper evaluation can, however, prove to be more economical particularly if the treatment system under consideration was tested previously on similar water. Assessment of historical water-quality data, water-quality measurements obtained during previous studies, and the use of experience and rigorous simulation modeling, along with cost considerations, constitute a paper evaluation and may prove to be valuable in many applications.

The logical steps that might be followed when planning a complete treatability study program are:

1. Conduct a paper evaluation of options using existing information to reduce or eliminate subsequent studies. There is little reason to conduct pilot studies for options which are not likely to be cost effective.

2. In certain cases it may be reasonable to move directly from the paper evaluation step directly to continuous-flow studies. An example would be the evaluation of a change in filter construction or operation. (Filter studies are not conducive to bench-scale studies.)

3. After bench-scale studies there are three possible routes:
   - Proceed with a secondary paper evaluation.

- Proceed directly with plant-scale prototype operations.
- Eliminate further studies and proceed directly to full plant-scale implementation.

4. When continuous-flow pilot-scale studies are necessary, a secondary paper evaluation should be performed. This will help to eliminate non-cost-effective options and focus the pilot study work on the more important issues.

5. The options after pilot studies are to proceed with either plant-scale prototype operations or full plant-scale implementation of the most cost-effective pilot-tested operations.

The important considerations when contemplating the use of a pilot plant are:

- Obtain and use the data and experience of others to minimize the extent and cost of pilot studies.
- Proceed with a pilot study only after it has been determined that such study has the potential of reducing costs greater than the cost of the study itself.
- Under some circumstances pilot studies may be required to "prove" the effectiveness of a treatment process although no cost savings could be expected for the new process.
- Since continuous pilot studies are expensive, develop a reliable bench-scale program which limits the use and extent of continuous flow studies.

17.5.4   Establishing Design Criteria

Based on the developed design philosophy, the water-quality objectives, and results of the treatability studies to evaluate alternative treatment technologies, the specific design criteria for a treatment system can be selected. These criteria are used to develop a design and then to construct the treatment system. In selecting design criteria, the focus should be on selecting treatment alternatives that satisfy the design objectives at a minimal life-cycle cost, taking into consideration reliability and operability. In addition to this objective, the selection process should also include considerations for future expansion of the treatment system, from both quantity and quality considerations.

17.6   BASIC TREATMENT ELEMENTS

This section provides a brief discussion of the basic processes for water-treatment systems. Since the unit processes discussed here are traditional water-treatment-unit processes and are the subject of almost all textbooks on water treatment, only a brief discussion is provided here. A more detailed discussion of any of these treatment processes could be found in any textbook relating to water treatment, e.g., *Water Treatment Plant Design* (ASCE/AWWA, 1990); *Water Quality and Treatment* (AWWA, 1990); *Water Treatment Principles and Design* (James M. Montgomery, 1985); *Water Clarification Processes* (Hudson, 1981); *Physicochemical Processes for Water Quality Control* (Weber, 1972); *Water Supply and Pollution Control* (Clark, et al., 1977); and *Water Treatment Plant Design* (Sanks 1982).

17.6.1 Coagulation and Flocculation

Coagulation and *flocculation* are integral parts of most water-treatment systems. The objective is to prepare the water for efficient removal of particulate contaminants in subsequent treatment processes such as filtration. During the process of coagulation and flocculation, naturally occurring particulate matter is rendered susceptible to aggregation and subsequent removal during the sedimentation and filtration process (Stumm and Morgan, 1962). The removal of particulate contaminants, however, is not the only objective achieved through coagulation and flocculation. Many particulate contaminants are microbiological in nature, while many other microbes are transported in water systems attached to particulate contaminants. Therefore, the removal of particulate contaminants through coagulation and subsequent processes also includes the removal of a vast number of pathogenic organisms (Ali-ani, et al., 1985). In addition to removing particulate contaminants, an optimized coagulation process can also effectively prepare the water for removal of dissolved contaminants such as NOM (Edwards and Amirtharajah, 1985), and other specific contaminants such as arsenic (Chowdhury, et al., 1994).

Particulate matters in a turbulent flow or a mixed reactor constantly come in contact with each other. Due to the natural stability of particles arising from the mineral layers of particulate matter or absorbed NOM, the aggregation of these particles is very slow (Morel, 1983). The probability that a collision between particles results in aggregation depends on the nature and stability of the particles. In a coagulation process, the chemicals added reduce the stability of the particles, increasing the chance of aggregation when suspended particles are transferred near one another. This process results in larger aggregates containing the suspended particles, which can be more easily settled or filtered out of the water. Detailed analyses of the mechanisms of coagulation can be found in Amirtharajah and O'Melia (1990).

Coagulation in a water-treatment system is often achieved through the addition of a metal salt and/or a synthetic polymer. Upon being added to water, the metal salt undergoes hydrolysis producing hydrolyzed metal coagulants which can interact with the suspended particles in three ways:

1. Adsorb on the particle surface and destabilize by charge neutralization
2. Form intermediate size precipitates that create a bridge between particles
3. Form larger size precipitate which enmeshes particles into flocs

Synthetic polymeric coagulants work either on the basis of particle destabilization or by bridging between the particles.

The process of coagulation, i.e., the destabilization of suspended particles or the formation of precipitates, usually takes place within a few seconds (Stumm and O'Melia, 1968). Adequate dispersion of coagulants into the body of the water is an essential part of effective coagulation. The more uniformly the coagulant chemical is dispersed into the water, the more effective is the coagulation process, particularly if charge neutralization is the predominant coagulation mechanism. If enmeshment or *sweep flocculation* is the primary coagulation mechanism, poor mixing may prove to be better, at least during the formation of the precipitates.

Flocculation, or the aggregation of the destabilized particles into larger flocs, often takes several minutes. In a conventional treatment plant the rapid mix process is on the order of 1 to 2 minutes, whereas the flocculation step is on the order of 10 to 30 minutes. The dosages of coagulants are directly related to the concentration of suspended particles. Large doses of metal coagulants (several 10s of mg/L) are generally used for high-turbidity source waters, whereas small doses (less than 10 mg/L) of metal coagulants or polymeric coagulants are used for low-turbidity source waters. In many applications, a combination of several coagulants is used to facilitate the process. In many instances where particle concentrations are relatively small, the coagulant dose is often determined by the amount of NOM present in the water.

17.6.2 Softening

Hardness in water is caused by the presence of polyvalent metal ions (cations) which generally interfere with the cleaning action of soap by preventing foam formation and may cause scaling problems. Most abundant of the polyvalent metal ions found in source waters are those of calcium and magnesium. Corresponding major anions associated with the calcium and magnesium ions are carbonates (CO_3^{-2}) and sulfates (SO_4^{-2}).

While hard water is not known to cause any adverse health effects, relatively softer water enhances consumer acceptability of water. There is no well-defined classification of hard water and soft water. In general, hardness values of less than 75 mg/L as $CaCO_3$ represents soft water, and values above 150 mg/L as $CaCO_3$ represents hard water (Sawyer and McCarty, 1967). Perception of hard water varies among people and geographical locations.

The removal of hardness from waters, termed *softening,* can be achieved through:

1. Chemical reactions which precipitate out the excess calcium and magnesium ions

2. *Ion exchange,* in which calcium and magnesium ions are replaced with sodium ions

3. Membrane filtration, in which the polyvalent ions are physically removed

Chemical precipitation for water softening involves shifting the equilibrium of calcium and magnesium solubility by increasing the pH of the water and by adding a source of carbonate. During precipitative softening, the calcium ion is normally removed as calcium carbonate precipitate and the magnesium ion is normally removed as magnesium hydroxide precipitate. Alkalinity and pH of the source waters play important roles during precipitative softening for calcium and magnesium removal.

During *precipitative softening,* hydroxide ions (added as $Ca(OH)_2$) are added to shift the carbonate specification to CO_3^{-2} which then facilitates the precipitation of $CaCO_3$. $Ca(OH)_2$ is often produced on site by *slaking lime* (CaO). This process generally takes place at a pH of 10. Additional removal of the calcium ion may be effected by adding a source of carbonate (soda ash) to the water. Removal of magnesium hardness is caused by the precipitation of $Mg(OH)_2$ (s) which generally takes place at a pH of 11 or higher. This process of chemical softening is often termed *lime–soda ash softening.* Depending on the characteristics of the source water and the relative concentration of calcium and magnesium ions and carbonate and noncarbonate ions, the amounts of lime and soda ash required for precipitative softening may vary, although quantities can be easily calculated based on the stoichiometry of chemical reactions. Details of the softening calculations and engineering considerations for precipitative softening can be found in Benefield and Morgan (1990).

Water softening using ion exchange is the most common application of the process. In this process the source water containing high levels of hardness is passed through an ion exchange resin in which a chemical exchange takes place between the sodium ions present in the resin and the calcium and magnesium ions present in the source water. As the operation progresses, the resin material becomes saturated with the calcium and magnesium ions and loses the softening capacity. At that time the resin is regenerated using sodium chloride solution. During the regeneration process a strong solution of sodium chloride (brine) is passed through the resin which removes the adsorbed calcium and magnesium ions and replaces the sites with sodium ions, thus restoring the capacity of the resin for softening. Ion-exchange processes are used for both individual applications and municipal use (Clifford, 1990). In many groundwater systems ion-exchange facilities are constructed near the well site to provide softening treatment prior to distribution.

Membrane processes remove the polyvalent ions by physical processes. The characteristics of the membranes used in these processes determine the rejection capacity of the membranes for the removal of hardness-causing ions. In general, nanofiltration and reverse osmosis processes are highly efficient in removing hardness. Engineering considerations for membrane processes are described further in Sec. 17.7.

17.6.3   Sedimentation

Sedimentation is a process used in water-treatment systems to remove easily settleable flocs formed during the flocculation process, thereby reducing the solids loading on the subsequent granular-media filters. Effective removal of larger flocs by settling during sedimentation helps produce better-quality filtered waters and facilitates the operation of filters. In a sedimentation process, water flows through large basins, which provide a relatively quiescent condition for allowing the large flocs to settle out by gravity. As the particles settle within a sedimentation basin, the concentration increases at greater depths of the basin resulting in a change in the settling behavior of the particles.

Depending on the concentration and behavior of particles in the settling environment, sedimentation is often classified into four basic types:

1. Discrete settling
2. Flocculent settling
3. Hindered settling
4. Compression settling

In a *discrete settling* mode the particle concentrations are low and the settling of one particle is often unaffected by the others. In *flocculent settling*, the particles continue to aggregate as the settling process continues, thereby increasing the settling velocity at lower depths. In *hindered* or *zone settling*, the particle concentrations are generally high and the settling of one particle often affects the settling of others. *Compression settling* usually takes place at the bottom of a sedimentation basin where the particle concentrations are very high, and continuous accumulation of particles usually compresses the particle density in the lower layers.

Design criteria for sedimentation basins are generally specified in terms of overflow rate, an estimate of the settling velocity of the smallest size particle expected to be removed by settling within the basin. *Overflow rate* is also a measure of treatment system flow per unit surface area of the sedimentation basin. The customary unit for overflow rate is gallons per day per square feet (gpd/ft^2) or meters per hour (m/h). The range of overflow rate for a conventional water treatment system may range from 500 to 4000 gpd/ft^2 (1 to 7 m/h). Lower overflow rates are normally used for difficult-to-settle waters such as highly colored waters, whereas the higher overflow rates are representative of the easily settleable flocs.

Sedimentation basins in water treatment systems are designed as rectangular or circular tanks providing detention times in the range of one to four hours or more. The influent to a sedimentation basin is generally designed with baffles to uniformly distribute the flow throughout the cross section of the basin with a minimal degree of turbulence. Rectangular sedimentation basins are usually designed with a length several times that of the width. Several parallel basins are usually constructed within a treatment system, rather than building larger basins for the ease of construction and to minimize the effects of wind-induced turbulence. Circular sedimentation basins are commonly designed as radial flow units (flows from the center to the circumference).

Sedimentation basins could also be constructed to be multilayered or fitted with plate or tube settlers to decrease the overflow rate. Plate or tube settlers use a large number of parallel plates or tubes placed at an angle inside the sedimentation basin. These tubes or plates reduce the distance that a settling particle needs to travel before reaching a surface, thereby increasing settling efficiency. Plate or tube settlers, however, increase the potential for attached growth, which poses maintenance problems. Other types of clarifiers used in water treatment include *upflow clarifiers* (water flows against gravity) and *sludge blanket clarifiers* (water flows through a blanket of sludge at the bottom of the sedimentation chamber). More details about the theory and design of sedimentation basins can be found in Gregory and Zabel (1990).

17.6.4 Filtration

Filtration is a process of physically straining flocculated particles by a granular medium. Water-treatment systems receiving waters with high concentrations of particulate contaminants often use sedimentation basins (or other types of clarifiers) prior to filtration to reduce the load of suspended particles on the filters (conventional treatment system). Systems with relatively low particulate concentrations in the source waters may apply the flocculated waters directly to the filters. These types of treatment systems are called *direct filtration plants*. Both conventional and direct filtration treatment systems use a properly designed filter bed consisting of one or more types of granular media.

Depending on the rate of filtration, filters are classified into *slow sand filters* and *rapid sand filters*. Rapid sand filters are often found in newer treatment systems which are dependent on chemical conditioning of the water applied to the filters. On the other hand, slow sand filters can be found in older installations which have little or no chemical pretreatment. The removal of particulate matter from the filter influent water may take place on the surface of the filters (cake filtration) or inside the depth of the filter media (depth filtration). Lower-rate filters often experience cake filtration whereas higher-rate filtration is more prone to depth filtration.

Slow sand filters are typically designed with a filtration rate of less than 2 gpm/ft^2 (5 m/h), whereas rapid sand filters are typically designed with a filtration rate of 2 to 10 gpm/ft^2 (5 to 25 m/h). Filtration rates higher than 10 gpm/ft^2 (25 m/h) are also being used by several water-treatment systems. For example, the direct filtration plant in Los Angeles, California uses a filtration rate of 13.5 gpm/ft^2 (33 m/h). Slow sand filters are usually constructed to have a very large surface area and a great depth of sand. Rapid sand filters, on the other hand, are often built as modular units of smaller concrete boxes containing a shallower layer of filter media.

Typically, rapid sand filter beds are made up of a combination of sand and anthracite coal (dual-media filter). However, *mono-medium filter* designs are often found in water-treatment systems, and *dual-media filters* with GAC as one of the media are increasingly used. In a mono-medium filter a single granular medium (e.g., anthracite coal) is placed on the support structure. Mono-medium filters are often constructed to have greater depths of granular medium. In a dual- or tri-media filter two or three different types of media are placed in layers. Different types of media with varying specific gravity are often chosen (e.g., sand and anthracite coal or sand and GAC). Due to the variation in the specific gravity of the media, larger particles of the lighter media (e.g., anthracite or GAC) can be placed in the top layer while the smaller particles of the heavier media (e.g., sand) can be placed in the bottom layers. Due to this gradation in the filter media a better efficiency is often achieved while maintaining the assurance of high-quality filtered water. The use of GAC as the top media in a dual-media filter has the added advantage of removal of biodegradable organic matter if proper conditions are maintained.

Slow sand filters, like rapid sand filters, are operated semi-continuously. As the cake build-up on slow sand filters increases, the rate of filtration is lowered. Eventually the filter is taken out of service and the top layer of the filter media is removed or replaced to revive the filtration rate. In a rapid sand filter, the media is back-washed after certain time intervals to remove the entrapped suspended particles from the filter media and to revive the filtration capabilities. The time interval is determined by either the quality of filtered water, the head loss build-up across the filter media, or a regular program of backwashing. In general, higher rate filters need more frequent backwashing compared to the lower rate filters. Details about the physical principles of filter design and operation are discussed by Cleasby (1990).

17.6.5 Disinfection

The primary goal of the disinfection process is to kill (or inactivate) the pathogenic organisms in the source waters. This objective is achieved through the addition of disinfectants to water. Inactivation within a treatment system is often termed *primary disinfection. Secondary* or *residual disinfection,* on the other hand, is the process of maintaining a disinfectant residual within the water-distribution system to provide a disinfecting environment and to combat accidental contamination with pathogens. Disinfectants are primarily oxidizing agents which react with the extracellular enzymes or the cellular material of the pathogenic microorganisms, thereby inactivating them. Several disinfectants commonly used in water-treatment systems as primary disinfectants include chlorine, chlorine dioxide, ozone, chloramines, and ultraviolet light. Except for ultraviolet light, disinfection is achieved through the addition of chemical agents. Chlorine and chloramines are the most commonly found secondary or residual disinfectants.

The process of primary disinfection involves contacting a disinfecting agent with the source water for a specified time within a disinfection chamber. The degree of microbial inactivation has been found to be roughly proportional to the product of disinfectant concentration and *contact time* (Hoff, 1986). This product is termed CT and is used for assessing the performance of the disinfection process. Depending on the potency of the disinfectants, the level of CT required for disinfection varies as a function of the type of disinfectant. Details about the theory of disinfection and the engineering considerations can be found in Haas (1990).

In the United States, the regulatory requirements for disinfection are contained in the Surface Water Treatment Rule (SWTR) and complemented by the *Guidance Manual for Compliance with the Filtration and Disinfection Requirements for Public Water Systems Using Surface Water Sources* (USEPA, 1990). The SWTR sets the minimum treatment requirements (mandatory), whereas the Guidance Manual suggests design, operating, and performance criteria for specific surface-water-quality conditions to provide the optimum protection from microbiological contaminants. These recommendations are presented as advisory guidelines only (not mandatory). According to the SWTR, all community and noncommunity public water systems which use a surface-water source or a groundwater under the direct influence of a surface water must achieve a minimum of 99.9 percent (3-log) removal and/or inactivation of *Giardia* cysts, and a minimum of 99.99 percent (4-log) removal and/or inactivation of viruses. Filtration plus disinfection or disinfection alone may be utilized to achieve these performance levels, depending on the source-water quality and site-specific conditions.

As with most strong chemical oxidants, disinfectants react with various organic (e.g., NOM) and inorganic (e.g., bromide) constituents present in the source water, resulting in the formation of disinfection by-products (DBPs). Many of these DBPs have been identified to date and have been characterized as probable carcinogens to humans (Symons, et al., 1975; Rook, 1976; Krasner, 1989; and Glaze, 1993). While the chronic effects of DBPs are a major concern for the drinking-water industry, the primary objective of the disinfection process is to make the water safer for consumers by inactivating pathogens, an objective that should not be compromised at the cost of reducing chronic carcinogenic effects of DBPs.

In most water-treatment systems chlorine is applied in the form of a gas (in larger systems) or as liquid sodium hypochlorite (in smaller systems). Most water-treatment systems maintain multiple locations for applying chlorine such as application of chlorine to the plant influent, after chemical addition for coagulation flocculation, after sedimentation, and after filtration. The biocidal efficiency of chlorine is expected to be better when applied to water containing smaller concentrations of particulate matter (such as after filtration). Moreover, delaying the application of chlorine in the process stream has the added advantage of producing fewer DBPs such as trihalomethanes and haloacetic acids. In some treatment systems ammonia is added after primary disinfection, thereby producing chloramines, to maintain a residual concentration in the water distribution system. The advantage of such combination of chlorine and chloramine is that this technique provides adequate primary and residual disinfection while reducing the formation of DBPs, as significantly fewer DBPs are formed with chloramines compared to chlorine.

In water-treatment systems using chlorine dioxide for primary disinfection, the disinfecting agent is generated on-site using a combination of sodium chlorite and chlorine gas. Chlorine dioxide is normally applied at the plant effluent or after sedimentation. As chlorine dioxide is a strong oxidant, smaller doses are required to achieve the desired microbial inactivation; however, the use of chlorine dioxide results in the formation of chlorite, a DBP. Both chlorine dioxide and chlorite are believed to have adverse effects on human health and are regulated at very low concentration. As a result of the quick decay of chlorine dioxide when applied to water, this chemical is not appropriate for maintaining residual disinfectant within the water-distribution system. For this reason, systems using chlorine dioxide for primary disinfection also use chlorine or chloramine for residual disinfection in the water-distribution system.

Some water-treatment plants in the United States and many others in Europe use ozone as a primary disinfectant. *Ozone* is a very strong oxidizing agent and is very effective for microbial inactivation. For some of the more resistant pathogenic organisms, ozone serves as an effective biocide. For use as a primary disinfectant, ozone is generated on-site at the treatment facility. The generators use either ambient air or liquid oxygen for generating a gaseous ozone stream which is applied (by bubbling) to the water in ozone contactors. *Ozone contactors* are generally completely enclosed to capture and destroy the off-gas after the disinfection process. By-products of the ozonation process are often fewer than chlorination or chlorine dioxide application; however, for waters containing bromide, ozonation produces bromate—a DBP which is regulated at very low concentration. *Ozonation* also increases the biodegradability of the NOM present in the source water, which can lead to increased fouling of the basins and filters within a treatment system. If the biodegradable fraction of the NOM generated by ozonation is not removed within a treatment system, this fraction can cause a biological regrowth problem in the water distribution system. As with chlorine dioxide, ozone can only be used as a primary disinfectant. Either chlorine or chloramine is used as a residual disinfectant in systems where ozone is used for primary disinfection. Additional discussion of ozone application in water-treatment systems is provided in Sec. 17.7.

With a view to minimizing the formation of DBPs, many water-treatment systems are currently in the process of modifying their disinfection practices so that water can be adequately disinfected in a manner which results in the lowest possible level of DBP formation. In most cases, this can best be accomplished by reducing the organic content of the water prior to disinfection.

17.6.6 Dissolved Air Flotation

Dissolved air flotation (DAF) is an alternative clarification process for removing lightweight flocs. DAF often replaces the traditional sedimentation process in removing low-density particles formed during coagulation and flocculation of waters containing algae and other NOM along with low turbidity. The DAF process involves the addition of a small stream of water supersaturated with air, at high pressure, immediately after flocculation. This supersaturated stream is normally added at the bottom of the DAF chamber near the influent. The pressure of the added stream is suddenly reduced at the entrance to the DAF chamber and the dissolved air is released, moving upward along with the low-density flocs present in the coagulated water. Clarified water is removed from the DAF chamber from the bottom at the effluent and the sludge accumulates at the top, where it is removed by continuous or intermittent mechanical scrappers or by flooding. Successful attachment between flocculated particles and air bubbles depends on proper destabilization of the suspended particles during coagulation and flocculation, which are often controlled by the coagulant dose, pH, and raw-water-quality conditions.

The pressurized supersaturated stream is often produced using the finished water from the treatment process and a saturation device such as a packed column or ejector. For this reason, this flow is sometimes termed the *recycle stream*. Typical recycle flows range between 6 and 12 percent of the system flow, and are often pressurized up to 70 to 80 psig (480 to 550 kPa) for producing effective supersaturation. Overflow rates of commonly used DAF units in water-treatment applications range between 5 and 15 m/h and nominal detention times between 5 and 15 minutes. Flocculation producing particles in the 10 to 30 μm range often produces the best result in the subsequent DAF units. More details of the DAF process considerations can be found in Gregory and Zabel (1990) and in the proceedings of the IAWQ-IWSA-AWWA Joint Specialized Conference, *Flotation Processes in Water and Sludge Treatment* (1994).

17.6.7 Precoat Filtration

Precoat filtration is most commonly used in smaller water-treatment systems to remove suspended particles from low- to moderate-turbidity source waters. This process relies on a thin layer of deposited granular medium for physical removal of particulate materials. The deposited medium is pumped into the filtration system to coat a rigid support called a *septum* prior to filtration of water. As raw water is pumped into the precoat filtration system the suspended particles are retained on the surface of the precoat medium. Additional filter medium is pumped into the system along with the raw water, which helps retain the permeability of the filter medium by intermingling the suspended particles removed from the raw water with the filter medium. This additional feed of filter medium is called the *body feed*. The absence of body feed will result in a quick increase of head loss across the medium as the filtration occurs at the surface. Intermingling of the filter medium from the body feed with the suspended particles results in a relatively lower head loss across the filter media. Nonetheless, continued operation with renewed layers of filter media ultimately results in a significant head loss across the filter. At that time, the flows are reversed and the deposited filter medium and the suspended particles are washed off.

The septum is commonly made of stainless steel wire mesh, which allows the deposition of precoat, although suspended particles in the raw water can easily pass through. The most common filter medium used in precoat filtration is *diatomaceous earth* (DE), which is composed of fossilized water plants called *diatoms*. At the fossilized state, DE is mostly silica. DE is commonly available in various grades according to the need for removal of different sizes of suspended particles from raw water. The pore sizes for different grades of DE filter media range from 5 to 17 μm. Details of precoat filter design can be found in AWWA (1988).

DE filters are often operated at a filtration rate of 0.5 to 2.0 gpm/ft^2 (1.25 to 5 m/h). Most installations of DE filters are small, with the largest being a 20 mgd plant in San Gabriel, California (although New York City has a very large plant under design). Most applications of DE filters treat surface waters and provide excellent removal of suspended particulates and organisms.

DE filters rely solely on physical straining of particles and often require little operator intervention. The cost of treatment is often significantly lower than conventional coagulation sedimentation and granular media filtration. Although DE filters provide an excellent barrier for particulate materials, they are not always adequate for meeting the turbidity MCL. Primary disinfection followed by DE filters are often required.

17.6.8 Corrosion Control

Until recently, corrosion-control practices were typically designed to improve aesthetics, protect marginal hydraulic capacity, and/or reduce long-term pipeline maintenance. Although these objectives remain worthwhile, in the United States the Lead and Copper Rule has essentially redefined corrosion control solely on the basis of public health impacts. The objective of current corrosion-control methods is to minimize the concentration of lead and copper in drinking water, without compromising other health-related water-quality goals.

Corrosion-control treatment can be characterized by two general approaches to inhibiting lead and copper dissolution: (1) forming a precipitate in the potable supply which deposits onto the pipe wall to create a protective coating; or (2) causing the pipe material and the potable supply to interact in such a way that metal compounds are formed on the pipe surface, creating a film of less soluble material. The most effective corrosion-control treatment may actually rely on some combination of these two mechanisms. In practice, corrosion-control treatment technology will involve one or more of the following:

Carbonate Passivation. Modification of pH and/or alkalinity (as a surrogate for dissolved inorganic carbonate) to induce the formation of less soluble compounds with the targeted pipe materials.

Calcium Carbonate Precipitation. Adjustment of the calcium carbonate system equilibrium such that a tendency for calcium carbonate precipitation results.

Corrosion Inhibitors. Application of specially formulated chemicals characterized by their ability to form metal complexes and thereby reduce corrosion. The most common corrosion inhibitors include orthophosphate, polyphosphates, poly-orthophosphate blends, and silicates.

A wide variety of proprietary chemicals have evolved to control pipeline and valve deterioration, eliminate "dirty water" complaints, reduce laundry staining, and so on. Some of these "corrosion inhibitor" chemicals can also help reduce lead and copper levels in drinking water, although many *will not,* and some can even increase lead concentrations. Comparison of corrosion inhibitors is often controversial because of the proprietary nature of the specific chemical formulations. This issue is further complicated by a lack of understanding by many users about the differences between chemical products (e.g., ortho- and polyphosphates) and their relationship to the formation of metallic precipitates and protective films in potable waters.

The 1990s represent a new era in corrosion control, and drinking-water suppliers must exercise caution in selecting technology which is consistent with opposing water-quality objectives. Selection of the most suitable method requires not only careful evaluation of water chemistry to minimize lead dissolution, but site-specific assessments of how candidate alternatives can affect disinfection efficacy, disinfection by-product formation, and other water quality constraints.

17.6.9 Arsenic Removal

Arsenic is one of the common constituents of soil and is often found in natural surface waters and groundwaters. The concentration of arsenic in source waters for water-treatment systems depends on local geology and the geology of the watershed. The most common forms of arsenic found in natural waters are present as pentavalent arsenate ($HAsO_4^{-2}$) and trivalent arsenite ($H_2AsO_3^-$) anions. The removal of these anions can be achieved through coagulation using alum or ferric chloride salts, ion exchange using anion exchange resins, or membrane filtration using nanofiltration or reverse osmosis membranes. The ion-exchange process proves to be very useful if the target finished-water arsenic concentrations are relatively high, in the range of 5 to 50 µg/L. However, ion-exchange resins are not very effective for finished-water arsenic concentrations in the <5 µg/L range. For this type of low arsenic concentration, membrane filtration or enhanced coagulation (optimized in terms of coagulant dose and coagulation pH) using alum or ferric chloride may prove to be the most suitable treatment technologies.

17.6.10 Nitrate Removal

Presence of nitrate in ingested water can cause severe acute health problems, particularly in infants. Nitrate when ingested can be reduced to nitrite which reacts with blood hemoglobin resulting in methemoglobinemia. Due to the particular physiological condition of infants, the health effects in infants are much more severe than in adults (Tate, 1990). For this reason, a water-quality standard for nitrate has been in existence since 1962 in the United States. The current MCL in the United States is 10 mg/L as N. Most of the source waters used for producing potable water, however, have nitrate concentrations much lower than the MCL. Occurrences of high nitrate concentration in individual groundwater wells are often reported. In fact, exceedence of nitrate concentration is reported to be the most common reason compelling the shutdown of small community water-supply wells (AWWA 1985). The most common treatment for nitrate removal in water-treatment systems includes the application of anion-exchange process at individual well heads (Clifford, 1990). In this process the contaminated water is passed through an anion exchange resin in which the nitrate concentration is exchanged for chloride concentration. Other treatment processes for nitrate removal include membrane processes such as reverse osmosis and electrodialysis.

17.7 ELEMENTS OF ADVANCED TREATMENT

Several processes used in water-treatment systems, such as enhanced coagulation, GAC adsorption, membrane filtration, and advanced oxidation with ozone, have been the focus of much research in recent years. The search for alternative treatment processes to meet the multiple objectives of today's water treatment (microbiologically safe water with minimum chronic health effects) is leading the water industry to make significant improvements in these treatment processes. The use of these processes for treatment-system applications has been increasing in recent years consistent with an objective to produce higher-quality water. This section describes some of the recent advances in water-treatment applications of these processes.

17.7.1 Enhanced Coagulation

As discussed earlier in this chapter, the coagulation process has been traditionally used in water-treatment systems to prepare for the removal of particulate material from source waters in subsequent treatment processes. The mechanism of the coagulation process, however, can also be applied to increase the removal of various organic and inorganic constituents from water. In recent years the coagulation process has been finding applications in water-treatment systems with the objective of maximizing removal of dissolved constituents such as NOM and arsenic. Depending on the alkalinity and the NOM concentrations of source waters, the coagulation process can remove a significant fraction of the NOM present in the source water. The coagulant dosage for optimized removal of NOM increases as the source-water alkalinity increases or the NOM concentration decreases.

Optimized removal of NOM through coagulation is normally observed at significantly higher dosages of coagulants (often greater than 25 mg/L as alum) and/or lower coagulation pH (often less than 7.0). As a result, the amount of residual solids produced from optimized coagulation is expected to be much higher compared to the conventional coagulation process.

17.7.2 Granular Activated Carbon (GAC)

In the *GAC adsorption process*, dissolved organics in the water diffuse into the grain and adsorb onto the internal surface of the GAC. The adsorption phenomenon may be physical adsorption, chemical adsorption, or a combination of both. Organic compounds, when adsorbed, are concentrated onto the carbon due to the extensive internal surface area produced during the activation portion of the GAC manufacturing process.

The raw materials for GAC production include bituminous coal, peat, lignite, petrol coke, wood, and coconut shells, each of which produces a unique internal structure. GAC is produced by pyrolytic carbonization of the raw material with subsequent or parallel activation. Carbonization removes the volatile compounds from the carbon and activation produces its internal porous structure. GAC for drinking-water treatment is generally produced by physical or thermal activation of carbonized char. Activating agents, such as steam, carbon dioxide, and air, are contacted with the char at temperatures typically ranging between 800 to 1000°C.

GAC has one or more of the following potential roles in drinking water treatment:

- Reduction of tastes and odors
- Reduction of *synthetic organic contaminants* (SOCs)
- Adsorption of NOM and DBP precursors
- Reduction of DBPs
- Providing the medium for biological degradation of NOM
- Gas-phase adsorption of off-gases from air-stripping units

A decision on the role of GAC may be based upon one of the above, but its use will normally accomplish multiple benefits. The actual benefits will be influenced by the characteristics of the water placed in contact with the GAC.

17.7.2.1 Taste and Odor Reduction. Tastes and odors result from a variety of agents, mostly organic in nature. Decayed vegetation, the growth and decay of algae, microorganisms, and industrial chemicals are the most common. Two particularly troublesome compounds are geosmin and methylisoborneol (MIB), both produced from algae and responsible for musty odors in water. Activated carbon, either powdered (PAC) or granular (GAC), is effective for the removal of many tastes and odors. When incidence is infrequent, the use of PAC may be significantly less costly than GAC. Where tastes and odors are pervasive over longer periods, GAC may be cost-effective. Geosmin and MIB are strongly adsorbed onto GAC. Typical GAC replacement and regeneration frequencies for taste and odor control have been reported to range between one and five years.

17.7.2.2 Reduction of SOCs. Most of the SOCs of health concern in drinking water are adsorbed on GAC to different degrees, depending upon the solubility, molecular size, and potency of the adsorbate. In general, SOCs of low molecular weight and high solubility are not easily removed by GAC, and those of very high molecular weight are readily adsorbed. SOCs frequently found in drinking-water supplies and readily adsorbable onto GAC include benzene, toluene, PCBs, phenol, chlorophenols, herbicides, carbon tetrachloride, and pesticides.

17.7.2.3 Adsorption of NOM. The adsorption properties of NOM (which serves as DBP precursors) on GAC vary greatly. In many waters DBP precursors can be effectively removed by GAC adsorption. The removal can be as high as 80 to 90 percent. Operating time required to reach steady state increases with *empty bed contact time* (EBCT) from about 6 weeks at 5 minutes EBCT to about 20 weeks at 20 minutes EBCT. Pilot-scale studies using the Ames Salmonella assay indicate that GAC is also effective at removing precursors of chlorine-produced mutagenicity.

17.7.2.4 *Reduction of Disinfection By-Products.*

Although GAC can effectively remove THMs to very low levels, the removal efficiency deteriorates rapidly due to competitive adsorption of background organic material, resulting in early breakthrough after only a few weeks. For this reason, GAC is not likely to be cost-effective for reducing DBPs.

17.7.2.5 *Medium for Biological Treatment.*

NOM can support microbial growth in treatment processes and in components of the distribution system even in very low concentrations. The parameter AOC can be used as a measure of the amount of NOM in water that can be consumed by specific strains of microorganisms. At higher levels of AOC, microorganisms can proliferate in treatment units and distribution systems. When AOC levels are lower, less food is available for microorganism growth. Bacterial growth in a distribution system may facilitate pipeline corrosion, produce tastes and odors, and increase the amount of disinfectant required to maintain a residual throughout the distribution system.

Biological treatment, a process frequently used in parts of Europe, can be an important process in future drinking-water treatment in the United States. It can reduce AOC and other biodegradable compounds, resulting in a biologically stable water. It can also remove certain trace SOCs by direct biochemical reaction. Biological treatment may be essential following ozonation to reduce the AOC of the water before distribution.

An attractive approach for incorporating a biological process into water-treatment systems would be in the filters (or filter/adsorbers) and/or post-filtration GAC adsorbers, ahead of disinfection. The irregular surfaces of GAC provide better protection for initial biofilm growth and reduce biofilm shear loss compared to smooth-surfaced material such as sand. The use of GAC may also enhance biological activity by concentrating organic matter, which serves as food to the microorganisms. GAC, either in filters or contactors, is therefore preferable to sand as a support medium for microbial growth.

One method to reduce the concentration of AOC delivered to the distribution system is to maximize the production of AOC within the treatment plant, and then provide for a biological treatment step. The formation of AOC can be enhanced by ozonation or an advanced oxidation process (AOP) such as ozone/hydrogen peroxide. The oxidation renders more of the NOM biodegradable, resulting in greater biological activity on the GAC instead of in the distribution system.

Any disinfectant residual in water applied to a GAC adsorber will be quickly consumed, resulting in a waste of chemicals and inhibition of biological activity on the GAC. Therefore, primary disinfection with chlorine should be provided after contact with GAC. This is an ideal location in the treatment train to obtain primary disinfection since it is the "cleanest" water available and will result in the least disinfectant demand and lowest DBP formation potential. On the other hand, if ozone is used as a disinfectant, it is advisable to use before the GAC process as ozonation is known to greatly enhance the biodegradability of NOM, thereby resulting in better NOM removal.

Under the conditions where an active biofilm accumulates on GAC, the associated biodegradation of organic matter can result in the freeing of previously filled adsorption sites. This process can be referred to as *bioregeneration.* Data on these phenomena are limited and much additional study is needed.

Although most SOCs have been shown to be nonbiodegradable, recent research has indicated that under specific environments a few, such as trichloroethene, chloroform, and phenols, can possibly be biodegraded. These SOCs are generally consumed as a secondary source of food, a process known as *cometabolism.* Biodegradation of

SOCs may result in longer bed lives for nonbiodegradable SOCs as compared with the adsorptive process.

17.7.2.6 Filter Medium. GAC is an effective filter medium for removal of turbidity and suspended solids. It provides surface rather then depth filtration, and its filtration characteristics are likely to be more like those of a sand medium rather than dual media.

17.7.2.7 Multiple Benefit Role. It is the multiple benefit role provided by the use of GAC as part of a treatment system that is, perhaps, the most interesting. As the regulatory process evolves, the potential benefits of GAC need to be carefully considered. The increased removal of both natural and synthetic organics will be an increasingly important strategy to meet new regulations and provide a superior quality of water.

17.7.3 Membrane Processes

Membrane processes can be used to remove a wide variety of materials from water, ranging from suspended particles to ions. While these processes all use a membrane barrier for separation of the material from the water, specific membrane applications can be as different as rapid sand filtration and ion exchange.

The advantages and limitations of the different types of membrane processes are important to understand. Applying sound engineering and testing principles, the appropriate membrane process can be selected to provide effective treatment for reducing many regulated and unregulated materials in water. The membrane process or processes to be used in a given application depends on the raw-water quality and the finished-water-treatment objectives. More specifically, it is necessary to determine what materials require removal, to what level they must be removed, and what effect other materials in the water may have on membrane performance.

Some potential applications for membrane processes are:

- Desalination (inorganic ion removals)
- Softening
- Removal of organic materials
- Particle removal
- Disinfection

Most membrane processes currently used in the water industry are pressure-driven. They include *hyperfiltration,* commonly called *reverse osmosis* (RO), *nanofiltration* (NF), which is commonly known as membrane softening, *ultrafiltration* (UF), and *microfiltration* (MF). Presented in ascending order with respect to the size of contaminant removed, RO membranes remove small ionic and organic materials while MF membranes are designed to remove large suspended particles such as bacteria, *Giardia* and *Cryptosporidium* cysts and colloidal particles. In addition to pressure-driven membrane processes, there are membrane processes that depend on an electrical potential (*electrodialysis,* ED; and *electrodialysis reversal,* EDR), concentration gradients, and other driving forces.

The rejection characteristics of pressure-driven membrane processes are normally identified by the size of material retained by the membrane. This can be in terms of *molecular weight cut off* (MWCO) for membrane desalination processes such as reverse osmosis, ultrafiltration, and nanofiltration, or in terms of *nominal pore size* (NPS) for membrane filtration processes such as microfiltration membranes.

Relative rejection properties of the generic classes of membranes for various dissolved and suspended materials are presented in the separation spectrum in Fig. 17.4. Rejection for a conventional media filter, without chemical conditioning, is also presented as a reference. While this figure can be used as a general guide, actual rejection of materials by a membrane can be dependent on factors other than MWCO and NPS. For example, microfiltration systems may provide effective removal of some virus species, even though the virus may be ten times smaller than the NPS of the membrane.

While the majority of existing membrane applications at the industrial scale have been used for desalting and softening, future applications are expected to rival conventional organic material and solids removal processes. This is due in part to improvements in membrane performance and longevity, which have significantly reduced membrane-process costs over the last ten years.

The performance of membranes can be described by both the quality of the filtrate (permeate) and the head loss (pressure drop) across the membrane. Using a lower MWCO membrane will normally provide the best permeate quality, but requires a larger driving pressure to operate. Looser membranes, those with larger MWCOs or NPSs, require lower driving pressures but are less effective at removing small materials, such as sodium. The performance of membranes can be quantified by determining the percent of salt passage and salt rejection. Percentages of *salt passage* and *rejection* through a membrane are defined as follows:

$$\% \text{ Salt passage} = \frac{\text{product quality}}{\text{feed quality}} \times 100\%$$

$$\% \text{ Rejection} = 100\% - \% \text{ salt passage}$$

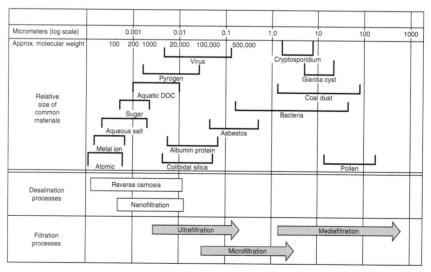

FIGURE 17.4 Separation spectrum.

Algae, *Giardia* and *Cryptosporidium* cysts, bacteria, and some viruses can be removed from water using microfiltration membranes which operate with driving pressures of 5 to 30 psig (35 to 200 kPa). Without pretreatment, dissolved organic materials will not be removed by microfiltration membranes. Ultrafiltration membranes are capable of removing the colloidal materials and some macromolecular material at driving pressures of 15 to 60 psig (100 to 400 kPa). Substantial improvements in performance can be realized when these membranes are used in conjunction with various chemical pretreatments.

A summary of operating conditions of the major types of membranes is presented in Table 17.1.

Nanofiltration membranes are capable of removing significant portions of the DOC in waters and operating at driving pressures of 80 to 120 psig (550 to 830 kPa). Divalent ions such as calcium are also rejected by nanofiltration membranes. Therefore, nanofiltration membranes are often referred to as softening membranes. However, the rejection of scale-causing ions such as calcium can also lead to the precipitation of these materials on the membrane. Precipitation typically results in irreversible fouling of the membrane surface which must be avoided by adding an antiscaling chemical and/or acid to the feed water. In addition, nanofiltration membranes normally require pretreatment for particle removal. Scaling considerations may also limit the maximum recovery for the system.

Reverse osmosis membranes reject much of the DOC and dissolved inorganic materials. Driving pressures range from 225 psig (1550 kPa), for low pressure reverse osmosis membranes, up to 1200 psig (8300 kPa) for seawater desalination. Typical rejections for sodium, calcium, chloride, and sulfate are in the range of 94 to 99+ percent. RO membranes will not, however, reject all contaminants in the source water. Nitrates are poorly rejected by most RO membranes with removals of 85 to 95 percent. Most nonpolar organics are effectively rejected by RO membranes but rejection may be as low as 40 to 60 percent for polar organics such as urea.

17.7.4 Ozonation

Ozone, a strong oxidant, has the potential to radically change the characteristics of a given water. Ozone applicators in water treatment have, for the most part, focused on providing disinfection while minimizing the formation of disinfection by-products. Other applications include:

TABLE 17.1 Common Operating Conditions

Membrane process	Operating pressures	Recovery	Flux	Primary application
Reverse osmosis	200–2,000 psig (1,400–14,000 kPa)	50–85%	3–20 gfd (5–33 Lh^{-1}m^{-2}	Desalting Trace contaminant removal
Nanofiltration	80–120 psig (550–830 kPa)	70–90%	15–25 gfd (25–42 Lh^{-1}m^{-2}	Softening NOM removal
Ultrafiltration	15–60 psig (105–415 kPa)	80–95%	20–300 gfd (42–500 Lh^{-1}m^{-2}	Disinfection Particulate removal
Microfiltration	5–30 psig (35–210 kPa)	95–98%	100–1,000 gfd (167–$1,670$ Lh^{-1}m^{-2}	Particulate removal

- Disinfection
- Increasing biodegradability
- Aiding coagulation (microflocculation)
- Taste and odor reduction
- Color removal
- Oxidation of iron and manganese

The ozone reaction mechanisms and the roles of ozone in drinking-water treatment have been summarized by Westerhoff, et al. (1991).

17.7.4.1 Disinfection. Molecular ozone, a very strong disinfectant, achieves disinfection by reacting with the contents of living cells, thereby destroying them. The most important factor in controlling disinfection reactions is the concentration of molecular ozone present in solution, although other factors affecting the disinfection efficiency of ozone are pH and temperature. The presence of dissolved and suspended inorganic and organic contaminants also impacts the ability of ozone to act as a disinfecting agent. These agents enhance the decomposition of ozone and promote the decomposition to hydroxyl radicals.

17.7.4.2 Formation of Disinfection By-Products. As stated previously, ozone has been proposed as a treatment process to reduce the formation of DBPs while maintaining adequate disinfection. In combination with chloramines as a secondary disinfectant (for maintaining a detectable residual in the distribution system), ozone can dramatically reduce the formation of trihalomethanes (THMs). Since THMs are the only DBP currently regulated, the ability to reduce THM formation has significant implications from a regulatory perspective.

Ozonation prior to chlorination, however, can change the characteristics of natural organic matter (NOM) towards either the reduction or formation of chlorination by-products such as THMs. The pH range at which the THM precursor enhancement occurs is dependent on the specific characteristics of the NOM and the pH buffering capacity of the water.

Ozonation of natural waters can lead to the formation of a number of other by-products. Among the by-products of ozonation for which occurrence data are most available are aldehydes, bromoform (a THM), and bromate. Other by-products of ozonation include H_2O_2, organic peroxides, carboxylic acids, ketones, phthalates, alkanes, and phenols (Glaze, 1989). At the dosages normally applied in drinking-water treatment, the concentration of the above mentioned DBPs are on the order of 2 to 5 µg/L. In the presence of bromide, ozonation of natural water can produce several µg/L of bromoform and bromate. This occurs through the oxidation of bromide to bromate or hypobromous acid. Hypobromous acid can subsequently react with NOM to produce bromoform. The formation of brominated by-products as a result of ozonation also can be controlled by the addition of ammonia prior to ozonation (Langlais, et al., 1991). Under these circumstances, ammonia combines with the hypobromous acid and prevents the reaction between the hypobromous acid and NOM.

In summary, ozone and chloramines as primary and secondary disinfectants, respectively, have the potential to reduce the formation of regulated DBPs, particularly THMs. For this reason, ozone is being designed and implemented at many utilities throughout the country. Other DBPs are formed, however, and the extent of this by-product formation and the epidemiology of these by-products are still the subjects of considerable research.

17.7.4.3 Enhanced Biodegradability. Ozonation tends to produce smaller fractions of organic materials from large macromolecules of NOM. As a result, this process of oxidizing NOM renders it more biodegradable after ozonation. In general, ozonation increases the biodegradability of NOM at dosages up to 1 mg/mg C. The degradable fraction typically represents 10 to 20 percent of the NOM on a carbon basis (as measured by TOC).

Although this is undesirable from the standpoint of potential microbiological regrowth in the distribution system, enhancing biodegradability can be a benefit if the treated water can be stabilized prior to distribution. In this manner, the overall product is a treated water lower in organic carbon that has a reduced food source for microorganisms in the distribution system.

One manner in which controlled biological growth can be implemented is to provide a granular medium following ozone addition. Microorganisms grow on the medium, and stabilize the assimilable organic fraction generated by ozonation. GAC also promotes biological activity as a result of its increased surface area compared to other granular media. Therefore, many applications use ozone ahead of GAC filter caps in rapid sand filters, or ahead of separate GAC contactors.

It is understood that nonpathogenic bacteria reproduce in the filter medium as substrate becomes available. These bacteria slough off the medium, and elevated *heterotrophic plate counts* (HPCs) are periodically found in filter effluents from biologically active filters. Research indicates, however, that HPC bacteria can be inactivated at contact times as short as 1 minute in the presence of free chlorine (Daniel, 1990).

17.7.4.4 Ozone as a Coagulant Aid. There is conflicting evidence in the literature about the ability of ozone to enhance coagulation demonstrated by improved particle removal. Several full-scale studies (Hodges, et al., 1979; Gerval, 1980; Jekel, 1983) reported improvement in particle removal as a result of ozonation. Laboratory scale studies, on the other hand, often failed to observe the coagulant aid behavior of ozone. In fact, some studies suggested that ozonation prior to coagulation may hinder particle removal (Reckhow and Singer, 1984). Although there is evidence in the field that under some conditions ozone facilitates the coagulation process, resulting in lower coagulant demand and lower cost for chemical coagulation, the mechanisms of this process are not well understood.

17.7.4.5 Reduction in Taste and Odor. Taste and odor originates from various sources, including microbial growth in the source water, industrial discharges, and chemical reactions within the treatment process. Metabolism of various organisms such as *Actinomycetes* and blue-green algae releases compounds (such as Geosmin and 2-methyl isoborneol, MIB) with objectionable taste and odor characteristics. Taste and odor can also be generated in the distribution system of potable water, mainly through biological activity. Metabolites of various microorganisms usually have very low threshold concentrations (in the nanogram per liter range).

Oxidation and biological treatment have been found to be effective on the removal of taste and odor compounds. Ozonation alone or ozonation assisted by UV light or H_2O_2 are reported as effective means for the reduction of taste- and odor-causing compounds (Glaze 1990, McGuire et al., 1989). Unsaturated compounds imparting taste and odor can be readily oxidized by ozone alone. Oxidation of saturated odor-causing compounds is enhanced by the formation of free hydroxyl radical promoted by the addition of H_2O_2 or UV light, particularly in the presence of low levels of radical scavengers such as HCO_3^- and CO_3^{-2}.

17.7.4.6 Color Removal.   Color in natural water is mostly derived from humic substances and is caused by dissolved organic matter which absorbs light at wavelengths of 400 to 800 nm. Most of the color-causing compounds are associated with aromatic organic structures. This group of compounds is easily oxidized with the application of ozone. Ozone dosages of 1 to 3 mg/mg C are adequate for most color removal applications (Langlais, et al., 1991). Ozonation followed by activated carbon filtration enhances color reduction, resulting in up to 95 percent removal (Langlais, et al., 1991).

17.7.4.7 Oxidation of Iron and Manganese.   Soluble forms of iron and manganese impart an unpleasant aesthetic quality to water. These compounds oxidize slowly in the presence of oxygen (atmospheric), causing a color change. The soluble forms of iron and manganese are readily oxidizable by ozone to insoluble precipitates. The kinetics of iron oxidation are much faster than those of manganese. For this reason, at low ozone dosages manganese oxidation does not take place until all the iron oxidation is complete. At high dosages of ozone (2.2. mg O_3/mg Mn) manganese is oxidized to permanganate which imparts a pink color to the water.

17.7.4.8 Oxidation of Specific Micropollutants.   Ozone is capable of oxidizing a large number of synthetic organic compounds. Oxidation normally occurs through the direct pathway, as described previously. Theoretical rate constants for direct oxidation of a number of micropollutants are summarized elsewhere (Langlais, et al., 1991). In advanced oxidation involving ozone and H_2O_2 or UV light, the free-radical formation is enhanced and the oxidation of some SOCs becomes more efficient under these conditions. For example, Glaze (1989) showed that the oxidation of TCE in groundwater can be enhanced by the use of H_2O_2 with ozone.

17.7.4.9 Points of Application in the Treatment Train.   The point of ozone application in the treatment train should be matched to the process objective. For example, if ozone is to be used as a coagulant aid, it should be added at the proper dosage to raw water prior to coagulation. On the other hand, disinfection is most efficient when ozone is added to the "cleanest" water possible; i.e., after some portion of particulate and organic matter has been removed. From a disinfection standpoint, it would be ideal to add ozone as the final unit process prior to secondary disinfection and distribution.

The impact of ozone on increased biodegradability of organic carbon, however, should be taken into consideration when selecting an application point. Ozone should not be added to treated water if there is not a granular medium barrier between the point of application and the distribution system to provide stabilization of the biodegradable fraction. A short period of free chlorination is required after the barrier.

17.8 RESIDUALS MANAGEMENT

Water-treatment processes essentially reverse some part of the geologic weathering process that erodes rocks and other geologic formations into the bodies of water. The suspended and dissolved contaminants found in natural-water sources are removed during water treatment to produce potable water. Suspended contaminants that are removed during the treatment process, as well as the dissolved contaminants which are converted to suspended form and removed, end up in the waste streams from the treatment facilities.

These waste streams or residuals consist primarily of the sludge from the sedimentation process and the backwash water from the filtration process. Most of the chemical agents added during the process of water treatment also end up in these waste streams. Heightened concerns in modern-day water treatment for the reduction of dissolved and suspended contaminants can often increase the amounts of residuals generated during the treatment process, e.g., the anticipated requirement for additional NOM removal through enhanced coagulation is expected to increase the sedimentation sludge as well as sludge in the filter-backwash water. With a potential increase in residuals production, the need for proper understanding and handling of residuals at water-treatment systems is an important consideration.

17.8.1 Treatment Plant Waste Streams

All water treatment plants produce one or more waste streams. The two major sources of these streams are the continuous or periodic removal of sludge from the sedimentation basins and the backwashing of filters. The waste streams within a treatment plant consist of various streams of varying solid concentrations originating at different locations within the plant.

Prior to the development of strict discharge limitations for water-treatment residuals, many treatment facilities practiced surface-water discharge of treatment-plant-water streams. However, this practice is changing as a result of classifying water-treatment plants as industries in 1972, thereby limiting treatment-plant discharges as industrial discharges. Surface-water discharges of treatment-plant residuals is therefore diminishing and alternate disposal of all or part of the water streams and the residuals is on an increase.

The most common method of residuals disposal is in a *landfill*. Most landfills require the waste to be disposed to meet certain classifications regarding the hazardous nature of the waste. Although water-treatment sludges are often classified as nonhazardous, the classification depends on the type of contaminants removed during water treatment and also on the chemical additives during the treatment process. For example, if a water-treatment system is optimized to remove high arsenic concentrations from source water through alum coagulation, the arsenic content of the sludge may be sufficient for a hazardous classification.

17.8.2 Residuals Reduction

The objective of a *residuals-dewatering facility* could be twofold: to produce a waste stream which has a more solid consistency, and to recover and recycle water associated with the residual. Filter-backwash water often contains much less solids than sludge produced in the sedimentation basins. For this reason, filter-backwash water is often collected in holding tanks so that the solids contained in the backwash water may settle out and the supernatant water be recycled back to the head of the treatment plant. This form of backwash-water recovery can take place with or without chemical addition. Backwash-water recovery processes, although recovering a large portion of the backwash water, pose a threat of increasing the concentration of particles and organisms such as *Giardia* and *Cryptosporidium*. The addition of coagulant chemicals to improve the efficiency of suspended-solids removal in the filter backwash recovery process can effectively reduce this threat.

Treatment of water streams from the sedimentation process ranges from water evaporation in drying beds to filter presses. The characteristics of the residual sludge often determine the dewaterability of the solids. Solids that are generated by treating highly turbid water are often easier to dewater, whereas solids rich in metal hydroxide (of aluminum or iron) and NOM are more difficult to dewater. Therefore, the additional water-treatment objective to remove more NOM from surface sources is expected to increase the difficulty of dewatering of water-treatment sludges.

17.8.3 Residuals Disposal

The difficult aspect of handling waste streams is the ultimate disposal of the residual solids. The most logical approach in engineering a system for handling the waste streams is to determine the option available for disposal of the residuals and then design a system which most efficiently prepares the residuals for such disposal. Disposal of residuals, after dewatering, in landfills will probably continue to be the predominant method. There is renewed interest in the development of beneficial uses for the residuals, including coagulant recovery and agricultural and horticultural uses. The engineering aspects for minimizing waste streams and dewatering techniques are described in numerous textbooks and publications, including those by the AWWA Research Foundation (1991); Westerhoff, et al. (1978; 1981); and in the proceedings of the AWWA/ASCE Joint Residuals Conference (1993).

17.9 FUTURE OF WATER TREATMENT

Author M. N. Baker (1949) in *The Quest for Pure Water* traces records of people's desire to improve water quality to as early as 2000 B.C. Since that time, advances in water treatment have continuously been made. Water treatment similar to that which we see today began in this country around 1900, with filtration followed by disinfection with chlorine. This history of modern water treatment is reflected in the experiences of Cincinnati, Ohio. Typhoid, the most dreaded of the waterborne diseases, was widespread in Cincinnati, averaging 42 to 56 deaths a year per 100,000 population during the 1880s, and more than tripling during the epidemic of 1887. This spurred efforts to obtain water treatment for the City, and two decades later, in 1907, the first rapid sand filtration plant was completed. The effects were apparent immediately: The death rate began decreasing in 1908, reaching a low of about 1 per 100,000 by 1940.

Today, as in the past, the basic objectives of water treatment are to produce an affordable water, safe for human consumption and appealing to the customer. The quest for pure water has been turned, more appropriately, into a quest for safe water. The elusive ingredient in the quest has been to define *safe*.

The definition of safe water has been, and will continue to be for some time, a blend of forces of science, technology, public interest, perception and expectations, state and federal regulatory efforts, media, and politics. The effects of these forces are resulting in profound changes in water treatment. Although the changes have already begun largely as a result of the 1986 Amendments to the Safe Drinking Water Act, they are likely to continue well into the beginning of the twenty-first century.

Involved in defining safe water are three predominant issues:

1. Minimizing the near-term health risk from infectious organisms

2. Minimizing the long-term health risks associated with compounds formed during treatment, e.g., disinfection by-products

3. Minimizing the long-term health risks associated with various and numerous chemical compounds introduced into the water supplies

During the extended period of time required to adequately define safe water, water treatment will undergo continual efforts to improve the already high quality of water being produced. A number of utilities have shifted their efforts from just meeting the continuously more stringent regulations being promulgated to developing and implementing longer-range plans designed to move themselves ahead of the regulations by producing a much higher quality of water than at present. This trend seems destined to gain momentum. As utilities move in this direction, they are likely to consider several major areas.

17.9.1 Process Considerations

Historically, coagulation/filtration plants were designed and operated to remove particles in the most cost-effective manner and provide basic disinfection to destroy pathogens. The most cost-effective manner usually meant using the least amount of coagulant which would adequately prepare the water for efficient filtration. Settling and filtration rates at some plants have been increased to increase plant capacities while conserving capital costs, with the primary concern being the ability to maintain acceptable filtration treatability. Except where color was a problem, there was little attention to the dissolved natural organics in the raw-water supply or attention to its removal.

The modern filter plant must be much more cognizant of the dissolved organics (natural and synthetic) in the raw water, and attention must be paid to design and operation of a facility which optimizes their removal. In addition, the basic disinfection practices of the past are, in most cases, no longer suitable to achieve the necessary inactivation of infectious organisms and to minimize production of undesirable disinfection residuals and/or disinfection by-products.

For most plants, future treatment processes will include:

- Chemical conversion of a significant amount of dissolved organics to particulates which can then be effectively removed by the subsequent coagulation/sedimentation/filtration processes.

- A disinfection strategy which assures maximum inactivation of infectious organisms and minimizes production of undesirable disinfection by-products.

- The use of advanced treatment processes such as membranes, ozonation, and/or GAC to effect a further removal of dissolved organics (natural or synthetic).

17.9.2 Distribution Systems

While most treatment facilities do not consider the distribution system (i.e., pipes, pumping systems, and storage facilities) to be part of the treatment system, this perception is about to change. With an emphasis on producing higher-quality water at the exit from the treatment plant, there is the potential that much of this effort will be impacted by the continued processing which occurs in the distribution system up to the point of use by the customer.

Future treatment facilities will consider at least the following potential changes in water quality within the distribution system:

- Potential changes resulting from the adsorption of chemicals from deposits within and materials or construction of the distribution system.
- Potential changes resulting from chemical reactions which continue in the distribution system.
- Potential changes resulting from biological activity in the distribution system.

17.9.3   System Reliability

Many water-treatment systems are being operated at the edge, with minimum redundancy. Typically this has occurred gradually as budgets were constrained and demands increased. For plants more than ten years old, it is prudent to perform a reliability audit to identify the sensitivity of each facility and essential pieces of equipment to assure continued satisfactory plant performance. New plants will need to include adequate provisions to assure system reliability.

17.9.4   Instrumentation and Control

New treatment systems are becoming more like industrial processing facilities with sophisticated, more sensitive operations, higher operating costs, increased need for reliability, and modern operator-friendly instrumentation and control systems. In addition to assisting in plant operations, instrumentation and control systems need to assist in the decision-making functions necessary to operate the plant. Thus they should be capable of detecting trends, providing historical inventory, and optimizing costs. The truly modern facility will even consider "smart" or "intelligent" systems to further assist in operations.

An early warning system for events impacting the raw-water supply as part of the treatment system is also becoming essential.

17.9.5   Management, Operations, and Maintenance

The future water-treatment system and its requirements will be much more complex and sensitive than those now in operation. Like an industrial processing facility, successful efficient operations will depend on its management, operations, and maintenance.

The key activity in the water-treatment plant of the future will be the decision-making processes. Many systems now have three distinct functional operating areas affecting treatment decisions:

1. Monitoring and data gathering
2. Laboratory control and analysis
3. Plant operations

It is surprising how many large systems have separated these functions into three distinct areas of responsibility, with inadequate communication between areas and unclear overall responsibility for input of all information in making daily operating decisions.

The future water-processing system of supply, treatment, and distribution will need an integrated plan for management, operation, and maintenance.

REFERENCES

Ali-Ani, M., J. M. McElroy, C. P. Hibler, and D. W. Hendricks, "Filtration of Giardia Cysts and Other Substances, Vol. 3, Rapid Rate Filtration," EPA/600/2-85-027, U.S. Environmental Protection Agency, 1985.

American Society Civil Engineers and American Water Works Association, *Water Treatment Plant Design,* 2d ed., McGraw-Hill, New York, 1990.

American Water Works Association, *Water Quality and Treatment—A Handbook of Community Water Supplies,* 4th ed., McGraw-Hill, New York, 1990.

American Water Works Association, AWWA/ASCE Joint Residuals Conference Proceedings, December 5–8, Phoenix, Ariz., 1993.

American Water Works Association Inorganics Contaminants Committee, "An AWWA Survey of Inorganic Contaminants in Water Supplies," *Journal American Water Works Association,* 77(5), p. 67, 1985.

American Water Works Association, "Precoat Filtration," *AWWA Manual M30,* 1988.

American Water Works Association Research Foundation, *Alum Sludge in the Aquatic Environment,* AWWARF, Denver Colo., 1991.

Amirtharaja, A., and C. R. O'Melia, "Coagulation Processes: Destabilization, Mixing and Flocculation," in American Water Works Association, ed., *Water Quality and Treatment—A Handbook of Community Water Supplies,* 4th ed., McGraw-Hill, New York, 1990.

Baker, M. N., *The Quest for Pure Water,* American Water Works Association, New York, 1949.

Benefield, L. D., and J. S. Morgan, "Chemical Precipitation," in American Water Works Association, ed., *Water Quality and Treatment—A Handbook of Community Water Supplies,* 4th ed., McGraw-Hill, New York, 1990.

Chowdhury, Z. K., S. Papadimas, M. Moran, and J. Chaffin, "Optimization of NOM and Arsenic Removal by Enhanced Coagulation," presented at the AWWA Annual Conference, New York, June, 1994.

Clark, J. W., W. Viessman, and M. J. Hammer, *Water Supply and Pollution Control,* 3d ed., IEP, New York, 1977.

Cleasby, J. L., "Filtration," in American Water Works Association, ed., *Water Quality and Treatment—A Handbook of Community Water Supplies,* 4th Edition, McGraw-Hill, New York, 1990.

Clifford, D. A., "Ion Exchange and Inorganic Adsorption," in American Water Works Association, ed., *Water Quality and Treatment—A Handbook of Community Water Supplies,* 4th ed., McGraw-Hill, New York, 1990.

Consumer Reports, "Is the Water Safe to Drink, Parts I–III," *Consumer Reports,* June, July, Aug., 1974, pp. 436, 538, 623.

Daniel, Phillipe, Environmental Engineer; Camp Dresser and McKee, Walnut Creek, Calif., personal communications, November 1990.

Edwards, G. A., and A. Amirtharajah, "Removing Color Caused by Humic Acids," *Journal American Water Works Association,* 77(3):50, 1985.

Flotation Processes in Water and Sludge Treatment, IAWQ-IWSA-AWWA Proceedings of Joint Specialized Conference, April 26–28, Orlando, Fla., 1994.

Glaze, W. H., et al., "Evaluating the Formation of Brominated DBPs During Ozonation," *Journal American Water Works Association,* 85(1), p. 96, 1993.

Glaze, W., "Ozonation By-Products, 2. Improvement of an Aqueous-Phase Derivitization Method for the Detection of Formaldehyde and Other Carboxyl Compounds Formed by the Ozonation of Drinking Water," *Environmental Science and Technology,* 23(7), pp. 838–847, 1989.

Gregory, R., and T. F. Zabel, "Sedimentation and Flotation," in American Water Works Association, ed., *Water Quality and Treatment—A Handbook of Community Water Supplies,* 4th ed., McGraw-Hill, New York, 1990.

Haas, C. N., "Disinfection," in American Water Works Association, ed., *Water Quality and Treatment—A Handbook of Community Water Supplies,* 4th ed., McGraw-Hill, New York, 1990.

Hoff, J. C., "Inactivation of Microbial Agents by Chemical Agents," EPA/600/2-86/067, U.S. Environmental Protection Agency, 1986.

Hudson, H. E., *Water Clarification Processes—Practical Design and Evaluation,* Van Nostrand Reinhold, New York, 1981.

James M. Montgomery, Consulting Engineers, Inc., *Water Treatment Principles and Design,* John Wiley, New York, 1985.

Jekel, M., "The Benefits of Ozone Treatment Prior to Flocculation Processes," *Ozone Science and Engineering,* 5, pp. 21–35, 1983.

Knoppert, P. L., "European Communities Drinking Water Standards: Corporation Implementation and Comments," *Proceedings AWWA Annual Conference,* Atlanta, Ga., June 1980.

Krasner, S. W., et al., "The Occurrence of Disinfection By-Products in U.S. Drinking Water," *Journal American Water Works Association,* 81(8), p. 41, August 1989.

Langlais, B., D. Reckhow, and D. Brink, eds., for AWWARF and Compagnie Generale des Eaux, *Ozone in Water Treatment Applications and Engineering,* Lewis Publishers, Chelsa Mich., 1991.

Morel, F. M. M., *Principles of Aquatic Chemistry,* Wiley-Interscience, New York, 1983.

Pontius, Frederick, *SDWA Advisor: Regulatory Update Service,* American Water Works Association, New York, 1994.

Rook, J. J., "Haloforms in Drinking Water," *Journal American Water Works Association,* 63(3), p. 168, March 1976.

Safe Drinking Water Act, 93rd Congress P.L. 93-523, December 6, 1974.

Safe Drinking Water Act Amendments, P.L. 99-339, June 19, 1986.

Sanks, R. L., *Water Treatment Plant Design for the Practicing Engineer,* Ann Arbor Science Publishers, Mich., 1982.

Sawyer, C. N., and P. L. McCarty, *Chemistry for Sanitary Engineers,* 2d ed., McGraw Hill, New York, 1967.

Stumm, W., and C. R. O'Melia, "Stoichiometry of Coagulation," *Journal American Water Works Association,* 60(5), p. 514, 1968.

Stumm, W., and J. J. Morgan, "Chemical Aspects of Coagulation," *Journal American Water Works Association,* 54(8), p. 971, 1962.

Symons, et al., "National Organics Reconaissance Survey for Halogenated Organics," *Journal American Water Works Association,* November 1975.

Symons, G. E. "That GAO Report," *Journal American Water Works Association,* 66(5), p. 275, 1974.

Tate, C., and K. Fox, "Health and Aesthetic Aspects of Water Quality," in American Water Works Association, ed., *Water Quality and Treatment—A Handbook of Community Water Supplies,* 4th ed., McGraw-Hill, New York, 1990.

U.S. Environmental Protection Agency, *Guidance Manual for Compliance with the Filtration and Disinfection Requirements for Public Water Systems Using Surface Water Sources,* Washington, D.C., October 1990

U.S. Public Health Service, "Drinking Water Standards," *Federal Register,* pp. 2152–2155, March 6, 1962.

U.S. Public Health Service, *Community Water Supply Study: Analysis of National Survey Findings,* NTIS PB214982, Springfield, Va., 1970.

Weber, W. J., *Physicochemical Processes for Water Quality Control,* John Wiley, New York, 1972.

Westerhoff, G. P., Z. K. Chowdhury, and D. M. Owen, "The Roles of Ozone in Drinking Water Treatment," *Proceedings AWWA Annual Conference,* June 23–27, Philadelphia, Pa., 1991.

Westerhoff, G. P. and D. A. Cornwell, "Lime Softening Sludge Treatment and Disposal," *Report of the Research Committee on Sludge Disposal,* presented at the American Water Works Association Conferences, St. Louis Mo., June, 1981.

Westerhoff, G. P. and D. A. Cornwell, "A New Approach to Alum Recovery," *Journal American Water Works Association,* 70(12), pp. 709–714, Dec., 1978.

Westerhoff, G. P., "Water Treatment Plant Sludges—An Update of The State-of-the-Art," *Journal American Water Works Association,* Part I, 70(9), pp. 498–503, Sept. 1978 and Part II, 70(10), pp. 548–554, Oct. 1978.

Westerhoff, G. P., "Treating Waste Streams: New Challenge To The Water Industry," *Civil Engineering,* Aug., 1978.

Westerhoff, G. P., "Minimization of Water Treatment Plant Sludges," presented at the American Water Works Association Conference, Atlantic City, N.J., June, 1978.

World Health Organization, *Guidelines for Drinking Water Quality,* vol. 1, Geneva, 1984.

CHAPTER 18
WATER DISTRIBUTION

Thomas M. Walski, Ph.D., P.E.
Wilkes University
Wilkes-Barre, Pennsylvania

18.1 OVERVIEW

18.1.1 Background

The term *water distribution* can refer to any system that transports water from a source to a user. In this chapter, it will refer to systems of closed pipes flowing full and the associated appurtenances that convey potable or near potable water to municipal, industrial, and fire-protection customers. Such systems vary greatly in size and complexity from systems with only a handful of pipes to those with millions of components and from pipes only a few inches in diameter to pipes that a human can easily walk through.

Water systems are constructed to consistently provide water in sufficient quantity to users at an acceptable pressure and quality as economically as possible. Because individual components of a system can fail, water-distribution systems are designed with a great deal of redundancy so that the system can perform adequately even when some individual components (e.g., pipes, valves, or pumps) are out of service.

Water distribution systems are constructed and operated by water utilities, which may be public or private. Systems may be owned by a public entity but operated by private firms under some contractual arrangement.

In most instances, design and operation is regulated by federal and state governmental agencies (e.g., Health Department, Environmental Resources Department, Environmental Protection Agency). In addition, fire insurance rating organizations evaluate distribution systems with regard to fire-suppression capability.

While water-source development and water treatment may be more visible components of an overall water-supply system, the water-distribution system often represents the largest capital investment for a utility. In an EPA study (Clark, Gillean, and Adams, 1977), water distribution was on the average the largest cost category for the 12 utilities surveyed. On average, distribution systems generally account for more than 80 percent of the capital associated with water-utility investment.

18.1.2 Water-Distribution Components

The major portion of water-distribution systems is buried underground and is not fully understood or appreciated by customers who simply know that when they open the tap water comes out. However, there are numerous components that must work together to deliver water.

Piping represents the largest single component. Treated-water-storage reservoirs are the most visible components, especially elevated tanks. Pumping stations are the largest energy users in most water systems. Valves serve a variety of functions, from isolating system components, to regulating pressures and flows, controlling surges, controlling the direction of flow, releasing air, breaking vacuums, or relieving excess pressure.

The upstream end of the distribution system is the clearwell of the water-treatment plant, where treated water is stored prior to delivery, or the well pump for a groundwater system. On the downstream end, the dividing line between the distribution system and the customer's plumbing varies with the utility but may be at the corporation tap at the main, the customer's property line, the customer's curb stop, or the customer's meter. Service lines will be only briefly mentioned in this chapter.

18.2 HYDRAULICS

18.2.1 Units of Measure

While the general principles of hydraulics are covered in Chap. 2, there are some features of basic hydraulics that need to be reviewed in the context of water-distribution systems. The two most important quantities in water-distribution systems are *flow* and *pressure.*

Flow is the volume of water moving through a pipe in a unit of time and for water distribution systems is usually given in units of gallons per minute (gpm) or liters per second (L/s). These units are not the "correct" English or metric units of cubic feet per second or cubic meters per day, but are convenient quantities that water-utility professionals have become accustomed to using.

Flow can be related to the average velocity of water in a pipe according to the relationship

$$Q = A V \qquad (18.1)$$

where Q = flow
A = cross section
V = average velocity

In most cases, the diameter in inches or millimeters is known rather than the area so that Eq. (18.1) can be more conveniently written as:

$$Q = k V D^2 \qquad (18.2)$$

where D = diameter
k = constant depending on units

When the diameter is in inches, the velocity in feet per second, and the flow is in gallons per minute, $k = 2.44$. When the diameter is in millimeters, the velocity in meters per second, and the flow is given in liters per second, $k = 0.0785$.

The second important quantity is that of *pressure,* which is the force exerted by the water per unit area over which the force is exerted. While the correct English and metric units for pressure are pounds per square foot and Pascal (Newton per square meter), the water industry uses units of pounds per square inch (psi) and kilopascal (kPa). Often individuals will incorrectly use the term "pounds of pressure" when they really mean psi.

In water distribution, pressure is expressed as gauge pressure, i.e., pressure in excess of atmospheric pressure, as opposed to absolute pressure, which is pressure above a perfect vacuum. A vacuum would correspond to a gauge pressure of −14.7 psi or −101 kPa.

Pressure can also be related to the height of a water column that would exert a given pressure. Pressure can then be referred to in units such as feet or meters of water, even though feet and meters are not true pressure units. As one moves down a water column, the pressure increases as shown in Fig. 18.1. The conversion factors relating water column height to pressure are:

$$h = 2.31\ P \tag{18.3}$$

where h is in feet and P is in psi, or

$$h = 0.102\ P$$

where h is in meters and P is in kPa.

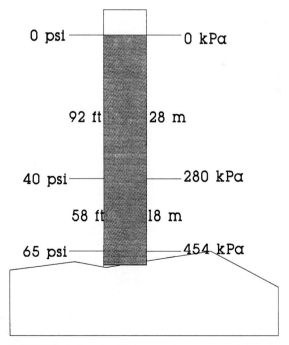

FIGURE 18.1 Relationship between pressure and head.

18.2.2 Continuity Principle

The two basic principles that are used to solve most water-distribution hydraulic problems are *continuity* and *energy*. The *continuity principle* is based on the law of conservation of mass. The mass of water moving into any control volume is equal to the mass flowing out plus any increase in storage of mass in that volume. For problems other than water hammer, water can be considered an incompressible fluid and the volume can be used instead of mass in writing the continuity equation as:

$$\frac{dS}{dt} = \sum Q_i \tag{18.4}$$

where S = volume of water in storage
t = time
Q_i = flow into control volume through ith pipe

In solving water-distribution-network problems, the continuity equation is usually written for each junction (i.e., intersections of two or more pipes) in the system. For junctions which do not have tanks associated with them, the left side of Eq. (18.4) can be set to zero because there is no storage at such points.

18.2.3 Energy Principle

While the continuity equation tracks the mass of water, the *energy equation* keeps track of the amount of energy in the water. There are three forms of energy that are important in water-distribution systems: kinetic, pressure, and potential (elevation). In water-distribution problems, energy is usually converted into units of energy per unit weight of the water which results in units of length (vertical distance). The energy at a point in a pipe can be given by:

$$\text{Energy (length units)} = z + \frac{P}{\gamma} + \frac{V^2}{2g} \tag{18.5}$$

where z = elevation corresponding to pressure p
g = acceleration due to gravity
γ = specific weight of water

The three terms on the right are the elevation (*potential energy*), pressure head (*pressure energy*), and velocity (*kinetic energy*) head, respectively.

The energy given by Eq. (18.5) can be related to a point in space above the pipe. The elevation of this point is referred to as the *energy head* and the set of these points along a pipe is called the *energy grade line*.

In most water-distribution-system problems, the kinetic energy term $V^2/2g$ is unimportant in comparison to the other terms because variations in kinetic energy are usually one or more orders of magnitude smaller than variations in other forms of energy in the distribution system. Equation (18.5) without the kinetic energy term gives the *static head* and the set of these points along a pipe is called the *hydraulic grade line*. The hydraulic grade line corresponds to the height water would rise in a column attached to the pipe.

The energy equation determines the direction of flow because water flows from areas of higher energy to lower energy, and the energy is used up in moving the water. This loss of energy is referred to as *head loss* and is the amount by which the hydraulic grade line drops between two points, as shown in Fig. 18.2. Energy can also be added to the flow by pumps or removed by turbines. The energy equation can then be written for flow from point 1 to point 2 along a pipe as:

$$z_1 + \frac{P_1}{\gamma} + \frac{V_1^2}{2g} + h_p = z_2 + \frac{P_2}{\gamma} + \frac{V_2^2}{2g} + h_L + h_t \qquad (18.6)$$

where h_p = head added by pump
       h_L = head loss
       h_t = head used by turbine

18.2.4 Head Loss

The head loss is a function of the length, diameter, velocity (or flow), and some parameter related to the roughness or carrying capacity of the pipe. The two most commonly used equations for calculating head loss are the *Darcy-Weisbach* and *Hazen-Williams equations*. The Darcy-Weisbach equation is usually written:

$$h_L = f \frac{L}{D} \frac{V^2}{2g} \qquad (18.7)$$

where f = dimensionless friction factor
       $L, D, h_L,$ and $V^2/2g$ are in the same units

While the Darcy-Weisbach equation is the most general head-loss equation, determination of the friction factor may involve several steps and practicing engi-

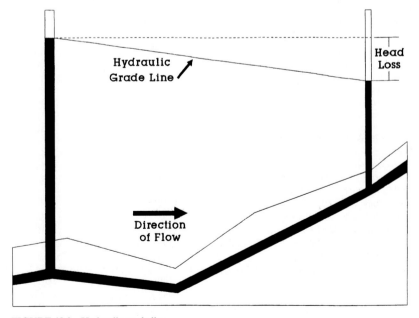

FIGURE 18.2 Hydraulic grade line.

neers have generally favored use of the Hazen-Williams equation which is appropriate only for turbulent flow of water. The Hazen-Williams equation can be written as:

$$h_L = \frac{10.46L}{D^{4.87}} \left(\frac{Q}{C} \right)^{1.85} \tag{18.8}$$

where D is in inches, L and h_L are in feet, and Q is in gpm.

C is the Hazen-Williams C-factor, which can range from a high of 160 for straight, smooth, large-diameter pipe to values as low as 20 for old, tuberculated, cast-iron pipe and depends mostly on pipe roughness, which is a function of pipe material, age, and water quality. Some typical values for C are given in Table 18.1. The effect of corrosive water on bare cast-iron pipe can vary greatly based on the corrosivity of the water. Lamont (1981) gave a set of C-factors depending on what he called the "severity of the attack." Only the minimum and maximum values he presented are given in Table 18.1.

18.2.5 Minor Losses

In addition to head losses that occur in straight runs of pipe, additional head losses occur in bends, fittings, meters, valves, and changes in diameter. These losses are referred to as *minor losses* and for most water-distribution-system problems are small in comparison to pipe losses. They often can be ignored or accounted for by minor adjustments in C-factors or pipe lengths. However, in some situations, such as the internal piping in pumping stations, minor losses can be significant.

TABLE 18.1 C-Factors for Some Pipe Materials

Type of pipe	6 in (150 mm)	48 in (1200 mm)
Uncoated, smooth, new cast iron	125	134
Coated, smooth, new cast iron	133	141
30-yr-old cast iron, mild water	106	120
30-yr-old cast iron, corrosive water	50	73
60-yr-old cast iron, mild water	102	112
60-yr-old cast iron, corrosive water	48	62
100-yr-old cast iron, mild water	95	104
100-yr-old cast iron, corrosive water	39	51
Newly scraped mains	121	127
Newly brushed mains	108	115
Coated, new, smooth spun cast iron	142	148
Smooth, new galvanized iron	133	
Smooth, new wrought iron	142	
Smooth, new coated steel	142	148
Smooth, new uncoated steel	145	150
Coated asbestos cement	149	
Uncoated asbestos cement	145	
Spun cement lined	149	153
Smooth PVC, copper, PE	149	153
Wavy PVC	145	150
New prestressed concrete		150

The formula for calculating head losses due to minor losses in distribution system piping are given as follows:

$$h_m = K \frac{V^2}{2g}$$

where h_m = minor loss
        K = minor loss coefficient

Some typical minor loss coefficients are given in Table 18.2. For more coefficients see Miller (1978).

TABLE 18.2 Minor Loss Coefficients

Source of minor loss	K
Gate valve open	0.39
Gate valve ¾ open	1.10
Gate valve ½ open	4.8
Gate valve ¼ open	27
Globe valve open	10
Angle valve open	4.3
Check valve conventional	4.0
Check valve clearway	1.5
Check valve ball	4.5
Butterfly valve	1.2
Corporation cock	0.5
Foot valve hinged	2.2
Foot valve poppet	12.5
90° elbow	0.9
45° elbow	0.45
Standard T flow through run	0.6
Standard T flow through branch	1.8
Return bend	1.5
90° radius/diameter = 2	0.3
90° bend $r/d = 8$	0.4
90° bend $r/d = 20$	0.5
Miter bend 90°	1.8
Miter bend 60°	0.75
Miter bend 30°	0.25
Expansion* $d/D = 0.75$	0.18
Expansion* $d/D = 0.50$	0.75
Expansion* $d/D = 0.25$	0.88
Contraction* $d/D = 0.75$	0.18
Contraction* $d/D = 0.50$	0.33
Contraction* $d/D = 0.25$	0.43
Entrance projecting	0.78
Entrance sharp	0.50
Entrance well rounded	0.04
Exit	1.0

* For expansion and contraction, velocities refer to the smaller pipe.

18.2.6 Pump Hydraulics

In almost all cases pumps in water-distribution systems are centrifugal pumps in which a motor turns an impeller which adds energy to the water as the water passes through the pump. For a given pump, with a given impeller, operating at a given speed, the increase in head depends on the flow through the pump. The relationship between the flow and the head added (also called *total dynamic head*) is referred to as the *pump head characteristic curve*. A typical curve is shown in Fig. 18.3.

The *power* (energy per unit time) added by the pump to the water can be given as:

$$\text{Water power} = k\gamma Q h_p \qquad (18.9)$$

where k is the appropriate unit conversion factor
h_p is total dynamic head of pump

The *efficiency* of a pump is the fraction of the motor power that is transferred to the water and is given by:

$$\text{Pump efficiency} = \frac{\text{water power}}{\text{motor power}} \times 100\% \qquad (18.10)$$

The pump efficiency depends on the flow through the pump and is given in pump manufacturer literature. However, in field conditions it is usually difficult to measure motor power (i.e., power delivered by the motor to the pump). Instead, all that can be measured is the power (usually electrical) used by the motor. Using the power input to the motor instead of power from the motor in determining efficiency gives the *wire-to-water* (or overall) *efficiency* which can be written as

$$\text{Wire-to-water efficiency} = \frac{\text{water power}}{\text{electric power}} \times 100\% \qquad (18.11)$$

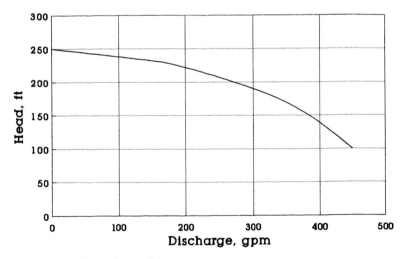

FIGURE 18.3 Pump characteristic curve.

Each pump has a flow rate at which it is most efficient, called the *best efficiency point*. A pump should be selected so that it will operate near its best efficiency point.

The actual point at which a pump will operate (i.e., operating point) is determined by combining the characteristics of the pump with the hydraulic characteristics of the system in which it is installed. For a given installation, the operating point will vary somewhat over time as water level in nearby tanks fluctuates, water consumption varies with time, valves are adjusted, and other pumps are turned on and off. The key to good pump selection is choosing a pump that will operate near its best efficiency point over the range of conditions it will experience over its life.

The characteristics of the piping system in which the pump is installed can be represented by the *system head curve* which is a function giving the head required to force water into the system versus the flow rate. An example of a system head curve is given in Fig. 18.4. The intersection of the system head curve and the pump head characteristic curve is the pump operating point. There is not a single system head curve for a distribution system but rather a band of system head curves depending on the water levels in tanks and water usage rates in the system at any point in time. (For more about system head curves see Walski and Ormsbee [1989] and Tullis [1993].)

18.2.7 Pipe Networks

Models view the water-distribution system as being made up of links which are components with different heads at each end (e.g., pipes, pumps, regulating valves) and nodes or junctions which have a single head (e.g., water users, changes in diameter, Ts and crosses, tanks). Most water-distribution-system hydraulic problems involve determining the flow in each link and the head at each node in the distribution system. This is done by setting up one energy equation for each link (or in some cases each loop) and one continuity equation for each node and solving the equations simultaneously.

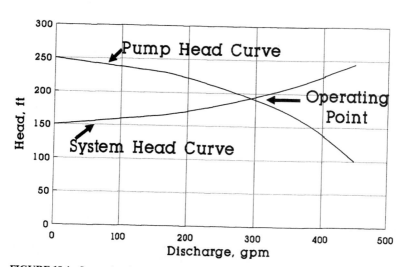

FIGURE 18.4 System head curve and operating point.

In systems that are branched with only one source, this can be done easily by applying the continuity equation to each node until all flows are known and then applying the energy equation to each link until the heads are known at each node. However, when a distribution system contains loops or more than one tank, it is not possible to separate the energy and continuity equations in this manner, and all the energy and continuity equations must be solved simultaneously.

The number of equations to be solved can be reduced significantly by noting that the head loss around a loop is zero. Therefore, instead of setting up one equation for each pipe, it is possible to set up one (somewhat longer) equation for each independent loop. The problem then becomes one of solving a set of nonlinear, simultaneous, algebraic equations. Numerous computer programs have been written to solve this problem and are available commercially. The techniques for solving these equations are presented in Bhave (1991) and Jeppson (1976). Walski (1984); AWWA Manual M32 (1989); Stephenson (1989); and Cesario (1995) describe how to use these models.

These computer programs for solving pipe-network problems have advanced from cumbersome mainframe programs with long output tables to microcomputer programs with graphical user interfaces. (Figure 18.5 shows a model representation of a distribution system while Fig. 18.6 shows how results can be presented.) Modeling of flows in pipe networks has advanced from being the domain of a few hydraulic engineers to being conducted by even small water utilities.

There are three types of hydraulic water-distribution-system models: (1) *steady-state models* which give the flows and pressures in a system at a given point in time, (2) *extended-period models* which trace the performance of a system over a period of time, and (3) *optimization models* which attempt to determine an optimal set of pipe sizes to provide adequate pressure at least cost or optimal set of pipe roughnesses or water-use distribution to match actual field data. Steady-state hydraulic models are most useful for pipe-sizing problems, pump selection, and evaluation of existing systems for hydraulic capacity. Extended-period models are most useful for tank location and sizing and determining energy consumption. Optimization models are specially written to solve particular problems such as pipe sizing and calibration.

Recently, water-quality models have emerged as an effective tool for tracking water quality through a distribution system under certain specific conditions (Rossman, et al., 1993; Rossman, 1994). Interest in this capability has grown with the passage of Safe Drinking Water Act (SDWA) Amendments requiring that drinking-water standards be met at the customer's tap and not only at the water-treatment plant. This act establishes *maximum contaminant levels* (MCL) or action levels for a number of critical water quality parameters associated with adverse health effects. Monitoring requirements as well as MCLs are promulgated under these regulations.

In addition to solving the hydraulic aspects of the water-distribution problem (i.e., determining flows and pressures) water-distribution-system models must also track the concentration of one or more water-quality parameters. This is done by incorporating mass balances for tracking fluid properties and kinetic models for determining changes in those properties over time into the model. Reactions affecting concentration of water-quality parameters occur both in the bulk fluid and the pipe wall. The U.S. Environmental Protection Agency model, EPANET (Rossman, 1994), is an explicit dynamic water-quality modeling algorithm based on a mass balance relation in pipes that considers both advective transport and reaction kinetics. A discrete volume element method (DVEM) is used for the tracking of dissolved substances (Rossman, et al., 1993). EPANET can be obtained on the Internet through World Wide Web browsers. One can also contact:

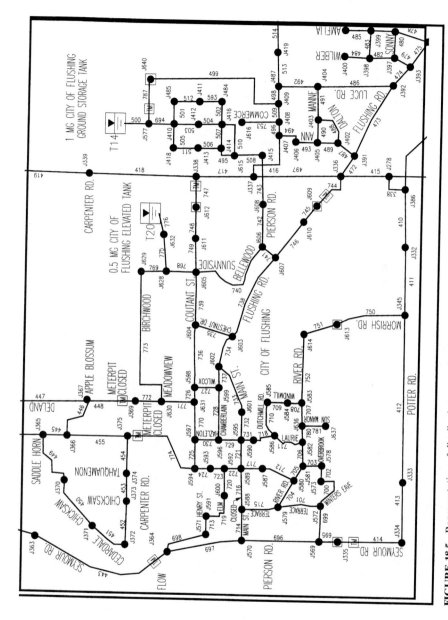

FIGURE 18.5 Representation of distribution system in pipe network model. (*Drawing courtesy of Haestad Methods, Inc.*)

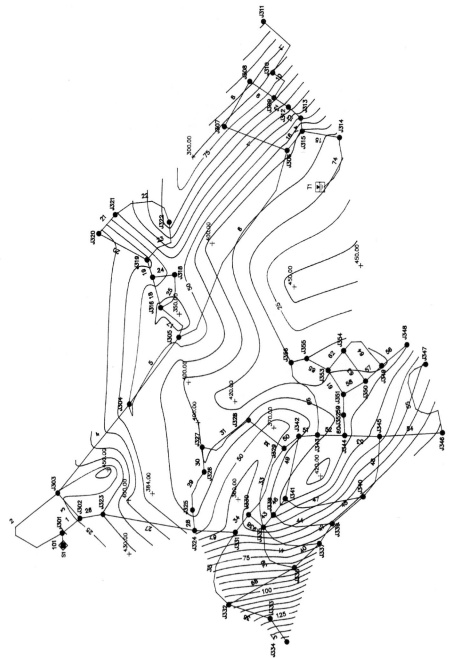

FIGURE 18.6 Model representation of pressure contours. (*Courtesy of Haestad Methods, Inc.*)

Risk Reduction Engineering Laboratory
Office of Research and Development
U.S. Environmental Protection Agency
Cincinnati, Ohio 45268

Water-quality models have a wide range of uses, including understanding the effect of system operation on water quality, tracking historical water-quality problems, calibrating hydraulic networks, sizing and locating rechlorination facilities, identifying monitoring locations, and developing flushing programs (AWWA Research Foundation, 1991). The EPANET model has received rather wide acceptance by municipalities. Water-quality modeling of distribution storage has also received attention with the development of new models (see Mau, et al., 1995, and Rossman, et al., 1995).

18.3 PIPING

18.3.1 Piping Materials

Pipes make up the largest capital investment of a water utility. Depending on the preference of the utility's engineers, water mains can be made of cast or ductile iron, steel, reinforced or prestressed concrete cylinders, asbestos cement, polyvinyl chloride, polyethylene, or fiberglass. Standards for pipe construction, installation, and performance are published by the American Water Works Association (continually updated) in its C-series standards.

Through the middle of the 1900s, cast iron was the primary type of pipe used in distribution systems. Ductile-iron pipe was developed in the years following World War II and gradually began replacing cast-iron pipe in water distribution systems since the 1960s. Today ductile iron (Fig. 18.7), which is stronger and less brittle than cast iron, has replaced cast iron for most water-distribution applications.

While it has been used in some places for small diameter mains, steel pipe has for the most part been used primarily for large diameter transmission mains. Because steel is very strong in tension, very thin pipes can be used and thus steel pipe is often found in above-ground installation. In small diameters the wall thickness of steel pipe may actually be controlled by external loading rather than internal pressure, which is limiting for most other kinds of pipe.

Most metal pipes used in water distribution are lined with cement mortar or other coatings to prevent internal corrosion. Water will tend to pull iron from unlined ferrous-metal pipes which can result in the formation of tubercles, which are lumps of oxidized iron, along the pipe wall (Fig. 18.8).

Metal pipes also need protection from corrosive environments outside of the pipe. Steel pipes may have some type of bonded coating or wrapped coating (Fig. 18.9) and/or cathodic protection while ductile-iron pipes are generally loosely wrapped (Fig. 18.10) in polyethylene in corrosive environments.

Reinforced or prestressed concrete cylinder pipe (Fig. 18.11) is made up of a thin steel cylinder wrapped with reinforcing or prestressing wire which provides the strength to resist internal pressures. This layer is coated with a thick layer of concrete which provides resistance to external loads. The pipe is lined with cement mortar to prevent corrosion of the cylinder by the water. Bends and fittings are made of lined and coated steel. Tapping or repairing concrete cylinder and other composite pipes in the field is more difficult than performing similar work on iron or plastic pipe.

FIGURE 18.7 Ductile iron pipe with polyethylene encasement. *(Photo courtesy of American Cast Iron Pipe Company.)*

Concrete cylinder pipe has been used primarily in large-diameter transmission mains. It can be economical in sizes 16 in (400 mm) and larger.

Asbestos cement pipe, made up of asbestos fibers, silica, and cement, has been used in smaller diameters, 6 to 20 in (150 to 500 mm). In recent years manufacture of asbestos cement pipe has decreased because of competition from plastic pipes and health concerns for those involved with pipe manufacture.

Polyvinyl chloride (PVC) is presently used extensively in diameters up to 36 in (900 mm). It requires no corrosion protection and is easy to handle. The price of PVC pipe correlates closely with the price of oil used in its production.

Polyethylene pipe (PE) has not been used extensively in water-distribution mains although it is widely used in water-service lines. Fiberglass or reinforced plastic mortar pipe can be economical for large-diameter transmission mains. The methods for design and construction of this pipe have changed since it was first introduced in the 1960s. Like concrete cylinder pipe, tapping or repairing the pipe in the field generally involves installing new sections of pipe with closure pieces.

FIGURE 18.8   Tuberculation in iron pipe. *(Photo courtesy of Knapp Polly Pig, Inc.)*

FIGURE 18.9   Wrapped steel pipe used for stream crossing.
(Photo by Walski.)

FIGURE 18.10 Polyethylene encasement on iron pipe. *(Photo courtesy of American Cast Iron Pipe Company.)*

FIGURE 18.11 Concrete cylinder pipe. *(Photo courtesy of Price Brothers Company.)*

Most pipes made of cast or ductile iron, PVC (AWWA C-900 standard), and asbestos cement are made with identical outer diameters so that they can be connected together in the field. This means that the internal diameter of these pipes will decrease as the wall thickness (and pressure rating) of the pipe increases. Ductile-iron pipe generally provides the greatest internal diameter for a given nominal diameter. For the pressures involved in water distribution actual internal diameters are usually greater than the nominal diameter of the pipe. For example, a 6 in (150 mm) diameter Class 50 ductile-iron pipe has an outside diameter of 6.90 in (175 mm) and an internal diameter of 6.40 in (162 mm).

For steel, fiberglass, and concrete cylinder pipe, the nominal and internal diameter are generally the same, and the external diameter varies with the pressure rating. As pipes get larger differences between internal and nominal diameters become less significant. In some instances, however, the nominal diameter may be the outer diameter of the pipe.

Most pipes described above are available in sections with a female bell end and a male spigot end. An O-ring rubber gasket provides the seal between the two pipe sections. Some metal pipes are also available with mechanical seals which force and hold the O-ring in place. Older metal pipe may be connected with lead or leadite joints. Steel pipe can also be butt welded or lap welded. Polyethylene pipe is usually butt welded. Plain ends can be connected with a variety of couplings (Fig. 18.12).

O-ring gasket joints are good for most buried applications but are not appropriate by themselves near bends and fittings where forces will tend to pull sections of

FIGURE 18.12 Pipe fittings. *(Photo by Walski.)*

pipe apart. Thrust blocks or restrained joints are used to keep these pipes from pulling apart. In above-ground installations, flanged joints are used to connect pipe. In buried applications, numerous pipe manufacturers have developed proprietary methods for restraining pipes.

18.3.2 Distribution-System Layout

Water-distribution systems are intended to deliver water at all times. However, because individual components can fail, a great deal of redundancy is provided in most systems to maintain service to customers even if individual components fail.

To accomplish this, water-distribution systems are generally laid out so that every customer can be fed from at least two directions and that isolating a segment of the system for repair will only place that individual segment of the system out of water. A *segment* is defined as the smallest portion of a distribution system that can be isolated by closing valves.

The reliability of a water-distribution system can be increased by installing more loops, installing more isolation valves, using larger-diameter pipe, using thicker-walled pipe, properly maintaining isolation valves, providing standby pumps and standby power for pumping operation, and keeping a large volume of water in elevated storage, to name a few. Walski (1993) describes methods for increasing water-distribution-system reliability. Mays (1989) presents methods for quantitatively accounting for reliability in water-distribution-system design.

In the pre–World War II era, most streets were laid out in rectangular grids that made it easy to provide looped systems. Today most land developments are laid out with numerous dead ends and cul-de-sacs, which makes it difficult to provide looping needed for reliability. The engineer is faced with decisions of whether to provide larger pipes so that hydrants can be placed on dead-end lines or smaller pipes sized only for domestic customers.

18.3.3 Pipe Sizing

Selecting pipe diameter in a water distribution system involves a great deal of judgement because of the complexity of water-distribution systems and the difficulty in predicting future water-consumption patterns. Ideally if one knows the head available at the source, the head required by the water user, the length of pipe between the two, and the flow to be delivered, one could use one of the head-loss equations to determine the optimal diameter (or combination of diameters because the optimal diameter most likely will not be a commercially available diameter).

The problem is much more complicated, however, because:

1. Water use varies daily, seasonally and over the expected life of the pipe.
2. Projecting future water-use patterns involves a great deal of uncertainty.
3. The sizing of an individual pipe cannot be done without considering the impacts of other pipes in the distribution system.
4. Pumps can be installed to raise the head at points in the system.
5. Tanks can be operated to change water-flow patterns during peak versus non-peak use times.
6. Water systems are usually not built in one construction project but evolve over long periods of time.

7. Water systems must be able to work when some individual components are out of service.

In spite of all these problems, engineers have been able to design workable systems for many years using essentially trial-and-error solutions in which they would select candidate pipe sizes, test the performance of the system using several model runs, and change the pipe sizes to eliminate bottlenecks and excess capacity. The advent of easy-to-use computer programs have made this job easier because now engineers can quickly examine numerous alternatives under a variety of different scenarios.

Pipe sizing is an iterative process generally involving:

1. Projecting demands

2. Developing a possible solution

3. Determining the critical condition (e.g., fire flow, peak hour in future years) for the pipe

4. Simulating the distribution performance for this worst case time (usually involves checking to determine if pressures are adequate and velocities are reasonable when meeting demand)

5. For the alternatives that are workable, testing their performance during other conditions over the design life; for unworkable alternatives, determining why they didn't work and formulating improved alternatives

6. Developing cost estimates for workable solutions

7. Selecting the best solution

Computer models have made it much easier to perform these steps. However, when using a computer model, it is essential that the model be calibrated to be certain that the model is an accurate representation of the existing system. Calibration involves comparing the model results with data collected from the field and adjusting the model input data (or correcting bad field data) so that the discrepancies between the two are acceptable.

The approach to pipe sizing varies depending on whether the sizing is done for a utility master plan, sizing of an individual transmission main, adding a new major customer, layout of a subdivision, or rehabilitation of an existing system (Walski, Gessler, and Sjostrom, 1990; Walski, 1994). In general, the sizing of large mains with terminal storage depends on peak day demands; the sizing of large mains without terminal storage depends on instantaneous (or peak hour) demands; while the sizing of neighborhood distribution mains depends primarily on fire flow requirements.

As a general rule of thumb, as velocities approach 10 ft/s (3 m/s), it becomes economical to move up to a larger pipe size. If the velocity at peak-flow periods is less than 5 ft/s (1.5 m/s), it may be possible to reduce the size of the pipe to realize cost savings without adversely affecting performance. This does not mean that one can simply look at the results of a computer model run at peak hourly demands and downsize pipes with low velocities. A given pipe may have as its purpose filling a reservoir during off-peak hours or providing an alternative feed when a given pumping station is out of service. It is important to consider the function of a pipe in the overall system.

A computer model using optimization can help steer the user toward good solutions but cannot be relied upon as the final word in selecting pipe sizes (Walski, et al., 1987). Optimization models are still not commonly used for pipe sizing in most design work but improvements to such models may make them practical in the future (Walski, 1885; Goulter, 1992; Walters and Cembrowicz, 1993).

18.3.4 Pressure Zones

In flat terrain, the overflow levels (i.e., maximum water level) of all distribution-system storage tanks can be set to the same elevation (i.e., a single pressure zone) and it should be possible to provide adequate pressure to all customers if the pipes are adequately sized. However, in hilly areas, a single water tank could not serve all customers without providing low pressure to the highest customers and excessive pressure to the lowest customers. Multiple tanks at different elevations can be used, but only if they can be isolated from one another in separate zones connected by pumps or pressure-regulating valves. These zones, which will have tanks with different overflow (i.e., different hydraulic grade lines), are referred to as pressure zones (or pressure planes).

Hydraulic grade line in a pressure zone is generally controlled to fall in a specific limited range to insure that the pressures are not too high or low for customers. Great care is needed in establishing tank overflow elevation (and, hence, hydraulic grade line elevation) in a pressure zone. All subsequent pipe sizing decisions are dependent on the hydraulic grade line in the zone. One needs to determine the range of ground elevations that can be served by a certain hydraulic grade line.

In general, the *nominal head* of a pressure zone (usually tank overflow elevation) should differ by 50 to 120 ft (15 to 40 m) between successive pressure zones. Too small of a step results in excessive tanks and inefficient piping while too large of a step results in an excessive range in pressures provided to customers.

Water is raised from a lower-pressure zone to a higher one by pumping and is passed from a high-pressure zone to a lower-pressure zone through pressure-reducing valves. In general a utility will waste energy if it pumps water to a higher zone only to have it flow to a lower zone through a pressure-reducing valve. The key is to minimize the amount of water pumped to each zone.

18.3.5 Demand Projections

One of the problems in laying out water-distribution systems lies in anticipating where and when the demands will actually occur. The vagaries of the local economy and associated land development and the ability of the utility to afford capital investment all complicate pipe-sizing decisions. Water conservation through demand management or reduction in loss also can affect pipe sizing.

For large existing systems future demands are generally based on historical records of per capita consumption corrected for projected trends in water conservation or increased water use. Forecasts of development in the utility service area or changes in the service area are usually made independently of water-distribution planning by another agency. The challenge for the water-system planner is to take these population and land-use projections and assign the corresponding water use to individual nodes in the water-distribution model for hydraulic analysis.

In addition to projecting domestic demands, the utility must project fire flow requirements. Usually for system planning, flows on the order of 1000 gpm (60 L/s) are used for residential areas while larger flows are used for commercial and industrial areas. Usually the utility makes a policy decision on the upper limit of fire protection it will provide and those wanting higher flows would need to provide their own system or reduce flow requirements by installing sprinkler systems, fire walls, or fire-retardant materials. Fire flows are addressed in AWWA Manual M31 (1989).

18.3.6 Staged Development

The water utility must balance between providing sufficient capacity and maintaining reasonable water rates. Part of the problem is that there are large economies of scale in water-distribution construction. Marginal extra units of capacity are relatively inexpensive. For example, it is much less expensive to install one 48 in (1200 mm) pipe rather than four 24 in (600 mm) pipes to carry roughly the same amount of water. However, the cost of a 48 in (1200 mm) pipe may be excessive in the year the pipe is first required and long detention time in the pipe may result in water-quality problems.

18.4 PUMPING STATIONS

While some water utilities are fortunate enough to have most of their water flow by gravity from mountain reservoirs, most utilities must pump water to their customers. Pumps are the most significant energy consumer in most utilities and as such account for a major portion of the operation cost of these utilities. Pump stations are the most complicated distribution facilities in any utility.

18.4.1 Types of Pumps

There are numerous ways of classifying pumps, depending on mechanical configuration, material pumped, orientation, bearings, and seals, to name a few. Some of the more commonly used classifications of water-distribution pumps are discussed in the following.

The most general classification of pumps is whether they are positive-displacement or kinetic pumps. *Positive-displacement pumps* include such pumps as piston, plunger, screw, diaphragm, progressive cavity, lobe, gear, and screw pumps. Most water-distribution pumps are *kinetic pumps* which are made up primarily of centrifugal and turbine pumps. Centrifugal pumps are most commonly used in water-distribution applications because of their low cost, simplicity, and reliability in the range of flows and heads encountered in water-distribution systems. Hydraulic Institute Standards (1983) contain a description of the nomenclature of pumps.

Among the kinetic-principle pumps, *turbine pumps* are used primarily when pumping from wells. Most other water-distribution pumps are *centrifugal pumps.* The impeller of the pump rotates inside of the pump casing and imparts energy to the fluid. Pumps are classified by whether the water leaves the impeller in the radial or axial direction with respect to the axis around which the impeller rotates. In the range of heads in most water-distribution systems, radial-flow pumps are most appropriate. Axial-flow pumps find use in high-flow, low-head applications.

The next classification is the orientation of the pump axis. In *vertical pumps,* the motor is mounted above the pump while in *horizontal pumps* (see Fig. 18.13), the motor is placed alongside the pump. Horizontal pumps are easier to reach for maintenance but use up more floor space. Vertical pumps see most use where floor space is at a premium, such as below-grade pumping stations.

Pumps are usually separately coupled, which means that the pump and motor shafts are two separate pieces of equipment which must be aligned and connected. In some instances, especially in smaller pumps, the pump and motor may be close-coupled, which means that the impeller is mounted at the end of the motor shaft.

FIGURE 18.13   Horizontal pump. *(Photo courtesy of Fairbanks Morse, Inc.)*

The point at which the pump shaft enters the pump casing is a potential location where water can leak out of the pump due to the high pressure in the pump. A seal is maintained using either packing, which is a braided fiber impregnated with lubricant, or a mechanical seal, which relies on a spring mechanism to hold the seal. Packing is less expensive, but allows for some leakage of water which is acceptable in most water-distribution applications.

Bearings are needed in a pump to carry the forces imposed on the shaft and impeller as it spins in the casing. The two types of arrangements used are overhung-impeller bearings, in which the two sets of bearings are located between the motor and pump (usually found on end suction pumps), and the impeller-between bearings, in which the impeller turns between the two bearings (usually found on horizontal pumps).

18.4.2   Types of Motors

In addition to the many types of pumps, there are also many types of motors which can be used to drive pumps. *Electric motors* are by far the most common type of prime mover for pumps, although some utilities will have one or more gas- or diesel-driven pumps for emergency standby service.

In water-distribution applications, most motors are constant speed because the pumps need not instantaneously match demands due to the existence of system storage. *Variable-speed motors* require significantly more expensive drives, are more expensive to maintain, and have somewhat lower overall efficiencies than constant-speed motors.

The voltage of the motor should be selected at the upper end of the voltages that can be used for a given motor. 480 V is the most common voltage for water-distribution pumping, although 4160 V and even 13.8 kV are desirable for very large pumps. Use of low-voltage pumps requires extra transformer capacity with associated transformer energy losses and larger-capacity cables for the extra amperage required. Three-phase power should be used for all but the smallest pumps.

18.4.3 Motor Speeds

Most pumps will operate at virtually any reasonable speed. A given pump curve corresponds to a single motor speed listed on the pump curve. As the speed changes the pump performance changes according to the pump affinity laws described in the following.

For a pump which produces flow Q_1 and head h_1 and uses horsepower HP_1 at the motor reference speed N_1, the flow, head, and horsepower at a different speed N_2 are given, respectively, by:

$$Q_2 = Q_1 \left(\frac{N_2}{N_1} \right)$$

$$h_2 = h_1 \left(\frac{N_2}{N_1} \right)^2$$

$$HP_2 = HP_1 \left(\frac{N_2}{N_1} \right)^3$$

18.4.4 Pump-Station Design

Pump stations are used to house the pumps and motors and auxiliary equipment which includes valves, meters, transformers, controls, monitoring equipment, surge control equipment, and any other ancillary operations the utility wishes to locate at the site.

The two principle types of stations are *in-line booster stations* and stations which pump from a tank. In-line booster stations should generally be avoided because if there is excessive head loss in the suction piping cavitation damage can occur in the pump or negative pressures can exist in the suction piping. (*Cavitation* is the creation and collapse of water-vapor bubbles due to very low pressure on the suction side of the pump impeller.) In general, pumps—especially in-line booster pumps—should be located at as low a level as economically feasible to minimize the chance that cavitation can occur.

Water-distribution pumping stations should contain at least two pumps and should have sufficient capacity such that the full design flow can be delivered when one pump is out of service. A larger number of pumps provides both extra reliability and operational flexibility, albeit at extra cost.

Valves should be included such that operators can isolate individual pumps for maintenance. There is usually a suction header pipe and a discharge header pipe. Between these two large pipes one would place, in order, an isolating valve on the suction side, the pump, a check valve, and an isolating valve on the discharge side (Fig. 18.14). If individual pumps are metered, the meter must be placed in this line. Taps for measuring pressure should be included on both the suction and discharge lines. If there is any chance of air getting into the pump, the pump should be equipped with an air-release valve.

In designing the pump station, it is essential to provide means for removing pumps. For all but the smallest stations, an overhead crane should be installed with sufficient clearance between the crane and the other pumps so that a pump can be moved to the access door.

Sump pumps are needed to protect stations from flooding both from external sources and from internal pipe breaks or leaks.

Ideally, the pump station will be located on a site with extra room for future expansion. Header pipes should be laid out so that one can simply add extra units should the need arise. If the expansion is fairly imminent, an extra pump bay should be included in the building while if additional pumps are just a long term possibility, there should be room on the property to expand the building.

The electrical-distribution grid should be investigated during the design of a pump station. If possible, the pump station should have two alternative sources from significantly different portions of the electrical grid. If not possible, a standby generator or internal combustion engine pumps should be provided.

For additional information on pump stations, the reader is referred to Hicks and Edwards (1971); Karassik, et al. (1976); or Sanks (1989).

18.5 WATER STORAGE

18.5.1 Purpose

Distribution-system storage is needed to equalize pump discharge near an efficient operating point in spite of varying demands, to provide supply during outages of individual components, to provide water for fire fighting, and to dampen out hydraulic transients.

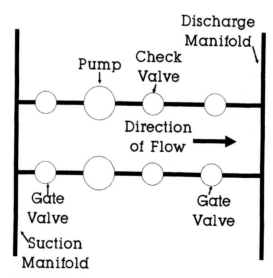

FIGURE 18.14 Pump station layout.

Water use varies with time but most water-treatment plants and their associated pump stations put out a fairly constant discharge. Storage tanks can equalize this difference between pump discharge and water consumption, thus preventing pumps from continually moving up and down their head characteristic curve as they would with no downstream storage. In addition, storage enables water-treatment-plant operators to operate their plants at a constant rate and reduce the changes in chemical feed rates.

Outages can occur in water-distribution systems due to power loss to pumping stations and pipe breaks. Water kept in storage can supply customers until the system is made operational again.

Fires can place short-term demands on a distribution system in excess of the capacity of the pumping equipment. Without storage, a significant amount of excess pumping capacity would be required, especially at smaller pumping stations. Storage tanks decrease the peak velocities and head losses in pipes carrying water from the source to the perimeter of the distribution system, thus resulting in reduced pipe sizes and costs for piping.

When velocities in pipes are changed quickly due to pump starts and stops, valve operation, or pipe breaks, hydraulic transient pressure fluctuation (*water hammer*) is created that can damage water-distribution systems. Storage tanks can dampen out the effects of these transients.

18.5.2 Sizing of Storage

Storage-tank sizing is generally controlled by regulations which may require on the order of 100 gal of storage per capita served. Fire-insurance-rating systems may also specify minimum storage requirements. The underlying requirements for storage are that the storage must be capable of dampening out fluctuations in daily water use, must provide water for fire fighting, and must have storage to meet emergency demands. These three types of storage are referred to as equalization, fire, and emergency storage.

Most utilities calculate the storage required for each type of storage and add them up to determine tank size. They then compare this value with values specified by regulatory or fire-rating standards. A significant amount of judgment is required, especially in determining the amount of emergency storage desired.

Distribution-system storage can have an adverse effect on water quality. As water stays in a tank, disinfectant residual drops and microbial regrowth can occur. The larger the tank, the longer the residence time and hence the more likely it is that disinfection residual of water in the tank will be low. Distribution-system operating strategies which promote significant water-level fluctuations over the course of a day can minimize this problem, but excessively large tanks will make even this difficult.

18.5.3 Types of Storage

Storage can be classified first based on whether the water level in the tank is the same as the hydraulic grade line of the system outside the tank. In this case water can be easily exchanged with the system, and the tank is said to *float on the system.* No energy is required to move water from the tank. Other tanks have water levels that are below the hydraulic grade line and water must be pumped from them into the distribution system. Many tanks near the borders of pressure zones will float on the lower-pressure zone and serve as a suction tank for pump-station pumping to the next-higher-pressure zone. In general, operation is much easier if all tanks float on the system, although there may be instances (e.g., to prevent freezing of tanks in very cold climates; reducing structural costs in seismically active areas) where tanks cannot be placed at an elevation to float on the system.

Storage in tanks can be classified as effective or ineffective. *Effective storage* is water above that elevation where the highest customer can be served. For example, if the highest customer to be served from a tank is located at elevation 930 ft (283 m), and requires water at 20 psi (14 m), any water stored below elevation 976 ft (298 m) cannot be delivered at adequate pressure and, hence, that storage is referred to as *ineffective.* All storage that does not float on the system is also ineffective.

18.5.4   Types of Tanks

Tanks can be classified by their shape, location with respect to the ground, and material of construction. Tanks can be buried, ground level, or elevated. Ground-level tanks which are considerably taller than their diameter are referred to as *standpipes.* Buried tanks tend to have a rectangular footprint to take advantage of building-lot size. Ground-level tanks and standpipes tend to be cylindrical while elevated tanks are generally cylindrical, round, or toroidal.

Buried tanks are generally constructed of concrete while elevated tanks and standpipes are made of steel. Ground-level tanks can be either steel or concrete. Steel tanks usually have lower initial costs but require cathodic protection and occasional coating. Some typical elevated storage tanks are shown in Fig. 18.15.

18.6   APPURTENANCES

In addition to the major components, pipes, pumps, and tanks, there are numerous other appurtenances necessary to make a distribution system function properly.

18.6.1   Hydrants

Hydrants are the most ubiquitous, visible water-distribution-system component. Hydrants are used primarily to provide water for fire fighting. They are also used for system testing and water-main flushing.

The two main types of hydrants are wet barrel and dry barrel. *Wet-barrel hydrants* can only be used in climates where freezing does not occur. Wet-barrel hydrants stay full of water at all times. The operating mechanism for controlling hydrant flow is located in the outlet of the hydrant.

Dry-barrel hydrants have their operating mechanism in the hydrant boot (i.e., the bottom of the hydrant). In this way the barrel stays dry and will not freeze, provided drains located in the base of the barrel allow the hydrant to drain.

In the United States, hydrant outlet nozzles are either 4½ in (114 mm) or 2½ in (63.5 mm). Fire hydrants generally have a thread pattern called National American Standard, although some cities actually have their own thread patterns.

Hydrants are controlled by turning the operating nut, which is usually a five sided nut, although other designs of operating nuts may be used to reduce theft of water. Hydrant outlet caps also have the same type of nut.

The ability of a fire hydrant to deliver water to a fire is usually expressed in terms of the flow it can deliver at 20 psi (138 kPa). To check the capacity of a hydrant at 20

FIGURE 18.15 Elevated steel tanks. *(Photo courtesy of Chicago Bridge and Iron, Inc.)*

psi, a hydrant-flow test is conducted. To conduct a hydrant-flow test, read the pressure at the hydrant (called the *residual hydrant*) during normal operating conditions, open up a nearby hydrant and measure the flow, and read the pressure at the residual hydrant when the second hydrant is flowing. Use the following equation to determine the flow at 20 psi:

$$Q_{20} = Q_t \left(\frac{P_r - P_{20}}{P_r - P_t} \right)^{0.54}$$ (18.12)

where Q_{20} = discharge at 20 psi, gpm (L/s)
  Q_t = discharge during flow test, gpm (L/s)
  P_r = residual pressure with no flow, psi (kPa)
  P_t = pressure during flow test, psi (kPa)
  P_{20} = 20 psi (138 kPa)

The discharge from a hydrant can be measured using a Pitot gage (Fig. 18.16) with the tip inserted into the flowing stream. The discharge can be related to the pressure reading by

$$Q = 29.8 D^2 C_D \sqrt{p}$$ (18.13)

FIGURE 18.16 Pitot gauge. *(Photo by Walski.)*

where Q = discharge, gpm
        D = nozzle diameter, in
        C_D = discharge coefficient
        p = pressure, psi

The 29.8 will change for different units than those given here. The discharge coefficient for a typical 2½-in outlet under typical pressure is approximately 0.90.

Smaller hydrants with only a 2½-in outlet can be used strictly for line flushing or for fire fighting on rural systems with limited distribution-system capacity. Standard fire hydrants found in systems providing fire protection are connected to mains at least 6 in (150 mm) in diameter.

An isolating valve, usually a gate valve, is placed in the hydrant lateral between the main and the hydrant. By closing this valve a hydrant can be repaired without the need to shut down the entire main. This is especially important for wet-barrel hydrants, which can discharge at a very high rate when struck by a vehicle.

18.6.2 Isolation Valves

Isolation valves enable a utility to shut down a segment of the distribution for maintenance or repair. A *segment* is the smallest portion of the system that can be isolated using valves.

Isolation valves in water-distribution systems are usually gate valves (Fig. 18.17) although in larger diameters butterfly (Fig. 18.18) valves have been used. Plug and ball valves can also be used. Ball valves are especially good when throttling is required.

Most isolation valves are not intended to be used for throttling but are intended to be used in either the fully open or fully closed position. A valve that is mistakenly left in the throttled position can give the utility the impression that the valve is open because the head loss is small during normal operation. However, when fire flows are required, the throttled valve can severely restrict flows.

Isolation valves are usually placed such that the operation of three or four valves can isolate a distribution segment. This usually corresponds to three valves at every cross-type intersection and two valves on every T. If the pipe diameters are different at a cross or T, it is economical to place the valves on the smaller-diameter pipes.

Because they are seldom used, isolation valves are often neglected. They can become frozen in place, have their valve boxes filled with debris or paved over, or simply be forgotten. It is essential that these valves be routinely exercised so that the mechanism will not freeze up, the valve boxes can be cleaned and raised during paving, and the utility personnel can become familiar with the valve locations.

18.6.3 Control Valves

Control valves (as shown in Fig. 18.19) are intended to regulate the flow or pressure in the distribution system. Most control valves have a globe-type body with the regulating mechanism placed on top of the valve.

FIGURE 18.17   Gate valves. *(Photo by Walski.)*

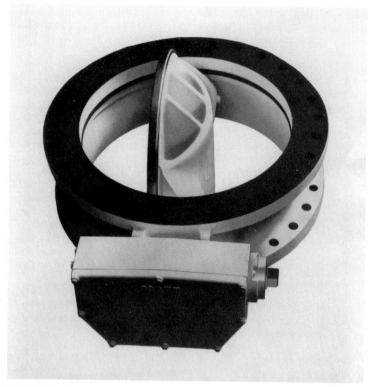

FIGURE 18.18   Butterfly valve. *(Photo courtesy of Henry Pratt Company.)*

The most commonly used control valve is the *pressure-reducing (regulating) valve* (PRV). PRVs are often placed at pressure-zone boundaries to reduce pressure to customers at lower elevations. Most regulating valves have a spring-operated pilot valve which in turn controls the main-valve mechanism.

PRVs are controlled by the downstream pressure. A pressure setting is established by operating a nut or screw in the valve. When the downstream pressure drops below that setting the valve opens to increase the pressure. As soon as the pressure rises, the valve throttles itself. If the downstream pressure does not drop below the pressure setting, the valve will not open. If the upstream pressure is below the pressure setting, the valve will remain wide open in an attempt to raise the downstream pressure.

A *pressure-sustaining valve* (PSV) (or back-pressure valve) operates similar to a pressure-reducing valve except that it monitors pressure at the upstream side of the valve. A PSV attempts to maintain a minimum pressure setting on the upstream side of the valve. It is used at pressure-zone boundaries or at large customers where the lower-pressure zone or the large customer has the potential to adversely affect the pressure in the main-pressure zone.

The pressure-reducing and -sustaining features can be combined into a single valve (a *pressure reducing–sustaining valve*) that maintains a set pressure downstream while sustaining pressure above a minimum pressure upstream.

DISCHARGE PRESSURE CONTROL PILOT
TURN HANDWHEEL CLOCKWISE TO RAISE
DISCHARGE PRESSURE

NEEDLE VALVE CONTROLS
CLOSING SPEED

STRAINER

STOP VALVE
NORMALLY OPEN

STOP VALVE NORMALLY OPEN
(MAY BE USED TO CONTROL SPEED)

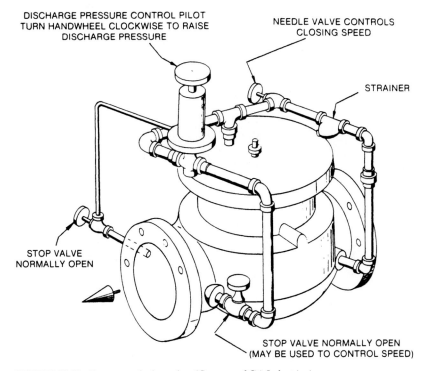

FIGURE 18.19   Pressure reducing valve. *(Courtesy of GA Industries.)*

A *flow-control valve* senses the flow rate and throttles the valve to maintain the flow at a preset rate. Flow-control valves are not used extensively in water-distribution systems.

The described valves can use the pressure of water to regulate the position of the control valve. These valves have been widely used for many years. It is also possible to install a wide variety of electrically controlled valves to perform similar functions. These valves require a power supply but have the ability to more selectively review signals before adjusting themselves. Valves attached to programmable logic controllers offer the greatest capability in responding to a widest variety of situations.

18.6.4   Direction-Control Valves

In some instances it is necessary to allow flow of water in only one direction. One such case is in pump-discharge lines where flow can pass backwards through a pump and damage the pump or motor if the discharge line is not equipped with a check valve which allows flow in only one direction.

Four types of *check valves* include swing check, rubber-flapper check, slanting-disc check, and double-door check. Swing check valves (Fig. 18.20) can be equipped with an external arm which can be used to indicate position of the valve and adjust the response of the valve to a reversal in pressure.

FIGURE 18.20 Swing check valve. *(Photo courtesy of Henry Pratt Company.)*

A second type of valve which controls the direction of flow is the *backflow-preventer valve* (Fig. 18.21). This type of valve is used in customer services to prevent backflow or backsiphonage from a customer's water system into the distribution system. A single check valve is not considered acceptable protection if the customer's service can contain something that can make the water nonpotable.

Backflow-prevention devices consist of two spring-loaded check valves and taps for testing the operation of the valves in the field. The highest level of protection is provided by reduced-pressure backflow preventers which have a relief valve between the two check valves which dumps water to the atmosphere if the pressure between the two check valves exceeds the pressure upstream of the valve.

18.6.5 Air-Release/Vacuum-Breaker Valves

The valves discussed previously have been main-line valves which control flow in the main. There are also valves which can allow air or water into or out of the main to eliminate undesirable situations.

Air can build up at high points in water-distribution systems to the extent that it can restrict the carrying capacity of the line. An air-release valve (Fig. 18.22) placed at the high point along the line can bleed off air. Air-release valves can also be used for bleeding off air when a pipe is initially filled, although fire hydrants and customers taps can also aid in releasing air from the system.

FIGURE 18.21 Reduced pressure backflow preventer valve. *(Photo courtesy of Febco.)*

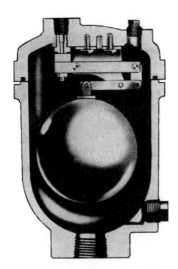

FIGURE 18.22 Air release valve. *(Courtesy of Crispin Valves Inc.)*

Air-release valves rely on a float that keeps the valve closed when the pipe is filled with water. However, when air gets into the valve the float is no longer buoyant and drops, thus allowing fluid (air) to escape. When the float is again submersed in water, it floats and seals the opening.

The mechanism in a *vacuum-breaker valve* prevents vacuums from forming at the valve. This is essential in some large-diameter steel or plastic pipes which may not be able to bear external loads when not full of water or under vacuum. When pressure drops below atmospheric, the valve mechanism opens and allows air into the pipe to break the vacuum. Once the pressure in the pipe returns to above atmospheric the air is released through an air-release valve. In water-distribution systems vacuum-breaker valves should ideally never open because the pressure should always be positive. The inlet for a vacuum-breaker valve should not be located anywhere it can be submerged.

18.6.6 Surge Control/Protection

Surges (hydraulic transients or, specifically in closed pipes, water hammer) occurs whenever the velocity of flowing water is changed (Chaudhry, 1979; Wylie and Streeter, 1978; Betamio de Almeida and Koelle, 1992). This can create large pressure fluctuations which have the potential to damage piping systems. For every 1 ft/s (0.3 m/s) change in velocity, there may be up to a 100 ft (30 m) change in head.

Methods to prevent surge involve using slow-closing valves and opening discharge valves slowly when pumps are turned on or off. Installing pumps with large moments of inertia or flywheels also reduces the rate of change in velocity of flowing water.

If a pump should stop suddenly, there will be a negative pressure wave that proceeds downstream from the pumps at roughly 3000 ft/s (1000 m/s) depending on pipe material and diameter. This negative wave will be reflected when it reaches a tank and will die out due to friction. If a typical water-storage tank is very far from the site where the water hammer forms, it may be possible to place a small pressurized tank, partly full of air, near the source of the surge to dampen out the surge.

If the pressure drops to the vapor pressure of water, the water will vaporize and then the vapor pocket will collapse suddenly and violently in a phenomena known as *column separation*, which can burst pipes. If the negative pressure is sufficiently low, a vacuum breaker can introduce air into the line to offset the negative pressure.

Pressure-release valves, at low points of the distribution systems, can protect the distribution system from very high pressures by opening to release pressure. There are also a variety of electrically controlled surge-control valves.

In most water-distribution systems, there are enough tanks, customer taps, and looping of pipes, and slow enough changes in velocity, such that water hammer is usually not a serious problem. However, water hammer can be a serious problem in long distribution pipelines and a surge analysis should be performed for major pipeline projects.

18.7 FLOW MEASUREMENT

Three overall categories of metering used in water systems are *main meters, hydrant-discharge gauges,* and *customer meters*. Customer meters are highly accurate meters which display cumulative flow and are usually displacement meters, although turbine meters are sometimes used in large services. Water-main metering involves a wide array of metering devices.

18.7.1 Change in Head Devices

When water flows through a restriction in a pipe, the pressure at that restriction changes and that change in pressure is a function of the flow rate. Three devices, *venturi meters, orifice plates,* and *nozzles,* rely on this principle to measure flow. These devices are simple, reliable, and do not cause excessive head loss. These devices require a differential-pressure meter to be attached to two points on the device and a method to relate the differential pressure to flow. The devices do not require any power, although power may be required by the differential-pressure transducer or transmitter to convey the reading to a remote location.

The velocity in the pipe can be related to the differential head by:

$$V = C_D\sqrt{2g\Delta h} \tag{18.14}$$

where V = velocity, ft/s
   C_D = discharge coefficient for that meter
   g = gravitational acceleration, ft/s^2
   Δh = differential head between pressure points, ft

The principle drawback to these devices is that they are accurate over a limited range of flows because of the limited accuracy of differential-pressure devices and the fact that the velocity is a function of the square root of the differential pressure. (For example, a factor of 4 change in velocity results in a factor of 16 change in differential pressure.) These devices are therefore used in locations such as treatment plants which run at a fairly constant flow rate or an individual pump discharge line because a pump will generally operate at a fairly constant flow when it is running. It will not be used at a customer meter or a pump station discharge because these flows can vary widely over time.

Orifice plates are the least expensive of this type of meter while venturi meters cause the least head loss. Flow tubes are variations on venturi meters which can produce very low head losses. These devices measure instantaneous flows and require some type of totalizer to yield cumulative flow.

18.7.2   Electromagnetic Meters

When a conducting substance flows through a magnetic field, a current is generated. This principle is used in *electromagnetic meters* to generate an electrical signal that can be displayed or recorded.

Most successful electromagnetic meters are spool pieces which are inserted into pipes, usually with flanged fittings. Some small sensors which can be inserted into taps in the pipe wall are also available. Electromagnetic flow meters create very little head loss, are accurate over a very wide range, but require a power supply.

18.7.3   Ultrasonic Meters

The nature of sound waves crossing a moving fluid is affected by the velocity of the fluid. This effect can be correlated to the average velocity of the fluid.

Ultrasonic meters can be insertion meters which are inserted into the pipe through taps or clamp-on meters which are merely clamped to the outside of the pipe. Clamp-on meters are sensitive to the pipe-wall material and uncertainty can be introduced in the reading to the extent that pipe-wall condition is uncertain. Clamp-on meters work best on steel or ductile-iron pipe.

Ultrasonic meters create very little head loss but are very sensitive to the length of straight pipe upstream of the sensor. They require as much as 20 pipe diameters of straight pipe upstream to be accurate. They require power and a totalizer if cumulative flow is desired.

18.7.4   Propeller or Turbine Meters

Propeller or *turbine meters* rely on the force of the flowing water to rotate a device inserted into the flow. These devices may occupy the full cross section of the pipe or may be smaller devices which measure velocity at a small area in the cross section and relate that velocity to the average velocity in the pipe. The smaller meters are very sensitive to the length of straight pipe upstream of the meter.

These meters are usually attached to a counter and the readout is usually in terms of cumulative flow. Rate of flow can be determined by measuring the cumulative flow over a known time interval and dividing by the time interval. Some meters have internal logic which can calculate flow rate and data loggers to store this information.

Propeller or turbine meters may also be used in larger-sized customer meters but are not as accurate over as wide a range of flows as displacement meters.

18.7.5 Pitot Rods and Gauges

A *Pitot tube* is the general name given to an L-shaped tube which can be inserted in a flowing stream to determine the total head at that point in the stream. The pressure reading at the gauge attached to the top of a Pitot tube gives the total pressure:

$$P_t = P_s + k\,\frac{v^2}{2g} \tag{18.15}$$

where P_t = reading of gauge (total pressure), psi (kPa)
  P_s = static pressure in fluid, psi (kPa)
  v = velocity at point in fluid, ft/s (m/s)
  k = 0.46 if p in psi and V in ft/s
    9.81 if p in kPa and V in m/s

If one knows the static pressure at the tip of the Pitot tube and the reading at the gauge attached to the Pitot, it is possible to determine velocity. In a pressure pipe, it is difficult to measure the static pressure at the Pitot tube tip and the first term in Eq. (18.15) would be much larger than the second term. The result is that a small error in determining either pressure will result in a large error in determining v.

This problem was solved by John Cole, who invented the *Pitot rod* (Fig. 18.23), which has two sensors connected to a differential-pressure-measuring device (e.g., manometer, differential-pressure meter, differential-pressure transducer). By measuring the differential pressure instead of two individual pressures, a much greater accuracy is possible, and Eq. (18.15) can be reorganized to

$$v = C\sqrt{\frac{2g\Delta P}{k}}$$

$$\Delta P = P_t - P_s \tag{18.16}$$

where C = coefficient for shape of Pitot tip.

A Pitot rod can be inserted into a pipe under pressure using a corporation cock. Unlike the other flow-measuring devices described above, a Pitot rod measures velocity at a point in the fluid, not the average velocity. Pitot rods can be very helpful in determining the velocity profile within a pipe. To determine the average flow in the pipe, the velocity profile must be integrated (Walski, 1984).

Another application for the Pitot tube principle in water-distribution systems is measurement of discharge from a fire hydrant outlet. In this case, because the stream is discharging to the atmosphere, the static pressure term in Eq. (18.15) can be neglected and the velocity can be given by

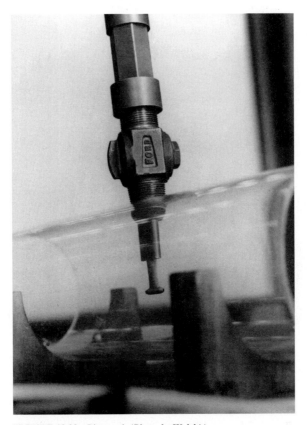

FIGURE 18.23 Pitot rod. *(Photo by Walski.)*

$$V = C_D\sqrt{P_t} \tag{18.17}$$

where C_D = discharge coefficient for the device. Equation (18.16) is another way of writing this equation.

Hydrant Pitot gauges can be hand held, clamped on to the hydrant, or installed in a diffuser device which directs and diffuses the flow. If a diffuser-type device is used (as shown in Fig. 18.24), the static pressure at the Pitot tip must be zero or some correction must be applied (Walski and Lutes, 1990).

18.7.6 Displacement Meters

Displacement meters are used in most customer meters. They rely on the water displacing some type of piston or chamber (Fig. 18.25). These meters are accurate over a fairly wide range of flows but would be prohibitively expensive to construct in sizes for large distribution-system meters. AWWA standards govern acceptable accuracy and head loss in customer meters.

FIGURE 18.24 Hydrant pitot diffuser. *(Photo by Walski.)*

FIGURE 18.25 Displacement meter. *(Photo courtesy of ABB.)*

FIGURE 18.26   Compound meter. *(Photo courtesy of ABB.)*

18.7.7   Compound Meters

Even though displacement meters are accurate over a wide range of flows, some larger customers may have an even wider range of flow. To overcome this, meter manufacturers can connect several different size meters in parallel as shown in Fig. 18.26. During low-use periods a check valve on the large meter remains closed and the small meter measures flow. As the downstream demand for water increases, the check valve opens and the larger meter comes on line. These meters are called *compound meters* and find use in larger installations which may use large quantities of water during normal operation, small quantities of water in off-peak periods, or very large quantities of water during emergencies.

18.8   OPERATION AND MAINTENANCE

Water-distribution systems are designed to operate with little operator intervention and even less maintenance. However, what operation-control and maintenance work that is required is very important for the proper operation of a water-distribution system (AWWA, 1986).

18.8.1   Operational Decisions

The most fundamental decision in water-system operation is which pumps to be running at any given time. This is controlled for the most part by water levels in distribution-system-storage tanks. There are actually three competing goals for a water system:

1. Maximize reliability, which is achieved by keeping the maximum amount of water in storage in case of emergencies such as pipe breaks and fires.

2. Minimize energy costs, which is achieved by operating pumps against as low of a head as possible (minimize water in storage) near the best efficiency point for the pump.

3. Meeting water-quality standards, which involves minimizing the time the water is in the distribution system and storage tanks, which is achieved by having storage-tank levels fluctuate as much as possible.

These three goals conflict with each other and water-system operation involves tradeoffs between these competing goals. Most systems try to maintain storage-tank levels in the upper half of the tanks with the tanks nearly full before the peak-water-use hours and roughly half full after the peak-water-use hours. However, operating rules for pumps and reservoirs vary widely between utilities.

In distribution systems without storage that floats on the system, pump-operation decisions are usually triggered by a pressure sensor somewhere in the distribution system. If the pressure drops below a certain value a pump is turned on or the speed is increased for a variable-speed pump.

18.8.2 Types of Control

Control of pumping operations can be as simple as having an individual manually operate a pump or control valve to highly sophisticated control systems which try to optimize system operation in real time. The first decision in establishing operation is whether control of equipment should be manual or automatic control, while the second decision is whether control should be remote or local.

Manual local control is usually limited to less-developed countries or short-term situations. Manual remote control is accomplished from a centralized control room where data is received from the system sensors and decisions are sent out to equipment. Today there is usually some sort of computer at the heart of this type of system, which is usually referred to as *SCADA* (*Supervisory Control and Data Acquisition*). Some systems may consist only of telemetry, in which control of equipment is at the equipment site but information about status and operating conditions is transmitted back to the control room.

Automatic remote control is usually accomplished using control units at the equipment location, which can range from float switches to computerized controllers called *RTU*s (*remote terminal units*) or *PLC*s (*programmable logic controllers*). Locating control units near the equipment it is controlling reduces problems with communications but limits the ability to operate the system as a unified whole. Real-time computer control from a central control room using a SCADA system offers promise for energy-cost savings with minimal loss of reliability (Ormsbee, 1991).

Figure 18.27 shows a traditional pump-control room with numerous individual gauges and charts. Modern controls are more likely to have a few computer monitors from which the entire distribution system can be checked.

18.8.3 Control Valves/Pressure-Zone Boundaries

Setting of pressure-regulating valves and adjustment of pressure-zone boundaries is another way that operators impact the performance of distribution systems. The goal

FIGURE 18.27   Pump station control panel. *(Photo by Walski.)*

in setting valves and adjusting pressure zones is to provide customers with pressures that are not too low or too high. Usually static pressures should be on the order of 40 to 80 psi (280 to 560 kPa).

In flat terrain, maintaining a good range of pressures is fairly easy, but in hilly areas establishing correct control-valve settings and pressure-zone boundaries is an ongoing process based mostly on experience with some assistance from pipe-network modeling.

18.8.4   Routine Maintenance Activities

Routine maintenance activities involve pump mechanical and electrical maintenance, valve exercising, leak detection and repair, tank coating, fire hydrant testing (although this may also be done by the fire department), main flushing, checking of telemetry equipment, and responding to customer concerns.

Water-distribution-system equipment is very reliable and can function for long periods of time without maintenance. It is therefore tempting to defer maintenance until failure of the distribution component. Utility maintenance managers must continually balance the need for maintenance with limited budgets.

A frequently ignored activity is valve exercising which benefits the utility by making valves easier to turn, keeping valve boxes cleaned, identifying and correcting valves which have been paved over, and educating utility employees as to valve location. Figure 18.28 shows a typical key used to operate underground valves. Because a valve may only be needed once every few years, it is easy to defer valve exercising. However, when a valve is needed in an emergency, it may not be available and a minor pipe break can become a major water outage.

FIGURE 18.28 Operating distribution valve. *(Photo by Walski.)*

Water-main flushing projects are undertaken to remove undesirable water from a portion of the distribution system, usually in response to problems with taste and odor, dirty water, or low disinfectant residuals. Mains are flushed using hydrants or blowoffs until the quality of the water flowing from the system is of acceptable quality. Directional flushing generally results in the biggest improvement in quality with the least use of water.

Leak-detection equipment is based on locating the sound made by water as it leaks from a pipe. Modern equipment (as shown in Fig. 18.29) can pinpoint leaks to within several feet (meters) in most instances. This makes it possible to locate and repair leaks before they erode the soil and become a major break or a street collapse.

Recently improvements in maintenance-management software have made it much easier to track preventive maintenance activities to the extent that work orders and supply inventories can be managed by the software. Maintenance-management activities have also been combined with spatial information in geographical information systems (GIS) to provide graphical information to help locate facilities.

18.8.5 Emergency Maintenance Activities

Maintenance personnel must be available at all times to respond to emergencies such as pipe breaks. A break can drain the entire distribution system in a matter of

FIGURE 18.29   Locating distribution leak. *(Courtesy of Heath Consultants.)*

hours unless isolated and repaired. Breaks are isolated using isolation valves (usually gate valves). Many breaks can be repaired with clamps. For larger breaks sections of pipe may need to be replaced and the pipe closed with closure pieces or couplings.

The keys to prompt repair are trained personnel who can be called out quickly and an adequate inventory of needed spare parts.

18.8.6   Rehabilitation/Replacement

Water-distribution-system components generally have very long lives. Eventually, however, some begin to fail at such a high rate that replacement or rehabilitation is economical. Decisions to repair, replace, or rehabilitate a pipe or pump should be based on the cost of the alternatives and present worth of future failures. However, in many cases budget constraints limit the number of projects that can be undertaken at any time even if they are justifiable.

Pipe-rehabilitation methods usually involve cleaning the inside of pipes using pigs (flexible plugs) or scrapers. The pipe may then be lined, usually with cement mortar, or a slightly smaller pipe may be pulled through the pipe. Another option is to pull a mole through the pipe, which bursts the existing pipe. A new pipe is then pulled behind the mole to replace the old pipe. Sock-type liners may also be used in low-pressure lines. Additional information about rehabilitation methods and decision making can be found in U.S. Department of Housing and Urban Development (1984); and Walski (1987).

REFERENCES

American Water Works Association, *Introduction to Water Distribution,* AWWA, Denver, Colo., 1986.

American Water Works Association, *Distribution Requirements for Fire Protection,* AWWA Manual M31, AWWA, Denver, Colo., 1989.

American Water Works Association, *Distribution Network Analysis for Water Utilities,* AWWA Manual M32, AWWA, Denver, Colo., 1989.

American Water Works Association, *AWWA Standards,* AWWA, Denver, Colo., continually updated.

AWWA Research Foundation, *Water Quality Modeling in Distribution Systems,* AWWA, Denver, Colo., 1991.

Betamio de Almeida, and E. Kolle, *Fluid Transients in Pipe Networks,* Computational Mechanics Publications, Boston, 1992.

Bhave, P. R., *Analysis of Flow in Water Distribution Systems,* Technomic, Lancaster, Pa., 1991.

Cesario, A. L., *Water Distribution System, Modeling, Analysis and Design,* American Water Works Association, Denver, Colo., 1995.

Chaudhry, M. H., *Applied Hydraulic Transients,* Van Nostrand Reinhold, New York, 1979.

Clark, R. M., J. I. Gillean, and W. K. Adams, "The Cost of Water Supply and Water Utility Management," U.S. Environmental Protection Agency Municipal Environmental Research Lab, Cincinnati, Ohio, 1977.

Goulter, I.C., "Systems Analysis in Water Distribution Network Design: From Theory to Practice," *Journal of Water Resources Planning and Management,* 118(3):238, 1992.

Hicks, T. G., and T. W. Edwards, *Pump Applications Engineering,* McGraw-Hill, New York, 1971.

Hydraulic Institute, *Hydraulic Institute Standards for Centrifugal, Rotary and Reciprocating Pumps,* Hydraulic Institute, Cleveland, Ohio, 1983.

Insurance Services Office, *Fire Flow Tests,* New York, 1973.

Jeppson, R. W., *Analysis of Flow in Pipe Networks,* Ann Arbor Science Publishers, Ann Arbor, Mich., 1976.

Karassik, I. J., et al., eds., *Pump Handbook,* McGraw-Hill, New York, 1976.

Lamont, P. A., "Common Pipe Flow Formulas Compared with the Theory of Roughness," *Journal American Water Works Association,* 73(5):274, 1981.

Mau, R. E., P. F. Boulos, R. M. Clark, W. M. Grayman, R. J. Tekippe, and R. R. Trussell, "Explicit Mathematical Models of Distribution Storage Water Quality," *Journal of Hydraulic Engineering,* ASCE, 121(10):699–709, October 1995.

Mays, L. W., ed., *Reliability Analysis of Water Distribution Systems,* American Society of Civil Engineers, New York, 1989.

Miller, D. S., *Internal Flow Systems,* BHRA Fluid Engineering, U.K., 1978.

O'Day, D. K., et al., *Water Main Evaluation for Replacement/Rehabilitation,* AWWA Research Foundation, Denver, Colo., 1986.

Ormsbee, L. E., ed., "Energy Efficient Operation of Water Distribution Systems," University of Kentucky Research Report, UKCE9104, 1991.

Ormsbee, L. E., and T. M. Walski, "Developing System Head Curves for Water Distribution Systems," *Journal American Water Works Association,* 81(7), 1989.

Rossman, L. A., "EPANET User's Manual," Risk Reduction Engineering Lab, U.S. Environmental Protection Agency, Cincinnati, Ohio, 1994.

Rossman, L. A., P. F. Boulos, and T. Altman, "Discrete Volume Element Method for Network Water Quality," *Journal of Water Resources Planning and Management,* 119(5):505, 1993.

Rossman, L. A., J. G. Uber, and W. M. Grayman, "Modeling Disinfectant Residuals in Drinking Water Storage Tanks," *Journal of Environmental Engineering,* ASCE, 121(10):752–755, October 1995.

Sanks, R. L., ed., *Pumping Station Design,* Butterworths, Stoneham, Mass., 1989.

Stephenson, G., *Network Analysis—A Code of Practice,* WRC, Swindon, U.K., 1989.

Tullis, P. A., "Selection and Use of Control Devices in Water Distribution Systems," in E. Cabrera and F. Martinez, eds., *Water Supply Systems: State-of-the-Art and Future Trends,* Computational Mechanics Publications, Boston, 1993.

U.S. Department of Housing and Urban Development, *Utility Infrastructure Rehabilitation,* Washington, D.C., 1984.

Walski, T. M., "State-of-the-Art Pipe Network Optimization," *Computer Applications in Water Resources,* American Society of Civil Engineers, Buffalo, N.Y., 1985.

Walski, T. M., *Analysis of Water Distribution Systems,* Kreiger, Melbourne, Fla., 1984.

Walski, T. M., "Practical Aspects of Providing Reliability in Water Distribution Systems," *Reliability Engineering and Systems Safety,* Elsevier, New York, 1993.

Walski, T. M., "Battle of the Network Models: Epilogue," *Water Resources Planning and Management,* 113(2):191, 1987.

Walski, T. M., ed., *Water Supply System Rehabilitation,* American Society of Civil Engineers, New York, 1987.

Walski, T. M., J. Gessler, and J. W. Sjostrom, *Water Distribution Systems: Simulation and Sizing,* Lewis, Chelsea, Mich., 1989.

Walski, T. M., "Optimization and Pipe Sizing Decisions," *Journal of Water Resources Planning and Management,* 121(4):340, 1995.

Walski, T. M., and T. L. Lutes, "Accuracy of Hydrant Flow Tests Using a Pitot Diffuser," *Journal American Water Works Association,* 82(7):58, 1990.

Walters, G., and R. G. Cembrowicz, "Optimal Design of Water Distribution Networks," in E. Cabrera and F. Martinez, eds., Water Supply Systems: State-of-the-Art and Future Trends, Computation Mechanics Publication, Boston, 1993.

Wylie, E. B., and V. L. Streeter, *Fluid Transients,* McGraw-Hill, New York, 1978.

CHAPTER 19
WASTEWATER-COLLECTION SYSTEMS

Franklin L. Burton, P.E.
Burton Environmental Engineering
Los Altos, California

19.1 INTRODUCTION

Wastewater-collection systems perform the essential function of conveying wastewater produced from residences, institutions, and commercial and industrial establishments, together with surface runoff from storms, to points of treatment or disposal. In many cases, collection systems also convey significant amounts of groundwater that enters the system through leaking pipes, connections, and manholes.

Because such systems must function properly without creating nuisances, the principles involved in their design and construction need to be understood. In this chapter, the important fundamentals of hydraulic design are reviewed and the design of gravity-flow sewer systems, alternative (pressure and vacuum) sewer systems, and pumping stations are discussed.

19.1.1 Overview

Wastewater-collection systems usually consist of a network of pipes and pumping stations for transporting wastewater to its final destination. The piping systems may be gravity-flow, the vast majority of the systems, or may consist of small-diameter pressure or vacuum sewers. Pumped systems are used where there are variations in topography or where deep excavations preclude construction of gravity sewers.

19.1.2 Definitions

Many of the frequently-used terms in wastewater-collection systems are defined here.

Sanitary wastewater. A combination of the liquid and solid wastes from residences, commercial buildings, industrial plants, institutions, and recreational facilities.

Stormwater. Runoff resulting from rainfall or snow melt.

Sewer. A pipe or conduit that carries wastewater or stormwater.

Infiltration. The water entering a sewer system and service connections from the ground, through such means as defective pipes, pipe joints, connections, or manhole walls. Infiltration does not include, and is distinguished from, inflow.

Inflow. The water discharged into a sewer system, including service connections, from such sources as roof leaders; cellar, yard, and area drains; foundation drains; cooling water discharges; drains from springs and swampy areas; manhole covers; cross connections from storm sewers; catch basins; storm sewers; surface runoff; street wash waters; or drainage. Inflow does not include, and is distinguished from, infiltration.

Infiltration/Inflow (I/I). The total quantity of water from both infiltration and inflow without distinguishing the source.

Sanitary sewer. A sewer intended to carry only sanitary and industrial wastewaters from residences, commercial buildings, industrial plants, and institutions.

Storm sewer. A sewer intended to carry only stormwaters, surface runoff, street wash waters, and drainage.

Combined sewer. A sewer intended to serve as a sanitary sewer and a storm sewer, or as an industrial sewer and storm sewer.

Building sewer. The extension from the building drain to the public sewer; it is frequently called a house connection.

Lateral sewer. A sewer that receives wastewater from building sewers and discharges into a main or trunk sewer.

Branch sewer. A sewer that receives wastewater from a relatively small area and discharges into a main sewer serving more than one branch sewer area.

Main sewer. The principal sewer into which lateral sewers discharge; it is also termed a trunk sewer.

Interceptor sewer. A sewer that receives dry-weather flow from a number of sewers and frequently additional quantities of stormwater (from a combined system) and conducts such flows to a treatment facility.

Force main. A pipeline that carries wastewater under pressure, usually from a pumping station.

19.2 QUANTITY OF SANITARY WASTEWATER

An important step in the design of sanitary sewers is estimating the design-flow rates, including the variations in flow rates. Sources of wastewater may include domestic and industrial wastewater as well as infiltration and inflow. Peak- and low-flow rates are important considerations for designing sewer systems and pumping stations. Ample capacity for peak-flow rates needs to be provided so sewers do not backup and overflow thereby causing public health problems and property damage. Low-flow conditions are also important in design so that deposition of solids and odor generation in the sewer system can be minimized.

19.2.1 Domestic Sources and Flow Rates

The principal sources of domestic wastewater in a community are the residential areas and commercial districts; other important sources include institutional and recreational facilities. For existing developments, flow-rate data should be obtained by directly measuring flows in the sewer system. Wastewater-flow rates for small systems (systems with 1000 people and less) may differ significantly from larger systems. The ratio of peak- to average-flow rates in small systems may be much higher than in larger systems.

19.2.1.1 Residential Areas.
For many residential areas, wastewater-flow rates are commonly determined on the basis of population density and the average per capita contribution of wastewater. Ranges and typical flow-rate values are given in Table 19.1. For large residential areas, it is often advisable to develop flow rates on

TABLE 19.1 Typical Wastewater-Flow Rates from Residential Sources

Source	Unit	Flow, gal/unit · d	
		Range	Typical
Apartment			
High rise	Person	35–75	50
Low rise	Person	50–80	65
Hotel	Guest	30–55	45
Individual residence			
Typical home	Person	45–90	70
Better home	Person	60–100	80
Luxury home	Person	75–150	95
Older home	Person	30–60	45
Summer cottage	Person	25–50	40
Motel			
With kitchen	Unit	90–180	100
Without kitchen	Unit	75–150	95
Trailer park	Person	30–50	40

Note: gal × 3.7854 = L
Source: Adapted in part from Salvato (1982).

the basis of land-use areas and anticipated population densities. Where possible, these rates should be based on actual flow data from selected similar residential areas, preferably in the same locale. In some communities, regulatory agencies mandate the design-flow rates to be used.

Wastewater-flow rates may also be estimated based on plumbing fixtures, if they are known. Typical per capita flow data from conventional domestic devices are given in Table 19.2 without and with the use of water-conservation devices.

Where flow-rate data do not exist, wastewater-flow rates may be estimated based on water consumption. As a rule of thumb, average wastewater-flow rates are about 70 percent of domestic water use. The ratio of wastewater to water consumption depends on factors such as climate, socioeconomic conditions, and conservation programs. More detailed information on estimating wastewater-flow rates from water supply data is found in Metcalf & Eddy (1991).

19.2.1.2 Commercial Districts.
Commercial wastewater-flow rates are generally expressed in gal/acre · d (m^3/ha · d) and are based on existing or anticipated future development or comparative data. Average unit-flow-rate allowances for commercial developments normally range from 800 to 1500 gal/acre · d (7.5 to 14

TABLE 19.2   Per Capita Wastewater-Flow Rates from Plumbing Fixtures

	Wastewater flow, gal/capita · d	
Device	Without conservation devices	With conservation devices
Bathtub faucet	7	6–7
Dishwashers	2	2
Clothes washing machine	16	13–14
Faucets	7–9	7–8
Showers	12–16	8–12
Toilet	22	14–19
Toilet leakage	4	4

Source: Adapted from Metcalf & Eddy (1975); and U.S. HUD (1984).

TABLE 19.3   Typical Wastewater-Flow Rates from Commercial Sources

		Flow, gal/unit · d	
Source	Unit	Range	Typical
Airport	Passenger	2–4	3
Automobile service station	Vehicle served	7–13	10
	Employee	9–15	12
Bar	Customer	1–5	3
	Employee	10–16	13
Department store	Toilet room	400–600	500
	Employee	8–12	10
Hotel	Guest	40–56	48
	Employee	7–13	10
Industrial building (sanitary waste only)	Employee	7–16	13
Laundry (self-service)	Machine	450–650	550
	Wash	45–55	50
Office	Employee	7–16	13
Restaurant	Meal	2–4	3
Shopping center	Employee	7–13	10
	Parking space	1–2	2

Source: Adapted in part from Geyer and Lentz (1962).

m^3/ha · d). Because unit-flow rates can vary widely for commercial facilities, every effort should be made to obtain records from similar facilities. Estimates for certain commercial sources may also be made from the data in Table 19.3.

19.2.1.3   Institutional Facilities.   Some typical flow rates from institutional facilities, which are essentially domestic in nature, are shown in Table 19.4. Again, it is stressed that flow rates vary with the region, climate, and type of facility. The actual records of institutions are the best sources of flow data for design purposes.

19.2.1.4   Recreational Facilities.   Wastewater-flow rates from many recreational facilities are highly seasonal. Typical data on wastewater-flow rates from recreational facilities are presented in Table 19.5.

TABLE 19.4 Typical Wastewater-Flow Rates from Institutional Sources

		Flow, gal/unit · d	
Source	Unit	Range	Typical
Hospital, medical	Bed	125–240	165
	Employee	5–15	10
Hospital, mental	Bed	75–140	100
	Employee	5–15	10
Prison	Inmate	75–150	115
	Employee	5–150	10
Rest home	Resident	50–120	85
School, day			
With cafeteria, gym, and showers	Student	15–30	25
With cafeteria only	Student	10–20	15
Without cafeteria and gym	Student	5–17	11
School, boarding	Student	50–100	75

Source: Adapted in part from Geyer and Lentz (1962).

TABLE 19.5 Typical Wastewater-Flow Rates from Recreational Facilities

		Flow, gal/unit · d	
Facility	Unit	Range	Typical
Apartment, resort	Person	50–70	60
Cabin, resort	Person	8–50	40
Cafeteria	Customer	1–3	2
	Employee	8–12	10
Campground (developed)	Person	20–40	30
Cocktail lounge	Seat	12–25	20
Coffee shop	Customer	4–8	6
	Employee	8–12	10
Country club	Member present	60–130	100
	Employee	10–15	13
Day camp (no meals)	Person	10–15	13
Dining hall	Meal served	4–10	7
Dormitory, bunkhouse	Person	20–50	40
Hotel, resort	Person	40–60	50
Store, resort	Customer	1–4	3
	Employee	8–12	10
Swimming pool	Customer	5–12	10
	Employee	8–12	10
Theater	Seat	2–4	3
Visitor center	Visitor	4–8	5

Source: Adapted in part from Salvato (1982).

19.2.2 Industrial Sources and Flow Rates

Wastewater-flow rates from industrial sources vary with the type and size of the facility, the degree of water reuse, and the on-site pretreatment methods used, if any. Typical design values for estimating the flows from industrial areas that have no or little wet-process-type industries is 1000 to 1500 gal/acre · d (9 to 14 m³/ha · d) for light industrial developments and 1500 to 3000 gal/acre · d (14 to 28 m³/ha · d) for medium industrial developments. Alternatively, for estimating industrial-flow rates where the nature of the industry is known, data such as those reported in Metcalf & Eddy (1991) can be used. For industries without internal recycling or reuse programs, it can be assumed that about 85 to 95 percent of the water used in the various operations and processes will become wastewater. For large industries with pretreatment facilities or internal water-reuse programs, separate estimates must be made. Average domestic (sanitary) wastewater contributed from industrial activities may vary from 8 to 25 gal/capita · d (30 to 95 L/capita · d).

19.2.3   Infiltration/Inflow

Extraneous flows in sewers are described as infiltration and inflow and are defined as follows:

Infiltration.   Water entering a sewer system, including sewer-service connections, from the ground through such means as defective pipes, pipe joints, connections, or manhole walls.

Steady Inflow.   Water discharged from cellar and foundation drains, cooling-water discharges, and drains from springs and swampy areas. This type of inflow is steady and is identified and measured with infiltration.

Direct inflow.   Those types of inflow that have a direct stormwater-runoff connection to the sanitary sewer and cause an almost immediate increase in wastewater flows. Possible sources are roof leaders, yard and areaway drains, manhole covers, cross connections from storm drains and catch basins, and combined sewers.

Total inflow.   The sum of the direct inflow at any point in the system plus any flow discharged from the system upstream through overflows, pumping station bypasses, and the like.

Delayed inflow.   Stormwater that may require several days or more to drain through the sewer system. This category of inflow can include the discharge of sump pumps from cellar drainage as well as the slowed entry of surface water through manholes in ponded areas.

Correcting infiltration/inflow problems in the collection system provides benefits by: (1) reducing overloaded or surcharged sewers and the associated problems of wastewater backups and overflows, (2) increasing the efficiency of operation of wastewater-treatment facilities, and (3) reducing pumping costs in systems that use pumping stations.

Detailed procedures for making an analysis of infiltration/inflow, including an example cost-effectiveness analysis, are provided in Metcalf & Eddy (1981). Because an understanding of the effects of infiltration/inflow is important in determining flow rates, a discussion of excessive infiltration/inflow is included in this section.

19.2.3.1   Infiltration into Sewers.   One portion of the rainfall in a given area runs quickly into the storm sewers or other drainage channels; another portion evaporates or is absorbed by vegetation; and the remainder percolates into the ground, becoming groundwater. The proportion of the rainfall that percolates into the ground depends on the character of the surface and soil formation and on the rate and distribution of the precipitation according to seasons. Any reduction in permeability, such as that due to buildings, pavements, or frost, decreases the opportunity for precipitation to become groundwater and increases the surface runoff correspondingly.

The rate and quantity of infiltration depend on the length of sewers, the area served, the soil and topographic conditions, and to a certain extent, the population density (which affects the number and total length of house connections).

The amount of groundwater flowing from a given area may vary from a negligible amount for a highly impervious district or a district with a dense subsoil, to 25 or 30 percent of the rainfall for a semipervious district with a sandy subsoil permitting rapid passage of water into it. The percolation of water through the ground from rivers or other bodies of water sometimes has considerable effect on the groundwater table, which rises and falls continually, especially in coastal areas affected by tides.

High groundwater results in leakage into sewers and in an increase in the quantity of wastewater, resulting in increased costs for conveyance, treatment, and disposal. The amount of flow that can enter a sewer from groundwater, or infiltration, may range from 100 to 10,000 gal/d · in · mi (0.0094 to 0.94 m^3/d · mm · km) or more. The number of inch · miles (millimeter · kilometers) in a wastewater collection system is the sum of the products of sewer diameters, in inches (millimeters), times the lengths, in miles (kilometers), of sewers of corresponding diameters. Expressed another way, infiltration may range from 20 to 3000 gal/acre · d (0.2 to 28 m^3/ha · d). During heavy rains, when there may be leakage through manhole covers, or inflow, as well as infiltration, the rate may exceed 50,000 gal/acre · d (470 m^3/ha · d). Infiltration/inflow is a variable part of the wastewater, depending on the quality of the material and workmanship in constructing the sewers and building connections, the character of the maintenance, and the elevation of the groundwater compared with that of the sewers.

Old sewers may receive comparatively large quantities of groundwater, while sewers built later may receive relatively smaller quantities of groundwater. Newer sewers may be built at higher elevations and the quality of the sewer-pipe materials and construction is better. In arid areas such as the southwestern United States, groundwater tables may be low so infiltration will not be a problem. For a developing community, an increase in the percentage of area that is paved or built over results in: (1) an increase in the percentage of stormwater that is conducted rapidly to the storm sewers and watercourses, and (2) a decrease in the percentage of the stormwater that can percolate into the earth and tend to infiltrate the sanitary sewers.

Modern sewer design calls for the use of high-quality pipe with dense walls, precast manhole sections, and joints sealed with rubber or synthetic gaskets. Use of these improved materials has greatly reduced infiltration into newly constructed sewers, and it is expected that the increase of infiltration rates with time will be much slower than has been the case with the older sewers. Improved construction of house laterals, often significant sources of leaks, will also reduce the amount of infiltration.

19.2.3.2 Inflow into Sewers. As described previously, the type of inflow that causes a steady flow cannot be identified separately and so is included in the measured infiltration. The direct inflow can cause an almost immediate increase in flow rates in sanitary sewers.

19.2.4 Flow Rates for Design

Flow rates used in design are per capita flow rates that establish the average daily flow rates based on the tributary population and peak hourly flow rates that are generally determined by applying a peaking factor to the average daily flow rate.

19.2.4.1 Per Capita Flow Rate. In general, new sewer systems should be designed on the basis of an average daily per capita flow of not less than 100 gal/d (0.38 m^3/d). This figure is assumed to cover normal infiltration, but an additional allowance should be made where unfavorable conditions exist such as high ground-water. For existing sewer systems, additional per capita allowances should be made where the average flow exceeds this value and no immediate remedial measures are proposed (Great Lakes–Upper Mississippi River Board of State Sanitary Engineers, 1978). Alternative design standards suggested by the New England Interstate Water Pollution Control Commission (1980) where flow data are not available include an average per capita flow value of 70 gal/d (0.27 m^3/d) plus an allowance for infiltration. An infiltration allowance of 250 to 500 gal/d · in · mi (0.24 to 0.48 m^3/cm of pipe diameter · km · d) is suggested as a normal range of infiltration.

19.2.4.2 Peaking Factor. Sanitary sewers should be designed based on peak hourly flow rates using one of the following methods:

1. The ratio of peak to average daily flow rate as determined from peaking factor curves similar to the one shown in Fig. 19.1. Many large cities or public agencies have developed peaking factor curves specially for use in their areas. Some regulatory agencies also mandate the peaking factors to be used.

2. Values established from a sewer-system-evaluation survey that has determined inflow/infiltration characteristics of existing tributary sewers.

19.3 QUANTITY OF STORMWATER

Stormwater runoff is the portion of precipitation that flows over the ground surface during and after a precipitation event. The design of storm sewers and combined sewers requires the estimation of design-stormwater-flow rates. There are many methods available for computing stormwater-flow rates including several computer-based models such as the public domain Stormwater Management Model (SWMM),

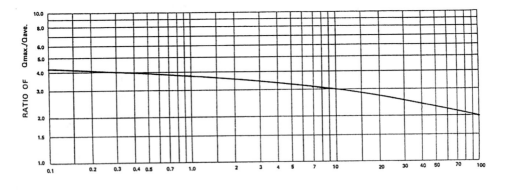

| Qmax: | Maximum rate of wastewater flow (Peak hourly flowrate) |
| Qave: | Average daily wastewater flowrate |

FIGURE 19.1 Peaking factors for the design of sewers. (*Great Lakes, 1978.*)

and the Corps of Engineers STORM and Hydrologic Simulation Program—Fortran (HSPF). Summaries of these three programs may be found in Nayyar (1992). The SCS technique, originally developed by the Soil Conservation Service of the U.S. Department of Agriculture, is also used to develop runoff hydrographs based on land use characteristics. This program and other computer simulation techniques are summarized in McGhee (1991). The procedure described in this section, however, is limited to the rational method, the simplest procedure in common use. Refer to Chaps. 24, 26, and 28 for more details.

19.3.1 The Rational Method

The equation for the rational method is

$$Q = XCiA \tag{19.1}$$

where Q = peak rate of runoff, ft³/s (m³/s)
  C = runoff coefficient
  A = area of drainage basin, acres (hectares)
  i = rainfall intensity, in/h (cm/h)
  X = 1.01 (0.0278 for SI units)

In computing the stormwater runoff rates using the rational method, the following data are required:

1. The time–intensity rainfall relation to be used as the basis of design.
2. The probable future condition of the drainage area.
3. The runoff coefficient, relating the peak rate of runoff at any location to the average rate of rainfall during the concentration time (see item 6) for that location.
4. The probable time required for water to flow over the surface of the ground to the first inlet, called the *inlet time.*
5. The area tributary to the sewer at the location at which the size is to be determined.
6. The time required for the water to flow in the sewer from the first inlet to the location in question that, added to the inlet time, gives the *time of concentration* at that location. The time of concentration is the length of time a rainstorm will persist for design purposes.

Several of these important factors are covered in the following discussion.

19.3.2 Time of Concentration

The rational method assumes that the maximum flow rate will occur at the time when all of the runoff flows from the contributing watersheds reach the outlet.

The inlet time will vary from 5 to 10 minutes for high-density areas with closely spaced street inlets, from 10 to 20 minutes for well-developed areas with relatively flat slopes, and from 20 to 30 minutes for residential areas with widely spaced street inlets.

19.3.3 Rainfall Frequency

The cost is usually prohibitive to construct storm sewers capable of handling the largest conceivable storms. Common practice is to use storm events having a frequency of occurrence of once every 2 to 10 years for the design of storm sewers in residential areas, and a frequency of occurrence of 10 to 30 years for commercial and high-value districts.

19.3.4 Rainfall Intensity-Duration-Frequency Relationships

The rainfall characteristics that must be known for storm sewer design are presented in a concise manner by intensity-duration-frequency curves, which can be prepared from a long record of precipitation at a given weather station. An example set of curves is shown in Fig. 19.2 for the northeastern area of Illinois for the years 1901 to

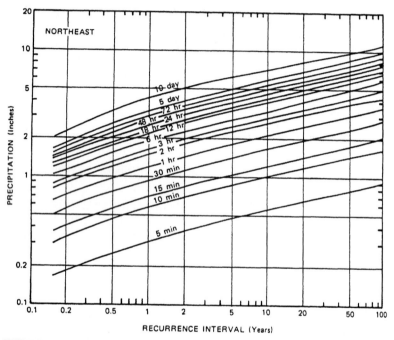

FIGURE 19.2 Example intensity-duration-frequency curves used for storm sewer design (*Huff and Angel, 1989*).

1983 (Huff and Angel, 1989). Use of the curve is exemplified as follows: for a system to be designed with a time of concentration of 30 minutes and a storm recurrence interval of 5 years, the total precipitation (from Fig. 19.2) is 1.3 in (33 mm). This means that the average precipitation of 1.3 in will be equaled or exceeded once every 5 years. The rainfall intensity term i is the precipitation (in/h) divided by the duration (expressed in hours) or $1.3/0.5 = 2.6$ in/h. Similar data for other localities are published by the U.S. Weather Bureau.

19.3.5 Runoff Coefficient

The runoff coefficient C in the rational method equation is difficult to estimate precisely. Whereas the use of the runoff coefficient implies there is a constant ratio of rainfall to runoff, the actual ratio will vary over the course of a storm due to the condition of the area and the variability of the rainfall patterns. A common practice is to use average coefficients for various types of areas and assume the coefficients will be constant throughout the duration of the storm. Typical values of C are given in Table 19.6.

TABLE 19.6 Rational Equation Runoff Factors

Type of area	C factors
Business	
Downtown	0.70–0.95
Neighborhood	0.50–0.70
Residential	
Single family	0.30–0.50
Multiunits, detached	0.40–0.60
Multiunits, attached	0.60–0.75
Residential (suburban)	0.50–0.70
Apartment	0.50–0.70
Industrial	
Light	0.50–0.70
Heavy	0.60–0.90
Parks, cemeteries	0.10–0.25
Playgrounds	0.20–0.35
Railroad yard	0.20–0.35
Unimproved	0.10–0.30
Pavement	
Asphaltic or concrete	0.70–0.95
Brick	0.70–0.85
Roofs	0.75–0.95
Lawns, sandy soil	
Flat, 2%	0.05–0.10
Average, 2–7%	0.10–0.15
Steep, >7%	0.15–0.20
Lawns, heavy soil	
Flat, 2%	0.13–0.17
Average, 2–7%	0.18–0.22
Steep, >7%	0.25–0.35
Gravel driveways and walks	0.15–0.30

Source: ASCE and WPCF (1969).

When estimating stormwater runoff quantities by the rational method, the average rainfall intensity is assumed to be the value that corresponds to a duration equal to the time of concentration.

19.4 HYDRAULICS OF SEWERS

It is normal practice to design sewers as open channel flow, i.e., to flow with a free-water surface. The advantages are: (1) the free-water surface will allow ventilation of the sewers, which prevents the accumulation of toxic gases, and (2) the velocities at lower flows can be kept reasonably high to facilitate self-cleaning of the sewers.

19.4.1   Flow Equations

The two equations most commonly used in gravity sewer and force-main design are the Manning equation and the Hazen-Williams equation. Refer to Chaps. 2 and 25 for more details.

19.4.1.1   Manning Equation.   The most widely used formula in calculating open-channel uniform flow is the Manning equation. The Manning equation is:

$$V = \frac{1.486}{n} R^{2/3} S^{1/2} \qquad \text{U.S. customary units} \qquad (19.2)$$

$$V = \frac{1}{n} R^{2/3} S^{1/2} \qquad \text{SI units} \qquad (19.2a)$$

where V = velocity, ft/s (m/s)
   n = dimensionless coefficient of roughness
   R = hydraulic radius, ft (m)
   S = slope of the energy grade line, ft/ft (m/m)

The hydraulic radius R is defined as

$$R = \frac{\text{cross-sectional area of flow, ft}^2 \text{ (m}^2)}{\text{wetted perimeter, ft (m)}} \qquad (19.3)$$

For a pipe flowing full, the hydraulic radius is

$$R = \frac{(\pi/4)\,(D^2)}{\pi D} = \frac{D}{4} \qquad (19.4)$$

where D = ft (m).

   The Manning equation can be transformed to determine the flow rate in a circular pipe *running full* at a particular slope as follows:

$$Q = \frac{0.463}{n} D^{8/3} S^{1/2} \qquad \text{U.S. customary units} \qquad (19.5)$$

$$Q = \frac{0.312}{n} D^{8/3} S^{1/2} \qquad \text{SI units} \qquad (19.5a)$$

where Q = ft³/s (m³/s).

   Typical n values for various types of pipe are presented in Table 19.7. In general, values ranging from 0.13 to 0.15 are used in sewer design.

19.4.1.2   Hazen-Williams Equation.   For the flow of water in pipes under pressure, such as wastewater-force mains, the Hazen-Williams equation is the one most commonly used. The Hazen-Williams equation is

TABLE 19.7   Ranges of Manning's n

Pipe material	Manning's n
Asbestos cement	0.011–0.015
Brick	0.013–0.017
Vitrified clay	0.011–0.015
Concrete	0.011–0.015
Cast iron	
Coated	0.011–0.013
Uncoated	0.012–0.015
Cement-lined	0.010–0.013
Corrugated metal	
Plain	0.022–0.026
Paved invert	0.018–0.022
Spun-asphalt lined	0.011–0.015
Plastic (smooth)	0.011–0.015
Steel	
Welded	0.010–0.014
Riveted	0.013–0.017

Source: Adapted from ASCE and WPCF (1969); Corbitt (1990); and Metcalf & Eddy (1981).

$$V = 1.318\ CR^{0.63}S^{0.54} \qquad \text{U.S. customary units} \qquad (19.6)$$

$$V = 0.849\ CR^{0.63}S^{0.54} \qquad \text{SI units} \qquad (19.6a)$$

where V = velocity, ft/s (m/s)
       C = coefficient of roughness (C decreases with roughness)
       R = hydraulic radius, ft (m)
       S = slope of the energy grade line ft/ft (m/m)

If $D/4$ is substituted for the hydraulic radius R, the Hazen-Williams equation written in terms of the flowrate Q is

$$Q = 0.432\ CD^{2.63}S^{0.54} \qquad \text{U.S. customary units} \qquad (19.7)$$

$$Q = 0.278\ CD^{2.63}S^{0.54} \qquad \text{SI units} \qquad (19.7a)$$

where Q = flowrate, ft³/s (m³/s).
Typical C values for various types of pipe are given in Table 19.8.

TABLE 19.8   Ranges of Hazen-Williams C

Pipe material	Hazen-Williams C
Asbestos cement	120–140
Clay	100–130
Concrete	100–140
Cast Iron*	100–130
Plastic	130–140
Steel	110–120

* Upper range is for lined pipe.
Source: Corbitt (1990).

19.4.2 Pipe Sizes

In the United States, it is standard practice to express pipe sizes in inches. There are differences when U.S. customary units are converted to metric as compared to standard metric sizes. For convenience of use, a comparison of U.S. customary to metric sizes is given in Table 19.9.

TABLE 19.9 Sewer Pipe Sizes in U.S. Customary and Metric Units

Standard U.S. pipe sizes, in	U.S. pipe sizes converted to metric, mm	Standard metric sizes, mm
4	101.6	100
6	152.4	150
8	203.2	200
10	254.0	250
12	304.8	300
15	381.0	375
18	457.2	450
20	508.0	500
21	533.4	525
24	609.6	600
27	685.8	675
30	762.0	750
36	914.4	900
42	1066	1050

19.4.3 Design Nomographs

To aid in the solution of flow equations, various design charts and tables may be used. Many have been developed for sewers that are flowing full. For the Manning equation, a nomograph for the solution of circular pipes flowing full in U.S. customary and SI units is given in Fig. 19.3. For solution of the Hazen-Williams equation, a similar nomograph is presented in Fig. 19.4. Software is also available for computer-aided design of sewers.

In general, sewers are designed to flow 90 percent full at maximum flow. Because of the variety of problems that occur in sewer design, it is necessary to estimate the velocity and flow rate when a sewer is flowing partially full. The relationships between the hydraulic elements for flow at full depth and at other depths in circular sewers, computed according to the Manning equation, are shown in Fig. 19.5. The hydraulic elements for a circular sewer, shown in Fig. 19.5 are the hydraulic radius R, the cross-sectional area of the flow A, the average velocity V, and the rate of discharge Q. The roughness factor, Manning's n, varies with depth and the curves in Fig. 19.5 indicate how the assumption for n affects the velocity and flow-rate calculations.

19.4.4 Minor Losses

Head losses in sewer systems where the magnitude and direction of flow changes are called *minor losses*. These losses can usually be expressed as functions of the squares

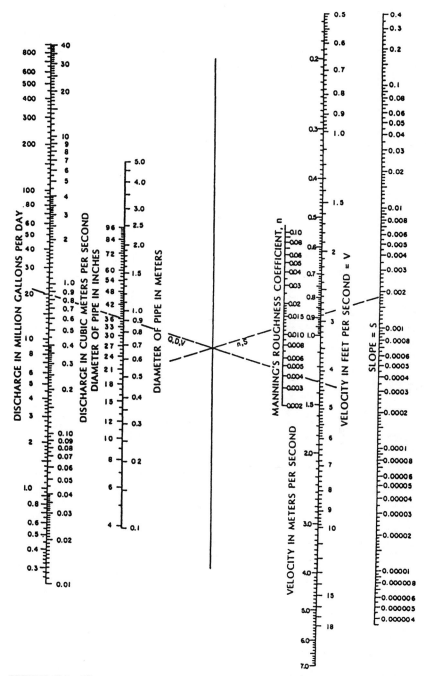

FIGURE 19.3 Alignment chart for flow in pipes, Manning formula (*ASCE and WPCF, 1982*).

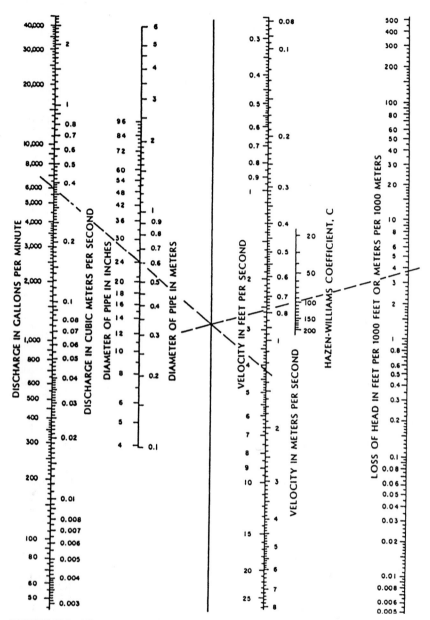

FIGURE 19.4 Alignment chart for flow in pipes, Hazen-Williams formula (*ASCE and WPCF, 1982*).

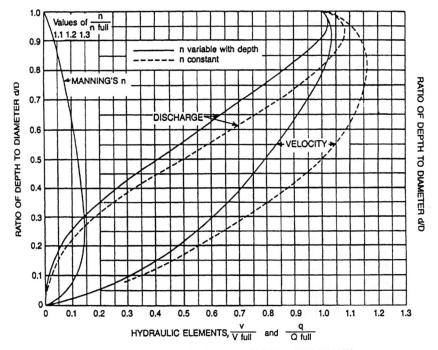

FIGURE 19.5 Hydraulic properties of circular sewers (*ASCE and WPCF, 1982*).

of the fluid velocities. Minor losses in closed conduits are somewhat different than those in open channels.

19.4.4.1 Minor Losses in Closed Conduits.
Most minor losses in a closed conduit system can be expressed as a multiple of the velocity head immediately upstream, within, or immediately downstream of the fitting or pipe configuration causing the loss. The general equation is

$$h_m = K \frac{V^2}{2g} \tag{19.8}$$

where h_m = minor loss, ft (m)
K = minor loss coefficient
$V^2/2g$ = velocity head, ft (m)
g = acceleration due to gravity of 32.2 ft/s^2 (9.81 m/s^2)

Minor loss coefficients for commonly used fittings and appurtenances are given in Table 19.10. Coefficients are applied to the upstream velocity head, except as noted.

19.4.4.2 Minor Losses in Open Conduits.
Transitions in open conduits include changes in direction, slope, or cross section that produce a change in the state of flow. Changes in direction that produce losses may be approximated using Eq. (19.8) and the loss coefficients given in Table 19.11.

TABLE 19.10 Typical Minor Loss Coefficients
for Pipe Fittings and Appurtenances

Fitting	K
90° bend—standard	0.25
90° bend—long radius	0.18
45° bend	0.18
180° bend	0.4
Tee—flow straight through	0.3
Tee—branch flow to mainstream	0.75
Reducer—conical	0.04*
Entrance (flush—sharp-edged)	0.5
Entrance—bellmouth	0.05
Exit	1.0
Gate valve—resilient seat	0.3
Check valve—swing type, fully open	0.6–2.2

   * Applied to exit velocity.
   Source: Adapted from Metcalf & Eddy (1981); and
Sanks, et al. (1989).

TABLE 19.11 Loss Coefficients in Open
Channels

Type of change	$K*$
Smooth 90° bend	0.6 b/r
Smooth 45° bend	0.4 b/r
Smooth 22½° bend	0.3 b/r
Angular deflection	$(\theta/90)^{0.5}$

   * b = channel width; r = radius of curvature; θ =
angular deflection.
   Source: McGhee (1991).

19.5 DESIGN OF SEWERS

Nearly all of the sanitary and storm sewer systems designed employ the concept of gravity flow. Gravity flow offers the advantages of low maintenance, low operating cost, and limited need for highly skilled personnel. The principal disadvantage of gravity-flow sewers is relatively high construction cost due to deep trench excavation, care in maintaining uniform slopes, and the construction of manholes at changes in direction of flow or at pipe intersections.

In small rural communities having low population density or hilly terrain, construction of conventional gravity-flow sewers is prohibitively expensive. To provide affordable alternatives for wastewater collection, alternative sewer-system designs such as pressure or vacuum sewers have been developed. The design aspects of conventional gravity-flow and alternative sewer systems are discussed in this section.

19.5.1 Design Criteria for Gravity-Flow Sewers

Sewer designs are usually governed by design standards in many areas. Some of the commonly used criteria are listed as follows (ASCE and WPCF, 1969; 1982; Metcalf & Eddy, 1975).

Design period. Lateral and branch sewers should be designed for the ultimate population density to be expected for the area served. Larger sewers are commonly designed to handle flow rates for 25 to 50 years in the future.

Minimum size for gravity sewers. 8 in (200 mm) in diameter.

Minimum depth. Sufficient to receive wastewater from basements and to prevent freezing.

Minimum slope. As required to produce minimum velocities when flowing full or half-full.

Maximum velocity for gravity sewers. 10 ft/s (3 m/s) unless special provisions for abrasion, turbulence, and thrust are made.

Alignment. Sewers less than 24 in (600 mm) in diameter should be straight between manholes. Curved sewers should be laid with radii not less than 100 ft (30 m) although many cities consider 200 ft (60 m) as the normal minimum.

19.5.2 Design of Gravity-Flow Sanitary Sewers

Important factors in the design of gravity-flow sanitary sewers are the location to receive wastewater flow from residential, commercial, and industrial contributors; depth of bury sufficient to permit gravity entry (in most cases); structural strength suitable to carry backfill, impact, and live loads; material of construction resistant to corrosion; and slopes sufficient to avoid deposition of solids. Design criteria commonly used for gravity-flow sanitary sewers are summarized in Table 19.12. Minimum slopes given in Table 19.13 have proven to be satisfactory for sewers with diameters ranging from 8 in (200 mm) to 36 in (900 mm).

TABLE 19.12 Commonly Used Design Criteria for Gravity-Flow Sanitary Sewers

Parameter	Criteria
Minimum velocity	2.0 ft/s (0.6 m/s), however, a minimum of 2.5 ft/s (0.76 m/s) is recommended to prevent deposition of mineral matter such as sand and grit.
Minimum depth	Top of sewer not less than 3 ft (0.9 m) below basement floor or invert not less than 6 ft (1.8 m) below house foundation. If water lines are located nearby, sewer depth should be below water line elevation.
Manhole spacing	Maximum spacing of 300 to 400 ft (90 to 120 m). For large sewers, spacing of 500 ft (150 m) may be used.
Construction at manholes	Crowns of pipes entering and leaving a manhole should be at nearly the same elevation so that gases are not trapped.

Source: Adapted from ASCE and WPCF (1969).

TABLE 19.13 Minimum Slopes of Gravity-Flow Sanitary Sewers

Size		Slope, ft/ft (m/m)	
in	mm	$n = 0.013$	$n = 0.015$
8	200	0.0040	0.0044
10	250	0.0028	0.0033
12	300	0.0022	0.0026
15	375	0.0015	0.0019
18	450	0.0012	0.0015
21	525	0.0010	0.0012
24	600	0.0008	0.0010
27	675	0.0007	0.0009
30	750	0.0006	0.0008
36	900	0.0005	0.0006

Source: Great Lakes–Upper Mississippi River Board of State Sanitary Engineers (1978); and Metcalf & Eddy (1981).

It is also important to streamline the flow through manholes, junction structures, and other structures to minimize turbulence and head loss. Minimizing turbulence is particularly important because turbulence can release hydrogen sulfide and accelerate corrosion of the pipe and concrete surfaces. Design considerations for corrosion control are discussed later in this section.

Common materials used in sanitary sewers are concrete, vitrified clay, plastic, and ductile iron (for installations where high structural strength is required or for pressure sewers). For concrete pipes subject to corrosion, special plastic linings are used. Types of plastic pipe used are thermoplastic pipe, i.e., acrylonitrile-butadiene-styrene (ABS), ABS composite, polyethylene (PE), and polyvinyl chloride (PVC); and thermoset plastic pipe, i.e., reinforced thermosetting resin (RTR) and reinforced plastic mortar (RPM) (ASCE and WPCF, 1982). Asbestos cement has been used extensively in the past for sewer pipe but its use has diminished significantly due to health hazards associated with its manufacture. Steel pipe is rarely used for sewers. Descriptions of pipe materials and their sensitivity to corrosion are discussed in Sec. 19.5.6.

19.5.3 Design of Gravity-Flow Storm Sewers

Design concepts for storm sewers have changed considerably in recent years. In the past, sewers were frequently designed to drain local areas rapidly with little consideration given to the cumulative effect on an urban area. The results of such development have included flooding because of increased runoff, and loss of groundwater recharge because of both increased imperviousness (due to more paved areas) and decreased detention time. It is particularly important now to consider how the peak flows can be reduced to mitigate potential flooding problems either to the receiving streams or to downstream sections of the storm sewer system. Techniques used for reducing the quantity of runoff and peak flows are summarized in Table 19.14 (McGhee, 1991).

Storm drainage systems should be considered to consist of two components, sometimes called the *minor system* and the *major system*. The minor system consists

TABLE 19.14 Commonly Used Techniques for Reducing Storm Flows

Parameter	Mitigation measure
Roof drains	Discharge to grassed areas rather than paved areas.
Surface grading	Use contour grading that maintains natural drainage patterns. Where possible use detention basins or ponds that cause deliberate ponding of runoff that can be discharged after the storm subsides.
Paving	Use porous pavements of asphalt, interlocking paving blocks, or gravel.
Ditches	Use grassed ditches instead of curbed streets.
Subsurface disposal	Use perforated storm sewers, infiltration trenches, or dry wells.

Source: Adapted from McGhee, (1991).

of channels and sewers designed to accommodate a storm of moderately short recurrence interval, on the order of 2 to 5 years, depending on the character of the area. The purpose of this system is to prevent flooding of roadways and adjoining areas by moderate storms that occur relatively frequently. The major system is the path followed by the flow during runoff events that exceed the capacity of the minor system. The components of the major system include the streets and adjoining land, drainage rights of way for surface flow, and natural drainage channels that are pre-served during development. The major system should be adequate to carry flows resulting from storms of long recurrence intervals—at least 25 years. Commonly used design criteria for major and minor systems are listed in Table 19.15.

19.5.3.1 Minor Systems. Because the minor system is intended to accommodate storms of relatively low intensity, a high degree of precision is not required. Subarea flows can be determined using the rational method, synthetic hydrographs, or simple simulation programs. The only difference in using the rational method as compared to the other methods is determining the design flows for the individual sewers.

The steps in designing individual storm sewers are outlined in Table 19.16 using the rational method for design as outlined in Sec. 19.3. The locations of the sewers are usually in streets and follow the natural slope of the ground. In storm-sewer design, the tributary area to each inlet must next be determined based on the ground contours. Additional tabular procedures used in computing sizes for a network of sewers are described in McGhee (1991).

TABLE 19.15 Commonly Used Design Criteria for Gravity-Flow Storm Sewers

Parameter	Criteria
Minimum velocity	3.0 ft/s (1.0 m/s).
Depth below grade	Sufficient to receive water from street inlets and other tribu-tary drains.
Depth of flow	Full depth or slightly surcharged at design flow.
Manhole spacing	Maximum spacing of 300 to 400 ft (90 to 120 m). For large sewers, spacing of 500 ft (150 m) may be used.
Construction at manholes	Crowns of pipes entering and leaving manhole should be at nearly the same elevation. At changes in direction of flow, drop the invert of the leaving sewer by about 1 in (30 mm).

Source: Adapted from ASCE and WPCF (1969); and Metcalf & Eddy (1981).

TABLE 19.16   Steps in Designing Gravity-Flow Storm Sewers

---

1. Determine sewer location.
2. Determine tributary area to sewer A.
3. Select runoff coefficient C for the tributary area (see Table 19.6).
4. Compute equivalent area, i.e., CA.
5. Select recurrence interval for storm, e.g., 2 years.
6. Select or compute the time of concentration. The time of concentration consists of inlet time or the time required for the runoff at the upper end of the area to reach the nearest inlet, plus the time of flow in the sewer from this inlet to the point being considered.
7. Using information from steps 5 and 6 and the intensity-duration-frequency curves (see Fig. 19.2), compute the precipitation and rainfall intensity i (precipitation/time of concentration).
8. Compute peak rate of runoff Q using Eq. (19.1).
9. Determine a preliminary slope of the sewer (usually based on ground slope).
10. Select an initial pipe size (Fig. 19.3).
11. Compute velocity and capacity of sewer.
12. Establish invert elevations of the sewer and check depth of cover.
13. Repeat steps 9 through 12 if necessary if slope, depth of cover, and velocity requires modification.

19.5.3.2   *Major Systems.* The major system, together with the minor system, is expected to prevent flooding of houses and other valuable property resulting from storms of relatively low frequency. The design storm will have a recurrent interval of 25 or more years. The major system may also incorporate features to reduce the peak flow of the tributary area such as stormwater-detention basins that have a controlled discharge. Flow calculations for the major system must be based on simulation models, some of which are listed in Sec. 19.3. The rational method is not sufficient.

19.5.4   Alternative Sewer Systems

In small and rural communities where soil conditions are not suitable for individual septic systems, sewer systems and treatment facilities are required. For many systems, the principal component of cost is constructing the sewer system. Alternative sewer systems have been developed to reduce the community-collection-system capital costs to affordable levels. The three types of alternative sewer systems used are pressure sewers, vacuum sewers, and small-diameter gravity sewers (Water Environment Federation, 1986).

19.5.4.1   *Pressure Sewers.* In pressure sewer systems, wastewater from individual residences or buildings is collected and discharged into a septic tank or holding tank and then pumped to a pressure or gravity-flow collector. The main components of a pressure-sewer system are shown in Fig. 19.6a. Where a septic tank is used to remove solids from the wastewater before it is pumped, the system is referred to as a *septic tank effluent pumping* (STEP) system. Where a holding tank is used, wastewater is discharged periodically into a pressure main by means of a grinder pump that can reduce the size of the solids in the wastewater. A septic tank or holding tank and pump are required at each inlet point to the pressure main. The advantages of a pressure-sewer system are: (1) it is especially adaptable to small communities, (2) small-diameter pipe laid at shallow depths can be used, and (3) several small pump-

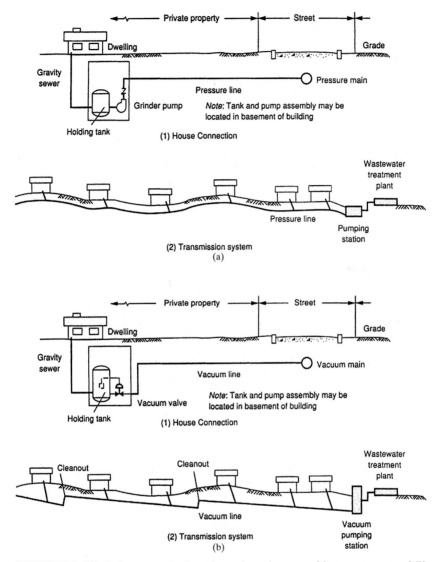

FIGURE 19.6 Principal components of pressure and vacuum sewers: (*a*) pressure sewer and (*b*) vacuum sewer. (*Metcalf & Eddy, 1981.*)

ing stations can be eliminated. The principal disadvantage is the operation and maintenance of several household pumping systems.

19.5.4.2 Vacuum Sewers. A modern vacuum-sewer system has three major subsystems: central collection station, collection network, and on-site facilities. The principal features are shown in Fig. 19.6*b*. In these systems wastewater from a building flows by gravity to an on-site vacuum ejector that consists of a holding tank and vacuum valve of special design. When a given amount of wastewater accumulates in the holding tank, the valve is programmed to open and close after the wastewater enters the system as a liquid plug and passes through the network to the central collection station. Vacuum pumps in a central collection station maintain vacuum in the system. The principal disadvantage of a vacuum system is the high potential for air to be captured through unsealed manholes. An advantage is that odor generation is minimized.

Plastic pipe is used throughout a vacuum-sewer system. A gravity house-sewer pipe is usually 4 in (100 mm) in size and mains range from 4 to 10 in (100 to 250 mm) depending on the flow and layout. Several mains may be served by a single collection station.

The advantages of vacuum sewers are similar to pressure sewers. Vacuum sewers differ from pressure sewers in that individual household pumps are not required but a central vacuum station is. The principal disadvantage of a vacuum sewer system is that it may not be as cost-effective as STEP or grinder pump pressure sewer systems.

19.5.4.3 Small-Diameter Gravity Sewers.

Small-diameter gravity sewers are used mostly in small, rural communities. Frequently small-diameter sewers are used to transport septic tank effluent. The advantages of small-diameter gravity sewers over conventional sewers include lower capital cost because of reduced pipe and installation costs, cleanouts can be used instead of manholes, and reduced inflow/infiltration resulting in smaller quantities of flow that need to be pumped and treated. The disadvantages are: Small-diameter sewers are prone to clog if oversize material is discharged to the piping system; where septic tanks are used, they have to be maintained and cleaned to prevent solids carryover; and odors have to be controlled where the wastewater transported is septic tank effluent.

19.5.5 Design Considerations for Sulfide and Corrosion Control

Most wastewaters naturally contain sulfate; it is present in many water supplies. Sulfate is reduced biologically to sulfide under anaerobic conditions that occur frequently in collection systems. Sulfide combines with hydrogen to form hydrogen sulfide (H_2S), which is released to the air space above the liquid surface in the sewer. Accumulated H_2S can be oxidized biologically to sulfuric acid that is corrosive to sewer pipes.

Most of the sulfate reduction occurs in the biological slime layer on the pipe wall or in the deposits of silt and sludge in the bottom of the pipe. These slime layers are a combination of filamentous organisms and gelatinous material embedding smaller bacteria. The slime layers are typically 0.01 to 0.04 in (0.3 to 1.0 mm) thick, depending on the velocity and abrasive content of the wastewater. Factors that contribute to the growth of slime layers and affect sulfide generation and corrosion are summarized in Table 19.17.

The release of hydrogen sulfide from wastewater can be controlled by: (1) maintaining aerobic conditions by adding air, oxygen, or hydrogen peroxide to the wastewater, (2) reducing anaerobic growth by adding disinfectants such as chlorine or changing the pH by the addition of sodium hydroxide, (3) oxidizing hydrogen sulfide by chemical addition such as potassium permanganate, or (4) reducing turbulence, especially in manholes. In large concrete sewers, plastic linings can be installed in the pipes to safeguard against corrosion. Additional information on the control of hydrogen sulfide in sewers can be found in ASCE (1989); Metcalf & Eddy (1981); and U.S. EPA (1985; 1992).

TABLE 19.17   Factors Affecting Sulfide Generation and Corrosion in Sewers

Factor	Effect
Wastewater characteristics	
Dissolved oxygen (DO)	Low DO favors proliferation of anaerobic bacteria and subsequent sulfide generation. Sulfate reduction usually occurs when the dissolved oxygen in the wastewater is in the range of 0.1 to 1.0 mg/L.
Concentration of organic material and nutrients	Organic material and nutrients diffuse into the slime layer resulting in increased growth of the microorganisms and buildup of the slime layer. High biochemical oxygen demand (BOD) encourages microbial growth and DO depletion and increases sulfide generation.
Sulfate concentration	If sulfate is abundant in the wastewater, the sulfide generation rate will be proportional to the organic matter and/or nutrient concentrations. If sulfate is limiting, sulfide production will be proportional to the sulfate concentration.
pH	Sulfate-reducing bacteria can exist in a pH range of 5.5 to 9.0. Low pH favors a shift to dissolved hydrogen sulfide gas.
Temperature	Higher temperatures accelerate the biological activity of sulfate producers. The rate of sulfide production increases 7% per each 1°C rise in temperature up to 30°C.
Humidity in sewer	Biological growth requires high humidity levels.
Metals concentration	Metals such as iron, zinc, copper, lead, and cadmium form insoluble metallic sulfides upon reaction with dissolved sulfide.
Sewer system characteristics	
Slope and velocity	Slope and velocity affect the degree of reaeration, solids deposition, and the release of hydrogen sulfide.
Turbulence	Turbulence, while improving reaeration, will increase the release of hydrogen sulfide.
Intermittent surcharging	Reduces oxygen transfer and promotes sulfide generation. May inhibit acid production by flushing and dilution.
Sewer pipe materials	Corrosion resistance of pipe materials varies widely.
Concrete alkalinity	Higher alkalinity concrete reduces the corrosion rate.
Accumulated grit and debris	Slows wastewater flow, traps organic solids, increases detention time, and increases sulfide generation.

Source: Adapted from U.S. EPA (1985; 1992).

19.5.6   Pipe-Materials Selection

The materials of which sewers are most commonly constructed are vitrified clay, concrete, plastic, asbestos cement, and ductile iron. Description of the pipe materials and their sensitivity to corrosion are reported in Table 19.18.

19.5.7   Pipe Linings and Coatings

Linings are differentiated from coatings in that the former are fixed to the interior wall of a pipe by projections embedded in the concrete at the time of manufacture, whereas the latter are fixed to the interior wall by adhesion. A typical thickness of a liner is on the order of 65 mils (1.65 mm), while a coating thickness on the order of 40 to 125 mils (1 to 3.18 mm) is required (ASCE, 1989).

TABLE 19.18 Pipe Materials Used in Sewer Construction

Material	Description	Sensitivity to acid corrosion
Vitrified clay pipe (VCP)	Manufactured in standard and extra-strength classifications. Widely used especially in smaller sizes. VCP is available in sizes ranging from 4 to 36 in (100 to 900 mm) and in laying lengths from 1 to 10 ft (0.3 to 3 m). VCP is brittle and requires careful handling in transport and installation.	VCP is virtually immune to acid attack.
Concrete pipe	Available in either reinforced or unreinforced classifications. Concrete pipe is widely used especially in larger sizes. Reinforced concrete pipe is available in sizes from 12 to 144 in (300 to 3600 mm) in diameter in five strength classifications. Unreinforced concrete pipe is available in sizes from 4 to 36 in (100 to 900 mm) in diameter.	Susceptible to acid corrosion. Concrete can be fortified against attack by using calcareous aggregate, increasing the cement content, or both, which provides additional alkalinity and acid neutralizing capacity. Plastic linings can also be installed.
Plastic pipe	Types of plastic pipe used in collection systems include polyvinyl chloride (PVC), acrylonitrile-butadiene-styrene (ABS), and polyethylene (PE). Materials offer low friction characteristics and light weight. Plastic pipes are generally less rigid and require proper bedding and lateral support.	Materials are resistant to acid corrosion although subject to strain corrosion in the presence of some materials such as detergents, organic solvents, and fats and oils.
Asbestos cement pipe (ACP)	ACP has been used widely in sewer construction but has been limited in current use because of health concerns in its manufacture. ACP is available in sizes up to 42 in (1000 mm).	Susceptible to acid corrosion. Due to its higher cement content, it corrodes at a slower rate than granitic concrete, although this attribute is offset generally by a thinner pipe wall.
Steel pipe	Used primarily for force mains and in locations where high-strength pipe is required.	Susceptible to corrosion by sulfuric acid and hydrogen sulfide and the normal oxidation of iron.
Ductile iron pipe (DIP)	Used primarily for force mains and in locations where high-strength pipe is required.	Also susceptible to corrosion by sulfuric acid and hydrogen sulfide. Generally lasts longer than steel.

Source: Adapted from ASCE (1989).

Liners for new concrete pipe generally come in two forms: sheets, in which sheets of PVC or PE are mechanically anchored to the pipe wall, or thin-walled pipe such as smaller-diameter PVC, high-density PE, or glass-reinforced polyester which are inserted into a larger-diameter conduit.

Two principal groups of coatings have been employed on new pipes: (1) thermoplastic coatings such as asphaltic/coal tar coatings and polyethylene that are applied hot and cured by cooling, and (2) thermosetting coatings such as polyurethane and epoxy that set by chemical reaction caused by a curing agent which induces molecular cross-linking. For a list of coatings and their performance in a corrosive environment, refer to ASCE (1989).

19.5.8 Sewer Rehabilitation

A number of methods have been used to rehabilitate wastewater-collection systems. These methods include open-trench excavation for pipeline repair, chemical grouting, and various types of structural renovation such as reinforced-shotcrete lining and slip lining. A summary of structural renovation techniques is provided in Table 19.19.

TABLE 19.19 Structural Renovation Techniques for Pipelines

Technique	Applicability		Diameter range, in	Comment
	Pressure lines	Gravity lines		
Inversion lining	Yes	Yes	up to 120	Low cross-sectional area loss. Access at existing manholes. Service lateral connections relatively easily remade.
Segment lining	No	Yes	up to 36	Hand installed. Access via existing manholes. Reconnections not difficult.
Pipe lining:				
Standard slip lining	Yes	Yes	As required	Loss of capacity due to smaller cross-sectional area. Connections difficult to remake. Access required to line. Grouting of annular space between pipe and lining.
Short-length lining	No	Yes	As required	Loss of capacity due to smaller cross-sectional area. Connections difficult to remake. Access by existing manholes. Grouting of annular space between pipe and lining.
Modified slip lining	Yes	No	4 to 20	Reduces cross-sectional area loss and eliminates grouting. Requires access to line.
Spiral winding	No	Yes	3 to 36	Minimizes cross-sectional area loss. Access via existing manholes. Uses PVC strip to form lining.
Spraying:				
Gunite	No	Yes	up to 72	Adaptable to variable cross sections. Access via manholes.
Ferrocement	Yes	Yes	up to 36	As above with increased strength and pressure capacity.

Source: Adapted from Fedotoff, et al. (1991).

19.6 SEWER APPURTENANCES

Many appurtenances are essential to the proper functioning of sewer systems. Appurtenances may include manholes, bends, and special structures for all types of sewers; service connections, cleanouts, and inverted siphons (depressed sewers) for sanitary sewers; and inlets and catch basins for storm sewers. Many municipal agencies have developed their own standards for design. The following discussion outlines the essential features of many common appurtenances. Additional details may be found in ASCE and WPCF (1982).

19.6.1 Standard Manholes

Manholes are located at junctions of sewers and at changes of grade, size, or alignment, except for curved sewers. Street intersections are common location points. Manholes should be installed at distances not greater than 400 ft (120 m) for sewers 15 in (375 mm) in diameter or less, and 500 ft (150 m) for sewers 18 to 30 inches (460 to 760 mm) (Great Lakes–Upper Mississippi River Board of State Sanitary Engineers, 1978). Typical manholes for small sewers are shown in Fig. 19.7. Greater spacing may be used for larger sewers up to a maximum of 600 ft (180 m). A manhole or terminal cleanout is used at the upper end of a sewer for convenience in flushing or cleaning.

19.6.2 Drop Manholes

Drop-type manholes should be provided for a sewer entering a manhole at an elevation of 24 in (0.6 m) or more above the manhole invert. Where the difference in elevation between the incoming sewer and the manhole is less than 24 in, the invert should be filleted to prevent solids deposition (Great Lakes–Upper Mississippi River Board of State Sanitary Engineers, 1978).

Typical drop manholes are constructed with either an internal or external drop connection (see Fig. 19.8), but an external connection is preferred for structural reasons. Problems with drop manholes usually relate to stoppages in the drop pipe; cleaning drop manholes while wastewater is spilling out of the upper pipe is objectionable and may be hazardous to the health and safety of the sewer maintenance personnel.

19.6.3 Bends

Where bends in sewers are required, they should be of long radius and constructed with a high degree of workmanship to produce channels that are smooth, with uniform sections, radius, and slope. The radius for optimum performance commonly recommended is three times the pipe diameter or channel width. Reasonably satisfactory conditions can usually be determined if the radius is not less than 1.5 times the diameter (ASCE and WPCF, 1982). If the velocity is supercritical, surface turbulence and energy losses arise even with long-radius bends.

19.6.4 Service Connections

Service connections, also called house connections, are branches between the street sewer and the property or curb lines. Sizes used are 4, 5, or 6 in diameter, although 5

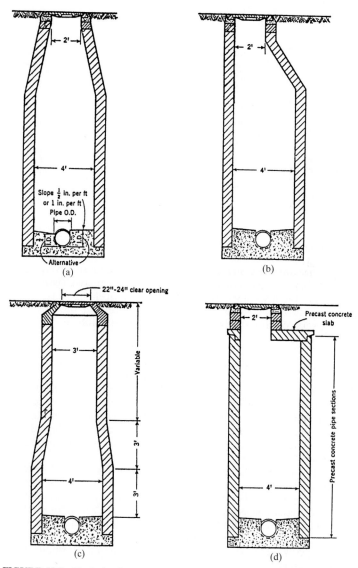

FIGURE 19.7 Manholes for small sewers (Note: ft × 0.3 = m; in × 2.54 = cm).
(*ASCE and WPCF, 1969.*)

in is seldom used in new construction. The slope of the service connection is usually
2 percent. Materials of construction, joints, and workmanship for the service connection should be equal to those of the street sewer to minimize infiltration and root
penetration. For deep sewers, "chimneys," vertical sections of pipe, should be provided in the street sewer to facilitate the service connections. A typical chimney is
shown in Fig. 19.9.

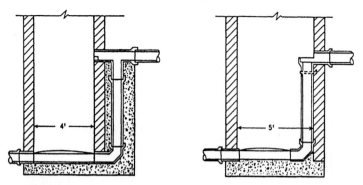

FIGURE 19.8   Drop manholes (Note: ft × 0.3 = m; in × 2.54 = cm). (*ASCE and WPCF, 1969.*)

19.6.5   Terminal Cleanouts

Terminal cleanouts are sometimes used at the upstream ends of sewers, in particular lateral sewers, where there is a relatively short run to a manhole. Their purpose is to provide a means for inserting cleaning rods or for flushing. A typical terminal cleanout is shown in Fig. 19.10. Many regulations restrict the use of terminal cleanouts to within 150 to 200 ft (45 to 60 m) of the next manhole.

19.6.6   Depressed Sewers (Inverted Siphons)

The term *inverted siphon* is commonly used for any dip or sag in the sewer profile that is necessary for the sewer to pass under an obstruction or to cross a waterway. The dip in the line is not really a siphon but is more appropriately termed a *depressed sewer;* however, this term is seldom used. Because the pipe depression is below the hydraulic grade line, it is always full of water under pressure. A typical depressed sewer is shown in Fig. 19.11.

Because of possible blockages, minimum diameters of depressed sewers are usually about the same as standard sewers: 8 in (200 mm) for sanitary sewers and 12 in (300 mm) for storm sewers. For minimizing deposition of material in the depressed sewers, velocities should be as high as practicable: 3 ft/s (0.9 m/s) for sanitary sewers and 4 to 5 ft/s (1.25 to 1.5 m/s) for storm sewers. To maintain reasonable velocities at all times, multiple pipes are used and arranged so that additional pipes are brought into service progressively as the flows increase. A minimum of two pipes should be used at a depressed sewer installation.

19.6.7   Inlets and Catch Basins

Inlets are structures through which stormwater enters sewers. Catch basins are a type of inlet except that a sump is provided for the collection of solids to prevent their discharge into the storm sewer. In modern practice, it is more common to design sewers with grades that are adequate to maintain self-cleansing velocities and to employ inlet structures that will not accumulate solids.

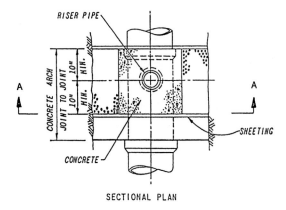

SECTIONAL PLAN

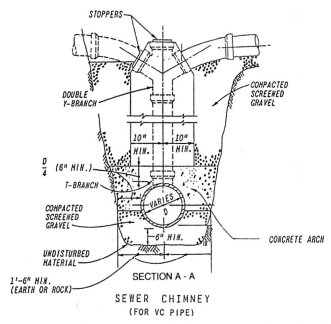

FIGURE 19.9 Typical sewer chimney for service connection (Note: ft × 0.3 = m; in × 2.54 = cm). (*Metcalf & Eddy, Inc., 1981.*)

19.6.7.1 *Inlets.* Stormwater that drains into street gutters is removed through inlets that connect to catch basins or the subsurface storm sewers. Street inlets are susceptible to clogging at the opening because of debris such as leaves, sticks, and wastepaper that may be present in the gutter. Their design depends on the extent that stormwater will be allowed to flood the streets.

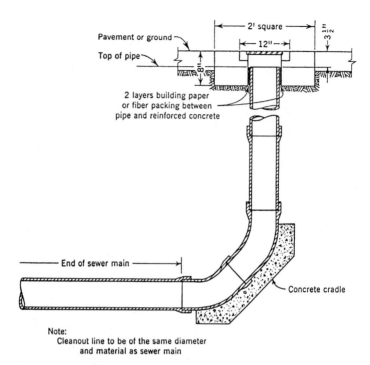

FIGURE 19.10 Terminal cleanout (Note: ft × 0.3 = m; in × 2.54 = cm). (*ASCE and WPCF, 1969.*)

Three general types of stormwater inlets are used: (1) curb inlets, (2) gutter inlets, and (3) combination inlets. A *curb inlet* is a vertical opening at the curb through which the flow passes. A *gutter inlet* is an opening in the gutter beneath one or more grates through which the gutter flow falls. A *combination inlet* includes both a curb opening and a gutter opening with the gutter opening directly in front of the curb opening. A typical street inlet is shown in Fig. 19.12.

19.6.7.2 Catch Basins. Catch basins are used where large quantities of sand and grit are likely to be washed into the inlet. Catch basins limit the amount of gritty material that might be washed into the sewer system and be difficult to remove. Catch basins require regular maintenance so that the inlet is not clogged thereby causing localized flooding. Catch basins that accumulate material and are not cleaned frequently may create offensive odors and provide breeding places for mosquitoes. A typical catch basin is shown in Fig. 19.13.

19.6.8 Ventilation

Natural ventilation from manholes, building vents, and flow variations of the wastewater is normally adequate to provide oxygen to the air space in the sewers. Forced ventilation of a sanitary sewer is considered only as a special application where large quantities of corrosive or odorous gases are present. When forced ven-

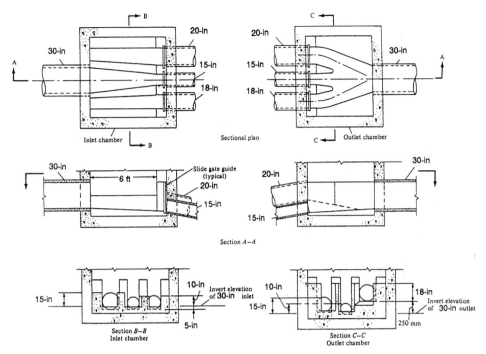

FIGURE 19.11 Definition sketch for a depressed sewer (Note: ft × 0.3 = m; in × 25.4 = mm). (*Metcalf & Eddy, 1981.*)

tilation is required, special airtight or pressure manhole covers must be used and the air exhausted to a high stack or an odor-control system.

19.7 PUMPING STATIONS AND FORCE MAINS

In wastewater-collection systems, pumping stations are often required to pump: (1) untreated wastewater, (2) stormwater, (3) a combination of untreated wastewater and stormwater, and (4) septic tank effluent (for small community systems). The characteristics of the wastewater to be pumped have a large influence on the design of the pumping facilities and the features that should be included. Sanitary wastewater typically does not contain large solids [more than 3 in (75 mm) in diameter] or large quantities of grit but can contain rags and stringy material that can clog pumps. Stormwater typically contains large solids, stringy material, and large quantities of grit. Both wastewater and stormwater pumps often require screens or other protective devices to remove solids or stringy material that may clog pumps. Other important design considerations are the source of energy (electricity, natural gas, or diesel), type of operation such as attended or remotely or automatically operated, and type of construction, i.e., wet-pit, dry-pit, custom designed, or factory assembled. In this section, some of the common types of designs of pumping stations will be discussed. Important features of force-main design will also be highlighted. For addi-

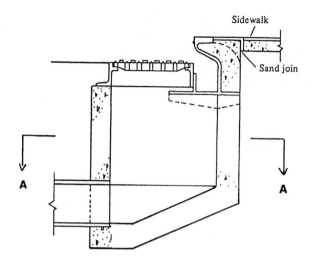

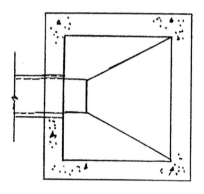

SECTION A - A

FIGURE 19.12 Typical street inlet for stormwater. (*Metcalf &
Eddy, Inc., 1981.*)

tional details on the design of pumping stations, the reader is referred to Hydraulic
Institute (1983); Metcalf & Eddy (1981); Sanks, et al. (1989); and Water Environment
Federation (1993a).

19.7.1 Types of Pumping Stations

Pumping stations that pump stormwater operate intermittently and usually are high
volume–low head type as they are required to move large volumes of water to

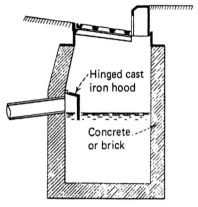

FIGURE 19.13 Typical catch basin. (*McGhee, 1991.*)

prevent flooding and overflows. Storm-water-pumping stations, in most cases, are custom-designed for their specific application.

Wastewater-pumping stations nearly always operate continuously but should also meet wide variations in flow that occur during a normal 24-hour period or are due to inflow/infiltration into the sewer system. Typical pumping stations include conventional wet-pit or dry-pit designs or factory-assembled units. Factory-assembled stations can also be designed for wet-pit or dry-pit applications. Factory-assembled stations are fabricated in modules with all of the equipment and components mounted and connected permanently within the module. Capacities of factory assembled units generally range from 100 to greater than 1600 gal/min (0.006 to >0.1 m³/s). Conventional pumping station capacities range from 300 to more than 10,000 gal/min (0.02 to >0.65 m³/s). The general features of conventional and factory-assembled pumping stations are summarized in Table 19.20 and typical installations are shown in Figs. 19.14 and 19.15.

19.7.2 Types of Pumping Equipment

Most of the pumps used in the pumping of wastewater or stormwater are centrifugal pumps. Centrifugal pumps are generally classified as radial-flow, mixed-flow, and axial-flow. Radial-flow and mixed-flow are most commonly used for pumping wastewater and stormwater as they are capable of passing small solids without clogging. Axial-flow pumps may be used for pumping storm drainage that is not mixed with wastewater. Inclined Archimedes screw pumps are also used for pumping wastewater or stormwater. The commonly used types of pumps are described in the following (Water Environment Federation, 1993a).

19.7.2.1 Vertical Pumps. Vertical pumps, the type most frequently used in stormwater-pumping stations, are available in many capacities and discharge heads (see Fig. 19.16a). Single-stage propeller pumps are used for low heads; mixed-flow pumps are used for higher heads. Two-stage propeller pumps are available that will approximately double the capacity of single-stage pumps.

19.7.2.2 Submersible Pumps. Submersible pumps originally found application as sump pumps but are now manufactured in many sizes. The impellers are of the nonclog design and the pumps are used for pumping untreated (raw) wastewater both for lift stations in the collection system and for raw-wastewater pumping at treatment plants (see Fig. 19.16b). With submersible pumps, no dry pit is required and minimal superstructure is needed. Access openings are required so pumps can be hoisted from wet well for maintenance. Provisions should also be made for periodic flushing of accumulated solids in the wet well to prevent pump clogging.

19.7.2.3 Centrifugal Pumps. A special type of centrifugal pump is used in dry-pit wastewater-pumping stations. The pumps are designed to handle small solids and

TABLE 19.20   General Features of Conventional and Factory-Assembled Pumping Stations

Item	Conventional	Factory-assembled
Station construction	Reinforced concrete substructure; superstructure may be of masonry, reinforced concrete, wood, or metal panels.	Steel, fiberglass.
Wet well	Pump protection equipment, including bar racks and comminutors, is often installed in the wet well of large stations. Entrance to wet well must be from outdoors, preferably by stairs.	Precast concrete manholes are often used as wet wells in small stations.
Dry well	Pumps are located on lower floor; motors and control panels should be located on upper floors for protection against flooding.	Pumps are mounted on station floor with motors mounted directly on pumps. Control panels are wall mounted. Dehumidifier usually provided to control corrosion.
Instrumentation	Includes automatic and manual controls for pumps, high and low wet well level alarms, and flow metering.	Same as conventional stations except if flow metering is required, it has to be located outside of pumping station.
Power source	For reliability, two power sources are required. Dual electric feed lines may be provided or one feed line and an emergency engine-generator may be provided.	Usually a single power source is used with a connection for a feed from a portable engine generator. A permanent engine-generator installation may also be provided.
Heating and ventilation	Depending on climate, heating of wet well and dry well may be required to prevent freezing. Both require ventilation for removal of noxious gases.	Heating and ventilating of dry well only is included in most cases.
Miscellaneous provisions	Lifting eye and hoist or a monorail may be provided for removal of pumps or motors. Access is also required for removal of screens or comminutors. Any electrical equipment in wet well must be explosion proof.	Access to equipment is provided through station entrance tube. Entrance to wet well is through a standard manhole cover.

Source: Adapted in part from Metcalf & Eddy (1981).

are of end-suction, nonclog design furnished in a horizontal configuration. Enclosed impellers and a handhole on the casing permit inspection and the removal of coarse material. Self-priming centrifugal pumps (see Fig. 19.16c) may also be used in small stations where the pumps are mounted above grade on the wet well cover.

19.7.2.4   Volute or Angle-Flow Pumps.   The volute or angle-flow pump (also called a dry-pit angle-flow pump or a single-suction centrifugal mixed-flow pump) usu-

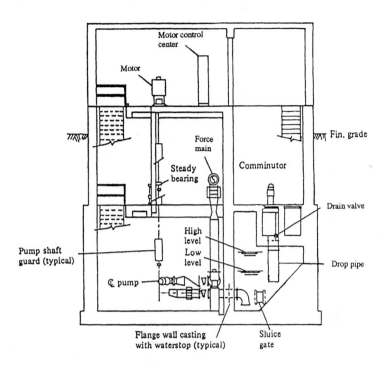

Section through pumping station

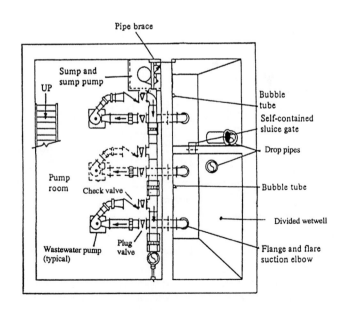

Basement plan

FIGURE 19.14 Conventional wastewater pumping station. (*Metcalf & Eddy, 1981.*)

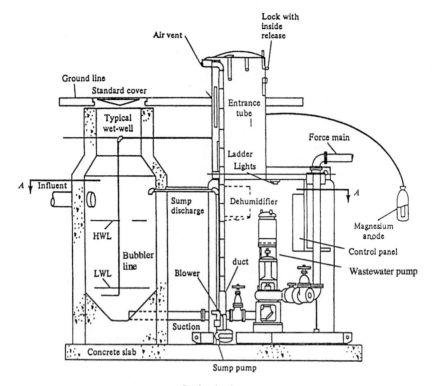

Section A - A

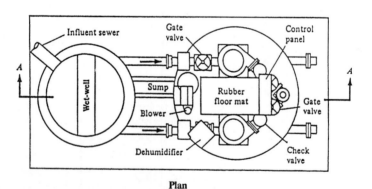

Plan

FIGURE 19.15   Factory-assembled wastewater pumping station. (*Metcalf & Eddy, 1981.*)

ally is mounted in the vertical position with the motor above the pump room operating floor (see Figs. 19.14 and 19.16*d*). Vertically mounted pumps have been used extensively in wastewater pumping. Pumps are also available in a horizontal configuration.

19.7.2.5   *Screw Pumps.* Screw pumps are primarily used for applications with low lift, high capacity, and nonclog requirements and where pressurized transfer is not

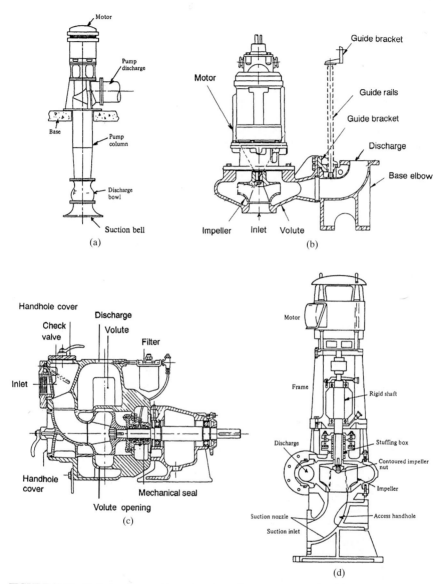

FIGURE 19.16 Typical pumps used in wastewater and stormwater pumping: (*a*) vertical (*Metcalf & Eddy, Inc.*); (*b*) submersible (*Water Environment Federation, 1992*); (*c*) self-priming centrifugal (*Sanks, et al., 1989*); and (*d*) volute (*Metcalf & Eddy, Inc.*).

required. They are manufactured in both open and closed configurations (see Fig. 19.17). Screw pumps are used both for wastewater and stormwater pumping as their performance is largely unaffected by variations of incoming flowrates. They range in size from a minimum of 12 to 144 in (0.3 to 3.7 m) in diameter and have a capacity range of 100 to 70,000 gal/min (6.3 to 4,400 L/s). In wastewater applications, screw

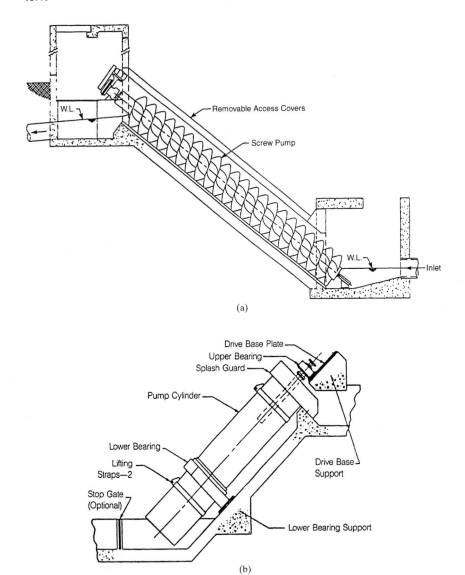

FIGURE 19.17 Screw pumps: (*a*) open and (*b*) enclosed. (*Water Environment Federation, 1993a, 1992.*)

pumps are generally used at treatment plants and not for lift stations in the collection system. Unlike centrifugal pumps, screw pumps require no wet well or suction or discharge piping. Bar racks may not be required because any solids capable of passing through the flights are readily conveyed through the pump. Proper design of the bottom bearing is very important to efficient operation and maintenance.

19.7.3 Pumping System Hydraulics

The analysis of centrifugal pumping systems, the most common system used in wastewater pumping, is discussed in this section. In these systems, energy must be expended to transport water from one point to another. The amount of energy expended depends on capacity (flow rate), head (the pressure the pump must develop to overcome the static lift and friction losses), and the efficiency of the pumps and pump drivers. These terms are defined further.

The *capacity* of a pump is the volume of liquid pumped per unit of time and is usually expressed as gallons per minute or million gallons per day (liters per second or cubic meters per second).

Head is the elevation of the free-water surface above or below a reference datum. Head can refer to both pump head and system head. *Pump head* refers to the height, measured in feet (meters), to which a pump can raise a liquid being pumped at a given flow rate. *System head* is the height to which the liquid must be raised to overcome the resistance of the piping system at a given flow rate. A definition sketch for the head on a pump is shown in Fig. 19.18. Terms specifically used in the analysis

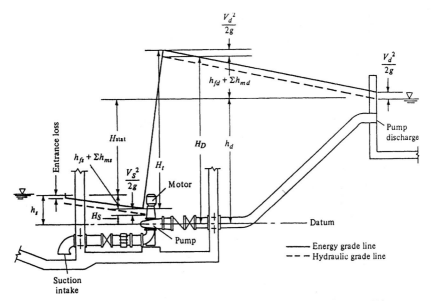

FIGURE 19.18 Definition sketch for the head on a pump. (*Metcalf & Eddy, Inc., 1981.*)

of pumps and pump systems are (1) static suction head h_s, (2) static discharge head h_d, (3) static head H_{stat}—the difference in elevation between the static discharge and static suction levels $(h_d - h_s)$, (4) friction head h_f, (5) velocity head—the kinetic energy contained in a fluid in motion, (6) minor head loss h_m usually due to valves and fitting, and (7) total dynamic head H_t. *Minor losses* and *velocity head* were defined in Sec. 19.4.4. *Total dynamic head* is the head against which the pump must work. It is determined by adding the static suction and discharge head (with respect to signs), frictional head losses, velocity heads, and minor losses. The equations for determining total dynamic head are (see Fig. 19.18):

$$H_t = H_D - H_S + \frac{V_d^2}{2g} - \frac{V_s^2}{2g} \tag{19.9}$$

$$H_D = h_d + h_{fd} + \Sigma h_{md} \tag{19.10}$$

$$H_S = h_s - h_{fs} - \Sigma h_{ms} - \frac{V_s^2}{2g} \tag{19.11}$$

where
H_t = total dynamic head, ft (m)
H_D (H_S) = discharge (suction) head measured at discharge (suction) nozzle of pump referenced to the centerline of the pump impeller, ft (m)
V_d (V_s) = velocity in discharge (suction) nozzle ft/s (m/s)
g = acceleration due to gravity, 32.2 ft/s² (9.81 m/s²)
h_d (h_s) = static discharge (suction) head, ft (m)
h_{df} (h_{fs}) = frictional headloss in discharge (suction) piping, ft (m)
h_{md} (h_{ms}) = minor fitting and valve losses in discharge (suction) piping, ft (m). Entrance loss is included in computing minor losses in suction piping.

The power requirement for pump operation is defined as the pump output divided by the pump efficiency. The power input determines the size of the driver required (electric motor or engine). The equations for power input are:

$$P_s = \frac{qH_t}{3960E_p} \qquad \text{U.S. customary units} \tag{19.12}$$

$$P_s = \frac{QH_t}{102E_p} \qquad \text{SI units} \tag{19.12a}$$

where
P_s = input shaft power, hp (kW)
q = flow rate, gal/min
Q = flow rate, L/s
H_t = total dynamic head, ft (m)
E_p = pump efficiency (as a decimal), dimensionless

19.7.4 Pump Selection

A step-by-step procedure for the selection of pumps is described in Table 19.21. The selection procedure applies mainly to centrifugal pumps but is useful in other types of pumping systems.

19.7.5 Design of Wastewater-Pumping Stations

Wastewater-pumping stations provide challenges to the design engineer because of the many considerations such as siting, space requirements, nature of the waste-

TABLE 19.21 Summary Procedure for Pump Selection

Step	Description
1.	Identify design point as well as all other operating points for which the pump must be selected.
2.	Define any other operating conditions.
3.	Compute and plot the station hydraulic characteristics for the design and extreme conditions.
4.	Select the number and type of pumps to be used.
5.	Obtain manufacturers' data for pumps that operate in the desired head-capacity range.
6.	Evaluate pump-performance curves including relationship of best efficiency point (BEP) to operating range. Select an appropriate pump speed. For wastewater pumps, select lower operating speeds to minimize abrasive wear (maximum speeds over 1200 r/min are not recommended).
7.	Plot pump and station head-capacity curves to ensure pump curves meet all station conditions (see Fig. 19.19).
8.	Determine if multiple-speed operation will be required to cover the operating range, especially at low-flow conditions.
9.	Evaluate pipe-friction losses using Hazen-Williams C-factors for initial (new pipe) and design conditions.
10.	Check power requirements for all operating conditions to ensure the selected motor size is not overloaded.
11.	To ensure that cavitation does not occur, check the net positive suction head (NPSH) to ensure the required NPSH does not exceed the available NPSH. If in doubt, obtain manufacturer's recommendations.
12.	If practicable, provide for possible future increase in capacity by: (1) using less than the maximum impeller size, (2) sizing suction and discharge piping conservatively to accommodate substituting future larger pumps, and (3) providing additional space for a future pump.

Source: Adapted in part from Sanks, et al. (1989).

water, operation and maintenance requirements, and aesthetics. This section will provide an overview of important design considerations; for additional details, Hydraulic Institute (1983); Metcalf & Eddy (1981); Sanks, et al. (1989); and Water Environment Federation (1993a) should be consulted.

The essential features of typical pumping stations were described previously in Table 19.20. Many of the design considerations, in addition to pump selection, relate to the size and configuration of the wet well, requirements for screening, sizing of the suction and discharge piping, instrumentation and controls, ventilation, and odor control.

19.7.5.1 Wet Wells. Wet wells are required in pumping stations to store wastewater before it is pumped. The storage volume depends on the type of pump operation, either constant- or variable-speed drive. If constant-speed operation is selected, the volume must be adequate to prevent frequent starting and stopping of the pumps. The time between starts is normally on the order of 15 to 20 min but should be confirmed by the pump and motor supplier. Large motors, over 250 hp (200 kW), may require longer rest periods between starts. The volume of the wet well between start and stop elevations for a single-speed pump or a single-speed control step for multiple-speed operation is given by the following equation:

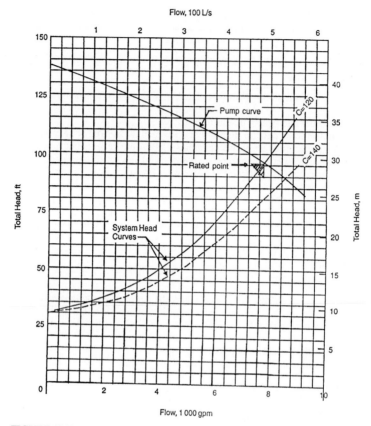

FIGURE 19.19 Typical pump and system head-capacity curves (Note: ft × 0.3048 = m; gal/min × 6.309 × 10⁻² = L/s). (*Water Environment Federation, 1993a.*)

$$V = \frac{qt}{4} \tag{19.13}$$

where V = storage volume, gal (m³)
 q = pump discharge capacity, gal/min (m³/min) or increment in pumping capacity where one pump is already operating and a second pump is started, or where speed is increased
 t = minimum time in minutes of one pumping cycle (time between successive starts or changes in speed of a pump operating over the control range)

Many public agencies specify a maximum storage time in a wet well to reduce the potential for odors and septic wastewater. A maximum storage retention time of 30 min at average flow is a common requirement (Water Environment Federation, 1993a).

Wet wells should be configured to prevent the accumulation of settleable solids that could occur independent of the pump-cycling time. Sloped floors are frequently used. In larger pumping stations, the wet wells are compartmentalized to allow cleaning and flushing.

19.7.5.2   Screens.   Screening devices such as bar racks or comminutors should be provided if coarse matter is expected in the wastewater that could cause clogging of the pump. Comminutors, mechanical devices that shred coarse wastewater solids, are preferred in unattended pumping stations so that manual handling of the screenings is avoided. In large stations, mechanically cleaned bar screens may be used. For protection of the pumping equipment, the clear opening between the bars in bar screens is one-third the size of the maximum solid that the pump can discharge (Metcalf & Eddy, 1981). The smallest clear opening is 1 in (100 mm). Velocities through the clear openings should be between a maximum of 4 ft/s (1.2 m/s) and a minimum of 1.2 ft/s (0.4 m/s).

19.7.5.3   Suction and Discharge Piping.   The velocity of the wastewater at the pump suction and discharge nozzles ranges from 10 to 14 ft/s (3 to 4.25 m/s). If the velocity is more or less than this, a better pump could probably be selected. Pumps with higher discharge velocities are necessary for heads of 100 ft (30 m) or more. Suction piping is usually one or two pipe sizes larger than the suction nozzle and discharge piping is generally one size larger than the discharge nozzle (Metcalf & Eddy, 1981). Common velocities in suction and discharge piping range from 4 to 8 ft/s (1.2 to 2.4 m/s). Suction-line velocities in the lower end of the range are generally recommended. Adequate submergence, as recommended by the pump manufacturer, must always be provided to prevent air from being drawn into the pump suction by vortexing action.

At the pump discharge, a check valve should be provided, mounted in the horizontal position to prevent clogging with debris. Isolation valves, either wedge-type gate valves or plug valves (see Sec. 19.7.8), should be provided at the pump suction and discharge so the pump can be removed from service for maintenance.

The joints in piping within a pumping station should almost always be bolted flanges. Adjacent to pumps, grooved-end couplings or flexible-sleeve-type couplings should be used to permit disassembly or to allow for misalignment.

19.7.5.4   Instrumentation and Controls.   Instruments for pumping stations should be selected to provide long life, low maintenance, and high reliability in damp, corrosive environments. In unattended stations, backup systems should be provided as failure of the pump operation may cause flooding, property damage, and potential health hazards. Typical instruments used in pumping stations include wet well level indicators, suction and discharge-pressure gauges, flow meter on either the influent or effluent, pump-status lights, running-time meters, and alarms. Typical types of control systems for pump operation include conventional control relays, programmable logic controllers (PLCs), single-loop digital controllers, proprietary controllers, and "smart" remote terminal units (RTUs). For detailed descriptions of the various control systems available, Water Environment Federation (1993a; 1993b) should be consulted.

Many pumping stations operate on-off based on wet well level. One or more pumps may be used to control the level in the wet well, with pumps starting and stopping at different preset levels. The capacity of the pumps may also be varied using eddy-current clutches, mechanical variable drives, or adjustable-speed drives. Adjustable-speed drives are the type most commonly used. Adjustable-speed drives may be controlled by a signal based on level, pressure, or flow. Most of the fully automatic control systems use a bubbler or an ultrasonic measurement for level detection. The two major methods of adjustable-speed control are variable-frequency controllers with induction motors and secondary energy-recovery units with wound-rotor motors. These systems are popular because (1) their efficiencies exceed 85 percent throughout their operating range and (2) starting current is low, which reduces the size of standby generators. Variable-frequency controllers are generally used on motors less than 1000 horsepower (746 kW), although they are available for sizes up to 4000 hp (3000 kW). Isolation transformers are usually required to prevent harmonic feedback and to protect the controller. For motors larger than 1000 horsepower that require voltages higher than 600 VAC, wound-rotor secondary energy-recovery systems may be more practical (Water Environment Federation, 1993b).

19.7.5.5 Electrical Equipment. Motor starters and controls should be located in factory-assembled, free-standing control centers located at ground level in a clean, dry area. Large stations should have a separate electrical room containing the motor starters, switch gear, and meters and instruments. All electrical equipment and lights located in the wet well must be of explosion-proof construction because of the potential danger of explosion of vapors or gases in the incoming wastewater.

19.7.5.6 Ventilation. Ventilating systems in the wet- and dry-well sides of the station must be entirely separated, and all openings for pipes or electrical conduits sealed gas-tight. Heating and ventilating equipment in the wet well must be of explosion-proof construction.

Intermittent ventilation is usually provided at small, unattended stations. Continuous ventilation is provided at large attended stations or where there are mechanically cleaned screens or other equipment that could be damaged by the moist and corrosive environment. Recommended wet well ventilation rates are a minimum of 30 air changes per hour for intermittent operation and 12 air changes per hour for continuous ventilation. The dry well should be positively ventilated with either supply or exhaust fans. Recommended dry well ventilation rates are a minimum of 15 air changes per hour for intermittent operation and at least 6 air changes per hour for continuous ventilation (Great Lakes–Upper Mississippi River Board of State Sanitary Engineers, 1978).

19.7.5.7 Odor Control. Odors from septic wastewater or industrial wastes may be a health and safety hazard to operating personnel and a nuisance to nearby residents. Most odor-producing compounds in domestic wastewater result from anaerobic biological activity. Hydrogen sulfide is the most common malodorous gas found in collection systems. As described in Sec. 19.5.5, hydrogen sulfide results from the anaerobic decomposition of sulfur present in human excreta and sulfates found in most water supplies. Odor control may require several design and operating measures for effective results. These may include odor prevention by limiting the accumulation of organic matter; limiting turbulence that releases gases; odor containment; ventilation and odor treatment by chemical scrubbers, activated carbon filters, or biological soil/compost filters; injection of chemicals in the wastewater such as chlorine, sodium hydroxide, oxygen, hydrogen peroxide, or iron salts; or odor masking with pleasant-smelling masking agents. In each case, the source and characteristics of the odors must be identified so that proper remedial measures can be developed. For detailed information on odor control methods, ASCE (1989); Metcalf & Eddy (1981; 1991); and Water Environment Federation (1993a) should be consulted.

19.7.6 Design of Stormwater-Pumping Stations

Stormwater pumping typically requires large pumps that operate at relatively low heads. A typical stormwater-pumping station includes the following features: (1) inlet piping or an approach channel, (2) trash racks, (3) wet well, (4) drainage pumps, (5) discharge piping, (6) an outlet structure, and (7) ancillary facilities. A typical large stormwater-pumping station is shown in Fig. 19.20.

19.7.6.1   Inlet Piping and Approach Channel.   Entrance to the pumping station depends on the siting of the pumping station and the conveyance system for the stormwater. In stations that use vertical pumps, special attention to the inlet conditions is required in order to ensure hydraulic conditions favorable to pump operation.

19.7.6.2   Trash Racks.   Trash racks should be provided at the entrance to the pumping station to catch debris and other objects that could harm the pumps or affect pumping operations. Because storm flows are intermittent, most stations are equipped with manually cleaned bar racks. Depending on the size and type of pumps used, clear openings between the bars usually range from 2 to 3 in. Stations that handle combined wastewater and stormwater flows may use mechanically cleaned bar screens.

19.7.6.3   Sump Design.   The design of the pump sump is important, particularly when large vertical axial- or mixed-flow pumps are used. If the water is not evenly distributed to the pump suction, unequal and inefficient pump performance could result. The ideal intake design is a channel going directly to the pumps. Any turn or obstruction causes eddy currents and tends to initiate deep-cored vortices. If possible, water should not flow past one pump to reach another. If pumps must be placed in the line of flow, it may be necessary to provide an open cell in front of each pump or to install turning vanes under the pump to deflect the water upward. Streamlining should be used to reduce the tail of alternating vortices in the wake of the pump or of other obstructions in the stream flow (Hydraulic Institute, 1983).

For pumps in the capacity range of 3,000 to 300,000 gal/min (0.19 to 19 m³/s), general guidelines for sump dimensions are shown in Figs. 19.21 and 19.22 (Hydraulic Institute, 1983). All of the dimensions shown in these figures are based on the rated

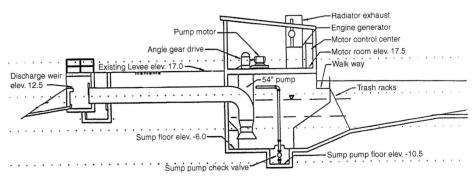

FIGURE 19.20   Typical large stormwater pumping station.

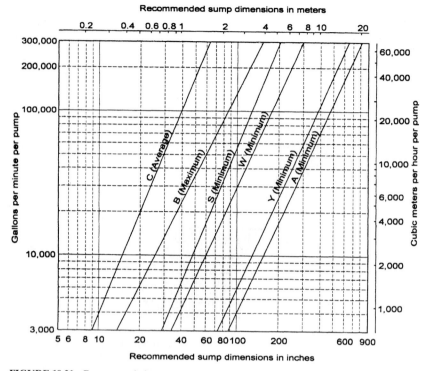

FIGURE 19.21　Recommended pump sump dimensions (Use with Figure 19.22; Note: in × 25.4 = mm; gal/min × 6.309 × 10^{-5} = m³/s; ft/s × 0.3048 = m/s). (*Courtesy of the Hydraulic Institute.*)

capacity of the pump at design head. If the pump is required to operate for significant periods at a higher capacity, the higher capacity should be used as a basis for determining dimensions. For specific pumping applications, sump dimensions should be verified by the pump supplier.

19.7.6.4 Stormwater Pumps.　The main types of pumps used for stormwater pumping include vertical (propeller or mixed-flow), submersible, centrifugal (horizontal or vertical nonclog), volute or angle-flow, or screw. Depending on the location, the pumps may be driven by electric motors or diesel or gas engines. Axial- or mixed-flow pumps are used most commonly because they have large unobstructed passages, are capable of pumping large quantities of water, and are relatively efficient. Screw pumps are also used because they are capable of pumping large volumes of water over a wide range of flow rates without clogging.

19.7.6.5 Discharge Piping and Outlet Structure.　Depending upon the length of the run of piping from the pump to the point of discharge, the discharge line may be of the same size as the pump discharge. For large discharge lines, steel pipe is normally used. Flap gates may be used in lieu of check valves depending on the type of discharge structure (see Table 19.24). If a discharge structure is used, provisions to dissipate the energy from the pump discharge need to be included.

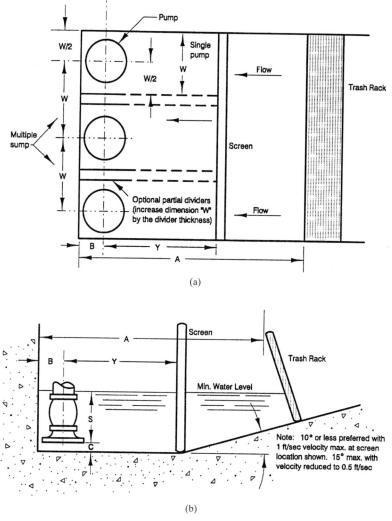

FIGURE 19.22 Recommended pump sump layout and dimensions: (*a*) plan and (*b*) section. (*Courtesy of the Hydraulic Institute.*)

19.7.6.6 Ancillary Facilities. For reliable operation, stormwater pumping stations have either gas- or diesel-engine drives or a second source of electric power such as an emergency engine-generator. Other facilities at pumping stations may include a crane for the removal of pumps and motors, removable roof hatches for lifting the pumps if a permanent crane is not provided, emergency fuel storage for engine drives, emergency lighting, automatic controls, a telemetry system for unattended stations, and a load bank for exercising the engine-generator. The station may also be equipped with a recirculation line from the discharge box to the wet well for exercising the pumps if there is an extended period when the station is not used.

19.7.7   Factory-Assembled Pumping Stations

Factory-assembled pumping stations (see Fig. 19.15) are common in collection systems with low flow rates and where the need to protect the pumps from clogging is minimal. Factory-assembled stations are shipped from the point of manufacture to the site in modules with all of the equipment and components mounted and connected permanently within the module. When the station components arrive at the site, the module base is mounted on a concrete foundation pad, the modules are interconnected, and all external piping and power connections are made. The station is ready to operate in a relatively short period. A manhole constructed of sections of precast concrete pipe usually serves as the wet well.

19.7.8   Design of Force Mains

Some of the commonly used criteria used in force-main design are listed in the following (Metcalf & Eddy, 1981).

Design period.   Commonly designed to handle flow rates for 25 to 50 years in the future.

Minimum depth.   Minimum depth of earth cover of 3 ft (0.9 m) is generally used to minimize the impact of live loads. For shallower depths of cover, concrete encasement should be used.

Slope.   Force mains can be laid on a flat slope, but good design practice consists of installing the force main on a gradual slope from the pump at the low end to the discharge point at the high end. High points in the line should be equipped with air and vacuum valves to remove entrapped air or gases or to permit the entry of air during draining. Low points should be provided with blowoff connections so the line can be drained and flushed to remove accumulated solids.

Velocity.   Desirable force-main velocities range from 3 to 5 ft/s (0.9 to 1.5 m/s). Higher velocities may be used, especially if there is a wide range in flow rates.

Pressure.   The force main and piping on the discharge side of the pump need to be pressure rated to withstand the maximum hydraulic head on the system, including abnormal pressures that may be produced by hydraulic transients such as water hammer and surge pressures.

Anchorage.   Force mains must be anchored to resist thrust at bends, branches, and plugs in the pipe. Anchorage may be concrete thrust blocks, restrained joints, or anchor blocks.

An important consideration in the design of force mains and pumping stations is water hammer that results from rapid changes in pressure, usually due to pump startup, pump shutdown, or power failure. The identification and calculation of pressures, velocities, and other abnormal behavior resulting from water hammer require special analysis that is discussed in other texts (Metcalf & Eddy, 1981; Sanks, et al., 1989; Water Environment Federation, 1993a). Computer modeling is also available. It is important to note that the objective of water hammer control is to limit pressure changes in the force main within allowable ranges by limiting the rate of change of velocity. Control systems generally used to limit water hammer pressures in wastewater-pumping stations and force mains are listed in Table 19.22.

TABLE 19.22   Common Controls for Limiting Water Hammer Pressures in Force Mains

---

- Swing check valve on pump discharge with outside lever and weight to assist closure.
- Spring-loaded check valve on pump discharge.
- Swing check valve either spring loaded or having an outside lever and weight together with a high-pressure (surge) relief valve.
- Positively controlled valve on pump discharge set to open at a preset pressure during startup and to close at a predetermined rate after power failure.
- Positively controlled discharge valve, together with one or more bypass surge relief valves, set to open at preset pressures and close at predetermined rates.
- Open surge tower or a pressurized surge tank with a bladder.
- Air and vacuum valves at the pumping station and high points in the force main to limit subatmospheric pressures.

Source: Adapted from Metcalf & Eddy (1981).

In transporting wastewater in long force-main runs and in warm climates, hydrogen sulfide gas may be generated that causes corrosion in the pipeline and appurtenances and unpleasant odors. Common methods of controlling sulfide in force mains are listed in Table 19.23.

19.7.9   Valves and Gates

Pumping stations use valves to isolate a pump, prevent backflow through the pump, isolate the valve that prevents backflow, and surge control. Force mains use valves for isolating segments of line for servicing, removal of trapped air and gases, and for the blowdown of accumulated solids. Gates such as slide gates and sluice gates are also used in place of valves at large openings for diverting flow or for isolating a compartment at the pumping station. The types of valves and gates that are commonly used in pumping stations and force mains are listed in Table 19.24.

TABLE 19.23   Common Sulfide Control Measures for Force Mains

---

- Addition of air or oxygen to the force main to reduce or eliminate anaerobic conditions in the wastewater and in part of the slime layer on the pipe.
- Addition of chemicals such as chlorine and hydrogen peroxide at the start of the force main to suppress the slime layer biological activity.
- Addition of oxidants at or near the end of the force main to oxidize the sulfide that has formed in the pipe.
- Addition of metal salts (primarily iron) to raise the oxidation–reduction potential to minimize the reduction of sulfate to sulfide and react with the sulfide generated to form insoluble metal sulfide compounds.
- Use of caustic wash on an intermittent basis to raise the pH to 12.5 or more for about 20 min to kill the organisms producing sulfide.

Source: ASCE (1989).

TABLE 19.24 Valve and Gate Types Commonly Used in Pumping Stations and Force Mains

Valve type	Application
Gate valve	Used to isolate specific piping sections. Wedge type is generally used in wastewater applications. Buried gate valves are usually of the nonrising-stem type. Where considerable solids are present and could cause clogging, such as in unscreened wastewater applications, plug valves should be used instead of gate valves.
Plug valve	Used in pump suction and discharge piping to isolate pumps for maintenance. Not susceptible to clogging by solids in wastewater. Valves are generally not used to regulate flowrates.
Butterfly valve	Should only be used in applications where the water is relatively free of debris and solids. The valve disc is susceptible to the accumulation of rags, sticks, and other debris.
Check valve	Used to prevent backflow in pump discharge lines. Backflow (or reversed flow) could cause the pump to run backwards, potentially causing damage, or could cause flooding of the wet well. Check valves should be installed in the horizontal position to prevent the accumulation of debris that could affect valve closure. Valves should also be equipped with an outside lever and weight to minimize water hammer.
Air-release and vacuum valve	Used to prevent formation of vacuums at the pumping stations and at the high points in force mains. Valves also vent air and gases that accumulate at high points in the pipe line. Valves are susceptible to clogging in solids-bearing wastewater and may require frequent maintenance.
Blowoff valve	Used to remove solids that accumulate in low points in a force main. Either gate or plug valves are used for blowoff service.
Sluice gate	Used to isolate wet well compartments for pump or wet well maintenance. Sluice gates are also used mainly for the diversion of gravity flows.
Flap gate	Used to prevent backflow at pipe outlet where discharges are to receiving streams, channels, or rivers.

REFERENCES

American Society of Civil Engineers, and Water Pollution Control Federation, *Design and Construction of Sanitary and Storm Sewers,* ASCE Manual of Practice no. 37 and WPCF Manual of Practice no. 9, 1969.

American Society of Civil Engineers, and Water Pollution Control Federation, *Gravity Sanitary Sewer Design and Construction,* ASCE Manual of Practice no. 60, New York, and WPCF Manual of Practice no. FD-5, Alexandria, Va., 1982.

American Society of Civil Engineers, and Water Pollution Control Federation, *Existing Sewer Evaluation and Rehabilitation,* ASCE Manual of Practice no. 62 and WPCF Manual of Practice no. FD-6, 1983.

American Society of Civil Engineers, *Sulfide in Wastewater Collection and Treatment Systems,* ASCE Manuals and Reports on Engineering Practice no. 69, 1989.

Corbitt, R. A., *Standard Handbook of Environmental Engineering,* McGraw-Hill, New York, 1990.

Fedotoff, R. C., P. H. Bellows, R. W. Frutchey, and J. Thomson, "Trenchless Techniques Improve Pipelines," *Water Environment and Technology,* April 1991, pp. 48–52.

Geyer, J. C., and J. J. Lentz, "Evaluation of Sanitary Sewer System Designs," The Johns Hopkins School of Engineering, 1962.

Great Lakes–Upper Mississippi River Board of State Sanitary Engineers, "Recommended Standards for Sewage Works," Albany, N.Y., 1978 ed.

Huff, F. A., and J. R. Angel, "Frequency Distributions and Hydroclimatic Characteristics of Heavy Rainstorms in Illinois," Illinois State Water Survey, 1989.

Hydraulic Institute, *Hydraulic Institute Standards for Centrifugal, Rotary & Reciprocating Pumps,* 14th ed., Hydraulic Institute, Cleveland, Ohio, 1983.

McGhee, T. J., *Water Supply and Sewerage,* 6th ed., McGraw-Hill, New York, 1991.

Metcalf & Eddy, Inc., "Report to National Commission on Water Quality on Assessment of Technologies and Costs for Publicly Owned Treatment Works," vol. 2, Boston, 1975.

Metcalf & Eddy, Inc., *Wastewater Engineering: Collection and Pumping of Wastewater,* McGraw-Hill, New York, 1981.

Metcalf & Eddy, Inc., *Wastewater Engineering: Treatment, Disposal and Reuse,* 3d ed., McGraw-Hill, New York, 1991.

Nayyar, M. L., *Piping Handbook,* 6th ed., McGraw-Hill, New York, 1992.

New England Interstate Water Pollution Control Commission, "Guidelines for the Design of Wastewater Treatment Works," Boston, 1980.

Salvato, J. A., *Environmental Engineering and Sanitation,* 3d ed., Wiley, New York, 1982.

Sanks, R. L., et al., *Pumping Station Design,* Butterworths, Boston, 1989.

U.S. Department of Housing and Urban Development, "Residential Water Conservation Projects," Summary Report, Washington, D.C., June 1984.

U.S. Environmental Protection Agency, "Odor and Corrosion Control in Sanitary Sewerage Systems and Treatment Plants," EPA/625/1-85/018, Cincinnati, Ohio, October 1985.

U.S. Environmental Protection Agency, "Detection, Control, and Correction of Hydrogen Sulfide Corrosion in Existing Wastewater Systems," EPA 832-R-92-001, September 1992.

Water Environment Federation, *Alternative Sewer Systems,* Manual of Practice FD-12, Alexandria, Va., 1986.

Water Environment Federation, *Design of Wastewater and Stormwater Pumping Stations,* Manual of Practice FD-4, Alexandria, Va., 1993a.

Water Environment Federation, *Design of Municipal Wastewater Treatment Plants,* Manual of Practice No. 8, Alexandria, Va., 1992.

Water Environment Federation, *Instrumentation in Wastewater Treatment Facilities,* Manual of Practice 21, Alexandria, Va., 1993b.

CHAPTER 20
WASTEWATER TREATMENT*

George Tchobanoglous
Department of Civil and Environmental Engineering
University of California at Davis
Davis, California

The design and construction of municipal wastewater-treatment facilities has resulted from (1) public concern about the impacts on the environment caused by the discharge of untreated or partially treated wastewater and (2) the passage of federal and state water-pollution-control legislation, the earliest being the Rivers and Harbors Act passed in 1889, which prohibited the deposition of solid wastes into navigable waters. In 1972 Congress enacted Public Law 92-500, the Federal Water Pollution Control Act Amendments of 1972. This act departed from previous pollution-control legislation in a number of significant ways. It expanded the federal role in water-pollution control, increased the level of federal funding for construction of publicly owned waste-treatment works, elevated planning to a new level of significance, opened new avenues for public participation, and created a regulatory mechanism requiring uniform technology-based effluent standards, together with a national permit system for all point-source dischargers as the means of enforcement. The passage of this law also marked a change in water-pollution-control philosophy. No longer was the classification of the receiving streams of ultimate concern as it was before. It was decreed in Public Law 92-500 that the quality of the nation's waters is to be improved by the imposition of specific effluent limitations.

The principal methods now used for wastewater treatment and their application in the removal of the contaminants of concern are introduced and reviewed in this chapter. The topics considered in this chapter include:

1. Wastewater characteristics
2. Wastewater treatment methods and standards
3. Physical treatment methods
4. Chemical treatment methods

---

* Adapted from R. K. Linsley, J. B. Franzini, D. Freyberg, and G. Tchobanoglous, *Water Resources Engineering,* 4th ed., McGraw-Hill, New York, 1992 and G. Tchobanoglous and F. L. Burton (Written and revised by), *Wastewater Engineering: Collection, Treatment, Disposal,* 3d ed., Metcalf & Eddy, Inc., McGraw-Hill, New York, 1991.

5. Biological treatment methods
6. Nutrient removal
7. Advanced treatment methods
8. Treatment and disposal of sludge and biosolids
9. Natural treatment systems
10. Wastewater reclamation and reuse
11. Effluent disposal
12. Treatment plant planning and design
13. Decentralized wastewater management systems

20.1   WASTEWATER CHARACTERISTICS

The selection and design of wastewater-treatment facilities is based on an analysis of: (1) the physical, chemical, and biological characteristics of the wastewater, (2) the quality that must be maintained in the environment to which the wastewater is to be discharged or for the reuse of the wastewater, and (3) the applicable environmental standards or discharge requirements that must be met. The principal physical, chemical, and biological impurities found in wastewater are reported in Table 20.1. For clarity, the chemical characteristics of wastewater are considered in two classifications, inorganic and organic. Some tests used to characterize the organic matter in wastewater involve the use of microorganisms, and are often classified as biochemical. Further, because of their special importance, *priority pollutants* and *volatile organic compounds* (VOCs) are often considered separately. Contaminants of concern in wastewater and the reasons for concern are summarized in Table 20.2. Of the contaminants listed in Table 20.2, suspended solids, biodegradable organics, and pathogenic organisms are of major importance, and most wastewater-treatment plants are designed to accomplish their removal. Although the other contaminants are also of concern, the need for their removal must be considered on a case-to-case basis.

20.1.1   Physical Characteristics

The principal physical characteristics of a wastewater, as reported in Table 20.1, are its solids content, color, odor, and temperature. The *total solids* in a wastewater consist of the insoluble or suspended solids and the soluble compounds dissolved in the water. The *suspended solids* content is found by drying and weighing the residue removed by filtering the sample. When this residue is ignited the *volatile solids* are burned off. Volatile solids are presumed to be organic matter, although some organic matter will not burn and some inorganic solids break down at high temperatures. From 40 to 75 percent of the solids in an average wastewater are suspended. Some of the suspended solids settle quite rapidly, but those of colloidal size settle slowly or not at all. *Settleable solids,* expressed as mL/L, are those that can be removed by sedimentation. The standard test consists of placing a wastewater sample in a 1-L conical glass container (Imhoff cone) and noting the volume of solids in milliliters that settles within the detention period of the particular plant. Usually about 60 percent of the suspended solids in a municipal wastewater are settleable.

   Color is a qualitative characteristic that can be used to assess the general condition of wastewater. If light brown in color, the wastewater is usually less than 6 hours old. A light- to medium-grey color is characteristic of wastewaters that have under-

TABLE 20.1 Physical, Chemical, and Biological Characteristics of Wastewater and Their Sources

Characteristic	Sources
Physical properties	
Color	Domestic and industrial wastes, natural decay of organic materials
Odor	Decomposing wastewater, industrial wastes
Solids	Domestic water supply, domestic and industrial wastes, soil erosion, inflow/infiltration
Temperature	Domestic and industrial wastes
Chemical constituents	
Organic	
Carbohydrates	Domestic, commercial, and industrial wastes
Fats, oils, and grease	Domestic, commercial, and industrial wastes
Pesticides	Agricultural wastes
Phenols	Industrial wastes
Proteins	Domestic, commercial, and industrial wastes
Priority pollutants	Domestic, commercial, and industrial wastes
Surfactants	Domestic, commercial, and industrial wastes
Volatile organic compounds	Domestic, commercial, and industrial wastes
Other	Natural decay of organic materials
Inorganic	
Alkalinity	Domestic wastes, domestic water supply, groundwater infiltration
Chlorides	Domestic wastes, domestic water supply, groundwater infiltration
Heavy metals	Industrial wastes
Nitrogen	Domestic and agricultural wastes
pH	Domestic, commercial, and industrial wastes
Phosphorus	Domestic, commercial, and industrial wastes; natural runoff
Priority pollutants	Domestic, commercial, and industrial wastes
Sulfur	Domestic water supply; domestic, commercial, and industrial wastes
Gases	
Hydrogen sulfide	Decomposition of domestic wastes
Methane	Decomposition of domestic wastes
Oxygen	Domestic water supply, surface-water infiltration
Biological constituents	
Eucaryotes	
Animals	Open watercourses and treatment plants
Plants	Open watercourses and treatment plants
Protists	
Algae	Surface-water infiltration, treatment plants
Fungi	Domestic wastes, surface-water infiltration
Protozoa	Domestic wastes, surface-water infiltration
Eubacteria	Domestic wastes, surface-water infiltration, treatment plants
Archaebacteria	Domestic wastes, surface-water infiltration, treatment plants
Viruses	Domestic wastes

gone some decomposition or that have been in the collection system for some time. If the color is dark grey or black, the wastewater typically is septic, having undergone extensive bacterial decomposition under *anaerobic* (in the absence of oxygen) conditions. The blackening of wastewater is often due to the formation of various sulfides, particularly ferrous sulfide. The formation of sulfides occurs when hydrogen sulfide produced from the reduction of sulfate under anaerobic conditions combines with a divalent metal, such as iron, which may be present.

TABLE 20.2 Contaminants of Concern in Wastewater Treatment

Contaminants	Reason for concern
Suspended solids	Sludge deposits and anaerobic conditions.
Biodegradable organics	Depletion of natural oxygen resources and the development of septic conditions.
Dissolved inorganics (e.g., total dissolved solids)	Inorganic constituents added by usage. Reclamation and reuse applications.
Heavy metals	Metallic constituents added by usage. Many metals are also classified as priority pollutants.
Nutrients	Growth of undesirable aquatic life; eutrophication.
Pathogens	Communicable diseases.
Priority organic pollutants	Suspected carcinogenicity, mutagenicity, teratogenicity, or high acute toxicity. Many priority pollutants resist conventional treatment methods (known as refractory organics).

The determination of *odor* has become increasingly important as the public has become more concerned with the proper operation of wastewater-treatment facilities. The odor of fresh wastewater is usually not offensive, but a variety of odorous compounds are released when wastewater is decomposed biologically under anaerobic conditions. The principal odorous compound is hydrogen sulfide (the smell of rotten eggs). Other compounds, such as indole, skatole, and mercaptans, form under anaerobic conditions and may cause odors that are more offensive than that of hydrogen sulfide. Detection thresholds for various odorous compounds are reported in Table 20.3. Because of public concern, special care is called for in the design of wastewater-treatment facilities to avoid conditions that will allow the development of odors.

The *temperature* of wastewater commonly is higher than that of the water supply because of the addition of warm water from municipal use. The measurement of temperature is important because most wastewater-treatment schemes include biological processes that are temperature dependent. The temperature of wastewater will vary from season to season and also with geographic location. In cold regions the temperature will vary from about 45 to 65°F (7 to 8°C) while in warmer regions the temperature will vary from 55 to 75°F (13 to 24°C).

TABLE 20.3 Odor Thresholds of Odorous Compounds Associated with Untreated Wastewater

Odorous compound	Chemical formula	Odor threshold, ppmV*
Ammonia	NH_3	46.8
Chlorine	Cl_2	0.314
Dimethyl sulfide	$(CH_3)_2S$	0.0010
Diphenol sulfide	$(C_6H_5)_2S$	0.0047
Ethyl mercaptan	CH_3CH_2SH	0.001
Hydrogen sulfide	H_2S	0.00047
Indole	C_8H_7N	0.0001
Methyl amine	CH_3NH_2	21.0
Methyl mercaptan	CH_3SH	0.0021
Skatole	C_9H_9NH	0.019

* Parts per million by volume.

20.1.2 Inorganic Chemical Characteristics

The principal chemical tests include free ammonia, organic nitrogen, nitrites, nitrates, organic phosphorus, and inorganic phosphorus. Nitrogen and phosphorus are important because these two nutrients have been identified most commonly as being responsible for the growth of aquatic plants. Other tests, such as chloride sulfate, pH, and alkalinity, are performed to assess the suitability of reusing treated wastewater and in controlling the various treatment processes.

Trace elements, which may include some heavy metals, are not determined routinely, but trace elements may be a factor in the biological treatment of wastewater. All living organisms require varying amounts of one or more trace elements, such as iron, copper, zinc, and cobalt, for proper growth. Heavy metals may also produce toxic effects; therefore, determination of the amounts of heavy metals is especially important where the further use of treated effluent or sludge is to be evaluated. Many of the metals are also classified as priority pollutants (see subsequent discussion). Specific inorganic constituents are determined to assess the presence or absence of priority pollutants and to determine if any potential treatment or disposal problems will develop (e.g., toxics in sludge).

Measurements of gases, such as hydrogen sulfide, oxygen, methane, and carbon dioxide, are made to help in the operation of the system. The presence of hydrogen sulfide is determined not only because it is an odorous gas but also because it is important in the maintenance of long sewers on flat slopes, as it can cause corrosion. Measurements of dissolved oxygen are made to monitor and control aerobic biological treatment processes. Methane and carbon dioxide measurements are used in connection with the operation of anaerobic digesters.

20.1.3 Organic Chemical Characteristics

The organic matter in wastewater typically consists of proteins, carbohydrates, and fats. Over the years, a number of different tests have been developed to determine the organic content of wastewaters. In general, the tests may be divided into those used to measure gross concentrations of organic matter greater than about 1 mg/L and those used to measure trace concentrations in the range of 10^{-12} to 10^0 mg/L. Laboratory methods commonly used today to measure gross amounts of organic matter (>1 mg/L) in wastewater include: (1) *biochemical oxygen demand* (BOD), (2) *chemical oxygen demand* (COD), and (3) *total organic carbon* (TOC). Trace organics are determined using instrumental methods including gas chromotography and mass spectroscopy. Specific organic compounds are determined to assess the presence of priority pollutants. Because of their special importance in the design and operation of wastewater-treatment plants, BOD, COD, and TOC are considered separately in the following.

20.1.3.1 Priority Pollutants.

The Environmental Protection Agency (EPA) has identified more than 120 priority pollutants in 65 classes to be regulated by categorical discharge standards. Priority pollutants (both inorganic and organic) were selected on the basis of their known or suspected carcinogenicity, mutagenicity, teratogenicity, or high acute toxicity. Many of the organic priority pollutants are also classified as *volatile organic compounds* (VOCs).

Two types of standards are used to control pollutant discharges to *publicly owned treatment works* (POTWs). The first, *prohibited discharge standards,* apply to all commercial and industrial establishments that discharge to POTWs. Prohibited standards restrict the discharge of pollutants that may create a fire or explosion hazard in sewers or treatment works, are corrosive (pH < 5.0), obstruct flow, upset treatment processes, or increase the temperature of the wastewater entering the plant to above 40°C. *Categorical standards* apply to industrial and commercial discharges in 25 industrial categories (categorical industries) and are intended to restrict the discharge of the identified priority pollutants. It is anticipated that the number of compounds listed will continue to expanded in the future.

20.1.3.2 *Volatile Organic Compounds (VOCs).* Organic compounds that have a boiling point <100°C and/or vapor pressure >1 mm Hg at 25°C are generally considered to be VOCs. Volatile organic compounds are of great concern because: (1) once such compounds are in the vapor state they are much more mobile, and therefore, more likely to be released to the environment, (2) the presence of some of these compounds in confined work areas and in the atmosphere may pose a significant public health risk, and (3) they contribute to a general increase in reactive hydrocarbons in the atmosphere, which can lead to the formation of photochemical oxidants. The release of these compounds in sewers and at treatment plants, especially at the headworks, are of particular concern with respect to the health of collection systems and wastewater-treatment plant workers.

20.1.4 Characterization of Organic Matter in Wastewater

The principal measures (often called parameters) used to characterize the gross amounts of organic matter found in wastewater include BOD_5 (5-day biochemical oxygen demand), COD (chemical oxygen demand), and TOC (total organic carbon). Complimenting these laboratory tests is the *theoretical oxygen demand* (ThOD), determined from the chemical formula of the compound.

20.1.4.1 *Biochemical Oxygen Demand.* The *BOD test* is the most common test used in the field of wastewater treatment. If sufficient oxygen is available, the aerobic biological decomposition of an organic waste, as depicted in Fig. 20.1, will continue until all of the waste is consumed. Three more or less distinct activities occur. First, a portion of the waste is oxidized to end products to obtain energy for cell maintenance and the synthesis of new cell tissue. Simultaneously, some of the waste is converted into new cell tissue, using part of the energy released during oxidation. Finally, when the organic matter is used, the new cells begin to consume their own cell tissue to obtain energy for cell maintenance. This third process is called *endogenous respiration*. Using the term COHNS (which represents the elements carbon, oxygen, hydrogen, nitrogen, and sulfur) to represent the organic waste and the term C_5HNO_2 to represent cell tissue, the three processes shown in Fig. 20.1 are defined by the following generalized chemical reactions:

Oxidation

$$COHNS + O_2 + bacteria \rightarrow CO_2 + H_2O + NH_3 + \text{other end products} + energy$$

$$(20.1)$$

Synthesis

$$COHNS + O_2 + bacteria + energy \rightarrow C_5H_7NO_2$$ $$(20.2)$$

New cell tissue

Endogenous respiration

$$C_5H_7NO_2 + 5O_2 \rightarrow 5CO_2 + NH_3 + 2H_2O$$ $$(20.3)$$

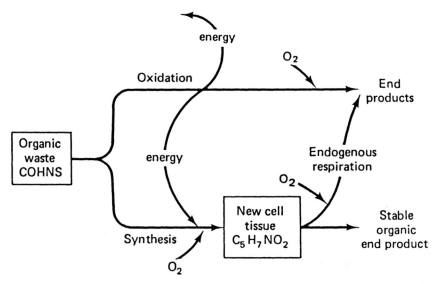

FIGURE 20.1 Schematic representation of the aerobic biological conversion of organic matter in wastewater to end products, to new cells, and ultimately to stable organic and other end products.

If only oxidation of the organic carbon that is present in the waste is considered, the ultimate BOD is the oxygen required to complete the three reactions. This oxygen demand is known as the *ultimate carbonaceous* or *first-stage* BOD and is usually denoted as BOD_u. In the standard test for BOD, a small sample of the wastewater to be tested is placed in a BOD bottle (300 mL). The bottle is then filled with dilution water saturated in oxygen and containing the nutrients required for biological growth. Before stoppering the bottle, the oxygen concentration in the bottle is measured. After incubating the bottle for 5 days at 20°C, the dissolved-oxygen concentration is measured again. The BOD of the sample is the difference in the dissolved-oxygen concentration values, expressed in mg/L, divided by the decimal fraction of sample used. The computed BOD value is known as the 5-day, 20°C biochemical oxygen demand. The exertion of the BOD with time is illustrated in Fig. 20.2.

20.1.4.2 Modeling of BOD Reaction. The rate of BOD exertion is modeled based on the assumption that the amount of organic material remaining at any time t is governed by a first order function, as given below.

$$L_t = L(10^{-K_1 t}) \tag{20.4}$$

where K_1 = first-order reaction rate constant, days
L_t = amount of waste remaining at time t (days) expressed in oxygen equivalents, mg/L
L = the total or/ultimate carbonaceous BOD, mg/L

Thus the BOD exerted up to time t is given by

$$BOD_t = L - L_t = L - L(10^{-K_1 t}) = L(1 - 10^{-K_1 t}) \tag{20.5}$$

Equation (20.5) is the standard expression used to define the BOD for wastewater. The value of K_1 is generally about 0.1 d^{-1} (base 10). The range is from 0.05 to 0.2 d^{-1} for effluents from biological treatment processes. For a given wastewater, the value of K_1 at 20°C can be determined experimentally by observing the variation with time of the dissolved oxygen in a series of incubated samples. If K_1 at 20°C is equal to 0.10 d^{-1}, the 5-day oxygen demand is about 68 percent of the ultimate first-stage demand. It has been found that K_1 varies with temperature as:

$$K_{1_T} = K_{1_{20}} 1.047^{T-20}$$

(20.6)

Equation (20.6), along with Eq. (20.5), makes it possible to convert test results from different time periods and temperatures to the standard 5-day 20°C test. Values of the 5-day 20°C BOD of municipal wastewater varies from about 100 to 500 mg/L. Although the BOD_5 test is commonly used, it suffers from several serious deficiencies. The most serious one is that the test has no stoichiometric validity. That is, the arbitrary 5-day period usually does not correspond to the point where all of the waste is consumed. Thus, it is not known where the 5-day BOD value falls along the curve (see Fig. 20.2). The 5-day value is used because the test was developed in England where the maximum time of flow of most rivers from headwaters to the ocean is about 4.8 days.

20.1.4.3 *Nitrification.*

An oxygen demand can also result from the biological oxidation of ammonia [see Eqs. (20.1) and (20.2)]. The reactions that define the nitrification process are as follows:

Conversion of ammonia to nitrite by *Nitrosomonas*

$$NH_3 + O_2 \rightarrow HNO_2 + H_2O$$

(20.7)

Conversion of nitrite to nitrate by *Nitrobacter*

$$HNO_2 + O_2 \rightarrow HNO_3$$

(20.8)

Overall conversion of ammonia to nitrate

$$NH_3 + 2O_2 \rightarrow HNO_3 + H_2O$$

(20.9)

The oxygen required for the conversion of ammonia to nitrate is known as the NOD (*nitrogenous oxygen demand*). Typically, the oxygen demand due to nitrification will occur from 5 to 8 days after the start of a conventional BOD test; however, it can occur sooner if sufficient nitrifying organisms are present initially (see Fig. 20.2).

20.1.4.4 *Carbonaceous Biochemical Oxygen Demand (CBOD).*

Nitrification is a common problem with the BOD test. The effects of nitrification can be overcome either by using various chemicals to suppress the nitrification reactions or by treating the sample to eliminate the nitrifying organisms. Pasteurization and chlorination are two methods that can be used. When the nitrification reaction is suppressed, the resulting BOD is known as the *carbonaceous biochemical oxygen demand* (CBOD). In effect, the CBOD is a measure of the oxygen demand exerted by the oxidizable carbon in the sample. The CBOD test should only be used on samples that contain little or no organic carbon (e.g., treated effluent). Large errors will occur in the measured BOD values (up to 20 percent) when the CBOD test is used on wastewater containing significant amounts of organic matter such as untreated wastewater.

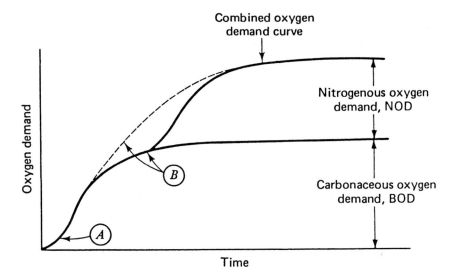

Combined oxygen demand curve

Nitrogenous oxygen demand, NOD

Carbonaceous oxygen demand, BOD

Oxygen demand

Time

(A) Lag period often occurs until microorganisms become acclimated

(B) Nitrification usually is observed to occur from 5 to 8 days after the start of the incubation period. In warm climates where nitrifying organisms may be present in sufficient numbers the combined oxygen demand curve will be approximated by the dotted curve

FIGURE 20.2 Definition sketch for the excretion of the carbonaceous and nitrogenous oxygen demand in the biological oxidation of the organic matter in wastewater. (*Tchobanoglous and Schroeder, 1985.*)

20.1.4.5 Chemical Oxygen Demand. The *COD test* is used to measure the oxygen equivalent of the organic material in wastewater that can be oxidized chemically using dichromate in an acid solution. Although it would be expected that the value of the ultimate BOD would approximate the COD, this is seldom the case. Some of the reasons for the observed differences are as follows: (1) many organic substances can be oxidized chemically but cannot be oxidized biologically, (2) inorganic substances that are oxidized by the dichromate increase the apparent organic content of the sample, (3) certain organic substances may be toxic to the microorganisms used in the BOD test, and (4) high COD values may occur because of the presence of interfering substances. From an operational standpoint, one of the main advantages of the COD test is that it can be completed in about 2.5 hours (compared to 5 or more days for the BOD test). To further reduce the time, a rapid COD test, which takes only about 15 min, has been developed.

20.1.4.6 Total Organic Carbon. The *TOC test*, done instrumentally, is used to determine the total organic carbon in an aqueous sample. The TOC of a wastewater can be used as a measure of its pollutional characteristics, and in some cases it has been possible to relate TOC to BOD and COD values. The TOC test is also gaining in favor because it takes only 5 to 10 min to complete. If a valid relationship can be established between results obtained with the TOC test and the results of the BOD test for a given wastewater, use of the TOC test for process control is recommended.

20.1.4.7 Comparison of Measures of Organic Matter. Typical values for the ratios of BOD_5/TOC and BOD_5/COD for municipal wastewater are reported in Table 20.4. If the ratio of BOD_5 to COD for untreated wastewater is 0.5 or greater, the waste is considered to be easily treatable by biological means. If the ratio is below about 0.3, either the waste may have some toxic components or acclimated microorganisms may be required in its stabilization.

20.1.5 Biological Characteristics

The principal groups of organisms found in surface water and wastewater include bacteria, fungi, algae, protozoa, plants, animals, and viruses. These organisms can be classified as *eucaryotes, eubacteria,* and *archaebacteria* (see Table 20.5). As reported in Table 20.5, most bacteria are classified as eubacteria. The category *protista*, contained within the eucaryote classification, includes algae, fungi, and protozoa. Plants, including seed plants, ferns, and mosses, are classified as multicellular eucaryotes. Invertebrates and vertebrates are classified as multicellular eucaryotic animals (Stanier, et al., 1986). Viruses are classified separately according to the host infected. Because of the extensive and fundamental role played by bacteria in the decomposition and stabilization of organic matter, both in nature and in treatment plants, their characteristics, functions, metabolism, and synthesis must be understood. Coliform bacteria are also used as an indicator of pollution by human wastes.

20.1.5.1 Pathogenic Organisms. Pathogenic organisms found in wastewater may be discharged by human beings who are infected with disease or who are carriers of a particular disease. The major categories of pathogenic organisms found in wastewater are reported in Table 20.6. The usual bacterial pathogenic organisms that may be excreted by humans cause diseases of the gastrointestinal tract, such as typhoid and paratyphoid fever, dysentery, diarrhea, and cholera. Because these organisms are highly infectious, they are responsible for many thousands of deaths each year in areas with poor sanitation, especially in the tropics.

20.1.5.2 Use of Indicator Organisms. Because the numbers of pathogenic organisms present in wastes and polluted waters are few and difficult to isolate and identify, the coliform organism, which is more numerous and more easily tested for, is commonly used as an indicator organism. The human intestinal tract

TABLE 20.4 Comparison of Ratios of Various Parameters Used to Characterize Municipal Wastewater

Type of wastewater	BOD_5/TOC	BOD_5/COD
Untreated	1.20–1.50	0.50–0.65
After primary settling	0.90–1.10	0.40–0.55
Final effluent	0.25–0.50	0.10–0.25

TABLE 20.5 Classification of Microorganisms

Group	Cell structure	Characterization	Representative members
Eucaryotes	Eucaryotic*	Multicellular with extensive differentiation of cells and tissue	Plants (seed plants, ferns, mosses) Animals (vertebrates, invertebrates)
		Unicellular or coenocytic or mycelial; little or no tissue differentiation	Protists (algae, fungi, protozoa)
Eubacteria	Procaryotic[†]	Cell chemistry similar to eucaryotes	Most bacteria
Archaebacteria	Procaryotic[†]	Distinctive cell chemistry	Methogens, halophiles, thermacidophiles

* Contain true nucleus.
[†] Contain no nuclear membrane.
Source: Stanier, et al. (1986).

contains countless rod-shaped bacteria known as *coliform organisms*. Each person discharges from 100 to 400 billion coliform organisms per day, in addition to other kinds of bacteria. Thus, the presence of coliform organisms is taken as an indication that pathogenic organisms may also be present, and the absence of coliform organisms is taken as an indication that the water is free from disease-producing organisms.

Either the *multiple-tube fermentation technique* or the *membrane-filter technique* is used to enmuerate the coliform group. The complete multiple-tube fermentation procedure for total coliform involves three test phases identified as the presumptive, confirmed, and completed test (Greenberg, et al., 1992). A similar procedure is available for the fecal coliform group as well as for other bacterial groups. The multiple-tube fermentation technique is based on the principle of dilution to extinction as illustrated in Fig. 20.3. Concentrations of total coliform bacteria are most often reported as the most probable number per 100 mL (MPN/100 mL). The MPN is based on the application of the Poisson distribution for extreme values to the analysis of the number of positive and negative results obtained when testing multiple portions of equal volume and in portions constituting a geometric series. It is emphasized that the MPN is not the absolute concentration of organisms that are present but only a statistical estimate of that concentration.

The membrane-filter technique is also used to determine the number of coliform organisms. The determination is accomplished by passing a known volume of water sample through a membrane filter that has a very small pore size. Bacteria are retained on the filter because they are larger than the size of the pores of the membrane filter. The membrane filter containing the bacteria is then contacted with an agar that contains nutrients necessary for the growth of the bacteria. After incubation, the coliform colonies can be counted and the concentration in the original water sample determined. The membrane-filter technique has the advantage of being faster than the MPN procedure and of giving a direct count of the number of coliforms. Both methods are subject to limitations in interpretation (Greenberg, et al., 1992).

TABLE 20.6   Infectious Agents Potentially Present in Raw Domestic Wastewater

Organism	Disease	Remarks
Bacteria		
Escherichia coli (enteropathogenic)	Gastroenteritis	Diarrhea
Leptospira (spp.)	Leptospirosis	Jaundice, fever (Weil's disease)
Salmonella typhi	Typhoid fever	High fever, diarrhea, ulceration of small intestine
Salmonella (~1700 spp.)	Salmonellosis	Food poisoning
Shigella (4 spp.)	Shigellosis	Bacillary dysentery
Vibrio cholerae	Cholera	Extremely heavy diarrhea, dehydration
Yersinia enterolitica	Yersinosis	Diarrhea
Viruses		
Adenovirus (31 types)	Respiratory disease	
Enteroviruses (67 types, e.g., polio, echo, and coxsackie viruses)	Gastroenteritis, heart anomalies, meningitis	
Hepatitis A virus	Infectious hepatitis	Jaundice, fever
Norwalk agent	Gastroenteritis	Vomiting
Parovirus (2 types)	Gastroenteritis	
Rotavirus	Gastroenteritis	
Protozoa		
Balantidium coli	Balantidiasis	Diarrhea, dysentery
Cryptosporidium	Cryptosporidiosis	Diarrhea
Entamoeba histolytica	Amebiasis (amoebic dysentery)	Prolonged diarrhea with bleeding, abscesses of the liver and small intestine
Giardia lamblia	Giardiasis	Mild to severe diarrhea, nausea, indigestion
Helminths*		
Ascaris lumbricoides	Ascariasis	Roundworm infestation
Enterobius vericularis	Enterobiasis	Pinworm
Fasciola hepatica	Fascioliasis	Sheep liver fluke
Hymenolepis nana	Hymenolepiasis	Dwarf tapeworm
Taenia saginata	Taeniasis	Beef tapeworm
T. solium	Taeniasis	Pork tapeworm
Trichuris trichiura	Trichuriasis	Whipworm

* The helmiths listed are those with a worldwide distribution
Source: Adapted from Feacham, et al. (1983); and Stanier, et al. (1986).

20.1.6   Toxicity Tests

Toxicity tests are used to: (1) assess the suitability of environmental conditions for aquatic life, (2) establish acceptable receiving-water concentrations for conventional parameters (such as DO, pH, temperature, salinity, or turbidity), (3) study the effects of water-quality parameters on wastewater toxicity, (4) assess the toxicity of wastewater to a variety of fresh, estuarine, and marine test species, (5) establish relative sensitivity of a group of standard aquatic organisms to effluent as well as standard

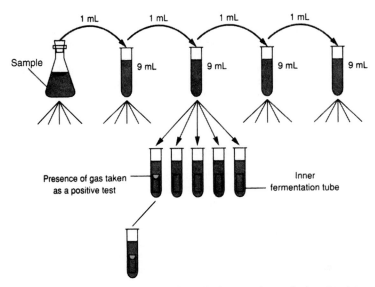

FIGURE 20.3 Illustration of the multiple-tube fermentation method used to determine bacterial densities. As shown, the multiple-tube method is based on the principal of dilution to extinction.

toxicants, (6) assess the degree of wastewater treatment needed to meet water-pollution-control requirements, (7) determine the effectiveness of wastewater-treatment methods, (8) establish permissible effluent-discharge rates, and (9) determine compliance with federal and state water-quality standards and water-quality criteria associated with NPDES permits. Such tests provide results that are useful in protecting human health and aquatic life from impacts caused by the release of contaminants into surface waters.

20.1.6.1 Toxicity Terminology. The following terms are used in evaluating the effects of contaminants on living organisms:

Acute toxicity. Exposure will result in significant response shortly after exposure (typically a response is observed within 48 or 96 hours).

Chronic toxicity. Exposure will result in sublethal response over a long term, often 1/10 of the life span or more.

Lethal toxicity. Exposure results in death.

Sublethal toxicity. Exposure will damage organism, but not cause death.

Cumulative toxicity. The effects on an organism caused by successive exposures.

20.1.6.2 Toxicity Testing. Toxicity tests are classified according to: (1) duration: short-term, intermediate, and/or long-term, (2) method of adding test solutions: static, recirculation, renewal, or flow-through, and (3) purpose: NPDES permit requirements, mixing zone determinations, and so forth. Toxicity testing has been widely validated in recent years. Even though organisms vary in sensitivity to effluent toxicity, EPA has documented that: (1) toxicity of effluents correlate well with toxicity measurements in the receiving waters when effluent dilution is measured, and (2) predictions of impacts from both effluent and receiving-water toxicity tests compare favorably with ecological community responses in the receiving waters. The EPA has conducted nationwide tests with freshwater, estuarine, and marine ecosystems. Methods include both acute as well as chronic exposures. Typical short-term chronic toxicity test methods are reported in Table 20.7. Detailed contemporary testing protocols are summarized in U.S. EPA (1985a; 1985b; 1988; 1989; 1991).

20.1.7 The Composition of Wastewater

Because the physical and chemical characteristics of wastewater vary throughout the day, frequent tests of the untreated wastewater (influent) and of the treated wastewater leaving the plant (effluent) are necessary for adequate characterization. An adequate determination of the waste characteristics will result only if the sample tested is representative. Hence, composite samples made up of portions of samples collected at regular intervals during a day are used. The amount used from each sample is proportional to the rate of flow at the time the sample was collected. Typical data on the composition of untreated domestic wastewater are reported in Table 20.8.

TABLE 20.7 Typical Examples of Short-Term Chronic Toxicity Test Methods Using Various Freshwater and Marine/Estuarine Aquatic Species

Species/common name	Test duration	Test endpoints
Freshwater species		
Cladoceran, *Ceriodaphnia dubia*	Approximately 7 days (until 60% of control have 3 broods)	Survival, reproduction
Fathead minnow, *Pimephales promelas*	7 days	Larval growth, survival
	9 days	Embryo-larval survival, percent hatch, percent abnormality
Freshwater algae, *Selenastrum capricomutum*	4 days	Growth
Marine/estuarine species		
Sea urchin, *Arbacia punctulata*	1.5 hours	Fertilization
Red macroalgae, *Champia parvula*	7–9 days	Cystocarp production (fertilization)
Mysid, *Mysidopsis bahia*	7 days	Growth, survival, fecundity
Sheepshead minnow, *Caprinodon variegatus*	7 days	Larval growth, survival
	7–9 days	Embryo-larval survival, percent hatch, percent abnormality
Inland silverside, *Menidia beryijina*	7 days	Larval growth, survival

Source: U.S. EPA (1991).

TABLE 20.8 Typical Composition of Untreated Domestic Wastewater*

Contaminants	Unit	Concentration		
		Weak	Medium	Strong
Solids, total (TS)	mg/L	350	720	1200
Dissolved, total (TDS)	mg/L	250	500	850
Fixed	mg/L	145	300	525
Volatile	mg/L	105	200	325
Suspended solids (SS)	mg/L	100	220	350
Fixed	mg/L	20	55	75
Volatile	mg/L	80	165	275
Settleable solids	mL/L	5	10	20
Biochemical oxygen demand, 5-day, 20°C (BOD_5, 20°C)	mg/L	110	220	400
Total organic carbon (TOC)	mg/L	80	160	290
Chemical oxygen demand (COD)	mg/L	250	500	1000
Nitrogen (total as N)	mg/L	20	40	85
Organic	mg/L	8	15	35
Free ammonia	mg/L	12	25	50
Nitrites	mg/L	0	0	0
Nitrates	mg/L	0	0	0
Phosphorus (total as P)	mg/L	4	8	15
Organic	mg/L	1	3	5
Inorganic	mg/L	3	5	10
Chlorides*	mg/L	30	50	100
Sulfate*	mg/L	20	30	50
Alkalinity (as $CaCO_3$)	mg/L	50	100	200
Grease	mg/L	50	100	150
Total coliform	no./100 mL	10^{-6}–10^{-7}	10^{-7}–10^{-8}	10^{-7}–10^{-9}
Nonspecific coliphage	no./100 mL	10^{-2}–10^{-4}	10^{-3}–10^{-4}	10^{-3}–10^{-5}
Volatile organic compounds (VOCs)	mg/L	<100	100–400	>400

* Values should be increased by amount present in domestic water supply.

In general, the per capita BOD and suspended solids contributions to municipal wastewater will vary as shown in Table 20.9. These values are sometimes used to compute the population equivalent of various waste discharges. In fact, many cities base their charges for treatment of industrial wastes at least partly on the computed population equivalent. For example, if the contribution of BOD_5 is 0.20 lb/capita · d (0.09 kg/capita · d), the population equivalent of an industrial discharge containing 20,000 lb BOD_5/d is 100,000 persons.

20.2 WASTEWATER-TREATMENT METHODS AND STANDARDS

Before considering the individual physical, chemical, and biological methods used for the treatment of wastewater, it is appropriate: (1) to review current effluent standards and regulations, (2) to review the classification and application of treatment methods, and (3) to define the various levels of wastewater treatment.

TABLE 20.9 Unit Waste Loading Factors

	Value, lb/capita · d	
Constituent	Range	Typical
Normal domestic wastewater without contribution from ground kitchen wastes		
BOD_5	0.13–0.24	0.18
SS	0.13–0.25	0.20
Nutrients*		
Ammonia nitrogen	0.004–0.008	0.007
Organic nitrogen	0.013–0.026	0.020
Total Kjeldahl nitrogen	0.020–0.031	0.026
Organic phosphorus	0.002–0.004	0.003
Inorganic phosphorus	0.004–0.007	0.006
Total phosphorus	0.007–0.011	0.008
Normal domestic wastewater with contribution from ground kitchen wastes*		
BOD_5	0.18–0.26	0.22
SS	0.20–0.33	0.26

* Values for nutrients are approximately the same as those shown for wastewater without contribution from ground kitchen wastes.
   Note: lb × 453.59 = g

20.2.1 Standards and Regulations

Pursuant to Section 304(d) of Public Law 92-500, the U.S. Environmental Protection Agency published its definition of secondary treatment. This definition, originally issued in 1973, was amended in 1985 to allow additional flexibility in applying the percent removal requirements of pollutants to treatment facilities serving separate sewer systems. The current definition of secondary treatment is reported in Table 20.10. The definition of secondary treatment includes three major effluent parameters: 5-day BOD, suspended solids, and pH. The substitution of 5-day carbonaceous BOD ($CBOD_5$) for BOD_5 may be made at the option of the NPDES permitting authority. Special interpretations of the definition of secondary treatment are permitted for publicly owned treatment works (1) served by combined sewer systems, (2) using waste-stabilization ponds and trickling filters, (3) receiving industrial flows, or (4) receiving less-concentrated influent wastewater from separate sewers. The secondary-treatment regulations were amended further in 1989 to clarify the percent removal requirements during dry periods for treatment facilities served by combined sewers (*Federal Register,* 1988; 1989).

Regulations are developed to implement legislation. For this reason, regulations are always subject to change as more information becomes available regarding the characteristics of wastewater, effectiveness of treatment processes, and environmental effects. It is anticipated that the focus of future regulations will be on the implementation of the Water Quality Act of 1987. Control of the pollutional effects of stormwater and nonpoint sources, toxics in wastewater (priority pollutants), and, as noted above, the overall management of biosolids including the control of toxic substances will receive the most attention. Nutrient removal, the control of pathogenic organisms, and the removal of organic and inorganic substances such as VOCs and total dissolved solids will also continue to receive attention in specific applications.

20.2.2 Classification and Application of Treatment Methods

The contaminants in wastewater are removed by physical, chemical, and biological means. Treatment methods in which the application of physical forces predominate are known as *physical unit operations* (*P*). Because most of these methods evolved directly from humans' first observations of nature, they were the first to be used for wastewater treatment. Treatment methods in which the removal or conversion of contaminants is brought about by the addition of chemicals or by other chemical reactions are known as *chemical unit processes* (*C*). Precipitation, adsorption, and disinfection are the most common examples used in wastewater treatment. Treatment methods in which the removal of contaminants is brought about by biological activity are known as *biological unit processes* (*B*). Biological treatment is used primarily to remove the biodegradable organic substances (colloidal or dissolved) in wastewater. A typical flow diagram for the treatment of wastewater to meet EPA secondary standards (see Table 20.10) is shown in Fig. 20.4. Although these operations and processes occur in a variety of combinations in treatment systems such as shown in Fig. 20.4, it has been found advantageous to study their scientific basis separately because the principles involved do not change.

20.2.3 Levels of Treatment

The treatment of wastewater involves the grouping together of several unit operations and processes to achieve various levels of treatment. Typical levels of wastewater treatment are identified in Table 20.11. Although the terms identified in Table 20.11 are in common use, a more rational approach would be to establish the level of contaminant removal (treatment) required before the wastewater can be reused or discharged to the environment. The required unit operations and processes necessary to achieve that required degree of treatment can then be grouped together on the basis of fundamental considerations without regard to the level of treatment.

TABLE 20.10 Minimum National Standards for Secondary Treatment*

Characteristic of discharge	Unit of measurement	Average 30-day concentration[†]	Average 7-day concentration[†]
BOD_5	mg/L	30[‡]	45
Suspended solids	mg/L	30[‡]	45
Hydrogen-ion concentration	pH units		Within the range of 6.0 to 9.0 at all times[§]
$CBOD_5$[¶]	mg/L	25[†]	40

* Present standards allow stabilization ponds and trickling filters to have higher 30-day average concentrations (45 mg/L) and 7-day average concentrations (65 mg/L) BOD/suspended solids performance levels as long as the water quality of the receiving water is not adversely affected. Exceptions are also permitted for combined sewers, certain industrial categories, and less-concentrated wastewater from separate sewers. For precise requirements of exceptions, the Federal Register (1988) should be consulted.

[†] Not to be exceeded

[‡] Average removal shall not be less than 85%

[§] Only enforced if caused by industrial wastewater or by in-plant inorganic chemical addition.

[¶] May be substituted for BOD_5 at the option of the NPDES permitting authority.

Source: *Federal Register* (1988; 1989).

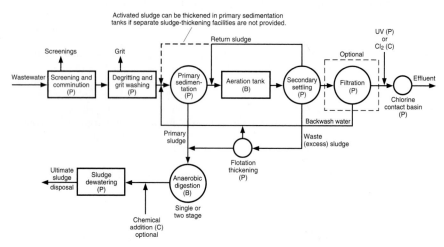

FIGURE 20.4 Typical flow diagram for the treatment of domestic wastewater with the activated-sludge process to meet EPA secondary requirements. The various physical (P), chemical (C), and biological (B) operations and processes are also identified.

TABLE 20.11 Levels of Wastewater Treatment

Treatment level	Description
Preliminary	Removal of wastewater constituents that may cause maintenance or operational problems with the treatment operations, processes, and ancillary systems.
Primary	Removal of a portion of the suspended solids and organic matter from the wastewater.
Advanced primary	Enhanced removal of suspended solids and organic matter from the wastewater. Typically accomplished by chemical addition or filtration.
Secondary	Removal of biodegradable organics and suspended solids. Disinfection is also typically included in the definition of conventional secondary treatment.
Secondary with nutrient removal	Removal of biodegradable organics, suspended solids, and nutrients (nitrogen, phosphorus, or both nitrogen and phosphorus).
Tertiary	Removal of residual suspended solids, usually by granular medium filtration. Disinfection is also typically a part of tertiary treatment. Nutrient removal is often included in this definition.
Advanced	Removal of dissolved and suspended materials remaining after normal biological treatment when required for water reuse or for the control of eutrophication in receiving waters.

20.3 PHYSICAL TREATMENT METHODS

Physical treatment methods used for the treatment of wastewater are reported in Table 20.12. Although flow metering is not, in the strict sense, a physical treatment method, it is a critical factor in the control and monitoring of wastewater-treatment plants regardless of size. Each of these methods is considered briefly in the following. Additional details on the physical methods of treatment described here may be found in Tchobanoglous and Burton (1991); and Water Environment Federation (1991).

20.3.1 Flow Metering

The correct application, selection, and maintenance of flow-metering devices is critical to the efficient operation of a modern wastewater-treatment facility. A complete flow-measurement system consists of two elements: (1) a sensor or detector and (2)

TABLE 20.12 Applications of Physical Unit Operations of Wastewater Treatment Arranged Alphabetically

Operation	Application
Adsorption	Removal of organics not removed by conventional chemical and biological treatment methods; also used for dechlorination of wastewater before final discharge of treated effluent.
Comminution	Grinding of coarse solids to a more or less uniform size.
Disinfection	Selective destruction and or inactivation of disease-causing organisms, usually with UV radiation or heat.
Filtration	Removal of fine residual suspended solids remaining after biological or chemical treatment. Thickening of sludge (e.g., belt filters).
Flocculation	Promotes the aggregation of small particles into larger particles to enhance their removal by gravity sedimentation.
Flotation	Removal of finely divided suspended solids and particles with densities close to that of water; also thickens biological sludges.
Flow equalization	Equalization of flow and mass loadings of BOD and suspended solids.
Flow metering	Measurement of wastewater-flow rates.
Gas transfer	Addition and removal of gases; removal of volatile substances; gas stripping.
Grit removal	Removal of grit to limit the accumulation of grit in subsequent treatment units.
Microscreening	Same as filtration; also removal of algae from stabilization pond effluents.
Mixing	Mixing of chemicals and gases with wastewater, and maintaining solids in suspension.
Screening	Removal of coarse and settleable solids by interception (surface straining).
Sedimentation	Removal of settleable solids and thickening of sludges.

a converter device. The sensor or detector is exposed to or affected by the flow; the converter is the device used to translate the signal or reading from the sensor into a flow reading.

20.3.2 Comminution

Comminutors (or *shredders*) are devices used to grind or cut waste solids to about 0.25 in (6 mm) size. In one type of comminutor, the wastewater enters a slotted cylinder within which another similar cylinder with sharp-edged slots rotates rapidly (see Fig. 20.5). As the solids are reduced in size, they pass through the slots of the cylinders and move on with the liquid to the treatment plant. Comminutors eliminate the problem of disposal of screenings by reducing the solids to a size that can be processed elsewhere in the plant.

20.3.3 Grit Removal

Specially designed *grit chambers* are used to remove inorganic particles (specific gravity about 1.6 to 2.65) such as sand, gravel, eggshells, and bone 0.2 mm or larger in size to prevent damage to pumps and to prevent the accumulation of this material in treatment facilities and sludge digesters. In the past, gravity-type grit chambers were used extensively. Today, however, spiral-flow aerated and free and forced-vortex grit chambers are most commonly used (see Fig. 20.6). The detention time in

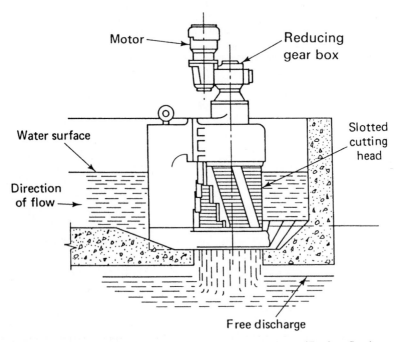

FIGURE 20.5 Example of circular comminutor with free discharge. (*Ecodyne Corp.*)

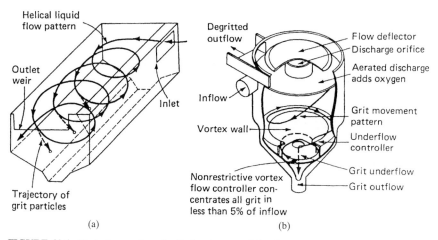

FIGURE 20.6 Typical examples of grit chambers used for the removal of grit from untreated wastewater: (*a*) aerated grit chamber, and (*b*) vortex type (teacup) grit separator. (*Eutek, Inc.*)

such units is 3 to 5 min at peak flow. Grit may be used for fill or hauled away for disposal in a landfill if it does not contain too much organic material.

20.3.4 Screening

Coarse screens or *bar racks* with 2-in (50-mm) openings or larger are used to remove large floating objects from wastewaters. A typical installation of the fixed-bar type is shown in Fig. 20.7*a*. They are installed ahead of pumps to prevent clogging. The material removed usually consists of wood, rags, and paper that will not putrefy and may be disposed of by incineration, burial, or dumping. Coarse screens have openings ranging from about ½ to 1½ in (12 to 40 mm). Coarse and medium screens should be large enough to maintain a velocity of flow through their openings under 3 ft/s (1 m/s). Limiting the head loss through the screens reduces the opportunity for screenings to be pushed through the openings. According to the method used to clean them, bar racks and screens are designated as hand-cleaned or mechanically cleaned. Screenings usually contain about 80 percent moisture by weight and will not burn without predrying. In most incinerators, the fresh screenings are dried by the heat of the fire before they enter the firebox. Fuel gas, oil, and sludge gas from the treatment plant may be used as fuel for the incinerator.

Fine screens with openings of ⅛ to ¹⁄₁₆ in (3.2 to 1.6 mm) are often used to pretreat industrial wastewater or to relieve the load on sedimentation basins at municipal plants where heavy solids are present. They will remove as much as 20 percent of the suspended solids in wastewaters. A fine screen should ordinarily be preceded by a medium screen (see Figs. 20.7*b,c*) or shredder to remove the larger particles. Very fine wire-mesh screens with openings varying from 0.001 to 0.003 in (25 to 75 μm) have been used as a replacement for conventional primary sedimentation (see Fig. 20.7*d*). Fine screens known as microscreens have also been used for the removal of algae. The screenings usually contain considerable organic material which may putrefy and become offensive and must be disposed of by incineration or burial.

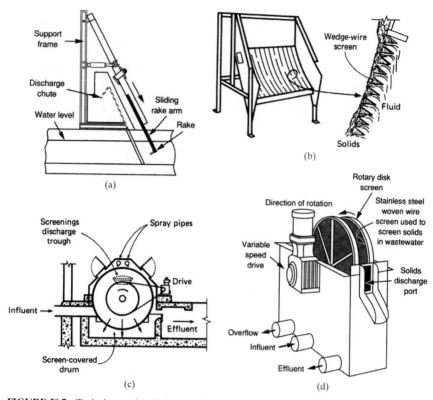

FIGURE 20.7 Typical screening devices used for the removal of solids from domestic wastewater: (*a*) inclined mechanically cleaned bar screen; (*b*) inclined fixed screen; (*c*) rotary drum screen; (*d*) rotary disk screen.

20.3.5 Flow Equalization

The variations that are observed in the influent-wastewater-flow rate and strength at almost all wastewater-treatment facilities were discussed in the previous chapter. Flow equalization involving on- or off-line storage is used to overcome the operational problems caused by flow-rate variations, to improve the performance of the downstream processes, and to reduce the size and cost of downstream treatment facilities. In effect, flow equalization dampens flow-rate variations so that a constant or nearly constant flow rate is achieved. This technique can be applied in a number of different situations, depending on the characteristics of the collection system.

20.3.6 Mixing

Mixing is an important unit operation in many phases of wastewater treatment including: (1) the mixing of one substance completely with another, (2) the mixing of liquid suspensions, (3) the blending of miscible liquids, (4) flocculation, and (5) heat transfer. The time range for rapid mixing is from a fraction of a second to about 30 seconds. The rapid mixing of chemicals in a liquid mixing can be carried out in a number of different ways, including: (1) in hydraulic jumps in open channels, (2) in Venturi flumes, (3) in pipelines, (4) by pumping, (5) with static mixers, and (6) with mechanical mixers. In the first four of these ways, mixing is accomplished as a result of turbulence that exists in the flow regime. In static mixers, turbulence is induced through the dissipation of energy. In mechanical mixing, turbulence is induced through the input of energy by means of rotating impellers, such as propellers, turbines, and paddles. Typical devices used for mixing in wastewater-treatment plants are shown in Fig. 20.8. Typical mixing intensities expressed in terms of the velocity gradient G are reported in Table 20.13.

20.3.7 Flotation

Flotation is a unit operation used to separate solid or liquid particles from a liquid phase. Separation is brought about by introducing fine gas (usually air) bubbles into the liquid phase. The bubbles attach to the particulate matter, and the buoyant force of the combined particle and gas bubbles is great enough to cause the particle to rise to the surface. Particles that have a higher density than the liquid can thus be made to rise. The rising of particles with lower density than the liquid can also be facilitated (e.g., oil suspension in water).

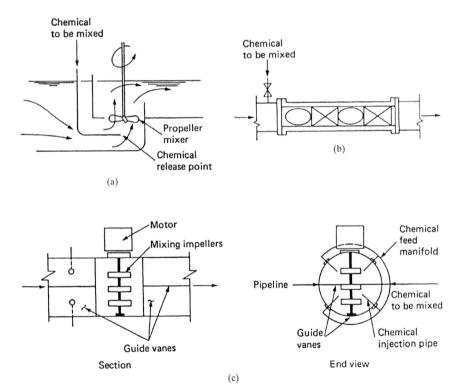

FIGURE 20.8 Typical devices used to mix chemicals with wastewater with mixing times equal to or less than 1 s: (*a*) mechanical mixer; (*b*) in-line static mixer; (*c*) in-line mechanical mixer.

TABLE 20.13 Typical Velocity Gradient G and Detention Time Values for Various Wastewater-Treatment Processes

Process	Range of values	
	Detention time	G value, s^{-1}
Mixing		
Typical rapid mixing operations in wastewater treatment	5–30 s	500–1500
Rapid mixing for effective initial contact and dispersion of chemicals	≤1 s	1500–6000
Rapid mixing of chemicals in contact filtration processes	<1 s	2500–7500
Flocculation		
Typical flocculation processes used in wastewater treatment	30–60 min	50–100
Flocculation in direct filtration processes	2–10 min	25–150
Flocculation in contact filtration processes	2–5 min	25–200

20.3.8 Gas Transfer

Gas transfer may be defined as the process by which gas is transferred from one phase to another, usually from the gaseous to the liquid phase. It is a vital part of a number of wastewater-treatment processes. For example, the functioning of aerobic processes, such as activated-sludge biological filtration and aerobic digestion, depends on the availability of sufficient quantities of oxygen. Chlorine, when used as a gas, must be transferred to solution in the water for disinfection purposes. Oxygen is often added to treated effluent after chlorination (postaeration). One process for removing nitrogen compounds consists of converting the nitrogen to ammonia and transferring the ammonia gas from the water to air. Typical aeration devices are shown in Fig. 20.9.

20.3.9 Sedimentation

Sedimentation is the separation from water, by gravitational settling, of suspended particles that are heavier than water. It is one of the most widely used unit operations in wastewater treatment. The terms *sedimentation* and *settling* are used interchangeably. A sedimentation basin may also be referred to as a sedimentation tank, settling basin, or settling tank. Sedimentation is used for grit removal, particulate-matter removal in the primary settling basin, biological-floc removal in the activated-sludge settling basin (see Fig. 20.10), and chemical-floc removal when the chemical-coagulation process is used. It is also used for solids concentration in sludge thickeners. In most cases, the primary purpose is to produce a clarified effluent, but it is also necessary to produce sludge with a solids concentration that can be easily handled and treated.

On the basis of the concentration and the tendency of particles to interact, four types of settling can occur in a nonaccelerated flow field: discrete particle, flocculant, hindered (also called zone), and compression. These types of settling phenomena are described in Table 20.14. During a sedimentation operation, it is common to have more than one type of settling occurring at a given time, and it is possible to have all four occurring simultaneously. The major function of a plain sedimentation basin in wastewater treatment is to remove the larger suspended material from the incoming wastewater. The material to be removed is high in organic content (50 to 75 percent) and has a specific gravity of 1.2 or less. The settling velocity of these organic particles is commonly as low as 4 ft/hr (1.25 m/h). A sloping bottom facilitates removal of the sludge. To get satisfactory performance from a sedimentation basin, the inlet must be designed to cause a uniform velocity distribution in the basin (see Fig. 20.11). Because large amounts of scum usually accumulate on the surface of sedimentation tanks, scum-removal facilities must be provided. A properly designed sedimentation basin will remove from 50 to 70 percent of the suspended solids in untreated wastewater. As noted in Table 20.14, gravity setting can also occur in an accelerated flow field, such as found in some centrifugal flow grit removal units (see Fig. 20.6b).

20.3.10 Granular-Medium Filtration

Although filtration is one of the principal unit operations used in the treatment of potable water, the filtration of effluents from wastewater-treatment processes is a relatively recent practice. Filtration is now used extensively for achieving supplemental removals of suspended solids (including particulate BOD) from wastewater effluents of biological and chemical treatment processes (see Fig. 20.12). Filtration is

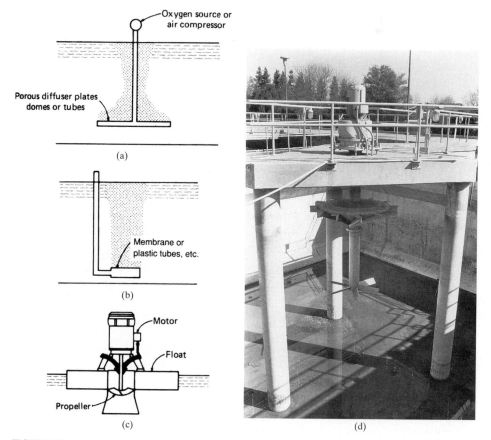

FIGURE 20.9   Typical devices used for the transfer of oxygen: (a) fine-bubble diffused air; (b) medium-bubble diffused air; (c) high-speed floating aerator; (d) low-speed turbine.

(a)

(b)

FIGURE 20.10 Typical examples of sedimentation tanks: (*a*) circular type used for the primary sedimentation of wastewater, and (*b*) circular type shown empty, used for the removal of solids following the activated-sludge process.

TABLE 20.14 Types of Settling Phenomena Involved in Wastewater Treatment

Type of settling phenomenon	Description	Application/occurrence
Discrete particle (type 1)	Refers to the settling of particles in a suspension of low solids concentration by gravity in a constant acceleration field. Particles settle as individual entities, and there is no significant interaction with neighboring particles.	Removes grit and sand particles from wastewater.
Flocculant (type 2)	Refers to a rather dilute suspension of particles that coalesce, or flocculate, during the settling operation. By coalescing, the particles increase in mass and settle at a faster rate.	Removes a portion of the suspended solids in untreated wastewater in primary settling facilities, and in upper portions of secondary settling facilities. Also removes chemical floc in settling tanks.
Hindered, also called zone (type 3)	Refers to suspensions of intermediate concentration, in which interparticle forces are sufficient to hinder the settling of neighboring particles. The particles tend to remain in fixed positions with respect to each other, and the mass of particles settles as a unit. A solids–liquid interface develops at the top of the settling mass.	Occurs in secondary settling facilities used in conjunction with biological treatment facilities.
Compression (type 4)	Refers to settling in which the particles are of such concentration that a structure is formed, and further settling can occur only by compression of the structure. Compression takes place from the weight of the particles, which are constantly being added to the structure by sedimentation from the supernatant liquid.	Usually occurs in the lower layers of a deep sludge mass, such as in the bottom of deep secondary settling facilities and in sludge-thickening facilities.
Accelerated gravity settling	Removal of particles in suspension by gravity settling in an acceleration field.	Removes grit and sand particles from wastewater.

also used to remove chemically precipitated phosphorus. The use of rapid granular medium filters for effluent polishing following secondary treatment is gaining in popularity, especially as the EPA published a definition of secondary treatment (see Table 20.10). Both gravity and pressure filters have been used. Slow sand filters are sometimes used for final or advanced treatment following secondary or other treatment processes, such as lagoons and stabilization ponds. Such filters are often called *polishing filters.* Wastewater is applied continuously at about 10 gal/ft^{2}d (40 L/m^{2}d) and the straining action of the sand and the biological mat that forms on the sand is relied upon to remove most of the remaining suspended solids in the wastewater.

20.3.11 Physical Disinfection

Ultraviolet (UV) light and the application of heat are the two most common physical means used for the disinfection of wastewater. Although the application of heat

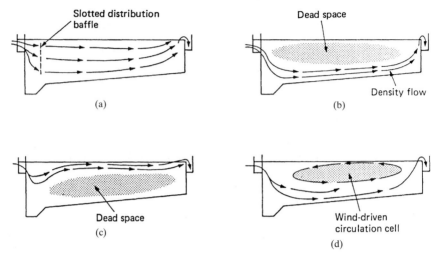

FIGURE 20.11   Flow patterns observed in rectangular sedimentation tanks: (*a*) ideal flow; (*b*) effect of density flow or thermal stratification; (*c*) effect of thermal stratification; (*d*) formation of wind-driven circulation cell.

is effective, this method is too costly given the large volumes of wastewater involved. With the proper dosage, UV radiation has been shown to be an effective bacterio-cide and virocide, while not contributing to the formation of any toxic compounds. More than 1000 wastewater-treatment facilities now use UV radiation for the disin-fection of wastewater.

20.3.11.1   Effectiveness of UV Radiation.   Ultraviolet light is a physical rather than a chemical disinfecting agent. Radiation with a wavelength of around 254 nm penetrates the cell wall of the microorganism and is absorbed by cellular materials including DNA and RNA, which either prevents replication or causes death of the cell to occur. Because the only ultraviolet radiation effective in destroying bacteria is that which reaches the bacteria, the water must be relatively free from turbidity (e.g., suspended solids) and chemical constituents that can absorb the ultraviolet energy and shield the bacteria. In effluents with minimal suspended solids, the inac-tivation of bacteria by UV radiation can be described, for practical purposes, using first-order kinetics.

20.3.11.2   Source of UV Radiation.   Low-pressure lamps and the newer, medium-pressure mercury arc lamps are, at present, the principal means of generat-ing UV energy used for disinfection. With low-pressure mercury lamps, about 85 percent of the light output is monochromatic at a wavelength of 253.7 nm, which is within the optimum range (250 to 270 nm) for germicidal effects. The output from medium-pressure lamps covers a broader range of wavelengths. To produce UV energy, the lamp, which contains mercury vapor, is charged by striking an electric arc. The energy generated by the excitation of the mercury vapor contained in the lamp results in the emission of UV light. Because UV radiation is not a chemical agent, toxic residuals are not produced. However, certain chemical compounds may be altered by the ultraviolet radiation. Based on the evidence to date, it appears that

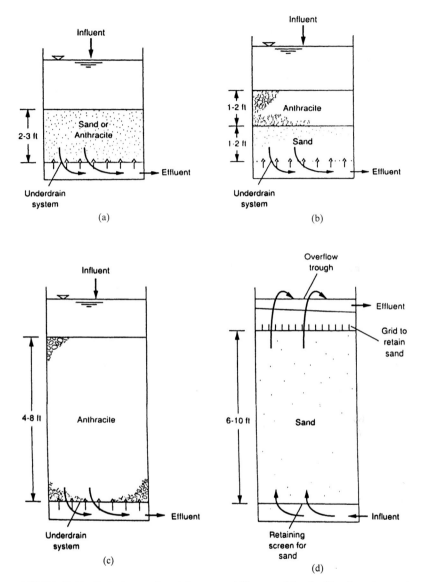

FIGURE 20.12 Typical examples of the types of filters used for the filtration of treated wastewater: (*a*) conventional mono-medium downflow; (*b*) conventional dual-medium downflow; (*c*) conventional mono-medium deep-bed downflow; (*d*) deep-bed upflow.

the compounds are broken down into a more innocuous form, but additional investigation into this occurrence is still warranted.

Operationally, low-pressure UV lamps, encased in quartz tubes to prevent cooling effects on the lamps, are suspended in the liquid to be treated (see Fig. 20.13). Because the distance over which ultraviolet light is effective is limited, the spacing

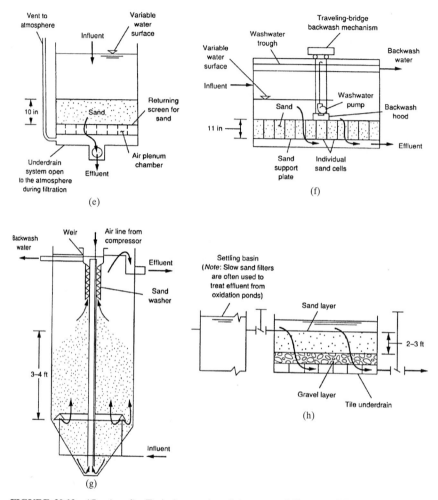

FIGURE 20.12 *(Continued)* Typical examples of the types of filters used for the filtration of treated wastewater: (*e*) pulsed-bed filter; (*f*) traveling-bridge filter; (*g*) continuous-backwash deep-bed upflow filter; (*h*) slow sand filter. Note filters (*e*) and (*g*) have also been used successfully for the filtration of settled primary effluent.

between lamps is typically 3 in (75mm). Both horizontal (parallel to flow) and vertical (perpendicular to flow) UV lamp configurations are used. To achieve effective inactivation of the microorganisms in wastewater, multiple banks of UV lights are installed in plug-flow channels. With medium-pressure lamps, the flow geometry around the lamp is fixed. Exposure time with both low- and medium-pressure UV lamps is controlled by the rate of flow past the lamps. The UV dosage range for unfiltered secondary effluent will vary from 20 to 100 mW-s/cm², depending on the effluent quality and the applicable inactivation requirements. In reclamation applications, where the effluent MPN value, based on total or fecal coliform, must be equal to less than 2.2 org/100mL, the required dosage is typically in the range from

100 to 140 mW-s/cm^2. Additional details on the application of UV disinfection for wastewater may be found in Darby, et al., 1995.

20.3.12 Adsorption

Adsorption, in general, is the process of collecting soluble substances that are in solution on a suitable interface. The interface can be between the liquid and a gas, a solid, or another liquid. In the past, the adsorption process has not been used extensively in wastewater treatment, but demands for a better quality of treated wastewater effluent have led to an intensive examination and use of the process of adsorption on activated carbon. Activated-carbon treatment of wastewater is usually thought of as a polishing process for water that has already received normal biological treatment. The carbon in this case is used to remove a portion of the remaining dissolved organic matter. Depending on the means of contacting the carbon with the water, the particulate matter that is present may also be removed.

20.4 CHEMICAL TREATMENT METHODS

Chemical methods used for the treatment of wastewater are reported in Table 20.15. The principal chemical unit processes used for wastewater treatment are *chemical precipitation* and *chemical disinfection.* A number of treatment plants use a combination of chemical processes (usually precipitation) and physical operations to achieve complete treatment. Such processes are identified as *physical-chemical*

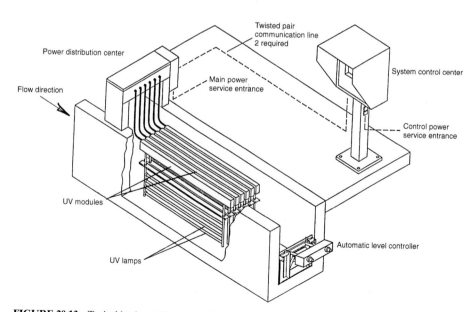

FIGURE 20.13 Typical horizontal-lamp parallel-flow ultraviolet (UV) facility used for the inactivation of pathogenic microorganisms (*Trojan Technologies*).

TABLE 20.15 Applications of Chemical Unit Processes in Wastewater Treatment Arranged Alphabetically

Process	Application
Chemical precipitation	Removal of phosphorus and enhancement of suspended solids removal in primary sedimentation facilities used for physical-chemical treatment.
Dechlorination	Removal of chlorine residual that exists after chlorination.
Disinfection	Selective destruction of disease-causing organisms, usually with chlorine, chlorine products, or ozone.
Others	Various other chemicals can be used to achieve specific objectives in wastewater treatment.

treatment processes. Additional details on the chemical methods of treatment described in the following may be found in Tchobanoglous and Burton (1991); and Water Environment Federation (1991).

20.4.1 Chemical Precipitation

Chemical precipitation in wastewater treatment involves the addition of chemicals to alter the physical state of dissolved and suspended solids and facilitate their removal by sedimentation. In some cases the alteration is slight, and removal is effected by entrapment within a voluminous precipitate consisting primarily of the coagulant itself. Another result of chemical addition is a net increase in the dissolved constituents in the wastewater. Chemical processes, in conjunction with various physical operations, have been developed for the complete secondary treatment of untreated wastewater, including the removal of either nitrogen or phosphorus, or both (Soap and Detergent Association, 1988; Tchobanoglous and Burton, 1991; and Water Environment Federation, 1991). Other chemical processes have also been developed to remove phosphorus by chemical precipitation and are designed to be used in conjunction with biological treatment.

Alum and ferric chloride are two commonly used coagulants in wastewater treatment. Lime is often added as an auxiliary chemical to improve the action of the coagulant. Chemical sedimentation is successful only if the chemicals and wastewater are mixed properly. Chemicals are introduced either in dry or in solution form. Rapid mixing is usually accomplished by mechanical agitation (see Fig. 20.8a). Rapid mixing is followed by about 10 to 20 min of gentle agitation in a flocculating basin before the wastewater is introduced to the sedimentation basin. Chemical sedimentation is particularly advantageous where there is a large seasonal variation in wastewater flow or as an emergency measure to increase the capacity of an overloaded plain sedimentation tank. The main disadvantages of chemical sedimentation are the increased costs and the considerable increase in sludge volume.

20.4.2 Chemical Disinfection

Chemical agents that have been used as disinfectants include: (1) chlorine and its compounds, (2) bromine, (3) iodine, (4) ozone, (5) phenol and phenolic compounds, (6) alcohols, (7) heavy metals and related compounds, (8) dyes, (9) soaps and synthetic detergents, (10) quaternary ammonium compounds, (11) hydrogen peroxide, and (12) various alkalies and acids. Of these, the most common disinfectants are the oxidizing chemicals, and chlorine is the one most universally used. Bromine and iodine have also been used for wastewater disinfection. Ozone is a highly effective disinfectant, and its use is increasing even though it leaves no residual. Highly acid or alkaline water can also be used to destroy pathogenic bacteria, because water with a pH greater than 11 or less than 3 is relatively toxic to most bacteria.

Chlorination may be used as a final step in the treatment of wastewater when an effluent low in bacterial content is necessary. Such use of chlorine is known as *postchlorination.* The disinfecting properties of chlorine reduce the bacterial count, and the oxidizing characteristics can reduce the BOD. Prechlorination before the wastewater enters the sedimentation tank helps to control odors, may prevent flies in trickling filters, and assists in grease removal. In proper doses, chlorine will destroy the bacteria which break down the sulfur compounds in the wastewater and produce hydrogen sulfide. For this reason, chlorine is sometimes put into the main collecting sewers to prevent the destructive action of hydrogen sulfide on concrete pipe. The usual chlorine dose varies considerably but is always much greater than in water purification. Prechlorination doses may be as high as 25 mg/L; postchlorination usually requires at least 3 mg/L (dosages observed in the field vary from 3 to 20 mg/L). Chlorine is sometimes added both at the beginning and end of the treatment process in what is known as *split chlorination.*

As a result of the handling and safety requirements set forth in the Uniform Fire Code enacted in 1991, the use of chlorine for wastewater disinfection is now relatively expensive. Because the cost of UV disinfection equipment has come down significantly in the past two years, UV radiation is now generally equal to or cheaper than chlorine for flows up to 100 mgd and is essentially the same for flows greater than 100 mgd. However, it has been shown that chlorine applied to control microorganisms may combine with some of the residual organic compounds in treated wastewater to form compounds that may be toxic to aquatic forms, so dechlorination facilities may also be required. Dechlorination can be accomplished by adding sulfur dioxide or by passing the chlorinated effluent through a bed of granular activated carbon.

20.5 BIOLOGICAL TREATMENT METHODS

The importance of the biological methods for the treatment of wastewater cannot be overstressed. They are basic components of almost all secondary treatment schemes. The principal types of biological treatment processes are identified in Table 20.16. Simply stated, biological treatment involves: (1) the conversion of the dissolved and colloidal organic matter in wastewater to biological cell tissue [Eq. (20.2)] and to end products [Eq. (20.1)], and (2) the subsequent removal of the cell tissue, usually by gravity settling. Referring to Eq. (20.2), it is clear that if the cell tissue produced is not removed by settling, the cell tissue in the wastewater will still exert a BOD and the treatment will be incomplete. Thus, the design and operation of the sedimentation facilities that follow the biological-treatment process must be considered carefully.

20.5.1 Classification of Biological Treatment Processes

From a practical standpoint, the major concerns in biological wastewater treatment are with the creation of the optimum environmental and physical conditions to bring

TABLE 20.16 Major Biological Treatment Processes Used for Wastewater Treatment

Type	Common name	Use*
Aerobic processes		
Suspended growth	Activated-sludge processes	Carbonaceous BOD removal
	Conventional (plug flow)	(nitrification)
	Complete-mix	
	Step aeration	
	Pure oxygen	
	Sequencing batch reactor	
	Contact stabilization	
	Extended aeration	
	Oxidation ditch	
	Deep tank (90 ft)	
	Deep shaft	
	Suspended-growth nitrification	Nitrification
	Aerated lagoons	Carbonaceous BOD removal,
		(nitrification)
	Aerobic digestion	Stabilization, carbonaceous
	Conventional air	BOD removal
	Pure oxygen	
Attached growth	Trickling filters	Carbonaceous BOD removal,
	Low-rate	nitrification
	High-rate	
	Roughing filters	Carbonaceous BOD removal
	Rotating biological contactors	Carbonaceous BOD removal,
		nitrification
	Packed-bed reactors	Carbonaceous BOD removal,
		nitrification
Combined suspended	Trickling filter/activated sludge;	Carbonaceous BOD removal,
and attached growth	activated sludge/trickling filter	nitrification
processes		
Anoxic processes		
Suspended growth	Suspended-growth denitrification	Denitrification
Attached growth	Fixed-film denitrification	Denitrification
Anaerobic processes		
Suspended growth	Anaerobic digestion	
	Standard-rate, single-stage	Stabilization, carbonaceous
		BOD removal
	High-rate, single-stage	Stabilization, carbonaceous
		BOD removal
	Two-stage	Stabilization, carbonaceous
		BOD removal
	Anaerobic contact processes	Carbonaceous BOD removal
Attached growth	Anaerobic filter	Carbonaceous BOD removal,
		waste stabilization
		(denitrification)
Combined aerobic, anoxic,		
and anaerobic processes		
Suspended growth processes,	Single- or multistage	Carbonaceous BOD removal,
Various proprietary processes		nitrification, dentrification,
Combined suspended and		phosphorus removal
attached growth	Single- or multistage processes	Carbonaceous BOD removal,
		nitrification, denitrification,
		and phosphorus removal

TABLE 20.16 Major Biological Treatment Processes Used for Wastewater Treatment *(Continued)*

Type	Common name	Use*
Pond processes		
Aerobic ponds		Carbonaceous BOD removal
Maturation (tertiary) ponds		Carbonaceous BOD removal, nitrification
Facultative ponds—type 1		Carbonaceous BOD removal
Facultative ponds—type 1		Carbonaceous BOD removal
Anaerobic-ponds		Carbonaceous BOD removal, waste stabilization

* Major uses are presented first; other uses are identified in parentheses.

about the rapid and effective conversion of organic matter to cell tissue and its subsequent removal. Biological conversion can be accomplished both *aerobically* (in the presence of oxygen) and *anaerobically* (in the absence of oxygen). The microorganisms responsible for the conversion can be maintained in suspension or attached to a fixed or moving medium. Such biological treatment processes are known as aerobic *suspended-growth* or *attached-growth* processes. The *activated-sludge* process, discussed in what follows, is the best-known example of an aerobic suspended-growth biologic treatment process. Where aerobic attached-growth processes are used, a suitable fixed or moving medium must be provided for these organisms to grow on. The *trickling filter,* and its variations, is the most common attached-growth process. The *rotating biological disk* process, in which the medium to which the microorganisms are attached is moving, is a variant.

20.5.2 Activated-Sludge Process

In the activated-sludge process as illustrated in Figs. 20.4 and 20.14, untreated or settled wastewater is mixed with 20 to 50 percent of its own volume of return activated sludge. The mixture enters an aeration tank (see Fig. 20.14) where the organisms and wastewater are mixed together with a large quantity of air. Under these conditions, the organisms oxidize a portion of the waste organic matter to carbon dioxide and water and synthesize the other portion into new microbial cells [Eqs. (20.1) and (20.2)]. The mixture then enters a settling tank where the flocculant microorganisms settle and are removed from the effluent stream. The settled microorganisms, or *activated sludge,* are then recycled to the head end of the aeration tank to be mixed again with wastewater. New activated sludge is continuously being produced in this process, and the excess sludge produced each day (waste activated sludge) must be disposed of together with the sludge from the primary-treatment facilities. The effluent from a properly operated activated-sludge plant is of high quality, usually having BOD_5 and suspended-solids concentrations varying from 10 to 20 mg/L for both constituents.

Design parameters commonly used for the sizing of the activated-sludge process include the *food to microorganism ratio* (F/M) and the *mean cell residence time* (MCRT). The F/M ratio, expressed as pounds BOD applied per pound of mixed liquor suspend solids (MLSS) per day, represents the mass of substrate (e.g., BOD, COD) applied to the aeration tank each day versus the mass of suspended solids (microorganisms) in the aeration tank. The MCRT, expressed in days, is a measure of the average amount of time the biological solids remain in the aeration tank. The total concentration of biological solids maintained in the aeration tank normally varies between 1000 and 4000 mg/L. Typically, 60 to 85 percent of the total suspended solids are volatile. Typical design values for various activated-sludge processes are given in Table 20.17. Additional design details on the activated-sludge process may be found in Eckenfelder and Grace (1992); Tchobanoglous and Burton (1991); and Water Environment Federation (1991).

The oxygen necessary for the biological process can be supplied by using air or pure oxygen. Three methods of introducing oxygen to the contents of the aeration tank are used commonly: (1) injection of diffused air, (2) mechanical aeration, and (3) injection of high-purity oxygen. The injection of diffused air involves introducing air under pressure into the aeration tank through diffusion plates or other suitable devices. The air injected into the reactor serves to keep the contents of the reactor well mixed. In mechanical aeration, rotating devices are used to mix the contents of the aeration basin and to introduce oxygen into the liquid by dispersing fine water droplets in the air so the oxygen can be adsorbed (Fig. 20.9). The use of high-purity oxygen requires the use of covered aeration tanks. High-purity oxygen generated at the site is injected into the aeration tank.

One of the most serious problems with the activated-sludge process is the phenomenon known as *bulking,* in which the sludge from the aeration tank will not settle. Where extreme bulking exists, a portion of the suspended solids from the aerator will be discharged in the effluent. Bulking can be caused by: (1) the growth of filamentous organisms (primarily *Sphaerotilus*) that will not settle, or (2) the growth of microorganisms that incorporate large volumes of water into their cell structure, making their density near that of water, thus causing them not to settle. In addition to the discharge of biological solids in the effluent, the large volume of sludge that must be handled is another problem. Foaming or frothing is also encountered with the activated-sludge process. Foaming is caused most often by the excessive growth of the organism *Nocardia.*

Control of filamentous organisms has been accomplished in a number of ways, including the addition of chlorine or hydrogen peroxide to the return waste activated sludge; alteration of the dissolved-oxygen concentration in the aeration tank; alteration of points of waste addition to the aeration tank to alter the F/M ratio; the addition of major nutrients (i.e., nitrogen and phosphorus); the addition of trace metals, nutrients, and growth factors; and more recently, the use of a selector (see Fig. 20.14). A selector is a small tank in which the incoming wastewater is mixed with the return sludge under anoxic or anaerobic conditions. The high substrate concentration in the selector favors the growth of nonfilamentous microorganisms. If the wastewater were added directly to a complete-mix reactor, the low substrate concentration in the reactor would favor the growth of filamentous organisms. Because the use of a biological selector has proved to be so successful for the control of filamentous microorganisms, this method is now favored by most designers. Another important advantage of this control method is the elimination of the need for return sludge chlorination facilities. The control of *Nocardia* can be accomplished by lowering the solids concentration in the reactor. Spraying *Nocardia* foam with a dilute chlorine spray has also proved to be an effective control measure (Jenkins, 1995).

20.5.3 Trickling-Filter Process

The effluent from primary sedimentation generally contains about 60 to 80 percent of the unstable organic matter originally present in the wastewater. The trickling-

TABLE 20.17 Typical Values for the Design Parameters for the Activated-Sludge Processes

Process modification	θ_c, d	F/M, lb BOD$_5$/lb MLVSS · d	Volumetric loading, lb BOD$_5$/10^3 ft^3 · d	MLSS, mg/L	V/Q, h	Qr/Q
Conventional plug flow	3–15	0.2–0.5	20–40	1000–3000	4–8	0.25–0.75
Complete-mix	0.9–15	0.2–1.0	50–120	800–6500	3–5	0.25–1.0
Step-feed	3–15	0.2–0.5	40–60	1500–3500	3–5	0.25–0.75
Modified aeration	0.2–0.5	1.5–5.0	75–150	200–1000	1.5–3	0.05–0.25
Contact stabilization	5–15	0.2–0.6	60–75	(1000–3000)* (4000–9000)†	(0.5–1.0)* (3–6)†	0.5–1.50
Extended aeration	20–30	0.05–0.15	10–25	1500–5000	18–36	0.5–1.50
High-rate aeration	5–10		100–1000	3000–6000	2–4	1.0–5.0
Kraus process	5–15	0.3–0.8	40–100	2000–3000	4–8	0.5–1.0
High-purity oxygen	3–10	0.25–1.0	100–200	3000–8000	1–3	0.25–0.5
Oxidation ditch	10–30	0.05–0.30	5–30	1500–5000	8–36	0.75–1.50
Sequencing batch reactor	N/A	0.05–0.30	5–15	1500–5000§	12–50	N/A
Deep shaft reactor	NI	0.5–5.0	NI	NI	0.5–5	NI
Single-stage nitrification	8–20	0.10–0.25 (0.02–0.15)‡	5–20	1500–3500	6–15	0.50–1.50
Separate-stage nitrification	15–100	0.05–0.20 (0.04–0.15)‡	3–9	1500–3500	3–6	0.50–2.00

* Contact unit.
† Solids-stabilization unit.
‡ TKN/MLVSS.
§ MLSS varies depending on the portion of the operating cycle.
Note: lb/10^3 ft^3 · d × 0.0160 = kg/m^3 · d; lb/b d = kg/kg · d; N/A = not applicable; NI = no information.

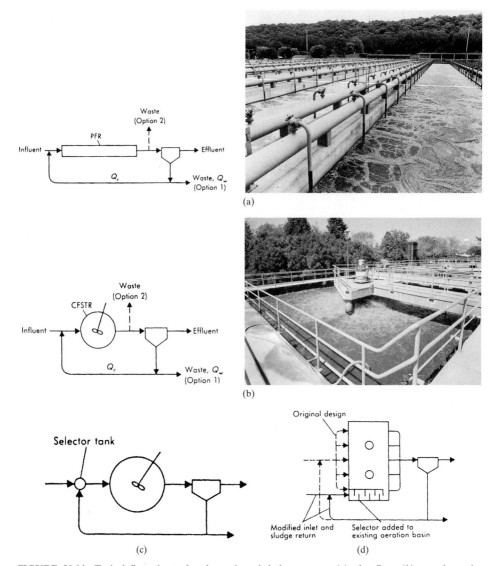

FIGURE 20.14 Typical flow sheets for the activated-sludge process: (*a*) plug-flow; (*b*) complete-mix; (*c*) complete-mix with added external selector tank; (*d*) complete-mix with selector constructed within existing aeration tank. The selector shown in (*c*) and (*d*) is used to control the growth of filamentous microorganisms. (*Tchobanoglous and Schroeder, 1985.*)

filter process is one method of oxidizing this putrescible matter remaining after primary treatment. A conventional trickling filter (see Fig. 20.15*a*) consists of a bed of crushed rock, slag, or gravel whose particles range from about 2 to 4 in (50 to 100 mm) in size. The bed is commonly 6 to 9 ft (2 to 3 m) deep, although shallower beds are sometimes used. Conventional filters are usually classified as low, intermediate, and high rate. The tower trickling filter (see Fig. 20.15b) is a modification of the conventional trickling-filter process in which specially designed high-porosity plastic modules are used as the fixed medium to which the microorganisms are attached. The specific surface area of these plastic media is about 30 ft²/ft³ (100 m²/m³) of volume. Such filters are built from 15 to 40 ft (4.5 to 12 m) high. They are often used in conjunction with conventional activated-sludge facilities to reduce high seasonal loadings resulting from canneries and similar activities. When used to reduce seasonal loadings, they are known as *roughing filters*. Typical design values for various types of trickling filters are given in Table 20.18.

Operationally, wastewater is applied to the surface of the filter intermittently by one or more rotary distributors and percolates downward through the bed to underdrains, where it is collected and discharged through an outlet channel. A gelatinous biological film forms on the filter medium, and the fine suspended, colloidal, and dissolved organic solids collect on this film where biochemical oxidation of the organic matter is accomplished by aerobic bacteria. The film eventually becomes quite thick with accumulated organic matter and will slough off (or unload) from time to time and be discharged with the effluent. Therefore, effluent from trickling filters requires sedimentation to remove the solids that pass the filter. Continuous sloughing can be achieved with proper control of the hydraulic application rate. Recirculation of trickling-filter effluent or effluent from secondary-settling facilities is a common feature of trickling filters. The rates of recirculation are generally adjusted as the wastewater flow changes to maintain approximately constant flow through the filters.

(a)

(b)

FIGURE 20.15 Typical examples of trickling filters: (*a*) conventional rock-filled low-rate filter, and (*b*) tower-type high-rate filter filled with a plastic medium.

TABLE 20.18 Typical Design Information for Various Types of Trickling Filters

Item	Intermediate low-rate	Super-rate	High-rate	High-rate	Roughing	Two-stage
Filter medium	Rock, slag	Rock, slag	Rock	Plastic	Plastic, redwood	Rock, plastic
Hydraulic loading, gal/ft² · min	0.02–0.06	0.06–0.16	0.16–0.64	0.2–1.20	0.8–3.2	0.16–0.64
Mgal/acre · d	1–4	4–10	10–40	15–90	50–200*	10–40*
BOD_5 loading, lb/10³ ft³ · d	5–25	15–30	30–60	30–100	100–500	60–120
Depth, ft	6–8	6–8	3–6	10–40	15–40	6–8
Recirculation ratio	0	0–1	1–2	1–2	1–4	0.5–2
Filter flies	Many	Some	Few	Few or none	Few or none	Few or none
Sloughing	Intermittent	Intermittent	Continuous	Continuous	Continuous	Continuous
BOD_5 removal efficiency, %	80–90	50–70	65–85	65–80	40–65	85–95
Effluent	Well nitrified	Partially nitrified	Little nitrification	Little nitrification	No nitrification	Well nitrified

* Does not include recirculation.
Note: ft × 0.3048 = m; gal/min × 58.6740 = m³/m² · d; Mgal/acre · d × 0.9354 = m³/m² · d; lb/10³ · d × 0.0160 = kg/m³ · d.

20.5.4 Rotating Biological Contactors

In the rotating biological contactor (RBC) process, a number of circular plastic disks are mounted on a central shaft (Fig. 20.16). These disks are partially submerged (from 40 to 80 percent) and rotated in a tank containing the wastewater to be treated. The microorganisms responsible for treatment become attached to the disks and rotate into and out of the wastewater. The oxygen necessary for the conversion of the organic matter adsorbed from the liquid is obtained by adsorption from the air as the slime layer on the disk is rotated out of the liquid. In some designs, air is added to the bottom of the tank to provide oxygen and to rotate the disks when the disks are equipped with air capture cups. Conceptually and operationally the biodisk process is similar to the trickling-filter process.

20.5.5 Stabilization Ponds and Aerated Lagoons

If an adequate area of flatland is available, secondary treatment of settled wastewater can be accomplished by retaining the partially treated wastewater in oxidation or stabilization ponds where organic matter is stabilized through the combined action of algae and other microorganisms. Stabilization ponds are usually classified according to the nature of the biological activity that is taking place as aerobic, anaerobic, or aerobic/anaerobic. Stabilization ponds have been used singly or in various combinations to treat both domestic and industrial wastes. *Aerobic ponds* are used primarily for the treatment of soluble organic wastes and effluents from wastewater-treatment plants. *Aerobic maturation ponds* are used to further treat (polish) effluents from conventional biological-treatment processes. *Aerobic/anaerobic ponds* (also known as a *facultative ponds*), are the most common type and have been used to treat domestic wastewater and a wide variety of industrial wastes. *Anaerobic ponds* are especially effective in bringing about rapid stabilization of strong organic wastes. Usually, anaerobic ponds are used in series with aerobic-anaerobic ponds to provide complete treatment.

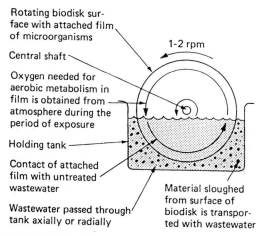

FIGURE 20.16 Definition sketch for the rotating biological contactor (RBC).

In an aerobic/anaerobic stabilization pond a symbiotic relationship (see Fig. 20.17a) exists between algae and microorganisms, such as bacteria and protozoa that oxidize organic matter. Algae, which are microscopic plants, produce oxygen while growing in the presence of sunlight. Oxygen is used by other microorganisms for oxidation of waste organic matter. End products of the process are carbon dioxide, ammonia, and phosphates that are required by the algae to grow and produce oxygen. The result of stabilization-pond treatment is the oxidation of the original organic matter and the production of algae that are discharged with the effluent to the stream. This results in a net reduction in BOD_5 as the algae are more stable than the original matter in wastewater and degrade slowly in the stream. Stabilization ponds should not be constructed upstream of lakes or reservoirs as the algae discharged from the ponds may settle in the reservoirs and cause anaerobic conditions and other water-quality problems. Properly operated ponds (Fig. 20.17b) may be as effective as trickling filters in reducing the BOD_5 of wastewater. Because of low loading rates, relatively large areas of land must be available. The banks of the ponds should be kept clean to avoid mosquito breeding, and the ponds should be located sufficiently far from residential areas to avoid complaints about odors. Typical design values for various pond processes are given in Table 20.19.

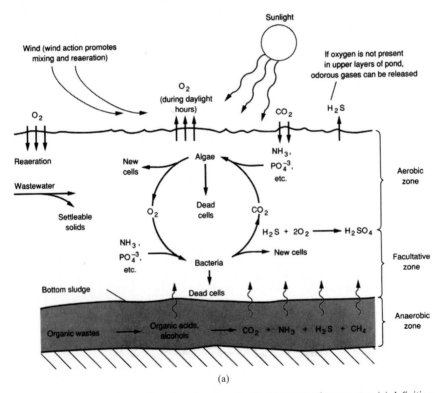

(a)

FIGURE 20.17 Typical examples of pond systems for the treatment of wastewater: (a) definition sketch for the operation of an aerobic/anaerobic (facultative) pond.

(b)

(c)

FIGURE 20.17 *(Continued)* Typical examples of pond systems for the treatment of wastewater: (b) view of aerobic/anaerobic pond; (c) view of aerated pond.

TABLE 20.19 Typical Design Parameters for Various Types of Stabilization Ponds

Parameter	Type of pond					
	Aerobic low-rate*	Aerobic high-rate	Aerobic maturation	Aerobic/anaerobic facultative†	Anaerobic pond	Aerated lagoon
Flow regime	Intermittently mixed	Intermittently mixed	Intermittently mixed	Mixed surface layer	Mixed surface layer	Completely mixed
Pond size, acres	<10 multiples	0.5–2	2–10 multiples	2–10 multiples	0.5–2 multiples	2–10 multiples
Operation‡	Series or parallel	Series	Series or parallel	Series or parallel	Series	Series or parallel
Detention time,‡ d	10–40	4–6	5–20	5–30	20–50	3–10
Depth, ft	3–4	1–1.5	3–5	4–8	8–16	6–20
pH	6.5–10.5	6.5–10.5	6.5–10.5	6.5–8.5	6.5–7.2	6.5–8.0
Temperature range, °C	0–30	5–30	0–30	0–50	6–50	0–30
Optimum temperature, °C	20	20	20	20	30	20
BOD$_5$ loading,§ lb/acre · d	60–120	80–160	≤15		200–500	
BOD$_5$ conversion, %	80–95	80–95	60–80	80–95	50–85	80–95
Principal conversion products	Algae, CO$_2$, bacterial cell tissue	Algae, CO$_2$, bacterial cell tissue	Algae, CO$_2$, NO$_3$, bacterial cell tissue	Algae, CO$_2$, CH$_4$, bacterial cell tissue	CO$_2$, CH$_4$, bacterial cell tissue	CO$_2$, bacterial cell tissue
Algal concentration, mg/L	40–100	100–260	5–10	5–20	0–5	0–5
Effluent SS,¶ mg/L	80–140	150–300	10–30	40–60	80–160	80–250

* Conventional aerobic ponds designed to maximize the amount of oxygen produced rather than the amount of algae produced.
† Pond includes supplemental aeration. For ponds without supplemental aeration, typical BOD$_5$ loadings are about one-third of those listed.
‡ Depends on climatic conditions.
§ Typical values. Much higher values have been applied at various locations. Loading values are often specified by state regulatory agencies. Values are based on an influent soluble BOD$_5$ of 200 mg/L and, with the exception of the aerobic ponds, an influent suspended solids of 200 mg/L.
¶ Includes algae, microorganisms, and residual suspended solids.
Note: acre × 0.4047 = ha; ft × 0.3048 = m; lb/acre · d × 1.1209 = kg/ha · d.

An *aerated lagoon* is essentially a pond system for the treatment of wastewaters in which oxygen is introduced by mechanical aerators rather than relying on photosynthetic oxygen production (Fig. 20.17c). The ponds are deeper than stabilization ponds and less detention time is required. Aeration lagoons require only 5 to 10 percent as much land as stabilization ponds. Treatment efficiencies of 60 to 90 percent can be obtained with detention times of 4 to 10 days. Aerated lagoons are used frequently for the treatment of industrial wastes.

One problem with both stabilization ponds and aerated lagoons and other similar treatment methods is that their effluent will not meet the definition of secondary treatment given in Table 20.10, especially with respect to suspended solids. To meet these requirements, a variety of solids-separation methods have been used or are under study. Among them are the use of conventional and large earthen settling basins in series, chemical precipitation, fine screening, intermittent sand filters, and rock filters. Because the performance of these separation methods has been variable, their application should be assessed on a case-by-case basis.

20.6 NUTRIENT REMOVAL

Because both nitrogen and phosphorus can impact receiving-water quality, the discharge of one or both of these constituents must often be controlled. Nitrogen may be present in wastewaters in various forms (e.g., organic, ammonia, nitrites, or nitrates). Most of the available nitrogen in municipal wastewater is in the form of organic or ammonia nitrogen at an average total concentration of around 25 to 35 mg/L. About 20 percent of the total nitrogen settles out during primary sedimentation. During biological treatment, a major portion of the organic nitrogen is converted to ammonia nitrogen, a portion of which is incorporated into biological cells that are removed from the treated wastewater stream before discharge, removing another 20 percent of the incoming nitrogen. The remaining 60 percent is normally discharged to the receiving waters.

Phosphorus is present in municipal wastewaters in organic form, as inorganic orthophosphate, or as complex phosphates at a total average concentration of about 6 to 10 mg/L. The complex phosphates represent about one-half of the phosphates in municipal wastewater and result from the use of these materials in synthetic detergents. Complex phosphates are hydrolyzed during biological treatment to the orthophosphate form (PO_4^{-3}). Of the total average phosphorus concentration of about 8 mg/L present in municipal wastewater, about 10 percent is removed as particulate material during primary sedimentation and another 10 to 20 percent is incorporated into bacterial cells during biological treatment. The remaining 70 percent of the incoming phosphorus normally is discharged with secondary-treatment-plant effluents.

20.6.1 Biological Nitrogen Removal

The two principal mechanisms for the biological removal of nitrogen are by *assimilation* and by *nitrification/denitrification*. Because nitrogen is a nutrient, microbes present in the treatment processes will assimilate ammonia nitrogen and incorporate it into cell mass. Nitrogen can be removed from the wastewater by removing cells from the system. However, in most wastewater, more nitrogen is available than can be assimilated into cell tissue, given the amount of organic carbon available for the production of cell tissue. In nitrification/denitrification, the removal of nitrogen is accomplished in two conversion steps. In the first step, *nitrification,* ammonia is oxidized biologically to nitrate. In the second step, *denitrification,* nitrate is converted under anoxic (without oxygen) conditions to nitrogen gas, which is vented from the system. In the past, the conversion process was often identified as *anaerobic denitrification.* However, the principal biochemical pathways are not anaerobic but rather a modification of aerobic pathways; therefore, the use of the term *anoxic* in place of *anaerobic* is considered appropriate. The denitrifying bacteria obtain energy for growth from the conversion of nitrate to nitrogen gas but require a source of carbon for cell synthesis. Because nitrified effluents are usually low in carbonaceous matter, an external source of carbon is required. In most biological denitrification systems, the incoming wastewater or cell tissue is used to provide the needed carbon. In the treatment of agricultural wastewaters, deficient in organic carbon, methanol and other organic compounds have been used as a carbon source. Industrial wastes that are poor in nutrients but contain organic carbon have also been used.

The principal *nitrification processes* may be classified as *suspended-growth* and *attached-growth processes* (Randall, et al., 1992; Soap and Detergent Association, 1988; Tchobanoglous and Burton, 1991; Water Environment Federation, 1991). In the suspended-growth process, nitrification can be achieved either in the same reactor used in the treatment of the carbonaceous organic matter or in a separate suspended-growth reactor following a conventional activated-sludge treatment process. Nitrification can also be achieved in the same attached-growth reactor used for carbonaceous organic matter removal or a separate reactor. Trickling filters, rotating biological contactors, and packed towers can be used for nitrifying systems. As with nitrification, the principal denitrification processes may also be classified as suspended-growth and attached-growth. Suspended-growth denitrification is usually carried out in a plug-flow type of activated-sludge system (i.e., following any process that converts ammonia and organic nitrogen to nitrates (nitrification).

20.6.2 Biological Phosphorus Removal

Microbes utilize phosphorus for cell synthesis and energy transport. As a result, 10 to 30 percent of the influent phosphorus is removed during secondary biological treatment. Under certain operating conditions, more phosphorus than is needed may be taken up by the microorganisms. Biological phosphorus removal is accomplished by sequencing and producing the appropriate environmental conditions in the reactor(s). Under anaerobic conditions, certain organisms respond to *volatile fatty acids* (VFAs) that are present in the influent wastewater by releasing stored phosphorus. When an anaerobic zone is followed by an aerobic (oxic) zone, the microorganisms exhibit phosphorus uptake above normal levels. Phosphorus not only is utilized for cell maintenance, synthesis, and energy transport but also is stored for subsequent use by the microorganisms. The sludge containing the excess phosphorus is either wasted (Fig. 20.18*a*) or removed and treated in a side stream to release the excess phosphorus (Fig. 20.18*b*). Release of phosphorus occurs under anoxic conditions. Thus, biological phosphorus removal requires both anaerobic and aerobic reactors or zones within a reactor. Currently, a number of proprietary processes take advantage of one of these mechanisms (Randall, et al., 1992; and Soap and Detergent Association, 1988).

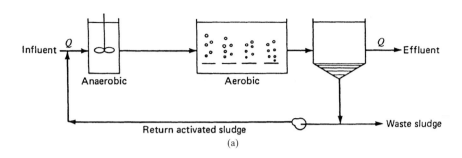

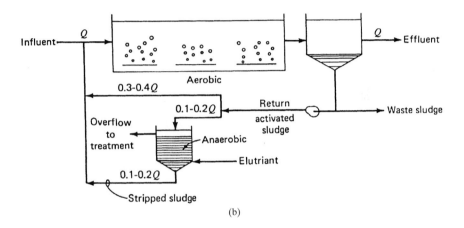

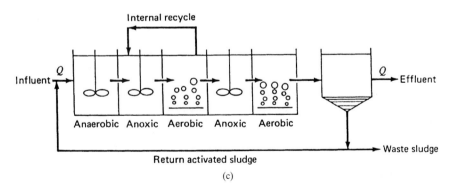

FIGURE 20.18 Typical flow diagrams used for the removal of nutrients from wastewater: (*a*) A/O process and (*b*) Phostrip process for the removal of phosphorus, and (*c*) Bardenpho process for the combined removal of nitrogen and phosphorus. (*Adapted from Randall, et al., 1992; and Soap and Detergent Association, 1988.*)

20.6.3 Combined Biological Nitrogen and Phosphorus Removal

Of fundamental importance in the biological removal of nutrients, as discussed previously, is the development of a cyclic sequence in the treatment process (e.g., anaerobic, anoxic, aerobic). For example, where both nitrogen and phosphorus are to be removed, combination processes such as the *Bardenpho process* (see Fig. 20.18c) are used. In the Bardenpho process, a sequence of anaerobic, anoxic, and aerobic steps are used to achieve both nitrogen and phosphorus removal. Nitrogen is removed by nitrification/denitrification. Phosphorus is removed by wasting sludge from the system.

20.6.4 Chemical Removal of Phosphorus

Compounds of phosphorus can be removed by addition of coagulants, such as alum, lime, ferric chloride, or ferrous sulfate. The chemicals may be added prior to primary sedimentation, alum and soil salts may be added to the aeration tank during the activated-sludge process, or chemicals may be added in a third stage following biological treatment. When added at the primary-treatment stage, a large portion of the organic material is removed as well as phosphorus so that a reduced loading on the biological treatment process results. However, a larger quantity of sludge is produced. When chemicals are added directly to the aeration tank of an activated-sludge plant, chemical treatment and biological treatment take place together, and little additional equipment is required. Chemical precipitation, especially using lime, is sometimes practiced in a third stage after biological treatment, both to remove the phosphorus and to increase the pH of the effluent in preparation for a process of ammonia-nitrogen removal.

20.7 TERTIARY/ADVANCED WASTEWATER TREATMENT

The terms *tertiary* or *advanced* waste treatment refer to the use of additional operations and processes beyond those used conventionally to prepare wastewater for direct reuse for industrial, agricultural, and municipal purposes. During a cycle of use for municipal purposes, the concentration of organic and inorganic materials in water is increased. Most of the readily biologically degradable organic material is removed during conventional treatment, but from 40 to 100 mg/L of dissolved and biologically resistant or *refractory* organic material remains in the effluent. These materials may be end products of normal biological decomposition or artificial products, such as synthetic detergents, pesticides, and/or organic industrial wastes. Although the removal of nitrogen and phosphorus represents treatment beyond the conventional, the removal of these constituents has become so common that it is no longer considered advanced treatment (see Table 20.11). As noted in Table 20.11, tertiary treatment typically involves the removal of residual suspended solids whereas advanced treatment generally involves the removal of dissolved constituents.

20.7.1 Removal of Residual Suspended Solids and Turbidity

Urban uses of reclaimed wastewater include landscape irrigation, groundwater recharge, recreational impoundment, and industrial cooling. For these uses, the health risk associated with the pathogens that may be present in the treated wastewater is an important issue. To achieve efficient bacterial and viral pathogen inactivation and removal, two key operating criteria must be met: (1) the treated effluent must be low in suspended solids and turbidity prior to disinfection to reduce shielding of pathogens and chlorine usage, and (2) a sufficient disinfectant dose and contact time must be provided for reclaimed wastewater. *Granular-medium filtration* is used most frequently to remove residual suspended solids found in secondary effluents that may interfere with subsequent disinfection, to reduce the concentration of organic matter that can react with a disinfectant, and to improve the esthetic quality of the reclaimed wastewater by reducing its turbidity. Typical flow diagrams employing granular-medium filtration to produce a high-quality effluent for reuse from treated wastewater are shown in Fig. 20.19.

20.7.2 Removal of Inorganic Salts

Inorganic salts can be removed from treated wastewaters by the processes used for the desalting of water. *Ion exchange, electrodialysis,* and *reverse osmosis* (RO), are also feasible. Fouling of the RO membranes by the refractory organic material in the treated effluent may be a problem. The most effective method for removing this refractory organic material is passage through columns containing granular acti-

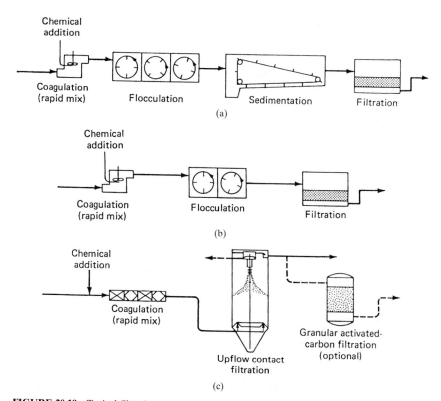

FIGURE 20.19 Typical filtration systems for wastewater-reclamation applications: (*a*) complete treatment; (*b*) direct filtration; (*c*) in-line contact filtration.

vated carbon. Processes have been developed for using activated carbon efficiently and for regenerating the carbon when its adsorptive capacity decreases.

20.8 TREATMENT AND DISPOSAL OF SLUDGE AND BIOSOLIDS

The processing and disposal of the solids and sludge resulting from the treatment of wastewater is one of the most difficult aspects of wastewater management. It is difficult because: (1) the sludge contains much of the material that was offensive in the incoming wastewater and, when biological treatment is used, waste biological sludge is also organic and will decay, and (2) only a small part of the sludge is solid matter. The principal methods now used for the processing of sludge are summarized in Table 20.20. The unit operations and chemical and biological processes used to accomplish each of these methods are also indicated. Alternative methods for the processing of sludge from wastewater-treatment plants are illustrated in Fig. 20.20.

20.8.1 Types and Sources of Solids and Sludge

The principal types and sources of solids and sludge from conventional wastewater-treatment plants include: (1) coarse solids, such as rags and twigs from screening facilities, (2) grit and scum from grit-removal facilities, (3) scum from preaeration or grease-removal facilities, (4) sludge and scum from primary settling tanks, (5) biological sludge produced during the treatment process and scum from the secondary settling facilities, and (6) sludge and ashes that require final disposal from the sludge-processing facilities.

20.8.2 Quantities of Sludge

The overall quantity of dry sludge that must be disposed of from a wastewater-treatment plant will vary from about 55 to 75 percent of the total incoming suspended solids on a dry basis. Within the treatment process, the quantities of sludge will vary with the removal conversion efficiencies and the thickening capabilities of the operations and processes. For example, about 60 to 70 percent of the incoming suspended solids will be removed in a primary settling tank and the solids content of the sludge will vary from 4 to 8 percent. The net yield in most biological-treatment processes will vary from about 0.3 to 0.6 lb cells/lb BOD_5 (0.3 to 0.6 kg/kg) added. The solids content of settled activated and trickling-filter sludge from secondary settling facilities will vary from 0.4 to 1 and 0.6 to 1.5 percent, respectively. Knowing the specific gravity and the solids content, the volume of sludge can be estimated as follows:

$$\text{Vol} = \frac{W_s}{sg(s/100)\gamma} \qquad (20.10)$$

where   Vol = volume, ft^3 (m^3)
   W_s = weight of sludge or dry sludge solids, lb (kg)
   sg = specific gravity of sludge or dry solids
   s = solids content of the sludge, percent
   γ = specific weight of water, 62.4 lb/ft^3 (1000 kg/m^3)

It will be noted that if the term $(s/100)$ is omitted, the resulting expression can be used to determine the volume of any substance.

TABLE 20.20   Sludge-Processing Methods

Method	Description	Representative unit operation, unit process, or treatment method
Preliminary operations	Used to improve the treatability of the sludge (such as blending and grinding to make it more uniform), to reduce wear of pumps (degritting), or to reduce the required capacity of other processing facilities (storage)	Sludge grinding Sludge degritting Sludge blending Sludge storage
Thickening	Used to reduce the volume of sludge	Gravity thickening Flotation thickening Centrifugation
Stabilization	Used to reduce the organic content of the sludge (digestion) or to alter its characteristics so that it will not cause nuisance conditions (oxidation and lime treatment)	Chlorine oxidation Lime stabilization Heat treatment Anaerobic digestion Aerobic digestion
Conditioning	Used to make the sludge more manageable in subsequent processing steps (heat treatment) or to reduce chemical requirements (elutriation)	Chemical conditioning Elutriation Heat treatment
Dewatering	Used to reduce the volume of sludge	Vacuum filtration Filter pressing Horizontal belt filtration Centrifugation Drying (bed) Lagooning
Disinfection	Used to control the bacterial activity of microorganisms in the sludge	Lime stabilization Heat treatment Composting Drying
Drying	Used to further reduce the volume of sludge by evaporating water	Flash drying Spray drying Rotary drying Multiple hearth drying
Composting	Used to produce a humus-like material that can be applied to land as a soil amendment	Composting (sludge only) Cocomposting with solid wastes or other organic materials
Thermal reduction	Used to reduce the volume of prethickened sludge by means of thermal reduction	Multiple-hearth incineration Fluidized-bed incineration Flash combustion Coincineration with solid wastes Copyrolysis with solid wastes Wet air oxidation

20.8.3   Anaerobic Sludge Digestion

Of the methods listed in Table 20.20, anaerobic digestion is the most common method used for the stabilization of sludge. Sludge digestion is a biological process in which anaerobic and facultative bacteria convert approximately 40 to 60 percent of the organic solids in sludge to carbon dioxide and methane gases. The organic

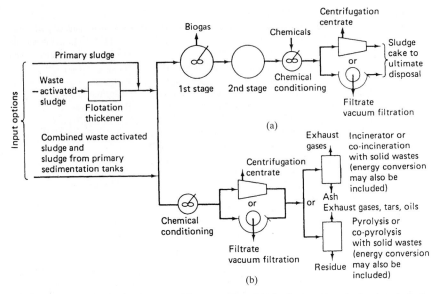

FIGURE 20.20 Typical flow diagram of the processing of sludge from wastewater-treatment plants: (*a*) option with anaerobic digestion (energy recovery may also be included), and (*b*) option with dewatering and thermal conversion.

matter that remains is chemically stable and practically odorless and contains 90 to 95 percent moisture. In older digesters where mixing was not used, detention times of 30 to 60 days were required for proper stabilization of the sludge; with modern high-rate digesters with proper mixing, detention times of 15 to 25 days are generally sufficient. Digestion is accomplished in sludge-digestion tanks, or digesters (see Fig. 20.21) which are ordinarily cylindrical or egg-shaped and are equipped with mixing and sampling devices, temperature recorders, and meters for measuring gas production.

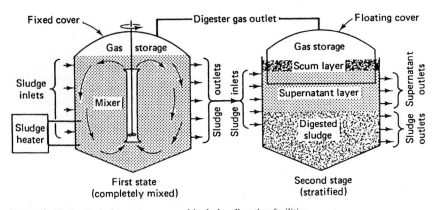

FIGURE 20.21 Typical two-stage anaerobic sludge digestion facilities.

Digesters are usually covered to retain heat and odors, to maintain anaerobic conditions, and to permit collection of gas. Digestion proceeds most rapidly at temperatures of 90 to 95°F (32 to 35°C), and hence digesters require some means of heating. The most common method is to preheat the sludge and continuously circulate digester liquid from the digester to an external heat exchanger. With this system, some mixing of the sludge in the digester is accomplished by currents set up in the tank by the circulating liquid. For more positive mixing in high-rate digesters, gas may be recirculated under pressure and discharged near the bottom of the tank. Mechanical stirrers may also be provided within the tank. Mixing brings the bacteria and sludge solids together for more rapid digestion and also maintains digester capacity by breaking up surface scum layers which form.

The digestion process is thought to occur in three steps. The first step in the process involves the *enzyme-mediated transformation* (*hydrolysis*) of higher-molecular-mass compounds into compounds suitable as a source of energy and cell carbon. In the second step, a group of organisms collectively identified as *acidogens*, or *acid formers*, converts the compounds resulting from the first step into identifiable lower-molecular-mass intermediate compounds such as formate and acetate. In the third step, a group of organisms collectively identified as *methanogens*, or methane formers, convert the intermediate compounds produced in the second step into simpler end products, principally methane and carbon dioxide. Currently, it is known that methanogens use the following substrates: $CO_2 + H_2$, formate, acetate, methanol, methylamines, and carbon monoxide in the production of methane. In an anaerobic digester, the two principal pathways involved in the formation of methane are: (1) the conversion of carbon dioxide and hydrogen to methane and water, and (2) the conversion of acetate to methane, carbon dioxide, and water. The methanogens and the acidogens form a syntrophic (mutually beneficial) relationship in which the methanogens convert fermentation end products such as hydrogen, formate, and acetate to methane and carbon dioxide. In effect, the methanogenic bacteria remove compounds that would inhibit the growth of acidogens. It should be noted that the material remaining after anaerobic digestion is now termed *biosolids*.

The gas evolved in a digester contains about 50 to 60 percent methane, 40 to 50 percent carbon dioxide, and traces of nitrogen, hydrogen sulfide, and other gases. About 15 ft³ of gas will be produced per pound of volatile solids (0.9 m³/kg) destroyed in the digester. Sludge gas has a fuel value of about 500 to 600 Btu/ft³ (18,300 to 22,000 kJ/m³) and may be used for heating digestion tanks and treatment plant buildings or in gas engines to drive pumps, generators, or air compressors. Sludge gas has also been mixed with natural gas for domestic consumption. Special gas holders are sometimes provided to store gas, or it can be stored in a digestion tank equipped with a floating, gastight cover. A floating cover rises and falls as the volume of sludge, liquid, and gas in the tank changes. Thus, constant gas pressure is maintained and storage is provided to even up production. Additional design details on the anaerobic digestion process may be found in Malina and Pohland (1992); Tchobanoglous and Burton (1991); and Water Environment Federation (1991).

20.8.4 Sludge Dewatering

Dewatering is a physical (mechanical) unit operation used to reduce the moisture content of sludge for one or more of the following reasons: (1) costs for trucking sludge to the ultimate disposal site become substantially lower when sludge volume is reduced by dewatering, (2) dewatered sludge is generally easier to handle than thickened or liquid sludge, (3) dewatering is required normally prior to the combustion of the sludge to increase the calorific value by removal of excess moisture, (4) dewatering is required before composting to reduce the requirements for supplemental bulking agents or amendments, (5) removal of the excess moisture may be required to render the sludge totally odorless and nonputrescible, and (6) sludge dewatering is required prior to landfilling in monofills to reduce leachate production at the landfill site.

Dewatering devices use a number of techniques for removing moisture. Some rely on natural evaporation and percolation to dewater the solids. Mechanical dewatering devices use mechanically assisted physical means to dewater the sludge more quickly. The physical means include filtration, squeezing, capillary action, vacuum withdrawal, and centrifugal separation and compaction. Selection of the dewatering device is determined by the type of sludge to be dewatered, characteristics of the dewatered product, and the space available. For smaller plants where land availability is not a problem, drying beds or lagoons are generally used. Conversely, for facilities situated on constricted sites, mechanical-dewatering devices are often chosen. Some sludges, particularly aerobically digested sludges, are not amenable to mechanical dewatering. These sludges can be dewatered on sand beds with good results. When a particular sludge must be dewatered mechanically, it is often difficult or impossible to select the optimum dewatering device without conducting bench-scale or pilot studies. Trailer-mounted full-size equipment is available from several manufacturers for field-testing purposes.

20.8.5 Conveyance of Sludges and Residues

One of the final steps in the management of sludge and other residues from treatment plants involves the conveyance of this material to a location for further processing or to its final resting point. Often overlooked in the design of wastewater-treatment plants, the means of conveyance can be a critical factor, especially in situations where the treatment plant is located in an urban area and the disposal sites are located some distance away. Methods that have been used to convey sludge include transport by pipeline, truck, barge, and rail. Of these, transport by truck is used most commonly.

20.8.6 Final Disposal of Residues and Biosolids

The number of final-disposal options for residues and biosolids from treatment plants is limited. The principal methods now used involve: (1) some form of land application, such as landfilling or spreading, (2) composting (see Fig. 20.22) followed by land application, and (3) incineration with ash disposal in municipal landfills, dedicated monofills, or hazardous-waste-disposal sites. Where properly evaluated and controlled, land application is the preferred method for the disposal of biosolids (U.S. EPA, 1983; 1984a). In land-application systems, biosolids can be applied to: (1) agricultural land, (2) forest land, (3) disturbed land, and (4) dedicated land disposal sites. Reuse (e.g., the use of composted biosolids), while a desirable objective, has yet to be realized on a large scale. Because of the many problems associated with the treatment and disposal of biosolids and other plant residues, more than 50 to 60 percent of the total cost of providing wastewater-treatment facilities is spent on the

FIGURE 20.22 Composting of digested sludge using the windrow method.

solids-management portion of the facilities. As a consequence, special attention must be devoted to the disposal of sludge when planning new or expanded waste-water-management facilities.

20.9 NATURAL TREATMENT SYSTEMS

Natural treatment systems are designed to take advantage of the physical, chemical, and biological processes that occur in nature to provide wastewater treatment. The processes involved in natural systems include all of those used in mechanical or in-plant treatment systems: sedimentation, filtration, gas transfer, adsorption, ion exchange, chemical precipitation, chemical oxidation and reduction, and biological conversion and degradation, plus others unique to natural systems, such as photo-synthesis, photooxidation, and plant uptake. In natural systems the processes occur at "natural" rates and tend to occur simultaneously in a single "ecosystem reactor," as opposed to mechanical systems in which processes occur sequentially in separate reactors or tanks at accelerated rates as a result of energy input.

20.9.1 Soil-Based Systems

Natural treatment systems include: (1) *soil-based systems*—slow rate, rapid infiltra-tion, and overland flow (U.S. EPA, 1984b) and (2) *aquatic-based systems*—con-structed and natural wetlands and aquatic-plant treatment systems. All forms of natural treatment systems are preceded by some form of mechanical pretreatment. For wastewater, a minimum of fine screening or primary sedimentation is necessary to remove gross solids that can clog distribution systems and lead to nuisance condi-tions. The need to provide preapplication treatment beyond some minimum level will depend on the system objectives and regulatory requirements. The most commonly used natural-treatment systems are described in what follows.

20.9.1.1 Slow Rate. Slow-rate treatment, the predominant natural treatment process in use today, involves the application of wastewater to vegetated land to provide treatment and to meet the growth needs of the vegetation. The applied water either is consumed through evapotranspiration or percolates vertically and horizontally through the soil profile (see Fig. 20.23a). Any surface runoff is usually collected and reapplied to the system. Treatment occurs as the applied water percolates through the soil profile. Slow-rate systems are classified as Type 1 when the principal objective is wastewater treatment and the hydraulic-loading rate is not controlled by the water requirements of the vegetation, but by the limiting design parameter of soil permeability or constituent loading. The design objective for Type 2 systems is water reuse through crop production or landscape irrigation.

20.9.1.2 Rapid Infiltration. In rapid-infiltration systems, wastewater that has received some preapplication treatment is applied on an intermittent schedule to shallow infiltration or spreading basins (Fig. 20.23b). Application of wastewater by high-rate sprinkling is also practiced. Vegetation is usually not provided in infiltration basins but is necessary for sprinkler application. Because loading rates are relatively high, evaporative losses are a small fraction of the applied water, and most of the applied water percolates through the soil profile where treatment occurs. Design objectives for rapid-infiltration systems include: (1) treatment followed by groundwater recharge to augment water supplies or prevent saltwater intrusion, (2) treatment followed by recovery using underdrains or pumped withdrawal, and (3) treatment followed by groundwater flow and discharge into surface waters.

20.9.1.3 Overland Flow. In overland flow, pretreated wastewater is distributed across the upper portions of carefully graded, vegetated slopes and allowed to flow over the slope surfaces to runoff-collection ditches at the bottom of the slopes (Fig. 20.23c). Overland flow is normally used at sites with relatively impermeable surface soils or subsurface layers, although the process has been adapted to a wide range of soil permeabilities because the soil surface tends to seal over time. Percolation through the soil profile is therefore a minor hydraulic pathway, and most of the applied water, less that lost to evaporation, is collected as surface runoff. Systems are operated using alternating application and drying periods, with the lengths of the periods depending on the treatment objectives.

20.9.2 Wetlands

Wetlands are inundated land areas with water depths typically less than 2 ft (0.6 m) that support the growth of emergent plants such as cattail, bulrush, reeds, and sedges. The vegetation provides surfaces for the attachment of bacteria films, aids in the filtration and adsorption of wastewater constituents, transfers oxygen into the water column, and controls the growth of algae by restricting the penetration of sunlight. Both natural and constructed wetlands have been used for wastewater treatment, although the use of natural wetlands is generally limited to the polishing or further treatment of secondary- or tertiary-treated effluent. Additional design details on constructed wetlands may be found in Moshiri (1993); Reed, et al. (1995); and U.S. EPA (1988).

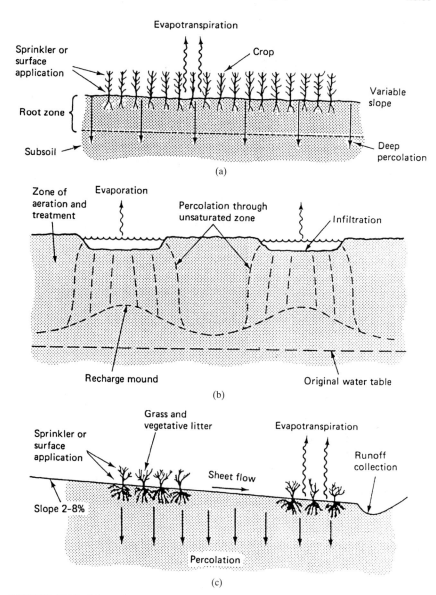

FIGURE 20.23 Schematic representation of the three principal land-treatment processes: (*a*) slow-rate irrigation; (*b*) rapid infiltration; (*c*) overland flow.

20.9.2.1 Natural Wetlands. Natural wetlands are usually considered to be receiving waters from a regulatory standpoint. Consequently, discharges to natural wetlands, in most cases, must meet applicable regulatory requirements that typically stipulate secondary or some form of advanced treatment. Furthermore, the principal objective when discharging to natural wetlands should be enhancement of existing habitat. Modification of existing wetlands to improve treatment capability is often very disruptive to the natural ecosystem and, in general, should not be attempted.

20.9.2.2 Constructed Wetlands. Constructed wetlands offer all of the treatment capabilities of natural wetlands without the constraints associated with discharging to a natural ecosystem. Two types of constructed-wetland systems have been developed for wastewater treatment: (1) free-water-surface (FWS) systems, and (2) subsurface-flow systems (SFS). When used to provide a secondary or advanced level of treatment, FWS systems (see Fig. 20.24*a, b*) typically consist of parallel basins or channels with relatively impermeable bottom soil or subsurface barrier, emergent vegetation, and relatively shallow water depths of 0.5 to 2 ft (0.15 to 0.6 m). Pretreated wastewater is normally applied continuously to such systems, and treatment occurs as the water flows slowly through the stems and roots of the emergent vegetation. Subsurface-flow systems are designed with an objective of secondary or advanced levels of treatment. These systems have also been called *root zone* or *rock-reed* filters and consist of excavated beds or trenches with relatively impermeable bottoms filled with sand or rock media to support emergent vegetation (see Fig. 20.24*c, d*).

20.9.3 Treatment of Systems Using Floating Aquatic Plants

Treatment systems using floating aquatic plants are similar in concept to FWS wetlands systems except that the plants are floating species such as water hyacinth (see Fig. 20.25*a*) and duckweed (see Fig. 20.25*b*). Water depths are typically deeper than wetlands systems, ranging from 1.6 to 3.0 ft (0.5 to 0.9 m). Supplementary aeration has been used with floating-plant systems to increase treatment capacity and to maintain aerobic conditions necessary for biological control of mosquitoes. Both water hyacinth and duckweed systems have been used to remove algae from lagoon and stabilization-pond effluents, while water hyacinth systems have been designed to provide secondary and advanced levels of treatment.

20.10 WASTEWATER RECLAMATION AND REUSE

Wastewater reclamation is the treatment or processing of wastewater to make it reusable. Wastewater reuse involves the beneficial use of treated wastewater in applications such as irrigation and industrial cooling. In response to continued population growth, contamination of both surface- and groundwaters, uneven distribution of water resources, and periodic droughts, the use of highly treated wastewater effluent, now discharged to the environment from municipal wastewater-treatment plants, is receiving more attention as an alternative source of water. In many parts of

(a)

(b)

FIGURE 20.24 Typical examples of constructed wetlands: (*a*) free-water system used for secondary treatment following aerobic/anaerobic ponds; (*b*) free-water system used for habitat development.

(c)

(d)

FIGURE 20.24 *(Continued)* Typical examples of constructed wetlands: (*c*) subsurface-flow system used for secondary treatment before vegetation has been planted; (*d*) subsurface-flow system used for secondary treatment at a resort on a Greek island.

(a)

(b)

FIGURE 20.25 Typical examples of treatment systems using floating aquatic plants: (*a*) water hyacinth system used for secondary treatment, and (*b*) duckweed system used to achieve enhanced removals of suspended solids in stabilization ponds. The water hyacinths are harvested using a truck-mounted articulated-arm open screen type clamshell. The person in back of the truck is redistributing the water hyacinths remaining in the pond. The floating plastic dividers shown in (*b*) are used to contain the duckweed to prevent it from being blown onto the pond banks. Duckweed is harvested using a specially designed harvestor such as shown.

the country, wastewater reuse is already an important element in water-resources planning (Pettygrove and Asano, 1985; and U.S. EPA, 1992).

20.10.1 Wastewater-Reuse Categories

The principal categories of municipal wastewater reuse are reported in Table 20.21 in descending order of projected volume of use. Potentially large quantities of reclaimed municipal wastewater can be used in the first four categories. Agricultural and landscape irrigation, the largest current and projected use of water, offers significant opportunities for wastewater reuse. The second major use of reclaimed municipal wastewater is in industrial activities, primarily for cooling and process

TABLE 20.21 Categories of Municipal-Wastewater Reuse and Potential Constraints Arranged in Descending Order of Projected Volume of Use

Wastewater-reuse categories	Potential constraints
Agricultural irrigation: crop irrigation, commercial nurseries Landscape irrigation: park, school yard, freeway median, golf course, cemetery, greenbelt, residential	Surface and groundwater pollution if not properly managed; marketability of crops and public acceptance; effect of water quality, particularly salts, on soils and crops; public health concerns related to pathogens, bacteria, viruses, and parasites; use-area control including buffer zone; may result in high user costs
Industrial recycling and reuse: cooling, boiler feed, process water, heavy construction	Constituents in reclaimed wastewater related to scaling, corrosion, biological growth, and fouling; public health concerns, particularly aerosol transmission of pathogens in cooling water
Groundwater recharge: groundwater replenishment, salt water intrusion control, subsidence control	Organic chemicals in reclaimed wastewater and their toxicologic effects; total dissolved solids, nitrates, and pathogens in reclaimed wastewater
Recreational/environmental uses: lakes and ponds, marsh enhancement, streamflow augmentation, fisheries, snowmaking	Health concerns of bacteria and viruses; eutrophication due to nitrogen and phosphorus in receiving water; toxicity to aquatic life
Nonpotable urban uses: fire protection, air conditioning, toilet flushing	Public health concerns on pathogens transmitted by aerosols; effects of water quality on scaling, corrosion, biological growth, and fouling; cross connection
Potable reuse: blending in water-supply reservoir, pipe-to-pipe water supply	Constituents in reclaimed wastewater, especially trace organic chemicals and their toxicologic effects; aesthetics and public acceptance; health concerns about pathogen transmission, particularly viruses

needs. Industrial uses vary greatly and to provide adequate water quality, additional treatment is often required beyond conventional secondary wastewater treatment. The third reuse application for reclaimed wastewater is groundwater recharge, either via spreading basins or direct injection to groundwater aquifers. A fourth use of reclaimed wastewater is characterized as miscellaneous subpotable uses, such as for recreational lakes, aquaculture, and toilet flushing. These subpotable uses are minor reclaimed-wastewater applications, at present, accounting for less than 5 percent of total wastewater reuse (Pettygrove and Asano, 1985; Tchobanoglous and Burton, 1991; and U.S. EPA, 1992).

20.10.2 Planning Wastewater-Reuse Applications

In effective planning for wastewater reclamation and reuse, the objectives should be defined clearly and there should be a well-defined basis for conducting the necessary planning studies. The optimum wastewater reclamation and reuse project is best achieved by integrating both wastewater-treatment and water-supply needs into one plan. This integrated approach is somewhat different from planning for conventional wastewater-treatment facilities where planning is done only for conveyance, treatment, and disposal of municipal wastewater. In the implementation of wastewater reclamation and reuse, the reuse application will usually govern the wastewater treatment needed, and the degree of reliability required for the treatment processes and operations. Because wastewater reclamation entails the provision of a continuous supply of water with consistent water quality, the reliability of the existing or proposed treatment processes and operations must be evaluated carefully in the planning stage. Typical flow diagrams for reclamation processes designed to produce a water of drinking-water quality are shown in Figure 20.26.

20.11 EFFLUENT DISPOSAL

After treatment, wastewater is either reused, as discussed earlier, or disposed of in the environment, where it re-enters the hydrologic cycle. Disposal can thus be viewed as the first step in a very indirect and long-term reuse. The most common means of treated-wastewater disposal is by discharge and dilution into ambient waters. Another means of disposal is land application, where the wastewater is made to seep into the ground and recharge underlying groundwater aquifers. Part of the wastewater destined for infiltration also evaporates and, in desert areas, this evaporated fraction can be substantial.

A fundamental element of wastewater disposal is the associated environmental impact. Numerous environmental regulations, criteria, policies, and reviews now ensure that the environmental impacts of treated-wastewater discharges to ambient waters are acceptable. This regulatory framework not only affects the selection of discharge locations and outfall structures, but also influences the level of treatment required. Treatment and disposal are thus linked and cannot be considered independently. For example, to achieve environmental acceptability, a choice may be available between enhanced treatment for one or several wastewater constituents or increased effluent dilution by, for example, moving the discharge further offshore or using a multiport-diffuser outfall. Another means is source reduction, whereby indi-

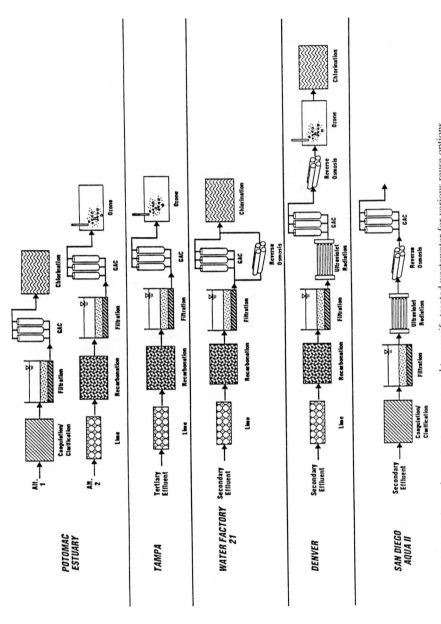

FIGURE 20.26 Advanced-treatment processes used to repurify treated wastewater for various reuse options including groundwater recharge and water-supply augmentation. (*Courtesy Montgomery/Watson.*)

vidual dischargers are required to decrease their contribution of specific contaminants to the sewers by process changes or pretreatment.

The emphasis of environmental-impact evaluations of wastewater discharges used to be on dissolved oxygen. The assimilative capacity of receiving waters, representing the amount of BOD_5 that can be assimilated without excessively taxing ambient dissolved oxygen levels, was of major concern. This emphasis on dissolved oxygen led to requirements for secondary treatment of wastewaters. Recently, the attention has broadened to a wider range of wastewater constituents, including nutrients, toxic compounds, and a variety of organic compounds. The environmental impacts of these constituents are diverse and often complex. The first element in evaluating these impacts, however, is the determination of the distribution and fate of these constituents in the water column and bottom sediments. Frequently, environmental criteria or standards exist regulating concentrations directly. In some cases, particularly for large discharges, additional environmental analyses are required; but the starting point is the distribution of constituent concentrations.

20.12 WASTEWATER-TREATMENT-PLANT DESIGN

The steps involved in the design of wastewater-treatment plants include: (1) conduct of flow and waste characterization studies, (2) preliminary process selection, (3) the conduct of pilot-plant studies, (4) the synthesis of alternative-treatment-flow diagrams, (5) the selection of design criteria and the sizing of units, (6) the layout of the physical facilities, (7) the preparation of hydraulic profiles, (8) the preparation of solids balances, and (9) the preparation of construction drawings, specifications, and cost estimates. Flow- and waste-characterization studies and process selection are considered further in the following.

Flow- and waste-characterization studies are of fundamental importance in the design of wastewater-treatment plants because the process design is based on this information. The rated capacity of wastewater-treatment plants is normally based on the average annual daily flow rate at the design year. As a practical matter, however, wastewater-treatment plants have to be designed to meet a number of conditions that are influenced by flow rates, wastewater characteristics, and a combination of both (mass loading). Peaking conditions also have to be considered, including peak hydraulic-flow rates and peak process mass-loading rates. Peak hydraulic-flow rates are important so that the unit operations and processes and their interconnecting conduits can be sized appropriately to handle the applied flow rates. Peak process loading rates are important in sizing the process units and their support systems so that treatment-plant-performance objectives can be achieved consistently and reliably. Typical flow rate and mass-loading factors that are important in the design and operation of wastewater-treatment facilities are described in Table 20.22.

The various combinations of unit operations and processes in a treatment plant work as a system, therefore, the designer must use a *systems approach* in the facilities design. The major part of the selection process is the evaluation of various combinations of unit operations and processes and their interactions. Part of this selection process may include consideration of flow equalization in reducing loadings on the treatment units. The evaluation process is not limited to the wastewater-

TABLE 20.22   Effect of Flow Rates and Constituent Mass Loadings on the Selection and Sizing of Secondary-Treatment-Plant Facilities

Unit operation or process	Critical design factor(s)	Sizing criteria	Effects of design criteria on plant performance
Wastewater pumping and piping	Maximum hour flow rate	Flow rate	Wet well may flood, collection system may surcharge, or treatment units may overflow if peak rate is exceeded.
Screening	Maximum hour flow rate	Flow rate	Head losses through bar rack and screens increase at high flow rates.
	Minimum hour flow rate	Channel approach velocity	Solids may deposit in approach channel at low flow rates.
Grit removal	Maximum hour flow rate	Overflow rate	At high flow rates, grit removal efficiency decreases in flow-through-type grit chambers causing grit problems in other processes.
Primary sedimentation	Maximum hour flow rate	Overflow rate	Solids-removal efficiency decreases at high overflow rates; increases loading on secondary-treatment system
	Minimum hour flow rate	Detention time	At low flow rates, long detention times may cause the wastewater to be septic.
Activated sludge	Maximum hour flow rate	Mean cell residence time	Solids washout at high flow rates; may need effluent recycle at low flow rates.
	Maximum organic load	Food/micro-organism ratio	High oxygen demand may exceed aeration capacity and cause poor treatment performance.
Trickling filters	Maximum hour flow rate	Hydraulic loading	Solids washout at high flow rates may cause loss of process efficiency.
	Maximum hour flow rate	Hydraulic and organic loading	Increased recycle at low flow rates may be required to sustain process.
	Maximum organic load	Mass loading/medium volume	Inadequate oxygen during peak load may result in loss of process efficiency and cause odors.
Secondary sedimentation	Maximum hour flow rate	Overflow rate or detention time	Reduced solids-removal efficiency at high overflow rates or short detention times.
	Minimum hour flow rate	Detention time	Possible rising sludge at long detention time.
Chlorine contact tank	Maximum hour flow rate	Detention time	Reduced bacteria kill at reduced detention time.
UV disinfection	Maximum day or week flow rate	Dose	Increased bacterial and virus inactivation at reduced flow rates

TABLE 20.23　Important Factors that Must Be Considered When Evaluating and Selecting Unit Operations and Processes

Factor	Comment
1. Process applicability	The applicability of a process is evaluated on the basis of past experience, data from full-scale plants, published data, and from pilot-plant studies. If new or unusual conditions are encountered, pilot-plant studies are essential.
2. Applicable flow range	The process should be matched to the expected range of flow rates. For example, stabilization ponds are not suitable for extremely large flow rates.
3. Applicable flow variation	Most unit operations and processes have to be designed to operate over a wide range of flow rates. Most processes work best at a relatively constant flow rate. If the flow variation is too great, flow equalization may be necessary.
4. Influent wastewater characteristics	The characteristics of the influent wastewater affect the types of processes to be used (e.g., chemical or biological) and the requirements for their proper operation.
5. Inhibiting and unaffected constituents	What constituents are present and may be inhibitory to the treatment processes? What constituents are not affected during treatment?
6. Climatic constraints	Temperature affects the rate of reaction of most chemical and biological processes. Temperature may also affect the physical operation of the facilities. Warm temperatures may accelerate odor generation and also limit atmospheric dispersion.
7. Reaction kinetics and reactor selection	Reactor sizing is based on the governing reaction kinetics. Data for kinetic expressions usually are derived from experience, published literature, and the results of pilot plant studies.
8. Performance	Performance is usually measured in terms of effluent quality, which must be consistent with the effluent-discharge requirements.
9. Treatment residuals	The types and amounts of solid, liquid, and gaseous residuals produced must be known or estimated. Often, pilot-plant studies are used to identify and quantify residuals.
10. Sludge processing	Are there any constraints that would make sludge processing and disposal infeasible or expensive? How might recycle loads from sludge processing affect the liquid-unit operations or processes? The selection of the sludge-processing system should go hand-in-hand with the selection of the liquid-treatment system
11. Environmental constraints	Environmental factors, such as prevailing winds and wind directions and proximity to residential areas, may restrict or affect the use of certain processes, especially where odors may be produced. Noise and traffic may affect selection of a plant site. Receiving waters may have special limitations, requiring the removal of specific constituents such as nutrients.
12. Chemical requirements	What resources and what amounts must be committed for a long period of time for the successful operation of the unit operation or process? What effects might the addition of chemicals have on the characteristics of the treatment residuals and the cost of treatment?
13. Energy requirements	The energy requirements, as well as probable future energy cost, must be known if cost-effective treatment systems are to be designed.
14. Other resource requirements	What, if any, additional resources must be committed to the successful implementation of the proposed treatment system using the unit operation or process being considered?

TABLE 20.23 Important Factors that Must Be Considered When Evaluating and Selecting Unit Operations and Processes *(Continued)*

Factor	Comment
15. Personnel requirements	How many people and what levels of skills are needed to operate the unit operation or process? Are these skills readily available? How much training will be required?
16. Operating and maintenance requirements	What special operating or maintenance requirements will need to be provided? What spare parts will be required and what will be their availability and cost?
17. Ancillary processes	What support processes are required? How do they affect the effluent quality, especially when they become inoperative?
18. Reliability	What is the long-term reliability of the unit operation or process being considered? Is the operation or process easily upset? Can it stand periodic shock loadings? If so, how do such occurrences affect the quality of the effluent?
19. Complexity	How complex is the process to operate under routine or emergency conditions? What levels of training must the operators have to operate the process?
20. Compatibility	Can the unit operation or process be used successfully with existing facilities? Can plant expansion be accomplished easily?
21. Land availability	Is there sufficient space to accommodate not only the facilities currently being considered but possible future expansion? How much of a buffer zone is available to provide landscaping to minimize visual and other impacts?

treatment units alone; the interaction of the liquid with the sludge-processing alternatives must be done as an integral part of the evaluation. The mass-balance analysis then becomes a critical element of the evaluation. The most important factors that must be considered in evaluating and selecting unit operations and processes are identified in Table 20.23. Additional details on wastewater treatment plant design may be found in Eckenfelder and Grau (1992); Tchobanoglous and Burton (1991); and Water Environment Federation (1991).

20.13 DECENTRALIZED WASTEWATER-MANAGEMENT SYSTEMS

Decentralized wastewater management (DWWM) may be defined as the collection, treatment, and disposal/reuse of wastewater from individual homes, clusters of homes, or isolated communities, industries, or institutional facilities at or near the point of origin. Although most decentralized systems maintain both the solid and liquid fractions of the wastewater near their point of origin, the liquid portion can be transported to a centralized point for further treatment and reuse. Alternative wastewater-management technologies for unsewered areas are reported in Table 20.24. Although a variety of on-site systems have been used, the most common system consists of a septic tank for the partial treatment of the wastewater and a subsurface-soil disposal field for final treatment and disposal of the septic

TABLE 20.24 Technologies Used for Decentralized Wastewater Management Systems

Collection and transport of wastewater	Wastewater treatment and/or containment	Wastewater disposal
House drains and building sewers	Primary treatment Septic tank (individual, home, home cluster, and community)	Subsurface soil disposal Leachfields Conventional
Conventional gravity sewers	Imhoff tank	Shallow trench Shallow sand-filled Seepage beds
Pressure sewers (with nongrinder pumps)	Advanced primary treatment Septic tank with attached growth reactor element	Mound systems Evapotranspiration/ percolation beds Drip application
Pressure sewers (with grinder pumps) Small-diameter variable-slope gravity sewers	Secondary treatment Aerobic/anaerobic units Aerobic units Anaerobic units Intermittent sand filter	Evaporation systems Evapotranspiration bed Evaporation pond
Vacuum sewers	Recirculating granular medium filter (sand, crushed glass)	Constructed wetlands (marsh)
Combinations of the above	Peat filter Constructed wetlands	Land disposal Surface application
	Recycle treatment systems Toilet flushing	Drip application
	Landscape watering and toilet flushing	Discharge to water bodies
		Combinations of the above
	On-site containment Holding tank Privy	

tank effluent (see Fig. 20.27). Because conventional disposal fields cannot be used in some locations, many alternative systems have been developed. The most successful of these include intermittent and recirculating granular-medium filters. Intermittent sand filters have become quite popular in many parts of the country for single family residences because of their excellent performance, reliability, and relatively lower cost. Recirculating granular-medium filters are used for larger flows. Complete-recycle systems have been developed for commercial buildings. Holding tanks are used where an acceptable on-site disposal system cannot be installed.

The treatment component of cluster and community systems will vary with the size of the installation. Typically, a large septic tank will be used for a cluster of homes. Imhoff tanks, commonly used in the past, are rarely used today because of their relatively high cost. In some communities, septic tanks may be used for the separation of settleable solids and greases and oils. Recirculating granular-medium filters are used in conjunction with septic tanks where a higher level of treatment is required. Pre-engineered and constructed package plants and individually designed plants are used where the flows are larger. The methods used for effluent disposal will also vary with the size of the system. For small installations serving a cluster of

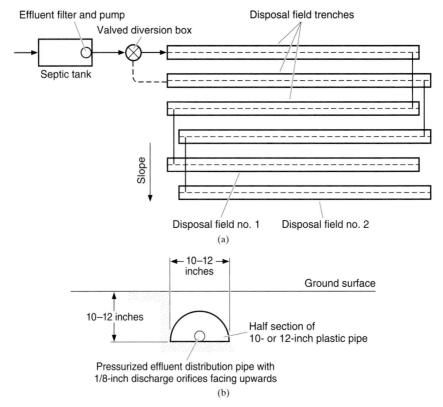

FIGURE 20.27   Typical subsurface soil-absorption disposal field: (*a*) valved pressure-dosed shallow-leachfield effluent-disposal system; and (*b*) cross section through shallow leachfield.

homes, effluent disposal is most commonly accomplished using disposal fields. As the size of the system increases, the methods used for the disposal of effluent are, as shown in Table 20.24, essentially the same as those used for larger systems. Additional details on decentralized systems may be found in Tchobanoglous and Burton (1991); and U.S. EPA (1980).

REFERENCES

Darby, J., et al., "Comparison of UV Irradiation to Chlorination: Guidance for Achieving Optimal UV Performance," Project 91-WWD-1, Water Environment Research Foundation, Alexandria, Va., 1995.

Eckenfelder, W. W., and P. Grau, eds., *Activated Sludge Process Design and Control: Theory and Practice,* Water Quality Management Library, Vol. 1, Technomic, Lancaster, Pa., 1992.

Feachem, R. G., D. J. Bradley, H. Garelick, and D. D. Mara, *Sanitation and Disease: Health Aspects of Excreta and Wastewater Management,* published for the World Bank by John Wiley, New York, 1983.

Federal Register, "Secondary Treatment Regulation," 40 CFR Part 133, Washington, D.C., July 1, 1988.

Federal Register, "Amendment to the Secondary Treatment Regulations: Percent Removal Requirements During Dry Weather Periods for Treatment Works Served by Combined Sewers," 40 CFR Part 133, Washington, D.C., January 27, 1989.

Greenberg, A. E., L. S. Clesceri, and A. D. Eaton, eds., *Standard Methods for the Examination of Water and Waste Water,* 18th ed., American Public Health Association, Washington, D.C., 1992.

Jenkins, D., personal communication, 1995.

Malina, J. F., Jr., and F. G. Pohland, eds., *Design of Anaerobic Processes for the Treatment of Industrial and Municipal Wastes,* Water Quality Management Library, Vol. 7, Technomic, Lancaster, Pa., 1992.

Moshiri, G. A., ed., *Constructed Wetlands for Water Quality Improvement,* Lewis, Boca Raton, Fla., 1993.

Pettygrove, G. S., and T. Asano, eds., *Irrigation with Reclaimed Wastewater—A Guidance Manual,* Lewis, Chelsea, Mich., 1985.

Randall, C. W., J. L. Barnard, and H. D. Stensel, eds., *Design and Retrofit of Wastewater Treatment Plants for Biological Nutrient Removal,* Water Quality Management Library, Vol. 5, Technomic, Lancaster, Pa., 1992.

Reed, S. C., E. J. Middlebrooks, and R. W. Crites, *Natural Systems for Waste Management and Treatment,* 2d ed., McGraw-Hill, New York, 1995.

Soap and Detergent Association, *Principles and Practice of Nutrient Removal from Municipal Wastewater,* New York, 1988.

Stanier, R. Y., J. L. Ingraham, M. L. Wheelis, and P. R. Painter, *The Microbial World,* 5th ed., Prentice-Hall, Englewood Cliffs, N.J., 1986.

Tchobanoglous, G., and F. L. Burton, *Wastewater Engineering: Treatment, Disposal, Reuse,* 3rd ed., Metcalf & Eddy, Inc., McGraw-Hill, New York, 1991.

Tchobanoglous, G., and E. D. Schroeder, *Water Quality: Characteristics, Modeling, Modification,* Addison-Wesley, Reading, Mass., 1985.

U.S. Environmental Protection Agency, "Design Manual: Onsite Wastewater Treatment and Disposal Systems," EPA 625/1-80-012, Office of Water Program Operations, Washington, D.C., 1980.

U.S. Environmental Protection Agency, "Process Design Manual for Land Application of Municipal Sludge," EPA 625/1-83-016, Cincinnati, Ohio, 1983.

U.S. Environmental Protection Agency, "Environmental Regulations and Technology, Use and Disposal of Municipal Wastewater Sludge," EPA 625/10-84-003, Cincinnati, Ohio, 1984a.

U.S. Environmental Protection Agency, "Process Design Manual for Land Treatment of Municipal Wastewater: Supplement on Rapid Infiltration and Overland Flow," EPA 625/1-81-013a, Cincinnati, Ohio, 1984b.

U.S. Environmental Protection Agency, "Design Manual for Constructed Wetlands and Floating Aquatic Plant Systems for Municipal Wastewater Treatment," EPA 625/1-88-022, Cincinnati, Ohio, 1988.

U.S. Environmental Protection Agency, "Methods for Measuring the Acute Toxicity of Effluents to Freshwater and Marine Organisms," U.S. EPA Environmental Monitoring and Support Laboratory, EPA-600/4-85/013, Cincinnati, Ohio, 1985a.

U.S. Environmental Protection Agency, "Short Term Methods for Estimating Chronic Toxicity of Effluents and Receiving Waters to Freshwater Organisms," EPA-660/4-85/014, Cincinnati, Ohio, 1985b.

U.S. Environmental Protection Agency, "Short Term Methods for Estimating the Chronic Toxicity of Effluents and Receiving Waters to Marine and Estuarine Organisms," EPA-600/4-88/028, Cincinnati, Ohio, 1988.

U.S. Environmental Protection Agency, "Short Term Methods for Estimating Chronic Toxicity of Effluents and Receiving Waters to Freshwater Organisms," EPA-660/2d ed., Cincinnati, Ohio, 1989.

U.S. Environmental Protection Agency, "Technical Support Document for Water Quality-Based Toxics Control," U.S. EPA Office of Water, EPA-505/2-90-001, Washington, D.C., 1991.

U.S. Environmental Protection Agency, "Guidelines for Water Reuse," EPA/625/R-92/004, Washington, D.C., 1992.

Water Environment Federation, *Design of Municipal Wastewater Treatment Plants*, Vol. I: Chaps. 1–12 and Vol. II, Chaps. 13–20, WEF Manual of Practice no. 8, Alexandria, Va., 1991.

CHAPTER 21
WATER RECLAMATION AND REUSE

James Crook
Black & Veatch
100 CambridgePark Drive
Cambridge, Mass.

21.1 INTRODUCTION

Reclamation and reuse of municipal wastewater for beneficial purposes is a well-established practice in the United States and elsewhere. Water reuse is now recognized as an important, integral component of water-resources management, although many of the early projects were implemented as a least-cost means of wastewater disposal. Reclaimed water is used for many purposes, ranging from pasture irrigation to augmentation of potable water supplies. Although many regulatory agencies do not advocate the use of reclaimed water for potable purposes, planned indirect potable reuse via groundwater recharge and surface-water augmentation is occurring and will likely increase in the future.

Depending on the intended use of reclaimed water, considerations may include health protection, user water quality and quantity requirements, irrigation effects, environmental effects, aesthetics, and public and/or user perception of the reuse concept. Making reclaimed water suitable and safe for reuse applications is achieved by eliminating or reducing the concentrations of pathogenic microorganisms and chemical constituents of concern through wastewater treatment and/or by limiting public or worker exposure to the water via design or operational controls.

The impacts or constraints on reuse from physical parameters, e.g., pH, color, temperature, and particulate matter, and chemical constituents, e.g., chlorides, sodium, heavy metals, and some trace organic compounds, are well known and recommended limits have been established for many of these constituents (National Academy of Sciences–National Academy of Engineering, 1973; U.S. EPA, 1981; WPCF, 1989; Wescot and Ayers, 1981). The health risks associated with pathogenic microorganisms are more difficult to assess.

Water-reuse regulations vary widely among the states in the United States. Some states have regulations or guidelines directed at land treatment or land application as a means for disposal or further wastewater treatment rather than regulations oriented to the intentional beneficial use of reclaimed water. Several states have no regulations or guidelines, and others disallow reuse altogether. The U.S. Environmental Protection Agency (EPA) published water-reuse guidelines in 1992 that are intended to provide guidance to states that have not developed their own criteria or guidelines (U.S. EPA, 1992).

This chapter does not provide detailed information on specific water-reuse facilities, wastewater-treatment processes, and their constituent removal effectiveness, operation and management, and research activities. Such information is adequately documented and summarized in the open literature, including publications by the American Water Works Association (AWWA) (1994), Crook, et al. (1994), U.S. EPA (1992), and Water Pollution Control Federation (WPCF) (1989).

21.2 STATUS OF REUSE

Historically, the largest-volume uses of reclaimed water were those that do not require a high quality water, e.g., pasture irrigation, and were often perceived as a method of wastewater disposal. Reclaimed water is now valued as a resource and, in recent years, the trend has shifted toward higher level uses such as urban irrigation, toilet flushing, industrial uses, and groundwater recharge. Table 21.1 lists many of the currently practiced types of water reuse.

TABLE 21.1   Uses of Reclaimed Water

Landscape irrigation	Nonpotable urban (other than irrigation)
Parks	Toilet and urinal flushing
Cemeteries	Fire protection
Golf courses	Air conditioner cooling water
Roadway rights-of-way	Vehicle washing
School grounds	Street cleaning
Greenbelts	Decorative fountains
Residential lawns	Impoundments
Agricultural irrigation	Ornamental
Food crops	Recreational
Fodder, fiber, and seed crops	Environmental
Nurseries	Stream augmentation
Sod farms	Marshes
Silviculture	Wetlands
Frost protection	Fisheries
Industrial	Miscellaneous
Cooling	Aquaculture
Boiler feed	Snow-making
Stack scrubbing	Soil compaction
Process water	Dust control
Groundwater recharge	Equipment washdown
Recharge aquifers	Livestock watering
Saltwater-intrusion control	
Potable-water supply augmentation	
Groundwater recharge	
Surface-water augmentation	

Agricultural-irrigation water and cooling water for thermoelectric-power generation each represent about 40 percent of the freshwater used in the United States (Van der Leeden, et al., 1991). The use of reclaimed water for agricultural irrigation and cooling provides significant opportunities for reuse, particularly at sites near urban areas where large volumes of reclaimed water are generated. Although commercial and domestic water use constitutes only about 10 percent of the total water demand, water reuse is more likely to be cost-effective in or near urban areas where reclaimed water is used to conserve or replace existing potable water for various applications.

While up-to-date statistical information on water reuse in the United States has not been compiled, it has been estimated that the use of reclaimed water approached 1.5 bgd (5.7×10^6 m³/d) in 1990 (Crook, 1992). An average of 240 mgd (0.93×10^6 m³/d) of municipal wastewater was reclaimed in California in 1987, representing about 12 percent of the total wastewater produced in the state. Sixty-three percent of the reclaimed water was used for agricultural irrigation, 14 percent for groundwater recharge, and 13 percent for landscape irrigation (California State Water Resources Control Board, 1990). In Florida, 290 mgd (1.1×10^6 m³/d), or about 30 percent of the state's municipal wastewater, was reused in 1992. Thirty-eight percent of the reclaimed water was used for landscape irrigation, 30 percent for agricultural irrigation, and 14 percent for groundwater recharge (Florida Department of Environmental Regulation, 1992).

21.3 WATER-QUALITY CONCERNS

The acceptability of reclaimed water for any particular use is dependent on the physical, chemical, and microbiological quality of the water. Factors that affect the quality of reclaimed water include source-water quality, wastewater-treatment processes and treatment effectiveness, treatment reliability, and distribution-system design and operation. Industrial source-control programs can limit the input of chemical constituents that may adversely affect biological-treatment processes and subsequent reclaimed-water quality. Assurance of treatment reliability is an obvious, yet sometimes overlooked, quality-control measure. Distribution-system design and operation is important to assure that the reclaimed water is not degraded prior to use and not subject to misuse. Open storage may result in water-quality degradation by microorganisms, algae, or particulate matter, and may cause objectionable odor or color in the reclaimed water.

21.3.1 Microorganisms

The potential transmission of infectious disease by pathogenic agents is the most common concern associated with reuse of treated municipal wastewater. While there is no reliable epidemiological evidence that the use of appropriately treated wastewater for any of the applications identified in Table 21.1 has caused a disease outbreak in the United States, the potential spread of infectious diseases through water reuse remains a public health concern.

The principal infectious agents that may be found in raw municipal wastewater can be classified into three broad groups: bacteria, parasites (protozoa and helminths), and viruses. Table 21.2 lists many of the infectious agents potentially present in raw municipal wastewater, and some of the important waterborne pathogens are discussed below.

TABLE 21.2   Infectious Agents Potentially Present in Raw Sewage

Pathogen	Disease
Bacteria	
Campylobacter jejuni	Gastroenteritis
Escherichia coli (enteropathogenic)	Gastroenteritis
Legionella pneumophila	Legionnaire's disease
Leptospira (spp.)	Leptospirosis
Salmonella typhi	Typhoid fever
Salmonella (2400 serotypes)	Salmonellosis
Shigella (4 spp.)	Shigellosis (dysentery)
Vibrio cholerae	Cholera
Yersinia enterocolitica	Yersiniosis
Protozoa	
Balantidium coli	Balantisiasis (dysentery)
Cryptosporidium	Cryptosporidiosis, diarrhea, fever
Entamoeba histolytica	Amebiasis (amebic dysentery)
Giardia lamblia	Giardiasis
Helminths	
Ancylostoma duodenale (hookworm)	Ancylostomiasis
Ascaris lumbricoides (roundworm)	Ascariasis
Echinococcus granulosis (tapeworm)	Hydatidosis
Enterobius vermicularis (pinworm)	Enterobiasis
Necator americanus (roundworm)	Necatoriasis
Schistosoma (spp.)	Schistosomiasis
Strongyloides stercoralis (threadworm)	Strongyloidiasis
Taenia (spp.) (tapeworm)	Taeniasis, cysticercosis
Trichuris trichiura (whipworm)	Trichuriasis
Viruses	
Adenovirus (51 types)	Respiratory disease, eye infections
Astrovirus (5 types)	Gastroenteritis
Calicivirus (2 types)	Gastroenteritis
Coronavirus	Gastroenteritis
Enteroviruses (72 types) (polio, echo, coxsackie, new enteroviruses)	Gastroenteritis, heart anomalies, meningitis, others
Hepatitis A virus	Infectious hepatitis
Norwalk agent	Diarrhea, vomiting, fever
Parvovirus (3 types)	Gastroenteritis
Reovirus (3 types)	Not clearly established
Rotavirus (4 types)	Gastroenteritis

Source: Adapted from Hurst, et al. (1989); and Sagik, et al. (1978).

21.3.1.1 Bacteria.   One of the most common pathogens found in municipal wastewater is the genus *Salmonella*. The *Salmonella* group contains a wide variety of species that can cause disease in humans and animals. The most severe form of salmonellosis is typhoid fever, caused by *Salmonella typhi*. A less common genus of bacteria in wastewater is *Shigella,* which produces an intestinal disease known as *bacillary dysentery* or *shigellosis.* Waterborne outbreaks of shigellosis have been reported from recreational swimming and where wastewater has contaminated wells used for drinking water (National Communicable Disease Center, 1969; 1973).

Other bacteria isolated from raw wastewater include *Vibrio, Mycobacterium, Clostridium, Leptospira,* and *Yersinia* species. *Vibrio cholerae* is the disease agent for cholera, which is not common in the United States but is still prevalent in other parts of the world. Humans are the only known hosts, and the most frequent mode of transmission is through water. *Mycobacterium tuberculosis* has been found in municipal wastewater (Greenberg and Kupka, 1957), and outbreaks have been reported among persons swimming in sewage-contaminated water (California Department of Health and Cooper, 1975).

Waterborne gastroenteritis of unknown cause is frequently reported, with the suspected agent being bacterial. One potential source of this disease is certain gram-negative bacteria normally considered to be nonpathogenic. These include entero-pathogenic *Escherichia coli* and certain strains of *pseudomonas,* which may affect the newborn. Waterborne enterotoxigenic *E. coli* have been implicated in gastrointestinal disease outbreaks (National Communicable Disease Center, 1975).

Campylobacter jejuni has been identified as the cause of a form of bacterial diarrhea in humans. While it has been well established that this organism causes disease in animals, it has also been implicated as the etiological agent in human waterborne disease outbreaks (Craun, 1988).

Coliform bacteria are commonly used as indicators of fecal contamination and the potential presence of pathogenic species. While coliforms generally respond similarly to environmental conditions and treatment processes as many bacterial pathogens, coliform bacteria determinations by themselves do not adequately predict the presence or concentration of pathogenic viruses or protozoa.

21.3.1.2 Protozoa.   Several pathogenic protozoan parasites have been detected in municipal wastewater. One of the most important of the parasites is the protozoan *Entamoeba histolytica,* which is responsible for amoebic dysentery and amoebic hepatitis. The diseases are found worldwide, but in the United States, *Entamoeba histolytica* has not been an important disease agent since the 1950s.

Waterborne disease outbreaks around the world have been linked to the proto-zoans *Giardia lamblia* and *Cryptosporidium* in drinking water and recreational water, and, although no giardiasis or cryptosporidiosis cases related to water-reuse projects have been reported, giardiasis and cryptosporidiosis are emerging as major waterborne diseases. Infection is caused by ingestion of *Giardia* cysts or *Cryptosporidium* oocysts. The cysts and oocysts are present in most wastewaters and are more difficult to inactivate by chlorination than are bacteria and viruses. Cryptosporidiosis can be fatal to immunocompromised individuals.

21.3.1.3 Helminths.   The most important helminthic parasites that may be found in wastewater are intestinal worms, including the stomach worm *Ascaris lumbricoides,* the tapeworms *Taenia saginata* and *Taenia solium,* the whipworm *Trichuris trichirar,* the hookworms *Ancylostoma duodenia* and *Necator americanus* and the threadworm *Strongyloides stercoralis.* The infective stage of some helminths is either the adult organism or larvae, while the eggs or ova of other helminths constitute the infective stage. The free-living nematode larvae stages are not pathogenic to human beings. The eggs and larvae are resistant to environmental stresses and may survive usual wastewater-disinfection procedures, although eggs can be removed by commonly used wastewater-treatment processes such as sedimentation, filtration, or stabilization ponds.

21.3.1.4 *Viruses.*

Over 100 different types of enteric viruses capable of producing infection or disease are excreted by humans. Enteric viruses multiply in the intestinal tract and are released in the fecal matter of infected persons.

The most important human enteric viruses are the enteroviruses (polio, echo, and coxsackie), Norwalk virus, rotaviruses, reoviruses, caliciviruses, adenoviruses, and hepatitis A virus (Hurst, et al., 1989; WPCF, 1989). The reoviruses and adenoviruses, which are known to cause respiratory illness, gastroenteritis, and eye infections, have been isolated from wastewater. Of the viruses that cause diarrheal disease, only the Norwalk virus and rotavirus have been shown to be major waterborne pathogens (Rose, 1986). There is no evidence that the human immunodeficiency virus (HIV), the pathogen that causes the acquired immunodeficiency syndrome (AIDS), can be transmitted via a waterborne route (Gover, 1993; Riggs, 1989).

It has been reported that viruses and other pathogens in wastewater used for crop irrigation do not readily penetrate fruits or vegetables unless the skin is broken (Bryan, 1974). In one study where soil was inoculated with poliovirus, viruses were detected in the leaves of plants only when the plant roots were damaged or cut (Shuval, 1978). Although absorption of viruses by plant roots and subsequent acropetal translocation has been reported (Murphy and Syverton, 1958), it probably does not occur with sufficient regularity to be a mechanism for the transmission of viruses. Therefore, the likelihood that pathogens would be translocated through trees or vines to the edible part of crops is extremely low.

21.3.2 Presence and Survival of Pathogens

The occurrence and concentration of pathogenic microorganisms in raw municipal wastewater depend on a number of factors, and it is not possible to predict with any degree of assurance what the general characteristics of a particular wastewater will be with respect to infectious agents. These factors include the sources contributing to the wastewater, the general health of the contributing population, the existence of disease carriers in the population, and the ability of infectious agents to survive outside their hosts under a variety of environmental conditions.

The occurrence of viruses in municipal wastewater fluctuates widely. Virus concentrations are generally highest during the summer and early autumn months. Viruses as a group are generally more resistant to environmental stresses than many of the bacteria, although some viruses persist for only a short time in municipal wastewater. The infectious doses of selected pathogens and the concentration ranges of some microorganisms in untreated municipal wastewater are presented in Tables 21.3 and 21.4, respectively.

Under favorable conditions, pathogens can survive for long periods of time on crops or in water or soil. While various pathogens exhibit a wide range of survival characteristics, environmental factors that affect survival include soil organic-matter content (presence of organic matter aids survival), temperature (longer survival at low temperatures), humidity (longer survival at high humidity), pH (bacteria survive longer in alkaline soils than in acid soils), amount of rainfall, amount of sunlight (solar radiation is detrimental to survival), protection provided by foliage, and competitive

TABLE 21.3 Infectious Doses of Selected Pathogens

Organism	Infectious dose*
Escherichia coli (enteropathogenic)	10^6–10^{10}
Clostridium perfringens	1–10^{10}
Salmonella typhi	10^4–10^7
Vibrio cholerae	10^3–10^7
Shigella flexneri 2A	180
Entameoba histolytica	20
Shigella dysentariae	10
Giardia lamblia	<10
Cryptosporidium	1–10
Ascaris lumbricoides	1–10
Viruses	1–10

* Some of the data for bacteria are given as ID_{50}, which is the dose that infects 50 percent of the people given that dose. People given lower doses also could become infected.
Source: Adapted from Feacham, et al. (1981; 1983).

microbial fauna and flora. Survival times for any particular microorganism exhibit wide fluctuations under differing conditions. Typical ranges of survival times for some common pathogens on crops and in water and soil are presented in Table 21.5.

At low temperatures (below 4°C) some microorganisms can survive in the underground for months or years. One study indicated that the die-off rate was approximately doubled with each 10°C rise in temperature between 5 and 30°C (Gerba and Goyal, 1985). Keswick, et al. (1982) reported a 10-fold decrease in poliovirus titer every 5 days at groundwater temperatures of 5 to 13°C, whereas Jansons, et al. (1989) found that the same decrease in virus titer required 26 days at groundwater temperatures of 15 to 18°C.

TABLE 21.4 Microorganism Concentrations in Raw Wastewater

Organisms	Concentration (number/100 mL)
Total coliforms	10^7–10^{10}
Clostridium perfringens	10^3–10^5
Enterococci	10^4–10^5
Fecal coliforms	10^4–10^9
Fecal streptococci	10^4–10^6
Pseudomonas aeroginosa	10^3–10^4
Protozoan cysts	10^3–10^5
Shigella	1–10^3
Salmonella	10^2–10^4
Helminth ova	10–10^3
Enteric virus	10^2–10^4
Giardia lambia cysts	10–10^4
Entamoeba histolytica cysts	1–10
Cryptosporidium oocysts	10–10^3

Source: Adapted from various sources.

TABLE 21.5 Typical Pathogen Survival Times at 20 to 30°C

Pathogen	Survival time (days)		
	Freshwater and sewage	Crop	Soil
Viruses*			
Enteroviruses[†]	<120 but usually <50	<60 but usually <15	<100 but usually <20
Bacteria			
Fecal coliforms*	<60 but usually <30	<30 but usually <15	<70 but usually <20
Salmonella spp.*	<60 but usually <30	<30 but usually <15	<70 but usually <20
Shigella spp.*	<30 but usually <10	<10 but usually <5	
Vibrio cholerae[‡]	<30 but usually <10	<5 but usually <2	<20 but usually <10
Protozoa			
E. histolytica cysts	<30 but usually <15	<10 but usually <2	<20 but usually <10
Helminths			
A. lumbricoides eggs	Many months	<60 but usually <30	Many months

* In seawater, viral survival is less, and bacterial survival is very much less, than in freshwater.
[†] Includes polio, echo, and coxsackie viruses.
[‡] V. cholerae survival in aqueous environments is a subject of current uncertainty.
Source: Adapted from Feacham, et al. (1983).

Viruses have been isolated in groundwater after various migration distances by a number of investigators examining a variety of recharge operations. Depending on soil conditions, bacteria and larger organisms associated with wastewater can be effectively removed by soil after percolation through a short distance of the soil mantle.

21.3.3 Aerosols

The concentration of pathogens in aerosols caused by spraying of wastewater is a function of their concentration in the applied wastewater and the aerosolization efficiency of the spray process. During spray irrigation, the amount of water that is aerosolized can vary from less than 0.1 percent to almost two percent, with a mean aerosolization efficiency of one percent or less (Bausum, et al., 1983; Camann, et al., 1988; Johnson, et al., 1980a; 1980b). Infection or disease may be contracted indirectly by aerosols deposited on surfaces such as food, vegetation, and clothes. The infective dose of some pathogens is lower for respiratory tract infections than for infections via the gastrointestinal tract; thus, for some pathogens, inhalation may be a more likely route for disease transmission than either contact or ingestion (Hoadley and Goyal, 1976; Sobsey, 1978).

In general, bacteria and viruses in aerosols remain viable and travel farther with increased wind velocity, increased relative humidity, lower temperature, and lower solar radiation. Other important factors include the initial concentration of pathogens in the wastewater and droplet size. Aerosols can be transmitted for several hundred meters under optimum conditions. Some types of pathogenic organisms, e.g., enteroviruses and Salmonella, appear to survive the wastewater-aerosolization process much better than the indicator organisms (Teltsch, et al., 1980). Bacteria and viruses have been found in aerosols emitted by spray-irrigation systems using untreated and poorly treated wastewater (Camann and Guentzel, 1985; Camann and Moore, 1988; Teltsch, et al., 1980).

Studies in the United States directed at residents in communities subjected to aerosols from sewage treatment plants have not detected any definitive correlation between exposure to aerosols and disease (Camann, et al., 1980; Fannin, et al., 1980; Johnson, et al., 1980a). There have not been any documented disease outbreaks resulting from spray irrigation with disinfected reclaimed water, and studies indicate that the health risk associated with aerosols from spray-irrigation sites using disinfected reclaimed water is low (Pahren and Jakubowski, 1980). The general practice is to limit exposure to aerosols produced from reclaimed water that is not highly disinfected through design or operational controls.

Studies (Adams, et al., 1978; 1980) indicate that cooling-tower aerosols can include substantial numbers of bacteria. Some state regulatory agencies require that reclaimed water used in cooling towers be essentially free of measurable levels of pathogens. Typical requirements include secondary treatment followed by filtration and high-level disinfection to achieve reclaimed water containing less than 2.2 total coliform organisms per 100 mL or no detectable fecal coliform organisms per 100 mL.

21.3.4 Disease Incidence Related to Water Reuse

Epidemiological investigations directed at wastewater-contaminated drinking-water supplies, the use of raw or minimally treated wastewater for food-crop irrigation, health effects on farmworkers who routinely come in contact with poorly treated wastewater used for irrigation, and the health effects of aerosols or windblown spray emanating from spray-irrigation sites using undisinfected wastewater have all provided evidence of infectious disease transmission from such practices (Fannin, 1980; Lund, 1980; Sepp, 1971; Shuval, et al., 1986).

The majority of documented disease outbreaks have been the result of contamination by bacteria or parasites. Several incidences of typhoid fever were reported in the early 1900s, and a major outbreak of cholera in Jerusalem in 1970 was reportedly caused by food-crop irrigation with undisinfected wastewater (Shuval, et al., 1986). Ascaris and hookworm infections have been attributed to the irrigation of vegetables and salad crops with untreated wastewater in Germany and India (Sepp, 1971; Shuval, et al., 1986). Human infection with the adult stage of the beef tapeworm *Taenia saginata,* due to ingestion of the cyst form, has resulted from irrigation of grazing land with raw and settled sewage. Evidence of tapeworm transmission via infected cattle and sheep in Europe, Australia, and elsewhere has been well documented (Sepp, 1971).

Excluding the use of raw sewage or primary effluent on sewage farms in the late nineteenth century, there have not been any confirmed cases of infectious disease resulting from reclaimed-water use in the United States. In developing countries, on the other hand, the irrigation of market crops with poorly treated wastewater is a major source of enteric disease (Shuval, et al., 1986).

Although pathogen-free water is not needed for all reclaimed-water applications, the general practice is to provide water of a quality appropriate for the highest use of the water in a community. Since regulatory agencies generally require essentially pathogen-free reclaimed water for the highest level nonpotable uses, e.g., residential-landscape irrigation, toilet flushing, and the irrigation of parks, in most cases all reclaimed water distributed throughout a community meets this requirement. Wastewater treated to this level would not present unreasonable risks of infectious disease from infrequent, inadvertent ingestion (Asano, et al., 1992; Rose and Gerba, 1991; Yanko, 1993).

21.3.5 Chemical Constituents

The chemical constituents potentially present in municipal wastewater generally are not a major health concern for urban uses of reclaimed water but may affect the acceptability of the water for uses such as food-crop irrigation, industrial applications, and indirect potable reuse. With the exception of the possible inhalation of volatile organic compounds from indoor exposure, there are minimal health concerns associated with chemical constituents where reclaimed water is not intended to be consumed or used for food-crop irrigation. Chemical constituents may be of concern when reclaimed water percolates into potable-groundwater aquifers as a result of irrigation, groundwater recharge, or other uses. Some inorganic and organic constituents and their potential significance in water reclamation and reuse are discussed as follows.

Biodegradable Organics. Biodegradable organics can create aesthetic and nuisance problems. Organics provide food for microorganics, adversely affect disinfection processes, make water unsuitable for some industrial or other uses, consume oxygen, and may cause acute or chronic health effects if reclaimed water is used for potable purposes.

Stable Organics. Some organic constituents tend to resist conventional methods of wastewater treatment. Some organic compounds are toxic in the environment, and their presence may limit the suitability of reclaimed water for irrigation or other uses.

Nutrients. Nitrogen, phosphorus, and potassium are essential nutrients for plant growth, and their presence normally enhances the value of the water for irrigation. When discharged to the aquatic environment, nitrogen and phosphorus can lead to the growth of undesirable aquatic life. When applied at excessive levels on land, the nitrate form of nitrogen will readily leach through the soil and may cause groundwater concentrations to exceed drinking-water standards.

Hydrogen Ion Concentration. The pH of wastewater affects disinfection efficiency, coagulation, metal solubility, and alkalinity of soils. Normal pH range in municipal wastewater is 6.5 to 8.5, but industrial wastes may have pH characteristics well outside of this range.

Heavy Metals. Some heavy metals accumulate in the environment and are toxic to plants and animals. Their presence in reclaimed water may limit the suitability of the water for irrigation or other purposes.

Dissolved Inorganics. Excessive salinity may damage some crops. Specific ions such as chloride, sodium, and boron are particularly toxic to some crops. Sodium may pose soil permeability problems.

Residual Chlorine. Excessive amounts of free available chlorine (>0.05 mg/L) may cause leaf-tip burn and damage some sensitive crops. However, most chlorine in reclaimed water is in a combined form, which does not cause crop damage (Asano, et al., 1985). The reaction of chlorine with organics in water creates a wide range of byproducts, some of which may be harmful to health when ingested over the long term. Inadvertent, infrequent ingestion of highly treated reclaimed water intended for nonpotable uses would present minimal health risks.

Suspended Solids. Organic constituents, heavy metals, and so forth are adsorbed on particulates. Suspended matter can shield microorganisms from disinfectants. Suspended solids can lead to sludge deposits and anaerobic conditions if discharged to the aquatic environment. Excessive amounts of solids cause clogging in irrigation systems or accumulate in soil and affect permeability.

The concentrations of inorganic constituents in reclaimed water depend mainly on the nature of the water supply, source of wastewater, and degree of treatment provided. Residential use of water typically adds about 300 mg/L of dissolved inorganic solids, although the amount added can range from approximately 150 mg/L to more than 500 mg/L (Metcalf & Eddy, 1991). The presence of total dissolved solids, nitrogen, phosphorus, heavy metals, and other inorganic constituents may affect the suitability of reclaimed water for different reuse applications. Wastewater treatment can reduce many trace elements to below recommended maximum levels for irrigation and other uses with existing conventional technology.

The health effects related to the presence of organic constituents are of primary concern with regard to potable reuse, and the release of volatile organic compounds (VOCs) may be a problem. Both organic and inorganic constituents need to be considered where reclaimed water is utilized for food-crop irrigation, where reclaimed water from irrigation or other beneficial uses reaches potable-groundwater supplies, or where organics may bioaccumulate in the food chain, e.g., in fish-rearing ponds.

21.4 WATER-QUALITY CONSIDERATIONS FOR REUSE APPLICATIONS

Information on water quality achievable via wastewater-treatment-unit processes is provided in Chap. 20. Recommended reclaimed-water microbiological limits for various applications are presented in Sec. 21.5 of this chapter.

21.4.1 Irrigation

Guidelines for evaluating irrigation-water quality are summarized in Table 21.6. EPA's recommended limits for heavy metals in irrigation water are presented in Table 21.7. The recommended maximum concentrations for long-term continuous use, i.e., more than 20 years, on all soils are set conservatively, to include sandy soils that have low capacity to react with the element in question. The criteria for short-term use, i.e., up to 20 years, are recommended for fine-textured neutral and alkaline soils with high capacities to remove the different pollutant elements (National Academy of Sciences–National Academy of Engineering, 1973).

Trace elements in reclaimed water normally occur in concentrations of less than a few mg/L, with usual concentrations less than 100 µg/L (Page and Chang, 1985). Some are essential for plants and animals, but many can become toxic at elevated concentrations or doses (Tanji, 1980). The mechanisms of potential food contamination from irrigation with reclaimed water include: physical contamination, where evaporation and repeated application may result in a build-up of contaminants on crops; uptake through the roots from the applied water or the soil; and foliar intake. Some chemical constituents are known to accumulate in particular crops, thus presenting potential health hazards to both grazing animals and/or humans. The concentrations of heavy metals and other trace elements in reclaimed water generally are much less than those in biosolids from wastewater-treatment plants, which also may be applied to agricultural land.

The elements of greatest concern at elevated levels are cadmium, copper, molybdenum, nickel, and zinc. Cadmium, copper, and molybdenum can be harmful to animals at concentrations too low to affect plants. Cadmium is of particular concern because it can accumulate in the food chain. It does not adversely affect ruminants in the small amounts they ingest. Most milk and beef products are unaffected by livestock ingestion of cadmium because it is stored in the liver and kidneys of the animal rather than the fat or muscle tissues. Copper is not toxic to monogastric ani-

TABLE 21.6 Guidelines for Interpretation of Water Quality for Irrigation

			Degree of restriction on use	
Potential irrigation problem	Units	None	Slight to moderate	Severe
Salinity (affects crop water availability)				
EC_w	dS/m	<0.7	0.7–3.0	>3.0
TDS	mg/L	<450	450–2000	>2000
Permeability (affects infiltrate rate of water into the soil. Evaluate using EC_w and SAR together)				
SAR = 0–3	and ECw =	<0.7	0.7–0.2	>0.2
= 3–6	=	<1.2	1.2–0.3	>0.3
= 6–12	=	<1.9	1.9–0.5	>0.5
= 12–20	=	<2.9	2.9–1.3	>1.3
= 20–40	=	<5.0	5.2–2.9	>2.9
Specific ion toxicity (affects sensitive crops)				
Sodium (Na)				
Surface irrigation	SAR	<3	3–9	>9
Sprinkler irrigation	mg/L	<70	>70	
Chloride (Cl)				
Surface irrigation	mg/L	<140	140–350	>350
Sprinkler irrigation	mg/L	<100	>100	
Boron (B)	mg/L	<0.7	0.7–3.0	>3.0
Miscellaneous effects (affects susceptible crops)				
Nitrogen (Total-N)	mg/L	<5	5–30	>30
Bicarbonate (HCO_3) (overhead sprinkling only)	mg/L	<90	90–500	>500
pH			Normal range 6.5–8.4	
Residual chlorine (overhead sprinkling only)	mg/L	<1.0	1.0–5.0	>5.0

Source: Adapted from Ayers and Westcot (1985); and University of California Committee of Consultants (1974).

mals, but may be toxic to ruminants; however, their tolerance to copper increases as available molybdenum increases (U.S. EPA, 1981). Molybdenum can also be toxic when available in the absence of copper. Nickel and zinc are a lesser concern than cadmium, copper, and molybdenum because they have visible adverse effects in plants at lower concentrations than the levels harmful to animals and humans. Zinc and nickel toxicity decrease as pH increases.

Crop uptake of certain pesticides has been studied (National Academy of Sciences–National Academy of Engineering, 1973; Palazzo, 1976). and uptake of polychlorinated biphenyls by root crops has been demonstrated under field conditions (Iwata and Gunther, 1976). Uptake of organic compounds is affected by: solubility, size, concentration, and polarity of the organic molecules; organic content, pH, and microbial activity of the soil; and climate. It has been postulated that most trace organic compounds are too large to pass through the semipermeable membrane of plant roots (U.S. EPA, 1981).

TABLE 21.7 Recommended Limits for Heavy Metals in Irrigation Water

Constituent	Long-term use, mg/L	Short-term use, mg/L	Remarks
Aluminum	5.0	20	Can cause nonproductivity in acid soils, but soils at pH 5.5 to 8.0 will precipitate the ion and eliminate toxicity.
Arsenic	0.10	2.0	Toxicity to plants varies widely, ranging from 12 mg/L for Sudan grass to less than 0.05 mg/L for rice.
Beryllium	0.10	0.5	Toxicity to plants varies, ranging from 5 mg/L for kale to 0.5 mg/L for bush beans.
Boron	0.75	2.0	Essential to plant growth, with optimum yields for many obtained at a few tenths mg/L in nutrient solutions. Toxic to many sensitive plants (e.g., citrus) at 1 mg/L. Usually sufficient quantities in reclaimed water to correct soil deficiencies. Most grasses relatively tolerant at 2.0 to 10 mg/L.
Cadmium	0.01	0.05	Toxic to beans, beets, and turnips at concentrations as low as 0.1 mg/L in nutrient solution. Conservative limits recommended.
Chromium	0.1	1.0	Not generally recognized as essential growth element. Conservative limits recommended due to lack of knowledge on toxicity to plants.
Cobalt	0.05	5.0	Toxic to tomato plants at 0.1 mg/L in nutrient solution. Tends to be inactivated by neutral and alkaline soils.
Copper	0.2	5.0	Toxic to a number of plants at 0.1 mg/L in nutrient solution.
Fluoride	1.0	15.0	Inactivated by neutral and alkaline soils.
Iron	5.0	20.0	Not toxic to plants in aerated soils, but can contribute to soil acidification and loss of essential phosphorus and molybdenum.
Lead	5.0	10.0	Can inhibit plant cell growth at very high concentrations.
Lithium	2.5	2.5	Tolerated by most crops at up to 5 mg/L; mobile in soil. Toxic to citrus at low doses—recommended limit is 0.075 mg/L.
Manganese	0.2	10.0	Toxic to a number of crops at a few tenths to a few mg/L in acid soils.
Molybdenum	0.01	0.05	Nontoxic to plants at normal concentrations in soil and water. Can be toxic to livestock if forage is grown in soils with high levels of available molybdenum.
Nickel	0.2	2.0	Toxic to a number of plants at 0.5 to 1.0 mg/L; reduced toxicity at neutral or alkaline pH.
Selenium	0.02	0.02	Toxic to plants at low concentrations and to livestock if forage is grown in soils with low levels of added selenium.
Tin, tungsten, and titanium	—	—	Effectively excluded by plants; specific tolerance levels unknown.
Vanadium	0.1	1.0	Toxic to many plants at relatively low concentrations.
Zinc	2.0	10.0	Toxic to many plants at widely varying concentrations; reduced toxicity at increased pH (6 or above) and in fine-textured or organic soils.

Source: Adapted from National Academy of Sciences–National Academy of Engineering (1973).

Free chlorine residual at concentrations less than 1 mg/L usually poses no problem to plants (U.S. EPA, 1992). However, some sensitive crops may be damaged at levels as low as 0.05 mg/L. Some woody crops may accumulate chlorine in the tissue to toxic levels. Excessive chlorine has a similar leaf-burning effect as sodium and chloride when sprayed directly on foliage. Chlorine at concentrations greater than 5 mg/L causes severe damage to most plants.

While excessive nitrogen stimulates vegetative growth in most crops, it may also delay maturity and reduce crop quality and quantity. Excessive nitrate in forages can cause an imbalance of nitrogen, potassium, and magnesium in grazing animals and is a concern if forage is used as a primary feed source for livestock. The addition of potassium with reclaimed water has little effect on the crop.

Aside from public health concerns associated with reclaimed-water reuse, salinity is one of the most important parameters in determining the suitability of water for irrigation. Salinity may influence the soil's osmotic potential, specific ion toxicity, and degradation of soil physical conditions. These conditions may result in reduced plant-growth rates, reduced yields, and, in severe cases, total crop failure. Highly saline water applied via overhead sprinklers results in direct sorption of sodium and/or chloride and can cause leaf injury, particularly during periods of high temperature and low humidity.

Landscape irrigation involves the irrigation of golf courses, parks, cemeteries, school grounds, freeway medians, residential lawns, and similar areas. The concern for pathogenic microorganisms is somewhat different than for agricultural irrigation in that landscape irrigation frequently takes place in urban areas where control over the use of the reclaimed water is more critical. Depending on the area being irrigated, its location relative to populated areas, and the extent of public access or use of the grounds, the microbiological requirements and operational controls placed on the system may differ. Except for TDS, the chemical composition of the irrigation water is not usually limiting. Both agricultural and landscape irrigation with reclaimed water is well-accepted and widely practiced in the United States.

Operational and seasonal storage are important factors in reclaimed-water systems, particularly irrigation systems and dual-water systems where there may be a variety of intermittent or seasonal uses. Often, on-site ponds, such as golf course water hazards, are used to reduce operational-storage requirements, but open-pond storage degrades water quality prior to reuse of the water. Where evaporation rates are high and rainfall is low, shallow ponds with a high area-to-volume ratio show significant increases in dissolved solids concentrations (Chapman and French, 1991).

In St. Petersburg, Florida, the absence of seasonal storage for reclaimed water used in the dual-water system results in a decreased ability to meet peak seasonal demands and permits a reuse commitment of less than 50 percent of the available reclaimed water (Crook and Johnson, 1991). At the Irvine Ranch Water District in Orange County, California, seasonal storage of reclaimed water is provided in large open reservoirs. Prior to introduction into the reclaimed-water distribution network, the water is filtered and chlorinated. Diurnal variations in demand are met by covered tanks near the reservoirs.

Where storage is intended to prevent any surface discharge of the water, the storage capacity needed depends on several factors, including climatic conditions and types and quantities of reclaimed-water usage. Storage periods exceeding 100 days may be required to prevent surface discharges (Clow, 1992; Mullarkey and Hall, 1990; U.S. EPA, 1992). Subsurface storage of reclaimed water in nonpotable aquifers is an attractive method of seasonal storage because many sites are available, evaporaton losses are eliminated, and degradation of the water is not likely to occur.

21.4.2   Dual Systems

Increasing use of reclaimed water in urban areas has resulted in the development of several large dual-water systems that distribute two grades of water to the same service area—one potable and the other nonpotable reclaimed water. Many regulatory agencies prohibit hose bibbs on reclaimed-water-distribution systems and, at use areas that receive both potable and reclaimed water, require backflow-prevention devices on the potable-water supply line to each site. A dual-water system can provide water for most of the uses listed in Table 21.1. Detailed information on the planning, design, construction, operation, and management of dual-water systems is available in several publications (California-Nevada Section AWWA, 1992; Crook, et al., 1994; U.S. EPA, 1992; WPCF, 1989).

Dual-water systems that make reclaimed water available throughout a community for irrigation and other uses where significant portions of the population will be exposed to the reclaimed water should be microbiologically safe such that inadvertent contact or ingestion does not constitute a health hazard. Chemical constituents in tertiary-treated reclaimed water are generally not a problem for most types of nonpotable urban reuse, but, if necessary, can be removed by specific advanced wastewater-treatment-unit processes.

21.4.3   Industrial Reuse

Typical industrial uses of reclaimed water are listed in Table 21.1. Some of the principal considerations regarding the use of reclaimed water in industrial applications are discussed in the following sections.

21.4.3.1   Cooling Water.   A health concern associated with industrial reuse is pathogens in the water that may pose hazards to workers and to the public in the vicinity of cooling towers. Aerosols produced in the workplace or from cooling towers also may present hazards from the inhalation of volatile organic chemicals, although little definitive research has been done in this area. Closed-loop cooling systems using reclaimed water present minimal health concerns unless there is inadvertent or intentional misuse of the water. Table 21.8 lists suggested chemical water-quality criteria for cooling-water supplies.

Cooling water should not lead to the formation of scale, which reduces the efficiency of the heat exchange. The principal causes of scaling are calcium (as carbonate, sulfate, and phosphate) and magnesium (as carbonate and phosphate) deposits. Scale control for reclaimed water is achieved through chemical means and sedimentation. Acidification or addition of scale inhibitors can control scaling. Acids (sulfuric, hydrochloric, and citric acids and acid gases such as carbon dioxide and sulfur dioxide) and other chemicals (chelants such as EDTA and polymeric inorganic phosphates) are often added to increase the water solubility of scale-forming constituents, such as calcium and magnesium (Strauss and Puckorius, 1984). Lime softening removes carbonate hardness and soda ash removes noncarbonate hardness. Other methods used to control scaling are alum treatment and sodium-ion exchange.

High levels of dissolved solids, ammonia, and heavy metals in reclaimed water can cause increased corrosion rates (Puckorius and Hess, 1991). TDS in reclaimed water can increase electrical conductivity and promote corrosion. Ammonia is very aggressive to copper alloys. Dissolved gases and certain metals with high oxidation states also promote corrosion. Heavy metals, particularly copper, can plate out on

TABLE 21.8　Recommended Cooling-Water-Quality Criteria for Make-Up Water to Recirculating Systems

Parameter*	Recommended limit
Cl	500
TDS	500
Hardness	650
Alkalinity	350
pH	6.9–9.0
COD	75
TSS	100
Turbidity	50
BOD	25
Organics[†]	1.0
NH_4-N	1.0
PO_4	4
SiO_2	50
Al	0.1
Fe	0.5
Mn	0.5
Ca	50
Mg	0.5
HCO_3	24
SO_4	200

* All values in mg/L except pH.
[†] Methylene blue active substances.
Source: Adapted from Goldstein, et al. (1979); and WPCF (1989).

mild steel, causing severe pitting. Corrosion also may occur when acidic conditions develop in cooling water. Corrosion inhibitors such as chromates, polyphosphates, zinc, and polysilicates can be used to reduce the corrosion potential of cooling water. These substances may have to be removed from the blowdown prior to discharge. An alternative to chemical addition is ion exchange or reverse osmosis (Strauss and Puckorius, 1984).

The moist environment in cooling towers is conducive to biological growth. Microorganisms can significantly reduce the heat-transfer efficiency, reduce water flow, and in some cases generate corrosive by-products (California State Water Resources Control Board, 1980; Fannin, et al., 1980; Troscinski and Watson, 1970). Sulfide-producing bacteria and sulfate-reducing bacteria are the most common corrosion-causing organisms in cooling systems using reclaimed water (Puckorius and Hess, 1991). These anaerobic sulfide producers cause pitting corrosion that is most severe on mild and stainless steels. Serious corrosion is caused by thiobaccillus bacteria, which convert sulfides to sulfuric acid. Similarly, nitrifying bacteria can convert ammonia to nitric acid, thus causing pH depression, which increases corrosion on most metals (Puckorius and Hess, 1991).

Legionella pneumophila, the bacterial agent that causes Legionnaire's Disease, is known to proliferate in air-conditioning cooling-water systems under certain conditions. There is no indication that reclaimed water is more likely to contain *Legionella* bacteria than waters of nonsewage origin.

Chlorine is the most common biocide used to control biological growth and is used as a disinfectant to reduce potential pathogens in reclaimed water. Nonoxidizing microbiocides are often needed in addition to chlorine because of the high nutrient content typically found in wastewater. Since most scale inhibitors and dispersants are anionic, either anionic or nonionic biocides are usually used. Low-foaming, nonionic surfactants enhance microbiological control by allowing microbiocides to penetrate the biological slimes (Puckorius and Hess, 1991).

Fouling is controlled by preventing the formation and settling of particulate matter. Chemical coagulation and filtration during the phosphorus-removal treatment phase significantly reduce levels of contaminants that can lead to fouling. Chemical dispersants are also used as required.

21.4.3.2 Boiler-Feed Water.   The use of reclaimed water differs little from the use of conventional public supplies for boiler-feed water; both usually require extensive additional treatment. Quality requirements for boiler-feed makeup water are also dependent on the pressure at which boilers are operated, as shown in Table 21.9. Generally, the higher the pressure, the higher the quality of water required. Very high pressure boilers require makeup water of distilled quality (Selby and Brooke, 1990). High alkalinity may contribute to foaming, resulting in deposits in super-

TABLE 21.9   Recommended Industrial Boiler-Feed Water-Quality Criteria

Parameter*	Low pressure, <150 psig	Intermediate pressure, 150–700 psig	High pressure, >700 psig
Silica	30	10	0.7
Aluminum	5	0.1	0.01
Iron	1	0.3	0.05
Manganese	0.3	0.1	0.01
Calcium	†	0.4	0.01
Magnesium	†	0.25	0.01
Ammonia	0.1	0.1	0.1
Bicarbonate	170	120	48
Sulfate	†	†	†
Chloride	†	†	†
Dissolved solids	700	500	200
Copper	0.5	0.05	0.05
Zinc	†	0.01	0.01
Hardness	350	1.0	0.07
Alkalinity	350	100	40
pH, units	7.0–10.0	8.2–10.0	8.2–9.0
Methylene blue active substances	1	1	0.5
Carbon tetrachloride extract	1	1	0.5
Chemical oxygen demand	5	5	1.0
Hydrogen sulfide	†	†	†
Dissolved oxygen	2.5	0.007	0.0007
Temperature, °F	†	†	†
Suspended solids	10	5	0.5

* Recommended limits in mg/L except for pH (units) and temperature (degrees Fahrenheit).
† Accepted as received (if meeting other limiting values); has never been a problem at concentrations encountered.
Source: Adapted from U.S. EPA (1980).

heaters, reheaters, and turbines. Bicarbonate alkalinity, under the influence of boiler heat, may lead to the release of carbon dioxide, which is a source of corrosion in steam-using equipment.

In general, both potable water and reclaimed water used for high-pressure-boiler water must be treated to reduce hardness to nearly zero. Removal or control of insoluble salts of calcium and magnesium and control of silica and aluminum are required, since these are the principal causes of scale build-up in boilers. Depending on the characteristics of the reclaimed water, lime treatment (including flocculation, sedimentation, and recarbonation) may be required, possibly followed by multi-media filtration, carbon adsorption, and nitrogen removal. High-purity boiler-feed water for high-pressure boilers also may require treatment by reverse osmosis or ion exchange (Meyer, 1991).

21.4.3.3 *Industrial-Process Water.*

The suitability of reclaimed water for use in industrial processes depends on the particular use. Table 21.10 presents industrial-process water-quality requirements for a variety of industries.

Use of reclaimed water in the paper and pulp industry is a function of the grade of paper produced. The higher the quality of the paper, the more sensitive it is to water quality. Impurities found in water, particularly certain metal ions and color bodies, can cause paper to change color with age. Biological growth can cause clogging of equipment and odors and can affect the texture and uniformity of the paper. Corrosion and scaling of equipment may result from the presence of silica, aluminum, and hardness. Discoloration of paper may occur due to iron, manganese, or microorganisms. Suspended solids may decrease the brightness of the paper (Camp Dresser & McKee, 1982).

Reclaimed water is used in the manufacture of a wide variety of paper products, ranging from kraft pulp newsprint to high-quality paper for stationery and wrappings. Reclaimed water used in the manufacture of paper products used as food wrap or beverage containers would have to be pathogen free and not contain any health-hazardous contaminants that could leach into the consumable products.

TABLE 21.10 Industrial-Process Water-Quality Requirements

| | Pulp and paper | | | | | Textile | | |
Parameter*	Mechanical pulping	Chemical, unbleached	Pulp and paper, bleached	Chemical	Petroleum and coal	Sizing suspension	Scouring, bleach and dye	Cement
Cu					0.05	0.01		
Fe	0.3	1.0	0.1	0.1	1.0	0.3	0.1	2.5
Mn	0.1	0.5	0.05	0.1		0.05	0.01	0.5
Ca		20	20	68	75			
Mg		12	12	19	30			
Cl	1,000	200	200	500	300			250
HCO$_3$				128				
NO$_3$				5				
SO$_4$				100				250
SiO$_2$		50	50	50				35
Hardness		100	100	250	350	25	25	
Alkalinity					125			400
TDS				1,000	1,000	100	100	600
TSS		10	10	5	10	5	5	500
Color	30	30	10	20		5	5	
pH	6–10	6–10	6–10	6.2–8.3	6–9			6.5–8.5
CCE								1

* All values in mg/L except color and pH.
Source: WPCF (1989).

Water used in textile manufacturing must be nonstaining; hence, it should be low in turbidity, color, iron, and manganese. Hardness causes curds to deposit on textiles and causes problems in some of the processes that use soap. Nitrates and nitrites may cause problems in dyeing.

21.4.4 Groundwater Recharge

The purposes of groundwater recharge using reclaimed water can include establishing saltwater-intrusion barriers in coastal aquifers, providing further soil-aquifer treatment (SAT) for future reuse, providing storage of reclaimed water, controlling or preventing ground subsidence, and augmenting potable or nonpotable aquifers.

21.4.4.1 Surface Spreading. In most cases wastewater receives at least secondary treatment and disinfection, and often tertiary treatment, prior to surface spreading, although primary effluent has been successfully used in soil-aquifer treatment systems at some spreading sites where the extracted water is to be used for nonpotable purposes (Carlson, et al., 1982; Rice and Bouwer, 1984). A disadvantage of using primary effluent is that infiltration-basin hydraulic-loading rates may be lower.

Algae can clog the soil surface of spreading basins and reduce infiltration rates. Algae further aggravate soil clogging by removing carbon dioxide, which raises the pH, causing precipitation of calcium carbonate. Reducing the detention time of standing water within the basins minimizes algal growth. Infiltration basins should be shallow enough to avoid compaction of the clogging layer (Bouwer and Rice, 1989). Scarifying, rototilling, or discing the soil following the drying cycle can help alleviate clogging potential, although scraping the bottom to remove the clogging layer is more effective.

Contaminants in the subsurface environment are subject to biodegradation by microorganisms, adsorption, filtration, ion exchange, volatilization, dilution, chemical oxidation and reduction, and chemical precipitation and complex formation (Roberts, 1980; U.S. EPA, 1989). For surface spreading operations, most of the removals of both chemical and microbiological constituents occur in the top 6 ft (2 m) of the vadose zone at the spreading site.

Particles larger than the soil pores are strained off at the soil–water interface. Particulate matter, including some bacteria, is removed by sedimentation in the pore spaces of the media during filtration. Viruses are removed mainly by adsorption. The accumulated particles gradually form a layer restricting further infiltration. Suspended solids that are not retained at the soil–water interface may be effectively removed by infiltration and adsorption in the soil profile. As water flows through passages formed by the soil particles, suspended and colloidal solids too small to be retained by straining are intercepted and adsorbed onto the surface of the stationary soil matrix through hydrodynamic actions, diffusion, impingement, and sedimentation.

Some inorganic constituents such as chloride, sodium, and sulfate are unaffected by ground passage, but there can be substantial removal of other inorganic constituents. Iron and phosphorus removals in excess of 90 percent have been achieved by precipitation and adsorption in the underground (Idelovitch, et al., 1980; Sontheimer, 1980), although the ability of the soil to remove these and other constituents may decrease over time. Heavy-metal removal varies widely for the different elements, ranging from 0 to more than 90 percent, depending on speciation of the influent metals.

Some trace metals, e.g., silver, chromium, fluoride, molybdenum, and selenium, are strongly retained by soil (Chang and Page, 1979; John, 1972). There are indications that once metals are adsorbed, they are not readily desorbed, although desorption depends, in part, on buffer capacity, salt concentrations, and reduction/oxidation potential (Sontheimer, 1980). Boron, which is mainly in the form of undissociated boric acid in soil solutions, is rather weakly adsorbed and, given sufficient amounts of leaching water, most of the adsorbed boron is desorbed (Rhoades, et al., 1979).

For surface spreading operations where an aerobic zone is maintained, ammonia is effectively converted to nitrates, but subsequent denitrification is dependent, in part, on anaerobic conditions during the flooding cycle and is often partial and fluctuating unless the system is carefully managed. Adsorption of organic constituents retards their movement and attenuates concentration fluctuations. The degree of attenuation increases with increasing adsorption strength, increasing distance from the recharge point, and increasing frequency of input fluctuation (Roberts, 1980). Some chemical constituents can desorb and move chromatographically in the underground.

21.4.4.2 Injection. Direct injection involves pumping reclaimed water directly into the groundwater zone, which is usually a confined aquifer. Injection requires water of higher quality than surface spreading to prevent clogging because of the absence of soil-matrix treatment afforded by surface spreading, and the potential requirement to have the injection water meet drinking-water standards or match or exceed the quality of the groundwater supply. Treatment processes beyond secondary treatment that may be used prior to injection include disinfection, filtration, air stripping, ion exchange, granular activated carbon, and reverse osmosis or other membrane-separation processes. With various subsets of these processes in appropriate combinations, it is possible to satisfy the full range of water-quality requirements for injection.

Clogging of injection wells can be caused by accumulation of organic and inorganic solids, biological and chemical contaminants, and dissolved air and gases from turbulence. Concentrations of suspended solids of 1 mg/L or greater can clog an injection well. Low concentrations of organic contaminants can cause clogging due to bacteriological growth near the point of injection.

21.4.5 Recreational and Environmental Uses

Impoundments may serve a variety of functions from aesthetic, noncontact uses, to boating, fishing, and swimming. As with other uses of reclaimed water, the level of treatment required will vary with the intended use of the water. Required treatment levels increase as the potential for human contact increases. The appearance of the reclaimed water is important when it is used for impoundments, and treatment for nutrient removal may be required. Without nutrient control there is a potential for algae blooms, resulting in odors, an unsightly appearance, and eutrophic conditions.

Stream augmentation is different from surface-water discharge in that augmentation seeks to accomplish a beneficial end, whereas surface discharge is a disposal alternative. As with impoundments, the water-quality requirements for stream augmentation are based on the designated use of the stream and maintenance of required water-quality standards. In addition, there may be an emphasis on creating a product that improves existing stream quality to sustain or enhance aquatic life. To achieve aesthetic goals, both nutrient removal and high-level disinfection are often needed. Dechlorination may be required to protect aquatic wildlife where chlorine is used as the wastewater disinfectant.

The primary intent in applying reclaimed water to wetlands is to provide additional treatment of effluent prior to discharge, although wetlands are sometimes created solely for environmental enhancement. Information on reclaimed-water use in wetlands is provided in Chap. 13.

21.4.6 Indirect Potable Reuse

Indirect potable reuse, where treated wastewater is discharged into a water course, a reservoir, or an underground aquifer and withdrawn downstream or downgradient at a later time for potable purposes, is widely practiced, if often inadvertently. Ultimately, the water must meet all physical, chemical, radiological, and microbiological drinking-water standards, which are discussed in Chap. 8. In recognition of the public health considerations associated with potable reuse, brief descriptions of relevant planned indirect potable-reuse projects are given below.

21.4.6.1 Groundwater Recharge. Reclaimed water has been used to recharge potable-groundwater aquifers by surface spreading since 1962 in Los Angeles County, California. In 1993, 45 mgd (170,000 m³/d) of highly disinfected tertiary (secondary treatment plus filtration) effluent from three wastewater-reclamation plants operated by the Sanitation Districts of Los Angeles County were spread at Whittier Narrows in the Montebello Forebay area of Los Angeles County. Reclaimed water accounts for more than 30 percent of the inflow into the basin (Crook, et al., 1990). Local stormwater runoff and imported surface water are also used for recharge.

A five-year health-effects study initiated in 1978 did not demonstrate any measurable adverse effects on the area's groundwater or the health of the population ingesting the water. The reclaimed water and groundwater complied with all federally prescribed drinking-water standards, and no pathogenic organisms were detected in either the reclaimed water or groundwater (Nellor, et al., 1984). The cancer-related epidemiological study findings are somewhat weakened by the fact that the minimum observed latency period for human cancers that have been linked to chemical agents is about 15 years. Due to the relatively short time period that groundwater containing reclaimed water had been consumed, it is unlikely that any positive effects of exposure to the reclaimed water would have been detected (State of California, 1987).

As of 1994 there were two projects in the United States involving groundwater recharge of reclaimed water by injection. Since 1976, reclaimed water produced at the Orange County Water District's Water Factory 21 in Fountain Valley, California, has been injected into a series of 23 injection wells to form a seawater-intrusion barrier. The 15 mgd (60,000 m³/d) facility receives secondary effluent and provides additional treatment by lime clarification, recarbonation, mixed-media filtration, granular activated carbon (two-thirds of the flow), reverse osmosis (one-third of the flow), and chlorination. COD concentrations averaged 8 mg/L in 1988, and TOC averaged 2.6 mg/L. The average turbidity of filter effluent was 0.22 FTU and did not exceed one FTU during 1988. Virus sampling indicated that the effluent is essentially free of measurable levels of viruses (McCarty, et al., 1982). The product water is blended 2:1 with deep-well water prior to injection. Once underground, some of the injected water flows toward the ocean forming the seawater barrier, but the majority of the water flows into the groundwater basin to augment the potable-groundwater supply (Argo and Cline, 1985).

A direct-injection project has been in operation in El Paso, Texas, since 1985, supplying more than 5 mgd (20,000 m³/d) of reclaimed water to the underground. The treatment processes at the Fred Hervey Water Reclamation Plant are similar to those at Water Factory 21. Ultimately, the treated wastewater returns to the city's potable-water system. While the reclaimed water currently recharged represents a small percentage of the total aquifer volume, the long-term goal is to provide 25 percent of El Paso's future water needs with reclaimed water (Knorr, et al., 1988).

21.4.6.2   Surface-Water Augmentation.   The Upper Occoquan Sewage Authority (UOSA) water-reclamation plant in Virginia discharges its effluent to Bull Run for indirect potable reuse. The discharge point is about 20 miles upstream of the water-supply intake. The treatment unit processes include primary and secondary treatment followed by five advanced waste-treatment processes: chemical clarification and two-stage recarbonation with intermediate settling; multimedia filtration; activated-carbon adsorption; ion exchange for nitrogen removal; and breakpoint chlorination. Typical constituent concentrations in the UOSA plant product water are as follows: BOD <1 mg/L; COD 5.1 to 17 mg/L; TOC 2.0 to 5.4 mg/L; turbidity 0.15 to 2.0 NTU; and total reactive phosphorus 0.01 to 0.35 mg/L (Robbins, 1993; Robbins and Ehalt, 1985).

Indirect potable reuse via surface-water augmentation has been evaluated at Tampa, Florida. Source water to the pilot facilities was undisinfected, denitrified, filtered, secondary effluent. Additional treatment included preaeration, two-stage lime clarification and recarbonation, filtration, granular-activated-carbon adsorption, and disinfection using ozone. The product water met National Primary and Secondary Drinking Water Regulations for inorganic and chemical parameters. Less than one coliform organism per 100 mL was detected in the product water, and *Giardia* and *Cryptosporidium* were not detected. No viruses were detected after granular-activated-carbon treatment and ozonation. All samples subjected to toxicological testing were negative (CH2M Hill, 1993).

San Diego, California, investigated the feasibility of using an innovative wastewater-treatment system utilizing channels containing water hyacinths for secondary treatment, followed by AWT to upgrade the secondary effluent to a quality that could be sufficient to serve as a raw-water source. The AWT demonstration plant included chemical coagulation, filtration, reverse osmosis, air stripping, carbon adsorption, and disinfection by ultraviolet radiation. A health-effects study indicated that the water quality and health risk associated with the use of AWT as a raw-water supply was less than or equal to that of the existing city raw water. The concentrations of parameters for which drinking-water standards exist were below those standards in the AWT effluent (Western Consortium for Public Health, 1992).

While it may be technically possible to produce reclaimed water of almost any desired quality, some health authorities and others have been reluctant to allow or support the augmentation of groundwater or surface-water supplies with reclaimed water and generally subscribe to the concept of using natural waters derived from the most protected source as raw-water supplies. Public health issues notwithstanding, indirect potable reuse is receiving increasing attention.

21.4.7   Direct Potable Reuse

Although direct potable reuse is not practiced anywhere in the United States, extensive research focusing on direct potable reuse was conducted at Denver, Colorado, over a 10-year period. The treatment process train used during the Denver Direct Potable Reuse Demonstration Project included undisinfected secondary effluent as the source water to the AWT pilot facilities, followed by high pH lime clarification, recarbonation, multimedia filtration, ultraviolet-radiation disinfection, granular-activated-carbon adsorption, reverse osmosis, air stripping, ozonation, and chloramination. The resulting product water was reported to be of a higher quality than that drawn from many sources of potable water in the region (Lauer, 1993). Toxicological testing did not reveal any adverse health effects (Rogers and Lauer, 1986).

21.5 WATER RECLAMATION AND REUSE CRITERIA

There are no federal regulations governing water reclamation and reuse in the United States; hence, the regulatory burden rests with the individual states. This has resulted in widely differing standards among states that have developed reuse criteria. No states have regulations that cover all potential uses of reclaimed water, but several states have extensive regulations or guidelines that prescribe requirements for a wide range of uses. Other states have regulations or guidelines that focus on land treatment of wastewater effluent, emphasizing additional treatment or effluent disposal rather than beneficial reuse, even though the effluent may be used for irrigation of agricultural sites, golf courses, or public-access lands. In 1992, 18 states had some form of water-reuse regulations, 18 states had guidelines, and 14 states had neither regulations nor guidelines (U.S. EPA, 1992).

Water-quality criteria are based on a variety of considerations:

Public health protection.   Reclaimed water should be safe for the intended use. Most existing reclaimed-water regulations are directed principally at public health protection, and many address only microbiological concerns. As of 1994, no state reclaimed-water microbiological-quality criteria were based on rigorous risk-assessment determinations.

Use requirements.   Many industrial uses and some other applications have specific physical and chemical water-quality requirements that are not related to health considerations. The physical, chemical, and microbiological quality may all limit the acceptability of reclaimed water for specific uses.

Irrigation effects.   The effect of individual constituents or parameters on crops or other vegetation, soil, and groundwater or other receiving water should be evaluated for potential reclaimed-water-irrigation applications.

Environmental considerations.   The natural flora and fauna in and around reclaimed-water-use areas and receiving waters should not be adversely impacted by the reclaimed water.

Aesthetics.   For high level uses, e.g., urban irrigation and toilet flushing, reclaimed water should be nonstaining and no different in appearance than potable water, i.e., clear, colorless, and odorless. For recreational impoundments, reclaimed water should not promote algal growth.

Public and/or user perception.   The water should be perceived as being safe and acceptable for the intended use. This could result in the imposition of conservative reclaimed-water-quality limits by regulatory agencies.

States which have water-reuse regulations or guidelines typically have set standards for reclaimed-water quality and/or specified minimum-treatment requirements. The most common parameters for which water-quality limits are imposed are biochemical oxygen demand (BOD), turbidity or total suspended solids (TSS), total or fecal coliform bacteria, nitrogen, and chlorine residual and contact time.

Where reclaimed water is used in an urban setting, most states require a high degree of treatment and disinfection. Where there is likely to be public contact with the reclaimed water, tertiary treatment to produce finished water that is essentially pathogen-free is typically required (Florida Department of Environmental Regulation, 1990; State of Arizona, 1987; State of California, 1978).

Reclaimed water used inside buildings for toilet and urinal flushing or for fire protection presents cross-connection control concerns. Although such uses do not result in frequent human contact with the water, regulatory agencies usually require that the reclaimed water be pathogen-free to reduce health hazards upon inadvertant cross-connection to potable-water systems (Florida Department of Environmental Regulation, 1990; Pawlowski, 1992).

Several states with active reuse programs, e.g., Arizona, California, Florida, and Texas, have comprehensive regulations that prescribe wastewater-treatment process and/or reclaimed-water-quality requirements according to the end use of the water and include treatment-reliability requirements, operational requirements, and use-area requirements. There are differences among these states' requirements, such as: California uses total coliform as the indicator organism, while the other three states use fecal coliform; Florida is the only one of the four states that requires monitoring for total suspended solids to determine particulate levels—the other four states use turbidity; California and Florida prescribe treatment processes in addition to water-quality limits, while Arizona and Texas do not specify treatment processes, although Arizona is considering the inclusion of treatment-process requirements in the state's reuse regulations (Pawlowski, 1992); and Arizona and California permit the use of reclaimed water for spray irrigation of food crops eaten raw, while such use is prohibited in Florida and Texas. EPA's *Guidelines for Water Reuse* (1992) summarize all state reuse criteria, including use-area requirements addressing items such as setback distances, cross-connection control, and public notification.

21.5.1 EPA Guidelines

EPA, in conjunction with the U.S. Agency for International Development, published *Guidelines for Water Reuse* in 1992. The guidelines address all important aspects of water reuse, including recommended treatment processes, reclaimed-water-quality limits, monitoring frequencies, setback distances, and other controls for various reuse applications. The recommended treatment, water quality, monitoring, and setbacks are presented in Table 21.11.

Both wastewater treatment and reclaimed-water-quality limits are recommended, for the following reasons: water-quality criteria involving surrogate parameters do not adequately characterize reclaimed-water quality; a combination of treatment and quality requirements known to produce reclaimed water of acceptable quality obviate the need to monitor the finished water for certain constituents; expensive, time consuming, and in some cases, questionable monitoring for pathogenic microorganisms is eliminated without compromising health protection; and treatment reliability is enhanced (U.S. EPA, 1992).

The guidelines include limits for fecal coliform organisms but do not include parasite or virus limits. Parasites have not been shown to be a problem at reuse operations in the United States at the treatment levels and reclaimed-water limits recommended in the guidelines, although definitive information on the presence and significance of the parasites *Giardia* and *Cryptosporidium* in reclaimed water is lacking. While viruses are a concern in reclaimed water, virus limits are not recommended in the guidelines for the following reasons: a significant body of information

TABLE 21.11 EPA Suggested Guidelines for Reuse of Municipal Wastewater

Types of reuse	Treatment	Reclaimed-water quality[a]	Reclaimed-water monitoring	Setback distance[b]	Comments
Urban reuse All types of land-scape irrigation (e.g., golf courses, parks, cemeteries) —also vehicle washing, toilet flushing, use in fire-protection systems and commercial air conditioners, and other uses with similar access or exposure to the water	Secondary[c] Filtration[d] Disinfection[e]	pH = 6–9 ≤10 mg/L BOD[f] ≤2 NTU[g] No detectable fecal coli/100 mL[h,i] 1 mg/L Cl$_2$ residual (min)[j]	pH—weekly BOD—weekly Turbidity—continuous Coliform—daily Cl$_2$ residual—continuous	50 ft (15 m) to potable-water-supply wells	Consult recommended agricultural (crop) limits for metals. At controlled-access irrigation sites where design and operational measures significantly reduce the potential of public contact with reclaimed water, a lower level of treatment, e.g., secondary treatment and disinfection to achieve ≤14 fecal coli/100 mL, may be appropriate. Chemical (coagulant and/or polymer) addition prior to filtration may be necessary to meet water-quality recommendations. The reclaimed water should not contain measurable levels of pathogens.[k] Reclaimed water should be clear, odorless, and contain no substances that are toxic upon ingestion. A higher chlorine residual and/or a longer contact time may be necessary to assure that viruses and parasites are inactivated or destroyed. A chlorine residual of 0.5 mg/L or greater in the distribution system is recommended to reduce odors, slime, and bacterial regrowth. Provide treatment reliability.
Restricted access area irrigation Sod farms, silviculture sites, and other areas where public access is prohibited, restricted, or infrequent	Secondary[c] Disinfection[e]	pH = 6–9 ≤30 mg/L BOD[f] ≤30 mg/L SS ≤200 fecal coli/100 mL[h,l,m] 1 mg/L Cl$_2$ residual (min)[j]	pH—weekly BOD—weekly SS—daily Coliform—daily Cl$_2$ residual—continuous	300 ft (90 m) to potable-water-supply wells 100 ft (30 m) to areas accessible to the public (if spray irrigation)	Consult recommended agricultural (crop) limits for metals. If spray irrigation, SS less than 30 mg/L may be necessary to avoid clogging of sprinkler heads. Provide treatment reliability.

21.25

TABLE 21.11 EPA Suggested Guidelines for Reuse of Municipal Wastewater (*Continued*)

Types of reuse	Treatment	Reclaimed-water quality[a]	Reclaimed-water monitoring	Setback distance[b]	Comments
Agricultural reuse —Food crops not commercially processed[a] Surface or spray irrigation of any food crop, including crops eaten raw	Secondary[c] Filtration[d] Disinfection[e]	pH = 6–9 ≤10 mg/L BOD[f] ≤2 NTU[g] No detectable fecal coli/100 mL[h,i] 1 mg/L Cl_2 residual (min)[j]	pH—weekly BOD—weekly Turbidity—continuous Coliform—daily Cl_2 residual—continuous	50 ft (15 m) to potable-water-supply wells	Consult recommended agricultural (crop) limits for metals. Chemical (coagulant and/or polymer) addition prior to filtration may be necessary to meet water quality recommendations. The reclaimed water should not contain measurable levels of pathogens.[k] A higher chlorine residual and/or a longer contact time may be necessary to assure that viruses and parasites are inactivated or destroyed. High nutrient levels may adversely affect some crops during certain growth stages. Provide treatment reliability.
Agricultural reuse —Food crops commercially processed[a] surface irrigation of orchards and vineyards	Secondary[c] Disinfection[e]	pH = 6–9 ≤30 mg/L BOD[f] ≤30 mg/L SS ≤200 fecal coli/100 mL[h,l,m] 1 mg/L Cl_2 residual (min)[j]	pH—weekly BOD—weekly SS—daily Coliform—daily Cl_2 residual—continuous	300 ft (90 m) to potable-water-supply wells 100 ft (30 m) to areas accessible to the public	Consult recommended agricultural (crop) limits for metals. If spray irrigation, SS less than 30 mg/L may be necessary to avoid clogging of sprinkler heads. High nutrient levels may adversely affect some crops during certain growth stages. Provide treatment reliability.
Agricultural reuse —Nonfood crops Pasture for milking animals; fodder, fiber, and seed crops	Secondary[c] Disinfection[e]	pH = 6–9 ≤30 mg/L BOD[f] ≤30 mg/L SS ≤200 fecal coli/100 mL[h,l,m] 1 mg/L Cl_2 residual (min)[j]	pH—weekly BOD—weekly SS—daily Coliform—daily Cl_2 residual—continuous	300 ft (90 m) to potable-water-supply wells 100 ft (30 m) to areas accessible to the public (if spray irrigation)	Consult recommended agricultural (crop) limits for metals. If spray irrigation, SS less than 30 mg/L may be necessary to avoid clogging of sprinkler heads. High nutrient levels may adversely affect some crops during certain growth stages. Milking animals should be prohibited from grazing for 15 days after irrigation ceases. A higher level of disinfection, e.g., to achieve ≤14 fecal coli/100 mL, should be provided if this waiting period is not adhered to. Provide treatment reliability.

Types of reuse	Treatment	Reclaimed water quality	Reclaimed water monitoring	Setback distances	Comments
Recreational impoundments Incidental contact (e.g., fishing and boating) and full body contact with reclaimed water allowed	Secondary[c] Filtration[d] Disinfection[e]	pH = 6–9 ≤10 mg/L BOD[f] ≤2 NTU[g] No detectable fecal coli/100 mL[h,i] 1 mg/L Cl$_2$ residual (min)[j]	pH—weekly BOD—weekly Turbidity—continuous Coliform—daily Cl$_2$ residual—continuous	500 ft (150 m) to potable-water-supply wells (minimum) if bottom not sealed	Dechlorination may be necessary to protect aquatic species of flora and fauna. Reclaimed water should be nonirritating to skin and eyes. Reclaimed water should be clear, odorless, and contain no substances that are toxic upon ingestion. Nutrient removal may be necessary to avoid algae growth in impoundments. Chemical (coagulant and/or polymer) addition prior to filtration may be necessary to meet water-quality recommendations. The reclaimed water should not contain measurable levels of pathogens.[k] A higher chlorine residual and/or a longer contact time may be necessary to assure that viruses and parasites are inactivated or destroyed. Fish caught in impoundments can be consumed. Provide treatment reliability.
Landscape impoundments Aesthetic impoundment where public contact with reclaimed water is not allowed	Secondary[c] Disinfection[e]	≤30 mg/L BOD[f] ≤30 mg/L SS ≤200 fecal coli/100 mL[l,m,n] 1 mg/L Cl$_2$ residual (min)[j]	pH—weekly SS—daily Coliform—daily Cl$_2$ residual—continuous	500 ft (150 m) to potable-water-supply wells (minimum) if bottom not sealed	Nutrient removal processes may be necessary to avoid algae growth in impoundments. Dechlorination may be necessary to protect aquatic species of flora and fauna. Provide treatment reliability.
Construction uses Soil compaction, dust control, washing aggregate, making concrete	Secondary[c] Disinfection[e]	≤30 mg/L BOD[f] ≤30 mg/L SS ≤200 fecal coli/100 mL[l,m,n] 1 mg/L Cl$_2$ residual (min)[j]	BOD—weekly SS—daily Coliform—daily Cl$_2$ residual—continuous		Worker contact with reclaimed water should be minimized. A higher level of disinfection, e.g., to achieve ≤14 fecal coli/100 mL, should be provided where frequent worker contact with reclaimed water is likely. Provide treatment reliability.

TABLE 21.11 EPA Suggested Guidelines for Reuse of Municipal Wastewater (*Continued*)

Types of reuse	Treatment	Reclaimed-water quality[a]	Reclaimed-water monitoring	Setback distance[b]	Comments
Industrial reuse Once-through cooling	Secondary[c]	$pH = 6\text{–}9$ ≤ 30 mg/L BOD[f] ≤ 30 mg/L SS ≤ 200 fecal coli/100 mL[h,l,m] 1 mg/L Cl_2 residual (min)[j]	pH—weekly BOD—weekly SS—weekly Coliform—daily Cl_2 residual—continuous	300 ft (90 m) to areas accessible to the public	Windblown spray should not reach areas accessible to users or the public.
Recirculating cooling towers	Secondary[c] Disinfection[e] (chemical coagulation and filtration[d] may be needed)	Variable, depends on recirculation ratio		300 ft (90 m) to areas accessible to the public May be reduced if high level of disinfection is provided	Windblown spray should not reach areas accessible to the public. Consult recommended water-quality limits for make-up water. Additional treatment by user is usually provided to prevent scaling, corrosion, biological growths, fouling, and foaming. Provide treatment reliability.
Environmental reuse Wetlands, marshes, wildlife habitat, stream augmentation	Variable Secondary[c] and disinfection[e] (min)	Variable, but not to exceed: ≤ 30 mg/L BOD[f] ≤ 30 mg/L SS ≤ 200 fecal coli/100 mL[h,l,m]	BOD—weekly SS—daily Coliform—daily Cl_2 residual—continuous		Dechlorination may be necessary to protect aquatic species of flora and fauna. Possible effects on groundwater should be evaluated. Receiving-water-quality requirements may necessitate additional treatment. The temperature of the reclaimed water should not adversely affect ecosystem. Provide treatment reliability.
Groundwater recharge By spreading or injection into nonpotable aquifers	Site-specific and use-dependent Primary[c] (min) for spreading Secondary[c] (min) for injection	Site-specific and use-dependent	Depends on treatment and use	Site-specific	Facility should be designed to ensure that no reclaimed water reaches potable-water-supply aquifers. For injection projects, filtration and disinfection may be needed to prevent clogging. Provide treatment reliability.

Types of reuse	Treatment	Reclaimed water quality	Reclaimed water monitoring	Setback distances	Comments
Indirect potable reuse Groundwater recharge by spreading into potable aquifers	Site-specific Secondary[c] and disinfection[e] (min) May also need filtration[d] and/or advanced wastewater treatment[o]	Site-specific Meet drinking-water standards after percolation through vadose zone	Includes, but not limited to, the following: pH—daily Coliform—daily Cl_2 residual—continuous Drinking-water standards—quarterly Other[p]—depends on constituent	2000 ft (600 m) to extraction wells. May vary depending on treatment provided and site-specific conditions	The depth to groundwater (i.e., thickness of the vadose zone) should be at least 6 ft (2 m) at the maximum groundwater-mounding point. The reclaimed water should be retained underground for at least 1 year prior to withdrawal. Recommended treatment is site-specific and depends on factors such as type of soil, percolation rate, thickness of vadose zone, native-groundwater quality, and dilution. Monitoring wells are necessary to detect the influence of the recharge operation on the groundwater. The reclaimed water should not contain measurable levels of pathogens after percolation through the vadose zone.[k] Provide treatment reliability.
Groundwater recharge by injection into potable aquifers	Secondary[c] Filtration[d] Disinfection[e] Advanced wastewater treatment[o]	Includes, but not limited to, the following: pH = 6.5–8.5 ≤2 NTU[g] No detectable fecal coli/100 mL[h,i] 1 mg/L Cl_2 residual (min)[j] Meet drinking-water standards	Includes, but not limited to, the following: Turbidity—continuous pH—daily Coliform—daily Cl_2 residual—continuous Drinking-water standards—quarterly Other[p]—depends on constituent	2000 ft (600 m) to extraction wells. May vary depending on site-specific conditions	The reclaimed water should be retained underground for at least 1 year prior to withdrawal. Monitoring wells are necessary to detect the influence of the recharge operation on the groundwater. Recommended-quality limits should be met at the point of injection. The reclaimed water should not contain measurable levels of pathogens at the point of injection.[k] A higher chlorine residual and/or a longer contact time may be necessary to assure virus inactivation. Provide treatment reliability.

TABLE 21.11 EPA Suggested Guidelines for Reuse of Municipal Wastewater (*Continued*)

Types of reuse	Treatment	Reclaimed-water quality[a]	Reclaimed-water monitoring	Setback distance[b]	Comments
Indirect potable reuse Augmentation of surface supplies	Secondary[c] Filtration[d] Disinfection[e] Advanced wastewater treatment[o]	Includes, but not limited to, the following: pH = 6.5–8.5[g] ≤2 NTU[g] No detectable fecal coli/100 mL[h,i] 1 mg/L Cl_2 residual (min)[j] Meet drinking-water standards	Includes, but not limited to, the following: pH—daily Turbidity—continuous Coliform—daily Cl_2 residual—continuous Drinking-water standards—quarterly Other[p]—depends on constituent	Site-specific	Recommended level of treatment is site-specific and depends on factors such as receiving-water-quality, time and distance to point of withdrawal, dilution and subsequent treatment prior to distribution for potable uses. The reclaimed water should not contain measurable levels of pathogens.[k] A higher chlorine residual and/or a longer contact time may be necessary to assure virus inactivation. Provide treatment reliability.

[a] Unless otherwise noted, recommended-quality limits apply to the reclaimed water at the point of discharge from the treatment facility.

[b] Setbacks are recommended to protect potable-water-supply sources from contamination and to protect humans from unreasonable health risks due to exposure to reclaimed water.

[c] Secondary-treatment processes include activated-sludge processes, trickling filters, rotating biological contactors, and many stabilization-pond systems. Secondary treatment should produce effluent in which both the BOD and SS do not exceed 30 mg/L.

[d] Filtration means the passing of wastewater through natural undisturbed soils or filter media such as sand and/or anthracite.

[e] Disinfection means the destruction, inactivation, or removal of pathogenic microorganisms by chemical, physical, or biological means. Disinfection may be accomplished by chlorination, ozonation, other chemical disinfectants, UV radiation, membrane processes, or other processes.

[f] As determined from the 5-day BOD test.

[g] The recommended turbidity limit should be met prior to disinfection. The average turbidity should be based on a 24-hour time period. The turbidity should not exceed 5 NTU at any time. If SS is used in lieu of turbidity, the average SS should not exceed 5 mg/L.

[h] Unless otherwise noted, recommended coliform limits are median values determined from the bacteriological results of the last 7 days for which analyses have been completed. Either the membrane filter or fermentation tube technique may be used.

[i] The number of fecal coliform organisms should not exceed 14/100 mL in any sample.

[j] Total chlorine residual after a minimum contact time of 30 minutes.

[k] It is advisable to fully characterize the microbiological quality of the reclaimed water prior to implementation of a reuse program.

[l] The number of fecal coliform organisms should not exceed 800/100 mL in any sample.

[m] Some stabilization-pond systems may be able to meet this coliform limit without disinfection.

[n] Commercially processed food crops are those that, prior to sale to the public or others, have undergone chemical or physical processing sufficient to destroy pathogens.

[o] Advanced wastewater-treatment processes include chemical clarification, carbon adsorption, reverse osmosis and other membrane processes, air stripping, ultrafiltration, and ion exchange.

[p] Monitoring should include inorganic and organic compounds, or classes of compounds, that are known or suspected to be toxic, carcinogenic, teratogenic, or mutagenic and are not included in the drinking-water standards.

Source: Adapted from U.S. EPA (1992).

exists indicating that viruses are inactivated or removed to low or immeasurable levels via appropriate wastewater treatment (Crook, 1989; Engineering-Science, 1987; Sanitation Districts of Los Angeles County, 1979); the identification and enumeration of viruses in wastewater are hampered by relatively low virus-recovery rates; there are a limited number of facilities having the personnel and equipment necessary to perform the analyses; laboratory analyses employing commonly used identification and enumeration techniques can take as long as four weeks to complete; there is no consensus among public-health experts regarding the health significance of low levels of viruses in reclaimed water; and there have not been any documented cases of viral disease resulting from the reuse of wastewater in the United States.

Whereas the water-quality requirements for nonpotable-water uses are tractable and not likely to change significantly in the future, the number of water-quality constituents to be monitored in drinking water—and, hence, reclaimed water intended for potable reuse—will increase and quality requirements may become more restrictive. Consequently, the authors of the guidelines concluded that it would not be prudent to suggest a complete list of reclaimed-water-quality limits for all constituents of concern (U.S. EPA, 1992). The guidelines provide some general and specific information to indicate the extensive treatment and water-quality requirements that are likely to be imposed where indirect potable reuse is contemplated. The guidelines do not advocate direct potable reuse and do not include recommendations for such use.

21.6   SUMMARY

The abbreviated presentation in this handbook does not cover all of the important aspects of water reclamation and reuse. Only highlights have been presented, with emphasis on reclaimed-water-quality considerations and water reclamation and reuse criteria and guidelines. Reference sources, some already cited in this chapter, that are useful sources of additional information include: the U.S. Environmental Protection Agency (1992) for evaluating the requirements and potential benefits of water reuse, planning and implementing a water-reuse system, identifying major technical and nontechnical issues, documenting water-reuse projects, and summaries of state regulations and guidelines; the World Health Organization (1989) for recommended guidelines for wastewater use in agriculture and aquaculture in developing countries; the American Water Works Association (1994) and California-Nevada Section AWWA (1992) for the planning, design, construction, operation, and management of community dual-water distribution systems; the National Academy of Science–National Academy of Engineering (1973) and the California State Water Resources Control Board (1985) for irrigation water quality limits and practices; Richard, et al. (1992) for water reclamation costs; and Crook et al. (1991) for reviewing existing information and scientific data relative to water reclamation and reuse and identifying specific areas having information shortfalls that warrant further research.

REFERENCES

Adams, A. P., M. Garbett, H. B. Rees, and B. G. Lewis, "Bacterial Aerosols from Cooling Towers," *Journal Water Pollution Control Federation* 50(10):2362–2369, 1978.

Adams, A. P., M. Garbett, H. B. Rees, and B. G. Lewis, "Bacterial Aerosols Produced from a Cooling Tower Using Wastewater Effluent as Makeup Water," *Journal Water Pollution Control Federation,* 52(3):498–501, 1980.

American Water Works Association, *Dual Water Systems,* Manual of Practice M24, Denver, Colo., 1994.

Argo, D. G., and N. M. Cline, 1985. "Groundwater Recharge Operations at Water Factory 21, Orange County, California," in T. Asano, ed., *Artificial Recharge of Groundwater,* Butterworth, Stoneham, Mass., 1985, pp. 359–395.

Asano, T., R. G. Smith, and G. Tchobanoglous, "Municipal Wastewater: Treatment and Reclaimed Water Characteristics," in G. S. Pettygrove and T. Asano, eds., *Irrigation with Reclaimed Municipal Wastewater—A Guidance Manual,* Lewis, Chelsea, Mich., 1985, pp. 2-1–2-26.

Asano, T., Y. C. Leong, M. G. Rigby, and R. H. Sakaji, "Evaluation of the California Wastewater Reclamation Criteria using Enteric Virus Monitoring Data," *Water Science Technology,* 26(7/8):1513–1524, 1992.

Ayers, R. S. and D. W. Westcot, *"Water Quality for Agriculture,"* FAO Irrigation & Drainage Paper 29, Rev. 1, U.N. Food and Agriculture Organization, Rome, 1985.

Bausum, H. T., S. A. Schaub, R. E. Bates, H. L. McKim, P. W. Schumacher, and B. E. Brockett, "Microbiological Aerosols From a Field-Source Wastewater Irrigation System," *Journal Water Pollution Control Federation,* 55(1):65–75, 1983.

Bouwer, H., and R. C. Rice, "Effect of Water Depth in Groundwater Recharge Basins on Infiltration Rate," *Journal of Irrigation and Drainage Engineering, ASCE,* 115:556–568, 1989.

Bryan, F. L. "Diseases Transmitted by Foods Contaminated by Wastewater," in *Wastewater Use in the Production of Food and Fiber,* EPA-660/2-74-041, U.S. Environmental Protection Agency, Washington, D.C., 1974, pp. 16–45.

California Department of Health, and R. C. Cooper, "Wastewater Contaminants and Their Effect on Public Health," in *A "State-of-the-Art" Review of Health Aspects of Wastewater Reclamation for Groundwater Recharge,* State of California Department of Water Resources, Sacramento, Calif., 1975, pp. 39–95.

California-Nevada Section AWWA, "Guidelines for Distribution of Nonpotable Water," California-Nevada Section, American Water Works Association, San Bernardino, Calif., 1992.

California State Water Resources Control Board, "California Municipal Wastewater Reclamation in 1987," California State Water Resources Control Board, Office of Water Recycling, Sacramento, Calif., 1990.

California State Water Resources Control Board, "Evaluation of Industrial Cooling Systems Using Reclaimed Municipal Wastewater," California State Water Resources Control Board, Office of Water Recycling, Sacramento, Calif., 1980.

California State Water Resources Control Board, G. S. Pettygrove and T. Asano, eds., Irrigation with Reclaimed Municipal Wastewater—A Guidance Manual. Lewis, Chelsea, Mich., 1985.

Camann, D. E., and M. N. Guentzel, "The Distribution of Bacterial Infections in the Lubbock Infection Surveillance Study of Wastewater Spray Irrigation," in *Proceedings of the Water Reuse Symposium III,* August 2–7, 1984, Denver, AWWA Research Foundation, Denver, Colo., 1985, pp. 1470–1495.

Camann, D. E., and B. E. Moore, 1988. "Viral Infections Based on Clinical Sampling at a Spray Irrigation Site," in *Proceedings of Water Reuse Symposium IV,* August 2–7, 1987, Denver, AWWA Research Foundation, Denver, Colo., 1988, pp. 847–863.

Camann, D. E., D. E. Johnson, H. J. Harding, and C. A. Sorber, "Wastewater Aerosol and School Attendance Monitoring at an Advanced Wastewater Treatment Facility: Durham Plant, Tigard, Oregon," in H. Pahren and W. Jakubowski, eds., *Wastewater Aerosols and Disease,* EPA-600/9-80-028, U.S. Environmental Protection Agency, Cincinnati, Ohio, 1980, pp. 160–179.

Camann, D. E., B. E. Moore, H. J. Harding, and C. A. Sorber, "Microorganism Levels in Air Near Spray Irrigation of Municipal Wastewater: The Lubbock Infection Surveillance Study," *Journal Water Pollution Control Federation,* 60(11):1960–1970, 1988.

Camp Dresser & McKee Inc., "Water Recycling in the Pulp and Paper Industry in California," report prepared for the California State Water Resources Control Board, Office of Water Recycling, Sacramento, Calif., 1982.

Carlson, R. R., K. D. Lindstedt, E. R. Bennett, and R. B. Hartman, "Rapid Infiltration Treatment of Primary and Secondary Effluents," *Journal Water Pollution Control Federation,* 54(3):270–280, 1982.

Chang, A. C., and A. L. Page, "Fate of Inorganic Micro-Contaminants during Groundwater Recharge," in T. Asano and P. V. Roberts, eds., *Water Reuse for Groundwater Recharge,* California State Water Resources Control Board, Office of Water Recycling, Sacramento, Calif., 1979, pp. 118–136.

Chapman, J. B., and R. H. French, "Salinity Problems Associated with Reuse Water Irrigation of Southwestern Golf Courses," in P. A. Krenkel, ed., *Environmental Engineering: Proceedings of the 1991 Specialty Conference,* sponsored by Environmental Engineering Division of the American Society of Civil Engineers, July 8–10, Reno, Nevada, American Society of Civil Engineers, New York, 1991, pp. 230–235.

CH2M Hill, "Tampa Water Resource Recovery Project Summary Report," CH2M Hill, Denver, Colo., 1993.

Clow, B. D., "Sizing Irrigation Reservoirs for Treated Domestic Wastewater," in *Proceedings of the Urban and Agricultural Water Reuse Specialty Conference,* June 28–July 1, Orlando, Fla., Water Environment Federation, Alexandria, Va., 1992, pp. 73–88.

Craun, G. F., "Surface Water Supplies and Health," *Journal of the American Water Works Association,* 80(2):40–52, 1988.

Crook, J., "Water Reclamation," in *Encyclopedia of Physical Science and Technology, Vol. 17,* Academic Press, San Diego, Calif., 1992, pp. 559–589.

Crook, J., "Viruses in Reclaimed Water," in *Proceedings of the 63rd Annual Technical Conference,* sponsored by the Florida Section American Water Works Association, Florida Pollution Control Association, and Florida Water & Pollution Control Operators Association, Nov. 12–15, St. Petersburg Beach, Fla., 1989, pp. 231–237.

Crook, J., T. Asano, and M. Nellor, "Groundwater Recharge with Reclaimed Water in California," *Water Environment and Technology,* 2(8):42–49, 1990.

Crook, J., and W. D. Johnson, "Health and Water Quality Considerations with a Dual Water System," *Water Environment and Technology,* 3(8):13–14, 1991.

Crook, J., D. A. Okun, and A. B. Pincince, "Water Reuse," Project 92-WRE-1, Water Environment Research Foundation, Alexandria, Va., 1994.

Engineering-Science, "Monterey Wastewater Reclamation Study for Agriculture: Final Report," prepared for the Monterey Regional Water Pollution Control Agency, Engineering-Science, Berkeley, Calif., 1987.

Fannin, K. F., K. W. Cochran, D. E. Lamphiear, and A. S. Monto, "Acute Illness Differences with Regard to Distance from the Tecumseh, Michigan Wastewater Treatment Plant," in H. Pahren and W. Jakubowski, eds., *Wastewater Aerosols and Disease,* EPA-600/9-80-028, U.S. Environmental Protection Agency, Health Effects Research Laboratory, Cincinnati, Ohio, 1980, pp. 117–135.

Feachem, R. G., D. J. Bradley, H. Garelick, and D. D. Mara, *Health Aspects of Excreta and Sullage Management: A State-of-the-Art Review,* The World Bank, Washington, D.C., 1981.

Feachem, R. G., H. Bradley, H. Garelick, and D. D. Mara, *Sanitation and Disease—Health Aspects of Excreta and Wastewater Management,* published for the World Bank, John Wiley, Chichester, England, 1983.

Florida Department of Environmental Regulation, "Reuse of Reclaimed Water and Land Application," Chap. 17-610, Florida Administrative Code, Florida Department of Environmental Regulation, Tallahassee, Fla., 1990.

Florida Department of Environmental Regulation, "1992 Reuse Inventory," Florida Department of Environmental Regulation, Tallahassee, Fla., 1992.

Gerba, C. P., and S. M. Goyal, "Pathogen Removal from Wastewater during Groundwater Recharge," in T. Asano, ed., *Artificial Recharge of Groundwater,* Butterworth, Boston, Mass., 1985, pp. 283–317.

Goldstein, D. J., I. Wei, and R. E. Hicks, "Reuse of Municipal Wastewater as Make-Up to Circulating Cooling Systems," in *Proceedings of the Water Reuse Symposium, Vol. 1,* Mar. 25–30, Washington, D.C., AWWA Research Foundation, Denver, Colo., 1979, pp. 371–397.

Gover, N., "HIV in Wastewater Not a Threat," *Water Environment & Technology,* 5(12):23, 1993.

Greenberg, A. E., and E. Kupka, "Tuberculosis Transmission by Wastewater—A Review," *Sewage and Industrial Wastes,* 29(5):524–537, 1957.

Hoadley, A. W., and S. M. Goyal, "Public Health Implications of the Application of Wastewater to Land," in R. L. Sanks and T. Asano, eds., *Land Treatment and Disposal of Municipal and Industrial Wastewater*, Ann Arbor Science, Ann Arbor, Mich., 1976, p. 1092.

Hurst, C. J., W. H. Benton, and R. E. Stetler, "Detecting Viruses in Water," *Journal of the American Water Works Association*, 81(9):71–80, 1989.

Idelovitch, E., R. Terkeltoub, and M. Michall, "The Role of Groundwater Recharge in Wastewater Reuse: Israel's Dan Region Project," *Journal of the American Water Works Association*, 72(7):391–400, 1980.

Iwata, Y., and F. A. Gunther, "Translocation of the Polychlorinated Biphenyl Oroclor 1254 from Soil into Carrots under Field Conditions," *Archive of Environmental Contamination and Toxicity*, 4(1):44–59, 1976.

Jansons, J., L. W. Edmonds, B. Speight, and M. R. Bucens, "Survival of Viruses in Groundwater," *Water Research*, 23(3):301–306, 1989.

John, M. K., "Cadmium Adsorption Maxima of Soils as Measured by Langmuir Isotherm," *Canadian Journal of Soil Science*, 52:343–350, 1972.

Johnson, D. E., D. E. Camaan, D. T. Kimball, R. J. Prevost, and R. E. Thomas, "Health Effects from Wastewater Aerosols at a New Activated Sludge Plant—John Egan Plant, Schaumburg, Ill.," in H. Pahren and W. Jakubowski, eds., *Wastewater Aerosols and Disease*, EPA-600/9-80-028, U.S. Environmental Protection Agency, Cincinnati, Ohio, 1980a, pp. 136–159.

Johnson, D. E., D. E. Camaan, J. W. Register, R. E. Thomas, C. A. Sorber, M. N. Guentzel, J. M. Taylor, and W. J. Harding, "The Evaluation of Microbiological Aerosols Associated With the Application of Wastewater to Land: Pleasanton, CA," Report no. EPA-600/1-80-015, U.S. Environmental Protection Agency, Cincinnati, Ohio, 1980b.

Keswick, B. H., C. P. Gerba, S. L. Secor, and I. Sech, "Survival of Enteric Viruses and Indicator Bacteria in Groundwater," *Journal of Environmental Science Health*, A17:903–912, 1982.

Knorr, D. B., J. Hernandez, and W. M. Copa, "Wastewater Treatment and Groundwater Recharge: A Learning Experience at El Paso, TX," in *Proceedings of Water Reuse Symposium IV*, Aug. 2–7, 1987, Denver, AWWA Research Foundation, Denver, Colo., 1988, pp. 211–232.

Lauer, W. C., "Denver's Direct Potable Water Reuse Demonstration Project—Final Report," Denver Water Department, Denver, Colo., 1993.

Lund, E., "Health Problems Associated with the Re-Use of Sewage: I. Bacteria, II Viruses, III. Protozoa and Helminths," working papers prepared for WHO Seminar on Health Aspects of Treated Sewage Re-Use, 1–5 June, Algers, Algeria, 1980.

McCarty, P. L., M. Reinhard, N. L. Goodman, J. W. Graydon, G. D. Hopkins, K. E. Mortelmans, and D. G. Argo, "Advanced Treatment for Wastewater Reclamation at Water Factory 21," Technical Paper no. 267, Department of Civil Engineering, Stanford University, Stanford, Calif., 1982.

Metcalf & Eddy, Inc., *Wastewater Engineering: Treatment, Disposal, Reuse*, McGraw-Hill, New York, 1991.

Meyer, R., "Preparing Water for Industrial Boilers," *Ind. Water Treatment*, 23(2):30–32, 1991.

Mullarkey, N., and M. Hall, "Feasibility of Developing Dual Water Distribution Systems for Non-potable Reuse in Texas," in *Proceedings of CONSERV 90*, Aug. 12–16, Phoenix, Ariz., National Water Well Association, Dublin, Ohio, 1990, pp. 59–63.

Murphy, W. H., and J. T. Syverton, "Absorption and Translocation of Mammalian Viruses by Plants. II. Recovery and Distribution of Viruses in Plants," *Virology*, 6(3):623–636, 1958.

National Academy of Sciences–National Academy of Engineering, "Water Quality Criteria 1972," EPA/R3/73/033, prepared by the Committee on Water Quality Criteria, National Academy of Sciences–National Academy of Engineering, for the U.S. Environmental Protection Agency, Washington, D.C., 1973.

National Communicable Disease Center, *Shigella Surveillance Second Quarter*, National Communicable Disease Center, Report 20, Atlanta, Ga., 1969.

National Communicable Disease Center, "Morbidity and Mortality, Weekly Report," 24(3):21, 1973.

National Communicable Disease Center, "Morbidity and Mortality, Weekly Report," 24(31):261, 1975.

Nellor, M. H., R. B. Baird, and J. R. Smyth, "Health Effects Study—Final Report," County Sanitation Districts of Los Angeles County, Whittier, Calif., 1984.

Page, A. L., and A. C. Chang, "Fate of Wastewater Constituents in Soil and Groundwater: Trace Elements," in G. S. Pettygrove and T. Asano, eds., *Irrigation with Reclaimed Municipal Wastewater—A Guidance Manual,* Lewis, Chelsea, Mich., 1985, pp. 13-1–13-16.

Pahren, H., and W. Jakubowski, eds., "Wastewater Aerosols and Disease. Proceedings of a Symposium," Sept. 19–21, 1979, Report EPA-600/9-80-028, U.S. Environmental Protection Agency, Health Effects Research Laboratory, Cincinnati, Ohio, 1980.

Palazzo, A. J., "The Effects of Wastewater Applications on the Growth and Chemical Composition of Forages," Report 76-9, U.S. Army Corps of Engineers, Cold Regions Research and Engineering Laboratory, Hanover, N.H., 1976.

Pawlowski, S., "Rules for the Reuse of Reclaimed Water," paper presented at the Salt River Project Water Reuse Symposium, Nov. 2, Tempe, Ariz., 1992.

Puckorius, P. R., and R. T. Hess, "Wastewater Reuse for Industrial Cooling Water Systems," *Industrial Water Treatment,* 23(5):43–48, 1991.

Rhoades, J. D., R. D. Ingvalson, and J. T. Hatcher, "Laboratory Determination of Leachable Soil Boron," *Soil Science Society of America Proceedings,* 34:871–875, 1979.

Rice, R. C., and H. Bouwer, "Soil-Aquifer Treatment Using Primary Effluent," *Journal Water Pollution Control Federation,* 56(1):84–88, 1984.

Richard, D., T. Asano, and G. Tchobanoglous, "The Cost of Wastewater Reclamation in California," Department of Civil and Environmental Engineering, University of California, Davis, Calif., 1992.

Riggs, J. L., "AIDS Transmission in Drinking Water: No Threat," *Journal of the American Water Works Association,* 81(9):69–70, 1989.

Robbins, M. H., Jr., "Supplementing a Surface Water Supply with Reclaimed Water," in *Proceedings of the 1993 AWWA Conference,* June 6–10, San Antonio, Tex., American Water Works Association, Denver, Colo., 1993, pp. 211–226.

Robbins, M. H., Jr., and C. G. Ehalt, "Operation and Maintenance of the UOSA Water Reclamation Plant," *Journal Water Pollution Control Federation,* 57(12):1122–1127, 1985.

Roberts, P. V., "Water Reuse for Groundwater Recharge: An Overview," *Journal of the American Water Works Association,* 72(7):375–379, 1980.

Rogers, S. E., and W. C. Lauer, "Denver's Demonstration of Potable Water Reuse: Final Report," paper presented at the Water Reuse Association Potable Reuse Technical Workshop, May 13–14, 1993, Newport Beach, Calif., 1986.

Rose, J. B., "Microbial Aspects of Wastewater Reuse for Irrigation," *CRC Critical Reviews in Environ. Control,* 16(3):231–256, 1986.

Rose, J. B., and C. P. Gerba, "Assessing Potential Health Risks from Viruses and Parasites in Reclaimed Water in Arizona and Florida, U.S.A.," *Water Science Technology,* 23:2091–2098, 1991.

Sagik, B. P., B. E. Moore, and C. A. Sorber, "Infectious Disease Potential of Land Application of Wastewater," in *State of Knowledge in Land Treatment of Wastewater, Vol. 1, Proceedings of an International Symposium,* U.S. Army Corps of Engineers, Cold Regions Research and Engineering Laboratory, Hanover, N.H., 1978, pp. 35–46.

Sanitation Districts of Los Angeles County, "Pomona Virus Study: Final Report," California State Water Resources Control Board, Sacramento, Calif., 1977.

Selby, K. A., and J. M. Brooke, "Introduction, Purpose and Objectives of a Boiler Water Treatment Program," *Industrial Water Treatment,* 23(1):17–19, 1990.

Sepp, E., "The Use of Sewage for Irrigation—A Literature Review," California Department of Public Health, Bureau of Sanitary Engineering, Berkeley, Calif., 1971.

Shuval, H. I., "Land Treatment of Wastewater in Israel," in *State of Knowledge in Land Treatment of Wastewater, Vol. 1. Proceedings of an International Symposium,* U.S. Army Corps of Engineers, Cold Regions Research and Engineering Laboratory, Hanover, N.H., 1978, pp. 429–436.

Shuval, H. I., A. Adin, B. Fattal, E. Rawitz, and P. Yekutiel, "Wastewater Irrigation in Developing Countries—Health Effects and Technical Solutions," World Bank Technical Paper no. 51, The World Bank, Washington, D.C., 1986.

Sobsey, M., "Public Health Aspects of Human Enteric Viruses in Cooling Waters," Report to NUS Corporation, Pittsburgh, Pa., 1978.

Sontheimer, H., "Experience with Riverbank Filtration along the Rhine River," *Journal of the American Water Works Association*, 72(7):386–390, 1980.

State of Arizona, *Regulations for the Reuse of Wastewater*, Arizona Administrative Code, Chap. 9, Art. 7, Arizona Department of Environmental Quality, Phoenix, Ariz., 1987.

State of California, *Wastewater Reclamation Criteria*, California Administrative Code, Title 22, Division 4, California Department of Health Services, Sanitary Engineering Section, Berkeley, Calif., 1978.

State of California, "Report of the Scientific Advisory Panel on Groundwater Recharge with Reclaimed Wastewater," prepared for the State Water Resources Control Board, Department of Water Resources, and Department of Health Services, Sacramento, Calif., 1987.

Strauss, S. D., and P. R. Puckorius, "Cooling Water Treatment for Control of Scaling, Fouling, Corrosion," *Power*, June 1984, pp. 1–24.

Tanji, K. K., ed., *Agricultural Salinity Assessment and Management*, American Society of Civil Engineers, New York, 1990.

Teltsch, B., S. Kidmi, L. Bonnet, Y. Borenzstajn-Roten, and E. Katzenelson, "Isolation and Identification of Pathogenic Microorganisms at Wastewater-Irrigated Fields: Ratios in Air and Wastewater," *Applied Environ. Microbiol.*, 39:1184–1195, 1980.

Troscinski, E. S. and R. G. Watson, "Controlling Deposits in Cooling Water Systems," *Chem. Engrg.*, March 9, 1970.

University of California Committee of Consultants, "Guidelines for Interpretation of Water Quality for Agriculture," Memo Report, 1974.

U.S. Environmental Protection Agency, "Guidelines for Water Reuse," Report EPA-600/8-80-036, U.S. Environmental Protection Agency, Municipal Environmental Research Laboratory, Cincinnati, Ohio, 1980.

U.S. Environmental Protection Agency, "Process Design Manual: Land Treatment of Municipal Wastewater," Report EPA/625/1-81-013, U.S. Environmental Protection Agency, Center for Environmental Research Information, Cincinnati, Ohio, 1981.

U.S. Environmental Protection Agency, "Transport and Fate of Contaminants in the Subsurface," Report EPA/625/4-89/019, EPA Center for Environmental Research Information, Cincinnati, Ohio, 1989.

U.S. Environmental Protection Agency, "Guidelines for Water Reuse," Report EPA/625/R-92/004, U.S. Environmental Protection Agency, Center for Environmental Research Information, Cincinnati, Ohio, 1992.

van der Leeden, F., F. L. Troise, and D. K. Todd, *The Water Encyclopedia*, 2d ed., Lewis, Chelsea, Mich., 1991.

Water Pollution Control Federation, *Water Reuse*, 2d ed., Manual of Practice SM-3, Water Pollution Control Federation, Alexandria, Va., 1989.

WateReuse Association of California, " '93 Survey: Future Water Recycling Potential," City of Los Angeles, Office of Water Reclamation, Los Angeles, Calif., 1993.

Westcot, D. W., and R. S. Ayers, "Irrigation Water Quality," in G. S. Pettygrove and T. Asano, eds., *Irrigation with Reclaimed Municipal Wastewater—A Guidance Manual*, Lewis, Chelsea, Mich., 1985, pp. 3-1–3-37.

Western Consortium for Public Health, "City of San Diego Total Resources Recovery Project—Health Effects Study—Final Report," Western Consortium for Public Health, Richmond, Calif., 1992.

World Health Organization, "Health Guidelines for the Use of Wastewater in Agriculture and Aquaculture," Technical Report Series 778, World Health Organization, Geneva, Switzerland, 1989.

Yanko, W. A., "Analysis of 10 Years of Virus Monitoring Data from Los Angeles County Treatment Plants Meeting California Wastewater Reclamation Criteria," *Water Environ. Research*, 65(3):221–226, 1993.

CHAPTER 22
IRRIGATION SYSTEMS

John A. Replogle
Albert J. Clemmens
U.S. Water Conservation Laboratory
Phoenix, Arizona

Marvin E. Jensen
Consultant
Fort Collins, Colorado

22.1 CURRENT STATUS OF IRRIGATION

In the decade preceding 1980, irrigation expanded rapidly worldwide. After 1980, the rate of expansion drastically slowed. In the 1990s, the general worldwide emphasis is on rehabilitation and management improvement of existing projects. Changes are occurring in the techniques used, in the organization for managing projects, and in the financing of rehabilitation efforts. Despite the recent change in emphasis and the slowed growth, there are indications that irrigated agriculture in developing countries will undergo significant change, including renewed expansion. These changes may reproportion the geographic distribution of irrigated areas (Higgins, et al., 1988).

World irrigation development has historically paralleled world population growth. Worldwide, irrigation is still expanding, but at a slightly slower rate than predicted. The United Nations Food and Agricultural Organization had estimated in 1977 that 630 million acres (255 million ha) would be under development by 1994 (FAO, 1993). Various other estimates vary from these somewhat (Field, 1990), but the general trends are similar, and indicate that the earlier estimate is about on schedule.

In the United States alone, about 57 million acres (23 million ha) are irrigated according to Shank (1992), or 47 million acres (19 million ha) according to U.S. census figures, as reported in FAO (1993) and USDA (1992). Both, while not agreeing on the total area, perhaps because of double cropping in one set of data, show decreases of about 9 percent over the last decade, counter to world trend. Each year some areas are withdrawn from irrigation, other areas are added. On the world scene, while salinity problems reduce the area irrigated in some locations, irrigation is constantly being developed at other locations.

The authors of this chapter assess recent trends and consider their effects on the community of water scientists, engineers, and others likely to be asked to furnish information, services, and creative ideas.

22.2 NEED FOR IRRIGATION

In terms of agricultural production, about 17 percent of the cropped area of the world is irrigated and contributes more than one third of the total world food production (Field, 1990). In the United States about 12 percent of the cropped area is irrigated and contributes about 25 percent of the total value of U.S. crops (USDA, 1982). Almost 60 percent of the production of major cereals in the developing world, rice and wheat, is derived from irrigated land.

Indirect benefits of irrigation are well known to agriculturalists in irrigated areas but not to the general public (Jensen, 1980). Many orchard crops could not survive the occasional freezing weather without frost protection provided by irrigation water. Livestock that graze dry rangeland during part of a season are sustained to graze again by feed produced on irrigated lands. In the United States, herds that would have perished during the severe drought of 1977 in Southern California, for example, survived on hay produced under irrigation in Montana and rainfed lands in Minnesota. In many arid and semiarid regions of the world, the use of limited forage on marginal to nonproductive rangeland depends on a supplemental supply of feed from irrigated areas. Thus the irrigated areas increase the net productivity of the nonirrigated areas.

For millennia, irrigation enabled civilizations to develop permanent residential sites in desert and semidesert areas. Nomadic operation was no longer required to acquire food for people and their animal herds. In our modern times irrigation continues to lessen the risks to food production due to the vagaries of climate and weather, and helps insure an uninterruptable food supply from the world agricultural community (FAO, 1977).

Despite this reliance on irrigation for stable food production, irrigation schemes worldwide, existing and planned, have been subjected to increasing adverse publicity (Jensen, 1980). The environmental effects of both the development and the abandonment of irrigation are frequently adverse. It is the adverse conditions that attract much attention from many areas. Often the critics are unable or unwilling to speculate on the probable negative effects that failure to irrigate might have on a region. According to Worthington (1976), this publicity emphasizes potential adverse effects that can accompany irrigation development while nearly ignoring the primary purpose of supplying water for improved living standards of millions of people.

Some concerns are based on land-damage estimates, real or imagined. Other concerns involve public health. Diseases transmitted by mosquitos, Simulium fly, tsetse fly, snails, and freshwater crustaceans are all of concern (Jensen, et al., 1990b). Many problems are aggravated by the concentration of population in an area that formerly had a dispersed population. Special precautions are necessary when converting to a more concentrated, irrigation-supported society, such as proper sanitary facilities and assured clean drinking water. Socioeconomic consequences need careful consideration. Always to be considered are the consequences of not irrigating, which may include hunger and political unrest.

In the United States, environmental concerns, particularly water-quality degradation in terms of salinity, nitrates, pesticides, and trace elements, are the major focuses for criticism of irrigated agriculture. In arid environments, water-quality degradation is a problem due to the concentration of salts and trace elements, particularly to wildlife habitats. However, many critics of irrigated agriculture ignore the increase in wildlife habitat provided by the irrigation development. Development of irrigation in a nonfarmed arid area changes the hydrology. In some areas, naturally occurring toxic elements such as selenium have been moved into return flows. Clearly, a better coordinated approach to planning is needed to enhance the positive impacts and reduce the negative effects on the environment. In more humid areas, irrigated and nonirrigated agriculture often face the same criticisms, primarily related to the movement of nitrate nitrogen and agricultural chemicals into groundwater and the runoff of sediments and pesticides into surface waters.

Because worldwide the most suitable irrigation areas have already been developed, planners must look carefully to determine the benefits of upgrading and improving already committed irrigation areas, as opposed to opening new projects on new lands.

Increasing irrigation to achieve economic development can also increase social costs for a region (Jensen, 1980). Extensive investments of capital are needed to accommodate increased population growth and density. Increased energy for irrigation is frequently needed. The hydropower lost due to diversions of irrigation water frequently must be replaced by a more expensive source (Whittlesey, et al., 1978).

22.3 NEED FOR DRAINAGE

Most of the world's projects appear to have been developed on the assumption, stated or unstated, that drainage was not justified because the benefits would be negligible, that is, requirements would either not materialize or drainage systems could be delayed indefinitely. This sometimes assumed that the drainage system could be afforded by the inherent increase in the economic growth in the project. For whatever reason, the cost of declining yield with increasing waterlogging and salinity was not adequately considered. Now, 20 to 30 years later, the drainage problems surface as waterlogged areas or downstream water-quality degradation problems that are becoming critical. Too frequently those economies never materialized for installing drainage systems due to many diverse sociopolitical and technical reasons. Often drainage problems occur sooner than anticipated due to canal seepage and deep percolation on irrigated lands. Thus, a two-pronged approach is usually needed: (1) reduce the excess water input to approach the natural drainage rate in an area, and (2) install artificial drainage when the natural-drainage rate is insufficient.

22.3.1 Salt Balance

All natural waters contain dissolved solids or salts. Salts remain in the soil as pure water is evaporated. These salt accumulations are unavoidable and predictable, and ultimately have to be removed (moved away from plant roots) by drainage of some sort, either natural or artificial. The amount of salt to be removed is related to: (1) the quality of the source water, (2) the volume evaporated, and (3) the amount of salt that can be stored in the soil root zone. It is not related to irrigation efficiency. In some geographic settings, additional salts are leached from the soil and rock. Here excess deep-percolation water adds additional salt to the drainage water, and thus excessive water application contributes more directly to the salinity problem.

The amount of salt that needs to be removed is related to the sensitivity of plants to root-zone salinity. Each crop has a different tolerance to each chemical. Threshold soil salinity values have been developed to guide management of the crop and soil. These threshold values are described in terms of the electrical conductivity of the soil-water extract, EC_e. While published values exist for the common chemicals and plants, these are rough estimates which likely vary with soils and climatic conditions. However, practical experience indicates that such handbook values are reasonable in most cases (see Tanji, 1990).

The salinity of the soil water at the bottom of the root zone is approximately the same as the salinity of the deep-percolation water. Thus, the soil salinity is dependent on the relative amount of deep percolation. The amount of water that needs to be leached in order to keep the soil salinity below some threshold value is called the *leaching requirement*, LR, which is generally expressed as a fraction of water which evapotranspires. Theoretical values of LR can be very small for high-quality water and nonsensitive crops (e.g., 1 percent) and as large as 15 or 20 percent for very poor quality water and sensitive crops. The leaching fraction is that amount of water that actually passes through the root zone, relative to the consumed water.

The above calculated leaching ratios are, in general, not sufficient to bring soil-salinity levels down to the prescribed level because of soil nonuniformity and nonuniform downward flow. Because of nonuniform flow, the water applied for leaching is not 100 percent effective in removing accumulated salt. The effectiveness of leaching is determined by local soil conditions and is generally not under the irrigator's control. The theoretical LR also does not consider salt precipitation in the root zone, which can reduce LR and the needed deep-percolation water. All of this makes determining real field LR as much art as science.

Because of the inherent nonuniformity of irrigation and soils, some areas receive insufficient water and gradually build up high soil-salinity levels, while other areas receive too much water and are excessively leached. Where natural drainage cannot remove excess deep percolation, soil becomes waterlogged. Waterlogged areas also tend to build up high salinity levels because of continual evaporation from the soil surface and insufficient net downward movement of water to remove accumulated salts. Even when not waterlogged, areas of high water tables may result in upward movement of water and salinity into the root zone, if there is not adequate downward movement of water. High-saline water-table conditions make management of water application difficult—too little water results in upward movement of saline groundwater into the root zone while too much water results in a rise in the water table.

Some areas do not drain to an ocean or saline sink. Salts are simply pushed to depths below the root zone. This can be practiced for decades in some areas with deep vadose zones. However, this practice has time limits in that the saline water will eventually reach the underlying groundwater. The time scale may be less than 20 years in some cases to more than a century in others. Typically, deep percolation will saturate about 10 times as much subsurface material as the percolated water depth. That is, an extra unit depth of seepage will occupy about 10 times that depth of gravel or sand above the saturated zone. If there is no natural drainage, the time line may be short.

22.3.2 Drainage Water Removal

To avoid increasing root-zone salinity, adequate leaching must be provided and drainage water must be removed. Where natural drainage is insufficient, constructed drainage systems are required. Methods of drainage include field-installed buried tile or plastic-tube drains, groundwater pumping, and deep-cut open ditches. Drainage outflows require a disposal location, such as a saline sink or the ocean. Sometimes, as discussed above, storage of drainage water occurs. Of particular immediate concern are subsurface-drainage systems in regions with poor-quality shallow groundwater that may collect and discharge water with potentially adverse environmental or local effects.

22.3.3 Drainage Water Treatment

Various approaches have been proposed to use or dispose of drainage waters. When not too saline, drainage water can be used to irrigate salt-tolerant crops. When not usable, and an outlet to a natural drain or the sea is not available, evaporation ponds may be used. Long-term storage for the residual salts accumulating in the ponds must be found. Stockpiling of such salts can cause problems due to wind erosion. Toxic elements in the ponds, concentrated by evaporation, also pose an environmental risk, such as adversely affecting waterfowl using the ponds.

A classic case is that of the Western San Joaquin Valley of California. Drainage water was stored in the Kesterson Reservoir while more permanent drainage facilities were being planned. In 1983 it was discovered that selenium concentrations in the Kesterson Reservoir were causing selenium (Se) toxicosis in waterfowl (Tanji, 1993). Since then studies have reported this and other problems in several other areas. Various approaches to removing the selenium have been suggested (Gates and Grismer, 1989; Tanji, 1990; Grismer, 1993; Engberg and Sylvester, 1993).

Some physical- and chemical-treatment technologies may have potential in some instances. An evapotranspiration method, *agroforestry,* is to plant certain varieties of trees or other crops that are saline resistant (Tanji and Karajeh, 1993). The Imperial Valley of California drains to the Salton Sea, which acts as an evaporation disposal system at the expense of increasing the salinity of those waters. Removal of the water by evaporation ponds (Tanji, et al., 1993) or evapotranspiration, however, still leaves the problem salts behind, which ultimately must be stockpiled in some less-harmful format. Drainage to the ocean seems to be the most viable long-term solution in most cases where salinity problems are severe, and that was the natural destination for the river water anyway.

22.4 IRRIGATION INFRASTRUCTURE (PROJECT LEVEL)

22.4.1 Water Supply Infrastructure

Water supplies for irrigation depend on: (1) managing precipitation runoff, either directly diverted from streams or from reservoir storage, (2) existing groundwater sources, or (3) reclaimed water from municipalities, either directly or as recharge to groundwater. Thus, most of the water supplies associated with irrigation require the management of reservoirs or groundwater aquifers. Only a limited area is irrigated using run-of-the-river systems with no storage.

All infrastructures require institutional arrangements. Within the irrigation infrastructure there are many types of organizations. These include various types of districts, such as irrigation, drainage, water-reclamation, conservancy, and water-storage districts; mutual canal and ditch companies, both public and private; commercial companies; and others organized for various purposes. These organizations frequently furnish both irrigation water and drainage service to agricultural lands, and some distribute electric power and other services (ASCE, 1991).

In the United States *districts* are public agencies organized under state laws that operate within specified geographic boundaries defined as legal boundaries, often providing services besides irrigation and drainage, such as electric-power distribution. The organization of a district is initiated by petition and majority vote. Many districts have taxing power, may issue bonds, may enter into contracts, may sue and be sued. They may acquire property by condemnation and may exercise other powers (ASCE, 1991). These legal districts probably outnumber all other types of irrigation-servicing organizations combined.

Mutual companies are private cooperative organizations organized under state laws to provide water or drainage service at cost to the members. These companies offer a potential advantage in keeping operation and maintenance costs low. However, if the companies are not properly managed, the operation and maintenance functions may not be performed efficiently, with resulting long-term higher costs (ASCE, 1991).

Commercial companies are organized under state laws for constructing and operating irrigation and drainage systems for profit. These companies usually select beneficiaries and establish relationships by contract. Rates usually are subject to public regulations. These enterprises, unless effectively managed, may have difficulty showing a profit (ASCE, 1991). In addition to institutional arrangements, physical structures are required.

22.4.1.1 Reservoirs. Surface waters are usually stored in reservoirs for multiple uses, including irrigation, recreation, power, municipal and industrial supplies, flood control, and in-stream concerns for wildlife. Groundwater and reclaimed municipal water are sometimes accumulated in surface reservoirs. Many reservoirs were constructed to supply irrigation water and for power generation. However, as the needs of society have expanded, these reservoirs have been adapted for many of the listed purposes when use is variable or is displaced from the wet season in order to accommodate constant-pumping rates. Reclaimed municipal water may require storage for similar reasons.

In irrigation practice, reservoirs may serve as the main storage reservoir at the head of an irrigation system or as regulatory reservoirs within the irrigation system. The latter are particularly useful in managing large irrigation systems with long canals and pipelines.

22.4.1.2 Well Fields. Single wells and groups of wells, called *well fields*, are widely used for both irrigation and municipal water supplies. Well fields present several concerns to water-resource planners. Among these is the conservation of the groundwater resource and its protection from external contamination by pollutants. Also, energy-load management and its relationship to electric-power or fossil-fuel-resource conservation must be considered (Gilley and Supalla, 1983).

22.4.1.3 River Diversions. For some irrigation systems, water supplies are based on run-of-the-river flows with no reservoir involved. Frequently small gravel dams are constructed in streams after the spring floods have passed to raise the summer flows high enough to divert the uncontrolled flow into canals. These systems do not attempt to match supply with demand, but are justified because of local surpluses of water that can be obtained for low investment of time, money and equipment.

Diversion methods range from these gravel dams to permanent structures with mechanized gates. River diversions are also used to supply water from upstream reservoirs to irrigation canals in a much more controlled manner (i.e., to regulate supply to match downstream demand).

22.4.2 Water Conveyance and Distribution Infrastructure

Where groundwater supplies are used for irrigation, water is most frequently pumped from individual farm wells. In areas where groundwater supplies are limited or depleted, groundwater-management districts are often established (typically under state laws) to regulate water use by individuals. Management of the water supply consists of managing power and the total seasonal volume pumped (Gilley and Supalla, 1983). For districts with surface-water supplies, including some groundwater districts and conjunctive surface-groundwater districts (which combined constitute more than half of irrigated acreage in the United States), water supplies are held in common to be distributed to users according to some established rules. Getting water from supply facilities to use areas requires a network of canals and pipelines to convey and distribute water to users. This infrastructure consists of various combinations of diversion structures, main and lateral canals, or pipelines, including the appropriate water-control structures or valves and regulating reservoirs. Open canals are often used to convey and distribute water because they are usually more economical in terms of capital costs, and in many cases maintenance costs, for the larger flows (e.g., >35 cfs [1 m³/s]). For smaller flows (<3 cfs [100 L/s]), pipelines are frequently more economical.

22.4.2.1 Canals. Many canals in conveyance-canal systems are unlined. The seepage from these unlined canals depends in part on the local water-table conditions and the permeability of the soil materials. Canals constructed in sandy soils, but accompanied by a high-water-table situation may have small seepage losses. High water tables near the canal system usually encourage the growth of unwanted vegetation that evapotranspires water diverting it from beneficial uses, and that hampers maintenance operations. Canals are lined for many reasons, including the facilitation of maintenance operations and the reduction of seepage. Frequently, seepage is cited as the justification for lining; but typically, once the local water table is raised further canal losses are diminished (Bouwer, 1969, 1978). This does not address the local problems such as waterlogging of adjacent farm lands caused by the presence of the higher water table, however.

Canal linings may vary from compacted earth to reinforced concrete. Canals that have been lined with ordinary clay, bentonite clay, soil cement, or membrane linings of plastic or rubber compounds pose a maintenance problem when trying to use heavy equipment to remove sediments. Many of these more vulnerable linings are covered with 1 ft (30 cm) of soil as a buffer against cleaning machinery.

22.4.2.2 High-Pressure Pipelines. *High-pressure pipeline delivery systems* for irrigation are often found in mountainous areas. These are constructed and operated much like a municipal delivery system, i.e., a demand irrigation system. Sometimes pressure regulators are used to control pressure fluctuations at the delivery point. Typical pipeline sizes are 12 in (300 mm) or less, but sizes over 4 ft (over 1 m) in diameter are in use. In mountainous areas, the elevation of the reservoir source supplies the needed pressure; otherwise pumps are used. High-pressure pipeline systems are well matched to support various farm sprinkler and microirrigation systems. Without pressure regulators at the farm, flow-rate fluctuations can be significant. The rates of flow delivered from such delivery systems are frequently too small to support efficient surface-irrigation methods.

22.4.2.3 Low-Head Pipelines.   *Low-pressure gravity pipelines* are often used for conveyance and distribution of flow rates typically on the order of 1 to 7 cfs (30 to 200 L/s). These systems have stand pipes that are open to the atmosphere and operate much like upstream-controlled canals—i.e., the rate of flow into the pipeline is set and that water is either delivered to offtakes or spilled (back pressure has little influence on the inflow into the pipeline in normal operations). Such systems are much less expensive than high-pressure pipelines and typically are designed for a higher-flow-rate capacity. For example, many low-head pipelines are made of nonreinforced concrete. Flow-regulating structures are also much less expensive. Operating pressures are typically less than 33 ft (10 m) of water head.

22.4.2.4 Semiclosed Pipelines.   For semiclosed pipeline systems, constructed with low-pressure pipe, float valves are used to limit fluctuations in delivery rate, and are claimed to be economically attractive for many applications that are frequently reserved for canal systems and high-pressure pipeline systems (Merriam, 1987a). Typical designs seldom exceed 30 to 50 ft (10 to 15 m) of head in any portion of the system (Merriam, 1987b). Float valves are used to reduce and control pressure in steps to limit the pressure in the pipe at any point. This allows the use of nonreinforced concrete pipe, or other low-pressure pipeline material, such as thin-walled plastic pipe.

This is an alternative approach to automated canals and high-head pipelines that are the usually suggested solutions for demand irrigation. Semiclosed pipelines are attractive alternatives because: (1) they are less expensive than high-head pipelines, (2) they do not have large pressure variations with changing flow rates, (3) they can usually avoid expensive and complex mechanical and electrical components frequently associated with automated canal systems, and (4) they support demand-type irrigation at the farm level. Disadvantages include a practical limit on flow rate because the economics of lowcost pipelines presently favors 12- to 15-in (305- to 380-mm) pipe diameters and float valves, with flow rates less than 10 cfs (350 L/s). This may inconvenience those farmers who irrigate with level basin systems by limiting the width of basins and thus increasing the labor needed to service smaller basins.

22.4.2.5 Structures.   Permanent structures of concrete, treated wood, or metal are common in most irrigation districts. These have several functions, as follows:

1. Inlet structure to control the entry of water from a supply source into the beginning of the canal.
2. Drop structure to control energy gradient in steep areas.
3. Check structures, sometimes called cross-regulators in international literature, that are intended to simply block flow for delivery through some upstream outlet, or are intended to maintain a water level in a reach of canal while passing some flow downstream.
4. Division boxes, which are used to divide the flow into two or more smaller canals. When splitting the flow into two parts, not necessarily equal, it is sometimes referred to as a *bifurcation*.
5. Measurement structures are used to quantify flow rates. These are usually flumes, weirs, or orifices. Infrequently a control structure can serve as a measuring struc-

ture, but usually the functions are separated to maintain accuracy and ease of control (Bos, 1989; Bos, et al., 1991)

Structures disrupt the natural flow of water in a canal so if sediment is being transported, it may settle and cause a maintenance problem. This maintenance requirement must be understood and plans and budget appropriately provided if the system is to maintain desired function.

22.4.2.6 *Operations.* The rules of water-delivery service and physical limitations of the system can restrict the ability of farm irrigation systems to operate effectively. There are several categories of water-delivery service used in irrigation-water distribution: continuous flow, rotation, arranged, and demand. There are several subcategories for each depending upon how the rate, frequency, and duration of a delivery are specified. For example, most high-pressure pipeline systems deliver water with a limited-rate demand schedule. However, if users can take water on demand only when it is their turn, then this becomes a type of arranged schedule. Delivery flexibility (i.e., less restriction of rate, frequency, and duration) usually requires either more investment in infrastructure or more labor (higher operating costs). The benefits are improved crop productivity and reduced water use and frequently more flexibility in crop selection. The chosen schedule is a compromise between the increased delivery cost and the benefits of more flexible service. The delivery schedule in use in a particular district may reflect an appropriate historical compromise, but may not reflect current conditions. Consequently, there is currently considerable emphasis in improving delivery flexibility (Hoffman, et al., 1990; also see ASCE, 1991; and Merriam, 1987a).

22.4.3 Project Economics

Most irrigation districts derive a majority of their revenue through the sale of water and delivery services to agricultural producers. Because these are frequently non-profit, quasi-governmental entities, charges are set to balance expenses. Farmers, through their board of directors, usually participate in deciding the charges for water and the level of service provided. Irrigation district finances do not necessarily reflect the economic health of the irrigation project because they are frequently buffered from the variations of farm income and profit.

22.4.3.1 *Capital Investment.* The capital investment devoted to irrigation systems varies widely. Logically, the more easily developed sites around the world are already operating. Thus, typically, new irrigation development is becoming ever more costly. Capital costs for new irrigation-water-distribution systems (not including costs of supply facilities and farm-irrigation systems) typically range from $100 to $500 per acre ($40 to $200 per ha). Capital costs of irrigation projects are paid in full by the water users. If funds are borrowed from commercial financial institutions repayment includes interest at prevailing rates. Some districts have the authority to issue bonds and repay them through property-tax assessments.

Some districts have received loans through the U.S. government, at reduced interest rates. These reduced interest rates were provided to promote the development of water resources and to promote rural development in the arid West. Repayment has been structured in several ways. The cost has been included in the unit cost of water. It has also been included as a charge per unit land area for delivery service or it has been charged as an annual land assessment (e.g., as part of county property taxes).

22.4.3.2   Operating Costs.   As with real estate, location is most frequently the controlling advantage. Some irrigation districts are so located that water is abundant and gravity flow is possible and pumping costs are small. Ownership rights and obligations among district farmers customarily require that operating and maintenance costs be shared. These costs are sometimes less than $5 per acre ft ($4 per 1000 m³) of water for some districts, and over $50 per acre ft ($40 per 1000 m³) of water for others (1000 m³ = 0.81 acre ft). Extremes are reported that greatly extend this usual range. Usually the more expensive water has been associated with high pumping costs, imported water from distant sites, or areas with extremely limited water supplies available for development.

22.4.3.3   Water Pricing.   Charges for water-delivery service are a mix of the cost of water and the cost of service. Some districts charge a fixed rate based on land area served, irrespective of the amount of water delivered. This allows the district to charge directly for services, which represent a fixed cost. Others charge for volume delivered, which encourages conservation but makes district financial management more difficult if the volume changes significantly from year to year. Where pumping is a significant part of water costs, such water pricing is important. Still others use a mix of the two, charging a fixed amount per unit land area that entitles the user to a certain volume per unit land area. Additional water delivered is charged according to volume delivered. This promotes a certain amount of conservation and also provides the district with a relatively fixed income that doesn't vary widely with irrigation demand.

22.4.3.4   Maintenance and Project Sustainability.   Irrigation project infrastructure maintenance is a major challenge. The pressure to keep operating costs low has often resulted in the degradation of the irrigation infrastructure. The problem is much worse in developing countries than in the United States. Procedures and recommendations for maintenance programs are readily available (e.g., ASCE, 1991), but not commonly practiced. Generally, proper attention is not rendered until long after delivery service starts to degrade, when the problems are already likely to be serious and expensive to correct.

22.5   FARM INFRASTRUCTURE

Irrigation is only one component of a farm enterprise. In arid areas, the enterprise heavily depends on irrigation, while in other areas irrigation is supplementary—one of many optional inputs. In the former, the decision to irrigate and the decision to farm are usually one and the same (e.g., you can't decide not to irrigate and still farm). The value of the land may be directly related to the availability of irrigation water. While in the latter, the decision to irrigate is simply financial—you can decide not to irrigate this year, or this crop, and still have a farming enterprise.

The success of a farm enterprise depends on many factors besides the cost and performance of the irrigation system. While in some cases, good crop management, good irrigation management and efficient irrigation systems go hand-in-hand, in other cases farms are financially successful even when they have relatively poor irrigation performance. In the following sections, we describe field-irrigation methods and farm-management and economic issues that can influence irrigation system performance.

There are three basic types of irrigation methods: *surface irrigation, sprinkler irrigation,* and *microirrigation.* Some irrigation methods are hard to categorize because they cross the traditional boundaries of these three general types. Surface-irrigation methods are so named because water is distributed across the field by flowing over the field surface. Thus soil infiltration, topography, and soil and crop flow resistance have a major influence on the distribution of water. By contrast, under sprinkler and microirrigation, most of the water distribution is accomplished in closed pipelines. In closed pipelines distribution is affected by changes in pressure and by variations in outlet properties (e.g., sprinkler head or emitter manufacturing variations). Once the water leaves the pipeline, distribution of water over the area served by the outlet occurs in a variety of ways depending on the system and the crop. With sprinklers, wind drift can have a significant effect on water distribution.

22.5.1 Surface Irrigation

Surface-irrigation methods can be grouped in the following categories: Continuous flood (paddy), basin, border strip, and furrow. Even within these categories, there are some further distinctions that affect how such systems are designed and operated.

Continuous-flood or *paddy irrigation* is commonly associated with rice production. Here, small basins are flooded essentially during the entire growing season. Often, water is supplied to one end of a series of basins, and water flows through each basin in turn. With other surface-irrigation methods, water is applied to the soil intermittently, with a frequency that depends on the soil, crop, crop growth stage, and so forth (Fig. 22.1).

Basin irrigation is a method of confining water to a given area. Water supplied to the basin advances across and then ponds over the entire surface. It differs from continuous-flood irrigation in that water does not remain ponded for long times (usu-

FIGURE 22.1   Contour paddy irrigation of rice in Southeast Asia.

ally less than one day). The basin may be level in all directions, may have a small downhill slope in the direction of water flow, or may be only roughly leveled. Under *border-strip irrigation*, water is applied to one end of a rectangular strip of land that is sloping. The water advances down the slope and either runs off the end or ponds behind a dike. The inflow is generally cut off before water reaches the end to reduce runoff or depth of ponding.

Furrow irrigation provides a means for controlling and guiding water on steep land. Furrows are created between crops that are planted in rows. Furrow irrigation is practiced on both steep undulating land and on very level land. Furrows can be oriented downslope, on the contour, or somewhere between. When land-grading equipment is used to smooth out the slope for furrow irrigation, it is often called *graded-furrow irrigation,* implying only sloping furrows. *Level-furrow irrigation* is also practiced, which is distinguished from furrows in level basins.

Level-basin irrigation, involving basins and border strips, has no soil-surface slope in any direction. Furrows may be imposed on these zero-sloped basins (Fig. 22.2). Water runs across the soil surface because of the slope of the water surface caused by introducing flow at one end of the basin or furrows. These basins are most usually constructed using laser-controlled grading equipment (Clemmens and Dedrick, 1982; Dedrick, 1986; 1989; 1990).

Uniformity of water distribution within level basins depends on the rate of application and the final intake rate of the soil. For example, if it requires three hours for the water to progress to the far end of a basin then the infiltration opportunity time at the beginning will be three hours longer than at the far end. The uniformity of water distribution across the basin would then depend on the uniformity of the soils, the accuracy of leveling, and whether the basin infiltration time was a few hours or

FIGURE 22.2 Furrows imposed onto level basins. Typical application rate is 2 L/s per m width of basin. An application time of 1.5 h provides 54 mm depth of water over a basin that is 200 m long. (3 h provides 108 mm.)

several hours. Drainage requirements to handle rainfall events are usually small. The rain stays and infiltrates where it falls and does not accumulate in the nonexistent low spots (Dedrick and Reinink, 1987a; 1987b).

Surge flow in surface irrigation is another technique that has been introduced in the past decade. It is a method of decreasing the infiltration rates of soils so that the irrigation water advances faster over the surface (Bishop, et al., 1981). This can improve the water distribution on sandy soils and other soils that have high enough infiltration rates so that they are difficult to irrigate with long furrows (Yonts and Eisenhauer, 1993). The process is to apply water to the head of a furrow for a period, then to stop and wait until that flow has infiltrated completely. Without undue delay restart the flow for a second surge, and so on, until the satisfactory completion of the irrigation. The effect is that the wetted and dewatered surface tends to consolidate and seal and thus reduces further infiltration so that subsequent applied water speeds more quickly across the field. The dewatering part appears to be the important stage of the mechanism inasmuch as the decrease in infiltration rate does not decrease so dramatically under continuous flow. The restart time appears not to be critical as long as complete dewatering has occurred and the soil surface has not dried out. The concept is usually more appropriate for sandy soils than clay soils. The process supplements or replaces other methods to control infiltration such as furrow packing.

Reuse-irrigation systems collect irrigation-runoff water (tailwater) from a field and make it available for reuse. They consist of collection ditches or diked areas at the lower end of a field, an open channel or pipe drain which directs the collected water to a storage area, and a means for returning the collected water to the same field or delivering it to a different field. The return-delivery system may include a pump and a pipeline (called a pumpback system) or open channels. Reuse systems are particularly applicable where legal constraints require that water which is delivered to a farm must be used on that farm. This is often the case in groundwater control districts. It is desirable when surface drainage might otherwise cause inundation and consequently damage to neighboring lands, and where water is pumped from aquifers, because repumping the collected water requires less energy. Reuse systems are most commonly used with sloping-furrow-irrigation systems. The reuse system allows large streams to be used throughout the irrigation without excessive loss of water. Provided the economics are favorable, most surface-irrigation systems might benefit from a tailwater-reuse system (Jensen, 1980).

Surface-irrigation systems are particularly applicable where investments in infrastructure and total crop values are low. Where major land leveling is not required, surface-irrigated systems usually require the lowest capital investment. Surface systems are also applicable where surface water is available from rivers, streams, and reservoirs. Although surface-irrigation systems function best with a well-controlled water supply, they are capable of operating with whatever supply of water is presented. Thus, irrigation projects that primarily use open canals will usually find that canal-flow fluctuations are best tolerated by surface-irrigation systems.

Surface-irrigation systems are most applicable in arid areas where evaporative demand is high, little rainfall occurs, large amounts of water are required, and water costs are low. Surface irrigation is best applied where soils are uniform, because nonuniform soils will reduce efficiency and uniformity of crop growth.

22.5.2 Sprinkler Irrigation

Sprinkler-irrigation systems are most applicable where irrigation water is supplemental to rainfall in meeting crop water needs. While sprinkler irrigation is used successfully in extremely arid conditions, it is generally not the most common method. Where the topography is steep or highly undulating, particularly when topsoil is shallow, sprinklers are generally preferred.

The main distinction between sprinkler-irrigation methods is whether the (usually rotating) sprinklers are on structures that move in the field during the irrigation event. Under each category of fixed or moving, there are several subcategories.

Some sprinkler systems are fixed in place for their useful life. These are generally referred to as *permanent, solid-set sprinklers.* These can be fixed on risers coming up from buried lines or on lines suspended above the crop, such as over-tree sprinklers. Another category of fixed sprinklers are the *hand-move sprinklers,* which are disassembled, moved and reassembled between irrigations, Fig. 22.3. There are several categories of fixed sprinklers that can be moved periodically without being disassembled. For these systems, the lateral lines move sideways on side-roll wheels. Nonmovable, fixed sprinklers are applicable to many orchard crops, where spray patterns are often a minor consideration. While fixed sprinklers are common for landscaping applications (turf) and are used some in permanent pastures, in general they are not the most efficient or economical sprinkler system for field crops such as small grains (such as wheat and barley). The hand-move systems are most economical for short-period needs, such as germination of vegetables, which might be inefficient with surface methods. In humid areas, they are used where irrigation is not normally practiced to reduce the impact of drought. The high labor requirements make them unattractive for most large-scale field crops, unless labor costs are low relative to equipment costs. The movable, fixed sprinklers (e.g., side roll) have been successfully used for field crops such as wheat and low-growing forage. Side-roll and side-move systems are used only on low-growing crops so that the lateral line will pass above the crop. In sandy soils, they offer some advantage over surface-irrigation methods. Generally, such moving systems need rectangular field shapes and can handle only mildly rolling topography and no obstructions.

For *continuous-move-sprinkler systems,* the system moves continuously during the irrigation. This tends to greatly increase the uniformity of application, because water application in the direction of movement is not significantly affected by the

FIGURE 22.3   Hand-moved-sprinkler irrigation of carrots in Portugal.

sprinkler pattern. The *traveling-gun* and *rotating-boom systems* are considered moving-sprinkler-irrigation systems. Under these systems, a high-pressure rotating sprinkler nozzle or sprinkler nozzles on a rotating boom are dragged along while water is supplied from a flexible hose. These systems can remain stationary, but more typically they move during the irrigation.

Center-pivot-irrigation systems are the most common moving-irrigation system. Here water is supplied at a central point and the lateral line rotates around this center. The linear-move system is similar to the center-pivot in size and style, Fig. 22.4. Here, however, the lateral moves in a straight line, with a supply source along one side of the field. Water-supply pickup can be from an open channel, a flexible hose, or a buried pipeline with automatic moving riser connectors.

The continuous-move-sprinkler irrigation systems developed within the last half of this century have been a major factor in the large increases in irrigated acreage in the semiarid and humid areas of the United States over the last thirty years. The center-pivot and linear-move systems are applicable to large areas and require little, although more highly skilled, labor. Center pivots can be run on extremely undulating topography, although such terrain may reduce the overall uniformity of application and aggravate soil traction problems. Linear-move systems are more restricted in their areas of application. They are practical only on relatively smooth terrain because of the increased difficulty in guiding the towers. They have been adapted more to arid conditions and specialty, high-value crops, such as vegetables, for example, where surface irrigation has traditionally been applied.

There are several types of sprinkler outlets used with the different methods described here. The most common types for agricultural purposes are the *spray nozzle* and the *rotating-head sprinklers* where the opening rotates in a circular path. The energy crisis of the mid-1970s caused a shift from high-pressure-sprinkler systems (60 lb/in^2 [400 kPa] and above) to lower-pressure-sprinkler systems (e.g., below

FIGURE 22.4 Linear-move system irrigating cotton with water supplied from a concrete-lined canal, Arizona.

45 lb/in^2 [300 kPa]). Low-pressure-spray nozzles (10 to 30 lb/in^2 [70 to 200 kPa]) are commonly used on continuous-move systems.

Some low-rate sprinklers are sometimes used for microirrigation (see following). The *low-energy precision-application* (LEPA) *system* is another example of sprinkler irrigation, but is sometimes classed as microirrigation. For this system, a continuous-move-sprinkler system (e.g., center pivot or linear move) is adapted so that low-rate-emission devices are lowered to the soil surface. Emission devices can be small sprayers, emitters, open tubes, or trailing perforated tubes.

22.5.3 Microirrigation

Microirrigation is a general category that includes various types of low-emission-rate devices. These systems include drip irrigation, trickle irrigation, subsurface irrigation, bubbler irrigation, and may include some of the moving LEPA systems. The idea is not to wet the entire soil surface or volume, but only that portion which needs to be wetted for the particular crop.

There are several basic types of microirrigation that relate to the way in which water is distributed over the field. As with sprinkler irrigation, there are fixed systems, which are applied to orchards and row crops, and moving systems, which are typically applied to row crops. Some of the earliest developments of microirrigation were for the watering of orchard crops, such as citrus. Individual emission devices were devised to deliver water to individual trees at low rates of flow, such that no surface ponding of water developed. These emitters can be large enough that they are connected in series by lengths of tubing, or can be small enough to be directly inserted into the soft plastic tubing by piercing the wall with a barblike connection, Fig. 22.5. *Line-source microirrigation* is another common system, in which flexible plastic tubing is constructed with small holes, slits, or perforations at fixed intervals. These lines are either laid on the ground surface or buried from 8 to 20 in (0.2 to 0.5 m) below the soil surface, the latter being called *subsurface irrigation. Bubbler irrigation* is a form of microirrigation where higher rates of flow are used, and is particularly useful in orchards. Water is delivered to small basins constructed around each tree to hold the water that ponds there.

Microirrigation is particularly applicable to orchards and vineyards, or other perennial crops for which the capital investment can be justified. Generally individual drip emitters, bubblers or microspray systems are used. Microirrigation is common in greenhouses and for ornamental crops where uniform and high crop quality is essential. Microirrigation is also well adapted to high-value crops that are sensitive to moisture and salinity stress; for example, sugar cane, strawberries, melons, tomatoes and other vegetables. For these, line-source emission is used, frequently underground. Subsurface microirrigation and LEPA systems have also been used on row crops such as cotton and corn with good success in terms of yield, but mixed success in terms of economics. Fixed microsystems require a relatively high value crop to make an adequate return on investment, an investment that is usually higher for microirrigation than for sprinkler irrigation.

Microirrigation generally requires that the water supply be of good quality. Most natural waters still need treatment to remove particles, to keep dissolved salts in solution, and to keep microbial activity from plugging emitters or emission points. Emitter clogging is one of the main limitations of microirrigation and has caused the failure of many installations. In some areas where water is saline, subsurface irrigation has been applied successfully because it keeps salts away from plant roots. This does require careful management.

FIGURE 22.5 Typical microirrigation lateral with emitters that irrigate grapes in Arizona.

Microirrigation is suitable to soils and topographies that are difficult to adapt to other methods. Microirrigation can be adapted to extremely sandy soils, or to rocky hillsides. Microirrigation is frequently used in landscaping settings where other methods might interfere with the landscape.

Microirrigation is also applied where water is scarce or expensive. In general, microirrigation is expected to be the most efficient method, although this does not always happen in practice. Where labor is expensive or scarce, microirrigation offers a good alternative. While microirrigation typically uses less labor than many other systems, it requires a more specialized training.

22.5.4 Considerations in Irrigation Method Selection

The factors that influence selection are so many and so varied that it is extremely difficult to make specific recommendations on which method is preferred above all others for a particular setting. Rather, it is usually appropriate to compare several alternatives before a final selection is made. The final selection is usually up to the discretion of the owner or financier of the irrigation system and not the designer or consultant. Also, new techniques or refinements of old techniques continue to be made, thus potentially expanding the ranges of suitable conditions for each method.

For the individual investor, it is important to define the objectives and expectations of the irrigation system before examination of selection conditions. Net economic return is usually the primary consideration. Other considerations might cause one to accept a system with other than maximum expected profit. All farming enterprises have an element of risk and the risk characteristics of the decision maker or system selector may alter the selection based on perceived risks. Future energy prices are always uncertain, and selection may depend on the expectation of these prices. Environmental goals, both from a personal and a societal perspective, may cause one to choose a particular design over another. Environmental assessments are becoming the rule rather than the exception in new irrigation development. The security of water supplies is always an issue, particularly with increasing pressures from municipal users and fish and wildlife needs, which may influence investment decisions.

The selection process can be divided into three main areas of consideration: physical site conditions, social or institutional considerations, and economic considerations. General site conditions that should be considered are listed in Table 22.1. A more thorough discussion of these issues is given in Clemmens and Dedrick (1994); also see Merriam and Keller (1978).

TABLE 22.1 Physical Site Conditions to Consider in Irrigation System Selection

Crops	Land	Climate
Crops grown	Field shape	Precipitation
Crop rotation	Obstructions	Temperature
Crop height/root volume	Topography	Frost conditions
Cultural practices	Soil	Humidity
Disease potential	Texture	Wind
Pests	Uniformity	
Water requirements	Depth	Energy
Climate modification	Intake rate	Availability
	Water-holding capacity	Reliability
Water supply	Erodibility	Cost
Source	Salinity	
Quantity	Drainability	
Quality	Bearing strength	
Salinity	Flood hazard	
Sodicity	Water table	
Sediments		
Organics		
Reliability		
Delivery		
Schedule		
Frequency		
Rate		
Duration		
Water cost		

Source: Adapted from Keller and Bliesner (1990) and Blair, et al. (1991).

An important issue in the selection of irrigation methods is the familiarity of the owner, management, and labor force with the method chosen. Choosing appropriate technology is not strictly a matter of having equipment, parts, and labor available, but is also related to local knowledge and support infrastructure.

One must also consider the legal issues related to water rights and laws. What responsibilities does the manager have for measuring and recording water use, and how does this affect the type of system and layout? How certain is the legal right to the water supply, or how senior is the right under prior appropriation laws? These questions need to be addressed when examining the desired longevity of the system chosen. If an irrigation system is currently in place, the cost of adopting a new system will be influenced by existing facilities.

Economic considerations play an important role in irrigation-system selection. The economic life of various irrigation-system components plus the annual maintenance costs as a percentage of initial cost should be considered in addition to capital cost and annual operating costs. The availability of money for investment in irrigation improvements may limit the growers' ability to select a system, which on paper is economically optimum. If lenders are not willing to invest in certain levels of technology, then other systems will likely be chosen.

There is always considerable debate on the *potential application efficiencies* (PAE, defined in a following section) of the different methods. Poor management and poor design complicate the issue and cause the range of actual performance to be extremely wide for all methods. What is relevant here is what performance is possible for these systems if they are reasonably well designed and managed. A range of PAE values for a number of selected methods is shown in Table 22.2 (adapted from Blair, et al., 1991; Keller and Bliesner, 1990; USDA, 1991; Dedrick, 1984; and others). In general such efficiencies are only achieved when management has some incentive to be efficient, for example, high water cost or limited supplies. Field-observed application efficiencies are usually within these ranges for these various systems (i.e., these are not based on theoretical considerations alone).

TABLE 22.2 Typical Potential Application Efficiencies (PAE) for Well-Designed and -Managed Irrigation Systems

Irrigation system	PAE
Furrow	50–70
Furrow: modern	60–80
Border strip	55–80
Basin (with or without furrows)	65–90
Basin: paddy	40–60
Bubbler: low head	80–90
Microspray	85–90
Micro: point source	85–90
Micro: line source	85–90
LEPA	80–90
Linear move	75–90
Center pivot	75–90
Traveling gun	60–75
Side roll	65–85
Hand move	65–85
Solid set	70–85

Finally, the selector must consider the length of the investment period and the long-term sustainability of the irrigation enterprise. Will the system selected help to maintain soil tilth and maintain crop productivity? More recently, the selector must also consider external factors such as environmental impact. For example, how will the irrigation system selected degrade the downstream water supply? Will drainage from the irrigated area be restricted in the future for environmental reasons? These questions can only be answered with a thorough investigation of specific site conditions.

22.6 IRRIGATION-SYSTEM PERFORMANCE

The performance of irrigated agriculture can be assessed from a number of different perspectives—rural development, economic return, environment, and so forth. Similarly, irrigation-system performance can be described from different perspectives. From a crop-production perspective, the irrigation-system performance is often described by the *irrigation efficiency,*

$$E_I = \frac{\text{volume of irrigation water beneficially used}}{\text{volume of irrigation water supplied}} \times 100\% \qquad (22.1)$$

This equation describes water use in terms of benefit for agricultural production. Beneficial uses include crop evapotranspiration (ET), water needed for removing excess salt from the soil, climate control, soil preparation, and weed control. Some of these uses actually consume water, while others do not. Other efficiencies for various parts of irrigation systems are discussed in several publications (Jensen, 1967; ASCE, 1978; Bos and Nugteren, 1978; Bos, 1979; Heermann, et al., 1990).

From the perspective of a hydrologic balance, the use of water by agriculture can be described by the *consumptive use coefficient*

$$C_{cu} = \frac{\text{volume of irrigation water consumptively used}}{\text{volume of irrigation water supplied}} \times 100\% \qquad (22.2)$$

This relationship includes uses that are not directly beneficial for agricultural production, such as phreatophyte ET and sprinkler and reservoir evaporation. These two definitions of water use are not the same, but are related. The relationship between them can be deduced from the illustration in Fig. 22.6.

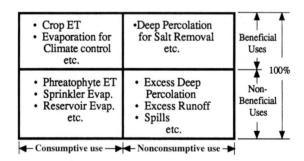

FIGURE 22.6   The division between consumptive and nonconsumptive uses differs from the division between beneficial and nonbeneficial uses.

Several points are worth emphasizing:

1. Nonconsumptive beneficial uses, particularly for maintaining salt balance in the root zone, are essential for sustained agricultural production and cannot be totally eliminated.

2. Reduction in the volume of nonconsumptive uses does not create new water supplies because this water remains in liquid form and may already be part of a reuse scheme (unless it flows to a saline sink).

3. Reductions in nonconsumptive uses do not reduce the total amount of salt that must be removed, because the major amount of salt to be removed is the amount of salt that was contained in the water consumed. Additional salts may be leached from lower soil and rock layers (salt loading). Here salt loading is influenced by the volume of deep-percolation water.

4. For most crops, production is directly related to crop consumptive use of water. Thus, reductions in crop consumptive use are frequently related to reductions in crop yield.

5. Reducing volume of nonconsumptive uses will usually have an associated cost and must produce a significant benefit to be justified.

The components in the earlier equations (and Fig. 22.6) are frequently difficult to determine with satisfactory accuracy. This results from the large spatial distribution of conditions for the large scale and measurement difficulties for the small scale. These equations can be applied to any scale of interest if the boundaries are clearly defined. Few efficiency studies satisfactorily define boundaries that allow accurate determination of all inflows and outflows.

The main components to determine are the irrigation inflow, rain, outflows, crop consumptive use, and water required for leaching. Crop water requirements for non-water-stressed crops can be estimated for most common agricultural crops in the United States with reasonable accuracy using weather-based calculations (Jensen, et al., 1990a). However, most fields have areas where water or salinity stress occurs due to nonuniformity of soils, topography, and so on. For large projects, variations in weather between stations can be significant. There also may be uncertainty in determining cropped areas. Thus estimates of actual beneficial consumptive uses are typically uncertain. Nonbeneficial evaporative losses are difficult to determine, among other water uses that are hard to quantify. The result is that accurate knowledge on agricultural-irrigation performance is lacking. These estimates are more difficult in areas where rainfall is significant because the amount of rainfall is not accurately known.

These problems are beyond the scope of needed assessments of field-irrigation systems. As a result, different terms are often used for defining the performance of the irrigation system itself (i.e., separated from crop management). Common terms are the *application efficiency* E_a and *low-quarter distribution uniformity,* DU_{lq}, defined in terms of that 25 percent of the field receiving the least depth of infiltrated water (ASAE, 1993):

$$E_a = \frac{\text{average depth of water stored in the root zone}}{\text{average depth applied}} \tag{22.3}$$

$$DU_{lq} = \frac{\text{average low quarter depth of water infiltrated}}{\text{average depth of water infiltrated}} \tag{22.4}$$

The numerator of the first equation refers to water which is available for consumptive use by the plant, and is eventually used for that purpose. This includes water added to root-zone soil moisture and water that evaporates instead of transpiring. In this sense, it considers a potential use of water—one that the irrigation system was trying to supply. These equations assume that the need for water is uniform over the field. However, due to soil differences and crop growth differences (some of which results from poor irrigation uniformity), this is not the case.

All irrigation systems apply water nonuniformly to a varying degree. Thus, for each irrigation event the farm manager is faced with a tradeoff between the total amount of water to apply and the amount of soil-water deficit to allow. Most farmers do not make such decisions based on analysis, but rather develop rules of thumb or use judgment to suit their particular needs. Such analysis or judgment must evaluate what effect each irrigation has on total yield. While some attempt to express this tradeoff in terms of seasonal water applied and yield (i.e., a simple economic model), the process is usually not that simple. First, there is uncertainty about the relationship between water applied and yield. Some of this uncertainty is due to uncertainty about irrigation uniformity, some due to uncertainty about yield potential for a given year, some due to uncertainty about the effects of irrigation timing. In the face of uncertainty, when water is plentiful and costs are low, farmers tend to apply more water than would be justified by simple economic analysis based on certainty. The costs associated with the excess are usually not considered by individual farmers. Also, yield reductions associated with excess water applications are less than that associated with water deficits.

Part of the difficulty with evaluating irrigation uniformity is the number of variables that contribute to *nonuniformity of water application*. Such factors include:

Surface irrigation.   Differences in opportunity time for infiltration caused by advance and recession, spatial variability of soil-infiltration properties, variations and undulations in soil-surface topography (including side fall), variations in flow rate at the supply end, and variations in application time for different parts of a field.

Sprinkler irrigation.   Variations in pressure caused by pipe friction and topography, sprinkler pattern effects, differences in wind drift, runoff due to excessive application rates, variation in timing, and field edge effects.

Microirrigation.   Variations in pressure caused by pipe friction and topography, variations in hydraulic properties of emitters or emission points (from manufacturing or from clogging), variations in soil wetting from emission points (e.g., line source along crop row), and variations in timing.

Many of these factors are difficult to quantify, even in a research setting, which adds to uncertainty. If these factors can be quantified, techniques are available for determining the influence of each on overall irrigation uniformity (Clemmens, 1991).

22.7   GOVERNMENT ASSISTANCE PROGRAMS

Government programs that benefit farmers result from various policies and priorities of governments. For example, in the United States low-interest loans were provided to develop water resources in the west at the beginning of the twentieth century. These programs were intended to help provide the needed infrastructure to populate western lands. The loans for water-supply development resulted in the development of hydroelectric power. In many cases, the irrigation districts paid for the development of both water and power and now have access to inexpensive power. If they had paid the full interest charges at that time, current costs would be higher. For districts that have paid back their loans to the federal government, the current low cost of water and power results in low operation and maintenance costs.

Inexpensive power was also made available for the development of irrigation projects where pumping was required. This inexpensive power made irrigation feasible in areas where otherwise it might have been uneconomical. These power rates, however, were part of the contract with the irrigation developers to justify their investment. Also, some irrigation districts still receive water from federally built and owned facilities at low cost, even though they did not pay for any of that infrastructure.

Many of the agricultural programs have been aimed at conserving soil and water resources. The Dust Bowl of the Great Depression caused changes in land-use practices. For many of these programs, the government was willing to pay part of the cost of a conservation-improvement project that might not provide an economic return to the farm, but is in the general interest of conservation and protection of resources. While much progress has been made, the need to improve water quality has resulted in a continuing need for some of these programs.

Variations in weather and crop prices in the past have caused large fluctuations in farm income. This unstable situation caused difficulty in sustaining an individual agricultural venture. Large diversified farms can generally withstand these fluctuations. However, small family farms that rely on one main crop (i.e., one that is most profitable due to local climate and soils) cannot withstand these cycles. Programs were established to protect these family farms by stabilizing crop prices and providing inexpensive crop insurance. In return for these guarantees, the farmers agree to various conservation measures and reductions in cropped acres (where low prices are caused by excess production).

To a certain extent, these programs have all been successful at fulfilling the desired objectives. While some now view these as wasteful, unnecessary subsidies, they were provided to fulfill some objective of society, not strictly to increase the profit to a few individuals. However, there are side effects to these programs that we now are having to consider. They have caused the development of some irrigation projects that may not be self-sustaining without the development incentives, price supports, or subsidies.

An example in supplemental-irrigation areas is the expansion of irrigated acreage in the western Great Plains in the 1950s and 1960s, which peaked in 1969 to 1978. Wheat-price supports caused the development of irrigation in dryland wheat areas above the Ogallala aquifer. This irrigation mined a valuable groundwater supply to produce surplus wheat—considered by most to have been a national-policy disaster. However, does the fault lie with the government policy or the agricultural entrepreneur? These issues in arid irrigated agriculture are more difficult. Removing support for irrigation may remove the underpinning for a rural society heavily relying on it. Thus, it works toward free markets but against rural development.

22.7.1 Research and Extension

Another form of support is through government-provided research and extension activities beneficial to irrigated agriculture. Agricultural research and extension programs were developed more than a century ago to improve the production of food and fiber. They have been a dramatic success and made the United States a world leader in agricultural production. These programs continue today and are focusing on issues other than production, such as water quality. These extension programs are also geared toward urban consumers—nutrition, landscaping, home economics, and so forth. These programs were aimed at improving the standard of living for the nation as a whole to stabilize food prices. It is argued that these programs have contributed heavily to the relatively low consumer food prices in the United States. These irrigation-support activities are generally handled through land grant universities (universities that were originally granted land and funds from the federal government provided that they teach agriculture and engineering, but are now mostly state supported), the U.S. Department of Agriculture, and the U.S. Bureau of Reclamation. Other agencies are also involved, including non-land-grant universities, the Bureau of Indian Affairs, the Bureau of Land Management, the Environmental Protection Agency, and the Department of Defense. A few of the activities supported are research and extension, or technology transfers, concerning irrigation-water requirements for crops (Erie, et al., 1982); flow-measurement techniques (Bos, et al., 1991; Bos, 1989); and groundwater-quality impacts of irrigation (Bouwer, et al., 1990).

22.8 WATER-RIGHTS ISSUES

22.8.1 Western Water Law and Irrigation History in the American West

Ownership rights to water and to its use varies by state, region within a state, and whether it is groundwater or surface water. Even this latter division is clouded. A well in gravel very near a stream that is essentially pumping from the stream might be classified as either in some states in the United States. Also see Chap. 5.

The doctrine of riparian water rights, used mostly in the eastern United States, grants the owner of the land adjacent to the water supply the right to its use. That is, the water right is attached to the land ownership. The doctrine follows the legal theory of reasonable use with the general guideline that a given water use must be compatible with other uses of the same source. A riparian owner whose water use is not materially affected by a reduction in stream flow has no basis for legal complaint. Some states now interpret the riparian rights doctrine to include diffuse surface water that is flowing across the ground surface before reaching a defined channel. Much of this water is the source for recharge for many ponds in the southeastern United States used for supplemental irrigation. Several eastern states do not recognize the doctrine of riparian rights, but use a modified form of the prior appropriation doctrine or other type of priority or permit system.

In most of the irrigated western United States the concept of *prior appropriation doctrine* prevails. These appropriative rights are usually translated as "first in time, first in right" to use the water beneficially, and applies primarily to water diverted from a stream. Pumping rights may follow the same patterns, but in most states these have been given special protection that differs from the surface-water rights. Because "beneficial use" is not rigorously and uniformly defined, this aspect of appropriative rights to surface waters is becoming a concern to present water users who anticipate that political power shifts may cause redefinition of beneficial use and reallocate existing supplies away from current uses. Water-resources specialists and society as a whole will probably be asked to deal with the shifts in an equitable and socially beneficial way. Again the definition of social benefit is a moving target. In some states, the public-trust doctrine is being used to protect environmental and ecosystem water uses (NRC, 1992).

Through the prior appropriation doctrine, a property right to water use is established that can be transferred or sold, subject to local laws and regulations. In those cases where a farmer has a specific quantity of water each year, his objective is usually to allocate the specific quantity among his various crops within a cropping season, which is currently considered to be beneficial use. If the farmer reduced water consumption through efficiencies or crop changes, his savings are not his but may be reallocated by the governing authority. This is a "use or lose" case, and it is readily argued that it does not encourage conservation. Thus, institutional planners usually attempt economic incentives to encourage efficiency and conservation measures.

Another example of institutional influence concerns a groundwater law in Arizona that has had a strong impact on irrigation. The law specifies that farmers must achieve certain levels of water conservation by given target dates. The objective is to conserve dwindling groundwater supplies. The result has been that farmers have strong incentives to adopt new equipment and water-management procedures as compared to the farmer that has a perpetual right to an annual given water quantity. The downside is that immediate extreme economic hardship may be experienced by the farmer and the supporting community. The counterargument to this is that it merely displaces the hardship in time, because the water, which is being mined, will be gone eventually and then no sustained community of any size is probable.

The long-term prospects concerning water institutions are that the appropriative doctrine prevalent in the water-scarce western United States can be expected to increase water sales by farmers to other users, and will encourage, if not require, more efficient water use in irrigated agriculture (Weatherford and Ingram, 1984). However, water-right sales involve mainly the portion that is consumed and not the total diverted source. The excess is usually the supply for a downstream user. Efficient water use in irrigated agriculture, in terms of production per unit of water consumed, will become increasingly important, with higher priority placed on improved strategies for decision making.

22.8.2 Water Transfers

Transfers of water from agriculture to municipal and industrial users will reduce the water supply available to irrigated agriculture. The net result is likely to increase the year-to-year variation in the supply to irrigation. This will require irrigators to incorporate more sophisticated risk strategies into equipment selection and irrigation management. In some areas, farmers are permitted to sell water on a year-to-year basis when supplies are low. A National Research Council committee recently completed a comprehensive assessment of water transfers in the western United States (NRC, 1992).

Interbasin and intrabasin transfers of water have been used extensively in the past. However, future schemes of water-transfer projects worldwide are facing formidable economic, political, and environmental constraints. Federally funded projects in most developed countries have drastically declined. New irrigation projects are still possible in some developing countries in Africa, Latin America, the Near East, and Asia. Adverse publicity and problems of increasing sediment loads in existing rivers, due in part to erosion caused by population pressures in existing food-production areas, severely limit the life of new projects that might reduce those pressures. The Tarbella Dam in Pakistan is estimated to have a half-life of 60 years because of sedimentation (Rangeley, 1986).

Water pricing has been frequently proposed by advocates of water conservation as a tool to force efficient use of water. The various proposals have gathered emotional advocates and opponents. Not least among the problems is the vesting of power to set the prices, the legal ownership of the water in the first place, and the depository and use of funds collected. Proponents of free marketing of water are gaining supporters. Moves in that direction are slow but appear to be steady. The local community impacts of moving water, called *third-party impacts,* must be considered against the end use at the purchasing community.

22.8.3 Conjunctive Uses of Water

The term *conjunctive use* originated in California and originally referred to the storing of surplus surface water in the rainy season for later use during periods of drought (Bradley, 1986). Subsequently it has evolved to mean coordinating the use of groundwater and surface water in time and space (Bradley, 1986; Hoffman, et al., 1990). The name applies to several practices and processes of coordinating the management of groundwater and surface waters for maximum social benefits from each (Bradley, 1986). When surface water is scarce, groundwater is pumped to make up the deficit. A conjunctive-use system may also recharge the groundwater supplies in wet periods with surplus surface water and may use aquifers for long-term storage (Goldman, 1986).

Despite the many advantages of conjunctive-use systems that make them attractive for water-resources projects, they are scarce because of their complexity (Goldman, 1986). Major advantages of conjunctive-use systems are the optimization of water uses. For example, excess river water can be diverted and stored in an aquifer and then withdrawn during periods of surface-water shortage. These systems can increase the usable yield of water that might otherwise be lost. This is particularly true for coastal cities where the loss would usually be to the ocean.

Other advantages for conjunctive-use schemes include stability of the water supply and the reduction of required capacity of the reservoirs. Because excess surface water is stored underground and can be used to augment the reservoir-water supply during drought periods, the reservoirs may maintain a smaller capacity than otherwise required. This can leave more capacity for flood control (Coe, 1990).

Greater flexibility in irrigation techniques may also be obtained from a conjunctive-use system. In some places water from one source may be high in salt content. Most plants are sensitive to salts, particularly during early growth stages, and irrigating with saline water can reduce crop yield. However, by applying the more-saline source during less-sensitive growth periods for the particular crop, saline waters can be effectively used (Hoffman, et al., 1990).

Further advantages of conjunctive use include proximity of the water users to stored groundwater for easy use in emergencies when the surface-water supply is interrupted and reduced need to line canals if seepage in the canal recharges the intended aquifer. Conjunctive use reduces the major problems associated with using either surface water or groundwater when they are used separately. That is, drainage problems from surface water are reduced because the excess water is infiltrated to the groundwater, and overdraft of groundwater is lessened because the surface water is used for recharge (Coe, 1990).

Conjunctive use cannot be considered a panacea for semiarid lands. Even though these systems offer benefits that make them attractive to many people, the physical, modeling, management, and legal drawbacks associated with conjunctive-use systems prevent them from being a cure-all. In areas where severe overdraft of groundwater has occurred, conjunctive use may not be able to restore the water supply or storage capacity. A safe, continuous, annual yield may be eventually obtainable from a conjunctive-use system, but the initial overdraft may not be replaceable because of irreversible consolidation of the aquifer during the initial removal of water.

One concept is to conjunctively nonuse water rather than to conjunctively use water (Bradley, 1986). Under this concept, water is stored permanently in the aquifer once all other water commitments have been met. This expands the goal of conjunctive use to include not only increasing the yield of a water-resource system but also reducing the amount of overdraft in aquifers. In general, most systems will require careful planning and modeling. In many places of the world, wells have been drilled by individual farmers to supplement shortages in surface water due to droughts without the benefit of central planning. Many of these are not true conjunctive-use systems, because they have no planned recharge component. With effective and efficient planning and management, a conjunctive-use system can optimize the yield of the total water resource.

22.9 CURRENT ISSUES

Current issues in water-resources planning related to irrigation include most of the points discussed previously in one form or the other.

22.9.1 Water Supply Competition

There are numerous emerging demands for water (U.S. Water Resources Research Council, 1978). Many citizens today seem highly concerned for environmental protection, consider irrigation unfriendly to the environment, and thus oppose expanding irrigation. Increasing emphasis is being placed on preserving free-flowing streams and preserving natural conditions for fish and wildlife. However, in some areas, free-flowing rivers would become dry without irrigation-return flows. Direct competition for water use from irrigation and hydroelectric power has become a major issue in the Pacific Northwest. Irrigation, as the largest consumer of water, is receiving increasing scrutiny and will likely be required to transfer supplies to other elements of societal use.

22.9.2 Some Effects of Changing Irrigation Uses to Other Uses

One of the effects of transferring water to urban uses is the increase in sewage-water flow from the urban area that can be an irrigation-water supply when properly treated (Pettygrove and Asano, 1985). Thus, it is likely that as irrigated areas upstream or within an urban area are populated, a new area of irrigated agriculture will form downstream of the city (Bouwer, 1993). Also, domestic users will have first use of water and irrigation will become a second user of water.

22.9.3 Need to Decommission Abandoned Systems

Many communities that expanded because of mining groundwater for a few years of crop production are now going through reduction because of uneconomical pumping lifts or complete water depletion. These areas need some sort of adjustment plan to aid in the return to nonirrigated agriculture or to their natural condition. These plans and the funding and research questions concerning effective implementation remain an open issue.

Parties associated with some existing irrigation systems today face arguments that the systems are not economically feasible and do not repay the construction costs, and therefore should be closed. Alternative arguments say that it is economically a sunk cost that cannot be recovered by closure, and that continued operation returns more productive economic value, even if not enough, than would be returned if closed. The validity of these arguments depends on the site-specific opportunities for alternate uses of the land, labor, and water resources. A primary consideration is the operating cost. If the project can cover the operating and maintenance costs, and there is no ready alternate use of the water, then it is argued that a project should be maintained and operated even if the capital loan expenses cannot be met and are defaulted. Again this is site specific and guidelines vary depending on social, economic, and political climate.

In practice, some projects are still operating that are not meeting maintenance, capital replacement, or loan costs. They are meeting only day-to-day operating costs of salary and emergency repairs, and will eventually stop operating when the system becomes essentially nonfunctional. This meets economic goals of recovering as much investment as possible from aging operating equipment and closing the plant. However, the real problems of environmental protection need to be addressed when systems are decommissioned.

22.9.4   Rehabilitation

Subsidized loans for development and rehabilitation of irrigation-delivery systems have been practiced for several decades. This has come about because many projects were constructed that failed to generate funds, for one reason or another, to support continual maintenance. This problem is more serious in the developing countries than in the United States. As a result, these projects deteriorated to the point that additional external loans, frequently government subsidized, were sought for rehabilitation. As these rehabilitation loans become more difficult to obtain, emphasis has turned to improving operation and maintenance, which requires increased collection of funds from the end users.

If the project cannot deliver water in proper amounts and at proper timing, agricultural profits are reduced and sufficient funds are not available for maintenance. The result is a generally declining system, and the eventual need for rehabilitation. However, it is important that these projects not be rehabilitated to the original design, if rehabilitation is justified, but that they be rehabilitated to a level that can be sustained by the anticipated agricultural production.

22.9.5   Global Commodity Competition

Developed countries appear to be eternally dealing with crop surpluses. World hunger seems to be more a global-distribution and political artifact than actual supply shortage. This food surplus picture may be only a temporary phenomenon and may soon be overrun by growing world population. Recent trends in developing countries toward smaller families seem to be happening faster than predictions of only a few years ago. Thus projections of population trends are hard to make because of the changing social and governmental factors. However, most projections see a world population of 8 to 10 billion by about the year 2025 before a leveling. This will place a high dependance on irrigated agriculture for a stable world food supply.

22.9.6   Return on Investment to a Nation

Economists frequently speak of production and services. Agricultural production is part of the wealth of a nation, and is considered basic to providing the foundation for improving living standards and economic growth. It extends beyond services. Modern agriculture in developed nations requires from 2 to 4 percent of the population working in direct agriculture. Estimates of indirect food-chain workers may increase this to 35 percent. Developing nations typically have over 75 percent of the population involved in acquiring food, frequently only at subsistence levels. The importance of agriculture needs to be considered in the context of overall national infrastructure. The infrastructure makes efficient agricultural production and marketing possible, while the relatively low consumer involvement (cost) allows attention to effective infrastructure.

22.9.7 Changes in Farm Types and Crop Diversification

Farming operations and management techniques are dynamic elements in the economy. Trends to larger farm units are market driven by relatively low farm-commodity prices and technological innovations that somehow still seem to provide a profit to the larger units and more efficient managers. The trends toward larger farms, larger machines, minimum-tillage techniques, and large capital investments in effective machinery is likely to continue. In places where water costs are increasing in the face of low commodity prices, farm abandonments are likely. In the United States, the consumer desire for fresh produce is likely to increase, and more farming operations will be needed to service this need. Most of this is practical under irrigation techniques that allow good quality control. Thus irrigation of fresh produce is likely to increase, while irrigation of animal foodstuff is likely to decrease.

22.9.8 Water Quality and Salinity

Water quality will be a continuing issue. The concentration of salts resulting from crop water consumption is a physical reality, independent of irrigation practices, which must be dealt with. Runoff from irrigation can be laced with insecticides or other chemicals. Excess infiltrated water that percolates below the root zone can contain these salts and chemicals and in some locations leach additional chemicals from underlying soil and rock. This becomes a problem if it reaches a recoverable groundwater source intended for other uses. Special management techniques are required (Gates and Grismer, 1989; Grismer, 1993).

22.9.9 Impact on Water and Other Resources

Water is becoming a limited resource. The involvement of interdisciplinary planning will be required to effectively make the institutional changes and management changes needed to optimally use the water resource and the attendant resources that depend on its presence.

22.10 SUMMARY

This chapter interprets some irrigation trends and their effects on the community of water scientists, engineers, and others likely to be asked to furnish information, services, and creative ideas.

In the United States alone, about 57 million acres (23 million ha) are irrigated, representing about 9 percent of the world's irrigated area. In terms of agricultural production, about 17 percent of the cropped area of the world is irrigated and contributes more than one third of the total world food production. In the United States about 12 percent of the cropped area is irrigated and contributes about 25 percent of the total value of U.S. crops. Almost 60 percent of the production of major cereals in the developing world, rice and wheat, is produced on irrigated land.

Most of the world's projects appear to have been developed on the assumption that drainage requirements will not materialize or will be afforded by the inherent increase in the economic growth in the project. Often drainage problems occur sooner than anticipated due to canal seepage and low farm-irrigation efficiencies that result in large deep-percolation losses. Two-pronged approaches are usually needed: (1) reduce the seepage losses to approach the natural drainage rate to an area, and (2) install artificial drainage when the natural-drainage rate is insufficient.

Within the irrigation infrastructure the many types of organizations include various types of districts, such as irrigation, drainage, water-reclamation, conservancy, and water-storage districts; mutual canal and ditch companies, both public and private; commercial companies; and others organized for various purposes. These organizations frequently furnish both irrigation water and drainage service to agricultural lands, and some distribute electric power and other services.

Most irrigation districts derive a majority of their revenue through the sale of water and delivery services to agricultural producers. Because these are frequently nonprofit, quasi-government entities, charges are set to balance expenses. Farmers, through their water organization board of directors, usually participate in deciding the charges for water and the level of service to be provided. Irrigation is only one component of a farm enterprise. In arid areas, farming is not possible, or at least not practical, without irrigation. In other areas, irrigation is supplemental and is only one of many optional inputs, and the decision to irrigate is simply financial.

There are three basic types of irrigation methods: surface irrigation, sprinkler irrigation, and microirrigation. Some irrigation methods are hard to categorize because they cross the traditional boundaries of these three general types. Surface-irrigation methods are so named because water is distributed across the field by flowing over the field surface. Thus soil infiltration and soil and crop flow resistance and topography have a major influence on the distribution of water. By contrast, under sprinkler and microirrigation, most water distribution is accomplished in closed pipelines.

Selecting an irrigation system is complex. The factors that influence system selection are so many and so varied that it is difficult to make firm recommendations for a particular setting. Rather, it is usually appropriate to compare several alternatives before a final selection is made.

Several points should be emphasized regarding sources and destinations of applied irrigation water:

- Nonconsumptive beneficial uses, particularly for maintaining salt balance in the crop root zone, are essential for sustained agricultural production and cannot be totally eliminated.

- Reduction in the volume of nonconsumptive uses does not create new water supplies because this water remains in liquid form and already may be committed to a reuse.

- Reduction in nonconsumptive uses does not reduce the total amount of salt that must be removed, because the major amount of salt to be removed is the amount of salt contained in the water consumed.

- For most crops, production is directly related to consumptive use of water. Thus, reductions in consumption usually reduce crop yield.

22.10.1 Current Issues in Water Resources Planning Related to Irrigation

Water-supply competition caused by the many emerging demands for water involve environmental and wildlife protection and alternate uses. These demands will likely result in decreased irrigation in the United States. However, changes in water use often result in increased sewage water from urban areas that can become an irrigation-water supply when properly treated.

Needs to decommission abandoned systems have become an issue because many irrigation systems have simply been abandoned and have caused undesirable environmental effects. Thus the real problems of environmental protection need to be addressed when systems go out of business.

Rehabilitation requirements of projects worldwide are a more serious problem in the developing countries than in the United States. These projects have deteriorated to the point that additional external loans are sought for rehabilitation. If rehabilitation is justified, consider rehabilitating the project to a reduced level that can be supported by current projected agricultural production.

Agricultural production is part of the wealth of a nation, and is considered basic to providing the foundation for cultural advances and extends beyond services. Developing nations typically have over 75 percent of the population involved in acquiring food, frequently only at subsistence levels. The importance of agriculture needs to be considered in the context of overall national infrastructure.

The trends toward larger farms, larger machines, minimum-tillage techniques, and large capital investments in effective machinery is likely to continue. In the United States the consumer desire for fresh produce is likely to increase, and more farming operations will be needed to service this need. Most of this is practical under irrigation techniques that allow good quality control. Thus irrigation of fresh produce is likely to increase, while irrigation of animal foodstuff is likely to decrease.

Water quality will be a continuing issue. Runoff from irrigation can contain high levels of insecticides or other chemicals. Deep seepage can thus become a problem if it reaches a recoverable groundwater source.

Water is becoming a limited resource. The involvement of interdisciplinary planning will be required to effectively make the institutional changes and management changes needed to optimally use the water resource and the attendant resources that depend on its presence.

REFERENCES

American Society of Agricultural Engineers, "Soil and Water Engineering Terminology," *Standard ASAE S526,* American Society of Agricultural Engineers, St. Joseph, Mich., 1993.

American Society of Civil Engineers, "Describing Irrigation Efficiency and Uniformity," *Journal of Irrigation and Drainage Division,* ASCE, 104(IR1):35–41, 1978.

American Society of Civil Engineers, "Management, Operation and Maintenance of Irrigation and Drainage Systems," *ASCE Manuals and Reports on Engineering Practice* no. 57, American Society of Civil Engineers, New York, 1991.

Bishop, A. A., W. R. Walker, N. L. Allen, and G. J. Poole, "Furrow Advance Rates under Surge Flow Systems," *Journal, Irrigation and Drainage Division,* ASCE, 107(IR3):257–264, 1981.

Blair, A. W., R. D. Bliesner, and J. L. Merriam, eds., "Selection of Irrigation Methods for Irrigated Agriculture," ASCE On-Farm Irrigation Committee Report (Draft), American Society of Civil Engineers, New York, 1991.

Bos, M. G., and J. Nugteren, *On Irrigation Efficiencies,* 2d ed., International Institute for Land Reclamation and Improvement/ILRI, Wageningen, Netherlands, 1978.

Bos, M. G., ed., *Discharge Measurement Structures,* pub. 20, 3d ed., International Institute for Land Reclamation and Improvement/ILRI, Wageningen, The Netherlands, 1989.

Bos, M. G., "Standards for Irrigation Efficiencies of ICID," *Journal of Irrigation and Drainage Division,* ASCE, 105(IR1):37–43, 1979.

Bos, M. G., J. A. Replogle, and A. J. Clemmens, *Flow Measuring Flumes for Open Channel Systems,* American Society of Agricultural Engineers, St. Joseph, Mich. (republication of John Wiley, 1984 ed.), 1991.

Bouwer, H., "Design Considerations for Earth Linings for Seepage Control," *Ground Water,* 20(5):531–537, 1982.

Bouwer, H., *Groundwater Hydrology,* McGraw-Hill, New York, 1978.

Bouwer, H., "Theory of Seepage from Open Channels," in V. T. Chow, ed., *Advances in Hydroscience,* Chap. 2, Academic Press, New York, 1969, pp. 121–170.

Bouwer, H., A. R. Dedrick, and D. B. Jaynes, "Irrigation Management for Groundwater Quality Protection," *Irrigation and Drainage Systems, An International Journal,* 4:375–383, 1990.

Bouwer, H., "Urban and Agricultural Competition for Water, and Water Reuse," in Richard G. Allen, ed., *Management of Irrigation and Drainage Systems, Integrated Perspectives,* American Society of Civil Engineers National Conference on Irrigation and Drainage, 21–23 July, Park City, Utah, 1993, pp. 79–84.

Bradley, M. D., "The Time Has Come to Define the Conjunctive Management of Water Resources," *Ground Water,* 24:551–552, 1986.

Clemmens, A. J., "Irrigation Uniformity Relationships for Irrigation System Management," *Journal of Irrigation and Drainage Division,* ASCE, 117(5):682–699, 1991.

Clemmens, A. J., and A. R. Dedrick, "Limits for Practical Level Basin Design," *Journal of Irrigation and Drainage Division,* ASCE, 108(IR2):127–141, 1982.

Clemmens, A. J., and Dedrick, A. R., "Irrigation Techniques and Evaluations," in K. K. Tanji and B. Yaron, eds., *Management of Water Use in Agriculture,* Advanced Series in Agricultural Sciences, vol. 22, Chap. 4, Springer-Verlag, Berlin, 1994, pp. 64–103.

Coe, J. J., "Conjunctive Use—Advantages, Constraints, and Examples," *Journal of Irrigation and Drainage Division,* ASCE, 116(3):427–443, 1990.

Dedrick, A. R., "Cotton Yields and Water Use on Improved Furrow Irrigation Systems," in *Water Today and Tomorrow,* Proc. Specialty Conference of the Irrigation and Drainage Division, American Society of Civil Engineers, New York, 1984, pp. 175–182.

Dedrick, A. R., "Control Requirements and Field Experience with Mechanized Level Basins," *Transactions, ASAE,* 29(6):1679–1684, 1986.

Dedrick, A. R., "Improvements in Design and Installation Features for Mechanized Level Basin Systems," *Applied Engineering in Agriculture, ASAE,* 5(3):372–378, 1989.

Dedrick, A. R., "Level-Basin Irrigation—An Update," in *Visions of the Future,* Proc. 3d Nat. Irrig. Symp., 28 Oct.–1 Nov., Phoenix, Ariz., 1990, p. 34–39.

Dedrick, A. R., and Y. Reinink, "Precipitation and Irrigation on Level Basins," in *Transactions of the Thirteenth International Congress on Irrigation and Drainage, Proc. Thirteenth Int. Congress on Irrigation and Drainage,* Morocco, Sept., 1-B:1387–1398, 1987a.

Dedrick, A. R., and Y. Reinink, "Water Ponding on Level Basins Caused by Precipitation," *Transactions, ASAE,* 30(4):1057–1064, 1987b.

Engberg, R. A., and M. A. Sylvester, "Concentrations, Distribution, and Sources of Selenium from Irrigated Lands in Western United States," *Journal of Irrigation and Drainage Engineering,* ASCE, 119(IR3):522–536, 1993.

Erie, L. J., O. F. French, D. A. Bucks, and K. Harris, "Consumptive Use of Water for the Major Crops in the Southwest," USDA-ARS Conservation Research Report no. 29, 1982.

Field, William P., "World Irrigation," *Irrigation and Drainage Systems,* 4(2):91–107, 1990.

Food and Agriculture Organization, "Water for Agriculture," *Food and Agriculture Organization of the United Nations, UN Water Conference,* Mar Del Plata, March, 1977.

Food and Agriculture Organization, "Production 1992," *FAO Yearbook,* FAO Statistics Series 112, vol. 46, UN, Rome, 1993.

Gates, Timothy K., and Mark E. Grismer, "Irrigation and Drainage Strategies in Salinity-Affected Regions," *Journal of Irrigation and Drainage Engineering,* ASCE, New York, 115(IR2):255–284, 1989.

Gilley, J. R., and R. J. Supalla, "Economic Analysis of Energy Saving Practices in Irrigation," *Transactions of the ASAE,* 26(6):1784–1792, 1983.

Goldman, J. C., "Conjunctive Use and Managed Groundwater Recharge: Engineering and Politics in Arid Lands," *Water Forum '86: World Water Issues in Evolution,* Proc. of Conf. sponsored by American Society of Civil Engineers, New York, Long Beach, Calif., Aug. 4–6, 1:261–265, 1986.

Grismer, M. E., "Subsurface Drainage System Design and Drain Water Quality," *Journal of Irrigation and Drainage Engineering,* ASCE, 119(IR3):537–543, 1993.

Heermann, D. F., W. W. Wallender, and M. G. Bos, "Irrigation Efficiency and Uniformity," in G. J. Hoffman, T. A. Howell, and K. H. Solomon, eds., *Management of Farm Irrigation Systems,* American Society of Agricultural Engineers, St. Joseph, Mich., 1990, pp. 125–149.

Higgins, G. M., P. J. Dieleman, and C. L. Abernethy, "Trends in Irrigation Development, and Their Implications for Hydrologists and Water Resources Engineers," *Hydrological Sciences-Journal-des Sciences Hydrologiques,* 33:43–59, 1988.

Hoffman, G. J., T. A. Howell, and K. H. Solomon, eds., *Management of Farm Irrigation Systems,* American Society of Agricultural Engineers, St. Joseph, Mich., 1990.

Jensen, M. E., ed., "Evaluating Irrigation Efficiencies," *Journal of Irrigation and Drainage Division,* ASCE, 93(IR1):83–98. Discussions: 93(IR4):149–151; 94(IR2):279. Closure: 95(IR1):213–214, 1967.

Jensen, M. E., *Design and Operation of Farm Irrigation Systems,* ASAE Monograph no. 3, American Society of Agricultural Engineers, St. Joseph, Mich., 1980.

Jensen, M. E., R. D. Burman, and R. G. Allen, eds., "Evaporation and Irrigation Water Requirements," *ASCE Manuals and Reports on Engineering Practice* no. 70, American Society of Civil Engineers, New York, 1990a.

Jensen, M. E., W. R. Rangeley, and P. J. Dieleman, "Irrigation Trends in World Agriculture," in B. A. Stewart and D. R. Nielsen, eds., *Irrigation of Agricultural Crops,* Agronomy no. 30, American Society of Agronomy, Madison, Wis., 1990b.

Keller, J., and Bliesner, R. D., *Sprinkle and Trickle Irrigation,* Van Nostrand Reinhold, New York, 1990.

Merriam, J. L., 1987a. "Pipelines for Flexible Deliveries," in D. D. Zimbelman, ed., *Proc. Symposium on Planning, Operation, Rehabilitation and Automation of Irrigation Water Delivery Systems,* Proc. of American Society of Civil Engineers Specialty Conference, 28–30 July, Portland, Oreg., 1987a, pp. 208–214.

Merriam, J. L., 1987b. "Design of Semi-closed Pipeline Systems," in D. D. Zimbelman, ed., *Planning, Operation, Rehabilitation and Automation of Irrigation Water Delivery Systems,* Proc. of American Society of Civil Engineers Specialty Conference, 28–30 July, Portland, Oreg., 1987b, pp. 224–236.

Merriam, J. L., and Keller, J., *Farm Irrigation System Evaluation: A Guide for Management,* Agricultural and Irrigation Engineering Department, Utah State University, Logan, Utah, 1978.

National Research Council, *Water Transfers in the West: Efficiency, Equity, and the Environment,* Academic Press, Washington, D.C., 1992.

Pettygrove, G. S., and T. Asano, "Municipal Wastewater—A Guidance Manual," University of California, Davis, Department of Land, Air and Water Resources, for the California State Water Resources Control Board, Lewis, 1985.

Rangeley, W. R., "Global Water Issues," *Civil Engineering,* 96(12):60–62, 1986.

Shank, Bruce F., 1992. "1991 Irrigation Survey," *Irrigation Journal,* 42(1):19–26, 1992.

Tanji, K. K., ed., *Agricultural Salinity Assessment and Management,* ASCE Manuals and Reports on Engineering Practice no. 71, 1990.

Tanji, K. K., "Prognosis on Managing Trace Elements," *Journal of Irrigation and Drainage Engineering,* ASCE, 119(IR3):577–583, 1993.

Tanji, K. K., and F. F. Karajeh, "Saline Drain Water Reuse in Agroforestry Systems," *Journal of Irrigation and Drainage Engineering,* ASCE, 119(IR1):170–180, 1993.

Tanji, K., S. Ford, A. Tito, J. Summers, and L. Willardson, "Evaporation Ponds: What Are They; Why Some Concerns," in Richard G. Allen, ed., *Management of Irrig. and Drain. Systems, Integrated Perspectives,* American Society of Civil Engineers National Conference on Irrigation and Drainage, 21–23 July, Park City, Utah, 1993, pp. 573–579.

U.S. Department of Agriculture, *Farm Irrigation Rating Index (FIRI): A Method for Planning, Evaluating and Improving Irrigation Management,* USDA, Soil Conservation Service, Western National Technical Center, Portland, Oreg., 1991.

U.S. Department of Agriculture, "Food—From Farm to Table," *1982 Yearbook of Agriculture,* U.S. Government Printing Office, Washington, D.C., 1982.

U.S. Department of Agriculture, *Agricultural Statistics—1992,* U.S. Government Printing Office, Washington, D.C., 1992.

U.S. Water Resources Research Council, *The Nations Water Resources 1975–2000, vol. 1: Summary; vol. 2: Water Quantity, Quality, and Related Land Considerations; vol. 3: Analytical Data; vol. 4: Water Resources Regional Reports,* Second National Water Assessment by the U.S. Water Resources Council, 052-045-00051-7, Superintendent of Documents, U.S. Government Printing Office, Washington, D.C., 1978.

Weatherford, G. D., and H. M. Ingram, "Legal Institutional Limitations on Water Use," in N. K. Whittlesey, ed., *Water Scarcity Impacts on Western Agriculture,* Westview Press, Boulder, Colo., 1984, pp. 73–100.

Whittlesey, N. K., K. C. Gibbs, and W. R. Butcher, "Social Overhead Capital Costs of Irrigation Development in Washington State," *Water Resources Bulletin,* 14(3):663–678, 1978.

Worthington, E. B., ed., *Arid Land Irrigation in Developing Countries: Environmental Problems and Effects,* Pergamon Press, New York, 1976.

Yonts, D. C., and D. E. Eisenhauer, "Impact of Surge Irrigation on Furrow Water Advance," in Richard G. Allen and Christopher M. U. Neale, eds., *Management of Irrigation and Drainage Systems: Integrated Perspectives,* American Society of Civil Engineers, New York, 1993, pp. 206–213.

CHAPTER 23
WATER DEMAND ANALYSIS

Benedykt Dziegielewski
Southern Illinois University–Carbondale
Carbondale, Illinois

Eva M. Opitz
Planning and Management Consultants, Ltd.
Carbondale, Illinois

David Maidment
University of Texas at Austin
Austin, Texas

23.1 WATER-USE DATA

23.1.1 Definitions of Water-Use Categories

From the hydrologic perspective, *water use* can be defined as all water flows that are a result of human intervention within the hydrologic cycle. This perspective was used in establishing the national system of water-use accounting of the National Water Use Information Program (NWUI Program) which is conducted by the United States Geological Survey (USGS). In the series of water-use circulars which are prepared by USGS at five-year intervals, this accounting system distinguishes among the following seven water-use flows: (1) water withdrawals for offstream purposes, (2) water deliveries at point of use or quantities released after use, (3) consumptive use, (4) conveyance loss, (5) reclaimed wastewater, (6) return flow, and (7) instream flow (Solley, et al., 1993). Figure 23.1 shows the relationships among these human-made flows at various points of measurement.

At the most general level, water use can be classified as *offstream use* or *instream use.* Conveyance losses and consumptive use account for differences in flows at various points of measurement. The USGS definitions of 10 water-use accounts are given in Table 23.1. Figure 23.2 shows a diagram developed by the USGS which tracks the sources, uses, and disposition of freshwater using the hydrologic accounting system (as outlined in Fig. 23.1). The measurements of water quantities in million gallons per day (mgd) in each block account for all freshwater withdrawals in 1990 in the United States. Approximately 338,400 mgd (89.4 million cubic meters per day)

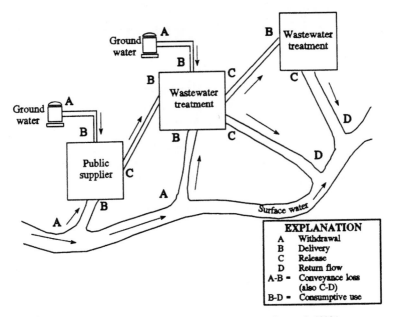

FIGURE 23.1   Definitions of water-use flows and losses. *(Solley, et al., 1993.)*

were withdrawn from freshwater sources during that year. More than three-fourths (i.e., 76.5 percent) of total withdrawals came from surface-water sources. Water disposition blocks in Fig. 23.2 indicate that 72.2 percent of all water withdrawn was released as return flow and thus made available for further use. The balance of 27.8 percent was evaporated, transpired, or incorporated into products or crops and was classified as consumptive use.

A more restrictive definition of water use refers to water that is actually used for a specific purpose. The USGS water-use circulars distinguish 10 categories of off-stream use which are grouped into four broad classes on Fig. 23.2. Table 23.2 contains definitions of the 10 uses. Water used in the generation of electricity at plants where the turbine generators are driven by falling water is classified as an instream use. This type of use (sometimes called *nonwithdrawal use* or *in-channel use*) also includes such purposes as navigation, water-quality improvement, fish propagation, and recreation. The USGS's National Water Use Information Program maintains an account for only one type of instream use—hydroelectric power use. Additional accounts of instream use will be added by the USGS in the future.

23.1.2   Measurement and Estimation of Water Use

Measurements of water use are reported as water volumes per unit of time. The volumetric units include cubic meters, cubic feet, gallons and liters, and their decimal multiples. In some cases, composite volumetric units such as acre foot or units of water depth such as inches of rain may be used. The time periods used include a second, minute, day, and year. Because the annual volumes of water use usually involve large numbers, annual water-use totals are often reported as the average daily usage

TABLE 23.1 Definitions of Water Use Terms

Term	Definition
Consumptive use	That part of water withdrawn that is evaporated, transpired, incorporated into products or crops, consumed by humans or livestock, or otherwise removed from the immediate water environment.
Conveyance loss	The quantity of water that is lost in transit from a pipe, canal, conduit, or ditch by leakage or evaporation.
Delivery and release	The amount of water delivered to the point of use and the amount released after use.
Instream use	Water that is used, but not withdrawn, from a ground- or surface-water source for such purposes as hydroelectric-power generation, navigation, water-quality improvement, fish propagation, and recreation.
Offstream use	Water withdrawn or diverted from a ground- or surface-water source for public water supply, industry, irrigation, livestock, thermoelectric-power generation, and other uses.
Public supply	Water withdrawn by public or private water suppliers and delivered to users.
Return flow	The water that reaches a ground- or surface-water source after release from the point of use and thus becomes available for further use.
Reclaimed wastewater	Wastewater-treatment-plant effluent that has been diverted for beneficial use before it reaches a natural waterway or aquifer.
Self-supplied water	Water withdrawn from a surface- or groundwater source by a user rather than being obtained from a public supply.
Withdrawal	Water removed from the ground or diverted from a surface-water source for offstream use.

Source: Adapted from Solley, et al. (1993).

rates. Two popular units are thousand cubic meters per day (km^3/d) and million gallons per day (mgd).

In order to make the estimates of water use easy to comprehend and to make meaningful comparisons of water use for various purposes (and various users), the annual or daily quantities are divided by some measures of size for each purpose of use. The result is an average rate of water use such as gallons per capita per day (gcd), gallons per employee per day (ged), or other unit-use coefficients.

The reported quantities of water use can be in the form of direct measurements obtained from water meters which register the volume of flow (such as displacement meters or Venturi meters) or they may be estimates. Estimates of water use that are derived from the measurements of water levels in storages or from pumping logs are generally more accurate than those derived from related data on the volume of water-using activity. For example, the estimates of water use for hydroelectric-power generation may be obtained by multiplying the amount of generated power by a water-use coefficient. For example, in 1990 the ratio of water use to power generated in the United States was 4016 gal/kWh (Solley, et al., 1993).

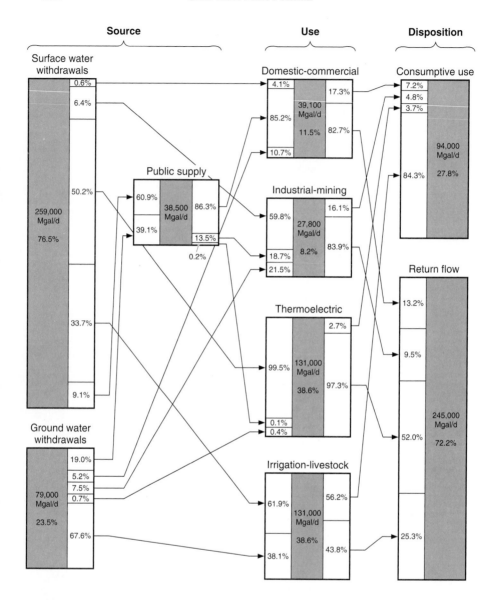

FIGURE 23.2   Source, use, and disposition of freshwater in the United States, 1990. *(Solley, et al., 1993.)*

23.1.3   Sampling of Water Users

Water-supply agencies and regional or state regulatory bodies usually have the ability to monitor the water use of all users or entire classes of users. In statistical terms,
studies involving all users would represent the use of entire populations. However, a
complete enumeration or inventory of all users may not always capture the "entire

TABLE 23.2 Major Purposes of Water Use

Water-use purpose	Definition
Domestic use	Water for household needs such as drinking, food preparation, bathing, washing clothes and dishes, flushing toilets, and watering lawns and gardens (also called residential water use).
Commercial use	Water for motels, hotels, restaurants, office buildings, and other commercial facilities and institutions.
Irrigation use	Artificial application of water on lands to assist in the growing of crops and pastures or to maintain vegetative growth in recreational lands such as parks and golf courses.
Industrial use	Water for industrial purposes such as fabrication, processing, washing, and cooling.
Livestock use	Water for livestock watering, feed lots, dairy operations, fish farming, and other on-farm needs.
Mining use	Water for the extraction of minerals occurring naturally and associated with quarrying, well operations, milling, and other preparations customarily done at the mine site or as part of a mining activity.
Public use	Water supplied from a public water supply and use for such purposes as firefighting, street washing, municipal parks, and swimming pools.
Rural use	Water for suburban or farm areas for domestic and livestock needs which is generally self-supplied.
Thermoelectric power use	Water for the process of the generation of thermoelectric power.

Source: Solley, et al. (1993).

population" because in addition to "populations" defined in terms of users, which can be viewed as finite (or delimited), some studies may require expanded definitions of populations. For example, a study population can be defined as the monthly water-use quantities of all users over a time horizon. Because such a definition includes future water use, historical records of withdrawals constitute only a part (or a sample) of the total population. Also, in many cases, a study of an entire population must limit the number of measurements of each unit (due to cost constraints) and may not be capable of producing answers to some research questions. Because of these considerations, knowledge of water use is almost invariably based on samples or fragments of total populations. Sampling has many advantages over a complete enumeration (or inventory) of the population under study. These advantages include reduced cost, greater speed of obtaining information, and a greater scope of information that can be obtained. In addition, a greater precision of measurement can be secured by employing trained personnel to take the necessary measurements and analyze the data.

Scientific sampling designs specify methods for sample selection and estimation of sample statistics that follow *the principle of specified precision at the minimum cost;* i.e., they provide, at the lowest possible cost, estimates that are precise enough for the study objectives. *Probability sampling* refers to any sampling procedure that relies on random selection and is amenable to the application of sampling theory to validate the measurements obtained through sampling. This requires that, within the sampled population, one is able to define a set of distinct samples (where each sample consists of sampling units) with known and equal probabilities of being selected. One of these samples is then selected through a random process. In practice, the sample is most commonly constructed by specifying probabilities of inclusion for the individual units, one by one or in groups, and then selecting a sample of desired size and type. *Nonprobability sampling* refers to sampling procedures that do not include the element of random selection because samples are restricted only to a part of the population that is readily accessible, selected haphazardly without prior planning, or they consist of "typical" units or volunteers. The only way of examining how good the nonprobability sample may be is to know parameters for the entire population or to compare it with the probability sample statistics taken from the same population. For more information on sampling procedures, see Dziegielewski, et al. (1993); Cochran (1963); Kish (1965); or Fowler (1988).

23.1.3.1 Types of Sampling Plans.   There are many ways of constructing a probability sample of water users. *Simple random sampling* refers to a method of selecting n sampling units out of a population of size N, such that every one of the distinct samples (where each sample consists of n sampling units) has an equal chance of being drawn. In *stratified sampling,* the sampled population of N units is first divided into several nonoverlapping subpopulations. These subpopulations are called *strata* because they divide a heterogeneous population into homogeneous subpopulations. If a simple random sample is taken from each stratum, then the sampling procedure is described as *stratified random sampling.* In order to design a stratified random-sampling plan, it is necessary to determine: (1) which population characteristic (i.e., variable) should be used in stratification, (2) how to construct the strata (i.e., how many strata to use and where to set the stratification boundaries), and (3) what sample sizes should be obtained from each stratum. The statistical theory of stratified sampling offers some methods for selecting the optimal number of strata, strata boundaries, and sample sizes in advance (Cochran, 1963). However, it is usually necessary to collect and examine some data before designing a good sampling plan.

Systematic sampling is often the most expeditious way of obtaining the sample and may be used in situations where time is critically constrained. The units in the population sampled are first numbered from 1 to N in some order. To select a sample of n units, one should take the first unit at random from the first k units and every kth unit thereafter. The selection of the first unit determines the whole sample, which is often called an every kth systematic sample.

If individual sampling units are arranged according to some characteristic or variable (e.g., water use), then the systematic sample is equivalent to a stratified sample in which one sampling unit is taken from each stratum. Constructing a list of sampling units can be avoided by dividing a geographic area into areal units such as county or river basin. This sampling plan, called *cluster sampling,* can result in significant cost savings. For example, a simple random sample of 600 industrial users may cover a state more evenly than 20 counties containing a sample of 30 plants each, but it will cost more because of the time devoted to travel and finding individual establishments. However, cluster sampling creates a greater risk of obtaining a nonrepresentative sample.

The *size of the sample* depends on the precision of measurement that is required and the variance in the parameters to be estimated. The precision of an estimate refers to the size of the deviations from the mean of all sample measurements obtained by repeated application of the sampling procedure. In contrast, the term *accuracy* is usually applied to indicate the deviations of the sample measurements from the true values in the population. For example, a simple random sample can be used to estimate average daily water use and variance in water use of all single-family houses in an urban area during a given year. According to sampling theory, the mean water use $\bar{y}$ obtained from the simple random sample is an unbiased estimate of the average water use $\bar{Y}$ for all houses (i.e., the population mean). Also, $\hat{Y} = N\bar{y}$ is an unbiased estimate of total water use of the population (N customers).

The *standard error* of $\bar{y}$, which describes the precision of the estimated mean value, is

$$\sigma_{\bar{y}} = \left(\sqrt{\frac{N-n}{N}}\right)\frac{S}{\sqrt{n}} \tag{23.1}$$

where S is obtained from population variance S^2 (by taking its square root). Because in practice S^2 may not be known, it must be estimated from the sample data using the formula:

$$S^2 = \frac{\sum\limits_{i}^{n}(y_i - \bar{y})^2}{n-1} \tag{23.2}$$

which provides an unbiased estimate of S^2 and where n is the sample size. Usually, with a population having a mean $\bar{Y}$ and a simple random sample having mean $\bar{y}$, control of the following probability condition is desired:

$$Pr\left(\left|\frac{\bar{y}-\bar{Y}}{\bar{Y}}\right| \geq r\right) = \alpha \tag{23.3}$$

where α is a small probability (e.g., 0.05) and r is relative error expressed as a fraction of the true population mean. By multiplying both sides of the parenthetical expression in the above equation by $\bar{Y}$, the same condition can be restated as:

$$Pr(|\bar{y}-\bar{Y}| \geq r\bar{Y}) = \alpha \tag{23.4}$$

If, instead of the relative error r, control of the absolute error d (i.e., the absolute value of the difference between the sample mean and the population mean) in $\bar{Y}$ is desired, the formula can be written as:

$$Pr(|\bar{y}-\bar{Y}| \geq d) = \alpha \tag{23.5}$$

It is usually assumed that $\bar{y}$ is normally distributed about the population mean $\bar{Y}$, and given its standard error from Eq. (23.1), the product $r\bar{Y}$ is:

$$r\bar{Y} = t\sigma_{\bar{y}} = t\sqrt{\frac{N-n}{N}} * \frac{S}{\sqrt{n}} \tag{23.6}$$

where t is the value of the normal deviate corresponding to the desired confidence probability. This value is 1.64, 1.96, and 2.58 for confidence probabilities 90, 95, and 99 percent, respectively. Solving the preceding equation for n gives

$$n = \frac{\left(\frac{tS}{r\bar{Y}}\right)^2}{\left[1 + \frac{1}{N}\left(\frac{tS}{r\bar{Y}}\right)^2\right]} \tag{23.7}$$

The expression in brackets represents a finite population correction, and it should be used when n/N is appreciable. Without this correction, we can take the first approximation of the desired sample size n_o as

$$n_o = \left(\frac{tS}{r\overline{Y}}\right)^2 \tag{23.8}$$

According to the preceding equation, n_o depends on the coefficient of variation (the ratio $S/\overline{Y}$) of the population that is often more stable and easier to guess in advance than S itself. It also depends on the error r that can be tolerated and the confidence level that is needed as captured by the value of t. For the absolute error specification as in Eq. (23.5), the preceding equation is changed into

$$n_o = \left(\frac{tS}{d}\right)^2 \tag{23.9}$$

The preceding sample size relationships are illustrated in the following example.

23.1.3.2 Example of Sample Size Determination for Continuous and Proportional Data.

A water-supply agency serves 80,000 customers. The analysis of billing frequencies for the entire fiscal year indicates that average daily use per customer is 250 gal and the standard deviation is 180 gal. Using simple random sampling, how many customers must be taken to be 95 percent confident of estimating average daily use within 2 percent of the true value?

Solution: $N = 80,000$; $S = 180$ gal; $\overline{Y} = 250$ gal; $\alpha = 0.05$; $t = 1.96$; and $r = 0.02$. Substituting these values into Eq. (23.8):

$$n_o = \frac{t^2 S^2}{r^2 \overline{Y}^2} = \frac{(1.96)^2 (180)^2}{(0.02)^2 (250)^2} = 4979$$

where n = sample size (n_o is the first approximation)
        N = population size
        S = population standard deviation
        Y = population mean
        t = confidence probability (t-statistic)
        r = relative error

Because n_o/N is not negligible, we need to take the finite population correction, found from Eqs. (23.7) and (23.8):

$$n = \frac{n_o}{1 + \dfrac{n_o}{N}} = \frac{4979}{1 + \dfrac{4979}{80,000}} = 4687$$

The results indicate that if the average water use is unknown, to be 95 percent confident of estimating it by sampling billing records with an error of 2 percent (or 5 gal), a sample of 4687 single-family homes would be required.

In some cases, it may be necessary to obtain estimates of the percent of water users who possess a certain characteristic (e.g., use groundwater as their source of supply) by surveying a sample of users. The sampling problem in this case is referred to as *sampling for proportions*, where the respondents are classified into two classes: groundwater users and users of other sources. In order to determine the required sample size, we must decide the margin of error d in the estimated proportion p of users who rely on groundwater and the risk α that the actual error will be larger than d. Therefore, control of the following probability condition is desired:

$$Pr\left(|p - P| \geq d\right) = \alpha \tag{23.10}$$

where P is the true proportion of users of groundwater. Assuming simple random sampling and a normal distribution of p, the standard error of p, σ_p, is given by:

$$\sigma_p = \sqrt{\frac{N-n}{N-1}} \sqrt{\frac{PQ}{n}} \tag{23.11}$$

where
N = population size
n = number of respondents in the sample
P = proportion of groundwater users in the population
Q = proportion of users of other sources in the population (i.e., $Q = 1 - P$)

The formula for the *desired degree of precision* is:

$$d = t \sqrt{\frac{N-n}{N-1}} \sqrt{\frac{PQ}{n}} \tag{23.12}$$

where t is the critical value of the t distribution corresponding to the desired confidence probability (i.e., the abscissa of the normal curve that cuts off an area of α at the tails). Solving Eq. (23.12) for n gives:

$$n = \frac{\left(\dfrac{t^2 PQ}{d^2}\right)}{\left[1 + \dfrac{1}{N}\left(\dfrac{t^2 PQ}{d^2} - 1\right)\right]} \tag{23.13}$$

If n/N is negligible because N is large, we can take the first approximation of n_o by using an advanced estimate p for P (and q for Q) from the formula

$$n_o = \frac{t^2 pq}{d^2} \tag{23.14}$$

After obtaining n_o, we can introduce the *finite population correction* to sample size from the following formula:

$$n = \frac{n_o}{1 + (n_o/N)} \tag{23.15}$$

The sampling plans and sample size determinations are important elements of the process of collecting water-use data. Because water users are unlikely to form a homogeneous group, stratified random sampling is the most useful procedure for obtaining representative samples of water users.

23.1.4 Development of Data Sets

Water-use data are usually collected for the purpose of monitoring water use. Many states require that water users submit annual (and/or monthly) records of their water withdrawals or discharges as a part of their permitting process. These data can be used for statistical analysis of water-use trends as well as for the development of water-use models. The latter purpose would also require data on variables which influence water use such as weather, price, employment, land use, and others.

In economics, data on economic activities (or variables) are collected at micro or macro levels. Observations on individual households, families, or firms are referred to as *microdata*. National-level accounts and observations on entire industries are called *macrodata*. In water-use modeling and analysis, the corresponding types of data are sometimes referred to as *disaggregate* and *aggregate* data. Levels of aggregation may vary from the end-use level (e.g., toilet flushing, lawn watering) or the municipal level (e.g., total production or total metered use) to total withdrawals in a region or state.

In developing water-use models, it is necessary to distinguish among different types of variables and their levels of measurement. The latter are also referred to as *scales*. In a mathematical sense, a variable is a quantity or function that may assume any given value or a set of values, as opposed to a constant that does not or cannot change or vary. Depending on the character (or type) of values that can be assumed by a variable, the following can be distinguished: (1) continuous variables, (2) discrete variables, and (3) random variables. A continuous variable can assume any value within the interval where it exists (e.g., monthly water use in a demand area). A discrete variable can assume only discrete values such as the number of customers in a demand area. Finally, a random (or stochastic) variable can take any set of values (positive or negative) with a given probability. Random variables can be discrete or continuous.

23.1.4.1 *Data Scales.*   The data on a variable can be measured on different kinds of scales. Such scales are used to describe the data and variables used in the analysis. The data can be nominal (either ordinal or nonordinal) or internal (including ratio data).

Nominal or *categorical data* are measurements that contain sufficient information to classify and count objects. Nominal data can be classified according to ordinal or nonordinal scales. *Ordinal* scales rank data according to the *value* of the variable that is being analyzed. Objects in an ordinal scale are characterized by relative rank so that a typical relationship is expressed in terms such as *higher, greater,* or *preferred to.* Ordinal ranking of data commonly occurs in the use of surveys. For example, survey data of household income is usually ranked (or classified ordinally) by income range, such as 1 = <\$25,000; 2 = \$25,000 to \$49,999; 3 = \$50,000 to \$74,999; and so on. Notice then that ordinal rankings are hierarchical. *Nonordinal* data are ranked by variable type and therefore cannot be ranked hierarchically on a numeric scale. An example of nonordinal data comes from a survey of residential landscapes in Southern California. Survey teams were asked to classify turf landscapes as 1 = Bermuda grass; 2 = other warm-season grasses; 3 = tall fescue; and 4 = cool-season grasses. Unlike the example of ordinal ranking given above, category 2 is not in any sense greater than category 1, because the categorization is based on landscape type without reference to numeric measurement of the distance between ranked objects. Finally, there are *interval* data that contain numerical values from a continuous scale. Variables with interval data are therefore most frequently called *continuous variables.* Because a continuous variable can assume an infinite number of values, continuous variables can theoretically be measured only over an interval, hence, the name interval data. If the continuous measurement scale contains a true theoretical zero (e.g., water use or temperature), then it is called a *ratio scale.*

Another classification of variables is used in statistical or econometric modeling. In constructing statistical relationships, an attempt is usually made to predict or explain the effects of one variable by examining changes in one or more other variables that are known or expected to influence the former variable. In mathematics, the variable to be predicted is called the *dependent* variable, while the variables that influence it are called *independent*. However, in the analytical literature from various disciplines, a number of alternative terms are often used to describe and classify variables. Table 23.3 contains a list of such terms.

TABLE 23.3 Alternative Terms for Dependent and Independent Variables

Dependent	Independent
Explained	Explanatory
Predicted	Predictor
Regressed	Regressor
Response	Causal
Endogenous	Exogenous
Target	Control

23.1.4.2 Data Arrangements. For modeling purposes, data (or observations) on water use and related variables can be obtained and arranged in several ways. Depending on which type of arrangement is used, four types of data configurations can be distinguished: (1) time series data, (2) cross-sectional data, (3) pooled time series and cross-sectional data, and (4) panel (or longitudinal) data. In *time series* data, observations on all variables in the data set are taken at regular time intervals (e.g., daily, weekly, monthly, annually). In *cross-sectional* data, observations are taken at one time (either point in time or time interval) but for different units, such as individuals, households, sectors of water users, cities, or counties. *Pooled* data sets combine both time series and cross-sectional observations to form a single data matrix. Finally, *panel* data represent repeated surveys of the same cross-sectional sample at different periods of time. Databases can be built from records of water use by supplementing them with data from other sources such as random samples of water users. The possible data configurations that can be constructed from water-use histories of individual water users include:

1. Time series of measurement period water-use data for individual water users
2. Time series of measurement period water-use data for all users in the sample
3. Cross-sectional water-use data for the same measurement period extending across all customers in the sample
4. Cross-sectional water-use data aggregated for two or more measurement periods representing seasonal or annual use extending across all users in the sample
5. Pooled time series cross-sectional data for all measurement periods and all water users
6. Pooled time series cross-sectional data aggregated over two or more measurement periods for all water users

In mathematical terms, we can describe each data configuration by designating water use of user i during measurement period t as q_{it}. If n and m represent, respectively, the number of users in the sample and the number of measurement periods, we can describe the six data configurations as:

1. *Customer time series* data representing total water use of user i in each time period t

$$q_{it}; \quad \text{where} \quad i = \text{constant}, t = 1 \ldots m$$

2. *Aggregate time series* data representing total water use of all users in the sample in each time period t

$$\sum_{i=1}^{n} q_{it}; \quad \text{where} \quad t = 1 \ldots m$$

3. *Cross-sectional billing-period* data, representing total water use of *each* user i during billing period t

$$q_{it}; \quad \text{where} \quad i = 1 \ldots n \quad \text{and} \quad t = \text{constant}$$

4. *Seasonal (or annual) cross-sectional* data, representing total water use of each user i during seasonal or annual period k

$$\sum_{t=1}^{k} q_{it}$$

where $i = 1 \ldots n$ and k = number of billing periods in each season

5. *Pooled time series cross-sectional data*, representing water use of each user i in each time period t

$$q_{it}; \quad \text{where} \quad i = 1 \ldots n \quad \text{and} \quad t = 1 \ldots m$$

6. *Pooled time series cross-sectional data with seasonal (or annual) aggregations*, representing water use of each user i in each season comprising k billing periods

$$\sum_{t=1}^{k} q_{it}$$

where $i = 1 \ldots n, t = 1 \ldots k$, and k = number of billing periods in each season

If observations on water use q_{it} are supplemented with data on variables that are believed to be predictors of water demand, then regression analysis can be applied to any of the above data configurations. Section 23.2.3 describes various water-use modeling techniques.

23.2 PUBLIC-SUPPLY WATER USE

National data on water withdrawals for public-supply purposes include public and private water systems that furnish water to at least 25 people, or that have a minimum of 15 hookups (Solley, et al., 1993). According to the American Water Works Association (AWWA), there were 203,500 public water-supply systems in the United States (AWWA 1988). Nearly 60,000 of these systems represented community water supplies serving approximately 220 million people. Table 23.4 shows the distribution of the community systems by the size of population served. Approximately 85 percent of the population is served by 5210 systems which deliver water to communities with more than 5000 persons. While 80 percent of public water-supply systems rely on groundwater, more than one-half of the larger systems use surface water as their principal source of supply. The data and techniques for analyzing water demand in a public water-supply system are the subject of this section.

TABLE 23.4 Public Water-Supply Systems in the United States, 1985

System size, population served	Number of systems	Percent of systems	Population served, thousands	Percent of population
25–100	19,717	33.2	1,089	0.5
101–500	18,321	30.9	4,550	2.1
501–1,000	6,254	10.5	4,646	2.1
1,001–2,500	6,432	10.8	10,496	4.8
2,501–3,300	1,432	2.4	4,158	1.9
3,301–5,000	1,980	3.3	7,831	3.6
5,001–10,000	2,128	3.6	15,351	7.0
10,001–25,000	1,782	3.0	28,496	13.0
25,001–50,000	697	1.2	25,034	11.4
50,001–75,000	230	0.4	14,063	6.4
75,001–100,000	94	0.2	8,099	3.7
Over 100,000	279	0.5	95,409	43.5
Total	59,346	100.0	219,220	100.0

Source: Adapted from U.S. EPA, as reported in van der Leeden, Troise, and Todd (1990).

In any public water-supply system, water-use records can be characterized with respect to the relative needs of various customer groups (e.g., single-family residential, hotels, food-processing plants), the purposes for which water is used (e.g., end uses such as sanitary needs, lawn watering, cooling), and the seasonal variation in water use. The analysis of water use can be expanded to also include the development of information on water users in the service area. Data on housing stock, household characteristics, business establishments, and other demographic and economic statistics are important because such characteristics are major determinants of water use.

23.2.1 Water-Use and Service-Area Data

Water-use data from water-supply agency (i.e., water utility) records can be used for examining historical trends in water use and disaggregating total use into seasons, sectors, and specific end uses within each sector.

23.2.1.1 Water-Use Data. *Water-production records* are a good source of data on total water demands in the area served by a public system. Water utilities usually have one or more production meters that are generally read at least daily. These production meters are typically maintained for accuracy and therefore usually produce highly reliable measurements of water flows into the distribution system. The water-treatment plants or pumping stations usually employ continuous metering of the flow of finished water to the distribution system. The data may be recorded on paper

recording charts which can be used to generate a time series of total production (or production from various supply sources) at daily or hourly time intervals. These data can be used for deriving temporal characteristics of aggregate water use (e.g., peak day, peak hour, day of week). The usefulness of the production data for water-demand analysis may include, but is not limited to: (1) the analysis of unaccounted water use (comparing production with water-sales data), (2) the measurement of the aggregate effect of emergency conservation campaigns on water use, or (3) the analysis of the relationship between water production and weather variability.

Customer billing data can be used for disaggregating water use into customer sectors. Typically, retail-water agencies maintain individual computer records of monthly, bimonthly, quarterly (or, less often, semiannual or annual) water-consumption records for all metered customers. Active computer files usually retain up to twelve or fifteen past meter readings for each customer. Depending on the length of the billing cycle, the active file records can contain a 12- to 36-month history of water use. This is a valuable source of water-use data that is often not exploited to its full potential. However, billing data can suffer from the following problems: (1) unequal billing periods, (2) a lack of correspondence between billing periods and calendar months, (3) estimated meter readings or incorrect meter readings due to meter misregistration, (4) unusual usage levels, (5) meter replacements and manual adjustments to meters, and (6) changes in customer occupancy. Because of these problems, sole reliance on customer billing records necessarily limits the usefulness of these data for various measurements. Although billing data are becoming more readily available on electronic media, this is still not routine with many utilities.

Special metering is sometimes undertaken in order to obtain water-use data for research purposes. In recent years, water utilities have begun to experiment with new meter-reading technologies that greatly reduce the cost of monitoring water use of individual customers. However, the initial investment costs of adopting these new technologies can be prohibitively high. The new technologies include automatic meter-reading (AMR) devices and electronic remote meter-reading (ERMR) devices (see Schlenger, 1991). Automatic meter-reading devices are carried by meter readers on their routes and plugged into the site meters for the automatic meter reading. The data from the automatic meter readings are stored in the AMR and then can be downloaded to central computer systems. The electronic remote meter-reading devices can be used to read site meters without actually visiting the meter location. Various methods of remote meter reading include: (1) telephone dial-outbound, (2) telephone dial-inbound, (3) telephone scanning, (4) cable television, (5) radio frequency, and (6) power-line carrier.

These new technologies may permit water agencies to obtain daily or weekly meter readings from individual accounts. More frequent readings can improve the precision of water use measurements. However, the most useful measurements can be obtained by devices that monitor individual pulses of water use on the service line. These pulses can be correlated with water flows through individual fixtures and appliances on customer premises (such as toilet flushes and shower flows) thus permitting accurate measurements of all end uses of water.

Currently, these new technologies are used only by a few water agencies. However, as the technology becomes more frequently adopted, there is great potential for new information sources for water-conservation planning. Unobtrusive metering technologies that permit accurate measurements of end uses of water would be particularly useful for water-conservation planning.

23.2.1.2 *Service-Area Data.*

Because water use is a function of demographic, economic, and climatic factors, an accurate description of the characteristics of the resident population, housing stock, economic activities, and weather patterns in the service area can serve as a basis for the analysis of water demands. This section discusses the types of information that can be used to characterize the service area and identify potential data sources.

Accurate *service-area maps* of the service area are indispensable. Often water-service areas do not follow political (i.e., city or county) boundaries. However, demographic and socioeconomic data are most readily available by political boundaries or by census-designated boundaries (i.e., census tracts). Therefore, in order to relate water-service-area data to demographic and socioeconomic characteristics for planning purposes, it is often necessary to determine the relationship between service-area boundaries and political or census-designated boundaries. A service-area map with overlays for census tract, zip code, or other political boundaries is most useful for this purpose. Service-area maps with geographic or land-use overlays have usefulness in many other planning activities. For example, only parts of the service area may be targeted for specific activities (e.g., the service area might be disaggregated spatially into pressure zones or rate zones for the purpose of water-use forecasting and/or facility planning).

Usually, a set of maps can be obtained from the facility planning or engineering department of the water agency. These maps should indicate the historical growth of the service territory and potential future additions. Also, the maps should note areas within a given community that are partially served or are served by other water-supply agencies. Maps denoting political boundaries and demographic characteristics can often be obtained from local and regional planning agencies. Mapping work could be greatly simplified by Geographic Information Systems (GIS) technology, which is a computerized mapping technique. The GIS system may have the service area divided into separate units (e.g., census tracts, pressure zones) and have several information bases about each separate unit (e.g., water-use characteristics, land use, socioeconomic characteristics).

Population and housing characteristics (i.e., household income, lot size, persons per household, home value) are determinants of residential water use. Therefore, it is important to obtain information on these characteristics as well as to understand their impact on water use.

The conventional method of estimating "population served" by multiplying "total service connections" by "persons per connection" is not very accurate. Although this method may be acceptable to estimate the population served in the residential sector (assuming an accurate measurement of persons per connection), this is not an accurate method for total service-area population served because of the confounding effect of commercial, institutional, governmental, and industrial accounts.

Knowledge of the number and type of housing units in the service area is very useful in water-demand analysis because water-use patterns differ among housing types. On both a per-housing-unit and a per-capita basis, water use in the multifamily sector tends to be lower than in the single-family sector. This is the result of different household composition and the fact that residents in multifamily housing, on a per-unit basis, have less opportunity for outdoor water-use practices.

The total number of residential *accounts* served by a water-supply system is not a good indication of the number of housing *units* in the service area because of the varying number of units served by multifamily accounts. However, the number of single-family customer accounts is typically a good indication of the number of single-family housing units served by the water system. Unless the water agency maintains records on the number of multifamily living units per multifamily account, housing-count data must be obtained from the demographic data developed by: (1)

U.S. Bureau of the Census, (2) State departments of finance or economic development, (3) regional associations of governments, and (4) local county and city planning agencies.

Again, deviations of the water-service boundaries from political boundaries must be considered when using these data. In addition to population and housing counts, some of the agencies listed above can also provide data on population characteristics (e.g., family size, age, income) and data on local housing (i.e., number of homes by type, new construction permits, vacancy rates, etc.). Table 23.5 gives examples of demographic and housing data that are included in the 1990 census files. Two primary questionnaires were used in the collection of census data—the short form which included questions that were asked of all persons and housing units (i.e., the 100 percent component) and the long form which targeted only about 20 percent of the population on additional subject items (i.e., the sample component). These data are presented by geographic and political subdivisions (i.e., states, counties, cities). For major urban areas, census data are further disaggregated into census tracts, city blocks, and block groups (but not for individual dwellings).

Information on *commercial, institutional,* and *industrial activities* in the service area is helpful in analyzing nonresidential water demands. There is a great diversity of purposes for which water is used in the commercial, institutional, and industrial (manufacturing) sectors. The uses of water may include sanitary, cooling and condensing, boiler feed, landscape irrigation, and others. The type of business activity conducted in a commercial or industrial establishment can provide useful information regarding the purposes for which water is used and therefore the types of conservation measures that might be applicable. Furthermore, data on square feet of

TABLE 23.5 Selected Examples of 1990 U.S. Census Data

Population	Housing
100 percent component	
Household relationship	Number of units in structure
Sex	Number of rooms in unit
Race	Tenure (owned or rented)
Age	Value of home or monthly rent paid
Ethnic/racial origin	Congregate housing (meals included in rent)
	Vacancy characteristics
20 percent sample component	
Social characteristics	
Education	Year moved into residence
Ancestry	Number of bedrooms
Migration	Plumbing and kitchen facilities
Language spoken at home	Telephone in unit
	Heating fuel
Economic characteristics	Source of water and method of sewage disposal
Labor force	Year structure built
Place of work and journey to work	Condominium status
Year last worked	Farm residence
Occupation, industry, and class	Shelter costs, including utilities
of worker	
Work experience in 1989	
Income in 1989	

floor space, land acreage, number of employees, number of rooms (for hotels and schools), and financial performance can also be useful information in predicting commercial and industrial water use. However, some of this information is not readily available for individual establishments or aggregated into political or census-designated boundaries. Some of the previously listed agencies maintain data on local economic activities including the number of establishments, employment, and financial performance of businesses (e.g., sales) disaggregated by industry type as denoted by the Department of Commerce Standard Industrial Classification (SIC) codes.

Additional establishment or employment data can be obtained from local and regional planning commissions, local chambers of commerce, or purchased from private firms. Some private vendors (e.g., Dun's Marketing Services, Contacts Influential) can provide customized computer databases containing information on large samples of businesses in designated geographical areas. Establishment and employment data can be analyzed to determine types of business establishments that represent a major portion of nonresidential water use either because of large employment or because of large water requirements for processing or other needs. Business types can be cross-checked with agency billing records and used for disaggregating water use into specific groups of nonresidential users.

Water used in parks, cemeteries, school playgrounds, and highway medians can account for a significant portion of total use and offer a potential for conservation. Data on *public and government facilities* can be obtained from city and regional planning departments, city park districts, street departments, and the Department of Transportation (for highway medians). Information that might be obtained include land-use data for various purposes (in square feet or acres) as well as the number of employees in various facilities.

23.2.2 Components of Water Demand

The quantities of water delivered by a public water-supply system can be disaggregated by user sector and season. Table 23.6 gives an example of such decomposition of water demands in the urban area of Southern California. This section describes analytical methods for disaggregating total urban water use.

23.2.2.1 Sectors of Water Users. Customer billing records can be used to obtain estimates of total metered use of water and to determine the distribution of total water use among several homogeneous classes of water users. A disaggregation of total metered use into major user sectors (such as residential, commercial, industrial, and others) can be developed from customer-billing records by using one of the following four methods: (1) analysis of available premise (user-type) categories, (2) distribution of meter sizes, (3) sampling of billing files, and (4) development of premise code data.

If individual customer files contain *customer premise categories* identifying the type of customers, then a simple computer program may be used to produce annual billing summaries by customer type. Each premise code can be assigned to one of the following homogeneous sectors of water users: (1) single-family residential, (2) multifamily complexes and apartment buildings, (3) commercial sector, (4) government and public sector, (5) manufacturing (industrial) sector, and (6) unaccounted uses. Water use in these sectors can be disaggregated into seasons and end uses if necessary.

In cases where the customer-billing file does not contain premise code (or customer-type) information, an approximate separation of users by residential, com-

TABLE 23.6 Sectoral and Seasonal Disaggregation of Urban Water Use in Southern California (Most Likely Ranges Given In Parentheses)

User sector subsector	Disaggregation of urban sectors	Seasonal disaggregation				Components of outdoor use		
		Nonseasonal (base use)	Seasonal (peak use)	Indoor use	Outdoor use	Irrigation	Cooling (AC)	Other
Residential sector								
Single-family	34.4	68.7 (65–75)	31.3 (25–35)	65.4 (60–70)	34.6 (30–40)	30.8	0.0	3.8
Multifamily	25.0	83.9 (80–90)	16.1 (10–20)	82.2 (75–85)	17.8 (15–25)	16.1	0.4	1.3
Total residential	59.4	72.1 (70–80)	27.9 (20–30)	69.6 (65–75)	30.4 (25–35)	27.2	0.2	3.0
Nonresidential sector								
Commercial	18.8	74.9 (70–80)	25.1 (20–30)	71.3 (70–75)	28.7 (25–30)	21.8	6.9	0.0*
Industrial	6.0	79.5 (75–90)	20.5 (10–25)	79.5 (75–90)	20.5 (10–25)	12.3	8.2	0.0*
Public	5.1	46.2 (30–50)	53.8 (50–70)	46.2 (30–50)	53.8 (50–70)	53.8	0.0†	0.0†
Other	1.1	58.0 (30–60)	42.0 (40–70)	58.0 (30–60)	42.0 (40–70)	42.0	0.0†	0.0†
Irrigation	0.4	34.0 (10–50)	66.0 (50–90)	0.0	100.0	100.0	0.0	0.0
Total nonresidential	31.4	70.0	30.0	67.4	32.6	26.9	5.7	0.0
Unaccounted use	9.2	100.0	0.0	100.0	0.0	—	—	—
Total urban use	100.0	74.0	26.0	71.7	28.3	24.6	1.9	1.8

Percent of total annual use

* Other uses in these sectors are included under landscape irrigation.
† Cooling and other uses are included under landscape irrigation.
Source: Dziegielewski, et al. (1990).

mercial, and industrial categories can be performed based on the distribution of *meter sizes* with a manual classification of the largest users. For example, single-family homes and small businesses are usually serviced by ⅝- or ¾-in meters. The problem with this method is that some meter sizes, particularly the larger meters (e.g., 1-in) may overlap several customer types thus decreasing the precision of the disaggregation of water use into customer classes.

In cases where water-use data by customer class are not available and the distribution of water use by meter size produces unreliable estimates of water use by customer class, disaggregation can be accomplished by taking a *random sample of customer accounts*. The number of accounts in the sample (i.e., sample size) will depend on the desired accuracy of water use in different customer classes. Sampling efficiency (or precision) can be improved by taking a stratified random sample of users with complete enumeration of the large water-using customers. Possible stratified sampling procedures include: (1) taking a random sample of customer accounts from each meter-size category (sample size within each stratum can be proportional to the total water use within each stratum); (2) using the same approach as in (1) except excluding the upper strata (e.g., meter sizes greater than 2 in) from the sampled population and performing complete enumeration of the largest users; or (3) separating all accounts into two categories based on meter size, with the smallest meter sizes representing the residential sector and the remaining meters representing the nonresidential sector, and then taking a random sample from each category.

The samples of the customer accounts can then be assigned manually into customer classes by visual inspection of customer record (account name) and/or by telephone verification of customer accounts. Depending upon the sample sizes, this can be a time- and resource-intensive exercise.

Regardless of which sampling approach is selected, analysis of the sample should produce the following two estimates: (1) proportion (or percent) of total customers by customer class, and (2) proportion (or percent) of total metered water use by customer class. It is also desirable to calculate the precision of the estimates based on the sample variance.

If sufficient time and resources are available, the classification of all customer accounts into appropriate user types is a worthwhile undertaking. The classification would require: (1) adding a data field (or using existing unassigned fields) to the customer computer file, (2) developing a set of nonoverlapping customer classes and precisely defining each class, (3) determining customer class for each existing customer by classifying all customers during meter reading or surveying (independently) all customers and requesting them to classify their premises on water bills, and (4) adding customer-classification categories to the application forms for new connections.

23.2.2.2 Seasonal and Nonseasonal Components.

23.2.2.2 Seasonal and Nonseasonal Components. Within each user sector, water use can be separated into its seasonal and nonseasonal components. *Seasonal use* can be defined as an aggregate of end uses of water, such as lawn watering or cooling, that varies from month to month in response to changing weather conditions (or due to other influences that are seasonal in nature). *Nonseasonal use,* on the other hand, can be defined as an aggregate of end uses of water, such as toilet flushing or dishwasher use, that remain relatively constant from month to month because these uses are not sensitive to weather conditions or other seasonal influences. Often, seasonal and nonseasonal components of water use are taken to represent the outdoor (or exterior) and indoor (or interior) water uses, respectively. Such an assumption is imprecise because some uses that occur inside the buildings can be seasonal (e.g., humidifier use or evaporative cooler), and some outdoor uses can be nonseasonal (e.g., car washing in warmer climates). The difficulties in classifying various end uses into outdoor and indoor categories must be kept in mind when water use is divided into seasonal and nonseasonal components.

Monthly water-use data can be used to derive estimates of seasonal and nonseasonal water use. The terms *seasonal* and *nonseasonal* relate to the method of characterizing a monthly time series of water-use records. This method is sometimes referred to as the *minimum-month* method because it uses the month of lowest use to represent the nonseasonal component of water use. With the minimum-month method, the percent of annual use in a given year that is considered seasonal is calculated from the formula

$$S_P = 100 - (M_P \cdot 12) \tag{23.16}$$

where S_P = percent of annual use that is seasonal
         M_P = percent of annual use during the minimum month

The best representation of how much water is used during a given calendar month is the aggregation of daily pumpage information which records how much water enters the distribution system every day. Monthly water-use information for customer groups is more difficult to obtain because of the effects of monthly and bimonthly billing cycles. When water utilities summarize the amount of water sold in a given month (both in aggregate or by customer group), this information typically represents the amount of water billed in a given month rather than the amount of water used. Bimonthly billing cycles (which indicate the amount of water used over a two-month period) further confounds calendar-month water use. Differences between the amount of water billed in a given month and the actual water consumed occurs because: (1) accounts are read in different months (e.g., some accounts are read in January, March, and May and others are read in February, April, and June), and (2) meter readings are typically recorded on any given day within a month.

In order to get a better representation of monthly water-use patterns, it is necessary to allocate water use from monthly and bimonthly billing-cycle records into water use during specific calendar months. Thus, the primary purpose of the allocation (or smoothing) techniques is to adjust water-consumption-billing records which are read on either a monthly or bimonthly basis (e.g., even accounts read on even months and odd accounts read on odd months in the bimonthly case) into calendar-month consumption for purposes of further analysis. Data smoothing procedures are performed on two levels. First, the information provided by water utilities on the amount of water billed to a customer group in a given month is smoothed to represent the amount of water actually consumed by a customer group in a given month. The smoothing procedure varies depending on whether a bimonthly or monthly billing cycle is in effect. Second, account-level water-use records are smoothed so that estimates of calendar-month water use can be determined. Both type of procedures are described in the following.

Figure 23.3 presents a graphical representation of the procedure used to smooth aggregate sales data which utilize a *monthly billing cycle*. A monthly billing practice involves reading water meters of individual customers in approximately one-month-long time intervals. Meters are read every working day, and all meters read during, for example, the month of February, are billed and recorded as the February water use Q_{Feb}^b In reality, only a portion of the billed water use Q_{Feb}^b actually occurred during the calendar month of February. Theoretically, Q_{Feb}^b 4 represents water use of individual customers during n monthly periods (where n is the number of billed customers) ending between the first and last meter reading date in February. Therefore, for individual customers, one-month-long periods of water use would fall between January 1 and February 28.

Assuming that all users in a given customer group (e.g., single-family, commercial, and so forth) are relatively homogeneous with respect to water use and that the effects of weather on water use during the two consecutive calendar months are not substantially different (i.e., that water use of the individual customers is evenly distributed throughout the period between meter reading dates), the calendar-month water use during month N can be estimated as:

$$Q_N^c = 0.5Q_N^b + 0.5Q_N^b + 1 \tag{23.17}$$

where Q^c = the amount of water used during the calendar month
Q^b = the amount of water billed during the calendar month

This equation indicates that water actually used during the calendar month of, for example, March Q_{Mar}^c would comprise one-half of the consumption billed in March Q_{Mar}^b and one-half of the consumption billed in April Q_{Apr}^b.

Whereas the above equation allows the calculation of calendar-month water use, a variant of this equation can be used to calculate *average per-account water use* in a given month when the number of billed accounts varies between the two consecutive months:

Monthly cycle billing records:

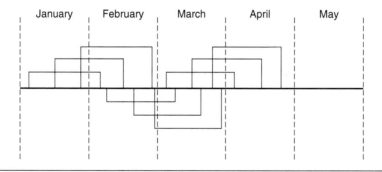

Estimate of calendar month water use (Q):

February: $Q_{Feb}^c = x\ Q_{Feb}^b + y\ Q_{Mar}^b$

March: $Q_{Mar}^c = x\ Q_{Mar}^b + y\ Q_{Apr}^b$

N-th month: $Q_N^c = x\ Q_N^b + y\ Q_{N+1}^b$

Legend:

▢ = beginning and end of customer's billing period

Q_{mar}^b = water use billed to customer during the month of March

x and y = proportions of Q_N^b and Q_{N+1}^b billed water use allocated to Q_N^c and Q_{N+1}^c

FIGURE 23.3 Allocation of monthly billing records of water use into calendar monthly consumption.

$$q_N = \frac{Q_N}{A_N} \cdot \frac{A_N}{A_N + A_{N+1}} + \frac{Q_{N+1}}{A_{N+1}} \cdot \frac{A_{N+1}}{A_N + A_{N+1}} \qquad (23.18)$$

where q_N = average per-account water use in any given month
        Q_N = amount of water billed in any given month
        A_N = number of accounts billed in any given month

Figure 23.4 presents the procedure that was used to allocate aggregate monthly sales data produced as a result of a bimonthly meter-reading cycle. In this procedure, although individual meters are read every two months, customer billing is performed during each calendar month. As a result, all customers within a given sector are divided into two groups, which can be referred to as Group A and Group B. Meters of customers in Group A are read and billed during odd months (i.e., January, March, and May), while in Group B, meters are read and billed during even months (i.e., February, April, and June).

Because some accounts are billed in any given month, aggregate monthly sales data will reflect, in any given month, only approximately one-half of the true number of accounts. For example, water use billed and recorded as the March consumption Q_{Mar}^b includes only customers of Group A whose individual two-month consumption period falls between January 1 and March 31. By analogy, water use billed and recorded as the April consumption Q_{Apr}^b includes only customers of Group B whose individual two-month consumption period falls between February 1 and April 30.

Again, assuming that water users in both groups are homogeneous with respect to water use (because they represent the same user sector) and that weather effects during the consecutive calendar months are not drastically different (i.e., that water use of the individual customer is evenly distributed throughout the two-month consumption period), we can estimate total water use during any calendar month using the following formula:

$$Q_N^c = 0.25 Q_N^b + 0.5 Q_{N+1}^b + 0.25 Q_{N+2}^b \qquad (23.19)$$

This relationship is derived from Fig. 23.4. It indicates that water actually used during the calendar month of March would comprise one-fourth of consumption billed in March, one-half of that billed in April, and one-fourth of that billed in May.

Whereas the above equation allows the calculation of calendar-month water use given bimonthly billing records, a variant of this equation can be used to calculate average per-account water use in a given month:

$$q_N = \left(0.25 \frac{Q_N}{A_N} + 0.25 \frac{Q_{N+2}}{A_{N+2}} \right) \cdot \frac{0.5 A_N + 0.5 A_{N+2}}{0.5 A_N + A_{N+1} + 0.5 A_{N+2}}$$

$$+ 0.5 \frac{Q_{N+1}}{A_{N+1}} \cdot \frac{A_{N+1}}{0.5 A_N + A_{N+1} + 0.5 A_{N+2}} \qquad (23.20)$$

$$= 0.25 \left(\frac{Q_N}{A_N} + \frac{Q_{N+2}}{A_{N+2}} \right) \cdot \frac{A_N + A_{N+2}}{A_N + 2 A_{N+1} + A_{N+2}} + 0.5 \frac{Q_{N+1}}{A_{N+1}} \cdot \frac{2 A_{N+1}}{A_N + 2 A_{N+1} + A_{N+2}}$$

$$(23.21)$$

where q_N = average per-account water use in any given month

Bimonthly cycle billing records:

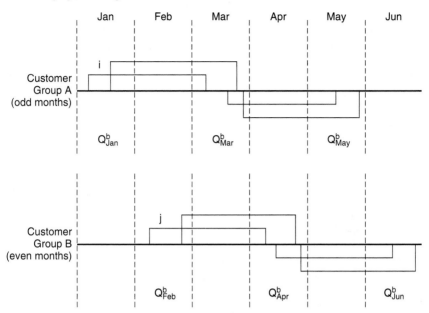

Legend:

i

☐ = beginning and end of consumption period i-th customer in Group A
     (billed during odd months)

j

☐ = beginning and end of consumption period j-th customer in Group B
     (billed during even months)

Q^b_{Mar} = billed water use during the month of March (includes only customers in Group A)

Q^b_{Apr} = billed water use during the month of April (includes only customers in Group B)

Estimate of calendar month water use (Q^c_N):

March: $Q^c_{Mar} = 0.25Q^b_{Mar(A)} + 0.50Q^b_{Apr(B)} + 0.25Q^b_{May(A)}$

April: $Q^c_{Apr} = 0.25Q^b_{Apr(A)} + 0.50Q^b_{May(B)} + 0.25Q^b_{Jun(A)}$

N-th month: $Q^c_N = 0.25Q^b_N + 0.50Q^b_{N+1} + 0.25Q^b_{N+2} = ((Q_N + Q_{N+2})/2 + Q_{N+1})/2$

FIGURE 23.4 Allocation of bimonthly billing records of water use into calendar monthly consumption.

In the case where *both monthly and bimonthly billing cycles* exist within a water utility, total monthly water use in a given month can be obtained by adding smoothed calendar water use from the monthly smoothing procedure to smoothed calendar water use from the bimonthly smoothing procedure. In the case of determining average per-account water use in any given month given the existence of both monthly and bimonthly billing cycles, the following weighting procedure can be used:

$$q_{N,\text{avg}} = q_{N,b} \cdot \text{WF}_b + q_{N,m} \cdot \text{WF}_m \tag{23.22}$$

where $q_{N,\text{avg}}$ = average per-account water use in any given month
$\quad\quad q_{N,b}$ = average per-account water use in any given month from bimonthly billing cycle
$\quad\quad q_{N,m}$ = average per-account water use in any given month from monthly billing cycle
$\quad\quad \text{WF}_b$ = weight factor for per-account water use from bimonthly billing cycle [see Eq. (23.23)]
$\quad\quad \text{WF}_m$ = weight factor for per-account water use from monthly billing cycle [see Eq. (23.24)]

$$\text{WF}_b = \frac{0.5\, A_{N,b} + A_{N+1,b} + 0.5\, A_{N+2,b}}{(0.5\, A_{N,b} + A_{N+1,b} + 0.5\, A_{N+2,b}) + (0.5\, A_{N,m} + 0.5\, A_{N+1,m})}$$

$$= \frac{A_{N,b} + 2A_{N+1,b} + A_{N+2,b}}{(A_{N,b} + 2A_{N+1,b} + A_{N+2,b}) + (A_{N,m} + A_{N+1,m})} \tag{23.23}$$

$$\text{WF}_m = \frac{0.5\, A_{N,m} + 0.5\, A_{N+1,m}}{(0.5\, A_{N,b} + A_{N+1,b} + 0.5\, A_{N+2,b}) + (0.5 A_{N,m} + 0.5 A_{N+1,m})}$$

$$= \frac{A_{N,m} + A_{N+1,m}}{(A_{N,b} + 2A_{N+1,b} + A_{N+2,b}) + (A_{N,m} + A_{N+1,m})} \tag{23.24}$$

In addition to the allocation of aggregate billing data into calendar month, water-consumption records for individual customers can also be allocated to calendar months. For example, if a single account's meter is read on March 15 and then again on May 15, a smoothing procedure can be used to standardize individual account billing-cycle data into calendar-month use. The meter-reading cycle can be monthly, bimonthly, or even trimonthly. The composition of the equation applied to this type of smoothing is as follows:

$$Q_n^c = \sum_{i=(n-x)}^{i=(n+y)} N_i \overline{Q_i^b} \tag{23.25}$$

where n = the nth calendar month
$\quad\quad x$ = the number of months prior to month n that fall within the billing period
$\quad\quad y$ = the number of months beyond month n that fall within the billing period
$\quad\quad i$ = summation index (the ith month)
$\quad\quad Q_n^c$ = quantity of water allocated to calendar month n

N_i = number of days in the ith month that are proportioned to consumption in month n

Q_i^b = average daily water consumption for the billing period which represents the ith month

This smoothing procedure is applied to the water-billing histories for each account. In this procedure, each account will be smoothed in accordance with the current read date and the prior read date. That is, the smoothing procedure must first look at the current and prior read dates. Next, consumption for each record is allocated into calendar months by the fraction of water use that belongs to each month encountered in the bill period [see Eq. (23.25)]. Finally, consumption is summed for each account by month.

23.2.2.3 *End Uses of Water.*   A meaningful assessment of the efficiency of water use cannot be made without breaking down the seasonal and nonseasonal uses of water into specific end uses. Precise measurements of the quantities of water used for showering, toilet flushing, and other purposes require installation of flow-recording devices on each water outlet found on customer premises. Because such measurements are very costly, engineering estimates are often used. However, such estimates are of limited validity because they tend to rely on many assumptions, and often ignore the physical and behavioral settings in which water use takes place. Analytical methods for quantifying the significant end uses of water are described Sec. 23.2.4

23.2.3 Water-Use Relationships

Reasonably precise estimates of water use can be obtained by disaggregating the total delivery of water to urban areas into two or more classes of water use and determining separate average rates of water use for each class. The *disaggregate estimation of water use* can be represented as the product of the number of users (or the demand driver count) and a constant average rate of usage:

$$Q_t = \sum_c N_{t,c} q_c \tag{23.26}$$

where $N_{t,c}$ represents the number of customers in a homogeneous user sector c at time t, and q_c is the unit use coefficient (or average rate of water use per customer) in that sector.

Gains in accuracy of disaggregated estimates are possible because the historical records often show that the variance of average rate of water use within some homogeneous sectors of water users is smaller than the variance of the aggregate use. Although in some nonresidential sectors the variance of q_c is large, it is often more than offset by the smaller variance in the residential sectors with a disproportionately greater number of customers.

The sectoral disaggregation of total urban demands may also be extended spatially and temporally. With the added dimensions of disaggregation, Eq. (23.26) would be expanded to the form:

$$Q_t = \sum_c \sum_s \sum_g N_{t,c,s,g} q_{c,s,g} \tag{23.27}$$

where s denotes the disaggregation of water use according to its seasonal variations (e.g., annual, seasonal, and monthly) and g represents the spatial disaggregation of water use into various geographical areas, such as pressure districts or land use units, which are relevant for planning purposes. An example of a unit-use coefficient $q_{c,s,g}$ could be average use in the single-family residential sector during the summer season in a pressure district.

The level of disaggregation is limited by the availability of data on average rates of water use in various sectors and the ability to obtain accurate estimates of driver counts N for each disaggregate sector. The latter are typically obtained from planning agencies who maintain data on population, housing, and employment for local areas.

The last three decades have produced numerous studies of the determinants of urban water demand. The advancements in theory were followed by the development of water-use models which recognized that both the level of average daily water use and its seasonal variation can be explained adequately by selected demographic, economic, and climatic characteristics of the study area. Such advanced models retain a high level of disaggregation; however, they allow the average rates of water use and the number of users (i.e., drivers) to change in response to changes in their determining factors:

$$Q_t = \sum_c \sum_s \sum_g (N_{t,c,s,g} q_{t,c,s,g}) \tag{23.28}$$

where

$$N_{t,c,s,g} = f(Z_i) \tag{23.29}$$

and

$$q_{t,c,s,g} = f(X_j) \tag{23.30}$$

where Z_i and X_j are, respectively, determinants of the number of water users (e.g., residents, housing units, or employees) and determinants of the average rates of water use (such as average per-person, per-household, or per-employee water use). Table 23.7 gives examples of determinants of the number of users and their average rates of water use.

TABLE 23.7   Determinants of Urban Water Demand

Determinants of demand drivers Z_i
Natural birth rates
Net migration
Family formation rates
Availability of affordable housing
Economic growth and output
Labor participation rates
Urban growth policies
Determinants of average rates of use X_j
Air temperature and precipitation
Type of urban landscapes
Housing density (average parcel size)
Water-use efficiency
Household size and composition
Median household income
Price of water service
Price of wastewater disposal
Industrial productivity

The remainder of this section describes the available data on the average rates of water use and reviews a number of empirical water-use models.

23.2.3.1 Average Rates of Water Use. The first step in the analysis of water demands in an area served by a public water-supply system is to determine average annual rates of water use. The simplest rate is the gross per-capita water use which is determined by dividing the total annual amount of water delivered to the distribution system by the estimated population served. Other rates are obtained by dividing the metered water use by various urban sectors by the number of users or customers in each sector. In addition to the average annual use, average rates during high-use and low-use seasons can be estimated.

The average rates of use are sometimes compared among service areas in order to assess the relative efficiency of water use. However, the aggregate nature of these rates precludes any meaningful comparisons among various service areas. These rates vary from city to city as a result of differences in local conditions that are unrelated to the efficiency in water use. For example, per-capita use may range from 50 to 500 gcd (189 to 1893 L/day). Table 23.8 shows the distribution of per-capita rates among 392 water-supply systems serving approximately 95 million people in the United States. The mean per-capita use in this sample was 175 gcd (662 L/day) with a standard deviation of 72 gcd (273 L/day) (AWWA 1986).

TABLE 23.8 Average Per-Capita Rates of Water Use

Range of per-capita use, gcd	Number of systems	Percent of systems
50–99	30	7.7
100–149	132	33.7
150–199	133	33.9
200–249	51	13.0
250–299	19	4.8
300+	27	6.9
Total	392	100.0

Source: AWWA (1984).

Generally, high per-capita rates are found in water-supply systems servicing large industrial or commercial sectors. Therefore, more meaningful comparisons would require the disaggregation of total use into homogeneous sectors of water users. Table 23.9 shows average-use rates per housing unit in the residential sector for selected cities. Nonresidential water-use rates in gallons per employee per day (ged) are shown in Table 23.10.

23.2.3.2 Modeling of Water Use. *Water-use models* are usually obtained by fitting theoretical functions to the data sets described in Sec. 23.1.4. The selection of appropriate data and estimation techniques depends on the desired characteristics of the final model. For example, time series models of aggregate data can be used for developing models of aggregate use for near-term forecasting. Generally, time series aggregate models can be expected to provide predictions of water use that are less reliable than those obtained from pooled time series cross-sectional observations on water use of individual customers from homogeneous groups of users.

TABLE 23.9 Average Rates of Water Use in Single-Family and Multiple-Unit Buildings

		Average use per dwelling unit, gpd	
Area/state	Year	Multiple-unit buildings	Single-family homes
Metered sales data			
National City, Calif.	1987	226	309
San Diego, Calif.	1990	261	—
Santa Monica, Calif.	1988	231	376
Torrance, Calif.	1987	183	383
Anaheim, Calif.	1987	284	526
Beverly Hills, Calif.	1987	241	917
Camarillo, Calif.	1987	246	437
Fullerton, Calif.	1988	246	566
Los Angeles, Calif.	1987	264	408
Cape Coral, Fla.	1990	123	—
Boston, Mass.	1990	167	—
Framingham, Mass.	1992	152	—
Newton, Mass.	1992	199	—
Las Vegas, Nev.	1989	290	—
Samples of buildings			
New York City, N.Y.	1990	248	—
Seattle, Wash.	1988	107	—
Baltimore, Md.	1965	218	—
Washington, D.C.	1981	197	—
Springfield, Ill.	1985	117	—
Los Angeles, Calif.	1976	165	—
San Francisco Bay, Calif.	1976	183	—
Central Valley, Calif.	1975	144	—
San Diego, Calif.	1991	137	—
Pasadena, Calif.	1991	187	—
Newton, Mass.	1992	155	—
Bailment, Mass.	1990	121	—
Framingham, Mass.	1992	164	—

The most commonly used technique for developing water-use models is *regression analysis*. A simple regression model that captures the relationship between two variables can be written as:

$$y_i = \alpha + \beta X_i + \epsilon_i \tag{23.31}$$

where y_i = water use of customer i

α = intercept term of the equation and the component of the effect of X upon y that is constant regardless of the value of X

β = slope coefficient of the equation and the component of the effect of X upon y that changes depending upon the value of X

ϵ_i = error term for the ith customer, and $i = 1, 2 \ldots n$, which measures the difference between the estimated value of y and the true observed value of y

This model also assumes that X, the independent variable, influences y, the dependent variable, while the dependent variable does not influence the independent variable in any way. Eq. (23.31) decomposes water use y_i into explained and unexplained

TABLE 23.10 Average Rates of Nonresidential Water Use from Establishment Level Data

Category	SIC code	Use rate, gal/employee · day	Sample size
Construction	—	31	246
General building contractors	15	118	66
Heavy construction	16	20	30
Special trade contractors	17	25	150
Manufacturing	—	164	2790
Food and kindred products	20	469	252
Textile mill products	22	784	20
Apparel and other textile products	23	26	91
Lumber and wood products	24	49	62
Furniture and fixtures	25	36	83
Paper and allied products	26	2614	93
Printing and publishing	27	37	174
Chemicals and allied products	28	267	211
Petroleum and coal products	29	1045	23
Rubber and miscellaneous plastics products	30	119	116
Leather and leather products	31	148	10
Stone, clay, and glass products	32	202	83
Primary metal industries	33	178	80
Fabricated metal products	34	194	395
Industrial machinery and equipment	35	68	304
Electronic and other electrical equipment	36	95	409
Transportation equipment	37	84	182
Instruments and related products	38	66	147
Miscellaneous manufacturing industries	39	36	55
Transportation and public utilities	—	50	226
Railroad transportation	40	68	3
Local and interurban passenger transit	41	26	32
Trucking and warehousing	42	85	100
U.S. postal service	43	5	1
Water transportation	44	353	10
Transportation by air	45	171	17
Transportation services	47	40	13
Communications	48	55	31
Electric, gas, and sanitary services	49	51	19
Wholesale trade	—	53	751
Wholesale trade—durable goods	50	46	518
Wholesale trade—nondurable goods	51	87	233
Retail trade	—	93	1044
Building materials and garden supplies	52	35	56
General merchandise stores	53	45	50
Food stores	54	100	90
Automotive dealers and service stations	55	49	198
Apparel and accessory stores	56	68	48
Furniture and homefurnishings stores	57	42	100
Eating and drinking places	58	156	341
Miscellaneous retail	59	132	161
Finance, insurance, and real estate	—	192	238
Depository institutions	60	62	77
Nondepository institutions	61	361	36

TABLE 23.10 Average Rates of Nonresidential Water Use from Establishment Level Data (*Continued*)

Category	SIC code	Use rate, gal/employee · day	Sample size
Security and commodity brokers	62	1240	2
Insurance carriers	63	136	9
Insurance agents, brokers, and service	64	89	24
Real Estate	65	609	84
Holding and other investment offices	67	290	5
Services	—	137	1878
Hotels and other lodging places	70	230	197
Personal services	72	462	300
Business services	73	73	243
Auto repair, services, and parking	75	217	108
Miscellaneous repair services	76	69	42
Motion pictures	78	110	40
Amusement and recreation services	79	429	105
Health services	80	91	353
Legal services	81	821	15
Educational services	82	117	300
Social services	83	106	55
Museums, botanical, zoological gardens	84	208	9
Membership organizations	86	212	45
Engineering and management services	87	58	5
Services, NEC	89	73	60
Public administration	—	106	25
Executive, legislative, and general	91	155	2
Justice, public order, and safety	92	18	4
Administration of human resources	94	87	6
Environmental quality and housing	95	101	6
Administration of economic programs	96	274	5
National security and international affairs	97	112	2

Source: Planning and Management Consultants, Ltd. (1994; unpublished data).

components, where the explained component is expressed as a function of a systematic force X. The unexplained component is expressed as random noise. In other words, in Eq. (23.31), $\alpha + \beta X$ is the deterministic component of y, and ϵ is the stochastic or random component.

In *ordinary least-squares* (OLS) regression analysis, the parameters α and β are estimated by fitting a regression line to water-use data so that the sum of squared residuals of y, $\Sigma \epsilon_i^2$, away from the line is minimized. The method of least squares dictates that one choose the regression line where the sum of the squared deviations of the points from the line is a *minimum*, resulting in a line that fits the data as well as possible.

In order for OLS to yield valid results, the residuals (ϵ) must meet the five assumptions of simple linear regression:

1. *Zero mean.* $E(\epsilon_i) = 0$ for all i, or the expected value of mean error is zero. In other words, the errors are expected to fluctuate randomly about zero and, in a sense, cancel each other out.

2. *Common (or constant) variance.* Var $(\epsilon_i) = \sigma^2$ for all i, which states that each error term has the same variance for each customer.

3. *Independence.* ϵ_i and ϵ_j are independent for all $i \neq j$.

4. *Independence of* X_j. ϵ_i and X_j are independent for all i and j, which says that the distribution of ϵ does not depend on the value of X.

5. *Normality.* ϵ_i are normally distributed for all i. This also implies that ϵ_i are independently and normally distributed with mean zero and a common variance σ^2. The concept of normality is needed for inferences on parameters, but is not required to find least square estimates.

When the five basic assumptions of the regression model are satisfied, OLS provides unbiased estimates of the regression coefficients α and β, which have minimum variance among all unbiased estimates. In other words, the least-squares estimators $\hat{\alpha}$ and $\hat{\beta}$ indeed yield the estimated straight line that has a smaller residual sum of squares than any other straight line. For this reason, OLS estimates are referred to as *best linear unbiased estimates* (or BLUE). Any violation of these assumptions can reduce the validity of the OLS method. The greater the departure of the model from this set of assumptions, the less reliable is OLS. In such situations, one must use alternative estimation procedures depending on the type of violation of these assumptions. One alternative estimation technique called *generalized least squares* (GLS) is described later in this section.

Multiple regression techniques should be used to model a dependent variable instead of simple regression because: (1) the dependent variable can be predicted more accurately if more than one independent variable is used, and (2) if the dependent variable depends on more than one independent variable, a simple regression on a single independent variable may result in a biased estimate of the effect of this independent variable on the dependent variable. The theoretical model of multiple regression is basically the same as in simple regression. The only difference is that the dependent variable is assumed to be a linear function of more than one independent variable. For example, if there are three independent variables, the model is

$$y = \alpha + \beta_1 X_1 + \beta_2 X_2 + \beta_3 X_3 + \epsilon \qquad (23.32)$$

where $X_1, X_2, X_3 =$ independent variables assumed to affect the dependent variable y
$\qquad \epsilon =$ random error term
$\qquad \alpha, \beta_1, \beta_2, \beta_3 =$ estimated coefficients

Just as in the case of simple regression, the coefficients α and β_i are estimated by finding the value of each that minimizes the sum of the squared deviations for the observed values of the dependent variable from the values of the dependent variable predicted by the regression equation. Furthermore, in order to obtain least-square estimates, multiple regression must follow each of the five assumptions required by simple regression, with two added conditions. The first condition is that none of the independent variables can be an exact linear combination of any of the other independent variables. In other words, no one variable can be an exact multiple (or linear combination) of any other independent variable. For example, X_1 cannot be written as aX_2. This situation is called *multicollinearity*. The second condition is related to degrees of freedom. Specifically, the number of observations N must exceed the number of coefficients being estimated. In practice, the sample size should be considerably larger than the number of coefficients to be estimated in order to obtain meaningful information about the underlying relationship.

A *time series analysis of monthly data* on volumes of water sold in consecutive billing periods can be used to estimate water-use models. However, the reliability of such models will depend on: (1) the ability to disaggregate sales data into classes of similar users (e.g., single-family residential, multiunit residential, small commercial, large industrial, and so on), (2) the ability to separate (or account for) the seasonal effects and weather effects in the time series data, and (3) the ability of the estimation technique to deal with nonconstant error variance and correlation of model errors through time. A theoretical time series model can be written as

$$y_t = a + \sum_{i=1}^{N} b_i S_{i,t} + \sum_{j=1}^{M} c_j W_{j,t} + \sum_{k=1}^{P} d_k X_{k,t} + \sum_{r=1}^{R} e_r C_{r,t} + \epsilon_t \qquad (23.33)$$

where

y_t = aggregate volume of water sold to a homogeneous class of customers during a monthly or bimonthly billing period t where $t = 1 \dots T$

a = model intercept

S_i = a set of N seasonal variables that capture the seasonal variability of water use, $i = 1 \dots N$

W_j = a set of M weather variables that capture the effect of actual weather conditions on water use, $j = 1 \dots M$

X_k = a set of P trend forming variables that capture changes in water use unrelated to seasonal and weather effects, $k = 1 \dots P$

C_r = a set of R conservation variables, $r = 1 \dots R$

ϵ_t = error term

b_i, c_j, d_k, e_r = coefficients to be estimated

The selection and definition of variables to represent the four types of systematic forces that affect aggregate water use over time are very important. The seasonal component in water-use data can be captured in many ways. Three possible specifications used in modeling time series water use data include: (1) a seasonal index, (2) a discrete step function, and (3) a Fourier series of sine and cosine terms.

A *seasonal index* is usually expressed as the average fraction of total annual water use to be expected during a given calendar month. This fraction can be estimated using the time series data on water use. For example, the value of the index in July can be obtained by dividing water use during the month of July by total annual use for each calendar year and then calculating the average value of the index for all years in the data set. The process is repeated for each calendar month until all twelve values of the seasonal index are obtained. The seasonal index is then used as a simple variable to capture the seasonal component of water use in Eq. (23.33). A *discrete step function* can be represented by twelve indicator variables corresponding to individual calendar months (e.g., $M_1, \dots M_{12}$, where $M_1 = 1$, if the month in the data is January, and $M_1 = 0$ elsewhere). When bimonthly data are modeled, six indicator variables would be created, one for each bimonthly period. In order to avoid multicollinearity, only $M - 1$ indicators should be specified, where M denotes the number of monthly or bimonthly periods. A *Fourier series* of sine and cosine terms is a harmonic function that can be applied to the data to generate a smooth sinusoidal cycle of seasonal effects. In the case of monthly data, the Fourier series may include six sine and cosine harmonics that can be written as

$$\sum_{h=1}^{6} \left(a_h \sin \frac{2\pi hm}{12} + b_h \cos \frac{2\pi hm}{12} \right) \qquad (23.34)$$

where a_h, b_h = coefficients to be estimated

m = calendar month ($m = 1$ for January, $m = 2$ for February, and so on)

The cycle corresponding to $h = 1$ has a 12-month period. The cycles corresponding to $h = 2$ are harmonics of the 6-month period. All six harmonics represent the seasonal cycle of water use, which is periodic but not directly sinusoidal. Because the lower harmonics tend to explain most of the seasonal fluctuations, in most situations, it may be possible to omit higher frequency harmonics in Eq. (23.34), thus representing the seasonal component as

$$\sum_{i=1}^{4} b_i \, S_{i,t} = b_1 \, \text{SIN}(1) + b_2 \, \text{COS}(1) + b_3 \, \text{SIN}(2) + b_4 \, \text{COS}(2) \qquad (23.35)$$

where b_1, b_2, b_3, b_4 = coefficients to be estimated
$\text{SIN}(1) = \sin (2\pi m/12)$
$\text{COS}(1) = \cos (2\pi m/12)$
$\text{SIN}(2) = \sin (4\pi m/12)$
$\text{COS}(2) = \cos (4\pi m/12)$

The significance of each cycle is usually tested first. The cycles with insignificant amplitudes (i.e., b_i) can then be deleted from the equation.

Air temperature and rainfall are usually used to capture the effects of weather on water use. In most cases, these two variables will be correlated with the seasonal variables. Therefore, the weather variables should be measured as deviations from their normal values for each month (or billing period). Also, lagged weather variables can be used to take into account: (1) the fact that the recorded consumption in any given month represents water use which took place during the current and the previous month, and (2) the short-term memory in water use (e.g., water use in month t is affected by rainfall in month $t-1$). The following weather variables can be included in the specification of the weather effects in Eq. (23.33): (1) deviation of monthly rainfall from monthly norms, (2) deviation of monthly average of maximum daily temperatures from monthly norms, (3) deviation of the number of days with precipitation greater than 0.01 inches from monthly norms, and (4) deviation of cooling degree days from monthly norms.

These deviations can be specified both as contemporaneous and lagged measurements. The "normal" values can be calculated for the period of the time series data or for weather data extending up to 30 years back (i.e., long-term averages).

In addition to properly measuring the seasonal and weather effects, it is necessary to include the effects of variables such as the number of customers, the price of water, the cost of wastewater disposal, and other factors such as income in residential sectors and productivity in nonresidential sectors. The changes in the *number of customers* can be incorporated by expressing the dependent variable in terms of water use per customer (by dividing total volume of water by the number of customers billed). The *price of water* and *wastewater disposal* should be included in the model. In modeling aggregate water-use data, it is difficult to determine what measure of price should be used. The relevant measure is the marginal price faced by an individual customer. This price is the same for all customers when a uniform rate structure is used. If increasing block rates are used, then average price determined for an average consumption level can be used, so that only actual increases in the price of water and wastewater are captured by the price variable. All nominal values of price should be converted into constant dollars using the Consumer Price Index (CPI) for all items. *Median household income* should also be included among the variables of residential models if the data are available and expressed in constant dollars. Usually, household-income statistics can be obtained for each quarter from the Internal Revenue Service. Monthly values of income can be obtained by interpolating the quarterly data. Several other variables that are known to influence water use can be omitted if the changes over time are minimal (e.g., average number of persons per customer connection, average lot size). If changes in such variables are significant, then these variables should be included in the model.

Finally, the effects of the conservation programs can be accounted for in the time series model by including an indicator variable which separates the data into pre- and postprogram periods. This variable takes on the value of 0 for all months before the program, and the value of 1 for the months after program implementation. The effects of other passive and active conservation measures such as a conservation-oriented plumbing code should be included in the model. This can be accomplished by introducing a variable that measures the cumulative number of new service connections sold after the code went into effect. The *error term* can be specified as additive or multiplicative. A logarithmic transformation of the water-use variable will result in a multiplicative error term. Such a transformation will often produce a better fit of the model than untransformed water use.

The potential problems with estimating the parameters of the time series regression model are *endogeneity of the price variable* (i.e., the price is related to the quantity of water used), *nonconstant error variance* (also known as *heteroskedasticity*), and *autocorrelation*. Most regression software packages have routines that attempt to correct for the problem of autocorrelation. Nonconstant error variance and endogeneity are problems best suited for alternative regression methods such as generalized least squares.

A time series *analysis of daily production* records can be used to develop a daily water-use model for the service area. Maidment et al. (1985) have developed a very sensitive model that explains the daily variability in water use in terms of maximum daily temperature, rainfall events, and the delayed response of water users to weather. This model, referred to as *WATFORE* (*wat*er use *fore*casting), can be used to reconstruct a time series of daily water use and for forecasting. The WATFORE model is described in Sec. 23.2.5.

Shaw and Maidment (1987; 1988) used the WATFORE model to measure the impacts of water-use restrictions in two Texas cities, Austin and Corpus Christi. Their analysis was based on a formulation of a time series with a single intervention variable (similar to a dummy indicator variable) as proposed by Hipel, et al. (1975):

$$y(t) = y^*(t) + N(t) \tag{23.36}$$

where $y^*(t)$ = the dynamic response of the process $y(t)$ to an intervention event
$N(t)$ = a stochastic background noise

The *intervention model* used by Shaw and Maidment (1987) was

$$W_s(t) = \overline{W}_s + \frac{\omega_{01}}{1 - \delta_{11}B} T(t) + \frac{\omega_{02} - \omega_{12}B}{1 + \delta_{12}B} R(t)$$

$$+ \sum_{i=1}^{v} y^*(t) + \frac{1}{1 - \phi_1 - \phi_2 B^2 - \phi_7 B^7} a(t) \tag{23.37}$$

where $W_s(t)$ = short-term memory water use
$\overline{W}_s$ = mean level component of short-memory series
T = daily maximum air temperature
R = previous day's seasonal water-use level for rainfall days
a = random shock input
ω, δ = transfer function coefficients
Φ = autoregressive coefficients of the noise model
B = backshift operator

and

$$y^*(t) = \frac{\omega(B)}{\delta(B)} B^b I(t)^T \tag{23.38}$$

where $y^* =$ intervention response term [v in Eq. (23.37) indicates the number of interventions affecting the series]
$I(t) =$ a pulse function input representing the event, $I(t) = 0$ when the event is not occurring

This model and the procedure used by Shaw and Maidment (1987; 1988) can be followed in order to estimate the reduction in average-day and maximum-day water use resulting from the intervention in the form of a drought emergency program or programs aimed at reducing maximum-day water use. For more information about intervention analysis and the use of transfer functions to model daily water use, see also Box and Tiao (1975); and Maidment, et al. (1985).

If customer-level monthly (or billing-period) data on water use for a period of two to four years can be obtained and supplemented with information on customer characteristics, and external factors such as price of water and weather, then a *pooled time series cross-sectional* (TSCS) data set can be constructed and used to estimate the parameters of a multiple regression model. The theoretical form of the model may be written as

$$y_{it} = \beta_1 + \sum_{k=2}^{K} \beta_k X_{k,it} + \epsilon_{it} \tag{23.39}$$

where $y_{it} =$ monthly water use of customer i in month t (instead of months, billing periods may be used)
$\beta_1 =$ the intercept term
$\beta_k =$ regression coefficients
$X_k =$ a set of independent variables that represent all possible systematic forces which affect that use
$\epsilon_{it} =$ the error term

The types of independent variables of residential water-use models that may be used in Eq. (23.39), are illustrated in Table 23.11. Measurements of these variables (or as many variables as possible) should be obtained for *each* customer and *each* time period using such sources of information as: (1) telephone or mail surveys of customers in the sample, (2) real estate and tax-assessor records, (3) aerial photographs, (4) driveby surveys of customer premises, (5) water and wastewater prices and rate structures, and (6) meteorological stations. Many variables will have values that are constant over time or will have only one observation in time that is available. In the latter case, their values can be assumed constant over the period for which water-use data is obtained.

Once the pooled time series cross-sectional database is complete, an appropriate form of the functional relationship between water use and its determinants must be selected. Also, in the context of the structure of systematic forces, the analyst should consider the appropriate structure of the model error. Residential water-demand models are often estimated using one of the following functional forms:

TABLE 23.11 Important Explanatory Variables for Residential Water-Use Models

Category/variable	Category/variable
Family characteristics	Frequency of outdoor uses
Family size (number of persons)	Lawn and landscape watering per
Number of children under 18	week
Household income	Car washing per week
Ownership of residence	Hosing of concrete (blacktop)
Household fixtures and appliances	surfaces
Number of showers	Price
Number of toilets	Marginal price of water
Washing machine (presence of)	Rate structure
Dishwasher (presence of)	Wastewater charge
Garbage disposal	Weather variables
Frequency of appliance use	Monthly average of maximum daily
Laundry loads per week	temperatures
Dishwasher loads per week	Total monthly precipitation
Outdoor features	Number of days with precipitation
Lot size	greater than 0.01 inches
Lawn size	Cooling degree days
Total irrigated area	Other characteristics
Automatic sprinkling system	Age of the house
Swimming pool	Type of sewerage system

1. Linear model

$$y_{it} = \beta_1 + \sum_{k=2}^{K} \beta_k X_{k,it} + \epsilon_{it} \tag{23.40}$$

in which the error is additive

2. Log-linear (or double-log) model

$$\log(y_{it}) = \log(\beta_1) + \sum_{k=2}^{K} \beta_k (\log X_{k,it}) + \log(\epsilon_{it}) \tag{23.41}$$

with multiplicative error

3. Exponential model

$$\log(y_{it}) = \beta_1 + \sum_{k=2}^{K} \beta_k X_{k,it} + \epsilon_{it} \tag{23.42}$$

with multiplicative (and/or exponential) error

Good examples of a pooled time series cross-sectional multivariate regression model can be found in Dziegielewski and Opitz (1988); and Dziegielewski, et al. (1993).

Most frequently, pooled time series cross-sectional analyses use the linear functional form of the model and estimate parameters using the *ordinary least-squares* (OLS) estimation technique. However, the nature of pooled time series cross-sectional data often results in the violation of one or more of the basic regression assumptions. Two common problems that have been encountered in practical research are heteroskedasticity and autocorrelation. Heteroskedasticity usually arises from cross-sectional components of the data. The variance of the error term ϵ is not constant across all observations. When this occurs, the scale of the dependent variable and the explanatory power of the OLS model tend to vary across observations. Autocorrelation is usually found in time series data. The disturbances ϵ_i are not independent of each other; they are correlated. Time series data often display a "memory" such that variation is not independent from one period to the next. For example, an earthquake or flood may affect water use in a particular community for many periods following the actual event. Note, however, that it does not always take such a large disturbance to produce autocorrelated errors. If either one of these problems exists, the coefficient estimates of the OLS model no longer have minimum variance among all linear unbiased estimators. In other words, the OLS estimators are no longer the best linear unbiased estimates (BLUE).

Several diagnostic tests for heteroskedasticity and autocorrelation are commonly available in statistical software packages. Perhaps the simplest diagnostic check is to plot the residuals. If heteroskedasticity and/or autocorrelation are detected, it is advisable to specify an alternative to the OLS model. One such model, the generalized least-squares (GLS) regression model, can be shown to provide the best linear unbiased estimators under these conditions. Instead of minimizing the sum of squared residuals as in OLS estimation, the GLS procedure produces a more efficient estimator by minimizing a *weighted* sum of the squared residuals. Observations whose residuals are expected to be large because the variances of their associated disturbances are known to be large are given a *smaller* weight. Observations whose residuals are expected to be large because other residuals are large are also given smaller weights (Kennedy 1985). In order to produce coefficient estimates using GLS, the variance-covariance matrix of the disturbance terms must be *known* (at least to a factor of proportionality). In actual estimating situations, however, this matrix is usually not known. A procedure called EGLS (*estimated generalized least squares*) can then be employed to estimate the variance-covariance matrix of the disturbances. EGLS estimators are no longer linear or unbiased, but because they account for the effects of heteroskedastic and autocorrelated errors, they are thought to produce better coefficient estimates.

The EGLS estimation technique is used to estimate what are called *error component* models of pooled time series and cross-sectional data (see Kennedy, 1985). The general specification for the random effects model can be written as

$$y_{it} = \beta_0 + \sum_{i=1}^{K} \beta_k X_{k,it} + \epsilon^* \tag{23.43}$$

where $\epsilon^* = u_i + v_t + \epsilon_{it}$.

Notice that this model has an overall intercept and an error term that consists of three components. The u_i represents the extent to which the ith cross-sectional units intercept differs from the overall intercept. The v_t represents the extent to which the tth time period's intercept differs from the overall intercept. The u_i and v_t are each assumed to be independently and identically distributed with a mean of zero and variance of σ_u^2 and σ_v^2, respectively. The third component, ϵ_{it}, represents the traditional error term that is unique to each observation. All three error components are assumed to be mutually independent. The extent to which the intercept coefficients differ across cross-sectional units and across time is assumed to be randomly distributed. Because of this, the error components model is sometimes referred to as the *random effects model.* A good example of an error components model can be found in Chesnutt and McSpadden (1990).

23.2.3.3 Conditional Demand Models. Econometric end-use models have recently been introduced in two analyses of household demand for electricity (McCollister and Hesterberg, 1986; EPRI, 1984). In the electric industry, end-use models are referred to as *microeconometric conditional demand models.* The general specification of such models was originally formulated by Parti and Parti (1980). *Conditional-demand analysis* (CDA) permits the disaggregation of total household demand for electricity into the component demand functions for electricity of particular appliances, even though no direct observations on the energy use of specific appliances exist.

The component-demand functions can be used to estimate monthly and annual energy use of each appliance, as well as the corresponding price and income elasticities. By using a treatment/control evaluation design, the end-use model also allows the analyst to single out changes in energy consumption stemming from external factors such as the introduction of a new conservation technology or conservation program.

In its application to end-use modeling of water consumption, a straightforward but expensive method for disaggregating total household water use would be to observe a sample of households in which water meters were attached to specific water appliances and fixtures. The volume of water used by these appliances during a month or year could be regressed upon various explanatory variables, including household characteristics and weather.

The conditional demand methodology allows the analyst to obtain such end-use demand functions even though no observations on water use of the appliances are available. The conditional analysis rests on the fact that the total water use in a household is the sum of water used through all fixtures, appliances, and facilities such that

$$Q = Q_0 + \sum_{i=1}^{n} Q_i \tag{23.44}$$

where Q_i ($i = 1 \ldots n$) is water used through a set of *specified* fixtures and appliances and Q_0 is water used through a set of *unspecified* fixtures and appliances.

For each Q_i, we can write

$$Q_i = f_i(X_j) \tag{23.45}$$

where Q_i is the conditional household demand for water use through the ith appliance, and X is the vector of j explanatory variables of this function.

Equation (23.45) can be rewritten by adding a dummy variable A_i, which takes on a value of 1 for those households possessing the ith appliance and 0 otherwise. Equation (23.45) then becomes:

$$Q_i = f_i(X_j) \cdot A_i \tag{23.46}$$

Water use in the unspecified appliances can also be assumed to be a function of the explanatory variables such that:

$$Q_0 = f_0(X_j) \tag{23.47}$$

If Eqs. (23.46) and (23.47) are linear, Eq. (23.44) can be rewritten

$$Q = \sum_{i=0}^{n} \sum_{j=0}^{m} b_{ij} (X_j A_i) \tag{23.48}$$

where each b_{ij} $(i = 0 \ldots n; j = 0 \ldots m)$ is the coefficient of the jth exogenous variable in the ith conditional demand function, the X_j $(j = 0 \ldots m)$ are the exogenous variables in the conditional demand functions, and X_o and A_o are unity. This framework allows for the possibility that the conditional demand functions are in a semilog or other nonlinear form. Equation (23.48) can be estimated by using linear regression techniques. The estimated coefficients for this equation are estimates of the parameters of the conditional water-demand functions and of the demand for water through the unspecified group of appliances.

A useful feature of the conditional demand analysis technique is that it can be used to provide econometric estimates of the average water-use levels of individual appliances without the necessity of directly metering end use (which is costly and perhaps unrealistic for any large sample or population). The *average* water use through the ith appliance in a sample of households can be computed as

$$\bar{q}_i = b_{io} + \sum_{j=1}^{M} b_{ij} (\bar{X}_{ij}); \qquad i = 0 \ldots N \tag{23.49}$$

where $\bar{X}_{ij}$ are the average values of the M exogenous variables in the households that possess the ith appliance. If one assumes that total water use is equal to the sum of average water use of each appliance and the deviations away from this average use, Eq. (23.48) can be written as

$$Q = \sum_{i=1}^{N} b_{io} A_i + \sum_{i=0}^{N} \sum_{j=1}^{M} b_{ij} \bar{X}_{ij} A_i$$

$$+ \sum_{i=0}^{N} \sum_{j=1}^{M} b_{ij} (X_j - \bar{X}_{ij}) A_i \tag{23.50}$$

or, using Eq. (23.49)

$$Q = \sum_{i=1}^{N} \bar{q}_i [(A_i)] + \sum_{i=0}^{N} \sum_{j=1}^{M} b_{i,j} [(X_j - \bar{X}_{i,j}) A_i] \tag{23.51}$$

An ordinary least-squares regression method can be used to regress Q on the terms in brackets. The estimated coefficients for the appliance dummy variables can be taken to represent average water use through those appliances by households possessing the appliances.

The number of end uses modeled by this technique can vary. However, costs of data collection typically rise with the number of end uses modeled. It may be appropriate to model household water use as the sum of a set of unspecified appliances (indoor water use) and a set of specified outdoor appliances or fixtures such as garden hoses and swimming pools. Water use through the unspecified appliances depends on explanatory variables such as price, weather, and a variety of household socioeconomic characteristics that can be obtained from a household survey. Outdoor water use depends on variables such as price, weather, type of landscape, and landscape area. Landscape information can be obtained through either a household survey or a driveby type of survey where trained personnel observe and rate the landscapes of sample households.

Although this method permits estimation of the parameters of the conditional demand functions, it requires a large number of explanatory variables, many with common components. For example, a strict interpretation of Eq. (23.48) requires the estimation of $n + 1$ price coefficients. This suggests that there may be a high degree of correlation among the independent variables, and that the identification of all the appliance-specific parameters might be impossible unless certain parameter restrictions are imposed. Usually a number of restrictions are imposed to alleviate this problem.

23.2.4 Estimation of Water-Conservation Savings

A precise measurement of water savings that can be attributed to various demand-management programs is difficult because the observed water use often shows great variability among different users and it also significantly varies over time for the same user. For example, the amounts of water used inside and outside a residential home can vary substantially from month to month and from household to household. This variability is caused by many factors, including conservation practices. Because of the great variability in water use, the observed changes in water use over time, or differences in use between individual customers or groups of customers may be caused by influences unrelated to the customer participation in a conservation program. Therefore, the most important consideration in measuring water-conservation savings is the design of a measurement procedure that is capable of correctly measuring not only the changes in water use but also separating these changes into those caused by the program and those caused by changes in weather, prices, economic factors, and other confounding factors. The precision of the measurements of water savings depends on whether the study design was capable of isolating and controlling for: (1) the characteristics of the conservation program that could significantly influence the results of the estimation of water savings, (2) the characteristics of the customer groups targeted by the program that could also influence the results, and (3) the characteristics that are external to both the conservation program design and the targeted customer groups. Research in evaluation designs identified a number of factors referred to as outside effects or externalities (Dziegielewski, et al. 1993).

Once a conservation practice is adopted, the baseline demand that represents water use without the practice cannot be directly measured, and the unaltered demand has to be reconstructed somehow. In practice, all study designs employ comparisons of water-use behavior (and other customer characteristics in some cases) over time and/or between groups of customers. Possible types of comparisons are illustrated by Fig. 23.5. Implementation of a conservation program divides the time continuum into two periods, namely, pretreatment conditions and posttreatment conditions. It also divides the water users into two groups—the control group of nonparticipants and the treatment group of program participants. The conditions of a valid study design are achieved by a careful selection of a sample of water users and the use of proper methods of data analysis.

There are two basic approaches for estimating water-conservation savings: statistical techniques and leveraged approaches. These two approaches are discussed in the following.

23.2.4.1 Statistical Estimation of Savings. *Statistical comparison methods* produce estimates of conservation savings by comparing water use between a partici-

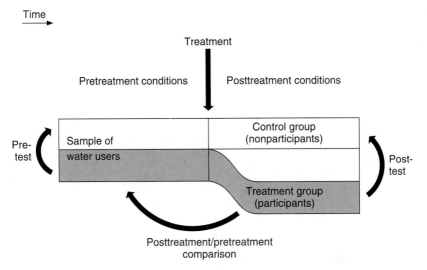

FIGURE 23.5 Evaluation designs for measuring water-conservation savings.

pant group and a control group (or changes in water use before and after the program). The *comparison of means method* is derived from the statistical theory of randomized controlled experiments which utilizes a treatment/control design. Conservation savings are estimated as the difference in the mean level of water use between the treatment group and the control group, i.e.:

$$d = \bar{q}_t - \bar{q}_c \qquad (23.52)$$

where

$$\bar{q}_t = \frac{1}{n_1} \sum_{t=1}^{n_1} q_t \qquad (23.53)$$

and

$$\bar{q}_c = \frac{1}{n_2} \sum_{c=1}^{n_2} q_c \qquad (23.54)$$

where d = conservation effect (difference between means)
$\bar{q}_t$ = mean water use in the treatment sample
$\bar{q}_c$ = mean water use in the control sample
q_t = water use of customer t in the treatment sample
q_c = water use of customer c in the control sample
n_1 = number of customers in the treatment sample
n_2 = number of customers in the control sample

The data can be used to test whether the observed difference d can be attributed to chance, or whether it is indicative that the two samples come from populations of unequal means. Given that the parent population distribution of the differences in means is unknown or not normal, and that the population standard deviation of water use in each group (σ_t and σ_c) are unknown but assumed equal, the *sampling* distribution of the differences in mean water use should follow a t-distribution for large samples.* A calculated t-statistic can be used to test the hypothesis that the differences between mean water use of the treatment and control samples is zero. In order to calculate a t-statistic, the standard error of the difference between the two means must be determined using the following formula:

$$S_d = \left(\frac{(n_t - 1)\, s_t^2 + (n_c - 1)s_c^2}{n_t + n_c - 2} \right) \left(\frac{n_c + n_t}{n_c\, n_t} \right) \tag{23.55}$$

where S_d = estimated standard error of the conservation effect d
         s_t = standard deviation of water use in the treatment sample
         s_c = standard deviation of water use in control sample
         n_t = number of customers in the treatment sample
         n_c = number of customers in the control sample

The t-statistic is then calculated as

$$t = \frac{\bar{q}_t - \bar{q}_c}{S_d} = \frac{d}{S_d} \tag{23.56}$$

Using the properties of the t-distribution, one can test the null hypothesis stating that the true population mean of the treatment group is equal to the population mean of the control group, or the alternative hypothesis stating that the population means of the treatment and control groups are different. The value of the t-statistic calculated by Eq. (23.56) uses the sample estimates to infer whether the difference d is large enough to reject the null hypothesis that the true difference in means is equal to zero. To test the null hypothesis, the resultant value of t must be compared to statistical tables of the t-distribution. These tables may be found in any standard statistics textbook. Depending on the type of assertion to be made about the difference between means, one may use either a one-tail or two-tail (also called one-sided or two-sided) test of significance.

The two-tail test should be used to test the hypothesis that the difference between means d is zero, against the alternative hypothesis that the difference in means is positive or negative. The one-sided test of significance should be used to test the null hypothesis of no difference in means against the alternative hypothesis that, after the treatment, the mean water use in the treatment group is *lower* than mean water use in the control group.

In order to produce reliable estimates, the comparison of means method must satisfy two requirements: (1) the two random variables representing water use in the treatment and control group must be drawn from the same population distribution, and (2) the distribution is normal. The first assumption is often violated. In the classic experimental design on which the comparison of means method is based, the experimenter has careful control over all factors that might affect the variable under consideration. Therefore, by carefully designing the samples to be used in the exper-

---

* The Central Limit theorem (and the concept of repeated sampling) implies that for large sample sizes, the sampling distribution of the difference in means between groups will approach a normal distribution. This allows statistical inference about population parameters when the population distribution is unknown or not normal. However, in order for the sampling distribution of the difference in two means to be approximated by the normal distribution, the population standard deviations (σ_t and σ_c) must be known. If σ_t and σ_c are not known but are assumed equal, one may use the sample estimates of σ_t and σ_c (s_t and s_c) and the t-distribution for statistical inference. For large sample sizes, the t-distribution will approximate the normal distribution.

iment, any difference between the treatment and control groups can be attributed solely to the treatment. However, the water planner is not likely to have complete control over the confounding factors that affect household water use. Although statistical theory suggests that randomly assigning households into treatment and control groups will result in groups that are less likely to be systematically different in terms of water use, this does not ensure that the two groups are identical with respect to household income, average number of persons per household, yard size, and many other factors. Therefore, unless a great deal of matching or sampling work is done, water use cannot be considered a random variable. It is related, if not caused, by the uncontrolled-for factors that differ between the treatment and control groups. In other words, one runs the risk of incorrectly attributing observed changes in water use to the treatment (e.g., a retrofit), when in fact they are caused by the different average values of external factors in each group. With respect to the second assumption, it must be stressed that empirical distributions of water use have most often been found not to follow a normal distribution. Typically, distributions of water use show a long right-hand tail, and thus do not conform to the symmetric bell-shaped appearance associated with the normal distribution. The violation of this assumption is not a fatal flaw, however, if a normalizing transformation of the data is used. For example, taking the log of water use should at least pull in the right-hand tail and minimize the leverage of "contaminated" or outlying data points or they can be screened out by rejecting values $> \bar{x} + 3s$.

When adhering to a strict experimental design, the comparison of means method is more likely to produce meaningful and reliable results in situations where: (1) the expected conservation effect is large when compared to mean water use, (2) the variance in water use is small compared to the conservation effect, (3) the mean and variance in water use are very similar (in terms of size) for both groups prior to treatment, and (4) the sample sizes in the treatment and control groups are large. The comparison of means method can produce reliable and informative results if used in conjunction with experimental designs and large sample sizes.

Multivariate regression models represent the most sophisticated method of comparing water use data over time or between groups of customers while controlling for the effects of a large number of external factors. One can choose from a variety of regression methods depending on the types of available data and the acceptable level of estimation complexity. The statistical models described in Sec. 23.2.3.2 can be used for this purpose.

23.2.4.2 Time Series Analysis of Conservation Effects. A time series of the volumes of water sold in consecutive billing periods can be used to measure conservation effects of full-scale programs while controlling for external influences other than the conservation program. The reliability of the estimates will depend on: (1) the ability to disaggregate sales data into classes of similar users (e.g., single-family residential, small commercial), (2) the ability to separate (or account for) the seasonal effects and weather effects in the time series data, and (3) the ability of the estimation technique to deal with nonconstant error variance and correlation of model errors through time. The effects of a conservation program under investigation can be measured by including an indicator variable which separates the time series data into pre- and postprogram periods. However, it is also important to capture the effects of other passive and active conservation measures which are adopted by water customers independently of the program under evaluation.

If customer-level monthly (or billing-period) data on water use for a period of two to four years can be supplemented with information on customer characteristics, and such external factors as price of water and weather conditions, then a pooled time series cross-sectional data set can be constructed and used to estimate the parameters of a multiple regression model. The important explanatory variables will depend on the customer class. In the residential sectors they usually include information on family characteristics, household features and appliances, frequency of appliance use, outdoor features, and frequency of outdoor uses. This measurement technique can produce very accurate estimates of actual water savings for some programs, especially those targeting the residential sector.

The major drawback of the multivariate regression models is that they are relatively expensive and time consuming. They require large amounts of data and a lot of expertise from the analyst. They also need large sample sizes and are less appropriate for nonresidential water users.

23.2.4.3   End-Use Accounting System.
The accumulating experience in the evaluation of conservation programs indicates that it is very difficult to obtain measurements of water savings with a high level of precision using a single best method. The most precise estimates can be achieved by taking advantage of the strong features of the statistical methods with engineering methods. The known strengths, weaknesses, and biases of each approach can be used to narrow down the confidence bands surrounding the actual water savings.

Engineering (or mechanical) estimates are obtained using laboratory measurements or published data on water savings per device or conservation practice. These data can be combined with assumptions regarding the magnitude of factors expected to impact on the results of the conservation program in order to generate estimates of program savings. However, the resultant estimates can be very sensitive to the underlying assumptions and relationships. For example, the savings resulting from the installation of an ultra-low-flush toilet replacing a standard toilet will depend on assumptions regarding flushing volumes and frequency of flushing. The resultant savings can range from 19.5 gal per person per day (3.9 gal $*$ 5 flushes per person per day) to 7.6 gal per person per day (1.9 gal $*$ 4 flushes per person per day). The high estimate is almost 3 times greater than the low estimate. The validity of the assumptions used in the above example can easily come under attack, since they rely on subjective conclusions and a great deal of professional judgment of the engineer or analyst.

Despite their obvious shortcomings, engineering estimates may be considered appropriate for providing preliminary estimates of potential savings when field measurements are not available. They can also be used to verify statistical estimates by setting limits on a possible range of savings. However, they become most useful in leveraged techniques where they can be used to augment and strengthen statistical models.

The most promising method of leveraging information involves the use of information from one approach within the procedures of another. For example, engineering estimates or special metering measurements can be used as independent variables in statistical models. Statistical models of urban water demands do not accommodate the needs of planning and evaluation of demand-management programs because of an inability to disaggregate water demands down to the end-use level. Because many demand-side programs target specific end uses, the absence of end-use water-demand models severely impairs the development of effective demand-management policies. Without adequate end-use models, the effects of various demand-management programs cannot be measured precisely. In order to enhance the ability of water planners to formulate, implement, and evaluate various demand-management alternatives, it is necessary to disaggregate the usually observed sectoral demands during a defined season of use into the applicable end uses. Only such a high level of disaggregation will permit water planners to make all necessary determinations in estimating water savings of various programs. A good example of a leveraged approach for estimating water savings is the *end-use accounting system* (EAS) described in Dziegielewski, et al. (1993) and is also presented in Sec. 23.2.5 in the discussion of the IWR-MAIN model.

23.2.5 Water-Use Forecasting

A forecast of urban water use usually forms a basis of one or more water planning and management activities. Depending on the activity, the forecasting horizon may range from several days to several decades. Operations of water systems, fiscal-year budgeting, and financial-management activities often rely on short-term forecasts of daily, weekly, or monthly water use for periods of up to one year. The development of plans for improving the system's distribution and treatment facilities as well as monitoring the performance of demand-side management programs require medium (or intermediate) term forecasts for periods extending from 1 to 5 or 10 years with an annual, monthly, or, sometimes, daily time step. Finally, the development of long-range policies and plans for managing water resources, including the expansion of capacity of water-supply sources and related facilities and the development of long-term demand-management programs, requires long-term forecasts for periods ranging from 10 to 100 years.

The most desirable characteristic of a water-use forecast is its accuracy. Since forecast accuracy can only be determined in retrospect, it must be prospectively assessed from the correlates of accuracy in forecasting approaches. For example, absence of disaggregation may result in inaccurate forecasts if significant trends or relationships are concealed.

23.2.5.1 Forecasting Approaches.   Although there are many types of forecasting approaches, they can be reduced to several prototypical methods which differ with respect to the level of disaggregation and the structural complexity of water-use equations. The simplest technique, known as *time extrapolation,* extrapolates the average change in past water-use records into the future. The forecasting equation can be written as:

$$Q_t = f(Q_{h1} \ldots Q_{hm}, t) \tag{23.57}$$

where Q_t = aggregate water use in forecast year t
$Q_{h1} \ldots Q_{hm}$ = historical time series of water use

The change in water use over time may be assumed to be linear, exponential, logistic, or other. The linear extrapolation uses the following relationship:

$$Q_t = Q_o + gT \tag{23.58}$$

where Q_o = intercept of the historical time series of water use $Q_{h1} \ldots Q_{hm}$
g = average change in water use per time period (i.e., the slope of the trend line fitted to historical data)
T = the number of time periods between the beginning of the historical record and the forecast year t

The historical water use Q_h can be disaggregated by customer classes, geographical areas, or both, without changing the basic premise of the time extrapolation approach, i.e., that water use is a function of time.

A distinctive class of forecasting approaches introduces a simple water-use relationship in which total water use is represented as the product of the number of users and an average rate of water use:

$$Q_t = N_t q_t \tag{23.59}$$

where Q_t = total water use in forecast year t
N_t = number of water users in year t
q_t = average rate of water use per customer (or unit use coefficient) in year t

Examples of this relationship were described in Sec. 23.2.3. In the case of time extrapolation, water use can be disaggregated by season, sector, and geographical area. A disaggregate version of the Eq. (23.27) would be:

$$Q_t = \sum_c \sum_s \sum_g (N_{t,c,s,g} q_{t,c,s,g}) \tag{23.60}$$

where the subscripts denote the three dimensions of disaggregation [see Eq. (23.60) in Sec. 23.2.3].

The third class of approaches includes methods that express the N_t and q_t components of the above equation as functions of their respective determinants. Table 23.7 contains examples of such determinants. The high level of disaggregation and structural relationships make all forecast assumptions very explicit; however, this approach is necessarily complex and it requires more data than the previous approaches.

Although many other forecasting approaches are conceivable, they generally fall between or within the simplest and the most complex classes of approaches, if the level of disaggregation of water use and the inclusion of determinants are used as the classifying factors. The following sections describe two forecasting models which illustrate the use of the forecasting approaches.

23.2.5.2 WATFORE Model.
WATFORE (i.e., WATer FOREcasting) is a computer model designed to forecast daily water use in cities for periods of a few days up to a few months ahead. The program requires daily data on the total pumpage, the daily rainfall and the maximum air temperature, as well as a prediction of expected weather conditions. The program generates a prediction of daily water needs and a probability estimate of exceeding specified usage limits.

The program requires values of 23 coefficients to describe water-use patterns in the study area in terms of: (1) water-use trends from year to year, (2) the seasonal variation of water use within a year, (3) the day-to-day response of water usage to prevailing weather conditions, and (4) an error correcting routine that continually adjusts the current forecast to account for errors in previous forecasts. Information needed to determine a particular set of coefficients includes the last 5 years of daily records and 10 years of monthly records of water use, rainfall, and daily maximum air temperatures.

Table 23.12 contains an example of a forecast generated by WATFORE. Prior to the forecast period which starts on June 15, the model corrects itself by forecasting one day ahead then observing the actual usage. During the forecast period, the actual water use is not known and the self-correction subroutine does not operate. Actual usage is shown in Table 23.12 just for visual purposes. However, actual rainfall and air-temperature data are used in the example although in practice they are not known at the time of the forecast. The model contains a subroutine for generating exceedence probabilities of specified usage limits for any given day. For example, the probabilities of actual water use exceeding 130 mgd for the first 5 days of the forecast period are 0, 2, 13.67, 0.01, and 0.05 percent. However, these probabilities are only valid in the event of no rainfall during the forecast period and an accurate temperature forecast.

According to the theoretical model employed by WATFORE, daily water use is made up of base and seasonal use, both exhibiting trends through time. Seasonal use has two components, one which varies smoothly over the year with normal air temperature and another which represents the short memory residuals:

$$W(t) = \hat{W}_b(t) + g(t) [\hat{W}_p(t) + W_s(t)] \qquad (23.61)$$

where W = daily water use
$\hat{W}_b$ = estimated base use
$\hat{g}$ = trend coefficient for peak seasonal use
$\hat{W}_p$ = estimated potential water use, a function of normal air temperature
W_s = short-memory water use
t = daily time index from beginning of series

A *transfer function noise model* is formulated for the short-memory series:

$$W_s(t) = \overline{W}_s + \sum_{i=1}^{2} \frac{\omega_0^{(T_i)}}{1 - \delta_1^{(T_i)}B} T_i(t) + \sum_{i=1}^{2} \frac{\omega_0^{(R_i)} - \omega_1^{(R_i)}B}{1 - \delta_1^{(R_i)}B} R_i(t)$$

$$+ \frac{1}{1 - \phi_1 B - \phi_2 B^2 - \phi_7 B^7} a(t) \qquad (23.62)$$

where $\overline{W}_s$ = level component of short-memory series model
T = transformed daily average air temperature
R = daily rainfall or a substitute variable for rainfall effects
a = independent normal random variable of zero mean and variance σ_a^2
i = index for the season of the year or the range of a variable
$\omega_0, \omega_1, \delta_1$ = transfer function coefficients
ϕ_1, ϕ_2, ϕ_7 = autoregressive coefficients of noise model
B = backshift operator

Trend Equations. Trends in both the base and seasonal components are visible in the monthly water-use data. They result from increasing population served by the water system and changes in water-use habits over time. Various methods are available for estimating these trends in water use including regressions using population, number of connections to the distribution system, water price, and household income as independent variables. (Maidment and Parzen, 1984a; 1984b). In this study, a fairly simple approach based on linear regression of maximum and minimum monthly water use against time was used as illustrated in Fig. 23.6.

Monthly average base water use $W_b(m)$ is identified from the lowest monthly water use. A polynomial function of time is fitted

$$\hat{W}_b(m) = a_0 + a_1 m + a_2 m^2 + \dots \qquad (23.63)$$

For Austin, Texas the lowest month is January, so $m = 1, 13, 25, \dots$, in Eq. (23.63). Typically only a_0 and a_1 are statistically significant, although their values may vary from one set of years to another as a result of population growth changes or major changes in the city's water-supply system.

TABLE 23.12 Example of a Water-Use Forecast Generated by the WATFORE Model

Date	Rain	W_{RAIN}	T_{NML}	T_{MAX}	W_{TEMP}	Base	W_{POT}	W_{ERR}	Fcast	Actl	Error
5-26 Sat	0.00	−3.7	87.0	96.0	20.8	69.9	21.1	10.3	121.0	118.3	−2.7
5-27 Sun	0.00	−2.8	87.3	90.0	12.2	69.9	21.4	4.3	110.4	105.0	−5.3
5-28 Mon	0.00	−2.2	87.6	80.0	3.8	69.9	21.8	11.6	97.9	105.0	7.1
5-29 Tue	0.00	−1.6	87.9	80.0	−0.8	70.0	22.1	24.4	99.8	114.0	14.3
5-30 Wed	0.00	−1.3	88.2	86.0	−1.2	70.0	22.5	31.0	111.0	121.0	9.9
5-31 Thu	0.00	−1.0	88.4	88.0	−0.8	70.0	22.8	37.2	118.1	128.3	10.1
6-01 Fri	0.00	−0.7	88.7	90.0	0.0	70.0	23.2	42.9	124.9	135.4	10.5
6-02 Sat	0.00	−0.6	89.0	91.0	2.1	70.0	23.5	44.4	132.3	139.5	7.3
6-03 Sun	0.22	−21.5	89.3	88.0	0.7	70.0	23.8	31.5	111.5	104.6	−6.9
6-04 Mon	0.64	−37.2	89.6	91.0	2.3	70.0	24.2	25.9	88.0	85.2	−2.7
6-05 Tue	0.34	−38.1	89.8	95.0	10.2	70.0	24.5	16.2	90.5	82.8	−7.8
6-06 Wed	0.00	−31.4	90.1	91.0	7.1	70.0	25.3	14.1	87.2	85.1	−2.1
6-07 Thu	0.00	−24.0	90.3	92.0	6.8	70.1	26.8	10.9	93.8	90.6	−3.3
6-08 Fri	0.01	−23.0	90.6	92.0	6.2	70.1	28.4	14.0	93.3	95.6	2.3
6-09 Sat	0.00	−19.3	90.8	92.0	5.4	70.1	29.8	18.5	100.1	104.5	4.4
6-10 Sun	0.00	−14.0	91.1	92.0	4.5	70.1	31.3	20.8	109.1	112.7	3.5
6-11 Mon	0.28	−23.1	91.3	89.0	−0.2	70.1	32.7	19.7	98.5	99.1	0.6
6-12 Tue	0.05	−32.5	91.5	91.0	−1.1	70.1	34.0	14.6	88.5	85.2	−3.3
6-13 Wed	0.00	−28.9	91.7	92.0	−0.1	70.1	35.3	19.1	90.0	95.5	5.4
6-14 Thu	0.00	−22.0	92.0	93.0	1.8	70.1	36.5	18.5	103.2	105.0	1.8
6-15 Fri	0.00	−16.7	92.1	93.0	2.5	70.1	37.7	16.7	110.3	112.9	0.0
6-16 Sat	0.00	−12.7	92.3	95.0	6.1	70.2	38.8	15.5	117.8	117.5	0.0
6-17 Sun	0.00	−9.7	92.5	95.0	7.7	70.2	39.9	14.5	122.6	125.8	0.0
6-18 Mon	0.12	−23.3	92.7	90.0	−0.6	70.2	40.9	13.6	100.8	117.3	0.0
6-19 Tue	0.00	−25.5	92.8	96.0	5.3	70.2	41.9	12.5	104.4	118.3	0.0
6-20 Wed	0.00	−19.5	93.0	96.0	8.2	70.2	42.8	11.9	113.6	136.7	0.0
6-21 Thu	0.00	−14.9	93.1	97.0	11.3	70.2	43.6	11.2	121.5	139.1	0.0
6-22 Fri	0.00	−11.4	93.3	97.0	12.7	70.2	44.4	10.6	126.6	144.1	0.0
6-23 Sat	0.00	−8.7	93.4	100.0	18.6	70.2	45.2	10.0	135.4	142.0	0.0
6-24 Sun	0.00	−6.6	93.5	102.0	25.1	70.2	45.9	9.4	144.1	144.3	0.0
6-25 Mon	0.00	−5.1	93.6	102.0	28.5	70.2	46.6	8.9	149.1	152.2	0.0
6-26 Tue	0.00	−3.9	93.7	103.0	31.8	70.3	47.2	8.4	153.8	146.8	0.0
6-27 Wed	0.01	−22.2	93.8	88.0	9.7	70.3	47.8	7.9	113.5	106.9	0.0
6-28 Thu	0.00	−23.8	93.9	95.0	7.2	70.3	48.4	7.5	109.5	116.7	0.0
6-29 Fri	0.00	−15.2	94.0	98.0	11.0	70.3	48.9	7.0	122.1	136.2	0.0
6-30 Sat	0.00	−9.7	94.1	98.0	12.9	70.3	49.4	6.6	129.5	137.8	0.0
7-01 Sun	0.00	−6.2	94.2	98.0	13.8	70.3	49.9	6.3	134.0	139.8	0.0
7-02 Mon	0.00	−4.0	94.2	97.0	12.4	70.3	50.4	5.9	134.9	137.4	0.0
7-03 Tue	0.00	−2.6	94.3	98.0	13.2	70.3	50.8	5.6	137.4	137.4	0.0
7-04 Wed	0.00	−1.7	94.4	98.0	13.6	70.3	51.3	5.3	138.7	137.5	0.0
7-05 Thu	0.00	−1.1	94.5	100.0	17.2	70.4	51.7	5.0	143.1	150.0	0.0
7-06 Fri	0.00	−0.8	94.5	99.0	17.2	70.4	52.1	4.8	143.7	152.1	0.0
7-07 Sat	0.00	−0.5	94.6	99.0	17.1	70.4	52.6	4.5	144.0	150.8	0.0
7-08 Sun	0.00	−0.3	94.7	98.0	15.1	70.4	53.0	4.3	142.5	148.6	0.0
7-09 Mon	0.00	−0.2	94.7	99.0	15.7	70.4	53.5	4.1	143.4	148.7	0.0
7-10 Tue	0.00	−0.2	94.8	100.0	17.7	70.4	53.9	3.9	145.7	147.8	0.0
7-11 Wed	0.00	−0.1	94.9	98.0	15.1	70.4	54.4	3.7	143.4	143.1	0.0
7-12 Thu	0.00	−0.1	95.0	98.0	13.5	70.4	54.8	3.5	142.2	142.3	0.0
7-13 Fri	0.00	−0.1	95.1	99.0	14.3	70.4	55.3	3.3	143.3	136.9	0.0
7-14 Sat	0.00	−0.0	95.1	99.0	14.6	70.4	55.8	3.2	143.9	137.5	0.0

Key to Columns:

T_{MAX} = maximum temperature, °F
Rain = rainfall, in
W_{TEMP} = water use due to temperature excess, mgd
W_{POT} = potential use (due to normal temperature), mgd
Fcast = forecasted water use, mgd

Error = forecast error, mgd
T_{NML} = normal maximum temperature for that day, °F
W_{RAIN} = water use due to rainfall, mgd
Base = base use, mgd
W_{ERR} = water use due to previous forecast errors, mgd
Actl = actual water use, mgd

The peak monthly use is estimated in a similar manner. The expected base use $\hat{W}_b(m)$ is subtracted from the data in the months typically exhibiting the highest water use (July and August in Austin, so $m = 7, 8, 19, 20, \ldots$), and the remaining seasonal water use $S_p(m)$ is regressed against time, monthly rainfall $R(m)$, and monthly average temperature $T(m)$:

$$\hat{S}_p(m) = b_0 + b_1 m + b_2 m^2 + \ldots + b_R\,[R(m) - \overline{R}(m)]$$

$$+ b_T\,[T(m) - \overline{T}(m)] \tag{23.64}$$

where $\overline{R}(m)$ and $\overline{T}(m)$ are the long-term average values of $R(m)$ and $T(m)$, respectively, in Eq. (23.64). Typically only b_0, b_1, b_R, and b_T are statistically significant. A weather-corrected estimate of peak monthly use is obtained by setting $T(m) - \overline{T}(m)$ and $R(m) - \overline{R}(m)$ to zero in Eq. (23.64). Equivalent daily equations for estimating the trends through time of base use $\hat{W}_b(t)$ and peak monthly seasonal use $\hat{S}_b(t)$ can readily be obtained from Eq. (23.63) and the truncated version of Eq. (23.64).

Equations (23.63) and (23.64) are employed to eliminate the base use from the data and convert the seasonal use component to a seasonally stationary time series

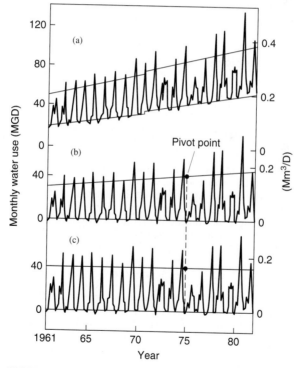

FIGURE 23.6 Seasonal variations. Procedure for isolating and detrending monthly seasonal water use: (a) Linear trend lines are fitted to base use; (b) base use is subtracted and a trend line fitted to maximum monthly seasonal use; (c) maximum trend line is rotated about a pivot point to produce a stationary seasonal water-use pattern.

(Fig. 23.6). First, actual seasonal use $S(t)$ is produced by subtracting base use from the total use $W(t)$:

$$S(t) = W(t) - \hat{W}_b(t) \qquad (23.65)$$

Second, a pivot time t_0 is selected, which is a reference date for standardizing the data. The growth coefficient $g(t) = \hat{S}_p(t)/\hat{S}_p(t_0)$ is computed and employed to produce a detrended, seasonally stationary time series $S_d(t)$ of the seasonal water use:

$$S_d(t) = \frac{S(t)}{g(t)} \qquad (23.66)$$

The transformations from Eqs. (23.65) and (23.66) are shown graphically in Figs. 23.6a and 23.6b, respectively.

Heat Function. The year-to-year variations in seasonal water use (Fig. 23.6c) arise from the collective action of hundreds of thousands of people responding to rainfall and heat conditions in their decisions regarding watering their lawns and gardens. Thus the process being modeled is partly physical and partly psychological, and the use of physical variables and equations to describe the collective response is an approximation of a very complex underlying process that involves millions of individual decisions. Therefore the climatic variables (rainfall and air temperature) used in describing this process can be regarded only as indicator variables.

In modeling the seasonal variation exhibited in Fig. 23.6 it is assumed that there is a functional relationship between water use and air temperature which is valid in the absence of recent rainfall. Figure 23.7 suggests that this function can be glimpsed during dry periods, but is obscured during wet periods. Thus it is logical to screen the data and select only those periods during which there are minimal rainfall effects in order to study the relationship between water use and air temperature. In Austin, weekly average values of water use and air temperature are used in estimating a heat function $H(T)$. During the summer (April–October), the data are selected such that there is no rain in the two previous weeks and in the week under consideration; for the remainder of the year, 3-day periods with no rain prior to the specified week are sufficient to define a dry period. The selected data for observed weekly average values of the seasonal water use are then plotted against weekly average air temperature T, (Fig. 23.8). The heat function was also studied using daily and monthly data, but weekly time intervals yielded the best compromise between stability and sensitivity to change. The heat function is estimated as:

$$H(T) = c_0 + c_1 T \qquad (23.67)$$

where c_0 and c_1 are coefficients that depend upon which of the three ranges of T the observed air temperature lies within. The values of the coefficients and the location of the break points are obtained by iterative application of linear regression. The observed data in Fig. 23.8 indicate that there are three seasons of the year: a summer season (April–October), a transition season (March and November), and a winter season (December–February).

A daily potential water use $W_p(t)$ may be estimated by substituting the normal daily air temperature estimated from long-term records $T_N(t)$ into the heat function for each day of the year.

$$W_p(t) = H[T_N(t)] \qquad (23.68)$$

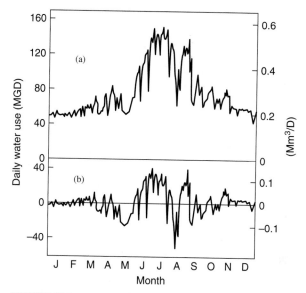

FIGURE 23.7 Daily water use in Austin, Texas: (*a*) 1980, and (*b*) the short-term memory series produced by detrending and deseasonalizing.

The short-memory effects of rainfall, air temperature, and random errors are then isolated as

$$W_s(t) = S_d(t) - W_p(t) \qquad (23.69)$$

In studying the short-memory effects of air temperature using Eq. (23.62), the residual air temperature is henceforth employed:

$$T_r(t) = T(t) - T_N(t) \qquad (23.70)$$

The rationale for using this deseasonalizing approach rather than more conventional methods such as fitting Fourier series is that changes in air temperature can either increase or reduce water usage, but a rainfall occurrence can only reduce usage. A Fourier-series model of historical seasonal water use may not represent these effects adequately as it contains rainfall as well as temperature effects. Because it must indicate increased water usage when there is no rainfall, such a model is inconsistent with the basic postulates cited earlier.

The short-memory series $W_s(t)$ obtained by Eq. (23.69) is illustrated in Fig. 23.7*b* for 1980. It fluctuates about zero, being positive when observed air temperature is greater than normal and negative following rainfall or lower than normal air temperature. The rainfall and air temperature effects are dynamic; that is, a change in rainfall or temperature influences water-use values both on that day and for a number of subsequent days. The functional form of a Box-Jenkins type transfer-function model which represents water usage is given by Eq. (23.62).

Identification of the appropriate model and estimation of the parameters can be accomplished by standard methods (see Box and Jenkins, 1976) provided that the

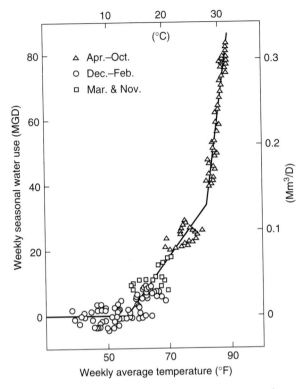

FIGURE 23.8 Heat function relates seasonal water use and average air temperature. Data points shown are weekly average values during rainless periods.

coefficients are constant through time and independent of the levels of the variables $W_s(t)$, $T_i(t)$, and $R_i(t)$. Unfortunately, the daily water-use data in the study do not satisfy these assumptions: the parameters are time varying because the response of water use to climatic variables changes seasonally; there are nonlinearities because the response of water use per unit change in air temperature or per unit of rainfall depends on the magnitude of the temperature and the amount of rainfall which occurs; there is an interaction effect between the responses to successive days of rainfall in that the water-use response to the last day's rain is less than to the first; also, air temperature and rainfall are correlated because the occurrence of rainfall lowers air temperatures. These difficulties necessitate the following modifications of a standard transfer function model.

1. *Time variation.* In order to allow for time variations in the parameters, the year is divided into the same three seasons employed for the heat function [December–February (winter), November–March (transition), and April–October (summer)]. Different parameter sets are estimated for each season in some cases. This particular division into three seasons is somewhat arbitrary but yields empirically satisfactory results.

2. *Nonlinearity.* The response variables are modeled as linear functions of nonlinearly transformed independent variables.

3. *Interaction.* An interaction effect exists in that the response of water use to a rainfall one day depends on the previous day's seasonal water use $S_d(t-1)$. If $S_d(t-1)$ is high, there is a large potential response; if it is lower due to rainfall, the potential response is less. As an alternative to rainfall depth, $S_d(t-1)$ is employed in Eq. (23.62) in conjunction with zero-one indicator variables associated with different magnitudes of $R(t)$. This modification to the model has the advantage that the actual water use cannot be driven below base use but instead approaches base use after several days of sustained rainfall. Empirical observations from the data support this hypothesis.

4. *Correlation.* The dependence of air temperature on rainfall requires the application of a prewhitening procedure in the model-identification phase. A transfer-function model is developed of daily air temperature using daily rainfall as the explanatory variable, and the residuals of this model are substituted in Eq. (23.62) for $T_i(t)$.

Least-squares estimates of the parameters of the short-memory model are obtained using Marquardt's algorithm (Marquardt, 1963). Diagnostic checking of the short-memory model is performed by testing the residual series for independence, stationarity, and normality. The cross-correlation matrix of the parameter estimates is also computed.

23.2.5.3 *IWR-MAIN Model.*

The IWR-MAIN Water-Demand Analysis Software (i.e., the most recent version 6.0) is a computer model for developing long-term forecasts of water use. The model translates demographic, housing, and business statistics (for cities, counties, or service areas) into estimates of existing water demands, and using the official projections of population, housing, and employment, it generates baseline forecasts of water use. The model can also be used to analyze the existing and projected water demands at the end-use level in order to generate estimates of conservation savings from passive, active, and emergency demand-reduction measures as well as their costs and benefits. Figure 23.9 shows the functional components of the model.

The model relies heavily on the existing information on water-use rates and relationships found in urban areas. This information is built into the model as knowledge-base libraries. The economic growth models operate on forecast drivers including the number of housing units, employment, and user-specified drivers. The residential model accounts for growth in housing stock and it maintains constant share allocation of housing units among residential subsectors. The nonresidential growth model maintains constant share of employment among industries. Both models use adjustments for seasonal variation in housing occupancy and employment.

The water-demand models disaggregate total urban water demands into customer classes, time periods, and study areas. Examples of the various dimensions of disaggregation are shown in Fig. 23.10. Disaggregation by purposes (i.e., end uses of water such as landscape irrigation, toilet flushing, and others) is performed by end-use efficiency models.

Within each disaggregate category of demand, average rates of water use are predicted using the following theoretical model:

$$Q_{isrt} = \alpha_{isr} \cdot [d^1_{isrt} (I^{\beta_{1,isrt}}_{it})] \cdot [d^2_{isrt} (MP^{\beta_{2,isrt}}_{is})] \cdot [d^3_{isrt} (e^{BD_{is} \cdot \beta_{3,isrt}})]$$
$$\cdot [d^4_{isrt}(H^{\beta_{4,isrt}}_{it})] \cdot [d^5_{isrt}(HD^{\beta_{5,isrt}}_{it})] \cdot T^{\beta_{6,isrt}}_{st} \cdot R^{\beta_{7,isrt}}_{st} \tag{23.71}$$

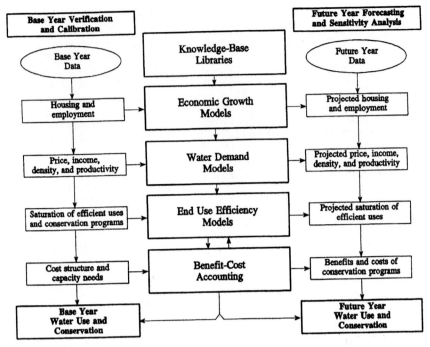

FIGURE 23.9 Major components of IWR-MAIN model.

where Q = predicted water use, in gal/unit · day
             I = median household income, $1000
             MP = effective marginal price, $/kgal
             BD = bill difference, $
             H = mean household size, persons/household
             HD = housing density
             M = average maximum daily temperature, °F
             R = total seasonal rainfall, in
             d = value distribution factor for the corresponding independent
                 variables
             α = intercept, gal/unit · day
   $\beta_1 \ldots \beta_7$ = elasticity values for each independent or explanatory variable

and subscripts

        i = residential subsector
        s = season (winter or summer)
        r = dimension (indoor or outdoor)
        t = base or projection year
        t' = base or long-term weather year (long-term used for all forecast years
            beyond base year)

The α's and β's vary by season (winter/summer), water-use dimension (indoor/outdoor), and by residential subsector.

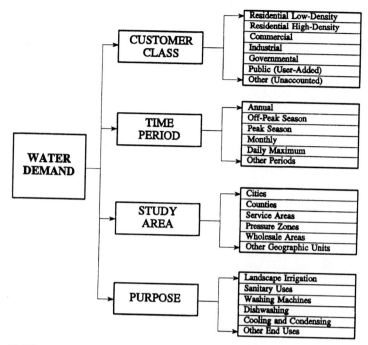

FIGURE 23.10   IWR-MAIN disaggregation of urban water use.

The distribution factors d are computed from the following formula:

$$d_k = \sum_{j=1}^{m} \frac{E_{jk}^{\beta_k} N_{jk}}{E_{\text{median}}^{\beta_k} N}$$
(23.72)

where
d_k = value distribution factor for kth explanatory variable
E = explanatory variable
j = value class of explanatory variable where $1 \le j \le m$
m = total number of classes
N_{jk} = number of units in jth class for kth variable that fits E
N = number of units in entire range
E_{median} = median value of explanatory variable E
β_k = elasticity for kth explanatory variable

The functional form of the model allows users to estimate water demands for each residential sector by indoor and outdoor water-use components and by summer and winter seasons. The functional residential water-demand model can be considered to be causal because it conforms to the economic theory of demand and the explanatory variables can be said to cause the demand. For example, income measures the consumer's ability to pay for water and price influences the amount of water the consumer is willing to purchase. The default elasticities of the explanatory variables of residential seasonal (summer and winter seasons) household water use are derived from econometric studies of water demand through a meta-analysis of empirical literature. An elasticity is a dimensionless measure of the relationship between quantity (water use) and any explanatory variable (e.g., price and income). The elasticity is interpreted as the percent change in quantity (e.g., water use) that is expected from a 1 percent change in the explanatory variable. For example, an elasticity of +0.4 on income in a water-demand equation indicates that a 1 percent increase in income will cause a 0.4 percent increase in water use. Approximately 60 studies of residential water demand which contained almost 200 empirically estimated water-use equations were integrated in order to derive a set of unbiased estimates of long-term elasticities for each explanatory variable.

The residential water-use models contained within IWR-MAIN are designed to estimate water use by: (1) residential subsector, (2) summer and winter seasons, and (3) indoor/outdoor water-use dimensions. It should be noted that because water-use information is not currently readily available regarding the disaggregation of summer and winter water uses for indoor/outdoor water-use purposes, the residential water-use models currently contained within IWR-MAIN assume the 50 percent of summer/winter season water use is used for indoor purposes and 50 percent of summer/winter season water use is used for outdoor purposes. Therefore, for a given residential subsector, season, and time period, the indoor and outdoor water-use estimates will be equal. If the user has more specific information regarding the indoor/outdoor residential water-use patterns of their service area, the intercepts in the residential coefficient library can be adjusted.

The forecast procedure also includes calculations for annual sewer contributions for each residential subsector. Annual sewer contributions for each residential subsector are a product of: (1) indoor water-use estimates, and (2) percent of homes connected to the sewer system. The default values in IWR-MAIN assume that 100 percent of the residential customers are connected to the sewer system. Therefore, for the residential subsector, it is assumed that all indoor residential water use of homes connected to the sewer system will result in sewer flows. Prior to utilizing the sewer contribution estimations generated by the forecast procedure, users should verify that the indoor water use estimates are appropriate for their study area.

The theoretical equation for the nonresidential sector calculates per-employee water use for each designated SIC category or industry group as:

$$Q_{ist} = \alpha_{is} PR_{ist}^{\beta_{1,is}} MP_{is}^{\beta_{2,is}} CDD_{st'}^{\beta_{3,is}} OTH_{ist}^{\beta_{4,is}} \tag{23.73}$$

where
- Q = water use, gal/employee · day
- PR = labor productivity
- MP = marginal price, $/kgal
- CDD = cooling degree days, number of days
- OTH = other (user added)
- α = model intercept, gal/employee · day
- $\beta_1 \ldots \beta_4$ = elasticities for dependent (explanatory) variables

and subscripts

- i = SIC category or industry group
- s = season (winter or summer)
- t = forecast year
- t' = base year or long-term weather year

Although this theoretical model is fully operational within nonresidential forecast procedures of the software, there are no currently available econometric (and generally applicable) models that contain model elasticities for price, productivity, cooling degree days, or the other variable. In this case, IWR-MAIN was designed to accommodate the model specification once data were available regarding the responsiveness of nonresidential water use to such variables.

In order to enhance the ability of water planners to formulate, implement, and evaluate various demand-management alternatives, IWR-MAIN disaggregates the observed sectoral demands during a defined season of use into the applicable end uses. Only such a high level of disaggregation permits water planners to make all necessary determinations in estimating water savings of various programs. The first step is to disaggregate the observed water demands into their specific components or end uses. Figure 23.11 illustrates how water demand of a homogeneous sector of water users can be disaggregated into its seasonal and end-use components. A rational representation of each end use is made using a *structural end-use equation* of the following form:

$$q = [(M_1S_1 + M_2S_2 + M_3S_3) * U + K * F] * A \qquad (23.74)$$

where
q = average quantity of water in a given end use
M_{1-3} = efficiency class of the end use (design parameter)
S_{1-3} = fraction of end uses within efficiency class
U = usage rate (or intensity of use)
K = average flow rate of leaks
F = fraction of end uses with leaks (incidence of leaks)
A = presence of end use in a given sector of users

A graphical representation of this structural end-use relationship is given in Fig. 23.12. An application of Eq. (23.74) to the toilet end use in the residential sector would require the knowledge of all end-use parameters. An example of the uses of this equation for analyzing the toilet end use is presented in Table 23.13. Other end uses and effects of improvements in their efficiency can be estimated using similar parameters and data.

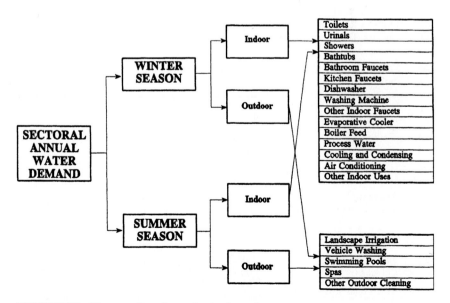

FIGURE 23.11   Disaggregation of annual water demands.

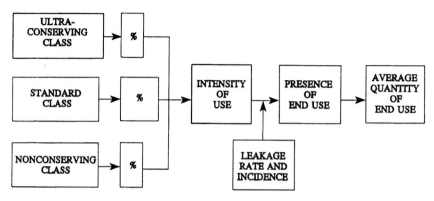

FIGURE 23.12 Structural end-use relationships.

The structural end-use relationship (Eq. 23.74) is dictated by the need to distinguish between changes in water demand caused by active and passive demand-management programs from the changes caused by other factors. The structure of end-use demands which exists in the service area at any point in time will not remain constant over the planning horizon. It will change in response to changes in the determinants of water use such as income, household size, and housing density. The effects of various interventions of demand-management programs must be counted relative to the baseline forecasts of water demands, which capture the effects of the relevant external factors. The parameters of each end use are affected by the external factors. For example, changes in price will cause a decrease in the incidence of leaks in the short run and will affect the distribution of end uses among the classes of efficiency in the long run. The other two parameters in the end-use equation (i.e., intensity and presence) also will be affected by changes in price.

For example, changes in price will cause a decrease in the incidence of leaks in the short run and will affect the distribution of end uses among the classes of efficiency in the long run. The other two parameters of the end-use equation (i.e., intensity and

TABLE 23.13 Example of Estimating Water Savings with End-Use Accounting System (Toilet End Use)

End-use parameter	Before change	After change	Net effect (savings)
Inefficient class rate	5.50	5.50	—
Inefficient class fraction	0.35	0.25	−0.10
Standard class rate	3.50	3.50	—
Standard class fraction	0.55	0.50	−0.05
Efficient class rate	1.60	1.60	—
Efficient class fraction	0.10	0.25	+0.15
Intensity of use, fpd	14.00	14.00	—
Presence of end use	1.00	1.00	—
Leakage rate, gpd	20.00	20.00	—
Incidence of leaks	0.15	0.15	—
Average quantity, gpd	59.10	52.35	−6.75

presence) also will be affected by changes in price. The ideal forecasting model would be capable of predicting the parameters of the structural end-use equations as a function of such influencing variables as price, income, household size, housing density, and weather. For example, in the case of lawn irrigation end use, the parameters of Eq. (23.71) should be modeled as functions of the explanatory variables of Eq. (23.74):

$$S_i = f(I, P, FC) \tag{23.75}$$

$$U = f(I, T, R, P, FC) \tag{23.76}$$

$$F = f(I, P, FC) \tag{23.77}$$

$$A = f(I, P, FC) \tag{23.78}$$

where the symbols are previously defined. These relationships would be used to generate forecasts of baseline end-use demands without demand-management interventions. The functional form of the end-use equations must be retained in order to predict the impacts of efficiency improvement programs on average levels of each end use.

The structure of the end-use equation allows the planner to estimate the net effects of long-term conservation programs by tracking the values of end-use parameters over time. The effects of the demand reduction can be passive or active depending on the nature of the demand-management program under consideration. Passive conservation effects are the result of natural shifts toward higher efficiency classes (i.e., from the *inefficient* 5.5-gal toilet, to the *standard* 3.5-gal toilet, and toward the *efficient* 1.6-gal toilet). The movements of end uses toward higher classes of efficiency are brought about primarily by efficient plumbing codes where (1) newly built structures *comply* with the codes and increase the fraction of end uses in the efficiency class, (2) standard and inefficient fixtures are *replaced* with efficient fixtures during remodeling, and (3) inefficient end uses are *retrofitted* to standard efficiency levels. Because these effects occur outside of the interventions by water utility they can be termed *passive*. The rate of compliance with plumbing codes accounts for changes in the proportions of customer pools that occur from new construction. Though most new construction is in compliance with plumbing codes, there is always some portion of new construction that does not conform to the code. The sources of noncompliance are varied as special exceptions to code granted by local governments to simple unsanctioned noncompliance. Regardless of the source, it may be assumed that this rate is constant for particular end uses. Since all end uses are not subject to plumbing codes, compliance rates may instead define whether new units will be standard or efficient. An example is the proportion of the housing units that purchase an efficient washing machine and those that purchase a standard washing machine. Natural replacement of a particular end use is dependent on the natural life of the item. The natural life can be expected to differ across end uses, as is the case when comparing the expected life of a washing machine to a faucet or a showerhead. Natural replacement also occurs due to remodeling or demolition. The effect of natural replacement will be to move customers to a higher-efficiency pool for the end use being replaced. The third form of passive shift in efficiency is from customer-initiated retrofit. These conservation actions often take the form of showerhead replacement, displacement devices placed in toilet tanks, or faucet aerators. Unlike natural replacement, these efforts are not dependent on the age of the end-use item being retrofitted, but depend primarily on customers' conservation attitudes. Evidence of this form of shift is often found in the surveys that have been performed in areas that have had a retrofit program. One of the most common reasons for not installing utility-provided retrofit devices were that similar devices were already in place. The active efficiency-improvement programs, which are conducted by water utilities, target one or more end uses and retrofit or replace inefficient or standard end uses of those customers who participate in the programs. The active programs will achieve water savings by changing the distribution of end uses among classes of efficiency.

The end-use accounting system accommodates the handling of such issues as interaction and overlapping of multiple programs, customer-initiated conservation effects, and the relationships between long-term and short-term (e.g., drought emergency) programs. Each of these effects creates problems in measuring the effectiveness of efficiency-improvement programs. The problems may be reduced by the tracking of the pools of customers. Interaction effects exist when conservation programs are subsequently applying percentage reductions over time in water use to an already decreased amount of water use. The reductions from previous programs would thereby limit the reductions from subsequent programs. Measuring the effectiveness of programs consequently becomes less accurate as programs are initiated, because the reduction effects of each program are ameliorated by the reduction in customers that have not already been affected by a previous program. Additionally, programs may overlap because portions of customer groups are targeted by multiple programs. These portions reduce the actual coverage of individual programs. The other form of overlap is one that occurs through customer-initiated retrofit. These customers are the group that would have implemented their own efficiency-improvement changes without intervention from active utility programs. Since most of these customers would have already made efficiency improvements the net effect of the utility programs is lower.

Short-term programs also introduce their own set of problems. Since they are unplanned events, they occur in addition to long-term programs that are planned and estimated to achieve a goal of forestalling supply-capacity expansion. The short-term program is a response to supply curtailment and therefore has motives beyond the improvement of efficiency. Most long-term programs are targeted toward improving the mechanical efficiency of end uses, whereas short-term programs go beyond this by affecting use behavior. This means that in addition to effects from shifting customers to higher-efficiency classes there are also changes in the other parameters of water use. All of these effects can be tracked using the end-use accounting system. As customers are targeted for long-term programs, the pool of lower-efficiency users is reduced. The rates of compliance, natural replacement, and customer-initiated retrofit are also tracked as they change the proportions of users in each efficiency class. Finally, the changes in both efficiency class and behavior are accounted for with short-term changes in both efficiency class and change in behavior from restriction programs. The effects of water-rationing plans can be estimated by resetting the usage intensities and the incidence of leaks to lower values for end uses that are rationed. Selected, less essential, end uses may be restricted or banned. A total ban of an end use can be captured by setting the presence parameter A in Eq. (23.74) to zero. During drought emergencies, water utilities often conduct emergency retrofit or replacement programs. Additional reduction of demand can be achieved by such programs, provided the potential for retrofits and replacements of inefficient or standard classes of end use is not preempted by previously implemented long-term programs.

A detailed description of the IWR-MAIN model can be found in Strus, et al. (1994).

REFERENCES

American Water Works Association, "1984 Water Utility Operating Data," Denver, Colo., 1986.

Box, G. E. P., and G. M. Jenkins, *Time Series Analysis, Forecasting and Control,* Holden-Day, San Francisco, Calif., 1976.

Box, G. E. P., and G. C. Tiao, "Intervention Analysis with Applications to Economic and Environmental Problems," *Journal of American Statistical Association,* 70(70), 1975.

Chesnutt, T. W., and C. N. McSpadden, "A Model-Based Evaluation of the Westchester Water Conservation Program," Metropolitan Water District of Southern California, Los Angeles, Calif., 1990.

Cochran, W. G., *Sampling Techniques,* John Wiley, New York, 1963.

Dziegielewski, B., "Development of a High Density Residential Water Use Forecast for New York City: Memorandum Report," Planning and Management Consultants, Ltd., Carbondale, Ill., 1994.

Dziegielewski, B., and E. Opitz, "Phoenix Emergency Retrofit Program: Impacts on Water Use and Consumer Behavior," Phoenix Water and Wastewater Department, Phoenix, Ariz., 1988.

Dziegielewski, B., E. Opitz, and D. Rodrigo, "Seasonal Components of Urban Water Use in Southern California," Metropolitan Water District of Southern California, Los Angeles, Calif., 1990.

Dziegielewski, B., E. M. Opitz, J. C. Kiefer, and D. D. Baumann, "Evaluation of Urban Water Conservation Programs: A Procedures Manual," American Water Works Association, Denver, Colo., 1993.

Electric Power Research Institute, "Measuring the Impact of Residential Conservation," vol. 1–3, EA-3606, Palo Alto, Calif., 1984.

Fowler, F. J. Jr., *Survey Research Methods,* Sage Publications, Newbury Park, Calif., 1988.

Hipel, K. W., et al., "Intervention Analysis in Water Resources," *Water Resources Research,* 11(6):855, 1975.

Kennedy, P., *A Guide to Econometrics,* MIT Press, Cambridge, Mass., 1985.

Kish, L., *Survey Sampling,* John Wiley, New York, 1965.

Maidment, D. R., and E. Parzen, "Time Patterns of Water Use in Six Texas Cities," *Journal of Water Resources Planning and Management Division,* ASCE, 110(1):909–106, 1984.

Maidment, D. R., S-P. Miaou, and Melba M. Crawford, "Transfer Function Models of Daily Urban Water Use," *Water Resources Research,* 21(4):425–32, 1985.

Maidment, D. R., and D. T. Shaw, "WATFORE Daily Municipal Water Forecasting User's Manual," Center of Research in Water Resources, Bureau of Engineering Research, Austin, Tex., 1985.

Maidment, D. R., S-P. Miaou, "Daily Water Use in Nine Cities," *Water Resources Research,* 22(6):845–851, 1986.

Marquardt, D. W., "An Algorithm for Least-Square Estimation of Nonlinear Parameters," *J. Soc. Ind. Appl. Math.,* 11(2):431–441, 1963.

McCollister, G. M., and B. C. Hesterberg, *A Model of Residential Consumption and Appliance Ownership,* Spectrum Economics, Mountain View, Calif., 1986.

Parti, M., and C. Parti, "The Total and Appliance-Specific Conditional Demand for Electricity in the Household Sector," *Bell Journal of Economics,* Spring:309–21, 1980.

Schlenger, D., "Current Technologies in Automatic Meter Reading," *Waterworld News,* 7(3):14–17, 1991.

Shaw, D. T., and D. R. Maidment, "Effects of Conservation on Daily Water Use," *Journal of the American Water Works Association,* 80(9):71–77, 1988.

Shaw, D. T., and D. R. Maidment, "Intervention Analysis of Water Use Restrictions, Austin, Texas," *Water Resources Bulletin,* 23(6):1037–1046, 1987.

Solley, W. B., R. R. Pierce, and H. A. Perlman, "Estimated Use of Water in the United States in 1990," U.S. Geological Survey Circular 1081, Washington, D.C., 1993.

Strus, C. A., E. M. Opitz, B. Dziegielewski, J. R. M. Steinbeck, and R. C. Hinckley, "IWR-MAIN, Version 6.0 Water Demand Analysis Software: User's Manual and System Description," Planning and Management Consultants, Carbondale, Ill., 1994.

van der Leeden, F., F. L. Troise, and D. K. Todd, *The Water Encyclopedia,* Lewis, Chelsea, Mich., 1990.

P · A · R · T · 4

WATER RESOURCES EXCESS MANAGEMENT

CHAPTER 24
HYDROLOGY FOR WATER-EXCESS MANAGEMENT

Robert W. Hinks
Larry W. Mays
Department of Civil and Environmental Engineering
Arizona State University, Tempe, Arizona

24.1 INTRODUCTION

This chapter describes the concepts and techniques of hydrology that provide a technical framework for the policies and procedures of water-excess management. Analysis and modeling of the processes that govern the transformation of precipitation falling on a watershed into runoff provides information on the watershed's surface flow characteristics. This information can be used to predict the occurrence and magnitude of a water-excess event that results from any precipitation event on the watershed. Books that cover hydrology as the main subject area include: Bedient and Huber (1992); Bras (1990); Chow (1964); Chow, et al. (1988); Linsley, et al. (1986); Maidment (1993); McCuen (1989); Ponce (1989); and Viessman, et al. (1989).

24.1.1 Streamflow

Determining characteristics of flow in rivers and other surface channels is a central task of surface-water hydrology, and is *the* critical hydrologic component in the analysis, modeling, and prediction of water-excess events. Streamflow is produced by precipitation during storms and by groundwater entering surface channels, with the former being the dominant contributor during high-flow periods. The hydrologic component processes involved in the transformation of rainfall into streamflow form part of the hydrologic cycle (Sec. 1.3.3), and are illustrated in the block diagram representation of Fig. 24.1 and the time distribution schematic of Fig. 24.2. These figures illustrate the spatial and temporal conceptual relationships that define the precipitation–streamflow transformation, and which yield the four quantitative aspects of streamflow important to water-excess management, namely:

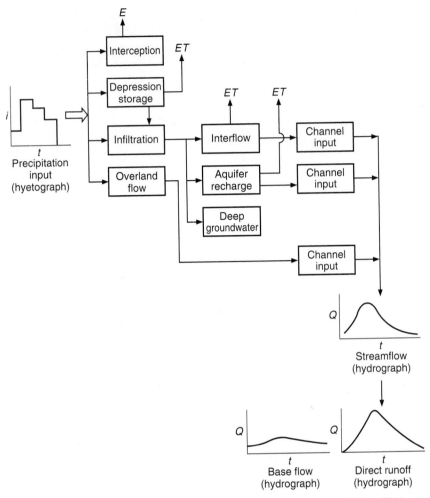

FIGURE 24.1 Distribution of precipitation input. *(Modified from Viessman and Welty, 1985.)*

1. Streamflow time series or peaks (hydrograph)
2. Flow volume time series or totals
3. Channel or reservoir water depth time series or maximums
4. Probabilities (frequencies) of extreme flow rate, volume, or depth magnitudes

24.1.2 Excess Rainfall and Direct Runoff

The *direct runoff* (DRO) component of the single-event discharge hydrograph shown as system output in Fig. 24.1 represents the response of a *watershed* (catchment, drainage basin) to *excess* (effective, net) *rainfall*, and is obtained by deduct-

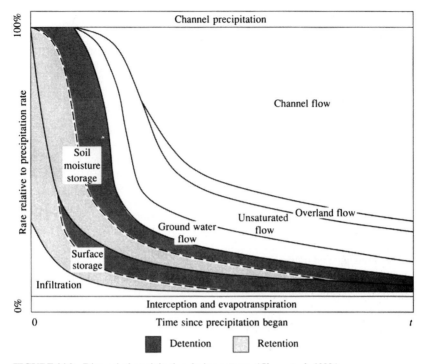

FIGURE 24.2 Disposal of precipitation during a storm. *(Chow, et. al., 1988.)*

ing all infiltration and storage losses from the total storm precipitation (the system input of Fig. 24.1). The area of the DRO part of the hydrograph represents the total volume of excess rainfall and constitutes most of the streamflow volume resulting from storm events in which overland flow and direct-channel precipitation predominate.

Whereas subsurface-flow velocities are normally so low that baseflow cannot contribute a significant amount of storm precipitation directly to streamflow, the *rate of infiltration* is a significant factor in determining the total flow volume and peak discharge of water-excess events. The other *hydrologic abstractions* shown in Fig. 24.1, *interception* and *depression storage,* represent that part of precipitation that may never contribute to runoff, but returns to the atmosphere by way of the processes of direct evaporation (E) and evapotranspiration (ET). Interception and depression storage abstractions are estimated based on the nature of both vegetation and ground surface, or are assumed to be negligible in a large storm.

The remainder of this chapter is devoted to a survey of the more widely used methods of analysis of rainfall–runoff transformation processes. Those processes define the critical hydrologic characteristics of extreme flow events and consequently define the necessary and desirable elements of models that simulate those processes and the resulting water-excess events. For greater detail of many of the principles and techniques discussed the reader is referred to the recent texts, especially Chow, et al. (1988), and Mays and Tung (1992), which serve as the principal references for this chapter.

24.1.3 Floodplain Hydrologic and Hydraulic Analysis

The hydrologic and hydraulic analysis of floods is required for the planning, design, and management of many types of facilities including hydrosystems within a floodplain or watershed. These analyses are needed for determining potential flood elevations and depths, areas of inundation, sizing of channels, levee heights, right of way limits, design of highway crossings and culverts, and many others. The typical requirements include (Hoggan, 1989):

1. *Floodplain information studies.*   Development of information on specific flood events such as the 10-, 100-, and 500-year frequency events.

2. *Evaluations of future land-use alternatives.*   Analysis of a range of flood events (different frequencies) for existing and future land uses to determine flood-hazard potential, flood damage, and environmental impact.

3. *Evaluation of flood-loss reduction measures.*   Analysis of a range of flood events (different frequencies) to determine flood damage reduction associated with specific design flows.

4. *Design studies.*   Analysis of specific flood events for sizing facilities to assure their safety against failure.

5. *Operation studies.*   Evaluation of a system to determine if the demands placed upon it by specific flood events can be met.

The methods used in hydrologic and hydraulic analysis are determined by the purpose and scope of the project and the data availability. Figure 24.3 is a schematic of hydrologic and hydraulic analysis for floodplain studies. The types of hydrologic analysis are to perform either a rainfall–runoff analysis or a flood-flow frequency analysis. If an adequate number of historical annual instantaneous peak discharges (*annual maximum series*) are available then a flood-flow frequency analysis can be

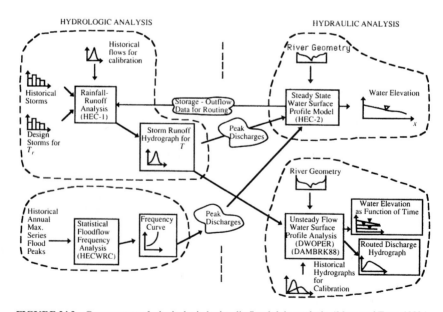

FIGURE 24.3   Components of a hydrologic-hydraulic floodplain analysis. *(Mays and Tung, 1992.)*

performed to determine peak discharges for various return periods. Otherwise, a rainfall–runoff analysis must be performed using a historical storm or design storm for a particular return period to develop a storm-runoff hydrograph.

Determination of water-surface elevations can be performed using a steady-state water-surface profile analysis (see Chap. 23) if only peak discharges are known or one can select the peak discharges from generated storm-runoff hydrographs. For a more detailed and comprehensive analysis, an unsteady-flow analysis based upon a hydraulic-routing model (see Chap. 23) and requiring the storm-runoff hydrograph can be used to more accurately define maximum water-surface elevations. The unsteady-flow analysis also provides more detailed information such as the routed-discharge hydrographs at various locations throughout a river reach.

24.2 DETERMINATION OF THE DESIGN STORM

A *design storm* is a precipitation event used as the basis of design for a hydrologic system. The key factors that define the design storm are the amount (volume) of precipitation and its distribution, both temporally and spatially over the watershed. Records of actual precipitation events, if available, usually contain information that can be used to establish relationships between precipitation intensity, duration, and frequency of occurrence. When a planning decision has been made on the desired design storm frequency and duration, the relationships can be used to determine precipitation intensity and volume. Construction of a *design-storm hyetograph* requires determination of the time sequence of rainfall, which is critical information in establishing peak runoff magnitude and timing in all but very short duration storms. A catchment's runoff response is also dependent on the areal distribution of rainfall, and even the orientation of a storm as it moves across the catchment. For small areas it is generally assumed that precipitation is uniformly distributed, however, this assumption is unrealistic for large watersheds.

24.2.1 Intensity-Duration-Frequency Analysis

The most common approach to establishing a design-storm volume involves use of a relationship between rainfall *intensity* (or depth), *duration,* and the *frequency* or return period appropriate for the catchment or site location. In many cases, the hydrologist is able to use standard *intensity-duration-frequency (IDF)* curves available for the location and does not have to perform this analysis. Typical IDF curves are shown in Fig. 24.4.

Care must be taken in the use of IDF curves. For example, they do not represent time histories of actual precipitation events but rather conditional probabilities of average rainfall intensities; also, the duration is not necessarily the duration of an actual storm but more typically represents an interval within a longer storm.

IDF curves have also been expressed as equations. Wenzel (1982) provided coefficients for a number of cities in the United States for an equation of the form

$$i = \frac{c}{T_d^e + f} \tag{24.1}$$

where i is the design-rainfall intensity and T_d is the duration. Table 24.1 shows values of the constants c, e, and f for 10-year return-period intensities at various locations in the United States.

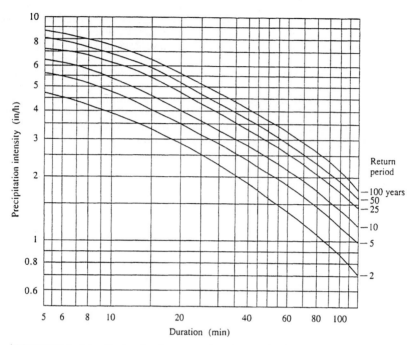

FIGURE 24.4 Intensity-duration-frequency of maximum rainfall curves for Chicago. *(Chow, et al., 1988.)*

24.2.2 Rainfall Distribution

Although the oldest and most empirical methods of analysis, such as the rational method, only consider uniform intensity of rainfall and compute the peak-discharge rate, most hydrologic modeling requires a time distribution of the design-storm rainfall for generation of the design hydrograph. There are a number of standard empir-

TABLE 24.1 Constants for Rainfall Eq. (24.1)

Location	c	e	f
Atlanta	97.5	0.83	6.88
Chicago	94.9	0.88	9.04
Cleveland	73.7	0.86	8.25
Denver	96.6	0.97	13.90
Houston	97.4	0.77	4.80
Los Angeles	20.3	0.63	2.06
Miami	124.2	0.81	6.19
New York	78.1	0.82	6.57
Santa Fe	62.5	0.89	9.10
St. Louis	104.7	0.89	9.44

Constants correspond to i in inches per hour and T_d in minutes.

Source: Wenzel, 1982; © American Geophysical Union.

ical techniques used to determine the shape of the design-storm hyetograph once the design precipitation depth and duration are known. Huff and the U.S. Department of Agriculture, Soil Conservation Service, have developed time distribution relationships that are widely used in hydrologic design.

Huff's (1967) storm patterns, which represent smoothed average rainfall distributions with time, are based on the analysis of heavy storms on areas ranging up to 400 mi² (1036 km²) in Illinois. Figure 24.5a shows the probability distribution of the most severe storms analyzed (first-quartile storms), and Figure 24.5b shows selected histograms of first-quartile storms for 10, 50, and 90 percent cumulative probabilities of occurrence, each illustrating the percentage of total storm rainfall for 10 percent increments of the storm duration. The 50-percent histogram represents a cumulative rainfall pattern that should be exceeded in about half the storms, and the 90-percent histogram can be interpreted as a storm distribution that is equaled or exceeded in 10 percent or less of the storms.

The Soil Conservation Service synthetic rainfall distributions are for storms of 6- and 24-hour durations. Four 24-hour storms were developed: Types I and IA are applicable in the wet winter and dry summer climates of the Pacific coast of the United States; Type III is applicable to the Gulf of Mexico and the Atlantic coastal region of the U.S., where tropical storms result in large 24-hour rainfall amounts;

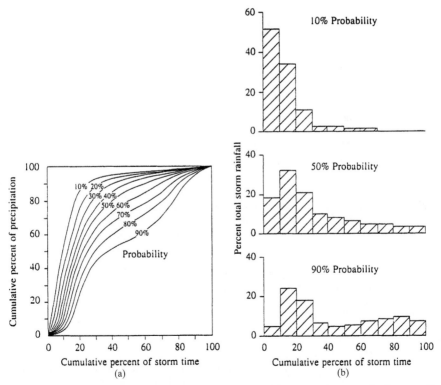

FIGURE 24.5   (a) Time distribution for first-quartile storms; (b) selected histograms for first-quartile storms. (Huff, 1967; © American Geophysical Union.)

Type II distribution is applicable to other areas of the nation. Table 24.2 and Fig. 24.6 show the cumulative hyetographs.

TABLE 24.2   SCS Rainfall Distributions

		24-h storm				6-h storm		
			P_t/P_{24}					
Hour t	$t/24$	Type I	Type IA	Type II	Type III	Hour t	$t/6$	P_t/P_6
0	0	0	0	0	0	0	0	0
2.0	0.083	0.035	0.050	0.022	0.020	0.60	0.10	0.04
4.0	0.167	0.076	0.116	0.048	0.043	1.20	0.20	0.10
6.0	0.250	0.125	0.206	0.080	0.072	1.50	0.25	0.14
7.0	0.292	0.156	0.268	0.098	0.089	1.80	0.30	0.19
8.0	0.333	0.194	0.425	0.120	0.115	2.10	0.35	0.31
8.5	0.354	0.219	0.480	0.133	0.130	2.28	0.38	0.44
9.0	0.375	0.254	0.520	0.147	0.148	2.40	0.40	0.53
9.5	0.396	0.303	0.550	0.163	0.167	2.52	0.42	0.60
9.75	0.406	0.362	0.564	0.172	0.178	2.64	0.44	0.63
10.0	0.417	0.515	0.577	0.181	0.189	2.76	0.46	0.66
10.5	0.438	0.583	0.601	0.204	0.216	3.00	0.50	0.70
11.0	0.459	0.624	0.624	0.235	0.250	3.30	0.55	0.75
11.5	0.479	0.654	0.645	0.283	0.298	3.60	0.60	0.79
11.75	0.489	0.669	0.655	0.357	0.339	3.90	0.65	0.83
12.0	0.500	0.682	0.664	0.663	0.500	4.20	0.70	0.86
12.5	0.521	0.706	0.683	0.735	0.702	4.50	0.75	0.89
13.0	0.542	0.727	0.701	0.772	0.751	4.80	0.80	0.91
13.5	0.563	0.748	0.719	0.799	0.785	5.40	0.90	0.96
14.0	0.583	0.767	0.736	0.820	0.811	6.00	1.0	1.00
16.0	0.667	0.830	0.800	0.880	0.886			
20.0	0.833	0.926	0.906	0.952	0.957			
24.0	1.000	1.000	1.000	1.000	1.000			

Source: U.S. Department of Agriculture, Soil Conservation Service (1973; 1986).

The simplest design-storm hyetograph has the shape of a triangle, in which the base length represents the storm duration and the area of the triangle is the total depth of precipitation, expressed as depth over the catchment area (Fig. 24.7). Both storm duration and total depth are *a priori* planning decisions. The height of the triangle represents the peak precipitation intensity and is easily calculated using the formula for the area of a triangle. The time to peak intensity t_a is given by the *storm advancement coefficient r*. A value of r of 0.5 corresponds to the peak intensity occurring in the middle of the storm, while a value less than 0.5 will have the peak earlier and a value greater than 0.5 will have the peak later than the midpoint. A suitable value of r is obtained by computing the ratio of time-to-peak intensity to storm duration for a series of observed storms of various durations. Generally, r is less than 0.5 (Table 24.3).

The *alternating block method* is a simple way of generating a design hyetograph from an intensity-duration-frequency curve. After selection of the design storm duration T_d the rainfall intensity is read from the IDF curve for each of the durations $\Delta t, 2\Delta t, 3\Delta t, \ldots, n\Delta t = T_d$, and the corresponding precipitation depth found as the

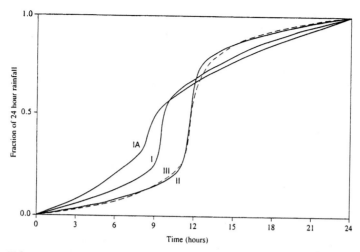

FIGURE 24.6   SCS 24-hour cumulative hyetographs. *(U.S. Department of Agriculture, Soil Conservation Service, 1986.)*

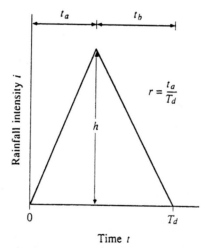

FIGURE 24.7   Generalized triangular design hyetograph. *(Chow, et al., 1988.)*

product of intensity and duration. By taking differences between successive precipitation depth values, the amount of precipitation to be added for each additional unit of time Δt is found. These increments, or blocks, are reordered into a time sequence with the maximum intensity occurring at the center of the required time duration T_d, and the remaining blocks arranged in descending order alternately to the right and left of the central block to form the design hyetograph. Table 24.4 and Fig. 24.8 illustrate the method.

TABLE 24.3 Typical Values of the Storm Advancement Coefficient r for Various Locations

Location	r	Reference
Baltimore	0.399	McPherson (1958)
Chicago	0.375	Keifer and Chu (1957)
Chicago	0.294	McPherson (1958)
Cincinnati	0.325	Preul and Papadakis (1973)
Cleveland	0.375	Havens and Emerson (1968)
Gauhati, India	0.416	Bandyopadhyay (1972)
Ontario	0.480	Marsalek (1978)
Philadelphia	0.414	McPherson (1958)

Source: Wenzel, 1982; © American Geophysical Union.

TABLE 24.4 Design Hyetograph Developed in 10-min Increments Using the Alternating Block Method

Column:	1 Duration, min	2 Intensity, in/h	3 Cumulative depth, in	4 Incremental depth, in	5 Time, min	6 Precipitation, in
	10	4.158	0.693	0.693	0–10	0.024
	20	3.002	1.001	0.308	10–20	0.033
	30	2.357	1.178	0.178	20–30	0.050
	40	1.943	1.296	0.117	30–40	0.084
	50	1.655	1.379	0.084	40–50	0.178
	60	1.443	1.443	0.063	50–60	0.693
	70	1.279	1.492	0.050	60–70	0.308
	80	1.149	1.533	0.040	70–80	0.117
	90	1.044	1.566	0.033	80–90	0.063
	100	0.956	1.594	0.028	90–100	0.040
	110	0.883	1.618	0.024	100–110	0.028
	120	0.820	1.639	0.021	110–120	0.021

Source: Chow, et al. (1988).

24.2.3 Estimated Limiting Storms

The estimated limiting values (ELV's) commonly employed in analysis for water-excess management are the *probable maximum precipitation* (PMP), the *probable maximum storm* (PMS), and the *probable maximum flood* (PMF). The PMP provides a depth of precipitation, and the PMS defines its time distribution. The PMS can be employed as input to a catchment-runoff model, which may then be used to develop a PMF for the design of water-excess management structures. Estimated limiting values have been found to be useful in the design of major water-excess management projects, such as the spillways of large dams, whose failure could lead to excessive downstream damage and loss of life.

The PMP is defined as the analytically estimated greatest depth of precipitation for a given duration that is physically possible and reasonably characteristic over a particular geographic region at a certain time of year. Because of uncertainties and limitations of data and knowledge it cannot be perfectly estimated and its probability of occurrence is unknown. Judgment must be used in setting its value.

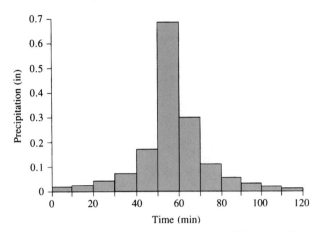

FIGURE 24.8   Design hyetograph developed from Table 24.4. *(Chow, et. al., 1988.)*

The procedure for determining the probable maximum precipitation has five important elements: (1) *depth-area-duration curves*, which specify the PMP for a specified storm area and duration, (2) a *standard isohyetal pattern* distributing the precipitation spatially in the form of an ellipse, (3) an *orientation adjustment factor*, which reduces the PMP estimates if the standard isohyetal pattern's longitudinal axis is not oriented in the direction of normal atmospheric moisture flow in the region, (4) a *critical storm area*, which generates the largest PMP over the catchment, and (5) an *isohyetal adjustment factor*, which specifies the percentage of the PMP depth that applies on each contour of the standard isohyetal pattern.

In probability-based point-precipitation estimates, there are three variables to be considered: intensity (or depth), duration, and frequency of occurrence. For determination of probable maximum precipitation, the frequency of occurrence is replaced by the storm area as the third variable. For a given location, depth-area-duration graphs are available as generalized charts that show plots of depth versus storm area for various storm durations.

24.2.3.1   Generalized PMP Charts.

Estimates of PMP may be made either for individual basins or for large basins encompassing numerous subbasins of various sizes. In the latter case, the estimates are called *generalized estimates* and are usually displayed as isohyetal maps that depict the regional variation of PMP for some specified duration, basin size, and annual or seasonal variation. These maps are commonly known as *generalized PMP charts*.

The most widely used generalized PMP charts in the United States are those contained in U.S. National Weather Service Hydrometeorological Report no. 51, commonly known as HMR 51, for the United States east of the 105th meridian. These maps specify the probable maximum precipitation depth for any time of the year (referred to as an *all-season estimate*) as a function of storm area ranging from 10 to 20,000 mi^2 (26 to 51,800 km^2) and storm duration ranging from 6 to 72 h. Figure 24.9 gives two examples of all-season PMP charts. For regions west of the 105th meridian, the National Academy of Sciences has prepared the diagram shown in Fig. 24.10, which specifies the appropriate National Weather Service manual from which PMP data may be obtained.

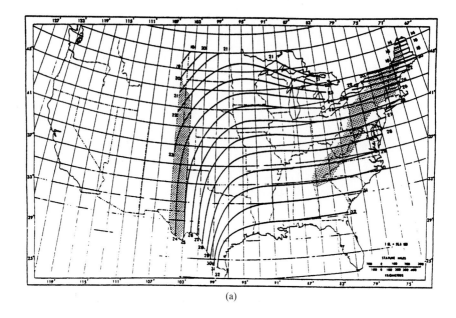

(a)

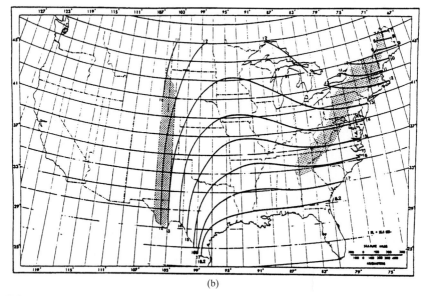

(b)

FIGURE 24.9   All-season PMP, (in) for 6 h: (*a*) for 10 mi²; (*b*) for 1000 mi².

If records of actual storms are available they may be maximized to obtain PMP values by increasing the observed storm precipitation by the ratio of the maximum moisture theoretically available over the catchment to the actual average moisture inflow to each storm. It is possible to transpose storms from other areas to the project catchment if the storms could have occurred in the catchment.

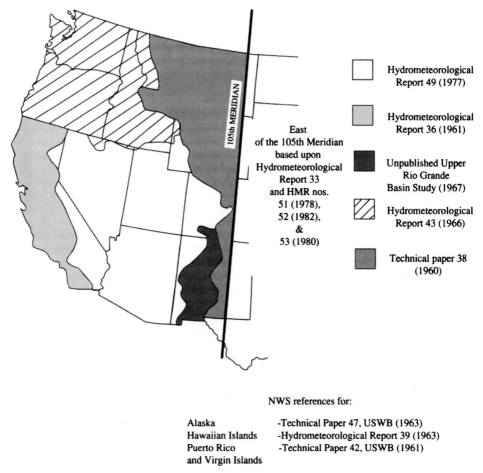

East
of the 105th Meridian
based upon
Hydrometeorological
Report 33
and HMR nos.
51 (1978),
52 (1982),
&
53 (1980)

☐ Hydrometeorological Report 49 (1977)

▨ Hydrometeorological Report 36 (1961)

■ Unpublished Upper Rio Grande Basin Study (1967)

▨ Hydrometeorological Report 43 (1966)

▨ Technical paper 38 (1960)

NWS references for:

Alaska	-Technical Paper 47, USWB (1963)
Hawaiian Islands	-Hydrometeorological Report 39 (1963)
Puerto Rico	-Technical Paper 42, USWB (1961)
and Virgin Islands	

FIGURE 24.10   Sources of information for probable maximum precipitation computation in the United States. *(National Academy of Science, 1983.)*

Precipitation depth estimates that could occur under very extreme circumstances are represented by the equation $P = 422T_d^{0.475}$, which approximates the world's greatest rainfall depths. P is the precipitation depth in mm and T_d is the precipitation duration in h. The equation was obtained by fitting data from observed extreme rainfall events at many locations for durations ranging from one minute to several months.

24.2.3.2 Standard Isohyetal Pattern.   A standard elliptical storm pattern adopted in HMR 52 is shown in Fig. 24.11. The ratio of the lengths of the major and minor axes of the ellipses shown is 2.5 to 1. If *a* and *b* are the lengths of the semimajor and semiminor axes, respectively, in Fig. 24.11, the area of the ellipse is given by $A = \pi ab$.

24.2.3.3 Orientation Adjustment Factor.   The orientation of the isohyetal pattern most likely to be conducive to a PMP event for any place in the United States east of the 105th meridian is indicated in Fig. 24.12. North is considered as 0° and the angles shown are measured clockwise from this direction. This figure is based on

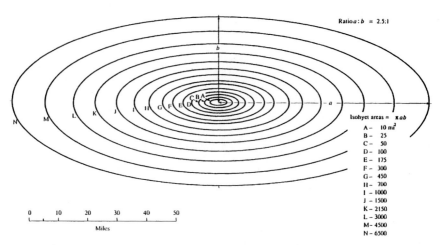

FIGURE 24.11 Standard isohyetal pattern for spatial distribution of PMP east of the 105th meridian. *(Hansen, et al., 1982.)*

averages of isohyetal orientations for major storms, which predominantly have moisture flowing from the south or west (180 to 270°). However, it is important to note that moisture flow for the PMP at any location is not restricted to any one orientation, and it is reasonable to expect the moisture flow to occur over a range of orientations centered on the values given in Fig. 24.12. If the longitudinal axis of the storm pattern is oblique to the recommended orientation, the PMP estimate is reduced by a percentage depending on the angle between the two directions. The adjustment factor for orientation differences is given in Fig. 24.13.

24.2.3.4   Critical Storm Area.   The critical storm area is that for which the precipitation from the depth-area-duration curve yields the greatest PMP over the catchment, taking into account the fact that the catchment area is not shaped elliptically like the standard isohyetal pattern.

24.2.3.5   Isohyetal Area Factor.   The value of the PMP represents the average precipitation depth over a specified area for a given duration. When a storm is represented by the standard isohyetal pattern there will be regions of greater precipitation depth near the center of the pattern and of lesser depth near the edges. Figure 24.14 shows, for the most intense 6-h storm interval, the percentage of the specified PMP depth to be applied to each contour of the standard isohyetal pattern to yield the correct spatial distribution. For example, using the PMP for an area of 10 mi^2 (26 km^2), according to the graph, contour A (area 10 mi^2) gets 100 percent of the PMP depth, and contour B [area 25 mi^2 (65 km^2)] gets 64 percent. Since the 6-h PMP depth for 10 mi^2 at Chicago, for example, is 26 in (66 cm) (Fig. 24.9a), this storm would have a depth of $26.0 \times 0.64 = 16.6$ in (42.3 cm) over 25 mi^2, which represents a volume of 25 mi$^2 \times 16.6$ in $= 415$ mi$^2 \cdot$ in (2730 km$^2 \cdot$ cm) of precipitation, of which 10 mi$^2 \times 26$ in $= 260$ mi$^2 \cdot$ in (1710 km$^2 \cdot$ cm), or 63 percent, occurs in the most intense 10 mi^2 area.

The PMS, or probable maximum storm, involves the *temporal* distribution of rainfall. To develop the hyetograph of a PMS, one needs to know the spatial and temporal distribution of the PMP. One procedure to determine the PMS is outlined in the National Weather Service Hydrometeorological Report no. 52 (HMR 52) for the nonorographic areas of the United States east of the 105th meridian. The time

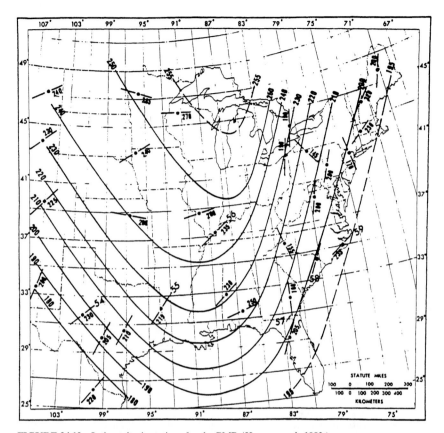

FIGURE 24.12   Isohyetal orientations for the PMP. *(Hansen, et al., 1982.)*

sequence of PMP increments is critical in modeling maximum runoff, and a commonly adopted sequence is the most advanced time distribution, beginning with the highest amount and continuing with decreasing increments.

The probable maximum flood, PMF, is the greatest flood event to be expected, assuming complete coincidence of all factors that would produce the heaviest rainfall and maximum runoff. Because the PMF is derived from the PMP, its frequency cannot be determined exactly. Hence, a pragmatic approach for many design situations is not to define the design rainfall-excess event as an estimated limiting value, but to scale it downward by a certain percentage depending on the type of structure and the hazard if it fails. In practice, the design rainfall-excess event is commonly called the *standard project flood* (SPF). The SPF is estimated using rainfall–runoff modeling by applying the unit hydrograph method to the *standard project storm* (SPS), which is the greatest storm that may be reasonably expected. The SPS can be derived from a detailed analysis of storm patterns and transpositions of storms to a position that would give maximum runoff. In general, the SPS estimate should represent the most severe flood-producing depth-area-duration relationship and isohyetal pattern of any storm that is considered reasonably characteristic of the region, and should consider the effect of any water-control structures in the catchment, and the effect of snow melt if it may constitute a substantial volume of runoff

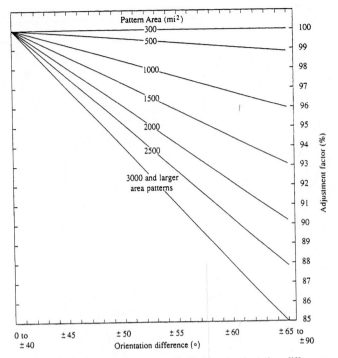

FIGURE 24.13 Adjustment factors for isohyetal orientation difference. *(Hansen, et al., 1982.)*

to the SPF hydrograph. Past estimates have indicated that SPS magnitudes and SPF discharges are generally in the range of 40 to 60 percent of the ELV for the same basin.

The U.S. Army Corps of Engineers Hydrologic Engineering Center has a computer program called HMR 52 which computes the basin-average precipitation for a probable maximum storm in accordance with the criteria in HMR 52. Program input includes PMP estimates from HMR 51, and user-specified storm centering and duration information. The program selects the storm area size and orientation that produces the maximum basin-average precipitation. The program can be used in conjunction with computer model HEC-1 to compute the probable maximum flood. Catchment-runoff model HEC-1 is discussed in Chap. 28.

24.3 DETERMINATION OF THE DESIGN-STORM HYDROGRAPH

24.3.1 Hydrologic Losses

The direct-runoff (DRO) hydrograph that is the central component of the study of water-excess events results from the transformation of excess rainfall into channel flow measured at the outlet of a drainage basin, or some subpart of the basin. The

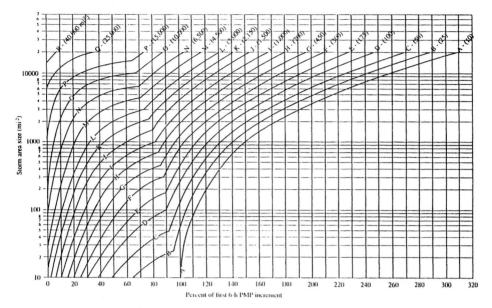

FIGURE 24.14 Nomograph for determining isohyet precipitation values from the PMP estimate for a given storm area. *(Hansen, et al., 1982.)*

graph of excess rainfall versus time, the excess-rainfall hyetograph, that is the primary input to determining the DRO hydrograph, is derived from the observed or design-storm hyetograph by subtracting the hydrologic abstractions, or losses, that represent the part of precipitation that does not contribute to runoff. Determination of a reasonable DRO hydrograph depends, therefore, on a good estimation of those abstractions.

The largest hydrologic abstraction is usually *infiltration,* which is the process of water penetrating the ground surface into the soil. Other losses (e.g., interception and surface storage) are usually estimated based on the catchment's vegetation and ground-surface characteristics.

There are many factors that influence infiltration, the most important of which are the condition of the soil surface and its vegetative cover, the properties of the soil such as its porosity and hydraulic conductivity, and the current moisture content of the soil. The *infiltration rate f,* expressed in in/h or cm/h, is the rate at which water enters the soil at the surface. Under conditions of ponded water on the surface, infiltration occurs at the *potential infiltration rate.* Most infiltration equations describe the potential rate. The *cumulative infiltration F* is the accumulated depth of water infiltrated during a specified time interval and is the integral of the infiltration rate over that time.

$$F(t) = \int_0^t f(\tau)d\tau \tag{24.2}$$

Clearly, the infiltration rate is the time derivative of the cumulative infiltration.

Three of the more widely used and accepted infiltration equations (Green-Ampt, Horton, and SCS) are presented in Table 24.5. The interrelationships of rainfall, infiltration rate, and cumulative infiltration are depicted in Fig. 24.15.

TABLE 24.5 Infiltration Equations

Cumulative infiltration, $F_{t+\Delta t}$	Infiltration rate, $F_{t+\Delta t}$	Comments
Green-Ampt equation $$F_{t+\Delta t} = F_t + K\Delta t + \psi\Delta\theta \ln\left[\frac{F_{t+\Delta t} + \psi\Delta\theta}{F_t + \psi\Delta\theta}\right]$$	$$f_{t+\Delta t} = K\left(\frac{\psi\Delta\theta}{F_{t+\Delta t}} + 1\right)$$	Hydraulic conductivity K Wetting front soil suction head ψ Change in moisture content $\Delta\theta$ $\Delta\theta = \eta - \theta_i$ Porosity η Initial moisture content θ_i See Table 24.6 for the infiltration parameters for various soil classes
Horton's equation $$F_{t+\Delta t} = F_t + f_c\Delta t + (f_t - f_c)\frac{(1 - e^{-k\Delta t})}{k}$$	$$f_{t+\Delta t} = f_t - k\,(F_{t+\Delta t} - F_t - f_c\Delta t)$$	Constant infiltration rate f_c Decay constant k
SCS method $$F_{t+\Delta t} = \frac{S(P_{t+\Delta t} + I_a)}{P_{t+\Delta t} - I_a + S} + I_a$$	$$f_{t+\Delta t} = \frac{S^2\dfrac{dP_{t+\Delta t}}{dt}}{(P_{t+\Delta t} - I_a + S)^2}$$	Potential maximum retention S $$S = \frac{1000}{CN} - 10$$ Dimensionless curve number CN $0 \leq CN \leq 100$ (see Table 24.7 for values of CN) Initial abstraction I_a $I_a = 0.2S$ Precipitation P

Source: Chow, et al. (1988).

24.3.2 The Unit Hydrograph Approach

A *unit hydrograph* is the direct runoff hydrograph resulting from 1 in or 1 cm of excess rainfall generated uniformly over a drainage area at a constant rate for an effective duration. The unit hydrograph is a simple linear model that can be used to derive the hydrograph resulting from any amount of excess rainfall. The following basic assumptions are inherent in the unit hydrograph approach:

1. The excess rainfall has a constant intensity within the effective duration.
2. The excess rainfall is uniformly distributed throughout the entire drainage area.
3. The base time of the DRO hydrograph (i.e., the duration of direct runoff) resulting from an excess rainfall of given duration is constant.
4. The ordinates of all DRO hydrographs of a common time base are directly proportional to the total amount of direct runoff represented by each hydrograph.
5. For a given drainage basin, the hydrograph resulting from a given excess rainfall reflects the unchanging characteristics of the basin.

Under natural conditions, these assumptions cannot be perfectly satisfied. However, when the hydrologic data to be used are carefully selected so that they come close to meeting the assumptions, the results obtained by the unit hydrograph model are generally acceptable for practical purposes. Although the model was originally

TABLE 24.6   Green-Ampt Infiltration Parameters for Various Soil Classes

Soil class	Porosity, η	Effective porosity, θ_e	Wetting front soil suction head, ψ (cm)	Hydraulic conductivity, K (cm/h)
Sand	0.437	0.417	4.95	11.78
	(0.374–0.500)	(0.354–0.480)	(0.97–25.36)	
Loamy sand	0.437	0.401	6.13	2.99
	(0.363–0.506)	(0.329–0.473)	(1.35–27.94)	
Sandy loam	0.453	0.412	11.01	1.09
	(0.351–0.555)	(0.283–0.541)	(2.67–45.47)	
Loam	0.463	0.434	8.89	0.34
	(0.375–0.551)	(0.334–0.534)	(1.33–59.38)	
Silt loam	0.501	0.486	16.68	0.65
	(0.420–0.582)	(0.394–0.578)	(2.92–95.39)	
Sandy clay loam	0.398	0.330	21.85	0.15
	(0.332–0.464)	(0.235–0.425)	(4.42–108.0)	
Clay loam	0.464	0.309	20.88	0.10
	(0.409–0.519)	(0.279–0.501)	(4.79–91.10)	
Silty clay loam	0.471	0.432	27.30	0.10
	(0.418–0.524)	(0.347–0.517)	(5.67–131.50)	
Sandy clay	0.430	0.321	23.90	0.06
	(0.370–0.490)	(0.207–0.435)	(4.08–140.2)	
Silty clay	0.479	0.423	29.22	0.05
	(0.425–0.533)	(0.334–0.512)	(6.13–139.4)	
Clay	0.475	0.385	31.63	0.03
	(0.427–0.523)	(0.269–0.501)	(6.39–156.5)	

The numbers in parentheses with each parameter are one standard deviation around the parameter value given.
Source: Rawls, Brakensiek, and Miller, 1983.

devised for large basins, it has been found applicable to small basins—that is, to areas as small as 1 acre (0.4 ha).

The following discrete *convolution equation* is used to compute direct runoff Q_n given excess rainfall P_m and the unit hydrograph U_{n-m+1}:

$$Q_n = \sum_{m=1}^{n \le M} P_m U_{n-m+1} \qquad (24.3)$$

The reverse process, called *deconvolution,* is used to derive a unit hydrograph given data on P_m and Q_n. Suppose that there are M pulses of excess rainfall and N pulses of direct runoff in the storm considered; then N equations can be written for $Q_n, n = 1, 2, 3, \ldots, N$, in terms of $N - M + 1$ unknown values of the unit hydrograph, as shown in Table 24.8.

Once the unit hydrograph has been determined, it may be applied to find the direct runoff and streamflow hydrographs. A design-rainfall hyetograph is determined, the abstractions estimated, and the excess-rainfall hyetograph calculated. The time interval used in defining the excess-rainfall hyetograph ordinates must be the same as that for which the unit hydrograph was specified. The discrete convolution Eq. (24.3) may then be used to yield the direct runoff hydrograph. By adding an estimated baseflow to the direct-runoff hydrograph, the streamflow hydrograph is obtained.

TABLE 24.7   Runoff Curve Numbers for Selected Agricultural, Suburban, and Urban Land Uses (Antecedent Moisture Condition II, $I_a = 0.2S$)

Land use description		Hydrologic soil group			
		A	B	C	D
Cultivated land*					
Without conservation treatment		72	81	88	91
With conservation treatment		62	71	78	81
Pasture or range land					
Poor condition		68	79	86	89
Good condition		39	61	74	80
Meadow, good condition		30	58	71	78
Wood or forest land					
Thin stand, poor cover, no mulch		45	66	77	83
Good cover[†]		25	55	70	77
Open spaces, lawns, parks, golf courses, cemeteries, etc.					
Good condition, grass cover on 75% or more of the area		39	61	74	80
Fair condition, grass cover on 50 to 75% of the area		49	69	79	84
Commercial and business areas (85% impervious)		89	92	94	95
Industrial districts (72% impervious)		81	88	91	93
Residential[‡]					
Average lot size	Average % impervious[§]				
⅛ acre or less	65	77	85	90	92
¼ acre	38	61	75	83	87
⅓ acre	30	57	72	81	86
½ acre	25	54	70	80	85
1 acre	20	51	68	79	84
Paved parking lots, roofs, driveways, etc.[¶]		98	98	98	98
Streets and roads					
Paved with curbs and storm sewers[¶]		98	98	98	98
Gravel		76	85	89	91
Dirt		72	82	87	89

* For a more detailed description of agricultural land use curve numbers, refer to Soil Conservation Service, 1972, Chap. 9.

[†] Good cover is protected from grazing and litter and brush cover soil.

[‡] Curve numbers are computed assuming the runoff from the house and driveway is directed towards the street with a minimum of roof water directed to lawns where additional infiltration could occur.

[§] The remaining pervious areas (lawn) are considered to be in good pasture condition for these curve numbers.

[¶] In some warmer climates of the country a curve number of 95 may be used.

The concept of an *instantaneous unit hydrograph* (IUH) considers a unit amount of excess rainfall for an infinitely small duration (see Chow, et al., 1988, and Maidment, 1993). The rainfall excess for an IUH is applied to the drainage area in zero time. This theoretical concept cannot be realized for actual watersheds, but is useful to characterize a watershed's response to rainfall without reference to the rainfall duration. The IUH can be related to watershed geomorphology (Rodriguez-Iturbe and Valdes, 1979; Gupta, et al., 1980).

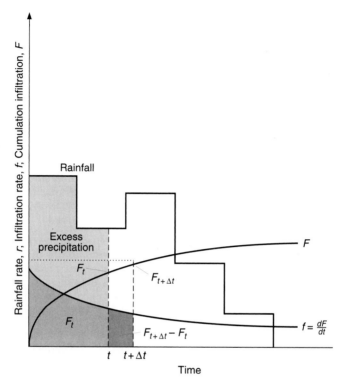

FIGURE 24.15 Interrelationships of rainfall, infiltration rate, and cumulative infiltration. *(Chow, et al., 1988.)*

24.3.3 Synthetic Unit Hydrograph

The unit hydrograph developed from rainfall and streamflow data on a drainage basin applies only for that basin and for the point on the stream where the stream-flow data were measured. *Synthetic unit hydrograph* procedures are used to develop unit hydrographs for other locations on the stream in the same basin or for nearby basins of a similar character.

TABLE 24.8 The Set of Equations for Discrete Time Convolution

$n = 1,2, \ldots ,N$

$$Q_1 \quad = P_1 U_1$$
$$Q_2 \quad = P_2 U_1 + P_1 U_2$$
$$Q_3 \quad = P_3 U_1 + P_2 U_2 \quad + P_1 U_3$$
$$\ldots$$
$$Q_M \quad = P_M U_1 + P_{M-1} U_2 + \quad + P_1 U_M$$
$$Q_{M+1} = \quad 0 \quad + P_M U_2 \quad + \ldots + P_2 U_M + P_1 U_{M+1}$$
$$\ldots$$
$$Q_{N-1} = \quad 0 + \quad 0 \quad + \ldots + \quad 0 \quad + \quad 0 \quad + \ldots + P_M U_{N-M} + P_{M-1} U_{N-M+1}$$
$$Q_N \quad = \quad 0 + \quad 0 \quad + \ldots + \quad 0 \quad + \quad 0 \quad + \ldots + \quad 0 \quad + P_M U_{N-M+1}$$

There are three types of synthetic unit hydrographs: (1) those relating *hydrograph* characteristics (such as peak flow rate, time to peak flow, and time base) to basin characteristics, (2) those based on the concept of a *dimensionless* unit hydrograph, and (3) those based on models of basin storage. Types 1 and 2 are described in the following.

Type 1: Snyder's Synthetic Unit Hydrograph. Snyder's unit hydrograph is based on synthetic relationships developed from a study of drainage basins in the eastern United States. From these relationships, five characteristics of a required unit hydrograph for a given excess rainfall duration can be calculated: the peak discharge per unit of basin area q_{pR}, the basin lag t_{pR} (the time difference between the centroid of the excess rainfall hyetograph and the unit hydrograph peak), the time base t_b, and the widths W (in time units) of the unit hydrograph at 50 and 75 percent of the peak discharge. Using these characteristics, the required unit hydrograph may be drawn. The variables are illustrated in Fig. 24.16.

Snyder defined a standard unit hydrograph as one whose rainfall duration t_r is related to the basin lag by

$$t_p = 5.5 \, t_r \tag{24.4}$$

For the *standard unit hydrograph* Snyder found that:

1. The *basin lag* is

$$t_p = C_1 C_t (LL_C)^{0.3} \tag{24.5}$$

where t_p is in h; L is the length of the main stream in mi (km) from the outlet to the upstream divide; L_C is the distance in mi (km) from the basin outlet to a point on the main stream channel nearest the centroid of the basin area; $C_1 = 1.0$ (0.75 for the SI system); and C_t is a coefficient derived from gauged basins in the same region.

2. The peak discharge per unit drainage area in cfs/mi^2 (m^3/s · km^2) of the standard unit hydrograph is

$$q_p = \frac{C_2 C_p}{t_p} \tag{24.6}$$

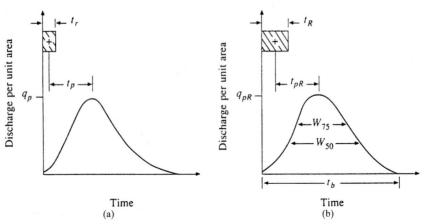

FIGURE 24.16 Snyder's synthetic unit hydrograph: (*a*) standard unit hydrograph ($t_p = 5.5 \, t_r$); (*b*) required unit hydrograph ($t_{pR} \neq 5.5 \, t_R$).

where $C_2 = 640$ (2.75, SI) and C_p is a coefficient derived from gauged basins in the same region.

To compute C_t and C_p for a gauged basin, the values of L and L_c are measured from the basin map. From a derived unit hydrograph of the basin are obtained values of its effective duration t_R in h, its basin lag t_{pR} in h, and its peak discharge per unit drainage area, q_{pR}, in cfs/mi^2 · in (m^3/s · km^2 · cm). If $t_{pR} = 5.5t_R$, then $t_R = t_r$, $t_{pR} = t_p$, and $q_{pR} = q_p$, and C_t and C_p are computed by Eqs. (24.5) and (24.6). If t_{pR} is quite different from $5.5t_R$, the standard basin lag is

$$t_p = t_{pr} + \frac{t_r - t_R}{4} \tag{24.7}$$

and Eqs. (24.4) and (24.7) are solved simultaneously for t_r and t_p. The values of C_t and C_p are then computed from Eqs. (24.5) and (24.6) with $q_{pR} = q_p$ and $t_{pR} = t_p$.

When an ungauged basin appears similar to a gauged basin, the coefficients C_t and C_p for the gauged basin can be used in the preceding equations to derive the required synthetic unit hydrograph for the ungauged basin.

3. The relationship between q_p and the peak discharge per unit area q_{pR} of the required unit hydrograph is

$$q_{pR} = \frac{q_p t_p}{t_{pR}} \tag{24.8}$$

4. The time base t_b in h of the unit hydrograph can be determined using the fact that the area under a unit hydrograph is always equivalent to a direct-runoff volume of 1 in or 1 cm uniformly distributed over the basin. Assuming a triangular shape for the unit hydrograph, the time base may be estimated by

$$t_b = \frac{C_3}{q_{pR}} \tag{24.9}$$

where $C_3 = 1290$ (5.56, SI)

5. The width in h of a unit hydrograph at a discharge equal to a certain percent of the peak discharge q_{pR} is given by

$$W = C_w q_{pR}^{-1.08} \tag{24.10}$$

where $C_w = 440$ (1.22, SI) for the 75-percent width and 770 (2.14, SI) for the 50-percent width. Usually one-third of this width is distributed before the unit hydrograph time peak and two-thirds after the peak.

Type 2: SCS Dimensionless Hydrograph. The SCS dimensionless hydrograph is a synthetic unit hydrograph in which the discharge is expressed by the ratio of discharge q to peak discharge q_p and the time by the ratio of the time t to the time of rise of the unit hydrograph T_p. Given the peak discharge and lag time for the duration of excess rainfall, the unit hydrograph can be estimated from the synthetic dimensionless hydrograph for the given basin. Figure 24.17a shows such a dimensionless hydrograph prepared from the unit hydrographs of a number of basins. The values of q_p and T_p may be estimated using a simplified model of a triangular unit hydrograph, as shown in Figure 24.17b, where the time is in h and the discharge in cfs/in (m^3/s · cm).

The Soil Conservation Service suggests the time of recession may be approximated as $1.67 T_p$. As the area under the unit hydrograph should be equal to 1 in or 1 cm of direct runoff, it can be shown that

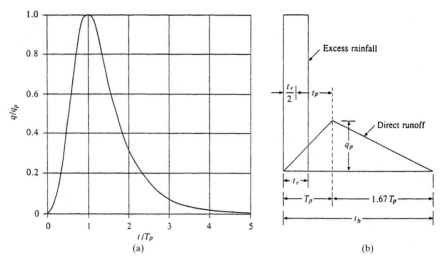

FIGURE 24.17 Soil Conservation Service synthetic unit hydrographs: (*a*) dimensionless hydrograph and (*b*) triangular unit hydrograph.

$$q_p = \frac{CA}{t_p} \qquad (24.11)$$

where $C = 483.4$ (2.08, SI) and A is the drainage area in mi^2 (km^2).

A study of many unit hydrographs obtained from large and small rural basins indicates the basin lag $t_p \cong 0.6\, T_c$, where T_c is the time of concentration of the basin. As shown in Fig. 24.17*b*, the time of rise T_p can be expressed in terms of lag time t_p and the duration of excess rainfall t_r

$$T_p = \frac{t_r}{2} + t_p \qquad (24.12)$$

24.3.4 Continuous-Simulation Models

The unit hydrograph approach described in the previous section is used to model *discrete* rainfall–runoff events, often in catchment-runoff simulation models such as HEC-1. *Event models* emphasize infiltration and surface runoff components with the objective of determining direct runoff, and are applicable to excess-water flow calculations in cases where direct runoff is the major contributor to stream-flow. *Continuous-simulation models* explicitly account for all runoff components, including surface flow and *indirect* runoff such as interflow and baseflow. Continuous models account for the overall moisture balance of the basin, including moisture accounting between storm events, and are well-suited to *long-term* runoff forecasting.

The three better-known continuous-simulation computer models are: (1) the Streamflow Synthesis and Reservoir Regulation (SSARR) catchment-runoff model developed by the U.S. Army Corps of Engineers North Pacific Division (1986); (2) the Stanford Watershed Model (SWM) developed at Stanford University

(Crawford and Linsley, 1966); and (3) the Sacramento Model developed by the Joint Federal–State River Forecast Center, the U.S. National Weather Service, and the State of California Department of Water Resources (Burnack, et al., 1973). The SSARR model is discussed in Chap. 28 and the Sacramento model is outlined in more detail in the remainder of this section.

The Sacramento soil-moisture accounting model divides the soil vertically into two soil-moisture accounting zones: the *upper zone,* which accounts for interception storage at the surface, water held by surface tension to the soil matrix, and water freely passing through the upper soil layer; the *lower zone* accounts for the bulk of the soil moisture and groundwater-storage capacity and includes that part of the free upper-zone water which percolates below the water table (the upper boundary of saturated soil).

The Sacramento Model generates five components of channel flow:

1. Direct runoff resulting from precipitation applied to impervious and temporarily impervious areas
2. Surface runoff resulting from excess precipitation (i.e., precipitation falling at a rate greater than the upper-zone infiltration rate
3. Interflow, which is lateral drainage from the upper-zone free-water storage
4. Primary baseflow, which is drainage from the lower free-water primary storage
5. Supplemental baseflow, which is drainage from the lower-zone free-water supplemental storage

The runoff-channel flow for each time interval is the sum of the above flow components. A modified version of the Sacramento Model that was incorporated into the U.S. National Weather Service River Forecast System uses 6-hour computational time intervals for calibration simulations and operational forecasts.

24.3.5 Distributed Event-Based Models

One of the more popular distributed approaches has been overland flow models based on *kinematic wave routing.* Kinematic wave routing is based on simplifications (ignoring the local and convective acceleration and pressure terms) of the Saint Venant equations (see Sec. 24.4.3 and Chap. 25). Reviews of kinematic wave models have been given by Overton and Meadows (1976), Woolhiser (1982), and Stephenson and Meadows (1986).

One of the newest models is KINEROS developed by Woolhiser, et al. (1990). This is a Hortonian model that simulates hillslope infiltration, performs unsteady routing of overland flow and channel flow using the kinematic assumption, and simulates channel losses and sediment transport. KINEROS characterizes a watershed as a dendritic network of overland flow planes, channels, and ponds. Over each plane or channel element the rainfall pattern, routing parameters, and initial soil water content are considered to be spatially uniform, but these may vary from plane to plane and element to element.

KINEROS uses a Newton-Raphson technique to solve the kinematic wave equations based on an implicit finite difference scheme (see Chap. 25). This approach allows simultaneous solution of the infiltration and routing equations at each time step and allows the simulated runoff to infiltrate as it travels downslope, even after rainfall has ceased. KINEROS was developed specifically for semiarid regions but should be applicable under any conditions.

Michaud and Sorooshian (1994) have applied KINEROS to the Walnut Gulch experimental watershed in southeastern Arizona. They also applied a simple lumped model based on the SCS method and a simple distributed model based on the SCS method. Results of this comparison showed that the lumped model performed very poorly. After calibration, the accuracy of the KINEROS model was similar to that of the simple distributed model. None of the three models accurately simulated peak flows or runoff volumes; however, they did a much better job in predicting times to peak and peak-to-volume ratios.

24.4 FLOW ROUTING

Broadly defined, *flow routing* is an analytical procedure intended to trace the flow of water through a hydrologic system, given some runoff event as input. The procedure determines the flow hydrograph at a point on a stream channel, from known or assumed hydrographs at one or more points upstream. If the flow is a water-excess event such as a flood then the procedure is specifically known as *flood routing*. Fread (1993) presented an excellent overview of various flood-routing techniques that have been developed. Also see Chap. 25.

Routing by lumped system methods is called *hydrologic routing*. These methods calculate the flow as a function of time alone. Routing by distributed system methods is called *hydraulic routing*, and the flow is calculated as a function of both space and time throughout the system.

For hydrologic routing the input $I(t)$, output $Q(t)$, and storage $S(t)$ are related by the continuity equation:

$$\frac{dS}{dt} = I(t) - Q(t) \tag{24.13}$$

If an inflow hydrograph $I(t)$ is known, Eq. (24.13) cannot be solved directly to obtain the outflow hydrograph $Q(t)$, because both Q and S are unknown. A second relationship, the *storage function*, is required to relate I, S, and Q. Coupling the continuity equation with the storage function provides a solvable combination of two equations and two unknowns.

The specific form of the storage function to be employed in hydrologic routing depends on the nature of the system being analyzed. In *reservoir routing* by the *level-pool method*, storage is a nonlinear function of Q only, $S = f(Q)$, and the function $f(Q)$ is determined by relating reservoir storage and outflow to reservoir water level. In *channel routing* by the *Muskingum method*, storage is linearly related to I and Q.

24.4.1 Lumped Reservoir Routing

Level-pool routing is a procedure for calculating the outflow hydrograph from a reservoir, assuming a horizontal water surface, given its inflow hydrograph and storage-outflow characteristics. When a reservoir has a horizontal water surface, its storage is a function of its water-surface elevation, or depth in the pool. Likewise, the outflow is a function of the water-surface elevation, or head on the outlet works. Combining these two functions yields the invariable, single-valued function $S = f(Q)$.

Integration of the continuity equation Eq. (24.13) over discrete time intervals provides an expression for the change in storage over the jth time interval Δt, $S_{j+1} - S_j$:

$$\int_{S_j}^{S_{j+1}} dS = \int_{j\Delta t}^{(j+1)\Delta t} I(t)\, dt - \int_{j\Delta t}^{(j+1)\Delta t} Q(t)dt \tag{24.14}$$

which can be rewritten as:

$$S_{j+1} - S_j = \frac{I_j + I_{j+1}}{2}\Delta t - \frac{Q_j + Q_{j+1}}{2}\Delta t \tag{24.15}$$

The inflow values at the beginning and end of the jth time interval are I_j and I_{j+1}, respectively, and the corresponding outflow values are Q_j and Q_{j+1}. The values of I_j and I_{j+1} are known because they are prespecified (i.e., they are inflow hydrograph ordinates). The values Q_j and S_j are known at the jth time interval. Hence Eq. (24.15) contains two unknowns, Q_{j+1} and S_{j+1}, which are isolated by multiplying Eq. (24.15) by $2/\Delta t$ and rearranging the result to produce:

$$\left(\frac{2S_{j+1}}{\Delta t} + Q_{j+1}\right) = (I_j + I_{j+1}) + \left(\frac{2S_j}{\Delta t} - Q_j\right) \tag{24.16}$$

In order to calculate the outflow Q_{j+1} from Eq. (24.16), a *storage-outflow function* relating $2S/\Delta t + Q$ and Q is needed. The method of developing this function using elevation-storage and elevation-outflow relationships is shown in Fig. 24.18.

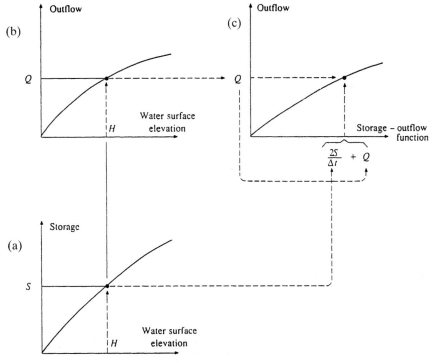

FIGURE 24.18 Development of the storage-outflow function for level pool routing. *(Chow, et al., 1988.)*

The relationship between water-surface elevation and reservoir storage can be obtained using topographic maps or field surveys. The elevation-discharge relationship is derived from hydraulic equations relating head and discharge for various types of spillways and outlet works. The value of Δt is the same as the time interval of the inflow hydrograph.

For a given value of water-surface elevation, the values of storage S and discharge Q are determined (parts a and b of Fig. 24.18). Then, the value of $2S/\Delta t + Q$ is calculated and plotted against Q (Fig. 24.18c). In routing the flow through the jth time interval, all terms on the right-hand side of Eq. (24.16) are known, and so the value of $2S_{j+1}/\Delta t + Q_{j+1}$ can be computed. The corresponding value of Q_{j+1} can be determined from the storage-outflow function $2S/\Delta t + Q$ versus Q. To set up the data for the next time interval, the value of $2S_{j+1}/\Delta t - Q_{j+1}$ is calculated by

$$\left(\frac{2S_{j+1}}{\Delta t} - Q_{j+1}\right) = \left(\frac{2S_{j+1}}{\Delta t} + Q_{j+1}\right) - 2Q_{j+1} \tag{24.17}$$

The computation is repeated iteratively for subsequent routing periods.

24.4.2 Lumped River Routing

A widely used hydrologic method is the *Muskingum method* for routing flows in channels. It models the storage volume of flooding in a channel reach by a combination of *wedge* and *prism* storage (Fig. 24.19). When the flood wave is advancing,

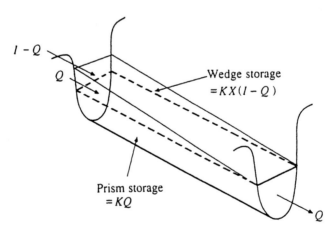

FIGURE 24.19 Prism and wedge storage in a channel reach. *(Chow, et al., 1988.)*

inflow exceeds outflow and a positive wedge of storage is produced. When the flood is receding, outflow exceeds inflow and a negative wedge results. In addition, a prism of storage is formed by a volume of (approximately) constant cross-section along the length of the channel reach.

Assuming that the cross-sectional area of the flood flow is directly proportional to the discharge at the section, the volume of prism storage is KQ where K is a coefficient of proportionality. The volume of wedge storage is assumed to be equal to

$KX(I-Q)$, where X is a weighting factor having the range $0 \leq X \leq 0.5$. The total storage is defined as the sum of the two storage components $S = KQ + KX(I-Q)$, which can be rearranged to give the linear storage function for the Muskingum method

$$S = K[XI + (1-X)Q] \tag{24.18}$$

The value of X depends on the shape of the modeled wedge storage. $X = 0$ corresponds to reservoir (level-pool) storage and Eq. (24.18) reduces to $S = KQ$. $X = 0.5$ for a full wedge. In most natural stream channels, X is between 0 and 0.3, with a mean value near 0.2. Great accuracy in determining X is usually not necessary because the Muskingum method is not sensitive to this parameter. The parameter K is the time of travel of the flood wave through the reach. Although techniques exist that allow for the values of K and X to vary according to flow rate and channel characteristics (e.g., the *Muskingum-Cunge method*), for hydrologic routing the values of K and X are assumed to be specified and constant throughout the range of flow.

The values of storage at times j and $j + 1$ can be written as

$$S_j = K[XI_j + (1-X)Q_j] \tag{24.19}$$

and

$$S_{j+1} = K[XI_{j+1} + (1-X)Q_{j+1}] \tag{24.20}$$

The change in storage over the time interval Δt is

$$S_{j+1} - S_j = K\{[XI_{j+1} + (1-X)Q_{j+1}] - [XI_j + (1-X)Q_j]\} \tag{24.21}$$

The change in storage can also be expressed as Eq. (24.15)

$$S_{j+1} - S_j = \frac{I_j + I_{j+1}}{2} \Delta t - \frac{Q_j + Q_{j+1}}{2} \Delta t \tag{24.15}$$

Combining Eqs. (24.21) and (24.15) yields the Muskingum routing equation

$$Q_{j+1} = C_1 I_{j+1} + C_2 I_j + C_3 Q_j \tag{24.22}$$

where

$$C_1 = \frac{\Delta t - 2KX}{2K(1-X) + \Delta t} \tag{24.23}$$

$$C_2 = \frac{\Delta t + 2KX}{2K(1-X) + \Delta t} \tag{24.24}$$

$$C_3 = \frac{2K(1-X) - \Delta t}{2K(1-X) + \Delta t} \tag{24.25}$$

Note that $C_1 + C_2 + C_3 = 1$.

The values of K and X are determined using observed inflow and outflow hydrographs in the river reach. By assuming various values of X and using known values of inflow and outflow, successive values of K can be computed using

$$K = \frac{0.5\Delta t[(I_{j+1} + I_j) - (Q_{j+1} + Q_j)]}{X(I_{j+1} - I_j) + (1-X)(Q_{j+1} - Q_j)} \tag{24.26}$$

Equation 24.26 is derived from Eqs. (24.15) and (24.21). The computed values of the numerator and denominator of Eq. (24.26) are plotted for each time interval Δt, with the numerator on the vertical axis and the denominator on the horizontal axis. This usually produces a graph in the form of a loop. The value of X that produces a loop closest to a single line is taken to be the correct value for the reach. The parameter K, from Eq. (24.26), is the slope of the line for this value of X.

24.4.3 Distributed Routing

Procedures for distributed-flow routing are popular for water-excess routing because they compute flow rate and water level as functions of both space and time (see Chap. 25). The methodologies are based upon the Saint-Venant equations of one-dimensional flow. In contrast, the lumped hydraulic-routing procedures discussed in previous sections compute flow rate as a function of time alone.

Unsteady flow is described by the *conservation form* of the Saint-Venant equations. This form provides the flexibility to simulate a wide range of flows from gradual long-duration flood waves in rivers to abrupt waves similar to those caused by dam failure. The equations below are derived in detail by Chow, et al. (1988). The continuity equation is

$$\frac{\partial Q}{\partial x} + \frac{\partial (A + A_0)}{\partial t} - q = 0 \tag{24.27}$$

and the momentum equation is

$$\frac{1}{A}\frac{\partial Q}{\partial t} + \frac{1}{A}\frac{\partial \left(\frac{\beta Q^2}{A}\right)}{\partial x} + g\left(\frac{\partial h}{\partial x} + S_f + S_e\right) - \beta q v_x + W_f B = 0 \tag{24.28}$$

where
- x = longitudinal distance along the channel or river
- t = time
- A = cross-sectional area of flow
- A_0 = cross-sectional area of off-channel dead storage
- q = lateral inflow per unit length along the channel
- h = water-surface elevation
- v_x = velocity of lateral flow in the direction of channel flow
- S_f = friction slope
- S_e = eddy loss slope
- B = width of the channel at the water surface
- V_f = wind shear force
- β = momentum correction factor
- g = acceleration due to gravity

The Saint-Venant equations operate under the following assumptions:

1. The flow is one-dimensional with depth and velocity varying only in the longitudinal direction of the channel. This implies that the velocity is constant and the water surface is horizontal across any section perpendicular to the longitudinal axis.

2. There is gradually varied flow along the channel so that hydrostatic pressure prevails and vertical accelerations can be neglected.

3. The longitudinal axis of the channel is approximated as a straight line.

4. The bottom slope of the channel is small and the bed is fixed, resulting in negligible effects of scour and deposition.

5. Resistance coefficients for steady uniform turbulent flow are applicable, allowing for the use of Manning's equation to describe resistance effects.

6. The fluid is incompressible and of constant density throughout the flow.

The momentum equation consists of terms for the physical processes that govern low momentum. When the water or flow rate is changed at a certain point in a channel with a subcritical flow, the effects of these changes propagate back upstream. These backwater effects can be incorporated into distributed routing methods through the local acceleration, convective acceleration, and pressure terms.

Hydrologic (lumped) routing methods may not perform well in simulating the flow conditions when backwater effects are significant and the river slope is mild, because these methods have no hydraulic mechanisms to describe upstream propagation of changes in the flow momentum.

24.5 HYDROLOGIC FREQUENCY ANALYSIS

This section presents methods of frequency analysis that quantify the uncertainty inherent in hydrologic data, by presenting the data in a standard probabilistic framework. The emphasis in this overview is on frequency analysis of water-excess flows in stream channels.

The objective of frequency analysis of hydrologic data is to relate the magnitude of extreme events, such as those of water excess, to their frequency of occurrence, or *recurrence interval T.* The recurrence interval, which is also called the *return period,* is defined as the *average* interval of time within which a hydrologic event of given magnitude is expected to be equaled or exceeded exactly once. Although the words *frequency* and *recurrence interval* are often used interchangeably in hydrology, no regular or stated *deterministic* interval of recurrence is implied by their use.

The application of theoretical probability distributions to a statistical population in hydrology centers on the analysis of observed data that represent a random sample of that population and extrapolation of the statistical summaries of that data to the entire population. Clearly, the reliability of the extrapolation depends on the degree to which the sample represents the population. In hydrology, the observed data typically represent only a small subset of the population; it is possible, therefore, that the available data may not be very representative. Hence, frequency analysis in hydrology must be applied to design problems with care.

The hydrologic data subject to frequency analysis are assumed to be independent and identically distributed in each time interval. Often this means in hydrology that *annual* values of the variable of interest (e.g., annual peak streamflow rate) are selected on the assumption that successive observations of this variable from year to year will be independent.

24.5.1 Flood-Flow Frequency Analysis

There are several types of theoretical probability distributions that have been successfully applied to hydrologic data. In flood-flow frequency determination the most popular have been the *lognormal, log-Pearson Type III* and *Gumbel* distributions. The first two of these distributions determine the probability of a hydrologic event by representing the magnitude of that event x_T as the mean $\bar{x}$ plus a departure of the variate x from the mean. This departure, Δx_T, is equal to the product of the standard deviation s_x and a *frequency factor* K_T. Parameters Δx_T and K_T are functions of the type of distribution being used and the recurrence interval of event x_T. For a given distribution, tables are available that relate K_T and T.

Chow proposed the following frequency factor equation

$$x_T = \bar{x} + K_T s_x \qquad (24.29)$$

When the variable being analyzed is $y = \log x$, as in the lognormal and log-Pearson Type III distributions, then the same method is applied to the statistics of the logarithms of the data, using

$$y_T = \bar{y} + K_T s_y \qquad (24.30)$$

and the required value of x_T is found by taking the antilog of y_T.

24.5.1.1   *The lognormal distribution.*   In general, a random variable has a lognormal distribution if the log of the variable is distributed normally. That is, if $y = \log x$, and y is distributed normally, then x is distributed lognormally. An example of a lognormal probability density function (PDF) is shown in Fig. 24.20. The shape of the function, its lower bound of zero, positive skewness, and ease of use through its relationship with the normal distribution, make it a useful distribution in hydrology.

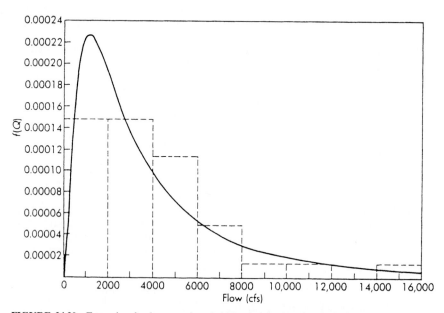

FIGURE 24.20   Example of a lognormal probability density function in hydrology. (*Bedient and Huber, 1992.*)

Care needs to be taken in transforming the statistics of the observations (x) into their log values (y). Equations are available relating the means, $\bar{x}$ and $\bar{y}$, and standard deviations, s_x and s_y. Alternatively, the observations can be transformed into their log values (base 10 or e), and statistical tables relating the (transformed) variable to cumulative probability for the *standard normal distribution* are used to compute the frequency factor used in the transformed Eq. (24.30).

24.5.1.2 The Log-Pearson Type III Distribution. For this distribution, the first step is to compute the logs of the hydrologic data, $y = \log x$. Usually logarithms to base 10 are used. The mean $\bar{y}$, standard deviation s_y, and coefficient of skewness C_{sy} are calculated. The frequency factor K_T is a function of T and C_{sy}. That is, whereas the lognormal distribution assumes that the logs of the observations are not skewed, L-P III includes skewness. Table 24.9 lists values of K_T for the log-Pearson Type III distribution (and Pearson Type III) for various values of the recurrence interval (return period) and coefficient of skewness.

24.5.1.3 The Gumbel Distribution. The Gumbel, or Extreme Value Type I, distribution has a shape similar to that shown in Fig. 24.20 for the lognormal distribution, but is unbounded on both the lower and upper ends. However, the (small) probability associated with negative or very large values of the variable is not usually a concern in its application in hydrology. As is the case with the lognormal and log-Pearson Type III distributions, the cumulative density function (CDF), rather than the PDF, is used in frequency analysis. The CDF for Gumbel's distribution is

$$F(x) = \exp\left[-\exp\left(-\frac{x-u}{\alpha}\right)\right] \qquad -\infty \le x \le \infty \qquad (24.31)$$

where parameters α and u are related to the variable mean $\bar{x}$ and standard deviation s_x by

$$\alpha = \frac{\sqrt{6}s_x}{\pi} \qquad (24.32)$$

and

$$u = \bar{x} - 0.5772\alpha \qquad (24.33)$$

It can be shown from Eqs. (24.31), (24.32), and (24.33) that a frequency factor K_T may readily be derived:

$$K_T = -0.7797\left\{0.5772 + \ln\left[\ln\left(\frac{T}{T-1}\right)\right]\right\} \qquad (24.34)$$

where T is the recurrence interval of interest.

24.5.2 U.S. Water Resources Council Guidelines

The former U.S. Water Resources Council (WRC) recommended in 1981 (Interagency Advisory Committees on Water Data, 1982) that the log-Pearson Type III distribution be used as a base distribution for flood-flow frequency studies. Although this decision was subjective to some extent (because no distribution is best for all sit-

TABLE 24.9 K_T Values for Log-Pearson Type III and Pearson Type III Distributions

Skew coefficient, C_s or C_w	Return period, years						
	2	5	10	25	50	100	200
	Exceedence probability						
	0.50	0.20	0.10	0.04	0.02	0.01	0.005
Positive skew							
3.0	−0.396	0.420	1.180	2.278	3.152	4.051	4.970
2.9	−0.390	0.440	1.195	2.277	3.134	4.013	4.909
2.8	−0.384	0.460	1.210	2.275	3.114	3.973	4.847
2.7	−0.376	0.479	1.224	2.272	3.093	3.932	4.783
2.6	−0.368	0.499	1.238	2.267	3.071	3.889	4.718
2.5	−0.360	0.518	1.250	2.262	3.048	3.845	4.652
2.4	−0.351	0.537	1.262	2.256	3.023	3.800	4.584
2.3	−0.341	0.555	1.274	2.248	2.997	3.753	4.515
2.2	−0.330	0.574	1.284	2.240	2.970	3.705	4.444
2.1	−0.319	0.592	1.294	2.230	2.942	3.656	4.372
2.0	−0.307	0.609	1.302	2.219	2.912	3.605	4.298
1.9	−0.294	0.627	1.310	2.207	2.881	3.553	4.223
1.8	−0.282	0.643	1.318	2.193	2.848	3.499	4.147
1.7	−0.268	0.660	1.324	2.179	2.815	3.444	4.069
1.6	−0.254	0.675	1.329	2.163	2.780	3.388	3.990
1.5	−0.240	0.690	1.333	2.146	2.743	3.330	3.910
1.4	−0.225	0.705	1.337	2.128	2.706	3.271	3.828
1.3	−0.210	0.719	1.339	2.108	2.666	3.211	3.745
1.2	−0.195	0.732	1.340	2.087	2.626	3.149	3.661
1.1	−0.180	0.745	1.341	2.066	2.585	3.087	3.575
1.0	−0.164	0.758	1.340	2.043	2.542	3.022	3.489
0.9	−0.148	0.769	1.339	2.018	2.498	2.957	3.401
0.8	−0.132	0.780	1.336	1.993	2.453	2.891	3.312
0.7	−0.116	0.790	1.333	1.967	2.407	2.824	3.223
0.6	−0.099	0.800	1.328	1.939	2.359	2.755	3.132
0.5	−0.083	0.808	1.323	1.910	2.311	2.686	3.041
0.4	−0.066	0.816	1.317	1.880	2.261	2.615	2.949
0.3	−0.050	0.824	1.309	1.849	2.211	2.544	2.856
0.2	−0.033	0.830	1.301	1.818	2.159	2.472	2.763
0.1	−0.017	0.836	1.292	1.785	2.107	2.400	2.670
0.0	0	0.842	1.282	1.751	2.054	2.326	2.576
Negative skew							
−0.1	0.017	0.846	1.270	1.716	2.000	2.252	2.482
−0.2	0.033	0.850	1.258	1.680	1.945	2.178	2.388
−0.3	0.050	0.853	1.245	1.643	1.890	2.104	2.294
−0.4	0.066	0.855	1.231	1.606	1.834	2.029	2.201
−0.5	0.083	0.856	1.216	1.567	1.777	1.955	2.108
−0.6	0.099	0.857	1.200	1.528	1.720	1.880	2.016
−0.7	0.116	0.857	1.183	1.488	1.663	1.806	1.926
−0.8	0.132	0.856	1.166	1.448	1.606	1.733	1.837
−0.9	0.148	0.854	1.147	1.407	1.549	1.660	1.749
−1.0	0.164	0.852	1.128	1.366	1.492	1.588	1.664
−1.1	0.180	0.848	1.107	1.324	1.435	1.518	1.581
−1.2	0.195	0.844	1.086	1.282	1.379	1.449	1.501

TABLE 24.9 K_T Values for Log-Pearson Type III and Pearson Type III Distributions (*Continued*)

Skew coefficient, C_s or C_w	Return period, years						
	2	5	10	25	50	100	200
			Exceedence probability				
	0.50	0.20	0.10	0.04	0.02	0.01	0.005
	Negative skew						
−1.3	0.210	0.838	1.064	1.240	1.324	1.383	1.424
−1.4	0.225	0.832	1.041	1.198	1.270	1.318	1.351
−1.5	0.240	0.825	1.018	1.157	1.217	1.256	1.282
−1.6	0.254	0.817	0.994	1.116	1.166	1.197	1.216
−1.7	0.268	0.808	0.970	1.075	1.116	1.140	1.155
−1.8	0.282	0.799	0.945	1.035	1.069	1.087	1.097
−1.9	0.294	0.788	0.920	0.996	1.023	1.037	1.044
−2.0	0.307	0.777	0.895	0.959	0.980	0.990	0.995
−2.1	0.319	0.765	0.869	0.923	0.939	0.946	0.949
−2.2	0.330	0.752	0.844	0.888	0.900	0.905	0.907
−2.3	0.341	0.739	0.819	0.855	0.864	0.867	0.869
−2.4	0.351	0.725	0.795	0.823	0.830	0.832	0.833
−2.5	0.360	0.711	0.771	0.793	0.798	0.799	0.800
−2.6	0.368	0.696	0.747	0.764	0.768	0.769	0.769
−2.7	0.376	0.681	0.724	0.738	0.740	0.740	0.741
−2.8	0.384	0.666	0.702	0.712	0.714	0.714	0.714
−2.9	0.390	0.651	0.681	0.683	0.689	0.690	0.690
−3.0	0.396	0.636	0.666	0.666	0.666	0.667	0.667

Source: Interagency Advisory Committee on Water Data (1982).

uations), it created a consistent approach to flood-flow frequency calculations used in federal studies involving water resources.

24.5.2.1 Determination of the Coefficient of Skewness.

The skew coefficient used in the log-Pearson Type III distribution is sensitive to the size of the data sample and it can be difficult to estimate accurately from small samples. The Water Resources Council recommended using a generalized estimate of C_w based on the equation

$$C_w = WC_s + (1 - W)C_m \qquad (24.35)$$

where W is a weighting factor, C_s is the skewness coefficient computed using the sample data, and C_m is a map skew, values of which are found in Fig. 24.21. The weighting factor W is calculated so as to minimize the variance of C_w. The weighting procedure adopted is a function of the variance of the sample skew and the variance of the map skew.

The estimates of the sample skew coefficient and the map skew coefficient in Eq. (24.35) are assumed to be independent with the same mean and different variances, $V(C_s)$ and $V(C_m)$. The variance, or mean square error, of the weighted skew $V(C_w)$ can be expressed as

$$V(C_w) = W^2 V(C_s) + (1 - W)^2 V(C_m) \qquad (24.36)$$

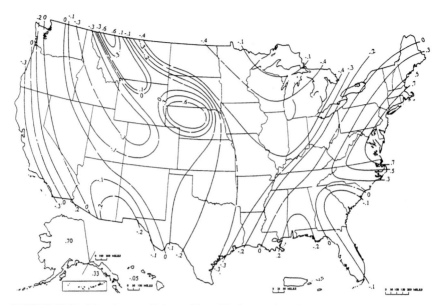

FIGURE 24.21 Map skew coefficients of logarithmic annual maximum streamflow. *(Interagency Advisory Committees on Water Data, 1982.)*

The skew weight that minimizes $V(C_w)$ can be determined by differentiating Eq. (24.36) with respect to W and solving $d[V(C_w)]/dW = 0$ for W to obtain

$$W = \frac{V(C_m)}{V(C_s) + V(C_m)} \tag{24.37}$$

The second derivative is greater than zero, confirming that the weight given by Eq. (24.37) minimizes $V(C_w)$.

Determination of W using Eq. (24.37) requires knowledge of $V(C_m)$ and $V(C_s)$. $V(C_m)$ can be estimated from the map of skew coefficients for the United States as 0.3025. Alternatively, $V(C_m)$ could be derived from a regression study relating the skew to physiographical and meteorological characteristics of the basin.

The weighted skew C_w can be determined by substituting Eq. (24.37) into Eq. (24.35)

$$C_w = \frac{V(C_m)C_s + V(C_s)C_m}{V(C_m) + V(C_s)} \tag{24.38}$$

The variance of the sample skew $V(C_s)$ can be approximated with sufficient accuracy as

$$V(C_s) = 10^{A - B \log (N/10)} \tag{24.39}$$

where

$$A = -0.33 + 0.08|C_s|, \qquad \text{if} \qquad |C_s| \le 0.90 \tag{24.40a}$$

or

$$A = -0.52 + 0.30|C_s|, \quad \text{if} \quad |C_s| > 0.90 \qquad (24.40b)$$

$$B = 0.94 - 0.26|C_s|, \quad \text{if} \quad |C_s| \leq 1.50 \qquad (24.40c)$$

or

$$B = 0.55 \qquad\qquad\qquad \text{if} \quad |C_s| > 1.50 \qquad (24.40d)$$

in which $|C_s|$ is the absolute value of the sample skew and N is the record length in years.

24.5.2.2 Testing for Outliers. *Outliers* are data points that depart significantly from the trend of the remaining data. Their retention (or deletion) can significantly affect the magnitude of statistical parameters computed from the data, especially for small samples. Procedures for treating outliers require judgment involving both hydrologic and mathematical considerations. According to the WRC, if the station (sample) skew is greater than +0.4, tests for high outliers are considered first; if the station skew is less than –0.4, tests for low outliers are considered first. Where the station skew is between ±0.4, tests for both high- and low-valued outliers should be applied before eliminating any outliers from the data set.

The following frequency equation can be used to detect high outliers

$$y_H = \bar{y} + K_N s_y \qquad (24.41)$$

where y_H is the high outlier threshold in log units and K_N is as given in Table 24.8 for sample size N. Use of the K_N values in Table 24.10 is equivalent to a one-sided test that detects outliers at the 10-percent level of significance. The K_N values are based on a normal distribution for the detection of single outliers. If the logs of peak values in a data sample are greater than y_H in Eq. (24.41), then they are considered high outliers. Flood-flow peaks considered high outliers should be compared with historic flood data and any flood-flow information at nearby sites. According to the

TABLE 24.10 Outlier Test K_N Values

Sample size N	K_N	Sample size N	K_N	Sample size N	K_N	Sample size N	K_N
10	2.036	24	2.467	38	2.661	60	2.837
11	2.088	25	2.486	39	2.671	65	2.866
12	2.134	26	2.502	40	2.682	70	2.893
13	2.175	27	2.519	41	2.692	75	2.917
14	2.213	28	2.534	42	2.700	80	2.940
15	2.247	29	2.549	43	2.710	85	2.961
16	2.279	30	2.563	44	2.719	90	2.981
17	2.309	31	2.577	45	2.727	95	3.000
18	2.335	32	2.591	46	2.736	100	3.017
19	2.361	33	2.604	47	2.744	110	3.049
20	2.385	34	2.616	48	2.753	120	3.078
21	2.408	35	2.628	49	2.760	130	3.104
22	2.429	36	2.639	50	2.768	140	3.129
23	2.448	37	2.650	55	2.804		

Source: Interagency Committee on Water Data (1982).

WRC, if information is available that indicates a high outlier is the maximum flow over an extended period of time, the outlier is treated as historic flood data and excluded from the analysis. If useful historic information is not available to compare to high outliers, then they should be retained as part of the record represented by the data set.

The following frequency equation can be used to detect low outliers

$$y_L = \bar{y} - K_N s_y \qquad (24.42)$$

where y_L is the low outlier threshold in log units. Flood-flow peaks considered low outliers are deleted from the record and a conditional probability adjustment described in WRC is applied.

24.6 HYDROLOGIC DESIGN

This section contains a discussion of some of the more common examples of design in water-excess management, with the emphasis being on urban stormwater systems. From an engineering viewpoint, the stormwater-management problem can be split into two parts: runoff prediction (hydrology) and system design (hydraulics). Runoff hydrographs generated in the first part provide input to the hydraulic design of storm sewers, detention basins, main channels, and facilities (such as discharge works) in water bodies that eventually receive the flow.

24.6.1 Design Flows—Storm Sewers

The Rational Method. The rational method, which has its origins in empirical design methods used in England in the mid-nineteenth century, is the most widely used method for designing drainage facilities for small urban and rural watersheds. The peak flow is estimated from the rational formula

$$Q = CiA \qquad (24.43)$$

where Q = the peak rate of runoff in cfs; C is a runoff coefficient (see Table 19.6 in Chap. 19); i is the average rainfall intensity in in/h, lasting for a critical period of time = t_c, the time of concentration of the watershed; and A is the size of the drainage area in acres. The product Ci is the net rainfall intensity in in/h at time $t = t_c$.

In urban areas, the drainage area usually consists of subareas of substantially different surface characteristics. In such cases the runoff coefficient is a weighted composite of the C values for the different subareas. If there are j subareas then the peak runoff is computed from

$$Q = i \sum_{j=1}^{m} C_j A_j \qquad (24.44)$$

where m is the number of subareas drained by the sewer system.

The rainfall intensity i, an average rate in in/h, is selected on the basis of design rainfall duration and design frequency of occurrence. The design duration and frequency may be specified in design standards or chosen by the engineer as a design parameter. The time of concentration t_c is the time, measured from the beginning of the storm, at which all parts of the drainage area are contributing to flow at the basin outlet. It is possible that runoff may reach a peak before all subareas are contributing to flow, and a trial-and-error procedure may be necessary to determine the critical t_c. In an urban basin, the time of concentration is the sum of the inlet time t_o and the flow time t_f where the flow time is given by

$$t_f = \sum_{j=1}^{n} \frac{L_j}{V_j} \qquad (24.45)$$

in which L_j is the length of the jth pipe along the flow path, and V_j is the average flow velocity in the pipe. The inlet time t_o is the longest time for the overland flow of water in the catchment to reach the storm-sewer inlet draining the system.

In the rational method, each sewer in the system is designed individually and independently (except for the computation of sewer-flow time), and the corresponding rainfall intensity i is computed repeatedly for the area drained by the sewer. As the design progresses to the downstream sewers, the drainage area and time of concentration increase. An increasing t_c in turn gives a decreasing i, which should be applied to the entire area drained by the sewer.

24.6.2 Design Volumes—Stormwater-Detention Basins

One of the common consequences of urban development is an increase in peak-runoff rates and volumes. An effective management method to minimize this detrimental impact is the use of *detention basins* to hold the excess water for a short period of time, and subsequent release of the stored water in a controlled manner. Detention basins do not significantly reduce the total volume of surface runoff, but simply reduce peak rates of flow by redistributing the runoff over time as shown in Fig. 24.22. A stormwater-detention basin can range from simply allowing the water to back up behind a highway culvert, to a reservoir with regulated outlet works, including spillways, valves, and possibly siphons. Detention storage can be provided at one or a combination of locations: (1) on or near the sites where the precipitation occurs, (2) in storm-sewer systems, or (3) in downstream impoundments. Several methods of stormwater detention exist, including basins and ponds on the ground surface, parking-lot storage, rooftop detention, and underground storage.

There are several major design determinations involved in the engineering design of stormwater-detention facilities. They include: (1) the selection of the design-rainfall event, (2) the volume of storage needed, (3) the maximum permitted release rate, (4) pollution-control requirements and opportunities, and (5) identification of practical detention methods and techniques for the specific projects.

Detention-Pond Sizing—Modified Rational Method. The modified rational method is an extension of the rational method for rainfalls lasting longer than the time of concentration. The method was developed so that the concept of the rational method could be used to develop hydrographs for *storage volume* design, rather than just peak discharges for storm-sewer capacity. It is often used for the preliminary design of detention storage for watersheds of up to 30 acres.

The shape of the hydrograph produced by the modified rational method is a trapezoid constructed by setting the rising and receding limbs equal to the time of concentration t_c and computing the peak discharge rate assuming various rainfall durations. Typical stormwater hydrographs for the modified rational method with various rainfall durations are shown in Fig. 24.23a. T_D represents the time of duration of rainfall and the area under the trapezoid (or triangle for $T_D = t_c$) represents the total volume of runoff. The peak discharge rate Q_p decreases as T_D increases

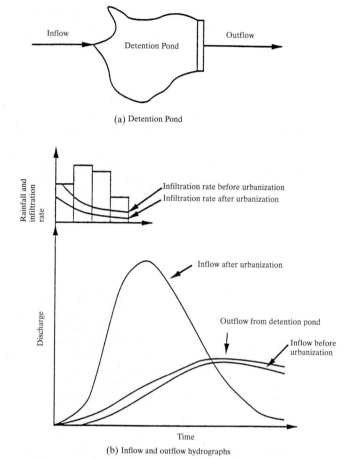

(a) Detention Pond

(b) Inflow and outflow hydrographs

FIGURE 24.22   Hydrologic effect of urbanization on storm runoff and function of detention ponds: (*a*) detention pond; (*b*) inflow and outflow hydrographs.

because, as IDF curves show (Sec. 24.2.1), longer duration rainfalls are associated with lower rainfall intensities.

If the maximum allowable rate of discharge out of the detention pond Q_A is known, then the required detention storage for each rainfall duration can be calculated as the volume difference between inflow and outflow hydrographs, as shown in Fig. 24.23*b*. The critical duration of the design storm is the one that yields the greatest required detention-storage volume.

Considering a general equation relating rainfall intensity and duration for a given frequency to be a variation of Eq. (24.1), namely

$$i = \frac{a}{(T_D + b)^c} \tag{24.46}$$

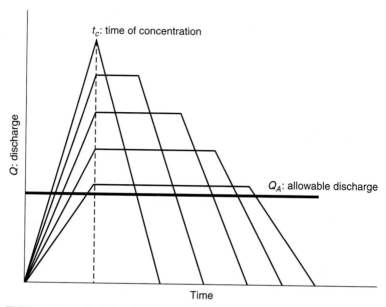

FIGURE 24.23a Typical modified rational method hydrographs.

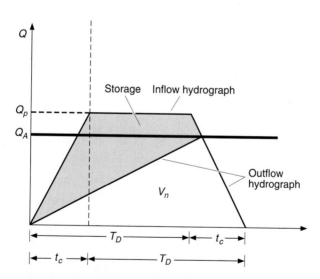

t_c: time of concentration (minutes)
T_D: time of duration (minutes)
Q_p: peak flows (cfs)
Q_A: maximum allowable release rate (cfs)

FIGURE 24.23b Inflow and outflow hydrographs for storage volume design.

where i is the average rainfall intensity for rainfall duration of T_D, and a, b, and c are constants, then using the rational equation the peak discharge rate can be expressed as

$$Q_p = CiA \qquad (24.47)$$

The inflow hydrograph volume V_i (Fig. 24.23b) is

$$V_i = 60(0.5)Q_p[(T_D - t_c) + (T_D + t_c)] \qquad (24.48)$$

and the outflow hydrograph volume is

$$V_o = 60(0.5)Q_A(T_D + t_c) \qquad (24.49)$$

and the required detention storage volume is

$$V_s = V_i - V_o = 60Q_p T_D - 30Q_A(T_D + t_c) \qquad (24.50)$$

The rainfall duration T_D for the maximum retention volume is determined by differentiating Eq. (24.50) with respect to T_D and setting the derivative equal to zero. This yields

$$\frac{T_D(1-c)+b}{(T_D+b)^{c+1}} - \frac{Q_A}{2CA} = 0 \qquad (24.51)$$

Eq. (24.51) can be solved for T_D by using Newton's iteration technique in which the iterative equation is

$$T_{D_{i+1}} = T_{D_i} - \frac{F(T_{D_i})}{F'(T_{D_i})} \qquad (24.52)$$

where

$$F(T_{D_i}) = \frac{[T_D(1-c)+b]}{(T_D+b)^{c+1}} - \frac{Q_A}{2CA} = 0 \qquad (24.53)$$

and

$$F'(T_{D_i}) = \frac{d[F(T_{D_i})]}{dT_D} = \frac{[T_D(1-c)+b](c+1)}{(T_D+b)^{c+2}} + \frac{1-c}{(T_D+b)^{c+1}} \qquad (24.54)$$

24.6.3 Risk-Based Design

Proposed water-excess management solutions are subject—as are most solutions to engineering problems—to an element of uncertainty. The uncertainty inherent in storm-sewer design derives from both hydraulic and hydrologic aspects of the problem. Recommended references on risk-based design include Ang and Tang (1975; 1984); Chow, et al. (1988); Harr (1987); Kapur and Lamberson (1977); and Yen (1986). Also see Chap. 7.

The key question is the ability of the proposed sewer design to accommodate the surface runoff generated by a storm. Although a factor of safety SF is inherent in the choice of a design frequency, the relationship between the sewer capacity Q_C and the storm runoff Q_L can also be *explicitly* considered: that is, $Q_C = SF \times Q_L$. Using a reliability analysis, a probability of failure $P(Q_L > Q_C)$ can be calculated for selected frequencies and safety factors. The corresponding risks and safety factors for each return period (recurrence interval) can be plotted to derive the risk–safety factor relationship for each return period. The procedure is as follows:

1. Select the return period T.
2. Use a rainfall–runoff model (such as the rational method) and perform an uncertainty analysis to compute the mean loading on the sewer $\bar{Q}_L$ (the mean surface runoff), and its coefficient of variation Ω_{QL}, where

$$\bar{Q}_L = \bar{C}\,\bar{i}\,\bar{A} \tag{24.55}$$

in which $\bar{C}$ is the mean runoff coefficient; $\bar{A}$ is the mean basin area in acres; $\bar{i}$ is the mean rainfall intensity in in/hr; and

$$\Omega_{QL}^2 = \Omega_c^2 + \Omega_i^2 + \Omega_A^2 \tag{24.56}$$

in which Ω_c, Ω_i, and Ω_A are the coefficients of variation of C, i, A, respectively.

3. Select a pipe diameter $\bar{d}$ and compute $\bar{Q}_c$ and Ω_{Q_c}, where $\bar{Q}_c$ is the mean capacity of the sewer and Ω_{Q_c} its coefficient of variation. The mean capacity can be obtained from a modified form of the Manning equation

$$\bar{Q}_c = \frac{0.463}{\bar{n}}\,\bar{S}_o^{1/2}\,\bar{d}^{8/3} \tag{24.57}$$

in which $\bar{n}$ is the mean Manning roughness coefficient and $\bar{S}_o$, is the mean sewer slope.

The coefficient of variation Ω_{Q_c} is computed from

$$\Omega_{Q_c}^2 = \Omega_n^2 + \frac{1}{4}\,\Omega_{S_o}^2 + \frac{64}{9}\,\Omega_d^2 \tag{25.58}$$

in which Ω_n, Ω_{S_o}, and Ω_d are the coefficients of variation of n, S_o, and d, respectively.

4. Compute the risk and safety factor.
5. Repeat Step 3 for other diameters.
6. Repeat Step 2 for other rainfall durations.
7. Return to Step 1 for each return period to be considered.

REFERENCES

Ang, A. H.-S., and W. H. Tang, *Probability Concepts in Engineering Planning and Design,* vol. I, *Basic Principles,* and vol. II, *Decision, Risk and Reliability,* John Wiley, New York, 1975, and 1984, respectively.

Bandyopadhyay, M., "Synthetic Storm Pattern and Runoff for Gauhati," *Journal of the Hydraulics Division,* ASCE, 98(HY5), pp. 845–857, 1957.

Bedient, P. B., and W. C. Huber, *Hydrology and Floodplain Analysis,* 2d ed., Addison-Wesley, Reading, Mass., 1992.

Bras, Rafael I., *Hydrology,* Addison-Wesley, Reading, Mass., 1990.

Burnach, J. C., R. L. Ferral, and R. A. McGuire, "A Generalized Streamflow Simulation System. Conceptual Modeling for Digital Computers," Joint Federal–State River Forecast Center, U.S. Department of Commerce, NOAA National Weather Service, and State of California Dept. Water of Resources, Mar. 1973.

Chow, V. T., ed., *Handbook of Applied Hydrology*, McGraw-Hill, New York, 1964.

Chow, V. T., D. R. Maidment, and L. W. Mays, *Applied Hydrology*, McGraw-Hill, New York, 1988.

Crawford, N. H., and R. K. Linsley, "Digital Simulation in Hydrology: Stanford Watershed Model IV," Technical Report no. 39, Department of Civil Engineering, Stanford University, Stanford, Calif., July 1966.

Federick, R. H., V. A. Myers, and E. P. Auciello, "Five to 60-Minute Precipitation Frequency for the Eastern and Central United States," NOAA Technical Memo NWS HYDRO-35, National Weather Service, Silver Spring, Md., June 1977.

Fread, D. L., "Flow Routing" in D. R. Maidment, ed., *Handbook of Hydrology*, Chap. 10, McGraw-Hill, New York, 1993.

Gupta, V. K., E. Waymire, and C. T. Wang, "A Representation of an Instantaneous Unit Hydrograph from Geomorphology," *Water Resources Research,* 16(5), pp. 855–862, 1980.

Hansen, E. M., L. C. Schreiner, and J. F. Miller, "Application of Probable Maximum Precipitation Estimates—United States East of the 105th Meridian," NOAA Hydrometeorological Report no. 52, National Weather Service, Washington, D.C., Aug. 1982.

Harr, M. E., *Reliability-based Design in Civil Engineering*, McGraw-Hill, New York, 1987.

Havens and Emerson, Consulting Engineers, "Master Plan for Pollution Abatement, Cleveland, Ohio," July 1968.

Hershfield, D. M., "Rainfall Frequency Atlas of the United States for Durations from 30 Minutes to 24 Hours and Return Periods from 1 to 100 Years," Tech. Paper 40, U.S. Department of Commerce, Weather Bureau, Washington, D.C., May 1961.

Hoggan, D. H., *Computer Assisted Floodplain Hydrology and Hydraulics*, McGraw-Hill, New York, 1989.

Huff, F. A., "Time Distribution of Rainfall in Heavy Storms," *Water Resources Research,* 3(4):1007–1019, 1967.

Interagency Advisory Committee on Water Data (formerly U.S. Water Resources Council), "Guidelines for Determining Frequency," Bulletin 17B, U.S. Department of the Interior, U.S. Geological Survey, Office of Water Data Coordination, Reston, Va., 1982.

Kapur, K. C., and L. R. Lambersen, *Reliability in Engineering Design*, John Wiley, New York, 1977.

Keifer, C. J., and H. H. Chu, "Synthetic Storm Pattern for Drainage Design," *Journal of the Hydraulics Division,* ASCE, 84(HY4), pp. 1–25, 1957.

Linsley, R. K., M. A. Kohler, and J. L. H. Paulhares, *Hydrology for Engineers,* McGraw-Hill, New York, 1986.

Maidment, D. R., ed., *Handbook of Hydrology,* McGraw-Hill, New York, 1993.

Marsalek, J., "Research on the Design Storm Concept," Technical memo 33, ASCE, Urban Water Resources Research Program, New York, 1978.

Mays, L. W., and Y. K. Tung, *Hydrosystems Engineering and Management,* McGraw-Hill, 1992.

McCuen, R. H., *Hydrologic Analysis and Design,* Prentice-Hall, Englewood Cliffs, N.J., 1989.

McPherson, M. B., "Discussion of Synthetic Storm Pattern for Drainage Design," *Journal of the Hydraulics Division,* ASCE, 84(HY1), pp. 49–60, 1958.

Michaud, J., and S. Sorooshian, "Comparison of Simple Versus Complex Distributed Runoff Models on a Midsized Semiarid Watershed," *Water Resources Research,* 30(3), pp. 593–605, March 1994.

Miller, J. F., R. H. Frederick, and R. J. Tracey, *Precipitation-Frequency Atlas of the Coterminous Western United States (by States),* NOAA Atlas 2, 11 vols., National Weather Service, Silver Spring, Md., 1973.

National Academy of Science, *Safety of Existing Dams: Evaluation and Improvement,* National Academy Press, Washington, D.C., 1983.

Overton, D. E., and M. E. Meadows, *Stormwater Modeling,* Academic Press, New York, 1976.

Ponce, V. M., *Engineering Hydrology: Principles and Practice,* Prentice-Hall, Englewood Cliffs, N.J., 1989.

Preul, H. D., and C. N. Papadukiz, "Development of Design Storm Hyetographs for Cincinnati, Ohio," *Water Resources Bulletin,* AWRA, 9(2), pp. 291–300, 1973.

Rawls, W. J., D. L. Brakensiek, and N. Miller, "Green-Ampt Infiltration Parameters from Soils Data," *Journal of Hydraulic Division,* ASCE, 109(1), pp. 62–70, 1983.

Rodriguez-Iturbe, I., and J. B. Valdes, "The Geomorphologic Structure of Hydrologic Response," *Water Resources Research,* 15(6), pp. 1409–1420, 1979.

Schreiner, L. C., and J. T. Riedel, "Probable Maximum Precipitation Estimates, United States East of the 105th Meridian," NOAA Hydrometeorological Report no. 51, National Weather Service, Washington, D.C., June 1978.

Stephenson, D., and M. E. Meadows, *Kinematic Hydrology and Modeling,* Developments in Water Science 26, Elsevier, Amsterdam, 1986.

U.S. Army Corps of Engineers Hydrologic Engineering Center, "Probable Maximum Storm (Eastern United States)," HMR 52, User's Manual, CPD-46, Mar. 1984.

U.S. Army Corps of Engineers, North Pacific Division, "Program Description and User Manual for SSARR Model, Stream Flow Synthesis and Reservoir Regulation," Portland, Oreg., 1986.

U.S. Department of Agriculture Soil Conservation Service, "A Method for Estimating Volume and Rate of Runoff in Small Watersheds," Tech. Paper 149, Washington, D.C., Apr. 1973.

U.S. Department of Agriculture Soil Conservation Service, "Urban Hydrology for Small Watersheds," Tech. Release no. 55, June 1986.

U.S. Department of Commerce, "Probable Maximum Precipitation Estimates, Colorado River and Great Basin Drainages," Hydrometeorological Report no. 49, NOAA, National Weather Service, Silver Spring, Md., Sept. 1977.

U.S. Department of Commerce, "Seasonal Variation of 10-Square-Mile Probable Maximum Precipitation Estimates, United States East of the 105th Meridian," Hydrometeorological Report no. 53, NOAA, National Weather Service, Silver Spring, Md., Apr. 1980.

U.S. Weather Bureau, "Seasonal Variation of the Probable Maximum Precipitation East of the 105th Meridian," Hydrometeorological Report no. 33, Washington, D.C., 1956.

U.S. Weather Bureau, "Rainfall-Intensity-Frequency Regime, Part 2—Southeastern United States," Tech. Paper no. 29, Mar. 1958.

U.S. Weather Bureau, "Generalized Estimates of Probable Maximum Precipitation West of the 105th Meridian," Tech. Paper no. 38, Washington, D.C., 1960.

U.S. Weather Bureau, "Generalized Estimates of Probable Maximum Precipitation and Rainfall-Frequency Data for Puerto Rico and Virgin Islands," Tech. Paper no. 42, Washington, D.C., 1961.

U.S. Weather Bureau, "Probable Maximum Precipitation in the Hawaiian Islands," Hydrometeorological Report no. 39, Washington, D.C., 1963a.

U.S. Weather Bureau, "Probable Maximum Precipitation Rainfall-Frequency Data for Alaska," Tech. Report no. 47, Washington, D.C., 1963b.

U.S. Weather Bureau, "Two- to Ten-Day Precipitation for Return Periods of 2 to 100 Years in the Contiguous United States," Tech. Paper 49, Washington, D.C., 1964.

U.S. Weather Bureau, "Meteorological Conditions for the Probable Maximum Flood on the Yukon River above Rampart, Alaska," Hydrometeorological Report no. 42, Environmental Science Services Administration, Washington, D.C., 1966a.

U.S. Weather Bureau, "Probable Maximum Precipitation, Northwest States," Hydrometeorological Report no. 43, Washington, D.C., 1966b.

U.S. Weather Bureau, "Interim Report—Probable Maximum Precipitation in California," Hydrometeorological Report no. 36, Washington, D.C., Oct. 1961; revised Oct. 1969.

Viessman, W., Jr., G. L. Lewis, and J. W. Knapp, *Introduction to Hydrology,* 3d. ed., Harper & Row, New York, 1989.

Viessman, W., and C. Welty, *Water Management: Technology and Institutions,* Harper & Row, New York, 1985.

Wenzel, H. G., "Rainfall for Urban Stormwater Design," in David F. Kibler, ed., *Urban Storm Water Hydrology,* Water Resources 7, American Geophysical Union, Washington, D.C., 1982.

Woolhiser, D. A., "Physically Based Models of Watershed Runoff," in V. P. Singh, ed., *Rainfall-Runoff Relationship,* Proc. International Symposium on Rainfall-Runoff Modeling, Water Resources Publications, Littleton, Colo., pp. 189–202, 1982.

Woolhiser, D. A., R. E. Smith, and D. C. Goodrich, "A Kinematic Runoff and Erosion Model: Documentation and User Manual," ARS 77, U.S. Dept. of Agric., Washington, D.C., 1990.

Yen, B. C., ed., *Stochastic and Risk Analysis in Hydraulic Engineering,* Water Resources Publications, Littleton, Colo., 1986.

CHAPTER 25
HYDRAULICS FOR EXCESS WATER MANAGEMENT

Ben Chie Yen
Professor of Civil Engineering
University of Illinois
Urbana, Illinois

25.1 INTRODUCTION

Excess water from natural sources and human uses is disposed of through channels, overland surfaces, and pipes into receiving water bodies. Groundwater flow is presented in Chaps. 2, 11, and 16. Hydraulics of pipe flow is given in Chaps. 2, 18, and 19. Presented in this chapter is the hydraulics of open-channel flow for water management. *Open-channel flow* is defined as the flow of a fluid in a channel in which the flow shares a free surface with a lighter fluid or vacuum above. For excess-water management, in most cases the fluid in the channel is water, the lighter fluid is air, and the free surface is subject to atmospheric pressure.

Water-engineering projects can be classified according to one of the following three purposes:

1. *Planning.* For which a long term, and often large spatial scale planning of water management is desired in advance without requiring details of the hydraulic system

2. *Design.* For which dimensions of the hydraulic structure are to be determined

3. *Operation.* For which the hydraulic structure already exists or its dimensions have already been determined; the flow through the hydraulic structure is to be simulated for the purposes of real-time flood or pollution control, real-time forecasting or performance evaluation

The appropriate level of hydraulics to be used in a project varies according to the purpose of the project in addition to data, personnel, and other factors. Planning projects usually require little use of hydraulics. Operational projects, especially those involving legal cases, demand a great deal of hydraulics for accuracy. On the other hand, a moderate level of hydraulics is often—but not always—sufficient for design projects, particularly when the commercially available sizes, rather than computed required dimensions, are used.

Water excess occurs in both urban and rural settings, in humid as well as arid regions, overland and in small ditches or large rivers, and in tropical as well as cold climates. The objectives of excess-water management are flood control, pollution control, or both, in urban and rural areas and in rivers. Irrespective of the project objectives, hydraulically the flows follow the same basic equations of continuity, momentum, and energy that will be described in Sec. 25.3.

25.2 CLASSIFICATION OF FLOW

Flow in channels and on the overland is usually expressed in terms of the *discharge, velocity, depth,* and *cross-sectional area* of the flow. They are related mathematically as follows:

$$V = \frac{Q}{A} \tag{25.1}$$

$$A = \int_{y_b}^{y_b + Y} B \, dy \tag{25.2}$$

$$Q = \frac{1}{A} \int_A \bar{u} \, dA \tag{25.3}$$

where (with reference to the coordinate system in Fig. 25.1)

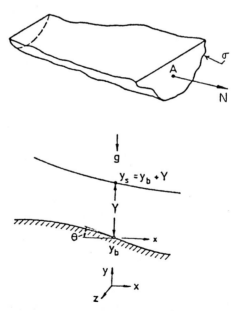

FIGURE 25.1 Gravity-oriented coordinates with depth measured vertically.

$A(x,t)$ = flow cross-sectional area, L^2
$B(x,y,t)$ = channel width, L
$Q(x,t)$ = discharge through A, L^3/T
$u(x,y,z,t)$ = local point velocity along the direction normal to A, L/T
$V(x,t)$ = cross-sectional averaged velocity, L/T
$Y(x,t)$ = depth of flow measured from channel bottom, L
$y_b(x,t)$ = elevation of channel bottom, L

The parentheses after a variable indicate the variable varies with time t; longitudinal location of the cross section or point x; vertical location y; and lateral location z. The dimension of the variable is given at the end, in which L represents length and T time. A bar above a variable indicates average over turbulent fluctuation. The water-surface elevation is $y_s = Y + y_b$ where the subscripts b and s represent channel bed and water surface, respectively.

Flow classification for channel and overland flow may be made according to the following criteria:

According to its variation with time:	Steady flow (time invariant)
	Unsteady flow (transient flow)
According to its variation with space:	Uniform flow (parallel streamlines)
	Nonuniform flow
	Gradually varied
	Rapidly varied
According to the viscosity effect:	Laminar flow
	Turbulent flow
According to the gravity effect:	Subcritical flow
	Supercritical flow

As mentioned in Chap. 2, the criterion to differentiate turbulent flow from laminar flow is the *Reynolds number*

$$\mathcal{R} = \frac{VR}{\nu} \qquad (25.4)$$

where R is the hydraulic radius defined as the cross-sectional area A divided by the wetted perimeter of A; and ν is the kinematic viscosity of the fluid. For channel and overland flows, the critical value of $\mathcal{R}$ is usually taken as 500 which corresponds to 2000 for full-pipe flow. Below this value, the flow is considered as laminar and above it turbulent. Actually, this critical value is lower for flow changing from turbulent to laminar, for spatially decelerating flow, and for flow with lateral inflow.

The criterion to differentiate subcritical flow from supercritical flow is the *Froude number* $\mathcal{F}$ which is usually defined as

$$\mathcal{F} = \frac{V}{\sqrt{gD}} \qquad (25.5)$$

where $D = A/B$ is called the *hydraulic mean depth*. This number is essentially the ratio of the flow velocity to the *celerity* that a disturbance would travel. The critical value of the Froude number is unity. When $\mathcal{F} < 1$, the flow velocity travels slower than the disturbance-wave celerity; hence, the disturbance will be able to propagate in both upstream and downstream directions with the former propagating slower than the latter, and the flow is subcritical. When $\mathscr{F} > 1$, the flow velocity is faster than the celerity, and hence, the disturbance waves will be swept downstream imposing no effect on the upstream; the flow is called *supercritical*. In the definition of Eq. (25.5), the effect of velocity distribution over the depth is not considered. If this effect is included, a correction factor should be included as will be discussed in the next section.

For fast free-surface flow, the drag from the channel bed may cause instability on the surface resulting in repeated roll waves on the water surface. The criterion of this instability is the *Vedernikov number* $\mathscr{V}$ which can be defined as the ratio of the kinematic wave velocity to the dynamic wave velocity. Both wave velocities are functions of velocity and pressure distributions over the cross-sectional area A. If uniform velocity and piezometric pressure distributions are assumed, the Vedernikov number can be expressed approximately as

$$\mathscr{V} = \frac{dQ/dA}{V + \sqrt{gD}} \tag{25.6}$$

When $\mathscr{V} < 1$ the flow is stable, whereas when $\mathscr{V} > 1$ the flow is unstable and roll waves occur. For a two-dimensional, plane, turbulent flow, by ignoring the velocity and pressure distributions, the marginal stability Froude number $\mathscr{F} = 2.0$ when $\mathscr{V} = 1$. The value of this marginal Froude number decreases for increasingly curved velocity distributions (Chen, 1995).

25.3 HYDRAULIC EQUATIONS

The basic hydraulic equations are derived from the physical principles of conservation of mass, momentum, and energy. In deriving the equations, two orthogonal coordinate systems can be employed (Yen, 1975). One is a *gravity-oriented coordinate system* in which y is vertical, x is horizontal along the channel longitudinal direction, and z is horizontal along the transverse direction. The other is a *natural coordinate system* in which y is normal to the channel bed making an angle θ with the vertical, x is the longitudinal direction along the channel and normal to y, and z is horizontal along the transverse direction. In field situations, depth is measured along the plumb or vertical direction, and the topographic and water-surface information is given in terms of elevation or stage. Therefore, the gravity-oriented orthogonal coordinate system is used here, and the *bed slope* is $S_o = \tan \theta$. The natural coordinate system is more suitable for laboratory flumes or human-made channels for which the bed slope $S_o = \sin \theta$ is usually constant and the cross-sectional shape is regular and prismatic. Open-channel flow equations using this natural coordinate system can be found elsewhere (e.g., Yen, 1973; 1975).

Occasionally, one of the following two nonorthogonal coordinate systems is employed carelessly with flow equations having incomplete or approximate terms. One is the coordinate system with the depth and y-direction along the vertical, while x is along the channel bed, not horizontal. The other is the system with the depth and y along the direction normal to the bed, while x is in the horizontal direction. The complete exact flow equations for these nonorthogonal coordinate systems are fairly complicated. Therefore, these coordinate systems should be avoided.

By using the gravity-oriented coordinates with flow depth measured vertically and the directional normal of the flow cross section along the x direction, the one-dimensional flow equations for a flow cross section can be derived from the principles of conservation of mass, momentum, and energy. Several such derivations appear in the literature. Those obtained by Yen (1975) are listed in Table 25.1 for the *continuity equation* in both discharge-area form and velocity-depth forms, and the cross-section averaged *one-dimensional momentum and energy equations* are given in Table 25.2. In these equations:

A = cross-sectional flow area
B = water-surface width of channel cross section
$D = A/B$ = hydraulic mean depth
$\mathscr{F}$ = Froude number of channel flow
g = gravitational acceleration
H_B = true cross-section average total head of flow (Table 25.2)
H_c = a reference head as defined in Table 25.2
H_L = piezometric head of lateral flow when joining channel flow, measured with respect to the horizontal reference datum
i,j = orthogonal coordinate directions $i,j = 1,2,3$ for x,y,z directions, respectively
K = piezometric pressure correction factor or local potential energy flux correction factor
K' = ambient piezometric pressure correction factor
m_q = normalized momentum input from lateral flow
N = directional normal of a surface, positive outward
P = local piezometric pressure with respect to channel bottom
$p = \tau_{ii}/3$, local pressure intensity
Q = discharge through cross-section A
q = time rate of lateral flow per unit length of σ
r = normal displacement of σ with respect to space or time projected on a plane parallel to A, positive outward
S_e = dissipated energy gradient
S_f = friction slope
$S_{oi} = \partial y_b/\partial x_i$, channel slope along the x_i direction
t = time
U = velocity of lateral flow when joining channel flow
u_i = instantaneous local (point) velocity component along x_i direction
u_i' = turbulent fluctuation with respect to $\bar{u}_i$
$V = V_x$ = x-component of channel flow velocity average over cross section
x = longitudinal coordinate

TABLE 25.1 Cross-Section Averaged One-Dimensional Continuity Equations for Incompressible Fluid Flow

Flow condition	In discharge-area form	In velocity-depth form
Unsteady with lateral flow	$\dfrac{\partial A}{\partial t} + \dfrac{\partial Q}{\partial x} = \int_{\sigma} \bar{q}\, d\sigma$	$\dfrac{\partial Y}{\partial t} + v\dfrac{\partial Y}{\partial x} + \dfrac{VD}{B}\dfrac{\partial B}{\partial x} + D\dfrac{\partial V}{\partial x} = \dfrac{1}{B}\int_{\sigma} \bar{q}\, d\sigma$
Steady with lateral flow	$\dfrac{dQ}{dx} = \int_{\sigma} \bar{q}\, d\sigma$	$\dfrac{dY}{dx} + \dfrac{D}{B}\dfrac{\partial B}{\partial x} + \dfrac{D}{V}\dfrac{dV}{dx} = \dfrac{1}{BV}\int_{\sigma} \bar{q}\, d\sigma$
Steady with no lateral flow	$\dfrac{dQ}{dx} = 0$	$\dfrac{dY}{dx} + \dfrac{D}{B}\dfrac{\partial B}{\partial x} + \dfrac{D}{B}\dfrac{dV}{dx} = 0$

Note: For prismatic channel, $\partial B/\partial x = 0$; for rectangular cross section, $D = Y$.

TABLE 25.2 Cross-Section Averaged One-Dimensional Dynamic Equations for Incompressible Homogeneous Fluid

Flow condition	Momentum equation in discharge-area form	Momentum equation in velocity-depth form	Energy equation
Unsteady flow	$\dfrac{1}{gA}\dfrac{\partial Q}{\partial t} + \dfrac{1}{gA}\dfrac{\partial}{\partial x}\left(\dfrac{\beta}{A}Q^2\right) + \dfrac{\partial}{\partial x}(KY)$ $+ (K-K')Y\dfrac{\partial A}{A\,\partial x} = S_o - S_f + \dfrac{1}{gA}\displaystyle\int_\sigma \bar{U}_x\bar{q}\,d\sigma$	$\dfrac{1}{g}\dfrac{\partial V}{\partial t} + (2\beta-1)\dfrac{V}{g}\dfrac{\partial V}{\partial x} + \dfrac{V^2}{g}\dfrac{\partial\beta}{\partial x}$ $+ \left[(\beta-1)\dfrac{V^2}{g} + (K-K')Y\right]\cdot\left[\dfrac{1}{D}\dfrac{\partial Y}{\partial x} + \dfrac{1}{B}\dfrac{\partial B}{\partial x}\right]$ $+ \dfrac{\partial}{\partial x}(KY) = S_o - S_f + \dfrac{1}{gA}\displaystyle\int_\sigma(\bar{U}_x - V)\bar{q}\,d\sigma$	$\dfrac{\partial H_B}{\partial t} + (H_c - H_B)\dfrac{1}{A}\dfrac{\partial Q}{\partial x} + \dfrac{Q}{A}\dfrac{\partial H_c}{\partial x} - W$ $- \dfrac{1}{A}\displaystyle\int_\sigma (H_L - H_B)\bar{q}\,d\sigma + \dfrac{Q}{A}S_e = \zeta\dfrac{\partial Y}{\partial t}$
Steady nonuniform flow	$\left[(K-K')Y - \dfrac{\beta Q^2}{gA^2}\right]\dfrac{1}{A}\dfrac{\partial A}{\partial x} + \dfrac{Q^2}{gA^2}\dfrac{d\beta}{dx}$ $+ \dfrac{d}{dx}(KY) = S_o - S_f + \dfrac{1}{gA}\displaystyle\int_\sigma\left(\bar{U}_x - 2\beta\dfrac{Q}{A}\right)\bar{q}\,d\sigma$	$\dfrac{d}{dx}(KY) + \left[(K-K')Y - \beta\dfrac{V^2}{g}\right]\left[\dfrac{1}{D}\dfrac{dY}{dx} + \dfrac{1}{B}\dfrac{\partial B}{\partial x}\right]$ $+ \dfrac{V^2}{g}\dfrac{d\beta}{dx} = S_o - S_f + \dfrac{1}{gA}\displaystyle\int_\sigma(\bar{U}_x - 2\beta V)\bar{q}\,d\sigma$	$\dfrac{dH_c}{dx} - \dfrac{1}{Q}\displaystyle\int_\sigma (H_L - H_c)\bar{q}\,d\sigma$ $= -S_e + \dfrac{A}{Q}W$
Steady uniform flow, no lateral flow	$S_o = S_f$	$S_o = S_f$	$S_o = S_e$

$\beta = \dfrac{A}{Q^2}\displaystyle\int_A \bar{u}_x^2\,dA$ $\qquad S_f = \dfrac{-1}{\gamma A}\displaystyle\int_\sigma [\bar{\tau}_{ij}]_\sigma N_j\,d\sigma$ $\qquad H_B = \beta_B\dfrac{V^2}{2g} + KY + y_b$ $\qquad \beta_B = \dfrac{1}{V^2 A}\displaystyle\int_A \bar{u}_i\bar{u}_i\,dA$ $\qquad S_e = \dfrac{1}{\gamma AV}\displaystyle\int_A \bar{\tau}_{ij}\dfrac{\partial\bar{u}_i}{\partial x_j}\,dA$

$K = \dfrac{1}{\gamma AY}\displaystyle\int_A \bar{P}\,dA$ $\qquad P = p + \Upsilon(y - y_b)$ $\qquad H_c = \alpha\dfrac{V^2}{2g} + \eta Y + y_b$ $\qquad \alpha = \dfrac{1}{V^2 Q}\displaystyle\int_A \bar{u}_i\bar{u}_i\bar{u}_x\,dA$ $\qquad \bar{\tau}_{ij} = \mu\left(\dfrac{\partial\bar{u}_i}{\partial x_j} + \dfrac{\partial\bar{u}_j}{\partial x_i}\right) + \rho\overline{u_i'u_j'}$

$K' = \left(\gamma Y\dfrac{\partial A}{\partial x}\right)^{-1}\displaystyle\int_\sigma\left(\bar{P}\dfrac{\partial\bar{r}}{\partial x} + p'\dfrac{\overline{\partial\bar{r}'}}{\partial x}\right)d\sigma$ $\qquad W = \dfrac{1}{\gamma A}\left(\dfrac{\partial}{\partial x}\displaystyle\int_A \bar{u}_i\bar{\tau}_{ix}\,dA + \displaystyle\int_\sigma [\bar{u}_x\bar{\tau}_{ij}]_\sigma N_j\,d\sigma\right)$ $\qquad \zeta = \dfrac{\partial Y}{\partial t} - \dfrac{1}{\gamma A}\displaystyle\int_A \dfrac{\partial\bar{p}}{\partial t}\,dA$ $\qquad \eta = \dfrac{1}{\gamma QY}\displaystyle\int_Z \bar{P}\bar{u}\,dA$

x_i = coordinate along ith direction
Y = depth of flow measured vertically
y = vertical coordinate
y_b = channel bed elevation above reference datum
y_s = water surface elevation above reference datum
α = convective kinetic energy flux correction factor
β = momentum flux correction factor
β_B = local kinetic energy flux correction factor
γ = specific weight of water
ζ = unsteady pressure correction factor
η = convective potential energy flux correction factor
θ = angle between channel bottom and horizontal plane
μ = dynamic viscosity of fluid
ρ = mass density of fluid
σ = perimeter bounding cross-sectional area A
$\bar{\tau}_{ij}$ = temporal mean stresses at a point

In the momentum equations in Table 25.2, the effect of the force due to internal stresses $\bar{\tau}_{ij}$ acting on the cross-sectional area A is neglected. Otherwise the term $(\partial T/\partial x)/\gamma A$ should be included, where

$$T = \int_A \left[2\mu \left(\frac{\partial \bar{u}_x}{\partial x} \right) - \rho \overline{u_x'^2} \right] dA \tag{25.7}$$

This internal-stress force term is zero for steady-uniform flow without lateral flow. Usually, this term is indeed relatively negligible; it is only for flow with rapid expansion and flow separation that one needs to check its significance.

In solving a problem, it is important to select the proper equations that can adequately account for the flow condition suitable for the available data and compatible with the accuracy required but not overly sophisticated and unnecessarily expensive. Understanding the physical meaning and variations of the coefficients and the terms of the dynamic equations listed in Table 25.2 would help. Some special values of the correction coefficients are listed in Table 25.3. For example, for gradually varying flow, the piezometric pressure distribution is almost constant over the cross section; hence, $K = K' = 1$ is a good approximation. Accordingly, the nonuniform-flow equations in Table 25.2 can be simplified for steady flow such as the momentum equations shown in Table 25.4 and energy equations in Table 25.5.

The rate of change of the *velocity distribution correction factors* β and α with x is usually small unless there is rapid contraction or expansion of the flow along the channel. In most rivers, the value of β lies between 1.05 and 1.10 whereas α lies between 1.15 and 1.25. Chow (1959) related β and α to the ratio of the maximum local velocity to the cross-sectional mean velocity, from which the approximate relationship between α and β is

$$\alpha = 1 + (\beta - 1)\left[2(2 - \beta) + \frac{1}{4\beta}\left(1 + \frac{1}{2\beta} \right)^2 \right] \tag{25.8}$$

Traditionally, the most commonly used flow equations for unsteady flow simulation are the *Saint-Venant equations* which can be written in velocity-depth (nonconservation) form as

$$\frac{\partial Y}{\partial t} + \frac{\partial}{\partial x}(VY) = 0 \tag{25.9}$$

TABLE 25.3 Special Values of Correction Factors in One-Dimensional Momentum and Energy Equations with Gravity-Oriented Coordinates for Incompressible Homogeneous Fluid

Correction factor	Special value	Condition
K	1	Constant piezometric pressure distribution over cross, $P = \gamma y + p$, sectional area A
K	$\cos^2 \theta$	Hydrostatic pressure distribution over rectangular cross section
K'	1	Constant piezometric pressure over A and σ
α	1	Uniform velocity distribution over A
α	1.5429	Parabolic velocity distribution over depth in rectangular cross section
α	2	Linearly varying velocity distribution over depth in rectangular cross section
α	$\dfrac{m}{m+3}\left(1+\dfrac{1}{m}\right)^3$	Power-law velocity distribution $\overline{u}_x/\overline{u}_{x,\, y\,=\,Y} = (y/Y)^{1/m}$ in rectangular cross section
β	1	Uniform velocity distribution over A
β	1.200	Parabolic velocity distribution over depth in rectangular cross section
β	1.333	Linearly varying velocity distribution over depth in rectangular cross section
β	$\dfrac{m}{m+2}\left(1+\dfrac{1}{m}\right)^2$	Power-law velocity distribution in rectangular cross section
β_B	1	Uniform velocity distribution over A
ζ	0	Steady flow
ζ	1	Constant piezometric pressure over A
ζ	$\cos^2 \theta$	Hydrostatic pressure distribution over rectangular cross section
η	1	Constant piezometric distribution over A

TABLE 25.4 Cross-Section Averaged One-Dimensional Momentum Equations for Special Cases of Steady Flow of Incompressible Homogeneous Fluid

Prismatic channel	$\left[K + (K - K')\dfrac{Y}{D} - \mathscr{F}^2\right]\dfrac{dY}{dx} = S_o - S_f - \dfrac{V^2}{g}\dfrac{d\beta}{dx} - Y\dfrac{dK}{dx} + m_q$
Constant piezometric pressure distribution $K = K' = 1$	$(1 - \mathscr{F}^2)\dfrac{dY}{dx} - S_o = (1 - \mathscr{F}^2)\dfrac{dy_s}{dx} - \mathscr{F}^2 S_o$ $= -S_f + \mathscr{F}^2\dfrac{D}{B}\dfrac{\partial B}{\partial x} - \dfrac{V^2}{g}\dfrac{d\beta}{dx} + m_q$
β = constant $K = K' = 1$ Prismatic or wide channel	$\dfrac{dY}{dx} = \dfrac{(S_o - S_f + m_q)}{(1 - \mathscr{F}^2)}$ or $\dfrac{dy_s}{dx} = \dfrac{(\mathscr{F}^2 S_o - S_f + m_q)}{(1 - \mathscr{F}^2)}$
	$\mathscr{F} = \dfrac{V}{\sqrt{gD/\beta}}$ $\qquad m_q = \dfrac{1}{gA}\int_\sigma (\overline{U}_x - 2\beta V)\overline{q}\, d\sigma$

TABLE 25.5   Cross-Section Averaged One-Dimensional Equations from Energy Principle for Special Cases of Steady Flow of Incompressible Homogeneous Fluid

Prismatic channel	$\left(\eta - \dfrac{\alpha V^2}{gD}\right)\dfrac{dY}{dx} = S_o - S_e - \dfrac{V^2}{2g}\dfrac{d\alpha}{dx} - Y\dfrac{d\eta}{dx} + \dfrac{W}{V}$ $\qquad\qquad + \dfrac{1}{Q}\displaystyle\int_\sigma \left(\dfrac{\overline{U_iU_i}}{2g} + \dfrac{\overline{P_L}}{\gamma} - \dfrac{3\alpha}{2g}V^2 - \eta Y\right)d\sigma$
$\eta = 1$ Constant piezometric pressure	$\left(1 - \dfrac{\alpha V^2}{gD}\right)\dfrac{dY}{dx} = S_o - S_e + \dfrac{\alpha V^2}{gA}\dfrac{\partial A}{\partial x}\bigg|_{Y=\text{constant}} - \dfrac{V^2}{2g}\dfrac{d\alpha}{dx} + \dfrac{W}{V} + E_q$
$\eta = 1$ $\alpha = $ constant Prismatic channel and W/V negligible	$\dfrac{dY}{dx} = \dfrac{(S_o - S_e + E_q)}{\left(1 - \dfrac{\alpha V^2}{gD}\right)}$
	$E_q = \dfrac{1}{Q}\displaystyle\int_\sigma \left(\dfrac{\overline{U_iU_i}}{2g} + \dfrac{\overline{P_L}}{\gamma} - \dfrac{3\alpha}{2g}V^2 - Y\right)d\sigma$

$$\frac{1}{g}\frac{\partial V}{\partial t} + \frac{V}{g}\frac{\partial V}{\partial x} + \frac{\partial Y}{\partial x} = S_o - S_f \qquad\qquad (25.10)$$

or in discharge-area (conservation) form as

$$\frac{\partial A}{\partial t} + \frac{\partial Q}{\partial x} = 0 \qquad\qquad (25.11)$$

$$\frac{1}{gA}\frac{\partial Q}{\partial t} + \frac{1}{gA}\frac{\partial}{\partial x}\left(\frac{Q^2}{A}\right) + \frac{\partial y_s}{\partial x} = -S_f \qquad\qquad (25.12)$$

No lateral flow is considered in these equations. For Eq. (25.9), the channel is prismatic or wide rectangular. For the momentum equations [Eqs. (25.10) and (25.12)], the additional simplifying assumptions are (1) uniform velocity distribution of $\overline{u}_x$ over A so that $\beta = 1$ and (2) constant piezometric pressure distribution over A so that $K = K' = 1$.

The hydraulic equations presented so far in this section are for one-dimensional open-channel flow. A technique suggested by Preissmann (Cunge and Wegner, 1964) permits the use of these equations for pressurized full-conduit unsteady flows. The idea is to transform the full-conduit flow into a conceptual open-channel flow by hypothetically introducing a free surface to the flow. This is achieved by introducing a longitudinally continuous, hypothetical, narrow, piezometric slot attached to the top of the conduit at its lower side and open to air at its top (Fig. 25.2). A

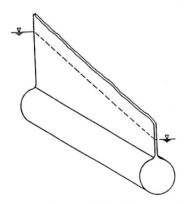

FIGURE 25.2   Hypothetical Preissmann piezometric open slot.

theoretical basis for the determination of the width of the slot b is to consider the surge celerity in the conduit accounting for the compressibility of water and elasticity of the pipe, making the width of the slot such that the wave-propogation celerity in the slot is the same as that of the actual elastic conduit. For more details on determination of the slot width see Yen (1986).

For certain local flow problems, simulation is desired on a depth-averaged, two-dimensional basis instead of cross-sectional averaged, one-dimensional basis. Such two-dimensional continuity and momentum equations are given in Tables 25.6 and 25.7, respectively.

TABLE 25.6   Depth-Averaged Two-Dimensional Continuity Equations for Incompressible Fluid

Unsteady with lateral flow	$\dfrac{\partial Y}{\partial t} + \dfrac{\partial}{\partial x}(YV_x) + \dfrac{\partial}{\partial z}(YV_z) = \overline{q}_Y + \overline{q}_b$
Unsteady, no lateral flow	$\dfrac{\partial Y}{\partial t} + \dfrac{\partial}{\partial x}(YV_x) + \dfrac{\partial}{\partial z}(YV_z) = 0$
Steady, no lateral flow	$\dfrac{\partial}{\partial x}(YV_x) + \dfrac{\partial}{\partial z}(YV_z) + 0$
$V_i = \dfrac{1}{Y}\displaystyle\int_{y_b}^{y_b + Y} \overline{u}_i\, dy$	

25.4   FLOW RESISTANCE

The *flow resistance* in the equations in the preceding section, expressed in terms of S_f according to the momentum concept and S_e from the energy concept, is conventionally related to the flow velocity by using a resistance coefficient. The most commonly used relationships are the *Manning formula*

$$V = \frac{K_n}{n}\, R^{2/3} S^{1/2} \tag{25.13}$$

the *Darcy-Weisbach formula*

$$V = \sqrt{\frac{8g}{f}}\, R^{1/2} S^{1/2} \tag{25.14}$$

and the *Chezy formula*

$$V = C\sqrt{RS} \tag{25.15}$$

in which $R = A/P$ is the hydraulic radius where P is the wetted perimeter of the flow cross section; g is gravitational acceleration; C is the *Chezy coefficient; f* is the *Weisbach coefficient; n* is the *Manning coefficient; K_n* is a constant depending on the measurement units used in the equation; and the slope S denotes either S_f or S_e, depending on whether the momentum or energy equation is used. The values of n, f, or C for steady uniform flow are usually used in these equations as approximations.

TABLE 25.7 Depth-Averaged Two-Dimensional Momentum Equations for Incompressible Homogeneous Fluid

	x-component	z-component
Unsteady flow	$\dfrac{1}{gY}\left[\dfrac{\partial}{\partial t}(V_xY) + \dfrac{\partial}{\partial x}(\beta_{xx}V_x^2Y) + \dfrac{\partial}{\partial z}(\beta_{xy}V_xV_zY)\right] + (2K - K_x')\dfrac{\partial Y}{\partial x}$ $= S_{ox} - S_{fx} - Y\dfrac{\partial K}{\partial x} + \dfrac{1}{\gamma Y}\left(\dfrac{\partial T_{xx}}{\partial x} + \dfrac{\partial T_{xz}}{\partial z}\right) + \dfrac{1}{gY}M_{qx}$	$\dfrac{1}{gY}\left[\dfrac{\partial}{\partial t}(V_zY) + \dfrac{\partial}{\partial x}(\beta_{xz}V_xV_zY) + \dfrac{\partial}{\partial x}(\beta_{zz}V_z^2Y)\right] + (2K - K_z')\dfrac{\partial Y}{\partial z}$ $= S_{oz} - S_{fz} - Y\dfrac{\partial K}{\partial z} + \dfrac{1}{\gamma Y}\left(\dfrac{\partial T_{xz}}{\partial x} + \dfrac{\partial T_{zz}}{\partial z}\right) + \dfrac{1}{gY}M_{qz}$
Steady flow	$(2\beta_{xx} - \beta_{xz})\dfrac{V_x}{g}\dfrac{\partial V_x}{\partial x} + \beta_{xz}\dfrac{V_z}{g}\dfrac{\partial V_x}{\partial z} + \dfrac{V_x^2}{g}\dfrac{\partial \beta_{xx}}{\partial x} + \dfrac{V_xV_z}{g}\dfrac{\partial \beta_{xz}}{\partial z}$ $+ \left[2K - K_x' + (\beta_{xx} - \beta_{xz})\dfrac{V_x^2}{gY}\right]\dfrac{\partial Y}{\partial x} = S_{ox} - S_{fx} - Y\dfrac{\partial K}{\partial x}$ $+ \dfrac{1}{\gamma Y}\left(\dfrac{\partial T_{xx}}{\partial x} + \dfrac{\partial T_{xz}}{\partial z}\right) + \dfrac{1}{gY}[M_{qx} - \beta_{xz}V_x(\bar{q}_Y + \bar{q}_b)]$	$(2\beta_{zz} - \beta_{xz})\dfrac{V_z}{g}\dfrac{\partial V_z}{\partial x} + \beta_{xz}\dfrac{V_x}{g}\dfrac{\partial V_z}{\partial z} + \dfrac{V_z^2}{g}\dfrac{\partial \beta_{zz}}{\partial z} + \dfrac{V_xV_z}{g}\dfrac{\partial \beta_{xz}}{\partial x}$ $+ \left[2K - K_z' + (\beta_{zz} - \beta_{xz})\dfrac{V_z^2}{gY}\right]\dfrac{\partial Y}{\partial z} = S_{oz} - S_{fz} - Y\dfrac{\partial K}{\partial z}$ $+ \dfrac{1}{\gamma Y}\left(\dfrac{\partial T_{xz}}{\partial x} + \dfrac{\partial T_{zz}}{\partial z}\right) + \dfrac{1}{gY}[M_{qz} - \beta_{xz}V_z(\bar{q}_Y + \bar{q}_b)]$
Steady flow with uniform velocity distribution over depth and $K = K' = 1$	$\dfrac{V_x}{g}\dfrac{\partial V_x}{\partial x} + \dfrac{V_z}{g}\dfrac{\partial V_x}{\partial z} + \dfrac{\partial y_s}{\partial x} = -S_{fx} + \dfrac{1}{\gamma Y}\left(\dfrac{\partial T_{xx}}{\partial x} + \dfrac{\partial T_{xz}}{\partial z}\right)$ $+ \dfrac{1}{gY}[M_{qx} - V_x(\bar{q}_Y + \bar{q}_b)]$	$\dfrac{V_x}{g}\dfrac{\partial V_z}{\partial x} + \dfrac{V_z}{g}\dfrac{\partial V_z}{\partial z} + \dfrac{\partial y_s}{\partial z} = -S_{fz} + \dfrac{1}{\gamma Y}\left(\dfrac{\partial T_{xz}}{\partial x} + \dfrac{\partial T_{zz}}{\partial z}\right)$ $+ \dfrac{1}{gY}[M_{qz} - V_z(\bar{q}_Y + \bar{q}_b)]$

$$\beta_{ij}V_iV_jY = \int_{y_b}^{y_b+Y}\bar{u}_i\bar{u}_j\,dy$$

$$K = \frac{1}{\gamma Y^2}\int_{y_b}^{y_b+Y}\bar{P}\,dy$$

$$M_{qi} = (\bar{q}_Y U_{Yx_i} + \bar{q}_b U_{bx_i})$$

$$K_i' = \frac{-1}{\left(\gamma Y\dfrac{\partial Y}{\partial x_i}\right)}\left[\overline{P_Y\frac{\partial \bar{y}_s}{\partial x}} + \overline{P_Y'\frac{\partial \bar{y}_s'}{\partial x_i}} + \overline{P_b\frac{\partial \bar{y}_b}{\partial x_i}} + \overline{P_b'\frac{\partial \bar{y}_b'}{\partial x_i}}\right]$$

$$T_{ij} = \int_{y_b}^{y_b+Y}\bar{\tau}_{ij}\,dy$$

From Eqs. (25.13), (25.14), and (25.15), the resistance coefficients f, n, and C are related as follows:

$$\sqrt{\frac{8}{f}} = \frac{K_n}{\sqrt{g}} \frac{R^{1/6}}{n} = \frac{C}{\sqrt{g}} \tag{25.16}$$

Weisbach's f has the advantage of being dimensionless. Manning's n has the advantage of being nearly independent of flow depth and velocity for fully developed turbulent flow over rough, rigid, impervious boundaries. On the other hand, the only advantage Chezy's C offers is that it is a simple velocity-slope expression as Eq. (25.15), and it is now seldom used in the United States. For a comprehensive view of open channel resistance, refer to Yen (1991).

Values of f for steady uniform flow in rigid, impervious, circular pipes are obtained from the *Moody diagram* which is given in Chap. 2. The corresponding f values for open channels are slightly modified and affected by the cross-sectional geometry. The modified Moody diagram proposed by Yen (1991) is reproduced as Fig. 25.3, in which the Reynolds number is defined as in Eq. (25.4). In the turbulent flow regime, the Moody diagram can be represented by the *Colebrook-White formula*

$$\frac{1}{\sqrt{f}} = -c_1 \log\left(\frac{k_s}{c_2 R} + \frac{c_3}{4\mathcal{R}\sqrt{f}}\right) \tag{25.17}$$

in which k_s is a boundary roughness measurement with a dimension of length, and c_1, c_2, and c_3 are coefficients. Some proposed values of c_1, c_2, and c_3 are given in Table 25.8. To avoid a trial-and-error process to find f using Eq. (25.17), the *Churchill (1973)-Barr (1972; 1977) formula* can be used for steady uniform flow in rigid pipes,

$$f = \frac{1}{4}\left[-\log\left(\frac{k_s}{14.8R} + \frac{5.76}{(4\mathcal{R})^{0.9}}\right)\right]^{-2} \tag{25.18}$$

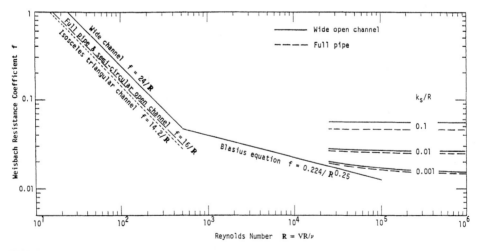

FIGURE 25.3 Weisbach resistance coefficient as function of Reynolds number for steady uniform flow in open channel with impervious rigid boundary (*Yen, 1991*).

TABLE 25.8 Coefficients of Colebrook-White-Type Equations for Steady Uniform Flow in Open Channels with Rigid Impervious Boundary

Channel geometry	Reference	c_1	c_2	c_3	Remarks
Full circular pipe	Colebrook (1938)-White	2.0	14.83	2.52	
Wide channel	Keulegan (1938)	2.03	11.09	3.41	
Wide channel	Rouse (1946, p. 214)	2.03	10.95	1.70	
Wide channel	Thijsse (1949)	2.03	12.2	3.033	
Wide channel	Sayre and Albertson (1961)	2.14	8.888	7.17	
Wide channel	Henderson (1966)	2.0	12.0	2.5	
Wide channel	Graf (1971, p. 305)	2.0	12.9	2.77	
Wide channel	Reinius (1961)	2.0	12.4	3.4	
Rectangular	Reinius (1961)	2.0	14.4	2.9	Width/depth = 4
Rectangular	Reinius (1961)	2.0	14.8	2.8	Width/depth = 2
Rectangular	Zegzhda (1938)	2.0	11.55	0	Dense sand

Yen (1991) proposed a similar formula for two-dimensional flow in wide, rigid, impervious, open channels for $\mathcal{R} > 30,000$ and $k_s/R < 0.05$

$$f = \frac{1}{4}\left[-\log\left(\frac{k_s}{12R} + \frac{1.95}{\mathcal{R}^{0.9}}\right)\right]^{-2} \qquad (25.19)$$

For narrower rectangular channels, the coefficients would be slightly larger than 12 and smaller than 1.95, respectively. Similar adjustments on the coefficients are needed for open channels of other shapes.

In open channels, Manning's n is far more popular than Weisbach's f because for fully developed turbulent flow over an impervious rigid boundary, n is almost invariant with the depth, while f varies with R even for high-Reynolds-number flows (Yen, 1991). In using Eq. (25.13), traditionally the constant K_n is taken as 1.486 for English units and 1.0 for SI units. A recently improved version is to take K_n as $\sqrt{g}$ such that the equation is dimensionally homogeneous and more compatible from the fluid mechanics perspective (Yen, 1992). A summary of these options is given in Table 25.9.

TABLE 25.9 Values and Units of Coefficient K_n in Eq. (25.14)

Value of K_n	Units of K_n	Units used in equation	Units of parameters			Values of n
			V	R	n	
$\sqrt{g}$ or	$\dfrac{m^{1/2}}{s}$	SI	$\dfrac{m}{s}$	m	$m^{1/6}$	Table 2.7 multiplied by $\sqrt{9.81} = 3.13$
	$\dfrac{ft^{1/2}}{s}$	English	fps	ft	$ft^{1/6}$	From Table 2.7 multiplied by 3.82
1	$\dfrac{m^{1/2}}{s}$	Metric	$\dfrac{m}{s}$	m	$m^{1/6}$	Table 2.7
1.486	$\dfrac{ft^{1/3} - m^{1/6}}{s}$	English	fps	ft	$m^{1/6}$	Table 2.7

The values of n as conventionally used with $K_n = 1$ m$^{1/2}$/s for SI units and $K_n = 1.486$ ft$^{1/3}$ − m$^{1/6}$/s for English units are given in Table 2.7 in Chap. 2. The n values for the improved version with $k_n = \sqrt{g}$ can be obtained by multiplying those in Table 2.14 by a constant $\sqrt{g} = 3.132$ for SI units or 3.189 for English units.

Determination of the flow resistance for sediment-laden movable-bed channels is far more involved than that for rigid-boundary channels. It depends on the bed-form, the channel geometry, sediment properties, whether the flow is steady and the sediment transport is in equilibrium, and other factors. Methods to estimate the *flow resistance of sediment-laden channels* can be divided into two groups: (1) those lin-early separating the resistance into two parts, and (2) those estimating the nonlinear resistance relationship without separation. In the linear separation group, the sepa-ration is usually based on wall resistance and form resistance or by plain-bed resis-tance and bedform resistance. A summary of selected linear separation methods is given in Table 25.10, whereas a summary of selected nonlinear methods is given in Table 25.11. Not all methods express the resistance directly in terms of resistance coefficients. Some of them are in terms of shear stress. In these two tables, the sym-bols that have not been previously defined are:

d_s = representative sediment size
d_x = equivalent sediment diameter with x percent by weight of the sediment in the sample being finer in size
h_{bedform} = height of bedform
L_{bedform} = length of bedform
R_b = hydraulic radius of channel bed
S_w = water-surface slope
$u_* = \sqrt{\tau_o/\rho}$, shear velocity
u_{*c} = critical shear velocity for incipient sediment motion
V_T = terminal fall velocity of sediment
δ = thickness of viscous sublayer
ρ_s = mass density of sediment
$\sigma_g = 0.5[(d_{84}/d_{50}) + (d_{50}/d_{16})]$
τ_o = average bed shear stress

A typical example of the linear separation approach in determining the resis-tance coefficient is that of Kennedy, Lovera, and Alam (Lovera and Kennedy, 1969; Alam and Kennedy, 1969) listed in Table 25.10. The Weisbach f is computed as $f' + f''$ where f' is the *plane-bed resistance* from Fig. 25.4, and f'' is the *bedform resistance* from Fig. 25.5.

Perhaps one of the easiest to use in determination of resistance coefficient is the set of nonlinear equations proposed by Camacho and Yen (1991) based on regres-sion of more than seven thousand laboratory and field data. They gave four equa-tions for four different ranges of the flow Froude number,

$$f = 10.7 \frac{T^{*0.35}}{\mathcal{R}^{0.38}} \qquad \text{for} \qquad \mathcal{F} < 0.4 \qquad (25.20)$$

$$f = 0.023 \left(\frac{d_{50}}{R}\right)^{0.26} \frac{\mathcal{R}^{0.10}}{\mathcal{F}^{1.76}} \qquad \text{for} \qquad 0.4 \leq \mathcal{F} < 0.7 \qquad (25.21)$$

$$f = \frac{0.22}{\mathcal{F}^{0.3}} \left(\frac{d_{50}}{R}\right)^{0.33} \qquad \text{for} \qquad 0.7 \leq \mathcal{F} < 1 \qquad (25.22)$$

$$f = \frac{0.231}{\mathcal{F}^{0.887}} \left(\frac{d_{50}}{R}\right)^{0.332} \qquad \text{for} \qquad 1 < \mathcal{F} < 2 \qquad (25.23)$$

TABLE 25.10   Selected Bed-Shear-Based Linear Separation Approaches to Resistance in Sediment-Laden Channels

Investigator	Dependent variables	Independent variables	Data used	Knowledge of bed required for application — Plane	Knowledge of bed required for application — Bedform	Remarks
		$R = R' + R''$				
Einstein and Barbarossa (1952)	V/u_*'	$\dfrac{d_{65}}{R'}, \dfrac{d_{65}u_*'}{11.6\,v}$	F	Yes		Assume log velocity distribution valid
	V/u_*''	$\dfrac{\Delta\rho_s}{\rho}\dfrac{d_{35}}{R'S}$			No	
Vanoni and Brooks (1957)	V/u_*'	$\dfrac{V^3}{gvS}, \dfrac{V}{\sqrt{gk_sS}}$		Yes		Modification of Einstein and Barbarossa on plane bed
	V/u_*''	Same as Einstein and Barbarossa	—		No	
Shen (1962)	V/u_*'	Same as Einstein and Barbarossa	L, F	Yes		Modification of Einstein and Barbarossa on V/u_*''
	V/u_*''	$\dfrac{\Delta\rho_s}{\rho}\dfrac{d_{35}}{R'S}, \dfrac{d_{50}V_T}{v}$			No	
		$S = S' + S''$				
Engelund and Hansen (1967) Engelund (1966)	V/u_*'	$\dfrac{\Delta\rho_s}{\rho}\dfrac{d_{65}}{R}$	L	No		
	V/u_*''	$\mathscr{F}^2\left(\dfrac{Y}{L_{\text{dune}}}\right)\left(\dfrac{H_{\text{dune}}}{Y}\right)^2$			No	
Simons and Richardson (1966)	$\dfrac{C'}{\sqrt{g}} = \dfrac{V}{\sqrt{gRS'}}$	$\dfrac{d_{85}}{Y}$	L	Yes		
	$\dfrac{C''}{\sqrt{g}} = \dfrac{V}{\sqrt{gRS''}}$	$\dfrac{d_{85}}{Y}, \dfrac{\Delta\tau_o}{\tau_o}$			Yes	
Vanoni and Hwang (1967)	f'	$\mathscr{R}, \dfrac{d_{50}}{R}$	L	Yes		
	f''	(Moody diagram) R/(modified dune height)			Yes	
Kennedy, et al. (Lovera and Kennedy, 1968; Alam and Kennedy, 1969)	f'	$\mathscr{R}, \dfrac{R}{d_{50}}$	L, F	Yes		
	f''	$\dfrac{V}{\sqrt{gd_{50}}}, \dfrac{d_{50}}{R}$			No	
Acaroglu (1972)	f'	$\mathscr{R}, \dfrac{d_{50}}{R}$ (Moody diagram)	L, F	No		
	f''	$\dfrac{R}{d_{50}},$ $\phi = C_s\mathscr{F}\dfrac{R}{d_{50}}\dfrac{1}{\sqrt{\Delta\rho_s/\rho}}$			No	

Note: For data used, L = laboratory, F = field
Source: Yen (1991).

TABLE 25.11 Selected Nonlinear Methods for Resistance in Sediment-Laden Channels

Investigator	Dependent variables	Independent variables	Data used	Knowledge of bedform required for application
Based on resistance				
Strickler (1923)	n	d_s	L, F	No
Yen and Liou (1969)	f	$\mathscr{F}, \dfrac{d_{50}}{R}, \mathscr{R}$	L	No
Mostafa and McDermid (1971)	$C_m = \dfrac{\sqrt{g}\,n}{K_n d_{50}^{1/6}}$	$\mathscr{F}, \dfrac{d_{50}}{\delta}$	F, L	No
Griffiths (1981)	f	d_{50}/R or $V/\sqrt{g d_{50}}$	F	Yes (moving bed or not)
Camacho and Yen (1991)	f	$\mathscr{R}, \mathscr{F}, \dfrac{d_{50}}{R}$ or $T^* = S_w \dfrac{R/d_{50}}{\Delta \rho_S/\rho}$	F, L	No
Based on shear				
Raudkivi (1967)	$\dfrac{V}{\sqrt{u_*^2 - u_{*c}^2}}$	$\dfrac{u_*^2}{\dfrac{\Delta \rho_S}{\rho} g d_{50}}$	None	Yes (sediment size)
Yalin (1977)	$\dfrac{C}{\sqrt{g}}$	$\dfrac{u_*^2}{\dfrac{\Delta \rho_S}{\rho} g d_s}, \dfrac{u_*^2 Y}{g d_s^2}$	L	No
Brownlie (1983)	$\dfrac{RS}{d_{50}}$	$\dfrac{VR}{\sqrt{g d_{50}^3}}, S, \sigma_g$	L, F	Yes
van Rijn (1984)	$\dfrac{C}{\sqrt{g}}$	$\dfrac{d_{90}}{R_b}, \dfrac{h_{bedform}}{R_b}, \dfrac{h_{bedform}}{L_{bedform}}$	L, F	Yes
Based on velocity				
Garde and Ranga Raju (1966)	$\dfrac{V}{\sqrt{\dfrac{\Delta \rho_S}{\rho} g d_{50}}}$	$\dfrac{R}{d_{50}}, \dfrac{S}{\Delta \rho_S/\rho}$	L, F	Yes
Based on mixed momentum and energy				
White, et al. (1980; 1987)	f	$\dfrac{\Delta \rho_S}{\rho} \dfrac{g}{v^2} d_{35}^3, \sqrt{\dfrac{\Delta \rho_S}{\rho} g d_s}$	L	No

Note: For data used, L = laboratory, F = field.
Source: Yen (1991).

The parameter T^* in Eq. (25.20) is defined in Table 25.11. The corresponding equations for n in Eq. (25.13) are

$$\frac{n}{R^{1/6}} = \frac{1.16}{c_n} \frac{T^{*0.175}}{\mathscr{R}^{0.19}} \qquad \text{for} \qquad \mathscr{F} < 0.4 \qquad (25.24)$$

$$\frac{n}{d_{50}^{1/6}} = \frac{0.054}{c_n} \left(\frac{d_{50}}{R}\right)^{-0.04} \frac{\mathscr{R}^{0.05}}{\mathscr{F}^{0.88}} \qquad \text{for} \qquad 0.4 \le \mathscr{F} < 0.7 \qquad (25.25)$$

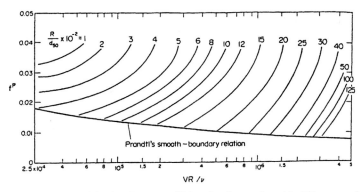

FIGURE 25.4 Weisbach resistance coefficient for plane sediment bed (*Lovera and Kennedy, 1969*).

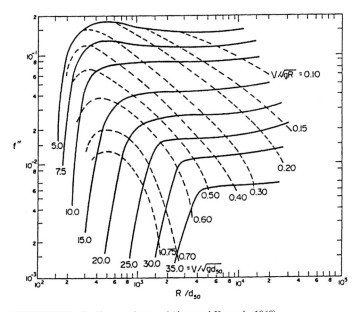

FIGURE 25.5 Bedform resistance (*Alam and Kennedy, 1969*).

$$\frac{n}{d_{50}^{1/6}} = \frac{0.17}{c_n \mathscr{F}^{0.15}} \qquad \text{for} \qquad 0.7 \le \mathscr{F} < 1 \qquad (25.26)$$

$$\frac{n}{d_{50}^{1/6}} = \frac{0.17}{c_n \mathscr{F}^{0.45}} \qquad \text{for} \qquad 1 < \mathscr{F} < 2 \qquad (25.27)$$

in which the dimensionless constant $c_n = 1$ for $K_n = \sqrt{g}$; $c_n = 3.132$ for SI units with $K_n = 1$ m$^{1/2}$/s; and $c_n = 3.819$ for English units with $K_n = 1.486$ ft$^{1/3}$–m$^{1/6}$/s and d_{50} in ft.

TABLE 25.12 Equations for Composite Roughness and Compound Channels

n_c	Concept	Equations	Reference
		Assumptions	
$= \dfrac{PR^{5/3}}{\sum \dfrac{P_i R_i^{5/3}}{n_i}}$	Total discharge is sum of subarea discharges	$S = S_i$ $Q = VA = \sum (V_i A_i)$	Lotter (1933)
$= \dfrac{\sum (n_i P_i R_i^{1/3})}{PR^{1/3}}$	Total shear force $P\sqrt{\gamma RS}$ is sum of subarea shear forces	$S = S_i$ $\sqrt{\gamma RS}\, P = \sum (\sqrt{\gamma R_i S}\, P_i)$ $\dfrac{V}{\sqrt{gRS}} = \dfrac{K_n}{\sqrt{g}}\dfrac{R^{1/6}}{n}$ $\dfrac{V_i}{V} = \left(\dfrac{R_i}{R}\right)^{1/2}$	
$= \dfrac{\sum (n_i P_i / R_i^{1/6})}{P/R^{1/6}}$	Total shear force $P\sqrt{\gamma RS}$ is sum of subarea shear forces	Same as above except $V_i/V = 1$	
$= \left[\dfrac{1}{P}\sum (n_i^2 P_i)\right]^{1/2}$	Total resistance force F is sum of subarea resistance forces ΣF_i	$S = S_i$ $F = \dfrac{\gamma n^2}{K_n^2 R^{1/3}} V^2 P$ $\dfrac{V^2}{R^{1/3}} = \dfrac{V_i^2}{R_i^{1/3}}$	Pavlovskii (1931)
$= \left[\dfrac{1}{P}\sum (n_i^{3/2} P_i)\right]^{2/3}$	Total cross section mean velocity equal to subarea mean velocity	$S = S_i$ $V = V_i$ $A = \Sigma A_i$	Horton (1933) Einstein (1934)
$= \dfrac{P}{\sum (P_i/n_i)}$		Special case of Lotter's Equation with $R_i/R = 1$	Felkel (1960)
$= \dfrac{\sum (n_i P_i)}{P}$	Contribution of component roughness is linearly proportional to wetted perimeter	$n_c P = \sum (n_i P_i)$	
$= \exp\left[\dfrac{\sum P_i h_i^{3/2} \ln n_i}{\sum P_i h_i^{3/2}}\right]$	Logarithmic velocity distribution over depth h for wide channel	$S = S_i,\ Q = \Sigma Q_i$ $\dfrac{Q_i}{2.5\sqrt{gS}} = h_i^{3/2} P_i\left[\ln\left(\dfrac{10.93 h_i}{k_i}\right)\right]$ $\dfrac{Q}{2.5\sqrt{gS}} = \sum h_i^{3/2} P_i\left[\ln\left(\dfrac{10.93 h_i}{k}\right)\right]$ $n = 0.0342\, k$	Krishnamurthy and Christensen (1972)
$= \dfrac{\sum n_i A_i}{A}$			U.S. Army Corps of Engineers Los Angeles District method; see Cox (1973)
$= \left[\dfrac{\sum (n_i^{3/2} A_i)}{A}\right]^{2/3}$	Same as Horton and Einstein's equation but derived erroneously		Colebatch (1941)

Most natural streams are actually channels with composite roughness or with floodplains. Suggested values of n for some composite and compound channels are given in Table 2.14. However, if the constituting component roughness of the local boundary n_i can be determined, the cross-sectional *composite or compound channel* resistance coefficient n_c can be determined by

$$n_c = \int_\sigma w_i n_i \, d\sigma \qquad (25.28)$$

in which w_i is the weighting function. Based on this concept, a number of equations have been proposed, and they are summarized in Table 25.12. A comprehensive comparison of these equations using laboratory and field data has not been performed. Because of the large depth difference in different subsections, and hence differences in local relative roughness, Reynolds number, and Froude number, the flow-geometry effect is more predominant in compound channels than in composite channels. Hence, those equations which involve the parameters R or A are more applicable to compound channels than those not involving these parameters.

25.5 STEADY UNIFORM FLOW

The simplest flow condition shown in Table 25.2 is *steady uniform flow* without lateral flow. For this kind of flow, the water-surface profile is parallel to the channel-bed profile, i.e., $S_o = S_f = S_e = S_w$. Since there is no need to compute the surface profile, steady uniform-flow problems usually involve solving for the discharge, velocity, or depth of the flow, designing for the dimensions of the cross section, or calibration of the resistance coefficient or slope. These different types of problems are listed in Table 25.13. In this table, the question-marked parameter is the parameter to be solved for, whereas the checked parameters are those for which the information is known for the problem.

Steady uniform-flow analysis utilizes the continuity relationship [Eq. 25.1] together with $S_f = S_o$ or $S_e = S_o$. In computations, the cross-sectional area A and hydraulic radius R are both functions of flow depth Y and channel geometry. The relationships between some geometric elements and the depth for some common cross sections are given in Table 25.14. If the flow depth, cross-sectional shape, resis-

TABLE 25.13 Types of Steady Uniform-Flow Problems

Discharge Q	Velocity V	Slope S	Resistance coefficient n or f	Depth Y	Cross section geometry and dimensions
?	—	✓	✓	✓	✓
✓	—	✓	✓	?	?
✓	—	✓	✓	✓	?
✓	—	✓	✓	?	✓
✓	—	✓	?	✓	✓
✓	—	?	✓	✓	✓
—	?	✓	✓	✓	✓
—	✓	?	✓	✓	✓
—	✓	✓	?	✓	✓
—	✓	✓	✓	?	✓

TABLE 25.14   Geometric Elements of Channel Cross Sections

Cross section	Area A	Wetted perimeter P	Hydraulic radius R	Top width B
Rectangle	bY	$b + 2Y$	$\dfrac{bY}{b+2Y}$	b
Trapezoid	$(b+zY)Y$	$b + 2Y\sqrt{1+z^2}$	$\dfrac{(b+zY)Y}{b+2Y\sqrt{1+z^2}}$	$b + 2zY$
Triangle	zY^2	$2Y\sqrt{1+z^2}$	$\dfrac{zY}{2\sqrt{1+z^2}}$	$2zY$
Circle	$\dfrac{1}{8}(\theta - \sin\theta)d_o^2$	$\dfrac{1}{2}\theta d_o$	$\dfrac{1}{4}\left(1-\dfrac{\sin\theta}{\theta}\right)d_o$	$(\sin\tfrac{1}{2}\theta)d_o$ or $2\sqrt{Y(d_o-Y)}$
Parabola $y = cx^2$	$\dfrac{2}{3}BY$	$B + \dfrac{8}{3}\dfrac{Y^2}{B}$ †	$\dfrac{2B^2Y}{3B^2+8Y^2}$ †	$\dfrac{3}{2}\dfrac{A}{Y}$
Power $y = cx^{1/m}$	$\dfrac{BY}{(m+1)}$	$\dfrac{2Y}{m}\sqrt{\dfrac{m^2}{c^{2m}}Y^{2m-2}+1} + \Sigma$ §	$\dfrac{A}{P}$	$2\left(\dfrac{Y}{c}\right)^m$
Round-cornered rectangle $(y>r)$	$\left(\dfrac{\pi}{2}-2\right)r^2 + (b+2r)Y$	$(\pi-2)r + b + 2Y$	$\dfrac{(\pi/2-2)r^2 + (b+2r)Y}{(\pi-2)r + b + 2Y}$	$b + 2r$
Round-bottomed triangle	$\dfrac{B^2}{4z} - \dfrac{r^2}{z}(1 - z\cot^{-1}z)$	$\dfrac{B}{z}\sqrt{1+z^2}$ $-\dfrac{2r}{z}(1 - z\cot^{-1}z)$	$\dfrac{A}{P}$	$2[z(Y-r)$ $+r\sqrt{1+z^2}]$

* For the section factor Z_c, the energy correction factor α or momentum correction factor β are assumed equal to unity. Otherwise, $Z_c = Q/\sqrt{g/\alpha}$ or $Z_c = Q/\sqrt{g/\beta}$.

† Satisfactory approximation for the interval $0 < x \le 1$, where $x = 4Y/b$. When $x > 1$, use the exact expression $P = (B/2)[\sqrt{1+x^2} + 1/x \ \ln(x+\sqrt{1+x^2})]$.

‡ Approximations from Straub as quoted by French (1986). For trapezoid, approximate Y_c valid for $0.1 < Q/b^{2.5} < 0.4$; when $Q/b^{2.5} < 0.1$, use rectangular formula.

tance coefficient and channel slope are known, the discharge can be computed by using the Manning equation

$$Q = \frac{K_n}{n} A R^{2/3} S_o^{1/2} \tag{25.29}$$

in which the product $AR^{2/3}$ for selected geometric shapes is given in Table 25.14 as a function of the depth Y. On the other hand, if the discharge, resistance coefficient n, and channel slope S_o are known, the flow depth Y can be determined by solving for

Hydraulic mean depth D	Uniform flow section factor $AR^{2/3} = \dfrac{Qn}{K_a S^{0.5}}$	Critical flow section factor* $Z_c = Q/\sqrt{g} = A_c\sqrt{D_c}$	Critical depth Y_c
Y	$\left[\dfrac{b^5 Y^5}{(b+2Y)^2}\right]^{1/3}$	$bY^{1.5}$	$\left(\dfrac{Z_c}{b}\right)^{2/3}$
$\dfrac{(b+zY)Y}{b+2zY}$	$\left[\dfrac{(b+zY)^5 Y^5}{(b+2Y\sqrt{1+z^2})^2}\right]^{1/3}$	$\dfrac{[(b+zY)Y]^{1.5}}{\sqrt{b+2zY}}$	$0.81\left(\dfrac{Z_c^4}{z^{1.5}b^{2.5}}\right)^{0.135} - \dfrac{b}{30z}$ ‡
$\dfrac{Y}{2}$	$\left[\dfrac{z^5 Y^8}{4(1+z^2)}\right]^{1/3}$	$\dfrac{\sqrt{2}}{2}zY^{2.5}$	$\left(\dfrac{\sqrt{2}Z_c}{z}\right)^{0.4}$
$\dfrac{1}{8}\left(\dfrac{\theta-\sin\theta}{\sin\frac12\theta}\right)d_o$	$\dfrac{1}{16}\left[\dfrac{(\theta-\sin\theta)^5 d_o^3}{2\theta^2}\right]^{1/3}$	$\dfrac{\sqrt{2}}{32}\dfrac{(\theta-\sin\theta)^{1.5}}{(\sin\frac12\theta)^{0.5}}d_o^{2.5}$	$\dfrac{1.01}{d_o^{0.26}}Z_c^{0.5}$ ‡ for $0.02 \le Y_c/d_o \le 0.85$
$\dfrac{2}{3}Y$	$\dfrac{2}{3}\left[\dfrac{4B^7 Y^5}{(3B^2+8Y^2)^2}\right]^{1/3}$ †	$\dfrac{2\sqrt{6}}{9}BY^{1.5}$	$0.958(c^{0.5}Z_c)^{0.5}$
$\dfrac{Y}{m+1}$	$\left[\dfrac{A^5}{P^2}\right]^{1/3}$	$\dfrac{BY^{1.5}}{(m+1)^{1.5}}$	$\left[\dfrac{c^m}{2}(m+1)^{1.5}Z_c\right]^{1/(m+1.5)}$
$\dfrac{(\pi/2-2)r^2}{b+2r}+Y$	$\left[\dfrac{A^5}{P^2}\right]^{1/3}$	$\dfrac{[(\pi/2-2)r^2+(b+2r)Y]^{1.5}}{\sqrt{b+2r}}$	
$\dfrac{A}{B}$	$\left[\dfrac{A^5}{P^2}\right]^{1/3}$	$A\sqrt{\dfrac{A}{B}}$	

$$\S\ \sum = \left(1-\frac{1}{m}\right)Y\sum_{k=0}^{\infty}\frac{\left(\frac12\right)_k\left(\frac{1}{2m-2}\right)_k\left[-\frac{m}{c^m}Y^{m-1}\right]^{2k}}{\left(1+\frac{1}{2m-2}\right)_k k!},\ \text{where } (w)_k = w(w+1)\cdots(w+k-1),\, k=1,2,\cdots,\, (w)_{k=o}=1$$

Y in the *uniform flow section factor* $AR^{2/3}$ which equals the constant $Qn/K_n S_o^{1/2}$. This solution is usually obtained by trial and error. The steady uniform flow for a specified Q in a given channel is referred to as the *normal flow* for that discharge, and the corresponding depth is called *normal depth*.

Among the different combinations of dimensions of a given cross-sectional geometric shape that can carry a specified discharge, the one with the smallest cross-sectional area is called the *best hydraulically efficient section,* whereas the most economic one is the one having least total cost. The latter is an optimization problem involving costs of various kinds and will not be discussed here. Only the former is considered in the following.

Freeboard is commonly used in channel design. The total cross-sectional area of the channel, including the freeboard, is

$$A_t = A + A_F \tag{25.30}$$

in which A is the flow cross-sectional area and A_F is the area above the freeboard. Both A and A_F, and hence A_t, are functions of the geometric parameters x_i, such as the depth Y, bottom width b, side slope z, and freeboard height F for a trapezoidal channel. The best hydraulic section is obtained by solving $dA_t/dx_i = 0$ and $dQ/dx_i = 0$ together with the relationship from the Manning formula, Eq. (25.29),

$$\frac{A^5}{P^2} = \frac{Qn}{K_n S_o^{1/2}} = \text{constant} \tag{25.31}$$

For a trapezoidal channel with specified side slope z and freeboard F, the relationship between the bottom width b and flow depth Y is

$$\frac{b}{Y} = \frac{1}{5 + 2\dfrac{h}{F}} \left\{ (2\sqrt{1+z^2} - 3z)\frac{h}{F} - (3\sqrt{1+z^2} + 2z) + \left[(5z^2 - 4z\sqrt{1+z^2} + 4)\left(\frac{h}{F}\right)^2 \right. \right.$$

$$\left. \left. + 2(-14z^2 + 19z\sqrt{1+z^2} - 6)\left(\frac{h}{F}\right) - 7z^2 + 32z\sqrt{1+z^2} + 9 \right]^{0.5} \right\} \tag{25.32}$$

For a rectangular cross section ($z = 0$) and no freeboard ($F = 0$), $b/Y = 2$. Chow (1959) gave the geometric parameters of the best hydraulic section without freeboard for six different cross-sectional shapes, and they are listed in Table 25.15. However, it should be checked in design that the best hydraulically efficient cross section does not violate the constraints of soil stability slope, critical tractive force, or maximum permissible velocity.

In open channels, consideration should be given to the erosion problem, especially for soft soils. One approach is to specify a *permissible maximum velocity* in

TABLE 25.15   Best Hydraulically Efficient Sections Without Freeboard

Cross section	Area A	Wetted perimeter P	Hydraulic radius R	Top width B	Hydraulic depth D	$AR^{2/3}$
Trapezoid, half of a hexagon	$\sqrt{3}Y^2$	$2\sqrt{3}Y$	$\frac{1}{2}Y$	$\frac{4}{3}\sqrt{3}Y$	$\frac{3}{4}Y$	$\sqrt{3}\left(\dfrac{Y^8}{4}\right)^{1/3}$
Rectangle, half of a square	$2Y^2$	$4Y$	$\frac{1}{2}Y$	$2Y$	Y	$(2Y^8)^{1/3}$
Triangle, half of a square	Y^2	$2\sqrt{2}Y$	$\frac{1}{4}\sqrt{2}Y$	$2Y$	$\frac{1}{2}Y$	$\frac{1}{2}Y^{8/3}$
Semicircle	$\frac{\pi}{2}Y^2$	πY	$\frac{1}{2}Y$	$2Y$	$\frac{\pi}{4}Y$	$\frac{\pi}{2}(2Y^8)^{1/3}$
Parabola, $B = 2\sqrt{2}Y$	$\frac{4}{3}\sqrt{2}Y^2$	$\frac{8}{3}\sqrt{2}Y$	$\frac{1}{2}Y$	$2\sqrt{2}Y$	$\frac{2}{3}Y$	$\frac{2\sqrt{2}}{3}(2Y^8)^{1/3}$
Hydrostatic catenary	$1.39586Y^2$	$2.9836Y$	$0.46784Y$	$1.917532Y$	$0.72795Y$	$0.84122Y^{8/3}$

channel design. Recommended permissible velocities for selected channel beds are listed in Table 25.16. These values are mostly taken from the 1975 Field Manual of the U.S. Soil Conservation Service and the U.S. Army Corps of Engineers' "Hydraulic Design of Flow Control Channels," Report EM 1110-2-1601, 1970. Another approach for sediment-laden channel beds is to consider the critical shear stress at incipient sediment motion and use a shear stress–sediment size diagram such as the Shields diagram. An alternate, dimensional diagram is given in Fig. 25.6, after Lane (1955). Chien (1956) presented a similar diagram showing a band of tractive force–sediment size curves which are located mostly between those of Kramer (1935) and DuBoys-Straub.

TABLE 25.16   Recommended Permissible Velocities

		Permissible velocity			
	Range of channel slope,	Erosion-resistant soil		Easily eroded soil	
Channel bed cover	%	fps	m/s	fps	m/s
Grass-line earth		(Silt clay)		(Sandy silt)	
	0–5	8	2.5	6	1.8
Bermuda grass	5–10	7	2.0	5	1.5
	>10*	6	1.8	4	1.2
Bahia					
Buffalo grass					
Kentucky bluegrass	0–5	7	2.0	5	1.5
Smooth brome	5–10	6	1.8	4	1.2
Blue grama	>10*	5	1.5	3	0.9
Tall fescue					
Grass mixtures	0–5	5	1.5	4	1.2
Reed canary grass	5–10	4	1.2	3	0.9
Lespedeza sericea					
Weeping lovegrass					
Yellow bluestem					
Redtop					
Alfalfa	0–5	3.5	1.0	2.5	0.8
Red fescue					
Common *Lespedeza*					
Sudan grass					
Earth	0–5	3.5 (Silt clay) 6 (Clay)	1.0 1.8	2.0	0.6
Sand	0–5	4 (Coarse)	1.2	2.0 (Fine)	0.6
Poor rock					
Soft shale		3.5	1.0		
Soft sandstone		8	2.5		
Good rock (usually igneous or hard metamorphic)		20	6		

* Only for vegetated slopes with concrete or stone center section

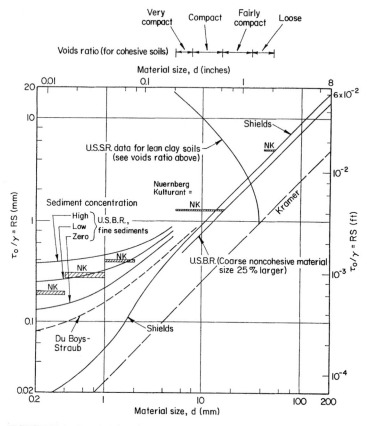

FIGURE 25.6 Permissible tractive force in erodible bed materials.

25.6 NONUNIFORM FLOW WATER-SURFACE PROFILE

Computation of nonuniform flow *backwater-surface profile* involves the following preparational steps before computation actually starts. For a given discharge and channel,

1. Determine the normal flow depth Y_n.
2. Determine the critical depth Y_c.
3. Determine whether the channel is hydraulically mild or steep.
4. Determine the control section(s).
5. Sketch the backwater profile.
6. Determine the direction to initiate the computation: for subcritical flow, the computation proceeds upstream; for supercritical flow, the computation proceeds downstream.

The *normal depth* is determined by solving Eq. (25.1) together with Eqs. (25.13), (25.14), or (25.15) with $S = S_o$. If the Manning formula is used,

$$AR^{2/3} = \frac{Qn}{K_n S_o^{1/2}} \qquad (25.33)$$

in which the right-hand side of the equation is a constant. The term $AR^{2/3}$ is a function of depth Y_n as shown in Table 25.14 for several cross-sectional shapes.

The *critical depth* is determined by

$$\frac{\beta V_c^2}{g} = D_c \qquad (25.34)$$

or

$$D_c^3 = \frac{\beta Q^2}{g B_c^2} \qquad (25.35)$$

in which the hydraulic mean depth D is a function of depth Y (Table 25.14). Accordingly, Y_c can be computed from D_c. For a rectangular channel,

$$Y_c = \sqrt[3]{\frac{\beta q^2}{g}} \qquad (25.36)$$

where $q = Q/B$ is the discharge per unit channel width. For a given Q, the channel is hydraulically mild if $Y_n > Y_c$ and steep if $Y_n < Y_c$.

Control sections are the locations where a unique relationship between depth and discharge exists. Control sections could be a free fall [critical depth, Eq. (25.35)], a dam and reservoir (horizontal or specified water surface at downstream), a sluice gate (critical depth), a weir, an obstacle, change in channel slope or roughness, or change in channel geometry. For subcritical flow $(Y > Y_c)$, the control section is sought at the downstream end to initiate the computation. For supercritical flow $(Y < Y_c)$, an upstream control section is sought. When the flow changes from subcritical to supercritical (*hydraulic drop*) the critical depth acts also as a control section; whereas from supercritical to subcritical (*hydraulic jump*), the equations of hydraulic jump height and length act as the control.

A sketch of the backwater profile is based on the *types of backwater profile* classified according to whether the channel is *mild (M), steep (S), critical (C), horizontal (H)*, or *adverse (A)*, and if Y is greater or less than Y_n and Y_c. The twelve types of backwater curves are shown in Fig. 25.7. These curves are named according to the channel slope and the depth zone. In zone 1, $Y > Y_n$ and $Y > Y_c$; hence $\mathcal{F} < 1$ and S_e or $S_f < S_o$, implying in zone 1 $dY/dx > 0$. In zone 2, Y is between Y_n and Y_c; hence S_e or $S_f > S_o$ and $\mathcal{F} < 1$ for mild, horizontal and adverse slopes, and S_e or $S_f < S_o$ and $\mathcal{F} > 1$ for steep slopes, yielding in zone 2 $dY/dx < 0$. In zone 3, $Y < Y_n$ and $Y < Y_c$; hence $\mathcal{F} > 1$ and S_e or $S_f > S_o$, thus $dY/dx > 0$.

The *sequent depths* before and after a hydraulic jump, Y_1 and Y_2 respectively, can be determined by solving the steady nonuniform flow momentum equation in Table 25.2 taking a control volume bounded by two cross sections having approximately hydrostatic pressure distribution with the jump in between. For a rectangular prismatic channel with a constant slope S_o, the relationship is

$$\frac{Y_2}{Y_1} = \frac{1}{2}\left(\sqrt{1 + 8G_1^2} - 1\right) \qquad (25.37)$$

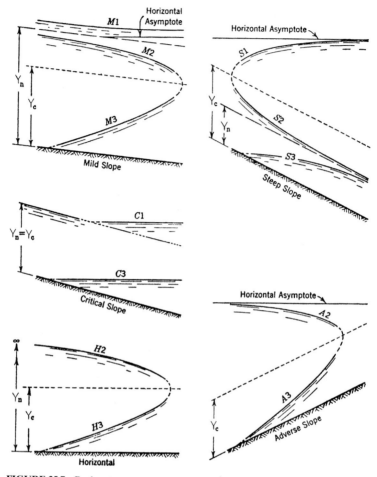

FIGURE 25.7 Backwater curve types.

in which

$$G_i^2 = \frac{1}{\cos^2 \theta} \left[\mathcal{F}_1^2 + \frac{1}{Y_1^2} \int_{x_1}^{x_2} \frac{p_b(x)}{\gamma} \sin \theta \, dx - \frac{1}{Y_1^2} \int_{x_1}^{x_2} \frac{\tau(x)}{\gamma} \cos \theta \, dx \right] \quad (25.38)$$

where p_b is the local pressure normal to channel bed, τ is the bed shear stress, and $\mathcal{F}_1$ is the upstream supercritical flow Froude number.

The pressure term can be approximated as

$$\int_{x_1}^{x_2} \frac{p_b(x)}{\gamma} \sin \theta \, dx = \int_{x_1}^{x_2} Y(x) \cos^3 \theta \, S_o \, dx \quad (25.39)$$

and the bed shear term can be approximated as

$$\int_{x_1}^{x_2} \frac{\tau(x)}{\gamma} \cos\theta \, dx = \int_{x_1}^{x_2} Y(x) \cos^2\theta \, S_o \, dx \qquad (25.40)$$

In most cases, the sum of the two integral terms in Eq. (25.38) is rather small relative to $\mathscr{F}_1^2$. If this is the case, or when the bed shear is neglected and the channel bed is horizontal, Eq. (25.37) can be reduced to

$$\frac{Y_2}{Y_1} = \frac{1}{2} \left(\sqrt{1 + 8\mathscr{F}_1^2} - 1 \right) \qquad (25.41)$$

The *length of the jump* L_j can be estimated by

$$L_j = \frac{10(Y_2 - Y_1)}{\mathscr{F}_1^{0.16}} \qquad (25.42)$$

After the preparational steps 1 to 6 for backwater computation are completed, the actual computation can be done by using one of the many methods that is most suitable for the problem. A number of such methods can be found in Chow (1959) and other references. One of the most direct methods is through integration of the simplified momentum equation given in Table 25.4,

$$\frac{dY}{dx} = \frac{S_o - S_f}{1 - \mathscr{F}^2} \qquad (25.43)$$

However, this method is not very practical in field conditions because the direct integration involves assumptions of simple cross-sectional geometry of the prismatic channel. Those who are interested in using this method should refer to Chow (1959, pp. 252–262).

The computation method that is most popular for field conditions is the so-called *standard step method*. Consider a short reach of the channel of length Δx between sections 1 and 2 as shown in Fig. 25.8. Following the *Bernoulli principle*, equating the total heads at the two sections

$$\frac{\alpha_2 V_2^2}{2g} + (Y_2 + y_{b2}) = \frac{\alpha_1 V_1^2}{2g} + (Y_1 + y_{b1}) + h_e \qquad (25.44)$$

in which the subscripts 1 and 2 denote sections 1 and 2, respectively; $Y + y_b$ is the stage of the water surface where the channel bed stage at section 1 is y_{b1} and that at section 2 is $y_{b2} = y_{b1} + S_o\Delta x$; the energy loss $h_e = S_e\Delta x$ where S_e is the slope of the energy line. If there is any other energy loss h_ℓ, it should also be added to the right-hand side of the equation.

The standard step method compares the cross section 2 total head computed from the left-hand side of Eq. (25.44), H_2, with the total head computed from the right-hand side of the equation, H_{12}, through trial and error until $H_2 = H_{12}$. In hand calculation, this is best done in tabular form. An example table is given in Table 25.17. In this table, H_2 is shown in column 6 which is computed as the sum of column 2 and column 5, and H_{12} in column 14 is computed as columns 12 and 13 added to column 14 of the previous station. The *energy-loss slope* S_e in column 9 is computed according to the Manning formula,

$$S_e = \frac{n^2 V^2}{K_n^2 R^{4/3}} \qquad (25.45)$$

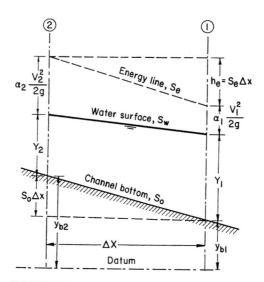

FIGURE 25.8 Channel reach for step method computation.

In the trial process, a likely river stage is assumed, as shown in the second line of numbers in the table. The geometry parameters A, P and R are obtained from the parameter-stage relationships predetermined for the channel at specified sections. Other values in the table are computed accordingly. Line 2 of computational numbers in Table 25.17 is crossed out because in the first trial $H_2 < H_{12}$, and hence a smaller river stage is assumed for the second trial at this station as shown in line 3 of the table.

For an experienced person, the trial required for the standard step method is rather minimal. It may be noted that in the trial-and-error process the true value of the total head H is much closer to H_{12} in column 14 than H_2 in column 6. Nevertheless, the computation is preferably done by using a computer. A computational algorithm is shown in Fig. 25.9 to aid programming. Since the method is applicable to nonprismatic channels, many existing computer programs, such as HEC-2 and WSPRO described in Chap. 28, use this method.

25.7 APPROXIMATE MODELS FOR UNSTEADY-FLOW COMPUTATION

25.7.1 Dynamic Wave Approximation

Mechanics-based unsteady-flow computation is commonly done by using either the Saint-Venant equations, Eqs. (25.9) and (25.10) or (25.11) and (25.12), or the corresponding momentum equation with the velocity distribution correction factor β included, as given in the following in discharge and velocity forms,

TABLE 25.17 Backwater Profile Computation by Standard Step Method

Station x, ft (1)	River stage $(Y+z_b)$, ft (2)	Area A, ft² (3)	Velocity $V=Q/A$, fps (4)	$\frac{\alpha V^2}{2g}$, ft (5)	Total head $H=(2)+(5)$, ft (6)	Wetted perimeter P, ft (7)	Hydraulic radius R, ft (8)	Energy slope S_e, in 10^{-3} (9)	Average $\overline{S}_e$ over reach, in 10^{-3} (10)	Length of reach Δx, ft (11)	Energy loss in reach h_f, ft (12)	Other losses h_L, ft (13)	Total head $H=(14)_1+(12)_2+(13)_2$, ft (14)
0	60.000	200.0	0.500	0.0043	60.004	38.28	5.22	0.0078	…	…	…	…	60.004
1000	*60.200*	*179.5*	*0.557*	*0.0053*	*60.205*	*36.30*	*4.94*	*0.0104*	*0.0091*	*1000*	*0.0091*	*0*	*60.013*
1000	60.008	174.0	0.575	0.0056	60.014	35.76	4.87	0.0113	0.0096	1000	0.0096	0	60.014
3000	60.041	127.3	0.786	0.0106	60.052	30.76	4.14	0.0263	0.0188	2000	0.0376	0	60.052
…	…	…	…	…	…	…	…	…	…	…	…	…	…
…	…	…	…	…	…	…	…	…	…	…	…	…	…
…	…	…	…	…	…	…	…	…	…	…	…	…	…

Trapezoidal channel, $b = 10$ ft, $z = 1$, $S_o = 0.0009$, $\alpha = 1.1$, $n = 0.095$ ft$^{1/6}$ with $K_n = \sqrt{g}$ in Eq. (25.13) (equivalent to $n = 0.025$ m$^{1/6}$
for $K_n = 1.486$ ft$^{1/3} - m^{1/6}$/sec) $Q = 100$ cfs, $Y_n = 1.43$ ft, $Y_c = 2.78$ ft, $Y_c = 1.43$ ft. Initial depth at $x = 0$ is $Y = 10.00$ ft, M1 type profile

25.29

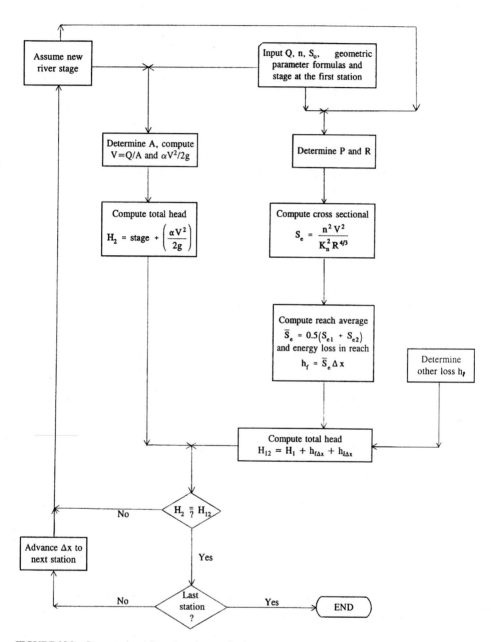

FIGURE 25.9 Computational flow chart for standard step method.

$$\frac{1}{gA}\frac{\partial Q}{\partial t}+\frac{1}{gA}\frac{\partial}{\partial x}\left(\frac{\beta}{A}Q^2\right) \qquad\qquad +\frac{\partial Y}{\partial x}\quad -S_o+S_f=0 \quad (25.46)$$

dynamic wave

quasi-steady dynamic wave

noninertia

kinematic wave

$$\frac{1}{g}\frac{\partial V}{\partial t}+(2\beta-1)\frac{V}{g}\frac{\partial V}{\partial x}+(\beta-1)\frac{V^2}{gA}\frac{\partial A}{\partial x}+\frac{\partial Y}{\partial x}\quad -S_o+S_f=0 \quad (25.47)$$

These equations are also referred to as full (or complete) *dynamic wave equations* (but not exact) because they account for all the inertia, pressure, body force, and resistance dynamic forces. They are also referred to as the shallow-water wave equations in contrast to deep-water wave equations for oceans. Because the *pressure distribution correction factors K and K'* in Table 25.2 are ignored (assumed equal to unity), these equations are unreliable when applied to highly curvilinear flow, such as flow with the Froude number near unity (hydraulic jump and hydraulic drop) and the initial stage of dam-break flow.

Mathematically, the Saint-Venant equations are a pair of first-order partial differential equations of hyperbolic type. No analytical solutions are known for these equations except for a few very special cases for which the equations are considerably simplified and often with linearization. In general, they are solved numerically with differential terms approximated by finite differences of selected grid points on a time-and-space domain such as the one shown in Fig. 25.10. Substitution of the finite differences into a partial differential equation transforms it into an algebraic equation. Thus, the original set of differential equations can be transformed into a set of simultaneous discretized algebraic equations for solution.

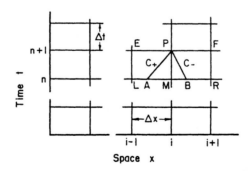

FIGURE 25.10 Computation grid for finite difference schemes.

The dynamic wave equations are rather complicated and it is not a simple task to obtain their numerical solutions. Therefore, various approximations have been proposed to provide simpler but acceptable solutions. Mathematically, a simplification is to linearize the equations. This approach is not very useful for unsteady-flow routing because in the linearization process, small perturbations are often assumed. In practice, if the time variation of the flow is small, the much simpler steady-flow approximation can be used. Another direction of simplification is by dropping the relatively less important terms in the dynamic wave equation before a solution is attempted. This approach leads to the kinematic wave, noninertia, and quasi-steady dynamic wave approximations as shown in Eqs. (25.46) and (25.47).

25.7.2 Initial and Boundary Conditions

Solution of the dynamic wave equations or their approximations can be obtained numerically only if the proper initial and boundary conditions are specified. The set of *initial conditions* of the flow consists of the discharge $Q(x,0)$ or velocity $V(x,0)$ paired with the flow area $A(x,0)$ or depth $Y(x,0)$ specified for the entire channel length at the initial time $t = 0$. Two *boundary conditions* are required for the full dynamic wave equations; they specify the time variations of the discharge, velocity, flow area, or depth at the two ends of the channel. For a subcritical flow, one boundary condition must be specified at the upstream end of the channel, whereas the other must be at the downstream end to reflect the backwater effect. For a supercritical flow, both boundary conditions are specified at the upstream end.

25.7.3 Kinematic Wave Approximation

Referring to Eqs. (25.46) or (25.47), the simplest approximation of the Saint-Venant momentum equation is by keeping only the last two slope terms and ignoring other terms representing the effect of inertia and pressure. This is the *kinematic wave approximation* for which

$$\frac{\partial A}{\partial t} + \frac{\partial Q}{\partial x} = 0 \tag{25.48}$$

$$S_o = S_f \tag{25.49}$$

where S_f can be related to the flow velocity or discharge by using the Manning formula [Eq. (25.13)] or Darcy-Weisbach formula [Eq. (25.14)].

This set of nonlinear kinematic wave differential equations requires only one boundary condition in addition to the initial condition to be properly posed for unique solution. The lone boundary condition is specified at the upstream end of the channel, usually the inflow hydrograph, and hence it permits solution to proceed from upstream towards downstream irrespective of the downstream conditions. On the other hand, it is unreliable for subcritical flow when the downstream backwater effect is important, such as the case of short channels in a network.

Neglecting the inertial and pressure terms in the momentum equation also eliminates the mechanism for flood wave peak attenuation. Theoretically, the continuity equations permits translation and some distortion of the flow hydrograph as the flood propagates downstream (Lighthill and Whitham, 1955). However, in seeking solution through a finite difference numerical procedure, inevitably numerical damping is introduced. This numerical attenuation usually acts advantageously in the same direction as the actual flood peak attenuation, and it is often misunderstood by those who, mistakenly thinking that small computational steps will give more accurate results, reduce the sizes of computational steps in hope of getting more attenuation. In fact, they are incurring more computational expense to obtain a solution that is closer to what the kinematic wave equations represent—no attenuation—and usually farther away from the actual physical condition. In other words, the numerical solution, if converged, gives a value that depends on the grid size Δx and Δt used. There have been attempts to match the numerical attenuation with the real hydraulic attenuation through properly selecting the sizes of Δx and Δt by relating them to the flood and channel properties, e.g., see Koussis (1980); Ponce and Theurer (1982). However, such matching, even carefully executed, is only approximate due to (among other factors) the fact that the flow changes with time and space, whereas in a computation, usually a fixed set of Δx and Δt is used.

Despite its deficiency of not accounting for the downstream backwater effect, the kinematic wave approximation, because of its relative simplicity, is the most extensively studied among the dynamic wave equation and its various approximations. Readers should refer to Bettess and Price (1976); Weinmann and Laurenson (1979); Koussis (1980); Smith (1980); and others for information on the various kinematic wave methods. Among them the most noticeable is the modification suggested by Cunge (1969), who showed that the hydrologic routing Muskingum method can be regarded as a special case of the kinematic wave approximation. The Muskingum coefficients can be computed from the channel width, reach length, discharge, kinematic wave celerity, and slope of the flow. Solution procedures of the *Muskingum-Cunge* method can be found in Fread (1993) and Ponce (1989).

25.7.4 Quasi-Steady Dynamic Wave Approximation

In Eq. (25.46), if only the local acceleration term is dropped, the approximation is a *quasi-steady dynamic wave* for which the continuity and momentum equations in discharge-area form are

$$\frac{\partial A}{\partial t} + \frac{\partial Q}{\partial x} = 0 \tag{25.48}$$

$$\frac{1}{gA} \frac{\partial}{\partial x} \left(\frac{\beta}{A} Q^2 \right) + \frac{\partial y_s}{\partial x} + S_f = 0 \tag{25.50}$$

Solution to this approximation requires two boundary conditions as for the full dynamic wave. For gradually varied unsteady flow in a truly or nearly prismatic channel, the local and convective acceleration terms are usually of the same order in magnitude but have opposite signs. Therefore, for such flow, neglecting only one of them gives worse results than neglecting both.

25.7.5 Noninertia Approximation

The *noninertia* (misnomer diffusion wave) *approximation* is perhaps the most useful among the approximations of the dynamic wave equation because it offers a balance between accuracy and simplicity to a large number of field situations. It ignores both *inertia terms* (local and convective accelerations) in Eq. (27.46) but retains the pressure, body force, and resistance terms. Thus,

$$\frac{\partial A}{\partial t} + \frac{\partial Q}{\partial x} = 0 \tag{25.48}$$

$$\frac{\partial Y}{\partial x} = S_o - S_f \tag{25.51}$$

or, in terms of water-surface elevation y_s,

$$\frac{\partial y_s}{\partial x} + S_f = 0 \tag{25.52}$$

Inclusion of the pressure term requires two boundary conditions. Hence, it accounts for downstream backwater effects and permits peak attenuation, distortion, and translation of the flood hydrograph, just as for the full and quasi-dynamic wave models. Because the inertia terms are ignored, it does not have computational problems when the flow passes through $\mathscr{F} = 1$ from supercritical to subcritical or vice versa, a difficulty that is often encountered in the full or quasi-dynamic wave models. For gradually varied flow in nearly or truly prismatic channels, it is more accurate than the quasi-dynamic wave approximation because the two inertia terms have opposite signs and are similar in magnitude. Readers may refer to Akan and Yen (1977); Strelkoff and Katopodes (1977); and Hromadka and Yen (1987) for versions of the application of the noninertia approximation to open-channels flows.

25.8   FINITE DIFFERENCE NUMERICAL SCHEMES FOR FLOW ROUTING

Hydraulic routing of unsteady open-channel flow is solved by applying a suitable numerical scheme directly to the full dynamic wave equations [such as Eqs. (25.11) and (25.12)] or their simplifications, or alternatively to a set of transformed ordinary differential equations known as the *characteristic equations* (see, for example, Lai, 1986; Wylie and Streeter, 1983). For instance, the characteristic equations transformed from Eqs. (25.9) and (25.10) are

$$\left(\frac{dV}{dt}\right) + \left(\frac{gB}{A}\right)^{1/2}\left(\frac{dY}{dt}\right) + g(S_f - S_o) = 0 \tag{25.53}$$

$$\left(\frac{dx}{dt}\right) = \left[V + \left(\frac{gA}{B}\right)^{1/2}\right] \tag{25.54}$$

$$\left(\frac{dV}{dt}\right) - \left(\frac{gB}{A}\right)^{1/2}\left(\frac{dY}{dt}\right) + g(S_f - S_o) = 0 \tag{25.55}$$

$$\left(\frac{dx}{dt}\right) = \left[V - \left(\frac{gA}{B}\right)^{1/2}\right] \tag{25.56}$$

The finite difference numerical schemes can be classified into two groups: explicit schemes and implicit schemes. Extensive discussion of these numerical schemes can be found in Lai (1986); Abbott, (1979); Mahmood and Yevjevich (1975); Cunge, et al. (1980); Abbott and Basco (1990); Chaudhry (1993); and a large number of references quoted in these books.

25.8.1   Explicit Schemes

In the *explicit schemes,* the differential equations are transferred into finite difference algebraic equations which are arranged such that the unknown parameters are each expressed explicitly as a function of known quantities, allowing them to be solved directly. As an example, consider the continuity equation for a rectangular channel for which the width B is constant, and there is no lateral flow. Furthermore, for simplicity in demonstration, assume the unlikely case where V is constant. Accordingly, the continuity equation in velocity form in Table 25.1 can be simplified as

$$\frac{\partial Y}{\partial t} + V\frac{\partial Y}{\partial x} = 0 \qquad (25.57)$$

Referring to the computational grid shown in Fig. 25.10, the flow variables (i.e., depth Y at all x in this example) are known initially at the grid points L, M, R, and so on, at the time level $n\Delta t$. The unknowns to be found are the depths at the next time level $(n + 1)\Delta t$. Suppose the depth at grid point P, Y_P is sought. By taking time and space backward differences to approximate the derivatives of the depth, and using three grid points, P, M, and L,

$$\left.\frac{\partial Y}{\partial t}\right|_P \approx \left.\frac{\Delta Y}{\Delta t}\right|_P \approx \frac{Y_P - Y_M}{\Delta t} \qquad (25.58)$$

$$\left.\frac{\partial Y}{\partial x}\right|_P \approx \left.\frac{\Delta Y}{\Delta x}\right|_P \approx \left.\frac{\Delta Y}{\Delta x}\right|_M \approx \frac{Y_M - Y_L}{\Delta x} \qquad (25.59)$$

Substituting Eqs. (25.58) and (25.59) into Eq. (25.57) and solving for Y_P yields

$$Y_P = Y_M - \frac{\Delta t}{\Delta x}V(Y_M - Y_L) \qquad (25.60)$$

Thus, the unknown Y_P is solved explicitly in terms of known quantities on the right-hand side of the equation.

Depending on the number and position of the grid points used in expressing the finite difference to approximate the derivatives, there are many different explicit schemes that can solve for the unknown depths, some of them are listed in Table 25.18 according to Lai (1986). Not listed in the table is the *MacCormack* (1969) *scheme* which is a two-step predictor-corrector scheme of second-order accuracy that recently has been applied to open-channel flow problems. In general, explicit schemes are relatively easy to understand, easy to formulate, and easy to program. But they are also computationally highly inefficient because of numerical stability and consistency problems. The solution varies with the value of $\Delta t/\Delta x$ used. For example, using $\Delta t/\Delta x = 0.5$, and assuming $V = 1$, Eq. (25.60) yields $Y_P = 0.5(Y_M + Y_L)$. Likewise, by using $\Delta t/\Delta x = 1$, $Y_P = Y_L$; and with $\Delta t/\Delta x = 2$, $Y_p = Y_M + 2Y_L$. One can easily see that for a given input stage hydrograph and an initial depth profile throughout the channel, the solution for the depths at subsequent time steps is different, depending on the values used for $\Delta t/\Delta x$. Usually, the true solution is unknown, and one would wonder which solution is acceptable and reliable. In this simple example, the case of $\Delta t/\Delta x = 1$ preserves the flood as it moves down the channel, it happens to yield the true solution for this assumed simple example. The case of $\Delta t/\Delta x = 0.5$ introduces artificially numerical damping of the flood peak, whereas the case of $\Delta t/\Delta x = 2$ is numerically unstable. To minimize the numerical instability, the computational grids are usually selected to satisfy the *Courant criterion*,

$$\frac{\Delta x}{\Delta t} \geq V + \sqrt{g\frac{A}{B}} \qquad (25.61)$$

TABLE 25.18 Explicit Finite-Difference Schemes

	Unstable	Diffusive	L-shaped (upstream)	Leap-frog	Lax-Wendroff
Computational grid-point structure (●) Unknown (o, x) Known					
$\Delta x = x_{j+1} - x_j$ $= x_j - x_{j-1}$ $\dfrac{\partial W}{\partial x} \approx \dfrac{W_{j+1}^k - W_{j-1}^k}{2\,\Delta x}$	$\dfrac{\partial W}{\partial x} \approx \dfrac{W_{j+1}^k - W_{j-1}^k}{2\,\Delta x}$ $\dfrac{\partial^2 W}{\partial x^2}$	$\dfrac{W_{j+1}^k - W_{j-1}^k}{2\,\Delta x}$	$\dfrac{W_{j+1}^k - W_j^k}{\Delta x}$ or $\dfrac{W_j^k - W_{j-1}^k}{\Delta x}$	$\dfrac{W_{j+1}^k - W_{j-1}^k}{2\,\Delta x}$	Depends on PDEs used; commonly includes $\dfrac{W_{j+1}^k - W_{j-1}^k}{2\,\Delta}$ $\dfrac{W_{j+1}^k - 2W_j^k + W_{j-1}^k}{(\Delta x)^2}$
$\Delta t = t_{k+1} - t_k$ $= t_k - t_{k-1}$ $\dfrac{\partial W}{\partial t} \approx \dfrac{W_j^{k+1} - W_j^k}{\Delta t}$	$\dfrac{\partial W}{\partial t} \approx \dfrac{W_j^{k+1} - W_j^k}{\Delta t}$	$W_j^{k+1} - \dfrac{W_{j+1}^k + W_{j-1}^k}{2}$ Δt	$\dfrac{W_j^{k+1} - W_j^k}{\Delta t}$	$\dfrac{W_j^{k+1} - W_j^{k-1}}{2\,\Delta t}$	
$W \approx W_j^k$	$W \approx W_j^k$	$\dfrac{W_{j+1}^k + W_{j-1}^k}{2}$ or W_j^k	W_j^k	$\dfrac{W_{j+1}^k + W_{j-1}^k}{2}$ or W_j^k	W_j^k
Discretization expressions	$O[\Delta^2]$	$O[\Delta^2]$	$O[\Delta^2]$	$O[\Delta^3]$	$O[\Delta^3]$
Discretization error	$O[\Delta^2]$	$O[\Delta^2]$	$O[\Delta^2]$	$O[\Delta^3]$	$O[\Delta^3]$

Source: Lai (1986).

25.36

Accordingly, the required Δt is often so small that the computation becomes very costly, yet the solution accuracy cannot be assured. Hence, from a practical viewpoint, the usefulness of the explicit schemes is rather limited.

25.8.2 Implicit Schemes

The *implicit schemes* express the unknown parameters of the differential equations implicitly by a set of simultaneous finite difference algebraic equations and then solve for them using an appropriate solution technique. Again, consider the simple example of Eq. (25.57) and refer to the computational grid shown in Figure 25.10. The time derivative is taken as in Eq. (25.58), while the space derivative is taken as

$$\frac{\partial Y}{\partial x}\bigg|_P \approx \frac{Y_F - Y_E}{2\Delta x} \tag{25.62}$$

Thus, Eq. (25.57) becomes

$$\frac{Y_P - Y_M}{\Delta t} + V\frac{Y_F - Y_E}{2\Delta x} = 0 \tag{25.63}$$

which contains three unknowns Y_P, Y_E and Y_F. Similar equations can be written for other unknown grid points at the same time step $n + 1$. Equation (25.63) and similar equations for other grid points at the same time step form a set of algebraic equations that can be solved simultaneously for the unknowns Y_E, Y_P, Y_F, and so on, with a specified boundary condition and a known initial condition at the time step n.

A popular implicit formulation utilizes four grid points in a box. Consider the points F, P, M and R between i and $i + 1$ in space and n and $n + 1$ in time in the computation grid shown in Fig. 25.10. The time and space derivatives of a variable (or function) y can be approximated as

$$\frac{\partial y}{\partial t} \approx \frac{1}{\Delta t}\{[w_x y_{i+1,n+1} + (1 - w_x)y_{i,n+1}] - [w_x y_{i+1,n} + (1 - w_x)y_{i,n}]\} \tag{25.64}$$

$$\frac{\partial y}{\partial x} \approx \frac{1}{\Delta x}[w_t(y_{i+1,n+1} - y_{i,n+1}) + (1 - w_t)(y_{i+1,n} - y_{i,n})] \tag{25.65}$$

in which w_x and w_t are weighting factors with their respective values between zero and unity; the first subscript of y denotes the space step and the second subscript represents the time step. A special case of Eqs. (25.64) and (25.65) is the *Preissmann* (1961) *scheme* for which $w_x = \frac{1}{2}$. It is also assumed that

$$y(x,t) = \frac{w_t}{2}(y_{i+1,n+1} + y_{i,n+1}) + \frac{1 - w_t}{2}(y_{i+1,n} + y_{i,n}) \tag{25.66}$$

where the value of w_t lies between 0.5 and 1.0. For $w_t = 1$, both y and its space derivative are expressed in terms of the unknowns at the new time step $n + 1$, and this is called a *full implicit scheme*. Substitution of Eqs. (25.64) to (25.66) for variables Q and Y into the Saint-Venant equations, knowing the channel geometry $A = A(Y)$, yields two algebraic equations written symbolically as

$$C_i(Q_{i+1,n+1}, Q_{i,n+1}, Y_{i+1,n+1}, Y_{i,n+1}) = 0 \tag{25.67}$$

$$M_i(Q_{i+1,n+1}, Q_{i,n+1}, Y_{i+1,n+1}, Y_{i,n+1}) = 0 \tag{25.68}$$

where C denotes the finite difference equation from the continuity equation and M the finite difference equation from the momentum equation. Similar algebraic equations are formulated for other grid points (channel reaches) of the channel for the same time step $n + 1$. This set of equations is solved simultaneously for the unknown depth Y and discharge Q at the grid points of the time level $(n + 1)\Delta t$ using known conditions at time step n and specified boundary conditions.

Four different types of implicit schemes are listed in Table 25.19. The implicit schemes are relatively more difficult to formulate and program, but if done properly and carefully, they can be computationally very efficient and stable. The finite difference computational grid sizes Δx and Δt can be chosen independently. Implicit schemes are often quoted as numerically inherently unconditionally stable. This statement is based on the linearized differential equation. For nonlinear differential equations, the numerical stability is, strictly speaking, not unconditional. However, for open channels, it has been found using a $\Delta t / \Delta x$ equal to 20 times the Courant criterion often imposes no serious problems.

25.8.3   Remarks on Numerical Schemes

There are many different versions of the implicit and explicit numerical schemes just mentioned, depending on how the time-space grid is formed, on the number and location of the grid points that are used in the formulation of the finite differences to approximate the derivatives, and on how the nonlinear terms (e.g., convective acceleration) are approximated. Undoubtedly, still more new versions of the schemes will be formed in the future.

There are three general considerations that are useful in selecting a potentially suitable numerical scheme for a channel flow problem. They are computational *stability, convergence,* and *compatibility* (or *consistency*) to the solution of the differential equations, and the computer capacity and time requirements.

Three types of instability problems may be involved in obtaining numerical solutions of the unsteady-flow differential equations. The first type is the natural phenomena of *hydrodynamic instabilities,* such as the transitions between supercritical and subcritical flows and roll waves for high-Froude-number flows. If the hydraulic equations truly represent the flow, under certain situations the solution may become unstable, reflecting faithfully the true physical situation, provided the numerical scheme is also reliable. The second type of instability comes from the assumptions made in deriving the differential equations to represent the physical phenomenon. For instance, the Saint-Venant equations are approximations of the exact equations in Tables 25.1 and 25.2. The assumptions may cause solution instability, although the numerical scheme itself is stable, and the real-world flow is also stable.

The third type of instability is the *numerical instability* coming from numerical operations, irrespective of the other two types of instabilities, although often they interact and make the situation worse. A finite-difference scheme is considered stable if the small truncation and round-off errors do not amplify in successive computations. The explicit schemes have been found conditionally stable, subject to proper selection of Δx and Δt, whereas most of the implicit schemes are regarded as unconditionally stable, and hence, Δt and Δx can be chosen independently, subject to the restrictions imposed by convergence considerations. The stability problem of explicit schemes appears to be less serious when applied to the characteristic equations than directly to the Saint-Venant equations.

TABLE 25.19 Implicit Finite-Difference Schemes

Computational grid-point structure	Box	Rectangle	Wide flange	Tee
(•) Unknown (○) Known $\Delta x = x_{j+1} - x_j$ $\Delta t = t_{k+1} - t_k$				
Discretization expressions	$\dfrac{\partial W}{\partial x} \approx \theta \dfrac{W_{j+1}^{k+1} - W_j^{k+1}}{\Delta x} + (1-\theta)\dfrac{W_{j+1}^k - W_j^k}{\Delta x}$	$\theta \dfrac{W_{j+1}^{k+1} - W_{j-1}^{k+1}}{x_{j+1} - x_{j-1}} + (1-\theta)\dfrac{W_{j+1}^k - W_{j-1}^k}{x_{j+1} - x_{j-1}}$	The same as rectangle schemes	$\dfrac{W_{j+1}^{k+1} - W_{j-1}^{k+1}}{2\Delta x}$
	$\dfrac{\partial W}{\partial t} \approx \psi \dfrac{W_{j+1}^{k+1} - W_{j+1}^k}{\Delta t} + (1-\psi)\dfrac{W_j^{k+1} - W_j^k}{\Delta t}$	$q\dfrac{W_{j+1}^{k+1} - W_{j+1}^k}{\Delta t} + (1-q)\dfrac{W_{j-1}^{k+1} - W_{j-1}^k}{\Delta t}$	$\dfrac{W_j^{k+1} - W_j^k}{\Delta t}$	$\dfrac{W_j^{k+1} - W_j^k}{\Delta t}$
	$W \approx \chi[\psi W_{j+1}^{k+1} + (1-\psi)W_j^{k+1}]$ $+ (1-\chi)[\psi W_{j+1}^k + (1-\psi)W_j^k]$	$\chi[qW_{j+1}^{k+1} + (1-q)W_{j-1}^{k+1}]$ $+ (1-\chi)[qW_{j+1}^k + (1-q)W_{j-1}^k]$	The same as rectangle schemes	W_j^k or W_j^{k+1}
	in which $0.5 \leq \theta \leq 1.0; 0 < \psi < 1.0, 0 < \chi < 1.0$	in which $q = \dfrac{x_j - x_{j-1}}{x_{j+1} - x_{j-1}}$		

Note: Discretization error varies with values and combinations of weighting factors. Aside from stability and other numerical and physical considerations, better accuracy is usually obtained by a more symmetrical arrangement.

Source: Lai (1986).

There exist no universal criteria to determine the sizes of Δx and Δt that would guarantee the stability of an explicit scheme. The Courant criterion, Eq. (25.61), is often used. However, since the velocity and depth of the flow vary with time and location within a channel, selection of the representative velocity and depth that would preserve numerical stability without requiring excessive computation is still a challenge.

The *selection of* Δt is often an unhappy compromise of three criteria. The first criterion is the physically significant time required for the flow to pass through the computational reach. Consider a typical range of channel length between 1 and 100 km and flow velocity of 0.5 to 3 m/s, a suitable computational time interval would be approximately 0.2 to 6 h. For a slowly varying unsteady flow, this criterion is not important and a larger computational Δt will suffice. For a rapidly varying unsteady flow, this criterion should be taken into account to ensure the computation is physically meaningful. The second criterion is a sufficiently small Δt to ensure numerical stability. An often used guide is the Courant criterion, Eq. (25.61). The third criterion is the time interval of the available input data. It is rare to have river-flow data with a time interval less than 1 h. Values for Δt smaller than the data time interval can only be interpolated. This criterion becomes important if the in-between values cannot be reliably interpolated. In a realistic application, all three criteria should be considered. Unfortunately, in many computations only the second, numerical stability, is considered.

A numerical computation without stability problems does not guarantee the solution to be correct. This solution-accuracy problem can easily be demonstrated by examining the simple example explicit solutions of Eq. (25.60) with $\Delta t / \Delta x = 1$ and 0.5. The case of $\Delta t / \Delta x = 0.5$ converges to a solution of the algebraic equation which is not compatible with the correct solution of the original differential equation. The term *compatible* (or *consistent*) in numerical solutions usually refers to the ability of the solution of the finite-difference scheme using finite grid size to approach the solution of the partial differential equation at the point being considered (Smith, 1978), as in the case of Eq. (25.60) with $\Delta t / \Delta x = 1$. From this definition, compatibility or consistency may loosely be considered as an indication of accuracy. However, since the true solutions of the differential equation such as the Saint-Venant equations are usually unknown, satisfactory criteria to assess the accuracy of numerical schemes have yet to be developed. Consequently, a true convergence and compatibility (consistency) test usually cannot be performed, and some indirect means must be adopted to assess the solution reliability. These means include: (1) checking with simple cases for which analytical solutions are known, (2) the capability of the numerical scheme to reproduce the initial condition for the steady state boundary conditions with or without an impulse or disturbance, and (3) the ability in conserving the volume of the flow throughout the computation as used by Sevuk and Yen (1973).

The third consideration on the selection of a numerical scheme is its computer requirements, including the memory storage and processing unit sizes, computation time and costs, difficulty in programming, and transferability of the program. No generally agreed criteria have been established concerning this aspect. Potentially useful yardsticks are also changing with the rapid changes in computer technology in recent years. Loosely, it may be stated that in programming, explicit schemes are easier than implicit schemes. Explicit schemes, because of their numerical-stability constraints, usually require considerably more computer time than implicit schemes. However, implicit schemes' advantage in computer time diminishes for short-duration rapidly varying flows because of the time interval limitation imposed by the accuracy consideration.

25.9 CHANNEL NETWORK

There are two issues that are important in simulation of flow in channel networks. One is the effect of channel junctions. The other is the sequence and technique in handling the computation of the components of the network.

25.9.1 Channel Junctions

The precise hydraulic description of the flow at *channel junctions* is rather complicated and difficult because of the high degree of flow mixing, separation, turbulence, and energy loss. Yet correct representation of the junction hydraulics is important in realistic and reliable computation of flow in river networks. In addition to the continuity relationship, the dynamic relationship can be represented by either the energy or the momentum equations. The momentum equations are vector equations, and the pressure acting on the junction boundaries should be included. The energy equation is usually expressed in one of the following simplified forms:

$$\sum Q_i \left(\frac{V_i^2}{2g} + Y_i + y_{bi} - h_{ei} \right) = Q_o \left(\frac{V_o^2}{2g} + Y_o + y_{bo} \right) \tag{25.69}$$

$$\frac{V_i^2}{2g} + Y_i + y_{bi} - h_{ei} = \frac{V_o^2}{2g} + Y_o + y_{bo} \tag{25.70}$$

in which Q is flow into or from the junction with a velocity V; Y is depth of flow; y_b is elevation of channel bed at the junction; and h_e is loss of energy head. The subscript i indicates the ith inflow channel at the junction; the subscript o represents the outflow channel. The head loss h_{ei} is not generally available, and hence assumptions are necessary to facilitate solutions. For practical purposes, the junctions can be classified into point type and reservoir type depending on whether the junction storage capacity is negligible relative to the volume of the flow (Yen, 1986).

25.9.1.1 Point-Type Junction. Junctions with insignificant storage capacity can be considered as a *point-type junction* which is assumed to be represented by a single confluence point without storage. The net discharge into the junction is therefore zero at all times. Hence,

$$\sum Q_i = Q_o \tag{25.71}$$

For a subcritical flow in an inflowing channel, the flow discharges freely into the junction only when a free fall exists over a nonsubmerged drop at the end of the channel. Otherwise, the subcritical flow in the inflowing channel is subject to backwater effects from the junction. Since the junction is considered as a point, the energy compatibility condition can be represented by a common water surface at the junction for all the joining channels. Thus,

$$Y_i = Y_{ic} \qquad \text{if} \qquad Z_i + Y_{ic} > Y_o + Z_o \tag{25.72}$$

$$Y_i + Z_i = Y_o + Z_o \qquad \text{otherwise} \tag{25.73}$$

in which Y_i = depth of flow (measured vertically) of the ith inflowing channel at the junction; Y_{ic} = critical depth corresponding to the instantaneous flow rate Q_i; Z_i = height of the drop of the ith inflowing channel; and Y_o and Z_o = depth and drop, respectively, of the outflowing channel. Flow in the outflow channel may be either subcritical or supercritical. In the latter case, Y_o in Eq. (25.72) is equal to the critical flow depth Y_{oc}, corresponding to the instantaneous flow rate Q_o.

Flow in the inflow channels can also be supercritical, discharging freely into the junction, provided the flow at the downstream end of the channel is not submerged by the backwater in the junction. For such cases, the discharge of the inflowing channels into the junction can be computed without considering the flow condition in the junction. Subsequently, the discharge into the outflowing channel Q_o can be computed directly by using Eq. (25.71).

25.9.1.2 Reservoir-Type Junction. The *reservoir-type junction* has a relatively large storage capacity in comparison to the flow. Consequently, it can be assumed to behave like a reservoir with a horizontal water surface and capable of absorbing and dissipating all the kinetic energy of the inflows. The net discharge into the junction is equal to the time rate of change of storage in the junction, i.e.,

$$\sum Q_i - Q_o = \frac{ds}{dt} \tag{25.74}$$

in which s = water stored in the junction. The stage of water surface H in the junction is assumed equal to the total energy head of the flow at the entrance of the outflowing channel, i.e.,

$$H = Y_o + \frac{V_o^2}{2g} + Z_o \tag{25.75}$$

Since the kinetic energy of the inflows is assumed lost at the junction, for subcritical flow in the inflow channels,

$$Z_i + Y_i = H \qquad \text{if} \qquad Z_i + Y_{ic} < H \tag{25.76}$$

$$Y_i = Y_{ic} \qquad \text{otherwise} \tag{25.77}$$

If the flow in the outflow channel is supercritical, critical flow condition exists at its entrance, and hence H in Eq. (25.76) should be replaced by the minimum specific energy corresponding to the instantaneous flow rate Q_o.

As in the case of point-type junctions, supercritical flow in the inflowing channels discharges freely into the reservoir-type junction provided the inflow is not submerged by the backwater from the junction, and the discharge from the inflowing channels into the junction can be computed without considering the existing flow conditions in the junction or outflowing channel.

25.9.2 Simulation of Flow in a Channel Network

A *channel network* can be considered as a number of nodes joined together by a number of links. The nodes are the junctions and network outlets. The links are the channels. Depending on their locations in the network, the nodes and links can be classified as *exterior* or *interior*. The exterior links are the most upstream channels or the last channels having the network exits at their respective downstream end. An exterior channel has only one end connected to the other channels. Interior links are the channels inside the network that have both ends connected to other channels. Exterior nodes are the junctions connected to the upstream end of the most upstream channels, or the exit node of the network. An exterior node has only one link connected to it. Interior nodes inside the network have more than one link connected to each node.

A systematic numeric representation of the nodes or links is important for computer simulation of a network. One approach is to number the links (channels) according to the branches and the order of the channels in the branch, similar to Horton's numbering of river networks. Another approach is to identify the links by the node numbers at the two ends of the link.

When applying the flow equations to a channel network, each channel between junctions is divided into a number of reaches. The continuity and momentum equations are applied to each reach and the channels are related through the interior node conditions. However, the flow conditions at the interior nodes are usually a part of the solution being sought. Consider, for example, the subcritical flow in a channel that is divided into two computational reaches and three computational stations as shown in Fig. 25.11. There are six equations for this channel, two continuity

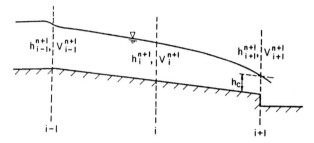

FIGURE 25.11 Subcritical flow in a channel with two computational reaches.

(identified as $C1$ and $C2$ for the first and second reaches, respectively), two momentum ($M1$ and $M2$) and two boundary conditions, one at the channel upstream entrance (BC1) and the other at the channel exit (BC2). An implicit solution of these equations in the form of Eqs. (25.67) and (25.68) using the gaussian elimination technique requires solving a banded matrix like the one given in Eq. (25.78),

$$
\begin{bmatrix}
\dfrac{\partial BC1}{\partial Y_{i-1}^{n+1}} & \dfrac{\partial BC1}{\partial V_{i-1}^{n+1}} & 0 & 0 & 0 & 0 \\[3mm]
\dfrac{\partial C1}{\partial Y_{i-1}^{n+1}} & \dfrac{\partial C1}{\partial V_{i-1}^{n+1}} & \dfrac{\partial C1}{\partial Y_{i}^{n+1}} & \dfrac{\partial C1}{\partial V_{i}^{n+1}} & 0 & 0 \\[3mm]
\dfrac{\partial M1}{\partial Y_{i-1}^{n+1}} & \dfrac{\partial M1}{\partial V_{i-1}^{n+1}} & \dfrac{\partial M1}{\partial Y_{i}^{n+1}} & \dfrac{\partial M1}{\partial V_{i}^{n+1}} & 0 & 0 \\[3mm]
0 & 0 & \dfrac{\partial C2}{\partial Y_{i}^{n+1}} & \dfrac{\partial C2}{\partial V_{i}^{n+1}} & \dfrac{\partial C2}{\partial Y_{i+1}^{n+1}} & \dfrac{\partial C2}{\partial V_{i+1}^{n+1}} \\[3mm]
0 & 0 & \dfrac{\partial M2}{\partial Y_{i}^{n+1}} & \dfrac{\partial M2}{\partial V_{i}^{n+1}} & \dfrac{\partial M2}{\partial Y_{i+1}^{n+1}} & \dfrac{\partial M2}{\partial V_{i+1}^{n+1}} \\[3mm]
0 & 0 & 0 & 0 & \dfrac{\partial BC2}{\partial Y_{i+1}^{n+1}} & \dfrac{\partial BC2}{\partial V_{i+1}^{n+1}}
\end{bmatrix}
\begin{bmatrix}
\Delta Y_{i-1}^{n+1} \\[3mm]
\Delta V_{i-1}^{n+1} \\[3mm]
\Delta Y_{i}^{n+1} \\[3mm]
\Delta V_{i}^{n+1} \\[3mm]
\Delta Y_{i+1}^{n+1} \\[3mm]
\Delta V_{i+1}^{n+1}
\end{bmatrix}
=
\begin{bmatrix}
-BC1_r \\[3mm]
-C1_r \\[3mm]
-M1_r \\[3mm]
-C2_r \\[3mm]
-M2_r \\[3mm]
-BC2_r
\end{bmatrix}
$$

(25.78)

in which $C1_r, C2_r, M1_r, M2_r, BC1_r,$ and $BC2_r$ are the numeric iteration residuals of the corresponding equations, and the time level is denoted as superscripts whereas the space level as the subscripts. For an exterior channel, one of the boundary conditions is specified, but the other is actually unknown, shared through the junction with the other channels joining the same junctions and to be solved together. For an interior channel, both boundary conditions are unknowns to be shared and solved together with other channels.

Thus, for a simple network of a single three-way junction having each channel divided into two computational reaches, the matrix of the equations to be solved for the 18 unknowns is shown in Fig. 25.12. In general, if there are N channels in the network and each channel is divided into M reaches, there are $N(2M + 2)$ algebraic equations for the $N(2M + 2)$ unknowns.

It is obvious that the number of unknowns and equations to be handled can easily become large for a network consisting of many channels. Consequently, the method of solution becomes important in achieving an efficient solution of the entire network. The network solution techniques can be classified into four groups as follows (Yen, 1979).

25.9.2.1 Cascade Method.

In this method, the solution is sought channel by channel starting from the most upstream channels. Each channel is solved for the entire duration of the flood before moving downstream to solve for the immediately following channel; i.e., the entire inflow hydrograph is routed through the channel before the immediate downstream channel is considered. For the noninertia, quasi-steady dynamic wave and dynamic wave equations, this cascading routing is possible only for the following two conditions: (1) the flows in all the channels are supercritical throughout, or (2) the exit-flow condition of each channel is specified, independent of the downstream junction condition, e.g., a free fall over a drop of sufficient height at the downstream end of each channel, such that BC2 in Eq. (25.78) is specified, and the solution for the channel can be obtained.

However, in some channel models, the downstream condition of the channel is arbitrarily assumed, e.g., using a forward or backward difference and approaching normal flow, so that the cascading routing computation can proceed, and the solution, of course, bears no relation to reality. Conversely, in the kinematic wave approximation, only one boundary condition is needed, and it is usually the inflow discharge or depth hydrograph of the channel. At the junction, only the downstream channel dynamic equation and junction continuity equations are used. The junction dynamic equations for the upstream channels are ignored. Thus, the computation can proceed downstream channel by channel in a cascading manner, completely ignoring the downstream backwater effects. This method of solving each channel individually for the entire hydrograph before proceeding to solve for the next downstream channel is relatively simple and inexpensive to execute. But it is inaccurate if the downstream backwater effect is significant.

25.9.2.2 One-Sweep Explicit Solution Method.

In this method, the flow equations of the channels and junctions are formulated by using an explicit finite difference scheme such that the flow depth, discharge, or velocity at a given computation station $x = i\Delta x$, and current time level $t = n\Delta t$ can be solved explicitly from the known information at the previous locations $x < i\Delta x$ at the same time level as well as known information at the previous time level $t = (n-1)\Delta t$. Thus, the solution is sought reach by reach, channel by channel, and junction by junction, individually from upstream to downstream, over a given time level for the entire network before progressing to the next time level for another sweep of individual solutions of the channels and junc-

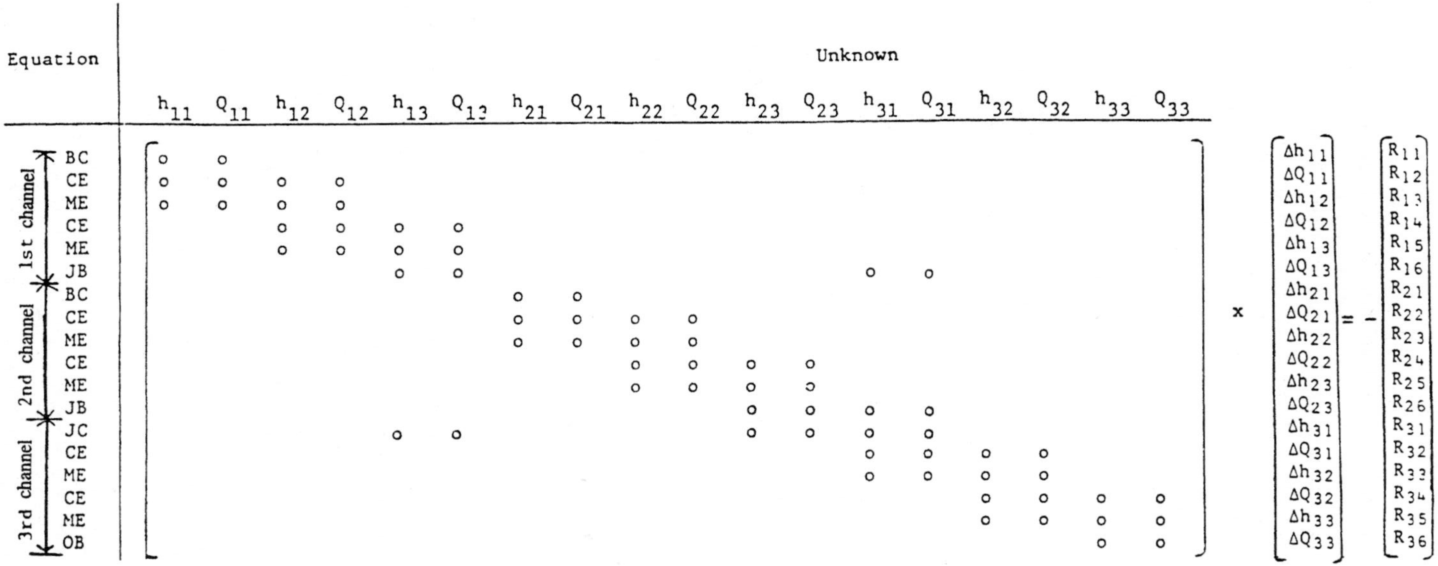

Equation | Unknown

Columns (Unknown): h_{11} Q_{11} h_{12} Q_{12} h_{13} Q_{13} h_{21} Q_{21} h_{22} Q_{22} h_{23} Q_{23} h_{31} Q_{31} h_{32} Q_{32} h_{33} Q_{33}

Equations (rows):

1st channel: BC, CE, ME, CE, ME, JB

2nd channel: BC, CE, ME, CE, ME, JB

3rd channel: JC, CE, ME, CE, ME, OB

Unknown vector:
$\Delta h_{11}, \Delta Q_{11}, \Delta h_{12}, \Delta Q_{12}, \Delta h_{13}, \Delta Q_{13}, \Delta h_{21}, \Delta Q_{21}, \Delta h_{22}, \Delta Q_{22}, \Delta h_{23}, \Delta Q_{23}, \Delta h_{31}, \Delta Q_{31}, \Delta h_{32}, \Delta Q_{32}, \Delta h_{33}, \Delta Q_{33}$

Residual vector:
$R_{11}, R_{12}, R_{13}, R_{14}, R_{15}, R_{16}, R_{21}, R_{22}, R_{23}, R_{24}, R_{25}, R_{26}, R_{31}, R_{32}, R_{33}, R_{34}, R_{35}, R_{36}$

BC=upstream inflow boundary condition; CE=continuity equation; ME=momentum equation;

JB=interior junction boundary condition; JC=interior junction continuity equation; OB=outlet boundary condition; R=residual.

FIGURE 25.12 Implicit scheme matrix configuration for three-way junction open channels.

tions of the entire network. In this method, only one or a few equations are solved for each station, avoiding the large matrix manipulation as in the implicit simultaneous solution of the entire network, and computer programming is relatively direct and simple compared to the simultaneous solution and overlapping segment methods to be discussed later. Nevertheless, this one-sweep method bears the drawback of computational stability and accuracy problems of explicit schemes that have been discussed previously. Variations of this one-sweep approach do exist. For example, each reach of a channel can be solved explicitly, or the channel is solved implicitly using Eq. (25.70) or similar equations, and then the junction flow condition is solved explicitly. Examples of slightly improved variations can be found in Harris (1968); Larson, et al. (1971); Sevuk (1973); Murota, et al. (1973); and Hsie, et al. (1974).

25.9.2.3 Simultaneous Solution Method. When the implicit difference formulations of the dynamic wave, quasi-steady dynamic wave, or noninertia equations and the corresponding junction equations are applied to the entire channel network and solved for the unknowns of flow variable together, simultaneous solution is ensured. The simultaneous solution usually involves solving a matrix similar to that in Fig. 25.11, but much bigger for most networks. There are numerically stable and efficient solution techniques for the sparse matrices of implicit schemes. Nonetheless, since the matrix is not banded, solution of the sparse matrix may still require a large amount of computation and large computer capacity if the network is large.

25.9.2.4 Overlapping Segment Method. To avoid the costly implicit simultaneous solution for large networks and still preserve the advantages of stability and numerical efficiency of the implicit schemes, Yen (1973) and Sevuk (1973) proposed the use of a technique called the *overlapping segment method*. Similar to the well-known Hardy Cross method for solving flow in distribution networks, the method is a successive iteration technique. Unlike the Hardy Cross method, which applies only to looped networks, the overlapping segment method can be applied to dendritic as well as loop-type networks. It decomposes the network into a number of small, overlapped, subnetworks or segments, and solutions are sought for the segments in succession.

A simple, single-step overlapping segment example of a network consisting of three segments is shown in Fig. 25.13. Each segment is formed by a junction together with all the channels joined to it. Thus, except for the most upstream and downstream channels, each interior channel belongs to two segments—as a downstream channel for one segment and then as an upstream channel for the sequent segment, i.e., "overlapped." Each segment is solved as a unit.

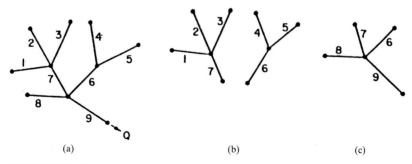

(a)   (b)   (c)

FIGURE 25.13   Network solution procedure by overlapping segment method.

The flow equations are first applied to each of the branches of the most upstream segment for which the upstream boundary condition is known and solved numerically and simultaneously with appropriate junction equations for all the channels in the segment. If the flow is subcritical and the boundary condition at the exit of the downstream channel of the segment is unknown, the forward or backward differences, depending on the numerical scheme, are used as an approximate substitution. Simultaneous numerical solution is obtained for all the channels and junctions of the segment for each time step, repeating until the entire flow duration is completed.

For example, for the network shown in Fig. 25.13a, solutions are first sought for each of the two segments shown in Figure 25.13b. Since the downstream boundary condition of the segment is assumed, the solution for the downstream channel is doubtful and discarded, whereas the solutions for the upstream channels are retained. The computation now proceeds to the next immediate downstream segment (e.g., the segment shown in Fig. 25.13c). The upstream channels of this new segment were the downstream channels of each of the preceding segments for which solutions have already been obtained. The inflows into this new segment are given as the outflow from the junctions of the preceding segments. With the inflows known, the solution for this new segment can be obtained. This procedure is repeated successively, segment by segment, going downstream until the entire network is solved. For the last (most downstream) segment of the network, the prescribed boundary condition at the network exit is used.

The method of overlapping segments reduces the computer size and time requirements when solving for large networks. It accounts for downstream backwater effect and simulates flow reversal, if it occurs. Its accuracy and practical usefulness have been demonstrated by Sevuk (1973) and Yen and Akan (1976). Unless the length of an interior channel is very short, the computed result using the overlapping segment method is practically the same as the simultaneous solution of the entire network.

Solution by the single-step overlapping segment method accounts for the downstream backwater effect of subcritical flow only for the adjacent upstream channels of the junction, but it cannot reflect the backwater effect from the junction to channels farther upstream if such case occurs. However, by considering the length to depth ratio of actual channels, in most cases, the effect of backwater beyond the immediate upstream branches is small, and hence, this imposes no significant error in routing of channel flows. For the rare case of two junctions being closely located, the overlapping segment method can be modified to perform double iteration by recomputing the upstream segment after the approximate junction condition is computed from the first iteration of the downstream segment. Alternatively and perhaps better, a two-step overlapping can be achieved by making a segment containing two neighboring junctions, forming the segment with three levels of channels instead of two and retaining only the solutions of the top level channels for each iteration. Thus, interior channels are iterated twice. The overlapping segment method can be modified to account for divided channels or loop networks in addition to tree-type networks.

25.10 BASIC PRINCIPLES FOR SELECTION OF HYDRAULIC LEVEL OF EQUATIONS FOR SOLVING PROBLEMS

With so many models and methods of different hydraulic levels available for the analysis of flow in channels, it is important in solving a problem to select the proper hydraulic equations that are not unnecessarily complicated and sophisticated requiring excessive amounts of computation or are incompatible with the available data. Conversely, the equations should not be too simple for the problem such that they cannot yield the scope or accuracy of information needed in solving the problem.

The appropriate level of hydraulic analysis to be selected for a water problem depends on a number of factors, including:

- Objective and approach taken—structural or nonstructural
- Information sought
- Accuracy required
- Tools available
- Flow conditions

In most problems, one-dimensional analysis solving for channel cross-sectional averaged quantities suffices. Sometimes when local information is sought, such as analyses for local scour and pollutant disposal and transport, or the variation of water surface in a channel bend, a local two-dimensional analysis is warranted, with flow averaged over the depth but variable along and across the channel (equations in Tables 25.6 and 25.7), or averaged across the channel to study the variation along the channel and over the depth. Two-dimensional simulation should be performed only to the spatial extent that is necessary, whereas the one-dimensional simulation beyond this local region provides the boundary conditions for the two-dimensional computation. Only in rare cases is three-dimensional simulation required.

Unsteady-flow analysis is far more complicated than the steady-flow analysis in the same channel. If the flow variation within Δt is not small (e.g., exceeding 5 percent), unsteady-flow routing may be justified. Conversely, for a slowly varying flow, a quasi- (stepwise) steady-flow routing with a Δt satisfying the time required to travel through Δx is sufficient to provide the required accuracy with far less computational effort and fewer numerical problems. This argument holds for the dynamic wave as well as noninertial and kinematic wave models.

If unsteady-flow analysis is desired, the exact and complete equations given in Table 25.2 are seldom required. They are useful for spatially and temporally rapidly varying flow, such as flow passing through a critical depth, and flow with rapid expansion and separation. For spatially gradually varied flow, the pressure distribution correction factors K and K' are essentially constant and near unity, and the Saint-Venant equations are good approximations.

Among the terms in the full dynamic wave equation, the acceleration terms are usually one or more order of magnitude smaller than the pressure term, which in turn is smaller than the two slope terms (Xia, 1992). Thus, for a majority of unsteady-flow cases, the noninertial approximation, which accounts for downstream backwater effects, suffices. The kinematic wave model is a good approximation only when the downstream backwater effect is small, a rather severe constraint for flow in channel networks. Ponce, et al. (1978) suggested that the kinematic wave approximation is applicable to channel flow when $T_r S_o V/Y > 85$ where T_r is the time-of-rise of the inflow hydrograph. Woolhiser and Liggett (1967) recommended a criterion $S_o L/F^2 Y_e > 20$ for flow on overland planes with a length L and equilibrium flow depth Y_e at the outlet. Xia (1992) provided values on terms in the dynamic wave equations that are helpful in selecting the approximation models. Table 25.20 summarizes major hydraulic capabilities of these approximations.

Undoubtedly, there are a few who think that with today's computational tools, the dynamic wave approximation is preferred because it is worthwhile to pay for the

TABLE 25.20 Theoretical Comparison of Approximations to Dynamic Wave Equation

	Kinematic wave	Noninertia	Quasi-steady dynamic wave	Dynamic wave
Boundary conditions required	1	2	2	2
Account for downstream backwater effect and flow reversal	No	Yes	Yes	Yes
Damping of flood peak	No	Yes	Yes	Yes
Account for flow acceleration	No	No	Only convective acceleration	Yes

computational penalty in order to gain a small theoretical accuracy over the more practical noninertial approximation. However, inadequate information on junction losses and data errors cause uncertainties far exceeding the small accuracy that might be gained by using the dynamic wave model. Moreover, the numerical solution of the dynamic wave model is often less robust than that of the noninertial model. Besides, in routing flow through a hydraulic jump or hydraulic drop, both models are approximations; the dynamic wave may have numerical difficulties while the noninertia will not.

Restrictions on the selection of the computational grid sizes Δx and Δt also affect the selection of the hydraulic equations. Consider, for instance, open-channel flow in a 300-ft- (90-m-) long sewer pipe having a velocity $V = 5$ fps (1.5 m/s). The time for the water to travel through the entire length is 300/5 = 60 seconds. Thus, the time step Δt used in the unsteady-flow routing should be in the order of a minute or less. If the channel length is subdivided into several computational reaches, Δt is even shorter. Measured rainfall and runoff data rarely have such short time resolution. If an explicit numerical scheme is used, the computational stability requirement usually limits Δt to seconds. Accordingly, the travel time requirement is satisfied but a penalty is paid for the large amount of computations. If an implicit scheme, such as the Preissmann four-point scheme, is used, often the model employs a Δt much larger than one minute, violating the travel time Δt constraint, making the computation physically improper. However, if a small Δt compatible with the travel time is used, the computational penalty is high.

25.11 DETERMINATION OF CHANNEL CARRYING CAPACITIES

On many occasions, it is desirable to know the capacity of the flow that a channel can carry. In drainage problems, how much water a channel can carry—the capacity—is a hydraulic problem, whereas how much water the channel is supposed to carry is a hydrology problem. The *channel capacity* can be defined as the maximum flow that the channel can carry with the water being confined within the channel without spilling over the banks into the floodplains and compatible with the downstream stage or discharge constraint, if any. Unsteady flow in a channel is a function of the inflow hydrograph, and the discharge is not a unique function of the upstream and downstream stages of the channel reach; hence, there is no unique unsteady-flow capacity of a channel. Thus, the channel capacity is for the steady-flow condition; in most cases, it is a good indication of unsteady-flow channel capacity.

If the flow is steady and uniform, it is rather easy to compute the channel capacity as described in Sec. 25.5, Eq. (25.29). However, for many channels, the slopes of the banks are not necessarily parallel to the channel bottom. Or, the flow in the channel or channel network is constrained by the water-surface elevation at the outlet. In such cases, the channel capacity is determined with the nonuniform flow profile that carries the maximum discharge satisfying the constraints. A recently developed channel hydraulic-performance graph method (Yen, 1987; Yen and Gonzalez, 1994) permits fast, direct determination of channel capacity.

Hydraulic-performance graph is a set of curves of different constant Q's; each curve indicates the pairs of water stages at the upstream and downstream ends of the channel corresponding to the given Q. An example is shown in Fig. 25.14. The hydraulic performance curve is an extension and refinement of the delivery curve originally introduced by Bakhmeteff (1932) to study nonuniform flow in a channel between two reservoirs.

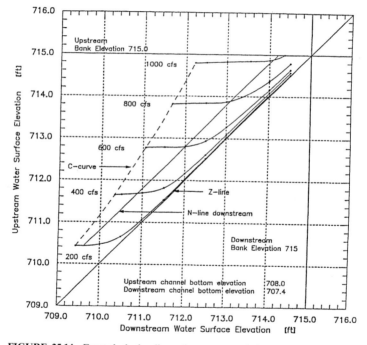

FIGURE 25.14 Example hydraulic performance graph for Boneyard Creek at Urbana, Illinois.

Figure 25.14 shows the hydraulic performance graph for a reach of the Boneyard Creek between stations 9052 and 9256 in Urbana, Illinois. The N-line in the graph corresponds to steady uniform flow (or normal flow) for different discharges in the channel. The Z-line represents the situation of the same water elevation at both ends of the channel, i.e., horizontal water surface and hence, no flow. Both N- and Z-lines are 45° lines. The horizontal distance between these two lines is equal to $S_o \Delta x$. The C-curve represents critical depth of the flow at the downstream end of the channel.

For the mild slope region between the *C*- and *Z*-curves, the backwater profiles between *Z* and *N* are *M*1 type, and those between *N*- and *C*-curves are *M*2 type.

For the example channel reach shown in Fig. 25.14, the bankfull steady uniform flow capacity is 1035 cfs at the intersection point of the *N*-line with upstream bankfull stage at 715.0 ft. The maximum channel capacity is 1045 cfs at the intersection of the *C*-curve with the upstream bank elevation 715.0, i.e., upstream end at bankfull stage of 715.0 ft, and downstream end stage at 712.1 ft corresponding to critical depth of 4.7 ft at the exit.

An expanded set of four-quadrant nondimensional curves summarizing the backwater information for open-channel flow in circular pipes is shown in Fig. 25.15 in which *D*, *L*, and S_o are, respectively, the diameter, length, and slope of the pipe; Q_f is the steady, uniform just-full pipe discharge, and Y_n is the normal flow depth for the specified *Q*. The subscripts *us* and *ds* of *Y* denote the depths at the upstream and downstream ends of the pipe, respectively. Use of these curves eliminates the cumbersome backwater profile computation in nonuniform flow channel-capacity determination.

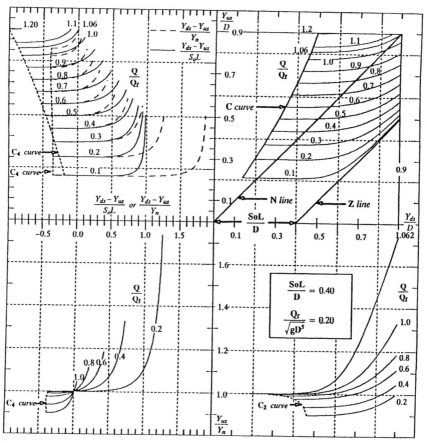

FIGURE 25.15 Nondimensional hydraulic performance graph for prismatic open channel with circular cross section.

REFERENCES

Abbott, M. B., *Computational Hydraulics: Elements of the Theory of Free Surface Flows,* Pitman, London, 1979.

Abbott, M. B., and D. R. Basco, *Computational Fluid Dynamics,* John Wiley, New York, 1990.

Acaroglu, E. R., "Friction Factors in Solid Material Systems," *J. Hydraulics Division,* ASCE, 98(HY4):681–699, 1972.

Akan, A. O., and B. C. Yen, "A Nonlinear Diffusion-Wave Model for Unsteady Open-Channel Flow," *Proceedings,* 17th Cong. Int. Assoc. Hydraul. Res., Baden Baden, Germany, 2:181–190, 1977.

Alam, A. M. Z., and J. F. Kennedy, "Friction Factors for Flow in Sand-Bed Channel," *Hydraulics Division,* ASCE, 95(HY6):1973–1992, 1969.

Bakhmeteff, B. A., *Hydraulic of Open Channels,* McGraw-Hill, New York, 1932.

Barr, D. I. H., "New Forms of Equations for the Correlation of Pipe Resistance Data," *Proceedings,* Institution of Civil Engineers (London), 53 part 2:383–390, 1972.

Barr, D. I. H., Discussion of "Accurate Explicit Equation for Friction Factor," *J. Hydraulics Division,* ASCE, 103(HY3):334–337, 1977.

Bettess, R., and R. K. Price, "Comparison of Numerical Methods for Routing Flow Along a Pipe," Report no. IT162, Hydraulics Research Station, Wallingford, England, 1976.

Brownlie, W. R., "Flow Depth in Sand-Bed Channels," *J. Hydraulic Eng.,* ASCE 109(7): 959–990, 1983.

Camacho, R., and B. C. Yen, "Nonlinear Resistance Relationships for Alluvial Channels," in B. C. Yen, ed., *Channel Flow Resistance: Centennial of Manning's Formula,* Water Resources Publications, Highlands Ranch, Colo. 1991, pp. 186–194.

Chaudhry, M. H., *Open-Channel Flow,* Prentice-Hall, Englewood Cliffs, N.J., 1993.

Chen, C. L., "Role of Velocity Distribution in Open-Channel Flow Stability Analysis," *J. Hydraulic Eng.,* ASCE, 121(9), 1995.

Chien, N., "The Present Status of Research on Sediment Transport," *Transactions,* ASCE, vol. 121:833–868, 1956.

Chow, V. T., *Open-Channel Hydraulics,* McGraw-Hill, New York, 1959.

Churchill, S. W., "Empirical Expressions for the Shear Stress in Turbulent Flow in Commercial Pipe," *J. AIChE,* 19(2):375–376, 1973.

Colebatch, G. T., "Model Tests on the Lawrence Canal Roughness Coefficients," *Journal, Institution of Civil Engineers* (Australia), 13(2):27–32, 1941.

Colebrook, C. F., "Turbulent Flow in Pipes with Particular Reference to the Transition Region Between the Smooth and Rough Pipe Laws," *Journal, Institution of Civil Engineers* (London), 11:133–156, 1938–1939.

Cox, R. G., "Effective Hydraulic Roughness for Channels Having Bed Roughness Different from Bank Roughness," *Misc. Paper,* H-73-2, U.S. Army Corps of Engineers Waterways Experiment Station, Vicksburg, Miss., Feb. 1973.

Cunge, J. A., "On the Subject of a Flood Propagation Computation Method," *J. Hydraulic Res.,* 7:205–230, 1969.

Cunge, J. A., and M. Wegner, "Intégration Numérique des Équations d'Écoulement de Barré de Saint-Venant par un Schéma Implicite de Différences Finies: Application au Cas d'Une Galerie Tantôt en Charge, Tantôt é Surface Libre," *La Houille Blanche,* 1:33–39, 1964.

Cunge, J. A., F. M. Holly, and A. Verwey, *Practical Aspects of Computational River Hydraulics,* Pitman, London, 1980.

Einstein, H. A., "Der Hydraulische oder Profil-Radius," *Schweizerische Bauzeitung,* Zurich, vol. 103(8):89–91, 1934.

Einstein, H. A., and N. L. Barbarossa, "River Channel Roughness," *Transactions,* ASCE, 117:1121–1146, 1952.

Engelund, F., "Hydraulic Resistance of Alluvial Streams," *J. Hydraulics Division,* ASCE, 92(HY2):315–326, 1966.

Engelund, F., "Hydraulic Resistance of Alluvial Streams," *J. Hydraulics Division,* ASCE, 93(HY4):287–297, 1967.

Engelund, F., and E. Hansen, *A Monograph on Sediment Transport in Alluvial Streams,* Teknisk Vorlag, Copenhagen, Denmark, 1967.

Felkel, K., "Gemessene Abflüsse in Gerinnen mit Weidenbewuchs," *Mitteilungen der BAW,* Heft 15, Karlsruhe, Germany, 1960.

Fread, D. L., "Flow Routing," in D. R. Maidment, ed., *Handbook of Hydrology,* Chap. 10 McGraw-Hill, New York, 1993, pp. 10.1–10.36.

French, R. H., *Open-Channel Hydraulics,* McGraw-Hill, New York, 1986.

Garde, R. J., and K. G. Ranga Raju, "Resistance Relationship for Alluvial Channel Flow," *J. Hydraulics Division,* ASCE, 92(HY4):77–99, 1966.

Graf, W. H., *Hydraulics of Sediment Transport,* McGraw-Hill, New York, 1971.

Griffiths, G. A., "Flow Resistance in Coarse Gravel Bed Rivers," *J. Hydraulics Division,* ASCE, 107(HY7):899–918, 1981.

Harris, G. S., "Development of a Computer Program to Route Runoff in The Minneapolis-St. Paul Interceptor Sewers," Memo M121, St. Anthony Falls Hydraulic Laboratory, University of Minnesota, Minneapolis, Minn., 1968.

Henderson, F. M., *Open Channel Flow,* Macmillian, New York, 1966.

Horton, R. E., "Separate Roughness Coefficients for Channel Bottoms and Sides," *Engineering News-Record,* 111(22):652–653, 1933.

Hromadka, T. V., and C. C. Yen, "A Diffusion Hydrodynamic Model," *Water Resources Investigations Report* 87-4137, U.S. Geological Survey, 1987.

Hsie, C. H., V. T. Chow, and B. C. Yen, "The Evaluation of a Hydrodynamic Watershed Model (IHW Model IV)," *Civil Engineering Studies, Hydraulic Engineering Series no. 28,* University of Illinois at Urbana-Champaign, Urbana, Ill., 1974.

Keulegan, G. H., "Laws of Turbulent Flow in Open Channels," *Journal, National Bureau of Standards,* Research Paper, 1151, 21:707–741, Washington, D.C., Dec. 1938.

Kramer, H., "Sand Mixtures and Sand Movement in Fluvial Models," *Transactions,* ASCE, 100:798–878, 1935.

Koussis, A. D., "Comparison of Muskingum Method Difference Schemes," *J. Hydraulics Division,* ASCE, 108(HY5):925–929, 1980.

Krishnamurthy, M., and B. A. Christensen, "Equivalent Roughness for Shallow Channels," *J. Hydraulics Division,* ASCE, 98(HY12):2257–2263, 1972.

Lai, C., "Numerical Modeling of Unsteady Open-Channel Flow," in B. C. Yen, ed., *Advances in Hydroscience,* vol. 14, Academic Press, Orlando, Fla., 1986, pp. 162–333.

Lane, E. W., "Design of Stable Channels," *Transactions,* ASCE, 120:1234–1260, 1955.

Larson, C. L., T. C. Wei, and C. E. Bowers, "Numerical Routing of Flood Hydrographs Through Open Channel Junctions," Water Resources Research Center Bulletin, no. 40, University of Minnesota, Minneapolis, Minn., 1971.

Lighthill, M. J., and G. B. Whitham, "On Kinematic Waves: I. Flood Movements in Long Rivers," *Proceedings, Royal Soc. London,* 229A(1178):281–316, 1955.

Lotter, G. K., "Soobrazheniia k Gidravlischeskomu Raschetu Rusel s Razlichnoi Scherokhovatostiiu Stenok," (Considerations on Hydraulic Design of Channels with Different Roughness of Walls), *Izvestiia Vsesoiuznogo Nauchno-Issledovatel'skog Instituta Gidrotekhniki* (Trans. All-Union Sci. Res. Inst. Hydraulic Eng.), Leningrad, 9:238–341, 1933.

Lovera, F., and J. F. Kennedy, "Friction Factors for Flat-Bed Flows in Sand Channels," *J. Hydraulics Division,* ASCE, 95(HY4):1227–1234, 1969.

MacCormack, R. W., "The Effect of Viscosity in Hypervelocity Impact Crating," AIAA Paper, 69-354, Cincinnati, Ohio, 1969.

Mahmood, K., and V. Yevjevich, eds., *Unsteady Flow in Open Channels,* vol. 1, Water Resources Publ., Highlands Ranch, Colo., 1975.

Martin, C. S., and J. J. Zovne, "Finite-Difference Simulation of Bore Propagation," *J. Hydraulics Division,* ASCE, 97(HY7):993–1010, 1971.

Mostafa, M. G., and R. M. McDermid, "Discussion of 'Sediment Transportation Mechanics: F. Hydraulic Relations for Alluvial Streams', by the Task Committee for Preparation of Sediment Manual," *J. Hydraulics Division,* ASCE, 97(HY10):1777–1780, 1971.

Murota, A., T. Kanda, and T. Eto, "Flood Routing for Urban River Network," *Proceedings International Symposium on River Mechanics,* Int. Assoc. Hydraul. Res., Bangkok, Thailand, 3:339–350, 1973.

Pavlovskii, N. N., "K Voporosu o Raschetnoi dlia Ravnomernogo Dvizheniia v Vodotokahk s Neodnorodnymi Stenkami," (On a Design Formula for Uniform Flow in Channels with Nonhomogeneous Walls), *Izvestiia Vsesoiuznogo Nauchno-Issledovatel'skog Instituta Gidrotekhniki* (Trans. All-Union Sci. Res. Inst. Hydraulic Eng.), Leningrad, 3:157–164, 1931.

Ponce, V. M., *Engineering Hydrology,* Prentice-Hall, Englewood Cliffs, N.J., 1989.

Ponce, V. M., R. M. Li, and D. B. Simons, "Applicability of Kinematic and Diffusion Models," *J. Hydraulic Division,* ASCE, 104(HY3):353–360, 1978.

Ponce, V. M., and F. D. Theurer, "Accuracy Criteria in Diffusion Routing," *J. Hydraulics Division,* ASCE, 108(HY6):747–757, 1982.

Preissmann, A., "Propagation des Intumescences dans les Canaux et Rivères," *Proc. Cong. de l'Assoc. Francaise de Calcul (1960),* 1st Grenoble, France, 1961, pp. 433–442.

Raudkivi, A. J., "Analysis of Resistance in Fluvial Channels, *J. Hydraulics Division,* ASCE, 93(HY5):2084–2093, 1967.

Reinius, E., "Steady Uniform Flow in Open-Channels," *Bulletin 60,* Division of Hydraulics, Royal Institute of Technology, Stockholm, Sweden, 1961.

Rouse, H., *Elementary Mechanics of Fluids,* John Wiley, New York, 1946.

Sayre, W. W., and M. L. Albertson, "Roughness Spacing in Rigid Open Channels," *J. Hydraulics Division,* ASCE, 87(HY3):121–150, 1961.

Sevuk, A. S., "Unsteady Flow in Sewer Networks," Ph.D. thesis, Department of Civil Engineering, University of Illinois at Urbana-Champaign, Urbana, Ill., 1973.

Sevuk, A. S., and B. C. Yen, "A Comparative Study on Flood Routing Computation," *Proceedings Int. Symp. on River Mech.,* Int. Assoc. Hydraul. Res., Bangkok, Thailand, 3:275–290, 1973.

Shen, H. W., "Development of Bed-Roughness in Alluvial Channels," *J. Hydraulics Division,* ASCE, 88(HY3):45–58, 1962.

Smith, A. A., "A Generalized Approach to Kinematic Flood Routing," *J. Hydrol.,* 45:71–87, 1980.

Smith, G. D., *Numerical Solution of Partial Differential Equations,* 2d ed., Oxford University Press, Oxford, UK, 1978.

Simons, D. B., and E. V. Richardson, "Resistance to Flow in Alluvial Channels," *Professional Paper 442-J,* U.S. Geological Survey, Washington D.C., 1966.

Strelkoff, T., and N. D. Katopodes, "Border Irrigation Hydraulics with Zero-Inertia," *J. Irrigation and Drainage Division,* ASCE, 103(IR3):325–342, 1977.

Strickler, A., "Beiträge zur Frage der Geschwindigkeitsformel und der Rauhigkeitszahlen für Ströme, Kanäle und geschlossene Leitungen," *Mitteilungen des Eidgenössischen Amtes für Wasserwirtschaft* 16, Bern, Switzerland, 1923 (Translated as "Contributions to the Question of a Velocity Formula and Roughness Data for Streams, Channels, and Closed Pipelines," by T. Roesgan and W. R. Brownie, Translation T-10, W.M. Keck Lab of Hydraulics and Water Resources, California Institute of Technology, Pasadena, Calif., January 1981).

Thijsse, J. T., "Formulae for the Friction Head Loss Along Conduit Walls Under Turbulent Flow," *Proceedings,* IAHR Third Congress, Grenoble France, III-4:1–11, 1949.

van Rijn, L. C., "Sediment Transport, Part III: Bed Forms and Alluvial Roughness," *J. Hydraulics Division,* ASCE, 110(12):1733–1754, 1984.

Vanoni, V. A., and N. H. Brooks, "Laboratory Studies of the Roughness and Suspended Load of Alluvial Streams," Report E-68, Sedimentation Lab., California Institute of Technology, Pasadena, Calif., 1957.

Vanoni, V. A., and L-S. Hwang, "Relation Between Bed Forms and Friction in Streams," *J. Hydraulics Division,* ASCE, 93(HY3):121–144, 1967.

Weinmann, P. E., and E. M. Laurenson, "Approximate Flood Routing Methods: A Review," *J. Hydraulics Division,* ASCE, 105(HY12):1521–1536, 1979.

White, W. R., E. Paris, and R. Bettess, "The Frictional Characteristics of Alluvial Streams: A New Approach," *Proceedings,* Institution of Civil Engineers (London), 69(2):737–750, 1980.

White, W. R., E. Paris, R. Bettess, and S. Wang, "Frictional Characteristics of Alluvial Streams in the Lower and Upper Regimes," *Proceedings,* Institution of Civil Engineers (London), vol. 83 part 2:685–700, 1987.

Woolhiser, D. A., and J. A. Liggett, "Unsteady One-Dimensional Flow Over A Plane: The Rising Hydrograph," *Water Resources Research,* 3(3):753–771, 1967.

Wylie, E. B., and V. L. Streeter, *Fluid Transients,* FEB Press, Ann Arbor, Mich., 1983.

Xia, R., "Sensitivity of Flood-Routing Models to Variations of Momentum Equation Coefficients and Terms," Ph.D. thesis, Department of Civil Engineering, University of Illinois at Urbana-Champaign, Urbana, Ill., 1992.

Yalin, M. S., *Mechanics of Sediment Transport,* 2d ed., Pergamon Press, Oxford, UK, 1977.

Yen, B. C., "Open-Channel Flow Equations Revisited," *J. Engineering Mechanics Division,* ASCE, 99(EM5):979–1009, 1973.

Yen, B. C., "Further Study on Open-Channel Flow Equations," *Sonderforschungsbereich 80,* Report no. SFB80/T/49, University of Karlsruhe, Karlsruhe, Germany, 1975.

Yen, B. C., "Unsteady Flow Mathematical Modeling Techniques," in H. W. Shen, ed., *Modeling of Rivers,* Chap. 13, Wiley-Interscience, New York, 1979, pp. 13.1–13.33.

Yen, B. C., "Hydraulics of Sewers," in B. C. Yen, ed., *Advances in Hydroscience,* vol. 14, Academic Press, Orlando, Fla., 1986, pp. 1–122.

Yen, B. C., "Urban Drainage Hydraulics and Hydrology: From Art to Science," (Joint Keynote at 22d IAHR Congress and 4th International Conference on Urban Storm Drainage, EPF-Lausanne, Switzerland), *Urban Drainage Hydraulics and Hydrology,* Water Resources Publications, Highlands Ranch, Colo., 1987, pp. 1–24.

Yen, B. C., "Hydraulic Resistance in Open Channels," in B. C. Yen, ed., *Channel Flow Resistance: Centennial of Manning's Formula,* Water Resources Publications, Highlands Ranch, Colo., 1991, pp. 1–135.

Yen, B. C., "Dimensionally Homogeneous Manning's Formula," *J. Hydraulic Eng.,* ASCE, 118(9):1326–1332, 1992.

Yen, B. C., and A. O. Akan, "Flood Routing Through River Junctions," *Rivers '76,* ASCE, 1:212–231, 1976.

Yen, B. C., and J. A. Gonzalez, "Determination of Boneyard Creek Flow Capacity by Hydraulic Performance Graph," Research Report 219, Water Resources Center, University of Illinois at Urbana-Champaign, Urbana, Ill., 1994.

Yen, B. C., and Y. C. Liou, "Hydraulic Resistance in Alluvial Channels," Research Report 22, Water Resources Center, University of Illinois at Urbana-Champaign, Urbana, Ill., 1969.

CHAPTER 26
URBAN STORMWATER MANAGEMENT

G. V. Loganathan, D. F. Kibler, and T. J. Grizzard
Dept. of Civil Engineering, Virginia Tech
Blacksburg, Virginia

26.1 INTRODUCTION

Urbanization of undeveloped areas increases imperviousness and creates channelization of runoff, thereby enhancing the hydraulic conveyance properties of the watershed. These hydraulic modifications tend to increase the runoff volume and peak flow and decrease the time to peak and recharge to groundwater storage. Most municipal stormwater ordinances require that the postdevelopment peak flows for certain designated return periods must not exceed predevelopment flow peaks. This requires in turn that channel improvements and flood-control detention ponds be installed to reduce the increased postdevelopment flows. An associated problem is that falling rain can capture airborne pollutants, as well as dislodge particulate matter and soluble pollutants which have settled on the ground. Surface runoff can dissolve, as well as transport, the adsorbed pollutants downstream. The water-quality impacts of stormwater discharges on receiving lakes and rivers are quite significant, and are regulated by the Environmental Protection Agency's (EPA) *stormwater-discharge permits* in the National Pollutant Discharge Elimination System (NPDES) program. Civil action may be brought by the EPA against developments or large municipalities which do not comply with the permit requirements.

The above description addresses one of the most frequent drainage problems within an urban area. The control measures put in place to alleviate this problem constitute the *minor drainage system* consisting of street gutters, inlets, culverts, roadside ditches, swales, small channels, and pipes. A less frequent but very significant flood-damage problem arises when an urban area is located along a watercourse draining a bigger basin. The issue here is controlling the floodplain use and providing for proper drainage of floodwaters. The Federal Emergency Management Agency (FEMA) issues guidelines with regard to floodplain management under the National Flood Insurance Program (NFIP). The components involved in this latter problem are the natural waterways, large human-made conduits, including the trunkline sewer system which receives the stormwater from the minor system, large water impoundments and other flood-protection devices. These components together are designated as the *major system.*

26.2 ESTIMATION OF URBAN STORMWATER QUANTITY

In this section methods of determining runoff quantity, which is the primary design input for both the minor and major systems, are reviewed. The methods include *direct measurement* and *indirect estimation* from rainfall, including computation of runoff depth, time distribution of runoff, and peak flow. Jennings (1982) and Wanielista (1990) provide good discussions on runoff-measuring devices. A design flow may be selected from the measured data by frequency analysis, provided the sample size is large enough. Because gauged streamflow records are rare in the urban environment the engineer is usually forced to apply an ungauged flow frequency method (see Sauer, et al., 1983) or a design-storm procedure. It may also be noted that a runoff measurement is associated with an areal measure representing the upstream area draining at the measurement point while the rainfall measurement is a point measurement. An areal average of rainfall from gauges located within the area has to be used in estimating the runoff for that area.

26.2.1 Design-Storm Approach

The wide availability of rainfall data compared to runoff data has led to methods of estimating runoff from design storms of a certain return period. An *intensity-duration-frequency* (IDF) *curve* provides rainfall intensity for a specified time duration at a chosen return period. These curves are obtained from the observed annual maximum intensities for a gauging station by fitting a conditional probability distribution for a given duration. If several stations are involved, a regional curve may be obtained by spatially averaging the rainfall intensities at chosen return periods (quantiles). The IDF curves can be fitted by a generic equation for a particular return period with the parameters of the equation dependent on location (Wenzel, 1982). The time distribution of rainfall can also be generated synthetically by following methods such as the Soil Conservation Service (SCS) and Huff distributions (Wenzel, 1982; Kibler, 1982). Further information is given in Chap. 24.

26.2.2 Rainfall–Runoff Relationship

In this section methods for estimating runoff volume only, time distribution of runoff (hydrograph), and peak flow only are outlined. Kibler, et al. (1995) provide a detailed description of hydrological processes in an urban area.

26.2.2.1 Runoff Depth. Rainfall excess is the amount which is in excess of the abstractions such as interception, infiltration, and depression storage. Rainfall excess is equal to the direct runoff in volume but the latter is distributed over a longer time period. The direct-runoff depth (volume) QR, in, is computed by the SCS procedure as

$$QR = \frac{(PR - 0.2SC)^2}{(PR + 0.8SC)} \tag{26.1}$$

where   PR = rainfall, in
         SC = (1000/CN) − 10
         CN = curve number (see Table 24.7 for values of CN), dependent on land
              use and soil type

Schueler (1987) presents an equation for runoff coefficient as

$$RC = 0.05 + 0.009 \text{ IMP} \tag{26.2}$$

in which IMP = percent imperviousness. The runoff depth QR is computed by

$$QR = RC * PR \tag{26.3}$$

26.2.2.2 Hydrograph Generation.   The *unit hydrograph* is defined as the direct-runoff hydrograph resulting from 1 in or 1 cm of rainfall excess of a particular duration (see Chap. 24). This duration is used to split arbitrary rainfall excess events into various subblocks of equal duration. The unit hydrograph is posted under each subblock with its ordinates multiplied by each subblock rain amount. These lagged, scaled hydrographs are added ordinate by ordinate to produce the *direct-runoff hydrograph* corresponding to an arbitrary rainfall-excess event. The unit hydrograph approach produces flood hydrographs only at a selected outlet point. To generate a hydrograph at any other point along a stream or an overland flow plane, the analyst must apply an auxiliary rating equation combined with the *kinematic wave-routing method* (see Chap. 25). The continuity equation is

$$\frac{\partial A}{\partial t} + \frac{\partial Q}{\partial x} = I \tag{26.4}$$

and the rating auxiliary equation

$$Q = \eta A^{\xi} \tag{26.5}$$

where A = flow area
         Q = discharge dependent on space and time
         I = lateral inflow per unit length
         η, ξ = parameters

Bedient and Huber (1992) provide details of this method. In selected cases, analytical solution is possible. The method produces a $Q(x,t)$ surface as opposed to the unit hydrograph approach where the spatial location x must be fixed.

26.2.2.3 Synthetic Hydrographs for Urban Areas.   When a unit hydrograph (UH) is not available from measured data, one may resort to a synthetic UH (see Chap. 24). Various UHs resembling the skewed bell-shaped hydrograph have been proposed. These include the gamma distribution, Snyder's synthetic unit hydrograph, SCS-dimensionless unit hydrograph, and the Espey 10-minute unit hydrograph. Viessman, et al. (1989) and Chap. 24 provide a good discussion of these methods. These methods in general provide equations for the peak flow, time to peak and time base of the hydrograph. Since the initial flow and the end flow are zeros, these three quantities together with the requirement of unit area, are sufficient to establish the full UH for a particular drainage system.

26.2.2.4 Peak Flow Estimation.   United States Geological Survey (USGS) regression equations (Sauer, et al., 1983) may be used to estimate peak flows at different return periods. Specifically these equations help to revise the rural flows as impacted by urbanization. For example, the urban 100-year flow UQ_{100} may be estimated by

$$UQ_{100} = 7.7\, A^{0.15}\, (13 - BDF)^{-0.32}\, RQ_{100}^{0.82} \qquad (26.6)$$

where UQ_{100} = urban 100 year peak flow, cfs
    A = basin area, mi^2, with the range 0.2 to 100
    BDF = basin development factor (dimensionless) with the range 0 to 12
    RQ_{100} = rural 100 year flow, cfs

The *basin development factor (BDF)* is estimated as follows: The basin is divided into thirds along the drainage channels as upper, middle, and lower by area. Within each third, four aspects of drainage namely: (1) channel improvements such as straightening, enlarging, deepening and clearing, (2) channel linings (with impervious material), (3) storm drains, and (4) curb-and-gutter streets are considered. If 50 percent or more of the channels are improved a code value of one for this particular aspect is assigned; otherwise the code value is set to zero. Similarly, if 50 percent or more lengths of channels are lined, a code value of one is assigned; otherwise the code value is zero. If at least 50 percent of the area is sewered or urbanized the respective codes will be set to unity; otherwise zero. The BDF value is the sum of the zero/one code values for all the three subdivisions and the four aspects.

Also the *rational formula* (see Chaps. 19 and 24) is widely used to estimate the peak flow by

$$Q_p = CIA \qquad (26.7)$$

where Q_p = peak flow, cfs
    I = rainfall intensity, in/h, and should have duration equal to the time of concentration
    A = area, acres, generally taken to be less than 200 acres
    C = runoff coefficient (dimensionless) dependent on land use

Singh (1992) and Chap. 19 provide values for the runoff coefficient. Additional information on the preceding methods is presented by Kibler, et al. (1995).

26.3 DRAINAGE OF URBAN AREAS

Removing rainwater from urban streets is accomplished by collecting overland flow in gutters and intercepting the gutter flow at inlets to the storm sewers. The underground part of an inlet may have a raised outlet to prevent heavy debris from entering the sewer pipe. This design is called a *catch basin*. In other parts of an urban area the stormwater runoff is concentrated in natural watercourses or artificial open channels. Both types of watercourses may involve control structures, such as culverts and detention ponds, channel improvements for quick passage of water, flow-retarding features such as riprap and vegetative lining. Some part of the overland flow itself may be trapped with the aid of porous pavements and infiltration trenches. The flow-controlling devices are also considered to be candidates for pollutant removal in stormwater. The following discussion on gutters and inlets is based on the Federal Highway Administration Study (Johnson and Chang, 1984).

26.3.1 Flow, Spread, and Depth in a Gutter

The purpose of a gutter, which is the depression along a roadway curb, is to collect overland flow along its length and concentrate it as channel flow. Following Burke (1981) (see Fig. 26.1) elemental gutter flow dQ_G through an elemental cross-section of width dx of the gutter is written as

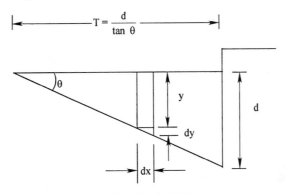

FIGURE 26.1 Gutter flow. *(Burke, 1981.)*

$$dQ_G = V\,dx\left(y + \frac{dy}{2}\right) \tag{26.8}$$

where V = velocity of flow within the elemental strip
   y = flow depth

The longitudinal slope for the gutter is S_o; the overland flow plane (or cross-slope) has a slope S_T; the spread of water measured laterally from the curb is T; the (maximum) flow depth in the gutter is d. The velocity in Eq. (26.8) is estimated by the Manning equation as

$$V = \frac{1.49}{n}\left(\frac{y\,dx + \frac{1}{2}\,dy\,dx}{ds}\right)^{2/3} S_o^{1/2} \tag{26.9}$$

where $ds = \sqrt{dx^2 + dy^2} = dx\sqrt{1 + S_T^2}$ where $S_T = \dfrac{dy}{dx} = \dfrac{d}{T}$
   n = Manning roughness coefficient

For small S_T, Eq. (26.9) becomes

$$V = \frac{1.49}{n}\left(y + \frac{dy}{2}\right)^{2/3} S_o^{1/2} \tag{26.10}$$

and therefore by ignoring the higher order terms Eq. (26.8) is written as

$$dQ_G = \frac{1.49}{n}\,y^{5/3}\,\frac{dy}{S_T}\,S_o^{1/2} \tag{26.11}$$

By integrating Eq. (26.11) across the cross section of the gutter as y goes from 0 to d we obtain the gutter flow in terms of depth as

$$Q_G = \frac{0.56}{n} \frac{S_o^{1/2}}{S_T} d^{8/3} \tag{26.12}$$

or the gutter flow in terms of water spread is

$$Q_G = \frac{0.56}{n} S_o^{1/2} S_T^{5/3} T^{8/3} \tag{26.13}$$

Also, note that the gutter flow Q_G must equal any flow from upstream plus the overland flow. Assuming for the time being that there is no upstream flow, then $Q_G = CIA$ by the rational formula with C the runoff coefficient, I the rainfall intensity, and A the area of the overland flow plane. Using the rational formula estimate of Q_G in Eqs. (26.12) and (26.13) one can choose acceptable depth and spread values.

26.3.2 Types of Street Inlets

There are four types of street inlets as shown in Fig. 26.2. These are the *grate, curb opening, slotted drain,* and the *combination* inlets. A *grate inlet* is made up of a metal plate with rectangular openings cut into it which are kept parallel to the gutter flow. A *curb opening inlet* is a vertical opening in the curb, typically covered by a top slab. A *slotted drain* is a buried pipe with openings cut at the top of the pipe wall along its longitudinal axis called *slot openings* kept in mesh with the road surface parallel to the gutter flow. A *combination inlet* usually consists of both a grate inlet and a curb opening inlet placed together side by side as shown in Fig. 26.2. Further classification is made depending upon whether the inlet is on a *unidirectional grade* or in a *sag* made up of rising curves from a low point. The interception capacity of an inlet on a *uniform grade* is dependent on the overland flow elements: namely the cross-slope S_T, and to a lesser extent its roughness; the gutter longitudinal slope, total gutter flow; the inlet configuration with its length, width, and cross-bar arrangement for grate inlets; and length of inlet for curb opening and slotted drains. All inlets located in a sag operate as a weir at low depths of flow. As depth increases a transition flow occurs which is followed by orifice flow at greater depths. Johnson and Chang (1984) give the pertinent empirical equations and charts based on experimental data. The inlets on uniform grade usually let a portion of the gutter flow bypass the inlet. The inlets in sag locations or sumps intercept all of the water due to ponding.

The inlets alter the rate at which stormwater is admitted into the sewer system. This flow rate and the control structure underneath the inlet will determine the downstream sewer size. In all inlet selections, possible clogging should be considered. A factor of safety as high as two is recommended when clogging is expected. A design return period of 10 years is typically used. The recommended maximum spread or encroachment is about half the pavement width for streets with parking, with a maximum gutter depth of 0.35 ft. The recommended minimum longitudinal slope for the gutter is 0.3 percent and the transverse slope of the pavement is 2 percent. The aforementioned criteria are some usual guidelines, but should be changed depending on the nature of traffic.

26.3.2.1 *Grate Inlets.* Woo (1984) emphasizes the three basic elements which govern the grate design: (1) hydraulic efficiency, (2) traffic (vehicle, bicycle, and pedestrian) safety, and (3) debris plugging. Other elements include structural strength to withstand wheel load, minimum cost, and protection against vandalism and rattling. Grate inlets are quite effective where debris is not a problem. The grate inlets are designated as P—1—7/8 and so on with the letter P for parallel bar grates

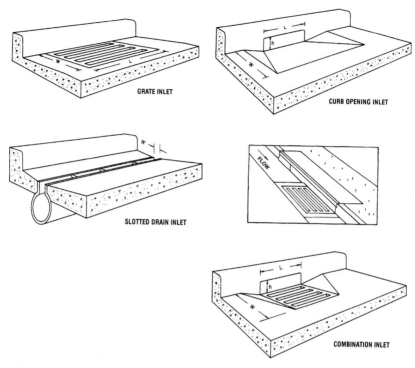

FIGURE 26.2 Types of inlets *(Johnson and Chang, 1984).*

and the number for the bar spacing; CV for curved vane and reticuline for honeycomb pattern (see Fig. 26.3; Johnson and Chang, 1984). The hydraulic analysis for *grates on uniform grade* is as follows. The amount of water from the gutter that can be removed as *frontal flow* which passes over the upstream end of the grate is limited by the width of the grate. If the velocity of flow in the gutter V exceeds a threshold value V_o, a part of the frontal flow would skip over the grate, called *splash-over.* The part of the gutter flow which goes around the grate when the spread T is larger than the width of grate W is called the *side flow.* The *capture efficiency* of side flow depends on the cross slope S_T, the length of the grate L, and the gutter-flow velocity V. The *frontal-flow ratio* E_o is given by

$$E_o = \frac{Q_W}{Q_G} = 1 - \left(1 - \frac{W}{T}\right)^{2.67}$$ (26.14)

where Q_W = frontal flow, m³/s (cfs)
Q_G = gutter flow, m³/s (cfs)
W = width of grate, m (ft)
T = spread of water, m (ft)

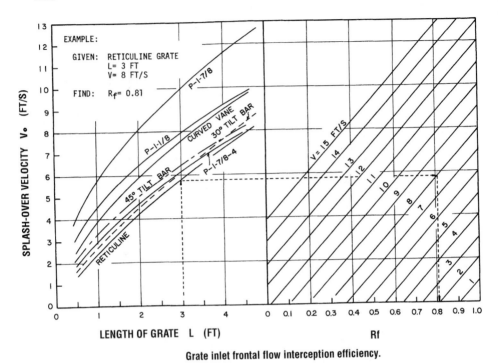

Grate inlet frontal flow interception efficiency.

FIGURE 26.3   Grate inlet interception nomograph. *(Johnson and Chang, 1984.)*

and therefore the *side-flow ratio* E_S is given as

$$E_S = 1 - E_O \tag{26.15}$$

When the gutter-flow velocity exceeds the threshold value, the *frontal-flow capture ratio* R_f, is given by

$$R_f = 1 - 0.09 \, (V - V_O) \tag{26.16}$$

where V = velocity of flow in gutter, m/s (ft/s) = Q_G/A_G; A_G = area of gutter flow; V_O = threshold gutter-flow velocity at which frontal flow skips over the grate without being captured totally, called *splash-over*. Use Fig. 26.3 to compute R_f which accounts for the various grates. The *side-flow capture ratio* R_S is given by

$$R_S = 1 / \left[1 + \frac{0.15 \, V^{1.8}}{S_T L^{2.3}} \right] \tag{26.17}$$

where S_T = cross slope
L = length of the grate, m (ft)

Therefore, the *interception efficiency of the grate E* is

$$E = R_f \, E_O + R_S \, E_S \tag{26.18}$$

and the *amount of intercepted inlet flow* Q_I is

$$Q_I = EQ_G \qquad (26.19)$$

The hydraulic analysis for *grates on sag (sump)* is given as follows. First of all note that in a sump condition, because water ponds, eventually all the water will be removed by the inlet. However, the rate of withdrawal from the street into the sewer is controlled by the inlet. For depths less than 0.3 ft the inlet acts as a weir; and for depths above 0.4 ft the inlet behaves as an orifice (Burke, 1981).

The *intercepted weir flow* $Q_{W,I}$ is given by

$$Q_{W,I} = C_W P d^{1.5} \qquad \text{for} \qquad d \le 0.3 \text{ ft} \qquad (26.20)$$

where P = perimeter of the grate, m (ft), disregarding bars and the side against the curb, i.e., $P = 2W + L$ (with curb) or $P = 2W + 2L$ (without curb); d = depth of flow in the gutter at the curb face, m (ft); $C_W = 1.66$ m^3/s (3.0 cfs).

The *intercepted orifice flow* $Q_{O,I}$ is given by

$$Q_{O,I} = C_O A (2gd)^{0.5} \qquad \text{for} \qquad d \ge 0.4 \text{ ft} \qquad (26.21)$$

where A = clear opening area, m^2 (ft^2)
  $C_O = 0.67$
  $g = 9.81$ m/s^2 (32.2 ft/s^2)

26.3.2.2 Curb Opening Inlets.

Curb opening inlets admit debris more easily because of the large opening and offer little interference to traffic. These inlets do not function well on steep slopes as it is difficult to direct the flow into the inlet. These are much longer in length than the grates. The hydraulic analysis for *curb opening on uniform grade* is given as follows. The *full interception curb opening length* L_{100} is given by

$$L_{100} = KQ_G^{0.42} S_O^{0.3} \left(\frac{1}{nS_T} \right)^{0.6} \qquad (26.22)$$

where $K = 0.076$ m^3/s (0.6 cfs)
  Q_G = gutter flow, m^3/s (cfs)
  S_O = longitudinal gutter slope
  n = Manning roughness coefficient
  S_T = cross-slope

For the provided curb length L_C the *interception efficiency* E is given by

$$E = 1 - \left(1 - \frac{L_C}{L_{100}} \right)^{1.8} \qquad (26.23)$$

in which L_C = installed curb opening length, m (ft). The efficiency E is used in Eq. (19) to obtain intercepted flow.

For *curb opening in a sag* the *intercepted weir flow* Q_W is given by

$$Q_W = C_W L_C d^{1.5} \qquad \text{for} \qquad d \le h \qquad (26.24)$$

where h = opening size (height of curb opening), m (ft)
C_W = 1.25 m³/s (2.3 cfs)
$d = S_T T$ = depth at curb

and when the opening is drowned the intercepted orifice flow Q_O is given by

$$Q_O = C_O h L_C (2g d_O)^{0.5} \tag{26.25}$$

where C_O = 0.67
d_O = vertical distance from the water surface to the mid point of the opening, m (ft)

26.3.2.3 Slotted Drains. *Slotted drains (inlets)* can be used on curbed and uncurbed sections. They do not interfere with the traffic and are much longer in length than the curb openings. The hydraulic functioning of *slotted inlets on uniform grade* is given by the same equations as that of the curb openings on uniform grade. The *slotted inlet in a sag* performs as a weir up to 0.2 ft depth of flow; a transition flow occurs between 0.2 and 0.4 ft; orifice flow occurs for depth d greater than 0.4 ft. Fig. 26.4 gives the amount of intercepted flow when slot width W is 1.75 in. For other slot widths multiply the answer from Fig. 26.4 by the ratio (W, in/1.75).

26.3.2.4 Combination Inlets. The *combination inlet* is mainly used to overcome the problems caused by debris. The hydraulic functioning of a combination inlet located on *uniform grade* is described as follows. The interception capacity of a combination inlet consisting of a grate and curb opening all of its length is that of the grate alone. However, a combination inlet with a curb opening upstream of the grate called a *sweeper* (See Fig. 26.2) has a capacity equal to the sum of the two inlets with the recognition that the frontal flow for the grate should be reduced by the amount of interception by the curb opening ahead. When a combination inlet is situated in a sag, during weir flow its capacity equals that of the grate alone. However, during orifice flow its capacity equals the sum of the capacities of both the grate and curb opening inlets.

26.3.2.5 Inlet Spacing. Inlet locations are often dictated by the practical street geometrical conditions rather than the spread-of-water computations. According to Johnson and Chang (1984), the inlets should be placed at all low points in the gutter grade, at median breaks, intersections, cross-walks, and on streets at intersections where runoff could enter the highway pavement. Wherever formation of ice is expected on pavements, the runoff should be removed prior to flowing over with appropriately placed inlets. On a continuous slope without hazardous ponding and icing conditions, an inlet spacing computation may be performed to drain runoff. This computation considers inlet capacity and how far away it should be located from its upstream inlet to drain the sheet flow generated between the successive two inlets and a permissible bypass flow.

26.4 STORM-SEWER LAYOUT AND DESIGN

A *storm-sewer system* is a pipe system receiving stormwater from the inlets and conveying it under free-surface conditions. Typically all the water that arrives at an inlet is assumed to enter the sewer pipe at the same rate if no storage is involved. The pipe size is determined using the peak discharge for each pipe and the Manning formula

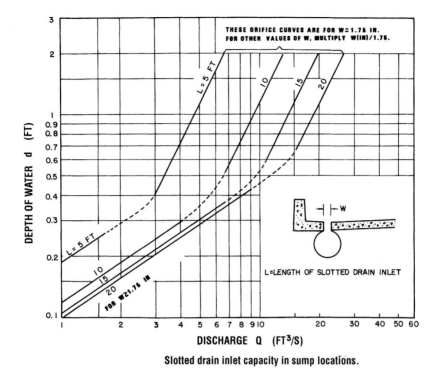

THESE ORIFICE CURVES ARE FOR W=1.75 IN.
FOR OTHER VALUES OF W, MULTIPLY W(IN)/1.75.

L=LENGTH OF SLOTTED DRAIN INLET

DISCHARGE Q (FT³/S)

Slotted drain inlet capacity in sump locations.

FIGURE 26.4   Slotted drain interception nomograph. *(Johnson and Chang, 1984.)*

assuming full-pipe-flow condition. Storm sewers are one of the physically separated underground utilities which include water distribution systems, gas mains, telephone line, television cable, and sanitary sewers. Sometimes, some of these may be placed within the same trench for the purpose of minimizing utility easements (National Association of Home Builders, 1974). The *right-of-way width* is the sum of the pavement width, sidewalk width, utility easement, planting strip or snow-storage width, and local or future-purpose width. A pipe-system layout typically follows the street layout because of the right of way.

As the first step in the detailed design of a sewer system, a comprehensive survey of the area including soil and topographic details, the location of existing sewers, manholes, and inlets, and the drainage boundaries of subbasins draining into the inlets must be conducted. From a set of potential layouts for the expansion of an existing system or a new system an optimal layout is selected. On this layout manholes and inlets should be located; a manhole is assigned wherever there is a change in direction, size, and slope of sewers and at selected intervals of 300 to 600 ft along straight runs (Velon and Persaud, 1993). Chow, et al. (1988) and Velon and Persaud (1993) point out the following practical considerations in sewer design. The sewers must be of 8-in minimum diameter and must be placed at a depth sufficient to protect them from freezing, about 6 ft, and be able to drain basements. They must be structurally strong enough to withstand the applied loadings, including traffic. Normally, the sewers are joined at junctions such that the crown elevation of the upstream pipe is not lower than that of the downstream pipe. To prevent deposition of solid material within the pipe, a minimum cleansing velocity of about 2.5 ft/s should be maintained. It is also recommended that velocities should be kept below about 15 ft/s. Any sewer downstream from a junction or manhole should not be smaller than any of the upstream sewers at that junction. The sewer system is generally a dendritic (tree) network (no loops) converging in the downstream direction. Sewers 24 in or less in diameter should be placed straight between manholes. Sewers should be placed at least 10 ft away from water mains. When they cross, an 18-in minimum vertical separation should be kept. If the stormwater has to be pumped to a higher location, the delivery end pressure should equal the expected maximum internal pressure. These are only general guidelines and one should consult local building codes and government requirements.

The sewer pipe diameter D, ft, is computed from the Manning equation by

$$D = \left[\frac{2.16 \, n \, Q}{\sqrt{S_O}} \right]^{3/8} \tag{26.26}$$

where Q = peak flow entering that pipe, cfs; n = Manning roughness coefficient; S_O = slope, ft/ft, obtained as the ratio of upstream and downstream invert (pipe bottom) elevation difference and the pipe length. If the computed diameter D from Eq. (26.26) is less than the regulatory minimum diameter, that minimum diameter is used. Also, adjustments should be made to the pipe slope to conform with velocity restrictions and topography. Ideally, the pipe slope will approximate the land slope in the direction of pipe travel in order to minimize excavation. If a detention-storage estimate is needed to limit downstream flow, the inlet hydrographs from the storm sewer should be routed through the trial detention basin until maximum routed discharge is less than or equal to the downstream specified, limited discharge. Also, note that if peak flows from various subareas are employed, the design of storm sewers differs from that of sanitary sewers primarily in the demand computation. By modifying the demand-computation part of the sanitary-sewer design programs (Gray and Quaranta, 1992), one may adapt them for storm-sewer design as well. However, modern practice dictates that in larger systems it is essential to employ unsteady-flow principles based on full hydrograph routing. Gupta (1989) presents comprehensive examples with practical details. Mays and Tung (1992) provide a detailed example of optimization of storm-sewer design by dynamic programming. Gray and Quaranta (1992) provide details for accommodating realistic costs in sewer design with due consideration to practical matters. Kibler, et al. (1995) provide detailed examples by hand computation as well as computer runs. Chapter 28 contains a description of currently available computer models. For structural design, construction, and other topics related to urban drainage the recent ASCE manual "Design and Construction of Urban Stormwater Management Systems," 1992, is quite comprehensive. The hydraulics of flow in sewers is fully discussed in Yen (1986) and Chapter 25 provides details on the principles of open channel flow.

26.4.1 Culvert Design

A *culvert* is a short closed conduit which conveys streamflow through a roadway embankment or some other type of flow obstruction. It may flow full over all its length or partly full. The full flow in a culvert is governed by the pipe-flow principles and is called *pressure flow*. The partly full flow or free-surface flow is governed by the principles of *open-channel flow*. The pressure flow situation may be rare. The culvert design calls for a minimum diameter which will carry the design flow without overtopping the road surface. In general, *certain freeboard,* which is the empty space between the water surface in the headwater pool and the top of the road surface, should be provided. This freeboard is accommodated by specifying a permissible headwater depth HW_p. If the design headwater depth HW is less than HW_p, the chosen minimum diameter is adopted. In general, the analysis of free-surface flow requires backwater-flow calculations to determine HW. The Federal Highway Administration (Norman, et al., 1985) recommends a semiempirical method based on the location of the control section which is either at the inlet or at the outlet. The design headwater depth is the larger of the two provided by the inlet and outlet control analyses. The method is described as follows.

Under *inlet control* a culvert has a shallow, high-velocity supercritical flow immediately downstream of the inlet; a hydraulic jump may form downstream depending upon the tailwater conditions. The culvert never flows full throughout its length. Inlet control means that the design flow is completely determined by the headwater elevation and the shape of the entrance regardless of other factors (which is in contrast to the outlet control where the energy losses within the culvert play a significant role). Because of this characteristic, a culvert under inlet control functions like an orifice when it is submerged and performs as a weir when it is not submerged. The (submerged) *orifice* discharge is computed from

$$\left[\frac{HW}{D}\right] = C\left[\frac{Q}{AD^{0.5}}\right]^2 + Y + Z \quad \text{for} \quad \left[\frac{Q}{AD^{0.5}}\right] \geq 4.0 \qquad (26.27)$$

where   HW = headwater depth above inlet control section invert, ft
   D = interior height of culvert barrel, ft
   Q = discharge, ft³/s
   A = full cross-sectional area of culvert barrel, ft²
   S_O = culvert barrel slope, ft/ft
   C, Y = constants from Table 26.1
   $Z = -0.5S_O$ in general
   $Z = +0.7S_O$ for mitered inlets

The (unsubmerged) *weir* discharge is computed from

$$\left[\frac{HW}{D}\right] = \left[\frac{H_c}{D}\right] + K\left[\frac{Q}{AD^{0.5}}\right]^M + Z \quad \text{for} \quad \left[\frac{Q}{AD^{0.5}}\right] \leq 3.5 \qquad (26.28)$$

where $H_c = d_c + V_c^2/2g$, ft
   d_c = critical depth, ft
   V_c = critical velocity, ft/s
   K, M = constants from Table 26.1

Equations (26.27) and (26.28) are implemented by assuming a culvert diameter D and using it on the right-hand side. The headwater depth HW is obtained by multiplying D with the computed HW/D ratio from Eqs. (26.27) and (26.28). Typically an inlet control nomograph is used to compute the headwater depth due to inlet control. A sample nomograph is shown in Fig. 26.5. The appropriate HW/D ratio is obtained by joining the culvert diameter and discharge. Note that a horizontal line is drawn after the first HW/D scale.

Under *outlet control* a culvert has either subcritical flow or full-pipe flow. The design discharge is determined by energy balance. For full flow, considering entrance

TABLE 26.1 Parameters for Inlet Control

Shape and material	Nomograph scale	Inlet edge designation	Unsubmerged K	M	Unsubmerged C	Y
Circular	1	Square edge w/headwall	0.0098	2.0	0.0398	0.67
Concrete	2	Groove end w/headwall	.0078	2.0	.0292	.74
	3	Groove end projecting	.0045	2.0	.0317	.69
Circular	1	Headwall	.0078	2.0	.0379	.69
CMP	2	Mitered to slope	.0210	1.33	.0463	.75
	3	Projecting	.0340	1.50	.0553	.54
Circular	A	Beveled ring, 45° bevels	.0018	2.50	.0300	.74
	B	Beveled ring, 33.7° bevels	.0018	2.50	.0243	.83
Rectangular	1	30 to 75° wingwall flares	.026	1.0	.0385	.81
Box	2	90 and 15° wingwall flares	.061	0.75	.0400	.80
	3	0° wingwall flares	.061	0.75	.0423	.82

Source: FHWA (1985).

loss H_e, friction loss (by Manning equation) H_f, and exit loss H_o, the total head loss H can be written as

$$H = \left[1 + K_e + \left(\frac{29n^2L}{R^{1.33}}\right)\right]\frac{V^2}{2g}$$ (26.29)

where K_e = entrance loss coefficient
n = Manning roughness coefficient
R = hydraulic radius (ratio of area over wetted perimeter of the full culvert barrel)
V = velocity

Let HW_o be the headwater depth above the outlet invert and TW the tailwater depth above the outlet invert. It is noted that for full flow TW $\geq D$. For partly full flow, the head loss should be computed from a backwater analysis. However, an empirical equation for this condition is suggested for the head loss which is

$$H = HW_o - h_o$$ (26.30)

where h_o = maximum $[TW, (D + d_c)/2]$.

To compute the *outlet-controlled headwater depth,* the following procedure is used. From field observations or backwater computations the tailwater depth measured above the outlet invert TW is obtained. Using Eq. (26.29) for full-flow conditions the head loss H is obtained. With the help of Eq. (26.30), the required outlet-controlled headwater elevation HW is obtained as

$$HW = H + h_o - LS_O$$ (26.31)

A sample outlet-control nomograph is shown in Fig. 26.6. The diameter and pipe length (on appropriate entrance loss scale) are connected to fix the point on the turning line which is in turn connected with the discharge to intersect the head loss H scale. Equation (26.31) is implemented once the head loss H is known. For different Manning n_1 value from that of the outlet nomograph, a modified length L_1 is used on the length scale given by

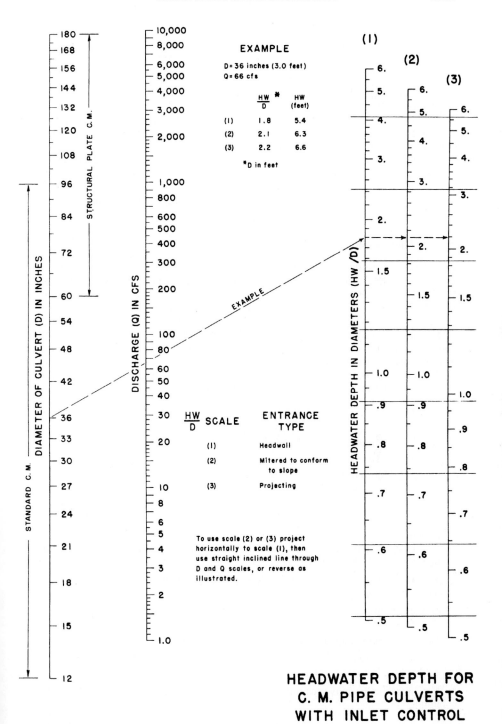

FIGURE 26.5 Culvert inlet control nomograph. *(FHWA, 1985.)*

$$L_1 = L\left(\frac{n_1}{n}\right)^2 \tag{26.32}$$

where L = actual culvert length
         n_1 = desired Manning n
         n = Manning n for the chart

The larger of the headwater elevations obtained from the inlet- and outlet-control calculations is adopted as the design headwater elevation. If the design headwater elevation exceeds the permissible headwater elevation (based on certain freeboard), a new culvert configuration must be selected and the process is repeated. For outlet control, an enlarged barrel will be necessary since inlet improvements are of limited benefit only. In the case of a very large size culvert, one may install multiple culverts with the new design discharge taken as the ratio of the original discharge to the number of culverts.

26.4.2 Open Channels with Flexible Linings

In general rock riprap and vegetative linings are considered to be *flexible linings* in the sense that they are able to conform to change in channel shape as opposed to the *rigid linings* such as concrete, grouted riprap, stone masonry, and asphalt which do not. Flexible linings also permit infiltration and exfiltration; filter out contaminants; provide greater energy dissipation; allow flow conditions that provide better habitat opportunities for local flora and fauna; and are less expensive. These flexible-lining characteristics are considered advantageous for stormwater conveyance. The following design procedure (Chen and Cotton, 1988; Kouwen, et al., 1969; Bathurst, et al., 1981; and Wang and Shen, 1985) is suitable for selecting appropriate flexible lining for given design flow, channel geometry and slope.

Step 1.   Choose a flexible lining from Table 26.2 and note its permissible shear stress τ_p.

Step 2.   Assume an appropriate flow depth d.

Step 3.   Use Table 26.3 for nonvegetative lining to find the Manning n. For vegetative lining use Table 26.4 to determine the appropriate retardance class. The Manning n is given by the general equation

$$n = k_1/(a_c + k_2) \tag{26.33}$$

in which $k_1 = R^{1/6}$ where R = hydraulic radius, ft; $k_2 = 19.97 \log (R^{1.4}S_0^{0.4})$ where S_O = channel longitudinal slope ft/ft; a_c = 15.8, 23.0, 30.2, 34.6, and 37.7 for retardance classes A, B, C, D, E respectively.

Step 4.   Calculate the computed flow depth d_{comp} from the Manning equation using the n from Step 3.

Step 5.   Compare d and d_{comp}. If they are not close enough replace d based on d_{comp} for a new d and go to Step 3.

Step 6.   Compute shear stress for the design condition by

$$\tau_{\text{des}} = \gamma \, dS_O \tag{26.34}$$

If $\tau_{\text{des}} < \tau_p$ the lining is acceptable. Otherwise go to Step 1 and choose a different lining.

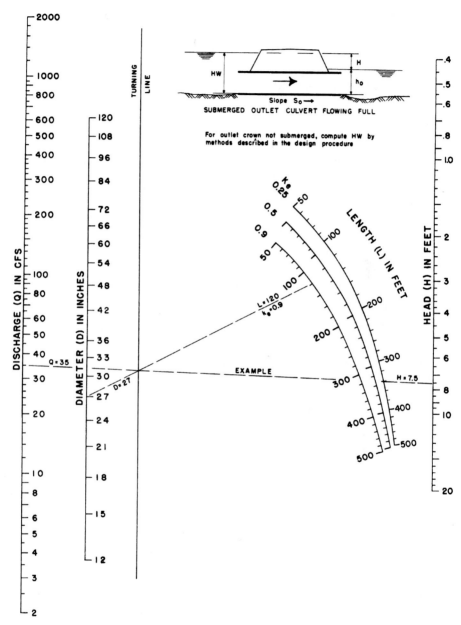

HEAD FOR STANDARD C. M. PIPE CULVERTS FLOWING FULL n = 0.024

FIGURE 26.6 Culvert outlet control nomograph. *(FHWA, 1985.)*

TABLE 26.2 Permissible Shear Stresses for Lining Materials

		Permissible shear stress	
Lining category	Lining type	lb/ft^2	kg/m^2
Temporary	Woven paper net	0.15	0.73
	Jute net	0.45	2.20
	Fiberglass roving		
	Single	0.60	2.93
	Double	0.85	4.15
	Straw with net	1.45	7.08
	Curled wood mat	1.55	7.57
	Synthetic mat	2.00	9.76
Vegetative	Class A	3.70	18.06
	Class B	2.10	10.25
	Class C	1.00	4.88
	Class D	0.60	2.93
	Class E	0.35	1.71
Gravel riprap	1-in	0.33	1.61
	2-in	0.67	3.22
Rock riprap	6-in	2.00	9.76
	12-in	4.00	19.52

Source: Chen and Cotton (1988).

TABLE 26.3 Manning's Roughness Coefficients

		n = value		
		Depth ranges		
Lining category	Lining type	0–0.5 ft (0–15 cm)	0.5–2.0 ft (15–60 cm)	>2.0 ft (>60 cm)
Rigid	Concrete	0.015	0.013	0.013
	Grouted riprap	0.040	0.030	0.028
	Stone masonry	0.042	0.032	0.030
	Soil cement	0.025	0.022	0.020
	Asphalt	0.018	0.016	0.016
Unlined	Bare soil	0.023	0.020	0.020
	Rock cut	0.045	0.035	0.025
Temporary	Woven paper net	0.016	0.015	0.015
	Jute net	0.028	0.022	0.019
	Fiberglass roving	0.028	0.021	0.019
	Straw with net	0.065	0.033	0.025
	Curled wood mat	0.066	0.035	0.028
	Synthetic mat	0.036	0.025	0.021
Gravel riprap	1-in (2.5-cm) D_{50}	0.044	0.033	0.030
	2-in (5-cm) D_{50}	0.066	0.041	0.034
Rock riprap	6-in (15-cm) D_{50}	0.104	0.069	0.035
	12-in (30-cm) D_{50}	—	0.078	0.040

Source: Chen and Cotton (1988).

TABLE 26.4 Classification of Vegetal Covers by Degree of Retardance

Retardance class	Cover	Condition
A	Weeping lovegrass	Excellent stand, tall [average 30 in (76 cm)]
	Yellow bluestem *Ischaemum*	Excellent stand, tall [average 36 in (91 cm)]
B	Kudzu	Very dense growth, uncut
	Bermuda grass	Good stand, tall [average 12 in (30 cm)]
	Native grass mixture (little bluestem, bluestem, blue gamma, and other long and short midwest grasses)	Good stand, unmowed
	Weeping lovegrasses	Good stand, tall [average 24 in (61 cm)]
	Lespedeza sericea	Good stand, not woody, tall [average 19 in (48 cm)]
	Alfalfa	Good stand, uncut [average 11 in (28 cm)]
	Weeping lovegrass	Good stand, unmowed [average 13 in (33 cm)]
	Kudzu	Dense growth, uncut
	Blue gamma	Good stand, uncut [average 13 in (28 cm)]
C	Crabgrass	Fair stand, uncut [10 to 48 in (25 to 120 cm)]
	Bermuda grass	Good stand, mowed [average 6 in (15 cm)]
	Common lespedeza	Good stand, uncut [average 11 in (28 cm)]
	Grass-legume mixture—summer (orchard grass, redtop, Italian ryegrass, and common lespedeza)	Good stand, uncut [6 to 8 in (15 to 20 cm)]
	Centipedegrass	Very dense cover [average 6 in (15 cm)]
	Kentucky bluegrass	Good stand, headed [6 to 12 in (15 to 30 cm)]
D	Bermuda grass	Good stand, cut to 2.5-in height (6 cm)
	Common lespedeza	Excellent stand, uncut [average 4.5 in (11 cm)]
	Buffalo grass	Good stand, uncut [3 to 6 in (8 to 15 cm)]
	Grass-legume mixture—fall, spring (orchard grass, redtop, Italian ryegrass, and common lespedeza)	Good stand, uncut [4 to 5 in (10 to 13 cm)]
	Lespedeza sericea	After cutting to 2-in height (5 cm) Very good stand before cutting
E	Bermuda grass	Good stand, cut to 1.5-in height (4 cm)
	Bermuda grass	Burned stubble

Source: SCS (1954).

26.5 STORAGE, TREATMENT, AND OVERFLOWS

The structural measures intended to control urban-runoff quantity are also assuming the role of pollutant trapping units under the name structural *best management practices (BMPs)*. According to Wanielista and Yousef (1993) a structural BMP requires a structural change in the flow-transport system. Nonstructural BMPs are preventive non-point-source control measures. The nonstructural BMPs include adoption of appropriate site development codes, minimization of impervious area in new development, public education on proper use of household chemicals, street sweeping and detection and elimination of illicit discharges into storm sewers, spill prevention and containment (Urbonas, 1993). The following description is based on Roesner (1993) (see also Kenel, et al., 1992). The U.S. EPA requires that all cities and urbanized counties with populations more than 100,000 apply for a permit to discharge stormwater. The permitting program requires the following elements: (1) General Information, (2) Adequate Legal Authority, (3) Source Identification, (4) Discharge Characterization, (5) Proposed Management Program, (6) Assessment of Controls, and (7) Fiscal Analysis.

The *General Information* identifies the municipality. *Adequate Legal Authority* refers to the creation and enforcement of ordinances to control pollutants by the municipality itself or through interagency agreements. *Source Identification* is the mapping of all outfalls greater than 36 in in diameter; all outfalls of 12 in and greater for industrial activity; and all National Pollutant Discharge Elimination System (NPDES) permitted point-source discharges to the storm-sewer system. It also requires the drainage area for each outfall; identification of land use and soil type; location of sanitary landfills, industrial facilities, open dumps, municipal incinerators, and Resource Conservation and Recovery Act (RCRA) hazardous-waste facilities; and a description of existing stormwater practices and structural controls. A 10-year projection of population and development activities considering high-, medium-, and low-growth scenarios should also be in place. *Discharge characterization* focuses on pollutant characteristics. From field screening, outfalls which carry dry-weather flows resulting from illicit connections of nonstormwater discharges to the storm-sewer system are identified. By land-use type monitor selected outfalls to characterize stormwater pollutants. Three storms with at least 0.1 in of rainfall and 72 h preceding dry period must be used. The spacing between storms must be at least one month. *Event mean concentration (EMC)* is computed by flow-weighted composite samples. A list of these pollutants is given in (Roesner, 1993). Based upon this analysis annual pollutant load for the entire drainage system as well as by each individual outfall are computed. Also, a long-term monitoring program should be set in place. *Proposed Management Program* includes the following elements: source controls and removal of illicit connections; installation and modification of structural treatment controls (BMPs) for new development; sediment and pollution control on construction sites. These controls are intended to achieve the reduction of pollutant load in stormwater discharges to the "*maximum extent practicable.*" *Assessment of Controls* provides an estimate of pollutant-load reduction achieved as a consequence of the proposed management program. *Fiscal Analysis* requires that for each of the five years of the permit period a capital- and operation- and maintenance-cost budget be provided as well as the identification of sources of funding for these expenditures.

26.5.1 Detention Ponds

These are ponds installed in the flow path to reduce peak flow and enhance trapping of pollutants (Whipple, et al., 1987; and Dorman, et al., 1988). Comprehensive analyses are given in Haan, et al. (1994) and Kibler, et al. (1995).

26.5.1.1 Peak Flow Reduction. The *sizing of a detention pond* for a specified reduction in peak flow for a chosen design-inflow hydrograph is accomplished by applying a reservoir routing procedure repeatedly. A description of *reservoir routing* is given in Chap. 24. The key equation is the discretized continuity equation given by

$$\frac{S_2 - S_1}{\Delta t} = \frac{I_1 + I_2}{2} - \frac{O_1 + O_2}{2} \tag{26.35}$$

An initial estimate of the water surface elevation h_2 at the end of time interval Δt is made. The storage S_2 and outflow O_2 are computed for this elevation from storage–elevation and outflow–elevation relationships respectively and substituted in Eq. (26.35) along with the known inflows I_1 and I_2 and beginning of time period, known storage S_1. The difference between the left-hand side and right-hand side quantities of Eq. (26.35) is the error. If the error is not small, revise h_2; if the error is small, update S_2 to be new S_1 and O_2 to be new O_1 and march one step ahead with the next inflow value and apply Eq. (26.35). This is an alternative to storage-indication routing described in Chap. 24.

A simplified procedure that avoids the repeated application of reservoir routing to choose an approximate pond size for a given reduction in peak flow is based on *dimensionless routing* (Akan, 1989; Kessler and Diskin, 1991; and McEnroe, 1992). Soil Conservation Service (1986) also provides such an end result, but based on numerous routings for many structures. The dimensionless routing procedure is developed as follows. The outflow $q = \eta\, h^u$, in which η is a constant which depends on the shape and size of the outlet. The exponent u depends on type of outlet and h is the water-surface elevation above the outlet. The detention unit water-storage volume varying with h is $V(h) = p\, h^w$.

Using a dimensionless hydrograph (e.g., SCS, 1986; McEnroe, 1992) of the form $f(t^*)$ the inflow hydrograph is denoted by $i(t) = I_p f(t^*)$ where I_p = peak inflow rate; $t^* = t/T_p$; T_p = time to inflow peak. The continuity equation involving inflow, outflow, and storage in dimensionless form becomes

$$R\frac{d}{dt^*}(q^*)^m + q^* = f(t^*) \tag{26.36}$$

in which $q^* = q(t)/I_p$; $R = p\eta^{-m} I_p^{m-1} T_p^{-1}$; $m = wu^{-1}$; and Kessler and Diskin (1991) and McEnroe (1992) recommend $f(t^*) = (t^*)^{N-1} \exp[(N-1)(1-t^*)]$ in which N is a parameter. Using the initial condition that $b^* = b/V_i$ where b = initial storage, V_i = inflow volume one may solve Eq. (26.36) by a numerical procedure such as the Runge-Kutta method. Using $Q^* = \max\{q^*(t^*)\}$ one may compute the dimensionless pond volume $V^* = R(Q^*)^m V_i^{-1} I_p T_p$ where Q^* is the ratio of peak outflow to peak inflow; and the pond volume $V_{pond} = V^* V_i$ and $V_* = \max\{V(h)/V_i\}$. It has been observed by both Kessler and Diskin (1991) and McEnroe (1992) that the parameter N does not play a crucial role in relating the dimensionless peak factor Q^* to the dimensionless maximum storage V^*. However, the parameter m which relates to the outlet type and storage growth does play a major role. Kessler and Diskin (1991) derive for level pool routing with $b = 0$, for a spillway outlet

$$V^* = 0.932 - 0.792\, Q^* \quad \text{for} \quad m = 0.666 \quad \text{and} \quad 0.2 \le Q^* \le 0.9 \tag{26.37}$$

and for a culvert outlet

$$V^* = 0.872 - 0.861\, Q^* \quad \text{for} \quad m = 2.0 \quad \text{and} \quad 0.2 \le Q^* \le 0.9 \tag{26.38}$$

It is emphasized that such equations let the designer choose the type of outlet desired. Also, note that the culvert outlet requires less storage over a spillway for the same amount of Q^*. As shown by Akan (1989) staged outlets can also be incorporated in the continuity equation Eq. (26.36).

26.5.1.2 Pollutant Trap Efficiency. As far as water quality is concerned the pollutant removal within these ponds is mostly by physical settling of particulate matter (Randall, et al., 1983; Haan, et al., 1994). *Drawdown time* is defined as the time required to empty an initially full pond. Many local governments use a drawdown time of 24 to 40 hours for the design to permit settling of particulate matter. Because of the sequential nature of runoff events, it is very likely that the detention unit will only be partly full at the onset of a runoff event. The actual detention time therefore, does not equal the drawdown time and is less than that. For detention ponds to work as *best management practices* (*BMPs*) and prevent pollutant loads from being transported downstream, two goals must be accomplished: (1) capture the pollutant load from the runoff by the pond, and (2) prevent the pollutant load from leaving the pond. The first goal is accomplished by providing sufficient volume within the pond to capture the runoff carrying the pollutant load. The second goal is accomplished by detaining the runoff in the pond for an extended time to allow the pollutant load to settle within the basin. These two goals are inseparable and in direct conflict. In the spirit of Akan (1990), Whipple, et al. (1987), and Wanielista and Yousef (1993), one can use the BMP pool volume meant as water-quality storage, as the initial condition to perform the flood routing for peak-flow control.

The methods of designing detention ponds generally fall into three categories: (1) *design-storm approach,* (2) *continuous-simulation modeling,* and (3) *statistical methods* which incorporate interevent times. The design-storm approach uses a single extreme event to size the basin to meet either a peak reduction or satisfy a drawdown-time requirement. Delleur and Padmanabhan (1981) and Goforth, et al. (1983) point out that, due to the sequential occurrence of runoff events, the available volume (empty space) for capturing an event is random and single-storm approaches do not account for it in the design. The second approach, involving continuous simulation, essentially duplicates the natural occurrence of runoff events and is very useful for analyzing the long-term performance of a given detention-basin configuration. By considering various alternative configurations, the engineer can select an appropriate design. However, continuous simulation can be time-consuming and data-intensive and for planning-stage calculations a simplified procedure may be sufficient. The third approach, involving statistical methods, also considers interevent times and accounts for the net empty space available between events. Because they are aimed towards developing simplified probability-based equations, statistical methods often incorporate certain assumptions. As a result, statistical methods are considered to be planning-level tools. The statistical planning methodologies by Howard (1976), Di Toro and Small (1979), and Loganathan, et al. (1985) concentrate on the fraction of untreated-runoff volumes leaving a detention basin. EPA (1986) and Driscoll (1989) interpret the results of Di Toro and Small for pollutant settling within a wet pond. The fraction of untreated overflow is interpreted as the fraction of pollutant that has not settled (Dorman, 1991). Haan, et al. (1994), Urbonas and Stahre (1993), and Wanielista and Yousef (1993) provide numerical examples of EPA (1986) methodology due to Driscoll (1989). Goforth, et al. (1983) carry out an extensive performance analysis of a detention basin using the EPA SWMM (Storm Water Management Model) computer program (Huber, et al., 1980). Chapter 28 has a description of SWMM.

Segarra and Loganathan (1992) and Loganathan, et al. (1994) present methods which provide an explicit, closed-form solution for the expected pollutant load and the *expected detention time* $E(D)$ under a random sequence of runoff events. Furthermore, to interpret this detention-time estimator in a usable way, a relationship between pollutant-settling efficiency η_{BMP} and expected detention time $E(D)$ is provided. The expected detention time is given as

$$E(D) = \exp\left[-\frac{\gamma}{a}(b-c)\right]\left\{\frac{c-b}{a} - \frac{1}{\gamma} + \left(\frac{m}{a} + \frac{k}{a}\right)(b-c) + \frac{m}{(\gamma-\beta)}\right.$$

$$\left. + \frac{k}{(\alpha a + \gamma)}\left(1 + \frac{\alpha a}{\gamma}\left(1 - \exp\left[-\frac{b}{a}(\alpha a + \gamma)\right]\right)\right)\right\} \quad (26.39)$$

$$- \frac{m}{(\gamma-\beta)}\exp\left[-\frac{\beta}{a}(b-c)\right] - \frac{k}{(\alpha a + \gamma)}\exp\left[-\alpha c - \frac{b}{a}\gamma\right] + \frac{1}{\gamma}$$

where α, β, γ = reciprocals of average runoff volume, duration, and interevent time respectively

c = empty space left over at the end of the previous runoff event = δb for $0 < \delta < 1$

a = outflow rate

b = detention pond size

$k = \beta\gamma/[(\alpha a + \beta)(\alpha a + \gamma)]$

$m = \alpha\gamma a/[(\alpha a + \beta)(\gamma - \beta)]$

The statistic α can be obtained from rainfall depth by using the runoff coefficient such as Eqs. (26.1) and (26.2). Typically, β and γ for rainfall and runoff are taken to be the same (Urbonas and Stahre, 1993). Eq. (26.39) can be used to evaluate upper- and lower-bound estimates for the detention time $E(D)$ by setting $c = 0$ and $c = b$ respectively, which are quite useful because these in turn set limit on the settling efficiency η_{BMP}. A series of SWMM simulations can be run to provide settling efficiency η_{BMP} versus $E(D)$ relation under the complete-mixing option. This relationship is given in Fig. 26.7 along with the settling column data curve developed by Randall, et al. (1983) at Virginia Tech's Occoquan Lab and reported by Schueler (1987). The overall efficiency is clearly dependent on how much runoff is captured and how well the captured pollutant load is treated (or removed by settling). The *capture efficiency* η_{FL} is given by Loganathan, et al. (1985) as

$$\eta_{FL} = 1 - k\left(\exp(-\alpha c) + \frac{\alpha a}{\gamma}\exp\left[-b\left(\alpha + \frac{\gamma}{a}\right) + \frac{\gamma}{a}c\right]\right) \quad (26.40)$$

The *overall efficiency* η is given by

$$\eta = (\text{Flow capture efficiency } \eta_{FL})\,[\text{BMP (settling) efficiency } \eta_{BMP}] \quad (26.41)$$

in which η_{FL} = fraction of volume captured; and η_{BMP} = fraction of pollutant concentration reduced by the detention unit as dictated by the detention time $E(D)$ given in Fig. 26.7. A crucial observation is that η_{FL} and η_{BMP} are not independent. This is because both are controlled by the design variables a and b as well as the runoff statistics. For a relatively small pond with a small withdrawal rate, the BMP efficiency η_{BMP} can be significant, but the capture efficiency η_{FL} will be so low that the resulting overall efficiency η may not be acceptable. Loganathan, et al. (1994) provide numerical examples.

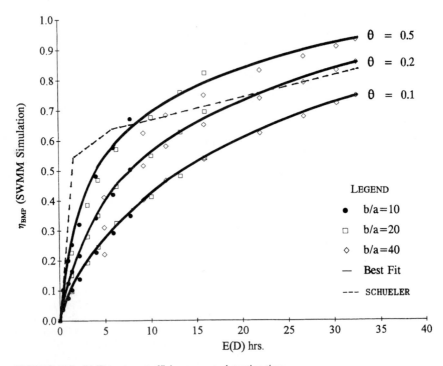

FIGURE 26.7   BMP treatment efficiency versus detention time.

26.5.2 Infiltration Trenches

The use of an infiltration trench is to collect surface runoff and to allow gradual infiltration to the surrounding soil. Of course the trench should have sufficient void space to store the stormwater. The Northern Virginia BMP handbook (1992) (see also "Design and Construction of Urban Stormwater Management Systems," ASCE, 1992; and "Stormwater and Nonpoint Source Control . . . ," 1992) recommends the following simplified procedure for the trench design. It is assumed that most of the infiltration into the surrounding soil takes place through the bottom of the trench as opposed to its sides; and the infiltration rate equals the vertical saturated hydraulic conductivity. Therefore by mass balance for a trench, the rate at which the void space is emptied must equal the product of the infiltration rate and the bottom area of the trench. That is, for a rectangular trench

$$\text{LEN} * \text{WID} * \text{DEP} * \text{porosity} = \text{LEN} * \text{WID} * K_S * \text{TIME} \qquad (26.42)$$

where   LEN = trench length
        WID = width
        DEP = depth
        K_S = saturated vertical hydraulic conductivity
      TIME = drainage time

Equation (26.42) yields the depth of the trench. The porosity and K_S must be based on soil type. The design void-space volume is taken to be the volume of a design-runoff event. It should be noted that Eq. (26.40) can be applied for a chosen level of the flow-capture efficiency to design the trench. An example problem of this kind is provided in Wanielista and Yousef (1993). The infiltration trenches are suitable only for small drainage areas. They do provide recharge to groundwater, but they are also highly susceptible to clogging by sediment.

26.6 ESTIMATING URBAN STORMWATER POLLUTANT LOADS

The diverse array of pollutants carried in urban runoff has been described by Roesner (1982) and Grizzard, et al. (1986). The discussion in this section is directed at various procedures for quantifying the pollutant load generated on urban watersheds.

26.6.1 Metropolitan Washington COG Method

This section presents an empirical method, called *MWCOG*, for estimating pollutant export from urban-development sites. The MWCOG method is empirical in nature, and utilizes the extensive database obtained in the Washington, D.C. area NURP (National Urban Runoff Program) study, as well as the NURP data analysis. The MWCOG method is versatile in that it predicts pollutant loadings under a variety of planning conditions. It can also be used to estimate the probability that pollutant concentrations exceed a given threshold level. The coverage presented here is taken from Schueler (1987).

The MWCOG method is primarily intended for use on developing sites less than a square mile in area. Despite its simplicity, the MWCOG method is sufficiently precise to make reasonable non-point-pollution management decisions at the site-planning level. *Storm pollutant export L,* lb, from a development site for any time can be determined by solving the following equation:

$$L = [(P) (Pj) (Rv)/12] (C) (A) (2.72) \qquad (26.43)$$

where P = rainfall depth, in, over the desired time interval
Pj = factor that corrects P for storms that produce no runoff
Rv = runoff coefficient, which expresses the fraction of rainfall which is converted to runoff
C = flow-weighted mean concentration of the pollutant in urban runoff, mg/L
A = area of development site, acres
12, 2.72 are unit conversion factors.

The user need only define five parameters, each of which is readily determined from site plan data, or are constants.

26.6.2 P (Depth of Rainfall)

The value of P selected depends on the time interval over which loading estimates are desired. For a normal year of rainfall, P will be about 40 in in the Washington, D.C. area. Values of 30 and 50 in can be used to characterize extremely dry and wet years, respectively. Long-term rainfall records from National Weather Service (NWS) stations should be used to estimate P in other regions of the country. If a load estimate is desired for a specific design storm or year of record, then the user can supply the relevant value of P.

26.6.3 *Pj* (Correction Factor)

The value of *Pj* is used to account for the fraction of annual or seasonal rainfall that does not produce any measurable runoff. Approximately 50 percent of the storms each year drop less than 0.2 in of precipitation. Storms of this size usually produce little or no runoff. An analysis of the Washington, D.C. area rainfall–runoff patterns indicates that only 90 percent of rainfall events produce runoff. Therefore, *Pj* should be set to 0.9 for annual and seasonal calculations. For individual storms, *Pj* should be set to 1.0 to avoid double counting.

26.6.4 *Rv* (Runoff Coefficient)

Rv is the measure of site response to rainfall events, and is calculated as:

$$Rv = \frac{r}{p} \tag{26.44}$$

where r = storm runoff, in
         p = storm rainfall, in

The *Rv* for a site depends on the nature of the soils, topography, and cover. However, the primary influence on the *Rv* is the degree of watershed imperviousness. The relation between the mean *Rv* and the degree of watershed imperviousness for 47 small urban catchments monitored throughout the region and the nation is shown in Eq. (26.45). Although some scatter was evident in the data plot, watershed imperviousness *I* does appear to be a reasonable predictor of the *Rv*. The following equation [presented previously as Eq. (26.2)] represents the best-fit line through the dataset (adjusted $R^2 = 0.71$):

$$Rv = 0.05 + 0.009\ (I) \tag{26.45}$$

where I = percent imperviousness of site.

Values for I are readily obtained from site plans or accompanying hydrological computations. This is done by summing the area of the site covered by structures, sidewalks, driveways, parking lots, roads, patios, and other impermeable areas (by planimetry or square counting) and dividing it by the total site area.

26.6.5 EPA/University of Florida Method

Previous discussions on the use of simplified desk-top procedures for *non-point-source (NPS) pollutant assessment* have been presented by Lakatos and Johnson (1979), Zison (1977), and by the EPA (1976). The discussion here is taken from an earlier report by Kibler (1982). Typical functions for estimating annual wet-weather and dry-weather flows are shown below. Annual average pollutant-loading rates also can be computed by the method shown below. The relationships shown below (Heaney, et al., 1977) were developed for EPA by the University of Florida on the basis of data collected in 248 standard metropolitan statistical areas as part of a nationwide evaluation of combined sewer overflows and stormwater discharges.

26.6.6 Annual Stormwater and Dry-Weather Quantity Prediction

The following equations for total annual storm runoff were developed by Heaney, et al. (1977):

$$AR = \left(\frac{0.15 + 0.75\,I}{100} \right)P - 5.23(DS)^{0.5957} \tag{26.46}$$

where AR is the annual runoff, in in/y.

$$I = 9.6PD_d^{0.573 \,-\, 0.039\,\log\,PD_d} \tag{26.47}$$

where I is the imperviousness, percent; PD_d is the population density in developed portion of the urbanized area, persons/acre; P is the annual precipitation, in/y, and

$$DS = 0.25 - 0.1875 \left(\frac{I}{100} \right) \qquad 0 \le I \le 100 \tag{26.48}$$

where DS is the depression storage, in $(0.005 \le DS \le 0.30)$.

For annual dry-weather flow, the following equation applies (Heaney, et al., 1977):

$$DWF = 1.34\,PD_d \tag{26.49}$$

where DWF is the annual dry-weather flow, in/y; and PD_d is the developed population density, persons/acre.

26.6.7 Annual Pollutant Loading Prediction

The following equations may be used to predict *annual average loading rates* as a function of land use, precipitation and population density (see Heaney, et al., 1977; and U.S. EPA, 1976):

Separate sewer areas

$$M_S = \alpha(i, j) \cdot P \cdot f_2(PD_d) \cdot \gamma \frac{lb}{acre \cdot y} \tag{26.50}$$

Combined sewer areas

$$M_C = \beta(i, j) \cdot P \cdot f_2(PD_d) \cdot \gamma \frac{lb}{acre \cdot y} \tag{26.51}$$

where
M = pounds of pollutant j generated per acre of land use i per year
P = annual precipitation, in/y
PD_d = developed population density, persons/acre
α, β = factors given in Table 26.5
γ = street-sweeping effectiveness factor
$f_2\,(PD_d)$ = population density function

Land uses

$i = 1$ residential

$i = 2$ commercial

$i = 3$ industrial

$i = 4$ other developed, e.g., parks, cemeteries, schools (assume $PD_d = 0$)

Pollutants

$j = 1$ BOD_5, total

$j = 2$ suspended solids (SS)

$j = 3$ volatile solids, total (VS)

$j = 4$ total PO_4 (as PO_4)

$j = 5$ total N

Population function

$$i = 1 \qquad f_2(PD_d) = 0.142 + 0.218 \cdot PD_d^{0.54}$$

$$i = 2,3 \qquad f_2(PD_d) = 1.0 \tag{26.52}$$

$$i = 4 \qquad f_2(PD_d) = 0.142$$

Factors α and β for separate sewers α and combined sewers β have units lb/acre · in. To convert to kg/ha · cm, multiply by 0.422. See Table 26.5.

Street sweeping factor γ is a function of street sweeping intervals N_S, days:

$$\gamma = N_S/20 \quad \text{if} \quad 0 \le N_S \le 20 \text{ days} \tag{26.53}$$

$$\gamma = 1.0 \quad \text{if} \quad N_S > 20 \text{ days}$$

As an example of the above methodology, annual stormwater runoff and pollutant loadings have been estimated for the Calder Alley basin in State College, Pennsylvania treated as a separate storm-drain system. The estimated average annual runoff quantity has been computed by equation (26.46), as shown in Table 26.6. The computation of average annual pollutant loads and concentrations is shown in Table 26.7 based upon equation (26.50). This example is taken from Kibler, et al. (1982).

TABLE 26.5 Pollutant Loading Factors α and β for Separate and Combined Sewer Areas

	Land use i	Pollutant*				
		BOD_5	SS	VS	PO_4	N
Separate areas, α	Residential	0.799	16.3	9.4	0.0336	0.131
	Commercial	3.200	22.2	14.0	0.0757	0.296
	Industrial	1.210	29.1	14.3	0.0705	0.277
	Other	0.113	2.7	2.6	0.0099	0.060
Combined areas, β	Residential	3.290	67.2	38.9	0.1390	0.540
	Commercial	13.200	91.8	57.9	0.3120	1.220
	Industrial	5.000	120.0	59.2	0.2910	1.140
	Other	0.467	11.1	10.8	0.0411	0.250

* Loading factors for each pollutant have units of lb/acre · in.
Source: EPA (1976) and Heaney, et al. (1977).

TABLE 26.6   Estimated Annual Stormwater Runoff from Calder Alley System

Land use type	Area, acres*	Impervious percent*	Depression storage, in[†]	Annual Runoff, in[‡]
Residential	93	30	.19	12.30
Commercial	134	53	.15	19.11
Industrial	—	—	—	—
Other	—	—	—	—

* From land use maps, State College, Pa.
[†] Computed from Eq. (26.48).
[‡] Computed from Eq. (26.46).
Source: Kibler, et al. (1982).

26.6.8   U.S. Geological Survey Storm Load Functions

The three-variable models for predicting storm loads by the USGS method developed by Driver and Tasker (1988) are summarized in Table 26.8. The explanation of individual terms in the USGS models follows.

$\hat{\beta}_0$ is the regression coefficient that is the intercept in the regression model; TRN is total storm rainfall, in; DA is total contributing drainage area in mi^2; IA is impervious area as percent; BCF is bias correction factor; COD is chemical oxygen demand in storm-runoff load, in lb; I is Region I representing areas that have mean annual rainfall less than 20 in; II is Region II representing areas that have mean annual rainfall of 20 to less than 40 in; III is Region III representing areas that have mean annual rainfall equal to or greater than 40 in; SS is suspended solids in storm-runoff load, in lb; TN is total nitrogen in storm-runoff load, in lb; TKN is total ammonia plus organic nitrogen as nitrogen in storm-runoff load, in lb; TP is total phosphorus in storm-runoff load, in lb; DP is dissolved phosphorus in storm-runoff load, in lb; Cd is total recoverable cadmium in storm-runoff load, in lb; Cu is total recoverable copper in storm-runoff load, in lb; Pb is total recoverable lead in storm-runoff load, in lb; Zn is total recoverable zinc in storm-runoff load, in lb; asterisk (*) indicates that the explanatory variable is not significant at the 5 percent level. The general equation form is:

$$Y = \hat{\beta}_0 \times X_1^{\beta_1} \times X_2^{\beta_2} \ldots X_n^{\beta} n \times B C F] \qquad (26.54)$$

TABLE 26.7   Estimated Average Annual Pollutant Loads and Concentrations for Calder Alley Storm Drain System

Land use	Area, acres	Pop. factor	Annual runoff, acre · in*	BOD₅, lbs[†]	BOD₅ mg/L[‡]	Suspended solids, lbs[†]	(mg/L[‡])	Phosphates, lbs[†]	(mg/L[‡])	Nitrogen, lb[†]	(mg/L[‡])
Residential	93	1.24	1144	3501	14	71,429	276	147	1	574	2
Commercial	134	1.00	2561	16294	28	113,042	195	385	1	1507	3
Industrial	—	—	—	—	—	—	—	—	—	—	—
Other	—	—	—	—	—	—	—	—	—	—	—

* Volume of annual runoff obtained by multiplying runoff depth in Table 26.6 by contributing area for respective land use.
[†] Pollutant load in pounds computed from loading factors in Table 26.5 and Eq. (26.50) with $\gamma = 1$ for street sweeping interval > 20 days and average annual precipitation = 38 in.
[‡] Pollutant concentrations computed by dividing respective pollutant loads by annual runoff volume and multiplying by 4.42. This constant is the number of mg/L per lb/acre · in.
Source: Kibler, et al. (1982).

TABLE 26.8 Summary of USGS Storm-Runoff Pollutant Load Models

Responsible variable and region	Regression coefficients						Standard error of estimate		Number of storms
	$\hat{\beta}_0$	TRN, in	DA, mi^2	IA + 1, %	BCF	R^2	%	log	
COD I	407	0.626	0.710	0.379	1.518	0.62	116	0.403	216
COD II	151	.823	.726	.564	1.451	.67	106	.376	793
COD III	102	.851	.601	.528	1.978	.46	186	.531	567
SS I	1,778.0	.867	.728	.157	2.367	.52	251	.613	176
SS II	812.0	1.236	.436	.202	1.938	.60	173	.512	964
SS III	97.7	1.002	1.009	.837	2.818	.53	290	.651	528
DS I	20.70	.637	1.311	1.180	1.249	.93	75	.293	175
DS II	3.26	1.251	1.218	1.964	1.434	.86	101	.367	281
TN I	20.20	.825	1.070	.479	1.258	.92	72	.286	121
TN II	4.04	.936	.937	.692	1.373	.77	97	.353	574
TN III	1.66	.703	.465	.521	1.845	.31	178	.518	617
TKN I	13.90	.722	.781	.328	1.722	.65	129	.431	188
TKN II	3.89	.944	.765	.556	1.524	.75	107	.381	858
TKN III	3.56	.808	.415	.199	1.841	.32	184	.529	613
TP I	1.725	.884	.826	.467	2.130	.56	184	.529	186
TP II	.697	1.008	.628	.469	1.790	.62	120	.411	1,091
TP III	1.618	.954	.789	.289	2.247	.50	210	.565	639
DP I	.540	.976	.795	.573	2.464	.55	161	.492	248
DP II	.060	.991	.718	.701	1.757	.63	121	.412	467
DP III	2.176	1.003	.280	−.448	2.254	.35	193	.542	247
Cd I	0.00001	0.886	.821	2.033	1.425	.72	101	0.365	65
Cd II	.021	1.367	1.062	.328	1.469	.62	109	.386	47
Cu I	.072	.746	.797	.514	1.675	.58	134	.440	212
Cu II	.013	.504	.585	.816	1.548	.55	123	.417	298
Cu III	.026	.715	.609	.642	2.819	.35	263	.625	464
Pb I	.162	.839	.808	.744	1.791	.59	166	.500	239
Pb II	.150	.791	.426	.522	1.665	.43	135	.442	943
Pb III	.080	.852	.857	.999	2.826	.54	228	.586	418
Zn I	.320	.811	.798	.627	1.639	.60	146	.465	224
Zn II	.046	.880	.808	1.108	1.813	.51	166	.500	357
Zn III	.024	.793	.628	1.104	2.533	.43	200	.551	591

Source: Driver and Tasker (1988).

26.7 POLLUTANT SETTLEABILITY

One of the first attempts to relate pollutant removal to detention time was that by Randall, et al. (1982) who conducted a series of column-settling experiments and established the removal efficiency curves still in use today. This data has been supplemented by experimental settling-column data reported by Grizzard, et al. (1986), Driscoll (1986), and Whipple and Hunter (1981). In each study, urban runoff was introduced into 4- to 6-ft-deep plexiglass chambers and the change in pollutant concentration over time was measured at sampling ports located at different depths in the column. In addition, a number of field monitoring studies on pollutant removal by detention ponds have been reported (MWCOG, 1983; OWML, 1986). More recently, Yu, et al. (1993) have reported typical pollutant removal for dry ponds ranging from 0 to 20 percent; while for extended detention ponds the removal efficiency can be in the 40 to 70 percent range for particulate forms (Yu and Kaign, 1992; Yu, et al., 1993). The Northern Virginia Planning District Commission (1992) has further confirmed the importance of detention time in the pollutant-removal process and has documented the disadvantages of the dry pond in this regard. The affinity of particulate pollutants for suspended sediment, including metals and organic constituents, has been documented by Tai (1991) and by Wanielista and Yousef (1993). Yu, et al. (1994) have reported pollutant removals of 50 percent for total suspended solids, 40 percent for total phosphorus, 35 percent for COD, and 30 percent for total zinc in an actual field-scale test of a retrofit extended detention pond in the Charlottesville, Virginia area. Excellent summaries of average removal efficiencies have been compiled by Schueler (1987), and most recently by Ostrowski (1994), for dry ponds, extended detention ponds, and wet ponds.

26.7.1 Pollutant Removal by Extended Detention Ponds

As pointed out by Schueler (1987), extending the detention time of dry or wet ponds can be an effective, low-cost means of removing particulate pollutants and controlling increases in downstream bank erosion. If stormwater is detained for 24 hours or more, as much as 90 percent removal of particulate pollutants is possible. However, extended detention only slightly reduces levels of soluble phosphorus and nitrogen found in urban runoff. Removal of these pollutants can be enhanced if the normally inundated area of the pond is managed as a shallow marsh or a permanent pool.

Extended detention ponds can also reduce the frequency of erosive floods, depending on the quantity of stormwater detained and the time over which it is released. According to Schueler (1987), extended detention is quite cost-effective, with construction costs seldom more than 10 percent above those reported for conventional dry ponds. Improved methods of estimating detention time in BMP or extended detention ponds has already been discussed in Sec. 26.5.1.2. Again, reference is made to the practical manual for planning urban BMPs by Schueler and the Metropolitan Washington Council of Governments (1987). Additional references on urban stormwater-quality modeling and management include ASCE Manual no. 77, (1992) and the recent text on municipal stormwater management by Debo and Reese (1995).

REFERENCES

Akan, O., "Detention Pond Sizing for Multiple Return Periods," *Journal of Hydraulic Engineering,* 115(5):650–664, 1989.

Akan, O., "Single-outlet Detention-Pond Analysis and Design," *Journal of Irrigation and Drainage Engineering,* 116(4):527–536, 1990.

American Society of Civil Engineers, "Design and Construction of Urban Stormwater Management Systems," *ASCE Manual and Report of Engineering Practice no. 77, WEF Manual of Practice FD-20,* New York, 1992.

Bathurst, J. C., R. M. Li, and D. B. Simons, "Resistance Equation for Large-Scale Roughness," *Journal of the Hydraulic Division, ASCE,* 107(HY12):1593–1613, 1981.

Bedient, P. B., and W. C. Huber, *Hydrology and Floodplain Analysis,* Addison-Wesley, Reading, Mass., 1992.

Burke, C. B., *County Storm Drainage Manual,* Highway Extension and Research Project for Indiana Counties, Purdue University—Engineering Experiment Station, 1981.

Chen, Y. H., and G. K. Cotton, "Design of Roadside Channels with Flexible Linings," Hydraulic Engineering Circular 15, FHWA-IP-87-7, Federal Highway Administration, McLean, Va., 1988.

Chow, V. T., D. R. Maidment, and L. W. Mays, *Applied Hydrology,* McGraw Hill, New York, 1988.

Debo, T. N., and A. J. Reese, *Municipal Storm Water Management,* Lewis Publishers, Boca Raton, Fla., 1995.

Delleur, J. W., and G. Padmanabhan, "An Extended Statistical Analysis of Synthetic Nonpoint Urban Quantity and Quality Data," *Proc. Internat. Symp. on Urban Hydrology, Hydraulics, and Sediment Control,* University of Kentucky, Lexington, 1981, pp. 229–238.

Dendrou, S. A., C. I. Moore, and R. S. Taylor, "Coastal Flooding Hurricane Storm Surge Model," Technical Report to FEMA, Camp, Dresser, and McKee, Inc., Annandale, Va., 1985.

Di Toro, D. M., and M. J. Small, "Stormwater Interception and Storage," *Journal of Environmental Engineering Division,* ASCE, 105(EE1):43–54, 1979.

Dorman, M. E., "A Methodology for the Design of Wet Detention Basins for Treatment of Highway Stormwater Runoff," M.S. thesis, Virginia Tech, 1991.

Dorman, M. E., J. Hartigan, F. Johnson, and B. Maestri, "Retention, Detention, and Overland Flow for Pollutant Removal from Highway Stormwater Runoff," FHWA/RD-87/056, McLean, Va., 1988.

Driscoll, E. D., "Lognormality of Point and Non-point Source Pollutant Concentrations," in B. Urbonas and L. Roesner, eds., *Urban Runoff Quality: Impact and Quality Enhancement Technology,* American Society of Civil Engineers, New York, 1986.

Driscoll, E. D., "Long Term Performance of Water Quality Ponds," *Design of Urban Runoff Quality Controls,* American Society of Civil Engineers, New York, 1989.

Driver, N. E., and G. D. Tasker, "Techniques for Estimation of Storm-Runoff Loads, Volumes and Selected Constituents Concentrations in Urban Watersheds in the U.S.," Open File Report 88-191, U.S. Geological Survey, Denver, 1988.

Federal Emergency Management Agency, "Guidelines and Specifications for Study Contractors," FEMA 37, Washington, D.C., 1992.

Federal Highway Administration, "Hydraulic Design of Highway Culverts," Hydraulic Design Series no. 5, FHWA-IP-85-15, 1985.

Feldman, A., "HEC Models for Water Resources System Simulation: Theory and Experience," in V. T. Chow, ed., *Advances in Hydroscience* vol. 12, Academic Press, San Diego, Calif., 1981.

Goforth, G. F., J. P. Heaney, and W. C. Huber, "Comparison of Basin Performance Modeling Techniques," *Journal of Environmental Engineering Division,* ASCE, 1983, pp. 1082–1098.

Gray, D. D., and J. D. Quaranta, GSDPM3—Gravity Sewer Design Program: Users Manual, Department of Civil Engineering, West Virginia University, Morgantown, W.Va., 1992.

Grizzard, T. J., C. W. Randall, B. L. Weand, and K. L. Ellis, "Effectiveness of Extended Retention Ponds," *Urban Runoff Quality,* American Society of Civil Engineers, New York, 1986.

Gupta, R. S., *Hydrology and Hydraulic Systems,* Prentice-Hall, Englewood Cliffs, N.J., 1989.

Haan, C. T., B. J. Barfield, and J. C. Hayes, *Design Hydrology and Sedimentology for Small Catchments,* Academic Press, San Diego, Calif., 1994.

Heaney, J. P., W. C. Huber, and S. J. Nix, "Storm Water Management Model: Level I Preliminary Screening Procedures," U.S. Environmental Protection Agency, Report EPA-600/2-76-275, MERL, Cincinnati, Ohio, 1976.

Heaney, J. F., et al., "Nationwide Evaluation of Combined Sewer Overflows and Urban Stormwater Discharges," vol. II, Cost Assessment and Impacts, Report EPA-600/2-77-064, Environmental Protection Agency, Washington, D.C., 1977.

Howard, C. D., "Theory of Storage and Treatment-Plant Overflows," *Journal of Environmental Engineering Division,* ASCE, 1976, pp. 709–722.

Huber, W. C., J. P. Heaney, and S. J. Nix, "Stormwater Management Model User's Manual—Version III," EPA Draft Report, National Environmental Research Center, Cincinnati, Ohio, 1980.

Huber, W. C., and R. E. Dickinson, "Storm Water Management Model, Version 4: User's Manual," Department of Environmental Engineering Sciences, University of Florida, Gainesville, Fla., Aug. 1988.

Hydrologic Engineering Center, *HEC-2 Water Surface Profiles,* U.S. Army Corps of Engineers, Davis, Calif., 1982.

Jennings, M. E., "Data Collection and Instrumentation," in D. Kibler, ed., *Urban Stormwater Hydrology,* Chap. 7, Water Resources Monograph 7, American Geophysical Union, Washington, D.C., 1982.

Johnson, F. L., and F. F. M. Chang, "Drainage of Highway Pavements," Hydraulic Engineering Circular 12, FHWA-TS-84-202, McLean, Va., 1984.

Kenel, P. P., J. M. Mussman, and H. O. Andrew, "Norfolk Develops A Comprehensive Stormwater Program," *Public Works,* 1992, pp. 44–46.

Kessler, A., and M. H. Diskin, "The Efficiency Function of Detention Reservoirs in Urban Drainage Systems," *Water Resources Research,* 27(3):253–258, 1991.

Kibler, D. F., *Urban Stormwater Hydrology,* Water Resources Monograph 7, American Geophysical Union, Washington, D.C., 1982.

Kibler, D. F., A. O. Akan, G. Aron, M. W. Glidden, and R. H. McCuen, "Urban Hydrology," in *Hydrology Manual 28,* Chap. 9 American Society of Civil Engineers, New York, 1995.

Kouwen, N., T. E. Unny, and H. M. Hill, "Flow Retardance in Vegetated Channel," *Journal of the Irrigation and Drainage Division,* ASCE, 95(IR2):329–342, 1969.

Kuo, C. Y., "Urban Stormwater Management in Coastal Areas," *Proceedings of the National Symposium,* ASCE, New York, 1980.

Lakatos, D. F., and G. M. Johnson, "Desk-top Analysis of Nonpoint Source Pollutant Loads," paper presented at Sixth International Symposium on Urban Runoff, University of Kentucky, Lexington, July 1979.

Loganathan, G. V., E. W. Watkins, and D. F. Kibler, "Sizing Stormwater Detention Basins for Pollutant Removal," *Journal of Environmental Engineering,* ASCE, 120(6):1380–1399.

Loganathan, G. V., J. W. Delleur, and R. I. Segarra, "Planning Detention Storage for Stormwater Management," *Journal of Water Resources Planning and Management Division,* ASCE, 111(4):382–398, 1985.

Mays, L. W., and Y. K. Tung, *Hydrosystems Engineering and Management,* McGraw-Hill, New York, 1992.

McEnroe, B. M., "Preliminary Sizing of Detention Reservoirs to Reduce Peak Discharges," *Journal of Hydraulic Engineering,* ASCE, 118(11):1540–1549, 1992.

Metropolitan Washington Council of Governments (MWCOG), Urban Runoff in the Metropolitan Washington Area, Final Report to U.S. Environmental Protection Agency, 1983.

Myers, V. A., "Joint Probability Method of Tide Frequency Analysis Applied to Atlantic City and Long Beach Island, NJ," ESSA Technical Memorandum II, U.S. Department of Commerce, Office of Hydrology, Silver Spring, Md., 1970.

National Association for Home Builders, *Land Development Manual,* Washington, D.C., 1974.

New Jersey Department of Environmental Protection and Energy, "Stormwater and Nonpoint Source Pollution Control—Best Management Practices Manual," Trenton, N.J., 1992.

Normann, J. M., R. J. Houghtalen, and W. J. Johnston, "Hydraulic Design of Highway Culverts," Hydraulic Design Series no. 5, FHWA-IP-85-15, 1985.

Northern Virginia Planning District Commission, "BMP Handbook for the Occoquan Watershed," Annandale, Va., Aug. 1987.

Northern Virginia Planning District Commission and Engineers and Surveyors Institute, "Northern Virginia BMP Handbook," Annandale, Va., 1992.

Occoquan Watershed Monitoring Lab (OWML), "Report on the Loudon Commons Extended Detention Pond Monitoring Study," prepared for Virginia Department of Conservation and Recreation, Richmond, Va., 1986.

Ostrowski, T. J., "Quantification of Urban Stormwater Pollutant Removal Efficiency of Detention Basins and Wetlands," Master of Science thesis, Department of Civil Engineering, Penn State University, University Park, Pa., 1994.

Randall, C. W., K. Ellis, T. J. Grizzard, and W. R. Knocke, "Urban Runoff Pollutant Removal by Sedimentation," *Proc. Stormwater Detention Facilities,* New England College, New Hampshire, Aug. 2–6, 1983.

Roesner, L. A., "Overview of Federal Law and USEPA Regulations for Urban Runoff," in J. Marsalek and H. C. Torno, eds., *Proceedings of the Sixth International Conference on Urban Storm Drainage,* Niagara Falls, Canada, 1993.

Sauer, V. B., W. O. Thomas, Jr., V. A. Strieker, and K. V. Wilson, "Flood Characteristics of Urban Watersheds in the United States," U.S. Geological Survey, *Water Supply Paper 2207,* Reston, Va., 1983.

Schueler, T. B., Controlling Urban Runoff: A Practical Manual for Planning and Designing Urban BMP's, Washington Metropolitan Water Resources Planning Board, 1987.

Segarra, R. I. and G. V. Loganathan, "Stormwater Detention Storage Design Under Random Pollutant Loading," *Journal of Water Resources Planning and Management,* ASCE, 118(5):475–491, 1992.

Singh, V. P., *Elementary Hydrology,* Prentice-Hall, Englewood Cliffs, N.J., 1992.

Soil Conservation Service (SCS), *Handbook of Channel Design for Soil and Water Conservation,* SCS-TP-61, Stillwater, Okla., 1954.

Soil Conservation Service (SCS), *Urban Hydrology for Small Watersheds,* U.S. Department of Agriculture, Washington, D.C., 1986.

Tai, Y. L., "Physical and Chemical Characteristics of Street Dust and Dirt from Urban Areas," M.S. thesis, Department of Civil Engineering, Penn State University, University Park, Pa., 1991.

Terstriep, M. L., and J. B. Stall, The Illinois Urban Drainage Area Simulator, ILLUDAS, Illinois State Water Survey, 1974.

Urbonas, B., "Assessment of BMP Use and Technology Today," in J. Marsalek and J. C. Torno, eds., *Proceedings of the Sixth International Conference on Urban Stormwater Drainage,* Niagara Falls, Canada, 1993.

Urbonas, B., and P. Stahre, *Stormwater,* Prentice-Hall, Englewood Cliffs, N.J., 1993.

Urbonas, B. R., and L. A. Roesner, "Hydrologic Design for Urban Drainage and Flood Control," in D. R. Maidment, ed., *Handbook of Hydrology,* Chap. 28, McGraw-Hill, New York, 1993.

U.S. Environmental Protection Agency, "Areawide Assessment Procedures Manual," vol. 1, Report EPA-600/9-76-004, Washington, D.C., July 1976.

U.S. Environmental Protection Agency, "Methodology for Analysis of Detention Basins for Control of Urban Runoff Quality," EPA 440/5-87-001, Office of Water, Nonpoint Source Branch, Washington D.C., 1986.

Velon, J. P., and R. J. Persaud, "Wastewater Conveyance and Treatment," in V. J. Zipparro, and H. Hasen, eds., *Davis' Handbook of Applied Hydraulics,* sect. 28, McGraw-Hill, New York, NY, 1993.

Viessman, Jr., W., G. L. Lewis, and J. W. Knapp, *Introduction to Hydrology,* Harper & Row, New York, 1989.

Wang, S. Y. and H. W. Shen, "Incipient Sediment Motion and Riprap Design," *Journal of the Hydraulics Division,* ASCE, 111(3):521–538, 1985.

Wanielista, M., *Hydrology and Water Quantity Control,* John Wiley, New York, 1990.

Wanielista, M. P., and Y. A. Yousef, *Stormwater Management,* Wiley Interscience, New York, 1993.

Wenzel, H. G., "Rainfall for Urban Stormwater Design," in D. F. Kibler, ed., *Urban Stormwater Hydrology,* Water Resources Monograph 7, American Geophysical Union, Washington, D.C., 1982.

Whipple, Jr., W., and J. V. Hunter, "Settleability of Urban Runoff Pollution," *J. Water Pollution Control Federation,* 53(1):1726–1732, 1981.

Whipple, W., N. S. Grigg, T. Grizzard, C. W. Randall, R. P. Shubinski, and L. S. Tucker, *Stormwater Management in Urbanizing Areas,* Prentice-Hall, Englewood Cliffs, N.J., 1983.

Whipple, Jr., W., R. Kropp, and S. Burke, "Implementing Dual Purpose Stormwater Detention Program," *Journal of Water Resources Planning and Management,* ASCE 113(6):779–792, 1987.

Woo, D. C., "Inlets in Stormwater Modeling," in P. Balmer, P. A. Malmqvist, and A. Sjobert, eds., *Proceedings of the Third International Conference on Urban Storm Drainage* Goteborg, Sweden, 1984.

Yen, B. C., ed., *Advances in Hydroscience,* vol. 14, Academic Press, San Diego, Calif., 1986.

Yu, S. L., and R. J. Kaighn Jr., "VDOT Manual of Practice for Planning Stormwater Management," Virginia Transportation Research Council, Charlottesville, Va., 1992.

Yu, S. L., S. L. Barnes, and V. W. Gerde, "Testing of Best Management Practices for Controlling Highway Runoff," Virginia Transportation Research Council, Charlottesville, Va., 1993.

Yu, S. L., S. L. Barnes, R. J. Kaighn, Jr., and S. L. Liao, "Pond Modification to Improve Water Quality," *Proceeding of the Symposium on Stormwater Runoff and Quality Management,* Penn State University, University Park, Pa., 1994.

Zison, S. W., Water Quality Assessment: "A Screening Method for Non-designated 208 Areas," Report EPA-600/9-77-023, Environmental Protection Agency, Washington, D.C., Aug. 1977.

CHAPTER 27

FEDERAL PERSPECTIVE FOR FLOOD-DAMAGE-REDUCTION STUDIES

U.S. Army Corps of Engineers
Hydrologic Engineering Center

This chapter was adapted from U.S. Army Corps of Engineers (USACE) guidance on hydrologic engineering requirements for performing flood-damage-reduction studies. It represents the federal perspective. The guidance was funded and prepared under direction of the Hydrologic Engineering Center (HEC), U.S. Army Corps of Engineers. Under separate contracts to the Hydrologic Engineering Center, Bechtel Corporation developed the initial materials and Dr. David Ford, Consulting Engineer, drafted the majority of the text.

27.1 INTRODUCTION

27.1.1 The Flood-Damage-Reduction Planning Problem

27.1.1.1 Overview. The Federal objective in flood-damage-reduction planning is to identify a plan that will reduce the flood-damage problem and ". . . contribute to national economic development consistent with protecting the Nation's environment, pursuant to national environmental statutes, applicable executive orders, and other Federal planning requirements [U.S. Water Resources Council (WRC), 1983]." Typically, this is accomplished by formulating a set of likely solutions, evaluating each in terms of the national economic development and other standards, comparing the results, and identifying the recommended plan from amongst the set.

27.1.1.2 Basis for Comparison. The measure of a flood-damage-reduction plan's contribution to *national economic development* (*NED*) is the *net benefit* of the plan. This is computed as the sum of location benefit, intensification benefit, and flood-inundation-reduction benefit, less the total cost of implementing, *operating, maintaining, repairing, replacing, and rehabilitating* (*OMRR&R*) the plan. *Location benefit* is the increased net income of additional floodplain development due to a plan. *Intensification benefit* is the increased net income of existing floodplain activities. *Inundation-reduction benefit* is the plan-related reduction in physical economic damage, income loss, and emergency cost.

27.1.1.3 Plan Components. A *flood-damage-reduction plan* includes one or more of the flood-damage-reduction measures listed in Table 27.1. The planning study determines which of these measures to include in the plan, where to locate the measures, what size to make the measures, and how to operate the measures. According to WRC guidelines, a study proceeds by formulating, evaluating, and comparing "various alternative plans . . . in a systematic manner." That is, candidate combinations of measures, with various locations, sizes, and operating schemes, are proposed. Each alternative is evaluated with the criteria described in the previous section. Of those formulated and evaluated, the alternative that reasonably yields the greatest NED contribution is referred to colloquially as the NED plan.

Subsequent sections in this chapter provide guidance on selecting appropriate locations, sizes, and operation policies and describe how the inundation-reduction benefit due to each of the measures can be estimated.

27.1.1.4 Standards. In addition to yielding maximum NED contribution, the flood-damage-reduction plan recommended for implementation must

- Protect the environment, consistent with the National Environmental Policy Act (NEPA) and other laws, orders, and requirements
- Be complete, efficient, effective, and acceptable (U.S. WRC, 1983), consistent with regulations, orders, and other legal requirements

27.1.2 Procedure for Finding a Solution to the Planning Problem

The federal approach to solving the flood-damage-reduction problem is through a sequential process that involves planning, design, construction, and operation. Planning or feasibility studies are performed in two phases.

In the first phase, the *reconnaissance phase,* alternative plans are formulated and evaluated in a preliminary manner. The goal is to determine if at least one plan exists that: (1) has positive net benefit, (2) is likely to satisfy the environmental-protection and performance standards, and (3) is acceptable to local interests. In this phase, the goal is to perform detailed hydrologic engineering and flood-damage analyses for the existing without-project condition if possible (USACE, 1988a). If a solution can

TABLE 27.1 Flood-Damage-Reduction Measures*

Measures that reduce damage by reducing discharge	Measures that reduce damage by reducing stage	Measures that reduce damage by reducing existing damage susceptibility	Measures that reduce damage by reducing future damage susceptibility
Reservoir Diversion Watershed management	Channel improvement	Levee or floodwall Floodproofing Relocation Flood warning and preparedness planning	Land-use and construction regulation Acquisition

* In general, not a detailed specification.

be identified, and if a local sponsor is willing to share the cost, the search for the recommended plan continues to the second phase, the feasibility phase.

In the *feasibility phase,* the set of feasible alternatives is refined and the search narrowed. The plans are nominated with specific locations and sizes of measures and operating policies as illustrated by Table 27.2. Detailed studies for all conditions are performed to establish channel capacities, structure configurations, levels of protection, interior flood-control requirements, residual or induced flooding, and so on. Then, the economic objective function is evaluated, and satisfaction of the performance, environmental, and other standards are tested. Feasible solutions are retained, inferior solutions are abandoned, and the cycle continues. The NED and locally preferred plans are identified from the final array. The process concludes with a recommended plan for design and implementation.

In the design or *preconstruction engineering and design (PED) stage,* necessary design documents and plans and specifications for implementation of the proposed plan are prepared. These further refine the solution to the point that construction can begin. Engineering-during-construction permits further refine the proposed plan and allow for design of those elements of the plan not initially implemented or constructed. Likewise, the engineering-during-operations stage permits fine tuning of OMRR&R decisions.

27.2   COMMON REQUIREMENTS

27.2.1   Overview

This section defines requirements for formulating and evaluating economically efficient flood-damage-reduction plans that will satisfy performance and environmental-protection standards. Some measures that may be included in a plan have unique

TABLE 27.2   Plan Formulation-Evaluation for Feasibility-Phase Studies

Nominate range of plans (1)	Iteratively screen and refine plans (2)	Develop final array of feasibility plans (3)
Plan A	Plan A	Plan A
Plan B		
Plan C	Plan C	
Plan D		
Plan E	Plan E	Plan E*
.	.	
.	.	Plan G
.	.	
.	.	Plan I†
Plan M	Plan M	

Column 1   Wide range of potential plans each consisting of one or more measures.
Column 2   Continuous screening and refining of plans with increasing detail.
Column 3   Each plan must have positive net benefits and meet specified performance, environmental, and other standards.

* Plan that maximizes National Economic Development (NED).
† Locally preferred plan.

requirements for formulation and evaluation. Others have some common requirements. Common requirements are summarized in Table 27.3.

27.2.2 Study Setup and Layout

Technical information is required to support the tasks of problem definition, plan formulation, and plan evaluation. The specific information needed and commensurate level of detail is dependent on the nature of the problem, the potential solutions, and the sensitivity of the findings to the basic information. Actions performed to set up and lay out the study are preliminary to the detail analysis. They include: defining the study scope and detail, field data collection and presence, review of previous studies and reports, and assembly of needed maps and surveys. Although this process involves more information gathering than analysis, it helps scope the study, lend credibility to the subsequent analysis, and provide insights as to potential solutions.

27.2.3 Requirements for Evaluating the NED Contribution

27.2.3.1 Benefit Evaluation Standard. The *economic efficiency* of a proposed flood-damage-reduction alternative is defined as

$$NB = (B_L + B_I + B_{IR}) - C \qquad (27.1)$$

in which NB = net benefit; B_L = location benefit; B_I = intensification benefit; B_{IR} = inundation-reduction benefit; and C = total cost of implementing, operating, maintaining, repairing, replacing, and rehabilitating the plan (the OMRR&R cost). The *inundation-reduction benefit* may be expressed as

$$B_{IR} = (D_{without} - D_{with}) \qquad (27.2)$$

in which $D_{without}$ = economic flood damage without the plan; and D_{with} = economic flood damage if the plan is implemented.

TABLE 27.3 Summary of Common Requirements

Objective or Standard	Requirement	Method/model
Economic objective	Develop discharge-frequency function and uncertainty	Frequency analysis or ungauged catchment methods
	Develop stage-discharge function and uncertainty	Observation or fluvial and alluvial process models
	Develop stage-frequency function and uncertainty	Statistical and system-accomplishment models
Performance standard	Expected capacity-exceedance probability	Risk-based analysis procedures
	Expected lifetime exceedance probability	Hydrologic risk binomial distribution
	Design exceedance consequences	Depends on measures
	Perform reliability evaluation	Risk-based analysis procedures
Environmental protection	Assess impact	May require runoff, fluvial, alluvial, and statistical process models

The random nature of flooding complicates determination of the inundation-reduction benefit. For example, a flood-damage-reduction plan that eliminates all inundation damage one year may be too small to eliminate all damage in an extremely wet year and much larger than required in an extremely dry year. WRC guidelines address this problem by calling for use of expected annual flood damage. Expected damage accounts for the risk of various magnitudes of flood damage each year, weighing the damage caused by each flood by the probability of occurrence. Combining Eqs. 27.1 and 27.2, and rewriting them in terms of expected values yields

$$NB = B_L + B_I + E\,[D_{\text{without}}] - E\,[D_{\text{with}}]) - C \qquad (27.3)$$

in which $E\,[.]$ denotes the expected value. For urban flood damages, this generally is computed on an annual basis because significant levels of flood damage are limited to annual recurrence. For agricultural flood damages, it may be computed as the expected damage per flood, as more than one damaging flood may occur in a given year. The NED plan then is the alternative plan that yields maximum net benefit, accounting for the full range of hydrologic conditions that might occur.

The so-called "without-project" condition in Eq. 27.3 represents existing and future system conditions in the absence of a plan. It is the base condition upon which alternative plans are formulated; from which all benefits are measured; against which all impacts are assessed (USACE, 1989).

27.2.3.2 *EAD Computation.*

The most widely used approach to computing the *expected value of annual damage (EAD)* is the frequency technique, which is illustrated in Fig. 27.1. To compute EAD with this technique, the *annual-damage-frequency function* is derived and integrated. This damage-frequency function commonly is derived from the *annual maximum-discharge-frequency function* (Fig. 27.1a), transformed with a *stage-discharge (rating) function* (Figure 27.1b), and a *stage-damage function* (Figure 27.1c). This stage-damage function may represent a single structure or it may be an aggregated function that represents many structures, their contents, and other damageable property. Dynamic catchment, channel, or economic conditions are accounted for by adjusting the appropriate functions and deriving and integrating the damage-frequency function to compute EAD for the present and for each future year. The resulting EAD values can be averaged over project life, with discounting if appropriate. The transforming, integrating, and discounting computations can be performed by the Hydrologic Engineering Center's (HEC) EAD program (USACE, 1989a).

27.2.3.3 *Risk-Based Analyses.*

The procedure illustrated in Fig. 27.1 ignores uncertainty in the functions. *Uncertainty* is due to measurement errors and the inherent variability of complex physical, social, and economic situations (USACE, 1994a). Traditionally, uncertainty has been accounted for by employing factors of safety, such as levee freeboard. However, the state of the art of risk analysis has advanced sufficiently as of the early 1990s to permit explicit accounting for uncertainty. Consequently, federal agencies are now adopting a policy where flood-damage-reduction studies use *risk-based analysis.* Figure 27.2 illustrates the analysis strategy.

The risk-based-analysis procedure seeks to quantify the uncertainty in the discharge-frequency function, stage-discharge function, and stage-damage function and to incorporate this analysis of the economic efficiency of alternatives. This is accomplished with Monte Carlo simulation, a numerical-analysis tool that yields the traditional estimate of the expected damage reduced, accounting explicitly for the

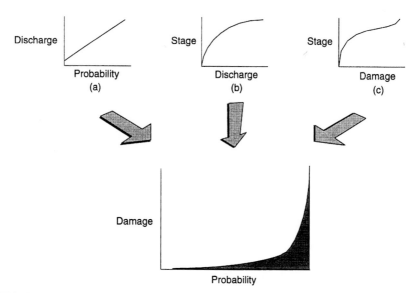

FIGURE 27.1 Derivation of damage-frequency function from hydrologic, hydraulic, and economic information.

errors in defining the discharge-frequency function, rating function, and stage-damage function. In addition, the Monte Carlo simulation procedure provides an assessment of the project performance as described in Sec. 27.2.3. Performance indicators derived are the true or expected capacity-exceedance probability and reliability of a flood-damage-reduction plan. The expected capacity-exceedance probability is the chance of flooding in any given year. Respectively, this is an index of the frequency with which the plan performs as designed. For example, in analysis of a proposed levee sized to contain the 1-percent-chance event, this procedure would estimate the probability that the levee would, in fact, contain the 1-percent-chance and other events, should these occur.

27.2.3.4 Discharge-Frequency Function Definition. The manner in which the discharge-frequency function is defined depends on the data available. For the existing, without-plan condition, if a sample of annual maximum discharge is available for the appropriate stream, the frequency function can be developed by fitting a statistical distribution with the sample. The procedures follow the guidelines proposed by the Water Resources Council (Interagency Advisory Committee, 1982). These procedures serve as the technical basis for the HEC-FFA computer program (USACE, 1992a; 1994b).

If a sample of annual discharge for an existing condition is not available and for future and with-project conditions, the discharge-frequency function must be developed with one of the procedures listed in Table 27.4. For special cases, such as regulated flows, different methods are required and must normally be augmented with modeling studies (USACE, 1994a).

The uncertainty in the discharge-frequency function varies depending on the physical characteristics of the stream, quality and nature of the available data, and other factors. With-project conditions uncertainty of the discharge-frequency func-

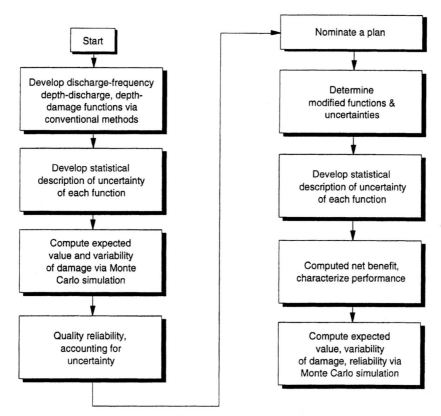

FIGURE 27.2 Risk-based analysis procedure.

tion may be less or greater than the without-project conditions. Future-conditions functions are almost always less certain.

27.2.3.5 Stage-Discharge Function Definition. The stage-discharge function, or *rating curve,* for the without-project, existing condition may be defined either by observations or with model studies. For cases that modify the function, the stage-discharge function must be defined with model studies. With-project conditions uncertainty may be less (concrete channel) than existing without-project conditions. Future-conditions uncertainty will most likely be greater. Alluvial streams involving mobile boundaries, ice, debris, and flow bulking from land-surface erosion can significantly add to the uncertainty of the stage-discharge function estimates.

Publications of the World Meteorological Organization (WMO, 1980; 1981) describe procedures for measuring stage and discharge to establish empirically the stage-discharge function for existing conditions. In most cases, stage-discharge relationships provided by the U.S. Geological Survey (USGS) for gauged sites are used.

Gradually Varied, Steady-Flow, Rigid-Boundary Conditions. Physical and numerical models may be used to establish stage-discharge functions for existing, future, without-project, or with-project conditions. Commonly, a numerical model of

TABLE 27.4 Procedures for Estimating Annual Maximum-Discharge-Frequency Function Without Discharge Sample

Method	Summary of procedure
Transfer	Frequency function is derived from discharge sample at nearby stream. Quantiles are extrapolated or interpolated for the location of interest.
Regional estimation of quantiles or frequency-function parameters	Quantiles or frequency functions are derived from discharge samples at nearby gauged locations. Frequency-function parameters are related to measurable catchment, channel, or climatic characteristics via regression analysis. The parameter-predictive equation is used for the location of interest.
Empirical equations	Peak discharge for specified probability event is computed from precipitation with a simple empirical equation. Typically, the probability of discharge and precipitation are assumed equal.
Hypothetical frequency events	Unique discharge hydrographs due to storms of specified probabilities and temporal and areal distributions are computed with a rainfall–runoff model. Results are calibrated to observed events or frequency relations at gauged locations so that probability of peak hydrograph equals storm probability.
Continuous simulation	Continuous record of discharge is computed from continuous record of precipitation with rainfall–runoff model, and annual discharge peaks are identified. Frequency function is fitted to series of annual hydrograph peaks, using statistical analysis procedures.

Source: Adapted from U.S. WRC (1981).

gradually varied, steady-flow, rigid-boundary (GVSF) in an open channel is used. Solution of the GVSF equations (see Chaps. 2 and 25) yields an estimate of stage at locations along a stream reach for a specified steady-flow rate. To solve the equations, the channel geometry and hydraulic-loss model parameters for the condition of interest must be defined. The geometry may be measured and parameters estimated for the existing-channel condition or defined as part of the proposal for a flood-damage-reduction plan. One commonly used GVSF model is the program HEC-2 (USACE, 1991a).

Erosion and Deposition. Channel-bed, channel-bank, and land-surface erosion and deposition complicate evaluation of the stage-discharge function. Mobilization of bed and bank materials in alluvial channels alters the channel shape. If that happens, stage at a channel cross section is not a unique, time-invariant function of discharge, channel geometry, and energy losses. Instead, the stage depends on material properties and the time history of discharge, and a movable-boundary hydraulics model is required to define the relationship for EAD computation. Two such models are HEC-6 (USACE, 1993a) and TABS-2 (Thomas and McAnally, 1985).

Mobilization and subsequent deposition of the sediment may cause other complications if not anticipated. For example, construction of a reservoir will alter a stream's natural gradient, but the flow and sediment load moving in the channel upstream of the reservoir are not changed. As the stream reaches the reservoir, velocity decreases significantly. The response of the stream is to deposit the bed load and decrease the gradient immediately upstream of the reservoir. This effect moves upstream as more sediment is deposited. This can induce flood damages upstream of the reservoir. Downstream, the effect is to scour the channel and erode the banks due to the relatively clear releases of the reservoir. Continuous downstream migration of the instability problem is likely over time.

Similarly, a channel straightening can alter the natural alluvial processes. Straightening increases the energy gradient while other conditions remain unchanged. This change can lead to increased erosion upstream of the realignment and increased deposition downstream. After some time, erosion of the channel banks and bed may occur.

Likewise, land-surface erosion increases the sediment load on a stream resulting in bulking of the flows. Also, if significant watershed construction accompanied by removal of vegetation occurs, the sediment runoff will increase during the construction period. Unless proper precautions are taken for these conditions, this sediment may move into adjacent channels, where it will be deposited. This, in turn, reduces the channel cross-section area, increases the stage for a given discharge, and induces damage. The National Pollution Discharge Elimination System (NPDES) General Permit for Storm Water Discharge from construction sites is the regulation governing EPA requirements for this condition.

Ice and Debris Impacts.   Ice and debris accumulation adversely alter accomplishments of flood-damage-reduction measures by restricting the flow in channels and conduits and by increasing pressure or forces on the measures. In cold regions, ice formation, build up, and break up must be anticipated, the impact must be evaluated, and project features must be adjusted to insure proper performance. With some measures, such as channel-lining improvements, this translates to an increase in project dimensions so the measures can withstand impacts of floating ice. Likewise, if ice is likely to form on a reservoir surface, the dam design must be altered to withstand the increased overturning moment due to the added force on the dam (USACE, 1992).

Debris.   The effect of debris is similar to that of ice; it can significantly reduce channel conveyance and constrict flows at obstacles. Examples are more volume associated with runoff, constrictions at bridges, and accumulation of urban trash and waste in channels. If debris is mobilized and subsequently redeposited, it may adversely affect performance of pumps, gates, and other plan features. Proper maintenance measures should be included as a component of any plan to avoid these problems.

27.2.3.6 *Stage-Frequency Definition.*

If flood inundation results from a flooding river, storm surges along a lake or ocean, wind-driven waves (run-up), a filling reservoir, or combinations of these events, a stage-frequency function is more appropriate for derivation of the damage-frequency function. Statistical-analysis procedures are similar to those used for fitting a frequency function with a discharge sample. For future conditions and other cases, the function must be defined with model studies. The model used depends on the condition to be analyzed. For example, if reservoir-operation changes are proposed to reduce flood damage due to reservoir pool-elevation rise, a reservoir-operation simulation model might be used to estimate the modified time series of lake levels.

27.2.4 Requirements for Satisfying Performance Standard

Selecting the alternative that maximizes NED contribution provides for efficient investment of public funds, but it does not guarantee that a plan will perform as effectively as the public has a right to expect. Two plans may yield the same net benefit, but one may be less vulnerable and thus more desirable. For example, consider two hypothetical alternatives: a levee plan and a channel-improvement plan, both sized and located to protect a floodplain from events less than the best estimate of the 1-percent-chance event. When a slightly larger event occurs, the levee will be overtopped and may be breached, causing catastrophic results. If this same event occurs with the channel plan, flow will be out of bank. However, the consequences of this out-of-bank flow likely will be less significant than those associated with a levee breach. The channel project is less vulnerable.

Performance indicators are used in determining the validity of the project and for comparing alternatives based on long-term project operational stability, public safety, and determining potential catastrophic-damage locations. They include defining the flood risk for the project life, determining the true or expected stage-exceedance probability, estimating the project reliability, describing the operation for a range of events and key assumptions, and defining the consequences of capacity-exceedance events of each plan. To ensure that flood-damage-reduction plans satisfy the performance standard, functioning as anticipated, the hydrologic engineer must participate actively in plan formulation. The performance indicators are described in more detail in subsequent paragraphs.

27.2.4.1 Expected Annual-Exceedance Probability.

The *expected annual-exceedance probability* is a key element of defining the performance of a given plan. It is the probability that the specified capacity or target stage will be exceeded in any given year. The value is determined from the risk-based analysis study that includes the uncertainties of the various functions. The target stage is normally that associated with the start of significant damage. For a levee or floodwall, the stage may be the stage where overtopping occurs. For a channel or nonstructural measures, the target stage may be that where flooding of the structures begins. Although variable for plans that modify the stage-damage function, the target stage should be consistent among plans that don't modify the stage-damage functions.

27.2.4.2 Expected Lifetime-Exceedance Probability.

The probability that one or more flood events will occur within a specified time period, normally the project life, is a means of indicating performance. The calculations may be made directly using the binomial distribution (USACE, 1993). Figure 27.3 graphically shows the relationships. The threat may be similar for all structures, such as behind a levee or floodwall, or variable depending on the elevation of individual structures, such as for a channel. For a channel example, a house located with the ground floor at the 1-percent-chance flood level (the so-called 100-year flood level), the probability of one or more exceedances is approximately 0.40, or about one chance in 2.5 over a 50-year project life. If the house is located with the ground floor at the 0.5-percent-chance level (the 200-year flood), the probability of one or more exceedances is 0.22. For a levee with an expected stage-exceedance probability of 1 percent there is a .40 probability of one or more event exceedances during the 50-year project life for all the protected structures.

27.2.4.3 Operation for Range of Events and Key Assumptions.

Each plan should be evaluated for performance against a range of events and key assumptions. Evaluation based solely on a specific design event is not a valid performance indicator by itself. For example, a pumping station must be configured to operate satisfactorily for a range of events, not simply designed for the 4-percent-chance event. The analysis should be for a range of frequent and rare events including those that exceed the project capacity.

Analysis of the sensitivity of the operation of the project to critical assumptions is required to assist in determining the stability of the project over its project life. An

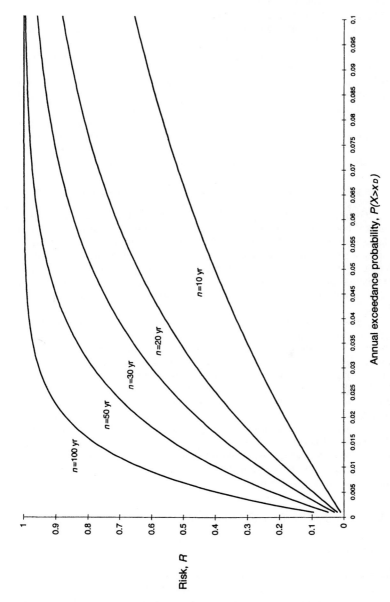

FIGURE 27.3 Probability of exceedance during project life.

example is that there is a somewhat high likelihood of future encroachment of the natural storage associated with an interior system although it was not assumed as part of the plan assumptions. The sensitivity of the encroachment on the project performance should be evaluated. Similarly, the sensitivity of future-development scenarios, erosion, debris, sediment, operation and maintenance, and other assumptions that are critical to having the project perform as planned and designed must be evaluated.

27.2.4.4 Consequences of Capacity-Exceedance Events. The project performance for one or more capacity-exceedance events is required. Analyses to determine the extent, depth, and velocities of flooding and warning times for each event are conducted by hydrologic engineers. Additional data to support definition of the population at risk, warning dissemination, and emergency response actions from the technical, social, and institutional perspectives for various times of the day are also required. The studies to determine the consequences of the capacity exceedance events may vary significantly depending on the plan. Plans such as levees and floodwalls normally require the most detail because of the significant loss potential.

27.2.5 Requirements for Satisfying Environmental-Protection Standard

27.2.5.1 Policy. The federal policy of flood-damage reduction is to develop, control, maintain, and conserve the nation's water resources in accordance with the laws and policies established by Congress and the administration, including those laws designed to protect the environment. The National Environmental Policy Act (NEPA) is the nation's broadest environmental law. It requires that federal agencies prepare an *environmental impact statement* (EIS) for proposed legislation or other major actions that would affect the environment significantly in most cases.

27.2.5.2 General Procedures. For federal actions, except those categorically excluded from NEPA requirements, the *environmental assessment (EA)* is developed to determine if the action will have a significant impact on environmental quality. The EA presents the alternatives and defines the environmental impacts of each. In the event of a finding of no significant impact, no further action is necessary. Otherwise, an EIS will be prepared. An EIS is normally prepared for feasibility reports for authorization and construction of major projects, for changes in projects which increase size substantially or incorporate additional purposes, and for major changes in the operation and/or maintenance of completed projects (USACE, 1989).

NEPA requires that an EIS include components shown on Table 27.5. Much of the scientific and engineering information required to develop these components is identical to or an expansion or extension of information otherwise required for economic and performance assessment.

27.3 WITHOUT-PROJECT CONDITIONS

27.3.1 Overview

The existing and future without-project conditions represent the base conditions for determining the economic value, performance, and environmental/social impacts of flood-damage-reduction measures and plans. Base conditions should be established in final detail as early in the process as possible to provide a stable basis of information and plan comparison. Table 27.6 presents a checklist that summarizes critical requirements for without-project conditions analysis. This and checklist sections are

TABLE 27.5 Technical Components of EIS

---

1. Description of the alternatives considered, including at least the "no-action" alternative, the Corps' preferred alternative, and the "environmentally preferable" alternative.
2. Presentation of the environmental impacts of each alternative.
3. Explanation of why any alternatives were eliminated from further consideration.
4. Delineation of the affected environment.
5. Assessment of the environmental consequences of each alternative, including: (1) direct effects, (2) foreseeable indirect effects, (3) cumulative effects from the incremental impact of the alternative plus other past, present, and foreseeable future actions, and (4) other effects, including unavoidable effects, irreversible or irretrievable commitments of resources, effect on urban quality, effect on historical and cultural quality.
6. Actions that may be taken to mitigate adverse impacts, including: (1) avoiding the impact by not implementing the plan, (2) minimizing the impact by limiting the plan, (3) rectifying the impact by repair, rehabilitation, or restoration, (4) reducing or eliminating the impact over time by preservation or maintenance, or (5) compensating by replacement or substitution of resources.

---

included as an aid to insure that nothing is left to chance. Some items, however, are listed just as a reminder to insure that details will not be overlooked.

27.3.2 Layout

The layout for the existing without-project conditions is crucial to the overall study. Preliminary efforts define the study limits, review available information, and establish a field presence. These activities assist with development of the initial definition

TABLE 27.6 Checklist for Without-Project Conditions

Study components	✓	Issues
Layout		Review/assemble available information
		Conduct field reconnaissance for historic flood data and survey specification
		Establish local contacts
		Assist in establishing study limits, damage reaches
Economic studies		Determine existing and future without-project conditions discharge-frequency and associated uncertainty
		Determine existing and future with-project conditions stage-discharge and associated uncertainty
Performance		Determine expected capacity-exceedance probability
		Determine expected lifetime-exceedance probability
		Evaluate existing project operation/stability for range of events and key assumptions
		Describe consequences of capacity exceedances
		Perform reliability analyses
Environmental and social		Evaluate without-project riparian impacts
		Evaluate without-project social impacts

of potential measures and plans to evaluate. Subbasins are delineated based on stream topology, gauge sites, runoff characteristics, and locations of existing and potential measures. Assistance is provided in estimating the maximum extent of flooding for structure inventories and defining damage reaches.

The reader should note that this and subsequent checklists will need to be expanded for a specific study. Others include financial, legal, geotechnical, and political study components.

27.3.3 Technical Analyses

Flood-damage-reduction investigations develop information that define the flood characteristics used in the economic analysis and determination of the performance and environmental/social impacts of the existing system. Information to be generated includes discharge-frequency, stage-discharge, flood-inundation boundaries, warning times, and the variability of flooding (shallow or deep, swift or slow, debris- and sediment-laden, ice, and so on). The information is developed using previously described conventional studies. Uncertainty of the discharge, stage, and damage functions are determined for the existing without-project conditions. These relationships form the basis of estimating uncertainties for the future without-project and with-project conditions. Risk-based analyses are then performed to determine economic and performance information. The nature of flooding and determination of the magnitude of major damage locations provide insights as to the type and range of costs of potential flood-damage-reduction measures.

27.4 RESERVOIRS

27.4.1 Applicability

Reservoirs reduce damage by reducing discharge directly. Table 27.7 is a checklist that summarizes critical requirements for reservoirs.

A reservoir is well suited for damage reduction in the following cases:

1. Damageable property is spread over a large geographic area downstream from the reservoir site with several remote damage centers and relatively small local-inflow areas between them.
2. A high degree of protection, with little residual damage (D_{with} of Eq. 27.2), is desired.
3. A variety of property, including infrastructure, structures, contents, and agricultural property is to be protected.
4. Water impounded may be used for other purposes, including water supply, hydropower, and recreation.
5. Sufficient real estate is available for location of the reservoir at reasonable economic, environmental, and social cost.
6. The great economic value of damageable property protected will justify the great cost of constructing the reservoir.

27.4.2 Reservoir Operation Overview

Figure 27.4 illustrates a multiple purpose reservoir. A reservoir reduces flood-inundation damage by temporarily holding excess runoff and then releasing that

TABLE 27.7 Checklist for Reservoir

Study components	✓	Issues
Layout		Consider alternative sites based on drainage area versus capacity considerations
		Delineate environmentally sensitive aquatic and riparian habitat
		Identify damage centers, delineate developed areas, define land uses for site selection
		Determine opportunities for system synergism due to location
Economics		Determine with-project modifications to downstream frequency function for existing and future conditions
		Quantify uncertainty in frequency function
		Formulate and evaluate range of outlet configurations for various capacities using risk-based analysis procedures
Performance		Determine expected annual-exceedance probability
		Determine expected lifetime-exceedance probability
		Describe operation for range of events and analyze sensitivity of critical assumptions
		Describe consequences of capacity exceedances
		Determine reliability for range of events
		Conduct dam-safety evaluation
Design		Formulate and evaluate preliminary spillway and outlet configurations
		Conduct pool sedimentation analysis
		Evaluate all downstream hydrologic and hydraulic impacts
		Formulate preliminary operation plans
Environmental and social		Evaluate with-project riparian habitat
		Evaluate aquatic and riparian habitat impact and identify enhancement opportunities
		Anticipate and identify incidental recreation opportunities

water downstream to the channel, either through the normal outlet system or over the emergency spillway for rare events, at a lesser rate over a longer period of time. This permits a reduction in peak-flow rate, resulting in lower stage and less damage. The rate of release depends on the characteristics of the outlet works and spillway. Note that in the illustration, the outlet serves two purposes: It limits the release of water during a flood event, and it provides a method of emptying the reservoir flood-control pool after the events.

Detention storage systems are simpler flood storage systems normally implemented in urban settings as shown in Fig. 27.5. They function in a similar manner to major reservoirs by modifying flood releases downstream of the project. The releases are typically uncontrolled such as shown in Fig. 27.6. In this figure, the

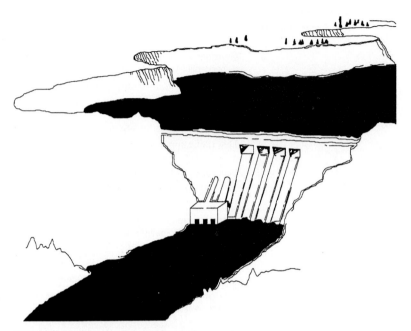

FIGURE 27.4   Multipurpose flood-control reservoir.

existing-condition, without-project peak discharge from a small catchment is 186 m³/s. This rate exceeds the maximum nondamaging discharge for the downstream reach, 113 m³/s, which is denoted "target flow" in the figure. To reduce the damage, storage is provided. The volume of water represented by the shaded area in the figure is held and released gradually at a rate that does not exceed the target. The total volume of the inflow and outflow hydrographs is the same, but the time distribution is altered by the storage.

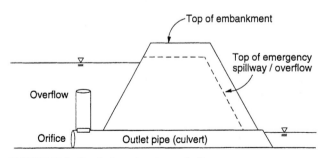

FIGURE 27.5   Simple detention storage facility.

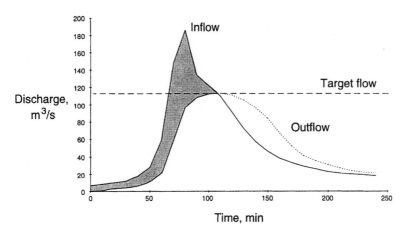

FIGURE 27.6   Impact of storage.

27.4.3   Discharge-Reduction Assessment

The primary effect of storage is reduction of discharge, and this is modeled for individual runoff events with the routing models (USACE, 1994b). Outflow from an impoundment that has a horizontal water surface can be computed with the so-called level-pool routing model (also known as modified Puls routing model, see Chap. 24). The reduction in discharge peak for individual events will translate, over the long term, into modification in the discharge-frequency function. This, in turn, yields a reduction in expected damage. The modified discharge-frequency function can be found by either: (1) evaluating reservoir operation with a long series of historical inflows and estimating regulated discharge probability from frequency of exceedance of magnitudes of the simulated reservoir outflow series, or (2) evaluating operation for a limited number of historical or hypothetical events. In this case, the probability of the unregulated inflow peak commonly is assigned to the peak of the corresponding computed outflow hydrograph. This is repeated for a range of runoff events to define adequately the modified discharge-frequency function. Hypothetical runoff events may be developed from rainfall–runoff analysis with rain depths of known probability, or from discharge duration-frequency analysis. In the first case, storms of specified probability are developed and the corresponding runoff hydrographs computed (USACE, 1994b). The runoff hydrographs are inflow to the reservoir. The peak outflows commonly are assigned probabilities equal to the corresponding storm probabilities. In the second case, a balanced inflow hydrograph is developed. This balanced hydrograph has volumes for specified durations consistent with established volume-duration-frequency relations. For example, a 0.10-probability balanced hydrograph is developed so the peak 1-hour volume equals the volume with probability 0.10 found through statistical analysis of runoff volumes. Likewise, the hydrograph's 24-hour volume equals the volume with probability 0.10. With either of the hypothetical inflow events, reservoir operation is simulated and the outflow peak is assigned the same probability as the inflow hydrograph. This is repeated for a range of hypothetical rainfall events to define adequately the modified discharge-frequency function.

Figure 27.7 shows typical modifications to the discharge-frequency function due to a reservoir. In this figure, the solid line represents the inflow and the without-project outflow discharge-frequency function. Note that the straight line shown here and in subsequent figures is a simplification for illustration. Discharge-frequency functions are not always straight lines when plotted on normal probability paper. Q_1 represents a target flow; this may be the channel capacity downstream, the flow corresponding to the maximum stage before damage is incurred, or any other target selected for a particular floodplain. Ideally, a reservoir would be designed and operated to maintain releases less than or equal to this target. If the inflow peak is less than the target, the reservoir need not exercise any control. If the inflow peak exceeds the target, the reservoir should restrict outflow to the target rate. Consequently, the with-project frequency function, which is shown as a dotted line, is equal to the without-project frequency function for events of exceedance probability greater than P_1 (events with discharge less than Q_1). For inflow events of exceedance probability less than P_1, release is limited to Q_1. However, regardless of the reservoir capacity, some extreme inflow events with peak greater than Q_2 and probability less than P_2 will exceed the capability of the reservoir to limit the outflow to Q_1. The reservoir may reduce flow somewhat, but as the magnitude of the events increase (and the probability decreases), the regulated outflow peaks will approach the inflow peaks. The reservoir will have less and less impact. Finally, for an event with inflow peak equal to Q_3, the reservoir will have negligible impact, and the without-project and with-project frequency functions will be identical.

27.4.4 Performance Considerations

The performance of a reservoir depends on its capacity, configuration, and location; and on its operation rules.

27.4.4.1 Capacity, Configuration, and Location. Table 27.8 suggests steps for evaluating reservoir alternatives. Additional guidance is available from the U.S.

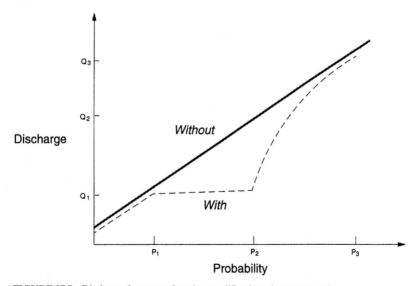

FIGURE 27.7 Discharge-frequency function modifications due to reservoir.

Army Corps of Engineers (1990), from the Bureau of Reclamation (1977), and from ASCE/WEF (1992).

27.4.4.2 *Operation Rules.*

For a simple uncontrolled reservoir, discharge reduction, and hence damage reduction, depends on the hydraulic characteristics of the structure. The computations for these systems can be done with a specialized computer program, such as HEC-1. For a reservoir with gates and valves that can be controlled, the damage reduction depends also on operation rules. Operation rules specify how and when the gates and valves are to be opened. Typically *flood-control operation rules* define the release to be made in the current time period as a function of one or more of the following: current storage in the reservoir, forecasted inflow to the reservoir, current and forecasted downstream flow, and current storage in and forecasted inflow to other reservoirs in a multiple-reservoir system.

Operation rules must be defined for controlled reservoirs as a component of any plan that includes such a reservoir. Computer program HEC-5 (USACE, 1982b) is designed for simulation of flood-control reservoir operation.

The effects of absence of personnel to regulate a reservoir, misoperation, and interruptions in communications during extreme events must be considered. For proper comparison of alternative plans, this cannot be simply an acknowledgment that these events may occur. A qualitative assessment must be made. For example, the discharge reduction possible should be defined for various events and the rate of pool rise defined.

27.4.4.3 *Other Considerations.*

To ensure proper performance of reservoirs for flood-damage reduction, the hydrologic engineer must consider also the following:

1. *Impact of debris/trash.* A complete plan must include features that will minimize adverse impacts of outlet plugging due to debris.
2. *Safety features.* A complete plan must include features to protect public safety at the reservoir site, particularly when the project is operating at capacity.

TABLE 27.8 Steps in Evaluating Proposed Storage Alternatives

1. Define a set of without-project inflow hydrographs. These should cover the range of likely events, including frequent small events, infrequent large events, major historical events, and so forth.
2. Identify a target for reliability analysis. This may be the channel capacity downstream, the flow corresponding to the maximum stage before damage is incurred, or any other target appropriate for a particular floodplain.
3. Select a trial reservoir location, capacity, and outlet configuration. Develop the elevation-area-discharge functions required for reservoir routing for this alternative.
4. For each inflow hydrograph in turn, compute the corresponding outflow hydrograph and all hydrographs of runoff downstream of the reservoir.
5. Compute the flood damage corresponding to the hydrograph peak.
6. Compare the outflow peak to the target to determine if the regulated flow or stage exceeds the target. Determine cost and net benefit of the alternative.
7. Repeat steps 3 to 5 for the range of inflow events. Determine the expected flood damage and the overall reliability of the alternative, defined as the frequency of meeting the target. Determine cost and net benefit of the alternatives.
8. Repeat steps 2 to 6 for all reservoir alternatives.
9. Compare the economic efficiency and the reliability of the alternatives to select a recommended plan.

TABLE 27.9   Impact of Reservoir on Stream-System Morphology

1. Rise in streambed at upper reservoir pool due to sediment deposition.
2. Change in channel and bed stability downstream from the dam due to degradation of the channel bed.

3. *Sedimentation.* A description of sedimentation problems due to reservoirs are shown in Table 27.9. Eventually, all reservoirs will fill with sediment. Therefore analyses must be conducted to identify the problems and remedial features (USACE, 1989).

27.4.5 Dam Safety Evaluation

The *discharge-reduction benefit* of a reservoir is accompanied by the hazard of dam failure. U.S. Army Corps of Engineers policy (1991b) is that ". . . a dam failure must not present a hazard to human life . . ." Accordingly, the analysis must: (1) formulate any reservoir plan to comply with this safety requirement, and (2) evaluate the impact of catastrophic failure of any proposed reservoir plan to confirm that this performance constraint is satisfied.

27.4.5.1 Formulation to Minimize Catastrophic Consequences When Capacity Is Exceeded.   Four design standards, depending on the type of dam and risk to life, are described in Table 27.10. Selection of the standard appropriate for plan formulation and to insure that the standard is used for all project features is required.

TABLE 27.10   Design Standards for Dam Safety (U.S. ACE, 1991b)

Standard	Application
1	Applies to dams located such that human life is at risk. In that case, the dam must be designed to pass safely a flood event caused by the probable maximum precipitation (PMP) occurring over the catchment upstream of the reservoir. PMP is a ". . . quantity of precipitation that is close to the physical upper limit for a given duration over a particular basin" (WMO, 1983). U.S. Army Corps of Engineers studies will use PMP amounts developed by the Hydrometeorological Section of the National Weather Service.
2	Applies to dams "where relatively small differentials between headwater and tailwater elevations prevail during major floods." These structures must be able to pass safely major floods typical of the region, without incurring excessive damage downstream and without sustaining damage that would render the dam inoperable.
3	Applies to dams "where failure would not jeopardize human life nor create damage beyond the capabilities of the owner to recover." These structures should be planned so failure related to hydraulic capacity will result in no measurable increase in population at risk and in a negligible increase in property damage over nonfailure damage.
4	Applies to small recreational and agricultural water-supply reservoirs. The design in this case is ". . . usually based on rainfall–runoff probability analysis and may represent events of fairly frequent occurrence." The decision likely will be based on economic considerations: Does the cost of a more reliable structure exceed the expected cost of repair or replacement?

27.4.5.2   *Failure Evaluation.* The impact of dam failure can be estimated with hydraulics models or routing models (USACE, 1993b and 1994b). Three aspects of dam failure must be considered: (1) formation of a breach, an opening in the dam as it fails, (2) flow of water through this breach, and (3) flow in the downstream channel. For analysis, the reservoir outflow hydrograph is computed conventionally. However, the operating characteristics of the reservoir change with time as the breach grows. For convenience in analysis, a breach commonly is assumed to be triangular, rectangular, or trapezoidal, and to enlarge at a linear rate. At each instant that the breach dimensions are known, the flow of water through the breach can be determined with principles of hydraulics. Subsequent movement of the outflow hydrograph through the downstream channel is modeled with a hydrodynamics model.

27.4.6   Environmental Impacts

Construction of a reservoir can have significant environmental and social impacts, and information provided by the hydrologic engineer can be critical in evaluation of these impacts. Table 27.11 illustrates this.

One particular serious environmental issue is preservation of wetlands. 40 CFR 230.41(a)(1) defines wetlands as ". . . those areas that are inundated or saturated by surface or groundwater at a frequency and duration sufficient to support, and that under normal circumstances do support, a prevalence of vegetation typically adapted for life in saturated soil conditions" (ASCE/WEF, 1992). The study specialists must identify any such areas to permit protection as required under Section 404 of the Federal Water Pollution Control Act.

27.5   DIVERSIONS

27.5.1   Applicability

This section presents special requirements for formulating and evaluating flood-damage-reduction diversion measures. *Diversions* reduce damage by reducing discharge directly. Table 27.12 is a checklist that summarizes critical requirements for diversions.

TABLE 27.11   Information Required to Assess Environmental Impacts

Potential impact	Hydrologic engineering information required to assess impact
Loss of wildlife habitat due to ponding	Inundation due to and duration of ponding
Loss of vegetation in ponded area	Inundation due to and duration of ponding
Inundation of archeological sites	Extent of and depth of inundation
Increased in-stream temperature, increased turbidity, reduced dissolved oxygen downstream of reservoir	With-plan discharge frequency, results of water-quality simulation
Improved recreational opportunities due to pond	Pond stage frequency
Change of downstream recreation due to reduced discharge	Discharge frequency, stage frequency

TABLE 27.12 Checklist for Diversion

Study components	✓	Issues
Layout		Delineate environmentally sensitive aquatic and riparian habitat
		Identify damage centers, delineate developed areas, define land uses for site selection
		Determine right-of-way restriction
		Identify infrastructure/utility-crossing conflicts
Economics		Determine with-project modifications to downstream frequency function for existing and future conditions
		Quantify uncertainty in frequency function
		Formulate and evaluate range of outlet configurations for various capacities using risk-based analysis procedures
Performance		Determine expected annual-exceedance probability
		Determine expected lifetime-exceedance probability
		Describe operation for range of events and analyze sensitivity of critical assumptions
		Describe consequences of capacity exceedances
		Determine reliability for range of events
Design		Formulate and evaluate preliminary control structure configurations
		Conduct diversion channel sedimentation analysis
		Evaluate all downstream hydrologic and hydraulic impacts
		Formulate preliminary operation plans
Environmental and social		Evaluate aquatic and riparian habitat impact and identify enhancement opportunities

A diversion is well-suited for damage reduction in the following cases:

1. Damageable property is spread over a large geographic area with relatively minor local inflows for diversions removing water from the system.
2. A high degree of protection, with little residual damage, is desired.
3. A variety of property, including infrastructure, structures, contents, and agricultural property is to be protected.
4. Sufficient real estate is available for location of the diversion channel or tunnel at reasonable cost.
5. The great value of damageable property protected will economically justify the great cost of the diversion.

27.5.2 Diversion Operation Overview

Figure 27.8 is a sketch of a diversion. This diversion includes a diversion channel and a control structure that is a broad-crested side-overflow weir. Alternatively, this control structure might be a conduit through an embankment or a gated, operator-controlled weir, and a pipe or other conduit might be used instead of the open diversion channel. For the design illustrated, when the discharge rate in the main channel reaches a predetermined threshold, the stage at the overflow is sufficient to permit water to flow into the diversion channel. This, in turn, reduces discharge in the main channel, thus eliminating or reducing damage to the downstream property. Downstream of the protected area, the bypass and the main channel may join. A plan view of this is shown in Figure 27.9.

27.5.3 Discharge-Reduction Assessment

As a diversion alters discharge for individual flood events, it will eventually alter the discharge-frequency function. Figure 27.10 shows typical modifications due to a diversion. The solid line represents the without-project discharge-frequency function at a location downstream of the diversion control structure. Q_1 represents a target flow at that point; as with a reservoir, this may be the channel capacity downstream, the flow corresponding to the maximum stage before damage is incurred, or any other target selected for a particular alternative. If the main-channel discharge is less than the target, no water need be diverted. When the main-channel discharge exceeds the target, the excess is diverted, limiting main-channel discharge to the target. Consequently, the with-project frequency function, which is shown as a dotted line, is equal to the without-project frequency function for events with exceedance probability greater than P_1 and discharge less than Q_1. The with-project function has flows equal to Q_1 when the main-channel discharge exceeds this target. However, regardless of the design, some extreme event of probability P_2 will cause the bypass channel to reach its capacity. Then the diversion will no longer be capable of limiting main-channel flow to Q_1. Of course, the diversion may reduce main-channel discharge somewhat. However, as the magnitude of the events increase (and the probability decreases), the with-project main-channel discharge will approach the without-project discharge. Finally, for an event in which the without-project peak

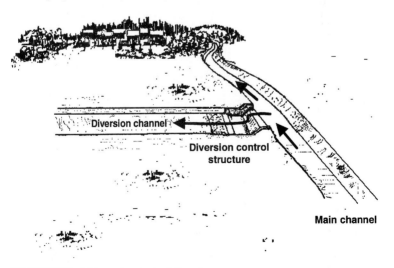

FIGURE 27.8 Major components of diversion.

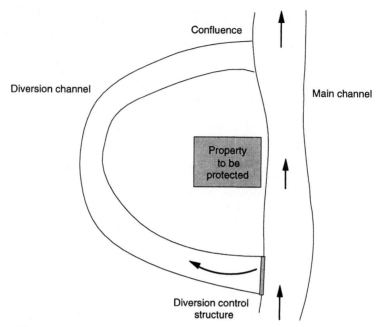

FIGURE 27.9   Plan view of diversion with downstream confluence.

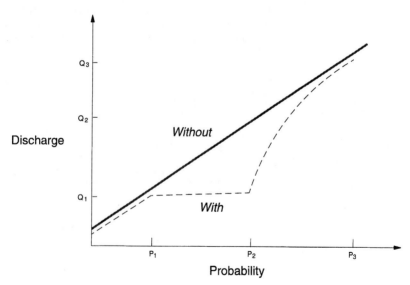

FIGURE 27.10   Discharge-frequency function modifications due to reservoir.

discharge equals Q_3, the diversion will have negligible impact, and the without-project and with-project frequency functions will be identical.

As with a reservoir, the impact of a diversion on the discharge-frequency function can be evaluated via *period-of-record analysis* or simulation of selected events. With the period-of-record analysis, the historical discharge time series is analyzed to estimate channel flow when the proposed diversion operates. The resulting modified main-channel discharge time series is analyzed with statistical procedures to define the frequency function. Otherwise, operation of the diversion with selected historical or hypothetical runoff hydrographs is simulated, and the resulting discharge peaks are assigned probabilities equal to the probabilities of the peaks without the diversion.

The behavior of a diversion can be modeled with the routing models. At the control structure, a hydraulics model estimates the distribution of discharge into the diversion channel and discharge in the main channel. This model may be as simple as a diversion-channel-flow versus main-channel-flow rating curve derived with a one-dimensional model or as complex as the two-dimensional models. Passage of flow in the diversion channel and in the main channel is modeled with a routing model, or, for more detailed analysis of the behavior, with a one-dimensional or gradually varied unsteady-flow model, or even a multidimensional flow model.

27.5.4 Technical Considerations

To ensure proper performance of a diversion the following potential problems must be considered: channel stability, deposition, and safety during operation.

1. *Channel stability.* A plan that includes a diversion must take care to insure channel stability in both the diversion and main channels.

2. *Deposition.* Deposition is a common problem at diversions. Consequently, a sedimentation analysis estimate of the magnitude of this problem and remedial actions must be conducted. This may include adjustments to the design to minimize deposition, or it might be limited to guaranteeing sufficient funds for continuous OMRR&R.

3. *Safety during operation.* A diversion such as that shown in Fig. 27.8 is an attractive nuisance. When the main channel reaches the design level, and water is discharged into the bypass, the public will be attracted. Care must be taken to provide for public safety.

Further, under normal circumstances, a diversion channel is dry and so is subject to unwise temporary or permanent use. If main-channel flows rise quickly, the diversion may begin to function with little advance notice, and the bypass channel will fill. Precautions should be taken to minimize damage within the channel or risk to life if the bypass channel is accessible to the public.

27.6 CHANNEL MODIFICATIONS

27.6.1 Applicability

This section describes the impact of channel modifications (sometimes called *channel improvements*) and the requirements for planning these modifications to reduce flood damage. A checklist of the requirements is presented as Table 27.13.

TABLE 27.13   Checklist for Channel Modification

Study components	✓	Issues
Layout		Determine right-of-way restriction
		Delineate environmentally sensitive aquatic and riparian habitat
		Identify damage centers, delineate developed areas, define land uses for site selection
		Identify infrastructure/utility-crossing conflicts
Economics		Determine with-project modifications to stage-discharge function for all conditions
		Determine any downstream effects due to frequency function changes due to loss of channel storage
		Quantify uncertainty in stage-discharge function
		Formulate and evaluate range of channel configurations using risk-based analysis procedures
Performance		Determine expected annual-exceedance probability
		Determine expected lifetime-exceedance probability
		Describe operation for range of events and analyze sensitivity of critical assumptions
		Describe consequences of capacity exceedances
		Determine reliability for range of events
Design		Account for ice and debris, erosion, deposition, and sediment transport, and high velocities
		Evaluate straightening effects on stability
		Evaluate all impact of restrictions or obstructions
Environmental and social		Evaluate aquatic and riparian habitat impact and identify enhancement opportunities
		Anticipate and identify incidental recreation opportunities

Channel modifications are effective flood-damage-reduction measures in the following cases:

1. Damageable property is locally concentrated.

2. A high degree of protection, with little residual damage, is desired.

3. A variety of property, including infrastructure, structures, contents, and agricultural property is to be protected.

4. Sufficient real estate is available for location of the channel modification at reasonable economic, environmental, and social cost.

5. The great economic value of damageable property protected will justify the great cost of constructing the reservoir.

27.6.2 Channel Overview

Stage in the floodplain is a function of: (1) the channel discharge rate; (2) the channel geometry, including invert slope, cross-sectional area, wetted perimeter, length, and alignment; and (3) the energy "lost" as water is conveyed in the channel. This description focuses on measures that reduce out-of-bank stage (and hence, damage) by modifying the geometry or by reducing the energy loss.

27.6.2.1 Channel Geometry Modification. The out-of-bank stage can be reduced for a given discharge rate if the channel is modified to increase the effective cross-sectional area. Figure 27.11 shows such a modification. In this elevation versus station plot, the original boundary is shown. When the material represented by the shaded polygons is removed, the new boundary is as shown with the dashed line. Now the total cross-sectional area beneath the water surface shown is greater than the without-plan area.

In the simplest case (steady, one-dimensional flow), discharge rate is directly proportional to cross-sectional area. Thus, if all else remains equal, the improved channel shown in Fig. 27.11 will convey a greater discharge with water surface at the same elevation or the same discharge at a reduced water-surface elevation.

The natural-channel geometry modifications may take place due to erosion and deposition or to bank instability. In either case, these will affect future with-plan and without-plan conditions. For example, if land-surface erosion increases as a consequence of development in a catchment, this sediment may be deposited in the channel. Without maintenance, this deposition will reduce the cross-sectional area over time, increasing stage for a specified discharge, and increasing EAD for the without-project condition ($E[D_{\text{without}}]$ in Eq. (27.3)). Similarly, scour may cause bank failure, thereby decreasing the effective flow area. This, too, may increase stage and the resulting EAD.

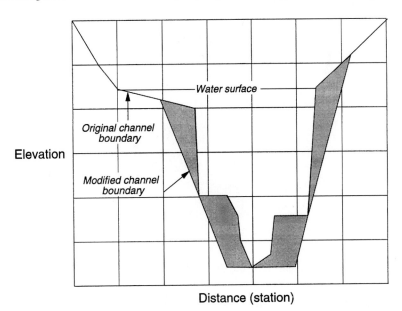

FIGURE 27.11 Illustration of channel geometry modification.

27.6.2.2   Energy Loss Reduction.   As water is conveyed in a channel, energy is converted from one form to another or "lost." As this loss of energy results in increased stage, stage may be reduced by reducing the energy loss. This may be accomplished by smoothing the channel boundary, straightening the channel, or minimizing the impact of obstructions in the channel.

The variation of water-surface elevation along a stream is largely a function of the boundary roughness and the stream energy required to overcome friction losses (USACE, 1993c). If all else remains the same, smoothing the channel to reduce the roughness will reduce the energy loss, which will in turn reduce stage and EAD.

The total energy loss due to friction between two points on a stream is the product of the energy loss per unit length and the distance between the points. Clearly if the stream distance can be reduced, the energy loss and stage may be reduced. Figure 27.12 illustrates how this may be accomplished. The original channel alignment is shown with the gray boundary. The boundary of the realigned channel is dotted. In this case, the energy loss in the improved channel is less, and the stage and damage will be reduced.

Although water-surface profiles are mostly influenced by friction forces, changes in the energy grade line and the corresponding water-surface elevations can also result from significant changes in stream velocity between cross sections. These velocity changes may be the result of natural or artificial expansions or contractions in channel width or of bridge crossings in which discharge is forced through an opening smaller than the upstream and downstream channels. To avoid the increase in stage, transitions must be designed carefully. Similarly, if restricted bridge openings cause stage increases, removal or modification of the bridges should be considered as a feature of the flood-damage-reduction plan.

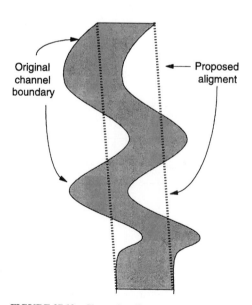

FIGURE 27.12   Channel realignment.

27.6.3 Stage-Reduction Assessment

The intended impact of a channel modification is reduction of stage for a given discharge, as illustrated by Fig. 27.13. In this figure, the existing, without-plan rating function is shown in gray, and the with-plan function is dotted. The modified rating function here shows a lower stage for all discharge values.

The impact of a channel modification can be evaluated with river-hydraulics models. These conceptual models have physically based parameters that the hydrologic engineer can modify to reflect the modifications. For example, the HEC-2 computer program (USACE, 1982a) includes a model of gradually varied steady flow (GVSF). The program uses the physical dimensions of the channel and Manning's n (an index of channel roughness) directly to estimate stage. To evaluate the impact of a proposed channel widening, for example, the hydrologic engineer will simply modify the program input to reflect the changes. Repeated solution of the GVSF equations for selected discharge rates yields the stage-discharge function for a proposed channel configuration. Likewise, if the proposed plan includes channel smoothing, the hydrologic engineer can change Manning's n to reflect this, rerun the program, and estimate the modified-condition rating function.

Channel modifications can also affect the discharge-frequency function. In many cases, the modifications will increase velocity in the improved section. Downstream, where no improvements have been made, this will yield greater discharge, and hence an increase in frequency-function quantiles. Further, the channel modifications may eliminate some of the natural storage in the channel. This natural storage, like the storage in a reservoir, would reduce flood peaks. In its absence, the downstream peaks may increase, and this too yields an increase in frequency-function quantiles.

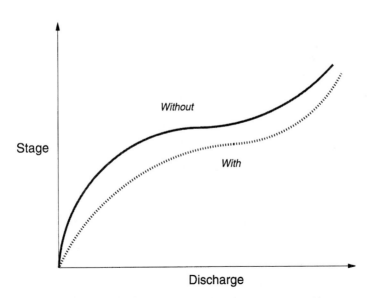

FIGURE 27.13 Stage-discharge function modifications due to channel improvement.

27.6.4 Incidental Impact of Channel Modifications

Channel modifications may also alter the discharge-frequency function if the modifications significantly reduce the timing of the hydrograph through the channel reach. For example, the channel realignment illustrated in Fig. 27.12 reduces the timing between the upstream and downstream cross sections by reducing its length. This reduction, in turn, may result in an increase in the downstream discharge peak for an event of specified probability. Major channel modifications cause an increase in cross section, as illustrated in Fig. 27.11, may increase the storage capacity, and consequently reduce the downstream peak for an event of specified probability. These incidental impacts that change the timing and storage should be investigated.

27.6.5 Technical Considerations

To ensure that channel modifications yield the damage reduction anticipated while formulating and evaluating alternative plans, careful consideration must be given to identification and solution of erosion and deposition problems, design for stability (especially if high velocities are anticipated), protection from ice and debris, and provision for ongoing OMRR&R.

27.6.5.1 Erosion and Deposition. When channel modifications are implemented, some of these problems may worsen. For example, if roughness is decreased, velocity increases, and the likelihood of erosion increases. If deposition was occurring in the without-plan condition, it may or may not continue. Similarly, if a channel is straightened, as shown in Fig. 27.12, the stream slope increases, and the potential for deposition increases where the improved reach rejoins the natural alignment downstream, and the potential for scour increases at the transition from the natural reach upstream. A study to identify these and other related performance problems must be conducted.

The following solutions should be considered as part of the plan if necessary to ensure proper performance: (1) stabilizers constructed of grouted or ungrouted rock, sheet piling, or a concrete sill, placed normal to the channel centerline, traversing the channel invert, and designed to limit channel degradation; (2) drop structures designed to reduce channel slopes, thus yielding nonscouring velocities; and (3) debris basins and check dams to trap and store bed-load sediments.

Channels that convey high velocity (supercritical) flow require special attention. High-velocity channel design must account for the effects of air entrainment, cross waves, superelevation at channel bends, and increased erosion potential.

27.6.5.2 Ice and Debris. Channels in cold regions and channels that carry floating debris (logs and vegetation) can cause special flooding problems. The formation of ice jams and the collection of floating debris at flow constrictions, like bridge crossings, can cause flooding upstream, as the bridge behaves like a dam. The formation of ice jams and the collection of floating debris at flow constrictions also may cause excessive scour due to a local increase in velocity. With such a buildup, the flood discharge must pass through an area that is constricted both laterally and vertically. This leads to increased velocity, which in turn leads to erosion of bed material near the constriction. Likewise, the channel bank in this area might be undermined and ultimately fail. The analysis should evaluate system behavior when it does occur, and must design an OMRR&R plan to minimize the likelihood of ice and debris problems.

27.6.5.3 OMRR&R. A local flood-protection project (including channel improvements) should include an OMRR&R plan to ensure that the modifications continue to function and provide protection as designed. This feature should provide for continuing inspection of the channel to identify evidence of scour damage to bank protection, significant erosion or deposition of sediment in the channel, and growth of vegetation that will increase resistance, thus increasing stage. The cost of this inspection and the anticipated cost of OMRR&R must be included as a component of the total plan cost.

27.6.6 Capacity-Exceedance Analysis

As with all proposed flood-damage-reduction plans, the hydrologic engineer must evaluate the impact of channel-capacity exceedance. In the case of channel improvements, this may be accomplished with the appropriate river-hydraulics model, using a steady flow or hydrograph with peak that exceeds the selected capacity. To minimize the extra effort required here, the hydrologic engineer should ensure that topographic data that are assembled for formulation and evaluation includes sufficient description of the floodplain outside the channel banks.

27.6.7 Environmental Impact

Channel modifications can have significant environmental impacts. For example, certain fish species depend on a pool-riffle aquatic environment typical of low flow in a meandering channel. If such a channel is straightened, the habitat will be disrupted, and the change may lead to reduction in the fish population. The hydrologic engineer should take care to identify such impacts. This will require consultation with environmental specialists. Similarly, consideration to the environmental impact of increased turbidity during construction activities must be given. Potential sources of fine-grained sediment should be identified, and a construction plan should be developed to control runoff from the construction site and to minimize the increase in sensitive areas of the stream.

27.7 LEVEES AND FLOODWALLS

27.7.1 Applicability

This section describes the impact of and requirements for planning levees and floodwalls. It also describes interior-area facilities and the requirements for planning those. A checklist of the requirements for formulating and properly evaluating plans is presented as Table 27.14. Because of their unique layout, sizing, and design requirements, a separate checklist is provided for interior areas as Table 27.15.

Levees and floodwalls are effective damage-reduction measures in the following circumstances:

1. Damageable property is clustered geographically.
2. A high degree of protection, with little residual damage, is desired.
3. A variety of property, including infrastructure, structures, contents, and agricultural property is to be protected.

TABLE 27.14 Checklist for Levees and Floodwalls

Study components	✓	Issues
Layout		Minimize contributing interior runoff areas (flank levees, diversions, collector system)
		Minimize area protected to reduce potential future development
		Investigate levee setback versus height tradeoffs
		Determine right-of-way available for levee or wall alignment
		Minimize openings requiring closure during flood events
Economics		Determine with-project modifications to stage-discharge function for all existing and future conditions
		Quantify uncertainty in stage-damage function
		Formulate and evaluate range of levee and interior area configurations for various capacities using risk-based analysis procedures
		Determine expected capacity- and stage-exceedance probability
Performance		Determine expected annual-exceedance probability
		Determine expected lifetime-exceedance probability
		Describe operation for range of events and sensitivity analysis of critical assumptions
		Describe consequences of capacity exceedances
		Determine reliability (expected probability of exceeding target) for range of events
Design		Design for levee or floodwall superiority at critical features (such as pump stations, high-risk damage centers)
		Design overtopping locations at downstream end, remote from major damage centers
		Provide levee height increments to accommodate settlement, wave run-up
		Design Levee exterior erosion protection
		Develop flood-warning-preparedness plan for events that exceed capacity
Environmental and social		Evaluate aquatic and riparian habitat impact and identify enhancement opportunities
		Anticipate and identify incidental recreation opportunities

TABLE 27.15 Checklist for Interior Areas

Study components	✓	Issues
Layout		Define hydraulic characteristics of interior system (storm-drainage system, outlets, ponding areas, and so forth)
		Delineate environmentally sensitive aquatic and riparian habitat
		Identify damage centers, delineate developed areas, define land uses for site selection
Economics		Determine with-project modifications to interior stage-frequency function for all conditions
		Quantify uncertainty in frequency function
		Formulate and evaluate range of pond, pump, outlet configurations for various capacities using risk-based analysis procedures
Performance		Determine expected annual-exceedance probability
		Determine expected lifetime-exceedance probability
		Determine operation for range of events and sensitivity analysis of critical assumptions
		Describe consequences of capacity exceedances
		Determine reliability (expected probability of exceeding target) for range of events
Design		Formulate and evaluate preliminary inlet and outlet configurations for facilities
		Formulate preliminary operation plans

4. Sufficient real estate is available for levee construction at reasonable economic, environmental, and social cost.

5. The great economic value of damageable property protected will justify the great cost of constructing the levee.

27.7.2 Levee and Floodwall Overview

A *levee* is an [earthen] embankment whose primary purpose is to furnish flood protection from seasonal high water and which is therefore subject to water loading for periods of only a few days or weeks a year. Figure 27.14 shows a cross section of a simple levee. A floodwall serves the same purpose under similar circumstances, differing only in the method of construction. It is subject to hydraulic loading on the one side which is resisted by little or no earth loading on the other side (USACE, 1948). Figure 27.15 shows a variety of floodwalls.

27.7.3 Flood-Damage-Reduction Assessment

Levees and floodwalls (hereafter referred to as levees for brevity) reduce damage by reducing flood stage in the protected area. They do so by blocking overflow from the

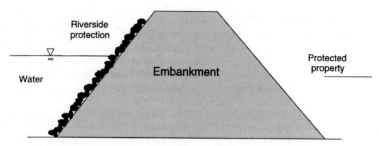

FIGURE 27.14   Cross section of simple levee.

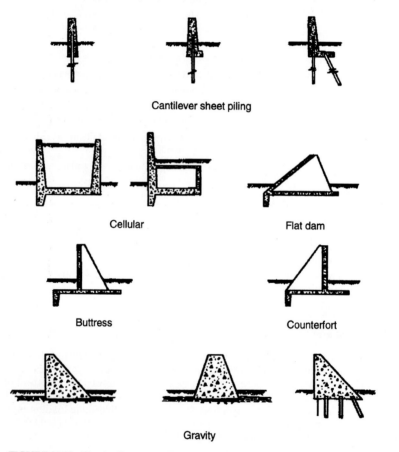

FIGURE 27.15   Floodwall types. In all cases, water to left, protected area to right. *(U.S. ACE, 1948.)*

channel onto the floodplain. This is represented by a modification to the stage-damage function, as shown in Fig. 27.16. S_1 represents the minimum stage, without the levee, at which damage is incurred. The solid curve represents the remainder of this without-levee function. With the levee in place, the stage at which damage is initially incurred rises to an elevation equal the height of the levee. This is designated S_2 in the figure. If the water stage rises above this, the levee is overtopped. Then the damage incurred, designated $\$_2$ on the figure, will equal or exceed the without-levee damage.

A levee may also modify the discharge-frequency function and the stage-discharge relationship. The levee restricts flow onto the floodplain, thus eliminating the natural storage provided by the floodplain. This may increase the peak discharge downstream of the levee for large events that would flow onto the floodplain without the levee. Further, as the natural channel is narrowed by the levee, the velocity may increase. This too may increase the peak discharge downstream for larger events. Modifications to the discharge-frequency function due to a levee are identified with the river-hydraulics models or with routing models (USACE, 1993b; 1994b). These model the impact of storage on the discharge hydrograph and will reflect the loss of this storage. Historical or hypothetical runoff hydrographs can be routed with the selected model to determine discharge peaks with the proposed levee. For example, the modified Puls routing model (USACE, 1994b, Chap. 24) uses a relationship of channel discharge to channel storage with the continuity equation to determine the channel outflow hydrograph. A levee will reduce storage for discharge magnitudes that exceed the channel capacity, so the impact will be reflected.

Introduction of a levee alters the effective channel cross section, so the levee alters the stage-discharge relationship. The impact of this change can be determined with the river-hydraulics models. As with channel alteration, the impact of a levee can be determined by modifying the parameters which describe the channel dimensions. Repeated application of the model with various discharge magnitudes yields the stage-discharge rating function for a specified levee configuration.

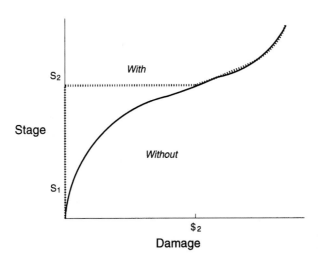

FIGURE 27.16   Stage-damage function modification due to levee or floodwall.

27.7.4 Interior-Area Protection

Figure 27.17 shows an area protected from riverine flooding by a levee. Such a levee (or *floodwall*) is referred to commonly as the *line of protection.* In this case, the line of protection is constructed so natural high ground integrates with the levee to provide the protection; elevation contours shown in the figure illustrate this. The elevation contours also illustrate a problem. The line of protection excludes floodwater, but it also blocks the natural flow path of runoff to the river. The protected area, which was formerly flooded by the slow-rising river is now flooded by local runoff, with little warning. This flooding may be only nuisance flooding, or in some cases, it may be flooding that is as dangerous or more dangerous than the riverine flooding.

27.7.4.1 Solutions to Interior Flooding Problem. To accommodate local runoff, some or all of the facilities shown in Fig. 27.18 may be provided. The interior-area runoff is passed through the line of protection by a gravity outlet when the interior water level is greater than the exterior level. This outlet may have a gate valve and a flap gate that close to prevent flow from the river into the interior area during high stage. When the exterior stage exceeds the interior stage, interior floodwater is stored in the interior pond and pumped over or through the line of protection. This is referred to as a *blocked gravity condition.*

27.7.4.2 Minimum Facility. Some portion of the interior-area components must be included as a part of any levee plan proposed; these are designated the *minimum interior facility.* This minimum facility should provide flood protection such that during gravity condition, the local storm-conveyance system functions essentially as it did without the line of protection in place, for floods less than the storm-sewer-design event. Consequently, the minimum facility often will consist of natural storage and gravity outlets sized to meet local drainage-design criteria. If no local

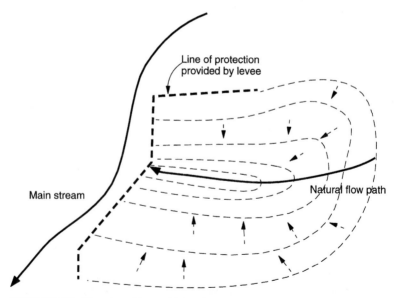

FIGURE 27.17   Plan view of levee with interior area.

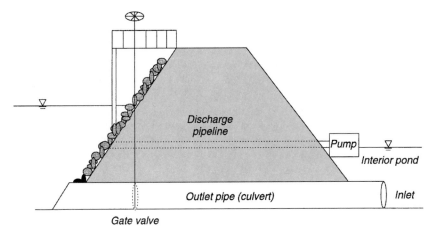

FIGURE 27.18   Components of interior-area protection system.

storm-sewer system exists, but one is planned, the anticipated design criteria are used for planning the minimum facility.

27.7.4.3 Analysis. Hydrologic analysis of interior-area behavior is complex because of the interaction of the interior and exterior waters.

Three interior-area analysis methods are summarized in Table 27.16. The method is chosen normally based on available resources, available data, and technical knowledge.

The HEC-IFH computer program (USACE, 1992b), is specifically designed for the simulation required for interior-area analysis.

27.7.5 Capacity Exceedance Analysis

The principle causes of levee failure are: (1) internal erosion, known as *piping;* (2) slides within the levee embankment or the foundation soils; (3) overtopping; and (4) surface erosion. The hydrologic engineer must work with geotechnical engineers to guard against failures due to piping and slides; flow nets may be required to provide sufficient information for proper design.

The likely locations of and impact of levee overtopping must be addressed by the hydrologic engineer. This is a particularly difficult task, because the hydraulics problem created by levee overtopping is a multidimensional, unsteady-flow problem. Further, when a levee is overtopped, it may breach, so complete analysis also includes the components of a dam-failure analysis. Nevertheless, information on the impact of the failure, including estimates of extent of the inundated area, warning time, and property and lives at risk should be determined. An unsteady fluvial-process model may provide information necessary for this analysis.

Surface erosion cannot be eliminated completely, but if proper precautions are taken, the likelihood of levee failure due to this can be minimized. Table 27.17 summarizes this guidance.

TABLE 27.16 Interior-Area Analysis Alternative

Method	Summary
Continuous record simulation	Simulate without-project and with-project conditions with continuous records of exterior and interior hydrology. These records may be historical flows or flows defined with streamflow-generation techniques. Use runoff-routing models with recorded rainfall if necessary to estimate discharge. Simulate pond, outlet, pump operation for period. Develop necessary stage-frequency functions, duration estimates for economic analysis.
Discrete historical or hypothetical event simulation	Develop stage-frequency function for exterior with flood events that have an effect on interior flooding when interior flooding occurs coincidentally. Simulate without-plan and with-plan conditions for interior area with discrete historical or hypothetical events for low exterior stages that do not affect interior flooding. Develop interior stage-frequency function. Combine the two stage-frequency functions using the joint-probability theorem.
Coincident frequency analysis	For situations in which occurrence of exterior and interior flooding is independent, apply total probability theorem to define stage-frequency functions. To do so, develop exterior stage-frequency function, simulate system performance to develop interior frequency function for various exterior stages, combine functions.

27.7.6 Other Technical Considerations

Most levee projects and some interior-area protection schemes are designed to operate automatically and only require surveillance of operation during floods. A complete plan will include provisions for this surveillance and for flood-fighting activities, which involve special precautions to ensure the safety and integrity of levees. It is important that managers of water-control systems be properly appraised of the status of levee projects in conjunction with the overall control of a water-resource system (USACE, 1987). This will ensure that gates are opened or closed

TABLE 27.17 Methods of Protecting Levee Riverside Slopes

1. If duration of flooding is brief, and velocities low, provide grass protection, unless currents or waves act against levee.
2. Provide additional protection if embankment materials are fine-grained soils of low plasticity (or silts), as these are most erodible.
3. If severe wave attack and currents are expected, shield riverside slope with timber stands and wide space between riverbank and levee.
4. Take care to accommodate scour due to flow constrictions and turbulence caused by bridge abutments and piers, gate structures, ramps, and drainage outlets.
5. To minimize turbulence and susceptibility to scour, avoid short-radius bends and provide smooth transitions where levees meet high ground or structures.
6. Depending on degree of protection needed and relative costs, provide slope protection with grass cover, gravel, sand-asphalt paving, concrete paving, articulated concrete mat, or riprap.

properly, pumps are turned on or off as necessary, and access openings in the levee or floodwall are closed properly in anticipation of rising floodwater.

27.8 OTHER MEASURES THAT REDUCE EXISTING-CONDITION DAMAGE SUSCEPTIBILITY

27.8.1 Overview

Existing-condition damage susceptibility, and hence EAD, can be reduced with so-called *nonstructural* measures described in this section. The measures include flood-proofing, relocation, and flood-warning-preparedness (FWP) plans. Requirements for the measures are summarized on Table 27.18.

27.8.2 Requirements for Floodproofing

27.8.2.1 Applicability. *Floodproofing* measures are appropriate for damage reduction for some single-story residential structures. In special cases, these measures have been used for other structures, but the economic and physical feasibility of such applications is limited. Floodproofing does not reduce damage to utilities, infrastructure, lawns, and other exterior property. These measures are limited generally to property frequently flooded. Floodproofing is generally less disruptive to the environment than other measures that require significant construction.

TABLE 27.18 Checklist for Measures that Reduce Existing-Condition Damage Susceptibility

Study components	✓	Issues
Layout		Based on qualification of flood hazard, identify structures for which measures are appropriate
Economics		Determine with-project modifications to stage-damage function for all existing and future conditions
		Quantify uncertainty in stage-damage function
		Formulate and evaluate range of floodproofing, relocation, and/or FWP plans, using risk-based analysis procedures
Performance		Determine expected annual-exceedance probability
		Determine expected lifetime-exceedance probability
		Determine operation for range of events and sensitivity analysis of critical assumptions
		Describe consequences of capacity exceedances
		Determine reliability for range of events
Design		Develop, for all these measures, FWP plans
Environmental and social		Evaluate aquatic and riparian habitat impact and identify enhancement opportunities
		Anticipate and identify incidental recreation opportunities

27.8.2.2   Overview of Floodproofing.   Floodproofing includes: (1) use of closures and small walls to keep out floodwaters and (2) raising existing structures in
place to reduce damage. The measures are spatially distributed, so they do not provide the same uniform protection possible with, for example, a reservoir. Floodproofing reduces damage to existing individual structures or parcels of land by
altering damage susceptibility.

Closures, like those shown in Fig. 27.19, reduce damage by keeping the floodwater out of the structure. This figure shows window closures, but similar closures can
be provided for doors and other openings. Closures may be temporary or permanent. The figure shows temporary closures; these are bolted into place during a flood
threat and removed afterwards. In addition to the closures, depending on site conditions, the following may be required: a waterproofing sealant applied to the walls
and floors to reduce seepage; a floor drain and sump pump to accommodate seepage; and a valve to eliminate flooding in the structure due to sewer backflow.

Similar damage reduction can be achieved with a small wall or levee built around
one or several structures. Such a wall is designed for compatibility with local landscaping and aesthetics, and generally is less than one meter high. Walls may be brick,
stone, concrete, or some other material designed to withstand lateral and uplift
forces associated with floodwaters. As with a major levee, runoff in the interior area
must be managed; often a small pump is adequate.

Figure 27.20 shows an existing structure after it was raised in-place to reduce
damage. The hazard is not eliminated here, but the damage is reduced. Now when a
flood occurs, the depth of water at the site, relative to the original ground level, is the
same, but the depth of flooding in the structure is less. In Figure 27.20, the structure
is a single-story wooden-frame residential structure that was constructed originally

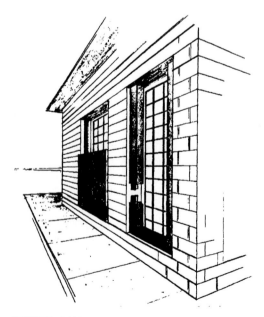

FIGURE 27.19   Floodproofing with closures. *(From
Sheaffer, et al., 1967.)*

FIGURE 27.20   Floodproofing by raising an existing structure in-place. *(From U.S. Department of Housing and Urban Development, 1977.)*

with a crawl space and no basement. Specific actions required to raise this structure are listed in Table 27.19. While it is possible to raise almost any structure, raising a structure such as that illustrated is most likely to be economically justified and physically feasible. Note that in the figure, fill was used to raise the car-parking pad.

27.8.2.3   Flood-Damage-Reduction Assessment.   Floodproofing alters the stage-damage relationship for structures. The manner in which it does so depends on the measures used. Figure 27.21 illustrates the alteration when a closure or small wall is used. The existing condition, without project stage-damage function is shaded gray; the modified function is the dotted curve. Without the closure or wall, damage begins when stage reaches S_1, as shown in the figure. With the closure or wall in place, the onset of damage is raised to stage S_2. Of course, if the stage exceeds S_2, the closure or wall is overtopped, and damage is essentially that which would be incurred without the measure. In the figure, this is represented by the sharp increase in with-project damage for stage greater than S_2.

TABLE 27.19   Actions Required to Raise a Structure In-Place

---

1. Disconnect all plumbing, wiring, and utilities that cannot be raised with the structure.
2. Place steel beams and hydraulic jacks beneath the structure and raise to desired elevation.
3. Extend existing foundation walls and piers or construct new foundation.
4. Lower the structure onto the extended or new foundation.
5. Adjust walks, steps, ramps, plumbing, and utilities. Regrade site as desired.
6. Reconnect all plumbing, wiring, and utilities.
7. Insulate exposed floors to reduce heat loss and protect plumbing, wiring, utilities, and insulation from possible damage.

Source: USACE (1978).

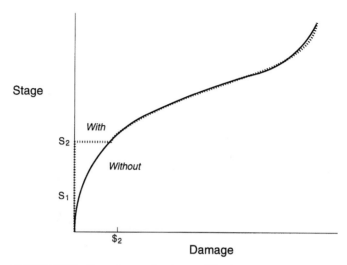

FIGURE 27.21 Stage-damage function modification due to floodproofing with closure wall.

Figure 27.22 illustrates the alteration when an existing structure is raised in-place. Again, the existing condition, without project stage-damage function is a solid curve, and the modified function is the dotted curve. For the existing, without-project condition, damage begins when stage reaches S_1. When the structure is raised, the stage-damage function is shifted upward a distance equal to the increased elevation, but the function retains essentially the same shape. Thus the onset of damage is raised to stage S_2, and the damage incurred at all stages equals the damage previously incurred at that stage less the distance the structure was raised.

27.8.2.4 Technical Considerations. Reports from HEC (USACE, 1978; 1985) describe various nonstructural measures in detail and identify critical technical considerations for formulating plans that include these measures. Some of the important considerations identified there are summarized in Table 27.20.

A complete plan that incorporates floodproofing must include an emergency evacuation plan. This can only be formulated properly with significant input from the hydrologic engineer. He or she must identify inundated areas for identifying escape routes and estimate flow velocities for evaluating the safety of the evacuation routes. For example, if a small two-foot-high levee is proposed for a group of residences, the hydrologic engineer should estimate velocities associated with flows corresponding to depths greater than two feet and evaluate the likelihood of evacuation by foot, vehicle, or boat in that case.

27.8.3 Requirements for Relocation

27.8.3.1 Applicability. Relocating contents within an existing structure at its current location is effective in any case, but the damage reduction possible is limited. The residual damage is likely to be great. Permanently removing the contents or the structure and contents from a flood-hazard area similarly reduces damage in any

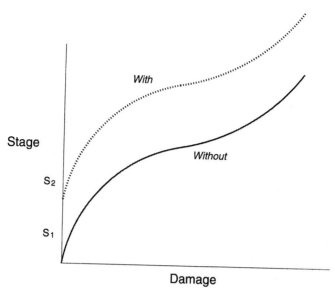

FIGURE 27.22 Stage-damage function modification due to floodproofing by raising in-place.

TABLE 27.20 Performance Requirements for Floodproofing

Floodproofing Method	Performance Requirement
Window or door closure	Provide adequate forecasting and warning to permit installation of closures.
	Identify *all* openings for closure, including fireplace cleanouts, weep holes, and so forth.
	Ensure structural adequacy to prevent failure due to hydrostatic pressure or floating of structure.
	Ensure watertightness to minimize and drainage to accommodate leakage.
	Arrange adequate, ongoing public training to ensure proper operation.
Small wall or levee	Requirements similar to major levee, but on a smaller scale, including: (1) providing for closure of openings in wall or levee, (2) ensuring structural stability of levee or wall, and (3) providing for proper interior drainage.
	Arrange adequate, ongoing public training to ensure proper operation.
	Plan for emergency access to permit evacuation if protected area is isolated by rising floodwaters.
Raising in-place	Protect beneath raised structure, as hazard is not eliminated.
	Ensure structural stability of raised structure.
	Plan for emergency access to permit evacuation if protected area is isolated by rising floodwaters.

case, but is likely to be costly and thus economically feasible only for higher-value structures. Permanent relocation is physically feasible for a limited class of structures (USACE, 1978).

27.8.3.2 Overview of Relocation. The term *relocation*, as used in this manual, means moving property so it is less susceptible to damage. This may be accomplished by: (1) relocating contents within an existing structure at its current location or (2) removing the contents or the structure and contents from a flood-hazard area.

Examples of relocation of contents within a structure are shown in Table 27.21. These are relatively simple measures that can be undertaken by any property owner. The relocation can be temporary or permanent. Effectiveness depends on the type of contents and flood hazard.

Removing contents or a relocation of flood-threatened structures are effective, if costly, solutions to the flood-damage problem in any circumstance. To accomplish this, a building site outside the flood-hazard area must be located and purchased or leased. In the case of moving a structure, the new site must be prepared; the structure must be raised, transported, and installed at the new site; contents must be moved; and the old site restored. For relocating contents only, a structure outside the hazard area must be built or leased, and contents must be moved.

27.8.3.3 Flood-Damage-Reduction Assessment. Relocation reduces flood damage by reducing the damage incurred at a given stage. In the extreme, if all structures and their contents are moved from the flood-hazard area, the stage-damage function is reduced to zero damage for all stages in the range of practical interest. More practically, if selected structures or contents are relocated, the stage-damage function will be modified to reflect the lowered value of property that would be inundated at a given elevation. In general, the damage for a specified stage will be reduced; the exact form of the modified with-project stage-damage function depends on location and value of property relocated.

27.8.4 Requirements for Flood-Warning-Preparedness Plans

27.8.4.1 Applicability. A FWP program may be implemented as: (1) a stand-alone measure when other measures are not feasible, (2) an interim measure until others are in place, or (3) as a component to other measures. FWP as a stand-alone measure provides only minimum damage reduction. Even with the most efficient forecast and best-planned response system, the possibility of catastrophic damage

TABLE 27.21 Examples of Relocation

1. Protecting HVAC equipment, appliances, shop equipment by raising off floor
2. Relocating property to higher floors
3. Relocating commercial and industrial products, merchandise, equipment to higher floor or higher building
4. Relocating finished products, materials, equipment, other moveable items now located outside to higher ground
5. Protecting electrical equipment by raising on pedestal, table, platform
6. Anchoring property that might be damaged by floodwater movement

Source: USACE (1978).

continues to exist in a managed manner. A FWP plan has no significant environmental impact in most cases.

27.8.4.2 Overview of Flood-Warning-Preparedness Plans.

A *flood-warning-preparedness plan* (*FWP plan*) reduces flood damage by providing the public with an opportunity to act before flood stages increase to damaging levels. The savings due to a FWP plan may be due to reduced inundation damage, reduced cleanup costs, reduced cost of disruption of services due to opportunities to shut off utilities and make preparations, and reduced costs due to reduction of health hazards. Further, FWP plans may reduce social disruption and risk to life of floodplain occupants.

A FWP plan is a critical component of other flood-damage-reduction measures, as pointed out elsewhere in this manual. In addition, federal *Flood Plain Management Services* (*FPMS*) staff may provide planning services in support of local agency requests for assistance in implementation of a FWP plan; this is authorized by Section 206 of the Flood Control Act of 1960.

27.8.4.3 Flood-Damage-Reduction Assessment.

A FWP plan reduces inundation damage by permitting the public to relocate property, close openings, close backflow valves, turn on sump pumps, and take other actions that will lower the damage incurred when water reaches a specified stage. Estimating the form of the modified, with-project stage-damage function requires estimating the accuracy of a forecast and how the public will respond to a warning. Day (1973) suggested a method for estimating the benefit. The estimates should be based on the specific study circumstances.

27.8.4.4 Technical Considerations.

Table 27.22 shows the components of a complete FWP plan. If the plan is to function properly, it must include each of these components. Likewise formulation of the emergency-response plan requires flood-hazard information including delineation of inundated areas and identification of escape routes. USACE (1986) provides guidance.

TABLE 27.22 Components of a FWP System

Component	Purposes
Flood-threat-recognition subsystem	Collection of data and information; transmission of data and information; receipt of data and information; organization and display of data and information; prediction of timing and magnitude of flood events.
Warning-dissemination subsystem	Determination of affected areas; identification of affected parties; preparation of warning messages; distribution of warning messages.
Emergency-response subsystem	Temporary evacuation; search and rescue; mass care center operations; public-property protection; flood fight; maintenance of vital services.
Post-flood-recovery subsystem	Evacuee return; debris clearance; return of services; damage assessment; provisions for assistance.
Continued system management	Public-awareness programs; operation, maintenance, and replacement of equipment; periodic drills; update and arrangements.

Source: Adapted from USACE (1988).

27.9 MEASURES THAT REDUCE FUTURE-CONDITION DAMAGE SUSCEPTIBILITY

27.9.1 Overview

Future-condition damage may be reduced through land-use and construction regulation or by acquisition. Although neither is used commonly in flood-damage-reduction plans, both are potentially components of a complete plan in which costs are shared with local partners. Consequently, requirements for these measures are described in this section. The checklist included in Sec. 27.8 describes requirements for measures described in this section.

27.9.2 Requirements for Construction and Land-Use Regulation

27.9.2.1 Overview. *Construction and land-use regulation* includes building codes, zoning ordinances, or subdivision regulations. These measures decrease future damage by reducing susceptibility of future development.

Figure 27.23 illustrates the result of one form of construction regulation. In this case, the building code requires that the lowest floor of new construction is above the 1-percent-chance flood stage. To comply, this structure is built on timber posts. This type of construction, of course, does not control the flood stage, but it does reduce the damage incurred. Construction on concrete walls, on steel, concrete, or masonry posts, piles, or piers, or on earth fill will have similar impact.

Damage susceptibility of new structures can be reduced also by regulating construction materials and practices. Table 27.23 lists typical requirements that may be included in such regulations.

Finally, future damage susceptibility can be reduced with *land-use regulations* that insure that future use of floodplains is compatible with the hazard there. *Zoning permits* district-by-district regulation of "... what uses may be conducted in flood hazard areas, where specific uses may be conducted, and how uses are to be

FIGURE 27.23   Illustration of construction per regulations to reduce damage susceptibility. *(From U.S. Department of Housing and Urban Development, 1977.)*

TABLE 27.23   Typical Requirements for New Construction to Reduce Damage
Susceptibility

Location	Requirement
Basement	Install drains, valves to equalize water pressure
	Use permeable backfill
	Use water-resistant flooring
	Use moisture-tolerant paints and paneling
	Provide ceiling drains to permit drywall drainage
	Provide anchored, water-resistant cabinets
	Construct stairways sufficiently wide for relocation of basement contents
First floor	Use water-resistant paints, paneling, flooring
	Provide cabinets, bookshelves, furnishings that are moisture tolerant
	Provide stairways sufficiently wide for relocation of first-floor contents
Exterior	Anchor tanks to prevent floatation, vent above first floor to prevent fuel escape
	Provide manually operated sewer-backflow valves
	Use nonabsorbent, exterior-grade materials and treated lumber
Electrical, heating, cooling system	Provide duct drains
	Separate electrical circuits to allow selective shut off
	Slope gas piping, fit with drains

Source: USACE (1978).

constructed or carried out" (U.S. WRC, 1971). Subdivision regulations ". . . guide division of large parcels of land into smaller lots for the purpose of sale of building developments . . . [they] often (1) require installation of adequate drainage facilities, (2) require that location of flood hazard areas be shown on the plat, (3) prohibit encroachment in floodway areas, (4) require filling of a portion of each lot to provide a safe building site at elevation above selected flood heights or provide for open support elevation to achieve the same ends, and (5) require the placement of streets and public utilities above a selected flood protection elevation" (U.S. WRC, 1971).

27.9.2.2   Flood-Damage-Reduction Assessment: Future, With-Project Evaluation.   Section 27.2 explains the computation of EAD and describes how, if conditions change over time, EAD is to be computed annually and discounted to determine an equivalent annual value over the life of a plan. Land-use and construction regulation will yield changes in the future-condition stage-damage function, thus reducing this equivalent annual value. This is illustrated by Fig. 27.24. This figure shows EAD computed over a period of 50 years. Without regulations, the value continues to increase each year as the value of development subject to flood damage increases. If construction and land-use regulations are imposed in 1999, however, the EAD stops increasing. Due to the regulations, the value of property exposed to flood damage does not increase beyond the 1999 level. In fact, if regulations prohibit new construction that is susceptible to flood damage, the EAD may decrease as structures and contents reach the end of their useful life and are replaced with structures and contents less susceptible to flood damage.

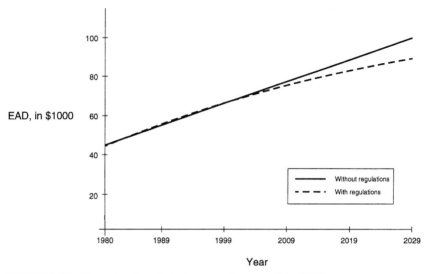

FIGURE 27.24   Illustration of regulation impact on future-condition EAD.

27.9.2.3 Technical Considerations. To some degree, construction and land-use regulation are applicable in all floodplains. To ensure success, the hydrologic engineer must participate actively in delineating the hazard area and characterizing the flooding. The delineation is necessary to identify property to which regulations should apply, and the characterization is necessary to determine the nature of the regulations.

27.9.3 Requirements for Acquisition

27.9.3.1 Overview. *Public acquisition* of floodplain property is another method by which the government, either federal or local, can ensure proper use, thus reducing damage susceptibility. Title to the property can be acquired, or a land-use easement can be acquired. In the first case, ownership of the property shifts to the public, so uses with high risk of damage can be abandoned. Instead, the property can be dedicated to use as a park or wildlife preserve. Acquisition of a land easement leaves property in the hands of private owners, but permits restriction of use. For example, building or filling within an easement can be prohibited.

27.9.3.2 Flood-Damage-Reduction Assessment: Future, With-Project Evaluation. Acquisition has an impact similar to that of construction and land-use control: It reduces future damage. Figure 27.24 might well illustrate the EAD with acquisition of floodplain property in 1999, as this too will reduce susceptibility to damage, and hence EAD, thereafter.

27.10 SYSTEM ANALYSIS

27.10.1 Plan Evaluation

Plans for reducing flood damage are comprised of one or more types of measures. For example, a mix of channel modifications, detention storage, floodproofing, regulatory policies, and flood-warning preparedness may be one plan for reducing flood damage throughout the study area. Another plan may have similar mixes of measures but be sized differently and may be used at different locations. Other plans

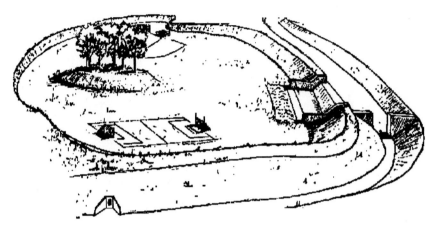

FIGURE 27.25 Example of flood-damage-reduction system. *(From drawing furnished by U.S. Army Engineer District, Tulsa, Okla.)*

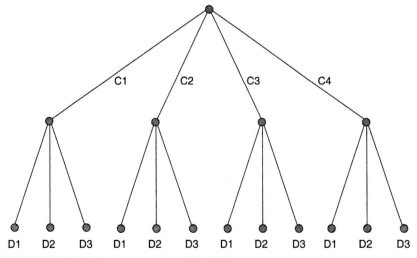

FIGURE 27.26 Decision tree for system of Fig. 27.25.

TABLE 27.24   Impacts of Flood-Damage-Reduction Measures

Measures	Impact of measure		
	Modifies discharge-frequency function	Modifies stage-discharge function	Modifies stage-damage function
Reservoir	Yes	Maybe, if stream and downstream channel erosion and deposition due to change in discharge	Maybe, if increased development in floodplain
Diversion	Yes	Maybe, if channel erosion and deposition due to change in discharge	Maybe, if increased development in floodplain
Channel improvement	Maybe, if channel affects timing and storage altered significantly	Yes	Not likely
Levee or floodwall	Maybe, if floodplain storage no longer available for flood flow	Yes	Yes
Floodproofing	Not likely	Not likely	Yes
Relocation	Not likely	Maybe, if flow obstructions removed	Yes
FWP plan	Not likely	Not likely	Yes
Land-use and construction regulation	Not likely	Maybe, if flow obstructions removed	Yes
Acquisition	Not likely	Maybe, if flow obstructions removed	Yes

may be completely different sets of measures and actions. The plan formulation and evaluation process is summarized in Sec. 27.2.

The total economic accomplishment, performance, and environmental impact of a flood-damage-reduction plan is not simply the sum of the output of the individual measures. Instead, a well-formulated plan can yield greater benefit, perform better, and have less adverse impact through synergism. For example, if land-use regulation is combined with a reservoir, the regulations will reduce damage susceptibility and the size of a reservoir may be reduced. Consequently, the same damage reduction may be achieved at less cost and, perhaps, with less adverse environmental impact. This interaction means that the components of a plan cannot be formulated and evaluated independently. Instead the interdisciplinary planning team must view a flood-damage-reduction plan as a system and must evaluate explicitly the interactions of the measures. These interactions will affect the economic benefit, performance, and environmental impact of the plan.

27.10.2   Economic-Objective Evaluation for System

The impact of interaction of plan components can be illustrated with the example in Fig. 27.25. For this example, development upstream has led to increased runoff that, in turn, causes flood damage downstream. The planning team has proposed a channel modification to reduce the downstream stage and corresponding damage. Based on engineering judgement and experience, several alternative sizes and configurations were proposed. In response to concern over the environmental impact of excavation required for larger channels, an upstream detention storage basin also has been proposed. This detention basin, configured as shown, reduces the flow to the channel, thus reducing the required capacity and the necessary excavation. Again based on engineering judgment and experience, several alternative sizes and configurations were proposed for the detention basin.

To evaluate the net benefit of each alternative, the hydrologic engineer must consider explicitly the interaction of the channel and the detention basin, since the channel impacts the stage-discharge function and detention storage the discharge-frequency function. To do so systematically, a decision tree like that shown in Fig. 27.26 might be constructed to identify the plans. In this illustration, four channel sizes and configurations are formulated; these are labelled C1, C2, C3, and C4. Three detention storage alternatives, labelled D1, D2, and D3, are proposed. Each branch in the decision tree represents an alternative plan with one of the proposed channel configurations and one of the proposed detention storage alternatives. Evaluation of the with- and without-project conditions frequency and stage relationships using procedures are described in Section 27.2. The expected annual-damage analyses are performed as illustrated in Fig. 27.1.

The process must take care to identify both planned and incidental changes to the discharge-frequency, stage-discharge, and stage-damage functions. Table 27.24 summarizes both for various flood-damage-reduction measures described in this manual, but the list is not universal.

REFERENCES

American Society of Civil Engineers/Water Environment Federation, *Design and Construction of Urban Stormwater Management Systems*, ASCE-WEF, New York, 1992.

Barkau, R. L., "A Mathematical Model of Unsteady Flow Through a Dendritic Network," Ph.D. dissertation, Department of Civil Engineering, Colorado State University, Ft. Collins, Colo., 1985.

Bureau of Reclamation, *Design of Small Dams*, U.S. Department of the Interior, Washington, D.C., 1977.

CECW-E/CECW-P/CECW-L, "Planning, Engineering, and Design Process, General Design Memoranda, and Reevaluation Reports," Memorandum, 5 Nov 1991.

Chow, V. T., *Open-Channel Hydraulics*, McGraw-Hill, New York, 1959.

Davis, D. W., "Optimal Sizing of Urban Flood Control Systems," TP-42, Hydrologic Engineering Center, Davis, Calif., 1974.

Day, H. J., (1973). "Benefit and Cost Analysis of Hydrological Forecasts: A State-of-the-Art Report," Operational Hydrology Report no. 3, WMO-no. 341, World Meteorological Organization, Geneva, Switzerland, 1973.

DeVries, J. J., and T. V. Hromadka, "Computer Models for Surface Water," in David R. Maidment, ed., *Handbook of Hydrology*, McGraw-Hill, New York, 1993.

Ford, D. T., and A. Oto, "Floodplain Management Plan Enumeration," *Journal of Water Resources Planning and Management Division*, ASCE, 115(6):472–485, 1989.

Ford, D. T., and D. W. Davis, "Hardware-Store Rules for System-Analysis Applications," *Closing the Gap Between Theory and Practice, Proceedings of the IAHS Symposium*, IAHS Publication 180, Baltimore, Md., 1989.

Interagency Advisory Committee on Water Data, "Guidelines for Determining Flood Flow Frequency," Bulletin 17B, U.S. Department of the Interior, U.S. Geological Survey, Office of Water Data Coordination, Reston, Va., 1982.

Jackson, T. L., "Application and Selection of Hydrologic Models," in C. T. Haan, et al., eds., *Hydrologic Modeling of Small Watersheds,* American Society of Agricultural Engineers, St. Joseph, Mo., 1982.

James, L. D., et al., "Integrating Ecological and Social Considerations into Urban Flood Control Programs," *Water Resources Research,* AGU, 14(2):177–184, 1978.

Larson, C. L., et al., "Some Particular Watershed Models," in C. T. Haan, et al., eds., *Hydrologic Modeling of Small Watersheds,* American Society of Agricultural Engineers, St. Joseph, Mo., 1982.

Loucks, D. P., et al., *Water Resources Systems Planning and Analysis,* Prentice-Hall, Englewood Cliffs, N.J., 1981.

Renard, K. G., W. J. Rawls, and M. M. Fogel, "Currently Available Models," in C. T. Haan, et al., eds., *Hydrologic Modeling of Small Watersheds,* American Society of Agricultural Engineers, St. Joseph, Mo., 1982.

Sheaffer, J. R., et al., "Introduction to Flood Proofing: An Outline of Principles and Methods," Center for Urban Studies, University of Chicago, Chicago, Ill., 1967.

Thomas, W. A., and W. H. McAnally, "TABS-2: Open-Channel Flow and Sedimentation, User's Manual." WES, Vicksburg, Miss., 1985.

U.S. Army Corps of Engineers, "Wall design: Flood walls," *Engineering Manual 1110-2-2501,* 1948.

U.S. Army Corps of Engineers, "Flood Control System Component Optimization: HEC-1 Capability." TD-9, Hydrologic Engineering Center, Davis, Calif., 1977.

U.S. Army Corps of Engineers, "Physical and Economic Feasibility of Nonstructural Flood Plain Management Measures," RD-11, Hydrologic Engineering Center, Davis, Calif., 1978.

U.S. Army Corps of Engineers, "HEC-2: Water Surface Profiles, User's Manual," CPD-2A, Hydrologic Engineering Center, Davis, Calif., 1982a.

U.S. Army Corps of Engineers, "HEC-5: Simulation of Flood Control and Conservation Systems, User's Manual," CPD-5A, Hydrologic Engineering Center, Davis, Calif., 1982b.

U.S. Army Corps of Engineers, "Ice Engineering," *Engineering Manual 1110-2-1612,* 1982c.

U.S. Army Corps of Engineers, "HMR-52: Probable Maximum Storm (Eastern United States), User's Manual," CPD-46, Hydrologic Engineering Center, Davis, Calif., 1984.

U.S. Army Corps of Engineers, "Engineering and Economic Considerations in Formulating Nonstructural Plans," TP-103, Hydrologic Engineering Center, Davis, Calif., 1985.

U.S. Army Corps of Engineers, "Proceedings of a Seminar on Local Flood Warning-Response System," Hydrologic Engineering Center, Davis, Calif., 1986.

U.S. Army Corps of Engineers, "Branch-and-Bound Enumeration for Reservoir Flood Control Plan Selection," RD-35, Hydrologic Engineering Center, Davis, Calif., 1987a.

U.S. Army Corps of Engineers, "Hydrologic Analysis of Interior Areas," *Engineering Manual 1110-2-1413,* 1987b.

U.S. Army Corps of Engineers, "Management of Water Control Systems," *Engineering Manual 1110-2-3600,* 1987c.

U.S. Army Corps of Engineers, "Flood Damage Reduction Reconnaissance-Phase Studies," Proceeding of a seminar, Hydrologic Engineering Center, Davis, Calif., 1988a.

U.S. Army Corps of Engineers, "FDA: Flood-Damage Analysis Package, User's Manual," CPD-59, Hydrologic Engineering Center, Davis, Calif., 1988b.

U.S. Army Corps of Engineers, "General Guidelines for Comprehensive Flood Warning/Preparedness Studies," RD-30, Hydrologic Engineering Center, Davis, Calif., 1988c.

U.S. Army Corps of Engineers, "Digest of Water Resources Policies and Authorities," *Engineering Pamphlet 1165-2-1,* 1989a.

U.S. Army Corps of Engineers, "EAD: Expected Annual Flood Damage Computation, User's Manual," CPD-30, Hydrologic Engineering Center, Davis, Calif., 1989b.

U.S. Army Corps of Engineers, "Sedimentation Investigations of Rivers & Reservoirs," *Engineering Manual 1110-2-4000,* 1989c.

U.S. Army Corps of Engineers, "SID: Structure Inventory for Damage Analysis, User's Manual," CPD-41, Hydrologic Engineering Center, Davis, Calif., 1989d.

U.S. Army Corps of Engineers, "HEC Data Storage System User's Guide and Utility Program Manuals," CPD-45, Hydrologic Engineering Center, Davis, Calif., 1990a.

U.S. Army Corps of Engineers, "HEC-1: Flood Hydrograph Package, User's Manual," CPD-1A, Hydrologic Engineering Center, Davis, Calif., 1990b.

U.S. Army Corps of Engineers, "Hydraulic Design of Spillways," *Engineering Manual 1110-2-1603,* 1990c.

U.S. Army Corps of Engineers, "HECLIB, Vol. 2: HEC-DSS Subroutines, Programmer's Manual," CPD-57, Hydrologic Engineering Center, Davis, Calif., 1991a.

U.S. Army Corps of Engineers, "Inflow Design Floods for Dams and Reservoirs." *Engineering Regulation 1110-8-2,* 1991b.

U.S. Army Corps of Engineers, "HEC-FFA: Flood Frequency Analysis, User's Manual," CPD-59, Hydrologic Engineering Center, Davis, Calif., 1992a.

U.S. Army Corps of Engineers, "HEC-IFH: Interior Flood Hydrology Package, User's Manual," CPD-31, Hydrologic Engineering Center, Davis, Calif., 1992b.

U.S. Army Corps of Engineers, "HEC-6: Scour and Deposition in Rivers and Reservoirs, User's Manual," CPD-6, Hydrologic Engineering Center, Davis, Calif., 1993a.

U.S. Army Corps of Engineers, "Hydrologic Frequency Analysis," *Engineering Manual 1110-2-1415,* 1993b.

U.S. Army Corps of Engineers, "River Hydraulics," *Engineering Manual 1110-2-1416,* 1993c.

U.S. Army Corps of Engineers, "UNET: One-dimensional Unsteady Flow Through a Full Network of Open Channels, User's Manual," CPD-66, Hydrologic Engineering Center, Davis, Calif., 1993d.

U.S. Army Corps of Engineers, "Risk Analysis for Evaluation of Hydrology/Hydraulics and Economics in Flood Damage Reduction Studies," *Engineering Circular 1105-2-205,* 1994a.

U.S. Army Corps of Engineers, "Flood-Runoff Analysis," *Engineering Manual 1110-2-1417,* 1994b.

U.S. Department of Housing and Urban Development, "Elevated Residential Structures," Federal Insurance Administration, Washington, D.C., 1977.

U.S. Water Resources Council, "Regulation of Flood Hazard Areas to Reduce Flood Losses," U.S. Government Printing Office, Washington, D.C., 1971.

U.S. Water Resources Council, "Estimating Peak Flow Frequency for Natural Ungaged Watersheds—A Proposed Nationwide Test," U.S. Government Printing Office, Washington, D.C., 1981.

U.S. Water Resources Council, "Economic and Environmental Principles and Guidelines for Water and Related Land Resources Implementation Studies," U.S. Government Printing Office, Washington, D.C., 1983.

World Meteorological Organization, "Intercomparison of Conceptual Models Used in Operational Hydrological Forecasting," Operational Hydrology Report no. 7, WMO-no. 429, Geneva, Switzerland, 1975.

World Meteorological Organization, "Manual on Stream Gauging, vol. I and II," Operational Hydrology Report no. 13, WMO-no. 519, Geneva, Switzerland, 1980.

World Meteorological Organization, "Guide to Hydrologic Practices, vol. I: Data Acquisition and Processing," WMO-no. 168. Geneva, Switzerland, 1981.

World Meteorological Organization, "Guide to Hydrologic Practices, Vol. II: Analysis, Forecasting and Other Applications," WMO-no. 168, 4th ed., Geneva, Switzerland, 1983.

CHAPTER 28
COMPUTER MODELS FOR WATER-EXCESS MANAGEMENT

David Ford
Consulting Hydrologic Engineer
Sacramento, California

Douglas Hamilton
Consulting Engineer
Tustin, California

28.1 INTRODUCTION

28.1.1 What Is a Water-Excess Management Computer Model?

This chapter describes *water-excess management models.* (These are also known as *stormwater, hydrologic engineering,* or *flood-control models.*) The models simulate critical processes to provide information for:

- Planning and designing new water-control facilities
- Operating such facilities
- Preparing for and responding to floods
- Regulating floodplain activities

For clarity, we make a distinction herein between mathematical models, computer models (also called programs), and applications. A *mathematical model* is a symbolic representation of the behavior of a system. For example, the combination of the continuity and momentum equations described in Chap. 2 is a mathematical model of flow in an open channel. To yield information, the equations of a mathematical model must be solved. If the equations are relatively simple, they may be solved with pencil and paper and electronic calculator. For example, the equations of the unit hydrograph model described in Chap. 24 can be solved in this fashion to predict runoff from a simple rainstorm. On the other hand, if the equations included in the model are too numerous or too complex to solve with pencil, paper, and calculator, they may be solved instead by translating the equations and an appropriate equation solver into computer code. The result is a *computer model* or computer program. When the equations of a mathematical model are solved with site-specific initial and boundary conditions and parameters, the model simulates the processes and predicts what will happen to the particular system. This solution with specified conditions is an *application* of the model. An application may use a computer model, or it may use the mathematical model with solution with pencil, paper, and calculator.

28.1.2 Selecting a Water-Excess Management Computer Model

28.1.2.1 Selection Problem and Solution. Ford and Davis (1989) write that water-resources planning and management is similar to home improvement: in both, the appropriate tool must be selected to solve the problems at hand. In the case of home improvement, the decision is what hand tool to use: Should it be a hand saw or a chain saw? In the case of water management, the decision is what computer tool or model to use. Jackson (1982) suggests that to select the best model, one should follow the procedure illustrated by Fig. 28.1. In the case of water-excess management, the information identified in Step 1 of this procedure typically includes:

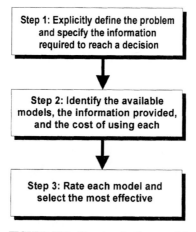

FIGURE 28.1 Steps in selecting a model.

- Stream-discharge time series or peaks
- Volume time series or totals
- River or reservoir water-depth time series or maximums
- Probabilities (frequencies) of extreme discharge, volume, or depth magnitudes
- Inundated-area geometry
- Landform changes due to erosion or deposition
- Economic, social, or environmental costs and benefits of any of these items

The remainder of this chapter is devoted to Step 2: identifying available models that can provide this information.

28.1.3 A Sample of Water-Excess Management Computer Models

Computer models that provide the required information are described in theses and dissertations, in project reports, and in a number of journals, including AGU's *Water Resources Research,* ASCE's *Journal of Hydraulic Engineering,* ASCE's *Journal of Water Resources Planning and Management,* and AWRA's *Water Resources Bulletin.* DeVries and Hromadka (1993); Renard, et al. (1982); Larson, et al. (1982); WMO (1975); Clarke (1973), and others have reviewed, summarized, and compared mathematical and computer models for water-excess management. Here we follow their lead, except that we focus on the information provided, rather than on the specific details of the mathematical formulation and solution techniques. We have selected a sample of the available computer models, using the following criteria:

Common usage. The water-excess management computer models described herein are restricted to those used commonly by engineers other than the model developers, as this chapter is written. Further, the models described are those used in a variety of applications, rather than in a single setting. Consequently, some models described in recent literature are omitted, and some models used predominately by a single entity or in a single application are omitted.

General application. The set of computer models reviewed includes only generalized computer models. A *generalized computer model* is one in which the characteristics of the system (initial and boundary conditions, model parameters, and system stimulus) are defined by the user's input. For example, with computer model HEC-1, which is described in Sec. 28.2, the user defines, via input, all pertinent catchment characteristics and model parameters, plus the temporal and spatial distribution of the rainfall. Thus runoff from any catchment due to any rainfall event can be estimated. Computer models written for site-specific or problem-specific applications typically do not have such flexibility. Although these special-purpose computer models may embody the same mathematical models, the characteristics and parameters typically are *hard-wired.* That is, they are fixed in the computer code. An example of such a site-specific model is a FORTRAN program created by an operating agency for repeated forecasting of flood runoff from a single catchment. In that case, the catchment characteristics and model parameters will not change, so these are included directly as a part of the code. This makes for efficient analysis, but it limits the general utility of the computer model.

Accessibility. The models reviewed are limited to those that: (1) are readily available to the reader from noncommercial sources, (2) will execute on commonly available hardware, and (3) have documentation sufficiently detailed to permit use without specialized training or consulting.

Table 28.1 lists the computer models included in the sample and the source of information on each.

28.1.4 Classification of the Computer Models

The information provided by a computer model is correlated directly with the processes modeled. For excess-water management, the critical processes include those listed in Table 28.2. Some computer models focus not on the processes but on system performance, so *performance models* are included in this chapter as an additional sixth classification. Performance models may simulate critical processes as a sec-

TABLE 28.1 Computer Models Described in This Chapter

Computer model	What is modeled (*primary use*)	Whom to contact for information	What is available
HEC-1	Catchment runoff	U.S. Army Corps of Engineers (USACE) Hydrologic Engineering Center (HEC) 609 Second St. Davis, CA 95616 USA *or* National Technical Information Service U.S. Dept of Commerce 5285 Port Royal Rd. Springfield, VA 22161 USA	Program in FORTRAN; source is available. PC-version available. Detailed user's manual is available. HEC technical reports, training documents, and project reports available. HEC distributes to U.S. government users. Vendors provide to others for a fee, but the program is public domain. Program and documentation are also available from U.S. National Technical Information Service (NTIS). Training is available from HEC (federal) and universities (other users).
TR-20, TR-55	Catchment runoff	Soil Conservation Service (SCS) P.O. Box 2890 Washington, DC 20013 USA	PC versions of both TR-20 and TR-55 are available from offices of the Soil Conservation Service throughout the U.S., along with documentation of the software. Both programs are public-domain so are available from a variety of other sources, including vendors and academic institutions. Technical reports describing math. models are available from NTIS
SSARR	Catchment runoff	USACE North Pacific Division Attn: CENPD-EN-WM-HES P.O. Box 2870 Portland, OR 97208-2870 USA	The program was developed for and is used by the Corps on mainframe computers. PC/workstation versions are available. Program user's manuals and project reports are available.
HSPF	Catchment runoff	Environmental Protection Agency (EPA) Environmental Research Laboratory Athens, GA 30613 USA	A PC version of HSPF is available, but as the program is written in FORTRAN, it may be compiled and executed on any computer for which a compiler is available. Documentation is available from EPA.
HEC-2	Fluvial	HEC (see above)	Same as for HEC-1 (see above).
WSPRO	Fluvial	U.S. Geological Survey (USGS) National Water Information Service Water Resources Division Reston, VA 22092 USA	FORTRAN source code, user's manual, and support documents are available. A PC version is available. WSPRO is available also from vendors.

Name	Type	Organization	Availability
UNET	Fluvial	HEC (see above)	Same as for HEC-1 (see above).
DWOPER	Fluvial	NOAA National Weather Service Hydrologic Research Laboratory 1325 East-West Highway Silver Spring, MD 20910 USA	FORTRAN source code, user's manuals, and reports of applications are available. A PC version is available from various vendors. Training in application of DWOPER is available from various universities in the U.S.
HEC-6	Alluvial	HEC (see above)	Same as for HEC-1 (see above).
GSTARS	Alluvial	U.S. Bureau of Reclamation Denver Federal Center, Bldg. 67 Denver, CO 80225 USA	Program source code, user's manual, and support documents are available. Additional application-support documents are available.
TABS-2	Alluvial	USACE Waterways Experiment Station (WES) P.O. Box 61 Vicksburg, MS 39180 USA	Program in FORTRAN; source is available. PC-version available. Detailed user's manual is available. WES technical reports, training documents, and project reports available. PC version available to public from vendors.
SWMM	Pressure flow	EPA (see above)	SWMM is available for a variety of platforms, including the PC. It is available from EPA and from a variety of vendors. Complete documentation is available from EPA and the vendors.
HYDRA	Pressure flow	McTrans Center University of Florida 512 Weil Hall Gainesville, FL 32611-2083 USA	A PC version of the program is available, along with full documentation. HYDRA is a component of HYDRAIN, which is also available for the PC.
HEC-FFA	Statistical	HEC (see above)	Same as for HEC-1 (see above).
J407	Statistical	USGS (see above)	J407 is programmed in FORTRAN, so it is available for any computer for which a compiler is available. A PC version and documentation are available.
EAD	Performance	HEC (see above)	Same as for HEC-1 (see above).
HEC-5	Performance	HEC (see above)	Same as for HEC-1 (see above).

TABLE 28.2   Processes Modeled by Water-Excess Management Models

Process	Description
Catchment runoff	These are the processes that govern how precipitation that falls on a catchment runs off that catchment. Runoff processes include evaporation, transpiration, infiltration, percolation, interflow, overland flow, and baseflow. Modeling these processes provides information on stream-discharge time series or peaks, and volume time series or totals.
Fluvial	These are the processes that govern fluid flow in an open channel when that fluid is subjected to external forces. Modeling these processes provides information on river or reservoir depth time series or maximums, and inundated-area geometry.
Alluvial	These are the processes that govern the erosion and deposition of sediment due to flow in an open channel. Modeling these processes provides information on landform changes due to erosion or deposition, river or reservoir water-depth time series or maximums, and inundated-area geometry.
Pressure flow	These are the processes that govern how water flows under pressure in closed conduits. For water-excess management in urban settings, the facilities are often planned to function as pressure conduits for the design flow or greater events (ASCE/WEF, 1992).
Statistical processes	Physical, chemical, or biological processes exhibit randomness and variability that cannot be accounted for with models of the behavior of a system (Hirsch, et al., 1993). Models of statistical processes recognize this and seek to describe the randomness and variability by establishing an empirical relationship between probability and magnitude. A statistical-process model yields information on probabilities associated with extreme discharge, volume, or depth magnitudes.

ondary function, but their primary function is to use information from such a simulation to evaluate economic, social, or environmental benefits and costs.

To provide further detail, the computer models reviewed herein are described, when appropriate, in terms of additional characteristics listed in Table 28.3.

28.2 RUNOFF-PROCESS COMPUTER MODELS

28.2.1 HEC-1

28.2.1.1 Overview. HEC-1 is a *single-event model* that estimates runoff from precipitation with a spatially- and temporally-lumped description of a catchment (USACE, 1990b). HEC-1 incorporates a variety of conceptual or quasi-conceptual mathematical models; the user specifies through input which of these are used. Parameters for the various mathematical models also are specified by user input. HEC-1 includes the capability to estimate parameters of most of the runoff models included, if proper hydrometeorological data are available. HEC-1 provides stream-discharge time series and peaks, and volume totals for decision making.

TABLE 28.3 Additional Characteristics for Model Classification

Characteristic	Description
Event or continuous model?	This distinction applies primarily to models of catchment-runoff processes. An event model represents a single flood event that occurs over a period ranging from minutes to days. A continuous model operates over a long period, predicting response during and between rainfall events.
Measured-parameter or fitted-parameter model?	This distinction applies to all models. A measured-parameter model is one in which parameters can be determined from system characteristics, either by direct measurement or by estimation from measurements. For example, channel cross sections for a fluvial-process model can be determined by measurement. A fitted-parameter model, on the other hand, includes parameters that cannot be measured. Instead, these must be found by fitting the model with observed values of the hydrometeorological phenomena. For example, the unit hydrograph models described in Chap. 24 have no explicit physical significance, so cannot be measured. Instead, for a given catchment, a unit graph is found by proposing model parameters, estimating runoff due to observed rainfall, and comparing the computed with observed runoff. If the values do not agree, the model parameters are adjusted to achieve an acceptable fit.
Distributed or lumped model?	This distinction applies to all models. A distributed model is one in which the spatial variations of characteristics and processes are considered explicitly, while in a lumped model, these spatial variations are averaged or ignored. For example, the UNET model (Sec. 28.3.3) accounts for variations of momentum and mass along a stream reach, while many of the simple routing models in HEC-1 (Sec. 28.2.1) account only for the variation of mass between the points of inflow to and outflow from a channel reach.
Complete or partial model?	This is related to the distinction between distributed and lumped models. A complete model represents in detail, more or less, all pertinent process. A partial model represents only a portion of the process. For example, HSPF (Sec. 28.2.4) attempts to represent all aspects of soil moisture, while the loss model of TR-55 (Sec. 28.2.2) accounts only for the distribution of rainfall volume between runoff and loss.
Conceptual or empirical model?	This distinction is based on characteristics of the mathematical models included in the computer model. A conceptual model is formulated from consideration of physical, chemical, and biological processes acting on the system input to produce the system output. An empirical model, on the other hand, is formulated from observation of input and output, without seeking to represent explicitly the process of conversion. The flood-frequency models of Chap. 24 are examples of the latter, while HSPF is an example of the former. The frequency models do not consider the physical process, while HSPF attempts to represent mathematically the relationship of input (rainfall) to output (discharge).

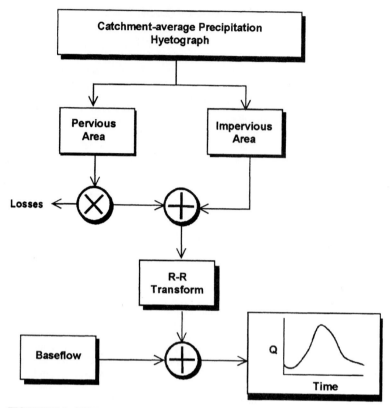

FIGURE 28.2 HEC-1 representation of runoff process.

28.2.1.2 Mathematical Models Included in the Computer Model. The runoff process, as represented in HEC-1, is illustrated by Fig. 28.2. The mathematical models incorporated include the following:

Loss models. To account for infiltration, depression storage, and other reductions in volume of runoff from pervious areas in a catchment, HEC-1 offers the alternatives listed in column 1 of Table 28.4. The user may select any one of these

TABLE 28.4 Runoff and Fluvial Process Models in HEC-1

Loss models	Runoff transforms	Routing models
Initial loss and uniform rate	Unit hydrograph	Muskingum
Soil Conservation Service (SCS) curve number (CN)	Clark's	Kinematic wave
	Snyder's	Modified Puls (level pool)
4-parameter exponential	SCS	Muskingum-Cunge
Holtan's	User specified	
Green and Ampt	Kinematic wave	

for a catchment. For complex catchments that are subdivided for analysis, the user can select combinations of the loss models.

Snowfall and snowmelt models. These models simulate snowfall formation and accumulation and estimate runoff volumes due to snowmelt. The snowfall model permits division of a catchment into elevation zones. The user specifies a time series of temperatures for the lowest-elevation zone, and the model estimates temperatures for all others with a lapse rate. Precipitation is assumed to fall as snow if the zone temperature is less than a user-defined freezing threshold. Melt occurs when the temperature exceeds a user-defined melting threshold. Snowfall is added to and snowmelt is subtracted from the snowpack in each zone. Snowmelt may be computed with either a degree-day model or an energy-budget model.

Runoff transforms. Runoff volumes may be transformed to runoff hydrographs in program HEC-1 with either a UH model or via solution of the kinematic-wave simplification of the St. Venant equations.

Baseflow model. HEC-1 incorporates a single model of baseflow, which is based on the assumption that drainage of water added to catchment storage (as soil moisture) can be modeled well with an exponential-decay function.

In addition to *runoff-process models,* HEC-1 includes the *fluvial-process models* shown in column 3 of Table 28.4 for routing hydrographs. The user may select any one appropriate for a given stream reach. As with other mathematical models included in HEC-1, any combination of these may be used. Parameters are defined with user input.

With the runoff- and fluvial-process models used in combination, large catchments in which parameters or precipitation vary spatially can be analyzed. To do so, the catchment is subdivided, the runoff-process models are used to compute runoff at various locations, and the routing models are used to account for flow in stream channels to common points. Figure 28.3 illustrates this approach. First runoff is computed for subcatchment 1 with the runoff-process models. The resulting hydrograph represents the flow at control point A. This hydrograph is routed from A to B with a fluvial-process model. The hydrograph of runoff from subcatchment 2 is computed and added to the routed hydrograph. This yields an estimate of total runoff, accounting for spatial variation in rainfall and catchment characteristics.

28.2.1.3 *Computer Model Input Requirements.* To estimate catchment runoff with HEC-1, the user must provide the following input:

Precipitation. The precipitation may be provided as catchment average depth or as depths observed at gauges. The user must provide a temporal distribution of precipitation; this may be the historical observation at a gauge, or it may be a design-storm distribution (see Chap. 24).

Catchment physical characteristics, including characteristics of water-control facilities. The user must delineate catchment boundaries and define, via input, the catchment area. If the catchment is subdivided for analysis, the user must define, through the sequence of input, how the system is schematized for modeling. If the stream system includes water-control facilities, such as detention ponds, the performance-governing characteristics of these must also be specified.

Model parameters. The user must specify all appropriate loss model, runoff-transform model, baseflow model, and routing model parameters.

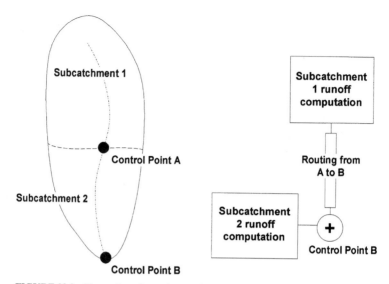

FIGURE 28.3   Illustration of complex catchment modeling by subdivision.

Simulation specification.   HEC-1 was designed for maximum flexibility; therefore, it relies on the user to specify the time step and duration of the simulation, subject to constraints imposed by the available computer memory.

28.2.1.4   Computer Model Output.   Output from HEC-1 includes the following: (1) a summary report of the user's input, (2) for each subcatchment, a report of the average-precipitation depth, the loss, and the excess for each simulation step, plus a report showing the computed runoff hydrograph ordinates, (3) for each stream reach modeled, a report of the outflow (downstream) hydrograph ordinates, and (4) various summary output tables that show the discharge peaks and times of peak at system control points.

28.2.1.5   Applications.   The Hydrologic Engineering Center (HEC) and U.S. Army Corps of Engineers (USACE) offices nationwide have used HEC-1 extensively; results of these applications are described in reports available from these offices. Additional applications have been described by Russell, et al. (1979); Bedient, et al. (1985); Bhaskar (1988); and Melching, et al. (1991).

28.2.1.6   Utility Programs and Specialized Versions.   The HEC has developed utility programs that simplify use of HEC-1 or provide additional capabilities. Several are identified in Table 28.5.

28.2.2   TR-20 and TR-55

28.2.2.1   Overview.   Programs TR-20 and TR-55 are single-event computer models that estimate design-event discharge time series, peaks, and volumes (USDA, 1983; 1986). Both programs use runoff-process models developed by the Soil Conservation Service; these mathematical models are described in the National Engineering Handbook (USDA, 1971). Program TR-55 computes the peaks, volumes,

TABLE 28.5   Utility Programs for and Specialized Versions of HEC-1

Program	Description
HEC-DSS	This is a time-series database management system (DBMS). It creates specially formatted random-access files, with a hierarchical system of record names to expedite storage and retrieval of data in the files. Data in the DBMS may be accessed through a set of front-end utility programs that permit data entry, reporting, charting, and database housekeeping. Further, the data can be accessed via a FORTRAN library of routines that read, write, and otherwise interact with database files. It is through this library that HEC-1 (and many other models from HEC) retrieves data from and files data in the database.
HMR-52	This program computes spatial and temporal distribution of probable maximum precipitation (PMP), using criteria established by the National Weather Service (NWS) for catchments east of the 103d meridian in the United States. The storm may be used in turn as input to HEC-1 to estimate the probable maximum flood (PMF) runoff. This extreme discharge is the basis for dam-safety analysis.
HEC-1F	This specialized version of HEC-1 is designed for real-time flow forecasting. It includes a subset of the runoff-process models of HEC-1, plus algorithms for updating model parameters during an event and for adjusting computed hydrographs to match observed discharge at the time a forecast is made. A companion program, PRECIP, estimates catchment-average precipitation for the forecasts.

and time series for a single catchment. Program TR-20 uses identical procedures to compute peaks, volumes, and time series, plus fluvial-process models to route and to combine catchment-runoff hydrographs, in a manner similar to that illustrated by Fig. 28.3.

28.2.2.2   Mathematical Models Included in the Computer Model.   The runoff process, as represented in TR-20 and in TR-55, is essentially the same as illustrated by Fig. 28.1. A catchment-average precipitation hyetograph is specified, losses are computed, and a runoff transform converts the runoff volume to discharge. TR-20 and TR-55 do not include a baseflow model. The features of the mathematical models are summarized in Table 28.6.

TABLE 28.6   Mathematical Models in TR-20 and TR-55

Component	Details of mathematical model
Loss model	SCS curve number model. Cumulative runoff volume is a function of cumulative rainfall volume. Model has two user-supplied parameters: initial abstraction and maximum retention. Programs include empirical relationships to estimate parameters from catchment characteristics.
Runoff transform	SCS one-parameter unit hydrograph. Programs include empirical relationship to estimate parameter from catchment travel time plus quasi-physically based models to estimate travel time.
Routing	TR-20 includes a linear channel-routing model and a storage-routing model for detention. TR-55 does not include a routing model.

28.2.2.3 Computer Model Input Requirements. To estimate catchment runoff with either TR-20 or TR-55, the user must provide the following input:

Precipitation. TR-20 and TR-55 are intended as tools for estimating runoff from rainfall. Consequently, both require specification of a catchment-average hyetograph. The SCS design storms are incorporated in both models, either directly in the code or as input files.

Catchment physical characteristics and/or runoff model parameters. The user must define the catchment area, the catchment-average rainfall, and the parameters of the loss model and the unit hydrograph model. Relationships are incorporated in both models to estimate the parameters from catchment characteristics; if those are to be used, the user must provide information on catchment soils, land use, overland flow surfaces, and channels.

Fluvial model parameters. If the routing models of TR-20 are required, the user must provide model parameters and channel descriptions. If the stream system includes detention structures, the characteristics of these must be provided also.

28.2.2.4 Computer Model Output. Depending on user preferences and the program version used, output from programs TR-20 and TR-55 may include the following: (1) a summary of user's input, (2) a report of estimated model parameters if the SCS empirical relationships are used, and (3) a report of catchment-runoff hydrographs and peaks.

28.2.2.5 Applications. Due to their simplicity and availability, programs TR-20 and TR-55 are used widely in the United States for planning and design. Engineers of local SCS offices use the models extensively and will provide information on various applications. McCuen (1987) describes applications. DeBarry and Carrington (1990) describe integration of a geographic information system (GIS) and TR-55 to estimate runoff with existing and future-condition land use.

28.2.3 SSARR

28.2.3.1 Overview. SSARR, the *Streamflow Synthesis and Reservoir Regulation model,* was developed by the USACE for planning, designing, and regulating water-control projects in the Columbia River basin, (USACE, 1991c). SSARR is primarily a runoff-process model, but it includes also fluvial-process and *performance-evaluation models.* SSARR is a continuous-accounting model, with fitted parameters. The model is a *spatially distributed model* in the sense that it permits the user to subdivide a basin to represent spatial variation of characteristics and processes, again as illustrated by Fig. 28.3.

28.2.3.2 Mathematical Models Included in the Computer Model. The mathematical models included in SSARR are a combination of conceptual and empirical models. These models simulate the following:

Snowmelt, snow accumulation. Precipitation is defined by the user at intervals from 0.1 to 24 h. If this can occur as snow, the user indicates a base temperature for snow formation. The snowpack can be defined by either elevation band or a snow-cover depletion function. SSARR computes snowmelt from the accumulated snowpack with either a temperature-index approach or USACE generalized snowmelt equations.

Soil moisture accumulation and evapotranspiration.   When rainfall and snowmelt volumes are determined, SSARR estimates runoff volume with an empirical relationship in which runoff is a function of an *accumulated soil-moisture index* (SMI) and rainfall intensity. Evapotranspiration is estimated as a function of either mean monthly or daily evaporation data; the SMI is adjusted to account for this evapotranspiration.

Baseflow.   Total runoff is determined with the SMI relationship. The runoff volume then is distributed to either baseflow or surface/subsurface flow with an empirical *baseflow infiltration index* (*BII*) relationship. The baseflow contribution is routed with a storage-routing model to estimate contribution to total streamflow.

Surface and subsurface flow.   Runoff volume that does not contribute to baseflow is considered surface or subsurface flow. An empirical relationship, the surface/subsurface separation curve, defines the contribution to each. Both are routed with storage-routing models, and the results are added to routed baseflow to estimate the total streamflow.

The interaction of these processes in the SSARR model is illustrated by Fig. 28.4.

28.2.3.3 Computer Model Input Requirements.   The input required for the SSARR model is similar to that required for the other runoff-process models (such as hydrometeorological data and stream-system configuration). In addition the user must define the SMI relationship, BII relationship, and *surface/subsurface separation curve,* the initial conditions for the SMI and BII, and the routing model parameters. Because of the empirical nature of these, they are best determined via fitting with recorded streamflow and precipitation.

28.2.3.4 Computer Model Output.   The SSARR model forecasts streamflow time series at locations throughout a basin, so the primary output is a tabulation of the forecasted discharge. Various report formats and plots of results may be selected by the user.

28.2.3.5 Applications.   The USACE has used SSARR continuously and successfully since its development for operational forecasting in the 622,000 km² Columbia basin; a variety of reports on application are available from the North Pacific Division of USACE. Larson, et al. (1982) describe other applications to catchments ranging in size from 12.6 to 370,000 km².

28.2.4   HSPF

28.2.4.1 Overview.   *HSPF,* the *Hydrologic Simulation Program—FORTRAN,* is a recent incarnation of Stanford Watershed Model (Crawford and Linsley, 1966). HSPF, developed for the U.S. Environmental Protection Agency (EPA) by Hydrocomp, Inc. in 1980, has been revised extensively by a variety of contractors and currently is available in its tenth release (Bicknell, et al., 1992). HSPF is a continuous, quasi-conceptual, distributed, fitted-parameter model of all aspects of catchment response to rainfall. In addition to the runoff-process model embodied in HSPF, this computer model includes mathematical models of fluvial, alluvial, chemical, and biological processes critical to the work of the EPA. HSPF provides information on stream-discharge time series or peaks, volume time series or totals, river or reservoir water-depth time series or maximums, and landform changes due to erosion or deposition.

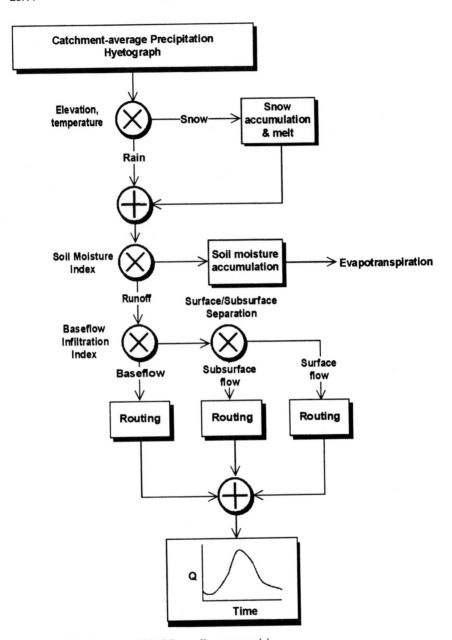

FIGURE 28.4   Structure of SSARR runoff-process model.

28.2.4.2 Mathematical Models Included in the Computer Model.

Like SSARR, HSPF computes catchment-runoff volume by accounting continuously for the spatial distribution of system moisture, as illustrated by Fig. 28.5. The computer model

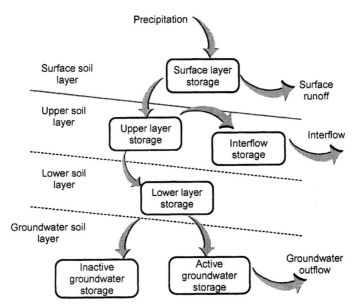

FIGURE 28.5   Structure of HSPF runoff-process model.

employs a combination of empirical and conceptual mathematical models in this accounting. For example, simulation of infiltration is based on the conceptual model developed by Philip, while overland flow, interflow, and groundwater flow are modeled with empirical storage-outflow relationships. The HSPF user's manual describes the various mathematical models and their parameters in detail.

28.2.4.3 Computer Model Input Requirements and Output.

Like other catchment-runoff models, HSPF requires input of hydrometeorological time series. In the case of a continuous-simulation model, these series are lengthy, so data management could be a formidable task. To minimize data management efforts and problems, HSPF includes a time series database-management system. In addition to time series input, the HSPF user must define values of the approximately 20 parameters of the various mathematical models. As many of these are parameters of empirical models, they are estimated from experience or via calibration with observations of precipitation and runoff. HSPF output includes hydrographs at system locations plus an accounting of moisture storage and flux throughout the hydrologic system.

28.3 FLUVIAL-PROCESS MODELS

28.3.1 HEC-2

28.3.1.1 Overview.

HEC-2 solves the equations of one-dimensional, steady, gradually varied flow to predict water-surface elevation along a natural or constructed open channel (USACE, 1982a). Water-surface profiles in either the subcritical or supercritical regime can be computed. HEC-2 also incorporates conceptual and empirical models that allow analysis necessary for common designing, planning, and regulating problems. These special capabilities are summarized in Table 28.7.

28.3.1.2 *Mathematical Models Included in the Computer Model.* Given a complete description of the geometric boundaries which contain the flow in an open channel, HEC-2 estimates the average flow depth and velocity in the prescribed cross sections by solving the one-dimensional energy equation. This formulation relies on the assumptions shown in Table 28.8. (Violation of one or more of these assumptions does not necessarily mean that results of analysis with HEC-2 are wrong. Instead, it means that the user must evaluate the relative effect of these assumptions upon the results of a particular application.)

By computing the energy loss between a river cross section with a known water-surface elevation and an adjacent cross section, the water-surface elevation at the adjacent section can be determined. For subcritical flow, the computations start with

TABLE 28.7   Special Capabilities of HEC-2 Model

Capability	Description
Treatment of effective flow areas	Several options are available to restrict flow to certain portions of a given cross section. This is often required because of sediment deposits, floodplain encroachments, oxbow lakes, and so forth.
Analysis of bridge and culvert losses	The energy loss due to bridge piers and culverts can be estimated.
Analysis of channel encroachments	Six methods of specifying floodplain encroachments are available. The equal conveyance reduction method is used to determine the floodway boundaries for a flood insurance study.
Evaluation of channel improvement	Natural river cross-section data may be modified simply with the channel-improvement option. This allows simulation of the effects of excavating a compound trapezoidal channel section into the natural section.
Calibration to high-water marks	When high-water marks are known for a specified discharge, HEC-2 can estimate the effective Manning's n value necessary to reproduce this observed elevation.
Development of storage-outflow function	HEC-2 includes the capability to develop a storage volume—discharge relationship for a river reach. This can, in turn, be used for streamflow routing with the modified Puls and other simple fluvial-process models.
Analysis of split flow	For flow splits (such as at diversion structures and levee overtoppings) HEC-2 balances the energy grade line elevations at the split and downstream confluence. Weir flow, normal depth, or a diversion rating curve may describe the hydraulics of the split.
Simulation of flow in ice-covered streams	Water-surface profiles with a stationary, floating ice cover can be estimated. The user must provide the thickness and effective n value of the ice cover.

TABLE 28.8   Assumptions of HEC-2

Flow is steady and gradually varied, with localized rapidly varied flow, such as at weirs or culvert inlets.

Flow is turbulent, and fully rough, with viscous forces playing a minor role.

Flow is homogeneous, with constant fluid density throughout the flow field.

Flow can be adequately characterized by movement in a single direction.

Pressure distribution at a cross section is hydrostatic.

---

a known relationship between discharge and water-surface elevation at the downstream boundary of the fluvial system and proceed in an upstream direction until the water-surface elevation is computed at each cross section. For supercritical flow, the computations start with a known water-surface elevation at the upstream boundary and proceed in a downstream direction.

HEC-2 estimates the total energy loss between two adjacent sections as the sum of frictional energy loss due to channel roughness; form energy loss due to expansion and contraction; and energy loss due to flow-through structures, such as bridges, culverts, or weirs. The frictional energy loss is the product of the average energy grade line slope and the distance between cross sections. This energy grade line slope at a section is computed with Manning's equation. Several schemes are available in HEC-2 for determining the average energy grade line slope between two cross sections: arithmetic, geometric, or harmonic mean energy slope at adjacent cross sections, or the average conveyance at adjacent cross sections. HEC-2 includes a contraction/expansion energy-loss model that estimates that loss as a function of the difference in velocity head between two cross sections.

28.3.1.3   Computer Model Input Requirements.   HEC-2 is a generalized computer program. The user must therefore provide all stream characteristics and boundary conditions via input. For a simple application, the basic input requirements are as listed in Table 28.9.

28.3.1.4   Computer Model Output.   A variety of output data may be selected by the user. The basic output includes a report of computed water-surface elevation, velocity, and other pertinent characteristics of flow at each channel cross section. HEC-2 will prepare an electronic file with the computed results for subsequent access by a graphing utility.

28.3.1.5   Applications.   HEC-2 is claimed to be the most commonly used water-excess management model in the United States. In most communities, HEC-2 is the model used to delineate the inundated area for establishing the 100-year floodplain. This floodplain, in turn, serves as the basis for most floodplain land-use controls. Reports of the delineation are available from the Flood Insurance Administration. Additional reports on HEC-2 application are available from the HEC.

28.3.2   WSPRO

28.3.2.1   Overview.   WSPRO is a one-dimensional steady-state backwater model developed by the U.S. Geological Survey (USGS), under contract to the Federal Highway Administration for the evaluation and design of bridge waterways (Sher-

TABLE 28.9 Input Required for HEC-2 Model

Input item	Description
Flow regime	The user must assess the location of normal depth relative to critical depth for each application. For a subcritical flow regime, cross-section data are specified progressing upstream. For a supercritical flow regime, data are specified progressing downstream. For unknown or mixed regimes, multiple input data sets are prepared and results combined, as discussed in the HEC-2 user's manual.
Starting boundary condition	HEC-2 solves the one-dimensional energy equation for a given stream, so the starting water-surface elevation must be specified. This can be input directly or estimated by the program.
Discharge	The steady-flow discharge must be specified for each stream segment. This may change along the profile in order to include effects of tributaries, diversions, and so forth.
Energy-loss coefficients	For a basic application of HEC-2, the user must specify Manning's n for the main channel. Manning's n are specified for the left and right overbanks, contraction-loss coefficient, and expansion-loss coefficient.
Cross-section geometry	Boundary geometry for the analysis is provided by a series of elevation—station coordinate points at each cross section. Cross sections are required at representative locations throughout the reach, but especially where slope, conveyance, or roughness change significantly.
Reach length	The distance between cross sections must be specified to permit computation of the turbulent energy loss due to boundary roughness. HEC-2 allows input of separate reach lengths for the main channel, left and right overbanks to describe curved channels, river meanders, and so forth.

man, 1986; 1988). Federal Highway Administration policy on design of floodplain encroachments is to consider the impact of encroachment alternatives on the floodplain, rather than to determine the bridge opening size necessary to pass a given design discharge. WSPRO is formulated for such analysis.

28.3.2.2 Mathematical Models Included in the Computer Model. WSPRO solves a one-dimensional steady-flow approximation of the momentum and continuity equations. In addition to the backwater solution, WSPRO includes empirical models for determining the effect of culvert, spur dike, and bridge abutments.

28.3.2.3 Computer Model Input Requirements and Output. Input requirements for WSPRO are similar to those of HEC-2. Special input describes typical highway bridge structures, including spur dikes, road grades, culverts, and bridges with multiple openings. For situations where the flow regime changes along a given profile, WSPRO has the capability to change from subcritical to supercritical flow computation schemes. Output from WSPRO includes computed conveyance, water-surface elevation, and velocity at each channel cross section, and results of computation for flow through bridge openings and other highway structures.

28.3.3 UNET

28.3.3.1 Overview. UNET simulates one-dimensional, unsteady flow through either a simple open channel, a dendritic system of open channels, or a network of open channels (USACE, 1993). This permits analysis of diversions and confluences in a looped system, including systems in which the direction of flow may reverse. UNET also has the capability to model flow in lakes, bridges, culverts, weirs, and gated spillways, using mathematical models that are essentially the same as those included in HEC-2.

28.3.3.2 Mathematical Models Included in the Computer Model. UNET solves a linearized finite difference approximation of the full one-dimensional, unsteady-flow equations (Barkau, 1985). The solution algorithm employs sparse-matrix techniques with Gaussian reduction.

28.3.3.3 Computer Model Input Requirements and Output. The input required for UNET is similar to that required for HEC-2. Additional input is required to describe the interconnection of stream segments, and location of lakes and storage elements. UNET uses the HEC-DSS described in Table 28.5 to store boundary conditions, such as rating curves and hydrographs.

Unsteady-flow models typically produce large reports of computational results, and UNET is not an exception. The model computes and reports depths, velocities, and other pertinent flow characteristics at each cross section for each time step of the discretized solution of the flow equations. These results may be filed with the HEC-DSS and subsequently plotted with DSPLAY, the graphing program of the database-management system.

28.3.3.4 Applications. UNET has become the USACE standard for analyses in which an unsteady flow model is required; consequently, project reports are readily available from offices nationwide. (See, for example, USACE, 1990c.)

28.3.4 DWOPER

28.3.4.1 Overview. The Dynamic Wave Operational model (DWOPER) was developed by the National Weather Service for use in riverine flood forecasting (Fread, 1987). DWOPER can be used for analysis of flow in a single channel, or a dendritic or bifurcated (looped) river system. Optional features include the simulation of locks and dams, lateral inflows, pressurized flows in storm drains, and wind effects. An automatic calibration feature is also available to estimate model parameters from observations of flow and water-surface elevation.

28.3.4.2 Mathematical Models within the Computer Model. DWOPER uses a four-point implicit finite difference scheme to approximate the complete one-dimensional unsteady-flow equations. The resulting nonlinear algebraic equations are solved using the Newton-Raphson method, along with Gaussian elimination with sparse-matrix techniques.

28.3.4.3 Computer Model Input Requirements and Output. Input for DWOPER includes channel cross-section geometry, expressed as elevation versus top width; a description of the network configuration; energy-loss model parameters, descriptions of structures, and flow hydrographs. Extensive reports of discharge, velocity, and depth at channel cross sections are available. Computed water-surface elevations can be filed electronically and graphed with utility programs.

28.3.4.4 Applications. Because of its widespread availability, DWOPER is in common use. Fread (1987) describes several applications.

28.4 ALLUVIAL-PROCESS MODELS

28.4.1 HEC-6

28.4.1.1 Overview. HEC-6 models the effects of *river sediment transport* and resulting changes in the flow boundaries with a one-dimensional representation of the open-channel flow (USACE, 1991b). The program computes changes in river bed profiles for a single flood event or for a long-term sequence of flows. It provides information on depths and landform changes due to erosion or deposition. Thus HEC-6 can be used to evaluate the lateral movement of a stream.

28.4.1.2 Mathematical Models Included in the Computer Model. HEC-6 solves a one-dimensional steady-flow approximation of the momentum and continuity equations. HEC-6 does not include the empirical models for bridge and culvert energy losses, but it does allow for the specification of an internal elevation-discharge boundary condition, the development of which can be accomplished using HEC-2. Transport calculations are made for a control volume defined using the cross-section locations and an assumed depth of alluvial deposits. The computed energy slope, depth, velocity, and shear stress at each cross section are used to compute the sediment-transport capacity at each cross section. These rates, along with sediment supply rate and armoring potential, are used for volumetric accounting of sediment movement through the system. The amount of scour or deposition is computed by dividing the surface area of the mobile boundary into the change in sediment volume. A new water-surface profile is then computed for the updated channel geometry.

Sediment-transport rates in HEC-6 are computed for 20 different grain-size categories ranging from clay (less than 0.004 mm) through silt (less than 0.063 mm) up to large boulders (2048 mm). A variety of sediment-transport equations, based on either cohesive or noncohesive theory, can be selected for the transport-capacity calculations. Mathematical models of incipient motion, channel-bed armoring, grain-size sorting, and particle entrainment are also included in HEC-6.

To account for unsteady flow, a hydrograph is discretized into a series of steady flows, and a water-surface profile is computed using a standard step backwater approach. This procedure is repeated until the entire flow hydrograph has been routed.

Due to the one-dimensional formulation, HEC-6 does not represent the multidimensional nature of sand bar formation, secondary flow currents, and stream bank failure.

28.4.1.3 Computer Model Input Requirements. HEC-6 also requires all of the information necessary for a one-dimensional fluvial model, including a complete description of the geometric boundaries of the channel that contains the flow, definition of the flow regime, and specification of energy-loss coefficients. In addition, the information shown in Table 28.10 must be developed and provided by the user.

TABLE 28.10   Input Required for HEC-6 Model

Category	Input requirement
Sediment grain-size distribution	Sediment grain sizes transported by streams vary over a wide range. HEC-6 considers the transport of sediment in the size range between clay-size particles and 2048 mm boulders. The user must stipulate the grain-size distribution of both the streambed material and the inflow sediment load. Vanoni (1985) provides guidance on preparation of these data.
Sediment properties	In addition to the grain size, the user must specify the specific gravity, shape factor, unit weight of deposits, and fall velocity of the sediment.
Other boundary conditions	HEC-6 computes the changes in a stream bed over time, so a time series of discharge must be provided. The discharge may be modified along the application reach to account for tributaries or diversions.

28.4.1.4   Computer Model Output.   HEC-6 provides reports of both hydraulic and sediment-transport calculations. The basic level of output data includes a report of initial conditions, hydraulic calculations, sediment-transport calculations, accumulated sediment volumes, and overall bed-elevation changes.

28.4.1.5   Applications.   Vanoni (1977) and Fan (1988) report on applications of this and other alluvial process models. In addition, a number of project reports are available from the HEC.

28.4.2   GSTARS

28.4.2.1   Overview.   The *Generalized Stream Tube Model for Alluvial River Simulation, GSTARS,* is a one-dimensional *alluvial-process model* developed by the USGS (Molinas and Yang, 1986). GSTARS provides information on river water depths, inundated-area geometry, and landform changes.

28.4.2.2   Mathematical Models Included in the Computer Model.   Like HEC-6, GSTARS solves the one-dimensional energy equation to determine the stream hydraulics at a given cross section for a given discharge, but it uses the concept of *stream tubes* to describe differences in the lateral distribution of conveyance along the stream cross section. Since the stream-tube-flow distribution is based on the computed horizontal water-surface elevation, the model does not explicitly include lateral stream processes, such as sand bar formation, secondary currents and bank failure. The stream tube approach is useful however for certain applications, such as simultaneous occurrence of deposition and scour at a cross section.

28.4.2.3   Computer Model Input Requirements and Output.   Input requirements for GSTARS are similar to those of HEC-6. Unlike HEC-6, GSTARS permits specification of lateral and vertical differences in the initial bed-material gradations for each cross section if the data is available. GSTARS output includes reports of the hydraulic and sediment-transport computations. It includes also convenient plots of streambed elevation versus time.

28.4.3 TABS-2

28.4.3.1   Overview.   TABS-2 is a fully *two-dimensional model* of sediment movement in an open channel (Thomas and McAnally, 1985). Consequently, it will model *sand bar formation, secondary currents,* and other flow and transport cases not well modeled with the one-dimensional representations.

28.4.3.2   Mathematical Models Included in the Computer Model.   TABS-2 solves the two-dimensional, depth-averaged momentum and continuity equations, for either steady or unsteady flow. TABS-2 uses a *finite-element technique* and computes, for each node of the finite-element representation, flow depth and longitudinal and lateral velocities. The sedimentation component of the model then computes the transport capacity using the two-dimensional *convection-diffusion equation* with bed source terms. The actual transport is based on sediment availability. TABS-2 can handle both *cohesive* and *noncohesive sediment transport.*

28.4.3.3   Computer Model Input Requirements and Output.   In addition to the grid network data, each element requires information on initial bed-material sizes. As with the other alluvial-process models, the inflowing sediment load and hydrograph must be specified by the user. TABS-2 will provide detailed reports of all computations. To aid the user in digesting this mass of output, TABS-2 includes also a postprocessor that displays the results of computations graphically. This graphical output includes velocity vector plots, contour plots of scour/deposit depths, and shear stress variations.

28.4.3.4   Applications.   This model is readily available in the United States, so is widely used by USACE, various other government agencies, and consultants. Thomas and Heath (1983) describe applications of TABS-2.

28.5   PRESSURE-FLOW PROCESS MODELS

Flow of water in modern drainage systems typically consists of both open-channel and closed-conduit flow. Open-channel flow always has a free surface, and is modeled with the fluvial-process computer models of Sec. 28.3. Water in a closed conduit, on the other hand, may flow with a free surface, or it may flow under pressure. This section describes models that are appropriate for simulation of the pressure flow. In fact, most computer models in this category account for both free-surface and pressure flow. In the first case, the computer models use mathematical models identical to those incorporated in the fluvial-process models. In the latter case, the computer models include additional mathematical models to account for energy losses due to pipe friction, entrances and inlets, bends, expansions, contractions, manholes, and valves, meters, and other appurtenances.

28.5.1 SWMM

28.5.1.1   Overview.   The *Storm Water Management Model, SWMM,* is a comprehensive program developed by the EPA for analysis of water quantity and quality in a combined-sewer system (Huber and Dickinson, 1988). For water-excess management, SWMM provides information on discharge time series and peaks, and on hydraulic grade elevation series or maximums.

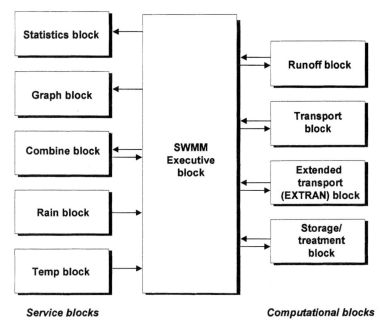

FIGURE 28.6 Structure of Storm Water Management Model (SWMM).

SWMM is constructed of computational "blocks," as shown in Fig. 28.6. The Runoff block uses nonlinear storage routing to estimate runoff due to specified rainfall. The Transport block solves a kinematic-wave simplification of the St. Venant equations to account for open channel flow. The EXTRAN block routes inlet hydrographs through a network of pipes, junctions, and flow-diversion structures. The Storage/Treatment block models flow-control devices. The Service blocks manipulate data, report results, and provide for statistical analysis of results. The blocks are linked through an executive routine. Each of the components may be used independently. In particular, the EXTRAN block may be used to simulate flow in a system "... whenever it is important to represent severe backwater conditions and special flow devices such as weirs, orifices, pumps, storage basins, and tide gates" (Roesner, et al., 1988).

28.5.1.2 Mathematical Models Included in the Computer Model. EXTRAN is a distributed, conceptual model of open-channel and closed-conduit flow. With it, a drainage system is represented in link-node form. This representation may include parallel pipes, looped systems, weirs, orifices, pumps, diversions due to surcharging, manholes, and storage facilities.

EXTRAN solves the gradually varied, one-dimensional form of the unsteady-flow equations for the system. For the link-node representation of the system, it uses the momentum equation in the links and a special lumped form of the continuity equation in the nodes. A finite difference solution scheme is used, with solution time step governed by the wave celerity in shorter channels or conduits in the system.

28.5.1.3 Computer Model Input Requirements and Output. To use the EXTRAN block to simulate flow in a system of closed conduits, the user must provide the following: (1) a link-node description of the system, (2) geometric description of the components of the system, including pipe sizes, shapes, slopes, and locations and dimensions of inlets, diversions, and overflows, and (3) the system inflow hydrographs, computed either with SWMM or a catchment-runoff model.

Output from the EXTRAN block includes reports and graphs of discharge hydrographs and velocities in selected conduits and flow depths and water-surface elevation at selected junctions. A data interface with other SWMM blocks permits transfer of the hydrographs to those blocks for subsequent analysis.

28.5.1.4 Applications. Jewell, et al. (1977) prepared a detailed SWMM application guide. Huber, Heaney, and Cunningham (1988) published a summary of SWMM applications. A SWMM users group meets annually to share information on applications.

28.5.2 HYDRA

28.5.2.1 Overview. HYDRA was developed for the U.S. Federal Highway Administration (FHWA) for analysis and design of storm drain, sanitary sewage, and combined-flow systems (FHWA, 1993). It is a component of the FHWA HYDRAIN integrated drainage design system. HYDRA will analyze flow in an existing or proposed system of closed conduits and/or open channels. If that system is overloaded (flow exceeds capacity), HYDRA will identify flow-reduction solutions. HYDRA also will select pipe size, slope, and invert elevations to satisfy user-specified design criteria.

28.5.2.2 Mathematical Models Included in the Computer Model. HYDRA is a complete design and analysis package that includes catchment-process, fluvial-process, and pressure-flow process models. The pressure-flow model is ". . . derived from the EXTRAN module of SWMM . . ." (FHWA, 1993). However, the model has been expanded to include water-control facilities common in highway drainage.

28.5.2.3 Computer Model Input Requirements and Output. HYDRA is designed for maximum flexibility and generality, thus permitting analysis of almost any reasonable drainage system. The user-prepared input file consists of commands, dimensions, and parameters to describe the drainage system and how it is to be modeled. As with EXTRAN, a link-node scheme is employed to describe system layout. The system's tailwater boundary condition is specified by the user. For pressure-flow computations, the user must identify each conduit's roughness and dimensions.

HYDRA computes the hydraulic gradeline throughout the system. To do so, it computes major and minor losses within the system.

28.6 STATISTICAL-PROCESS COMPUTER MODELS

Chapters 7 and 24 of this handbook describe statistical models that provide information for decision making. All these models are empirical, fitted-parameter models. When these are employed in water-excess management studies, either of two approaches is common: (1) special-purpose computer software is created to satisfy the needs of the particular study, perhaps using a library of statistical analysis routines, or (2) the water-management problem is formulated in general terms and one of the readily available general-purpose statistical analysis software packages is used. The exception is *annual maximum-discharge frequency analysis,* in which the probability of the annual maximum discharge exceeding a specified magnitude is estimated. Because of the critical role of and recurrent need for such analysis, and due to the widespread use of a single statistical model proposed by the U.S. Water Resources Council, generalized computer models have been developed. This section describes two of those models.

28.6.1   HEC-FFA and Program J407

28.6.1.1   Overview.   In the United States, a number of federal agencies conduct annual maximum-discharge frequency analysis for decision making. Until 1967, each agency established its own methods and procedures for the analysis, leading to occasional differences in estimates of quantiles or probabilities. To promote a consistent approach, a multiagency committee of the U.S. Water Resources Council (WRC) studied alternatives and recommended the log-Pearson Type III distribution for use by U.S. federal agencies (Interagency Advisory Committee, 1982). The committee also recommended procedures for treating small samples, outliers, zero flows, broken and incomplete records, and historical flood information. The USACE and USGS developed computer models to implement the guidelines. The USACE computer model is designated HEC-FFA (USACE, 1992), and the USGS model is designated J407. These models are essentially identical, with minor differences to meet the unique needs of the agencies.

28.6.1.2   Mathematical Models Included in the Computer Model.   HEC-FFA and J407 fit a Pearson Type III statistical model (distribution) to logarithms of an observed flood series, using modified method-of-moments parameter estimators. Bulletin 17B and Chap. 24 of this handbook describe the statistical model and fitting procedures in more detail. Table 28.11 lists the analysis procedures used by the models in fitting the distribution.

28.6.1.3   Computer Model Input Requirements.   Both HEC-FFA and J407 provide information on probabilities (frequencies) of extreme discharge magnitudes. To do so, both require as input a sample series of unregulated, annual-maximum flows that is free of climatic trends, representative of constant watershed conditions, and from a common parent population. In addition to the systematic time series, HEC-FFA and J407 require the following:

Model execution specifications.   The user may select various plotting positions for visually inspecting the goodness of fit, and various reports and plots of results.

Model parameters.   Both computer models estimate the log-Pearson Type III parameters from sample statistics. The sample statistics are computed from the input series. However, if desired, the user may specify the sample statistics, thus overriding the computation. Further, the user must specify the regional skew coefficient if the weighting scheme of Bulletin 17B is to be used.

Historical data.   If historical flow data are available, the user must identify these.

28.6.1.4   Computer Model Output.   Output from HEC-FFA and J407 includes the following: (1) a summary report of the user's input, (2) computed sample statis-

TABLE 28.11 HEC-FFA and J407 Features

Feature	Analysis procedure
Parameter estimation	Estimate parameters with method of moments; this assumes sample mean, standard deviation, skew coefficient = parent population mean, standard deviation, and skew coefficient. To account for variability in skew computed from small samples, use weighted sum of station skew and regional skew.
Outliers	These are observations that ". . . depart significantly from the trend of the remaining data." Models identify high and low outliers. If information available indicates that a high outlier is maximum in the extended time period, it is treated as historical flow. Otherwise, they are treated as part of a systematic sample. Low outliers are deleted from the sample, and conditional probability adjustment is applied.
Zero flows	If the annual maximum flow is zero (or below a specified threshold), the observations are deleted from the sample. The model parameters are estimated with the remainder of the sample. The resulting probability estimates are adjusted to account for the conditional probability of exceeding a specified discharge, given that a nonzero flow occurs.
Historical flood information	If information is available indicating that an observation represents the greatest flow in a period longer than that represented by the sample, model parameters are computed with "historically" weighted moments.
Broken record	If observations are missing due to ". . . conditions not related to flood magnitude," different sample segments are analyzed as a single sample with the size equal to the sum of the sample sizes.
Expected probability adjustment	This adjustment is made to the model results ". . . to incorporate the effects of uncertainty in application of the [frequency] curve."

tics and estimated model parameters, (3) a report of the computed frequency function, showing selected quantiles, and (4) plots of the frequency function.

28.7 PERFORMANCE MODELS

The computer models described in earlier sections of this chapter provide information on *system behavior;* they simulate processes by which a system input is transformed to a system output. But for informed water-resources planning, we often need information on *system performance:* the consequence of a particular system output or a particular state of the system. Several of the models described include the capability to assess performance. For example, HEC-1 and TR-20 include routines to model detention-structure performance, given hydrographs computed with the runoff process models they include. But for more detailed analysis, computer models designed especially for evaluation are available. Two are described here: EAD, a flood-damage evaluation model; and HEC-5, a reservoir-system evaluation model.

28.7.1   EAD

28.7.1.1   Overview.   The objective of the HEC *EAD* (*Expected Annual Flood Damage*) program is to compute inundation damage and inundation-reduction benefit, thus permitting evaluation of existing flood hazard and of the anticipated performance of proposed damage-reduction measures (USACE, 1984a). The *Principles and Guidelines* (U.S. WRC, 1983), a document that provides the rules for federal water resource planning in the United States, stipulates that the economic benefit of a flood-damage-reduction project is the sum of location, intensification, and inundation-reduction benefit. *Location benefit* is associated with addition of activity to a floodplain, while *intensification benefit* is a consequence of modified operation of existing activity in a floodplain due to the protection provided. *Inundation-reduction benefit* is the difference between damage due to flooding without and with the project. The *Principles and Guidelines* further require that the damage estimates should be ". . . potential average annual dollar damages to activities affected by flooding . . . [estimated] using standard damage-frequency integration techniques . . ."

28.7.1.2   Mathematical Models Included in the Computer Model.   *Average annual damage,* also properly called the *expected annual damage,* is computed by integrating the cumulative distribution function (cdf) of annual damage. In the simplest application, EAD uses a numerical integration scheme to integrate a user-provided damage-frequency function and reports the results. These computations can be performed for various damage categories for any number of *reaches* (subdivisions of the floodplain). Damage-frequency functions are not commonly available, but are derived from statistical, fluvial, and economic data or models, as illustrated in Fig. 28.7. The functions may represent the existing without-project, existing with-project, future without-project, and/or future with-project state of the floodplain. EAD will perform this manipulation for any alternative conditions defined by the user.

The functions shown in Fig. 28.7 may change with time. EAD includes the appropriate discounting formulas as required by the *Principles and Guidelines* to ". . . convert future monetary values to present values."

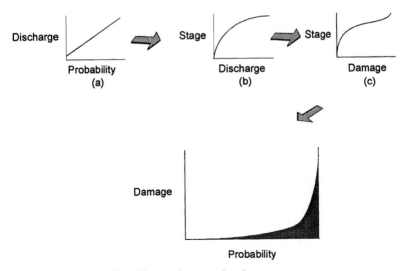

FIGURE 28.7   Derivation of damage-frequency function.

TABLE 28.12 Input Required for HEC EAD Model

Category	Input requirement
Job specification	User must define discount rate, period of analysis.
Statistical function	User must define either discharge—probability, stage—probability, or damage—probability function.
Other functions	Depending on the form of the statistical function provided, user must provide other functions necessary to derive a damage versus probability function. These may include stage-damage and/or stage-discharge functions.

28.7.1.3 Computer Model Input and Output. Table 28.12 lists the input required for EAD. EAD output includes a report of the derived damage-frequency functions, sorted by reach, for each damage category, plus the aggregate function, for the existing, without-project condition, and for each alternative condition defined by the user. It includes also a report of the computed average annual damage, sorted by reach, for each damage category, plus the aggregate function. The *inundation-reduction benefit* of each with-project condition is displayed.

28.7.1.4 Utility Programs. The HEC has developed utility programs that simplify use of EAD or provide additional capabilities. The *SID* program provides data-management capabilities for the numerous stage-damage functions typical of a major flood-control study (USACE, 1989). It yields input in the format required for EAD. The *FDA package* is a complete ensemble of flood-damage analysis models (USACE, 1988). It includes EAD, SID, and utility programs that permit linkage with statistical- and fluvial-process models through the HEC-DSS.

28.7.2 HEC-5

28.7.2.1 Overview. Program HEC-5 models the performance of a reservoir or system of reservoirs operated to manage excess water (USACE, 1982b). Other computer models, including HEC-1 and TR-20, can simulate the operation of a detention structure in which the performance is a function of the properties of the outlet works. HEC-5, however, simulates performance that is a function of both the properties of the outlet works and an operator's specification of the manner in which the reservoirs should be operated. With HEC-5, storage in each reservoir in a system is divided into zones, as illustrated by Fig. 28.8. Within each zone, the user defines indexed storage levels. The model will simulate operation to meet specified system constraints and to keep system reservoirs in balance, with each at the same index level. System constraints that may be modeled are summarized in Table 28.13.

In addition to modeling reservoir *flood-control operation,* HEC-5 includes algorithms for modeling *reservoir-system* operation for conservation purposes.

28.7.2.2 Mathematical Models Included in the Computer Model. HEC-5 includes various simplified models for streamflow routing and a reservoir storage routing model. For reservoirs with hydroelectric power-generation facilities, an energy-production model is included.

28.7.2.3 Computer Model Input Requirements. Table 28.14 lists the input required for HEC-5 for analysis of performance of a flood-control reservoir system.

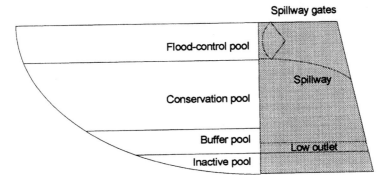

FIGURE 28.8   HEC-5 flood-control operation rules.

28.7.2.4 Computer Model Output. HEC-5 output includes the following: (1) a summary of the user's input, (2) for each reservoir, a summary of inflows, releases, and storages for the period of analysis, (3) for each system control point, a summary of flows for the period of analysis, and (4) if flood-damage relationships are provided, a summary of damage at each location. HEC-5 also includes links to HEC-DSS, the database-management system described in Sec. 28.2.1. Thus flood hydrographs can be computed and filed in the database by a catchment-process model, then retrieved for reservoir-performance analysis with HEC-5.

TABLE 28.13 HEC-5 Flood-Control Operation Rules

Constraint on release made	Condition
Release to draw storage to top of conservation pool without exceeding channel capacity at reservoir or downstream points for which reservoir is operated	Storage is between top of conservation pool and top of flood-control pool
Release equal to or greater than minimum desired flow	Storage greater than top buffer storage
Release equal to minimum desired flow	Storage between top inactive and top of buffer pool
No release	Storage below top of inactive pool
Release required to satisfy hydropower requirement	If that release is greater than controlling desired or required flows for above conditions
Release limited to user-specified rate of change	Unless reservoir is in surcharge operation
No release that will contribute to flooding downstream	If flood storage available
Release to maintain downstream flow at channel capacity	If operating for flood control
Release from reservoir at greatest level	If two or more reservoirs on parallel streams operate for common downstream point
Release to bring upper reservoir to same index level as downstream reservoir	If two reservoirs in tandem

TABLE 28.14 Input Required for HEC-5 Model

Category	Input requirement
Job specification	Output requirements
System hydrological data	Reservoir inflows and intermediate-area runoff, reservoir-evaporation data
Fluvial-model parameters	Routing coefficients
System layout	Description of individual reservoirs and physical relationship of reservoirs, channels, and so forth
Operating policy	Reservoir storage zones and levels

28.8 WARNINGS

28.8.1 Rapid Changes in Technology

Scott McNeally, chairman of Sun Microsystems, suggests that ". . . the shelf life of biscuits and technology is about the same" (*New York Times,* 27 March 1993). Accordingly, the reader is cautioned that the state-of-practice in water-excess modeling changes rapidly. He or she should contact appropriate local, state, and federal government agencies and review the current water-resources literature for information on computer model updates or new computer models before selecting for application one of the models described in this chapter.

28.8.2 Pitfalls in Analysis

Biswas (1979), Ford and Davis (1989), and many others have written about the potential problems in applying models for water-resource decision making. They note that, although mathematics and the computer are now essential ingredients of decision making, great care must be exercised to insure that results provide the information necessary to make the decisions. Quade (1980) provides a list of common pitfalls in formulation and modeling; these are summarized in Table 28.15. The reader is encouraged to consider these when applying the models described herein.

REFERENCES

American Society of Civil Engineers (ASCE)/Water Environment Federation, *Design and Construction of Urban Stormwater Management Systems,* Manuals and Reports of Engineering Practice no. 77. ASCE, New York, 1992.

Barkau, R. L., "A Mathematical Model of Unsteady Flow Through a Dendritic Network," Ph.D. dissertation, Department of Civil Engineering, Colorado State University, Ft. Collins, Colo., 1985.

Bedient, P., A. Flores, S. Johnson, and P. Pappas, "Floodplain Storage and Land Use Analysis at The Woodlands, Texas," *Water Resources Bulletin,* 21(4):543–551, 1985.

Bhaskar, N. R., "Projection of Urbanization Effects on Runoff Using Clark Instantaneous Unit Hydrograph Parameters," *Water Resources Bulletin,* 24(1):113–124, 1988.

TABLE 28.15 Pitfalls in Formulation and Modeling

Pitfall	Description
Insufficient attention to formulation	Modeler gets in hurry to get on with "real work" and pays insufficient attention to formulation. Ends up working on problem that has little relation to real issue.
Unquestioning acceptance of stated goals and constraints	Uncritical acceptance of client's or policymaker's statement of objectives and constraints on actions to be considered.
Measuring achievement by proxy	Danger is that measure of achievement is a product of the modeler's judgment of the decision maker's values, rather than the decision maker's judgment of values.
Misjudging the difficulties	Analysis is, necessarily, based on an abstraction of reality. Attempts to include "everything" result in ambiguity, confusion, and intractability.
Bias	Beware hidden bias introduced by data used, alternatives evaluated, and model selected.
Equating modeling with analysis	Models are only one ingredient and modeling only one step in policy analysis; searching out the right problem, designing a better alternative for consideration, and skillfully interpreting the computations from the model and relating them to the decision maker's problem are equally significant.
Improper treatment of uncertainties	Impact of unpredictabilities in factors that affect outcome of a course of action must be considered, including uncertainties that are not stochastic in nature.
Attempting to simulate reality	Heed the principle that the problem (the question being asked), as well as the process being modeled, determines what should be modeled.
Belief that a model can be proved correct	Models can, at best, be invalidated. Aim of validation tests is to increase degree of confidence that events inferred from application of model will, in fact, occur under the conditions assumed. Modeler should also check data.
Neglecting the by-products of modeling	The results of computation with model are not the only valuable output. "Building" a model is valuable because of what one learns about the problem and because of the guidance that a model can provide to the judgement and intuition of the analyst and policy maker.
Overambition	Large-scale models designed for many purposes are likely to serve none in the end.
Seeking academic, rather than policy, goals	Applications or development of models should focus on decision-making needs (relevance, reliability, usability, cost-effectiveness), rather than on tenure-granting needs (nontriviality, power, computational efficiency, elegance).
Internalizing the policy maker	Model cannot be made so comprehensive that it captures the preferences and constraints on the policymakers, so that it can designate the best alternative in a credible and acceptable manner.
Not keeping the model relevant	Modeler must work to insure that the model's "knobs" represent the policy variables under the control of the policymaker.
Not keeping the model simple	Model should be no more complicated than needed, and analyst should be able to exhibit in simplified form the key structrural elements of the model and their most important relationships. Thus client can understand both how results came about and how to incorporate logic of analysis into continued thinking about the problem.
Capture of the user by the developer	Full documentation is critical. Otherwise user is at mercy of developer.

Source: Quade (1980).

Bicknell, B. R., et al., *Hydrologic Simulation Program—FORTRAN: User's Manual for Release 10,* Environmental Research Laboratory, Environmental Protection Agency, Athens, Ga., 1992.

Biswas, A. K., "Models for Water Resources Decision-Making: Problems and Prospects," *Water Supply & Management,* 3:1–7, 1979

Clarke, R. T. "A Review of Some Mathematical Models Used in Hydrology with Observations on Their Calibration and Use," *Journal of Hydrology,* 19:1–20, 1973.

Crawford, N. H., and R. K. Linsley, "Digital Simulation in Hydrology: Stanford Watershed Model IV" Technical Report no. 39, Department of Civil Engineering, Stanford University. Stanford, Calif., 1966.

DeBarry, P. A., and J. T. Carrington, "Computer Watersheds," *Civil Engineering,* American Society of Civil Engineers, New York, 1990, pp. 68–70.

DeVries, J. J., and T. V. Hromadka, "Computer Models for Surface Water," in David R. Maidment, ed., *Handbook of Hydrology,* McGraw-Hill, New York, 1993.

Fan, S-S., ed., "Twelve Selected Computer Stream Sedimentation Models Developed in the United States," Federal Energy Regulatory Commission, Washington, D.C., 1988.

Federal Highway Administration, "HYDRAIN—Integrated Drainage Design Computer System. Vol. III. HYDRA—Storm Drains," Structures Division, Federal Highway Administration, Washington, D.C., 1993.

Ford, D. T., and D. W. Davis, "Hardware-Store Rules for System-Analysis Applications," *Closing the Gap Between Theory and Practice,* Proceedings of the IAHS Symposium, Baltimore, MD, IAHS Pub. 180, 1989.

Fread, D. L., "NWS Operational Dynamic Wave Model," *Proceedings, ASCE 25th Annual Hydraulics Division Specialty Conference,* American Society of Civil Engineers, New York, 1978, pp. 455–464.

Fread, D. L., *National Weather Service Operation Dynamic Wave Model, Appendix A: User's Manual,* National Weather Service, NOAA, Silver Springs, Md., 1987.

Huber, W. C., J. P. Heaney, and B. A. Cunningham, "Storm Water Management Model (SWMM) Bibliography," EPA/600/3-85/077, Environmental Protection Agency, Athens, Ga., 1988.

Huber, W. C., and R. E. Dickinson, Storm Water Management Model, Version 4: User's Manual, Environmental Research Laboratory, Environmental Protection Agency, Athens, Ga., 1988.

Interagency Advisory Committee on Water Data, "Guidelines for Determining Flood Flow Frequency," Bulletin 17B, U.S. Department of the Interior, U.S. Geological Survey, Office of Water Data Coordination, Reston, Va., 1982.

Jackson, T. L., "Application and Selection of Hydrologic Models," in C. T. Haan, et al., eds., *Hydrologic Modeling of Small Watersheds,* American Society of Agricultural Engineers, St. Joseph, Mo., 1982.

Jewell, T. K., et al., "SWMM Application Study Guide," in F. A. Digiano, et al. eds., Short Course Proceedings: Applications of Stormwater Management Models, EPA-600/2-77-065, Environmental Protection Agency, Cincinnati, Ohio, 1977.

Larson, C. L., et al., "Some Particular Watershed Models," in C. T. Haan, et al., eds., *Hydrologic Modeling of Small Watersheds,* American Society of Agricultural Engineers, St. Joseph, MO, 1982.

McCuen, R., *A Guide to Hydrologic Analysis Using SCS Methods,* Prentice-Hall, Englewood Cliffs, N.J., 1982.

Melching, C. S., et al., "Output Reliability as Guide for Selection of Rainfall-Runoff Models," *Journal of Water Resources Planning and Management,* 117(3):383–398, 1991

Molinas, A., and C. T. Yang, Computer Program User's Manual for GSTARS (Generalized Stream Tube Model for Alluvial Rivers Systems), U.S. Department of Interior, Bureau of Reclamation, Engineering and Research Center, Denver, Colo., 1986.

Quade, E. S., "Pitfalls in formulation and modeling," in G. Majone and E. S. Quade, eds., *Pitfalls of Analysis,* John Wiley, New York, 1980.

Renard, K. G., W. J. Rawls, and M. M. Fogel, "Currently Available Models," in C. T. Haan, et al., eds., *Hydrologic Modeling of Small Watersheds,* American Society of Agricultural Engineers, St. Joseph, Mo., 1982.

Roesner, L. A., et al., "Storm Water Management Model User's Manual Version 4: EXTRAN Addendum," EPA/600/3-88/001b. Environmental Research Laboratory, Environmental Protection Agency, Athens, Ga., 1988.

Russell, S. O., et al., "Estimating Design Flows for Urban Drainage," *Journal of Hydraulics Division,* ASCE, 105(HY1):43–52, 1979.

Sherman, J. O., et al., "Bridge Waterways Analysis Model," FHWA Report no. FHWA/RD86-108, Federal Highway Administration, Washington, D.C., 1986.

Sherman, J. O., "User's Manual for WSPRO: A Model for Water Surface Profile Computations," Federal Highway Administration, Washington, DC, 1988.

Thomas, W. A., and R. E. Heath, "Application of TABS-2 to Greenville Reach, Mississippi River," *River Meandering,* Proceedings of ASCE Conference on Rivers '83, New Orleans, La., 1983.

Thomas, W. A., and W. H. McAnally, Open-Channel Flow and Sedimentation TABS-2: User's Manual, Waterways Experiment Station, Vicksburg, Miss., 1985.

U.S. Army Corps of Engineers, "HEC-2: Water Surface Profiles, User's Manual," Hydrologic Engineering Center, Davis, Calif., 1982a.

U.S. Army Corps of Engineers, "HEC-5: Simulation of Flood Control and Conservation Systems, User's Manual," Hydrologic Engineering Center, Davis, Calif., 1982b.

U.S. Army Corps of Engineers, "Expected Annual Flood Damage Computation—EAD, User's Manual," Hydrologic Engineering Center, Davis, Calif., 1984a.

U.S. Army Corps of Engineers, "HMR-52: Probable Maximum Storm (Eastern United States), User's Manual," Hydrologic Engineering Center, Davis, Calif., 1984b.

U.S. Army Corps of Engineers, "FDA: Flood-Damage Analysis Package, User's Manual," Hydrologic Engineering Center, Davis, Calif., 1988.

U.S. Army Corps of Engineers, "SID: Structure Inventory for Damage Analysis, User's Manual," Hydrologic Engineering Center, Davis, Calif., 1989.

U.S. Army Corps of Engineers, "HEC Data Storage System User's Guide and Utility Program Manuals," Hydrologic Engineering Center, Davis, Calif., 1990a.

U.S. Army Corps of Engineers, "HEC-1: Flood Hydrograph Package, User's Manual," Hydrologic Engineering Center, Davis, Calif., 1990b.

U.S. Army Corps of Engineers, "Red River of the North UNET Application," Project Report no. 91-01, Hydrologic Engineering Center, Davis, Calif., 1990c.

U.S. Army Corps of Engineers, "HECLIB, Vol. 2: HECDSS Subroutines, Programmer's Manual," CPD-57, Hydrologic Engineering Center, Davis, Calif., 1991a.

U.S. Army Corps of Engineers, "HEC-6: Scour and Deposition in Rivers and Reservoirs: User's Manual," Hydrologic Engineering Center, Davis, Calif., 1991b.

U.S. Army Corps of Engineers, "SSARR Model—Streamflow Synthesis and Reservoir Regulation: User Manual," North Pacific Division, Portland, Oreg., 1991c.

U.S. Army Corps of Engineers, "HEC-FFA: Flood Frequency Analysis: User's Manual," Hydrologic Engineering Center, Davis, Calif., 1992.

U.S. Army Corps of Engineers, "UNET: One-Dimensional Unsteady Flow Through a Full Network of Open Channels, User's Manual," Hydrologic Engineering Center, Davis, Calif., 1993.

U.S. Department of Agriculture, *SCS National Engineering Handbook, Sec. 4, Hydrology,* Soil Conservation Service, Washington, D.C., 1971.

U.S. Department of Agriculture, "Computer Program for Project Formulation—Hydrology," Technical Release no. 20, 2d ed., Soil Conservation Service, Washington, D.C., 1983.

U.S. Department of Agriculture, "Urban Hydrology for Small Watersheds," Technical Release no. 55, Soil Conservation Service, Washington, D.C., 1986.

U.S. Water Resources Council, Economic and Environmental Principles and Guidelines for Water and Related Land Resources Implementation Studies, U.S. Government Printing Office, Washington, DC, 1983.

Vanoni, V. A., ed., *Sedimentation Engineering.* Manuals and Reports of Engineering Practice no. 54, American Society of Civil Engineers, New York, 1977.

World Meteorological Organization, "Intercomparison of Conceptual Models Used in Operational Hydrological Forecasting," WMO-no. 429, WMO Operational Hydrology Report no. 7, Geneva, Switzerland, 1975.

P · A · R · T · 5

WATER RESOURCES FOR THE FUTURE

CHAPTER 29

GLOBAL CLIMATE CHANGE: EFFECT ON HYDROLOGIC CYCLE

Dennis P. Lettenmaier
Department of Civil Engineering
University of Washington
Seattle, Washington

Gregory McCabe
U.S. Geological Survey
Lakewood, Colorado

Eugene Z. Stakhiv
Institute for Water Resources
U.S. Army Corps of Engineers
Fort Belvoir, Virginia

29.1 INTRODUCTION

The *long-term energy budget of the earth* as seen from space (i.e., for the flux of energy through an artificial control surface at the top of the atmosphere) is a balance between two components: (net) incoming shortwave radiation from the sun, and outgoing longwave radiation from the earth. The difference in wavelength of the incoming and outgoing radiation is a consequence of the much higher temperature of the sun than of the earth. The temperature of the earth is determined by how much of the incoming solar radiation is absorbed by the atmosphere and the earth's surface, and by the characteristics (emissivity) of the matter that absorbs the radiation.

The earth's temperature is also affected by the fact that part of the longwave radiation emitted by the surface is absorbed by the atmosphere, and then re-emitted as longwave radiation in either the upward or downward direction (Fig. 29.1). The net result of the internal radiation balance of the atmosphere is the so-called *greenhouse effect*, which results in both the atmosphere and the earth's surface being several tens of °C warmer than would otherwise be the case. The magnitude of the effect depends on the composition of the atmosphere, and most importantly, on the concentrations of water vapor, carbon dioxide, and, to a lesser extent, on certain other trace gases, such as methane.

It is well known that global concentrations of carbon dioxide have increased, essentially monotonically, by about 25 percent since the Industrial Revolution, and by over 10 percent since the beginning of long-term instrumental records in the late 1950s. Depending on the scenario used for economic growth and fossil-fuel use, current global concentrations of CO_2 are projected to double within about the next 80 years (see Fig. 29.2). Concerns about the consequences of global warming are motivated by paleoclimatic studies, which show that global temperatures closely follow variations in atmospheric carbon dioxide concentrations, and by modeling studies using *general circulation models (GCMs)* of the atmosphere (see Sec. 29.1.1). Most modeling studies have tested the sensitivity of climate to an equivalent doubling of CO_2. (The term *equivalent* means that the effects of changes in concentrations of trace gases other than CO_2 are expressed as changes in CO_2 that would have the same effect.) Generally, the modeling studies show temperature increases would be greater at the poles than at the equator, and that global precipitation would increase. The models, which operate on spatial scales of hundreds of km, are much less able to allow interpretation of the magnitude, or even direction, of climate change on regional or local scales (see Sec. 29.2.3).

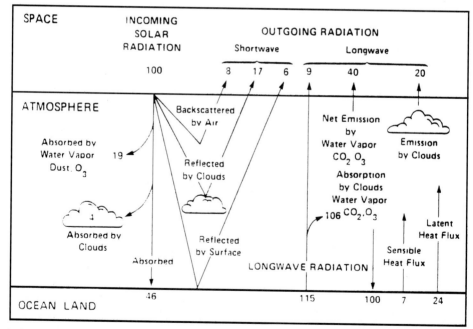

FIGURE 29.1　Globally averaged atmospheric energy balance, expressed as a percentage of incoming solar radiation (*MacCracken and Luther, 1985*).

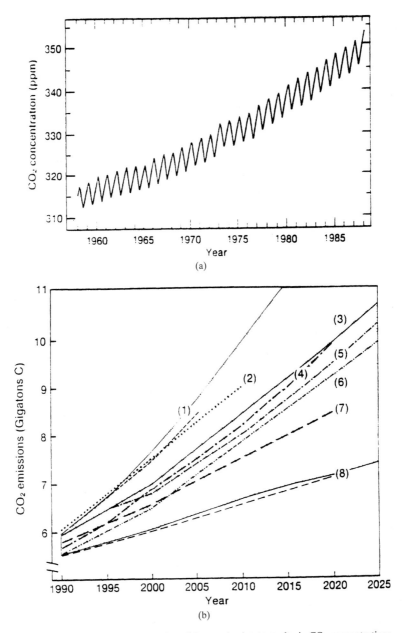

FIGURE 29.2   Observed and projected future rise in atmospheric CO_2 concentrations and emissions: (*a*) observed atmospheric CO_2 concentrations at Mauna Loa Observatory, Hawaii (*from Mitchell, 1989*); (*b*) projected Intergovernmental Panel on Climate Change (IPCC) range of annual global CO_2 emissions in intermediate term (1990 to 2025) (*from Leggett, et al., 1992*).

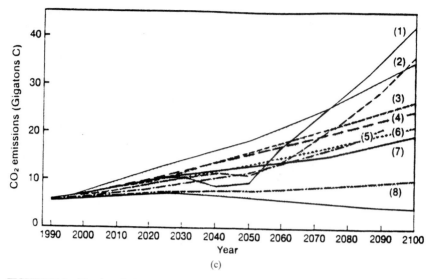

(c)

FIGURE 29.2 *(Continued)* Observed and projected future rise in atmospheric CO_2 concentrations and emissions: *(c)* projected IPCC range of annual global CO_2 emissions in long term (1990 to 2100) *(from: Mitchell, 1989; Leggett, et al., 1992).*

29.1.1 General Circulation Models and Climate-Change Prediction

Most predictions of global warming are based on computer simulations using general circulation models (GCMs) of the atmosphere. GCMs solve, in a discrete space-time space, the conservation equations that describe the geophysical fluid dynamics of the atmosphere (see, e.g., Washington and Parkinson, 1986). These models have the same general structure as the *numerical weather-prediction models (NWPMs)* widely used for weather forecasting; the main difference is that NWPMs are run for relatively short periods (several days) using prescribed initial conditions. While the horizontal resolution of most current NWPMs is now less than 50 km, GCMs used for climate simulation typically use spatial discretizations of several degrees of latitude and longitude in the horizontal. Vertically, GCMs partition the atmosphere into layers, usually numbering 5 to 20, of unequal thickness (with more layers close to the surface). The temporal discretization is defined by numerical-stability criteria, and is typically considerably less than 1 h. However, a distinction should be made between the computational time step of GCMs, and the physically realistic time step at which results can be interpreted. For instance, some GCMs do not represent the diurnal cycle, so while a time step of, for instance, 20 min might be used to assure numerical stability, intepretation of results as (e.g., averages over) periods of less than a day is not physically meaningful.

The main consideration determining spatial resolution of GCMs is computation time. Most GCMs are run on supercomputers. The current generation of GCMs, such as those used in the 1995 IPCC Second Assessment Report (Viner, et al., 1994), require one to several days of CPU time per year of simulation, depending on the particulars of the model. Doubling the spatial resolution (halving the grid-mesh size) can be expected to increase the computational time by roughly an order of magnitude for most GCMs.

Early GCMs prescribed as boundary conditions the physical characteristics of the earth's surface. In particular, for the ocean, sea-surface temperatures were prescribed, and for the land surfaces, surface temperature and soil wetness were prescribed. Subsequently, interactive surface models, in particular for land–atmosphere coupling, have been developed, and coupled ocean–atmosphere models are now run by some of the major GCM modeling centers. However, there remain a number of weaknesses with GCMs, many of which are related to the spatial resolution at which they must operate. Among the most serious of these are difficulties in reproducing feedbacks between the ocean and the atmosphere, especially those that result in important teleconnections in weather patterns, inability to represent the subgrid characteristics of clouds and their effects on the surface radiation budget, and inadequate representation of the land-surface hydrology. In general, GCMs do a better job of representing large-scale features of the atmosphere, such as the evolution of major storm fronts, than they do of reproducing surface processes, such as precipitation and streamflow. Likewise, while they often do a reasonable job of reproducing seasonal variations in temperature, biases of several °C are not uncommon (Grotch, 1988).

Because GCMs represent the *radiation balance* of the atmosphere, it is a relatively straightforward matter to perform simulations in which the atmospheric chemical composition is assumed to have changed. GCM simulations performed in the mid- and late 1980s, such as those used in the Intergovernmental Panel on Climate Change First Assessment Report (IPCC, 1990) generally predicted that an equivalent doubling of the concentration of CO_2 would result in global average changes in temperature in the range 4 to 6 °C. These simulations did not, however, consider various mitigating effects. These include recent work showing that anthropogenic aerosols may cause a significant masking in the short term of the surface warming that would otherwise result from increases in concentrations of the greenhouse gases (Charlson, et al., 1992). A second major effect is the thermal inertia of the oceans, which is ignored by steady-state GCM runs.

GCM simulations performed for the IPCC Second Assessment Report (Viner, et al., 1994) are based on transient runs of coupled ocean–atmosphere models, in which the atmospheric concentrations of CO_2 are allowed to increase from current conditions, according to projections of global emissions. The resulting temperature changes, for the time of global equivalent CO_2 doubling (predicted to occur in the second half of the twenty-first century by many emissions models) are generally smaller than those used in the 1990 IPCC report. Although a comprehensive comparison has not been performed of the 1995 IPCC transient scenarios as of the time of this writing, preliminary analyses suggest that typical values for northern hemisphere land areas at the time of global equivalent CO_2 doubling are in the range of 2 to 4 °C warmer than present.

From the standpoint of *climate-effects assessments*, there is a strong consensus among GCMs that global warming will occur, and most GCMs agree on the direction of temperature change on a regional basis as well. For other variables, there is much less agreement. For the reasons discussed in the next section, most GCMs agree that on a global-average basis, precipitation will increase. However, there is little consensus at the regional level. Further, and more disturbing, many GCMs do not reproduce current-climate precipitation at all well, especially over the land surfaces. Other variables of hydrologic interest, such as runoff, are generally reproduced poorly, even at the scale of large continental rivers (Miller and Russell, 1992). For this reason, interpretations of the hydrologic effects of climate change at present are unable to rely directly on GCM simulations at the regional scale, and instead must make use of interim strategies to provide scenarios of changes in the most hydrologically important surface variables, such as precipitation.

29.1.2 Hydrologic Effects of Climate Change

The cycling of water through the atmosphere and the earth surface can be viewed as an *engine*. The residence time of water on the land surface and the oceans is much greater than in the atmosphere, hence the speed of the engine is limited by the rate at which water can be supplied to the atmosphere by evaporation.

About 70 percent of the earth's surface is ocean, over which evaporation is limited only by the water-holding capacity of the atmosphere, and the energy available for evaporation. Over the land surface, evaporation is limited by the rate at which water can be transferred to the atmosphere, as well as soil moisture.

On a global basis, it can be expected that a general warming would result in increased evaporation and increased precipitation. Unfortunately, such a global prediction is of little help in understanding how the hydrology of a river basin might respond to climate change. In addition to the problem of spatial scale, interpreting climate-change effects over land is more difficult than over the ocean, because of the complicating effect of the dependence of evaporation on soil moisture. In the southern hemisphere, where less of the earth's surface is covered by land, most precipitation is derived from ocean evaporation. Over the continents of the northern hemisphere, especially in the interior of North America and Eurasia, the land surface plays a larger role, especially in summer, when soil moisture limits evaporation.

From a water-supply perspective, the most important impacts of global warming would be those associated with changes in runoff and groundwater recharge. In areas with rain-dominated hydrology, it is possible, using simple water-balance models, to estimate the sensitivity of runoff to changes in precipitation and evaporation (see, for example, Schaake, 1990; Nemec and Schaake, 1982). These studies show that on an average basis (e.g., using mean annual precipitation, evaporation, and runoff) the *elasticity of runoff* with respect to precipitation (ratio of percentage change in streamflow to percentage change in precipitation) is greater than one, and is larger for arid than for humid climates. In a crude sense, the elasticity tends to be inversely related to the *long-term runoff ratio* (ratio of mean runoff to mean precipitation), which is less in arid than in humid climates. The elasticity of runoff with respect to potential evaporation tends to be lower than the precipitation elasticity, but it too is generally greater than one, and seems to be inversely related to the runoff ratio.

In river basins where the hydrology is snowmelt-dominated, such as the western United States, the sensitivity of runoff to temperature and potential evaporation is complicated by the changes in the seasonal runoff pattern that would result from a warmer climate. For instance, although *potential evapotranspiration (PET)* would increase with warmer temperatures, warmer temperatures also shift the runoff hydrograph toward the winter from the spring (Lettenmaier and Gan, 1990), when PET is seasonally lower. As a result, Lettenmaier and Gan found, for several California catchments, that the change in annual runoff in response to temperature increases of around 4°C would be small, even though there were major changes in the seasonal runoff patterns.

Prediction of the *temperature sensitivity of runoff* is complicated, even for snow-free catchments, by the dependence of PET on variables other than temperature. Potential evaporation depends on net radiation, which is strongly affected by cloud cover, as well as vapor-pressure deficit and surface wind. It is not known how any of these variables would change in a warmer climate. For instance, if the water cycle "engine" does in fact turn faster with warmer temperatures, increased precipitation might well be accompanied by greater cloudiness, which would reduce net radiation and hence potential evaporation. Increased temperature increases the water-holding capacity of the atmosphere, which would increase the vapor-pressure deficit at the same absolute humidity, but whether absolute humidity would increase or decrease is unknown. Finally, changes in surface wind under altered climate are a major unknown which can have a strong effect on potential evaporation. All of the above combine to result in a highly uncertain picture for runoff if significant global warming occurs.

The *sensitivity of groundwater recharge* to precipitation and temperature changes is even more problematic. If one assumes that the groundwater system is in equilibrium (no net long-term storage change), and that the groundwater and surface-water basin boundaries are coincident (an assumption that is often not justified) then in the long term groundwater recharge must balance stream discharge of groundwater (baseflow) plus pumping. By partitioning runoff into direct and baseflow components, one can, at least conceptually, attribute the baseflow (hence recharge) to infiltration. In a warmer climate, infiltration would be decreased by increased potential evaporation (due to greater capture by evapotranspiration of precipitation intercepted by the vegetation canopy, and increased evaporation of ponded and near-surface soil moisture following storm events). However, infiltration would be strongly affected by the pattern and intensity of storms, which would probably change in an altered climate in a manner not presently understood.

Further, in areas with strong precipitation gradients (such as the mountainous western United States), much of the infiltration occurs in the upper reaches of the major river basins, and groundwater discharge to streams, and/or pumping, occurs in the lower reaches. In these cases, a seasonal shift in recharge from snowmelt dominated to winter-rainfall dominated might cancel some of the effect on recharge of higher PET. One of the few studies that has directly addressed the climate sensitivity of groundwater recharge is reported by Vaccaro (1992). He coupled a daily stochastic weather generator with a soil-water balance model for part of the semi-arid Columbia Plateau, Washington. He found that the sensitivity of recharge to a warmer (CO_2 doubled) climate was relatively small and depended on land use. For present land use, which is dominated by irrigated agriculture, recharge decreased by about 16 percent for the current land use, but increased by about 10 percent under predevelopment conditions. The reason for the difference was that the altered climate simulations include an increase in winter precipitation, which is the dominant source of recharge under predevelopment conditions. Under present land use, recharge is greatly affected by infiltration of irrigation water, more of which would be lost to evapotranspiration in a warmer climate.

29.1.3 Water-Resource-Systems Effects

The impact of potential climate change on water-resources management has been the topic of many recent studies (e.g., Kirshen and Fennessey, 1993; Lettenmaier and Sheer, 1991; Arnell and Reynard, 1995; Nash and Gleick, 1991). A number of these studies are reviewed in Chap. 14 of the Intergovernmental Panel on Climate Change (IPCC) Second Assessment Report (Kaczmarek, et al., 1996). However, for the reasons discussed in Sec. 29.1.1, predictions of the regional incidence, timing, and magnitude of the physical hydrologic changes that would accompany global-climate change are highly uncertain, and all such studies are best interpreted as sensitivity analyses. The methods of climate-change effects interpretation described in Secs. 29.2 to 29.4 should be viewed in the same context. It is impossible at present to predict climate-change effects accurately enough on a regional scale to view the results of such studies as other than evaluations of the consequences of alternative climate scenarios.

Such studies have nonetheless proved useful in moving beyond the traditional retrospective analyses that have been the basis for most water-resource-systems design and operations studies, to an explicit assessment of system robustness. However, attempts to transform the primary physical effects into water-resources-management impacts (availability, use, distribution, operation) and then to socio-economic and environmental impacts encounter a hierarchy of cascading uncertainty. For instance, Lettenmaier and Sheer (1991) used a three-level modeling approach (consisting of precipitation–runoff modeling for four index catchments, a stochastic model to relate index-catchment streamflows to major inflows to the combined California State Water Project–Central Valley Project reservoir system, and a water-resources-system model to describe the operation of the reservoir system) to assess the possible effects of climate change on the Sacramento–San Joaquin River basin.

Stakhiv, et al. (1992) argue that such multilevel model analyses, while unavoidable given the current state of the art of regional climate change prediction, greatly complicate the problem of identifying possible response strategies. While the resultant uncertainty in prediction of water-resource-system response to climate change can in principle be evaluated, at least for the present climate, it is bound to hamper an analytical evaluation of preferred responses. It is difficult to prescribe a set of practical or reasonable steps, actions and precautions that a water-resources manager, decision maker, or policy analyst should be considering, short of simply ignoring the problem until better information is available. There is a very large gap between believing that global warming will someday result in serious adverse consequences and in mobilizing policies and preparing for large investments in anticipation of trends whose specific outcomes are highly uncertain.

Water-resources management, by its nature, is intended to ameliorate the extremes in climate variability. It stands to reason, then, that managed watersheds, i.e., those under some form of water control and distribution, are apt to experience less-dramatic climate impacts than those without such water-management systems even though the natural hydrologic response may be similar. Clearly, climate change could exacerbate an increasingly complex resource-management problem. Stakhiv, et al. (1992) argue that water-resources management opportunities for adapting to climate change are perhaps better understood and offer more realistic promise to mitigate whatever adverse consequences that may materialize than our collective ability to predict, understand and manage future socioeconomic changes. On the other hand, Lettenmaier, et al. (1993) have conducted sensitivity studies showing that, in the case of a modestly sized multiple-purpose reservoir in a snowmelt-dominated western United States stream, more efficient system operation would be unable to mitigate water-supply failures caused by a shift in the seasonal hydrograph resulting from a warmer climate.

Reservoir reallocation has been one of the traditional ways of adapting to changing water demands and uses. The U.S. Army Corps of Engineers has engaged in a number of studies of reallocation of storage in its approximately 600 reservoirs (Stakhiv, et al., 1992). Many of these studies have been driven by the growing demand for municipal and industrial water supply by urban areas. Reallocation of existing storage is an ongoing process, both formally as a result of congressional authorizations or as part of other related comprehensive studies, where reallocation was one among many possible alternatives to more efficient water-resources management.

The specification of climate-change scenarios is at the heart of all attempts to assess the hydrologic implications of climate change. Therefore, the following section evaluates various approaches that have been used in past studies. Then Secs.

29.3 and 29.4, respectively, discuss hydrologic and water-resource-systems modeling approaches that can be used in conjunction with climate-change scenarios to perform integrated assessments.

29.2   CLIMATE-CHANGE SCENARIOS

To assess the potential effects of predicted climatic change on water resources, scenarios of changes in atmospheric variables that affect surface and subsurface hydrology, such as temperature, precipitation, and evapotranspiration, are needed. Methods of climatic-change scenario development range from prescribing hypothetical climatic conditions to the use of predictions from complex mathematical models such as general circulation models (see Sec. 29.1.1). This section discusses three methods of climate scenario development, and their advantages and disadvantages. The three methods are: (1) historical and spatial analogues, (2) output from general circulation models, and (3) prescribed climate-change scenarios.

29.2.1   Historical Analogues

In many parts of the world long climatological records have been collected, in some cases in excess of 100 years. This is especially the case in North America and Europe. These data can be used to develop climate-change scenarios that are based on direct analogues between the past and predicted doubled-CO_2 conditions (i.e. historical analogues).

Historical analogues include *instrumental analogues* and *proxy analogues*. The variables with the longest instrumental records are usually temperature, precipitation, and atmospheric pressure. The advantage of using instrumental records is that the sequences actually occurred at the location of interest, so uncertainties introduced by transferring data from another location, or model error, are avoided. One drawback to the use of instrumental records is that they usually are no more than a century in length, especially in the United States, so the climate variability over the period of record may not be large enough to capture the magnitude of climate change predicted by GCMs. To compensate for this problem, attempts have been made to use only the most extreme years from instrumental records (Williams, 1980; Jager and Kellogg, 1983; Smith and Tirpak, 1989a; 1989b). The drawback of this approach is that single extreme years, or even decade-long periods, are transitory events and do not represent equilibrium conditions under the specified climatic conditions (Jager and Kellogg, 1983). In addition, different periods with similar mean climates are often caused by different processes and are associated with different spatial distributions of climate parameters (Robock, et al., 1993).

Proxy records include climate information derived from recorded historical events (e.g. diaries; agricultural records) or from paleoclimatic studies (e.g., sea-floor and ice cores; dendrochronologies) (Fig. 29.3). Some of these records provide information from past periods when the atmospheric concentration of CO_2 was close to the level predicted for a doubling of preindustrial levels of CO_2. For example, the Holocene optimum (approximately 6000 to 5000 years B.P.) may be a suitable analogue for a 1°C warming; the last interglacial period (approximately 125,000 years B.P.) could be an analogue for a 2 to 2.5°C warming; and the Pliocene climatic optimum (approximately 3 to 4 million years B.P.) represents a warming of 3 to 4°C (Gleick, 1989; Budyko, 1991). Another positive attribute of proxy records is that they often include biological or ecological information related to the climatic data, indicating the effects of past climatic changes on biological and ecological processes. A drawback to the use of proxy analogues is that the climatic information provided is for time steps which often are too large for most water-resource sensitivity analyses (e.g., years, decades, or centuries). Another drawback is that the driving forces of past climatic conditions most likely were different than the driving forces of predicted climatic conditions resulting from anthropogenically increased concentrations of atmospheric CO_2 (MacCracken and Kutzbach, 1991).

29.2.2 Spatial Analogues

Spatial analogues are based on the transfer of data from a location that has a current climate similar to that expected for future conditions (e.g., doubled CO_2) at the loca-

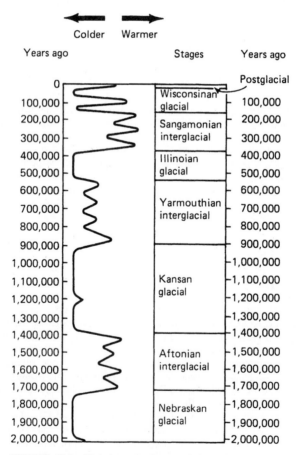

FIGURE 29.3 Global temperature variations determined by the study of small fossils in deep-sea sediments (*Ericson and Wollin, 1968*).

tion of interest. For example, the current climatic conditions of Atlanta, Georgia might be used to represent a doubled CO_2 climate at New York City. GCM predictions for doubled CO_2 conditions are often used to guide the choosing of locations to serve as the analogues (Smith and Tirpak, 1989a). The advantage of this approach is that observed climate records can be used to perform sensitivity analyses. A significant drawback is that many environmental factors contribute to the climate of an area, such as topography, vegetation, position relative to storm tracks, and position relative to sources of atmospheric moisture. These local factors are clearly not represented when data are transferred spatially.

29.2.3 Climate-Change Scenarios Based on GCM Simulations

GCMs currently provide the only physically based predictions of how climate might change as a result of increasing concentrations of atmospheric CO_2. GCMs are mathematical representations of the earth's climate system and simulate atmospheric processes at a field of grid points that cover the surface of the earth. The resolution of GCMs, or grid spacing, has improved as supercomputers have become faster; for long-term climate simulations (runs of multiple decades) most of the major GCM centers are now running at grid resolutions of about 5 degrees of latitude by 5 degrees of longitude, and long-term runs within the next few years at double the current resolution are likely.

29.2.3.1 Scenarios Based Directly on GCM Simulations. Although GCMs are able to reproduce the general spatial and temporal distributions of most climatic variables on a global scale, GCM estimates of climate on regional scales (10^4 to 10^6 km^2) can vary markedly (Grotch and MacCracken, 1990). One problem is that coarse-grid meshes used by GCMs to represent global-climate processes (Fig. 29.4) preclude accurate representation of regional or local climatic and hydrologic variables such as temperature, precipitation, evapotranspiration, soil moisture, and runoff (Manabe and Stouffer, 1980; Gleick, 1987; Lamb, 1987; Grotch and Mac-Cracken, 1990). Large errors exist in GCM estimates of current climate for many regions of the world (Grotch, 1988). In addition, Miller and Russell (1992) showed, by routing GCM grid-node runoff predictions through a channel network for the largest of the world's largest rivers (such as the Amazon, Yangtze, and Mississippi), that large errors resulted even in the model-based estimates of the mean annual flow for the current climate. Therefore, direct use of GCM predictions for areas the size of most watersheds is not recommended.

29.2.3.2 Ratio and Difference Methods. A popular use of GCMs in water-resources effects assessments is to compute the change in climate from current to doubled-CO_2 conditions. This approach makes use of a pair of GCM simulations, one of which is the model's best representation of current climate, and the other in which only the atmospheric CO_2 concentration has changed. The difference between the two runs, expressed, for instance, as a ratio or difference of the temporal means of the doubled-CO_2 run and the current (control) run, is used to adjust an observed series of climate variables. The motivation for this approach is that although GCMs may not accurately estimate the local statistics of regional climate variables, their internal consistency and strong physical basis may provide plausible estimates of relative changes in climatic variables (Gates, 1985). The advantages of this approach are that it is simple and involves direct use of observed records, thus avoiding the introduction of additional model error when applying GCM results to the local level

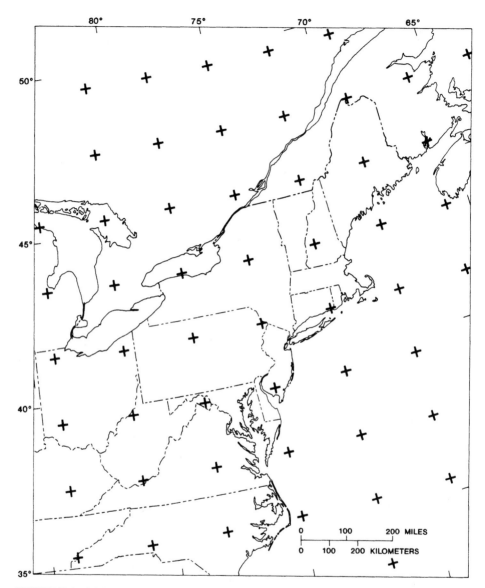

FIGURE 29.4 Grid points for the northeastern United States from the Geophysical Fluid Dynamics Laboratory (GFDL) general circulation model (GCM). These grid points are from the R30 version of the GFDL GCM and the grid-point spacing is approximately 2.25° of latitude by 3.75° of longitude (*Ayers, et al., 1993*).

(for instance, if multiple precipitation stations are used in a hydrological simulation, the temporal and spatial correlations of the historical observations are preserved). The disadvantage is that the historical temporal sequencing of events is retained. For instance, if, in addition to changes in precipitation amounts, the timing of storms (storm inter-arrival times) changes with increases in atmospheric concentrations of

CO_2, this will not be reflected in the simulated amounts. Also, there is the problem of determining the appropriate temporal and spatial averaging for the GCM results. If the GCM simulation period is short (many early climate assessments were based on GCM simulations of much less than 10 years length), the estimate of the temporal mean will be subject to considerable error, hence the estimate of the ratio or difference between a CO_2-doubled and control run may be poor. This problem has been alleviated as faster supercomputers have made GCM runs of as long as several hundred years possible.

In the ratio and difference methods, the appropriate number of GCM grid cells to average must be selected as well. There is a temptation to use either the nearest GCM grid node, or even to interpolate spatially to the location of interest. However, most GCMs use a *spectral representation* (meaning that the GCMs solve for discrete spatial Fourier transforms of the predicted variables, which are in turn inverse transformed to the spatial domain). This means that when the predicted variables are represented on a spatial grid mesh the fundamental (*Nyquist*) frequency, which is the highest resolution resolved by the models, is represented by two grid nodes in each dimension. Therefore, von Storch, et al. (1993) argue that GCMs provide no information at spatial resolutions higher than twice the distance between grid cells, and that the minimum effective spatial resolution is defined by four or more grid nodes (at least two in each direction). This implies that rather than interpolating, a spatial average of at least the four nearest grid nodes should be taken as the basis for computing the ratio or difference of altered- and base-climate simulations.

29.2.3.3 Scenarios Based on GCM Atmospheric Circulation Patterns.

Recent research suggests that GCMs may be better able to simulate synoptic-scale weather patterns than local surface conditions (Hay, et al., 1992; Hewitson and Crane, 1992; McCabe and Legates, 1992). Methods have been developed to use weather patterns, such as frontal passages and high-pressure systems, to apply GCM estimates of future climatic conditions to small areas and thus overcome the spatial incompatibility between GCMs and watersheds (Hay, et al., 1992; Hewitson and Crane, 1992). Variations in weather and climate can be represented by changes in atmospheric circulation. Types of atmospheric circulation patterns can be grouped into classes of similar characteristics (*synoptic weather patterns*), thus stratifying the mean climatic conditions of a region into classes of weather patterns that determine the climate. The frequencies and characteristics of the different classes of weather patterns relate to changes in climatic conditions such as temperature, precipitation, and flood and drought frequency. In addition, the spatial size of weather patterns (typically about 10 degrees of latitude by 10 degrees of longitude) is large enough to be reasonably predicted by GCMs, and knowledge of weather-pattern frequencies and characteristics in and around a watershed may provide adequate information for the prediction of regional climatic variables (Fig. 29.5). In addition to using synoptic weather patterns, another avenue of scenario development is the use of *atmospheric pressure anomalies* to develop links between atmospheric circulation and surface climatic variations. Atmospheric pressure anomalies are especially useful for analyses of drought and flood frequencies (Namias, 1983; Hirschboeck, 1987; Knox and Lawford, 1990).

Using the weather-pattern approach, climate-change scenarios are developed by determining empirical relations between atmospheric circulation patterns and temperature and precipitation from the historical record, and then determining the change in weather-pattern or weather-anomaly frequencies predicted by GCMs for future climatic conditions. The predicted frequencies in weather patterns or weather anomalies are used in conjunction with the empirical relations between the weather patterns and temperature and precipitation to generate regional precipitation and

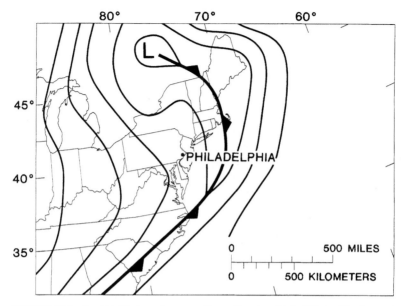

FIGURE 29.5   A mid-latitude cyclone over the northeastern United States. Weather patterns, such as mid-latitude cyclones and high-pressure systems, are large enough to be reasonably simulated by general circulation models.

temperature estimates for future climatic conditions. Use of empirical relationships between atmospheric circulation and surface climate variables, such as temperature and precipitation, may provide a realistic mechanism for describing the regional effects of climate change. The primary drawback of this approach are the assumptions that: (1) relations between atmospheric circulation and surface-climate variables, such as temperature and precipitation, are properly simulated by GCMs, (2) these relations will not change appreciably in a climate-changed world, and (3) changes in surface climate in a climate-changed world primarily will be the result of changes in atmospheric circulation.

29.2.3.4 Stochastic Downscaling Models.   Stochastic models have traditionally been used to solve practical problems of hydrologic data simulation (Lettenmaier, 1993), but also have been found useful for the development of climate time series for use in climatic-change studies. Except for the direct use of GCM output and the direct use of historical and spatial analogues, climate-change scenarios generally only provide changes in mean climate conditions. Changes in mean climate conditions need to be evaluated with respect to natural climate variability. Stochastic equations can be used to include natural climate variability in climate-change scenarios. Stochastic equations include parameters that describe mean climate conditions as well as natural climate variability. Climate-change scenarios can be used to alter the parameters of stochastic equations. The stochastic equations then can be used to generate multiple time series of climate variables for specified scenarios. Stochastic equations can be used to study the effects of steady-state climatic changes, as well as long-term transient changes in climate on water resources (McCabe and Wolock, 1992).

In general, stochastic models of climate variables (i.e. temperature and precipitation) have been statistical descriptions of observed climate data that do not incorporate any additional meteorological information (Foufoula-Georgiou, 1985; Kavvas and Delleur, 1981; Smith and Karr, 1985; Woolhiser and Roldan, 1986). More recently, with interest in developing realistic climate-change scenarios, investigators have explored the development of stochastic models in the context of large-scale atmospheric circulation. Climatologists historically have viewed precipitation and temperature as just two components of atmospheric conditions, and have extensively studied the relation between surface-climate variables such as temperature and precipitation and regional weather patterns (Muller and Wax, 1977; Barry, et al., 1981; Yarnal, 1985; Faiers, 1988; Yarnal and Leathers, 1988). Regional weather patterns, or synoptic weather types, are aggregate representations of a number of meteorological variables. In an effort to bridge the gap between the hydrologist's and the climatologist's view of surface-climate processes, stochastic models founded in climatology have been developed. These models are developed with the objective of generating realistic time series of temperature and precipitation by using statistical relations between atmospheric circulation patterns and temperature and precipitation (Hay, et al., 1991; Bardossy and Plate, 1991; 1992; Wilson, et al., 1991; 1992).

To develop such models, the daily weather patterns affecting a region need to be classified into a set of weather patterns that represent the range of weather conditions affecting an area. Weather-pattern classification procedures range from subjective manual techniques to highly statistical techniques that use combinations of principal components and cluster analyses (Muller, 1977; Kalkstein, et al., 1987; McCabe, 1990; Yarnal, 1993). Once the daily weather patterns are classified into discrete groups, historical statistics that describe the probability of occurrence of specific weather patterns, the sojourn times of each weather pattern, and the transition probabilities from one weather pattern to another are used to develop a stochastic model that simulates temporal sequences of the weather patterns. In addition, for each weather pattern, historical statistical probabilities of temperature and precipitation are used to simulate time series of temperature and precipitation (Hay, et al., 1991; Wilson, et al., 1991; 1992). Such models have been found to reliably simulate the statistics of daily precipitation and temperature.

The use of *stochastic downscaling methods*, such as those described in this section, to produce scenarios for use in hydrologic and water-resources effects studies has two major advantages in comparison with ratio and difference methods. First, GCMs have been found to simulate atmospheric circulation more reliably than they simulate surface temperature and precipitation (Hewitson and Crane, 1992; McCabe and Legates, 1992). Thus, methods of simulating local surface variables that are conditioned on GCM simulations of large-scale features are arguably drawing from the strength of GCMs. Second, runoff production is highly sensitive to the timing of precipitation, which is modeled explicitly with stochastic methods, and therefore can infer changes in the timing of precipitation that might accompany climate change. The ratio method, on the other hand, assumes that the timing of precipitation will be identical to that observed in the historical record.

29.2.3.5 Nested Models.   Another use of GCMs for regional climatic-change scenario development is the use of *nested models*. Several recent studies have used output from GCMs to initialize mesoscale or regional-climate models (Giorgi, 1990; Giorgi and Mearns, 1991). The mesoscale models provide better estimates of regional climate for current conditions than GCMs, and are being evaluated for use in climatic-change studies. Studies are being performed that use coarse-gridded GCM output as input to a mesoscale model, and subsequently the output from the mesoscale model is used as input to regional precipitation and temperature models to generate more accurate estimates of climatic conditions for small areas such as watersheds. An example of the use of the nested model approach is a study of the Gunnison River, Colorado, by the U.S. Geological Survey. In 1988, the U.S. Geological Survey began developing a set of nested models to evaluate the effects of climate change on the water resources of the Gunnison River basin in southwestern Colorado (Hay, et al, 1992). The nested model approach (Fig. 29.6) was chosen for the development of climate-change scenarios for this study because accurate estimates of spatial and temporal distributions of winter precipitation in the complex mountainous terrain of the Gunnison River basin were required (Hay, et al., 1992). The simulation of future climatic conditions in the Gunnison River basin begins with simulations from the *National Center for Atmospheric Research (NCAR) Community Climate Model* (version CCM1), a full scale GCM (Giorgi, 1990). The output from CCM1 is used to determine boundary conditions for a mesoscale atmospheric model. The *mesoscale atmospheric model* used in the Gunnison River basin study is an augmented version of the Pennsylvania State University/NCAR mesoscale model MM4 (Anthes, et al., 1987) with grid resolution 60 km (Hostetler and Giorgi, 1993). Output from MM4 provides information describing the state of the atmosphere which is used as input to the Rhea-Colorado State University (RHEA-CSU) orographic precipitation model (Rhea, 1977). The RHEA-CSU model simulates the interaction of air layers with the underlying topography by allowing vertical displacement of the air column while keeping track of the resulting condensate or evaporation. The RHEA-CSU orographic precipitation model provides estimates of daily precipitation for 2.5-by-2.5-km grid cells covering the Gunnison River basin. Simulations from the RHEA-CSU model are subsequently used as input to a precipitation–runoff model (Hay, et al., 1992; Leavesley, et al., 1992; Parker, et al., 1992).

The primary benefit of these approaches is that they provide relatively accurate, physically based estimates of climatic conditions. The drawback is that they require large amounts of computer time and generate a limited number of scenarios. An additional drawback is that the boundary conditions of the mesoscale atmospheric model depend on simulations from a low-resolution GCM, therefore the high-resolution simulations from the mesoscale model may be contaminated by errors in boundary conditions simulated by the GCM (Robock, et al., 1993).

29.2.4 Prescribed Climatic Change Scenarios

Prescribed climatic changes provide a suite of scenarios that cover ranges of changes in climatic variables (e.g., temperature, precipitation, and evapotranspiration). Prescribed changes can be applied uniformly throughout the year, or variable changes can be made on a monthly or seasonal basis. Appropriate ranges of changes for each climatic variable can be found from a search of current climatic change literature, or can be directed by predictions from GCMs. Generally, increases in temperature are used to reflect the effects of global warming. Because of the large uncertainty regarding the direction and magnitude of changes in precipitation, both increases and decreases in precipitation are employed. Changes in precipitation can be simple changes in monthly or annual precipitation totals, or can include changes in storm duration, storm intensity, and interstorm period (Wolock and Hornberger, 1991) (Table 29.1). In addition to changes in temperature and precipitation, changes in other variables often are used. For example, because of evidence that increasing levels of atmospheric CO_2 may directly affect plants and transpiration rates, prescribed scenarios often include changes in stomatal resistance to transpiration (Rosenberg, et al., 1990; Wolock and Hornberger, 1991).

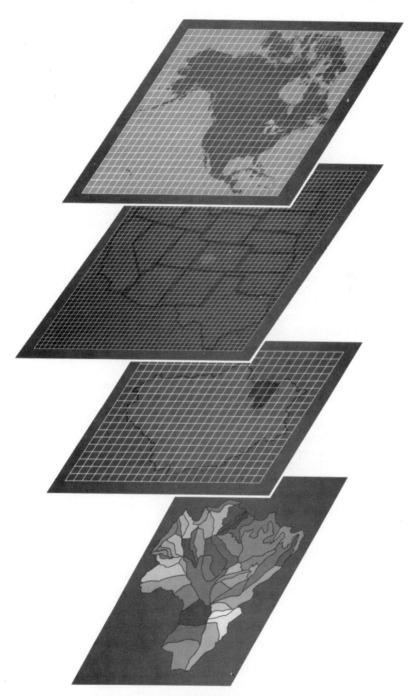

FIGURE 29.6   Nesting of a coarse-gridded general circulation model, a mesoscale general circulation model, and a regional precipitation model, and a catchment-scale hydrologic model to generate high-resolution spatial precipitation and runoff estimates for the Gunnison River, Colorado (*Leavesley, et al., 1992.*)

TABLE 29.1 Range of Prescribed Hypothetical Changes in Temperature and Precipitation Tested in Selected Climate-Change Studies.

Study	ΔT^*	$\Delta P^\dagger$
Nemec and Schaake (1982)	+1°C; +3°C	+10%; +25%
Gleick (1987)	+2°C; +4°C	0%; +10%; +20%
McCabe and Ayers (1989)	0°C; +2°C; +4°C	0%; +10%; +20%

* Change in mean annual temperature.
† Change in mean annual precipitation.

Prescribed hypothetical climatic changes can be used to evaluate the effects of steady-state changes in climate, as well as the effects of long-term transient changes. Prescribed climatic changes also can be used to evaluate the effects of changes in mean climatic conditions, or can be used with stochastic equations to evaluate the effects of prescribed climatic changes amid natural climatic variability. This is an important aspect of climatic change because for a given expected future climate, natural variability creates a wide range of climatic conditions that may actually occur. For example, the normalized frequency distributions of the percentage change in the 7-day low flow in the Delaware River basin shown in Fig. 29.10 illustrates how the range in climatic conditions that may be realized due to natural variability can mask the effects of long-term changes in climate (McCabe and Wolock, 1991; 1992).

The primary benefit of prescribed climatic-change scenarios is that they provide a range of climatic conditions and permit detailed sensitivity analyses to be performed. In some cases, it may be possible to use a carefully chosen range of prescribed changes to form a response surface, which may be amenable to economic analysis. This method sidesteps the scale incompatibility problems associated with using low-resolution climate estimates from GCMs, and is arguably a reasonable way to proceed, given the present uncertainties in GCMs. A drawback is that prescribed scenarios are not necessarily physically consistent, since they fail to account in any way for the dynamics of the land-ocean-atmosphere system. Although prescribed scenarios are hypothetical, they have been widely used in sensitivity analyses and have provided important insights as to the effects of climatic change on water resources (Gleick, 1989; Lettenmaier, et al., 1988; McCabe and Ayers, 1989; Rosenberg, et al., 1990). In general, prescribed scenarios permit a large number of sensitivity analyses to be performed in a short period of time, at very little cost.

29.2.5 Combined Methods

Perhaps the most useful scenarios of future climatic conditions are those developed by using a combination of methods (Fig. 29.7). For example, although GCM estimates of future climatic conditions for regional areas may be uncertain, these estimates can be useful to guide the development of seasonal and annual prescribed scenarios. Information from GCMs also is useful to describe changes in atmospheric circulation and major climate zones for future conditions (Mather and Feddema, 1986). This information can be used to direct the development of analogue or prescribed scenarios. Historical or proxy data also can be used to guide the development of prescribed scenarios (Williams, 1980; Jager and Kellogg, 1983). Figure 29.7 presents three methods of climate scenario development using combinations of methods. The first example (Fig. 29.7a) illustrates the use of estimates from general circulation models to guide the selection of prescribed hypothetical scenarios (Gleick, 1987; McCabe and Ayers, 1989; McCabe and Wolock, 1991; 1992). The second example (Fig. 29.7*b*) shows the use of observed data to develop empirical relations between surface climate and atmospheric circulation, and using these relations in conjunction with general circulation model estimates of future atmospheric circulation to develop a climate-change scenario (Hay, et al., 1991; Bardossy and Platte, 1991; 1992; Hewitson and Crane, 1992; Wilson, et al., 1991; 1992). The third example (Fig. 29.7*c*) illustrates the use of general circulation model estimates to select historical and/or spatial analogues for use as climate-change scenarios (Williams, 1980; Jager and Kellogg, 1983; Smith and Tirpak, 1989b).

29.2.6 Evaluation of Scenario Development Approaches

Each of the methods discussed in this section have some strengths and weaknesses. The method chosen for a particular analysis should depend on the temporal and spatial scales of interest, and the type of climatic information needed for the analysis. Figure 29.8 shows the conceptual ranges of temporal and spatial scales of temperature and precipitation estimates for various methods of climatic-change scenario development.

Several of the methods have a wide temporal range, but a limited spatial range. Estimates from GCMs are limited to both large temporal and spatial scales. Proxy analogues are most useful for local and regional spatial scales and for temporal

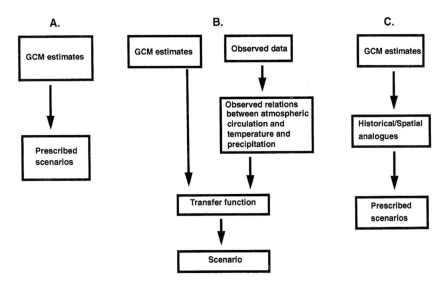

FIGURE 29.7   Examples of combinations of methods to develop climate-change scenarios: (*a*) illustration of the use of estimates from general circulation models to guide the selection of prescribed hypothetical scenarios; (*b*) illustration of use of observed data to develop empirical relationships between surface climate and atmospheric circulation, and combination of these relationships with general circulation model estimates of future atmospheric circulation to develop a climate-change scenario; and (*c*) illustration of use of general circulation model estimates to select historical and/or spatial analogue climate-change scenarios.

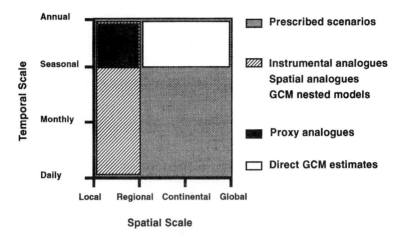

FIGURE 29.8   Conceptual ranges of temporal and spatial scales of temperature and precipitation estimates for various methods of climate-change scenario development.

scales longer than a season. Instrumental analogues, spatial analogues, and estimates from GCM nested models are useful for a wide range of temporal scales and for local and regional spatial scales. Only prescribed scenarios are applicable to all temporal and spatial scales, but as indicated in Sec. 29.2.4, their use implies greater accuracy of GCM precipitation and temperature predictions than appears justifiable. Climate variables other than temperature and precipitation may be associated with different sets of temporal and spatial scales for each of the methods. For example, GCM estimates of temperature and precipitation have limited usefulness for temporal scales smaller than a season; however, GCM simulations of atmospheric pressure may be reliable for time steps as short as one day (Hewitson and Crane, 1992).

The method of scenario development employed for particular sensitivity analysis is dependent on the requirements of the specific analysis, such as the temporal scale on which the analysis is performed (i.e., daily, monthly, annually), the spatial scale of the area of interest, the climatic information needed (e.g., simple descriptive statistics, magnitude and frequency of climatic anomalies, or time series of climatic variables), and the type of analysis (e.g., broad sensitivity analysis or more focused predictions). Eventually, we expect that the best option will be the use of fully coupled land-atmosphere-ocean GCMs, perhaps using nested strategies. However, the accuracy of predictions by both GCMs and mesoscale of the surface processes required to drive atmospheric models is not currently at the point where this strategy is generally viable. For this reason, stochastic downscaling methods, such as those described in Sec. 29.2.3, appear to offer the most defensible method of climate scenario development for water-resources studies in the near term.

29.3 USE OF HYDROLOGICAL MODELS FOR CLIMATE ASSESSMENT STUDIES

Generally, at least two modeling steps are required to interpret how a water-resources system might perform if the climate were to change. First, the hydrologic conditions that would accompany altered land-surface climatic conditions (characterized by a climate scenario, as discussed in Sec. 29.2) must be determined. Where historical analogue scenarios are used, it may be possible in some situations to base climate-change effects interpretations on historical sequences of streamflow corresponding, for instance, to abnormally warm periods. However, in most locales historical records do not include lengthy sequences with temperature anomalies as large as those projected to occur, for instance, under CO_2 doubling. Likewise, although GCMs simulate large-scale runoff, direct predictions of streamflow from GCMs are not presently feasible, due to differences in the spatial scale of the GCMs and the river basins for which effects are to be interpreted, because of the highly simplified representation of land-surface hydrology in GCMs, and due to the large errors in GCM precipitation. This means that in most cases modeling will be required to produce hydrological scenarios (e.g., streamflow) corresponding to scenarios of surface atmospheric variables.

Even in those unusual cases where modeling is not required to produce a hydrological scenario, a second modeling step, to determine water-resources effects, will be needed. Water demand is continually changing, as are the configurations and operating policies of water-resource systems. Therefore, interpreting how a water-resources system might perform given the hydrologic conditions that would accompany an altered-climate scenario will virtually always require modeling of the water-resources system. Section 29.3.1 discusses hydrological modeling issues for climate-effects assessments; water-resource-systems modeling aspects are described in Sec. 29.3.2.

29.3.1 Hydrological-Process Modeling

The selection of an appropriate hydrological-modeling strategy for climate-effects assessment represents a tradeoff between model complexity and data availability. In addition, if the climate scenario is based on a GCM simulation, the hydrological model also should account for the need to downscale from a GCM grid cell to a more hydrologically meaningful spatial scale.

The purpose of a hydrological model, in the context of climate-change effects studies, is to predict streamflow, given precipitation and other surface meteorological conditions. Because streamflow is an areally integrating variable, and climate scenarios are based on large-scale conditions, it would seem logical that the hydrological model inputs would be spatial averages of the surface meteorological variables. In practice, however, hydrological models must be calibrated and verified using historical meteorological observations, which at present imply point measurements. Eventually this may not be the case, in particular as lengthy observational records of radar precipitation become available, but at present operational hydrological models are driven by point observations, and climate scenarios seek to represent the same points (e.g., station locations) under alternative climate conditions. The state of the art of *time-continuous precipitation–runoff modeling,* using point meteorological inputs, is well developed (see, for instance, Francini and Pacciani, 1991, for a review). The accepted approach is to partition concurrent historical records of precipitation and other surface meteorological inputs into calibration and verification periods. Parameters are estimated using the calibration period, and model performance is validated using the independent verification period. Calibration is typically performed by minimizing some metric of the difference between simulated and observed streamflows. This has historically been accomplished using trial-and-error procedures, although recently powerful automated methods, such as the *shuffled complex evolution method* of Duan, et al. (1992) have evolved. All of these procedures are applicable to the first step in climate-assessment studies, which is to implement a hydrologic model for the current climate (base case) using historical observed data.

In the last few years, the state of the art in precipitation–runoff modeling has evolved beyond the spatially lumped, conceptual models such as those reviewed by Francini and Pacciani (1991) to account explicitly for spatial variations in surface conditions such as terrain, soils, and vegetation within a watershed have seen an explosion in distributed hydrological models. The popularity of these models can be traced to the ready availability of digital terrain data, and the increased computational power of desktop workstations which can retain in memory images of land-surface characteristics in addition to elevation, such as other geometrical attributes (slope, aspect) and soil and vegetation characteristics. Nonetheless, accurate spatial representation of precipitation, usually the most important boundary condition, is generally lacking. This situation may be resolved in some areas as radar precipitation products become more available, but for the present, spatial variations in precipitation usually must be interpolated from point data. Therefore, there may be little to distinguish spatially distributed and spatially lumped models in the context of climate-effects assessments. For instance, Lettenmaier, et al. (1993) report on an assessment of the effects of climate change in the American River, a Cascade Mountains catchment using *Topmodel* (Beven and Kirkby, 1979), a quasi-distributed model that maps runoff response characteristics determined using a digital elevation model on the basis of a topographic index. However, although Topmodel captures the spatial distribution of runoff production, the input (precipitation or snowmelt) is spatial average precipitation, which is represented by a single point, or station.

Arguably, from the standpoint of climate-effects interpretation, the knowledge of the spatial distribution of runoff production is only important if it results in better streamflow predictions. In the case of the Lettenmaier, et al. study of the American River, the streamflow simulations produced using a spatially lumped model (the Sacramento soil-moisture accounting model of Burnash, et al., 1973) were at least as good as those produced using Topmodel.

From a practical standpoint, what appears to be more important than the spatial scale of a hydrological model (assuming that the sole goal is accurate simulation of runoff) is the time step. Even although water-resources systems may respond to relatively long time steps, particularly in cases when the reservoir storage is large compared to the mean flow of a river, it is important that the hydrological model operate at a time step short enough to capture the stochastic structure of storm events (wet and dry periods). Generally, this means a time step no longer than one day, and quite possibly shorter. As a rough rule, one might argue that the modeling time step should be no longer than several times the time of concentration of the catchment. It is usually the case that the streamflow simulations produced by a hydrological model will have to be aggregated to provide the input to a water-resources-systems model. For instance, although Lettenmaier, et al. used a daily time step for their hydrological simulations, the simulated streamflow was aggregated to a weekly time step for reservoir simulation.

29.3.2 Spatial Extrapolation of Hydrological Simulations

Hydrological simulation approaches that are based on point meteorological data are effectively limited to catchments of modest size (e.g,, a few hundred to a few thousand km^2) unless the model is extended by application to a large number of subbasins. Although such an exhaustive approach has proved successful, and is used operationally for purposes such as water-supply forecasting in large river basins such as the Colorado and Columbia, it results in a massive amount of data handling and model calibration. The effort required may well be inconsistent with the exploratory nature of climate-effects studies. An alternative approach is to simulate relatively small index catchments, and then to extend the index-catchment results using stochastic modeling methods. This approach was followed by Lettenmaier and Gan (1990) and Lettenmaier and Sheer (1991) in a study of climate sensitivity of the California State Water Project. Four headwater catchments of size 500 to 1000 km² were modeled with a daily time step, using the Sacramento soil-moisture accounting model, and the temperature index snow accumulation and ablation model of Anderson (1973). The daily simulations were aggregated to monthly flows. A simple stochastic disaggregation model which explicitly preserved only the lag-zero correlations of the monthly flow logarithms was used.

Miller and Brock (1989) utilized a more sophisticated stochastic disaggregation model based on the work of Grygier and Stedinger (1988) to simulate streamflows at 42 reservoir inflow nodes of a model of the Tennessee Valley Authority (TVA) reservoir system as part of a study of the climate sensitivity of the TVA system. The TVA study conceptual design was similar to the California State Water Project climate sensitivity analysis of Lettenmaier and Sheer (1991); specifically, a conceptual streamflow-simulation model was used to simulate daily streamflow at six index catchments given sequences of daily precipitation and temperature, and the daily index-site flows were aggregated to a weekly time step.

Application of the disaggregation procedure for climate-effects assessments requires a few decisions not faced by the analyst in more conventional applications. First, the covariances from which the parameter matrices are determined can be estimated either from observed historic flows at the index sites or from flows simulated using the conceptual precipitation–runoff simulation model applied to historical meteorological observations. The recommended approach is to use the simulated index-site flows, because the disaggregation procedure can then compensate for some of the model error incurred in the precipitation–runoff transformation. The choice might be different if the disaggregation model parameters could be re-estimated for the altered-climate scenario. However, although a set of index-site flows will be produced for the altered climate, the node flows are of course unknown (otherwise the entire disaggregation procedure would be unnecessary). This necessitates use of the same coefficient matrices for both present and altered climate, which in turn introduces another problem.

The disaggregation procedure is a form of regression. Regression procedures have a property sometimes termed *regression to the mean*, which essentially means that anomalies in the independent variable (in this case, the index-site flows) are reduced in magnitude when viewed as anomalies in the dependent variable (node flows). For altered-climate simulations, this means that if a set of index-site flows (simulated, for instance, using a precipitation–runoff model with meteorological forcings corresponding to a warmer climate) are used to produce scenarios of node flows that will be used to drive a reservoir-system simulation model, the disaggregation procedure will tend to attenuate the climate "signal" as seen in the resulting simulated node flows. Sias and Lettenmaier (1994) circumvented this problem by computing the means of the base- and altered-climate flows at each site and time (week), then adjusted the altered-climate index-site simulated flows to have the same weekly means as the historical climate. The adjusted flows were then disaggregated, and the simulated node flows were adjusted by the means ratio at the associated index site.

This procedure implicitly assumes that climate change would affect only the mean, and not the higher moment of the flows. The procedure could be generalized by scaling both means and variances. The necessity for such an adjustment procedure does point out a critical deficiency in index-site simulation strategies, and suggests why direct-simulation approaches, currently under development (see, for instance, Nijssen, et al., 1995; Lettenmaier, et al, 1996a; 1996b) may be preferred. The *index site method,* coupled with stochastic disaggregation, is probably appropriate for preliminary studies, and has the advantage of simplicity and reliance on well-established methods.

29.4 RESERVOIR MODELING FOR CLIMATE-CHANGE ASSESSMENT

Most highly developed water-resources systems exploit some combination of surface and subsurface storage. This storage serves to buffer both intra- and interannual variations in streamflow. One of the major issues to be addressed in climate-sensitivity studies of water-resources systems is how well existing water-resource systems could perform in buffering the hydrologic variability that would be associated with altered climate. It is not surprising, then, that most of the studies of water-resources effects of climate change reviewed in the IPCC Second Assessment Report (Kaczmarek, et al., 1995) involved some aspect of water-resource-system simulation and/or performance assessment. In fact, one of the questions that should be asked in climate-effects interpretations is the extent to which changes in the system operating policy can compensate for climate change. Reservoir simulation and optimization are addressed in detail in texts such as Loucks, et al. (1981), so this section addresses only those aspects peculiar to climate-change assessments.

29.4.1 Choice of Operating Rules

The methods for simulating streamflows under altered-climate scenarios outlined in Secs. 29.2 and 29.3 can result in as many as four classes of reservoir inflows:

1. Observed historical flows for some base period
2. Simulated streamflows based on observed meteorological sequences for a period that may be different from the base period
3. Simulated streamflows corresponding to a GCM current climate condition
4. Simulated streamflows corresponding to one or more alternative climates

In many cases, a simulation model exists for purposes of system operation and/or planning, and is based on the historical flows. For the base period, some process has usually been employed to determine operating rules that are appropriate for the base period. For instance, an analysis may have been conducted to determine the critical period during the base period, given some particular set of demands and reservoir system configuration (e.g., the current physical system, which may not have been in place during the entire base period). For reservoir inflow classes 2 and 3, the operating rules based on the observed historical streamflows may result in inferior system performance. The question then arises as to whether the historical operating rules are appropriate for streamflow sequences other than those from which they were derived. In some cases, sequence-independent operating rules are used, in which case the problem is avoided. For instance, Lettenmaier, et al. (1993) evaluated the performance of a simple hypothetical reservoir using a sequence-independent "fill and spill" operating rule for a base case and two alternative-climate scenarios. On the other hand, Sheer and Randall (1989) evaluated the performance of the combined California State Water Project and Central Valley Project for a base run and several alternative-climate scenarios, each of length 57 years. In all cases, the operating rules were tailored to the particular streamflow sequence.

29.4.2   Optimization versus Simulation

Reservoir-system performance can be characterized either by optimization or simulation models. Simulation models are widely used for both planning and system operation. Their advantages are that they are relatively easy to develop, they are not computationally intensive (although this advantage has become less important with the advent of high-performance microprocessors), and they can be made easily understandable. Optimization models have the advantage of providing optimal system performance, subject to the model assumptions and with respect to the objective function used. Their disadvantages are: (1) they can be computationally intensive, especially for multiple-reservoir systems, (2) it may be difficult to identify the "true" objective function, hence they may not be optimal in a practical sense, and (3) it may be difficult for water-resources managers to understand (and hence accept) the basis for the algorithm's decisions.

From the standpoint of climate-change assessment, one can argue that it does not make much difference which approach is used, so long as it is applied consistently to the base case and altered-climate scenarios. Most assessments of water-resource-system sensitivity to climate change (e.g., Lettenmaier and Sheer, 1991; Nash and Gleick, 1991; Croley, 1990) have used simulation models, probably because of their simplicity. Lettenmaier, et al. (1993) used both simulation and optimization models for a hypothetical single water-supply reservoir. They explored the question of whether optimal operation could mitigate the loss of water-supply reliability resulting from a seasonal shift in the snowmelt-dominated hydrograph of the American River, Washington, that would accompany global warming. They found that the optimal-operation algorithm in some cases was able to improve the objective function (which was taken to be total hydropower revenues). However, system water-supply reliability, which was treated as a constraint, was more often violated with optimal operation than with the simple simulation algorithm. In general, the ability of the system to meet water-supply demand seemed to be strongly controlled by the size of the reservoir system, and much less so by the system operating policy.

29.4.3   Quantitative Measures of Climate-Change Effects

Various means can be used to quantify the effect of climate change on water-resource-system performance. Depending on the purpose of the system, a variety of performance indices may be defined. In the case of water-supply systems, these might include the magnitude, number, and length of water-supply deficits. For flood-control systems, flood damage is the obvious performance index. For navigation performance, the length of periods with channel depth below a critical value might be used. For hydropower systems constituting a small part of a combined hydrothermal system, the total seasonal or annual hydropower avoidance cost (cost that would have been incurred in the absence of hydropower generation) could be used. For base-loaded hydropower systems (such as the Columbia River system), performance measures similar to those used for water supply, such as the length, magnitude, and duration of deficits from "firm power" might be appropriate. Since most systems are operated for multiple purposes, more than one performance measure will often be needed.

A convenient method of comparing system performance for alternative climate scenarios is the empirical probability distribution of the annual values of the performance measure. These are plotted in the same manner as an empirical probability distribution of an annual flood series. If the sequence $\{Y_j, j = 1, \ldots N\}$ are the values of a performance index for N years, ranked from smallest to largest, the empirical probability distribution is formed by plotting Y_j against P_j, the empirical cumulative probability. For screening purposes, the Weibull plotting position formula

$$P_j = \frac{j}{N+1}$$

is usually adequate. The scale for P_j can be distorted for better visual discrimination. A normal probability scale works well for this purpose even if the data are not normal. Figure 29.9 shows the empirical probability distribution of California State Water Project deliveries for a base climate and several alternative GCM CO_2-doubled climates plotted in this manner. Figure 29.10 shows changes in the 7-day low flow in the Delaware River basin, summarized as frequency distributions for simulations with no climate change and with a gradual 3°C warming. Fifty independent 60-year simulations were used to estimate the frequency distributions. Because of natural climatic variability the range of changes in 7-day low flow for the two climatic conditions cause the distributions of changes in 7-day low flow for the two climate scenarios to overlap.

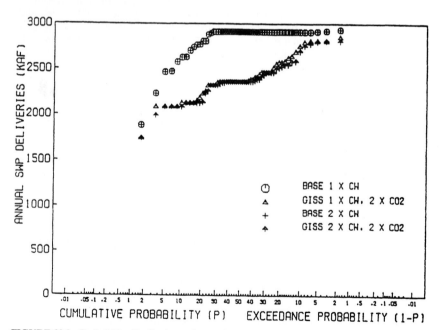

FIGURE 29.9 Probability distributions of annual California State Water Project water deliveries under present climate and three CO_2-doubled climate scenarios (*Lettenmaier and Sheer, 1991*).

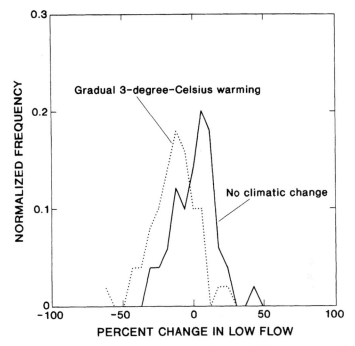

FIGURE 29.10 Normalized frequency distribution of the percentage of change in the 7-day low flow in the Delaware River basin for simulations with no climate change and with a gradual 3°C warming, both with current precipitation amounts and variability. The distributions were developed from fifty 60-year simulations. (*Ayers, et al., 1993.*)

REFERENCES

Anderson, E. A., "National Weather Service River Forecast System—Snow Accumulation and Ablation Model," NOAA Technical Memorandum NWS-HYDRO-17, 1973.

Anthes, R. A., E. Y. Hsie, and Y. H. Kuo, "Description of the Penn State/NCAR Mesoscale Model Version 4 (MM4)," NCAR Technical Note, NCAR/TN-282+STR, 1987.

Arnell, N. W., and N. S. Reynard, "Impact of Climate Change on River Flow Regimes in the United Kingdom," in review, *Water Resources Research,* 1995.

Ayers, M. A., D. M. Wolock, G. J. McCabe, L. E. Hay, and G. D. Tasker, "Sensitivity of Water Resources in the Delaware River Basin to Climate Variability and Change," U.S. Geological Survey Open-File Report 92-52, 1993, 68 pp.

Barry, R., G. Kiladis, and R. S. Bradley, "Synoptic Climatology of the Western United States in Relation to Climatic Fluctuations During the Twentieth Century," *Journal of Climatology,* 1:97–113, 1981.

Bardossy, A., and E. J. Plate, "Modeling Daily Rainfall Using a Semi-Markov Representation of Circulation Pattern Occurrence," *Journal of Hydrology,* 122:33–47, 1991.

Bardossy, A., and E. J. Plate, "Space-Time Model for Daily Rainfall Using Atmospheric Circulation Patterns," *Water Resources Research* 28:1247–1259, 1992.

Beven, K. J., and M. J., Kirkby, "A Physically Based, Variable Contributing Area Model of Basin Hydrology," *Hydrological Sciences Bulletin,* 24:43–69, 1979.

Budyko, M. I., "The Analogue Method of Estimating Future Climate Changes," *Meteorology and Hydrology,* 39–50, 1991.

Burnash, R. J. C., R. L. Ferral, and R. A. McGuire, "A Generalized Streamflow Simulation System, Conceptual Modeling for Digital Computers," U.S. National Weather Service, Sacramento, Calif., 1973.

Charlson, R. J., S. E. Schwartz, J. M. Hales, R. D. Cess, J. A. Oakley, J. E. Hansen, and D. J. Hofmann, "Climate Forcing by Anthropogenic Aerosols," *Science,* 255(5043):423–430, 1992.

Croley, T. E., "Laurentian Great Lakes Double CO_2 Climate Change Hydrological Impacts," *Climatic Change,* 17:27–47, 1990.

Duan, Q., S. Sorooshian, and V. Gupta, "Effective and Efficient Global Optimization for Conceptual Rainfall-Runoff Models," *Water Resources Research,* 28(4):1015–1031, 1992.

Ericson, D. B., and G. Wollin, "Pleistocene Climates and Chronology in Deep-Sea Sediments," *Science,* 157: 1233, 1968.

Faiers, G., "A Synoptic Weather Type Analysis of January Hourly Precipitation at Lake Charles, Louisiana," *Professional Geographer,* 9:223–230, 1988.

Foufoula-Georgiou, E., "Discrete-Time Point Process Models for Daily Rainfall," Water Resources Technical Report no. 93, University of Washington, Seattle, Wash., 1985.

Francini, M., and M. Pacciani, "Comparative Analysis of Several Conceptual Rainfall-Runoff Models," *Journal of Hydrology,* 122:161–219, 1991.

Gates, W. L., "The Use of General Circulation Models in the Analysis of Ecosystem Impacts of Climatic Change," *Climatic Change,* 7:267–284, 1985.

Giorgi, F., "On the Simulation of Regional Climate Using a Limited Area Model Nested in a General Circulation Model," *Journal of Climate,* 3:941–963, 1990.

Giorgi, F., and L. O. Mearns, "Approaches to the Simulation of Regional Climate Change: A Review," *Reviews of Geophysics,* 29(2):191–216, 1991.

Gleick, P. H., "Regional Hydrologic Consequences of Increases in Atmospheric CO_2 and Other Trace Gases," *Climatic Change,* 10:137–161, 1987.

Gleick, P. H., "Climate Change, Hydrology, and Water Resources," *Reviews of Geophysics,* 7(3):329–344, 1989.

Grotch, S. L., "Regional Intercomparisons of General Circulation Model Predictions and Historical Data," Technical Note DOE/NBB-0084, U.S. Department of Energy, 1988.

Grotch, S. L., and M. C. MacCracken, "The Use of General Circulation Models to Predict Regional Climatic Change," *Journal of Climate,* 4:286–303, 1990.

Grygier, J. C., and J. R. Stedinger, "Condensed Disaggregation Procedures and Conservation Corrections for Stochastic Hydrology," *Water Resources Research,* 24(10):1574–1584, 1988.

Hay, L. E., M. D. Branson, and G. H. Leavesley, "Simulation of Precipitation in the Gunnison River Basin Using an Orographic-Precipitation Model," *Managing Water Resources During Global Change, Proceedings of the American Water Resources Association,* Reno, Nev., 1992, pp. 651–660.

Hay, L. E., G. J. McCabe, D. M. Wolock, and M. A. Ayers, "Simulation of Precipitation by Weather Type Analysis," *Water Resources Research,* 27:493–501, 1991.

Hewitson, B., and R. G. Crane, "Regional Climates in the GISS GCM: Synoptic Scale Circulation," *Journal of Climate,* 5:1002–1011, 1992.

Hirschboeck, K. K., "Catastrophic Flooding and Atmospheric Anomalies," in L. Mayer and D. Nash, eds., *Catastrophic Flooding,* Allen and Unwin, Boston, 1987, pp. 23–56.

Hostetler, S. W., and F. Giorgi, "Use of Output from High-Resolution Atmospheric Models in Landscape-Scale Hydrologic Models: An Assessment," *Water Resources Research,* 29:1685–1695, 1993.

Houghton, J. T., G. J. Jenkins, and J. J. Ephraums, eds., *Climate Change: The IPCC Scientific Assessment,* Cambridge University Press, Cambridge, U.K., 1990.

Intergovernmental Panel on Climate Change, *Climate Change: The IPCC Scientific Assessment,* Cambridge University Press, New York, 1990.

Jager, J., and W. W. Kellogg, "Anomalies in Temperature and Rainfall During Warm Arctic Seasons as a Guide to the Formulation of Climatic Scenarios," *Climatic Change,* 5:39–60, 1983.

Kaczmarek, Z., N. W. Arnell, and E. Z. Stakhiv, "Water Resources Management," in Intergovernmental Panel on Climate Change Second Assessment Report, Chap. 10, Cambridge University Press, 1996.

Kalkstein, L. S., G. Tan, and J. A. Skindlov, "An Evaluation of Three Clustering Procedures for Use in Synoptic Climatological Classification," *Journal of Climate and Applied Meteorology,* 26:717–730, 1987.

Kavvas, M. L., and J. W. Delleur, "A Stochastic Cluster Model of Daily Rainfall Sequences," *Water Resources Research,* 17:1151–1160, 1981.

Kirshen, P. H., and N. M. Fennessey, "Potential Impacts of Climate Change upon the Water Supply of the Boston Metropolitan Area," U.S. Environmental Protection Agency, 1993.

Knox, J. L., and R. G. Lawford, "The Relationship Between Canadian Prairie Dry and Wet Months and Circulation Anomalies in the Mid-Troposphere," *Atmosphere-Ocean,* 28(2): 189–215, 1990.

Lamb, P. J., "On the Development of Regional Climatic Scenarios for Policy-Oriented Climatic-Impact Assessment," *Bulletin of the American Meteorological Society,* 68:1116–1123, 1987.

Leavesley, G. H, M. D. Branson, and L. E. Hay, "Investigation of the Effects of Climate Change in Mountainous Regions using Coupled Atmospheric and Hydrologic Models," in *Managing Water Resources During Global Change, Proceedings of the American Water Resources Association,* Reno, Nev., 1992, pp. 691–700.

Leggett, J., W. J. Pepper, and R. J. Swart, "Emissions Scenarios for the IPCC: An Update," *Climate Change 1992: The Supplementary Report to the IPCC Scientific Assessment,* J. T. Houghton, B. A. Callander, and S. K. Varney (eds.), Cambridge University Press, New York, 1992.

Lettenmaier, D. P., and T. Y. Gan, "Hydrologic Sensitivities of the Sacramento-San Joaquin River Basin, California, to Global Warming," *Water Resources Research,* 26:69–86, 1990.

Lettenmaier, D. P., K. L. Brettmann, L. W. Vail, S. B. Yabusaki, and M. J. Scott, "Sensitivity of Pacific Northwest Water Resources to Global Warming," *Northwest Environmental Journal,* 8(2):265–283, 1993.

Lettenmaier, D. P., and D. P. Sheer, "Climatic Sensitivity of California Water Resources," *Journal of Water Resources Planning and Management,* 117(1):108–125, 1991.

Lettenmaier, D. P., T. Y. Gan, and D. R. Dawdy, "Interpretation of Hydrologic Effects of Climate Change in the Sacramento-San Joaquin River Basin, California," Water Resources Technical Report no. 110, Department of Civil Engineering, Environmental Engineering and Science, University of Washington, Seattle, Wash., 1988.

Lettenmaier, D. P., "Applications of Stochastic Modeling in Climate Change Impact Assessment," in *Proceedings, International Conference on Water Resource Systems,* Waterloo, Ontario, 1993.

Lettenmaier, D. P., D. Ford, B. Nijssen, J. P. Hughes, S. P. Millard, and S. M. Fisher, "Water Management Implications of Global Warming: 4. The Columbia River System," Report to U.S. Army Corps of Engineers, Institute for Water Resources, Ft. Belvoir, Va., 1996a.

Lettenmaier, D. P., D. Ford, E. F. Wood, and S. M. Fisher, "Water Management Implications of Global Warming: 5. The Missouri River System," Report to U.S. Army Corps of Engineers, Institute for Water Resources, Ft. Belvoir, Va., 1996b.

Loucks, D. P., J. R. Stedinger, and D. A. Haith, *Water Resource Systems Planning and Analysis,* Prentice-Hall, Englewood Cliffs, N.J., 1981.

MacCracken, M. C., and F. M. Luther, eds., "Detecting the Climatic Effects of Increasing Carbon Dioxide," Report DOE/ER-0235, U.S. Department of Energy, Washington, D.C., 1985.

MacCracken, M. C., and J. Kutzbach, "Comparing and Contrasting Holocene and Eemian Warm Periods with Greenhouse-Gas-Induced Warming," in M. E. Schlesinger, ed., *Greenhouse-Gas-Induced Climatic Change: A Critical Appraisal of Simulations and Observations,* Elsevier, New York, 1991, pp. 17–34.

Manabe, S., and R. J. Stouffer, "Sensitivity of a Global Climate Model to an Increase of CO_2 Concentration in the Atmosphere," *Journal of Geophysical Research*, 85:5529–5554, 1980.

Mather, J. R., and J. Feddema, "Hydrologic Consequences of Increases in Trace Gases and CO2 in the Atmosphere," in *Effects of Changes in Stratospheric Ozone and Global Climate, Vol. 3— Climate Change*, U.S. Environmental Protection Agency, Washington, D.C., 1986, pp. 251–271.

McCabe, G. J., and M. A. Ayers, "Hydrologic Effects of Climate Change in the Delaware River Basin," *Water Resources Bulletin*, 25(6):1231–1242, 1989.

McCabe, G. J., "A Conceptual Weather-Type Classification Procedure for the Philadelphia, Pennsylvania Area," Water-Resources Investigations Report 89-4183, U.S. Geological Survey, Reston, Va., 1990.

McCabe, G. J., and D. M. Wolock, "Detectability of the Effects of a Hypothetical Temperature Increase on the Thornthwaite Moisture Index," *Journal of Hydrology*, 125:25–35, 1991.

McCabe, G. J., and D. R. Legates, "General Circulation Model Simulations of Winter and Summer Sea-Level Pressures Over North America," *International Journal of Climatology*, 12:815–827, 1992.

McCabe, G. J., and D. M. Wolock, "Effects of Climatic Change and Climatic Variability on the Thornthwaite Moisture Index in the Delaware River Basin," *Climatic Change*, 20:143–153, 1992.

Miller, B. A., and W. G. Brock, "Potential Impacts of Climate Change on the Tennessee Valley Authority Reservoir System," in The Potential Impacts of Global Climate Change on the United States, Appendix A: Water Resources, U.S. EPA, Washington, D.C., 1989.

Miller, J. R., and G. C. Russell, "The Impacts of Global Warming on River Runoff," *Journal of Geophysical Research*, 97:2757–2764, 1992.

Mitchell, J. F. B., "The 'Greenhouse' Effect and Climate Change," *Reviews of Geophysics*, 27(1):115–139, 1989.

Muller, R. A., "A Synoptic Climatology for Environmental Baseline Analysis: New Orleans," *Journal of Applied Meteorology*, 16:20–32, 1977.

Muller, R. A., and C. L. Wax, "A Comparative Synoptic Climatic Baseline for Coastal Louisiana," *Geoscience and Man*, 18:121–129, 1977.

Namias, J., "Some Causes of United States Drought," *Journal of Climate and Applied Meteorology*, 22:30–39, 1983.

Nash, L. L., and P. H. Gleick, "Sensitivity of Streamflow in the Colorado Basin to Climatic Changes," *Journal of Hydrology*, 125:221–241, 1991.

Nemec, J., and J. C. Schaake, "Sensitivity of Water Resource Systems to Climate Variation," *Journal of Hydrological Sciences*, 27:327–343, 1982.

Nijssen, B., D. P. Lettenmaier, F. Abdulla, E. F. Wood, and S. W. Wetzel, "Simulation of Runoff of Continental-Scale River Basins Using a Grid-Based Land Surface Scheme," in review, *Water Resources Research*, 1995.

Parker, R. S., G. Kuhn, L. E. Hay, and J. G. Elliot, "Effects of Potential Climate Change on the Hydrology and the Maintenance of Channel Morphology in the Gunnison River Basin, Colorado," *Proceedings of the Workshop on the Effects of Global Climate Change on Hydrology and Water Resources at the Catchment Scale*, Tsukuba-shi, Japan, 1992, pp. 399–410.

Rhea, J. O., "Orographic Precipitation Model for Hydrometeorological Use," Ph.D. dissertation, Colorado State University, Department of Atmospheric Science, Fort Collins, Colo., 1977.

Robock, A., R. P. Turco, M. A. Harwell, T. P. Ackerman, R. Andressen, H. Chang, and M. V. K. Sivakumar, "Use of General Circulation Model Output in the Creation of Climate Change Scenarios for Impact Analysis," *Climatic Change*, 23:293–335, 1993.

Rosenberg, N. J., B. A. Kimball, P. Martin, and C. F. Cooper, "From Climate and CO_2 Enrichment to Evapotranspiration," in P. E. Waggoner, ed., *Climate Change and U.S. Water Resources*, John Wiley, New York, 1990, pp. 151–175.

Schaake, J. C., "From Climate to Flow," in P. E. Waggoner, ed., *Climate Change and U.S. Water Resources*, chap. 8, John Wiley, New York, 1990.

Sheer, D. P., and D. Randall, "Methods for Evaluating the Potential Impacts of Global Climate Change: Case Studies of the State of California and Atlanta, Georgia," in J. B. Smith and D. A. Tirpak, eds., "The potential effects of global climate change on the United States, Appendix A: Water Resources," Report EPA-230-05-89-051, U.S. Environmental Protection Agency, 1989, pp. 2-1–2-28.

Sias, J. C., and D. P. Lettenmaier, "Potential Effects of Climatic Warming on the Water Resources of the Columbia River Basin," Water Resources Series Technical Report 142, Department of Civil Engineering, University of Washington, Seattle, Wash., 1994.

Smith, J. A., and A. F. Karr, "Statistical Inference for Point Process Models of Rainfall," *Water Resources Research,* 21(1):73–80, 1985.

Smith, J. B., and D. A. Tirpak, "The Potential Effects of Global Climate Change on the United States," Report EPA-230-05-89-050, U.S. Environmental Protection Agency, Office of Policy, Planning, and Evaluation, Washington, D.C., 1989a.

Smith, J. B., and D. A. Tirpak, "The Potential Effects of Global Climate Change on the United States: Appendix A—Water Resources," Report EPA-230-05-89-051, U.S. Environmental Protection Agency, Office of Policy, Planning, and Evaluation, Washington, D.C., 1989b.

Stakhiv, E., I. Shiklomanov, and H. Lins, "Hydrology and Water Resources," in W. J. McG Tegart and G. W. Sheldon, eds., *Climate Change 1992, Supplementary Report to IPCC Impacts Assessment,* Chap. VI Australian Government Publishing Service, Canberra, 1992.

Vaccaro, J. J., "Sensitivity of Groundwater Recharge Estimates to Climate Variability and Change, Columbia Plateau, Washington," *Journal of Geophysical Research,* 97 (D3): 2821–2833, 1992.

Viner, D., M. Hulme, and S. C. B. Raper, "Climate Change Scenarios for the IPCC Working Group II Impacts Assessment," Climate Impacts LINK Technical Note 6, Climate Research Unit, Norwich, U.K., 1994.

von Storch, H., E. Zorita, and U. Cubasch, "Downscaling of Global Climate Change Estimates to Regional Scales: An Application to Iberian Rainfall in Wintertime," *Journal of Climate,* 6:1161–1171, 1993.

Washington, W. M., and C. L. Parkinson, *An Introduction to Three-Dimensional Climate Modeling,* University Science Books and Cambridge University Press, 1986.

Williams, J., "Anomalies in Temperature and Rainfall During Warm Arctic Seasons as a Guide to the Formulation of Climate Scenarios," *Climatic Change,* 2:249–266, 1980.

Wilson, L. L., D. P. Lettenmaier, and E. F. Wood, "Simulation of Daily Precipitation in the Pacific Northwest Using a Weather Classification Scheme," in E. F. Wood., ed., *Land-Surface-Atmosphere Interactions for Climate Modeling: Observations, Models, and Analysis, Surveys of Geophysics,* 12:127–142, 1991.

Wilson, L. L., D. P. Lettenmaier, and E. Skyllingstad, "A Hierarchical Stochastic Model of Large-Scale Atmospheric Circulation Patterns and Multiple Station Daily Precipitation," *Journal of Geophysical Research,* 97:2791–2809, 1992.

Wolock, D. M., and G. M. Hornberger, "Hydrological Effects of Changes in Levels of Atmospheric Carbon Dioxide," *Journal of Forecasting,* 10:105–116, 1991.

Woolhiser, D. A., and J. Roldan, "Seasonal and Regional Variability of Parameters for Stochastic Daily Precipitation Models: South Dakota, U.S.A.," *Water Resources Research,* 22:965–978, 1986.

Yarnal, B., "A 500mb Synoptic Climatology of Pacific Northwest Winters in Relation to Climatic Variability, 1948–1949 to 1977–1978," *Journal of Climatology,* 5:237–252, 1985.

Yarnal, B., and D. J. Leathers, "Relationships between Interdecadal and Interannual Climatic Variations and Their Effect on Pennsylvania Climate," *Annals of the Association of American Geographers,* 78:624–641, 1988.

Yarnal, B., *Synoptic Climatology in Environmental Analysis,* Belhaven Press, London, 1993.

CHAPTER 30

ECOLOGICAL EFFECTS OF GLOBAL CLIMATE CHANGE ON FRESHWATER ECOSYSTEMS WITH EMPHASIS ON STREAMS AND RIVERS

Stuart G. Fisher
Nancy B. Grimm
Department of Zoology
Arizona State University
Tempe, Arizona

30.1 GLOBAL CLIMATE CHANGE AND FRESHWATERS: BOUNDARIES AND SCENARIOS

Scientific opinions vary about future climate change driven by continued increased emission of greenhouse gases; however, many scientists agree that a global temperature rise of 1 to 2°C by 2050 is likely (Levine, 1992; Houghton, et al. 1990). Precise knowledge of global average-temperature change is of little predictive value, especially with respect to aquatic ecosystems and water resources. Temperature change will vary regionally and seasonally, but so too will changes in precipitation and runoff, and the direction of such changes is by no means clear. Hydrologic changes are likely to influence patterns of water storage and exchange among aquifers, streams, rivers and lakes. Climate change also can have a significant effect on the organisms living in these aquatic systems.

Our objective in this chapter is to discuss the potential effects of climate change on the ecology of aquatic organisms and the structure and functioning of aquatic ecosystems. We will emphasize running waters, which are better known to us. We will not enter the debate about whether global warming will occur, but rather will discuss potential ecological effects should it happen. Few papers have focused on this problem (but see Firth and Fisher, 1992; Carpenter, et al., 1992); thus we will derive predictions from papers written for other reasons and will attempt to meld this sometimes-specific information with the broader specter of global climate change.

30.1.1 Climate Change Scenarios

Studies at global and other very large scales, primarily using *general circulation models (GCM)* that predict the climatic effects of a doubling of CO_2, have indicated temperature rise in the northern hemisphere coupled with decreased precipitation in midlatitudes and increased precipitation in northern latitudes (e.g., Wigley, et al., 1980; Manabe and Wetherald, 1987; Levine, 1992). Dickinson (1989) has reviewed uncertainties inherent in GCMs; examples include both positive (Ravel and Ramanathan, 1989) and negative (Mitchell, et al., 1989) feedback effects of clouds and effects of orogeny (Manabe and Broccoli, 1990).

At regional scales, several investigators have used GCM output, hypothetical scenarios, and empirical data on extremes to generate more specific predictions. Global models rather poorly apply to this scale (Coleman, 1988; Cushman and Spring, 1989; Hostetler, 1994) and do not predict seasonal responses well. Although development of regional models is progressing rapidly, these models, like global models, are fraught with uncertainty and their utility in impact assessment is low (Giorgi, et al., 1994; Hostetler, 1994). It remains necessary, therefore, to generate predictions from combinations of empirical studies, regional models, and comparative ecosystem studies. The problem of regional or local climate change is compounded if one is interested in predicting changes in hydrology, since numerous interactions among temperature, plant physiological response, precipitation, and runoff confound simple interpretation of effects of a temperature increase (Skiles and Hanson, 1994). See Chap. 29 for more details on GCMs.

30.1.2 Future Distributions of Freshwaters

Forecasts of future precipitation and evaporation are uncertain, but the consensus is that greenhouse warming will increase both (Schneider, 1992). River runoff may increase or decrease, depending upon how these parameters change. Thus some areas of the Earth will become wetter while others become drier, thereby shifting the global distribution of streams, lakes, and wetlands (Waggoner, 1990).

Future distribution of fresh surface water and groundwater will depend in large part on altered patterns of runoff associated with climate change. Numerous hydrologic studies have appeared in the last decade that predict changes in runoff over the next century if global warming of 2 to 4°C occurs. These regional studies of large basins apply water-balance models (Thornthwaite and Mather, 1955) by selecting independent variables of precipitation, soil moisture, and potential evapotranspiration. Generalizations that emerge from these efforts are: (1) the precipitation–runoff relationship is nonlinear (Wigley and Jones, 1985), (2) under reduced-precipitation scenarios (predicted for continental interiors; Wigley, et al., 1980), reductions in streamflow are likely, particularly in western U.S. basins (Williams, 1989), (3) arid basins are more sensitive to precipitation changes than are humid basins (Wigley and Jones, 1985; Karl and Riebsame, 1989; Mimikou, et al., 1991; Cohen, 1991), and (4) seasonal shifts in streamflow distributions are probably more significant than changes in total annual runoff, particularly in cold or mountainous regions where winter precipitation now occurs as snow (Gleick, 1987; McCabe and Ayers, 1989;

Lettenmaier and Gan, 1990; Mimikou, et al., 1991). A recent modeling study (Lettenmaier and Gan, 1990) simulated large increases in flood maxima and a shift in flood timing from spring to winter.

Substantial uncertainty in hydrologic modeling concerns complex responses of plants to elevated CO_2 which has the net effect of reducing evapotranspiration (Idso and Brazel, 1984; Wigley and Jones, 1985; Verbyla, 1990). Most water-balance-based predictions do not consider these responses because of their uncertainty (McCabe and Ayers, 1990); however, the net effect of lower evapotranspiration is increased stream flow. A recent analysis of USGS stream flow data at 559 sites in the conterminous United States for the period 1941 to 1988 indicates a significant increase in stream flow, especially in autumn and winter (Lins, 1994). This occurred despite no significant change in precipitation or temperature and is attributed to increased cloudiness and reduced evapotranspiration. Analysis of 1009 stream sites by Lettenmaier, et al. (1994) similarly found evidence for increased streamflow for most sites between 1941 and 1988.

In arid regions such as the Great Basin, drier conditions may decrease water-supply rates below current demand by the human population (Flaschka, et al., 1987). Lakes and wetlands might then be expected to contract in area. For example, Poiani and Johnson (1993) modeled impacts of various climate change scenarios on a prairie wetland, and predicted drying of the wetland, reduction in maximum depth, and change in vegetation. In moist regions, drier conditions will decrease stream flows, lake levels, and water supplies (Cohen, 1987; Richey, et al., 1989). In heavily populated areas, the magnitude of change is uncertain because effects of climate change on water demand for irrigation, energy, and cooling are difficult to anticipate (Cohen, 1987). Although certain regions may experience wetter conditions as a result of global climate change (Schneider, et al., 1990; Giorgi, et al., 1994), consequences of increased water supply have received less attention.

30.2   EFFECTS ON STREAMS AND RIVERS

30.2.1   Overview: Aquatic Ecosystems

Ecologists working with streams, rivers, and lakes view these ecosystems in two ways. Organismal, population, and community ecologists focus on organisms either alone or in groups of similar or diverse species. The aquatic environment is thus seen as a complex of habitat factors, and response variables are reckoned as presence or absence of one or a number of species, physiological or behavioral responses of individuals, and population sizes of individual species relative to others in the same environment. Clearly climatic factors such as temperature and seasonal regimes of precipitation are important drivers of these biologic responses. Ecosystem ecologists, on the other hand, view streams and lakes as integrated complexes of physical, chemical, and biological components that interact to produce a collective response; e.g., a rate of total primary production or nutrient retention. These *ecosystem properties* also are sensitive to climatic drivers but are less dependent on the identity of individual organisms present. For example, the African Savannah and North American grasslands may function similarly in terms of energy flow and nutrient cycling yet have no common species. A consideration of the ecological effects of climate change must consider responses of individuals, species, and whole ecosystems (see also Carpenter, et al., 1992; Grimm 1993). Responses of these different entities may be neither correlative nor additive. In the sections following, we will consider a broad spectrum of potential ecological responses to climate change. A comprehensive assessment of ecological impact must incorporate this broad view.

30.2.2 Organismal Life History and Biogeography

Stream water temperature depends upon the temperatures of source (e.g., snowmelt) and groundwater, the percentage contributions of groundwater and discharge, heating from the atmosphere (conduction), and insolation (Ward, 1985; Burton and Likens, 1982). Evaporative cooling may be important in arid-zone rivers (Weatherly, 1967). Based upon data from New Zealand, Mosley (1982) found that 73 percent of the variation in stream water temperature could be explained by elevation and latitude. For most streams, an increase in air temperature will be translated directly into an increase in water temperature; however, this may be modified by other elements of climate which influence amount and seasonality of runoff, cloudiness, and amount of shading by riparian vegetation. For example, higher runoff will project the influence of source water farther downstream; low discharge will amplify the influence of (relatively) constant temperature groundwater (Ward, 1985). Thus while global warming will result in elevated water temperature on average, thermal regimes may also be altered in several ways that are difficult to predict.

Temperature controls many life-history features of aquatic invertebrates. Variable and unpredictable patterns of water temperature in New Zealand (Winterbourn, et al., 1981) and Australia (Lake, 1982) have generated a general view of invertebrate communities there as having flexible, poorly synchronized life cycles in contrast to those of more predictable parts of the world, such as temperate North America (Vannote and Sweeney, 1980). Such life-history features as size and fecundity (Ward and Cummins, 1979), egg hatching (Sweeney and Schnack, 1977), growth rate (Cudney and Wallace, 1980), emergence (Danks 1978; Harper, et al., 1970), and voltinism (Ward and Stanford, 1982) are sensitive to temperature. Adult longevity, and presumably time available for oviposition, are related to water temperature during development (Macan, 1970) and atmospheric temperature and humidity encountered by emergent adults (Jackson, 1988).

The *thermal equilibrium hypothesis* of Vannote and Sweeney (1980) states that temperature determines the range of a species with optimal conditions near the center of that range. Life-history characters such as growth, reproductive potential, and generation time vary as one moves from center to periphery. When closely related species co-occur, temperature appears to be a critical factor maintaining synchronization of life cycles and minimizing competition. Temperature may similarly account for the longitudinal array of species in a river system, again limiting competition to zones of overlap (Hildrew and Edington, 1979).

Climate change, through altered temperature, will likely have a profound effect on the distribution and abundance of aquatic insects in rivers and streams. Effects will first be seen at the edges of species ranges and the demise of certain taxa may result proximally from competition or predation. Adjustments to climate change will depend upon genetic variability and ability to disperse. Unfortunately, genetic variability may be least where projected changes are greatest (e.g., northeastern North America for mayflies), and dispersal is limited in these fragile species with short-lived adults (Sweeney, et al., 1992).

While climate change is likely to influence invertebrate life histories through altered temperature, other elements of climate affect invertebrates as well. Stanley and Short (1988) found that the thermal equilibrium hypothesis of Vannote and Sweeney (1980) did not apply to dynamics of invertebrates in streams of semiarid west Texas. Flood disturbance and intermittency rather than competition selected for organisms with rapid life cycles, many generations, and asynchronous, overlapping life cycles. Towns (1983) similarly noted the prominence of abiotic over biotic control of life-cycle characteristics among aquatic insects of New Zealand. In Arizona, flash floods and drying appear to be strong selective factors influencing insect life history, yet three major strategies exist. Large, mobile, long-lived organisms such as belastomatids (Hemiptera) resist flood disturbance or fly to nearby permanent water in time of drought. Beetles, which are susceptible to floods as larvae but not as adults, restrict reproduction to the period of the year (spring and fall) when the probability of floods is lowest (Gray, 1980). Small, short-lived mayflies and chironomids continuously reproduce and complete life cycles extremely rapidly (8 to 13 days) at far fewer degree days than comparable species elsewhere (Gray, 1981). Many of these "fast" taxa are present as aerial adults for less than a day and, as flash floods rise and fall within hours, rapidly recolonize flood-denuded substrates (Fisher, et al., 1982; Jackson, 1988).

All of these life-history characteristics are related to the disturbance regime, which is highly sensitive to small changes in weather patterns, runoff amount and timing, and the rate at which storms move across the desert southwest. Species that adjust reproduction to extant disturbance regimes are likely to change rapidly if this regime is altered. However, even the most generalized, small, fast mayflies will be imperiled if floods occur too rapidly in succession or if flood recession is slowed, thus interfering with timely oviposition (Jackson, 1988).

30.2.3 Disturbance and Hydroperiod

Many climate-change scenarios project an increase in extreme events (Schneider, 1992). These include floods, droughts, and thermal extremes that cause abrupt change in ecosystem structure or process rates. Extreme events are of substantial ecological significance because temporal variability in streams and lakes is keyed by cycles of hydrologic events and recruitment of keystone species (Grimm and Fisher, 1992; Carpenter, 1988). Such cycles typically have periods of one to many years, measured at the spatial scale of a particular stream or lake ecosystem. Disturbance characteristics are regionally distributed. A regional analysis of flow characteristics of 78 streams of the United States showed them to be separable by cluster analysis into 9 types based on flow characteristics (Poff and Ward, 1989). This separation distinguishes between the perennial streams of the east where the kinds and numbers of organisms are controlled by temperature and by biotic interactions (Vannote and Sweeney, 1980; Poff and Ward, 1990) and the flashy, intermittent streams of the Southwest where abiotic factors such as flash floods and drought are prominent (Fisher, et al., 1982). Both temperature and hydrology are part of the physical habitat template (Poff and Ward, 1990) which determines the structure of biologic communities. Both will be altered by global climate change, will affect streams of different regions in distinct ways, and will likely alter the geographic distribution of stream types. Biologic communities in streams that experience a shift from one control type to another will likely change in composition, provided that the regional species pool contains species appropriate to these new conditions. Loss of species without replacement will generate a new, simplified community of uncertain functioning.

30.2.3.1 Perturbations of Riparian Zones and Floodplains.   The riparian zone influences and is influenced by hydrologic disturbance regimes of flood and drought. In streams of forested watersheds, the stream channel–riparian complex can be viewed as a mosaic of geomorphic surfaces generated by valley landforms, geomorphology, hydrologic processes, and forest succession (Gregory, et al., 1991). Fluvial geomorphology changes slowly owing to the long turnover time of both live and fallen trees and the slow decomposition of wood; however, hydrologic events of annual return can influence within- and near-channel patchiness and the habitat structure for both aquatic organisms and riparian vegetation. Combined, these factors result in smaller, younger vegetation near the stream channel and older vegetation in the more distal floodplain. Not only is age structure shaped by hydrologic events; because the channel is a complex mosaic, species diversity is thereby enhanced when reckoned on a scale of several hundred meters (Gregory, et al., 1991).

Rare catastrophic events such as mass wasting of hillslopes remove soil, debris, and riparian vegetation from affected areas and expose bedrock surfaces. These rare events affect only a small fraction of the watershed but may influence up to 10 percent of the channel network as material enters and dams recipient streams (Swanson, et al., 1987). The geomorphic effect of these slumps and earthflows can persist for thousands of years (Dietrich, et al., 1982); however, biota of recipient streams may recover to predisturbance levels in 1 to 2 years (Lamberti, et al., 1991). Rate of recovery of riparian vegetation is somewhere between these extremes.

Alluvial streams and rivers of arid and semiarid regions of the west also support a dynamic riparian zone that is greatly influenced by hydrologic regime. Indeed, some riparian tree species depend upon scarification of seeds for germination (Stromberg, et al., 1991). Baker (1990) studied the age structure of longleaf cottonwood of the Animas River, Colorado and attributed observed patterns to the interplay of processes of seedling establishment and flood removal. Conditions favorable for the establishment of cottonwood seedlings occur every 4.3 years on average; however, recruitment of large trees occurred only every 11 years due to high mortality associated with bank and channel bar erosion during infrequent but severe flash floods. As is the case with Oregon streams disturbed by earthflows, recovery of in-stream processes after flood or drought in hot desert streams is exceedingly rapid (Fisher, et al., 1982); however, the riparian zone is likely to respond more slowly. Because of the strong interaction of aquatic and riparian components even in aridland streams where riparian vegetation is sparse, it is unlikely that either subsystem reaches any semblance of equilibrium given that time constants for effective disturbance events differ markedly and disturbance-recovery cycles are always out of phase.

In arid regions, riparian vegetation is very sensitive to availability of water, which in turn depends upon amount and distribution of precipitation. Conversion of chaparral to grass in a small Arizona watershed increased water yield and converted a previously intermittent stream to permanent flow. Riparian vegetation increased in density and extent (Ingebo, 1971). Similar oscillations may occur in unmanipulated desert streams in response to year-to-year changes in weather or longer-term climate change (Grimm and Fisher, 1992). Climatic shifts to drier conditions pose a serious threat to aridland riparian ecosystems owing to the combined influence of reduced or altered streamflow and increased human demand for scarce water resources.

DeCamps, et al. (1988) have similarly shown that riparian vegetation of the Garonne River in France participates in a cycle of establishment and destruction due to existing patterns of hydrologic disturbance and recruitment characteristics of riparian trees, especially willows and alder. At places or during times where disturbance interval is lengthened to beyond approximately 10 years, transition to a relatively stable climax association of native hardwoods such as oaks occurs. This community is relatively immune to subsequent floods.

The floodplain is the zone of maximum interaction between a river and its valley (Chauvet and DeCamps, 1989), depending upon regular inundation for nutrient and organic-matter supply and provision of breeding ground for fishes and their food organisms. Inundation influences river-water chemistry, especially during initial stages (Welcomme, 1988). Inundation extent and duration are sensitive to climate, so changes in precipitation and runoff amount and timing will influence flood-plain–river interactions. The inundation period both waters and stresses riparian plants and controls colonization by exotics, depending upon duration (Bravard, et al., 1986). There is thus a substantial potential for climate change to alter riparian vegetation in this and similar drainages.

In the Mississippi River, floods of less than annual-return frequency may be significant disturbances which influence the distribution of macrophytes and primary production through erosion and sedimentation (Sparks, et al., 1990). By influencing the transport of materials, the effect of floodplains in this system extends far downstream. By comparison, local flood-control structures have a less pervasive effect. Similarly, dependence upon allochthonous inputs increases longitudinally with river size in the blackwater Ogeechee River of Georgia due to lateral floodplain contributions (Meyer and Edwards, 1990). Floodplain contribution to the river's energy budget has been linked to seasonal flooding, which provides a bimodal pulse of organic matter to stream consumers (LeCren and Lowe-McConnell, 1981; Merritt and Lawson, 1979).

30.2.3.2 Hydrologic Disturbance of Streams.

The most obvious and immediate effects of global climate change on stream ecosystems will involve changes in hydrologic patterns. Increased or decreased total precipitation, extreme rainfall events, and changes in seasonality will affect the amount, variability and timing of flow. Changes in total amount or timing of runoff may result in increased or decreased incidence of intermittency (Ward, et al., 1992; Poff, 1992), while altered amount, variability, and extremes of runoff may produce changes in timing or increased frequency and severity of flash flooding (Gleick, 1987). Such changes in magnitude and temporal distribution of extreme events may disrupt ecosystems more than will changes in mean conditions (Walker, 1991; Poff, 1992; Grimm and Fisher, 1992). A complex set of relationships governs how climate influences runoff, which in turn interacts with geology and vegetation to modify geomorphology, soil chemistry, and hydrology to exert both independent and interactive effects on stream processes (Minshall, 1988; Ward, et al., 1992). Thus, as Poff (1992) states, regional changes in climate will produce an *integrated catchment response* that includes the stream response.

The role of disturbance in lotic communities and ecosystems is currently a major focus of stream ecology research (Peckarsky, 1983; Resh, et al., 1988; Power, et al., 1988; Poff and Ward, 1989; Yount and Niemi, 1990; Niemi, et al., 1990). Despite the attention it has received, there is no general consensus as to the importance of disturbance in determining structure and functioning (but see Resh, et al., 1988), although a strong case can be made for expecting regional variation (Poff and Ward, 1989; 1990; Fisher, 1990). Poff (1992) suggests that in a warmer, drier climate many perennial-runoff and intermittent-runoff streams in the continental United States might experience increased intermittency because of their high flow variability, while groundwater-fed streams would be buffered against such changes. He further indicates that streams fed by snowmelt, which have highly predictable flood and flow regimes, might become less predictable with winter warming (increased rain-on-snow events).

Altered frequency of disturbance under drier climate scenarios also has been considered by other authors. Ward, et al. (1992) calculated $7Q_2$ and $7Q_{10}$ flows (discharge for 7-day periods of lowest flow with recurrence intervals of 2 and 10 years, respectively) under current and reduced-precipitation climates for Alabama streams differing in parent geology. They predicted that flows would be reduced (or cease during some times of the year), but to a lesser extent in carbonate terrains than in sandstone or quartzite terrains. Thus the geologic setting would influence stream-flow response to precipitation change. The nonlinear response of streamflow in arid and semiarid regions to changes in rainfall discussed previously is relevant to the question of hydrologic disturbance, because increasing aridity may render flows of many more streams unpredictable (e.g., Grimm and Fisher, 1992). One possible explanation for this type of response, for which there is mounting evidence (e.g., Dahm and Molles, 1992), is that soils develop a hydrophobic layer during periods of low rainfall that increases the likelihood of overland flow.

Response to Spates and Drought. Rather than attempt a comprehensive review, we focus on a few studies to illustrate how stream ecosystems respond to and recover from hydrologic disturbance, both at the wet (spates) and dry (drought or stream drying) ends of the hydrologic spectrum. The following effects might arise from increased frequency of drought or drying conditions: community shifts to species with resistant stages or short life cycles; heightened biotic interactions; concentration of pollutants; loss of riparian vegetation; increased frequency of anoxia or hypoxia in sediments; and loss of hydrologic connections between surface and subsurface waters. Increased frequency or severity of flash floods (or decreased predictability; Poff and Ward, 1989) might: favor species with behavioral or life-history adaptations to variable regimes; reduce importance of biotic interactions; decrease habitat or substrate stability; prevent establishment of high biotic standing crops in stream or riparian zones; increase hydrologic linkage with subsurface systems.

Studies in temporary streams have shown that faunas are dominated by organisms with resistant life stages (Williams and Hynes, 1977; Harrison 1966; Boulton, et al., 1989) or by insects with aerial adults that recolonize newly wetted habitats from permanent ones (Gray, 1981; Boulton and Suter, 1986). Streams in the desert Southwest subject to severe flash flooding also are inhabited by insects with short life cycles (Gray, 1981) that recolonize after floods by oviposition of aerial adults (Gray and Fisher, 1981). Native fishes in these systems exhibit behaviors in laboratory flood simulations that presumably allow them to avoid decimation by flash floods (Meffe, 1984). Rates of recolonization and recovery of biotic communities decimated by flash flooding depend upon both characteristics of the disturbance and life histories of the organisms (Hanson and Waters, 1974; Seigfreid and Knight, 1977; Fisher, et al., 1982; Power and Stewart, 1987; Molles, 1985; Grimm and Fisher, 1989; Lake and Barmuta, 1986; Niemi, et al., 1990).

As streams dry, mobile organisms are concentrated and biotic interactions surely increase (Taylor, 1983; Stanley and Fisher, 1992). These interactions or physiological stresses may differentially affect species or size classes of organisms (e.g., Resh, 1982). Competitive interactions are seasonally important in a California stream, but persisted through the winter in a drought year, when winter caddisfly densities were not reduced by the usual high winter flows (Feminella and Resh, 1990). Year-to-year variation in competition was attributed in another study to variation in timing and intensity of winter floods (Hemphill, 1991).

Importance of Disturbance Regime. The previous discussion is somewhat unrealistic in that it assumes a single outcome of global climate change (i.e., increased flooding or increased drought). In reality, complex changes to the hydrologic *regime* of streams should be expected, with pronounced temporal and/or spatial variation in controls of system state (e.g., flooding, drying, biotic interaction; Grimm and Fisher, 1992; Grimm, 1993).

Prediction of stream ecosystem response to changes in frequency, timing, or magnitude of disturbances is very speculative at this stage, and must be based upon comparison of the same system in different years (year-to-year variation) or different systems with similar biotas but subject to different disturbance regimes. The latter is fraught with difficulties (Resh, et al., 1988), but has been suggested as an application of space-for-time substitution to the problem of global climate change (Grimm and Fisher, 1992). They suggested that both spatial variation in flash-flood magnitude (across the United States; Baker, 1977) and gradients in seasonality of precipitation (in the Southwest) could be exploited for comparative studies.

Poff and Ward (1989) suggested that the degree to which biotic interactions shape community structure in any stream is a function of the disturbance regime. This is a reasonable model when such different streams as highly variable, flashy grassland systems and groundwater-fed streams are compared. It is likely that in most streams, however, there is potential for year-to-year and spatial variation in the temporal alternation of disturbance and biotic controls of community structure and ecosystem functioning (Grimm and Fisher, 1992; Grimm, 1993). The vulnerability to climate change of stream ecosystems may be related to their history of exposure to hydrologic variability (Poff, 1992); Ford and Thornton (1992) have suggested that intrinsically variable systems might provide early warning of ecological change, but might also be the least at risk.

A few studies have considered how disturbance and biotic interactions can shift temporally and spatially in controlling ecosystem functioning. Grimm (1993) compared the 5 highest versus the 5 lowest runoff years of the 30-year streamflow record for Sycamore Creek, a Sonoran Desert stream. State frequency analysis, with alternate states being *successional* (<30 d since spate), *drying* (>200 d since spate) or *biotic* (remainder of year, when biotic interactions are presumed to control structure and functioning), suggested that dry years had few successional days but many drying days, wet years exhibited opposite state frequencies, and the biotic state was constant across these hydrologically different years. Conclusions were based entirely upon hydrologic data, but an earlier study comparing this system with a mesic, forest stream (Fort River) suggested that flash-flood disturbance explained most of the variance in algal biomass in Sycamore Creek (but not Fort River) at annual time scales (Fisher and Grimm, 1988). Shorter-term variation in algal biomass (e.g., during successional sequences) is more strongly related to nitrogen availability, which is controlled by biotic uptake processes (Grimm and Fisher, 1986; Grimm, 1992).

Flood-induced redistribution of predators interact with strong three-trophic-level interactions to control algal biomass distribution in an Oklahoma stream (Power and Matthews, 1983; Power, et al., 1985). Stream pools studied by Power and colleagues exhibited pronounced spatial variation in trophic interactions: some pools lacked bass and algae were sparse because of effects of the grazer *Campostoma anomalum,* while others were algae-choked because bass effectively suppressed the algivorous fish. Bass distribution was determined by movements among pools during floods and isolation within pools as flows decreased.

A series of experiments in Rattlesnake Creek, a southern California stream that experiences high flows during a winter wet season and prolonged periods of low flow, supports the notion of temporally alternating periods of disturbance and biotic control of invertebrate community structure. High winter flows reduce algal and invertebrate standing crops (Dudley, et al., 1986). Early spring establishment of blackflies occurs on substrata exposed by scour, but these animals are competitively displaced by hydropsychid caddisflies and maintain high densities only with repeated substratum disturbance (Hemphill and Cooper, 1983; Hemphill, 1991). During low-flow periods, fish and invertebrate predators strongly regulate composition and abundance of lower trophic levels (Cooper 1984; Hemphill and Cooper, 1984), and grazers can influence macroalgae, especially during establishment on newly opened substrata (Dudley and D'Antonio, 1991).

In streams such as Rattlesnake Creek and the Eel River studied by Power (1990), storms and flooding are highly seasonal and periods of low flow, during which biotic interactions can predominate (Power, 1992), are predictable in time. Altered seasonality of both rainfall and spates is a likely consequence of global warming and this would clearly influence such seasonal ecosystems. Grimm (1993) illustrated the effect of different seasonal flash-flood regimes on state frequency (successional, drying, biotic, as described previously) by comparing two years with similar annual runoff but different temporal distributions of spates. One year had only two spates and was in the drying state 80 percent of the time, whereas the second year had 9 small spates and was in succession 55 percent but in drying only 20 percent of the year.

30.2.4 Terrestrial–Aquatic Interactions

The land–water interface is a zone of intense biogeochemical activity that expands and contracts with fluctuations in water supply to landscapes (Wetzel 1990). This interface is especially important in streams because of their large perimeter to volume ratio. Because of the dependence of riparian systems on the integrity of streams, riparian influences such as shading, leaf production, and bank stabilization are at risk if climate change occurs sufficiently rapidly or drying is so severe that riparian vegetation is lost. Stream ecologists have paid close attention to the influence of terrestrial vegetation on freshwaters (Hynes, 1963; Fisher and Likens, 1973; Hynes, 1975). Utility of terrestrial leaf litter to stream consumers varies greatly among species (Peterson and Cummins, 1974) and the combination of species within individual leaf accumulations or *packs* (Benfield, et al., 1977). Change in climate will likely alter the total biomass, productivity, and species composition of the riparian community and, in turn, the supply of energy and its utility to aquatic consumers (Meyer and Pulliam, 1992). Mockernut hickory leaves added to streams outside the natural range of this tree species decomposed more slowly than in native streams (Minshall, et al., 1982). In general the importance of exotic species in the riparian zone may increase with climate change (Sweeney, et al., 1992). These exotics may thus alter ecosystem metabolism even if input rates are unchanged.

Nitrogen content of leaves has been shown to decrease under conditions of elevated CO_2 (Williams, et al., 1986). This may decrease the rate of consumption of terrestrial leaves (Fajer, et al., 1989), increasing the proportion of leaf production that enters streams and slowing in-stream decomposition (Meyer and Pulliam, 1992). In the arctic, projected increased cloudiness, water stress, and winter browsing are likely to increase foliar nitrogen and decrease secondary compounds (Oswood, et al., 1992); however, enhanced summer grazing can cause the reverse. While experiments have shown that shredder caddisflies prefer high-nitrogen, low-tannin leaves from fertilized trees over unfertilized controls (Irons, et al., 1988), the direction and magnitude of changes in litter quality in response to climate change in this region are complex and the net outcome remains uncertain (Oswood, et al., 1992).

Specificity of consumers for native litter species may also extend to different reaches of the same stream in the same watershed. For example, McArthur, et al. (1985) demonstrated a distinct preference of bacteria from headwater streams in the Konza Prairie of Kansas for dissolved organic substrates derived from grasses while those of downstream reaches in gallery forests were adapted to extracts of characteristic trees such as oak. Because of the amplified relationship between precipitation and runoff, especially in arid and semiarid regions (Wigley and Jones, 1985), a slight change in precipitation and consequent runoff can move this grassland-gallery forest riparian boundary substantially in an up- or downstream direction—more rapidly than the grassland-forest boundary in nearby upland regions. This change in the riparian zone will likely force marked alterations in aquatic communities and the nature and rates of ecosystem processes. While there is little question that adjustment of the aquatic sector can keep pace with the slower changing riparian vegetation, consequences of this boundary shift may be translated far downstream through transport.

A surrogate for climate change in riparian forest species composition was provided by the American chestnut blight of the first half of this century which removed a dominant species (>50 percent of forest canopy) from forests of eastern North America and altered the composition of leaf input to streams. Smock and Mac-Gregor (1988) suggested that this shift from chestnut to oak- or hickory-dominated forests may have had several effects on streams. For example, increased input of wood may have increased the frequency of organic dams, which are important organic-matter storage sites in streams (Bilby and Likens, 1980). This would decrease organic matter spiralling length and enhance decomposition (Newbold, et al., 1982). On the other hand, chestnut leaves decompose faster than oaks, and are a higher-quality food for and enhance fecundity and secondary production of insects which consume leaves directly (Smock and MacGregor, 1988). This effect is muted somewhat in the southern range of chestnut where hickory, a species with food quality similar to chestnut, was the predominant replacement species.

In the American Southwest, saltcedar (*Tamarisk* spp) provides an analog for change in composition of riparian vegetation that might be induced by climate. Tamarisk has rapidly spread through the Colorado River system by virtue of moderated discharge and reduced flooding—not from changing climate, but by flow regulation (Stanford and Ward, 1986). Tamarisk provides a poorer-quality habitat for riparian species such as birds (Johnson, et al., 1977), but in dense stands, it also traps sediments, constricts river channels, and increases the incidence of overbank flows (Graf, 1978). Future climate-related changes in either natural flow regime or temperature may extend the northern limits of this now widely distributed and dominant riparian species.

30.2.5 Ecosystem Metabolism

Stream ecosystem metabolism can be described as whole-system primary production and respiration, import and export of organic matter, and dynamics of intrasystem organic-storage pools (Fisher and Likens, 1973). Both temperature and hydrology influence these variables. Temperature directly affects rate of photosynthesis and respiration in streams, but it does so differentially; thus as temperature rises, *respiration R* tends to increase faster then *primary production P* and the P/R ratio falls (Busch and Fisher, 1981; McIntire and Phinney, 1965), assuming ample organic substrates are present in storage. Other factors being equal, a drop in P/R renders streams more sinklike than sourcelike for organic matter and will decrease export to larger rivers, lakes, reservoirs, or estuaries.

This simple model of stream metabolism assumes other inputs and outputs remain constant, but they may not. Litter input from the riparian zone may change in quantity and quality as temperature rises. Dissolved organic carbon entering streams via interflow may decrease as soil respiration rises (Meyer and Pulliam, 1992). Decomposition of organic matter in situ may change as well if shredder (leaf-consuming) macroinvertebrate populations are enhanced or decline as a result of increased temperature or altered flow regimes.

In some streams and rivers, wood is an important component of organic matter storage and serves as a retention device for leaves and fine particulate organics (Bilby and Likens, 1980; Smock and MacGregor, 1988). Ecosystem respiration can rise if organic storage is enhanced at the expense of export (Fisher, 1977). Furthermore, wood provides an important, although rare, habitat in sand-bottom rivers and supports large populations of filter feeders which reduce seston in transport (Benke, et al., 1984). Wood also alters stream geomorphology, generally increasing channel length and area (Sedell and Froggatt, 1984; Minckley and Rinne, 1985) and increasing the area available for both photosynthesis and respiration. Amount of wood present depends upon riparian dynamics, forest successional state (Bilby and Likens, 1980), and recent history of watershed burning (Minshall, et al., 1989). While wood derived from conifers and most hardwoods decomposes slowly, some species such as alder, typical of earlier successional stages, decompose rapidly (Dudley and Anderson, 1987). Streams draining watersheds in early successional stages thus are starved for wood and organic-matter storage (and metabolism) is reduced. As previously noted, climate change can result in altered fire frequency (Overpeck, et al., 1990), riparian composition, and invertebrate communities; thus there is reasonable expectation that climate will also affect the dynamics of stored organic matter.

Time since flood disturbance and flood magnitude were the most significant variables explaining algal standing crop in Sycamore Creek, Arizona. By inference, integral primary production also is more closely related to disturbance than to temperature or any other factor (Fisher and Grimm, 1988). Downstream trends in algal abundance in Salmon River, Idaho, are partially attributable to scouring (Cushing, et al., 1983). Biggs (1988), in a survey of 66 streams in New Zealand, reported that disturbance by flooding was the most significant factor shaping photosynthetic capacity. Alteration in magnitude and frequency of disturbance events thus can greatly alter the annual pattern and cumulative rate of primary production in streams. This does not hold everywhere; for example, eastern U.S. streams show little response in state variables to floods of a magnitude which occurs several times each year (Fisher and Grimm, 1988; Skinner and Arnold, 1988).

Floods may alter organic matter accumulations and export leaves when they are most abundant in the system (Fisher, 1977; Bilby and Likens, 1980). Shredder macroinvertebrates are positively correlated with abundance of leaf litter (Smock, et al., 1989) and may experience food limitation in late summer just prior to leaf fall (Grafius and Anderson, 1979). Since shredders break down coarse organic particles and ecosystem respiration is an inverse function of particle size (Naiman and Sedell, 1979), climate change resulting in flows which diminish preautumn standing crops of detritus can restrict ecosystem respiration through the winter.

Disturbance can also remove entire ecosystem subsystems. Grimm and Fisher (1984) showed that the hyporheic zone can be an important contributor to ecosystem respiration, reducing the P/R ratio in proportion to hyporheic volume. Storm events of 1 to 5 year return can alter hyporheic volume by export of sand and shift the P/R ratio upward (Fisher, 1990; Grimm and Fisher, 1992). In this manner, altered hydrology due to changing climate can convert a system which is largely a processor of organic matter ($P/R < 1$) to one which is primarily an exporter ($P/R > 1$).

Organic matter budgets for heterotrophic streams of coniferous forests are equally sensitive to hydrologic regime, and streams are far less efficient as processors of organic matter during wet than dry years (Sedell, et al., 1974; Cummins, et al., 1983). Spiralling distance, an index of processing efficiency (Newbold, et al., 1981), also is increased by increased discharge. This affects not only spiralling of organic matter (Minshall, et al., 1983) but inorganic substances as well, since organic substrates are especially active in nutrient retention (Munn and Meyer, 1990).

30.2.6 Geomorphology

Fluvial geomorphological processes are strongly influenced by climate, thus the geomorphic setting of lotic ecosystems will change as climate changes. Erosion rates and sediment transport have been linked to climate (Judson and Ritter, 1964; Wilson, 1973), with highest rates associated with regions having highly variable precipitation and runoff (Harlin, 1980; Baker, 1977). Paleoclimatic data combined with stratigraphic analysis suggested that Holocene arid phases were characterized by rare, high-magnitude flash floods while floods in humid phases were more frequent but less intense (Kochel and Baker, 1982). The generalization that arid and semiarid regions experience amplified runoff response to precipitation change (Wigley and Jones, 1985; Dahm and Molles, 1992), higher streamflow variability (Poff, 1992), and more severe flash floods than more humid regions (Baker, 1977; Graf, 1988) led several authors to suggest that these systems might provide early warnings of global change (Dahm and Molles, 1992; Grimm and Fisher, 1992; Poff, 1992). The significance of geomorphic change and its interaction with hydrologic regime to stream ecosystem structure and function is clear, as both factors strongly influence the physical template against which ecological processes are set (Minshall, 1988; Poff, 1992; Ward, et al., 1992).

Climatic events with decade- to century-scale recurrence intervals also influence stream geomorphology. A late 1800s episode of accelerated erosion in the American Southwest drained *ciénegas* (desert marshes), leaving steep-walled, channelized, and incised arroyos (Hastings, 1959; Hastings and Turner, 1965). Although the cattle versus climate issue was hotly debated for many years, it is now generally agreed that both degradation by overgrazing and climatic change led to arroyo cutting in southern Arizona (Cooke and Reeves, 1976; Graf, 1979). In this case, increased frequency of intense rains and resulting severe flash floods converted complex, diverse, and stable (in terms of bank sediments) streams and ciénegas [see Hendrickson and Minckley (1984) for a description] to much more simple, highly erosive ones.

Channel widening induced by high-magnitude floods following droughts is exacerbated by sparse riparian and channel vegetation and has been shown to continue for decades, until periods of increased precipitation permit reestablishment of vegetation (Schumm and Lichty, 1963; Burkham, 1972; Martin and Johnson, 1987). This type of geomorphic response to precipitation variation at decade scales has parallels in longer-term arid-humid cycles in the Holocene in Australia (Schumm, 1968) and Wisconsin (Knox, 1972), when arid phases were characterized by sparse vegetation, high peak flows, and wide, straight channels, compared to the vegetated, sinuous, more moderated flows of humid phases.

In forested streams of the Pacific Northwest, large particulate detritus (wood) is an important geomorphic agent with a long residence time. Large woody debris in streams influences channel morphology (Keller and Swanson, 1979; Triska, 1984; Harmon, et al., 1986) and stream ecosystem processes (Bilby and Likens, 1980), but can be removed by debris flows (Benda, 1990). Because they are triggered by landslides and earthflows (Swanson, et al., 1987; Benda, 1990) which result from extreme rainstorms, frequency of debris flows will no doubt change as precipitation variability and timing changes.

30.2.7 Hyporheic and Groundwater Ecosystems

Landscape units including streams and lakes are intimately connected by a continuous aquatic thread. Streams and rivers receive water from overland flow and groundwater, transport water down channels, and deliver it to receiving systems such as lakes, estuaries, or—particularly in arid regions—aquifers. Streams support a rich benthic community at the sediment–water interface and an interactive water-column community as well. In the past decade, ecological studies have revealed that an equally rich, taxonomically distinct, and scientifically poorly known community exists below stream sediments (Pennak and Ward, 1986; Strayer, 1988; Strommer and Smock, 1989; Danielopol, 1989; Williams, 1989). This hyporheic meiofaunal community is composed of many invertebrate taxa and may achieve high densities. Furthermore, this community often extends far beyond the stream margin, under the riparian zone, into distal stream terraces, and in many cases, into groundwater (Stanford and Ward, 1988; Vervier and Gilbert, 1991). Community composition of the hyporheic fauna varies as a function of its position relative to groundwater flux, chemical characteristics of the hyporheic environment, and degree of hydrologic connectance with surface streams (Boulton, et al., 1992; Dole-Olivier and Marmonier, 1992). Recent studies of the extent and role of this hyporheic environment have revolutionized our view of flowing-water ecosystems.

During floods, the hydrologic linkage between streams and hyporheic or groundwater zones is accentuated (Marmonier and Creuze des Chatelliers, 1991). As surface flow declines during drying, the hyporheic zone assumes more importance in its effects on surface-water chemistry and biota (Stanley and Valett, 1992), including input of low-oxygen and high-nutrient water (Valett, et al., 1994). The ultimate cessation of hydrologic connection between these subsystems (Stanley and Valett, 1992), however, can result in significant stresses to biota in these underground habitats.

Climate, especially hydrology, is critical to the continued existence of this unique fauna and to the bacterial processes which occur in hyporheic environments. These organisms and these functions continue to exist in southwestern U.S. stream beds without surface flow in summer; however, the extent of this tolerance is unknown at present. Climate change in a direction which increases water in channels is likely to accentuate this habitat and the role of hyporheic organisms to whole ecosystem functioning. Dewatering will imperil this group and these important functions.

30.3 EFFECTS ON LAKE ECOSYSTEMS

In temperate lake ecosystems, climate change may alter seasonal metabolism through effects on length of the ice-free or stratified periods, thermal structure and water temperature, and surface irradiance. Changes in water-renewal time and trophic structure, which can also affect ecosystem metabolism, act at multiyear time scales and were addressed earlier.

In temperate lakes, global warming is expected to increase the heat content and the durations of the ice-free and stratified seasons (McCormick, 1990). Analyses across a latitude gradient from the equator to 75°N showed that primary production was directly related to mean air temperature and length of the growing season (Brylinsky and Mann, 1973). Trophic transfer efficiencies across the same gradient suggest that secondary and tertiary production are also related inversely to latitude (Brylinsky and Mann, 1973). The seasonal variability of primary production within the year increased with latitude across a similar gradient (Melack, 1979). An empirical analysis of responses of aquatic systems to temperature change suggested that global warming would lead to a general increase in lake productivity at all trophic levels (Regier, et al., 1990).

Changes in cloud cover are among the critical uncertainties in projections of future climate (Dickinson, 1989). Increased frequency of cloudy days has significant implications for deep-dwelling phytoplankton that account for substantial primary production in oligotrophic and mesotrophic lakes during summer stratification (Moll and Stoermer, 1982). In Lake Michigan, for example, up to 70 percent of the primary production occurs in the deep algal layer (Moll and Stoermer, 1982). Deep algal layers exploit high rates of nutrient supply by eddy diffusion from the hypolimnion, but can be limited by grazers, low temperatures, or low light (St. Amand and Carpenter, 1993). If global warming reduces vertical mixing (McCormick, 1990), the metalimnion will become a larger, even more important habitat for phytoplankton. Increased cloud cover may reduce surface irradiance to the point where algae cannot maintain a positive carbon balance in deep, nutrient-rich water. If photosynthesis were restricted to nutrient-poor surface waters, primary production would decrease by factors of two to three (Moll and Stoermer, 1982; St. Amand and Carpenter, 1993).

The net effects of global change on lake ecosystem metabolism are not yet resolved. Longer growing seasons, warmer temperatures, and decreased water-renewal rates should stimulate production. On the other hand, lower surface irradiance may dramatically lower production in certain lakes. Increased water-renewal rates in some regions should decrease production. Changes in community structure can have powerful effects on the utilization of production at higher trophic levels and feedbacks that can alter ecosystem metabolism (Carpenter, et al., 1985; Carpenter, et al., 1992; Schindler, 1990; Schindler and Bayley, 1990).

30.4   WATER QUALITY

In both lakes and streams, changes in water supply and temperature may have profound effects on water quality, even in the absence of compensatory adjustments in water use as the human population responds to the stresses of climate warming. Water quality of rivers will reflect changes in both in-stream and watershed processes, since the chemical content of a river is shaped by large areas of landscape. Changes in lake-water quality will depend upon internal processes but also will respond to alteration of water-renewal rate.

In lotic environments, increased concentration of pollutants can result from reduced flows (Larimore, et al., 1959; Towns, 1985; Boulton and Lake, 1990). Much of what is known about water-quality changes associated with extended drought comes from rivers receiving sewage effluent (Slack, 1977; Walling and Foster, 1978; Chessman and Robinson, 1987). Chessman and Robinson (1987), for example, reported poor water quality in an Australian river associated with record-low flows. These water-quality changes included reduced oxygen concentration, a likely consequence of reduced or intermittent flow even in unimpacted streams (Stanley and Valett, 1992). Because rivers often are exploited for their "purification" capabilities, the combined stresses of reduced flow and increased human population are likely to result in negative water-quality effects similar to those documented for droughts. For this reason, assessment of the likelihood of increased or extended extreme low flows in streams influenced by pollutant discharges is a high priority (see also Ford and Thornton, 1992; Meyer and Pulliam, 1992).

Even without reduced streamflow, increased concentrations of nutrients and pollutants are evident in many rivers (Smith, et al., 1991). Turner and Rabalais (1991) documented increases in nitrogen and phosphorus and decreases in silicate in the Mississippi River during the 20th century, which they attributed to increased fertilizer use. Nitrate concentration in river water in Iowa increased as expected during drought, but also during wet periods as nitrate stored on agricultural fields or in groundwater was released to streams (Lucey and Goolsby, 1993). Salinization (increased salt content) due to irrigation of drylands is a global problem (Williams, 1987) that is likely to increase with desertification (Verstraete and Schwartz, 1991) and increased water demand attending warming.

Water-quality changes in lakes will result from the balance of allochthonous inputs, utilization or storage (sedimentation) within the lake, and the rate at which pollutants or nutrients are flushed from the lake. Increased precipitation might be expected to increase allochthonous inputs to lakes, but the detrimental effect on water quality might be offset by an increase in the flushing rate. In general, concentration of nutrients or contaminants varies inversely with the water-renewal rate (Schindler, 1978; Reckhow and Chapra, 1983). Schindler, et al. (1990) reported increased concentrations of chemical constituents in boreal lakes of western Ontario associated with decreased precipitation and increased evaporation during 1969 to 1989. They attributed these water-quality changes to decreased water-renewal rate and effects of watershed fires (i.e., allochthonous inputs). Internal processes also played a role in water-quality changes: increased nutrient concentration enhanced primary production and this reduced hypolimnetic oxygen (Schindler, et al., 1990).

Temperature changes in lakes (and to a lesser degree, streams) can directly affect water quality. Solubility of oxygen and other gases decreases with increasing temperature, and at the same time elevated temperature results in higher microbial respiration, which can further reduce oxygen. Changes in seasonality of lake turnover and onset of stratification discussed earlier can affect timing and extent of nutrient release from sediments.

Consideration of water-quality changes should recognize interaction of groundwater with surface aquatic ecosystems. Where groundwater feeds streams, climate-induced changes in its chemical or thermal characteristics may alter stream water quality. For example, Meisner, et al. (1988) proposed this scenario: Because of altered vegetative cover as well as elevated surface air temperature, groundwaters may warm. Warmer groundwater could be associated with depressed oxygen concentration at discharge zones, owing to reduced oxygen solubility at higher temperature and increased groundwater microbial metabolism. Their analysis considered effects of such a scenario on salmonid fishes (Meisner, et al., 1988). An interesting case of groundwater impact is presented by lakes that derive most of their chemical inputs from rainwater. A delicate balance exists between acid-neutralizing groundwater inputs and acidic precipitation in some softwater lakes; drought tips the balance toward precipitation inputs, resulting in lake acidification (Webster, et al., 1990). Finally, the quality of groundwater is affected by surface processes and water use prior to recharge; water-quality changes discussed above therefore will ultimately have an effect on groundwaters.

30.5 CONSERVATION ISSUES

The potential for alteration of ecosystem structure and functioning in response to global warming is substantial, and adjustments will be made to some extent via shifts in community composition, species replacements, and changes in fluvial geomorphology. Some of these adjustments may result in substantial changes in extant communities and in severe cases, the eradication of entire species. Davis (1989) discusses the poleward movement of terrestrial vegetation that will be necessitated by global warming. She predicts that north-temperate vegetation will be required to move northward 100 km per °C of warming, or about 300 km per century. This rate of movement is historically unprecedented and may lead to the extinction of many species with limited powers of dispersal.

Aquatic organisms will experience similar pressures to extend their ranges poleward. In Minnesota, streamwater temperatures are projected to increase by 2.4 to 4.7°C with a doubling of CO_2 (Stefan and Sinokrot, 1993). Simultaneous removal of riparian vegetation will add another 6°C to this. This will move the northern boundaries for largemouth bass 200 km (McCauley and Kilgour, 1990) and will result in northward extentions of yellow perch and smallmouth bass as well (Shuter and Post, 1990). Studies of brook trout in two Ontario streams indicate a reduction of available summer habitat by 30 to 42% with a 4°C temperature rise and a general headwater shift of these species in these drainages (Meisner, 1990). Striped bass are likely to be extirpated in southern extremities of their range and to expand northward. Combined effects of warming, for example deoxygenation of deep waters of the Chesapeake Bay, may eliminate striped bass from areas where they are now abundant and commercially important (Coutant, 1990). Paleoecological analogues were used to project Chinook Salmon abundance in the Columbia River near Yakima under a 2°C temperature increase. This species is predicted to decrease by at least 50 percent, genetic diversity of this small population will likely be reduced, and commercial losses will approach $20 million basin wide (Chatters, et al., 1991).

The imperative of poleward migration presents a challenge to aquatic and semiaquatic organisms. Many lakes are hydrologically isolated or exist in limited drainages. Stream and river organisms are similarly restricted. Those of north–south flowing drainages such as the McKenzie and Mississippi in North America, the Nile and Murray-Darling in Australia, and the Ob-Irtysh and Mekong in Asia may adjust by poleward range adjustments within the drainage. In order for movement in response to temperature change to be successful, other environmental requirements must also be met; for example, suitable substrates for spawning and an adequate food supply. East–west oriented rivers such as the Amazon, the Zaire, the Yangtze, and the Huang, which subtend only 5 to 8 degrees of latitude, will provide little opportunity for longitudinal adjustment, and many species may be lost. In the Zaire, for example, 80 percent of fish species are endemic to this single river system. Fortunately, latitudinal shifts are likely to be less pronounced in the tropics. Obviously, adjustments to temperature change by longitudinal shifts require that the river be navigable by the organisms in question. High dams and water projects on many systems will prohibit this process.

Organisms of even small streams of east and west continental margins will be unable to adjust to temperature by migration within the river corridor. Range shifts are likely to be pronounced and local endemics are subject to extirpation. Endemism is high in salmonines of the Pacific Northwest, the Sierra Nevada, and the Rocky Mountains (Benke and Zarn, 1976; Moyle, 1976). Many of the endangered fishes of western North America are restricted to one or a few localities or are already highly stressed by habitat modification associated with water projects (Mathews and Zimmerman, 1990; Minckley, 1973). Simplification of the world's ichthyofauna is a likely consequence of global warming.

Riparian vegetation is similarly threatened, especially in arid regions where tree distribution is restricted to watercourses and where water diversions and flow regulation have already greatly reduced suitable habitat. On a national basis, nearly two-thirds of riparian vegetation has been lost in the past 100 years. In the western United States, this figure is nearer 80 percent and up to 98 percent in heavily used drainages such as the Sacramento in California (Swift, 1984). This is an especially crucial environment considering the high percentage of terrestrial species that are riparian obligates—up to 60 percent of vertebrates of the American Southwest, for example (Olendorf, et al., 1989). It may be difficult for riparian species to adjust to changing climate by range extensions. Riparian plants, especially those in an arid matrix, are nearly as restricted to the watercourse as are fish—and their rate of movement is substantially slower.

Conservation efforts by such organizations as The Nature Conservancy have been focused on the identification and preservation of wetland areas of unique composition but of limited size. The specter of climate change casts doubt on the wisdom of this approach. In many cases, the combination of thermal and hydrologic factors that have made these preserves so unique are likely to abandon them in their pole-ward march, leaving behind a community of little distinction.

30.6 FRESHWATER SYSTEMS AS FEEDBACKS ON CLIMATE CHANGE

Not only are freshwater ecosystems influenced by climate change, they also play a role as drivers or moderators through positive and negative feedbacks. *Feedbacks* occur through three principal mechanisms: (1) greenhouse trace-gas production, (2) alteration of carbon storage and consequent uptake or mineralization of atmospheric carbon, and (3) direct moderation or control of local and/or regional climate by large aquatic systems. In addition, a fourth indirect effect on climate occurs through anthropogenic alteration of existing feedback systems (e.g., drainage of wetlands, groundwater pumping, and river channelization).

Freshwater ecosystems are important sites of methanogenesis (Schindler and Bayley, 1990). Methane is an important greenhouse gas because its potency is 50 times that of CO_2 (Rodhe, 1990). On a global basis, wetlands contribute 22 percent of the annual methane flux to the atmosphere (Harriss and Frolking, 1992). Polar and tropical regions are especially active sites of methanogenesis (Harriss and Frolking, 1992) because waterlogged soils and anoxia (required for the anaerobic process of methanogenesis) are so prevalent there. Regional warming is expected to be greatest at the poles (Mitchell, et al., 1989), thus northern wetlands are of particular interest. Increased temperature will thaw permafrost and increase active-layer depth, which may increase the extent of wetlands and dry those that currently exist (Roots, 1989). The balance of these two processes will determine whether northern latitude regions increase or decrease as methane producers. Long-term changes in extent of wetlands anywhere on the globe are predictable only if precipitation and temperature change are accurately known. Paleoclimatic evidence of low atmospheric methane concentration during the cold Younger Dryas was attributed to freezing of northern wetlands and drying of tropical wetlands (Chappellaz, et al.,

1990). A reversal of these conditions would indicate that wetland methane emission is a positive feedback to climate warming.

Over the short term, direct temperature effects on methane production rates have been modeled by Harriss and Frolking (1992). Methane emission rates are much higher in summer than in winter (Whalen and Reeburgh, 1988), thus their model compared rates for hypothetical hot versus cold summers. They showed a 58 percent increase from the cold to the hot summer, and concluded that methane emission was highly sensitive to temperature. In contrast, modeled methane flux from a subtropical blackwater river's floodplain was more sensitive to precipitation change (which influences inundation patterns) than to temperature increase (Meyer and Pulliam, 1992). In the latter case, predictions of a drier climate (20 percent reduction in precipitation) would translate to 40 to 60 percent reduction in methane flux. Expansion of aridity in some regions caused by climate change coupled with desertification could thus act as a negative feedback on climate change in that methane production of aridland streams is less than 1 percent of rates reported for the Southeast (Jones, et al., in press).

Riverine wetlands associated with large rivers in tropical regions are highly productive ecosystems (Welcomme, 1979; Lowe-McConnell, 1975) and often support extensive anaerobic zones (Richey, et al., 1988). Methane production is very high in Amazon floodplain wetlands (Devol, et al., 1988). Floodplains are connected to rivers via seasonal cycles of flood and drought (Welcomme, 1979; Ward, 1989). Few speculations with respect to the influence of climate change on these ecosystems have appeared in the literature, but it is likely that disruptions of the hydrologic cycle will profoundly influence seasonal patterns of inundation and thereby affect anaerobic processes.

One of the largest uncertainties in global climate feedbacks concerns alteration of carbon storage. Will increased CO_2 stimulate productivity sufficiently to counterbalance greenhouse effects? It is likely that increased temperature will result in thawed permafrost. Large quantities of carbon stored as peat will then be subject to increased erosion and decomposition. Aquatic systems are particularly important in that carbon export in arctic rivers is derived from erosion of peat (Peterson, et al., 1986; Oswood, et al., 1992). This erosion is a substantial percentage of terrestrial carbon accumulation (Peterson, et al., 1986), and should increase as temperatures rise. In Alaska, where temperature increase over the past century has already increased the depth of the active layer (= depth to permafrost; Lachenbruch and Marshall, 1986), CO_2 emission from aquatic ecosystems (rivers, lakes, and thaw ponds) may be a significant carbon source. Kling, et al. (1991) estimate that annual CO_2 flux from aquatic ecosystems balances carbon accumulation on the Alaskan coastal plain.

Large freshwater ecosystems moderate regional climates through effects on the hydrologic cycle and on ocean circulation (Carpenter, et al., 1992). Water evaporated from extensive tropical wetlands falls as rain over a much wider area (e.g., all of South America plus the southwestern Atlantic Ocean for Amazon basin evaporation; Melack, 1992); drying or expansion of these areas will thus influence climate over large regions. Changes in discharge from the world's large rivers may alter ocean circulation and thus influence world climate (Broeker and Denton, 1989; Broeker, et al., 1990). Melack (1992) summarized paleooceanographic evidence for a major climatic impact on North Atlantic regions of cool glacial meltwater discharged by the Mississippi River into the Gulf of Mexico. North-flowing rivers like the MacKenzie can move enough heat into arctic regions to mitigate regional climate (Schindler and Bayley, 1990). Glacier melting due to global warming, in addition to having a large effect as a catastrophic disturbance to stream ecosystems (Oswood, et al., 1992), may also contribute to changes in oceanic circulation and thus climate.

30.7 INTERACTIONS OF CLIMATE CHANGE AND ANTHROPOGENIC CHANGE

In addition to the anthropogenic climate feedback introduced by fossil-fuel burning, numerous other anthropogenic changes will strongly alter the climate-feedback system. In most developed nations, there has already been a substantial drainage and filling of wetland environments (NRC, 1992) which are important to trace gas budgets. Channel simplification of most major rivers (and even smaller ones; Benke, 1990) in North America and Europe has reduced the extent of floodplains and associated wetlands. This habitat modification includes manipulations such as channelization, levee building, impoundment, flow diversion, and snag removal. Several historical reconstructions have indicated that snag removal and channelization efforts to "clean up waterways" for navigation purposes grossly simplified once complex, heterogeneous riverine environments (Sedell and Froggatt, 1984; Triska, 1984; Bravard, et al., 1986), much to the detriment of fisheries productivity (Sedell, et al., 1990). Because they have been so drastically modified, we simply do not know the historical importance of temperate-zone floodplain wetlands to climate feedbacks. A related human influence was beaver trapping that reduced populations of this important vertebrate species to near extinction in North America. Naiman, et al. (1991) calculate that 1 percent of the rise in atmospheric methane over the past 50 years can be attributed indirectly to population increase of beaver, since these animals impound streams and increase the extent of anaerobic zones (Coleman and Dahm, 1990) and thus rates of methane emission (Ford and Naiman, 1988; Naiman, et al., 1991).

One major feedback to climate change often overlooked is that between water quality and both the activity and size of the human population. Assured, naturally controlled, high-quality water supplies for municipal and agricultural uses obviate the need for massive economic and industrial investments in water development, water treatment, desalinization, fertilizer application, health care, flow diversion, and stream restoration which inevitably follow from abuses of water supply. All of these industrial activities require combustion of fossil fuels and the release of CO_2 and other greenhouse gases as byproducts.

In sum, history has shown that humans can egregiously degrade aquatic ecosystems, foul water supplies, and eradicate aquatic species without the assistance of global warming. Global warming effects are uncertain as we enter the twenty-first century. Considered apart from other assaults on the aquatic environment, global warming is cause for serious concern; but other practices such as dam construction, irrigation practices, groundwater pumping, channelization, water diversion, release of toxic chemicals, and introduction of exotic organisms are more immediate and urgent in the short run. The problem is that global warming, superimposed on other impacts of a burgeoning human population on water resources, will accelerate the kinds of water projects that have been so deleterious in the past. The question then is not so much one of what climate change will do to aquatic ecosystems, but what humans will do to aquatic ecosystems in trying to avert its effects.

REFERENCES

Baker, V. R., "Stream-Channel Response to Floods, with Examples from Central Texas," *Geol. Soc. Amer. Bull.*, 88:1057–1071, 1977.

Baker, W. L., "Climatic and Hydrologic Effects on the Regeneration of *Populus angustifolia James* along the Animas River, Colorado," *Journal of Biogeography*, 17:59–73, 1990.

Benda, D. J., "The Influence of Debris Flows on Channels and Valley Floors in the Oregon Coast Range, U.S.A., *Earth Surf. Proc. Landforms*, 15:457–466, 1990.

Benfield, E. F., D. S. Jones, and M. F. Patterson, "Leaf Pack Processing in a Pastureland Stream," *Oikos*, 29:99–103, 1977.

Benke, A. C., Jr., T. C. Van Arsdall, D. M. Gillespie, and F. K. Parrish, "Invertebrate Productivity in a Subtropical Blackwater River: The Importance of Habitat and Life History," *Ecological Monographs*, 54:25–63, 1984.

Benke, A. C., "A Perspective on America's Vanishing Streams," *Journal of the North American Benthological Society*, 9:77–88, 1990.

Benke, R. J., and M. Zarn, "Biology and Management of Threatened and Endangered Western Trouts," USDA Forest Service Technical Report RM-28. Rocky Mountains Forest and Range Experimental Station, Fort Collins, Colo., 1976.

Biggs, B. J. F., "Algal Proliferations in New Zealand's Shallow Stony Foothills-Fed Rivers: Toward a Predictive Model," *Verh. Internat. Verein. Limnol.*, 23:1405–1411, 1988.

Bilby, R. E., and G. E. Likens, "Importance of Organic Debris Dams in the Structure and Function of Stream Ecosystems," *Ecology*, 61:1107–1113, 1980.

Boulton, A. J., "Oversummering Refuges of Aquatic Macroinvertebrates in Two Intermittent Streams in Central Victoria," *Trans. Roy. Soc. S. Australia*, 113:23–34, 1989.

Boulton, A. J., and P. S. Lake, "The Ecology of 2 Intermittent Streams in Victoria, Australia. 1. Multivariate Analyses of Physicochemical Features," *Freshwat. Biol.*, 24:123–141, 1990.

Boulton, A. J., and P. J. Suter, "Ecology of Temporary Streams—An Australian Perspective," in P. DeDeckker and W. D. Williams, eds., *Limnology in Australia*, CSIRO/Junk Publishers, Melbourne and the Hague, 1986, pp. 313–327.

Boulton, A. J., H. M. Valett, and S. G. Fisher, "Spatial Distribution and Taxonomic Composition of the Hyporheos of Several Sonoran Desert Streams," *Archiv Fur Hydrobiologie*, 125:37–61, 1992.

Bravard, J. P., C. Amoros, and G. Pautou, "Impact of Civil Engineering Works on the Successions of Communities in a Fluvial System," *Oikos*, 47:92–111, 1986.

Broecker, W. S., and G. H. Denton, "The Role of Ocean-Atmosphere Reorganizations in Glacial Cycles," *Geochim. Cosmochim. Acta*, 53:2465–2501, 1989.

Broecker, W. S., T. H. Peng, J. Jouzel, and G. Russell, "The Magnitude of Global Fresh-Water Transports of Importance to Ocean Circulation," *Clim. Dynam.*, 4:73–79, 1990.

Brylinsky, M., and K. H. Mann, "An Analysis of Factors Governing Productivity in Lakes and Reservoirs," *Limnol. Oceanogr.*, 18:1–14, 1973.

Burkham, D. E., "Channel Changes of the Gila River, Safford Valley, Arizona, 1846–1970," U.S. Geological Survey Prof. Paper 655-G, 1972.

Burton, T. M., and G. E. Likens, "The effect of Strip-Cutting on Stream Temperatures in the Hubbard Brook Experimental Forest, New Hampshire," *BioScience*, 23:433–435, 1973.

Busch, D. E., Fisher, S. G., "Metabolism of a Desert Stream," *Freshwat. Biol.*, 11:301–307, 1981.

Carpenter, S. R., ed., *Complex Interactions in Lake Communities*, Springer-Verlag, New York, 1988.

Carpenter, S. R., S. G. Fisher, N. B. Grimm, and J. R. Kitchell, "Global Climate Change and Freshwater Ecosystems: Lakes and Streams," *Annual Review of Ecology and Systematics*, 23:119–139, 1992.

Carpenter, S. R., J. F. Kitchell, and J. R. Hodgson, "Cascading Trophic Interactions and Lake Productivity," *Bioscience*, 35:634–639, 1985.

Chappelaz, J., J. M. Barnola, D. Raynaud, Y. S. Korotkevich, and C. Lorius, "Ice-core Record of Atmospheric Methane over the Past 160,000 Years." *Nature*, 345:127–131, 1990.

Chatters, J. C., D. A. Nietzel, M. J. Scott, and S. A. Shankle, "Potential Impacts of Global Climate Change on Pacific Northwest Spring Chinook Salmon (*Oncorhynchus tshawytscha*): An Exploratory Case Study," *The Northwest Environmental Journal*, 7:71–92, 1991.

Chauvet, E., and H. DeCamps, "Lateral Interactions in a Fluvial Landscape: The River Garonne, France," *J. No. Am. Benthol. Soc.,* 8:9–17, 1989.

Chessman, B. C., and D. P. Robinson, "Some Effects of the 1982–1983 Drought on Water Quality and Macroinvertebrate Fauna in the Lower LaTrobe River, Victoria," *Austr. J. Mar. Freshwat. Res.,* 38:288–299, 1987.

Cohen, S. J., "Influences of Past and Future Climates on the Great Lakes Region of North America," *Water International,* 12:163–169, 1987.

Cohen, S. J., "Possible Impacts of Climatic Warming Scenarios on Water Resources in the Saskatchewan River Sub-Basin, Canada," *Climatic Change,* 19:291–317, 1991.

Coleman J. M., "Climatic Warming and Increased Summer Aridity in Florida USA," *Climatic Change,* 12:165–178, 1988.

Coleman, R. L., and C. N. Dahm, "Stream Geomorphology: Effects on Periphyton Standing Crop and Primary Production," *Journal of the North American Benthological Society,* 9:293–302, 1990.

Cooke R. U., and R. W. Reeves, *Arroyos and Environmental Change in the American Southwest,* Oxford University Press, London, 1976.

Cooper, S. D., "The Effects of Trout on Water Striders in Stream Pools," *Oecologia,* 63:376–379, 1984.

Coutant, C. C., "Temperature-Oxygen Habitat for Freshwater and Coastal Striped Bass in a Changing Climate," *Trans. Amer. Fish. Soc.,* 119:240–253, 1990.

Cudney, M. D., and J. B. Wallace, "Life cycles, Microdistribution and Production Dynamics of Six Species of Net-Spinning Caddisflies in a Large Southeastern (USA) River," *Holarctic Ecol,* 3:169–182, 1980.

Cummins, K. W., J. R. Sedell, F. J. Swanson, G. W. Minshall, S. G. Fisher, C. E. Cushing, R. C. Petersen, and R. L. Vannote, "Organic Matter Budgets for Stream Ecosystems: Problems in Their Evaluation," in J. R. Barnes and G. W. Minshall, eds., *Stream Ecology: Application and Testing of General Ecological Theory,* Plenum, New York, 1983, pp. 299–353.

Cushing, C. E., K. W. Cummins, G. W. Minshall, and R. L. Vannote, "Periphyton, Chlorophyll a, and Diatoms of the Middle Fork of the Salmon River, Idaho," *Holarctic Ecol,* 6:221–227, 1983.

Cushman, R. M., and P. N. Spring, "Differences among Model Simulations of Climate Change on the Scale of Resource Regions," *Environmental Management,* 13:789–795, 1989.

Dahm, C. N., and M. C. Molles, Jr., "Streams in Semiarid Regions as Sensitive Indicators of Global Climate Change," in P. Firth and S. G. Fisher, eds., *Climate Change and Freshwater Ecosystems,* Springer-Verlag, New York, 1992, pp. 250–260.

Danielopol, D. L., "Groundwater Fauna Associated with Riverine Aquifers," *Journal of the North American Benthological Society,* 8:18–35, 1989.

Danks, H. V., "Some Effects of Photoperiod, Temperature and Food on Emergence in Three Species of Chironomidae (Diptera)," *The Canadian Entomologist,* 110:289–300, 1978.

Davis, M. B., "Lags in Vegetation Response to Greenhouse Warming," *Climatic Change,* 15:75–82, 1989.

DeCamps, H., M. Fortune, F. Gazelle, and G. Pautou, "Historical Influence of Man on the Riparian Dynamics of a Fluvial Landscape," *Landscape Ecology,* 1:163–173, 1988.

Devol, A. H., J. E. Richey, W. Clark, S. King, and L. Martinelli, "Methane Emissions to the Troposphere from the Amazon Floodplain," *J. Geophys. Res.,* 93:1583–1592, 1988.

Dickinson, R. E., "Uncertainties of Estimates of Climate Change: A Review," *Climate Change,* 15:5–14, 1989.

Dietrich, W. E., T. Dunne, N. F. Humphrey, and L. M. Reid, "Construction of Sediment Budgets for Drainage Basins," in F. J. Swanson, R. J. Janda, T. Dunne, and D. N. Swantson, eds., "Sediment Budgets and Routing in Forested Drainage Basins," USDA Forest Service Technical Report PNW-141, Portland, Oreg., 1982, pp. 5–23.

Dole-Olivier, M. J., and P. Marmonier, "Patch Distribution of Interstitial Communities— Prevailing Factors," *Freshwater Biology,* 27:177–191, 1992.

Dudley, T. L., and N. H. Anderson, "The Biology and Life Cycles of *Lipsothrix* spp. (Diptera: Tipulidae) Inhabiting Wood in Western Oregon Streams," *Freshwater Biology,* 17:437–451, 1987.

Dudley, T. L., S. D. Cooper, and N. Hemphill, "Effects of Macroalgae on a Stream Invertebrate Community," *J. No. Amer. Benthol. Soc.,* 5:93–106, 1986.

Dudley, T. L., and C. M. D'Antonio, "The Effects of Substrate Texture, Grazing and Disturbance on Macroalgal Establishment in Stream Riffles," *Ecology,* 72:297–309, 1991.

Fajer, E. D., M. D. Bowers, and F. A. Bazazz, "The Effects of Enriched Carbon Dioxide Atmospheres on Plant-Insect Herbivore Interactions," *Science,* 243:1198–1200, 1989.

Feminella, J. W., and V. H. Resh, "Hydrologic Influences, Disturbance, and Intraspecific Competition in a Stream Caddisfly Population," *Ecology,* 71:2083–2094, 1990.

Firth, P., and S. G. Fisher, eds., *Climate Change and Freshwater Ecosystems,* Springer-Verlag, New York, 1992.

Fisher, S. G., "Organic Matter Processing by a Stream-Segment Ecosystem: Fort River, Massachusetts, USA," *Internationale Revue der Gesamten Hydrobiologie,* 62:701–727, 1977.

Fisher, S. G., "Recovery Processes in Lotic Ecosystems: Limits of Successional Theory," *Environ. Manage.,* 14:725–736, 1990.

Fisher, S. G., L. J. Gray, N. B. Grimm, and D. E. Busch, "Temporal Succession in a Desert Stream Ecosystem Following Flash Flooding," *Ecol. Monogr.,* 52:93–110, 1982.

Fisher, S. G., and N. B. Grimm, "Disturbance as a Determinant of Structure in a Sonoran Desert Stream Ecosystem," *Verh. Internat. Verein Limnol.,* 23:1183–1189, 1988.

Fisher, S. G., and G. E. Likens, "Energy Flow in Bear Brook, New Hampshire: An Integrative Approach to Stream Ecosystem Metabolism," *Ecological Monographs,* 43:421–439, 1973.

Flaschka, I. M., C. W. Stockton, and W. R. Boggess, "Climatic Variation and Surface Water Resources in the Great Basin Region," *Wat. Res. Bull.,* 23:47–57, 1987.

Ford, D. E., and K. W. Thornton, "Water Resources in a Changing Climate," in P. Firth and S. G. Fisher, eds., *Climate Change and Freshwater Ecosystems,* Springer-Verlag, New York, 1992, pp. 26–47.

Ford, T. E., and R. J. Naiman, "Alteration of Carbon Cycling by Beaver: Methane Evasion Rates from Boreal Forest Streams and Rivers," *Canadian Journal of Zoology,* 66:529–533, 1988.

Giorgi, F., C. S. Brodeur, and G. Bates, "Regional Climates Change Scenarios over the United States Produced with a Nested Regional Climate Model," *Journal of Climate,* 7:375–399, 1994.

Gleick, P. H., "Regional Hydrologic Consequences of Increases in Atmospheric CO_2 and Other Trace Gases," *Clim. Change,* 10:127–161, 1987.

Graf, W. L., "Fluvial Adjustments to the Spread of Tamarisk in the Colorado Plateau Region," *Bulletin of the Geological Society of America,* 89:1491–1501, 1978.

Graf, W. L., "The Development of Montane Arroyos and Gullies," *Earth Surface Processes,* 4:1–14, 1979.

Graf, W. L., *Fluvial Processes in Dryland Rivers,* Springer-Verlag, New York, 1988.

Grafius, E., and N. H. Anderson, "Population Dynamics, Bioenergetics and Role of *Lepidostoma quercina Ross* (Trichoptera: Lepidostomatidae) in an Oregon Woodland Stream," *Ecology,* 60:433–441, 1979.

Gray, L. J., "Recolonization Pathways and Community Development of Desert Stream Invertebrates," PhD dissertation, Arizona State University, Tempe, Ariz., 1980.

Gray, L. J., "Species Composition and Life Histories of Aquatic Insects in a Lowland Sonoran Desert Stream," *American Midland Naturalist,* 106:229–242, 1981.

Gray, L. J., and S. G. Fisher, "Postflood Recolonization Pathways of Macroinvertebrates in Lowland Sonoran Desert Stream," *American Midland Naturalist,* 106:249–257, 1981.

Gregory, S. V., F. J. Swanson, W. A. McKee, and K. W. Cummins, "An Ecosystem Perspective of Riparian Zones," *BioScience,* 41:540–551, 1991.

Grimm, N. B., "Biogeochemistry of Nitrogen in Arid-Land Stream Ecosystem," *J. Ariz.-Nev. Acad. Sci.,* 26:130–146, 1992.

Grimm, N. B., "Implications of Climate Change for Stream Communities," in P. Kareiva, J. Kingsolver, and R. Huey, eds., *Biotic Interactions and Global Change.* Sinauer Associates, Sunderland, Mass., 1993, pp. 293–314.

Grimm, N. B., and S. G. Fisher, "Exchange between Interstitial and Surface Water: Implications for Stream Metabolism and Nutrient Cycling," *Hydrobiologia,* 111:219–228, 1984.

Grimm, N. B., and S. G. Fisher, "Nitrogen Limitation in a Sonoran Desert Stream," *J. No. Amer. Benthol. Soc.,* 5:2–15, 1986.

Grimm, N. B., and S. G. Fisher, "Stability of Periphyton and Macroinvertebrates to Disturbance by Flash Floods in a Desert Stream," *J. No. Amer. Benthol. Soc.,* 8:293–307, 1989.

Grimm, N. B., and S. G. Fisher, "Responses of Arid-Land Streams to Changing Climate," in P. Firth and S. G. Fisher, eds., *Climate Change and Freshwater Ecosystems,* Springer-Verlag, New York, 1992, pp. 211–233.

Hanson, D. L., and T. F. Waters, "Recovery of Standing Crop and Production Rate of a Brook Trout Population in a Flood-Damaged Stream," *Transactions of the American Fisheries Society,* 103:431–439, 1974.

Harlin, J. M., 1980. "The Effect of Precipitation Variability on Drainage Basin Morphometry," *Amer. J. Sci.,* 280:812–825, 1980.

Harmon, M. E., J. F. Franklin, F. J. Swanson, P. Sollins, S. V. Gregory, "Ecology of Coarse Woody Debris in Temperate Ecosystems," *Adv. Ecol. Res.,* 15:133–302, 1986.

Harper, P. P., and J. G. Pilon, "Annual Patterns of Emergence of Some Quebec Stoneflies (Insecta: Plecoptera)," *Canadian Journal of Zoology,* 48:681–694, 1970.

Harrison, A. D., "Recolonization of a Rhodesian Stream after Drought," *Archiv fur Hydrobiologie,* 62:405–421, 1966.

Harriss, R. C., and S. E. Frolking, "The Sensitivity of Methane Emissions from Northern Freshwater Wetlands to Global Warming," in P. Firth and S. G. Fisher, eds., *Climate Change and Freshwater Ecosystems,* Springer-Verlag, New York, 1992, pp. 48–67.

Hastings, J. R., "Vegetation Change and Arroyo Cutting in Southeastern Arizona," *Journal of Arizona Academy of Science,* 60–67, 1959.

Hastings, J. R., and R. M. Turner, *The Changing Mile,* University of Arizona Press, Tucson, Ariz., 1965.

Hemphill, N., "Disturbance and Variation in Competition between Two Stream Insects," *Ecology,* 72:864–872, 1991.

Hemphill, N., and S. D. Cooper, "The Effect of Physical Disturbance on the Relative Abundance of Two Filter-Feeding Insects in a Small Stream," *Oecologia,* 58:378–82, 1983.

Hemphill, N., and S. D. Cooper, "Differences in the Community Structure of Stream Pools Containing or Lacking Trout," *Verh. Internat. Verein. Limnol.,* 22:1858–1861, 1984.

Hendrickson, D. A., and W. L. Minckley, "Cienegas—Vanishing Climax Communities of the American Southwest," *Desert Plants,* 6:131–175, 1984.

Hildrew, A. G., and J. M. Edington, "Factors Facilitating the Coexistence of Hydropsychid Caddis Larvae (Trichoptera) in the Same River System," *J. Anim. Ecol.,* 48:557–576, 1979.

Hostetler, S. W., "Hydrologic and Atmospheric Models: The (Continuing) Problem of Discordant Scales," *Climatic Change,* 27:345–350, 1994.

Houghton, J. T., G. J. Jenkins and J. J. Ephraums, eds., *Climate Change: The IPCC Scientific Assessment,* Cambridge University Press. Cambridge, 1990.

Hynes, H. B. N., "Imported Organic Matter and Secondary Productivity in Streams," *Proceedings of the International Congress of Zoology,* 16:324–329, 1963.

Hynes, H. B. N., "The Stream and its Valley," *Verh. Internat. Verein. Limnol.,* 19:1–15, 1975.

Idso, S. B., and A. J. Brazel, "Rising Atmospheric Carbon Dioxide Concentrations May Increase Streamflow," *Nature,* 312:51–53, 1984.

Ingebo, P. A., "Suppression of Channelside Chaparral Cover Increases Streamflow," *J. Soil Wat. Conserv.,* 26:79–81, 1971.

Irons, J. G., M. W. Oswood, and J. P. Bryant, "Consumption of Leaf Detritus by a Stream Shredder: Influence of Tree Species and Nutrient Status," *Hydrobiologia,* 160:53–61, 1988.

Jackson, J. K., "Diel Emergence, Swarming and Longevity of Selected Adult Aquatic Insects from a Sonoran Desert Stream," *American Midland Naturalist,* 119:344–352, 1988.

Johnson, R. R., L. T. Haight, and J. M. Simpson, 1977. "Endangered Species vs. Endangered Habitats: A Concept," in R. R. Johnson and D. A. Jones, eds., "Importance, Preservation, and Management of Riparian Habitat: A Symposium," USDA Forest Service General Technical Report RM-43, Tucson, Ariz., 1977, pp. 68–79.

Jones, J. B., Jr., R. M. Holmes, S. G. Fisher, N. B. Grimm, and D. M. Greene, "Methanogenesis in Sonoran Desert Stream Ecosystems," *Biogeochemistry:* in review.

Judson, S., and D. F. Ritter, "Rates of Regional Denudation in the United States," *J. Geophys. Res.,* 69:3395–3401, 1964.

Karl, T. R., and W. E. Riebsame, "The Impact of Decadal Fluctuations in Mean Precipitation and Temperature on Runoff: A Sensitivity Study over the United States," *Clim. Change,* 15:423–447, 1989.

Keller, E. A., and F. J. Swanson, "Effects of Large Organic Material on Channel Form and Fluvial Processes," *Earth Surf. Proc.,* 4:361–380, 1979.

Kling, G. W., G. W. Kipphut, and M. C. Miller, "Arctic Lakes and Streams as Gas Conduits to the Atmosphere: Implications for Tundra Carbon Budgets," *Science,* 251:298–301, 1991.

Knox, J. C., "Valley Alluviation in Southwestern Wisconsin," *Annals of the Association of American Geographers,* 62:401–410, 1972.

Kochel, R. C., and V. R. Baker, "Paleoflood Hydrology," *Science,* 215:353–361, 1982.

Lamberti, G. A., S. V. Gregory, L. R. Ashkenas, R. C. Wildman, and K. M. S. Moore, "Stream Ecosystem Recovery Following a Catastrophic Debris Flow," *Can. J. Fish. Aquat. Sci.,* 48:196–208, 1991.

Lachenbruch, A. H., and B. V. Marshall, "Changing Climate: Geothermal Evidence from Permafrost in the Alaskan Arctic," *Science,* 234:689–696, 1986.

Lake, P. S., "The 1981 Jolly Award Address: Ecology of the Macroinvertebrates of Australian Upland Streams—A Review of Current Knowledge," *Bull. Austr. Soc. Limnol.,* 8:1–15, 1982.

Lake, P. S. and L. A. Barmuta, "Stream Benthic Communities: Persistent Presumptions and Current Speculations," in P. De Decker and W. D. Williams, eds., *Limnology in Australia,* CSIRO/Dr. W. Junk, Melbourne, Australia and Dordrecht, The Netherlands, 1986, pp. 263–276.

Larimore, R. W., W. F. Childers, and C. Heckrotte, "Destruction and Re-establishment of Stream Fish and Invertebrates Affected by Drought," *Trans. Amer. Fish. Soc.,* 88:261–285, 1959.

LeCren, E. D., and R. H. Lowe-McConnell, eds., *Functioning of Freshwater Ecosystems,* Cambridge University Press, London, 1981.

Lettenmaier, D. P., and T. Y. Gan, "Hydrologic Sensitivities of the Sacramento-San Joaquin River Basin, California, to Global Warming," *Wat. Resour. Res.,* 26:69–86, 1990.

Lettenmaier, D. P., E. F. Wood, and J. R. Wallis, "Hydro-Climatological Trends in the Continental United States, 1948–1988," *Journal of Climate,* 7:586–607, 1994.

Levine, J. S., "Global Climate Change," in P. Firth and S. Fisher, eds., *Climate Change and Freshwater Ecosystems,* Springer-Verlag, New York, 1992, pp. 1–25.

Lins, H. F., "Increasing US Streamflow Linked to Greenhouse Forcing," *EOS, Transactions,* 75:283–285, 1994.

Lowe-McConnell, R. H., "Fish Communities of Tropical Freshwaters," Longmans, London, 1975.

Lucey, and Goolsby, "Effects of Climate Variations over 11 Years on Nitrate-Nitrogen Concentrations in the Racoon River, Iowa," *Journal of Environmental Quality,* 22:38–46, 1993.

Macan, T. T., *Biological Studies of the English Lakes,* American Elsevier, New York, 1970.

Manabe, S., and A. J. Broccoli, "Mountains and Arid Climates of Middle Latitudes," *Science,* 247:192–195, 1990.

Manabe, S., and R. T. Wetherald, "Large-Scale Changes of Soil Wetness Induced by an Increase in Atmospheric Carbon Dioxide," *Journal of the Atmospheric Sciences,* 44:1211–1235, 1987.

Marmonier, P., and M. C. D. Chatelliers, "Effects of Spates on Interstitial Assemblages of the Rhone River: Importance of Spatial Heterogeneity," *Hydrobiologia*, 210:243–251, 1991.

Martin, C. W., and W. C. Johnson, "Historical Channel Narrowing and Riparian Vegetation Expansion in the Medicine Lodge River Basin, Kansas, 1871–1983," *Ann. Assoc. Am. Geog.*, 77:436–449, 1987.

Matthews, W. J., and E. G. Zimmerman, "Potential Effects of Global Warming on Native Fishes of the Southern Great Plains and the Southwest," *Fisheries*, 15:26–32, 1990.

McArthur, J. V., G. R. Marzolf, and J. E. Urban, "Response of Bacteria Isolated from a Pristine Prairie Stream to Concentration and Source of Soluble Organic Carbon," *Appl. Environ. Microbiol.*, 499:238–241, 1985.

McCabe, G. J., and M. A. Ayers, "Hydrologic Effects of Climate Change in the Delaware River Basin," *Water Resources Bulletin*, 25:1231–1242, 1989.

McCabe, G. J., and M. A. Ayers, "Hydrologic Effects of Climate Change in the Delaware River Basin: Reply," *Water Resources Bulletin*, 26:833–834, 1990.

McCauley, R. W. and D. M. Kilgour, "Effect of Air Temperature on Growth of Largemouth Bass in North America," *Trans. Amer. Fish. Soc.*, 119:276–281, 1990.

McCormick, M. J., "Potential Changes in Thermal Structure and Cycle of Lake Michigan due to Global Warming," *Trans. Am. Fish. Soc.*, 119:183–194, 1990.

McIntire, C. D., and H. K. Phinney, "Laboratory Studies of Periphyton Production and Community Metabolism in Lotic Environments," *Ecological Monographs*, 35:237–258, 1965.

Meffe, G. K., "Effects of Abiotic Disturbance on Coexistence of Predator-Prey Fish Species," *Ecology*, 65:1525–1534, 1984.

Meisner, J. D., "Potential Loss of Thermal Habitat for Brook Trout, Due to Climate Warming, in Two Southern Ontario Streams," *Trans. Amer. Fish. Soc.*, 119:282–291, 1990.

Meisner, J. D., J. S. Rosenfeld, and H. A. Regier, "The Role of Groundwater in the Impact of Climate Warming on Stream Salmonines," *Fisheries*, 13:2–8, 1988.

Melack, J. M., "Temporal Variability of Phytoplankton in Tropical Lakes," *Oecologia*, 44:1–7, 1979.

Melack, J. M., "Reciprocal Interactions Among Lakes, Large Rivers, and Climate," in P. Firth and S. G. Fisher, eds., *Climate Change and Freshwater Ecosystems*, Springer-Verlag, New York, 1992, pp. 68–87.

Merritt, R. W., and D. L. Lawson, "Leaf Litter Processing in Floodplain and Stream Communities," in R. R. Johnson and J. F. McCormick, eds., "Strategies for Protection and Management of Floodplain Wetlands and Other Riparian Ecosystems," USDA Forest Service General Technical Report WO-a2, 1979, pp. 93–105.

Meyer, J. L., and R. T. Edwards, "Ecosystem Metabolism and Turnover of Organic Carbon along a Blackwater River Continuum," *Ecology*, 71:668–677, 1990.

Meyer, J. W., and W. M. Pulliam, "Modification of Terrestrial-Aquatic Interactions by a Changing Climate," in P. Firth and S. G. Fisher, eds., *Climate Change and Freshwater Ecosystems*, Springer-Verlag, New York, 1992.

Mimikou, M., Y. Kouvopoulos, G. Cavadias, and N. Vayianos, "Regional Hydrological Effects of Climate Change," *Journal of Hydrology*, 123:119–146, 1991.

Minckley, W. L., *Fishes of Arizona*, Arizona Game and Fish Department. Phoenix, Ariz., 1973.

Minckley, W. L., and J. N. Rinne, "Large Organic Debris in Hot Desert Streams: An Historical Review," *Desert Plants*, 7:142–153, 1985.

Minshall, G. W., "Stream Ecosystem Theory: A Global Perspective," *J. No. Amer. Benthol. Soc.*, 7:263–288, 1988.

Minshall, G. W., J. T. Brock, T. W. LaPoint, "Characterization and Dynamics of Benthic Organic Matter and Invertebrate Functional Feeding Group Relationships in the Upper Salmon River, Idaho (USA)," *Int. Rev. Ges. Hydrobiol.*, 67:793–820, 1982.

Minshall, G. W., J. T. Brock, and J. D. Varney, "Wildfires and Yellowstone's Stream Ecosystems," *BioScience*, 39:707–715, 1989.

Minshall, G. W., R. C. Petersen, K. W. Cummins, T. L. Bott, J. R. Sedell, "Interbiome Comparison of Stream Ecosystem Dynamics," *Ecol. Monogr.,* 53:1–25, 1983.

Mitchell, J. F. B., C. A. Senior, and W. J. Ingram, "CO_2 and Climate: A Missing Feedback?" *Nature,* 341:132–134, 1989.

Moll, R. A., and E. F. Stoermer, "A Hypothesis Relating Trophic Status and Subsurface Chlorophyll Maxima of Lakes," *Arch. Hydrobiol.,* 94:425–440, 1982.

Molles, M. C., Jr., "Recovery of a Stream Invertebrate Community From a Flash Flood in Tesuque Creek, New Mexico," *Southwestern Naturalist,* 30:279–287, 1985.

Mosley, M. P., "New Zealand River Temperature Regimes," *Water and Soil Misc Publ. no. 36.,* Wellington, N.Z., 1982.

Moyle, P. B., *Inland Fishes of California,* University of California, Berkeley, Calif., 1976.

Munn, N. L., and J. L. Meyer, "Habitat-Specific Solute Retention in Two Small Streams: An Intersite Comparison," *Ecology,* 71:2069–2082, 1990.

Naiman, R. J., T. Manning, and C. A. Johnston, "Beaver Population Fluctuations and Tropospheric Methane Emissions in Boreal Wetlands," *Biogeochemistry,* 12:1–15, 1991.

Naiman, R. J., and S. R. Sedell, "Benthic Organic Matter as a Function of Stream Order in Oregon," *Arch. fur Hydrobiologie,* 87:404–422, 1979.

Newbold, J. D., J. W. Elwood, R. V. O'Neill, and W. Van Winkle, "Measuring Nutrient Spiralling in Streams," *Can. J. Fish. Aquat. Sci.,* 38:860–863, 1981.

Newbold, J. D., P. J. Mulholland, J. W. Elwood, and R. V. O'Neill, "Organic Carbon Spiralling in Stream Ecosystems," *Oikos,* 38:266–272, 1982.

Niemi, G. J., P. Devore, N. Detenbeck, D. Taylor, and A. Lima, "Overview of Case Studies on Recovery of Aquatic Systems from Disturbance," *Environ. Manage.,* 14:571–587, 1990.

Olendorf, R. R., D. D. Bibles, M. T. Dearn, J. R. Haugh, and M. N. Kochert, "Raptor Habitat Management under the U.S. Bureau of Land Management Multiple-Use Mandate," *Raptor Research Report,* 8:1–80, 1989.

Oswood, M. W., A. M. Milner, and J. G. Irons III, "Climate Change and Alaska Rivers and Streams," in P. Firth and S. G. Fisher, eds., *Climate Change and Freshwater Ecosystems,* Springer-Verlag, New York, 1992, pp. 192–210.

Overpeck, J. T., D. Rind, and R. Goldberg, "Climate-Induced Changes in Forest Disturbance and Vegetation," *Nature,* 343:51–53, 1990.

Peckarsky, B. L., "Biotic Interactions or Abiotic Limitations? A Model of Lotic Community Structure," in T. D. Fontaine III, and S. M. Bartell, eds., *Dynamics of Lotic Ecosystems,* Ann Arbor Science, Ann Arbor, Mich., 1983, pp. 303–324.

Pennak, R., and J. V. Ward, "Interstitial Fauna Communities of the Hyporheic and Adjacent Groundwater Biotopes of a Colorado Mountain Stream," *Archiv fur Hydrobiologie Supplement,* 74:356–396, 1986.

Petersen, R. C., Cummins, K. W., "Leaf Processing in a Woodland Stream," *Freshwat. Biol.,* 4:343–368, 1974.

Peterson, B. J., J. E. Hobbie, and T. L. Corliss, "Carbon Flow in a Tundra Stream Ecosystem," *Can. J. Fish. Aquat. Sci.,* 43:1259–1270, 1986.

Poff, N. L., "Regional Hydrologic Response to Climate Change: An Ecological Perspective," in P. Firth and S. G. Fisher, eds., *Climate Change and Freshwater Ecosystems,* Springer-Verlag, New York, 1992, pp. 88–115.

Poff, N. L., and J. V. Ward, "Implications of Streamflow Variability and Predictability for Lotic Community Structure: A Regional Analysis of Streamflow Patterns," *Can. J. Fish. Aquat. Sci.,* 46:1805–1818, 1989.

Poff, N. L., and J. V. Ward, "Physical Habitat Template of Lotic Systems: Recovery in the Context of Historical Pattern of Spatiotemporal Heterogeneity," *Environmental Management,* 14:629–645, 1990.

Poiani, K. A., and W. C. Johnson, "Potential Effects of Climate Change on a Semi-Permanent Prairie Wetland," *Climatic Change,* 24:213–232, 1993.

Power, M. E., "Effects of Fish in River Food Webs," *Science,* 250:811–814, 1990.

Power, M. E., "Hydrologic and Trophic Controls of Seasonal Algal Blooms in Northern California Rivers," *Arch. Hydrobiol.,* 125:385–410, 1992.

Power, M. E., and W. J. Matthews, "Algae-Grazing Minnows (*Campostoma anomalum*), Pisciv-orous Bass (*Micropterus* spp.), and the Distribution of Attached Algae in a Small Prairie-Margin Stream," *Oecologia,* 60:328–332, 1983.

Power, M. E., W. J. Matthews, and A. J. Stewart, "Grazing Minnows, Piscivorous Bass and Stream Algae: Dynamics of a Strong Interaction," *Ecology,* 66:1448–1456, 1985.

Power, M. E., and A. J. Stewart, "Disturbance and Recovery of an Algal Assemblage Following Flooding in an Oklahoma Stream," *The American Midland Naturalist,* 117:333–345, 1987.

Power, M. E., R. J. Stout, C. E. Cushing, P. P. Harper, and F. R. Hauer, "Biotic and Abiotic Con-trols in River and Stream Communities," *J. No. Amer. Benthol. Soc.,* 7:456–479, 1988.

Raval A., and V. Ramanathan, "Observational Determination of the Greenhouse Effect," *Nature,* 342:758–761, 1989.

Reckhow, K., and S. C. Chapra, *Engineering Approaches for Lake Management. Vol. 1. Data Analysis and Empirical Modeling,* Butterworth's Boston, 1983.

Regier, H. A., J. A. Holmes, and D. Pauly, "Influence of Temperature Change on Aquatic Ecosystems: An Interpretation of Empirical Data," *Trans. Am. Fish. Soc.,* 119:374–389, 1990.

Resh, V. H., "Age Structure Alteration in a Caddisfly Population after Habitat Loss and Recov-ery," *Oikos,* 38:280–284, 1982.

Resh, V. H., A. V. Brown, A. P. Covich, M. E. Gurtz, and H. W. Li, "The Role of Disturbance in Stream Ecology," *J. No. Amer. Benthol. Soc.,* 7:433–455, 1988.

Richey, J. E., C. Nobre, and C. Deser, "Amazon River Discharge and Climate Variability, 1903–1985," *Science,* 246:101–103, 1989.

Rodhe, H., "A Comparison of the Contribution of Various Gases to the Greenhouse Effect," *Science,* 248:1217–1219, 1990.

Roots, E. F., "Climate Change: High Latitude Regions," *Clim. Change,* 15:223–253, 1989.

Schindler, D. W., "Factors Regulating Phytoplankton Production and Standing Crop in the World's Freshwaters," *Limnol. Oceanogr.,* 23:478–486, 1978.

Schindler, D. W., "Experimental Perturbations of Whole Lakes as Tests of Hypotheses Con-cerning Ecosystem Structure and Function," *Oikos,* 57:25–41, 1990.

Schindler, D. W., and S. E. Bayley, "Fresh Waters in Cycle," in C. Mungall and D. J. McLaren, eds., *Planet Under Stress,* Oxford University Press, Toronto, 1990, pp. 149–167.

Schindler, D. W., K. G. Beaty, E. J. Fee, D. R. Cruikshank, and E. R. DeBruyn, "Effects of Cli-matic Warming on Lakes of the Central Boreal Forest," *Science,* 250:967–70, 1990.

Schneider, S. H., "Scenarios of Global Warming," in P. M. Kareiva, J. G. Kingsolver, and R. B. Huey, eds., *Biotic Interactions and Global Change,* Sinauer Associates, Sunderland, Mass., 1992, pp. 9–23.

Schneider, S. H., P. H. Gleick, and L. O. Mearns, "Prospects for Climate Change," in P. E. Wag-goner, ed., *Climate Change and U.S. Water Resources,* John Wiley, New York, 1990, pp. 41–73.

Schumm, S. A., "River Adjustment to Altered Hydrologic Regimen—Murrumbidgee River and Paleochannels, Australia," U.S. Geological Survey Prof. Paper 598, 1968.

Schumm, S. A., and R. W. Lichty, "Time, Space, and Causality in Geomorphology," *American J. Science,* 263:110–119, 1965.

Sedell, J. R., and J. L. Froggatt, "Importance of Streamside Forests to Large Rivers: The Isola-tion of the Willamette River, Oregon, U.S.A., from its Floodplain by Snagging and Streamside Forest Removal," *Verh. Internat. Verein. Limnol.,* 22:1828–1834, 1984.

Sedell, J. R., G. H. Reeves, F. R. Hauer, J. A. Stanford, and C. P. Hawkins, "Role of Refugia in Recovery from Disturbances of Modern Fragmented and Disconnected River Systems," *En-viron. Manage.,* 14:711–724, 1990.

Sedell, J. R., F. J. Triska, J. D. Hall, N. H. Anderson, and J. H. Lyford, "Sources and Fates of Organic Inputs in Coniferous Forest Streams," in R. H. Waring and R. L. Edmonds, eds., *Inte-*

grated Research in the Coniferous Forest Biome, Bulletin #5, U.S. International Biological Program, 1974.

Shuter, B. J., and J. R. Post, "Climate, Population Variability, and the Zoogeography of Temperate Fishes," *Trans. Am. Fish. Soc.,* 119:314–336, 1990.

Siegfried, C. A., and A. W. Knight, "The Effects of Washout in a Sierra Foothill Stream," *American Midland Naturalist,* 98:200–207, 1977.

Skiles, J. W., and J. D. Hanson, "Responses of Arid and Semiarid Watersheds to Increasing Carbon Dioxide and Climate Change as Shown by Simulation Studies," *Climatic Change,* 26:377–397, 1994.

Skinner, W. D., and D. E. Arnold, "Absence of Temporal Succession of Invertebrates of Pennsylvania Streams," *Bull. N. Amer. Benthol. Soc.,* 51:63, 1988.

Slack, J. G., "River Water Quality in Essex During and After the 1976 Drought," *Effluent and Water Treatment Journal,* November: 575–578, 1977.

Smith, M. E., C. T. Driscoll, B. J. Wyskowski, C. M. Brooks, and C. C. Cosentini, "Modification of Stream Ecosystem Structure and Function by Beaver (*Castor canadensis*) in the Adirondack Mountains, New York," *Canadian Journal of Zoology,* 69:55–61, 1991.

Smith, R. A., R. B. Alexander, and M. G. Wolman, "Water Quality Trends in the Nations Rivers," *Science,* 235:1607–1615, 1987.

Smock, L. A., and C. M. MacGregor, "Impact of the American Chestnut Blight on Aquatic Shredding Macroinvertebrates," *J. N. Am. Benthol. Soc.,* 7:212–221, 1988.

Smock, L. A., G. M. Metzler, and J. E. Gladden, "Role of Debris Dams in the Structure and Functioning of Low-Gradient Headwater Streams," *Ecology,* 70:764–775, 1989.

Sparks, R. E., P. B. Bayley, S. L. Kohler, and L. L. Osborne, "Disturbance and Recovery of Large Floodplain Rivers," *Environmental Management,* 14:699–709, 1990.

St. Amand, A., Carpenter, S. R., "Metalimnetic Phytoplankton Dynamics," in S. R. Carpenter and J. F. Kitchell, eds., *Trophic Cascades in Lakes,* Cambridge University Press, London, 1993.

Stanford, J. A., and J. V. Ward, "The Colorado River System," in B. R. Davies and K. F. Walker, eds., *The Ecology of River Systems,* D. W. Junk, Dordrecht, 1986, pp. 353–374.

Stanford, J. A., and J. V. Ward, "The Hyporheic Habitat of River Ecosystems," *Nature,* 335:64–66, 1988.

Stanley, E. H., and S. G. Fisher, "Intermittency, Disturbance, and Stability in Stream Ecosystems," in R. D. Roberts and M. L. Bothwell, eds., "Aquatic Ecosystems in Semiarid Regions: Implications for Resource Management," NHRI Symposium Series 7, Environment Canada, Saskatoon, Sask., 1992, pp. 271–280.

Stanley, E. H., and R. A. Short, "Temperature Effects on Warmwater Stream Insects: A Test of the Thermal Equilibrium Hypothesis," *Oikos,* 52:313–320, 1988.

Stanley, E. H., and H. M. Valett, "Interaction between Drying and the Hyporheic Zone of a Desert Stream Ecosystem," in P. Firth and S. G. Fisher, eds., *Climate Change and Freshwater Ecosystems,*" Springer-Verlag, New York, 1992.

Stefan H. G. and B. A. Sinokrot, "Projected Global Climate Change Impact on Water Temperatures in Five North Central U.S. Streams," *Climatic Change,* 24:353–381, 1993.

Strayer, D., "Crustaceans and Mites (Acari) from Hypoheic and Other Underground Waters in Southeastern New York," *Stygologia,* 4:192–207, 1988.

Stromberg, J. C., D. T. Patten, and B. D. Richter, "Flood Flows and Dynamics of Sonoran Riparian Forests," *Rivers,* 2:221–235, 1991.

Strommer, J. L., and L. A. Smock, "Vertical Distribution and Abundance of Invertebrates within the Sandy Substrate of a Low-Gradient Headwater Stream," *Freshwater Biology,* 22:263–274, 1989.

Swanson, F. J., L. E. Benda, S. H. Duncan, G. E. Grant, and W. F. Megahan, "Mass Failures and Other Processes of Sediment Production in Pacific Northwest Forest Landscapes," in E. O. Salo and T. W. Cundy, eds., *Streamside Management: Forestry and Fishery Interactions,* Institute of Forest Resources, University of Washington, Seattle, 1987.

Sweeney, B. W., J. K. Jackson, J. D. Newbold, and D. H. Funk, "Climate Change and the Life Histories and Biogeography of Aquatic Insects in Eastern North America," in P. Firth and S. G. Fisher, eds., *Climate Change and Freshwater Ecosystems,* Springer-Verlag, New York, 1992, pp. 143–176.

Sweeney, B. W., and J. A. Schnack, "Egg Development, Growth, and Metabolism of Sigara Alternata (Say) (Hemiptera: Corixidae) in Fluctuating Thermal Environments," *Ecology,* 58:265–277, 1977.

Swift, B. L., "Status of Riparian Ecosystems in the United States," *Water Resources Bulletin,* 20:223–228, 1984.

Taylor, R. C., "Drought-Induced Changes in Crayfish Populations along a Stream Continuum," *Amer. Midl. Natur.,* 110:286–98, 1983.

Thornthwaite, C. W., and J. R. Mather, *The Water Balance,* Publications in Climatology, vol. 8, no. 1, Drexel Institute of Technology, 1955.

Towns, R. D., "Limnological Characteristics of a South Australian Intermittent Stream, Brown Hill Creek," *Aust. J. Mar. Freshwat. Res.,* 36:821–837, 1985.

Towns, D. R., "Life History Patterns of Emergence of Six Species of Leptophlebiidae (Ephemeroptera) in a New Zealand Stream and the Role of Interspecific Competition in their Evolution," *Hydrobiologia,* 99:37–50, 1983.

Triska, F. J., "Role of Wood Debris in Modifying Channel Geomorphology and Riparian Areas of a Large Lowland River under Pristine Conditions: A Historical Case Study," *Verh. Internat. Verein. Limnol.,* 22:1876–1892, 1984.

Turner, R. E., and N. N. Rabalais, "Changes in Mississippi River Water Quality This Century," *Bioscience,* 41:140–147, 1991.

Valett, H. M., S. G. Fisher, N. B. Grimm, and P. Camill, "Vertical Hydrologic Exchange and Ecological Stability of a Desert Stream Ecosystem," *Ecology,* 75:548–560, 1994.

Vannote, R. L., and B. W. Sweeney, "Geographic Analysis of Thermal Equilibria: A Conceptual Model for Evaluating the Effect of Natural and Modified Thermal Regimes on Aquatic Insect Communities," *American Naturalist,* 115:667–695, 1980.

Verbyla, D. L., "Hydrologic Effects of Climate Change in the Delaware River Basin," *Water Resources Bulletin,* 26:831–832, 1990.

Verstraete, M. M., and S. A. Schwartz, "Desertification and Global Change," *Vegetation,* 91:3–13, 1991.

Vervier, P., and J. Gibert, "Dynamics of Surface Water-Groundwater Ecotones in a Karstic Aquifer," *Freshwater Biology,* 26:241–250, 1991.

Waggoner, P. E., ed., 1990. *Climate Change and U.S. Water Resources,* Wiley, New York, 1990.

Walker, B. H., "Ecological Consequences of Atmospheric and Climate Change," *Clim. Change,* 18:301–316, 1991.

Walling, D. E., and I. D. L. Foster, "The 1976 Drought and Nitrate Levels in the River Exe Basin," *Journal of the Institute of Water Engineering Sciences,* 32:341–352, 1978.

Ward, A. K., G. M. Ward, J. Harlin, and R. Donahoe, "Geological Mediation of Stream Flow and Sediment and Solute Loading to Stream Ecosystems Due to Climate Change," in P. Firth and S. G. Fisher, eds., *Climate Change and Freshwater Ecosystems,* Springer-Verlag, New York, 1992, pp. 116–142.

Ward, G. M., and K. W. Cummins, "Effects of Food Quality on Growth of a Stream Detritivore, *Paratendipes albimanus* (Meigen)(Diptera: chironomidae)," *Ecology,* 60:57–64, 1979.

Ward, J. V., "Thermal Characteristics of Running Waters," *Hydrobiologia,* 125:31–46, 1985.

Ward, J. V., "The Four-Dimensional Nature of Lotic Ecosystems," *Journal of the North American Benthological Society,* 8:2–8, 1989.

Ward, J. V., and J. A. Stanford, "Thermal Responses in the Evolutionary Ecology of Aquatic Insects," *Annual Review of Entomology,* 27:97–117, 1982.

Ward, J. V., H. J. Zimmermann, and L. D. Cline, "Lotic Zoobenthos of the Colorado System," in B. R. Davies and K. F. Walker, eds., *The Ecology of River Systems,* Dr. W. Junk, Dordrecht, The Netherlands, 1986.

Weatherly, A. H., *Australian Inland Waters and their Fauna,* Australian National University Press, Canberra, 1967.

Webster, K. E., A. D. Newell, L. A. Baker, and P. L. Brezonik, "Climatically Induced Rapid Acidification of a Softwater Seepage Lake," *Nature,* 347:374–376, 1990.

Welcomme, R. L., *Fisheries Ecology of Floodplain Rivers,* Longmanns, London, 1979.

Welcomme, R. L., "Concluding Remarks I. On the Nature of Large Tropical Rivers, Flood-plains, and Future Research Directions," *J. No. Am. Benthol. Soc.,* 7:525–526, 1988.

Wetzel, R. G., "Land-Water Interfaces: Metabolic and Limnological Regulators," *Verh. Internat. Verein. Limnol.,* 24:6–24, 1990.

Whalen, S., and W. Reeburgh, "A Methane Flux Time Series for Tundra Environments," *Global Biogeochemical Cycles,* 2:399–409, 1988.

Wigley, T. M. L., and P. D. Jones, "Influences of Precipitation Changes and Direct CO_2 Effects on Streamflow," *Nature,* 314:149–152, 1985.

Wigley, T. M. L., P. D. Jones, and P. M. Kelly, "Scenario for a Warm, High-CO_2 World," *Nature,* 283:17–21, 1980.

Williams, D. D., "Towards a Biological and Chemical Definition of the Hyporheic Zone in Two Canadian Rivers," *Freshwater Biology* 22:189–208, 1989.

Williams, D. D., and H. B. N. Hynes, "The Ecology of Temporary Streams. II. General Remarks on Temporary Streams," *Internationale Revue der Gesamten Hydrobiologie,* 62:53–61, 1977.

Williams, P., "Adapting Water Resources Management to Global Climate Change," *Clim. Change,* 15:83–93, 1989.

Williams, W. D., "Salinization of Rivers and Streams: An Important Environmental Hazard," *Ambio,* 16:180–185, 1987.

Williams, W. E., K. Garbutt, F. A. Bazzaz, and P. M. Vitousek, "The Response of Plants to Elevated CO_2. IV. Two Deciduous Forest-Tree Communities," *Oecologia,* 69:454–459, 1986.

Wilson, L., "Variations in Mean Annual Sediment Yield as a Function of Mean Annual Precipitation," *Amer. J. Sci.,* 273:335–349, 1973.

Winterbourn, M. J., J. S. Rounick, and B. Cowie, "Are New Zealand Stream Ecosystems Really Different?" *New Zealand Journal of Marine and Freshwater Research,* 15:321–328, 1981.

Yount, J. D., and G. J. Niemi, "Recovery of Lotic Communities and Ecosystems from Disturbance: A Narrative Review of Case Studies," *Environmental Management,* 14:547–569, 1990.

CHAPTER 31
ENERGY AND WATER

Benjamin F. Hobbs
Professor of Geography and Civil Engineering
The Johns Hopkins University
Baltimore, Maryland

Richard L. Mittelstadt
Hydropower Engineer
North Pacific Division, U.S. Army Corps of Engineers
Portland, Oregon

Jay R. Lund
Associate Professor of Civil and Environmental Engineering
University of California
Davis, California

Energy production is a major and growing use of water. Falling water drives electric turbines, and contributes about 20 percent of the world's electricity production. No water is consumed directly by this process, but the timing of streamflows is altered and water is evaporated from the reservoirs built to store water and create a hydraulic head. Water is also consumed for various purposes as a result of thermal power and synthetic fuel production (Table 31.1). For instance, water is a process input for production of synthetic fuels and oil derived from shale. Water is also used to dispose of waste materials. But the largest consumptive use results from carrying away waste heat generated by fossil-fuel and nuclear steam-cycle power plants. The most efficient steam plants convert only 40 percent of the heat content of fuel into power; the rest must be rejected into the atmospheric or water environment. On average, U.S. thermal power plants evaporate about 2.5 percent of the 130,000 Mgal of fresh water that they withdraw daily. This evaporation accounts for approximately 3.6 percent of all consumptive use of fresh water in the United States, and a much larger fraction than that in some regions (Solley, et al., 1993).

Power demands in the United States are expanding by 2 to 3 percent each year, while load growth in developing countries can be as high as 10 percent a year. Thus, it can be expected that the energy sector's demand for water will also increase, perhaps causing conflict with other users. However, energy production can also contribute to the solution of water-supply problems. Profits from hydroelectricity production can

TABLE 31.1 Typical Water Consumption in Energy-Related Activities, acre · ft/10^{15} Btu of Product

Activity	Process water	Evaporative cooling	Waste water	Total
Nuclear power stations	—	537,200	55,600	592,800*
Fossil-fueled power stations	—	358,100	37,100	395,200†
Coal gasification	2,800	36,700	17,500	57,000
Coal gasification, high Btu	32,500	68,100	2,800	103,400
Oil shale conversion	21,700	32,000	8,000	61,700
Nuclear fuel processing	—	37,400	3,900	41,300
Coal-slurry pipeline	—	—	—	34,000
Oil refining	—	16,000	6,200	22,200
Underground coal mining	—	7,700	—	7,700
Strip coal mining, with revegetation	—	3,400	—	3,400
Strip coal mining, no revegetation	—	1,800	—	1,800

* Equivalent to 0.66 gal/kWh
† Equivalent to 0.44 gal/kWh
Source: Buras (1980).

make possible the construction of reservoirs that can be devoted to other purposes. Indeed, subsidization of irrigation by power revenues has until recently been a guiding principle of the "reclamation" of the U.S. Southwest by the U.S. Bureau of Reclamation. Future contributions by energy to solving water problems, at least in the United States, are more likely to take the form of desalinization of brackish or salt-waters as a byproduct of thermal power production. Such facilities are a fact of life in a few water-short regions outside of the United States (*Power*, 1993).

The purpose of this chapter is to present an overview of the technology of water use in energy production and associated planning issues. The focus is on the two major energy uses of water: hydropower production and waste-heat disposal. Section 31.1 reviews the principles of hydropower production, along with methods for estimating hydro project energy output. In Sec. 31.2, our attention shifts to cooling-water supply for thermal power facilities. Cooling technologies are discussed, as are planning issues and approaches. Finally, Sec. 31.3 closes this chapter with a case study that illustrates the complexity of water–energy interrelationships: the management of the Columbia River Basin in the northwestern United States.

31.1 HYDROPOWER SYSTEMS

31.1.1 Introduction

Extracting energy from falling water is one of the oldest tools used by humankind for performing useful work. Primitive water wheels have been in use for thousands of years, for lifting irrigation water, for grinding grain, and for other simple applications. The industrial revolution was powered to a large extent by machinery driven by falling water. During the early years of the electrification of America, hydropower was the energy source of choice for many communities. Hydropower remains the major source of electric power in some parts of the world. In other places, where power systems are now dominated by thermal power generation, hydropower is used to produce peaking power.

By extracting 80 to 90 percent of the potential energy in falling water, hydropower is by far the most efficient of the major sources of electric-power generation. Hydropower is a renewable and a relatively clean power source. There are no air pollution or radioactive-waste problems associated with hydropower, and it makes almost no contribution to global warming. Hydropower is also a valuable byproduct at projects constructed for irrigation, municipal water supply, navigation, or flood control. Unfortunately, operations that are best for hydropower sometimes conflict with other river uses, and the construction and operation of hydropower projects can have a major adverse impact on the river environment.

A knowledge of the basic principles of hydropower engineering is important for any engineer involved in the planning, design, and management of water-resource projects. In some parts of the world, new hydropower projects are being planned as key elements for meeting future electric-power needs. Activity is also under way in parts of the world where river development is essentially complete, such as the United States and much of Europe. Older plants are being redeveloped to make them operate more efficiently, and generating facilities are being added at existing dams that currently do not have powerplants. In addition, the demands on our reservoir systems are continually changing. Operational plans must be periodically updated to meet current needs and environmental requirements.

31.1.2 Terms and Definitions

31.1.2.1 Types of Hydroelectric Plants. The major types of hydro projects are *run-of-river, pondage, storage, reregulating,* and *pumped storage.*

A pure *run-of-river* project has no usable storage. Power output at any time is strictly a function of inflow. Typical run-of-river projects include: (1) navigation projects where the pool must be maintained at a constant elevation, (2) irrigation diversion dams, and (3) projects where the topography does not allow for storage. Powerplants on irrigation canals and water-supply pipelines can also be classified as run-of-river projects.

Some run-of-river projects have a small amount of storage capability, which can be used to shape discharges to follow the daily and weekly load patterns. Daily or weekly storage is referred to as *pondage,* and the use of pondage permits a project to serve peaking power loads.

The term *storage* generally refers to projects that have seasonal regulation capability. A project with power storage can be used to regulate seasonal streamflows in order to follow the seasonal power-demand pattern. While regulation of power storage will increase generation at the storage project itself, it can increase production at downstream power projects as well. Some storage reservoirs have been constructed in this country primarily for power generation, but most are multipurpose projects.

Reregulating reservoirs are designed to receive fluctuating discharges from large hydroelectric peaking plants and release them downstream in a smooth pattern that meets downstream flow criteria. Reregulating projects (also sometimes known as *afterbay reservoirs*) may be constructed to operate with a conventional hydro peaking plant or a pump-back installation.

Pumped-storage projects are designed to convert low value off-peak energy to high value on-peak energy. Low-cost energy is used to pump water to an upper reservoir at nights and on weekends. Water from the upper reservoir is then released during high-demand hours to generate peaking power. There are two basic types of pumped-storage projects: *pump-back* and *off-stream.* At *pump-back* projects, water is moved back and forth between two reservoirs on the same stream, while an *off-stream* project uses an adjacent reservoir to store water.

31.1.2.2   Components of Hydroelectric Plants.   Four elements are necessary to generate power from water: (1) a means of creating head, (2) an intake structure, (3) a conduit to convey the water, and (4) a powerplant (see Fig. 31.1).

The *dam* creates the head or elevation differential necessary to move the turbines and impounds the storage used to maintain the daily or seasonal flow-release pattern. The *reservoir* is the water impoundment behind the dam. The *forebay* is that portion of the reservoir immediately upstream of the intake structure.

Intake structures direct water from the reservoir into the penstock or power conduit. *Gates* or *valves* are used to shut off the flow of water to permit emergency shutdown or turbine and penstock maintenance. *Racks* or *screens* on the intake structure prevent trash and debris from entering the turbine units. When the powerhouse is integral with the dam, the intake is part of the dam structure.

A *conduit* or *water passage* conveys water from the intake structure to the powerhouse and can take many configurations, depending upon the project layout. At some projects, an *open canal* or *low-pressure tunnel* may be used to convey water part of the distance from the forebay to the powerhouse. The remainder of the water passage, where most of the drop in elevation occurs, is a pressurized pipe or tunnel called the *penstock*. A *surge tank* is an air-filled chamber that is sometimes constructed at the head of long penstocks to reduce the momentum changes due to water-hammer effects.

The *powerhouse* encloses the *turbines, generators, control* and *auxiliary equipment,* and sometimes erection and service areas. The powerhouse discharges water into the *tailrace* or *tailwater.* The *draft tube* conveys the water from the discharge side of the turbine to the tailrace.

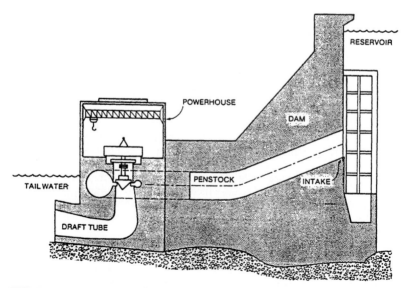

FIGURE 31.1   Elements of a typical hydroelectric plant. *(From USACE, 1985.)*

31.1.2.3 Power System Terms and Definitions.

Energy refers to that which is capable of doing work. Mechanical energy is expressed in foot-pounds, while electrical energy is expressed in *kilowatt-hours* (1 kWh = 2,656,000 ft · lb). The energy output of a hydroelectric plant is also called *generation.*

Power is the rate of energy production or consumption, expressed in either horsepower or kilowatts. While this is the technically correct definition of power, the term is often used in a broad sense to describe the commodity of electricity, which includes both energy and power.

Load is demand for electricity. Load can be expressed in terms of energy demand (average power demand), or capacity demand (peak power demand). For planning purposes, capacity demand is measured in terms of the expected maximum annual capacity demand (annual peak load). Energy demand is normally measured in terms of an average annual energy requirement.

Resources are sources of electrical power. A system's power resources could include both generating plants and imports from adjacent power systems. A utility must have sufficient resources available to meet the expected load and to provide a *reserve margin* of 15 to 20 percent to cover unplanned outages of powerplants.

Load factor is the ratio of average power demand to peak power demand for the period being considered. Load factor can be computed on a daily, weekly, monthly, or annual basis. For example,

$$\text{Daily load factor} = \frac{\text{average power demand per day}}{\text{peak power demand per day}} \qquad (31.1)$$

Hydroelectric energy is the electrical energy produced by converting the potential energy of flowing water by means of a hydraulic turbine connected to a generator. Firm, average annual, and secondary energy are of interest in hydropower studies.

Average annual energy is an estimate of the average amount of energy that could be generated by a hydro project in a year, based on examination of a long period of historical streamflows.

Firm energy, from a marketing standpoint, is electrical energy that is available on an assured basis. For hydroelectric energy to be marketable as firm energy, the streamflow used to generate it must also be available on an assured basis. Thus, hydroelectric firm energy (also called *primary energy*) is based on a project's energy output over the most adverse sequence of flows (*critical period*) in the existing streamflow record. Where a hydro plant or hydro system carries a large portion of a power system's load, the hydro plant's firm energy output must closely follow the seasonal-demand pattern.

Secondary energy is that energy generated in excess of a project's or system's firm energy output. At least some secondary energy is produced most years, and it is often concentrated in the high-runoff season.

Capacity is the maximum amount of power that a generating unit or power plant can deliver at any given time, expressed in kW. The *rated capacity* of a generating unit is the capacity that it is designed to deliver. *Overload capacity* refers to a level of output that a generator can deliver in excess of the normal rated capacity (or *nameplate capacity*) for a limited amount of time. In the past, generators at hydro plants were typically purchased with an overload capacity 15 percent greater than rated or nameplate capacity. However, in recent years the practice of specifying dual ratings has been discontinued.

Installed capacity is the nominal capacity of a powerplant and is the aggregate of the rated (or nameplate) capacities of all of the units in the plant. *Dependable capacity* (or *firm capacity*) is the amount of capacity that a powerplant can reliably contribute towards meeting peak power demands. It has been traditionally defined as the load-carrying ability of a powerplant under adverse load and flow conditions.

Hydraulic capacity is the maximum flow that a hydroelectric plant can use for power generation. Hydraulic capacity varies with head, and is a maximum at rated head. Above rated head, it is limited by generator capacity, and below rated head it is limited by the full gate discharge at that head.

Plant factor is the ratio of the average load on a plant for the time period being considered to its aggregate rated capacity (installed capacity). For example, the *average annual plant factor* is defined as follows:

$$\text{annual plant factor} = \frac{\text{average annual energy}}{(8760 \text{ h})(\text{installed capacity})} \tag{31.2}$$

where the average annual energy is expressed in kWh and the installed capacity is in kW.

The demand for electricity varies over the course of the day, ranging from a minimum in the early hours of the morning to peak loads in late morning or early evening. Power system operators divide the load into three segments: (1) the continuous 24-hour-a-day *base load,* (2) the *peak load,* which is the highest portion of the load occurring for only a few hours a day, and (3) the *intermediate load,* which represents the portion between the base load and the peak load. Hydropower can be used in a power system for peaking, for meeting intermediate loads, for base-load operation, or for meeting a combination of these loads.

Hydroelectric energy has a very low variable cost, because it does not require the burning of fuel. The optimum loading of hydropower would be to use it for carrying peak load, because it would displace the system's most expensive fuel-burning powerplants. Many hydro projects are used in this manner. However, some projects are constrained from peaking operation by operating limits designed to protect the environment and other project purposes. Others are constrained by limited storage or pondage.

31.1.3   Estimating Energy Potential

31.1.3.1   The Water Power Equation.   The amount of power that a hydraulic turbine can develop is a function of the discharge, the net hydraulic head across the turbine, and the efficiency of the turbine. This relationship is expressed by the *water power equation:*

$$\text{hp} = \frac{QHe_t}{8.815} \tag{31.3}$$

where   hp = the theoretical horsepower available
Q = the discharge, ft^3/s (cfs)
H = the net available head, ft
e_t = the turbine efficiency

This equation can also be expressed in terms of kilowatts of electrical output:

$$\text{kW} = \frac{QHe}{11.81} \tag{31.4}$$

In this equation, the turbine efficiency e_t has been replaced by the overall efficiency e, which is the product of the generator efficiency e_g and the turbine efficiency e_t. The efficiency of turbine-generator units varies with both head and discharge. For preliminary studies, an overall efficiency of 80 to 85 percent is sometimes used. Equation 31.4 can be simplified by incorporating an 85 percent overall efficiency as follows:

$$kW = 0.072\, QH \tag{31.5}$$

In order to convert a project's power output to energy, this equation must be integrated over time.

$$kWh = \frac{1}{11.81} \int eQ(t)\, H(t)\, dt \tag{31.6}$$

The integration process is accomplished using either the sequential streamflow-routing procedure or by flow-duration curve analysis.

The *discharge* value in the water power equation is the flow that is available for power generation during the time period being considered. Where the sequential streamflow-routing method is used to compute energy, discrete flows must be used for each time increment in the period being studied. In a nonsequential analysis, the series of expected flows is represented by a flow-duration curve. In either case, the streamflow used must represent the usable flow available for power generation. This flow must reflect at-site or upstream storage regulation; leakage and other losses; nonpower water usage for fish passage, lockage, and so on; and limitations imposed by turbine characteristics (minimum and maximum discharges).

Gross or *static head* is determined by subtracting the water-surface elevation at the tailwater of the powerhouse from the water-surface elevation of the forebay. At most hydropower projects, the forebay and tailwater elevations do not remain constant, so the head will vary with project operation.

Net head represents the actual head available for power generation and should be used in calculating energy. Head losses due to intake structures, penstocks, and outlet works are deducted from the gross head to establish the net head. Hydraulic losses between the entrance to the turbine and the draft-tube exit are accounted for in the turbine efficiency.

31.1.3.2   *Input Data.* The most important type of hydrologic data required for a hydropower feasibility study is the long-term streamflow record that represents the flow available for power production. Other important hydrologic data includes tailwater rating curves, reservoir storage-elevation tables, evaporation losses and other types of losses, and downstream-flow requirements.

Streamflow records are the backbone of the hydropower study. The U.S. Geological Survey (USGS) is the principal source of streamflow records in this country. For sequential routing studies, historical streamflow records are normally used. Thirty years of historical streamflow data is generally considered to be the minimum necessary to ensure statistical reliability. However, for many sites, considerably less than 30 years of record is available. The record can be extended using correlation techniques or stochastic hydrology.

A daily time interval should be used with the duration-curve method. For the sequential streamflow-routing method, a weekly or monthly routing interval is normally used, although some studies may require the modeling of short periods using the daily or hourly interval.

The tailwater elevation is a function of the total project discharge, outlet-channel geometry, and backwater effects. This elevation can be represented either by a *tailwater rating curve* or a constant elevation based on a weighted average-tailwater value.

Not all of the streamflow passing a dam site may be available for power generation. *Consumptive losses* might include reservoir-surface evaporation losses and diversions for irrigation or water supply. *Nonconsumptive losses* could include navigation-lock requirements, water to operate fish-passage facilities, and leakage through or around the dam, around spillway gates, and through turbine wicket gates.

Downstream-flow requirements are sometimes established to ensure that the range of project discharges produced by power operations do not adversely impact other river uses. Examples of streamflow uses which might establish flow requirements include navigation, recreation, water quality, maintenance of fish and wildlife habitat, and flood-control discharge limitations.

31.1.3.3 Methods for Estimating Energy Potential.

Two basic approaches are used in determining the energy potential of a hydropower site: (1) the nonsequential or flow-duration curve method, (2) and the sequential streamflow-routing (SSR) method.

The *flow-duration curve method* uses a duration curve developed from observed or estimated streamflow conditions as the starting point. Streamflows corresponding to selected percentage exceedance values are applied to the water power equation [Eq. (31.4)] to obtain a power-duration curve. Forebay and tailwater elevations must be assumed to be constant or to vary only with discharge. For this reason, the effects of storage operation at reservoir projects cannot be taken into account. A fixed average-efficiency value or a value that varies with discharge may be used. When specific power installations are being examined, unit operating characteristics must be applied to limit generation to that which can actually be produced by that installation. These parameters include minimum single-unit turbine discharge, minimum turbine operating head, and generator installed capacity. The area under the power-duration curve provides an estimate of the plant's energy output.

The flow-duration curve method has the advantage of being relatively simple and fast once the basic flow-duration curve has been developed. For this reason it can be used economically for computing power output using daily streamflow data. However, it cannot accurately simulate the use of power storage to increase energy output, and it cannot handle projects where head varies independently of flow. Also, it cannot be used to analyze systems of projects.

With the *sequential streamflow-routing method,* the energy output is computed sequentially for each interval in the period of analysis. The method uses the continuity equation [Eq. (31.7)] to route streamflows through the project. This makes it possible to account for the variations in reservoir elevation resulting from reservoir regulation. This method can be used to simulate reservoir operation for hydropower as well as nonpower objectives (such as flood control and water supply).

The advantages of SSR are that it can be used to examine projects where head varies independently of streamflow, it can be used to model the effects of reservoir regulation, and it can be used to investigate projects that are operated as a part of a system. The primary disadvantage of SSR is its complexity.

Of the two methods, the sequential streamflow method is perhaps the most widely used for evaluating major projects. It will be discussed in more detail in the following section. The duration-curve method is limited to the analysis of small hydro projects and for conducting preliminary studies for other types of projects. For further details on the duration-curve method, reference should be made to standard hydropower engineering handbooks such as Gulliver and Arndt (1991); Warnick (1984); USACE (1979); and USACE (1985).

31.1.4 Sequential Streamflow-Routing Method

31.1.4.1 Introduction. The sequential streamflow-routing procedure was developed primarily for evaluating storage projects and systems of storage projects, and it is based on the *continuity equation:*

$$\Delta S = I - O - L \tag{31.7}$$

where ΔS = change in reservoir storage
I = reservoir inflow
O = reservoir outflow
L = losses (evaporation, diversion, and so on)

Energy output can then be estimated by applying the reservoir outflow values to the water power equation. At storage projects, head and efficiency as well as flow may be affected by the operation of the conservation equation, through the ΔS component.

A key element in evaluating the energy potential of a storage project is the amount of storage available for regulation. In the case of some new reservoirs, the size can be established by conducting firm yield studies (USACE, 1975). For existing projects, the amount of storage that can be used for power regulation is usually defined by a combination of physical factors and other project requirements. The usable power storage would be the reservoir storage between the minimum pool and normal full-pool elevations. The normal full-pool elevation is determined by establishing a dam height and deducting freeboard requirements and exclusive flood-control storage requirements (if any). The minimum power-pool elevation is usually defined by the minimum head at which the turbines can operate.

A number of different storage regulation strategies may be used to optimize hydropower benefits, some of which require the consideration of other project purposes such as flood control, irrigation, and recreation. However, to illustrate the mechanics of storage regulation for hydropower, the regulation of a single-purpose power-storage project to maximize firm energy will be examined. A similar approach is followed when using other operating strategies.

31.1.4.2 Maximizing Firm Energy Production. In determining the energy output of a project where maximization of firm energy output is the primary objective, the following general steps would be undertaken:

- Identify the critical period.
- Make a preliminary estimate of the firm energy potential.
- Make one or more critical period SSR routings to determine the actual firm energy capability and to define operating criteria that will guide year-by-year reservoir operation.
- Make an SSR routing for the total period of record to determine average annual energy.
- If desired, make additional period-of-record routings using alternative operating strategies to determine which one optimizes power benefits.

Each of these operations may be done automatically using a computerized SSR routing model such as HEC-5, as described in USACE (1983).

The objective of the reservoir operation is to maximize the firm yield of the basin. This is accomplished by operating the project such that reservoir storage is fully utilized to supplement natural streamflows within the most adverse sequence of streamflows. Fully utilizing this storage means that at some point during this adverse streamflow period the usable storage will have been completely drafted, leaving the reservoir empty. This adverse streamflow period, which is called the *critical period*, is identified by examining the historical streamflow record.

The use of the term critical period varies somewhat from region to region. It always refers to the most adverse streamflow period, and by definition it always begins at a point in time when the reservoir is full. In some power systems, the end of the critical period is defined as the point when the reservoir is empty. In other systems, the end of the critical period is the point when the reservoir has refilled following the drought period. Here, the period ending with the reservoir empty will be identified as the *critical drawdown period*, while the term *critical period* will refer to the complete cycle, ending with the reservoir full.

The larger the amount of reservoir storage, the higher the firm yield that can be sustained at a given site. Increasing the amount of reservoir storage also increases the length of the critical period, sometimes even changing the critical period to a completely different sequence of historical streamflows.

Identification of the critical period can be accomplished in several ways. Creager and Justin (1950) describe the mass curve, which has been used historically as the manual technique for identifying the critical period. However, since computers have become available for performing SSR studies, standard practice is to perform a series of period-of-record SSR studies using alternative firm energy requirements. The first iteration is based on a preliminary estimate of firm energy. This is then adjusted by trial and error until a level of firm energy is achieved which will completely utilize the available storage once during the period of record.

31.1.4.3 Preliminary Firm Energy Estimate. The preliminary firm energy estimate is made by applying an estimated firm flow, an estimated average head, and an average efficiency of 85 percent to the power equation. One way of making a first estimate of the firm flow would be to add the usable storage to the average flow in the driest year in the historical streamflow period. The average head can be estimated as the head with one-quarter to one-third of the storage withdrawn.

31.1.4.4 The Sequential Routing Procedure. The basis for the sequential streamflow-routing analysis is the continuity equation [Eq. (31.7)]. The reservoir outflow component must include powerplant discharge plus outflow not available for generation, e.g., spill, leakage, and project water requirements (station service, navigation lock and fish ladder operation, and so on). Reservoir inflows are obtained from streamflow records. Losses refer to the loss in reservoir storage as a result of evaporation (adjusted to account for precipitation falling on the reservoir) plus any withdrawals from the reservoir for water supply or irrigation.

For purposes of illustrating the application of the continuity equation to a storage project, a single-purpose power reservoir will be examined using monthly flows. Because the first objective in the regulation process is to determine the project's firm energy output, the initial regulations are limited to the critical period. The objective in each monthly time increment is to determine how reservoir storage can be used to ensure that the monthly firm energy demand will be met. In periods of high reservoir inflow, inflow may be greater than the required discharge for power, and the excess water is stored if possible. In low-flow periods, storage is drafted to supplement inflow. The task then is to solve the continuity equation for change in storage (ΔS) in each interval during the critical period.

Expanding Eq. (31.7) to include all categories of losses and all outflow components, the continuity equation, expressed in ft³/s, becomes:

$$\Delta S = I - (Q_P + Q_L + Q_S) - (E + W) \tag{31.8}$$

where ΔS = change in reservoir storage during the routing interval
     I = reservoir inflow
     Q_P = power discharge
     Q_L = leakage and nonconsumptive water requirements
     Q_S = spill
     E = net evaporation losses (evaporation from minus precipitation onto
        the reservoir surface)
     W = withdrawals for water supply, irrigation, and so forth.

Also, the ΔS for a given time increment can be further defined as

$$\Delta S = \frac{(S_2 - S_1)}{C_S} \tag{31.9}$$

where S_1 = start-of-period storage, acre · ft
     S_2 = end-of-period storage, acre · ft
     C_S = discharge-to-storage conversion factor*

Substituting Eq. (31.9) into Eq. (31.8) and rearranging the terms, the following equation is obtained:

$$S_2 = S_1 + C_S (I - Q_P - Q_L - Q_S - E - W) \tag{31.10}$$

This equation is expressed in acre · ft and is used to solve for the principal unknown, the end-of-period storage S_2. In the critical period, spill Q_S would be zero.

The first iteration through the critical period is based on a preliminary estimate of the project's monthly firm energy capability. The Q_P for each month is computed using the power equation [Eq. (31.6)] and the monthly firm energy requirement for that month. This computation is repeated for each period in turn, using the end-of-period storage of the previous period as the start-of-period storage.

If the preliminary energy estimate is correct, the energy requirements will be met in all months, all of the power storage will be used at one point in the critical period, and the project will be able to refill in the following months. If the power storage is not fully used, the firm energy estimate must be increased and the computations performed again. If the reservoir goes empty and cannot meet the firm energy requirements for one or more months, the firm energy estimate must be decreased and additional iterations performed.

31.1.4.5 Determining Average Annual Energy.   Once the firm energy has been calculated, the next step is to determine the project's average annual energy output. To determine the average annual energy, a sequential routing is made for the entire period of record using the monthly firm energy requirements derived from the critical period routing. The project's average annual energy is the average of the annual energy production values for all of the years in the period of record. The average annual secondary energy is the difference between the average annual energy and the annual firm energy.

---

* For example, if the routing interval is a 30-day month, the conversion factor C_S would be 59.50 acre · ft/ft³ · s⁻¹ · month.

Several alternative strategies are available for operating in better-than-critical streamflow conditions. The simplest strategy is to operate to just meet the firm energy requirements. Secondary energy will be produced only when the reservoir is at the normal full pool and net reservoir inflow exceeds the discharge required to meet firm energy requirements. However, this approach permits no flexibility of operation during periods of better-than-critical streamflow. In order to permit better use of secondary energy and also more flexibility in using storage for nonpower river uses, rule curves may be developed to govern reservoir regulation.

A *rule curve* is a guideline for reservoir operation, and it is generally based on detailed sequential analysis of various critical combinations of hydrologic conditions and demands. Rule curves do not fall exclusively under the domain of hydropower. For example, rule curves may be developed to govern use of conservation storage for irrigation, for navigation, or for municipal water supply. Rule curves may also be derived to guide flood-control operation, and composite rule curves may be developed to regulate operation for a combination of purposes. The development and use of rule curves for both hydropower and multipurpose operation are discussed in Kuiper (1965); USACE (1975); USACE (1977), USACE (1985); and Warnick (1984).

31.1.4.6 Alternative Power Operation Strategies.

The power-regulation procedures described in the preceding sections are designed to ensure that firm energy capability will be provided in all years in the period of record. While this is the classic approach to regulation of power storage, it applies mainly to power systems where hydropower is the dominant resource. In such systems, the main objective is to maximize the amount of energy production that can be assured under even the most adverse water conditions. This is because no other source of energy is available to back up the hydro system during drought years.

Where a hydro project is operated in a thermal-based power system, a lower level of energy output is acceptable in the occasional drought year, because generation from thermal plants can make up the shortfall. In a thermal-based power system, several alternative strategies might be considered in regulating power storage (Gulliver and Arndt, 1991; USACE, 1985).

Maximize average annual energy.   Storage is regulated to maximize energy production under average water conditions.

Maximize energy benefits.   If the value of energy varies from month to month, specific values can be assigned to the energy output in each month, and successive iterations made to develop operating rules that maximize dollar benefits.

Maximize dependable capacity.   The objective in this case is to maintain the reservoir at or above the rated head; this is to insure that the project's full rated capacity is available at all times (or at least during all of the peak-demand months).

Variable draft.   Storage is drafted for energy production based on the market value of energy at the time; this approach is now being used either explicitly or implicitly in several U.S. hydropower systems.

31.1.4.7 System Analysis.

The analysis of a system of hydropower projects follows the same basic principles as for a single hydro-storage project. The major difference is that analysis of a hydropower system is more complex, and, when the system is operated for multiple purposes, the analysis is even more complex. For analysis of systems, computerized SSR models become a necessity.

The basic problem in operating a system of reservoir projects is determining the order of drafting storage from the various reservoirs that will maximize power output. One approach for determining the draft sequence is the storage effectiveness technique, which is described in USACE (1985). As a practical matter, determining the optimum draft sequence is only the first step in developing a system operation plan. When all factors are considered, it is more often the case that drafts must be applied somewhat equitably to all reservoirs in the system, rather than in a manner that optimizes economic benefits.

31.1.4.8 Computer Models for Hydropower Studies. Sequential streamflow routing can require considerable data manipulation. Models have been developed which are capable of handling automatic optimization of firm energy production, evaluation of multiproject systems, and operation of projects for flood control and other functions simultaneously with power production. The HEC-5 model is one widely used general-purpose model, but the unique requirements of some river systems have required that customized models also be developed (USACE, 1985). Many of these are deterministic models. Linear and dynamic programming techniques are also being applied to a variety of hydropower planning and operation problems, both to single projects and to systems of projects (Gulliver and Arndt, 1991; Loucks, et al., 1981; Mays and Tung, 1992; Wurbs, 1995).

31.1.5 Complements and Conflicts with Other River Uses

31.1.5.1 Introduction. Section 31.3 discusses the continuing conflicts between hydropower and other river uses on the Columbia River system, with a focus on the attempts to develop an operational plan that will meet the demands of all interests. At the time of this writing, the Corps of Engineers, the states, and other entities are involved in similar reexaminations of reservoir operations on the upper Missouri and on the Alabama-Coosa-Tallapoosa and Apalachicola-Chattahoochee-Flint river basins. Similar studies are under way elsewhere. Some current issues are the relicensing of existing hydroelectric projects and the protection of fish and wildlife under the Endangered Species Act. Others are the increasing recreational use of hydropower reservoirs and the reallocation of storage from hydropower to water and other uses. Addressing the requirements of the different water uses is even more important in planning new projects, because failure to satisfy all interested parties will almost certainly lead to organized efforts to halt the project.

Attempting to meet multiple objectives does not always result in conflict, however. A careful analysis often leads the analyst to discover that diverse water uses can complement rather than conflict with one another.

A key element in the analysis of multiple water-resource objectives is the simulation of alternative operating plans for the reservoirs and river systems. These studies require that the projects be operated to meet a combination of sometimes conflicting requirements. The following paragraphs describe some of the more common multiple-use problems and how they are dealt with in reservoir modeling.

31.1.5.2 Storage Zones. Many hydroelectric projects with storage also provide space for flood-control regulation. At some projects, the storage meets other water needs in addition to power production. The storage at a multipurpose project can typically be divided into functional *storage zones,* as shown in Fig. 31.2. The top zone is the *flood-control zone,* which is kept empty except when regulating floods.

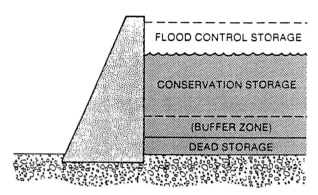

FIGURE 31.2   Reservoir storage zones. *(From USACE, 1985.)*

Below the flood-control zone is the *conservation storage zone.* This space stores water to be used to serve various at-site and downstream water uses, which could include power generation, irrigation, municipal and industrial water supply, navigation, water quality, fish and wildlife, and recreation. The term *power storage* is sometimes used instead of conservation storage when discussing power operation (as in Sec. 31.1.4), but conservation storage is the term most often used when describing multipurpose operation.

Below the conservation zone is the *dead storage zone,* which is kept full at all times to provide minimum head for power generation and sedimentation storage space.

The conservation storage zone is often subdivided into two or more zones based on the level of service that can be provided with the amount of available storage. A common division is into an upper zone, where releases can be made in excess of those required to meet firm or minimum requirements, and a lower zone (sometimes called a *buffer zone*), where releases are made only to meet firm or minimum requirements. The division between the upper and lower zone may vary seasonally. The power rule curve shown in Fig. 31.3 is an example of a seasonally varying division. At this project, drafts during the summer months (points C to D) would be made only to meet firm power requirements. During the remainder of the year additional drafts could be made if the reservoir were above the rule curve.

31.1.5.3   Operation with Fixed Flood-Control Zone.   The simplest flood-control configuration is that where a fixed amount of storage space is maintained above the top of the conservation pool the year around. This approach is followed in basins where large floods can be expected at any time of the year, such as in the South Atlantic coastal basins. The reservoir is normally maintained at or below the top of conservation pool, with the flood-control space being filled only to control floods. Following a flood, this space is evacuated as quickly as possible within the limits of downstream channel capacity.

31.1.5.4   Operation with Joint-Use Storage.   In many river basins, major floods are concentrated in one season of the year. This permits establishment of a joint-use storage zone, which can be used for flood regulation during part of the year and conservation storage in the remainder of the year (USACE, 1976). This is illustrated by Fig. 31.4. Such an allocation requires less total reservoir storage than does providing

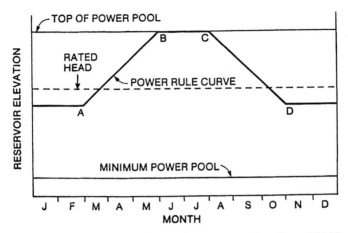

FIGURE 31.3 Example of rule curve for power operation. *(From USACE, 1985.)*

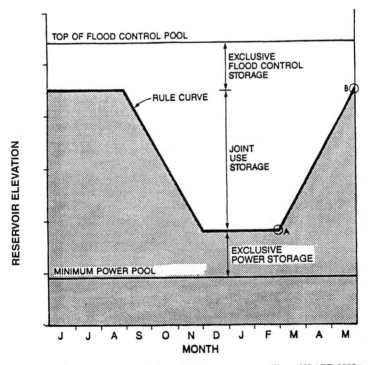

FIGURE 31.4 Rule curve for project with joint-use storage. *(From USACE, 1985.)*

separate exclusive storage zones for flood control and conservation. For this reason, the use of joint-use storage should be considered wherever hydrologic conditions permit.

Because the joint-use zone must be evacuated annually, not all of the conservation storage may contribute to the project's firm energy capability. The refill curve (A–B on Fig. 31.3) is defined by a careful balancing of the probability of floods in each interval within the refill period against the probability of sufficient runoff to permit refill.

31.1.5.5 Joint-Use Storage with Snowmelt Runoff.

In the mountainous river basins of the western United States, much of the runoff is from snowmelt, and the magnitude of that runoff can be forecasted several months in advance with some degree of confidence. This makes it possible to manage joint-use storage space more efficiently. Drafts for flood control are scheduled to ensure that sufficient flood-control space is provided to maintain the required level of protection. At the same time, sufficient conservation storage must be maintained to meet firm energy requirements in all years and permit refill in most years (USACE, 1976).

31.1.5.6 Nonpower Conservation Requirements.

At most projects having power storage, releases must also be scheduled to meet other downstream uses, which might include navigation, irrigation, municipal and industrial water supply, fish and wildlife, water quality, and recreation. In some cases, these requirements may be determined independently of the reservoir-regulation study. Examples are minimum flows required to maintain navigation downstream, minimum releases to maintain fish populations, and the water-supply requirements of a downstream community. Often these requirements have been mandated as having a higher priority than power generation, so the reservoir operating plan must meet these requirements first and then maximize power benefits within whatever regulation capabilities remain.

31.1.5.7 Multipurpose Operational Studies.

Making an operational SSR study to determine the energy output of a project serving multiple purposes is basically the same as making a study for a single-purpose power-storage reservoir. The difference is that the requirements of other functions would be superimposed on the power regulation. In some periods, it may not be possible to meet all requirements. This requires a set of operating rules that establish priorities, and it is sometimes necessary to make alternative studies with different priority orders to identify the best overall plan.

Although it is impossible to specify a priority system that applies in all cases, some general guidelines can be identified. Operation for the safety of the structure usually has the highest priority. Of the functional purposes, flood control must have a high priority, particularly where downstream levees, bridges, or other vital structures are threatened. For example, it is not unusual that releases for power generation be curtailed entirely during periods of flood regulation.

Among the conservation purposes, municipal and industrial water supply, irrigation, and hydroelectric-power generation are often given a high priority, particularly where alternative supplies are not readily available. High priority is also usually assigned to flows required for fish and wildlife. In fact, flow requirements established under the Endangered Species Act may have the highest priority of any conservation purpose.

Navigation may receive a lower priority, and water-quality management and other low-flow augmentation priorities can be somewhat lower yet, because temporary shortages are usually not disastrous. Finally, recreation and aesthetic considerations would normally have the lowest priority, although political considerations often dictate that they be given greater consideration. There can be marked exceptions to the priorities listed above. Regional differences in needs and in legal and institutional factors may greatly affect priorities.

31.1.6 Related Considerations

The preceding sections deal primarily with the methodology for determining the energy output of hydro projects and systems. Related issues are the determination of installed capacity, the selection of turbines, economic analysis, and the unique characteristics of pumped-storage hydro projects. Space does not permit detailed discussion of these topics, but a number of useful references are cited.

- Selecting plant capacity: Warnick (1984) and USACE (1985)
- Turbine selection: ASCE (1989); Creager and Justin (1950); Gulliver and Arndt (1991); Warnick (1984); USACE (1979); and U.S. DOI (1976)
- Economic and financial analysis: Goodman (1983); Gulliver and Arndt (1981); James and Lee (1971); and U.S. DOE (1979)
- Pumped storage: Public Service (1976); Karadi (1974); USACE (1985); Warnick (1984); and Harza (1990)

31.2 WATER SUPPLY FOR THERMAL POWER GENERATION

In this section, we summarize power-plant cooling technologies, issues associated with the energy sector's demand for water supply, and approaches to including water in power system planning.

31.2.1 Technology

By far, the largest consumptive use of water by the energy sector is for thermal power production. For a review of water requirements for both thermal and nonthermal power plants and for fossil-fuel production, see Gleick, 1994. Nuclear and fossil steam plants need boiler makeup water, condenser-cooling water, potable water, and plant-service water. In addition, coal plants often require water for ash handling and flue-gas desulfurization (Shorney, 1983). The most important use is cooling; the typical steam plant must reject about two-thirds of the heat value of its fuel into the environment. This is usually accomplished by passing the steam through the turbines and then through condensers, where heat is rejected to a secondary water loop. This secondary loop then transfers the heat either directly to the atmosphere by evaporation, radiation, or conduction, or to a water body by conduction. Water consumed in cooling generally accounts for 70 to 90 percent of the total water consumption by a power plant.

For a plurality of U.S. generation capacity, heat rejection is accomplished by once-through cooling. In *once-through cooling,* water is withdrawn from the source, passed through the condensers, and then immediately returned to the source at an elevated temperature. This method has several advantages. It is the least costly to construct, imposes the smallest capacity penalty (i.e., loss of power due to fan and pump requirements and turbine backpressure), requires less water treatment, and evaporates less water than evaporative cooling towers or cooling ponds. Once-through cooling evaporates water because it elevates the temperature of the receiving water body. Under an ambient wet-bulb temperature of 50°F, once-through cooling consumes about 130 gal/10^6 Btu of rejected heat, while mechanical draft-cooling towers and cooling ponds evaporate 180 to 240 gal/10^6 Btu (Probstein and Gold, 1978). Once-through cooling consumes less because, compared to towers and ponds, a greater fraction of the heat ultimately reaches the atmosphere via convection and radiation.

However, once-through cooling has a major drawback: thermal pollution. Concerns in the United States over the impacts of heating on aquatic ecosystems contributed to the passage of the 1972 Federal Water Pollution Control Act Amendments. That law mandates, with a few exceptions, that all new power plants adopt the "best available control technology" (BACT) for thermal pollution. BACT was subsequently defined by the U.S. Environmental Protection Agency as closed cycle evaporative-cooling towers.

Consequently, most new plants in the United States use evaporative-cooling systems. Some existing power plants are also being retrofitted with cooling towers because of the damage caused by their thermal discharges (Lander and Christensen, 1993). Most evaporatively cooled facilities use cooling towers rather than ponds, because ponds require more land (1 to 5 acres/MW) and consume more water. But if land and water are cheap, or if water storage is desired for other purposes, then the low capital and maintenance costs of ponds can make them attractive (Probstein and Gold, 1978). Cooling ponds tend to be located in the southern United States because higher evaporation rates there minimize the amount of land required (Sonnichsen, et al., 1982).

In *evaporative cooling-tower systems*, water in the secondary cooling loop is passed through an air stream, cooled, and recycled to the condensers (Fig. 31.5). The movement of air may be induced by fans, as in Fig. 31.5, or by the chimney effect created by building a 200-ft or taller hyperbolic shell. Figure 31.6 presents four different types of mechanical- and natural-draft cooling towers. The choice among the types of cooling towers depends on the relative cost of capital (more being required for natural-draft towers) and energy (more of which is consumed by mechanical towers), and on whether climatic conditions are favorable for natural-draft towers. The actual costs of either type of system depend strongly on climate. Generally, their capital cost is determined by the following three factors (Probstein and Gold, 1978; Cheremisinoff and Cheremisinoff, 1981; Mitchell, 1989; Adams and Stevens, 1993):

1. The design ambient temperature and humidity, representing the most severe environmental conditions under which the plant is expected to operate (in practice, the wet-bulb temperature that is exceeded only 1 to 5 percent of the time).
2. The desired cooling-water temperatures at the tower inlet and outlet.
3. The rate at which water is recirculated.

Because evaporative towers cannot cool water to below the wet-bulb temperature, they exact an efficiency penalty during hot periods compared to cooling ponds and once-through cooling, which instead are limited by water temperatures.

The water consumption resulting from expanded use of evaporative cooling presents problems in both the arid western United States and the humid east. It is an unhappy coincidence that in the West, rainfall and streamflows are relatively small precisely in those regions that contain the bulk of the nation's oil shale, tar sands, and thick low-sulfur coal seams. The region's streamflows are often over-appropriated, meaning that increased water diversions to energy will necessarily be

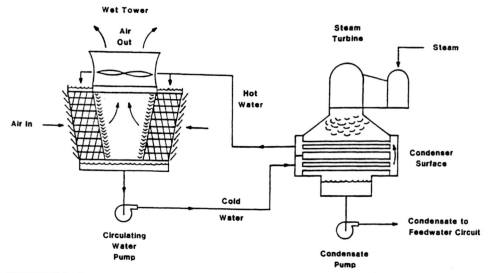

FIGURE 31.5   Conventional mechanical-draft wet cooling-tower system. *(From Mitchell, 1989.)*

at the expense of other users. Several states concerned about the social changes that would result from such reallocations have thrown legal barriers in the way of water rights transfers—even though the economic value of water to irrigated agriculture is at least two orders of magnitude smaller than its worth to energy producers. In contrast, in the relatively wet eastern United States, there is plenty of extra water available, on average. But specific subregions are chronically water short. This is not because of a lack of reservoir sites, but because social and environmental concerns have made dam construction very difficult. Water-supply problems are usually worse in exactly those basins whose proximity to energy consumers make them attractive for power development.

As a result, utilities are increasingly turning to dry and hybrid wet/dry cooling systems in order to increase siting flexibility. These systems reject more heat to the atmosphere via conduction and radiation than do evaporative systems. In 1980, there was only one dry-cooled plant of significant size in the United States and no wet/dry capacity. But by 1994, that number had grown to 36 dry-cooled plants in the United States, representing 3200 MW of power, and 460 plants worldwide, providing 21,000 MW (Bartz, 1994). Wet/dry systems have also become more popular.

Figure 31.7 shows two alternative dry-cooling configurations, one based on mechanical draft and the other using a natural-draft tower. *Direct dry-cooling systems* pass the condensing steam from the turbines directly through heat exchangers, while the indirect systems in the figure contain a secondary loop that picks up heat at the condensers and rejects it at the towers.

Dry cooling, while easing water problems, is unfortunately costly—three to five times as much as evaporative cooling (Bartz and Maulbetsch, 1981). The incremental cost compared to evaporative cooling is about $3 to $10 per 1000 gal of water saved (Allemann, et al., 1987; Hobbs, 1984). A third to a half of the cost of dry cooling is due to the severe capacity penalties that result from its inability to lower the temperature of cooling water below the dry-bulb temperature; as much as 15 percent of a plant's capacity may be lost as a result. Most of the remainder of the cost is for the huge heat

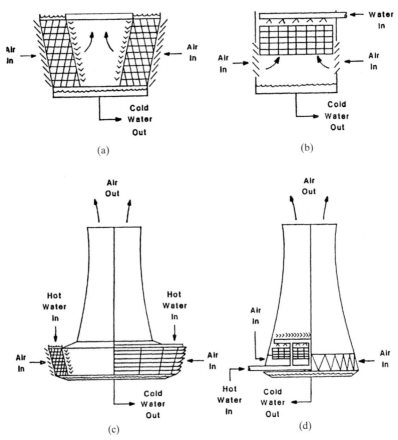

FIGURE 31.6 Different types of wet-cooling towers: (*a*) mechanical-draft crossflow; (*b*) mechanical-draft counterflow; (*c*) natural-draft crossflow; (*d*) natural-draft counterflow. *(From Mitchell, 1989.)*

exchangers made necessary by the relative inefficiency of radiation and convection compared to evaporation in removing heat. As in evaporative systems, these exchangers have to be sized to accommodate extreme design-temperature conditions. This is true even if far smaller exchangers would suffice for most of the year.

Because of dry cooling's high cost, it is most commonly adopted by mine-mouth power plants, where avoiding coal-transport costs more than makes up for the expense of dry cooling. An example is the huge 3960 MW Matimba complex in South Africa (*Modern Power Systems,* 1991). In the United States, dry cooling is primarily used by nonutility generators (NUGs). NUGs value siting flexibility and maintaining good relations with neighbors, meaning that they are willing to pay a premium to avoid discharges of water vapor (Guyer and Bartz, 1991). Cogeneration and waste-to-energy plants, which are favored by many NUGs, face particularly severe siting constraints. Some NUGs have constructed natural-gas-fired combined-cycle plants with dry cooling, in part because the thermal efficiencies of about 45 percent for such facilities mean that less heat rejection is needed.

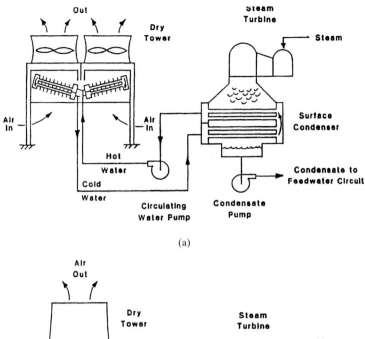

(a)

(b)

FIGURE 31.7 Indirect dry cooling-tower system: (*a*) with surface condenser and mechanical-draft tower; (*b*) with direct-contact condenser and natural-draft tower. (*From Mitchell, 1989.*)

The expense of dry cooling and the need to conserve water have motivated development of *combined wet/dry cooling towers* (Bartz and Maulbetsch, 1981; Mitchell, 1989). One such system has two sets of cooling-tower sections, one dry and one wet (Fig. 31.8). The dry section is sized so that it can handle the heat load for much of the year, but not so large as to be able to reject all of the heat under the hottest conditions. During warm periods, some of the thermal load is diverted to the wet sections. Only as many wet sections as are needed for maintaining the desired water temperature are turned on. Other wet/dry systems involve *deluging,* in which water can be sprayed on the dry tower's heat exchangers. In another design, air can

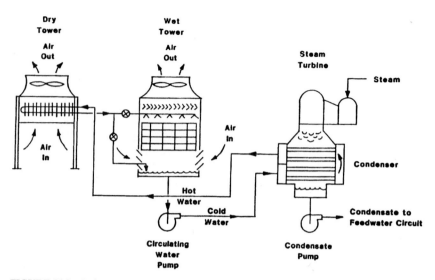

FIGURE 31.8   Series-connected wet/dry cooling-tower system with separate wet and dry towers. *(From Mitchell, 1989.)*

be presaturated with water before it contacts heat exchangers. Yet another type of wet/dry system is called the *binary tower;* in it, water can be diverted either inside or outside of plastic heat exchangers, depending on environmental conditions. In all of these systems, the ability to cool by evaporation during hot periods allows the dry heat exchangers to be much smaller than for 100 percent dry-cooled plants, resulting in considerable capital cost savings.

A convenient way to characterize wet/dry systems is to describe them in terms of the amount of water they consume in an average year compared to a 100-percent evaporative system. The smaller the amount of water consumed, the higher the power plant's cost becomes because of efficiency losses and the need for larger dry heat exchangers. However, water-acquisition expenses will decrease at the same time. Figure 31.9 shows that the optimal level of water conservation depends strongly on the cost of water. Although the exact results depend on the climate and type of power plant, the figure implies that if water costs about $3/1000 gal in that region, then 30 percent wet/70 percent dry cooling may be advantageous compared to evaporative cooling. But at $14/1000 gal, 97 percent dry cooling would be justified.

Research and demonstration efforts have been directed at lowering these costs. For instance, anhydrous ammonia has been substituted for water in the secondary loop in the Kern plant in California (Mitchell, 1989). Nonmetallic heat exchangers (Dynatech R&D Company, 1984) and capacitive-cooling systems (Allemann, et al., 1987) also hold promise. In capacitive systems, part of the waste heat is rejected to a water-storage tank during the hottest period of the day. The water in the tank is then cooled at night by either direct air cooling or an ammonia dry-cooling system.

Over the next decade, however, conventional evaporative, wet/dry, and dry-cooling systems will be the technologies that most new power plants will choose. Power companies will have to weigh the incremental cost of saving water via dry or wet/dry cooling against the expense and political uncertainty of water supply.

31.2.2 Cooling-Water Issues

Two groups of issues are reviewed here: water demands and water supplies.

31.2.2.1 *Cooling-Water-Demand Issues.*

In the 1970s, it was feared that energy's appetite for water would grow exponentially, impacting the environment and causing conflict with existing users. For example, the U.S. Water Resources Council (1978) projected that electric utilities' share of water consumption in the United States would grow from 1 to as much as 8 percent in the year 2000. Massive synfuel and coal developments in the arid west were being planned in regions where water supplies were already oversubscribed.

However, those fears were not realized for several reasons. First, electricity-demand growth rates slackened from the 7 percent/y levels common in the early 1970s to 2 to 3 percent/y today. Most of this decrease was due to electricity price increases, but some credit can also be given to the energy conservation and load-management programs that some utilities have energetically promoted. The consequence of lower growth is that not as many power plants are needed. Second, oil prices fell to levels which, in real terms, had not been experienced since before the 1973 embargo. This killed off President Carter's synfuels-development plans.

The third reason why the projections were wrong is that electrical-generation technology has been evolving in a way that lessens the pressure it places on scarce water supplies. Economies of scale in power production have largely disappeared; it is no longer necessary to build power plants 1000 MW in size or larger to maxi-

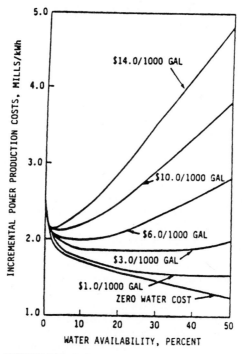

FIGURE 31.9 Influence of water cost on the cost of wet/dry cooling. *(From Alleman, et al., 1987.)*

mize efficiency. Smaller plants have shorter lead times and more flexibility, which more than compensates for any scale economies in boiler and generator size. This is especially true for combustion turbines and combined-cycle plants burning natural gas; their optimal size is, at most, a few hundred MW. These facilities are becoming more efficient, too; in the near future, the best combined-cycle facilities will have efficiencies in the range of 50 percent. As a result, these plants need less water per MWh, and dry-cooling technologies become more affordable. Renewable energy represents another technological change. Some renewables, such as wind turbines, have become popular in California and require no water at all. A final trend is so-called *distributed generation,* which may meet as much as 20 to 30 percent of utilities' load growth (Weinberg, et al., 1993). These are small units that are strategically sited near growing local-load centers; one of their purposes is to decrease the amount of necessary transmission and distribution investment. The technologies involved, such as combustion turbines, diesels, and fuel cells, use radiators rather than evaporative cooling.

Thus, the energy sector's demand for water will not reach the apocalyptic levels once anticipated. However, in regions (especially in the developing world) where demand growth remains robust and water is scarce, conflicts with other water users will still arise. Even in the United States, utilities anticipate difficulties obtaining even the reduced supplies they need. The Electric Power Research Institute recently surveyed 40 major U.S. utilities located in areas where water shortages or siting constraints make alternative cooling technologies potentially viable (Guyer and Bartz, 1991). Forty percent of those utilities foresee dry cooling as potentially becoming a future requirement for their facilities, for the most part between the years 2000 and 2010. Another 20 percent of the utilities polled see water-resource issues as very important, although they doubt that dry cooling will become a requirement for them.

31.2.2.2 Cooling-Water-Supply Issues.

The main problem of wet/dry and dry cooling is its expense; the difficulties of water-supply acquisition, in contrast, are mainly political and institutional. In much of the United States, there are more demands for water than supply, when instream uses are included among the demands. To obtain a reliable water supply, energy industries will often have to divert water from these other uses.

The resulting conflicts with agricultural, municipal, environmental, and other users of water can be resolved by four different institutions: (1) the marketplace and (2) the legislative, (3) judicial, and (4) administrative branches of government (Abbey and Lucero, 1980; Trelease, 1980; Weatherford, et al., 1982). For example, a utility may buy a water right from an irrigator and obtain the necessary approval from the state engineer, only to find itself sued by a downstream user that believes that its property rights to the return flows are being compromised. The state legislature may become a party to the controversy by passing a law restricting the circumstances under which water can be transferred to industrial use. Therefore, hydrologic availability alone is not enough; permission from one or perhaps more of these institutions is required.

Chapters 4 and 5 review water-allocation institutions in detail. One issue of particular concern are policies that define utility uses of water as being lower in priority than other uses. For instance, legislatures in California, Montana, and North Dakota have effectively banned transfers of water rights from irrigated agriculture to industry. As another example, the Delaware River Basin Commission has stated that:

[t]o continue to evaporate waters of the Delaware River Basin in large quantities in the cooling of electrical generating stations appears to be inconsistent with the doctrine of equitable apportionment. Therefore, it has been assumed for purposes of this staff report that no new quantities of water will be available for the electric utility industry beyond that which will be required for power installations to be operational by the year 1982 (DRBC, 1975, pp. 1–66).

Such prioritizations may prevent transfers of water from low-value uses to high-value uses, or force utilities to resort to alternative, high-cost supplies. Some other particular water-supply issues that can impact energy planning include (Trelease, 1980; Weatherford, et al., 1982a):

- Unsettled Indian rights
- Whether water rights are implicitly reserved for national forests and other federal lands
- The extent to which instream uses will be subject to increased protection

However, some observers (e.g., Abbey and Lucero, 1980) maintain that the water-for-energy literature overemphasizes governmental forums for resolving water conflicts and the constraints they present, while failing to note the many opportunities that the market presents for securing water supplies. In general, utilities have often found clever solutions to water-supply problems, even in highly constrained areas. These opportunities, which are reviewed by Hobbs (1988), were overlooked by many water-for-energy studies. The main reason for this oversight is the difficulty faced in national-level studies in obtaining data on a uniform basis on water sources other than streamflows (e.g., Dobson and Shepherd, 1979; Sonnichsen, 1982).

Many of the utilities' solutions have involved purchases of water rights. Weatherford, et al. (1982b) argued that although formal market mechanisms are rudimentary in most areas of the West, markets that promote economically efficient reallocations of water supply will nevertheless evolve. Indeed, the last decade has seen the emergence of markets in Colorado, California, and elsewhere. Abbey and Loose (1979) list 12 transfers of water rights to energy firms in the West, totaling 86 billion gal/y. Most of this water was sold by irrigators, although some consisted of sewage effluent provided by cities. At least two of these transfers involved groundwater rights. Transfers have also been occurring in the East. For instance, in the early 1980s, Pennsylvania utilities negotiated with the U.S. Army Corps of Engineers to reallocate storage in a flood-control reservoir to water supply.

Another water-supply issue is climate change (see Chap. 29). There is the prospect that climate warming will simultaneously decrease hydropower production, increase power demands by air conditioners, and make water supplies for new generation scarcer. As an example, an analysis of the Tennessee Valley Authority found that generation losses would result from actions taken to avoid violations of environmental and safety constraints, such as plant deratings, cooling-tower usage, and/or nuclear plant shutdowns (Miller, et al., 1992). Additional reductions would occur because plant efficiency deteriorates at higher temperatures. These reductions would total between 1 and 2 percent of TVA's total generation. However, a study of other U.S. utilities found that reductions in thermal power generation due to increased water temperatures or decreased availability is likely to be small compared to temperature-induced increases in power demands and, for some utilities, hydropower impacts (Linder, et al., 1989).

31.2.3 Planning for Cooling-Water Supplies

Cooling-water issues are considered at three different stages in the electric utility planning process. First, they can be a concern in corporate planning when comparing different methods of meeting growing demands for power. Second, once a plant type and size has been chosen, then economic and environmental tradeoffs between different water sources and cooling methods must be weighed. Finally, economic and environmental considerations come into play when choosing a final design for a condenser-cooling system.

31.2.3.1 Utility Resource Planning.

Resource planning is the evaluation of alternative energy-supply and demand-side (conservation and load management) resources for meeting power demands over a multidecade time horizon. Until the 1970s, the resource planner's task was just to determine the best size, timing, and type of large central-station generation plants to meet ever-growing electric loads. The problem today is different, and considerably more complex (Hirst and Goldman, 1991). First, there are more options. These now include demand-side management, cogeneration of power and heat for industrial processes or space conditioning, power purchases and sales in an increasingly deregulated market, and small-scale and renewable-power sources.

A second reason why resource planning is more complex today is that the utility's objectives have expanded beyond cost. Other important objectives now include customer value maximization, promotion of economic development, preservation of environmental quality, and equity. In the United States and elsewhere, this expansion is a result of an increase in competitive pressures and, at the same time, deeper involvement by government and various interest groups in the planning process.

Previously, water was only implicitly considered in resource planning to the extent that the expense of supply and cooling affected the relative cost of different energy-supply options. However, because of the increased prominence of government regulation and environmental considerations, water impacts now receive explicit consideration on occasion. This sometimes occurs through the environmental impact statement process, in which a utility must project the impacts of its resource plan upon air, water, and other environmental resources.

Another way water impacts are considered in resource planning is through the use of *externality adders.* Several state regulatory commissions now require utilities to include monetary estimates of external environmental costs in their comparisons of alternative means of meeting demands for electric service. For instance, the Nevada Public Service Commission has recently required utilities to add $1500/ton for every ton of SO_2 that a proposed energy source would emit, $7000 for each ton of NO_x, and $22 per ton of CO_2 (Wiel, 1991). Adders are also defined for other environmental impacts, including water. When a utility proposes a new supply source, it must add these values to the internal capital and operating costs, and then recommend the least costly resource. These emissions penalties are not actually paid by the utility or its ratepayers; they are instead an accounting device designed to tip the planning scale towards less-polluting options, including energy conservation and renewable sources.

However, the overwhelming emphasis of present externality adder systems is upon air emissions. For instance, under a system adopted in New York State, a proposed coal plant would be subjected to a 1.405 ¢/kWh adder—of which only 0.005 ¢ is for water impacts (Putta, 1990). Obviously, such a number is meaningless, since a facility sited on the shore of one of the Great Lakes will have a vastly different water impact than one placed within the water-short Delaware River basin. The failure of adder systems to appropriately account for impacts that are site specific is a major failing.

Water impacts sometimes also receive consideration in multiobjective resource-planning exercises. In the multiobjective approach, the economic, environmental, reliability, and other attributes of each alternative resource portfolio are explicitly enumerated. Water impacts can be one of the attributes considered. The results are then displayed so that participants in the process understand the tradeoffs involved in choosing one plan over another. Participants might also be asked to make numerical statements of value judgments in the form, for example, of importance weights for the various objectives. This was done, for instance, in a study at Seattle City Light where water consumption for evaporative cooling was one of 12 attributes considered in an evaluation of alternative resource plans (Hobbs and Meier, 1994). Utility planners and interest group representatives chose weights for the attributes. On average, they assigned less than 2 percent of the total weight to water consumption.

With so little importance assigned to water in resource-planning studies, it is unlikely that water considerations will significantly impact resource choices. However, the relative importance of water could increase in the future. One reason would be the growing scarcity of water. Another is that air emissions are increasingly subject to market-based regulatory systems that convert air pollution into just another cost concern. An example is the U.S. SO_2 allowance system; since the national number of allowances is fixed, a decision to decrease emissions at one location will just result in more emissions elsewhere, leading to negligible net environmental impact. Consequently, the utility only needs to consider the cost of acquiring allowances in its planning and can disregard SO_2's environmental effects. As air emissions become of less concern as an environmental issue, water resources will become more important. This is especially true in the Pacific Northwest, where water is perhaps the most contentious environmental issue (see Sec. 31.3, following).

31.2.3.2 Water Supply and Conservation Planning by Utilities. Once a utility has decided on a supply and demand-side resource plan, then it must consider how its water requirements will be met. An integrated water-supply study for a particular facility or set of facilities can be conducted using either a cost-minimizing approach or a multiple-objective framework. An engineering economy study would pick the mix of supply sources and water conservation measures that yields the lowest possible expense of water provision and use. Figure 31.9 presented a simple example of such a study, where the cost of water is balanced against the incremental expense of water conservation by wet/dry or dry cooling. But even for a profit-making corporation, consideration of cost alone is generally not enough. Environmental and social impacts are also important, if only because more severe impacts increase the chance of lawsuits and delay in project completion. Important, too, are legal and institutional uncertainties concerning whether water supplies will really be available in the amounts and at the costs assumed.

In general, tradeoffs among these objectives should be considered. For instance, the option that minimizes expected cost for a particular plant might consist of the purchase of water rights from farmers. But that alternative might also be risky, in that the state engineer might disapprove the sale or third parties may sue to prevent it. [For instance, the Wheatland power station experienced expensive construction delays as a result of court suits over water rights (Abbey and Lucero, 1980).] A less risky but more costly option in that instance might be to install dry or wet/dry cooling towers instead. As pointed out above, private power developers are especially eager to avoid siting and regulatory hassles and so have proven willing to incur the expense of dry cooling in order to ensure that projects will be completed on time.

Increasingly, systems-analysis tools are being used to quantify the reliability of different sources, and to identify optimal mixes of water supply and conservation options (see Chaps. 6 and 7). For instance, Shaw, et al. (1981) analyzed the situation in which a utility purchases water from different users, depending on the level of streamflow. An example of this would be where a utility has obtained the option to use a farmer's senior water rights during exceptionally dry years. The method considers the resulting randomness in the cost of water, the expense of on-site storage, and the possibilities of evaporative and wet/dry towers and cooling ponds. Palmer and Lund (1988) apply another such method. Their approach includes the option of water storage, along with purchases during droughts of either power from other utilities or water rights. Lence, et al. (1992) consider how reservoir releases can be coordinated with thermal generation downstream in order to comply with constraints on river temperature while meeting energy demands.

As an example of such a system study, Hobbs (1984) used systems analysis to evaluate the following options for meeting the water needs of power plants in the Texas-Gulf region in the years 2000 and 2030:

- *Surface supplies,* including excess firm yields of reservoirs, excess safe yield of aquifers, transferable water rights, and sewage plant effluent. The costs of these options were described by supply curves for each region within each river basin.

- *Water transport,* including rivers, existing canals, and potential pipelines. Legal restrictions on interbasin transport were explicitly recognized.

- *Water conservation* by either conventional or ammonia-based wet/dry cooling systems. The expense of conservation was portrayed using demand curves for water in each region. A demand curve shows the marginal worth of water to the power industry as a function of the amount of water provided, assuming a fixed amount of generation capacity. [It is also possible to use this systems framework to consider plant location as another option to lower water costs; see, e.g., Lall and Mays (1981).]

A linear program was used to calculate the supply-demand equilibrium in each region, which yielded the least-cost combination of the above options. A purpose of this study was to assess the market potential for conventional and ammonia-based wet/dry systems. The study concluded that the availability of the less-expensive ammonia systems would lower costs by about $1 million per year for the utilities in that region, while yielding 28,400 acre · ft/y of water savings.

31.2.3.3 Economic and Environmental Design of Cooling Systems.

Once a decision has been made to install a type of cooling system at a particular facility, then the next step is detailed system design.

The optimization of heat-exchanger size, cooling-water flow rates, tower fill, number of fans, and other variables in order to minimize the cost of a cooling tower at a particular facility can be quite complex (Allemann, et al., 1987). For instance, choosing the design wet-bulb temperature for an evaporative-cooling tower represents a balance between facility reliability (the probability that the design temperature will be exceeded, resulting in a derating of the facility and increased power costs) and the expense of larger heat exchangers (Adams and Stevens, 1993). For the case of dry-cooling towers, mathematical programming methods have been used to obtain cost-minimizing tower designs (Buys and Kroger, 1989).

Environmental considerations further complicate the design process. These include (Mirsky, et al., 1992):

Discharge water composition. Water circulated through cooling systems is usually treated with chemicals to prevent biofouling. Further, dissolved minerals are concentrated and the temperature is raised. Therefore, cooling system materials and hardware must be resistant to corrosion, and *blowdown* (water discharged from the facility to the environment) must be monitored and perhaps treated. Some facilities are designed for *zero-discharge* operation, in which case any blowdown must be evaporated on site.

Plume effects. Fogging, icing, and drift can result from vapor plumes from cooling towers. Visibility can be threatened at nearby roads and airports, and pavements may be subject to icing during winter. Drift containing salt and minerals can harm local vegetation, soils, and even cars and equipment, especially if the cooling towers use salt water. Extensive modeling studies are often made to assess the impact of alternative cooling-tower designs (Policastro, et al., 1994). Design options include drift eliminators, fan-stack extensions to elevate plumes, heating discharge air to lower plume visibility, or, in extreme cases, switching to dry or wet/dry towers.

Discharge temperature. Although this is of most concern for once-through facilities, it can occasionally be a problem for evaporative-cooling systems. An easy (yet perhaps costly) approach is to operate the plant at lower levels during warm conditions to avoid exceeding discharge temperature limitations.

Noise generation. Noise in cooling towers results from splashing water and mechanical equipment and can annoy neighbors. There are a variety of mitigation options, such as mufflers, anti-splash devices, slow-speed fans, and planting of shrubbery.

Stack gas interactions. Cooling-tower plumes and exhaust gases can interact. This can sometimes be taken advantage of, such as the use of the buoyant tower plume to pull up flue gases that have been cooled by scrubbers.

31.2.4 Conclusions

Alarmist projections that water shortages would stand in the way of necessary energy development now appear to be overly pessimistic. Technological improvements in dry and wet/dry cooling offer the prospect of less costly water conservation, and energy industries have proven adept at developing and using nontraditional sources of water, such as groundwater, wastewater, and rights transfers. There nevertheless remain political and institutional uncertainties that can make water acquisition expensive and frustrating. Integrated analysis of the options available, combined with explicit recognition of institutional uncertainties, can help identify cost-effective and politically feasible alternatives.

31.3 WATER AND POWER FROM THE COLUMBIA RIVER

The previous two sections have summarized principles involved in designing and operating water-for-energy systems. In this section, a case study is presented which illustrates many of the conflicts and issues that arise in using water to produce energy.

The hydroelectric system in the Columbia River basin is unique in many ways but typifies many contemporary problems and alternatives for regional hydropower-system management. While the Columbia system has great hydropower importance, larger than any other large U.S. river system, it has not escaped the tradeoffs among multiple objectives, pluralistic water management, and changing societal objectives for water-resources management that characterize most contemporary water projects. Also, like many other large reservoir systems, the response to these problems has benefitted from considerable creativity, thought, and effort in the development, testing, and implementation of management measures.

31.3.1   The Columbia River System

The Columbia River basin is the fourth-largest river basin in North America, and the second-largest river basin in the United States (Fig. 31.10). The basin covers 259,000 mi^2, including parts of seven states and Canadian British Columbia. The river's main stem is 1214 miles long.

Inflow to this system derives largely from snowmelt during the spring and early summer months, with lower flows during the fall and winter. Average annual runoff from the Columbia basin is over 180,000 cfs or roughly 130 million acre · ft (maf) per year. Roughly 102 maf of this runoff occurs in the seven months between January and the end of July. The historical droughts on the system occurred between 1928 to 1932 (4 y), 1943 to 1945 (2 y), and 1977 to 1978 (1 y). Major floods for the system occurred in 1861, 1894, 1948, 1956, and 1973, with the greatest flood impacts occurring below Bonneville toward the lower end of the system at and below Portland.

The reservoirs on this system have a total storage capacity of over 55 maf, of which 42 maf is available for coordinated system operations. This implies a systemwide ratio of storage to mean annual flow of about 0.33, a system with relatively little overyear storage. This compares to storage-to-annual-inflow ratios of over 3.0 for the main-stem Missouri River system and the Colorado River. Reservoirs on the system tend to refill most years, typically by the end of July. However, there is enough overyear storage on the system to reduce the effects of occasional overyear droughts. Storage on the system is divided among over 212 water projects, with the 17 largest projects having over 36 maf of available storage (87 percent of total available storage).

Operating purposes for reservoirs on the Columbia River system include hydropower, recreation, fish, wildlife, irrigation, flood control, navigation, and water supply. Hydropower on the Columbia system is of unique multiregional importance, as discussed later. Operation for some purposes often conflicts with operation for other purposes, subjects of eternal controversy.

The intraregional tradeoffs in the operation of the Columbia River system are similar in many ways to tradeoffs in other river basins, such as the conflicts between environmental and flood control, hydropower, irrigation, and recreation operations. The case of Columbia River hydropower is unique in its dominance of regional electric-power supplies and its major role in the operation of power systems as far away as southern California via power-system interties.

31.3.2   System Governance

Like most U.S. river basins, the Columbia River system is governed by a complex web of private, local, state, and federal institutions interacting through federal, state, and treaty legislation and a variety of regulations, contracts, and other forms of coordination. As it is an international river system, Canadian institutions at private, local,

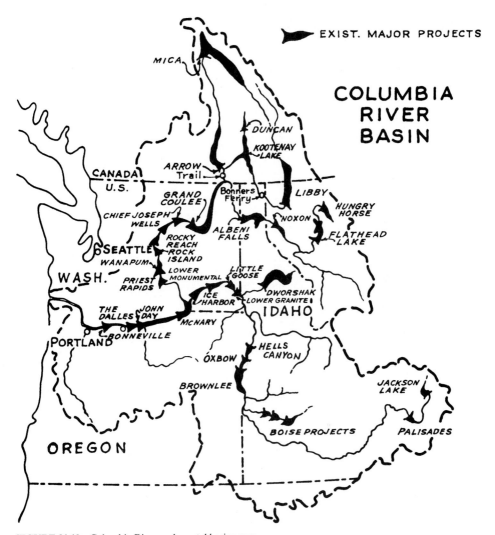

FIGURE 31.10 Columbia River and coastal basins map.

provincial, and federal levels also have significant roles in system water management. With over 20 of the system's 42 maf of available storage located in Canada, a variety of U.S.–Canadian treaties, agreements, and other forms of coordination are vital to the system's operation. Some of the parties involved are identified in Table 31.2.

The enormous and central role of electric-power production from such a large system further complicates system management. Integrated hydropower production requires the coordinated operation of many reservoir-operating federal, Canadian, and private agencies, local electric-power utilities, and power-transmission authorities.

TABLE 31.2 Sampling of Major Institutions Involved in Governing the Columbia River System

Type of Participant	Interest	
	Hydropower	Other
U.S. federal	Bonneville Power Authority Army Corps of Engineers Federal Energy Regulatory Comm. Bureau of Reclamation	Army Corps of Engineers Bureau of Reclamation National Marine Fisheries Service U.S. Fish and Wildlife Service National Weather Service Soil Conservation Service
Canadian federal		Canadian Federal Government
U.S. states		Washington DOE Oregon Fish and Wildlife Idaho Fish and Game Montana DNRC
Canadian Provincial	B.C. Hydro	B.C. provincial agencies
U.S. local and interagency governments	NW Power Planning Council Grant County PUD Douglas County PUD Chelan County PUD Seattle City Light Many others	Irrigation districts Urban water supplies Countless others
Canadian local governments		Local towns
Native American tribes		19 tribes
U.S. private	Idaho Power Puget Sound Power & Light Washington Water Power Co. Montana Power Co. Portland General Electric Power Co. Pacific Gas and Electric Co. Power customers	Navigation interests Recreation interests Countless others

Coordination, sales, and distribution of electric power from federal U.S. Army Corps of Engineers and Bureau of Reclamation dams is administered by the Bonneville Power Administration (BPA). BPA is required to give priority in sales to publicly owned utilities and Pacific Northwest power customers, but also sells power to private utility companies and directly to some industries, such as aluminum producers.

The Pacific Northwest Coordination Agreement (PNCA) is a major instrument for system coordination, integrating the operation of 120 hydropower projects on U.S. portions of the Columbia River system (U.S. DOE, 1993). The 17 parties to the PNCA include three federal agencies, five private power companies, three municipalities, six public utility districts, and one aluminum company subsidiary. The intent of the agreement is to operate hydropower production as if it were operated by a single entity. The benefits of this coordinated operation are then distributed among PNCA members. Since all parties have hydropower production facilities downstream of storage controlled by other parties, agreement members have incentive to cooperate.

The International Joint Commission (IJC) and the Columbia River Treaty oversee operation of reservoirs which cross the U.S.–Canada border and international development and operation of the Columbia River system. A major aspect of these agreements is the development and operation of Canadian water-storage projects to enhance hydropower and flood-control operations on U.S. portions of the Columbia River.

In addition to hydropower interests, a host of other interests are involved in the governance of the Columbia River system from perspectives of flood control, irrigation, fisheries, recreation, navigation, and Native American tribal interests. A wide variety of state regulatory and enabling legislation and federal treaties, regulations, and project-authorizing legislation defines the activities and limitations of government agencies. In addition, there are almost countless private and public economic interests regarding each aspect of system operation. Some of these are also listed in Table 31.2.

31.3.3 Regional and Interregional Power Flows

The Columbia River system provides up to 75 percent of the electrical-power generation in the Pacific Northwest. Total hydropower production for the Columbia system averages 18,500 MW, equivalent in total power to roughly 20 nuclear power plants running full-time. Large amounts of surplus hydropower are sold throughout the western United States and Canada as far south as Los Angeles and Arizona. In an average year, the Pacific Northwest provides about 2000 to 3000 MW of average power capacity to the Southwest. Bonneville Power Authority (BPA) also purchases lesser amounts of power from outside the region (Canada and the Southwest) to allow storage of hydropower energy in Columbia River reservoirs.

Figure 31.11 shows the power demands and supply for the coordinated portion of the Columbia River system for July 1991 to July 1992, a relatively dry year. The gray portion of the diagram shows that portion of power demands served from general streamflows. Unlike most electric-power grids, the Columbia basin power grid is base-loaded largely with hydropower. In the spring and summer, most power demand can be met by typical streamflows. During the fall, winter, and early spring months, with higher power demands for heating (during winter) and lower runoff, power demands must be supplied from energy stored in reservoirs and from thermal power plants located within and outside of the basin. Most of this production comes from water stored in reservoirs during the previous year's spring and summer freshet. Since almost all peaking for the Columbia Basin can be provided by hydropower, additional base-load power can be supplied from relatively efficient thermal-electric plants.

The Pacific Northwest–Southwest power transmission intertie takes advantage of the difference in power-demand peaks in the Northwest, with a winter heating peak, and the Southwest, with a summer air-conditioning peak. Excess Southwestern baseload thermal-electric capacity can help satisfy Northwest demands in the winter, allowing storage of energy in reservoirs and maintenance of hydropower heads. Then, during the spring and summer freshet, surplus hydropower generation in the Northwest is available to cool the Southwest. During most months, there is substantial two-way flow of energy on the intertie, varying hourly to help balance both systems, with a net annual flow of energy southward. The presence of an intertie substantially diversifies the power production available to accommodate unusual local power demands or drought-induced shortages in hydropower production at both ends of the intertie.

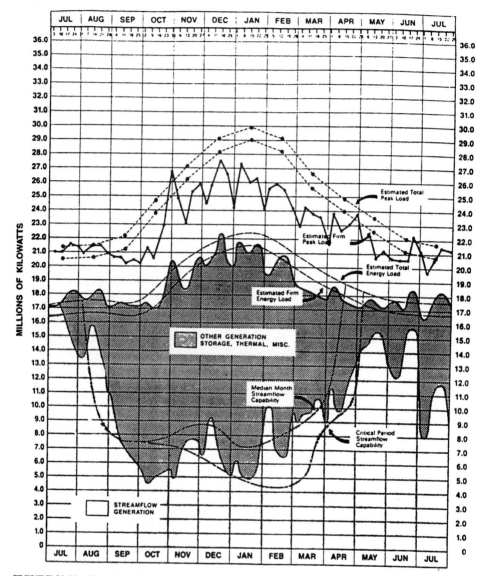

FIGURE 31.11   Coordinated system—loads and resources.

31.3.4 Power Struggles

Significant controversies have arisen over time regarding the operation of the Columbia River system (Broches and Spranger, 1985). Beyond these, there are often significant conflicts between various nonhydropower purposes.

Fish versus Hydropower.   Conflicts between fish and hydropower interests are the most controversial aspect of managing the Columbia River system. These disputes are long-standing. In recent decades, salmon populations have declined drastically from the abundant populations originally found in the basin. Much of the decline of these species is attributed to the existence and operation of dams, primarily for hydropower. The effects of these dams range from excluding or inhibiting access of spawning salmon to spawning grounds, impeding the migration of young salmon to the sea, sheltering predators of young salmon, altering the hydrologic regime to disturb both upstream and downstream migration of salmon, and reduction in the availability of salmon spawning beds. Other factors not related to the reservoir system have also been linked to the decline of salmon populations, including overfishing, alteration of land use, changes in stream sedimentation and chemistry, degradation of spawning areas, and introduction of exotic species. A wide variety of structural and nonstructural measures for addressing this conflict have been implemented, planned, and proposed (Collins, 1992). The potential of fishery operations to disrupt hydropower operations also implies disruptions to other reservoir uses compatible with traditional hydropower operations (such as navigation and recreation).

Irrigation versus Hydropower. Irrigation diversions from the Columbia River average roughly 10 million acre · ft/y over about 7.8 million acres of irrigated land in the basin. Most irrigation use is for large irrigation projects at relatively high elevations. This implies that water diverted for irrigation, even considering return flows, reduces potential hydropower generation. Irrigation diversions also are generally pumped in this region, resulting in additional hydropower demands (GAO, 1986). Hydropower losses have been estimated to be between $100/y per additional acre of irrigated land for lost hydropower alone and $200/y · acre if subsidies for pumping and irrigation water distribution are also included (McCarl and Ross, 1985). Water markets and interruptible irrigation demands have been proposed to resolve this water-use conflict in an economical way (McCarl and Parandvash, 1988; Hamilton, et al., 1989).

Recreation versus Hydropower. In most years, recreation and hydropower uses of Columbia River reservoirs are quite compatible. Hydropower operations deplete reservoir storage during the late fall, winter, and early spring months, when recreation demand is at its lowest, refilling in the peak recreation season during the summer. However, unusual drawdown of reservoirs during drought for maintaining hydropower production can harm recreation activities. Recreational fishing can conflict with hydropower, due to increased drafting for fish flows downstream.

Navigation versus Hydropower. Navigation requirements for water on the lower Columbia and Snake Rivers are relatively minor and quite compatible with hydropower production. Typical hydropower operation keeps storage in these lower reservoirs full, maximizing hydropower head in these high-flow locations. Minimum flows for navigation below Bonneville and below Hells Canyon on the Snake River benefit other water uses as well. Such operations for hydropower support navigation well.

Flood Control versus Hydropower. Flood control is generally compatible with hydropower operations. Hydropower operations require drawing down system reservoirs in the fall, winter, and early spring to capture streamflow during the late spring and summer peak-runoff season. This operation substantially reduces peak annual-flood flows and creates substantial storage capacity for regulating most large and all but localized small-flood flows. (The reduction in peak flows also hinders the migration of juvenile salmon downstream.)

The uncertainty of the timing and magnitude of flood flows increases the amount of annual drawdown in several cases, however. The timing of this additional flood-control drawdown and the likelihood that drawdown will be larger than that needed for flood control (reducing refill probability) can misallocate and reduce hydropower production.

31.3.5 Current Operations Strategy

Until recent years, the operation of the Columbia River system was based largely on hydropower operations. In recent years, the listing of several species as "threatened" or "endangered" has resulted in significant changes in system operation, based on the Biological Opinion issued by the National Marine Fisheries Service (NMFS). One result for reservoir operations is that several major reservoirs in the system are no longer likely to refill during the summer. The details of operating strategy under this new situation are still subject to change and reflect the situation of changing project purposes common to many of the world's hydropower facilities. Following, the traditional operation of the Columbia River system is presented.

Current operation of the Columbia River system typically involves three seasons: (1) fixed drawdown, (2) variable drawdown, and (3) reservoir refill. The fixed-drawdown season runs from August until December, when generally full reservoirs are drawn down to create hydropower and flood-storage space. The variable-drawdown season, from January until March, creates additional hydropower and storage space for controlling floods from spring snowmelt, the region's season of greatest flooding. The amount of draft in the variable-drawdown season varies with runoff forecasts for the season, which are made monthly. The refill season, from April to July, is one where reservoirs are replenished with spring and summer snowmelt, storing water for winter power generation, recreation, and late-summer irrigation (U.S. DOE, 1991b).

While the system has storage equivalent to only about a third of mean annual runoff and there is significant seasonal variation in runoff, the Columbia River system can mitigate for droughts longer than one year. This is done by drawing down storage not usually employed in normal or wet years.

To assist with the downstream migration of juvenile salmon, about 4 maf of water from reservoir storage is made available to increase flow through the system in late spring. The magnitude and use of this so-called *water budget* for salmon has become a significant feature of seasonal operations in recent years.

Detailed operation of the system relies heavily on the coordinated use of computer models by several agencies and coordinating bodies.

31.3.6 System Models for Planning and Management

Modeling is a fundamental basis for the operation of the Columbia River system, with many computer models used by different agencies for examining the planning and operation of the Columbia River system. Some models are specifically related to power generation, flood control, fisheries, or other specific operating purposes. Other models are more multipurpose, with generally less detailed representations of several operating purposes (U.S. DOE, 1991a; 1992). Models also vary in their spatial and temporal scale, with different time-step lengths and levels of geographic simplification of this extensive river system. The existence of multiple system models, each serving a different purpose, is common for large multiple-reservoir and hydropower systems. This section reviews the use of computer models for system operation and planning.

31.3.6.1   Major Operational and Planning Models.   Six computer models are typically used to simulate alternatives and scenarios for operation and planning of the Columbia River system (USDOE, 1992). These models and their role in system operation and planning are described in the following.

"Monthly" Models.   The first three models operate with a monthly time-step, except for splitting April and August into two periods each, for a total of 14 time-steps per year. The major hydropower and multipurpose reservoir planning models for the system (HYSSR, HYDROSIM, and HYDREG) all incorporate some representation of regional and interregional thermal power generation and power demand.

HYSSR (Hydro System Seasonal Regulation), developed and maintained by the Corps of Engineers (COE), is the oldest Columbia River system model, developed originally in 1958 for comprehensive system planning. The model includes 65 individual reservoir projects incorporating hydropower, flood-control, and target streamflow operations. HYSSR and HYDROSIM are used as the basis for most reservoir annual-operation plans, particularly refill studies, and long-term planning for reservoir operation strategies.

HYDROSIM (Hydro Simulator) is a BPA model for seasonal hydropower planning. The model can be run for 36- or 80-reservoir representations of the system.

HYDREG, originally a BPA model, is now maintained by the Northwest Power Pool. This program is used to establish seasonal operating guidelines and interchange rights, obligations, and headwater benefits for individual parties to the Pacific Northwest Coordination Agreement (PNCA). This model includes 150 individual projects and is run as often as weekly throughout the year.

Daily Simulation.   SSARR (Stream Flow Synthesis and Reservoir Regulation) is a Corps of Engineers model for flood-control operations and daily river forecasting. SSARR can estimate flood stages at selected locations.

Hourly Simulation.   Hourly time-step models are also employed. HLDPA (Hourly Load Distribution and Pondage Analysis) is a COE model for examining hourly peaking performance. The model is typically run on a weekly basis. The BPA's Pondage Model also examines hourly peaking performance for a weekly period. In addition to their use for hydropower peaking coordination, these models provide information for evaluating environmental impacts of peaking operations.

31.3.6.2   Other Models.   A variety of other computer models are employed in the planning and management of the Columbia River system. Some of these models are specific to particular project purposes and particular locations or parts of the system.

Other Multipurpose Models.   The Upper Snake Model is used by the Bureau of Reclamation for modeling operation of reservoirs on the Upper Snake River. HEC-PRM (Hydrologic Engineering Center Prescriptive Reservoir Model) is a recent experimental COE model for exploring optimal reservoir operations based on network flow optimization. The HEC-PRM model currently aggregates the system into 19 reservoirs, including 14 with significant water-storage capacity (USACE, 1993).

Hydropower.   Several other models support hydropower aspects of system modeling. These include PMDAM (Power Market Decision Analysis Model) and SAM. SAM is a monthly BPA power-marketing and economic-simulation model which includes Canadian and Californian interactions with the power system. SAM is a companion model to HYDROSIM.

Fish.   Models of fish populations in the Columbia River system typically address either anadromous fish, which migrate through the reservoir system, or resident fish, which remain within a single reservoir or river reach. Two multireservoir models for anadromous fish are CRiSP (Columbia River Salmon Passage) developed by the University of Washington, Seattle, which models mostly fish passage between dams, and SLCM (Stochastic Life-Cycle Model) developed by Resources for the Future, which models population dynamics of anadromous fish. A model for locally resident fish in Libby and Hungry Horse reservoirs and downstream has been developed by BPA and the State of Montana.

Water Quality. HEC-5Q is a COE model for analyzing the effects of reservoir operations on temperature and chemical concentrations at different locations in the system.

31.3.6.3 Model Runs. The models identified previously, particularly the monthly planning models, such as HYSSR, can be run in various ways for various purposes. Model use for planning and operations of the Columbia River system is intensive. Roughly 500 runs are made annually of HYSSR alone.

Critical Period. Critical-period studies assess the operation of a system over a particularly severe operating period, typically severe drought. For the Columbia River system, the 1928 to 1932 and the 1943 to 1945 droughts are commonly used. Critical-period studies are used to assess how much firm load can be supplied by the hydropower system.

50-Year Continuous Runs. When run for the entire period of hydrologic record, a model's results can be used to represent the likely range and distribution of operations over the long-term future. This approach is useful for assessing the long-term performance of new operating strategies or facilities.

Refill Studies, 50 1-Year Runs. Refill studies are typically used for the development of annual-operation plans or other short-term operation studies and involve using current reservoir capacities and forecast flows followed by historical flows for the remainder of the planning period. Typically, this would be done 50 times, once for each year of the hydrologic record. This approach provides information on the performance of a proposed set of operating decisions under a wide range of circumstances.

31.3.7 Evolving System Management

The operation and management of the Columbia River hydropower system has changed since its construction. In the beginning the Columbia River system was envisioned as a hydropower, irrigation, navigation, and flood-control system for stimulating economic development in the region. However, the passage of time has diminished the relative importance of many of these system uses to the region.

While hydropower is of undisputed importance for sustaining the region's inexpensive power supplies, cheap hydropower is no longer seen as being synonymous with regional prosperity. The success of most of the region's manufacturing base is much more dependent on the creativity of its firms (such as Boeing and Microsoft) and the quality of its labor force than the price of a relatively minor energy input. Agriculture, and therefore irrigation, is also of less relative importance to the region's economy. Expansion of irrigated agriculture would be to less suitable locations and so would suffer diminishing and perhaps negative economic marginal returns. Navigation, which depends on agriculture and heavy industries for its benefits, is similarly limited in its prospects for growth. And flood control, limited to relatively small land areas within the basin, is unlikely to experience great increases in benefits relative to the region's economy.

In contrast to the relatively limited nature of benefits from traditional system-operating purposes, other system purposes have seen growth in real and perceived importance. Recreation on the Columbia River system has grown exponentially in recent decades, and is likely to continue growing faster than the region's population. Fish and wildlife purposes of operation, particularly preservation of anadromous fish, are also of greater importance (at least perceived, if not real importance) to the region than would have been envisioned in the past.

The management of the system has not been unresponsive to these changes. Decades of work and tens of millions of dollars have been spent to expand recreation opportunities on the system and to improve the survival of fish species (particularly salmon) passing upstream and downstream through many of the dams and reservoirs. Still, these changes are unrelenting and seem unsatisfied.

The most recent effort to respond to this situation is the current (1994) System Operational Review (SOR). This review of 14 federal hydropower projects is reexamining the operation of most of the largest storage projects on U.S. portions of the Columbia River system. This reexamination is from a multipurpose perspective, with special motivation arising from recent endangered species listing of several of the basin's anadromous fish species. The SOR also examines prospects and alternatives for renewal and renegotiation of PNCA and Canadian treaty provisions. Reflecting this multipurpose perspective, the SOR process has technical-level involvement from 11 federal agencies, state agencies from 4 basin states, electric utilities, Native American tribes, universities, and dozens of interest groups and individuals (U.S. DOE, 1992a). The initial screening-level studies examined 90 alternative operating plans for the affected reservoirs. Each alternative was evaluated by multiparty work groups representing anadromous fish, resident fish, recreation, irrigation, flood-control, water-quality, wildlife, power, navigation, and cultural-resources perspectives. Each work group also was involved in the design of alternatives. The results of the screening evaluation are being used to suggest and support a recommendation for improving system operations.

The SOR is just one more episode in the continuing evolution of the operation of the Columbia River system. Continued changes in legal regulations, regional economic demands on the basin's water resources, and social expectations of resource management are all likely to keep management of the Columbia River system in the public eye and place demands on the institutions which manage the basin. Such water-resource problems are likely to never be solved.

31.3.8 Implications for Other Hydropower Systems

The Columbia River hydropower system, while in many ways unique, offers several lessons for the management of other contemporary hydropower systems.

1. Changes in the relative value of different system-operating purposes have significant implications for the operation of hydropower systems. This is particularly true for newly recognized environmental purposes, reinforced by Endangered Species Act designations. The environmental impacts of hydropower facilities and operations have been particularly severe in terms of habitat fragmentation from the existence of dams, modification of the natural seasonal hydrologic regime from seasonal storage, and rapid changes in physical habitat, from peaking releases.

2. Hydropower-system management also cannot be divorced from operation and planning of other parts of the electric-utility system, including thermal-electric generation, system interties, and demandside management. Alternatives for operation of hydropower systems can take advantage of opportunities to modify all aspects of this system.

3. Hydropower systems exist with a complex system of electric-power and water-resource institutions, particularly in the United States. Each of these institutions has different and often conflicting interests. The development of changes in hydropower-system operations, or the defense of unchanged operations, rests on an ability on the part of the project operator to communicate with a wide range of relevant and interested parties and develop and evaluate management alternatives in this context.

4. Computer models are essential for developing, refining, and evaluating alternatives for the improving operation of hydropower systems in the joint context of other water uses and the larger electric-power system. With computer models, options can be developed and examined quickly and inexpensively. The development of computer models also is an opportunity to develop a consistent, agreed-upon technical basis for system studies.

5. The physical and institutional diversity of hydropower systems, while imposing significant management problems, also provides a wide range of opportunities for addressing problems and developing integrated alternatives. The planning and operation of hydropower systems can involve a wide and integrated range of structural and nonstructural measures in both the water-resources and electric-power systems, including demand management and market transfers for both water and power demands.

Acknowledgments

Cindy Henriksen is thanked for her insightful comments on recent developments in Columbia River system operations.

REFERENCES

Abbey, D., and V. Loose, "Water Supply and Demand in a Large Scale Optimization Model of Energy Supply," Report DOE/EV/10180-2, U.S. Department of Energy, Washington, D.C., 1979.

Abbey, D., and F. Lucero, "Water Related Planning and Design at Energy Firms," Report DOE/EV/10180-1, U.S. Department of Energy, Washington, D.C., 1980.

Adams, S., and J. Stevens, "Strategies for Improved Cooling-Tower Economy," *Power*, Jan: 49–50, 1993.

Allemann, R. T., B. M. Johnson, and E. V. Werry, "Wet-Dry Cooling Demonstration: A Transfer of Technology," Report EPRI CS-5016, Electric Power Research Institute, Palo Alto, Calif., 1987.

American Society of Civil Engineers, *Civil Engineering Guidelines for Planning and Designing Hydroelectric Developments*, Vol. 3, Chap. 2, ASCE, New York, 1989.

American Society of Civil Engineers, *Compendium of Pumped-Storage Plants in the United States*, ASCE, New York, 1993.

Bartz, J. A., "New Developments in Cooling Towers," *Power Engineering* 98(6):23–25, 1994.

Bartz, J. A., and J. S. Maulbetsch, "Are Dry-Cooled Power Plants a Feasible Alternative?," *Mechanical Engineering*, 103(10):34–41, 1981.

Broches, C. F., and M. S. Spranger, *The Politics and Economics of Columbia River Water,* Washington Sea Grant, University of Washington, Seattle, Wash., 1985.

Buras, N., "Role of Water in Energy Development," in Y. Y. Haimes and J. Kindler, eds., *Water and Related Land Resource Systems, Proceedings,* International Federation for Automatic Control, Cleveland, Ohio, 1980.

Buys, J. D., and D. G. Kroger, "Cost-Optimal Design of Dry Cooling Towers through Mathematical Programming Techniques," *Transactions of the ASME,* 111(May):322–327, 1989.

Cheremisinoff, N. P., and P. N. Cheremisinoff, *Cooling Towers—Selection, Design and Practice,* Ann Arbor Science, Ann Arbor, Mich., 1981.

Collins, B., "Salmon Were an Afterthought to 136 Columbia River Dams," *High Country News,* July 13, 1992, pp. 15–21.

Creager, W. P., and J. D. Justin, *Hydroelectric Handbook,* John Wiley, New York, 1950.

Delaware River Basin Commission, *Water Management of the Delaware River Basin,* Trenton, N.J., 1975.

Dobson, J. E., and A. D. Shepherd, "Water Availability for Energy in 1985 and 1990," Report ORNL/TM-6777, Oak Ridge National Laboratory, Oak Ridge, Tenn., 1979.

Dynatech R&D Company, "Nonmetallic Heat Exchangers: A Survey of Current and Potential Designs for Dry-Cooling Systems," Report EPRI CS-3454, Electric Power Research Institute, Palo Alto, Calif., 1984.

Gleick, P. H., "Water and Energy," *Annual Review of Energy and Environment,* 19:267–299, 1994.

Goodman, A. S., *Principles of Water Resources Planning,* Prentice-Hall, Englewood Cliffs, N.J., 1983.

Gulliver, J. S., and R. E. A. Arndt, *Hydropower Engineering Handbook,* McGraw-Hill, New York, 1991.

Guyer, E. C., and J. A. Bartz "Dry Cooling Moves into the Mainstream," *Power Engineering,* 96(8):29–32, 1991.

Hamilton, J. R., N. K. Whittlesey, and P. Halverson, "Interruptible Power Markets in the Pacific Northwest," *American Journal of Agricultural Economics,* Feb.:64, 1989.

Harza Engineering Co., *Pumped-Storage Planning and Engineering Guide,* EPRI GS-6669, Electric Power Research Institute, Palo Alto, Calif., 1990.

Hirst, E., and C. Goldman, "Creating the Future: Integrated Resource Planning for Electric Utilities," *Annual Review of Energy,* 16, 1991.

Hobbs, B. F., "Water Supply for Power in the Texas-Gulf Region," *Journal of Water Resources Planning and Management,* 110(4):373–391, 1984.

Hobbs, B. F., "Cooling Water Supply for Energy Production," in P. Cheremisinoff, et al., eds., *Civil Engineering Practice,* vol. 4, Technomic, Lancaster, Pa., 1988.

Hobbs, B. F., and P. M. Meier, "Multicriteria Methods for Resource Planning: An Experimental Comparison," *IEEE Transactions on Power Systems,* 9(4):1811–1817, 1994.

James, L. D., and R. R. Lee, *Economics of Water Resources Planning,* McGraw-Hill, New York, 1971.

Karadi, G. M., *Pumped-Storage Development and Its Environmental Effects,* Project GK-31683, National Science Foundation, Washington, D.C., 1974.

Kuiper, E., *Water Resources Development,* Butterworths, London, 1965.

Lall, U., and L. W. Mays, "Model for Planning Water-Energy Systems," *Water Resources Research,* 17(4):853–865, 1981.

Lander, J., and G. Christensen, "Add Helper Cooling Towers to Control Discharge Temperatures," *Power,* April:137–139, 1993.

Lence, B. J., M. I. Latheef, and D. H. Burn, "Reservoir Management and Thermal Power Generation," *Journal of Water Resources Planning and Management,* 118(4):388–405, 1992.

Linder, K. P., M. J. Gibbs, and M. R. Inglis, "Potential Impacts of Climate Change on Electric Utilities," Report EPRI EN-6249, Electric Power Research Institute, Palo Alto, Calif., 1989.

Loucks, D. P., J. R. Stedinger, and D. A. Haith, *Water Resource Systems Planning and Analysis,* Prentice-Hall, Englewood Cliffs, N.J., 1981.

McCarl, B. A., and G. H. Parandvash, "Irrigation Development versus Hydroelectric Generation: Can Interruptible Irrigation Play a Role?," *Western Journal of Agricultural Economics,* 13(2):267–276, 1988.

McCarl, B. A., and M. Ross, "The Cost Borne by Electricity Consumers under Expanded Irrigation from the Columbia River," *Water Resources Research,* 21(9):1319–1328, 1985.

Mays, L. W., and Y.-K. Tung, *Hydrosystems Engineering and Management,* McGraw-Hill, New York, 1992.

Mitchell, R. D., "Survey of Water-Conserving Heat Rejection Systems," Report EPRI GS-6252, Electric Power Research Institute, Palo Alto, Calif., 1989.

Miller, B. A., et al., "Impacts of Changes in Air and Water Temperature on Thermal Power Generation," in *Managing Water Resources During Global Change,* American Water Resources Association, Bethesda, Md., 1992, pp. 439–448.

Mirsky, G. R., J.-P. Libert, and K. Bryant, "Designing Cooling Towers to Accommodate the Environment," *Power,* May:95–99, 1992.

Modern Power Systems, "Dry Cooling Towers at Biggest Coal Fuel Power Station," 11(7):29–32, 1991.

Palmer, R. N., and J. R. Lund, "Drought and Power Production II: Risk Analysis Planning," *Journal of Water Resources Planning and Management,* 114, 1988.

Policastro, A. J., W. E. Dunn, and R. A. Carhart, "A Model for Seasonal and Annual Cooling Tower Impacts," *Atmospheric Environment,* 28(3):379–395, 1994.

Power, "Desalination Can Produce New Revenue Stream," Aug.: 9–12, 1993.

Probstein, R. F., and H. Gold, *Water in Synthetic Fuel Production: The Technology and Alternatives,* MIT Press, Cambridge, Mass., 1978.

Public Service Electric & Gas Corp., *An Assessment of Energy Storage Systems Suitable for Use by Electric Utilities,* Vol. III, EPRI EM-264, Electric Power Research Institute, Palo Alto, Calif., 1976.

Putta, S., "Valuing Externalities in Bidding in New York," *Electricity Journal,* 3(6):42–47, 1990.

Shaw, J. J., E. E. Adams, D. R. F. Harleman, and D. H. Marks, "A Methodology for Assessing Alternative Water Acquisition and Use Strategies for Energy Facilities in the American West," Energy Laboratory Report no. MIT-EL 81-051, Massachusetts Institute of Technology, Cambridge, Mass., 1981.

Shorney, F. L., "Water Conservation and Reuse at Coal-Fired Power Plants," *Journal of Water Resources Planning and Management,* 109(4):345–359, 1983.

Solley, W. B., R. R. Pierce, and H. A. Perlman, "Estimated Use of Water in the United States in 1990," U.S. Geological Survey Circular 1081, U.S. Government Printing Office, Washington, D.C., 1993.

Sonnichsen, J. C., "An Assessment of the Need for Dry Cooling, 1981 Update," Report HEDL-TME 81-47, UC-11, 12, Hanford Engineering Development Laboratory, Richland, Wash., 1982.

Sonnichsen, J. C., et al., "Steam-Electric Power Plant Cooling Handbook," Report HEDL-TME 81-53, UC-12, 12, Hanford Engineering Development Laboratory, Richland, Wash., 1982.

Trelease, F. J., "Legal Problems in the Allocation and Transfer of Water to Electric Utilities," in *Proceedings: Workshop on Water Supply for Electric Energy,* Report EPRI WS-79-237, Electric Power Research Institute, Palo Alto, Calif., 1980.

U.S. Army Corps of Engineers, *Reservoir Yield,* Hydrologic Engineering Methods for Water Resources Development, Vol. 8, U.S. Army Corps of Engineers, The Hydrologic Engineering Center, Davis, Calif., 1975.

U.S. Army Corps of Engineers, *Flood Control by Reservoirs,* Hydrologic Engineering Methods for Water Resources Development, Vol. 7, U.S. Army Corps of Engineers, The Hydrologic Engineering Center, Davis, Calif., 1976.

U.S. Army Corps of Engineers, *Reservoir System Analysis for Conservation,* Hydrologic Engineering Methods for Water Resources Development, Vol. 9, U.S. Army Corps of Engineers, The Hydrologic Engineering Center, Davis, Calif., 1977.

U.S. Army Corps of Engineers, *Feasibility Studies for Small Scale Hydropower Additions,* U.S. Army Corps of Engineers, The Hydrologic Engineering Center, Davis, Calif., 1979.

U.S. Army Corps of Engineers, *Application of the HEC-5 Hydropower Routines,* Training Document no. 12, U.S. Army Corps of Engineers, The Hydrologic Engineering Center, Davis, Calif., 1983.

U.S. Army Corps of Engineers, "Hydropower Engineering and Design," EM 1110-2-1701, U.S. Army Corps of Engineers, Washington, D.C., 1985.

U.S. Army Corps of Engineers, "Columbia River Reservoir System Analysis: Phase II," Technical Report PR-21, Hydrologic Engineering Center, U.S. Army Corps of Engineers, Davis, Calif., 1993.

U.S. Department of Energy, "Hydroelectric Power Evaluation," DOE/FERC-0031, U.S. Department of Energy, Federal Energy Regulatory Commission, Washington D.C., 1979.

U.S. Department of Energy, *The Columbia River System Operation Review Scoping Document,* U.S. Department of Energy, Bonneville Power Administration, U.S. Army Corps of Engineers, North Pacific Division, U.S. Bureau of Reclamation, Pacific Northwest Region, Portland, Oreg., 1991a

U.S. Department of Energy, *The Columbia River System: The Inside Story,* U.S. Department of Energy, Bonneville Power Administration, U.S. Army Corps of Engineers, North Pacific Division, U.S. Bureau of Reclamation, Pacific Northwest Region, Portland, Oreg., 1991b.

U.S. Department of Energy, *Screening Analysis: A Summary,* U.S. Department of Energy, Bonneville Power Administration, U.S. Army Corps of Engineers, North Pacific Division, U.S. Bureau of Reclamation, Pacific Northwest Region, Portland, Oreg., 1992a.

U.S. Department of Energy, *Modeling the System: How Computers are used in Columbia River Planning,* U.S. Department of Energy, Bonneville Power Administration, U.S. Army Corps of Engineers, North Pacific Division, U.S. Bureau of Reclamation, Pacific Northwest Region, Portland, Oreg., 1992b.

U.S. Department of Energy, *Power System Coordination: A Guide to the Pacific Northwest Coordination Agreement,* U.S. Department of Energy, Bonneville Power Administration, U.S. Army Corps of Engineers, North Pacific Division, U.S. Bureau of Reclamation, Pacific Northwest Region, Portland, Oreg., 1993.

U.S. Department of Interior, "Selecting Hydraulic Reaction Turbines," Engineering Monograph no. 20, U.S. Department of Interior, Bureau of Reclamation, Denver, Colo., 1976.

U.S. General Accounting Office "Water Resources Issues Concerning Expanded Irrigation in the Columbia Basin Project," Report no. GAO/RCED-86-82BR, Jan., U.S. General Accounting Office, Washington, D.C., 1986.

U.S. Water Resources Council, *The Nation's Water Resources, 1975–2000,* 4 vols., U.S. Government Printing Office, Washington, D.C., 1978.

Warnick, C. C., *Hydropower Engineering,* Prentice-Hall, Englewood Cliffs, N.J., 1984.

Weatherford, G., et al., *Acquiring Water for Energy: Institutional Aspects,* Water Resources Publications, Littleton, Colo., 1982a.

Weatherford, G., L. Brown, H. Ingram, and D. Mann, eds., *Water and Agriculture in the Western U.S.: Conservation, Reallocation, and Markets,* Westview Press, Boulder, Colo., 1982a.

Weinberg, C. J., J. J. Iannucci, and M. M. Reading, "The Distributed Utility: Technology, Customer, and Public Policy Changes Shaping the Electrical Utility of Tomorrow," *Energy Systems and Policy,* 15:307–322, 1993.

Wiel, S., "The New Environmental Accounting: A Status Report," *The Electricity Journal,* 4(9):46–54, 1991.

Wurbs, R. A., *Water Management Models: A Guide to Software,* Prentice-Hall, Englewood Cliffs, N.J., 1995.

CHAPTER 32
WATER-USE MANAGEMENT: PERMIT AND WATER-TRANSFER SYSTEMS

J. Wayland Eheart
Department of Civil Engineering
University of Illinois at Urbana-Champaign
Urbana, Illinois

Jay R. Lund
Department of Civil and Environmental Engineering
University of California, Davis
Davis, California

32.1 INTRODUCTION

This chapter reviews two approaches to managing the use of water, *water-use permit systems* and *water-transfer* or *water-marketing systems*. Permit systems typically imply a level of central administration and control, while transferable-use systems require a level of decentralized management. The two systems often operate in tandem, since transferable water-use systems usually require the firm establishment of property rights provided by a permit system.

The first sections of the chapter focus exclusively, if somewhat arbitrarily, on water-withdrawal permits programs. It is divided into nine sections following this introduction. Section 32.2 reviews the legal foundations upon which water regulations are or might be based. Section 32.3 reviews various types of permit systems. In Sec. 32.4 program objectives are introduced and discussed. Section 32.5 discusses options and decisions for management programs and associated tradeoffs among the program objectives. Section 32.6 presents some technical aspects that must be addressed for any system of transferable or nontransferable permits. The review of water-transfer systems begins with Sec. 32.7. This discussion focuses primarily on the transfer of water to cities, but the principles remain applicable to other contexts of water transfers. This section contains some historical and economic background on the subject. Section 32.8 reviews various forms of water-transfer arrangements. Section 32.9 discusses various issues for implementing water transfers, including the role of government in supporting and regulating water transfers. Section 32.10 summarizes the discussion and implications of all the preceding sections and offers some concluding remarks on water-use management.

It should be noted that there are other forms of water-use management which are beyond the scope of this chapter. These include the use of plumbing codes, land-use regulation and incentives, and general water-conservation regulations and incentives now common in many cities in the western United States.

32.2 FOUNDATIONS OF WATER-WITHDRAWAL REGULATIONS

In the so-called humid regions of the world, rainfalls and streamflows have historically been sufficient to supply nearly all human needs. In those regions a set of common-law precedents known as the *riparian doctrine* has often evolved to govern surface-water use.

In the arid regions of the western United States, a different type of water-use doctrine evolved in recognition that in arid or semiarid environments the worth of land was intimately tied to the amount of water that could be used with it. The *appropriative doctrine* that was developed in that region grants each water user a clear right to a certain quantity of water. Associated with that right is a priority rank relative to all other users on the same stream that establishes the order of forfeiture of water in times of shortage.

The riparian doctrine is rather imprecise compared to the appropriative doctrine and many riparian areas lack a strong, comprehensive set of water-use regulations (Linsley and Franzini, 1972; Dixon and Cox, 1985). In the days when the technical capacity to use water was limited and abstractions from streams were needed only for a few cattle or a small vegetable plot, the riparian doctrine functioned adequately. In recent years, however, problems of concentrated and massive use have become more common and severe in humid regions. Aquifer levels and aquatic stream habitats have been threatened by concentrated withdrawals as cities have expanded, and irrigation, which is highly consumptive, has increased (e.g., Goering and Cekay, 1988).

32.3 WATER-USE PERMIT SYSTEMS

With the recognition that the common-law riparian doctrine is ill-suited to manage water withdrawals in the current era, many riparian areas have begun to pass more detailed water-use laws (e.g., the 1972 Florida Water Resources Act, Florida Statutes, Chap. 373; Virginia Water Use Act, 1991).

While such water laws constitute great progress in setting the general principles under which regulatory programs must operate, they have generally not attained the specificity necessary for day-to-day regulation of water withdrawals. These tasks are usually implicitly or explicitly left to local regulatory agencies to implement as they best see fit. Saarinen and Lynn (1993), for example, discuss the problems of achieving economic efficiency under Florida's statute, even though such efficiency is a stated goal of the statute. While discussion of comprehensive management programs has occurred (e.g., Mack and Peralta, 1987; Walker, et al., 1983), implementation has not been widespread.

There is thus a need for state, provincial, regional, or basin management agencies in traditionally riparian areas to develop day-to-day administrative systems or water-management programs. Ideally, such programs should routinely regulate the withdrawal of water efficiently (in both the bureaucratic and economic sense) and fairly and involve the courts only in exceptional cases such as interstate or international disputes.*

There is a corresponding need, albeit a lesser one, for a review of administrative systems in areas traditionally governed by the appropriative doctrine. In many such cases, ad hoc and even inconsistent procedures have become ensconced, so that revisions are appropriate.

In essence, such programs should have two overall purposes, although they may have several competing objectives. First, they should serve as water-allocation instruments in times of drought. We define drought, in this context, as an absence or shortage of water *when the availability of more is expected*. The lack of water in Death Valley, California is not a drought, as no one expects Death Valley to be a verdant paradise. The existence of water users, their water-using capacity, and their wants define the drought as much as the water shortage (WSTB, 1986). The second overall purpose of such programs is to prevent problems attending concentrated withdrawals. Such concentrated uses may cause problems with maintenance of minimum streamflows (i.e., aquatic habitats) and aquifer levels in times of normal rainfall as well as drought.

There are many options for such programs and a useful precursor to their development is an assessment and comparison of these options, with a view toward providing guidance to agencies engaged in their development. This chapter provides such a comparison. Alternative decisions involved in creating a water-withdrawal permit program are compared qualitatively in the context of six management objectives. Certain alternatives are identified as currently appearing more attractive than others, pending further research and public dialogue. No specific recommendations are made but an attempt is made to provide guidance to decision makers without imposing values on them.

32.3.1 Water-Management Programs

It is assumed here that a water-management agency (referred to as the *agency*) has been charged with developing and administering a water-withdrawal management program (although these two activities need not be undertaken by the same entity). The program is to consist of a set of uniform rules governing water withdrawal. A set of target minimum streamflows and aquifer levels (or allowable drawdown rates) is assumed already set. These external requirements may vary with time, monotonically changing or oscillating on an annual frequency. It is also assumed that the agency has been given statutory authority to impose any of the management programs discussed here on raw-water users. Obviously, if the statute precludes any of these options, they must be excluded from consideration.

There are four major philosophical approaches toward restricting water use, discussed as follows. They are not all mutually exclusive and may be combined in a program. Most programs will probably concentrate on one approach, possibly with a second filling in the gaps.

---

* Work is ongoing to develop methods to effect water sharing among states in a smooth manner.

1. *Ad hoc restrictions.* This is essentially the absence of a management program. These restrictions simply disallow certain types of uses under certain conditions. Examples of such restrictions are the classical municipal prohibitions of lawn and garden watering and car washing during times of shortage. On a larger scale, one might envision restrictions on agricultural or industrial use in times of drought. Such emergency restrictions may be satisfactory in cases where supply shortfalls are brief and infrequent, but are ill-suited for long-term maintenance of water resources once the crisis has passed. They often are not economically efficient. Water restrictions may be imposed just when they can be accommodated least easily (e.g., restrictions on irrigation just when crops are in greatest need of water). WSTB (1986) presents the results of a workshop devoted to a thorough analysis of the problems and options of ad hoc restrictions.

2. *Water charges.* A second approach is the requirement that users pay a charge for using water. This is not simply a registration fee; it must be proportional to the amount of water used and high enough to induce conservation. Under such a program, installation or inspection of flow meters, monitoring of flows (to establish the required payments), and collection of payments will be necessary agency activities. A strength of such a program is that it tends to induce an economically efficient use of water. An important weakness, however, is that the charge must constantly be readjusted to keep the water-use rate within the limits required to maintain groundwater resources and streamflows. The program controls not the supply of water, but its unit cost to the users. Any user may take as much water as desired as long as the charge is paid. The charge program generates revenue for the government and forces the users to pay for the public resource they receive. It may be argued, as Lyon (1982) does for sales of transferable permits, that attempts to address the financial burden issue with refunds to users will render the program ineffective. If the refunds vary with the amount of water used, the users will incorporate that knowledge in their decision making and the outcome will not be economically efficient.

3. *Subsidies.* A variation of the charge program is its opposite, i.e., a program of subsidies for water conservation. An analogous example is the federal construction grants program in controlling water pollution. These programs have the same weaknesses as the charge program, plus some of their own. First, they require money to be transferred to the users from the government and must be financed by the taxpayers. Second, users may receive subsidies for conservation measures they might have undertaken anyway.

4. *Water-Use Permits.* A fourth approach is a program under which each user receives a permit allowing a rate of withdrawal that depends, generally, on ambient conditions. Legally, the permit may be a deed to the water itself (implying a property right), or it may be a license (implying temporary permission to use the resource).* The agency must decide exactly what the permit should entitle its holder to do, and under what circumstances. It is customary to think of it as entitling its holder to withdraw a certain volumetric flow rate (e.g., L/s) of water, but an operational definition must be more complete and must say by whom, where, and for how long the water may be withdrawn and what happens when there is not enough water to supply all permits. Other agency decisions are on what basis permits should be distributed initially; how many should be distributed; whether or not they should be transferable and, if so, under what circumstances or restrictions; and so on.

---

* No decision in that regard is rendered here; throughout this document the word *permit* denotes an explicit legal tender to water, however limited. The word *right* denotes a more general claim, whether legally explicit or legally or culturally implicit.

Without offering a quantitatively buttressed argument, it is asserted here that of these four approaches, the latter, permits, is most useful as the mainstay of a water-management program. Some ad hoc processes may be necessary to address unforeseen circumstances, and charges or subsidies may prove useful in certain cases. Permit systems, however, have the intuitive appeal of being revenue-neutral to the agency while, conceivably at least, offering a foundation for the goal of day-to-day operation of the management program stated previously.

32.4   PERMIT PROGRAM OBJECTIVES

The objectives in designing a water-management program can be grouped into six major categories. The categories are not mutually exclusive and some overlap may occur among them.

32.4.1   Ease of Implementation, Administration, and Enforcement

The first objective is to maximize the ease on the agency's part in setting up the program initially (implementation), operating it routinely (administration), and insuring compliance with it (enforcement). This objective could be included in the cost-efficiency objective following, but is separated because many agencies will consider their own costs separately from those to society at large.

Two important components of this objective are simplicity and comprehensibility. Regulations that are simple and easy to understand are usually easy to implement, administer, and enforce as well. Simple regulations are more likely to be followed because regulatees understand more clearly what is required of them and cannot as successfully use a confusion defense if caught in a violation.

32.4.2   Equity

The second objective is maintaining equity among water users. Equity is very much in the eye of the beholder, and the decision about what is equitable or inequitable is arbitrary to some extent. Moreover, each party affected by regulations can usually be counted on to view the most equitable rule as the one which benefits him or her the most. Consensus over equity issues may be difficult to achieve, but a compromise may be acceptable. Through negotiation, it should be possible at least to avoid a program that is agreed by all to be inequitable.

32.4.3   Effectiveness in Protecting Water Resources

The third objective is that for which the program was originally devised, viz., the maintenance of streamflows or groundwater resources. Some programs address these needs directly, others only indirectly. Better protection of resources may require greater effort of administration and enforcement or less economic efficiency.

32.4.4   Robustness and Flexibility

In devising any program to control water withdrawals, there will inevitably be errors in data collection and in predicting the outcome of the program. Hence, an objective in program design is that the social benefit of the program be robust, or insensitive to errors in the data upon which its design depends. Robustness is not necessarily a static ability to withstand unforseen changes but may instead be a dynamic correction in response to such changes.

32.4.5 Economic Efficiency

A fifth social objective, already touched upon, is economic efficiency of water use. Different management programs will generally result in different distributions of water among users, implying different aggregate levels of economic benefit (efficiency) from water use. Given the finite supply of water, only one distribution will maximize benefit, but some other programs may fare better with respect to other program objectives.

Generally, a lack of regulations results in an economically inefficient distribution because those to whom water is physically most available take a larger share than is efficient. On the other hand, regulations that restrict water use without consideration of economic efficiency may erect barriers to voluntary redistribution that would increase efficiency.

32.4.6 Political and Legal Feasibility

The final objective is political and legal feasibility. No matter how well designed a program is with respect to the other five objectives, it will fail if it is not accepted politically or if it cannot operate within prevailing laws. Most other objectives are fortunately coincident with, if not constituent of, this objective. However, a well-designed program may require small changes in prevailing law, and such programs may be rendered politically infeasible if misconceptions about them gain widespread acceptance through the actions of pressure groups or lobbyists.

32.5 PERMIT PROGRAMS FOR WATER-WITHDRAWAL CONTROL

The agency's choice of options determines how the water-permits program fares with respect to program objectives. It will not generally be possible to find a set of choices that maximizes all objectives, and the agency will be forced to trade one objective off against others. There is no optimal or "best" tradeoff point among those objectives, only one that is most acceptable in the agency's judgment. The following paragraphs discuss several options and the tradeoffs among the objectives for each.

Although it is important in the design of a program, no distinction will be drawn for the moment between water withdrawal and consumption; these terms will be used interchangeably, along with the term *water use*. It is assumed that all use is consumptive and that water, once withdrawn from a watercourse, disappears forever. The importance of return flows, their effects, and the options for incorporating them in the regulatory program are discussed near the end of this section.

32.5.1 Geographical and Temporal Configuration of Programs

The agency must decide the geographical and temporal extent of water-control programs and how the activities of smaller-scale programs should be delineated and coordinated to achieve overall agency goals at the larger scale (e.g., state or province). Conceivably, limits for surface-water programs would best follow basin boundaries, but those boundaries do not necessarily correspond to the natural lines of cleavage for groundwater. Moreover, there is a frequent reluctance of state or provincial and local authorities to delegate the necessary authority to multistate or multiprovince river-basin commissions, especially when nonwater interstate issues are at stake. Thus, the central coordination might most feasibly take place at the state or provincial level.

For both surface- and groundwater, there is the question of whether individual ad hoc programs should be developed for each local problem area, whether each new program should be extended to the smallest political boundary that contains the problem area (e.g., a city, township, county, or multicounty unit), or whether a more comprehensive program should be developed for a state, province, or river basin. A comprehensive uniform state- or province-wide program is more difficult to administer but has no greater effectiveness than individual programs designed for problem areas. Nevertheless, if the water problems of several individual areas, each with a local governing authority, expand to their jurisdictional boundaries, each local authority may act to protect its local interest. There may then be a need for a higher authority to intercede to distribute water among them. With local authorities already in place, political realities may require the central authority to administer through the local authorities. This may be less efficient than if more central regulation were in operation from the outset. Thus, local programs are likely to be effective only as long as an adequate plan for their future coordination and, if need be, their future subvention to central authority, is in place at the outset.

32.5.2 Permit Definition Basis

The manner of determining how much water the permit holder may take is referred to as the *definition basis* of the permit. The two major permit definition bases are discussed briefly here and in more detail by Eheart and Lyon (1983). Other alternatives are discussed by Eheart (1989); for brevity those are omitted here.

Let the amount of water (e.g., L/s) available for distribution to the permit holders be referred to as the *total allowable withdrawal* (*TAW*). For surface water the TAW equals the streamflow minus the minimum flow requirement for instream flow needs. For groundwater, the TAW is set considering target aquifer levels or drawdown rates. For now, assume that when the agency sets the TAW it also distributes it among users according to some "fair" formula. (The complexities of channel geometry, discussed later, will complicate the formula for rivers.) There are then three bases, discussed as follows, for defining permits, i.e., relating the allowable withdrawal to the TAW.

1. *Constant use basis.*   Under this basis, each user is allowed a certain constant use rate. The rate may vary with time, possibly following an annual schedule, but does not depend on immediate stream conditions unless the streamflow is insufficient to satisfy all claims. The chief drawback of this basis is that it does not spell out what happens during drought.

2. *Prioritized permit basis.*   Traditionally this basis has been used to allocate surface water in the western United States under the appropriative doctrine. A set of priorities among users is established. Any given user is allowed a certain constant rate of water withdrawal as long as the TAW is enough to satisfy him or her and all other users with a higher priority (including instream needs). When the TAW is insufficient to satisfy all users, they forego withdrawals according to their priorities.

3. *Flexible permit bases.*   Under these bases, users' allowable withdrawals increase and decrease continuously with the immediate streamflow. The simplest of these bases is the fractional basis, under which each user is allotted a constant percentage of the TAW. Thus, as the TAW fluctuates, so does the amount of water allotted to each user and no user is ever entirely deprived of water. This type of permit has the advantage of homogeneity over prioritized permits; no assignment of priority need be made to individual users since no user's permit has priority over any other's. There may, however, be some perceived equity and administrative disadvantages to this basis. First, there is no easy way to issue free permits to newcomers, although they may easily buy their way in and may be accommodated through staggered limited-duration permits (see following). If entering users are to share available water with existing users who are already using it all, then the existing users must give up some. Forcing them to do so might be viewed as a confiscation of property without compensation. By contrast, the prioritized-permit basis may be structured to accommodate newcomers by assigning them lowest priority. The second disadvantage is that it is more difficult to account for the geometric complexity of real river systems under this type of permit-definition basis. Ways of addressing this problem also exist, however, and are discussed in greater detail following.

32.5.3 Allocation Basis

The agency must decide the basis for distributing permits among the users, i.e., deciding what size permit (and the priority, if appropriate) each user receives. The size of permit is the rate of allowable withdrawal for the prioritized-permit system; for the fractional-permit system, it is the fraction of the TAW the user is allowed to withdraw.

In the West, priority allocations were established by the rule of "first in time, first in right." Size allocations were based on the agency's judgment about the amount of water each user needed or deserved, based in part on past use or anticipated future use. These allocations were made on a case-by-case basis, and there was no guarantee of consistency from one time period or location to another.

It would be difficult to apply appropriative rights in riparian areas. First, since no water-rights system existed historically, it would be difficult to establish priorities among users. Second, the notion of priorities runs counter to the legal precedent of the riparian doctrine and has been legally rejected in some riparian areas (see, e.g., for Illinois, *Bliss v. Kennedy,* 1867).

One alternative approach is a formula that considers the historical use of water by each prospective recipient as a measure of need. Presumably, the size allocations would be roughly proportional to some measure of past-use rate. Provisions could be incorporated to avoid rewarding those who had used water wastefully in the past, and certainly to dissuade users from profligate use in the present for the sole purpose of receiving a higher allocation in the future. This poses the problem of estimating past withdrawal rates that may have been unmeasured. Most municipal and industrial withdrawals are gauged, but many agricultural withdrawals are not.

An alternative basis rests on the theory that the right to a certain amount of water is internalized in the worth of riparian land. In this context, the issuance of a permit by the agency may be viewed as an attempt to separate and grant to the riparian landowner the water right that was historically bound up with the land right. Under this approach, the size of the fractional permit issued to a given surface-water user might be proportional to such a parameter as the length of riparian streamfront or the area of riparian land. (An equivalent rule for the distribution of groundwater permits for irrigation might be in proportion to the area of overlying land.) Such a basis of allocation might be considered equitable for agricultural users, but the agency may deem it desirable to set aside a portion of the water for other uses, to be distributed among them on a different basis (e.g., proportional to population equivalent for municipalities and earnings, taxes, or employees for industries).

Another alternative is simply an ad hoc allocation basis that seeks to use no formula in distributing the available water among users but, rather, leaves such decisions to the judgment of the agency. While such an approach has the benefit of flexibility, it risks potential challenges of arbitrary or capricious behavior on the part of the agency.

While priority allocation to individuals is inconsistent with the riparian doctrine, as noted, some states in the United States have embraced a priority allocation among use *types*, placing so-called *natural needs* above *artificial wants* in priority (e.g., Clark, 1985). Even in appropriative regions, such priorities often take precedence over the first-in-time rule. Thus, a regulatory system might serve all municipal requirements before satisfying the requests of industrial, power, or agricultural uses. It may be, however, that only the portion of municipal withdrawal that constitutes natural needs is felt to be eligible to be set aside first, so that the remaining portion of the municipal claim should have to compete with other claims.

32.6   TECHNICAL DETAILS OF PERMIT SYSTEMS

Several detailed aspects must be addressed for establishing any permit system.

32.6.1   Duration of Permits and Accommodation of Newcomers

The agency faces a dilemma in deciding the length of time a permit is valid. As noted by Eheart and Lyon (1983) and Young (this volume, Chap. 3), permits that are valid for only a short time may not allow the users sufficient time to pay off capital equipment and may thus result in economically inefficient decisions. (For example, the user may purchase less-expensive equipment that uses water inefficiently.) A long permit validity will present difficulty in accommodating newcomers, and there is a risk of overcommitting the resource and being unable to reverse the allocation process except by repurchase. Eheart and Lyon (1983) note that one way of addressing this dilemma is a system of staggered permits of n-year duration under which the agency may lower the total number of permits by as much as $1/n$ per year simply by not reissuing expired permits.

There is a potential problem in initiating such a system or accommodating new users, since users prefer long-term to short-term permits. Nevertheless, it may be possible to initiate the system if the agency acts at an early stage, before demands become significant in comparison to supplies. If $1/n$ or less of the target number of permits are currently held and less than $1/n$ additional will be claimed in aggregate by all new users each year, the agency may issue n-year permits to all new applicants.

A staggered system of limited-duration permits could be structured to include newcomers in the allocation, granting them the same status as existing users and enabling them to acquire an increasingly large number of permits each year. Thus, for example, under a 5-year staggered system, a newcomer who, as an existing user, would have claim to 15 percent of the permits would, like existing users, be allocated

15 percent of the 20 percent of new permits (3 percent) that are reissued each year. Requiring a user to wait five years for the full allocation might not be practical; depending on the water-use application, operating at reduced production capacity may not be economical. In such cases the staggered system might be effective when combined with a spot market, so that the user could rent permits for the interim.

32.6.2 Averaging Periods

It is not possible for water users to restrict their withdrawals to a certain flowrate at all times, nor is it always desirable from the agency's perspective for them to do so. Crops need only be irrigated when they undergo moisture deficit, and municipal demands fluctuate according to the weather and the incidence of fire. It is therefore appropriate for the agency to grant users some flexibility by restricting their time-averaged, rather than their instantaneous, withdrawals. The question then becomes one of choosing the averaging period. The larger the averaging period, the more flexibility the users have but the greater the opportunity that exists for them to over-exploit the water resource, at least temporarily.

If dewatering of an unimpounded stream is to be prevented, the averaging period will usually have to be from less than a day to a few days, depending on the size of the stream. Since most aquifers recharge only at certain times of the year, a 1-year averaging period might be appropriate for groundwater withdrawals.

In its monitoring activities, the agency will calculate the average periodically, since it is likely to be impossible for it to keep a running tab on the time average of every user's withdrawal continuously. The necessity of performing these calculations at certain times presents incentives for perverse behavior, however. Toward the beginning of a period (i.e., just after averaging) there is an incentive for users who put less stock in future worth to increase withdrawals, perhaps hoping that it will rain more toward the end of the averaging period. Toward the end of the averaging period (i.e., just before averaging) there is an incentive for users who originally husbanded their water to increase withdrawals because they will lose what they don't take. (If they don't lose what they don't take, the averaging period is effectively longer.) Thus, around the time that the calculations are performed, there may be increases in total withdrawals. To address this problem, it may be worthwhile to stagger the times of calculating averages for different users so that the increased withdrawals are spread evenly over time. Under prioritized permits, carryover from the previous period could be assigned a low priority, thus making it available to the user, but not with the same value.

32.6.3 Large-Scale Groundwater Restrictions

There are several issues related to selecting an aggregate allowable rate of removal from groundwater aquifers. One is how much water should be saved for the future. When aggregate withdrawal exceeds aggregate recharge, aquifer mining is said to be taking place. The word *mining,* although commonly used, may be a misnomer; unlike coal or oil, groundwater is a renewable resource and will eventually return to higher levels when aggregate pumping stops or is reduced. Nevertheless, many years' worth of recharge may be removed in a very short time, and future generations may have to wait some time before enjoying the aquifer levels of their forebears. Sometimes it is difficult to document mining because aquifer levels usually undergo annual fluctuations that may mask long-term decreases.

Apart from this issue, aquifer mining implies a continual depression of the piezo-metric head and affects most users drawing from a groundwater unit. It requires periodic redrilling of wells or lowering of pumps as a matter of course, as well as a continuous increase in pumping costs. Thus, even if aquifer mining has the endorse-ment of the agency, it is reasonable to require that a substantial number of users in a region be similarly disposed toward mining and be willing to cope with its cost and inconvenience in order to reap the benefit of (temporarily) increased allowable pumping rates.

32.6.4 Complexity of Surface-Water Programs Using the Flexible-Permit Basis

There is a problem with adopting the flexible-permit basis for surface water in humid regions. If it happens that all water flows from an area where none is used, through a central point from which it is physically available to all users, fractional permits could have the clear meaning of restricting each user to its fraction of the flow rate at that point. But where sources and sinks of water are geographically dis-persed, and the amounts that are physically available to be shared are not the same, such fractional assignments among users lose their meaning.

Two methods are proposed here of indexing the allowable withdrawal to the observed streamflow. Under one, a gauge upstream of each user determines its allowable withdrawal; under the other, the stream-gauge reading controls users' withdrawals upstream of it. The former, termed the upstream-gauged system is administratively easier, but requires a large number of stream gauges to be used effectively. The latter, termed the downstream-gauged system, is administratively difficult but uses a smaller number of gauges. They are discussed in greater detail as follows.

32.6.4.1 Upstream-Gauged Systems. Under these systems, a group of users located in a stream reach will share a TAW for that reach equal to the streamflow at the nearest upstream gauge minus minimum streamflow and an amount set aside for users downstream of their reach. There are two variations of this system, distin-guished according to how the amount of water a user must pass through to down-stream users is determined. Only the fixed pass-through system is practicable; a discussion of the variable pass-through system is important to aid conceptual under-standing of the issue.

32.6.4.2 Variable Pass-Through. Under the variable pass-through system, a set of contingency arrangements based on readings at upstream gauges determines the withdrawal rate to which any permit holder is entitled. Assuming fractional permits as the implementation of flexible permits, each user is assigned a number propor-tional to its share of all the water available in any basin that contains it, minimized over all basins that contain it. As the basin becomes larger, more water is coming into it, but there are more users with whom the user has to share. For example, a user located on a small tributary to the Mississippi River in Iowa might be entitled to the minimum of: half of the water shared by it and its nearest two neighbors; 5 percent of the water shared by those users and those in the basin upstream of the next downstream gauge; and so on, to 0.00015 percent of all the available water in the Mississippi basin. Unfortunately, this system is impracticable as it requires an infeasible amount of information processing and assumes a zero residence time in the channels.

32.6.4.3 Fixed Pass-Through. An alternative that avoids this problem is to impose fixed obligations to downstream users, so that curtailment orders to a given group of users are in no way dependent on stream-gauge readings downstream of them, regardless of what actually happens to the downstream users' water supply. This will allow indexing a user's allowable withdrawal to the nearest upstream gauge. Under this system, users are divided into groups residing between adjacent stream gauges and each group of users is required to forego a fixed fraction of its incoming TAW for downstream users. An important question is how that fraction should be determined. Conceivably, it should be chosen in consideration of equity and economic efficiency, and should ideally take account of the number and type of downstream users as well as their potential for supplies from other tributaries. One possibility is to use an estimate of the average value of the aggregate downstream claim (as a fraction of incoming TAW) under the variable pass-through system. Harrison (1991) has studied this method for a limited data set and has concluded that it has high economic efficiency.

Permit transfers would be administratively convenient for the fixed pass-through system; the size of the transferred permit would simply be assigned to the new reach. As an example, consider an upstream user that currently is allowed to take 20 percent of the TAW at its upstream gauge, 45 percent being allocated to other users in the group and 35 percent passed to downstream users. If that user wishes to sell its permit to a downstream user, the remaining users in the upstream group will be required to pass 20 + 35 = 55 percent to the downstream users. The downstream neighbors of the buyer will each have a smaller percentage of their incoming TAW, but that TAW will be commensurately larger. For a transfer of a permit in the upstream direction, the system operates the same, except that in the example, no more than 55 percent could be moved to the upstream group.

32.6.4.4 Downstream-Gauged Systems. Under these systems, the permit is actually a set of different permits, each contingent on a different gauge constraining or limiting withdrawals. In essence, the system operates like the upstream-gauged system except that an attempt is made to infer the streamflow at an ungauged point upstream of the users from the observed streamflow downstream of them. The allowable withdrawal by a user is its fraction of the TAW at a stream gauge, as shared with all other users upstream of the stream gauge. A correction must be made to the TAW to account for current use of water; thus, assuming 100 percent consumption, the sum of current uses is added to the observed flow rate at the gauge, the minimum flow requirement is subtracted from that quantity, and the resulting TAW is distributed among the users between that gauge and the nearest upstream gauge. While this system is practicable, it may not be practical, since it essentially uses a feedback control mechanism in that a user's withdrawal influences the streamflow at a gauge, which in turn influences the user's allowable withdrawal. This may lead to problems of stability if return flows are not adequately accounted for, but some alteration of the curtailment order process based on experience and trial-and-error by the agency may address this problem in some cases, especially if the basin is small.

32.6.5 Complexity of Surface-Water Programs Using Prioritized Permits

For prioritized permits there is little difference between upstream variable pass-through and downstream-gauged systems. Curtailment orders are issued to upstream user's in decreasing order of seniority whenever a senior user's withdrawal is unavailable. The gauges are used primarily to determine where curtailment occurs and to insure that minimum streamflows are met.

The fixed pass-through system could be used for such permits if a fixed estimate is used to represent downstream senior claims. For example, if the claim of a senior user is 15 m³/s, and the average flow available to that user from tributaries with no users is 10 m³/s, one possible fixed pass-through system would require an upstream junior user to pass the first 5 m³/s of its incoming TAW to the downstream senior user.

This system does not use a feedback mechanism, but it could result in a reversal of allocation compared to the variable pass-through system. If the upstream user's stream has a high flow and the other tributaries of the downstream senior user have low flows, the upstream junior user might be completely satisfied while only the first 5 m³/s of the downstream senior user's claim would be fulfilled.

32.6.6   Withdrawal versus Consumption and Accounting for Return Flows

As noted later, if all withdrawal permits are defined in terms of consumption, some, but not all, of the problems of third-party impacts may be avoided. The avoided problems are the middle-user impairment problem (Anderson, 1983; 1983b; Johnson, et al., 1981; Tregarthen, 1983) and exacerbation of the feedback problem for downstream-gauged systems. Definition in terms of consumption also serves as a greater incentive for water conservation since users would be required to hold permits for only the water they consume.

There are some drawbacks to consumptive-use permit definition, however. It requires assuming that there is always adequate additional flow in the stream for each user's pass-through use, that such use is returned near the point of abstraction, and that the return flow rate is accurately known. Unfortunately, return flow rates or consumptive fractions are not always constant, may change seasonally and unpredictably, and may be difficult to determine (especially for irrigation, whose return flows are geographically dispersed). Thus, while it may be more fair to base permit definition on consumption, it is administratively easier to base it on withdrawal.

One approach to this dilemma is to operate the program as though all withdrawal is 100 percent consumptive unless the user can document his or her return flow accurately, and to choose a program structure that is robust to the operational uncertainty caused by this assumption. Assuming 100 percent consumption, the return flows could be thought of as tributaries of uncertain and unpredictable magnitude. Downstream-gauged systems would not be robust to this uncertainty because they require accurate predictions of return flow, as do upstream-gauged, variable pass-through systems. For fixed pass-through upstream-gauged systems, however, uncertain return flows show up at nearby gauges to add to the flow at the next reach. Under that system, therefore, while the assumption of 100 percent consumption might be viewed as inequitable by users with significant but difficult-to-measure return flows, it would not cause operational problems.

32.6.7   Interactions between Ground and Surface Waters

If an aquifer is hydraulically connected to a stream, it may be possible for withdrawals from one medium to deplete the other, interfering with proper accounting of both. The ideal accounting method is to determine for a given user what portion of his or her withdrawal comes ultimately from each source and to require a permit of that quantity to be held. Unfortunately, it is difficult to determine a priori what each portion is, because it depends on a complex set of natural parameters that vary with time (e.g., the relative fractions of rainfall that infiltrate and run off), as well as the actions of the user and other users.

32.7   VOLUNTARY WATER-TRANSFER SYSTEMS

One feature that the agency may wish to consider is to allow permits to be transferred voluntarily among users, either on a permanent or temporary basis. Water transfers are a common component of many regional water systems and are being increasingly considered for meeting growing water demands and for managing the impacts of drought. Water transfers can take many forms and can serve a number of different purposes in the planning and operation of water-resource systems. However, to be successful, water transfers must be carefully integrated with traditional water-supply augmentation and demand-management measures as well as with the institutional systems which regulate water use. This integration requires increased cooperation among different water-use sectors and resolution of numerous technical and institutional issues, including impacts to third parties. This section identifies the many forms that water transfers can take, some of the benefits they can generate, and the difficulties and constraints which must be overcome in their implementation.

The most frequently cited argument in favor of this approach is an economic one (see, e.g., Anderson, 1983a; 1983b; Eheart and Lyon, 1983; Wong and Eheart, 1983; Enright and Lund, 1989). Greater economic efficiency will accrue if permit trading is allowed; the approach is also flexible, robust, and does not require strong intervention by the agency. For example, when newcomers buy permits, there is an automatic redistribution of water use toward greater efficiency.

Historically, advances in water-system management have been motivated by socioeconomic and environmental considerations. Since the 1970s, the increasing expense and environmental impact of developing traditional water supplies (e.g., reservoirs) have encouraged innovative use of existing facilities (e.g., conjunctive use and pumped-storage schemes) and have led to expanded demand-management efforts. In recent years, growth in water demands and environmental concerns have caused even these innovations to yield diminishing marginal returns. These economic and environmental conditions, combined with recent droughts, have spurred further efforts to improve traditional supply-augmentation and demand-management measures and have motivated the recent consideration and use of water transfers. The use of water transfers in many parts of the country, especially in the West, can be seen as a natural development of the water resources profession seeking to explore and implement new approaches in water management. We begin with brief reviews of recent water-transfer activity in California (Lund and Israel, 1995a; 1995b; Lund, et al., 1992). Following that, some of the more relevant issues for water managers and planners contemplating the use of water transfers are reviewed.

32.7.1   Existing Examples of Water Transfers

Water transfers and water marketing have existed in one form or another in many parts of the United States since early in this century. Many metropolitan areas have some form of water market in operation, usually involving a single large seller, typically a large central city or utility company, selling water to numerous large and small suburban cities and water districts. These sales arise from the economies of scale of urban water-supply acquisition, conveyance, and treatment and the historical legacy of central cities being the first to acquire most of the better, larger, and least expensive water supplies in many regions. Both central city and suburban parties to these transfers and sales accrue significant advantages from this arrangement, in the form of lower water-supply costs, higher supply reliabilities, and greater capability and certainty in regional water-supply planning. Still, there is often some degree of controversy and conflict between parties to these transfers (Lund, 1988).

Water marketing and transfers within agricultural regions is a still more ancient practice. Maass and Anderson (1978) describe a very effective water-marketing arrangement that has been in effect in one area of Spain since the fifteenth century. In addition, there are almost countless water trades and sales between farmers throughout much of the western United States. The majority of these transfers occur within mutual irrigation companies. These companies are typically informally constituted cooperatives of farmers, without governmental status. Each farmer has a share of the total amount of water available to the company. Water is then transferred by rental or sale of these shares to other farmers within the venture (Hartman and Seastone, 1970). It has been estimated that there are roughly 9200 such mutual water companies in the western United States (Revesz and Marks, 1981).

Other examples of existing water transfers are presented by MacDonnell (1990). This review found that almost 6000 water-right change applications were filed in six western states between 1975 and 1984, primarily in Colorado, New Mexico, and Utah. The vast majority of these applications were approved by state authorities. There are untold additional cases where transfers have been effected without legal need for state approval. For example, water transfers within the Bureau of Reclamation's Central Valley Project (CVP) generally do not require state review, since the Bureau is the holder of very general and flexible water rights. Between 1981 and 1988, CVP contractors were involved in over 1200 short-term transfers involving over 3700 Mm^3 (3 million acre · ft; Gray, 1990).

Most of the transfers described above are confined to specific water sectors and within individual metropolitan areas or irrigation systems. However, contemporary interest in water transfers has broadened the scope of traditional transfers to include transfers between different water-use sectors, e.g., agriculture to urban, often over larger geographical distances. These transfers often involve many parties with diverse views, facilities, and water demands which are more geographically separated. They may also require the use of conveyance and storage systems controlled by parties who are neither selling nor purchasing water. The controversies and complexities of effecting water transfers under these conditions may have initially deterred water managers from pursuing this option. However, with the changing economic and policy environment of water management and the absence of other attractive choices, water transfers can offer engineers a cost-effective alternative for enhancing the performance and flexibility of their systems (Lund, et al., 1992).

32.7.2 Economic Theory of Water Transfers

There is vast literature on the merits of water markets and voluntary water transfers (Milliman, 1959; Hartman and Seastone, 1970; Howe, et al., 1986; Brajer, et al., 1989; Eheart and Lyon, 1983). One important question addressed by some of this literature is the magnitude of the potential efficiency gains from trading. In several studies, it was estimated by computer simulation of markets in irrigation water to be significant. Wong and Eheart (1983) report an improvement of about 13 percent over nontransferable permits for surface-water permits from the Little Wabash

River in Illinois. Enright and Lund (1991) report only around 1 percent for a simple demand system exposed to different hydrologies, the Mad River in California and Tionesta Creek in Pennsylvania. Eheart and Barclay (1990) report an improvement ranging between 3 and 86 percent, depending on the amount of water available and the predictability of weather and crop-yield response. Other investigators (e.g., Tregarthen, 1983) have indirectly confirmed these findings.

Improved economic efficiency is not the only important advantage of such a program. Transfers also provide an incentive to develop and adopt ways of using water more efficiently through recycling and waste reduction. It is widely believed that many of the historically documented cases of wasteful use of water under the appropriative system might have been avoided by allowing transfers (Anderson, 1983b).

With regard to political feasibility, transfers of both water and pollution permits have received endorsements from policy analysis organizations of various political persuasions (Tietenberg, 1985; Bandow, 1986; Stavins, 1989). Permit transfers are a part of the most recent version of the Clean Air Act (1990), following a report commissioned by the U.S. Congress that endorsed a host of market-based incentives for environmental protection (U.S. Congress, 1988).

The additional administrative costs imposed by transfers are expected to be modest. The agency must maintain a registry of permits. Registration of trades must be sufficiently formal, and enforcement adequate, to prevent users from simply transferring permits from one to the other just before the enforcement agent arrives to check for violations. Transfer restrictions must be decided upon and administered. No cost data need be collected, however, and no cost optimization need be done by the agency. The agency may opt to set up a brokering operation or may let a private concern do so, but transfers are generally voluntary and need not be brokered by anyone.

In addition to these potential strengths of permit-transfer systems, there are some issues that must be addressed before an agency would wish to embark on the development of a system of such transfers.

32.7.3 Imperfect Markets

While water-market transfers are often desirable, the economic efficiency of water markets is usually imperfect when compared to ideal market performance. The conditions required for a perfect market are difficult to attain for a commodity such as water. Some problems include (Howe, et al., 1986; Brajer, et al., 1989):

- Water rights are often poorly defined.
- Water transfers can have high transaction costs.
- Water markets will often consist of relatively few buyers and/or sellers.
- Water is often costly to convey between willing buyers and sellers.
- Communication between buyers and sellers may be difficult.
- In humid regions, the dispersed nature of water sources and sinks may make definition of water rights difficult.

These difficulties commonly exist for other goods and services provided with great success through market mechanisms, and are not barriers to the use of water markets. The political appeal of market in offering trading opportunities and incentives for innovation is undiminished by these concerns. However, in appraising water transfers, planners, engineers, and policymakers should consider that trading activity may not be as lively as originally anticipated.

32.7.4 Third-Party and Environmental Impacts

The transfer of water can significantly affect third parties not directly involved in the transfer. The neighbors of the buyer may be impaired and the neighbors of the seller may receive a windfall benefit under permit transfers. For example, a freely transferable permit to pump a given amount of groundwater could impair the neighbors of the buyer with additional drawdowns and pumping costs in their wells. Furthermore, permits for irrigation will tend to be transferred toward farms whose soils have low moisture-retention capacities and these farms are often close together.

Another type of *third-party impairment* associated with prioritized water permits defined as withdrawals (Anderson, 1983a; 1983b; Johnson, et al., 1981) is the forfeiture of third-party rights that may attend a permit transfer. Consider, for example, a low-priority, third-party user situated downstream of the seller of a high-priority permit and upstream of its buyer. If the seller historically has a large return flow, the middle user has become dependent on that return flow, and the transfer requires the middle user to curtail or forego withdrawal to preserve a streamflow adequate to satisfy the downstream user.

The greatest challenge for implementing water transfers in the future may lie in properly identifying the affected parties and adequately mitigating these impacts. Many interests can be affected by water transfers, as noted in Table 32.1. Impacts can be direct, as with reduced instream flows below the diversion point for a transfer; or secondary, as represented by the loss of farm-related jobs in an agricultural region when farmers choose to transfer their water supplies. More detailed discussions of the third-party impacts of water transfers appear in Eheart and Lyon (1983); National Research Council (1992); Howe, et al. (1990); and Little and Greider (1983).

TABLE 32.1   Some Potential Third Parties to Water Transfers

Urban
  Downstream urban uses
  Landscaping firms and employees
  Retailers of lawn and garden supplies

Rural
  Farm workers
  Farm service companies and employees
  Rural retailers and service providers
  Downstream farmers
  Local governments

Environmental
  Fish and wildlife habitat
  Those affected by potential land subsidence, overdraft, and well interference
  Those affected by potential groundwater-quality deterioration

General
  Taxpayers

Paradoxically, water transfers might aid members of a group in one region while harming other members of the same group in another region. Water transfers from one farming region to another will lower farm employment and demand for farming services in the selling region and increase them in the purchasing region. Similarly, transfers of surface water from farms to cities can both help and harm fish and wildlife. By reducing application of water to farms, water quality downstream of the farm might improve, to the benefit of fish and perhaps other downstream water users. Also, there is a likely reduction in fish kills at the farm intake pumps because of the decreased withdrawals. Yet, where the on-farm application of water serves as habitat for migrating waterfowl, the removal of this water could harm bird populations.

Several mechanisms have been suggested to ameliorate the impact of or compensate groups harmed by water transfers. These mechanisms include (National Research Council, 1992; California Action Network, 1992):

- Taxing transfers to compensate harmed third parties
- Requiring transferors to provide additional water for environmental purposes
- State compensation to help economic transitions in water-selling regions
- Requiring public review and regulatory and third-party approval of transfers
- Requiring prior evaluation of third-party impacts of transfers, similar to an environmental impact report
- Requiring formal monitoring of third party impacts
- Restricting transfers to "surplus" waters
- In certain cases, redefining water rights to prevent third-party effects

Trading restrictions may undermine efficiency gains and may not be effective in protecting third parties anyway. Eheart and Barclay (1990) found in simulations of water-permit markets for irrigation in Kankakee County, Illinois that the return of water was sufficiently homogenous among users that no user bought more than 40 percent of his or her original allotment even without the imposition of transfer restrictions. At the same time, Cravens, et al. (1989) found that seasonal drawdown associated with irrigation is very significant for the same aquifer, even without transfers.

Third-party compensation has been endorsed by some researchers (e.g., Coase, 1960). Others (e.g., Baumol and Oates, 1988) note potential problems of strategic behavior if third parties are given final authority on whether a transfer is allowed.

Direct impacts on third parties can be reduced through legislation. Potential third-party effects from changes in return-flow quantity are commonly eliminated by state regulation allowing the transfer of only consumptive water use (Gray, 1989). Nevertheless, difficulties in assessing consumptive use may cause impacts to users of return flows (Ellis and DuMars, 1978). Likewise, the relative magnitude of secondary impacts is often difficult to determine accurately, but their presence is undeniable. Under ideal economic conditions of full employment and perfect labor and materials markets, such secondary impacts should be self-canceling in the aggregate. However, the common presence of significant unemployment, imperfect labor and capital mobility, and potentially important equity impacts raise these secondary economic impacts of water transfers to prominence.

32.7.5 Nonuser Market Players

Fourth, if permits are initially given away rather than sold and are also transferable, there is a potential for certain other kinds of inequities, if only perceived ones. A user who never intended to exercise a permit could, upon receiving it free from the agency, sell it and reap a windfall reward. There thus may be a need, if only a political one, for the agency to scrutinize requests to insure that claimants are bona fide potential users of water and not just speculators attempting to acquire windfall permits to sell later. On the other hand, some would argue that anyone who meets the established requirement to receive a permit is entitled to do so whether or not his or her doing so is for speculative reasons.

32.7.6 Market Thinness

A fifth issue related to water transfers is that because of the usually small number of participants, one or a few parties may be able to manipulate the permits market to their advantage. Many other markets are vulnerable to such manipulation, and while manipulation may pay off to an individual, trading will always improve the overall economic efficiency of the outcome. No exchange, no matter how it is manipulated, will take place unless both parties have an interest in it. Several researchers (e.g., Hahn, 1984; Saleth, et al., 1991) have studied this problem and have estimated that the potential for market dominance is small, but exceptions are possible, and would occur where one user is much larger in size than any other and can exercise monopoly or monopsony strength. Such cases pose problems of both efficiency and equity.

32.7.7 Multiple-Forum Origins of Transfers

A sixth issue is that water transfers can emerge from various forums: bipartisan or multilateral negotiations, several forms of brokerage and bidding, and other means (Hartman and Seastone, 1970; Saleth, et al., 1989). There is of course the potential to mix the use of different forums in the water-transfer process, using one forum to set a price and quantity, with other forums performing technical and legal review of transfer proposals. The forum or institutional mechanism under which water transfers are developed, reviewed, and approved can substantially affect the number, type, and details of transfers that actually take place, and is particularly important for the consideration of third-party impacts (Nunn and Ingram, 1988; Little and Greider, 1983; Eheart and Lyon, 1983).

32.8 TYPES OF WATER-TRANSFER ARRANGEMENTS

Water transfers can take many forms, as presented in Table 32.2. The specific needs of the purchasing and selling parties may dictate the type of transfer sought and the forum through which transfer arrangements are made. However, existing legislation and recent transfer experiences will also be important in selecting the most appropriate form of transfer. Each transfer form can have a different use in system operation and has different advantages and disadvantages for water buyers, water sellers, and other groups (Lund, et al., 1992). The various uses and associated benefits of water transfers are summarized in Table 32.3. Additionally, water transfers, like many forms of water-source diversification, increase the flexibility of a water system's operation, particularly in responding to drought. This flexibility allows new forms of operation that could not be accomplished without transfers and in many cases allows modification of system operations on a rapid time-scale. The following discussion on transfer types focuses on the possible uses and associated benefits of each.

TABLE 32.2 Major Types of Water Transfers

Permanent transfers
Contingent transfers/dry-year options
  Long-term
  Intermediate-term
  Short-term
Spot market transfers
Water banks
Transfer of reclaimed, conserved, and surplus water
Water wheeling or water exchanges
  Operational wheeling
  Wheeling to store water
  Trading waters of different qualities
  Seasonal wheeling
  Wheeling to meet environmental constraints

TABLE 32.3 Major Benefits and Uses of Transferred Water

Directly meet demand and reduce costs
  Use transferred water to meet demand, either permanently or during drought
  Use purchased water to avoid higher cost of developing new sources
  Use purchased water to avoid increasingly costly demand-management measures
  Seasonal storage of transferred water to reduce need for peaking capacity
  Use drought-contingent transfers to reduce need for overyear storage facilities
  Wheeling low-quality water for high-quality water to reduce treatment costs
Improve system reliability
  Direct use of transferred water to avoid depletion of storage
  Overyear storage of transferred water to maintain storage reserves
  Drought-contingent contracts to make water available during dry years
  Wheeling water to make water available during dry years
Improve water quality
  Trading low-quality water for higher-quality water to reduce water-quality concerns
  Purchase water to reduce agricultural runoff
Satisfy environmental constraints
  Purchasing water to meet environmental constraints
  Wheeling water to meet environmental constraints
  Using transferred water to avoid environmental impacts of new supply capacity

32.8.1 Permanent Transfers

A permanent transfer of water involves the acquisition of water rights and a change in ownership of the right. Permanent transfers are a form of supply augmentation and serve many of the same needs as capacity-expansion projects, including direct use to meet demands and improved system reliability. In some instances, the direct use of permanently transferred water can delay the implementation of increasingly costly demand-management measures or the need for system expansion, which in turn has the advantage of avoiding or at least delaying potential environmental impacts associated with construction (Table 32.3).

The majority of permanent transfers involve the purchase of agricultural water rights by urban interests. These transfers can involve reversion of the farmland to dryland agriculture, the immediate or gradual fallowing of farmland, the replacement of the farm's water supplies with an alternate supply source of possibly lower quality (from an urban-use perspective), or the lease of the transferred water back to the farmer in wet years when urban supplies are plentiful. Another form of permanent water transfer, common in Arizona, is for the developer to acquire groundwater rights associated with recently developed, formerly agricultural suburban lands. Some Arizona cities have made the provision of such rights to the urban water supplier a prerequisite for annexation of new suburban developments to urban water systems (MacDonnell, 1990). This ties permanent changes in water use to changes in land use and does not require water rights to be severed from the land, a political and legal difficulty in some cases.

32.8.2 Contingent Transfers/Dry-Year Options

In many cases, potential buyers of water are less interested in acquiring permanent supplies than in increasing the reliability of their water-supply system during drought, supply interruptions due to earthquake, flooding, contamination, or mechanical failure, or during periods of unusually great demand. For these cases temporary transfers contingent on water shortages may be desirable. The appropriate time horizon and conditions for a contingent transfer agreement will depend somewhat on the particular source of unreliability that the buyer would like to eliminate. For example, the timing of the call mechanism for earthquake supply interruptions would likely be very different from the call mechanism for responding to drought. Regardless, drought-contingent contracts for water are probably best made with holders of senior water rights, since they are the least likely to be shorted during drought. However, the increased reliability of water from senior rights tends to raise its market value (Lund, et al., 1992; National Research Council, 1992). An important benefit of contingent transfers is that longer-term arrangements allow for a more thorough analysis and mitigation of potential third-party impacts.

The time horizon of contingent transfers is important. Contingent-transfer agreements can be established to cover a period of several decades. This provides each party long-term assurance of the terms and conditions of water availability. Such long-term agreements can help an urban water utility modify release rules for reservoir storage to maintain less drought storage than would otherwise be desired or reduce the need for new source development. Long-term arrangements also can provide flexibility where future water demands may not meet expectations. However, long-term leasing of water does entail risk for water buyers if water demands meet or exceed current forecasts. Long-term leasing or contingent contracts allow water-right owners to retain long-term investment flexibility in anticipation of potentially greater future values for water leasing or sale of a water right.

Intermediate-term (3 to 10 year) contingent-transfer contracts might be employed to help reduce the susceptibility of the buyer's system to drought during periods prior to the construction or acquisition of new supplies. Short-term (1 to 2 year) contingent-transfer contracts might be utilized in the midst of a drought by a water agency with depleted storage, preparing for the possibility that the drought might last a year or two longer. This type of short-term contingent-transfer contract would enable the buyer to have committed water supplies when their system might be extremely vulnerable.

Advantages of contingent transfers for the seller, typically agricultural interests, are the immediate infusion of cash when the contract is made, the infusion of additional revenues if the contingent-transfer option is called, and an increased ability to predict the conditions and timing of transfers, rather than relying on the vagaries of timing, price, and quantity of a water spot market.

The potential sale of water by farmers during drought affects the need for groundwater management (if available as an alternate supply of water) and the special operation of conveyance and storage facilities. The ability of farmers to sell water might also affect the operation-rule curves used by agricultural water suppliers for allocating water from storage to farmers over multiyear droughts. Perhaps additional hedging or overyear storage by agricultural water suppliers will increase farm incomes more than adherence to current reservoir operating rules, by creating a greater scarcity of water and higher water-transfer incomes during drought years. Similar issues relate to the overyear use of groundwater storage.

32.8.3 Spot Market Transfers

It may be desirable for the agency to allow short-term transfers in a water-rental or water-futures market. Spot market transfers are short-term transfers or leases. Typically spot market transfers are agreed to and completed within a single water year. However, for large systems, there is a possibility to establish spot futures markets for water for seasonal or overyear periods.

Spot transfers may garner improvements in economic efficiency beyond those from long-term permits. Eheart and Barclay (1990) and Wong and Eheart (1983) estimated that for irrigation such improvements are small (less than 2 percent), as long as trading of long-term permits is allowed. For other types of uses, they could conceivably be more significant, but the seasonal variation of the value of water is usually greater for irrigation than other types of water use.

Spot market transfers are typically established by some sort of bidding process, often with some of the conditions for transfer being fixed (e.g., price and quantity). However, spot market transfers can arise from negotiations between individuals or groups of buyers and sellers. A wide variety of bargaining rules for the operation of spot markets have been examined on a theoretical basis and through the use of simulation (Saleth, et al., 1991). These results illustrate the importance of bargaining rules when the numbers of buyers and sellers are small, less than about a dozen participants. For large spot markets, the effects of particular bargaining rules are quickly overshadowed by competition among buyers and sellers.

Spot market purchases can be advantageous in both dry or wet years. During periods of drought, short-term transfers may be sought to directly meet demands, especially demands still not met after implementation of drought water-conservation and traditional supply-augmentation measures. As with permanent transfers, temporary transfers used to meet demands directly can have the advantage of delaying or avoiding the costs of developing new supply sources or implementing more stringent demand-management measures.

In wet years, water purchased through a spot market can be stored in reservoirs or aquifers as overyear storage. This enhances the yield of the system during drought years by increasing the amount of stored water available upon entering a drought. Overyear storage of transferred water is particularly well suited to acquiring water from junior water-rights holders. Junior water rights are typically less expensive than senior water rights, although they may only be available during relatively wet years. However, storage of transferred water during wet years may require additional surface or groundwater-storage capacity, and is subject to evaporative and seepage losses and any costs associated with storage. This approach may also work for within-year storage.

32.8.4 Water Banks

Water banks are a relatively constrained form of spot market operated by a central banker. Here, users sell water to the bank for a fixed price and buy water from the bank at a higher fixed price. The difference in prices typically goes to covering the bank's administrative and technical costs. Each user's response to the bank and involvement in the market is largely restricted to the *quantity* of water it is willing to buy or sell at the fixed price.

The California Drought Emergency Water Banks, beginning in 1991, are examples of water banks or spot markets where the terms and price of transfer were relatively fixed, with the state acting as a banker (California DWR, 1992; Howitt, et al., 1992). A similar, but smaller water bank was established in Solano County, California (Lund, et al., 1992). In agricultural regions, it is common for water banks or pools to exist within large irrigation systems. For many existing water pools, sellers avoid only the cost of purchasing unneeded water from the system. Water buyers in these pools pay the system normal wholesale water prices, plus some administrative cost (National Research Council, 1992; Gray, 1990).

Where spot market or water bank transfers have become established, as in California, agencies of all types are likely to plan on these markets being available for either buying or selling water (Lund, et al., 1992; Israel and Lund, 1995). The existence of spot markets and water banks during droughts provides incentives for urban water suppliers to rely somewhat less on more expensive forms of conventional water-supply capacity expansion and urban water conservation in planning, and also may encourage different designs for new facilities and modified operation of existing facilities. For agricultural water districts, the existence of water banks and spot markets during drought has implications for the wording of water-supply contracts and the management of water and cropland during a drought.

32.8.5 Wheeling and Exchanges

In the electric-power industry, power is often *wheeled* through the transmission system between power companies and electric generation plants to make power less expensive and more reliable. Water can similarly be wheeled or exchanged through water-conveyance and storage facilities to improve water-system performance. Again, such movements of water involve the institutional transfer of water among water users and agencies. There are a number of forms of wheeling water or water exchanges (Lund, et al., 1992).

Sometimes the cost of conveying water or the losses inherent in water conveyance can be reduced by wheeling water through conveyance and storage systems controlled by others. An example would be the use of excess capacity in a parallel lined canal owned by another agency, rather than use an agency's own unlined canal to convey water. Differences in pumping efficiencies might also motivate operational wheeling between conveyance facilities. Similar considerations might apply to decisions on where to store water during a drought when different reservoirs have different seepage or evaporation rates (Kelly, 1986) or if the distribution of hydropower heads is considerable for different storage options.

Seasonal wheeling of water is common in agricultural regions where different subareas have complementary demands for water over time. This can provide opportunities for one water user to exchange water to another user during the low-demand season, with repayment coming in the form of additional water during the user's high-demand season.

Also, by paying farmers not to use their rights to water, the consumptive use foregone becomes available for in-stream demands downstream. This mechanism is particularly applicable to riparian rights which cannot be legally transferred for use away from the riparian lands (Lund, et al., 1992). Another application of wheeling to meet environmental constraints could involve the use of storage facilities to release water when desired for in-stream flows while meeting demands before this time from other reservoirs or groundwater.

In many cases, historical happenstance has left agricultural users with rights to high-quality water for irrigation while new urban development is left with remaining water sources of lesser quality. In such cases the additional costs of treating low-quality water for urban use is usually much greater than the costs from slightly lower crop yields from use of the lower-quality water. Given reasonable conveyance costs, it therefore becomes desirable for water-quality-based trades between agricultural and urban users. Urban users can often afford to make these trades on an uneven basis, trading more low-quality water for less high-quality water or providing a monetary inducement for a volumetrically even trade of water. Lesser-quality waters might also be traded for environmental uses of aquifer recharge or habitat maintenance (Lund, et al., 1992).

32.8.6 Transfer of Reclaimed, Conserved, and Surplus Water

Although not always recognized as such, the purchase of water made available by reclamation or reductions in water demand is a form of water transfer. Numerous urban water utilities have become involved in purchasing water back from their retail customers. Such schemes usually involve rebates to customers for installing low-flow toilets or removing relatively water-intensive forms of landscaping (California DWR, 1988). Some cities have developed clever schemes where water transfers are made within their customer base. For instance, Morro Bay, California has a program whereby developers can receive water utility hook-up permits if they cause a more than equivalent reduction in existing water demand through plumbing retrofits, landscaping, or other measures (Laurent, 1992).

Urban areas have taken an interest in financing the conservation of irrigation water to make additional water available for urban supplies. This has primarily been accomplished through the lining of irrigation canals. For example, the transfer of water between the Imperial Irrigation District (IID) and the Metropolitan Water District of Southern California (MWD) involves a 35-year contract for MWD payments for canal lining and other system improvements in IID's irrigation infrastructure in exchange for the water saved by these improvements. The savings are estimated at 123.3 Mm^3/y (100,000 acre · ft/y) from IID's Colorado River water supplies (Gray, 1990; Sergent, 1990). This approach can have additional benefits where agricultural seepage and drainage water has led to water-quality problems or high water tables, but can create additional problems where canal seepage is used to recharge groundwater.

32.9 IMPLEMENTING WATER TRANSFERS

Perhaps the most important implication of water-transfer planning is the need to increase integration and cooperation among diverse water users. Since for economic reasons most water for water transfers will probably come from agricultural users and much of this water will go to urban and perhaps environmental users, any planning for water transfers implicitly integrates urban, agricultural, and environmental water supplies. As the tendency to seek and implement water transfers continues, it will become less possible, and less desirable, for individual urban or agricultural water districts or regions to plan and operate their water supplies independently. This necessary coordination of planning and operations between functionally diverse water agencies will imply potentially protracted and probably controversial negotiations, at least for long-term transfer arrangements.

Additionally, if intersectoral and interregional water transfers are to become significant long-term components of water-resources planning, they must be integrated with traditional water-supply augmentation and demand-management measures. Given the complex nature of many water-resource systems and the wide variety of different possible water-transfer designs, it seems apparent that some form of water-supply-system computer modeling will be required to achieve this integration of water transfers with other water-management measures.

Most major water-supply agencies already possess significant conventional water-modeling capability. However, most models are specific to individual water systems, in accordance with the needs of traditional water-supply and water-conservation measures which can be implemented by a single system. The integration of water transfers will likely require significant modifications to these single-system models to allow explicit examination of long- and short-term water transfers and exchanges. Water transfers also encourage more explicit consideration of the economic nature of water-supply operations in system modeling. System models for examining water-transfer options together with supply source and water-conservation expansions and modifications might usefully provide economic measures of performance (component and net costs) in addition to traditional technical measures of performance (e.g., yields and shortages). Various agencies and academic researchers have already begun such efforts (Lund and Israel, 1995a; Smith and Marin, 1993).

The economic nature of the design of water transfers and their integration with other water-supply-management measures encourages the use of optimization models, where the model itself suggests promising combinations of water transfers, construction, and water conservation. While technically more difficult and still somewhat inexact, optimization modeling can aid in identifying promising solutions, which can then be examined in more detail with simulation models. Performance of economically based optimization (or simulation) of water-resource systems with water transfers requires technical studies estimating the value of and the willingness to pay for different water uses and different water quantities.

During California's recent drought in which water transfers were actively pursued and implemented, both traditional supply infrastructure and demand-management strategies continued to have important, albeit modified, roles in water management (Israel and Lund, 1995). Some hints of how the integration should take and several specific areas of concern for implementing water transfers are discussed below.

32.9.1   Legal Transferability of Water

The legal transferability of water is a major consideration in designing water transfers. Legislation pertaining to the transferability of water will vary between states and can vary within a given state over time as a state's water law evolves. California has strong statutory directives to promote water transfers (Gray, 1989; Sergent, 1990), yet legal constraints still pose a significant threat to water-transfer activity. Legal considerations are particularly important when a proposed transfer involves changes in conditions stipulated by the original water right, such as changes in type of use, place of use, or timing of withdrawals. The type of water right to be transferred is also an important consideration. Riparian rights, for instance, are generally nontransferable from their initial location of use, and the transferability of groundwater rights varies substantially by state.

Also, different types of water contracts impose different transferability requirements. In California, many water contracts stipulate that any water not used by the contractor reverts to the contractee, while others may stipulate that water cannot be transferred outside of a district and can only be transferred within a district at cost. These types of provisions reduce the ability and incentive of contractors to sell surplus or conserved water (Sergent, 1990; Gray, 1989). Short-term, emergency water transfers may be able to gain relatively easy approval and rapid implementation, given sufficient flexibility in the conveyance and storage system and sufficient professional flexibility and readiness on the part of water managers. Legislation often exists which reduces or eliminates barriers to transfers during drought or other emergency conditions. This was certainly the case for the 1991 and 1992 California Drought Emergency Water Banks (California DWR, 1992). On the other hand, long-term, planned transfers, such as dry-year option contracts and permanent water transfers typically face more difficult legal and economic constraints. Many of the longer-term transfers that require the storage of surplus water during wet years also involve complex legal issues (Getches, 1990), particularly for groundwater storage (Kletzing, 1988). The costs, delays, and risks involved in overcoming these constraints can induce agencies not to consider or participate in water transfers.

32.9.2 Real versus Paper Water

Where water transfers are motivated by real water shortages, the transfer of water by contract must correspond closely with the transfer of water in the field. This is sometimes known as the distinction between *real* and *paper* water. Associating quantities of paper water to real water is a difficult technical problem. In the case of transfers from farms, farmers typically do not know with certainty how much water they use or how much real water would become available if land were to be fallowed or cropping patterns altered (Ellis and DuMars, 1978). Even where such flow measurements are made, they are often inexact.

As water moves through a complex conveyance and storage system, there are seepage and evaporation losses, withdrawals by or return flows from other users, and natural accretions downstream. All these factors complicate the estimation of how much water is physically available to the receiver of a water transfer, given that the sender has relinquished use of a given amount. Another problem with linking paper water to real water is establishing the hydrologic independence or interdependence of water sources. This is a common problem where pumped groundwater may induce recharge from nearby surface water.

Particularly where there are many potential buyers and sellers of water, there would seem to be some need for standards or governmental involvement in tying real water to paper water transfers (Blomquist, 1992). Without such standard accounting, amounts of paper water are likely to exceed amounts of wet water available, leading to excessive withdrawals by water users to the detriment of downstream users and those not party to transfers. This will be true for transfers of water for both consumptive and in-stream uses. Litigation and calls for greater regulation of water transfers would be the likely result.

32.9.3 Conveyance, Storage, and Treatment

The mere purchase of water is usually insufficient to effect a water transfer. Transferred water must typically be conveyed and pumped to a new location, often stored, and commonly treated. Since both emergency short-term transfers and long-term transfers may require modifying the operation of existing water infrastructure, considerable work may be required to coordinate the use of conveyance, storage, and treatment systems. This can be particularly challenging because these facilities are often designed for very different operations. Occasionally, canals must be run backwards, water must flow backwards through pumps, and treatment plants must treat waters of a quality different from their design specifications. Construction of additional conveyance interties or other facilities may be required in some cases.

The difficulties encountered by San Francisco illustrate well the traditional engineering limitations and concerns with the use of water transfers in system operations and planning (Lougee, 1991; Lund, et al., 1992). San Francisco purchased 62 Mm^3 (50,000 acre · ft) from the 1991 California Drought Water Bank, but their water-treatment plant was unable to accept more than a limited rate of transferred water from the Sacramento-San Joaquin Delta. Delta water is of lower quality than San Francisco's normal Sierra supply and the mixing of waters in the treatment plant beyond certain ratios increased the likelihood of trihalomethane formation. This limitation forced much of the transferred water to be stored in state-owned facilities and slowly released into San Francisco's treatment plant. California's East Bay Municipal Utility District (EBMUD) faced similar quality limitations on the treatability of transferred water which, combined with other difficulties in effecting transfers, led EBMUD not to use transferred water and to rely more on urban water-conservation measures.

Water transfers are likely to be more successful in regions with an extensive system of conveyance and storage facilities and well-coordinated operations. Locations with restricted conveyance and storage infrastructure are likely to have limited potential for effecting water transfers unless creative operations or new conveyance and storage facilities can be developed. The coordination and physical completion of water transfers will be more difficult, and perhaps impossible, if agencies controlling major components of a region's water conveyance and storage system choose not to participate in transfers, are legally restrained from participating, or participate only in a limited way.

32.9.4 Contracts and Agreements

The legal transfer of water is typically effected by contracts which must specify a number of logistical and financial conditions of the transaction. Among the logistical and fiscal details that must be specified are: the location and timing of water pickup from the seller; the fixed or variable price of the water; the fixed or variable quantity of water; and potentially the quality of the water. The responsibilities for contract execution and liabilities for failure to completely execute the contract might also be included.

Where transferred water cannot be conveyed directly between the buyer and seller, agreements are often required with other entities, either to make use of their conveyance facilities (pumps or aqueducts) or to coordinate the conveyance of transferred water through natural waterways, within environmental limitations (Lougee, 1991). Similarly, facilities owned or operated by entities not directly involved in the transfer may be necessary to store transferred water until it can be used. This will often require agreements or contracts for the storage of water with agencies which oversee storage facilities. When water is stored in aquifers, recharge and pumping facilities will be required, and legal arrangements with overlying landowners are common (Kletzing, 1988). Likewise, contractual arrangement may be required for the treatment of transferred water.

32.9.5 Price, Transaction Costs, and Risks

As demonstrated by the 1991 and 1992 California Drought Water Banks, both sellers and buyers can be quite sensitive to the price established for water (Lund, et al., 1992). At lower prices, there are fewer willing sellers and greater demand for water from agricultural users. Higher prices encourage sellers but tend to exclude most potential agricultural buyers. The price set by the market, through negotiations, or by a governmental water bank has important implications for the character and number of resulting transfers.

The cost of water to a user includes more than its purchase price. As noted above, much of the work in establishing successful transfers of water lies in arranging for the conveyance, storage, and perhaps treatment of the transferred water. In some cases, the costs of these activities may exceed the cost of the water itself. For example, in 1991 San Francisco purchased 18.5 Mm^3 (15,000 acre · ft) from Placer County at a price of $36,500/Mm^3 ($45/acre · ft). However, total costs including wheeling charges through state and federal facilities and storage costs were between $200,000 and $300,000/Mm^3 ($250 to $350/acre · ft). Also, the final delivery cost of water purchased by San Francisco from the Water Bank was nearly double the purchase price of $142,000/Mm^3 ($175/acre · ft). Water transfers are also subject to numerous other transaction costs, including legal fees, costs of public agency review, costs of required technical studies, and costs involved in settling claims from third parties. MacDonnell's survey (1990) found that transaction costs averaged several hundred dollars per acre · ft of transferred perpetual water right, with averages of $309,000/Mm^3 ($380/acre · ft) of perpetual right in Colorado and $150,000/Mm^3 ($184/acre · ft) in New Mexico. These transaction costs can add substantially to the purchase price of water, which in Colorado and New Mexico ranges from $200,000 to $1.2 million per Mm^3 ($243 to $1,500 per acre · ft). The unit costs for transactions commonly decrease for larger transfers and increase with the controversy of a transfer. Still, transaction costs are highly variable between transfers.

The risks of a transfer not being completed may also dissuade potential partners in transfers. The risk of a proposed transfer being stopped entirely is particularly palpable where a substantial part of the transaction costs must be expended before a transfer agreement is finally approved, or if there are high costs to delaying implementation of other water-supply alternatives while transfers are being negotiated. This would be the case where large expenditures for technical and legal work must be made before final approval of a transfer is in place (Lund, 1993).

32.9.6 Evaluation of Impacts to Third Parties

Evaluating the third-party impacts of water transfers can be formidable and inexact, involving difficult ecological and economic studies (National Research Council, 1992). There is currently little technical work quantifying physical, environmental, economic, and social impacts from water transfers (Howe, et al., 1990; Agricultural Issues Center, 1993). Less is known about how these impacts would vary with different specific transfer cases and mechanisms and how effective different approaches to mitigating third-party impacts might be.

Some of the technical issues in managing third-party impacts are illustrated by the case of Yolo County, California. Farms in Yolo County contributed about 185 Mm^3 (150,000 acre · ft) of water to the 1991 California Drought Water Bank. Some of this water came from fallowing farmland and transferring the surface-water rights. However, most of the surface water was replaced by increased groundwater pumping. Yet the county does not employ a water engineer or groundwater specialist dedicated to countywide water-supply problems who could assess and manage the long-term impacts of these transfers (Jenkins, 1992). There is also little legal authority for counties to assume this role. Furthermore, rural county governments may lack expertise to estimate the economic impacts of different types of transfers. Without an understanding of the economic and physical effects of water transfers, water-exporting regions are likely to be suspicious of and somewhat resistant to water transfers.

This same lack of a technical basis for assessing and managing impacts of water transfers takes on a more important role at the statewide level where water-transfer policies are made. Technical studies are needed to support policies and perhaps specific cases should be investigated of when and how water transfers are made and how any third-party impacts should be managed (Howitt, et al., 1992). Of course, as noted earlier, there is a possibility of avoiding certain kinds of third-party impacts by defining water rights in terms of consumption.

32.9.7 Roles for Government in Water Transfers

The role of state and federal government is so important in many cases that it must be considered part of the system engineering. In California, for instance, a significant part of state and federal involvement in water transfers is due to the technical role required by their ownership and operation of major conveyance and storage facilities and their requirements and responsibilities under various environmental regulations.

A number of roles for federal, state, and local governments can be identified for facilitating water transfers, some of which may require modification of existing regulations, legislation, and local agency enabling legislation. Perhaps the most appropriate role for government in water transfers is that of an arbiter of technical and third-party disputes and a regulator of the market. This role is needed to ensure a close tie between trades of paper water and real water and the coordination of the movement of transferred water with environmental regulations (Blomquist, 1992). State or regional governments would also seem to have a useful role in the collection and analysis of data for monitoring and resolving external and third-party impacts. Regional governments can also act as bankers in the formation of regional water markets, taking advantage of the regional hierarchy of governmental water jurisdictions commonly found in water management.

Government involvement can improve the prospects for water transfers by:

1. Improving information regarding transfers and transfer impacts
2. Establishing a process for managing third-party impacts
3. Reducing the transaction costs of arranging and implementing water transfers
4. Increasing the probability that efforts between parties to arrange a water transfer will be successful and reducing the risks to parties from involvement with transfers

Of course, as noted previously, government must maintain the registry of permits and may act as broker (but need not necessarily). Individual agencies have their own agendas and will continue to pursue short- and long-term contracts regardless of the existence of government-sponsored water banks. However, government involvement can greatly accelerate the development of water-transfer agreements by initial sponsorship of transfers through the establishment of water banks or by other means. The development of transfers as part of a larger water-resource system is likely to continue after government sponsorship of water banks has ended.

32.10 SUMMARY AND IMPLICATIONS FOR REGULATORY PROGRAMS AND WATER TRANSFERS

Against the backdrop of the sections on water regulatory programs, it is possible to single out several options that currently appear more attractive than others and that should be considered for a water-permits program in a humid region. This is not to be construed as a recommendation, but a suggestion that these options are appropriate to consider first, pending further research and public dialogue for each specific situation.

The flexible-permit basis seems to be more consistent, on an equity basis, with the riparian doctrine than prioritized permits or nonpermit approaches such as ad hoc restrictions, subsidies, or charges. The western first-in-time rule for prioritization may not be legally feasible in riparian areas. Fractional permits are also more consistent with legal precedents for groundwater regulation. The constant use basis for permits is simply a physically intractable means of allocating water from a watercourse whose flow rate fluctuates with time.

One way or another, the agency must grapple with the equity issue of deciding how the resource must be distributed among the participants. The historical precedent or riparian equivalent bases are administratively easier and will probably be regarded as more equitable than either the historical-use or ad hoc bases, even though the latter may be more flexible.

It should be possible to set up a system of regulation that requires little administrative effort until supplies become constraining and that automatically invokes controls at that point. In areas where demand is currently light compared to supply, users could be issued permits which entitle them to their current normal withdrawal. By issuing the same number of new permits of limited duration every year, the agency could eventually achieve a staggered permit system.

Given the host of potential problems associated with alternative approaches, the upstream-gauged, fixed pass-through system seems to be the most administratively tractable method for addressing the complexity of river systems. It enables the use of fractional permits without ambiguity of individual users' allowable withdrawal. The disadvantage of this approach is that it requires a larger number of gauges and therefore entails a higher administrative cost.

Averaging periods could be set at any level initially, as long as there is a proviso that they may be lowered later. To prevent substantial streamflow depletion, averaging periods for free-flowing streams should usually be from hours to a week. Lakes and reservoirs could use a longer averaging period, perhaps months, depending on the size of the water body. Aquifer averaging periods could be as large as a year or more. The agency should consider staggering the time of calculation of averages over users so that the incentives for increased pumping near that time do not occur simultaneously.

Allowing transfers of water rights among users has advantages and disadvantages. The principal advantage is that economic efficiency may be improved by voluntary trading and the allowance of such trading is considered equitable by the users. This improvement in efficiency is robust to data errors and does not require data collection or planning by the agency. The principal disadvantage is that the agency relinquishes some control over where withdrawals occur. This may undermine its effectiveness in protecting streamflows and aquifers and may lead to some third-party impairment problems. Most third-party problems are solved when permits are defined in terms of consumption, however.

Water transfers have far-reaching implications for water-resource planning and management. In addition to contributing to the bag of tricks available to water managers, transfers require a broader conceptualization of water-management problems. Unlike traditional supply-augmentation and demand-management measures, which can typically be accomplished by a single water agency, water transfers require coordinated planning and operations between both groups party to the transfer. Also, water transfers often require the use of storage and conveyance facilities belonging to or operated by entities not directly involved in the buying or selling of water. The evaluation of transfers demands a more explicit economic perspective on the purposes of water-resource systems and more detailed economic measures of operation performance. The water acquired by transfers can serve a variety of operational, environmental, and economic purposes. Overall, the multiple forms of water transfers and their flexibility, combined with legal, third-party, and technical issues in implementing transfers, make water transfers one of the more promising, yet complex techniques for improving water management.

As traditional forms of water-resource development become more difficult and expensive, the profession must turn to the management of water use, including the management of water allocations and water demands. This chapter has reviewed the use of permit systems and water-transfer systems for managing water allocations. Notwithstanding significant difficulties, both approaches have been increasingly employed in recent years and show promise for the future.

Acknowledgments

The authors gratefully acknowledge the support of the Energy and Environmental Affairs Division of the Illinois Department of Energy and Natural Resources under Project SENR WR-26, "Analysis of Options for Water Use Regulation for the State of Illinois," for the study of permit systems and funding provided by the U.S. Army Corps of Engineers Hydrologic Engineering Center and University of California Water Resources Center for the study of water-transfer systems. While the authors assume full responsibility for any errors or omissions, appreciation is extended to the following students and colleagues who have helped conduct this research or whose viewpoints have contributed to its planning and the interpretation of its results: John Barclay, Jean Bowman, John Braden, Gary Clark, Christopher Enright, Inês Ferreira, Paul Hutton, Morris Israel, Mimi Jenkins, Jon Liebman, Karen Miller, Rezaur Rahman, Robert Reed, Maria Saleth, Maureen Sergent, other current and former students, and the many professionals interviewed for this research for their comments on work on these subjects. The findings reported in this paper are those of the authors and do not necessarily reflect the views of the Illinois Department of Energy and Natural Resources or the U.S. Army Corps of Engineers.

REFERENCES

Agricultural Issues Center, "California Water Transfers: Gainers and Losers in Two Northern Counties," University of California, Davis, 1993.

Anderson, T. L., ed., *Water Rights,* Pacific Institute for Public Policy Research, San Francisco, 1983a.

Anderson, T. L., *Water Crisis: Ending the Policy Drought,* Cato Institute and Johns Hopkins University Press, Baltimore, Md., 1983b.

Bandow, D., *Critical Issues—Protecting the Environment: a Free-Market Strategy,* The Heritage Foundation, Washington, D.C., 1986.

Baumol, W. J., and W. E. Oates, *The Theory of Environmental Policy,* Prentice-Hall, Englewood Cliffs, N.J., 1988.

Bliss v. Kennedy (43 Ill. 67) 1867.

Blomquist, W., *Dividing the Waters: Governing Groundwater in Southern California,* ICS Press, San Francisco, 1992.

Brajer, V., A. L. Church, R. Cummings, and P. Farah, "The Strengths and Weaknesses of Water Markets as They Affect Water Scarcity and Sovereignty Interests in the West," *Natural Resources Journal,* 29(Spring): 489–509, 1989.

California Action Network, "Open Letter from California Action Network Regarding Water Marketing and Public Policy," Davis, Calif., August 11, 1992.

California Department of Water Resources, *Landscape Water Conservation Guidebook No. 8,* Sacramento, Calif., 1988.

California Department of Water Resources, *The 1991 Drought Water Bank,* Sacramento, Calif., 1992.

Clark, G. R., "Illinois Groundwater Law[:] The Rule of 'Reasonable Use,' " paper prepared by the State of Illinois, Department of Transportation, Division of Water Resources for presentation to the Illinois Groundwater Association semiannual meeting, Joliet, Illinois, October 8, 1985.

Cravens, S. J., S. D. Wilson, and R. C. Barry, "Summary Report: Regional Assessment of the Ground-Water Resources in Eastern Kankakee and Northern Iroquois Counties," SWS Contract Report 456, by the Illinois State Water Survey to the Illinois Department of Energy and Natural Resources and the Illinois Department of Transportation Division of Water Resources, April 1989.

Coase, R. H., "The Problem of Social Costs," *Journal of Law and Economics,* 3:1–44, 1960.

Dixon, W. D., and W. E. Cox, "Minimum Flow Protection in Riparian States," *Journal of Water Resources Planning and Management,* ASCE, 111 (2):149–156, 1985.

Eheart, J. W., and J. P. Barclay, "Economic Aspects of Groundwater Withdrawal Permit Transfers, *Journal of Water Resources Planning and Management,* ASCE, 116(2):282–303, Mar/Apr, 1990.

Eheart, J. W. and R. M. Lyon., "Alternative Structures for Water Rights Markets." *Water Resources Research* 19(4):887–894, 1983.

Eheart, J. W., "Programs for Management of Water Withdrawals in Illinois," Final Report, Project SENR WR-26 Illinois Department of Energy and Natural Resources, Energy and Environmental Affairs Division, Springfield, Ill., 1989.

Ellis, W. H., and C. T. DuMars, "The Two-Tiered Market in Western Water," *Nebraska Law Review,* 57(2):333–367, 1978.

Enright, C., and J. R. Lund, "Alternative Water District Organization: A Screening Level Analysis," *Journal of Water Resources Planning and Management,* ASCE, 117(1):86–107, 1991.

Florida Water Resources Act, Florida Statutes, Chap. 373, 1972.

Getches, D. H., *Water Law,* West Publishing, St. Paul, Minn., 1990.

Gray, Brian E., "Water Transfers in California: 1981–1989," in MacDonnell, Lawrence J. (principal investigator), "The Water Transfer Process As A Management Option for Meeting Changing Water Demands," vol. II, USGS Grant Award No. 14-08-0001-G1538, Natural Resources Law Center, University of Colorado, Boulder, 1990.

Goering, Laurie, and Thomas Cekay, "Irrigation Pulls Plug on Residents' Water," Chicago *Tribune,* July 25, 1988.

Gray, Brian E., "A Primer on California Water Transfer Law," *Arizona Law Review,* 31:745–781, 1989.

Hahn, R. W., "Market Power and Transferable Property Rights," *The Quarterly Journal of Economics,* 99(4):754–765, 1984.

Harrison, K. W., "The Allocation of Flowing Water in Humid Regions," M.S. thesis, University of Illinois, Urbana, Ill., 1991.

Hartman, L. M., and D. Seastone, *Water Transfers: Economic Efficiency and Alternative Institutions,* Johns Hopkins Press, Baltimore, Md., 1970.

Howe, C. W., D. R. Schurmeier, and W. D. Shaw, Jr., "Innovative Approaches to Water Allocation: The Potential for Water Markets," *Water Resources Research,* 22(4):439–445, 1986.

Howe, C. W., J. K. Lazo, and K. R. Weber, "The Economic Impacts of Agriculture-to-Urban Water Transfers on the Area of Origin: A Case Study of the Arkansas River Valley in Colorado," *American Journal of Agricultural Economics,* Dec.:1200–1204, 1990.

Howitt, R., N. Moore, and R. T. Smith, "A Retrospective on California's 1991 Emergency Drought Water Bank," March, California Department of Water Resources, Sacramento, Calif., 1992.

Israel, M. and J. R. Lund, "Recent California Water Transfers: Implications for Water Management," *Natural Resources Journal,* 35:1–32, Winter 1995.

Jenkins, Mimi, "Yolo County California's Water Supply System: Conjunctive Use without Management," Masters report, Department of Civil and Environmental Engineering, University of California, Davis, Calif., 1992.

Johnson, R. N., M. Gisser, and M. Werner, "The Definition of a Surface Water Right and Transferability," *Journal of Law and Economics,* 24:273–288, Oct. 1981.

Kelly, K. F., "Reservoir Operation During Drought: Case Studies," Hydrologic Engineering Center, U.S. Army Corps of Engineers, Research Document no. 25, Davis, Calif., 1986.

Kletzing, R., "Imported Groundwater Banking: The Kern County Water Bank—A Case Study," *Pacific Law Journal,* 19(4):1225–1266, 1988.

Laurent, M. L., "Overview New Development Process/Water Allocations/Conservation," City of Morro Bay, Calif., 1992.

Linsley, R. K., and J. B. Franzini, *Water-Resources Engineering,* Chap. 6, McGraw-Hill, New York, 1972, pp. 148–160.

Little, R. L., and T. R. Greider, *Water Transfers from Agriculture to Industry: Two Utah Examples,* Research Monograph no. 10, Institute for Social Science Research on Natural Resources, Utah State University, Logan, Utah, 1983.

Lougee, N. H., "Uncertainties in Planning Inter-Agency Water Supply Transfers," in J. L. Anderson, ed., *Water Resources Planning and Management and Urban Water Resources,* American Society of Civil Engineers, New York, 1991, pp. 601–604.

Lund, J. R., "Metropolitan Water Market Development: Seattle, Washington, 1887–1987," *Journal of Water Resources Planning and Management,* ASCE, 114(2):223–240, 1988.

Lund, J. R., "Transaction Risk versus Transaction Costs in Water Transfers," *Water Resources Research,* 29(9):3103–3107, 1993.

Lund, J. R., and M. Israel, "Optimization of Water Transfers in Urban Water Supply Planning," *Journal of Water Resources Planning and Management,* ASCE, 121(1):41–48, 1995a.

Lund, J. R., and M. Israel, "Water Transfers in Water Resource Systems," *Journal of Water Resources Planning and Management,* ASCE, 121(2):193–204, 1995b.

Lund, J. R., M. Israel, and R. Kanazawa, "Recent California Water Transfers: Emerging Options in Water Management," Center for Environmental and Water Resources Eng. Report 92-1, Dept. of Civil and Env. Eng., University of California, Davis, 1992.

Lyon, R. M., "Auctions and Alternative Procedures for Allocating Pollution Rights," *Land Economics,* 58(1):16–32, 1982.

Mack, L. E., and A. W. Peralta, "Water Allocation: Benevolent Czar or Crystal Pitcher Approach," in *Proceedings of the Seventeenth Mississippi Resources Conference, March 25–27, 1987,* Arkansas Water Research Center, Fayetteville, Ark., 1987.

Maass, A., and R. Anderson, . . . *And the Desert Shall Rejoice: Conflict, Growth, and Justice in Arid Environments,* MIT Press, Cambridge, Mass., 1978.

MacDonnell, L. J. (principal investigator), "The Water Transfer Process As A Management Option for Meeting Changing Water Demands," vol. I, USGS Grant Award No. 14-08-0001-G1538, Natural Resources Law Center, University of Colorado, Boulder, 1990.

Milliman, J. W., "Water Law and Private Decision-Making: A Critique," *Journal of Law and Economics,* 2(Oct.):41–63, 1959.

National Research Council, *Water Transfers in the West: Efficiency, Equity, and the Environment,* National Academy Press, Washington, D.C., 1992.

Nunn, S. C., and H. M. Ingram, "Information, the Decision Forum, and Third Party Effects in Water Transfers," *Water Resources Research,* 24(4):473–480, 1988.

Revesz, R. L., and D. H. Marks, "Local Irrigation Agencies," *Journal of the Water Resources Planning and Management Division,* ASCE, 107(WR2):329–338, 1981.

Saarinen, P. P., and G. D. Lynn, "Getting the Most Valuable Water Supply Pie: Economic Efficiency in Florida's Reasonable-Beneficial Use Standard," *Journal of Land Use and Environmental Law,* 8(2):491–520, 1993.

Saleth, R. M., J. B. Braden, and J. W. Eheart, "Bargaining Rules for a Thin Spot Water Market," *Land Economics,* 67(3):326–339, 1991.

Sergent, M. E., "Water Transfers: The Potential for Managing California's Limited Water Resources," Masters thesis, Civil Engineering Department, University of California, Davis, 1990.

Smith, M. G., and C. M. Marin, "Analysis of Short-Run Domestic Water Supply Transfers Under Uncertainty," *Water Resources Research,* 29(8):2909–2916, 1993.

Stavins, R. N., "Harnessing Market Forces to Protect the Environment," *Environment,* 31(1):5, 1989.

Tietenberg, T. H., *Emissions Trading: an Exercise in Reforming Pollution Policy,* Resources For the Future, Washington, D.C., 1985.

Tregarthen, T. D., "Water in Colorado: Fear and Loathing of the Marketplace," in T. L. Anderson, ed., *Water Rights,* Pacific Institute for Public Policy Research, San Francisco, 1983.

U.S. Congress, "Project 88—Harnessing Market Forces to Protect our Environment: Initiatives for the New President," a public policy study sponsored by Senators Timothy E. Wirth of Colorado and John Heinz of Pennsylvania, December, 1988.

Virginia Water Use Act of 1991.

Walker, W. R., W. E. Cox, and M. S. Hrezo, eds., "Legal and Administrative Systems for Water Allocation and Management: Planning in the Southeastern States," OWRT B-123-VA(1), NTIS # PB187120, March 1983.

Water Science and Technology Board, National Research Council, Commission on Engineering and Technical Systems, Commission on Physical Sciences, Mathematics, and Resources, "Drought Management and Its Impact on Public Water Systems," Report on a Colloquium Sponsored by The Water Science and Technology Board, September 5, 1985, National Academy Press, Washington D.C., 1986.

Wong, B. D. C., and J. W. Eheart, "Market Simulations for Irrigation Water Rights." *Water Resources Research* 19(5):1127–1138, 1983.

CHAPTER 33

DECISION SUPPORT SYSTEMS (DSS) FOR WATER-RESOURCES MANAGEMENT

Rene F. Réitsma
Research Associate, CADSWES/CEAE
University of Colorado,
Boulder, Colorado

Edith A. Zagona
Research Associate, CADSWES/CEAE
University of Colorado,
Boulder, Colorado

Steven C. Chapra
Professor, CADSWES/CEAE
University of Colorado,
Boulder, Colorado

Kenneth M. Strzepek
Associate Professor, CADSWES/CEAE
University of Colorado,
Boulder, Colorado

33.1 INTRODUCTION: THE DSS CONCEPT

In a recent review of *decision support systems* (DSS) and DSS research, Konsynski, et al. (1992) remark that "a significant amount of soul searching has taken place regarding definitions of the DSS field." The term DSS is indeed used for many different kinds of computer-based information and modeling systems. And although there seems to exist some consensus as to the purpose of these systems, namely to support decision making in more or less complex (i.e., multiobjective or semistructured) situations, a single, clear, and unambiguous definition is lacking.

Mittra (1986) defines DSS as "... a computer-based information system that helps a manager make decisions by providing him or her with all the relevant data in an easy to understand form."

Similarly, Keen (1986) states the purpose of DSS is "... to help improve the effectiveness and productivity of managers and professionals." Similar definitions are applied by Sprague (1986) and Keen and Scott Morton (1978).

Where Mittra and Keen and Scott Morton emphasize support for managerial decision making, Sprague and Watson (1986), McLean and Sol (1986), and Guariso and Werthner (1989) emphasize the type of problems (ill-structured and semistructured problems) for which DSS offers a solution. Fedra, et al. (1986) emphasize the latter as well:

> ... there is a class of (decision) problem situations that are not well understood by the group of people involved. Such problems cannot be properly solved by a single systems analysis effort or a highly structured computerized decision aid. They are neither unique—so that a one-shot effort would be justified given the problem is big enough—nor do they recur frequently enough in sufficient similarity to subject them to rigid mathematical treatment. Due to the mixture of uncertainty in the scientific aspects of the problem, and the subjective and judgmental elements in its sociopolitical aspects, there is no wholly objective way to find a best solution.

Guariso and Werthner (1989) also recognize the various types of decision making that environmental DSS needs to support. They adopt Anthony's (1965) three-tiered taxonomy of decision making (strategic planning, management control, and operational control). In water-resources management, these types of decision-making problems are clearly present. Managing reservoirs and river systems, for instance, implies both short-term operational decision making (daily and hourly operations of the dams and power plants) as well as more infrequent long-term planning decisions (new construction, long-term environmental planning, and so on). This distinction between short-term operational and long-term planning decision making in water-resources management is tightly coupled with the ways in which the management organization works and conducts its business. It is internalized by the organization and manifests itself in how the organization is composed, how its authority and responsibilities are arranged, and how it distributes information. DSS for water-resources management need to support this complex combination of operational decision making and long-term planning.

In essence, this problem relates to the dual objective of increasing both the efficiency and effectiveness of organizations or decision-making processes (Mallach, 1994). Typically, efficiency increases are achieved through automation whereas increases in effectiveness or productivity are achieved by reorganization of the decision-making or production system itself. In the context of water-resources management, efficiency-based objectives tend to relate to operational decision making, whereas effectiveness-based objectives relate more to long-term policymaking. Although DSS are meant to support both operational and policy decision making, there exists an inverse relationship between the nature of these two objectives and the formalisms needed to express them on a computer. Where operational activities tend to be relatively well defined in terms of operational procedures, guidelines, constraints, and responsibilities, long-term policymaking is a much less well defined, much more unruly process (Klosterman, 1987). Where operational decision making assumes a predefined set of objectives as well as a set of rules and procedures for how to achieve these objectives, strategic planning, for instance, questions the objectives themselves, thereby opening up the process for external effects from differential interests associated with the resources under question. And although some of the previous definitions seem to indicate that DSS exhibits its major strength in the domain of ill-structured long-term planning, both the logical foundation of current-day computing and our lack of formalisms for accurately representing complex decision-making situations cause those aspects also to be the hardest to model on a computer.

In practice, though, every water-resources-management organization deals with actual, real problems. Some of these are of an operational nature, some involve long-term planning issues. The goal of DSS is to support both.

Unlike definitions based on the objectives or activities supported by DSS, we propose a definition based on the architectural structure or components found in most DSS for water resources (Fig. 33.1):

> Decision support systems are computer-based systems which integrate state information, dynamic or process information, and plan evaluation tools into a single software implementation.

- *State information* refers to data which represent the water resource or environmental system's state at any point in time.
- *Process information* represents first principles governing the resource's behavior over time.
- *Evaluation tools* refer to utility software for transforming raw system data into information relevant for decision making.

A complementary functional definition of DSS can then be formulated as follows:

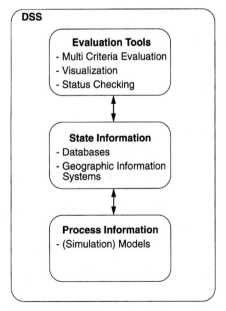

FIGURE 33.1   DSS structure and components.

Decision support systems provide information on the historic, current, and future states of an environmental resource where future states are computed by one or more simulation and/or optimization models. This information is communicated in forms which are directly useful for operational decision making and/or long-term planning.

The remainder of this chapter consists of five sections. First, we present an example of a DSS for water-resources management. We use this example, a system developed for the Tennessee Valley Authority (TVA) and the Electric Power Research Institute (EPRI), throughout the rest of the chapter for purposes of illustration. This brief presentation is followed by a general discussion of the three main components of a DSS: (1) state representation, (2) dynamics, and (3) evaluation. In this section we only briefly present some of the technical components of DSS, such as geographic information systems, relational databases, and programming tools. As such, technical aspects of DSS occupy only a small part of this chapter. Although computer technologies provide the ultimate tools with which a DSS is developed, experience has shown that the real challenges for designing, developing, and implementing these systems only marginally relate to problems of a technical nature. For additional information on any of the technical tools discussed in this chapter, we refer to the extensive literature on nearly all of these subjects. In a short section on Tools and Toys (Sec. 33.5), however, some of the most recent technical developments are briefly discussed. The main section of this chapter contains a discussion on the process of actually developing and fielding DSS. This process is laid out in various phases and the major activities and concerns are presented for each phase. Finally, some concluding remarks and expectations for water-resources management DSS in the future are presented.

33.2 TERRA: EXAMPLE AND WORKING HYPOTHESIS

The TVA Environment and River Resource Aid (TERRA, Fig. 33.2) was designed and developed to support everyday management of the TVA river, reservoir, and power resources (CADSWES, 1993).

As a water-resources management agency, the TVA has a long history of managing interests and utilizations associated with the water provided by its Tennessee and Cumberland rivers and tributaries. The TVA controls a watershed which contains more than 40 reservoirs and a variety of hydroelectric, fossil-fuel, and nuclear power plants. Managing such a system on often very narrow time scales (1 to 6 h) requires close cooperation and communication among the various departments within the organization, as well as very current and sound information on the status of the TVA river and power system. TERRA is a system which provides the various geographically distributed TVA departments with historic, (near) real time, and (estimated) future information on the status and trajectory of the TVA water and power systems. The following are some of the features of the TERRA system.

- TERRA is constructed around a geo-relational database which is accessed by users, by applications run by users, and by automated procedures for monitoring and checking the current and expected future health of the TVA river and power system.

- The system is planned to function as the central data-storage and retrieval system for operational information necessary to operate the TVA water and power

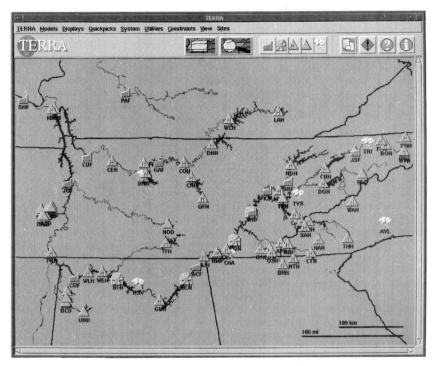

FIGURE 33.2   TVA Environmental and River Resources Aid (TERRA).

system on a daily basis. Although many of TVA's departments use additional data and models, TERRA contains the data and information most commonly used among these various groups. As such, it serves as a conduit for sharing information.

- TERRA is equipped with facilities to continually update its own database with new information, including real-time observed data, model results, or complete text-based TVA memoranda or directives. It keeps a log of all data transfers so that a complete record of the TERRA information flow is available for possible future analysis.

- To support operational decision making and planning, several (resident) models are tightly coupled with the system. Through the use of specialized software known as *data management interfaces* (*DMI*s) they maintain connections with TERRA's database. The DMIs allow models to initialize themselves with the most recent system data by retrieving them from the database, and allow models to leave their results behind in the database for other applications (such as models, visualizations, and report generators) to be used as input.

- Various visualization tools are integrated in the system. These range from simple historical overviews and bibliographies to complex graphics showing past model performance and historic trajectories as well as the current values of many system variables and parameters. All visualization tools interact with the TERRA database through their own set of DMIs.

- Data entering the database or data from both resident and nonresident models are checked against various sets of operational constraints (environmental, recreational, special/emergency, navigational, and so forth). If, for instance, measured (observed) values approach or exceed the values listed in these constraints, TERRA users are automatically notified through the posting of these violations on the system's bulletin board.

TERRA was developed as a tool to fit into an existing organization. It was laid out and equipped so that it fits the organization's requirements and daily activities. As such, TERRA contains numerous *utilities:* small software applications which, although hardly or not at all integrated with any of the main components of a DSS, provide many of the little extras which make a system work well within an organization. Examples of these utilities in TERRA are facilities which connect the user to other TVA computer systems; programs for computing temperature and humidity data; electronic mail; and facilities for ad hoc querying of the TERRA databases.

TERRA contains the three essential components of a DSS: (1) management of the state information of the TVA river basins, (2) the models for conducting simulations and optimizations, and (3) a comprehensive set of reporting and visualization tools for studying, analyzing and evaluating current and forecasted states of the river system. Moreover, TERRA was designed and developed with a well-defined organizational context in mind. Although such *tuning* of a DSS to the characteristics of the organization cannot be considered a property of a DSS per se, it represents an aspect of system design which is of vital importance to the success of a DSS. We will return to this latter aspect later in the chapter.

33.3   DSS: A CLOSER LOOK

In this section we take a closer look at each of the main components of a DSS. Brief discussions of the role and function of these components are presented, together with an overview of recent developments relevant for DSS construction and use.

33.3.1   State Representation

State representation by means of databases forms the heart of today's DSS for water-resources and environmental management. Whereas only a few years ago DSS developers had to develop their own tools or systems for data management, today's relational and spatial database-management systems offer comprehensive facilities for storing, retrieving, and manipulating data of the water resource under question.

In DSS for water-resources management, it is customary to make a distinction between *relational* and *spatial database-management systems*. The reason for this is that although both the attribute and spatial characteristics of a water resource are important, they are managed differently. *Geographic information systems* (GIS) are database-management systems which have been developed for the storage, retrieval, and management of spatial data (Burrough, 1986; Scholten and Stillwell, 1990; Star and Estes, 1990). Unlike most attribute data, spatial data have a well-defined structure and are composed of simple building blocks such as points, lines, and polygons. Furthermore, spatial data are limited enough in their dimensional extent to be made subject to equally well-defined mathematical manipulations. This implies that with relatively simple spatial building blocks complex structures and manipulations of those structures can be achieved. Examples of more complex structures are digital-elevation models, stream networks, or finite-element meshes. Examples of spatial computations are the calculation of area, the length of a route through a topological network, the intersection of a group of superimposed polygons, or the aggregation of values from irregularly shaped surfaces into a single composite value.

General attribute data, on the other hand, do not conform to such strict formalisms. As a result, more general data models have been developed to cope with attribute data (Date, 1985). The most popular of these is the *relational data model* (Codd, 1983; Date, 1986). The relational model relates units of information (e.g., a reservoir's identification and its storage capacity) in a tabular manner, where tables can be manipulated using the rules of relational algebra. Examples of this technique are presented in Sec. 33.5, Of Tools and Toys. Although the relational model is not a panacea for all data representation (spatial and temporal data, for instance, do not lend themselves easily to storage in a relational form), it has both intuitive and implementational* elegance. Hence, many relational database-management systems are available in today's market, together with various tools and facilities for querying these databases (either from query shells or directly from inside application programs), maintenance of database integrity, triggering other subsequent processes, networking facilities, and so on.

In water-resources modeling, both the spatial and attribute information associated with the water resource being modeled are important. As such, simulation and optimization models need to access both types of information. This can be accomplished by the creation of a *geo-relational data model* accompanied by a set of facilities for storing and retrieving both the spatial and attribute data in a coordinated manner. The geo-relational data model does not combine the spatial and relational models into a higher, more general data model. Instead, it preserves both data models while providing a link or common *key* between them. For instance, by assigning objects' identical identifiers in both data models or by specifying a mapping of identifiers between data models, application programs can associate both spatial and attribute data with the environmental resource under study.

The TERRA system employs the geo-relational data model. Point structures in the spatial database are associated with attribute data in the relational database. As such, each designated point on a map can be treated as a reference to a much larger complex of data stored in the relational database. And since the relational database is maintained on a (near) real-time basis, both historic and current information on the various reservoirs, dams, power plants, or meteorological stations is easily accessible through the map of the TVA region (Fig. 33.2).

33.3.2 State Transitions

Although many types of models can be employed in DSS, this section focuses on those used for *simulation;* that is, those models that represent a water-resource's or environmental system's behavior and can serve as simplified representations of the system under analysis.

Table 33.1 contains a brief taxonomy of models used to support decision making in water-resources management. They range from purely physical quantity and qual-

---

* Unlike some other data models which also have intuitive appeal, the relational model lends itself to efficient implementation on a computer.

ity models to those which simulate economic and societal impacts. Regardless of the domain, they all provide decision makers with a means to assess the impacts of their decisions by simulating outcomes as a function of actions.

Unfortunately, early computing environments often limited the utility of these models in managerial contexts. Most early models were written in procedural languages (e.g., FORTRAN) and were designed for a batch mode of operation. Input was via punched cards or text files and output usually consisted of tables of numbers. As depicted in Fig. 33.3, the analyst traditionally was at the center of the information processing. This configuration was dictated in large part by deficiencies of early hardware and software and exacerbated by the accompanying lack of integration of computing tools. Several harmful impacts on the analysis and evaluation process resulted:

- The process was inefficient and hence costly. Because the analyst served as the intermediary among the tools and products—the flow of data between models was accomplished manually—inefficiencies arose from the effort expended in data processing and display.

TABLE 33.1 Taxonomy of Models in Water-Resources Management

Model type	Description
Watershed models	Relate meteorological phenomena to the runoff quantity and/or quality from a land area. They vary from simple empirical relationships like the rational formula to complex physically deterministic models that take into account land and vegetation types, antecedent moisture, evapotranspiration, infiltration, solar radiation, and other details.
Surface-water quantity models	Describe the movement of water on the earth's surface, including rivers, lakes and artificial channels. They include representations of structures such as dams, power plants, and control structures, as well as laws or policies such as operating procedures and water rights. In mathematical complexity they range from simple mass balance to three-dimensional partial differential equations of fully dynamic flow. They may also include boundary conditions with the atmosphere and/or the groundwater. In addition, some address the problem of sediment transport.
Surface-water quality models	These include quantity as required, but focus on the fate of constituents carried by the water. Modeled variables include temperature, dissolved oxygen, plant biomass, metals, toxic organics, sediments and dissolved solids. Water-quality models have grown in importance over the past two decades in managing water resources within environmental regulations.
Groundwater models	Simulate quantity and/or quality of water below the surface of the ground. They may include infiltration, pumping, movement through geological formations, aquifer management, contaminant plume tracking, and conjunctive management of surface and groundwater.
Economic models	These do not explicitly model physical processes, but consider the economic aspects of water-resources development and management. These aspects include cost-benefit analysis of projects or policies, regional economic impacts of water-resources projects, costs of pollution control, and risk-benefit analysis.
Social models	Consider the social impacts of water-resources projects and policies, including migration, displacement, farming practices, education, and the uneven distribution of benefits among various population groups.

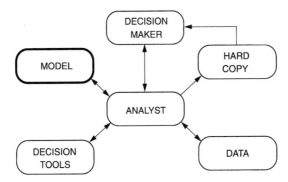

FIGURE 33.3 Flows of information in traditional water-resources management.

- The analyst often required considerable expertise in both the computing and environmental domains. As a consequence, the analyst usually served as a costly intermediary between the tools and the decision maker.
- Multiple evaluations were unfeasible because of the large time expenditures in developing model runs. Therefore, only a portion of the decision space could be explored.
- Because of the difficulty in processing and transmission, only a portion of the data was utilized in model applications and evaluations.
- In many cases, simplified models were adopted in an attempt to overcome the inefficiencies in the process. Although this sometimes expedited implementation, the lack of realism in the simplified models detracted from the credibility of the final analysis.
- In many instances, models became institutionalized because of common usage. This often related to the fact that models were so difficult to modify that updates were discouraged. Thus, in many cases, a gap opened between state-of-the-art modeling and model use.
- Because of the sectorial nature of models themselves, problems were often attacked in a piecemeal fashion. For example, a drainage-basin model might be run to generate boundary conditions for river routing and/or groundwater models. At best, because separate models had to be run without electronic linkage, the process was extremely cumbersome. At worst, models would be underutilized or not used at all.
- Lack of visualization tools meant that erroneous results could occur without being detected. Also, nuances of results were lost because most output was numerical.

As a consequence, although considerable resources were expended on the information evaluation process, the impact of the models on the ultimate management decision was diminished.

Although model applications in the past have been less than optimal, it should be noted that some successful cases have occurred. For example, in the 1960s, the Delaware estuary in the eastern United States was the site of a concentrated data collection and modeling effort. These were combined with optimization algorithms to develop waste-load allocations to remediate the estuary's water quality (Thomann and Sobel, 1964; Thomann, 1972). Another example relates to efforts to reduce phosphorus loadings to ameliorate Great Lakes eutrophication (Loehr, et al., 1980). In this case, several mathematical models were used to determine phosphorus-reduction goals that became part of the 1977 Great Lakes Water Quality Agreement between the United States and Canada. Examples in water-resources management are also described in Maass, et al. (1962) and Major and Lenton (1979).

These successes were marked by the fact that the problems were important enough to marshal the substantial economic and political resources needed to address them appropriately. For example, they capitalized on the development of a variety of new models, generated considerable expenditures for data collection, and involved the top expertise in the development, application, and interpretation of model results.

Although other examples of successful model applications exist, there is no question that the use of models for more routine environmental decision making has been hampered by past technological impediments.

From these experiences it can be inferred that *electronic integration* dictates the role of models in DSS. Electronic integration dramatically changes the structure and interactions represented in Fig. 33.3. Although users will still drive the process, the DSS rather than its user will serve as the primary conduit for information flowing between the different system components. This means that modeling will no longer be the primary focus of the information evaluation effort. Although all the components will be more in balance, data and data management will, in fact, dominate. Beyond this general observation, the following specific changes can be expected:

- Models will be closely integrated with the rest of the system. No longer will translation barriers impede access between models, data, and plan-evaluation tools.

- Electronic integration will stimulate closer integration of the different modeling tools themselves. For example, a drainage-basin analysis package could integrate all the tools outlined in Table 33.1 under a single framework.

- Because of advances in both hardware and software, models can be as complex as the problem and data permit. For example, three-dimensional spatial characterizations and nonlinear formulations would be feasible. Thus, by providing more realistic simulations, decisions will be grounded on a more credible foundation.

- New types of models can be more easily employed because of advances in programming languages and the speed of hardware. For example, stochastic approaches could be more easily implemented, allowing decision makers to assess the reliability of model results. Another example might relate to real-time simulations which can be designed to inform and expedite operational applications.

- Improvements in input/output will ameliorate both the quality and quantity of modeling efforts and their use in decision making. By allowing all data and simulation results to be used and visualized, model calibration should be more reliable and comprehensive. Further, by providing quick model runs and visualizations, more of the decision space will be explored.

- Modern DSS allow more intelligence and information to be embedded in the system and be made readily accessible to the analyst. Consequently, less expertise (particularly computer science expertise) will be required by the analyst. In some instances, it could be possible to allow direct interaction between the decision maker and the system. In both cases, the gap between analysis and implementation or the lack of communication between modelers and policy makers (U.S. Office of Technology Assessment, 1982) can be reduced.

- New types of data not traditionally used for modeling can now be used to extend model capabilities. The prime example are detailed spatial data. New models could be developed to exploit the more detailed spatial characterizations provided by GIS systems.

In summary, automation and integration should provide a revolutionary change in both the models themselves and the effectiveness with which they inform the decision-making process.

TERRA supports models in both a resident and nonresident manner. Resident models are models which are tightly coupled with the TERRA DSS. This implies that the models can be run from within the DSS and that their data, both input and output, are managed through the TERRA database and a set of DMIs. In practice this means that upon starting a model of, for instance, a power plant scheduling problem, the model's DMI accesses the TERRA database so that it initializes with the most current data of the system. After further setting the decision variables and running the model, the model's output goes back into the database where it can be used by other components of the system such as visualization tools, constraint checkers, report generators, and so on.

Nonresident models are models which are run and maintained elsewhere in the organization but output of which can be incorporated into the TERRA database where it can be subject to the same operations as all other model results. This concept of nonresidency is both simple and attractive. It is simple because the only facilities needed to incorporate foreign model results are a data interchange format and an incorporation procedure. The attractiveness comes from the fact that regardless of how well a DSS's functionality matches the structure of an organization, there is always a need for ad hoc information to be incorporated and cross-referenced with other information in the system. Also, it is unfeasible to tightly couple many existing computerized models with a new DSS. As such, loose coupling in a nonresident fashion constitutes an attractive alternative.

33.3.3 Plan Evaluation

Plan evaluation comprises an important component of DSS. Plan-evaluation tools can be defined as those DSS facilities which present system data in a form which accommodates the assessment or appraisal of a policy or plan (Yeh and Becker, 1982; Fedra and Zhao, 1988; DeSanctis and Gallupe, 1987; Yakowitz, 1992). Plan-evaluation tools are those tools that filter, modify, and present parts of the system's data in a form suited for alternative evaluation and decision making, thereby transforming a system's *data* into *information*. At least two rather different forms of plan evaluation can be recognized: *formal* and *informal* tools.

Formal plan evaluation involves the use of mathematical techniques for the systematic evaluation of various choice alternatives. Policymaking frequently requires the choice of a particular alternative over one or more other alternatives. Although often the selection is motivated by political rather than analytical considerations, a careful comparative analysis of the alternatives can be of great value for the quality of the decision.

Multicriterian evaluation (MCE) techniques allow such comparative analysis by casting the choice problem in a mathematical framework also known as the *decision space* (Nijkamp and Spronk, 1981; Wierzbicky, 1979; Winkelbauer and Markstrom, 1990; Nijkamp, et al., 1990; Yakowitz and Lane, 1992).

Utility functions are at the heart of MCE techniques. A utility function is a mathematical expression which combines the various characteristics of alternatives (e.g., reservoir yield and cost variables) into a single measure of attractiveness or utility. It does this by combining three factors:

1. The value an alternative takes for a particular dimension or criterion
2. The weight or priority attached to that criterion
3. A *combination rule* specifying how the various criteria are combined into a single utility value

An example of such a rule, a *weighted additive rule,* is represented in Eq. (33.1).

$$\hat{U}_x = \sum_{i=1}^{I} w_i x_i \qquad (33.1)$$

where $\hat{U}_x$ = estimated utility of alternative x
       w_i = priority or weight of attribute i
       x_i = score of alternative x on attribute i

Various MCE approaches exist. In the *context-free* approach, alternatives are ranked on the basis of their individual characteristics alone. In other words, the results from the utility functions are the end-result of each alternative's evaluation and alternatives are ranked in order of their utility values. In the *reference-point* approach alternatives are evaluated not only on the basis of their own characteristics, but are evaluated against a target or desired distribution of criteria.

Unlike context-free approaches, the reference-point approach is better suited for negotiation situations because it allows exploration of various goal states. It allows adjustments of both the weight set and the desired distribution to more realistically reflect the feasibility and acceptability of the alternatives' criteria scores. Additional sensitivity analysis provides clues as to the robustness of the computed evaluations and provides information on the critical differences and similarities between the various alternatives.

MCE techniques are useful for any situation in which a number of alternatives need to be evaluated for implementation. Although MCE techniques allow the inclusion of some political variables by means of setting weights or priorities, their main virtue lies in the fact that they allow for impartial, systematic investigation of alternatives and objectives. Also, MCE techniques are useful in that they offer a framework for combining characteristics and variables which can otherwise hardly be integrated into a single evaluation framework.

Unlike formal plan-evaluation tools, informal ones do not rely on mathematical formalism to synthesize the overall utility of choice alternatives. Instead, they are aimed at representing the various relevant characteristics of the options whereas the combination of these into utility or attractiveness judgments are left to the human interpreter. Although informal plan evaluation was always available in the form of more or less comprehensive visualization of system and model data, recent objections against MCE techniques emphasize the importance of informal plan evaluation (Hendriks and Van der Smagt, 1988; Reitsma, 1990; Reitsma and Behrens, 1991). Critics argue, for instance, that algebraic combination rules do not allow for noncompensatory factors, i.e., those which must meet minimum requirements. Moreover, they argue that often the utility of choice alternatives is determined by complex, conditional relationships among the combined characteristics of the alternatives and the objective a decision maker has in mind. These conditional relationships cannot be modeled with an equation. Yet another problem with utility theoretical models is that different objectives imply different perspectives of the same situation. As such, different interest groups may have very different perspectives or *views* of the same environmental resource. Views then are representations of a system's data in the context of one or more objectives. In other words, a view represents a system's data relative or with reference to one or more objectives or interests. Views can contain various types of information:

- Variables, either basic or derived
- Scalings and categorizations of those variables
- Display methods
- Active components which monitor the behavior of one or more variables and notify users if a threshold is violated
- Evaluation models that summarize the effects of decisions on predefined sets of criteria.*

Where simulation and plan evaluation nicely fit the three-component model of DSS presented earlier, optimization integrates the behavioral and evaluation components of a DSS into a single numerical analysis component. Whereas simulation models represent a system's behavior over time using the dynamic relationships between a system's variables, its boundary conditions, and its forcing functions, optimization implies searching for a solution which maximizes an objective function, subject to a set of constraints. Two types of constraints exist: physical constraints and policy constraints. Physical constraints define physical boundary conditions, the behavioral envelope within which the system needs to operate. For instance, a reservoir cannot contain more water than its maximum capacity and its power plant cannot take water in from below its minimum power-pool level. Policy constraints, on the other hand, reduce the degrees of freedom for finding a solution by imposing various policy-induced limitations. For instance, lake levels might be constrained to certain elevations for recreational or environmental purposes and in-stream flows might be constrained above certain flow thresholds to protect fish habitat. In short, where state transitions in simulation models are a consequence of specific causal relationships embedded in the model's dynamics, optimization models derive states by finding solutions given an objective function and constraints. And since the objective functions and constraints represent implicit criteria for operating the environmental resource, no explicit evaluation step is necessary. Optimization models for water resources have been developed and refined primarily in academia, but water-management agencies such as the Tennessee Valley Authority have developed dedicated optimization models (e.g., Gilbert and Shane, 1982) to assist in managing reservoir and hydropower systems.

The role and place of optimization in DSS for water-resources management is open for debate. Although integration of simulation and evaluation in an optimization method seems attractive, many water-resources management problems do not lend themselves for optimization. The complex configurations of interests associated with a common resource, the conditional relationships between objectives, and uncertainty as to the future development of both objectives and constraints often limit optimization applications.

---

* MCE applied within the context of one set of interests escapes some of the critiques mentioned earlier, provided that there exists no (semantic) interaction between variables.

33.4 DSS: DESIGN AND DEVELOPMENT

Figure 33.1 depicts the general structure of a DSS. But since it defines a DSS as a composite of functions, it does not contain information as to how such functionality can be achieved.

In this section we discuss the design and development activities necessary to take a DSS from its inception as an idea to the final stages of fielding, testing, training, and long-term maintenance. Although many methodologies for information analysis, system design, software design, development, and implementation exist, each with their own way of structuring the design and development process (e.g., Nijssen, 1976; Ross and Schoman, 1977; Fairly, 1985; Gould, 1988), we feel that eight phases can be recognized in the DSS design, development, and fielding process:

1. Needs analysis
2. High-level design
3. Detailed functional design
4. Detailed software design
5. Development
6. Initial fielding
7. Testing, training, and fine tuning
8. Maintenance and further development

Although these phases can be clearly recognized in most comprehensive environmental DSS design and development efforts, the actual process frequently behaves in a less linear fashion. Often, initial design decisions are revoked or modified based on new understanding of the workings of the organization, the nature of the associated data, or changes in the available budgets or the development team. And although a structured design and development process considerably reduces the likelihood of failures, misunderstandings, and temporary setbacks, DSS development is a complex process susceptible to all the usual political, financial, and sociopsychological factors associated with organizational change, interdisciplinary problems, different interests, and stubborn, seemingly unwilling computers. The eight phases of DSS design and development discussed in the remainder of this session, therefore, are to be considered as those aspects to which the development process tends to gravitate along its way.

33.4.1 Phase 1: Needs Analysis

The primary goal of a needs analysis is to arrive at a systematic inventory of the functions the system is to support. Based on the results of this inventory, a high-level system design, focused on the overall system architecture and its various components, can take place. Although various techniques for arriving at this initial inventory of needs and requirements exist, we briefly discuss a technique known as *problem-centered design* (Lewis, et al., 1991; Polsen, et al., 1992; Lewis and Rieman, 1993). This software design technique, originally a program design methodology only, concentrates on the everyday actions of workers in an existing organization (scenarios), rather than attempting to elicit functional specifications directly. It is the task of the DSS developers to abstract from these often very diverse scenarios a smaller set of generic system requirements or functions.

Approaching the needs analysis by means of problem scenarios rather than through asking future users for a list of functional specifications not only forces the developers to continually conceptualize the system as it would function within the everyday practice of the organization, it also involves future users in the DSS design and development process. This, in its turn, stimulates discussion and anticipation regarding the new system, recognizes individual roles in the organization, and stimulates users to actually (re)evaluate current practice, efficiency, and effectiveness. When conducted correctly, problem-centered design can act as a valuable catalyst, rallying people in the organization behind the new system.

Problem-centered design, however, is not a panacea. Its main problems are that it does not guarantee a comprehensive, exhaustive design and that since the design adheres closely to everyday practice in the organization, there is an inherent danger of retaining or even reinforcing existing inefficiencies and ineffectiveness. Regarding the problem of completeness, we can say that, although we have never attempted to quantify these variables, in practice even a small number (three or four) of carefully designed scenarios will represent almost all of the generic needs of the system. Some minor adjustments may have to be made later on, but the main character and functions of the system will be captured appropriately. As to the problem of retaining inefficiencies, it can be noted that although we know of no quantitative underpinning of this problem, the problem-centered design approach seems to be very successful at truly engaging users in the needs analysis. During scenario analysis, they often reflect on the current practices of the organization and detect and communicate inefficiencies or situations in need of improvement.

The acquisition of a DSS may imply some significant organizational changes and/or adaptations. Although the problem-centered design process tries to adhere as much as possible to the reality of everyday tasks and organizational structure, changes due to the introduction of a more or less comprehensive DSS are inevitable. These changes may be rather modest such as a different format used for presenting data, but can also have repercussions for people's jobs, the way people conduct these jobs, and the ownership of models or data. A good example of this is the creation of an organization-wide central database to support DSS application software. The construction of such a database, as well as its long-term maintenance, most likely alters the distribution of ownership of data. Traditionally, much of this ownership reflects the organizational structure (departments, sections, and so on). As a consequence, people's responsibilities and authorities regarding the data, or at least their perception of such ownership, might be affected. Although this is not a DSS design or development issue in itself, it needs to be taken into account during the early phases of system design.

The products of a needs analysis can be many. Quite common is a matrix which combines DSS functions with the models and data required to achieve such functionality. The needs analysis can also include inventories of required data, their ownership, format, availability, and nature (constant or variable data, time series or parameter data, and so on).

When a problem-centered design is applied, the needs analysis also yields as comprehensive a set of scenarios as possible. These scenarios are used in almost all subsequent phases of the design and development process for reference and evaluation of system components, user interfaces, and overall capability of the system.

In some cases, the needs analysis is part of a larger initial effort to explore the feasibility of a large-scale DSS. Such a feasibility study typically contains a complete needs analysis as well as some of the components of the high-level design discussed in the next paragraph. Whether or not a feasibility study is warranted depends mostly on the magnitude and political feasibility of a proposed DSS. For instance, in 1992 the Colorado Department of Natural Resources, in cooperation with that state's Water Conservation Board and its Division of Water Resources, commissioned a feasibility study to explore the possibility of developing a DSS for monitoring, analyzing and managing the Colorado River. A feasibility study was issued because it was felt that the magnitude and political significance of such a system were sufficient to spend some money up front to see if the project was indeed feasible. Based on the study results (Dames and Moore, 1993), the state's legislature decided to initiate the project.

33.4.2 Phase 2: High-Level (Component) Design

33.4.2.1 System Architectures.   Perhaps the first and foremost decision to be made during this phase of the design is the selection of an appropriate system architecture. In environmental DSS, at least four types of architecture can be recognized. The differences are mainly a reflection of how the various functions are integrated. The choice of architecture has profound consequences for such aspects as modality, long-term maintenance, extensibility, and performance (Table 33.2).

Dedicated Models.   One approach to building DSS for water-resources and environmental management concentrates almost exclusively on dedicated models. Dedicated models are models which are developed for a specific case. Examples of this approach are the Colorado River Simulation Model (CRSM) developed by the U.S. Bureau of Reclamation (Schuster, 1987), the Delaware Estuary Model (DOI, 1966), or the Potomac River Simulation Model (Hetling, 1966). In the implementation of such models, representations of physical processes and those of the attributes of the system being modeled are not separated. In other words, the attributes and characteristics (constants in the state representation) of, for instance, a river basin are *hard-wired* into the source code of the DSS.

This entirely dedicated architecture has the advantage that the resultant DSS can be both highly effective and efficient if applied to a well-defined domain. However,

TABLE 33.2   DSS Architecture Evaluation Criteria

Criterion	Explanation
Extensibility—state	Ease with which an architecture can accommodate an extension of its database or can be applied to another case, such as another watershed, stream, or river basin.
Extensibility—process	Extensibility with respect to models. Concerns the efforts needed to extend an existing system with new versions of already integrated models or the addition of new models.
Maintainability	Operation and maintenance of a particular architecture. Systems mainly based on third-party software are easier to maintain than custom-developed software.
Performance	Execution speed of a system.
Modularity of implementation	Availability of intermediate or by-products during system development. Some architectures allow for parts of the system/functionality to become available during development, whereas others only allow products to be brought on line after complete development is finished.
Development costs	Costs to implement the architecture.
Extensibility costs	Costs to extend an existing system.

it takes great effort (and cost) when the system needs to be applied to a different problem, e.g., a new river basin or a new set of operating policies, or when new functionality needs to be added. Another shortcoming is that case-specific implementations cannot evolve over time and adjust to the availability of new sources of data, newly available models, or simply new functions required by users. Therefore, case-specific systems score poorly on both extensibility criteria. The modularity of case-specific systems is rated as poor because they tend to be ready for use only after development is completed. Little functionality becomes available during the development process. These systems are moderately expensive to develop and tend to be very expensive to extend.

Dedicated DSS. An approach developed more recently and perfected by such developers as the Advanced Computer Applications (ACA) group at the International Institute for Applied Systems Analysis (IIASA) in Laxenburg, Austria (Fedra, 1991; Fedra, et al., 1993) has a more hybrid character in that it marries generic models with dedicated data. Generic models are models which, unlike dedicated ones, do not contain any case-specific information about the domain they are applied to. Instead, they contain facilities which allow users to create separate input representing the system being modeled. In a dedicated DSS, one or more generic models are carefully prepared for operation on a very specific situation. The generic models are coupled with preprocessors which allow users to interactively define model input, often through the use of graphical user interfaces. This preparation may involve some hard-wiring of the data and the connections between the data and the model, but this hard-wiring is not nearly as extensive as in the case of fully dedicated approaches.

This alternative for system architecture has specific advantages in that one has the availability of generic models while the input and output operations have been tuned to the specific application domain. Dedicated DSS tend to be very efficient in their coding and are characterized by high performance (speed). The disadvantage of this approach is that embedded pre- and postprocessors are often of a dedicated nature, making it hard to apply the system to a slightly different domain, such as a geographically different study location. Also, dedicated DSS architectures are not *open.* Thus, when new data sources become available, the preprocessor may have to be rebuilt. Likewise, when new models or new versions of existing models become available, resources must be spent on redeveloping both the pre- and postprocessor parts of the system. Although dedicated DSS architectures offer moderate opportunities for extensibility, they are rated poor on modularity. The reason is that, like case-specific systems, little functionality becomes available during the development process.

Data-Centered DSS. Recently, more data-driven approaches to the development of DSS have been developed (Reitsma and Sieh, 1992; CADSWES, 1993; Reitsma, et al., 1994). Data-driven architectures avoid adherence to either completely generic or dedicated models. Although the application of generic models in a dedicated DSS constitutes an improvement, there is still a significant amount of work involved in reapplying such a model to new cases. If an architecture could be more adaptive to application to another case, that would be a benefit. Another reason for looking at more data-driven approaches originates from organizations which on the one hand have a definite interest in decision support, but on the other hand have invested substantially in the maintenance of large-scale databases containing data about their organizational domain. The last 10 to 15 years have shown a very strong increase in the amounts of data that organizations have been collecting on their domains. This data constitutes a rich resource for modeling, analysis, and decision making. Moreover, since collecting and maintaining this data is a costly endeavor it stands to reason that organizations want to use that data whenever possible. As a result, DSS architectures have evolved to arrange applications around a central database. These applications are equipped with specialized data-accessing software or data management interfaces (DMIs) which facilitate interaction with the database for retrieval or storage of data. The TERRA system provides an example of such a data-driven DSS.

Data-centered architectures allow addition and extension in both the application domain and the modeling in the system. As such, data-centered architectures score "good" on extensibility criteria. Just like purely generic systems, however, extensibility is somewhat traded off against performance. Data-centered systems are expensive to develop. However, they offer high extensibility and modular development, thus allowing various functions to become available during the course of development.

GIS-Based Systems. GIS-based systems are an outcome of recent developments in the field of geographic information systems. GIS systems are specialized in storing, retrieving, modifying, and generating spatially referenced data. Some of the commercially available GIS packages provide extended functionality and have been applied successfully for various operational and long-term planning studies. Recent developments in modeling in GIS (NCGIA, 1991; 1993) suggest that GIS can be extended even further into other domains of modeling, e.g., water resources. This type of architecture does offer certain advantages in that it makes use of sophisticated software for management and evaluation of spatial data. A distinct problem, however, is that although rapid improvements are being made in the integration of GIS and modeling (NCGIA, 1991; 1993), the full integration of all three components of DSS in GIS is, to say the least, problematic. For these reasons GIS-based systems perform poorly on model extensibility and moderately on data extensibility. Development in GIS systems is relatively cost-effective, however, because of the commercial availability of these systems. Extensions, though, can become very expensive, if possible at all.

Once an appropriate DSS architecture has been selected, a conceptual, high-level design of its components and infrastructure can be developed (Fig. 33.4).

During the high-level design phase, developers identify the database-management system, the hardware and software platforms for DSS implementation, and the software tools to use (programming languages, programming tools, graphics libraries, and so on). The high-level design also lists all required models, their availability, source codes, and compatibilities, and at least a large part of the supporting data. Finally, the high-level design should contain an indication of the duration and costs of the project (these will be specified in much more detail during the detailed design), as well as the organizational issues involved in the introduction of the DSS.

33.4.2.2 *Selection/Design of Models.*

As part of the high-level design, decisions as to which models to include in the system need to be made. This implies choices as to whether to include existing models or to develop new ones. The choice depends on what models are available, who owns them, the importance of using generally accepted models, the costs of modifying an existing model to fit the requirements of the DSS, and the cost of new model development. Here we offer some selection criteria for models to be included in a DSS. These criteria can be considered design criteria if a new model is to be developed.

Representation. The design of a DSS requires that the data, the user interface, the models, and the analysis tools support decision making in an integrated, concerted manner. Hence, the foremost selection criterion for inclusion of a model in a DSS is that the model represents the relationships, processes, and objectives which permit exploration of the decision space. A river-basin-management DSS should contain a model of the river basin with the relevant features, e.g., routing water through the system, hydrologic inflows, hydropower plants, and so on. If the decision requires

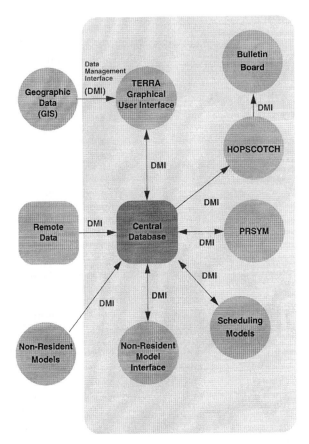

FIGURE 33.4   TERRA architecture.

optimization of hydropower benefits, the model must include both hydropower simulation and the economics (costs and benefits) of the power produced.

Accuracy and Flexibility.   The value of a model as a DSS tool lies in its ability to portray the significant interactions of the environmental or water-resources system at the appropriate level of accuracy. Many DSS are designed to support various types of decisions regarding the subject system. Hence, models must not only perform their computations at the appropriate levels of spatial, temporal, and domain granularity, but also must be flexible in terms of time-step size, time horizon, spatial resolution, and selection of decision variables and outputs. Often, existing models do not have the needed flexibility, but can sometimes be enhanced to meet these requirements.

Type of Model.   A model selected for a DSS must be appropriate to the type of decision it needs to support. For example, multiobjective models and stochastic models may be effectively employed as screening models for identifying a set of feasible alternatives or for eliminating inferior plans. Deterministic optimization models provide a single optimal solution and sometimes a sensitivity analysis of the results. These models can be used to identify a good design or operating plan, but because optimization often requires many simplifications in the representation of the system under study, the results should always be evaluated using more precise simulation models (Loucks, et al., 1981). In a review of the use of models in planning, Klosterman (1987) observes that "Mathematical programming techniques have proven to be extremely powerful for examining well-structured problems with a specified number of calculable variables, clearly defined goals and firmly established technical solutions, but largely inappropriate for the majority of planning problems which lack all of these characteristics." This observation is consistent with the argument by Fedra (1986) mentioned earlier for systems which are developed with the semistructured nature of many decision problems in mind.

Robustness. Many complicated models are known for their difficulty of use. A model may lack robustness because it does not allow straightforward expressions of a well-formulated problem. Simulation models which include the numerical solution of differential equations may have convergence problems as a result of inappropriate boundary conditions or badly specified convergence criteria. When a model is embedded in a DSS it is often intended to provide analysis capabilities to users who may not otherwise have the expertise to use these models independently. Also, some DSS are designed to be used in situations where finding a solution in a reasonable length of time is essential. Other DSS are used for real-time operating decisions. In these cases it is essential that the DSS and the model be designed such that users are given enough additional computational facilities to be able to identify modeling problems and eventually find a solution.

Propriety. Ownership of existing models is often an issue in their inclusion in a DSS, either because of the cost of licensing or because modifications to the model might be necessary. Often, the source codes of proprietary models are not available, making it very difficult to adapt the model to the requirements of the DSS. If the DSS developers own models which were custom built for the application, thought should be given to whether source codes must be accessible to users, and whether or not the developers will continue to maintain and debug the code in the future.

Acceptability. Much of the modeling, planning, and policy formulation in water-resources management today involve compliance with regulatory thresholds and limits. Although in many instances specific models may not be explicitly required in any regulations, certain models have been traditionally used and accepted. For example, the QUAL2E water quality model is accepted by the U.S. EPA (Brown and Barnwell, 1987), whereas HEC-2, a steady-state hydraulic model (U.S. Army Corps of Engineers, 1982), is a commonly accepted tool for defining floodplain boundaries. Many other models likewise have credibility in the water-resources domain. A new model may have credibility problems and will have to undergo rigorous testing. Some organizations have developed their own models for particular applications. Often, much time and resources have been invested in these models, and the organization knows and trusts them. In such cases it may not make economic sense to develop a new model, even if the existing model was developed with antiquated coding practices. The selection of a known, accepted model which must be adapted to the use of the DSS, or a new, flexible, and perhaps more powerful model will depend largely on the use of the DSS and the importance of credibility as judged by those whose decisions depend on it.

33.4.3 Phase 3: Detailed Functional Design

Once a high-level design has been completed and accepted by the sponsoring organization, a more detailed functional design, followed by a detailed software design can be developed. The detailed functional design departs from the results of the high-level design and describes in detail all components and all infrastructure of the DSS from a functional point of view.

A detailed description of all user interfaces comprises an important part of the detailed design phase. In working out the user interface design, all generic system functions need to be specified in detail and must be mapped onto a set of user interfaces which support these functions in an efficient, consistent, and user-friendly manner. User-interface design has two aspects: control flow and dialog elements. The control flow specifies the order in which user interface actions need to occur and thus the order-dependencies between the various interface elements. The user interface dialogs contain the actual interface components such as prompts (command line interfaces), forms (form-based interfaces), or graphical user interface (GUI) elements.

Similar designs must be developed for the models, their data, their interconnections, the various visualization tools and all system utilities. Note that the scenarios developed during the earlier phases of the project remain important in that all detailed specifications need to be validated against them. After all, the scenarios contain many instances of the functions the system needs to support. Also, scenario refinement continues during detailed design.

An important component of the detailed functional design is a development schedule and an associated budget. Especially for large DSS, continuation of the project typically involves substantial amounts of funding, as well as the organizational repercussions mentioned earlier. The decision to continue the project beyond the phase of detailed design implies major commitments from both the developers and the sponsoring organization. The developers commit to actually developing and fielding the system as it is specified in the detailed functional design. The sponsoring organization commits the funding and the organizational support to make this happen.

33.4.4 Phase 4: Detailed Software Design

Two activities transform the detailed functional design into actual DSS source codes: detailed software design and the actual development or implementation* of the software. The detailed software design takes the specifications from the detailed functional design and translates them into software representations (data structures, algorithms, program functions, function calls and parameters, and so on). The software design also outlines other complementary issues and conventions needed during the actual software development process. Examples of these are coding standards, in-line documentation standards, conventions for function naming, variable and function declarations, specification of libraries, and so forth. If databases are to be part of the DSS, detailed software design also includes the specification of the database schema or data model. The data model specifies how data are to be stored in the database and how the various data relate to each other. Section 33.5, Of Tools and Toys, contains examples of some of these activities in the context of relational databases.

33.4.5 Phase 5: Development

Different groups of developers follow different methodologies in conducting the software design and the actual coding. Some groups choose to commit to a compre-

---

* *Implementation* is sometimes referred to as the actual fielding of software. Here, *development* and *implementation* are used interchangeably and refer to the construction of a system's source codes.

hensive software design before any coding can take place, others follow a more liberal approach where coding of certain components can start before others have been designed completely. Naturally, the system's architecture determines to a large extent whether or not such freedom is allowed. Generally, the more modular a system's architecture, the more freedom developers have to accept a liberal design and coding methodology. However, even the most modular architectures require a fair amount of software design before actual coding can commence.

At this point a comment on the technique of *rapid prototyping* is in order. Rapid prototyping refers to a design and development method where the developers, after having gained some insight into the organization's needs for a DSS, quickly produce a first, initial version of a system (Gray and Took, 1984; Budde and Bacon, 1992; Lowell, 1992). The rationale underlying this approach is that it is often much easier for users to react to a proposal from the developers than to try and come up with proposals of their own. As such, rapid prototyping implies an almost evolutionary development of a system, allowing for redirection and changes during the development process, rather than putting all your eggs in one very carefully but possibly incorrectly designed basket.

Rapid prototyping, however, is not all it seems, especially when it comes to complex systems which are likely to have their effect on an organization's daily practice. In our experience, we have learned that it often takes substantial amounts of time to learn about the sponsoring organization's problems, about the reasons that this organization thinks it needs a DSS, and the processes and tasks this DSS is supposed to support. At least an equally strong effort is needed to find out about an organization's models and data and the flow of information through the organization. The complexity of many water-resources management problems is such that careful study, design and organizational reorientation seem both welcome and inevitable requirements. Furthermore, systems which have clear operational components must be designed and developed correctly, simply because there is so much at stake. Evolutionary development introduces too much uncertainty, resulting in a too complicated, nonstandardized mixture of features retained during the various rounds of the development-evaluation process. However, some developers as well as sponsoring organizations are comfortable with the technique of rapid prototyping.

Coding the DSS is the stage that most developers are eager to reach, even though eagerness to code, or *keyboard-happiness,* can be a serious problem for developers. During this phase there is intensive contact with the sponsoring organization's representatives. Data and model codes need to be acquired, myriads of small, unforeseen problems need to be resolved, and many decisions need to be made. Periodic reviews and progress reports are in order, since this is the stage which consumes most of the sponsor's budget, while the developers have maximum freedom in putting the system together. Keeping the sponsoring organization up to date on progress, good and bad news, and the plans for the next 30 days of development are good project-management practices.

33.4.6   Phase 6: Initial Fielding

Once development is almost finished, it is a good idea to subject the DSS to an *alpha-testing* or an initial fielding phase. This amounts to installing the system on the sponsor's machine(s) and making sure that all the data and functional components as well as the connecting infrastructure work appropriately. In the case of a single-user DSS, which is completely installed on a single machine and is to be used by a single person at a time, this fielding phase might be rather insignificant. However, in case of a system such as TERRA, initial field testing implies the installation of the software on various machines, connected with each other through a corporate wide-area network (WAN). The database servers need to be installed and each individual application must connect to the database. In such a situation it is likely that logistic bottlenecks will occur. These need to be traced and resolved. Also, systems such as TERRA need to automatically incorporate (near) real-time data at certain times and time intervals. Initial fielding implies that all these functions are tested by the developers while running the system on the sponsor's machines/network.

33.4.7   Phase 7: Testing, Training, and Fine Tuning

Once the developers are confident that at least all the basic functions of the DSS are in place, *beta-testing* and user training can take place. Since most DSS are not products in the sense that they can be sold as-is to a large group of customers or clients, it often makes sense to use the training sessions not only as an opportunity to make the organization's workers familiar with the idiosyncrasies of the DSS, but to also use those sessions for software-testing purposes. It is a well-known fact that no matter how well versed DSS developers are in the daily problems of the sponsor, and no matter how well they have designed the user interfaces to comply with the nature of the problem at hand, developers will only have tested a small subset of the actual situations a DSS must be able to support. Moreover, since the developers are intimately aware of the internal logic of the system, there are a great number of actions they will never even think of attempting. During a first testing and training session, therefore, users will typically be able to break the system on many more places and occasions than the developers ever thought possible. There is an old Dutch proverb that says: "One fool can ask more questions than ten wise people can answer." During initial testing and training, developers often reflect on who the fools are.

Proper system design and appropriate coding styles, however, facilitate the fixing of bugs and the realization of small changes to, for instance, the user interfaces. This allows for many fixes and/or flaws in the DSS to be corrected in, often, a few hours. More serious mistakes take longer to rectify. However, by allowing several weeks between testing/training sessions, sufficient time for updates is available.

33.4.8   Phase 8: Maintenance and Further Development

Once a DSS is *on-line,* i.e., fielded and functioning within the sponsoring organization, both immediate and long-term system maintenance and version-control policies become relevant. Unlike new releases of third-party software which seem to be completely induced by the software's manufacturer and, for the bulk of its users, simply happen, once an organization has a comprehensive DSS in place it needs to think about continued bug fixing, version-control strategies, database updates and archival, new releases of supporting software (!), newly emerging hardware, and the extension of the DSS with new capabilities. Unfortunately, too often these considerations are left out of the initial design phases of the DSS and as a result, software is often "thrown over the wall" by the developers, thus leaving the receiving organization unprepared for the management of a complex software system.

Various models exist to avoid this type of situation. Obviously, long-term system maintenance should be an issue dealt with during the early phases of the project. Not only does maintenance have a technical dimension (changes need recoding), but substantial costs and personnel commitments may be required. As part of the Colorado River DSS feasibility study mentioned earlier, for instance, it was estimated that the total annual cost for maintaining and extending the system's databases and software and replacement of hardware, amounts to about 5 percent of the initial development cost.*

One possibility for long-term system maintenance is for the sponsoring organization to assume this task. Although this is a definite possibility, it requires the organization to invest in this solution even during the initial development process; for example, by stationing one or more of its designated system administrators with the developers during the developing process. Another possibility is to have a third party involved in the entire development process where it is known from the start that it will maintain the fielded DSS on a contractual basis. A third possibility is that the developers conduct long-term maintenance. The latter solution, however, requires the developers to have the necessary organizational infrastructure in place. Long-term software maintenance and version control do not simply occur naturally.

33.5   OF TOOLS AND TOYS

Many of the tools by means of which DSS are developed have evolved quickly over the last ten years or so. Perhaps the most important difference between today's and yesterday's DSS concerns the way in which state information is maintained in terms of the storage, retrieval, and overall management of a system's data. Where systems such as TERRA are built on a data-centered architecture with a comprehensive geo-relational database at their center, early DSS developers did not have such comprehensive tools for data management available. Instead, they developed their own data-storage facilities and their own data-storage and retrieval functions. With the recent advance of relational database management and GIS systems, however, DSS developers have been given a number of most effective tools for the management of their often very complex data.

33.5.1   Relational Databases

The power of, for instance, the relational model for representing attribute data is remarkable, especially in light of its simplicity. In relational databases, data is represented as tables, not unlike traditional data matrices, i.e., variables as columns, observations as rows. However, when different tables share one or more columns, complex search or join operations can be executed. For instance, assume that

- Table *A* holds the ID, name, and generating capacity of each power plant in the system
- Table *B* holds power-plant IDs and generation values as measured over the last two weeks
- Table *C* holds IDs and discharges for all reservoirs in the system, including those equipped with power plants

---

* Total initial development costs were estimated at $5 million, not including costs for data and core-model source-code development.

One can then search for all generating (overall) efficiencies by joining tables *A, B,* and *C* and dividing generation by discharge. The join operation results in a table containing all records which have the ID field in common, and for those, efficiencies are computed by dividing the associated generations by the release values. Structured query languages such as SQL (Date, 1989) provide the tools for conducting these types of operations from both command-level SQL interpreters and from inside programs (embedded SQL):

```
select plant_name,generation/discharge
from A,B,C
where A.id = B.id and B.id = C.id and
B.date_time = C.date_time and discharge > 0;
```

The associated low-level data storage, searching, joining, and computing are carried out by the relational database software.

Although this example shows some of the power of the relational model and associated querying facilities, it also illustrates one of its main problems: its inefficacy to handle time-series data. Although, strictly speaking, the relational model does not forbid manipulation of complex data types such as time-series as units of information, relational operations can only be conducted on the level of those units. Therefore, if one needs the database to be able to select from a time series, the criteria for searching that time series need to be stored in the database as well. As a result, time series often need to be stored as records, where each record holds both the value and a time stamp or an index into a table which maps the indices on time stamps. In addition, each of those records needs to contain the ID of the object with which the time series is associated. Although from a relational point of view this does not constitute a redundancy in information—in relational terms; a violation of *normality*—semantic redundancy exists in the sense that all that is needed to define a time series is a vector of values and an associated time definition consisting of a start time, a frequency, and the length of the vector (StatSci, 1991) rather than a vector of values and an associated vector of time stamps.

The TERRA system was built exclusively around a relational database. This database holds practically all information which can be queried through the TERRA interface or accessed by an application's DMI. From a functional point of view, information in the TERRA database can be grouped into the following seven categories:

1. Time series data
2. Historical data
3. Physical attribute data
4. Operational constraints
5. Model data
6. Security data
7. Meta data

The meta data comprises data about how all other data are organized. This allows DMIs not having to know where data are located in the database. Instead, they only have to query the meta data tables which will then tell them where the actual data can be found. Using meta data greatly improves the portability of a system to other cases.

Modern relational databases are also equipped with elaborate security facilities for managing security-based data access. For complex organizations such as TVA it is very feasible that not all information stored in the databases or all functions provided through the TERRA system are available to everybody. The database's facilities for implementing security can then be used to protect some of the system's data and functions.

A measure of complexity of a relational database is its number of tables. For a water-resources management application, the TERRA database can be considered of medium complexity; it has about 150 tables. The table structure of a relational database is also known as the *data model*. Modern relational-database-management systems are equipped with sophisticated tools to build these data models from scratch. One of these is the entity relationship diagram or *ERD*. ERDs are diagrammatic representations of how data are organized in the database, i.e., what tables consist of and how they relate to each other, both relationally and semantically. Figure 33.5, for example, contains a small portion ($\approx$3 percent) of TERRA's ERD. It represents that there exist three types of power plants: hydro, nuclear, and fossil; and two more abstract groupings of these, thermal plants and power plants in general. It also represents that, for instance, fossil-plant compliance summaries are associated with the fossil plants. The crow foot for that relationship implies that there is a one-to-many relationship between the fossil-plant table and the compliance summary table, meaning that one plant can have multiple compliance summaries associated with it.

Also note the relationships between power plants, streams (TSTREAM), and zebra mussel sites. According to the ERD, streams have multiple instances of both mussel sites and power plants associated with them. Hence, both the mussel-site and power-plant tables contain a column for the associated ID of the stream they are associated with. In case we need to know which mussel sites are within five (stream) miles from a particular hydroelectric power plant and what the associated stream is, we simply select that information from the joined *mussels, hydros* and *streams* tables:

```
select hydro_plant_id,hydro_plant_name,
       mussel_id,
       stream_id,stream_name
from hydros, mussels, streams
where hydros.stream_id = mussels.stream_id and
      hydros.stream_id = streams.stream_id and
      hydro_plant_id = 3 and
      abs (hydros.stream_mile - mussels.stream_mile) < 5;
```

33.5.2 Geographic Information Systems

Where relational-database-management systems increased both the productivity and amenities of a developer's life regarding various types of attribute data, GIS have supported the management and manipulation of complex spatial data sets. By coupling both data models, today's developers have the availability of powerful geo-relational data models which serve as data managers for complex DSS applications.

The use of a geo-relational data model not only allows efficient and effective storage of both types of information, it also suggests some interesting opportunities for (partly) automated generation of models of the system.

To illustrate this, consider a model of a river basin consisting of three components:

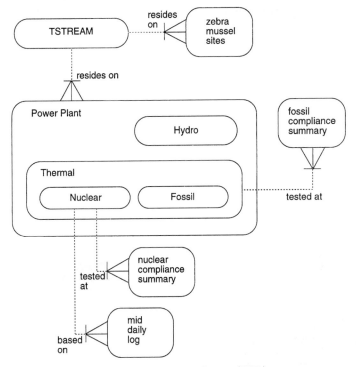

FIGURE 33.5 TERRA entity-relationship diagram (ERD).

1. A representation of the system's configuration in the form of network topology (Labadie, et al., 1986; CADSWES, 1993a)

2. Attributes of the basin's components (e.g., head/area · volume relationships, historical flows, elevation data, and so on)

3. Mathematical expressions representing the system's behavior (e.g., evaporation, power generation, and mass balance)

The first two components represent the state information of a model, whereas the third component, the process information, contains the *state transition rules* under which the system operates. And because the state information is comprised of both spatial and attribute information, it is possible to generate a model's (initial) state information directly from a supporting geo-relational database. Recent attempts at object-oriented modeling of river basins use this technique for the automated generation of river-basin models (CADSWES, 1993a; Reitsma, et al., 1994).

The use of GIS for water resources and, in particular, hydrologic modeling is not limited to object-oriented approaches, however. Recent international conferences on GIS and environmental modeling (NCGIA, 1991; 1993), have shown a rapidly expanding repository of modeling techniques which are implemented directly on top of a GIS and its functions for the storage, retrieval, and manipulation of spatial data.

TERRA uses GIS only to the extent that TERRA's spatial information was generated using a standard GIS commercial product. TERRA software accesses that information and uses it for mapping and for maintaining a relationship between the spatial and attribute information.

33.5.3 Object Orientation

The advance of *object-oriented programming* (OOP) (Ellis and Stroustrup, 1990; Lippman, 1991; Coplien, 1992) caused another revolution in both modeling and coding of complex computer-based systems. Object-oriented techniques discretize a modeling or programming problem in self-contained domains of action, the so-called *objects*. Objects maintain both data and exhibit behavior in a true stimulus-response manner; i.e., an object's state changes as a function of the application of a dynamic. The dynamic itself is triggered as a result of receiving a message from another object or as the result of a (previous) state change. Applied to a model of a river basin, objects could be reservoirs, power plants, diversions, or reaches, each of which maintains its own state by managing its own data and applying its own dynamics whenever necessary. The objects communicate with each other by sending each other water or by requesting water. Although each of the objects acts autonomously, collectively their behavior models the behavior of the system as a whole.

With the new object-oriented programming techniques, it suddenly became possible to address many programming problems in ways coders had always wanted (true modularity, encapsulation of data, class hierarchies and inheritance, and so on). At the same time, new compilers and new complete-programming environments facilitated the management and coding of large complicated programs and eliminated a lot of tedious and often difficult code testing. New and continuing improvements in language standards keep supporting this process.

TERRA makes only limited use of object orientation. Most of its programming was done in the C language, although a few object-oriented (C++) libraries are used for conducting certain types of operations. Although object-orientation provides many advantages, some of its overhead can sometimes be traded off against the more efficient and somewhat simpler structures of traditional programming.

33.5.4 Graphical Standards and Interface Tools

Another major leap in the development of DSS was caused by the coming of age of graphics-development tools such as graphics libraries, user-interface builders, graphics standards, and of late, platform-independent development-tool kits. Although it is still possible for developers to write their own graphics libraries and, if so desired, their own complete windowing systems, writing graphics applications has become increasingly more efficient and accessible through the availability of these high-level tool kits.

Advances in graphics systems have occurred hand-in-hand with the emergence of windowing systems and mouse-driven interface systems. Both tools have dramatically increased the opportunities for both data visualization and the development of user interfaces.

33.5.5 Networks

Productivity and comfort levels of DSS developers were further boosted by the emergence and rapid growth of national and international computer networks. The networks and their sheer limitless supply of public program libraries, utilities, tools, news, stories, and pictures provide valuable resources for DSS developers. World-wide networks, together with such institutions as the Open Software Foundation (OSF) and clever constructs such as *share-ware,* have created a very large market where it is a pleasure to shop for other people's intellectual gems.

Networks not only facilitate the work of DSS developers, they also open new opportunities for putting DSS technology to work. In case of the TERRA system, for instance, people throughout the TVA system can simultaneously log onto the TERRA database server using the corporate wide-area network (WAN) (Fig. 33.6). TERRA itself is equipped with data-collection programs which, at specified times, use the network to see if new data for incorporation is available at various sites within the organization. If so, the data is copied over, checked for operational con-straint violations, and incorporated into the database. Violations are logged on a bul-letin board and users throughout the organization are notified about the new postings to their bulletin board.

33.5.6 Interpreters

A number of less dramatic innovations have further strengthened a DSS developer's arsenal of tools and toys. A very recent one, for instance, is the (re)emergence of interpreters, written on top of conventional programming languages such as C, which allow for the creation of complex interpreted code segments. Some of these languages, such as "Tcl" (Ousterhout, 1990) or Tcl-based tools such as "expect" (Libes, 1991), allow for the modification and extension of DSS without having to recompile the system's source codes. As such, developers can field a system while continuing to work on, for instance, the library of visualization tools. Every time a new tool is completed it can be added to the system without having to go through the often complicated process of version release and control. Similar benefits are nowa-days being reaped from interpreter-based plotting tools such as ACE/Xmgr (Turner, 1993) or statistical data analysis and visualization packages such as S-PLUS (StatSci, 1991; Becker, et al., 1988).

33.6 DSS: COMING OF AGE OR FOREVER A PROMISE?

The early pioneers of DSS technology envisioned DSS as integrated, computer-based systems which provide a logical rather than analog view of an environment or water resource, to be manipulated at will so we can try and find optimal, acceptable, or at least defendable operational and long-term planning policies. Although this idea was a promising one, preciously few of these systems have come to fruition and are actually used on a regular basis. Perhaps the single most important reason for this is that most DSS were poorly designed, both from the points of view of func-tionality and software development.

For too long, developers of environmental DSS concentrated almost exclusively on the rather mechanical aspects of DSS (modeling, user interfaces, database con-nections, and so on), without paying much attention to the organizational context in which these systems were supposed to function. For too long, developers did not address the often difficult to control and rather mundane aspects of organization, use, and long-term maintenance and training. Naive expectations by early develop-

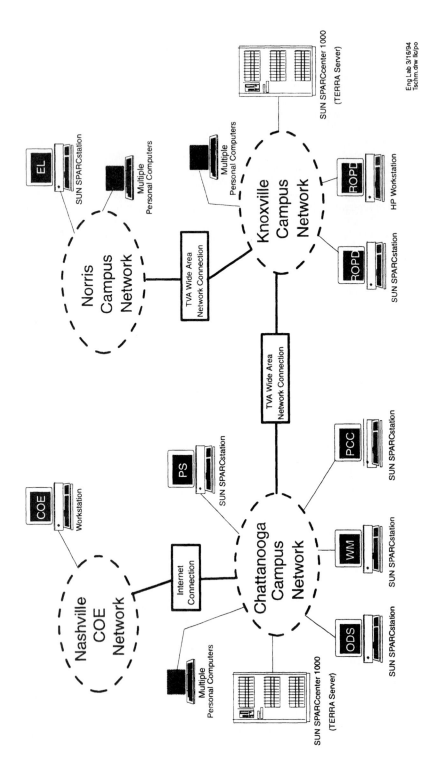

FIGURE 33.6 TERRA user network.

Eng Lab 3/16/94
Tschm.drw llc/po

SUN SPARCcenter 1000
(TERRA Server)

Multiple
Personal Computers

EL

SUN SPARCstation

Multiple
Personal Computers

Knoxville
Campus
Network

ROPD

HP Workstation

ROPD

SUN SPARCstation

Norris
Campus
Network

TVA Wide Area
Network Connection

TVA Wide Area
Network Connection

PS

SUN SPARCstation

PCC

SUN SPARCstation

COE

Workstation

Nashville
COE
Network

Internet
Connection

Chattanooga
Campus
Network

WM

SUN SPARCstation

Multiple
Personal Computers

SUN SPARCcenter 1000
(TERRA Server)

ODS

SUN SPARCstation

ers as to the type of users DSS technology should be targeted at illustrate this point. Some of the DSS definitions mentioned at the beginning of this chapter, for example, imply that DSS were to be developed for managers—for actual decision makers. Indeed, some developers still argue that DSS's greatest virtue lies in putting large amounts of data, models, and management information directly into the hands of high-level managers (Fedra, et al., 1993). Too often, however, this objective degraded into unacceptable simplifications of the problem represented in a DSS, or to the simplification of complex model inputs and outputs into lumped, predefined configurations of values or fuzzy, intractable categories (e.g., small, medium, and large). These simplifications were based on the assumption that high-level managers do not deal with the technical details of the models and associated data. And although the latter might be the case—surprisingly little research is available to support this claim—it was frequently ignored that decision makers often rely on highly trained technical-support staff who do need access to detailed data.

Evidently, the type of user a DSS is meant for holds implications for the type of tasks the system needs to support. The distinction between managers and technical-support staff, for instance, might be valid in the context of long-term planning and strategic decision making. For short-term operational decision making, however, this distinction becomes less clear. On the operational level, decision authority tends to be distributed among groups of people, each with specific task sets and with limited decision-making authority. On the level of operational decision making, therefore, decision makers may very well address the DSS themselves, whereas on the level of long-term policymaking, the user group is mainly composed of support staff.

In order for environmental DSS to fulfill their promise, they need to support the existing information flow in an organization and the tasks and responsibilities of its workers, resolving occasional inefficiencies in the process. Furthermore, they need to provide policymakers' support staffs with information and facilities that allow them to better argue why various plan alternatives need to be accepted, put on hold, or rejected.

Although the record for DSS in environmental decision making is all but a clean one, DSS technology, in both its technological and organizational aspects, has come of age and is, although somewhat late, delivering on its promise. DSS such as TERRA illustrate that through close cooperation between developers and sponsoring organizations many mistakes made in earlier attempts can be avoided. Problem-centered design methodologies augmented with software design and adequate project management help make DSS technology deliver on its promise. Multidisciplinary development teams safeguard against oversimplification of the problem to be solved, while bringing the necessary diverse skill sets to its solution. In addition, developments in hardware—but especially software engineering, graphics, data bases and programming environments—have significantly enriched the developer's repository of tools for building DSS. As a result, developers can spend more time on the organizational aspects of DSS, thereby overcoming some of the deficiencies characterizing earlier work.

Notwithstanding these advancements in technology and methodology, DSS development remains a precarious endeavor. Too often, technical aspects are emphasized at the cost of more mundane issues related to the actual needs of future users. This predisposition towards technical rather than utility-based evaluations of DSS becomes clear, for instance, when attending water-resources-management conferences. Whereas developers are eager to exhibit their technical accomplishments with demonstrations and colorful pictures of the systems they developed, preciously few presentations address the resultant utility of those systems, whether or not they are used for what they were meant for, or what the long-term ramifications for the organization of such systems are. Although idle speculation, this might be because most developers have a desire to continue developing rather than to address the difficult issue of evaluating whether or not their products make any difference. But if the late Sir Karl Popper (1965; 1968) was right, and only from our mistakes can we learn, the next few years will be most interesting.

REFERENCES

Anthony, R. N., *Planning and Control Systems: A Framework for Analysis,* Harvard University Graduate School of Business Administration, Cambridge, Mass., 1965.

Becker, R. A., J. M. Chambers, and A. R. Wilks, *The New S Language,* Cole Computer Science Series, Wadsworth & Brooks, Pacific Grove, Calif., 1991.

Brown, L. C., and T. O. Barnwell, "The Enhanced Stream Water Quality Models QUAL2E and QUAL2E-UNCAS: Documentation and User Manual," Environmental Research Laboratory, Office of Research and Development, U.S. Environmental Protection Agency, Athens, Ga., 1987.

Budde, R., and P. Bacon, *Prototyping: an Approach to Evolutionary System Development,* Springer-Verlag, Berlin and New York, 1992.

Burrough, P. A., *Principles of Geographical Information Systems for Land Resources Assessment,* Oxford Science Publications, 1986.

CADSWES, *TVA Environmental and River Resources Aid (TERRA) Design Document for 1993 Prototype Implementation,* CADSWES/Department of CEAE, University of Colorado, Boulder, Colo., 1993.

CADSWES, *Power and Reservoir System Model (PRSYM) Design Document for 1993 Implementation,* CADSWES/Department of CEAE, University of Colorado, Boulder, Colo., 1993a.

Codd, E. F., "A Relational Model of Data for Large Shared Data Banks," *Communications of the AMC,* 13:377–387, 1983.

Coplien, J., *Advanced C++. Programming Styles and Idioms,* Addison-Wesley, Reading, Mass., 1992.

Dames, and Moore, "Feasibility Study Report for a Colorado River Decision Support System," submitted to the Colorado Department of Natural Resources, Colorado Water Conservation Board and Colorado Division of Water Resources, Colorado Department of Natural Resources, Denver, Colo., 1993.

Date, C. J., *An Introduction to Database Systems, Vol. 1 and 2,* Addison-Wesley, Reading, Mass., 1985.

Date, C. J., ed., *Relational Databases: Selected Writings,* Addison-Wesley, Reading, Mass., 1986.

Date, C. J., *The SQL Standard,* 2d ed., Addison-Wesley, Reading, Mass., 1989.

DeSanctis, G., and R. B. Gallupe, "A Foundation for the Study of Group Decision Support Systems," *Management Science,* 33:589–609, 1987.

Department of the Interior, "Delaware Estuary Comprehensive Study. Preliminary Report and Findings," Federal Water Pollution Co. Inc., Redwood, Calif., 1966.

Ellis, M. A., and B. Stroustrup, *The Annotated C++ Reference Manual,* Addison-Wesley, Reading, Mass., 1990.

Fairly, R. E., *Software Engineering Concepts,* McGraw-Hill Series in Software Engineering and Technology, McGraw-Hill, New York, 1985.

Fedra, K., *A Computer-Based Approach to Environmental Impact Assessment,* IIASA RR-91-13, IIASA, Laxenburg, Austria, 1991.

Fedra, K., E. Weigkricht, and L. Winkelbauer, "A Hybrid Approach to Information and Decision Support Systems: Hazardous Substances and Industrial Risk Management," in *IFAC Proceedings of the Conference on Economy and Artificial Intelligence,* Aix-en-Provence, France, 1986, pp. 169–175.

Fedra, K., and C. Zhao, "Multi-Criteria Data Evaluation Based on DIDASS/DISCRET," in *IIASA/ACA (1988) Expert Systems for Integrated Development; A Case Study of Shanxi Province, The People's Republic of China: Final Report, vol. 1: General Systems Documentation,* Laxenburg, Austria, 1988, pp. 209–218.

Fedra, K., E. Weigkricht, and L. Winkelbauer, *Decision Support and Information Systems for Regional Development Planning,* IIASA RR-93-13, IIASA, Laxenburg, Austria, 1993.

Gilbert, K. C., and R. M. Shane, "TVA Hydro Scheduling Model: Theoretical Aspects," *Journal of Water Resources Planning and Management,* ASCE, 108(WR1):21–36, 1982.

Guariso G., and H. Werthner, *Environmental Decision Support Systems,* Ellison Horwood Series in Computers and their Applications, Ellison Horwood Limited, Chichester, U.K., 1989.

Gould, J., "How to Design Usable Systems," in M. Helander, ed., *Handbook of Human-Computer Interaction,* North-Holland, New York, 1988, pp. 757–789.

Gray, P., and R. Took, eds., *Proceedings of the Working Conference on Prototyping, Namur, October 1983,* sponsored by the Commission of the European Communities, Springer-Verlag, Berlin and New York, 1984.

Hendriks, P. H. J., and A. G. M. van der Smagt, "Definition of Choice Sets and Choice Set Attributes, Some Problems Inherent in Decompositional Modeling," in R. G. Golledge, and Timmermans, eds., *Behavioral Modeling in Geography and Planning,* Croom Helm, New York, 1988, pp. 27–37.

Hetling, L. J., "A Mathematical Model for the Potomac—What it Has Done and What it Can Do," CB-SRPB Technical Paper no. 8, Federal Water Pollution Control Administration, Washington, D.C., 1966.

Keen, P. G. W., "Value Analysis. Justifying Decision Support Systems," in R. Sprague and H. J. Watson, eds., *Decision Support Systems. Putting Theory into Praxis,* Prentice-Hall, Englewood Cliffs, N.J., 1986, pp. 48–64.

Keen, P. G. W., and M. S. Scott Morton, *Decision Support Systems: An Organizational Perspective,* Addison-Wesley Series on Decision Support, Addison-Wesley, Reading, Mass., 1978.

Klosterman, R. E., "The Politics of Computer-Aided Planning, *Town Planning Review,* 58:441–451, 1987.

Konsynski, B. R., E. A. Stohr, and J. McGee, "Review and Critique of DSS," in E. A. Stohr, and B. R. Konsynski, eds., *Information Systems and Decision Processes,* IEEE Computer Society Press, 1992, pp. 7–26.

Labadie, J. W., D. A. Bode, and A. M. Pineda, "Network Model for Decision-Support in Municipal Raw Water Supply," *Water Resources Bulletin,* 22:927–940, 1986.

Lewis, C., J. Rieman, and B. Bell, "Problem-Centered Design for Expressiveness and Facility in a Graphical Programming System," *Human-Computer Interaction,* 6:319–355, 1991.

Lewis, C., and J. Rieman, "Task-Centered User Interface Design; A Practical Introduction," ftp.cs.colorado.edu (ftp address), Boulder Colo., 1993.

Libes, D., "Scripts for Controlling Interactive Processes," *Computing Systems,* 4, 1991.

Lippman, S. B., *C++ Primer,* Addison-Wesley, Reading, Mass., 1991.

Loehr, R. C., C. S. Martin, and W. R. Rast, *Phosphorus Management Strategies for Lakes,* Ann Arbor Science, Ann Arbor, Mich., 1980.

Loucks, D. P., J. R. Stedinger, and D. H. Haith, *Water Resource Systems Planning and Analysis,* Prentice-Hall, Englewood Cliffs, N.J., 1981.

Lowell, A. J., *Rapid Evolutionary Development: Requirements, Prototyping & Software Creation,* John Wiley, New York, 1992.

Maass, A., M. M. Hufschmidt, R. Dorfman, H. A. Thomas, Jr., S. A. Marglin, and G. M. Fair, *Design of Water-Resource Systems,* Harvard University Press, Cambridge, Mass., 1962.

Major, D. C., and R. L. Lenton, *Applied Water Resources Systems Planning,* Prentice-Hall, Englewood Cliffs, N.J., 1979.

Mallach, E. G., *Understanding Decision Support Systems and Expert Systems,* Irwinn Inc., Burr Ridge, Ill., 1994.

McLean, E. R., and H. Sol, eds., "Decision Support Systems: A Decade in Perspective," *Proceedings of the IFIP WG 8.3 Working Conference on Decision Support Systems: A Decade in Perspective, Noordwijkerhout, The Netherlands, 16–18 June 1986,* Elsevier Science, Amsterdam, 1986.

Mittra, S. S., *Decision Support Systems, Tools and Techniques,* John Wiley, New York, 1986.

National Center for Geographic Information and Analysis, *Proceedings of the First International Conference on GIS and Environmental Modeling, August, 1991,* Boulder, Colo., 1991.

National Center for Geographic Information and Analysis, *Proceedings of the Second International Conference on GIS and Environmental Modeling, September, 1993,* Breckenridge, Colo., 1993.

Nijkamp, P., and J. Spronk, eds., *Multiple Criteria Analysis: Operational Methods,* Aldershot, Gower, 1981.

Nijkamp, P., H. Voogd, and H. Rietveld, *Multicriteria Evaluation in Physical Planning,* Elsevier, New York, 1990.

Nijssen, G. M., *Modeling in Database Management Systems,* North Holland, Amsterdam, 1976.

Ousterhout, J., "Tcl: An Embeddable Command Language," *Proceedings of the Winter 1990 USENIX Conference, Jan. 22–26,* Washington, D.C., 1990.

Polsen, P. G., C. H. Lewis, J. Rieman, and C. Wharton, "Cognitive Walkthroughs: A Method for Theory-Based Evaluation of User Interfaces," *Intern. Journal of Man-Machine Studies,* 36:741–773, 1992.

Popper, K. R., *Conjectures and Refutations: The Growth of Scientific Knowledge,* Harper & Row, New York, 1968.

Popper, K. R., *Logic of Scientific Discovery,* Harper & Row, New York, 1965.

Reitsma, R. F., *Functional Classification of Space, Aspects of Site Suitability Assessment in a Decision Support Environment,* IIASA RR-90-2, Laxenburg, Austria, 1990.

Reitsma, R. F., and J. Behrens, "Integrated River Basin Management (IRBM): A Decision Support Approach," in R. E. Klosterman, ed., *Proceedings of Second International Conference on Computers in Urban Planning and Urban Management, Oxford, U.K., July 6–8, 1991,* 1991, pp. 29–41.

Reitsma, R. F., and D. Sieh, "Bootstrapping Models Using Existing Databases and Object-Orientation," *Proceedings of the Eighth Conference on Computing in Civil Engineering,* Dallas, Tex., 1992.

Reitsma, R. F., A. M. Sautins, and S. C. Wehrend, "RSS: A Construction Kit for Visual Programming of River Basin Models," *Journal of Computing in Civil Engineering,* ASCE, 8:378–384, 1994.

Ross, D. T., and K. E. Schoman, "Structured Analysis for Requirements Definition," *IEEE Transactions on Software Engineering,* SE-3:6–15, 1977.

Scholten, H. J., and J. C. H. Stillwell, *Geographical Information Systems for Urban and Regional Planning,* Kluwer Academic Publishers, Dordrecht, Boston, and London, 1990.

Schuster, R. J. "Colorado River Simulation System Documentation. System Overview," U.S. Department of the Interior, United States Bureau of Reclamation, Denver, Colo., 1987.

Sprague, R., and H. J. Watson, eds., *Decision Support Systems. Putting Theory into Praxis,* Prentice-Hall, Englewood Cliffs, N.J., 1986.

Star, J., and J. Estes, *Geographic Information Systems: An Introduction,* Prentice-Hall, Englewood Cliffs, N.J., 1990.

StatSci, *S-PLUS User Manual, Vol. 1,* Statistical Sciences, Seattle, Wash., 1991.

Thomann, R. V., *Systems Analysis and Water Quality Management,* Environmental Research and Applications, New York, 1972.

Thomann, R. V., and M. J. Sobel, "Estuarine Water Quality Management and Forecasting," *San. Engr. Div. ASCE,* 92(SA5):6–36, 1964.

Turner, P., *ACE/gr User's Manual: Graphics for Exploratory Data Analysis,* Software Documentation Series, SDS3, 91-3, 1993.

U.S. Army Corps of Engineers, *HEC-2 Water Surface Profiles,* Hydrologic Engineering Center, Davis, Calif., 1982.

Wierzbicki, A., *A Methodological Guide to Multi-Objective Optimization,* IIASA WP-79-122, Laxenburg, Austria, 1979.

Winkelbauer, L., and S. Markstrom, "Symbolic and Numerical Methods in Hybrid Multi-Criteria Decision Support," *Expert Systems with Applications,* 1:345–358, 1990.

Yakowitz, D. S., and L. J. Lane, "A Multi-Attribute Decision Tool for Ranking a Finite Number of Alternatives," Working Paper #891, Southwest Watershed Research Center, Tuscon, Ariz., 1992.

Yakowitz, D. S., "A Decision Support System for Water Quality Modeling," in M. Karamouz, ed., *Water Resources Planning and Management, Proceedings of the Water Resources Sessions at Water Forum '92,* American Society of Civil Engineers, New York, 1992.

Yeh, W. G., and L. Becker, "Multi-Objective Analysis of Multi-Reservoir Operations," *Water Resources Research,* 18:1326–1336, 1982.

Zigurs, I., E. V. Wilson, A. M. Sloane, R. F. Reitsma, and C. Lewis, "Simulation Models and Group Negotiation: Problems of Task Understanding and Computer Support," in R. Sprague, and J. F. Nunamaker, Jr., eds., *Proceedings of the 27th Hawaii International Conference on System Sciences,* vol. 4, IEEE Computer Society Press, Los Alamitos, Calif., 1994.

INDEX

FE